CHILTON®
LABOR GUIDE
2006 EDITION

Imported Vehicles

ASIAN VEHICLES

Acura
Daewoo
Daihatsu
Honda
Hyundai
Infiniti
Isuzu
Kia
Lexus
Mazda
Mitsubishi
Nissan
Scion
Subaru
Suzuki
Toyota

EUROPEAN VEHICLES

Alfa Romeo
Audi
BMW
Jaguar
Land Rover
Mercedes-Benz
Mini
Peugeot
Porsche
Renault
Saab
Sterling
Volkswagen
Volvo
Yugo

THOMSON
DELMAR LEARNING

Australia • Canada • Mexico • Singapore • Spain • United Kingdom • United States

Chilton®
Labor Guide
2006 Edition

Imported Vehicles

Vice President
Technology Professional Business Unit:
Gregory L. Clayton

Publisher,
Professional Business Unit:
David Koontz

Production Director:
Mary Ellen Black

Marketing:
Beth A. Lutz

Marketing Specialist:
Brian McGrath

Publishing Coordinator:
Paula Baillie

Sr. Production Editor:
Elizabeth C. Hough

Editorial Assistant:
Christine Wade

Editors:
Kirk Fuhrman
Steven D. Junker
Tracy Junker

COPYRIGHT 2006 by Delmar Learning, a division of Thomson Learning, Inc. Thomson Learning™ is a trademark used herein under license.

Printed in the United States of America
1 2 3 4 5 6 XX 07 06 05

For more information contact
Delmar Learning
Executive Woods
5 Maxwell Drive, PO Box 8007,
Clifton Park, NY 12065-8007
Or find us on the World Wide Web at :
www.trainingbay.com
or, www.chiltonsonline.com

ALL RIGHTS RESERVED. No part of this work covered by the copyright hereon may be reproduced in any form or by any means—graphic, electronic, or mechanical, including photocopying, recording, taping, Web distribution, or information storage and retrieval systems—without the written permission of the publisher.

For permission to use material from the text or product, contact us by
Tel.   (800) 730-2214
Fax   (800) 730-2215
www.thomsonrights.com

ISBN: 1-4018-1537-7

NOTICE TO THE READER

Publisher does not warrant or guarantee any of the products described herein or perform any independent analysis in connection with any of the product information contained herein. Publisher does not assume, and expressly disclaims, any obligation to obtain and include information other than that provided to it by the manufacturer.

The reader is expressly warned to consider and adopt all safety precautions that might be indicated by the activities herein and to avoid all potential hazards. By following the instructions contained herein, the reader willingly assumes all risks in connection with such instructions.

The publisher makes no representation or warranties of any kind, including but not limited to, the warranties of fitness for particular purpose or merchantability, nor are any such representations implied with respect to the material set forth herein, and the publisher takes no responsibility with respect to such material. The publisher shall not be liable for any special, consequential, or exemplary damages resulting, in whole or part, from the readers' use of, or reliance upon, this material.

EDITORIAL POLICY

Comprehensive Information

Every effort is made to gather current data from the Original Equipment Manufacturers (OEMs) when they publish it. Different OEMs choose to release their information at different times of the year. As a result, some models are more current than others when each edition of this guide is published.

Although information in this guide is based on industry sources and is as complete as possible at the time of publication, some vehicle manufacturers may make changes which cannot be included here. While striving for total accuracy, the publisher cannot assume responsibility for any errors, changes, or omissions that may occur in the compilation of this data.

Safety Notice

Proper service and repair procedures are vital to the safe, reliable operation of all motor vehicles, as well as the personal safety of those performing the repairs. Standard safety procedures should be followed at all times to eliminate the possibility of personal injury or improper service that could damage the vehicle or compromise its safety.

Repair procedures, tools, parts, and technician skill and experience vary widely. It is not possible to anticipate all conceivable ways or conditions under which vehicles may be serviced, or to provide cautions for all possible hazards that may result. Standard and accepted safety precautions and equipment should be used when handling toxic or flammable substances, and safety goggles or other protection should be used during any process that may cause sparking, material removal or projectiles. Some procedures require the use of tools specially designed for a specific purpose. Before substituting another tool or procedure you must be completely satisfied that neither your personal safety nor the performance of the vehicle will be endangered.

LOCATING AND USING THE INFORMATION

Locating Information

To find where a particular model-specific section is located, look in the Table of Contents or on the Tab Index preceding each major manufacturer section. Once you have found the vehicle model you are looking for, refer to the System Index and Operations Index.

Overlapping Labor Times

Labor times shown, unless noted otherwise, consider all necessary steps required in performing the operation. However, when multiple operations are performed on a vehicle, and have overlapping labor times, a reasonable deduction should be made from the total times given. Your judgment is the best guide in these situations.

This guide helps professional technicians and service writers compute fair labor times, but it is not intended to be unconditional. There are many variations between the way different service providers are equipped, or perform the same job. Labor times provided are averages based on a typical blend of technician skills and experience found in the aftermarket.

Severe Service Times

Extra labor time should be considered when the vehicle exhibits advanced age, high mileage, lack of maintenance, abuse, or has operated in a climate that creates extra effort needed to free up seized bolts, replace broken fasteners, remove rust, degrease and clean parts, etc. Additional time should also be added when more than normal road testing is required. Suggested Severe Service Times are displayed for your use as needed.

Technician's Skill Level Codes

(A) Highly Skilled: A technician with multiple certifications, usually a master technician, who has thorough knowledge of operating systems and has strong diagnostic skills.

(B) Skilled: A technician with one or more certification(s) who has knowledge of operating systems and can make accurate diagnosis using test equipment.

(C) Semi-Skilled: An apprentice technician working toward certification.

Acknowledgements & Copyright Notice

No part of this product may be reproduced, transmitted or stored in any form or by any means, electronic or mechanical, including photocopying, recording, or by information storage or retrieval system, without prior written permission from the publisher.

TABLE OF CONTENTS

VOLUME 2—IMPORTED VEHICLES

ASIAN VEHICLES   **PAGE**

ACURA:
CL : Integra : Legend : NSX : RL : RSX : TL : TSX : VIGOR ACU-1
MDX : SLX ACU-19

DAEWOO:
Lanos : Leganza : Nubira DAE-1

DAIHATSU:
Charade : Rocky DAI-1

HONDA:
Accord : Civic : Civic del Sol : CRX : Insight : Prelude : S2000 .... HON-1
CR-V : Element : Odyssey : Passport : Pilot : Ridgeline HON-23

HYUNDAI:
Accent : Elantra : Excel : Pony : Santa Fe : Scoupe : Sonata : Stellar : Tiburon : Tuscon :
   XG300 : XG350 .... HYU-1

INFINITI:
G20 : G35 : I30 : I35 : J30 : M30 : M45 : Q45 .... INF-1
FX35 : FX45 : QX4 : QX56 .... INF-29

ISUZU:
I-Mark : Impulse : Stylus ISU-1
Amigo : Ascender : Axiom : Hombre : Oasis : Pickup : Rodeo : Rodeo Sport : Trooper :
   VehiCROSS ISU-15

KIA:
Amanti : Optima : Rio : Sephia : Spectra KIA-1
Sedona : Sorento : Sportage KIA-11

LEXUS:
ES250 : ES300 : ES330 : GS300 : GS400 : GS430 : IS300 : LS400 : LS430 : SC300 : SC400 : SC430 .... LEX-1
GX470 : LX450 : LX470 : RX300 : RX330 .... LEX-23

MAZDA:
323 : 626 : 929 : GLC : Mazda3 : Mazda6 : Miata : Millenia : MX-3 : MX-6 : Protege : Protege5 :
   RX-7 : RX-8 .... MAZ-1
B-Series : MPV : Navajo : Tribute MAZ-49

MITSUBISHI:
3000GT : Cordia : Diamante : Eclipse : Galant : Lancer : Lancer Evolution : Mirage : Precis :
   Sigma : Starion : Tredia MITSU-1
Endeavor : Expo : Expo LRV : Mighty Max : Montero : Montero Sport : Outlander : Van MITSU-35

NISSAN:
200SX : 210 : 240SX : 280ZX : 300ZX : 310 : 350Z : 510 : 810 : Altima : Maxima : NX : Pulsar :
   Pulsar NX : Sentra : Stanza NI-1
720 Pickup : Armada : Axxess : D21 Pickup : Frontier : Murano : Pathfinder : Pickup : Quest :
   Titan : Van : Xterra NI-47

SCION:
TC : XA : XB SC-1

SUBARU:
Baja : Brat : DL : Forester : GL : GL-10 : GLF : Impreza : Justy : Legacy : Loyale : Outback :
   RX : Standard : SVX : XT : XT6 .... SUB-1

SUZUKI:
Aerio : Esteem : Forenza : Grand Vitara : Samurai : Sidekick : Swift : Verona : Vitara : X-90 : XL-7 ...... SUZ-1

TOYOTA:
Avalon : Camry : Celica : Corolla : Cressida : Echo : Matrix : MR2 : MR2 Spyder : Paseo : Prius :
   Solara : Starlet : Supra : Tercel TOY-1
4Runner : Highlander : Land Cruiser : Pickup : Prerunner : Previa : RAV4 : Sequoia : Sienna :
   T100 : Tacoma : Tundra : Van : Van Wagon TOY-57

EUROPEAN VEHICLES	PAGE
ALFA ROMEO:	
164 : Milano	AR-1
AUDI:	
80 : 90 : 100 : 200 : 4000 : 5000 : A4 : A6 : A8 : Cabriolet : Coupe : S4 : S6 : TT : V8 Quattro	AUD-1
BMW:	
3 Series : 524TD : 5 Series : 6 Series : 7 Series : 8 Series : Alpina Roadster : M3 : M5 : X3 : X5 : Z3 : Z4 : Z8	BMW-1
JAGUAR:	
S-Type : Super V8 : Vanden Plas : XJ12 : XJ6 : XJ8 : XJR : XJS : XK8 : XKR : X-Type	JAG-1
LAND ROVER:	
Defender 90 : Discovery : Freelander : LR3 : Range Rover	LR-1
MERCEDES BENZ:	
190 Series : 240D : 260E : 280 Series : 300 Series : 380 Series : 400 Series : 420SEL : 500 Series : 560 Series : 600 Series : C Series : CL Series : CLK Series : E Series : S Series : SL Series : SLK Series	MB-1
G Series : ML Series	MB-35
MINI:	
Cooper	MIN-1
PEUGEOT:	
405 : 504 : 505 : 604	PEU-1
BOXSTER:	
911 : Boxster	POR-1
924 : 928 : 944 : 968	POR-17
RENAULT:	
Alliance : Encore : Fuego : Lecar : R181	REN-1
SAAB:	
900 : 9000 : 9-2X : 9-3 : 9-5	SAA-1
STERLING:	
825 : 827	STR-1
VOLKSWAGEN:	
Beetle : Cabrio : Cabriolet : Corrado : Dasher : Fox : Golf : Jetta : Passat : Quantum : Rabbit : Rabbit Convertible : Rabbit Pickup : Scirocco	VW-1
Eurovan : Vanagon	VW-29
VOLVO:	
240 : 242 : 244 : 245 : 262 : 264 : 265 : 740 : 745 : 760 : 780 : 850 : 940 : 960 : C70 : S40 : S60 : S70 : S80 : S90 : V40 : V70 : V90 : XC70 : XC90	VLV-1
YUGO:	
Cabrio : GV : GVL : GVS : GVX	YUG-1

ASIAN VEHICLES

Acura ACU-1
Daewoo DAE-1
Daihatsu DAI-1
Honda HON-1
Hyundai HYU-1
Infiniti INF-1
Isuzu ISU-1
Kia...KIA-1
Lexus LEX-1
Mazda MAZ-1
Mitsubishi MITSU-1
Nissan NI-1
Scion SC-1
Subaru SUB-1
Suzuki...SUZ-1
Toyota...TOY-1

ASIAN VEHICLES

Acura	ACU-1	Lexus	LEX-1
Daewoo	DAE-1	Mazda	MAZ-1
Daihatsu	DAI-1	Mitsubishi	MITSU-1
Honda	HON-1	Nissan	NI-1
Hyundai	HYU-1	Scion	SC-1
Infiniti	INF-1	Subaru	SUB-1
Isuzu	ISU-1	Suzuki	SUZ-1
Kia	KIA-1	Toyota	TOY-1

ACU

CL : Integra : Legend : NSX : RL : RSX : TL : TSX : VIGOR

SYSTEM INDEX

MAINTENANCE	ACU-2
Maintenance Schedule	ACU-2
CHARGING	ACU-2
STARTING	ACU-2
CRUISE CONTROL	ACU-3
IGNITION	ACU-3
EMISSIONS	ACU-3
EXHAUST	ACU-5
ENGINE COOLING	ACU-5
ENGINE	ACU-6
Assembly	ACU-6
Cylinder Head	ACU-7
Camshaft	ACU-8
Crank & Pistons	ACU-8
Engine Lubrication	ACU-9
CLUTCH	ACU-10
MANUAL TRANSAXLE	ACU-10
AUTO TRANSAXLE	ACU-10
SHIFT LINKAGE	ACU-11
DRIVELINE	ACU-11
BRAKES	ACU-12
FRONT SUSPENSION	ACU-13
REAR SUSPENSION	ACU-13
STEERING	ACU-14
HEATING & AC	ACU-15
WIPERS & SPEEDOMETER	ACU-16
LAMPS & SWITCHES	ACU-16
BODY	ACU-17

OPERATIONS INDEX

A
- AC Hoses ACU-15
- Air Bags ACU-14
- Air Conditioning ACU-15
- Alignment ACU-13
- Alternator (Generator) ACU-2
- Antenna ACU-16
- Anti-Lock Brakes ACU-12

B
- Back-Up Lamp Switch ACU-16
- Ball Joint ACU-13
- Battery Cables ACU-3
- Bleed Brake System ACU-12
- Blower Motor ACU-16
- Brake Disc ACU-12
- Brake Hose ACU-12
- Brake Pads and/or Shoes ACU-12

C
- Camshaft ACU-8
- Catalytic Converter ACU-5
- Coolant Temperature (ECT) Sensor ... ACU-4
- Crankshaft ACU-8
- Crankshaft Sensor ACU-3
- Cruise Control ACU-3
- CV Joint ACU-12
- Cylinder Head ACU-7

D
- Differential ACU-11
- Distributor ACU-3
- Drive Belt ACU-2

E
- EGR ACU-4
- Electronic Control Module (ECM/PCM) ... ACU-4
- Engine ACU-6
- Engine Lubrication ACU-9
- Engine Mounts ACU-6
- Evaporator ACU-15
- Exhaust ACU-5
- Exhaust Manifold ACU-5
- Expansion Valve ACU-16

F
- Flywheel ACU-10
- Fuel Vapor Canister ACU-4

G
- Gear Selector Lever ACU-11
- Generator ACU-2

H
- Halfshaft ACU-11
- Headlamp ACU-16
- Heater Core ACU-16
- Horn ACU-16

I
- Idle Air Control (IAC) Valve ... ACU-5
- Ignition Coil ACU-3
- Ignition Module ACU-3
- Ignition Switch ACU-3
- Inner Tie Rod ACU-15
- Intake Air Temperature (IAT) Sensor ... ACU-4

K
- Knock Sensor ACU-4

L
- Lower Control Arm ACU-13

M
- Maintenance Schedule ACU-2
- Manifold Absolute Pressure (MAP) Sensor ... ACU-4
- Master Cylinder ACU-12
- Muffler ACU-5

N
- Neutral Safety Clutch Switch ... ACU-16

O
- Oil Pan ACU-9
- Oil Pump ACU-9
- Outer Tie Rod ACU-15
- Oxygen Sensor ACU-4

P
- Parking Brake ACU-12
- Pistons ACU-9
- Positive Crankcase Ventilation (PCV) Valve ... ACU-4

R
- Radiator ACU-5
- Radiator Hoses ACU-5
- Radio ACU-16
- Rear Main Oil Seal ACU-9

S
- Shock Absorber/Strut, Front ... ACU-13
- Shock Absorber/Strut, Rear ... ACU-14
- Spark Plug Cables ACU-3
- Spark Plugs ACU-3
- Spring, Front Coil ACU-13
- Spring, Rear Coil ACU-13
- Starter ACU-3
- Steering Wheel ACU-15

T
- Thermostat ACU-5
- Throttle Body ACU-5
- Throttle Position Sensor (TPS) ... ACU-4
- Timing Belt ACU-8
- Timing Chain ACU-8
- Torque Converter ACU-10

U
- Upper Control Arm ACU-13

V
- Valve Body ACU-10
- Valve Cover Gasket ACU-7
- Valve Job ACU-8

W
- Water Pump ACU-6
- Wheel Balance ACU-2
- Window Regulator ACU-17
- Windshield Washer Pump ACU-16
- Windshield Wiper Motor ACU-16

ACU-1

ACU-2 CL : INTEGRA : LEGEND : NSX : RL : RSX : TL : TSX : VIGOR

	LABOR TIME	SEVERE SERVICE
MAINTENANCE		
Air Cleaner Filter Element, Replace (C)		
1986-05 (.2)	.3	.3
Chassis Lubrication, Change Oil & Filter (C)		
Includes: Correct fluid levels.		
1986-05 (.3)	.5	.5
Drive Belt Replace (B)		
Air conditioning (.5)	.8	.9
Alternator		
exc. below (.3)	.5	.6
1990-01 Integra (.4)	.6	.7
1986-90 Legend (.5)	.7	.8
Power steering		
exc. below (.5)	.8	.9
2.2CL, 2.3CL, RSX (.3)	.5	.6
3.0CL	.6	.7
Integra		
1986-89 (1.0)	1.4	1.5
1990-01 (.2)	.3	.3
Fuel Filter, Replace (B)		
Exc. below (.3)	.5	.5
2.2CL, 2.3CL (.6)	.9	1.0
RSX, TSX (.4)	.6	.7
1999-05 TL (.6)	1.0	1.1
Halogen Headlamp Bulb, Replace (B)		
Exc. below (.2)	.3	.3
NSX each (.4)	.6	.6
2005 RL each (.4)	.6	.6
2004-05 TL each (.5)	.7	.7
Oil & Filter, Change (C)		
Includes: Correct fluid levels.		
1986-05 (.3)	.5	.5
Timing Belt, Replace (B)		
2.2CL, 2.3CL, 2.5TL (2.3)	3.3	3.7
3.0CL, 3.2CL, 3.2TL (3.0)	4.3	4.6
Integra (2.0)	2.9	3.2
Legend (3.0)	4.3	4.6
NSX (3.7)	5.2	5.5
RL		
1996-04 (2.5)	3.7	4.0
2005 (2.1)	3.0	3.3
Vigor (2.3)	3.5	4.0
Replace		
cam seal each add	.5	.5
water pump add	.6	.7
Tire, Replace (C)		
Includes: Dismount old tire and mount new tire to rim.		
Each (.4)	.6	.7
Tires, Rotate (C)		
1986-05 (.4)	.6	.7
Wheel, Balance (B)		
Each (.3)	.5	.5

	LABOR TIME	SEVERE SERVICE
SCHEDULED MAINTENANCE INTERVALS		
If necessary, refer to appropriate Chilton maintenance service information.		
7,500 Mile Service (C)		
All Models	.9	.9
Inspect front brake pads		
NSX add	.1	.1
15,000 Mile Service (B)		
2.2CL, 3.0CL	2.1	2.1
2.5TL	2.2	2.2
Integra, NSX	1.8	1.8
1996-04 RL	1.8	1.8
22,500 Mile Service (C)		
All Models	.9	.9
Inspect front brake pads		
NSX add	.1	.1
30,000 Mile Service (B)		
2.2CL, 3.0CL	3.1	3.1
2.5TL	4.3	4.3
3.2CL, 3.2CL, 3.2TL	2.9	2.9
Integra	3.9	3.9
NSX	3.5	3.5
1996-04 RL	3.1	3.1
w/AT add	.5	.5
37,500 Mile Service (C)		
All Models	.9	.9
Inspect front brake pads		
NSX add	.1	.1
45,000 Mile Service (B)		
2.2CL, 3.0CL, 3.2CL, 3.2TL	2.8	2.8
2.5TL	2.5	2.5
Integra	2.4	2.4
NSX	2.9	2.9
1996-04 RL	2.5	2.5
52,500 Mile Service (C)		
All Models	.9	.9
Inspect front brake pads		
NSX add	.1	.1
60,000 Mile Service (B)		
2.2CL, 3.0CL	5.1	5.1
2.5TL	5.6	5.6
3.2CL, 3.2TL, Integra	5.4	5.4
NSX	5.7	5.7
1996-04 RL	5.1	5.1
w/AT add	.5	.5
67,500 Mile Service (C)		
All Models	.9	.9
Inspect front brake pads		
NSX add	.1	.1
75,000 Mile Service (B)		
2.2CL, 3.0CL, 3.2CL, 3.2TL	2.8	2.8
2.5TL	2.5	2.5
Integra	2.4	2.4
NSX	2.9	2.9
1996-04 RL	2.5	2.5

	LABOR TIME	SEVERE SERVICE
82,500 Mile Service (C)		
All Models	.9	.9
Inspect front brake pads		
NSX add	.1	.1
90,000 Mile Service (B)		
2.2CL, 3.0CL	7.2	7.2
2.5TL	8.5	8.5
3.2CL, 3.2TL	7.4	7.4
Integra	7.6	7.6
NSX	7.1	7.1
1996-04 RL	4.6	4.6
w/AT add	.5	.5
97,500 Mile Service (C)		
All Models	.9	.9
Inspect front brake pads		
NSX add	.1	.1
CHARGING		
Alternator Circuits, Test (B)		
Includes: Test component output.		
1986-05 (.3)	.6	.6
Alternator, R&R and Recondition (B)		
Includes: Complete disassembly, inspect, test, replace parts as required, reassemble.		
Exc. below (1.5)	2.2	2.5
2.2CL, 2.3CL, NSX (1.1)	1.6	1.8
RSX (2.1)	3.0	3.2
Alternator Assy., Replace (B)		
Includes: Pulley transfer.		
2.2CL, 2.3CL, NSX (.5)	.8	.9
2.5TL, Vigor (.6)	.9	1.0
3.0CL, 3.2CL (.7)	1.1	1.2
3.2TL		
1996-03 (.9)	1.2	1.3
2004-05 (.5)	.8	.9
Integra, Legend (.9)	1.4	1.5
RL (1.1)	1.6	1.7
RSX (1.5)	2.2	2.3
Alternator Voltage Regulator, Replace (B)		
2.2CL, 2.3CL, NSX (.7)	1.1	1.2
2.5TL, Vigor (.8)	1.2	1.3
3.0CL, 3.2CL (.9)	1.3	1.4
3.2TL, Integra (1.1)	1.6	1.7
Legend		
1986-90 (1.5)	2.2	2.3
1991-95 (1.1)	1.7	1.8
1996-04 RL (1.1)	1.6	1.7
RSX (1.6)	2.3	2.4
STARTING		
Starter Draw Test (On Car) (B)		
1986-05 (.3)	.6	.6
Battery, Replace (B)		
Includes: Testing battery.		
1986-05 (.4)	.6	.7

	LABOR TIME	SEVERE SERVICE
Battery Cables, Replace (C)		
Exc. below		
positive (.4)	.6	.7
negative (.2)	.3	.4
NSX		
positive (1.5)	2.2	2.3
negative (.5)	.8	.9
RSX		
positive (1.2)	1.7	1.8
negative (.2)	.3	.4
Starter, R&R and Recondition (B)		
Includes: Turn down armature.		
Exc. below (1.4)	2.1	2.3
1996-98 3.2TL (3.0) ..	4.3	4.5
1990-93 Integra (1.8) ..	2.7	2.9
1991-95 Legend (3.0) ..	4.4	4.6
1996-04 RL (4.0)	5.8	6.0
RSX, TSX (2.0)	3.0	3.2
Starter Assy., Replace (B)		
Add draw test if performed.		
Exc. below (.7)	1.1	1.3
2.5TL, Vigor (.9)	1.3	1.5
1996-98 3.2TL (2.5) ..	3.7	3.9
1991-95 Legend (2.5) .	3.7	3.9
RL		
1996-04 (3.5)	5.1	5.3
2005 (.5)	.8	1.0
RSX, TSX (1.5)	2.2	2.4

CRUISE CONTROL

	LABOR TIME	SEVERE SERVICE
Diagnose Cruise System Component Each (A)		
1986-05 (.6)	.9	.9
Control Actuator Cable, Adjust (B)		
1986-05 (.3)	.5	.5
Actuator Assy., Replace (B)		
1986-05 (.3)	.5	.5
Control Actuator Valve, Replace (B)		
1986-05 (.3)	.5	.5
Control Check Valve, Replace (B)		
1986-04 (.2)	.3	.3
Control Controller (Module), Replace (B)		
1986-04		
exc. NSX (.3)	.5	.5
NSX (.6)	.8	.8
Control Sensor, Replace (B)		
1986-05 (.5)	.8	.9
Control Switch, Replace (B)		
Clutch or brake		
exc. below (.3)	.5	.5
Legend coupe (.5)	.8	.8
Steering wheel		
exc. below (.2)	.3	.3
1999-00 RL (.5)	.8	.8

IGNITION

	LABOR TIME	SEVERE SERVICE
Diagnose Ignition System Component Each (A)		
1986-05 (.6)	.8	.8
Retrieve Fault Codes (A)		
1986-05 (.3)	.4	.4
Ignition Timing, Reset (B)		
1986-05	.6	.6
Crankshaft Position (CKP), Cylinder Position, Top Dead Center (TDC) or Crankshaft Speed Fluctuation (CKF) Sensors, Replace (B)		
Distributor mounted ...	1.1	1.2
Engine or cam cover mounted		
2.2CL, 2.3CL	3.8	4.3
2.5TL, Vigor	2.9	3.3
3.0CL, 3.2CL	4.6	5.2
3.2TL		
1996-98	3.5	4.0
1999-03	4.6	5.2
Integra		
1986-93	.6	.7
1994-01	2.3	2.6
Legend		
1986-90	4.6	5.2
1991-95	3.5	4.0
NSX	5.6	6.4
1996-04 RL	3.5	4.0
RSX	.3	.3
Distributor, Replace (B)		
Includes: Reset ignition timing.		
Exc. below (.3)	.5	.6
2.5TL, Vigor (.5)	.7	.8
1986-90 Legend (.7) ..	1.0	1.1
Distributor Cap and/or Rotor, Replace (B)		
Exc. below (.2)	.3	.3
2.5TL, Vigor (.5)	.7	.7
Electronic Ignition Control Unit, Replace (B)		
2.2CL, 2.3CL,		
3.0CL (.3)	.5	.5
2.5TL (.6)	.8	.8
3.2CL, 3.2TL		
1996-98 (.6)	.8	.8
1999-03 (.4)	.6	.6
Integra, NSX (.3)	.5	.5
Legend (.4)	.6	.6
1996-04 RL	.6	.6
RSX (.2)	.3	.3
Vigor (.6)	.8	.8
w/Legend coupe add	.2	.2
Igniter Unit, Replace (B)		
Exc. below (.2)	.3	.3
2.2CL, 2.3CL, 3.0CL,		
Integra (.4)	.6	.6

	LABOR TIME	SEVERE SERVICE
Ignition Coil, Replace (B)		
Exc. below (.2)	.3	.3
2.2CL, 2.3CL,		
Integra (.4)	.6	.6
2.5TL, Vigor (.3)	.5	.5
1986-90 Legend (.5)	.7	.7
Ignition Switch, Replace (B)		
Exc. below	1.2	1.2
1991-95 Legend	1.5	1.5
1996-04 RL	1.5	1.5
Ignition Switch (Electrical Portion), Replace (B)		
Exc. below	.6	.6
2.2CL, 2.3CL, 2.5TL,		
3.0CL	1.2	1.2
1996-98 3.2TL	1.2	1.2
1986-93 Integra	1.0	1.0
1986-90 Legend	1.0	1.0
2005 RL (.7)	1.1	1.1
Pulse Generator, Replace (B)		
CL	.8	1.0
Integra	1.8	2.0
Legend		
1986-90	1.8	2.0
1991-95	.6	.8
1996-04 RL	.6	.8
1995-03 TL	.6	.8
Vigor	.5	.7
Spark Plug (Ignition) Cables, Replace (B)		
Exc. below	.3	.4
1986-90 Legend	.7	.8
Spark Plugs, Replace (B)		
Exc. below	.8	.9
2.2CL, 2.3CL,		
Integra, RSX	.5	.6
2.5TL, Vigor	.6	.7
NSX	1.2	1.4
Vacuum Advance Unit, Replace (B)		
Includes: Reset timing.		
1986-88		
Integra	.8	.9
Legend	1.3	1.4

EMISSIONS

The following operations do not include testing. Add time as required.

	LABOR TIME	SEVERE SERVICE
Diagnose Emission Control System Component Each (A)		
1986-05 (.6)	.9	.9
Reprogram Electronic Control Module (ECM) (A)		
All Models (.3)	.5	.5
Retrieve Fault Codes (A)		
All Models (.3)	.5	.5

ACU-4   CL : INTEGRA : LEGEND : NSX : RL : RSX : TL : TSX : VIGOR

	LABOR TIME	SEVERE SERVICE
Barometric Manifold Absolute Pressure Sensor, Replace (B)		
Exc. below	.3	.3
1996-98 3.2TL	.6	.6
1986-89 Integra	.7	.7
Legend	.7	.7
1996-04 RL	.6	.6
Barometric Pressure Sensor, Replace (B)		
2.2CL, 2.3CL, 3.0CL	.5	.5
2.5TL	.8	.8
3.2CL, 3.2TL	.7	.7
1986-96 Integra	.5	.6
1986-90 Legend	.5	.5
Vigor	.8	.8
Bypass Control Solenoid Valve, Replace (B)		
Exc. below	.3	.3
1999-03 3.2CL, 3.2TL	.6	.6
Integra, NSX, RSX	.6	.6
Coolant Temperature (ECT) Sensor, Replace (B)		
1986-05 (.4)	.6	.6
EGR Control Solenoid Valve, Replace (B)		
Exc. below	.3	.3
3.2TL, NSX	.6	.6
1990-93 Integra	.6	.6
1991-95 Legend	.6	.6
1996-04 RL	.6	.6
EGR Valve, Replace (B)		
Exc. below	.5	.7
1986-90 Legend	.9	1.0
Electronic Control Module (ECM/PCM), Replace (B)		
Exc. below	.5	.5
2.5TL, Vigor	.8	.8
3.2CL, Legend	.6	.6
3.2TL		
1996-98	.8	.8
1999-03 (.4)	.6	.6
RL		
1996-04	.6	.6
2005 (.6)	.8	.8
Fuel Vapor Canister, Replace (B)		
Exc. below	.5	.5
2.5TL	2.3	2.5
3.2CL, RSX	.5	.5
3.2TL		
1996-98	2.3	2.5
1999-03 (.6)	.9	1.0
RL		
1996-99	2.3	2.5
2000-04 (.6)	.9	1.0
2005 (1.9)	2.7	2.9
TSX (.6)	.9	.9
Intake Air Temperature (IAT) Sensor, Replace (B)		
1986-05 (.2)	.3	.3

	LABOR TIME	SEVERE SERVICE
Knock Sensor, Replace (B)		
2.2CL, 2.3CL, RSX	.6	.8
2.5TL, Integra, Vigor	.9	1.1
3.2CL, 3.2TL		
1996-98	4.3	4.5
1999-03	1.1	1.3
Legend	4.3	4.5
NSX (1.3)	1.9	2.1
RL		
1996-04	4.3	4.5
2005 (1.5)	2.2	2.4
Oxygen Sensor, Replace (B)		
1986-05 (.3)	.5	.5
Replace front sensor 2.5TL add	.8	.8
Positive Crankcase Ventilation (PCV) Valve, Replace (B)		
Exc. below	.3	.3
1996-98 3.2TL	.6	.6
Legend	.6	.6
1996-04 RL	.6	.6
Purge Control Solenoid Valve, Replace (B)		
Exc. below	.3	.3
2.2CL, 2.3CL, 3.0CL	.5	.5
NSX, RSX	.5	.5
Throttle Position Sensor (TPS), Replace (B)		
Exc. below	.7	.7
2.5TL, Vigor	1.0	1.0
1996-98 3.2TL	1.0	1.0
Legend		
1986-90	1.4	1.4
1991-95	1.0	1.0
NSX	1.2	1.2
1996-04 RL	1.0	1.0
Timing Adjuster, Replace (B)		
1998-03 3.2TL (.4)	.6	.8
Legend, NSX	.6	.8
Vigor	.3	.5
Two-Way Valve, Replace (B)		
Exc. below	.6	.7
2.5TL, NSX, Vigor	1.8	1.9
1996-98 3.2TL	1.6	1.7
Integra		
1986-89	1.6	1.7
1990-93	2.7	2.8
Legend	1.6	1.7
RSX, TSX (.4)	.6	.7
Vapor Separator, Replace (B)		
Integra	1.6	1.6
Legend	1.6	1.6

FUEL
DELIVERY

	LABOR TIME	SEVERE SERVICE
Fuel Filter, Replace (B)		
Exc. below (.3)	.5	.5
2.2CL, 2.3CL (.6)	.9	1.0
RSX, TSX (.4)	.6	.7
1999-05 TL (.6)	1.0	1.1

	LABOR TIME	SEVERE SERVICE
Fuel Gauge (Dash), Replace (B)		
Exc. below	1.4	1.4
1996-04 RL	1.9	1.9
Vigor	1.9	1.9
Fuel Gauge (Tank), Replace (B)		
Exc. below	1.0	1.1
1986-89 Integra	1.4	1.5
RSX, TSX (.4)	.6	.7
2004-05 TL (.4)	.6	.7
Fuel Pump, Replace (B)		
Includes: testing		
2.2CL, 2.3CL, 3.0CL	1.6	1.8
2.5TL, Vigor	1.6	1.8
3.2CL, 3.2TL		
1996-98	1.6	1.8
1999-03	1.0	1.1
Integra		
1986-93	1.6	1.8
1994-01	1.0	1.1
Legend		
1986-90	1.6	1.8
1991-95	1.0	1.1
NSX	2.1	2.3
1996-04 RL	1.0	1.1
RSX, TL, TSX, (.4)	.6	.7
Fuel Tank, Replace (B)		
Includes: Drain and refill.		
Exc. below	1.3	1.5
3.2CL	4.0	4.2
NSX, RSX	1.9	2.1
TL		
1999-03	4.0	4.2
2004-05 (3.0)	4.4	4.6
TSX (3.0)	4.4	4.6
Fuel Tank Pressure Sensor, Replace (B)		
Exc. below	.3	.3
Integra, NSX, RSX	.6	.6
Intake Manifold and/or Gasket, Replace (B)		
Includes: Adjustments.		
2.2CL, 2.3CL, Integra	3.4	3.6
2.5TL, Vigor	3.5	3.7
3.0CL	.8	1.0
3.2CL, 3.2TL		
1996-98	3.6	3.8
1999-03	.8	1.0
Legend		
1986-90	2.7	2.9
1991-95	3.6	3.8
NSX, RSX	1.9	2.1
1996-04 RL	3.6	3.8
TSX (1.0)	1.4	1.6
Replace injector plate RSX add	.3	.3

INJECTION

	LABOR TIME	SEVERE SERVICE
Idle Speed Adjust, Replace (B)		
1986-05 (.2)	.3	.3

CL : INTEGRA : LEGEND : NSX : RL : RSX : TL : TSX : VIGOR **ACU-5**

	LABOR TIME	SEVERE SERVICE
Fast Idle Valve, Replace (B)		
1986-05 (.4) 6		.6
Fuel Injection Relay, Replace (B)		
1986-05 (.4) 6		.6
Fuel Injector Resistor, Replace (B)		
1986-05 (.2) 3		.3
Fuel Injectors, Replace (B)		
Exc. below 9		1.0
3.0CL, 3.2CL 2.3		2.6
1986-90 Legend 1.1		1.2
TL		
1999-03 .... 2.3		2.6
2004-05 (.9) 1.3		1.4
w/second bank add		
96-98 3.2TL 4		.5
Legend		
86-90 .... 1.0		1.2
91-95 4		.5
NSX 4		.5
1996-04 RL 4		.5
Fuel Pressure Regulator, Replace (B)		
Exc. below 5		.6
3.0CL 1.4		1.5
2005 RL (.7) 1.1		1.2
RSX TSX (.4) 6		.7
TL		
1999-03 (1.2) 1.7		1.8
2004-05 (.4) 6		.7
Idle Air Control (IAC) Valve, Replace (B)		
Exc. below 7		.8
1999-03 3.2TL 1.2		1.4
CL, RSX 1.2		1.4
Throttle Body Assy., Replace (B)		
Exc. below 8		.9
2.5TL, 3.2TL, Vigor ... 1.1		1.2
Legend		
1986-90 .... 1.4		1.6
1991-95 .... 1.1		1.2
NSX 1.3		1.5
RL		
1996-04 .... 1.1		1.2
2005 (1.0) 1.4		1.6

EXHAUST

	LABOR TIME	SEVERE SERVICE
Catalytic Converter, Replace (B)		
Exc. below 8		.9
2004-05 TL (1.1) 1.6		1.7
Exhaust Extension Pipe (Tailpipe), Replace (B)		
1986-05 (.2) 3		.4
Exhaust Manifold or Gasket, Replace (B)		
2.2CL, 2.3CL 8		1.1
2.5TL, Vigor 2.2		2.5
3.0CL		
one side 1.9		2.2
both sides 3.5		4.0

	LABOR TIME	SEVERE SERVICE
3.2CL, 3.2TL		
1996-98		
one side 2.9		3.2
both sides 5.3		5.9
1999-03		
one side 1.9		2.2
both sides 3.5		4.0
Integra		
1986-89 8		1.1
1990-01 .... 1.1		1.4
Legend		
one side 2.9		3.2
both sides 5.3		5.9
NSX		
front 2.2		2.5
rear 1.6		1.8
1996-04 RL		
one side 2.9		3.2
both sides 5.3		5.9
RSX, TSX (.4) 6		.7
w/6-spd MT w/1 pc. front		
exh man NSX front add 1.8		1.8
Exhaust Pipe or Crossover Pipe Flange Gasket or Seal, Replace (B)		
1986-05 one (.6) 1.0		1.1
Front Exhaust Pipe, Replace (B)		
Exc. NSX (.6) 1.0		1.1
NSX (1.4) 2.2		2.5
Rear Exhaust Pipe, Replace (B)		
1986-05 (.6) 1.0		1.1
Muffler, Replace (B)		
1986-05 (.5) 8		.9

ENGINE COOLING

	LABOR TIME	SEVERE SERVICE
Pressure Test Cooling System (C)		
1986-05 3		.3
AC Condenser Cooling Fan Relay, Replace (B)		
1986-05 (.2) 3		.5
Bypass Hoses, Replace (B)		
1986-05 one (.4) 6		.8
Coolant Thermostat, Replace (B)		
2.2CL, 2.3CL, Integra 8		.9
2.5TL, Vigor 5		.6
3.0CL 6		.7
3.2CL, 3.2TL, Legend,		
RSX, TSX (.8) 1.2		1.4
NSX 1.4		1.7
1996-04 RL (.8) 1.2		1.4
Cooling Fan Switch, Replace (B)		
Exc. NSX 7		.7
NSX 1.2		1.2
Cooling Fan Relay, Replace (B)		
Exc. below 3		.3
2005 RL (.3) 4		.4

	LABOR TIME	SEVERE SERVICE
Idler Pulley, Replace (B)		
3.2CL 5		.7
Integra		
1986-89 7		.9
1990-93 9		1.1
1994-99 5		.7
Legend 7		.9
Vigor 5		.7
1996-04 RL 7		.7
1995-03 TL 5		.7
Radiator Assy., R&R or Replace (B)		
Includes: Refill with proper coolant mix.		
2.2CL, 2.3CL,		
Integra, Vigor 2.1		2.4
2.5TL, 3.0CL 2.1		2.4
3.2CL, 3.2TL 2.1		2.4
Legend 2.4		2.7
NSX, RSX, TSX (1.2) . 1.8		2.1
RL		
1996-04 .... 2.1		2.4
2005 (1.8) 2.6		2.9
w/AC Integra only		
86-87 add 7		.7
88-01 add 3		.3
w/AT add 3		.3
Radiator Fan and/or Fan Motor, Replace (B)		
2.2CL, 2.3CL, 3.0CL 8		.9
2.5TL 1.5		1.6
3.2CL, 3.2TL		
1996-98 .... 1.5		1.6
1999-03 8		.9
Integra 1.2		1.3
Legend, RSX, Vigor .. 1.5		1.6
NSX 5		.6
RL		
1996-04 .... 1.5		1.6
2005 (1.2) 1.7		1.8
TSX (.6) 9		1.0
Radiator Hoses, Replace (B)		
Includes: Refill with proper coolant mix.		
Exc. below		
upper or lower 8		.9
both 1.3		1.4
NSX		
front each 8		.9
middle each 1.4		1.5
rear each 1.2		1.4
2005 RL		
upper (.5) 8		.9
lower (.7) 1.0		1.1
Temperature Control Assy., Replace (B)		
Includes: adjust cables when necessary		
Exc. below 1.4		1.4
1990-93 Integra 2.1		2.1
RSX 3		.3
2004-05 TL (.7) 9		1.0
TSX (.5) 6		.6

ACU-6 CL : INTEGRA : LEGEND : NSX : RL : RSX : TL : TSX : VIGOR

	LABOR TIME	SEVERE SERVICE
Temperature Gauge (Dash), Replace (B)		
Exc. below	1.4	*1.4*
1996-04 RL	1.9	*1.9*
Vigor	1.9	*1.9*
Temperature Gauge (Engine), Replace (B)		
Includes: coolant refill and bleed		
1986-05	.6	*.7*
Water Control Valve, Replace (B)		
Exc. below (1.0)	1.4	*1.5*
2005 RL (.5)	.7	*.8*
RSX, TSX (.5)	.7	*.8*
2004-05 TL (.5)	.7	*.8*
Water Pump and/or Gasket, Replace (B)		
Includes: Refill with proper coolant mix.		
2.2CL, 2.3CL	4.0	*4.2*
2.5TL, Vigor	3.5	*3.7*
3.0CL	5.0	*5.2*
3.2CL, 3.2TL		
1996-98	7.6	*7.8*
1999-03	5.0	*5.2*
Integra		
1986-89	3.1	*3.3*
1990-01	3.5	*3.7*
Legend		
1986-90	3.6	*3.8*
1991-95	7.6	*7.8*
NSX	6.2	*6.4*
RL		
1996-04 (5.0)	7.6	*7.8*
2005 (2.4)	3.4	*3.6*
RSX, TSX (2.0)	3.0	*3.2*

ENGINE
ASSEMBLY
Times shown are for OEM assemblies. Time to replace assemblies from aftermarket rebuilder may vary.

Cylinder Block, Replace (B)

	LABOR TIME	SEVERE SERVICE
2.2CL, 2.3CL	19.9	*22.5*
2.5TL, Vigor	22.3	*25.3*
3.0CL	26.9	*30.4*
3.2CL, 3.2TL		
1996-98	27.7	*31.3*
1999-03	26.9	*30.4*
Integra		
1986-89	17.9	*20.3*
1990-93	16.9	*19.1*
1994-01	17.8	*20.1*
Legend	27.7	*31.3*
NSX, RSX (17.5)	22.8	*23.3*
RL		
1996-04 RL	27.7	*31.3*
2005 (17.9)	23.3	*23.8*
TSX (17.1)	22.2	*22.7*

Engine Assy., R&I (B)
Does not include parts or component transfer.

	LABOR TIME	SEVERE SERVICE
2.2CL, 2.3CL	8.3	*8.3*
2.5TL	8.0	*8.0*
3.0CL	10.7	*10.7*
3.2CL	11.3	*11.3*
Integra		
1986-89	6.2	*6.2*
1990-01	8.5	*8.5*
Legend		
1986-90	10.7	*10.7*
1991-95	14.8	*14.8*
1996-04 RL	11.3	*11.3*
1996-03 3.2TL	11.3	*11.3*
Vigor	8.0	*8.0*

Engine Assy., R&R and Recondition (A)
Includes: Replacing rings, rod and main bearings, cylinder head reconditioning and engine tune-up.

	LABOR TIME	SEVERE SERVICE
2.2CL, 2.3CL	23.1	*23.1*
2.5TL	32.0	*32.0*
3.0CL	41.4	*41.4*
3.2CL, 3.2TL		
1996-98	42.2	*42.2*
1999-03	41.4	*41.4*
Integra		
1986-89	26.7	*26.7*
1990-01	25.6	*25.6*
Legend	42.2	*42.2*
1996-04 RL	42.2	*42.2*
Vigor	32.0	*32.0*

Engine Assy. (Complete), Replace (B)
Includes: Component transfer and engine tune-up.

	LABOR TIME	SEVERE SERVICE
2.2CL, 2.3CL	10.8	*12.2*
2.5TL, Vigor	15.7	*17.7*
3.0CL, 3.2CL, NSX (9.0)	12.5	*13.0*
3.2TL		
1996-98	19.3	*21.8*
1999-03	13.7	*15.5*
Integra		
1986-89	8.5	*8.5*
1990-01	10.0	*11.4*
Legend		
1986-90	13.7	*13.7*
1991-95	19.3	*21.8*
RL		
1996-04	19.3	*21.8*
2005 (7.9)	11.1	*11.6*
RSX (8.7)	12.1	*12.6*
TSX (8.4)	11.7	*12.2*

Engine Assy. (Short Block), Replace (B)
Assembly consists of engine block, piston assemblies, crankshaft, camshaft, timing chain and gears. Does not include cylinder heads. Operation includes: R&R engine, transfer necessary parts and all necessary adjustments.

	LABOR TIME	SEVERE SERVICE
2.2CL, 2.3CL	14.6	*16.5*
2.5TL, Vigor	18.7	*21.2*
3.0CL, 3.2CL	22.0	*24.9*
3.2TL		
1996-03	22.3	*25.3*
2004-05 (14.7)	19.3	*20.2*
Integra	14.3	*16.2*
Legend	22.3	*25.3*
NSX	20.5	*23.2*
RL		
1996-04	22.3	*25.3*
2005 (12.3)	16.5	*17.4*
RSX (15.0)	21.8	*22.7*
TSX (14.7)	19.3	*20.2*

Recondition valves add

	LABOR TIME	SEVERE SERVICE
4 cyl.	4.9	*5.5*
5 cyl.	5.8	*6.5*
6 cyl.	7.9	*8.9*

Engine Mounts, Replace (B)

	LABOR TIME	SEVERE SERVICE
2.5TL, Vigor		
front or right side	.8	*.9*
left side	3.0	*3.4*
rear	1.1	*1.3*
trans. side	1.0	*1.1*
all	4.3	*4.9*
3.2TL		
1996-98		
one	.7	*.9*
all	2.4	*2.7*
1999-03		
exc. rear	.8	*.9*
rear	2.7	*3.1*
all	3.2	*3.6*
CL		
exc. rear	.8	*.9*
rear	2.7	*3.1*
all	3.2	*3.6*
Integra		
1986-89 one	.7	*.9*
1990-93		
exc. rear	.7	*.9*
rear	1.5	*1.7*
1994-01		
one	.7	*.9*
all	2.4	*2.7*
Legend, RSX, TSX		
one (.5)	.7	*.9*
all (1.5)	2.2	*2.4*

CL : INTEGRA : LEGEND : NSX : RL : RSX : TL : TSX : VIGOR ACU-7

	LABOR TIME	SEVERE SERVICE
NSX		
exc. left or right side	.8	.9
left or right side	1.3	1.4
all	2.9	3.2
RL		
1996-04		
one (.5)	.7	.9
all (1.5)	2.2	2.4
2005		
front (1.2)	1.8	2.0
left or right side (.3)	.5	.7
rear (1.0)	1.5	1.7
trans. side (4.1)	5.9	6.1

CYLINDER HEAD
Compression Test (B)
2.2CL, 2.3CL	.7	.9
2.5TL	.8	1.0
3.2CL, 3.2TL	1.3	1.5
Integra	.7	.9
Legend	1.3	1.5
RL		
1996-04 (.4)	.5	.7
2005 (.7)	1.1	1.3
Vigor	.8	1.0
Cylinder leak test add	.5	.7

Valve Clearance, Adjust (B)
2.2CL, 2.3CL	1.2	1.4
2.5TL, Vigor	1.5	1.7
3.0CL 3.2CL, 3.2TL	2.9	3.3
Integra, RSX, TSX (.8)	1.2	1.4
NSX (3.1)	4.7	5.3

Camshaft and/or Rocker Arms (One Shaft), Replace (B)
2.2CL, 2.3CL	4.3	4.8
2.5TL, Vigor	4.6	5.2
3.0CL	6.4	7.2
3.2CL, 3.2TL		
1996-98	3.8	4.3
1999-03	6.4	7.2
Integra, RSX, TSX (2.5)	3.7	4.0
Legend		
1986-90	5.0	5.1
1991-95 (2.5)	3.7	4.0
NSX (11.6)	15.7	16.0
RL		
1996-04 (2.5)	3.7	4.0
2005		
one (1.8)	2.6	2.7
all (4.0)	5.6	5.9

Cylinder Head (w/Valves), Replace (B)
Includes: Parts transfer, & valve adjustment.
2.2CL, 2.3CL	7.0	7.9
2.5TL, Vigor	9.9	11.2
3.0CL		
one	9.1	10.3
both	10.6	12.0
3.2CL, 3.2TL		
1996-98		
one	8.5	9.6
both	11.6	13.1
1999-03		
one	9.1	10.3
both	10.6	12.0
2004-05		
one (6.0)	8.6	9.1
both (8.0)	11.2	11.7
Integra		
1986-89	8.5	9.6
1990-93	8.1	9.1
1994-01	6.1	6.9
Legend		
1986-90		
one	9.1	9.6
both	11.8	12.3
1991-95		
one	8.5	9.6
both	11.6	13.1
NSX		
one	16.3	18.4
both	17.8	20.1
RL		
1996-04		
one	8.5	9.6
both	11.6	13.1
2005		
front (5.2)	7.2	7.7
rear (6.1)	8.4	8.9
RSX (6.7)	10.2	10.7
TSX (6.2)	8.8	9.3

Cylinder Head (w/o Valves), Replace (A)
Includes: Parts transfer and adjustments.
2.2CL, 2.3CL	8.2	9.3
2.5TL, Vigor	11.6	13.1
3.0CL	10.7	11.2
one	10.8	12.2
both	13.8	15.7
3.2CL, 3.2TL		
1996-98		
one	10.2	11.5
both	15.7	17.7
1999-03		
one	10.8	12.2
both	13.8	15.7
2004-05		
one (7.1)	10.0	10.5
both (9.9)	13.6	14.1
Integra		
1986-93	9.7	10.2
1994-01	8.7	9.8
Legend		
1986-90		
one	10.7	11.2
both	13.8	14.3
1991-95		
one	10.2	11.5
both	15.7	17.7
NSX		
one	19.3	21.8
both	20.8	23.6
RL		
1996-04		
one	10.2	11.5
both	15.7	17.7
2005		
front (8.8)	11.8	12.7
rear (9.8)	12.9	13.8
both (15.6)	19.5	20.4
RSX (8.4)	11.7	12.2
TSX (7.9)	11.1	11.6

Cylinder Head Gasket, Replace (B)
2.2CL, 2.3CL	6.2	7.1
2.5TL, Vigor	9.1	10.3
3.0CL		
one side	8.4	9.5
both sides	9.9	11.2
3.2CL, 3.2TL		
1996-98		
one side	7.8	8.8
both sides	10.8	12.2
1999-03		
one side	8.4	9.5
both sides	9.9	11.2
2004-05		
one (5.5)	7.9	8.4
both (8.5)	11.8	12.3
Integra	5.8	6.1
Legend		
1986-90		
one side	8.3	8.6
both sides	11.0	11.3
1991-95		
one side	7.8	8.8
both sides	10.8	12.2
NSX		
one side (10.7)	14.6	15.1
both sides (11.7)	15.8	16.3
RL		
1996-04		
one side	7.8	8.8
both sides	10.8	12.2
2005		
front (5.2)	7.2	7.7
rear (6.1)	8.4	8.9
both (8.3)	11.1	11.6
RSX (6.3)	8.9	9.4
TSX (5.8)	8.3	8.8

Valve Cover Gasket, Replace (B)
2.2CL, 2.3CL	.5	.6
2.5TL, Vigor	.8	.9
3.0CL, 3.2CL, 3.2TL		
one	1.2	1.3
both	1.8	2.0

ACU-8   CL : INTEGRA : LEGEND : NSX : RL : RSX : TL : TSX : VIGOR

	LABOR TIME	SEVERE SERVICE
Valve Cover Gasket, Replace (B)		
Integra, RSX, TSX (.3)	.5	.6
Legend		
1986-90 one	.5	.6
1991-95		
one	.9	1.0
both	1.8	2.1
NSX		
one (.6)	.9	1.0
both (1.6)	2.4	2.8
RL		
1996-04		
one (.6)	.9	1.0
both (1.2)	1.8	2.0
2005 one (1.1)	1.6	1.8
w/rear cover and/or gasket NSX add	.9	1.0
Valve Job (A)		
Includes: Valve adjustment.		
2.2CL, 2.3CL	10.6	12.0
2.5TL, Vigor	15.0	17.0
3.0CL	15.7	17.7
3.2CL, 3.2TL		
1996-98	20.1	22.7
1999-03	15.7	17.7
Integra		
1983-93	12.1	12.6
1994-01	9.9	11.2
Legend		
1986-90	18.7	19.2
1991-95	20.1	22.7
NSX (16.0)	23.0	23.9
1996-04 RL (13.2)	17.3	18.2
RSX (9.2)	12.7	13.6
TSX (8.7)	11.7	12.6
Valve Springs and/or Oil Seals, Replace (B)		
Includes: Cylinder head R&R.		
3.0CL		
one head	11.2	11.5
both heads	13.5	13.8
3.2CL, 3.2TL		
1996-98		
one head	10.6	10.9
both heads	15.3	15.6
1999-03		
one head	11.2	11.5
both heads	13.5	13.8
Legend		
1986-90		
one head	12.1	12.3
both heads	15.2	15.4
1991-95		
one head	10.6	10.8
both heads	15.3	15.5
1996-04 RL		
one head	10.6	10.9
both heads	15.3	15.6

	LABOR TIME	SEVERE SERVICE
CAMSHAFT		
Timing Belt, Adjust (B)		
2.2CL, 2.3CL (.7)	1.1	1.2
2.5TL, Vigor (.9)	1.3	1.4
1996-03 3.2TL (.6)	.9	1.0
Integra		
1986-89 (.3)	.5	.6
1990-01 (.7)	1.1	1.2
Legend (.6)	.9	1.0
NSX (1.0)	1.5	1.6
1996-04 RL (.6)	.9	1.0
Camshaft, Replace (B)		
2.2CL, 2.3CL	4.3	4.8
2.5TL, Vigor	4.6	5.2
3.0CL	6.4	7.2
3.2CL, 3.2TL		
1996-98	3.8	4.3
1999-03	6.4	7.2
Integra, RSX, TSX (2.5)	3.7	4.0
Legend		
1986-90	5.0	5.1
1991-95 (2.5)	3.7	4.0
NSX (11.6)	15.7	16.0
RL		
1996-04 (2.5)	3.7	4.0
2005		
one (2.7)	3.8	4.1
all (5.2)	7.2	7.5
Camshaft Seal, Replace (B)		
2.2CL, 2.3CL	1.7	1.8
2.5TL	1.7	1.8
3.0CL		
one	2.3	2.4
both	3.1	3.2
3.2CL, 3.2TL		
1996-98		
one	3.8	3.9
both	6.5	6.6
1999-03		
one	2.3	2.4
both	3.1	3.2
Integra		
1986-89		
one	1.6	1.7
both	1.9	2.0
1990-01		
one	1.3	1.4
both	1.6	1.7
Legend		
1986-90		
one	1.6	1.7
both	1.9	2.0
1991-95		
one	3.8	3.9
both	6.5	6.6
NSX (10.7)	14.0	14.5

	LABOR TIME	SEVERE SERVICE
RL		
1996-04		
one (2.5)	3.8	3.9
both (4.3)	6.5	6.6
2005		
one (1.2)	1.7	1.9
both (1.9)	2.7	2.9
Vigor	1.7	1.8
Timing Belt, Replace (B)		
2.2CL, 2.3CL, 2.5TL (2.3)	3.3	3.7
3.0CL, 3.2CL, 3.2TL (3.0)	4.3	4.6
Integra (2.0)	2.9	3.2
Legend (3.0)	4.3	4.6
NSX (3.7)	5.2	5.5
RL		
1996-04 (2.5)	3.7	4.0
2005 (2.1)	3.0	3.3
Vigor (2.3)	3.5	4.0
Replace		
cam seal each add	.5	.5
water pump add	.6	.7
Timing Chain, Replace (B)		
RSX, TSX (2.8)	4.1	4.4
Timing Chain Tensioner, Replace (B)		
RSX, TSX (.4)	.6	.7
CRANK & PISTONS		
Connecting Rod Bearings, Replace (A)		
Includes: Check all bearing clearances.		
2.2CL, 2.3CL	3.1	3.6
2.5TL, Vigor	12.2	13.8
3.0CL, RSX, TSX (2.3)	3.4	3.9
3.2CL, 3.2TL		
1996-98	16.3	18.4
1999-03	3.5	4.0
Integra	2.9	3.4
Legend		
1986-90	3.8	4.3
1991-95	16.3	18.4
NSX	16.0	18.1
1996-04 RL (10.7)	14.6	15.5
Add engine removal time if necessary.		
Crankshaft and Main Bearings, Replace (A)		
Includes: Engine R&R, check bearing clearances.		
2.2CL, 2.3CL	13.1	14.8
2.5TL, NSX, Vigor (17.5)	22.8	23.7
3.0CL	25.7	29.1
3.2CL, 3.2TL		
1996-98	27.4	31.0
1999-03	25.7	29.1
Integra		
1986-89	11.7	13.2
1990-93	12.1	13.8
1994-01	11.2	12.7

CL : INTEGRA : LEGEND : NSX : RL : RSX : TL : TSX : VIGOR — ACU-9

	LABOR TIME	SEVERE SERVICE
Legend		
1986-90	26.4	27.3
1991-95	27.4	28.3
RL		
1996-04 (18.0)	23.4	24.3
2005 (14.2)	18.7	19.6
RSX (11.7)	15.8	16.7
TSX (11.3)	15.3	16.2
Crankshaft Front Oil Seal, Replace (A)		
2.2CL, 2.3CL, Integra	3.8	4.3
2.5TL, Vigor	7.3	8.3
3.0CL	5.0	5.7
3.2CL, 3.2TL		
1996-98	4.3	4.8
1999-03	5.0	5.7
Legend		
1986-90	4.6	5.2
1991-95	4.3	4.8
NSX (4.2)	6.1	6.6
RL		
1996-04 (2.8)	4.1	4.6
2005 (2.4)	3.5	4.0
RSX, TSX (2.0)	3.0	3.5
Main Bearings, Replace (A)		
2.2CL, 2.3CL	3.3	3.8
2.5TL, Vigor	14.7	15.6
3.0CL	5.3	5.8
3.2CL, 3.2TL		
1996-98	20.1	20.6
1999-03	5.3	5.8
Integra		
1986-89	3.3	3.8
1990-01	4.4	4.9
Legend		
1986-90	5.6	6.1
1991-95	20.1	20.6
NSX		
engine		
installed (4.0)	5.8	6.3
removed (5.3)	7.6	8.1
1996-04 RL (13.2)	17.6	18.1
RSX, TSX (11.7)	15.8	16.3

Add engine removal time if necessary.

Pistons or Connecting Rods, Replace (A)
Includes: Ridge reaming, cylinder wall deglazing, installing new rings and rod bearings, engine tune-up.

2.2CL, 2.3CL	12.8	14.4
2.5TL	21.4	24.3
3.0CL	18.8	21.3
3.2CL, 3.2TL		
1996-98	23.1	25.5
1999-03	18.8	21.3
Integra		
1986-93	11.2	12.2
1994-01	12.3	13.9

	LABOR TIME	SEVERE SERVICE
Legend		
1986-90	18.6	19.5
1991-95	23.1	25.5
NSX (17.5)	22.8	23.7
1996-04 RL	23.1	25.5
RSX (10.5)	16.0	18.1
TSX (9.6)	13.2	14.1
Vigor	19.7	21.5
Rear Main Oil Seal, Replace (A)		
Includes: Trans. R&R. when necessary.		
2.2CL, 2.3CL		
AT	9.0	10.1
MT	7.9	8.9
2.5TL	10.2	11.5
3.0CL	8.7	9.8
3.2CL, 3.2TL		
1996-98	9.4	10.7
1999-03	8.7	9.8
2004-05		
AT (8.2)	11.5	12.0
MT (7.6)	10.6	11.1
Integra		
1986-89		
AT	5.0	5.6
MT	4.6	5.2
1990-01		
AT	6.2	7.1
MT	4.9	5.5
Legend		
1986-90		
AT	6.7	7.3
MT	6.2	6.8
1991-95		
AT	9.9	11.2
MT	9.4	10.7
NSX		
AT (5.3)	7.6	8.1
MT (5.0)	7.2	7.7
RL		
1996-04 (6.2)	8.8	9.3
2005 (8.1)	11.3	11.8
RSX, TSX (5.2)	7.5	8.0
Vigor		
AT	10.9	12.4
MT	10.2	11.5

Rings, Replace (A)
Includes: Ridge reaming, cylinder wall deglazing, installing new rings, engine tune-up.

2.2CL, 2.3CL	12.0	13.6
2.5TL	20.7	23.4
3.0CL	17.6	20.0
3.2CL, 3.2TL		
1996-98	21.3	24.1
1999-03	17.6	20.0
Integra	11.6	13.1
Legend		
1986-90	16.8	17.7
1991-95	21.3	24.1

	LABOR TIME	SEVERE SERVICE
NSX (17.5)	22.8	23.7
1996-04 RL (14.0)	18.5	19.4
RSX (10.0)	13.7	14.6
TSX (9.6)	13.2	14.1
Vigor	18.2	20.6

ENGINE LUBRICATION
Engine Oil Cooler, Replace (B)

2.5TL, NSX, Vigor	1.5	1.7
1996-03 3.2TL (.5)	.8	.9
Legend, RSX (.5)	.8	.9
1996-04 RL (.5)	.8	.9

Engine Oil Pressure Switch (Sending Unit), Replace (B)

Exc. below (.3)	.5	.6
NSX (.4)	.6	.7
2005 RL (.6)	.8	.9

Oil Pan and/or Gasket, Replace (B)

2.2CL, 2.3CL, 3.0CL	1.9	2.2
2.5TL, Vigor	6.6	6.9
3.2CL, 3.2TL		
1996-98	7.2	7.5
1999-03	1.9	2.2
Integra (1.2)	1.8	2.1
Legend		
1986-90	1.9	2.2
1991-95	7.2	7.5
NSX (1.2)	1.8	2.1
RL		
1996-04 (4.5)	6.5	6.8
2005 (1.3)	1.9	2.2
RSX, TSX (1.7)	2.5	2.8
w/6-Speed MT add		
Integra, Legend	5.2	5.4
NSX w/1 piece front		
exhaust. manifold	1.8	2.0

Oil Pump, Replace (B)

2.2CL, 2.3CL, 3.0CL	6.8	7.7
2.5TL, Vigor	9.7	11.0
3.2CL, 3.2TL		
1996-98	12.0	13.6
1999-03	6.8	7.7
Integra		
1986-89	5.3	5.8
1990-93	5.8	6.3
1994-01	5.3	6.0
Legend		
1986-90	7.5	8.0
1991-95	12.0	13.6
NSX (5.9)	8.4	8.9
RL		
1996-04 (11.4)	15.4	15.9
2005 (3.7)	5.4	5.9
RSX, TSX (2.0)	3.0	3.5
w/AC 2.2CL, 3.2TL,		
Integra, Legend add	.5	.5

ACU-10   CL : INTEGRA : LEGEND : NSX : RL : RSX : TL : TSX : VIGOR

	LABOR TIME	SEVERE SERVICE
CLUTCH		
Clutch Hydraulic System, Bleed (B)		
1986-03 (.4)	.6	.6
Clutch Pedal Free Play, Adjust (B)		
1986-03 (.3)	.5	.6
Clutch Switch, Adjust (B)		
All Models (.2)	.3	.3
Clutch Assy., Replace (B)		
2.2CL, 2.3CL	6.1	6.9
3.2CL	8.1	9.1
Integra		
1986-89	5.5	5.7
1990-93	4.0	4.2
1994-01	3.6	4.1
Legend		
1986-90	6.1	6.9
1991-95	5.8	6.5
NSX, RSX (4.5)	6.5	7.0
2004-05 TL (7.1)	10.0	10.5
TSX (5.0)	7.2	7.7
Vigor	5.8	6.5
Replace rear main seal add		
exc. below	1.7	1.8
91-95 Legend, NSX, Vigor	.3	.4
04-05 TL	3.1	3.2
Clutch Control Cable, Replace (B)		
1986-93 Integra	.7	.8
Clutch Master Cylinder, Replace (B)		
Includes: System bleeding.		
Exc. below (.8)	1.2	1.3
1986-90 Legend (1.0)	1.4	1.5
RSX (.6)	.9	1.0
Clutch Release Bearing or Fork, Replace (B)		
2.2CL, 2.3CL	5.9	6.4
3.2CL, RSX, TSX (4.8)	6.9	7.4
Integra		
1986-89	5.2	5.7
1990-93	3.7	4.2
1994-01	3.4	3.9
Legend	5.8	6.3
NSX (4.5)	6.5	7.0
2004-05 TL (7.3)	10.3	10.8
Vigor	5.4	5.9
Clutch Slave Cylinder, Replace (B)		
Includes: System bleeding.		
Exc. below (.8)	1.3	1.4
1986-90 Legend	1.6	1.8
NSX (1.0)	1.5	1.7
TSX (.6)	.9	1.0
Clutch Switch, Replace (B)		
All Models (.3)	.5	.5
Flywheel, Replace (B)		
2.2CL, 2.3CL	6.7	7.2
3.2CL	8.8	9.3
Integra	4.3	4.8
Legend, Vigor	6.7	7.2

	LABOR TIME	SEVERE SERVICE
NSX (4.7)	6.8	7.3
RSX, TSX (5.2)	7.5	8.0
2004-05 TL (7.2)	10.2	10.7
Replace pilot bearing add	.2	.3
MANUAL TRANSAXLE		
Transaxle Assy., R&R and Recondition (A)		
2.2CL, 2.3CL	11.8	12.7
3.2CL	13.6	14.5
Integra		
1986-89	8.3	9.2
1990-93	9.9	10.8
1994-01	9.3	10.2
Legend		
1986-90	9.9	10.8
1991-95	12.3	13.2
NSX (7.0)	9.9	10.8
RSX, TSX (8.4)	11.7	12.6
2004-05 TL (9.5)	13.1	14.0
Vigor	11.5	12.4
Recondition differential assembly add	2.2	2.7
Transaxle Assy., R&R and Reseal (B)		
2.2CL, 2.3CL	6.8	7.3
Integra		
1986-89	6.6	7.1
1990-93	5.0	5.5
1994-99	4.5	5.0
Legend		
1986-90	8.1	8.6
1991-95	7.0	7.5
Vigor	6.1	6.6
Transaxle Assy., R&R or Replace (B)		
2.2CL, 2.3CL	5.8	6.3
3.2CL	7.5	8.0
Integra		
1986-89	5.2	5.7
1990-93	3.6	4.1
1994-01	3.2	3.7
Legend		
1986-90	6.7	7.2
1991-95	5.9	6.4
NSX (4.2)	6.1	6.6
RSX, TSX (4.6)	6.6	7.1
2004-05 TL (6.8)	9.6	10.1
Vigor	5.0	5.5
Replace assembly add	.5	.7
AUTOMATIC TRANSAXLE		
SERVICE TRANSAXLE INSTALLED		
Diagnose Electronic Transmission Component Each (A)		
1986-05 (.6)	1.0	1.0
Retrieve Fault Codes (A)		
1986-05 (.3)	.4	.4
Pressure Test (B)		
1986-05 (1.0)	1.5	1.5

	LABOR TIME	SEVERE SERVICE
3-4 Shift Switch, Replace (B)		
1990-93 Integra	.7	.7
1986-90 Legend	.7	.7
External Solenoid or Sensor, Replace (B)		
Exc. 3.0CL each (.4)	.7	.7
3.0CL each (.5)	.8	.8
Internal Solenoid or Sensor, Replace (B)		
One or all (1.2)	1.9	2.1
Lock-Up Control Solenoid Valve, Replace (B)		
Exc. below (.3)	.5	.5
1996-98 3.2TL (.5)	.7	.7
NSX (.4)	.6	.6
Shift Control Solenoid Valve, Replace (B)		
2.2CL, 2.3CL	.5	.5
2.5TL, Vigor (1.2)	1.6	1.6
3.0CL	.7	.7
3.2CL, 3.2TL		
1996-98	1.6	1.6
1999-05 (.5)	.7	.7
Integra, RSX (.3)	.5	.5
Legend		
1986-90	.5	.5
1991-95	1.6	1.6
NSX (.4)	.6	.6
1996-04 RL (1.2)	1.6	1.6
Speed Sensor, Replace (B)		
Exc. below (.4)	.6	.7
1996-04 RL (.6)	.9	1.0
Transmission Control Module (TCM/PCM), Replace (B)		
2.2CL, 2.3CL, 3.0CL, Integra, Legend	.5	.5
2.5TL, NSX, TSX, Vigor (.5)	.8	.8
3.2CL, 3.2TL		
1996-98 (.5)	.8	.8
1999-05 (.4)	.6	.6
1996-04 RL	.5	.5
RSX (.2)	.3	.3
Vacuum Modulator, Replace (B)		
1990-93 Integra	.6	.6
Valve Body, R&R and Recondition (A)		
All Models (2.0)	3.0	3.5
Valve Body, Replace (B)		
1991-04 (1.0)	1.5	1.7
SERVICE TRANSAXLE REMOVED		
Transaxle R&R included unless otherwise noted.		
Torque Converter, Replace (B)		
2.2CL, 2.3CL	7.8	8.3
2.5TL, Vigor	10.4	10.9
3.0CL, NSX, RSX (5.4)	8.6	9.1

CL : INTEGRA : LEGEND : NSX : RL : RSX : TL : TSX : VIGOR ACU-11

	LABOR TIME	SEVERE SERVICE
3.2CL, 3.2TL		
1996-98	9.6	10.1
1999-03	8.8	9.3
2004-05 (6.8)	9.6	10.1
Integra	5.4	5.9
Legend		
1986-90	7.0	7.5
1991-95	9.6	10.1
RL		
1996-04 (6.0)	8.6	9.1
2005 (7.6)	10.6	11.1
Transaxle Assy., R&I (B)		
2.2CL, 2.3CL	7.4	7.9
2.5TL, Vigor	9.8	10.3
3.0CL, NSX, RSX	8.2	8.7
3.2CL, 3.2TL		
1996-98	9.1	9.6
1999-03	8.3	8.8
2004-05 (6.8)	9.6	10.1
Integra		
1989-93	5.4	5.9
1994-01	5.2	5.7
Legend	6.9	7.4
1986-90	6.7	7.2
1991-95	9.1	9.6
RL		
1996-04 (6.0)	8.6	9.1
2005 (7.6)	10.6	11.1
TSX (4.4)	6.4	6.9
Transaxle Assy., R&R and Recondition (B)		
2.2CL, 2.3CL	13.9	14.8
2.5TL, Vigor	16.5	17.4
3.0CL, 3.2CL (9.3)	12.8	13.7
3.2TL		
1996-98	15.7	16.6
1999-03 (9.3)	12.8	13.7
2004-05 (10.5)	14.4	15.3
Integra		
1986-89	13.2	14.1
1990-01	11.5	12.4
Legend		
1986-90	14.7	15.6
1991-95	15.7	16.6
NSX (10.0)	13.7	14.6
RL		
1996-04 (9.8)	13.4	14.3
2005 (12.9)	17.2	18.1
RSX, TSX (9.0)	12.5	13.4
Recondition differential assy. add		
exc. NSX	2.4	2.7
NSX	4.0	4.3
Transaxle and Torque Converter, Replace (B)		
2.2CL, 2.3CL	7.8	8.3
2.5TL, Vigor	10.4	10.9
3.0CL, NSX, RSX	8.6	9.1

	LABOR TIME	SEVERE SERVICE
3.2CL, 3.2TL		
1996-98	9.6	10.1
1999-03	8.8	9.3
2004-05 (6.8)	9.6	9.1
Integra	5.4	5.9
Legend		
1986-90	7.0	7.5
1991-95	9.6	10.1
RL		
1996-04 (6.0)	8.6	9.1
2005 (7.6)	10.6	11.1
TSX (4.4)	6.4	6.9

SHIFT LINKAGE
AUTOMATIC TRANSAXLE
Shift Control Cable, Adjust (B)

Exc. below (.5)	.8	.9
1992-05 2.5TL, 3.2CL, 3.2TL, Vigor (.6)	.9	1.0
1996-04 RL (1.6)	2.4	2.5
1990-93 Integra	1.5	1.6
1991-95 Legend	2.4	2.5

Throttle Lever Control Cable, Adjust (B)

2.2CL, 2.3CL	.3	.4
Integra	.3	.4
Legend	.3	.4

Shift Control Cable, Replace (B)

2.2CL, 2.3CL, RSX, TSX (1.0)	1.5	1.7
2.5TL, Vigor	1.4	1.6
3.0CL (1.9)	2.6	2.8
3.2CL, 3.2TL		
1996-98	1.4	1.6
1999-03	2.9	3.1
2004-05 (1.2)	1.8	2.0
Integra		
1986-89	1.4	1.6
1990-93	1.9	2.1
1994-01	1.5	1.7
Legend		
1986-90	1.4	1.5
1991-95	2.9	3.3
NSX (2.0)	3.0	3.2
RL		
1996-04 (1.9)	2.8	3.0
2005 (1.3)	1.9	2.1

Throttle Lever Control Cable, Replace (B)

2.2CL, 2.3CL	.5	.6
Integra	.5	.6
Legend	.5	.6

MANUAL TRANSAXLE
Gearshift Lever, Replace (B)

Exc. below (1.0)	1.5	1.7
2.2CL, 2.3CL, 3.2CL (1.2)	1.8	2.0
1994-01 Integra (1.2)	1.8	2.0

	LABOR TIME	SEVERE SERVICE
NSX (1.5)	2.2	2.4
RSX (.6)	.9	1.1
2004-05 TL (.8)	1.0	1.2
Shift Control Cable, Replace (B)		
2.2CL, 2.3CL, TSX (1.4)	2.0	2.2
NSX (2.0)	3.0	3.2
RSX (1.0)	1.5	1.7
2004-05 TL (1.2)	1.7	1.9
Shift Lever Bushing, Replace (B)		
Exc. NSX (.8)	1.2	1.3
NSX (1.5)	2.2	2.3
Shift Rod, Replace (B)		
1986-01 (.6)	1.0	1.1
Torque Rod Bushing, Replace (B)		
1986-01 (.5)	.8	.9

DRIVELINE
DRIVE AXLE
Differential, Drain and Refill (B)

1986-05	.7	.7

Differential Assy., R&R and Recondition (B)

Front (9.0)	12.5	13.4

Differential Assy., R&R or Replace (B)

Front		
2.5TL, Vigor	5.1	5.6
1996-98 3.2TL (3.2)	5.1	5.6
Legend 1991-95	5.8	6.3
1996-04 RL (3.6)	5.3	5.8
Rear NSX	1.4	1.6

Differential Oil Seal, Replace (B)

Front each (.9)	1.4	1.6
Rear each (1.4)	2.2	2.5

Pinion Shaft Oil Seal, Replace (B)

Front		
2.5TL, 3.2TL	6.7	7.6
Legend, Vigor		
AT	6.7	7.6
MT	5.9	6.7
1996-04 RL	6.7	7.6

HALFSHAFTS
Halfshaft, R&R or Replace (B)

Front exc. below		
one side	1.1	1.3
both sides	1.9	2.1
intermediate	1.4	1.6
2005 RL		
one side (1.0)	1.5	1.7
Rear		
one side (1.5)	2.2	2.4
both sides (2.5)	3.7	3.9
intermediate (1.7)	2.5	2.7
Replace support bearing add	.5	.5

ACU-12 CL : INTEGRA : LEGEND : NSX : RL : RSX : TL : TSX : VIGOR

	LABOR TIME	SEVERE SERVICE
Inboard CV Joint and/or Boot, Replace (B)		
Front		
one side	1.6	1.8
both sides	2.7	2.9
Rear one (1.5)	2.2	2.4
Intermediate Shaft Support Bearing, Replace (B)		
Exc. below (1.2)	1.8	2.0
1991-05 NSX (1.7)	2.5	2.7
Outboard CV-Joint and/or Boot, Replace (B)		
Front		
one side	1.4	1.6
both sides	2.3	2.5
Rear one (1.5)	2.2	2.4

BRAKES
ANTI-LOCK
The following operations do not include testing. Add time as required.

Diagnose Anti-Lock Brake System Component Each (A)
1990-05 (.6) .... .8 .8
Retrieve Fault Codes (A)
1986-05 (.3) .... .4 .4
Bleed ABS System (B)
1990-05 (1.0) 1.5 1.6
Accumulator Assy., Replace (B)
Exc. NSX 1.9 2.1
NSX (.9) 1.3 1.5
Control Module, Replace (B)
2.2CL, 2.3CL, 3.0CL .... .5 .5
2.5TL, Vigor 1.0 1.0
3.2CL, 3.2TL
  1996-98 .... 1.0 1.0
  1999-03 .... 1.4 1.4
Integra
  1990-93 .... .6 .6
  1994-01 .... 1.0 1.0
Legend, NSX .... .8 .8
31996-04 RL .... .8 .8
High Pressure Hoses, Replace (B)
Exc. below 1.3 1.4
1990 Legend .... .8 .9
NSX (.7) 1.0 1.1
Low Pressure Hoses, Replace (B)
Exc. below 1.4 1.5
1991-95 Legend .... .6 .7
NSX (.8) 1.2 1.3
Modulator, Replace (B)
Exc. below (1.0) 1.4 1.4
2005 RL (1.2) 1.7 1.7
Power Unit, Replace (B)
Exc. 1990 Legend 1.6 1.6
1990 Legend 1.1 1.1
Pressure Switch, Replace (B)
1990-05 (1.0) 1.4 1.4

Wheel Sensor, Replace (B)
2.5TL, 3.2TL, Vigor (.5) .. .8 .9
CL, RSX, TSX (.3) ..... .5 .6
Integra
  front .... .5 .6
  rear 1.1 1.2
Legend, RL
 1990
  front .... .6 .7
  rear 1.6 1.7
 1991-04
  front
   right .... .8 .9
   left 1.3 1.4
  rear .... .8 .9
2005 one (.5) .... .8 .9
NSX (.4) .... .6 .7

SYSTEM
Bleed Brakes (B)
Includes: Add fluid.
1986-05 (.5) .... .8 .9
Brake System, Flush and Refill (B)
All models 1.3 1.3
Brake Failure Relay, Replace (B)
1986-05 (.2) .... .3 .3
Brake Hose (Flexible), Replace (B)
Includes: System bleeding.
Exc. below one (.5) ..... .7 .8
2005 RL
 front or rear one (.7) .. .7 .8
 each addl. add .... .3 .4
Brake Warning Switch, Replace (B)
1986-03 .... .5 .5
Master Cylinder, R&R and Recondition (B)
Includes: System bleeding.
1986-96 .... 2.8 3.1
Master Cylinder, Replace (B)
Includes: System bleeding.
Exc. below (1.2) 1.7 1.9
2005 RL (1.4) 2.0 2.2
Power Booster Vacuum Assy., Replace (B)
Includes: System bleeding.
Exc. below (1.3) 1.9 2.1
2005 RL (2.1) 3.0 3.2
Power Booster Vacuum Check Valve, Replace (B)
1986-05 (.2) .... .3 .3
Proportioning Valve, Replace (B)
Exc. below (1.0) 1.4 1.5
RSX, TSX (.5) .... .8 .9
2004-05 TL (.5) .... .8 .9

SERVICE BRAKES
Brake Pads, Replace (B)
1986-05
 front or rear (.7) 1.1 1.3
 four wheels 2.1 2.3

COMBINATION ADD-ONS
Replace
 brake hose add each ... .3 .3
 caliper add .... .3 .3
 disc rotor add each ... .2 .2
 master cylinder add .... .7 .7
Resurface rotor
 add each .... .5 .5
Caliper Assy., R&R and Recondition (B)
Includes: System bleeding.
Exc. below
 one (.8) 1.3 1.5
 both (1.7) 2.6 2.8
NSX
 one (1.0) 1.5 1.7
 both (2.1) 3.2 3.4
2005 RL
 one (1.0) 1.5 1.7
 both (2.1) 3.2 3.4
Caliper Assy., Replace (B)
Includes: System bleeding.
Front
 one (.5) .... .8 .9
 both (1.1) 1.6 1.8
Rear
 exc. below
  one (.7) 1.0 1.1
  both (1.3) 1.9 2.1
 NSX
  one (.8) 1.1 1.2
  both (1.5) 2.2 2.4
 2005 RL
  one (.8) 1.1 1.2
  both (1.5) 2.2 2.4
Disc Brake Rotor, Replace (B)
Front
 exc. below
  one (.8) 1.1 1.2
  both (1.5) 2.2 2.4
 1997 2.2CL, 2.3CL, 3.0CL
  one (1.2) 1.8 1.9
  both (2.4) 3.5 3.7
Rear
 exc. NSX
  one (.7) 1.0 1.1
  both (1.3) 1.9 2.1
 NSX
  one (.8) 1.1 1.2
  both (1.5) 2.2 2.4
Replace parking brake
 drum add .... .2 .3

PARKING BRAKE
Parking Brake Cable, Adjust (C)
1986-05 (.4) .... .6 .7
Parking Brake Cable, Replace (B)
Front
 exc. below (.6) 9 1.1
 2005 RL (1.4) 2.0 2.2

CL : INTEGRA : LEGEND : NSX : RL : RSX : TL : TSX : VIGOR **ACU-13**

	LABOR TIME	SEVERE SERVICE
Rear		
exc. below		
one (1.0)	1.6	1.8
both (1.5)	2.4	2.7
2005 RL		
one (2.2)	3.3	3.5
both (3.0)	4.4	4.6
Parking Brake Shoes, Replace (B)		
One side (.8)	1.3	1.4
Both sides (1.6)	2.6	2.8

FRONT SUSPENSION

Unless otherwise noted, time given does not include alignment.

Front & Rear Alignment, Check (A)
- 1986-05 (1.2) 1.9   2.2

Front Suspension Height, Adjust (A)
- Integra
  - one side 2   .3
  - both sides 3   .4

Front Toe, Adjust (A)
- 1986-05 (.6) 8   1.0

Ball Joint Boot (Tie Rod), Replace (B)
- 1986-02 6   .8

Engine Cradle (Frame), Replace (A)
- 2.2CL, 2.3CL, 3.0CL, TSX (2.5) 3.7   3.9
- 2.5TL, Vigor (2.3) 3.4   3.6
- 3.2CL (4.3) 6.2   6.4

Front Crossmember, Replace (B)
- 2.2CL, 2.3CL, 3.0CL 2.4   2.6
- 3.2CL (4.0) 5.8   6.0
- 3.2TL
  - 1999-03 (4.3) 6.2   6.4
  - 2004-05 (3.8) 5.5   5.7
- 1990-93 Integra (2.5) 4.0   4.2
- Legend (3.1) 4.5   4.7
- RL
  - 1996-04 (3.1) 5.0   5.2
  - 2005 (4.6) 6.6   6.8
- RSX (1.0) 1.5   1.7
- TSX (2.5) 3.7   3.9

Front Radius Rod and/or Bushings, Replace (B)
- One side (.6) 1.0   1.1
- Both sides (1.2) 1.9   2.2

Front Strut Damper, Replace (B)
- Exc. below
  - one side (.8) 1.2   1.3
  - both sides (1.6) 2.4   2.6
- 1986-89 Integra
  - one side 1.0   1.1
  - both sides 1.9   2.1
- 2005 RL
  - one side (1.3) 1.9   2.0
  - both sides (2.5) 3.7   3.9

	LABOR TIME	SEVERE SERVICE
RSX		
one side (.5)	.8	.9
both sides (1.0)	1.6	1.8

Front Torsion Bar, Replace (B)
- 1986-89 Integra
  - one side 1.8   2.0
  - both sides 2.8   3.0

Hub Bearings, Replace (B)
- One side
  - w/ABS 1.8   2.0
  - w/o ABS 1.5   1.7
- Both sides
  - w/ABS 3.6   3.8
  - w/o ABS 3.0   3.2
- NSX one (1.0) 1.6   1.8

Lower Control Arm, Replace (B)
Includes: Reset toe.
- Exc. below
  - one side (.6) 9   1.0
  - both sides (1.2) 1.8   2.0
- 1996-98 3.2TL
  - one side 1.3   1.4
  - both sides 2.6   2.9
- Legend
  - one side 1.3   1.4
  - both sides 2.6   2.9
- NSX
  - one side (2.2) 3.3   3.4
  - both sides (4.4) 6.4   6.6
- RL
  - 1996-04
    - one side (.8) 1.2   1.3
    - both sides (1.6) 2.4   2.6
  - 2005
    - one side (1.1) 1.6   1.7
    - both sides (1.8) 2.7   2.9
- RSX
  - one side (.4) 6   .7
  - both sides (.8) 1.3   1.5
- *Replace bushings add each side* 3   .5

Rear Crossmember, Replace (B)
- 2.2CL, 2.3CL, 3.0CL 4.0   4.5
- 3.2CL 6.9   7.7
- 3.2TL
  - 1999-03 (4.3) 6.2   6.4
  - 2004-05 (3.2) 4.7   4.9
- Integra
  - 1986-93 (4.5) 6.5   6.7
  - 1994-01 (1.7) 2.5   2.7
- Legend (3.4) 5.0   5.2
- NSX (1.0) 1.5   1.7
- 1996-04 RL (3.4) 5.0   5.2
- TSX (2.5) 3.7   3.9

Stabilizer Bar, Replace (B)
- Exc. below 1.1   1.3
- 1991-95 Legend 3.0   3.2
- NSX (.8) 1.2   1.4
- 2005 RL (4.1) 5.9   6.1

	LABOR TIME	SEVERE SERVICE
Stabilizer Bar Bushings or Link, Replace (B)		
1986-05 (.4)	.6	.7

Steering Knuckle, Replace (B)
- Exc. NSX, RSX
  - one (1.1) 1.8   1.9
  - both (2.2) 3.5   3.7
- NSX one (2.5) 4.0   4.1
- RSX
  - one (.8) 1.3   1.4
  - both (1.6) 2.6   2.8
- *Replace ball joint add each* 5   .5

Strut Coil Spring, Replace (B)
- Exc. RSX
  - one side (.8) 1.3   1.4
  - both sides (1.6) 2.4   2.6
- RSX
  - one side (.5) 8   .9
  - both sides (1.0) 1.6   1.8

Upper Control Arm, Replace (B)
Includes: Reset toe.
- Exc. below
  - one side (.6) 1.0   1.1
  - both sides (1.2) 1.8   2.0
- NSX
  - one side (2.2) 3.3   3.4
  - both sides (4.4) 6.4   6.6
- 2005 RL one side (1.3) 1.9   2.1
- *Replace bushings add each side* 5   .7

REAR SUSPENSION

Unless otherwise noted, time given does not include alignment.

Rear Suspension (Complete), Align (A)
- 1986-05 (1.2) 1.8   2.0

Rear Toe, Check and Adjust (A)
- 1986-05 (.5) 8   1.0

Coil Spring, Replace (B)
- 1986-89 Integra
  - one 8   1.0
  - both 1.8   2.0
- 1986-90 Legend
  - one 7   .9
  - both 1.6   1.8

Control Arm, Replace (B)
- One 6   .7
- Both 1.3   1.4

Hub and/or Bearings, Replace (B)
- Exc. NSX
  - one side (.6) 1.0   1.1
  - both sides (1.2) 1.8   2.0
- NSX
  - one side (1.0) 1.5   1.6
  - both sides (2.0) 3.0   3.2

ACU-14 CL : INTEGRA : LEGEND : NSX : RL : RSX : TL : TSX : VIGOR

	LABOR TIME	SEVERE SERVICE
Lower Control Arm (A or B), Replace (B)		
Exc. NSX, RSX		
one (.5)	.8	1.0
both (1.0)	1.6	1.8
NSX		
one (2.0)	3.0	3.1
both (4.0)	5.8	6.0
RSX		
one (.9)	1.4	1.6
both (1.8)	2.9	3.2
Replace bushings add		
each side	.3	.5
Panhard Rod Bushing, Replace (B)		
1986-89 Integra		
one or both (.5)	.8	1.0
Radius Rod or Bushings, Replace (B)		
Includes: Rear wheel alignment.		
Legend		
one side (.9)	1.4	1.6
both sides (1.8)	2.9	3.2
Rear Wheel Knuckle Assy., Replace (B)		
Exc. below		
one side (.9)	1.4	1.6
both sides (1.8)	2.7	2.9
Legend, RL		
1986-90		
one side	2.2	2.5
both sides	4.5	5.0
1991-04		
one side (1.2)	1.8	1.9
both sides (2.4)	3.5	3.7
NSX		
one side (2.1)	3.1	3.2
both sides (4.2)	6.1	6.3
Shock Absorbers, Replace (B)		
1986-89 Integra		
one side	.8	.9
both sides	1.8	1.9
1986-90 Legend		
one side	.8	.9
both sides	1.8	1.9
Stabilizer Bar, Replace (B)		
Exc. below (.7)	1.1	1.2
2005 RL (.5)	.6	.7
RSX (.5)	.6	.7
Stabilizer Bar Bushings or Link, Replace (B)		
1986-05 (.4)	.6	.7
Strut Coil Spring or Damper, R&R or Replace (B)		
Exc. below		
one (.8)	1.2	1.3
both (1.6)	2.6	2.8
2005 RL		
one (.7)	1.1	1.2
both (1.1)	1.6	1.8

	LABOR TIME	SEVERE SERVICE
RSX		
one (.5)	.8	.9
both (1.1)	1.6	1.8
Trailing Arm or Link, Replace (B)		
Exc. below		
one side (.6)	1.0	1.1
both sides (1.2)	1.9	2.2
3.2CL, 3.2TL, TSX		
one side (.4)	.6	.8
both sides (.8)	1.2	1.4
Integra		
one side	1.6	1.7
both sides	3.2	3.4
Upper Control Arm, Replace (B)		
Exc. NSX		
one side (.4)	.6	.8
both sides (.8)	1.3	1.4
NSX		
one side (2.2)	3.3	3.4
both sides (4.4)	6.4	6.6

STEERING
AIR BAGS
The following operations do not include testing. Add time as required.

	LABOR TIME	SEVERE SERVICE
Diagnose Air Bag System Component Each (A)		
1990-05 (.6)	.9	.9
Retrieve Fault Codes (A)		
1990-05 (.3)	.4	.4
Air Bag Assy., Replace (B)		
1990-05		
driver side (.3)	.5	.5
passenger side		
exc. below (.5)	.7	.7
NSX (.7)	.9	.9
3.2TL, TSX (.8)	1.0	1.0
side one		
exc. below (1.3)	1.6	1.6
2005 RL (.6)	.8	.8
RSX (.8)	1.0	1.0
2004-05 TL (.6)	.8	.8
Cable Reel, Replace (B)		
1986-05 (.9)	1.2	1.2
Electronic Control Unit, Replace (B)		
Exc. below (.6)	.8	.8
2.5TL, Vigor (1.5)	2.1	2.1
3.2CL (1.7)	2.3	2.3
3.2TL		
1996-03 (1.7)	2.3	2.3
2004-05 (.8)	1.0	1.0
RL		
1996-04 (1.2)	1.6	1.6
2005 (1.6)	2.0	2.0
TSX (.8)	1.0	1.0
Sensor, Replace (B)		
Front or dash (.5)	.8	.8
Seat pad (1.0)	1.4	1.4

	LABOR TIME	SEVERE SERVICE
Side impact		
front		
3.2CL (.6)	.8	.8
3.2TL		
1996-03 (.6)	.8	.8
2004-05 (.3)	.5	.5
1996-04 RL (1.0)	1.4	1.4
RSX, TSX (.3)	.5	.5
rear		
2004-05 TL (.5)	.6	.6
TSX (.5)	.6	.6

ELECTRIC STEERING

	LABOR TIME	SEVERE SERVICE
Diagnose Electric Steering System Component Each (A)		
NSX (.6)	.8	.8
Steering Gearbox, Adjust (B)		
NSX (.3)	.5	.5
Control Unit, Replace (A)		
NSX (.6)	.8	.8
Power Unit, Replace (A)		
NSX (1.0)	1.4	1.5
Steering Column, Replace (B)		
NSX (1.4)	1.9	1.9
Steering Gearbox, Replace (B)		
Includes: Reset toe.		
NSX (3.5)	5.3	6.0
Steering Wheel, Replace (B)		
NSX (.5)	.7	.7
Tie Rod, Tie Rod End and/or Boot, Replace (A)		
Includes: Reset toe.		
One side	2.9	3.2
Both sides	3.8	4.3

POWER RACK & PINION

	LABOR TIME	SEVERE SERVICE
Pump Pressure Check (B)		
1986-05 (.4)	.6	.6
Rack & Pinion Assy., Adjust (B)		
1986-05 (.5)	.8	.9
Control Unit, Replace (B)		
2.2CL, 2.3CL, 3.0CL	4.3	4.4
2.5TL	4.3	4.4
3.2CL, 3.2TL	4.7	4.8
Integra	4.3	4.4
Legend		
1986-90	4.3	4.4
1991-95	4.7	4.8
RL		
1996-04	4.7	4.8
2005 (.7)	1.0	1.1
Vigor	4.7	4.8
Power Steering Switch Assy., Replace (B)		
Exc. 2005 RL (.3)	.4	.5
2005 RL (.6)	.9	1.0
Power Steering Reservoir, Replace (B)		
Exc. below (.2)	.5	.5
2005 RL (.4)	.6	.6

CL : INTEGRA : LEGEND : NSX : RL : RSX : TL : TSX : VIGOR **ACU-15**

	LABOR TIME	SEVERE SERVICE
Rack & Pinion Assy., R&R or Replace (B)		
Includes: Reset toe.		
Exc. below (3.0)	4.4	4.6
1991-95 Legend (3.1)	4.5	4.7
RL		
1996-04 (3.1)	4.5	4.7
2005 (3.8)	5.5	5.7
RSX (2.6)	3.8	4.0
Rack & Pinion Assy., Reseal (A)		
Includes: Alignment.		
2.2CL, 2.3CL, 2.5TL	4.9	5.4
3.0CL	4.9	5.4
3.2CL (4.0)	5.8	6.3
3.2TL		
1996-03 (4.0)	5.8	6.3
2004-05 (3.1)	4.5	5.0
Integra, Legend	4.9	5.2
RL		
1996-04 (4.0)	5.8	6.3
2005 (4.9)	7.1	7.6
RSX (3.1)	4.5	5.0
TSX (3.3)	4.8	5.3
Vigor	5.3	5.9
Steering Column, Replace (B)		
Exc. below (1.2)	1.6	1.6
2005 RL (1.7)	2.1	2.1
RSX (.8)	1.1	1.1
2004-05 TL (.6)	.8	.8
TSX (.6)	.8	.8
Steering Pump, R&R and Recondition (A)		
2.2CL, 2.3CL	1.2	1.4
2.5TL, Vigor	1.8	2.1
3.0CL, RSX, TSX (.8)	1.2	1.4
3.2CL, 3.2TL		
1996-98	1.8	2.1
1999-05 (.8)	1.2	1.4
Integra		
1986-89	2.1	2.4
1990-01	1.2	1.4
Legend, RL		
1986-90	2.4	2.8
1991-04 (1.2)	1.8	2.1
Steering Pump, R&R or Replace (B)		
2.2CL, 2.3CL	.5	.7
2.5TL, Vigor	1.2	1.4
3.0CL, RSX	.6	.8
3.2CL, 3.2TL		
1996-98	1.2	1.4
1999-03	.6	.8
Integra		
1986-89	1.5	1.7
1990-01	.5	.7
Legend, RL		
1986-90	1.8	2.1
1991-03	1.2	1.4

	LABOR TIME	SEVERE SERVICE
Steering Pump Hoses, Replace (B)		
Does not include purging.		
Output		
exc. below (1.0)	1.4	1.5
Integra		
1986-89	3.0	3.4
1990-01	2.3	2.6
Legend	3.0	3.4
2005 RL (1.2)	1.7	1.8
RSX, TSX (.4)	.6	.7
2004-05 TL (.6)	.9	1.0
Return		
exc. below (1.0)	1.4	1.5
RSX, TSX (.3)	.8	.9
2004-05 TL (.3)	.8	.9
Steering Shaft Coupler, Replace (B)		
1986-02	.8	.9
Steering Wheel, Replace (B)		
All Models (.5)	.7	.7
Tie Rod, Tie Rod End and/or Boot, Replace (A)		
Includes: Reset toe.		
Exc. below		
one or both (1.0)	1.4	1.6
2005 RL		
inner		
one side (1.0)	1.6	1.7
both sides (1.7)	2.4	2.6
outer		
one side (.6)	.9	1.0
both sides (1.1)	1.6	1.8
RSX, TSX		
one side (.8)	1.2	1.3
both sides (1.0)	1.6	1.8
2004-05 TL		
one side (.8)	1.2	1.3
both sides (1.0)	1.6	1.8

HEATING & AIR CONDITIONING

When more than one component requires replacement where evacuation, recovery and recharging is already included, deduct 1.0 hour for each additional component from the time given.

	LABOR TIME	SEVERE SERVICE
Evacuate/Recover and Recharge System (B)		
Exc. 2005 RL (.5)	.7	.7
2005 RL (.7)	1.0	1.0
AC Hoses, Replace (B)		
Includes: Evacuate/recover and recharge.		
Exc. below each (.9)	1.4	1.6
NSX each (1.2)	1.8	2.0
RSX each (1.5)	2.3	2.5

	LABOR TIME	SEVERE SERVICE
Compressor Assy., Replace (B)		
Includes: Transfer parts.		
Evacuate/recover and recharge.		
Exc. below (1.5)	2.3	2.5
2005 RL (3.0)	4.3	4.5
RSX (3.0)	4.3	4.5
2004-05 TL (1.8)	2.6	2.8
TSX (1.8)	2.6	2.8
Compressor Clutch Assy., Replace (B)		
Includes: Evacuate/recover and recharge.		
Exc. RSX (1.8)	2.8	3.0
RSX (3.3)	4.7	4.9
Compressor Shaft Seal, Replace (B)		
Includes: Transfer parts.		
Evacuate/recover and recharge.		
1986-02	3.1	3.3
Condenser Assy., Replace (B)		
Includes: Evacuate/recover and recharge.		
2.2CL, 2.3CL, 3.0CL	1.6	1.7
2.5TL, Vigor	2.8	2.9
3.2CL, 3.2TL		
1996-98	2.8	2.9
1999-03	1.6	1.7
2004-05 (1.0)	1.4	1.5
Integra		
1986-89	2.0	2.1
1990-01	1.6	1.7
Legend, RL		
1986-90	1.9	2.0
1991-04	1.8	1.9
2005 (2.1)	3.0	3.1
NSX each (2.1)	3.0	3.1
RSX (1.3)	2.0	2.1
TSX (1.0)	1.4	1.5
Condenser Fan Motor, Replace (B)		
Exc. below	.9	1.0
NSX each (1.2)	1.8	2.1
RSX (1.0)	1.5	1.7
Condenser Fan Motor Relay, Replace (B)		
1986-05 (.1)	.3	.3
Evaporator Coil, Replace (B)		
Includes: Evacuate/recover and recharge.		
2.2CL, 2.3CL, 3.0CL	2.3	2.5
2.5TL, Vigor	7.6	7.8
3.2CL, 3.2TL		
1996-98	7.6	7.8
1999-03	2.3	2.5
2004-05 (2.3)	3.3	3.5
RL (2.5)	3.8	4.0
Integra	2.3	2.4
Legend		
1986-90	2.3	2.5
1991-95	7.6	7.8
NSX (6.2)	8.5	8.7
1996-04 RL (2.5)	3.8	4.0
RSX (1.6)	2.4	2.6

ACU-16 CL : INTEGRA : LEGEND : NSX : RL : RSX : TL : TSX : VIGOR

	LABOR TIME	SEVERE SERVICE
Expansion Valve, Replace (B)		
Includes: Evacuate/recover and recharge.		
2.2CL, 2.3CL, 3.0CL	2.3	2.5
2.5TL, Vigor	7.6	7.8
3.2CL, 3.2TL		
1996-98	7.6	7.8
1999-03	2.3	2.5
2004-05 (2.4)	3.4	3.6
Integra	2.3	2.5
Legend		
1986-90	2.3	2.5
1991-95	7.6	7.8
NSX (6.5)	8.9	9.1
RL		
1996-04 (2.5)	3.8	4.0
2005 (2.8)	4.0	4.2
RSX (1.6)	2.4	2.6
Heater Blower Motor, Replace (B)		
Exc. below (.3)	.5	.6
2.5TL, Vigor	6.1	6.2
1996-98 3.2TL	6.1	6.2
1986-89 Integra	2.5	2.6
1991-95 Legend	2.5	2.6
NSX (.7)	.9	1.0
Heater Blower Motor Resistor, Replace (B)		
1986-05	.8	.8
Heater Blower Motor Switch, Replace (B)		
Integra		
exc. below (1.0)	1.4	1.4
1990-93 (1.4)	2.1	2.1
Heater Core, R&R or Replace (B)		
Exc. below (4.5)	6.8	7.0
NSX (5.5)	8.4	8.6
2005 RL (5.0)	7.6	7.8
RSX (5.0)	7.6	7.8
2004-05 TL (6.5)	8.1	8.3
TSX (7.1)	8.9	9.1
Heater Hoses, Replace (B)		
Exc. NSX each (.5)	.8	.9
NSX		
front any (.5)	.8	.9
middle (.9)	1.4	1.5
rear (.8)	1.2	1.4
Mode Selector Switch, Replace (B)		
2.2CL, 2.3CL, 3.0CL	1.6	1.6
2.5TL, 3.2CL, 3.2TL	.3	.3
Vigor	.5	.5
Receiver/Drier Assy., Replace (B)		
Includes: Evacuate/recover and recharge.		
Exc. below (1.0)	1.5	1.7
2.2CL, 2.3CL, 3.0CL	2.4	2.6
NSX (1.5)	2.3	2.5
RSX (1.3)	2.0	2.2
TSX (1.1)	1.6	1.8

	LABOR TIME	SEVERE SERVICE
Temperature Control Assy., Replace (B)		
Includes: Adjust cables when necessary.		
Exc. below (1.0)	1.4	1.4
1990-93 Integra	2.1	2.1
RSX (.2)	.3	.3
2004-05 TL (.7)	.9	.9
TSX (.5)	.6	.6

WIPERS & SPEEDOMETER

	LABOR TIME	SEVERE SERVICE
Antenna, Replace (B)		
1986-03		
manual (.4)	.6	.6
power motor and mast (.6)	.8	.8
Electronic Wiper/Washer Control Unit, Replace (B)		
Exc. NSX (.5)	.7	.7
NSX (.7)	.9	.9
Instrument Panel Cluster, Replace (B)		
RSX (.5)	.7	.7
2004-05 TL (.3)	.4	.4
Radio, R&R (B)		
Exc. RSX (1.0)	1.4	1.4
RSX (.3)	.5	.5
Rear Window Wiper Motor, Replace (B)		
Integra, RSX (.6)	.9	1.0
Speedometer Cable & Casing, Replace (B)		
Integra		
1986-89	.9	.9
1990-01	1.8	1.8
Legend	1.8	1.8
Speedometer Cable (Inner), Replace or Lubricate (B)		
Integra	.5	.5
1986-90 Legend	.5	.5
Speedometer Head, R&R or Replace (B)		
Exc. below	1.4	1.4
2.5TL, Vigor	1.9	1.9
1996-04 RL	1.9	1.9
1996-98 3.2TL	1.9	1.9
Windshield or Rear Window Washer Motor, Replace (B)		
Exc. below (.7)	1.0	1.0
1986-93 Integra	.5	.5
NSX (.9)	1.4	1.5
2005 RL (.4)	.6	.6
2004-05 TL (.4)	.6	.6
Windshield Wiper & Washer Switch, Replace (B)		
Exc. below (.7)	1.0	1.0
2005 RL (.3)	.4	.4

	LABOR TIME	SEVERE SERVICE
Windshield Wiper Linkage, Replace (B)		
Exc. NSX (.6)	.9	1.0
NSX (.7)	1.1	1.2
Windshield Wiper Motor, Replace (B)		
Exc. below (.6)	.9	1.0
NSX (.7)	1.1	1.2
2005 RL (.7)	1.1	1.2

LAMPS & SWITCHES

	LABOR TIME	SEVERE SERVICE
Back-Up Lamp Bulb, Replace (C)		
One (.2)	.3	.3
Back-Up Lamp Switch, Replace (B)		
1986-05 MT only (.3)	.5	.5
Brake Lamp Switch, Replace (B)		
1986-05 (.3)	.5	.5
Clutch Start Switch, Replace (B)		
Exc. Legend (.3)	.5	.5
Legend	.7	.7
Dashlamp Dimmer Switch, Replace (B)		
Exc. NSX (.3)	.5	.5
NSX (.6)	.8	.8
Halogen Headlamp Bulb, Replace (B)		
Exc. below (.2)	.3	.3
NSX each (.4)	.6	.6
2005 RL each (.4)	.6	.6
2004-05 TL each (.5)	.7	.7
Hazard Warning Switch, Replace (B)		
Exc. below (.2)	.3	.3
3.2CL, RSX (.5)	.7	.7
1996-03 3.2TL (.5)	.7	.7
NSX (.7)	.9	.9
2005 RL (.7)	.9	.9
Headlamps, Aim (B)		
Two	.5	.5
Four	.8	.8
High Mount Stop Lamp Bulb, Replace (C)		
1986-05 (.2)	.3	.3
Horn, Replace (B)		
Exc. NSX (.4)	.6	.6
NSX (.9)	1.2	1.2
Horn Relay, Replace (B)		
1986-05 (.1)	.2	.2
License Lamp Bulb, Replace (C)		
1986-05 one or all (.2)	.3	.3
Park & Turn Signal Lamp Bulb or Lens, Replace (C)		
Each (.2)	.3	.3
Parking Brake Warning Lamp Switch, Replace (B)		
Exc. below (.3)	.5	.5
NSX, 2005 RL (.4)	.6	.6
Side Marker Lamp Bulb, Replace (C)		
1986-05 each (.2)	.3	.3
Tail Lamp Lens or Bulb, Replace (C)		
Exc. below one	.6	.6
NSX one	.3	.3
RSX one	.8	.8

	LABOR TIME	SEVERE SERVICE
Turn Signal & Headlamp Dimmer Switch, Replace (B)		
Exc. below (.7)	1.0	1.0
2005 RL (.3)	.4	.4
RSX (.5)	.6	.6
2004-05 TL (.5)	.6	.6
Turn Signal or Hazard Warning Flasher, Replace (B)		
1986-05 (.1)	.2	.2

SEAT BELTS

	LABOR TIME	SEVERE SERVICE
Diagnose Seat Belt System Component Each (A)		
1990-05 (.4)	.6	.6
Retrieve Fault Codes (A)		
1990-05 (.3)	.4	.4
Anchor Rail Assy., Replace (B)		
1990-95 Integra one	1.7	1.7

	LABOR TIME	SEVERE SERVICE
Automatic Shoulder Belt Control Unit, Replace (B)		
1990-95 Integra	.5	.5
Retractor-Receiver Assy., Replace (B)		
1990-95 Integra one	.8	.8

BODY

	LABOR TIME	SEVERE SERVICE
Door Window Regulator, Replace (B)		
Exc. below (1.0)	1.4	1.5
NSX (1.6)	2.0	2.1
RSX, TSX (.5)	.7	.8
2004-05 TL (.7)	1.0	1.1
Front Door Latch, Replace (B)		
Exc. below (.8)	1.1	1.3
NSX (1.4)	1.9	2.1
RSX, TSX (.5)	.7	.9
2004-05 TL (.5)	.7	.9
Hood Release Cable, Replace (B)		
Exc. below (.8)	1.3	1.3
2005 RL (1.3)	1.6	1.6

	LABOR TIME	SEVERE SERVICE
Lock Striker Plate, Replace (B)		
1986-05 (.3)	.5	.5
Rear Door Latch, Replace (B)		
Exc. 3.2TL (.8)	1.1	1.2
3.2TL (.5)	.7	.8
Tailgate, Hatch Support, Replace (B)		
Exc. below		
one (.2)	.3	.3
both (.4)	.6	.6
2005 RL one (.4)	.6	.6
Trunk, Tailgate or Hatch Lock Cylinder, Replace (B)		
1986-05 (.3)	.5	.5
Window Regulator Motor, Replace (B)		
Exc. below (1.0)	1.4	1.6
RSX (.5)	.7	.9
2004-05 TL (.5)	.7	.9

NOTES

MDX : SLX — ACU

SYSTEM INDEX

MAINTENANCE ACU-20
   Maintenance Schedule .. ACU-20
CHARGING ACU-20
STARTING ACU-20
CRUISE CONTROL ACU-20
IGNITION ACU-20
EMISSIONS ACU-21
FUEL ACU-21
EXHAUST ACU-21
ENGINE COOLING ACU-22
ENGINE ACU-22
   Assembly ACU-22
   Cylinder Head ACU-22
   Camshaft ACU-22
   Crank & Pistons ACU-22
   Engine Lubrication ACU-23
AUTO TRANSMISSION ... ACU-23
TRANSFER CASE ACU-23
SHIFT LINKAGE ACU-23
DRIVELINE ACU-23
BRAKES ACU-24
FRONT SUSPENSION ACU-24
REAR SUSPENSION ACU-25
STEERING ACU-25
HEATING & AC ACU-26
WIPERS &
  SPEEDOMETER ACU-26
LAMPS & SWITCHES ACU-26
BODY ACU-27

OPERATIONS INDEX

A
AC Hoses. ACU-26
Air Bags ACU-25
Air Conditioning ACU-26
Alignment................. ACU-24
Alternator (Generator) ACU-20
Antenna ACU-26
Anti-Lock Brakes ACU-24

B
Ball Joint ACU-25
Battery Cables............ ACU-20
Bleed Brake System ACU-24
Blower Motor ACU-26
Brake Disc ACU-24
Brake Hose ACU-24
Brake Pads and/or Shoes .. ACU-24

C
Camshaft ACU-22
Camshaft Sensor......... ACU-20
Catalytic Converter ACU-21
Coolant Temperature (ECT)
  Sensor ACU-21
Crankshaft ACU-23
Crankshaft Sensor........ ACU-21
Cruise Control ACU-20
CV Joint ACU-24
Cylinder Head ACU-22

D
Differential ACU-23
Drive Belt................. ACU-20
Driveshaft................. ACU-23

E
EGR....................... ACU-21
Electronic Control Module
  (ECM/PCM) ACU-21
Engine ACU-22
Engine Lubrication........ ACU-23
Engine Mounts........... ACU-22
Evaporator ACU-26
Exhaust ACU-21
Exhaust Manifold......... ACU-21
Expansion Valve ACU-26

F
Flexplate ACU-23
Fuel Injection ACU-21
Fuel Pump................ ACU-21
Fuel Vapor Canister....... ACU-21

G
Gear Selector Lever ACU-23
Generator................. ACU-20

H
Halfshaft.................. ACU-24
Headlamp ACU-26
Heater Core ACU-26
Horn ACU-26

I
Idle Air Control (IAC) Valve . ACU-21
Ignition Coil ACU-21
Ignition Module ACU-21
Ignition Switch ACU-21
Inner Tie Rod ACU-25
Intake Air Temperature (IAT)
  Sensor ACU-21
Intake Manifold ACU-21

K
Knock Sensor ACU-21

L
Lower Control Arm........ ACU-25

M
Maintenance Schedule ACU-20
Manifold Absolute Pressure
  (MAP) Sensor.......... ACU-21
Mass Air Flow (MAF)
  Sensor ACU-21
Master Cylinder ACU-24
Muffler ACU-21
Multifunction Lever/Switch .. ACU-26

O
Oil Pan ACU-23
Oil Pump ACU-23
Outer Tie Rod ACU-25
Oxygen Sensor ACU-21

P
Parking Brake ACU-24
Pistons ACU-23
Positive Crankcase
  Ventilation (PCV) Valve ... ACU-21

R
Radiator ACU-22
Radiator Hoses ACU-22
Radio ACU-26
Rear Main Oil Seal ACU-23

S
Shock Absorber/Strut,
  Front.................. ACU-24
Shock Absorber/Strut,
  Rear................... ACU-25
Spark Plug Cables........ ACU-21
Spark Plugs ACU-21
Spring, Front Coil......... ACU-24
Spring, Rear Coil.......... ACU-25
Starter ACU-20
Steering Wheel ACU-25

T
Thermostat............... ACU-22
Throttle Body............. ACU-21
Throttle Position Sensor
  (TPS) ACU-21
Timing Belt............... ACU-22
Torque Converter......... ACU-23

U
U-Joint ACU-23
Upper Control Arm........ ACU-25

V
Valve Body................ ACU-23
Valve Cover Gasket....... ACU-22
Valve Job................. ACU-22
Vehicle Speed Sensor ACU-21

W
Water Pump.............. ACU-22
Wheel Balance........... ACU-20
Window Regulator........ ACU-27
Windshield Washer Pump .. ACU-26
Windshield Wiper Motor.... ACU-26

ACU-20 MDX : SLX

	LABOR TIME	SEVERE SERVICE
MAINTENANCE		

Air Cleaner Filter Element, Replace (C)
  1996-05 (.2) 3   .4
Chassis Lubrication, Change Oil & Filter (C)
Includes: Correct fluid levels.
  1996-05 (.3) 5   .5
Drive Belt, Adjust (B)
  Alternator
  1996-99 (.2) 3   .4
Drive Belt, Replace (B)
  A/C
  1996-97 (.4) 6   .7
  1998-99 (.2) 3   .4
  2001-05 (.5) 8   .9
  Alternator
  1996-97 (.5) 8   .9
  1998-05 (.3) 5   .5
  P/S
  1996-05 (.4) 6   .7
Fuel Filter, Replace (B)
  1996-99 (.3) 5   .6
  2001-05 (1.0) 1.6   1.7
Halogen Headlamp Bulb, Replace (B)
  1996-05 each (.2) 3   .3
Oil & Filter, Change (C)
Includes: Correct fluid levels.
  1996-05 (.3) 5   .5
Timing Belt, Replace (B)
  1996-97 .... 4.1   4.4
  1998-99 .... 3.6   3.9
  2001-05 (3.0) 4.3   4.6
Tire, Replace (C)
Includes: Dismount old tire and mount new tire to rim.
  One (.4) 6   .7
  each addl. add 6   .7
Wheel, Balance (B)
  One (.3) 5   .5
  each addl. add 5   .5

SCHEDULED MAINTENANCE INTERVALS
If necessary, refer to appropriate Chilton maintenance service information.

7,500 Mile Service (C)
  1996-97 .... 2.1   2.1
  1998-99 .... 2.7   2.7
  2001-05 .... 1.5   1.5
15,000 Mile Service (B)
  1996-97 .... 4.1   4.1
  1998-99 .... 4.4   4.4
  2001-05 .... 1.2   1.2
22,500 Mile Service (C)
  1996-97 .... 2.1   2.1
  1998-99 .... 2.7   2.7
  2001-05 .... 1.0   1.0

30,000 Mile Service (B)
  1996-97 .... 6.9   6.9
  1998-99 .... 7.0   7.0
  2001-05 .... 2.0   2.0
37,500 Mile Service (C)
  1996-97 .... 2.1   2.1
  1998-99 .... 2.7   2.7
  2001-05 .... 1.0   1.0
45,000 Mile Service (B)
  1996-97 .... 3.3   3.3
  1998-99 .... 3.8   3.8
  2001-05 .... 3.6   3.6
52,500 Mile Service (C)
  1996-97 .... 2.1   2.1
  1998-99 .... 2.7   2.7
  2001-05 .... 1.0   1.0
60,000 Mile Service (B)
  1996-97 .... 9.0   9.0
  1998-99 .... 9.5   9.5
  2001-05 .... 2.0   2.0
67,500 Mile Service (C)
  1996-97 .... 2.1   2.1
  1998-99 .... 2.7   2.7
  2001-05 .... 1.0   1.0
75,000 Mile Service (B)
  1996-97 .... 3.2   3.2
  1998-99 .... 5.2   5.2
  2001-05 .... 2.5   2.5
82,500 Mile Service (C)
  1996-97 .... 2.1   2.1
  1998-99 .... 2.7   2.7
  2001-05 .... 1.0   1.0
90,000 Mile Service (B)
  1996-97 .... 6.9   6.9
  1998-99 .... 7.0   7.0
  2001-05 .... 2.6   2.6
97,500 Mile Service (C)
  1996-97 .... 2.1   2.1
  1998-99 .... 2.7   2.7
  2001-05 .... 1.0   1.0
105,000 Mile Service (C)
  1998-99 .... 3.8   3.8
  2001-05 .... 7.7   7.7
Replace water pump
  01-05 add 5   .5
112,500 Mile Service (C)
  1998-99 .... 2.7   2.7
  2001-05 .... 1.0   1.0
120,000 Mile Service (C)
  2001-05 .... 2.0   2.0

CHARGING

Alternator Circuits, Test (B)
Includes: Test component output.
  1996-05 (.3) 5   .5
Alternator, R&R and Recondition (B)
  1996-99 (1.5) 2.2   2.4
  2001-05 (2.0) 3.0   3.2

Alternator Assy., Replace (B)
Includes: Pulley transfer.
  1996-99 (.5) 7   .9
  2001-03 (1.4) 2.0   2.1
Voltage Regulator, Replace (B)
  2001-05 (1.6) 2.3   2.5
Voltmeter, Replace (B)
  1996-99 (1.2) 1.6   1.7

STARTING

Starter Draw Test (On Car) (B)
  1996-05 (.3) 5   .5
Battery, Replace (B)
  1996-05 (.4) 6   .7
Battery Cables, Replace (C)
  1996-05
    positive (.4) 6   .7
    negative (.2) 3   .4
Starter, R&R and Recondition (B)
  1996-99 (1.4) 2.0   2.2
  2001-03 (1.2) 1.7   1.9
Starter Assy., Replace (B)
  1996-99 (.5) 7   .8
  2001-03 (.7) 1.0   1.1
Starter Relay, Replace (B)
  1996-05 (.2) 3   .3

CRUISE CONTROL

Diagnose Cruise System Component Each (A)
  1996-05 (.6) 9   .9
Actuator Assy., Replace (B)
  1996-05 (.4) 6   .6
Control Actuator Cable, Replace (B)
  1996-05 (.3) 5   .5
Control Main Switch, Replace (B)
  1996-99 (.8) 1.1   1.1
  2001-05 (.2) 3   .3
Control Pump, Replace (B)
  1996-97 (.4) 6   .6
Control Relay, Replace (B)
  1996-05 (.2) 3   .3
Control Sensor, Replace (B)
  1996-05 (.6) 8   .9
Controller Module, Replace (B)
  1996-97 (.8) 1.1   1.1
  1998-99 (.3) 5   .5

IGNITION

Diagnose Ignition System Component Each (A)
  1996-05 (.6) 8   .8
Retrieve Fault Codes (A)
  1996-05 (.3) 4   .4
Camshaft Position Sensor, Replace (B)
  1996-97 (.2) 3   .4
  1998-99 (1.5) 2.2   2.3
  2001-05 (3.0) 4.3   4.4

MDX : SLX **ACU-21**

	LABOR TIME	SEVERE SERVICE
Crankshaft Angle Sensor, Replace (B)		
1996-99 (.4)	.6	.7
2001-05 (3.0)	4.3	4.4
Ignition Coil, Replace (B)		
1996-05 (.2)	.3	.3
Ignition Module, Replace (B)		
1996-97 (.2)	.3	.3
Ignition Switch, Replace (B)		
1996-05 (.9)	1.3	1.3
Key Warning Buzzer or Chime, Replace (B)		
1996-99 (.2)	.3	.3
Spark Plug (Ignition) Cables, Replace (B)		
1996-99	.7	.9
Spark Plugs, Replace (B)		
1996-05 (.6)	.9	1.0

EMISSIONS

The following operations do not include testing. Add time as required.

	LABOR TIME	SEVERE SERVICE
Diagnose Emission Control System Component Each (A)		
1996-05 (.6)	.9	.9
Reprogram Electronic Control Module (ECM) (A)		
All Models (.3)	.5	.5
Retrieve Fault Codes (A)		
All Models (.3)	.5	.5
Dynamometer Test (B)		
1996-99	.4	.4
Coolant Temperature (ECT) Sensor, Replace (B)		
1996-97 (1.3)	1.8	1.8
1998-03 (.4)	.6	.6
ECI Control Relay, Replace (B)		
1996-99 (.3)	.5	.5
EGR Pressure Feedback Transducer, Replace (B)		
1996-97 (.3)	.5	.5
EGR Valve, Replace (B)		
1996-97 (.4)	.6	.6
1998-05 (.3)	.5	.5
Electronic Control Module (ECM/PCM), Replace (B)		
1996-99 (.8)	1.1	1.1
2001-05 (.4)	.6	.6
Electronic Spark Control Module, Replace (B)		
1996-97 (.2)	.3	.3
Fuel Tank Pressure Sensor, Replace (B)		
1996-97 (.2)	.3	.4
1998-05 (.5)	.7	.8
Fuel Vapor Canister, Replace (B)		
1996-05 (.6)	.8	.9

	LABOR TIME	SEVERE SERVICE
Intake Air Temperature (IAT) Sensor, Replace (B)		
1996-05 (.3)	.5	.5
Knock Sensor, Replace (B)		
1996-97 (2.0)	2.9	3.0
1998-99 (1.3)	1.9	2.0
2001-05 (.7)	1.1	1.2
Manifold Absolute Pressure (MAP) Sensor, Replace (B)		
1996-05 (.2)	.3	.3
Mass Air Flow (MAF) Sensor, Replace (B)		
1996-99 (.2)	.3	.3
Oxygen Sensor, Replace (B)		
One (.3)	.5	.6
Positive Crankcase Ventilation (PCV) Valve, Replace (B)		
1996-05 (.2)	.3	.3
Throttle Position Sensor (TPS), Replace (B)		
1996-97 (.6)	.8	.8
1998-99 (.2)	.3	.3
2001-05 (.5)	.7	.7
Vacuum Control Valve, Replace (B)		
1996-99 (.3)	.5	.5
Vacuum Switching Valve, Replace (B)		
1996-99 (.4)	.6	.6
Vapor Pressure Control Valve, Replace (B)		
1996-99 (.3)	.5	.5
2001-03 (1.2)	1.6	1.6
Vehicle Speed Sensor, Replace (B)		
1996-99 (1.0)	1.4	1.5
2001-05 (.5)	.7	.8

FUEL

DELIVERY

	LABOR TIME	SEVERE SERVICE
Fuel Pump, Test (B)		
1996-05	.3	.3
Fuel Gauge (Dash), Replace (B)		
1996-05 (1.0)	1.3	1.3
Fuel Gauge (Tank), Replace (B)		
1996-99 (.4)	.6	.8
2001-05 (1.0)	1.6	1.8
Fuel Pump, Replace (B)		
1996-99 (.8)	1.1	1.3
2001-05 (1.0)	1.6	1.8
Fuel Pump Relay, Replace (B)		
1996-05 (.2)	.3	.3
Fuel Tank, Replace (B)		
Includes: Drain and refill.		
1996-05 (1.0)	1.6	1.8
Intake Manifold and/or Gasket, Replace (B)		
Includes: Adjustments.		
1996-97 (1.8)	2.8	3.1
1998-99 (1.1)	1.7	2.0
2001-05 (.5)	.8	1.0

	LABOR TIME	SEVERE SERVICE
INJECTION		
Idle Speed, Adjust (B)		
1996-05 (.3)	.5	.5
Air Regulator, Replace (B)		
1996-99 (.3)	.5	.5
Fuel Injector Resistor, Replace (B)		
1996-99 (.2)	.3	.3
Fuel Injectors, Replace (B)		
1996-99 (2.2)	3.1	3.3
2001-05 (1.5)	2.2	2.4
Fuel Pressure Regulator Replace (B)		
1996-97 (.4)	.6	.7
1998-99 (.6)	.8	.9
2001-05 (.3)	.5	.6
Idle Air Control (IAC) Valve, Replace (B)		
1996-97 (.2)	.3	.3
1998-99 (.5)	.7	.7
2001-05 (.7)	.9	.9
Throttle Body Assy., Replace (B)		
1996-05 (.5)	.8	.9

EXHAUST

	LABOR TIME	SEVERE SERVICE
Catalytic Converter, Replace (B)		
1996-99		
three-way converter		
right side	.7	.8
left side	1.0	1.1
warm up converter	1.4	1.5
2001-05 (.5)	.8	.9
Exhaust Manifold or Gasket, Replace (B)		
1996-97		
right side	3.6	3.9
left side	4.9	5.2
both sides	6.5	6.8
1998-99		
right side	2.5	2.8
left side	2.0	2.3
both sides	4.4	4.7
2001-05		
one side (1.2)	1.8	2.1
both sides (2.2)	3.3	3.6
Exhaust Pipe, Replace (B)		
1996-99 rear (.4)	.6	.7
2001-05		
front or rear (.6)	1.0	1.1
Muffler and Tailpipe Assy., Replace (B)		
1996-99	.7	.8
2001-05		
muffler (.5)	.8	.9
tailpipe extension	.3	.3

ENGINE COOLING

Pressure Test Cooling System (C)
  1996-05 3   .3
Coolant Bypass Hose, Replace (B)
  2001-05 one (.4) 6   .7
Coolant Thermostat, Replace (B)
  1996-97 (.3) 5   .7
  1998-99 (1.3) 1.8   1.9
  2001-05 (.8) 1.2   1.4
Fan Clutch Assy., Replace (B)
  1996-99 (.5) 8   .9
Radiator Assy., R&R or Replace (B)
Includes: Refill with proper coolant mix.
  1996-99 (1.2) 1.6   1.7
  2001-05 (1.7) 2.6   2.9
Radiator Fan and/or Fan Motor, Replace (B)
  2001-05 (.5) 8   .9
Radiator Fan Temperature Switch, Replace (B)
  2001-05 (.5) 8   .9
Radiator Fan Relay, Replace (B)
  2001-05 (.1) 2   .2
Radiator Hoses, Replace (B)
Includes: Refill with proper coolant mix.
  1996-99 each (.3) 5   .6
  2001-05 each (.7) . . . 1.1   1.2
Temperature Gauge (Dash), Replace (B)
  1996-05 (1.0) 1.3   1.3
Water Pump and/or Gasket, Replace (B)
Includes: Includes: Refill with proper coolant mix.
  1996-97 (3.2) 4.7   5.0
  1998-99 (2.6) 3.8   4.1
  2001-05 (3.3) 4.8   5.1

ENGINE

ASSEMBLY
Times shown are for OEM assemblies. Time to replace assemblies from aftermarket rebuilders may vary.
Engine Assy., R&I (B)
Does not include parts or component transfer.
  1996-99 . . . . 11.3   11.3
Engine Assy., R&R and Recondition (A)
Includes: Replacing rings, rod and main bearings, cylinder head reconditioning and engine tune-up.
  1996-99 . . . . 32.2   32.2
Engine (New or Rebuilt Complete), Replace (B)
Includes: Component transfer and engine tune-up.
  1996-05 (9.5) 14.4   16.3

Engine Assy. (Short Block), Replace (B)
Assembly consists of cylinder block, piston assemblies, crankshaft, camshaft, timing chain and gears. Does not include cylinder heads. Operation Includes: R&R engine, transfer necessary parts and all necessary adjustments.
  1996-05 (15.0) 22.8   25.8
Engine Mounts, Replace (B)
  1996-99
    front
      right side (.8) 1.3   1.4
      left side (.9) 1.4   1.5
    rear (.5) 7   .9
  2001-05
    front (1.7) 2.7   2.8
    rear (.5) 8   .9
    side (right or left) (.5) . .8   .9
    all (2.0) 3.0   3.1

CYLINDER HEAD
Compression Test (B)
  1996-05 . . . . 1.1   1.2
Cylinder Head, Replace (B)
Includes: Parts transfer, adjustments.
  1996-99
    3.2L
      one side 12.6   13.1
      both sides 17.1   17.6
    3.5L
      one side 14.7   15.2
      both sides 18.8   19.3
  2001-05
    one side (6.0) 8.6   9.1
    both sides (7.0) . . . 9.9   10.4
Cylinder Head Gasket, Replace (B)
  1996-99
    one side 10.1   10.4
    both sides 12.3   12.6
  2001-05
    one side (5.5) 7.9   8.4
    both sides (6.5) 9.2   9.7
Rocker Arms or Shafts, Replace (B)
  1996-99
    right side 7.6   7.8
    left side 5.0   5.2
  2001-05 (4.2) 6.1   6.3
Valve/Cam Cover and/or Gasket, Replace (B)
  1996-99
    3.2L
      right side 4.5   4.6
      left side 2.0   2.1
      both sides 6.5   6.6
    3.5L
      right side 2.0   2.1
      left side 2.3   2.4
      both sides 4.3   4.4

  2001-03
    one side (.8) 1.2   1.3
    both sides (1.1) . . . 1.7   1.8
Valve Job (A)
Includes: Valve adjustment.
  1996-99
    3.2L 21.4   21.9
    3.5L 22.9   23.4
  2001-05 (10.3) 13.6   14.1
Valve Lifters, Replace (B)
  1996-99
    right side 8.2   8.4
    left side 5.6   5.8
Valve Springs and/or Oil Seals, Replace (B)
  1996-99
    one side (5.7) 8.2   8.7
    both sides (7.3) . . . 10.3   10.8

CAMSHAFT
Timing Belt, Adjust (B)
  1996-99 (.3) 5   .7
Camshaft, Replace (B)
  1996-99
    3.2L
      right side 8.2   8.4
      left side 5.3   5.5
      both sides 9.7   9.9
    3.5L
      one side 5.1   5.3
      both sides 8.6   8.8
  2001-05 (4.2) 6.1   6.4
Camshaft Seal, Replace (B)
  1996-99
    one (1.8) 2.7   2.8
    both (2.2) 3.3   3.4
  2001-05
    one or both (3.4) . . . 5.0   5.7
Replace timing belt
  96-99 add 1.0   1.2
Timing Belt, Replace (B)
  1996-97 . . . . 4.1   4.6
  1998-99 . . . . 3.6   4.1
  2001-05 (3.0) 4.3   4.6
Timing Belt Tensioner, Replace (B)
  1996-97 . . . . 4.1   4.3
  1998-99 . . . . 3.6   3.8
Timing Cover and/or Gasket, Replace (B)
  1996-99 . . . . 3.1   3.3

CRANK & PISTONS
Connecting Rod Bearings, Replace (A)
Includes: Check bearing oil clearance.
  1996-99 . . . . 12.1   12.6
  2001-05 (2.3) 3.5   4.0

MDX : SLX ACU-23

	LABOR TIME	SEVERE SERVICE

Crankshaft and Main Bearings, Replace (A)
Includes: Engine R&R 1996-99.
Does not include: Cylinder head removal 2001-03.
  1996-99 . . . . **23.2** *23.7*
  2001-05 (17.4) **22.6** *23.1*

Crankshaft Front Oil Seal, Replace (B)
  1996-99 (2.9) **4.4** *4.6*
  2001-05 (3.3) **4.8** *5.0*

Crankshaft Pulley or Damper, Replace (B)
  1996-99 . . . . **3.9** *4.0*

Pistons and Connecting Rods, Replace (B)
Includes: Ridge reaming, cylinder wall deglazing, installing new rings and rod bearings, engine tune-up.
  1996-99 . . . . **24.2** *25.1*
  2001-05 (12.4) **16.6** *17.5*

Rear Main Oil Seal, Replace (B)
  1996-99 (7.7) **11.7** *12.2*
  2001-05 (6.9) **9.7** *10.2*

Rings, Replace (A)
Includes: Ridge reaming, cylinder wall deglazing, installing new rings, engine tune-up.
  1996-99 (14.8) **19.4** *20.3*
  2001-03 (11.6) **15.7** *16.6*

ENGINE LUBRICATION

Engine Oil Cooler, Replace (B)
  1996-97 (.6) **1.0** *1.1*

Engine Oil Cooler Coolant Hose, Replace (B)
  1996-97 (.4) . . . .**6** *.7*

Engine Oil Pressure Switch (Sending Unit), Replace (B)
  1996-05 (.4) . . . .**6** *.6*

Oil Pan and/or Gasket, Replace (B)
  1996-99 (5.0) **7.2** *7.5*
  2001-05 (1.2) **1.8** *2.1*

Oil Pressure Gauge (Dash), Replace (B)
  1996-99 (1.2) **1.6** *1.6*

Oil Pump, Replace (B)
  1996-99
    3.2L **11.4** *11.6*
    3.5L **12.1** *12.3*
  2001-05 (4.5) **6.5** *6.7*

Oil Pump Relief Valve, Replace (B)
  2001-05 (1.3) **1.9** *2.2*

Sub-Oil Pan or Gasket, Replace (B)
  1996-99 (.5) **8** *1.0*

AUTOMATIC TRANSMISSION
SERVICE TRANSMISSION INSTALLED

Diagnose Electronic Transmission Component Each (A)
  1996-05 (.6) . . . .**9** *.9*

Drain and Refill Unit (B)
  1996-99 . . . . **1.4** *1.4*

Pressure Test (B)
  1996-99 (.3) . . . .**5** *.5*
  2001-05 (1.0) **1.5** *1.6*

Retrieve Fault Codes (A)
  1996-05 (.3) . . . .**4** *.4*

Oil Pan and/or Gasket, Replace (B)
  1996-99 (1.0) **1.4** *1.4*

Speedometer Driven Gear and/or Seal, Replace (B)
  1996-99 . . . .**7** *.8*

Transmission Control Module (TCM/PCM), Replace (B)
  1996-05 (.4) . . . .**6** *.6*

Transmission Speed Sensor, Replace (B)
  1996-99 (.6) **1.0** *1.0*
  2001-05 (.3) . . . .**5** *.5*

Valve Body Assy., R&R and Recondition (A)
  1996-99 . . . . **3.8** *4.1*

Valve Body Assy., Replace (B)
  1996-99 . . . . **1.9** *2.1*

SERVICE TRANSMISSION REMOVED

Transmission R&R included unless otherwise noted.

Flywheel (Flexplate), Replace (B)
  1996-99 . . . . **12.7** *12.9*

Front Oil Pump, Replace (B)
  1996-99 . . . . **13.1** *13.3*

Torque Converter, Replace (B)
  1996-99 . . . . **12.6** *12.8*
  2001-05 (6.5) **9.2** *9.7*

Transmission, Recondition (A)
Includes: Replace Torque Converter.
  1996-99 . . . . **18.3** *19.2*
  2001-05 (10.3) **14.1** *15.0*

Transmission Assy., R&R (B)
  1996-99 . . . . **12.4** *12.6*
Replace assembly add . . . **.5** *.7*

TRANSFER CASE

4WD Actuator Motor, Replace (B)
  1998-99 (.8) **1.2** *1.3*

4WD Control Module, Replace (B)
  1997-05 (.6) **9** *1.0*

Transfer Case, R&R and Recondition (B)
  1996-99 (4.3) **6.2** *7.1*
  2001-05 (2.8) **4.1** *5.0*

Transfer Case, R&R or Replace (B)
  1996-99 (3.1) **4.5** *4.7*
  2001-05 (.8) **1.3** *1.5*

Transfer Case Oil Seal, Replace (B)
  1996-99
    front or rear (1.2) . . . **1.8** *1.9*

SHIFT LINKAGE
AUTOMATIC TRANSMISSION

Shift Control Cable, Adjust (B)
  2001-05 (.5) . . . .**7** *.7*

Shift Control Cable, Replace (B)
  1996-99 (1.0) **1.4** *1.4*
  2001-05 (1.9) **2.8** *3.0*

Gear Selector Lever, R&R and Recondition (B)
  2001-05 (1.7) **2.5** *2.7*

Gear Selector Lever, Replace (B)
  2001-05 (1.5) **2.2** *2.3*

DRIVELINE
DRIVESHAFT

Driveshaft, Replace (B)
  1996-99
    front or rear (.7) **1.1** *1.2*
  2001-05 rear (.4) . . . .**6** *.7*
Recondition or replace
  U joints add each . . . .**3** *.5*

DRIVE AXLE

Axle Shaft, Replace (B)
  1996-99
    front (3.6) **5.3** *5.5*
    rear (1.2) **1.8** *2.0*

Axle Shaft Bearing and/or Oil Seal, Replace (B)
  1996-99 rear (1.2) **1.8** *2.0*

Axle Shaft Oil Seal, Replace (B)
  2001-05
    front
      right (.7) **1.1** *1.2*
      left (.9) **1.3** *1.4*
    rear (.7) **1.1** *1.2*

Differential Assy., R&R and Recondition (B)
  1996-99
    front (6.0) **8.6** *9.1*
    rear (3.4) **5.1** *5.6*

Differential Assy., Replace (B)
  1996-99
    front (4.1) **5.9** *6.4*
    rear (1.9) **2.8** *3.3*
  2001-05 rear (2.0) **3.0** *3.5*

ACU-24 MDX : SLX

	LABOR TIME	SEVERE SERVICE
Differential Carrier, R&R or Replace (B)		
1996-99		
front	6.2	6.7
rear	2.9	3.4
Replace assembly add	.5	.7
Differential Side Bearings, Replace (B)		
1996-99		
front	7.0	7.2
rear	3.6	3.8
Front Axle Shaft Actuator, Replace (B)		
1996-99 (1.0)	1.5	1.6
Intermediate Axle Shaft, Replace (B)		
2001-05 (.9)	1.4	1.6
Pinion Shaft Oil Seal, Replace (B)		
1996-99		
front	1.3	1.4
rear	1.6	1.7
2001-05		
rear (.7)	1.1	1.2
TCS/Rear Lock Control Unit., Replace (B)		
2001-05 (.5)	.7	.7

HALFSHAFTS

	LABOR TIME	SEVERE SERVICE
CV Joint and/or Boots, Replace (B)		
2001-05		
inner (1.1)	1.6	1.7
outer (.9)	1.3	1.4
Halfshaft, Replace (B)		
2001-05 each (.7)	1.1	1.3

BRAKES
ANTI-LOCK

The following operations do not include testing. Add time as required.

	LABOR TIME	SEVERE SERVICE
Diagnose Anti-Lock Brake System Component Each (A)		
1996-05 (.6)	.8	.8
Retrieve Fault Codes (A)		
1996-05 (.3)	.4	.4
Anti-Lock Valve, Replace (B)		
1996-97 (.8)	1.1	1.1
Combination Valve, Replace (B)		
1996-97 (.9)	1.2	1.2
Control Module, Replace (B)		
ABS/TCS/VSA		
1996-99 (.4)	.6	.6
2001-05 (1.0)	1.4	1.4
Exciter Ring, Replace (B)		
1996-99 (2.3)	3.4	3.4
G Sensor, Replace (B)		
1996-99 (.6)	.8	.8
Hydraulic Assy. (Modulator), Replace (B)		
1996-05 (1.0)	1.5	1.7

	LABOR TIME	SEVERE SERVICE
Wheel Speed Sensor, Replace (B)		
1996-98		
2 wheel system (.3)	.5	.5
4 wheel system		
front one (.6)	.8	.8
rear one (.3)	.5	.5
2001-05 each (.3)	.5	.5

SYSTEM

	LABOR TIME	SEVERE SERVICE
Bleed Brakes (B)		
Includes: Add fluid.		
1996-05 (.4)	.6	.6
Brake System, Flush and Refill (B)		
1996-05	1.3	1.3
Brake Hose (Flexible), Replace (B)		
Includes: System bleeding.		
One	1.0	1.1
each addl. add	.3	.4
Combination Valve, Replace (B)		
Includes: System bleeding.		
1996-99	1.4	1.4
Master Cylinder, Replace (B)		
Includes: System bleeding.		
1996-99 (1.0)	1.4	1.5
2001-05 (1.2)	1.7	1.8
Power Booster Unit, Replace (B)		
Includes: Adjust push rod and system bleeding.		
1996-99 (1.0)	1.4	1.5
2001-05 (1.3)	1.9	2.0
Power Booster Vacuum Check Valve, Replace (B)		
1996-05 (.2)	.3	.3
Proportioning Valve, Replace (B)		
1996-05 (1.0)	1.4	1.5

SERVICE BRAKES

	LABOR TIME	SEVERE SERVICE
Brake Pads, Replace (B)		
Front or rear (1.0)	1.4	1.4
Four wheels	2.7	2.8
Caliper Assy., R&R and Recondition (B)		
Includes: System bleeding.		
1996-99		
front		
one	1.2	1.5
both	2.3	2.6
rear		
one	1.5	1.8
both	2.9	3.2
2001-05		
one (.8)	1.3	1.4
both (1.6)	2.6	2.9
Caliper Assy., Replace (B)		
Includes: System bleeding.		
Front or rear		
one (.6)	1.0	1.1
both (1.2)	1.9	2.2

	LABOR TIME	SEVERE SERVICE
Disc Brake Rotor, Replace (B)		
1996-99		
front		
one	1.5	1.6
both	2.9	3.0
rear		
one	1.0	1.1
both	1.9	2.0
2001-05		
one (.7)	1.1	1.3
both (1.2)	2.2	2.5

PARKING BRAKE

	LABOR TIME	SEVERE SERVICE
Parking Brake Cable, Adjust (C)		
1996-05 (.4)	.6	.7
Parking Brake Apply Actuator, Replace (B)		
1996-05 (.8)	1.3	1.3
Parking Brake Apply Warning Indicator Switch, Replace (B)		
1996-99 (.6)	.8	.8
2001-05 (.3)	.5	.5
Parking Brake Cable, Replace (B)		
1996-99 (1.7)	2.6	2.7
2001-05		
front (.6)	1.0	1.1
rear		
one (.8)	1.3	1.4
both (1.5)	2.4	2.5
Parking Brake Shoes, Replace (B)		
1996-99		
one side	.7	.8
both sides (1.0)	1.4	1.5
2001-05		
one side (1.0)	1.4	1.5
both sides (2.0)	2.6	2.9

FRONT SUSPENSION

Unless otherwise noted, time given does not include alignment.

	LABOR TIME	SEVERE SERVICE
Align Front End (A)		
1996-99 (1.0)	1.5	1.6
2001-05 (.6)	1.0	1.1
Front & Rear Alignment, Check & Adjust (A)		
2001-05 (1.2)	1.9	2.0
Front Toe, Adjust (B)		
1996-99 (.6)	1.0	1.1
Torsion Bar, Adjust (B)		
One side	.5	.7
Both sides	.7	.9
Front Damper Spring, Replace (B)		
MDX		
one (.8)	1.2	1.4
both (1.6)	2.4	2.6
Front Strut or Shock Absorbers, Replace (B)		
1996-99		
one	.5	.6
both (1.0)	1.5	1.6

MDX : SLX **ACU-25**

	LABOR TIME	SEVERE SERVICE
2001-05		
one (.6)	1.0	*1.1*
both (1.2)	1.9	*2.2*
Hub Oil Seal, Replace (B)		
1996-99 (.6)	.8	*.9*
Locking Hub Cam and/or Clutch, Replace (B)		
1996-99	.7	*.9*
Lower Ball Joint, Replace (B)		
1996-99 (.7)	1.1	*1.2*
2001-05		
one (1.4)	2.2	*2.3*
both (2.7)	4.0	*4.2*
Lower Control Arm Assy., Replace (B)		
1996-99 (2.0)	3.1	*3.2*
2001-05		
one (.6)	1.0	*1.1*
both (1.2)	1.9	*2.1*
Replace bushings		
96-99 add	.5	*.7*
Stabilizer Bar, Replace (B)		
1996-99 (1.0)	1.5	*1.6*
2001-05 (.7)	1.1	*1.2*
Stabilizer Bar Bushings or Link, Replace (B)		
1996-05 (.4)	.6	*.7*
Steering Knuckle, Replace (B)		
1996-99		
one (2.0)	3.1	*3.2*
both (4.0)	5.8	*6.0*
2001-05		
one (1.1)	1.6	*1.7*
both (2.2)	3.3	*3.5*
Recondition knuckle		
96-99 add	.5	*.6*
Torsion Bar, Replace (B)		
Includes: Height adjustment.		
1996-99 (.7)	1.0	*1.3*
w/skid plate add	.3	*.3*
Upper Ball Joint, Replace (B)		
1996-99 (.6)	.8	*1.0*
Upper Control Arm Assy., Replace (B)		
1996-99 (2.0)	3.1	*3.3*
Upper Control Arm Bushings and/or Shaft, Replace (B)		
1996-99 (2.3)	3.4	*3.6*
Wheel Bearings, Replace or Clean & Repack (B)		
1996-99		
2WD		
one side (1.0)	1.6	*1.7*
both sides (2.0) ..	3.1	*3.3*
4WD		
one side (1.5)	2.2	*2.3*
both sides (2.5) ..	3.7	*3.9*
2001-05		
one (1.1)	1.6	*1.7*
both (2.2)	3.3	*3.7*

	LABOR TIME	SEVERE SERVICE
REAR SUSPENSION		
Center Link, Replace (B)		
1996-99 (.4)	.6	*.7*
Replace bushings		
add each	.2	*.3*
Lateral Link, Replace (B)		
1996-99 (.6)	.6	*.7*
Replace bushings		
add each	.2	*.2*
Rear Coil Spring, Replace (B)		
1996-99 (.5)	.8	*.9*
2001-05		
one (.5)	.8	*.9*
both (1.0)	1.5	*1.7*
Rear Crossmember, Replace (B)		
2001-05 (3.0)	4.4	*4.6*
Rear Stabilizer Bar & Bushings, Replace (B)		
1996-99 (.7)	1.1	*1.2*
Replace bushings add ...	.2	*.2*
Shock Absorbers or Bushings, Replace (B)		
1996-99		
one (.5)	.7	*.8*
both (1.0)	1.5	*1.7*
2001-05		
one (.3)	.5	*.5*
both (.6)	1.0	*1.1*
Trailing Arm or Link, Replace (B)		
1996-99 (.6)	.6	*.7*
2001-05		
one (.6)	1.0	*1.1*
both (1.2)	1.9	*2.2*
Replace bushings		
add each	.2	*.2*
STEERING		
AIR BAGS		
Diagnose Air Bag System Component Each (A)		
1996-05 (.6)	.9	*.9*
Retrieve Fault Codes (A)		
1996-05 (.3)	.4	*.4*
Air Bag Assy., Replace (B)		
Driver side (.3)	.5	*.5*
Passenger side		
1996-99 (1.5)	2.2	*2.2*
2001-05 (.5)	.6	*.6*
Side		
2001-05 (1.3)	1.6	*1.6*
Coil, Replace (B)		
1996-99 (.9)	1.2	*1.2*
Occupant Position Detection System Module, Replace (B)		
2001-05 (1.3)	1.6	*1.6*
Sensing & Diagnostic Module (SDM), Replace (B)		
1996-99 (.8)	1.1	*1.1*

	LABOR TIME	SEVERE SERVICE
Sensor, Replace (B)		
Seat pad		
2001-05 (1.0)	1.3	*1.3*
Side Impact		
2001-05 one (.6)	.8	*.8*
SRS Control Unit, Replace (B)		
2001-05 (.6)	.8	*.8*
LINKAGE		
Idler Arm, Replace (B)		
1996-99 (.5)	.7	*1.0*
Pitman Arm, Replace (B)		
1996-99	.7	*1.0*
Relay Rod (Center Link), Replace (B)		
Includes: Reset toe.		
1996-99	1.6	*1.9*
Tie Rod or Tie Rod Ends, Replace (B)		
Includes: Reset toe.		
1996-99		
one	.7	*.9*
each addl. add	.3	*.5*
2001-05		
one side (.8)	1.3	*1.4*
both sides (1.0)	1.6	*1.8*
POWER WORM & SECTOR		
Troubleshoot Power Steering (A)		
1996-05 (.3)	.5	*.5*
Gear, Adjust (On Vehicle) (B)		
1996-99 (.7)	1.0	*1.1*
Gear Assy., R&R and Recondition (B)		
Includes: Reset toe.		
1996-99 (2.1)	3.0	*3.2*
2001-05 (4.4)	6.2	*6.4*
Gear Assy., Replace (B)		
Includes: Reset toe.		
1996-99 (1.4)	2.2	*2.4*
2001-05 (4.0)	5.6	*5.8*
Reservoir, Replace (B)		
1996-05 (.3)	.5	*.6*
Steering Column, R&R or Replace (B)		
1996-05 (1.1)	1.6	*1.7*
Steering Pump, Replace (B)		
1996-99 (.8)	1.3	*1.4*
2001-05 (.4)	.6	*.7*
Steering Pump Hoses, Replace (B)		
Does not include purging.		
1996-99 each (.3)	.5	*.6*
2001-05 each (1.0) ...	1.4	*1.5*
Steering Wheel, Replace (B)		
1996-05 (.5)	.6	*.6*

HEATING & AIR CONDITIONING

When more than one component requires replacement where evacuation/recovery and recharging is already included, deduct 1.0 for each additional component from the time given.

	LABOR TIME	SEVERE SERVICE
Evacuate/Recover and Recharge System (B)		
1996-99 (1.0)	1.4	1.4
2001-05 (.5)	.7	.7
AC Hoses, Replace (B)		
Includes: Evacuate/recover and recharge.		
1996-99		
discharge (2.0)	2.9	3.0
suction (1.7)	2.4	2.5
2001-05 each (1.4)	2.0	2.1
Blower Motor, Replace (B)		
1996-99 (.3)	.5	.5
2001-05		
front (.3)	.5	.5
rear (.7)	.9	.9
Blower Motor Relay, Replace (B)		
1996-99 (.2)	.3	.3
2001-05		
front (.1)	.2	.2
rear (.6)	.8	.8
Blower Motor Resistor, Replace (B)		
1996-99 (.3)	.5	.5
2001-05		
front or rear (.6)	.8	.8
Blower Motor Switch, Replace (B)		
1996-99	1.3	1.3
Compressor Assy., Replace (B)		
Includes: Parts transfer, evacuate/recover and recharge.		
1996-99 (2.0)	3.0	3.1
2001-05 (1.5)	2.3	2.4
Compressor Clutch and/or Pulley, Replace (B)		
Includes: Compressor R&I.		
1996-99 (2.3)	3.3	3.4
2001-05 (1.8)	2.6	2.7
Compressor Shaft Seal, Replace (B)		
Includes: Parts transfer, evacuate/recover and recharge.		
1996-99	4.1	4.3
Condenser Assy., Replace (B)		
Includes: Evacuate/recover and recharge.		
1996-99 (2.1)	3.1	3.4
2001-05 (1.5)	2.2	2.4
Condenser Pressure Switch, Replace (B)		
1996-99 (.7)	1.1	1.1
2001-05 (.2)	.3	.3

	LABOR TIME	SEVERE SERVICE
Evaporator Coil, Replace (B)		
Includes: Evacuate/recover and recharge.		
1996-99 (5.5)	7.6	7.8
2001-05		
front (1.5)	2.3	2.5
rear (3.5)	4.9	5.1
Expansion Valve, Replace (B)		
Includes: Evacuate/recover and recharge.		
1996-99 (5.5)	7.6	7.8
2001-05		
front (1.5)	2.3	2.5
rear (3.5)	4.9	5.1
Heater Core, R&R or Replace (B)		
1996-99 (4.5)	6.3	6.5
2001-05		
front (6.0)	8.3	8.5
rear (.9)	1.3	1.5
Heater Hoses, Replace (B)		
1996-99 each (.5)	.7	.8
2001-05		
front (.5)	.7	.8
rear (2.0)	2.9	3.0
Receiver/Drier Assy., Replace (B)		
Includes: Evacuate/recover and recharge.		
1996-99 (2.0)	2.9	3.0
2001-03 (1.3)	1.9	2.0
Temperature Control Assy., Replace (B)		
1996-99 (1.0)	1.4	1.4
2001-05 each (.3)	.5	.5

WIPERS & SPEEDOMETER

	LABOR TIME	SEVERE SERVICE
Antenna Assy., Replace (B)		
1996-99		
manual	.7	.7
power (1.0)	1.4	1.4
2001-05 (1.0)	1.4	1.4
Intermittent Wiper Control Unit, Replace (B)		
1996-99	.5	.5
Radio, R&R (B)		
1996-99 (1.0)	1.4	1.4
2001-05 (.3)	.5	.5
Rear Window Washer Pump Motor, Replace (B)		
1996-05 (.6)	.9	1.0
Rear Window Wiper/Washer Switch, Replace (B)		
1996-05 (.8)	1.1	1.1
Rear Wiper Motor, Replace (B)		
1996-05 (.6)	.9	1.0
Speedometer Head, R&R or Replace (B)		
1996-99 (1.0)	1.3	1.3
2001-05 (.5)	.7	.7

	LABOR TIME	SEVERE SERVICE
Windshield Washer Pump Motor, Replace (B)		
1996-05 (.6)	.9	1.0
Windshield Wiper Motor, Replace (B)		
1996-99 (.6)	.8	.9
2001-05 (.9)	1.4	1.5
Windshield Wiper Switch, Replace (B)		
1996-05 (.7)	1.0	1.0
Windshield Wiper Transmission, Replace (B)		
1996-99 (1.2)	1.7	1.8
2001-05 (.9)	1.3	1.4
Wiper Relay, Replace (B)		
1996-99 (.3)	.5	.5
2001-05 (.1)	.2	.2

LAMPS & SWITCHES

	LABOR TIME	SEVERE SERVICE
Back-Up Lamp Bulb, Replace (C)		
Each (.2)	.3	.3
Combination Switch Assy., Replace (B)		
1996-05 (.7)	1.0	1.0
Front Combination Lamp Bulb, Replace (C)		
One (.2)	.3	.3
each addl. add	.1	.1
Front Combination Lamp Lens, Replace (B)		
One (.2)	.3	.3
each addl. add	.1	.1
Halogen Headlamp Bulb, Replace (B)		
1996-05 each (.2)	.3	.3
Hazard Warning Switch, Replace (B)		
1996-99 (.8)	1.0	1.0
2001-05 (.4)	.6	.6
Headlamps, Aim (B)		
Two (.3)	.5	.5
Four (.4)	.6	.6
High Mount Stop Lamp Assy., Replace (B)		
1996-05 (.3)	.5	.5
High Mount Stop Lamp Bulb, Replace (C)		
1996-05 (.2)	.3	.3
Horn, Replace (B)		
One (.4)	.6	.6
each addl. add	.3	.3
License Lamp Assy., Replace (B)		
1996-05 (.2)	.3	.3
License Lamp Lens or Bulb, Replace (C)		
1996-05 (.2)	.3	.3
Rear Combination Lamp Assy., Replace (B)		
1996-99 each (.2)	.3	.3
2001-05 each (.4)	.6	.6

	LABOR TIME	SEVERE SERVICE
Rear Combination Lamp Bulb, Replace (C)		
Each (.2)	.3	*.3*
Side Marker Lamp Bulb, Replace (C)		
One (.2)	.3	*.3*
each addl. add	.1	*.1*
Side Marker Lamp Lens, Replace (B)		
One (.2)	.3	*.3*
each addl. add	.1	*.1*
Stop Lamp Switch, Replace (B)		
1996-05 (.3)	.5	*.5*
Turn Signal or Hazard Warning Flasher, Replace (B)		
1996-05 (.2)	.3	*.3*

BODY

	LABOR TIME	SEVERE SERVICE
Door Latch, Replace (B)		
1996-99		
front or rear (.8)	1.0	*1.1*
tailgate (.3)	.5	*.6*
2001-05		
front (.7)	.9	*1.0*
rear (.4)	.6	*.7*
tailgate (.8)	1.0	*1.1*
Door Lock Cylinder, Replace (B)		
1996-05 (.6)	.8	*.9*
Door Window Regulator, Replace (B)		
Front or rear (1.0)	1.4	*1.6*

	LABOR TIME	SEVERE SERVICE
Hood Hinge, Replace (B)		
1996-99 (.5)	.7	*.8*
Hood Lock, Replace (B)		
1996-05 (.3)	.5	*.6*
Hood Release Cable, Replace (B)		
1996-99 (.3)	.5	*.6*
2001-05 (.8)	1.1	*1.2*
Lock Striker Plate, Replace (B)		
1996-05 (.3)	.5	*.6*
Window Regulator Motor, Replace (B)		
Front or rear (1.0)	1.4	*1.6*

ACU-28 MDX : SLX

NOTES

DAE

Lanos : Leganza : Nubira

SYSTEM INDEX

- MAINTENANCE DAE-2
- CHARGING DAE-2
- STARTING DAE-2
- IGNITION DAE-2
- EMISSIONS DAE-2
- FUEL DAE-2
- EXHAUST DAE-2
- ENGINE COOLING DAE-3
- ENGINE DAE-3
  - Assembly DAE-3
  - Cylinder Head DAE-3
  - Camshaft DAE-3
  - Crank & Pistons DAE-4
  - Engine Lubrication DAE-4
- CLUTCH DAE-4
- MANUAL TRANSAXLE ... DAE-4
- AUTO TRANSAXLE DAE-4
- SHIFT LINKAGE DAE-4
- DRIVELINE DAE-4
- BRAKES DAE-5
- FRONT SUSPENSION DAE-5
- REAR SUSPENSION DAE-6
- STEERING DAE-6
- HEATING & AC DAE-6
- WIPERS & SPEEDOMETER DAE-6
- LAMPS & SWITCHES DAE-7
- BODY DAE-7

OPERATIONS INDEX

A
- AC Hoses DAE-6
- Air Bags DAE-6
- Air Conditioning DAE-6
- Alignment DAE-5
- Alternator (Generator) ... DAE-2
- Antenna DAE-6
- Anti-Lock Brakes DAE-5

B
- Back-Up Lamp Switch DAE-7
- Ball Joint DAE-5
- Battery Cables DAE-2
- Bleed Brake System DAE-5
- Blower Motor DAE-6
- Brake Disc DAE-5
- Brake Drum DAE-5
- Brake Hose DAE-5
- Brake Pads and/or Shoes .. DAE-5

C
- Camshaft DAE-3
- Coolant Temperature (ECT) Sensor DAE-2
- Crankshaft DAE-4
- CV Joint DAE-4
- Cylinder Head DAE-3

D
- Differential DAE-4
- Drive Belt DAE-2

E
- EGR DAE-2
- Electronic Control Module (ECM/PCM) DAE-2
- Engine DAE-3
- Engine Lubrication DAE-4
- Engine Mounts DAE-3
- Evaporator DAE-6
- Exhaust DAE-2
- Exhaust Manifold DAE-2

F
- Flywheel DAE-4
- Fuel Injection DAE-2
- Fuel Pump DAE-2
- Fuel Vapor Canister DAE-2

G
- Gear Selector Lever DAE-4
- Generator DAE-2

H
- Halfshaft DAE-4
- Headlamp DAE-7
- Heater Core DAE-6
- Horn DAE-7

I
- Ignition Coil DAE-2
- Inner Tie Rod DAE-6
- Intake Manifold DAE-2

K
- Knock Sensor DAE-2

L
- Lower Control Arm DAE-5

M
- Master Cylinder DAE-5
- Muffler DAE-2
- Multifunction Lever/Switch .. DAE-7

N
- Neutral Safety Switch DAE-7

O
- Oil Pan DAE-4
- Oil Pump DAE-4
- Outer Tie Rod DAE-6
- Oxygen Sensor DAE-2

P
- Parking Brake DAE-5
- Pistons DAE-4
- Positive Crankcase Ventilation (PCV) Valve ... DAE-2

R
- Radiator DAE-3
- Radiator Hoses DAE-3
- Radio DAE-6
- Rear Main Oil Seal DAE-4

S
- Shock Absorber/Strut, Front DAE-5
- Shock Absorber/Strut, Rear DAE-6
- Spark Plug Cables DAE-2
- Spark Plugs DAE-2
- Spring, Front Coil DAE-5
- Spring, Rear Coil DAE-6
- Starter DAE-2
- Steering Wheel DAE-6

T
- Thermostat DAE-3
- Throttle Body DAE-2
- Throttle Position Sensor (TPS) DAE-2
- Timing Belt DAE-3

V
- Valve Body DAE-4
- Valve Cover Gasket DAE-3
- Valve Job DAE-3
- Vehicle Speed Sensor DAE-2

W
- Water Pump DAE-3
- Wheel Balance DAE-2
- Wheel Cylinder DAE-5
- Window Regulator DAE-7
- Windshield Washer Pump .. DAE-7
- Windshield Wiper Motor ... DAE-7

DAE-1

DAE-2 LANOS : LEGANZA : NUBIRA

	LABOR TIME	SEVERE SERVICE

MAINTENANCE

Air Cleaner Filter Element, Replace (C)
1999-02 .... .5 .6

Drive Belt, Adjust (B)
1999-02 .... .6 .7

Drive Belt, Replace (B)
SOHC .... .7 .8
DOHC .... .6 .7

Fuel Filter, Replace (B)
Lanos .... .6 .8
Leganza, Nubira .... .5 .7

Halogen Headlamp Bulb, Replace (C)
Lanos, Leganza
 one .... .3 .3
 both .... .7 .7
Nubira
 one .... .3 .3
 both .... .5 .5

Oil & Filter, Change (C)
Includes: Correct all fluid levels.
1999-02 .... .6 .6

Timing Belt, Replace (B)
SOHC 1.8 2.1
DOHC
 Lanos 1.5 1.8
 Leganza, Nubira ... 1.8 2.1

Tire, Replace (C)
Includes: Dismount old tire and mount new tire to rim.
One .... .5 .5

Wheel, Balance (B)
One .... .3 .3
each addl. add .... .2 .2

CHARGING

Alternator Circuits, Test (A)
Includes: Test component output.
1999-02 .... .3 .3

Alternator Assy., Replace (B)
Includes: Pulley transfer.
SOHC 1.1 1.3
DOHC
 Lanos 1.5 1.7
 Leganza 1.1 1.3
 Nubira 1.0 1.2

STARTING

Starter Draw Test (On Car) (B)
1999-02 .... .5 .5

Battery Cables, Replace (C)
1999-02 .... .5 .6

Battery, Replace (C)
1999-02 .... .3 .4

Starter Assy., Replace (B)
Lanos .... .8 .9
Leganza, Nubira .... .7 .8

Starter Drive, Replace (B)
Includes: Starter R&R.
Lanos 1.1 1.3
Leganza, Nubira 9 1.1

Starter Solenoid, Replace (B)
Lanos 1.3 1.5
Leganza, Nubira 1.2 1.4

IGNITION

Diagnose Ignition System (A)
1999-02 .... .8 .8

Ignition Coil, Replace (B)
Lanos .... .3 .3
Leganza, Nubira .... .5 .5

Spark Plug (Ignition) Cables, Replace (B)
1999-02 .... .3 .4

Spark Plugs, Replace (B)
1999-02 .... .5 .6

EMISSIONS

The following operations do not include testing. Add time as required.

Diagnose Emission Control System (A)
1999-02 .... .8 .8

Canister Purge Solenoid Valve, Replace (B)
1999-02 .... .3 .3

Coolant Temperature (ECT) Sensor, Replace (B)
1999-02 .... .5 .5

EGR Valve, Replace (B)
1999-02 .... .5 .7

Electronic Control Module (ECM/PCM), Replace (B)
1999-02 .... .7 .7

Fuel Vapor Canister, Replace (B)
Lanos, Leganza .... .5 .5
Nubira .... .6 .6

Knock Sensor, Replace (B)
1999-02 .... 2.3 2.5

Oxygen Sensor, Replace (B)
1999-02 .... .5 .7

Positive Crankcase Ventilation (PCV) Valve, Replace (B)
1999-02 .... .3 .3

Throttle Position Sensor (TPS), Replace (B)
1999-02 .... .3 .3

Vehicle Speed Sensor, Replace (B)
4T40E .... .5 .7
40LE .... .3 .5

FUEL
DELIVERY

Fuel Pump, Test (B)
1999-02 .... .5 .5

Fuel Filter, Replace (B)
Lanos .... .8 .8
Leganza, Nubira .... .7 .7

Fuel Gauge, (Dash), Replace (B)
Lanos .... .7 .7
Leganza .... .8 .8
Nubira .... .6 .6

Fuel Gauge, (Tank), Replace (B)
Lanos, Nubira .... .6 .8
Leganza 1.1 1.3

Fuel Pump, Replace (B)
Lanos, Nubira .... .6 .8
Leganza 1.1 1.3

Fuel Tank, Replace (B)
Includes: Drain and refill.
Lanos 1.4 1.7
Leganza, Nubira 1.6 1.9

Intake Manifold and/or Gasket, Replace (B)
Includes: Adjustments.
SOHC 2.8 3.1
DOHC
 Lanos 2.2 2.5
 Leganza 1.9 2.2
 Nubira 2.5 2.8

INJECTION

Diagnose Fuel Injection System (A)
1999-02 .... 1.0 1.0

Fuel Rail Assy., Replace (B)
SOHC .... .6 .8
DOHC
 Lanos, Nubira .... .4 .6
 Leganza .... .6 .8

Fuel Injectors, Replace (B)
1999-02 .... .7 .9

Fuel Pressure Regulator, Replace (B)
1999-02 .... .5 .7

Throttle Body and/or Gasket, R&R or Replace (B)
SOHC .... .6 .8
DOHC 1.2 1.4

EXHAUST

Exhaust Manifold or Gasket, Replace (B)
Lanos, Leganza .... .6 .8
Nubira 8 1.0

Exhaust Pipe, Replace (B)
1999-02 .... .7 .9

Muffler, Replace (B)
1999-02 .... .7 .9

LANOS : LEGANZA : NUBIRA DAE-3

	LABOR TIME	SEVERE SERVICE
ENGINE COOLING		
Pressure Test Cooling System (B)		
1999-02	3	.3
Coolant Thermostat, Replace (B)		
Lanos, Nubira	.5	.7
Leganza	.7	.9
Engine Coolant Temp. Sending Unit, Replace (B)		
1999-02	.5	.7
Radiator Assy., R&R or Replace (B)		
Lanos	1.1	1.3
Leganza, Nubira	1.5	1.7
w/AT add	.3	.3
Radiator Fan Motor and/or Fan, Replace (B)		
One	.6	.8
Both	.7	.9
Radiator Hoses, Replace (B)		
Includes: Refill with proper coolant mix.		
Lanos		
one	.6	.8
both	.7	.9
Leganza, Nubira		
upper	.6	.8
lower	1.1	1.3
both	1.3	1.5
Water Pump and/or Gasket, Replace (B)		
Includes: Refill with proper coolant mix.		
SOHC	1.9	2.1
DOHC		
Lanos	2.8	3.0
Leganza, Nubira	1.9	2.1
ENGINE		
ASSEMBLY		
Times shown are for OEM assemblies. Time to replace assemblies from aftermarket rebuilders may vary.		
Engine Assy., R&I (B)		
Does not include parts or component transfer.		
SOHC		
AT	6.2	6.7
MT	5.0	5.5
DOHC		
Lanos		
AT	5.6	6.1
MT	9.4	9.9
Leganza	6.4	6.9
Nubira	6.1	6.6
w/AC add	1.1	1.3

	LABOR TIME	SEVERE SERVICE
Engine Assy., R&R and Recondition (A)		
Includes: Replacing rings, rod and main bearings, cylinder head reconditioning and engine tune-up.		
SOHC		
AT	9.6	10.1
MT	10.0	10.5
DOHC		
Lanos		
AT	9.6	10.1
MT	13.2	13.7
Leganza	10.5	11.0
Nubira	10.1	10.6
w/AC add	1.1	1.3
Engine Assy. (New or Rebuilt Unit Complete), Replace (B)		
Includes: Component transfer and engine tune-up.		
SOHC		
AT	7.1	7.6
MT	5.9	6.4
DOHC		
Lanos		
AT	6.7	7.2
MT	10.3	10.8
Leganza	7.3	7.8
Nubira	7.0	7.5
w/AC add	1.1	1.3
Engine Mounts, Replace (B)		
Each	1.2	1.4
CYLINDER HEAD		
Compression Test (B)		
1999-02	.6	.7
Camshaft Followers, Replace (B)		
SOHC		
one	.7	.9
all	1.4	1.6
Camshaft Housing Cover Gasket, Replace (B)		
SOHC	.5	.7
DOHC	.7	.9
Cylinder Head and/or Gasket, Replace (B)		
Includes: Parts transfer and adjustments.		
SOHC	5.9	6.4
DOHC		
Lanos	5.2	5.7
Leganza, Nubira	5.9	6.4
w/PS add	.6	.6
Valve Job (A)		
SOHC	9.4	9.9
DOHC		
Lanos	9.7	10.2
Leganza	12.1	12.6
Nubira	12.6	13.1
w/PS add	.6	.6

	LABOR TIME	SEVERE SERVICE
Valve Lifters, Replace (B)		
DOHC		
Lanos	3.2	3.7
Leganza	2.6	3.1
Nubira	2.8	3.3
Valve Springs and/or Oil Seals, Replace (B)		
SOHC	5.6	6.1
DOHC		
Lanos	3.2	3.7
Leganza	3.3	3.8
Nubira	2.6	2.1
w/PS add	.6	.8
CAMSHAFT		
Camshaft, Replace (A)		
SOHC	3.3	3.8
DOHC		
Lanos		
one	3.5	4.0
both	4.0	4.5
Leganza, Nubira		
one	2.0	2.5
both	2.5	3.0
w/PS add	.6	.8
Camshaft Oil Seals, Replace (B)		
SOHC	1.8	2.3
DOHC		
Lanos		
one	3.2	3.7
both	3.5	4.0
Leganza, Nubira		
one	1.6	2.1
both	1.9	2.4
w/PS add	.6	.8
Camshaft Sprocket, Replace (B)		
SOHC	1.6	2.1
DOHC		
Lanos		
one	2.3	2.8
both	2.6	3.1
Leganza, Nubira		
one	1.5	2.0
both	1.8	2.4
w/PS add	.6	.8
Timing Belt, Replace (B)		
SOHC	1.8	2.3
DOHC		
Lanos	1.5	2.0
Leganza, Nubira	1.8	2.3
Timing Belt Tensioner, Replace (B)		
SOHC	1.5	2.0
DOHC	1.8	2.3

DAE-4 LANOS : LEGANZA : NUBIRA

	LABOR TIME	SEVERE SERVICE
CRANK & PISTONS		
Connecting Rod Bearings, Replace (A)		
Includes: Check bearing oil clearance.		
SOHC	3.1	3.6
DOHC		
Lanos	2.9	3.4
Leganza	4.9	5.4
Nubira	4.4	4.9
Crankshaft and Main Bearings, Replace (A)		
Includes: Engine R&R, check bearing oil clearance.		
SOHC	11.7	12.2
DOHC		
Lanos	15.3	15.8
Leganza	11.4	11.9
Nubira	9.6	10.1
w/AC add	1.1	1.3
Crankshaft Pulley, Replace (B)		
SOHC	.7	.9
DOHC		
Lanos	.6	.9
Leganza, Nubira	.5	.7
Crankshaft Front Oil Seal, Replace (B)		
SOHC	2.2	2.4
DOHC	2.5	2.7
Crankshaft Sprocket, Replace (B)		
SOHC	1.4	1.6
DOHC	1.7	1.9
Pistons, Connecting Rods or Rings, Replace (A)		
Includes: Ridge reaming, cylinder wall deglazing, installing new rings and rod bearings, engine tune-up.		
SOHC	9.6	10.1
DOHC		
Lanos	9.7	10.3
Leganza	10.5	11.0
Nubira	10.5	11.0
w/PS add	.6	.8
Rear Main Oil Seal, Replace (B)		
SOHC		
AT	5.3	5.8
MT	6.2	6.7
DOHC		
Lanos		
AT	5.8	6.3
MT	6.5	7.0
Leganza		
AT	6.3	6.8
MT	6.2	6.7
Nubira		
AT	6.0	6.5
MT	5.9	6.4

	LABOR TIME	SEVERE SERVICE
ENGINE LUBRICATION		
Engine Oil Pressure Switch (Sending Unit), Replace (B)		
1999-02	.9	.9
Oil Pan and/or Gasket, Replace (B)		
SOHC	1.8	2.0
DOHC		
Lanos	1.8	2.0
Leganza	1.5	1.7
Nubira	1.1	1.3
Oil Pump, Replace (B)		
SOHC	3.8	4.3
DOHC		
Lanos	3.5	4.0
Leganza	3.0	3.5
Nubira	2.7	3.2
CLUTCH		
Clutch Hydraulic System, Bleed (B)		
1999-02	.5	.5
Clutch Assy., Replace (B)		
Lanos	4.7	4.9
Leganza, Nubira	4.9	5.1
Clutch Master Cylinder, Replace (B)		
Includes: System bleeding.		
1999-02	1.0	1.2
Clutch Release Bearing, Replace (B)		
1999-02	4.7	4.9
Clutch Slave Cylinder, Replace (B)		
Includes: System bleeding.		
1999-02	.7	.8
Flywheel, Replace (B)		
Lanos	5.1	5.3
Leganza, Nubira	5.3	5.5
MANUAL TRANSAXLE		
Transaxle Assy., R&R and Recondition (A)		
Includes: Recondition transmission only.		
1999-02	11.6	12.1
Transaxle Assy., R&R or Replace (B)		
1999-02	4.7	4.9
AUTOMATIC TRANSAXLE		
SERVICE TRANSAXLE INSTALLED		
Diagnose Transaxle (A)		
1999-02	.8	.8
Check Unit for Oil Leaks (C)		
1999-02	.5	.5
Input Shaft Speed Sensor, Replace (B)		
40LE	.8	.8
Oil Pan and/or Gasket, Replace (B)		
4T40E, ZF	.8	.8
Oil Filter, Replace (B)		
4T40E	1.0	1.5
40LE	6.9	7.4
ZF	1.1	1.6

	LABOR TIME	SEVERE SERVICE
Output Shaft Speed Sensor, Replace (B)		
40LE	.3	.3
Valve Body, Replace (B)		
40LE	1.4	1.9
ZF	1.2	1.7
SERVICE TRANSAXLE REMOVED		
Transaxle R&R included unless otherwise noted.		
Transaxle Assy., R&I (B)		
4T40E	7.4	7.9
40LE	5.9	6.4
ZF	4.7	5.2
Transaxle Assy., R&R and Recondition (A)		
Includes: Recondition transmission only. Road test.		
4T40E	14.7	15.2
40LE	10.9	11.4
ZF	14.6	15.1
SHIFT LINKAGE		
AUTOMATIC		
Shift Lever, Replace (B)		
1999-02	.7	.7
MANUAL		
Gear Selector Lever, Replace (B)		
1999-02	.7	.7
DRIVELINE		
DRIVE AXLE		
Differential Assy., R&R and Recondition (B)		
1999-02	4.7	5.2
Differential Assy., Replace (B)		
1999-02	2.7	3.2
Differential Side Bearings, Replace (B)		
1999-02	3.2	3.7
Ring Gear and Pinion, Replace (B)		
1999-02	3.8	4.3
HALFSHAFTS		
Halfshaft, R&I or Recondition (B)		
Lanos, Nubira		
one side	1.1	1.3
both sides	1.9	2.1
Leganza		
one side	.8	1.0
both sides	1.5	1.7
R&R CV boot add each	.2	.3
R&R CV joint add each	.5	.5
Replace shaft add each	.5	.5

LANOS : LEGANZA : NUBIRA **DAE-5**

	LABOR TIME	SEVERE SERVICE

BRAKES
ANTI-LOCK
The following operations do not include testing. Add time as required.

Diagnose Anti-Lock Brake System (A)
- 1999-02 .... 1.0 *1.0*

Antilock Modulator, Replace (B)
- Lanos, Nubira 1.0 *1.2*

Antilock Solenoid, Replace (B)
- Lanos, Nubira
  - one .... .3 *.5*
  - all .... .5 *.7*

Electronic Control Module, Replace (B)
- 1999-02 .... .5 *.7*

Hydraulic Control Assy., Replace (B)
- Leganza 1.1 *1.3*

Hydraulic Modulator Assy., Replace (B)
- Lanos, Nubira 1.0 *1.2*

Motor Pack, Replace (B)
- Lanos, Nubira 1.1 *1.3*

Speed Sensor Assy. Replace (B)
- Front
  - one .... .5 *.7*
  - both .... .6 *.8*
- Rear
  - one 8 *1.0*
  - both 1.0 *1.2*

Valve Assy., Replace (B)
- Lanos, Nubira 1.2 *1.4*

SYSTEM

Bleed Brakes (B)
Includes: Add fluid.
- 1999-02 .... .6 *.6*

Flush and Refill Brake System, (B)
- 1999-02 .... .8 *.8*

Brake Hose (Flexible), Replace (B)
Includes: System bleeding.
- Front
  - one side 1.1 *1.3*
  - both sides 1.5 *1.7*
- Rear
  - one side 1.5 *1.7*
  - both sides 2.3 *2.5*

Master Cylinder, R&R and Rebuild (B)
Includes: System bleeding.
- 1999-02 .... 1.0 *1.2*

Master Cylinder, Replace (B)
Includes: System bleeding.
- 1999-02 .... .7 *.9*

Power Booster Unit, Replace (B)
Includes: System bleeding.
- 1999-02 .... 1.2 *1.4*

Proportioning Valve, Replace (B)
Includes: System bleeding.
- 1999-02 .... .8 *1.0*

SERVICE BRAKES

Brake Drum, Replace (B)
Includes: Repack wheel bearings.
- One .... .8 *.9*
- Both 1.2 *1.3*

Caliper Assy., Replace (B)
Includes: System bleeding.
- One 1.0 *1.2*
- Both 1.2 *1.4*

Disc Brake Rotor, Replace (B)
- One .... .8 *.9*
- Both 1.1 *1.2*

Pads and/or Shoes, Replace (B)
Includes: Adjust service and parking brake. System bleeding.
- Front disc .... .6 *.8*
- Rear disc 1.1 *1.3*
- Rear drum 1.6 *1.8*
- Four wheels
  - Drum brakes 1.8 *2.0*
  - Disc brakes 1.4 *1.6*

COMBINATION ADD-ONS
Replace
- brake drum add each ...3 *.3*
- brake hose add each ...3 *.3*
- disc rotor add each .....3 *.3*
- wheel cylinder add .....3 *.3*

Resurface
- brake drum add .... .5 *.5*
- disc rotor add .... .5 *.5*

Wheel Cylinder, Replace (B)
Includes: System bleeding.
- One side 1.5 *1.7*
- Both sides 2.2 *2.4*

PARKING BRAKE

Parking Brake Cable, Adjust (C)
- Lanos .... .8 *.8*
- Leganza, Nubira ..... .5 *.5*

Parking Brake Apply Actuator, Replace (B)
- 1999-02 .... .6 *.6*

Parking Brake Cable, Replace (B)
- Front .... .7 *.9*
- Rear
  - Lanos 1.1 *1.3*
  - Leganza, Nubira ... 2.7 *2.9*

FRONT SUSPENSION
Unless otherwise noted, time given does not include alignment.

Align Front End (A)
- 1999-02 .... 1.4 *1.6*

Toe In, Adjust (B)
- 1999-02 .... .8 *1.0*

Ball Joint, Replace (B)
- Lanos
  - one side 1.8 *2.0*
  - both sides 3.0 *3.2*
- Leganza, Nubira
  - one side 1.0 *1.2*
  - both sides 1.6 *1.8*

Front Spring, Replace (B)
- Lanos
  - one side 2.2 *2.4*
  - both sides 4.1 *4.3*
- Leganza, Nubira
  - one side 1.4 *1.6*
  - both sides 2.5 *2.7*

Lower Control Arm, Replace (B)
- Lanos, Leganza
  - one side 1.4 *1.6*
  - both sides 2.6 *2.8*
- Nubira
  - one side 1.1 *1.3*
  - both sides 1.8 *2.0*

Lower Control Arm Bushings, Replace (B)
- Lanos
  - one side 1.6 *1.8*
  - both sides 3.5 *3.7*

Stabilizer Bar, Replace (B)
- Lanos, Nubira 1.2 *1.4*
- Leganza 1.6 *1.8*

Stabilizer Bar Link, Replace (B)
- One side .... .6 *.8*
- Both sides 8 *1.0*

Steering Knuckle, Replace (B)
- Lanos
  - one side 2.8 *3.0*
  - both sides 5.1 *5.3*
- Leganza, Nubira
  - one side 1.8 *2.0*
  - both sides 3.2 *3.4*

Strut Assy., Replace (B)
- Lanos
  - one side 1.7 *1.9*
  - both sides 2.5 *2.7*
- Leganza, Nubira
  - one side .... .7 *.9*
  - both sides 1.3 *1.5*

Wheel Hub and/or Bearings, Replace (B)
- Lanos
  - one side 2.1 *2.3*
  - both sides 3.8 *4.0*
- Leganza, Nubira
  - one side 1.8 *2.0*
  - both sides 3.2 *3.4*

DAE-6 LANOS : LEGANZA : NUBIRA

	LABOR TIME	SEVERE SERVICE

REAR SUSPENSION

Unless otherwise noted, time given does not include alignment.

Parallel Link, Replace (B)
Leganza, Nubira
  front or rear
    one side 5   .7
    both sides 7   .9
Strut Assy., Replace (B)
Leganza
  one side 1.4   1.6
  both sides 2.2   2.4
Nubira
  one side 1.1   1.3
  both sides 1.9   2.1
Rear Spring, Replace (B)
Lanos
  one side 1.0   1.2
  both sides 1.1   1.3
Leganza
  one side 5   .7
  both sides 7   .9
Nubira
  one side 1.1   1.3
  both sides 1.9   2.1
Shock Absorber, Replace (B)
Lanos
  one side 7   .9
  both sides 1.0   1.2
Trailing Link, Replace (B)
Leganza, Nubira
  one side 6   .8
  both sides 8   1.0
Stabilizer Bar, Replace (B)
Lanos 1.1   1.3
Leganza, Nubira 6   .8
Stabilizer Bar Bushings, Replace (B)
1999-02 3   .5
Rear Wheel Bearing, Replace (B)
Lanos, Nubira
  one side 1.0   1.2
  both sides 1.6   1.8

STEERING

AIR BAGS
The following operations do not include testing. Add time as required.
Diagnose Air Bag System (A)
1999-02 .... 1.2   1.2
Control Unit, Replace (B)
1999-02 7   .7
Module Assembly, Replace (A)
Driver side
  Lanos 5   .5
  Leganza, Nubira 7   .7
Passenger side
  Lanos 2.6   2.6
  Leganza, Nubira ... 2.7   2.7

STEERING GEAR
Troubleshoot Power Steering Component Each (A)
1999-02 5   .5
Power Steering Pump Hoses, Replace (B)
Does not include purging.
Pressure
  Lanos, Nubira 1.0   1.2
  Leganza 7   .9
Return
  Lanos 7   .9
  Nubira 6   .8
Suction
  Lanos, Nubira 6   .8
  Leganza 5   .7
Steering Column, R&I (B)
1999-02 .... 1.4   1.6
w/air bags add 1   .1
Steering Gear, Replace (B)
Manual steering 1.4   1.6
Power steering
  Lanos, Leganza 2.3   2.5
  Nubira 1.6   1.8
Steering Pump, Replace (B)
SOHC 1.2   1.4
DOHC
  Lanos 2.5   2.7
  Leganza 1.2   1.4
  Nubira 2.1   2.3
Steering Wheel, Replace (B)
Lanos, Leganza 5   .5
Nubira 6   .6
w/air bags add
  Lanos, Leganza 1   .1
  Nubira 3   .3
Tie Rod Ends, Replace (B)
Includes: Alignment.
One 1.4   1.6
each addl. add 2   .3

HEATING & AIR CONDITIONING

When more than one component requires replacement where evacuation/recovery and recharging is already included, deduct 1.0 hour for each additional component from the time given.

Evacuate/Recover and Recharge System (B)
1999-02 .... 1.0   1.0
AC Control Assy., Replace (B)
Lanos 6   .6
Leganza, Nubira 5   .5
AC Hoses, Replace (B)
One 1.7   1.9
each addl. add 5   .6
ATC Control Switch, Replace (B)
Leganza 5   .7
Blower Motor, Replace (B)
Lanos 5   .7
Leganza, Nubira 7   .9
Blower Motor Switch, Replace (B)
Lanos 5   .5
Leganza, Nubira 8   .8
Blower Motor Resistor, Replace (B)
1999-02 6   .8
Compressor Assy., Replace (B)
Does not include evacuate/recover and recharge.
Lanos 1.8   2.0
Leganza, Nubira 2.3   2.5
Condenser Assy., Replace (B)
Does not include evacuate/recover and recharge.
1999-02 .... 2.5   2.7
Evaporator Core, Replace (B)
Does not include evacuate/recover and recharge.
Lanos 5.6   5.8
Leganza 6.2   6.3
Nubira 5.9   6.1
Heater Core, Replace (B)
Does not include evacuate/recover and recharge.
Lanos 5.8   6.0
Leganza 5.3   5.5
Nubira 5.0   5.2
Heater Hoses, Replace (B)
1999-02 5   .7
Inside Temperature Sensor, Replace (B)
Leganza 5   .5
Outside Temperature Sensor, Replace (B)
Leganza 3   .3
Nubira 6   .6
Receiver Drier Assy., Replace (B)
Does not include evacuate/recover and recharge.
1999-02 .... 1.0   1.2
Sun Sensor, Replace (B)
Leganza 3   .3
Temperature Control Cable, Replace (B)
Lanos 5   .5
Leganza, Nubira 7   .7
Vacuum Distributor, Replace (B)
Leganza, Nubira 6   .6

WIPERS & SPEEDOMETER

Antenna Assy., Replace (B)
Antenna to connector . 2.1   2.3
Radio to connector ... 1.2   1.4
Radio, R&R (B)
1999-02 5   .5

LANOS : LEGANZA : NUBIRA **DAE-7**

	LABOR TIME	SEVERE SERVICE
Speedometer Head, R&R or Replace (B)		
Lanos	1.1	*1.1*
Leganza	.8	*.8*
Nubira	.6	*.6*
Windshield Washer Pump, Replace (B)		
Lanos, Leganza	.7	*.7*
Nubira	.5	*.5*
Windshield Wiper Linkage, Replace (B)		
1999-02	.8	*1.0*
Windshield Wiper Motor, Replace (B)		
1999-02	.8	*1.0*
Windshield Wiper (Combination) Switch, Replace (B)		
1999-02	.6	*.6*

LAMPS & SWITCHES

	LABOR TIME	SEVERE SERVICE
Back-Up Lamp Switch, Replace (B)		
1999-02	.5	*.5*
Combination (Multifunction) Switch, Replace (B)		
1999-02	.6	*.6*
Halogen Headlamp Bulb, Replace (C)		
Lanos, Leganza		
one	.3	*.3*
both	.7	*.7*
Nubira		
one	.3	*.3*
both	.5	*.5*
Headlamps, Aim (B)		
1999-02	.5	*.5*
Horn, Replace (C)		
1999-02	.3	*.3*
Horn Relay, Replace (B)		
1999-02	.3	*.3*

	LABOR TIME	SEVERE SERVICE
Neutral Safety Switch, Replace (B)		
1999-02	.5	*.5*
Stop Light Switch, Replace (B)		
1999-02	.3	*.3*
Parking Brake Warning Switch, Replace (C)		
1999-02	.5	*.5*
Turn Signal or Hazard Warning Flasher, Replace (C)		
1999-02	.3	*.3*

BODY

	LABOR TIME	SEVERE SERVICE
Door Window Regulator, Replace (B)		
Lanos, Nubira each	1.1	*1.4*
Leganza each	.8	*1.1*

NOTES

Charade : Rocky — DAI

SYSTEM INDEX

MAINTENANCE	DAI-2
CHARGING	DAI-2
STARTING	DAI-2
IGNITION	DAI-2
EMISSIONS	DAI-2
FUEL	DAI-2
EXHAUST	DAI-3
ENGINE COOLING	DAI-3
ENGINE	DAI-3
Assembly	DAI-3
Cylinder Head	DAI-3
Camshaft	DAI-3
Crank & Pistons	DAI-3
Engine Lubrication	DAI-4
CLUTCH	DAI-4
MANUAL TRANSAXLE	DAI-4
MANUAL TRANSMISSION	DAI-4
AUTO TRANSAXLE	DAI-4
TRANSFER CASE	DAI-4
DRIVELINE	DAI-4
BRAKES	DAI-5
FRONT SUSPENSION	DAI-5
REAR SUSPENSION	DAI-6
STEERING	DAI-6
HEATING & AC	DAI-6
WIPERS & SPEEDOMETER	DAI-7
LAMPS & SWITCHES	DAI-7
BODY	DAI-7

OPERATIONS INDEX

A
- AC Hoses DAI-6
- Air Conditioning DAI-6
- Alignment DAI-5
- Alternator (Generator) DAI-2
- Antenna DAI-7

B
- Back-Up Lamp Switch DAI-7
- Ball Joint DAI-5
- Battery Cables DAI-2
- Bleed Brake System DAI-5
- Blower Motor DAI-6
- Brake Disc DAI-5
- Brake Drum DAI-5
- Brake Hose DAI-5
- Brake Pads and/or Shoes ... DAI-5

C
- Camshaft DAI-3
- Catalytic Converter DAI-3
- Crankshaft DAI-3
- CV Joint DAI-5
- Cylinder Head DAI-3

D
- Differential DAI-4
- Distributor DAI-2
- Drive Belt DAI-2
- Driveshaft DAI-5

E
- EGR DAI-2
- Engine DAI-3
- Engine Lubrication DAI-4
- Engine Mounts DAI-3
- Evaporator DAI-6
- Exhaust DAI-3
- Exhaust Manifold DAI-3

F
- Flywheel DAI-4
- Fuel Injection DAI-2
- Fuel Pump DAI-2
- Fuel Vapor Canister DAI-2

G
- Generator DAI-2

H
- Halfshaft DAI-5
- Headlamp DAI-7
- Heater Core DAI-6
- Horn DAI-7

I
- Ignition Coil DAI-2
- Ignition Module DAI-2
- Ignition Switch DAI-2
- Inner Tie Rod DAI-6
- Intake Manifold DAI-2

L
- Lower Control Arm DAI-5

M
- Master Cylinder DAI-5
- Muffler DAI-3

N
- Neutral Safety Switch DAI-7

O
- Oil Pan DAI-4
- Oil Pump DAI-4
- Outer Tie Rod DAI-6

P
- Parking Brake DAI-5
- Pistons DAI-4

R
- Radiator DAI-3
- Radiator Hoses DAI-3
- Radio DAI-7
- Rear Main Oil Seal DAI-4

S
- Shock Absorber/Strut, Front .. DAI-5
- Shock Absorber/Strut, Rear ... DAI-6
- Spark Plug Cables DAI-2
- Spark Plugs DAI-2
- Spring, Front Coil DAI-5
- Spring, Leaf DAI-6
- Starter DAI-2
- Steering Wheel DAI-6

T
- Thermostat DAI-3
- Throttle Body DAI-2
- Timing Belt DAI-3
- Torque Converter DAI-4

U
- U-Joint DAI-5
- Upper Control Arm DAI-6

V
- Valve Cover Gasket DAI-3
- Valve Job DAI-3
- Vehicle Speed Sensor DAI-2

W
- Water Pump DAI-3
- Wheel Balance DAI-2
- Wheel Cylinder DAI-5
- Window Regulator DAI-7
- Windshield Washer Pump ... DAI-7
- Windshield Wiper Motor DAI-7

DAI-2   CHARADE : ROCKY

	LABOR TIME	SEVERE SERVICE

MAINTENANCE

Air Cleaner Filter Element, Replace (C)
1988-92 3   .4
Drive Belt, Adjust (B)
1988-92 3   .4
Drive Belt, Replace (B)
1988-92 5   .6
Fuel Filter, Replace (B)
1988-92 5   .5
Halogen Headlamp Bulb, Replace (C)
One 3   .3
Oil & Filter, Change (C)
Includes: Correct all fluid levels.
1988-92 3   .4
Timing Belt, Replace (B)
Charade 2.0   2.5
Rocky 2.5   3.0
Tire, Replace (C)
Includes: Dismount old tire and mount new tire to rim.
One 5   .5
Wheel, Balance (B)
One 3   .3
each addl. add 2   .2

CHARGING

Alternator Circuits, Test (B)
Includes: Test component output.
1988-92 5   .5
Alternator, R&R and Recondition (B)
Charade 1.5   2.0
Rocky 1.8   2.3
Alternator, Replace (B)
Includes: Pulley transfer.
Charade 8   1.0
Rocky 1.1   1.3
Alternator Voltage Regulator, Replace (B)
1988-92 .... 1.7   1.9
Front Alternator Bearing, Replace (B)
1988-92 .... 1.5   1.7
Voltmeter and/or Ammeter Gauge, Replace (B)
Rocky 4   .4

STARTING

Starter Draw Test (On Car) (B)
1988-92 3   .3
Battery Cables, Replace (C)
Charade 4   .5
Rocky each 5   .6
Starter, R&R and Recondition (B)
Charade
  1.0L 2.2   2.7
  1.3L 3.2   3.7
Rocky 1.9   2.4

Starter Assy., Replace (B)
Charade
  1.0L 6   .8
  1.3L 1.6   1.8
Rocky 9   1.1
Starter Drive Assy., Replace (B)
Includes: Starter R&R.
Charade
  1.0L 1.2   1.5
  1.3L 2.1   2.4
Rocky 1.5   1.8
Starter Relay, Replace (B)
1988-92 5   .5
Starter Solenoid and/or Switch, Replace (B)
Includes: Starter R&R.
Non-reduction starter
 Charade
  1.0L 6   .8
  1.3L 1.6   1.8
 Rocky 1.2   1.4
Reduction starter
  1.0L 9   1.1
  1.3L 1.6   1.8

IGNITION

Diagnose Ignition System Component Each (A)
1988-92 8   .8
Ignition Timing, Reset (B)
1988-92 5   .5
Distributor Armature, Replace (B)
1988-92 4   .7
Distributor, R&R and Recondition (B)
Includes: Reset base ignition timing.
Charade 1.7   1.9
Rocky 1.2   1.4
Distributor, Replace (B)
Includes: Reset base ignition timing.
1988-92 7   .9
Distributor Cap and/or Rotor, Replace (B)
1988-92 4   .4
Electronic Ignition Control Unit, Replace (B)
1988-92 7   .7
Igniter Unit, Replace (B)
Charade 5   .5
Rocky 1.1   1.1
Ignition Coil, Replace (B)
1988-92 3   .3
Ignition Switch, Replace (B)
1988-92 7   .7
Spark Plug (Ignition) Cables, Replace (B)
1988-92 5   .6

Spark Plugs, Replace (B)
1988-92 4   .5
Vacuum Advance, Replace (B)
Includes: Reset base ignition timing.
Rocky 9   1.0

EMISSIONS

The following operations do not include testing. Add time as required.
Diagnose Emission Control System Component Each (A)
1988-92 8   .8
EGR Modulator, Replace (B)
1988-92 4   .5
EGR Control Valve, Replace (B)
1988-92 5   .7
EGR Switching Valve, Replace (B)
Rocky 4   .4
Fuel Vapor Canister, Replace (B)
1988-92 6   .6
Vehicle Speed Sensor, Replace (B)
1988-92 6   .6

FUEL

DELIVERY

Fuel Filter, Replace (B)
1988-92 5   .5
Fuel Gauge (Dash), Replace (B)
1988-92 .... 1.0   1.0
Fuel Gauge (Tank), Replace (B)
1988-92 .... 1.5   2.0
Fuel Pump (In Tank), Replace (B)
Rocky 7   1.0
Fuel Tank, Replace (B)
1988-92 .... 1.8   2.1
Intake Manifold and/or Gasket, Replace (B)
Includes: Adjustments.
1988-92 .... 1.8   2.1

INJECTION

Cold Start Injector, Replace (B)
Charade 1.0L 4   .4
Fuel Injectors, Replace (B)
Charade
  1.0L 1.3   1.5
  1.3L 9   1.1
Rocky 1.1   1.4
Fuel Injector Start Time Switch, Replace (B)
Charade 4   .4
Fuel Pressure Regulator, Replace (B)
Rocky 1.0   1.2
Throttle Body Assy., Replace (B)
1988-92 8   .8
Vacuum Pump, Replace (B)
Charade 1.4   1.4

CHARADE : ROCKY DAI-3

	LABOR TIME	SEVERE SERVICE

EXHAUST

Catalytic Converter, Replace (B)
- Rocky .... .8 .9

Exhaust Manifold or Gasket, Replace (B)
- Charade
  - 1.0L 1.5 1.8
  - 1.3L 1.9 2.2
- Rocky 2.0 2.3

Front Exhaust Pipe, Replace (B)
- Charade .... .7 .8
- Rocky 1.2 1.3

Muffler, Replace (B)
- 1988-92 .... .6 .7

Tail Pipe, Replace (B)
- 1988-92 .... .7 .8

ENGINE COOLING

Pressure Test Cooling System (B)
- 1988-92 .... .3 .3

Coolant Temperature Sensor Switch, Replace (B)
- 1988-92 .... .3 .3

Coolant Thermostat, Replace (B)
- 1988-92 .... .8 1.0

Fan Blade, Replace (B)
- Rocky .... .6 .8
- *Replace fluid coupling add* .... .1 .2

Radiator Assy., R&R or Replace (B)
- 1988-92 .... 1.1 1.2

Radiator Fan Motor and/or Fan, Replace (B)
- 1988-92 .... .8 1.0

Radiator Fan Motor Switch (Coolant Temp.), Replace (B)
- 1988-92 .... .6 .6

Radiator Hoses, Replace (B)
Includes: Refill with proper coolant mix.
- Charade
  - 1.0L each .... .6 .7
  - 1.3L each .... .8 .9
- Rocky .... .6 .7

Temperature Gauge (Dash), Replace (B)
- 1988-92 .... 1.1 1.1

Temperature Gauge (Engine), Replace (B)
- 1988-92 .... .5 .5

Water Pump and/or Gasket, Replace (B)
Includes: Refill with proper coolant mix.
- Charade 2.2 2.5
- Rocky 1.7 2.0

ENGINE
ASSEMBLY
Times shown are for OEM assemblies. Time to replace assemblies from aftermarket rebuilders may vary.

Engine Assy., R&I (B)
Does not include parts or component transfer.
- Charade 6.1 6.6
- Rocky 7.0 7.5

Engine Assy., R&R and Recondition (A)
Includes: Replacing rings, rod and main bearings, cylinder head reconditioning and engine tune-up.
- Charade
  - 1.0L 17.4 17.9
  - 1.3L 20.0 20.5
- Rocky 20.5 21.0

Engine Assy. (Short) Replace (B)
Assembly consists of cylinder block, piston assemblies, crankshaft, camshaft, timing chain and gears. Does not include cylinder heads. Operation Includes: R&R engine, transfer necessary parts and all necessary adjustments.
- Charade 10.0 10.5
- Rocky 10.5 11.0
- *Recondition cylinder head add* 3.3 3.8

Engine Mounts, Replace (B)
- 1988-92 .... .9 1.1

CYLINDER HEAD
Valve Clearance, Adjust (B)
- 1988-92 .... 1.1 1.4

Cylinder Head, Replace (B)
Includes: Parts transfer, adjustments.
- Charade
  - 1.0L 4.7 5.2
  - 1.3L 5.1 5.6
- Rocky 4.8 5.3

Rocker Arms or Shafts, Replace (B)
- Charade
  - 1.0L 1.8 2.0
  - 1.3L 1.4 1.6
- Rocky 1.4 1.6

Valve/Cam Cover Gasket, Replace or Reseal (B)
- 1988-92 .... .7 .8

Valve Job (A)
- Charade
  - 1.0L 7.8 8.3
  - 1.3L 9.4 9.9
- Rocky 10.7 11.2

Valve Springs and/or Oil Seals, Replace (B)
- Charade
  - 1.0L 2.9 3.4
  - 1.3L 5.4 5.9
- Rocky 8.6 9.1

CAMSHAFT
Balance Shaft, Replace (B)
- Charade 1.0L 7.7 8.2
- *Replace bearings add* .... .4 .9

Balance Shaft Gear Cover, R&R or Replace (B)
- Charade 1.0L 4.7 5.2

Camshaft, Replace (B)
- Charade
  - 1.0L 3.2 3.7
  - 1.3L 2.5 3.0
- Rocky 3.0 3.5

Camshaft Seal, Replace (B)
- Charade 1.7 2.2

Camshaft Timing Gear, Replace (B)
- Charade 1.8 2.3

Timing Belt, Replace (B)
- Charade 2.0 2.5
- Rocky 2.5 3.0

Timing Belt/Chain Cover, Replace (B)
- Charade 1.8 2.3
- Rocky 2.2 2.7

Timing Belt Tensioner, Replace (B)
- Charade 1.9 2.4
- Rocky 2.4 2.9

CRANK & PISTONS
Connecting Rod Bearings, Replace (A)
Includes: Check bearing oil clearance.
- Charade
  - 1.0L 6.5 7.0
  - 1.3L 7.0 7.5
- Rocky 7.5 8.0

Crankshaft and Main Bearings, Replace (A)
Includes: Engine R&R.
- Charade
  - 1.0L 8.6 9.1
  - 1.3L 9.0 9.5
- Rocky 12.5 13.0

Crankshaft Front Oil Seal, Replace (A)
- Charade
  - 1.0L 2.5 2.7
  - 1.3L 2.2 2.4
- Rocky 2.4 2.6

Crankshaft Pulley, Replace (B)
- Charade 1.2 1.4
- Rocky 1.7 1.9

Crankshaft Sprocket, Replace (B)
- Charade 1.9 2.1

DAI-4 CHARADE : ROCKY

	LABOR TIME	SEVERE SERVICE
Pistons or Connecting Rods, Replace (A)		
Includes: Ridge reaming, cylinder wall deglazing, installing new rings and rod bearings, engine tune-up.		
Charade		
1.0L	7.5	8.0
1.3L	8.0	8.5
Rocky	8.7	9.2
Rear Main Oil Seal, Replace (B)		
Charade		
1.0L	6.8	7.3
1.3L	7.3	7.8
Rocky	7.2	7.7
Rings, Replace (A)		
Includes: Ridge reaming, cylinder wall deglazing, installing new rings, engine tune-up.		
Charade		
1.0L	6.5	7.0
1.3L	7.0	7.5
Rocky	7.5	8.0
w/AC add	1.0	1.0

ENGINE LUBRICATION

	LABOR TIME	SEVERE SERVICE
Engine Oil Cooler, Replace (B)		
Charade	.9	1.1
Rocky	1.3	1.5
Oil Pan and/or Gasket, Replace (B)		
Charade	2.1	2.3
Rocky	2.9	3.1
Oil Pressure Gauge (Dash), Replace (B)		
1988-92	1.0	1.0
Oil Pressure Gauge (Engine), Replace (B)		
Charade	.4	.4
Rocky	.7	.7
Oil Pump, R&R and Recondition (B)		
Charade		
1.0L	2.7	3.2
1.3L	3.1	3.6
Rocky	5.5	6.0
Oil Pump, Replace (B)		
Charade		
1.0L	2.4	2.7
1.3L	2.8	3.1
Rocky	5.2	5.5
Oil Pump Chain, Replace (B)		
Charade 1.0L	4.7	5.2

CLUTCH

	LABOR TIME	SEVERE SERVICE
Clutch Pedal Free Play, Adjust (B)		
1988-92	.3	.4
Clutch Assy., Replace (B)		
Charade	5.3	5.8
Rocky	6.5	7.0
Clutch Release Bearing, Replace (B)		
Charade	.9	1.1
Rocky	1.4	1.6

	LABOR TIME	SEVERE SERVICE
Flywheel, Replace (B)		
Charade		
1.0L	5.6	6.1
1.3L	6.2	6.7
Rocky	7.0	7.5

MANUAL TRANSAXLE

	LABOR TIME	SEVERE SERVICE
Axle Shaft Oil Seal, Replace (B)		
One	1.2	1.4
Both	1.7	1.9
Differential Case Bearings, Replace (B)		
1988-92	6.5	7.0
Differential Assy. and Transaxle, R&R and Recondition (A)		
1988-92	6.7	7.2
Ring Gear & Pinion Assy., Replace (B)		
1988-92	6.5	7.0
Speedometer Driven Gear and/or Seal, Replace (B)		
1988-92	.4	.4
Transaxle Assy., R&R and Recondition (A)		
Includes: Transmission recondition only.		
1988-92	7.5	8.0
Transaxle Assy., R&R or Replace (B)		
1988-92	4.7	5.2

MANUAL TRANSMISSION

	LABOR TIME	SEVERE SERVICE
Extension Housing and/or Gasket, Replace (B)		
Rocky	.7	.9
Transmission and Transfer Case Assy., R&I (B)		
Rocky	5.9	6.4
Transmission and Transfer Case Assy., R&R and Recondition (A)		
Rocky	10.7	11.2

AUTOMATIC TRANSAXLE

SERVICE TRANSAXLE INSTALLED

	LABOR TIME	SEVERE SERVICE
Check Unit For Oil Leaks (C)		
1988-92	.5	.5
Drain & Refill Unit (B)		
1988-92	1.0	1.2
Axle Shaft Oil Seals, Replace (B)		
One	1.3	1.5
Both	1.7	1.9
Oil Pan and/or Gasket, Replace (B)		
1988-92	1.0	1.2
Throttle Lever Valve Control Cable, Replace (B)		
Includes: Cable adjustment.		
1988-92	1.5	1.7
Transaxle Control Computer, Replace (B)		
1988-92	.7	.7

	LABOR TIME	SEVERE SERVICE
Transaxle Solenoid Assy., Replace (B)		
1988-92	1.7	1.9

SERVICE TRANSAXLE REMOVED

	LABOR TIME	SEVERE SERVICE
Differential Case Bearings, Replace (B)		
Includes: Transaxle R&R.		
1988-92	6.1	6.6
Differential Assy., Recondition (A)		
Includes: Transaxle R&R.		
1988-92	6.7	7.2
Torque Converter or Converter Seal, Replace (B)		
1988-92	5.2	5.4
Transaxle Assy. (Complete), R&R and Recondition (A)		
1988-92	10.0	10.5
Transaxle Assy., R&R or Replace (B)		
1988-92	5.1	5.3

TRANSFER CASE

	LABOR TIME	SEVERE SERVICE
Front Cover Gasket, Replace (B)		
Rocky	5.2	5.7
Output Flange Oil Seal, Replace (B)		
Rocky	1.6	1.8
Transfer Case Drive Chain, Replace (B)		
Rocky	3.8	4.3
Transfer Case Shift Lever, Replace (B)		
Rocky	.7	.7

DRIVELINE

DRIVE AXLE

	LABOR TIME	SEVERE SERVICE
Axle Shaft, Replace (B)		
Rear axle		
Rocky	4.4	4.6
Differential Carrier, R&R or Replace (B)		
Front axle		
Rocky	4.7	5.2
Rear axle		
Rocky	3.7	4.2
Differential Case, Replace (B)		
Rear axle		
Rocky		
standard	6.0	6.5
limited slip	6.5	7.0
Differential Cover Gasket, Replace (B)		
Front axle		
Rocky	1.1	1.3
Differential Side Bearings, Replace (B)		
Front axle		
Rocky	5.5	6.0
Rear axle		
Rocky	4.8	5.3

CHARADE : ROCKY — DAI-5

	LABOR TIME	SEVERE SERVICE
Pinion Bearings, Replace (B)		
Front axle		
Rocky	6.1	6.6
Pinion Shaft Oil Seal, Replace (B)		
Front or rear axle		
Rocky	1.8	2.0
Rear Wheel Bearings & Cups, Replace (B)		
Rear axle		
Rocky one side	4.3	4.5
Ring & Pinion Gear Set, Replace (B)		
Front axle		
Rocky	6.4	6.9
Rear axle		
Rocky	7.0	7.5

DRIVESHAFT

	LABOR TIME	SEVERE SERVICE
Driveshaft, Replace (B)		
Rocky one piece	1.0	1.2
Recondition U-joint		
add each	.3	.5

HALFSHAFTS

	LABOR TIME	SEVERE SERVICE
CV Joint, Replace or Recondition (B)		
One side	1.8	2.3
Halfshaft, R&R or Replace (B)		
One	1.5	1.7
Both	1.9	2.1
Halfshaft Assy., R&R or Replace (B)		
Rocky	1.9	2.1
Recondition CV-joint		
add each	.4	.6
Replace boot add	.1	.2
Halfshaft Flange or Seal, Replace (B)		
Rocky		
outer	3.9	4.1
inner	2.6	2.8
Halfshaft Oil Seal, Replace (B)		
One	1.3	1.5
Both	1.8	2.0

BRAKES
SYSTEM

	LABOR TIME	SEVERE SERVICE
Bleed Brakes (B)		
Includes: Add fluid.		
1988-92	.6	.6
Brakes, Adjust (B)		
Rear wheels	.5	.5
Brake Hose (Flexible), Replace (B)		
Includes: System bleeding.		
Charade		
one	1.1	1.3
both	1.4	1.6
Rocky one	.9	1.0
each addl. add	.3	.4
Brake Proportioning Valve, Replace (B)		
1988-92	1.2	1.4

	LABOR TIME	SEVERE SERVICE
Master Cylinder, R&R and Rebuild (B)		
1988-92	2.1	2.6
Master Cylinder, Replace (B)		
1988-92	1.7	1.9
Power Booster Unit, Replace (B)		
1988-92	2.2	2.4
Power Booster Vacuum Check Valve, Replace (B)		
1988-92	.3	.3

SERVICE BRAKES

	LABOR TIME	SEVERE SERVICE
Brake Drum, Replace (B)		
Charade		
one	.9	1.0
both	1.1	1.2
Rocky		
one	1.2	1.3
both	1.5	1.6
Caliper Assy., R&R and Recondition (B)		
Includes: System bleeding.		
Charade		
one	1.5	1.8
both	2.2	2.5
Rocky		
one	1.8	2.1
both	2.9	3.2
Caliper Assy., Replace (B)		
Includes: System bleeding.		
One	1.4	1.6
Both	2.0	2.2
Disc Brake Rotor, Replace (B)		
Charade		
one	.8	.9
both	1.2	1.3
Rocky		
one	1.4	1.5
both	2.1	2.2
Pads and/or Shoes, Replace (B)		
Includes: Adjust service and parking brake. System bleeding.		
Charade		
front disc	1.3	1.5
rear drum	1.5	1.7
Rocky		
front disc	1.4	1.6
rear drum	2.4	2.6

COMBINATION ADD-ONS

	LABOR TIME	SEVERE SERVICE
Replace		
caliper add	.2	.2
brake drum add	.2	.2
wheel cylinder add	.2	.2
brake hose add	.3	.3
Resurface		
disc rotor add	.5	.5
brake drum add each	.5	.5

	LABOR TIME	SEVERE SERVICE
Wheel Cylinder, R&R and Rebuild (B)		
Includes: System bleeding.		
Charade		
one	1.8	2.1
both	2.4	2.7
Rocky		
one	1.9	2.2
both	2.8	3.1
Wheel Cylinder, Replace (B)		
Includes: System bleeding.		
Charade		
one	1.6	1.8
both	2.0	2.2
Rocky		
one	1.7	1.9
both	2.4	2.6

PARKING BRAKE

	LABOR TIME	SEVERE SERVICE
Parking Brake Cable, Adjust (C)		
1988-92	.2	.3
Parking Brake Apply Actuator, Replace (B)		
1988-92	.8	.8
Parking Brake Apply Warning Indicator Switch, Replace (B)		
1988-92	.3	.3
Parking Brake Cable, Replace (B)		
Charade		
one	1.1	1.3
both	2.1	2.3
Rocky each	1.8	2.1

FRONT SUSPENSION

Unless otherwise noted, time given does not include alignment.

CHARADE

	LABOR TIME	SEVERE SERVICE
Align Front End (B)		
1988-92	1.0	1.2
Front Toe, Adjust (B)		
1988-92	.5	.7
Front Strut Coil Spring, Replace (B)		
One	1.5	1.7
Both	2.2	2.4
Lower Control Arm and/or Ball Joint, Replace (B)		
One	1.1	1.3
Both	1.5	1.7
Stabilizer Bar, Replace (B)		
1988-92	1.3	1.5
Steering Knuckle, Replace (B)		
One	1.8	2.0
Both	2.7	2.9
Strut, Replace (B)		
One	1.4	1.6
Both	2.0	2.2
Wheel Bearings, Replace (B)		
One	1.8	1.8
Both	2.7	2.7

CHARADE : ROCKY

	LABOR TIME	SEVERE SERVICE
ROCKY		
Align Front End (B)		
1990-92	1.0	1.2
Toe, Front, Adjust (B)		
1990-92	.5	.7
Free Wheeling Hub, Replace (B)		
1990-92 one	1.0	1.2
Front Axle Hub Oil Seal, Replace (B)		
1990-92 one	1.7	1.9
Front Hub Bearing, Replace (B)		
1990-92 one	1.8	2.0
Front Hub Seal, Replace (B)		
1990-92 one side	1.8	2.0
Front Shock Absorbers, Replace (B)		
1990-92		
one	1.1	1.3
both	1.8	2.0
Lower Control Arm Assy., Replace (B)		
1990-92	2.0	2.2
Stabilizer Bar, Replace (B)		
1990-92	1.1	1.3
Steering Knuckle, Replace (B)		
1990-92 one	2.5	2.7
Torsion Bar, Replace (B)		
Includes: Height adjustment.		
1990-92 one side	1.3	1.5
Upper Control Arm Assy., Replace (B)		
1990-92 one	1.4	1.6

REAR SUSPENSION

	LABOR TIME	SEVERE SERVICE
Rear Control Arm, Replace (B)		
Charade one	1.2	1.4
Rear Spring, Replace (B)		
One	1.8	2.0
Both	2.6	2.8
Rear Spring Eye Bushing, Replace (B)		
Rocky		
one side	1.9	2.1
both sides	2.9	3.1
Rear Stabilizer Bar & Bushings, Replace (B)		
1988-92	1.0	1.2
Shackle and/or Bushing, Replace (B)		
Rocky		
one side	.8	1.0
both sides	1.2	1.4
Shock Absorbers or Bushings, Replace (B)		
Charade		
one	1.6	1.8
both	2.3	2.5
Rocky		
one	1.0	1.2
both	1.7	1.9

	LABOR TIME	SEVERE SERVICE
Stub Axle, Replace (B)		
Charade		
one	2.2	2.4
both	3.3	3.5
Strut Rod, Replace (B)		
Charade one side	1.0	1.2
Wheel Bearings, Replace (B)		
Charade		
one	1.2	1.4
both	1.5	1.7

STEERING

Unless otherwise noted, time given does not include alignment.

LINKAGE

	LABOR TIME	SEVERE SERVICE
Pitman Arm, Replace (B)		
Rocky	1.7	2.0
Relay Rod (Center Link), Replace (B)		
Includes: Reset toe.		
Rocky	2.8	3.1
Tie Rod or Tie Rod Ends, Replace (B)		
Includes: Reset toe.		
Rocky one	1.5	1.7
Inner Tie Rod Assy., Replace (B)		
Includes: Reset toe.		
Charade		
one	1.7	1.9
both	2.2	2.4
Outer Tie Rod End, Replace (B)		
Includes: Reset toe.		
Charade		
one	1.0	1.2
both	1.3	1.5
Rack & Pinion Assy., R&R and Recondition (B)		
Charade	3.7	4.2
Rack & Pinion Assy., Replace (B)		
Charade	2.1	2.3
Rack Boots, Replace (B)		
Charade		
one	1.3	1.5
both	1.5	1.7
Steering Shaft U-Joint, Replace (B)		
Charade	1.0	1.2
Steering Wheel, Replace (B)		
Charade	.6	.6

POWER STEERING

	LABOR TIME	SEVERE SERVICE
Gear Assy., R&R or Replace (B)		
Rocky	2.5	3.0
Rack & Pinion Assy., R&R or Replace (B)		
Includes: Alignment.		
Charade	3.7	4.2
Steering Pump, R&R or Replace (B)		
1988-92	1.5	1.7
Rocky	1.3	1.5

	LABOR TIME	SEVERE SERVICE
Steering Pump Hoses, Replace (B)		
Does not include purging.		
Charade each	1.1	1.3

HEATING & AIR CONDITIONING

When more than one component requires replacement where evacuation/recovery and recharging is already included, deduct 1.0 hour for each additional component from the time given.

	LABOR TIME	SEVERE SERVICE
Evacuate/Recover and Recharge System (B)		
1988-92	1.0	1.0
AC Hoses, Replace (B)		
Includes: Evacuate/recover and recharge.		
Suction hose		
1.0L	1.6	1.8
1.3L	1.7	1.9
Hose assy.	1.5	1.7
Amplifier, Replace (B)		
1988-92	.4	.4
Compressor Assy., Replace (B)		
Includes: Parts transfer. Evacuate/recover and recharge.		
Charade		
1.0L	2.2	2.4
1.3L	2.7	2.9
Rocky	2.0	2.2
Condenser Assy., Replace (B)		
Includes: Evacuate/recover and recharge.		
Charade		
1.0L	2.5	2.7
1.3L	2.9	3.1
Rocky	2.3	2.5
Evaporator Coil, Replace (B)		
Includes: Evacuate/recover and recharge.		
Charade	4.2	4.4
Rocky	3.2	3.4
Heater Blower Motor, Replace (B)		
1988-92	.9	.9
Heater Blower Motor Relay, Replace (B)		
1988-92	.7	.7
Heater Blower Motor Resistor, Replace (B)		
1988-92	.5	.5
Heater Control Assy., Replace (B)		
Charade	.6	.6
Rocky	1.4	1.4
Heater Core, R&R or Replace (B)		
Charade	4.1	4.3
Rocky	3.7	3.9
Heater Hoses, Replace (B)		
Each	.4	.5

CHARADE : ROCKY DAI-7

	LABOR TIME	SEVERE SERVICE
Receiver/Drier Assy., Replace (B)		
Includes: Evacuate/recover and recharge.		
1988-92	1.6	1.8
Thermostatic Control Switch, Replace (B)		
1988-92	.9	.9

WIPERS & SPEEDOMETER

	LABOR TIME	SEVERE SERVICE
Antenna, Replace (B)		
Charade	2.0	2.2
Rocky	.8	.8
Radio, R&R (B)		
1988-92	.7	.7
Rear Window Wiper Motor, Replace (B)		
1988-92	.6	.8
Rear Windshield Washer Pump, Replace (B)		
Charade	1.0	1.0
Rocky	.5	.5
Speedometer Cable & Casing, Replace (B)		
1988-92	.5	.5
Speedometer Head, R&R or Replace (B)		
Charade	1.0	1.0
Rocky	.8	.8
Window Wiper & Washer Rear Switch, Replace (B)		
1988-92	.4	.4
Windshield Wiper & Washer Switch, Replace (B)		
1988-92	.7	.7
Windshield Wiper Interval Relay, Replace (B)		
1988-92	.4	.4

	LABOR TIME	SEVERE SERVICE
Windshield Wiper Linkage, Replace (B)		
1988-92	1.0	1.2
Windshield Wiper Motor, Replace (B)		
1988-92	.9	1.1

LAMPS & SWITCHES

	LABOR TIME	SEVERE SERVICE
Back-Up Lamp/Neutral Safety Switch, Replace (B)		
1988-92	.5	.5
Clearance Lamp Assy., Replace (B)		
One	.3	.3
Clearance Lamp Lens or Bulb, Replace (C)		
One	.2	.2
Clutch Start Switch, Replace (B)		
1988-92	.5	.5
Halogen Headlamp Bulb, Replace (C)		
One	.3	.3
Hazard Warning Switch, Replace (B)		
Rocky	.5	.5
Headlamps, Aim (B)		
Two	.4	.4
Four	.6	.6
High Mount Stop Light Assy., Replace (B)		
1988-92	.3	.3
Horn, Replace (B)		
1988-92	.4	.4
Horn Relay, Replace (B)		
1988-92	.7	.7
License Lamp Assy., Replace (B)		
1988-92	.3	.3
Parking Brake Warning Lamp Switch, Replace (B)		
1988-92	.4	.4
Rear Combination Lamp Assy., Replace (B)		
Charade	1.0	1.0
Rocky	.3	.3

	LABOR TIME	SEVERE SERVICE
Rear Combination Lamp Lens, Replace (B)		
One	1.0	1.0
Side Marker Lamp Assy., Replace (B)		
One	.2	.2
Side Marker Lamp, Front, Replace (B)		
One	.2	.2
Stop Light Switch, Replace (B)		
1988-92	.5	.5
Turn Signal & Headlamp Dimmer Switch, Replace (B)		
1988-92	.9	.9
Turn Signal or Hazard Warning Flasher, Replace (B)		
1988-92	.2	.2
Turn Signal Lamp Lens or Bulb, Replace (B)		
One	.3	.3

BODY

	LABOR TIME	SEVERE SERVICE
Door Window Regulator, Replace (B)		
Charade	.9	1.1
Rocky	1.6	1.8
Front Door Lock, Replace (B)		
1988-92	.6	.8
Front Door Lock Cylinder, Replace (B)		
1988-92	.5	.7
Front Door Window Motor, Replace (B)		
1988-92	1.1	1.3
Hood Lock, Replace (B)		
1988-92	.4	.4
Hood Release Cable, Replace (B)		
1988-92	.9	.9
Rear Door Lock, Replace (B)		
Rocky	.7	.7

NOTES

HON

Accord : Civic : Civic del Sol : CRX : Insight : Prelude : S2000

SYSTEM INDEX

- **MAINTENANCE** HON-2
  - Maintenance Schedule .. HON-2
- **CHARGING** HON-2
- **STARTING** HON-3
- **CRUISE CONTROL** HON-3
- **IGNITION** HON-3
- **EMISSIONS** HON-4
- **FUEL** HON-5
- **EXHAUST** HON-6
- **ENGINE COOLING** HON-7
- **ENGINE** HON-7
  - Assembly HON-7
  - Cylinder Head HON-8
  - Camshaft HON-9
  - Crank & Pistons HON-10
  - Engine Lubrication ... HON-10
- **CLUTCH** HON-11
- **MANUAL TRANSAXLE** ... HON-11
- **AUTO TRANSAXLE** HON-12
- **TRANSFER CASE** HON-13
- **SHIFT LINKAGE** HON-13
- **DRIVELINE** HON-13
- **BRAKES** HON-13
- **FRONT SUSPENSION** ... HON-15
- **REAR SUSPENSION** HON-16
- **STEERING** HON-17
- **HEATING & AC** HON-19
- **WIPERS & SPEEDOMETER** HON-20
- **LAMPS & SWITCHES** HON-21
- **BODY** HON-21

OPERATIONS INDEX

A
- AC Hoses HON-19
- Air Bags HON-17
- Air Conditioning HON-19
- Alignment HON-15
- Alternator (Generator) HON-3
- Antenna HON-20
- Anti-Lock Brakes HON-13

B
- Back-Up Lamp Switch HON-21
- Ball Joint HON-15
- Battery Cables HON-3
- Bleed Brake System HON-14
- Blower Motor HON-19
- Brake Disc HON-14
- Brake Drum HON-14
- Brake Hose HON-14
- Brake Pads and/or Shoes .. HON-15

C
- Camshaft HON-9
- Catalytic Converter HON-6
- Coolant Temperature (ECT) Sensor HON-4
- Crankshaft HON-10
- Crankshaft Sensor HON-3
- Cruise Control HON-3
- CV Joint HON-13
- Cylinder Head HON-8

D
- Differential HON-13
- Distributor HON-4
- Drive Belt HON-2
- Driveshaft HON-13

E
- EGR HON-5
- Electronic Control Module (ECM/PCM) HON-5
- Engine HON-7
- Engine Lubrication ... HON-10
- Engine Mounts HON-8
- Evaporator HON-19
- Exhaust HON-6
- Exhaust Manifold HON-7
- Expansion Valve HON-19

F
- Flexplate HON-12
- Flywheel HON-11
- Fuel Injection HON-6
- Fuel Pump HON-6
- Fuel Vapor Canister .. HON-5

G
- Gear Selector Lever HON-13
- Generator HON-3

H
- Halfshaft HON-13
- Headlamp HON-21
- Heater Core HON-20
- Horn HON-21

I
- Idle Air Control (IAC) Valve . HON-6
- Ignition Coil HON-4
- Ignition Switch HON-4
- Inner Tie Rod HON-17
- Intake Air Temperature (IAT) Sensor HON-5
- Intake Manifold HON-6

K
- Knock Sensor HON-5

L
- Lower Control Arm HON-15

M
- Maintenance Schedule HON-2
- Manifold Absolute Pressure (MAP) Sensor HON-5
- Master Cylinder HON-14
- Muffler HON-7

N
- Neutral Safety Clutch Switch HON-21

O
- Oil Pan HON-11
- Oil Pump HON-11
- Outer Tie Rod HON-17
- Oxygen Sensor HON-5

P
- Parking Brake HON-15
- Pistons HON-10
- Positive Crankcase Ventilation (PCV) Valve ... HON-5

R
- Radiator HON-7
- Radiator Hoses HON-7
- Radio HON-20
- Rear Main Oil Seal HON-10

S
- Shock Absorber/Strut, Front HON-16
- Shock Absorber/Strut, Rear HON-17
- Spark Plug Cables HON-4
- Spark Plugs HON-4
- Spring, Front Coil ... HON-15
- Spring, Leaf HON-16
- Spring, Rear Coil ... HON-16
- Starter HON-3
- Steering Wheel HON-17

T
- Thermostat HON-7
- Throttle Body HON-6
- Throttle Position Sensor (TPS) HON-5
- Timing Belt HON-9
- Timing Chain HON-9
- Torque Converter ... HON-12

U
- U-Joint HON-13
- Upper Control Arm ... HON-16

V
- Valve Body HON-12
- Valve Cover Gasket HON-9
- Valve Job HON-9
- Vehicle Speed Sensor ... HON-5

W
- Water Pump HON-7
- Wheel Balance HON-2
- Wheel Cylinder HON-15
- Window Regulator HON-21
- Windshield Washer Pump .. HON-20
- Windshield Wiper Motor ... HON-21

HON-1

HON-2   ACCORD : CIVIC : CIVIC DEL SOL : CRX : INSIGHT : PRELUDE : S2000

	LABOR TIME	SEVERE SERVICE
MAINTENANCE		
Air Filter Element, Replace (B)		
1981-05 (.2)	.3	.4
Chassis Lubrication, Change Oil & Filter (C)		
Includes: Correct fluid levels.		
1981-05 (.3)	.5	.5
Drive Belt, Adjust (B)		
All Models	.3	.4
Drive Belt, Replace (B)		
Air conditioning		
exc. below (.5)	.7	.8
Accord		
1981-85 (.2)	.3	.4
Civic		
2000-05 (.2)	.3	.4
Prelude		
1981-87 (.3)	.5	.6
1992-01 (.2)	.3	.4
S2000 (.3)	.5	.6
Air Pump		
1981-87 (.2)	.3	.4
Alternator		
1981-05		
4 cyl. (.3)	.5	.6
V6 (.5)	.8	.9
Power steering		
Accord		
4 cyl. (.3)	.5	.6
V6 (.4)	.6	.7
Civic, DelSol,		
Prelude (.3)	.5	.6
Replace alternator belt		
w/AC 81-00 add	.2	.2
w/PS Civic 81-00 add	.2	.2
w/top mount AC or FI interference add	1.2	1.2
Drive Belt Tensioner/Pulley, Replace (B)		
Accord, S2000 (.3)	.4	.5
Fuel Filter, Replace (B)		
Accord		
1981-97 (.3)	.5	.6
1998-02 (.6)	.9	1.0
2003-05 (.4)	.6	.7
Civic, CRX (.4)	.6	.7
del Sol	.5	.6
Insight, S2000 (1.5)	2.4	2.5
Prelude (.3)	.5	.6
Halogen Headlamp Bulb, Replace (B)		
1988-02		
exc. S2000 each (.2)	.3	.3
S2000 each (.3)	.5	.5
2003-05		
exc. below each (.2)	.3	.3
Accord		
right (.2)	.3	.3
left (.4)	.6	.7
S2000 each	.5	.5

	LABOR TIME	SEVERE SERVICE
Oil & Filter, Change (C)		
Includes: Correct fluid levels.		
1981-05 (.3)	.5	.5
Sealed Beam Headlamp, Replace (B)		
1981-92 each (.2)	.3	.3
w/CRX add	.3	.3
Timing Belt, Replace (B)		
Accord		
1981-83	2.0	2.3
1984-89	3.1	3.4
1990-02		
4 cyl.	3.5	3.8
V6	5.0	5.3
2003-05 (3.0)	4.3	4.6
Civic, CRX		
1981-85	1.7	2.0
1986-87	2.0	2.3
1988-00	2.7	3.0
2001-05 (2.0)	3.0	3.3
del Sol (1.5)	2.2	2.5
Prelude		
1981-82	2.0	2.3
1983-91	3.1	3.4
1992-96	3.5	3.8
1997-01 (2.5)	3.8	4.1
w/PS add	.2	.2
w/top mount AC add	1.2	1.2
w/twin cam add	.3	.3
Replace water pump add	.5	.7
Tire, Replace (C)		
Includes: Dismount old tire and mount new tire to rim. Wheel balance.		
One (.4)	.6	.7
each addl. add	.6	.7
Tires, Rotate (C)		
1981-05 (.3)	.5	.5
Wheel, Balance (B)		
One (.3)	.5	.5
each addl. add	.5	.5
SCHEDULED MAINTENANCE INTERVALS		
If necessary, refer to appropriate Chilton maintenance service information.		
7,500 Mile Service (C)		
All Models	1.2	1.2
15,000 Mile Service (B)		
All Models	2.1	2.1
Adjust valves add	1.1	1.1
22,500 Mile Service (C)		
All Models	1.2	1.2
30,000 Mile Service (B)		
All Models	3.4	3.4
Adjust valves add	1.1	1.1
Replace		
brake fluid add	.6	.6
spark plugs add	.5	.5
transaxle fluid add	.5	.5

	LABOR TIME	SEVERE SERVICE
37,500 Mile Service (C)		
All Models	1.2	1.2
45,000 Mile Service (B)		
All Models	2.1	2.1
Replace		
brake fluid add	.6	.6
coolant add	.7	.7
transaxle fluid add	.5	.5
52,500 Mile Service (C)		
All Models	1.2	1.2
60,000 Mile Service (B)		
All Models	3.6	3.6
Adjust valves add	1.1	1.1
Replace		
brake fluid add	.6	.6
fuel filter add	.3	.3
spark plugs add	.5	.5
transaxle fluid add	.5	.5
67,500 Mile Service (C)		
All Models	1.2	1.2
75,000 Mile Service (B)		
All Models	2.7	2.7
Adjust valves add	1.1	1.1
Replace		
coolant add	.7	.7
transaxle fluid add	.5	.5
82,500 Mile Service (C)		
All Models	1.2	1.2
90,000 Mile Service (B)		
All Models	3.6	3.6
Adjust valves add	1.1	1.1
Replace		
brake fluid add	.6	.6
spark plugs add	.5	.5
timing belt		
exc. V6 add	2.2	2.2
V6 add	3.5	3.5
transaxle fluid add	.5	.5
97,500 Mile Service (C)		
All Models	1.2	1.2
CHARGING		
Alternator Circuits, Test (B)		
Includes: Test component output.		
1981-05 (.4)	.6	.6
Alternator Assy., R&R and Recondition (A)		
Accord		
1981-89	2.6	2.8
1990-97	1.7	1.9
1998-05		
4 cyl. (1.1)	1.7	1.9
V6 (1.5)	2.2	2.4
Civic		
1981-00	2.3	2.5
2001-05		
exc. Si (1.5)	2.3	2.5
Si (1.8)	2.7	2.9
CRX, del sol (1.5)	2.3	2.5

ACCORD : CIVIC : CIVIC DEL SOL : CRX : INSIGHT : PRELUDE : S2000

	LABOR TIME	SEVERE SERVICE
Prelude		
1981-87	3.1	3.3
1988-01 (1.5)	2.3	2.5
S2000 (1.5)	2.3	2.5
Alternator Assy., Replace (B)		
Includes: Pulley transfer.		
Accord		
1981-89	1.4	1.6
1990-97	.8	1.0
1998-05		
4 cyl. (.5)	.7	.9
V6 (.8)	1.1	1.3
Civic		
1981-83	1.2	1.4
1984-00	1.4	1.6
2001-05		
exc. Si (.9)	1.4	1.6
Si (1.2)	1.8	2.1
CRX, del Sol (.9)	1.4	1.6
Prelude		
1981-87	2.3	2.5
1988-01 (.9)	1.4	1.6
S2000 (.9)	1.4	1.6
Alternator Voltage Regulator, Replace (B)		
Accord		
1981-89	2.4	2.6
1990-05 (.7)	1.1	1.3
Civic		
1981-83 (.3)	.5	.7
1984-05		
exc. Si (1.0)	1.6	1.8
Si (1.8)	2.6	2.8
CRX, del Sol (1.5)	2.3	2.5
Prelude		
1981-87	2.8	3.0
1988-01 (1.0)	1.6	1.8
S2000 (1.0)	1.6	1.8
Front Alternator Bearing, Replace (B)		
Accord, CRX, del Sol	2.0	2.2
Civic		
1981-83	1.5	1.7
1984-95	1.2	1.4
Prelude		
1981-87	2.6	2.8
1988-95	1.9	2.1

INTEGRATED MOTOR ASSIST SYSTEM
Battery Assembly, Replace (B)
Civic (.8)	1.2	1.4
Insight (1.2)	1.8	2.1

Battery Electronic Control Unit, Replace (B)
Civic, Insight (.5)	.8	.9

	LABOR TIME	SEVERE SERVICE
Commutation Sensor, Replace (B)		
Civic (3.2)	4.9	5.2
Insight (3.4)	5.2	5.5
DC-DC Converter, Replace (B)		
Accord (.7)	1.0	1.1
Civic, Insight (.5)	.8	.9
Motor Rotor, Replace (B)		
Accord (7.8)	10.9	11.4
Civic (3.8)	5.5	6.0
Insight (4.0)	5.8	6.3
Motor Stator, Replace (B)		
Accord (8.2)	11.5	12.0
Civic (3.5)	5.1	5.6
Insight (3.7)	5.4	5.9
Power Cable, Replace (B)		
Accord (3.1)	4.5	5.0
Civic (1.5)	2.2	2.7
Insight (2.9)	4.3	4.8
Power Inverter Module, Replace (B)		
Civic, Insight (.5)	.8	.9
Predriver Unit/Intelligent Power Unit, Replace (B)		
Civic, Insight (.5)	.8	.9

STARTING
Starter Draw Test (On Car) (B)
1981-05 (.3)	.5	.5

Battery Cables, Replace (C)
Positive (.4)	.6	.7
Negative (.2)	.3	.4
Replace positive cable w/four cylinder Accord 03-05 add	1.2	1.2

Starter Assy., R&R and Recondition (A)
Exc. below (1.2)	1.9	2.4
1985-91 Civic wagon	2.7	3.2
1981-87 Prelude	2.0	2.5
S2000 (2.1)	3.1	3.6

Starter Assy., Replace (B)
1981-99		
exc. Civic wagon	1.1	1.3
Civic wagon	1.8	2.0
2000-05		
exc. below (.7)	1.1	1.3
Civic Si 2.0L (1.5)	2.3	2.6
S2000 (1.6)	2.5	2.8

Starter Drive Assy., Replace (B)
Includes: Starter R&R.
Exc. below	1.4	1.7
Civic wagon	2.0	2.3
1988-95 Prelude	1.8	2.1

Starter Solenoid and/or Switch, Replace (B)
Includes: Starter R&R.
Exc. below	1.3	1.5
Civic wagon	1.9	2.1
1988-95 Prelude	1.7	1.9

	LABOR TIME	SEVERE SERVICE
### CRUISE CONTROL		
Diagnose Cruise System Component Each (A)		
1982-05 (.6)	.8	.8
Control Actuator Cable, Adjust (B)		
1982-05 (.2)	.3	.4
Actuator Assy., Replace (B)		
1982-05 (.3)	.5	.5
Control Actuator Valve, Replace (B)		
1982-05 (.3)	.5	.5
Control Brake or Clutch Switch, Replace (B)		
1982-05 (.3)	.5	.5
Control Check Valve, Replace (B)		
1982-05 (.2)	.3	.3
Control Controller Module, Replace (B)		
Exc. S2000 (.3)	.5	.5
S2000 (.6)	.9	1.0
Control On/Off Switch, Replace (B)		
1982-05 (.2)	.3	.3
Control Slip Ring, Replace (B)		
1986-96 (.7)	.9	.9
Control Steering Wheel Switch, Replace (B)		
1982-05 (.3)	.5	.5
Control Vehicle Speed Sensor, Replace (B)		
Exc. below	.8	.9
1986-89 Accord	1.8	1.8
1988-91 Civic, CRX, Prelude	1.8	1.8

IGNITION
Diagnose Ignition System Component Each (A)
1981-05 (.6)	.8	.8

Retrieve Fault Codes (A)
1981-05 (.3)	.4	.4

Ignition Timing, Reset (B)
1981-03 (.2)	.3	.3

Crankshaft Position (CKP), Cylinder Position, Top Dead Center (TDC) or Crankshaft Speed Fluctuation (CKF) Sensors, Replace (B)
Distributor mounted (.5)	.8	.9
Engine or cam cover mounted		
Accord		
1994-97		
exc. below	1.2	1.5
4 cyl. CKP/TDC	3.8	4.1
V6 CKP	2.3	2.6
1998-02		
4 cyl.	3.8	4.1
V6 (2.8)	4.0	4.3
2003-05		
4 cyl. (.2)	.3	.4
V6 (3.0)	4.3	4.6

HON-3

ACCORD : CIVIC : CIVIC DEL SOL : CRX : INSIGHT : PRELUDE : S2000

	LABOR TIME	SEVERE SERVICE
Crankshaft Position (CKP), Cylinder Position, Top Dead Center (TDC) or Crankshaft Speed Fluctuation (CKF) Sensors, Replace (B)		
Civic		
1996-00 (1.5)	2.3	2.6
2001-03		
1.3L, 2.0L	.3	.3
1.7L (1.5)	2.2	2.5
2000 Insight, S2000 (.2)	.3	.3
1997-01 Prelude	3.8	4.3
Distributor, R&R and Recondition (B)		
Includes: Reset base ignition timing.		
1981-93	1.9	2.2
1994-95		
4 cyl.	1.9	2.2
V6	1.4	1.5
w/FI 81-93 add	2.3	2.6
Distributor Assy., Replace (B)		
Includes: Reset base ignition timing.		
1981-93	.5	.7
1994-03	.5	.7
w/FI 81-93 add	.2	.2
Distributor Cap and/or Rotor, Replace (B)		
1981-03	.3	.3
Ignition Coil, Replace (B)		
1981-87	.3	.4
1988-05		
Accord (.4)	.6	.7
Civic (.4)	.6	.7
CRX, del Sol (.2)	.3	.4
Insight, S2000 (.2)	.3	.4
Prelude (.4)	.6	.7
w/coil mounted inside distributor Accord add	.3	.3
Igniter Unit, Replace (B)		
Accord, Civic, CRX, del Sol (.4)	.6	.7
Prelude		
1981-91	.3	.3
1992-01	.7	.7
Ignition Switch (Electrical Portion), Replace (B)		
1981-93	1.0	1.0
1994-97		
exc. Accord	1.0	1.0
Accord	1.4	1.5
1998-03	.6	.7
Ignition Switch and Lock Assy., Replace (B)		
1981-05 (.9)	1.4	1.5
Spark Plug (Ignition) Cables, Replace (B)		
1981-05	.3	.4

	LABOR TIME	SEVERE SERVICE
Spark Plugs, Replace (B)		
1981-93	.5	.6
1994-05		
4 cyl.	.5	.6
V6	.8	.9
Vacuum Advance Unit, Replace (B)		
Includes: Reset base timing.		
1981-88	.7	.7
w/FI add	.3	.3

EMISSIONS

The following operations do not include testing. Add time as required.

	LABOR TIME	SEVERE SERVICE
Diagnose Emission Control System Component Each (A)		
1985-05 (.6)	.8	.8
Evaporative Function Test (A)		
2001-02 Civic (.2)	.3	.3
Reprogram Electronic Control Module (ECM/PCM) (A)		
2000-05		
exc. below (.3)	.4	.4
2005 Accord (.5)	.6	.6
Retrieve Fault Codes (A)		
1985-05 (.3)	.3	.3
Dynamometer Test (B)		
1981-01	.5	.5
Air Bypass Solenoid Valve, Replace (B)		
1981-87	.5	.5
Air Bypass Valve, Replace (B)		
1981-87	.7	.7
Air Delay Valve, Replace (B)		
1981-87	.5	.5
Air Distribution Manifold, Replace (B)		
1981-87	.7	.7
Air Injection Nozzles, Replace (B)		
1981-87		
one	.3	.3
all	.9	.9
Air Injection Tube Check Valve Assy., Replace (B)		
1981-87	.3	.3
Air Pump Assy., Replace (B)		
1981-87	.7	.9
2000-05 S2000 (.6)	.9	1.0
Air Pump Current Sensor, Replace (B)		
S2000 (.6)	.9	1.0
Air Pump Relay, Replace (B)		
S2000 (.6)	.9	1.0
Air Pump Solenoid, Replace (B)		
S2000 (.6)	.9	1.0
Air Vacuum Switch, Replace (B)		
1981-87	.5	.5
Anti-Backfire Valve, Replace (B)		
1981-87	.3	.3

	LABOR TIME	SEVERE SERVICE
Barometric Pressure Sensor, Replace (B)		
Accord		
1986-89	.3	.3
1990-97		
4 cyl.	.5	.5
V6	.7	.7
1998-05 (.4)	.7	.7
Civic		
1987-91	.3	.3
1992-05 (.3)	.5	.5
CRX, del Sol (.3)	.5	.5
Prelude		
1987-91	.7	.7
1992-01 (.3)	.5	.5
Bypass Control Solenoid Valve, Replace (B)		
Accord		
1990-93	.7	.7
1994-97	.3	.3
1998-05 (4.)	.6	.7
Civic		
1996-05 (.4)	.6	.7
Insight, S2000 (.4)	.6	.7
Prelude		
1997-01 (.2)	.3	.3
Canister Purge Control Valve, Replace (B)		
1981-86	.5	.5
Constant Vacuum Regulator, Replace (B)		
1981-86	.7	.7
Coolant Temperature (ECT) Sensor, Replace (B)		
1987-97		
4 cyl.	.6	.7
V6	1.2	1.4
1998-05		
exc. below (.4)	.6	.7
Insight (.6)	.8	.9
S2000 (.7)	1.1	1.2
Deceleration Valve, Replace (B)		
1981-86	.3	.3
EFI Main Relay, Replace (B)		
Exc. Insight	.6	.7
Insight	.3	.3
EGR Control Frequency Solenoid Valve, Replace (B)		
1985-90	.7	.7
EGR Control Solenoid Valve, Replace (B)		
1986-95	.7	.7
1996-01 Accord, Prelude	.6	.7
1996-00 Civic	.3	.3

ACCORD : CIVIC : CIVIC DEL SOL : CRX : INSIGHT : PRELUDE : S2000

	LABOR TIME	SEVERE SERVICE
EGR Valve, Replace (B)		
Accord		
1981-89	.8	.9
1990-97		
4 cyl.	.5	.6
V6	.9	1.0
1998-05 (.3)	.5	.6
Civic		
1981-95	.7	.8
1996-05 (.2)	.3	.4
CRX	.7	.8
del Sol (.2)	.3	.4
Insight (.2)	.3	.4
Prelude	.8	.9
Electronic Control Module (ECM/PCM), Replace (B)		
Accord (.4)	.7	.7
Civic		
1985-00	.7	.7
2001-05		
1.3L, 1.7L (.5)	.8	.9
2.0L (.2)	.3	.3
CRX	.9	.9
del Sol (.3)	.7	.7
Insight, S2000 (.3)	.5	.5
Prelude		
1985-91	1.2	1.2
1992-01	.5	.6
Electronic Load Detector, Replace (B)		
1988-05 (.5)	.7	.7
Fuel Vapor Canister, Replace (B)		
1981-97	.3	.3
1998-05		
Accord (.6)	.9	1.0
Civic		
1998-00	.3	.3
2001-05 (.5)	.8	.9
Insight, S2000 (.6)	.9	1.0
Prelude	.3	.3
Intake Air Temperature (IAT) Sensor, Replace (B)		
1981-86	.8	.8
1987-03		
exc. Civic 1.3L (.2)	.3	.3
Civic 1.3L (.4)	.6	.7
Knock Sensor, Replace (B)		
Accord		
1994-02	.3	.5
2003-05		
4 cyl. (.4)	.6	.7
V6 (.7)	1.1	1.2
1996-05 Civic		
exc. Si. 2.0L (.2)	.3	.5
Si. 2.0L (.4)	.6	.7
Insight (.3)	.5	.7

	LABOR TIME	SEVERE SERVICE
Prelude		
1992-96	.6	.8
1997-01		
exc. SH	.3	.5
SH	4.6	5.2
S2000 (.2)	.3	.5
Main Relay, Replace (B)		
1987-05		
exc. below (.4)	.6	.7
2003-05 Accord, 2001-05 Civic (.2)	.3	.3
Insight (.2)	.3	.3
Manifold Absolute Pressure (MAP) Sensor, Replace (B)		
Accord (.2)	.3	.4
Civic (.4)	.6	.7
CRX, del Sol	.5	.5
Insight, S2000 (.2)	.3	.3
Prelude		
1987-91	.7	.7
1992-01	.3	.3
Oxygen Sensor, Replace (B)		
Accord		
1985-91	1.2	1.3
1992-05 (.3)	.5	.6
Civic (.3)	.5	.6
CRX	.7	.8
del Sol	.5	.6
Insight, S2000 (.3)	.5	.6
Prelude		
1985-91	1.2	1.3
1992-01	.5	.6
Positive Crankcase Ventilation (PCV) Valve, Replace (B)		
1981-05 (.2)	.3	.3
Pulse Generator, Replace (B)		
1981-87	.7	.7
Throttle Position Sensor (TPS), Replace (B)		
Accord		
1986-89	1.6	1.6
1990-97		
4 cyl.	.8	.9
V6	1.6	1.7
1998-05 (.5)	.8	.9
Civic		
1987-91	1.3	1.3
1992-05 (.5)	.8	.9
CRX, del Sol	.7	.7
Insight (.6)	.9	1.0
Prelude		
1987-91	1.6	1.6
1992-01	.8	.9
S2000 (.5)	.8	.9
Vacuum Holding Valve, Replace (B)		
1981-86	.5	.5
Vacuum Switch, Replace (B)		
1987-91 Prelude	.7	.7

	LABOR TIME	SEVERE SERVICE
Vehicle Speed Sensor, Replace (B)		
1986-05 Accord (.5)	.7	.7
1987-05 Civic (.5)	.7	.7
1987-91 CRX, del Sol	.8	.8
Insight (.2)	.3	.3
Prelude		
1987-91		
carburetor	1.9	1.9
FI	.8	.8
1992-01	.7	.7
S2000 (.5)	.7	.7

FUEL
CARBURETOR
Carburetor, Adjust (On Car) (B)
1981-91	.5	.5

Carburetor, R&R and Clean or Recondition (A)
Includes: Adjustments.
Accord, Civic	4.1	4.1
Prelude		
1983-86	4.5	4.5
1988-91	6.2	6.2

Carburetor, Replace (B)
Includes: Adjustments.
Accord, Civic	1.9	1.9
Prelude		
1983-86	3.1	3.1
1988-91	4.2	4.2

Carburetor Choke Cover, Replace (B)
1981-91		
rivet retained	1.2	1.2
screw retained	.7	.7

Choke Cable, Replace (B)
1981-88	.7	.7

Throttle Cable, Replace (B)
1981-91	.8	.8

DELIVERY
Fuel Pump, Test (B)
1981-05 (.2)	.3	.3

Fuel Filter, Replace (B)
Accord		
1981-97 (.3)	.5	.6
1998-02 (.6)	.9	1.0
2003-05 (.4)	.6	.7
Civic, CRX (.4)	.6	.7
del Sol	.5	.6
Insight (1.4)	2.2	2.3
Prelude (.3)	.5	.6
S2000 (1.5)	2.4	2.5

Fuel Gauge (Dash), Replace (B)
1981-00 Accord	1.6	1.6
Civic		
1981-87	1.2	1.2
1988-00	1.6	1.6

HON-5

HON-6 ACCORD : CIVIC : CIVIC DEL SOL : CRX : INSIGHT : PRELUDE : S2000

	LABOR TIME	SEVERE SERVICE
Fuel Gauge (Dash), Replace (B)		
CRX		
1984-87	1.2	1.2
1988-91	1.9	1.9
del Sol		
1993-95	2.8	2.8
1996-97	.5	.5
Prelude		
1981-87	2.3	2.3
1988-91	1.8	1.8
1992-01	1.4	1.4
Fuel Gauge (Tank), Replace (B)		
Accord		
1981-85	1.4	1.6
1986-02 (.6)	.9	1.1
2003-05 (.4)	.6	.8
Civic, CRX		
1981-87 (.9)	1.3	1.5
1988-05 (.6)	.9	1.1
del Sol	1.7	1.9
Insight (1.4)	2.0-	2.2
Prelude		
1981-85	1.3	1.5
1986	2.3	2.5
1987-91	1.7	1.9
1992-01	1.0	1.2
S2000 (1.5)	2.2	2.4
w/Accord wagon add	.2	.2
Fuel Cut-off Solenoid, Replace (B)		
1981-91 Accord	.7	.7
1981-91 Civic	.5	.5
CRX	.5	.5
1981-91 Prelude	.7	.7
Fuel Tank, Replace (B)		
Includes: Drain and refill.		
Accord		
1981-95 (.9)	1.3	1.6
1996-97 (.8)	1.2	1.5
1998-02		
disc brakes	4.0	4.3
drum brakes	4.6	4.9
2003-05 (3.1)	4.5	4.8
Civic (1.0)	1.5	1.8
CRX, del Sol	1.7	2.0
Insight (1.2)	1.8	2.1
Prelude		
1981-82	1.4	1.7
1983-87	2.3	2.6
1988-96	1.3	1.6
1997-01	2.1	2.4
S2000 (4.0)	5.8	6.1
w/NGV Civic GX add	.5	.5
Fuel Pump, Replace (B)		
Includes: Testing.		
1981-86		
electric	1.2	1.5
manual	.7	.9

	LABOR TIME	SEVERE SERVICE
Accord		
1987-89	.8	1.1
1990-97	1.6	1.8
1998-05 (.6)	1.0	1.3
Civic, CRX, del Sol		
1988-91	1.6	1.9
1992-05 (.6)	1.0	1.3
Civic HB, IMA (.4)	.6	.8
Insight (1.4)	2.2	2.5
Prelude	1.0	1.2
S2000 (1.5)	2.4	2.7
Fuel Pump Relay, Replace (B)		
1981-01	.5	.5
Intake Manifold and/or Gasket, Replace (B)		
Includes: Adjustments.		
Accord		
1981-97	3.8	4.3
1998-05		
4 cyl.	1.6	1.8
V6	.8	.9
Civic		
1981-87	3.6	3.9
1988-00	3.8	4.3
2001-05		
1.3L, 1.7L (.9)	1.4	1.5
2.0L (1.4)	2.1	2.4
CRX		
1984-87	3.0	3.3
1988-91	4.1	4.4
Insight (1.2)	1.8	2.1
del Sol, Prelude	3.8	4.3
S2000 (2.0)	3.1	3.4
Replace		
injector plate		
01-05 Civic add	.3	.3
manifold add	.5	.7

INJECTION

	LABOR TIME	SEVERE SERVICE
Fast Idle Valve, Replace (B)		
1985-01 (.4)	.7	.7
Fuel Injectors, Replace (B)		
Accord		
1985-93	.7	1.0
1994-03		
4 cyl. (.4)	.7	1.0
V6	2.3	2.6
Civic		
1985-00	.7	1.0
2001-03		
1.3L, 2.0L (.4)	.7	1.0
1.7L (.7)	1.1	1.2
CRX	1.7	2.0
del Sol	1.2	1.5
Insight, S2000 (.4)	.7	1.0
1985-01 Prelude	.7	1.0

	LABOR TIME	SEVERE SERVICE
Fuel Pressure Regulator, Replace (B)		
1985-05 (.3)	.5	.7
w/98-01 Accord V6 add	1.1	1.2
w/01-05 Civic 1.3L, 1.7L add	.2	.2
Idle Air Control (IAC) Valve, Replace (B)		
Accord		
1986-93	.5	.5
1994-02	1.2	1.4
2003-05 (.5)	.8	.9
Civic		
1987-95	.3	.3
1996-00		
exc. Si	.9	1.0
Si	.3	.3
2001-05 (.5)	.9	1.0
CRX, del Sol	.3	.3
Insight (.7)	1.1	1.2
Prelude	.5	.5
S2000 (.3)	.5	.5
Throttle Body Assy., Replace (B)		
Accord		
1985-89	1.8	1.9
1990-97		
4 cyl.	.8	.9
V6	1.6	1.7
1998-05 (.5)	.8	.9
Civic, CRX		
1985-91	1.5	1.6
1992-05 (.5)	.8	.9
del Sol	1.2	1.3
Insight (.6)	1.0	1.1
Prelude		
1985-91	1.8	1.9
1992-01	.8	.9
S2000 (.5)	.8	.9

EXHAUST

	LABOR TIME	SEVERE SERVICE
Catalytic Converter, Replace (B)		
Accord		
1981-85	1.2	1.3
1986-05 (.5)	.8	.9
Civic, CRX		
1981-87	1.2	1.3
1988-05 (.5)	.8	.9
del Sol	1.2	1.3
Insight (.5)	.8	.9
Prelude		
1981-87	1.2	1.3
1988-01 (.5)	.8	.9
S2000 (.5)	.8	.9
Center Muffler, Replace (B)		
1986-87	1.7	1.8

ACCORD : CIVIC : CIVIC DEL SOL : CRX : INSIGHT : PRELUDE : S2000   HON-7

	LABOR TIME	SEVERE SERVICE
Exhaust Manifold or Gasket, Replace (B)		
Accord		
1981-83	3.7	4.0
1984-89	1.5	1.8
1990-97	.8	1.1
1998-02		
4 cyl.	.8	1.1
V6		
one side	1.8	2.1
both sides	2.3	2.6
2003-05 (.4)	.6	.7
Civic, CRX		
1981-87	3.7	4.0
1988-91	1.2	1.5
1992-05 (.5)	.8	1.1
del Sol	1.2	1.5
Insight (.3)	.5	.8
Prelude		
1981-82	3.7	4.0
1983-87	1.7	2.0
1988-91	2.6	2.9
1992-01	.8	1.1
S2000 (.7)	1.1	1.3
Exhaust Pipe or Crossover Pipe Flange Gasket or Seal, Replace (B)		
Exc. Insight each (.6)	.9	1.0
Insight each (.3)	.5	.5
Front or Rear Exhaust Pipe, Replace (B)		
Each (.8)	1.2	1.3
Muffler, Replace (B)		
1981-05 (.5)	.8	.9
One-Piece Exhaust System, Replace (B)		
CRX (1.1)	1.7	1.8
Tail Pipe, Replace (B)		
1981-05 (.2)	.3	.4

ENGINE COOLING

	LABOR TIME	SEVERE SERVICE
Diagnose Cooling Fan System Component Each (B)		
1981-05 (.6)	.8	.8
Pressure Test Cooling System (C)		
1981-05	.3	.3
Cooling Fan Timer, Replace (B)		
1988-01 (.5)	.7	.7
Coolant Thermostat, Replace (B)		
Accord		
4 cyl.		
1976-02 (.5)	.8	.9
2003-05 (.8)	1.2	1.4
V6 (.8)	1.2	1.4
Civic, CRX		
1981-83	.8	.9
1984-87	1.2	1.3

	LABOR TIME	SEVERE SERVICE
1988-05		
exc. 2.0L (.5)	.8	.9
2.0L (.8)	1.2	1.4
del Sol	.7	.8
Insight (.5)	.8	.9
Prelude	.8	.9
S2000 (1.3)	2.0	2.2
Engine Coolant Temp. Sending Unit, Replace (B)		
1981-05 (.4)	.6	.7
Fan Blade, Replace (B)		
1981-01	.7	.9
Radiator Assy., R&R or Replace (B)		
Includes: Refill with proper coolant mix.		
Accord, Civic (1.2)	1.8	2.1
Insight (.8)	1.2	1.4
Prelude, S2000 (1.2)	1.8	2.1
w/AC add	.3	.3
w/AT add	.3	.3
Radiator Fan and/or Fan Motor, Replace (B)		
1981-05 (.5)	.8	.9
w/Civic 2.0L Si add	.5	.5
Radiator Fan Motor Relay, Replace (B)		
1981-05 (.2)	.3	.3
Radiator Fan Motor Switch (Coolant Temp.), Replace (B)		
1981-85	.5	.5
1986-05 (.5)	.7	.7
Radiator Hoses, Replace (B)		
Includes: Refill with proper coolant mix.		
Exc. S2000		
upper or lower (.5)	.8	.9
bypass (.4)	.6	.7
S2000		
upper (.9)	1.4	1.5
lower (1.2)	1.8	2.1
bypass (.4)	.6	.7
Temperature Gauge (Dash), Replace (B)		
Accord		
1981	1.2	1.2
1982-00	1.6	1.6
Civic		
1981-87	1.2	1.2
1988-00	1.6	1.6
CRX		
1984-87	1.2	1.2
1988-91	1.9	1.9
del Sol		
1993-95	2.8	2.8
1996-97	.5	.5
Prelude		
1981	1.2	1.2
1982-87	2.3	2.3
1988-91	1.8	1.8
1992-01	1.4	1.4

	LABOR TIME	SEVERE SERVICE
Water Pump and/or Gasket, Replace (B)		
Includes: Refill with proper coolant mix.		
Accord		
1981-83	1.8	2.3
1984-89	3.4	3.9
1990-02		
4 cyl.	3.5	4.0
V6	5.0	5.7
2003-05		
4 cyl. (1.7)	2.4	2.8
V6 (3.3)	5.0	5.7
Civic		
1981-87	2.3	2.8
1988-00	2.8	3.3
2001-05		
exc. below (.7)	1.1	1.2
1.7L, 2.0L Si (2.0)	3.0	3.4
CRX	3.1	3.6
del Sol	3.1	3.6
Insight (.8)	1.2	1.4
Prelude		
1981-87	3.2	3.7
1988-91	3.4	3.9
1992-01	4.0	4.5
S2000 (1.2)	1.8	2.1
w/AC add	.5	.5
w/top mount AC or alternator add	1.2	1.2

ENGINE

ASSEMBLY

Times shown are for OEM assemblies. Time to replace assemblies from aftermarket rebuilders may vary.

Engine Assy., R&I (B)
Does not include parts or component transfer.

	LABOR TIME	SEVERE SERVICE
Accord		
1981-85	6.2	6.7
1986-01		
4 cyl.	9.5	10.0
V6	11.9	12.4
Civic	7.9	8.4
Civic wagon	8.9	9.4
CRX, del Sol	6.2	6.7
Insight	6.8	7.3
Prelude		
1981-82	6.2	6.7
1983-01	9.5	10.0
S2000	9.0	9.5
w/AC add	.3	.3
w/AT add	1.2	1.2
w/Prelude type SH add	1.2	1.2
w/PS add	.2	.2
w/top mount AC add	1.2	1.2
w/twin cam add	.7	.7

HON-8 ACCORD : CIVIC : CIVIC DEL SOL : CRX : INSIGHT : PRELUDE : S2000

	LABOR TIME	SEVERE SERVICE
Engine Assy., R&R and Reconditioning (A)		
Includes: Replacing rings, rod and main bearings, cylinder head reconditioning and engine tune-up.		
Accord		
1981-85	23.2	24.1
1986-89	24.1	25.0
1990-01		
4 cyl.	26.2	27.1
V6	28.0	28.9
Civic	22.3	23.2
Civic wagon	25.6	26.5
CRX, del Sol	23.2	24.1
Insight	22.9	23.8
Prelude		
1981-82	23.2	24.1
1983-91	24.3	25.2
1992-01	25.2	26.1
S2000	29.3	30.2
w/AC add	.3	.3
w/AT add	1.2	1.2
w/Prelude type SH add	1.2	1.2
w/PS add	.2	.2
w/top mount AC add	1.2	1.2
w/twin cam add	.7	.7
Engine Assy. (New or Rebuilt Complete), Replace (B)		
Includes: Component transfer and engine tune-up.		
Accord		
1981-85	13.1	13.6
1986-02		
4 cyl. (incl. 98 V6)	10.3	10.8
V6 (exc. 1998)	13.2	13.7
2003-05		
4 cyl. (8.0)	12.2	12.7
V6 (9.0)	13.7	14.2
Civic		
exc. 2.0L Si,		
Civic HB (6.1)	8.8	9.3
2.0L Si,		
Civic HB (9.0)	13.2	13.7
Civic wagon	15.8	16.3
CRX, del Sol	13.1	13.6
Insight (6.3)	9.6	10.1
Prelude		
1981-85	13.1	13.6
1986-01		
exc. SH	10.8	11.3
SH	12.3	12.8
S2000 (8.2)	12.5	13.0
w/AC add	.5	.5
w/AT add	1.2	1.2
w/Prelude type SH add	1.2	1.2
w/PS add	.3	.3
w/top mount AC add	1.2	1.2
w/twin cam add	.8	.9

	LABOR TIME	SEVERE SERVICE
Engine Mounts, Replace (B)		
Accord		
1981-89	.8	1.0
1990-02		
front or side	.8	1.0
rear	2.7	2.9
2003-05 each (.5)	.8	1.0
Civic, CRX, del Sol		
each (.5)	.8	1.0
Insight each (.5)	.8	1.0
Prelude		
1981-91	.8	1.0
1992-01		
exc. rear each	.8	1.0
rear	2.7	2.9
S2000 each (.4)	.7	.9
CYLINDER HEAD		
Compression Test (B)		
1991-96		
4 cyl.	.8	.9
V6	1.2	1.2
1997-01	.8	.8
Cylinder leak test add	.5	.7
Valve Clearance, Adjust (B)		
1981-97	1.3	1.8
1998-05		
4 cyl. (.8)	1.2	1.4
V6 (1.9)	2.9	3.3
Cylinder Head, Replace (B)		
Includes: Parts transfer, adjustments.		
Accord		
1981-89	7.6	8.1
1990-93		
4 cyl.	8.4	8.9
V6		
one side	10.2	10.7
both sides	13.2	13.7
1994-02		
4 cyl.	6.8	7.3
V6		
one side	9.1	9.6
both sides	10.6	11.1
2003-05		
one head (6.2)	8.8	9.3
two heads (7.2)	10.2	10.7
Civic		
1981-85	7.1	7.6
1986-95	6.8	7.3
1996-05		
exc. 2.0L Si,		
Civic HB (3.5)	5.3	5.8
2.0L Si,		
Civic HB (6.7)	9.5	10.0
CRX		
1984-85	7.1	7.6
1986-91	6.8	7.3

	LABOR TIME	SEVERE SERVICE
del Sol	6.8	7.3
Insight (3.5)	5.3	5.8
Prelude		
1981-85	7.1	7.6
1983-91	7.6	8.1
1992-96	8.4	8.9
1997-01	6.8	7.3
S2000 (6.1)	8.7	9.2
w/AC add	.3	.3
w/top mount AC add	1.2	1.2
w/twin cam add	.8	.9
Cylinder Head Gasket, Replace (B)		
Accord		
1981-93		
4 cyl.	6.2	6.7
V6		
one side	8.4	8.9
both sides	10.6	11.1
1994-02		
4 cyl.	6.1	6.6
V6		
one side	8.4	8.9
both sides	9.9	10.4
2003-05		
one head (5.8)	8.8	9.3
two heads (6.8)	9.6	10.1
Civic, CRX, del Sol		
1981-85	5.5	6.0
1986-05		
exc. 2.0L Si,		
Civic HB (3.1)	4.5	5.0
2.0L Si,		
Civic HB (6.3)	8.9	9.4
Insight (3.3)	5.0	5.5
Prelude		
1981-82	5.5	6.0
1983-91	7.6	8.1
1992-01	6.2	6.7
S2000 (5.6)	8.0	8.5
w/AC add	.3	.3
w/top mount AC add	1.2	1.2
w/twin cam add	.8	.9
w/VTEC add	.5	.5
Replace cam and		
rockers add	1.2	1.4
Pre-chamber, R&R and Reconditioning (B)		
1981-85		
one	1.5	1.8
all	2.8	3.1
Rocker Arm Shaft Assy., Reconditioning (B)		
1981-85	2.4	2.9
1986-01		
12V	2.6	3.1
16V	3.1	3.6

ACCORD : CIVIC : CIVIC DEL SOL : CRX : INSIGHT : PRELUDE : S2000 — HON-9

	LABOR TIME	SEVERE SERVICE
Springs and/or Valve Seals, Replace (B)		
Includes: Cylinder head R&R.		
Accord		
1994-97 V6	15.2	15.4
1998-01 V6	13.5	13.7
Insight	6.2	6.5
Valve Cover Gasket, Replace (B)		
1981-87	.5	.6
1988-97		
4 cyl.	.5	.6
V6		
one	.5	.6
both	.9	1.0
1998-05		
4 cyl. (.5)	.8	.9
V6		
one (.8)	1.2	1.4
both (1.2)	1.7	1.9
Valve Job (B)		
Accord		
1981-93		
4 cyl.	10.7	11.6
V6	16.9	17.8
1994-02		
4 cyl.	10.6	11.5
V6		
one head	13.2	14.1
both heads	15.7	16.6
2003-05		
one head (8.7)	12.1	13.0
two heads (10.3)	14.1	15.0
Civic, CRX, del Sol		
1981-85	8.2	9.1
1986-05		
exc. 2.0L Si (6.0)	9.1	10.0
2.0L Si (9.2)	12.7	13.2
Insight (5.5)	7.9	8.8
Prelude		
1981-82	8.2	9.1
1983-91	12.5	13.4
1992-01	10.7	11.6
S2000 (8.6)	12.0	12.9
w/AC add	.3	.3
w/top mount AC add	1.2	1.2
w/twin cam add	.8	.9
Valve Springs and/or Oil Seals, Replace (B)		
One	1.7	1.9
each addl.	.5	.6
All		
single cam	2.4	2.6
dual cam	4.4	4.6
CAMSHAFT		
Timing Belt, Adjust (B)		
1981-91	.5	.7
1992-05 (.7)	1.1	1.2

	LABOR TIME	SEVERE SERVICE
Balance Shaft Seals, Replace (B)		
1990-01 Accord	3.5	3.7
1992-01 Prelude	3.5	3.7
Camshaft and/or Rocker Arms, Replace (A)		
Accord		
1981-82	2.8	3.0
1983-05		
4 cyl. (2.5)	3.8	4.3
V6		
1993-97	6.1	6.9
1998-02	9.1	10.3
2003-05 (4.3)	6.5	7.4
Civic, CRX, del Sol		
1981-87	3.0	3.2
1988-05 (2.5)	3.8	4.3
Insight, S2000 (2.5)	3.8	4.3
Prelude		
1981-82	2.8	3.0
1983-96	3.7	3.9
1997-01	3.8	4.3
w/12V add	.5	.5
w/AC add	.3	.3
w/top mount AC add	1.2	1.2
w/twin cam add	.5	.5
w/VTEC add	.3	.3
Recondition rocker arms add	1.2	1.2
Replace balance shaft belt add		
exc. below	1.2	1.2
94-05 Accord	1.5	1.7
96-05 Civic	.8	.9
97-01 Prelude	1.5	1.7
timing belt add	1.2	1.2
Camshaft Seal, Replace (B)		
Accord		
1983-85		
right side	.9	1.1
left side	1.3	1.5
1986-93	1.3	1.5
1994-02		
4 cyl.	3.8	4.0
V6	5.3	5.5
2003-05		
one (3.3)	5.0	5.2
both (3.7)	5.6	5.8
CRX	.9	1.1
del Sol	1.2	1.4
Civic		
1981-85		
right side	1.3	1.5
left side	.7	.9
1986-87	.9	1.1
1988-95		
one	1.1	1.3
both	1.4	1.6
1996-05 (2.2)	3.3	3.5

	LABOR TIME	SEVERE SERVICE
Prelude		
1983-85		
right side	.9	1.1
left side	1.3	1.5
1986-96		
one	1.6	1.8
both	1.8	2.0
1997-01	4.1	4.3
w/AC add	.3	.3
w/PS add	.2	.2
w/top mount AC add	1.2	1.2
w/twin cam add	.2	.2
Timing Belt, Replace (B)		
Accord		
1981-83	2.0	2.5
1984-89	3.1	3.6
1990-02		
4 cyl.	3.5	4.0
V6	5.0	5.5
2003-05 (3.0)	4.3	4.8
Civic, CRX		
1981-85	1.7	2.2
1986-87	2.0	2.5
1988-00	2.7	3.2
2001-05 (2.0)	3.0	3.5
del Sol	2.6	3.1
Prelude		
1981-82	2.0	2.5
1983-91	3.1	3.6
1992-96	3.5	4.0
1997-01	3.8	4.3
w/PS add	.2	.2
w/top mount AC add	1.2	1.2
w/twin cam add	.3	.3
Replace water pump add	.5	.7
Timing Belt Lower Cover, Replace (B)		
1981-85 Accord, Prelude	.9	1.2
1981-85 Civic	1.4	1.7
w/AC add	1.2	1.5
w/PS add	.2	.2
Timing Belt Upper Cover, Replace (B)		
Accord, Civic, CRX, del Sol	.6	.7
Prelude		
1981-85	.6	.7
1986-01	1.2	1.3
w/AC add	1.2	1.3
Timing Chain, Replace (B)		
Accord, Civic (2.8)	4.0	4.5
Insight (9.8)	13.2	13.7
S2000 (4.9)	7.4	7.9
Timing Chain Tensioner, Replace (B)		
Accord, Civic (.4)	.6	.7
Insight (9.8)	13.2	13.7
S2000 (5.1)	7.8	8.3

HON-10 ACCORD : CIVIC : CIVIC DEL SOL : CRX : INSIGHT : PRELUDE : S2000

	LABOR TIME	SEVERE SERVICE
CRANK & PISTONS		
Connecting Rod Bearings, Replace (A)		
Includes: Check bearing oil clearance.		
Accord		
1981-85	3.4	3.9
1986-02		
4 cyl.	2.6	3.1
V6	3.3	3.8
2003-05		
4 cyl.(2.2)	3.3	3.8
V6 (2.4)	3.6	4.1
Civic		
1981-85	3.1	3.6
1986-03		
exc. 2.0L Si (1.7)	2.6	3.1
2.0L Si (2.2)	3.3	3.8
Civic wagon	6.0	6.5
CRX	3.4	3.9
del Sol	2.9	3.4
Insight (7.8)	10.9	11.4
Prelude, S2000 (1.7)	2.6	3.1
Crankshaft and Main Bearings, Replace (A)		
Includes: Engine R&R.		
Does not include: Cylinder head removal.		
Accord		
1981-85	9.4	9.9
1986-05		
4 cyl.		
exc. 03-05 (8.3)	12.6	14.1
03-05 (11.0)	15.0	15.5
V6		
1994-97	26.0	26.5
1998-02	25.2	25.7
2003-05 (16.9)	22.0	22.5
Civic, CRX, del Sol		
1981-87	9.4	9.9
1988-05		
exc. 2.0L Si (7.3)	10.3	10.8
2.0L Si (11.7)	15.8	16.3
Civic wagon	13.1	13.6
Insight (10.8)	14.7	15.2
Prelude		
1981-82	9.4	9.9
1983-01	13.1	13.6
S2000 (11.2)	15.2	15.7
w/AC add	.5	.5
w/AT add	1.2	1.2
w/Prelude type SH add	1.2	1.2
w/PS add	.2	.2
w/top mount AC add	1.2	1.2
w/twin cam add	.7	.7
Crankshaft Front Oil Seal, Replace (A)		
Accord		
1981-87	2.5	2.7
1988-89	3.6	3.8

	LABOR TIME	SEVERE SERVICE
1990-02		
4 cyl.	3.3	3.5
V6	4.6	4.8
2003-05		
4 cyl. (2.0)	3.0	3.2
V6 (3.3)	4.8	5.0
Civic, CRX, del Sol		
1981-87	1.8	2.0
1988-05		
exc. 2.0L Si (1.7)	2.6	2.8
2.0L Si (2.0)	3.0	3.2
Insight (1.0)	1.5	1.7
Prelude		
1981-82	2.5	2.7
1983-01	3.8	4.0
S2000 (2.2)	3.3	3.5
w/AC add	.5	.5
w/top mount AC add	1.2	1.2
Crankshaft Pulley, Replace (B)		
1981-01	.5	.6
w/AC add	.2	.2
w/PS add	.2	.2
Pistons or Connecting Rods, Replace (A)		
Includes: Ridge reaming, cylinder wall deglazing, installing new rings and rod bearings, engine tune-up.		
Accord		
1981-85	11.2	12.1
1986-02		
4 cyl.	12.7	13.6
V6	18.4	29.3
2003-05		
4 cyl. (9.8)	14.9	15.8
V6 (12.4)	18.9	19.8
Civic		
1981-87	10.8	11.7
1988-05		
exc. 2.0L Si (7.3)	11.2	12.1
2.0L Si (10.5)	16.0	16.9
CRX, del Sol	12.0	12.9
Insight (11.6)	17.3	18.2
Prelude		
1981-87	11.2	12.1
1988-01	13.2	14.1
S2000 (10.9)	16.4	17.3
w/AC add	.5	.5
w/top mount AC add	1.2	1.2
w/twin cam add	.8	.9
w/VTEC add	.5	.5
Rear Main Oil Seal, Replace (A)		
Includes: Trans. R&R. when necessary.		
Accord		
1981-89	5.8	6.3
1990-93	6.3	6.8
1994-97		
4 cyl.	7.0	7.5
V6	7.8	8.3

	LABOR TIME	SEVERE SERVICE
1998-03		
4 cyl.	7.9	8.4
V6	8.7	9.2
Civic		
1981-95		
2WD	4.5	5.0
4WD	8.0	8.5
1996-05		
exc. 2.0L Si (3.4)	5.2	5.7
2.0L Si (5.2)	7.5	8.0
CRX, del Sol	5.8	6.3
Insight (4.2)	6.1	6.6
Prelude		
1981-82	5.8	6.3
1983-01	7.0	7.5
S2000 (5.6)	8.0	8.5
w/AT add	1.4	1.5
Rings, Replace (A)		
Includes: Ridge reaming, cylinder wall deglazing, installing new rings, engine tune-up.		
Accord		
1981-87	9.8	10.7
1988-89	11.8	12.7
1990-02		
4 cyl.	11.6	12.5
V6	17.2	18.1
2003-05		
4 cyl. (9.3)	14.1	15.0
V6 (11.6)	17.6	18.5
Civic		
1981-87	9.9	10.8
1988-05		
exc. 2.0L Si (6.8)	10.3	11.2
2.0L Si (10.0)	15.2	16.1
Civic wagon	13.3	14.2
CRX, del Sol	10.7	11.6
Insight (10.6)	16.1	17.0
Prelude		
1981-87	9.8	10.7
1988-91	13.7	14.6
1992-01	12.0	12.9
S2000 (10.0)	15.2	16.1
w/AC add	.5	.5
w/top mount AC add	1.2	1.2
w/twin cam add	.8	.9
w/VTEC add	.5	.5
ENGINE LUBRICATION		
Engine Oil Cooler, Replace (B)		
1985-01	.7	.9
Engine Oil Pressure Switch (Sending Unit), Replace (B)		
Accord, Civic		
1981-05 (.4)	.6	.6
CRX		
1984-87	.5	.5
1988-91	1.2	1.2
del Sol	.5	.5
Insight (.3)	.5	.5

ACCORD : CIVIC : CIVIC DEL SOL : CRX : INSIGHT : PRELUDE : S2000 — HON-11

	LABOR TIME	SEVERE SERVICE
Prelude		
1981	.5	.5
1982-87	1.6	1.6
1988-01	.5	.5
S2000 (.3)	.5	.5
Oil Pan and/or Gasket, Replace (B)		
Accord		
1981-85	1.2	1.7
1986-02	1.9	2.4
2003-05		
4 cyl. (1.7)	2.7	3.2
V6 (1.2)	1.9	2.4
Civic		
1981-85	1.8	2.2
1986-05		
exc. 2.0L Si (1.2)	1.9	2.4
2.0L Si (1.7)	2.7	3.2
Civic wagon	2.8	3.3
CRX		
1984-85	2.0	2.5
1986-91	2.8	3.3
del Sol	2.8	3.3
Insight (6.8)	9.6	10.1
Prelude		
1981-85	1.2	1.4
1986-01	1.9	2.2
S2000 (2.0)	3.2	3.6
Oil Pump, Replace (B)		
Accord		
1981	1.4	1.9
1982-83	2.0	2.5
1984-89	3.6	4.1
1990-02	7.2	7.9
2003-05		
4 cyl. (2.0)	3.2	3.7
V6 (4.5)	6.5	7.0
Civic		
1981-83	2.0	2.5
1984-87	4.2	4.7
1988-00	5.6	6.1
2001-05		
1.3L (2.8)	4.1	4.6
1.7L (3.5)	5.1	5.6
2.0L (2.0)	3.2	3.7
Civic wagon	7.8	8.3
CRX		
1984-87	4.2	4.7
1988-91	5.2	5.7
del Sol	5.2	5.7
Insight (10.3)	14.1	14.6
Prelude		
1981	1.4	1.9
1982	2.0	2.5
1983-87	3.7	4.2
1988-91	5.2	5.7
1992-01	7.2	7.7
S2000 (5.1)	7.3	7.8
w/AC add	.3	.3
Recondition pump add	.5	.8
Replace timing belt add	1.2	1.2

CLUTCH

	LABOR TIME	SEVERE SERVICE
Bleed Clutch Hydraulic System (B)		
1981-05 (.4)	.6	.7
Clutch Pedal Free Play, Adjust (B)		
1981-05 (.3)	.4	.5
Clutch Assy., Replace (B)		
Accord		
1990-97	6.4	6.9
1998-02	7.4	7.9
2003-05		
4 cyl. (5.0)	7.2	7.7
V6 (6.2)	8.8	9.3
Civic		
1981-95		
2WD	3.6	4.1
4WD	8.4	8.9
1996-05		
exc. 2.0L Si, Civic HB (2.6)	3.8	4.3
2.0L Si, Civic HB (5.0)	7.2	7.7
CRX, del Sol	5.4	5.9
del Sol	5.2	5.7
Insight (3.0)	4.4	4.9
Prelude		
1981-82	5.2	5.7
1983-87	5.8	6.3
1988-01	6.4	6.9
S2000 (5.2)	7.5	8.0
w/ATTS Prelude add	4.3	4.8
Clutch Control Cable, Replace (B)		
1981-87	.7	.9
1988-91	1.2	1.4
Clutch Master Cylinder, Replace (B)		
Includes: System bleeding.		
1990-05 Accord (.8)	1.3	1.5
1992-05 Civic (.8)	1.3	1.5
CRX, del Sol	1.4	1.6
Insight, S2000 (.8)	1.3	1.5
Prelude	1.3	1.5
Clutch Release Bearing, Replace (B)		
Accord		
1990-97	5.9	6.4
1998-02	6.9	7.4
2003-05		
4 cyl. (4.7)	6.8	7.3
V6 (5.9)	8.4	8.9
Civic		
1981-95		
2WD	3.1	3.6
4WD	7.9	8.4
1996-05		
exc. 2.0L Si, Civic HB (2.3)	3.4	3.9
2.0L Si, Civic HB (4.7)	6.8	7.3
CRX, del Sol	5.0	5.5
Insight (2.5)	3.7	4.2

	LABOR TIME	SEVERE SERVICE
Prelude		
1981-82	5.0	5.5
1983-01	5.9	6.4
S2000 (4.6)	6.6	7.1
w/ATTS Prelude add	4.0	4.5
Clutch Slave Cylinder, Replace (B)		
Includes: System bleeding.		
1981-99 (.8)	1.3	1.5
2000-05		
Accord, Civic (.8)	1.3	1.5
Insight, S2000 (.6)	1.0	1.2
Prelude (.8)	1.3	1.5
Flywheel, Replace (B)		
Accord		
1990-97	6.7	7.2
1998-02	7.7	8.2
2003-05		
4 cyl. (5.2)	7.5	8.0
V6 (6.4)	9.1	9.6
Civic		
1981-95		
2WD	3.9	4.4
4WD	8.8	9.3
1996-05		
exc. 2.0L Si, Civic HB (2.8)	3.8	4.3
2.0L Si, Civic HB (5.2)	7.5	8.0
CRX, del Sol	5.5	5.7
Insight (3.2)	4.7	5.2
Prelude		
1981-82	5.5	5.7
1983-01	6.7	7.6
S2000 (5.4)	7.7	8.2
w/ATTS Prelude add	4.0	4.5
Replace		
pilot bearing add	.2	.2
rear main seal add	.2	.2

MANUAL TRANSAXLE

	LABOR TIME	SEVERE SERVICE
Transaxle Assy. R&R and Recondition (A)		
Accord		
1990-97	11.8	12.7
1998-02	12.8	13.7
2003-05		
4 cyl. (8.4)	11.7	12.6
V6 (9.6)	13.2	14.1
Civic		
1981-91		
2WD	7.4	8.3
4WD	11.8	12.7
1992-00	9.3	10.2
2001-05		
exc. 2.0L Si, Civic HB (6.0)	8.6	9.5
2.0L Si, Civic HB (8.4)	11.7	12.6
CRX, del Sol	8.9	9.8
Insight (6.4)	9.1	10.0

HON-12 ACCORD : CIVIC : CIVIC DEL SOL : CRX : INSIGHT : PRELUDE : S2000

	LABOR TIME	SEVERE SERVICE
Transaxle Assy. R&R and Reconditon (A)		
Prelude		
1981-82	8.9	9.8
1983-91	9.7	10.6
1992-01	11.8	12.7
S2000 (8.2)	11.5	12.4
w/ATTS Prelude add	4.0	4.5
Recondition		
differential add	2.4	2.9
transfer case 81-95 add	2.8	3.3
Transaxle Assy., R&R or Replace (B)		
Accord		
1990-97	5.8	6.3
1998-02	6.7	7.2
2003-05		
4 cyl. (4.6)	6.6	7.1
V6 (5.8)	8.3	8.8
Civic		
1981-91		
2WD	4.5	5.0
4WD	9.1	9.6
1992-00	3.2	3.7
2001-05		
exc. 2.0L Si, Civic HB (2.4)	3.5	4.0
2.0L Si, Civic HB (4.6)	6.6	7.1
CRX, del Sol	5.0	5.5
Insight (2.6)	3.8	4.3
Prelude		
1981-82	5.0	5.5
1983-87	5.8	6.3
1988-91	6.8	7.3
1992-01	5.8	6.3
S2000 (4.4)	6.4	6.9
w/ATTS Prelude add	4.0	4.5
Transaxle Assy., Reseal (B)		
Accord, CRX, del Sol	7.9	8.4
Civic		
1981-91		
2WD	6.4	6.9
4WD	10.7	11.2
1992-01	5.2	5.7
Prelude		
1981-82	7.3	7.8
1983-87	7.8	8.3
1988-91	8.9	9.4
1992-01	7.6	8.1
Reseal transfer case add	2.3	2.6

AUTOMATIC TRANSAXLE
SERVICE TRANSAXLE INSTALLED

	LABOR TIME	SEVERE SERVICE
Diagnose Electronic Transmission Component Each (A)		
1988-05 (.6)	.8	.8
Retrieve Fault Codes (A)		
1988-05 (.3)	.4	.4
Drain & Refill Unit (B)		
1981-05	.5	.5
Throttle Control Cable, Adjust (B)		
1981-97	.3	.4
3-4 Shift Assy., Replace (B)		
1988-01	.5	.7
CVT Speed Sensor, Replace (B)		
Civic, Insight		
1996-05 one (.3)	.5	.6
CVT Valve Body Assembly, Replace (B)		
Civic, Insight		
1996-05 (1.0)	1.6	1.8
Governor Assy., R&R or Replace (B)		
1983-89 Accord, Prelude	7.8	8.0
1981-89 Civic, CRX	6.8	7.0
Lock-Up Control Solenoid Valve, Replace (B)		
1988-05 (.3)	.5	.7
Shift Control Solenoid Valve, Replace (B)		
1988-05 (.3)	.5	.7
w/94-05 Accord V6 add	.5	.5
Speed Pulser, Replace (B)		
1988-05 (.3)	.5	.6
Throttle Valve Control Cable, Replace (B)		
1981-93	1.0	1.1
1994-97	.5	.5
Transmission Control Computer, Replace (B)		
All Models (.5)	.8	.9

SERVICE TRANSAXLE REMOVED
Transaxle R&R included unless otherwise noted.

	LABOR TIME	SEVERE SERVICE
Torque Converter, Replace (B)		
Accord		
1983-89	7.8	8.3
1990-97		
4 cyl.	7.8	8.3
V6	8.6	9.1
1998-05		
4 cyl. (5.5)	7.9	8.4
V6 (6.6)	9.4	9.9
Accord IMA		
2005 (7.0)	9.9	10.4

	LABOR TIME	SEVERE SERVICE
Civic		
1981-91		
2WD	6.4	6.9
4WD	9.9	10.4
1992-05 (3.3)	4.8	5.3
1983-01 Prelude	7.8	8.3
Flush oil cooler and lines add	.5	.5
Transaxle Assy., R&I or Replace (B)		
Accord		
1983-89	6.9	7.2
1990-97		
4 cyl.	7.4	7.7
V6	8.2	8.5
1998-01		
4 cyl.	8.4	8.7
V6	9.1	9.4
Civic		
1981-91		
2WD	6.3	6.6
4WD	9.8	10.1
1992-01	5.0	5.3
Civic wagon	7.8	8.1
CRX, del Sol	5.8	6.1
1983-01 Prelude	7.4	7.7
Replace		
flexplate add	.3	.3
transaxle assy. add	1.2	1.5
Transaxle, Recondition (A)		
Includes: Replace torque converter.		
Accord		
1983-89	13.4	14.3
1990-97		
4 cyl.	13.9	14.8
V6	14.7	15.6
1998-02		
4 cyl.	14.9	15.8
V6	15.7	16.6
2003-05		
4 cyl. (8.3)	11.5	12.4
V6 (9.4)	13.0	13.9
Accord IMA		
2005 (12.1)	16.2	17.1
Civic		
1981-91	12.8	13.7
1992-03 (7.2)	10.2	11.1
Civic wagon	12.3	13.2
CRX	9.8	10.7
del Sol	14.2	15.1
Prelude		
1982-87	10.5	11.4
1988-01	13.9	14.8
Flush oil cooler and lines 81-93 add	1.2	1.2
Recondition		
differential add	2.4	2.7
transfer case 81-93 add	2.3	2.5

ACCORD : CIVIC : CIVIC DEL SOL : CRX : INSIGHT : PRELUDE : S2000   HON-13

	LABOR TIME	SEVERE SERVICE
Valve Body Assy., Replace (B)		
1981-87 Civic, CRX	9.3	9.5
1981-87 Prelude	10.0	10.2
Recondition assy. add	.5	.8

TRANSFER CASE
Control Unit, Replace (B)
1997-01 Prelude 5   .5
Transfer Case, R&R and Recondition (B)
1985-91 Civic 10.5   11.4
Transfer Case, R&R or Replace (B)
1985-91 Civic 2.3   2.5
1997-01 Prelude 4.0   4.5

SHIFT LINKAGE
AUTOMATIC TRANSAXLE
Shift Control Cable, Adjust (B)
1981-05 (.5) 8   .9
Gear Selector Lever, R&R and Recondition (B)
1981-93 (.7) 1.1   1.1
1994-05
　Accord
　　1994-97 (1.9) 2.4   2.5
　　1998-05 (1.2) 1.5   1.6
　Civic (1.0) 1.2   1.4
　Prelude (1.0) 1.2   1.2
Gear Selector Lever, Replace (B)
1990-05
　Accord
　　1990-93 (.8) 1.0   1.1
　　1994-97 (1.7) 2.1   2.2
　　1998-05 (1.0) 1.3   1.4
　Civic (.8) 1.0   1.1
　Prelude (.8) 1.0   1.1
Shift Control Cable, Replace (B)
1981-87 (.8) 1.3   1.4
1988-05 (1.0) 1.5   1.7
w/catalytic converter 81-87 add 5   .5

MANUAL TRANSAXLE
Gearshift Lever, Replace (B)
1981-87 (1.0) 1.4   1.5
1988-05
　exc. below (1.2) 1.7   1.8
　2001-05 Civic
　　1.3L, 1.7L (.6) 9   1.0
　S2000 (.3) 5   .5
w/catalytic converter 81-87 add 5   .5
Gearshift Lever Boot, Replace (B)
1981-93 8   .8
1994-97
　Accord 2.3   2.6
　Civic, Prelude 9   1.0

	LABOR TIME	SEVERE SERVICE
1998-05		
exc. below (.3)	.5	.5
Civic 1.3L, 1.7L (.6)	.9	1.0
Prelude (.6)	.9	1.0
S2000 (.2)	.3	.3

Shift Lever Bushing, Replace (B)
1981-87 (.6) 9   1.0
1988-01 (.8) 1.3   1.4
w/catalytic converter 81-87 add 5   .5
Shift Rod, Replace (B)
1981-87 7   .8
1988-00 9   1.0
w/catalytic converter 81-87 add 5   .5
Torque Rod Bushing, Replace (B)
1981-00 8   .9
w/catalytic converter 81-87 add 5   .5

4WD
Shift Cables, Adjust (B)
Civic wagon 4WD 5   .6
Shift Cables, Replace (B)
Civic wagon 4WD 8   .9
Shift Cover, Reseal or Recondition (B)
Civic wagon 4WD 2.6   2.8

DRIVELINE
DRIVESHAFT
Center Support Bearing, Replace (B)
Civic wagon 4WD
　one 8   1.0
　both 1.4   1.6
Driveshaft, R&R or Replace (B)
Civic wagon 4WD
　one 7   .9
　all 1.7   1.9
S2000 (.4) 6   .7
Recondition U-joint add each 2   .2
Driveshaft Support Bearing, Replace (B)
Civic wagon 4WD
　one 8   1.1
　both 1.6   1.9

DRIVE AXLE
Intermediate Shaft and/or Bearing, Replace (B)
1988-05 (.9) 1.4   1.6
Replace
　oil seal add 2   .2
　support bearing add 5   .5
Rear Axle Shaft Oil Seal, Replace (B)
Civic wagon 4WD
　one 7   .8
　both 1.3   1.4

	LABOR TIME	SEVERE SERVICE
Rear Differential Assy., Recondition (B)		
Civic wagon 4WD	5.8	6.3
S2000 (4.0)	5.8	6.7
Rear Differential Assy., Replace (B)		
S2000 (1.6)	2.6	2.9
Rear Wheel Bearings, Replace (B)		
Civic wagon 4WD		
one side	1.2	1.5
both sides	2.2	2.5

HALFSHAFTS
Halfshaft, R&R or Replace (B)
Exc. below (.7) 1.1   1.3
Civic wagon 4WD rear
　one 2.3   2.5
　both 3.1   3.3
w/4WD front add 5   .5
Replace
　CV boot Civic wagon
　　4WD rear add each 3   .5
　CV joint Civic wagon
　　4WD rear add each 7   .9
　support bearing exc. Civic
　　wagon 4WD add 3   .3
Halfshaft Oil Seal, Replace (B)
One side 1.4   1.6
Both sides 2.3   2.4
w/4WD add 5   .5
Inboard CV Joint and/or Boot, Replace (B)
1981-05 (1.0) 1.6   1.8
w/4WD add 5   .5
Outboard CV Joint and/or Boot, Replace (B)
Exc. below (.9) 1.4   1.6
S2000 each (.8) 1.3   1.5
w/4WD add 5   .5

BRAKES
ANTI-LOCK
Diagnose Anti-Lock Brake System Component Each (A)
1988-05 (.6) 8   .8
Retrieve Fault Codes (A)
1988-05 (.3) 4   .4
Bleed ABS System (B)
1991-97 8   .9
Accumulator Assy., Replace (B)
Accord
　1991-93 . . . . 1.6   1.6
　1994-97 . . . . 1.9   2.2
1992-95 Civic 1.6   1.6
1995 del Sol 8   .8
1988-96 Prelude 1.9   1.9

HON-14 ACCORD : CIVIC : CIVIC DEL SOL : CRX : INSIGHT : PRELUDE : S2000

	LABOR TIME	SEVERE SERVICE
Control Module, Replace (B)		
1988-99 (.3)	.5	.5
2000-05		
Accord, Civic, Prelude (.3)	.5	.5
Insight, S2000 (1.0)	1.6	1.8
High Pressure Hoses, Replace (B)		
1988-96 one (.8)	1.1	1.1
Low Pressure Hoses, Replace (B)		
1988-97 Accord, Prelude (.9)	1.4	1.6
1992-95 Civic, del Sol (.3)	.5	.5
Modulator, Replace (B)		
Exc. 2005 Accord IMA		
1988-05 (1.0)	1.6	1.8
2005 Accord IMA (1.4)	2.0	2.2
Power Unit, Replace (B)		
Accord		
1990-93	1.4	1.4
1994-97	1.9	2.2
1992-95 Civic, del Sol	1.4	1.4
1988-96 Prelude	1.6	1.6
Pressure Switch, Replace (B)		
1990-97 (1.0)	1.4	1.4
Wheel Speed Sensor, Replace (B)		
Accord each (.5)	.8	.9
Civic		
1992-95 each (.3)	.5	.5
1996-00		
front each (.3)	.5	.5
rear each (.7)	1.1	1.3
2001-05 each (.3)	.5	.5
Insight each (.3)	.5	.5
Prelude		
1992-96 each	.5	.5
1997-01		
front each	.5	.5
rear each	1.1	1.3
S2000		
front each (.3)	.5	.5
rear each (.7)	1.1	1.3

SYSTEM

	LABOR TIME	SEVERE SERVICE
Bleed Brakes (B)		
Includes: Add fluid.		
1981-05 (.4)	.6	.7
Brake System, Flush and Refill (B)		
1981-01	1.2	1.2
Brakes, Adjust (B)		
Includes: System bleeding.		
1981-01 rear wheels	.5	.5
Brake Adjuster, Replace (B)		
One	.7	.8
Both	1.2	1.3
Brake Failure Relay, Replace (B)		
1981-88	.3	.3

	LABOR TIME	SEVERE SERVICE
Brake Hose (Flexible), Replace (B)		
Includes: System bleeding.		
Front		
one	.9	1.0
both	1.6	1.7
Rear		
one	.7	.8
each addl.	.3	.4
Brake Proportioning Valve, Replace (B)		
Accord, Civic, Prelude		
one (.5)	.8	1.0
both (.8)	1.3	1.5
CRX, del Sol		
one (.7)	1.2	1.4
both (1.0)	1.5	1.7
Insight, S2000 (.5)	.8	1.0
Master Cylinder, R&R and Recondition (B)		
Includes: System bleeding.		
1981-01	2.8	3.1
Master Cylinder, Replace (B)		
Includes: Adjust Push Rod, System bleeding.		
1981-05 (1.2)	1.7	1.9
Power Booster Unit, Replace (B)		
Includes: System bleeding.		
Exc. below (1.4)	2.1	2.3
Accord		
1981-85 (1.9)	2.7	2.9
1986-89 (1.6)	2.3	2.5
Civic		
1981-91 (1.9)	2.7	2.9
1992-05 (1.6)	2.3	2.5
CRX, del Sol (1.6)	2.8	3.0
Power Booster Vacuum Check Valve, Replace (B)		
1981-05 (.2)	.3	.3

SERVICE BRAKES

	LABOR TIME	SEVERE SERVICE
Brake Drum, Replace (B)		
One (.3)	.5	.6
Both (.6)	1.0	1.1
Caliper Assy., R&R and Recondition (B)		
Includes: System bleeding.		
Front		
1981-91		
Accord, Civic, Prelude		
1981-85		
one	1.6	1.9
both	2.7	3.0
1986-91		
one	1.6	1.9
both	3.0	3.3
CRX		
one	1.8	2.1
both	3.0	3.3

	LABOR TIME	SEVERE SERVICE
1992-01		
one	1.1	1.4
both	2.2	2.5
Rear		
1987-91		
one	1.6	1.9
both	3.0	3.3
1992-01		
one	1.1	1.4
both	2.1	2.4
Caliper Assy., Replace (B)		
Includes: System bleeding.		
Front		
1981-91		
Accord		
1981-85		
one	.9	1.0
both	1.6	1.7
1986-91		
one	1.2	1.3
both	2.2	2.3
Civic, Prelude		
one	1.2	1.3
both	2.2	2.3
CRX		
1984-85		
one	.9	1.0
both	1.6	1.7
1986-91		
one	1.2	1.3
both	1.7	1.8
1992-05		
one (.5)	.8	.9
both (1.0)	1.6	1.8
Rear		
1987-91		
one	1.2	1.3
both	2.2	2.3
1992-05		
one (.6)	1.0	1.1
both (1.2)	1.9	2.2
Disc Brake Rotor, Replace (B)		
Front		
Accord		
1981		
one	2.0	2.1
both	3.1	3.2
1982-85		
one	1.2	1.3
both	1.7	1.8
1986-97		
one	1.8	2.0
both	3.5	4.0
1998-05		
one (.7)	1.1	1.3
both (1.4)	2.2	2.5
Civic, del Sol, Prelude		
one	1.1	1.3
both	2.2	2.5

ACCORD : CIVIC : CIVIC DEL SOL : CRX : INSIGHT : PRELUDE : S2000 — HON-15

	LABOR TIME	SEVERE SERVICE
CRX		
one	1.2	1.3
both	1.7	1.8
Insight, S2000		
one (.7)	1.1	1.3
both (1.4)	2.2	2.5
Rear		
1987-91		
one	.7	.8
both	1.6	1.7
1992-05		
one (.6)	1.0	1.1
both (1.2)	1.9	2.2

Front Pads & Rotors, Replace (B)
w/bolt on rotors (1.4) .. 2.2   2.5

Pads and/or Shoes, Replace (B)
Includes: Service and parking brake adjustment, bleed system.
Front or rear disc (.7)	1.1	1.3
Rear drum (.9)	1.4	1.6

COMBINATION ADD-ONS
Replace
brake drum add	.2	.2
brake hose add	.3	.3
caliper add	.3	.3
disc rotor add	.8	.8
disc rotor studs add	.2	.2
wheel cylinder add	.5	.5

Resurface
brake drum add	.5	.5
disc rotor add	.7	.7

Wheel Cylinder, R&R and Recondition (B)
Includes: System bleeding.
1981-91		
one	1.9	2.2
both	2.6	2.9
1992-01		
one	1.1	1.4
both	1.8	2.1

Wheel Cylinder, Replace (B)
Includes: System bleeding.
1981-91		
one	1.3	1.5
both	2.4	2.6
1992-05		
one (.6)	1.0	1.1
both (1.2)	1.9	2.2

PARKING BRAKE
Parking Brake Cable, Adjust (C)
1981-05 (.4) 6   .8

Parking Brake Apply Actuator, Replace (B)
1981-05 (.5) 8   .9
w/console mounted
  98-05 Accord add 5   .5

Parking Brake Cable, Replace (B)
1981-87		
one	.9	1.1
each addl.	.2	.3
1988-91		
one	1.6	1.8
each addl.	.5	.6
1992-05		
one (.8)	1.3	1.4
both (1.5)	2.4	2.7
w/catalytic converter add	.5	.5

Parking Brake Equalizer, Replace (B)
1981-01 7   .9

FRONT SUSPENSION
Unless otherwise noted, time given does not include alignment.

Align Front End (A)
1981-95 (1.1)	1.7	1.9
1996-05 (.6)	1.0	1.1

Front & Rear Alignment, Check & Adjust (A)
1981-95		
front	1.7	1.9
rear ()	1.2	1.4
front & rear	2.8	3.0
4WS	3.1	3.3
1996-05		
front (.6)	1.0	1.1
front & rear (1.2)	1.9	2.2

Front Suspension Height, Adjust (B)
1981-86 5   .7

Front Toe, Adjust (B)
1981-01 8   1.0

Front Hub Assy., Replace (B)
1981-87		
one	2.0	2.3
both	3.1	3.4
1988-05		
exc. below		
one side (1.2)	1.8	2.0
both sides (2.2)	3.5	4.0
2001-05 Civic		
one side (.8)	1.3	1.5
both sides (1.6)	2.6	2.9
Insight		
one side (.8)	1.3	1.5
both sides (1.6)	2.6	2.9

Front Radius Rod, Replace (B)
1981-85	1.6	1.8
1986-87	2.3	2.5
1988-05		
one side (.6)	1.0	1.2
both sides (1.2)	1.9	2.2

Front Radius Rod Bushings, Replace (B)
1981-85	1.6	1.8
1986-87	2.3	2.5
1988-05		
one side (.6)	1.0	1.2
both sides (1.2)	1.9	2.2

Front Strut Coil Spring, Replace (B)
1981-87		
one	1.7	1.9
both	3.1	3.3
1988-05		
exc. below		
one (.8)	1.3	1.5
both (1.6)	2.5	2.7
2001-05 Civic		
one side (.5)	.8	.9
both sides (1.0)	1.6	1.8

Front Strut Fork, Replace (B)
1988-01		
one	.7	.9
both	1.2	1.4

Front Torsion Bar, Replace (B)
1981-87		
one (1.2)	1.8	2.0
both (2.5)	2.8	3.0

Lower Ball Joint, Replace (B)
1988-05 each (.9) 1.3   1.5

Lower Control Arm, Replace (B)
Includes: Alignment on S2000.
1981-87	1.8	2.1
1988-05		
exc. below		
one (.6)	1.0	1.1
both (1.2)	1.9	2.1
2003-05 Accord		
one (.9)	1.3	1.4
both (1.6)	2.4	2.6
2001-05 Civic		
one (.4)	.6	.7
both (.8)	1.3	1.5
S2000		
one (1.1)	1.8	1.9
both (2.2)	3.5	3.7

Lower Control Arm Bushings, Replace (B)
Accord, Prelude		
1981-87	2.3	2.5
1988-01		
one side	1.7	1.9
both sides	3.2	3.4
Civic, CRX		
1981-83		
one side	1.7	1.9
both sides	3.2	3.4
1984-87		
one side	3.2	3.4
both sides	5.2	5.4
1988-01		
one side	1.7	1.9
both sides	3.2	3.4

ACCORD : CIVIC : CIVIC DEL SOL : CRX : INSIGHT : PRELUDE : S2000

	LABOR TIME	SEVERE SERVICE
Stabilizer Bar, Replace (B)		
1981-05		
exc. below (.7)	1.2	1.4
Accord		
2003-05 (2.1)	3.1	3.3
Stabilizer Bar Bushing and/or Bracket, Replace (B)		
1981-05 (.4)	.6	.8
Steering Knuckle, Replace (B)		
1981-87		
one	2.0	2.3
both	3.1	3.4
1988-05		
exc. below		
one side (1.2)	1.8	2.0
both sides (2.2)	3.5	4.0
2001-05 Civic		
one side (.8)	1.3	1.5
both sides (1.6)	2.6	2.9
Insight		
one side (.8)	1.3	1.5
both sides (1.6)	2.6	2.9
Strut Assy., R&R or Replace (B)		
Exc. below		
one (1.0)	1.5	1.7
both (1.7)	2.5	2.7
2001-05 Civic		
one side (.5)	.8	.9
both sides (1.0)	1.6	1.8
Strut Assy., R&R and Recondition (B)		
1981-87		
one	2.3	2.8
both	4.1	4.6
1988-95		
one	1.5	2.0
both	3.0	3.5
Upper Ball Joints, Replace (B)		
Includes: Reset toe-in.		
1986-01 Prelude, Accord		
one side	.5	.8
both sides	.9	1.2
Upper Control Arm, Replace (B)		
Includes: Reset toe.		
1986-05		
exc. below		
one side (.6)	1.0	1.1
both sides (1.2)	1.9	2.2
2003-05 Accord		
one side (1.0)	1.5	1.6
both sides (1.7)	2.5	2.8
Upper Control Arm Bushings, Replace (B)		
1988-01		
one side	1.6	1.7
both sides	2.9	3.0

	LABOR TIME	SEVERE SERVICE
Wheel Bearing and Hub Assy., Replace (B)		
1981-87		
exc. below		
one	2.0	2.3
both	3.1	3.4
1983-87 Prelude		
one side	2.6	2.8
both sides	3.7	3.9
1988-05		
exc. below		
one side (1.2)	1.8	2.0
both sides (2.2)	3.5	4.0
2001-05 Civic		
one side (.8)	1.3	1.5
both sides (1.6)	2.6	2.9
Insight		
one side (.8)	1.3	1.5
both sides (1.6)	2.6	2.9

REAR SUSPENSION

	LABOR TIME	SEVERE SERVICE
Rear Toe, Check and Adjust (B)		
1986-05 (1.2)	1.2	1.4
Lower "A" or "B" Control Arm or Bushings, Replace (B)		
Includes: Alignment on 2001-03 Civic.		
1981-85 bushings		
one side	2.0	2.2
both sides	3.0	3.2
1986-05		
exc. below		
one (.5)	.8	.9
both (1.0)	1.6	1.8
Civic, CRX, del Sol		
2001-05		
one (.9)	1.4	1.5
both (1.8)	2.7	2.9
Replace bushings add		
each side	.5	.7
Panhard Rod Bushing, Replace (B)		
1984-87 Civic	1.7	1.9
Radius Rod or Bushings, Replace (B)		
1981-85 Accord		
one side	2.0	2.2
both sides	2.6	2.8
1981-83 Civic		
one side	2.6	2.8
both sides	3.5	3.7
1981-87 Prelude		
one side	2.6	2.8
both sides	3.5	3.7
Rear Hub and/or Bearings, Replace (B)		
1981-87		
Civic		
one side (.9)	1.3	1.4
both sides (1.6)	2.4	2.6

	LABOR TIME	SEVERE SERVICE
1988-05		
exc. below		
one side (.7)	1.1	1.2
both sides (1.2)	1.9	2.2
S2000		
one side (1.1)	1.6	1.7
both sides (2.2)	3.5	4.0
2003-05 Accord		
one side (1.0)	1.5	1.6
both sides (1.9)	2.8	3.0
w/4WD add	.6	.6
Rear Knuckle, Replace (B)		
1986-05		
one (1.1)	1.6	1.7
both (1.8)	2.7	2.9
Rear Leaf Spring, Replace (B)		
1981-83 Civic wagon		
one side	1.7	2.0
both sides	3.0	3.3
Rear Strut Coil Spring, Replace (B)		
Accord, S2000		
1981-85		
one	2.2	2.3
both	3.3	3.4
1986-05		
one (.8)	1.2	1.3
both (1.6)	2.4	2.6
Civic		
1981-87		
one	2.2	2.3
both	3.3	3.4
1988-00		
one	1.3	1.4
both	2.6	2.9
2001-05		
one (.5)	.8	.9
both (1.0)	1.6	1.8
CRX		
1984-87		
one	2.2	2.3
both	3.3	3.4
1988-91		
one	1.9	2.0
both	3.2	3.3
del Sol		
one	1.9	2.0
both	3.2	3.3
Insight		
one (.5)	.8	.9
both (1.0)	1.6	1.8
Prelude		
1981-87		
one	2.2	2.3
both	3.3	3.4
1988-01		
one	1.3	1.4
both	2.6	2.9

ACCORD : CIVIC : CIVIC DEL SOL : CRX : INSIGHT : PRELUDE : S2000 **HON-17**

	LABOR TIME	SEVERE SERVICE
Shock Absorbers or Bushings, Replace (B)		
1981-83 Civic wagon		
one	.8	1.1
both	1.2	1.5
Shackle and/or Bushing, Replace (B)		
1981-83 Civic wagon		
one side	1.2	1.4
both sides	2.2	2.4
Stabilizer Bar, Replace (B)		
Exc. below (.7)	1.1	1.2
Civic		
1984-87 (1.1)	1.6	1.7
2001-05 (.4)	.6	.7
Stabilizer Bar Bushings, Replace (B)		
1981-05 (.5)	.7	.8
Strut Assy., R&R or Replace (B)		
Does not include alignment.		
Accord		
1981-85		
one	1.9	2.0
both	3.0	3.1
1986-01		
one	1.7	1.8
both	2.8	2.9
Civic, Prelude		
1981-87		
one	1.9	2.0
both	3.0	3.1
1988-01		
one	1.7	1.8
both	2.8	2.9
CRX		
1984-87		
one	2.0	2.1
both	3.0	3.1
1988-91		
one	1.7	1.8
both	2.8	2.9
del Sol		
one	1.7	1.8
both	2.8	2.9
Insight, S2000		
one	1.1	1.2
both	2.2	2.3
Strut Shock Absorber, Replace (B)		
Accord		
1981-85		
one	2.2	2.3
both	3.3	3.4
1986-02		
one (.8)	1.2	1.3
both (1.6)	2.4	2.6
2003-05		
one (1.1)	1.6	1.7
both (2.1)	3.1	3.3
Civic		
1981-87		
one	2.2	2.3
both	3.3	3.4

	LABOR TIME	SEVERE SERVICE
1988-00		
one	1.3	1.4
both	2.6	2.9
2001-05		
one (.5)	.8	.9
both (1.0)	1.5	1.7
CRX		
one	2.2	2.3
both	3.3	3.4
del Sol		
one	1.9	2.0
both	3.2	3.3
Insight		
one (.3)	.5	.6
both (.6)	.9	1.1
Prelude		
1981-87		
one	2.2	2.3
both	3.3	3.4
1988-01		
one	1.3	1.4
both	2.6	2.9
S2000		
one (.8)	1.2	1.3
both (1.6)	2.4	2.6
Toe Adjusting Bar, Replace (B)		
Does not include alignment.		
1988-01		
one side	1.2	1.5
both sides	2.0	2.3
Replace bushings		
add each	.5	.7
Trailing Arm or Link, Replace (B)		
Does not include alignment.		
Accord		
1986-97		
one side (.6)	1.0	1.1
both sides (1.2)	1.8	2.0
1998-05		
one side (.4)	.6	.7
both sides (.8)	1.3	1.4
1988-00 Civic		
one side (.9)	1.4	1.7
both sides (1.8)	2.9	3.2
Insight both (2.8)	4.1	4.3
1992-01 Prelude		
one side (.6)	1.0	1.1
both sides (1.2)	1.9	2.2
Replace bushings add		
each side	.5	.7
Upper Control Arm or Bushings, Replace (B)		
Does not include alignment.		
One side (.4)	.7	1.0
Both sides (.8)	1.3	1.6
Replace bushings		
add each	.5	.7

	LABOR TIME	SEVERE SERVICE
STEERING		
AIR BAGS		
Diagnose Air Bag System		
Component Each (A)		
1990-05 (.6)	.8	.8
Retrieve Fault Codes (A)		
1990-05 (.3)	.4	.4
Air Bag Assy., Replace (B)		
Driver (.3)	.5	.5
Passenger (.5)	.8	.8
Side		
Accord each		
1998-02 (1.3)	1.8	1.8
2003-05 (.8)	1.2	1.2
2001-05 Civic each	1.1	1.1
Side curtain		
2003-05		
Accord (1.0)	1.4	1.4
Cable Reel, Replace (B)		
1990-05 (.9)	1.4	1.4
Electronic Control Unit, Replace (B)		
Exc. below (.6)	.8	.8
Accord		
1998-02 (1.2)	1.5	1.5
2003-05 (.8)	1.1	1.1
Front Sensor, Replace (B)		
1990-05 any (.5)	.7	.7
Side Impact Sensor, Replace (B)		
Accord		
1998-02 (.6)	.8	.8
2003-05 (.3)	.5	.5
2001-05 Civic (.3)	.5	.5
ELECTRIC STEERING		
Diagnose Electric Steering		
System (A)		
2000-05 (.6)	.8	.8
Electric Steering Memorization		
Procedure (A)		
S2000 (.6)	.8	.8
Steering Gearbox, Adjust (B)		
Insight, S2000 (.5)	.7	.7
Control Unit, Replace (B)		
Exc. below (.3)	.5	.5
2005 Accord IMA (.8)	1.0	1.0
Steering Column, Replace (B)		
Insight, S2000 (1.0)	1.4	1.4
Steering Gearbox, Replace (B)		
Insight, S2000 (1.2)	1.8	2.1
Steering Wheel, Replace (B)		
2000-05 (.5)	.7	.7
Tie Rod, Tie Rod End and/or Boot, Replace (B)		
Includes: Reset toe.		
One (.8)	1.3	1.5
Both (1.0)	1.6	1.8

HON-18 ACCORD : CIVIC : CIVIC DEL SOL : CRX : INSIGHT : PRELUDE : S2000

	LABOR TIME	SEVERE SERVICE
MANUAL RACK & PINION		
Unless otherwise noted, time given does not include alignment.		
Gear Assy., Adjust (B)		
1981-00	.7	1.0
Flexible Coupling, Replace (B)		
1981-00	.8	1.0
Gear Assy., R&R (B)		
1981-00	2.9	3.4
Gear Assy., R&R and Recondition (A)		
1981-00	4.5	4.8
Rack Mounting Bushings, Replace (B)		
1981-00	1.4	1.7
Steering Column, Replace (B)		
1981-85 Accord	.8	.8
Civic		
1981-85	.8	.8
1986-00	1.6	1.6
CRX		
1984-85	.8	.8
1986-91	1.2	1.2
del Sol	1.2	1.2
1981-83 Prelude	.8	.8
Steering Wheel, Replace (B)		
1981-85 Accord	.7	.7
Civic, CRX		
1981-85	.7	.7
1986-91	.3	.3
1992-00	.7	.7
del Sol	.7	.7
1981-83 Prelude	.3	.3
Tie Rod, Tie Rod End and/or Boot, Replace (B)		
Includes: Reset toe.		
1981-00		
one (.8)	1.3	1.5
both (1.0)	1.6	1.8
POWER RACK & PINION		
Unless otherwise noted, time given does not include alignment.		
Pump Pressure Check (B)		
1981-05 (.3)	.5	.5
Rack & Pinion Assy., Adjust (B)		
1981-05 (.5)	.7	.9
Cut-Off Valve, Replace (B)		
1981-96 (.6)	.9	.9
Power Steering Reservoir, Replace (B)		
1981-91 (.5)	.7	.7
1993-05 (.2)	.3	.4
Pump Flow Control Valve, Replace (B)		
Accord		
1981-85	2.6	2.8
1986-96	1.4	1.6
1981-96 Civic	4.8	5.0
1993-96 del Sol	4.2	4.4

	LABOR TIME	SEVERE SERVICE
Prelude		
1981-85	2.6	2.8
1986-96	1.4	1.6
Pump Shaft Seal, Replace (B)		
1981-91	1.7	2.0
1992-01	1.3	1.6
Rack & Pinion Assy., R&R or Replace (B)		
Includes: Reset toe.		
Accord		
1981-93		
4 cyl.	3.7	4.2
V6	4.2	4.7
1994-97	4.1	4.6
1998-02		
4 cyl. (2.9)	4.6	4.8
V6 (3.2)	4.7	4.9
1998-05		
4 cyl. (2.9)	4.6	4.8
V6 (3.2)	4.7	4.9
Civic, del Sol (2.5)	4.1	4.6
1988-01 Prelude		
2WS	4.1	4.6
4WS	4.4	4.9
Insight (1.0)	1.5	1.9
S2000 (1.2)	1.8	2.1
Replace assembly add	.5	.7
Rack & Pinion Assy., Reseal (A)		
Includes: Alignment.		
Accord		
4 cyl.	5.3	5.9
V6	5.5	6.1
Civic, del Sol	5.3	5.9
1988-01 Prelude		
2WS	5.3	5.9
4WS	5.8	6.4
Rack Mounting Bushings, Replace (B)		
1981-01	1.4	1.7
Speed Sensor, Replace (B)		
Accord, Civic, del Sol	1.5	1.5
Prelude		
1981-87	1.5	1.5
1988-91	2.4	2.4
1992-01	.8	.8
Steering Column, Replace (B)		
Exc. below (1.0)	1.6	1.6
2003-05 Accord (.5)	.8	.8
w/air bags add	.2	.2
w/tilt wheel add	.2	.2
Steering Pump, R&R and Recondition (B)		
Accord		
1981-89 (1.6)	2.3	2.8
1990-05 (.7)	1.1	1.6
Civic, del Sol		
1981-87 (1.6)	2.5	3.0
1988-00 (1.3)	1.8	2.3
2001-05 (.8)	1.2	1.7

	LABOR TIME	SEVERE SERVICE
Prelude		
1981-91	1.3	1.8
1992-01	1.1	1.6
Steering Pump, R&R or Replace (B)		
Accord		
1981-85 (1.2)	1.7	1.9
1990-05 (.6)	.9	1.1
Civic		
1981-87 (1.2)	1.9	2.1
1988-95 (.6)	1.0	1.2
1996-05 (.4)	.6	.8
del Sol (.3)	.5	.7
Prelude		
1981-96	1.1	1.3
1997-01	.5	.7
Steering Pump Hoses, Replace (B)		
Does not include purging.		
Pressure		
Accord		
1981-89	3.1	3.2
1990-05		
4 cyl. (.6)	.9	1.0
V6 (.8)	1.2	1.4
Civic		
1981-00 (1.5)	2.3	2.6
2001-05 (.3)	.5	.5
del Sol	2.3	2.4
Prelude		
1981-87	1.7	1.8
1988-91	3.1	3.2
1992-01	1.2	1.3
Return		
1981-93	.7	.8
1994-05 (.5)	.8	.9
Steering Wheel, Replace (B)		
All Models (.5)	.7	.7
w/cruise control add	.1	.1
Tie Rod, Tie Rod End and/or Boot, Replace (B)		
Includes: Reset toe.		
1981-05		
one (.8)	1.3	1.5
both (1.0)	1.6	1.8
REAR STEERING 4WS		
Gearbox Performance Test (B)		
1988-94 4WS	1.7	2.0
Control Unit, Replace (B)		
1992-94 4WS	.5	.5
Driveshaft, Replace (B)		
1988-94 4WS	2.0	2.3
Gearbox, Replace (B)		
1988-94 4WS	2.3	2.6
Performance test add	1.2	1.2
Gearshift Dust Boot, Replace (B)		
1988-94 4WS	.8	.9
Main Sensor, Replace (B)		
1992-94 4WS	.7	.8

ACCORD : CIVIC : CIVIC DEL SOL : CRX : INSIGHT : PRELUDE : S2000   HON-19

	LABOR TIME	SEVERE SERVICE
Tie Rod, Tie Rod End and/or Boot, Replace (B)		
Includes: Reset toe.		
1988-94 4WS		
one side	1.3	1.5
both sides	1.6	1.8

HEATING & AIR CONDITIONING

When more than one component requires replacement where evacuation/recovery and recharging is already included, deduct 1.0 hour for each additional component from the time given.

Evacuate/Recover and Recharge System (B)
1981-05 (.7)	1.0	1.0

AC Hoses, Replace (B)
Includes: Evacuate/recover and recharge.
1981-85		
one	2.2	2.3
each addl.	.5	.6
1986-05		
exc. below each (1.1)	1.5	1.6
2003-05 Accord		
discharge (1.7)	2.4	2.5
2001-05 Civic Si 2.0L, Civic HB		
suction (1.9)	2.7	2.8
discharge (1.7)	2.4	2.5

AC On/Off Control Switch, Replace (B)
Accord		
1986-93	2.4	2.4
1994-02	1.6	1.6
2003-05 (.5)	.7	.7
Civic		
1986-95	2.4	2.4
1996-05 (.6)	.8	.8
Insight, S2000 (1.1)	1.4	1.4
1986-01 Prelude	1.4	1.4

Compressor Assy., Replace (B)
Includes: Parts transfer, evacuate/recover and recharge.
Accord		
1981-91	3.6	3.8
1992-02	2.8	3.0
2003-05		
4 cyl. (1.7)	2.4	2.6
V6 (2.0)	2.9	3.1
Civic		
1981-85	2.8	3.0
1986-91	3.7	3.9
1992-00	2.8	3.0

	LABOR TIME	SEVERE SERVICE
2001-05		
1.3L, 2.0L Si (3.2)	4.5	4.7
1.7L (2.3)	3.3	3.5
CRX, del Sol	2.8	3.0
Insight, S2000 (2.0)	2.8	3.0
Prelude		
1981-85	2.8	3.0
1986-91	3.7	3.9
1992-01	2.8	3.0
w/transfer clutch assy. add	.5	.7

Compressor Clutch Assy., Replace (B)
Accord		
1981-85	1.4	1.6
1986-91	3.1	3.3
1992-02	2.0	2.2
2003-05 (1.0)	1.5	1.7
Accord IMA (2.0)	2.9	3.1
Civic, CRX, del Sol		
1981-85	1.7	1.9
1986-87	4.2	4.4
1988-91	3.1	3.3
1992-00	2.0	2.2
2001-05		
1.3L, 2.0L Si (2.5)	3.6	3.8
1.7L (1.6)	2.4	2.6
Insight, S2000 (1.0)	1.5	1.7
Prelude		
1981-85	1.7	1.9
1986-87	4.2	4.4
1988-91	3.1	3.3
1992-01	2.0	2.2

Compressor or Fan Relay, Replace (B)
Exc. below (.2)	.3	.3
1994-97 Accord		
4 cyl. (.2)	.3	.3
V6 (.6)	.9	.9

Compressor Shaft Seal, Replace (B)
Includes: Parts transfer. Evacuate/recover and recharge.
Accord	4.1	4.4
Civic, CRX	3.0	3.3
Prelude		
1981-85	3.0	3.3
1986-87	3.7	4.0
1988-91	4.1	4.4

Condenser Assy., Replace (B)
Includes: Evacuate/recover and recharge.
Accord		
1981-02 (1.2)	1.7	1.9
2003-05 (1.7)	2.4	2.6
Civic, del Sol		
1981-00 (1.2)	1.7	1.9
2001-05 (1.4)	2.0	2.2
CRX	2.8	3.0
Insight (1.9)	2.7	2.9

	LABOR TIME	SEVERE SERVICE
Prelude		
1981-91	2.7	2.9
1992-96	2.0	2.2
1997-01	1.5	1.7
S2000 (1.2)	1.7	1.9

Evaporator Coil, Replace (B)
Includes: Evacuate/recover and recharge.
Accord, Civic, del Sol (1.8)	2.6	2.7
CRX	2.3	2.4
Insight, S2000 (1.7)	2.4	2.5
Prelude		
1981-82	2.3	2.4
1983-87	3.1	3.2
1988-01	2.7	2.8

Expansion Valve, Replace (B)
Includes: Evacuate/recover and recharge.
Accord, Civic, del Sol (1.8)	2.6	2.7
Accord IMA (2.9)	4.1	4.2
CRX	2.3	2.4
Insight, S2000 (1.7)	2.4	2.5
Prelude		
1981-82	2.3	2.4
1983-87	3.1	3.2
1988-01	2.7	2.8

Heater & AC Mode Switch, Replace (B)
Includes: Adjust cables when necessary.
Accord		
1981-85 (1.1)	1.3	1.4
1986-93 (1.5)	1.9	2.0
1994-97 (1.0)	1.3	1.4
1998-05 (.4)	.5	.6
Civic		
1981-87 (.7)	.9	1.0
1988-91 (1.5)	1.9	2.0
1992-00 (1.0)	1.3	1.4
2001-05 (.4)	.5	.6
CRX, del Sol (.6)	.8	.9
Prelude		
1981-91 (1.1)	1.4	1.5
1992-01 (.6)	.8	.9

Heater Blower Motor, Replace (B)
Accord		
1981-93	2.5	2.5
1994-05 (.3)	.5	.5
Civic		
1981-86	1.2	1.2
1987	.8	.8
1988-05 (.3)	.5	.5
CRX		
1984-86	1.2	1.2
1987	.8	.8
1988-91	1.5	1.5
del Sol (.2)	.3	.3
Insight, S2000 (.3)	.5	.5

ACCORD : CIVIC : CIVIC DEL SOL : CRX : INSIGHT : PRELUDE : S2000

	LABOR TIME	SEVERE SERVICE
Heater Blower Motor, Replace (B)		
Prelude		
1981-82	1.2	1.2
1983-87	1.7	1.7
1988-01	.5	.5
Add time to evacuate & recharge if necessary.		
Heater Blower Motor Resistor, Replace (B)		
Accord		
1981-89	.7	.7
1990-93	2.2	2.2
1994-05 (.5)	.7	.7
Civic, CRX		
1981-83	.8	.8
1984-87		
w/AC	1.2	1.2
w/o AC	.5	.5
1988-05 (.5)	.7	.7
del Sol		
1993-95	.7	.7
1996-97	.3	.3
Insight, S2000 (.5)	.7	.7
Prelude		
1981-83	.8	.8
1984-87	1.7	1.7
1988-01	.7	.7
Heater Blower Motor Switch, Replace (B)		
Accord		
1981-89	.7	.7
1990-97 (1.0)	1.6	1.6
Civic		
1981-91	.8	.8
1992-95	1.6	1.6
1996-05 (.5)	.8	.8
CRX, del Sol (.5)	.8	.8
Prelude		
1981-82	.7	.7
1983-87	1.9	1.9
1988-91	.7	.7
1992-01	1.6	1.6
Heater Control Assy., Replace (B)		
Includes: Adjust cables when necessary.		
Accord		
1981-87	1.9	1.9
1988-93	2.4	2.4
1994-97 (1.0)	1.6	1.6
2003-05 (.4)	.6	.6
Civic, CRX		
1981-87	.7	.7
1988-91	2.4	2.4
1992-95	1.6	1.6
1996-05 (.6)	.8	.8
del Sol (.6)	.8	.8
Insight, S2000 (1.0)	1.6	1.6
Prelude	1.6	1.6

	LABOR TIME	SEVERE SERVICE
Heater Core, R&R or Replace (B)		
Accord		
1981-97	6.8	7.3
1998-02 (5.0)	7.0	7.5
2003-05 (5.4)	7.5	8.0
Civic, CRX		
1981-87	2.8	2.3
1988-00	6.8	7.3
2001-05 (5.0)	7.0	7.5
del Sol		
1993-95	5.2	5.7
1996-97	4.2	4.7
Insight, S2000 (4.5)	6.3	6.8
Prelude		
1981-82	3.7	4.2
1983-87	1.7	2.2
1988-01	6.8	7.3
w/AC add	1.2	1.2
Heater Hoses, Replace (B)		
1981-87 each	.5	.6
1988-05 each	.8	.9
Heater Water Shut-Off Valve, Replace (B)		
1981-05 (.5)	.8	.9
Receiver/Drier Assy., Replace (B)		
Includes: Evacuate/recover and recharge.		
Accord, S2000		
1981-02 (1.2)	1.6	1.7
2003-05 (1.9)	2.7	2.8
Civic		
1981-00 (1.2)	1.6	1.7
2001-05 (1.5)	2.2	2.3
del Sol (1.0)	1.4	1.5
Prelude, Insight (1.5)	2.2	2.3
S2000 (1.2)	1.6	1.7

WIPERS & SPEEDOMETER

	LABOR TIME	SEVERE SERVICE
Power Antenna, Replace (B)		
Accord		
1986-89	2.4	2.5
1990-97	.8	.9
2002-05 Civic HB, Civic IMA (.2)	.3	.4
1998-96 Prelude	.8	.9
w/wagon add	.6	.7
Instrument Panel Cluster, Replace (B)		
1998-05 Accord (.4)	.7	.7
2001-05 Civic (.3)	.5	.5
Insight (.5)	.7	.7
S2000 (1.0)	1.4	1.4
Intermittent Wiper Control Unit, Replace (B)		
1986-00 (.5)	.7	.7
Intermittent Wiper Relay, Replace (B)		
1986-05 (.2)	.3	.3

	LABOR TIME	SEVERE SERVICE
Radio, R&R (B)		
Accord, Prelude		
1981-89	.7	.7
1990-02	1.6	1.6
2003-05 (.5)	.7	.7
Civic, CRX, del Sol		
1981-91	.7	.7
1992-95	1.6	1.6
1996-05 (.6)	.8	.8
Insight (.3)	.5	.5
S2000 (1.0)	1.3	1.3
Rear Window Wiper Motor, Replace (B)		
All Models (.6)	.9	1.0
Speedometer Cable & Casing, Replace (B)		
1981-85	1.3	1.3
Speedometer Driven Gear, Replace (B)		
Includes: Oil seal.		
1981-85	.5	.5
w/speed sensor add	.5	.5
Speedometer Head, R&R or Replace (B)		
Accord		
1981-97	1.8	1.8
1998-02	1.6	1.6
Civic		
1981-87	1.2	1.2
1988-91	1.4	1.4
1992-95	1.8	1.8
1996-00	.6	.6
CRX	1.9	1.9
del Sol	1.8	1.8
Prelude		
1981	1.2	1.2
1982-87	2.3	2.3
1988-01	1.8	1.8
Tachometer, R&R or Replace (B)		
Accord 1981-02	1.8	1.8
Civic		
1981-95	1.8	1.8
1996-00	.6	.6
CRX		
1984-87	1.2	1.2
1988-91	1.9	1.9
del Sol	1.8	1.8
Prelude		
1981	1.2	1.2
1982-87	2.3	2.3
1988-01	1.6	1.6
Windshield Washer Motor, Replace (B)		
1981-85	.5	.6
1986-05 one (.6)	.9	1.0
Windshield Wiper & Washer Switch, Replace (B)		
1981-85	.8	.8
1986-99	1.2	1.2

ACCORD : CIVIC : CIVIC DEL SOL : CRX : INSIGHT : PRELUDE : S2000   HON-21

	LABOR TIME	SEVERE SERVICE
2000-05		
Accord, Civic,		
Prelude (.7) 1.0		1.0
Insight, S2000 (.5) . . .7		.7
Windshield Wiper Linkage,		
Replace (B)		
All Models (.6) 9		1.0
Windshield Wiper Motor, Replace (B)		
Accord (.6) 9		1.0
Civic		
1981-83 . . . . 1.3		1.5
1984-05 (.6) 9		1.0
CRX		
1984-87 5		.7
1988-91 . . . . 1.2		1.4
del Sol 8		1.0
Insight, S2000 (.6) 9		1.0
Prelude (.6) 9		1.0

LAMPS & SWITCHES

Back-Up Lamp Bulb, Replace (C)
  One (.2) 3   .3
Back-Up Lamp Switch, Replace (B)
  All Models (.3) 5   .5
Brake Lamp Switch, Replace (B)
  All Models (.3) 5   .5
Brake/Tail Lamp Lens Housing,
Replace (B)
  1988-05 one ( ) 8   .9
  w/Accord wagon add 2   .2
  w/96-97 del Sol add 2   .2
Clutch Start Switch, Replace (B)
  1986-05 (.3) 5   .5
Halogen Headlamp Bulb, Replace (B)
  1988-02
    exc. S2000 each 3   .3
    S2000 each 5   .5
  2003-05
    exc. below each (.2) . .3   .3
    Accord
      right (.2) 3   .3
      left (.4) 6   .7
    S2000 each (.3) 5   .5
Hazard Warning Switch, Replace (B)
  1981-05
    exc. below (.2) 3   .3
    2003-05 Accord (.4) . . .5   .5
  w/81-91 Prelude add 3   .3
Headlamp Dimmer Switch,
Replace (B)
  1981-91 . . . . 1.7   1.7
Headlamp Switch, Replace (B)
  1981-91 5   .5
Headlamps, Aim (B)
  All Models (.3) 5   .5

High Mount Stop Lamp Assy.,
Replace (B)
  Accord
    1986-89 (.3) 5   .5
    1990-97 (.7) 1.0   1.0
    1998-05 (.3) 5   .5
  Civic
    1986-00 (.3) 5   .5
    2001-05 (.7) 1.0   1.0
  CRX, del Sol, Prelude . . .5   .5
  w/Civic coupe or
    sedan add 6   .6
High Mount Stop Lamp Bulb,
Replace (B)
  All Models (.2) 3   .3
Horn, Replace (B)
  Accord
    1981-85 (.5) 7   .7
    1986-93 (.7) 1.0   1.0
    1994-05 (.4) 6   .6
  Civic, CRX
    1981-87 (.5) 7   .7
    1988-95 (.7) 1.0   1.0
    1996-05 (.4) 6   .6
  del Sol
    1993-95 . . . . 1.2   1.2
    1996-97 3   .3
  Insight (.4) 6   .6
  Prelude
    1981-87 (.5) 7   .7
    1988-96 (.7) 1.0   1.0
    1997-01 (.4) 6   .6
  S2000 (.4) 6   .6
Horn Relay, Replace (B)
  All Models (.2) 3   .3
License Lamp Lens or Bulb,
Replace (C)
  1981-05 (.3) 4   .4
Parking Brake Warning Lamp
Switch, Replace (B)
  1981-05 (.4) 5   .5
Parking Lamp Lens or Bulb,
Replace (C)
  Each (.2) 3   .3
Sealed Beam Headlamp, Replace (B)
  1981-92 each (.2) 3   .3
  w/CRX add 3   .3
Side Marker Lamp Bulb, Replace (B)
  1981-85 (.1) 2   .2
  1986-05 one (.2) 3   .3
Tail Lamp Bulb, Replace (C)
  Each (.2) 3   .3
Tailgate Lamp Bulb, Replace (C)
  1990-97 Accord wagon . .2   .2
Tailgate Lamp Lens, Replace (B)
  1990-97 Accord wagon
    one 7   .7
    both 1.3   1.3

Turn Signal & Headlamp Dimmer
Switch, Replace (B)
  1981-85 7   .8
  1986-05 (.7) 9   1.0
Turn Signal or Hazard Relay,
Replace (B)
  1981-05 (.2) 3   .3
Turn Signal or Hazard Warning
Flasher, Replace (B)
  1981-05 (.3) 4   .4

SEAT BELTS

Diagnose Seat Belt System (A)
  1990-93 . . . . 1.0   1.0
Anchor Rail Assy., Replace (B)
  1990-93 . . . . 1.3   1.3
Control Module, Replace (B)
  1990-93 5   .5
Lap Belt Assy., Replace (B)
  1990-93 5   .5
Retractor-Receiver Assy.,
Replace (B)
  1990-93 7   .7

BODY

Door Latch, Replace (B)
  Accord
    1981-87 8   .8
    1988-97 . . . . 1.3   1.3
    1998-02
      front 9   .9
      rear 6   .6
    2003-05 (.5) 7   .7
  Civic
    1981-87 7   .7
    1988-95 . . . . 1.3   1.3
    1996-05 (.5) 7   .7
  CRX
    1984-90 8   .8
    1991 . . . . 1.8   1.8
  del Sol
    1993-95 . . . . 1.6   1.6
    1996-97 5   .5
  Insight, S2000 (.8) . . . 1.2   1.2
  Prelude
    1981-87 8   .8
    1988-01 . . . . 1.3   1.3
  w/power locks add 2   .2
Door Window Regulator, Replace (B)
  Front
    Accord
      1981-02 . . . . 1.6   1.8
      2003-05 (.5) 8   .9
    Civic
      1981-95 . . . . 1.3   1.5
      1996-05 (.5) 8   .9
    CRX 1.9   2.1

HON-22 ACCORD : CIVIC : CIVIC DEL SOL : CRX : INSIGHT : PRELUDE : S2000

	LABOR TIME	SEVERE SERVICE
Door Window Regulator, Replace (B)		
del Sol (1.0)	1.6	*1.8*
Insight (.7)	1.1	*1.2*
Prelude (1.0)	1.6	*1.8*
S2000 (1.0)	1.6	*1.8*
Rear		
Accord		
1981-02	1.6	*1.8*
2003-05 (.5)	.8	*.9*
Civic		
1981-95	1.6	*1.8*
1996-05 (.5)	.8	*.9*
Prelude (1.0)	1.7	*1.9*
w/power windows add	.5	*.5*
Gas Tank Lid Release Cable, Replace (B)		
Exc. below (1.0)	1.6	*1.7*

	LABOR TIME	SEVERE SERVICE
2001-05 Civic (.6)	.9	*1.0*
Hood Lock, Replace (B)		
1981-05		
exc. below (.3)	.5	*.5*
Accord (.5)	.7	*.7*
Hood Release Cable, Replace (B)		
Exc. below (.8)	1.2	*1.4*
Prelude		
1988-01 (1.1)	1.8	*1.9*
Lock Striker Plate, Replace (B)		
1981-05 (.3)	.4	*.4*
Tailgate/Hatch Latch, Replace (B)		
Accord		
exc. wagon (.7)	1.1	*1.1*
wagon (.5)	.7	*.7*

	LABOR TIME	SEVERE SERVICE
Civic		
1981-91 (.4)	.6	*.5*
1992-00 (.8)	1.1	*1.1*
Insight (.3)	.5	*.5*
Trunk Latch, Replace (B)		
1981-05 (.3)	.5	*.5*
Window Regulator Motor, Replace (B)		
Exc. below (1.1)	1.6	*1.8*
2003-05 Accord (.5)	.8	*.9*
1996-05 Civic (.5)	.8	*.9*
del Sol	2.5	*2.7*

HON

CR-V : Element : Odyssey : Passport : Pilot : Ridgeline

SYSTEM INDEX

MAINTENANCE	HON-24
Maintenance Schedule	HON-24
CHARGING	HON-24
STARTING	HON-24
CRUISE CONTROL	HON-25
IGNITION	HON-25
EMISSIONS	HON-25
FUEL	HON-26
EXHAUST	HON-27
ENGINE COOLING	HON-27
ENGINE	HON-28
Assembly	HON-28
Cylinder Head	HON-28
Camshaft	HON-29
Crank & Pistons	HON-30
Engine Lubrication	HON-30
CLUTCH	HON-31
MANUAL TRANSAXLE	HON-31
MANUAL TRANSMISSION	HON-31
AUTO TRANSAXLE	HON-31
AUTO TRANSMISSION	HON-32
TRANSFER CASE	HON-32
SHIFT LINKAGE	HON-32
DRIVELINE	HON-32
BRAKES	HON-33
FRONT SUSPENSION	HON-34
REAR SUSPENSION	HON-35
STEERING	HON-36
HEATING & AC	HON-37
WIPERS & SPEEDOMETER	HON-38
LAMPS & SWITCHES	HON-38
BODY	HON-39

OPERATIONS INDEX

A

AC Hoses	HON-37
Air Bags	HON-36
Air Conditioning	HON-37
Alternator (Generator)	HON-24
Antenna	HON-38
Anti-Lock Brakes	HON-33

B

Back-Up Lamp Switch	HON-38
Ball Joint	HON-34
Battery Cables	HON-24
Bleed Brake System	HON-33
Blower Motor	HON-37
Brake Disc	HON-33
Brake Drum	HON-33
Brake Hose	HON-33
Brake Pads and/or Shoes	HON-33

C

Camshaft	HON-29
Catalytic Converter	HON-27
Coolant Temperature (ECT) Sensor	HON-26
Crankshaft	HON-30
Crankshaft Sensor	HON-25
Cruise Control	HON-25
CV Joint	HON-32
Cylinder Head	HON-28

D

Differential	HON-32
Distributor	HON-25
Drive Belt	HON-24
Driveshaft	HON-32

E

EGR	HON-26
Electronic Control Module (ECM/PCM)	HON-26
Engine	HON-28
Engine Lubrication	HON-30
Engine Mounts	HON-28
Evaporator	HON-37
Exhaust	HON-27
Exhaust Manifold	HON-27
Expansion Valve	HON-37

F

Flywheel	HON-31
Fuel Injection	HON-27
Fuel Pump	HON-26
Fuel Vapor Canister	HON-26

G

Gear Selector Lever	HON-32
Generator	HON-24

H

Halfshaft	HON-32
Headlamp	HON-38
Heater Core	HON-38
Horn	HON-39

I

Idle Air Control (IAC) Valve	HON-27
Ignition Coil	HON-25
Ignition Module	HON-25
Ignition Switch	HON-25
Inner Tie Rod	HON-36
Intake Air Temperature (IAT) Sensor	HON-26
Intake Manifold	HON-26

K

Knock Sensor	HON-26

L

Lower Control Arm	HON-34

M

Maintenance Schedule	HON-24
Manifold Absolute Pressure (MAP) Sensor	HON-26
Mass Air Flow (MAF) Sensor	HON-26
Master Cylinder	HON-33
Muffler	HON-27

N

Neutral Safety Clutch Switch	HON-38
Neutral Safety Switch	HON-39

O

Oil Pan	HON-30
Oil Pump	HON-30
Outer Tie Rod	HON-36
Oxygen Sensor	HON-26

P

Parking Brake	HON-34
Pistons	HON-30
Positive Crankcase Ventilation (PCV) Valve	HON-26

R

Radiator	HON-27
Radiator Hoses	HON-27
Radio	HON-38
Rear Main Oil Seal	HON-30

S

Shock Absorber/Strut, Front	HON-34
Shock Absorber/Strut, Rear	HON-35
Spark Plug Cables	HON-25
Spark Plugs	HON-25
Spring, Front Coil	HON-34
Spring, Leaf	HON-35
Spring, Rear Coil	HON-35
Starter	HON-25
Steering Wheel	HON-36

T

Thermostat	HON-27
Throttle Body	HON-27
Throttle Position Sensor (TPS)	HON-26
Timing Belt	HON-29
Timing Chain	HON-30
Torque Converter	HON-31

U

U-Joint	HON-32
Upper Control Arm	HON-35

V

Valve Cover Gasket	HON-29
Valve Job	HON-29
Vehicle Speed Sensor	HON-26

W

Water Pump	HON-28
Wheel Balance	HON-24
Wheel Cylinder	HON-34
Window Regulator	HON-39
Windshield Washer Pump	HON-38
Windshield Wiper Motor	HON-38

HON-23

HON-24 CR-V : ELEMENT : ODYSSEY : PASSPORT : PILOT : RIDGELINE

	LABOR TIME	SEVERE SERVICE
MAINTENANCE		
Air Cleaner Filter Element, Replace (B)		
All Models (.2)	.3	.4
Chassis Lubrication, Change Oil & Filter (C)		
Includes: Correct fluid levels.		
All Models (.3)	.5	.5
Drive Belt, Adjust (B)		
All Models (.3)	.3	.4
Drive Belt, Replace (B)		
Air conditioning		
exc. below (.5)	.8	.9
Element, Ridgeline (.2)	.3	.3
Alternator (.3)	.5	.6
Power steering		
CR-V (.3)	.5	.6
Odyssey, Pilot (.4)	.6	.7
Passport (.2)	.3	.4
Replace alternator belt w/AC add		
CR-V, Odyssey	.2	.2
Passport	.2	.2
w/PS add		
Odyssey	.2	.2
Passport	.2	.2
Fuel Filter, Replace (B)		
CR-V, Element (.4)	.6	.7
Odyssey, Pilot		
1995-98	.5	.5
1999-04 (1.0)	1.6	1.8
Passport (.3)	.5	.5
Halogen Headlamp Bulb, Replace (C)		
All Models (.2)	.3	.3
Oil & Filter, Change (C)		
Includes: Correct fluid levels.		
All Models (.3)	.5	.5
Timing and Balance Shaft Belt, Replace (B)		
1997-01 CR-V (2.0)	3.1	3.6
Odyssey, Pilot		
1995-98	3.5	4.0
1999-05 (3.0)	4.6	5.2
Passport		
1994-97		
4 cyl.	3.0	3.5
V6	3.7	4.2
1998-02	3.7	4.2
w/AC Passport add	.3	.3
w/twin cam CR-V add	.3	.3
Replace water pump add		
CR-V	.5	.5
95-98 Odyssey	.5	.5
99-05 Odyssey, Pilot	.7	.7
Tire, Replace (C)		
Includes: Dismount old tire and mount & balance new tire to rim.		
One (.4)	.6	.7
each addl. add	.6	.7

	LABOR TIME	SEVERE SERVICE
Tires, Rotate (C)		
All Models (.3)	.5	.5
Wheel, Balance (B)		
One (.3)	.5	.5
each addl. add	.5	.5
SCHEDULED MAINTENANCE INTERVALS		
If necessary, refer to appropriate Chilton maintenance service information.		
7,500 Mile Service (C)		
CR-V, Odyssey	1.3	1.3
Passport	1.8	1.8
15,000 Mile Service (B)		
CR-V, Odyssey	1.4	1.4
Passport		
2.6L	4.7	4.7
3.2L	3.6	3.6
22,500 Mile Service (C)		
CR-V, Odyssey	1.3	1.3
Passport	1.8	1.8
30,000 Mile Service (B)		
CR-V, Odyssey	4.0	4.0
Passport		
2.6L	7.3	7.3
3.2L	3.6	3.6
37,500 Mile Service (C)		
CR-V, Odyssey	1.3	1.3
Passport	1.8	1.8
45,000 Mile Service (B)		
CR-V, Odyssey	1.8	1.8
Passport	4.1	4.1
2.6L	2.8	2.8
3.2L	1.9	1.9
52,500 Mile Service (C)		
CR-V, Odyssey	1.3	1.3
Passport	1.8	1.8
60,000 Mile Service (B)		
CR-V, Odyssey	3.9	3.9
Passport		
2.6L	8.8	8.8
3.2L	8.1	8.1
67,500 Mile Service (C)		
CR-V, Odyssey	1.3	1.3
Passport	1.8	1.8
75,000 Mile Service (B)		
CR-V, Odyssey	1.8	1.8
Passport		
2.6L	2.8	2.8
3.2L	1.9	1.9
82,500 Mile Service (C)		
CR-V, Odyssey	1.3	1.3
Passport	1.8	1.8
90,000 Mile Service (B)		
CR-V, Odyssey	6.9	6.9
Passport		
2.6L	7.3	7.3
3.2L	6.6	6.6

	LABOR TIME	SEVERE SERVICE
97,500 Mile Service (C)		
CR-V, Odyssey	1.3	1.3
Passport	1.8	1.8
105,000 Mile Service (B)		
CR-V, Odyssey	1.8	1.8
Passport	4.1	4.1
Replace timing belt add	2.4	2.4
112,500 Mile Service (C)		
CR-V, Odyssey	1.4	1.4
Passport	1.8	1.8
120,000 Mile Service (B)		
CR-V, Odyssey	4.0	4.0
Passport		
2.6L	7.4	7.4
3.2L	6.7	6.7
CHARGING		
Alternator Circuits, Test (B)		
Includes: Test component output.		
All Models (.4)	.6	.6
Alternator Assy., R&R and Recondition (B)		
CR-V, Passport		
exc. below (1.5)	2.3	2.8
2002-05 CR-V (2.1)	3.0	3.5
Element (1.8)	2.6	3.1
Odyssey		
1995-98	2.3	2.8
1999-04 (2.0)	3.0	3.5
Pilot (2.0)	3.0	3.5
Alternator Assy., Replace (B)		
Includes: Pulley transfer.		
CR-V		
1997-01 (.9)	1.3	1.5
2002-05 (1.5)	2.2	2.4
Element (1.2)	1.7	1.9
Odyssey		
1995-98 (.9)	2.3	2.5
1999-04 (1.4)	2.0	2.2
2005 (1.1)	1.6	1.8
Passport (.5)	.8	1.0
Pilot (1.3)	1.9	2.1
Ridgeline (.5)	.8	1.0
Alternator Voltage Regulator, Replace (B)		
CR-V (1.6)	2.3	2.4
Element (1.8)	2.6	2.7
Odyssey		
1995-98 (1.5)	2.2	2.3
1999-04 (2.0)	2.9	3.0
Pilot (1.6)	2.3	2.5
Voltmeter, Replace (B)		
1994-97 Passport	1.9	1.9
STARTING		
Starter Draw Test (On Car) (B)		
All Models (.3)	.5	.5
Battery Cables, Replace (C)		
Positive (.6)	.8	.9
Negative (.4)	.6	.7

CR-V : ELEMENT : ODYSSEY : PASSPORT : PILOT : RIDGELINE — HON-25

	LABOR TIME	SEVERE SERVICE
Starter, R&R and Recondition (B)		
CR-V		
1997-01 (1.2)	1.9	2.2
2002-05 (2.0)	3.0	3.3
Element (2.2)	3.1	3.3
Odyssey, Pilot (1.6)	2.3	2.6
Passport		
1994-97		
4 cyl. (1.9)	2.7	3.0
V6 (1.4)	2.6	2.9
1998-02 (1.4)	2.2	2.5
Starter Assy., Replace (B)		
CR-V		
1997-01	1.1	1.3
2002-05 (1.7)	2.3	2.6
Element (1.7)	2.3	2.5
Odyssey, Pilot,		
Ridgeline (.7)	1.1	1.3
Passport		
4 cyl.	1.6	1.8
V6	.8	1.0
Starter Drive Assy., Replace (B)		
Includes: Starter R&R.		
All Models	2.0	2.3
Starter Relay, Replace (B)		
Passport (.2)	.3	.3
Starter Solenoid and/or Switch, Replace (B)		
Includes: Starter R&R.		
All Models	1.3	1.6

CRUISE CONTROL

	LABOR TIME	SEVERE SERVICE
Diagnose Cruise System Component Each (A)		
1995-05 (.6)	.8	.8
Control Actuator Cable, Adjust (B)		
All Models (.2)	.3	.4
Actuator Assy., Replace (B)		
Exc. below (.3)	.5	.5
1994-97 Passport (.4)	.7	.7
Control Unit, Replace (B)		
1997-01 CR-V (.2)	.3	.3
1995-98 Odyssey (.3)	.5	.5
1994-97 Passport (.3)	.5	.5
Control Actuator Cable, Replace (B)		
Exc. Passport (.3)	.5	.7
Passport	.3	.5
Control Actuator Valve, Replace (B)		
1995-04 Odyssey (.3)	.5	.5
Control Brake or Clutch Switch, Replace (B)		
All Models (.3)	.5	.5
Control Check Valve, Replace (B)		
All Models (.2)	.3	.3
Control Combination Switch, Replace (B)		
1996-02 Passport	.7	.7

	LABOR TIME	SEVERE SERVICE
Control Main Switch, Replace (B)		
Exc. Passport (.2)	.3	.3
Passport (.8)	1.3	1.3
Control Pump, Replace (B)		
1994-97 Passport	.7	.7
Control Sensor, Replace (B)		
All Models (.5)	.8	.9
Control Steering Wheel Switch, Replace (B)		
All Models		
exc. Pilot (.2)	.3	.3
Pilot (.3)	.5	.5
Control Switch, Replace (B)		
1994-95 Passport	.7	.7
Control Vacuum Bottle and/or Hose, Replace (B)		
All Models (.2)	.3	.3

IGNITION

	LABOR TIME	SEVERE SERVICE
Diagnose Ignition System Component Each (A)		
1994-06 (.6)	.8	.8
Retrieve Fault Codes (A)		
1994-06 (.3)	.4	.4
Ignition Timing, Reset (B)		
1995-01	.3	.3
Crankshaft Position (CKP), Cylinder Position, Top Dead Center (TDC) or Crankshaft Speed Fluctuation (CKF) Sensors, Replace (B)		
Distributor mounted	.8	.9
Engine or cam cover mounted		
CR-V		
1997-01	2.3	2.6
2002-05 (.2)	.3	.3
Element (.2)	.3	.3
Odyssey, Pilot		
1995-98	3.8	4.3
1999-04 (3.0)	4.3	4.5
Passport (.4)	.6	.8
Distributor, Replace (B)		
Includes: Reset base timing.		
CR-V (.3)	.5	.7
1995-98 Odyssey (.3)	.5	.7
1994-97 Passport		
4 cyl. (.5)	.7	.9
Distributor Cap and/or Rotor, Replace (B)		
Exc. Element (.2)	.3	.3
Element (.3)	.5	.5
Ignition Coil, Replace (B)		
CR-V		
1997-01	.6	.7
2002-05 (.2)	.3	.3
Element (.2)	.3	.3
Odyssey, Pilot		
1995-98	.6	.7
1999-05 (.2)	.3	.3

	LABOR TIME	SEVERE SERVICE
Passport		
1994-97	.7	.7
1998-02	.3	.3
Ridgeline (.2)	.3	.3
Ignition Module, Replace (B)		
All Models (.4)	.7	.7
Ignition Switch, Replace (B)		
Exc. below (.9)	1.4	1.5
2005 Odyssey (.6)	.8	.9
Ridgeline (1.6)	2.0	2.1
Ignition Switch (Electrical Portion), Replace (B)		
Exc. below (.4)	.6	.7
Odyssey		
1995-04 (.7)	.9	1.0
2005 (.3)	.5	.6
Passport, Pilot		
Ridgeline (.9)	1.4	1.5
Spark Plug (Ignition) Cables, Replace (B)		
CR-V (.2)	.3	.4
1995-04 Odyssey (.2)	.3	.4
1994-97 Passport		
4 cyl. (.2)	.3	.4
V6 (.4)	.7	.8
Spark Plugs, Replace (B)		
Exc. below (.5)	.8	.9
CR-V (.3)	.5	.6
Element (.3)	.5	.6
Odyssey		
1995-98 (.3)	.5	.6

EMISSIONS

The following operations do not include testing. Add time as required.

	LABOR TIME	SEVERE SERVICE
Diagnose Emission Control System Component Each (A)		
1994-06 (.6)	.8	.8
Dynamometer Test (B)		
All Models	.4	.4
Air Injection Manifold, Replace (B)		
1994-97 Passport (.8)	1.3	1.3
Air Pump Assy., Replace (B)		
1994-97 Passport (.5)	.7	.7
Barometric Pressure Sensor, Replace (B)		
All Models	.7	.7
By-Pass Control Solenoid Valve, Replace (B)		
CR-V, Element (.4)	.6	.7
Odyssey, Pilot		
1995-98 (.2)	.3	.4
1999-04 (.4)	.6	.7
Passport		
1994-97 (.3)	.5	.6
Pilot (.4)	.6	.7

HON-26 CR-V : ELEMENT : ODYSSEY : PASSPORT : PILOT : RIDGELINE

	LABOR TIME	SEVERE SERVICE
Coolant Temperature (ECT) Sensor, Replace (B)		
Exc. Passport (.4) 6		.7
Passport		
1994-97		
4 cyl. (.3) 5		.5
V6 (1.3) 2.0		2.0
1998-02 (.3) 5		.5
EFI Main Relay, Replace (B)		
All Models (.5) 7		.7
EGR Back Pressure Transducer, Replace (B)		
1994-97 Passport 3		.3
EGR Control Solenoid Valve, Replace (B)		
1995-98 Odyssey (.4) ...7		.7
1994-97 Passport (.4) ...7		.7
2003-05 Pilot (.3) 5		.5
EGR Vacuum Control Valve, Replace (B)		
Odyssey 7		.7
EGR Valve, Replace (B)		
Exc. below (.3) 5		.7
Passport		
1994-97 (.6) 9		1.0
1998-02 (.2) 3		.5
Electronic Load Detector, Replace (B)		
All Models (.5) 7		.7
Electronic Spark Control Module, Replace (B)		
1994-97 Passport V6 ...3		.3
Electronic Control Module (ECM/PCM), Replace (B)		
CR-V (.4) 6		.7
Odyssey		
1995-04 (.4) 7		.7
2005 (.6) 9		.9
Passport		
1994-97 (.5) 7		.7
1998-99 (.8) 1.3		1.3
2000-02 (.3) 5		.5
Pilot (.4) 7		.7
Ridgeline (.6) 9		.9
Evaporator Purge Solenoid Valve, Replace (B)		
All Models (.3) 5		.5
Fuel Vapor Canister, Replace (B)		
CR-V		
1997-01 (.2) 3		.3
2002-05 (.5) 8		.9
Odyssey, Pilot		
1995-98 (.2) 3		.3
1999-05 (.5) 8		.9
Passport		
1994-95 (.2) 3		.3
1996-02 (.6) 8		.8
Ridgeline (.5) 8		.9

	LABOR TIME	SEVERE SERVICE
Intake Air Temperature (IAT) Sensor, Replace (B)		
Exc. Passport (.2) 3		.3
Passport (.3) 5		.5
Knock Sensor, Replace (B)		
2002-05 CR-V, Element (.4) 6		.7
Odyssey		
1999-04 (.7) 1.1		1.2
2005 (1.5) 2.2		2.3
Passport		
1994-97		
4 cyl. 3		.5
V6 ....3.0		3.2
1998-02 ....2.0		2.2
Pilot (.7) 1.1		1.2
Ridgeline (1.5) 2.2		2.3
Lockup Control Solenoid, Replace (B)		
All Models 5		.5
Main Relay, Replace (B)		
All Models 7		.7
Manifold Absolute Pressure (MAP) Sensor, Replace (B)		
Exc. below (.2) 3		.3
1995-98 Odyssey (.4) ...6		.6
Mass Air Flow (MAF) Sensor, Replace (B)		
Passport (.2) 3		.3
Oxygen Sensor, Replace (B)		
All Models (.3) 5		.7
Positive Crankcase Ventilation (PCV) Valve, Replace (B)		
All Models (.2) 3		.3
Throttle Position Sensor (TPS), Replace (B)		
Exc. below (.5) 8		.9
1998-02 Passport (.2) ...3		.3
Vehicle Speed Sensor, Replace (B)		
All Models (.6) 8		.9

FUEL
DELIVERY

	LABOR TIME	SEVERE SERVICE
Fuel Pump, Test (B)		
All Models (.2) 3		.3
Fuel Filter, Replace (B)		
CR-V, Element (.4) 6		.7
Odyssey, Pilot		
1995-98 (.3) 5		.5
1999-05 (1.0) 1.6		1.8
Passport (.3) 5		.5
Fuel Gauge (Dash), Replace (B)		
1997-01 CR-V (1.0) ..1.4		1.4
Odyssey		
1995-98 (1.0) 1.4		1.4
1999-04 (.5) 8		.9
Passport (1.2) 1.9		1.9

	LABOR TIME	SEVERE SERVICE
Fuel Gauge (Tank), Replace (B)		
CR-V		
1997-01 (.9) 1.4		1.5
2002-05 (.4) 6		.7
Element (.4) 6		.7
Odyssey, Pilot, Ridgeline (1.0) 1.6		1.7
Passport		
1994-97 6		.8
1998-02 ....1.1		1.3
Fuel Pump, Replace (B)		
CR-V		
1997-01 (.9) 1.4		1.5
2002-05 (.4) 6		.7
Element (.4) 6		.7
Odyssey, Pilot		
1995-05 (1.0) 1.6		1.8
Passport		
1994-97 (2.0) 2.7		2.9
1998-02 (.8) 1.1		1.3
Fuel Pump Relay, Replace (B)		
Odyssey, Ridgeline (.2) ..5		.5
Fuel Tank, Replace (B)		
Includes: Drain and refill.		
CR-V		
1997-01 (1.8) 2.7		2.9
2002-05 (1.0) 1.4		1.6
Element (.8) 1.2		1.4
Odyssey		
1995-04 (1.0) 1.6		1.8
2005 (1.4) 2.1		2.3
Passport (1.3) 1.8		2.0
Pilot (1.0) 1.4		1.6
Ridgeline (2.1) 3.1		3.3
w/4WD add		
CR-V 1.1		1.1
Eelment 6		.6
Intake Manifold and/or Gasket, Replace (B)		
Includes: Adjustments.		
CR-V		
1997-01 (2.0) 3.1		3.4
2002-05 (1.4) 2.1		2.4
Element (1.4) 2.1		2.4
Odyssey		
1995-98 (2.5) 3.6		3.9
1999-04 (.5) 8		.9
2005 (1.0) 1.4		1.7
Passport		
1994-97		
4 cyl. 4.7		5.0
V6 ....2.4		2.7
1998-02 ....1.5		1.8
Pilot (.5) 8		.9
Ridgeline (1.1) 1.6		1.9
Replace manifold add 5		.7

CR-V : ELEMENT : ODYSSEY : PASSPORT : PILOT : RIDGELINE **HON-27**

	LABOR TIME	SEVERE SERVICE
INJECTION		
Fast Idle Valve, Replace (B)		
All Models (.4)	.6	.6
Fuel Injector Resistor, Replace (B)		
1994-04 Odyssey, Passport (.2)	.3	.3
Fuel Injectors, Replace (B)		
CR-V (.4)	.6	.8
Element (.7)	1.0	1.2
Odyssey		
1995-98 (.4)	.6	.8
1999-04 (3.0)	4.3	4.5
2005		
one bank (.9)	1.3	1.5
all (1.1)	1.6	1.8
Passport		
1994-97		
4 cyl.	.8	1.0
V6	2.9	3.1
1998-02 (2.4)	3.4	3.6
Pilot (1.5)	2.2	2.4
Ridgeline		
one bank (.9)	1.3	1.5
all (1.2)	1.7	1.9
Fuel Pressure Regulator, Replace (B)		
Exc. Passport	.6	.7
Passport		
1994-97	.5	.7
1998-02	.8	1.0
Idle Air Control (IAC) Valve, Replace (B)		
CR-V		
1997-01	.3	.3
2002-05 (.6)	.9	1.0
Element (.6)	.9	1.0
Odyssey, Pilot		
1995-05 (.8)	1.3	1.4
Passport (.2)	.3	.3
Idle Control Solenoid Frequency Valve, Replace (B)		
1995-97 Odyssey	.7	.7
Throttle Body Assy., Replace (B)		
Exc. below (.5)	.8	.9
2005 Odyssey, Ridgeline (1.0)	1.4	1.5

EXHAUST

Catalytic Converter, Replace (B)		
Exc. below (.5)	.8	.9
2005 Odyssey		
front or rear (1.7)	2.5	2.6
Passport		
1994-95 (.3)	.5	.6
1996-99		
right side	.7	.8
left side	1.0	1.1
Ridgeline		
front or rear (1.2)	1.8	1.9

	LABOR TIME	SEVERE SERVICE
Center Exhaust Pipe, Replace (B)		
Passport		
1994-97		
one	.6	.7
both	.8	.9
1998-02	.8	.9
Exhaust Manifold or Gasket, Replace (B)		
CR-V, Element (.5)	.8	1.0
Odyssey		
1995-98 (.5)	.8	1.0
1999-04		
one side (1.2)	1.9	2.1
both sides (2.2)	3.3	3.5
2005 one side (1.4)	2.1	2.3
Pilot		
one side (1.2)	1.9	2.1
both sides (2.2)	3.3	3.5
Passport		
1994-97		
4 cyl.	1.3	1.6
V6		
right side	3.6	3.9
left side	4.8	5.1
both sides	6.5	6.8
1998-02		
right side	2.5	2.8
left side	2.0	2.3
both sides	4.4	4.7
Ridgeline		
front (1.0)	1.5	1.7
rear (1.4)	2.1	2.3
Front Exhaust Pipe, Replace (B)		
Exc. below (.6)	1.0	1.1
1994-97 Passport		
one (.5)	.7	.8
both (.6)	1.0	1.1
Intermediate Exhaust Pipe, Replace (B)		
1994-95 Passport	.7	.8
Muffler, Replace (B)		
All Models (.6)	.9	1.0
Rear Exhaust Pipe, Replace (B)		
Exc. Passport (.6)	.9	1.0
Passport (.3)	.5	.6
Tail Pipe, Replace (B)		
All Models (.2)	.3	.4

ENGINE COOLING

Diagnose Cooling Fan System Component Each (B)		
1995-05 (.6)	.8	.8
Pressure Test Cooling System (C)		
All Models (.2)	.3	.3
Coolant Thermostat, Replace (B)		
CR-V		
1997-01 (.5)	.8	.9
2002-05 (.8)	1.2	1.3
Element (.8)	1.3	1.4

	LABOR TIME	SEVERE SERVICE
Odyssey, Pilot		
1995-98 (.5)	.8	.9
1999-05 (.8)	1.3	1.4
Passport		
1994-97 (.3)	.5	.7
1998-02 (1.3)	2.0	2.1
Ridgeline (.6)	.9	1.0
Cooling Clutch and/or Pulley Fan, Replace (B)		
Passport (.5)	.7	.8
Replace fan clutch add	.1	.1
Cooling Fan Timer, Replace (B)		
Odyssey (.5)	.7	.7
Engine Coolant Temp. Sending Unit, Replace (B)		
All Models (.4)	.6	.7
Fan Blade, Replace (B)		
Odyssey		
1995-97 (.5)	.7	.9
2005 (1.3)	1.9	2.0
Ridgeline (.7)	1.0	1.1
Radiator Assy., R&R or Replace (B)		
Includes: Refill with proper coolant mix.		
CR-V (1.2)	1.8	2.0
Element (1.3)	1.9	2.1
Odyssey		
1995-98 (1.2)	1.8	2.0
1999-04 (2.0)	2.9	3.1
2005 (1.6)	2.3	2.5
Passport (1.1)	1.6	1.8
Pilot (1.7)	2.4	2.6
Ridgeline (1.5)	2.2	2.4
w/AT transmission add	.3	.3
Radiator Fan and/or Fan Motor, Replace (B)		
CR-V		
1997-01 (.5)	.8	.9
2002-05 (1.0)	1.4	1.5
Element (1.0)	1.4	1.5
Odyssey		
1995-98 (.5)	.8	.9
1999-04 (2.5)	3.6	3.7
2005 (1.5)	2.2	2.3
Pilot (.5)	.8	.9
Ridgeline (.8)	1.2	1.3
Radiator Fan Motor Relay, Replace (B)		
All Models (.2)	.3	.3
Radiator Fan Motor Switch (Coolant Temp.), Replace (B)		
All Models (.5)	.7	.7
Radiator Hoses, Replace (B)		
Includes: Refill with proper coolant mix.		
CR-V, Element		
upper or lower (.5)	.8	.9
both (.8)	1.3	1.4
bypass (.4)	.7	.8

HON-28 CR-V : ELEMENT : ODYSSEY : PASSPORT : PILOT : RIDGELINE

	LABOR TIME	SEVERE SERVICE
Radiator Hoses, Replace (B)		
Odyssey, Pilot		
1995-98		
upper or lower (.5)	.8	.9
both (.8)	1.3	1.4
bypass (.4)	.7	.8
1999-05		
upper or lower (.7)	1.1	1.2
both (1.0)	1.6	1.7
bypass (.4)	.7	.8
Passport		
upper or lower	.5	.6
both	.8	.9
Ridgeline		
upper (.4)	.7	.8
lower (.7)	1.1	1.2
Temperature Gauge (Dash), Replace (B)		
CR-V		
1997-05 (1.0)	1.4	1.4
Odyssey		
1995-98 (1.0)	1.4	1.4
1999-04 (.5)	.8	.9
Passport (1.2)	1.6	1.6
Water Pump and/or Gasket, Replace (B)		
Includes: Refill with proper coolant mix.		
CR-V		
1997-01 (2.6)	3.8	4.0
2002-05 (2.0)	3.0	3.2
Element (2.0)	3.0	3.2
Odyssey		
1995-98 (2.6)	3.8	4.0
1999-04 (3.3)	5.0	5.2
2005 (2.4)	3.5	3.7
Passport		
1994-97		
4 cyl.	3.0	3.2
V6	3.5	3.7
1998-02	3.5	3.7
Pilot (3.3)	5.0	5.2
Ridgeline (2.4)	3.5	3.7
w/AC add	.5	.5
w/PS add	.3	.3

ENGINE
ASSEMBLY
Times shown are for OEM assemblies. Time to replace assemblies from aftermarket rebuilders may vary.

Engine Assy., R&I (B)
Does not include parts or component transfer.

	LABOR TIME	SEVERE SERVICE
CR-V	6.7	6.7
Odyssey		
1995-98	8.2	8.2
1999-01	11.4	11.4
Passport		
4 cyl.	5.5	5.5
V6	7.9	7.9
w/4WD Passport add	1.2	1.2
w/AC add	.5	.5
w/AT Passport add	1.4	1.4

Engine Assy., R&R and Recondition (A)
Includes: Replacing rings, rod and main bearings, cylinder head reconditioning and engine tune-up.

CR-V, Odyssey	27.9	27.9
Passport		
4 cyl.	24.1	24.1
V6	30.2	30.2
w/4WD Passport add	1.2	1.2
w/AC add	.5	.5
w/AT Passport add	1.4	1.4

Engine Assy. (Complete), Replace (B)
Includes: Component transfer and engine tune-up.

CR-V		
1997-01 (6.8)	9.6	10.1
2002-05 (8.7)	12.1	12.6
Element (8.7)	12.1	12.6
Odyssey		
1995-98 (7.1)	10.0	10.5
1999-05 (9.0)	12.5	13.0
Passport		
1994-97		
4 cyl.	7.3	7.3
V6	10.2	10.2
1998-02	10.7	10.7
Pilot (9.5)	13.1	13.6
Ridgeline (9.0)	12.5	13.0
w/4WD Passport add	1.2	1.2
w/AC add	.5	.5
w/AT Passport add	1.4	1.4

Engine Assy. (Short Block), Replace (B)
Assembly consists of cylinder block, piston assemblies, crankshaft, camshaft, timing chain and gears. Does not include cylinder heads. Operation Includes: R&R engine, transfer necessary parts and all necessary adjustments.

CR-V, Pilot		
1997-01 (9.1)	12.6	13.5
2002-05 (15.0)	19.5	20.4
Element (15.0)	19.5	20.4
Odyssey		
1995-98 (9.1)	12.6	13.5
1999-05 (14.5)	19.0	19.9

	LABOR TIME	SEVERE SERVICE
Passport		
1994-97		
4 cyl.	18.6	18.6
V6	23.9	23.9
1998-02	24.3	24.3
Ridgeline (12.2)	16.3	17.2
w/4WD Passport add	1.2	1.2
w/AC add	.5	.5
w/AT Passport add	1.4	1.4
Engine Mounts, Replace (B)		
CR-V		
each (.5)	.8	1.0
all (1.5)	2.3	2.5
Element each (.5)	.8	1.0
Odyssey, Pilot		
1995-98		
front or side (.5)	.8	1.0
rear (1.7)	2.6	2.8
all (2.0)	3.0	3.2
1999-04		
front (1.7)	2.6	2.8
rear or side (.5)	.8	1.0
all (2.0)	3.0	3.2
Passport		
right side	1.4	1.6
left side	1.4	1.6
rear	.6	.8
Ridgeline		
front or rear (.8)	1.2	1.4
side left or right (.5)	.8	1.0
w/4WD Passport add	1.2	1.4

CYLINDER HEAD
Compression Test (B)

Exc. below (.6)	.9	1.0
Passport		
4 cyl.	.7	.8
V6	1.2	1.3

Valve Clearance, Adjust (B)

CR-V, Element, Pilot (.8)	1.3	1.5
Odyssey		
1995-98 (.8)	1.3	1.5
1999-04 (1.9)	2.7	2.9
2005 (2.2)	3.1	3.3
Passport		
1994-97	.8	1.1
1998-02 (4.8)	6.7	6.9
Ridgeline (2.1)	3.0	3.2
w/V6 add		
Element	1.2	1.2
Odyssey 99-04	2.7	2.7

Cylinder Head, Replace (B)
Includes: Parts transfer, adjustments.

CR-V, Element		
1997-01	6.1	6.6
2002-05 (6.7)	9.2	9.7

CR-V : ELEMENT : ODYSSEY : PASSPORT : PILOT : RIDGELINE — HON-29

	LABOR TIME	SEVERE SERVICE
Odyssey, Pilot		
1995-98	6.7	7.6
1999-04		
one side (6.0)	8.3	8.8
both sides (8.0)	10.8	12.3
2005		
one (5.4)	7.7	8.2
Passport		
1994-97		
4 cyl.	11.4	11.9
V6		
one side	12.1	12.6
both sides	17.0	17.5
1998-02		
one side	15.0	15.5
both sides	18.8	19.3
Ridgeline		
one (5.1)	7.3	7.8
both (6.9)	9.7	10.2
Replace balancer or timing belts		
exc. CR-V add	1.5	1.7
CR-V add	.8	.9
Cylinder Head Gasket, Replace (B)		
CR-V, Element		
1997-01	5.5	6.2
2002-05 (6.3)	8.6	9.1
Odyssey, Pilot		
1995-98	6.2	6.9
1999-04		
one side (5.5)	7.6	8.1
both sides (7.5)	10.2	10.7
2005		
one side (5.4)	7.7	8.2
one sides (7.7)	10.8	11.3
Passport		
1994-97		
4 cyl.	8.8	9.1
V6		
one side	10.2	10.5
both sides	12.6	12.9
1998-02		
one side	10.7	11.0
both sides	12.8	13.1
Ridgeline		
one (5.1)	7.3	7.8
both (6.9)	9.7	10.2
w/AC add	.5	.5
w/twin cam Pilot add	3.7	3.7
Rocker Arm Cover Gasket, Replace or Reseal (B)		
CR-V, Element (.3)	.5	.6
Odyssey, Pilot		
1995-98	.5	.6
1999-04		
one (.8)	1.2	1.3
both (1.1)	1.7	1.9
2005		
one (1.1)	1.7	1.8
both (1.5)	2.2	2.4

	LABOR TIME	SEVERE SERVICE
Passport		
1994-97		
4 cyl.	.7	.8
V6		
right side	4.0	4.1
left side	1.8	1.9
both sides	5.5	5.7
1998-02		
one side	2.1	2.2
both sides	3.8	4.0
Ridgeline		
front (.6)	.9	1.0
rear (1.0)	1.5	1.6
both (1.4)	2.0	2.2
Rocker Arms or Shafts, Replace (B)		
CR-V (2.5)	3.6	3.8
Odyssey		
1995-98	4.1	4.3
1999-04 (4.2)	6.7	6.9
Passport		
1994-97		
4 cyl.	1.4	1.6
V6 one		
right side	7.9	8.1
left side	5.2	5.4
Pilot (4.2)	6.7	6.9
Replace rocker arms add	.3	.5
Valve Job (A)		
CR-V, Element		
1997-01	9.9	10.8
2002-05 (9.2)	12.7	13.6
Odyssey		
1995-98	10.7	12.0
1999-04 (10.3)	14.1	15.0
Passport		
1994-97		
4 cyl.	12.5	13.0
V6	18.5	19.0
1998-02	23.6	24.1
Pilot (10.3)	14.1	15.0
Ridgeline (6.6)	9.4	10.3
w/AC add	.5	.5
Valve Lifters, Replace (B)		
1994-97 Passport V6		
one cyl.		
right side	7.6	7.8
left side	5.0	5.2
all cyls.	9.4	9.6
Valve Stem Oil Seal, Replace (B)		
CR-V	5.0	5.2
1999-01 Odyssey		
one side	11.4	11.6
both side	13.7	13.9
Passport		
4 cyl.		
one cyl.	3.1	3.3
all cyls.	7.6	7.8

	LABOR TIME	SEVERE SERVICE
V6		
one cyl.	8.7	8.9
all cyls.	11.0	11.2
Ridgeline all (4.6)	6.4	6.6
CAMSHAFT		
Camshaft, Replace (B)		
CR-V, Element (2.5)	3.6	3.8
Odyssey		
1995-98	4.1	4.3
1999-04 (4.2)	6.7	6.9
Passport		
1994-97		
4 cyl.	2.4	2.7
V6		
one side	7.8	8.1
both sides	9.4	9.7
1998-02		
one side	5.4	5.7
both sides	8.7	9.0
Pilot (4.2)	6.7	6.9
Camshaft Seal, Replace (B)		
CR-V		
one	3.3	3.8
both	3.9	4.1
Odyssey, Pilot		
1995-98	3.8	4.3
1999-03		
one	5.0	5.7
both	5.6	6.4
Passport		
one side	2.7	2.8
both sides	3.3	3.4
Timing and Balance Shaft Belt, Replace (B)		
1997-01 CR-V (2.0)	3.0	3.5
Odyssey		
1995-98	3.5	4.0
1999-04 (3.0)	4.6	5.2
2005 (2.1)	3.1	3.6
Passport		
1994-97		
4 cyl.	3.0	3.5
V6	3.7	4.2
1998-02	3.7	4.2
Pilot (3.0)	4.6	5.2
Ridgeline (2.1)	3.1	3.6
w/AC Passport add	.5	.5
w/twin cam CR-V add	.3	.3
Replace water pump add		
CR-V	.5	.5
95-98 Odyssey	.5	.5
99-05 Odyssey, Pilot	.7	.7

HON-30 CR-V : ELEMENT : ODYSSEY : PASSPORT : PILOT : RIDGELINE

	LABOR TIME	SEVERE SERVICE
Timing Belt/Chain Cover, Replace (B)		
Odyssey (.3)	.5	.7
Passport		
1994-97		
4 cyl.	1.8	2.0
V6	3.0	3.2
1998-02	3.0	3.2
Ridgeline (.4)	.6	.8
Timing Chain, Replace (B)		
CR-V, Element		
2002-05 (2.8)	4.0	4.5
Timing Chain Tensioner, Replace (B)		
CR-V, Element		
2002-05 (.4)	.6	.7
CRANK & PISTONS		
Connecting Rod Bearings, Replace (A)		
Includes: Check bearing oil clearance.		
CR-V		
1997-01 (1.7)	2.6	3.1
2002-05 (2.2)	3.3	3.8
Element (2.2)	3.3	3.8
Odyssey		
1995-98 (1.7)	2.6	3.1
1999-04 (2.3)	3.5	4.0
Passport		
1994-97		
4 cyl.	3.2	3.7
V6	6.4	6.9
1998-02 (7.0)	9.9	10.4
Pilot (2.3)	3.5	4.0
w/4WD Passport add	1.2	1.2
w/AC add	.5	.5
w/AT Passport add	1.4	1.4
Crankshaft and Main Bearings, Replace (A)		
Includes: Engine R&R.		
CR-V		
1997-01 (7.4)	11.4	11.9
2002-05 (11.7)	17.8	18.3
Element (11.7)	17.8	18.3
Odyssey		
1995-98 (8.6)	13.1	13.6
1999-04 (16.9)	22.0	22.5
2005 (13.4)	17.8	18.3
Passport		
1994-97		
4 cyl.	11.9	12.4
V6	19.7	20.2
1998-02 (13.1)	17.4	17.9
Pilot (17.4)	22.6	23.1
Ridgeline (14.4)	18.9	19.4
w/4WD Passport add	1.2	1.2
w/AC add	.5	.5
w/AT Passport add	1.4	1.4

	LABOR TIME	SEVERE SERVICE
Crankshaft Front Oil Seal, Replace (A)		
CR-V, Element (2.0)	3.0	3.3
Odyssey		
1995-98	3.8	4.1
1999-04 (3.0)	4.4	4.7
2005 (2.4)	3.5	3.8
Passport		
4 cyl.	3.3	3.6
V6	3.9	4.2
Pilot (3.3)	4.8	5.1
Ridgeline (2.4)	3.5	3.8
w/AC add	.5	.5
Crankshaft Pulley, Replace (B)		
Odyssey (.3)	.5	.6
w/AC add	.5	.5
Main Bearings, Replace (A)		
Includes: Bearing clearance check.		
CR-V		
1997-01 (2.2)	3.3	3.8
2002-05 (11.7)	15.8	16.3
Element (11.7)	15.8	16.3
Odyssey		
1995-98	3.3	3.8
1999-04 (2.4)	3.7	4.2
Passport (7.7)	10.8	11.3
Pilot (2.4)	3.7	4.2
w/4WD Passport add	1.5	1.5
w/AT Passport add	1.8	1.8
Pistons or Connecting Rods, Replace (A)		
Includes: Ridge reaming, cylinder wall deglazing, installing new rings and rod bearings, engine tune-up.		
CR-V		
1997-01	11.9	12.4
2002-05	16.0	16.5
Odyssey		
1995-98	12.8	13.5
1999-04	18.4	18.9
Passport		
1994-97		
4 cyl.	15.5	16.0
V6	24.1	24.6
1998-02	24.6	25.1
Pilot	18.8	19.3
w/AC add	.5	.5
w/AT Passport add	1.4	1.4
Rear Main Oil Seal, Replace (B)		
CR-V		
1997-01		
AT	5.9	6.4
MT	4.9	5.4
2002-05 (5.2)	7.5	8.0
Element (5.2)	7.5	8.0
Odyssey		
1995-98 (5.3)	7.6	8.1
1999-04 (6.3)	8.9	9.4
2005 (7.6)	10.6	11.1

	LABOR TIME	SEVERE SERVICE
Passport		
4 cyl.	6.4	6.9
V6	7.1	7.6
Pilot (6.9)	9.7	10.2
Ridgeline (7.3)	10.3	10.8
w/4WD add		
97-01 CR-V	2.3	2.6
02-05 CR-V	.8	.9
Element	.8	.9
Passport	1.5	1.5
w/AT add		
95-98 Odyssey	1.1	1.1
CR-V	1.3	1.3
Passport	1.9	1.9
Rings, Replace (A)		
Includes: Ridge reaming, cylinder wall deglazing, installing new rings, engine tune-up.		
CR-V		
1997-01	10.8	11.3
2002-05 (10.0)	13.7	14.2
Odyssey		
1995-98	12.0	12.5
1999-04 (11.3)	15.3	15.8
Passport		
1994-97		
4 cyl.	13.7	14.2
V6	22.3	22.8
1998-02	22.8	23.2
Pilot	17.6	18.1
w/AC add	.5	.5
w/AT Passport add	1.4	1.4
ENGINE LUBRICATION		
Engine Oil Pressure Switch (Sending Unit), Replace (B)		
Exc. below (.3)	.5	.5
Passport (.4)	.6	.6
2005 Odyssey, Ridgeline (.6)	.6	.6
Oil Pan and/or Gasket, Replace (B)		
CR-V		
1997-01	1.9	2.2
2002-05 (1.7)	2.5	2.8
Element (1.7)	2.5	2.8
Odyssey, Pilot, Ridgeline (1.3)	1.9	2.2
Passport		
4 cyl.	2.3	2.6
V6	6.1	6.4
w/4WD Passport add	1.2	1.2
Oil Pressure Gauge (Dash), Replace (B)		
1994-97 Passport	1.9	1.9
Oil Pump, Replace (B)		
CR-V		
1997-01 (3.5)	5.1	5.6
2002-05 (2.0)	3.0	3.5
Element (2.0)	3.0	3.5

CR-V : ELEMENT : ODYSSEY : PASSPORT : PILOT : RIDGELINE — HON-31

	LABOR TIME	SEVERE SERVICE
Odyssey		
1995-04 (4.5)	7.1	7.6
2005 (3.7)	5.4	5.9
Passport		
1994-97		
4 cyl.	2.0	2.5
V6	9.9	10.4
1998-02	10.7	11.2
Pilot (4.5)	6.5	7.0
Ridgeline (3.7)	5.4	5.9
w/4WD Passport add	1.2	1.2
Sub-Oil Pan or Gasket, Replace (B)		
Passport 4 cyl.	.8	1.0

CLUTCH

Bleed Clutch Hydraulic System (B)
All Models (.4) .... .6 / .7

Clutch Assy., Replace (B)
CR-V
  1997-01 (2.4) .... 3.5 / 4.0
  2002-05 (5.0) .... 7.2 / 7.7
Element (5.0) .... 7.2 / 7.7
Passport
  4 cyl. .... 5.5 / 5.7
  V6 .... 7.0 / 7.2
w/4WD add
  97-01 CR-V .... 3.0 / 3.0
  02-05 CR-V .... .8 / .9
  Passport .... 1.9 / 2.2

Clutch Master Cylinder, Replace (B)
Includes: System bleeding.
CR-V, Element (.8) ... 1.2 / 1.5
Passport (1.0) .... 1.5 / 1.8

Clutch Release Bearing, Replace (B)
CR-V
  1997-01 (2.1) .... 3.1 / 3.6
  2002-05 (4.7) .... 6.8 / 7.3
Element (4.7) .... 6.8 / 7.3
Passport
  4 cyl. .... 5.1 / 5.3
  V6 .... 6.5 / 6.7
w/4WD add
  97-01 CR-V .... 3.0 / 3.0
  02-05 CR-V, Element . .8 / .8
  Passport .... 1.9 / 1.9

Clutch Slave Cylinder, Replace (B)
Includes: System bleeding.
CR-V, Element, Passport (.8) ... 1.2 / 1.5

Flywheel, Replace (B)
CR-V
  1997-01 (2.6) .... 3.8 / 4.3
  2002-05 (5.2) .... 7.5 / 8.0
Element (5.2) .... 7.5 / 8.0
Passport
  4 cyl. .... 5.8 / 6.0
  V6 .... 7.3 / 7.5

---

w/4WD add
  97-01 CR-V .... 3.2 / 3.6
  02-05 CR-V, Element . .8 / .9
  Passport .... 1.9 / 2.2

MANUAL TRANSAXLE

Transaxle Assy., R&R and Recondition (A)
CR-V
  1997-01
    2WD .... 9.3 / 10.2
    4WD .... 11.7 / 12.6
  2002-05
    2WD (8.4) ...... 11.7 / 12.6
    4WD (8.9) ...... 12.4 / 13.3
Element
  2WD (8.4) .... 11.7 / 12.6
  4WD (8.9) .... 12.4 / 13.3
Recondition diff. add ... 2.4 / 2.7
Replace flywheel add ..... .3 / .3

Transaxle Assy., Replace (B)
CR-V
  1997-01
    2WD .... 3.2 / 3.6
    4WD .... 5.6 / 6.3
  2002-05
    2WD (4.6) ...... 6.6 / 7.1
    4WD (5.1) ...... 7.3 / 7.8
Element
  2WD (4.6) .... 6.6 / 7.1
  4WD (5.1) .... 7.3 / 7.8
Replace flywheel add ..... .3 / .3

MANUAL TRANSMISSION

Transmission Assy., R&I (B)
Passport
  4 cyl. .... 5.2 / 5.4
  V6 .... 6.5 / 6.7
w/4WD add .... 1.2 / 1.2

Transmission Assy., R&R and Recondition (A)
Passport
  4 cyl. .... 10.8 / 11.4
  V6 .... 12.6 / 13.2
w/4WD add .... 2.8 / 2.8

AUTOMATIC TRANSAXLE

SERVICE TRANSAXLE INSTALLED

Diagnose Electronic Transmission Component Each (A)
1994-06 (.6) .... .8 / .8

Drain & Refill Unit (B)
All Models .... .5 / .5

Throttle Valve Control Cable, Adjust (B)
1995-98 Odyssey ...... .3 / .4

Lock-Up Control Solenoid Valve, Replace (B)
Exc. below (.3) .... .5 / .6
2005 Odyssey (.7) ... 1.1 / 1.2
Ridgeline (1.7) .... 2.5 / 2.6

Shift Control Solenoid Valve, Replace (B)
All Models (.3) .... .5 / .5

Speed Sensor, Replace (B)
All Models (.3) .... .5 / .5

Throttle Valve Control Cable, Replace (B)
1995-98 Odyssey (.3) ... .5 / .6

Transmission Control Computer, Replace (B)
All Models (.4) .... .6 / .7

SERVICE TRANSAXLE REMOVED

Torque Converter, Replace (B)
CR-V
  1997-01
    2WD .... 4.8 / 5.3
    4WD .... 7.2 / 7.7
  2002-05
    2WD (5.2) ...... 7.5 / 8.0
    4WD (5.7) ...... 8.2 / 8.7
Element
  2WD (6.5) .... 9.2 / 9.7
  4WD (7.0) .... 9.9 / 10.4
Odyssey
  1995-98 (4.9) ...... 7.1 / 7.6
  1999-05 (6.8) ...... 9.6 / 10.1
Pilot (7.3) .... 10.4 / 10.9
Ridgeline (6.8) .... 9.6 / 10.1
w/V6 Element add .... .7 / .7

Transmission, Recondition (A)
Includes: Replace Torque converter and flush cooler lines.
CR-V, Element
  1997-01
    2WD .... 11.2 / 12.6
    4WD .... 13.6 / 15.3
  2002-05
    2WD (9.0) ...... 12.5 / 13.4
    4WD (9.5) ...... 13.1 / 14.0
Odyssey
  1995-98 (9.0) ...... 12.5 / 13.4
  1999-04 (9.8) ...... 13.4 / 14.3
  2005 (12.1) .... 16.2 / 17.1
Pilot (10.8) .... 14.7 / 15.6
Ridgeline (12.1) .... 16.2 / 17.1
Recondition diff. add ... 2.4 / 2.7

HON-32 CR-V : ELEMENT : ODYSSEY : PASSPORT : PILOT : RIDGELINE

	LABOR TIME	SEVERE SERVICE
Transmission and Torque Converter, R&R or Replace (B)		
Includes: Flush cooler lines.		
CR-V, Element		
1997-01		
2WD	5.9	6.4
4WD	8.3	8.8
2002-05		
2WD (5.2)	7.5	8.0
4WD (5.7)	8.2	8.7
Odyssey		
1995-98	8.2	8.7
1999-04 (6.0)	8.6	9.1
2005 (6.8)	9.6	10.1
Pilot (6.5)	9.2	9.7
Ridgeline (6.8)	9.6	10.1

AUTOMATIC TRANSMISSION
SERVICE TRANSMISSION INSTALLED

	LABOR TIME	SEVERE SERVICE
Shift Control Solenoid Valve, Replace (B)		
Passport one or all (1.2)	1.6	1.6
Shift Lock Cable, Replace (B)		
Passport (1.0)	1.4	1.5
Speed Sensor, Replace (B)		
Passport (.7)	1.0	1.0
Transmission Control Module, Replace (B)		
Passport		
1994-97 (.3)	.5	.5
1998-02 (.8)	1.1	1.1
Transmission Pan and/or Gasket, Replace (B)		
Passport (1.0)	1.4	1.4

SERVICE TRANSMISSION REMOVED

	LABOR TIME	SEVERE SERVICE
Transmission Assy., R&R or Replace (B)		
Passport		
2WD	11.9	12.1
4WD	13.4	13.6
Replace assembly add	.5	.7

TRANSFER CASE

	LABOR TIME	SEVERE SERVICE
Transfer Case, R&R and Recondition (B)		
CR-V, Element, Pilot (2.8)	4.1	5.0
Passport (4.3)	6.2	7.1
Replace add		
bearing and/or shaft	.7	.9
gear	.7	.9
shift shaft and/or fork	.8	1.0

	LABOR TIME	SEVERE SERVICE
Transfer Case, R&R or Replace (B)		
CR-V, Element, Pilot (.8)	1.3	1.8
Passport (3.1)	4.5	5.2
Ridgeline (1.2)	1.8	2.3
Transfer Case Oil Seal, Replace (B)		
Passport each (1.2)	1.9	2.1

SHIFT LINKAGE
AUTOMATIC

	LABOR TIME	SEVERE SERVICE
Shift Control Cable, Adjust (B)		
All Models (.5)	.8	.9
Gear Position Indicator, Replace (B)		
Exc. below (.5)	.7	.8
1996-02 Passport (.3)	.5	.6
Shift Control Cable, Replace (B)		
Exc. below (1.0)	1.6	1.7
2002-05 CR-V (.6)	.9	1.0
1998-02 Passport	1.0	1.0
Ridgeline (.8)	1.2	1.3
Gear Selector Lever, Replace (B)		
1995-01 CR-V, Odyssey (1.0)	1.5	1.7
2002-05		
exc. CR-V (1.0)	1.5	1.7
CR-V (.5)	.7	.8

MANUAL

	LABOR TIME	SEVERE SERVICE
Shift Control Cable, Adjust (B)		
CR-V, Element (.5)	.8	.9
Gear Shift Lever, Replace (B)		
CR-V, Element		
1997-01 (1.2)	1.8	2.1
2002-05 (.6)	.9	1.0
Passport (.6)	.8	.8
Shift Control Cable, Replace (B)		
CR-V, Element (1.0)	1.6	1.8

DRIVELINE
DRIVESHAFT

	LABOR TIME	SEVERE SERVICE
Center Support Bearing, Replace (B)		
1994-97 Passport (.6)	.9	1.0
Replace bearing or cushion add	.2	.2
Driveshaft, R&R or Replace (B)		
CR-V (.8)	1.1	1.2
Element, Pilot (.4)	.6	.7
Passport		
1994-97		
front	1.0	1.1
rear		
one	.7	.8
both	1.1	1.2
1998-02 single	1.0	1.1
Replace U-joint add each	.3	.3

DRIVE AXLE

	LABOR TIME	SEVERE SERVICE
Axle Shaft, Seal and/or Bearing, Replace (B)		
Rear axle		
Passport each	1.8	2.0
Axle Housing (Case), Replace (B)		
Rear axle		
Passport	9.7	10.2
Differential Assy., R&R or Replace (B)		
Front		
Passport	6.3	6.8
Rear		
CR-V, Element (1.5)	2.4	2.7
Passport	4.4	4.9
Pilot (.9)	1.4	1.6
w/Dana axle add	1.2	1.2
Differential Oil Seal, Replace (B)		
CR-V, Element, Pilot each (.7)	1.1	1.3
Pinion Shaft Oil Seal, Replace (B)		
CR-V, Element, Pilot (.7)	1.1	1.3
Passport		
front axle	1.4	1.6
rear axle	1.1	1.3

HALFSHAFTS

	LABOR TIME	SEVERE SERVICE
Halfshaft, R&R or Replace (B)		
Exc. Passport		
one (.7)	1.1	1.3
both		
exc. below (1.5)	2.2	2.4
2005 Odyssey, Ridgeline (1.2)	2.2	2.4
Passport		
one	5.2	5.4
both	10.0	10.4
Halfshaft Oil Seal, Replace (B)		
CR-V, Element, Odyssey, Pilot one side (.9)	1.4	1.6
Inboard CV Joint and/or Boot, Replace (B)		
Exc. below (1.1)	1.6	1.8
2005 Odyssey (1.2)	1.8	2.0
Passport (3.8)	5.5	5.7
Intermediate Shaft, Replace (B)		
CR-V, Element, Odyssey, Pilot (.9)	1.4	1.6
Replace bearing add	.5	.5
Outboard CV Joint and/or Boot, Replace (B)		
Exc. below (.9)	1.4	1.6
2005 Odyssey (1.2)	1.8	2.0
Passport (3.8)	5.5	5.7
Ridgeline (1.2)	1.8	2.0

CR-V : ELEMENT : ODYSSEY : PASSPORT : PILOT : RIDGELINE **HON-33**

BRAKES

ANTI-LOCK

Diagnose Anti-Lock Brake System Component Each (A)
	LABOR TIME	SEVERE SERVICE
1994-05 (.6)	.8	.8

Bleed ABS System (B)
1995-98 Odyssey	.8	.9

Retrieve Fault Codes (A)
1994-05 (.3)	.4	.4

Accumulator Assy., Replace (B)
1995-98 Odyssey (1.2)	1.9	2.2

Anti-Lock Valve, Replace (B)
1994-97 Passport	.8	.8
w/4WD add	.2	.2

Combination Valve, Replace (B)
1994-97 Passport	.8	.8

Control Module, Replace (B)
CR-V (.3)	.5	.6
1995-04 Odyssey (.3)	.5	.6
Pilot (1.0)	1.6	1.7

Exciter Ring, Replace (B)
Passport (2.3)	3.4	3.5

G Sensor, Replace (B)
1996-97 Passport	.8	.8

Low Pressure Hoses, Replace (B)
1995-98 Odyssey	1.4	1.6

Modulator or Modulator/Control Module, Replace (B)
Exc. Ridgeline (1.0)	1.6	1.8
Ridgeline (1.3)	1.9	2.0

Power Unit, Replace (B)
1995-98 Odyssey	1.9	2.2

Pressure Switch, Replace (B)
1995-98 Odyssey	1.6	1.8

Wheel Speed Sensor, Replace (B)
Exc. below one (.3)	.5	.6
Passport one (.5)	.8	.9
Ridgeline one (.6)	.9	1.0

SYSTEM

Bleed Brakes (B)
Includes: Add fluid.
All Models (.5)	.7	.8

Brake System, Flush and Refill (B)
All Models (.8)	1.2	1.2

Brakes, Adjust (B)
Includes: Refill master cylinder.
Rear wheels (.3)	.5	.5

Brake Adjuster, Replace (B)
Odyssey
one (.5)	.7	.8
both (.8)	1.2	1.3

Brake Hose (Flexible), Replace (B)
Includes: System bleeding.
CR-V
one (.6)	.9	1.0

Odyssey
1995-04
front
	LABOR TIME	SEVERE SERVICE
one	.9	1.0
both	1.6	1.7

rear
one	.7	.8

2005
front one (.7)	1.1	1.2
all (1.7)	2.5	2.6

Ridgeline
front one (.7)	1.1	1.2
all (1.7)	2.5	2.6
each addl. add	.3	.4

Master Cylinder, Replace (B)
Includes: System bleeding.
Exc. Passport (1.2)	1.7	1.9
Passport (.9)	1.2	1.4

Power Booster Unit, Replace (B)
Includes: System bleeding.
Exc. Passport (1.3)	1.9	2.1
Passport (1.1)	1.6	1.8

Power Booster Vacuum Check Valve, Replace (B)
All Models (.2)	.3	.3

SERVICE BRAKES

Brake Drum, Replace (B)
Exc. CR-V, Passport
one (.3)	.5	.6
both (.6)	1.0	1.1

CR-V
one (.6)	.9	1.0
both (.9)	1.3	1.4

Passport
one (.7)	1.1	1.2
both (1.2)	1.8	1.9

Brake Proportioning Valve, Replace (B)
Front (.6)	.9	1.0

Rear
1995-98 Odyssey (1.3)	2.1	2.3

Caliper Assy., R&R and Recondition (B)
Includes: System bleeding.
CR-V
front
one (1.3)	1.9	2.0
both (2.3)	3.4	3.6

rear
one (.8)	1.9	2.0
both (1.6)	3.4	3.6

Odyssey
1995-04
one (1.3)	1.9	2.0
both (2.3)	3.4	3.6

2005
front
	LABOR TIME	SEVERE SERVICE
one (1.5)	2.2	2.3
both (2.8)	5.0	5.2

rear
one (1.0)	1.5	1.6
both (1.5)	2.2	2.4

Passport
front
one (1.5)	2.2	2.3
both (2.7)	4.0	4.2

rear
one (1.2)	1.8	1.9
both (2.4)	3.5	3.7

Pilot
front or rear
one (.8)	1.2	1.3
both (1.6)	2.4	2.6

Ridgeline
front
one (1.5)	2.2	2.3
both (2.8)	5.0	5.2

rear
one (1.0)	1.5	1.6
both (1.5)	2.2	2.4

Caliper Assy., Replace (B)
Includes: System bleeding.
Exc. Passport
one (.5)	1.0	1.1
both (1.2)	1.9	2.1

Passport
one (1.0)	1.4	1.5
both (1.7)	2.5	2.6

Disc Brake Rotor, Replace (B)
Exc. below
one (.7)	1.1	1.2
both (1.4)	2.1	2.3

CR-V, Odyssey
front
one (1.4)	2.1	2.2
both (2.5)	3.7	3.9

Passport
front
one (1.4)	2.1	2.2
both (2.5)	3.7	3.9

rear
one (1.0)	1.5	1.6
both (1.7)	2.5	2.7

Pads and/or Shoes, Replace (B)
Includes: Service and parking brake adjustment. System bleeding.
Exc. below
front or rear
disc (.7)	1.1	1.3
rear drum (.9)	1.4	1.6

HON-34 CR-V : ELEMENT : ODYSSEY : PASSPORT : PILOT : RIDGELINE

	LABOR TIME	SEVERE SERVICE
Pads and/or Shoes, Replace (B)		
Odyssey, Passport		
front or rear		
disc (1.0)	1.5	1.6
rear drum (1.2)	1.8	1.9
COMBINATION ADD-ONS		
Resurface		
drum add	.5	.5
rotor add		
front	.9	.9
rear	.5	.5
Wheel Cylinder, Replace (B)		
Includes: System bleeding.		
CR-V, Passport		
one (1.1)	1.6	1.7
both (1.8)	2.7	2.9
Element		
one (.6)	1.0	1.2
both (1.2)	1.9	2.2
Odyssey		
1995-98		
one (.6)	1.0	1.2
both (1.2)	1.9	2.2
1999-05		
one (.8)	1.2	1.3
both (1.5)	2.2	2.4
PARKING BRAKE		
Parking Brake Cable, Adjust (C)		
All Models (.4)	.6	.8
Parking Brake Apply Actuator, Replace (B)		
CR-V		
1997-01 (.3)	.5	.5
2002-05 (.6)	1.0	1.1
Element (.3)	.5	.5
Odyssey		
1995-98 (.3)	.5	.5
1999-05 (.5)	.8	.9
Passport (.8)	1.2	1.3
Pilot (.8)	1.2	1.3
Ridgeline (.6)	.9	1.0
Parking Brake Apply Warning Indicator Switch, Replace (B)		
Exc. Passport (.6)	.8	.8
Passport (.3)	.5	.5
Parking Brake Cable, Replace (B)		
CR-V		
front (1.1)	1.8	2.0
rear		
one (1.0)	1.5	1.6
both (1.5)	2.2	2.4
Odyssey		
1995-04		
front (.6)	1.0	1.1
rear		
one (.6)	1.3	1.4
both (1.5)	2.2	2.4

	LABOR TIME	SEVERE SERVICE
2005		
front (.6)	1.0	1.1
rear		
left (1.6)	2.4	2.5
right (1.9)	2.8	2.9
both (2.8)	4.1	4.3
Passport rear each (1.5)	2.3	2.5
Pilot		
front (.6)	1.0	1.1
rear		
one (.6)	1.3	1.4
both (1.5)	2.2	2.4
Ridgeline		
front (1.1)	1.6	1.7
rear each (1.7)	2.5	2.6
both (2.3)	3.4	3.6
Parking Brake Shoes, Replace (B)		
Exc. Passport		
one side (.9)	1.3	1.4
both sides (1.6)	2.6	2.9
Passport		
one side (.8)	1.2	1.3
both sides (1.3)	1.9	2.1
FRONT SUSPENSION		
Unless otherwise noted, time given does not include alignment.		
Align Front End (A)		
All Models (.6)	1.0	1.1
Front & Rear Alignment, Check & Adjust (A)		
All models (1.2)	1.9	2.2
Front Axle Hub Oil Seal, Replace (B)		
Passport one (.6)	.8	1.1
Front Strut Coil Spring, Replace (B)		
CR-V, Odyssey		
one (.8)	1.3	1.4
both (1.6)	2.6	2.8
Element		
one (.5)	1.3	1.4
both (1.0)	2.6	2.8
Pilot		
one (.6)	.9	1.0
both (1.2)	1.9	2.1
Ridgeline one (.8)	1.2	1.3
Locking Hub, Replace or Recondition (B)		
1994-95 Passport	.7	1.0
Lower Ball Joint, Replace (B)		
CR-V		
one (.9)	1.3	1.4
both (1.8)	2.7	2.9
Element		
one (.8)	1.2	1.3
one (1.5)	2.2	2.4
1995-98 Odyssey		
one (.9)	1.3	1.4
both (1.8)	2.7	2.9
Passport one (.6)	1.0	1.3

	LABOR TIME	SEVERE SERVICE
Lower Control Arm (FWD), Replace (B)		
Exc. below		
one side (.6)	.9	1.0
both sides (1.2)	1.9	2.1
2002-05 CR-V, Element		
one side (.4)	.6	.7
both sides (.8)	1.2	1.4
2005 Odyssey		
one side (1.0)	1.5	1.7
both sides (1.8)	2.7	1.9
Ridgeline		
one side (.8)	1.2	1.3
both sides (1.4)	2.1	2.3
Lower Control Arm or Bushings (RWD), Replace (B)		
1994-97 Passport one	2.7	3.0
Replace bushings add	.3	.5
Radius Rod and/or Bushings, Replace (B)		
1995-98 Odyssey		
one side	1.0	1.1
both sides	1.9	2.2
Shock Absorbers, Replace (B)		
Passport		
one (.3)	.5	.6
both (.6)	.8	.9
Stabilizer Bar, Replace (B)		
Exc. below (.7)	1.1	1.2
2005 Odyssey (.8)	1.2	1.3
Ridgeline (.4)	.6	.7
Stabilizer Bar Bushing and/or Bracket, Replace (B)		
Exc. below one (.4)	.6	.7
Passport one (.2)	.3	.4
Ridgeline one (.7)	1.1	1.2
Steering Knuckle, Replace (B)		
Exc. below		
one side (1.1)	1.8	2.0
both sides (2.2)	3.5	4.0
2002-05 CR-V, Element		
one side (.8)	1.3	1.4
both sides (1.6)	2.6	2.9
2005 Odyssey, Ridgeline		
one side (1.6)	2.4	2.5
both sides (3.0)	4.4	4.6
Passport		
one side	2.1	2.4
both sides	4.1	4.4
Replace ball joint add each		
CR-V, Odyssey	.3	.5
Passport	1.2	1.4
Strut, R&R or Replace (B)		
CR-V, Element		
1997-01		
one	1.3	1.4
both	2.6	2.9
2002-05		
one (.5)	.8	.9
both (1.0)	1.6	1.8

CR-V : ELEMENT : ODYSSEY : PASSPORT : PILOT : RIDGELINE — HON-35

	LABOR TIME	SEVERE SERVICE
Odyssey		
1995-98		
one	1.3	1.4
both	2.6	2.9
1999-04		
one (.6)	1.0	1.1
both (1.2)	1.9	2.2
2005		
one (1.2)	1.8	1.9
both (2.2)	3.3	3.5
Pilot		
one (.6)	1.0	1.1
both (1.2)	1.9	2.2
Ridgeline		
one (1.2)	1.8	1.9
both (2.2)	3.3	3.5
Torsion Bar, Replace (B)		
Passport one (.6)	1.0	1.2
w/skid plate add	.2	.2
Upper Ball Joint, Replace (B)		
Passport one (.5)	.8	.9
Upper Control Arm (FWD), Replace (B)		
Includes: Reset toe.		
1997-01 CR-V		
one side (.6)	1.0	1.1
both sides (1.2)	1.9	2.2
1995-98 Odyssey		
one side (.6)	1.0	1.1
both sides (1.2)	1.9	2.2
Upper Control Arm or Bushings (RWD), Replace (B)		
Passport one	3.0	3.3
Replace bushings add	.5	.7
Wheel Bearing & Hub Assy., Replace (B)		
Exc. below		
one side	1.8	2.0
both sides	3.5	4.0
2002-05 CR-V, Element		
one side (.8)	1.3	1.4
both sides (1.6)	2.6	2.9
2005 Odyssey, Ridgeline		
one side (1.5)	2.2	2.3
both sides (2.9)	4.3	4.5
Wheel Bearings, Replace (B)		
Passport		
one side	1.4	1.4
both sides	2.7	2.7
w/4WD add each	.3	.3

REAR SUSPENSION

	LABOR TIME	SEVERE SERVICE
Rear Toe, Check and Adjust (B)		
All Models (.9)	1.3	1.5
Coil Spring, R&R or Replace (B)		
CR-V, Element		
1997-01		
one	1.3	1.4
both	2.6	2.9
2002-05		
one (.5)	.8	.9
both (1.0)	1.6	1.8
Odyssey, Pilot		
1995-98		
one	1.0	1.1
both	1.9	2.2
1999-05		
one (.5)	.8	.9
both (1.0)	1.6	1.8
Passport		
1998-02 one or both	.7	.9
Ridgeline		
one (.6)	.9	1.0
both (1.1)	1.7	1.9
Lateral Link, Replace (B)		
1998-02 Passport	.6	.8
Replace bushings add	.1	.3
Lower "A" or "B" Control Arm or Bushings, Replace (B)		
CR-V, Element		
1997-01		
one	.6	.7
both	1.3	1.4
2002-05		
one (.9)	1.4	1.6
both (1.8)	2.9	3.2
Odyssey		
1995-98		
one side	.8	.9
both sides	1.6	1.8
1999-04		
one side (.4)	.6	.9
both sides (.8)	1.3	1.4
2005		
one side (.9)	.6	.9
both sides (1.6)	1.3	1.4
Pilot		
one side (.4)	.6	.9
both sides (.8)	1.3	1.4
Ridgeline		
one side (.9)	.6	.9
both sides (1.6)	1.3	1.4
Rear Compensator Arm or Bushings, Replace (B)		
1997-01 CR-V		
one side	.6	.9
both sides	1.3	1.4
Rear Cradle or Crossmember, Replace (B)		
1999-04 Odyssey (2.0)	3.0	3.5
Pilot (3.0)	4.4	4.9
Ridgeline (3.5)	5.1	5.6

	LABOR TIME	SEVERE SERVICE
Rear Hub and/or Bearings, Replace (B)		
CR-V		
1997-01		
2WD		
one side	1.0	1.1
both sides	1.9	2.1
4WD		
one side	1.8	1.9
both sides	3.5	3.7
2002-05		
one side (.6)	1.0	1.1
both sides (1.2)	1.8	2.0
Element, Odyssey, Pilot		
one side (.7)	1.1	1.1
both sides (1.3)	1.9	2.1
Ridgeline		
one side (2.2)	3.3	3.4
both sides (4.3)	6.2	6.4
Rear Knuckle, Replace (B)		
CR-V, 1995-04 Odyssey, Pilot		
one side (.9)	1.4	1.6
both sides (1.8)	2.9	3.2
Rear Leaf Spring, Replace (B)		
1994-97 Passport one	.8	1.2
Replace bushings add each	.3	.5
Rear Strut, Replace (B)		
CR-V, Element		
1997-01		
one	1.3	1.4
both	2.6	2.9
2002-05		
one (.5)	.8	.9
both (1.0)	1.6	1.8
Odyssey		
1995-98		
one	.8	.9
both	1.6	1.8
1999-04		
one (.3)	.5	.6
both (.6)	1.0	1.1
2005		
one (.7)	1.1	1.2
both (1.3)	1.9	2.1
Pilot		
one (.3)	.5	.6
both (.6)	1.0	1.1
Ridgeline		
one (1.0)	1.5	1.6
both (1.9)	2.8	3.0
Rear Trailing Arm or Bushings, Replace (B)		
CR-V		
one side (.6)	1.0	1.1
both sides (1.2)	1.9	2.2

HON-36 CR-V : ELEMENT : ODYSSEY : PASSPORT : PILOT : RIDGELINE

	LABOR TIME	SEVERE SERVICE
Rear Trailing Arm or Bushings, Replace (B)		
Odyssey		
1995-04		
one side (.6)	1.0	1.1
both sides (1.2)	1.9	2.2
2005		
one side (2.0)	3.0	3.1
both sides (3.8)	5.5	5.7
Pilot		
one side (.6)	1.0	1.1
both sides (1.2)	1.9	2.2
Ridgeline		
one side (2.0)	3.0	3.1
both sides (3.8)	5.5	5.7
Shock Absorbers or Bushings, Replace (B)		
Passport		
one	.5	.6
both	.8	.9
Spring Shackle, Replace (B)		
1994-97 Passport	.7	.9
Replace bushings add	.2	.4
Stabilizer Bar, Replace (B)		
Exc. below (.7)	1.1	1.3
2002-05 CR-V, Element (.4)	.6	.8
Ridgeline (1.8)	2.7	2.9
Stabilizer Bar Bushings and or Link, Replace (B)		
Exc. Ridgeline (.4)	.6	.7
Ridgeline		
one (.6)	.9	1.0
both (1.0)	1.5	1.7
Stabilizer Bar Bushing Holder, Replace (B)		
Ridgeline		
one or both (1.7)	2.5	2.7
Toe Adjusting Bar, Replace (B)		
Add for alignment.		
Odyssey		
one side	1.2	1.5
both sides	2.0	2.3
Replace bushings add each	.5	.7
Trailing Link, Replace (B)		
1998-99 Passport		
upper one side	.7	1.0
lower one side	.6	.9
Replace bushings add each	.1	.3
Upper Control Arm or Bushings, Replace (B)		
One side (.4)	.6	.9
Both sides (.8)	1.3	1.4

	LABOR TIME	SEVERE SERVICE
STEERING		
AIR BAGS		
The following operations do not include testing. Add time as required.		
Diagnose Air Bag System Component Each (A)		
1994-05 (.6)	.8	.8
Retrieve Fault Codes (A)		
1994-05 (.3)	.4	.4
Air Bag Assy., Replace (B)		
Driver side (.3)	.5	.5
Passenger side		
exc. below (.5)	.8	.8
1998-02 Passport	2.1	2.1
Side (.6)	.9	.9
Cable Reel, Replace (B)		
All Models (.9)	1.2	1.2
Electronic Control Unit, Replace (B)		
Exc. Ridgeline (.6)	.8	.8
Ridgeline (.8)	1.0	1.0
Sensing & Diagnostic Module (SDM), Replace (B)		
Passport	1.1	1.1
POWER RACK & PINION		
Troubleshoot Power Steering System (A)		
1995-06 (.4)	.6	.6
Pump Pressure Check (B)		
All Models (.3)	.5	.5
Rack & Pinion Assy., Adjust (B)		
Exc. below (.5)	.8	.9
1998-02 Passport (.7)	1.1	1.2
Power Steering Reservoir, Replace (B)		
All Models (.4)	.6	.6
Rack & Pinion Assy., R&R or Replace (B)		
Includes: Reset toe.		
CR-V		
1997-01	4.1	4.5
2002-05 (2.6)	3.8	4.2
Element (2.0)	3.0	3.4
Odyssey		
1995-98	4.1	4.5
1999-04 (3.0)	4.4	4.8
2005 (4.2)	6.1	6.5
1998-02 Passport		
2WD	2.8	3.2
4WD	4.4	4.8
Pilot (3.5)	5.3	5.7
Ridgeline (4.2)	6.1	6.5
w/4WD Passport add	1.8	1.8
Rack & Pinion Assy., Reseal (A)		
Includes: Alignment.		
CR-V, Element (3.2)	4.8	5.7
Odyssey, Pilot		
1995-98 (3.2)	4.8	5.7
1999-04 (3.5)	5.1	6.0

	LABOR TIME	SEVERE SERVICE
Steering Column, Replace (B)		
CR-V, Element		
1997-01		
AT	2.0	2.0
MT	1.5	1.5
2002-05 (.6)	.8	.8
Odyssey		
1995-04 (1.4)	1.9	1.9
2005 (1.6)	2.3	2.3
Passport (1.0)	1.4	1.4
Pilot (1.1)	1.5	1.5
w/AT add		
CR-V	.3	.3
Element	.9	.9
95-04 Odyssey	.3	.3
Steering Flex Coupling, Replace (B)		
Exc. Element (.5)	.8	.9
Element (.7)	1.1	1.2
Steering Pump, R&R and Recondition (B)		
CR-V		
1997-01 (1.0)	1.4	1.9
2002-05 (.8)	1.2	1.7
Element (.5)	.7	1.2
Odyssey		
1995-98 (1.0)	1.4	1.9
1999-04 (.8)	1.2	1.7
Passport	1.6	2.0
Pilot (.8)	1.2	1.7
Steering Pump, R&R or Replace (B)		
CR-V (.6)	.5	.5
Element, Odyssey, Pilot (.4)	.6	.8
Passport, Ridgeline (.8)	1.1	1.3
Steering Pump Hoses, Replace (B)		
Does not include purging.		
Pressure		
CR-V (.6)	.9	1.0
Odyssey, Pilot		
1995-98	.9	1.0
1999-05 (.9)	1.4	1.5
Passport (.3)	.5	.6
Ridgeline (3.6)	5.1	5.2
Return		
exc. below (.5)	.8	.9
2005 Odyssey (.7)	1.1	1.2
Passport (.2)	.3	.4
Ridgeline (3.6)	5.1	5.2
Steering Wheel, Replace (B)		
All Models (.5)	.7	.7
Tie Rod, Tie Rod End and/or Boot, Replace (B)		
Includes: Reset toe.		
Exc. below		
one (.8)	1.2	1.3
both (1.0)	1.5	1.7
2005 Odyssey, Ridgeline		
one (.5)	.7	.8
both (.8)	1.2	1.4

CR-V : ELEMENT : ODYSSEY : PASSPORT : PILOT : RIDGELINE — HON-37

	LABOR TIME	SEVERE SERVICE
1998-02 Passport		
one	.7	.9
both	1.0	1.2

POWER WORM & SECTOR

Troubleshoot Power Steering System (A)
- 1994-97 Passport 5   .5

Steering Gear, Adjust (B)
- 1994-97 Passport 1.2   1.5

Flexible Coupling, Replace (B)
- 1994-97 Passport 6   .8

Gear Assy., R&R and Recondition (B)
- 1994-97 Passport
  - 2WD 2.8   3.3
  - 4WD 4.4   4.9

Gear Assy., Replace (B)
- 1994-97 Passport 1.9   2.2

Idler Arm, Replace (B)
- 1994-97 Passport 7   1.0

Pitman Arm, Replace (B)
- 1994-97 Passport 7   1.0

Relay Rod (Center Link), Replace (B)
Includes: Reset toe.
- 1994-97 Passport 1.6   1.9

Steering Column, R&R or Replace (B)
- 1994-97 Passport 1.4   1.4

Steering Pump, R&R or Replace (B)
- 1994-97 Passport 1.1   1.2
- *Reseal pump add* 5   .8

Steering Pump Hoses, Replace (B)
Does not include purging.
- 1994-97 Passport
  - pressure 5   .6
  - return 3   .4

Steering Wheel, Replace (B)
- 1994-97 Passport 7   .7

Tie Rod, End and/or Boot, Replace (B)
Includes: Reset toe.
- 1994-97 Passport
  - one side 7   .9
  - both sides 1.0   1.2

HEATING & AIR CONDTIONING

When more than one component requires replacement where evacuation, recovery and recharging is already included, deduct 1.0 hour for each additional component from the time given.

Evacuate, Recover and Recharge System (B)
- All Models (.7) 1.0   1.0

AC Hoses, Replace (B)
Includes: Evacuate, recover and recharge.
- Exc. below
  - each (1.1) 1.6   1.8
  - *each addl. add* 1.6   1.8
- 2002-05 CR-V
  - suction (1.5) 2.2   2.4
  - discharge (1.7) 2.4   2.6
- Element
  - suction (2.0) 2.9   3.1
  - discharge (2.1) 3.0   3.2
- 2005 Odyssey
  - suction (2.2) 3.1   3.3
  - discharge (2.4) 3.4   3.6
- Passport
  - high pressure
    - 4 cyl. 2.8   2.9
    - V6
      - 1994-97 ...... 2.7   2.8
      - 1998-02 ...... 2.2   2.3
  - low pressure
    - (liquid line) 3.8   3.9
    - suction or discharge 2.2   2.3
- Ridgeline
  - suction (2.3) 3.3   3.5
  - discharge (2.2) 3.1   3.3

AC On, Off Control Switch, Replace (B)
- CR-V
  - 1997-01 (1.0) 1.4   1.4
  - 2002-05 (.3) 4   .4
- Element, Pilot, Ridgeline (.6) 8   .8
- Odyssey
  - 1995-98 (1.0) 1.4   1.4
  - 1999-05 (.4) 6   .6
- Passport (1.0) 1.4   1.4

Blower Motor, Replace (B)
- Front
  - exc. below (.3) 5   .5
  - 2005 Odyssey, Ridgeline (.4) 6   .6
- Rear
  - Odyssey
    - 1995-98 (.6) 8   .8
    - 1999-05 (2.0) 2.7   2.7
  - Pilot (.7) 9   .9

Blower Motor Resistor, Replace (B)
- Front (.5) 7   .7
- Rear
  - Odyssey
    - 1995-98 (.4) 5   .5
    - 1999-04 (1.5) 1.9   1.9
  - Pilot (.6) 8   .8

Blower Motor Switch, Replace (B)
- CR-V
  - 1997-01 (.6) 8   .8
  - 2002-05 (.3) 5   .5

	LABOR TIME	SEVERE SERVICE
1995-98 Odyssey		
front (1.3)	1.4	1.4
rear (.2)	.3	.3
Passport (.7)	1.0	1.0

Compressor Assy., Replace (B)
Includes: Parts transfer, evacuate, recover and recharge.
- CR-V
  - 1997-01 (2.0) 2.9   3.1
  - 2002-05 (3.2) 4.5   4.7
- Odyssey
  - 1995-04 (2.0) 2.9   3.1
  - 2005 (2.8) 4.0   4.2
- Passport
  - 1994-95 (2.0) 2.9   3.1
  - 1996-02 (2.4) 3.4   3.6
- Pilot (1.7) 2.4   2.6
- Ridgeline (2.8) 4.0   4.2
- *Transfer clutch assy. add* 5   .7

Compressor or Fan Relay Replace (B)
- All Models (.2) 3   .3

Condenser Assy., Replace (B)
Includes: Evacuate, recover and recharge.
- CR-V
  - 1997-01 (1.2) 1.7   1.9
  - 2002-05 (1.5) 2.2   2.4
- Element, Ridgeline (2.2) 3.1   3.3
- Odyssey
  - 1995-04 (1.3) 1.9   2.1
  - 2005 (2.1) 3.0   3.2
- Passport (1.8) 2.6   2.8
- Pilot (1.3) 1.9   2.1

Evaporator Coil, Replace (B)
Includes: Evacuate, recover and recharge.
- CR-V (1.7) 2.4   2.6
- Element (1.8) 2.6   2.8
- Odyssey
  - front
    - 1995-04 (1.7) 2.4   2.6
    - 2005 (2.8) 4.0   4.2
  - rear
    - 1995-98 (1.2) 1.7   1.9
    - 1999-04 (2.7) 3.8   4.0
- Passport (5.2) 7.2   7.4
- Pilot
  - front (1.7) 2.3   2.5
  - rear (3.7) 4.3   4.5
- Ridgeline (2.8) 4.0   4.2

Expansion Valve, Replace (B)
Includes: Evacuate, recover and recharge.
- CR-V (1.7) 2.4   2.6

HON-38 CR-V : ELEMENT : ODYSSEY : PASSPORT : PILOT : RIDGELINE

	LABOR TIME	SEVERE SERVICE
Expansion Valve, Replace (B)		
Odyssey		
front (1.7) **2.4**		2.6
rear		
1995-98 (1.3) **1.9**		2.1
1999-05 (2.7) **3.8**		4.0
Passport (4.5) **7.2**		7.4
Pilot		
front (1.7) **2.4**		2.6
rear (1.0) **1.4**		1.6
Ridgeline (2.8) **4.0**		4.2
Heater & AC Mode Control Motor, Replace (B)		
CR-V (.8) **1.0**		1.0
Element (.9) **1.1**		1.1
Odyssey		
front (.6) .... **.8**		.8
rear heater		
1999-04 (2.0) **2.5**		2.5
Pilot		
front (.3) .... **.4**		.4
rear heater (.6) .... **.8**		.8
Ridgeline (.5) .... **.6**		.6
Heater Core, R&R or Replace (B)		
CR-V, Element		
front		
1997-01 (4.5) **6.3**		6.5
2002-05 (5.0) **7.0**		7.2
rear		
1997-05 (2.0) **3.0**		3.2
Odyssey		
1995-98 front (4.5) .. **6.3**		6.5
1999-05		
front (6.0) **8.3**		8.5
rear (2.0) **2.9**		3.1
Passport		
w/AC (4.5) **6.3**		6.5
w/o AC (4.5) **6.3**		6.5
Pilot		
front (6.0) **8.3**		8.5
rear (.9) **1.4**		1.5
Ridgeline (4.7) **6.5**		6.7
Add time to evacuate & recharge if necessary.		
Heater Hoses, Replace (B)		
Each (.6) .... **.8**		.9
Heater Water Shut-Off Valve, Replace (B)		
All Models (.6) .... **.8**		.9
Pressure Cycling Switch, Replace (B)		
Exc. below (.2) .... **.3**		.3
1999-04 Odyssey (.6) ... **.8**		.8
1996-02 Passport (.7) . **1.0**		1.0
Receiver/Drier Assy., Replace (B)		
Includes: Evacuate, recover and recharge.		
CR-V		
1997-01 (1.2) **1.6**		1.7
2002-05 (1.5) **2.2**		2.3

	LABOR TIME	SEVERE SERVICE
Element (1.2) **1.6**		1.7
Odyssey		
1995-98 (1.3) **1.9**		2.0
1999-04 (1.5) **2.2**		2.3
Passport (1.7) **2.4**		2.5
Pilot (1.5) **2.2**		2.3
Temperature Control Assy., Replace (B)		
CR-V		
1997-01 (1.0) **1.4**		1.4
2002-05 (.3) .... **.4**		.4
Element, Ridgeline (.5) .. **.7**		.7
Odyssey		
1995-98 (1.0) **1.4**		1.4
1999-05 (.4) .... **.5**		.5
Passport (1.0) **1.4**		1.4
Pilot (.6) .... **.8**		.8
Thermostat/Temperature sensor, Replace (B)		
Includes: Evacuate, recover and recharge.		
Exc. Passport (1.0) ... **1.4**		1.4
Passport (5.2) **7.2**		7.4
## WIPERS & SPEEDOMETER		
Antenna Assy., Replace (B)		
Exc. below (.4) .... **.6**		.6
2002-05 CR-V,		
Element (1.0) **1.4**		1.4
Instrument Panel Cluster (IPC), Replace (B)		
2002-05 CR-V, Element		
Pilot (.3) .... **.5**		.5
2005 Odyssey (.7) ... **1.0**		1.0
Intermittent Wiper Control Unit, Replace (B)		
All Models (.5) .... **.7**		.7
Intermittent Wiper Relay, Replace (B)		
All Models (.2) .... **.3**		.3
Radio, R&R (B)		
CR-V, Element		
1997-01 (1.0) **1.4**		1.4
2002-05 (.3) .... **.5**		.5
Odyssey, Pilot		
1995-98 (1.0) **1.4**		1.4
1999-05 (.4) .... **.6**		.6
Passport (1.0) **1.4**		1.4
Ridgeline (.5) .... **.7**		.7
w/Navi Pilot add .... **.3**		.3
Rear Window Wiper Motor, Replace (B)		
Exc. Element (.6) .... **.9**		1.0
Element (.4) .... **.6**		.7
Rear Window Wiper Switch, Replace (B)		
Passport **1.3**		1.3

	LABOR TIME	SEVERE SERVICE
Speedometer Head, R&R or Replace (B)		
CR-V (1.0) **1.4**		1.4
Odyssey		
1995-98 (1.0) **1.4**		1.4
1999-04 (.5) .... **.7**		.7
Passport (1.2) **1.6**		1.6
Washer Motor, Replace (B)		
CR-V, Element, Odyssey,		
Pilot each (.6) .... **.9**		1.0
Passport each		
1994-97 (.2) .... **.3**		.3
1998-02 (.4) .... **.6**		.6
Ridgeline (.4) .... **.6**		.6
Windshield Wiper Motor, Replace (B)		
CR-V, Element,		
Passport (.6) .... **.9**		1.0
Odyssey		
1995-04 (.9) **1.4**		1.5
2005 (.6) .... **.9**		1.0
Pilot (.9) **1.4**		1.5
Ridgeline (.7) **1.0**		1.1
Windshield Wiper Switch, Replace (B)		
All Models (.7) **1.0**		1.0
Windshield Wiper Transmission, Replace (B)		
Exc. below (.7) **1.0**		1.1
Odyssey		
1995-98 (1.2) **1.8**		2.1
2005 (.5) .... **.7**		.8
Passport (1.2) **1.8**		2.1
## LAMPS & SWITCHES		
Back-Up Lamp Bulb, Replace (C)		
Each (.2) .... **.3**		.3
Back-Up Lamp Switch, Replace (B)		
CR-V, Element (.3) ..... **.5**		.5
Passport (.5) .... **.7**		.7
Brake Lamp Switch, Replace (B)		
All Models (.3) .... **.5**		.5
Combination Lamp Switch, Replace (B)		
Exc. below (.5) .... **.7**		.7
1997-01 CR-V (.7) ... **1.0**		1.0
1995-04 Odyssey (.7) . **1.0**		1.0
Passport (.7) **1.0**		1.0
Clutch Start Switch, Replace (B)		
CR-V, Element,		
Passport (.3) .... **.5**		.5
Halogen Headlamp Bulb, Replace (B)		
All Models (.2) .... **.3**		.3
Hazard Warning or Turn Signal Flasher, Replace (B)		
All Models (.2) .... **.3**		.3
Hazard Warning or Turn Signal Relay, Replace (B)		
All Models (.2) .... **.3**		.3

CR-V : ELEMENT : ODYSSEY : PASSPORT : PILOT : RIDGELINE   **HON-39**

	LABOR TIME	SEVERE SERVICE
Hazard Warning Switch, Replace (B)		
Exc. below (.3)	.4	.4
Passport (.8)	1.0	1.0
Ridgeline (.5)	.6	.6
Headlamp Switch, Replace (B)		
Exc. Passport (.8)	1.0	1.0
1994-95 Passport (.5)	.7	.7
Headlamps, Aim (B)		
Exc. below		
one side (.3)	.4	.4
both sides (.5)	.7	.7
High Mount Stop Lamp Assy., Replace (B)		
All Models (.3)	.5	.5
High Mount Stop Lamp Bulb, Replace (C)		
All Models (.2)	.3	.3
Horn, Replace (B)		
Exc. below (.4)	.6	.6
1999-04 Odyssey (.8)	1.2	1.2
1996-02 Passport (.2)	.3	.3
each addl. add	.2	.2
License Lamp Lens or Bulb, Replace (B)		
Exc. Ridgeline one or all (.2)	.3	.3
Ridgeline one or all (.5)	.6	.6
Neutral Safety Switch, Replace (B)		
Passport	.7	.7

	LABOR TIME	SEVERE SERVICE
Park & Turn Signal Lamp Bulb or Lens, Replace (C)		
Each (.2)	.3	.3
Parking Brake Indicator Switch, Replace (B)		
Exc. below (.3)	.4	.4
2002-05 CR-V (.5)	.7	.7
Passport (.6)	.8	.8
Parking Lamp Lens or Bulb, Replace (C)		
Each (.2)	.3	.3
Side Marker Lamp Lens or Bulb, Replace (C)		
Each (.2)	.3	.3
Tail Lamp Lens or Bulb, Replace (C)		
Each (.4)	.6	.6
Turn Signal & Headlamp Dimmer Switch, Replace (B)		
All Models (.7)	1.0	1.0

BODY

Door Latch, Replace (B)		
CR-V, Element (.5)	.7	.7
Odyssey, Passport (.8)	1.1	1.1
Pilot		
front (.7)	.9	.9
rear (.4)	.6	.6
Ridgeline (.8)	1.1	1.1

	LABOR TIME	SEVERE SERVICE
Door Window Regulator, Replace (B)		
Exc. below (1.0)	1.3	1.4
CR-V (.7)	1.1	1.2
2005 Odyssey rear (1.3)	1.6	1.7
Gas Tank Lid Release Cable, Replace (B)		
Exc. Ridgeline (1.0)	1.5	1.6
Ridgeline (2.2)	3.1	3.2
Hood Lock, Replace (B)		
Exc. Ridgeline (.3)	.5	.5
Ridgeline (.6)	.9	.9
Hood Release Cable, Replace (B)		
Exc. below (.8)	1.2	1.4
2002-05 CR-V, Passport (.6)	.8	.8
Ridgeline (.9)	1.3	1.3
Lock Striker Plate, Replace (B)		
All Models (.3)	.5	.5
Window Regulator Motor, Replace (B)		
CR-V		
1997-01		
front (.7)	1.1	1.2
rear (.6)	.9	1.0
2002-05 each (.5)	1.1	1.2
Odyssey, Pilot (1.0)	1.5	1.7
Passport one (1.0)	1.4	1.6

NOTES

HYU

Accent : Elantra : Excel : Pony : Santa Fe : Scoupe : Sonata : Stellar : Tiburon : Tuscon : XG300 : XG350

SYSTEM INDEX

MAINTENANCE	HYU-2
Maintenance Schedule	HYU-2
CHARGING	HYU-3
STARTING	HYU-3
CRUISE CONTROL	HYU-4
IGNITION	HYU-4
EMISSIONS	HYU-5
FUEL	HYU-5
EXHAUST	HYU-7
ENGINE COOLING	HYU-7
ENGINE	HYU-8
Assembly	HYU-8
Cylinder Head	HYU-10
Camshaft	HYU-12
Crank & Pistons	HYU-14
Engine Lubrication	HYU-16
CLUTCH	HYU-17
MANUAL TRANSAXLE	HYU-17
MANUAL TRANSMISSION	HYU-18
AUTO TRANSAXLE	HYU-18
AUTO TRANSMISSION	HYU-20
SHIFT LINKAGE	HYU-20
DRIVELINE	HYU-20
BRAKES	HYU-21
FRONT SUSPENSION	HYU-23
REAR SUSPENSION	HYU-25
STEERING	HYU-26
HEATING & AC	HYU-28
WIPERS & SPEEDOMETER	HYU-29
LAMPS & SWITCHES	HYU-30
SEAT BELTS	HYU-30
BODY	HYU-30

OPERATIONS INDEX

A
AC Hoses	HYU-28
Air Bags	HYU-26
Air Conditioning	HYU-28
Alignment	HYU-23
Alternator (Generator)	HYU-3
Anti-Lock Brakes	HYU-21

B
Ball Joint	HYU-24
Battery Cables	HYU-3
Bleed Brake System	HYU-21
Blower Motor	HYU-28
Brake Disc	HYU-22
Brake Drum	HYU-22
Brake Hose	HYU-21
Brake Pads and/or Shoes	HYU-23

C
Camshaft	HYU-12
Camshaft Sensor	HYU-4
Catalytic Converter	HYU-7
Coolant Temperature (ECT) Sensor	HYU-5
Crankshaft	HYU-15
Crankshaft Sensor	HYU-4
Cruise Control	HYU-4
CV Joint	HYU-21
Cylinder Head	HYU-10

D
Differential	HYU-20
Distributor	HYU-4
Drive Belt	HYU-2
Driveshaft	HYU-20

E
EGR	HYU-5
Electronic Control Module (ECM/PCM)	HYU-5
Engine	HYU-8
Engine Lubrication	HYU-16
Engine Mounts	HYU-10
Evaporator	HYU-28
Exhaust	HYU-7
Exhaust Manifold	HYU-7
Expansion Valve	HYU-29

F
Flexplate	HYU-19
Flywheel	HYU-17
Fuel Injection	HYU-6
Fuel Pump	HYU-6
Fuel Vapor Canister	HYU-5

G
Gear Selector Lever	HYU-20
Generator	HYU-3

H
Halfshaft	HYU-21
Heater Core	HYU-29

I
Ignition Coil	HYU-4
Ignition Module	HYU-4
Ignition Switch	HYU-4
Inner Tie Rod	HYU-26
Intake Air Temperature (IAT) Sensor	HYU-5
Intake Manifold	HYU-6

K
Knock Sensor	HYU-5

L
Lower Control Arm	HYU-24

M
Maintenance Schedule	HYU-2
Master Cylinder	HYU-22
Muffler	HYU-7

N
Neutral Safety Switch	HYU-18

O
Oil Pan	HYU-16
Oil Pump	HYU-17
Outer Tie Rod	HYU-26
Oxygen Sensor	HYU-5

P
Parking Brake	HYU-23
Pistons	HYU-15
Positive Crankcase Ventilation (PCV) Valve	HYU-5

R
Radiator	HYU-8
Radiator Hoses	HYU-8
Rear Main Oil Seal	HYU-16

S
Shock Absorber/Strut, Front	HYU-24
Shock Absorber/Strut, Rear	HYU-25
Spark Plug Cables	HYU-4
Spark Plugs	HYU-4
Spring, Front Coil	HYU-23
Spring, Leaf	HYU-25
Spring, Rear Coil	HYU-25
Starter	HYU-3
Steering Wheel	HYU-26

T
Thermostat	HYU-7
Throttle Body	HYU-7
Throttle Position Sensor (TPS)	HYU-5
Timing Belt	HYU-13
Timing Chain	HYU-14
Torque Converter	HYU-19

U
U-Joint	HYU-20
Upper Control Arm	HYU-25

V
Valve Body	HYU-19
Valve Cover Gasket	HYU-11
Valve Job	HYU-11

W
Water Pump	HYU-8
Wheel Balance	HYU-2
Wheel Cylinder	HYU-23

HYU-1

HYU-2   ACCENT : ELANTRA : EXCEL : PONY : SANTA FE : SCOUPE : SONATA

MAINTENANCE

	LABOR TIME	SEVERE SERVICE
Air Cleaner Filter Element, Replace (C)		
1985-05	.3	.4
Composite Headlamp Bulb, Replace (B)		
1985-05 each	.4	.4
Drive Belt, Adjust (B)		
Exc. below	.3	.4
1992-95 Elantra	.5	.6
Pony, Stellar	.5	.6
1997-98 Sonata	.5	.6
w/PS add	.3	.4
Drive Belt, Replace (B)		
Alternator		
exc. below	.5	.6
Santa Fe		
2.4L	.6	.7
3.5L	.8	.9
XG300/350	.8	.9
Compressor		
Santa Fe	.8	.8
1999-05 Sonata	.6	.7
Tiburon, Tuscon	.8	.9
XG300/350	.6	.7
Steering pump		
Accent	.3	.4
Elantra		
1991-95	.5	.6
1996-05	.8	.9
Excel	.4	.5
Santa Fe, Sonata, Tuscon	.5	.6
Tiburon, XG300/350	.6	.7
w/AC alternator add		
Accent, Tiburon	.2	.2
Elantra	.3	.3
95-05 Sonata 4 cyl.	.2	.2
w/DOHC alternator add		
92-95 Elantra	.3	.3
95-98 Sonata 4 cyl.	.2	.2
w/DOHC compressor add		
92-95 Elantra	.5	.5
96-00 Elantra	.3	.3
95-98 Sonata	.3	.3
w/DOHC steering pump		
95-98 Sonata add	.6	.6
w/PS alternator add		
Accent	.2	.2
92-05 Elantra	.3	.3
Drive Belt Tensioner, Replace (B)		
Exc. below	.5	.5
Santa Fe V6, Tuscon V6, XG300/350	.8	.9
Fuel Filter, Replace (B)		
Accent		
1995-99	1.0	1.1
2000-05		
in-line	.5	.6
on fuel pump	.8	.9
Elantra	1.0	1.1
Excel		
1986-89	.5	.6
1990-94	.8	.9
Pony, Stellar	.3	.4
Santa Fe	1.0	1.1
Scoupe	.8	.9
Sonata		
1989-98	1.0	1.1
1999-05		
in-line	.5	.6
on fuel pump	.8	.9
Tiburon, Tuscon	1.0	1.1
XG300/350	.5	.6
Oil & Filter, Change (C)		
Includes: Correct fluid levels.		
1985-05	.4	.6
Sealed Beam Headlamp, Replace (B)		
1985-93 one	.3	.3
Timing Belt, Replace (B)		
Accent	2.3	2.6
Elantra	2.8	3.1
Excel		
1986-89	1.5	1.9
1990-94	2.6	2.9
Santa Fe		
2.4L	2.6	2.9
2.7L	2.9	3.2
3.5L	6.6	7.3
Scoupe	2.5	2.9
Sonata		
1989-98		
4 cyl.	3.1	3.5
V6	3.3	3.7
1999-05	2.8	3.1
Stellar 2.0L	1.8	2.2
Tuscon		
4 cyl.	2.3	2.8
V6	6.0	6.5
Tiburon	3.1	3.4
XG300/350	4.6	5.1
w/AC add		
Accent	.5	.5
Elantra	.6	.6
Sonata		
89-98	.6	.6
99-05	.2	.2
Tiburon, Tuscon	.3	.3
w/DOHC add		
Elantra	1.2	1.2
89-98 Sonata 4 cyl.	.6	.6
w/PS add		
Accent, Stellar	.3	.3
92-05 Elantra	.3	.3
Replace tensioner add	.2	.2
Tire, Replace (C)		
Includes: Dismount old tire and mount new tire to rim.		
1986-05	.5	.5
Tires, Rotate (C)		
1986-05	.5	.5
Wheel, Balance (B)		
1985-05		
one	.3	.3
each addl.	.2	.2

SCHEDULED MAINTENANCE INTERVALS

If necessary, refer to appropriate Chilton maintenance service information.

	LABOR TIME	SEVERE SERVICE
7,500 Mile Service (C)		
All Models	.6	.6
15,000 Mile Service (C)		
Accent, Elantra	.7	.7
Santa Fe, Tiburon, Tuscon, XG300/350	.7	.7
Sonata	.8	.8
w/AT add	.1	.1
22,500 Mile Service (C)		
All Models	.4	.4
30,000 Mile Service (B)		
Accent, Elantra	3.5	3.5
Santa Fe, Tiburon, Tuscon, XG300/350	3.5	3.5
Sonata	3.3	3.3
w/AT add	.6	.6
37,500 Mile Service (C)		
All Models	.4	.4
45,000 Mile Service (C)		
All Models	.8	.8
w/AT add	.1	.1
52,500 Mile Service (C)		
All Models	.8	.8
60,000 Mile Service (B)		
Accent, Elantra	4.4	4.4
Santa Fe, Tiburon, Tuscon, XG300/350	5.2	5.2
Sonata	5.6	5.6
w/3.0L Sonata add	.5	.5
w/AT add	.6	.6
67,500 Mile Service (C)		
All Models	.4	.4
75,000 Mile Service (C)		
All Models	.4	.4
w/AT add	.1	.1
82,500 Mile Service (C)		
All Models	.4	.4
90,000 Mile Service (B)		
Accent, Elantra	3.7	3.7
Santa Fe, Tiburon, Tuscon, XG300/350	3.7	3.7
Sonata	3.3	3.3
w/AT add	.6	.6

STELLAR : TIBURON : TUSCON : XG300 : XG350   HYU-3

	LABOR TIME	SEVERE SERVICE
97,500 Mile Service (C)		
All Models	.4	.4

CHARGING

Alternator Circuits, Test (B)
Includes: Test component output.
1985-05	.3	.3

Alternator, R&R and Recondition (A)
Accent		
1995-99	2.8	3.1
2000-05	2.3	2.6
Elantra		
1992-00	2.8	3.1
2001-05	2.3	2.6
Excel	2.8	3.1
Pony, Stellar	2.3	2.6
Santa Fe		
4 cyl.	3.3	3.6
V6	2.0	2.3
Scoupe	2.8	3.1
Sonata		
1989-98	2.9	3.2
1999-05	3.3	3.6
Tiburon		
4 cyl.	2.6	2.9
V6	3.2	3.5
Tuscon		
4 cyl.	2.8	3.1
V6	3.0	3.3
XG300/350	4.0	4.3
w/AC Excel add	.7	.7
w/PS add		
Excel	.7	.7
Pony, Scoupe, Stellar	.2	.2
89-94 Sonata	.2	.2

Alternator Assy., Replace (B)
Includes: Pulley transfer.
Accent		
1995-99	1.3	1.5
2000-05	.8	.9
Elantra		
1992-95	1.2	1.4
1996-05	.8	.9
Excel	1.3	1.5
Pony, Stellar	.5	.7
Santa Fe		
4 cyl.	2.0	2.2
V6	.6	.7
Scoupe	1.3	1.5
Sonata		
1989-98		
4 cyl.	1.3	1.5
V6	1.8	2.1
1999-05	1.8	2.1
Tiburon		
4 cyl.	1.1	1.2
V6	1.8	2.1
Tuscon		
4 cyl.	1.8	2.0
V6	2.0	2.2
XG300/350	2.6	2.9
w/AC Excel add	.7	.7
w/PS add		
Excel	.7	.7
Pony, Scoupe, Stellar	.2	.2
89-94 Sonata	.2	.2

Alternator Voltage Regulator, Replace (B)
Accent, Elantra	1.7	2.0
Excel, Scoupe	1.7	2.0
Pony, Stellar	.9	1.2
Santa Fe		
2.4L	1.5	1.8
2.7L	1.2	1.5
3.5L	3.8	4.1
Sonata		
1989-98		
4 cyl.	1.8	2.1
V6	2.3	2.6
1999-05		
4 cyl.	1.4	1.7
V6	2.3	2.6
Tiburon		
4 cyl.	2.0	2.3
V6	2.3	2.6
Tuscon		
4 cyl.	2.0	2.3
V6	2.3	2.6
XG300/350	3.1	3.4
w/AC Excel add	.7	.7
w/PS add		
Excel	.7	.7
Pony, Scoupe, Stellar	.2	.2
92-94 Sonata	.2	.2

Voltmeter and/or Ammeter Gauge, Replace (B)
Exc. Elantra	.8	.8
Elantra	1.1	1.1

STARTING

Starter Draw Test (On Car) (B)
1985-05	.5	.6

Battery, Replace (C)
1985-05	.5	.6

Battery Cables, Replace (C)
Accent each	.6	.7
Elantra		
1991-00 each	.8	.9
2001-05		
positive	1.7	1.9
negative	.5	.6
Excel each	.5	.6
Santa Fe		
4 cyl.	1.8	2.0
V6		
positive cable	1.8	2.0
cable assembly	2.6	2.8
Sonata each	.5	.6
Tuscon		
4 cyl cable assy.	2.0	2.2
V6 cable assy.	2.3	2.5
Tiburon		
4 cyl.	.7	.8
V6		
positive	1.7	1.9
negative	.5	.6
XG300/350		
positive	1.7	1.9
negative	.5	.6

Starter, R&R and Recondition (A)
Accent, Elantra, Excel	1.9	2.2
Pony, Stellar	1.9	2.2
Santa Fe	1.7	2.0
Scoupe	2.7	3.0
Sonata		
1989-98		
4 cyl.	2.4	2.7
V6	1.7	2.0
1999-05	1.7	2.0
Tuscon		
4 cyl.	3.0	3.3
V6	2.7	3.0
Tiburon	1.9	2.2
XG300/350	1.7	2.0

Starter Assy., Replace (B)
Accent, Elantra, Excel	.8	.9
Pony, Stellar	.5	.6
Santa Fe, Scoupe	.8	.9
Sonata		
1989-98		
4 cyl.	1.0	1.1
V6	.6	.7
1999-05	.8	.9
Tuscon		
4 cyl.	1.8	2.0
V6	1.6	1.8
Tiburon, XG300/350	.8	.9

Starter Drive Assy., Replace (B)
Includes: Starter R&R.
Accent	1.3	1.4
Elantra, Excel		
1992-95	1.7	2.0
1996-05	1.4	1.6
Pony, Stellar, Tiburon	1.4	1.6
Santa Fe	1.3	1.4
Scoupe	1.6	1.9
Sonata		
1994-98	1.7	2.0
1999-05	1.1	1.3
Tuscon		
4 cyl.	2.6	2.9
V6	2.4	2.7
XG300/350	1.1	1.3

HYU-4

ACCENT : ELANTRA : EXCEL : PONY : SANTA FE : SCOUPE : SONATA

	LABOR TIME	SEVERE SERVICE
Starter Solenoid and/or Switch, Replace (B)		
Includes: Starter R&R.		
Accent	1.0	1.2
Elantra	1.1	1.3
Excel		
1986-89	.9	1.1
1990-94	1.1	1.3
Pony, Stellar	.7	.8
Santa Fe	.8	.9
Scoupe, Tiburon	1.0	1.2
Sonata		
1989-98		
4 cyl.	1.4	1.7
V6	.8	.9
1999-05	.8	.9
Tuscon		
4 cyl.	2.0	2.3
V6	1.7	2.0
XG300/350	.8	.9

CRUISE CONTROL

	LABOR TIME	SEVERE SERVICE
Diagnose Cruise Control System Component Each (A)		
1989-05	.5	.5
Actuator Assy., Replace (B)		
1989-05	.5	.5
Control Controller Module, Replace (B)		
Elantra	.8	.8
Scoupe	.3	.3
Sonata		
1989-93	.3	.3
1994-05	.7	.7
Tiburon, Tuscon	.3	.3
Control Dash Switch (On/Off), Replace (B)		
Elantra		
1992-95	.3	.3
1996-05	.6	.6
Scoupe	.3	.3
1994-05 Sonata	.3	.3
Tiburon		
1997-01	1.1	1.1
2003-05	.3	.3
Tuscon	1.3	1.3
XG300/350	.3	.3
Control Cable and Core, Replace (B)		
1989-05	.3	.3

IGNITION

	LABOR TIME	SEVERE SERVICE
Diagnose Ignition System Component Each (A)		
1985-05	.9	.9
Camshaft Position Sensor, Replace (B)		
Accent	.5	.5
Elantra	.3	.3

	LABOR TIME	SEVERE SERVICE
Santa Fe		
2.4L	.5	.5
2.7L	.5	.5
3.5L	1.2	1.4
Sonata		
1994-98		
4 cyl.	3.3	3.7
V6	3.8	4.2
1999-05	.5	.5
Tiburon, Tuscon	.5	.5
XG300/350	1.2	1.4
Computer Control Relay, Replace (B)		
Exc. below	.3	.3
Accent, Elantra	.5	.5
1999-05 Sonata	1.7	1.9
XG300/350	.8	.9
Crankshaft Angle Sensor, Replace (B)		
1985-88	.5	.7
1992-95 Elantra	.7	.9
1989-98 Sonata 4 cyl.	.7	.9
Crankshaft Position Sensor, Replace (B)		
1985-88	.5	.7
Accent, Elantra	.5	.5
Santa Fe		
2.4L	3.5	3.9
2.7L	.5	.5
3.5L	2.4	2.7
Sonata		
4 cyl.	3.1	3.4
V6		
1994-98	1.5	1.7
1999-05	.5	.5
Tiburon, Tuscon	.5	.5
XG300/350	3.2	3.6
Distributor, R&R and Recondition (A)		
Includes: Reset base ignition timing.		
1986-89 Excel	1.7	2.1
Pony, Stellar	1.6	1.0
Distributor, Replace (B)		
Includes: Reset base ignition timing.		
Excel	.7	.9
Pony, Stellar	.5	.7
Scoupe	.7	.9
Sonata		
1989-93	.7	.9
1995-98 V6	.8	.9
w/12V add Excel, Pony, Scoupe, Stellar	.5	.7
Distributor Cap and/or Rotor, Replace (B)		
1985-98	.5	.5
Distributor Points & Condenser, Replace (B)		
Pony, Stellar	.5	.5
Electronic Ignition Control Unit, Replace (B)		
Exc below	.9	1.1
1999-2005 Sonata	2.4	2.6

	LABOR TIME	SEVERE SERVICE
Igniter Unit, Replace (B)		
1985-94	.7	.9
Ignition Coil, Replace (B)		
Exc. below	.5	.5
Santa Fe, Sonata	.8	.9
Tiburon, XG300/350	.8	.9
w/DOHC add	.2	.2
91-95 Elantra	.3	.3
95-98 Sonata	.3	.3
Ignition Lock Cylinder, R&R or Replace (B)		
Exc. below	.9	.9
1986-89 Excel	.7	.7
Pony, Stellar	.7	.7
Santa Fe	1.2	1.2
1989-95 Sonata	1.2	1.2
2003-05 Tiburon	1.1	1.1
Ignition Switch, Replace (B)		
Exc. below	1.2	1.2
Pony, Stellar wagon		
3 & 4 door	.5	.5
5 door	.9	.9
Ignition Timing, Reset (B)		
1985-98	.8	.8
Ignition Trigger Wheel, Replace (B)		
1985-05	.7	.9
Spark Plug (Ignition) Cables, Replace (B)		
Exc. below	.5	.6
Santa Fe, Tiburon V6		
right side	2.6	2.9
left side	.6	.7
both sides	3.1	3.4
1999-05 Sonata V6		
right side	2.6	2.9
left side	.5	.6
both sides	3.1	3.4
Tuscon V6		
right side	1.6	1.9
left side	.5	.6
both sides	2.1	2.4
XG300/350		
right side	2.6	2.9
left side	.5	.6
Spark Plugs, Replace (B)		
Accent, Elantra	.5	.6
Excel		
1986-89	.5	.6
1990-94	.3	.4
Pony, Stellar	.5	.6
Santa Fe		
4 cyl.	.5	.6
V6		
right side	2.9	3.2
left side	.8	.9
both sides	3.3	3.7
Scoupe	.3	.4

STELLAR : TIBURON : TUSCON : XG300 : XG350   HYU-5

	LABOR TIME	SEVERE SERVICE
Sonata		
1989-98	.6	.7
1999-05		
4 cyl.	.5	.6
V6		
right side	2.6	2.9
left side	.5	.6
both sides	2.9	3.2
Tiburon		
4 cyl.	.5	.6
V6		
right side	2.6	2.9
left side	.6	.7
both sides	2.9	3.2
Tuscon		
4 cyl.	.5	.6
V6		
right side	1.5	1.8
left side	.6	.7
both sides	2.0	2.3
XG300/350		
right side	2.6	2.9
left side	.5	.6
both sides	2.9	3.2
Steering & Ignition Lock Switch Assy., Replace (B)		
1985-05	.3	.3

EMISSIONS

The following operations do not include testing. Add time as required.

Dynamometer Test (A)
  1985-05 .... .5 .5

Air Cleaner Temperature Sensor, Replace (B)
  1985-05 .... .5 .5

Air Cleaner Vacuum Motor, Replace (B)
  1985-97 .... .3 .3

Canister Purge Control Valve, Replace (B)
  1985-05 .... .3 .3

Coolant Temperature (ECT) Sensor, Replace (B)
  All Models .... .8 .8

Coolant Vacuum Switch Valve, Replace (B)
  1986-89 Excel .... .3 .3
  Santa Fe .... .3 .3
  1999-05 Sonata .... .3 .3
  Stellar 2.0L .... .5 .5
  XG300/350 .... .3 .3

EGR Valve, Replace (B)
  Exc. below .... .5 .7
  1986-89 Excel .... .3 .5
  1989-98 Sonata 4 cyl. .6 .8
  XG300/350 .... 1.2 1.4

EGR Valve Solenoid, Replace (B)
  1985-05 .... .3 .3

Electronic Control Module (ECM/PCM), Replace (B)
  Accent, Excel .... .6 .6
  Elantra
    1992-95 .... .5 .5
    1996-05 .... .8 .8
  Pony, Scoupe, Stellar .5 .5
  Santa Fe 1.4 1.4
  Sonata
    1989-98 .... .5 .5
    1999-05 .... 2.4 2.4
  Tiburon
    1997-01 .... .8 .8
    2003-05 .... 1.2 1.2
  Tuscon, XG300/350 ... .9 .9

Fuel Vapor Canister, Replace (B)
  Exc. below .... .5 .5
  2003-05 Tiburon ..... .8 .9

High Altitude Compensator, Replace (B)
  1986-89 Excel .... .3 .3

Knock Sensor, Replace (B)
  Accent .... .5 .7
  1996-05 Elantra ..... .5 .7
  Santa Fe
    4 cyl .... .5 .7
    V6 .... 2.4 2.6
  1999-05 Sonata
    4 cyl. .... .5 .7
    V6 .... 2.3 2.5
  Tiburon, Tuscon
    4 cyl. .... .5 .7
    V6 .... 2.4 2.6
  XG300/350 .... 2.3 2.5

Oxygen Sensor, Replace (B)
  Accent
    front
      SOHC .... .3 .5
      DOHC 1.1 1.2
    rear
      1995-97 .... 1.7 1.9
      1998-99 .... .6 .8
      2000-05 .... .3 .5
  Elantra
    1992-95 each ..... .6 .8
    1996-05 .... .3 .5
  Excel 8 1.0
  Santa Fe .... .3 .3
    front .... .3 .5
    rear
      exc. below ..... .6 .8
      3.5L upper
        right side ... 4.4 4.6
        left side 1.1 1.3
  Scoupe 8 1.0
  Sonata
    1989-93
      4 cyl. 8 1.0
      V6 .... .5 .7
    1994-05 each ..... .6 .8

Tuscon
  4 cyl. .... .5 .7
  V6 lower each ...... .5 .7
    upper right side ... 1.2 1.5
    upper left side 8 1.3

Tiburon
  1997-01 each .... .3 .5
  2003-05 each .... .6 .8

XG300/350
  front .... .3 .5
  rear each .... .6 .8

w/turbo Scoupe add ..... .2 .2

Positive Crankcase Ventilation (PCV) Valve, Replace (B)
  1985-05 .... .3 .3

Purge Control Solenoid Valve, Replace (B)
  1985-05 .... .3 .3

Thermal Vacuum Valve, Replace (B)
  Elantra .... .6 .7
  Excel
    1986-89 .... .3 .3
    1990-94 .... .7 .8
  Pony, Stellar .... .5 .5
  Scoupe .... .7 .8
  1995-98 Sonata ..... .8 .9

Throttle Angle Sensor, Replace (B)
  1985-97 .... .3 .3
  w/turbo add .... .3 .3

Throttle Position Sensor (TPS), Replace (B)
  1989-05 .... .5 .5
  w/DOHC add .... .5 .5

FUEL
CARBURETOR

Automatic Choke Coil, Replace (B)
  1985-93 .... .7 .7

Automatic Choke, Replace (B)
  1985-93 .... .3 .3

Carburetor, R&R and Clean or Recondition (B)
Includes: Adjustments.
  1986-93 Excel 3.3 3.3
  Pony, Stellar 2.8 2.8

Carburetor, Replace (B)
Includes: Adjustments
  1986-93 Excel 1.4 1.4
  Pony, Stellar 1.3 1.3

Float or Needle and/or Seat, Replace (B)
  1986-93 Excel 1.4 1.4

DELIVERY

Fuel Pump, Test (B)
  1985-05 .... .5 .5

HYU-6   ACCENT : ELANTRA : EXCEL : PONY : SANTA FE : SCOUPE : SONATA

	LABOR TIME	SEVERE SERVICE
Fuel Filter, Replace (B)		
Accent		
1995-99	1.0	1.1
2000-05		
in-line	.5	.6
on fuel pump	.8	.9
Elantra	1.0	1.1
Excel		
1986-89	.5	.6
1990-94	.8	.9
Pony, Stellar	.3	.4
Santa Fe	1.0	1.1
Scoupe	.8	.9
Sonata		
1989-98	1.0	1.1
1999-05		
in-line	.5	.6
on fuel pump	.8	.9
Tiburon, Tuscon	1.0	1.1
XG300/350	.5	.6
Fuel Gauge (Dash), Replace (B)		
Accent	.6	.6
Elantra		
1992-00	1.1	1.1
2001-05	.7	.7
Excel		
1986-89	.9	.9
1990-94	1.0	1.0
Pony, Stellar	.8	.8
Santa Fe	.7	.7
Scoupe	1.5	1.5
Sonata		
1989-93	.8	.8
1994-05	.5	.5
Tiburon		
1997-01	1.4	1.4
2003-05	1.1	1.1
Tuscon	1.2	1.2
XG300/350	.6	.6
Fuel Gauge (Tank), Replace (B)		
Accent	.8	.9
Elantra, Excel	1.0	1.1
Santa Fe	1.0	1.1
Scoupe	.7	.9
Sonata		
1989-93	.8	1.0
1994-98	1.6	1.8
1999-05	1.1	1.3
Tiburon		
1997-01	.7	.9
2003-05	1.4	1.6
Tuscon	1.6	1.8
XG300/350	1.1	1.3
Fuel Pump, R&R and Recondition (B)		
Pony, Stellar	1.2	1.4
Fuel Pump, Replace (B)		
Electric		
Accent	.8	.9
Elantra		

	LABOR TIME	SEVERE SERVICE
1992-95	2.0	2.1
1996-05	.8	.9
Santa Fe	1.0	1.1
Scoupe, Excel	2.0	2.1
Sonata		
1989-98	1.7	1.8
1999-05	.6	.7
Tuscon	.9	1.0
Tiburon	.8	.9
XG300/350	.6	.7
Mechanical		
1986-93 Excel	1.6	1.6
Pony, Stellar	.7	.7
Fuel Tank, Replace (B)		
Includes: Drain and refill.		
Accent, Elantra	1.7	2.0
Excel		
1986-89	1.3	1.5
1990-94	1.7	1.9
Pony, Stellar	.9	1.1
Santa Fe, XG300/350	3.2	3.6
Scoupe, Tiburon	1.7	1.9
Sonata		
1989-93	1.4	1.6
1994-98	2.2	2.5
1999-05	3.2	3.6
Tuscon	2.3	2.5
Intake Manifold and/or Gasket, Replace (B)		
Includes: Adjustments.		
Accent	2.3	2.6
Elantra		
1992-00	3.2	3.5
2001-05	2.4	2.7
Excel	3.1	3.4
Pony, Stellar	2.0	2.3
Santa Fe		
2.4L, 2.7L	2.6	2.9
3.5L	2.9	3.2
Scoupe	3.1	3.4
Sonata		
1989-98		
4 cyl.	3.4	3.7
V6	2.6	2.9
1999-05	2.4	2.7
Tiburon		
4 cyl.	3.2	3.5
V6	2.6	2.9
Tuscon		
4 cyl.	2.4	2.9
V6	1.6	2.1
XG300/350	2.4	2.9
Replace manifold add	.5	.7

INJECTION

	LABOR TIME	SEVERE SERVICE
Fuel Accumulator, Replace (B)		
1990-94 Excel	.5	.5
Scoupe	.5	.5
Replace fuel feed tube add	.2	.2

	LABOR TIME	SEVERE SERVICE
Fuel Injectors, Replace (B)		
Accent	1.0	1.1
Elantra	.8	.9
Excel	.8	.9
Pony, Stellar	.9	1.1
Scoupe, Tiburon	.7	.9
Santa Fe		
right side or both	2.9	3.2
left side	.5	.5
Sonata		
1989-98		
4 cyl.	.7	.9
V6	1.4	1.6
1999-05		
4 cyl.	.8	1.0
V6		
right side or both	2.6	2.9
left side	.5	.5
Tiburon		
1997-01	.6	.7
2003-05		
4 cyl.	1.1	1.3
V6		
right side or both	2.8	3.1
left side	.5	.5
Tuscon		
4 cyl.	.8	1.0
V6		
right side or both	1.6	1.9
left side	.5	.5
XG300/350		
right side or both	2.6	2.9
left side	.5	.5
Fuel Pressure Regulator, Replace (B)		
Accent	.8	.9
Elantra		
1992-95	.6	.7
1996-05	.8	.9
Excel	.6	.7
Pony, Scoupe, Stellar	.7	.8
Santa Fe, Sonata	.6	.7
Tiburon		
1997-01	.5	.7
2003-05		
4 cyl.	1.1	1.2
V6	.6	.7
Tuscon	.8	.9
XG300/350	.6	.7
Fuel Rail, Replace (B)		
Accent		
1995-99	1.0	1.2
2000-05	.8	.9
Elantra	.8	.9
1990-94 Excel	.6	.8
Santa Fe		
4 cyl.	.8	.9
V6		
right side	2.9	3.2
left side	.5	.7
Scoupe	.7	.9

STELLAR : TIBURON : TUSCON : XG300 : XG350   **HYU-7**

	LABOR TIME	SEVERE SERVICE
Sonata		
1989-98		
4 cyl.	.8	.9
V6	1.4	1.5
1999-05		
4 cyl.	.8	.9
V6		
right side	2.6	2.9
left side	.5	.7
Tiburon		
4 cyl.	1.1	1.2
V6		
right side	2.6	2.9
left side	.5	.7
4 cyl.	.8	1.0
V6		
right side	1.4	1.6
left side	.5	.5
XG300/350		
right side	2.6	2.9
left side	.5	.7
w/turbo add	.2	.2
Surge Tank and/or Gasket, Replace (B)		
Exc. below	1.3	1.6
Santa Fe, Tiburon V6	2.7	3.0
1999-04 Sonata V6	2.4	2.7
XG300/350	2.4	2.7
w/turbo add	.2	.2
Throttle Body Assy., Replace (B)		
Accent	.8	.9
Elantra		
1992-95	1.1	1.2
1996-05	.8	.9
1990-94 Excel	1.0	1.1
Santa Fe	1.4	1.5
Scoupe	1.3	1.4
Sonata		
1994-98	1.1	1.2
1999-05		
4 cyl.	1.1	1.2
V6	1.4	1.5
Tiburon	.8	.9
Tuscon		
4 cyl.	.8	1.0
V6	1.4	1.6
XG300/350	1.4	1.5
w/turbo add	.2	.2
Throttle Control Solenoid, Replace (B)		
1989-05	.5	.5

TURBOCHARGER

	LABOR TIME	SEVERE SERVICE
Exhaust Valve Actuator, Replace (B)		
1993-95 Scoupe	2.4	2.6
Oil Feed Pipe, Replace (B)		
1993-95 Scoupe	.9	1.0
Turbocharger Assy., R&R or Replace (B)		
1993-95 Scoupe	2.3	2.8
Turbocharger Coolant Line, Replace (B)		
1993-95 Scoupe		
feed	.9	1.0
return	.5	.6
Turbo-to-Exhaust Manifold Gasket, Replace (B)		
1993-95 Scoupe	1.5	1.8

EXHAUST

	LABOR TIME	SEVERE SERVICE
Catalytic Converter, Replace (B)		
All Models	.6	.7
Center Muffler, Replace (B)		
All Models	.8	.9
Exhaust Manifold or Gasket, Replace (B)		
Accent	1.1	1.3
Elantra		
1992-95	1.4	1.7
1996-00	.8	1.1
2001-05	1.1	1.3
Excel	1.4	1.7
Pony, Stellar	.9	1.2
Santa Fe		
2.4L	1.6	1.8
2.7L		
right side	2.2	2.5
left side	1.6	1.8
3.5L		
right side	6.6	7.4
left side	3.5	3.9
Scoupe	1.4	1.7
Sonata		
1989-98		
4 cyl.	1.9	2.1
V6		
right side	1.1	1.3
left side	1.4	1.6
1999-05		
4 cyl.	1.6	1.9
V6		
right side	4.0	4.4
left side	1.6	1.9
Tiburon		
4 cyl.	1.4	1.6
V6		
right side	2.9	3.2
left side	1.6	1.8
Tuscon		
4 cyl.	1.6	1.9
V6		
left or right side	1.8	2.1
XG300/350		
right side	4.0	4.4
left side	2.1	2.3
w/DOHC add		
Elantra	.3	.3
Sonata	1.0	1.0

	LABOR TIME	SEVERE SERVICE
w/PS Elantra add	.7	.7
w/turbo add	.5	.5
Front Muffler, Replace (B)		
All Models	.8	.9
Rear Muffler, Replace (B)		
Exc. Pony, Stellar	.6	.7
Pony, Stellar	.8	.9

ENGINE COOLING

	LABOR TIME	SEVERE SERVICE
Pressure Test Cooling System (B)		
1986-05	.3	.3
Coolant Thermostat, Replace (B)		
4 cyl.		
exc. Scoupe	.6	.7
Scoupe	.7	.9
V6		
exc. below	.9	1.0
1990-98 Sonata	.6	.7
Electric Fan Thermo Switch, Replace (B)		
All Models	.6	.7
Engine Coolant Temp. Sending Unit, Replace (B)		
All Models	.5	.5
Electric Cooling Fan Assy., Replace (B)		
Accent	.6	.7
Elantra	.8	.9
Excel	.5	.8
Pony, Stellar	.5	.8
Santa Fe	2.1	2.4
Scoupe	.5	.8
Sonata		
1989-98	.6	.7
1999-05	.9	1.0
Tiburon, Tuscon		
XG300/350	.9	1.0
Fan Blade, Replace (B)		
Exc. below	.7	.8
Pony, Stellar	.5	.6
Tiburon	.8	.9
Fan Motor Resistor, Replace (B)		
Elantra	.5	.5
Excel	.3	.3
Pony, Stellar	.5	.5
Scoupe	.3	.3
Sonata	.5	.5
Fluid Fan Drive or Fan Blade, Replace (B)		
Pony, Stellar	.7	.9
Freeze Plugs (Water Jacket), Replace (B)		
Add access time as required.		
Exc. below each	.5	.7
Accent		
intake side	3.8	4.0
exhaust side	2.1	2.3

HYU-8
ACCENT : ELANTRA : EXCEL : PONY : SANTA FE : SCOUPE : SONATA

	LABOR TIME	SEVERE SERVICE
Freeze Plugs (Water Jacket), Replace (B)		
Elantra		
1992-95		
intake side	3.6	3.8
exhaust side	2.6	2.8
1996-00		
intake side	3.3	3.5
exhaust side	1.8	2.0
2001-05		
intake side	3.8	4.0
exhaust side	2.1	2.3
Santa Fe		
4 cyl.		
intake side	3.8	4.0
exhaust side	2.1	2.3
V6		
intake side	6.5	6.7
exhaust side		
2.7L	3.2	3.4
3.5L	5.4	5.6
Sonata		
1995-98		
4 cyl.		
intake side	3.6	3.8
exhaust side	2.6	2.8
V6		
intake side	6.4	6.6
exhaust side	2.8	3.0
1999-05		
4 cyl.		
intake side	3.3	3.5
exhaust side	2.1	2.3
V6		
intake side	5.6	5.8
exhaust side	2.9	3.1
Tiburon		
1997-01		
intake side	3.3	3.5
exhaust side	1.8	2.0
2003-05		
4 cyl.		
intake side	4.3	4.5
exhaust side	2.8	3.0
V6		
intake side	5.7	5.9
exhaust side	3.3	3.5
Tuscon		
4 cyl.		
intake side	3.1	3.4
exhaust side	2.0	2.3
V6		
intake side	6.0	6.5
exhaust side	2.9	3.1
XG300/350		
intake side	5.9	6.1
exhaust side	3.5	3.7
w/4WD Santa Fe V6 add	.9	.9

	LABOR TIME	SEVERE SERVICE
w/AC add		
intake side		
Accent	.9	.9
Elantra, Tiburon	.6	.6
95-05 Sonata	.9	.9
exhaust side		
95-05 Sonata V6	.6	.6
Tiburon V6	.8	.8
w/DOHC exhaust side add		
92-95 Elantra	.9	.9
95-98 Sonata 4 cyl.	1.4	1.4
Idler Pulley, Replace (B)		
All Models	.5	.5
Radiator Assy., R&R or Replace (B)		
Includes: Refill with proper coolant mix.		
Accent	.8	.9
Elantra	.9	1.0
Excel		
1986-89	1.2	1.3
1990-94	.8	.9
Pony, Stellar	.9	1.0
Santa Fe	2.3	2.6
Scoupe	.8	.9
Sonata		
1989-98	.8	.9
1999-05		
4 cyl.	.9	1.0
V6	1.2	1.4
Tiburon	1.1	1.2
Tuscon	1.4	1.6
XG300/350	1.2	1.4
w/AC add	.2	.2
w/AT add	.3	.3
Radiator Fan Motor Relay, Replace (B)		
Sonata	.3	.3
Radiator Hoses, Replace (B)		
Includes: Refill with proper coolant mix.		
Accent, Elantra		
upper	.5	.6
lower	.8	.9
Excel		
1986-89	.5	.6
1990-94		
upper	.3	.4
lower	.6	.7
Pony, Stellar	.5	.6
Santa Fe		
upper	.5	.6
lower	.8	.9
Scoupe		
upper	.3	.3
lower	.7	.8
Sonata		
1989-98		
upper	.3	.3
lower	.6	.7
1999-05		
upper	.5	.6
lower	.8	.9

	LABOR TIME	SEVERE SERVICE
Tiburon, Tuscon, XG300/350		
upper	.5	.6
lower	.8	.9
Temperature Gauge (Dash), Replace (B)		
Accent	.6	.6
Elantra	.8	.8
1986-94 Excel	1.0	1.0
Pony, Stellar	.8	.8
Santa Fe	.6	.6
Scoupe	1.5	1.5
Sonata		
1989-93	.3	.3
1994-05	.5	.5
Tiburon, Tuscon	1.4	1.4
XG300/350	.6	.6
Water Pump and/or Gasket, Replace (B)		
Includes: Refill with proper coolant mix.		
Accent	2.7	3.0
Elantra		
1991-95	3.0	3.4
1996-00	2.2	2.5
2001-05	2.9	3.2
Excel	2.9	3.3
Pony, Stellar	.9	1.3
Santa Fe	3.2	3.6
Scoupe	3.0	3.4
Sonata		
1989-98		
4 cyl.	3.8	4.3
V6	4.4	5.0
1999-05		
4 cyl.	3.3	3.7
V6	4.0	4.4
Tiburon		
1997-01	2.2	2.5
2003-05	3.2	3.6
Tuscon	2.8	3.3
XG300/350	4.9	5.5
w/AC add	.3	.3
w/DOHC 92-95 Elantra add	1.4	1.5
w/PS add	.3	.3

ENGINE
ASSEMBLY
Times shown are for OEM assemblies. Time to replace assemblies from aftermarket rebuilders may vary.

Engine Assy., R&I (B)
Does not include parts or component transfer.

	LABOR TIME	SEVERE SERVICE
Accent		
1995-99	6.3	6.8
2000-05	8.7	9.3

STELLAR : TIBURON : TUSCON : XG300 : XG350 — HYU-9

	LABOR TIME	SEVERE SERVICE
Engine Assy., R&I (B)		
Elantra		
1992-95	6.2	6.7
1996-05	9.2	9.7
Excel	6.3	6.8
Pony, Stellar	5.6	6.1
Scoupe	6.4	6.9
Sonata		
4 cyl.	7.3	7.8
V6	8.7	9.2
Stellar 2.0L	5.6	6.1
Santa Fe		
4 cyl.	9.2	9.7
V6	11.0	11.5
Tiburon		
4 cyl.	5.2	5.7
V6	9.6	10.1
Tuscon		
4 cyl.	7.7	8.3
V6	10.8	11.3
XG 300/350	14.8	15.3
w/AC add		
Accent	.6	.6
Elantra		
92-95	.7	.7
96-05	.3	.3
Excel, Pony, Scoupe, Stellar, Tiburon	.3	.3
Sonata, Santa Fe, Tuscon, XG 300/350		
4 cyl.	.5	.5
V6	.7	.7
w/AT add		
Accent, Excel	.6	.6
92-05 Elantra	.5	.5
Pony, Stellar	.1	.1
Scoupe, Sonata 4 cyl.	.5	.5
Stellar 2.0L, Tiburon	.5	.5
w/DOHC add		
92-95 Elantra	.5	.5
Sonata 4 cyl.	.3	.3
w/PS add		
Accent	.5	.5
92-95 Elantra	.5	.5
Excel, Pony, Stellar	.3	.3
Scoupe	.5	.5
Stellar 2.0L	.3	.3
w/turbo add	.5	.5
Engine Assy., R&R and Recondition (A)		
Includes: Replacing rings, rod and main bearings, cylinder head reconditioning and engine tune-up.		
Accent	21.5	23.8
Elantra		
1992-95	25.7	28.4
1996-05	23.8	24.3
Excel		
1986-89	23.1	23.1
1990-94	27.6	27.6
Pony, Stellar	15.8	15.8
Santa Fe		
4 cyl.	27.1	30.0
V6	35.7	39.5
Scoupe	27.6	27.6
Sonata		
1989-98		
4 cyl.	38.3	31.2
V6	29.4	32.6
1999-05		
4 cyl.	25.8	28.6
V6	35.1	38.9
Tiburon		
4 cyl.	22.9	25.4
V6	31.2	34.6
Tuscon		
4 cyl.	21.8	22.3
V6	31.3	31.8
XG300/350	43.1	47.7
w/4WD Santa Fe add	1.0	1.0
w/AC add		
Accent	.6	.6
Elantra		
92-95	1.0	1.0
96-05	.5	.5
Excel, Pony, Scoupe, Stellar	.3	.3
Sonata		
4 cyl.	.6	.6
V6	1.0	1.0
Tiburon		
97-01	.3	.3
03-05	1.7	2.0
w/AT add		
Accent	1.0	1.0
92-05 Elantra	.8	.8
Excel	.5	.5
Pony	.1	.1
Santa Fe	.8	.8
Scoupe	.5	.5
Sonata 4 cyl.	.8	.8
Stellar		
1.5L, 1.6L	.1	.1
2.0L	.5	.5
Tiburon	.8	.8
w/DOHC add		
Accent	1.1	1.3
Sonata 4 cyl.	.8	.8
w/PS add		
Accent	.8	.8
92-95 Elantra	.8	.8
Excel, Pony, Stellar	.3	.3
Scoupe	.5	.5
Engine Assy. (Short Block), Replace (B)		
Assembly consists of engine block, piston assemblies, crankshaft, camshaft, timing chain and gears. Does not include cylinder heads. Operation includes: R&R engine, transfer necessary parts and all necessary adjustments.		
Accent		
1995-99	13.1	14.6
2000-05	13.5	15.1
Elantra		
1992-95	13.8	15.4
1996-00	12.6	14.1
2001-05	13.8	15.4
Excel		
1986-89	13.3	13.3
1990-94	13.8	13.8
Pony	12.2	12.2
Santa Fe		
4 cyl.	16.1	17.9
V6		
2.7L	19.7	22.0
3.5L	24.0	26.7
Scoupe	14.1	14.1
Sonata		
1989-98		
4 cyl.	16.8	18.8
V6	18.2	20.3
1999-05		
4 cyl.	14.6	16.3
V6	19.1	21.2
Stellar		
1.5L, 1.6L	12.2	12.2
2.0L	9.8	9.8
Tiburon		
1997-01	12.6	14.1
2003-05		
4 cyl.	13.7	15.2
V6	17.9	19.9
Tuscon		
4 cyl.	12.0	12.5
V6	21.3	21.8
XG300/350	23.1	25.6
w/4WD Santa Fe add	1.0	1.0
w/AC add		
Accent		
95-99	.6	.6
00-05	1.0	1.0
Elantra		
92-95	1.0	1.0
96-05	.5	.5
Excel, Pony, Scoupe, Stellar	.3	.3
Sonata		
4 cyl.	.6	.6
V6	1.0	1.0

HYU-10 ACCENT : ELANTRA : EXCEL : PONY : SANTA FE : SCOUPE : SONATA

	LABOR TIME	SEVERE SERVICE
Engine Assy. (Short Block), Replace (B)		
Tiburon		
97-01	.3	.3
03-05	1.7	2.0
w/AT add		
Accent	1.0	1.0
92-05 Elantra	.8	.8
Excel	.5	.5
Pony	.1	.1
Santa Fe	.8	.8
Scoupe	.5	.5
Sonata 4 cyl.	.8	.8
Stellar		
1.5L, 1.6L	.1	.1
2.0L	.5	.5
Tiburon	.8	.8
w/DOHC add		
92-95 Elantra	1.9	2.1
Sonata 4 cyl.	.8	.8
w/PS add		
Accent	.8	.8
92-95 Elantra	.8	.8
Excel, Pony, Stellar	.3	.3
Scoupe	.5	.5
Engine Mounts, Replace (B)		
Right mount	.7	.9
Front roll stopper	.7	.9
Rear roll stopper		
Accent	.6	.8
Elantra		
1992-00	.9	1.0
2001-05	.4	.4
1986-89 Excel	.7	.9
Santa Fe		
2.4L	.8	.9
2.7L	1.9	2.1
3.5L	2.4	2.7
Scoupe	.9	1.1
Sonata		
1995-98	1.1	1.3
1999-05		
4 cyl.	3.2	3.4
V6	1.9	2.1
Tiburon		
1997-01	.7	.9
2003-05		
4 cyl.	.7	.9
V6	1.9	2.1
Tuscon		
4 cyl.	1.0	1.2
V6	1.6	1.8
XG300/350	2.5	2.7
Transmission mounts		
Accent	.6	.8
Elantra	1.0	1.2
Excel, Scoupe	.9	1.1
Santa Fe, Sonata, Tiburon, Tuscon, XG300/350	1.1	1.3

	LABOR TIME	SEVERE SERVICE
CYLINDER HEAD		
Compression Test (B)		
4 cyl.	.7	.8
V6	.8	.9
Valve Clearance, Adjust (B)		
Includes: Adjust jet valves.		
All Models	.9	1.0
Cylinder Head, Replace (B)		
Accent		
1995-99	7.9	8.4
2000-05	8.7	9.2
Elantra		
1992-95	10.1	10.6
1996-05	9.7	10.2
Excel		
1986-89	8.2	8.7
1990-94	9.5	10.0
Pony, Stellar	6.2	6.7
Santa Fe		
2.4L	11.5	12.0
2.7L		
right side	15.8	16.3
left side	12.5	13.0
both sides	18.3	18.8
3.5L		
right side	17.7	18.2
left side	16.9	17.4
both sides	22.2	22.7
Scoupe	9.5	10.0
Sonata		
1989-98		
4 cyl.	10.4	10.9
V6		
one side	11.0	11.5
both sides	13.0	13.5
1999-05		
4 cyl.	11.4	11.9
V6		
right side	14.7	15.2
left side	12.0	12.5
both sides	17.6	18.1
Tiburon		
1997-01	9.5	10.0
2003-05		
4cyl.	9.6	11.1
V6		
right side	14.7	15.2
left side	12.0	12.5
both sides	16.6	17.1
Tuscon		
4 cyl.	9.6	10.1
V6		
left side	10.1	10.6
right side	10.9	11.4
both sides	17.2	17.7

	LABOR TIME	SEVERE SERVICE
XG300/350		
right side	18.6	19.1
left side	15.6	16.1
both sides	22.7	23.2
w/AC 89-05 Sonata		
V6 add	.2	.2
w/DOHC add		
Accent	.8	.9
92-05 Elantra	.5	.5
Scoupe	.7	.7
89-05 Sonata 4 cyl.	.5	.5
w/turbo add	.3	.3
Cylinder Head Gasket, Replace (B)		
Accent		
1995-99	4.0	4.5
2000-05	5.7	6.2
Elantra		
1992-00	4.6	5.1
2001-05	6.0	6.5
Excel		
1986-89	3.7	4.2
1990-94	4.6	5.1
Pony, Stellar	4.3	4.9
Santa Fe		
2.4L	5.5	6.0
2.7L		
right side	6.7	7.2
left side	6.5	7.0
both sides	8.5	9.0
3.5L		
right side	12.8	13.3
left side	12.2	12.7
both sides	13.3	13.8
Scoupe	4.4	4.9
Sonata		
1989-98		
4 cyl.	5.5	6.0
V6		
one side	6.5	7.0
both sides	6.7	7.3
1999-05		
4 cyl.	5.5	6.0
V6		
right side	6.9	7.4
left side	6.2	6.7
both sides	7.7	8.2
Stellar 2.0L	3.7	4.2
Tiburon		
1997-01	4.5	5.0
2003-05		
4 cyl.	6.5	7.0
V6		
right side	8.4	8.9
left side	7.6	8.1
both sides	9.4	9.9
Tuscon		
4 cyl.	7.6	8.1

STELLAR : TIBURON : TUSCON : XG300 : XG350 HYU-9

	LABOR TIME	SEVERE SERVICE
Engine Assy., R&I (B)		
Elantra		
1992-95	6.2	*6.7*
1996-05	9.2	*9.7*
Excel	6.3	*6.8*
Pony, Stellar	5.6	*6.1*
Scoupe	6.4	*6.9*
Sonata		
4 cyl.	7.3	*7.8*
V6	8.7	*9.2*
Stellar 2.0L	5.6	*6.1*
Santa Fe		
4 cyl.	9.2	*9.7*
V6	11.0	*11.5*
Tiburon		
4 cyl.	5.2	*5.7*
V6	9.6	*10.1*
Tuscon		
4 cyl.	7.7	*8.3*
V6	10.8	*11.3*
XG 300/350	14.8	*15.3*
w/AC add		
Accent	.6	*.6*
Elantra		
92-95	.7	*.7*
96-05	.3	*.3*
Excel, Pony, Scoupe,		
Stellar, Tiburon	.3	*.3*
Sonata, Santa Fe		
Tuscon, XG 300/350		
4 cyl.	.5	*.5*
V6	.7	*.7*
w/AT add		
Accent, Excel	.6	*.6*
92-05 Elantra	.5	*.5*
Pony, Stellar	.1	*.1*
Scoupe, Sonata 4 cyl.	.5	*.5*
Stellar 2.0L, Tiburon	.5	*.5*
w/DOHC add		
92-95 Elantra	.5	*.5*
Sonata 4 cyl.	.3	*.3*
w/PS add		
Accent	.5	*.5*
92-95 Elantra	.5	*.5*
Excel, Pony, Stellar	.3	*.3*
Scoupe	.5	*.5*
Stellar 2.0L	.3	*.3*
w/turbo add	.5	*.5*
Engine Assy., R&R and Recondition (A)		
Includes: Replacing rings, rod and main bearings, cylinder head reconditioning and engine tune-up.		
Accent	21.5	*23.8*
Elantra		
1992-95	25.7	*28.4*
1996-05	23.8	*24.3*

	LABOR TIME	SEVERE SERVICE
Excel		
1986-89	23.1	*23.1*
1990-94	27.6	*27.6*
Pony, Stellar	15.8	*15.8*
Santa Fe		
4 cyl.	27.1	*30.0*
V6	35.7	*39.5*
Scoupe	27.6	*27.6*
Sonata		
1989-98		
4 cyl.	38.3	*31.2*
V6	29.4	*32.6*
1999-05		
4 cyl.	25.8	*28.6*
V6	35.1	*38.9*
Tiburon		
4 cyl.	22.9	*25.4*
V6	31.2	*34.6*
Tuscon		
4 cyl.	21.8	*22.3*
V6	31.3	*31.8*
XG300/350	43.1	*47.7*
w/4WD Santa Fe add	1.0	*1.0*
w/AC add		
Accent	.6	*.6*
Elantra		
92-95	1.0	*1.0*
96-05	.5	*.5*
Excel, Pony, Scoupe,		
Stellar	.3	*.3*
Sonata		
4 cyl.	.6	*.6*
V6	1.0	*1.0*
Tiburon		
97-01	.3	*.3*
03-05	1.7	*2.0*
w/AT add		
Accent	1.0	*1.0*
92-05 Elantra	.8	*.8*
Excel	.5	*.5*
Pony	.1	*.1*
Santa Fe	.8	*.8*
Scoupe	.5	*.5*
Sonata 4 cyl.	.8	*.8*
Stellar		
1.5L, 1.6L	.1	*.1*
2.0L	.5	*.5*
Tiburon	.8	*.8*
w/DOHC add		
Accent	1.1	*1.3*
Sonata 4 cyl.	.8	*.8*
w/PS add		
Accent	.8	*.8*
92-95 Elantra	.8	*.8*
Excel, Pony, Stellar	.3	*.3*
Scoupe	.5	*.5*

	LABOR TIME	SEVERE SERVICE
Engine Assy. (Short Block), Replace (B)		
Assembly consists of engine block, piston assemblies, crankshaft, camshaft, timing chain and gears. Does not include cylinder heads. Operation includes: R&R engine, transfer necessary parts and all necessary adjustments.		
Accent		
1995-99	13.1	*14.6*
2000-05	13.5	*15.1*
Elantra		
1992-95	13.8	*15.4*
1996-00	12.6	*14.1*
2001-05	13.8	*15.4*
Excel		
1986-89	13.3	*13.3*
1990-94	13.8	*13.8*
Pony	12.2	*12.2*
Santa Fe		
4 cyl.	16.1	*17.9*
V6		
2.7L	19.7	*22.0*
3.5L	24.0	*26.7*
Scoupe	14.1	*14.1*
Sonata		
1989-98		
4 cyl.	16.8	*18.8*
V6	18.2	*20.3*
1999-05		
4 cyl.	14.6	*16.3*
V6	19.1	*21.2*
Stellar		
1.5L, 1.6L	12.2	*12.2*
2.0L	9.8	*9.8*
Tiburon		
1997-01	12.6	*14.1*
2003-05		
4 cyl.	13.7	*15.2*
V6	17.9	*19.9*
Tuscon		
4 cyl.	12.0	*12.5*
V6	21.3	*21.8*
XG300/350	23.1	*25.6*
w/4WD Santa Fe add	1.0	*1.0*
w/AC add		
Accent		
95-99	.6	*.6*
00-05	1.0	*1.0*
Elantra		
92-95	1.0	*1.0*
96-05	.5	*.5*
Excel, Pony, Scoupe,		
Stellar	.3	*.3*
Sonata		
4 cyl.	.6	*.6*
V6	1.0	*1.0*

HYU-10 ACCENT : ELANTRA : EXCEL : PONY : SANTA FE : SCOUPE : SONATA

	LABOR TIME	SEVERE SERVICE
Engine Assy. (Short Block), Replace (B)		
Tiburon		
97-01	.3	.3
03-05	1.7	2.0
w/AT add		
Accent	1.0	1.0
92-05 Elantra	.8	.8
Excel	.5	.5
Pony	.1	.1
Santa Fe	.8	.8
Scoupe	.5	.5
Sonata 4 cyl.	.8	.8
Stellar		
1.5L, 1.6L	.1	.1
2.0L	.5	.5
Tiburon	.8	.8
w/DOHC add		
92-95 Elantra	1.9	2.1
Sonata 4 cyl.	.8	.8
w/PS add		
Accent	.8	.8
92-95 Elantra	.8	.8
Excel, Pony, Stellar	.3	.3
Scoupe	.5	.5
Engine Mounts, Replace (B)		
Right mount	.7	.9
Front roll stopper	.7	.9
Rear roll stopper		
Accent	.6	.8
Elantra		
1992-00	.9	1.0
2001-05	.4	.4
1986-89 Excel	.7	.9
Santa Fe		
2.4L	.8	.9
2.7L	1.9	2.1
3.5L	2.4	2.7
Scoupe	.9	1.1
Sonata		
1995-98	1.1	1.3
1999-05		
4 cyl.	3.2	3.4
V6	1.9	2.1
Tiburon		
1997-01	.7	.9
2003-05		
4 cyl.	.7	.9
V6	1.9	2.1
Tuscon		
4 cyl.	1.0	1.2
V6	1.6	1.8
XG300/350	2.5	2.7
Transmission mounts		
Accent	.6	.8
Elantra	1.0	1.2
Excel, Scoupe	.9	1.1
Santa Fe, Sonata, Tiburon, Tuscon, XG300/350	1.1	1.3

	LABOR TIME	SEVERE SERVICE
CYLINDER HEAD		
Compression Test (B)		
4 cyl.	.7	.8
V6	.8	.9
Valve Clearance, Adjust (B)		
Includes: Adjust jet valves.		
All Models	.9	1.0
Cylinder Head, Replace (B)		
Accent		
1995-99	7.9	8.4
2000-05	8.7	9.2
Elantra		
1992-95	10.1	10.6
1996-05	9.7	10.2
Excel		
1986-89	8.2	8.7
1990-94	9.5	10.0
Pony, Stellar	6.2	6.7
Santa Fe		
2.4L	11.5	12.0
2.7L		
right side	15.8	16.3
left side	12.5	13.0
both sides	18.3	18.8
3.5L		
right side	17.7	18.2
left side	16.9	17.4
both sides	22.2	22.7
Scoupe	9.5	10.0
Sonata		
1989-98		
4 cyl.	10.4	10.9
V6		
one side	11.0	11.5
both sides	13.0	13.5
1999-05		
4 cyl.	11.4	11.9
V6		
right side	14.7	15.2
left side	12.0	12.5
both sides	17.6	18.1
Tiburon		
1997-01	9.5	10.0
2003-05		
4cyl.	9.6	11.1
V6		
right side	14.7	15.2
left side	12.0	12.5
both sides	16.6	17.1
Tuscon		
4 cyl.	9.6	10.1
V6		
left side	10.1	10.6
right side	10.9	11.4
both sides	17.2	17.7

	LABOR TIME	SEVERE SERVICE
XG300/350		
right side	18.6	19.1
left side	15.6	16.1
both sides	22.7	23.2
w/AC 89-05 Sonata V6 add	.2	.2
w/DOHC add		
Accent	.8	.9
92-05 Elantra	.5	.5
Scoupe	.7	.7
89-05 Sonata 4 cyl.	.5	.5
w/turbo add	.3	.3
Cylinder Head Gasket, Replace (B)		
Accent		
1995-99	4.0	4.5
2000-05	5.7	6.2
Elantra		
1992-00	4.6	5.1
2001-05	6.0	6.5
Excel		
1986-89	3.7	4.2
1990-94	4.6	5.1
Pony, Stellar	4.3	4.9
Santa Fe		
2.4L	5.5	6.0
2.7L		
right side	6.7	7.2
left side	6.5	7.0
both sides	8.5	9.0
3.5L		
right side	12.8	13.3
left side	12.2	12.7
both sides	13.3	13.8
Scoupe	4.4	4.9
Sonata		
1989-98		
4 cyl.	5.5	6.0
V6		
one side	6.5	7.0
both sides	6.7	7.3
1999-05		
4 cyl.	5.5	6.0
V6		
right side	6.9	7.4
left side	6.2	6.7
both sides	7.7	8.2
Stellar 2.0L	3.7	4.2
Tiburon		
1997-01	4.5	5.0
2003-05		
4 cyl.	6.5	7.0
V6		
right side	8.4	8.9
left side	7.6	8.1
both sides	9.4	9.9
Tuscon		
4 cyl.	7.6	8.1

STELLAR : TIBURON : TUSCON : XG300 : XG350 **HYU-11**

	LABOR TIME	SEVERE SERVICE
V6		
left side	7.1	*7.6*
right side	7.9	*8.4*
both sides	13.2	*13.7*
XG300/350		
right side	9.9	*10.4*
left side	9.4	*9.9*
both sides	10.8	*11.2*
w/AC add		
89-05 Sonata V6	.2	*.2*
Tiburon	.5	*.5*
w/DOHC add		
92-05 Elantra	.6	*.6*
Scoupe	.7	*.7*
89-05 Sonata 4 cyl.	.5	*.5*
w/turbo add	.3	*.3*
Jet Valves, Replace (B)		
Excel	1.4	*1.6*
1989-98 Sonata 4 cyl.	1.3	*1.5*
Rocker Arm Cover Gasket, Replace or Reseal (B)		
Exc. below	.5	*.6*
Santa Fe		
2.7L		
right side	.5	*.6*
left side or both	1.3	*1.4*
3.5L		
right side	1.2	*1.4*
left side or both	2.6	*2.9*
Sonata V6		
1995-98		
right side	.3	*.4*
left side or both	3.3	*3.7*
1999-05		
right side	.5	*.6*
left side or both	1.3	*1.4*
Tiburon		
right side	.5	*.6*
left side or both	3.3	*3.7*
Tuscon		
4 cyl.	.6	*.7*
V6		
left side	.7	*.8*
right side or both	2.9	*3.4*
XG300/350		
right side	.8	*.9*
left side or both	2.1	*2.4*
w/turbo add	.2	*.2*
Rocker Arms or Shafts, Replace (B)		
Accent one side	1.3	*1.5*
Elantra	1.5	*1.7*
Excel		
1986-89	1.2	*1.4*
1990-94	1.5	*1.7*
Pony	1.4	*1.6*
Santa Fe	3.0	*3.4*
2.4L one side	2.8	*3.1*
3.5L		
right side	7.6	*8.5*
left side	8.2	*9.1*

	LABOR TIME	SEVERE SERVICE
Scoupe	1.5	*1.7*
Sonata		
1989-98		
4 cyl.	1.7	*1.9*
V6	2.8	*3.0*
1999-05 4 cyl.		
one side	2.6	*2.9*
Stellar	1.7	*1.9*
XG300/350		
right side	6.3	*7.0*
left side	6.8	*7.5*
w/DOHC add		
Elantra, Scoupe	.3	*.3*
89-98 Sonata 4 cyl	.3	*.3*
w/turbo add	.2	*.2*
Valve Job (B)		
Accent		
1995-99	7.9	*8.4*
2000-05	8.3	*8.8*
Elantra		
1992-95	9.9	*10.4*
1996-05	9.6	*10.1*
Excel		
1986-89	8.7	*9.2*
1990-94	8.3	*8.8*
Pony, Stellar	6.8	*7.3*
Santa Fe		
2.4L	11.5	*12.0*
2.7L		
right side	14.5	*15.0*
left side	13.0	*13.5*
both sides	19.6	*20.1*
3.5L		
right side	17.0	*17.5*
left side	16.7	*17.3*
both sides	22.4	*22.9*
Scoupe	8.3	*8.8*
Sonata		
1989-98		
4 cyl.	10.2	*10.7*
V6	10.8	*11.3*
1999-05		
4 cyl.	11.5	*12.0*
V6		
right side	13.9	*14.4*
left side	12.6	*13.1*
both sides	19.2	*19.7*
Tiburon		
4 cyl.	8.8	*9.3*
V6		
right side	13.1	*13.6*
left side	11.8	*12.3*
both sides	16.9	*17.4*
Tuscon		
4 cyl.	9.8	*10.3*
V6		
right side	10.1	*10.6*
left side	9.8	*10.3*
both sides	14.9	*15.4*

	LABOR TIME	SEVERE SERVICE
XG300/350		
right side	18.4	*18.9*
left side	15.7	*16.2*
both sides	23.4	*23.9*
w/AC add		
Sonata V6	.2	*.2*
03-05 Tiburon	.5	*.5*
w/DOHC add		
Accent	.9	*.9*
92-05 Elantra	.9	*.9*
Scoupe	.7	*.7*
89-05 Sonata 4 cyl.	.9	*.9*
w/turbo add	.3	*.3*
Valve Lash Adjuster, Replace (B)		
Accent one side	1.4	*1.9*
Elantra		
1992-95	1.4	*1.9*
1996-00	2.0	*2.5*
2001-05	3.8	*4.3*
Santa Fe		
2.4L one side	2.8	*3.3*
2.7L		
right side	3.2	*3.7*
left side	3.6	*4.1*
both sides	4.4	*4.9*
3.5L		
right side	7.7	*8.3*
left side	8.2	*8.7*
both sides	9.3	*9.8*
Scoupe	1.3	*1.8*
Sonata		
1989-98		
4 cyl.	1.4	*1.9*
V6		
right side	1.1	*1.6*
left side	2.3	*2.8*
both sides	2.8	*3.3*
1999-05		
4 cyl. one side	3.8	*4.3*
V6		
right side	6.9	*7.4*
left side	7.3	*7.8*
both sides	13.3	*13.8*
Tiburon		
4 cyl.	4.0	*4.5*
V6		
right side	3.3	*3.8*
left side	3.8	*4.3*
both sides	4.7	*5.2*
Tuscon		
4 cyl.	4.6	*4.5*
V6		
right side	5.3	*5.8*
left side	4.8	*5.3*
both sides	8.7	*9.2*

HYU-12 ACCENT : ELANTRA : EXCEL : PONY : SANTA FE : SCOUPE : SONATA

	LABOR TIME	SEVERE SERVICE
Valve Lash Adjuster, Replace (B)		
XG300/350		
right side	6.3	6.8
left side	6.8	7.3
both sides	7.6	8.1
w/DOHC each side add		
Accent, Scoupe, Sonata	.6	.6
Valve Springs and/or Oil Seals, Replace (B)		
Accent		
1995-99	5.8	6.3
2000-05	6.3	6.8
Elantra	6.8	7.3
Pony, Stellar	6.2	6.7
Santa Fe		
2.4L	7.5	8.0
2.7L		
right side	8.0	8.5
left side	7.1	7.7
both sides	9.2	9.7
3.5L		
right side	13.2	13.7
left side	12.8	13.3
both sides	15.4	15.9
Scoupe	6.9	7.4
Sonata		
1989-98		
4 cyl.	7.2	7.6
V6		
right side	7.5	8.0
left side	6.8	7.3
both sides	8.8	9.3
1999-05		
4 cyl.	7.5	8.0
V6	9.1	9.6
Tiburon		
1997-01	6.9	7.4
2003-05		
4 cyl.	7.4	7.9
V6		
right side	7.5	8.0
left side	7.5	8.0
both sides	8.8	9.7
Tuscon		
4 cyl.	6.4	6.9
V6		
right side	7.7	8.1
left side	6.5	7.0
both sides	9.8	10.3
XG300/350		
right side	11.8	12.3
left side	10.1	10.6
both sides	12.8	13.3
w/AC add		
99-05 Sonata V6	.2	.2
Tiburon	.5	.5

	LABOR TIME	SEVERE SERVICE
w/DOHC add		
Accent	.5	.5
Elantra	1.3	1.3
Scoupe	.3	.3
89-98 Sonata 4 cyl.	1.0	1.0
w/turbo add	.3	.3
CAMSHAFT		
Timing Belt, Adjust (B)		
Stellar 2.0L	.5	.7
Balance Shaft, Replace (B)		
1992-95 Elantra		
right side	9.5	10.6
left side	5.0	5.7
Santa Fe		
right side	10.9	12.1
left side	11.6	12.9
Sonata		
1989-98		
right side	9.5	10.6
left side	5.0	5.7
1999-05		
one side	9.5	10.6
Stellar each side	4.6	4.9
w/AC add		
Elantra	.5	.5
Sonata	.6	.6
Stellar	.2	.2
w/AT add		
Elantra	.3	.3
Santa Fe	.8	.8
89-98 Sonata	.3	.3
99-04 Sonata	.8	.8
w/DOHC add		
Elantra, Sonata	1.2	1.2
w/PS add Stellar	.2	.2
Balance Shaft Bearing, Replace (B)		
1992-95 Elantra		
right side	10.3	11.5
left side	5.5	6.2
Santa Fe		
right or left side	11.9	13.1
Sonata		
1989-98 4 cyl.		
right side	10.3	11.5
left side	5.5	6.2
1999-05 4 cyl.		
right or left side	10.5	11.6
w/AC add		
Elantra, Sonata, Santa Fe	.6	.6
Stellar	.2	.2
w/AT add		
Elantra, Sonata, Santa Fe	.8	.8
w/DOHC add		
Elantra, Sonata	.8	.8
w/PS add Stellar	.3	.3

	LABOR TIME	SEVERE SERVICE
Camshaft, Replace (B)		
Accent		
1995-99	1.8	2.3
2000-05		
right or left side	2.4	2.9
both sides	2.8	3.3
Elantra		
1992-95	2.8	3.3
1996-05	3.5	4.0
Excel		
1986-89	3.7	3.9
1990-94	2.8	3.0
Pony, Stellar	2.3	2.5
Scoupe	2.8	3.0
Santa Fe		
4 cyl.	2.6	2.9
V6		
2.7L		
right side	3.2	3.6
left side	3.6	4.1
both sides	4.4	4.9
3.5L		
right side	7.6	8.5
left side	8.2	9.1
both	9.3	10.4
Sonata		
1989-98		
4 cyl.	2.8	3.0
V6		
right side	3.5	3.9
left side	5.3	6.0
both sides	6.0	6.7
1999-05		
4 cyl.		
right or left side	2.6	2.9
both sides	2.8	3.1
V6		
right side	2.9	3.2
left side	3.3	3.7
both sides	4.3	4.7
Tiburon		
4 cyl.	3.8	4.2
V6		
right side	3.3	3.7
left side	3.8	4.2
both sides	4.7	5.2
Tuscon		
4 cyl.	3.4	3.9
V6		
right side	7.4	7.9
left side	6.9	7.4
XG300/350		
right side	6.2	6.8
left side	6.8	7.5
both sides	7.5	8.3
w/DOHC add		
Accent, Elantra	.3	.3
89-02 Sonata 4 cyl.	.3	.3
w/turbo add	.2	.2

STELLAR : TIBURON : TUSCON : XG300 : XG350 HYU-13

	LABOR TIME	SEVERE SERVICE
Camshaft Seal, Replace (B)		
Accent		
1995-99	1.8	2.1
2000-05	2.3	2.6
Elantra		
1992-95	1.5	1.7
1996-00	1.1	1.2
2001-05	2.6	2.9
Excel		
1986-89	1.2	1.3
1990-94	1.6	1.7
Santa Fe		
4 cyl.	2.1	2.4
V6		
2.7L		
right or left side	3.3	3.7
both sides	4.3	4.7
3.5L		
right or left side	8.8	9.7
both sides	9.6	10.7
Scoupe	1.6	1.7
Sonata		
1989-98		
4 cyl.	1.7	1.9
V6		
right side	2.9	3.2
left side	2.6	2.9
both sides	3.2	3.6
1999-05		
4 cyl.	2.0	2.2
V6		
one side	3.1	3.4
both sides	4.1	4.6
Stellar 2.0L	1.4	1.5
Tiburon	1.2	1.3
1997-01	1.1	1.2
2003-05		
4 cyl.	3.2	3.6
V6		
right or left side	3.1	3.4
both sides	4.1	4.6
Tuscon		
4 cyl.	2.4	2.9
V6		
right or left side	2.8	3.3
both sides	5.9	6.4
XG300/350		
right side	1.5	1.7
left side	1.1	1.2
both sides	3.1	3.4
w/AC add		
Sonata, Tiburon	.5	.5
w/DOHC add		
Elantra	.5	.5
89-98 Sonata 4 cyl.	.5	.5
w/turbo add	.2	.2
Camshaft Sprocket, Replace (B)		
Accent		
1995-99	2.0	2.2
2000-05	2.3	2.6

	LABOR TIME	SEVERE SERVICE
Elantra		
1992-95	1.2	1.4
1996-05	2.8	3.1
Excel		
1986-89	.7	.9
1990-94	1.3	1.5
Santa Fe		
4 cyl.	2.1	2.4
V6		
2.7L		
right or left side	2.0	2.2
both sides	3.2	3.6
3.5L		
right or left side	6.8	7.5
both sides	6.9	7.7
Scoupe	1.3	1.5
Sonata		
1989-98		
4 cyl.	1.4	1.6
V6		
right side	2.9	3.1
left side	2.5	2.7
both sides	3.1	3.3
1999-05		
4 cyl.	2.0	2.2
V6		
right or left side	2.8	3.1
both sides	3.1	3.4
Tiburon		
4 cyl.	2.2	2.7
V6		
right side	3.2	3.6
left side	2.8	3.1
both sides	3.3	3.7
Tuscon		
4 cyl.	2.8	3.3
V6		
right or left side	3.6	3.1
both sides	5.6	6.1
XG300/350		
right or left side	4.6	5.1
both sides	4.8	5.4
w/AC add		
Excel, Scoupe	.1	.1
w/DOHC add		
Accent, Elantra	.8	.8
89-05 Sonata 4 cyl.	.7	.7
Engine Front Cover and/or Gasket, Replace (B)		
Accent, Excel, Elantra	4.6	5.1
Santa Fe		
exc. 3.5L	5.7	6.4
3.5L	8.9	9.9
Scoupe	4.6	5.1
Sonata		
1989-98		
4 cyl.	5.0	5.5
V6	5.6	6.2
1999-05	5.1	5.7
Stellar 2.0L	3.9	4.2

	LABOR TIME	SEVERE SERVICE
Tiburon	4.8	5.4
1997-01	4.7	5.3
2003-05	5.3	5.9
Tuscon		
4 cyl.	5.0	5.5
V6	8.2	8.7
XG300/350	8.5	9.4
w/AC add		
Accent, Elantra, Stellar	.3	.3
Santa Fe		
4 cyl.	.3	.3
V6	.6	.6
89-05 Sonata	.6	.6
Tiburon V6	.6	.6
w/DOHC add		
Accent, Elantra	1.3	1.3
89-98 Sonata 4 cyl.	1.3	1.3
w/PS add		
Accent	.3	.3
92-95 Elantra	.3	.3
Timing Auto Tensioner, Replace (B)		
Elantra		
1996-00	4.1	4.3
2001-05	3.6	3.8
Santa Fe		
4 cyl.	2.0	2.2
V6		
2.7L	2.9	3.1
3.5L	6.6	6.8
Sonata		
1989-98 4 cyl.	4.4	4.6
1999-05		
4 cyl.	2.0	2.2
V6	2.8	3.0
2003-05 Tiburon		
4cyl.	4.0	4.2
V6	3.1	3.3
XG300/350	4.6	4.8
w/AC add		
91-95 Elantra	.6	.6
89-98 Sonata 4 cyl.	.5	.5
Tiburon	.3	.3
Timing Belt, Replace (B)		
Accent	2.3	2.6
Elantra	2.8	3.1
Excel		
1986-89	1.5	1.9
1990-94	2.6	2.9
Santa Fe		
2.4L	2.6	2.9
2.7L	2.9	3.2
3.5L	6.6	6.9
Scoupe	2.5	2.9
Sonata		
1989-98		
4 cyl.	3.1	3.5
V6	3.3	3.7
1999-05	2.8	3.1
Stellar 2.0L	1.8	2.2

HYU-14 ACCENT : ELANTRA : EXCEL : PONY : SANTA FE : SCOUPE : SONATA

	LABOR TIME	SEVERE SERVICE
Timing Belt, Replace (B)		
Tiburon	3.1	3.4
Tuscon		
4 cyl.	2.3	2.8
V6	6.0	6.5
XG300/350	4.6	5.1
w/AC add		
Accent	.5	.5
Elantra	.6	.6
Sonata		
89-98	.6	.6
99-05	.2	.2
Tiburon, Tuscon	.3	.3
w/DOHC add		
Elantra	1.2	1.2
89-98 Sonata 4 cyl.	.6	.6
w/PS add		
Accent, Stellar	.3	.3
92-95 Elantra	.3	.3
Replace tensioner add	.2	.2
Timing Belt Tensioner, Replace (B)		
Accent	2.3	2.6
Elantra	2.8	3.1
Excel		
1986-89	1.6	2.0
1990-94	2.5	2.9
Santa Fe		
4 cyl.	2.0	2.2
V6		
2.7L	2.9	3.2
3.5L	6.6	7.3
Scoupe	2.5	2.9
Sonata		
1989-98	3.1	3.5
1999-05		
4 cyl.	2.0	2.2
V6	2.8	3.1
Stellar 2.0L	1.9	2.3
Tiburon	3.1	3.4
Tuscon		
4 cyl.	1.8	2.3
V6	6.0	6.5
XG300/350	4.6	5.1
w/AC add		
Accent	.5	.5
92-95, 01-04 Elantra	.5	.5
89-05 Sonata	.5	.5
99-05 Sonata V6	.2	.2
Stellar, Tiburon	.3	.3
w/DOHC add		
Accent, Elantra	.8	.8
89-98 Sonata 4 cyl.	.9	.9
w/PS add		
Accent	.3	.3
92-95 Elantra	.3	.3
Timing Chain, Replace (B)		
Accent		
1995-99	1.9	2.1
2000-05	2.4	2.7

	LABOR TIME	SEVERE SERVICE
Elantra		
1996-00	1.8	2.3
2001-05	3.5	3.9
Pony, Stellar	5.2	5.7
Santa Fe	3.6	4.1
1999-05 Sonata V6	3.3	3.7
Tiburon		
1997-01	1.9	2.1
2003-05	4.1	4.6
w/AC Pony, Stellar add	.2	.2
w/AT Pony, Stellar add	.3	.3
Replace chain guide add	.1	.1
Timing Chain Cover and/or Gasket, Replace (B)		
Pony, Stellar	4.9	5.3
w/AC add	.2	.2
w/PS add	.3	.3
Timing Cover and/or Gasket, Replace (B)		
Accent, Elantra		
upper	.3	.3
lower	1.2	1.4
Excel		
1986-89		
upper	.3	.4
lower	.7	.9
1990-94		
upper	.3	.4
lower	1.4	1.6
Scoupe		
upper	.3	.4
lower	1.4	1.6
Sonata		
1989-98		
4 cyl.		
upper	.5	.5
lower	1.8	2.1
V6		
upper	.5	.5
lower	1.4	1.5
rear	2.6	2.9
1999-05		
upper		
4 cyl.	.3	.3
V6	.9	1.0
lower	1.5	1.7
Stellar 2.0L		
upper	.5	.6
lower	1.2	1.4
Tiburon		
1997-02		
upper	.3	.4
lower	1.2	1.4
2003-05		
upper		
4 cyl.	.3	.4
V6	1.1	1.2
lower	1.7	1.9

	LABOR TIME	SEVERE SERVICE
Tuscon		
4 cyl.	5.0	5.5
V6	8.2	8.7
XG300/350	4.6	5.1
w/AC add		
Accent, Excel, Scoupe	.5	.5
92-05 Elantra	.4	.4
Sonata		
exc. below	.2	.2
92-98 4 cyl.	.6	.6
Tiburon	.3	.3
w/DOHC add		
92-95 Elantra	.8	.8
w/PS add		
Accent, Excel	.3	.3
92-05 Elantra	.2	.2
Timing Cover Front Seal, Replace (B)		
Pony, Stellar	1.4	1.6
Stellar 2.0L	2.3	2.5
w/AC add	.2	.2
w/PS add	.3	.3
CRANK & PISTONS		
Connecting Rod Bearings, Replace (A)		
Includes: Check bearing oil clearance.		
Accent, Elantra	5.3	5.8
Excel	4.4	4.9
Pony, Stellar	9.2	9.7
Santa Fe		
4 cyl.	4.3	4.8
V6		
2.7L	4.6	5.1
3.5L	5.5	6.2
Scoupe	4.4	4.9
Sonata		
1989-98	4.7	5.3
1999-05	4.6	5.1
Stellar 2.0L	8.2	8.7
Tiburon		
1997-01	4.0	4.4
2003-05		
4 cyl.	4.6	5.1
V6	4.9	5.5
Tuscon		
4 cyl.	5.4	5.9
V6	9.2	9.7
XG300/350	5.2	5.8
w/AC Tiburon add	.5	.5
w/DOHC add		
89-98 Sonata 4 cyl.	.2	.2
Elantra	.2	.2

STELLAR : TIBURON : TUSCON : XG300 : XG350 HYU-15

	LABOR TIME	SEVERE SERVICE
Crankshaft and Main Bearings, Replace (A)		
Includes: Engine R&R.		
Accent	14.3	16.1
Elantra		
1992-00	13.7	15.2
2001-05	14.0	15.7
Excel		
1986-89	14.8	15.3
1990-94	13.5	14.0
Pony, Stellar	10.8	11.3
Santa Fe		
4 cyl.	15.6	17.5
V6		
2.7L	20.6	23.0
3.5L	25.9	28.8
Scoupe	13.5	14.0
Sonata		
1989-98		
4 cyl.	16.2	18.0
V6	18.0	20.0
1999-05		
4 cyl.	14.2	15.9
V6	20.1	22.4
Stellar 2.0L	11.4	11.9
Tiburon		
1997-01	13.2	14.8
2003-05		
4 cyl.	14.6	16.3
V6	20.5	22.8
Tuscon		
4 cyl.	15.4	15.9
V6	26.9	27.4
XG300/350	25.0	27.8
w/4WD add	.9	.9
w/AC add		
Accent		
95-99	.5	.5
00-05	.9	.9
Elantra		
92-95	.9	.9
96-05	.5	.5
Excel, Pony, Stellar	.3	.3
Scoupe, Stellar 2.0L	.3	.3
Sonata	.9	.9
Tiburon		
97-01	.5	.5
03-05	1.7	1.7
w/AT add		
Accent, Elantra	.9	.9
Excel	.5	.5
Pony, Stellar	.1	.1
Santa Fe, Tiburon	.8	.8
Scoupe, Stellar 2.0L	.5	.5
Sonata 4 cyl.	.8	.8
w/DOHC add		
92-95 Elantra	.3	.3
95-98 Sonata 4 cyl.	.8	.8

	LABOR TIME	SEVERE SERVICE
w/PS add		
Accent	.8	.8
92-95 Elantra	.8	.8
Excel, Pony, Stellar	.3	.3
Scoupe	.5	.5
w/turbo add	.5	.5
Crankshaft Front Oil Seal, Replace (B)		
Accent		
1995-99	2.6	2.8
2000-05	4.7	5.3
Elantra		
1992-95	3.2	3.6
1996-00	2.4	2.7
2001-05	4.7	5.3
Excel		
1986-89	1.9	2.1
1990-94	3.1	3.3
Santa Fe		
4 cyl.	5.8	6.5
V6		
2.7L	3.0	3.4
3.5L	9.3	10.4
Scoupe	3.0	3.2
Sonata		
1989-98	3.6	3.8
1999-05		
4 cyl.	5.5	6.2
V6	2.4	2.7
Tiburon		
1997-01	2.4	2.6
2003-05	5.5	6.2
Tuscon		
4 cyl.	5.2	5.7
V6	8.4	8.9
XG300/350	8.8	9.9
w/AC add		
Accent, Elantra	.3	.3
Sonata	.6	.6
Tiburon		
4 cyl.	.3	.3
V6	.6	.6
w/DOHC add		
92-95 Elantra	1.4	1.4
89-98 Sonata 4 cyl.	1.3	1.3
Replace		
balance shaft seal add		
Elantra, Sonata	.3	.3
Santa Fe 4 cyl.	.3	.3
cam seal add		
Accent, Excel, Scoupe	.2	.2
Elantra	.5	.5
Santa Fe	1.0	1.0
89-05 Sonata		
4 cyl.	.5	.5
V6	1.0	1.0
Tiburon	.3	.3
XG300/350	1.0	1.0

	LABOR TIME	SEVERE SERVICE
Crankshaft Pulley, Replace (B)		
Exc. below	.5	.6
Santa Fe V6 3.5L	1.0	1.1
XG300/350	.8	.8
w/AC add	.5	.5
w/DOHC add		
92-95 Elantra	.5	.5
89-98 Sonata 4 cyl.	.2	.2
Crankshaft Sprocket, Replace (B)		
Accent		
1995-05	2.2	2.5
Elantra		
1992-95	2.2	2.5
1996-00	1.7	2.0
2001-05	2.4	2.7
Excel		
1986-89	1.9	2.1
1990-94	2.2	2.4
Pony, Stellar	5.2	5.4
Santa Fe		
4 cyl.	2.1	2.3
V6		
2.7L	2.5	2.9
3.5L	7.2	8.1
Scoupe	2.2	2.4
Sonata		
1992-98		
4 cyl.	3.2	3.6
V6	2.2	2.5
1999-05	2.2	2.5
Tiburon		
1997-01	1.7	2.0
2003-05	2.7	3.0
Tuscon		
4 cyl.	1.8	2.3
V6	4.5	5.0
XG300/350	4.7	5.3
w/AC add		
Accent	.5	.5
Elantra	.5	.5
Excel, Pony, Scoupe, Stellar	.3	.3
92-98 Sonata 4 cyl.	.5	.5
99-05 Sonata V6	.3	.3
Tiburon	.5	.5
w/DOHC add		
92-95 Elantra	1.3	1.4
92-98 Sonata 4 cyl.	1.1	1.3
w/PS add		
Accent	.3	.3
Elantra	.3	.3
Pistons or Connecting Rods, Replace (A)		
Includes: Ridge reaming, cylinder wall deglazing, installing new rings and rod bearings, engine tune-up.		
Accent		
1995-99	8.4	8.9
2000-05	10.2	10.7

HYU-16 ACCENT : ELANTRA : EXCEL : PONY : SANTA FE : SCOUPE : SONATA

	LABOR TIME	SEVERE SERVICE
Pistons or Connecting Rods, Replace (A)		
Elantra		
1992-95	9.5	10.0
1996-05	10.3	10.8
Excel	9.7	10.2
Pony, Stellar	9.7	10.2
Santa Fe		
4 cyl.	10.6	11.1
V6		
2.7L	12.1	12.6
3.5L	16.1	16.6
Scoupe	9.7	10.2
Sonata		
1989-98		
4 cyl.	10.3	10.8
V6	13.4	13.9
1999-05		
4 cyl.	10.6	11.1
V6	11.6	12.1
Stellar 2.0L	9.3	9.8
Tiburon		
1997-01	10.3	10.8
2003-05		
4 cyl.	9.6	11.1
V6	14.4	14.9
Tuscon		
4 cyl.	5.0	5.5
V6	8.2	8.7
XG300/350	14.5	15.0
w/AC add		
89-98 Sonata V6	.2	.2
04 Tiburon	.5	.5
w/DOHC add		
92-95 Elantra	1.0	1.0
Scoupe	.7	.7
89-98 Sonata 4 cyl.	.5	.5
w/turbo add	.3	.3
Rear Main Oil Seal, Replace (B)		
Accent		
1995-99	5.3	5.8
2000-05	7.3	7.8
Elantra		
1992-95	4.6	5.1
1996-00	6.1	6.6
2001-05	7.3	7.8
Excel		
1986-89		
AT	5.3	5.8
MT		
4-Speed	5.0	5.5
5-Speed	5.2	5.7
1990-94		
AT	5.3	5.8
MT	4.6	5.1
Pony, Stellar		
AT	3.1	3.6
MT	2.9	3.4
Santa Fe		
4 cyl.	8.8	9.4
V6	10.5	11.0

	LABOR TIME	SEVERE SERVICE
Scoupe		
MT	4.6	5.1
AT	5.3	5.8
Sonata		
1989-98	6.0	6.5
1999-05		
4 cyl.	8.8	9.4
V6	9.7	10.3
Tiburon		
1997-01	6.1	6.6
2003-05	7.8	8.4
Tuscon		
4 cyl.	7.7	8.2
V6	9.2	9.7
XG300/350	12.1	12.6
w/4WD Santa Fe add	1.0	1.1
w/AT not listed add	1.0	1.1
Rings, Replace (A)		
Includes: Ridge reaming, cylinder wall deglazing, installing new rings, engine tune-up.		
Accent		
1995-99	7.9	8.4
2000-05	9.4	9.9
Elantra		
1992-95	8.5	9.0
1996-05	9.5	10.0
Excel	8.7	9.2
Pony, Stellar	8.6	9.1
Santa Fe		
4 cyl.	9.7	10.3
V6		
2.7L	11.3	11.8
3.5L	14.9	15.4
Scoupe	8.7	9.3
Sonata		
1989-98		
4 cyl.	10.3	10.8
V6	12.9	13.4
1999-05		
4 cyl.	9.7	10.3
V6	10.9	11.4
Stellar 2.0L	8.0	8.5
Tiburon		
1997-01	9.4	9.0
2003-05		
4 cyl.	9.0	9.5
V6	11.5	12.0
Tuscon		
4 cyl.	10.9	11.4
V6	14.5	15.0
XG300/350	13.7	14.2
w/AC add		
89-05 Sonata V6	.2	.2
03-05 Tiburon	.5	.5
w/DOHC add		
92-95 Elantra	1.0	1.0
Scoupe	.7	.7
89-98 Sonata 4 cyl.	.5	.5
w/turbo add	.3	.3

	LABOR TIME	SEVERE SERVICE
ENGINE LUBRICATION		
Oil Pan and/or Gasket, Replace (B)		
Accent	1.6	1.8
Elantra		
1992-95	1.6	1.8
1996-02	2.2	2.5
2003-05	1.7	2.0
Excel	1.6	1.8
Pony, Stellar	1.2	1.4
Scoupe	1.6	1.8
Santa Fe	1.3	1.4
4 cyl.		
upper	1.7	2.0
lower	1.1	1.3
V6		
upper	2.4	2.7
lower	1.4	1.6
Sonata		
1989-98		
4 cyl.	1.6	1.8
V6	2.7	3.0
1999-05		
4 cyl.		
upper	1.6	1.8
lower	1.1	1.3
V6		
upper	1.9	2.1
lower	1.3	1.4
Tiburon		
1997-01	2.2	2.5
2002-05		
4 cyl.	1.9	2.1
V6		
upper	2.1	2.4
lower	1.3	1.5
Tuscon		
4 cyl.	1.7	2.0
V6		
upper	2.0	2.3
lower	1.2	1.5
XG300/350		
upper	2.2	2.5
lower	1.3	1.4
Oil Pressure Gauge (Dash), Replace (B)		
Elantra		
1992-95	1.3	1.3
1996-00	1.0	1.0
Scoupe	1.3	1.3
1989-98 Sonata	.8	.8
1997-01 Tiburon	1.6	1.8
Oil Pressure Gauge (Engine), Replace (B)		
Includes: Pressure test.		
1985-89		
analog gauge	.5	.5
warning light	.3	.3
1990-05		
sending unit	.5	.5
switch assembly	.3	.3
w/turbo add	.2	.2

STELLAR : TIBURON : TUSCON : XG300 : XG350 **HYU-17**

	LABOR TIME	SEVERE SERVICE
Oil Pump, Replace (B)		
Accent, Elantra	5.0	5.5
Excel	4.8	5.3
Pony, Stellar	.8	1.3
Santa Fe	5.8	6.3
V6 3.5L	9.8	10.3
Scoupe	4.8	5.3
Sonata		
1989-98	5.0	5.5
1999-05	7.4	7.9
Stellar 2.0L	4.4	4.8
Tiburon		
1997-01	5.0	5.7
2003-05	5.7	6.3
Tuscon	5.0	5.5
XG300/350	9.0	10.0
w/AC add		
Accent	.5	.5
92-95 Elantra	.6	.6
96-05 Elantra	.3	.3
Sonata 4 cyl.	.6	.6
Sonata V6	.2	.2
Tiburon, Tuscon 4 cyl.	.3	.3
Tiburon, Tuscon V6	.6	.6
w/DOHC add		
Elantra	1.1	1.1
89-98 Sonata 4 cyl.	1.3	1.4
w/PS add		
00-05 Accent	.3	.3
92-95 Elantra	.3	.3
99-05 Sonata	.3	.3

CLUTCH

	LABOR TIME	SEVERE SERVICE
Bleed Clutch Hydraulic System (B)		
1985-05	.5	.5
Clutch Pedal Free Play, Adjust (B)		
1985-05	.3	.3
Clutch Assy., Replace (B)		
Accent		
1995-99	3.0	3.4
2000-05	6.0	6.7
Elantra		
1992-95	4.1	4.6
1996-00	5.2	5.8
2001-05	7.0	7.9
Excel		
1986-89		
4-Speed	4.4	4.6
5-Speed	4.8	5.0
1990-94	4.3	4.5
Pony, Stellar		
4-Speed	2.4	2.6
5-Speed	2.7	2.9
Santa Fe	8.5	9.6
Scoupe	4.3	4.5
Sonata		
1989-98	4.1	4.6
1999-05	8.5	9.6

	LABOR TIME	SEVERE SERVICE
Tiburon		
1997-01	5.2	5.8
2002-05		
4 cyl.	6.9	7.7
V6	7.3	8.2
Tuscon		
2WD	6.6	7.1
4WD	7.6	8.1
Clutch Cable, Replace (B)		
Pony, Stellar	.5	.7
Clutch Master Cylinder, R&R and Reconditon (A)		
Includes: System bleeding.		
Accent	1.1	1.3
Elantra		
1992-95	1.1	1.2
1996-00	1.8	2.1
2001-05	1.4	1.6
1990-94 Excel	1.2	1.5
Santa Fe	1.4	1.6
Scoupe		
1991-92	1.2	1.5
1993-95	1.6	1.9
Sonata	1.3	1.4
Tiburon		
1997-01	1.8	2.1
2002-05	1.4	1.6
Tuscon	1.4	1.6
Clutch Master Cylinder, Replace (B)		
Includes: System bleeding.		
Accent	1.0	1.1
Elantra		
1992-95	.7	.8
1996-00	1.7	1.8
2001-05	1.3	1.4
1990-94 Excel	.7	.8
Santa Fe	1.3	1.4
Scoupe		
1991-92	.9	1.0
1993-95	1.3	1.4
Sonata	1.0	1.1
Tiburon		
1997-01	1.7	1.8
2003-05	1.3	1.4
Tuscon	1.2	1.3
Clutch Release Bearing, Replace (B)		
Accent		
1995-99	3.1	3.4
2000-05	5.7	6.3
Elantra		
1992-95	4.1	4.6
1996-00	5.1	5.7
2001-05	7.0	7.9
Excel		
1986-89		
4-Speed	4.2	4.4
5-Speed	4.5	4.7
1990-94	3.8	4.0

	LABOR TIME	SEVERE SERVICE
Pony, Stellar		
4-Speed	2.0	2.2
5-Speed	2.4	2.6
Santa Fe	8.5	9.6
Scoupe	3.8	4.0
Sonata		
1989-98	4.1	4.6
1999-05	8.5	9.6
Tiburon		
1997-01	5.1	5.7
2003-05		
4 cyl.	6.9	7.7
V6	7.3	8.2
Tuscon		
2WD	6.6	7.1
4WD	7.6	8.1
Clutch Slave Cylinder, R&R and Reconditon (B)		
Includes: System bleeding.		
All Models	1.2	1.4
Clutch Slave Cylinder, Replace (B)		
Includes: System bleeding.		
Exc. below	.9	1.0
1992-95 Elantra	1.1	1.2
Tiburon	1.1	1.2
Flywheel, Replace (B)		
Accent		
1995-99	3.2	3.6
2000-05	6.9	7.7
Elantra		
1992-95	4.3	4.8
1996-00	5.3	6.0
2001-05	6.9	7.7
Excel		
1986-89		
4-Speed	4.8	5.0
5-Speed	5.2	5.4
1990-94	4.6	4.8
Santa Fe	9.0	10.0
Scoupe	4.3	4.5
Sonata		
1989-98	4.3	4.8
1999-05	8.7	9.7
Tiburon		
1997-01	5.3	6.0
2003-05		
4 cyl.	7.3	8.2
V6	7.8	8.7
Tuscon		
2WD	6.8	7.3
4WD	7.8	8.4

MANUAL TRANSAXLE

	LABOR TIME	SEVERE SERVICE
Speedometer Driven Gear and/or Seal, Replace (B)		
1986-05 (FWD)	.6	.6

HYU-18 ACCENT : ELANTRA : EXCEL : PONY : SANTA FE : SCOUPE : SONATA

	LABOR TIME	SEVERE SERVICE
Transaxle Assy., R&I (B)		
Includes: Reset toe.		
Accent		
1995-99	2.9	*3.1*
2000-05	6.9	*7.7*
Elantra		
1992-95	3.8	*4.0*
1996-05	6.9	*7.7*
Excel		
1986-89		
4-Speed	4.3	*4.5*
5-Speed	4.4	*4.6*
1990-94	3.8	*4.0*
Santa Fe	8.2	*9.2*
Scoupe	3.8	*4.0*
Sonata		
1989-98	3.8	*4.0*
1999-05	8.2	*9.2*
Tiburon	6.6	*7.4*
Tuscon		
2WD	6.1	*7.1*
4WD	7.3	*8.0*
Transaxle Assy., R&R and Recondition (A)		
Includes: Recondition transaxle only.		
Accent		
1995-99	9.0	*10.0*
2000-05	12.3	*13.8*
Elantra		
1992-95	10.0	*11.2*
1996-00	10.8	*12.0*
2001-05	12.5	*13.9*
Excel		
1986-89		
4-Speed	11.7	*12.3*
5-Speed	12.3	*12.9*
1990-94	9.3	*9.9*
Santa Fe	15.6	*17.5*
Scoupe	9.8	*10.4*
Sonata		
1989-98	10.0	*11.2*
1999-05	16.5	*18.4*
Tiburon		
1997-01	10.8	*12.0*
2002-05		
4 cyl.		
exc. below	12.3	*13.8*
MFA60 trans	13.1	*14.6*
V6		
M5GF5 trans	14.9	*16.6*
MFA60 trans	13.4	*15.0*
Tuscon		
2WD	14.0	*15.0*
4WD	15.0	*15.9*
Pull type clutch		
Elantra add	.3	*.3*

	LABOR TIME	SEVERE SERVICE
Transaxle Rear Cover, R&R or Replace (B)		
Accent	.8	*.9*
Elantra		
1992-95	.7	*.9*
1996-05	2.4	*2.7*
Excel, Scoupe	.7	*.9*
1989-98 Sonata	.7	*.9*
Tiburon		
1997-01	2.1	*2.3*
2003-05	2.5	*2.9*

MANUAL TRANSMISSION

	LABOR TIME	SEVERE SERVICE
Extension Housing and/or Gasket, Replace (B)		
Pony, Stellar	2.7	*2.9*
Extension Housing Oil Seal, Replace (B)		
Pony, Stellar	.7	*.9*
Transmission Assy., R&R and Recondition (A)		
Pony, Stellar		
4-Speed	6.3	*6.9*
5-Speed	7.1	*7.7*
Transmission Assy., R&R or Replace (B)		
Pony, Stellar		
4-Speed	2.0	*2.2*
5-Speed	2.4	*2.6*

AUTOMATIC TRANSAXLE
SERVICE TRANSAXLE INSTALLED

	LABOR TIME	SEVERE SERVICE
Check Unit for Oil Leaks (C)		
1986-05	.5	*.5*
Drain & Refill Unit (B)		
Accent	1.2	*1.4*
Elantra	1.1	*1.3*
Excel	1.0	*1.0*
Sonata	1.1	*1.3*
Tiburon, Tuscon, XG300/350	1.0	*1.1*
Kickdown Linkage, Adjust (B)		
1986-05	.3	*.3*
Oil Pressure Check (B)		
1986-05	1.0	*1.0*
Throttle Control Cable, Adjust (B)		
1986-89 Excel	.3	*.4*
ECT Switch, Replace (B)		
1986-05	.8	*.8*
Kickdown Switch or Solenoid, Replace (B)		
1986-05	.8	*.8*
Neutral Safety Switch, Replace (B)		
Exc. below	.5	*.5*
1996-00 Elantra	.8	*.8*
1997-05 Tiburon	.8	*.8*

	LABOR TIME	SEVERE SERVICE
Oil Pan and/or Gasket, Replace (B)		
Accent	1.3	*1.4*
Elantra	1.1	*1.2*
Excel	1.0	*1.0*
Sonata	1.1	*1.2*
Tiburon, Tuscon, XG300/350	1.0	*1.1*
Overdrive Switch, Replace (B)		
Accent, Elantra, Excel	1.3	*1.4*
Scoupe	1.3	*1.4*
1989-98 Sonata	1.7	*2.0*
Tiburon, Tuscon, XG300/350	1.1	*1.3*
Park Position Switch, Replace (B)		
Accent	.6	*.6*
Elantra, Tiburon, Tuscon, XG300/350	.8	*.8*
Excel, Sonata	.5	*.5*
Parking Pawl, Replace (B)		
1986-89	2.5	*2.7*
Pulse Generator, Replace (B)		
Accent	1.0	*1.1*
Elantra		
1992-95	.5	*.5*
1996-05	1.1	*1.3*
Excel, Scoupe	.5	*.5*
1989-98 Sonata	.5	*.5*
Tiburon, Tuscon, XG300/350	1.1	*1.3*
Solenoid Valve (All), Replace (B)		
Accent	2.1	*2.3*
Elantra		
1992-95	2.4	*2.7*
1996-00	1.9	*2.1*
2001-05	4.1	*4.6*
Excel		
1986-89	1.3	*1.5*
1990-94	2.4	*2.6*
Santa Fe	4.1	*4.6*
Scoupe	2.4	*2.6*
Sonata		
1989-98	2.4	*2.7*
1999-05	4.1	*4.6*
Tiburon		
1997-01	1.9	*2.1*
2002-05	4.1	*4.6*
Tuscon	3.6	*4.1*
XG300/350	4.1	*4.6*
Speedometer Driven Gear and/or Seal, Replace (B)		
1986-05	.6	*.6*
Throttle Lever Cable (Valve) Control, Replace (B)		
1986-89 Excel	1.3	*1.5*
Transaxle Control Unit, Replace (B)		
Accent	.6	*.7*
Elantra		
1992-95	1.3	*1.4*
1996-05	.8	*.9*

STELLAR : TIBURON : TUSCON : XG300 : XG350 HYU-19

	LABOR TIME	SEVERE SERVICE
Santa Fe	1.4	1.6
Sonata		
1989-98	1.0	1.1
1999-05 V6	2.9	3.2
Tiburon, Tuscon	1.0	1.1
XG300/350	1.4	1.6

Valve Body, R&R and Recondition (A)

Accent	4.0	4.4
Elantra		
1992-95	4.9	5.5
1996-00	3.6	4.1
2001-05	6.0	6.7
Excel, Scoupe	5.0	5.4
Santa Fe	6.0	6.7
Sonata		
1989-98	4.9	5.5
1999-05	6.0	6.7
Tiburon		
1997-01	3.6	4.1
2003-05	6.0	6.7
Tuscon	5.4	6.0
XG300/350	6.0	6.7

Transaxle Solenoid Assy., Replace (B)

1986-05	.8	.9

Valve Body, Replace (B)

Accent	1.7	2.0
Elantra		
1992-95	2.1	2.3
1996-00	1.6	1.8
2001-05	4.3	4.8
Excel	2.0	2.2
Santa Fe	4.3	4.8
Scoupe	2.0	2.2
Sonata		
1989-98	2.1	2.3
1999-05	4.3	4.8
Tiburon		
1997-01	1.6	1.8
2003-05	4.4	5.0
Tuscon	3.9	4.4
XG300/350	4.3	4.8

SERVICE TRANSAXLE REMOVED
Transaxle R&R included unless otherwise noted.

Flywheel (Flexplate), Replace (B)

Accent		
1995-99	4.9	5.5
2000-05	7.8	8.7
Elantra		
1992-95	4.6	5.1
1996-00	5.3	6.0
2001-05	7.8	8.7
Excel		
1986-89	5.2	5.4
1990-94	5.0	5.2

	LABOR TIME	SEVERE SERVICE
Santa Fe		
4 cyl.	10.5	11.7
V6	10.3	11.5
Scoupe	5.0	5.2
Sonata		
1989-98		
4 cyl.	4.9	5.5
V6	5.7	6.3
1999-05		
4 cyl.	8.5	9.6
V6	9.5	10.7
Tiburon		
1997-01	5.3	6.0
2003-05		
4 cyl.	7.3	8.2
V6	8.2	9.2
Tuscon		
2WD	6.7	7.5
4WD	8.5	9.3
XG300/350	11.8	13.2

Front Pump Oil Seal, Replace (B)

Accent		
1995-99	5.5	6.2
2000-05	7.9	8.9
Elantra		
1992-95	6.3	7.0
1996-00	5.7	6.3
2001-05	8.2	9.2
Excel		
1986-89	5.2	5.4
1990-94	5.9	6.1
Santa Fe		
exc. below	10.3	11.5
4WD ring seal	12.2	13.7
Scoupe	5.9	6.1
Sonata		
1989-98		
4 cyl.	6.3	7.0
V6	7.5	8.4
1999-05	10.3	11.5
Tiburon		
1997-01	5.7	6.3
2003-05		
4 cyl.	8.2	9.2
V6	9.0	10.0
Tuscon		
2WD	7.6	8.3
4WD	8.9	9.8
XG300/350	12.3	13.8

Governor Assy., Replace or Recondition (B)

1986-89 Excel	8.9	9.1
Recondition gov. add	.5	.7

Oil Pump and/or Gasket, Replace (B)

Accent		
1995-99	5.2	5.8
2000-05	8.1	9.0

	LABOR TIME	SEVERE SERVICE
Elantra		
1992-95	6.3	7.0
1996-00	5.7	6.3
2001-05	8.2	9.2
Excel		
1986-89	5.2	5.4
1990-94	6.2	6.4
Santa Fe	10.8	12.0
Scoupe	6.2	6.4
Sonata		
1989-98		
4 cyl.	6.2	7.0
V6	7.5	8.4
1999-05	10.3	11.5
Tiburon		
1997-01	5.7	6.3
2003-05	6.0	6.7
4 cyl.	8.2	9.2
V6	8.8	9.9
Tuscon		
2WD	7.7	8.4
4WD	9.0	9.9
XG300/350	12.8	14.3

Torque Converter, Replace (B)

Accent		
1995-99	4.7	5.3
2000-05	7.8	8.7
Elantra		
1992-95	4.6	5.1
1996-00	5.3	6.0
2001-05	7.3	8.2
Excel		
1986-89	4.7	4.9
1990-94	4.5	4.7
Santa Fe		
2WD	9.7	10.9
4WD	11.5	12.8
Scoupe	4.6	4.8
Sonata		
1989-98		
4 cyl.	4.6	5.1
V6	5.8	6.5
1999-05	9.5	10.7
Tiburon		
1997-01	5.3	6.0
2003-05		
4 cyl.	7.3	8.2
V6	7.9	8.9
Tuscon		
2WD	6.6	7.3
4WD	7.7	8.5
XG300/350	11.6	13.0

Transaxle Assy., R&R and Recondition (A)

Accent		
1995-99	11.9	13.3
2000-05	14.6	16.3

	LABOR TIME	SEVERE SERVICE
Transaxle Assy., R&R and Recondition (A)		
Elantra		
1992-95	**12.6**	*14.1*
1996-00	**13.2**	*14.8*
2001-05	**16.8**	*18.8*
Excel		
1986-89	**13.3**	*14.2*
1990-94	**12.9**	*13.8*
Santa Fe		
2WD	**18.8**	*21.0*
4WD		
exc. 3.5L	**21.8**	*24.3*
3.5L	**20.6**	*23.0*
Scoupe	**12.9**	*13.8*
Sonata		
1989-98		
4 cyl.	**12.6**	*14.1*
V6	**13.7**	*15.4*
1999-05	**18.9**	*21.1*
Tiburon		
1997-01	**13.2**	*14.8*
2003-05		
4 cyl.	**16.7**	*18.7*
V6	**17.4**	*19.4*
Tuscon		
2WD	**15.5**	*16.4*
4WD	**16.9**	*17.8*
XG300/350	**20.6**	*23.0*
Transaxle Assy., Replace (B)		
Accent		
1995-99	**4.7**	*5.3*
2000-05	**7.6**	*8.5*
Elantra		
1992-95	**4.4**	*5.0*
1996-00	**5.2**	*5.8*
2001-05	**7.3**	*8.2*
Excel		
1986-89	**4.6**	*4.8*
1990-94	**4.4**	*4.6*
Santa Fe		
2WD	**10.0**	*11.2*
4 WD	**11.5**	*12.8*
Scoupe	**4.4**	*4.6*
Sonata		
1989-98		
4 cyl.	**4.4**	*5.0*
V6	**5.7**	*6.3*
1999-05	**9.4**	*10.6*
Tiburon		
1997-01	**5.2**	*5.8*
2003-05		
4 cyl.	**7.3**	*8.2*
V6	**7.9**	*8.9*
Tuscon		
2WD	**6.6**	*7.4*
4WD	**8.4**	*9.2*
XG300/350	**11.5**	*12.8*

AUTOMATIC TRANSMISSION

SERVICE TRANSMISSION INSTALLED

	LABOR TIME	SEVERE SERVICE
Check Unit for Oil Leaks (C)		
Pony, Stellar	**.9**	*.9*
Drain and Refill Unit (B)		
Pony, Stellar	**.5**	*.5*
Oil Pressure Check (B)		
Pony, Stellar	**1.2**	*1.2*
Throttle Cable, Adjust (B)		
Pony, Stellar	**.3**	*.4*
Extension Housing Oil Seal, Replace (B)		
Pony, Stellar	**.8**	*1.0*
Governor Assy., R&R and Recondition (B)		
Pony, Stellar	**2.5**	*3.9*
Governor Assy., R&R or Replace (B)		
Pony, Stellar	**2.2**	*2.6*
Neutral Safety Switch, Replace (B)		
Pony, Stellar	**.5**	*.5*
Oil Pan Gasket, Replace (B)		
Pony, Stellar	**.8**	*.8*
Speedometer Driven Gear and/or Seal, Replace (B)		
Pony, Stellar	**.5**	*.5*
Throttle Cable, Replace (B)		
Pony, Stellar	**1.2**	*1.3*
Valve Body Assy., R&R and Recondition (A)		
Pony, Stellar	**3.7**	*4.1*
Valve Body Assy., Replace (B)		
Pony, Stellar	**2.0**	*2.2*

SERVICE TRANSMISSION REMOVED

Transmission R&R included unless otherwise noted.

	LABOR TIME	SEVERE SERVICE
Flywheel (Flexplate), Replace (B)		
Pony, Stellar	**3.1**	*3.3*
Front Oil Pump, Replace (B)		
Pony, Stellar	**3.7**	*3.9*
Recondition pump add	**.5**	*.8*
Front Pump Oil Seal, Replace (B)		
Pony, Stellar	**3.4**	*3.6*
Torque Converter, Replace (B)		
Pony, Stellar	**3.1**	*3.3*
Transmission Assy., R&R (B)		
Pony, Stellar	**3.1**	*3.3*
Transmission Assy., R&R and Recondition (A)		
Pony, Stellar	**9.4**	*10.3*
Stellar 2.0L	**9.8**	*10.7*

SHIFT LINKAGE

AUTOMATIC

	LABOR TIME	SEVERE SERVICE
Floor Shift Control Assy., Replace (B)		
Exc. Santa Fe	**1.1**	*1.1*
Santa Fe	**2.1**	*2.1*
Gearshift Control Cable or Rod, Replace (B)		
Accent	**1.7**	*2.0*
Elantra	**1.9**	*2.1*
Excel		
1986-89	**1.2**	*1.4*
1990-94	**1.4**	*1.6*
Santa Fe	**1.7**	*2.0*
Scoupe	**1.6**	*1.8*
Sonata, Tiburon, Tuscon, XG300/350	**2.2**	*2.5*

MANUAL

	LABOR TIME	SEVERE SERVICE
Gearshift Control Assy. (Floor Shift), R&R or Replace (B)		
1985-05	**.8**	*.8*
Transaxle Shift Cable, Replace (B)		
1986-05	**.5**	*.6*

DRIVELINE

DRIVESHAFT

	LABOR TIME	SEVERE SERVICE
Center Support Bearing, Replace (B)		
Pony, Santa Fe, Stellar, Tuscon	**1.4**	*1.6*
Center Yoke or Flange, Replace (B)		
Pony, Santa Fe, Stellar, Tuscon	**1.1**	*1.3*
Driveshaft R&R or Replace (B)		
Pony, Santa Fe, Stellar, Tuscon	**.8**	*.9*
U-Joint, Replace or Recondition (B)		
Pony, Santa Fe, Stellar, Tuscon	**1.4**	*1.5*

REAR DIFFERENTIAL

	LABOR TIME	SEVERE SERVICE
Axle Shaft Oil Seal, Replace (B)		
Pony, Stellar	**.8**	*1.0*
Differential Carrier, R&R and Recondition (B)		
Pony, Stellar	**4.8**	*5.3*
Santa Fe	**4.3**	*4.8*
Tuscon	**4.3**	*4.8*
w/limited slip add	**.3**	*.3*
Differential Carrier, R&R or Replace (B)		
Pony, Stellar	**2.9**	*3.4*
Santa Fe	**1.9**	*2.1*
Tuscon	**2.4**	*2.8*
w/limited slip add	**.3**	*.3*

STELLAR : TIBURON : TUSCON : XG300 : XG350   **HYU-21**

	LABOR TIME	SEVERE SERVICE
Differential Side Bearing, Replace (B)		
Pony, Stellar, Tuscon	3.1	3.6
Santa Fe		
right or left side	3.0	3.4
both sides	3.2	3.6
Pinion Shaft Oil Seal, Replace (B)		
Pony, Stellar	.9	1.0
Santa Fe	1.3	1.4
Tuscon	2.7	3.1
Rear Axle Shaft, Replace (B)		
Pony, Stellar one side	1.3	1.5
Santa Fe, Tuscon		
one side	1.9	2.1
Ring Gear & Pinion Assy., Replace (B)		
Pony, Stellar	4.4	4.9
Santa Fe, Tuscon	3.5	4.0
Replace pinion bearings add	.3	.3

HALFSHAFTS

Halfshaft, R&R or Replace (B)		
Exc. below	1.1	1.3
1999-05 Sonata	.6	.7
XG300/350	.6	.7
Inner CV Joint or Boot, Replace (B)		
Exc. below	1.3	1.4
Elantra, Tiburon	1.4	1.6
Sonata, XG300/350	1.0	1.1
w/Birfield joint add	.2	.2
Inner Shaft, R&R or Replace (B)		
Santa Fe	1.4	1.6
Sonata		
1989-98 V6	1.6	1.8
1999-05 V6	1.1	1.3
Scoupe, Tuscon	1.6	1.8
2003-05 Tiburon V6	1.6	1.8
XG300/350	1.3	1.4
Replace bearing or dust seal add	.2	.2
Outer CV-Joint, Replace (B)		
1995-05 Accent	1.3	1.4
Elantra	1.6	1.8
Excel, Santa Fe	1.3	1.4
Scoupe, Tuscon	1.2	1.4
Sonata		
1989-98	1.4	1.6
1999-05	1.0	1.1
Tiburon	1.6	1.8
XG300/350	1.0	1.1

BRAKES
ANTI-LOCK

The following operations do not include testing. Add time as required.

Diagnose Anti-Lock Brake System Component Each (A)		
1992-05	1.3	1.4
Anti-Lock Relay, Replace (B)		
Accent	.5	.5
Elantra	1.1	1.1
Sonata		
1989-93	.3	.3
1994-98	.8	.8
Tiburon, Tuscon, XG300/350	.9	.9
Anti-Lock Control Module, Replace (B)		
All Models	.6	.6
Front Sensor Assy., Replace (B)		
Accent		
one side	.6	.6
both sides	1.1	1.1
Elantra		
1992-95		
one side	.6	.6
both sides	1.1	1.1
1996-00		
one side	.9	.9
both sides	1.5	1.7
2001-05		
one side	.5	.5
both sides	1.0	1.1
Santa Fe		
one side	.6	.6
both sides	.9	.9
Sonata		
one side	.3	.3
both side	.5	.5
Tiburon		
1997-01	.9	1.0
2003-05		
one side	1.0	1.0
both sides	1.1	1.2
Tuscon, XG300/350		
one side	.5	.5
both sides	.8	.8
Hydraulic Modulator Assy., R&R or Replace (B)		
Includes: System bleeding.		
Accent	1.3	1.4
Elantra		
1992-95	4.3	4.7
1996-00	2.0	2.2
2001-05	1.4	1.6
Santa Fe	1.6	1.8
Sonata		
1989-93	2.4	2.7
1994-98	2.8	3.1
1999-05	1.4	1.6
Tiburon		
1997-01	2.0	2.2
2003-05	1.6	1.8
Tuscon, XG300/350	2.2	2.5
Rear Sensor Assy., Replace (B)		
Accent, Elantra		
one side	1.0	1.1
both sides	1.3	1.4

Santa Fe		
one side	1.9	2.1
both sides	2.5	2.9
Sonata		
1989-93		
one side	.5	.5
both sides	.7	.7
1994-98		
one side	.6	.6
both sides	1.1	1.1
1999-05		
one side	.8	.9
both sides	1.0	1.1
Tiburon, Tuscon, XG300/350		
one side	.5	.5
both sides	.8	.8

SYSTEM

Bleed Brakes (B)		
Includes: Add fluid.		
1985-05	.5	.5
Brake System, Flush and Refill (B)		
1985-05	1.2	1.2
Brakes, Adjust (B)		
Includes: System bleeding.		
1985-05 rear wheels	.6	.6
Brake Hose (Flexible), Replace (B)		
Includes: System bleeding.		
Exc. below	.8	.9
Pony, Stellar front	.9	1.0
2003-05 Tiburon		
front or rear	1.0	1.1
each addl. add	.3	.4
Combination Valve, Replace (B)		
Includes: System bleeding.		
Accent	.9	1.0
Elantra		
1992-95	1.4	1.5
1996-05	.9	1.0
Excel		
1986-89	.9	1.1
1990-94	1.2	1.4
Pony, Stellar	1.3	1.5
Santa Fe	1.1	1.2
Scoupe	1.4	1.6
Sonata		
1994-98	1.4	1.5
1999-05	1.1	1.2
Tiburon	.8	.9
Tuscon, XG300/350	1.2	1.4
Master Cylinder, R&R and Rebuild (B)		
Includes: System bleeding.		
Accent	1.7	1.9
Elantra		
1992-95	1.7	1.9
1996-05	1.2	1.4
Excel	1.5	1.8
Pony, Stellar	1.9	2.2
Santa Fe, Scoupe	1.5	1.7

HYU-22 ACCENT : ELANTRA : EXCEL : PONY : SANTA FE : SCOUPE : SONATA

	LABOR TIME	SEVERE SERVICE
Master Cylinder, R&R and Rebuild (B)		
Sonata		
1989-98	1.7	1.9
1999-05	1.1	1.2
Tiburon	1.5	1.7
Tuscon, XG300/350	1.1	1.2
Master Cylinder, Replace (B)		
Includes: System bleeding.		
Accent		
1995-99	1.7	1.9
2000-05	.8	.9
Elantra		
1992-95	1.4	1.5
1996-05	1.1	1.2
Excel	1.2	1.4
Pony, Scoupe, Stellar	1.6	1.8
Santa Fe	1.2	1.4
Sonata		
1989-98	1.2	1.4
1999-05	.9	1.0
Tiburon	1.2	1.4
Tuscon, XG300/350	.9	1.0
Power Brake Booster, Replace (B)		
Includes: System bleeding.		
Accent	1.8	2.1
Elantra	1.7	1.9
Excel		
1986-89	1.6	1.8
1990-94	.7	.9
Pony, Stellar	1.9	2.1
Santa Fe	1.5	1.7
Scoupe	.7	.9
Sonata		
1989-98	6.3	7.0
1999-05	1.4	1.5
Tiburon	1.7	1.9
Tuscon, XG300/350	1.4	1.5
w/AC add		
91-95 Elantra	2.6	2.6
Excel, Pony, Scoupe, Stellar	1.9	1.9
89-98 Sonata	1.4	1.4
Vacuum Check Valve, Replace (B)		
1985-05	.3	.3

SERVICE BRAKES

	LABOR TIME	SEVERE SERVICE
Brake Drum, Replace (B)		
1985-05	.6	.7
Caliper Assy., R&R and Recondition (B)		
Includes: System bleeding.		
Front		
Accent, Excel		
one side	1.1	1.3
both sides	1.7	1.9
Elantra		
one side	1.1	1.3
both sides	1.4	1.6

	LABOR TIME	SEVERE SERVICE
Pony, Stellar		
one side	1.2	1.5
Santa Fe		
one side	1.3	1.4
both sides	2.1	2.3
Scoupe		
one side	1.3	1.6
both sides	2.3	2.6
Sonata		
1989-98		
one side	1.4	1.5
both sides	2.0	2.2
1999-05		
one side	.8	.9
both sides	1.3	1.4
Tiburon, Tuscon		
one side	1.3	1.4
both sides	1.6	1.8
XG300/350		
one side	.8	.9
both sides	1.3	1.4
Rear		
Elantra		
1995		
one side	1.4	1.5
both sides	1.8	2.1
1996-00		
one side	1.7	1.9
both sides	2.3	2.6
2001-05		
one side	1.1	1.3
both sides	1.6	1.8
Santa Fe		
one side	1.1	1.2
both sides	1.5	1.7
Sonata		
1989-98		
one side	1.4	1.5
both sides	1.8	2.1
1999-05		
one side	1.0	1.1
both sides	1.4	1.6
Tiburon		
1997-01		
one side	1.7	1.9
both sides	2.3	2.6
2003-05		
one side	1.3	1.4
both sides	1.9	2.1
Tuscon, XG300/350		
one side	1.0	1.1
both sides	1.6	1.8
Caliper Assy., Replace (B)		
Includes: System bleeding.		
Front		
Accent, Elantra, Excel		
one side	.8	.9
both sides	1.3	1.4

	LABOR TIME	SEVERE SERVICE
Pony, Stellar		
one side	.7	.8
Santa Fe		
one side	.8	.9
both sides	1.3	1.4
Scoupe		
one side	.7	.8
both sides	1.3	1.4
Sonata		
1989-98		
one side	.9	1.0
both sides	1.5	1.7
1999-05		
one side	.5	.6
both sides	.8	.9
Tiburon		
one side	.8	.9
both sides	1.3	1.4
Tuscon, XG300/350		
one side	.5	.5
both sides	.8	.9
Rear		
Elantra		
1995		
one side	.9	1.0
both sides	1.3	1.4
1996-00		
one side	1.4	1.5
both sides	1.8	2.1
2001-05		
one side	.8	.9
both sides	1.3	1.4
Santa Fe, Sonata		
one side	.9	1.0
both sides	1.3	1.4
Tiburon		
1997-01		
one side	1.3	1.3
both sides	1.6	1.7
2003-05		
one side	1.0	1.1
both sides	1.4	1.6
Tuscon, XG300/350		
one side	.8	.9
both sides	1.1	1.3
Disc Brake Rotor, Replace (B)		
Front		
Accent		
1995-99	1.2	1.4
2000-05	.5	.6
Elantra		
1992-00	1.7	1.9
2001-05	.6	.7
Excel		
1986-89	1.4	1.5
1990-94	1.5	1.6
Pony, Stellar	.7	.8
Santa Fe	.6	.7
Scoupe	1.7	1.8
Sonata	.6	.7

STELLAR : TIBURON : TUSCON : XG300 : XG350 HYU-23

	LABOR TIME	SEVERE SERVICE
Tiburon	1.5	1.6
1997-01	1.7	1.9
2003-05	.6	.7
Tuscon, XG300/350	.5	.5
Rear		
All Models	.6	.7

Pads and/or Shoes, Replace (B)
Includes: Service and parking brake adjustment, system bleeding.
- Front disc
  - exc. below 9   1.0
  - Santa Fe 5   .5
  - 1999-05 Sonata 5   .5
  - XG300/350 6   .7
- Rear disc
  - Elantra, Sonata, Tuscon 9   1.0
  - Santa Fe, Tiburon, XG300/350 6   .7
- Rear drum
  - exc. below 1.3 1.4
  - 1995-99 Accent 1.7 1.9
  - 1992-95 Elantra ... 1.5 1.7
  - Excel 1.5 1.6
  - Pony, Stellar 1.9 2.0
  - Santa Fe, Scoupe .. 1.6 1.8

COMBINATION ADD-ONS
Repack wheel bearings
- two wheels add 7   .7

Replace
- brake drum add 2   .2
- brake hose add 3   .3
- caliper add 3   .3
- disc rotor add
  - FWD 3   .3
  - RWD 2   .2
- master cylinder add . .5   .5
- wheel cylinder add .. .2   .2

Resurface
- brake drum add 5   .5
- disc rotor add
  - FWD 5   .5
  - RWD 9   .9

Wheel Cylinder, R&R and Rebuild (B)
Includes: System bleeding.
- Exc. below
  - one side 1.3 1.4
  - both sides 2.2 2.5
- 1986-89 Excel
  - one side 1.7 2.0
  - both sides 3.1 3.4

Wheel Cylinder, Replace (B)
Includes: System bleeding.
- Exc. below
  - one side 1.1 1.3
  - both sides 1.9 2.1
- Pony, Stellar one side .. .7   .8

PARKING BRAKE
Parking Brake Cable, Adjust (C)
- 1985-05 3   .5

Parking Brake Apply Actuator, Replace (B)
- Accent, Excel 6   .6
- Elantra 8   .8
- Pony, Scoupe, Stellar .. .7   .7
- Santa Fe 7   .7
- Sonata 5   .5
- Tiburon, Tuscon 8   .8
- XG300/350 5   .5

Parking Brake Apply Warning Indicator Switch, Replace (B)
- All Models 6   .6

Parking Brake Cable, Replace (B)
- Accent
  - 1995-99
    - one side 1.5 1.7
    - both sides 2.1 2.4
  - 2000-05
    - one side 1.3 1.4
    - both sides 1.7 2.0
- Elantra
  - 1992-95
    - one side 1.5 1.7
    - both sides 2.1 2.3
  - 1996-05
    - one side 1.9 2.1
    - both sides 2.7 3.0
- Excel, Scoupe
  - one side 1.4 1.6
  - both sides 1.9 2.1
- Pony, Stellar
  - one side 9   1.1
- Santa Fe
  - one side 8   .9
  - both sides 1.2 1.4
- Sonata
  - 1989-98
    - one side 1.7 1.9
    - both sides 2.4 2.7
  - 1999-05
    - one side 1.6 1.8
    - both sides 1.9 2.1
- Tiburon
  - 1997-01
    - one side 2.0 2.2
    - both sides 2.6 2.9
  - 2003-05
    - one side 1.0 1.1
    - both sides 1.6 1.8
- Tuscon
  - one side 2.0 2.2
  - both sides 2.6 2.9
- XG300/350
  - one side 1.6 1.8
  - both sides 1.9 2.1

FRONT SUSPENSION
Unless otherwise noted, time given does not include alignment.

Align Front End (B)
Includes: Adjust caster (when possible), and center steering wheel.
- All FWD models 1.2 1.4
- Pony, Stellar 1.5 1.7

Front & Rear Alignment, Check & Adjust (B)
- All FWD models 2.0 2.2

Front Toe, Adjust (B)
- 1985-05 6   .8

Coil Spring, Replace (B)
- Pony, Stellar
  - control arm suspension
    - one side 2.7 2.9
    - both sides 5.0 5.2
  - strut suspension .. 1.3 1.5

Front Strut Coil Spring, Replace (B)
- Accent 1.0 1.1
- Elantra
  - 1991-95 .... 1.4 1.5
  - 1996-05 .... 1.1 1.3
- Excel 1.6 1.8
- Santa Fe 1.4 1.6
- Scoupe 1.4 1.6
- Sonata
  - 1989-98 .... 1.4 1.6
  - 1999-05 .... 1.1 1.3
- Tiburon, Tuscon 1.3 1.4
- XG300/350 .... 1.1 1.3

Front Cross Member, Replace (B)
- 2000-05 Accent 4.9 5.5
- 1996-05 Elantra 6.1 6.9
- 1986-89 Excel
  - No. 1 .... 1.8 2.3
  - No. 2 .... 3.1 3.6
- Santa Fe
  - 2WD 6.3 7.0
  - 4WD 6.7 7.5
- Sonata
  - 1989-98 .... 4.1 4.6
  - 1999-05 .... 6.4 7.2
- Tiburon 6.1 6.9
- Tuscon
  - 2WD 5.4 6.0
  - 4WD 5.9 6.6
- XG300/350 .... 6.4 7.2

Front Hub Assy., Replace (B)
- Accent
  - 1995-99 .... 1.8 2.1
  - 2000-05 .... 1.4 1.6
- Elantra
  - 1991-00 .... 2.0 2.2
  - 2001-05 .... 1.4 1.6
- Excel 2.0 2.2
- Scoupe 2.0 2.2

HYU-24 ACCENT : ELANTRA : EXCEL : PONY : SANTA FE : SCOUPE : SONATA

	LABOR TIME	SEVERE SERVICE
Front Hub Assy., Replace (B)		
Sonata		
1989-98	2.0	2.2
1999-05	1.6	1.8
Tiburon		
1997-01	1.8	2.1
2003-05	1.6	1.8
Santa Fe, Tuscon, XG300/350	1.6	1.8
Front Shock Absorbers, Replace (B)		
Stellar		
one side	.5	.6
both sides	.7	.8
Front Stabilizer Link, Replace (B)		
Exc. below		
one side	.3	.4
both sides	.6	.7
2003-05 Tiburon		
one side	.5	.6
both sides	.6	.7
Front Strut Assembly, Replace (B)		
Accent	1.0	1.1
Elantra		
1991-95	1.4	1.5
1996-05	1.1	1.3
Excel	1.2	1.4
Pony, Stellar	1.7	1.9
Santa Fe, Scoupe	1.4	1.6
Sonata		
1989-98	1.4	1.6
1999-05	1.1	1.3
Tiburon, Tuscon	1.3	1.4
XG300/350	1.1	1.3
Front Strut Shock Absorber, Replace (B)		
Pony, Stellar	2.3	2.5
Stellar 2.0L	1.7	1.9
Lower Ball Joint, Replace (B)		
Accent	1.0	1.1
Elantra		
1992-95	.6	.9
1996-05	1.3	1.4
Excel	.6	.9
Pony, Stellar	1.3	1.6
Santa Fe, Tuscon	1.7	2.0
Scoupe	.7	1.0
Sonata		
1989-98	1.0	1.1
1999-05	.5	.5
Stellar 2.0L	.7	1.0
Tiburon	1.3	1.4
XG300/350	.5	.5
Lower Control Arm, Replace (B)		
Accent	.8	.9
Elantra		
1992-95	1.6	1.8
1996-05	1.1	1.3
Excel		
1986-89	1.2	1.5
1990-94	1.4	1.7

	LABOR TIME	SEVERE SERVICE
Pony, Stellar		
control arm suspension	3.0	3.3
strut suspension	.8	1.1
Santa Fe, Tuscon	1.6	1.8
Scoupe	1.6	1.9
Sonata		
1989-98	1.6	1.8
1999-05	.8	.9
Tiburon	1.1	1.3
XG300/350	.8	.9
Lower Control Arm Strut (Tension Bar), Replace (B)		
1986-89 Excel	.7	1.0
Shock Absorber Strut Cartridge, Replace (B)		
1986-89 Excel	1.8	2.1
Stabilizer Bar, Replace (B)		
Accent	.8	.9
Elantra		
1992-95	1.4	1.6
1996-00	4.1	4.6
2001-05	1.3	1.4
Excel		
1986-89	.5	.7
1990-94	1.2	1.4
Pony, Stellar	.7	.9
Santa Fe		
2WD	4.9	5.5
4 WD	5.5	6.2
Scoupe	1.4	1.6
Sonata		
1989-98	1.4	1.6
1999-05	3.2	3.6
Tiburon		
1997-01	3.5	3.7
2003-05	2.1	2.3
Tuscon		
2WD	4.2	4.7
4 WD	4.7	5.3
XG300/350	3.2	3.6
Stabilizer Bar Bushings, Replace (B)		
Accent		
one side	.8	.9
both sides	1.0	1.1
Elantra		
one side	.6	.7
both sides	1.0	1.1
Excel		
1986-89	.5	.6
1990-94		
one side	.6	.7
both sides	.8	.9
Pony, Stellar		
one side	.3	.4
both sides	.5	.6

	LABOR TIME	SEVERE SERVICE
Santa Fe		
2WD one or both sides	4.7	5.3
4WD one or both sides	5.3	6.0
Scoupe, Tiburon		
one side	.5	.6
both sides	.8	.9
Sonata		
1989-98		
one side	.6	.7
both sides	1.0	1.1
1999-05		
one side	1.0	1.1
both sides	1.7	2.0
Tiburon		
1997-01		
one side	.5	.5
both sides	1.0	1.1
2003-05		
one side	1.0	1.1
both sides	1.4	1.6
Tuscon		
2WD	4.1	4.6
4 WD	4.6	5.2
XG300/350		
one side	1.0	1.1
both sides	1.6	1.8
Stabilizer Bar Links and/or Grommets, Replace (B)		
Stellar 2.0L	1.2	1.4
Steering Knuckle, Replace (B)		
Accent		
1995-99	1.7	2.0
2000-05	1.4	1.6
Elantra		
1991-95	2.1	2.3
1996-05	1.6	1.8
Excel	1.8	2.1
Santa Fe	1.4	1.6
Scoupe	2.0	2.3
Sonata		
1989-98	2.2	2.5
1999-05	1.6	1.8
Stellar	.8	1.1
Tiburon, Tuscon, XG300/350	1.6	1.8
Steering Knuckle Oil Seal, Replace (B)		
Inner		
Accent	.9	1.0
Elantra	1.3	1.4
Excel		
1986-89	.7	.8
1990-94	1.1	1.2
Scoupe	1.3	1.4
1989-98 Sonata	1.4	1.6

STELLAR : TIBURON : TUSCON : XG300 : XG350 **HYU-25**

	LABOR TIME	SEVERE SERVICE
Outer		
Accent	1.7	2.0
1992-95 Elantra	2.0	2.2
Excel		
1986-89	1.5	1.6
1990-94	1.8	1.9
Scoupe	2.0	2.1
1989-98 Sonata	2.1	2.3
Strut Rod, Replace (B)		
Pony, Stellar	.7	.9
Upper Ball Joint, Replace (B)		
1999-05 Sonata	.8	.9
XG300/350	.8	.9
Upper Control Arm Assy., Replace (B)		
1999-05 Sonata	.8	.9
Stellar	3.3	3.6
XG300/350	.8	.9

REAR SUSPENSION

Unless otherwise noted, time given does not include alignment.

Rear Toe, Check & Adjust (B)
  1986-05 .... .6 / .8

Lower Control Arm, Replace (B)
- Accent
  - 1995-99 .... 1.1 / 1.3
  - 2000-05 .... .8 / .9
- Elantra
  - 1996-00 .... 1.1 / 1.3
  - 2001-05 .... .5 / .6
- Sonata
  - 1994-98 .... 1.6 / 1.8
  - 1999-05 .... .6 / .7
- Stellar one side .... .8 / 1.1
- Tiburon .... 1.1 / 1.3
- XG300/350 .... .3 / .4

Rear Coil Spring, Replace (B)
- Accent
  - 1995-99
    - one side .... 1.8 / 2.1
    - both sides .... 2.8 / 3.1
  - 2000-05
    - one side .... 1.3 / 1.4
    - both sides .... 2.1 / 2.4
- Elantra
  - 1992-95
    - one side .... 1.1 / 1.2
    - both sides .... 1.8 / 2.1
  - 1996-05
    - one side .... 1.7 / 2.0
    - both sides .... 2.7 / 3.0
- Excel
  - 1986-89
    - one side .... .5 / .6
    - both sides .... .9 / 1.0
  - 1990-94
    - one side .... .6 / .9
    - both sides .... 1.0 / 1.1

- Santa Fe
  - one side .... .5 / .6
  - both sides .... .9 / 1.0
- Scoupe
  - one side .... .7 / 1.0
  - both sides .... 1.2 / 1.5
- Sonata
  - 1989-93
    - one side .... 1.2 / 1.4
    - both sides .... 1.8 / 2.1
  - 1994-99
    - one side .... 1.8 / 2.1
    - both sides .... 2.6 / 2.9
  - 2000-05
    - one side .... .8 / .9
    - both sides .... 1.4 / 1.6
- Stellar
  - one side .... .5 / .6
  - both sides .... .7 / .9
- Tiburon
  - one side .... 1.4 / 1.6
  - both sides .... 2.4 / 2.7
- Tuscon, XG300/350
  - one side .... .8 / 1.0
  - both sides .... 1.4 / 1.6

Rear Cross Member Assy., R&R or Replace (B)
- Elantra
  - 1996-00 .... 2.1 / 2.3
  - 2001-05 .... 1.4 / 1.6
- Santa Fe
  - 2WD .... 2.1 / 2.3
  - 4WD .... 2.5 / 2.7
- Sonata
  - 1994-98 .... 4.0 / 4.2
  - 1999-05 .... .9 / 1.1
- Tiburon
  - 1997-01 .... 2.2 / 2.4
  - 2003-05 .... 1.7 / 1.9
- Tuscon
  - 2WD .... 1.6 / 1.8
  - 4WD .... 2.0 / 2.2
- XG300/350 .... 1.4 / 1.6

Rear Leaf Spring, R&R and Recondition (B)
  Pony, Stellar one .... 1.7 / 2.2

Rear Leaf Spring, Replace (B)
  Pony, Stellar one .... 1.2 / 1.7

Rear Strut Assy., Replace (B)
- Accent
  - 1995-99
    - one side .... 1.8 / 2.1
    - both sides .... 2.8 / 3.1
  - 2000-05
    - one side .... 1.3 / 1.4
    - both sides .... 2.2 / 2.5
- Elantra
  - 1992-95
    - one side .... 1.1 / 1.2
    - both sides .... 1.9 / 2.1

- 1996-05
  - one side .... 1.8 / 2.1
  - both sides .... 2.7 / 3.0
- 1990-94 Excel
  - one side .... .6 / .8
  - both sides .... 1.0 / 1.2
- Santa Fe, Tuscon
  - one side .... .5 / .5
  - both sides .... .8 / .9
- Scoupe
  - one side .... .7 / .9
  - both sides .... 1.2 / 1.4
- Sonata
  - 1989-93
    - one side .... 1.2 / 1.4
    - both sides .... 1.8 / 2.0
  - 1994-05
    - one side .... 1.9 / 2.1
    - both sides .... 2.7 / 3.0
- Tiburon
  - one side .... 1.4 / 1.6
  - both sides .... 2.4 / 2.7
- XG300/350
  - 2000
    - one side .... .8 / .9
    - both sides .... 1.1 / 1.3
  - 2001-05
    - one side .... 1.9 / 2.1
    - both sides .... 2.7 / 3.0

Rear Suspension Arm Assy., R&R or Replace (B)
- 1992-95 Elantra .... 3.0 / 3.4
- Excel .... 2.8 / 3.2
- Scoupe .... 3.1 / 3.6
- 1989-98 Sonata .... 3.0 / 3.4

Rear Wheel Bearings and Seals, Replace (B)
- One side
  - Elantra .... 1.1 / 1.2
  - Excel .... 1.0 / 1.1
  - Santa Fe, Tuscon
    - 2WD .... .8 / .9
    - 4WD .... 1.7 / 1.8
  - Scoupe .... 1.0 / 1.1
  - Sonata .... 1.3 / 1.4

Shock Absorbers or Bushings, Replace (B)
- Excel
  - 1986-89
    - one side .... .5 / .6
    - both sides .... .7 / .8
  - 1990-94
    - one side .... .6 / .7
    - both sides .... 1.0 / 1.1
- Pony, Stellar
  - one side .... .5 / .6
  - both sides .... .7 / .8
- Scoupe
  - one side .... .7 / .8
  - both sides .... 1.2 / 1.3

HYU-26 ACCENT : ELANTRA : EXCEL : PONY : SANTA FE : SCOUPE : SONATA

	LABOR TIME	SEVERE SERVICE
Spring Shackle, Replace (B)		
Pony, Stellar each	.5	.8
Stabilizer Bar, Replace (B)		
Accent	.5	.7
Elantra		
1996-00	.8	.9
2001-05	.5	.5
Excel	3.6	3.8
1994-05 Sonata	.6	.8
Santa Fe, Tiburon, Tuscon, XG300/350	.6	.7
Upper Control Arm, Replace (B)		
Sonata		
1994-98		
one side	1.7	2.0
both sides	2.4	2.7
1999-05		
one side	.6	.7
both sides	1.0	1.1
Stellar one	.8	1.1
XG300/350		
one side	.6	.7
both sides	1.0	1.1
Wheel Hub, Replace (B)		
One side		
exc. below	.8	.9
4WD Santa Fe	1.7	1.9

STEERING
AIR BAGS
The following operations do not include testing. Add time as required.

	LABOR TIME	SEVERE SERVICE
Diagnose Air Bag System Component Each (A)		
1989-05	.8	.8
Air Bag Clock Spring, Replace (B)		
Accent	.9	.9
Elantra		
1991-95	.6	.6
1996-05	.8	.8
Santa Fe	.8	.8
Sonata		
1994-98	.6	.6
1999-05	.9	.9
Tiburon, Tuscon	.8	.8
XG300/350	.9	.9
Air Bag Assy., Replace (B)		
Driver side	.5	.5
Passenger side		
exc. below	3.3	3.3
2000-05 Accent	2.4	2.4
2001-05 Elantra	2.4	2.4
Santa Fe	2.7	2.7
1999-05 Sonata	2.4	2.4
2003-05 Tiburon	2.8	2.8
Tuscon	3.7	3.7
XG300/350	2.7	2.7
Right or left curtain		
Sonata	2.0	2.0
Tuscon	2.4	2.4

	LABOR TIME	SEVERE SERVICE
Module Assembly, Replace (A)		
Accent	.5	.5
Elantra		
1992-95	.5	.5
1996-05	.8	.8
Santa Fe	1.1	1.1
Sonata		
1989-98	.9	.9
1999-05	2.3	2.3
Tiburon	.8	.8
Tuscon, XG300/350	2.2	2.2

MANUAL RACK & PINION

	LABOR TIME	SEVERE SERVICE
Horn Contact, Replace (B)		
All Models	.5	.5
Outer Tie Rod End, Replace (B)		
Includes: Reset toe.		
1985-99 one side	2.7	3.0
2000-05 one side	1.0	1.1
Rack & Pinion Assy., R&R and Recondition (B)		
Includes: Reset toe.		
Accent		
1995-99	4.1	4.6
2000-05	3.5	3.9
1992-95 Elantra	4.1	4.6
Excel	4.2	4.7
Scoupe	3.9	4.4
Rack & Pinion Assy., Replace (B)		
Includes: Reset toe.		
Accent		
1995-99	2.9	3.2
2000-05	2.2	2.5
Elantra	2.9	3.2
Excel	2.5	2.8
Scoupe	2.5	2.8
Rack Boots, Replace (B)		
Includes: Reset toe.		
Accent		
1995-99		
one side	.6	.7
both sides	1.1	1.3
2000-05		
one side	1.0	1.1
both sides	1.3	1.4
Excel, Scoupe		
one side	.6	.8
both sides	1.0	1.2
Elantra		
one side	.6	.7
both sides	1.1	1.3
Stellar		
one side	.8	1.0
both sides	1.4	1.6
Steering Column, R&R and Recondition (B)		
1986-89 Excel	2.0	2.0
Stellar	2.2	2.2
Steering Column, Replace (B)		
Accent	1.1	1.2

	LABOR TIME	SEVERE SERVICE
Elantra	1.7	1.9
Excel		
1986-89	1.9	1.9
1990-94	1.6	1.6
Scoupe	1.6	1.6
Stellar	1.2	1.2
Steering Shaft U-Joint, Replace (B)		
Accent, Excel, Scoupe	1.3	1.5
Stellar	1.2	1.4
Steering Wheel, Replace (B)		
All Models	.5	.5

POWER RACK & PINION

	LABOR TIME	SEVERE SERVICE
Horn Contact or Canceling Cam, Replace (B)		
1985-05	.5	.5
Inner Tie Rod Assy., Replace (B)		
Includes: Reset toe.		
Accent	3.5	3.9
Elantra		
1992-95	4.0	4.4
1996-05	3.3	3.7
Excel		
1986-89	3.3	3.5
1990-94	3.6	3.8
Santa Fe, Tuscon		
2WD	5.7	6.3
4WD	6.0	6.7
Scoupe	3.4	3.6
Sonata		
1989-98		
4 cyl.	2.8	3.1
V6	4.1	4.6
1999-05	3.3	3.7
Tiburon	3.9	4.1
1997-01	4.0	4.4
2003-05	3.3	3.7
Stellar 2.0L	2.9	3.1
XG300/350	4.3	4.8
w/AC add		
91-95 Elantra	.3	.3
Excel	.3	.3
95-96 Sonata	.3	.3
w/AT 93-95 Scoupe add	.3	.3
w/ECPS 89-94 Sonata add	.3	.3
Outer Tie Rod Ends, Replace (B)		
Includes: Reset toe.		
Exc. below		
one side	.8	1.0
both sides	1.3	1.5
1999-05 Sonata		
one side	1.0	1.2
both sides	1.5	1.7
Tiburon		
one side	1.0	1.2
both sides	1.5	1.7

STELLAR : TIBURON : TUSCON : XG300 : XG350 — HYU-27

	LABOR TIME	SEVERE SERVICE
Power Steering Reservoir, Replace (B)		
1985-05	.5	.6
Rack & Pinion Assy., R&R and Recondition (A)		
Includes: Reset toe.		
Accent		
1995-99	4.3	4.8
2000-05	3.8	4.3
Elantra		
1992-95	5.0	5.7
1996-00	4.6	5.1
2000-05	4.0	4.4
Excel	4.9	5.4
Santa Fe, Tuscon		
2WD	6.4	7.2
4WD	6.9	7.7
Scoupe	4.5	5.0
Sonata		
1989-98		
4 cyl.	4.6	5.1
V6	5.5	6.2
1999-05	4.6	5.1
Stellar 2.0L	4.1	4.6
Tiburon		
1997-01	4.6	5.1
2003-05	4.0	4.4
XG300/350	5.5	6.2
w/AT Excel, Scoupe add	.3	.3
w/ECPS add	.3	.3
Rack & Pinion Assy., R&I (B)		
Includes: Reset toe.		
Accent	3.2	3.6
Elantra		
1992-00	3.6	4.1
2001-05	3.0	3.4
Excel	3.1	3.4
Santa Fe, Tuscon		
2WD	5.2	5.8
4WD	5.8	6.5
Scoupe	3.0	3.3
Sonata		
1989-98		
4 cyl.	2.4	2.7
V6	3.8	4.2
1999-05	3.2	3.6
Stellar 2.0L	2.6	2.9
Tiburon		
1997-01	3.8	4.3
2003-05	3.3	3.7
XG300/350	4.1	4.6
w/AT Excel, Scoupe add	.3	.3
w/ECPS add	.3	.3
Rack Boots, Replace (B)		
Includes: Reset toe.		
Accent		
one side	.8	.9
both sides	1.4	1.6

	LABOR TIME	SEVERE SERVICE
Elantra		
1992-95		
one side	.6	.7
both sides	1.1	1.3
1996-05		
one side	1.1	1.3
both sides	1.4	1.6
Excel		
one side	.7	.9
both sides	1.1	1.3
Santa Fe, Tuscon		
2WD		
one side	5.3	6.0
both sides	5.5	6.2
4WD		
one side	6.0	6.7
both sides	6.1	6.9
Scoupe		
one side	.7	.9
both sides	1.2	1.4
Sonata		
1989-98		
one side	.6	.7
both sides	1.1	1.3
1999-05		
one side	3.2	3.6
both sides	3.6	4.1
Stellar 2.0L		
one side	.8	1.0
both sides	1.4	1.6
Tiburon		
one side	1.3	1.4
both sides	1.7	2.0
XG300/350		
one side	4.3	4.8
both sides	4.4	5.0
Steering Column, R&R and Recondition (B)		
1986-89 Excel	2.0	2.0
Stellar	2.2	2.2
Steering Column, Replace (B)		
Accent	.9	1.0
Elantra	1.7	1.9
Excel	1.4	1.4
Santa Fe, Tuscon	1.5	1.7
Scoupe	1.6	1.6
Sonata		
1989-98	1.2	1.4
1999-05	1.5	1.7
Stellar	1.2	1.2
Tiburon		
1997-01	1.7	1.9
2003-05	1.2	1.4
XG300/350	1.7	1.9
Steering Pump, R&R and Recondition (B)		
Accent	1.5	1.7
Elantra		
1992-95	2.1	2.4
1996-05	1.4	1.5

	LABOR TIME	SEVERE SERVICE
Excel	1.8	2.3
Santa Fe, Tuscon	2.0	2.2
Scoupe		
1991-92	2.0	2.5
1993-95	1.6	2.1
Sonata	1.8	2.1
Tiburon	1.4	1.5
XG300/350	2.4	2.7
w/DOHC 89-98 Sonata add	.9	.9
Steering Pump, R&R or Replace (B)		
Accent	.9	1.0
Elantra		
1992-95	1.2	1.4
1996-05	.6	.7
Excel	1.1	1.3
Santa Fe, Tuscon	1.2	1.4
Scoupe		
1991-92	1.3	1.5
1993-95	.7	.9
Sonata	.9	1.0
Tiburon	.6	.7
XG300/350	1.5	1.7
w/DOHC 89-98 Sonata add	.7	.7
Steering Wheel, Replace (B)		
Exc. below	.5	.5
1999-05 Sonata	.8	.9
XG300/350	.8	.9
POWER WORM & SECTOR		
Gear Assy., Replace (B)		
Pony, Stellar	1.2	1.5
Gear Assy., R&R and Recondition (B)		
Pony, Stellar	2.3	2.8
Idler Arm, Replace (B)		
Pony, Stellar	1.4	1.6
Horn Contact, Replace (B)		
Pony, Stellar	.5	.5
Outer Tie Rod End, Replace (B)		
Includes: Reset toe.		
Pony, Stellar	.8	1.0
Pitman Arm, Replace (B)		
Includes: Reset toe-in.		
Pony, Stellar	.5	.8
Steering Column, R&R or Replace (B)		
Pony, Stellar	1.8	1.8
Steering Intermediate Shaft, Replace (B)		
Pony, Stellar	1.2	1.5
Steering Shaft Lower Coupling, Replace (Pot Joint) (B)		
Pony, Stellar	1.2	1.4
Steering Wheel, Replace (B)		
Pony, Stellar	.3	.3

HYU-28 ACCENT : ELANTRA : EXCEL : PONY : SANTA FE : SCOUPE : SONATA

HEATING & AIR CONDITIONING

When more than one component requires replacement where evacuation/recovery and recharging is already included, deduct 1.0 hour for each additional component from the time given.

	LABOR TIME	SEVERE SERVICE
Evacuate/Recover and Recharge System (B)		
1985-05	1.1	1.2
AC Hoses, Replace (B)		
Includes: Evacuate/recover and recharge.		
Exc. below suction or discharge	1.4	1.5
Santa Fe suction	2.2	2.3
XG300/350 V6 suction	2.3	2.4
Accumulator Assy., Replace (B)		
Includes: Evacuate/recover and recharge.		
1985-05	1.4	1.4
Ambient Temperature Sensor, Replace (B)		
All Models	.3	.3
Blower Motor, Replace (B)		
Exc. below	.5	.5
Pony, Stellar	.7	.7
1999-05 Sonata	.6	.6
XG300/350	.6	.6
Blower Motor Relay, Replace (B)		
1985-05	.3	.3
Blower Motor Resistor, Replace (B)		
Accent	.6	.6
Elantra	.6	.6
Excel, Pony, Stellar	.5	.5
Scoupe	.7	.7
Sonata, Tiburon	.5	.5
XG300/350	.5	.5
Blower Motor Switch, Replace (B)		
Exc. below	.6	.6
Excel		
1986-89	1.2	1.2
1990-94	.3	.3
Tiburon	.3	.3
Tiburon coupe	1.1	1.1
Compressor Assy., R&R and Recondition (A)		
Includes: Parts transfer, evacuate/recover and recharge.		
1995-99 Accent	3.1	3.4
Elantra		
1992-95	3.1	3.4
1996-05	2.6	2.9
Excel	3.1	3.6
Pony, Stellar	3.1	3.6

	LABOR TIME	SEVERE SERVICE
Scoupe	3.3	3.8
1989-98 Sonata		
4 cyl.	3.5	3.9
V6	3.1	3.4
Tiburon		
1997-01	2.4	2.7
2003-05	3.1	3.4
Compressor Assy., Replace (B)		
Includes: Parts transfer, evacuate/recover and recharge.		
Accent	2.1	2.4
Elantra		
1992-95	2.6	2.9
1996-05	2.0	2.2
Excel	2.7	2.9
Pony, Stellar	2.2	2.4
Santa Fe		
exc. 3.5L	2.1	2.4
3.5L	4.7	5.2
Scoupe	3.1	3.3
Sonata		
1989-98		
4 cyl.	3.1	3.4
V6	2.4	2.7
1999-05	2.0	2.2
Tiburon, Tuscon	2.4	2.6
XG300/350	3.2	3.6
Compressor Clutch Assy., Replace (B)		
Includes: Parts transfer, evacuate/recover and recharge when necessary.		
Accent		
1995-99	.6	.7
2000-05	2.3	2.4
Elantra		
1992-95	2.9	3.0
1996-00	.6	.7
2001-05		
exc. 1.6 ENG	2.3	2.4
1.6 ENG	.6	.7
Excel		
1986-89	2.0	2.1
1990-94	2.4	2.5
Pony, Stellar	1.2	1.3
Santa Fe		
4 cyl.	.5	.6
V6	4.5	4.6
Scoupe	2.0	2.1
Sonata		
1989-94		
4 cyl.	2.4	2.5
V6	1.9	2.0
1995-98		
4 cyl.	.6	.7
V6	2.8	2.9
1999-05	1.0	1.1

	LABOR TIME	SEVERE SERVICE
Tiburon		
1997-01	2.1	2.2
2003-05		
4 cyl.	2.2	2.3
V6	1.6	1.7
Tuscon	2.6	2.9
XG300/350	.6	.7
Condenser Assy., Replace (B)		
Includes: Evacuate/recover and recharge.		
Accent		
1995-99	1.8	2.1
2000-05	1.5	1.7
Elantra		
1992-00	2.3	2.6
2001-05	1.8	2.1
Excel	2.4	2.5
Pony, Stellar	2.8	2.9
Santa Fe	2.6	2.9
Scoupe	2.0	2.1
Sonata		
1989-98	3.1	3.4
1999-05	2.6	2.9
Tiburon		
1997-01	2.3	2.6
2003-05	1.8	2.1
Tuscon, XG300/350	2.6	2.9
w/DOHC add		
92-95 Elantra	.5	.5
89-98 Sonata	.5	.5
Condenser Fan Motor, Replace (B)		
Accent	.6	.7
Elantra		
1992-00	.6	.7
2001-05	.3	.3
Excel	.5	.6
Santa Fe, Tuscon	1.2	1.4
Scoupe	.5	.6
Sonata		
1989-98	.8	.9
1999-05	.3	.3
Tiburon	.5	.6
XG300/350	.3	.3
Evaporator Core, Replace (B)		
Includes: Evacuate/recover and recharge.		
Accent		
1995-99	2.6	2.6
2000-05	1.8	2.1
Elantra		
1992-95	3.1	3.1
1996-00	3.9	3.9
2001-05	5.4	6.0
Excel	3.3	3.3
Pony, Stellar	3.8	3.8
Santa Fe	5.7	5.7
Scoupe	3.3	3.3

STELLAR : TIBURON : TUSCON : XG300 : XG350   HYU-29

	LABOR TIME	SEVERE SERVICE
Sonata		
1989-98		
4 cyl.	6.3	6.3
V6	6.6	6.6
1999-01	4.7	5.2
2002-05	2.4	2.7
Tiburon		
1997-01	4.1	4.1
2003-05	6.3	7.0
Tuscon	6.5	7.2
XG300/350	4.7	5.2
Expansion Valve, Replace (B)		
Includes: Evacuate/recover and recharge.		
Accent		
1995-99	3.2	3.6
2000-05	1.4	1.5
Elantra		
1992-95	3.2	3.6
1996-00	4.4	4.9
2001-05	1.8	2.1
1986-89 Excel	3.6	3.6
Pony, Stellar	4.3	4.3
Santa Fe, Tuscon	1.8	2.1
Scoupe	2.9	2.9
Sonata		
1992-98	7.5	8.3
1999-05	1.4	1.5
Tiburon		
1997-01	4.6	5.1
2003-05	1.8	2.1
XG300/350	1.4	1.5
Heater Blower Motor Relay, Replace (B)		
1985-05	.3	.3
Heater Blower Motor Resistor, Replace (B)		
Accent		
1995-99	.6	.6
2000-05	.3	.3
Elantra		
1992-00	.6	.6
2001-05	.3	.3
Excel, Sonata, Tiburon	.5	.5
Scoupe	.7	.7
Heater Control Valve, Replace (B)		
1986-89 Excel	.7	.9
Pony, Stellar	.8	1.0
Heater Core, R&R or Replace (B)		
Accent		
1995-99	4.9	4.9
2000-05	4.1	4.1
Elantra		
1991-95	3.0	3.0
1996-00	4.3	4.3
2001-05	3.8	3.8
Excel		
1986-89	1.6	1.6
1990-94	3.5	3.5

	LABOR TIME	SEVERE SERVICE
Pony, Stellar	3.1	3.1
Santa Fe	5.1	5.1
Scoupe	3.0	3.0
Sonata		
1989-98	4.5	4.5
1999-05	4.3	4.3
Tiburon		
1997-01	3.9	3.9
2003-05	4.5	4.5
Tuscon	7.0	7.0
XG300/350	4.9	4.9
w/AC add		
95-99 Accent	.8	.8
00-05 Accent	1.0	1.0
Elantra, Scoupe, Tiburon, Tuscon	1.1	1.1
86-89 Excel	2.5	2.5
90-94 Excel	1.1	1.1
Pony, Stellar	2.1	2.1
89-98 Sonata	.9	.9
99-05 Sonata	1.1	1.1
XG300/350	.8	.8
Heater Hoses, Replace (B)		
Right or left side		
Accent		
1995-99	1.1	1.2
2000-05	.5	.6
Elantra		
1992-95	.8	.9
1996-05	.5	.6
Excel	.7	.8
Pony, Stellar	.5	.6
Santa Fe	.8	.9
Scoupe	.7	.8
Sonata		
1989-98	1.2	1.4
1999-05	.5	.6
Tiburon, Tuscon, XG300/350	.5	.6
In-Vehicle Sensor, Replace (B)		
1989-98 Sonata	.8	.8
2001-05 Elantra	.7	.7
Receiver/Drier Assy., Replace (B)		
Includes: Evacuate/recover and recharge.		
Accent	1.5	1.7
Elantra	1.7	1.9
1986-89 Excel	2.3	2.3
Pony, Stellar	1.6	1.6
Scoupe	1.7	1.7
1999-05 Sonata	2.0	2.2
Tiburon	1.7	1.9
Tuscon, XG300/350	1.7	1.9

WIPERS & SPEEDOMETER

	LABOR TIME	SEVERE SERVICE
Antenna, Replace (B)		
Manual	.3	.3
Power	.6	.6

	LABOR TIME	SEVERE SERVICE
Radio, R&R (B)		
Exc. below	.6	.6
Accent	.7	.7
2001-05 Elantra	.7	.7
Rear Window Washer Motor, Replace (B)		
1986-05	.3	.3
Rear Window Wiper Motor, Replace (B)		
Accent		
1995-99	.5	.8
2000-05	.8	.9
Elantra	.8	.9
Excel	.5	.8
Santa Fe, Tuscon	.6	.7
Tiburon	.5	.6
Rear Window Wiper Switch, Replace (B)		
Accent, Excel	.3	.3
1992-00 Elantra	.5	.5
Santa Fe, Tuscon	.5	.5
2003-05 Tiburon	.5	.5
Speedometer Cable & Casing, Replace (B)		
Accent	.6	.7
1992-00 Elantra	.9	1.0
Excel	.7	.8
Pony, Scoupe, Stellar	.8	.9
Sonata	.7	.8
Speedometer Head, R&R or Replace (B)		
Accent	.6	.6
Elantra		
1992-95	1.1	1.1
1996-05	.7	.7
Excel		
1986-89	.9	.9
1990-94	1.0	1.0
Pony, Stellar	.9	.9
Santa Fe, Tuscon	.7	.7
Scoupe	1.5	1.5
Sonata		
1989-93	1.3	1.3
1994-05	.5	.5
Tiburon	.5	.5
XG300/350	.6	.6
Windshield Washer Pump, Replace (B)		
1985-05	.6	.6
Windshield Wiper & Washer Switch, Replace (B)		
Exc. below	.6	.6
Excel		
1986-89	.7	.7
1990-94	1.0	1.0
Scoupe	1.4	1.4
1989-93 Sonata	1.6	1.6

HYU-30 ACCENT : ELANTRA : EXCEL : PONY : SANTA FE : SCOUPE : SONATA

	LABOR TIME	SEVERE SERVICE
Windshield Wiper Interval Relay, Replace (B)		
1985-05	.5	.5
Windshield Wiper Linkage, Replace (B)		
Exc. below	.8	.9
Accent, Santa Fe, Tuscon	.9	1.0
Sonata	1.2	1.4
Windshield Wiper Motor, R&R and Recondition (B)		
Pony, Stellar	.7	1.0
Windshield Wiper Motor, Replace (B)		
Exc. below	.8	.9
2000-05 Accent	.5	.5
1991-00 Elantra	.5	.5
1986-89 Excel	.9	1.0
Santa Fe, Tuscon	.9	1.0

LAMPS & SWITCHES

	LABOR TIME	SEVERE SERVICE
Back-Up Light Switch, Replace (B)		
1985-05	.5	.5
Combination Switch Assy., Replace (B)		
Accent	.9	.9
Elantra		
1992-95	1.1	1.1
1996-05	.8	.8
Excel	1.0	1.0
Pony, Stellar	.8	.8
Santa Fe, Tuscon	.9	.9
Scoupe	1.2	1.2
Sonata	1.1	1.1
Tiburon	.8	.8
XG300/350	1.1	1.1
Composite Headlight Bulb, Replace (B)		
1985-05 each	.3	.3
Hazard Warning Switch, Replace (B)		
Exc. below	.5	.5
2000-05 Accent	.6	.6
1996-05 Elantra	.6	.6
1997-01 Tiburon	1.1	1.1
Headlight Sealed Beam, Replace (B)		
1985-93 one	.3	.3
Headlight Switch, Replace (B)		
Exc. below	.6	.6
Excel	1.0	1.0
Scoupe	1.6	1.6

	LABOR TIME	SEVERE SERVICE
Headlights, Aim (B)		
Both sides	.5	.5
High Mount Stop Light Assy., Replace (B)		
Exc. below	.5	.5
2000-05 Accent	1.1	1.1
Scoupe	.7	.7
2002-05 Sonata	.7	.7
Horn, Replace (B)		
1985-05	.5	.5
Horn Relay, Replace (B)		
1985-05	.3	.3
License Lamp Assy., Replace (C)		
Exc. below both sides	.5	.5
1999-05 Sonata	1.7	1.9
2003-05 Tiburon	1.7	1.9
Parking Lamp Lens or Bulb, Replace (C)		
1985-05	.3	.3
Rear Combination Lamp Lens, Replace (B)		
1985-05	.5	.5
Side Marker Lamp Lens, Replace (B)		
Exc. below	.3	.3
1992-95 Elantra	.6	.6
1998-98 Sonata	.6	.6
Stop Light Switch, Replace (B)		
1985-05	.5	.5
Turn Signal Lamp Lens or Bulb, Replace (C)		
1985-05	.3	.3
Turn Signal or Hazard Warning Flasher, Replace (B)		
1985-05	.3	.3
Turn Signal Switch, Replace (B)		
Pony, Stellar	.8	.8

SEAT BELTS

	LABOR TIME	SEVERE SERVICE
Automatic Shoulder Belt Control Unit, Replace (B)		
1994-95 Scoupe	.7	.7
1989-95 Sonata	.7	.7
Automatic Shoulder Belt Motor, Replace (B)		
1989-95 Sonata	.9	.9
Lap Belt Assy., Replace (B)		
1985-98	.5	.5
Shoulder Belt Track, Replace (B)		
1989-98 Sonata	.8	.8

BODY

	LABOR TIME	SEVERE SERVICE
Door Latch Assy., Replace (B)		
Exc. below	.9	1.0
Elantra		
1992-95	1.1	1.2
1996-05	1.4	1.5
Santa Fe, Tuscon	1.4	1.5
Door Lock, Replace (B)		
Exc. below	.7	.7
1996-00 Elantra	.9	.9
Front Door Window Regulator and/or Motor, Replace (B)		
Accent	.9	1.0
Elantra		
1992-95	.9	1.0
1996-99	1.1	1.2
2001-05	1.5	1.7
Excel	.9	1.0
Pony, Stellar	.7	.9
Santa Fe, Tuscon	1.5	1.7
Scoupe, Sonata	.9	1.0
Tiburon		
1997-01	.7	.9
2003-05	1.1	1.2
XG300/350	1.4	1.5
Hood Hinge, Replace (B)		
Exc. below	.6	.7
95-98 Sonata	.9	1.0
Hood Lock, Replace (B)		
1985-05	.5	.5
Hood Release Cable, Replace (B)		
Exc. below	1.2	1.4
2003-05 Tiburon	1.4	1.5
Lock Striker Plate, Replace (B)		
1985-05	.5	.5
Rear Door Window Regulator and/or Motor, Replace (B)		
Exc. below	.8	.9
2001-05 Elantra	1.5	1.7
Santa Fe, Tuscon	1.2	1.4

INF

G20 : G35 : I30 : I35 : J30 : M30 : M45 : Q45

SYSTEM INDEX

MAINTENANCE	INF-2
Maintenance Schedule	INF-2
CHARGING	INF-2
STARTING	INF-3
CRUISE CONTROL	INF-3
IGNITION	INF-3
EMISSIONS	INF-4
FUEL	INF-5
EXHAUST	INF-6
ENGINE COOLING	INF-6
ENGINE	INF-7
Assembly	INF-7
Cylinder Head	INF-8
Camshaft	INF-10
Crank & Pistons	INF-10
Engine Lubrication	INF-11
CLUTCH	INF-11
MANUAL TRANSAXLE	INF-12
MANUAL TRANSMISSION	INF-12
AUTO TRANSAXLE	INF-12
AUTO TRANSMISSION	INF-13
TRANSFER CASE	INF-13
SHIFT LINKAGE	INF-13
DRIVELINE	INF-14
BRAKES	INF-15
FRONT SUSPENSION	INF-18
REAR SUSPENSION	INF-20
STEERING	INF-21
HEATING & AC	INF-23
WIPERS & SPEEDOMETER	INF-25
LAMPS & SWITCHES	INF-25
BODY	INF-26

OPERATIONS INDEX

A
- AC Hoses ... INF-24
- Air Bags ... INF-21
- Air Conditioning ... INF-23
- Alignment ... INF-18
- Alternator (Generator) ... INF-2
- Antenna ... INF-25
- Anti-Lock Brakes ... INF-15

B
- Back-Up Lamp Switch ... INF-25
- Ball Joint ... INF-18
- Battery Cables ... INF-3
- Bleed Brake System ... INF-16
- Blower Motor ... INF-23
- Brake Disc ... INF-17
- Brake Hose ... INF-16
- Brake Pads and/or Shoes ... INF-16

C
- Camshaft ... INF-10
- Camshaft Sensor ... INF-3
- Catalytic Converter ... INF-6
- Coolant Temperature (ECT) Sensor ... INF-4
- Crankshaft ... INF-10
- Crankshaft Sensor ... INF-3
- Cruise Control ... INF-3
- CV Joint ... INF-15
- Cylinder Head ... INF-8

D
- Differential ... INF-14
- Distributor ... INF-3
- Drive Belt ... INF-2
- Driveshaft ... INF-14

E
- EGR ... INF-4
- Electronic Control Module (ECM/PCM) ... INF-4
- Engine ... INF-7
- Engine Lubrication ... INF-11
- Engine Mounts ... INF-8
- Evaporator ... INF-24
- Exhaust ... INF-6
- Exhaust Manifold ... INF-6
- Expansion Valve ... INF-24

F
- Flywheel ... INF-12
- Fuel Injection ... INF-5
- Fuel Pump ... INF-5
- Fuel Vapor Canister ... INF-4

G
- Gear Selector Lever ... INF-13
- Generator ... INF-2

H
- Halfshaft ... INF-14
- Headlamp ... INF-25
- Heater Core ... INF-24
- Horn ... INF-25

I
- Idle Air Control (IAC) Valve ... INF-5
- Ignition Coil ... INF-3
- Ignition Module ... INF-4
- Ignition Switch ... INF-4
- Inner Tie Rod ... INF-22
- Intake Air Temperature (IAT) Sensor ... INF-4
- Intake Manifold ... INF-5

K
- Knock Sensor ... INF-4

L
- Lower Control Arm ... INF-19

M
- Maintenance Schedule ... INF-2
- Mass Air Flow (MAF) Sensor ... INF-4
- Master Cylinder ... INF-16
- Muffler ... INF-6
- Multifunction Lever/Switch ... INF-25

N
- Neutral Safety Switch ... INF-25

O
- Oil Pan ... INF-11
- Oil Pump ... INF-11
- Outer Tie Rod ... INF-22
- Oxygen Sensor ... INF-4

P
- Parking Brake ... INF-17
- Pistons ... INF-11
- Positive Crankcase Ventilation (PCV) Valve ... INF-5

R
- Radiator ... INF-7
- Radiator Hoses ... INF-7
- Radio ... INF-25
- Rear Main Oil Seal ... INF-11

S
- Shock Absorber/Strut, Front ... INF-19
- Shock Absorber/Strut, Rear ... INF-20
- Spark Plug Cables ... INF-4
- Spark Plugs ... INF-4
- Spring, Front Coil ... INF-18
- Spring, Rear Coil ... INF-20
- Starter ... INF-3
- Steering Wheel ... INF-23

T
- Thermostat ... INF-7
- Throttle Body ... INF-6
- Throttle Position Sensor (TPS) ... INF-5
- Timing Belt ... INF-10
- Timing Chain ... INF-10
- Torque Converter ... INF-12

U
- Upper Control Arm ... INF-20

V
- Valve Body ... INF-12
- Valve Cover Gasket ... INF-9
- Valve Job ... INF-9
- Vehicle Speed Sensor ... INF-5

W
- Water Pump ... INF-7
- Wheel Balance ... INF-2
- Window Regulator ... INF-26
- Windshield Washer Pump ... INF-25
- Windshield Wiper Motor ... INF-25

INF-1

INF-2 G20 : G35 : I30 : I35 : J30 : M30 : M45 : Q45

	LABOR TIME	SEVERE SERVICE
MAINTENANCE		
Air Cleaner Filter Element, Service (C)		
All Models	.3	.4
Chassis Lubrication, Change Oil & Filter (C)		
Includes: Correct fluid levels.		
All Models	.8	.8
Drive Belt, Adjust (B)		
One	.6	.6
each addl. add	.2	.2
Drive Belt, Replace (B)		
G20, G35		
1991-96		
one	.6	.7
two	1.1	1.2
1999-05		
one	.6	.7
two	.9	1.0
I30, I35		
one	.8	.9
two	1.1	1.2
J30		
one	.6	.7
two	.8	.9
three	1.1	1.2
M30		
one	.9	1.0
two	1.1	1.2
three	1.2	1.3
M45		
one	.3	.3
two	.8	.9
Q45		
one	.6	.7
two	.8	.9
three	1.1	1.2
four	1.2	1.3
Fuel Filter, Replace (B)		
In fuel tank		
G35	.8	.9
I30	.9	1.0
M45	.6	.7
2002-04 Q45	.8	.9
In fuel line		
G20	.6	.7
G35, M45	.8	.9
I30, I35	1.0	1.1
J30, M30	.5	.6
Q45		
1990-96	.5	.6
1997-01	1.1	1.2
Halogen Headlamp Bulb, Replace (B)		
G20, I30, I35, J30	.3	.3
G35	.8	.8
M30	.5	.5
M45	1.4	1.4
Q45		
1990-00	.7	.7
2001-05	1.6	1.6

	LABOR TIME	SEVERE SERVICE
Oil & Filter, Change (C)		
Includes: Correct all fluid levels.		
All Models	.8	.8
Timing Belt, Replace (B)		
J30	4.7	5.2
M30	4.9	5.4
w/AC M30 add	.3	.3
w/PS M30 add	.3	.3
Replace timing gears add	1.0	1.0
Tire, Replace (C)		
Includes: Dismount old tire and mount new tire to rim.		
Exc. below	.8	.8
M45, 2002-05 Q45	1.3	1.3
each addl. add	.6	.6
Tires, Rotate (C)		
1991-05	.5	.5
Wheel, Balance (B)		
Exc. below	.6	.6
G35	1.1	1.1
M30	.8	.8
each addl. add		
exc. M30	.2	.2
M30	.5	.5
SCHEDULED MAINTENANCE INTERVALS		
If necessary, refer to appropriate Chilton maintenance service information.		
7,500 Mile Service (C)		
All Models	.6	.6
15,000 Mile Service (C)		
G20, Q45	1.0	1.0
J30, I30	.9	.9
22,500 Mile Service (C)		
All Models	.4	.4
30,000 Mile Service (B)		
G20, I30, J30	1.7	1.7
Q45	1.6	1.6
37,500 Mile Service (C)		
All Models	.6	.6
45,000 Mile Service (C)		
G20, Q45	1.0	1.0
J30, I30	.9	.9
52,500 Mile Service (C)		
All Models	.6	.6
60,000 Mile Service (B)		
G20, I30, J30	7.3	7.3
Q45	7.9	7.9
67,500 Mile Service (C)		
All Models	.6	.6
75,000 Mile Service (C)		
G20, Q45	1.0	1.0
J30, I30	.9	.9
82,500 Mile Service (C)		
All Models	.6	.6

	LABOR TIME	SEVERE SERVICE
90,000 Mile Service (B)		
G20, I30, J30	2.2	2.2
Q45	2.3	2.3
97,500 Mile Service (C)		
All Models	.6	.6
CHARGING		
Alternator Circuits, Test (B)		
Includes: Test component output.		
All Models	.5	.5
Alternator, R&R and Recondition (A)		
G20, G35		
1991-96	3.2	3.7
1999-05	2.4	2.9
I30, I35	3.6	4.1
J30	4.0	4.5
M30	3.1	3.6
M45	2.6	3.1
Q45		
1990-96	5.0	5.5
1997-01	4.4	4.9
2002-05	2.3	2.8
Alternator Assy., Replace (B)		
Includes: Pulley transfer.		
G20		
1991-96	1.1	1.3
1999-02	1.5	1.7
G35, M30	1.4	1.6
I30, I35	2.0	2.2
J30	2.1	2.3
M45	1.7	1.9
Q45		
1990-96	4.8	5.0
1997-05	1.8	2.0
w/AC add		
G20	.9	.9
G35	1.2	1.2
M45	1.7	1.7
Alternator Relay, Replace (B)		
1990-96 Q45	1.1	1.1
Alternator Voltage Regulator, Replace (B)		
G20		
1991-96	2.4	2.7
1999-02	2.0	2.3
G35	2.3	2.6
I30, I35		
1996-99	2.1	2.4
2000-04	2.4	2.7
J30	2.9	3.2
M30, M45	2.1	2.4
Q45		
1990-96	5.6	5.9
1997-01	2.1	2.4
2002-05	1.8	2.2
w/AC add		
G20	.9	.9
G35	1.2	1.2
M45	1.7	1.7

G20 : G35 : I30 : I35 : J30 : M30 : M45 : Q45 **INF-3**

	LABOR TIME	SEVERE SERVICE

STARTING

Starter Draw Test (On Car) (B)
All Models .... .5 .5
Battery Cables, Replace (C)
G20
  1991-96 .... .6 .7
  1999-02 .... 1.8 1.9
  G35, I30, I35 ..... 1.4 1.5
  J30 .... 1.7 1.8
M30 .... .6 .7
M45 .... 2.0 2.1
Q45
  1990-96 .... 1.2 1.3
  1997-05 .... 2.0 2.1
Starter, R&R and Recondition (A)
G20 .... 2.4 2.7
G35 .... 1.8 2.1
I30, I35
  1996-99 .... 2.7 3.0
  2000-04 .... 2.1 2.4
J30 .... 2.1 2.4
M30, M45 .... 2.7 3.0
Q45
  1990-96 .... 3.8 4.1
  1997-01 .... 4.3 4.6
  2002-05 .... 2.1 2.4
Starter Assy., Replace (B)
G20 .... 1.4 1.6
G35 .... .8 1.0
I30, I35 .... 1.1 1.2
J30 .... .6 .8
M30 .... .7 .9
M45 .... 1.4 1.6
Q45
  1990-96 .... 1.9 2.1
  1997-01 .... 3.3 3.5
  2002-05 .... 1.1 1.3
Starter Drive Assy., Replace (B)
Includes: Starter R&R.
G20 .... 1.9 2.2
G35 .... 1.4 1.7
I30, I35 .... 1.6 1.9
J30 .... 1.0 1.3
M30 .... 1.1 1.4
M45 .... 1.9 2.2
Q45
  1990-96 .... 2.4 2.7
  1997-01 .... 3.8 4.1
  2002-05 .... 1.6 1.9
Starter Relay, Replace (B)
Exc. G35 .... .3 .3
G35 .... .8 .9
Starter Solenoid and/or Switch, Replace (B)
Includes: Starter R&R.
G20 .... 1.6 1.9
G35 .... .9 1.2
I30, I35 .... 1.4 1.7
J30 .... .7 1.0
M30 .... 1.0 1.3

M45 .... 1.6 1.9
Q45
  1990-96 .... 2.1 2.4
  1997-01 .... 3.5 3.8
  2002-05 .... 1.3 1.7

CRUISE CONTROL

Diagnose Cruise System Component Each (A)
All Models .... .6 .6
Actuator Assy., Replace (B)
Exc. below .... .5 .5
G35, Q45 .... .6 .6
Brake or Clutch Release Switch, Replace (B)
G20 .... .8 .8
G35, M45 .... .5 .5
I30, I35 .... .3 .3
J30, M30 .... .6 .6
Q45
  1990-01 .... 1.1 1.1
  2002-05 .... .5 .5
Control Controller Module, Replace (B)
Exc. below .... .8 .8
M30 .... 1.1 1.1
M45 .... .3 .3
1990-01 Q45 .... 1.1 1.1
Control Main Switch, Replace (B)
G20
  1991-96 .... .3 .3
  1999-02 .... .7 .7
G35, Q45 .... .6 .6
I30, I35, M45 ...... .5 .5
J30, M30 .... .3 .3
Control Sensor, Replace (B)
G20, G35, J30 ...... .5 .5
I30, I35 .... .6 .6
M30 .... 1.1 1.1
Q45
  1990-96 .... 1.4 1.4
  1997-05 .... .6 .6
Control Set/Cancel/Phone Switch, Replace (B)
Exc. G35 .... .5 .5
G35 .... .6 .6
Control Vacuum Pump, Replace (B)
All Models .... .5 .5

IGNITION

Diagnose Ignition System Component Each (A)
All Models .... .8 .8
Camshaft Position Sensor, Replace (B)
G35
  one .... .6 .7
  both 1.1 1.3

I30, I35
  one .... .5 .7
  both .... .6 .8
J30 .... .6 .8
M45 .... .3 .5
Q45
  1990-96 .... .8 1.0
  1997-01 .... 5.9 6.1
  2002-05 .... .3 .5
Crankshaft Position Sensor, Replace (B)
OBD
  G20, G35, I30,
   I35, M45 .... .5 .7
  J30 .... .6 .8
  Q45 .... .5 .7
Reference
  G35 .... 1.1 1.3
  I30, I35 .... .6 .8
  1997-01 Q45 ...... .3 .5
Distributor, Replace (B)
Includes: Reset base ignition timing.
G20, M30 .... .7 .9
Distributor Cap and/or Rotor, Replace (B)
G20, M30 .... .5 .5
Key Warning Buzzer or Chime, Replace (B)
G20
  1991-96 .... 1.2 1.2
  1999-02 .... .5 .5
G35 .... .7 .7
I30, I35, M45 ...... .6 .6
J30 .... .8 .8
M30 .... 1.4 1.4
Q45
  1990-96 .... 1.6 1.6
  1997-05 .... 1.1 1.1
Ignition Coil, Replace (B)
G20 .... .5 .5
G35 .... 1.1 1.1
I30 .... .6 .6
I35 .... .6 .6
  w/VQ35DE eng. 1.1 1.1
J30 .... .7 .7
M30, M45 .... .6 .6
Q45
  1990-96
   w/TCS 6.0 6.0
   w/o TCS 3.1 3.1
  1997-05 .... 1.2 1.4
Ignition Coil Relay, Replace (B)
G35 .... .8 .8
I30, M45 .... .3 .3
1990-96 Q45 .... .6 .6
Ignition & Steering Lock Switch Assy., Replace (B)
Exc. below 1.2 1.2
G35 .... 1.8 1.8
1997-01 Q45 .... 1.6 1.6

INF-4 G20 : G35 : I30 : I35 : J30 : M30 : M45 : Q45

	LABOR TIME	SEVERE SERVICE
Ignition Switch, Reset (B)		
Exc. below	.7	.7
G35	1.1	1.1
M45	.8	.8
Q45		
1990-96	1.2	1.2
Ignition Timing, Reset (B)		
All Models	.5	.5
Spark Plug (Ignition) Cables, Replace (B)		
G20	.5	.6
M30	.8	.9
Spark Plugs, Replace (B)		
G20	.6	.7
G35, I35	1.4	1.5
I30, J30	1.1	1.2
M30, M45	1.2	1.3
Q45		
1990-96		
w/TCS	6.5	6.6
w/o TCS	3.5	3.6
1997-05	1.4	1.5
Transistorized Ignition Control Unit, Replace (B)		
G20	.6	.6
J30	.5	.5
M30	.3	.3
1990-96 Q45		
w/TCS	1.4	1.4
w/o TCS	.5	.5

EMISSIONS

The following operations do not include testing. Add time as required.

	LABOR TIME	SEVERE SERVICE
Diagnose Emission Control System Component Each (A)		
All Models	.7	.7
Dynamometer Test (A)		
All Models	.5	.5
Reprogram Electronic Control Module (ECM) (A)		
All Models	.6	.6
Retrieve Fault Codes (A)		
All Models	.3	.3
Air Flow Meter, Replace (B)		
Exc. below	.5	.5
2000-04 I30, I35	.8	.8
M30	.6	.6
Air Injection Control Unit, Replace (B)		
G20	.7	.7
Anti-Backfire Valve, Replace (B)		
G20	.5	.5
Back Pressure Variable Transducer Valve, Replace (B)		
G20		
1991-96	.5	.5
1999-02	.3	.3
1996-97 I30	.7	.7

	LABOR TIME	SEVERE SERVICE
Q45		
1990-96	.5	.5
1997-01	.3	.3
Canister Purge Solenoid, Replace (B)		
Exc. below	.3	.3
I30, I35	.5	.5
Canister Vent Control Valve, Replace (B)		
G20	.3	.3
G35	.6	.6
I30, I35, M45	.5	.5
1997-05 Q45	.5	.5
Coolant Temperature (ECT) Sensor, Replace (B)		
G20	.6	.6
G35	2.0	2.0
I30, I35, J30, M30	.5	.5
M45	1.5	1.5
Q45		
1990-96	5.3	5.3
1997-01	.8	.8
2002-05	3.3	3.3
EGR Control Valve, Replace (B)		
G20		
1991-96	.5	.7
1999-02		
vacuum	.8	1.0
electric	1.5	1.7
I30, I35		
1996-99		
vacuum	1.0	1.2
electric	1.5	1.7
2000-04	1.8	2.0
J30	2.2	2.4
M30	.5	.7
Q45	.5	.7
EGR Solenoid Valve, Replace (B)		
G20	.6	.6
I30	.7	.7
J30, M30	.5	.5
1990-01 Q45	.6	.6
EGR Temperature Sensor, Replace (B)		
G20	.5	.5
I30, I35	.7	.7
J30, M30	.6	.6
1990-01 Q45	.6	.6
Electronic Control Module (ECM/PCM), Replace (B)		
G20	.8	.8
G35	.9	1.0
I30, I35		
1996-99	.6	.6
2000-04	1.6	1.6
J30	.6	.6
M30	.5	.5
M45	.9	1.0

	LABOR TIME	SEVERE SERVICE
Q45		
1990-01	.5	.5
2002-04	.8	.8
Evaporative Pressure Sensor, Replace (B)		
All Models	.5	.5
Fuel Vapor Canister, Replace (B)		
Exc. I30, I35	.6	.6
I30, I35		
carbon canister	1.5	1.5
vapor canister	.8	.8
Intake Air Temperature (IAT) Sensor, Replace (B)		
All Models	.5	.5
Knock Sensor, Replace (B)		
G20		
1991-96	1.3	1.5
1999-02	2.0	2.2
G35	1.7	1.9
I30, I35		
1996-99	.9	1.1
2000-04	3.8	4.0
J30	8.8	9.0
M30	.5	.7
M45	3.1	3.3
Q45		
1990-96	5.3	5.5
1997-05	3.6	3.8
Mixture Ratio Control Solenoid, Replace (B)		
G20	.5	.5
1990-96 Q45	.7	.7
Oxygen Sensor, Replace (B) n		
Front		
G20	.5	.7
G35		
one	.9	1.1
both	1.1	1.3
I30, I35		
1996-99		
one	.7	.9
both	1.0	1.2
1999-04		
one	.8	1.0
both	1.8	2.0
J30		
one	1.0	1.2
both	1.1	1.3
M30	.6	.8
M45		
one	.6	.8
both	1.0	1.2
Q45		
1990-96		
one	.6	.8
both	1.0	1.2
1997-01		
one	2.5	2.7
both	2.7	2.9

G20 : G35 : I30 : I35 : J30 : M30 : M45 : Q45   INF-5

	LABOR TIME	SEVERE SERVICE
2002-05		
one	.8	1.0
both	1.1	1.3
Rear		
G20, G35 one		
or both	.5	.7
I30, I35		
one	1.0	1.2
both	1.1	1.3
J30		
one	1.4	1.6
both	1.5	1.7
M45 one or both	.5	.5
Q45		
1990-96		
one	1.0	1.2
both	1.1	1.3
1997-05		
one or both	.5	.7
Positive Crankcase Ventilation (PCV) Valve, Replace (B)		
All Models	.5	.5
Purge Control Valve, Replace (B)		
G20, G35	.3	.3
I30, I35	.5	.5
1997-01 Q45	.3	.3
Throttle Position Sensor (TPS)/Switch, Replace (B)		
G20, G35		
1991-96	1.0	1.0
1999-04	.5	.5
I30, I35	.7	.7
J30, M30, M45	.5	.5
Q45	.6	.6
Vehicle Speed Sensor, Replace (B)		
Exc. below	.6	.6
M30	1.2	1.2
M45	.3	.3

FUEL
DELIVERY

	LABOR TIME	SEVERE SERVICE
Electric Fuel Pump, Replace (B)		
G20, G35		
1991-96	1.3	1.4
1999-05	.8	.9
I30, I35	1.0	1.1
J30	1.4	1.6
M30, M45	.8	.9
Q45		
1990-96	1.9	2.0
1997-01	2.4	2.5
2002-05	.6	.7
Fuel Filter, Replace (B)		
In fuel tank		
G35	.8	.9
I30	1.0	1.1
M45	.6	.7
2002-05 Q45	.8	.9

	LABOR TIME	SEVERE SERVICE
In fuel line		
G20	.6	.7
G35, M45	.8	.9
I30, I35	1.0	1.1
J30, M30	.5	.6
Q45		
1990-96	.5	.6
1997-01	1.1	1.2
Fuel Gauge (Dash), Replace (B)		
G20		
1991-96	1.2	1.2
1999-02	.8	.8
G35	1.2	1.2
I30, I35, J30	1.1	1.1
M30	1.5	1.5
M45	.8	.8
Q45		
1990-96	1.5	1.5
1997-01	1.2	1.2
2002-05	.6	.6
Fuel Gauge (Tank), Replace (B)		
G20		
1991-96	1.0	1.2
1999-02	.6	.8
G35	1.1	1.3
I30, I35	1.0	1.2
J30	1.9	2.1
M30, M45	.8	1.0
Q45		
1990-96	1.9	2.1
1997-01	2.4	2.6
2002-05	.8	1.0
Fuel Pump Control Module, Replace (B)		
Exc. 1990-96 Q45	.6	.6
1990-96 Q45	1.6	1.6
Fuel Pump Relay, Replace (B)		
G20	.3	.3
G35, M30, Q45	.8	.8
I30, I35, J30	.6	.6
M45	1.0	1.0
Fuel Tank, Replace (B)		
Includes: Drain and refill.		
G20	2.5	2.7
G35		
2WD	7.0	7.2
4WD	7.8	8.0
I30, I35	1.9	2.1
J30	1.7	1.9
M30	2.5	2.7
M45	1.7	1.9
Q45		
1990-96	3.5	3.7
1997-01	4.0	4.2
2002-05	1.3	1.5
Fuel Tank Temperature Sensor, Replace (B)		
G20, I30, I35	1.0	1.0
G35	.8	.8

	LABOR TIME	SEVERE SERVICE
J30, M45	.6	.6
2002-05 Q45	.6	.6
Intake Manifold and/or Gasket, Replace (B)		
Includes: Adjustments.		
G20		
1991-96	1.7	2.0
1999-02	6.5	6.8
G35		
upper	1.4	1.7
lower	2.0	2.3
I30		
1996-99		
upper	2.9	3.2
lower	3.6	3.9
2000-01		
upper	4.0	4.3
lower	4.1	4.4
I35		
upper	1.1	1.4
lower	2.0	2.3
J30		
upper	5.9	6.2
lower	9.8	10.1
M30		
upper	3.8	4.1
lower	4.1	4.4
M45		
upper	2.6	2.9
lower	3.3	3.6
Q45		
1990-96		
w/TCS		
upper	7.2	7.5
lower	9.3	9.6
w/o TCS		
upper	4.1	4.4
lower	6.2	6.5
1997-01		
upper	3.3	3.6
lower	3.4	3.7
2002-05		
upper	2.6	2.9
lower	3.3	3.6

INJECTION
The following operations do not include testing. Add time as required.

Diagnose Fuel Injection System Component Each (A)
  All Models 7   .7

Air Regulator, Replace (B)
  G20
    1991-96 6   .6
    1999-02 .... 2.1   2.1
  J30 8   .8
  M30 6   .6

	LABOR TIME	SEVERE SERVICE
Cylinder Head Temperature Sensor, Replace (B)		
M30	.7	.7
M45	1.5	1.5
Q45		
2002-05	1.1	1.1
EFI Relay, Replace (B)		
G20	.5	.5
G35	.8	.8
I30, I35	.3	.3
J30	.6	.6
M30	.5	.5
M45	.9	.9
Q45	.6	.6
Fuel Injectors, Replace (B)		
G20		
1991-96	1.1	1.2
1999-02	3.1	3.4
G35	2.1	2.4
I30		
1996-99	3.1	3.4
2000-01	3.5	3.9
I35	1.8	2.1
J30	5.2	5.4
M30, M45	3.3	3.7
Q45		
1990-96	5.1	5.3
1997-01	3.6	3.8
2002-05	2.8	3.1
Fuel Pressure Damper, Replace (B)		
G35, M45	.7	.7
I35	.6	.6
2002-05 Q45	.5	.5
Fuel Pressure Regulator, Replace (B)		
G20	.6	.8
G35	.8	1.0
I30, I35		
1996-99	.6	.8
2002-04	1.2	1.4
J30	.5	.7
M30, M45	.8	1.0
Q45		
1990-96	.9	1.1
1997-01	2.9	3.1
2002-05	.6	.8
Throttle Body Assy., Replace (B)		
G20	1.4	1.5
G35	.5	.6
I30, I35		
1996-99	1.4	1.5
2000-04	1.8	1.9
J30	.9	1.0
M30	1.7	1.8
M45	.5	.6
Q45		
1990-96		
w/TCS	3.1	3.2
w/o TCS	1.7	1.8
1997-01	2.0	2.1
2002-05	.6	.7

EXHAUST

	LABOR TIME	SEVERE SERVICE
Catalytic Converter, Replace (B)		
G20		
1991-96	.6	.7
1999-02	1.1	1.2
G35		
one	1.4	1.5
both	1.7	1.8
I30, I35	1.1	1.2
J30	1.6	1.7
M30	.5	.6
M45		
one	4.4	4.5
both	6.1	6.2
Q45		
1990-96	1.9	2.0
2002-05		
one	5.2	5.3
both	8.8	8.9
Center Exhaust & Pre-Muffler Pipe, Replace (B)		
G20, G35	.8	.9
I30, I35	.8	.9
J30	1.5	1.6
M30	.5	.6
M45	1.3	1.4
Q45	1.4	1.5
Exhaust Manifold or Gasket, Replace (B)		
G20		
1991-96	1.3	1.6
1999-02	2.4	2.7
G35		
right side	2.7	3.0
left side	3.0	3.3
I30, I35		
1996-99		
right side	6.6	6.9
left side	4.0	4.3
2000-04		
right side	6.7	7.0
left side	3.3	3.6
J30		
one side	3.8	4.1
M30		
right side	2.1	2.4
left side	3.1	3.4
M45		
right side	4.4	4.7
left side	3.2	3.5
Q45		
1990-96		
right side	4.8	5.1
left side	8.6	8.9
1997-01		
right side	5.7	6.0
left side	5.9	6.2

	LABOR TIME	SEVERE SERVICE
2002-05		
right side	5.2	5.5
left side	4.7	5.0
Front Exhaust Pipe, Replace (B)		
G20	1.3	1.4
G35	.6	.7
I30, I35	2.1	2.2
J30		
one	1.6	1.7
both	2.2	2.3
M30	.6	.7
M45		
one	1.3	1.4
both	2.1	2.2
Q45		
1990-96		
one	1.7	1.8
both	2.2	2.3
1997-01		
one	1.4	1.5
both	1.9	2.0
2002-05	1.1	1.2
Muffler, Replace (B)		
G20	.6	.7
G35	1.0	1.1
I30, I35, M45	.6	.7
J30	1.1	1.2
M30	.8	.9
Q45		
1990-01	1.1	1.2
2002-05	.6	.7
Tail Pipe, Replace (B)		
All Models	.5	.5

ENGINE COOLING

	LABOR TIME	SEVERE SERVICE
Pressure Test Cooling System (B)		
All Models	.3	.3
Bypass Hoses, Replace (B)		
G20	1.1	1.2
G35	3.3	3.7
I30, I35, J30	2.8	3.1
M30	1.9	2.1
M45	4.4	4.9
Q45		
1990-96	5.6	5.8
1997-01	3.4	3.6
2002-05	3.3	3.7
Coolant Temperature Sensor Switch, Replace (B)		
G20	.5	.5
G35	2.0	2.0
I30, I35, J30	.5	.5
M30	.6	.6
M45	1.5	1.5
Q45		
1990-96	5.3	5.3
1997-01	.5	.5
2002-05	3.3	3.3

G20 : G35 : I30 : I35 : J30 : M30 : M45 : Q45

	LABOR TIME	SEVERE SERVICE
Coolant Thermostat, Replace (B)		
G20	1.1	1.2
G35	1.4	1.5
I30, I35		
inlet/outlet	1.1	1.2
outlet/gasket	1.5	1.6
J30	2.4	2.5
M30	2.3	2.4
M45	1.4	1.5
Q45		
1990-96	1.1	1.2
1997-01	1.4	1.5
2002-05		
inlet/outlet	.9	1.0
outlet/gasket	2.0	2.1
Fan Blade, Replace (B)		
G20		
1991-96		
one	1.1	1.2
two	1.4	1.5
1999-02		
one	.3	.4
two	.5	.6
G35		
one	1.4	1.5
two	1.5	1.6
I30, I35		
1996-99		
one	.8	.9
two	1.2	1.3
2000-04		
one	1.4	1.5
two	1.7	1.8
J30	.6	.7
M30	1.1	1.2
M45	1.5	1.6
Q45		
1990-96	1.2	1.3
1997-01	.6	.7
2002-05	1.1	1.2
Radiator Assy., R&R or Replace (B)		
Includes: Refill with proper coolant mix.		
G20	1.4	1.6
G35	1.7	1.9
I30, I35	1.5	1.7
J30	1.8	2.0
M30	1.5	1.7
M45	2.3	2.5
Q45		
1990-01	1.8	2.0
2002-05	2.3	2.5
w/AT add		
G20, G35, M30	.2	.2
I30, I35	.3	.3

	LABOR TIME	SEVERE SERVICE
Radiator Fan and/or Fan Motor, Replace (B)		
Includes: R&I shroud as necessary.		
G20		
1991-96		
one	1.2	1.5
two	1.7	2.0
1999-02		
one	.5	.8
two	.6	.9
G35		
coupling	.8	1.1
motor		
one	1.4	1.7
two	1.5	1.8
I30, I35		
1996-99		
one	.9	1.2
both	1.2	1.5
2000-05		
one	1.4	1.7
two	1.5	1.8
J30 coupling	.6	.9
M30 coupling	1.1	1.4
1990-01 Q45 coupling	1.2	1.5
Radiator Hoses, Replace (B)		
Includes: Refill with proper coolant mix.		
Upper		
exc. below	.9	1.0
G35, M45	.5	.6
2002-05 Q45	.3	.4
Lower		
G20	.9	1.0
G35	1.4	1.5
I30, I35	1.2	1.3
J30	.8	.9
M30	1.1	1.2
M45	1.5	1.6
Q45		
1990-96	1.5	1.6
1997-01	.9	1.0
2002-05	1.2	1.3
Radiator Fan Motor Relay, Replace (B)		
G20, I30, I35	.3	.3
G35	.8	.8
1990-01 Q45	.5	.5
Temperature Gauge (Dash), Replace (B)		
G20		
1991-96	1.2	1.2
1999-02	.8	.8
G35	1.2	1.2
I30, I35, J30	1.1	1.1
M30	1.5	1.5
M45	.8	.8
Q45		
1990-96	1.5	1.5
1997-01	1.2	1.2
2002-05	.6	.6

	LABOR TIME	SEVERE SERVICE
Temperature Gauge (Engine), Replace (B)		
G20, J30, M30	.5	.5
G35	2.0	2.0
I30, I35	.5	.5
M30	.6	.6
M45	1.5	1.5
Q45		
1990-96	5.3	5.3
1997-00	.5	.5
2001-05	3.3	3.3
Water Pump and/or Gasket, Replace (B)		
Includes: Refill with proper coolant mix.		
G20	3.1	3.5
G35	2.1	2.5
I30, I35	4.4	4.8
J30	5.0	5.4
M30	4.9	5.3
M45	2.0	2.4
Q45		
1990-96	2.8	3.2
1997-01	.9	1.3
2002-05	1.4	1.8

ENGINE

ASSEMBLY

Times shown are for OEM assemblies. Time to replace assemblies from aftermarket rebuilders may vary.

Engine Assy., R&I (B)
Does not include parts or component transfer.

	LABOR TIME	SEVERE SERVICE
G20	11.4	11.4
I30	12.9	12.9
J30	15.5	15.5
M30	10.8	10.8
Q45		
1990-96	12.8	12.8
1997-01	11.8	11.8
w/AC G20 add	.5	.5
w/PS G20 add	.5	.5

Engine Assy., R&R and Recondition (A)
Includes: Replacing rings, rod and main bearings, cylinder head reconditioning and engine tune-up.

	LABOR TIME	SEVERE SERVICE
G20		
1991-96	30.1	40.0
1999-02	28.1	29.0
G35		
2WD	31.2	32.1
AWD	33.5	34.3
I30		
1996-99	34.6	35.5
2000-01	35.0	35.9
I35	33.9	34.8
J30	45.1	46.0
M30	31.6	32.5
M45	32.0	32.9

INF-7

INF-8   G20 : G35 : I30 : I35 : J30 : M30 : M45 : Q45

	LABOR TIME	SEVERE SERVICE
Engine Assy., R&R and Reconditioning (A)		
Q45		
1990-96	37.7	*38.6*
1997-01	38.5	*39.4*
2002-05	31.3	*32.2*
w/AC add		
91-96 G20	.8	*.8*
M30	.8	*.8*
w/PS add		
91-96 G20	.6	*.6*
M30	.6	*.6*
Engine (New or Rebuilt Complete), Replace (B)		
Includes: Component transfer and engine tune-up.		
G20		
1991-96	14.4	*14.9*
1999-02	15.3	*15.8*
G35		
2WD	14.5	*15.0*
AWD	17.3	*17.8*
I30	17.4	*17.9*
I35	16.8	*17.3*
J30	19.5	*20.0*
M30	13.9	*14.4*
M45	16.5	*16.8*
Q45		
1990-96	18.5	*19.0*
1997-01	17.3	*17.8*
2002-05	15.8	*16.3*
w/AC 91-96 G20 add	.8	*.8*
w/PS 91-96 G20 add	.6	*.6*
Engine Assy. (Short Block), Replace (B)		
Assembly consists of cylinder block, piston assemblies, crankshaft, camshaft, timing chain and gears. Does not include cylinder heads. Operation Includes: R&R engine, transfer necessary parts and all necessary adjustments.		
G20	18.8	*19.3*
G35		
2WD	19.8	*20.3*
AWD	22.2	*22.7*
I30	24.0	*24.5*
I35	24.3	*24.8*
J30	26.2	*26.7*
M30	17.9	*18.4*
M45	21.2	*21.7*
Q45		
1990-96	25.5	*26.0*
1997-01	22.9	*23.4*
2002-05	20.5	*21.0*
w/AC add		
91-96 G20	.8	*.8*
M30	.8	*.8*

	LABOR TIME	SEVERE SERVICE
w/PS add		
91-96 G20	.6	*.6*
M30	.6	*.6*
Engine Mounts, Replace (B)		
Front		
G20		
1991-96	.8	*1.0*
1999-02	1.2	*1.4*
G35		
2WD	5.0	*5.2*
AWD	6.7	*7.2*
I30, I35		
1996-99	1.1	*1.3*
2000-04	1.7	*1.9*
J30	3.1	*3.3*
M30	2.4	*2.6*
M45	3.6	*3.8*
Q45		
1990-96	3.5	*3.7*
1997-01	3.2	*3.4*
2002-05	1.9	*2.1*
Rear		
G20	1.3	*1.5*
G35	.6	*.7*
I30, I35	1.9	*2.1*
J30	1.1	*1.3*
M30, M45	.6	*.8*
Q45	.8	*1.0*
CYLINDER HEAD		
Compression Test (B)		
G20	1.2	*1.3*
I30	1.3	*1.4*
J30	1.5	*1.6*
M30	1.7	*1.8*
Q45		
1990-96		
w/TCS	7.3	*7.4*
w/o TCS	4.1	*4.2*
1997-05	1.8	*1.9*
Cylinder Head, Replace (B)		
Includes: Parts transfer, grind valves, adjustments. G20 does not include grind valves.		
G20		
1991-96	18.6	*19.1*
1999-02	13.6	*14.1*
G35		
2WD		
one side	19.9	*20.4*
both sides	25.3	*25.8*
AWD		
one side	21.8	*22.3*
both sides	27.2	*27.7*
I30, I35		
right side	23.8	*24.3*
left side	22.1	*22.6*
both sides	28.6	*29.1*

	LABOR TIME	SEVERE SERVICE
J30		
right side	20.4	*20.9*
left side	20.0	*20.5*
both sides	27.5	*28.0*
M30		
right side	13.6	*14.1*
left side	14.2	*14.7*
both sides	19.5	*20.0*
M45		
one side	21.8	*22.3*
both sides	25.8	*26.3*
Q45		
1990-96		
one side	30.5	*31.0*
both sides	37.1	*47.6*
1997-01		
right side	19.3	*19.8*
left side	20.6	*21.1*
both sides	26.0	*26.5*
2002-05		
one side	21.1	*21.6*
both sides	25.0	*25.5*
Cylinder Head Front Cover, Replace (B)		
G20	1.6	*1.8*
Cylinder Head Gasket, Replace (B)		
G20		
1991-96	12.0	*12.3*
1999-02	9.5	*9.8*
G35		
2WD		
one side	15.9	*16.2*
both sides	17.2	*17.5*
AWD		
one side	18.5	*18.8*
both sides	23.0	*23.3*
I30, I35		
right side	20.7	*21.0*
left side	19.3	*19.6*
both sides	23.0	*23.3*
J30		
right side	15.8	*16.1*
left side	14.1	*14.4*
both sides	16.2	*16.5*
M30		
right side	10.3	*10.6*
left side	10.9	*11.2*
both sides	12.6	*12.9*
M45		
one side	19.0	*19.3*
both sides	20.3	*20.6*
Q45		
1990-96		
one side	22.8	*23.1*
both sides	24.1	*24.4*
1997-01		
one side	14.5	*14.8*
both sides	16.4	*16.7*

G20 : G35 : I30 : I35 : J30 : M30 : M45 : Q45 INF-9

	LABOR TIME	SEVERE SERVICE
2002-05		
one side	18.4	*18.7*
both sides	19.7	*20.0*
w/AC 91-96 G20 add	.3	*.3*

Rocker Arm Cover Gasket, Replace or Reseal (B)

	LABOR TIME	SEVERE SERVICE
G20	1.0	*1.2*
G35		
one side	2.0	*2.2*
both sides	2.8	*3.0*
I30		
1996-99		
right side	2.8	*3.1*
left side	.6	*.7*
both sides	3.1	*3.4*
2000-01		
right side	4.0	*4.2*
left side	1.2	*1.4*
both sides	4.8	*5.0*
I35		
right side	2.1	*2.3*
left side	.9	*1.1*
both sides	2.6	*2.8*
J30		
one side	5.0	*5.2*
both sides	5.8	*6.0*
M30		
right side	1.4	*1.6*
left side	4.1	*4.3*
both sides	4.7	*4.9*
M45		
right side	.9	*1.1*
left side	1.4	*1.6*
both sides	2.0	*2.2*
Q45		
1990-96		
w/TCS		
right side	2.8	*3.0*
left side	7.9	*8.1*
both sides	10.1	*10.3*
w/o TCS		
right side	2.8	*3.0*
left side	5.0	*5.2*
1997-01		
right side	1.6	*1.8*
left side	3.1	*3.3*
both sides	3.2	*3.4*
2002-05		
right side	1.1	*1.3*
left side	1.7	*1.9*
both sides	2.4	*2.6*

Rocker Arms or Shafts, Replace (B)
Includes: R&R cams when required.

	LABOR TIME	SEVERE SERVICE
G20		
1991-96	4.4	*4.6*
1999-02	5.5	*5.7*
G35		
one side	12.8	*13.0*
both sides	13.9	*14.1*

	LABOR TIME	SEVERE SERVICE
I30, I35		
1996-99		
right side	13.5	*13.7*
left side	11.3	*11.5*
both sides	15.0	*15.2*
2000-04		
right side	13.6	*13.8*
left side	12.0	*12.2*
both sides	15.1	*15.3*
J30		
right side	11.5	*11.7*
left side	11.2	*11.4*
both sides	12.8	*13.0*
M30		
right side	2.1	*2.3*
left side	4.6	*4.8*
both sides	6.1	*6.3*
M45		
one side	16.7	*16.9*
both sides	17.6	*17.8*
Q45		
1990-96		
w/TCS		
right side	4.1	*4.3*
left side	11.5	*11.7*
both sides	12.3	*12.5*
w/o TCS		
right side	4.1	*4.3*
left side	6.2	*6.4*
both sides	9.7	*9.9*
1997-01		
right side	4.7	*4.9*
left side	6.2	*6.4*
both sides	9.1	*9.3*
2002-05		
one side	16.0	*16.2*
both sides	17.0	*17.2*

Valve Job (A)

	LABOR TIME	SEVERE SERVICE
G20		
1991-96	18.6	*19.1*
1999-02	13.6	*14.1*
G35		
2WD	25.8	*26.3*
AWD	27.5	*28.0*
I30, I35		
1996-99	24.2	*24.7*
2000-04	28.2	*28.7*
J30	27.5	*28.0*
M30	19.5	*20.0*
M45	26.5	*27.0*
Q45		
1990-96	37.1	*37.6*
1997-01	26.0	*26.5*
2002-05	25.7	*26.2*

Valve Springs and/or Oil Seals, Replace (B)

	LABOR TIME	SEVERE SERVICE
G20		
1991-96	7.6	*7.8*
1999-02	10.4	*10.6*

	LABOR TIME	SEVERE SERVICE
G35		
one side	13.1	*13.3*
both sides	14.5	*14.7*
I30, I35		
1996-99		
right side	14.7	*14.9*
left side	12.7	*12.9*
both sides	18.0	*18.2*
2000-04		
right side	14.8	*15.0*
left side	13.2	*13.4*
both sides	18.0	*18.2*
J30		
right side	16.9	*17.1*
left side	16.3	*16.5*
both sides	20.2	*20.4*
M30		
right side	3.8	*4.0*
left side	6.2	*6.4*
both sides	9.2	*9.4*
M45		
one side	17.6	*17.8*
both sides	19.5	*19.7*
Q45		
1990-96		
right side	9.4	*9.6*
left side	11.5	*11.7*
both sides	17.8	*18.0*
1997-01		
right side	5.8	*6.0*
left side	7.3	*7.5*
both sides	11.2	*11.4*
2002-05		
one side	17.0	*17.2*
both sides	18.8	*20.0*

Valve Timing Control (VTC) Solenoid Valves, Replace (B)

	LABOR TIME	SEVERE SERVICE
G35		
one	.5	*.8*
all	.6	*.9*
I35		
one	.3	*.6*
all	.5	*.8*
J30		
one	4.7	*5.0*
all	5.2	*5.5*
M45		
one	.6	*.9*
all	.9	*1.2*
Q45		
1990-96		
one	2.3	*2.6*
all	6.9	*7.2*
1997-01		
one	.5	*.8*
all	.7	*1.0*
2002-05		
one	.8	*1.1*
all	1.1	*1.4*

INF-10 G20 : G35 : I30 : I35 : J30 : M30 : M45 : Q45

	LABOR TIME	SEVERE SERVICE
CAMSHAFT		
Camshaft, Replace (A)		
G20		
1991-96		
one	3.1	3.3
both	4.1	4.3
1999-02		
one	3.5	3.7
both	4.1	4.3
G35		
2WD		
one side	12.8	13.0
both sides	14.2	14.4
AWD		
right side	17.9	18.1
left side	16.9	17.1
both sides	19.3	19.6
I30, I35		
1996-99		
right side	13.8	14.0
left side	11.6	11.8
both sides	15.4	15.6
2000-04		
right side	13.8	14.0
left side	12.3	12.5
both sides	15.6	15.8
J30		
one side	10.9	11.1
both sides	12.6	12.8
M30		
right side	11.2	11.4
left side	12.1	12.3
both sides	14.1	14.3
M45		
one side	16.6	16.8
both sides	17.3	17.5
Q45		
1990-96		
right side	6.3	6.5
left side	8.5	8.7
both sides	12.4	12.6
1997-01		
right side	3.6	3.8
left side	5.2	5.4
both sides	6.8	7.0
2002-08		
one side	15.9	16.1
both sides	16.7	16.9
Camshaft Seal, Replace (B)		
J30		
one side	5.9	6.1
both sides	6.5	6.7
M30		
one side	8.7	8.9
both sides	9.1	9.3
Camshaft Timing Gear, Replace (B)		
G20		
1991-96	1.9	2.1
1999-02	2.6	2.8

	LABOR TIME	SEVERE SERVICE
Timing Belt, Replace (B)		
J30	4.7	5.2
M30	4.9	5.4
w/AC M30 add	.3	.3
w/PS M30 add	.3	.3
Replace timing gears add	1.0	1.0
Timing Belt Cover, Replace (B)		
J30	4.6	5.1
M30	4.7	5.2
w/AC M30 add	.3	.3
w/PS M30 add	.3	.3
Timing Chain or Gear, Replace (B)		
Includes: Oil seal.		
G20		
1991-96	9.5	9.9
1999-02	9.9	10.3
G35		
2WD	9.2	9.6
AWD	13.0	13.3
I30, I35		
1996-99	9.6	10.0
2000-04	10.4	10.8
M45	15.7	16.1
Q45		
1990-96	23.6	24.0
1997-01	17.2	17.6
2002-05	15.0	15.4
Timing Chain Cover and/or Gasket, Replace (B)		
G20		
1991-96	8.2	8.5
1999-02	8.8	9.1
G35		
2WD	6.3	6.6
AWD	10.2	10.5
I30, I35		
1996-99	9.0	9.3
2000-04	9.6	9.9
M45	15.1	15.4
Q45		
1990-96	23.2	23.5
1997-01	17.2	17.5
2002-05	14.5	14.8
w/AC 91-96 G20 add	.5	.5
Timing Chain Tensioner, Replace (B)		
G20	.5	.8
G35		
2WD		
upper	14.2	14.5
lower	1.4	1.7
AWD		
upper	16.9	17.0
lower	1.4	1.7
I30, I35		
upper	13.3	13.6
lower	1.3	1.6
M45		
upper	1.4	1.7
lower	1.7	2.0

	LABOR TIME	SEVERE SERVICE
Q45		
1990-96	3.1	3.4
1997-01	17.3	17.6
2002-05		
upper	.8	1.1
lower	1.3	1.7
CRANK & PISTONS		
Connecting Rod Bearings, Replace (A)		
Includes: Check bearing oil clearance. R&R engine and oil pan.		
G20		
1991-96	21.0	21.5
1999-04	19.8	20.3
G35		
2WD	21.9	22.4
AWD	24.1	24.6
I30	24.2	24.7
I35	23.3	23.8
J30	30.1	30.6
M30, M45	22.0	22.5
Q45		
1990-96	27.7	38.2
1997-01	23.3	23.8
2002-05	21.4	21.9
Crankshaft & Main Bearings, Replace (A)		
Includes: Engine R&R.		
G20		
1991-96	21.1	21.6
1999-02	21.5	22.0
G35		
2WD	22.8	23.3
AWD	23.8	24.3
I30, I35	26.0	26.5
J30	30.6	31.1
M30	22.2	22.7
M45	22.8	23.3
Q45		
1990-96	28.4	28.9
1997-01	24.4	24.9
2002-05	22.1	22.6
w/AC add		
96-99 G20	.8	.8
M30	.8	.8
w/PS add		
96-99 G20	.6	.6
M30	.6	.6
Crankshaft Front Oil Seal, Replace (B)		
Includes: R&I oil pump assembly on J30, M30.		
G20, G35	1.6	1.8
I30, I35	1.4	1.6
J30	5.1	5.3
M30	4.8	5.0
M45	2.1	2.3

G20 : G35 : I30 : I35 : J30 : M30 : M45 : Q45

	LABOR TIME	SEVERE SERVICE
Q45		
1990-96	2.3	2.5
1997-01	1.6	1.8
2002-05	2.1	2.3
w/AC M30 add	.3	.3
w/PS M30 add	.3	.3
Crankshaft Pulley, Replace (B)		
G20, G35	1.6	1.8
I30, I35	1.4	1.6
J30	1.9	2.1
M30	2.2	2.4
M45	1.8	2.0
Q45		
1990-96	2.1	2.3
1997-01	1.4	1.6
2002-05	2.1	2.3

Pistons or Connecting Rods, Replace (A)
Includes: Ridge reaming, cylinder wall deglazing, installing new rings and rod bearings, engine tune-up.

	LABOR TIME	SEVERE SERVICE
G20		
1991-96	22.0	22.9
1999-02	21.4	22.3
G35		
2WD	22.9	23.8
AWD	25.2	26.9
I30	26.5	27.4
I35	27.5	26.4
J30	32.0	32.9
M30	23.5	24.4
M45	24.9	25.8
Q45		
1990-96	29.8	30.7
1997-01	26.1	27.0
2002-05	24.3	25.2
w/AC add		
96-99 G20	.8	.8
M30	.8	.8
w/PS add		
96-99 G20	.6	.6
M30	.6	.6

Rear Main Oil Seal, Replace (B)
Includes: R&R trans. when necessary.

	LABOR TIME	SEVERE SERVICE
G20		
1991-96		
AT	7.0	7.6
MT	6.7	7.3
1999-02		
AT	7.5	8.1
MT	7.3	7.9
G35		
AT		
2WD	5.0	5.6
AWD	6.0	6.6
MT	4.1	4.7

	LABOR TIME	SEVERE SERVICE
I30, I35		
1996-99		
AT	11.3	11.9
MT	10.9	11.5
2000-04	11.9	12.5
J30	6.1	6.7
M30, M45	5.8	6.3
Q45		
1990-96	9.5	10.1
1997-05	6.0	6.6

Rings, Replace (A)
Includes: Ridge reaming, cylinder wall deglazing, installing new rings, engine tune-up.

	LABOR TIME	SEVERE SERVICE
G20		
1991-96	22.0	22.9
1999-02	21.4	22.3
G35		
2WD	22.9	23.8
AWD	25.2	26.9
I30	26.5	27.4
I35	27.5	26.4
J30	32.0	32.9
M30	23.5	24.4
M45	24.9	25.8
Q45		
1990-96	29.8	30.7
1997-01	26.1	27.0
2002-05	24.3	25.2
w/AC add		
96-99 G20	.8	.8
M30	.8	.8
w/PS add		
96-99 G20	.6	.6
M30	.6	.6

ENGINE LUBRICATION

Engine Oil Pressure Switch (Sending Unit), Replace (B)

	LABOR TIME	SEVERE SERVICE
G20, G35		
1991-96	1.4	1.5
1999-05	.6	.7
I30, I35, J30, M30	.5	.6
M45	.8	.9
Q45		
1990-96	.5	.6
1997-05	.9	1.0

Oil Pan and/or Gasket, Replace (B)

	LABOR TIME	SEVERE SERVICE
G20		
1991-96		
upper	4.6	4.9
lower	1.7	1.9
1999-02		
upper	4.3	4.6
lower	1.0	1.2
G35		
2WD		
upper	7.8	8.1
lower	1.1	1.3

	LABOR TIME	SEVERE SERVICE
AWD		
upper	8.8	9.1
lower	1.1	1.3
I30, I35		
1996-99		
upper	5.0	5.3
lower	.8	1.0
2000-04		
upper	6.1	6.4
lower	1.3	1.5
J30	4.8	5.1
M30	3.1	3.4
M45	7.0	7.3
Q45		
1990-96	8.2	8.5
1997-01	4.1	4.4
2002-05	6.1	6.4

Oil Pump, Replace (B)

	LABOR TIME	SEVERE SERVICE
G20	8.8	9.3
G35	7.5	8.0
I30, I35		
1996-99	11.5	12.0
2000-04	12.6	13.4
J30	10.8	11.3
M30	7.9	8.4
M45	15.8	16.3
Q45		
1990-96	24.9	25.4
1997-01	18.8	19.3
2002-05	15.1	15.6

Oil Pump Chain, Replace (B)

	LABOR TIME	SEVERE SERVICE
Q45		
1990-96	24.9	25.5
1997-01	18.1	18.7

Oil Pump Chain Guides, Replace (B)

	LABOR TIME	SEVERE SERVICE
Q45		
1990-96	24.9	25.4
1997-01	18.1	18.6

CLUTCH

Clutch Pedal Free Play, Adjust (B)

	LABOR TIME	SEVERE SERVICE
All Models	.3	.3

Clutch Assy., Replace (B)

	LABOR TIME	SEVERE SERVICE
G20		
1991-96	6.3	6.5
1999-02	5.5	5.7
G35	3.5	3.7
I30, I35		
1996-99	6.3	6.5
2000-04	7.0	7.2

Replace release bearing add .3 .3

Clutch Control Cable, Replace (B)

	LABOR TIME	SEVERE SERVICE
G20	1.0	1.2

Clutch Master Cylinder, Replace (B)
Includes: System bleeding.

	LABOR TIME	SEVERE SERVICE
1999-05 G20, G35	1.3	1.5
I30, I35	1.6	1.8

INF-12 G20 : G35 : I30 : I35 : J30 : M30 : M45 : Q45

	LABOR TIME	SEVERE SERVICE
Clutch Slave Cylinder, Replace (B)		
Includes: System bleeding.		
1999-05 G20, G35	.8	1.0
I30, I35	1.4	1.6
Clutch Switch, Replace or Adjust (B)		
All Models	.5	.5
Flywheel, Replace (B)		
G20		
1991-96	6.6	6.8
1999-02	7.3	7.5
G35	4.0	4.2
I30, I35		
1996-99	6.6	6.8
2000-04	7.2	7.4

MANUAL TRANSAXLE

	LABOR TIME	SEVERE SERVICE
Differential Case, Replace (B)		
G20		
1991-96	8.7	9.2
1999-02	9.4	9.9
I30, I35		
1996-99	8.5	9.0
2000-04		
RS5F50A/V	9.5	10.0
RS6F51H	10.3	10.8
Differential Case, Recondition (A)		
G20		
1991-96	9.1	9.6
1999-02		
RS5F30/31/32A/V	9.9	10.4
RS5F70A	9.4	9.9
I30, I35		
1996-99	9.0	9.5
2000-04		
RS5F50A/V	10.0	10.5
RS6F51H	10.5	11.0
Differential Gear Cover, Replace or Reseal (B)		
G20, I30	.8	1.0
Differential Side Oil Seal, Replace (B)		
G20		
one side	1.4	1.6
both sides	2.5	2.7
I30, I35		
1996-99		
one side	1.6	1.8
both sides	2.7	2.9
2001-04		
one side	2.9	3.1
both sides	3.5	3.7
Transaxle Assy., R&R and Recondition (A)		
Includes: Differential case assy.		
G20		
1991-96	12.3	13.2
1999-02		
RS5F30/31/32A/V	13.1	14.0
RS5F70A	14.3	15.2
I30, I35		
1996-99	12.3	13.2
2000-04		
RS5F50A/V	13.4	14.3
RS6F51H	15.2	16.1
Transaxle Assy., Replace (B)		
G20		
1991-96	6.0	6.2
1999-02	6.9	7.1
I30, I35		
1996-99	6.0	6.2
2000-04	6.9	7.1

MANUAL TRANSMISSION

	LABOR TIME	SEVERE SERVICE
Rear Extension Oil Seal, Replace (B)		
G35	1.1	1.3
Transmission Assy., R&R and Recondition (A)		
G35	12.6	13.5
Transmission Assy., Replace (B)		
G35	3.6	3.8

AUTOMATIC TRANSAXLE
SERVICE TRANSAXLE INSTALLED

	LABOR TIME	SEVERE SERVICE
Drain & Refill Unit (B)		
All Models	1.0	1.0
Oil Pressure Check (B)		
All Models	1.0	1.0
Accumulator Assy., R&R or Replace (B)		
G20		
1991-96	2.5	2.7
1999-02	1.7	1.9
I30, I35	2.2	2.4
Band Servo, Replace (B)		
G20		
1991-96	1.0	1.2
1999-02	1.4	1.6
I30, I35	1.1	1.3
Differential Oil Seal (Halfshaft), Replace (B)		
G20		
one side	1.4	1.6
both sides	2.4	2.7
I30, I35		
one side	1.6	1.8
both sides	2.4	2.7
Governor Assy., R&R or Replace (B)		
G20	.8	1.0
Lock-Up Torque Converter Solenoid, Replace (B)		
G20		
1991-96	2.5	2.7
1999-02	1.7	1.9
I30, I35	2.2	2.4
Oil Pan and/or Gasket, Replace (B)		
All Models	1.1	1.3

	LABOR TIME	SEVERE SERVICE
Overdrive Cancel Solenoid, Replace (B)		
G20		
1991-96	2.5	2.7
1999-02	1.7	1.9
I30, I35	2.2	2.4
Parking Pawl, Replace (B)		
G20	1.9	2.1
I30, I35	8.1	8.3
Side Cover, Replace (B)		
G20	1.6	1.8
I30, I35	2.1	2.3
Shift Solenoid, Replace (B)		
G20		
1991-96	2.5	2.7
1999-02	1.7	1.9
I30, I35	2.2	2.4
Speedometer Driven Gear and/or Seal, Replace (B)		
G20	.5	.5
I30, I35	.7	.7
Valve Body, R&R and Recondition (A)		
G20		
1991-96	4.9	5.4
1999-02	4.0	4.5
I30, I35	4.6	5.1
Valve Body, Replace (B)		
G20		
1991-96	2.5	2.7
1999-02	1.7	1.9
I30, I35	2.2	2.4

SERVICE TRANSAXLES REMOVED
Transaxle R&R included unless otherwise noted.

	LABOR TIME	SEVERE SERVICE
Differential Case, Recondition (A)		
G20	9.3	9.8
I30, I35	9.7	10.2
Front Oil Pump, Replace or Recondition (B)		
G20	8.4	8.6
I30, I35	8.7	8.9
Torque Converter, Replace (B)		
G20	6.9	7.1
I30, I35	7.3	7.5
Transaxle Assy. (Complete), R&R and Recondition (A)		
G20	18.8	19.7
I30, I35	19.3	20.2
Flush oil cooler and lines add	.8	.8
Transaxle Assembly, Replace (B)		
G20	7.2	7.4
I30, I35	7.6	7.8
Transaxle Assy., Reseal (B)		
G20	12.6	13.2
I30, I35	12.9	13.5

G20 : G35 : I30 : I35 : J30 : M30 : M45 : Q45 **INF-13**

AUTOMATIC TRANSMISSION

SERVICE TRANSMISSION INSTALLED

	LABOR TIME	SEVERE SERVICE
Drain and Refill Unit (B)		
J30, M30	1.0	1.0
Q45		
1990-96	1.1	1.1
1997-01	.8	.8
2002-05	1.1	1.1
Oil Pressure Check (B)		
All Models	1.0	1.0
Accumulator Assy., Replace (B)		
J30, M30	2.4	2.6
Q45		
1990-96	2.4	2.6
1997-01	1.6	1.8
Downshift Solenoid, Replace (B)		
G35	1.9	2.1
M30	.8	1.0
M45	1.6	1.8
Q45		
1990-96	.8	1.0
2002-05	1.6	1.8
Extension Housing and/or Gasket, Replace (B)		
G35	2.1	2.3
J30	2.9	3.1
M30, M45	2.5	2.7
Q45		
1990-01	2.9	3.2
2002-05	2.4	2.6
Extension Housing Oil Seal, Replace (B)		
G35	2.9	3.1
J30	1.4	1.6
M30	1.1	1.3
1990-01 Q45	2.1	2.3
Lock-Up Torque Converter Solenoid, Replace (B)		
All Models	1.9	1.9
Oil Pan and/or Gasket, Replace (B)		
G35	1.4	1.6
J30, M30	1.0	1.2
M45, Q45	1.1	1.3
Overdrive Cancel Solenoid, Replace (B)		
G35	1.9	2.1
M30	.8	1.0
M45	1.6	1.8
Q45		
1990-96	.8	1.0
2002-05	1.6	1.8
Parking Pawl, Replace (B)		
G35	3.6	3.9
J30	2.9	3.1
M30	2.7	2.9
M45	2.4	2.6

	LABOR TIME	SEVERE SERVICE
Q45		
1990-96	3.0	3.2
1997-05	2.4	2.6
Speedometer Driven Gear and/or Seal, Replace (B)		
All Models	.6	.6
Valve Body Assy., R&R and Recondition (A)		
G35	2.5	3.0
J30	4.7	5.3
M30	4.9	5.5
M45	1.6	2.1
Q45		
1990-96	4.9	5.5
1997-01	3.0	3.6
2002-05	1.6	2.1
Valve Body Assy., Replace (B)		
G35, J30	2.5	2.7
M30	2.1	2.3
M45	1.6	1.8
Q45		
1990-96	2.0	2.2
1997-05	1.6	1.8
Vehicle Speed Sensor, Replace (B)		
G35	4.3	4.3
J30	.5	.5
M30	2.3	2.3
M45	2.9	2.9
Q45		
1990-01	.8	.8
2002-05	2.9	2.9

SERVICE TRANSMISSION REMOVED

Transmission R&R included unless otherwise noted.

	LABOR TIME	SEVERE SERVICE
Front Oil Pump, Replace (B)		
G35		
2WD	5.5	5.7
AWD	6.3	6.5
J30	6.4	6.6
M30	6.1	6.3
M45	6.4	6.6
Q45		
1990-96	9.4	9.6
1997-01	6.2	6.4
2002-05	6.6	6.8
Recondition pump add		
exc. below	1.4	1.6
M30	1.6	1.8
RE5R05A A/T	.6	.8
Torque Converter, Replace (B)		
G35		
2WD	5.0	5.2
AWD	5.8	6.0
J30	6.0	6.2
M30	5.5	5.7
M45	6.0	6.2

	LABOR TIME	SEVERE SERVICE
Q45		
1990-96	9.0	9.2
1997-01	5.6	5.8
2002-05	6.1	6.3
Transmission & Converter, R&R and Recondition (A)		
G35		
2WD	9.5	10.4
AWD	10.3	11.2
J30	14.9	15.8
M30	18.4	19.3
M45	10.5	11.4
Q45		
1990-96	17.7	18.6
1997-01	15.2	16.1
2002-05	10.6	11.5
Transmission Assy., Replace (B)		
G35		
2WD	5.7	5.9
AWD	6.4	6.6
J30	6.3	6.5
M30, M45	6.6	6.8
Q45		
1990-96	9.4	9.6
1997-05	6.7	6.9
Transmission Assy., Reseal (B)		
G35		
2WD	8.7	9.3
AWD	9.4	10.0
J30	10.8	11.4
M30	10.5	11.1
M45	9.4	10.0
Q45		
1990-96	14.0	14.6
1997-01	10.6	11.2
2002-05	9.5	10.1

TRANSFER CASE

	LABOR TIME	SEVERE SERVICE
Companion Flange, Replace (B)		
G35	1.9	2.1
Front Case Oil Seals, Replace (B)		
G35	1.9	2.1
Rear Case Oil Seals, Replace (B)		
G35	1.9	2.1
Transfer Case, R&R and Recondition (B)		
G35	4.4	5.3
Transfer Case, Replace (B)		
G35	2.5	2.8

SHIFT LINKAGE

AUTOMATIC TRANSAXLE

	LABOR TIME	SEVERE SERVICE
Gear Selector Lever, Replace (B)		
G20	1.2	1.3
I30, I35		
1996-99	2.4	2.5
2000-04	1.6	1.7

G20 : G35 : I30 : I35 : J30 : M30 : M45 : Q45

	LABOR TIME	SEVERE SERVICE
Manual Shaft &/or Parking Rod, Replace (B)		
G20		
1991-96	3.2	3.4
1999-02	2.2	2.4
I30, I35	2.7	2.9
AUTOMATIC TRANSMISSION		
Gear Selector Lever, Replace (B)		
Exc. 1990-96 Q45	.9	1.0
1990-96 Q45	1.2	1.3
Manual Shaft &/or Parking Rod, Replace (B)		
G35	4.0	4.2
J30, M30	2.2	2.4
M45	3.5	3.7
I30, I35		
1990-96	2.2	2.4
1997-01	1.6	1.8
2002-04	3.3	3.5
MANUAL TRANSAXLE		
Gearshift Control Rod, Replace (B)		
G20	.6	.7
I30, I35	1.2	1.3
Gearshift Lever Assy., Replace (B)		
G20	.8	.9
I30, I35	1.4	1.5
Gearshift Support Rod, Replace (B)		
All Models	1.4	1.5
MANUAL TRANSMISSION		
Gearshift Lever Assy., Replace (B)		
G35	.6	.7
DRIVELINE		
Axle Shaft Oil Seal, Replace (B)		
M30		
one	1.7	2.0
both	3.3	3.7
Center Support Bearing, Replace (B)		
G35	1.1	1.4
J30	1.7	2.0
M30	1.3	1.6
M45	1.7	2.0
Q45		
1990-96	2.0	2.3
1997-05	1.4	1.7
Differential Case Assy., R&R and Recondition (A)		
G35		
front axle	11.3	11.9
rear axle	4.1	4.7
J30	3.5	4.1
M30	4.6	5.2
M45	4.9	5.5
Q45		
1990-96	3.4	4.0
1997-01	4.3	4.9
2002-05	4.7	5.3

	LABOR TIME	SEVERE SERVICE
Differential Cover Gasket, Replace (B)		
G35		
front axle	6.7	7.0
rear axle	2.4	2.7
J30	2.5	2.8
M30	1.7	2.0
M45	3.2	3.5
Q45	2.9	3.2
Differential Side Oil Seal, Replace (B)		
G35		
front axle		
one	2.4	2.6
both	3.0	3.2
rear axle		
one	1.1	1.3
both	1.9	2.1
J30		
one	1.0	1.2
both	1.7	1.9
M30		
one	1.9	2.1
both	3.1	3.3
M45		
one	2.4	2.6
both	3.3	3.5
Q45		
1990-96		
one	1.0	1.2
both	1.7	1.9
1997-01		
one	.7	.9
both	1.1	1.3
2002-05		
one	1.9	2.1
both	2.9	3.1
Differential Assy., Replace (B)		
G35		
front axle	5.8	6.3
rear axle	2.5	3.0
J30	2.4	2.9
M30	3.3	3.8
M45	3.0	3.5
Q45	2.7	3.2
Differential Assy., R&R and Recondition (A)		
G35		
front axle	11.5	12.0
rear axle	6.0	6.6
J30	7.2	7.8
M30	8.2	8.8
M45	6.3	6.9
Q45		
1990-96	7.2	7.8
1997-05	6.1	6.9

	LABOR TIME	SEVERE SERVICE
Driveshaft Shaft, R&R or Replace (B)		
G35	1.0	1.2
AWD		
front	1.9	2.1
rear	1.0	1.2
J30	1.3	1.5
M30	.7	.9
M45	1.3	1.5
Q45		
1990-96	1.6	1.8
1997-05	1.0	1.2
Front Axle Side Flange, Replace (B)		
G35		
one	2.4	2.6
both	3.0	3.2
Halfshaft, Replace (B)		
G35		
front		
one	1.4	1.6
both	2.9	3.1
rear		
one	.8	1.0
both	1.4	1.6
J30		
one	1.9	2.1
both	3.3	3.5
M30		
one	1.6	1.8
both	3.0	3.2
M45		
one	2.1	2.3
both	2.7	2.9
Q45		
1990-96		
one	1.6	1.8
both	2.9	3.1
1997-01		
one	2.6	2.8
both	4.1	4.3
2002-05		
one	1.6	1.8
both	2.2	2.4
Pinion Shaft Oil Seal, Replace (B)		
G35		
front axle	2.1	2.3
rear axle	1.3	1.5
J30, M30	1.4	1.6
Q45	1.1	1.3
Rear Axle Housing, Replace (B)		
G35		
one side	4.1	4.6
both sides	5.8	6.4
J30		
one side	1.9	2.4
both sides	3.2	3.7
M45		
one side	3.5	4.0
both sides	5.2	5.7

	LABOR TIME	SEVERE SERVICE

Q45
  1990-96
    one side 2.6   3.1
    both sides 4.9   5.4
  1997-01
    one side 3.8   4.3
    both sides 5.5   6.0
  2002-05
    one side 2.2   2.7
    both sides 4.0   4.5
Replace
  bushings add
    each side 8   .9
  wheel hub or bearing
    add each side 5   .5
Rear Axle Shaft, Replace (B)
M30
  one 2.1   2.4
  both 4.0   4.3
Rear Axle Side Flange, Replace (B)
G35
  one 1.1   1.3
  both 2.1   2.3
J30
  one 7   .9
  both 1.1   1.3
M45
  one 2.2   2.4
  both 3.2   3.4
Q45
  1990-96
    one 7   .9
    both 1.1   1.3
  1997-01
    one 1.0   1.2
    both 1.4   1.6
  2002-05
    one 2.2   2.4
    both 3.2   3.4
Rear Wheel Hub Grease Seal, Replace (B)
G35
  one side 8   1.0
  both sides 1.4   1.6
J30
  one side 1.6   1.8
  both sides 2.9   3.1
M45
  one side 2.1   2.3
  both sides 2.2   2.4
Q45
  1990-96
    one side 1.6   1.8
    both sides 2.9   3.1
  1997-01
    one side 2.4   2.6
    both sides 4.0   4.2

Ring Gear & Pinion Assy., Replace (B)
G35
  front axle 10.2 10.7
  rear axle 4.3   4.8
J30 .... 5.8   6.3
M30 .... 6.9   7.4
M45 .... 5.3   5.8
Q45
  1990-96 .... 5.8   6.3
  1997-05 .... 5.2   5.7

FRONT HALFSHAFTS
CV Joint and Boot, Replace or Recondition (B)
Includes: R&R halfshaft.
G20
  one side 2.4   2.7
  both sides 4.3   4.6
G35
  AWD
    one side 1.4   1.6
    both sides 2.9   3.1
I30, I35
  one side 2.7   3.0
  both sides 4.4   4.7
Halfshaft, R&R or Replace (B)
G20
  one side 1.1   1.4
  both sides 1.9   2.2
G35
  AWD
    one side 1.4   1.6
    both sides 2.9   3.2
I30, I35
  one side 1.3   1.6
  both sides 2.1   2.4
Halfshaft Boots, Replace (B)
Inner and outer boots
  one side 2.4   2.7
  both sides 4.4   4.7
Outer boots only
  one side 1.9   2.2
  both sides 3.3   3.6
Intermediate Shaft Support Bearing, Replace (B)
Includes: R&I front halfshaft assembly.
G20
  1991-96 .... 2.1   2.4
  1999-02 .... 1.6   2.0
I30, I35
  1996-99 .... 2.5   2.8
  2000-05 .... 2.2   2.5

BRAKES
ANTI-LOCK
Diagnose Anti-Lock Brake System Component Each (A)
All Models 7   .7

Retrieve Fault Codes (A)
All Models 3   .3
Actuator Assy., Replace (B)
Includes: System bleeding.
G20 .... 2.1   2.2
G35 .... 2.9   3.0
I30, I35, M30 ..... 2.1   2.2
J30, M45 .... 1.7   1.8
Q45
  1990-96 .... 1.1   1.2
  1997-05 .... 1.7   1.8
Actuator Relay, Replace (B)
Exc. below 3   .3
G35 5   .5
2002-05 Q45 .... 1.5   1.5
Control Module, Replace (B)
G20 8   .8
G35 .... 2.9   2.9
I30, I35, M30 6   .6
J30 5   .5
M45 .... 1.4   1.4
Q45
  1990-96
    standard 6   .6
    LAN 1.2   1.2
  1997-05 9   .9
Front Sensor Rotor Assy., Replace (B)
G20
  one 1.3   1.5
  both 2.4   2.6
G35
  one 1.1   1.3
  both 2.1   2.3
I30, I35
  1996-99
    one 1.4   1.6
    both 2.5   2.7
  2000-04
    one 1.6   1.8
    both 3.2   3.4
J30, M30
  one 1.1   1.3
  both 1.9   2.1
M45
  one 1.7   1.9
  both 2.4   2.6
Q45
  1990-01
    one 1.1   1.3
    both 1.9   2.1
  2002-05
    one 6   .8
    both 1.0   1.2
Front ABS Sensor Assy., Replace (B)
G20
  one 8   .9
  both 1.1   1.2

INF-16 G20 : G35 : I30 : I35 : J30 : M30 : M45 : Q45

	LABOR TIME	SEVERE SERVICE
Front ABS Sensor Assy., Replace (B)		
G35		
one	.5	.6
both	.6	.7
I30, I35		
one	.8	.9
both	1.1	1.2
J30, M30		
one	.8	.9
both	1.3	1.4
M45		
one	1.1	1.2
both	1.6	1.7
Q45		
1990-01		
one	.6	.7
both	.8	.9
2002-05		
one	.8	.9
both	1.4	1.5
Rear Sensor Rotor Assy., Replace (B)		
G20		
1991-96		
one	1.3	1.4
both	2.1	2.2
1999-02		
one	.8	.9
both	1.3	1.4
G35		
one	1.1	1.2
both	2.1	2.2
I30, I35		
one	1.1	1.2
both	1.9	2.0
J30	1.3	1.4
M30	.6	.7
M45		
one	2.4	2.5
both	3.3	3.4
Q45		
1990-96	.6	.7
1997-01		
one	2.7	2.8
both	4.6	4.7
2002-05		
one	1.9	2.0
both	2.9	3.0
Rear ABS Sensor Assy., Replace (B)		
G20		
one	.8	.9
both	1.0	1.1
G35 one or both	.5	.6
I30, I35		
1996-99		
one	.6	.7
both	1.1	1.2
2000-04		
one	.5	.6
both	.6	.7

	LABOR TIME	SEVERE SERVICE
J30	.8	.9
M30	1.0	1.1
M45		
one	.5	.6
both	.6	.7
Q45		
1990-01 one or both	1.0	1.1
2002-05		
one	.5	.6
both	.6	.7
SYSTEM		
Bleed Brakes (B)		
Includes: Add fluid.		
Exc. 1999-02 G20	1.0	1.2
1999-02 G20	1.1	1.3
Brake Hose (Flexible), Replace (B)		
Includes: System bleeding.		
Exc. below.		
one	.8	1.0
both	1.1	1.3
G35, M45		
one	1.1	1.3
both	1.4	1.6
1997-05 Q45		
one	1.0	1.2
both	1.3	1.5
Proportioning (Load Sensing) Valve, Replace (B)		
Includes: System bleeding.		
G20		
1991-96	1.0	1.1
1999-02	1.4	1.5
I30		
1996-99	4.3	4.4
2000-04	2.7	2.8
J30	1.0	1.1
M45	1.1	1.2
2002-05 Q45	1.3	1.4
Master Cylinder, R&R and Recondition (A)		
Includes: System bleeding.		
G20		
1991-96	2.2	2.5
1999-02	1.6	1.9
G35	2.4	2.7
I30, I35	1.8	2.1
J30	2.1	2.4
M30, M45	1.7	2.0
Q45		
1990-01	1.7	2.0
2002-05	1.8	2.1
Master Cylinder, Replace (B)		
Includes: System bleeding.		
G20	1.5	1.7
G35	2.0	2.2
I30, I35, J30	1.4	1.6
M30, M45	1.2	1.4

	LABOR TIME	SEVERE SERVICE
Q45		
1990-01	1.1	1.3
2002-05	1.4	1.6
Power Brake Booster Unit, Replace (B)		
Includes: System bleeding.		
G20, I30, I35	2.3	2.5
G35	2.9	3.1
J30	5.6	5.8
M30	1.1	1.3
M45	1.8	2.0
Q45		
1990-96	2.3	2.5
1997-01	1.2	1.4
2002-05	1.8	2.0
Vacuum Check Valve, Replace (B)		
Exc. below	.3	.3
2002-05 Q45	1.8	2.1
SERVICE BRAKES		
Brake Pads, Replace (B)		
Front	1.1	1.3
Rear		
exc. below	1.3	1.5
G35	.6	.8
All four wheels	2.0	2.1
Resurface brake rotor add	.6	.6
Front Caliper Assy., Replace (B)		
Includes: System bleeding.		
G20		
1991-96		
one	1.1	1.2
both	1.9	2.0
1999-02		
one	.9	1.0
both	1.5	1.6
G35, I30, I35		
one	1.1	1.2
both	1.7	1.8
J30		
one	1.4	1.3
both	2.5	2.6
M30, M45		
one	1.1	1.2
both	1.7	1.8
Q45		
one	1.1	1.2
both	1.7	1.8
Front Caliper Assy. R&R and Recondition (A)		
Includes: Bleed system.		
G20		
one	1.6	1.9
both	3.0	3.3
G35		
one	2.2	2.5
both	3.8	4.1

	LABOR TIME	SEVERE SERVICE
I30, I35		
one	1.7	2.0
both	2.5	2.8
J30		
one	1.9	2.2
both	3.5	3.8
M30, M45		
one	1.4	1.7
both	2.2	2.5
Q45		
1990-01		
one	1.6	1.9
both	2.7	3.0
2002-05		
one	1.3	1.6
both	2.1	2.4

Front Disc Brake Rotor, Replace (B)
	LABOR TIME	SEVERE SERVICE
Exc. below		
one	.8	1.0
both	1.4	1.6
G35		
one	.6	.8
both	.8	1.0

Rear Caliper Assy., Replace (B)
Includes: System bleeding.
	LABOR TIME	SEVERE SERVICE
G20		
1991-96		
one	1.1	1.3
both	1.9	2.1
1999-02		
one	.9	1.1
both	1.4	1.6
G35		
one	1.1	1.3
both	1.4	1.6
I30, I35		
one	1.1	1.3
both	2.1	2.3
J30		
one	1.4	1.6
both	2.5	2.7
M30, M45		
one	1.1	1.3
both	1.9	2.1
Q45		
1990-01		
one	1.3	1.5
both	2.1	2.3
2002-05		
one	.8	1.0
both	1.3	1.5

Rear Caliper Assy., R&R and Recondition (A)
Includes: System bleeding.
	LABOR TIME	SEVERE SERVICE
G20		
1991-96		
one	1.7	2.0
both	2.9	3.2

	LABOR TIME	SEVERE SERVICE
1999-02		
one	1.4	1.7
both	2.4	2.7
G35		
one	1.6	1.9
both	2.5	2.8
I30, I35		
one	1.7	2.0
both	3.0	3.3
J30		
one	2.1	2.4
both	3.2	3.5
M30, M45		
one	1.6	1.9
both	2.7	3.0
Q45		
one	1.7	2.0
both	3.0	3.3

Rear Disc Brake Rotor, Replace (B)
Disc Type Parking Brake
	LABOR TIME	SEVERE SERVICE
G20		
one	.9	1.1
both	1.4	1.6
G35		
one	.5	.7
both	.8	1.0
I30, I35, J30		
one	.6	.8
both	1.1	1.3
M30, M45		
one	1.1	1.3
both	1.6	1.8
Q45		
1990-01		
one	.8	1.0
both	1.3	1.5
2002-05		
one	.6	.8
both	.8	1.0

Drum Type Parking Brake
	LABOR TIME	SEVERE SERVICE
G35, M30, M45		
one	1.1	1.3
both	1.6	1.8
J30		
one	.6	.8
both	1.1	1.3
Q45		
1990-96		
one	.9	1.1
both	1.6	1.8
1997-05		
one	.8	1.0
both	1.1	1.3

PARKING BRAKE

Parking Brake Cable, Adjust (C)
	LABOR TIME	SEVERE SERVICE
Exc. below	.5	.7
1999-02 G20	.8	1.0

Front Parking Brake Cable, Replace (B)
	LABOR TIME	SEVERE SERVICE
G20, I30, I35	1.3	1.5
G35		
hand lever type	.8	1.0
pedal type	1.3	1.5
J30	3.0	3.2
M30	1.4	1.6
M45	2.1	2.3
Q45	2.4	2.6

Parking Brake Apply Actuator, Replace (B)
	LABOR TIME	SEVERE SERVICE
G20, G35	.9	.9
I30, I35	1.1	1.1
J30	3.0	3.0
M30, M45	.8	.8
Q45		
1990-96	6.2	6.2
1997-01	2.6	2.6
2002-05	3.6	3.6

Parking Brake Apply Warning Indicator Switch, Replace (B)
	LABOR TIME	SEVERE SERVICE
Exc. below	.7	.7
1990-96 Q45	.9	.9

Parking Brake Equalizer, Replace (B)
	LABOR TIME	SEVERE SERVICE
G20	1.3	1.4
M30, Q45	.6	.7

Parking Brake Shoes, Replace (B)
	LABOR TIME	SEVERE SERVICE
G35	1.6	1.8
J30	1.9	2.1
M30	3.6	3.8
M45	1.9	2.1
Q45		
1990-96	3.6	3.8
1997-01	1.6	1.8
2002-05	1.3	1.5

Rear Parking Brake Cable, Replace (B)
	LABOR TIME	SEVERE SERVICE
G20, I30, I35, M30		
one	1.1	1.3
both	1.6	1.8
G35		
one	1.7	1.9
both	2.7	2.9
J30		
one	1.3	1.5
both	2.2	2.4
M45		
one	2.2	2.4
both	3.5	3.7
Q45		
1990-96		
one	1.3	1.5
both	1.7	1.9
1997-01		
one	1.4	1.6
both	2.2	2.4
2002-05		
one	1.9	2.1
both	3.2	3.4

INF-18 G20 : G35 : I30 : I35 : J30 : M30 : M45 : Q45

	LABOR TIME	SEVERE SERVICE
TRACTION CONTROL		
Actuator Assy., Replace (B)		
G35	2.9	*3.2*
1990-96 Q45	2.1	*2.4*
Control Unit, Replace (B)		
G35, I30, I35	.6	*.6*
M45	1.2	*1.2*
Q45		
1990-96		
standard model	.6	*.6*
LAN model	1.2	*1.2*
1997-06	.8	*.8*
Main Relay, Replace (B)		
Q45	.5	*.5*
Motor Assembly, Replace (B)		
Q45		
1990-96	.9	*1.0*
1997-01	.5	*.6*
Pump Assy., Replace (B)		
Q45		
1990-96	1.2	*1.3*
2002-05	2.0	*2.1*
Reservoir Tank, Replace (B)		
1990-96 Q45	.5	*.5*
Switch Assembly, Replace (B)		
Exc. below	.3	*.3*
1996-99 I30	.5	*.5*
Throttle Control Unit, Replace (B)		
Q45		
1990-96		
standard model	.6	*.6*
LAN model	1.8	*1.8*
1997-01	.6	*.6*

FRONT SUSPENSION
Unless otherwise noted, time given does not include alignment.

	LABOR TIME	SEVERE SERVICE
Front & Rear Alignment, Check & Adjust (A)		
1991-96 G20	1.3	*1.5*
G35	3.0	*3.2*
J30	1.7	*1.9*
M30	1.3	*1.5*
M45, Q45	1.7	*1.9*
Front Toe, Adjust (B)		
Exc. below	.8	*1.0*
G35	1.1	*1.3*
M45	1.3	*1.5*
Rear Toe, Adjust (B)		
1991-96 G20	1.1	*1.3*
G35	1.4	*1.6*
J30, M30	.8	*1.0*
M45	1.3	*1.5*
Q45	.8	*1.0*
Ball Joint, Replace (B)		
1996-99 I30		
one	1.6	*1.9*
both	2.4	*2.7*

	LABOR TIME	SEVERE SERVICE
Connecting Rod Sensor, Replace (B)		
1990-96 Q45		
one or both	1.3	*1.3*
Drain Filter, Replace (B)		
1990-96 Q45	4.0	*4.0*
Fan Motor, Replace (B)		
1990-96 Q45	6.7	*6.7*
Front Accumulator, Replace (B)		
1990-96 Q45	6.7	*6.7*
Front Actuator Assy., Replace (B)		
M30		
one	.6	*.7*
both	1.1	*1.2*
Q45		
full-active suspension		
1990-96		
one	5.7	*5.8*
both	6.7	*6.8*
2002-05 one or both	.3	*.4*
sonar suspension		
1997-05 one or both	.4	*.5*
Front Air Filter, Replace (B)		
1990-96 Q45		
one	.3	*.3*
both	.5	*.5*
Front Coil Spring, Replace (B)		
G20		
1991-96		
one	1.7	*2.0*
both	3.2	*3.5*
1999-02		
one	1.6	*1.9*
both	2.7	*3.0*
G35		
one	1.4	*1.7*
both	2.2	*2.5*
1990-96 Q45		
one	1.4	*1.7*
both	2.7	*3.0*
Front Control Valve, Replace (B)		
1990-96 Q45	7.9	*7.9*
Front Height Sensor, Replace (B)		
1990-96 Q45		
one	1.4	*1.4*
both	1.7	*1.7*
Front Hub Assy., Replace (B)		
G20		
1991-96		
one	1.4	*1.7*
both	2.7	*3.0*
1999-02		
one	2.2	*2.5*
both	3.5	*3.8*
G35		
2WD		
one	1.1	*1.4*
both	2.1	*2.4*

	LABOR TIME	SEVERE SERVICE
AWD		
one	1.6	*1.9*
both	2.1	*2.4*
1990-96 Q45		
one	1.4	*1.7*
both	2.7	*3.0*
Front Joint Assy., Replace (B)		
1990-96 Q45		
one	3.5	*3.9*
both	4.0	*4.4*
Front Shock Absorbers, Replace (B)		
G20		
1991-96		
one	1.6	*1.8*
both	2.9	*3.1*
1999-02		
one	1.4	*1.6*
both	2.4	*2.6*
G35		
one	1.1	*1.3*
both	1.4	*1.6*
1990-96 Q45		
one	1.3	*1.5*
both	2.4	*2.6*
Height Control Switch, Replace (B)		
1990-96 Q45	.3	*.3*
Hub Bearing or Seal, Replace (B)		
I30, I35		
one	2.9	*3.2*
both	4.9	*5.2*
J30		
one	1.5	*1.8*
both	2.6	*2.9*
M30		
one	1.8	*2.1*
both	3.3	*3.6*
M45		
one	1.9	*2.1*
both	2.4	*2.6*
J30		
one	1.6	*1.9*
both	2.6	*2.9*
Q45		
1997-01		
one	1.9	*2.2*
both	2.9	*3.2*
2002-05		
one	1.2	*1.5*
both	1.5	*1.8*
Insulator Cap, Replace (B)		
1990-96 Q45		
one	4.6	*4.6*
both	4.9	*4.9*
King Pins & Bearings, Replace (B)		
Includes: Reset toe.		
G20		
1991-96		
one side	1.4	*1.9*
both sides	2.5	*3.0*

G20 : G35 : I30 : I35 : J30 : M30 : M45 : Q45 **INF-19**

	LABOR TIME	SEVERE SERVICE
1999-02		
one side	1.9	2.4
both sides	3.0	3.5
1990-96 Q45		
one side	1.7	2.2
both sides	3.2	3.7

Multi-Valve Assy., Replace (B)
1990-96 Q45 .... 8.4   8.4

Oil Cooler, Replace (B)
1990-96 Q45 .... 6.7   6.7

Oil Tank, Replace (B)
1990-96 Q45 .... 4.4   4.4

Pump Accumulator, Replace (B)
1990-96 Q45 .... 6.3   6.3

Selector Switch, Replace (B)
M30 3   .3
Q45 5   .5

Sonar Suspension Control Unit, Replace (B)
M30 6   .6
Q45
  1997-01 .... 1.4   1.4
  2002-05 6   .6

Stabilizer Bar, Replace (B)
G20 8   1.0
G35 .... 1.1   1.3
I30, I35 .... 2.7   2.9
J30 .... 1.6   1.8
M30, M45 .... 1.3   1.5
Q45
  1990-96 .... 1.0   1.2
  1997-05 .... 1.4   1.6

Stabilizer Bar Bushings, Replace (B)
G20 7   .8
G35, I30, I35 .... 1.1   1.2
J30 .... 1.3   1.4
M30 6   .7
M45 .... 1.0   1.1
Q45
  1990-96 6   .7
  1997-01 .... 1.1   1.2
  2002-05 5   .6

Stabilizer Bar Connecting Rods and/or Bushings, Replace (B)
One 6   .7
Both 1.0   1.1

Steering Angle Sensor, Replace (B)
M30 .... 1.0   1.0
Q45
  1990-01 .... 1.4   1.4
  2002-05 8   .8

Steering Knuckle, Replace (B)
G20
  1991-96
    one side 2.4   2.7
    both sides 4.4   4.7
  1999-02
    one side 2.9   3.2
    both sides 4.7   5.0

G35
  one side 2.2   2.5
  both sides 3.8   4.1
1990-96 Q45
  one side 2.2   2.5
  both sides 4.3   4.6

Steering Knuckle Arm, Replace (B)
Includes: Reset toe.
M30
  one 1.0   1.2
  both 1.6   1.8

Steering Spindle, Replace (B)
Includes: Reset toe.
G20
  1991-96
    one side 2.4   2.6
    both sides 4.4   4.6
  1999-02
    one side 2.9   3.1
    both sides 4.7   4.9
G35
  2WD
    one side 2.2   2.4
    both sides 3.8   4.0
  AWD
    one side 1.9   2.1
    both sides 3.2   3.4
Q45
  1990-96
    one side 2.2   2.4
    both sides 4.3   4.5

Strut, R&R or Replace (B)
Includes: Alignment.
I30, I35
  one 1.7   1.9
  both 2.5   2.7
J30
  one 2.4   2.6
  both 4.4   4.6
M30
  one 3.2   3.4
  both 6.0   6.2
M45
  one 3.2   3.4
  both 4.7   4.9
Q45
  1997-05
    one 2.5   2.7
    both 4.3   4.5
Replace shock kit J30, M30, 97-01 Q45 add each 8   .8

Suspension Crossmember, Replace (B)
Includes: Alignment.
G20
  1991-96 .... 5.2   5.7
  1999-02 .... 4.7   5.2
G35 .... 2.4   2.9
I30, I35 .... 7.5   8.0
J30 .... 5.2   5.7
M30 .... 3.0   3.5

M45 .... 5.3   5.8
Q45
  1990-96 .... 3.2   3.7
  1997-01 .... 6.0   6.5
  2002-05 .... 4.4   4.9

Tension Rod and/or Bushing Replace (B)
Includes: Reset toe.
J30
  one 8   1.0
  both 1.0   1.2
M30
  one 8   1.0
  both 1.3   1.5
Q45
  1990-96
    one 8   1.0
    both 1.3   1.5
  1997-01
    one 1.3   1.5
    both 1.6   1.8

Tension Rod Bracket, Replace (B)
Includes: Reset toe.
M30
  one 8   1.0
  both 1.3   1.5
1990-96 Q45
  one 1.3   1.5
  both 1.6   1.8

Transverse Lower Links, Replace (B)
Includes: Alignment.
G20
  1991-96
    one side 1.4   1.7
    both sides 2.5   2.8
  1999-02
    one side 1.4   1.7
    both sides 2.1   2.4
G35
  one side 1.6   1.9
  both sides 1.9   2.2
I30, I35
  one side 2.9   3.2
  both sides 4.4   4.7
J30
  one side 1.1   1.4
  both sides 1.9   2.2
M30
  one side 2.0   2.3
  both sides 3.3   3.6
M45
  one side 2.4   2.7
  both sides 2.9   3.2
Q45
  1990-96
    one side 1.7   2.1
    both sides 2.4   2.7
  1997-05
    one side 1.6   1.9
    both sides 1.9   2.2

INF-20 G20 : G35 : I30 : I35 : J30 : M30 : M45 : Q45

	LABOR TIME	SEVERE SERVICE
Upper Links or Brackets, Replace (B)		
Includes: Reset toe.		
G20		
1991-96		
one side	1.3	1.8
both sides	2.2	2.7
1999-02		
one side	1.3	1.8
both sides	1.7	2.2
G35		
one	1.4	1.9
both	2.1	2.6
1990-96 Q45		
one	1.7	2.2
both	2.4	2.9

REAR SUSPENSION

Lower Link, Replace (B)		
Includes: Alignment.		
G20, J30		
one	.8	1.0
both	1.1	1.3
G35		
front		
one	1.7	1.9
both	1.9	2.1
rear		
one	1.4	1.6
both	1.9	2.1
M45		
front		
one	.8	1.0
both	1.4	1.6
rear		
one	2.2	2.4
both	2.9	3.1
Q45		
1990-96		
one	.7	.9
both	1.1	1.3
1997-01		
one	1.7	1.9
both	2.1	2.3
2002-05		
one	2.4	2.6
both	3.0	3.2
Parallel Rods, Replace (B)		
G20		
one	1.0	1.3
both	1.6	1.9
Rear Actuator Assy., Replace (B)		
Full-Active Suspension		
Q45		
1990-96		
one	5.5	5.5
both	6.7	6.7
2002-05		
one	1.7	1.7
both	1.9	1.9

	LABOR TIME	SEVERE SERVICE
Sonar Suspension		
M30		
one	.6	.6
both	1.1	1.1
Q45		
1997-01		
one	.5	.5
both	.6	.6
2002-05		
one or both	1.4	1.4
Rear Accumulator, Replace (B)		
1990-96 Q45	5.5	5.5
Rear Coil Spring, Replace (B)		
G20		
1991-96		
one	1.7	2.0
both	2.7	3.0
1999-02		
one	1.3	1.6
both	2.2	2.5
G35		
separate coil/seat		
one	1.4	1.7
both	1.7	2.0
shock mounted		
one	3.0	3.3
both	3.3	3.6
I30, I35		
one	1.3	1.6
both	2.2	2.5
J30		
one	1.4	1.7
both	2.5	2.8
M30		
one	1.6	1.9
both	3.0	3.3
M45		
one	2.1	2.4
both	2.5	2.8
Q45		
1990-96		
one	1.7	2.0
both	3.3	3.6
1997-01		
one	1.4	1.7
both	2.4	2.7
2002-05		
one	1.7	2.0
both	2.1	2.4
Rear Control Arm, Replace (B)		
Includes: Alignment. R&I axle shaft preload measurements.		
M30		
one	3.8	4.1
both	7.2	7.5

	LABOR TIME	SEVERE SERVICE
Rear Control Arm Bushing, Replace (B)		
Includes: Alignment.		
M30		
one side	2.5	2.8
both sides	4.7	5.0
Rear Control Valve, Replace (B)		
1990-96 Q45	6.3	6.3
Rear Height Sensor, Replace (B)		
1990-96 Q45		
one	1.3	1.3
both	1.6	1.6
Rear Joint Assy., Replace (B)		
1990-96 Q45		
one	5.5	5.7
both	6.6	6.8
Rear Joint Seats, Replace (B)		
1990-96 Q45		
one	5.8	6.0
both	6.7	6.9
Rear Lower Arm/Bushings, Replace (B)		
J30		
one	1.0	1.3
both	1.6	1.9
Q45		
1990-96		
one	1.0	1.3
both	1.7	2.0
1997-01		
one	1.9	2.2
both	2.7	3.0
Rear Shock Absorbers and Spring Assy., Replace (B)		
G35		
one	1.7	1.9
both	2.1	2.3
J30		
one	1.3	1.5
both	2.2	2.4
M30		
one	1.6	1.8
both	3.0	3.2
M45		
one	1.4	1.6
both	1.9	2.1
Q45		
1990-96		
one	1.7	1.9
both	3.2	3.4
1997-01		
one	1.5	1.7
both	2.4	2.6
2002-05		
one	1.7	1.9
both	2.1	2.3
Rear Stabilizer Bar, Replace (B)		
G20	.8	1.0
G35	.5	.7
J30	2.2	2.4

	LABOR TIME	SEVERE SERVICE
M30, M45	1.0	*1.2*
Q45		
1990-96	.7	*.9*
1997-01	1.1	*1.3*
2002-05	.5	*.7*

Rear Stabilizer Bar Bushings, Replace (B)
1990-96 G20	1.0	*1.2*
G35	.5	*.7*
J30	.8	*1.0*
M30	1.0	*1.2*
M45	.6	*.8*
Q45		
1990-01	1.0	*1.2*
2002-05	.6	*.8*

Rear Stabilizer Connecting Rods and/or Bushings, Replace (B)
G20, M30, M45		
one	.6	*.7*
both	1.0	*1.1*
G35		
one	.5	*.6*
both	.6	*.7*
J30		
one	.6	*.7*
both	1.1	*1.2*
Q45		
1990-01		
one	1.0	*1.1*
both	1.3	*1.4*
2002-05		
one	.5	*.6*
both	.6	*.7*

Rear Strut Shock Absorber, Replace (B)
1999-02 G20		
one	1.1	*1.3*
both	1.9	*2.1*
I30, I35		
one	1.1	*1.3*
both	1.9	*2.1*

Rear Strut, Replace (B)
1991-96 G20		
one	1.6	*1.8*
both	2.4	*2.6*

Rear Upper Arm/Bushings, Replace (B)
G35		
one	2.7	*3.0*
both	4.0	*4.3*
M45		
one	1.3	*1.6*
both	1.9	*2.2*
Q45		
2002-05		
one	2.5	*2.8*
both	3.3	*3.6*

Rear Wheel Knuckle Spindle, Replace (B)
G20		
one	1.4	*1.7*
both	2.5	*2.8*

Upper Link, Replace (B)
Front		
J30		
one side	.8	*1.0*
both sides	1.1	*1.3*
Q45		
1990-96		
one side	.7	*.9*
both sides	1.1	*1.3*
1997-01		
one	1.4	*1.6*
both	1.9	*2.1*
Rear		
G35, J30		
one side	.8	*1.0*
both sides	1.4	*1.6*
Q45		
1990-96		
one side	.6	*.8*
both sides	1.1	*1.3*
1997-01		
one	2.2	*2.4*
both	2.7	*2.9*

Wheel Bearings, Hubs or Oil Seals, Replace (B)
G20, I30, I35		
one side	1.1	*1.3*
both sides	1.9	*2.1*

STEERING
AIR BAGS

Diagnose Air Bag System Component Each (A)
All Models	.7	*.7*

Retrieve Fault Codes (A)
All Models	.3	*.3*

Air Bag Assy., Replace (B)
Includes: Diagnosis.
Driver side	.6	*.6*
Passenger side		
G20, J30	1.1	*1.1*
G35	1.4	*1.4*
I30, I35, M45	1.0	*1.1*
Q45		
1990-96	4.9	*4.9*
1997-01	1.1	*1.1*
2002-05	.8	*.9*
Side impact		
G20		
one side	.6	*.6*
both sides	.8	*.8*
G35		
one side	1.1	*1.1*
both sides	1.8	*1.8*
I30, I35		
one side	.7	*.7*
both sides	.9	*.9*
M45		
one side	1.2	*1.2*
both sides	2.2	*2.2*
Q45		
1997-01		
one side	.6	*.6*
both sides	.8	*.8*
2002-05		
one side	1.2	*1.2*
both sides	2.0	*2.0*
Curtain		
G35		
coupe		
one side	1.6	*1.6*
both sides	1.9	*1.9*
sedan		
one side	2.0	*2.0*
both sides	2.3	*2.3*
M45		
one side	2.7	*2.7*
both sides	3.0	*3.0*
Q45		
2002-05		
one side	2.7	*2.7*
both sides	3.0	*3.0*

Air Bag On/Off Switch, Replace (B)
All Models	.8	*.8*

Air Bag Side Sensor, Replace (B)
Includes: Diagnosis.
G20		
one side	.6	*.6*
both sides	.8	*.8*
G35		
one side	.7	*.7*
both sides	1.1	*1.1*
I30, I35		
one side	1.0	*1.0*
both sides	1.6	*1.6*
M45		
one side	.8	*.8*
both sides	1.2	*1.2*
Q45		
one side	.9	*.9*
both sides	1.5	*1.5*

Air Bag Spiral Cable, Replace (B)
Includes: Diagnosis.
Exc. below	.8	*.8*
M30	.9	*.9*
1990-96 Q45	2.0	*2.0*

Crash Sensor, Replace (B)
Includes: Diagnosis.
Center		
G20	1.7	*1.7*
G35	.8	*.8*
I30, I35	.5	*.5*
J30	1.7	*1.7*
M30	.9	*.9*

G20 : G35 : I30 : I35 : J30 : M30 : M45 : Q45

	LABOR TIME	SEVERE SERVICE
Crash Sensor, Replace (B)		
M45	.5	.5
Q45		
1990-96	1.2	1.2
2002-05	.6	.6
Side		
M30		
one	.9	.9
both	1.7	1.7
1990-96 Q45		
one or both	3.6	3.6
Electronic Control Unit, Replace (B)		
Includes: Diagnosis.		
1991-96 G20	1.9	1.9
G35, J30, M30	.8	.8
1990-96 Q45	.8	.8
Sensing and Diagnostic Module (SDM), Replace (B)		
Includes: Diagnosis.		
1999-02 G20	.9	.9
G35	.9	.9
I30, I35		
1996-99	.5	.5
2000-04	1.1	1.1
J30	.8	.8
M45	1.1	1.1
Q45		
1990-96	1.6	1.6
1997-05	.8	.8
Tunnel/Safing Sensor, Replace (B)		
Includes: Diagnosis.		
G20	1.7	1.7
G35	.9	.9
I30	2.8	2.8
J30, M30	.8	.8
M45	1.1	1.1
Q45	.9	.9

POWER RACK & PINION

	LABOR TIME	SEVERE SERVICE
Inner Tie Rod, Replace (B)		
Includes: Reset toe.		
G20, J30, M30		
one side	1.1	1.3
both sides	1.6	1.8
G35		
one side	1.3	1.5
both sides	1.7	1.9
I30, I35		
1996-99		
one side	1.1	1.3
both sides	1.6	1.8
2000-04		
one side	1.6	1.8
both sides	2.4	2.6
M45		
one side	1.9	2.1
both sides	2.5	2.7

	LABOR TIME	SEVERE SERVICE
Q45		
1990-01		
one side	1.1	1.3
both sides	1.6	1.8
2002-05		
one side	1.4	1.6
both sides	2.2	2.4
Outer Tie Rod, Replace (B)		
Includes: Reset toe.		
Exc. below		
one side	1.1	1.3
both sides	1.6	1.8
M45		
one side	1.6	1.8
both sides	1.9	2.1
Power Steering Hoses, Replace (B)		
Includes: System bleeding.		
G20	1.7	1.8
G35	1.8	1.9
I30, I35		
1996-99	1.3	1.4
2000-04	2.4	2.5
J30	2.5	2.6
M30	1.6	1.7
M45	2.7	2.8
Q45		
1990-96	2.4	2.5
1997-01	.8	.9
2002-05	2.5	2.6
Pressure Switch, Replace (B)		
Exc. below	.6	.6
M45	1.4	1.4
2002-05 Q45	1.0	1.0
Rack & Pinion Assy., R&R and Recondition (A)		
Includes: Reset toe.		
G20		
1991-96	6.1	6.7
1999-02	5.0	5.6
G35	4.0	4.6
I30, I35		
1996-99	7.2	7.8
2000-04	9.0	9.6
J30	4.3	4.9
M30	5.0	5.6
M45	4.6	5.2
Q45		
1990-96	4.6	5.2
1997-01	3.6	4.2
2002-05	4.7	5.3
Rack & Pinion Assy., R&R or Replace (B)		
Includes: All adjustments.		
G20		
1991-96	4.3	4.6
1999-02	2.8	3.1
G35	2.4	2.7
I30, I35		
1996-99	5.5	5.8
2000-04	7.3	7.6

	LABOR TIME	SEVERE SERVICE
J30	2.8	3.1
M30	3.3	3.6
M45	3.0	3.3
Q45		
1990-96	3.1	3.4
1997-01	2.2	2.5
2002-05	3.2	3.5
Rack Boots, Replace (B)		
Includes: Reset toe.		
G20		
one side	1.1	1.3
both sides	1.7	1.9
G35		
one side	1.3	1.5
both sides	1.7	1.9
I30, I35		
1996-99		
one side	1.0	1.2
both sides	1.6	1.8
2000-04		
one side	1.4	1.6
both sides	2.1	2.3
J30, M30		
one side	1.0	1.2
both sides	1.6	1.8
M30		
one side	1.0	1.2
both sides	1.4	1.6
M45		
one side	1.9	2.1
both sides	2.5	2.7
Q45		
1990-96		
one side	1.0	1.2
both sides	1.6	1.8
1997-01		
one side	1.0	1.2
both sides	1.2	1.4
2002-05		
one side	1.4	1.6
both sides	2.2	2.4
Steering Column, Replace (B)		
G20	1.6	1.6
G35	2.0	2.0
I30, I35		
1996-99	1.4	1.4
2000-04	3.1	3.1
J30	1.9	1.9
M30	3.5	3.5
M45	2.8	2.8
Q45		
1990-96	4.7	4.7
1997-05	2.8	2.8
Steering Column Lower Joint, Replace (B)		
G20		
1991-96	.8	1.0
1999-02	1.6	1.8

	LABOR TIME	SEVERE SERVICE
G35, M45	.6	.8
I30, I35	1.9	2.1
J30, M30, Q45	.6	.8
Steering Lock Assy., Replace (B)		
G20	1.4	1.4
G35	1.6	1.6
I30, I35		
1996-99	1.1	1.1
2000-04	1.5	1.5
J30	1.4	1.4
M30, M45, Q45	1.1	1.1
Steering Oil Cooler, Replace (B)		
I30	.8	.9
J30, Q45	1.4	1.5
M45	1.5	1.7
Steering Pump, R&R and Recondition (A)		
G20		
1991-96	3.2	3.7
1999-02	2.3	2.8
G35	2.8	3.3
I30, I35	3.2	3.8
J30	3.8	4.3
M30, M45	2.6	3.1
Q45		
1990-96		
w/active susp	6.1	6.6
w/o active susp	4.3	4.8
1997-05	2.9	3.4
Steering Pump, R&R or Replace (B)		
G20	2.0	2.2
G35	2.1	2.3
I30, I35	2.0	2.2
J30		
w/HICAS	3.6	3.8
w/o HICAS	2.4	2.6
M30	1.2	1.4
M45	1.8	2.0
Q45		
1990-96		
w/active susp	4.7	4.9
w/o active susp	3.1	3.3
1997-01	1.4	1.6
2002-05	2.1	2.3
Steering Pump Reservoir or Seals, Replace (B)		
Exc. below	.8	.9
1990-96 Q45	1.3	1.5
Steering Wheel, Replace (B)		
Exc. below	.6	.6
I30, I35	.7	.7

4 WHEEL STEERING

	LABOR TIME	SEVERE SERVICE
4WS Control Unit, Replace (B)		
1990-96 Q45	1.6	1.6
J30	.5	.5
4WS Control Valve, Replace (B)		
1990-96 Q45	1.9	1.9
J30	1.1	1.1
4WS Cut-Off Valve, Replace (B)		
1990-96 Q45	1.3	1.3
J30	1.1	1.1
4WS Power Cylinder, Replace (B)		
1990-96 Q45	1.1	1.1
J30	1.4	1.4
4WS Power Steering Tubes & Hose Set, Replace (B)		
1990-96 Q45	1.7	1.7
J30	1.1	1.1
4WS Return Hose, Replace (B)		
1990-96 Q45	1.0	1.0
J30	1.6	1.6
4WS Steering Sensor Assy., Replace (B)		
1990-96 Q45	.7	.7
J30	.5	.5
4WS Tie Rod End Boots, Replace (B)		
J30, Q45 rear		
one	.7	.7
both	1.0	1.0
4WS Tie Rod Ends, Replace (B)		
J30, Q45 rear		
one	.8	.8
both	1.3	1.3

HEATING & AIR CONDTIONING

When more than one component requires replacement where evacuation/recovery and recharging is already included, deduct 1.0 hour for each additional component from the time given.

	LABOR TIME	SEVERE SERVICE
Evacuate/Recover and Recharge System (A)		
R12	1.2	1.2
R134A		
exc. G35	.8	.8
G35	.9	.9
Ambient Sensor, Replace (B)		
Automatic AC		
exc. below	.5	.5
1990-96 Q45	.7	.7
Manual AC	.3	.3
Blower Motor, Replace (B)		
G20	.5	.5
G35	1.2	1.2
I30, I35	.8	.8
J30	1.1	1.1
M30	.6	.6
M45	.3	.3
Q45		
1990-96	1.2	1.2
1997-05	.5	.5

	LABOR TIME	SEVERE SERVICE
Compressor Assy., Replace (B)		
Includes: Parts transfer, evacuate/recover and recharge.		
G20		
1991-96	2.2	2.4
1999-02	2.1	2.3
G35	2.5	2.7
I30, I35	2.4	2.6
J30	2.7	2.9
M30	2.5	2.7
M45	3.0	3.2
Q45		
1990-96	4.0	4.2
1997-05	2.9	3.1
w/R12 add		
G20, J30	.4	.4
1996-99 I30	.4	.4
1990-96 Q45	.4	.4
Condenser Assy., Replace (B)		
Includes: Evacuate/recover and recharge.		
G20		
1991-96	2.9	3.0
1999-02	1.7	1.8
G35	3.3	3.4
I30, I35	3.0	3.1
J30	2.7	2.8
M30	4.3	4.4
M45	3.5	3.6
Q45		
1990-96	3.8	3.9
1997-01	4.9	5.0
2002-05	3.5	3.6
w/R12 add		
G20, J30	.4	.4
1996-99 I30	.4	.4
1990-96 Q45	.4	.4
Condenser Fan Motor, Replace (B)		
G20		
1991-96	1.2	1.4
1999-02	.5	.7
G35	.8	1.0
J30, M30	1.4	1.6
Q45		
1990-96	2.1	2.3
1997-01	.8	1.0
Condenser Fan Motor Relay, Replace (B)		
Exc. below	.3	.3
G35	.8	.8
Coolant Temperature Switch, Replace (B)		
I30	.5	.5
M30	.7	.7
Q45 1990-96	.8	.8

G20 : G35 : I30 : I35 : J30 : M30 : M45 : Q45

	LABOR TIME	SEVERE SERVICE
Evaporator Core, Replace (B)		
Includes: Evacuate/recover and recharge.		
G20	1.9	2.0
G35	2.7	2.8
I30, I35		
1996-99	1.8	1.9
2000-04	2.8	2.9
J30	2.1	2.2
M30	2.3	2.4
M45	7.6	7.7
Q45		
1990-96	2.2	2.3
1997-01	2.0	2.1
2002-05	5.7	5.8
w/R12 add		
G20, J30	.4	.4
1996-99 I30	.4	.4
1990-96 Q45	.4	.4
Expansion Valve, Replace (B)		
Includes: Evacuate/recover and recharge.		
G20		
1991-96	2.4	2.5
1999-02	1.7	1.8
G35	2.8	2.9
I30, I35		
1996-99	2.6	2.7
2000-04	3.0	3.1
J30	2.4	2.5
M30	3.1	3.2
M45	1.9	2.0
Q45		
1990-96	3.0	3.1
1997-01	1.9	2.0
2002-05	2.7	2.8
w/R12 add		
G20, J30	.4	.4
1996-99 I30	.4	.4
1990-96 Q45	.4	.4
Heater and AC Control Assy., Replace (B)		
G20		
1991-96	1.5	1.5
1999-02	.5	.5
G35	1.5	1.5
I30, I35	.8	.8
J30	.7	.7
M30	1.2	1.2
M45	.6	.6
Q45		
1990-96	.7	.7
1997-05	1.0	1.0
Heater Blower Motor Relay, Replace (B)		
Exc. below	.5	.5
2000-04 I30, I35	.7	.7
M45	.8	.8

	LABOR TIME	SEVERE SERVICE
Heater Core, R&R or Replace (B)		
Includes: R&I evaporator. Evacuate/recover and recharge.		
G20		
1991-96	5.3	5.6
1999-02	4.6	4.9
G35	5.9	6.2
I30, I35		
1996-99	5.0	5.3
2000-04	6.3	6.6
J30	5.1	5.4
M30	5.9	6.2
M45	6.8	7.1
Q45		
1990-96	7.8	8.1
1997-01	5.3	5.6
2002-05	6.1	6.4
w/AC add		
91-96 G20	.3	.3
G35	.2	.2
J30	.2	.2
Heater Hoses, Replace (B)		
G20, G35	1.2	1.4
I30, I35		
1996-99	.6	.8
2000-04	2.2	2.4
J30, M30	1.4	1.6
M45	2.0	2.2
Q45		
1990-96	2.2	2.4
1997-01	1.1	1.3
2002-05	1.7	1.9
High Pressure Hose, Replace (B)		
Includes: Evacuate/recover and recharge.		
G20		
1991-96	2.0	2.1
1999-02	1.7	1.8
G35	1.8	1.9
I30, I35	3.1	3.2
J30	2.9	3.0
M30	2.6	2.7
M45	3.2	3.3
Q45		
1990-96	1.8	1.9
1997-01	3.8	3.9
2002-05	2.6	2.7
w/R12 add		
G20, J30	.4	.4
1996-99 I30	.4	.4
1990-96 Q45	.4	.4
Intake Sensor, Replace (B)		
G20	1.0	1.0
G35	1.6	1.6
I30, I35		
1996-99	.5	.5
2000-01	1.8	1.8
J30	.9	.9
M30	1.4	1.4

	LABOR TIME	SEVERE SERVICE
M45	6.8	6.8
Q45		
1990-96	.8	.8
1997-01	1.6	1.6
2002-05	1.1	1.1
In-Vehicle Sensor, Replace (B)		
G20	.7	.7
G35, M45	.3	.3
I30, I35	.8	.8
J30	.5	.5
M30	1.1	1.1
Q45		
1990-96	1.1	1.1
1997-05	.5	.5
Low Pressure and/or Cycling Switch, Replace (B)		
Includes: Evacuate/recover and recharge.		
G20	1.7	1.7
G35	3.2	3.2
I30, I35, J30, M30	1.8	1.8
M45	1.3	1.3
Q45	1.5	1.5
w/R12 add		
G20, J30	.4	.4
1996-99 I30	.4	.4
1990-96 Q45	.4	.4
Low Pressure Hose, Replace (B)		
Includes: Evacuate/recover and recharge.		
G20		
1991-96	3.0	3.1
1999-02	1.3	1.4
G35	2.1	2.2
I30, I35	1.6	1.7
J30	2.1	2.2
M30	1.9	2.0
M45	2.7	2.8
Q45		
1990-01	1.4	1.5
2002-05	2.4	2.5
w/R12 add		
G20, J30	.4	.4
1996-99 I30	.4	.4
1990-96 Q45	.4	.4
Magnetic Clutch Assy., Replace (B)		
Includes: Evacuate/recover and recharge.		
G20		
1991-96	3.2	3.3
1999-02	2.2	2.3
G35	2.9	3.0
I30, I35	2.2	2.3
J30	3.3	3.4
M30	3.0	3.1
M45	1.9	2.0

G20 : G35 : I30 : I35 : J30 : M30 : M45 : Q45   **INF-25**

	LABOR TIME	SEVERE SERVICE
Q45		
1990-96	4.4	4.5
1997-01	3.3	3.4
2002-05	1.9	2.0
w/R12 add		
G20, J30	.4	.4
1996-99 I30	.4	.4
1990-96 Q45	.4	.4
Receiver/Drier Assy., Replace (B)		
Includes: Evacuate/recover and recharge.		
G20	1.6	1.7
G35	3.3	3.4
I30, I35, J30	1.7	1.8
M30	2.0	2.1
M45	1.4	1.5
Q45	1.9	2.0
w/R12 add		
G20, J30	.4	.4
1996-99 I30	.4	.4
1990-96 Q45	.4	.4

WIPERS & SPEEDOMETER

	LABOR TIME	SEVERE SERVICE
Power Antenna, Replace (B)		
Exc. G20	.6	.6
G20	.7	.7
Radio, R&R (B)		
G20	.7	.7
G35	1.6	1.6
I30, I35	1.1	1.1
J30, M30, M45	.8	.8
Q45		
1990-96	.7	.7
1997-01	1.1	1.1
2002-05	.7	.7
Speedometer Cable & Casing, Replace (B)		
G20	1.2	1.2
M30	1.5	1.5
Speedometer Head, R&R or Replace (B)		
G20		
1991-96	1.2	1.2
1999-02	.8	.8
G35, J30	1.2	1.2
I30, I35	1.1	1.1
M30	1.5	1.5
M45	.8	.8
Q45		
1990-01	1.5	1.5
2002-05	.6	.6
Windshield Washer Pump, Replace (B)		
G20	1.1	1.2
G35, I30, I35	.8	.9
J30	.6	.6
M30, M45	.9	1.0

	LABOR TIME	SEVERE SERVICE
Q45		
1990-96	3.2	3.2
1997-01	.6	.6
2002-05	1.1	1.2
Windshield Wiper & Washer Switch, Replace (B)		
Exc. below	.5	.5
1997-05 Q45	.6	.6
Windshield Wiper Interval Relay, Replace (B)		
G20, I30	.5	.5
G35		
coupe	.8	.8
sedan	.3	.3
J30, M45, Q45	.3	.3
M45	.8	.8
Windshield Wiper Motor, Replace (B)		
G20	.8	1.1
G35	.9	1.2
I30, I35		
1996-99	.5	.8
2000-04	.9	1.2
J30	1.1	1.4
M30, M45	.6	.9
Q45		
1990-96	1.7	2.0
1997-01	.5	.8
2002-05	.9	1.2

LAMPS & SWITCHES

	LABOR TIME	SEVERE SERVICE
Back-Up Lamp Bulb, Replace (C)		
Exc. below	.3	.3
Q45		
1990-96	.5	.5
1997-01	.8	.8
2002-05	.5	.5
Back-Up Lamp Switch, Replace (B)		
G20	.5	.5
G35	1.1	1.1
I30, I35	.6	.6
Combination Switch Assy., Replace (B)		
G20	.9	.9
G35	.7	.7
I30, I35		
1996-99	.7	.7
2000-04	1.2	1.2
J30	.8	.8
M30	.6	.6
M45	.3	.3
Q45		
1990-96	2.3	2.3
1997-01	.9	.9
2002-05	.3	.3
Cornering Lamp Bulb, Replace (C)		
Exc. below	.3	.3
G35	.6	.6
J30	.5	.5

	LABOR TIME	SEVERE SERVICE
Q45		
1997-05	.5	.5
Daylight Running Lamp Module, Replace (B)		
1995-05	.5	.5
Halogen Headlamp Bulb, Replace (B)		
G20, I30, I35, J30	.3	.3
G35	.8	.8
M30	.5	.5
M45	1.4	1.4
Q45		
1990-01	.7	.7
2002-05	1.1	1.1
Hazard Warning Switch, Replace (B)		
G20	.5	.5
G35, I30, I35	.3	.3
J30	.6	.6
M30, M45	.5	.5
Q45		
1990-96	1.2	1.2
1997-01	.5	.5
2002-05	.9	.9
Headlamp & Turn Signal Switch, Replace (B)		
All Models	.5	.5
Headlamps, Aim (B)		
Two	.4	.4
Four	.6	.6
High Mount Stop Lamp Assy., Replace (B)		
G20	.5	.5
G35		
coupe	1.6	1.6
sedan	.9	.9
I30, I35	.6	.6
J30, M30, M45	.5	.5
Q45		
1990-01	.7	.7
2002-05	.3	.3
Horn, Replace (B)		
Exc. below	.5	.5
I30, I35	1.2	1.2
1990-96 Q45	.9	.9
Horn Relay, Replace (B)		
Exc. below	.3	.3
G35	.7	.7
J30	.5	.5
Neutral Safety Switch, Replace (B)		
ATM		
Exc. below	.8	.8
G35	1.9	1.9
M45	1.6	1.6
2002-04 Q45	1.6	1.6
ATX		
G20	.6	.6
I30, I35	1.0	1.0
MT		
G35	.3	.3

G20 : G35 : I30 : I35 : J30 : M30 : M45 : Q45

	LABOR TIME	SEVERE SERVICE
Parking Brake Warning Lamp Switch, Replace (B)		
Exc. below	.8	.8
1990-96 Q45	1.1	1.1
Rear Combination Lamp Assy., Replace (B)		
G20		
1991-96		
one	1.5	1.5
both	2.3	2.3
1999-02		
one	.6	.6
both	.9	.9
G35, M45		
one	.3	.3
both	.5	.5
I30, I35		
1996-99		
one	1.0	1.0
both	1.4	1.4
2000-04		
one	.6	.6
both	1.1	1.1
J30		
one	.5	.5
both	.7	.7
M30		
one	.8	.8
both	1.1	1.1
Q45		
1990-96		
one	.8	.8
both	1.2	1.2
1997-01 one or both	.6	.6
2002-05		
one	.6	.6
both	.9	.9
Rear Combination Lamp Bulb, Replace (C)		
Exc. below	.3	.3
2002-04 Q45	.5	.5
each addl. add	.1	.1
Stop Lamp Switch, Replace (B)		
G20, G35	.5	.5
I30, I35	.6	.6
J30	1.0	1.0
M30, M45	.5	.5
Q45		
1990-96	1.0	1.0
1997-01	.3	.3
2002-05	.7	.7
Turn Signal or Hazard Warning Flasher, Replace (B)		
Exc. below	.5	.5
G35, M45	.7	.7
Q45		
1990-96	1.1	1.1
1997-05	.6	.6

SEAT BELTS

The following operations do not include testing. Add time as required.

	LABOR TIME	SEVERE SERVICE
Control Module, Replace (B)		
G20	.7	.7
Front Limit Switch, Replace (B)		
G20		
one	.3	.3
both	.7	.7
Rear Limit Switch, Replace (B)		
G20		
one	.5	.5
both	.8	.8
Automatic Seat Belt Assy., Replace (B)		
G20		
one	.8	.8
both	1.2	1.2
Seat Belt Guide Rails, Replace (B)		
G20		
one	1.2	1.2
both	2.4	2.4
Seat Belt Motor, Replace (B)		
G20		
one	1.2	1.2
both	2.4	2.4
Seat Slide Switch, Replace (B)		
G20	.5	.5

BODY

	LABOR TIME	SEVERE SERVICE
Door Lock Actuator, Replace (B)		
G20	1.2	1.2
G35, M45	1.1	1.1
I30, I35	1.1	1.1
J30, M30	.9	.9
Q45		
1990-01	1.1	1.1
2002-05	.7	.7
Front Door Lock, Replace (B)		
G20	1.1	1.1
G35, J30, M45	1.4	1.4
I30, I35		
1996-99	1.4	1.4
2000-04	1.0	1.0
M30	1.9	1.9
Q45		
1990-96	2.1	2.1
1997-01	1.2	1.2
2002-05	.9	.9
Front Door Lock Cylinder, Replace (B)		
G20	.7	.7
G35	1.1	1.1
I30, I35	1.2	1.2
J30	.6	.6
M30	1.6	1.6
M45	.8	.8

	LABOR TIME	SEVERE SERVICE
Q45		
1990-96	1.5	1.5
1997-01	1.0	1.0
2002-05	.8	.8
Front Door Window Regulator, Replace (B)		
G20	1.1	1.1
G35	1.2	1.2
I30, I35, J30	1.1	1.1
M30	2.3	2.3
M45	1.9	1.9
Q45		
1990-96	3.0	3.0
1997-01	1.4	1.4
2002-05	.8	.8
w/AC 91-96 G20 add	.2	.2
Front Door Window Regulator Motor, Replace (B)		
Exc. below	.9	.9
G35 coupe	1.2	1.2
M30	2.6	2.6
M45	1.9	1.9
Q45		
1990-96	2.0	2.0
1997-01	1.4	1.4
Hood Hinge, Replace (B)		
G20		
1991-96		
one	1.2	1.2
both	1.4	1.4
1999-02		
one	3.0	3.0
both	3.9	3.9
G35		
coupe		
one	2.3	2.3
both	3.1	3.1
sedan		
one	1.1	1.1
both	1.2	1.2
I30, I35		
one	1.5	1.5
both	1.8	1.8
J30		
one	1.8	1.8
both	2.8	2.8
M30		
one	.9	.9
both	1.1	1.1
Q45		
1990-96		
one	1.4	1.4
both	1.5	1.5
1997-05		
one	1.1	1.1
both	1.2	1.2
Hood Lock, Replace (B)		
Exc. below	.5	.5
2000-04 I30, I35	1.1	1.1
Q45	.6	.6

	LABOR TIME	SEVERE SERVICE
Hood Release Cable, Replace (B)		
G20	1.3	*1.4*
G35, M45	1.2	*1.3*
I30, I35		
1996-99	1.1	*1.2*
2000-04	1.8	*1.9*
J30	1.1	*1.2*
M30	1.8	*1.9*
Q45		
1990-01	1.3	*1.4*
2002-05	1.5	*1.6*
Lock Striker Plate, Replace (B)		
1990-05	.5	*.5*

	LABOR TIME	SEVERE SERVICE
Rear Door Lock, Replace (B)		
G20	1.1	*1.1*
G35	.7	*.7*
I30, I35	1.2	*1.2*
J30	.8	*.8*
M45	1.2	*1.2*
Q45		
1990-01	1.1	*1.1*
2002-05	.7	*.7*
Rear Door Window Regulator, Replace (B)		
1991-96 G20	1.7	*1.7*
G35 sedan	.8	*.8*
I30, I35	1.4	*1.4*
M45	2.0	*2.0*
Q45		
1990-96	2.0	*2.0*
1997-01	1.5	*1.5*
2002-05	.8	*.8*

	LABOR TIME	SEVERE SERVICE
Rear Door Window Regulator Motor, Replace (B)		
G20	1.1	*1.1*
G35	.9	*.9*
I30, I35, J30	1.1	*1.1*
M45	2.0	*2.0*
Q45		
1990-96	1.9	*1.9*
1997-05	1.3	*1.3*
Remote Door Lock Control Unit, Replace (B)		
G20	.6	*.6*
G35	.7	*.7*
I30, J30	.5	*.5*
M45	.8	*.8*
Q45		
1990-96	1.2	*1.2*
1997-05	.6	*.6*

NOTES

FX35 : FX45 : QX4 : QX56

SYSTEM INDEX

- MAINTENANCE INF-30
  - Maintenance Schedule . . . INF-30
- CHARGING INF-30
- STARTING INF-30
- CRUISE CONTROL INF-30
- IGNITION INF-30
- EMISSIONS INF-31
- FUEL INF-31
- EXHAUST INF-32
- ENGINE COOLING INF-32
- ENGINE INF-33
  - Assembly INF-33
  - Cylinder Head INF-33
  - Camshaft INF-34
  - Crank & Pistons INF-34
  - Engine Lubrication INF-35
- AUTO TRANSMISSION INF-35
- TRANSFER CASE INF-36
- SHIFT LINKAGE INF-36
- DRIVELINE INF-36
- BRAKES INF-37
- FRONT SUSPENSION INF-38
- REAR SUSPENSION INF-39
- STEERING INF-39
- HEATING & AC INF-40
- WIPERS &
  SPEEDOMETER INF-40
- LAMPS & SWITCHES INF-41
- BODY INF-41

OPERATIONS INDEX

A
- AC Hoses INF-40
- Air Bags INF-39
- Air Conditioning INF-40
- Alignment INF-38
- Alternator (Generator) INF-30
- Antenna INF-40
- Anti-Lock Brakes INF-37

B
- Ball Joint INF-39
- Battery Cables INF-30
- Bleed Brake System INF-37
- Blower Motor INF-40
- Brake Disc INF-38
- Brake Drum INF-38
- Brake Hose INF-37
- Brake Pads and/or Shoes . . . INF-38

C
- Camshaft INF-34
- Camshaft Sensor INF-30
- Catalytic Converter INF-32
- Coolant Temperature (ECT)
  Sensor INF-31
- Crankshaft INF-35

- Crankshaft Sensor INF-31
- Cruise Control INF-30
- Cylinder Head INF-33

D
- Differential INF-37
- Distributor INF-31
- Drive Belt INF-30
- Driveshaft INF-37

E
- EGR INF-31
- Electronic Control Module
  (ECM/PCM) INF-31
- Engine INF-33
- Engine Lubrication INF-35
- Engine Mounts INF-33
- Evaporator INF-40
- Exhaust INF-32
- Exhaust Manifold INF-32
- Expansion Valve INF-40

F
- Flexplate INF-36
- Fuel Injection INF-32
- Fuel Pump INF-31
- Fuel Vapor Canister INF-31

G
- Gear Selector Lever INF-36
- Generator INF-30

H
- Halfshaft INF-37
- Headlamp INF-41
- Heater Core INF-40
- Horn INF-41

I
- Ignition Coil INF-31
- Ignition Switch INF-31
- Inner Tie Rod INF-40
- Intake Air Temperature (IAT)
  Sensor INF-31
- Intake Manifold INF-31

K
- Knock Sensor INF-31

L
- Lower Control Arm INF-39

M
- Maintenance Schedule INF-30
- Manifold Absolute Pressure
  (MAP) Sensor INF-31
- Mass Air Flow (MAF)
  Sensor INF-31
- Master Cylinder INF-38
- Muffler INF-32
- Multifunction Lever/Switch . . . INF-41

N
- Neutral Safety Switch INF-41

O
- Oil Pan INF-35
- Oil Pump INF-35
- Outer Tie Rod INF-40
- Oxygen Sensor INF-31

P
- Parking Brake INF-38
- Pistons INF-35
- Positive Crankcase
  Ventilation (PCV) Valve . . . INF-31

R
- Radiator INF-32
- Radiator Hoses INF-32
- Radio INF-40
- Rear Main Oil Seal INF-35

S
- Shock Absorber/Strut,
  Front INF-38
- Shock Absorber/Strut,
  Rear INF-39
- Spark Plug Cables INF-31
- Spark Plugs INF-31
- Spring, Front Coil INF-38
- Spring, Rear Coil INF-39
- Starter INF-30
- Steering Wheel INF-40

T
- Thermostat INF-32
- Throttle Body INF-32
- Throttle Position Sensor
  (TPS) INF-31
- Timing Belt INF-34
- Timing Chain INF-34
- Torque Converter INF-36

U
- Upper Control Arm INF-39

V
- Valve Body INF-36
- Valve Cover Gasket INF-34
- Valve Job INF-34

W
- Water Pump INF-32
- Wheel Balance INF-30
- Wheel Cylinder INF-38
- Window Regulator INF-41
- Windshield Washer Pump . . . INF-41
- Windshield Wiper Motor INF-41

INF-30 FX35 : FX45 : QX4 : QX56

	LABOR TIME	SEVERE SERVICE
MAINTENANCE		
Air Cleaner Filter Element, Service (C)		
All Models	.3	.4
Chassis Lubrication, Change Oil & Filter (C)		
All Models	.7	.7
Drive Belt, Adjust (B)		
One	.6	.7
each addl. add	.2	.2
Drive Belt, Replace (B)		
One	.8	.9
each addl. add	.2	.2
Fuel Filter, Replace (B)		
FX35, FX45	1.1	1.2
QX4	.5	.6
QX56	2.1	2.2
Halogen Headlamp Bulb, Replace (B)		
FX35, FX45, QX56	.6	.6
QX4	.3	.3
Oil & Filter, Change (C)		
Includes: Correct all fluid levels.		
All Models	.7	.7
Timing Belt, Replace (B)		
1997-00 QX4	3.3	3.8
w/AC add	.2	.2
w/PS add	.5	.5
Replace gears add each	.3	.3
Tire, Replace (C)		
Includes: Dismount old tire and mount new tire to rim.		
FX35, FX45 one	.7	.9
QX4, QX56 one	1.1	1.3
each addl. add	.6	.6
Tires, Rotate (C)		
All Models	.5	.5
Wheel, Balance (B)		
One	.6	.6
each addl. add	.2	.2
Xenon Headlamp, Replace (B)		
FX35, FX45	2.0	2.0
QX56	1.5	1.5

SCHEDULED MAINTENANCE INTERVALS
If necessary, refer to appropriate Chilton maintenance service information.

	LABOR TIME	SEVERE SERVICE
7,500 Mile Service (C)		
All Models	.6	.6
15,000 Mile Service (B)		
All Models	2.1	2.1
22,500 Mile Service (C)		
All Models	.6	.6
30,000 Mile Service (B)		
All Models	5.1	5.1
37,500 Mile Service (C)		
1994-99	.6	.6
45,000 Mile Service (B)		
All Models	2.1	2.1
52,500 Mile Service (C)		
All Models	.6	.6
60,000 Mile Service (B)		
All Models	6.1	6.1
67,500 Mile Service (C)		
All Models	.6	.6
75,000 Mile Service (B)		
All Models	2.1	2.1
82,500 Mile Service (C)		
All Models	.6	.6
90,000 Mile Service (B)		
All Models	6.1	6.1
97,500 Mile Service (C)		
All Models	.6	.6
100,000 Mile Service (B)		
All Models Replace timing belt	3.1	3.1
105,500 Mile Service (B)		
All Models	2.1	2.1
113,000 Mile Service (C)		
All Models	.6	.6
120,000 Mile Service (B)		
All Models	6.1	6.1

CHARGING

	LABOR TIME	SEVERE SERVICE
Alternator Circuits, Test (B)		
Includes: Test component output.		
QX4	.3	.3
Alternator, R&R and Recondition (A)		
FX35, FX45	2.1	2.4
QX4	3.1	3.4
Alternator Assy., Replace (B)		
Includes: Pulley transfer.		
FX35	.9	1.0
FX45	1.2	1.3
QX4		
1997-00	1.1	1.2
2001-03	2.1	2.2
QX56	2.1	2.2
w/AC add		
FX35	.9	.9
FX45	1.1	1.1
Alternator Voltage Regulator, Replace (B)		
FX35	1.4	1.6
FX45	1.7	1.9
QX4		
1997-00	1.4	1.6
2001-03	2.4	2.6
w/AC add		
FX35	.9	.9
FX45	1.1	1.1

STARTING

	LABOR TIME	SEVERE SERVICE
Starter Draw Test (On Car) (B)		
All Models	.5	.5
Battery Cables, Replace (C)		
FX35	1.1	1.2
FX45	2.6	2.7
QX4, QX56	.8	.9
Starter, R&R and Recondition (A)		
FX35	1.8	2.0
FX45	2.6	2.8
QX4	2.0	2.2
Starter Assy., Replace (B)		
FX35	1.1	1.2
FX45	1.6	1.7
QX4	1.0	1.1
QX56	1.3	1.4
Starter Drive Assy., Replace (B)		
Includes: Starter R&R.		
FX35	1.6	1.7
FX45	2.1	2.2
QX4	1.4	1.5
Starter Relay, Replace (B)		
FX35	.8	.8
Starter Solenoid and/or Switch, Replace (B)		
Includes: Starter R&R.		
FX35	1.3	1.4
FX45	1.7	1.8
QX4	1.1	1.2

CRUISE CONTROL

	LABOR TIME	SEVERE SERVICE
Diagnose Cruise System Component Each (A)		
All Models	1.0	1.0
Actuator Assy., Replace (B)		
QX4	.5	.5
Brake or Clutch Release Switch, Replace (B)		
All Models	.3	.3
Control Controller (Module), Replace (B)		
FX35, FX45	.6	.6
QX4, QX56	.4	.4
Control Main Switch, Replace (B)		
QX4	.3	.3
QX56	.8	.8
Control On/Off Switch, Replace (B)		
FX35, FX45	.7	.7
QX4 1997	.3	.3
QX56	.8	.8
Control Speed Sensor, Replace (B)		
QX4	.5	.5
QX56	1.2	1.2
Control Vacuum Pump, Replace (B)		
QX4	.5	.5

IGNITION

	LABOR TIME	SEVERE SERVICE
Diagnose Ignition System Component Each (A)		
All Models	.5	.5
Ignition Timing, Reset (B)		
QX4	.5	.5
Camshaft Position Sensor, Replace (B)		
All Models each	.6	.6

FX35 : FX45 : QX4 : QX56 INF-31

	LABOR TIME	SEVERE SERVICE
Crankshaft Position Sensor, Replace (B)		
All Models	.6	*.6*
Distributor, Replace (B)		
Includes: Reset base ignition timing.		
QX4	.5	*.7*
Distributor Cap and/or Rotor, Replace (B)		
QX4	.5	*.5*
Ignition Coil, Replace (B)		
FX35	1.5	*1.6*
FX45	.9	*1.0*
QX4		
1997-00	.5	*.6*
2001-03	1.1	*1.2*
QX56	.6	*.7*
Ignition Coil Relay, Replace (B)		
FX35	.8	*.8*
Ignition & Steering Lock Switch Assy., Replace (B)		
FX35, FX45	.9	*1.0*
QX4	1.2	*1.3*
QX56	2.6	*2.7*
Ignition Switch, Replace (B)		
FX35, FX45	1.1	*1.1*
QX4, QX56	.8	*.8*
Key Warning Buzzer or Chime, Replace (B)		
FX35, FX45	.9	*.9*
QX4	.6	*.6*
QX56	1.1	*1.1*
Spark Plug (Ignition) Cables, Replace (B)		
QX4	.5	*.6*
Spark Plugs, Replace (B)		
FX35	2.1	*2.3*
FX45	.9	*1.1*
QX4		
1997-00	.5	*.7*
2001-03	1.2	*1.4*
QX56	.9	*1.1*

EMISSIONS

The following components do not include testing. Add time as required.

	LABOR TIME	SEVERE SERVICE
Diagnose Emission Control System Component Each (A)		
1997-05	.5	*.5*
Dynamometer Test (A)		
QX4	.4	*.4*
Reprogram Electronic Control Module (ECM) (A)		
All Models	.6	*.6*
Retrieve Fault Codes (A)		
1997-05	.3	*.3*
Air Flow Meter, Replace (B)		
FX35, FX45	.5	*.6*
QX4	.8	*.9*
QX56	.3	*.3*

	LABOR TIME	SEVERE SERVICE
Back Pressure Variable Transducer Valve, Replace (B)		
QX4	.5	*.5*
Barometric Manifold Absolute Pressure Sensor, Replace (B)		
QX4	.3	*.3*
Coolant Temperature (ECT) Sensor, Replace (B)		
FX35, FX45	.8	*.8*
QX4		
1997-00	.5	*.5*
2001-03	2.7	*2.7*
EGR Control Valve, Replace (B)		
QX4	.5	*.7*
EGR Solenoid Valve, Replace (B)		
QX4	.5	*.5*
EGR Temperature Sensor, Replace (B)		
QX4	.5	*.5*
Electronic Control Module (ECM/PCM), Replace (B)		
FX35, FX45	.8	*.9*
QX4, QX56	.6	*.7*
Fuel Vapor Canister, Replace (B)		
FX35, FX45, QX4	.6	*.7*
QX56	1.1	*1.2*
Intake Air Temperature (IAT) Sensor, Replace (B)		
Exc. below	.4	*.4*
2001-03 QX4	.6	*.6*
Knock Sensor, Replace (B)		
FX35	2.1	*2.3*
FX45	2.6	*2.7*
QX4		
1997-00	2.8	*3.0*
2001-03	3.6	*3.8*
QX56	1.8	*2.0*
Oxygen Sensor, Replace (B)		
FX35, FX45		
front		
one	.8	*1.0*
both	1.0	*1.2*
rear one or both	.6	*.8*
QX4		
1997-00		
front one or both	1.4	*1.6*
rear one or both	.6	*.8*
2001-03		
front		
one	2.2	*2.4*
both	2.7	*2.9*
rear		
one	.8	*1.0*
both	1.3	*1.5*
QX56		
front	.8	*1.0*
rear one or both	.6	*.8*
Positive Crankcase Ventilation (PCV) Valve, Replace (B)		
FX35, FX45	.6	*.6*

	LABOR TIME	SEVERE SERVICE
QX4		
1997-00	.3	*.3*
2001-03	3.1	*3.1*
QX56	.5	*.5*
Throttle Position Sensor (TPS)/Switch, Replace (B)		
FX35, FX45, QX4	.6	*.6*

FUEL

DELIVERY

	LABOR TIME	SEVERE SERVICE
Air Cleaner Filter Element, Service (C)		
All Models	.3	*.4*
Fuel Filter, Replace (B)		
FX35, FX45	1.1	*1.2*
QX4	.5	*.6*
QX56	2.1	*2.2*
Fuel Gauge (Dash), Replace (B)		
All Models	.8	*.8*
Fuel Level Gauge (Tank), Replace (B)		
Exc. below	1.1	*1.1*
QX56	2.1	*2.2*
Fuel Pump Dropping Resistor, Replace (B)		
FX35	1.0	*1.0*
FX45	.4	*.4*
Fuel Pump(Electric), Replace (B)		
FX35, FX45	1.1	*1.3*
QX4	.8	*1.0*
QX56	2.1	*2.3*
Fuel Pump Relay, Replace (B)		
FX35, FX45	.8	*.8*
QX4	.3	*.3*
QX56	.6	*.6*
Fuel Tank, Replace (B)		
FX35, FX45	7.6	*7.8*
QX4	2.5	*2.7*
QX56	3.8	*4.0*
Intake Manifold and/or Gasket, Replace (B)		
Includes: Adjustments.		
FX35		
upper	1.6	*1.9*
lower	2.4	*2.7*
FX45		
upper	2.4	*2.7*
lower	3.2	*3.5*
QX4		
1997-00		
upper	2.4	*2.7*
lower	2.8	*3.1*
2001-03		
upper	3.1	*3.4*
lower	3.6	*3.9*
QX56	1.8	*2.0*
Vacuum Delay Valve, Replace (B)		
1997-00 QX4	.3	*.3*

FX35 : FX45 : QX4 : QX56

	LABOR TIME	SEVERE SERVICE
INJECTION		
Diagnose Fuel Injection System (A)		
FX35, FX45	.6	.7
QX4		
1997-00	1.0	1.0
2001-03	2.2	2.2
QX56	1.4	1.4
Cylinder Head Temperature Sensor, Replace (B)		
FX45, QX56	.9	.9
EFI Relay, Replace (B)		
FX35	.8	.8
QX4	.3	.3
QX56	1.1	1.1
Fuel Injectors, Replace (B)		
FX35	2.1	2.4
FX45	2.9	3.2
QX4	3.3	3.6
QX56	1.4	1.7
Fuel Pressure Regulator, Replace (B)		
FX35, FX45	1.1	1.3
QX4		
1997-00	2.6	2.8
2001-03	.6	.8
Throttle Body Assy., Replace (B)		
FX35, FX45	.6	.7
QX4	1.4	1.5
QX56	1.1	1.2
EXHAUST		
Catalytic Converter, Replace (B)		
FX35, FX45		
one	1.1	1.2
both	1.4	1.5
QX4		
1997-00		
2WD		
one	1.6	1.7
both	1.9	2.0
4WD		
one	1.6	1.7
both	6.3	6.4
2001-03		
2WD		
one	1.4	1.5
both	1.9	2.0
4WD		
one	2.4	2.5
both	3.2	3.3
QX56		
one	4.6	4.7
both	6.3	6.4
Center Exhaust & Pre-Muffler Pipe, Replace (B)		
FX35, FX45	.8	.9
QX4		
1997-00	.8	.9
2001-03	1.3	1.4
QX56	1.0	1.1

	LABOR TIME	SEVERE SERVICE
Exhaust Manifold or Gasket, Replace (B)		
FX35		
right side	3.2	3.5
left side	2.7	3.0
FX45		
right side	4.6	4.9
left side	4.0	4.3
QX4		
1997-00		
2WD		
right side	3.0	3.3
left side	3.6	3.9
4WD		
right side	3.0	3.3
left side	4.7	5.0
2001-03		
2WD		
right side	4.6	4.9
left side	3.2	3.5
4WD		
right side	4.6	4.9
left side	3.6	3.9
QX56		
2WD		
right or left side	2.9	3.2
4WD		
right or left side	4.6	4.9
Muffler & Rear Pipe, Replace (B)		
Exc. below	.6	.7
QX56	1.0	1.1
ENGINE COOLING		
Pressure Test Cooling System (B)		
QX4	.3	.3
Coolant Temperature Sensor Switch, Replace (B)		
FX35, FX45	.9	.9
Coolant Thermostat, Replace (B)		
FX35	1.1	1.3
FX45		
inlet/outlet	.8	1.0
outlet/gasket	3.1	3.3
QX4		
1997-00	1.8	2.0
2001-03	1.4	1.6
QX56	.8	1.0
Engine Coolant Temp. Sending Unit, Replace (B)		
FX35, FX45	.9	.9
QX4		
1997-00	.5	.5
2001-03	2.7	2.7
Fluid Fan Clutch Assy., Replace (B)		
FX45	1.1	1.3
QX4		
1997-00	.8	1.0
2001-03	1.2	1.4
QX56	2.0	2.2

	LABOR TIME	SEVERE SERVICE
Freeze Plugs (Water Jacket), Replace (B)		
Add access time as required.		
FX35, FX45, QX4 each	.5	.7
Radiator Assy., R&R or Replace (B)		
FX35, FX45, QX4	1.5	1.6
QX56	2.1	2.2
w/AT QX4 add	.2	.2
Radiator Fan &/or Fan Motor, Replace (B)		
FX35 one	.6	.6
each addl. add	.2	.2
Radiator Fan Motor Relay, Replace (B)		
FX35	.8	.8
Radiator Hoses, Replace (B)		
Includes: Refill with proper coolant mix.		
FX35, FX45		
upper	.6	.7
lower	1.2	1.3
QX4		
1997-00		
upper	.5	.6
lower	.9	1.0
2001-03		
upper	1.1	1.2
lower	1.4	1.5
QX56		
upper	.8	.9
lower	1.1	1.2
Temperature Gauge (Dash), Replace (B)		
FX35	.8	.8
FX45	1.2	1.2
QX4		
analog	.8	.8
digital	.3	.3
QX56	.6	.6
Temperature Gauge (Engine), Replace (B)		
FX35, FX45	.9	.9
QX4		
1997-00	.5	.5
2001-03	3.1	3.1
Water Pump and/or Gasket, Replace (B)		
Includes: Refill with proper coolant mix.		
FX35, FX45	1.7	2.1
QX4		
1997-00	3.5	3.9
2001-03	2.4	2.8
QX56	2.1	2.5

FX35 : FX45 : QX4 : QX56 **INF-33**

	LABOR TIME	SEVERE SERVICE

ENGINE
ASSEMBLY
Times shown are for OEM assemblies. Time to replace assemblies from aftermarket rebuilders may vary.

Engine Assy., R&I (B)
Does not include parts or component transfer.
- 3.3L
  - 2WD 8.9 8.9
  - 4WD 10.8 10.8
- 3.5L
  - 2WD 14.9 14.9
  - 4WD 15.3 15.3
- w/AC 3.3L add .... .5 .5
- w/PS 3.3L add .... .3 .3

Engine Assy., R&R and Recondition (A)
Includes: Replacing rings, rod and main bearings, cylinder head reconditioning and engine tune up.
- FX35
  - 2WD 31.8 32.9
  - 4WD 34.0 35.1
- FX45 .... 34.9 36.0
- QX4
  - 1997-00
    - 2WD 22.4 23.3
    - 4WD 24.8 25.7
  - 2001-03
    - 2WD 35.1 36.2
    - 4WD 35.7 36.8
- QX56
  - 2WD 35.7 36.8
  - 4WD 37.7 38.8
- w/AC 97-00 QX4 add ..... .6 .6
- w/PS 97-00 QX4 add ..... .5 .5

Engine (New or Rebuilt Complete), Replace (B)
Includes: Component transfer and engine tune-up.
- FX35
  - 2WD 14.5 14.7
  - 4WD 16.0 16.2
- FX45 .... 16.9 17.1
- QX4
  - 1997-00
    - 2WD 12.3 12.5
    - 4WD 14.7 14.9
  - 2001-03
    - 2WD 17.2 17.4
    - 4WD 17.9 18.1
- QX56
  - 2WD 15.9 16.1
  - 4WD 17.6 17.8
- w/AC 97-00 QX4 add ..... .6 .6
- w/PS 97-00 QX4 add ..... .3 .3

Engine Assy. (Short Block), Replace (B)
Assembly consists of cylinder block, piston assemblies, crankshaft, camshaft, timing chain and gears. Does not include cylinder heads. Operation Includes: R&R engine, transfer necessary parts and all necessary adjustments.
- FX35
  - 2WD 20.3 20.8
  - 4WD 22.3 22.8
- FX45 .... 23.1 23.6
- QX4
  - 1997-00
    - 2WD 15.8 16.3
    - 4WD 18.4 18.9
  - 2001-03
    - 2WD 23.8 24.3
    - 4WD 24.3 24.8
- QX56
  - 2WD 21.9 22.4
  - 4WD 23.6 24.1
- w/AC 97-00 QX4 add ..... .6 .6
- w/PS 97-00 QX4 add ..... .3 .3

Engine Mounts, Replace (B)
- FX35
  - 2WD
    - front 5.0 5.2
    - rear .... .6 .8
  - 4WD
    - front 5.5 5.7
    - rear .... .6 .8
- FX45
  - front 2.5 2.7
  - rear .... .6 .8
- QX4
  - 1997-03
    - front 3.8 4.0
    - rear .... .6 .8
  - 2001-03
    - 2WD
      - front 2.2 2.4
      - rear 1.9 2.1
    - 4WD
      - front 5.0 5.2
      - rear 1.9 2.1
- QX56
  - front or rear 1.3 1.5

CYLINDER HEAD
Compression Test (B)
- 3.3L .... .6 .7
- 3.5L 1.3 1.4

Cylinder Head, Replace (B)
Includes: Parts transfer, grind valves, adjustments.
- FX35
  - 2WD
    - right side 20.7 21.2
    - left side 21.0 21.5
    - both sides 25.8 26.3
  - 4WD
    - right side 21.9 22.4
    - left sides 22.2 22.7
    - both sides 27.1 27.6
- FX45
  - one side 23.0 23.5
  - both sides 27.2 27.7
- QX4
  - 1997-00
    - 2WD
      - one side 10.7 11.2
      - both sides .. 14.2 14.7
    - 4WD
      - one side 11.2 11.7
      - both sides .. 14.5 15.0
  - 2001-03
    - 2WD
      - one side 23.3 23.8
      - both sides .. 26.8 27.3
    - 4WD
      - one side 23.7 24.2
      - both sides .. 27.2 27.7
- QX56
  - 2WD
    - one side 22.4 22.9
    - both sides 27.7 28.3
  - 4WD
    - one side 24.0 24.5
    - both sides 29.4 29.9

Cylinder Head Gasket, Replace (B)
- FX35
  - 2WD
    - right side 17.5 17.8
    - left side 18.0 18.3
    - both sides 19.9 20.2
  - 4WD
    - right side 18.8 19.1
    - left sides 19.3 19.6
    - both sides 21.1 21.4
- FX45
  - one side 20.0 20.3
  - both sides 21.5 21.8
- QX4
  - 1997-00
    - 2WD
      - one side 8.6   8.9
      - both sides .. 10.5 10.8
    - 4WD
      - one side 9.2   9.5
      - both sides .. 10.9 11.2

	LABOR TIME	SEVERE SERVICE

Cylinder Head Gasket, Replace (B)
 2001-03
  2WD
   one side 20.9 21.2
   both sides 21.9 22.2
  4WD
   one side 21.5 21.8
   both sides 22.5 22.8
 QX56
  2WD
   one side 19.6 20.1
   both sides 21.3 21.8
  4WD
   one side 21.1 21.6
   both sides 22.9 30.4

Rocker Arm Cover Gasket, Replace or Reseal (B)
 FX35
  one side 2.6 2.7
  both sides 3.2 3.3
 FX45
  one side 2.0 2.1
  both sides 3.5 3.6
 QX4
  1997-00
   right side .... .7 .8
   left side 3.1 3.2
   both sides 3.3 3.4
  2001-03
   one side 3.5 3.6
   both sides 4.0 4.1
 QX56
  one side .... .8 .9
  both sides 1.2 1.3

Valve Job (A)
 FX35
  2WD 26.4 26.9
  4WD 27.5 28.0
 FX45 .... 27.8 28.3
 QX4
  1997-00 .... 14.5 15.0
  2001-03
   2WD 26.8 27.3
   4WD 27.3 27.8
 QX56
  2WD 27.7 28.2
  4WD 29.4 29.9

Valve Springs and/or Oil Seals, Replace (B)
 FX35
  one side 20.3 20.5
  both sides 23.4 23.6
 FX45
  one side 21.3 21.5
  both sides 23.6 23.8
 QX4
  1997-00
   right side 2.5 2.7
   left side 4.7 4.9
   both sides 6.4 6.6

 2001-03
  one side 18.2 18.4
  both sides 20.9 21.1
 QX56
  2WD
   one side 20.0 20.2
   both sides 22.2 22.4
  4WD
   one side 21.7 21.9
   both sides 24.0 24.5

CAMSHAFT
Camshaft, Replace (A)
 FX35
  2WD
   one side 15.8 16.0
   both sides 16.9 17.1
  4WD
   one side 17.5 17.7
   both sides 18.4 18.6
 FX45
  one side 18.5 18.7
  both sides 19.7 19.9
 QX4
  1997-00
   2WD
    right side 4.4 4.6
    left side 8.6 8.8
    both sides 10.7 10.9
   4WD
    right side 4.9 5.1
    left side 8.8 9.0
    both sides 11.2 11.4
  2001-03
   2WD
    one side 15.6 15.8
    both sides 16.9 17.1
   4WD
    one side 17.2 17.4
    both sides 18.5 18.7
 QX56
  2WD
   one side 19.0 19.2
   both sides 20.4 20.6
  4WD
   one side 20.7 20.9
   both sides 22.0 22.2

Camshaft Seal, Replace (B)
 1997-00 QX4
  one side 3.6 3.8
  both sides 3.8 4.0

Timing Belt, Replace (B)
 1997-00 QX4 .... 3.3 3.8
 w/AC add .... .2 .2
 w/PS add .... .5 .5
 Replace gears add each .3 .3

Timing Chain, Replace (B)
 FX35 .... 8.5 9.0
 FX45 .... 17.6 18.1

 2001-03 QX4
  2WD 13.3 13.8
  4WD 15.1 15.6
 QX56
  2WD 16.7 17.2
  4WD 18.4 18.9

Timing Chain Tensioner, Replace (B)
 FX35
  upper
   2WD 14.8 15.1
   4WD 16.6 16.9
  lower 9 1.2
 FX45
  upper 2.0 2.3
  lower 1.5 1.8
 2001-03 QX4
  upper
   2WD 15.6 15.9
   4WD 17.3 17.6
  lower 1.4 1.7
 QX56
  upper or lower 9 1.2

Timing Cover and/or Gasket, Replace (B)
 FX35 .... 5.6 5.8
 FX45 .... 17.1 17.3
 QX4
  1997-00 .... 3.3 3.5
  2001-03
   2WD 10.9 11.1
   4WD 12.7 12.9
 QX56
  2WD 16.0 16.3
  4WD 17.8 18.1
 w/AC 97-00 QX4 add ..... .2 .2
 w/PS 97-00 QX4 add ..... .5 .5

CRANK & PISTONS
Connecting Rod Bearings, Replace (A)
Includes: Check bearing oil clearances.
 FX35
  2WD 20.2 20.7
  4WD 22.0 22.5
 FX45 .... 23.6 24.1
 QX4
  1997-00
   2WD 17.8 18.3
   4WD 18.4 18.9
  2001-03
   2WD 23.7 24.2
   4WD 24.2 24.7
 QX56
  2WD 22.9 23.4
  4WD 24.5 25.0

FX35 : FX45 : QX4 : QX56 INF-35

	LABOR TIME	SEVERE SERVICE
Crankshaft & Main Bearings, Replace (A)		
Includes: Engine R&R.		
FX35		
2WD	23.1	23.6
4WD	25.0	25.5
FX45	25.5	26.0
QX4		
1997-00		
2WD	15.2	15.7
4WD	17.0	17.5
2001-03		
2WD	26.7	27.2
4WD	27.2	27.7
QX56		
2WD	27.2	27.7
4WD	27.7	28.2
w/AC 97-00 QX4 add	.6	.6
w/PS 97-00 QX4 add	.5	.5
Crankshaft Front Oil Seal, Replace (B)		
FX35	1.4	1.6
FX45	2.1	2.3
QX4		
1997-00	3.2	3.4
2001-03	1.7	1.9
QX56	2.8	3.0
Crankshaft Pulley, Replace (B)		
FX35	1.4	1.6
FX45	1.9	2.1
QX4		
1997-00	1.2	1.4
2001-03	1.6	1.8
QX56	2.9	3.1
Pistons or Connecting Rods, Replace (A)		
Includes: Ridge reaming, cylinder wall deglazing, installing new rings and rod bearings, engine tune-up.		
FX35		
2WD	23.3	24.2
4WD	25.3	26.2
FX45	26.6	27.5
QX4		
1997-00		
2WD	18.9	19.8
4WD	22.3	23.2
2001-03		
2WD	27.0	27.9
4WD	27.5	28.0
QX56		
2WD	26.1	27.0
4WD	27.8	28.7
Rear Main Oil Seal, Replace (B)		
Includes: R&R trans. when necessary.		
FX35		
2WD	5.5	6.1
4WD	6.9	7.5
FX45	6.1	6.7

	LABOR TIME	SEVERE SERVICE
QX4		
1997-00		
2WD		
AT	6.7	7.3
MT	5.3	5.9
4WD		
AT	11.9	12.5
MT	10.9	11.5
2001-03		
2WD		
AT	9.0	9.6
MT	5.3	5.9
4WD		
AT	10.3	10.9
MT	10.9	11.5
QX56		
2WD	4.6	5.2
4WD	5.7	6.3
Rings, Replace (A)		
Includes: Ridge reaming, cylinder wall deglazing, installing new rings, engine tune-up.		
FX35		
2WD	23.3	24.2
4WD	25.3	26.2
FX45	26.6	27.5
QX4		
1997-00		
2WD	18.9	19.8
4WD	22.3	23.2
2001-03		
2WD	27.0	27.9
4WD	27.5	28.0
QX56		
2WD	26.1	27.0
4WD	27.8	28.7

ENGINE LUBRICATION

	LABOR TIME	SEVERE SERVICE
Oil Pressure Gauge (Engine), Replace (B)		
Includes: Pressure test.		
Exc. below	.6	.6
2001-03 QX4	.9	.9
Oil Pan and/or Gasket, Replace (B)		
FX35		
upper		
2WD	8.1	8.3
4WD	10.2	10.4
lower	1.1	1.3
FX45	8.1	8.3
QX4		
1997-00		
2WD	3.6	3.8
4WD	7.5	7.7
2001-03		
upper		
2WD	5.2	5.4
4WD	7.5	7.7
lower	.8	1.0

	LABOR TIME	SEVERE SERVICE
QX56		
upper		
2WD	14.9	15.1
4WD	16.7	16.9
lower	.8	1.0
Oil Pump, Replace (B)		
FX35	10.8	11.3
QX4		
1997-00	7.5	8.0
2001-03	11.8	12.3
QX56	16.6	17.1
Recondition pump add	.5	.5

AUTOMATIC TRANSMISSION

SERVICE TRANSMISSION INSTALLED

	LABOR TIME	SEVERE SERVICE
Accumulator Assy., Replace (B)		
QX4	1.9	2.1
Drain and Refill Unit (B)		
QX4	1.3	1.3
Oil Pressure Test (B)		
All Models	.9	.9
Electronic Control Unit, Replace (B)		
FX35	1.5	1.5
FX45	2.2	2.2
QX4, QX56	.5	.5
Extension Housing &/or Gasket, Replace (B)		
FX35		
2WD	2.2	2.4
4WD	3.6	3.8
FX45	3.6	3.8
QX4	1.6	1.8
QX56		
2WD	1.6	1.8
4WD	4.9	5.1
Extension Housing Oil Seal, Replace (B)		
FX35		
2WD	1.3	1.4
4WD	2.9	3.0
FX45	2.9	3.0
QX4	.8	.9
QX56		
2WD	.8	.9
4WD	4.3	4.5
Inhibitor Switch, Replace (B)		
FX35		
2WD	1.7	1.8
4WD	2.5	2.6
FX45	2.5	2.6
QX4	.8	.9
QX56	1.9	2.0
Oil Pan Gasket, Replace (B)		
All Models	1.3	1.4

INF-36 FX35 : FX45 : QX4 : QX56

	LABOR TIME	SEVERE SERVICE
Parking Pawl, Replace (B)		
FX35		
2WD	2.2	2.4
4WD	3.6	3.8
FX45	3.6	3.8
QX4	1.7	1.9
QX56		
2WD	1.6	1.8
4WD	4.9	5.1
Speedometer Driven Gear and/or Seal, Replace (B)		
QX4 2WD	.9	1.0
Valve Body Assy., R&R and Recondition (A)		
FX35		
2WD	1.7	2.2
4WD	2.5	3.0
FX45	2.5	3.0
QX4	2.7	3.2
QX56	1.9	2.4
Valve Body Assy., Replace (B)		
FX35		
2WD	1.7	2.0
4WD	2.5	2.8
FX45	2.5	2.8
QX4	1.7	2.0
QX56	1.9	2.1

SERVICE TRANSMISSION REMOVED
Transmission R&R included unless otherwise noted.

	LABOR TIME	SEVERE SERVICE
Flywheel (Flexplate), Replace (B)		
FX35		
2WD	5.5	5.7
4WD	6.9	7.1
FX45	6.0	6.2
QX4		
1997-00		
2WD	6.6	6.8
4WD	11.9	12.1
2001-03		
2WD	5.8	6.0
4WD	7.3	7.5
QX56		
2WD	4.3	4.5
4WD	5.5	5.7
Front Oil Pump, Replace (B)		
FX35		
2WD	6.0	6.2
4WD	7.5	7.7
FX45	6.3	6.5
QX4		
1997-00		
2WD	7.3	7.5
4WD	11.8	12.0
2001-03		
2WD	6.9	7.1
4WD	9.0	9.2

	LABOR TIME	SEVERE SERVICE
QX56		
2WD	4.9	5.1
4WD	6.9	7.1
Torque Converter, Replace (B)		
FX35		
2WD	5.5	5.7
4WD	7.0	7.2
FX45	5.8	6.0
QX4		
1997-00		
2WD	6.4	6.6
4WD	11.5	11.7
2001-03		
2WD	6.0	6.2
4WD	8.1	8.3
QX56		
2WD	4.4	4.6
4WD	6.3	6.5
Transmission Assy., R&R and Recondition (A)		
FX35		
2WD	9.9	10.8
4WD	11.3	12.2
FX45	10.3	11.2
QX4		
1997-00		
2WD	15.2	16.1
4WD	19.7	20.6
2001-03		
2WD	14.6	15.5
4WD	16.6	17.5
QX56		
2WD	10.2	11.1
4WD	12.1	13.0
Flush oil cooler and lines add	.5	.7
Transmission Assy., Replace (B)		
FX35		
2WD	6.0	6.2
4WD	7.5	7.7
FX45	6.3	6.5
QX4		
1997-00		
2WD	6.6	6.8
4WD	11.6	11.8
2001-03		
2WD	6.4	6.6
4WD	8.5	8.7
QX56		
2WD	4.3	4.5
4WD	6.3	6.5
Transmission Assy., Reseal (B)		
FX35		
2WD	9.0	9.6
4WD	10.5	11.1
FX45	9.4	10.0
QX4		
1997-00		
2WD	11.2	11.8
4WD	16.3	17.2

	LABOR TIME	SEVERE SERVICE
2000-03		
2WD	10.8	11.7
4WD	12.8	13.7
QX56		
2WD	8.7	9.6
4WD	10.6	11.5

TRANSFER CASE

	LABOR TIME	SEVERE SERVICE
Companion Flange, Replace (B)		
FX35, FX45	1.7	1.9
QX4, QX56	1.0	1.2
Front or Rear Cover Seals, Replace (B)		
FX35, FX45 one	2.1	2.3
QX4		
front	5.3	5.5
rear	1.0	1.2
QX56		
front	4.3	4.5
rear	.8	1.0
Speedometer Drive Gear, Replace (B)		
QX4	1.0	1.1
Transfer Case, R&R and Recondition (A)		
QX4	12.1	12.7
QX56	11.0	11.6
Transfer Case, Replace (B)		
FX35, FX45	2.7	2.9
QX4	5.2	5.4
QX56	4.3	4.5
Transfer Case Front Cover, R&R or Replace (B)		
QX4	5.3	5.5
QX56	4.3	4.5

SHIFT LINKAGE
AUTOMATIC TRANSMISSION

	LABOR TIME	SEVERE SERVICE
Gear Selector Lever, Replace (B)		
FX35, FX45	.9	.9
QX4	.6	.6
QX56	.5	.5
Manual Shaft &/or Parking Rod, Replace (B)		
FX35	3.3	3.5
FX45	4.0	4.2
QX4	1.9	2.1
QX56	2.4	2.6

DRIVELINE
DRIVESHAFT

	LABOR TIME	SEVERE SERVICE
Center Support Bearing, Replace (B)		
FX35, FX45	1.7	1.9
QX4	1.1	1.3
QX56	.5	.7

	LABOR TIME	SEVERE SERVICE

Driveshaft, R&R or Replace (B)
FX35, FX45
  front 2.1   2.3
  rear 1.3   1.5
QX4
  front 8   1.0
  rear 6   .8
QX56 8   1.0

DRIVEAXLE

Axle Shaft Oil Seal, Replace (B)
Rear axle
  one side 1.4   1.6
  both sides 2.2   2.4

Differential Assy., R&R and Recondition (A)
FX35, FX45
  front axle 9.4   10.3
  rear axle 6.1   7.0
QX4
  front axle 8.4   9.3
  rear axle
    standard 6.3   7.2
    limited slip 7.3   8.2
QX56
  front axle 6.4   7.3
  rear axle 8.8   9.7

Differential Assy., Replace (B)
FX35, FX45
  front axle 4.9   5.4
  rear axle 3.0   3.5
QX4
  front axle 4.7   5.2
  rear axle 3.6   4.1
QX56
  front axle 2.1   2.5
  rear axle 4.0   4.5

Differential Case, R&R and Recondition (A)
FX35, FX45
  front axle 7.2   7.7
  rear axle 4.9   5.4
QX4
  front axle 6.7   7.2
  rear axle
    standard 5.3   5.8
    limited slip 6.1   6.6
QX56
  front axle 4.9   5.4
  rear axle 7.3   7.8

Differential Cover Gasket, Replace (B)
Front axle
  FX35, FX45 .... 5.2   5.4
  QX4 .... 2.9   3.1
  QX56 .... 2.4   2.6
Rear axle
  FX35, FX45 .... 2.5   2.7
  QX56 .... 2.2   2.4

Pinion Shaft Oil Seal, Replace (B)
FX35, FX45
  front axle 4.1   4.3
QX4
  front axle 1.1   1.3
  rear axle 1.3   1.5
QX56
  front axle 1.4   1.6
  rear axle 1.9   2.1

Rear Axle/Half Shaft, Replace (B)
FX35, FX45
  one 8   1.0
  both 1.3   1.5
QX4
  one 2.1   2.3
  both 3.3   3.5
QX56
  one side 1.0   1.2
  both sides 1.6   1.8

Ring Gear & Pinion Set, Replace (B)
FX35, FX45
  front axle 8.4   8.9
  rear axle 4.7   5.2
QX4
  front axle 7.3   7.8
  rear axle 5.5   6.0
QX56
  front axle 4.9   5.4
  rear axle 8.4   8.9

Side Flange Oil Seals, Replace (B)
Front axle
  FX35, FX45
    one or both sides . 6.1   6.3
  QX4
    one side 2.4   2.6
    both sides 3.3   3.5
  QX56
    one or both sides . 2.2   2.4

FRONT HALFSHAFTS

Front Axle Shaft Boot, Replace (B)
FX35, FX45
  one side 3.2   3.4
  both sides 5.0   5.2
QX4
  one side 2.4   2.6
  both sides 4.1   4.3
QX56
  one side 2.1   2.3
  both sides 3.6   3.8

Front Halfshaft, R&R (B)
FX35, FX45
  one side 2.4   2.6
  both sides 3.2   3.4
QX4
  one side 1.6   1.8
  both sides 2.5   2.7
QX56
  one side 1.4   1.6
  both sides 2.4   2.6

BRAKES
ANTI-LOCK

Diagnose Anti-Lock Brake System Component Each (A)
All Models 7   .7

Retrieve Fault Codes (A)
All Models 3   .3

Actuator Assy., Replace (B)
Includes: System bleeding.
FX35, FX45 .... 2.1   2.1
QX4 .... 1.5   1.5
QX56 .... 2.2   2.2

Control Module, Replace (B)
FX35, FX45 .... 1.8   1.8
QX4 5   .5
QX56 .... 1.9   1.9

Front Sensor Rotor Assy., Replace (B)
FX35, FX45
  one 1.3   1.3
  both 2.1   2.1
QX4
  one 1.1   1.1
  both 1.9   1.9
QX56
  one 1.9   1.9
  both 3.2   3.2

Front Sensor Assy., Replace (B)
One 8   .8
Both 1.1   1.1

Rear Sensor Rotor Assy., Replace (B)
FX35, FX45
  one 1.0   1.0
  both 1.4   1.4
QX4
  one 1.3   1.3
  both 2.2   2.2
QX56
  one 1.7   1.7
  both 2.5   2.5

Rear Sensor Assy., Replace (B)
Exc. below
  one 5   .5
  both 6   .6
QX56
  one 1.6   1.6
  both 2.5   2.5

SYSTEM

Bleed Brakes (B)
Includes: Add fluid.
All Models 1.0   1.0

Brake Hose (Flexible), Replace (B)
Includes: System bleeding.
FX35, FX45
  one 1.0   1.1
  both 1.3   1.4

INF-38 FX35 : FX45 : QX4 : QX56

	LABOR TIME	SEVERE SERVICE
Brake Hose (Flexible), Replace (B)		
QX4		
one	.6	.7
both	1.0	1.1
QX56		
one	.8	.9
both	1.3	1.4
Load Sensing Valve, Replace (B)		
Includes: System bleeding.		
QX4	1.0	1.0
Master Cylinder, R&R and Recondition (A)		
Includes: System bleeding.		
FX35, FX45	2.0	2.3
QX4	1.5	1.8
QX56	1.7	1.9
Master Cylinder, Replace (B)		
Includes: System bleeding.		
FX35, FX45	1.6	1.8
QX4	.8	1.0
QX56	1.5	1.7
Power Booster Unit, Replace (B)		
FX35, FX45, QX4	2.1	2.3
QX56	3.3	3.5

SERVICE BRAKES

	LABOR TIME	SEVERE SERVICE
Brake Drum, Replace (B)		
One	.6	.7
Both	.8	.9
Caliper Assy., Replace (B)		
Includes: System bleeding.		
One	1.1	1.2
Both	1.7	1.8
Caliper Assy., R&R and Recondition (A)		
Includes: System bleeding.		
One	1.6	1.9
Both	2.7	3.0
Front Disc Brake Rotor, Replace (B)		
FX35, FX45		
one	.6	.7
both	1.0	1.1
QX4		
one	1.3	1.4
both	2.2	2.3
QX56		
one	1.0	1.1
both	1.4	1.5
Pads and/or Shoes, Replace (B)		
Front disc	1.0	1.1
Rear disc	1.1	1.1
Rear drum	1.4	1.5
All four disc wheels	1.8	1.8
All four disc/drum	2.3	2.4
Rear Disc Brake Rotor, Replace (B)		
FX35, FX45		
one	1.0	1.1
both	1.1	1.2
QX56		
one	1.1	1.2
both	1.4	1.5

	LABOR TIME	SEVERE SERVICE
COMBINATION ADD-ONS		
Replace		
caliper add	.3	.3
wheel cyl. add each	.2	.2
Resurface		
brake drum add each	.6	.6
disc rotor add each	.6	.6
Wheel Cylinder, R&R and Recondition (B)		
Includes: System bleeding.		
One	1.4	1.7
Both	2.2	2.5
Wheel Cylinder, Replace (B)		
Includes: System bleeding.		
One	1.3	1.4
Both	1.9	2.0

PARKING BRAKE

	LABOR TIME	SEVERE SERVICE
Parking Brake Apply Actuator, Replace (B)		
FX35, FX45	1.1	1.1
QX4, QX56	.7	.7
Parking Brake Apply Warning Indicator Switch, Replace (B)		
QX4, QX56	.5	.5
Parking Brake Cable, Adjust (C)		
All Models	.5	.6
Parking Brake Cable, Replace (B)		
FX35, FX45		
front	2.5	2.7
rear		
one	1.1	1.3
both	1.6	1.8
QX4		
front	.8	1.0
rear		
one	1.3	1.5
both	1.9	2.1
QX56		
front	2.9	3.1
rear		
one	2.2	2.4
both	3.6	3.8
Parking Brake Rotor, Replace (B)		
One	1.1	1.3
Both	1.4	1.5
Parking Brake Shoes, Replace (B)		
FX35, FX45	1.3	1.4
QX56	3.3	3.4

FRONT SUSPENSION

Unless otherwise noted, time given does not include alignment.

	LABOR TIME	SEVERE SERVICE
Front & Rear Alignment, Check & Adjust (A)		
FX35, FX45, QX56	1.9	2.1
Front Toe, Adjust (B)		
FX35, QX4	.8	.9
FX45, QX56	1.3	1.4
Rear Toe, Adjust (B)		
FX35, FX45, QX56	1.3	1.5

	LABOR TIME	SEVERE SERVICE
4WD Free Wheeling Hub, Replace (B)		
Includes: Grease seal and bearing.		
FX35, FX45		
one side	1.3	1.6
both sides	2.1	2.4
QX4		
one side	2.5	2.8
both sides	4.9	5.2
QX56		
one side	1.7	2.0
both sides	3.2	3.5
Coil Spring, Replace (B)		
Includes: Alignment.		
FX35, FX45		
one side	2.2	2.4
both sides	3.5	3.7
QX4		
one side	2.1	2.3
both sides	3.6	3.8
QX56		
one side	1.1	1.3
both sides	2.1	2.3
Front Actuator Assy., Replace (B)		
One	.3	.3
Both	.5	.5
Front Hub, Replace (B)		
FX35, FX45		
one side	1.3	1.5
both sides	2.1	2.3
QX4		
one side	1.7	1.9
both sides	3.2	3.4
QX56		
2WD		
one side	1.0	1.2
both sides	1.6	1.8
AWD		
one side	1.7	1.9
both sides	3.2	3.4
Front Hub Grease Seal, Replace (B)		
FX35, FX45		
one side	1.3	1.5
both sides	2.1	2.3
QX4		
one side	2.5	2.7
both sides	4.9	5.1
QX56		
one side	1.7	2.0
both sides	3.2	3.5
Front Shock Absorbers, Replace (B)		
QX56		
one	1.0	1.2
both	1.7	1.9
Front Spindle, Replace (B)		
FX35, FX45		
one side	2.2	2.5
both sides	3.3	3.6
QX4		
one side	2.2	2.5
both sides	4.0	4.3

FX35 : FX45 : QX4 : QX56 INF-39

	LABOR TIME	SEVERE SERVICE

QX56
  one side 4.6   4.9
  both sides 6.1   6.4
Lower Ball Joint, Replace (B)
Includes: Adjust toe.
  QX4
    one 1.6   1.8
    both 2.2   2.4
  QX56
    one 3.5   3.7
    both 4.1   4.3
Selector Switch, Replace (B)
  QX4 3   .3
Stabilizer Bar, Replace (B)
  Exc. below 1.4   1.6
  QX56 8   1.0
Stabilizer Bar Connecting Rods &/or Bushings, Replace (B)
  Exc. below
    one side 8   .9
    both sides 1.0   1.1
  QX56
    one or both sides 5   .6
Stabilizer Bar Bushings, Replace (B)
  FX35, FX45, QX4 . . . . 1.1   1.3
Strut Assy., Replace (B)
Includes: Alignment.
  FX35, FX45
    one 1.7   1.8
    both 2.5   2.6
  QX4
    one side 1.9   2.0
    both sides 3.3   3.4
Suspension Crossmember, Replace (B)
Includes: Alignment.
  FX35, FX45 . . . . 4.3   4.8
  QX4 . . . . 7.8   8.3
Transverse Links, Replace (B)
Includes: Alignment.
  FX35, FX45
    one 1.6   1.8
    both 2.1   2.3
  QX4
    one 1.1   1.3
    both 1.7   1.9
  QX56
    upper link
      one 1.1   1.3
      both 1.9   2.1
    lower link
      one 3.5   3.7
      both 4.1   4.3
Upper Ball Joint, Replace (B)
  QX56
    one 1.1   1.3
    both 1.9   2.1

REAR SUSPENSION
Lower Link, Replace (B)
  FX35, FX45
    front
      one 2.2   2.5
      both 2.9   3.2
    rear
      one 1.6   1.9
      both 1.9   2.2
  QX4
    one 6   .9
    both 1.0   1.3
  QX56
    front
      one 4.1   4.4
      both 5.0   5.3
    rear
      one 3.5   3.8
      both 4.0   4.3
Rear Coil Spring, Replace (B)
  FX35, FX45
    one side 1.6   1.8
    both sides 1.7   1.9
  QX4
    one side 3.0   3.2
    both sides 3.2   3.4
  QX56
    one side 8   1.0
    both sides 1.3   1.5
Shock Absorbers, Replace (B)
  FX35, FX45
    one side 2.1   2.2
    both sides 2.5   2.6
  QX4
    one side 6   .7
    both sides 1.0   1.1
  QX56
    one side 8   .9
    both sides 1.3   1.4
Stabilizer Bar, Replace (B)
  FX35, FX45 5   .7
  QX4 6   .8
  QX56 8   1.0
Stabilizer Connecting Rod, Replace (B)
  FX35, FX45
    one or both sides 5   .6
  QX4
    one side 6   .7
    both sides 8   .9
  QX56
    one or both sides 5   .6
Stabilizer Bar Bushings, Replace (B)
  All Models 6   .8
Upper Arm/Bushings, Replace (B)
  FX35, FX45
    one 2.7   3.0
    both 3.8   4.1
  QX56
    one or both 4.7   5.0

Upper Link, Replace (B)
  QX4
    one 5   .8
    both 8   1.1

STEERING
AIR BAGS
Diagnose Air Bag System Component Each (A)
  All Models 6   .6
Retrieve Fault Codes (A)
  All Models 3   .3
Air Bag Assy., Replace (B)
Includes: Diagnosis.
  FX35, FX45
    curtain
      one 3.2   3.2
      both 3.6   3.6
    driver 5   .5
    passenger 7   .7
    side
      one 1.1   1.1
      both 1.9   1.9
  QX4
    curtain
      one 3.1   3.1
      both 3.5   3.5
    driver side 5   .5
    passenger side 2.8   2.8
    side
      one 5   .5
      both 7   .7
  QX56
    curtain
      one 4.6   4.6
      both 4.9   4.9
    driver 6   .6
    passenger 3.5   3.5
    side
      one 1.5   1.5
      both 2.4   2.4
Air Bag On/Off Switch, Replace (B)
  QX4 8   .8
Air Bag Sensors, Replace (B)
  FX35, FX45
    center crash 6   .6
    side
      one 6   .6
      both 8   .8
  QX4, QX56 side
    center crash 6   .6
    side
      one 6   .6
      both 1.2   1.2
Air Bag Spiral Cable, Replace (B)
Includes: Diagnosis.
  Exc. below 7   .7
  QX56 . . . . 1.4   1.4

FX35 : FX45 : QX4 : QX56

	LABOR TIME	SEVERE SERVICE
Sensing & Diagnostic Module (SDM), Replace (B)		
Exc. below	.8	.8
QX56	1.2	1.2

POWER RACK & PINION
Gear Assy., Adjust (On Vehicles) (B)
- All Models 5   .7

Inner Tie Rod, Replace (B)
Includes: Reset toe.
- FX35, FX45
  - one side 1.9   2.1
  - both sides 2.5   2.7
- QX4
  - one side 1.1   1.3
  - both sides 1.9   2.1
- QX56
  - one side 1.9   2.1
  - both sides 2.9   3.1

Outer Tie Rod, Replace (B)
Includes: Reset toe.
- FX35, FX45
  - one side 1.6   1.8
  - both sides 1.7   1.9
- QX4
  - one side 7   .9
  - both sides 1.3   1.5
- QX56
  - one side 1.3   1.5
  - both sides 1.6   1.8

Power Steering Hoses, Replace (B)
Includes: System bleeding.
- Exc. below 2.0   2.2
- QX56 .... 2.7   2.9

Power Steering Switch Assy., Replace (B)
- All Models 6   .7

Rack & Pinion Assy., R&R and Recondition (A)
Includes: Reset toe.
- FX35, FX45, QX4 .... 4.6   5.1

Rack & Pinion Assy., Replace (B)
Includes: Alignment charges.
- FX35, FX45, QX4 .... 3.2   3.5
- QX56 .... 3.6   3.9

Rack Boots, Replace (B)
Includes: Toe adjustment.
- FX35, FX45
  - one side 1.7   1.9
  - both sides 2.2   2.4
- QX4
  - one side 1.0   1.2
  - both sides 1.6   1.8
- QX56
  - one side 1.9   2.1
  - both sides 2.9   3.1

Steering Column, Replace (B)
- FX35, FX45 .... 2.6   2.6
- QX4, QX56 .... 1.4   1.4

	LABOR TIME	SEVERE SERVICE
Steering Column Lower Joint, Replace (B)		
All Models	.5	.7
Steering Lock Assy., Replace (B)		
All Models	1.2	1.2
Steering Oil Cooler, Replace (B)		
FX35, FX45	1.2	1.3
QX56	.8	.9
Steering Pump, R&R and Recondition (A)		
FX35, FX45	2.3	2.8
QX4, QX56	2.6	3.1
Steering Pump, Replace (B)		
All Models	1.5	1.7
Steering Wheel, Replace (B)		
All Models	.7	.7

HEATING & AIR CONDITIONING

When more than one component requires replacement where evacuation/recovery and recharging is already included, deduct 1.0 hour for each additional component from the time listed.

Evacuate/Recover and Recharge System (B)
- All Models 1.0   1.0

AC Hoses Replace (B)
Includes: Evacuate/recover and recharge.
- High pressure
  - FX35, FX45 .... 1.8   2.0
  - QX4
    - 1997-00 .... 1.4   1.6
    - 2001-03 .... 2.4   2.6
  - QX56 .... 4.6   4.8
- Low Pressure
  - FX35, FX45, QX4 .. 1.7   1.9
  - QX56 .... 3.3   3.5

Ambient Sensor, Replace (B)
- All Models 5   .5

Blower Motor, Replace (B)
- All Models 5   .7

Compressor Assy., Replace (B)
Includes: Transfer parts. Evacuate, recover/recycle and charge system.
- FX35, FX45 .... 1.9   2.2
- QX4 .... 2.3   2.6
- QX56 .... 2.4   2.7

Condenser Assy., Replace (B)
Includes: Evacuate/recover and recharge.
- FX35, FX45, QX4 .... 2.6   2.7
- QX56 .... 3.1   3.4

Evaporator Core, Replace (B)
Includes: Evacuate/recover and recharge.
- FX35, FX45 .... 2.0   2.2

	LABOR TIME	SEVERE SERVICE
QX4	1.5	1.7
QX56	1.2	1.4
Expansion Valve, Replace (B)		
Includes: Evacuate/recover and recharge.		
FX35, FX45	1.9	2.0
QX4	1.6	1.7
QX56	7.4	7.5
Heater and AC Control Assy., Replace (B)		
All Models	.6	.6
Heater Blower Motor Relay, Replace (B)		
All Models	.6	.6
Heater Core, R&R or Replace (B)		
Includes: R&I evaporator. Evacuate/recover and recharge.		
FX35, FX45	4.5	4.8
QX4	4.1	4.4
QX56	7.0	7.3
Heater Hoses, Replace (B)		
FX35	1.2	1.4
FX45	2.1	2.4
QX4	.6	.7
QX56	1.4	1.6
In-Vehicle Sensor, Replace (B)		
FX35, FX45, QX4	.5	.5
QX56	1.4	1.4
Low Pressure and/or Cycling Switch, Replace (B)		
Includes: Evacuate/recover and recharge.		
FX35, FX45	.9	.9
QX4, QX56	1.2	1.2
Magnetic Clutch Assy., Replace (B)		
Includes: Evacuate/recover and recharge.		
FX35, FX45	2.1	2.3
QX4		
1997-00	4.4	4.6
2000-03	2.9	3.1
QX56	2.9	3.1
Receiver/Drier Assy., Replace (B)		
Includes: Evacuate/recover and recharge.		
FX35, FX45	1.2	1.4
QX4	1.7	1.9
QX56	3.1	3.3
Sunload Sensor, Replace (B)		
FX35, FX45, QX4	.3	.3

WIPERS & SPEEDOMETER
Power Antenna, Replace (B)
- QX4 .... 1.0   1.0

Radio, R&R (B)
- All Models 8   .8

Rear Window Washer Motor, Replace (B)
- QX4, QX56 3   .3

FX35 : FX45 : QX4 : QX56 INF-41

	LABOR TIME	SEVERE SERVICE
Rear Window Wiper Motor, Replace (B)		
All Models	.6	.8
Rear Window Wiper & Washer Switch, Replace (B)		
QX4, QX56	.5	.5
Speedometer Head, R&R or Replace (B)		
All Models	1.1	1.1
Windshield Washer Pump, Replace (B)		
FX35, FX45	1.6	1.9
QX4	.7	.9
QX56	.5	.7
Windshield Wiper & Washer Switch, Replace (B)		
All Models	.5	.5
Windshield Wiper Interval Relay, Replace (B)		
FX35, FX45	.8	.8
QX56	.5	.5
Windshield Wiper Motor, Replace (B)		
All Models	.8	1.0

LAMPS & SWITCHES

	LABOR TIME	SEVERE SERVICE
Back-Up Lamp Bulb, Replace (B)		
FX35, FX45, QX4	.3	.3
QX56	.6	.6
Combination Switch Assy., Replace (B)		
FX35, FX45	.3	.3
QX4	.7	.7
QX56	1.2	1.2
Cornering Lamp Bulb, Replace (C)		
All Models	.3	.3
Daylight Running Lamp Module, Replace (B)		
QX4	.5	.5
QX56	1.1	1.1

	LABOR TIME	SEVERE SERVICE
Halogen Headlamp Bulb, Replace (B)		
FX35, FX45, QX56	.6	.6
QX4	.3	.3
Hazard Warning Switch, Replace (B)		
All Models	.3	.3
Headlamp & Turn Signal Switch, Replace (B)		
FX35, FX45	.5	.5
QX4	.7	.7
QX56	1.2	1.2
Headlamps, Aim (B)		
Two	.4	.4
Four	.6	.6
High Mount Stop Lamp Assy., Replace (B)		
FX35, FX45, QX4	.5	.5
Horns, Replace (B)		
All Models	.5	.5
Horn Relay, Replace (B)		
All Models	.5	.5
Inhibitor Switch, Replace (B)		
FX35		
2WD	1.7	1.8
4WD	2.5	2.6
FX45	2.5	2.6
QX4	.8	.9
QX56	1.9	2.0
Parking Brake Warning Lamp Switch, Replace (B)		
QX4	.5	.5
Rear Combination Lamp Bulb, Replace (C)		
All Models	.3	.3
Stop Lamp Switch, Replace (B)		
All Models	.5	.5
Turn Signal or Hazard Warning Flasher, Replace (B)		
All Models	.5	.5

	LABOR TIME	SEVERE SERVICE
Xenon Headlamp, Replace (B)		
FX35, FX45	2.0	2.0
QX56	1.5	1.5

BODY

	LABOR TIME	SEVERE SERVICE
Front Door Lock, Replace (B)		
All Models	1.1	1.1
Hood Hinge, Replace (B)		
FX35, FX45		
one or both	1.2	1.2
QX4		
one	.7	.7
both	.9	.9
QX56		
one	2.4	2.4
both	3.3	3.3
Hood Lock, Replace (B)		
All Models	.5	.5
Hood Release Cable, Replace (B)		
FX35, FX45	1.5	1.7
QX4, QX56	1.1	1.2
Lock Striker Plate, Replace (B)		
All Models	.5	.5
Rear Door Lock, Replace (B)		
FX35, FX45	.9	.9
QX4, QX56	.7	.7
Window Regulator, Replace (B)		
FX35, FX45	.9	.9
QX4		
front	.8	.8
rear	1.2	1.2
QX56	1.1	1.1
w/AC QX4 add	.2	.2
Window Regulator Motor, Replace (B)		
FX35, FX45	.9	.9
QX4, QX56	1.2	1.4

NOTES

ISU

I-Mark : Impulse : Stylus

SYSTEM INDEX

- **MAINTENANCE** ISU-2
- **CHARGING** ISU-2
- **STARTING** ISU-2
- **CRUISE CONTROL** ISU-2
- **IGNITION** ISU-2
- **EMISSIONS** ISU-3
- **FUEL** ISU-3
- **EXHAUST** ISU-4
- **ENGINE COOLING** ISU-4
- **ENGINE** ISU-5
  - Assembly ISU-5
  - Cylinder Head ISU-5
  - Camshaft ISU-6
  - Crank & Pistons ISU-6
  - Engine Lubrication ISU-7
- **CLUTCH** ISU-7
- **MANUAL TRANSAXLE** ISU-8
- **MANUAL TRANSMISSION** .. ISU-8
- **AUTO TRANSAXLE** ISU-8
- **AUTO TRANSMISSION** ISU-8
- **TRANSFER CASE** ISU-8
- **SHIFT LINKAGE** ISU-9
- **DRIVELINE** ISU-9
- **BRAKES** ISU-9
- **FRONT SUSPENSION** ISU-10
- **REAR SUSPENSION** ISU-11
- **STEERING** ISU-11
- **HEATING & AC** ISU-12
- **WIPERS & SPEEDOMETER** ISU-12
- **LAMPS & SWITCHES** ISU-13
- **BODY** ISU-13

OPERATIONS INDEX

A
- AC Hoses ISU-12
- Air Conditioning ISU-12
- Alignment ISU-10
- Alternator (Generator) ... ISU-2
- Antenna ISU-12
- Anti-Lock Brakes ISU-9

B
- Ball Joint ISU-10
- Battery Cables ISU-2
- Bleed Brake System ISU-9
- Blower Motor ISU-12
- Brake Disc ISU-10
- Brake Drum ISU-9
- Brake Hose ISU-9
- Brake Pads and/or Shoes ISU-10

C
- Camshaft ISU-6
- Catalytic Converter ISU-4
- Coolant Temperature (ECT) Sensor ISU-3

Crankshaft ISU-6
Crankshaft Sensor ISU-2
Cruise Control ISU-2
CV Joint ISU-9
Cylinder Head ISU-5

D
- Differential ISU-9
- Distributor ISU-2
- Drive Belt ISU-2
- Driveshaft ISU-9

E
- EGR ISU-3
- Electronic Control Module (ECM/PCM) ISU-3
- Engine ISU-5
- Engine Lubrication ISU-7
- Engine Mounts ISU-5
- Evaporator ISU-12
- Exhaust ISU-4
- Exhaust Manifold ISU-4
- Expansion Valve ISU-12

F
- Flexplate ISU-8
- Flywheel ISU-8
- Fuel Injection ISU-4
- Fuel Pump ISU-3
- Fuel Vapor Canister ISU-3

G
- Generator ISU-2
- Glow Plug ISU-2

H
- Halfshaft ISU-9
- Headlamp ISU-13
- Heater Core ISU-12
- Horn ISU-13

I
- Idle Air Control (IAC) Valve ... ISU-4
- Ignition Coil ISU-2
- Ignition Module ISU-2
- Ignition Switch ISU-2
- Injection Pump ISU-4
- Inner Tie Rod ISU-11
- Intake Air Temperature (IAT) Sensor ISU-3
- Intake Manifold ISU-4

K
- Knock Sensor ISU-3

L
- Lower Control Arm ISU-10

M
- Manifold Absolute Pressure (MAP) Sensor ISU-3
- Mass Air Flow (MAF) Sensor ISU-3
- Master Cylinder ISU-9
- Muffler ISU-4
- Multifunction Lever/Switch ... ISU-13

N
- Neutral Safety Switch ISU-13

O
- Oil Pan ISU-7
- Oil Pump ISU-7
- Outer Tie Rod ISU-11
- Oxygen Sensor ISU-3

P
- Parking Brake ISU-10
- Pistons ISU-7

R
- Radiator ISU-4
- Radiator Hoses ISU-5
- Radio ISU-12
- Rear Main Oil Seal ISU-7

S
- Shock Absorber/Strut, Front .. ISU-10
- Shock Absorber/Strut, Rear .. ISU-11
- Spark Plug Cables ISU-2
- Spark Plugs ISU-2
- Spring, Front Coil ISU-10
- Spring, Rear Coil ISU-11
- Starter ISU-2
- Steering Wheel ISU-11

T
- Thermostat ISU-4
- Throttle Body ISU-4
- Throttle Position Sensor (TPS) ISU-3
- Timing Belt ISU-6
- Timing Chain ISU-6
- Torque Converter ISU-8

U
- Upper Control Arm ISU-11

V
- Valve Body ISU-8
- Valve Cover Gasket ISU-6
- Valve Job ISU-6
- Vehicle Speed Sensor ISU-3

W
- Water Pump ISU-5
- Wheel Balance ISU-2
- Wheel Cylinder ISU-10
- Window Regulator ISU-13
- Windshield Washer Pump ISU-13
- Windshield Wiper Motor ISU-13

ISU-1

ISU-2 I-MARK : IMPULSE : STYLUS

	LABOR TIME	SEVERE SERVICE
MAINTENANCE		
Air Cleaner Filter Element, Replace (C)		
1981-93	.2	.3
Drive Belt, Adjust (B)		
One	.4	.5
each addl. add	.1	.2
Drive Belt, Replace (B)		
One	.6	.7
each addl. add	.1	.2
Fuel Filter, Replace (B)		
I-Mark	.4	.4
Impulse, Stylus		
1983-89	.5	.5
1990-93	.7	.7
Halogen Headlamp Bulb, Replace (C)		
1988-93	.3	.3
Oil & Filter, Change (C)		
Includes: Correct all fluid levels.		
1981-93	.3	.4
Sealed Beam Headlamp, Replace (C)		
I-Mark	.4	.4
Impulse, Stylus	.3	.3
Tire, Replace (C)		
Includes: Dismount old tire and mount new tire to rim.		
1981-89	.5	.5
Wheel, Balance (B)		
One	.3	.3
each addl. add	.2	.2
CHARGING		
Alternator Circuits, Test (B)		
Includes: Test component output.		
1981-93	.5	.5
Alternator, R&R and Recondition (A)		
I-Mark		
FWD		
SOHC	2.6	3.1
DOHC	2.9	3.4
RWD	2.3	2.8
Impulse, Stylus	2.2	2.7
w/1.8L DOHC add	.2	.2
w/turbo add	.5	.5
Alternator Assy., Replace (B)		
Includes: Pulley transfer.		
I-Mark		
FWD		
SOHC	.6	.8
DOHC	.9	1.1
RWD	.9	1.1
Impulse, Stylus		
1983-89	.6	.8
1990-93	.9	1.1
w/1.8L DOHC add	.2	.2
w/turbo add	.5	.5

	LABOR TIME	SEVERE SERVICE
Front Alternator Bearing, Replace (B)		
1981-93	1.0	1.3
w/turbo add	.5	.5
Voltage Regulator, Replace (B)		
1981-93		
external	.5	.5
internal	1.7	2.0
w/turbo add	.5	.5
Voltmeter and/or Ammeter Gauge, Replace (B)		
I-Mark	.7	.7
Impulse, Stylus		
1983-89	1.9	1.9
1990-93	.9	.9
STARTING		
Starter Draw Test (On Car) (B)		
1981-93	.3	.3
Battery Cables, Replace (C)		
Positive	.6	.7
Negative	.4	.5
Starter, R&R and Recondition (A)		
1981-93	1.8	2.3
w/turbo add	.5	.5
Starter Assy., Replace (B)		
1981-93	1.0	1.2
w/turbo add	.5	.5
Starter Drive Assy., Replace (B)		
Includes: Starter R&R.		
1981-93	1.4	1.7
w/turbo add	.5	.5
Starter Relay, Replace (B)		
1981-93	.4	.6
Starter Solenoid and/or Switch, Replace (B)		
Includes: Starter R&R.		
1981-93	1.3	1.6
w/turbo add	.5	.5
CRUISE CONTROL		
Actuator Assy., Replace (B)		
1981-93	.5	.5
Control Actuator Cable, Replace (B)		
1981-93	.5	.5
Control Clutch Switch, Replace (B)		
Impulse, Stylus	.5	.5
Control Controller Module, Replace (B)		
1981-93	.5	.5
Control On/Off Switch, Replace (B)		
1981-93	1.9	1.9
Control Relay, Replace (B)		
I-Mark	.4	.4
Control Resume Solenoid, Replace (B)		
Impulse	1.9	1.9

	LABOR TIME	SEVERE SERVICE
Control Sensor, Replace (B)		
I-Mark	.9	.9
Impulse, Stylus	1.0	1.0
Control Switch, Replace (B)		
1981-93	.4	.4
IGNITION		
Ignition Timing, Reset (B)		
1981-93	.3	.3
w/turbo add	.1	.1
Crank Sensor, Replace (B)		
Impulse	.5	.7
Distributor, R&R and Recondition (A)		
Includes: Reset base ignition timing.		
1981-93	1.0	1.3
Distributor, Replace (B)		
Includes: Reset base ignition timing.		
1981-93	.5	.7
Distributor Cap and/or Rotor, Replace (B)		
1981-93	.3	.3
DIS Control Unit, Replace (B)		
Impulse turbo	.8	1.0
Glow Plug, Replace (B)		
Diesel		
one	.5	.6
all	.9	1.0
Glow Plug Relay, Replace (B)		
Diesel	.4	.5
Ignition Coil, Replace (B)		
1981-93	.5	.5
w/turbo add	.2	.2
Replace igniter add	.1	.3
Ignition Lock Cylinder and/or Buzzer Switch, Replace (B)		
I-Mark		
1983-87	.5	.5
1988-89	.8	.8
Impulse, Stylus		
1983-87	.5	.5
1988-89	1.6	1.6
1990-93	.8	.8
Ignition Module, Replace (B)		
Impulse turbo	.6	.6
Ignition Switch, Replace (B)		
I-Mark		
1981-87	.5	.5
1988-89	.7	.7
Impulse, Stylus		
1983-87	.5	.5
1988-93	1.6	1.6
Spark Plug (Ignition) Cables, Replace (B)		
1981-93	.6	.7
w/turbo add	.2	.2
Spark Plugs, Replace (B)		
1981-93	.6	.7

I-MARK : IMPULSE : STYLUS ISU-3

	LABOR TIME	SEVERE SERVICE

Vacuum Advance Unit, Replace (B)
Includes: Reset base ignition timing.
1981-87 .... .7 .7
1988-93 .... .5 .5

EMISSIONS

Accelerator Switch, Replace (B)
1983-87
  I-Mark, Impulse ...... .5 .5
Air Cleaner Vacuum Motor, Replace (B)
1981-89 .... .5 .5
Air Cleaner Temperature Sensor, Replace (B)
1981-93 .... .5 .5
Air Distribution Manifold, Replace (B)
1981-93 .... 1.5 1.5
Air Pump Assy., Replace (B)
1981-93 .... 1.0 1.3
Air Injection Check Valve, Replace (B)
1981-93 .... .5 .5
Barometric Switch, Replace (B)
1981-93 .... .4 .4
Check/Relief Valve, Replace (B)
1981-93 .... .4 .4
Coolant Temperature (ECT) Sensor, Replace (B)
  I-Mark .... .5 .5
  Impulse, Stylus .... .3 .3
Diverter and/or Switching Valve, Replace (B)
1981-93 .... .5 .5
EFE Heater, Replace (B)
1981-93 .... 1.3 1.3
EGR Valve, Replace (B)
  Gasoline
    I-Mark
      FWD .... .7 .9
      RWD 9 1.1
    Impulse, Stylus ...... .6 .8
  Diesel
    I-Mark, Impulse 8 1.0
  w/DOHC add .... .1 .1
  w/turbo add .... .5 .5
EGR Controller, Replace (B)
  Diesel .... .5 .5
EGR Pressure Feedback Transducer, Replace (B)
Does not include testing.
  I-Mark .... .6 .6
  Impulse, Stylus .... .5 .5
  w/turbo add .... .5 .5
Electronic Control Module (ECM/PCM), Replace (B)
1981-93 .... .7 .7
Engine Speed Sensor, Replace (B)
1983-87
  I-Mark, Impulse ...... .5 .5

Fast Idle Solenoid, Replace (B)
1981-93 .... .5 .5
Fuel Tank Pressure Control Valve, Replace (B)
1988-89 I-Mark .... .5 .5
Fuel Vapor Canister, Replace (B)
1981-93 .... .5 .5
w/turbo add .... .5 .5
Idle Compensator, Replace (B)
1981-93 .... .5 .5
Idle and Wide Open Throttle Switch, Replace (B)
1981-93 .... .5 .5
Knock Sensor, Replace (B)
  I-Mark .... .5 .7
  Impulse, Stylus
    1983-89 .... .5 .7
    1990-93 .... .7 .9
Manifold Absolute Pressure (MAP) Sensor, Replace (B)
1983-93 .... .5 .5
Mass Air Flow (MAF) Sensor, Replace (B)
  Impulse .... .5 .5
MAT Sensor, Replace (B)
1983-93 .... .5 .5
Mixture Control Valve, Replace (B)
1981-93 .... .7 .7
Oxygen Sensor, Replace (B)
  I-Mark .... .7 .9
  Impulse, Stylus .... .5 .7
PROM, Replace (B)
  I-Mark, Impulse turbo . 1.0 1.0
Solenoid and Vacuum Regulator, Replace (B)
1981-93 .... .5 .5
Temperature Switch, Replace (B)
1981-93 .... .3 .3
Thermal Vacuum Valve, Replace (B)
1981-93 .... .3 .3
Throttle Position Sensor (TPS), Replace (B)
  I-Mark
    SOHC .... .6 .6
    DOHC 1.0 1.0
Vacuum Delay Valve, Replace (B)
1981-93 .... .4 .4
w/turbo add .... .5 .5
Vacuum Motor, Replace (B)
1981-93 .... .5 .5
Vacuum Switching Valve, Replace (B)
  Diesel .... .5 .5
Vapor Separator, Replace (B)
  I-Mark .... .5 .5
  Impulse, Stylus .... .7 .7

Vehicle Speed Sensor, Replace (B)
  I-Mark 1.0 1.0
  Impulse, Stylus
    1983-89 .... 1.3 1.3
    1990-93 .... 1.1 1.1

FUEL
CARBURETOR
Carburetor, Adjust (On Vehicle) (B)
  I-Mark .... .3 .3
Carburetor, R&R and Clean or Recondition (B)
Includes: Adjustments.
  I-Mark 2.6 2.6
Carburetor, Replace (B)
Includes: Adjustments.
  I-Mark
    FWD 1.6 1.6
    RWD 1.2 1.2
Choke, Replace (B)
  I-Mark .... .9 .9
Float or Needle and Seat, Replace (B)
  I-Mark 1.0 1.0

DELIVERY
Fuel Pump, Test (B)
1981-93 .... .3 .3
Fuel Filter, Replace (B)
  I-Mark .... .4 .4
  Impulse, Stylus
    1983-89 .... .5 .5
    1990-93 .... .7 .7
Fuel Gauge (Dash), Replace (B)
  I-Mark 1.0 1.0
  Impulse, Stylus
    1983-89 .... 1.9 1.9
    1990-93 .... .9 .9
Fuel Gauge (Tank), Replace (B)
  I-Mark 7 1.0
  Impulse, Stylus 1.5 1.8
Fuel Pump, Replace (B)
  I-Mark
    FWD 1.7 2.0
    RWD .... .8 .8
  Impulse, Stylus
    1983-88 .... .9 .9
    1989-93 .... 1.6 1.9
Fuel Tank, Replace (B)
Includes: Drain and refill.
  I-Mark
    FWD 1.5 1.8
    RWD 1.4 1.7
  Impulse, Stylus 1.4 1.7

ISU-4 I-MARK : IMPULSE : STYLUS

	LABOR TIME	SEVERE SERVICE
Intake Manifold and/or Gasket, Replace (B)		
Includes: Adjustments.		
Gasoline		
I-Mark		
FWD		
SOHC	1.3	1.6
DOHC	2.9	3.2
RWD	3.1	3.4
Impulse, Stylus		
1983-89	1.8	2.1
1990-93		
SOHC	2.3	2.6
DOHC		
1.6L	3.9	4.2
1.8L	3.3	3.6
Diesel	1.5	1.8
w/turbo add	1.7	1.7

INJECTION (DIESEL)
Idle Speed, Adjust (C)		
1981-87	.5	.6
Injection Timing, Check & Adjust (B)		
1981-87	1.1	1.3
w/AC add	.2	.2
Fuel Injection Pump, Replace (B)		
1981-87	4.2	4.4
High Pressure Fuel Lines, Replace (B)		
One	.7	.8
All	1.1	1.3
Injector Nozzle and/or Seal, Replace (B)		
One	.8	.9
All	1.2	1.4
Clean nozzles add each	.3	.3
Vacuum Pump, Replace (B)		
1981-87	.8	1.0
Recondition pump add	.4	.7

INJECTION (GAS)
Idle Speed, Adjust (B)		
1981-93	.3	.4
Air Injection Manifold, Replace (B)		
1983-93	1.2	1.5
Air Regulator, Replace (B)		
Impulse	.9	.9
Electronic Control Unit, Replace (B)		
Impulse, Stylus	.5	.5
Fuel Injection Relay, Replace (B)		
Impulse, Stylus	.4	.4
Fuel Injectors, Replace (B)		
I-Mark	2.2	2.4
Impulse, Stylus		
1983-89	1.7	1.9

	LABOR TIME	SEVERE SERVICE
1990-93		
SOHC	1.4	1.6
DOHC	1.8	2.0
w/turbo add	1.5	1.5
Clean and test injectors add	.8	.8
Fuel Pressure Regulator, Replace (B)		
I-Mark	.6	.8
Impulse, Stylus		
1983-89	.5	.7
1990-93	.8	1.0
Fuel Rail, Replace (B)		
I-Mark	1.2	1.4
Impulse, Stylus		
1983-89	1.4	1.6
1990-93	1.2	1.4
Idle Air Control (IAC) Valve, Replace (B)		
I-Mark		
SOHC	.5	.5
DOHC	1.5	1.5
Impulse, Stylus	.5	.5
Throttle Body Assy., Replace (B)		
I-Mark	.9	.9
Impulse, Stylus		
1983-89	.8	.8
1990-93		
SOHC	.7	.7
DOHC	1.0	1.0
Replace seal or gasket add	.3	.3
Throttle Valve Switch, Replace (B)		
Impulse	.6	.6

TURBOCHARGER
Intercooler, Replace (B)		
1991-92 Impulse	.9	1.1
Stepping Motor, Replace (B)		
1986-89 Impulse	.7	.7
Stepping Motor Cable, Replace (B)		
1986-89 Impulse	.5	.6
TCS Controller, Replace (B)		
1986-89 Impulse	.5	.5
Turbo Boost Pressure Sensor, Replace (B)		
I-Mark	.5	.5
Turbocharger Assy., R&R or Replace (B)		
I-Mark	5.0	5.3
1986-92 Impulse	4.5	4.8
Wastegate Actuator, Replace (B)		
I-Mark	.9	1.1
1986-92 Impulse	.7	1.0
Wastegate Solenoid, Replace (B)		
I-Mark	.8	.8
Water Valve Assy., Replace (B)		
1986-89 Impulse	.5	.5

	LABOR TIME	SEVERE SERVICE
EXHAUST		
Catalytic Converter, Replace (B)		
1981-89	1.0	1.1
1990-93	1.3	1.4
Exhaust Manifold or Gasket, Replace (B)		
I-Mark		
gasoline	1.5	1.8
diesel	1.0	1.3
Impulse, Stylus		
SOHC	1.3	1.6
DOHC	1.7	2.0
w/turbo add	1.7	2.0
Exhaust Pipe & Resonator, Replace (B)		
I-Mark	1.2	1.3
Impulse	.8	.9
Front Exhaust Pipe, Replace (B)		
I-Mark	1.2	1.3
Impulse, Stylus		
1983-89	.8	.9
1990-93	.9	1.0
Intermediate Exhaust Pipe, Replace (B)		
I-Mark	1.2	1.3
Impulse, Stylus	.8	.9
Muffler, Replace (B)		
I-Mark		
FWD	.8	.9
RWD	1.0	1.1
Impulse, Stylus		
1983-89	1.0	1.1
1990-93	.8	.9

ENGINE COOLING
Coolant Thermostat, Replace (B)		
I-Mark		
gasoline	.6	.7
diesel	.9	1.0
Impulse, Stylus		
1983-89	.7	.8
1990-93	.4	.5
w/1.8L DOHC add	.2	.2
w/turbo add	.5	.5
Radiator Assy., R&R or Replace (B)		
I-Mark	1.0	1.1
Impulse, Stylus		
1983-89	1.6	1.7
1990-93	1.2	1.3
w/AT add	.2	.2
w/turbo add	.3	.3
Radiator Fan and/or Fan Motor, Replace (B)		
I-Mark	.8	1.0
Impulse, Stylus		
1983-87	1.0	1.2
1988-93	.7	.9
w/turbo add	.3	.3

I-MARK : IMPULSE : STYLUS ISU-5

	LABOR TIME	SEVERE SERVICE
Radiator Fan Motor Relay, Replace (B)		
1981-93	.4	.4
Radiator Fan Motor Switch (Coolant Temp.), Replace (B)		
1981-93	.5	.5
Radiator Hoses, Replace (B)		
Includes: Refill with proper coolant mix.		
Upper	.5	.5
Lower		
I-Mark, Stylus		
one	.6	.7
both	.9	1.0
Impulse		
one	.8	.9
both	1.1	1.2
w/turbo add	.3	.3
Temperature Gauge (Dash), Replace (B)		
I-Mark	.9	.9
Impulse, Stylus		
1983-89	1.9	1.9
1990-93	.8	.8
Temperature Gauge (Engine), Replace (B)		
1981-93	.6	.6
Water Pump and/or Gasket, Replace (B)		
Includes: Refill with proper coolant mix.		
I-Mark		
FWD	3.0	3.4
RWD		
gasoline	1.5	1.9
diesel	1.5	1.9
Impulse, Stylus		
1983-89	2.0	2.4
1990-93		
SOHC	1.6	2.0
DOHC		
1.6L	2.2	2.6
1.8L	2.9	3.3
w/AC add	.2	.2
w/PS add	.2	.2
w/turbo add	.5	.5

ENGINE

ASSEMBLY

Times shown are for OEM assemblies. Time to replace assemblies from aftermarket rebuilders may vary.

Engine Assy., R&I (B)
Does not include parts or component transfer.

	LABOR TIME	SEVERE SERVICE
Gasoline		
I-Mark		
1981-87	6.4	6.6
1988-89		
SOHC	6.6	6.8
DOHC	7.3	7.5
Impulse, Stylus		
1983-89	7.2	7.4
1990-93		
SOHC	4.9	5.1
DOHC		
1.6L	6.8	7.0
1.8L	7.5	7.7
Diesel	6.5	6.7
w/AC add	.3	.3
w/AT add	.3	.3
w/PS add	.3	.3
w/turbo add	.5	.5

Engine Assy., R&R and Recondition (A)
Includes: Replacing rings, rod and main bearings, cylinder head reconditioning and engine tune-up.

	LABOR TIME	SEVERE SERVICE
I-Mark		
SOHC	24.8	25.0
DOHC	27.9	28.1
Impulse, Stylus		
1983-89	25.7	25.9
1990-93		
SOHC	21.9	22.1
DOHC	24.4	24.6
1.6L	24.4	24.6
1.8L	25.8	26.0
w/AC add	.3	.3
w/AT add	.3	.3
w/PS add	.3	.3
w/turbo add	.5	.5

Engine Assy. (Short), Replace (B)
Assembly consists of cylinder block, piston assemblies, crankshaft, camshaft, timing chain and gears. Does not include cylinder heads. Operation Includes: R&R engine, transfer necessary parts and all necessary adjustments.

	LABOR TIME	SEVERE SERVICE
Gasoline		
I-Mark		
SOHC	12.9	13.1
DOHC	13.9	14.1
Impulse, Stylus		
1983-89	13.6	13.8
1990-93		
SOHC	9.8	10.0
DOHC		
1.6L	12.3	12.5
1.8L	12.9	13.1
Diesel	15.5	15.7
w/AC add	.3	.3
w/AT add	.3	.3
w/PS add	.3	.3
w/turbo add	.5	.5

Engine Mounts, Replace (B)

	LABOR TIME	SEVERE SERVICE
Gasoline		
front	1.0	1.2
right		
I-Mark		
FWD	.6	.8
RWD	.9	1.1
Impulse	.8	1.0
Stylus	.7	.9
left	.8	1.0
rear		
I-Mark RWD	.9	1.1
Impulse		
1983-89	.8	1.0
1990-92	.9	1.1
Diesel		
one	.6	.8
both	1.2	1.4
rear	.9	1.1
w/1.8L DOHC add	.2	.2
w/turbo add	.2	.2

CYLINDER HEAD

Compression Test (B)

	LABOR TIME	SEVERE SERVICE
Gasoline	.8	.9
Diesel	1.0	1.1

Valves, Adjust (B)

	LABOR TIME	SEVERE SERVICE
Diesel	1.3	1.5

Cylinder Head, Replace (B)
Includes: Parts transfer, adjustments.

	LABOR TIME	SEVERE SERVICE
Gasoline		
I-Mark		
1981-87		
FWD	5.2	5.4
RWD	4.9	5.1
1988-89	6.0	6.2
Impulse, Stylus		
1983-87	5.7	5.9
1988-89	6.5	6.7
1990-93		
SOHC	6.2	6.4
DOHC	11.6	11.8
Diesel	6.3	7.0
w/AC add	.3	.3
w/PS add	.3	.3
w/turbo add	1.0	1.0

Cylinder Head Gasket, Replace (B)

	LABOR TIME	SEVERE SERVICE
Gasoline		
I-Mark		
1981-87		
FWD	4.4	4.6
RWD	4.1	4.3
1988-89	4.7	4.9
Impulse, Stylus		
1983-87	4.9	5.2
1988-89	5.7	5.9
1990-93		
SOHC	2.9	3.1
DOHC	5.0	5.2

ISU-6 I-MARK : IMPULSE : STYLUS

	LABOR TIME	SEVERE SERVICE
Cylinder Head Gasket, Replace (B)		
Diesel	5.9	6.1
w/AC add	.3	.3
w/PS add	.3	.3
w/turbo add	1.0	1.0
Rocker Arms, Replace (B)		
Gasoline	1.7	1.9
Diesel	1.3	1.5
w/turbo add	.5	.5
Valve Adjuster, Replace (B)		
1981-93	.9	1.1
w/DOHC add	.5	.5
Valve Cover Gasket, Replace or Reseal (B)		
Gasoline	.5	.7
Diesel	.7	.9
w/DOHC add	.4	.4
w/turbo add	.5	.5
Valve Lash Adjuster, Replace (B)		
1988-92 Impulse		
one or all	3.5	3.7
Valve Job (A)		
I-Mark		
1981-87		
FWD	6.5	6.7
RWD	5.9	6.1
1988-89		
SOHC	6.7	6.9
DOHC	9.8	10.0
Impulse, Stylus		
1983-87	6.8	7.0
1988-89	7.6	7.8
1990-93		
SOHC	7.6	7.8
DOHC	14.5	14.7
w/AC add	.3	.3
w/PS add	.3	.3
w/turbo add	1.0	1.0
Valve Lifters, Replace (B)		
I-Mark	3.0	3.2
Impulse, Stylus	4.1	4.3
Valve Springs and/or Oil Seals, Replace (B)		
Gasoline		
1981-87		
one cyl.	1.3	1.5
each addl. add	.2	.3
1988-93		
SOHC	2.6	2.8
DOHC	5.2	5.4
Diesel		
one cyl.	1.7	1.9
each addl. add	.3	.4
w/turbo add	1.0	1.0
R&R cyl. head if necessary add	2.5	3.0

	LABOR TIME	SEVERE SERVICE
CAMSHAFT		
Camshaft, Replace (A)		
Gasoline		
1981-87	1.8	2.0
1988-93		
SOHC	2.5	2.7
DOHC	3.9	4.0
Diesel	2.8	3.0
w/turbo add	.5	.5
Replace front seal add	.2	.2
Camshaft Timing Gear, Replace (B)		
Diesel	4.5	4.7
Timing Belt and/or Tensioner, Replace (B)		
Gasoline		
I-Mark		
1986-87	8.4	8.6
1988-89	2.6	2.8
Impulse, Stylus		
1986-89	3.5	3.7
1990-93		
SOHC	1.7	1.9
DOHC		
1.6L	2.3	2.5
1.8L	2.9	3.1
Diesel	3.5	3.7
w/AC add	.3	.3
w/AT add	.3	.3
w/PS add	.3	.3
w/turbo add	.5	.5
Replace tensioner add	.3	.5
Timing Belt/Chain Cover, Replace (B)		
I-Mark		
1981-85	9.5	9.7
1986-87	7.8	8.0
1988-89	2.2	2.4
Impulse, Stylus		
1983-85	9.5	9.7
1986-89	3.2	3.4
1990-93		
upper	.4	.5
lower		
SOHC	.8	1.0
DOHC		
1.6L	1.2	1.4
1.8L	1.9	2.1
w/AC add	.3	.3
w/AT add	.3	.3
w/PS add	.3	.3
w/turbo add	.5	.5
Timing Chain, Gear and Tensioner, Replace (B)		
1981-85 I-Mark	10.5	10.7
1983-85 Impulse	11.4	11.6
w/AC add	.3	.3
w/AT add	.3	.3
Timing Case Cover Seal, Replace (B)		
1981-87		
I-Mark, Impulse	3.5	3.7

	LABOR TIME	SEVERE SERVICE
CRANK & PISTONS		
Connecting Rod Bearings, Replace (A)		
Includes: Check bearing oil clearance.		
Gasoline		
I-Mark		
1981-87		
FWD	7.7	7.9
RWD	7.4	7.6
1988-89		
SOHC	8.0	8.2
DOHC	9.8	10.0
Impulse, Stylus		
1983-89	8.1	8.3
1990-93	3.7	3.9
Diesel	7.3	7.5
w/1.8L DOHC add	.5	.5
w/AC add	.3	.3
w/AT add	.3	.3
w/PS add	.3	.3
w/turbo add	.5	.5
Crankshaft and Main Bearings, Replace (A)		
Includes: Engine R&R, check bearing oil clearance.		
Gasoline		
I-Mark		
1981-87	9.0	9.2
1988-89		
SOHC	9.3	9.5
DOHC	11.2	11.4
Impulse, Stylus		
1983-89	11.8	12.0
1990-93		
SOHC	7.3	7.5
DOHC		
1.6L	9.5	9.7
1.8L	10.6	10.8
Diesel	9.0	9.2
w/AC add	.3	.3
w/AT add	.3	.3
w/PS add	.3	.3
w/turbo add	.5	.5
Crankshaft Pulley or Balancer, Replace (B)		
Gasoline		
I-Mark		
1981-87	1.6	1.8
1988-89	1.9	2.1
Impulse, Stylus		
1983-89	1.6	1.8
1990-93		
SOHC	.9	1.1
DOHC		
1.6L	1.5	1.7
1.8L	1.8	2.0

I-MARK : IMPULSE : STYLUS **ISU-7**

	LABOR TIME	SEVERE SERVICE
Diesel 1.7		1.9
w/AC add 3		.3
w/AT add 3		.3
w/PS add 3		.3
w/turbo add 5		.5

Front Crankshaft Oil Seal, Replace (B)
Gasoline
  I-Mark
    1981-87 . . . . 2.4   2.6
    1988-89 . . . . 2.6   2.8
  Impulse, Stylus
    1983-87 . . . . 2.6   2.8
    1988-89
      1.9L 2.6   2.8
      2.0L 4.0   4.2
    1990-93
      SOHC 1.8   2.0
      DOHC
        1.6L 2.3   2.5
        1.8L 2.9   3.1
  Diesel 3.6   3.8
  w/AC add 3   .3
  w/AT add 3   .3
  w/PS add 3   .3
  w/turbo add 5   .5

Pistons or Connecting Rods, Replace (A)
Includes: Ridge reaming, cylinder wall deglazing, installing new rings and rod bearings, engine tune-up.
Gasoline
  I-Mark
    1981-87
      FWD 12.5   12.7
      RWD 13.2   13.4
    1988-89
      SOHC 11.3   11.5
      DOHC 12.7   12.9
  Impulse, Stylus
    1983-87 . . . . 16.1   16.3
    1988-89 . . . . 14.5   14.7
    1990-93
      SOHC 8.5   8.7
      DOHC
        1.6L 10.6   10.8
        1.8L 14.5   14.6
  Diesel 11.3   11.5
  w/AC add 3   .3
  w/AT add 3   .3
  w/PS add 3   .3
  w/turbo add 5   .5

Rear Main Oil Seal, Replace (B)
Gasoline
  I-Mark
    1981-87 . . . . 3.7   3.9
    1988-89 . . . . 3.9   4.1

	LABOR TIME	SEVERE SERVICE

Impulse, Stylus
  1983-89
    AT 5.8   6.0
    MT 5.5   5.7
  1990-93
    AT 5.4   5.6
    MT 4.7   4.9
Diesel
  AT 3.9   4.1
  MT 3.6   3.8
w/1.8L DOHC add 5   .5
w/turbo add 2.5   2.7

Rings, Replace (A)
Includes: Ridge reaming, cylinder wall deglazing, installing new rings, engine tune-up.
Gasoline
  I-Mark
    1981-87
      FWD 9.7   9.9
      RWD 10.4   10.6
    1988-89
      SOHC 10.1   10.3
      DOHC 11.5   11.7
  Impulse, Stylus
    1983-89 . . . . 13.3   13.5
    1990-93
      SOHC 7.3   7.5
      DOHC
        1.6L 9.4   9.6
        1.8L 10.9   11.1
  Diesel 10.1   10.3
  w/AC add 3   .3
  w/AT add 3   .3
  w/PS add 3   .3
  w/turbo add 5   .5

ENGINE LUBRICATION

Engine Oil Cooler, Replace (B)
  1990-93 6   .8
  w/turbo add 5   .5

Engine Oil Pressure Switch (Sending Unit), Replace (B)
  1981-93 5   .5

Oil Pan and/or Gasket, Replace (B)
Gasoline
  I-Mark
    1981-87 . . . . 6.5   6.7
    1985-89 . . . . 1.6   1.8
  Impulse, Stylus
    1983-89 . . . . 7.3   7.5
    1990-93 . . . . 2.1   2.3
  Diesel 6.5   6.7
  w/1.8L DOHC add 5   .5
  w/AC add 3   .3
  w/AT add 3   .3
  w/PS add 3   .3
  w/turbo add 5   .5

	LABOR TIME	SEVERE SERVICE

Oil Pressure Gauge (Dash), Replace (B)
  I-Mark 8   .8
  Impulse, Stylus
    1983-89 . . . . 1.8   1.8
    1990-93 9   .9

Oil Pump, Replace (B)
Gasoline
  FWD
    I-Mark 2.8   3.0
    Impulse, Stylus
      SOHC 3.5   3.7
      DOHC
        1.6L 4.3   4.5
        1.8L 4.8   5.0
  RWD
    I-Mark 6.8   7.0
    Impulse
      1983-87 . . . . . . 7.6   7.8
      1988-89 . . . . . . 7.8   8.0
  Diesel 3.6   3.8
  w/1.8L DOHC add 5   .5
  w/AC add 3   .3
  w/AT add 3   .3
  w/PS add 3   .3
  w/turbo add 5   .5

CLUTCH

Clutch Hydraulic System, Bleed (B)
  Impulse 5   .5

Clutch Pedal Free Play, Adjust (B)
  1981-93 3   .5

Clutch Assy., Replace (B)
  I-Mark FWD,
    Impulse, Stylus 5.0   5.2
  I-Mark RWD 3.8   4.0
  w/1.8L DOHC add 5   .5
  w/AWD add 3.3   3.5
  Replace pilot bearing add 1   .3

Clutch Control Cable, Replace (B)
  I-Mark
    FWD 6   .8
    RWD 9   1.1
  w/AWD add 4   .6

Clutch Master Cylinder, Replace (B)
Includes: System bleeding.
  Impulse 1.1   1.3

Clutch Release Bearing or Fork, Replace (B)
  I-Mark
    FWD 4.9   5.1
    RWD 3.7   3.9
  w/1.8L DOHC add 5   .5
  w/AWD add 3.3   3.5

Clutch Slave Cylinder, Replace (B)
Includes: System bleeding.
  Impulse 8   1.0

ISU-8   I-MARK : IMPULSE : STYLUS

	LABOR TIME	SEVERE SERVICE
Flywheel, Replace (B)		
I-Mark		
FWD	5.2	5.4
RWD	4.0	4.2
w/1.8L DOHC add	.5	.5
w/AWD add	3.3	3.3

MANUAL TRANSAXLE

	LABOR TIME	SEVERE SERVICE
Final Drive Pinion Case, Replace (B)		
Impulse AWD	1.9	2.1
Final Drive Pinion Oil Seal, Replace (B)		
Impulse AWD	2.0	2.2
Speedometer Driven Gear and/or Seal, Replace (B)		
1985-93	.5	.5
Transaxle Assy., R&R and Recondition (A)		
I-Mark	7.8	8.0
Impulse, Stylus	11.2	11.4
w/1.8L DOHC add	.5	.5
w/AWD add	3.3	3.5
Recondition diff. add	1.5	1.7
Transaxle Assy., R&R or Replace (B)		
I-Mark	4.7	4.9
Impulse, Stylus	4.7	4.9
w/1.8L DOHC add	.5	.5
w/AWD add	3.3	3.5
Transaxle Shift Cover and/or Gasket, Replace (B)		
1985-93	.5	.7

MANUAL TRANSMISSION

	LABOR TIME	SEVERE SERVICE
Extension Housing Oil Seal, Replace (B)		
I-Mark, Impulse	.9	1.1
Transmission Assy., R&R and Recondition (A)		
I-Mark	6.0	6.2
Impulse	6.0	6.2

AUTOMATIC TRANSAXLE

SERVICE TRANSAXLE INSTALLED

	LABOR TIME	SEVERE SERVICE
Drain & Refill Unit (B)		
I-Mark, Stylus	1.0	1.2
Impulse	1.2	1.4
Check Unit for Oil Leaks (B)		
1985-93	.5	.5
Oil Pressure Check (B)		
1985-93	.5	.5
Electronic Control Unit, Replace (B)		
1990-92 Impulse	.7	.7
Governor Assy., R&R or Replace (B)		
1985-93	.6	.9
Recondition assy. add	.2	.5
Intermediate Servo, R&R (B)		
I-Mark, Stylus	1.0	1.2
Recondition servo add	.2	.5

	LABOR TIME	SEVERE SERVICE
Kickdown Solenoid, Replace (B)		
I-Mark, Stylus	.5	.5
Oil Pan and/or Gasket, Replace (B)		
I-Mark, Stylus	1.0	1.2
1990-92 Impulse	1.2	1.4
Shift Lock Controller, Replace (B)		
1990-93	.7	.7
Shift Lock Solenoid & Switch, Replace (B)		
1990-93	.7	.7
Solenoid Valve, Replace (B)		
1990-92 Impulse	1.9	2.1
Speedometer Driven Gear and/or Seal, Replace (B)		
1985-93	.5	.5
Vacuum Modulator, Replace (B)		
I-Mark, Stylus	.6	.6
Valve Body, R&R and Recondition (A)		
I-Mark, Stylus	2.8	3.1
1990-92 Impulse	3.3	3.5
Valve Body, Replace (B)		
I-Mark, Stylus	1.3	1.5
1990-92 Impulse	1.7	1.9

SERVICE TRANSAXLE REMOVED
Transaxle R&R included unless otherwise noted.

	LABOR TIME	SEVERE SERVICE
Front Pump Oil Seal, Replace (B)		
I-Mark	3.6	3.8
Impulse, Stylus	5.5	5.7
Torque Converter, Replace (B)		
I-Mark	3.5	3.7
Impulse, Stylus	5.4	5.6
Torque Converter Drive Plate (Flexplate), Replace (B)		
I-Mark	3.5	3.7
Impulse, Stylus	5.6	5.8
w/1.8L DOHC add	.5	.5
Transaxle Assy., R&R or Replace (B)		
I-Mark	3.3	3.5
Impulse, Stylus	5.3	5.5
Replace assembly add	.5	.7
Transaxle and Converter Assy., R&R and Recondition (A)		
I-Mark	10.1	10.3
Impulse	10.5	10.7
Stylus	9.9	10.1
Recondition diff. add	1.5	1.7

AUTOMATIC TRANSMISSION

SERVICE TRANSMISSION INSTALLED

	LABOR TIME	SEVERE SERVICE
Check Unit for Oil Leaks (C)		
I-Mark, Impulse	.5	.5
Drain and Refill Unit (B)		
I-Mark, Impulse	.6	.6

	LABOR TIME	SEVERE SERVICE
Oil Pressure Check (B)		
I-Mark, Impulse	.5	.5
Extension Housing and/or Gasket, Replace (B)		
I-Mark, Impulse	2.0	2.2
Extension Housing Oil Seal, Replace (B)		
I-Mark, Impulse	1.0	1.2
Governor Assy., Replace or Recondition (B)		
I-Mark, Impulse	2.7	2.9
Manual Shaft Seal and/or Detent Lever, Replace (B)		
I-Mark, Impulse	3.8	4.1
Oil Pan and/or Gasket, Replace (B)		
I-Mark, Impulse	.6	.6
Parking Pawl, Replace (B)		
I-Mark, Impulse	3.5	3.7
Speedometer Driven Gear and/or Seal, Replace (B)		
I-Mark, Impulse	.5	.5
Throttle Valve & Kickdown Cam, Replace (B)		
I-Mark	2.3	2.5
Impulse	1.5	1.7
Valve Body Assy., R&R and Recondition (A)		
I-Mark	3.2	3.4
Impulse	3.9	4.1
Valve Body Assy., Replace (B)		
I-Mark	1.6	1.8
Impulse	2.2	2.4

SERVICE TRANSMISSION REMOVED
Transmission R&R included unless otherwise noted.

	LABOR TIME	SEVERE SERVICE
Front Oil Pump, Replace (B)		
I-Mark, Impulse	4.2	4.4
Replace front seal add	.2	.4
Torque Converter, Replace (B)		
I-Mark, Impulse	4.0	4.2
Torque Converter Drive Plate (Flexplate), Replace (B)		
I-Mark, Impulse	4.1	4.3
Transmission Assy., R&R (B)		
I-Mark, Impulse	3.8	4.0
Replace assembly add	.5	.8
Transmission and Converter, R&R and Recondition (A)		
I-Mark, Impulse	10.5	10.7

TRANSFER CASE

	LABOR TIME	SEVERE SERVICE
Pinion Shaft Seal, Replace (B)		
Impulse	1.3	1.5
Transfer Case, R&I (B)		
Impulse	4.0	4.2

I-MARK : IMPULSE : STYLUS ISU-9

	LABOR TIME	SEVERE SERVICE
Transfer Case, R&R and Reconditioning (A)		
Impulse	9.5	9.7

SHIFT LINKAGE
AUTOMATIC TRANSAXLE/TRANSMISSION
Shift Linkage, Adjust (B)
- 1981-93 .5 .6

Gearshift Control Cable or Rod, Replace (B)
- I-Mark .5 .7
- Impulse, Stylus .8 1.0

Gearshift Selector Lever, Replace (B)
- 1981-89 .5 .5
- 1990-93 .8 .8

Park Lock Cable, Replace (B)
- 1981-93 1.1 1.3

Shift Lever Seal, Replace (B)
- 1981-87 .6 .7

MANUAL TRANSAXLE/TRANSMISSION
Gearshift Lever, Replace (B)
- I-Mark .5 .5
- Impulse, Stylus .9 .9
- *console add* .1 .1

Gearshift Seal and/or Boot, Replace (B)
- 1981-89 .5 .5
- *w/console add* .1 .1

Transaxle Selector Cable, Replace (B)
- 1985-93 1.0 1.2
- *w/AWD add* .4 .4

Transaxle Shift Cable, Replace (B)
- 1985-93 1.1 1.3

DRIVELINE
DRIVESHAFT
Driveshaft, Replace (B)
- 1981-93 .5 .7

Driveshaft and Joints, Replace (B)
- Impulse AWD
  - front 1.5 1.7
  - center coupling 1.9 2.1
  - rear 1.5 1.7
- *Recondition center bearing add* .6 .8

Extension Housing Support, Replace (B)
- I-Mark, Impulse 1.5 1.7

Extension Shaft, R&R or Replace (B)
- I-Mark, Impulse 1.3 1.5
- *Replace*
  - rubber mount add .2 .2
  - shaft assy. add .2 .2
  - shaft bearing add .2 .2
  - shaft cushion bearing add .2 .2

Rear Companion Flange, Replace (B)
- I-Mark, Impulse .5 .7

U-Joints, Recondition (B)
- 1981-93 one .7 .9

DRIVE AXLE
Differential, Drain and Refill (B)
- I-Mark, Impulse .5 .5

Axle Housing Cover or Gasket, Replace (C)
- I-Mark, Impulse .5 .7

Axle Shaft and/or Bearing, Replace (B)
- I-Mark, Impulse
  - one 1.0 1.2
  - both 1.7 1.9

Differential Carrier Bearings, Replace (B)
- I-Mark, Impulse 3.0 3.2

Differential Side & Pinion Gear, Replace (B)
- I-Mark, Impulse 3.4 3.6
- *Replace case assy. add* 1.0 1.2

Final Drive, R&R and Recondition (A)
- Impulse AWD 5.3 5.5

Final Drive, R&R or Replace (A)
- Impulse AWD 2.8 3.0

Final Drive Mounting Bracket, Replace (B)
- Impulse AWD 2.1 2.3

Pinion Shaft Oil Seal, Replace (B)
- I-Mark, Impulse 1.6 1.8

Rear Axle Carrier and/or Gasket, Replace (B)
- I-Mark, Impulse 2.1 2.3

Rear Axle Housing, Replace (B)
- I-Mark, Impulse 5.6 5.8

Ring Gear & Pinion Set, Replace (B)
- I-Mark, Impulse 5.1 5.3
- *w/limited slip add* .4 .4

HALFSHAFTS
Halfshaft, R&R or Replace (B)
- AWD
  - one .8 1.0
  - both 1.3 1.5
- FWD 1.4 1.6
- *Replace*
  - CV-boot FWD add each .5 .7
  - inner CV-joint FWD add each .5 .7

Halfshaft Oil Seal, Replace (B)
- One .9 1.0
- Both 1.2 1.3

BRAKES
ANTI-LOCK
Control Unit, Replace (B)
- 1986-93 .6 .6

Front Sensor Assy., Replace (B)
- 1986-93 .8 .8

Hydraulic Assy., Replace (B)
- 1986-93 1.7 1.9

Load Sensing Proportioning Valve, Replace (B)
- 1986-93 4.0 4.2

Rear Sensor Assy., Replace (B)
- 1986-93 one .8 .8

SYSTEM
Bleed Brakes (B)
Includes: Add fluid.
- 1981-93 .5 .5

Brakes, Adjust (B)
Includes: Refill master cylinder.
- Rear wheels .3 .3

Brake Hose (Flexible), Replace (B)
Includes: System bleeding.
- Front
  - one .7 .8
  - both 1.1 1.3
- Rear
  - one .8 .9
  - each addl. add .4 .5

Brake Proportioning Valve, Replace (B)
- One or both .8 1.0

Combination Valve, Replace (B)
Includes: System bleeding.
- I-Mark RWD .9 1.1
- Impulse, Stylus .7 .9

Master Cylinder, R&R and Rebuild (A)
Includes: System bleeding.
- I-Mark RWD 1.6 1.8
- Impulse, Stylus 1.7 1.9

Master Cylinder, Replace (B)
Includes: System bleeding.
- I-Mark RWD 1.2 1.4
- Impulse, Stylus 1.3 1.5

Power Booster Unit, Replace (B)
- I-Mark
  - FWD 1.3 1.5
  - RWD 1.4 1.6
- Impulse, Stylus 1.5 1.7

Power Booster Vacuum Check Valve, Replace (B)
- 1981-93 .3 .3

SERVICE BRAKES
Brake Drum, Replace (B)
- One .5 .6
- Both .7 .8

ISU-10 I-MARK : IMPULSE : STYLUS

	LABOR TIME	SEVERE SERVICE
Caliper Assy., R&R and Reconditon (A)		
Includes: System bleeding.		
Front		
1981-87		
one	1.5	1.7
both	2.5	2.7
1988-89		
one	1.6	1.8
both	2.9	3.1
1990-93		
one	1.3	1.5
both	2.1	2.3
Rear		
one	1.3	1.5
both	1.9	2.1
Replace brake hose add	.4	.5
Caliper Assy., Replace (B)		
Includes: System bleeding.		
Front		
one	.9	1.1
both	1.6	1.8
Rear		
one	.9	1.1
both	1.5	1.7
Replace brake hose add	.4	.5
Disc Brake Rotor, Replace (B)		
Front		
1981-89		
one	.9	1.1
both	1.5	1.7
1990-93		
one	.5	.6
both	.8	.9
Rear		
one	.6	.7
both	.9	1.0
Pads and/or Shoes, Replace (B)		
Includes: Adjust service and parking brake. System bleeding.		
1981-87		
front disc	1.0	1.2
rear drum	1.5	1.7
four wheels	2.2	2.4
1988-93		
front disc	.9	1.1
rear drum	1.0	1.2
four wheels	1.6	1.8
COMBINATION ADD-ONS		
Replace		
brake drum add	.1	.1
wheel cylinder add	.2	.2
Resurface		
brake drum add	.5	.6
disc rotor add	.5	.6
Self-Adjusting Units, Free-Up or Replace (B)		
One	.5	.6
each addl. add	.3	.4

	LABOR TIME	SEVERE SERVICE
Wheel Cylinder, R&R and Rebuild (B)		
Includes: System bleeding.		
One	1.3	1.5
Both	2.2	2.4
Wheel Cylinder, Replace (B)		
Includes: System bleeding.		
One	.8	1.0
Both	1.5	1.7
PARKING BRAKE		
Parking Brake Cable, Adjust (B)		
1981-93	.3	.5
Parking Brake Apply Actuator, Replace (B)		
1981-93	.8	.8
Parking Brake Apply Warning Indicator Switch, Replace (B)		
I-Mark	.5	.5
Impulse, Stylus		
1983-89	.7	.7
1990-93	.5	.5
Parking Brake Cable, Replace (B)		
I-Mark		
FWD	1.2	1.4
RWD	1.5	1.7
Impulse, Stylus		
1983-87	1.5	1.7
1988-93		
front	1.4	1.6
rear		
one	1.3	1.5
both	2.2	2.4
Parking Brake Shoes, Replace (B)		
Impulse	.9	1.1
FRONT SUSPENSION		
Unless otherwise noted, time given does not include alignment.		
FRONT WHEEL DRIVE		
Align Front End (A)		
1985-93	.8	1.1
Front Strut Coil Spring, Replace (B)		
I-Mark		
one	2.1	2.3
both	3.7	3.9
Impulse, Stylus		
one	1.5	1.7
both	3.0	3.2
Front Torsion Bar, Replace (B)		
1985-93		
one	.6	.8
both	.8	1.0
Lower Ball Joint, Replace (B)		
I-Mark		
one side	.9	1.1
both sides	1.5	1.7
Impulse, Stylus		
one side	.7	.9
both sides	1.2	1.4

	LABOR TIME	SEVERE SERVICE
Lower Control Arm, Replace (B)		
I-Mark		
one side	1.8	2.0
both sides	3.1	3.3
Impulse, Stylus		
one side	1.1	1.3
both sides	1.9	2.1
Replace		
ball joint add each side	.3	.5
bushings add each side	.2	.4
Stabilizer Bar, Replace (B)		
I-Mark	.7	.9
Stylus, Impulse	1.3	1.5
Replace bushings add	.3	.5
Steering Knuckle, Replace (B)		
I-Mark		
one side	2.3	2.5
both sides	4.4	4.6
Impulse, Stylus		
one side	1.2	1.4
both sides	2.2	2.4
Strut, Replace (B)		
I-Mark		
one	2.2	2.4
both	3.9	4.1
Impulse, Stylus		
one	1.7	1.9
both	3.1	3.3
Wheel Bearings and Cups, Replace (B)		
I-Mark		
one side	1.2	1.4
both sides	2.3	2.5
Impulse, Stylus		
one side	1.1	1.3
both sides	2.0	2.2
REAR WHEEL DRIVE		
Align Front End (A)		
1981-89	1.0	1.2
w/AC add	.5	.5
Front Toe, Adjust (B)		
1981-89	.6	.8
Front Shock Absorbers, Replace (B)		
One	.6	.7
Both	.9	1.0
Front Spring, Replace (B)		
One	1.8	2.0
Both	3.5	3.7
Front Strut Bar, Replace (B)		
1981-89	.5	.7
Lower Ball Joint, Replace (B)		
One	1.1	1.3
Both	2.1	2.3

I-MARK : IMPULSE : STYLUS ISU-11

	LABOR TIME	SEVERE SERVICE
Lower Control Arm Assy., Replace (B)		
One	2.3	2.5
Both	4.4	4.6
Replace		
ball joint add each	.3	.5
bushings add		
each side	.3	.5
Stabilizer Bar, Replace (B)		
1981-89	.5	.7
Replace bushings add	.1	.3
Steering Knuckle, Replace (B)		
One	1.7	1.9
Both	3.3	3.5
Upper Ball Joint, Replace (B)		
One	1.2	1.4
Both	2.1	2.3
Upper Control Arm Assy., Replace (B)		
One	2.1	2.3
Both	3.7	3.9
Replace bushings add		
each side	.4	.6
Wheel Bearing and Cups, Replace (B)		
1981-89	1.5	1.7
Wheel Bearing Grease Seal, Replace (B)		
One	.5	.5
Both	.9	.9
Wheel Bearings, Clean & Pack (B)		
1981-89 both wheels	1.5	1.7

REAR SUSPENSION

	LABOR TIME	SEVERE SERVICE
Lateral Link, Replace (B)		
1990-93	1.0	1.2
Rear Axle Housing, R&R or Replace (B)		
1985-93	2.5	2.7
Replace bushings add	1.0	1.2
Rear Control Arm, Replace (B)		
I-Mark, Impulse		
one	.7	.9
both	1.0	1.2
Replace bushings		
add each	.3	.5
Rear Knuckle, Replace (B)		
1990-93	1.7	1.9
Rear Spring, Replace (B)		
I-Mark, Impulse		
one	.9	1.1
both	1.6	1.8
Rear Strut Coil Spring, Replace (B)		
1985-89 I-Mark		
one	.8	1.0
both	1.2	1.4
1990-92 Impulse		
one	1.6	1.8
both	3.0	3.2

	LABOR TIME	SEVERE SERVICE
1991-93 Stylus		
one	1.9	2.1
both	3.6	3.8
Rear Struts, R&R or Replace (B)		
1985-89 I-Mark		
one	.7	.9
both	1.1	1.3
1990-92 Impulse		
one	1.3	1.5
both	2.4	2.6
1991-93 Stylus		
one	1.6	1.8
both	3.0	3.2
Rear Wheel Spindle, Replace (B)		
1985-89 I-Mark	1.7	1.9
Shock Absorbers or Bushings, Replace (B)		
I-Mark, Impulse		
one	.7	.9
both	1.0	1.2
Stabilizer Bar & Bushings, Rear, Replace (B)		
I-Mark	.5	.7
Replace bushings add	.2	.4
Trailing Link, Replace (B)		
1990-93	.9	1.1
Upper Control Arm, Replace (B)		
I-Mark, Impulse		
one	.6	.8
both	1.0	1.2
Replace bushings		
add each	.3	.5
Wheel Bearing, Replace (B)		
1985-89 I-Mark		
one side	1.3	1.5
both sides	2.1	2.3
Wheel Hub, Replace (B)		
1990-93	.9	1.1

STEERING
MANUAL RACK & PINION

	LABOR TIME	SEVERE SERVICE
Column Lock Actuator, Replace (B)		
I-Mark	1.4	1.6
Impulse, Stylus		
1983-89	2.8	3.0
1990-93	2.1	2.3
Horn Contact, Replace (B)		
1981-93	.5	.7
Inner Tie Rod Assy., Replace (B)		
Includes: Reset toe.		
FWD	.8	1.0
RWD	1.3	1.5
Outer Tie Rod End, Replace (B)		
Includes: Reset toe.		
1981-93	.8	1.0

	LABOR TIME	SEVERE SERVICE
Rack & Pinion Assy., R&R or Replace (B)		
Includes: Adjustments.		
1981-87	1.2	1.4
1988-93	3.4	3.6
Steering Wheel, Replace (B)		
1981-93	.5	.7

POWER RACK & PINION

	LABOR TIME	SEVERE SERVICE
Column Lock Actuator, Replace (B)		
I-Mark	1.4	1.6
Impulse, Stylus		
1983-89	2.8	3.0
1990-93	2.1	2.3
Horn Contact or Canceling Cam, Replace (B)		
1981-93	.5	.7
Inner Tie Rod Assy., Replace (B)		
Includes: Reset toe.		
FWD	.8	1.0
RWD	1.3	1.5
Outer Tie Rod Ends, Replace (B)		
Includes: Reset toe.		
1981-93	.8	1.0
Rack & Pinion Assy., R&R and Recondition (A)		
Includes: Reset toe.		
1981-89	3.6	3.8
1990-93	4.7	4.9
Replace tie rod ends add	.3	.5
Rack & Pinion Assy., R&R or Replace (B)		
Includes: Adjustments.		
1981-89	2.4	2.6
1990-93	3.8	4.0
Steering Pump, R&R and Recondition (A)		
Impulse	2.2	2.4
Steering Pump, R&R or Replace (B)		
I-Mark	1.2	1.4
Impulse, Stylus		
1983-89	1.7	1.9
1990-93	1.2	1.4
w/1.6L add	.2	.2
Steering Pump Hoses, Replace (B)		
Does not include purging.		
High pressure	.6	.7
Low pressure	.7	.8
Steering Pump Reservoir, Replace (B)		
1981-87	.4	.5
1988-93	.7	.8

ISU-12 I-MARK : IMPULSE : STYLUS

	LABOR TIME	SEVERE SERVICE

HEATING & AIR CONDITIONING

When more than one component requires replacement where evacuation/recovery and recharging is already included, deduct 1.0 hour for each additional component from the time given.

Evacuate/Recover and Recharge System (B)
- 1981-93 .... 1.0   1.0

AC Hoses, Replace (B)
Includes: Evacuate/recover/ and recharge.
- 1981-89
  - one 1.7   1.9
  - each addl. add 3   .4
- 1990-93
  - liquid line
    - high pressure 1.5   1.7
    - low pressure 2.6   2.8
  - suction hose 1.4   1.6

Blower Motor, Replace (B)
- I-Mark 1.5   1.7
- Impulse, Stylus
  - 1983-87 .... 2.9   3.1
  - 1988-89 9   .9
  - 1990-93 6   .6

Blower Motor Resistor, Replace (B)
- I-Mark
  - FWD 5   .5
  - RWD 7   .7
- Impulse, Stylus 5   .5

Blower Motor Switch, Replace (B)
- I-Mark 5   .5
- Impulse 1.9   1.9

Compressor Assy., R&R and Recondition (A)
Includes: Evacuate/recover/ and recharge.
- I-Mark
  - FWD 2.6   2.8
  - RWD 3.9   4.1
- Impulse, Stylus
  - 1983-89 .... 2.9   3.1
  - 1990-93 .... 2.9   3.1

Compressor Assy., Replace (B)
- I-Mark
  - FWD 1.5   1.7
  - RWD 2.5   2.7
- Impulse, Stylus
  - 1983-89 .... 2.1   2.3
  - 1990-93 .... 1.7   1.9
- w/PS add 3   .3
- w/turbo add 5   .5

Compressor Clutch Assy., Replace (B)
- I-Mark
  - FWD 1.6   1.8
  - RWD 2.6   2.8
- Impulse, Stylus
  - 1983-89 .... 2.2   2.4
  - 1990-93 .... 1.8   2.0
- w/PS add 3   .3
- w/turbo add 5   .5
- Replace pulley bearing add 2   .5

Compressor Front Seal, Replace (B)
- I-Mark
  - FWD 1.7   1.9
  - RWD 2.7   2.9
- Impulse, Stylus
  - 1983-89 .... 2.3   2.5
  - 1990-93 .... 1.9   2.1
- w/PS add 2   .2
- w/turbo add 5   .5

Condenser Assy., Replace (B)
Includes: Evacuate/recover/ and recharge.
- I-Mark
  - FWD 1.6   1.8
  - RWD 2.8   3.0
- Impulse, Stylus
  - 1983-89 .... 2.3   2.5
  - 1990-93 .... 1.6   1.8
- w/turbo add 5   .5

Condenser Cooling Fan, Replace (B)
- Impulse, Stylus 5   .8
- I-Mark 6   .9
- Replace fan motor add 2   .5

Condenser Cooling Fan Relay, Replace (B)
- 1981-93 4   .4

Condenser Pressure Switch, Replace (B)
- I-Mark 1.5   1.7
- Impulse, Stylus 1.3   1.5

Evaporator Core, Replace (B)
Includes: Evacuate/recover/ and recharge.
- 1981-93 .... 3.4   3.6

Evaporator Thermostatic Switch, Replace (B)
- I-Mark
  - FWD 3.1   3.3
  - RWD 3.6   3.8
- Impulse, Stylus
  - 1983-89 .... 2.9   3.1
  - 1990-93 .... 2.1   2.3

Expansion Valve, Replace (B)
Includes: Evacuate/recover/ and recharge.
- 1981-89 .... 3.1   3.3
- 1990-93 .... 1.3   1.5

Fan Switch, Replace (B)
- I-Mark 5   .7

Heater/AC Relay, Replace (B)
- Impulse, Stylus 4   .4

Heater Blower Motor, Replace (B)
- I-Mark
  - gasoline 1.5   1.7
  - diesel 2.1   2.3
- Impulse, Stylus
  - 1983-89 9   .9
  - 1990-93 6   .6

Heater Blower Motor Resistor, Replace (B)
- I-Mark
  - FWD 5   .5
  - RWD 7   .7
- Impulse, Stylus 5   .5

Heater Blower Motor Switch, Replace (B)
- I-Mark 5   .5

Heater Core, R&R or Replace (B)
- I-Mark 5.2   5.4
- Impulse, Stylus
  - 1983-89 .... 5.8   6.0
  - 1990-93
    - w/AC 6.1   6.3
    - w/o AC 4.9   5.1

Heater Hoses, Replace (B)
- Each 7   .8

Receiver/Drier Assy., Replace (B)
Includes: Evacuate/recover/ and recharge.
- I-Mark
  - FWD 1.3   1.5
  - RWD 2.1   2.3
- Impulse, Stylus
  - 1983-89 .... 1.3   1.5
  - 1990-93 .... 1.6   1.8

Water Control Valve, Replace (B)
- I-Mark 5   .7
- Impulse, Stylus 6   .8

WIPERS & SPEEDOMETER

Antenna, Replace (B)
- 1990-93 9   .9

Radio, R&R (B)
- 1981-89 9   .9
- 1990-93 5   .5

Rear Window Washer Motor, Replace (B)
- I-Mark, Impulse 4   .4

Rear Window Wiper Motor, Replace (B)
- 1981-93 7   .9

I-MARK : IMPULSE : STYLUS ISU-13

	LABOR TIME	SEVERE SERVICE
Rear Window Wiper and Washer Switch, Replace (B)		
I-Mark		
FWD	.9	.9
RWD	.6	.6
Impulse, Stylus		
1983-89	1.9	1.9
1990-93	.5	.5
Speedometer Cable and Casing, Replace (B)		
I-Mark	1.0	1.2
Impulse, Stylus	.9	1.1
Speedometer Head, R&R or Replace (B)		
I-Mark	.9	.9
Impulse, Stylus		
1983-89	1.9	1.9
1990-93	.8	.8
Windshield Washer Motor, Replace (B)		
Impulse, Stylus	.5	.5
Windshield Washer Pump, Replace (B)		
1981-93	.5	.5
Windshield Washer Switch, Replace (B)		
I-Mark		
FWD	.6	.6
RWD	1.1	1.1
Impulse, Stylus		
1983-89	2.0	2.0
1990-93	.9	.9
Windshield Wiper Interval Relay, Replace (B)		
1983-89	1.9	1.9
1990-93	.5	.5
Windshield Wiper Motor, Replace (B)		
I-Mark		
FWD	.7	1.0
RWD	1.1	1.4
Impulse, Stylus		
1983-89	1.6	1.9
1990-93	1.1	1.4

	LABOR TIME	SEVERE SERVICE
Windshield Wiper Transmission, Replace (B)		
One side		
I-Mark	1.8	2.1
Impulse, Stylus	1.4	1.7

LAMPS & SWITCHES

	LABOR TIME	SEVERE SERVICE
Combination Switch Assy., Replace (B)		
I-Mark	1.2	1.2
Impulse, Stylus		
1988-93	2.3	2.3
Halogen Headlamp Bulb, Replace (C)		
1988-93	.3	.3
Headlamps, Aim (B)		
1981-93	.3	.3
Headlamp Switch, Replace (B)		
I-Mark	.5	.5
Impulse, Stylus		
1983-89	1.9	1.9
1990-93	.7	.7
Horn, Replace (B)		
1981-93		
one	.4	.4
each addl. add	.1	.1
Horn Relay, Replace (B)		
Impulse, Stylus	.4	.4
Neutral Safety Switch, Replace (B)		
FWD	.5	.5
RWD		
1981-87	.5	.5
1988-89	.6	.6
w/AWD FWD add	.4	.4
w/console RWD add	.1	.1
Parking Brake Warning Lamp Switch, Replace (B)		
1981-93	.5	.5
Parking Lamp Lens or Bulb, Replace (B)		
1981-93	.4	.4
Sealed Beam Headlamp, Replace (C)		
1981-87 I-Mark	.4	.4
Impulse, Stylus	.3	.3
Starter Safety Switch, Replace (B)		
1981-93	.5	.5

	LABOR TIME	SEVERE SERVICE
Stop Light Switch, Replace (B)		
1981-89	.5	.5
1990-93	.6	.6
Tail Lamp Lens or Bulb, Replace (C)		
1981-93	.2	.2
Turn Signal or Hazard Relay, Replace (B)		
1981-93	.3	.3
Turn Signal or Hazard Warning Flasher, Replace (B)		
1981-93	.2	.2
Turn Signal Lamp or Lens, Front, Replace (B)		
1981-93 one	.3	.3
Turn Signal Switch, Replace (B)		
Impulse, Stylus	1.9	1.9

BODY

	LABOR TIME	SEVERE SERVICE
Door Lock Remote Control, Replace (B)		
1981-93	.8	.8
Front Door Lock, Replace (B)		
I-Mark	1.3	1.5
Impulse, Stylus		
1983-89	1.7	1.9
1990-93	1.0	1.2
Front Door Window Regulator, Replace (B)		
I-Mark	1.2	1.4
Impulse, Stylus		
1983-89	1.6	1.8
1990-93		
manual	1.3	1.5
power	1.6	1.8
Hood Release Cable, Replace (B)		
1981-93	.5	.5
Lock Striker Plate, Replace (B)		
1981-93	.5	.5
Rear Door Lock, Replace (B)		
I-Mark, Stylus	.7	.7
Rear Door Window Regulator, Replace (B)		
I-Mark, Stylus	.9	1.1
Window Regulator Motor, Replace (B)		
1990-93	1.6	1.8

ISU-14 I-MARK : IMPULSE : STYLUS

NOTES

ISU

Amigo : Ascender : Axiom : Hombre : Oasis : Pickup : Rodeo : Rodeo Sport : Trooper : VehiCROSS

SYSTEM INDEX

MAINTENANCE ISU-16
   Maintenance Schedule ... ISU-16
CHARGING ISU-16
STARTING ISU-17
CRUISE CONTROL ISU-17
IGNITION ISU-18
EMISSIONS ISU-19
FUEL ISU-20
EXHAUST ISU-22
ENGINE COOLING ISU-23
ENGINE ISU-24
   Assembly ISU-24
   Cylinder Head ISU-26
   Camshaft ISU-29
   Crank & Pistons ISU-30
   Engine Lubrication ISU-32
CLUTCH ISU-33
MANUAL TRANSMISSION .. ISU-34
AUTO TRANSAXLE ISU-35
AUTO TRANSMISSION ISU-35
TRANSFER CASE ISU-37
SHIFT LINKAGE ISU-38
DRIVELINE ISU-38
BRAKES ISU-41
FRONT SUSPENSION ISU-43
REAR SUSPENSION ISU-45
STEERING ISU-46
HEATING & AC ISU-48
WIPERS &
  SPEEDOMETER ISU-50
LAMPS & SWITCHES ISU-50
BODY ISU-51

OPERATIONS INDEX

A
AC Hoses ISU-48
Air Bags ISU-46
Air Conditioning ISU-48
Alignment ISU-43
Alternator (Generator) ISU-17
Antenna ISU-50
Anti-Lock Brakes ISU-41

B
Back-Up Lamp Switch ISU-50
Ball Joint ISU-43
Battery Cables ISU-17
Bleed Brake System ISU-41
Blower Motor ISU-48
Brake Disc ISU-42
Brake Drum ISU-42
Brake Hose ISU-41
Brake Pads and/or Shoes .. ISU-42

C
Camshaft ISU-29
Camshaft Sensor ISU-18
Catalytic Converter ISU-22
Coolant Temperature (ECT)
 Sensor ISU-19
Crankshaft ISU-31
Crankshaft Sensor ISU-18
Cruise Control ISU-17
CV Joint ISU-40
Cylinder Head ISU-26

D
Differential ISU-39
Distributor ISU-18
Drive Belt ISU-16
Driveshaft ISU-38

E
EGR ISU-19
Electronic Control Module
 (ECM/PCM) ISU-19
Engine ISU-24
Engine Lubrication ISU-32
Engine Mounts ISU-26
Evaporator ISU-49
Exhaust ISU-22
Exhaust Manifold ISU-23
Expansion Valve ISU-49

F
Flexplate ISU-35
Flywheel ISU-34
Fuel Injection ISU-21
Fuel Pump ISU-21
Fuel Vapor Canister ISU-19

G
Gear Selector Lever ISU-38
Generator ISU-17
Glow Plug ISU-18

H
Halfshaft ISU-40
Headlamp ISU-51
Heater Core ISU-49
Horn ISU-51

I
Idle Air Control (IAC) Valve ... ISU-22
Ignition Coil ISU-18
Ignition Module ISU-18
Ignition Switch ISU-18
Inner Tie Rod ISU-47
Intake Air Temperature (IAT)
 Sensor ISU-20
Intake Manifold ISU-21

K
Knock Sensor ISU-20

L
Lower Control Arm ISU-43

M
Maintenance Schedule ISU-16
Manifold Absolute Pressure
 (MAP) Sensor ISU-20
Mass Air Flow (MAF)
 Sensor ISU-20
Master Cylinder ISU-42
Muffler ISU-23
Multifunction Lever/Switch ... ISU-51

N
Neutral Safety Clutch Switch . ISU-50
Neutral Safety Switch ISU-51

O
Oil Pan ISU-32
Oil Pump ISU-33
Orifice Tube/Valve ISU-49
Outer Tie Rod ISU-47
Oxygen Sensor ISU-20

P
Parking Brake ISU-43
Pistons ISU-31
Positive Crankcase
 Ventilation (PCV) Valve ... ISU-20

R
Radiator ISU-24
Radiator Hoses ISU-24
Radio ISU-50

S
Shock Absorber/Strut, Front .. ISU-43
Shock Absorber/Strut, Rear .. ISU-45
Spark Plug Cables ISU-19
Spark Plugs ISU-19
Spring, Front Coil ISU-43
Spring, Leaf ISU-45
Spring, Rear Coil ISU-45
Starter ISU-17
Steering Wheel ISU-48

T
Thermostat ISU-23
Throttle Body ISU-22
Throttle Position Sensor
 (TPS) ISU-20
Timing Belt ISU-30
Timing Chain ISU-30
Torque Converter ISU-36

U
U-Joint ISU-38
Upper Control Arm ISU-43

V
Valve Body ISU-36
Valve Cover Gasket ISU-27
Valve Job ISU-28
Vehicle Speed Sensor ISU-20

W
Water Pump ISU-24
Wheel Balance ISU-16
Wheel Cylinder ISU-43
Window Regulator ISU-51
Windshield Washer Pump . ISU-50
Windshield Wiper Motor .. ISU-50

ISU-15

ISU-16 AMIGO : ASCENDER : AXIOM : HOMBRE : OASIS : PICKUP

	LABOR TIME	SEVERE SERVICE
MAINTENANCE		
Air Cleaner Filter Element, Replace (C)		
All Models	.3	.4
Drive Belt, Adjust (B)		
All Models		
one	.2	.3
each addl. add	.1	.2
Drive Belt, Replace (B)		
Exc. below one	.5	.6
1992-97 Trooper		
compressor	.7	.9
each addl. add	.1	.2
w/AC Oasis add	.1	.2
Drive Belt Tensioner, Replace (B)		
All Models	.5	.6
Fuel Filter, Replace (B)		
All Models	.5	.5
Halogen Headlamp Bulb, Replace (B)		
1988-05	.3	.3
Oil & Filter, Change (C)		
Includes: Correct all fluid levels.		
All Models	.5	.5
Timing Belt, Replace (B)		
2.2L	1.2	1.4
2.3L	3.7	3.9
2.6L		
Amigo, Pick-Up, Trooper II	2.9	3.1
Rodeo/Sport	1.7	1.9
3.2L, 3.5L, 4.3L		
Amigo	2.5	2.7
Axiom (1.6)	2.2	2.4
Oasis	3.1	3.3
Rodeo/Sport		
1991-97	3.1	3.3
1998-04	2.5	2.7
Trooper		
1992-97		
SOHC	2.7	2.9
DOHC	2.4	2.6
1998-02	3.0	3.2
VehiCROSS	2.4	2.6
Diesel	3.1	3.3
Replace		
oil pump add	2.3	2.5
tensioner add	.2	.5
water pump add	.3	.5
Tire, Replace (C)		
Includes: Dismount old tire and mount new tire to rim.		
All Models	.5	.7
Wheel, Balance (B)		
All Models		
one	.4	.6
each addl. add	.2	.4

	LABOR TIME	SEVERE SERVICE
SCHEDULED MAINTENANCE INTERVALS		
If necessary, refer to appropriate Chilton maintenance service information.		
7,500 Mile Service (C)		
Amigo (2.1)	2.8	3.0
Hombre (1.6)	2.2	2.4
Oasis (1.6)	2.2	2.4
Rodeo (1.3)	1.8	2.0
Trooper (2.1)	2.8	3.0
15,000 Mile Service (B)		
Amigo (3.8)	5.1	5.3
Hombre (3.6)	4.9	5.1
Oasis (3.8)	5.1	5.3
Rodeo (4.1)	5.5	5.7
Trooper (3.8)	5.1	5.3
22,500 Mile Service (C)		
Amigo (2.1)	2.8	3.0
Hombre (2.3)	3.1	3.3
Oasis (2.3)	3.1	3.3
Rodeo (2.1)	2.8	3.0
Trooper (2.1)	2.8	3.0
30,000 Mile Service (B)		
Amigo (7.3)	9.9	10.1
Hombre (5.3)	7.2	7.4
Oasis (5.4)	7.3	7.5
Rodeo (5.3)	7.2	7.4
Trooper (7.3)	9.9	10.1
37,500 Mile Service (C)		
Amigo (2.1)	2.8	3.0
Hombre (2.3)	3.1	3.3
Oasis (2.3)	3.1	3.3
Rodeo (1.6)	2.2	2.4
Trooper (2.1)	2.8	3.0
45,000 Mile Service (B)		
Amigo (4.1)	5.5	5.7
Hombre (3.8)	5.1	5.3
Oasis (3.8)	5.1	5.3
Rodeo (3.3)	4.5	4.7
Trooper (4.1)	5.5	5.7
52,500 Mile Service (C)		
Amigo (2.3)	3.1	3.3
Hombre (2.3)	3.1	3.3
Oasis (2.5)	3.4	3.6
Rodeo (2.3)	3.1	3.3
Trooper (2.3)	3.1	3.3
60,000 Mile Service (B)		
Amigo (8.2)	11.1	11.3
Hombre (7.9)	10.7	10.9
Oasis (8.0)	10.8	11.0
Rodeo (8.0)	10.8	11.0
Trooper (8.2)	11.1	11.3
67,500 Mile Service (C)		
Amigo (1.6)	2.2	2.4
Hombre (1.3)	1.8	2.0
Oasis (1.6)	2.2	2.4
Rodeo (1.6)	2.2	2.4
Trooper (2.1)	2.8	3.0

	LABOR TIME	SEVERE SERVICE
75,000 Mile Service (B)		
Amigo (3.8)	5.1	5.3
Hombre (3.6)	4.9	5.1
Oasis (3.8)	5.1	5.3
Rodeo (3.3)	4.5	4.7
Trooper (3.8)	5.1	5.3
Replace timing belt		
Rodeo add	2.7	2.9
82,500 Mile Service (C)		
Amigo (2.1)	2.8	3.0
Hombre (1.3)	1.8	2.0
Oasis (1.6)	2.2	2.4
Rodeo (1.6)	2.2	2.4
Trooper (2.1)	2.8	3.0
90,000 Mile Service (B)		
Amigo (8.2)	11.1	11.3
Hombre (7.9)	10.7	10.9
Oasis (8.0)	10.8	11.0
Rodeo (5.3)	7.2	7.4
Trooper (8.2)	11.1	11.3
Replace timing belt		
Oasis add	2.5	2.7
97,500 Mile Service (C)		
Amigo (2.1)	2.8	3.0
Hombre (1.3)	1.8	2.0
Oasis (1.6)	2.2	2.4
Rodeo (1.6)	2.2	2.4
Trooper (2.1)	2.8	3.0
105,000 Mile Service (B)		
Amigo (4.1)	5.5	5.7
Hombre (3.6)	4.9	5.1
Oasis (3.8)	5.1	5.3
Rodeo (3.3)	4.5	4.7
Trooper (4.1)	5.5	5.7
112,500 Mile Service (C)		
Amigo (2.3)	3.1	3.3
Hombre (2.3)	3.1	3.3
Oasis (2.5)	3.4	3.6
Rodeo (2.3)	3.1	3.3
Trooper (2.3)	3.1	3.3
120,000 Mile Service (B)		
Amigo (8.2)	11.1	11.3
Hombre (7.9)	10.7	10.9
Oasis (8.0)	10.8	11.0
Rodeo (8.0)	10.8	11.0
Trooper (8.2)	11.1	11.3
CHARGING		
Alternator Circuits, Test (B)		
Includes: Test component output.		
All Models	.6	.6
Alternator, R&R and Recondition (B)		
4 cyl.		
2.2L, 2.3L, 2.6L	2.5	3.0
diesel	2.8	3.3
V6		
Amigo, Rodeo	2.2	2.7
Trooper	2.6	3.1

RODEO : RODEO SPORT : TROOPER : VEHICROSS   **ISU-17**

	LABOR TIME	SEVERE SERVICE
Alternator Assy., Replace (B)		
Includes: Pulley transfer.		
4 cyl.		
2.2L 16 valve	1.4	1.6
2.2L		
exc. below	1.0	1.2
2002-04 Rodeo/		
Sport (.3)	.5	.7
diesel	1.2	1.4
2.3L, 2.6L	.7	.9
L6 4.2L (.6)	.8	1.0
V6		
Amigo, Rodeo/Sport	.5	.7
Axiom	.5	.7
Ascender		
Trooper (.6)	.8	1.0
VehiCROSS	1.2	1.4
V8 5.3L (.6)	.8	1.0
Alternator Voltage Regulator, Replace (B)		
All Models		
internal	2.0	2.2
external	.7	.7
Front Alternator Bearing and/or Retainer & End Plate, Replace (B)		
4 cyl.		
2.2L, 2.3L, 2.6L	1.2	1.4
diesel	1.8	2.0
V6	1.3	1.5
Voltmeter, Replace (B)		
Amigo, Rodeo	1.2	1.2
Pick-Up	1.2	1.2
Trooper		
1984-91	.7	.7
1992-02	1.1	1.1

STARTING

	LABOR TIME	SEVERE SERVICE
Starter Draw Test (On Car) (B)		
All Models		
exc. Ascender	.3	.3
Ascender	.8	.9
Battery, Replace (C)		
All Models	.6	.7
Battery Cables, Replace (C)		
All Models		
exc. below each	.3	.4
Ascender, VehiCROSS		
each	.8	.9
Starter, R&R and Recondition (B)		
Amigo		
1989-94	2.3	2.8
1998-00	1.9	2.4
Hombre	2.3	2.8
Oasis	2.0	2.5
Pick-Up		
1981-87	2.6	3.1
1988-95	2.3	2.8

	LABOR TIME	SEVERE SERVICE
Rodeo/Sport		
4 cyl.	3.1	3.6
V6	1.9	2.4
Trooper		
1984-87	2.6	3.1
1988-91		
4 cyl.		
2WD	2.3	2.8
4WD	2.6	3.1
V6		
1992-02	1.7	2.2
Replace field coils add	.5	.8
Starter Assy., Replace (B)		
Amigo		
1989-94	.5	.7
1998-00		
2WD	.7	.9
4WD	1.4	1.6
Ascender		
L6 (.6)	.8	1.0
V8 (3.2)	4.3	4.5
Axiom		
2WD (.6)	.8	1.0
4WD (1.0)	1.4	1.6
Hombre		
4 cyl.	1.0	1.2
V6	1.2	1.4
Oasis	1.2	1.4
Pick-Up		
1981-87	1.0	1.2
1988-95	.5	.7
Rodeo/Sport		
4 cyl.	.6	.7
V6		
2WD	.7	.9
4WD	1.3	1.4
Trooper		
4 cyl.		
1984-87	1.0	1.2
1988-91		
2WD	1.2	1.4
4WD	1.5	1.7
V6		
1992-02	.8	1.0
VehiCROSS	.7	.9
Starter Drive Assy., Replace (B)		
Includes: Starter R&R.		
Amigo		
1989-94	.7	.9
1998-00	1.6	1.8
Hombre, Oasis	1.8	2.0
Pick-Up		
1981-87	1.4	1.6
1988-97	.7	.9
Rodeo/Sport		
4 cyl.	.7	.9
V6	1.6	1.8

	LABOR TIME	SEVERE SERVICE
Trooper		
1984-87	1.4	1.6
1988-99		
4 cyl.		
2WD	1.4	1.6
4WD	1.7	1.9
V6	1.2	1.4
Starter Relay, Replace (B)		
All Models	.3	.3
Starter Solenoid and/or Switch, Replace (B)		
Includes: Starter R&R.		
Amigo		
1989-94	.7	.9
1998-00	1.7	1.9
Hombre, Oasis	1.4	1.6
Pick-Up		
1981-87	1.4	1.6
1988-95	.7	.9
Rodeo		
4 cyl.	.8	1.0
V6	1.7	1.9
Trooper		
1984-87	1.4	1.6
1988-02		
4 cyl.		
2WD	1.4	1.6
4WD	1.7	1.9
V6	1.2	1.4

CRUISE CONTROL

	LABOR TIME	SEVERE SERVICE
Diagnose Cruise Control System Component Each (A)		
All Models	1.0	1.0
Actuator Assy. and/or Bracket, Replace (B)		
All Models	.5	.5
Brake or Clutch Switch, Replace (B)		
Oasis	.5	.5
Check Valve, Replace (B)		
Oasis	.3	.3
Control Actuator Cable, Adjust (B)		
1994-04	.3	.4
Control Actuator & Bracket, Replace (B)		
Axiom, Rodeo/Sport	.5	.5
Trooper	.7	.7
VehiCROSS	.5	.5
Control Actuator Cable, Replace (B)		
Exc. below	.3	.4
1991-01 Rodeo	.7	.8
1992-97 Trooper	.7	.8
Control Actuator Valve, Replace (B)		
Oasis	.5	.5
Control Check Valve, Replace (B)		
Oasis	.3	.3

ISU-18 AMIGO : ASCENDER : AXIOM : HOMBRE : OASIS : PICKUP

	LABOR TIME	SEVERE SERVICE
Control Unit/Module, Replace (B)		
Amigo		
1989-94	.5	.5
1998-00	.3	.3
Axiom (.3)	.5	.5
Oasis, Pick-Up	.5	.5
Rodeo/Sport		
1991-97	.5	.5
1998-04	.3	.3
Trooper		
1984-91	.7	.7
1992-96	1.3	1.3
1997-02	.7	.7
VehiCROSS	.3	.3
Control Dash Switch, Replace (On Off) (B)		
Amigo, Rodeo	.7	.7
Pick-Up	.7	.7
Control Main Switch, Replace (B)		
Amigo, Rodeo/Sport		
1991-99	.7	.7
2000-04	.5	.5
Ascender, Axiom (.7)	.9	.9
Oasis	.5	.5
Trooper		
1992-94	.3	.3
1995-99	.8	.8
2000-02	.5	.5
VehiCROSS	.5	.5
Control Relay, Replace (B)		
All Models	.3	.3
Control Sensor, Replace (B)		
Ascender, Axiom (1.0)	1.4	1.4
Amigo, Rodeo/Sport		
1991-97	1.2	1.2
1998-04	1.0	1.0
Oasis	1.8	1.8
Trooper	1.2	1.2
Steering Wheel Control Switch, Replace (B)		
Oasis	.3	.3

IGNITION

	LABOR TIME	SEVERE SERVICE
Diagnose Ignition System Component Each (A)		
All Models (.8)	1.1	1.1
Ignition Timing, Reset (B)		
1981-97 4 cyl.	.3	.3
1989-97 V6	.5	.5
Camshaft Sensor, Replace (B)		
Amigo, Rodeo/Sport		
1996-97		
2.2L	.7	.9
3.2L	.5	.7
1998-04		
2.2L	2.6	2.8
3.2L	1.8	2.0

	LABOR TIME	SEVERE SERVICE
Ascender		
L6 (.5)	.7	.9
V8 (2.1)	2.8	3.0
Axiom	1.8	2.0
Trooper		
1996-97	.3	.5
1998-02	2.3	2.5
VehiCROSS	2.6	2.8
Crankshaft Angle Sensor, Replace (B)		
Amigo	.8	1.0
Ascender		
L6 (.5)	.7	.9
V8 (3.3)	4.5	4.7
Axiom	.9	1.1
Hombre		
2.2L	.7	.9
4.3L	.8	1.0
Rodeo/Sport		
1991-97		
2.2L	.5	.7
3.2L	.9	1.1
1998-04	.9	1.1
Trooper		
1992-02	.5	.7
VehiCROSS	.8	1.0
Distributor, R&R and Recondition (B)		
Includes: Reset base ignition timing.		
4 cyl.	1.7	1.9
V6	1.9	2.1
Distributor, Replace (B)		
Includes: Reset base ignition timing.		
1981-97	.7	.9
Distributor Cap and/or Rotor, Replace (B)		
1981-97	.5	.5
Distributor Vacuum Control Unit, Replace (B)		
1981-95	.7	.8
ECI Control Relay, Replace (B)		
1995-97 Trooper	.3	.3
EFI Main Relay, Replace (B)		
Oasis	.7	.7
Electronic Spark Control Module, Replace (B)		
1992-95 Trooper	.5	.5
Electronic Spark Controller, Replace (B)		
1992-95 Trooper	.5	.5
Glow Plug Control Switch, Replace (B)		
Diesel	.3	.3
Glow Plug Relay, Replace (B)		
Diesel	.3	.3
Glow Plug, Replace (B)		
Diesel		
one	.5	.6
all	.8	1.0

	LABOR TIME	SEVERE SERVICE
Ignition Coil, Replace (B)		
Amigo		
1989-94	.5	.5
1998-00		
2.2L	.8	.9
3.2L one or all	.6	.7
Ascender one or all		
L6 (.5)	.7	.9
V8 (1.0)	1.4	1.6
Hombre, Axiom	.6	.7
Rodeo/Sport		
1991-95	.5	.6
1996-97	.6	.7
1998-04 one or all	.8	.9
Trooper		
1984-95	.5	.6
1996-99	.6	.7
2000-04	.5	.6
VehiCROSS	.5	.5
Replace igniter add	.1	.2
Ignition Coil and Module, Replace (B)		
1994-95 Rodeo		
3.2L V6	.5	.6
Ignition Lock Cylinder, Replace (B)		
Ascender (1.5)	2.0	2.2
Hombre	1.2	1.4
Trooper		
1992-94	.7	.9
1995-04	1.7	1.9
w/air bags add	.2	.4
Ignition Module, Replace (B)		
Ascender		
one or all (.6)	.8	1.0
Hombre	.7	.9
Pick-Up		
1988-93	.5	.7
Rodeo		
1991-93	.5	.7
1994-95	.5	.7
1996-04		
2.2L	.8	1.0
3.2L, 3.5L		
one or all	.5	.7
Trooper		
1988-93	.5	.7
1994-95	.5	.7
1996-04	.3	.5
Ignition Switch, Replace (B)		
Amigo		
1989-94	1.0	1.0
1998-99	.8	.8
2000	1.1	1.1
Ascender, Axiom (.9)	1.2	1.2
Hombre	1.0	1.0
Oasis	1.6	1.6
Pick-Up		
1981-87	.5	.5
1988-95	1.0	1.0

RODEO : RODEO SPORT : TROOPER : VEHICROSS   ISU-19

	LABOR TIME	SEVERE SERVICE
Rodeo/Sport		
1991-97	1.0	1.0
1998-99	.8	.8
2000-04	1.1	1.1
Trooper		
1984-87	.5	.5
1988-94	1.0	1.0
1995-02	1.7	1.7
VehiCROSS	1.3	1.3

Ignition Switch (Electrical Portion), Replace (B)
Oasis	1.2	1.2

Key Warning Buzzer or Chime, Replace (B)
Amigo		
1989-91	1.2	1.2
1992-94	.5	.5
Ascender (.9)	1.2	1.2
Hombre	.8	.8
Pick-Up		
1981-91	1.2	1.2
1992-95	.5	.5
Trooper		
1984-91	1.2	1.2
1992-97	.3	.3
1998-04	.8	.8
w/shift lock add	.1	.1

Spark Plug (Ignition) Cables, Replace (B)
1998-04 Amigo, Rodeo/Sport		
2.2L	.7	.9
3.2L, 3.5L	1.4	1.6
Ascender		
4.2L (.9)	1.2	1.4
5.3L (1.2)	1.4	1.6
Axiom	1.2	1.4
Trooper		
1991-95	.6	.8
1996-04	1.2	1.4
VehiCROSS	.7	.9

Spark Plugs, Replace (B)
1998-04 Amigo, Rodeo/Sport		
2.2L	.5	.7
3.2L, 3.5L	1.2	1.4
Ascender		
4.2L (.8)	1.1	1.2
5.3L (.9)	1.2	1.4
Axiom	1.1	1.2
Trooper		
1991-95	.5	.7
1996-02	1.2	1.4
VehiCROSS	.7	.9

EMISSIONS

The following operations do not include testing. Add time as required.

Diagnose Electronic Emission Control System Component Each (A)
	LABOR TIME	SEVERE SERVICE
1986-05 (1.0)	1.4	1.6

Dynamometer Test (A)
All Models	.4	.4

Air Cleaner Temperature Sensor, Replace (B)
All Models	.5	.5

Air Cleaner Vacuum Motor, Replace (B)
1981-97	.5	.5

Air Distribution Manifold, Replace (B)
Includes: Exhaust manifold R&R.
1981-90	1.8	2.0
1991-97	1.4	1.6

Air Flow Sensor, Replace (B)
1996-04 Amigo, Rodeo/Sport	.5	.5
1996-02 Trooper	.5	.5

Air Injection Check Valve, Replace (B)
1981-97	.5	.5

Air Pump Assy., Replace (B)
1981-97 4 cyl.	.8	1.0
1989-97 V6	1.2	1.4

Anti-Dieseling Solenoid, Replace (B)
1989-94 Amigo	.7	.7
Pick-Up	.7	.7

Check and Relief Valve, Replace (B)
1994-97	.3	.3

Coolant Temperature (ECT) Sensor, Replace (B)
Ascender, Axiom (.3)	.5	.7
Hombre, Oasis	.5	.7
Rodeo/Sport		
1992-97		
2.2L	.5	.7
3.1L	.7	.9
3.2L	2.0	2.2
1998-04	.5	.7
Trooper		
1992-97	1.7	1.9
1998-02 (1.5)	2.0	2.2
VehiCROSS	.5	.7
1998-00 Amigo	.5	.7

Crankcase Vent Cleaner Filter Element, Replace (B)
1981-97	.3	.3

ECI Control Relay, Replace (B)
1995-97 Trooper	.3	.3

EGR Back Pressure Transducer, Replace (B)
1991-97 Rodeo		
2.2L	.7	.9
3.2L	.5	.7
1992-02 Trooper	.7	.9

EGR Control Solenoid, Replace (B)
Oasis	.7	.9

EGR Control Valve, Replace (B)
	LABOR TIME	SEVERE SERVICE
Oasis	.5	.7

EGR Temperature Sensor, Replace (B)
1981-97	.7	.9

EGR Vacuum Control Solenoid, Replace (B)
Hombre	.5	.7
Rodeo	.5	.7

EGR Valve, Replace (B)
Exc. below	.8	1.0
1998-00 Amigo		
2.2L	.6	.8
3.2L	.7	.9
Ascender 5.3L (.3)	.5	.6
VehiCROSS	.7	.9

Electronic Control Module (ECM/PCM), Replace (B)
Amigo, Rodeo/Sport		
1998-99	.7	.7
2000-04	.3	.3
Ascender (.8)	1.1	1.1
Axiom	.3	.5
Hombre	.3	.3
Oasis	.7	.7
Trooper		
1992-99	.8	.8
2000-02	.3	.3
VehiCROSS	.8	.8

Electronic Load Detector, Replace (B)
Oasis	.8	.8

Evap Canister, Replace (B)
Exc. below (.3)	.5	.7
Axiom	.8	1.0
Hombre, Oasis	.3	.5
2002-04 Rodeo/Sport 3.2L, 3.5L	.8	1.0

Evap Canister Purge Solenoid, Replace (B)
1996-04 Amigo, Rodeo/Sport		
2.2L	1.2	1.4
3.2L, 3.5L	.3	.5
Ascender, Axiom (.3)	.5	.7
1996-02 Trooper	.3	.5
VehiCROSS	.3	.5

Evap Canister Purge Thermal Vacuum Switch, Replace (B)
Hombre	.5	.7

Evap Canister Vapor Separator, Replace (B)
All Models	.5	.7

Evaporative Control Valve, Replace (B)
1994-97	.5	.5

Fast Idle Coolant Temperature Switch, Replace (B)
1981-87 Pick-Up	.3	.5
1986-87 Trooper	.3	.5

ISU-20 AMIGO : ASCENDER : AXIOM : HOMBRE : OASIS : PICKUP

	LABOR TIME	SEVERE SERVICE
Intake Air Temperature (IAT) Sensor, Replace (B)		
Exc. below	.3	.5
Hombre, Oasis	.5	.7
Idle Compensator, Replace (B)		
1981-97	.5	.7
Knock Sensor, Replace (B)		
1998-00 Amigo		
2.2L	.7	.9
3.2L	2.4	2.6
Ascender front and/or both		
L6 (.3)	.5	.7
V8 (.8)	1.1	1.3
Axiom	2.4	2.6
Hombre		
2.2L	.5	.7
3.2L		
right side	1.0	1.2
left side	1.2	1.4
1991-95 Pick-Up	.8	1.0
Rodeo/Sport		
1991-95	.8	1.0
1996-97	2.6	2.8
1998-04		
2.2L	.6	.8
3.2L, 3.5L	2.8	3.0
Trooper		
1992-95	3.0	3.2
1996-99	2.4	2.6
2000-02	.3	.5
VehiCROSS	2.6	2.8
Knock Sensor Module, Replace (B)		
Hombre	.5	.5
Lockup Control Solenoid, Replace (B)		
Oasis	.5	.5
Manifold Air Temperature Sensor, Replace (B)		
Does not include testing.		
Rodeo	.3	.3
1992-97 Trooper	.3	.3
Manifold Pressure Sensor, Replace (B)		
Ascender, Axiom (.3)	.5	.5
Amigo		
1989-94	.5	.5
1998-00	.3	.3
Hombre	.5	.5
Oasis	.7	.7
Pick-Up	.5	.5
1992-04 Rodeo/Sport	.3	.3
1992-02 Trooper	.3	.3
VehiCROSS	.3	.3
Mass Air Flow (MAF) Sensor, Replace (B)		
1983-05 (.3)	.5	.5
Mixture Control Valve, Replace (B)		
1981-97	.7	.7
Oxygen Sensor, Replace (B)		
Amigo		

	LABOR TIME	SEVERE SERVICE
2.2L one or both	.5	.7
3.2L		
front each	.7	.7
rear each	.5	.7
Ascender each (.3)	.5	.7
Axiom		
front each		
2WD	.6	.8
4WD	.8	1.0
rear each	.5	.7
Hombre		
engine comp	.5	.7
exhaust pipe each	.3	.5
Oasis	.7	.9
Rodeo/Sport		
1991-95	.5	.7
1996-97		
2.2L		
one	.5	.7
both	.8	1.0
3.2L		
one	.5	.7
both	.7	.9
1998-04		
2.2L one or both	.6	.8
3.2L, 3.5L		
front each	.8	1.0
rear each	.5	.7
Trooper		
1992-95	.5	.7
1996-99		
right front	1.4	1.6
left front	.7	.9
right or left rear	.5	.7
pre, post catalytic converter	.5	.7
2000-02		
right of left side front or rear	.6	.8
VehiCROSS each	.5	.7
w/4WD Amigo, Rodeo add each	.5	.7
Positive Crankcase Ventilation (PCV) Valve, Replace (B)		
All Models	.3	.3
w/4.3L add	.3	.3
Thermal Vacuum Valve, Replace (B)		
1981-97	.5	.7
Thermostatic Vacuum Control Switch, Replace (B)		
Hombre	.5	.7
Throttle Position Sensor (TPS), Replace (B)		
Amigo, Ascender, Axiom (.5)	.7	.9
Pick-Up, Rodeo/Sport	.5	.7
Hombre	.3	.5
Oasis	.8	1.0
1996-02 Trooper	.5	.7
VehiCROSS	.5	.7

	LABOR TIME	SEVERE SERVICE
Two-Way Valve, Replace (B)		
Oasis	1.5	1.7
Vacuum Control Valve, Replace (B)		
Rodeo	.3	.3
1992-02 Trooper	.5	.5
Vacuum Switching Valve, Replace (B)		
Rodeo	.5	.7
1992-02 Trooper	.3	.5
Vehicle Speed Sensor, Replace (B)		
Amigo, Rodeo/Sport		
1994-97	.5	.7
1998-04		
AT	.7	.9
MT	.3	.5
Ascender (.5)	.7	.9
Axiom (.8)	1.1	1.5
Oasis	.8	1.0
VehiCROSS	.5	.7

FUEL
CARBURETOR

	LABOR TIME	SEVERE SERVICE
Carburetor, Adjust (On Vehicle) (B)		
1989-94 Amigo	.5	.5
Pick-Up	.5	.5
Air Cleaner Temperature Sensor, Replace (B)		
1989-94 Amigo	.5	.5
Pick-Up	.5	.5
Air Cleaner Vacuum Motor, Replace (B)		
1989-94 Amigo	.5	.5
Pick-Up	.5	.5
Anti Dieseling Solenoid, Replace (B)		
1989-94 Amigo	.7	.7
Pick-Up	.7	.7
Automatic Choke, Replace (B)		
1989-94 Amigo	.8	.8
Pick-Up	.8	.8
Carburetor, Replace (B)		
Includes: Adjustments.		
1989-94 Amigo	1.3	1.5
1981-94 Pick-Up	1.3	1.5
Carburetor, R&R and Clean or Recondition (B)		
Includes: Adjustments.		
1989-94 Amigo	3.2	3.4
Pick-Up	3.2	3.4
Carburetor Accelerator Pump, Replace (B)		
1989-94 Amigo	.8	1.0
Pick-Up	.8	1.0
Carburetor Float, Needle and/or Seat, Replace (B)		
1989-94 Amigo	1.2	1.4
Pick-Up	1.2	1.4
Fuel Filter, Replace (B)		
1989-94 Amigo	.3	.5
Pick-Up	.3	.5

RODEO : RODEO SPORT : TROOPER : VEHICROSS — ISU-21

	LABOR TIME	SEVERE SERVICE
DELIVERY		
Fuel Pump, Test (B)		
All Models	.5	.5
Diesel Fuel Injection Pump, Replace (B)		
Diesel	3.7	3.9
w/AC add	.3	.5
w/turbo add	.7	.9
Fuel Filter, Replace (B)		
All Models	.5	.5
Fuel Gauge (Dash), Replace (B)		
Amigo, Ascender (.9)	1.2	1.4
Axiom (.9)	1.2	1.4
Oasis	.7	.9
Pick-Up	1.2	1.4
Rodeo/Sport	1.2	1.4
Trooper		
1984-91	.7	.9
1992-02	1.1	1.3
VehiCROSS	.8	1.0
Fuel Gauge (Tank), Replace (B)		
Amigo		
1989-94	.7	.9
1998-00	1.5	1.7
Ascender (2.0)	2.8	3.0
Axiom	1.8	2.0
Hombre		
left tank	2.3	2.5
rear tank	1.9	2.1
Oasis	1.0	1.2
Pick-Up	.7	.9
Rodeo/Sport		
1991-97	.7	.9
1998-00	1.5	1.7
2001	1.5	1.7
w/short resin tank	1.4	1.6
w/long resin tank	1.9	2.1
2002-04	1.7	2.0
w/short resin tank	1.1	1.3
w/long resin tank	2.2	2.4
Trooper		
1984-97	.7	.9
1996-02	1.7	1.9
VehiCROSS	.5	.7
w/short wheelbase		
Pick-Up add	1.2	1.4
Fuel Pump, Replace (B)		
Electric		
Amigo		
1989-94	2.3	2.4
1998-00	1.5	1.7
Ascender (1.7)	2.3	2.5
Axiom	1.8	2.0
Hombre		
left tank	2.3	2.5
rear tank	1.7	1.9
Oasis, Pick-Up	2.3	2.5
Rodeo/Sport		
1991-95	2.9	3.1
1996-97	1.7	1.9

	LABOR TIME	SEVERE SERVICE
1998-00	1.5	1.7
2001	1.7	2.0
w/short resin tank	3.2	3.4
w/long resin tank	2.2	2.4
2002-04		
w/short resin tank	2.9	3.1
w/long resin tank	1.8	2.0
Trooper		
1984-91	2.3	2.5
1992-95	1.4	1.6
1996-02	1.7	1.9
VehiCROSS	3.7	3.9
Mechanical		
4 cyl.	2.0	2.2
V6	1.2	1.4
w/tank shield		
Ascender add	.5	.7
Fuel Pump Relay, Replace (B)		
1983-05 (.3)	.5	.5
Fuel Tank, Replace (B)		
Includes: Drain and refill.		
Amigo		
1989-94	2.0	2.2
1998-00	1.0	1.2
Ascender (2.0)	2.7	2.9
Axiom	1.8	2.0
Hombre		
left tank	2.0	2.2
rear tank	1.9	2.1
Oasis, Pick-Up	2.0	2.2
Rodeo/Sport		
1991-95	2.0	2.2
1996-97	1.6	1.8
1998 00	1.0	1.2
2001		
w/short resin tank	3.2	3.4
w/long resin tank	2.2	2.4
2002-04		
w/short resin tank	2.9	3.1
w/long resin tank	1.8	2.0
Trooper		
1984-91	2.0	2.2
1992-02	1.4	1.6
VehiCROSS	3.6	3.8
w/tank shield		
Ascender add	.5	.7
Intake Manifold and/or Gasket, Replace (B)		
Includes: Adjustments on all. R&R Fuel Injectors on 2.2L Amigo, Rodeo, Rodeo Sport. R&R Common Chamber on V6 Amigo, Axiom, Rodeo, Rodeo Sport.		
1998-00 Amigo	2.3	2.5
Ascender		
L6 (2.5)	3.4	3.6
V8 (1.9)	2.6	2.8
Axiom		
2002-03 (1.6)	2.2	2.4
2004 (.5)	.7	.9

	LABOR TIME	SEVERE SERVICE
Hombre		
2.2L		
1996-97	2.2	2.4
1998	1.7	1.9
1999-00 upper	1.0	1.2
4.3L		
upper	1.9	2.1
lower	4.1	4.3
Rodeo/Sport		
1998-04		
2.2L	2.2	2.3
3.2L	2.3	2.5
3.5L	.8	1.0
Trooper		
1992-97		
SOHC	2.3	2.5
DOHC	2.9	3.1
1998-02	2.9	3.1
VehiCROSS	2.3	2.5
w/AC 96-98 Hombre add	.2	.2
Replace upper manifold		
Hombre add	.2	.2
INJECTION		
Diesel Fuel Injection Timing, Check & Adjust (B)		
Diesel	1.0	1.2
w/AC add	.2	.2
w/turbo add	.7	.7
Idle Speed, Adjust (B)		
1983-98	.5	.5
Air Regulator, Replace (B)		
1983-91	1.6	1.8
1992-97 Rodeo	1.6	1.8
1992-97 Trooper	.5	.7
Central Multiport Fuel Injector, Replace (B)		
Hombre	1.6	2.0
Diesel Injection Nozzles, Replace (B)		
Diesel		
one	.7	.8
all	1.2	1.4
Clean nozzles add each	.2	.3
EFI Relay, Replace (B)		
Oasis	.7	.7
Electronic Control Unit, Replace (B)		
4 cyl.	.7	.7
V6		
2.8L, 3.1L	.5	.5
3.2L, 3.5L		
Trooper	.8	.8
Amigo, Rodeo	.5	.5
Fast Idle Valve, Replace (B)		
Oasis	.7	.7
Fuel Injection Lines, Replace (B)		
Diesel		
one	.5	.5
all	1.0	1.0
Fuel Injector Resistor, Replace (B)		
1983-97	.3	.3

AMIGO : ASCENDER : AXIOM : HOMBRE : OASIS : PICKUP

	LABOR TIME	SEVERE SERVICE
Fuel Injectors, Replace (B)		
4 cyl.		
2.2L one	.5	.7
2.2L 16 valve	1.2	1.4
2.6L	1.8	2.0
L6		
4.2L Ascender		
one or all (3.0)	4.1	4.3
V6		
2.8L, 3.1L	.8	.9
3.2L		
1992-01 Amigo, Rodeo	1.3	1.5
1992-97 Trooper		
SOHC	3.3	3.6
DOHC	2.8	3.1
1998-99	1.3	1.5
2000-04		
2WD	1.0	1.2
4WD	1.3	1.5
3.5L	1.2	1.4
V8		
5.3L one or all (1.2)	1.4	1.6
Fuel Pressure Regulator, Replace (B)		
4 cyl.		
2.2L	1.6	1.8
2.2L 16 valve	.5	.7
2.3L, 2.6L	.5	.7
L6		
4.2L Ascender (.6)	.8	1.0
V6		
2.8L, 3.1L	.7	.9
3.2L, 3.5L		
1992-97		
Rodeo	1.2	1.4
Trooper	.8	1.0
1998-04		
Amigo, Rodeo/Sport, Axiom (.3)	.5	.7
Trooper	.8	1.0
VehiCROSS	.8	1.0
4.3L		
Hombre	1.8	2.0
V8		
5.3L (.3)	.5	.7
Fuel Rail, Replace (B)		
1998-00 Amigo		
2.2L	1.3	1.5
3.2L	1.5	1.7
Ascender		
4.2L (3.0)	4.1	4.3
5.3L (1.1)	1.3	1.5
Axiom	2.3	2.6
Rodeo/Sport		
1994-97		
2.2L	1.8	2.1
3.2L	3.4	3.7
1998-04		
2.2L	1.3	1.6
3.2L, 3.5L	1.5	1.8

	LABOR TIME	SEVERE SERVICE
Trooper		
1996-97	3.0	3.3
1998-02	1.3	1.6
VehiCROSS	1.8	2.1
Idle Air Control Motor, Replace (B)		
Rodeo	.5	.5
Idle Air Control (IAC) Valve, Replace (B)		
Amigo, Rodeo/Sport		
1991-97	.5	.5
1998-04	.3	.3
Axiom	.3	.3
Hombre, Oasis	.7	.7
Idle Air Regulator, Replace (B)		
VehiCROSS	.3	.3
Throttle Body Assy., Replace (B)		
1985-04		
exc. below	1.2	1.2
Ascender (.5)	.7	.9
Axiom (.6)	.8	.9
Throttle Body Fuel Injector and/or Gaskets, Replace (B)		
Hombre		
one or both	.7	.7
Throttle Body Fuel Meter, Replace (B)		
Hombre	1.2	1.2
Vacuum Pump, Replace (B)		
Diesel	.7	.9

EXHAUST

	LABOR TIME	SEVERE SERVICE
Catalytic Converter, Replace (B)		
Amigo		
1989-94	1.0	1.2
1998-99		
4 cyl.	2.3	2.5
V6		
right side	1.7	1.9
left side	1.8	2.0
both sides	2.8	3.0
2000		
4 cyl.	1.8	2.0
V6		
right side	1.2	1.4
left side	1.1	1.3
both sides	1.8	2.0
Ascender		
L6 cyl. (1.0)	1.4	1.6
V8		
right side (1.2)	1.6	1.8
left side (1.0)	1.4	1.6
Axiom		
right or left side (1.2)	1.6	1.8
both sides (1.5)	2.0	2.2
Hombre, Oasis	1.2	1.4
Pick-Up	1.0	1.2

	LABOR TIME	SEVERE SERVICE
Rodeo/Sport		
1991-95		
three way	1.0	1.2
warm up	1.7	1.9
1996-97		
three way		
4 cyl.	1.0	1.2
V6		
right side	2.7	2.9
left side	1.8	2.0
both sides	3.3	3.5
warm up	1.7	1.9
1998-00		
4 cyl.	2.3	2.5
V6		
right side	1.7	1.9
left side	1.8	2.0
both sides	2.8	3.0
2001-04 three way		
4 cyl.		
AT	3.3	3.5
MT	2.1	2.3
V6		
right or left side	1.4	1.6
both sides	1.8	2.0
Trooper		
1984-91		
4 cyl.	1.0	1.2
V6	.7	.9
1992-02		
three way		
1992-95	1.2	1.4
1996-02		
right side	2.7	2.9
left side	1.8	2.0
both sides	3.3	3.5
warm up	1.3	1.5
VehiCROSS		
right side	2.0	2.2
left side	1.8	2.0
both sides	2.6	2.8
w/4WD add		
Axiom	.3	.3
01-04 Rodeo/Sport	.3	.3
Center Exhaust Pipe, Replace (B)		
1998-02 Amigo	1.7	1.9
Axiom, Pick-Up V6	.8	1.0
Rodeo/Sport		
1991-97		
4 cyl.	.7	.9
V6	1.9	2.1
1998-99	1.7	1.9
2000		
4 cyl.	.7	.9
2001-04		
V6 MT	.8	1.0
Trooper		
1984-91	1.0	1.2
1992-02	1.7	1.9

RODEO : RODEO SPORT : TROOPER : VEHICROSS ISU-23

	LABOR TIME	SEVERE SERVICE
Crossover Pipe, Replace (B)		
Hombre	1.8	2.0
Exhaust Manifold or Gasket, Replace (B)		
Amigo		
1989-94		
w/Carb	1.2	1.4
w/FI	1.6	1.8
1998-00		
4 cyl.	1.6	1.8
V6		
right side	3.8	4.0
left side	4.3	4.5
both sides	5.8	6.0
Ascender		
L6 (2.2)	3.0	3.2
V8		
right or left (1.9)	2.6	2.8
Axiom		
V6		
right side	3.8	4.0
left side	5.0	5.2
both sides	6.7	6.9
Hombre		
4 cyl.	1.4	1.6
V6		
right side	3.4	3.6
left side	1.9	2.1
Oasis	1.2	1.4
Pick-Up		
1981		
gasoline	2.2	2.4
diesel	1.4	1.6
1982-88		
gasoline		
w/Carb	1.2	1.4
w/FI	1.6	1.8
diesel	1.4	1.6
1989-95		
4 cyl.	1.6	1.8
V6		
right side	2.6	2.8
left side	1.5	1.7
both sides	3.7	3.9
Rodeo/Sport		
1991-97		
4 cyl.	1.6	1.8
V6		
3.1L		
right side	2.6	2.8
left side	1.5	1.7
both sides	3.8	4.0
3.2L		
right side	3.6	3.8
left side	4.8	5.0
both sides	6.2	6.4
1998-01		
4 cyl.	1.6	1.8

	LABOR TIME	SEVERE SERVICE
V6		
right side	3.8	4.0
left side	5.0	5.2
both sides	6.8	7.0
2002-04		
4 cyl.	1.4	1.6
V6		
right side	3.8	4.0
left side	4.3	4.5
both sides	5.8	6.0
Trooper		
1984-88		
gasoline	1.6	1.8
diesel	1.4	1.6
1989-91		
4 cyl.	1.6	1.8
V6		
right side	2.6	2.8
left side	1.5	1.7
both sides	3.8	4.0
1992-97		
right side	3.0	3.2
left side	4.1	4.3
both sides	6.4	6.6
1998-02		
right side	3.0	3.2
left side	3.4	3.6
both sides	5.0	5.2
VehiCROSS		
right side	3.3	3.5
left side	3.0	3.2
both sides	4.9	5.1
w/4WD add		
Axiom	.5	.5
01-04 Rodeo/Sport	.5	.5
Exhaust Pipe, Replace (B)		
Ascender		
L6 (.5)	.7	.9
Axiom		
front	.8	.9
rear	.5	.6
Oasis		
front	1.0	1.1
rear	.8	.8
both	1.8	1.8
Exhaust System (Complete), Replace (B)		
Ascender (2.7)	3.6	3.8
Hombre	1.6	1.7
Front Exhaust Pipe (To Converter), Replace (B)		
Hombre	.7	.8
Ascender		
left (1.0)	1.4	1.6
right (1.3)	1.8	2.0
Intermediate Exhaust Pipe, Replace (B)		
1989-94 Amigo	.5	.6
Hombre, Pick-Up	.5	.6
1991-97 Rodeo	.8	.9

	LABOR TIME	SEVERE SERVICE
Trooper		
1984-91	1.8	1.9
1992-96	.7	.8
Muffler, Replace (B)		
Amigo		
1989-94	1.0	1.1
1998-00	.7	.8
Ascender (1.9)	2.6	2.8
Axiom	.6	.7
Hombre	.7	.8
Oasis, Pick-Up	1.0	1.1
Rodeo/Sport		
1991-95	1.2	1.3
1996-97	.5	.6
1998-04	.7	.8
Trooper	1.2	1.3
VehiCROSS	.7	.8
Tail Pipe, Replace (B)		
1989-94 Amigo	.5	.6
Axiom, Hombre	.5	.6
Oasis	.3	.4
Pick-Up, Trooper	.7	.8
1991-97 Rodeo	.7	.8
2001-04 Rodeo/Sport	.5	.5
VehiCROSS	.5	.6

ENGINE COOLING

	LABOR TIME	SEVERE SERVICE
Pressure Test Cooling System (C)		
All Models	.3	.3
Cooling Fan Timer, Replace (B)		
Oasis	.8	.8
Coolant Thermostat, Replace (B)		
1981-97		
2.3L, 2.6L	.5	.7
2.2L, 2.8L, 3.1L	.8	1.0
3.2L	.5	.7
4.3L	.8	1.0
diesel	.8	1.0
1998-00 Amigo	1.1	1.3
Ascender		
4.2L, 5.3L (.9)	1.2	1.4
Axiom (1.5)	2.0	2.2
Rodeo/Sport		
2.2L	.5	.7
3.2L, 4.3L	2.0	2.2
1998-02 Trooper	2.1	2.3
1998-02 VehiCROSS	2.0	2.2
Engine Coolant Temp. Sending Unit, Replace (B)		
All Models	.5	.5
Fan Blade, Replace (B)		
4 cyl.		
2.2L, 2.3L, 2.6L	.7	.9
L6 4.2L	3.2	3.6
V6		
2.8L, 3.1L	1.7	1.9
3.2L, 3.5L, 4.3L	.7	.9
V8 5.3L	3.2	3.6

	LABOR TIME	SEVERE SERVICE

Fan Clutch Assy., Replace (B)
- 4 cyl.
  - 2.2L, 2.3L, 2.6L 7   1.0
  - L6 4.2L 3.2   3.6
- V6
  - 2.8L, 3.1L 5   .8
  - 3.2L, 3.5L, 4.3L 7   1.0
- V8 5.3L 3.2   3.6

Freeze Plugs (Water Jacket), Replace (B)
Add access time as required.
- All Models each 5   .7

Radiator Assy., R&R or Replace (B)
- 1981-97
  - 2.2L, 2.3L, 2.6L 1.5   1.7
  - 2.8L, 3.1L 1.7   1.9
  - 3.2L
    - Rodeo 1.5   1.7
    - Trooper 1.3   1.5
- 1998-04
  - Amigo, Rodeo/Sport
    - 2.2L, 3.2L (.7) 9   1.1
  - Ascender
    - 4.2L, 5.3L
      - w/AC (3.2) 4.3   4.5
      - w/o AC (2.5) 3.4   3.6
  - Axiom 1.1   1.3
  - Trooper, ... 1.2   1.4
  - VehiCROSS 1.2   1.4

Radiator Fan and/or Fan Motor, Replace (B)
- Amigo, Rodeo/Sport 5   .7
- Oasis 8   1.0

Radiator Fan Motor Relay, Replace (B)
- Amigo, Rodeo/Sport 3   .3
- Oasis 3   .3

Radiator Fan Motor Switch (Coolant Temp.), Replace (B)
- Oasis 8   .8

Radiator Hoses, Replace (B)
Includes: Refill with proper coolant mix.
- 1981-97
  - upper 5   .7
  - lower
    - 2.3L, 2.6L 5   .7
    - 3.1L 8   1.0
    - 3.2L 5   .7
  - 2.2L
    - upper or lower 7   .9
    - both 1.3   1.5
    - by-pass 7   .9
- 1998-04
  - Amigo, Rodeo/Sport
    - 2.2L
      - upper 3   .5
      - lower. 5   .7
    - 3.2L, 3.5L
      - upper 5   .7
      - lower 7   .9

Ascender
- 4.2L, 5.3L
  - upper (.4) 6   .8
  - lower (.9) 1.2   1.4
Axiom
- Upper (.6) 8   1.0
- lower (.9) 1.2   1.4
Hombre
- upper or lower ... 1.3   1.5
Trooper
- upper or lower 5   .7
VehiCROSS
- upper 3   .5
- lower 5   .7

Temperature Gauge (Dash), Replace (B)
- Amigo, Ascender (.9) . 1.2   1.2
- Axiom 1.2   1.2
- Oasis 7   .7
- Pick-Up, Rodeo/Sport . 1.2   1.2
- Trooper
  - 1984-91 7   .7
  - 1992-02 .... 1.2   1.2
- VehiCROSS 8   .8

Water Pump and/or Gasket, Replace (B)
Includes: Refill with proper coolant mix.
- 4 cyl.
  - 2.2L 2.0   2.2
  - 2.2L 16 valve 3.1   3.3
  - 2.3L 4.2   4.4
  - 2.6L 1.9   2.1
  - diesel 1.4   1.6
- L6 4.2L (2.2) 3.0   3.2
- V6
  - 2.8L, 3.1L 2.8   3.0
  - 3.2L, 3.5L 3.8   4.0
  - 4.3L 1.9   2.1
- V8 5.3L (2.6) 3.5   3.7
- w/AC 81-01 add 3   .3
- w/AT 81-01 add 5   .5
- w/PS 81-01 add 3   .3

ENGINE
ASSEMBLY
Times shown are for OEM assemblies. Time to replace assemblies from aftermarket rebuilders may vary.
Engine Assy., R&I (B)
Does not include parts or component transfer.
- Amigo
  - 1989-94
    - 2.3L
      - 2WD 6.5   6.7
      - 4WD 9.8   10.0
    - 2.6L
      - 2WD 8.9   9.1
      - 4WD 10.0   10.2

- 1998-00
  - 2.2L 4.8   5.0
  - 3.2L
    - 2WD 8.1   8.3
    - 4WD 9.7   9.9
- Ascender
  - 4.2L
    - 2WD (10.9) 14.7   14.9
    - 4WD (13.9) 18.8   19.0
  - 5.3L (9.3) 12.6   12.8
- Hombre 6.2   6.4
- Oasis 8.0   8.2
- Pick-Up
  - 2.3L
    - 2WD 6.5   6.7
    - 4WD 9.8   10.0
  - 2.6L
    - 2WD 8.9   9.1
    - 4WD 10.0   10.2
  - 3.1L 9.6   9.8
  - diesel
    - 2WD 7.7   7.9
    - 4WD 9.2   9.4
- Rodeo
  - 1991-97
    - 2.6L
      - 2WD 5.2   5.4
      - 4WD 7.7   7.9
    - 3.1L
      - 2WD 6.4   6.6
      - 4WD 8.9   9.1
    - 3.2L 9.3   9.5
  - 1998-04
    - 2.2L (3.6) 4.9   5.1
    - 3.2L, 3.5L
      - 2WD (6.0) 8.1   8.3
      - 4WD (7.2) 9.7   9.9
- Trooper
  - 1984-91
    - 2.3L
      - 2WD 6.5   6.7
      - 4WD 9.8   10.0
    - 2.6L 10.0   10.2
    - 3.2L 9.6   9.8
  - 1998-02
    - AT 10.3   10.5
    - MT 7.5   7.7
- w/AC add 3   .5
- w/C69 AC recover recharge
  - Ascender add 1.6   1.8
- w/AT add 3   .5
- w/PS add 3   .5
- w/tank shield
  - Ascender add 6   .8
- w/transfer case
  - Ascender add 5   .7

RODEO : RODEO SPORT : TROOPER : VEHICROSS

ISU-25

	LABOR TIME	SEVERE SERVICE
Engine Assy., R&R and Reconditon (A)		

Includes: Replacing rings, rod and main bearings, cylinder head reconditioning and engine tune-up.

1989-94 Amigo
- 2.3L
  - 2WD .... 24.9 / 25.1
  - 4WD .... 28.8 / 29.0
- 2.6L
  - 2WD .... 29.8 / 30.0
  - 4WD .... 28.7 / 28.9

Axiom
- 3.5L
  - 2WD .... 29.8 / 30.0
  - 4WD .... 36.2 / 36.4

Hombre .... 17.8 / 18.0
Oasis .... 27.9 / 28.1

Pick-Up
- 2.3L
  - 2WD .... 24.9 / 25.1
  - 4WD .... 28.4 / 29.0
- 2.6L
  - 2WD .... 27.1 / 27.3
  - 4WD .... 28.7 / 28.9
- diesel
  - 2WD .... 27.6 / 27.8
  - 4WD .... 29.1 / 29.3

1991-97 Rodeo/Sport
- 2.6L
  - 2WD .... 24.1 / 24.3
  - 4WD .... 26.4 / 26.6
- 3.1L
  - 2WD .... 26.8 / 27.0
  - 4WD .... 29.3 / 29.5
- 3.2L .... 28.4 / 28.6

2002-04 Rodeo/Sport
- 2.2L .... 29.0 / 29.2
- 3.2L, 3.5L
  - 2WD .... 27.5 / 27.7
  - 4WD .... 32.1 / 32.3

Trooper
- 1984-91
  - 2.3L
    - 2WD .... 24.9 / 25.1
    - 4WD .... 28.4 / 28.6
  - 2.6L
    - 2WD .... 27.1 / 27.3
    - 4WD .... 28.7 / 28.9
  - 2.8L .... 31.9 / 32.1
- 1992-97
  - SOHC .... 31.3 / 31.5
  - DOHC .... 35.9 / 36.1
- 1998-02
  - AT .... 33.5 / 33.7
  - MT .... 29.2 / 29.4
- w/AC add .... .3 / .5
- w/AT add .... .3 / .5
- w/PS add .... .3 / .5

Engine Assy. (Complete), Replace (B)

Includes: Engine R&R. Component transfer.

Amigo
- 1989-94
  - 2.3L, 2.6L
    - 2WD .... 6.8 / 7.0
    - 4WD .... 8.8 / 9.0
- 1998-00
  - 2.2L .... 6.3 / 6.5
  - 3.2L
    - 2WD .... 10.0 / 10.2
    - 4WD .... 11.6 / 11.8

Ascender
- 4.2L
  - 2WD (12.6) .... 17.0 / 17.2
  - 4WD (15.3) .... 20.7 / 20.9
- 5.3L (13.6) .... 18.4 / 18.6

Axiom
- 3.5L
  - 2WD (8.8) .... 11.9 / 12.1
  - 4WD (10.2) .... 13.8 / 14.0

Hombre .... 7.3 / 7.5
Oasis .... 10.2 / 10.4

Pick-Up
- 2.3L, 2.6L
  - 2WD .... 6.8 / 7.0
  - 4WD .... 8.8 / 9.0
- 3.1L .... 10.3 / 10.5

Rodeo/Sport
- 1991-97
  - 2.6L
    - 2WD .... 6.5 / 6.7
    - 4WD .... 9.1 / 9.3
  - 3.1L
    - 2WD .... 7.8 / 8.0
    - 4WD .... 10.0 / 10.2
  - 3.2L
    - 2WD .... 10.2 / 10.4
    - 4WD .... 12.0 / 12.2
- 1998-01
  - 2.2L .... 6.3 / 6.5
  - 3.2L
    - 2WD .... 10.0 / 10.2
    - 4WD .... 11.6 / 11.8
- 2002-04
  - 2.2L (5.4) .... 7.3 / 7.5
  - 3.2L, 3.5L
    - 2WD (8.3) .... 11.2 / 11.4
    - 4WD (9.7) .... 13.1 / 13.3

Trooper
- 1984-97
  - AT .... 14.5 / 14.7
  - MT .... 14.1 / 14.3
- 1998-02
  - AT .... 17.5 / 17.7
  - MT .... 13.2 / 13.4

	LABOR TIME	SEVERE SERVICE
VehiCROSS	13.4	13.6
w/A/C add	.3	.5
w/rear auxiliary A/C add	1.0	1.2
w/AT add	.5	.7
w/PS 81-01 add	.3	.5

Engine Assy. (Short), Replace (B)

Assembly consists of cylinder block, piston assemblies, crankshaft, camshaft, timing chain and gears. Does not include cylinder heads. Operation Includes: R&R engine, transfer necessary parts and all necessary adjustments.

Amigo
- 1989-94
  - 2.3L
    - 2WD .... 13.5 / 13.7
    - 4WD .... 17.1 / 17.3
  - 2.6L
    - 2WD .... 15.4 / 15.6
    - 4WD .... 17.1 / 17.3
- 1998-00
  - 2.2L .... 10.3 / 10.5
  - 3.2L
    - 2WD .... 16.2 / 16.4
    - 4WD .... 17.3 / 17.5

Axiom
- 2WD .... 19.7 / 19.9
- 4WD .... 20.7 / 20.9

Hombre .... 10.3 / 10.5
Oasis .... 13.4 / 13.6

Pick-Up
- 2.3L
  - 2WD .... 13.5 / 13.7
  - 4WD .... 17.1 / 17.3
- 2.6L
  - 2WD .... 15.4 / 15.6
  - 4WD .... 17.1 / 17.3
- 3.1L .... 19.3 / 19.5

Rodeo/Sport
- 1991-97
  - 2.6L .... 13.5 / 13.7
  - 3.1L .... 15.8 / 16.0
  - 3.2L .... 17.1 / 17.3
- 1998-01
  - 2.2L (7.6) .... 10.3 / 10.5
  - 3.2L
    - 2WD (12.0) .... 16.2 / 16.4
    - 4WD (12.8) .... 17.3 / 17.5
- 2002-04
  - 2.2L
    - AT (9.1) .... 12.3 / 12.5
    - MT (8.6) .... 11.6 / 11.8
  - 3.2L, 3.5L
    - 2WD
      - AT .... 19.7 / 19.9
      - MT .... 17.9 / 18.1
    - 4WD
      - AT .... 20.7 / 20.9
      - MT .... 18.9 / 19.1

ISU-26 AMIGO : ASCENDER : AXIOM : HOMBRE : OASIS : PICKUP

	LABOR TIME	SEVERE SERVICE
Engine Assy. (Short), Replace (B)		
Trooper		
1984-91		
2.3L	17.1	17.3
2.6L	10.0	10.2
2.8L	13.1	13.3
1992-97		
AT	23.1	23.3
MT	22.6	22.8
1998-02		
AT (19.3)	26.1	26.3
MT (16.0)	21.6	21.8
VehiCROSS	24.8	25.0
w/AC add	.3	.5
w/AT add	.3	.5
w/PS add	.3	.5
Engine Mounts, Replace (B)		
Amigo		
1989-94		
2.3L	.7	.9
2.6L		
front each	1.0	1.2
rear	1.2	1.4
1998-00		
2.2L		
front	1.0	1.2
rear	.7	.9
3.2L		
front 2WD		
right side	3.2	3.4
left side	3.8	4.0
front 4WD		
right side	4.6	4.8
left side	4.9	5.1
rear	1.3	1.5
Ascender		
4.2L, 5.3L		
one side (1.2)	1.6	1.8
both sides (1.6)	2.2	2.4
Axiom		
front		
2WD		
right side	3.8	4.3
left side	4.4	5.0
4WD		
right side	5.3	6.0
left side	5.7	6.3
rear	1.3	1.4
Hombre		
2.2L		
right side	.8	1.0
left side	1.0	1.2
both sides	1.7	1.9
3.2L, 4.3L		
right or left side	2.6	2.8
both sides	3.4	3.6
Oasis		
side or front	.8	1.0
rear	2.7	2.9
all	3.5	3.7

	LABOR TIME	SEVERE SERVICE
Pick-Up		
2.3L	.7	.9
2.6L		
front each	1.6	1.8
rear	1.2	1.4
3.1L		
front each	1.7	1.9
rear	.7	.9
Rodeo/Sport		
1991-97		
2.6L		
front		
2WD	1.7	1.9
4WD	3.1	3.3
3.1L, 3.2L		
front		
2WD	1.8	2.0
4WD	3.2	3.4
rear		
2WD	.7	.9
4WD	1.2	1.4
1998-04		
2.2L		
right or left side	1.0	1.2
rear	.6	.7
3.2L, 3.5L		
front 2WD		
right side	3.8	4.3
left side	4.4	5.0
front 4WD		
right side	5.3	6.0
left side	5.7	6.3
rear	1.3	1.4
Trooper		
1984-91		
2.3L	.7	.9
2.6L		
front each	1.0	1.2
rear	1.2	1.4
2.8L		
front each	1.7	1.9
rear	.7	.9
1992-97		
front	2.7	2.9
rear	.7	.9
1998-02		
right side	2.8	3.0
left side	2.6	2.8
rear	.6	.8
VehiCROSS		
right	4.1	4.3
left	3.4	3.6
rear	1.3	1.5
CYLINDER HEAD		
Compression Test (B)		
2.2L, 2.3L, 2.6L	.7	.9
2.8L, 3.1L, 3.2L	1.6	1.8
3.5L, 4.3L	1.5	1.7
Diesel	1.2	1.4

	LABOR TIME	SEVERE SERVICE
Valves, Adjust (B)		
4 cyl.		
2.2L, 2.6L	1.4	1.6
2.3L	1.2	1.4
V6		
2.8L, 3.1L	2.8	3.0
3.2L		
1992-01	4.5	4.7
2002-04	5.0	5.2
3.5L		
1998-01	4.5	4.7
2002-04	5.0	5.2
Diesel	1.2	1.4
Cylinder Head, Replace (B)		
Includes: Parts transfer, adjustments.		
Amigo		
1989-94		
2.3L	6.2	6.4
2.6L	9.9	10.1
1998-00		
2.2L	14.9	15.1
3.2L		
right side	11.0	11.2
left side	9.9	10.1
both sides	11.7	11.9
Ascender		
4.2L		
2WD (14.1)	19.8	21.0
4WD (16.7)	22.5	22.7
5.3L		
one side (8.3)	11.2	11.4
both sides (12.1)	16.3	16.5
Axiom		
right side	9.8	10.0
left side	8.8	9.0
both sides	11.3	11.5
Hombre	6.4	6.6
Oasis	8.3	8.5
Pick-Up		
2.3L	6.8	7.0
2.6L	9.9	10.1
3.1L		
one side	8.4	8.6
both sides	11.0	11.2
Rodeo/Sport		
1991-97		
2.6L	8.8	9.0
3.1L		
one side	8.4	8.6
both sides	10.7	10.9
3.2L		
right side	10.0	10.2
left side	7.1	7.3
both sides	13.1	13.3
1998-01		
2.2L	14.9	15.1
3.2L		
right side	11.0	11.2
left side	9.0	9.2
both sides	11.7	11.9

	LABOR TIME	SEVERE SERVICE
2002-04		
2.2L (4.3)	**5.8**	*6.0*
3.2L, 3.5L		
right side (7.1)	**9.6**	*9.8*
left side (6.3)	**8.5**	*8.7*
both sides (8.4)	**11.3**	*11.5*
Trooper		
1984-91		
2.3L	**6.2**	*6.4*
2.6L	**9.9**	*10.1*
2.8L		
one side	**8.3**	*8.5*
both sides	**11.6**	*11.8*
1992-95		
3.2L		
SOHC		
right side	**9.6**	*9.8*
left side	**8.6**	*8.8*
both sides	**11.6**	*11.8*
DOHC		
right side	**9.2**	*9.4*
left side	**8.6**	*8.8*
both sides	**9.9**	*10.1*
1996-97		
3.2L		
right side	**9.3**	*9.5*
left side	**5.4**	*5.6*
both sides	**11.0**	*11.2*
1998-02		
3.5L, 4.3L		
right side	**7.8**	*8.0*
left side	**7.9**	*8.1*
both sides	**11.1**	*11.3*
VehiCROSS		
right side	**12.4**	*12.6*
left side	**11.8**	*12.0*
both sides	**18.0**	*18.2*
Cylinder Head Gasket (B)		
Amigo		
1989-94		
2.3L	**4.4**	*4.6*
2.6L	**8.0**	*8.2*
1998-00		
2.2L.	**4.9**	*5.1*
3.2L		
right side	**8.6**	*8.8*
left side	**7.7**	*7.9*
both sides	**10.5**	*10.7*
Ascender		
4.2L		
2WD (13.2)	**17.8**	*18.0*
4WD (15.8)	**21.3**	*21.5*
5.3L		
one side (8.2)	**11.1**	*11.3*
both sides (11.9)	**16.1**	*16.7*
Axiom		
right side	**9.5**	*9.7*
left side	**8.5**	*8.7*
both sides	**11.3**	*11.5*
Hombre		
2.2L	**4.4**	*4.6*
3.2L, 4.3L		
right side	**6.3**	*6.5*
left side	**6.8**	*7.0*
both sides	**8.6**	*8.8*
Oasis	**6.0**	*6.2*
Pick-Up		
2.3L	**4.4**	*4.6*
2.6L	**8.0**	*8.2*
3.1L		
one side	**7.1**	*7.3*
both sides	**7.7**	*7.9*
Rodeo/Sport		
1991-97		
2.6L	**5.2**	*5.4*
3.1L		
one side	**7.4**	*7.6*
both sides	**9.5**	*9.7*
3.2L		
right side	**8.9**	*9.1*
left side	**5.8**	*6.0*
both sides	**10.5**	*10.7*
1998-01		
2.2L.	**4.9**	*5.1*
3.2L		
right side	**8.6**	*8.8*
left side	**7.8**	*8.0*
both sides	**10.5**	*10.7*
2002-04		
2.2L (4.0)	**5.4**	*5.6*
3.2L, 3.5L		
right side (7.0)	**9.4**	*9.6*
left side (6.3)	**8.5**	*8.7*
both sides (8.4)	**11.3**	*11.5*
Trooper		
1984-91		
2.3L	**4.4**	*4.6*
2.6L	**8.0**	*8.2*
2.8L		
right side	**6.4**	*6.6*
left side	**6.3**	*6.5*
both sides	**7.7**	*7.9*
1992-95		
3.2L		
SOHC		
right side	**8.6**	*8.8*
left side	**5.5**	*5.7*
both	**10.1**	*10.3*
DOHC sides		
right side	**8.2**	*8.4*
left side	**5.5**	*5.7*
both sides	**8.9**	*9.1*
1996-97		
3.2L		
right side	**8.3**	*8.5*
left side	**4.9**	*5.1*
both sides	**9.8**	*10.0*
1998-02		
3.5L, 4.3L		
right side	**6.9**	*7.1*
left side	**7.1**	*7.3*
both sides	**9.8**	*10.0*
VehiCROSS		
right side	**9.2**	*9.4*
left side	**8.8**	*9.0*
both sides	**11.3**	*11.5*
Rocker Arm Cover Gasket, Replace or Reseal (B)		
Amigo		
1989-94		
2.3L	**.7**	*.9*
2.6L	**1.0**	*1.2*
1998-00		
2.2L	**1.0**	*1.2*
3.2L		
right or left side	**1.1**	*1.3*
both sides	**2.1**	*2.3*
Ascender		
4.2L		
w/AC (4.7)	**6.3**	*6.5*
w/o AC (3.8)	**5.1**	*5.3*
5.3L		
one side (1.4)	**1.9**	*2.1*
both sides (1.9)	**2.6**	*2.8*
Axiom		
right or left side	**1.2**	*1.4*
both sides	**2.3**	*2.5*
Hombre		
2.2L	**.7**	*.9*
3.2L, 4.3L		
one side	**.7**	*.9*
both sides	**1.4**	*1.6*
Oasis	**.5**	*.7*
Pick-Up		
2.3L	**.7**	*.8*
2.6L	**1.0**	*1.2*
3.1L		
right side	**1.4**	*1.6*
left side	**.8**	*1.0*
both sides	**2.2**	*2.4*
Rodeo/Sport		
1991-97		
2.6L	**.8**	*1.0*
3.1L		
right side	**1.4**	*1.6*
left side	**.8**	*1.0*
both sides	**2.2**	*2.4*
3.2L		
right side	**4.4**	*4.6*
left side	**2.0**	*2.2*
both sides	**5.2**	*5.4*
1998-04		
2.2L (.8)	**1.1**	*1.3*
3.2L, 3.5L		
right or left side (.9)	**1.2**	*1.4*
both sides (1.7)	**2.3**	*2.5*

ISU-28 AMIGO : ASCENDER : AXIOM : HOMBRE : OASIS : PICKUP

	LABOR TIME	SEVERE SERVICE
Rocker Arm Cover Gasket, Replace or Reseal (B)		
Trooper		
1984-91		
2.3L	.7	.9
2.6L	1.0	1.2
2.8L		
right side	1.4	1.6
left side	.8	1.0
both sides	2.2	2.4
1992-97		
3.2L		
SOHC		
right side	4.4	4.6
left side	2.0	2.2
both sides	5.0	5.2
DOHC		
right side	5.0	5.2
left side	2.5	2.7
both sides	6.0	6.2
1998-02		
3.5L, 4.3L		
right side	1.2	1.4
left side	1.7	1.9
both sides	2.5	2.7
VehiCROSS		
right side	1.4	1.6
left side	1.7	1.9
both sides	3.0	3.2
Rocker Arm Shaft, Replace (B)		
Gasoline	1.4	1.6
Diesel	1.4	1.6
w/AT add	.2	.4
Replace rocker arms add each side	.3	.5
Rocker Arms, Replace (B)		
4 cyl.		
2.2L, 2.3L, 2.6L		
one cyl.	1.5	1.7
all cyls.	1.8	2.0
L6		
Ascender		
one cyl. (4.6)	6.2	6.4
all cyls. (5.1)	6.9	7.1
V6		
Hombre		
one side	1.3	1.5
both sides	2.2	2.4
Pick-Up		
one cyl.		
right side	1.8	2.0
left side	1.6	1.8
each addl. add	.2	.3
Rodeo/Sport		
3.1L		
one cyl.		
right side	1.8	2.0
left side	1.6	1.8
each addl. add	.2	.3

	LABOR TIME	SEVERE SERVICE
3.2L		
one cyl.		
right side	7.9	8.1
left side	5.9	6.1
each addl. add	.2	.3
Trooper		
SOHC		
right side	8.2	8.4
left side	5.9	6.1
V8		
Ascender		
one side (1.2)	1.6	1.8
both sides (1.9)	2.6	2.8
w/AC Ascender add	.9	1.0
Valve Job (A)		
Amigo		
1989-94		
2.3L	7.2	7.4
2.6L	9.5	9.7
1998-00		
2.2L	15.3	15.5
3.2L		
right side	11.6	11.8
left side	9.7	9.9
both sides	12.4	12.6
Ascender		
4.2L		
2WD (13.5)	18.2	18.4
4WD (16.1)	21.7	21.9
5.3L		
one side (7.6)	10.3	10.5
both sides (10.8)	14.6	14.8
Axiom		
right side	12.0	12.2
left side	9.9	10.1
both sides	12.7	12.9
Hombre	7.3	7.5
Oasis	10.2	10.4
Pick-Up		
2.3L	7.2	7.4
2.6L	9.5	9.7
3.1L		
one side	9.2	9.4
both sides	12.7	12.9
Rodeo/Sport		
1991-97		
2.6L	8.3	8.5
3.1L		
one side	8.9	9.1
both sides	11.2	11.4
3.2L		
right side	10.5	10.7
left side	7.1	7.3
both sides	14.2	14.4
1998-01		
2.2L	18.9	19.1
3.2L		
right side	11.0	11.2
left side	9.0	9.2
both sides	11.7	11.9

	LABOR TIME	SEVERE SERVICE
2002-04		
2.2L (14.6)	19.7	19.9
3.2L, 3.5L		
right side (8.9)	12.0	12.2
left side (7.3)	9.9	10.1
both sides (9.4)	12.7	12.9
Trooper		
1984-91		
2.3L	7.2	7.4
2.6L	9.5	9.7
2.8L		
one side	9.2	9.4
both sides	13.2	13.4
1992-97		
3.2L		
SOHC		
right side	10.5	10.7
left side	7.6	7.8
both sides	14.2	14.4
DOHC		
right side	9.8	10.0
left side	7.9	8.8
both sides	11.5	11.7
Valve Lifters, Replace (B)		
4 cyl.		
intake or exhaust		
one or all	2.1	2.3
L6		
Ascender		
4.2L		
intake or exhaust (4.7)	6.3	6.5
both (5.3)	7.2	7.4
V6		
Amigo		
one cyl.		
right side	1.8	2.0
left side	1.6	1.8
each addl. add	.3	.4
Axiom		
right side	1.8	2.0
left side	1.5	1.7
both sides	3.3	3.5
second camshaft each side add	.5	.7
Hombre 3.2L	4.2	4.4
Pick-Up		
one cyl.	4.5	4.7
each addl. add	.3	.4
Rodeo/Sport		
1991-97		
3.1L		
one cyl.	5.5	5.7
each addl. add	.3	.4
3.2L		
right side	6.3	6.5
left side	5.6	5.8

RODEO : RODEO SPORT : TROOPER : VEHICROSS ISU-29

	LABOR TIME	SEVERE SERVICE
1998-01		
one cyl.		
right side	1.8	2.0
left side	1.6	1.8
both sides	3.3	3.5
each addl. add	.3	.4
2002-04		
3.2L		
right side (1.2)	1.8	2.0
left side (1.1)	1.5	1.7
second camshaft		
each side add	.5	.5
Trooper		
1989-91		
one cyl.	4.5	4.8
each addl. add	.3	.4
1992-97		
right side	6.3	6.6
left side	5.6	5.9
VehiCROSS		
right side	2.3	2.6
left side	2.7	3.0
both sides	4.3	4.6
V8		
Ascender		
5.3L		
one side (6.8)	9.2	9.4
both sides (10.8)	14.6	14.8
w/skid plate Hombre add	.3	.3
Valve Pushrods, Replace (B)		
2.8L		
Trooper one		
right side	1.9	2.1
left side	1.5	1.7
each addl. add	.1	.2
3.1L		
Hombre		
one	1.0	1.2
all	1.5	1.7
Pick-Up		
one right side	1.9	2.1
one left side	1.5	1.7
each addl. add	.1	.2
Rodeo		
one right side	1.9	2.1
one left side	1.7	1.9
each addl. add	.1	.2
5.3L		
Ascender		
one side (2.2)	3.0	3.2
both sides (3.2)	4.3	4.5
Valve Springs and/or Oil Seals, Replace (B)		
4 cyl.		
Amigo		
1989-94	3.8	4.0
1998-00	2.7	2.9
Hombre	2.8	3.0
Oasis	4.5	4.7

	LABOR TIME	SEVERE SERVICE
Pick-Up	3.8	4.0
Rodeo/Sport	3.1	3.3
Trooper II	3.8	4.0
L6		
4.2L Ascender		
one cyl. (5.8)	7.8	8.0
all cyls. (7.9)	10.7	10.9
V6		
2.8L Trooper		
one cyl.		
right side (2.1)	2.8	3.0
left side (1.4)	1.9	2.1
all cyls. (5.3)	7.2	7.4
3.1L Pick-Up, Rodeo		
one cyl.		
right side (1.8)	2.4	2.6
left side (1.4)	1.9	2.1
all cyls. (5.0)	6.8	7.0
3.2L		
Amigo		
right side	8.3	8.5
left side	6.3	6.5
both sides	12.8	13.0
Rodeo/Sport		
right side	8.5	8.7
left side	6.3	6.5
both sides	10.7	10.9
each addl. add	.8	.8
Trooper		
1992-97 SOHC		
right side	7.8	8.0
left side	6.7	6.9
both sides	9.4	9.6
1992-97 DOHC		
one cyl. right side	6.3	6.5
one cyl. left side	5.2	5.4
all cyls.	10.0	10.2
1998-02		
right side	6.4	6.6
left side	5.3	5.5
3.5L		
Axiom		
right side	8.5	8.7
left side	6.3	6.5
both sides	10.7	10.9
each addl. add	.8	.8
VehiCROSS		
right side	11.8	12.0
left side	11.4	11.6
both sides	17.0	17.2
V8		
5.3L Ascender		
one side all (2.5)	3.4	3.6
both sides (4.3)	5.8	6.0

	LABOR TIME	SEVERE SERVICE
CAMSHAFT		
Timing Belt, Adjust (B)		
2.2L 16 valve	.5	.7
3.2L	.5	.7
Camshaft, Replace (A)		
Amigo		
1989-94		
2.3L	1.7	1.9
2.6L	3.1	3.3
1998-00		
2.2L		
intake or exhaust	2.4	2.6
both	2.9	3.1
3.2L		
right side one	1.8	2.0
left side one	1.8	2.0
Ascender		
4.2L		
intake or exhaust (4.9)	6.6	6.8
both (5.3)	7.2	7.4
5.3L (16.1)	21.7	21.9
Axiom		
3.5L		
one side		
intake or exhaust	2.1	2.3
both sides	3.2	3.4
second camshaft		
each side add	.3	.3
Hombre	6.4	6.6
Oasis	3.7	3.9
Pick-Up		
2.3L	1.8	2.0
2.6L	3.1	3.3
2.8L	10.0	10.2
diesel		
2WD	11.0	11.2
4WD	11.8	12.0
Rodeo/Sport		
1991-97		
2.6L	2.3	2.5
3.1L	10.0	10.2
3.2L		
right side	7.4	7.6
left side	5.2	5.4
both sides	9.1	9.3
1998-04		
2.2L		
intake or exhaust (1.8)	2.4	2.6
both (2.2)	3.0	3.2
3.2L, 3.5L		
one side		
intake or exhaust (1.3)	1.8	2.1
both sides (2.2)	3.0	3.2
second camshaft		
each side add	.3	.3

ISU-30   AMIGO : ASCENDER : AXIOM : HOMBRE : OASIS : PICKUP

	LABOR TIME	SEVERE SERVICE
Camshaft, Replace (A)		
Trooper		
1984-91		
2.3L	1.8	2.0
2.6L	2.3	2.5
2.8L	9.2	9.4
diesel	11.8	12.0
1992-97		
3.2L		
SOHC		
right side	7.4	7.6
left side	5.0	5.2
both sides	9.1	9.3
DOHC		
right side	5.5	5.7
left side	4.9	5.1
both sides	6.5	6.7
1998-02		
3.5L		
right side	1.6	1.9
left side	2.3	2.5
both sides	3.4	3.6
VehiCROSS		
right side		
intake or exhaust	1.8	2.0
both	2.0	2.2
left side		
intake or exhaust	2.2	2.4
both	2.5	2.7
both sides		
intake or exhaust	3.4	3.6
Camshaft Drive Gear, Replace (B)		
VehiCROSS		
right	4.2	4.4
left	4.8	5.0
both	5.8	6.0
Camshaft Idle Gear Shaft Oil Seal, Replace (B)		
VehiCROSS		
right or left	2.6	2.8
Camshaft Seal, Replace (B)		
Amigo, Rodeo		
2.2L one	1.9	2.1
each addl. add	.3	.4
Oasis	1.4	1.6
Camshaft Timing Pulley, Replace (B)		
Axiom		
one	2.3	2.5
both	4.9	5.1
Rodeo/Sport		
3.2L, 3.5L		
one	2.3	2.5
both	4.9	5.1
VehiCROSS		
right or left	2.5	2.7
Timing Belt, Replace (B)		
2.2L	1.2	1.4
2.3L	3.9	4.1
2.6L		

	LABOR TIME	SEVERE SERVICE
Amigo, Pick-Up,		
Trooper II	2.9	3.1
Rodeo	1.7	1.9
3.2L, 3.5L, 4.3L		
Amigo	2.5	2.7
Axiom	2.4	2.6
Oasis	3.1	3.3
Rodeo/Sport		
1991-97	3.1	3.3
1998-04	2.5	2.7
Trooper		
1992-97		
SOHC	2.8	3.0
DOHC	2.4	2.6
1998-02	3.0	3.3
VehiCROSS	2.4	2.6
Diesel	3.1	3.3
Replace		
oil pump add	2.3	2.5
tensioner add	.2	.4
water pump add	.3	.5
Timing Belt Case and/or Cover, Replace (B)		
Amigo		
1989-94		
2.3L	3.4	3.6
2.6L	2.6	3.0
1998-00		
2.2L	1.0	1.2
3.2L	2.0	2.2
Axiom	2.0	2.2
Oasis	.5	.7
Pick-Up		
2.3L	3.4	3.6
2.6L	2.8	3.0
Rodeo/Sport		
1991-97		
2.6L	1.4	1.6
3.1L, 3.2L	2.9	3.1
1998-04		
2.2L	1.0	1.2
3.2L	2.0	2.2
Trooper		
1984-91		
2.3L	3.4	3.6
1992-95	1.9	2.1
1996-02	.8	1.1
VehiCROSS	2.0	2.2
Timing Belt Idle Pulley, Replace (B)		
VehiCROSS	2.2	2.4
Timing Belt Tension Pulley, Replace (B)		
Axiom	2.4	2.6
Rodeo/Sport	1.1	1.3
VehiCROSS	2.2	2.4

	LABOR TIME	SEVERE SERVICE
Timing Chain, Replace (B)		
Includes: Engine R&R if necessary, adjustments.		
Ascender		
4.2L		
2WD (10.6)	14.3	14.5
4WD (14.1)	19.0	19.2
5.3L (8.6)	11.6	11.8
Trooper		
2.8L	5.2	5.4
w/AC add		
Ascender	1.0	1.2
Trooper	.3	.5
w/AT Trooper add	.3	.5
w/PS Trooper add	.3	.5
Timing Chain Cover and/or Gasket, Replace (B)		
Ascender		
4.2L		
2WD (10.2)	13.8	14.0
4WD (13.7)	18.5	18.7
5.3L (8.2)	11.1	11.3
Trooper		
2.8L	4.8	5.0
w/AC add		
Ascender	1.0	1.2
Trooper	.3	.5
w/AT Trooper add	.3	.5
w/PS Trooper add	.3	.5
Timing Gear, Replace (B)		
Diesel	4.4	4.6
CRANK & PISTONS		
Connecting Rod Bearings, Replace (A)		
Includes: Check bearing oil clearance.		
Amigo		
1989-94		
2.3L		
2WD	8.0	8.2
4WD	10.8	11.0
2.6L		
2WD	9.9	10.1
4WD	11.3	11.5
1998-00		
2.2L	5.5	5.7
3.2L		
2WD	9.6	9.8
4WD	14.7	14.9
Ascender		
4.2L		
2WD (13.2)	17.7	17.9
4WD (15.7)	21.2	21.4
5.3L		
2WD (5.2)	7.0	7.2
4WD (6.6)	8.9	9.1
Axiom		
2WD	9.4	9.6
4WD	14.3	14.5

RODEO : RODEO SPORT : TROOPER : VEHICROSS   ISU-31

	LABOR TIME	SEVERE SERVICE
Hombre	9.8	10.0
Oasis	3.4	3.6
Pick-Up		
2.3L		
2WD	8.0	8.2
4WD	10.8	11.0
2.6L		
2WD	9.9	10.1
4WD	11.3	11.5
3.1L	17.1	17.3
Rodeo/Sport		
1991-97		
2.6L	5.4	5.6
3.1L	6.8	7.0
3.2L		
2WD	7.8	8.0
4WD	8.7	8.9
1998-04		
2.2L	5.3	5.5
3.2L, 3.5L		
2WD	9.4	9.6
4WD	14.4	14.6
Trooper		
1984-91		
2.3L		
2WD	8.0	8.2
4WD	10.8	11.0
2.6L	11.0	11.2
2.8L	16.8	17.0
1992-02	14.9	15.1
VehiCROSS	19.9	20.1
w/AC add	.3	.5
w/AT add	.2	.4
w/PS add	.3	.5

Crankshaft and Main Bearings, Replace (A)
Includes: Engine R&R, check bearing oil clearance.

	LABOR TIME	SEVERE SERVICE
Amigo		
1989-95		
2.3L		
2WD	10.0	10.2
4WD	12.7	12.9
2.6L		
2WD	11.0	11.2
4WD	12.7	12.9
1998-00		
2.2L	14.3	14.5
3.2L		
2WD	17.5	17.7
4WD	19.9	20.1
Ascender		
4.2L		
2WD (16.3)	22.0	22.2
4WD (18.9)	25.5	25.7
5.3L		
2WD (21.7)	29.3	29.5
4WD (23.7)	32.0	32.2

	LABOR TIME	SEVERE SERVICE
Axiom		
2WD	17.1	17.3
4WD	18.9	19.1
Hombre	11.7	11.9
Oasis	12.3	12.5
Pick-Up		
2.3L		
2WD	10.0	10.2
4WD	12.7	12.9
2.6L		
2WD	11.0	11.2
4WD	12.7	12.9
3.1L	17.3	17.5
Rodeo/Sport		
1991-97		
2.6L	11.4	11.6
3.1L	12.3	12.5
3.2L		
2WD	15.5	15.7
4WD	17.1	17.3
1998-04		
2.2L	14.3	14.5
3.2L, 3.5L		
2WD	17.1	17.3
4WD	18.9	19.1
Trooper		
1984-91		
2.3L	12.7	12.9
2.6L	11.0	11.2
2.8L	17.3	17.5
1992-97	19.3	19.5
1998-02		
AT	24.9	25.1
MT	22.2	22.4
w/AC add	.3	.5
w/AT add	.6	.8
w/PS add	.3	.5

Crankshaft Front Oil Seal, Replace (B)

	LABOR TIME	SEVERE SERVICE
4 cyl.		
2.2L	3.7	3.9
2.2L DOHC	2.2	2.4
2.3L	4.2	4.4
2.6L	2.5	2.7
V6		
2.8L, 3.1L	3.8	4.0
3.2L	3.5	3.7
3.5L	3.3	3.5
V8		
5.3L	1.5	1.8
Diesel	3.7	3.9

Crankshaft Pulley or Damper, Replace (B)

	LABOR TIME	SEVERE SERVICE
4 cyl.		
Amigo		
1989-94	1.6	1.9
1998-00	.3	.4
Hombre		
pulley	1.0	1.2
balancer	1.4	1.6

	LABOR TIME	SEVERE SERVICE
Oasis	.5	.7
Pick-Up	.8	1.0
Rodeo/Sport		
1991-97	1.0	1.2
1998-04	.3	.5
Trooper	1.6	1.8
L6		
Ascender (2.8)	3.8	4.0
V6		
Amigo, Hombre	.8	1.0
Axiom, Pick-Up	.8	1.0
Rodeo/Sport		
1991-97	1.0	1.2
1998-04	.8	1.0
Trooper		
1989-91	.8	1.0
1992-97	2.8	3.0
1998-02	2.0	2.2
VehiCROSS	.7	.9
V8		
Ascender (1.1)	1.5	1.7
Diesel	1.7	1.9
w/AC add	.2	.2

Main Bearings, Replace (A)
Includes: Check bearing oil clearance.

	LABOR TIME	SEVERE SERVICE
Gasoline		
Ascender		
4.2L		
2WD (16.6)	22.4	22.6
4WD (19.3)	26.1	26.3
5.3L		
2WD (21.6)	29.2	29.4
4WD (22.4)	30.2	30.3
Axiom		
2WD	17.1	19.1
4WD	18.9	21.0
Hombre	9.4	10.6
Oasis	3.7	4.2
Diesel		
2WD	11.6	13.0
4WD	12.4	13.8
w/AC add	.8	.8
w/AT Axiom add	.6	.6
w/turbo add	.7	.7

Pistons and Connecting Rods, Replace (A)
Includes: Ridge reaming, cylinder wall deglazing, installing new rings and rod bearings, engine tune-up.

	LABOR TIME	SEVERE SERVICE
Amigo		
1989-94		
2.3L		
2WD	11.9	12.8
4WD	15.4	16.3
2.6L		
2WD	13.1	14.0
4WD	14.5	15.4

ISU-32 AMIGO : ASCENDER : AXIOM : HOMBRE : OASIS : PICKUP

	LABOR TIME	SEVERE SERVICE
Pistons and Connecting Rods, Replace (A)		
1998-00		
2.2L	13.2	14.1
3.1L		
2WD	16.3	17.2
4WD	17.9	18.8
Ascender		
4.2L		
2WD (17.9)	24.2	24.4
4WD (20.4)	27.5	27.7
5.3L		
2WD (19.1)	25.8	26.0
4WD (21.5)	29.0	29.2
Axiom		
2WD		
right side	15.5	17.3
left side	14.5	16.2
both sides	17.3	19.3
4WD		
right side	20.1	22.3
left side	19.2	21.3
both sides	21.9	24.3
Hombre	16.3	17.3
Oasis	14.1	15.0
Pick-Up		
2.3L		
2WD	11.9	12.8
4WD	15.4	16.3
2.6L		
2WD	13.1	14.0
4WD	14.5	15.4
3.1L	16.4	17.3
Rodeo/Sport		
1991-97		
2.6L	12.9	13.8
3.1L	15.3	16.2
3.2L	16.3	17.2
1998-99		
2.2L	13.2	14.1
3.1L		
2WD	16.3	17.2
4WD	17.9	18.8
2002-04		
2.2L	12.9	14.5
3.2L, 3.5L		
2WD		
right side	15.5	17.3
left side	14.5	16.2
both sides	17.3	19.3
4WD		
right side	20.1	22.3
left side	19.2	21.3
both sides	21.9	24.3
Trooper		
1984-91		
2.3L		
2WD	11.9	12.8
4WD	15.4	16.3

	LABOR TIME	SEVERE SERVICE
2.6L	14.5	15.0
2.8L	16.4	17.3
1992-97		
SOHC		
right side	14.5	15.4
left side	8.3	9.2
both sides	16.9	17.8
DOHC		
right side	14.1	15.0
left side	12.0	12.9
both sides	14.8	15.7
1998-02		
right side	12.8	13.7
left side	10.3	11.2
both sides	15.8	16.7
VehiCROSS		
right	17.2	18.1
left	16.8	17.7
both	19.6	20.5
w/AC 81-01 add	.3	.3
w/AT 81-01 add	.2	.2
w/PS 81-01 add	.3	.3
Rings, Replace (A)		
Includes: Ridge reaming, cylinder wall deglazing, installing new rings, engine tune-up.		
Amigo		
1981-95		
2.3L		
2WD	10.5	11.7
4WD	14.2	15.9
2.6L		
2WD	11.8	13.2
4WD	13.3	14.8
1998-00		
2.2L	13.0	14.6
3.2L		
2WD	16.1	17.9
4WD	21.2	23.5
Ascender		
4.2L		
2WD (16.9)	22.8	23.0
4WD (19.4)	26.2	26.4
5.3L		
2WD (18.1)	24.4	24.6
4WD (20.5)	27.7	27.9
Axiom		
2WD	15.9	17.7
4WD	20.6	22.8
Hombre	15.3	17.1
Oasis	12.9	14.5
Pick-Up		
2.3L		
2WD	10.5	11.7
4WD	14.2	15.9
2.6L		
2WD	11.8	13.2
4WD	13.3	14.8
3.1L	14.6	16.3

	LABOR TIME	SEVERE SERVICE
Rodeo/Sport		
1991-97		
2.6L	11.6	13.0
3.1L	13.3	14.8
3.2L	14.5	16.2
1998-01		
2.2L	14.6	16.3
3.2L		
2WD	14.9	16.6
4WD	15.9	17.7
2002-04		
2.2L	14.3	16.1
3.2L, 3.5L		
2WD	15.9	17.7
4WD	20.6	22.8
Trooper		
1984-91		
2.3L	14.2	15.9
2.6L	13.3	14.8
2.8L	14.6	16.3
1992-97		
SOHC	15.2	17.0
DOHC	14.9	16.6
1998-02	15.8	17.5
w/AC 81-01 add	.3	.3
w/AT 81-01 add	.2	.2
w/PS 81-01 add	.3	.3
ENGINE LUBRICATION		
Engine Oil Cooler, Replace (B)		
Hombre	.8	1.0
1991-97 Rodeo	1.2	1.4
1992-97 Trooper	1.2	1.4
Engine Oil Pressure Switch (Sending Unit), Replace (B)		
1998-00 Amigo		
2.2L	.3	.5
3.2L	.7	.9
Ascender		
4.2L (.8)	1.1	1.3
5.3L (.5)	.7	.9
Axiom	.3	.5
Pick-Up	.5	.7
1998-04 Rodeo/Sport		
2.2L	.3	.5
3.2L, 3.5L		
exc. below	.7	.9
2002-04	.3	.5
Trooper		
1984-97	.5	.7
1998-02	.5	.7
VehiCROSS	.5	.7
Oil Pan and/or Gasket, Replace (B)		
Amigo		
1989-94		
2WD	2.4	2.6
4WD	10.7	10.9

RODEO : RODEO SPORT : TROOPER : VEHICROSS ISU-31

	LABOR TIME	SEVERE SERVICE
Hombre	9.8	10.0
Oasis	3.4	3.6
Pick-Up		
2.3L		
2WD	8.0	8.2
4WD	10.8	11.0
2.6L		
2WD	9.9	10.1
4WD	11.3	11.5
3.1L	17.1	17.3
Rodeo/Sport		
1991-97		
2.6L	5.4	5.6
3.1L	6.8	7.0
3.2L		
2WD	7.8	8.0
4WD	8.7	8.9
1998-04		
2.2L	5.3	5.5
3.2L, 3.5L		
2WD	9.4	9.6
4WD	14.4	14.6
Trooper		
1984-91		
2.3L		
2WD	8.0	8.2
4WD	10.8	11.0
2.6L	11.0	11.2
2.8L	16.8	17.0
1992-02	14.9	15.1
VehiCROSS	19.9	20.1
w/AC add	.3	.5
w/AT add	.2	.4
w/PS add	.3	.5

Crankshaft and Main Bearings, Replace (A)
Includes: Engine R&R, check bearing oil clearance.

	LABOR TIME	SEVERE SERVICE
Amigo		
1989-95		
2.3L		
2WD	10.0	10.2
4WD	12.7	12.9
2.6L		
2WD	11.0	11.2
4WD	12.7	12.9
1998-00		
2.2L	14.3	14.5
3.2L		
2WD	17.5	17.7
4WD	19.9	20.1
Ascender		
4.2L		
2WD (16.3)	22.0	22.2
4WD (18.9)	25.5	25.7
5.3L		
2WD (21.7)	29.3	29.5
4WD (23.7)	32.0	32.2
Axiom		
2WD	17.1	17.3
4WD	18.9	19.1
Hombre	11.7	11.9
Oasis	12.3	12.5
Pick-Up		
2.3L		
2WD	10.0	10.2
4WD	12.7	12.9
2.6L		
2WD	11.0	11.2
4WD	12.7	12.9
3.1L	17.3	17.5
Rodeo/Sport		
1991-97		
2.6L	11.4	11.6
3.1L	12.3	12.5
3.2L		
2WD	15.5	15.7
4WD	17.1	17.3
1998-04		
2.2L	14.3	14.5
3.2L, 3.5L		
2WD	17.1	17.3
4WD	18.9	19.1
Trooper		
1984-91		
2.3L	12.7	12.9
2.6L	11.0	11.2
2.8L	17.3	17.5
1992-97	19.3	19.5
1998-02		
AT	24.9	25.1
MT	22.2	22.4
w/AC add	.3	.5
w/AT add	.6	.8
w/PS add	.3	.5

Crankshaft Front Oil Seal, Replace (B)

	LABOR TIME	SEVERE SERVICE
4 cyl.		
2.2L	3.7	3.9
2.2L DOHC	2.2	2.4
2.3L	4.2	4.4
2.6L	2.5	2.7
V6		
2.8L, 3.1L	3.8	4.0
3.2L	3.5	3.7
3.5L	3.3	3.5
V8		
5.3L	1.5	1.8
Diesel	3.7	3.9

Crankshaft Pulley or Damper, Replace (B)

	LABOR TIME	SEVERE SERVICE
4 cyl.		
Amigo		
1989-94	1.6	1.9
1998-00	.3	.4
Hombre		
pulley	1.0	1.2
balancer	1.4	1.6
Oasis	.5	.7
Pick-Up	.8	1.0
Rodeo/Sport		
1991-97	1.0	1.2
1998-04	.3	.5
Trooper	1.6	1.8
L6		
Ascender (2.8)	3.8	4.0
V6		
Amigo, Hombre	.8	1.0
Axiom, Pick-Up	.8	1.0
Rodeo/Sport		
1991-97	1.0	1.2
1998-04	.8	1.0
Trooper		
1989-91	.8	1.0
1992-97	2.8	3.0
1998-02	2.0	2.2
VehiCROSS	.7	.9
V8		
Ascender (1.1)	1.5	1.7
Diesel	1.7	1.9
w/AC add	.2	.2

Main Bearings, Replace (A)
Includes: Check bearing oil clearance.

	LABOR TIME	SEVERE SERVICE
Gasoline		
Ascender		
4.2L		
2WD (16.6)	22.4	22.6
4WD (19.3)	26.1	26.3
5.3L		
2WD (21.6)	29.2	29.4
4WD (22.4)	30.2	30.3
Axiom		
2WD	17.1	19.1
4WD	18.9	21.0
Hombre	9.4	10.6
Oasis	3.7	4.2
Diesel		
2WD	11.6	13.0
4WD	12.4	13.8
w/AC add	.8	.8
w/AT Axiom add	.6	.6
w/turbo add	.7	.7

Pistons and Connecting Rods, Replace (A)
Includes: Ridge reaming, cylinder wall deglazing, installing new rings and rod bearings, engine tune-up.

	LABOR TIME	SEVERE SERVICE
Amigo		
1989-94		
2.3L		
2WD	11.9	12.8
4WD	15.4	16.3
2.6L		
2WD	13.1	14.0
4WD	14.5	15.4

ISU-32 AMIGO : ASCENDER : AXIOM : HOMBRE : OASIS : PICKUP

	LABOR TIME	SEVERE SERVICE
Pistons and Connecting Rods, Replace (A)		
1998-00		
2.2L	13.2	14.1
3.1L		
2WD	16.3	17.2
4WD	17.9	18.8
Ascender		
4.2L		
2WD (17.9)	24.2	24.4
4WD (20.4)	27.5	27.7
5.3L		
2WD (19.1)	25.8	26.0
4WD (21.5)	29.0	29.2
Axiom		
2WD		
right side	15.5	17.3
left side	14.5	16.2
both sides	17.3	19.3
4WD		
right side	20.1	22.3
left side	19.2	21.3
both sides	21.9	24.3
Hombre	16.3	17.3
Oasis	14.1	15.0
Pick-Up		
2.3L		
2WD	11.9	12.8
4WD	15.4	16.3
2.6L		
2WD	13.1	14.0
4WD	14.5	15.4
3.1L	16.4	17.3
Rodeo/Sport		
1991-97		
2.6L	12.9	13.8
3.1L	15.3	16.2
3.2L	16.3	17.2
1998-99		
2.2L	13.2	14.1
3.1L		
2WD	16.3	17.2
4WD	17.9	18.8
2002-04		
2.2L	12.9	14.5
3.2L, 3.5L		
2WD		
right side	15.5	17.3
left side	14.5	16.2
both sides	17.3	19.3
4WD		
right side	20.1	22.3
left side	19.2	21.3
both sides	21.9	24.3
Trooper		
1984-91		
2.3L		
2WD	11.9	12.8
4WD	15.4	16.3

	LABOR TIME	SEVERE SERVICE
2.6L	14.5	15.0
2.8L	16.4	17.3
1992-97		
SOHC		
right side	14.5	15.4
left side	8.3	9.2
both sides	16.9	17.8
DOHC		
right side	14.1	15.0
left side	12.0	12.9
both sides	14.8	15.7
1998-02		
right side	12.8	13.7
left side	10.3	11.2
both sides	15.8	16.7
VehiCROSS		
right	17.2	18.1
left	16.8	17.7
both	19.6	20.5
w/AC 81-01 add	.3	.3
w/AT 81-01 add	.2	.2
w/PS 81-01 add	.3	.3
Rings, Replace (A)		
Includes: Ridge reaming, cylinder wall deglazing, installing new rings, engine tune-up.		
Amigo		
1981-95		
2.3L		
2WD	10.5	11.7
4WD	14.2	15.9
2.6L		
2WD	11.8	13.2
4WD	13.3	14.8
1998-00		
2.2L	13.0	14.6
3.2L		
2WD	16.1	17.9
4WD	21.2	23.5
Ascender		
4.2L		
2WD (16.9)	22.8	23.0
4WD (19.4)	26.2	26.4
5.3L		
2WD (18.1)	24.4	24.6
4WD (20.5)	27.7	27.9
Axiom		
2WD	15.9	17.7
4WD	20.6	22.8
Hombre	15.3	17.1
Oasis	12.9	14.5
Pick-Up		
2.3L		
2WD	10.5	11.7
4WD	14.2	15.9
2.6L		
2WD	11.8	13.2
4WD	13.3	14.8
3.1L	14.6	16.3

	LABOR TIME	SEVERE SERVICE
Rodeo/Sport		
1991-97		
2.6L	11.6	13.0
3.1L	13.3	14.8
3.2L	14.5	16.2
1998-01		
2.2L	14.6	16.3
3.2L		
2WD	14.9	16.6
4WD	15.9	17.7
2002-04		
2.2L	14.3	16.1
3.2L, 3.5L		
2WD	15.9	17.7
4WD	20.6	22.8
Trooper		
1984-91		
2.3L	14.2	15.9
2.6L	13.3	14.8
2.8L	14.6	16.3
1992-97		
SOHC	15.2	17.0
DOHC	14.9	16.6
1998-02	15.8	17.5
w/AC 81-01 add	.3	.3
w/AT 81-01 add	.2	.2
w/PS 81-01 add	.3	.3
ENGINE LUBRICATION		
Engine Oil Cooler, Replace (B)		
Hombre	.8	1.0
1991-97 Rodeo	1.2	1.4
1992-97 Trooper	1.2	1.4
Engine Oil Pressure Switch (Sending Unit), Replace (B)		
1998-00 Amigo		
2.2L	.3	.5
3.2L	.7	.9
Ascender		
4.2L (.8)	1.1	1.3
5.3L (.5)	.7	.9
Axiom	.3	.5
Pick-Up	.5	.7
1998-04 Rodeo/Sport		
2.2L	.3	.5
3.2L, 3.5L		
exc. below	.7	.9
2002-04	.3	.5
Trooper		
1984-97	.5	.7
1998-02	.5	.7
VehiCROSS	.5	.7
Oil Pan and/or Gasket, Replace (B)		
Amigo		
1989-94		
2WD	2.4	2.6
4WD	10.7	10.9

RODEO : RODEO SPORT : TROOPER : VEHICROSS — ISU-33

	LABOR TIME	SEVERE SERVICE
1998-00		
2.2L	1.3	1.5
3.2L		
2WD	2.8	3.0
4WD	7.9	8.1
Ascender		
4.2L		
2WD (3.9)	5.3	5.3
4WD (7.0)	9.4	9.6
5.3L	2.0	2.4
2WD (3.0)	4.1	4.3
4WD (4.9)	6.6	6.8
Axiom		
2WD	2.9	3.1
4WD	8.1	8.3
Hombre	7.3	7.5
Oasis	2.8	3.0
Pick-Up		
4 cyl.		
2WD		
2.3L	2.3	2.5
2.6L	9.8	10.0
4WD	10.9	11.1
V6	4.9	5.1
diesel		
2WD	8.2	8.4
4WD	9.2	9.4
Rodeo/Sport		
1991-97		
2.6L	4.2	4.4
3.1L	4.9	5.1
3.2L		
2WD	5.9	6.1
4WD	6.8	7.0
1998-04		
2.2L	1.3	1.5
3.2L, 3.5L		
2WD	2.9	3.1
4WD	8.1	8.3
Trooper		
1984-91		
2.3L, 2.6L	2.0	2.2
2.8L	10.7	10.9
diesel	9.2	9.4
1992-97	5.2	5.6
1998-99	7.1	7.3
2000-02		
2WD	5.0	5.2
4WD	7.1	7.3
VehiCROSS	7.6	7.8

Oil Pressure Gauge (Dash), Replace (B)
Exc. below	1.2	1.4
1984-91 Trooper	.7	.9

Oil Pump, Replace (B)
Amigo		
1989-94	2.0	2.5

	LABOR TIME	SEVERE SERVICE
1998-00		
2.2L	3.3	3.8
3.2L		
2WD	6.2	6.7
4WD	11.4	11.9
Ascender		
4.2L		
2WD (7.6)	10.3	10.5
4WD (10.4)	14.0	14.2
5.3L		
2WD (10.3)	13.9	14.1
4WD (12.3)	16.6	16.8
Axiom		
2WD	6.4	6.6
4WD	11.5	11.7
Hombre	7.6	7.8
Oasis	6.5	6.7
Pick-Up		
4 cyl.		
2.3L, 2.6L	2.0	2.2
3.1L	5.2	5.4
diesel		
2WD	8.3	8.5
4WD	9.3	9.5
Rodeo/Sport		
1991-97		
2.6L	2.0	2.2
3.1L		
2WD	2.8	3.0
4WD	5.0	5.2
3.2L		
2WD	9.8	10.0
4WD	10.5	10.77
1998-04		
2.2L	3.5	3.7
3.2L, 3.5L		
2WD	6.4	6.8
4WD	11.5	11.7
Trooper		
1984-91		
2.3L, 2.6L	2.0	2.2
2.8L	10.0	10.2
diesel	9.3	9.5
1992-02	7.5	7.7
VehiCROSS	10.5	10.7

Sub Oil Pan or Gasket, Replace (B)
2.2L	.8	1.0
3.2L, 3.5L	1.6	1.8

CLUTCH

Clutch Hydraulic System, Bleed (B)
All Models	.5	.7

Clutch Pedal Free Play, Adjust (B)
All Models	.3	.5

Clutch Actuator Cylinder, Replace (B)
Includes: System bleeding.
Hombre	1.3	1.5

	LABOR TIME	SEVERE SERVICE
Clutch Assy., Replace (B)		
1981-93		
4 cyl.		
2.3L, 2.6L		
2WD	5.3	5.5
4WD	6.8	7.0
V6		
3.1L		
2WD	8.0	8.2
4WD	8.9	9.1
3.2L		
2WD	6.5	6.7
4WD	8.4	8.6
1994-02 Amigo		
2.2L	3.8	4.0
3.2L		
2WD	6.4	6.6
4WD	8.6	8.8
1994-02 Hombre	4.5	4.7
1994-02 Pick-Up		
2.3L, 2.6L	5.2	5.4
3.1L	6.2	6.4
Rodeo/Sport		
2.2L		
1994-97	5.4	5.6
1998-04	3.8	4.3
3.1L, 3.2L		
1994-04		
2WD	6.4	7.2
4WD	8.4	9.4
Trooper		
1994-97	8.0	8.2
1998-02	6.8	7.0

Clutch Control Cable, Replace (B)
1981-97	1.2	1.4

Clutch Master Cylinder, R&R and Recondition (B)
Includes: System bleeding.
All Models	2.3	2.5

Clutch Master Cylinder, Replace (B)
Includes: System bleeding.
All Models	1.2	1.4

Clutch Release Bearing or Fork, Replace (B)
1981-93		
2.3L, 2.6L		
2WD	5.2	5.4
4WD	6.4	6.6
2.8L, 3.1L		
2WD	7.7	7.9
4WD	8.6	8.8
3.2L		
2WD	7.1	7.3
4WD	8.2	8.4

ISU-34 AMIGO : ASCENDER : AXIOM : HOMBRE : OASIS : PICKUP

	LABOR TIME	SEVERE SERVICE
Clutch Release Bearing or Fork, Replace (B)		
1994-02		
Amigo		
1994-95		
2.3L, 2.6L	5.0	5.2
1998-00		
2.2L	3.5	3.7
3.2L		
2WD	6.1	6.3
4WD	8.3	8.5
Hombre	4.2	4.4
Pick-Up		
2.3L, 2.6L	5.0	5.2
3.1L	5.8	6.0
Rodeo/Sport		
1998-04 2.2L	3.5	3.7
1994-97 2.6L	5.3	5.5
3.2L		
2WD	6.1	6.3
4WD	8.1	8.3
Trooper		
1994-97	7.3	7.5
1998-02	6.4	6.6
Clutch Slave Cylinder, R&R and Reconditon (B)		
Includes: System bleeding.		
All Models	1.2	1.4
Clutch Slave Cylinder, Replace (B)		
Includes: System bleeding.		
All Models	.7	.9
Flywheel, Replace (B)		
1981-93		
2.3L, 2.6L		
2WD	5.6	5.8
4WD	7.1	7.3
2.8L, 3.1L		
2WD	8.3	8.5
4WD	9.2	9.4
3.2L		
2WD	6.9	7.1
4WD	8.8	9.0
1994-02		
Amigo		
1994-95		
2.3L, 2.6L	5.5	5.7
1998-00		
2.2L	4.1	4.3
3.2L		
2WD	6.7	6.9
4WD	8.9	9.1
Hombre	4.9	5.1
Pick-Up		
2.3L, 2.6L	5.5	5.7
3.1L	6.3	6.5
Rodeo/Sport		
2.2L	4.0	4.2
2.6L	5.8	6.0

	LABOR TIME	SEVERE SERVICE
3.2L		
2WD	6.7	6.9
4WD	8.7	8.9
Trooper		
1994-97	8.2	8.4
1998-02	7.3	7.5

MANUAL TRANSMISSION

	LABOR TIME	SEVERE SERVICE
Extension Housing and/or Gasket, Replace (B)		
Amigo		
1989-95		
2WD	3.1	3.3
4WD	2.4	2.6
1998-00		
2WD	2.3	2.5
4WD	3.2	3.4
Pick-Up		
2WD	3.1	3.3
4WD	2.4	2.6
Rodeo/Sport		
2WD	2.4	2.6
4WD	3.2	3.4
1992-02 Trooper	3.4	3.6
Extension Housing Oil Seal, Replace (B)		
All Models	1.2	1.4
Speed Sensor, Replace (B)		
1998-00 Amigo	.3	.5
Hombre	.7	.9
Rodeo	.3	.5
1992-02 Trooper	.3	.5
Speedometer Driven Gear, Replace (B)		
All Models	.5	.7
Transmission Assy., R&R and Recondition (A)		
1981-93		
2.3L, 2.6L		
2WD	11.5	11.7
4WD	13.5	13.7
2.8L, 3.1L		
2WD	10.3	10.5
4WD	15.4	15.6
3.2L		
2WD	10.0	10.2
4WD	14.5	14.7
1994-04		
Amigo		
1994-95		
2WD		
2.3L	10.5	10.7
2.6L	11.5	11.7
4WD		
2.3L	10.5	10.7
2.6L	13.5	13.7
3.1L	14.1	14.3
1998-00		
2.2L	18.2	18.4

	LABOR TIME	SEVERE SERVICE
3.2L		
2WD	12.6	12.8
4WD	15.8	16.0
Hombre	7.6	7.8
Pick-Up		
2WD		
2.3L	10.5	10.7
2.6L	11.5	11.7
4WD		
2.3L	10.5	10.7
2.6L	13.5	13.7
3.1L	14.1	14.3
Rodeo/Sport		
2.2L	9.0	9.2
2.6L		
1994-97	10.3	10.5
3.2L		
1994-01		
2WD	18.2	18.4
4WD	21.5	21.7
2002-04		
2WD	12.3	12.5
4WD	15.5	15.7
Trooper		
1994-97	12.6	12.8
1998-02	11.6	11.8
Transmission Assy., R&R or Replace (B)		
1981-93		
2.3L, 2.6L		
2WD	5.2	5.4
4WD	6.4	6.6
2.8L, 3.1L	7.7	7.9
2WD	7.7	7.9
4WD	8.6	8.8
3.2L		
2WD	7.1	7.3
4WD	7.6	7.8
1994-99		
Amigo		
1994-95		
2.3L, 2.6L	5.0	5.2
1998-00		
2.2L	5.9	6.1
3.2L		
2WD	5.9	6.1
4WD	7.9	8.1
Pick-Up		
1994-95		
2.3L, 2.6L	5.0	5.2
3.1L	5.8	6.0
Hombre	4.2	4.4
Rodeo/Sport		
2.2L	3.5	3.7
2.6L		
1994-97	5.3	5.5
3.2L		
2WD	6.1	6.3
4WD	8.1	8.3

RODEO : RODEO SPORT : TROOPER : VEHICROSS ISU-35

	LABOR TIME	SEVERE SERVICE
Trooper		
1994-97	8.3	8.5
1998-02	7.0	7.2

AUTOMATIC TRANSAXLE

Flywheel (Flexplate), Replace (B)
Oasis 7.9 8.1
Shift Solenoid Valve, Replace (B)
Oasis .... .5 .7
Speed Pulser, Replace (B)
Oasis .... .5 .7
Transaxle Assy., R&R and Reconditon (A)
Oasis 15.2 15.3
Flush oil cooler and lines add 1.2 1.4
Recondition diff. add ... 1.7 1.9
Transaxle Assy., R&R or Replace (B)
Oasis 7.6 7.8
Replace assembly add . 1.2 1.4

AUTOMATIC TRANSMISSION

SERVICE TRANSMISSION INSTALLED

Check Unit for Oil Leaks (C)
All Models 9 1.1
Oil Pressure Check (B)
All Models .... .5 .7
Detent Cable, Adjust (B)
1981-87 .... .3 .5
Drain and Refill Unit (B)
Amigo
 1989-94 .... 1.6 1.8
 1998-00 (.8) 1.1 1.3
Ascender, Axiom (.8) .. 1.1 1.3
Pick-Up
 1981-87
  THM200 (.8) 1.1 1.3
  AW55 .... .7 .9
 1988-95 .... 1.6 1.8
Rodeo/Sport
 1991-97
  2.3L, 2.6L 1.8 2.0
  3.1L 1.7 1.9
 1998-04 (.8) 1.1 1.3
Trooper
 1984-87
  THM200 (.8) 1.1 1.3
  AW55 .... .7 .9
 1988-91
  2.3L, 2.6L 1.8 2.0
  2.8L 2.8 3.0
 1992-02 .... 1.6 1.8
VehiCROSS 1.6 1.8
Detent Cable, Replace (B)
1981-87 .... .7 .9

Extension Housing and/or Gasket, Replace (B)
1989-94 Amigo 3.2 3.4
Ascender (1.0) 1.4 1.6
Pick-Up
 1981-87 .... 2.5 2.7
 1988-95 .... 3.2 3.5
1991-97 Rodeo
 2.6L 2.8 3.0
Trooper
 1984-87 .... 2.5 2.7
 1988-91 .... 2.8 3.0
 1992-96 .... 2.8 3.0
Extension Housing Oil Seal, Replace (B)
1981-87 .... .8 1.0
1988-97 .... 1.5 1.7
Ascender 8 1.0
Governor Assy., R&R or Replace (B)
1981-87
 THM200 .... 1.2 1.4
 AW55 .... 2.7 2.9
1988-96 .... 3.1 3.3
Inhibitor Switch, Replace (B)
4 cyl.
 2.2L, 2.3L, 2.6L .... .5 .7
V6
 1989-91
  2.8L, 3.1L, 3.2L .. 1.2 1.4
 1992-97
  3.1L, 3.2L 1.6 1.8
 1998-04
  3.2L, 3.5L
   Amigo, Axiom ... 1.1 1.3
   Rodeo/Sport 1.1 1.3
   Trooper 1.6 1.8
   VehiCROSS ..... .7 .9
Intermediate Servo, R&R or Replace (B)
1981-87 .... 1.0 1.2
Manual Valve Lever and/or Seal, Replace (B)
1989-94 Amigo 3.9 4.1
Pick-Up 3.9 4.1
1991-97 Rodeo
 2.6L 5.0 5.2
Trooper
 1984-87 .... 3.9 4.1
 1988-97 .... 5.0 5.1
Oil Pan and/or Gasket, Replace (B)
Amigo
 1989-94 .... 1.6 1.8
 1998-00 .... 1.0 1.2
Ascender (.9) 1.3 1.5
Axiom 1.0 1.2
Pick-Up
 1981-87
  THM200 .... 1.0 1.2
  AW55 .... .7 .9
 1988-95 .... 1.6 1.8

Rodeo/Sport
 1991-97
  2.3L, 2.6L 1.8 2.0
  3.1L 1.7 1.9
 1998-04 .... 1.0 1.2
Trooper
 1984-87
  THM200 .... 1.0 1.2
  AW55 .... .7 .9
 1988-91
  2.3L, 2.6L 1.8 2.0
  2.8L 2.8 3.0
 1992-02 .... 1.6 1.8
VehiCROSS 1.6 1.8
Overdrive Control Switch, Replace (B)
2.2L, 2.3L, 2.6L ..... .3 .5
Parking Pawl, Replace (B)
1989-94 Amigo 3.9 4.1
Ascender (1.6) 2.2 2.4
Pick-Up
 1981-87
  THM200 .... 2.2 2.4
  AW55 .... 3.9 4.1
 1988-95 .... 3.9 4.1
1991-97 Rodeo 5.0 5.2
Trooper
 1984-87
  THM200 .... 2.2 2.4
  AW55 .... 3.9 4.1
 1988-95 .... 5.0 5.2
Pressure Regulator Valve, Replace (B)
1981-87 .... 1.7 1.9
Speedometer Driven Gear and/or Seal, Replace (B)
2.2L, 2.3L, 2.6L ..... .5 .7
2.8L, 3.1L,
 3.2L, 3.5L, 4.3L ..... .7 .9
Throttle Valve & Kickdown Cam, Replace (B)
1981-87 .... 1.4 1.6
1988-02 2.2L,
 2.3L, 2.6L 3.1 3.3
Transfer Case Oil Pan and/or Gasket, Replace (B)
1988-04 2.2L,
 2.3L, 2.6L 9 1.1
Transfer Case Valve Body, R&R or Replace (B)
1988-02 2.2L,
 2.3L, 2.6L 1.8 2.0
Transfer Case Valve Body, R&R and Recondition (A)
1988-04 2.2L,
 2.3L, 2.6L, 3.5L ... 2.9 3.1

ISU-36 AMIGO : ASCENDER : AXIOM : HOMBRE : OASIS : PICKUP

	LABOR TIME	SEVERE SERVICE
Transmission Control Module, Replace (B)		
1988-04		
2.2L, 2.3L, 2.6L	.5	.7
2.8L, 3.1L, 3.2L, 3.5L, 4.3L	.7	.7
Transmission Speed Sensor, Replace (B)		
4 cyl.		
2.2L, 2.3L, 2.6L		
1989-95	1.2	1.4
1996-04	.7	.9
L6, V8		
Ascender (.3)	.5	.7
V6		
2.8L, 3.1L, 3.2L, 3.5L, 4.3L		
Amigo, Hombre	.7	.9
Axiom	.8	1.0
Pick-Up	1.2	1.4
Rodeo/Sport		
1991-95	1.2	1.4
1996-04	.8	1.0
Trooper		
1989-91	1.2	1.4
1992-95	.3	.5
1996-97	.7	.9
1998-02	.5	.7
VehiCROSS	.5	.7
Valve Body Assy., R&R and Recondition (A)		
1998-00 Amigo		
main case	2.8	3.0
adapter case	2.2	2.4
Ascender (2.1)	2.8	3.0
Axiom		
main case	2.9	3.1
adapter case	2.2	2.4
Pick-Up		
1981-87	3.1	3.3
1988-95		
2.3L, 2.6L	4.9	5.1
3.1L	3.8	4.0
Rodeo/Sport		
1998-04		
main case	2.9	3.1
adapter case	2.2	2.4
Trooper		
1984-87	3.1	3.3
1988-97		
2.3L, 2.6L	4.9	5.1
2.8L	3.8	4.0
1998-02		
main case	2.5	2.7
adapter case	2.0	2.2
VehiCROSS		
main case	2.5	2.7
adapter case	2.0	2.2

	LABOR TIME	SEVERE SERVICE
Valve Body Assy., Replace (B)		
1998-00 Amigo		
main case	2.0	2.2
adapter case	1.7	1.9
Ascender (1.3)	1.8	2.0
Axiom		
main case	2.1	2.3
adapter case	1.8	2.0
Pick-Up		
1981-87		
THM200	1.9	2.1
AW55	1.8	2.0
1988-95		
2.3L, 2.6L	3.1	3.3
3.1L	1.9	2.1
Rodeo/Sport		
1998-04		
main case	2.1	2.3
adapter case	1.8	2.0
Trooper		
1984-87		
THM200	1.9	2.1
AW55	1.8	2.0
1988-97		
2.3L, 2.6L	3.1	3.3
2.8L	1.9	2.1
1998-02		
main case	1.8	2.0
adapter case	1.2	1.4
VehiCROSS		
main case	1.8	2.0
adapter case	1.2	1.4
Replace solenoids Pick-Up, Trooper add	.3	.5

SERVICE TRANSMISSION REMOVED

Transmission R&R included unless otherwise noted.

	LABOR TIME	SEVERE SERVICE
Flywheel (Flexplate), Replace (B)		
Amigo		
1989-94	7.7	7.9
1998-00		
2WD	5.9	6.1
4WD	7.4	7.6
Ascender		
2WD (4.3)	5.8	6.0
4WD (5.6)	7.6	7.8
Axiom		
2WD	6.1	6.3
4WD	7.6	7.8
Hombre	4.5	4.7
Pick-Up	7.7	7.9
Rodeo/Sport		
1991-97		
2.6L	7.1	7.3
3.1L	7.3	7.5
3.2L	10.3	10.5

	LABOR TIME	SEVERE SERVICE
1998-04		
2WD	6.1	6.3
4WD	7.6	7.8
Trooper		
1984-91		
2.3L, 2.6L	8.8	9.0
2.8L	9.1	9.3
1992-97	9.7	9.9
1998-02		
2WD	5.6	5.8
4WD	11.0	11.2
VehiCROSS	7.7	7.9
Front Oil Pump, Replace (B)		
Amigo		
1989-94	7.8	8.0
1998-00		
2WD	8.7	8.9
4WD	10.1	10.3
Ascender		
2WD (6.6)	8.9	9.1
4WD (7.4)	10.0	10.2
Axiom		
2WD	8.8	9.0
4WD	10.3	10.5
Hombre	6.7	6.9
Pick-Up	7.8	8.0
Rodeo/Sport		
1991-97		
2.6L, 3.1L	8.4	8.6
3.2L	9.7	9.9
1998-04		
2.2L	5.8	6.0
3.2L		
2WD	8.8	9.0
4WD	10.3	10.5
Trooper		
1984-91		
2, 3L, 2.6L	8.9	9.1
2.8L	9.2	9.4
1992-97	10.1	10.3
1998-02		
2WD	6.6	6.8
4WD	12.0	12.2
VehiCROSS	7.6	7.8
Replace front seal add	.1	.1
Reaction Gear Set, Replace (B)		
Hombre	7.0	7.5
Reverse & Input Clutch Input Drum, Replace (B)		
Hombre	6.9	7.4
Torque Converter, Replace (B)		
Amigo		
1989-94	7.6	7.8
1998-00		
2WD	5.8	6.0
4WD	7.3	7.5
Ascender		
2WD (5.6)	7.6	7.8
4WD (7.1)	9.6	9.8

RODEO : RODEO SPORT : TROOPER : VEHICROSS   ISU-37

	LABOR TIME	SEVERE SERVICE
Axiom		
2WD	6.0	6.2
4WD	7.5	7.7
Hombre	4.3	4.5
Pick-Up	7.6	7.8
Rodeo/Sport		
1991-97		
2.6L	7.2	7.4
3.1L	9.0	9.2
1998-04		
2.2L	4.3	4.5
3.2L		
2WD	6.0	6.2
4WD	7.5	7.7
Trooper		
1984-91		
2.3L, 2.6L	8.7	8.9
2.8L	9.0	9.2
1992-97	9.7	9.9
1998-02		
2WD	5.8	6.0
4WD	11.2	11.4
VehiCROSS	7.2	7.4
w/tank shield		
Ascender add	.6	.7

Transmission and Converter, R&R and Recondition (A)

	LABOR TIME	SEVERE SERVICE
Amigo		
1989-94		
2WD	18.7	18.9
4WD	30.5	30.7
1998-00		
2WD	17.6	17.8
4WD	19.9	20.1
Ascender		
2 WD (11.3)	15.3	15.5
4 WD (12.4)	16.7	16.9
Axiom		
2 WD	17.3	17.5
4 WD	19.3	19.5
Hombre	12.8	13.0
Pick-Up		
2 WD	18.7	18.9
4 WD	30.5	30.7
Rodeo/Sport		
1991-97		
2.6L	15.8	16.0
3.1L	15.8	16.0
3.2L	17.1	17.3
1998-04		
2WD	17.3	17.5
4WD	19.3	19.5
Trooper		
1984-91		
2.3L, 2.6L	29.8	30.0
2.8L	17.5	17.7
1992-97	19.4	19.6

	LABOR TIME	SEVERE SERVICE
1998-02		
2WD	15.3	15.5
4WD	20.9	21.1
VehiCROSS	15.2	15.4
w/tank shield		
Ascender add	.5	.7

Transmission Assy., R&R (B)

	LABOR TIME	SEVERE SERVICE
Amigo		
1989-94	4.8	5.0
1998-00		
2WD	5.6	5.8
4WD	7.0	7.2
Ascender		
2 WD (5.6)	7.6	7.8
4 WD (7.1)	9.6	9.8
Axiom		
2 WD	9.0	9.2
4 WD	11.2	11.4
Hombre	4.2	4.4
Pick-Up	7.4	7.6
Rodeo/Sport		
1991-97		
2.6L	6.8	7.0
3.1L	6.8	7.0
3.2L	10.0	10.2
1998-04		
2.2L		
2WD	6.0	6.2
4WD	8.2	9.2
3.2L		
2WD	9.8	10.0
4WD	11.2	11.4
Trooper		
1984-91		
2.3L, 2.6L	8.4	8.6
2.8L	8.8	9.0
1992-97	9.7	9.9
1998-02		
2WD	5.3	5.5
4WD	10.7	10.9
VehiCROSS	7.6	7.8
w/tank shield		
Ascender add	.5	.7
Replace assembly add	1.7	1.9

TRANSFER CASE

Rear Cover and/or Gasket, Replace (B)

	LABOR TIME	SEVERE SERVICE
1981-97	.5	.7

Transfer Case, R&R and Recondition (A)

	LABOR TIME	SEVERE SERVICE
Amigo		
1989-94	8.8	9.0
1998-00		
AT	9.1	9.3
MT	10.5	10.7
Ascender (5.0)	6.8	7.0
Axiom	5.7	5.9
Pick-Up	8.8	9.0

	LABOR TIME	SEVERE SERVICE
Rodeo/Sport		
1991-97		
AT	9.2	9.4
MT	12.2	12.4
1998-01		
AT	9.1	9.3
MT	10.5	10.7
2002-04		
AT	5.7	5.9
MT	11.5	11.7
Trooper		
1984-91	3.8	4.0
1992-97		
AT	9.9	10.1
MT	12.0	12.2
1998-02		
AT	9.9	10.1
MT	7.6	7.8
VehiCROSS	6.9	7.1
w/tank shield		
Ascender add	.5	.7

Transfer Case, R&R or Replace (B)

	LABOR TIME	SEVERE SERVICE
Amigo		
1989-94	6.9	7.1
1998-00		
AT	3.7	3.9
MT	11.6	11.8
Ascender	3.0	3.2
Axiom	3.8	4.0
Pick-Up	6.9	7.7
Rodeo/Sport		
1991-97		
AT	4.5	4.7
MT	10.5	10.7
1998-04		
AT	3.8	4.0
MT	11.9	12.1
Trooper		
1984-91	1.2	1.4
1992-97		
AT	5.0	5.2
MT	11.6	11.8
1998-02		
AT	4.3	4.5
MT	2.9	3.1
VehiCROSS	3.3	3.5
w/tank shield		
Ascender add	.5	.7

Transfer Case Oil Pan and/or Gasket, Replace (B)

	LABOR TIME	SEVERE SERVICE
1981-97	.7	.9

Transfer Case Oil Seal(s), Replace (B)

	LABOR TIME	SEVERE SERVICE
Amigo		
1998-00		
AT		
front	1.2	1.4
rear	.7	.9
MT	9.0	9.2

ISU-38 — AMIGO : ASCENDER : AXIOM : HOMBRE : OASIS : PICKUP

	LABOR TIME	SEVERE SERVICE
Transfer Case Oil Seal(s), Replace (B)		
Ascender		
Front (.6)	.8	1.0
rear (.4)	.6	.8
Axiom		
front	1.1	1.3
rear	.8	1.0
Pick-Up		
front	2.0	2.2
rear	1.7	1.9
Rodeo/Sport		
1991-97		
AT		
front	1.7	1.9
rear	1.8	2.0
1998-04		
AT		
front	1.2	1.4
rear	.8	1.0
MT		
1998-01	9.0	9.2
2002-04	9.9	10.1
Trooper		
1984-91		
front	2.0	2.2
rear	1.7	1.9
1992-97		
front	1.6	1.8
rear	1.2	1.4
1998-02		
front	1.2	1.4
rear	.8	1.0
VehiCROSS		
front		
input shaft	3.4	3.6
output shaft	1.2	1.4
rear	1.2	1.4

SHIFT LINKAGE
AUTOMATIC TRANSMISSION

	LABOR TIME	SEVERE SERVICE
Shift Control Rod, Adjust (B)		
All Models	.5	.7
Shift Linkage, Adjust (B)		
All Models	.7	.9
Throttle Cable, Adjust (B)		
All Models	.3	.5
Floor Shift Control Assy., Replace (B)		
Amigo	1.7	1.9
Ascender (.4)	.6	.8
Axiom	1.1	1.3
Rodeo/Sport		
1998-01	1.7	1.9
2002-04	1.1	1.3
Pick-Up	1.2	1.4

	LABOR TIME	SEVERE SERVICE
Trooper		
1984-97	1.2	1.4
1998-02	.8	1.0
VehiCROSS	1.2	1.4
Shift Control Cable, Replace (B)		
Amigo	1.4	1.6
Ascender (1.3)	1.8	2.0
Axiom, Rodeo/Sport	1.4	1.6
Oasis	1.6	1.8
Shift Control Lever, (Trans. Mount) (B)		
All Models	.5	.7
Shift Control Rod, Replace (B)		
All Models	.7	.9
Shift Lock Cable, Replace (B)		
1998-02 Trooper	.8	1.0
VehiCROSS	.8	1.0
Throttle Cable, Replace (B)		
1981-97	3.1	3.3
1998-04	.5	.7

MANUAL TRANSMISSION

	LABOR TIME	SEVERE SERVICE
Gearshift Control Lever, Replace (B)		
All Models	.6	.8
w/console add	.2	.4
Transfer Case Control Lever, Replace (B)		
All Models	.5	.7

TRANSFER CASE

	LABOR TIME	SEVERE SERVICE
Transfer Case Linkage, Adjust (B)		
All Models	.5	.7
Transfer Case Lever and/or Boot, Replace (B)		
Exc. below	.7	.7
Axiom (.8)	1.1	1.2
VehiCROSS	1.4	1.4

DRIVELINE
DRIVESHAFT

	LABOR TIME	SEVERE SERVICE
Center Support Bearing, Replace (B)		
1981-97	1.0	1.2
Driveshaft, Replace (B)		
Amigo		
1989-94		
front		
2.3L, 2.6L	.5	.7
1998-00		
front	1.6	1.8
rear		
2WD	.5	.7
4WD	1.2	1.4
Ascender		
front (.6)	.8	1.0
rear (.3)	.5	.7

	LABOR TIME	SEVERE SERVICE
Axiom		
front	1.6	1.8
rear		
2WD	.5	.7
4WD	1.6	1.8
Hombre	.5	.7
Pick-Up		
front		
4 cyl.	.5	.7
V6	1.5	1.7
Rodeo/Sport		
1991-97		
front		
2.6L, 3.1L	1.6	1.8
3.2L	1.8	2.0
rear		
one	.7	.9
each addl. add	.3	.4
1998-04		
front	1.6	1.8
rear		
2WD	.5	.7
4WD	1.2	1.4
Trooper		
1984-91		
front	.5	.7
rear	.8	1.0
1992-02		
front	1.8	2.0
rear	.8	1.0
VehiCROSS		
front	1.2	1.4
rear	.8	1.0
U-Joint, Replace (B)		
Amigo		
1989-94		
one	1.2	1.4
each addl. add	.5	.7
1998-02		
front		
one	1.8	2.0
each addl. add	.5	.7
rear		
2WD one	.8	1.0
4WD one	1.6	1.8
each addl. add	.3	.5
Ascender		
front (1.0)	1.4	1.6
rear (.8)	1.1	1.3
both (1.4)	1.9	2.1
Axiom		
front		
one	1.9	2.1
both	2.1	2.3
rear		
2WD one	1.0	1.1
4WD one	1.6	1.8
each addl. add	.3	.5

RODEO : RODEO SPORT : TROOPER : VEHICROSS ISU-39

	LABOR TIME	SEVERE SERVICE
Pick-Up		
one	1.2	1.4
each addl. add	.5	.7
Rodeo/Sport		
1991-97		
one	1.2	1.4
each addl. add	.5	.7
1998-04		
front		
one	1.9	2.1
each addl. add	.2	.3
rear		
2WD one	1.0	1.1
4WD one	1.6	1.8
each addl. add	.3	.5
Trooper		
one	1.2	1.4
each addl. add	.5	.7
VehiCROSS		
front	1.2	1.4
rear one	1.7	1.9
each addl. add	.7	.9

DRIVE AXLE

Axle Housing, Replace (B)
Front axle		
1992-01	5.9	6.1
2002-04		
Axiom	6.6	6.8
Rodeo/Sport	6.6	6.8
Rear axle		
integral type		
Amigo		
1998-02		
2WD	8.3	8.5
4WD	8.6	8.8
Ascender (5.4)	7.3	7.5
Axiom	8.4	8.6
Rodeo/Sport		
1998-04	8.4	8.6
VehiCROSS	5.3	5.5
removable type		
Pick-Up	5.2	5.4
Rodeo		
1991-97	5.2	5.4
Trooper		
1984-97	5.2	5.4
1998-99	5.3	5.5
semi-floating		
Hombre	5.8	6.0

Axle Housing Cover Gasket, Replace (B)
Amigo, Axiom, Rodeo/Sport	1.8	2.0
Ascender	1.3	1.5
Hombre	.7	.9

Axle Shaft, Replace (B)
Front axle		
Amigo		
1989-94	4.9	5.1
1998-00		
one side	3.3	3.5
both sides	6.0	6.2
Axiom		
left or right side	3.5	3.7
complete assy.	5.5	5.7
Ascender		
one side (1.5)	2.0	2.2
both sides (1.7)	2.3	2.5
Pick-Up	4.9	5.1
Rodeo/Sport		
1991-97	5.2	5.4
1998-04		
one side	3.5	3.7
both sides	6.4	6.6
complete assy.	5.5	5.7
Trooper		
1984-91	4.9	5.1
1992-97	5.3	5.5
1998-02	3.1	3.3
VehiCROSS		
left or right side	3.1	3.3
complete assy.	4.9	5.1
Rear axle		
integral type		
Amigo, Rodeo/Sport	1.8	2.0
Ascender		
right or left side	2.1	2.3
both	2.4	2.6
Axiom		
right of left side	1.7	1.9
VehiCROSS	1.8	2.0
removable type		
Trooper		
1992-97	1.8	2.0
1998-02	1.3	1.5
semi-floating		
Hombre	1.0	1.2

Axle Shaft and/or Bearing, Replace (B)
Rear axle integral type		
1989-94 Amigo	1.8	2.0
Ascender		
right or left (1.6)	2.2	2.4
both (2.2)	3.0	3.2
Axiom		
right or left	2.1	2.3
Rodeo/Sport		
1991-97	1.8	2.0
1998-04	2.1	2.3
1992-97 Trooper	1.9	2.1

Axle Shaft Bearing and/or Oil Seal, Replace (B)
Rear axle		
removable type		
1998-00 Amigo	2.0	2.2
Rodeo		
1991-97	1.8	2.0
1998-01	2.0	2.2
Trooper		
1992-97	2.0	2.2
1998-02	1.4	1.6
VehiCROSS	2.0	2.2
semi-floating		
Hombre	1.3	1.5

Axle Shaft Inner Oil Seal, Replace (B)
Rear axle removable type		
Trooper		
1992-97	2.0	2.2
1998-02	1.4	1.6

Axle Shaft, Seal and/or Bearing, Replace (B)
Front axle		
Pick-Up		
one side	1.9	2.1
both sides	3.2	3.4
Trooper		
one side	1.9	2.1
both sides	3.2	3.4
Rear axle		
1989-94 Amigo	1.8	2.1
Ascender		
right or left (1.6)	2.2	2.4
both (2.0)	2.7	2.9
Axiom		
right or left	2.1	2.3
Hombre	1.3	1.5
Pick-Up	1.5	1.7
1991-97 Rodeo	1.8	2.0
Trooper		
1984-91	1.5	1.7
1992-02	1.5	1.7

Differential Assy., R&R and Recondition (A)
Front axle		
1998-00 Amigo	9.1	9.3
Ascender	14.6	14.8
Axiom	9.1	9.3
Rodeo/Sport		
1991-04	9.1	9.3
Trooper		
1992-97	9.3	9.5
1998-02	9.1	9.3
VehiCROSS	8.0	8.2
Rear axle		
integral type		
Amigo		
1989-94	5.3	5.5

AMIGO : ASCENDER : AXIOM : HOMBRE : OASIS : PICKUP

	LABOR TIME	SEVERE SERVICE
Differential Assy., R&R and Reconditon (A)		
Ascender		
Standard (5.6)	7.6	7.8
Limited slip (7.1)	9.6	9.8
Axiom	7.5	7.7
Rodeo/Sport		
1991-97	5.3	5.5
1998-04	7.5	7.7
VehiCROSS	5.8	6.0
removable type		
1981-97	5.4	5.6
1998-02	6.4	6.6
semi-floating		
Hombre		
Standard	4.2	4.4
Limited slip	5.2	5.4
Differential Carrier, R&R or Replace (B)		
Front axle		
exc. Ascender	5.9	6.1
Ascender	11.8	12.0
Rear axle	3.2	3.4
Differential Case, R&I (B)		
Rear axle semi-floating		
Hombre		
w/Eaton case	3.1	3.3
w/o Eaton case	3.0	3.2
Differential Cover Gasket, Replace (B)		
Rear axle		
integral type		
Amigo		
1989-94	1.0	1.2
1998-00	1.8	2.0
Ascender (1.0)	1.4	1.6
Axiom	1.7	1.9
Rodeo/Sport		
1991-97	1.0	1.2
1998-04	1.7	1.9
removable type		
Pick-Up, Trooper	3.1	3.3
1993-94 Rodeo	1.2	1.4
semi-floating		
Hombre	.7	.9
Differential Side Bearings, Replace (B)		
Front axle		
exc. Ascender	7.2	7.4
Ascender	12.9	13.1
Rear axle		
integral type		
Amigo		
1989-94	3.1	3.3
1998-00	4.7	4.29
Axiom	4.9	5.1
Ascender	3.5	3.7

	LABOR TIME	SEVERE SERVICE
Rodeo/Sport		
1991-97	3.6	3.8
1998-04	4.9	5.1
VehiCROSS	4.1	4.3
removable type		
1981-97	3.8	4.0
1998-02	4.5	4.7
semi-floating		
Hombre	3.0	3.2
Limited Slip Rear Clutch Plates, Replace (B)		
Rear axle		
1996-97	2.0	2.2
Pinion Shaft, Gears and/or Side Gears, Replace (B)		
Front axle		
exc. Ascender	8.2	8.4
Ascender	12.1	12.3
Rear axle		
integral type		
Amigo		
1989-94	1.7	1.9
1998-00	5.4	5.6
Ascender, Axiom		
w/limited Slip	6.7	6.9
w/o limited Slip	1.6	1.8
Rodeo/Sport		
1991-97	1.7	1.9
1998-04	6.0	6.2
removable type		
1993-94 Rodeo	3.7	3.9
semi-floating		
Hombre		
w/limited slip	2.3	2.5
w/o limited slip	1.7	1.9
Ring Gear & Pinion, Replace (B)		
Front axle		
Axiom	6.7	6.9
Rodeo/Sport		
2002-04	6.7	6.9
Rear axle		
integral type		
Amigo		
1989-94	4.4	4.6
1998-00		
2WD	6.4	6.6
4WD	6.5	6.7
Ascender	7.0	7.2
Axiom	7.3	7.5
Rodeo/Sport		
1991-97	4.9	5.1
1998-01	6.5	6.7
2002-04	7.3	7.5
removable type		
1993-97 Rodeo	5.4	5.6
1994-97 Trooper	5.2	5.4
semi-floating		
Hombre	4.3	4.5

	LABOR TIME	SEVERE SERVICE
Pinion Shaft Oil Seal, Replace (B)		
Front axle		
exc. below	1.5	1.7
Ascender, Axiom	1.1	1.3
2002-04 Rodeo	1.1	1.3
Rear axle		
integral type		
Amigo		
1989-00	1.3	1.5
Ascender, Axiom	1.3	1.5
Rodeo/Sport		
1991-04	1.3	1.5
VehiCROSS	.8	1.0
removable type		
1981-97	1.2	1.4
1998-02	.8	1.0
semi-floating		
Hombre	1.2	1.4
HALFSHAFTS		
Halfshaft, R&R or Replace (B)		
Ascender		
right or left	2.1	2.3
both	3.5	3.7
Oasis		
one side	1.8	2.0
both sides	3.2	3.4
Replace support bearing add	.3	.5
Halfshaft Oil Seal, Replace (B)		
Ascender	2.2	2.4
Oasis		
one side	1.2	1.4
both sides	2.3	2.5
Inboard CV Joint and/or Boot, Replace (B)		
Ascender		
one side	2.8	3.0
both sides	4.7	4.9
Oasis		
one side	1.5	1.7
both sides	2.8	3.0
Intermediate Shaft and/or Bearing, Replace (B)		
Ascender	2.5	2.7
Oasis	2.3	2.5
Replace oil seal add	.2	.4
Outboard CV Joint and/or Boot, Replace (B)		
Ascender		
one side	2.7	2.9
both sides	4.7	4.9
Oasis	2.3	2.5

RODEO : RODEO SPORT : TROOPER : VEHICROSS ISU-41

	LABOR TIME	SEVERE SERVICE

BRAKES
ANTI-LOCK
The following operations do not include testing. Add time as required.

Diagnose Anti-Lock Brake System Component Each (A)
- 1988-04 8   *1.0*

Bleed ABS System (B)
Includes: Relieve accumulator pressure.
- 1988-05 7   *.9*

Accumulator Assy., Replace (B)
- Oasis 1.2   *1.4*

Anti-Lock Valve, Replace (B)
- Pick-Up 1.0   *1.2*
- 1994-97 Rodeo 7   *.9*
- Trooper
  - 1988-97 .... 1.0   *1.2*
  - w/cruise control add 2   *.4*

Control Module, Replace (B)
- Amigo
  - 1989-00 5   *.7*
- Ascender (.7) 9   *1.1*
- Axiom 5   *.7*
- Hombre
  - 2 wheel ABS 5   *.7*
  - 4 wheel ABS 1.6   *1.8*
- Oasis 7   *.9*
- 1988-95 Pick-Up 5   *.7*
- Rodeo/Sport
  - 1991-04 5   *.7*
- Trooper
  - 1988-91 5   *.7*
  - 1992-93 7   *.9*
  - 1994-02 6   *.8*

Enable Relay, Replace (A)
- Hombre 7   *.9*

Excitor Ring, Replace (B)
- 1998-00 Amigo
  - front
    - 2WD 8   *1.0*
    - 4WD 1.0   *1.2*
  - rear 3.4   *3.6*
- Axiom
  - front 1.1   *1.3*
  - rear 3.8   *4.0*
- 1998-04 Rodeo/Sport
  - front
    - 2WD 1.0   *1.2*
    - 4WD 1.3   *1.5*
  - rear
    - 1998-01 .... 3.4   *3.6*
    - 2002-04 .... 3.8   *4.0*
- Trooper
  - 1988-97 .... 3.4   *3.6*
  - 1998-02
    - front 1.7   *1.9*
    - rear 1.4   *1.6*

- VehiCROSS
  - front 1.3   *1.5*
  - rear 1.8   *2.0*

G Sensor, Replace (B)
- 1998-00 Amigo 1.1   *1.3*
- Axiom 1.1   *1.3*
- Rodeo/Sport
  - 1995-04 .... 1.1   *1.3*
- 1992-02 Trooper 7   *.9*
- VehiCROSS 5   *.7*

Hydraulic Assy., Replace (B)
- 1998-00 Amigo 1.1   *1.3*
- Axiom 1.1   *1.3*
- Rodeo/Sport
  - 1995-97 7   *.9*
  - 1998-04 .... 1.1   *1.3*
- 1992-02 Trooper 1.5   *1.7*
- VehiCROSS 1.2   *1.4*
- *Replace*
  - *gear add* 1   *.1*
  - *motor pack add* 1   *.1*
  - *solenoid add* 1   *.1*

Low Pressure Hoses, Replace (B)
- Oasis 1.5   *1.7*

Modulator, Replace (B)
- Oasis 7   *.9*

Power Unit, Replace (B)
- Oasis 1.2   *1.4*

Pressure Modulating Valve, Replace (B)
- Hombre 1.2   *1.4*

Speed Sensor Buffer, Replace (B)
- 1989-96 V6 5   *.7*

Wheel Speed Sensor, Replace (B)
- Exc. Ascender
  - 2 wheel ABS 5   *.7*
  - 4 wheel ABS
    - front one 8   *1.0*
    - rear one 5   *.7*
- Ascender
  - front
    - right or left 1.0   *1.2*
    - both 1.6   *1.8*

SYSTEM
Bleed Brakes (B)
Includes: Add fluid.
- All Models 6   *.8*
- w/ABS add 5   *.7*

Brake System, Flush and Refill (B)
- All Models 1.3   *1.5*

Brakes, Adjust (B)
Includes: Refill master cylinder.
- All Models rear wheels ..5   *.7*

Brake Adjuster, Replace (B)
- Oasis
  - front 7   *.9*
  - both 1.2   *1.4*

Brake Hose (Flexible), Replace (B)
Includes: System bleeding.
- Amigo
  - front
    - one 7   *.9*
    - both 1.3   *1.5*
  - rear
    - one 7   *.9*
    - each addl. add 3   *.5*
- Ascender
  - front or rear
    - one 1.4   *1.6*
    - both 1.7   *1.9*
- Axiom
  - front
    - one 6   *.8*
  - rear
    - one 5   *.7*
  - Center 6   *.8*
- Hombre
  - one 1.2   *1.4*
  - each addl. add 3   *.5*
- Oasis
  - front
    - one 1.0   *1.2*
    - both 1.6   *1.8*
  - rear
    - one 7   *.9*
    - each addl. add 3   *.5*
- Pick-Up, Rodeo/Sport
  - front
    - one 7   *.9*
    - both 1.3   *1.5*
  - rear
    - one 7   *.9*
    - each addl. add 3   *.5*
- Trooper
  - 1984-91
    - front
      - one 7   *.9*
      - both 1.3   *1.5*
    - rear
      - one 7   *.9*
      - each addl. add 3   *.5*
  - 1992-02
    - front 8   *1.0*
    - rear 5   *.7*
- VehiCROSS
  - front one 8   *1.0*
  - rear one 5   *.7*

Combination Valve, Replace (B)
Includes: System bleeding.
- Amigo
  - 1989-94 7   *.9*
  - 1998-00 9   *1.1*
- Axiom 9   *1.1*
- Hombre 1.4   *1.6*
- Pick-Up 7   *.9*
- Rodeo/Sport 9   *1.1*

ISU-42 AMIGO : ASCENDER : AXIOM : HOMBRE : OASIS : PICKUP

	LABOR TIME	SEVERE SERVICE
Combination Valve, Replace (B)		
Trooper		
1984-91	.7	.9
1992-02	1.4	1.6
Master Cylinder, Replace (B)		
Includes: System bleeding.		
Amigo		
1989-94	1.0	1.2
1998-00	.8	1.0
Ascender	1.4	1.6
Axiom	.9	1.1
Hombre	1.2	1.4
Oasis	1.7	1.9
Pick-Up	1.0	1.2
Rodeo/Sport		
1991-97	1.2	1.4
1998-04	.8	1.0
Trooper		
1984-91	1.0	1.2
1992-02	1.6	1.8
VehiCROSS	.8	1.0
Power Booster Unit, Replace (B)		
Amigo		
1989-94	1.2	1.4
1998-00	1.0	1.2
Ascender, Axiom	1.2	1.4
Hombre	1.6	1.8
Oasis	2.8	3.0
Pick-Up	1.2	1.4
Rodeo/Sport		
1991-97	1.3	1.5
1998-01	1.0	1.2
2002-04	1.7	1.9
Trooper		
1984-91	1.2	1.4
1992-02	2.0	2.2
VehiCROSS	2.0	2.2
Power Booster Vacuum Check Valve, Replace (B)		
All Models	.3	.5
Proportioning Valve, Replace (B)		
Includes: System bleeding.		
Oasis	2.5	2.7
SERVICE BRAKES		
Brake Drum, Replace (B)		
All Models one	.5	.7
Caliper Assy., R&R and Recondition (B)		
Includes: System bleeding.		
Exc. below		
front		
one	1.9	2.1
both	3.1	3.3
rear		
one	1.3	1.5
both	2.0	2.2

	LABOR TIME	SEVERE SERVICE
Amigo, Rodeo/Sport		
1998-04		
one	1.2	1.4
two	1.8	2.0
all four wheels	3.1	3.3
VehiCROSS		
front or rear		
one	1.8	2.0
both	3.1	3.3
Caliper Assy., Replace (B)		
Includes: System bleeding.		
Exc. below		
front		
one	1.2	1.4
both	1.9	2.1
rear		
one	1.3	1.5
both	1.9	2.1
four wheels	3.0	3.2
1998-04 Amigo, Rodeo/Sport		
one	.8	1.0
two	1.3	1.5
all four wheels	2.5	2.7
Axiom		
front		
right side	.8	1.0
left side	.5	.7
rear		
right or left side	.8	1.0
both sides	1.2	1.4
VehiCROSS		
front or rear		
one	1.2	1.4
both	1.8	2.0
Disc Brake Rotor, Replace (B)		
Exc. below		
front		
2WD		
one	1.4	1.6
both	2.4	2.6
4WD		
one	1.8	2.0
both	2.6	2.8
rear		
one	.8	1.0
both	1.5	1.7
four wheels	3.3	3.5
1998-04 Amigo, Rodeo/Sport		
front		
one	1.2	1.4
both	2.0	2.2
Ascender		
front or rear		
one	1.0	1.2
both	1.9	2.1

	LABOR TIME	SEVERE SERVICE
1992-02 Trooper		
front		
right side	1.8	2.0
left side	1.2	1.4
both sides	2.0	2.2
VehiCROSS		
front		
one	1.6	1.8
both	2.6	2.8
rear		
one	.5	.7
both	.8	1.0
Front Pads and/or Rotors, Replace (B)		
Includes: Service and parking brake adjustment, bleed system.		
Oasis	2.0	2.2
Pads and/or Shoes, Replace (B)		
Includes: Adjust service and parking brake. System bleeding.		
Exc. below		
front disc	.8	1.0
rear drum	1.9	2.1
four wheels	2.8	3.0
2002-04		
Ascender		
front or rear	1.6	1.8
four wheels	3.2	3.4
Axiom		
front or rear disc	1.1	1.3
rear drum	1.4	1.6
Rodeo		
front or rear disc	1.1	1.3
rear drum	1.4	1.6
COMBINATION ADD-ONS		
Repack wheel bearings		
add two wheels	.7	.7
Replace		
brake drum add each	.1	.1
brake hose add each	.3	.3
caliper add	.3	.3
disc rotor add each	.2	.2
master cylinder add	.5	.5
wheel cylinder add	.2	.2
Resurface		
brake drum add	.5	.5
brake rotor add	1.0	1.0
Self-Adjusting Units, Free-Up or Replace (B)		
All Models		
one	.7	.8
each addl. add	.5	.6
Wheel Cylinder, R&R and Rebuild (B)		
Includes: System bleeding.		
All Models		
one	1.4	1.6
both	2.2	2.4

RODEO : RODEO SPORT : TROOPER : VEHICROSS — ISU-43

	LABOR TIME	SEVERE SERVICE
Wheel Cylinder, Replace (B)		
Includes: System bleeding.		
All Models		
one	1.1	1.3
both	1.9	2.1
PARKING BRAKE		
Parking Brake Cable, Adjust (C)		
Exc. Ascender	.5	.7
Ascender	1.1	1.3
Parking Brake Apply Actuator, Replace (B)		
Amigo		
1989-94	.8	.9
1998-00	.7	.8
Ascender	1.2	1.3
Axiom	.6	.7
Hombre	1.6	1.7
Oasis	.8	.9
Pick-Up	.8	.9
Rodeo/Sport		
1991-97	.5	.6
1998-04	.7	.8
Trooper		
1984-91	.8	.9
1992-02	1.3	1.4
VehiCROSS	.7	.8
Parking Brake Apply Warning Indicator Switch, Replace (B)		
Amigo	.5	.5
Ascender	.6	.6
Axiom	.3	.3
Hombre	.7	.7
Oasis, Pick-Up	.5	.5
Rodeo/Sport	.5	.5
Trooper		
1984-91	.5	.5
1992-02	.8	.8
VehiCROSS	.5	.5
Parking Brake Cable, Replace (B)		
Amigo		
1989-94		
front	2.3	2.5
rear		
one	1.7	1.9
both	2.3	2.5
Center	1.0	1.2
1998-00		
AT	1.6	1.8
MT	1.3	1.5
Ascender		
right or left	1.3	1.5
both	1.6	1.8
Center	1.4	1.6
Axiom		
one	1.6	1.8
both	2.4	2.6

	LABOR TIME	SEVERE SERVICE
Hombre		
front	1.3	1.5
rear each	1.2	1.4
intermediate	.7	.9
Oasis		
one	1.0	1.2
both	2.0	2.2
Pick-Up		
front	2.3	2.4
rear		
one	1.7	1.9
both	2.3	2.5
Center	1.0	1.2
Rodeo/Sport		
1991-97		
front	2.3	2.5
rear		
one	1.7	1.9
both	2.3	2.5
Center	1.0	1.2
1998-04		
AT		
one	1.6	1.8
both	2.4	2.6
MT		
one	1.3	1.5
both	2.2	2.4
Trooper		
1984-91		
front	1.3	1.5
rear both	2.3	2.4
Center	1.0	1.2
1992-02		
rear one	2.0	2.2
VehiCROSS		
right or left	1.6	1.8
both	2.4	2.6
Parking Brake Equalizer, Replace (B)		
Hombre	.7	.9
Parking Brake Shoes, Replace (B)		
1998-00 Amigo		
one side	.7	.9
both sides	1.4	1.6
Ascender, Axiom		
both sides	1.9	2.1
Rodeo/Sport		
one side	.7	.9
both sides	1.4	1.6
1992-02 Trooper		
one side	.8	1.0
both sides	1.3	1.5
VehiCROSS		
right or left	.7	.9
both	1.3	1.5

	LABOR TIME	SEVERE SERVICE
FRONT SUSPENSION		
FRONT WHEEL DRIVE		
Unless otherwise noted, time given does not include alignment.		
Align Front End (A)		
Oasis	1.7	1.9
Front & Rear Alignment, Check & Adjust (A)		
Oasis		
front	1.7	1.9
rear	1.2	1.4
front & rear	2.8	3.0
Front Toe, Adjust (B)		
Oasis	.8	1.0
Front Coil Spring, Replace (B)		
Oasis		
one side	1.4	1.6
both sides	2.6	2.8
Front Hub Assy., Replace (B)		
Oasis		
one	2.2	2.4
both	4.1	4.3
Hub Bearings, Replace (B)		
Oasis		
one side	1.6	1.8
both sides	3.1	3.3
Lower Ball Joint, Replace (B)		
Oasis each	1.7	1.9
Lower Control Arm, Replace (B)		
Includes: Reset toe.		
Oasis		
one side	1.4	1.6
both sides	2.6	2.8
Radius Rod Bushings, Replace (B)		
Oasis		
one side	1.0	1.2
both sides	1.7	1.9
Stabilizer Bar, Replace (B)		
Oasis	1.2	1.4
Stabilizer Bar Bushings, Replace (B)		
Oasis	.7	.9
Steering Knuckle, Replace (B)		
Oasis		
one side	2.2	2.4
both sides	3.4	3.6
Replace lower ball joint add	.3	.5
Strut, Replace (B)		
Oasis		
one side	1.2	1.4
both sides	2.0	2.2
Tie Rod Boot (B)		
Oasis one	.7	.9
Upper Control Arm, Replace (B)		
Oasis		
one side	1.3	1.6
both sides	2.3	2.5

ISU-44   AMIGO : ASCENDER : AXIOM : HOMBRE : OASIS : PICKUP

	LABOR TIME	SEVERE SERVICE
Upper Control Arm Bushings, Replace (B)		
Oasis		
one side	1.6	1.8
both sides	2.9	3.1
REAR WHEEL DRIVE		
Align Front End (A)		
Exc. Ascender 4WD	1.5	1.7
Ascender 4WD	2.5	2.7
Front Toe, Adjust (B)		
All Models	.8	1.0
Axle Shaft Bearings, Replace (B)		
All Models		
one side	1.9	2.1
both sides	3.2	3.4
Coil Spring, Replace (B)		
Ascender		
one side	1.1	1.3
both sides	1.7	1.9
Hombre		
one side	1.3	1.5
both sides	2.3	2.5
Front Crossmember, Replace (B)		
1992-02 Trooper	.5	.7
VehiCROSS	.5	.7
Front Hub Assy., Replace (B)		
Ascender		
right or left	1.6	1.8
both	2.5	2.7
1992-02 Trooper	1.6	1.8
VehiCROSS		
right or left side	1.2	1.4
Front Hub Oil Seal, Replace (B)		
Exc. below	1.1	1.3
2002-04 Rodeo/Sport		
right or left side	1.4	1.6
Front Shock Absorbers, Replace (B)		
Amigo, Pick-Up	.5	.7
Ascender	1.0	1.3
Axiom, Hombre	.8	1.0
Rodeo/Sport	.8	1.0
Trooper		
1984-91	.5	.7
1992-02	.7	.9
VehiCROSS		
right or left	.8	1.0
Front Strut Bar, Replace (B)		
1981-98	.7	.9
Replace bushing add each	.1	.3
Locking Hub Cam and/or Clutch, Replace (B)		
Automatic	.7	.9
Manual	1.0	1.2

	LABOR TIME	SEVERE SERVICE
Lower Ball Joint, Replace (B)		
Amigo		
1989-94	.8	1.0
1998-00		
2WD	2.2	2.4
4WD	2.4	2.6
Ascender		
right or left	1.3	1.5
both	1.9	2.1
Axiom		
right or left	2.2	2.4
both	4.6	4.8
Hombre, Pick-Up	1.2	1.4
Rodeo/Sport		
1991-97	1.2	1.4
1998-04		
one	2.4	2.6
both	4.6	4.8
Trooper, VehiCROSS	1.2	1.4
Lower Control Arm Assy., Replace (B)		
Amigo		
1989-94	3.1	3.3
1998-99		
2WD	1.8	2.0
4WD	2.0	2.2
2000		
2WD or 4WD	3.0	3.2
Ascender		
right or left	1.3	1.5
both	1.9	2.1
Axiom		
right or left	3.2	3.4
Hombre	1.9	2.1
Pick-Up	3.0	3.2
Rodeo/Sport		
1991-97	3.0	3.2
1998-99		
2WD	1.8	2.0
4WD	2.0	2.2
2000-04		
2WD or 4WD	3.2	3.4
Trooper		
1984-91	3.0	3.2
1992-02	2.5	2.7
VehiCROSS	2.5	2.7
Replace bushings add	.5	.7
Stabilizer Bar, Replace (B)		
Amigo		
1989-94	.5	.7
1998-00	.7	.9
Ascender	1.1	1.3
Axiom	.6	.8
Hombre	1.0	1.2
Pick-Up	.5	.7
Rodeo/Sport	.7	.9
Trooper		
1984-91	.5	.7
1992-02	1.2	1.4
VehiCROSS	1.2	1.4

	LABOR TIME	SEVERE SERVICE
Steering Knuckle, Replace (B)		
Amigo		
1989-94	2.2	2.4
1998-99		
2WD	2.2	2.4
4WD	2.3	2.5
2000		
2WD	1.8	2.0
4WD	2.2	2.4
Ascender		
2WD		
right or left	1.7	1.9
both	2.8	3.0
4WD		
right or left	1.9	2.1
both	3.3	3.5
Axiom		
right or left	2.2	2.4
Hombre	2.0	2.2
Pick-Up		
2WD	1.8	2.0
4WD	2.0	2.2
Rodeo/Sport		
1991-97	2.0	2.2
1998-04	2.3	2.5
Trooper		
1984-91	2.0	2.2
1992-02	1.7	1.9
VehiCROSS		
each side	1.7	1.9
Rebuild knuckle add each	.5	.7
Stabilizer Bar Bushings, Replace (B)		
One	.5	.7
Both	.7	.9
Stabilizer Bar Links and/or Grommets, Replace (B)		
Amigo, Axiom, Rodeo/Sport		
each side	.5	.7
Steering Knuckle Inner Bearings, Replace (B)		
All Models		
one	1.5	1.7
both	2.8	3.0
2002-04		
Axiom, Rodeo/Sport		
right or left	2.5	2.7
Steering Knuckle Inner Oil Seal, Replace (B)		
1981-97		
one	1.4	1.6
both	2.6	2.8
1998-04 one	2.3	2.5
Torsion Bar, Adjust (B)		
All Models		
one side	.5	.7
both sides	.7	.9

RODEO : RODEO SPORT : TROOPER : VEHICROSS ISU-45

	LABOR TIME	SEVERE SERVICE
Torsion Bar, Replace (B)		
Includes: Height adjustment.		
Amigo		
1989-94	.7	.9
1998-00		
right side	.7	.9
left side	.5	.7
Axiom		
right side	.8	1.0
left side	.5	.7
Pick-Up	.7	.9
Rodeo/Sport		
1991-97	1.0	1.2
1998-04		
right side	.8	1.0
left side	.5	.7
Trooper		
1986-91	.7	.9
1992-02	1.3	1.5
VehiCROSS	.7	.9
Upper Ball Joint, Replace (B)		
Amigo		
1989-94	.5	.7
1998-00	1.4	1.6
Ascender		
right or left	1.1	1.3
both	1.7	1.9
Axiom		
right or left	1.4	1.6
Hombre	1.2	1.4
Pick-Up	.8	1.0
Rodeo		
1991-97	1.0	1.2
1998-99	1.4	1.6
Trooper	.8	1.0
VehiCROSS	.8	1.0
Upper Control Arm Assy., Replace (B)		
Amigo		
1989-94	2.6	2.8
1998-99	1.8	2.0
2000	3.1	3.3
Ascender		
right or left	1.1	1.3
both	1.7	1.9
Axiom		
right or left	3.2	3.4
Hombre	1.6	1.8
Pick-Up	2.0	2.2
Rodeo/Sport		
1991-97	2.0	2.2
1998-99	1.8	2.0
2000-04		
right or left	3.2	3.4
Trooper		
1984-91	2.0	2.2
1992-02	3.0	3.2
VehiCROSS	3.0	3.2
Replace bushings add	.5	.7

	LABOR TIME	SEVERE SERVICE
Upper Control Arm Bushings and/or Shaft, Replace (B)		
Amigo	2.3	2.4
Hombre	2.0	2.2
Rodeo	2.4	2.6
1992-02 Trooper	3.3	3.5
Wheel Bearing & Cups, Replace (B)		
Amigo		
1989-94		
one side	2.3	2.5
1998-00		
one side	1.2	1.4
Axiom		
right or left	1.1	1.3
both	2.2	2.4
Hombre one side	1.2	1.4
Pick-Up		
2 WD one	1.0	1.2
4 WD one	2.3	2.5
Rodeo/Sport		
1991-97		
2WD one	1.0	1.2
4WD one	1.5	1.7
1998-04		
one side	1.1	1.3
both	2.2	2.4
Trooper		
1984-91 one	2.3	2.4
1992-02 one	1.4	1.6
VehiCROSS		
one side	1.4	1.6
Wheel Bearing Grease Seal, Replace (B)		
Amigo		
1989-94		
2WD	1.0	1.2
4WD	.8	1.0
1998-00		
2WD	.8	1.0
4WD	1.0	1.2
Hombre one side	1.0	1.2
Pick-Up		
2WD	1.0	1.2
4WD	.8	1.0
Rodeo/Sport		
1991-97	1.2	1.4
1998-01		
2WD	.8	1.0
4WD	1.0	1.2
2002-04		
2WD	1.4	1.6
4WD	1.6	1.8
Trooper		
1984-91	.8	1.0
1992-02	1.3	1.5

	LABOR TIME	SEVERE SERVICE
REAR SUSPENSION		
Center Link, Replace (B)		
1992-02 Trooper	.7	.9
VehiCROSS	.7	.9
Replace bushings add each	.1	.3
Lateral Link, Replace (B)		
1998-04 Amigo, Rodeo/Sport	.8	1.0
Axiom, VehiCROSS	.8	1.0
1992-02 Trooper	.8	1.0
Replace bushings add each	.2	.4
Rear Coil Spring, Replace (B)		
Ascender, Axiom Rodeo/Sport		
one	1.1	1.3
both	1.4	1.6
Oasis one	.5	.7
1992-02 Trooper	.8	1.0
VehiCROSS		
right or left	.8	1.0
Rear Hub and/or Bearings, Replace (B)		
Oasis one	1.0	1.2
w/rear disc brakes add	.2	.4
Rear Leaf Spring, Replace (B)		
One side	1.2	1.4
Both sides	2.0	2.2
Recondition spring add each	.5	.7
Replace bushings add each	.3	.5
Rear Strut Shock Absorbers, Replace (B)		
1995-99 Oasis		
right side	.7	.9
left side	.5	.7
both sides	1.2	1.4
Shock Absorbers or Bushings, Replace (B)		
Exc. VehiCROSS		
one	.7	.9
both	.8	1.0
VehiCROSS one	.8	1.0
Spring Rear Eye Bushing, Replace (B)		
Hombre		
right front or rear	1.7	1.9
left front or rear	2.0	2.2
Replace fuel tank add	1.7	1.9
Spring Shackle, Replace (B)		
One side	.7	.9
Both sides	1.2	1.4
Replace bushings add each	.2	.4
Spring Shackle and/or Bushing, Replace (B)		
Hombre	1.4	1.6

ISU-46 AMIGO : ASCENDER : AXIOM : HOMBRE : OASIS : PICKUP

	LABOR TIME	SEVERE SERVICE
Stabilizer Bar, Replace (B)		
Amigo, Axiom, Hombre, Rodeo/Sport	1.0	1.2
Oasis, VehiCROSS	1.2	1.4
1994-02 Trooper	1.2	1.4
Replace bushings add	.1	.2
Stabilizer Bar Bushings, Replace (B)		
Axiom, Rodeo/Sport	.2	.4
Oasis	.7	.9
Trailing Link, Replace (B)		
1998-04 Amigo, Axiom, Rodeo/Sport		
lower		
right side	.5	.7
left side	.7	.9
upper		
right side	.5	.7
left side	.8	1.0
1992-02 Trooper lower		
right or left side	.7	.9
VehiCROSS lower		
right or left side	.7	.9
Replace bushings add each	.1	.2

STEERING

AIR BAGS

The following operations do not include testing. Add time as required.

	LABOR TIME	SEVERE SERVICE
Diagnose Air Bag System Component Each (A)		
1995-05	1.0	1.2
Air Bag Assy., Replace (B)		
Amigo		
1998-00		
drivers side	.5	.7
passenger side	1.1	1.3
Ascender		
drivers side	.7	.9
passenger side	1.4	1.6
Axiom		
drivers side	.3	.5
passenger side	1.0	1.2
Oasis		
drivers side	.5	.7
passenger side	.5	.7
Rodeo/Sport		
1995-97		
drivers side	.5	.7
passenger side	2.2	2.4
1998-99		
drivers side	.5	.7
passenger side	1.3	1.5
2000-04		
drivers side	.5	.7
passenger side	1.0	1.2
1995-02 Trooper		
drivers side	.5	.7
passenger side	1.6	1.8

	LABOR TIME	SEVERE SERVICE
VehiCROSS		
drivers side	.5	.7
passenger side	1.2	1.4
Arming Sensor, Replace (B)		
Hombre	.7	.9
Coil, Replace (B)		
1998-99 Amigo	1.3	1.5
Ascender	1.4	1.6
Axiom	1.1	1.3
Hombre	1.6	1.8
Rodeo/Sport		
1995-97	1.6	1.8
1998-04	1.3	1.5
1995-02 Trooper	1.6	1.8
VehiCROSS	.8	1.0
Dual Pole Arming Sensor, Replace (B)		
1995 Trooper	2.9	3.1
Electronic Control Unit, Replace (B)		
Oasis	.5	.7
Energy Reserve Module, Replace (B)		
Hombre	.5	.7
Forward Sensor, Replace (A)		
Hombre		
each	.7	.9
1995 Trooper		
one side	1.3	1.5
both sides	1.6	1.8
Inflatable Cushion Module, Replace (B)		
Hombre	.5	.7
Lower Instrument Panel Trim Pad, Replace (B)		
Hombre	.5	.7
Sensing & Diagnostic Module, Replace (SDM) (B)		
1998-00 Amigo	1.6	1.8
Ascender	.7	.9
Axiom	1.4	1.6
Rodeo/Sport		
1995-97	.7	.9
1998-04	1.4	1.6
1995-02 Trooper	.8	1.0
VehiCROSS	.7	.9

LINKAGE

	LABOR TIME	SEVERE SERVICE
Idler Arm, Replace (B)		
All Models	.7	.9
Pitman Arm, Replace (B)		
1989-94 Amigo	.5	.7
Hombre	.7	.9
Pick-Up	.5	.7
Rodeo, Trooper	.7	.9
VehiCROSS	.7	.9
Relay Rod (Center Link), Replace (B)		
Includes: Reset toe.		
1989-94 Amigo	1.6	1.8
Hombre	1.4	1.6
Pick-Up	1.6	1.8
Rodeo	1.3	1.5

	LABOR TIME	SEVERE SERVICE
Trooper		
1984-91	1.3	1.5
1992-02	1.6	1.8
VehiCROSS	1.4	1.6
Tie Rod or Tie Rod Ends, Replace (B)		
Includes: Reset toe.		
Amigo		
1989-94	.8	1.0
1998-02	1.0	1.2
Hombre	1.2	1.4
Oasis		
right side	1.4	1.6
left side	1.6	1.8
both sides	1.7	1.9
Pick-Up	.8	1.0
Rodeo	1.0	1.2
Trooper		
1984-91	.7	.9
1992-02	1.2	1.4
VehiCROSS	1.2	1.4

MANUAL WORM & SECTOR

	LABOR TIME	SEVERE SERVICE
Gear Assy., Adjust (On Vehicle) (B)		
1989-94 Amigo	.7	.9
Pick-Up	.7	.9
1991-95 Rodeo	.7	.9
Flexible Coupling, Replace (B)		
1989-94 Amigo	.7	.9
Pick-Up	.7	.9
1991-95 Rodeo	.7	.9
Gear Assy., R&R and Recondition (A)		
1989-94 Amigo	3.1	3.3
Pick-Up	3.1	3.3
1991-95 Rodeo	2.3	2.5
Gear Assy., Replace (B)		
1989-94 Amigo	1.3	1.5
Pick-Up	1.3	1.5
1991-95 Rodeo	1.3	1.5
Steering Column, R&R or Replace (B)		
1989-94 Amigo	1.5	1.7
Pick-Up	1.5	1.7
1991-95 Rodeo	1.7	1.9
Steering Wheel, Replace (B)		
1989-94 Amigo	.5	.7
Pick-Up	.5	.7
1991-95 Rodeo	.5	.7

POWER RACK & PINION

	LABOR TIME	SEVERE SERVICE
Power Steering Hoses, Replace (B)		
Includes: System bleeding.		
1998-02 Amigo		
pressure or return	.8	1.0
Ascender		
pressure or return	.8	1.0
Axiom		
pressure or return	.5	.7

RODEO : RODEO SPORT : TROOPER : VEHICROSS — ISU-47

	LABOR TIME	SEVERE SERVICE
Power Steering Hoses, Replace (B)		
Rodeo/Sport		
pressure	.5	.7
return	.3	.5
Oasis		
pressure	1.0	1.2
return	.8	1.0
VehiCROSS		
pressure	1.6	1.8
return	1.6	1.8
Power Steering Reservoir, Replace (B)		
1998-02 Amigo	.5	.7
Ascender (.8)	1.1	1.3
Axiom, Rodeo/Sport	.5	.7
Oasis	.7	.9
VehiCROSS	.5	.7
Pressure Switch, Replace (B)		
1998-02 Amigo, Axiom	.5	.7
Rodeo/Sport		
4 cyl.	.6	.8
V6	.5	.7
VehiCROSS	1.4	1.6
Rack & Pinion Assy., Adjust (B)		
1998-04 Amigo, Axiom Rodeo/Sport	.5	.7
Rack & Pinion Assy., R&R or Replace (B)		
Includes: Alignment.		
1998-04 Amigo, Axiom Rodeo/Sport		
2WD	1.4	1.6
4WD	1.8	2.0
Ascender	3.2	3.4
Oasis	3.5	3.7
Replace assembly add	.5	.7
Speed Sensor, Test & Replace (B)		
Oasis	.8	1.0
Steering Column, R&I (B)		
Amigo, Rodeo/Sport		
1998-99	1.9	2.1
2000-04	1.7	1.9
Ascender, Axiom	1.8	2.0
Oasis	1.4	1.6
VehiCROSS	1.8	2.0
w/air bags add	.2	.2
Replace assembly add	.3	.5
Steering Flex Coupling, Replace (B)		
Amigo, Ascender, Rodeo/Sport	.5	.7
Steering Pump, R&R or Replace (B)		
Amigo, Rodeo/Sport		
4 cyl.	.6	.8
V6	1.1	1.3
Ascender, Axiom	1.1	1.3
Oasis	1.1	1.3
Steering Unit Assy., Replace (B)		
VehiCROSS	1.8	2.0

	LABOR TIME	SEVERE SERVICE
Steering Unit Assy., R&R and Recondition (A)		
VehiCROSS	2.4	2.6
Steering Wheel, Replace (B)		
Amigo, Rodeo/Sport		
1998-99	.3	.5
2000-04	.5	.7
Ascender	.8	1.0
Axiom	.5	.7
Oasis	.7	.9
VehiCROSS	.5	.7
w/cruise control add	.1	.2
Tie Rod, Tie Rod End and/or Boot, Replace (B)		
Includes: Alignment.		
Amigo, Rodeo/Sport	1.0	1.2
Ascender, Axiom	.8	1.0
Oasis		
right side	1.4	1.6
left side	1.6	1.8
both sides	1.7	1.9
VehiCROSS	1.2	1.4
Replace boot add each	.1	.3

POWER WORM & SECTOR

	LABOR TIME	SEVERE SERVICE
Troubleshoot Power Steering (B)		
All Models	.5	.5
Gear Assy., Adjust (On Vehicle) (B)		
All Models	1.2	1.4
Flexible Coupling, Replace (B)		
All Models	.7	.9
Gear Assy., R&R and Recondition (A)		
1989-94 Amigo	3.0	3.2
Hombre	3.3	3.5
Pick-Up	3.0	3.2
1991-97 Rodeo	3.9	4.1
Trooper		
1984-91	3.0	3.2
1992-02	2.3	2.5
Gear Assy., Replace (B)		
All Models	2.2	2.4
Horn Contact, Replace (B)		
Hombre	1.4	1.6
w/air bags add	.2	.4
Oil Cooler, Replace (B)		
Hombre	1.4	1.6
Power Steering Switch, Replace (B)		
1991-97 Rodeo		
4 cyl.	.7	.9
V6	.5	.7
1992-02 Trooper V6	.5	.7
Remote Reservoir, Replace (B)		
All Models	.5	.7
Steering Column, R&I (B)		
1989-94 Amigo	1.5	1.7
Ascender (1.3)	1.8	2.0
Hombre	1.5	1.7
Pick-Up	1.5	1.7

	LABOR TIME	SEVERE SERVICE
Rodeo		
1991-95	1.7	1.9
1996-97	2.3	2.5
Trooper		
1984-91	1.0	1.2
1992-94	1.3	1.5
1995-02	2.5	2.7
w/air bags add	.2	.4
Recondition		
standard column add	1.2	1.4
tilt column add	1.7	1.9
Replace assy. add	.2	.2
Steering Column Lock Actuator Parts, Replace (B)		
Hombre	1.6	1.8
w/air bags add	.2	.2
w/cruise control add	.1	.1
Steering Pump, R&R and Recondition (B)		
1989-94 Amigo	1.8	2.0
Pick-Up	1.7	1.9
1991-97 Rodeo		
4 cyl.	1.7	1.9
V6	2.0	2.2
Trooper		
1986-91		
2.3L, 2.6L	1.7	1.9
2.8L	2.0	2.2
1992-02	1.8	2.0
Steering Pump, R&R or Replace (B)		
1989-94 Amigo	1.0	1.3
Ascender (1.1)	1.5	1.7
Hombre	.8	101
Pick-Up		
2.3L, 2.6L	1.3	1.5
3.1L	1.4	1.7
diesel	1.0	1.2
1991-97 Rodeo		
4 cyl.	1.2	1.4
V6	1.4	1.6
Trooper		
1984-91		
2.3L, 2.6L	1.3	1.5
2.8L	1.4	1.6
diesel	1.0	1.2
1992-02	1.5	1.7
Steering Pump Hoses, Replace (B)		
Does not include purging.		
1989-94 Amigo		
pressure	.7	.9
return	.5	.7
Ascender ea. hose (.5)	.7	.9
hydraulic hose assembly (2.3)	3.1	3.3
Hombre each	.5	.7
Pick-Up		
pressure	.7	.9
return	.5	.7

ISU-48 AMIGO : ASCENDER : AXIOM : HOMBRE : OASIS : PICKUP

	LABOR TIME	SEVERE SERVICE
Steering Pump Hoses, Replace (B)		
1991-97 Rodeo		
pressure	.7	.9
return	.5	.7
Trooper		
1984-91		
pressure	.7	.9
return	.5	.7
1992-02 each	1.6	1.8
Steering Wheel, Replace (B)		
All Models	.5	.5
w/air bags add	.2	.2

HEATING & AIR CONDITIONING

When more than one component requires replacement where evacuation, recovery and recharging is already included, deduct 1.0 hour for each additional component from the time given.

	LABOR TIME	SEVERE SERVICE
Evacuate, Recover and Recharge System (B)		
All Models	1.0	1.2
AC Hoses, Replace (B)		
Includes: Evacuate, recover and recharge.		
Amigo		
1989-94		
high or low pressure	1.9	2.1
evap. liquid line	3.1	3.3
suction or discharge	1.9	2.1
1998-00		
high pressure	1.3	1.5
low pressure	1.2	1.4
suction hose	1.6	1.8
Discharge hose	1.7	1.9
Ascender		
high pressure	1.1	1.3
low pressure	1.1	1.3
suction or discharge	.6	.8
Axiom		
high pressure	1.2	1.4
low pressure	1.5	1.7
suction/discharge	1.5	1.7
Hombre		
one	1.9	2.1
each addl. add	.5	.6
Oasis		
one	1.7	1.9
each addl. add	.5	.7
Pick-Up		
high pressure	1.9	2.1
low pressure	1.9	2.1
evap. liquid line	3.1	3.3

	LABOR TIME	SEVERE SERVICE
Rodeo/Sport		
1991-97		
high pressure	1.9	2.1
low pressure	1.9	2.1
evap. liquid line	3.1	3.3
suction or discharge	1.9	2.1
1998-04		
high pressure	1.3	1.5
low pressure	1.3	1.5
suction/discharge	1.7	1.9
Trooper		
1984-91		
high or low pressure	1.9	2.1
evap. liquid line	3.1	3.3
1992-97		
high pressure	1.2	1.4
low pressure	1.4	1.6
evap. liquid line	1.6	1.8
1998-02		
high pressure	1.2	1.4
low pressure	1.2	1.4
evap. liquid line	1.3	1.5
VehiCROSS		
high pressure liquid line	1.6	1.8
suction	1.4	1.6
Discharge	1.4	1.6
AC Switch, Replace (B)		
Rodeo/Sport 2002-04	1.2	1.4
VehiCROSS	.8	1.0
Accumulator Assy., Replace (B)		
Includes: Evacuate, recover and recharge.		
Ascender	.5	.7
Hombre	1.7	1.9
Blower Motor, Replace (B)		
1981-87	2.0	2.2
1988-94	.7	.9
1995-97		
Hombre	.8	1.0
Pick-Up, Rodeo, Trooper	.5	.7
1998-99		
Amigo, Rodeo	.3	.5
Oasis		
front	.5	.7
rear	1.0	1.2
Trooper	.3	.5
VehiCROSS	.3	.5
2000-04		
Amigo, Rodeo/Sport	.6	.8
Ascender		
A/C	.5	.7
auxiliary or rear	.8	1.0
Axiom, VehiCROSS	.3	.5
Blower Motor Relay, Replace (B)		
All Models	.3	.5

	LABOR TIME	SEVERE SERVICE
Blower Motor Resistor, Replace (B)		
Exc. Ascender		
front	.5	.7
rear	.7	.7
Ascender front or rear	.8	1.0
Blower Motor Switch, Replace (B)		
Hombre	.5	.7
Oasis rear	.3	.5
Trooper		
1994	.8	1.0
1995-02	1.2	1.4
Compressor or Fan Relay, Replace (B)		
1996-04	.3	.5
Compressor Assy., Replace (B)		
Includes: Parts transfer. Evacuate, recover and recharge.		
Amigo		
1989-94	1.7	1.9
1998-00	2.2	2.4
Ascender	1.8	2.0
Axiom	2.1	2.3
Hombre	1.8	2.0
Oasis	2.8	3.0
Pick-Up		
1981-87	2.3	2.4
1988-95	1.7	1.9
Rodeo/Sport		
1991-92	2.3	2.4
1993-97	3.1	3.3
1998-04	2.1	2.3
Trooper		
1986-97	2.3	2.5
1998-02	3.1	3.3
VehiCROSS	2.6	2.8
Replace receiver/drier add	.2	.4
Transfer clutch add	.5	.7
Compressor Clutch Plate & Hub Assy., Replace (B)		
1981-87	1.6	1.8
1988-97	2.6	2.8
Compressor Clutch Assy., Replace (B)		
Amigo, Axiom, Rodeo/Sport	2.0	2.2
1992-97 Trooper	2.7	2.9
VehiCROSS	2.9	3.1
Compressor Shaft Seal, Replace (B)		
Includes: Compressor R&R. Evacuate, recover and recharge.		
1981-87	1.9	2.1
1988-97	2.9	3.1
Condenser Cooling Fan, Replace (B)		
1993-97 3.2L	.5	.7
VehiCROSS	.8	1.0

RODEO : RODEO SPORT : TROOPER : VEHICROSS ISU-49

	LABOR TIME	SEVERE SERVICE
Condenser Pressure Switch, Replace (B)		
1998-02 Amigo	1.1	1.3
Rodeo/Sport		
1991-97	1.6	1.8
1998-04	1.1	1.3
Trooper		
1984-91	1.6	1.8
1992-02	1.2	1.4
VehiCROSS	1.2	1.4
Evaporator Core, Replace (B)		
Includes: Evacuate, recover and recharge.		
1981-87	5.2	5.4
1988-94	3.1	3.3
1995-97		
Hombre	2.8	3.0
Pick-Up, Rodeo	3.7	3.9
Trooper	2.5	2.7
1998-04		
Amigo, Axiom, Rodeo/Sport	4.3	4.5
Ascender		
front	5.4	5.6
rear	2.1	2.3
Oasis front or rear	2.3	2.5
Trooper	2.5	2.7
VehiCROSS	2.2	2.4
Evaporator Thermostatic Switch, Replace (B)		
1998-00 Amigo	4.1	4.3
Pick-Up	2.0	2.2
Rodeo/Sport		
1991-97	1.7	1.9
1998-04	4.1	4.3
Trooper		
1984-94	2.0	2.2
1995-02	2.2	2.4
VehiCROSS	2.2	2.4
Add time to recharge system if needed.		
Expansion Valve, Replace (B)		
Includes: Evacuate, recover and recharge.		
1998-00 Amigo	4.4	4.6
Ascender	1.5	1.7
Axiom	2.9	3.1
Hombre	3.7	3.9
Oasis		
front	2.3	2.5
rear	1.8	2.0
Pick-Up		
1981-87	5.2	5.4
1988-95	3.1	3.3
Rodeo/Sport		
1995-97	3.7	3.9
1998-04	4.4	4.6

	LABOR TIME	SEVERE SERVICE
Trooper		
1984-87	5.2	5.4
1988-94	3.1	3.3
1995-02	2.3	2.5
VehiCROSS	2.2	2.4
Heater Blower Motor Resistor, Replace (B)		
Exc. Ascender	.5	.7
Ascender	.9	1.1
Heater Blower Motor Switch, Replace (B)		
1998-00 Amigo	1.3	1.5
Hombre	.8	1.0
Oasis	2.0	2.2
Rodeo		
1994-97	.5	.7
1998-01	1.3	1.5
Heater Control Valve, Replace (B)		
Oasis	.8	1.0
Heater Core, R&R or Replace (B)		
Amigo		
1989-94		
w/AC	6.3	6.5
w/o AC	5.2	5.4
1998-00		
w/AC	7.6	7.8
w/o AC	6.4	6.6
Ascender		
AC (4.0)	5.4	5.6
auxiliary heater (1.4)	1.9	2.1
Axiom		
w/AC	6.8	7.0
w/o AC	5.7	5.9
Hombre		
w/AC	1.6	1.8
w/o AC	2.3	2.5
auxiliary	2.3	2.5
Oasis		
w/AC	9.3	9.5
w/o AC	8.3	8.5
Pick-Up		
w/AC	6.3	6.5
w/o AC	5.2	5.4
Rodeo/Sport		
1991-97		
w/AC	6.3	6.5
w/o AC	5.2	5.4
1998-01		
w/AC	7.6	7.8
w/o AC	6.4	6.6
2002-04		
w/AC	6.8	7.0
w/o AC	5.7	5.9
Trooper		
1984-97		
w/o AC	6.3	6.5
w/AC	5.2	5.4

	LABOR TIME	SEVERE SERVICE
1998-02		
w/AC	6.8	7.0
w/o AC	5.8	6.0
VehiCROSS	4.8	5.0
Heater Hoses, Replace (B)		
Amigo		
1989-94	.5	.7
1998-00		
4 cyl.	.5	.7
V6	.8	1.0
Axiom	1.2	1.4
Hombre	.8	1.0
Oasis each	.7	.9
Pick-Up	.5	.7
Rodeo/Sport		
1991-97	.5	.7
1998-04		
4 cyl.	.5	.7
V6	.8	1.0
Trooper		
1986-91	.5	.7
1992-02	1.0	1.2
VehiCROSS	1.2	1.4
Orifice Valve (Tube), Replace (B)		
Includes: Evacuate, recover and recharge.		
Hombre	1.6	1.8
Pressure Cycling Switch, Replace (B)		
Hombre	.3	.5
Receiver/Drier Assy., Replace (B)		
Includes: Evacuate, recover and recharge.		
1981-87	2.0	2.2
1988-97	1.6	1.8
1998-04		
exc. Rodeo/Sport	1.7	1.9
Rodeo/Sport	1.1	1.3
Temperature Control Assy., Replace (B)		
Ascender	.7	.9
Axiom, Hombre	1.1	1.3
Oasis	1.7	1.9
Rodeo/Sport		
1996-04	1.1	1.3
Trooper	1.0	1.2
VehiCROSS	.7	.9
Temperature Control Unit, Replace (B)		
Hombre	1.0	1.2
Pick-Up	1.0	1.2
1995-97 Rodeo	1.5	1.7
Trooper	1.2	1.4
Thermostat, Replace (B)		
Includes: Evacuate, recover and recharge.		
Oasis	2.3	2.5
Vacuum Selector Valve, Replace (B)		
Hombre	.8	1.0

ISU-50 AMIGO : ASCENDER : AXIOM : HOMBRE : OASIS : PICKUP

	LABOR TIME	SEVERE SERVICE
WIPERS & SPEEDOMETER		
Antenna, Replace (B)		
1981-98	.5	.7
1994-04		
exc. Ascender		
manual	.6	.8
power	.8	1.0
Ascender		
mast	.3	.5
base (.7)	.9	1.1
lead-in (1.2)	1.8	2.0
Instrument Cluster, R&I (B)		
Amigo, Rodeo/Sport	.8	1.0
Ascender	.7	.9
Axiom	1.2	1.4
Hombre	.7	.9
VehiCROSS	.5	.7
Intermittent Wiper Control Unit, Replace (B)		
Oasis	.5	.7
Pulse Wipe Control Module, Replace (B)		
Hombre	.5	.7
Radio, R&R (B)		
Amigo	.3	.5
Ascender, Axiom	.7	.9
Hombre	1.7	1.9
Oasis	1.2	1.4
Pick-Up	.5	.7
Rodeo/Sport		
1996-97	.7	.9
1998-04	.3	.5
Trooper		
1984-95	.5	.7
1996-02	.7	.9
VehiCROSS	.7	.9
Rear Window Washer Motor, Replace (B)		
1991-97 Rodeo	.3	.5
Ascender	.8	1.0
Trooper		
1984-91	.3	.5
1992-97	.7	.9
1998-02	.5	.7
Rear Window Wiper Motor, Replace (B)		
1998-00 Amigo	.6	.8
Axiom, Oasis	.5	.7
Ascender (.6)	.8	1.0
Rodeo/Sport		
1991-97	.5	.7
1998-04	.6	.8
Trooper	.7	.9
Rear Window Wiper Switch, Replace (B)		
1998-00 Amigo	.6	.8
Ascender	.8	1.0
Axiom, Pick-Up	.6	.8
Rodeo/Sport	.6	.8

	LABOR TIME	SEVERE SERVICE
Trooper		
1984-94	.5	.7
1995-02	.8	1.0
Speedometer Cable & Casing, Replace (B)		
All Models	.6	.8
Speedometer Cable (Inner), Replace or Lubricate (B)		
All Models	.7	.9
Speedometer Head, R&R or Replace (B)		
Amigo		
1989-94	1.2	1.4
1998-00	1.0	1.2
Ascender, Axiom (.9)	1.2	1.4
Oasis	.7	.9
Pick-Up	1.2	1.4
Rodeo/Sport		
1991-04	1.0	1.2
Trooper		
1984-91	.7	.9
1992-02	1.2	1.4
VehiCROSS	.8	1.0
Windshield Washer Pump, Replace (B)		
Amigo		
1989-94	.3	.5
1998-00	1.3	1.5
Ascender	.8	1.0
Hombre	.3	.5
Oasis	.8	1.0
Pick-Up		
1981-87	.5	.7
1988-95	.3	.5
Rodeo/Sport		
1991-97	.3	.5
1998-04	1.3	1.5
Trooper		
1984-87	.5	.7
1988-91	.3	.5
1992-02	.7	.9
VehiCROSS	.3	.5
Windshield Washer Pump Valve, Replace (B)		
All Models	.3	.5
Windshield Wiper Interval Relay, Replace (Rear) (B)		
1992-97 Trooper	.5	.7
Windshield Wiper Motor, Replace (B)		
Amigo		
1989-94	1.0	1.2
1998-00	1.1	1.3
Ascender, Axiom (.7)	.9	1.1
Hombre, Pick-up	1.0	1.2
Oasis	1.4	1.6
Rodeo/Sport		
1991-04	1.1	1.3

	LABOR TIME	SEVERE SERVICE
Trooper		
1984-91	1.0	1.2
1992-02	.5	.7
VehiCROSS	.7	.9
Windshield Wiper Switch, Replace (B)		
See combination switch also.		
1989-94 Amigo	1.0	1.2
Ascender, Axiom	.9	1.1
Hombre		
standard column	1.4	1.6
tilt column	1.8	2.0
Oasis	1.2	1.4
Pick-Up	1.0	1.2
1991-97 Rodeo	.8	1.0
Trooper		
1984-91	1.0	1.2
1992-94	.5	.7
1995-02	1.4	1.6
Windshield Wiper Transmission, Replace (B)		
Amigo		
1989-94	1.4	1.6
1998-00	1.3	1.5
Ascender (.7)	.9	1.1
Pick-Up		
Hombre	.8	1.0
Oasis	1.7	1.9
1981-87	1.7	1.9
1988-95	1.4	1.6
Rodeo	1.3	1.5
Trooper		
1984-87	1.7	1.9
1988-91	1.4	1.6
1992-02	.8	1.0
VehiCROSS	.8	1.0
LAMPS & SWITCHES		
Back-Up Lamp Bulb, Replace (C)		
Ascender	.3	.3
Oasis one	.2	.2
each addl. add	.1	.1
Back-Up Lamp Switch, Replace (B)		
All Models		
AT	.5	.6
MT	.3	.3
Clutch Start Switch, Replace (B)		
1986-04	.5	.5
Combination Switch Assy., Replace (B)		
Amigo		
1989-94	1.0	1.2
1998-00	1.0	1.2
Ascender	1.1	1.3
Axiom	.9	1.1
Oasis	1.2	1.4
Pick-Up	1.0	1.2

RODEO : RODEO SPORT : TROOPER : VEHICROSS   **ISU-51**

	LABOR TIME	SEVERE SERVICE
Rodeo/Sport		
1991-97	1.6	*1.8*
1998-04	1.0	*1.2*
Trooper		
1984-91	1.0	*1.2*
1992-94	.7	*.9*
1995-02	1.4	*1.6*
VehiCROSS	.8	*1.0*
Daylight Running Light Module, Replace (B)		
Ascender (2.1)	2.8	*3.0*
Hombre	.8	*1.0*
Halogen Headlamp Bulb, Replace (B)		
1988-05	.3	*.3*
Hazard Warning Switch, Replace (B)		
1998-00 Amigo	.6	*.8*
Axiom	.5	*.7*
1998-04 Rodeo/Sport	.6	*.8*
Trooper		
1992-94	.5	*.7*
1995-02	.8	*1.0*
VehiCROSS	.5	*.7*
Headlamp Dimmer Switch, Replace (B)		
1981-90	.7	*.9*
Hombre	1.0	*1.2*
Headlamp Sealed Beam Bulb, Replace (B)		
1981-97 each	.3	*.3*
Headlamp Switch, Replace (B)		
Ascender (.3)	.5	*.7*
Hombre	.7	*.9*
Pick-Up		
1981-87	.5	*.7*
1988-95	1.6	*1.8*
1994-95 Rodeo	.8	*1.0*
Trooper		
1984-94	.5	*.7*
1995-02	1.4	*1.6*
Headlamps, Aim (B)		
Two	.4	*.4*
Four	.6	*.6*
High Mount Stop Lamp Assy., Replace (B)		
1998-00 Amigo	.3	*.3*
Ascender, Axiom	.3	*.3*
Hombre, Oasis	.3	*.3*
1993-95 Pick-Up	.5	*.5*
1992-04 Rodeo/Sport	.3	*.3*
High Mount Stop Lamp Bulb, Replace (C)		
All Models	.2	*.2*
Horn, Replace (B)		
Exc. VehiCROSS	.3	*.3*
VehiCROSS	.8	*.8*
each addl. add	.1	*.1*
Horn Relay, Replace (B)		
All Models	.5	*.5*

	LABOR TIME	SEVERE SERVICE
License Lamp Assy., Replace (B)		
Amigo		
1989-94	.7	*.7*
1998-00	.3	*.3*
Ascender, Axiom	.3	*.3*
Hombre	.5	*.5*
Pick-Up, Rodeo/Sport	.3	*.3*
1984-02 Trooper	.3	*.3*
VehiCROSS	.3	*.3*
License Lamp Bulb, Replace (C)		
1994-05 one or all	.2	*.2*
Multifunction Switch Lever, Replace (B)		
Hombre	1.4	*1.6*
w/air bags add	.2	*.3*
Neutral Start/Safety Switch, Replace (B)		
All Models	.7	*.7*
Parking Brake Warning Lamp Switch, Replace (B)		
Amigo	.5	*.5*
Ascender, Hombre	.7	*.7*
Axiom	.3	*.3*
Oasis, Pick-Up	.5	*.5*
Rodeo/Sport	.5	*.5*
Trooper		
1984-91	.5	*.5*
1992-02	.8	*.8*
VehiCROSS	.5	*.5*
Parking Lamp Lens or Bulb, Replace (C)		
1998-00 Amigo one	.2	*.2*
Ascender	.5	*.5*
Hombre one	.3	*.3*
Oasis one	.2	*.2*
1994-02 Rodeo one	.2	*.2*
each addl. add	.1	*.1*
Side Marker Lamp Lens, Replace (C)		
All Models	.2	*.2*
Stop & Tail Lamp Bulb, Replace (C)		
All Models one	.3	*.3*
each addl. add	.1	*.1*
Stop Light Switch, Replace (B)		
1981-87	.3	*.3*
1988-04	.5	*.5*
Turn Signal or Hazard Relay, Replace (B)		
1986-91 Trooper	.3	*.3*
Turn Signal or Hazard Warning Flasher, Replace (B)		
All Models	.5	*.5*
Turn Signal & Hazard Warning Switch, Replace (B)		
Hombre	1.0	*1.0*
Pick-Up	.7	*.7*
Trooper		
1984-91	.7	*.7*
1992-95	1.6	*1.6*
1996-02	.7	*.7*

	LABOR TIME	SEVERE SERVICE
Turn Signal & Headlamp Dimmer Switch, Replace (B)		
Oasis	1.2	*1.2*
Turn Signal or Parking Lamp Lens or Bulb, Replace (C)		
All Models	.2	*.2*
Turn Signal Switch, Replace (B)		
Trooper		
1992-94	.7	*.7*
1995-97	1.4	*1.4*

BODY

	LABOR TIME	SEVERE SERVICE
Front Door Lock, Replace (B)		
Amigo	1.1	*1.3*
Ascender, Axiom	.7	*.9*
Pick-Up	.8	*1.0*
Rodeo/Sport	1.1	*1.3*
Trooper	.8	*1.0*
VehiCROSS	.7	*.9*
Front Door Window Regulator, Replace (B)		
Amigo		
1989-94	1.0	*1.2*
1998-02		
manual	1.0	*1.2*
power	1.1	*1.3*
Ascender, Axiom (.7)	.9	*1.1*
Hombre		
manual	1.0	*1.2*
power	1.2	*1.4*
Pick-Up	1.0	*1.2*
Rodeo/Sport		
manual	1.0	*1.2*
power	1.1	*1.3*
Trooper		
1984-91	1.0	*1.2*
1992-02		
manual	1.2	*1.4*
power	.8	*1.0*
VehiCROSS	.8	*1.0*
Hood Hinge, Replace (One) (B)		
All Models	.7	*.9*
2002-04		
Ascender (1.0)	1.4	*1.6*
Axiom, Rodeo/Sport	.5	*.7*
Hood Lock, Replace (B)		
Exc. Axiom	.5	*.7*
Axiom	1.1	*1.4*
Hood Release Cable, Replace (B)		
Amigo		
1989-94	.7	*.9*
1998-02	1.0	*1.2*
Ascender, Axiom	1.1	*1.4*
Hombre	.7	*.9*
Oasis	1.2	*1.4*
Pick-Up	.7	*.9*
Rodeo/Sport	1.0	*1.2*

ISU-52   AMIGO : ASCENDER : AXIOM : HOMBRE : OASIS : PICKUP

	LABOR TIME	SEVERE SERVICE
Hood Release Cable, Replace (B)		
Trooper		
1984-91	.7	.9
1992-02	1.3	1.5
Liftgate/Tailgate Lock, Replace (B)		
Ascender	2.3	2.5
Axiom	.5	.7
Hombre	.7	.9
Rodeo/Sport	.5	.7
Lock Striker Plate, Replace (B)		
All Models	.5	.7
Rear Door Lock, Replace (B)		
Ascender, Axiom,		
Hombre	.8	1.0
Rodeo/Sport		
1991-97	.7	.9
1998-04	.8	1.0
Trooper	.7	.9
w/power locks add	.2	.4

	LABOR TIME	SEVERE SERVICE
Rear Door Window Regulator, Replace (B)		
Ascender	.9	1.1
Axiom	.6	.8
Hombre		
manual	.8	1.0
power	1.0	1.2
Rodeo/Sport		
1991-97		
manual	1.3	1.5
power	1.4	1.6
1998-04	1.1	1.3
Trooper	.8	1.0
Window Regulator Motor, Replace (B)		
Front		
Amigo	1.1	1.3
Ascender, Axiom	.9	1.1
Hombre	1.3	1.5

	LABOR TIME	SEVERE SERVICE
Oasis	1.6	1.8
Pick-Up	1.7	1.9
Rodeo/Sport	1.1	1.3
Trooper		
1984-91	.7	.9
1992-02	1.2	1.4
VehiCROSS	.8	1.0
Rear		
Axiom	.6	.8
Hombre	1.7	1.9
Oasis	1.6	1.8
Rodeo		
1991-97	.8	1.0
1998-04	1.1	1.3
Trooper,	.8	1.0
VehiCROSS	.8	1.0

KIA

Amanti : Optima : Rio : Sephia : Spectra

SYSTEM INDEX

- MAINTENANCE KIA-2
  - Maintenance Schedule... KIA-2
- CHARGING KIA-2
- STARTING KIA-2
- IGNITION KIA-2
- EMISSIONS KIA-3
- FUEL KIA-3
- EXHAUST KIA-3
- ENGINE COOLING KIA-3
- ENGINE KIA-4
  - Assembly KIA-4
  - Cylinder Head KIA-4
  - Camshaft KIA-5
  - Crank & Pistons KIA-5
  - Engine Lubrication .. KIA-6
- CLUTCH KIA-6
- MANUAL TRANSAXLE KIA-6
- AUTO TRANSAXLE KIA-6
- SHIFT LINKAGE KIA-7
- DRIVELINE KIA-7
- BRAKES KIA-7
- FRONT SUSPENSION KIA-8
- REAR SUSPENSION KIA-8
- STEERING KIA-9
- HEATING & AC KIA-9
- WIPERS & SPEEDOMETER .. KIA-10
- LAMPS & SWITCHES KIA-10
- BODY KIA-10

OPERATIONS INDEX

A
- AC Hoses KIA-9
- Air Bags KIA-9
- Air Conditioning KIA-9
- Alignment KIA-8
- Alternator (Generator) KIA-2
- Antenna KIA-10
- Anti-Lock Brakes KIA-7

B
- Back-Up Lamp Switch ... KIA-10
- Ball Joint KIA-8
- Battery Cables KIA-2
- Bleed Brake System KIA-7
- Blower Motor KIA-9
- Brake Disc KIA-7
- Brake Drum KIA-7
- Brake Hose KIA-7
- Brake Pads and/or Shoes KIA-8

C
- Camshaft KIA-5
- Camshaft Sensor KIA-2
- Catalytic Converter ... KIA-3
- Coolant Temperature (ECT) Sensor KIA-3
- Crankshaft KIA-5
- Crankshaft Sensor KIA-2
- CV Joint KIA-7
- Cylinder Head KIA-4

D
- Distributor KIA-2
- Drive Belt KIA-2

E
- EGR KIA-3
- Electronic Control Module (ECM/PCM) KIA-3
- Engine KIA-4
- Engine Lubrication KIA-6
- Engine Mounts KIA-4
- Evaporator KIA-9
- Exhaust KIA-3
- Exhaust Manifold KIA-3
- Expansion Valve KIA-10

F
- Flywheel KIA-6
- Fuel Injection KIA-3
- Fuel Pump KIA-3
- Fuel Vapor Canister ... KIA-3

G
- Gear Selector Lever ... KIA-7
- Generator KIA-2

H
- Halfshaft KIA-7
- Headlamp KIA-10
- Heater Core KIA-10
- Horn KIA-10

I
- Idle Air Control (IAC) Valve .. KIA-3
- Ignition Coil KIA-2
- Ignition Module KIA-2
- Ignition Switch KIA-2
- Inner Tie Rod KIA-9
- Intake Manifold KIA-3

L
- Lower Control Arm KIA-8

M
- Maintenance Schedule .. KIA-2
- Mass Air Flow (MAF) Sensor KIA-3
- Master Cylinder KIA-7
- Muffler KIA-3
- Multifunction Lever/Switch ... KIA-10

N
- Neutral Safety Switch . KIA-10

O
- Oil Pan KIA-6
- Oil Pump KIA-6
- Outer Tie Rod KIA-9
- Oxygen Sensor KIA-3

P
- Parking Brake KIA-8
- Pistons KIA-6
- Positive Crankcase Ventilation (PCV) Valve KIA-3

R
- Radiator KIA-3
- Radiator Hoses KIA-4
- Radio KIA-10
- Rear Main Oil Seal KIA-6

S
- Shock Absorber/Strut, Front KIA-8
- Shock Absorber/Strut, Rear KIA-9
- Spark Plug Cables KIA-2
- Spark Plugs KIA-3
- Spring, Front Coil KIA-8
- Spring, Rear Coil KIA-8
- Starter KIA-2
- Steering Wheel KIA-9

T
- Thermostat KIA-3
- Throttle Body KIA-3
- Throttle Position Sensor (TPS) KIA-3
- Timing Belt KIA-5

V
- Valve Body KIA-7
- Valve Cover Gasket KIA-4
- Valve Job KIA-5
- Vehicle Speed Sensor .. KIA-3

W
- Water Pump KIA-4
- Wheel Balance KIA-2
- Wheel Cylinder KIA-8
- Window Regulator KIA-10
- Windshield Washer Pump KIA-10
- Windshield Wiper Motor KIA-10

KIA-1

KIA-2 AMANTI : OPTIMA : RIO : SEPHIA : SPECTRA

	LABOR TIME	SEVERE SERVICE
MAINTENANCE		
Air Cleaner Filter Element, Replace (C)		
All Models	.3	.4
Chassis Lubrication, Change Oil & Filter (C)		
Includes: Correct all fluid levels.		
Sephia, Spectra	.5	.5
Drive Belt, Adjust (B)		
One	.3	.4
each addl. add	.1	.2
Drive Belt, Replace (B)		
Exc. below	.5	.6
Amanti	.8	.8
w/AC add	.1	.1
Fuel Filter, Replace (B)		
Exc. below	.3	.3
Amanti	.9	.9
Halogen Headlamp Bulb, Replace (C)		
One	.3	.3
each addl. add	.2	.2
Oil and Filter, Change (C)		
Includes: Correct all fluid levels.		
All Models	.5	.5
Timing Belt, Replace (B)		
Exc. below	2.8	3.1
Amanti	4.6	4.9
Optima, Rio		
4 cyl.	2.6	2.9
w/AC add	.2	.2
w/DOHC 94-96		
Sephia add	.5	.5
w/PS add	.2	.2
Tire, Replace (C)		
Includes: Dismount old tire and mount new tire to rim.		
One	.5	.5
Tires, Rotate (C)		
Sephia, Spectra	.5	.5
Wheel, Balance (B)		
Two	.5	.5
each addl. add	.3	.3

SCHEDULED MAINTENANCE INTERVALS
If necessary, refer to appropriate Chilton maintenance service information.

	LABOR TIME	SEVERE SERVICE
7,500 Mile Service (C)		
All Models	.6	.6
15,000 Mile Service (C)		
All Models	.6	.6
22,500 Mile Service (C)		
All Models	.6	.6
30,000 Mile Service (B)		
All Models	2.1	2.1
37,500 Mile Service (C)		
All Models	.6	.6
45,000 Mile Service (C)		
All Models	.6	.6
52,500 Mile Service (C)		
All Models	.6	.6
60,000 Mile Service (B)		
Exc. California	4.1	4.1
California	2.3	2.3
67,500 Mile Service (C)		
All Models	.6	.6
75,000 Mile Service (C)		
All Models	.6	.6
82,500 Mile Service (C)		
All Models	.6	.6
90,000 Mile Service (B)		
All Models	2.4	2.4
97,500 Mile Service (C)		
All Models	.6	.6
105,000 Mile Service (B)		
All Models	.6	.6
Replace timing belt		
CA models add	2.0	2.0
112,500 Mile Service (C)		
All Models	.6	.6
120,000 Mile Service (B)		
Exc. California	4.1	4.1
California	2.3	2.3

CHARGING
	LABOR TIME	SEVERE SERVICE
Alternator Circuits, Test (A)		
Includes: Test component output.		
All Models	.5	.5
Alternator Assy., R&R and Recondition (A)		
Optima	3.3	3.7
Rio	1.8	2.1
Sephia, Spectra	2.4	2.7
Alternator Assy., Replace (B)		
Includes: Pulley transfer.		
Amanti, Optima	1.8	2.1
Rio	.8	.9
Sephia, Spectra	1.1	1.2
Voltage Regulator, Replace (B)		
Optima		
4 cyl.	1.4	1.5
V6	2.3	2.6
Rio	1.5	1.7
Sephia, Spectra	2.0	2.2

STARTING
	LABOR TIME	SEVERE SERVICE
Starter Draw Test (On Car) (B)		
All Models	.5	.5
Battery, Replace (C)		
All Models	.6	.6
Battery Cables, Replace (C)		
All Models	.6	.6
Keyless Entry Module, Replace (C)		
Exc. below	.6	.6
Amanti, Optima	1.1	1.1
Keyless Entry Transmitter, Replace (C)		
All Models	.3	.3
Starter Assy., R&R and Recondition (A)		
Optima, Rio	1.7	2.0
Sephia, Spectra	2.4	2.7
Starter Assy., Replace (B)		
All Models	1.0	1.1
Starter Drive, Replace (B)		
Includes: Starter R&R.		
Rio	1.3	1.4
Sephia, Spectra	2.1	2.3
Starter Solenoid, Replace (B)		
Optima, Rio	1.0	1.1
Sephia, Spectra	1.4	1.6

IGNITION
	LABOR TIME	SEVERE SERVICE
Diagnose Ignition System Component Each (A)		
All Models	.5	.5
Ignition Timing, Adjust (C)		
All Models	.5	.5
Camshaft Position Sensor, Replace (B)		
Exc. below	.5	.7
Amanti	1.2	1.4
Crankshaft Position Sensor, Replace (B)		
Exc. below	.5	.8
Amanti	3.2	3.5
Optima 4 cyl.	2.6	2.9
Distributor, R&R and Recondition (B)		
Includes: Reset base ignition timing.		
Optima, Sephia	1.5	1.7
Distributor, Replace (B)		
Includes: Reset base ignition timing.		
Optima, Sephia	.8	.9
Distributor Cap and/or Rotor, Replace (B)		
Optima, Sephia	.5	.5
Igniter Assy., Replace (B)		
Sephia, Spectra	.7	.8
Ignition Coil, Replace (B)		
All Models	.6	.7
Ignition Switch, Replace (B)		
Exc. below	.7	.8
Optima	1.1	1.2
Spark Plug (Ignition) Cables, Replace (B)		
Exc. below	.3	.4
Amanti	2.6	2.9
Optima		
4 cyl.	.5	.5
V6	2.6	2.9

AMANTI : OPTIMA : RIO : SEPHIA : SPECTRA — KIA-3

	LABOR TIME	SEVERE SERVICE
Spark Plugs, Replace (B)		
Exc. below	.5	.6
Amanti, Optima		
4 cyl.	2.6	2.9
V6		
right side	.5	.5
left side	2.6	2.9
both sides	2.9	3.2

EMISSIONS

The following operations do not include testing. Add time as required.

Diagnose Emission Control System Component Each (A)		
All Models	.5	.5
Retrieve Fault Codes (A)		
All Models	.3	.3
Air Bypass Check Valve, Replace (B)		
All Models	.5	.5
Air Flow Meter, Replace (B)		
Rio, Sephia, Spectra	.3	.3
Canister Purge Solenoid Valve, Replace (B)		
All Models	.3	.3
Coolant Temperature (ECT) Sensor, Replace (B)		
All Models	.5	.5
EGR Check Valve, Replace (B)		
All Models	.5	.7
EGR Control Valve, Replace (B)		
Exc. below	.5	.7
Amanti	1.2	1.4
EGR Modulator, Replace (B)		
Sephia, Spectra	.3	.3
EGR Solenoid, Replace (B)		
All Models	.3	.3
EGR Valve Pipe, Replace (B)		
All Models	.6	.7
Electronic Control Module (ECM/PCM), Replace (B)		
Exc. below	.5	.5
Amanti	.9	.9
Optima	2.4	2.4
Fuel Vapor Canister, Replace (B)		
Exc. below	.3	.3
Spectra 2.0L	1.2	1.4
Mass Air Flow (MAF) Sensor, Replace (B)		
All Models	.5	.5
Oxygen Sensor, Replace (B)		
All Models	.5	.7
Positive Crankcase Ventilation (PCV) Valve, Replace (B)		
All Models	.3	.3
Thermo Water Valve, Replace (B)		
All Models	.7	.7
Throttle Position Sensor (TPS), Replace (B)		
All Models	.6	.7

	LABOR TIME	SEVERE SERVICE
Vehicle Speed Sensor Assy., Replace (B)		
All Models	.5	.5

FUEL
DELIVERY

Fuel Pump, Test (B)		
All Models	.5	.5
Fuel Filter, Replace (B)		
Exc. below	.5	.6
Amanti	1.0	1.1
Fuel Gauge (Dash), Replace (B)		
All Models	.7	.7
Fuel Gauge (Tank), Replace (B)		
Exc. below	.6	.7
Amanti, Optima	1.1	1.3
Fuel Pump, Replace (B)		
All Models	.6	.7
Fuel Pump Relay, Replace (B)		
All Models	.5	.5
Fuel Tank, Replace (B)		
Includes: Drain and refill.		
Amanti, Optima	3.2	3.4
Rio, Sephia, Spectra	1.6	1.8
Spectra 2.0L	1.9	2.1
Intake Manifold and/or Gasket, Replace (B)		
Includes: Adjustments.		
Amanti, Optima	2.4	2.7
Rio	1.7	1.9
Sephia, Spectra	2.0	2.2
Spectra 2.0L	2.4	2.6

INJECTION

Diagnose Fuel Injection System Component Each (A)		
Sephia, Spectra	.7	.7
Bypass Air Control Valve, Replace (B)		
Sephia, Spectra	.5	.5
Fuel Injector Relay, Replace (B)		
All Models	.3	.3
Fuel Injectors, Replace (B)		
Exc. below	1.1	1.3
Amanti, Optima		
4 cyl.	.8	.9
V6		
right	.5	.5
left or both	2.6	2.9
Fuel Pressure Regulator, Replace (B)		
Exc. below	.5	.7
Spectra 2.0L	.6	.8
Fuel Pressure Solenoid, Replace (B)		
Sephia, Spectra one	.3	.3
each addl. add	.2	.2
Fuel Rail Assy., Replace (B)		
Exc. below	.9	1.0
Amanti, Optima		
4 cyl.	.8	.9

	LABOR TIME	SEVERE SERVICE
V6		
right	.5	.5
left or both	2.6	2.9
Idle Air Control (IAC) Valve, Replace (B)		
All Models	.6	.7
Throttle Body and/or Gasket, R&R or Replace (B)		
Exc. below	.8	.9
Amanti, Optima	1.4	1.5

EXHAUST

Catalytic Converter, Replace (B)		
All Models	.6	.7
Exhaust Manifold or Gasket, Replace (B)		
Amanti, Optima		
4 cyl.	1.6	1.8
V6		
right	4.0	4.4
left	1.6	1.8
both	4.7	5.3
Rio	1.3	1.4
Sephia, Spectra		
SOHC	1.6	1.8
DOHC	1.1	1.3
Spectra 2.0L	1.6	1.8
Front Muffler Assy. and/or Front Pipe, Replace (B)		
All Models	.8	.9
Intermediate Muffler Assy. or Inlet Pipe, Replace (B)		
All Models	.8	.9
Muffler, Replace (B)		
All Models	.8	.9
Resonator, Replace (B)		
All Models	.8	.9

ENGINE COOLING

Pressure Test Cooling System (B)		
All Models	.3	.3
Coolant Thermostat, Replace (B)		
All Models		
4 cyl.	.6	.8
V6	.8	1.0
Electric Fan Relay, Replace (B)		
All Models	.3	.3
Engine Coolant Temp. Sending Unit, Replace (B)		
All Models	.5	.5
Freeze Plugs (Water Jacket), Replace (B)		
Add access time as required.		
One	.5	.7
Radiator Assy., R&R or Replace (B)		
4 cyl.	.8	.9
V6	1.2	1.4
w/AC add	.2	.2
w/AT add	.2	.2

AMANTI : OPTIMA : RIO : SEPHIA : SPECTRA

	LABOR TIME	SEVERE SERVICE
Radiator Fan Motor and/or Fan, Replace (B)		
4 cyl.	.8	.9
V6	1.1	1.2
Radiator Hoses, Replace (B)		
Includes: Refill with proper coolant mix.		
All Models		
upper	.5	.6
lower	.8	.9
Radiator Supply Tank, Replace (B)		
All Models	.3	.3
Water Pump and/or Gasket, Replace (B)		
Includes: Refill with proper coolant mix.		
Optima		
4 cyl.	3.5	3.9
V6	3.8	4.2
Rio	2.9	3.2
Sephia, Spectra		
1994-98		
SOHC	2.6	3.1
DOHC	3.0	3.5
1999-05		
1.8L	3.6	4.1
2.0L	2.8	3.2
w/AC add	.3	.3
w/PS add	.2	.2

ENGINE
ASSEMBLY

Times shown are for OEM assemblies. Time to replace assemblies from aftermarket rebuilders may vary.

Engine Assy., R&I (B)
Does not include parts or component transfer.

	LABOR TIME	SEVERE SERVICE
1993-00		
Sephia		
AT	7.0	7.0
MT	6.4	6.4
Spectra		
AT	7.8	7.8
MT	7.0	7.0
w/AC add	.5	.5
w/DOHC 94-96		
Sephia add	.7	.7
w/PS add	.3	.3

Engine Assy., R&R and Recondition (A)
Includes: Replacing rings, rod and main bearings, cylinder head reconditioning and engine tune-up.

Amanti	43.1	44.1
Optima		
4 cyl.	25.8	26.8
V6	35.1	36.1
Rio	21.6	22.6
Sephia, Spectra		
1994-98	24.6	25.6
1999-05		
1.8L	24.2	25.2
2.0L	22.8	23.8
w/AC add	.3	.3
w/AT exc. Amanti add	.8	.9
w/DOHC 94-96		
Sephia add	.7	.7
w/PS exc. Optima add	.3	.3

Engine (New or Rebuilt Complete), Replace (B)
Includes: Component transfer and engine tune-up.

Amanti	21.6	22.5
Optima		
4 cyl.	12.5	13.4
V6	15.6	16.4
Rio	7.3	8.1
Sephia, Spectra		
1993-98	7.3	8.1
1999-05		
1.8L	8.1	9.0
2.0L	10.9	11.8
w/AC add	.3	.3
w/AT add	.8	.8
w/DOHC 94-96		
Sephia add	.7	.7
w/PS exc. Optima add	.3	.3

Engine (Short Block), Replace (B)
Assembly consists of cylinder block, piston assemblies, crankshaft, camshaft, timing chain and gears. Does not include cylinder heads. Operation Includes: R&R engine, transfer necessary parts and all necessary adjustments.

Amanti	23.1	24.2
Optima		
4 cyl.	17.0	18.1
V6	19.1	20.2
Rio	12.2	13.3
Sephia, Spectra		
1993-98	12.8	13.9
1999-05		
1.8L	18.9	21.0
2.0L	13.8	14.9
w/AC add	.5	.5
w/AT 01-03 exc. Amanti add	.8	.9
w/DOHC 94-96 Sephia add	.7	.7
w/PS exc. Amanti, Optima, Spectra 2.0L add	.3	.3

	LABOR TIME	SEVERE SERVICE
Engine Mounting Bracket, Replace (B)		
One		
exc. below	1.0	1.1
Amanti, Optima		
V6 No.1	1.9	2.0
each addl. add	1.0	1.0
Engine Mounting Rubber, Replace (B)		
One	.7	.9
each addl. add	.7	.7
Roll Stop, Replace (B)		
Amanti		
front bracket	.8	.9
rear bracket	1.9	2.1
Optima		
4 cyl. one	.8	.9
V6		
front bracket	.5	.6
rear bracket	1.9	2.1
each addl. Optima		
4 cyl. add	.8	.8

CYLINDER HEAD

Compression Test (B)
Sephia, Spectra	.6	.8

Valve Clearance, Adjust (B)
All Models	.9	1.0

Cylinder Head and/or Gasket, Replace (B)
Includes: Parts transfer and adjustments.

Exc. below	5.6	6.2
Amanti, Optima		
4 cyl.	5.1	5.7
V6		
right	6.5	7.1
left	5.7	6.3
both	7.6	8.2
w/AC add	.3	.3
w/DOHC 94-96		
Sephia add	.7	.7

Rocker Arm Cover Gasket, Replace or Reseal (B)
Exc. below	.6	.7
Amanti, Optima		
4 cyl.	.5	.6
V6		
right	.8	.9
left	2.1	2.3
both	2.3	2.5

Rocker Arms or Shafts, Replace (B)
Amanti		
right	6.6	6.9
left	7.0	7.3
both	7.9	8.2
Optima		
4 cyl.	3.1	3.4

AMANTI : OPTIMA : RIO : SEPHIA : SPECTRA — KIA-5

	LABOR TIME	SEVERE SERVICE
V6		
one	2.6	*2.9*
both	2.9	*3.2*
Rio	.9	*1.0*
Sephia, Spectra	1.5	*1.6*
Valve Job (A)		
Amanti		
right	18.4	*19.6*
left	15.7	*16.9*
both	23.4	*24.6*
Optima		
4 cyl.	11.5	*12.7*
V6		
right	13.9	*15.1*
left	12.6	*13.8*
both	19.2	*20.4*
Rio	10.7	*11.9*
Sephia, Spectra		
1993-98	11.9	*12.4*
1999-05		
1.8L	12.9	*13.8*
2.0L	9.6	*10.8*
w/AC add	.3	*.3*
w/DOHC 94-96		
Sephia add	.7	*.7*
Valve Lifters/Lash Adjustment Assy., Replace (B)		
Amanti		
right	6.3	*6.7*
left	6.8	*7.2*
both	7.6	*8.0*
Optima		
4 cyl.	2.9	*3.2*
V6		
right	3.2	*3.6*
left	4.8	*5.4*
both	5.7	*6.4*
Rio	3.3	*3.7*
Sephia, Spectra		
1.8L	4.1	*4.6*
2.0L	1.5	*2.0*
Valve Springs and/or Oil Seals, Replace (B)		
Amanti		
right	11.8	*12.3*
left	10.1	*10.6*
both	12.8	*13.3*
Optima		
4 cyl.	7.5	*8.0*
V6		
one side	6.8	*7.3*
both sides	8.8	*9.2*
Rio	5.4	*5.9*
Sephia, Spectra		
1.8L	9.8	*10.3*
2.0L	6.8	*7.3*
w/AC add	.2	*.2*
w/DOHC 94-96		
Sephia add	.7	*.7*

	LABOR TIME	SEVERE SERVICE
CAMSHAFT		
Camshaft, Replace (A)		
Amanti		
right	6.2	*6.7*
left	6.8	*7.3*
both sides	7.5	*8.0*
Optima		
4 cyl.	2.9	*3.4*
V6		
right	2.9	*3.4*
left	3.3	*3.8*
both sides	4.3	*4.8*
Rio	2.9	*3.4*
Sephia, Spectra	3.5	*4.0*
w/AC add	.2	*.2*
w/DOHC 94-96		
Sephia add	.7	*.7*
Camshaft Bearings, Replace (B)		
With cam removed		
Sephia, Spectra	.3	*.4*
Camshaft Oil Seals, Replace (B)		
Amanti, Optima		
4 cyl.	2.1	*2.4*
V6	4.1	*4.4*
Rio, Sephia, Spectra	2.6	*2.9*
w/AC add	.2	*.2*
w/DOHC 94-96		
Sephia add	.3	*.3*
Camshaft Sprocket, Replace (B)		
Amanti		
one side	4.6	*5.0*
both sides	4.8	*5.2*
Optima		
4 cyl.	2.8	*3.1*
V6		
one side	2.8	*3.1*
both sides	3.1	*3.4*
Rio	2.3	*2.6*
Sephia		
1993-98	1.4	*1.5*
1999-01	3.2	*3.6*
Spectra		
1.8L	3.2	*3.6*
2.0L	2.8	*3.2*
w/AC add	.2	*.2*
w/DOHC 94-96		
Sephia add	.7	*.7*
w/PS add	.2	*.2*
Timing Belt, Replace (B)		
Exc. below	2.8	*3.1*
Amanti	4.6	*4.9*
Optima, Rio		
4 cyl.	2.6	*2.9*
w/AC add	.2	*.2*
w/DOHC 94-96		
Sephia add	.5	*.5*
w/PS add	.2	*.2*

	LABOR TIME	SEVERE SERVICE
Timing Belt Idler Pulley, Replace (B)		
Exc. below	2.3	*2.6*
Amanti	4.6	*4.9*
Optima V6	2.8	*3.1*
Spectra 2.0L	2.8	*3.1*
Timing Belt Tensioner, Replace (B)		
Exc. below	2.0	*2.3*
Amanti	4.6	*4.9*
Optima V6	2.8	*3.1*
Spectra 2.0L	2.8	*3.1*
w/AC add	.3	*.3*
w/PS add	.3	*.3*
CRANK & PISTONS		
Connecting Rod Bearings, Replace (A)		
Includes: Check bearing oil clearance.		
Exc. below	4.3	*4.6*
Amanti	5.0	*5.3*
Optima	4.4	*4.7*
Crankshaft and Main Bearings, Replace (A)		
Includes: Engine R&R, check bearing oil clearance.		
Amanti	25.0	*25.9*
Optima		
4 cyl.	14.2	*15.1*
V6	20.1	*21.0*
Rio	13.1	*14.0*
Sephia, Spectra		
1.8L	16.1	*17.0*
2.0L	14.0	*14.9*
w/AC add	.3	*.3*
w/AT 94-00 add	.5	*.5*
w/DOHC 94-96		
Sephia add	.7	*.7*
w/PS add	.3	*.3*
Crankshaft Pulley, Replace (B)		
All Models	.8	*.9*
w/AC add	.2	*.2*
Crankshaft Sprocket, Replace (B)		
Amanti	4.6	*4.9*
Optima, Rio	2.3	*2.6*
Sephia, Spectra		
1993-98	1.9	*2.1*
1999-05	2.4	*2.7*
w/AC add	.3	*.3*
w/PS exc. Amanti, Optima, Spectra 2.0L add	.3	*.3*
Crankshaft/Front Oil Pump Seal, Replace (B)		
Amanti	8.8	*9.2*
Optima	5.5	*6.1*
Rio	4.0	*4.6*
Sephia, Spectra		
1993-98	2.3	*2.5*

KIA-6   AMANTI : OPTIMA : RIO : SEPHIA : SPECTRA

	LABOR TIME	SEVERE SERVICE
Crankshaft/Front Oil Pump Seal, Replace (B)		
1999-05		
1.8L	4.3	4.9
2.0L	4.7	5.3
w/AC add		
exc. below	.3	.3
Amanti, Optima	.6	.6
w/PS add	.3	.3
Pistons or Connecting Rods, Replace (A)		
Includes: Ridge reaming, cylinder wall deglazing, installing new rings and rod bearings, engine tune-up.		
Amanti	14.5	15.7
Optima		
4 cyl.	10.6	11.9
V6	11.6	13.0
Rio	11.5	12.8
Sephia, Spectra	10.3	11.5
w/AC add	.3	.3
w/DOHC 94-96		
Sephia add	.7	.7
Piston Rings, Replace (A)		
Includes: Ridge reaming, cylinder wall deglazing, installing new rings, engine tune-up.		
Amanti	13.7	14.9
Optima		
4 cyl.	9.7	10.9
V6	10.9	12.2
Rio	11.6	13.0
Sephia, Spectra	9.5	10.7
w/AC add	.3	.3
w/DOHC 94-96		
Sephia add	.7	.7
Rear Main Oil Seal, Replace (B)		
Amanti	12.1	13.2
Optima		
4 cyl.	8.8	9.9
V6	9.7	10.9
Rio	6.1	6.9
Sephia, Spectra		
1.8L	4.9	5.5
2.0L	7.3	7.9
w/AT exc. Amanti add	.3	.3

ENGINE LUBRICATION

	LABOR TIME	SEVERE SERVICE
Oil Pan and/or Gasket, Replace (B)		
Exc. below	2.1	2.3
Amanti, Spectra 2.0L	2.4	2.6
Optima 4 cyl.	1.6	1.9
Oil Pressure Switch (Sending Unit), Replace (B)		
All Models	.5	.5
Oil Pump, Replace (B)		
Amanti	9.0	9.6
Optima	5.5	6.1
Rio	4.4	5.0

	LABOR TIME	SEVERE SERVICE
Sephia, Spectra		
1993-98	3.8	4.3
1999-05		
1.8L	5.7	6.3
2.0L	5.0	5.6
w/AC add	.3	.3
w/DOHC 94-96		
Sephia add	.3	.3
w/PS exc. Amanti, Optima, Spectra 2.0L add	.3	.3

CLUTCH

	LABOR TIME	SEVERE SERVICE
Clutch Hydraulic System, Bleed (B)		
All Models	.5	.5
Clutch Pedal Free Play, Adjust (B)		
All Models	.3	.3
Clutch Assy., Replace (B)		
Amanti, Optima	8.5	9.6
Rio	4.3	4.8
Sephia, Spectra		
1993-98	4.0	4.4
1999-05		
1.8L	4.6	5.1
2.0L	7.0	7.6
Clutch Control Cable, Replace (B)		
All Models	.8	.9
Clutch Master Cylinder, R&R and Recondition (A)		
Includes: System bleeding.		
Exc. below	1.3	1.4
1993-03 Sephia, Spectra	1.9	2.0
Clutch Master Cylinder, Replace (B)		
Includes: System bleeding.		
Exc. below	.6	.7
Amanti, Optima	1.0	1.1
Spectra 2.0L	1.3	1.4
Clutch Release Bearing, Replace (B)		
Amanti, Optima	8.5	9.1
Rio	4.0	4.6
Sephia, Spectra		
1993-98	3.8	4.4
1999-05		
1.8L	4.4	5.0
2.0L	7.0	7.6
Clutch Slave Cylinder, R&R and Recondition (A)		
Includes: System bleeding.		
Optima	1.3	1.4
Sephia, Spectra	.6	.7
Clutch Slave/Release Cylinder Assy., Replace (B)		
Includes: System bleeding.		
Amanti, Optima, Spectra 2.0L	1.0	1.1
Sephia, Spectra	.5	.6
Flywheel, Replace (B)		
Amanti, Optima	8.7	9.3
Rio	4.6	5.1

	LABOR TIME	SEVERE SERVICE
Sephia, Spectra		
1993-98	3.8	4.2
1999-05		
1.8L	4.9	5.5
2.0L	6.9	7.4

MANUAL TRANSAXLE

	LABOR TIME	SEVERE SERVICE
Transaxle Assy., R&R and Recondition (A)		
Includes: Recondition transmission only.		
Amanti, Optima	16.5	18.4
Rio	7.5	8.4
Sephia, Spectra		
1.8L	8.8	9.7
2.0L	12.5	13.4
Recondition differential assy.		
94-97 add	2.5	2.5
Transaxle Assy., R&R or Replace (B)		
Amanti, Optima	8.2	9.2
Rio	4.0	4.4
Sephia, Spectra		
1993-98	4.0	4.4
1999-05		
1.8L	4.4	5.0
2.0L	6.9	5.4

AUTOMATIC TRANSAXLE

SERVICE TRANSAXLE INSTALLED

	LABOR TIME	SEVERE SERVICE
Diagnose Electronic Transaxle System Component Each (A)		
All Models	.5	.5
Check Unit for Oil Leaks (C)		
All Models	.5	.5
Downshift Solenoid, Replace (B)		
Amanti, Optima	4.1	4.6
Rio	2.2	2.5
Sephia, Spectra		
1994-98	1.1	1.4
1999-05		
1.8L	2.2	2.5
2.0L	4.1	4.6
ECAT Switch, Replace (B)		
Rio, Sephia, Spectra	.6	.7
Electronic Control Unit, Replace (B)		
Sephia, Spectra	.5	.5
Governor Assy., R&R and Recondition (A)		
Sephia, Spectra	.6	1.0
Governor Assy., Replace (A)		
Sephia, Spectra	.5	.8
Kickdown Switch, Replace (B)		
Rio, Sephia, Spectra	.8	.9
Oil Filter/Pan Magnet, Replace (B)		
Rio, Sephia, Spectra	1.0	1.1
Oil Pan and/or Gasket, Replace (B)		
Rio, Sephia, Spectra	1.1	1.1
Overdrive Switch, Replace (B)		
Sephia, Spectra	.7	.7

AMANTI : OPTIMA : RIO : SEPHIA : SPECTRA KIA-7

	LABOR TIME	SEVERE SERVICE
Pulse Generator, Replace (B)		
Rio, Sephia, Spectra	.5	.7
Speedometer Driven Gear, Replace (B)		
All Models	.6	.7
Throttle Valve Cable, Replace (B)		
Sephia, Spectra		
1994-97	.8	.9
2000-05	1.6	1.8
Transaxle Control Module, Replace (B)		
Sephia, Spectra	.5	.5
Transaxle Range/Inhibitor Switch Assy., Replace (B)		
All Models	.6	.7
Transaxle Temperature Sensor, Replace (B)		
Amanti, Optima, Spectra 2.0L	3.0	3.4
Sephia, Spectra	.5	.5
Valve Body, R&R and Recondition (A)		
Exc. below	2.9	3.2
Amanti, Optima, Spectra 2.0L	6.0	6.7
Valve Body, Replace (B)		
Exc. below	2.1	2.3
Amanti, Optima, Spectra 2.0L	4.3	4.8

SERVICE TRANSAXLE REMOVED
Transaxle R&R included unless otherwise noted.

	LABOR TIME	SEVERE SERVICE
Transaxle Assy., R&R (B)		
Exc. below	4.7	5.3
Amanti	4.1	4.7
Optima	11.5	12.1
Spectra 2.0L	7.3	7.9
Transaxle Assy., Recondition (A)		
Includes: Replace torque converter.		
Exc. below	11.3	12.7
Optima	18.9	21.1

SHIFT LINKAGE
AUTOMATIC
Shift Lever, Replace (B)
All Models 8   .8

MANUAL
Gear Selector Lever, Replace (B)
All Models 8   .8

DRIVELINE
HALFSHAFTS

	LABOR TIME	SEVERE SERVICE
CV Joint, Replace/Recondition (B)		
Inner or outer		
exc. below		
one side	1.4	1.6
both sides	2.5	2.7
Amanti		
one side	1.6	1.8
both sides	2.9	3.1
Rio		
one side	1.3	1.5
both sides	1.9	2.1
Halfshaft, R&R or Replace (B)		
Exc. below		
one side	1.1	1.3
both sides	1.6	1.8
Amanti, Spectra 2.0L		
one side	1.1	1.3
both sides	2.2	2.4
Halfshaft Boot Set, Replace (B)		
One side	1.4	1.6
Both sides	2.5	2.7

BRAKES
ANTI-LOCK
The following operations do not include testing. Add time as required.

Diagnose Anti-Lock Brake System Component Each (A)
Sephia, Spectra 7   .7

	LABOR TIME	SEVERE SERVICE
ABS Control Unit, Replace (B)		
Rio, Spectra	1.7	1.9
Sephia		
1994-99	.5	.5
2000-01	1.7	1.9
Wheel Speed Sensor Assy. Replace (B)		
Front		
exc. below		
one side	.6	.7
both sides	1.0	1.1
Optima		
one side	.3	.3
both sides	.5	.5
Rear		
exc. below		
one side	.8	.9
both sides	1.0	1.1
Amanti		
one side	.9	1.0
both sides	1.1	1.2

SYSTEM
Bleed Brakes (B)
Includes: Add fluid.
All Models 5   .5

	LABOR TIME	SEVERE SERVICE
Brakes, Adjust (B)		
Includes: Refill master cylinder.		
Rear wheels	.3	.3
Flush and Refill Brake System (B)		
All Models	.6	.7
Brake Hose (Flexible), Replace (B)		
Includes: System bleeding.		
Amanti, Optima, Rio		
one side	.6	.7
both sides	1.3	1.4
Sephia, Spectra		
one side	.6	.7
both sides	.8	.9
Master Cylinder, R&R and Recondition (A)		
Includes: System bleeding.		
Exc. below	1.7	1.9
Amanti, Optima	1.1	1.2
Master Cylinder, Replace (B)		
Includes: System bleeding.		
Exc. below	1.5	1.7
Amanti, Optima, Spectra 2.0L	.9	1.0
Power Booster Unit, Replace (B)		
Includes: System bleeding.		
Exc. below	2.0	2.2
Amanti, Optima	1.4	1.5
Proportioning Valve, Replace (B)		
Includes: System bleeding.		
Exc. below	.8	.9
Amanti, Optima	1.1	1.2

SERVICE BRAKES

	LABOR TIME	SEVERE SERVICE
Brake Drum, Replace (B)		
Includes: Repack wheel bearings.		
One side	.6	.7
Both sides	1.0	1.1
Caliper Assy., R&R and Recondition (A)		
Includes: System bleeding.		
Exc. below		
one side	1.3	1.4
both sides	1.9	2.1
Optima		
one side	1.0	1.1
both sides	1.4	1.6
Caliper Assy., Replace (B)		
Includes: System bleeding.		
Front or Rear		
one side	.8	1.1
both sides	1.3	1.6
Disc Brake Rotor, Replace (B)		
Front or Rear		
one side	.6	.7
both sides	1.0	1.1

KIA-8 AMANTI : OPTIMA : RIO : SEPHIA : SPECTRA

	LABOR TIME	SEVERE SERVICE
Pads and/or Shoes, Replace (B)		
Includes: Adjust service and parking brake. System bleeding.		
Front or rear disc	.8	.9
Rear drum	1.3	1.4
Four wheels		
Drum brakes	2.1	2.3
Disc brakes	1.6	1.8

COMBINATION ADD-ONS

Replace		
brake drum add each	.3	.3
brake hose add each	.3	.3
disc brake rotor add each	.3	.3
wheel cylinder add	.3	.3
Resurface		
disc rotor add	.5	.5
brake drum add	.5	.5
Wheel Cylinder, R&R and Reconditon (B)		
Includes: System bleeding.		
Exc. below		
one side	1.3	1.4
both sides	1.7	2.0
Optima		
one side	.6	.8
both sides	1.0	1.2
Wheel Cylinder, Replace (B)		
Includes: System bleeding.		
One side	1.1	1.3
Both sides	1.7	1.9

PARKING BRAKE

Parking Brake Cable, Adjust (C)		
All Models	.3	.3
Parking Brake Apply Actuator, Replace (B)		
All Models	.5	.5
Parking Brake Cable, Replace (B)		
Amanti		
front		
lever type	.6	.6
pedal type	1.4	1.6
rear		
one side	1.6	1.8
both sides	1.9	2.1
Optima rear		
one side	1.6	1.8
both sides	1.9	2.1
Rio		
front	.5	.5
rear		
one side	1.1	1.3
both sides	1.7	2.0
Sephia, Spectra		
front	1.0	1.1
rear		
one side	.8	.9
both sides	1.1	1.3

	LABOR TIME	SEVERE SERVICE
Spectra 2.0L		
one side	1.3	1.5
both sides	1.4	1.6
Parking Brake Shoes, Replace (B)		
All Models	1.1	1.2

FRONT SUSPENSION

Unless otherwise noted, time given does not include alignment.

Align Front End (A)		
All Models	1.6	1.8
Toe In, Adjust (B)		
Two wheels	.8	.9
Four wheels	1.6	1.8
Ball Joint, Replace (B)		
Amanti, Optima		
upper		
one side	.8	.9
both sides	1.3	1.4
Sephia, Spectra		
lower		
one side	.6	.7
both sides	1.1	1.3
Front Spring, Replace (B)		
Amanti, Optima		
one side	1.1	1.3
both sides	2.2	2.4
Rio		
one side	1.0	1.2
both sides	1.6	1.8
Sephia, Spectra		
one side	1.4	1.6
both sides	2.2	2.4
Spectra 2.0L		
one side	1.0	1.2
both sides	1.7	1.9
Front Wheel Outer Oil Seal, Replace (B)		
Rio, Sephia, Spectra		
one side	1.6	1.8
both sides	3.0	3.2
Lower Control Arm, Replace (B)		
Amanti, Optima		
one side	.8	.9
both sides	1.6	1.8
Rio		
one side	.8	.9
both sides	1.1	1.3
Sephia, Spectra		
one side	1.7	2.0
both sides	2.4	2.7
Spectra 2.0L		
one side	1.0	1.2
both sides	1.3	1.5
Lower Control Arm Bushings, Replace (B)		
Amanti, Optima		
one side	.8	.9
both sides	1.6	1.8

	LABOR TIME	SEVERE SERVICE
Rio		
one side	.8	.9
both sides	1.1	1.3
Sephia, Spectra		
one side	1.7	2.0
both sides	2.4	2.7
Spectra 2.0L		
one side	1.0	1.2
both sides	1.3	1.5
Stabilizer Bar, Replace (B)		
Amanti, Optima	3.2	3.6
Rio, Sephia, Spectra, Spectra 2.0L	1.4	1.6
Stabilizer Bar Link, Replace (B)		
One side	.3	.3
Both sides	.6	.7
Steering Knuckle, Replace (B)		
One side	1.6	1.8
Both sides	3.0	3.4
Strut Assy., Replace (B)		
Amanti, Optima		
one side	1.1	1.3
both sides	2.2	2.5
Rio		
one side	1.0	1.1
both sides	1.6	1.8
Sephia, Spectra		
one side	1.4	1.6
both sides	2.2	2.5
Spectra 2.0L		
one side	1.0	1.2
both sides	1.7	1.9
Wheel Hub, Replace (B)		
One side	1.6	1.7
Both sides	3.0	3.1
Wheel Bearings, Replace (B)		
One side	1.6	1.7
Both sides	3.0	3.1

REAR SUSPENSION

Unless otherwise noted, time given does not include alignment.

Toe In, Adjust (B)		
All Models	.7	.9
Lateral Link, Replace (B)		
Sephia, Spectra		
one side	1.0	1.1
both sides	1.3	1.4
Rear Spring, Replace (B)		
Amanti, Optima, Rio		
one side	1.1	1.3
both sides	1.4	1.6
Sephia, Spectra		
one side	1.9	2.1
both sides	2.7	2.9
Spectra 2.0L		
one side	1.4	1.6
both sides	2.5	2.7

AMANTI : OPTIMA : RIO : SEPHIA : SPECTRA **KIA-9**

	LABOR TIME	SEVERE SERVICE
Rear Wheel Hub and Bearing, Replace (B)		
One side	.8	.9
Both sides	1.4	1.6
Rear Wheel Oil Seal, Replace (B)		
Rio		
one side	1.9	2.1
both sides	3.5	3.9
Stabilizer Bar, Replace (B)		
All Models	.6	.7
Stabilizer Bar Bushings, Replace (B)		
All Models	.3	.3
Strut Assy., Replace (B)		
Amanti, Optima, Rio		
one side	1.1	1.3
both sides	1.4	1.6
Sephia, Spectra		
one side	1.9	2.1
both sides	2.7	2.9
Spectra 2.0L		
one side	1.4	1.6
both sides	2.5	2.7
Trailing Link, Replace (B)		
One side	1.0	1.1
Both sides	1.4	1.6

STEERING

AIR BAGS

The following operations do not include testing. Add time as required.

Diagnose Air Bag System Component Each (A)

	LABOR TIME	SEVERE SERVICE
Sephia, Spectra	1.0	1.0
Retrieve Fault Codes (A)		
All Models	.3	.3
Air Bag Control Module (A)		
Exc. below	1.4	1.4
Amanti, Optima	2.3	2.3
Spectra 2.0L	.8	.8
Air Bag Relay, Replace (B)		
Sephia, Spectra	.5	.5
Module Assembly, Replace (A)		
Driver side		
exc. below	.3	.3
2000-03 Sephia, Spectra	.7	.7
Passenger side		
Amanti	3.5	3.5
Optima	2.4	2.4
Rio	1.6	1.6
Sephia, Spectra	.8	.8
Spectra 2.0L	2.2	2.2
Side curtain assy.		
Amanti		
one side	1.9	1.9
both sides	2.2	2.4
Spectra 2.0L		
one side	1.6	1.6
both sides	1.8	1.8

	LABOR TIME	SEVERE SERVICE
Sensor, Replace (B)		
Right side	1.1	1.1
Left side	1.5	1.5

STEERING GEAR

	LABOR TIME	SEVERE SERVICE
Troubleshoot Power Steering (A)		
All Models	.5	.5
Steering Gear, Adjust (B)		
All Models	.7	.9
Inner Tie Rod, Replace (B)		
Includes: Alignment.		
Exc. below one	3.8	4.3
Amanti		
one side	4.3	4.9
both sides	4.6	5.2
Optima		
one side	3.3	3.9
both sides	3.8	4.3
Outer Tie Rod, Replace (B)		
Includes: Alignment.		
One side	1.1	1.3
Both sides	1.3	1.4
Power Steering Pressure Switch Assy., Replace (B)		
All Models	.5	.5
Power Steering Pump Hoses, Replace (B)		
Pressure	1.1	1.2
Suction	.5	.6
Steering Column, R&I (B)		
All Models	1.6	1.7
Steering Gear, Replace (B)		
Amanti	4.1	4.4
Optima		
manual	2.7	3.0
power	3.2	3.6
Rio		
manual	2.7	3.0
power	3.6	4.1
Sephia, Spectra		
manual	2.1	2.3
power	2.5	2.9
Spectra 2.0L	2.7	3.0
Steering Gear Rack Assy., R&R and Recondition (A)		
Exc. below		
manual	3.8	4.3
power	4.7	5.3
Amanti	5.3	5.9
Optima	4.6	5.1
Spectra 2.0L	3.8	4.3
Steering Ignition Lock Assy. Replace (B)		
Exc. below	.7	.7
Amanti, Optima, Spectra 2.0L	.9	.9
Steering Pump, R&R and Recondition (A)		
Exc. below	1.8	2.0
Sephia, Spectra	2.1	2.4

	LABOR TIME	SEVERE SERVICE
Steering Pump, Replace (B)		
Amanti, Optima, Rio, Spectra 2.0L	.9	1.0
Sephia, Spectra	1.5	1.6
Steering Wheel, Replace (B)		
Exc. below	.3	.3
Amanti, Optima, Spectra 2.0L	.7	.7

HEATING & AIR CONDITIONING

When more than one component requires replacement where evacuation/recovery and recharging is already included, deduct 1.0 hour for each additional component from the time given.

Evacuate/Recover and Recharge System (B)

	LABOR TIME	SEVERE SERVICE
All Models	1.0	1.0

AC Hoses, Replace (B)
Does not include evacuate/recover and recharge.

	LABOR TIME	SEVERE SERVICE
Discharge Hose		
Amanti, Optima, Spectra 2.0L	.3	.3
Rio, Sephia, Spectra	1.2	1.4
Drain Hose	.3	.3
Suction Hose		
exc. below	1.2	1.2
Spectra 2.0L	.3	.3
Blower Motor, Replace (B)		
A/C	.6	.7
Heater		
Optima	3.5	3.5
Rio	3.1	3.1
Sephia, Spectra	2.4	2.4
Spectra 2.0L	.5	.5
Blower Motor Switch, Replace (B)		
All Models	.7	.7
Blower Motor Resistor, Replace (B)		
Rio, Sephia, Spectra	.6	.6

Compressor Assy., Replace (B)
Does not include evacuate/recover and recharge.

	LABOR TIME	SEVERE SERVICE
Exc. below	2.1	2.4
Optima	.9	1.0

Condenser Assy., Replace (B)
Does not include evacuate/recover and recharge.

	LABOR TIME	SEVERE SERVICE
Exc. below	1.5	1.7
Sephia, Spectra	1.8	2.0

Evaporator Core, Replace (B)
Does not include evacuate/recover and recharge.

	LABOR TIME	SEVERE SERVICE
Amanti, Spectra 2.0L	4.4	4.6
Optima, Rio	3.8	4.0
Sephia, Spectra	2.9	3.2

KIA-10 AMANTI : OPTIMA : RIO : SEPHIA : SPECTRA

	LABOR TIME	SEVERE SERVICE
Expansion Valve, Replace (B)		
Does not include evacuate/recover and recharge.		
Amanti, Optima	.3	.3
Rio	3.5	3.9
Spectra 2.0L	.8	.9
Heater Assy., Replace (B)		
Amanti	4.6	4.9
Optima	4.2	4.5
Rio	3.5	3.8
Sephia, Spectra	2.3	2.6
Spectra 2.0L	3.1	3.4
Heater Core, Replace (B)		
Amanti, Optima	5.1	5.5
Rio, Spectra 2.0L	4.3	4.7
Sephia, Spectra	3.1	3.4
Heater Hoses, Replace (B)		
Amanti, Optima, Spectra 2.0L		
one side	.5	.6
both sides	.8	.9
Rio, Sephia, Spectra		
one side	.9	1.0
both sides	1.5	1.6
Pressure Switch, Replace (B)		
Sephia, Spectra	.5	.5
Receiver Drier Assy., Replace (B)		
Does not include evacuate/recover and recharge.		
Exc. below	1.2	1.4
Amanti	.6	.8
Spectra 2.0L	1.5	1.7
Temperature Control Assy., Replace (B)		
All Models	.7	.7
Temperature Control Cable, Replace (B)		
Rio, Sephia, Spectra	.7	.7

WIPERS & SPEEDOMETER

	LABOR TIME	SEVERE SERVICE
Antenna Assy., Replace (B)		
All Models	.7	.7
Instrument Panel Cluster Switch, Replace (C)		
All Models	.5	.5
Radio, R&R (B)		
All Models	.6	.6
Speedometer Cable & Casing, Replace (B)		
Optima, Rio	.3	.3
Sephia, Spectra	.7	.7
Speedometer Head, R&R or Replace (B)		
All Models	.7	.7
Windshield Washer Pump, Replace (B)		
Front or rear	.6	.6
Windshield Washer Switch, Replace (B)		
Sephia, Spectra	.7	.7
Windshield Wiper (Combination) Switch, Replace (B)		
Front		
exc. below	.8	.8
Optima	1.4	1.4
Rear	.3	.3
Windshield Wiper Motor, Replace (B)		
Front	.8	.9
Rear	.3	.3
Windshield Wiper Pivot Shaft & Link Assy., Replace (B)		
All Models	.8	.9

LAMPS & SWITCHES

	LABOR TIME	SEVERE SERVICE
Back-Up Lamp Switch, Replace (B)		
All Models	.5	.5
Combination (Multifunction) Switch, Replace (B)		
Exc. below	.7	.7
Optima	1.1	1.1
Halogen Headlamp Bulb, Replace (C)		
One	.3	.3
each addl. add	.2	.2
Headlamps, Aim (B)		
All Models	.5	.5
Horn, Replace (C)		
One side	.3	.3
Both sides	.5	.5
Horn Relay, Replace (B)		
All Models	.3	.3
Neutral Safety Switch, Replace (B)		
Rio, Sephia, Spectra	.3	.3
Stop Lamp Switch, Replace (B)		
All Models	.5	.5
Parking Brake Warning Switch, Replace (C)		
Exc. below	.3	.3
Spectra 2.0L	.6	.6
Turn Signal or Hazard Warning Flasher, Replace (C)		
All Models	.3	.3

BODY

	LABOR TIME	SEVERE SERVICE
Door Window Regulator, Replace (B)		
One side	.8	.8
Both sides	1.2	1.2
Power Window Motor, Replace (B)		
Exc. below		
one side	.6	.6
both sides	1.1	1.1
Optima		
one side	.9	.9
both sides	1.5	1.5
Power Window Switch, Replace (B)		
One side	.3	.3
Both sides	.5	.5

KIA

Sedona : Sorento : Sportage

SYSTEM INDEX

MAINTENANCE KIA-12
   Maintenance Schedule... KIA-12
CHARGING.............. KIA-12
STARTING.............. KIA-12
CRUISE CONTROL KIA-12
IGNITION.............. KIA-12
EMISSIONS............. KIA-13
FUEL.................. KIA-13
EXHAUST............... KIA-14
ENGINE COOLING........ KIA-14
ENGINE................ KIA-14
   Assembly........... KIA-14
   Cylinder Head...... KIA-15
   Camshaft........... KIA-15
   Crank & Pistons.... KIA-15
   Engine Lubrication. KIA-16
CLUTCH................ KIA-16
MANUAL TRANSMISSION . KIA-16
AUTO TRANSMISSION ... KIA-17
TRANSFER CASE KIA-17
SHIFT LINKAGE......... KIA-17
DRIVELINE............. KIA-17
BRAKES................ KIA-18
FRONT SUSPENSION..... KIA-19
REAR SUSPENSION KIA-19
STEERING.............. KIA-20
HEATING & AC.......... KIA-20
WIPERS &
  SPEEDOMETER........ KIA-21
LAMPS & SWITCHES KIA-21
BODY.................. KIA-21

OPERATIONS INDEX

A
AC Hoses KIA-20
Air Bags............... KIA-20
Air Conditioning........ KIA-20
Alignment.............. KIA-19
Alternator (Generator)... KIA-12
Antenna................ KIA-21
Anti-Lock Brakes....... KIA-18

B
Back-Up Lamp Switch... KIA-21
Ball Joint.............. KIA-19
Battery Cables......... KIA-12
Bleed Brake System.... KIA-18
Blower Motor........... KIA-20
Brake Disc............. KIA-18
Brake Drum............ KIA-18
Brake Hose............ KIA-18
Brake Pads and/or Shoes... KIA-18

C
Camshaft.............. KIA-15
Camshaft Sensor....... KIA-12
Catalytic Converter..... KIA-14

Coolant Temperature (ECT)
  Sensor............. KIA-13
Crankshaft............. KIA-16
Crankshaft Sensor..... KIA-12
Cruise Control......... KIA-12
CV Joint............... KIA-17
Cylinder Head......... KIA-15

D
Differential............ KIA-18
Drive Belt............. KIA-12
Driveshaft............. KIA-17

E
EGR................... KIA-13
Electronic Control Module
  (ECM/PCM)......... KIA-13
Engine................ KIA-14
Engine Lubrication..... KIA-16
Engine Mounts......... KIA-14
Evaporator............ KIA-21
Exhaust............... KIA-14
Exhaust Manifold...... KIA-14
Expansion Valve....... KIA-21

F
Flexplate.............. KIA-17
Flywheel............... KIA-16
Fuel Injection.......... KIA-13
Fuel Pump............. KIA-13
Fuel Vapor Canister.... KIA-13

G
Gear Selector Lever ... KIA-17
Generator............. KIA-12

H
Halfshaft.............. KIA-17
Headlamp............. KIA-21
Heater Core........... KIA-21
Horn.................. KIA-21

I
Idle Air Control (IAC) Valve.. KIA-13
Ignition Coil........... KIA-13
Ignition Module........ KIA-12
Ignition Switch........ KIA-13
Inner Tie Rod.......... KIA-20
Intake Manifold........ KIA-13

K
Knock Sensor KIA-13

L
Lower Control Arm..... KIA-19

M
Maintenance Schedule... KIA-12
Manifold Absolute Pressure
  (MAP) Sensor....... KIA-13
Mass Air Flow (MAF)
  Sensor............. KIA-13
Master Cylinder........ KIA-18
Muffler................ KIA-14
Multifunction Lever/Switch... KIA-21

N
Neutral Safety Switch .. KIA-21

O
Oil Pan................ KIA-16
Oil Pump.............. KIA-16
Outer Tie Rod......... KIA-20
Oxygen Sensor........ KIA-13

P
Parking Brake......... KIA-19
Pistons................ KIA-16
Positive Crankcase
  Ventilation (PCV) Valve.... KIA-13

R
Radiator............... KIA-14
Radiator Hoses........ KIA-14
Radio................. KIA-21
Rear Main Oil Seal KIA-16

S
Shock Absorber/Strut,
  Front............... KIA-19
Shock Absorber/Strut,
  Rear................ KIA-20
Spark Plug Cables KIA-13
Spark Plugs........... KIA-13
Spring, Front Coil...... KIA-19
Spring, Rear Coil...... KIA-19
Starter................ KIA-12
Steering Wheel........ KIA-20

T
Thermostat............ KIA-14
Throttle Body.......... KIA-13
Throttle Position Sensor
  (TPS).............. KIA-13
Timing Belt........... KIA-15
Torque Converter..... KIA-17

U
U-Joint................ KIA-17
Upper Control Arm..... KIA-19

V
Valve Body KIA-17
Valve Cover Gasket.... KIA-15
Valve Job............. KIA-15

W
Water Pump........... KIA-14
Wheel Balance........ KIA-12
Wheel Cylinder........ KIA-19
Window Regulator..... KIA-21
Windshield Washer Pump... KIA-21
Windshield Wiper Motor ... KIA-21

KIA-11

KIA-12 SEDONA : SORENTO : SPORTAGE

	LABOR TIME	SEVERE SERVICE

MAINTENANCE

Air Cleaner Filter Element, Replace (C)
- All Models .... .3 / .3

Chassis Lubrication, Change Oil & Filter (C)
Includes: Correct all fluid levels.
- All Models .... .5 / .5

Drive Belt, Adjust (B)
- One .... .3 / .3
- each addl. add .... .1 / .1

Drive Belt, Replace (B)
- All Models .... .5 / .5
- w/AC add .... .1 / .1

Drive Belt Tensioner, Replace (B)
- Sorento .... .5 / .5
- 2005 Sportage .... .5 / .5

Fuel Filter, Replace (B)
- Exc. below .... .5 / .5
- 2005 Sportage .... .7 / .7

Halogen Headlamp Bulb, Replace (C)
- One .... .3 / .3
- each addl. add .... .3 / .3

Oil & Filter, Change (C)
Includes: Correct all fluid levels.
- All Models .... .5 / .5

Timing Belt, Replace (B)
- Sedona .... 3.6 / 3.9
- Sorento .... 4.8 / 5.1
- Sportage
  - 1995-02 .... 3.7 / 4.0
  - 2005 .... 2.9 / 3.2
- w/4WD add .... .9 / .9
- w/AC add .... .3 / .3

Tire, Replace (C)
Includes: Dismount old tire and mount new tire to rim.
- All Models one .... .5 / .5

Tires, Rotate (C)
- Sportage .... .5 / .5

Wheel, Balance (B)
- Two wheels .... .5 / .5
- each addl. add .... .3 / .3

SCHEDULED MAINTENANCE INTERVALS

If necessary, refer to appropriate Chilton maintenance service information.

7,500 Mile Service (C)
- All Models .... .6 / .6

15,000 Mile Service (C)
- All Models .... 1.1 / 1.1

22,500 Mile Service (C)
- All Models .... .6 / .6

30,000 Mile Service (B)
- All Models .... 3.1 / 3.1

37,500 Mile Service (C)
- All Models .... .6 / .6

45,000 Mile Service (C)
- All Models .... 1.1 / 1.1

52,500 Mile Service (C)
- All Models .... .6 / .6

60,000 Mile Service (B)
- All Models .... 5.3 / 5.3

67,500 Mile Service (C)
- All Models .... .6 / .6

75,000 Mile Service (C)
- All Models .... 1.1 / 1.1

82,500 Mile Service (C)
- All Models .... .6 / .6

90,000 Mile Service (B)
- All Models .... 5.3 / 5.3

97,500 Mile Service (C)
- All Models .... .6 / .6

105,000 Mile Service (C)
- All Models .... 1.1 / 1.1

CHARGING

Alternator Circuits, Test (A)
Includes: Test component output.
- All Models .... .5 / .5

Alternator Assy., R&R and Recondition (B)
- Sedona .... 3.3 / 3.5
- Sorento .... 2.3 / 2.6
- Sportage
  - 1995-02 .... 1.6 / 1.8
  - 2005
    - 4 cyl. .... 2.2 / 2.4
    - V6 .... 2.7 / 2.9

Alternator Assy., Replace (B)
- Sedona .... 3.0 / 3.1
- Sorento .... 2.0 / 2.1
- Sportage
  - 1995-02 .... 1.3 / 1.4
  - 2005
    - 4 cyl. .... 1.7 / 1.9
    - V6 .... 2.3 / 2.4
- Transfer pulley add .... .2 / .2

Voltage Regulator, Replace (B)
- Sedona .... 3.3 / 3.5
- Sorento .... 1.7 / 1.9
- Sportage
  - 1995-02 .... 1.6 / 1.8
  - 2005
    - 4 cyl. .... 2.0 / 2.2
    - V6 .... 2.4 / 2.6

STARTING

Starter Draw Test (On Car) (B)
- All Models .... .5 / .5

Battery, Replace (C)
- All Models .... .5 / .5

Battery Cables, Replace (C)
- Exc. below .... .5 / .7
- 2005 Sportage
  - positive .... 2.4 / 2.5

Keyless Entry Module, Replace (C)
- All Models .... .5 / .5

Keyless Entry Transmitter, Replace (C)
- All Models .... .3 / .3

Starter Assy., R&R and Recondition (B)
- Sedona, Sorento .... 2.2 / 2.5
- 1995-02 Sportage .... 3.0 / 3.4

Starter Assy., Replace (B)
- Sedona .... .6 / .7
- Sorento .... 1.2 / 1.3
- Sportage
  - 1995-02 .... 1.4 / 1.5
  - 2005
    - 4 cyl. .... 1.2 / 1.3
    - V6 .... .6 / .7

Starter Drive, Replace (B)
Includes: Starter R&R.
- Sedona, Sorento .... 2.1 / 2.3
- 1995-02 Sportage .... 3.0 / 3.4

Starter Solenoid, Replace (B)
- Sedona, Sorento .... 1.0 / 1.1
- 1995-02 Sportage .... 1.7 / 2.0

CRUISE CONTROL

Diagnose Cruise Control System Component Each (A)
- All Models .... .5 / .5

Cruise Control Cable, Replace (B)
- All Models .... .7 / .7

Cruise Control Module, Replace (B)
- All Models .... .5 / .5

Cruise Control Switch, Replace (B)
- All Models .... .3 / .3

IGNITION

Diagnose Ignition System Component Each (A)
- All Models .... .5 / .5

Ignition Timing, Adjust (C)
- Sportage .... .7 / .7

Camshaft Position Sensor, Replace (B)
- All Models .... .5 / .5

Crankshaft Angle Sensor, Replace (B)
- All Models .... .5 / .5

Crankshaft Position Sensor, Replace (B)
- Sedona .... .5 / .6
- Sorento .... 4.3 / 4.4
- Sportage
  - 1995-02 .... .5 / .6
  - 2005 .... .9 / 1.0

Igniter Assy., Replace (B)
- Sportage .... .7 / .7

SEDONA : SORENTO : SPORTAGE **KIA-13**

	LABOR TIME	SEVERE SERVICE
Ignition Coil, Replace (B)		
Exc. Sorento		
one or all	.5	.6
Sorento		
right	2.4	2.5
left	.7	.8
all	2.7	2.8
Ignition Lock & Cylinder, Replace (B)		
Sedona	.8	.8
Sorento	1.0	1.0
Sportage		
1995-02	1.0	1.0
2005	.8	.8
Ignition Switch, Replace (B)		
Sedona	.8	.8
Sorento	1.0	1.0
Sportage		
1995-02	1.0	1.0
2005	.3	.3
Spark Plug (Ignition) Cables, Replace (B)		
Sedona, Sorento	2.7	2.9
1995-02 Sportage	.5	.6
Spark Plugs, Replace (B)		
Sedona, Sorento		
right side	.5	.5
left side	2.6	2.7
all	2.9	3.0
Sportage		
1995-02	1.4	1.5
2005		
4 cyl.	.5	.6
V6	1.7	1.8

EMISSIONS

The following operations do not include testing. Add time as required.

	LABOR TIME	SEVERE SERVICE
Diagnose Emission Control System Component Each (A)		
All Models	.5	.5
Retrieve Fault Codes (A)		
All Models	.3	.3
Throttle Position Sensor, Adjust (A)		
All Models	.5	.5
Air By-Pass Valve, Replace (B)		
Sportage	.5	.5
Air Temperature Sensor, Replace (B)		
All Models	.3	.3
Canister Purge Solenoid Valve, Replace (B)		
Exc. below	.3	.3
2005 Sportage	.9	.9
Coolant Temperature (ECT) Sensor, Replace (B)		
All Models	.4	.4
EGR Modulator, Replace (B)		
Sportage	.3	.3

	LABOR TIME	SEVERE SERVICE
EGR Solenoid, Replace (B)		
Exc. below	.5	.5
Sedona	1.1	1.1
1995-02 Sportage	1.1	1.1
EGR Valve, Replace (B)		
Sedona	1.1	1.2
Sorento, Sportage	.5	.5
EGR Valve Pipe, Replace (B)		
Sedona, Sportage	.3	.3
Electronic Control Module (ECM/PCM), Replace (B)		
Exc. below	.6	.6
2005 Sportage	.9	.9
Fuel Vapor Canister, Replace (B)		
Exc. below	.5	.5
2005 Sportage	.9	.9
Knock Sensor, Replace (B)		
Sedona	2.2	2.3
Sportage		
1995-02	.3	.3
2005		
4 cyl.	.4	.4
V6	1.9	2.0
Manifold Absolute Pressure (MAP) Sensor, Replace (B)		
Sedona, Sorento	.5	.5
Mass Air Flow (MAF) Sensor, Replace (B)		
All Models	.4	.4
Oxygen Sensor Replace (B)		
All Models	.5	.5
Positive Crankcase Ventilation (PCV) Valve, Replace (B)		
All Models	.3	.3
Throttle Position Sensor (TPS), Replace (B)		
All Models	.5	.5

FUEL

DELIVERY

	LABOR TIME	SEVERE SERVICE
Fuel Pump, Test (B)		
All Models	.5	.5
Fuel Filter, Replace (B)		
Exc. below	.5	.5
2005 Sportage	.7	.7
Fuel Gauge (Dash), Replace (B)		
All Models	.7	.7
Fuel Gauge (Tank), Replace (B)		
Exc. below	.6	.7
2005 Sportage	.9	1.0
Fuel Pump, Replace (B)		
Exc. below	.6	.7
2005 Sportage	.9	1.0
Fuel Pump Relay, Replace (B)		
All Models	.5	.5
Fuel Tank, Replace (B)		
Includes: Drain and refill.		
Exc. below	1.3	1.4
2005 Sportage	2.5	2.6

	LABOR TIME	SEVERE SERVICE
Intake Manifold and/or Gasket, Replace (B)		
Includes: Adjustments.		
Sedona	2.7	2.9
Sorento	3.0	3.2
Sportage		
1995-02	3.0	3.2
2005	1.7	1.9

INJECTION

	LABOR TIME	SEVERE SERVICE
Diagnose Fuel Injection System Component Each (A)		
All Models	.7	.7
Fuel Injector Relay, Replace (B)		
Sportage	.5	.5
Fuel Injectors, Replace (B)		
Sedona, Sorento		
right side	.8	.9
left side	2.1	2.2
both sides	2.4	2.5
Sportage		
1995-02	2.3	2.4
2005		
right side	.8	.9
left side	1.5	1.5
both sides	1.9	2.0
Fuel Pressure Regulator, Replace (B)		
Exc. below	.4	.4
2005 Sportage	.9	.9
Fuel Pressure Solenoid, Replace (B)		
One	.3	.3
each addl. add	.2	.2
Fuel Rail Assy., Replace (B)		
Sedona, Sorento		
right side	.8	.9
left side	2.1	2.2
both sides	2.4	2.5
Sportage		
1995-02	2.3	2.4
2005		
right side	.8	.9
left side	1.5	1.5
both sides	1.9	2.0
Idle Air Control (IAC) Valve, Replace (B)		
All Models	.6	.7
Pulsation Damper, Replace (B)		
Sedona, Sportage	.3	.3
Throttle Body and/or Gasket, R&R or Replace (B)		
Sedona, Sorento	.8	.9
Sportage		
1995-02	.9	1.0
2005		
4cyl.	1.1	1.2
V6	1.3	1.4

KIA-14 SEDONA : SORENTO : SPORTAGE

	LABOR TIME	SEVERE SERVICE
EXHAUST		
Catalytic Converter, Replace (B)		
Front		
Sedona		
right	3.0	3.1
left	1.5	1.6
Sorento	1.1	1.2
1995-02 Sportage	1.1	1.2
Rear		
exc. below	.8	.9
1995-02 Sportage	1.1	1.2
Center Muffler Assy., Replace (B)		
All Models	.7	.8
Exhaust Manifold or Gasket, Replace (B)		
Sedona		
right side	.7	.8
left side	1.3	1.4
Sorento		
right side	1.5	1.6
left side	1.9	2.0
Sportage		
4 cyl.		
1995-02	1.1	1.3
2005	1.6	1.7
V6 one	2.1	2.2
w/4WD add	.3	.3
Front Pipe Assy., Replace (B)		
All Models	1.1	1.3
Intermediate or Inlet Pipe, Replace (B)		
Sportage	.7	.8
Muffler, Replace (B)		
One	.8	.9
each addl. add	.6	.6
Resonator, Replace (B)		
Sportage	.6	.7
Tail Pipe, Replace (B)		
Sedona	.6	.7
Sorento	.7	.8
1995-02 Sportage	.7	.8
ENGINE COOLING		
Pressure Test Cooling System (B)		
All Models	.3	.3
Coolant Thermostat, Replace (B)		
Sedona	2.4	2.5
Sorento	.6	.7
Sportage		
4 cyl.	.6	.7
V6	.9	1.0
Electric Fan Relay, Replace (B)		
Sorento, Sportage	.3	.3
Engine Coolant Temp. Sending Unit, Replace (B)		
All Models	.3	.3

	LABOR TIME	SEVERE SERVICE
Freeze Plugs (Water Jacket), Replace (B)		
Add access time as required.		
One	.5	.7
Radiator Assy., Replace (B)		
Exc. below	1.5	1.7
2005 Sportage	2.4	2.6
w/AC add	.2	.2
w/AT add	.2	.2
Radiator Fan Motor and/or Fan, Replace (B)		
Sedona	1.1	1.2
Sorento, Sportage	.8	.9
Radiator Hoses, Replace (B)		
Includes: Refill with proper coolant mix.		
Upper	.5	.5
Lower	.8	.9
Radiator Supply Tank, Replace (B)		
All Models	.3	.3
Thermo Water Valve, Replace (B)		
Sportage	.7	.7
Water Pump and/or Gasket, Replace (B)		
Includes: Refill with proper coolant mix.		
Sedona	4.3	4.5
Sorento	5.1	5.3
Sportage		
1995-02	4.3	4.5
2005	3.0	3.2
w/4WD add	.9	.9
w/AC add	.2	.2
ENGINE		
ASSEMBLY		
Times shown are for OEM assemblies. Time to replace assemblies from aftermarket rebuilders may vary.		
Engine Assy., R&I (B)		
Does not include parts or component transfer.		
Sportage		
1995-02		
AT	7.3	7.3
MT	6.9	6.9
w/4WD add	.9	.9
w/AC add	.5	.5
w/PS add	.3	.3
w/SOHC 95-96 add	.7	.7
Engine Assy., R&R and Recondition (B)		
Includes: Replacing rings, rod and main bearings, cylinder head reconditioning and engine tune-up.		
Sedona	24.4	25.3
Sorento	29.5	30.4

	LABOR TIME	SEVERE SERVICE
Sportage		
4 cyl.		
1995-03	22.4	23.3
2005	20.5	21.4
V6	32.5	33.4
w/4WD add	.9	.9
w/AC add	.3	.3
w/AT add	.8	.8
w/SOHC 95-96 add	.7	.7
Engine (New or Rebuilt Complete) Assy., Replace (B)		
Includes: Component transfer and engine tune-up.		
Sedona	9.7	10.2
Sorento	9.5	10.0
Sportage		
4 cyl.	9.2	9.7
V6	9.5	10.0
w/4WD add	.9	.9
w/AC add	.3	.3
w/AT add	.5	.5
w/SOHC 95-96 add	.7	.7
Engine (Short block), Replace (B)		
Assembly consists of cylinder block, piston assemblies, crankshaft, camshaft, timing chain and gears. Does not include cylinder heads. Operation Includes: R&R engine, transfer necessary parts and all necessary adjustments.		
Sedona, Sorento	16.7	17.6
Sportage	16.1	17.0
w/4WD add	.9	.9
w/AC add	.3	.3
w/AT add	.5	.5
w/SOHC 95-96 add	.7	.7
Engine Mounts, Replace (B)		
Sedona		
right or left	.6	.7
front	.9	1.0
transmission	.8	.9
Sorento		
one	.7	.8
both	1.5	1.6
transmission	.9	1.0
Sportage		
1995-02		
one	.6	.7
both	1.2	1.3
transmission	.5	.6
2005		
right or left	.6	.7
front	.9	1.0
transmission	1.3	1.4

SEDONA : SORENTO : SPORTAGE — KIA-15

	LABOR TIME	SEVERE SERVICE
CYLINDER HEAD		
Compression Test (B)		
Sportage	.6	.8
Valve Clearance, Adjust (B)		
Sportage	1.2	1.5
Cylinder Head and/or Gasket, Replace (B)		
Includes: Parts transfer and adjustments.		
Sedona, Sorento		
right side	13.3	13.8
left side	10.6	11.1
both sides	15.3	15.8
Sportage		
1995-02	12.7	13.2
2005 V6		
right side	12.0	12.5
left side	10.9	11.4
both sides	15.7	16.2
w/AC add	.2	.2
w/SOHC 95-96 add	.7	.7
Hydraulic Lash Adjuster Assy., Replace (B)		
Sedona		
right side	3.8	4.1
left side	6.1	6.4
both sides	5.9	6.4
Sorento		
one side	9.7	10.0
both sides	10.5	11.0
Sportage		
1995-02 all	2.9	3.1
2005		
4 cyl. all	3.7	4.0
V6		
right side	4.8	5.1
left side	3.4	3.7
both sides	5.9	6.4
Rocker Arm Cover Gasket, Replace or Reseal (B)		
Sedona		
right side	.5	.6
left side	2.4	2.5
both sides	2.9	3.1
Sorento		
right side	2.6	2.8
left side	2.3	2.5
both sides	2.9	3.1
Sportage		
1995-02	.6	.7
2005		
4 cyl.	.6	.7
V6		
right side	2.4	2.6
left side	.6	.7
both sides	2.9	3.1

	LABOR TIME	SEVERE SERVICE
Rocker Arms or Shafts, Replace (B)		
Sedona		
one	3.3	3.6
both	3.6	3.9
Sorento all	10.4	10.7
Sportage		
one or both	2.9	3.2
Valve Job (A)		
Sedona		
right side	11.7	12.6
left side	10.4	11.3
both sides	17.4	18.3
Sorento		
right side	13.7	14.6
left side	13.0	13.9
both sides	15.1	16.0
Sportage		
1995-02	11.2	12.1
2005		
4 cyl.	8.0	8.9
V6		
right side	10.9	11.8
left side	9.9	10.8
both sides	14.4	15.3
w/4WD add	.9	.9
w/AC add	.3	.3
w/SOHC 95-96 add	.7	.7
Valve Lifters, Replace (B)		
Sportage	1.8	2.1
Valve Springs and/or Oil Seals, Replace (B)		
Includes: Valve grinding, valve adjustments, removing carbon from piston top and cylinder head.		
Sedona		
right side	10.2	10.4
left side	8.9	9.1
both sides	14.8	15.2
Sorento		
right side	12.6	12.8
left side	11.9	12.1
both sides	13.3	13.7
Sportage		
1995-02	8.1	8.3
2005		
4 cyl.	6.8	7.0
V6		
right side	7.9	8.1
left side	6.8	7.0
both sides	8.8	9.4
CAMSHAFT		
Camshaft, Replace (B)		
Sedona	5.9	6.1
Sorento	9.9	10.1
Sportage		
1995-02	4.3	4.5

	LABOR TIME	SEVERE SERVICE
2005		
4 cyl.	3.6	3.8
V6	5.7	5.9
w/4WD add	.9	.9
w/AC add	.3	.3
Camshaft Bearings (Cam Removed), Replace (B)		
Sportage	.3	.8
Camshaft Oil Seals, Replace (B)		
Sedona		
left side	5.3	5.9
right side	3.6	4.1
both sides	7.1	7.8
Sportage	3.8	4.2
w/SOHC 95-96 add	.7	.7
Camshaft Sprocket, Replace (B)		
Sedona		
right	3.4	3.6
left	4.9	5.1
all	6.5	6.7
Sorento		
all	6.0	6.2
Sportage all		
1995-02	3.6	3.8
2005 4 cyl.	4.4	4.6
w/SOHC 95-96 add	.7	.7
Timing Belt, Replace (B)		
Sedona	3.6	3.9
Sorento	4.8	5.1
Sportage		
1995-02	3.7	4.0
2005	2.9	3.2
w/4WD add	.9	.9
w/AC add	.3	.3
Timing Belt Tensioner, Replace (B)		
Sedona	3.3	3.5
Sorento	2.2	2.4
Sportage		
1995-02	2.2	2.4
2005	2.9	3.1
Timing Belt Idler Pulley, Replace (B)		
Sedona	3.3	3.5
Sorento	2.2	2.4
Sportage		
1995-02	2.2	2.4
2005	2.7	2.9
CRANK & PISTONS		
Connecting Rod Bearings, Replace (A)		
Includes: Check bearing oil clearance.		
Sedona	15.0	15.5
Sorento	5.0	5.5
Sportage		
1995-02	8.6	9.1
2005	4.4	4.9

KIA-16 SEDONA : SORENTO : SPORTAGE

	LABOR TIME	SEVERE SERVICE
Crankshaft and Main Bearings, Replace (A)		
Includes: Engine R&R, check bearing oil clearance.		
Sedona	17.2	17.7
Sorento	18.2	18.7
Sportage		
1995-02	12.5	13.0
2005		
4 cyl.	13.3	13.8
V6	18.9	19.4
w/4WD add	.9	.9
w/AC add	.3	.3
w/AT add	.5	.5
w/SOHC 95-96 add	.7	.7
Crankshaft Pulley, Replace (B)		
Exc. below	1.7	1.8
2005 Sportage	.6	.7
w/AC 05 Sportage add	.3	.3
Crankshaft/Front Oil Pump Seal, Replace (B)		
Sedona	6.2	6.4
Sorento	5.8	6.0
Sportage		
1995-02	5.9	6.1
2005		
4 cyl.	4.8	5.0
V6	5.7	5.9
w/4WD add	.9	.9
w/AC add	.2	.2
Crankshaft Sprocket, Replace (B)		
Sedona, Sorento	5.1	5.3
Sportage		
1995-02	4.8	5.0
2005		
4 cyl.	2.5	2.7
V6	3.1	3.3
w/4WD add	.9	.9
w/AC add	.2	.2
Pistons or Connecting Rods, Replace (A)		
Includes: Ridge reaming, cylinder wall deglazing, installing new rings and rod bearings, engine tune-up.		
Sedona	15.2	15.5
Sorento	15.8	16.1
Sportage		
4 cyl.	8.9	9.2
V6	16.5	16.8
w/4WD add	.9	.9
w/AC add	.3	.3
w/SOHC 95-96 add	.7	.7
Rear Main Oil Seal, Replace (A)		
Sedona	8.7	9.0
Sorento	4.5	4.8
Sportage		
1995-02	5.0	5.3

	LABOR TIME	SEVERE SERVICE
2005		
4 cyl.	7.3	7.6
V6	8.0	8.3
w/4WD add	.9	.9
Rings, Replace (A)		
Includes: Ridge reaming, cylinder wall deglazing, installing new rings, engine tune-up.		
Sedona	15.5	15.8
Sorento	13.6	13.9
Sportage		
1995-02	9.5	9.8
2005		
4 cyl.	8.3	8.6
V6	15.5	15.8
w/4WD add	.9	.9
w/AC add	.3	.3
w/SOHC 95-96 add	.7	.7

ENGINE LUBRICATION

	LABOR TIME	SEVERE SERVICE
Engine Oil Pressure Switch (Sending Unit), Replace (B)		
All Models	.3	.3
Oil Pan and/or Gasket, Replace (B)		
Sedona		
upper	2.5	2.7
lower	.7	.9
Sorento	3.3	3.5
Sportage		
1995-02	2.9	3.1
2005	1.9	2.1
w/AC add	.3	.3
Oil Pump, Replace (B)		
Sedona	5.1	5.3
Sorento	5.5	5.7
Sportage		
1995-02	5.5	5.7
2005		
4 cyl.	4.5	4.7
V6	4.8	5.0
w/4WD add	.9	.9
w/AC add	.3	.3

CLUTCH

	LABOR TIME	SEVERE SERVICE
Clutch Hydraulic System, Bleed (B)		
Sorento, Sportage	.5	.5
Clutch Pedal Free Play, Adjust (B)		
Sorento, Sportage	.5	.5
Clutch Assy., Replace (B)		
Sorento	8.3	8.8
Sportage	7.1	7.6
w/4WD add	.9	.9
Clutch Master Cylinder, R&R and Recondition (B)		
Includes: System bleeding.		
Sportage	1.1	1.3

	LABOR TIME	SEVERE SERVICE
Clutch Master Cylinder, Replace (B)		
Includes: System bleeding.		
Sorento	1.0	1.1
Sportage		
1995-02	.8	.9
2005	1.3	1.4
Clutch Release Bearing, Replace (B)		
Sorento	8.3	8.8
Sportage	7.1	7.6
w/4WD add	.9	.9
Clutch Slave Cylinder, R&R and Recondition (B)		
Includes: System bleeding.		
Sportage	.8	.9
Clutch Slave Cylinder, Replace (B)		
Includes: System bleeding.		
Sorento	.9	1.0
Sportage		
1995-02	.6	.7
2005	.9	1.0
Flywheel, Replace (B)		
Sorento	8.3	8.8
Sportage		
1995-02	6.9	7.4
2005	7.2	7.7

MANUAL TRANSMISSION

	LABOR TIME	SEVERE SERVICE
Extension Housing, Replace (B)		
Sorento	1.5	1.7
Sportage		
1995-02	1.5	1.7
2005	2.4	2.6
Speedometer Driven Gear, Replace (B)		
Sorento	.8	.9
Sportage		
1995-02	1.1	1.2
2005	.6	.7
Transmission Assy., R&R and Recondition (A)		
Includes: Recondition transmission only.		
Sorento	11.3	12.2
Sportage		
1995-02	11.3	12.2
2005	14.2	15.1
w/4WD add	.9	.9
Transmission Assy., R&R or Replace (B)		
Sorento	8.3	8.8
Sportage	6.8	7.3
w/4WD add	.9	.9
Replace clutch pilot bearing add	.9	.9

SEDONA : SORENTO : SPORTAGE **KIA-17**

	LABOR TIME	SEVERE SERVICE
AUTOMATIC TRANSMISSION		
SERVICE TRANSMISSION INSTALLED		
Diagnose Transmission (A)		
Sportage	1.2	1.2
Oil Pan and/or Gasket, Replace (B)		
All Models	1.4	1.6
Oil Filter, Replace (B)		
Sportage	1.1	1.1
Speed Sensor, Replace (B)		
All Models	.6	.7
Speedometer Drive Gear, Replace (B)		
All Models	.8	.9
Valve Body, R&R and Recondition (A)		
Sedona	5.8	6.7
Sorento	5.5	6.4
Sportage		
1995-02	4.0	4.9
2005	5.8	6.7
Valve Body, Replace (B)		
Sedona	4.3	4.8
Sorento	4.0	4.5
Sportage		
1995-02	2.7	3.2
2005	4.3	4.8
SERVICE TRANSMISSION REMOVED		
Transmission R&R included unless otherwise noted.		
Flywheel (Flexplate), Replace (B)		
Sedona	8.6	9.1
Sorento	8.3	8.8
Sportage		
1995-02	6.9	7.4
2005	7.3	7.8
Torque Converter Assy., R&R (A)		
Sedona	8.4	8.9
Sorento	7.3	7.8
Sportage		
1995-02	7.3	7.8
2005	7.7	8.2
w/4WD add	.9	.9
Replace		
drive plate assy. add	.3	.3
engine rear oil seal add	.3	.3
Transmission Assy., R&I (B)		
Sportage	4.7	4.9
Replace assembly add	.5	.7
Transmission Assy., R&R (A)		
Sedona	8.4	8.9
Sorento, Sportage	7.3	7.8
w/4WD add	.9	.9
Replace		
drive plate assy. add	.3	.3
engine rear oil seal add	.3	.3
transmission assy. add	.5	.5

	LABOR TIME	SEVERE SERVICE
Transmission Assy., R&R and Recondition (A)		
Sorento	14.8	16.5
Sportage	12.0	12.9
TRANSFER CASE		
Drive Chain, Replace (B)		
Sorento	5.9	6.4
Sportage	5.3	5.8
Shift Lever Cover Assy., Replace (B)		
Sorento, Sportage	.5	.5
Transfer Case, R&R and Recondition (B)		
Sorento	8.2	9.1
1995-02 Sportage	8.2	9.1
Transfer Case, R&R (B)		
Sorento	5.8	6.3
Sportage		
1995-02	4.4	4.9
2005	2.7	3.2
SHIFT LINKAGE		
AUTOMATIC		
Shift Cable, Adjust (B)		
All Models	.3	.4
Shift Lever, Replace (B)		
All Models	1.1	1.1
Shift Lock Control Cable, Replace (B)		
Sendona, Sorento	1.4	1.6
Sportage		
1995-02	.3	.5
2005	8.0	8.2
Throttle Valve Cable, Replace (B)		
All Models	1.4	1.6
MANUAL		
Shift Cable, Adjust (B)		
Sorento, Sportage	.5	.6
Gear Selector Lever, Replace (B)		
Sorento, Sportage	.6	.7
Shift Control Cable, Replace (B)		
2005 Sportage	7.9	8.1
w/AC add	.7	.7
DRIVELINE		
DRIVESHAFT		
4WD Coupling, R&R and Recondition (B)		
2005 Sportage	2.4	2.6
4WD Coupling, Replace (B)		
2005 Sportage	1.3	1.5
Center Bearing Assy., Replace (B)		
2005 Sportage	1.8	2.0
Center Support Bearing or Oil Seal, Replace (B)		
1995-02 Sportage	1.5	1.7

	LABOR TIME	SEVERE SERVICE
Driveshaft, Replace (B)		
Sorento, Sportage		
front	1.9	2.1
rear		
exc. below	.8	1.0
2005 Sportage	1.1	1.3
U-Joint Assy., Replace (B)		
Sorento, Sportage	2.2	2.4
DRIVE AXLES		
Axle Shaft, Replace (B)		
Front		
exc. below		
one side	1.6	1.8
both sides	2.9	3.2
2005 Sportage		
front or rear		
one side	.9	1.1
both sides	1.3	1.6
Intermediate		
Sedona, Sorento	1.9	2.0
2005 Sportage	1.3	1.4
Rear		
Sorento		
one side	1.5	1.6
both sides	2.5	2.7
Sportage		
1995-02		
one side	1.5	1.6
both sides	2.5	2.7
2005		
one side	.9	1.0
both sides	1.3	1.5
Companion/Pinion Flange, Replace (B)		
Sorento, Sportage		
front	.6	.7
rear		
exc. below	.9	1.0
2005 Sportage	2.7	2.8
w/4WD 05 Sportage		
rear add	.6	.6
CV Joint and/or Boots, Replace		
Inner or outer		
Sedona, Sorento		
one side	1.8	1.9
both sides	3.5	3.7
Sportage		
1995-02		
one side	1.5	1.6
both sides	2.8	3.0
2005		
front		
one side	1.2	1.3
both sides	1.9	2.1
rear		
one side	1.3	1.4
both sides	2.2	2.4

KIA-18   SEDONA : SORENTO : SPORTAGE

	LABOR TIME	SEVERE SERVICE
Differential Assy., R&R and Reconditon (B)		
Front		
Sedona	13.4	14.3
Sorento	6.0	6.9
Sportage	7.0	7.9
Rear		
Sorento, Sportage	5.1	6.0
Differential Assy., Replace (B)		
Front		
Sedona	11.2	12.5
Sorento	4.3	4.8
Sportage	4.7	5.3
Rear		
Sorento	3.8	4.3
Sportage	3.2	3.6
Differential Carrier, Replace (B)		
Front		
Sorento	5.0	5.7
Sportage	5.3	6.0
Rear		
Sportage	3.3	4.0
Differential Side Bearings, Replace (B)		
Front		
Sedona	11.6	13.0
Sorento	5.0	5.7
Sportage	5.7	6.3
Rear		
Sorento	4.3	4.8
Sportage	3.6	4.1
Free Wheeling Hub, R&R and Recondition (B)		
Sportage		
one side	.5	.7
both sides	.7	.9
Housing Assy., Replace (B)		
Front		
Sorento	5.5	6.0
Sportage	6.3	6.8
Rear	4.1	4.6
Pinion Seal, Replace (B)		
Sorento, Sportage		
front or rear	.7	.9
Replace spacer add	2.3	2.5
Ring Gear and Pinion, Replace (B)		
Front		
Sorento	4.7	5.3
Sportage	5.5	6.2
Rear		
Sorento	4.4	5.0
Sportage	3.8	4.3

BRAKES
ANTI-LOCK

The following operations do not include testing. Add time as required.

	LABOR TIME	SEVERE SERVICE
Diagnose Anti-Lock Brake System Component Each (A)		
All Models	1.0	1.0
Retrieve Fault Codes (A)		
All Models	.5	.5
ABS Control Unit, Replace (B)		
All Models	.5	.5
ABS Hydraulic Unit, Replace (B)		
Exc. below	2.2	2.3
2005 Sportage	1.5	1.6
ABS Main or Motor Relay, Replace (B)		
All Models	.3	.3
Wheel Speed Sensor, Replace (B)		
Front or rear		
exc. below		
one side	.6	.7
both sides	1.0	1.1
2005 Sportage rear		
one side	1.8	1.9
both sides	2.5	2.7

SYSTEM

	LABOR TIME	SEVERE SERVICE
Bleed Brakes (B)		
Includes: Add fluid.		
All Models	.5	.5
Brakes, Adjust (B)		
Includes: Refill master cylinder.		
Rear wheels	.3	.3
Flush and Refill Brake System (B)		
Sportage	.8	.8
Brake Hose (Flexible), Replace (B)		
Includes: System bleeding.		
Front		
one side	.9	1.0
both sides	1.1	1.2
Rear		
Sedona, Sportage		
one side	.8	.9
both sides	.9	1.0
Sorento		
one side	1.1	1.2
both sides	1.9	2.1
Master Cylinder, R&R and Recondition (B)		
Includes: System bleeding.		
All Models	1.2	1.4
Master Cylinder, Replace (B)		
Includes: System bleeding.		
All Models	.9	1.0
Power Booster Unit, Replace (B)		
Includes: System bleeding.		
Sedona, Sorento	4.1	4.3

	LABOR TIME	SEVERE SERVICE
Sportage		
1995-02	4.0	4.2
2005	1.5	1.7
Proportioning Valve, Replace (B)		
Includes: System bleeding.		
All Models	1.1	1.3

SERVICE BRAKES

	LABOR TIME	SEVERE SERVICE
Brake Drum, Replace (B)		
Includes: Repack wheel bearings.		
All Models		
one side	.6	.7
both sides	.8	.9
Caliper Assy., R&R and Recondition (B)		
Includes: System bleeding.		
Front		
one side	1.6	1.8
both sides	2.5	2.9
Caliper Assy., Replace (B)		
Includes: System bleeding.		
Front		
one side	.8	.9
both sides	1.2	1.4
Rear		
Sorento		
one side	.9	1.0
both sides	1.4	1.5
2005 Sportage		
one side	.9	1.0
both sides	1.4	1.5
Disc Brake Rotor, Replace (B)		
Front or Rear		
exc. below		
one side	.8	.9
both sides	1.2	1.4
2005 Sportage		
front		
one side	.6	.7
both sides	.9	1.1
rear		
one side	.5	.6
both sides	.8	1.0
Pads and/or Shoes, Replace (B)		
Includes: Adjust service and parking brake. System bleeding.		
Front or rear disc	1.1	1.3
Rear drum	1.3	1.4

COMBINATION ADD-ONS

	LABOR TIME	SEVERE SERVICE
Replace		
brake drum add each	.3	.3
brake hose add each	.3	.3
disc brake rotor add each	.3	.3
wheel cylinder add	.3	.3
Resurface		
disc rotor add	.5	.5
brake drum add	.5	.5

SEDONA : SORENTO : SPORTAGE

	LABOR TIME	SEVERE SERVICE
Wheel Cylinder, R&R and Reconditon (B)		
Includes: System bleeding.		
One side	1.4	1.6
Both sides	2.1	2.3
Wheel Cylinder, Replace (B)		
Includes: System bleeding.		
One side	1.1	1.3
Both sides	1.7	2.0

PARKING BRAKE

	LABOR TIME	SEVERE SERVICE
Parking Brake Cable, Adjust (C)		
All Models	.3	.5
Parking Brake Apply Actuator, Replace (B)		
All Models		
foot control	1.9	2.0
hand control	.8	.9
Parking Brake Cable, Replace (B)		
Front		
Sedona	.9	1.0
Sorento	1.8	1.9
1995-02 Sportage	.9	1.0
Rear		
Sedona, Sorento		
one side	1.3	1.4
both sides	2.1	2.3
Sportage		
1995-02		
one side	1.0	1.1
both sides	2.5	2.7
2005		
one side	1.3	1.4
both sides	1.9	2.1
Parking Brake Shoes, Replace (B)		
Sorento	.8	.9
Sportage	1.8	2.0

FRONT SUSPENSION

Unless otherwise noted, time given does not include alignment.

	LABOR TIME	SEVERE SERVICE
Align Front End (B)		
All Models	1.1	1.2
Toe In, Adjust (B)		
Two wheels	.8	.9
Four wheels		
Sedona, Sorento	1.0	1.2
Sportage	1.6	1.8
Front Spring, Replace (B)		
Exc. below		
one side	1.8	2.0
both sides	3.1	3.4
2005 Sportage		
one side	1.3	1.5
both sides	2.4	2.7
Front Wheel Oil Seal, Replace (B)		
Inner or outer		
one side	1.7	1.9
both sides	2.8	3.1

	LABOR TIME	SEVERE SERVICE
Lower Arm Ball Joint, Replace (B)		
Sedona		
one side	1.2	1.4
both sides	1.8	2.1
Sorento		
one side	1.6	1.8
both sides	2.8	3.1
Sportage		
1995-02		
one side	1.6	1.8
both sides	2.8	3.1
2005		
one side	1.2	1.4
both sides	1.6	1.9
Lower Control Arm, Replace (B)		
Exc. below		
one side	1.7	1.9
both sides	2.9	3.1
2005 Sportage		
one side	1.3	1.5
both sides	1.9	2.2
Lower Control Arm Bushings, Replace (B)		
Sorento		
one side	1.7	1.8
both sides	2.9	3.0
Sportage		
one side	1.6	1.7
both sides	2.2	2.3
Shock Absorber, Replace (B)		
Sportage		
one side	1.0	1.1
both sides	1.6	1.7
Stabilizer Bar, Replace (B)		
Sedona	2.7	2.9
Sorento	.9	1.1
Sportage		
1995-02	.9	1.1
2005	4.5	4.7
Stabilizer Bar Link, Replace (B)		
Sedona		
one side	1.3	1.4
both sides	2.2	2.4
Sorento		
one side	.7	.8
both sides	1.2	1.4
Sportage		
1995-02		
one side	1.3	1.4
both sides	2.2	2.4
2005		
one side	.3	.4
both sides	.6	.8
Steering Knuckle, Replace (B)		
One side	1.7	2.0
Both sides	2.9	3.2
Strut Assy., Replace (B)		
Sedona		
one side	1.6	1.8
both sides	2.8	3.1

	LABOR TIME	SEVERE SERVICE
Sorento, Sportage		
one side	1.3	1.5
both sides	2.4	2.7
Strut Fork, Replace (B)		
Sportage		
one side	1.0	1.1
both sides	1.6	1.8
Upper Control Arm, Replace (B)		
Sorento		
one side	1.6	1.8
both sides	2.8	3.1
1995-02 Sportage		
one side	1.6	1.8
both sides	2.8	3.1
Upper Control Arm Bushings, Replace (B)		
Sorento		
one side	1.6	1.8
both sides	2.8	3.1
1995-02 Sportage		
one side	1.6	1.8
both sides	2.8	3.1
Wheel Hub, Replace (B)		
One side	1.6	1.8
Both sides	2.7	3.0
Wheel Bearings, Replace (B)		
One side	1.6	1.8
Both sides	2.7	3.0

REAR SUSPENSION

Unless otherwise noted, time given does not include alignment.

	LABOR TIME	SEVERE SERVICE
Toe In, Adjust (B)		
All Models	.5	.5
Lateral Rod, Replace (B)		
All Models	.6	.7
Lower Arm Complete, Replace (B)		
Exc. below		
one side	.5	.6
both sides	.8	.9
Sedona		
one side	1.2	1.4
both sides	1.8	2.2
Rear Spring, Replace (B)		
Sedona, Sorento		
one side	1.0	1.1
both sides	1.6	1.8
1995-02 Sportage		
one side	1.0	1.1
both sides	1.6	1.8
Rear Wheel Bearing Assy., Replace		
Sedona, Sorento		
one side	1.6	1.7
both sides	2.7	2.9
Sportage		
1995-02		
one side	1.6	1.7
both sides	2.7	2.9

KIA-20 SEDONA : SORENTO : SPORTAGE

	LABOR TIME	SEVERE SERVICE
Rear Wheel Bearing Assy., Replace		
2005		
2WD		
one side	.7	.8
both sides	1.3	1.5
4WD		
one side	1.6	1.7
both sides	3.0	3.2
Rear Wheel Hub and Bearing Assy., Replace (B)		
Sedona, Sorento		
one side	1.9	2.0
both sides	3.4	3.6
Sportage		
2005		
2WD		
one side	.7	.8
both sides	1.3	1.5
4WD		
one side	1.6	1.7
both sides	3.0	3.2
Rear Wheel Oil Seal, Replace (B)		
Sedona, Sorento		
one side	2.1	2.3
both sides	3.6	4.1
Sportage		
one side	1.4	1.6
both sides	1.9	2.1
Rear Wheel Oil Seal (Axel Casing), Replace (B)		
Sorento		
one side	1.6	1.8
both sides	2.9	3.2
1995-02 Sportage		
one side	1.6	1.8
both sides	2.9	3.2
Shock Absorber, Replace (B)		
Sedona, Sorento		
one side	1.0	1.1
both sides	1.6	1.8
Sportage		
1995-02		
one side	.9	1.0
both sides	1.5	1.7
2005		
one side	.5	.6
both sides	.7	.9
Stabilizer Bar, Replace (B)		
All Models	.7	.9
Upper Arm Complete, Replace (B)		
One side	.6	.7
Both sides	.9	1.1

STEERING
AIR BAGS
The following operations do not include testing. Add time as required.

	LABOR TIME	SEVERE SERVICE
Diagnose Air Bag System Component Each (A)		
All Models	.7	.7
Air Bag Control Module, Replace (B)		
Sedona, Sorento	1.1	1.1
Sportage	.9	.9
Coil Spring Assy., Replace (B)		
All Models	.8	.8
Knee Module Assy., Replace (B)		
Sportage	.7	.7
Module Assembly, Replace (A)		
Driver side	.7	.7
Passenger side	.7	.7
Sedona, Sorento	2.4	2.4
Sportage	1.2	1.2
Side Curtain	2.4	2.4
Sensor, Replace (B)		
Front or side	.8	.8

STEERING GEAR

	LABOR TIME	SEVERE SERVICE
Troubleshoot Power Steering (A)		
Sportage	.5	.5
Steering Gear, Adjust (B)		
Sportage	.7	.9
Center Link, Replace (B)		
Sportage	1.2	1.3
Idler Arm, Replace (B)		
Sportage	.8	.9
Pitman Arm, Replace (B)		
Sportage	.8	.9
Power Steering Pump Hoses, Replace (B)		
Does not include purging.		
Sedona, Sorento		
pressure or return	1.5	1.7
suction	1.1	1.2
Sportage		
1995-02		
pressure	.6	.7
suction	.6	.6
2005		
pressure	.5	.6
suction	.3	.4
Power Steering Pressure Switch Assy., Replace (B)		
Sportage	.5	.5
Steering Column, R&I (B)		
Sedona, Sorento	2.5	2.7
Sportage		
1995-02	2.5	2.7
2005	1.3	1.5
Steering Gear, R&R and Recondition (B)		
Sedona	4.7	5.6
Sorento	3.2	4.1
Sportage		
1995-02	2.4	3.3
2005		
2WD	5.9	6.8
4WD	6.4	7.3
w/PS add	.5	.5
Steering Gear, Replace (B)		
Sedona	2.9	3.4
Sorento	2.1	2.6
Sportage		
1995-02	1.8	2.3
2005		
2WD	4.8	5.3
4WD	5.3	5.8
w/PS add	.5	.5
Steering Pump, R&R and Recondition (A)		
Sedona, Sorento	1.7	1.9
Sportage	2.0	2.2
Steering Pump, Replace (B)		
Sedona, Sorento	1.5	1.7
Sportage		
1995-02	1.5	1.7
2005	1.8	2.0
Transfer pulley add	.5	.5
Steering Wheel, Replace (B)		
All Models	.5	.5
Tie Rod and/or Tie Rod Ends, Replace (B)		
Includes: Toe-in and steering wheel off-center adjustment.		
One side	1.0	1.1
Both sides	1.4	1.6

HEATING & AIR CONDITIONING

When more than one component requires replacement where evacuation/recovery and recharging is already included, deduct 1.0 hour for each additional component from the time given.

	LABOR TIME	SEVERE SERVICE
Evacuate/Recover and Recharge System (B)		
All Models	1.1	1.2
AC Hoses, Replace (B)		
Does not include evacuate/recover and recharge.		
Exc. below one	1.4	1.5
2005 Sportage suction	.5	.6
each addl. add	.5	.6
Ambient Sensor, Replace (B)		
Sedona	.8	.9
Sorento	.5	.6
Sportage		
1995-02	.8	.9
2005	.5	.6
Blower Motor, Replace (B)		
All Models	.6	.7

SEDONA : SORENTO : SPORTAGE   **KIA-21**

	LABOR TIME	SEVERE SERVICE
Blower Motor Switch, Replace (B)		
All Models	.5	.5
Blower Motor Resistor, Replace (B)		
All Models	.5	.5
Compressor Assy., Replace (B)		
Does not include evacuate/recover and recharge.		
All Models	1.8	2.1
Condenser Assy., Replace (B)		
Does not include evacuate/recover and recharge.		
Sedona, Sorento	1.8	2.0
Sportage		
1995-02	1.7	1.9
2005	1.4	1.6
Condenser Fan Assy., Replace (B)		
Sedona, Sorento	1.0	1.1
1995-02 Sportage	.9	1.0
Evaporator Core, Replace (B)		
Does not include evacuate/recover and recharge.		
Sedona, Sorento	3.0	3.3
Sportage		
1995-02	3.4	3.7
2005	6.5	6.8
Expansion Valve, Replace (B)		
Does not include evacuate/recover and recharge.		
Sedona	3.0	3.2
Sorento	2.7	2.9
Sportage		
1995-02	2.7	2.9
2005	.6	.8
Heater Assy., Replace (B)		
Sedona	3.3	3.5
Sorento	3.4	3.6
Sportage		
1995-02	2.7	2.9
2005	7.0	7.2
Heater Core, Replace (B)		
Sedona	3.3	3.5
Sorento	3.4	3.6
Sportage		
1995-02	3.4	3.6
2005	7.2	7.4
Heater Hoses, Replace (B)		
Sedona, Sportage	.8	.9
Heater Pipe, Replace (B)		
One side	.8	.9
Both sides	1.4	1.5
In-Car Sensor, Replace (B)		
Sedona	1.3	1.4
Sorento	.8	.9
1995-02 Sportage	1.3	1.4
Pressure Switch, Replace (B)		
All Models	.5	.5

	LABOR TIME	SEVERE SERVICE
Receiver Drier Assy., Replace (B)		
Does not include evacuate/recover and recharge.		
All Models	1.4	1.5
Sun Sensor, Replace (B)		
Sedona	.3	.4
2005 Sportage	.3	.4
Temperature Control Assy., Replace (B)		
Exc. below	.8	.8
2005 Sportage	.4	.4
Temperature Control Cable, Replace (B)		
Sportage	1.1	1.1

WIPERS & SPEEDOMETER

	LABOR TIME	SEVERE SERVICE
Antenna Assy., Replace (B)		
Sedona	.4	.5
Sorento	.6	.7
Sportage		
1995-02	.4	.5
2005	1.0	1.1
Instrument Panel Cluster Assy., Replace (C)		
All Models	.5	.5
Radio, R&R (B)		
Sedona	.4	.5
Sorento	1.0	1.1
Sportage		
1995-02	1.0	1.1
2005	.4	.5
Rear Window Washer Pump, Replace (B)		
Sedona	.6	.7
Rear Window Washer/Wiper Switch, Replace (B)		
All Models	.5	.5
Rear Window Wiper Motor, Replace (B)		
All Models	.6	.7
Speedometer Cable & Casing, Replace (B)		
Sorento, Sportage	.7	.7
Speedometer Head, R&R or Replace (B)		
All Models	.7	.7
Windshield Washer Pump, Replace (B)		
All Models	.6	.7
Windshield Wiper Motor, Replace (B)		
All Models	.6	.7
Windshield Wiper (Combination) Switch, Replace (B)		
All Models	.8	.8
Wiper Pivot Shaft & Link Assy., Replace (B)		
Sportage	.8	.9

LAMPS & SWITCHES

	LABOR TIME	SEVERE SERVICE
Back-Up Lamp Switch, Replace (B)		
All Models	.3	.3
Combination (Multi-function) Switch, Replace (B)		
All Models	.8	.8
Halogen Headlamp Bulb, Replace (C)		
One	.3	.3
each addl. add	.3	.3
Headlamps, Aim (B)		
All Models	.5	.5
Horn, Replace (C)		
Exc. below	.3	.3
2005 Sportage	.9	.9
Horn Relay, Replace (B)		
All Models	.3	.3
Multifunction Switch Assy., Replace (B)		
All Models	.8	.8
Neutral Safety Switch, Replace (B)		
All Models	.7	.7
Stop Light Switch, Replace (B)		
All Models	.5	.5
Parking Brake Warning Switch, Replace (C)		
All Models	.6	.6
Turn Signal or Hazard Warning Flasher, Replace (C)		
All Models	.5	.5

BODY

	LABOR TIME	SEVERE SERVICE
Door Window Regulator, Replace (B)		
Exc. below		
one side	.8	.9
both sides	1.4	1.5
2005 Sportage		
power		
front		
one side	1.1	1.2
both sides	2.0	2.1
rear		
one side	.9	1.0
both sides	1.6	1.7
Door Handle, Replace (B)		
Inside	.4	.5
Outside		
front	.5	.6
rear		
exc. below	.6	.7
2005 Sportage	.9	1.0
Tailgate	.4	.5
Door Latch Assy., Replace (B)		
Exc. below		
one side	.8	.9
both sides	1.1	1.2

KIA-22   SEDONA : SORENTO : SPORTAGE

	LABOR TIME	SEVERE SERVICE
Door Latch Assy., Replace (B)		
2005 Sportage		
front		
one side	1.1	1.2
both sides	2.0	2.1
rear		
one side	.5	.6
both sides	.9	1.0
Tailgate	.5	.6
Electric Door Lock Actuator, Replace (B)		
All Models	.8	.9

	LABOR TIME	SEVERE SERVICE
Fuel Door Cable, Replace (B)		
Exc. below	1.5	1.6
Sportage		
1995-02	.5	.6
2005	.9	1.0
Fuel Door Latch, Replace (B)		
All Models	.4	.5
Hood Hinge, Replace (B)		
One	.5	.6
Both	.9	1.0
Hood Latch, Replace (B)		
All Models	.5	.6

	LABOR TIME	SEVERE SERVICE
Hood Release Cable, Replace (B)		
Sedona, Sorento	1.9	2.0
2005 Sportage	1.1	1.2
Hood Release Handle, Replace (B)		
All Models	.3	.3
Power Window Motor, Replace (B)		
All Models	.5	.6

LEX

ES250 : ES300 : ES330 : GS300 : GS400 : GS430 : IS300 : LS400 : LS430 : SC300 : SC400 : SC430

SYSTEM INDEX

MAINTENANCE	LEX-2
Maintenance Schedule	LEX-2
CHARGING	LEX-2
STARTING	LEX-3
CRUISE CONTROL	LEX-3
IGNITION	LEX-3
EMISSIONS	LEX-4
FUEL	LEX-5
EXHAUST	LEX-6
ENGINE COOLING	LEX-6
ENGINE	LEX-7
Assembly	LEX-7
Cylinder Head	LEX-7
Camshaft	LEX-9
Crank & Pistons	LEX-9
Engine Lubrication	LEX-10
CLUTCH	LEX-10
MANUAL TRANSAXLE	LEX-11
MANUAL TRANSMISSION	LEX-11
AUTO TRANSAXLE	LEX-11
AUTO TRANSMISSION	LEX-11
SHIFT LINKAGE	LEX-12
DRIVELINE	LEX-12
BRAKES	LEX-13
FRONT SUSPENSION	LEX-15
REAR SUSPENSION	LEX-16
STEERING	LEX-18
HEATING & AC	LEX-19
WIPERS & SPEEDOMETER	LEX-20
LAMPS & SWITCHES	LEX-21
BODY	LEX-21

OPERATIONS INDEX

A
- AC Hoses ... LEX-19
- Air Bags ... LEX-18
- Air Conditioning ... LEX-19
- Alignment ... LEX-15
- Alternator (Generator) ... LEX-2
- Antenna ... LEX-20
- Anti-Lock Brakes ... LEX-13

B
- Back-Up Lamp Switch ... LEX-21
- Ball Joint ... LEX-16
- Battery Cables ... LEX-3
- Bleed Brake System ... LEX-14
- Blower Motor ... LEX-19
- Brake Disc ... LEX-14
- Brake Hose ... LEX-14
- Brake Pads and/or Shoes ... LEX-14

C
- Camshaft ... LEX-9
- Camshaft Sensor ... LEX-3
- Catalytic Converter ... LEX-6
- Coolant Temperature (ECT) Sensor ... LEX-4
- Crankshaft ... LEX-9
- Cruise Control ... LEX-3
- CV Joint ... LEX-13
- Cylinder Head ... LEX-8

D
- Differential ... LEX-12
- Distributor ... LEX-3
- Drive Belt ... LEX-2
- Driveshaft ... LEX-12

E
- EGR ... LEX-4
- Electronic Control Module (ECM/PCM) ... LEX-4
- Engine ... LEX-7
- Engine Lubrication ... LEX-10
- Engine Mounts ... LEX-7
- Evaporator ... LEX-20
- Exhaust ... LEX-6
- Exhaust Manifold ... LEX-6
- Expansion Valve ... LEX-20

F
- Flexplate ... LEX-11
- Flywheel ... LEX-11
- Fuel Injection ... LEX-5
- Fuel Pump ... LEX-5
- Fuel Vapor Canister ... LEX-4

G
- Gear Selector Lever ... LEX-12
- Generator ... LEX-2

H
- Halfshaft ... LEX-13
- Headlamp ... LEX-21
- Heater Core ... LEX-20
- Horn ... LEX-21

I
- Idle Air Control (IAC) Valve ... LEX-6
- Ignition Coil ... LEX-3
- Ignition Module ... LEX-3
- Ignition Switch ... LEX-3
- Inner Tie Rod ... LEX-18
- Intake Manifold ... LEX-5

K
- Knock Sensor ... LEX-4

L
- Lower Control Arm ... LEX-16

M
- Maintenance Schedule ... LEX-2
- Mass Air Flow (MAF) Sensor ... LEX-4
- Master Cylinder ... LEX-14
- Muffler ... LEX-6
- Multifunction Lever/Switch ... LEX-21

N
- Neutral Safety Switch ... LEX-21

O
- Oil Pan ... LEX-10
- Oil Pump ... LEX-10
- Orifice Tube/Valve ... LEX-20
- Outer Tie Rod ... LEX-18
- Oxygen Sensor ... LEX-4

P
- Parking Brake ... LEX-14
- Pistons ... LEX-10
- Positive Crankcase Ventilation (PCV) Valve ... LEX-5

R
- Radiator ... LEX-6
- Radiator Hoses ... LEX-7
- Radio ... LEX-20
- Rear Main Oil Seal ... LEX-10

S
- Shock Absorber/Strut, Front ... LEX-16
- Shock Absorber/Strut, Rear ... LEX-17
- Spark Plug Cables ... LEX-4
- Spark Plugs ... LEX-4
- Spring, Front Coil ... LEX-15
- Spring, Rear Coil ... LEX-17
- Starter ... LEX-3
- Steering Wheel ... LEX-19

T
- Thermostat ... LEX-6
- Throttle Body ... LEX-6
- Throttle Position Sensor (TPS) ... LEX-5
- Timing Belt ... LEX-9
- Torque Converter ... LEX-11

U
- Upper Control Arm ... LEX-16

V
- Valve Body ... LEX-11
- Valve Cover Gasket ... LEX-8
- Valve Job ... LEX-8
- Vehicle Speed Sensor ... LEX-11

W
- Water Pump ... LEX-7
- Wheel Balance ... LEX-2
- Window Regulator ... LEX-22
- Windshield Washer Pump ... LEX-20
- Windshield Wiper Motor ... LEX-21

LEX-1

LEX-2   ES250 : ES300 : ES330 : GS300 : GS400 : GS430

	LABOR TIME	SEVERE SERVICE
MAINTENANCE		
Air Cleaner Filter Element, Service (B)		
1990-05	2	.3
Chassis Lubrication, Change Oil & Filter (C)		
Includes: Correct fluid levels.		
1990-05	.6	.6
Drive Belt, Adjust (B)		
1990-05	.3	.4
ES250, ES300	.4	.5
Drive Belt, Replace (B)		
Fan		
exc. below	.3	.4
ES250, ES300, ES330	.5	.6
Power steering		
ES250, ES300, ES330		
1990-98	1.2	1.3
1999-05	1.0	1.1
Halogen Headlamp Bulb, Replace (B)		
Exc. below	.3	.3
ES300, ES330		
2002-05	1.1	1.1
IS300	.8	.8
Oil & Filter, Change (C)		
Includes: Correct fluid levels.		
1990-05	.6	.6
Timing Belt, Replace (B)		
ES250	3.7	4.0
ES300, ES330		
1992-01	2.9	3.2
2002-05	3.6	3.9
GS300		
1994-97	3.2	3.5
1998	4.2	4.5
1999-05	3.0	3.3
GS400, GS430	4.0	4.3
IS300	2.9	3.2
LS400		
1990-94	4.6	4.9
1995-98	3.8	4.1
1999-00	4.5	4.8
LS430, SC300	3.1	3.4
SC400	4.8	5.1
SC430	3.8	4.1
Tire, Replace (C)		
Includes: Dismount old tire and mount new tire to rim.		
1990-05	.5	.5
Tires, Rotate (C)		
1990-05	.6	.7
Wheel, Balance (B)		
1990-05 one	.3	.3
each addl.	.2	.2

	LABOR TIME	SEVERE SERVICE
SCHEDULED MAINTENANCE INTERVALS		
If necessary, refer to appropriate Chilton maintenance service information.		
7,500 Mile Service (C)		
All Models	.4	.4
LS400 add	.2	.2
15,000 Mile Service (C)		
ES300, SC300	1.6	1.6
GS300/400	1.5	1.5
LS400	1.7	1.7
SC400	1.6	1.6
22,500 Mile Service (C)		
All Models	.4	.4
LS400 add	.2	.2
30,000 Mile Service (B)		
ES300, SC300	2.4	2.4
GS300/400	2.3	2.3
LS400	2.5	2.5
SC400	2.4	2.4
37,500 Mile Service (C)		
All Models	.4	.4
LS400 add	.2	.2
45,000 Mile Service (B)		
ES300, SC300	2.1	2.1
GS300/400	2.0	2.0
LS400	2.2	2.2
SC400	2.1	2.1
52,500 Mile Service (C)		
All Models	.4	.4
LS400 add	.2	.2
60,000 Mile Service (B)		
ES300, SC300	3.9	3.9
GS300/400	3.8	3.8
LS400	4.2	4.2
SC400	3.9	3.9
67,500 Mile Service (C)		
All Models	.4	.4
LS400 add	.2	.2
75,000 Mile Service (C)		
ES300, SC300	1.6	1.6
GS300/400	1.5	1.5
LS400	1.7	1.7
SC400	1.6	1.6
82,500 Mile Service (C)		
All Models	.4	.4
LS400 add	.2	.2
90,000 Mile Service (B)		
ES300, SC300	2.4	2.4
GS300/400	2.3	2.3
LS400	2.5	2.5
SC400	2.4	2.4
97,500 Mile Service (C)		
All Models	.4	.4
LS400 add	.2	.2

	LABOR TIME	SEVERE SERVICE
CHARGING		
Alternator Circuits, Test (B)		
Includes: Test component output.		
1990-05	.3	.3
Alternator, R&R and Recondition (A)		
ES250, ES300, ES330		
1990-98	1.8	2.1
1999-05	2.0	2.3
GS300, GS400, GS430	1.7	2.0
IS300	2.0	2.3
LS400, LS430		
1990-94	2.2	2.5
1995-05	1.9	2.2
SC300, SC400, SC430		
1992-98	1.9	2.2
1999-05	1.7	2.0
Alternator Assy., Replace (B)		
Includes: Pulley transfer.		
ES250, ES300, ES330	.6	.7
GS300, GS400, GS430		
1993-98	1.2	1.3
1999-05	.8	.9
IS300	.9	1.0
LS400, LS430		
1990-94	1.2	1.3
1995-98	.8	.9
1999-00	.7	.8
2001-05	.9	1.0
SC300		
1992-98	1.2	1.3
1999-00	.8	.9
SC400, SC430		
1992-05	.8	.9
Alternator Voltage Regulator, Replace (B)		
ES250, ES300, ES330	.9	1.1
GS300, GS430	1.2	1.4
GS400	1.5	1.7
IS300	1.3	1.5
LS400		
1990-94	1.5	1.7
1995-00	1.4	1.6
LS430, SC400, SC430	1.2	1.4
SC300	1.5	1.7
Front Alternator Bearing, Replace (B)		
ES250, ES300	1.5	1.8
GS300, GS400, GS430		
1993-98	2.0	2.3
1999-05	1.7	2.0
IS300	1.9	2.2
LS400, LS430		
1990-94	2.2	2.5
1995-05	1.9	2.2

	LABOR TIME	SEVERE SERVICE
SC300		
1992-00	2.0	2.3
SC400		
1992-98	1.9	2.2
1999-00	1.8	2.1

STARTING

Starter Draw Test (On Car) (B)
  1990-05 5   .5

Battery Cables, Replace (C)
Ground
  All Models 6   .7
Starter
  ES330 6   .7

Starter, R&R and Recondition (A)
ES250 .... 2.5   2.8
ES300, ES330
  1994-01 .... 1.7   2.0
  2002-05 .... 1.3   1.6
GS300, SC300
  1992-98 .... 1.7   2.0
  1999-05 .... 1.5   1.8
GS400, GS430 .... 4.4   4.7
IS300 .... 1.3   1.6
LS400
  1990-94 .... 5.6   5.9
  1995-98 .... 4.2   4.5
  1999-00 .... 4.4   4.7
LS430 .... 3.4   3.7
SC400 .... 5.6   5.9
SC430 .... 4.0   4.3

Starter Assy., Replace (B)
ES250 .... 1.7   1.9
ES300, ES330 8   1.0
GS300, SC300
  1992-05 9   1.1
GS400, LS400
  1990-94 .... 5.3   5.5
  1995-98 .... 3.7   3.9
  1999-00 .... 3.8   4.0
GS430 .... 4.0   4.2
LS430
  2001-03 9   1.0
  2004-05 .... 2.7   2.9
IS300 8   .9
SC400 .... 5.3   5.5
SC430 .... 2.8   3.0

Starter Solenoid and/or Switch, Replace (B)
Includes: Starter R&R.
ES250 .... 1.8   2.0
ES300
  1992-98 8   1.0
  1999-01 .... 1.2   1.4
GS300, SC300
  1992-98 .... 1.3   1.5
  1999-04 .... 1.1   1.3

	LABOR TIME	SEVERE SERVICE
GS400, GS430, LS400		
1990-94	5.2	5.4
1995-98	3.9	4.1
1999-05	4.1	4.3
LS430	2.8	3.0
IS300	.9	1.1
SC400		
1992-98	5.2	5.4
1999-00	5.3	5.5

CRUISE CONTROL

Diagnose Cruise System Component Each (A)
  1990-05 5   .5

Actuator Assy., Replace (B)
Exc. below 7   .7
SC300, SC400 8   .8
SC430 .... 1.5   1.5

Control Computer, Replace (B)
Exc. below 8   .8
ES300
  1997-01 5   .5
GS300, GS400
  1994-98 9   .9
LS400
  1990-94 9   .9
SC300, SC400 .... 1.7   1.7

Control Switch, Replace (B)
Exc. below 7   .7
ES300, ES330
  1992-96 .... 1.3   1.3
  2002-05 9   .9

IGNITION

Diagnose Ignition System Component Each (A)
  1990-05 9   .9

Retrieve Fault Codes (A)
  1990-05 3   .3

Camshaft Sensor, Replace (B)
1994-05 ES300,
  ES330 3   .4
1999-05 GS300 ...... 3.6   3.7
GS400, GS430 .... 1.1   1.2
IS300 .... 3.3   3.4
LS430 .... 1.1   1.2
SC430 .... 1.7   1.8

Distributor, Replace (B)
Includes: Reset base ignition timing.
ES250 .... 1.2   1.4
1992-93 ES300 ...... 1.3   1.5
GS300, SC300 .... 1.3   1.5
LS400
  1990-94 .... 1.6   1.8
  1995-00 .... 1.7   1.9
SC400 .... 1.1   1.2

	LABOR TIME	SEVERE SERVICE
Distributor Cap and/or Rotor, Replace (B)		
ES250, ES300	.3	.3
GS300, SC300	.5	.5
LS400		
1990-94	1.4	1.4
1995-00	1.6	1.6
SC400	.7	.7

Igniter Unit, Replace (B)
  1990-05 5   .5

Ignition Coil, Replace (B)
ES250, ES300, ES330
  1990-93
    one 5   .5
    both 8   .8
  1994-05
    either side 5   .5
GS300 5   .5
GS400, GS430
  right side 3   .3
  left side 5   .5
IS300 .... 1.4   1.4
LS400
  1990-94
    one 5   .5
    both 7   .7
  1995-00
    right side 3   .3
    left side 5   .5
LS430
  right side 6   .6
  left side 5   .5
SC300 5   .5
SC400
  1992-98
    one 5   .5
    both 7   .7
  1999-00
    either side 5   .5
SC430
  either side 9   .9

Ignition Switch, Replace (B)
ES250 5   .5
ES300, ES330
  1992-96 8   .8
  1997-05 6   .6
IS300 5   .5
GS300, GS400, GS430
  1994-00 9   .9
  2001-05 8   .8
LS400, LS430
  1990-94 .... 1.4   1.4
  1995-05 6   .6
SC300, SC400 .... 1.8   1.8
SC430 .... 2.5   2.5

Ignition Timing, Reset (B)
1990-95 ES250,
  ES300 .... 1.2   1.2
1992-95 GS300,
  SC300 .... 1.2   1.2

LEX-4 ES250 : ES300 : ES330 : GS300 : GS400 : GS430

	LABOR TIME	SEVERE SERVICE
Spark Plug (Ignition) Cables, Replace (B)		
ES250, ES300	.5	.6
GS300, SC300	1.6	1.7
IS300	1.3	1.4
LS400		
1990-94	1.7	1.8
1995-00	.8	.9
SC400	1.3	1.4
Spark Plugs, Replace (B)		
ES250	.8	1.0
ES300, ES330		
1992-93	.7	.9
1994-96	.8	1.0
1997-98	1.6	1.8
1999-01	1.8	2.0
2002-05	1.2	1.4
GS300	1.6	1.8
GS400, GS430	1.2	1.4
IS300	1.6	1.8
LS400		
1990-94	1.3	1.5
1995-00	.8	1.0
LS430	.9	1.1
SC300	2.0	2.2
SC400, SC430	1.2	1.4

EMISSIONS

The following operations do not include testing. Add time as required.

	LABOR TIME	SEVERE SERVICE
Diagnose Emission Control System Component Each (A)		
1990-05	.6	.6
Dynamometer Test (A)		
1990-05	.5	.5
Retrieve Fault Codes (A)		
1990-05	.3	.3
Air Distribution Manifold, Replace (B)		
Includes: R&R exhaust manifold when required.		
1993-94 LS400		
one side	6.3	6.5
both sides	7.6	7.8
Air Injection Check Valve, Replace (B)		
1993-94 LS400		
one side	5.2	5.2
both sides	5.2	5.2
Air Pump Assy., Replace (B)		
1993-94 LS400	5.0	5.0
Coolant Temperature (ECT) Sensor, Replace (B)		
All Models		
1990-98	1.2	1.3
1999-05 ES300, ES330, SC300	.8	.9
1999-05 GS300	.7	.8
1999-00 GS400, LS400	1.3	1.4

	LABOR TIME	SEVERE SERVICE
GS430	1.2	1.3
IS300	1.6	1.7
LS430	2.0	2.1
1999-00 SC400, SC430	.7	.8
Diverter and/or Switching Valve, Replace (B)		
1993-94 LS400	5.0	5.0
EGR Control Valve, Replace (B)		
1990-98	.7	.9
1999-01 ES300	.5	.7
1999-00 LS400	.8	1.0
1999-00 SC300, SC400	.7	.9
EGR Modulator, Replace (B)		
1990-02	.3	.3
EGR Switching Valve, Replace (B)		
1990-98	.5	.5
1999-02	1.2	1.2
Electronic Control Module (ECM/PCM), Replace (B)		
ES250	.7	.8
ES300, ES330		
1992-96	1.2	1.4
1997-05	.8	.9
GS300, GS400, GS430		
1993-98	1.4	1.6
1999-05	.8	.9
IS300	.8	.9
LS400		
1990-94	.9	1.0
1995-98	1.5	1.7
1999-00	1.2	1.4
LS430	.8	.9
SC300, SC400		
1992-98	1.2	1.4
1999-00	.8	.9
SC430	.6	.7
Fuel Vapor Canister, Replace (B)		
ES250, ES300, ES330	.8	.9
2002-05	2.3	2.6
GS300, GS400, GS430		
1993-97	.8	.9
1998-05	1.2	1.4
IS300	.8	.9
LS400		
1990-94	1.2	1.4
1995-00	.8	.9
LS430	1.5	1.7
SC300, SC400	.3	.3
SC430	2.0	2.2
Knock Sensor, Replace (B)		
Front		
1993-94		
ES300	2.8	3.1
GS300	.7	.9
LS400	4.8	5.0
LS430	2.4	2.7

	LABOR TIME	SEVERE SERVICE
SC300	1.3	1.5
SC400	4.6	4.8
each addl.	.1	.2
1995-05		
ES300, ES330	2.7	2.9
GS300	.7	.9
GS400	4.9	5.1
GS430	3.6	4.1
IS300	.8	.9
LS400	3.4	3.6
LS430	2.4	2.7
SC300	1.2	1.4
SC400	4.6	4.8
SC430	1.7	1.9
each addl.	.2	.2
Rear		
1993-05		
GS300	1.2	1.4
IS300	.6	.7
SC300	.3	.5
Mass Air Flow (MAF) Sensor, Replace (B)		
1990-97		
3.0L L6, 4.0L	.9	.9
2.5L, 3.0L V6	1.2	1.2
1998		
GS300	.3	.3
SC300	.9	.9
1999-05		
exc. below	.3	.3
LS400, LS430	.8	.8
Oxygen Sensor, Replace (B)		
Front		
1990-98		
3.0L L6, 2.5L, 3.0L V6	.7	.9
4.0L	.8	1.0
each addl.	.2	.3
1999-05		
3.0L L6, 4.0L	.8	.9
3.0L V6	.5	.7
3.3L V6	.9	1.0
4.3L V8 exc. below	.6	.7
2005 GS430	.8	.9
each addl.	.1	.2
Rear		
1990-94		
3.0L L6	1.7	1.9
2.5L, 3.0L V6		
ES250	.7	.9
ES300	1.2	1.4
4.0L		
LS400	3.0	3.2
SC400	1.4	1.6
each addl.	.2	.3
1995-98		
ES300	.8	1.0
GS300	1.2	1.4

IS300 : LS400 : LS430 : SC300 : SC400 : SC430 **LEX-5**

	LABOR TIME	SEVERE SERVICE
LS400		
one	1.8	2.0
both	2.8	3.0
SC300	1.5	1.7
SC400		
one	1.2	1.4
both	1.4	1.6
1999-05		
ES300	.7	.9
GS300, GS400,		
GS430	1.2	1.4
IS300		
one	1.4	1.9
both	1.7	2.2
LS400, LS430		
one	1.8	2.0
both	2.9	3.1
SC300	1.6	1.8
SC400		
one	1.3	1.5
both	1.5	1.7
SC430		
one	.5	.7
both	.8	1.0
Positive Crankcase Ventilation (PCV) Valve, Replace (B)		
1990-05	.5	.5
Thermal Vacuum Valve, Replace (B)		
ES250	1.2	1.2
1992-96 ES300	1.2	1.2
1990-94 LS400	.7	.7
Throttle Position Sensor (TPS), Replace (B)		
Exc. below	.7	.7
2002-05 ES300, ES330	1.7	1.9
2001-05 IS300, LS430	2.0	2.2
SC430	1.1	1.2
Vacuum Control Valve, Replace (B)		
1990-94 LS400	.3	.3
SC430	2.1	2.4

FUEL
DELIVERY

	LABOR TIME	SEVERE SERVICE
Fuel Pump, Test (B)		
1990-05	.3	.3
Fuel Filter, Replace (B)		
ES250	.8	.9
ES300, ES330		
1992-98	.8	.9
1999-05	1.0	1.1
GS300, GS400 GS430		
1994-97	.8	.9
1998	1.2	1.3
1999-05	1.3	1.4
IS300	1.0	1.1

	LABOR TIME	SEVERE SERVICE
LS400, LS430		
1990-94	.7	.8
1995-05	1.3	1.4
SC300, SC400		
1992-98	1.2	1.3
1999-00	.8	.9
SC430	1.7	2.0
Fuel Gauge (Dash), Replace (B)		
ES250	1.3	1.3
ES300		
1992-01	1.3	1.3
GS300, GS400, GS430	1.3	1.3
IS300	.6	.6
LS400	1.3	1.3
SC300, SC400	1.3	1.3
Fuel Gauge (Tank), Replace (B)		
ES250	.5	.7
ES300, ES330		
1992-96	.8	1.0
1997-05	1.0	1.2
GS300		
1993-00	1.2	1.4
2001-05	1.1	1.3
GS400, IS300	1.2	1.4
GS430	1.1	1.3
LS400, LS430		
1990-94	.5	.7
1995-05	1.3	1.5
SC300, SC400		
SC430	.8	1.0
Fuel Pump Control Module, Replace (B)		
GS300, GS400 GS430	.6	.7
1993-94 LS400	.6	.7
SC300, SC400	.9	1.0
Fuel Pump, Replace (B)		
ES250	2.8	3.0
ES300, ES330	1.0	1.2
GS300, GS400 GS430		
1994-97	2.0	2.2
1998-05	1.3	1.5
IS300	1.0	1.2
LS400		
1990-94	2.6	2.8
1995-00	1.6	1.8
LS430	1.3	1.5
SC300, SC400	1.2	1.4
SC430	1.9	2.1
Fuel Tank, Replace (B)		
Includes: Drain and refill.		
ES250	2.9	3.2
ES300, ES330	3.0	3.3
GS300, GS400 GS430		
1994-98	2.9	3.2
1999-05	3.1	3.4
IS300	3.7	4.0

	LABOR TIME	SEVERE SERVICE
LS400		
1990-94	2.9	3.2
1995-00	1.8	2.1
LS430	4.0	4.3
SC300, SC400	2.0	2.3
SC430	3.4	3.7
Intake Manifold and/or Gasket, Replace (B)		
ES250, ES300, ES330		
1990-05	2.6	2.8
GS300, GS400		
GS430	3.6	3.8
IS300	3.8	4.0
LS400		
1990-94	4.6	4.8
1995-00	3.3	3.5
LS430		
2001-03	2.3	2.5
2004-05	2.7	2.9
SC300		
1992-98	3.9	4.1
1999-00	4.2	4.4
SC400	4.6	4.8
SC430	2.3	2.5
Replace manifold add	.5	.8

INJECTION

	LABOR TIME	SEVERE SERVICE
Air Cleaner Filter Element, Service (B)		
1990-05	.2	.3
Air Intake Chamber or Gasket, Replace (B)		
ES250	1.7	1.9
ES300, ES330	2.4	2.6
GS300, SC300	1.5	1.7
GS400, GS430	2.4	2.6
IS300	2.1	2.3
LS400, LS430		
1990-94	4.9	5.1
1995-05	2.6	2.8
SC400		
1992-98	3.7	3.9
1999-00	5.0	5.2
SC430	2.1	2.3
Cold Start Injector, Replace (B)		
2.5L, 3.0L V6	.5	.5
4.0L	4.5	4.5
Fuel Injectors, Replace (B)		
ES250	2.5	2.7
ES300, ES330	2.7	2.9
2002-05	2.1	2.3
GS300, SC300		
1992-98	3.7	3.9
1999-00 SC300	3.8	4.0
1999-05 GS300	3.8	4.0
GS400, GS430	3.3	3.5
IS300	1.7	1.9

LEX-6 ES250 : ES300 : ES330 : GS300 : GS400 : GS430

	LABOR TIME	SEVERE SERVICE
Fuel Injectors, Replace (B)		
LS400		
1990-94	5.2	5.4
1995-98	3.1	3.3
1999-00	3.4	3.6
LS430	1.8	2.0
SC400	5.3	5.5
SC430	1.1	1.3
Fuel Pressure Regulator, Replace (B)		
ES250	1.2	1.4
ES300, ES330	.8	.9
GS300	.7	.8
1999-05	.5	.6
GS400, GS430	1.3	1.4
IS300	.8	.9
LS400		
1990-94	.7	.8
1995-00	2.6	2.8
LS430	1.2	1.4
SC300, SC400	.7	.8
SC430	1.7	1.9
Idle Air Control (IAC) Valve, Replace (B)		
ES250	1.6	1.6
ES300, ES330		
1992-96	1.2	1.2
1997-01	1.4	1.4
2002-05	.5	.5
GS300, SC300		
1992-00	.9	.9
LS400		
1990-94	1.2	1.2
1995-00	.9	.9
SC400	1.2	1.2
SC430	1.8	1.8
Start Injector Time Switch, Replace (B)		
ES250	.8	.8
1992-93 ES300	1.2	1.2
1990-94 LS400	1.2	1.2
SC400	1.2	1.2
Throttle Body Assy., Replace (B)		
ES250	1.4	1.5
ES300, ES330	1.2	1.3
2002-05	1.7	1.9
GS300, SC300		
GS400, GS430	1.7	1.9
IS300	1.5	1.7
LS400, LS430	1.8	2.1
SC400	1.5	1.7
SC430	2.0	2.2

EXHAUST

	LABOR TIME	SEVERE SERVICE
Catalytic Converter, Replace (B)		
1990-96 ES250, ES300	.9	1.0
GS300, SC300		
1992-02	.9	1.0
1999-03 SC300	.8	.9
GS430, LS400, LS430		
one	.6	.7
both	.8	.9
SC400, SC430		
one	.8	.9
both	1.1	1.3
Exhaust Manifold or Gasket, Replace (B)		
ES250		
right side	8.9	9.3
left side	2.3	2.6
ES300, ES330		
1994-01		
right side	8.6	8.9
left side	2.6	2.9
2002-05		
right side	1.1	1.3
left side	.8	.9
GS300	2.0	2.2
GS400, GS430		
right or left side	8.5	8.9
IS300	1.1	1.3
LS400		
1990-94		
right side	3.0	3.3
left side	2.4	2.7
1995-00		
right side	8.0	8.3
left side	8.2	8.5
LS430		
right or left side	6.5	6.8
SC300	1.7	2.0
SC400		
right or left side	6.0	6.3
SC430		
right	10.9	11.2
left	11.8	12.1
Front Exhaust Pipe, Replace (B)		
1990-98		
3.0L L6, 2.5L, 3.0L V6	.7	.8
4.0L	.9	1.0
1999-05 exc. below	.8	.9
GS300	2.1	2.3
Intermediate Exhaust Pipe, Replace (B)		
1990-05	.7	.8
Muffler & Tailpipe Assy., Replace (B)		
1990-91	.7	.8
1992-98		
one	.7	.8
both	.8	.9
1999-05		
one	.6	.7
both	.8	.9

ENGINE COOLING

	LABOR TIME	SEVERE SERVICE
Pressure Test Cooling System (C)		
1990-05	.3	.3
Bypass Hoses, Replace (B)		
3.0L L6	1.4	1.6
3.3LV6		
ES330	2.4	2.7
2.5L, 3.0L V6, 4.0L, 4.3L exc. below	.7	.9
ES300		
2002-03	2.4	2.7
LS430	1.4	1.6
SC430	.9	1.0
Coolant Thermostat, Replace (B)		
2.5L, 3.0L V6	1.3	1.4
3.0L L6	.9	1.0
3.3L V6	1.4	1.5
4.0L, 4.3L exc. below	.9	1.0
GS430, LS430, SC430 4.3L	1.4	1.5
Engine Coolant Temp. Sending Unit, Replace (B)		
Exc. below	.9	1.0
2001-03 GS430	1.2	1.4
2001-05 IS300	1.2	1.3
1990-94 LS400	.5	.5
2001-03 LS430	2.0	2.2
Fan Blade, Replace (B)		
3.0L L6	1.3	1.4
3.3L V6	1.7	1.8
4.0L	.8	1.0
4.3L	1.4	1.5
Fan Motor Module, Replace (B)		
ES300, LS430	.8	.9
SC400	1.2	1.4
SC430	.3	.3
Radiator Assy., R&R or Replace (B)		
Includes: Refill with proper coolant mix.		
ES250	1.8	2.0
ES300, ES330		
1992-96	2.2	2.4
1997-05	1.7	2.0
GS300, GS400		
GS430		
1992-97	2.2	2.4
1998-05	1.8	2.1
IS300	1.7	1.9
LS400		
1990-94	1.8	2.1
1995-98	1.9	2.2
1999-00	1.7	1.9
LS430	2.2	2.5
SC300		
1992-98	2.0	2.2
1999-00	1.8	2.1
SC400, SC430		
1992-05	1.8	2.1

IS300 : LS400 : LS430 : SC300 : SC400 : SC430   **LEX-7**

	LABOR TIME	SEVERE SERVICE
Radiator Hoses, Replace (B)		
Includes: Refill with proper coolant mix.		
ES250		
upper	.5	.5
lower	1.5	1.6
ES300, ES330		
upper	1.1	1.2
lower	1.4	1.5
GS300		
upper	.9	1.0
lower	1.2	1.3
GS400, GS430		
upper or lower	1.4	1.5
IS300		
upper	1.2	1.3
lower	1.4	1.5
LS400, LS430		
1990-94		
upper	1.2	1.3
lower	.9	1.0
1995-05		
upper or lower	1.4	1.5
SC300		
upper	.9	1.0
lower	1.2	1.3
SC400		
upper or lower	1.4	1.5
SC430		
upper	2.0	2.1
lower	1.7	1.8
Radiator Fan and/or Fan Motor, Replace (B)		
ES250	.7	1.0
ES300, ES330		
1992-98	2.5	2.8
1999-01	.7	1.0
2002-05	1.8	2.1
GS300, GS400, GS430,		
1999-05	1.7	1.9
IS300	1.4	1.5
LS430	2.0	2.2
SC300 1999-00	.7	1.0
SC400		
hydraulic motor	2.2	2.5
hydraulic pump	2.8	3.1
SC430	1.5	1.7
Radiator Fan Motor Relay, Replace (B)		
Exc. below	.2	.2
ES250	.3	.3
Temperature Gauge (Dash), Replace (B)		
ES300		
1992-98	1.3	1.3
1999-01	1.2	1.2
GS300, GS400, GS 430		
1994-97	1.3	1.3
1998	1.4	1.4
1999-05	1.2	1.2

	LABOR TIME	SEVERE SERVICE
LS400		
1990-94	1.3	1.3
1995-98	1.4	1.4
1999-00	1.2	1.2
SC300, SC400		
1992-98	1.3	1.3
1999-00	1.2	1.2
Water Control Valve, Replace (B)		
ES250	.9	1.1
ES300	1.3	1.5
GS300, GS400,		
GS430	1.2	1.4
LS400, LS430	1.2	1.4
SC300, SC400	1.3	1.5
Water Pump and/or Gasket, Replace (B)		
Includes: Refill with proper coolant mix.		
ES250	4.2	4.5
ES300, ES330	3.3	3.6
GS300, SC300		
1992-98	3.7	4.0
1999-00	3.6	3.9
2001-05	3.7	4.0
GS400, GS430	4.4	4.7
IS300	3.6	3.9
LS400		
1990-94	4.8	5.1
1995-98	4.2	4.7
1999-00	4.3	4.6
LS430	3.5	3.8
SC400, SC430		
1992-05	4.7	5.0

ENGINE
ASSEMBLY

Times shown are for OEM assemblies. Time to replace assemblies from aftermarket rebuilders may vary.

Engine Assy., R&I (B)
Does not include parts or component transfer.

	LABOR TIME	SEVERE SERVICE
ES250, ES300	14.6	14.6
GS300, GS400	16.2	16.2
LS400		
1990-94	12.3	12.3
1995-00	14.6	14.6
SC300	13.6	13.6
SC400	14.2	14.2

Engine Assy., R&R and Recondition (A)
Includes: Replacing rings, rod and main bearings, cylinder head reconditioning and engine tune-up.

	LABOR TIME	SEVERE SERVICE
ES250, ES300		
exc. below	46.2	47.1
2002-03	53.6	54.5
ES330	51.6	52.5
GS300	45.5	46.4
GS400	55.3	55.2

	LABOR TIME	SEVERE SERVICE
GS430	53.0	53.9
IS300	42.8	43.7
LS400	55.5	56.4
LS430		
2001-03	55.5	56.4
2004-05	53.4	54.3
SC300	43.9	44.8
SC400	55.3	56.2
SC430	53.2	54.1
Engine Assy. (Short Block), Replace (B)		
Assembly consists of cylinder block, piston assemblies, crankshaft, camshaft, timing chain and gears. Does not include cylinder heads. Operation Includes: R&R engine, transfer necessary parts and all necessary adjustments.		
ES250, ES300, ES330		
exc. below	20.2	21.1
2002-03	22.2	23.1
2004-05 ES330	20.9	21.8
GS300	18.6	19.5
GS400	21.3	22.2
GS430	20.3	21.2
IS300	17.4	18.3
LS400		
1990-94	23.2	24.1
1995-00	21.7	22.6
LS430	20.3	21.2
SC300	17.0	17.9
SC400, SC430	20.5	21.4
Engine Mounts, Replace (B)		
Front one side		
1990-97		
3.0L L6		
GS300	1.5	1.7
SC300	2.2	2.4
2.5L, 3.0L V6	.9	1.1
4.0L	1.3	1.4
1998-05		
ES300, ES330	.9	1.1
GS300	1.4	1.6
GS400, GS430		
LS400, LS430	1.3	1.4
IS300	1.7	1.9
SC300	2.2	2.4
SC400, SC430	1.3	1.4
Rear		
SC300	.7	.9

CYLINDER HEAD
Compression Test (B)

	LABOR TIME	SEVERE SERVICE
ES250	1.2	1.4
ES300		
1992-93	.8	1.0
1994-96	1.2	1.4
1997-01	1.8	2.0

LEX-8   ES250 : ES300 : ES330 : GS300 : GS400 : GS430

	LABOR TIME	SEVERE SERVICE
Compression Test (B)		
GS300	1.9	2.1
GS400	1.5	1.7
LS400		
1990-94	1.5	1.7
1995-00	1.2	1.4
SC300	2.3	2.5
SC400	1.5	1.7
Cylinder Head, Replace (B)		
Includes: Parts transfer, adjustment.		
ES250, ES300, ES330		
exc. below		
one side	13.3	13.8
both sides	20.5	21.0
2002-05		
one side	13.1	13.6
both sides	20.0	20.5
GS300	21.6	22.1
GS400, GS430		
one side	16.9	17.4
both sides	25.5	26.0
IS300	20.1	20.6
LS400		
1990-94		
one side	18.7	19.2
both sides	28.3	28.8
1995-00		
one side	18.2	18.7
both sides	27.8	28.3
LS430		
2001-05		
one side	14.3	14.8
both sides	21.6	22.1
SC300	24.6	25.1
SC400		
one side	18.3	18.8
both sides	23.7	24.2
SC430		
one side	13.7	14.2
both sides	20.1	20.6
Cylinder Head Gasket, Replace (B)		
ES250, ES300, ES330		
exc. below		
one side	8.6	9.1
both sides	11.2	11.7
2002-05		
one side	8.4	8.9
both sides	11.1	12.6
GS300	9.3	9.8
GS400		
one side	9.7	10.2
both sides	11.8	12.3
GS430		
one side	9.8	10.3
both sides	11.5	12.0
IS300	10.9	11.4
LS400		
1990-94		
one side	11.2	11.7
both sides	13.3	13.8

	LABOR TIME	SEVERE SERVICE
1995-00		
one side	9.7	10.2
both sides	11.8	12.3
LS430		
one side	7.6	8.1
both sides	8.8	9.3
SC300	10.1	10.6
SC400		
one side	10.7	11.2
both sides	12.8	13.3
SC430		
one side	7.2	7.7
both sides	9.7	10.2
Valve/Cam Cover Gasket, Replace or Reseal (B)		
ES250, ES300, ES330		
exc. below		
right side	2.6	2.8
left side	.7	.9
2002-05		
right side	2.1	2.4
left side	1.4	1.5
GS300, SC300		
1992-98	1.5	1.7
1999-00 SC300	1.4	1.6
1999-05 GS300	1.3	1.5
GS400, GS430		
right side	2.0	2.2
left side	.8	.9
IS300	2.0	2.2
LS400		
1990-94		
right side	3.0	3.2
left side	1.5	1.7
1995-00		
right side	2.2	2.4
left side	.7	.9
LS430		
right side	.9	1.0
left side	.8	.9
SC400		
right side	2.4	2.6
left side	1.2	1.4
SC430		
right side	.6	.7
left side	.8	.9
Valve Clearance, Adjust (B)		
ES250	4.9	5.2
ES300, ES330	5.6	6.2
GS300, SC300	4.6	5.1
GS400	5.8	6.1
GS430, LS430	6.2	6.8
IS300	4.6	5.1
LS400		
1990-94	7.0	7.3
1995-00	5.8	6.1

	LABOR TIME	SEVERE SERVICE
SC400		
1992-98	5.8	6.1
1999-00	5.3	5.6
SC430	4.1	4.6
Valve Job (A)		
ES250, ES300		
exc. below	27.7	28.6
2002-03	29.0	29.9
ES330	24.9	25.8
GS300	25.9	26.8
GS400	34.4	35.3
GS430	31.6	32.5
IS300	26.3	27.2
LS400		
1990-94	35.0	35.9
1995-00	34.3	35.2
LS430	30.8	31.7
SC300	28.9	29.8
SC400	36.2	37.1
SC430	28.0	28.9
Valve Lifters, Replace (B)		
1990-98		
3.0L L6	8.2	8.7
2.5L, 3.0L V6	9.1	9.6
4.0L	11.7	12.2
1999-05		
ES300, ES330	8.5	9.0
GS300	8.3	8.8
GS400, GS430	10.2	10.7
LS400	10.1	10.6
LS430	9.0	9.5
IS300	7.7	8.2
SC300	8.3	8.8
SC400	12.2	12.7
SC430	10.5	11.0
Valve Springs and/or Oil Seals, Replace (B)		
1990-98		
3.0L L6	15.2	15.7
2.5L, 3.0L V6	14.2	14.7
4.0L	18.2	18.7
1999-05		
ES300	15.6	16.1
ES330	13.2	13.7
GS300		
1993-00	14.9	15.4
2001-05	14.1	14.6
GS400, LS400	16.8	17.3
GS430	15.7	16.2
IS300	14.4	14.9
LS430	14.8	15.3
SC300	15.4	15.9
SC400	18.4	18.9
SC430	14.3	14.8

IS300 : LS400 : LS430 : SC300 : SC400 : SC430 **LEX-9**

	LABOR TIME	SEVERE SERVICE
CAMSHAFT		
Camshaft, Replace (A)		
ES250	9.2	*10.2*
ES300, ES330		
1992-98	9.6	*10.1*
1999-01	9.9	*10.4*
2002-05	8.6	*9.1*
GS300, SC300		
1992-97	8.6	*9.1*
1998-05	8.6	*9.1*
GS400, GS430	14.4	*14.9*
IS300	8.4	*8.9*
LS400		
1990-94	12.3	*12.8*
1995-98	10.5	*11.0*
1999-00	11.3	*11.8*
LS430	9.3	*9.8*
SC400		
1992-98	12.3	*12.8*
1999-00	12.9	*13.4*
SC430	10.3	*10.8*
Camshaft Timing Gear or Sprocket, Replace (B)		
ES250, ES300, ES330		
1990-98	3.0	*3.2*
1999-05	2.7	*3.0*
GS300, SC300		
1992-97	1.5	*1.7*
1998 SC300	1.5	*1.7*
1998 GS300	3.0	*3.2*
1999-05	1.3	*1.6*
GS400, GS430, LS400		
1990-94	3.8	*4.0*
1995-98	2.8	*3.0*
1999-05	2.7	*3.0*
IS300	3.0	*3.3*
LS430	3.7	*4.0*
SC400	2.6	*2.8*
SC430	4.4	*4.7*
Crankshaft Timing Gear or Sprocket, Replace (B)		
ES250, ES300, ES330		
exc. below	2.9	*3.2*
2002-05	3.6	*3.9*
GS300, SC300		
1992-97	3.3	*3.7*
1998 SC300	3.3	*3.7*
1999-00 SC300	3.2	*3.6*
1998 GS300	4.2	*4.5*
1999-05 GS300	3.0	*3.3*
GS400, GS430	4.0	*4.3*
IS300	2.9	*3.2*
LS400		
1990-94	4.6	*5.1*
1995-98	3.8	*4.2*
1999-00	4.5	*4.9*
LS430	3.3	*3.6*

	LABOR TIME	SEVERE SERVICE
SC400		
1992-98	5.2	*5.6*
1999-00	4.8	*5.4*
SC430	4.0	*4.3*
Timing Belt, Replace (B)		
ES250	3.7	*4.0*
ES300, ES330	2.9	*3.2*
2002-05	3.6	*3.9*
GS300		
1994-97	3.2	*3.5*
1998	4.2	*4.5*
1999-05	3.0	*3.3*
GS400, GS430	4.0	*4.3*
IS300	2.9	*3.2*
LS400		
1990-94	4.6	*4.9*
1995-98	3.8	*4.1*
1999-00	4.5	*4.8*
LS430, SC300	3.1	*3.4*
SC400	4.8	*5.1*
SC430	3.8	*4.1*
Timing Belt Tensioner, Replace (B)		
ES250, ES300, ES330		
exc. below	.5	*.7*
2002-05	3.4	*3.7*
GS300, SC300		
1992-98	.9	*1.0*
1999-05	.8	*.9*
GS400, GS430	.8	*.9*
IS300	3.2	*3.5*
LS400, LS430		
1990-05	.9	*1.2*
SC400, SC430		
1992-98	1.2	*1.5*
1999-00	.9	*1.0*
Timing Cover and/or Gasket, Replace (B)		
ES250	2.5	*2.8*
ES300	2.6	*2.9*
ES330	2.3	*2.6*
GS300		
1992-97	2.6	*2.9*
1998	3.6	*3.9*
1999-04	2.8	*3.1*
GS400, GS430	3.4	*3.7*
IS300	2.3	*2.6*
LS400		
1990-94	4.2	*4.5*
1995-98	3.3	*3.6*
1999-00	3.8	*4.1*
LS430	2.7	*3.0*
SC300		
1992-00	2.6	*2.9*
SC400	4.6	*4.9*
SC430	3.3	*3.6*

	LABOR TIME	SEVERE SERVICE
CRANK & PISTONS		
Connecting Rod Bearings, Replace (A)		
Includes: Check bearing oil clearance.		
ES250, ES300, ES330		
exc. below	7.6	*8.1*
2002-05	16.5	*17.0*
GS300	15.4	*15.9*
1998-05	15.2	*15.7*
GS400, GS430, LS400		
1990-94	15.3	*15.8*
1995-98	13.1	*13.6*
1999-00	13.7	*14.2*
2001-05	13.7	*14.2*
LS430	14.7	*15.2*
IS300, SC300	19.1	*19.6*
SC400	13.8	*14.3*
SC430	15.1	*15.6*
Crankshaft & Main Bearings, Replace (A)		
Includes: Engine R&R.		
ES250	21.2	*21.7*
ES300, ES330		
exc. below	23.2	*23.7*
2002-05	23.4	*23.9*
GS300	21.1	*21.6*
GS400, GS430	23.5	*24.0*
IS300	20.1	*20.6*
LS400		
1990-94	25.0	*25.5*
1995-00	24.6	*27.1*
LS430	23.7	*24.2*
SC300	19.3	*19.8*
SC400	23.8	*24.3*
SC430	24.1	*24.6*
Crankshaft Front Oil Seal, Replace (B)		
ES250	5.4	*5.6*
ES300, ES330		
1992-98	6.2	*6.4*
1999-01	3.0	*3.2*
2002-05	4.0	*4.4*
GS300		
1992-97	3.6	*3.8*
1998	4.2	*4.4*
1999-04	3.2	*3.6*
GS400, GS430	4.4	*4.9*
IS300	3.2	*3.6*
LS400		
1990-94	5.0	*5.2*
1995-00	4.2	*4.4*
LS430	3.5	*3.9*
SC300		
1992-97	3.6	*3.8*
1998-00	3.0	*3.2*
SC400	5.0	*5.2*
SC430	4.3	*4.7*

LEX-10   ES250 : ES300 : ES330 : GS300 : GS400 : GS430

	LABOR TIME	SEVERE SERVICE
Crankshaft Pulley, Replace (B)		
ES250, ES300, ES330		
1990-05	1.2	1.4
GS300, SC300	2.4	2.6
GS400, GS430	1.8	2.1
IS300	2.1	2.4
LS400	1.9	2.1
LS430, SC400	2.4	2.6
SC430	2.1	2.4

Pistons or Connecting Rods, Replace (A)
Includes: Ridge reaming, cylinder wall deglazing, installing new rings and rod bearings, engine tune-up.

ES250, ES300	18.5	19.0
2002-03	25.7	26.2
ES330	24.2	24.7
GS300	22.2	22.7
GS400, GS430	23.8	24.3
IS300	20.7	21.2
LS400, LS430	23.8	24.3
SC300	21.5	22.0
SC400, SC430	23.5	24.0

Rear Main Oil Seal, Replace (B)
Includes: R&R trans. when necessary.

ES250		
AT	9.8	10.3
MT	8.4	8.9
ES300	9.6	10.1
2001-03	13.6	14.1
ES330	13.0	13.5
GS300, SC300		
AT	6.9	7.4
MT	5.7	6.2
GS400, GS430, LS400		
1990-94	7.4	7.9
1995-98	6.2	6.7
1999-05	6.1	6.6
LS430	4.8	5.3
IS300		
AT	5.1	5.6
MT	8.7	9.2
SC400	7.6	8.1
SC430	6.2	6.7

Rings, Replace (A)
Includes: Engine, cylinder head(s) and oil pan R&I. Ridge reaming, cylinder wall deglazing, clean carbon.

ES250, ES300		
1990-97	16.2	16.7
1998-03	24.7	25.2
ES330	23.5	24.0
GS300	20.7	21.2
1998-05	19.9	20.4
GS400, GS430	22.4	22.9
IS300	19.8	20.3
LS400		
1990-94	24.2	24.7
1995-00	23.4	23.9
LS430	22.5	23.0

	LABOR TIME	SEVERE SERVICE
SC300	18.7	19.2
SC400	23.1	23.6
SC430	22.6	23.1

ENGINE LUBRICATION

Engine Oil Pressure Switch (Sending Unit), Replace (B)

ES250	.5	.5
ES300	.6	.7
ES330	.3	.3
GS300	2.0	2.2
GS400, GS430	.6	.7
IS300	.6	.7
LS400, LS430, SC400	.8	.9
SC300	1.3	1.3
SC430	.5	.5

Oil Pan and/or Gasket, Replace (B)

ES250	4.1	4.6
ES300		
1994-01		
No.1	5.1	5.6
No.2	3.2	3.7
2002-03		
No.1	4.1	4.6
No.2	3.0	3.5
ES330		
No.1	3.8	4.3
No.2	2.2	2.7
GS300		
1993-97		
No.1	12.4	12.9
No.2	3.3	3.8
1998-05		
No.1	12.0	12.5
No.2	3.4	3.9
GS400, GS430		
No.1	11.7	12.2
No.2	3.0	3.5
IS300		
No.1	13.0	13.5
No.2	3.4	3.9
LS400, LS430		
1990-94		
No.1	13.1	13.6
No.2	2.5	3.0
1995-00		
No.1	12.3	12.8
No.2	3.0	3.5
2001-05		
No.1	12.5	13.0
No.2	2.7	3.2
SC300		
No.1	9.7	10.2
No.2	3.0	3.5
SC400		
No.1	11.5	12.0
No.2	2.6	3.1
SC430		
No.1	13.1	13.6
No.2	1.5	2.0

	LABOR TIME	SEVERE SERVICE
Oil Pump, Replace (B)		
ES250	5.4	5.9
ES300, ES330		
1992-98	6.0	6.5
1999-05	7.5	8.0
GS300	13.8	14.3
GS400, GS430	14.3	14.8
IS300	13.4	13.9
LS400		
1990-94	6.2	6.7
1995-00	14.4	14.9
LS430	14.3	14.8
SC300	11.2	11.7
SC400	14.1	14.6
SC430	13.7	14.2
Recondition pump add	.3	.3

CLUTCH

Bleed Clutch Hydraulic System (B)
1990-05	.6	.7

Clutch Pedal Free Play, Adjust (B)
1990-05	.5	.5

Clutch Assy., Replace (B)
ES250	6.0	6.5
1992-93 ES300	7.2	7.7
IS300	8.4	8.9
SC300	5.3	5.8

Replace
pilot bearing add	.2	.2
rear main oil seal add	.3	.3
release bearing add	.2	.2

Clutch Master Cylinder, R&R and Recondition (A)
Includes: System bleeding.

ES250	1.7	2.0
1992-93 ES300	1.7	2.0
IS300	1.4	1.7
SC300		
1992-97	1.8	2.1
1998-00	1.6	1.9

Clutch Master Cylinder, Replace (B)
Includes: System bleeding.

ES250	1.4	1.6
1992-93 ES300	1.4	1.6
IS300	1.1	1.3
SC300	1.4	1.6

Clutch Slave Cylinder, R&R and Recondition (A)
Includes: System bleeding.

ES250	1.4	1.7
1992-93 ES300	1.4	1.7
IS300	1.1	1.3
SC300		
1992-97	1.6	1.9
1998-00	1.3	1.6

Clutch Slave Cylinder, Replace (B)
Includes: System bleeding.

ES250	.9	1.1
1992-93 ES300	.9	1.1

IS300 : LS400 : LS430 : SC300 : SC400 : SC430 **LEX-11**

	LABOR TIME	SEVERE SERVICE
IS300	1.0	1.2
SC300	1.2	1.4

Flywheel, Replace (B)
- ES250 .... 6.2   *6.7*
- 1992-93 ES300 ...... 7.5   *8.0*
- IS300 .... 8.4   *8.9*
- SC300 .... 5.8   *6.0*

Replace
- pilot bearing add 2   *.2*
- rear main oil seal add 3   *.3*
- release bearing add 2   *.2*

MANUAL TRANSAXLE

Differential Case, Replace (B)
- ES250 .... 10.3   *10.8*
- 1992-93 ES300 ..... 11.2   *11.7*

Differential Oil Seal, Replace (B)
- ES250 one 1.9   *2.1*
- 1992-93 ES300 one .. 2.2   *2.4*

Transaxle Assy. R&R and Reconditon (A)
- ES250 .... 14.2   *15.1*
- 1992-93 ES300 ..... 14.8   *15.7*

Transaxle Assy., R&I (B)
Includes: Reset toe.
- ES250 .... 5.8   *6.0*
- 1992-93 ES300 ...... 7.0   *7.2*

MANUAL TRANSMISSION

Extension Housing and/or Gasket, Replace (B)
- IS300 .... 9.1   *9.6*
- SC300 .... 6.0   *6.5*

Transmission Assy., R&R and Recondition (A)
- IS300 .... 14.6   *15.5*
- SC300 .... 9.9   *10.8*

Transmission Assy., R&R and Replace (B)
- IS300 .... 8.3   *8.8*
- SC300 .... 5.1   *5.6*

Transmission Mount, Replace (B)
- IS300 8   *.9*
- SC300 7   *.8*

AUTOMATIC TRANSAXLE

SERVICE TRANSAXLE INSTALLED

Check Unit For Oil Leaks (C)
- All Models 9   *.9*

Drain & Refill Unit (B)
- All Models 1.5   *1.5*

Oil Pressure Check (B)
- All Models 2.1   *2.1*

Axle Shaft Oil Seal, Replace (B)
- ES250 .... 2.0   *2.2*
- ES300, ES330
  - 1992-97 .... 2.3   *2.5*
  - 1998-05 .... 3.0   *3.2*

Oil Pan and/or Gasket, Replace (B)
- All Models 2.3   *2.3*

	LABOR TIME	SEVERE SERVICE

Speed Sensor, Replace (B)
Front
- ES250 .... 1.4   *1.4*
- ES300, ES330 7   *.7*

Valve Body, Replace (B)
Includes: R&R pan.
- ES250 .... 2.8   *3.2*
- ES300, ES330 .... 3.0   *3.4*

SERVICE TRANSAXLE REMOVED

Differential Pinion and/or Side Gear, Replace (B)
- ES250 .... 9.9   *10.4*
- ES300, ES330
  - 1992-97 .... 15.9   *16.8*
  - 2002-05 .... 13.7   *14.6*

Torque Converter, Replace (B)
- ES250 .... 6.2   *6.7*
- ES300, ES330
  - 1992-97 .... 8.7   *9.2*
  - 2002-05 .... 11.7   *12.2*

Transaxle Assy., R&R and Recondition (A)
- ES250 .... 19.2   *21.3*
- ES300, ES330 ..... 20.5   *21.4*

Transaxle Assy., R&R and Replace (B)
- ES250 .... 6.0   *6.5*
- ES300, ES330
  - 1992-97 .... 8.4   *8.9*
  - 2002-05 .... 11.3   *11.8*

Replace flexplate add 6   *.6*

AUTOMATIC TRANSMISSION

SERVICE TRANSMISSION INSTALLED

Check Unit for Oil Leaks (C)
- 1990-05 9   *.9*

Drain and Refill Unit (B)
- 1990-05 exc. below .. 1.3   *1.3*
- ES300, ES330 .... 1.6   *1.6*

Oil Pressure Check (B)
- 1990-05 .... 1.9   *1.9*

Extension Housing Gasket, Replace (B)
- GS300, GS400, GS430 .. 3.4   *3.7*
- IS300 .... 3.3   *3.6*
- LS400, SC400 .... 3.7   *4.0*
- LS430 .... 3.8   *4.1*
- SC300, SC430 .... 4.4   *4.7*

Extension Housing Oil Seal, Replace (B)
- GS300, GS400, GS430 .. 3.4   *3.6*
- IS300 .... 2.7   *2.9*
- LS400
  - 1990-94 .... 3.0   *3.2*
  - 1995-00 .... 3.7   *3.9*

	LABOR TIME	SEVERE SERVICE
LS430	3.8	*4.0*
SC300	2.7	*2.9*
SC400	3.1	*3.3*
SC430	4.0	*4.2*

Oil Pan and/or Gasket, Replace (B)
- Exc. below 1.9   *2.1*
- IS300, SC430 .... 2.4   *2.6*

Valve Body Assy., Replace (B)
- 1990-05 exc. below .. 2.4   *2.7*
- IS300, SC430 .... 2.7   *3.0*

Vehicle Speed Sensor, Replace (B)
Front
- ES300, ES330
  - 1994-96 3   *.3*
  - 1997-05 .... 1.3   *1.4*
- GS300, GS400, GS430, IS300 6   *.6*
- LS400
  - 1990-94 .... 1.2   *1.2*
  - 1995-00 .... 1.4   *1.4*
- SC300, SC400, SC430 7   *.8*
Rear
- 1990-05 7   *.8*

SERVICE TRANSMISSION REMOVED

Transmission R&R included unless otherwise noted.

Flywheel (Flexplate), Replace (B)
- ES330 .... 12.6   *13.1*
- GS300 .... 6.8   *7.3*
- GS400, GS430 .... 5.9   *6.4*
- IS300 .... 5.1   *5.6*
- LS400
  - 1990-94 .... 7.2   *7.7*
  - 1995-98 .... 6.9   *7.4*
  - 1999-03 .... 5.8   *6.3*
- LS430 .... 4.8   *5.3*
- SC300, SC400
  - 1992-98 .... 6.7   *7.2*
  - 1999-00 SC300 .... 6.4   *6.9*
  - 1999-00 SC400 .... 6.5   *7.0*
- SC430 .... 6.2   *6.7*

Front Pump Oil Seal, Replace (B)
- GS300, GS400, GS430
  - 1994-98 .... 6.7   *7.2*
  - 1999-05 .... 6.5   *7.0*
- IS300 .... 5.7   *6.2*
- LS430 .... 6.1   *6.6*
- S400
  - 1990-94 .... 7.2   *7.7*
  - 1995-98 .... 6.9   *7.4*
  - 1999-00 .... 6.0   *6.5*
- SC300, SC400
  - 1992-98 .... 6.7   *7.2*
  - 1999-00 .... 5.8   *6.3*
- SC430 .... 7.3   *7.8*

LEX-12 ES250 : ES300 : ES330 : GS300 : GS400 : GS430

	LABOR TIME	SEVERE SERVICE
Torque Converter, Replace (B)		
GS300, GS400, GS430	6.2	6.7
IS300	5.4	5.9
LS400		
1990-94	7.1	7.6
1995-98	6.8	7.3
1999-00	6.4	6.9
LS430	4.7	5.2
SC300, SC400, SC430	6.1	6.6
Transmission Assy., R&R and Recondition (A)		
GS300, GS400, GS430	13.6	14.5
IS300	16.6	17.5
LS400		
1990-94	15.0	15.9
1995-98	17.8	18.7
1999-00	19.3	20.2
LS430	15.7	16.6
SC300, SC400	15.3	16.2
SC430	13.4	14.3
Transmission Assy., R&R and Replace (B)		
GS300, GS400	6.8	7.0
LS400		
1990-94	6.9	7.1
1995-00	6.8	7.0
SC300, SC400	6.8	7.0

SHIFT LINKAGE
AUTOMATIC TRANSMISSION

	LABOR TIME	SEVERE SERVICE
Gear Selector Lever, Replace (B)		
ES250, ES300	.7	.7
GS300, GS400, GS430		
1994-05	1.2	1.2
IS300	.9	.9
LS400	1.7	1.7
LS430	2.6	2.6
SC300, SC400, SC430	2.0	2.0
Shift Control Cable, Replace (B)		
ES250 both	1.9	2.1
ES300, ES330 one or both		
1992-97	1.9	2.1
1998-05	5.9	6.1
IS300	.5	.7
Selector Lever Indicator Bulb, Replace (B)		
Exc. below	.5	.5
ES300, ES330		
1997-05	.9	.9
LS400		
1990-94	.7	.7
1995-00	1.4	1.4
LS430	2.2	2.2

	LABOR TIME	SEVERE SERVICE
Throttle Valve Cable, Replace (B)		
ES250	3.0	3.2
ES300	3.6	3.8
GS300, SC300	2.5	2.7
LS400		
1990-94	7.2	7.4
1995-00	6.8	7.0
SC400	6.9	7.1
Throttle Valve Cable or Rod, Adjust (B)		
1990-05	.5	.6

MANUAL TRANSMISSION

	LABOR TIME	SEVERE SERVICE
Gearshift Lever, Replace (B)		
ES250	1.3	1.3
1992-93 ES300	.8	.8
IS300, SC300	.7	.7

DRIVELINE
DRIVESHAFT

	LABOR TIME	SEVERE SERVICE
Center Support Bearing, Replace (B)		
GS300, GS400, GS430	2.9	3.1
IS300	2.4	2.6
LS400, LS430		
1990-94	2.6	2.8
1995-05	3.0	3.2
SC300, SC400	2.7	2.9
SC430	3.5	3.7
Driveshaft, Replace (B)		
GS300, GS400, GS430	2.5	2.9
LS400, LS430		
1990-94	2.4	2.6
1995-05	2.5	2.9
IS300	2.1	2.3
SC300	2.0	2.2
SC400	2.5	2.7
SC430	2.9	3.2

DRIVE AXLE

	LABOR TIME	SEVERE SERVICE
Differential, Drain & Refill (B)		
1990-05	.7	.7
Axle Shaft Bearings, Replace (B)		
ES300		
one side	1.6	1.9
both sides	2.4	2.7
GS300, GS400, GS430		
1994-98		
one side	3.1	3.4
both sides	4.6	4.9
1999-05		
one side	3.5	3.8
both sides	5.3	5.6
IS300		
one side	3.5	3.8
both sides	5.0	5.3

	LABOR TIME	SEVERE SERVICE
LS400		
1990-94		
one side	3.1	3.4
both sides	5.0	5.3
1995-00		
one side	3.6	3.9
both sides	5.8	6.1
SC300, SC400		
one side	3.6	3.9
both sides	5.5	5.8
SC430		
one side	2.7	3.0
both sides	4.3	4.6
Axle Shaft Oil Seal, Replace (B)		
GS300, GS400, GS430		
1994-98		
one side	3.1	3.4
both sides	4.6	4.9
1999-05		
one side	3.5	3.8
both sides	5.3	5.6
LS400		
1990-94		
one side	2.7	3.0
both sides	4.2	4.5
1995-00		
one side	3.6	3.9
both sides	5.8	6.1
SC300, SC400		
one side	3.6	3.9
both sides	5.5	5.8
Differential Carrier Assy., R&R and Recondition (A)		
GS300, GS400, GS430		
1994-98	9.2	10.1
1999-05	9.7	10.6
1999-00 LS400	9.6	10.5
LS430	9.7	10.6
IS300	8.4	9.3
SC300, SC400	9.3	9.2
SC430	10.4	11.3
Differential Carrier, R&I (B)		
GS300, GS400	3.1	3.6
LS400		
1990-94	2.6	3.1
1995-00	3.1	3.6
SC300, SC400	3.0	3.5
Pinion Shaft Oil Seal, Replace (B)		
GS300, GS400, GS430	5.4	5.6
1990-94 LS400	4.5	4.7
LS430	5.4	5.6
IS300	2.8	3.0
SC300, SC400	5.2	5.4
SC430	8.8	9.0

IS300 : LS400 : LS430 : SC300 : SC400 : SC430 LEX-13

	LABOR TIME	SEVERE SERVICE
Rear Wheel Hub Assy., Replace (B)		
ES300		
1992-96		
one side	1.2	1.5
both sides	1.6	1.8
1997-03		
one side	1.6	1.8
both sides	3.0	3.2
ES330		
one side	2.1	2.3
both sides	3.0	3.2
GS300, GS400		
one side	3.1	3.3
both sides	4.6	4.8
LS400, LS430		
1990-94		
one side	2.7	2.9
both sides	4.2	4.4
1995-05		
one side	3.1	3.3
both sides	5.4	5.6
SC300, SC400		
one side	3.6	3.8
both sides	5.4	5.6
Ring Gear & Pinion Set, Replace (B)		
GS300, GS400, GS430		
1993-97	8.6	9.1
1998-05	8.7	9.2
IS300	7.5	8.0
LS400		
1990-91	7.2	7.7
1992-00	8.8	9.3
LS430	8.7	9.2
SC300, SC400	8.4	8.9
SC430	10.3	10.8
Side Gear Shaft Oil Seal, Replace (B)		
GS300, GS400, GS430	1.7	1.9
IS300	1.7	1.9
LS400		
1990-94	3.0	3.2
1995-00	3.4	3.6
LS430	1.9	2.1
SC300, SC400	1.7	1.9
SC430	3.1	3.3

HALFSHAFTS
CV Joint and Boot, Replace or Recondition (B)

	LABOR TIME	SEVERE SERVICE
Front		
ES250		
one side	3.1	3.4
both sides	4.3	4.6
ES300, ES330		
one side	3.3	3.6
both sides	5.3	5.6

	LABOR TIME	SEVERE SERVICE
Rear		
GS300, GS400, GS430		
one side	1.9	2.1
both sides	3.5	3.7
IS300		
one side	2.5	2.7
both sides	3.8	4.0
LS400, LS430		
1990-94		
one side	2.0	2.3
both sides	3.4	3.7
1995-05		
one side	2.7	2.9
both sides	4.5	4.7
SC300, SC400		
one side	2.0	2.3
both sides	3.4	3.7
SC430		
one side	3.0	3.2
both sides	5.7	6.3
w/air suspension rear GS300, GS400 add each side	.2	.2
Halfshaft, R&R or Replace (B)		
Front		
ES250		
one	2.7	2.9
both	3.6	3.8
ES300		
one	1.7	1.9
both	2.4	2.6
ES330		
one side	3.0	3.2
both sides	4.1	4.3
Rear		
GS300, GS400, GS430		
one side	1.2	1.3
both sides	1.7	2.0
IS300		
one side	1.8	2.0
both sides	2.4	2.6
LS400, LS430		
1990-94		
one side	1.2	1.3
both side	1.8	2.0
1995-05		
one side	1.3	1.4
both sides	2.0	2.2
SC300, SC400		
one side	1.3	1.4
both sides	1.8	2.0
SC430		
one side	2.4	2.7
both sides	4.6	5.1
w/air suspension rear add 90-94 LS400	.2	.2
Recondition shaft add each GS300, GS400	.5	.7

	LABOR TIME	SEVERE SERVICE
BRAKES		
ANTI-LOCK		
Diagnose Anti-Lock Brake System Component Each (A)		
1990-05	.5	.5
Retrieve Fault Codes (A)		
1990-05	.3	.3
Actuator Assy., Replace (B)		
ES250	1.9	1.9
ES300, ES330	2.1	2.1
GS300, GS400		
1994-97 GS300		
w/TRAC	3.2	3.2
w/o TRAC	1.6	1.6
1998 GS300	3.7	3.7
1999-04 GS300, GS400, GS430	3.3	3.3
IS300	2.7	2.8
LS400		
1990-94	1.8	1.8
1995-00	1.8	1.8
LS430	2.6	2.6
SC300, SC400		
w/TRAC	3.4	3.4
w/o TRAC	1.4	1.4
w/TRAC add		
90-94 LS400	.2	.2
95-00 LS400	.7	.7
Control Module, Replace (B)		
ES300, ES330		
1992-96	1.1	1.2
1997-01	.7	.7
2002-05	1.1	1.2
GS300, GS400, GS430		
1994-97	.9	.9
1998-05	1.4	1.4
LS400		
1990-94	.9	.9
1995-00	1.3	1.3
LS430	.8	.9
SC300, SC400		
1992-98	1.1	1.2
1999-00	.8	.8
SC430	1.4	1.5
Wheel Speed Sensor, Replace (B)		
Front		
ES250		
one	.8	.8
both	1.3	1.5
ES300, SC300, SC400		
one	.9	1.0
both	1.5	1.7
ES330, SC430		
one	1.1	1.2
both	1.5	1.7
GS300, GS400, GS430		
one	.9	1.0
both	1.4	1.6

LEX-14   ES250 : ES300 : ES330 : GS300 : GS400 : GS430

	LABOR TIME	SEVERE SERVICE
Wheel Speed Sensor, Replace (B)		
IS300		
one	1.2	1.4
both	1.7	1.9
LS400		
1990-94		
one	1.2	1.4
both	1.7	1.7
1995-00		
one	.8	.8
both	1.4	1.6
LS430		
one side	1.8	2.1
both sides	2.4	2.7
Rear		
ES250		
one	.9	.9
both	1.5	1.5
ES300, ES330		
1992-97		
one	.8	.8
both	1.3	1.3
1998-05		
one side	2.8	2.8
both sides	3.3	3.3
GS300, GS400, GS430		
one	.8	.8
both	1.1	1.1
IS300		
one side	1.2	1.2
both sides	1.4	1.4
LS400		
one	.7	.7
both	1.2	1.2
LS430		
one	1.5	1.5
both	1.8	1.8
SC300, SC400		
one	.7	.7
both	1.2	1.2
SC430		
one side	1.2	1.2
both sides	1.5	1.5

SYSTEM

Bleed Brakes (B)
Includes: Add fluid.
One circuit	.8	.8
Both circuits	1.2	1.2

Brake System, Flush and Refill (B)
1990-05	1.4	1.4

Brake Hose (Flexible), Replace (B)
Includes: System bleeding.
Front or rear one		
exc. below	.9	1.0
ES300, ES330		
2002-05	1.4	1.5
SC430		
2002-05	1.5	1.7
each addl. add	.3	.4

	LABOR TIME	SEVERE SERVICE
Master Cylinder, R&R and Rebuild (B)		
Includes: System bleeding.		
ES250	1.9	2.2
ES300, ES330		
1992-98	1.9	2.2
1999-05	1.8	2.1
GS300, GS400, GS430	2.0	2.3
IS300	1.8	2.1
LS400, LS430		
1990-94	1.8	2.1
1995-05	1.7	1.9
SC300, SC400, SC430		
1992-98	1.9	2.2
1999-00	1.7	2.0
2001-05	4.7	5.0
w/TRAC GS300, GS400 add	.3	.3

Master Cylinder, Replace (B)
Includes: System bleeding.
ES250	1.6	1.8
ES300, ES330	1.4	1.6
GS300, GS400, GS430		
1994-97	1.6	1.8
1998-05	2.0	2.2
IS300	1.7	1.9
LS400, LS430		
1990-94	1.5	1.7
1995-05	1.4	1.6
SC300, SC400	1.4	1.6
SC430	4.4	4.9
w/TRAC GS300, GS400 add	.3	.3

Power Booster Unit, Replace (B)
ES250	3.4	3.6
ES300	2.9	3.2
2002-03	2.1	2.4
GS300, GS400	3.0	3.2
IS300	2.3	2.6
LS400		
1990-94	2.6	2.9
1995-00	2.3	2.6
LS430	2.3	2.6
SC300, SC400, SC430	2.6	2.9
w/TRAC add		
GS300, GS400	.7	.7
90-94 LS400	.7	.7
95-00 LS400	.5	.5

Vacuum Check Valve, Replace (B)
1990-05	.6	.6

SERVICE BRAKES

Brake Pads, Replace (B)
1990-05 exc. below		
front or rear	1.0	1.1
four wheels	1.7	2.0

	LABOR TIME	SEVERE SERVICE
IS300		
front	1.4	1.6
rear	1.6	1.8
four wheels	2.7	3.0
LS400, LS430		
front	1.4	1.6
rear	1.6	1.8
four wheels	2.5	2.9

COMBINATION ADD-ONS
Replace
brake hose add	.3	.3
caliper add	.2	.2
disc brake rotor add	.2	.2
Resurface disc rotor add	.5	.5

Brake Proportioning Valve, Replace (B)
ES250	2.0	2.2
1997-98 ES300	1.6	1.8
1990-94 LS400	2.0	2.2

Caliper Assy., R&R and Recondition (B)
Includes: System bleeding.
1990-05 one front or rear	1.4	1.6
w/LS400 add	.1	.1

Caliper Assy., Replace (B)
Includes: System bleeding.
1990-05 one front or rear exc. below	1.2	1.4
ES300, ES330	1.9	2.1
w/LS400 add	.5	.5

Disc Brake Rotor, Replace (B)
Front or rear		
one	.8	.9
two	1.4	1.6

PARKING BRAKE
Parking Brake Apply Actuator, Replace (B)
ES250	.7	.7
ES300		
1992-93	1.2	1.2
1994-03	1.6	1.6
ES330	1.1	1.1
GS300, GS400, GS430	1.8	1.8
IS300	.9	.9
LS400	1.7	1.7
LS430	1.3	1.3
SC300, SC400, SC430	1.4	1.3

Parking Brake Cable, Adjust (C)
1990-05	.7	.9

Parking Brake Cable, Replace (B)
Front		
ES250	.9	1.1

IS300 : LS400 : LS430 : SC300 : SC400 : SC430   **LEX-15**

	LABOR TIME	SEVERE SERVICE
ES300, ES330		
1992-96	1.3	*1.5*
1997-05	1.4	*1.6*
GS300, GS400, GS430	2.2	*2.4*
IS300	1.3	*1.4*
LS400		
1990-94	4.2	*4.4*
1995-00	1.8	*2.0*
LS430	4.4	*4.6*
SC300, SC400	1.6	*1.8*
SC430	3.6	*4.1*
Rear		
ES250	2.0	*2.2*
ES300, ES330	2.7	*3.0*
2002-05	1.1	*1.3*
GS300, GS400, GS430, IS300	1.9	*2.1*
LS400		
1990-94	4.4	*4.6*
1995-00	3.7	*3.9*
LS430	4.3	*4.5*
SC300, SC400	2.0	*2.2*
SC430	4.6	*4.8*

TRACTION CONTROL
ABS & TRAC Control Module, Replace (B)

ES250	.7	*.7*
ES300, ES330		
1992-96	1.1	*1.3*
1997-01	.6	*.7*
2002-05	1.1	*1.3*
GS300, GS400, GS430		
1994-05	1.2	*1.2*
LS400		
1990-00	1.3	*1.3*
LS430, SC300, SC400	.8	*.9*
SC430	1.4	*1.4*

Accumulator Assy., Replace (B)

GS300	1.5	*1.5*
1990-94 LS400	1.7	*1.7*
SC300, SC400	1.3	*1.3*

Actuator Assy., Replace (B)

1997-01 ES300	2.1	*2.1*
GS300, GS400		
1994-97	1.2	*1.2*
1998-02	3.1	*3.1*
LS400		
w/Trac	2.8	*2.8*
w/o Trac	1.8	*1.8*
SC300, SC400		
1992-98	3.2	*3.2*
1999-00		
w/Trac	3.4	*3.4*
w/o Trac	1.4	*1.4*

Main Relay, Replace (B)

1997-01 ES300	.2	*.2*
GS300, GS400, GS430	.3	*.3*
1990-94 LS400	.3	*.3*
SC300, SC400	.5	*.5*

Motor Relay, Replace (B)

1997-01 ES300	.2	*.2*
GS300	.2	*.2*
LS400		
1990-94	.5	*.5*
1995-00	.3	*.3*
SC300, SC400	.5	*.5*

Pump Assy., Replace (B)

GS300	1.6	*1.6*
LS400, SC300, SC400	1.7	*1.7*

Throttle Relay, Replace (B)

1990-91 LS400	.5	*.5*

FRONT SUSPENSION
Unless otherwise noted, time given does not include alignment.

Align Front End (A)

GS300, GS400	2.8	*3.0*
LS400	2.8	*3.0*
SC300, SC400	2.8	*3.0*

Front & Rear Alignment, Check & Adjust (A)

All Models	1.9	*2.1*
Adjust caster or camber front or rear, each add	.3	*.3*
ES300, ES330 front camber add	.8	*.8*

Front Toe, Adjust (B)

All exc. below	.2	*.2*
ES300	1.2	*1.4*
GS300, GS400, LS400, SC300, SC400	.5	*.6*

Coil Spring, Replace (B)

GS300, GS400, GS430		
one	1.9	*2.1*
both	2.7	*3.0*
IS300		
one	2.2	*2.5*
both	2.9	*3.2*
LS400, LS430		
1990-94		
one	1.3	*1.6*
both	1.9	*2.2*
1995-05		
one	2.1	*2.3*
both	2.9	*3.2*
SC300, SC400		
one	3.1	*3.4*
both	5.0	*5.3*
SC430		
one	1.3	*1.4*
both	2.2	*2.5*

Front Axle Hub, Replace (B)

ES300, ES330		
1999-01	2.6	*2.8*
2002-05	3.2	*3.6*
GS300, GS400, GS430	3.2	*3.6*
IS300	3.0	*3.4*
LS400		
1990-94	2.9	*3.1*
1995-00	2.7	*2.9*
LS430	1.9	*2.1*
SC300, SC400	3.0	*3.2*
SC430	2.5	*2.9*

Front Bearing, Replace (B)

1999-03 ES300		
one side	2.5	*2.7*
both sides	3.4	*3.6*
ES330		
one side	2.9	*3.2*
both sides	4.7	*5.0*
GS300, GS400, GS430		
1993-00		
one side	2.2	*2.4*
both sides	3.8	*4.0*
2001-05		
one side	3.0	*3.4*
both sides	4.3	*4.6*
IS300		
one side	2.7	*3.0*
both sides	3.8	*4.3*
LS400		
one side	2.7	*2.9*
both sides	3.8	*4.0*
SC300, SC400		
one side	2.2	*2.4*
both sides	3.8	*4.0*
SC430		
one side	2.4	*2.7*
both sides	3.6	*3.9*

Front Hub Seal, Replace (B)

ES250		
one side	3.0	*3.2*
both sides	4.2	*4.4*
ES300		
one side	2.6	*2.8*
both sides	3.7	*3.9*
GS300, GS400, GS430		
one side	3.2	*3.6*
both sides	4.6	*5.1*
IS300		
one side	2.9	*3.2*
both sides	4.1	*4.4*
SC430		
one side	2.5	*2.9*
both sides	3.8	*4.3*

LEX-16 ES250 : ES300 : ES330 : GS300 : GS400 : GS430

	LABOR TIME	SEVERE SERVICE
Front Shock Absorbers, Replace (B)		
GS300, GS400, GS430		
one	1.7	1.9
both	2.5	2.7
IS300		
one	2.1	2.3
both	2.7	3.0
LS400, LS430		
1990-94		
one	1.2	1.4
both	1.7	1.9
1995-05		
one	1.9	2.1
both	2.6	2.8
SC300, SC400		
one	3.0	3.2
both	4.6	4.8
SC430		
one	1.1	1.3
both	2.5	2.9
w/air suspension		
LS400, LS430 add	.1	.1
Lower Ball Joint, Replace (B)		
GS300, GS400, GS430		
1994-98	1.2	1.5
1999-05	1.0	1.1
IS300	1.6	1.8
LS400, LS430		
1990-94	1.2	1.5
1995-05	1.6	1.8
SC430	.8	.9
Lower Control Arm, Replace (B)		
ES250		
one side	2.2	2.5
both sides	2.9	3.2
ES300, ES330		
1992-98		
one side	2.1	2.3
both sides	3.0	3.2
1999-01		
one side	1.3	1.6
both sides	2.2	2.5
2002-05		
one side	12.3	13.8
both sides	12.6	14.1
GS300, GS400, GS430	7.0	7.9
IS300	2.4	2.7
LS400, LS430		
1990-94	2.9	3.2
1995-00	2.5	2.8
2001-05	1.9	2.1
SC300, SC400	3.1	3.4
SC430	2.2	2.5
w/air suspension		
LS400 add	.1	.1
Replace ball joint		
GS300, GS400 add	.3	.3

	LABOR TIME	SEVERE SERVICE
Stabilizer Bar, Replace (B)		
ES300	1.7	1.9
GS300, GS400, GS430	.8	.9
IS300	1.0	1.1
LS400	3.6	3.8
LS430	1.0	1.1
SC300, SC400, SC430		
1994-05	1.3	1.5
w/air suspension		
LS400 add	.5	.5
Stabilizer Bar Bushings, Replace (B)		
ES300		
one	.7	.9
all	.8	1.0
GS300, GS400, GS430		
one	.6	.7
all	.8	.9
IS300		
one	.6	.7
all	.8	.9
LS400, LS430		
1999-05		
one	.8	.9
all	.9	1.0
IS300		
one	.6	.7
all	.8	.9
SC430		
one	1.0	1.1
all	1.1	1.3
Steering Knuckle, Replace (B)		
ES250, ES300, ES330	3.2	3.6
GS300, GS400, GS430	3.2	3.6
IS300	2.9	3.2
LS400, LS430		
1990-94	3.0	3.4
1995-00	2.7	3.0
2001-05	2.1	2.3
SC300, SC400	3.0	3.4
SC430	1.9	2.1
Strut/Shock Absorber Replace (B)		
ES250		
one	2.3	2.5
both	3.0	3.4
ES300, ES330		
one	1.9	2.1
both	2.7	3.0
Strut Coil Spring, Replace (B)		
ES250		
one	2.4	2.6
both	3.2	3.4
ES300, ES330		
one	1.8	2.0
both	3.0	3.2

	LABOR TIME	SEVERE SERVICE
Upper Control Arm, Replace (B)		
GS300, GS400, GS430	1.6	1.8
IS300	1.9	2.1
LS400, LS430	2.3	2.6
SC300, SC400	3.8	4.1
SC430	1.1	1.3
Wheel Bearing Grease Seal, Replace (B)		
GS300, GS400, GS430	3.2	3.6
IS300	2.9	3.2
LS400		
1990-94	2.8	2.8
1995-00	2.5	2.5
SC300, SC400	3.0	3.0
SC430	2.5	2.9

REAR SUSPENSION

Unless otherwise noted, time given does not include alignment.

	LABOR TIME	SEVERE SERVICE
Rear Alignment, Adjust (A)		
ES300	1.5	1.7
GS300, GS400	2.0	2.2
LS400, SC300, SC400	2.0	2.2
Rear Control Arm No. 1, Replace (B)		
ES250		
one	.7	1.0
both	.9	1.2
ES300, ES330		
1992-01		
one	2.2	2.5
both	2.5	2.8
2002-05		
one	1.7	2.0
both	2.2	2.5
GS300, GS400, GS430		
one	1.4	1.6
both	2.1	2.3
IS300		
one	1.3	1.4
both	1.4	1.6
LS400, LS430		
1990-94		
one	1.3	1.6
both	1.6	1.9
1995-05		
one	1.4	1.6
both	1.7	2.0
SC300, SC400		
one	1.4	1.7
both	1.7	2.0
SC430		
one	.8	.9
both	1.3	1.4

IS300 : LS400 : LS430 : SC300 : SC400 : SC430 **LEX-17**

	LABOR TIME	SEVERE SERVICE
Rear Control Arm No. 2, Replace (B)		
ES250		
one	.8	1.1
both	1.2	1.4
ES300, ES330		
one	1.3	1.4
both	1.7	2.0
GS300, GS400, GS430		
one	1.6	1.8
both	2.2	2.5
IS300		
one	1.4	1.6
both	1.7	2.0
LS400, LS430		
1990-94		
one	1.3	1.4
both	1.5	1.8
1995-05		
one	1.4	1.6
both	1.6	1.8
SC300, SC400, SC430		
one	1.3	1.4
both	2.1	2.3
Rear Strut Coil Spring, Replace (B)		
ES250		
one	2.2	2.5
both	2.9	3.1
ES300, ES330		
one	2.4	2.7
both	3.3	3.6
GS300, GS400, GS430		
one	1.7	2.0
both	2.5	2.8
IS300		
one	2.2	2.5
both	2.9	3.2
LS400		
1990-94		
one	2.7	3.0
both	3.6	3.9
1995-00		
one	3.4	3.7
both	5.3	5.6
LS430		
one	2.7	3.0
both	3.3	3.6
SC300, SC400, SC430		
one	2.2	2.5
both	3.1	3.4
Rear Strut, Replace (B)		
ES250		
one	2.0	2.2
both	2.7	2.9
ES300, ES330		
one	1.8	2.0
both	2.6	2.8
2002-05		
one	2.4	2.6
both	3.3	3.5

	LABOR TIME	SEVERE SERVICE
GS300, GS400, GS430		
one	1.6	1.8
both	2.2	2.5
IS300		
one	1.9	2.1
both	2.4	2.6
LS400		
1990-94		
w/air suspension		
one	2.2	2.4
both	2.6	2.8
w/o air suspension		
one	2.5	2.7
both	3.1	3.3
1995-00		
w/air suspension		
one	2.9	3.1
both	3.8	4.0
w/o air suspension		
one	3.1	3.3
both	4.4	4.6
LS430		
one	2.5	2.9
both	3.0	3.4
SC300, SC400, SC430		
one	1.7	2.0
both	2.4	2.7
Strut Rod, Replace (One Side) (B)		
ES250		
one	.7	.9
both	.9	1.1
ES300, ES330		
one	.7	.9
both	.8	1.0
2002-05		
one	1.1	1.3
both	1.7	2.0
GS300, GS400, GS430		
one	1.0	1.1
both	1.1	1.3
LS400, LS430		
1990-94		
one	.8	1.0
both	1.2	1.4
1995-05		
one	1.3	1.5
both	1.4	1.6
SC300, SC400		
one	1.2	1.4
both	1.3	1.5
SC430		
one	1.3	1.4
both	2.1	2.3
Stabilizer Bar, Replace (B)		
Exc. below	.8	.9
2002-05 E300,		
ES330	1.4	1.6
SC430	1.4	1.6

	LABOR TIME	SEVERE SERVICE
Stabilizer Bar Bushings, Replace (B)		
One		
Exc. below	.7	.9
2002-05 SC430		
one	1.4	1.6
all	1.6	1.8
Upper Control Arm, Replace (B)		
GS300, GS400, GS430		
one	1.9	2.1
both	2.9	3.2
IS300		
one	2.7	3.0
both	3.6	4.1
LS400, LS430		
1990-94		
one	2.0	2.3
both	3.0	3.3
1995-05		
one	3.0	3.4
both	4.9	5.5
SC300, SC400		
one	1.8	2.1
both	2.8	3.1
SC430		
one	1.0	1.1
both	1.6	1.8
AIR SUSPENSION		
Compressor Assy., Replace (B)		
LS400, LS430		
1990-05	1.6	1.8
Control Switch, Replace (B)		
1997-03 ES300	.5	.5
ES330	.2	.2
LS400, LS430		
1990-94	.3	.3
1995-05	.5	.5
Electronic Control Module, Replace (B)		
1997-04 ES300,		
ES330	.6	.6
LS400, LS430	.8	.8
Front Height Control Sensor, Replace (B)		
LS400, LS430		
1990-94		
one	.7	.8
both	.9	1.0
1995-05		
one	1.1	1.3
both	1.7	2.0
Front Suspension Control Actuator, Replace (B)		
1997-01 ES300		
one	1.8	2.0
both	2.9	3.1

LEX-18 ES250 : ES300 : ES330 : GS300 : GS400 : GS430

	LABOR TIME	SEVERE SERVICE
Front Suspension Control Actuator, Replace (B)		
LS400, LS430		
1990-94		
one	.3	.5
both	.5	.7
1995-05		
one	.5	.7
both	.7	.9
Height Control Dryer, Replace (B)		
LS400, LS430	1.4	1.6
Height Control Relay, Replace (B)		
LS400	.7	.7
LS430	.2	.2
Rear Height Control Sensor, Replace (B)		
LS400, LS430		
1990-94		
one	.8	.8
both	1.2	1.2
1995-05		
one	.7	.7
both	.8	.8
Rear Suspension Control Actuator, Replace (B)		
1997-01 ES300		
one	2.2	2.4
both	3.1	3.3
LS400, LS430		
one	1.4	1.6
both	1.7	1.9
Steering Sensor, Replace (B)		
1997-01 ES300	.7	.7
ES330	.8	.8
LS400, LS430		
1990-94	1.7	1.7
1995-00	1.8	1.8
2002-05	.8	.8

STEERING

AIR BAGS

The following operations do not include testing. Add time as required.

	LABOR TIME	SEVERE SERVICE
Diagnose Air Bag System Component Each (A)		
1993-05	.5	.5
Retrieve Fault Codes (A)		
1993-05	.3	.3
Air Bag Assy., Replace (B)		
Passenger side		
1993-94	.8	.8
ES300		
1995-96	.8	.8
1997-01	.7	.7
2002-03	2.9	2.9
ES330	2.4	2.4
GS300, GS400, GS430		
1995-97	.8	.8
1998-05	4.6	4.6
IS300	2.0	2.0
LS400	4.4	4.4
LS430	5.0	5.0
SC300, SC400	.7	.7
SC430	3.5	3.5
Side impact		
GS300, GS400, GS430		
1998-04		
one side	6.8	6.8
both sides	6.9	6.9
ES300, ES330		
2002-05		
one	2.4	2.4
both	2.8	2.8
IS300		
2001-05		
one	3.8	3.8
both	3.9	3.9
Wagon		
one	4.2	4.2
both	4.3	4.3
1999-03 LS400		
one side	1.2	1.2
both sides	1.8	1.8
LS430		
one side	2.7	2.7
both sides	2.8	2.8
Center Sensor, Replace (B)		
ES250	1.2	1.2
ES300, ES330	.7	.7
2002-05	1.1	1.1
GS300, GS400, GS430	.8	.8
IS300	.9	.9
LS400, LS430		
1990-94	3.1	3.1
1995-05	1.5	1.5
SC300, SC400		
SC430	1.2	1.2
Airbag (Crash) Sensor, Replace (B)		
1997-01 ES300	.7	.7
1998-02 GS300, GS400	.8	.8
1995-00 LS400	1.4	1.4
Front Sensor, Replace (B)		
ES250, ES300, ES330		
one	.8	.9
both	1.1	1.2
GS300, GS400, GS430		
one	.8	.9
both	1.2	1.4
IS300		
one	.6	.7
both	.9	1.0
LS400, LS430		
one	.8	.9
both	.9	1.0

	LABOR TIME	SEVERE SERVICE
SC300, SC400		
one	1.2	1.2
both	1.7	1.7
SC430		
one	.8	.9
both	1.1	1.2
Side Sensor, Replace (B)		
one	.9	1.0
both	1.5	1.7

POWER RACK & PINION

	LABOR TIME	SEVERE SERVICE
Electronic Control Module, Replace (B)		
ES250	.7	.7
ES300	.7	.7
GS300, GS400, GS430	1.2	1.2
LS400	1.2	1.2
SC300, SC400	.8	.8
Horn Contact or Canceling Cam, Replace (B)		
1990-04	.5	.5
Inner Tie Rod Assy., Replace (B)		
Includes: Reset toe.		
ES250	4.1	4.3
ES300, ES330		
1992-96	3.8	4.0
1997-01	3.3	3.7
2002-05	5.8	6.5
GS300, GS400, GS430	3.2	3.6
IS300	3.6	4.1
LS400		
1990-94	3.1	3.3
1995-00	3.0	3.2
LS430	3.5	3.9
SC300, SC400, SC430	3.0	3.4
Outer Tie Rod, Replace (B)		
Includes: Reset toe.		
1990-04 exc. below	1.4	1.6
2001-05 IS300	2.1	2.3
2002-03 ES300	3.3	3.7
2002-05 SC430	2.4	2.7
Rack Boots, Replace (B)		
ES250	3.7	3.9
ES300, ES330	1.4	1.6
2002-05	5.7	6.3
GS300, GS400, GS430	1.4	1.6
IS300	1.7	2.0
LS400, LS430		
1990-94	3.7	3.9
1995-05	1.4	1.6
SC300, SC400	1.4	1.6
SC430	2.4	2.7

IS300 : LS400 : LS430 : SC300 : SC400 : SC430 **LEX-19**

	LABOR TIME	SEVERE SERVICE
Rack & Pinion Assy., R&R or Replace (B)		
ES250	3.6	4.1
ES300, ES330		
1992-96	3.5	3.9
1997-01	2.7	3.0
2002-05	5.3	6.0
GS300, GS400, GS430	2.9	3.2
IS300	3.3	3.7
LS400		
1990-94	2.9	3.4
1995-00	2.7	3.2
LS430	3.2	3.6
SC300, SC400	2.7	3.2
SC430	3.3	3.7
Steering Pump, R&R and Recondition (B)		
ES250	3.2	3.7
ES300, ES330		
1992-96	4.0	4.4
1997-01	3.5	3.9
2002-05	3.0	3.4
GS300, GS400, GS430, IS300	3.2	3.6
LS400		
1990-94	4.1	4.6
1995-00	4.9	5.4
w/TRAC add	.3	.3
LS430	3.5	3.9
SC300	3.0	3.5
SC400	3.3	3.8
SC430	2.4	2.7
Steering Pump, R&R or Replace (B)		
ES250	2.0	2.2
ES300, ES330		
1992-96	2.5	2.9
1997-05	2.0	2.2
GS300, GS400, GS430	1.7	1.9
IS300	2.0	2.2
LS400, LS430		
1990-94	2.5	2.7
1995-00	3.1	3.3
2001-05	2.1	2.4
w/TRAC add	.3	.3
SC400	2.0	2.2
SC300, SC430	1.6	1.8
Steering Pump Hoses, Replace (B)		
Does not include purging.		
1990-94		
Pressure		
ES250	2.0	2.2
ES300	1.4	1.6
GS300, LS400, SC300	1.5	1.7
SC400	1.7	1.9

	LABOR TIME	SEVERE SERVICE
Return		
ES250	.7	.9
ES300	1.3	1.5
GS300, SC300, SC400	1.4	1.6
LS400	.7	.9
1995-05		
Pressure		
ES300, ES330	1.4	1.6
2002-05	1.7	2.0
GS300, GS400	1.6	1.8
IS300, LS400, LS430, SC300	1.6	1.8
SC400	1.7	1.9
SC430	2.7	3.0
Return		
Exc. below	1.4	1.6
2002-05 LS430	1.9	2.1
Steering Wheel, Replace (B)		
1990-00	.5	.5
2001-05	.7	.7
w/LS400 99-00 add	.1	.1
Upper Mast Jacket Bearing, Replace (B)		
Includes: Insulators.		
1990-94 LS400	2.7	2.9

HEATING & AIR CONDTIONING

When more than one component requires replacement where evacuation/recovery and recharging is already included, deduct 1.0 hour for each additional component from the time given.

	LABOR TIME	SEVERE SERVICE
Evacuate/Recover and Recharge System (B)		
1990-05	1.5	1.5
AC Control Assy., Replace (B)		
Exc. below	.7	.7
ES250	1.2	1.2
SC300, SC400	.9	.9
AC Hoses, Replace (B)		
Includes: Evacuate/recover and recharge.		
ES250	1.9	2.1
ES300, ES300	1.8	2.1
GS300, GS400, GS430, IS300	2.0	2.2
LS400, LS430	4.8	5.5
SC300, SC400, SC430	1.8	2.1
Blower Motor, Replace (B)		
ES250	.7	.9
ES300, ES330	.5	.7
GS300, GS400, GS430	.7	.9
IS300	.3	.3

	LABOR TIME	SEVERE SERVICE
LS400, LS430		
1990-94	2.2	2.4
1995-05	.5	.5
SC300, SC400	1.2	1.4
SC430	.5	.5
Blower Motor Resistor, Replace (B)		
ES250	.5	.5
ES300	.7	.7
LS400		
1990-94	1.5	1.5
1995-00	.5	.5
SC300, SC400	.7	.7
Compressor Assy., Replace (B)		
Includes: Parts transfer, evacuate/recover and recharge.		
ES250	1.9	2.1
ES300, ES330	3.3	3.7
GS300, GS400, GS430	2.9	3.2
IS300	2.4	2.7
LS400, LS430	2.6	2.9
SC300	2.0	2.2
SC400	1.7	1.9
SC430	2.8	3.1
Compressor Clutch Assy., Replace (B)		
ES250	2.2	2.4
ES300, ES330		
1992-01	3.5	3.9
2002-05	3.1	3.4
GS300, GS400, GS430	3.2	3.6
IS300	2.3	2.6
LS400	2.4	2.6
LS430	2.6	2.9
SC300	2.6	2.9
SC400	2.2	2.4
SC430	2.8	3.1
Condenser Assy., Replace (B)		
Includes: Evacuate/recover and recharge.		
ES250	2.0	2.2
ES300, ES330		
1992-01	4.4	4.9
2002-05	2.8	3.1
GS300, GS400, GS430	5.1	5.7
IS300	2.0	2.2
LS400		
1990-94	3.7	3.9
1995-00	3.0	3.2
LS430	2.6	2.9
SC300, SC400	3.6	3.8
SC430	2.6	2.9

LEX-20 ES250 : ES300 : ES330 : GS300 : GS400 : GS430

	LABOR TIME	SEVERE SERVICE
Evaporator Coil, Replace (B)		
Includes: Evacuate/recover and recharge.		
ES250	2.5	2.5
ES300, ES330		
1992-01	2.5	2.5
2002-05	5.6	5.6
GS300, GS400, GS430		
1994-97		
w/TRAC	5.0	5.0
w/o TRAC	2.6	2.6
1998-05	2.9	2.9
IS300	5.4	5.4
LS400		
1990-94	3.4	3.4
1995-98	6.5	6.5
1999-00	6.3	6.3
LS430	8.9	8.9
SC300, SC400		
w/TRAC	5.5	5.5
w/o TRAC	3.6	3.6
SC430	5.7	5.7
Expansion Valve, Replace (B)		
Includes: Evacuate/recover and recharge.		
ES250	2.3	2.3
ES300, ES330		
1992-01	2.4	2.4
2002-05	5.4	5.4
GS300, GS400, GS430		
1994-97		
w/TRAC	5.0	5.0
w/o TRAC	2.6	2.6
1998-05	2.9	2.9
IS300	5.4	5.4
LS400		
1990-94	3.4	3.4
1995-98	5.9	5.9
1999-00	5.0	5.0
LS430	5.9	5.9
SC300, SC400		
w/TRAC	5.5	5.5
w/o TRAC	3.6	3.6
SC430	5.1	5.1
Heater Core, R&R or Replace (B)		
ES250	8.3	8.3
ES300, ES330		
1992-98	2.7	2.7
1999-01	1.8	1.8
2002-05	5.0	5.0
GS300, GS400, GS430		
1994-98	10.0	10.0
1999-05	8.1	8.1
IS300	4.6	4.6

	LABOR TIME	SEVERE SERVICE
LS400		
1990-94	12.8	12.8
1995-98	9.3	9.3
1999-00	7.3	7.3
LS430	8.4	8.4
SC300, SC400		
1992-98	19.3	19.3
1999-00	15.4	15.4
SC430	4.9	4.9
Heater Hoses, Replace (B)		
ES250	.9	1.1
ES300, ES330		
1992-01	1.2	1.4
2002-05	.3	.3
GS300, GS400, GS430	1.2	1.4
IS300	.9	1.0
LS400, LS430	1.1	1.2
SC300, SC400	1.3	1.5
SC430	.6	.7
Low Pressure Cut-Off Switch, Replace (B)		
Includes: Evacuate/recover and recharge.		
Exc. ES250	2.0	2.2
ES250	2.3	2.6
Pressure Regulator, Replace (B)		
GS300, GS400, GS430		
1994-97		
w/TRAC	3.1	3.1
w/o TRAC	1.7	1.7
1998-05	1.7	1.7
LS400	1.7	1.7
SC300, SC400		
w/TRAC	4.6	4.6
w/o TRAC	2.9	2.9
Receiver/Drier Assy., Replace (B)		
Includes: Evacuate/recover and recharge.		
ES250	1.6	1.6
ES300	3.2	3.6
GS300, GS400, GS430	2.0	2.2
IS300	2.1	2.4
LS400		
1990-94	3.3	3.7
1995-00	2.9	3.2
LS430	2.8	3.1
SC300, SC400	4.1	4.6
SC430	2.6	2.9

WIPERS & SPEEDOMETER

	LABOR TIME	SEVERE SERVICE
Antenna Assy., Replace (B)		
1990-05 Exc. below	.7	.7
IS300	2.1	2.4
Wagon	2.9	3.2
SC430	1.1	1.9

	LABOR TIME	SEVERE SERVICE
Radio, R&R (B)		
Exc. below	.6	.6
ES250	.3	.3
SC300, SC400	.7	.7
w/auxiliary amp add	.3	.3
Speedometer Cable & Casing, Replace (B)		
ES250	1.4	1.4
Speedometer Head, R&R or Replace (B)		
Exc. below	1.1	1.1
1995-00 LS400	1.4	1.4
GS300		
1994-98	.7	.7
1999-05	1.4	1.4
Windshield Washer Pump, Replace (B)		
ES250	.7	.7
ES300, IS300	.5	.5
ES330	1.1	1.1
GS300, GS400, GS430	.7	.7
LS400		
1990-94	1.3	1.3
1995-00	.7	.7
LS430	1.4	1.4
SC300, SC400	1.2	1.2
SC430	.6	.6
Windshield Wiper & Washer Switch, Replace (B)		
ES250	.9	.9
ES300, ES330		
1992-96	1.4	1.4
1997-01	.7	.7
2002-05	.3	.3
GS300, GS400, GS430		
1994-97	1.5	1.5
1998-05	.3	.3
IS300	.3	.3
LS400		
1990-94	2.2	2.2
1995-00	.3	.3
SC300, SC400		
1992-98	2.4	2.4
1999-00	2.0	2.0
Windshield Wiper Linkage, Replace (B)		
ES250	.8	1.0
ES300, ES330	1.2	1.4
2002-05	.6	.7
GS300, GS400, GS430	.7	.9
IS300	.6	.7
LS400, LS430	.7	.9
SC300, SC400	.8	1.0
SC430	.5	.5

IS300 : LS400 : LS430 : SC300 : SC400 : SC430   **LEX-21**

	LABOR TIME	SEVERE SERVICE
Windshield Wiper Motor, Replace (B)		
Exc. below	.6	.7
GS300, GS400, GS430, IS300	.8	.9
LS400, LS430	.8	.9

LAMPS & SWITCHES

Back-Up Lamp Assy., Replace (B)
ES250 .... .3 .3
Back-Up Lamp Switch, Replace (B)
ES250 .... .5 .5
ES300 .... .3 .3
SC300 .... .5 .5
Front Turn Signal Lamp or Lens, Replace (B)
ES250 .... .5 .5
ES300, ES330
  1997-05 .... .3 .3
GS300, GS430
  2001-05 .... .3 .3
IS300 .... 1.1 1.1
LS400
  1990-94 .... .3 .3
  1995-00 .... .5 .5
LS430 .... .5 .5
SC430 .... .3 .3
Halogen Headlamp Bulb, Replace (B)
Exc. below .... .3 .3
ES300, ES 330
  2002-05 .... 1.1 1.1
IS300 .... .8 .8
Headlamp Switch, Replace (B)
ES250 .... 1.2 1.2
ES300, ES330
  1992-96 .... 1.4 1.4
  1997-01 .... .7 .7
  2002-05 .... .3 .3
GS300, GS400, GS430
  1994-97 .... 1.5 1.5
  1998-05 .... .3 .3
IS300 .... .3 .3
LS400, LS430
  1990-94 .... 2.0 2.0
  1995-98 .... 1.2 1.2
  1999-03 .... .3 .3
SC300, SC400
  1992-98 .... 2.0 2.0
  1999-00 .... 2.6 2.6
SC430 .... 2.3 2.3
Headlamps, Aim (B)
Two .... .5 .5
Four .... .7 .7

High Mount Stop Lamp Assy., Replace (B)
Exc. below
GS300
  1994-97 .... 1.2 1.2
w/rear spoiler add
ES300, IS300 ..... .3 .3
GS300 .... .5 .5
Horn, Replace (B)
ES250 .... .5 .5
ES300, ES330 ..... .7 .7
  2002-05 .... .3 .3
GS300, GS400, GS430 .... .3 .3
IS300 .... .5 .5
LS400, LS430 ..... .3 .3
SC300, SC400 ..... .6 .7
SC430 .... .3 .3
License Lamp Lens or Bulb, Replace (C)
1990-05 .... .3 .3
Neutral Safety/Back-Up Lamp Switch, Replace (B)
ES250, ES300 ..... .7 .7
GS300, GS400 ..... 1.2 1.2
LS400
  1990-94 .... 1.3 1.3
  1995-00 .... 1.7 1.7
SC300, SC400 ..... 1.2 1.2
Parking Brake Warning Lamp Switch, Replace (B)
ES250, ES300, ES330 .. .7 .7
GS300, GS400, GS430 .... 1.8 1.8
IS300 .... 1.1 1.1
LS400
  1990-94 .... 1.7 1.7
  1995-00 .... 1.6 1.6
LS430 .... 1.1 1.1
SC300, SC400, SC430 .... .5 .5
Parking Lamp Lens or Bulb, Replace (B)
ES250 .... .3 .3
1992-96 ES300 .... .7 .7
2001-04 ES300 .... .2 .2
GS300, IS300 ..... .3 .3
LS400 .... .3 .3
LS430 .... .5 .3
SC300, SC400 ..... 1.4 1.4
SC430 .... .5 .5
Rear Combination Lamp Bulb, Replace (C)
Exc. below .... .5 .5
SC300, SC400 ..... .7 .7

Side Marker Lamp Lens, Replace (B)
ES300, ES330
  2002-05 .... .3 .3
IS300 .... .5 .5
LS400 .... .3 .3
SC300, SC400 ..... .7 .7
Stop Light Switch, Replace (B)
Exc. below .... .7 .8
ES250 .... .5 .5
LS400
  1990-94 .... 1.4 1.4
SC300, SC400 ..... .5 .5
Turn Signal or Hazard Warning Flasher, Replace (B)
Exc. below .... .5 .5
2002-03 ES300 .... .7 .7
Turn Signal Switch, Replace (B)
ES250 .... .8 .8
ES300, ES330
  1992-96 .... .9 .9
  1997-01 .... .6 .6
  2002-05 .... .9 .9
GS300, GS400, GS430 .... 1.1 1.1
IS300 .... .7 .7
LS400, LS430
  1990-94 .... 2.0 2.0
  1995-05 .... .8 .8
SC300, SC400 ..... 1.9 1.9
SC430 .... .5 .5

BODY

Deck Lid Lock, Replace (B)
1990-05 .... .6 .6
Deck Lid Torsion Bars, Replace (B)
ES300, ES330 ..... .3 .3
  2002-05 .... .6 .6
1990-94 LS400 .... .7 .7
IS300 .... .3 .3
Front Door Lock, Replace (B)
ES250 .... 1.2 1.2
ES300, ES330
  1992-96 .... 1.5 1.5
  1997-01 .... 1.3 1.3
  2002-05 .... .5 .5
GS300, GS400, GS430 .... 1.7 1.7
IS300 .... 1.1 1.1
LS400
  1990-94 .... 1.6 1.6
  1995-00 .... 1.3 1.3
LS430 .... 1.8 1.8
SC300, SC400 ..... 2.2 2.2
SC430 .... .7 .7
Hood Lock, Replace (B)
Exc. below .... .5 .5
ES330, SC430 ..... .7 .7
LS400
  1990-94 .... 2.0 2.0

LEX-22   ES250 : ES300 : ES330 : GS300 : GS400 : GS430

	LABOR TIME	SEVERE SERVICE
Hood Release Cable, Replace (B)		
ES250, ES300	1.3	1.3
2002-03 ES300	.8	.8
ES330	.8	.8
GS300, GS400, GS430	2.9	2.9
IS300	1.4	1.4
LS400		
1990-94	1.3	1.3
1995-00	2.6	2.6
LS430	.8	.8
SC300, SC400	2.0	2.0
SC430	1.2	1.4
Lock Striker Plate, Replace (B)		
1990-05	.5	.5
Rear Door Lock, Replace (B)		
ES250	.9	.9
ES300, ES330	1.2	1.2
2002-05	.8	.8

	LABOR TIME	SEVERE SERVICE
GS300, GS400, GS430	1.6	1.6
IS300	1.1	1.1
LS400		
1990-94	1.6	1.6
1995-00	1.3	1.3
LS430	1.8	1.8
Window Regulator and/or Motor, Replace (B)		
Front		
ES250	1.2	1.2
ES300		
1992-96	1.6	1.6
1997-03	1.3	1.3
ES330	.9	.9
GS300, GS400, GS430	1.2	1.2
IS300	1.2	1.2

	LABOR TIME	SEVERE SERVICE
LS400		
1990-94	1.7	1.7
1995-00	1.4	1.4
LS430	1.8	1.8
SC300, SC400	1.8	1.8
SC430	.9	.9
Rear		
ES250	1.2	1.4
ES300, ES330		
1994-96	1.4	1.6
1997-98	1.6	1.8
1999-05	1.1	1.1
IS300	1.2	1.2
GS300, GS400, GS430	1.5	1.7
LS400, LS430		
1990-94	1.7	1.9
1995-05	1.5	1.7

LEX

GX470 : LX450 : LX470 : RX300 : RX330

SYSTEM INDEX

MAINTENANCE LEX-24
CHARGING LEX-24
STARTING LEX-24
CRUISE CONTROL LEX-24
IGNITION LEX-24
EMISSIONS LEX-24
FUEL LEX-25
EXHAUST LEX-25
ENGINE COOLING LEX-25
ENGINE LEX-26
   Assembly LEX-26
   Cylinder Head LEX-26
   Camshaft LEX-26
   Crank & Pistons LEX-27
   Engine Lubrication LEX-27
AUTO TRANSAXLE LEX-27
AUTO TRANSMISSION LEX-28
TRANSFER CASE LEX-28
SHIFT LINKAGE LEX-28
DRIVELINE LEX-28
BRAKES LEX-29
FRONT SUSPENSION LEX-30
REAR SUSPENSION LEX-31
STEERING LEX-31
HEATING & AC LEX-32
WIPERS &
  SPEEDOMETER LEX-32
LAMPS & SWITCHES LEX-32
BODY LEX-33

OPERATIONS INDEX

A
AC Hoses LEX-32
Air Bags LEX-31
Air Conditioning LEX-32
Alignment LEX-30
Alternator (Generator) LEX-24
Antenna LEX-32
Anti-Lock Brakes LEX-29

B
Battery Cables LEX-24
Bleed Brake System LEX-29
Blower Motor LEX-32
Brake Disc LEX-30
Brake Hose LEX-29
Brake Pads and/or Shoes ... LEX-30

C
Camshaft LEX-26
Camshaft Sensor LEX-24
Catalytic Converter LEX-25
Coolant Temperature (ECT)
  Sensor LEX-24
Crankshaft LEX-27
Cruise Control LEX-24
CV Joint LEX-29
Cylinder Head LEX-26

D
Differential LEX-28
Distributor LEX-24
Drive Belt LEX-24
Driveshaft LEX-28

E
EGR LEX-24
Electronic Control Module
  (ECM/PCM) LEX-25
Engine LEX-26
Engine Lubrication LEX-27
Engine Mounts LEX-26
Evaporator LEX-32
Exhaust LEX-25
Exhaust Manifold LEX-25
Expansion Valve LEX-32

F
Flexplate LEX-27
Fuel Injection LEX-25
Fuel Pump LEX-25
Fuel Vapor Canister LEX-25

G
Gear Selector Lever LEX-28
Generator LEX-24

H
Halfshaft LEX-29
Headlamp LEX-32
Heater Core LEX-32
Horn LEX-33

I
Idle Air Control (IAC) Valve .. LEX-25
Ignition Coil LEX-24
Ignition Module LEX-24
Ignition Switch LEX-24
Inner Tie Rod LEX-31
Intake Manifold LEX-25

K
Knock Sensor LEX-25

L
Lower Control Arm LEX-30

M
Mass Air Flow (MAF)
  Sensor LEX-25
Master Cylinder LEX-30
Muffler LEX-25
Multifunction Lever/Switch ... LEX-33

N
Neutral Safety Switch LEX-33

O
Oil Pan LEX-27
Oil Pump LEX-27
Outer Tie Rod LEX-31
Oxygen Sensor LEX-25

P
Parking Brake LEX-30
Pistons LEX-27
Positive Crankcase
  Ventilation (PCV) Valve ... LEX-25

R
Radiator LEX-25
Radiator Hoses LEX-26
Radio LEX-32
Rear Main Oil Seal LEX-27

S
Shock Absorber/Strut,
  Front LEX-30
Shock Absorber/Strut,
  Rear LEX-31
Spark Plug Cables LEX-24
Spark Plugs LEX-24
Spring, Front Coil LEX-30
Spring, Rear Coil LEX-31
Starter LEX-24
Steering Wheel LEX-31

T
Thermostat LEX-25
Throttle Body LEX-25
Throttle Position Sensor
  (TPS) LEX-25
Timing Belt LEX-27
Timing Chain LEX-27
Torque Converter LEX-27

U
U-Joint LEX-29
Upper Control Arm LEX-31

V
Valve Body LEX-27
Valve Cover Gasket LEX-26
Valve Job LEX-26
Vehicle Speed Sensor LEX-25

W
Water Pump LEX-26
Wheel Balance LEX-24
Window Regulator LEX-33
Windshield Washer Pump ... LEX-32
Windshield Wiper Motor LEX-32

LEX-23

LEX-24 GX470 : LX450 : LX470 : RX300 : RX330

	LABOR TIME	SEVERE SERVICE
MAINTENANCE		
Air Cleaner Filter Element, Service (B)		
1996-05 exc. below	.2	.2
RX330	.5	.5
Chassis Lubrication, Change Oil & Filter (C)		
Includes: Correct all fluid levels.		
Exc. below	.8	.8
RX300, RX330	1.0	1.0
Drive Belt, Adjust (B)		
Air conditioning		
LX450, RX330	.2	.3
Fan		
1996-05	.3	.4
Drive Belt, Replace (B)		
Air conditioning		
LX450	.5	.6
RX330	.9	1.0
Fan		
1996-05	.5	.6
Power Steering		
1996-03	.6	.6
Fuel Filter, Replace (B)		
GX470	2.1	2.2
LX450, LX470	.8	.9
RX300	1.2	1.3
RX330	1.7	1.8
Halogen Headlamp Bulb, Replace (B)		
All Models	.3	.3
Oil & Filter, Change (C)		
Includes: Correct all fluid levels.		
Exc. below	.8	.8
RX300, RX330	1.0	1.0
Timing Belt, Replace (B)		
GX470	3.6	4.1
LX470	4.0	4.4
RX300, RX330	3.2	3.6
Tire, Replace (C)		
Includes: Dismount old tire and mount new tire to rim.		
1996-05	.6	.6
Tires, Rotate (C)		
1996-05	.8	.8
Wheels, Balance (B)		
one	.3	.3
each addl. add	.2	.2
CHARGING		
Alternator Circuits, Test (B)		
Includes: Test component output.		
1996-05	.7	.7
Alternator, R&R and Recondition (A)		
GX470	2.0	2.2
LX450	2.3	2.5
LX470	2.0	2.2
RX300, RX330	1.8	2.0

	LABOR TIME	SEVERE SERVICE
Alternator Assy., Replace (B)		
Includes: Pulley transfer.		
GX470	1.2	1.3
LX450, LX470, RX300	.8	.9
RX330	.5	.6
Alternator Voltage Regulator, Replace (B)		
LX450	1.2	1.4
LX470	2.0	2.2
RX300	.8	1.0
Front Alternator Bearing Replace (B)		
Exc. below	1.5	1.7
GX470	1.7	1.9
RX330	1.2	1.4
STARTING		
Starter Draw Test (On Car) (B)		
1996-05	.3	.3
Battery Cables, Replace (C)		
Positive	.6	.7
Negative	.5	.6
Starter, R&R and Recondition (A)		
GX470	3.7	3.9
LX450	1.6	1.8
LX470	3.6	3.8
RX300, RX330	1.7	1.9
Starter Assy., Replace (B)		
GX470	2.7	2.8
LX450	.7	.8
LX470	3.0	3.1
RX300, RX330	.9	1.0
Starter Solenoid and/or Switch, Replace (B)		
Includes: Starter R&R.		
GX470	3.1	3.3
LX450	.8	1.0
LX470	3.3	3.5
RX300	1.2	1.4
CRUISE CONTROL		
Diagnose Cruise Control System Component Each (A)		
1996-05	.8	.8
Actuator Assy., Replace (B)		
1996-03	.6	.6
Control Computer, Replace (B)		
All Models	.7	.7
Control Main Switch, Replace (B)		
All Models	.6	.6
IGNITION		
Diagnose Ignition System Component Each (A)		
1996-05	.8	.8
Retrieve Fault Codes (A)		
1996-05	.3	.3

	LABOR TIME	SEVERE SERVICE
Camshaft Sensor, Replace (B)		
RX300, RX330	.5	.6
GX470, LX470	1.1	1.2
Distributor, Replace (B)		
Includes: Reset base ignition timing.		
LX450	1.3	1.5
Distributor Cap and/or Rotor, Replace (B)		
LX450	.3	.3
Igniter Unit, Replace (B)		
LX450	.5	.5
Ignition Coil, Replace (B)		
GX470		
right or left side	.3	.3
LX450	.5	.5
LX470		
right or left side	.3	.3
RX300		
right side	1.5	1.5
left side	.3	.3
RX330		
right side	2.1	2.1
left side	.5	.5
Ignition Switch, Replace (B)		
1996-05	.7	.7
Spark Plug (Ignition Cables), Replace (B)		
LX450	.3	.5
Spark Plugs, Replace (B)		
Exc. below	.8	.9
RX300, RX330	2.4	2.7
Steering & Ignition Lock Switch Assy., Replace (B)		
GX470, LX470, RX300, RX330	.5	.5
LX450	.7	.7
EMISSIONS		
The following operations do not include testing. Add time as required.		
Diagnose Emission Control System Component Each (A)		
1996-05	.6	.6
Dynamometer Test (A)		
1996-05	.5	.5
Retrieve Fault Codes (A)		
1996-05	.3	.3
Coolant Temperature (ECT) Sensor, Replace (B)		
GX470, LX450	1.2	1.4
LX470	1.8	2.1
RX300, RX330	.9	1.0
EGR Check Valve, Replace (B)		
LX450	.3	.3
EGR Control Valve, Replace (B)		
LX450	1.2	1.5
EGR Modulator, Replace (B)		
LX450	.3	.3

GX470 : LX450 : LX470 : RX300 : RX330   **LEX-25**

	LABOR TIME	SEVERE SERVICE
EGR Temperature Sensor, Replace (B)		
LX450	.3	.3
EGR Thermal Vacuum Valve, Replace (B)		
LX450	1.3	1.3
Electronic Control Module (ECM/PCM), Replace (B)		
GX470	.5	.5
LX450	1.3	1.3
LX470	.6	.6
RX300, RX330	.7	.7
Fuel Vapor Canister, Replace (B)		
GX470, LX450, LX470	.5	.5
RX300	.8	.8
RX330	1.1	1.1
Knock Sensor, Replace (B)		
GX470	2.4	2.7
LX450		
front or rear	.7	.9
LX470 one	3.1	3.4
RX300, RX330 one	3.2	3.6
each addl. side add	.2	.2
Mass Air Flow (MAF) Sensor, Replace (B)		
All Models	.3	.3
LX450	1.2	1.2
Oxygen Sensor, Replace (B)		
GX470		
front, one	2.4	2.7
rear, one	.5	.5
LX450 front or rear	.8	1.0
LX470		
front or rear one	.5	.5
each addl. side add	.2	.2
RX300 front	.3	.3
each addl. side add	.2	.2
RX330		
front, one	.6	.6
rear, one	1.7	1.7
Positive Crankcase Ventilation (PCV) Valve, Replace (B)		
GX470, LX450, LX470	.3	.3
RX300	.5	.5
Throttle Position Sensor (TPS), Replace (B)		
All Models	.7	.7
Vehicle Speed Sensor, Replace (B)		
Exc. below	.7	.7
RX330	.9	.9

FUEL
DELIVERY
	LABOR TIME	SEVERE SERVICE
Fuel Pump, Test (B)		
1996-05	.3	.3
Fuel Filter, Replace (B)		
GX470	2.1	2.3
LX450, LX470	.8	.9
RX300	1.2	1.4
RX330	1.7	2.0
Fuel Gauge (Dash), Replace (B)		
LX450	.8	.8
Fuel Gauge (Tank), Replace (B)		
GX470	2.0	2.2
LX450	2.5	2.7
LX470	.5	.7
RX300	.9	1.1
RX330	1.7	1.9
Fuel Pump, Replace (B)		
GX470, LX450	2.1	2.3
LX470	.7	.9
RX300	1.3	1.5
RX330	1.9	2.1
Fuel Tank, Replace (B)		
Includes: Drain and refill.		
GX470	2.2	2.4
LX450, LX470	3.8	4.0
RX300		
2WD	3.2	3.4
4WD	3.8	4.0
RX300		
2WD	3.6	3.8
4WD	4.5	4.7
Intake Manifold Gasket, Replace (B)		
GX470	2.4	2.6
LX450	8.8	9.0
LX470	3.0	3.2
RX300	4.1	4.4
RX330	3.0	3.2
Replace manifold add	.5	.5

INJECTION
	LABOR TIME	SEVERE SERVICE
Air Cleaner Filter Element, Service (B)		
1996-05 exc. below	.2	.2
RX330	.5	.5
Fuel Injectors, Replace (B)		
GX470	1.7	1.9
LX450	3.4	3.6
LX470	2.2	2.4
RX300	3.0	3.2
RX330	2.6	2.8
Fuel Pressure Regulator, Replace (B)		
GX470	.5	.6
LX450	1.3	1.4
LX470	.6	.7
RX300	.7	.8
RX330	1.7	1.8
Idle Air Control (IAC) Valve, Replace (B)		
LX450	1.7	1.8
RX300	1.3	1.4
Throttle Body Assy., Replace (B)		
GX470	1.5	1.7
LX450, LX470	1.7	1.9
RX300	2.1	2.3
RX330	1.7	1.9

EXHAUST
	LABOR TIME	SEVERE SERVICE
Catalytic Converter, Replace (B)		
GX470	.6	.7
LX450	1.7	1.8
LX470	.8	.9
RX300	.6	.7
Center Exhaust Pipe, Replace (B)		
GX470	.6	.7
LX450	1.3	1.4
LX470	1.0	1.1
RX300, RX330	.8	.9
Exhaust Manifold or Gasket, Replace (B)		
GX470		
right side	2.0	2.2
left side	1.4	1.6
LX450	2.0	2.2
LX470		
right side	7.7	7.9
left side	7.9	8.1
RX300		
right side	7.5	7.7
left side	2.6	2.8
RX330		
right side	1.7	1.9
left side	1.2	1.4
Front Exhaust Pipe, Replace (B)		
GX470	.6	.7
LX450	1.2	1.3
LX470, RX300	.7	.8
RX330	1.3	1.4
Muffler & Tailpipe Assy., Replace (B)		
GX470, RX300, RX330	.7	.8
LX450, LX470	.8	.9

ENGINE COOLING
	LABOR TIME	SEVERE SERVICE
Pressure Test Cooling System (C)		
1996-02	.3	.3
By-Pass Hoses, Replace (B)		
Exc. below	.6	.7
GX470	1.2	1.4
RX330	3.1	3.3
Coolant Thermostat, Replace (B)		
GX470	1.4	1.5
LX450	1.2	1.4
LX470, RX300	.9	1.0
RX330	1.4	1.5
Engine Coolant Temp. Sending Unit, Replace (B)		
GX470	1.2	1.4
LX450	.9	1.0
LX470	1.8	2.1
RX300, RX330	.9	1.0
Radiator Assy., R&R or Replace (B)		
GX470	1.7	1.9
LX450	1.8	2.0
LX470	1.5	1.7
RX300	2.0	2.2
RX330	2.3	2.6

LEX-26 GX470 : LX450 : LX470 : RX300 : RX330

	LABOR TIME	SEVERE SERVICE
Radiator Hoses, Replace (B)		
Includes: Refill with proper coolant mix.		
GX470		
upper	1.2	1.4
lower	1.4	1.5
LX450		
upper	.8	.8
lower	1.3	1.4
LX470		
upper	1.1	1.2
lower	1.4	1.5
RX300		
upper	.9	1.0
lower	1.4	1.5
RX330		
upper or lower	1.1	1.2
Temperature Gauge (Dash), Replace (B)		
LX450	1.2	1.2
Water Pump and/or Gasket, Replace (B)		
Includes: Refill with proper coolant mix.		
GX470	3.6	3.8
LX450	2.2	2.4
LX470	4.0	4.2
RX300	2.7	2.9
RX330	3.8	4.0

ENGINE
ASSEMBLY
Times shown are for OEM assemblies. Time to replace assemblies from aftermarket rebuilders may vary.

Engine Assy., R&I (B)
Does not include parts or component transfer.
1996-05	13.8	14.3

Engine Assy., R&R and Recondition (A)
Includes: Replacing rings, rod and main bearings, cylinder head reconditioning and engine tune-up.
GX470	54.3	55.2
LX450	49.0	49.5
LX470	58.5	59.4
RX300		
2WD	44.9	45.8
4WD	45.1	46.0
RX330		
2WD	51.2	52.1
4WD	52.1	53.0

	LABOR TIME	SEVERE SERVICE
Engine Assy. (Short Block), Replace (B)		
Assembly consists of cylinder block, piston assemblies, crankshaft, camshaft, timing chain and gears. Does not include cylinder heads. Operation Includes: R&R engine, transfer necessary parts and all necessary adjustments.		
GX470	22.1	23.0
LX450	20.8	20.8
LX470	20.9	21.8
RX300		
2WD	22.6	25.1
4WD	24.5	25.0
RX330		
2WD	21.7	22.6
4WD	22.6	23.5
Recondition valves add		
GX470	23.0	23.0
LX450	18.2	18.2
LX470	20.0	20.0
RX300	17.5	17.5
RX330	18.3	18.3
Engine Mounts, Replace (B)		
Front exc. below	1.3	1.4
LX450, LX470	.8	.9

CYLINDER HEAD
Compression Test (B)
1996-05	1.2	1.4

Cylinder Head, Replace (B)
Includes: Parts transfer, adjustments.
GX470		
one side	17.8	18.3
both sides	28.0	28.5
LX450		
one side	20.5	21.0
LX470		
one side	15.7	16.2
both sides	24.8	25.3
RX300, RX330		
one side	14.7	15.2
both sides	21.7	22.2

Cylinder Head Gasket, Replace (B)
GX470		
one side	7.2	7.7
both sides	9.2	9.7
LX450		
one side	9.7	10.2
LX470		
one side	9.4	9.9
both sides	12.6	13.1
RX300		
one side	9.5	10.0
both sides	13.5	14.0
RX330		
one side	9.5	10.0
both sides	12.2	12.7

	LABOR TIME	SEVERE SERVICE
Valve/Cam Cover Gasket, Replace or Reseal (B)		
GX470		
left side	.5	.5
right side	.6	.7
LX450	1.3	1.5
LX470		
left side	.8	.9
right sides	.6	.7
RX300		
left side	.6	.7
right side	4.1	4.6
RX330		
left side	1.5	1.7
right side	2.9	3.2
Valve Clearance, Adjust (B)		
GX470	7.0	7.5
LX450, LX470	5.9	6.4
RX300, RX330	5.7	6.2
Valve Job (A)		
GX470	30.5	31.4
LX450	27.7	28.6
LX470	31.8	32.7
RX300	26.1	27.0
RX330	25.8	26.7
Valve Lifters, Replace (B)		
GX470	9.8	10.3
LX450	7.0	7.5
LX470	10.2	10.7
RX300	10.3	10.8
RX330	9.0	9.5
Valve Springs and/or Oil Seals, Replace (B)		
GX470	14.0	14.5
LX450	14.2	14.7
LX470	15.9	16.4
RX300	14.3	14.8
RX330	16.6	17.1

CAMSHAFT
Camshaft, Replace (A)
GX470	9.3	9.8
LX450	6.4	6.6
LX470	9.3	9.8
RX300	10.4	10.9
RX330	9.0	9.5

Camshaft Gear or Sprocket, Replace (B)
GX470	3.7	3.9
LX470	4.1	4.3
RX300, RX330	3.4	3.6

Crankshaft Gear or Sprocket, Replace (B)
GX470	3.8	4.2
LX450	16.2	18.0
LX470	4.1	4.6
RX300, RX330	3.3	3.7

GX470 : LX450 : LX470 : RX300 : RX330   LEX-27

	LABOR TIME	SEVERE SERVICE
Timing Belt, Replace (B)		
GX470	3.6	*4.1*
LX470	4.0	*4.4*
RX300, RX330	3.2	*3.6*
Timing Belt/Chain Cover and/or Gasket, Replace (B)		
GX470	3.3	*3.7*
LX450	15.9	*17.7*
LX470	3.6	*4.1*
RX300	2.9	*3.2*
RX330	1.4	*1.5*
Timing Belt Tensioner, Replace (B)		
GX470	3.7	*3.8*
LX470	.7	*.8*
RX300	.5	*.6*
RX330	3.0	*3.1*
Timing Chain, Gear & Tensioner, Replace (B)		
LX450	16.2	*18.0*

CRANK & PISTONS

Connecting Rod Bearings, Replace (A)
Includes: Check bearing oil clearance.

GX470	15.5	*16.0*
LX450	7.0	*7.5*
LX470	15.1	*15.6*
RX300	5.5	*6.2*
RX330		
2WD	17.0	*17.5*
4WD	17.8	*18.3*

Crankshaft & Main Bearings, Replace (A)
Includes: Engine R&R.

GX470	24.8	*25.3*
LX450	23.6	*24.1*
LX470	24.1	*24.6*
RX300	23.3	*23.8*
RX330		
2WD	24.4	*24.9*
4WD	25.4	*25.9*

Crankshaft Front Oil Seal, Replace (B)

GX470	3.8	*4.0*
LX450	3.4	*3.6*
LX470	4.3	*4.5*
RX300	3.3	*3.5*
RX330	3.7	*3.9*

Crankshaft Pulley, Replace (B)

GX470	1.3	*1.4*
LX450	3.0	*3.4*
LX470	2.1	*2.3*
RX300	1.1	*1.3*
RX330	1.4	*1.6*

	LABOR TIME	SEVERE SERVICE
Pistons or Connecting Rods, Replace (A)		

Includes: Ridge reaming, cylinder wall deglazing, installing new rings and rod bearings, engine tune-up.

GX470	29.1	*32.2*
LX450	16.9	*17.8*
LX470	26.8	*29.7*
RX300	18.5	*20.6*
RX330		
2WD	27.0	*29.8*
4WD	27.8	*30.7*

Rear Main Oil Seal, Replace (B)
Includes: Trans. R&R when necessary.

GX470, LX470	7.6	*8.1*
LX450	8.9	*9.5*
RX300	10.5	*11.0*
RX330		
2WD	12.5	*13.0*
4WD	13.3	*13.8*

Rings, Replace (A)
Includes: Engine, cylinder head(s) and oil pan R&I. Ridge reaming, cylinder wall deglazing, clean carbon.

GX470	27.9	*30.9*
LX450	16.9	*17.8*
LX470	25.5	*28.1*
RX300	18.5	*20.6*
RX330		
2WD	25.7	*28.4*
4WD	26.6	*29.4*

ENGINE LUBRICATION

Engine Oil Cooler, Replace (B)

GX470, LX450	1.4	*1.5*
LX470	1.7	*1.9*

Engine Oil Pressure Switch (Sending Unit), Replace (B)

Exc. below	.5	*.5*
RX330	.6	*.7*

Oil Pan and/or Gasket, Replace (B)

GX470		
No. 1	13.6	*14.1*
No. 2	3.6	*4.1*
LX450		
No. 1	5.2	*5.7*
No. 2	3.1	*3.6*
LX470		
No. 1	13.8	*14.3*
No. 2	12.8	*13.3*
RX300		
No. 1	4.7	*5.2*
No. 2	2.9	*3.4*
RX330		
No. 1	6.2	*6.7*
No. 2	3.3	*3.8*

	LABOR TIME	SEVERE SERVICE
Oil Pump, Replace (B)		
GX470	15.9	*16.4*
LX450	15.0	*15.5*
LX470	15.2	*15.7*
RX300	6.6	*7.1*
RX330	9.1	*9.6*

AUTOMATIC TRANSAXLE

SERVICE TRANSAXLE INSTALLED

Oil Pressure Test (B)

RX300	2.0	*2.0*

Check Unit for Oil Leaks (C)

RX300	.9	*.9*

Drain and Refill Unit (B)

RX300, RX330	.8	*.8*

Oil Pan and/or Gasket, Replace (B)

RX300, RX330	2.2	*2.5*

Manual Lever Shaft Seal, Replace (B)

RX300	1.1	*1.3*
RX330	2.4	*2.7*

Valve Body Assy., Replace (B)
Includes: R&R oil pan.

RX300, RX330	3.0	*3.4*

SERVICE TRANSAXLE REMOVED

Flywheel (Flexplate), Replace (B)

RX300		
2WD	10.6	*11.1*
4WD	10.9	*11.4*
RX330		
2WD	12.4	*12.9*
4WD	13.4	*13.9*

Torque Converter, Replace (B)
Includes: R&R transaxle.

RX300		
2WD	10.5	*11.0*
4WD	10.8	*11.3*
RX330		
2WD	12.7	*13.2*
4WD	14.0	*14.5*

Transaxle Assy., R&R and Recondition (A)

RX300		
2WD	23.0	*23.9*
4WD	23.3	*24.2*
RX330		
2WD	20.5	*21.4*
4WD	22.1	*23.0*

Transaxle Assy., R&I (B)

RX300		
2WD	10.3	*10.8*
4WD	10.5	*11.0*
RX330		
2WD	11.8	*12.3*
4WD	13.0	*13.5*

LEX-28 GX470 : LX450 : LX470 : RX300 : RX330

AUTOMATIC TRANSMISSION

SERVICE TRANSMISSION INSTALLED

	LABOR TIME	SEVERE SERVICE
Oil Pressure Test (B)		
1996-05	2.1	2.3
Check Unit for Oil Leaks (C)		
1996-05	.9	.9
Drain and Refill Unit (B)		
1996-05	1.3	1.4
Oil Pan and/or Gasket, Replace (B)		
GX470	8.0	8.3
LX450, LX470	2.8	3.1
Transmission Speed Sensor, Replace (B)		
GX470	.5	.5
LX450, LX470	.8	.9
Valve Body Manual Lever Shaft Seal, Replace (B)		
GX470	9.4	9.6
LX450, LX470	1.3	1.5
Valve Body Assy., Replace (B)		
GX470	2.7	3.0
LX450, LX470	3.4	3.7

SERVICE TRANSMISSION REMOVED

Transmission R&R included unless otherwise noted.

	LABOR TIME	SEVERE SERVICE
Flywheel (Flexplate), Replace (B)		
GX470	8.2	9.2
LX450	7.9	8.1
LX470	8.1	9.0
Front Pump Oil Seal, Replace (B)		
GX470	9.0	10.0
LX450, LX470	7.8	8.7
Torque Converter, Replace (B)		
GX470	8.2	9.2
LX450, LX470	7.6	8.5
Transmission Assy., R&R and Recondition (A)		
GX470	12.6	13.5
LX450	20.8	21.7
LX470	19.6	20.5
Transmission Assy., R&I (B)		
GX470	8.2	9.2
LX450, LX470	8.4	9.4

TRANSFER CASE

	LABOR TIME	SEVERE SERVICE
Transfer Case, R&R and Recondition (A)		
GX470	15.4	16.3
LX450, LX470	11.4	12.3
RX300	14.4	15.3
RX330	16.2	17.1
Transfer Case, R&I (B)		
GX470	7.2	7.7
LX450, LX470	3.3	3.5
RX300	10.8	11.3
RX330	13.3	13.8

SHIFT LINKAGE

	LABOR TIME	SEVERE SERVICE
Shift Indicator Plate, Replace (B)		
LX470	.4	.4
RX300, RX330	.8	.8
Shift Cable, Replace (B)		
GX470	1.3	1.5
RX300	1.9	2.1
RX330	7.7	7.9
Shift Lever, Replace (B)		
LX470, RX300	1.0	1.1
Throttle Cable, Adjust (B)		
LX450, LX470, RX300	.5	.7
Throttle Cable, Replace (B)		
LX450, LX470	3.3	3.5

DRIVELINE

DRIVESHAFT

	LABOR TIME	SEVERE SERVICE
Driveshaft, R&R or Replace (B)		
GX470		
front	1.0	1.1
rear	.8	.9
LX450		
front	.9	1.1
rear	1.2	1.4
LX470		
front or rear	1.0	1.1
RX300, RX330		
front	1.9	2.1

DRIVEAXLE

	LABOR TIME	SEVERE SERVICE
Axle Inner Shaft, Replace (B)		
Front axle		
LX450	1.7	2.0
Axle Shaft, Replace (B)		
Rear axle		
GX470	3.1	3.3
LX450		
semi-floating	1.9	2.1
full-floating	.7	.9
LX470	2.5	2.7
RX300	1.6	1.8
Axle Shaft Bearings, Replace (B)		
Rear axle		
GX470		
one side	3.0	3.2
both sides	4.1	4.3
LX450		
semi-floating	1.9	2.2
full-floating	1.2	1.5
LX470		
one side	2.1	2.3
both sides	3.1	3.3

	LABOR TIME	SEVERE SERVICE
RX300		
2WD		
one side	1.6	1.8
both sides	2.4	2.7
4WD		
one side	3.5	3.9
both sides	3.1	3.4
Axle Shaft Oil Seal, Replace (B)		
Rear axle		
GX470		
one side	2.5	2.7
both sides	3.4	3.6
LX450		
semi-floating	1.9	2.2
full-floating	.8	1.1
LX470		
one side	2.4	2.6
both sides	3.7	3.9
RX300		
one side	3.5	3.9
both sides	5.8	6.5
Differential Carrier Assy., R&R and Recondition (A)		
Rear axle		
GX470	8.2	9.1
LX450		
Standard	6.8	7.4
Locking	7.5	8.1
LX470		
Standard	6.5	7.4
Locking	6.9	7.8
RX300	9.3	10.4
RX330	12.0	12.9
Front axle		
GX470	6.4	7.3
LX450	8.2	8.8
LX470	8.9	9.8
Differential Carrier, R&R or Replace (B)		
Rear axle		
GX470	3.8	4.3
LX450	2.2	2.7
LX470	2.4	2.9
RX300	1.9	2.4
RX330	7.1	7.6
Front axle		
GX470	3.5	4.0
LX450	3.1	3.6
LX470	3.4	3.9
Differential Cover Gasket, Replace (B)		
Rear axle		
GX470	3.7	4.0
LX450, LX470	2.1	2.4
RX300	2.2	2.5
RX330	7.2	7.5

GX470 : LX450 : LX470 : RX300 : RX330 **LEX-29**

	LABOR TIME	SEVERE SERVICE
Differential Side Bearings, Replace (B)		
Rear axle		
Both sides		
GX470	6.5	7.0
LX450		
standard	3.4	3.7
locking	3.8	4.1
LX470		
standard	3.4	3.9
locking	3.8	4.3
RX300	6.3	6.8
RX330	10.8	11.3
Front axle		
GX470	5.4	5.9
LX450	4.5	5.0
LX470	7.2	7.7
Pinion Bearings, Replace (B)		
Rear axle		
GX470	6.6	7.1
LX450		
Standard	5.6	6.1
Locking	6.0	6.5
LX470		
Standard	5.5	6.0
Locking	5.9	6.4
RX300	6.8	7.3
RX330	10.9	11.4
Front axle		
GX470	6.6	7.1
LX450, LX470	7.1	7.6
Pinion Shaft Oil Seal, Replace (B)		
Rear axle		
GX470	1.7	1.9
LX450, LX470	1.9	2.1
RX300	1.7	1.9
RX330	3.3	3.5
Front axle		
Exc. below	1.9	2.1
LX450	2.3	2.5
Ring Gear & Pinion Set, Replace (B)		
Rear axle		
GX470	6.9	7.4
LX450		
Standard	6.2	6.7
Locking	6.3	6.8
LX470		
Standard	5.8	6.3
Locking	6.2	6.7
RX300	7.9	8.4
RX330	11.3	11.8
Front axle		
GX470	5.8	6.3
LX450	7.7	8.2
LX470	7.1	7.6
Universal Joint Flange, Replace (B)		
Rear axle		
GX470, LX470	1.6	1.8
RX300	1.3	1.5
RX330	3.3	3.5

	LABOR TIME	SEVERE SERVICE
Front axle		
GX470	1.5	1.7
LX450	2.2	2.4
LX470	1.8	2.0
Universal Spider Joint, Replace or Rebuild (B)		
GX470	1.6	1.8
LX450, LX470	2.1	2.3
HALFSHAFTS		
CV Joint and Boot, Replace or Recondition (B)		
GX470		
one	3.3	3.7
both	4.9	5.5
LX470		
one	2.2	2.5
both	3.5	3.9
RX300		
front		
one	2.5	2.9
both	4.4	5.0
rear		
one	3.5	3.9
both	5.5	6.2
RX330		
front		
one	3.5	3.9
both	5.5	6.2
rear		
w/air suspension		
one	4.1	4.6
both	5.7	6.3
w/o air suspension		
one	3.6	4.1
both	5.2	5.8
Halfshaft, R&R or Replace (B)		
Front		
GX470		
one	2.9	3.2
both	4.0	4.4
LX470		
one	1.7	2.0
both	2.4	2.7
RX300		
one	1.3	1.4
both	1.9	2.1
RX330		
2WD		
one	3.0	3.4
both	4.1	4.6
4WD		
one	3.2	3.6
both	4.7	5.3
Rear		
RX300		
one	2.1	2.3
both	2.5	2.9

	LABOR TIME	SEVERE SERVICE
RX330		
w/air suspension		
one	3.5	3.9
both	4.4	5.0
w/o air suspension		
one	3.0	3.4
both	4.0	4.4

BRAKES
ANTI-LOCK

	LABOR TIME	SEVERE SERVICE
Diagnose Anti-Lock Brake System Component Each (A)		
1993-05	.5	.5
Retrieve Fault Codes (A)		
1993-05	.3	.3
Actuator Assy., Replace (B)		
Exc. below	2.4	2.6
RX330	1.9	2.1
Control Module, Replace (B)		
GX470	.9	1.0
LX450	.8	.9
LX470	.6	.7
RX300	.5	.6
Wheel Speed Sensor, Replace (B)		
Exc. below		
Front or rear	.9	1.0
each addl. add	.3	.4
RX300		
front		
one side	1.1	1.2
both sides	1.7	1.8
rear		
one side	2.4	2.5
both sides	3.5	3.6

SYSTEM

	LABOR TIME	SEVERE SERVICE
Bleed Brakes (B)		
Includes: Add fluid.		
One circuit	.8	.8
Two circuits	1.5	1.5
Brake System, Flush and Refill (B)		
1996-05 exc. below	1.7	1.9
LX450	1.1	1.3
Brake Hose (Flexible), Replace (B)		
Includes: System bleeding.		
GX470		
front or rear	1.4	1.5
LX450, LX470, RX300		
front or rear	1.0	1.1
RX330		
front or rear	1.3	1.4
each addl. add	.3	.4
Load Sensing Valve, Replace (B)		
Includes: System bleeding.		
LX450, LX470	2.2	2.3

LEX-30   GX470 : LX450 : LX470 : RX300 : RX330

	LABOR TIME	SEVERE SERVICE
Master Cylinder, Replace (B)		
Includes: System bleeding.		
GX470	2.4	2.6
LX450, LX470	1.4	1.6
RX300	2.0	2.2
RX330	1.2	1.4
Power Booster Unit, Replace (B)		
LX450	2.0	2.2
RX300, RX330	2.9	3.1
Power Booster Vacuum Check Valve, Replace (B)		
LX450	.7	.8
RX300, RX330	.9	1.0

SERVICE BRAKES

	LABOR TIME	SEVERE SERVICE
Brake Pads, Replace (B)		
GX470		
front	.8	.9
rear	1.6	1.8
LX450, LX470		
front or rear	1.1	1.3
RX300		
front or rear	.9	1.0
RX330		
front or rear	.8	.9
Caliper Assy., R&R and Recondition (A)		
Includes: System bleeding.		
LX450, LX470		
front or rear	1.4	1.5
RX300		
front or rear	1.4	1.5
Caliper Assy., Replace (B)		
Includes: System bleeding.		
GX470		
front or rear, one	1.4	1.6
LX450, LX470		
front or rear, one	1.0	1.1
RX300		
front or rear, one	1.1	1.2
RX300		
front or rear, one	1.6	1.8
Disc Brake Rotor, Replace (B)		
GX470		
front one	1.6	1.8
rear one	.9	1.0
LX450, LX470		
front one	1.6	1.8
rear one	.8	.9
RX300		
front one	.6	.7
rear one	.8	.9
RX330		
front one	1.0	1.1
rear one	.6	.7

PARKING BRAKE

	LABOR TIME	SEVERE SERVICE
Parking Brake Cable, Adjust (C)		
1996-04	.8	.9
Parking Brake Apply Actuator, Replace (B)		
GX470	.6	.6
LX450	.7	.7
LX470	.9	.9
RX300	1.2	1.2
RX330	1.9	1.9
Parking Brake Cable, Replace (B)		
GX470	1.9	2.1
LX450	1.2	1.4
LX470	1.3	1.5
RX300		
front or rear cables	1.6	1.8
RX330		
2WD		
front	2.2	2.4
rear both	2.5	2.7
4WD		
front	2.2	2.4
rear both	3.7	4.6

FRONT SUSPENSION

Unless otherwise noted, time given does not include alignment.

	LABOR TIME	SEVERE SERVICE
Align Front End (A)		
1996-05	1.9	2.1
Front Toe, Inspect/Adjust (B)		
1996-05	2.1	2.3
Front Axle Hub Oil Seal, Replace (B)		
GX470	3.5	3.7
LX450 one	1.3	1.5
LX470 one	2.3	2.5
Front Hub, Replace (B)		
GX470 one	3.5	3.7
LX450 one	2.5	2.7
LX470 one	2.9	3.1
RX330 one	2.5	2.7
RX330 one	3.0	3.2
Front Hub Bearing, Replace (B)		
LX450 one	1.6	1.8
LX470 one	2.9	3.1
RX300 one	2.3	2.5
RX330 one	2.9	3.1
Front Coil Spring, Replace (B)		
GX450		
one side	2.1	2.4
both sides	2.7	3.0
LX450		
one side	1.2	1.5
both sides	1.3	1.6
Front Torsion Bar Spring, Replace (B)		
LX470		
one side	1.0	1.2
both sides	1.3	1.5
Leading Arm, Replace (B)		
LX450		
one side	.7	.9
both sides	.9	1.1

	LABOR TIME	SEVERE SERVICE
Lower Control Arm, Replace (B)		
GX470		
one side	1.4	1.7
both sides	1.7	2.0
LX470		
one side	1.7	2.0
both sides	2.5	2.8
RX300		
one side	1.3	1.4
both sides	2.2	2.5
RX330		
2WD		
one side	10.4	10.7
both sides	10.6	10.9
4WD		
one side	10.9	11.2
both sides	11.2	11.5
Strut/Shock Absorber Replace (B)		
GX470		
one side	1.7	1.9
both sides	2.1	2.4
LX450		
one side	.7	.8
both sides	1.2	1.3
LX470		
one side	1.1	1.2
both sides	1.4	1.5
RX300		
one side	2.3	2.6
both sides	2.9	3.2
RX330		
one side	1.7	2.0
both sides	2.4	2.7
Strut Coil Spring, Replace (B)		
RX300		
one side	2.5	2.9
both sides	3.3	3.7
RX330		
one side	1.9	2.1
both sides	2.7	3.0
Stabilizer Bar, Replace (B)		
GX470	.8	1.0
LX450	.7	.9
LX470	1.1	1.3
RX300	1.7	1.9
RX300		
2WD	3.0	3.2
4WD	3.5	3.7
Stabilizer Bar Bushings, Replace (B)		
GX470		
one side	.6	.7
both sides	.8	1.0
LX450		
one side	.5	.6
both sides	.7	.9
LX470		
one side	1.1	1.2
both sides	1.6	1.8

GX470 : LX450 : LX470 : RX300 : RX330 **LEX-31**

	LABOR TIME	SEVERE SERVICE
RX300		
one side	.6	.7
both sides	1.0	1.2
RX300		
one side	1.0	1.1
both sides	1.1	1.3
Steering Knuckle, Replace (B)		
GX470	2.8	3.2
LX450	3.7	4.0
LX470	2.7	3.0
RX300	2.4	2.7
RX330	1.9	2.2
Suspension Support, Replace (B)		
RX300		
one side	2.4	2.7
both sides	3.0	3.4
RX330		
one side	1.7	2.0
both sides	2.5	2.9
Upper Control Arm, Replace (B)		
GX470		
one side	1.4	1.6
both sides	1.7	2.0
LX470		
one side	1.3	1.4
both sides	1.9	2.1

REAR SUSPENSION

	LABOR TIME	SEVERE SERVICE
Coil Spring, Replace (B)		
LX450, LX470		
one side	1.3	1.4
both sides	1.4	1.6
RX300, RX330		
one side	1.4	1.6
both sides	2.2	2.5
Lower Control Arm, Replace (B)		
GX470, LX450, LX470	.5	.8
Rear Control Arm No. 1, Replace (B)		
RX300	2.1	2.4
RX330		
2WD	1.5	1.8
4WD	4.8	5.1
Rear Control Arm No. 2, Replace (B)		
RX300	1.2	1.4
RX330	2.2	2.4
Shock Absorbers or Bushings, Replace (B)		
GX470		
one side	.8	.9
both sides	1.1	1.3
LX450		
one side	.7	.8
both sides	.8	1.0
LX470		
one side	1.3	1.4
both sides	1.6	1.8

	LABOR TIME	SEVERE SERVICE
RX300, RX330		
one side	1.3	1.4
both sides	1.9	2.1
Stabilizer Bar, Replace (B)		
GX470, LX450, LX470	.8	1.0
RX300	.5	.7
RX330	1.4	1.6
Stabilizer Bar Bushings, Replace (B)		
GX470		
one	.6	.7
both	.8	.9
LX450, LX470		
one	.5	.5
both	.6	.7
RX300, RX330		
one	.3	.4
both	.5	.6
Strut Rod, Replace (B)		
RX300		
one side	.8	.9
both sides	1.0	1.1
RX330		
one side	1.4	1.6
both sides	2.1	2.3

STEERING
AIR BAGS

	LABOR TIME	SEVERE SERVICE
Diagnose Air Bag System Component Each (A)		
All Models	.5	.5
Retrieve Fault Codes (A)		
All Models	.3	.3
Air Bag Assy., Replace (B)		
GX470		
passenger side	2.7	2.7
drivers side	.6	.6
side impact	2.6	2.6
LX450		
passenger side	2.5	2.5
drivers side	.5	.5
LX470		
passenger side	2.4	2.4
drivers side	.3	.3
side impact	1.5	1.5
RX300		
passenger side	.5	.5
drivers side	.5	.5
side impact	2.6	2.6
RX330		
passenger side	2.4	2.4
drivers side	.6	.6
side impact	6.0	6.7
Air Bag Sensor, Replace (B)		
GX470	1.0	1.0
LX470	.7	.7
RX300	.9	.9
RX330	1.3	1.4

	LABOR TIME	SEVERE SERVICE
Center Air Bag Sensor, Replace (B)		
GX470	.8	.8
LX450	.5	.5
LX470	.7	.7
RX300	.9	.9
RX330	2.2	2.5

LINKAGE

	LABOR TIME	SEVERE SERVICE
Tie Rod, Replace (B)		
Includes: Reset toe.		
LX450	1.5	1.7
Tie Rod End, Replace (B)		
Includes: Reset toe.		
GX470	1.6	1.8
LX450	1.0	1.2
LX470	1.1	1.3
RX300, RX330	1.4	1.6
Opposite side add.	.5	.5

POWER WORM & SECTOR

	LABOR TIME	SEVERE SERVICE
Gear Assy., R&R and Recondition (B)		
LX450	2.9	3.4
Gear Assy., Replace (B)		
LX450	1.8	2.1
Power Steering Control Module, Replace (B)		
LX450	.7	.7
Steering Pump, R&R or Replace (B)		
LX450	2.0	2.2
Steering Pump Hoses, Replace (B)		
Does not include purging.		
Pressure		
LX450	1.4	1.6
Return		
LX450	1.4	1.5
Steering Wheel, Replace (B)		
1996-97	.7	.7

POWER RACK & PINION

	LABOR TIME	SEVERE SERVICE
Gear Assy., R&R and Recondition (B)		
GX470	5.3	6.2
LX470	4.8	5.7
RX330	5.7	6.6
Gear Assy., Replace (B)		
GX470	3.3	3.8
LX470	2.1	2.6
RX300	3.1	3.6
RX330	3.4	3.9
Steering Pump, R&R or Replace (B)		
GX470	1.5	1.8
LX470	3.1	3.4
RX300, RX330	1.7	2.1

LEX-32 GX470 : LX450 : LX470 : RX300 : RX330

	LABOR TIME	SEVERE SERVICE
Steering Pump Hoses, Replace (B)		
Does not include purging.		
Pressure		
GX470 . . . . 1.8		2.1
LX470 . . . . 1.4		1.6
RX300, RX330 . . . . 1.6		1.9
Return		
GX470 . . . . 1.8		2.0
LX470 . . . . 1.3		1.5
RX300 . . . . 1.2		1.4
RX330 . . . . 1.5		1.7
Steering Wheel, Replace (B)		
1998-05 7		.7

HEATING & AIR CONDITIONING

When more than one component requires replacement where evacuation/recovery and recharging is already included, deduct 1.0 hour for each additional component from the time given.

Evacuate/Recover and Recharge System (B)
  1996-05 . . . . 1.2   1.2

AC Control Switch, Replace (B)
  GX470, LX450 7   .7
  LX470, RX300 3   .3

AC Hoses, Replace (B)
Includes: Evacuate/recover and recharge.
  1996-05 . . . . 1.5   1.7

Blower Motor, Replace (B)
  GX470 3   .3
  LX450 7   .9
  LX470, RX330 5   .5
  RX300 3   .3

Blower Motor Resistor, Replace (B)
  GX470, LX450 3   .3

Compressor Assy., Replace (B)
Includes: Transfer parts as required, evacuate/recover and recharge.
  GX470 . . . . 2.5   2.9
  LX450 . . . . 1.8   2.1
  LX470 . . . . 2.1   2.4
  RX300 . . . . 3.1   3.4
  RX330 . . . . 2.5   2.9

Compressor Clutch Assy., Replace (B)
  GX470 . . . . 2.8   3.1
  LX450 . . . . 2.2   2.4
  LX470 . . . . 2.3   2.6
  RX300 . . . . 3.5   3.9
  RX330 . . . . 2.9   3.2

Condenser Assy., Replace (B)
Includes: Evacuate/recover and recharge.
  GX470 . . . . 2.1   2.4
  LX450 . . . . 2.0   2.2
  LX470 . . . . 1.5   1.7
  RX300, RX330 . . . . . . 2.6   2.9

Evaporator Coil, Replace (B)
Includes: Evacuate/recover and recharge.
  GX470 . . . . 6.0   6.2
  LX450 . . . . 1.7   1.9
  LX470 . . . . 3.3   3.5
  RX300, RX330 . . . . . . 5.6   5.8

Expansion Valve, Replace (B)
Includes: Evacuate/recover and recharge.
  GX470 . . . . 1.3   1.5
  LX450 . . . . 1.7   1.9
  LX470 . . . . 3.1   3.3
  RX300 . . . . 5.6   5.8
  RX330 . . . . 2.9   3.1

Heater & AC Control Assy., Replace (B)
  1996-05 exc. below 5   .5
    GX470 8   .8

Heater Control Valve, Replace (B)
  LX450, LX470 . . . . . . 1.2   1.4
  RX300 . . . . 1.8   2.0

Heater Core, R&R or Replace (B)
  GX470 . . . . 6.7   6.9
  LX450 . . . . 4.9   5.1
  LX470 . . . . 3.7   3.9
  RX300 . . . . 5.7   5.9

Heater Hoses, Replace (B)
  Exc. below 1.2   1.4
  RX330 . . . . 1.5   1.7

High or Low Pressure Switch, Replace (B)
  LX450, LX470,
  RX300 . . . . 1.8   1.8

Receiver/Drier Assy., Replace (B)
Includes: Evacuate/recover and recharge.
  GX470 . . . . 2.2   2.4
  LX450 . . . . 1.7   1.9
  LX470 . . . . 1.4   1.6
  RX300 . . . . 2.4   2.6

WIPERS & SPEEDOMETER

Antenna Assy., Replace (B)
  LX470, RX300 8   .9
  RX330 . . . . 3.9   4.0

Radio, R&R (B)
  All Models 6   .6

Rear Window Washer Motor & Pump, Replace (B)
  GX470 8   .9
  LX450, LX470 5   .5

Rear Window Wiper Motor, Replace (B)
  GX470, LX450,
  LX470 8   .9
  RX300, RX330 5   .7

Rear Windshield Wiper Switch, Replace (B)
  LX450 . . . . 1.3   1.3

Speedometer Head, R&R or Replace (B)
  GX470 7   .7
  LX450 . . . . 1.2   1.2
  LX470, RX300 . . . . . . 1.1   1.1
  RX330 5   .5

Washer Motor & Pump Assy., Replace (B)
  GX470 8   .9
  LX450, LX470 5   .5
  RX300 . . . . 1.2   1.4
  RX330 9   .9

Windshield Wiper Switch, Replace (B)
  GX470 3   .3
  LX450 . . . . 1.3   1.3
  LX470 3   .3
  RX300, RX330 5   .5

Windshield Wiper Linkage, Replace (B)
  GX470 6   .7
  LX450 9   1.0
  LX470 8   .9
  RX300 . . . . 1.1   1.2
  RX330 6   .7

Windshield Wiper Motor, Replace (B)
  GX470 6   .7
  LX450, LX470 5   .8
  RX300 . . . . 1.1   1.2
  RX330 6   .7

LAMPS & SWITCHES

Back-Up Lamp Bulb, Replace (C)
  Exc. below 3   .3
    GX470 6   .6
    RX330 5   .5

Halogen Headlamp Bulb, Replace (B)
  All Models 3   .3

Hazard Warning Switch, Replace (B)
  1996-05 5   .5

Headlamp Switch, Replace (B)
  Exc. below 3   .3
  LX450 . . . . 1.3   1.3

Headlamps, Aim (B)
  Two 5   .5
  Four 7   .7

High Mount Stop Lamp Assy., Replace (B)
  Exc. below 3   .3
    LX470 5   .5

GX470 : LX450 : LX470 : RX300 : RX330   LEX-33

	LABOR TIME	SEVERE SERVICE
Horn, Replace (B)		
1996-05	.3	.3
License Lamp Lens or Bulb, Replace (C)		
1996-05	.3	.3
Neutral Safety Switch, Replace (B)		
GX470, LX450, LX470	.9	1.0
RX300	.8	.9
RX330	1.1	1.2
Parking Brake Warning Lamp Switch, Replace (B)		
GX470	.6	.7
LX450	.5	.5
LX470	1.0	1.1
RX300	.5	.5
RX330	1.7	2.0
Parking Lamp Lens or Bulb, Replace (C)		
LX450, LX470, RX330	.3	.3
RX300	.5	.5
Rear Combination Lamp Lens, Replace (B)		
GX470, LX450, LX470	.3	.3
RX300, RX330	.5	.5
Stop Lamp Switch, Replace (B)		
GX470	.5	.5
LX450, LX470	.7	.7
RX300	.6	.6
RX330	.8	.8

	LABOR TIME	SEVERE SERVICE
Turn Signal Lamp Lens or Bulb, Replace (C)		
Exc. below	.3	.3
RX300	.8	.8
Turn Signal or Hazard Warning Flasher, Replace (B)		
GX470, LX450, LX470, RX330	.3	.3
RX300	.5	.5
Turn Signal Switch, Replace (B)		
Exc. below	.7	.7
LX450	.8	.8

BODY

	LABOR TIME	SEVERE SERVICE
Front Door Lock, Replace (B)		
Exc. below	.7	.7
LX450	.8	.8
Hood Hinge, Replace (B)		
GX470	.5	.5
LX450, LX470	.8	.9
RX300, RX330	.9	.9
Hood Lock, Replace (B)		
Exc. below	.3	.3
LX450	.7	.7
Hood Release Cable, Replace (B)		
GX470	.8	.9
LX450, LX470	.8	.9
RX300	1.2	1.4
RX330	.5	.5

	LABOR TIME	SEVERE SERVICE
Door Lock Striker Plate, Replace (B)		
1996-05	.5	.5
Rear Door Lock, Replace (B)		
GX470	.9	1.0
LX450	1.1	1.2
LX470	.6	.7
RX300, RX330	.8	.9
Window Regulator, Replace (B)		
Front		
GX470, LX450	.9	.9
LX470, RX300, RX330	.7	.8
Rear		
GX470	.8	.8
LX450	1.1	1.1
LX470	.7	.8
RX300, RX330	.9	.9
Window Regulator Motor, Replace (B)		
Front		
GX470	1.2	1.4
LX450	1.1	1.2
LX470	.8	.9
RX300, RX330	.8	.9
Rear		
GX470	.9	1.0
LX450	1.1	1.2
LX470	.8	.9
RX300, RX330	.9	1.0

NOTES

MAZ

**323 : 626 : 929 : GLC : Mazda3 : Mazda6 : Miata : Millenia :
MX-3 : MX-6 : Protege : Protege5 : RX-7 : RX-8**

SYSTEM INDEX

MAINTENANCE	MAZ-2
Maintenance Schedule	MAZ-2
CHARGING	MAZ-2
STARTING	MAZ-3
CRUISE CONTROL	MAZ-4
IGNITION	MAZ-4
EMISSIONS	MAZ-5
FUEL	MAZ-6
EXHAUST	MAZ-9
ENGINE COOLING	MAZ-10
PISTON ENGINE	MAZ-11
Assembly	MAZ-11
Cylinder Head	MAZ-13
Camshaft	MAZ-15
Crank & Pistons	MAZ-17
Engine Lubrication	MAZ-19
ROTARY ENGINE	MAZ-20
CLUTCH	MAZ-21
MANUAL TRANSAXLE	MAZ-22
MANUAL TRANSMISSION	MAZ-23
AUTO TRANSAXLE	MAZ-23
TRANSFER CASE	MAZ-26
SHIFT LINKAGE	MAZ-26
DRIVELINE	MAZ-27
BRAKES	MAZ-29
FRONT SUSPENSION	MAZ-32
REAR SUSPENSION	MAZ-36
STEERING	MAZ-39
HEATING & AC	MAZ-43
WIPERS & SPEEDOMETER	MAZ-44
LAMPS & SWITCHES	MAZ-45
BODY	MAZ-46

OPERATIONS INDEX

A
AC Hoses	MAZ-43
Air Bags	MAZ-39
Air Conditioning	MAZ-43
Alignment	MAZ-32
Alternator (Generator)	MAZ-3
Antenna	MAZ-44
Anti-Lock Brakes	MAZ-29

B
Back-Up Lamp Switch	MAZ-45
Ball Joint	MAZ-33
Battery Cables	MAZ-3
Bleed Brake System	MAZ-30
Blower Motor	MAZ-43
Brake Disc	MAZ-31
Brake Drum	MAZ-31
Brake Hose	MAZ-30
Brake Pads and/or Shoes	MAZ-31

C
Camshaft	MAZ-15
Catalytic Converter	MAZ-9
Coolant Temperature (ECT) Sensor	MAZ-5
Crankshaft	MAZ-17
Crankshaft Sensor	MAZ-4
Cruise Control	MAZ-4
CV Joint	MAZ-28
Cylinder Head	MAZ-13

D
Differential	MAZ-27
Distributor	MAZ-4
Drive Belt	MAZ-2

E
EGR	MAZ-5
Electronic Control Module (ECM/PCM)	MAZ-6
Engine	MAZ-11
Engine Lubrication	MAZ-19
Engine Mounts	MAZ-12
Evaporator	MAZ-43
Exhaust	MAZ-9
Exhaust Manifold	MAZ-9
Expansion Valve	MAZ-44

F
Flywheel	MAZ-22
Fuel Injection	MAZ-8
Fuel Pump	MAZ-7
Fuel Vapor Canister	MAZ-6

G
Gear Selector Lever	MAZ-26
Generator	MAZ-3

H
Halfshaft	MAZ-28
Headlamp	MAZ-45
Heater Core	MAZ-44
Horn	MAZ-46

I
Idle Air Control (IAC) Valve	MAZ-8
Ignition Coil	MAZ-4
Ignition Module	MAZ-5
Ignition Switch	MAZ-5
Inner Tie Rod	MAZ-40
Intake Air Temperature (IAT) Sensor	MAZ-6
Intake Manifold	MAZ-7

K
Knock Sensor	MAZ-6

L
Lower Control Arm	MAZ-34

M
Maintenance Schedule	MAZ-2
Mass Air Flow (MAF) Sensor	MAZ-6
Master Cylinder	MAZ-30
Muffler	MAZ-10
Multifunction Lever/Switch	MAZ-45

N
Neutral Safety Clutch Switch	MAZ-45
Neutral Safety Switch	MAZ-45

O
Oil Pan	MAZ-19
Oil Pump	MAZ-20
Outer Tie Rod	MAZ-40
Oxygen Sensor	MAZ-6

P
Parking Brake	MAZ-32
Pistons	MAZ-18
Positive Crankcase Ventilation (PCV) Valve	MAZ-6

R
Radiator	MAZ-10
Radiator Hoses	MAZ-11
Radio	MAZ-45
Rear Main Oil Seal	MAZ-18

S
Shock Absorber/Strut, Front	MAZ-35
Shock Absorber/Strut, Rear	MAZ-38
Spark Plug Cables	MAZ-5
Spark Plugs	MAZ-5
Spring, Front Coil	MAZ-33
Spring, Rear Coil	MAZ-36
Starter	MAZ-3
Steering Wheel	MAZ-40

T
Thermostat	MAZ-10
Throttle Body	MAZ-9
Throttle Position Sensor (TPS)	MAZ-6
Timing Belt	MAZ-16
Timing Chain	MAZ-17
Torque Converter	MAZ-24

U
U-Joint	MAZ-28
Upper Control Arm	MAZ-36

V
Valve Body	MAZ-24
Valve Cover Gasket	MAZ-14
Valve Job	MAZ-14
Vehicle Speed Sensor	MAZ-6

W
Water Pump	MAZ-11
Wheel Balance	MAZ-2
Wheel Cylinder	MAZ-32
Window Regulator	MAZ-46
Windshield Washer Pump	MAZ-45
Windshield Wiper Motor	MAZ-45

MAZ-2   323 : 626 : 929 : GLC : MAZDA3 : MAZDA6 : MIATA : MILLENIA

	LABOR TIME	SEVERE SERVICE

MAINTENANCE

Air Cleaner Filter Element, Replace (C)
All Models 3   .5

Chassis Lubrication, Change Oil & Filter (C)
Includes: Correct fluid levels.
All Models 5   .6

Drive Belt, Adjust (B)
Alternator 5   .6
Serpentine 9   1.0

Drive Belt Replace (B)
Air pump
 1981-85 626, GLC 3   .5
 1983-91 RX-7 5   .7
Alternator
 323, Protégé/5
  1986-94 5   .7
  1995-05 9   1.1
 626, MX-6 9   1.1
 929
  1988-91 8   1.0
  1992-95 ....1.2 1.4
 GLC, Mazda6 8   1.0
 Mazda3 5   .7
 Miata 6   .8
 Millenia 7   .9
 MX-3 8   1.0
 RX-7
  1981-91 5   .7
  1993-95 ....1.4 1.6
 RX-8 (.5) 7   .9
Compressor
 exc. below 7   .9
 Miata 5   .7
Fan
 1981-82 626 7   .9
 1988-95 929 7   .9
 1981-91 RX-7 7   .9
Steering pump
 323, Protégé/5 6   .8
 626, MX-6
  1981-92 5   .7
  1993-97
   4 cyl. 7   .9
   V6 ....1.2 1.4
  1998-02 7   .9
 929 7   .9
 GLC, Miata 5   .7
 Millenia 7   .9
 MX-3
  4 cyl. 5   .7
  V6 8   1.0
 RX-7
  1981-85 3   .5
  1986-91 ....1.2 1.4
  1993-95 7   .9
w/AC add
 air pump belt only add ..2 .2
 fan belt only ......1 .1

Fuel Filter, Replace (B)
323, Protégé/5
 1986-94 3   .5
 1995-05 9   1.1
626, MX-6
 1981-92 3   .5
 1993-02 9   1.1
929
 1988-91 3   .5
 1992-95 7   .9
GLC 3   .5
Mazda3 ....2.4 2.6
Mazda6 8   1.0
Miata
 1990-93 3   .5
 1994-05 (.8) 1.1 1.3
Millenia 7   .9
MX-3 8   1.0
RX-7
 1981-85 3   .5
 1986-91 5   .7
 1993-95 7   .9
RX-8 (.5) 7   .9

Halogen Headlamp Bulb, Replace (B)
1985-05 each 3   .3

Oil & Filter, Change (C)
Includes: Correct fluid levels.
All Models 5   .6

Sealed Beam Headlamp, Replace (B)
One side 5   .5

Timing Belt, Replace (B)
323, Protégé/5
 1986-94
  SOHC 2.4 2.6
  DOHC 3.4 3.6
 1995-05 (2.4) 3.2 3.3
626, MX-6
 1983-87 ....1.9 2.1
 1988-92 ....3.5 3.7
 1993-02
  4 cyl. (2.2) 3.0 3.2
  V6 (3.1) 4.2 4.4
929
 1988-91 ....3.1 3.3
 1992-95 ....3.9 4.1
Miata 3.1 3.2
Millenia
 KJ 5.8 6.0
 KL 4.1 4.3
MX-3
 4 cyl.
  1992-93 ....2.3 2.5
  1994-95 ....3.1 3.3
 V6 ....3.9 4.1
w/AC add ....2 .2
w/PS add ....1 .1

Tire, Replace (C)
Includes: Dismount old tire and mount new tire to rim.
All Models 5   .5

Wheel, Balance (B)
One 3   .3
each addl. add 2   .2

SCHEDULED MAINTENANCE INTERVALS

If necessary, refer to appropriate Chilton maintenance service information.

323, 626, MAZDA3, MAZDA6, MIATA, MILLENIA, MX-3, MX-6, PROTÉGÉ, RX-7, RX-8

7,500 Mile Service (C)
All Models 4   .6
15,000 Mile Service (C)
All Models 4   .6
22,500 Mile Service (C)
All Models 4   .6
30,000 Mile Service (B)
Exc. below 3.8 4.0
Miata 3.6 3.8
MX-3, Protégé 3.2 3.4
37,500 Mile Service (C)
All Models 4   .6
45,000 Mile Service (C)
All Models 4   .6
52,500 Mile Service (C)
All Models 4   .6
60,000 Mile Service (B)
Exc. below 6.6 6.8
Miata 7.1 7.3
MX-3 ....6.3 6.5
Protégé 6.8 7.0
67,500 Mile Service (C)
All Models 4   .6
75,000 Mile Service (C)
All Models 4   .6
82,500 Mile Service (C)
All Models 4   .6
90,000 Mile Service (B)
Exc. below 3.8 4.0
Miata 3.6 3.8
MX-3, Protégé 3.2 3.4
97,500 Mile Service (C)
All Models 4   .6

CHARGING

Alternator Circuits, Test (B)
Includes: Test component output.
All Models 7   .7

Alternator, R&R and Recondition (B)
323, Protégé
 1986-94 ....2.5 2.7
 1995-98 ....2.7 2.9
 1999-02 (1.8) 2.4 2.6
626, MX-6
 1983-92 ....2.3 2.5
 1993-02 (2.2) 3.0 3.2

MX-3 : MX-6 : PROTEGE : PROTEGE5 : RX-7 : RX-8 — MAZ-3

	LABOR TIME	SEVERE SERVICE
929		
1988-91	2.4	2.6
1992-95	2.8	3.0
GLC	2.5	2.7
Miata	2.4	2.6
Millenia	3.0	3.2
MX-3		
1992-93	2.6	2.8
1994-95		
4 cyl.	3.4	3.6
V6	3.2	3.4
RX-7		
1981-91	2.3	2.5
1993-95	3.7	3.9

Alternator Assy., Replace (B)
Includes: Pulley transfer.

	LABOR TIME	SEVERE SERVICE
323, Protégé/5		
1986-91	1.2	1.4
1992-94	.8	1.0
1995-98	1.4	1.6
1999-00	.8	1.0
2001-05 (1.0)	1.4	1.6
626, MX-6		
1983-92	.7	.9
1993-02	1.6	1.8
929		
1988-91	.7	.9
1992-95	1.4	1.6
GLC	1.2	1.4
Mazda3	.7	.9
Mazda6		
2.3L (1.0)	1.4	1.6
3.0L (1.5)	2.0	2.2
Miata	1.2	1.4
Millenia	1.6	1.8
MX-3		
1992-93		
4 cyl.	.9	1.1
1994-95		
4 cyl.	2.0	2.2
V6	1.7	1.9
RX-7		
1981-91	.5	.7
1993-95	2.3	2.5
RX-8 (.5)	.7	.9

Alternator Voltage Regulator, Replace (B)

	LABOR TIME	SEVERE SERVICE
323, Protégé/5		
1986-94	2.2	2.4
1995-98	2.6	2.8
1999-05 (1.6)	2.2	2.4
626, MX-6		
1983-92	1.7	1.9
1993-02 (2.1)	2.8	3.0
929		
1988-91	1.8	2.0
1992-95	2.3	2.5
GLC		
external	.5	.7
internal	2.2	2.4

	LABOR TIME	SEVERE SERVICE
Mazda3	1.9	2.1
Mazda6		
2.3L (1.9)	2.6	2.8
3.0L (2.4)	3.2	3.4
Miata		
1990-97	2.4	2.6
1999-05 (.9)	1.2	1.4
Millenia	2.9	3.1
MX-3		
1992-93 4 cyl.	2.5	2.7
1994-95	3.1	3.3
RX-7		
1981-91	1.7	1.9
1993-95	3.2	3.4
RX-8 (1.4)	1.9	2.1

Front Alternator Bearing, Replace (B)

	LABOR TIME	SEVERE SERVICE
323, Protégé	1.4	1.6
626, MX-6		
1981-82	.9	1.1
1983-93	1.2	1.4
929		
1988-91	.9	1.1
1992-95	1.6	1.8
MX-3		
front	1.2	1.4
rear	1.5	1.7
RX-7		
1981-91	.8	1.0
1993-95	2.6	2.8
Replace rear bearing add	.2	.4

STARTING

Starter Draw Test (On Car) (B)
All Models .3 / .3

Battery, Replace (C)
All Models .4 / .6

Battery Cables, Replace (C)
Positive .7 / .9
Negative .3 / .5

Starter, R&R and Recondition (B)

	LABOR TIME	SEVERE SERVICE
323, Protégé		
1986-94		
2WD	2.4	2.6
4WD	3.1	3.3
1995-98	2.5	2.7
1999-02	2.6	2.8
626, MX-6		
1981-92	2.5	2.7
1993-97		
4 cyl.	2.6	2.8
V6		
AT	3.8	4.0
MT	2.2	2.4

	LABOR TIME	SEVERE SERVICE
1998-02		
4 cyl.	2.6	2.8
V6		
AT	2.9	3.1
MT	2.2	2.4
929	2.7	2.9
GLC3	1.9	2.2
Miata	2.8	3.1
Millenia	2.6	2.9
MX-3		
4 cyl.	3.0	3.2
V6		
AT	2.8	3.0
MT	2.3	2.5
RX-7		
1981-85	2.0	2.2
1986-91	1.7	1.9
1993-95		
AT	2.7	2.9
MT	2.4	2.6

Starter Assy., Replace (B)

	LABOR TIME	SEVERE SERVICE
323, Protégé/5		
1986-94		
2WD	.7	.9
4WD	1.7	1.9
1995-05 (.6)	.8	1.0
626, MX-6		
1981-92	.8	1.0
1993-97		
4 cyl.	1.3	1.5
V6		
AT	2.6	2.8
MT	.7	.9
1998-02		
4 cyl.	.9	1.1
V6		
AT	1.4	1.6
MT	.7	.9
929	.9	1.1
GLC	.7	.9
Mazda3 (.6)	.8	1.0
Mazda6 (1.0)	1.4	1.6
Miata	1.7	1.9
Millenia	1.2	1.4
MX-3		
4 cyl.	1.4	1.6
V6		
AT	1.3	1.5
MT	.7	.9
RX-7		
1981-91	.7	.9
1993-95		
AT	1.2	1.4
MT	.8	1.0
RX-8 (.4)	.6	.8

MAZ-4 323 : 626 : 929 : GLC : MAZDA3 : MAZDA6 : MIATA : MILLENIA

	LABOR TIME	SEVERE SERVICE
Starter Drive Assy., Replace (B)		
Includes: R&R starter.		
323, Protégé/5		
1986-94		
2WD	2.0	2.2
4WD	2.9	3.1
1995-05 (1.9)	2.6	2.8
626, MX-6		
1981-82	1.9	2.1
1983-92	2.2	2.4
1993-97		
4 cyl.	2.5	2.7
V6		
MT	2.0	2.2
AT	3.5	3.7
1998-02		
4 cyl.	2.5	2.7
V6		
AT	2.7	2.9
MT	2.0	2.2
929	2.2	2.4
GLC	1.4	1.6
Mazda3 (1.0)	1.4	1.6
Mazda6 (1.9)	2.6	2.8
Miata		
1990-97	2.4	2.6
1999-05 (2.2)	3.0	3.2
Millenia	2.2	2.4
MX-3		
4 cyl.	2.5	2.7
V6		
AT	2.4	2.6
MT	1.7	1.9
RX-7		
1981-91	1.6	1.8
1993-95		
AT	2.3	2.5
MT	2.0	2.2
RX-8	1.3	1.5
Starter Solenoid, Replace (B)		
323, Protégé/5		
1986-94		
2WD	1.3	1.5
4WD	2.2	2.4
1995-05 (1.0)	1.4	1.6
626, MX-6		
1981-92	1.3	1.5
1993-97		
4 cyl.	1.3	1.5
V6		
AT	2.7	2.9
MT	.7	.9
1998-02		
4 cyl.	1.3	1.5
V6		
AT	1.6	1.8
MT	.7	.9
929	1.3	1.5
GLC	1.2	1.4
Mazda3	1.0	1.2

	LABOR TIME	SEVERE SERVICE
Mazda6	1.6	1.8
Miata		
1990-97	1.4	1.6
1999-05	1.9	2.1
Millenia	1.3	1.5
MX-3		
4 cyl.	1.6	1.8
V6		
AT	1.4	1.6
MT	.7	.9
RX-7		
1981-91	.7	.9
1993-95	1.5	1.7
RX-8 (.5)	.7	.9

CRUISE CONTROL

	LABOR TIME	SEVERE SERVICE
Diagnose Cruise Control System (A)		
All Models (.7)	1.1	1.1
Actuator Assy., Replace (B)		
All Models	.7	.9
Control Main Switch, Replace (B)		
Exc. below	.5	.7
929	.7	.9
Mazda3, Mazda6	.6	.8
MX-3	.8	1.0
RX-8	.6	.8
Control Safety Switch (Cut-Out), Replace (B)		
All Models	.5	.7
Control Switch, Replace (B)		
929	.8	1.0
1986-91 RX-7	.5	.7
Control Unit (Box), Replace (B)		
323, Protégé/5		
1986-94	.7	.9
1995-98	1.2	1.4
1999-05	.5	.7
626, MX-6	.5	.7
929	.5	.7
Miata	.8	1.0
Millenia	.7	.9
MX-3	.8	1.0
RX-7	.5	.7

IGNITION

	LABOR TIME	SEVERE SERVICE
Diagnose Ignition System Component Each (A)		
1983-05	.8	.8
Ignition Timing, Reset (B)		
All Models	.7	.9
Crankshaft Position Sensor, Replace (B)		
323, Protégé/5	.9	1.1
626, MX-6		
1993-98		
4 cyl.	.9	1.1
V6	1.6	1.8
1999-02	.8	1.0
929	1.6	1.8

	LABOR TIME	SEVERE SERVICE
Mazda3 (.7)	.9	1.1
Mazda6		
2.3L (1.5)	2.0	2.2
3.0L (.8)	1.1	1.3
Miata	.9	1.1
Millenia		
1995-98	1.2	1.4
1999-02 each side	.8	1.0
MX-3	1.2	1.4
RX-7		
1986-91	.8	1.0
1993-95	1.2	1.4
RX-8 (.5)	.7	.9
Distributor, R&R and Recondition (B)		
Includes: Reset base ignition timing.		
323, GLC	2.0	2.2
626, MX-6		
1981-82	2.0	2.2
1983-92		
w/turbo	1.4	1.6
w/o turbo	2.2	2.4
Distributor, Replace (B)		
Includes: Reset base ignition timing.		
All Models	.8	1.0
Distributor Cap and/or Rotor, Replace (B)		
All Models	.3	.3
Distributor Igniter Set, Replace (B)		
Exc. below	1.3	1.5
626, MX6		
1981-82	.5	.7
1983-92	1.4	1.6
1993-97	.7	.9
929		
1988-91	.5	.7
1992-95	1.2	1.4
Distributor Pickup Set, Replace (B)		
1986-89 323	.9	1.1
626, MX-6		
1981-85	1.3	1.5
1986-92	.8	1.0
GLC	1.4	1.6
1983-85 RX-7	1.4	1.6
Electronic Spark Control Module, Replace (B)		
1986-92 626, MX-6	.5	.5
Ignition Coil, Replace (B)		
323, Protégé/5	.5	.7
626, MX-6	.5	.7
929		
1988-91	.5	.7
1992-95	.9	1.1
GLC	.5	.7
Mazda3	.3	.5
Mazda6		
2.3L	.3	.5
3.0L		
right side	1.8	2.0
left side	.5	.7
both sides	2.0	2.2

MX-3 : MX-6 : PROTEGE : PROTEGE5 : RX-7 : RX-8 **MAZ-5**

	LABOR TIME	SEVERE SERVICE
Miata	.7	.9
Millenia		
right side	1.2	1.4
left side	.5	.7
both sides	1.4	1.6
MX-3	.7	.9
RX-7		
1981-91	.5	.7
1993-95	2.0	2.2
RX-8 (.4)	.6	.8

Ignition Coil Igniters, Replace (B)

| 1986-95 RX-7 | .7 | .9 |

Ignition Lock Switch Assy., Replace (B)

Exc. below	1.4	1.6
Mazda3	.6	.8
RX-8	.9	1.1

Ignition Module, Replace (B)

1986-95 323, Protégé	.8	1.0
626, MX-6		
1987-92	1.3	1.5
1993-97	.5	.7
929		
1988-91	.5	.7
1992-95	1.2	1.4
MX-3	.5	.7
1986-95 RX-7	.7	.9

Ignition Switch, Replace (B)

323, Protégé/5	.7	.9
626, MX-6		
1981-82	1.4	1.6
1983-87	.5	.7
1988-92	1.2	1.4
1993-02	.8	1.0
929		
1988-91	.7	.9
1992-95	1.2	1.4
GLC	.7	.9
Mazda3 (.4)	.6	.8
Mazda6 (.3)	.5	.7
Miata	.5	.7
Millenia	.7	9
MX-3	.9	1.1
RX-7	.8	1.0
RX-8 (.3)	.5	.7

Spark Plug (Ignition) Cables, Replace (B)

Exc. below	.5	.7
1992-95 929	1.9	2.1
1993-95 RX7	1.9	2.1

Spark Plugs, Replace (B)

323, Protégé/5	.5	.7
626, MX-6	.5	.7
929	.7	.9
GLC	.5	.7
Mazda3 (.4)	.6	.8
Mazda6		
2.3L (.4)	.6	.8
3.0L (1.6)	2.2	2.4
Miata	.5	.7

	LABOR TIME	SEVERE SERVICE
Millenia		
KJ	1.6	1.8
KL	.5	.7
MX-3		
4 cyl.	.5	.7
V6	.7	.9
RX-7		
1981-91	.5	.7
1993-95	.9	1.1
RX-8 (.4)	.6	.7

Vacuum Advance Unit, Replace (B)
Includes: Reset base ignition timing.

| All Models | .7 | .9 |

EMISSIONS

Diagnose Emissions System (A)

| All Models (.8) | 1.1 | 1.1 |

Dynamometer Test (A)

| All Models | .5 | .5 |

Air Control Valve, Replace (B)

1981-85 626	.5	.7
1992-95 MX-3	.5	.7
RX-7		
1983-91	1.3	1.5
1993-95	2.6	2.8

Air Flow Meter, Replace (B)

1986-94 323, Protégé	.7	.9
1983-87 626	.7	.9
929		
1988-91	.7	.9
1992-95	.9	1.1
Miata	.7	.9
Millenia	.8	1.0
MX-3	.5	.7

Air Injection Check Valve, Replace (B)

RX-7		
1983-85	.7	.9
1986-91	1.2	1.4
1993-95	2.7	2.9
RX-8 (.4)	.6	.8

Air Injection Nozzles, Replace (B)

| GLC | .9 | 1.1 |

Air Pump Assy., Replace (B)

1981-85 626, GLC	.7	.9
1981-95 RX-7	.8	1.0
RX-8	.5	.7
w/AC add	.2	.2

Air Temperature Sensor, Replace (B)

| 929 | .5 | .7 |
| Mazda6, Miata, Protégé/5 | .3 | .5 |

Altitude Compensator, Replace (B)

| 1983-85 RX-7 | .7 | .9 |

Anti-Backfire Valve, Replace (B)

| GLC | .5 | .7 |

	LABOR TIME	SEVERE SERVICE
Check Valve, Replace (B)		
Exc. below	.7	.9
1981-92 626, MX-6		
2WS	.8	1.0
4WS	1.4	1.6

Coolant Temperature (ECT) Sensor, Replace (A)

323, Protégé/5		
1986-95	.7	.9
1996-05 (.3)	.5	.7
626, MX-6		
1987-92	.7	.9
1993-02	.5	.7
929	.7	.9
Mazda3, Mazda6	1.1	1.3
MX-3 (.3)	.5	.7
Millenia		
KJ	4.5	4.7
KL	.5	.7
RX-7		
1986-91	.5	.7
1993-95	4.2	4.4
RX-8 (1.0)	1.4	1.6

Deceleration Valve, Replace (B)

| 1981-91 RX-7 | .7 | .9 |

Detonation Sensor, Replace (B)

RX-7		
1987-91	2.3	2.5
1993-95	3.2	3.4

EGI Control Unit, Replace (B)

Miata	.7	.9
RX-7		
1983-91	.7	.9
1993-95	.5	.7

EGR Modulator Valve, Replace (B)

626, MX-6		
1987-92	.5	.7
1993-97	.3	.5

EGR Valve, Replace (B)

323, Protégé/5	.8	1.0
626, MX-6		
1981-87	.8	1.0
1988-92	.7	.9
1993-97		
4 cyl.	2.0	2.2
V6	3.1	3.3
1998-02		
4 cyl.	.8	1.0
V6	2.5	2.7
929		
SOHC	.7	.9
DOHC	2.3	2.5
GLC	.7	.9
Mazda3	1.5	1.7
Mazda6		
2.3L	1.4	1.6
3.0L	.6	.8
Miata	.5	.7
Millenia	1.8	2.0

MAZ-5

MAZ-6 323 : 626 : 929 : GLC : MAZDA3 : MAZDA6 : MIATA : MILLENIA

	LABOR TIME	SEVERE SERVICE
EGR Valve, Replace (B)		
MX-3	2.0	2.2
RX-7		
1986-91	1.7	1.9
1993-95	2.0	2.2
Electronic Control Module (ECM/PCM), Replace (B)		
All Models	.8	1.0
Fuel Vapor Canister, Replace (B)		
Piston engine		
exc. below	.7	.9
Mazda3	1.9	2.1
Mazda6	3.0	3.2
Rotary engine	.5	.7
High Altitude Compensator, Replace (B)		
All Models		
piston engine	.5	.7
rotary engine	.7	.9
Injection Tube Check Valve Assy., Replace (B)		
1981-85 626	.5	.7
GLC	.5	.7
RX-7	1.3	1.5
Intake Air Temperature (IAT) Sensor, Replace (A)		
Exc. below	.3	.5
1988-95 929	.5	.7
RX-7		
1986-91	.5	.7
1993-95	2.3	2.5
Knock Sensor, Replace (B)		
1988-05 323, Protégé/5	.8	1.0
626, MX-6		
1988-92	.8	1.0
1993-97	3.1	3.3
1998-02		
4 cyl.	.7	.9
V6	3.0	3.2
929		
1990-91		
one side	1.8	2.0
both sides	2.9	3.1
1992-95		
one side	2.8	3.0
both sides	4.8	5.0
Mazda3	2.6	2.8
Mazda6		
2.3L (1.9)	2.6	2.8
3.0L (.4)	.6	.8
1999-05 Miata	.9	1.1
Millenia		
KJ	5.3	5.5
KL	3.3	3.5
MX-3	3.1	3.3
RX-7		
1987-91	1.8	2.0
1993-95	2.7	2.9
RX-8 (.3)	.5	.7

	LABOR TIME	SEVERE SERVICE
Manifold Air Temperature Sensor, Replace (B)		
1992-95 MX-3	.5	.7
Mass Air Flow (MAF) Sensor, Replace (B)		
Exc. below	.8	1.0
1986-91 RX-7	.5	.7
Oxygen Sensor, Replace (A)		
323, Protégé/5		
1 or 2	.7	.9
3 or 4	.8	1.0
626, MX-6		
1983-92		
1 or 2	.5	.7
1993-02		
1 or 2	.7	.9
3 or 4	.8	1.0
929	.7	.9
Mazda3		
1 or 2	.7	.9
3, or 4	.7	.9
Mazda6		
1 or 2	.9	1.1
3 or 4	1.4	1.6
Miata both	.6	.8
Millenia		
KJ		
1 or 2	1.3	1.5
3 or 4	1.8	2.0
KL		
1 or 2	.8	1.0
3 or 4	1.3	1.5
MX-3	.7	.9
RX-7		
1986-91	.5	.7
1993-95	2.2	2.4
RX-8		
1 or 2 (.3)	.5	.7
Positive Crankcase Ventilation (PCV) Valve, Replace (B)		
Exc. below	.3	.5
Mazda3	2.6	2.8
Mazda6 2.3L	2.8	3.0
Purge Control Solenoid Valve, Replace (B)		
Exc. below	.3	.5
Mazda3, Mazda6	.6	.8
RX-7		
1986-91	.5	.7
1993-95	2.4	2.6
Relay, Replace (B)		
1986-02 626, MX-6 all	.5	.7
Miata all	.5	.7
MX-3	.5	.7
Solenoid Valve, Replace (B)		
Exc. below	.5	.7
1993-95 RX-7	3.0	3.2
RX-8	1.8	2.0

	LABOR TIME	SEVERE SERVICE
Throttle Closing Dashpot, Replace (B)		
Exc. below	.5	.7
1983-85 RX-7	.7	.9
Throttle Position Sensor (TPS), Replace (A)		
323, Protégé/5		
1987-98	.7	.9
1999-05		
1.6L	.7	.9
1.8L, 2.0L	1.5	1.7
626, MX-6		
1983-92	.7	.9
1993-02		
4 cyl.	1.6	1.8
V6	.7	.9
929		
1988-91	.7	.9
1992-95	1.5	1.7
Mazda3, Mazda6	.5	.7
Miata		
1990-98	.7	.9
1999-05	1.7	1.9
Millenia		
KJ	1.2	1.4
KL	.7	.9
MX-3	.8	1.0
RX-7	.8	1.0
RX-8 (.3)	.5	.7
Vehicle Speed Sensor, Replace (B)		
1986-89 323, Protégé	1.4	1.6
1984-92 626, MX-6	1.4	1.6
1988-91 929	2.0	2.2
1995-02 Millenia	.5	.7

FUEL
CARBURETOR

	LABOR TIME	SEVERE SERVICE
Carburetor, Adjust (On Car) (B)		
323, 626, GLC, RX-7	.3	.5
Carburetor, R&R and Clean or Recondition (B)		
323, 626, MX-3	3.1	3.1
RX-7	3.3	3.5
Carburetor, Replace (B)		
Includes: Adjustments.		
323	1.7	1.9
626, MX-3	2.0	2.2
RX-7	2.3	2.5

DELIVERY

	LABOR TIME	SEVERE SERVICE
Fuel Pump, Test (B)		
All Models	.3	.5
Coolant Temperature Switch, Replace (B)		
626, MX-6	.7	.9
Fuel Gauge (Dash), Replace (B)		
323, Protégé	1.2	1.4
626, MX-6		
1981-82	1.2	1.4

MX-3 : MX-6 : PROTEGE : PROTEGE5 : RX-7 : RX-8 — MAZ-7

	LABOR TIME	SEVERE SERVICE
1983-85		
analog	1.8	2.0
electronic	2.0	2.2
1986-87	.8	1.0
1988-92	1.2	1.4
1993-97	.9	1.1
1998-02	1.3	1.5
929	1.4	1.6
GLC	1.5	1.7
Mazda3 (1.8)	2.4	2.6
Mazda6 (.5)	.7	.9
Miata	.8	1.0
Millenia	1.7	1.9
MX-3	1.3	1.5
RX-7	1.2	1.4
RX-8 (.6)	.8	1.0
Fuel Gauge (Tank), Replace (B)		
323, Protégé/5		
1986-95		
2WD	.5	.7
4WD	2.2	2.4
1996-05 (.4)	.6	.8
626, MX-6		
1981-82	.5	.7
1983-92		
main	.7	.9
sub	.3	.5
1993-97		
main	2.0	2.2
sub	.5	.7
1998-02	2.9	3.1
929		
1988-91	.8	1.0
1992-95		
w/CD changer	1.9	2.1
w/o CD changer	1.2	1.4
GLC	.5	.7
Mazda3 (1.6)	2.2	2.4
Mazda6 (.6)	.8	1.0
Miata	.9	1.1
Millenia		
one (main unit)	.5	.7
two	.7	.9
MX-3	.5	.7
RX-7		
1981-85	1.5	1.7
1986-91	.5	.7
1993-95	.8	1.0
RX-8 (.3)	.5	.7
Fuel Pump, Replace (B)		
Electric		
323, Protégé/5		
1986-98	.7	.9
1999-05 (1.0)	1.4	1.6
626, MX-6		
1983-92		
main	.7	.9
sub	3.1	3.3
1993-97	2.2	2.4
1998-02	2.9	3.1

	LABOR TIME	SEVERE SERVICE
929		
1988-91	.7	.9
1992-95		
w/CD changer	2.0	2.2
w/o CD changer	1.3	1.5
Mazda3 (1.6)	2.2	2.4
Mazda6 (.6)	.8	1.0
Miata	1.1	1.3
Millenia	1.8	2.0
MX-3	1.2	1.4
1986-95 RX-7	.7	.9
RX-8 (.3)	.5	.7
Mechanical		
626, GLC, RX-7	.8	1.0
Fuel Pump Relay, Replace (B)		
323, Protégé/5		
1986-98	.7	.9
1999-05	.3	.5
626, 929, MX-6	.5	.7
Mazda3, RX-8	.4	.6
Mazda6, Millenia	.3	.5
Miata	.5	.6
MX-3, RX-7	.7	.9
Fuel Tank, Replace (B)		
Includes: Drain and refill.		
323, Protégé/5		
1986-94		
2WD	1.4	1.6
4WD	2.0	2.2
1995-98	1.7	1.9
1999-05 (2.2)	3.0	3.2
626, MX-6		
1981-92		
2WS	1.5	1.7
4WS	3.1	3.3
1993-97	1.9	2.1
1998-02	2.9	3.1
929		
1988-91	1.4	1.6
1992-95		
w/CD changer	6.3	6.5
w/o CD changer	5.6	5.8
GLC		
sedan	1.5	1.7
wagon	1.4	1.6
Mazda3	2.6	2.8
Mazda6	2.9	3.1
Miata		
1990-93	6.4	6.6
1994-05 (3.5)	4.7	4.9
Millenia	2.5	2.7
MX-3	2.2	2.4
RX-7		
1981-91	1.4	1.6
1993-95	2.0	2.2
RX-8 (3.0)	4.1	4.3

	LABOR TIME	SEVERE SERVICE
Intake Manifold and/or Gasket, Replace (B)		
Includes: Adjustments.		
323, Protégé/5		
1986-94	2.5	2.7
1995-98		
1.5L	3.0	3.2
1.6L, 1.8L	3.5	3.7
1999-05		
1.6L	3.0	3.2
1.8L, 2.0L	3.3	3.5
626, MX-6		
1981-82	2.2	2.4
1983-92		
w/turbo	3.1	3.3
w/o turbo	2.8	3.0
1993-97		
4 cyl.	3.9	4.1
V6	3.1	3.3
1998-02		
4 cyl.	3.5	3.7
V6	3.4	3.6
929		
1988-91		
SOHC	3.1	3.3
DOHC	2.8	3.0
1992-95	3.0	3.2
GLC	3.1	3.3
Mazda3	2.7	2.9
Mazda6		
2.3L (2.1)	2.8	3.0
3.0L (1.8)	2.4	2.6
Miata	3.5	3.7
Millenia		
KJ		
right	4.6	4.8
left	3.1	3.3
both	4.9	5.1
KL	3.8	4.0
MX-3		
1992-93	3.4	3.6
1994-95		
4 cyl.	2.8	3.0
V6	2.5	2.7
RX-7		
1981-91	2.5	2.7
1993-95	9.5	9.7
RX-8		
ATX	9.9	10.1
MTX	10.3	10.5
w/ABS RX-7 add	.1	.1
w/EGI RX-7 add	.5	.5
w/turbo RX-7 add	1.7	1.9
Replace manifold add	.5	.7

MAZ-8 323 : 626 : 929 : GLC : MAZDA3 : MAZDA6 : MIATA : MILLENIA

	LABOR TIME	SEVERE SERVICE
INJECTION		
Bypass Air Control Valve, Replace (B)		
323, Protégé/5		
1986-94	1.7	1.9
1995-98	1.6	1.8
1999-05		
1.6L	1.6	1.8
1.8L, 2.0L	1.2	1.4
626, MX-6		
1988-92	1.2	1.4
1993-02	1.7	1.9
929		
1988-91	.5	.7
1992-95	.8	1.0
Mazda3		
1.6L	1.4	1.6
1.8L, 2.0L	1.1	1.3
Miata	1.8	2.0
Millenia	1.8	2.0
MX-3	1.7	1.9
1986-91 RX-7	.3	.5
Fuel Distributor, Replace (B)		
323, Protégé/5		
1986-94	.9	1.1
1995-98		
1.5L	2.4	2.6
1.6L, 1.8L	1.6	1.8
1999-05	1.2	1.4
626, MX-6		
1983-92	1.7	1.9
1993-02		
4 cyl.	1.3	1.5
V6	1.8	2.0
929		
1988-89		
one	1.8	2.0
both	2.0	2.2
1990-95		
SOHC	2.6	2.8
DOHC	2.3	2.5
Mazda3	.8	1.0
Mazda6		
2.3L (.7)	.9	1.1
3.0L (1.8)	2.4	2.6
Miata		
1990-97	1.6	1.8
1999-05 (1.8)	2.4	2.6
Millenia		
KJ	4.1	4.3
KL	2.0	2.2
MX-3		
4 cyl.	1.6	1.8
V6	2.0	2.2
RX-7		
1983-85	1.7	1.9
1986-91	.5	.7
1993-95	3.7	3.9
RX-8	2.4	2.6
w/turbo RX-7 add	1.2	1.2

	LABOR TIME	SEVERE SERVICE
Fuel Filter, Replace (B)		
323, Protégé/5		
1986-94	.3	.5
1995-05	.9	1.1
626, MX-6		
1981-92	.3	.5
1993-02	.9	1.1
929		
1988-91	.3	.5
1992-95	.7	.9
GLC	.3	.5
Mazda3	2.4	2.6
Mazda6	.8	1.0
Miata		
1990-93	.3	.5
1994-05 (.8)	1.1	1.3
Millenia	.7	.9
MX-3	.8	1.0
RX-7		
1981-85	.3	.5
1986-91	.5	.7
1993-95	.7	.9
RX-8 (.5)	.7	.9
Fuel Injectors, Replace (B)		
323, Protégé/5		
1986-94	1.4	1.6
1995-98		
1.5L	2.2	2.4
1.6L, 1.8L	1.6	1.8
1999-05	1.2	1.4
626, MX-6		
1983-92	1.9	2.1
1993-02		
4 cyl.	1.4	1.6
V6	2.3	2.5
929		
one side	2.0	2.2
both sides	2.5	2.7
Mazda3	.8	1.0
Mazda6		
2.3L (.7)	.9	1.1
3.0L (1.7)	2.3	2.5
Miata	2.4	2.6
Millenia		
KJ	4.1	4.3
KL	2.0	2.2
MX-3		
4 cyl.	1.5	1.7
V6	2.3	2.5
RX-7		
1983-85	1.7	1.9
1986-91		
primary or secondary	1.8	2.0
all	2.3	2.5
1993-95		
primary or secondary	3.8	4.0
all	4.2	4.4

	LABOR TIME	SEVERE SERVICE
RX-8	2.4	2.6
w/turbo RX-7 add	.7	.9
Fuel Pressure Regulator, Replace (B)		
323, Protégé/5		
1986-94	.7	.9
1995-05	1.3	1.5
626, MX-6		
1983-92	.7	.9
1993-02	1.3	1.5
929		
1988-91	1.2	1.4
1992-95	1.7	1.9
Miata		
1990-93	.7	.9
1994-05 (.8)	1.1	1.3
Millenia		
KJ	3.3	3.5
KL	1.3	1.5
MX-3	1.3	1.5
RX-7		
1983-85	1.7	1.9
1986-91	.5	.7
1993-95	3.7	3.9
w/turbo RX-7 add	.5	.5
Idle Air Control (IAC) Valve, Replace (B)		
1992-93 MX-3	1.9	2.1
1993-95 RX-7	2.0	2.2
Pressure Sensor, Replace (B)		
323, Protégé/5	.6	.8
1998-02 626	.7	.9
Miata	.6	.8
Millenia	.5	.7
Pulsation Damper, Replace (B)		
626, MX-6		
1983-92	.5	.7
929		
1988-91	1.2	1.4
1992-95	1.8	2.0
Mazda6		
2.3L	.8	1.0
3.0L	2.0	2.2
Miata		
1999-05		
No. 1	.9	1.1
No. 2 (on fuel distributor)	2.4	2.6
1999-02 Protégé/5		
1.6L	1.2	1.4
1.8L, 2.0L	.9	1.1
RX-7		
1983-91	1.7	1.9
1993-95	3.7	3.9
RX-8	2.4	2.6
w/turbo RX-7 add	1.2	1.4

MX-3 : MX-6 : PROTEGE : PROTEGE5 : RX-7 : RX-8

	LABOR TIME	SEVERE SERVICE
Surge Tank and/or Gasket, Replace (B)		
323, Protégé/5		
1986-94	1.5	1.7
1995-98	1.8	2.0
1999-05	3.3	3.5
1986-92 626, MX-6	1.7	1.9
1988-91 929		
SOHC	2.7	2.9
DOHC	2.2	2.4
Mazda6	2.0	2.2
Miata	2.0	2.2
Millenia	2.7	2.9
MX-3		
1992-93	2.8	3.0
1994-95	2.4	2.6
RX-7		
1983-85	1.5	1.7
1986-91	2.2	2.4
w/turbo RX-7 add	.2	.2
System Relay, Replace (B)		
929	.3	.3
Throttle Body Assy., Replace (B)		
323, Protégé/5		
1986-94	1.5	1.7
1995-05	1.8	2.0
626, MX6	1.6	1.8
929	1.4	1.6
Mazda3	1.5	1.7
Mazda6	1.4	1.6
Miata		
1990-93	.7	.9
1994-05	1.5	1.7
Millenia	1.8	2.0
1992-95 MX-3	1.7	1.9
RX-7		
1983-85	.9	1.1
1986-91	1.9	2.1
1993-95	1.3	1.5
RX-8 (.9)	1.2	1.4
Water Temperature Sensor, Replace (B)		
929	.5	.7

SUPERCHARGER
	LABOR TIME	SEVERE SERVICE
Supercharger, Replace (B)		
Millenia	5.0	5.2

TURBOCHARGER
	LABOR TIME	SEVERE SERVICE
Intercooler, Replace (B)		
1988-89 323	.5	.7
1991 MX-6	.8	1.0
RX-7		
1987-91	.7	.9
1993-95	1.2	1.4
Turbo Boost Pressure Sensor, Replace (B)		
1988-89 929	.5	.7
1986-95 RX-7	.5	.7

	LABOR TIME	SEVERE SERVICE
Turbocharger Assy., R&R or Replace (B)		
1988-89 323	3.1	3.3
1986-92 626, MX-6	3.9	4.1
Mazdaspeed Protégé	4.1	4.3
RX-7		
1987-90	4.2	4.4
1991	4.8	5.0
1993-95	6.8	7.0
w/AC add	.5	.7
Water Pipes and/or Gaskets, Replace (B)		
1988-90 323		
inlet	.7	.9
outlet	2.3	2.5
Mazdaspeed Protégé	1.4	1.6
1986-91 RX-7		
inlet	.7	.9
outlet	3.9	4.1

EXHAUST
	LABOR TIME	SEVERE SERVICE
Catalytic Converter, Replace (B)		
323, Protégé/5		
3-way converter	.7	.9
3-way warm up		
1995-98	2.4	2.6
1999-05 (1.0)	1.4	1.6
626, MX-6		
1981-82 each	1.2	1.4
1983-93	.8	1.0
1994-02		
4 cyl.		
3-way converter	.7	.9
3-way warm up	1.4	1.6
front type	.7	.9
V6		
right side	2.9	3.1
left side	2.0	2.2
both sides	3.6	3.8
929		
1988-91	1.3	1.5
1992-95	.9	1.1
GLC		
front	1.9	2.1
rear	1.6	1.8
Mazda3 (2.1)	2.8	3.0
Mazda6		
4 cyl.	.8	1.0
V6		
3-way converter		
one side	1.1	1.3
both sides	1.4	1.6
3-way warm up		
right side	2.9	3.1
left side	1.9	2.1
both sides	3.5	3.7
Miata		
3-way converter	.8	1.0
3-way warm up	2.4	2.6

	LABOR TIME	SEVERE SERVICE
Millenia		
3-way converter	.7	.9
3-way warm up		
right side	2.9	3.1
left side		
KJ	2.2	2.4
KL	1.8	2.0
both sides		
KJ	3.8	4.0
KL	3.4	3.6
MX-3		
3-way converter	.7	.9
3-way warm up		
4 cyl.	1.6	1.8
V6		
one	1.6	1.8
both	2.0	2.2
RX-7		
1981-91		
front or center	1.4	1.6
rear	1.6	1.8
1993-95		
front	2.2	2.4
rear	.8	1.0
RX-8 (.7)	.9	1.1
Center Exhaust Pipe, Replace (B)		
1988-91 323, Protégé	.5	.7
1983-92 626, MX-6	.8	1.0
1988-91 929		
1992-95 MX-3	.7	.9
1981-85 RX-7	.9	1.1
Exhaust Manifold and/or Gasket, Replace (B)		
323, Protégé/5		
1986-94		
SOHC	.9	1.1
DOHC	2.3	2.5
1995-98	2.6	2.8
1999-05	2.2	2.4
626, MX-6		
1981-82	1.5	1.7
1983-85	1.9	2.1
1986-92		
w/turbo	3.4	3.6
w/o turbo	1.8	2.0
1993-02		
4 cyl.	2.0	2.2
V6		
right side	2.9	3.1
left side	2.0	2.2
both sides	3.6	3.8
929		
1988-91		
right side	.9	1.1
left side	1.4	1.6
both sides	2.2	2.4
1992-95		
right side	2.0	2.2
left side	2.4	2.6
both sides	3.8	4.0

MAZ-10 323 : 626 : 929 : GLC : MAZDA3 : MAZDA6 : MIATA : MILLENIA

	LABOR TIME	SEVERE SERVICE
Exhaust Manifold and/or Gasket, Replace (B)		
GLC	2.3	2.5
Mazda3	2.7	2.9
Mazda6		
L4 (1.0)	1.4	1.6
V6		
right side (2.1)	2.8	3.0
left side (1.3)	1.8	2.0
both sides (2.8)	3.8	4.0
Miata		
1990-93	2.3	2.5
1994-97	1.9	2.1
1999-05	2.0	2.2
Millenia		
right side	2.9	3.1
left side	1.8	2.0
both sides	3.4	3.6
MX-3		
4 cyl.	1.6	1.8
V6		
one side	1.6	1.8
both sides	2.3	2.5
RX-7		
1981-85	.9	1.1
1986-91	1.2	1.4
1993-95	6.8	7.0
RX-8	2.7	2.9
w/turbo RX-7 add	2.8	3.0
Replace manifold add	.5	.7
Front Exhaust Pipe, Replace (B)		
323, Protégé/5	.8	1.0
626		
1981-82	1.2	1.4
1983-02	.8	1.0
1988-91 929		
one	.7	.9
both	.8	1.0
GLC	1.2	1.4
Mazda3	.6	.8
Mazda6	1.5	1.7
Miata	.9	1.1
Millenia		
exc. flexible pipe	.8	1.0
flexible pipe	.7	.9
MX-3	.9	1.1
RX-7		
1981-85	1.2	1.4
1986-91		
front	1.3	1.5
rear	.8	1.0
RX-8 (.7)	.9	1.1
Muffler, Replace (B)		
323, Protégé/5	.8	1.0
626, MX-6		
1981-82	.9	1.1
1983-05	.6	.8
929	.7	.9
GLC	.8	1.0

	LABOR TIME	SEVERE SERVICE
Mazda3	.9	1.1
Mazda6	.6	.8
Miata	.6	.8
Millenia	.7	.9
MX-3	.7	.9
RX-7		
1981-85	1.3	1.5
1986-91		
one	.8	1.0
both	1.2	1.4
1993-95	.7	.9
RX-8 (.7)	.9	1.1
Pre-Catalytic Converter, Replace (B)		
1992-95 929	.9	1.1
Pre-Muffler, Replace (B)		
1987-97 626, MX-6	.5	.7
Millenia, MX-3	.7	.9
1981-85 RX-7	1.2	1.4
Thermal Reactor, Replace (B)		
1981-85 RX-7	3.1	3.3

ENGINE COOLING

	LABOR TIME	SEVERE SERVICE
Pressure Test Cooling System (C)		
All Models	.3	.3
Bypass Hoses, Replace (B)		
323, Protégé/5		
1988-94	.5	.7
1995-05		
exc. 2.0L	1.1	1.3
2.0L	.9	1.1
626, 929, MX-6	.9	1.1
Mazda3	1.3	1.5
Mazda6, MX-3	1.2	1.4
Miata		
1991-93	.5	.7
1994-05 (.8)	1.1	1.3
1995-98 Millenia		
KJ	1.2	1.4
KL	.9	1.1
RX-7		
1981-91	.5	.7
1993-95		
main	1.3	1.5
return	1.8	2.0
both	2.0	2.2
RX-8 (1.4)	1.9	2.1
Condenser Fan Motor, Replace (B)		
323, Protégé/5		
1986-94	.3	.5
1995-98	.7	.9
1999-05	.3	.5
626, MX-6		
1983-92	1.2	1.4
1993-02	.7	.9
Miata	.6	.8
Millenia	.7	.9
MX-3	.7	.9
1981-91 RX-7	.5	.7

	LABOR TIME	SEVERE SERVICE
Coolant Thermostat, Replace (B)		
323, Protégé/5	1.2	1.4
626, MX-6		
1983-92	.7	.9
1993-97	1.7	1.9
1998-02		
4 cyl.	1.2	1.4
V6	1.7	1.9
929		
1988-91	.8	1.0
1992-95	2.0	2.2
GLC	.7	.9
Mazda3	1.5	1.7
Mazda6		
2.3L	1.7	1.9
3.0L	1.2	1.4
Miata	1.2	1.4
Millenia	1.4	1.6
MX-3	1.3	1.5
RX-7		
1981-91	.7	.9
1993-95	2.8	3.0
RX-8	1.6	1.8
Cooling Fan Control, Replace (B)		
1986-91 RX-7	.5	.7
Fan Blade, Replace (B)		
All Models	.5	.7
Fluid Fan Drive or Fan Blade, Replace (B)		
929	1.7	1.9
RX-7		
1981-91	.5	.7
1993-95		
AT	3.4	3.6
MT	3.1	3.3
Radiator Fan and/or Fan Motor, Replace (B)		
323, Protégé/5		
1986-94	.7	.9
1995-98	1.8	2.0
1999-05	.5	.7
626, MX-6	.7	.9
GLC	.8	1.0
Mazda3	.6	.8
Mazda6		
2.3L (1.9)	2.6	2.8
3.0L (2.3)	3.1	3.3
Miata	.6	.8
Millenia, MX-3	.7	.9
1993-95 RX-7	3.5	3.7
RX-8 (1.8)	2.4	2.6
Radiator Assy., R&R or Replace (B)		
Includes: Refill with proper coolant mix.		
323, Protégé/5		
1986-94	1.2	1.4
1995-05 (1.3)	1.8	2.0
626, MX-6		
1981-92	1.2	1.4
1993-02	1.7	1.9
929	1.9	2.1

	LABOR TIME	SEVERE SERVICE
GLC	1.2	1.4
Mazda3 (1.2)	1.6	1.8
Mazda6 (2.6)	3.5	3.7
Miata		
1990-93	.8	1.0
1994-05	1.7	1.9
Millenia	2.0	2.2
MX-3	1.4	1.6
RX-7		
1981-85	1.2	1.4
1986-91	1.7	1.9
1993-95	3.1	3.3
RX-8	2.4	2.5
w/AT add	.2	.2

Radiator Fan Motor Relay, Replace (B)

1986-05	.5	.7

Radiator Fan Motor Switch (Coolant Temp.), Replace (B)

626, MX-6	.8	1.0
MX-3	.5	.7

Radiator Hoses, Replace (B)
Includes: Refill with proper coolant mix.

323, Protégé/5		
1986-94 each	.5	.7
1995-98 each	1.4	1.6
1999-05		
upper	.9	1.1
lower	1.2	1.4
626, MX-6		
1981-92 each	.5	.7
1993-02		
upper	.8	1.0
lower	1.2	1.4
929		
1988-91		
upper	.8	1.0
lower	1.4	1.6
1992-02 each	1.2	1.4
GLC each	.5	.7
Mazda3 each	1.2	1.4
Mazda6		
upper	1.1	1.2
lower		
2.3L	1.7	1.9
3.0L	1.4	1.6
Miata		
1990-93 each	.5	.7
1994-05		
upper	.9	1.1
lower	1.2	1.4
Millenia		
upper	.9	1.1
lower		
KJ	1.4	1.6
KL	1.2	1.4
MX-3		
upper	.8	1.0
lower	1.2	1.4

	LABOR TIME	SEVERE SERVICE
RX-7		
1981-91 each	.5	.7
1993-95 each	1.3	1.5
RX-8 each	1.5	1.7

Temperature Gauge (Dash), Replace (B)

323, Protégé	1.4	1.6
626, MX-6		
1981-82	.7	.9
1983-87	1.7	1.9
1988-92	1.2	1.4
1993-97	.8	1.0
1998-02	1.3	1.5
929	1.4	1.6
GLC	.9	1.1
Miata	1.2	1.4
Millenia	1.2	1.4
MX-3	1.6	1.8
RX-7	1.2	1.4

Temperature Gauge (Engine), Replace (B)

Piston engine		
All Models	.5	.7
Rotary engine		
1981-91	.5	.7
1993-95	4.2	4.4

Water Pump and/or Gasket, Replace (B)
Includes: Refill with proper coolant mix.

323, Protégé/5		
1986-98		
1.5L	3.4	3.6
1.6L	3.8	4.0
1.8L	3.1	3.3
1999-05		
1.6L (3.4)	4.6	4.8
1.8L, 2.0L (2.5)	3.4	3.6
626, MX-6		
1981-82	1.7	1.9
1983-87	2.7	2.9
1999-92	3.4	3.6
1993-02		
4 cyl.	3.1	3.3
V6	4.2	4.4
929		
1988-91	3.8	4.0
1992-95	4.8	5.0
GLC	1.7	1.9
Mazda3	1.4	1.6
Mazda6		
2.3L (1.1)	1.5	1.7
3.0L (1.7)	2.3	2.5
Miata	4.3	4.5
Millenia		
KJ	5.3	5.5
KL	4.1	4.3

	LABOR TIME	SEVERE SERVICE
MX-3		
1992-93	3.1	3.3
1994-95		
4 cyl.	3.8	4.0
V6	4.8	5.0
RX-7		
1981-85	1.6	1.8
1986-91	2.4	2.6
1993-95	4.2	4.4
RX-8 (1.0)	1.4	1.6

Water Bypass Pipe and/or O Ring, Replace (B)

323, Protégé/5		
1986-94	.8	1.0
1995-98	1.2	1.4
1999-05 (1.1)	1.5	1.7
626, MX-6		
1981-92	.7	.9
1993-02		
4 cyl.	1.6	1.8
V6	2.9	3.1
929	1.6	1.8
Mazda6	1.8	2.0
Miata		
1990-93	.8	1.0
1994-05	1.2	1.4
Millenia		
KJ	1.3	1.5
KL	3.1	3.3
MX-3		
4 cyl.	1.2	1.4
V6	2.3	2.5

PISTON ENGINE
ASSEMBLY

Times shown are for OEM assemblies. Time to replace assemblies from aftermarket rebuilders may vary.

Engine Assy., R&I (B)
Does not include parts or component transfer.

323, Protégé		
1986-94		
AT	6.3	6.5
MT	5.9	6.1
1995-02	10.5	10.7
626, MX-6		
1981-82		
AT	5.6	5.8
MT	5.2	5.4
1983-92		
AT	7.0	7.2
MT	6.3	6.5
1993-97		
4 cyl.	9.3	9.5
V6	9.7	9.9
1998-02		
4 cyl.	10.0	10.2
V6	13.5	13.7

MAZ-12   323 : 626 : 929 : GLC : MAZDA3 : MAZDA6 : MIATA : MILLENIA

	LABOR TIME	SEVERE SERVICE
Engine Assy., R&I (B)		
929		
1988-91		
AT	7.7	7.9
MT	7.1	7.3
1992-95	13.0	13.2
GLC		
AT	6.7	6.9
MT	5.9	6.1
Mazda3	11.6	11.8
Mazda6		
2.3L	12.1	12.3
3.0L	13.3	13.5
Miata	7.3	7.5
Millenia		
KJ	11.2	11.4
KL	10.3	10.5
MX-3		
4 cyl.	9.1	9.3
V6	9.5	9.7
w/4WD 86-94 323, Protégé add	.5	.7
w/AC add		
86-94 323, Protégé	.5	.7
81-92 626, MX-6	.5	.7
93-97 626, MX-6	.3	.5
Miata	.2	.4
MX-3	.3	.5
95-02 Protégé	.3	.5
w/AT add		
93-97 626, MX-6	.1	.2
Miata	1.3	1.5
MX-3	.1	.2
w/DOHC add		
88-91 929	.5	.7
86-94 323, Protégé	.5	.7
w/PS add		
86-94 323, Protégé	.5	.7
81-92 626, MX-6	.2	.4
Miata	.2	.4
w/turbo 81-92 626, MX-6 add	1.2	1.4
Engine Assy., R&R and Recondition (A)		

Includes: Replacing rings, rod and main bearings, cylinder head reconditioning and engine tune-up.

	LABOR TIME	SEVERE SERVICE
323, Protégé/5		
1986-94		
2WD	26.0	26.2
4WD	27.0	27.2
1995-98	30.7	30.9
1999-05	27.5	27.7
626, MX-6		
1981-82		
AT	22.4	22.6
MT	21.7	21.9
1983-92	28.0	28.2
1993-02		
4 cyl.	29.9	30.1
V6	38.8	39.0
929		
1988-91		
AT	25.5	25.7
MT	25.0	25.2
1992-95	39.9	40.1
GLC		
AT	24.2	24.4
MT	23.3	23.5
Mazda3	27.3	27.5
Mazda6		
2.3L	29.0	29.2
3.0L	36.7	36.9
Miata		
1990-93	29.9	30.1
1994-05		
AT	28.6	28.8
MT	27.8	28.0
Millenia		
KJ	43.1	43.3
KL	39.4	39.6
MX-3		
4 cyl.	29.2	29.4
V6	36.0	36.2
w/4WD 86-94 323, Protégé add	.5	.7
w/AC add		
86-94 323, Protégé	.5	.7
81-92 626, MX-6	.5	.7
93-97 626, MX-6	.3	.5
Miata	.2	.4
MX-3	.3	.5
95-05 Protégé	.3	.5
w/AT add		
93-97 626, MX-6	.1	.2
Miata	1.3	1.5
MX-3	.1	.2
w/DOHC add		
86-94 323, Protégé	.5	.7
88-91 929	.5	.7
w/PS add		
86-94 323, Protégé	.5	.7
81-92 626, MX-6	.2	.4
Miata	.2	.4
w/turbo 81-92 626, MX-6 add	1.2	1.4
Engine Assy. (Short Block), Replace (B)		

Assembly consists of cylinder block, piston assemblies, crankshaft, camshaft, timing chain and gears. Does not include cylinder heads. Operation Includes: R&R engine, transfer necessary parts and all necessary adjustments.

	LABOR TIME	SEVERE SERVICE
323, Protégé/5		
1986-98	18.4	18.6
1999-05	16.7	16.9
626, MX-6		
1981-82		
AT	18.0	18.2
MT	17.4	17.6
1983-92	19.4	19.6
1993-02		
4 cyl.	17.3	17.5
V6	20.5	20.7
929		
1988-91		
AT	17.3	17.5
MT	16.4	16.6
1992-95	21.2	21.4
GLC		
AT	20.7	20.9
MT	20.1	20.3
Mazda3	17.6	17.8
Mazda6		
2.3L	18.8	19.0
3.0L	20.6	20.8
Miata		
1990-93	16.3	16.5
1994-05	17.1	17.3
Millenia		
KJ	25.2	25.4
KL	21.2	21.4
MX-3		
4 cyl.	18.7	18.9
V6	20.1	20.3
w/4WD 86-94 323, Protégé add	.5	.7
w/AC add		
86-94 323, Protégé	.5	.7
81-92 626, MX-6	.5	.7
93-97 626, MX-6	.3	.5
Miata	.2	.4
MX-3	.3	.5
95-02 Protégé	.3	.5
w/AT add		
93-97 626, MX-6	.1	.2
Miata	1.3	1.5
MX-3	.1	.2
w/DOHC add		
88-91 929	.5	.7
86-94 323, Protégé	.5	.7
w/PS add		
86-94 323, Protégé	.5	.7
81-92 626, MX-6	.2	.4
Miata	.2	.4
w/turbo 81-92 626, MX-6 add	1.2	1.4
Engine Mounts, Replace (B)		
Front		
323, Protégé/5		
1986-98		
No. 1	1.3	1.5
No. 2, 3, 4	.7	.9
1999-05		
No. 1, 2, 4	.9	1.1
No. 3	.6	.8

MX-3 : MX-6 : PROTEGE : PROTEGE5 : RX-7 : RX-8 — MAZ-13

	LABOR TIME	SEVERE SERVICE
626, MX-6		
1981-82		
one	.5	.7
both	.8	1.0
1983-92 each	.8	1.0
1993-02		
No. 1 or 2	1.4	1.6
No. 3 or 4	.8	1.0
929		
1988-91		
one	.8	1.0
both	1.2	1.4
1992-95		
right side	3.5	3.7
left side	3.8	4.0
both sides	5.4	5.6
GLC each	.7	.9
Mazda3		
No. 1	.5	.7
No. 3	.6	.8
No. 4	.9	1.1
Mazda6		
No. 1	.6	.8
No. 3		
2.3L	.8	1.0
3.0L	2.4	2.6
No. 4		
2.3L	.9	1.1
3.0L	2.6	2.8
Miata		
one	1.1	1.3
both	1.4	1.6
Millenia		
No. 1	1.8	2.0
No. 2, 4	.9	1.1
No. 3	.7	.9
MX-3		
No. 1	1.4	1.6
No. 2 or 3	.7	.9
No. 4	.9	1.1
Rear		
929, Miata, MX-3	.7	.9
w/AC Miata front add	1.2	1.4

CYLINDER HEAD
Compression Test (B)

	LABOR TIME	SEVERE SERVICE
1981-87 4 cyl.	.7	.9
323, Protégé/5	.9	1.1
626, MX-6		
1988-92	.5	.7
1993-02		
4 cyl.	.7	.9
V6	1.2	1.4
929		
1988-91	.8	1.0
1992-95	1.2	1.4
Mazda3	.8	.9
Mazda6		
2.3L	.8	1.0
3.0L	2.6	2.8

	LABOR TIME	SEVERE SERVICE
Miata	1.2	1.4
Millenia		
KJ	1.9	2.1
KL	1.2	1.4
MX-3		
4 cyl.	.7	.9
V6	.8	1.0

Valves, Adjust (B)

Exc. below	1.2	1.4
1988-02 V6	2.9	3.1
Mazda6	4.1	4.3
Miata, Protégé/5	2.0	2.2

Cylinder Head, Replace (B)
Includes: Transfer parts, adjustments.

323, Protégé/5		
1986-94		
8V		
SOHC	10.8	11.0
DOHC	13.3	13.5
16V	13.6	13.8
1995-98	11.9	12.1
1999-05	12.4	12.6
626, MX-6		
1981-82	8.2	8.4
1983-92	13.1	13.3
1993-02		
4 cyl.	13.1	13.3
V6		
one side	13.6	13.8
both sides	18.2	18.4
929		
1988-91		
SOHC		
one side	11.7	11.9
both sides	19.8	20.0
DOHC		
right side	12.4	12.6
left side	12.8	13.0
both sides	20.7	20.9
1992-95		
right side	13.1	13.3
left side	13.5	13.7
both sides	19.8	20.0
GLC	9.2	9.4
Mazda3	14.4	14.6
Mazda6		
4 cyl	14.8	15.0
V6		
right side	19.0	19.2
left side	20.4	20.6
both sides	25.0	25.2
Miata		
1990-93	13.5	13.7
1994-05	12.7	12.9
Millenia		
KJ		
right side	18.4	18.6
left side	17.9	18.1
both sides	25.3	25.5

	LABOR TIME	SEVERE SERVICE
KL		
one side	14.5	14.7
both sides	20.2	20.4
MX-3		
1992-93	11.9	12.1
1994-95		
4 cyl.	12.7	12.9
V6		
one side	12.3	12.5
both sides	17.8	18.0
w/AC add		
86-94 323, Protégé	.1	.1
95-02 323, Protégé	.1	.1
81-92 626, MX-6	.5	.5
Miata	.1	.1
w/PS add		
86-94 323, Protégé	.3	.3
81-92 626, MX-6	.2	.2
Miata	.2	.2
w/turbo 81-92 626, MX-6 add	.9	.9

Cylinder Head Gasket, Replace (B)

323, Protégé/5		
1986-94		
SOHC		
carbureted	7.0	7.2
fuel injected	6.4	6.6
DOHC	7.6	7.8
1995-98	8.3	8.5
1999-05	6.9	7.1
626, MX-6		
1981-82	5.5	5.7
1983-92	9.6	9.8
1993-02		
4 cyl.	7.4	7.6
V6		
one side	9.4	9.8
both sides	10.8	11.0
929		
1988-91		
SOHC		
right side	5.4	5.6
left side	5.8	6.0
both sides	7.4	7.6
DOHC		
right side	6.9	7.1
left side	7.3	7.5
both sides	9.0	9.2
1992-95		
right side	8.4	8.6
left side	9.1	9.3
both sides	10.2	10.4
GLC	6.5	7.0
Mazda3	9.3	9.5
Mazda6		
4 cyl	9.9	10.1
V6		
right side	15.7	15.9
left side	17.0	17.2
both sides	18.6	18.8

MAZ-14 323 : 626 : 929 : GLC : MAZDA3 : MAZDA6 : MIATA : MILLENIA

	LABOR TIME	SEVERE SERVICE
Cylinder Head Gasket, Replace (B)		
Miata		
1990-93	6.3	6.5
1994-97	7.0	7.2
1999-05	6.3	7.5
Millenia		
KJ		
right side	14.1	14.3
left side	13.3	13.5
both sides	16.9	17.1
KL		
right side	10.3	10.5
left side	10.2	10.4
both sides	12.6	12.8
MX-3		
1992-93	6.4	6.6
1994-95		
4 cyl.	7.8	8.0
V6		
one side	8.4	8.6
both sides	9.9	10.1
w/AC add		
86-94 323, Protégé	.1	.1
95-02 323, Protégé	.1	.1
81-92 626, MX-6	.5	.5
Miata	.1	.1
w/PS add		
86-94 323, Protégé	.3	.3
81-92 626, MX-6	.2	.2
Miata	.2	.2
w/turbo 81-92 626, MX-6 add	.9	.9
Cylinder Head, Retorque (B)		
1988-89 929	1.4	1.6
Miata	1.2	1.4
MX-3		
4 cyl.	1.2	1.4
V6	3.1	3.3
Cylinder Head Cover Gasket, Replace or Reseal (B)		
323, Protégé/5	.6	.8
626, MX-6		
1981-92	.5	.7
1993-02		
4 cyl.	.7	.9
V6		
right side	3.4	3.6
left side	.7	.9
both sides	3.7	3.9
929		
1988-91		
SOHC		
one side	1.4	1.6
both sides	2.3	2.5
DOHC		
right side	1.4	1.6
left side	1.9	2.1
both sides	2.7	2.9

	LABOR TIME	SEVERE SERVICE
1992-95		
right side	2.7	2.9
left side	3.0	3.2
both sides	3.3	3.5
GLC	.5	.7
Mazda3	.8	1.0
Mazda6		
4 cyl.	.6	.8
V6		
right side	2.9	3.1
left side	1.2	1.4
both sides	3.6	3.8
Miata		
1990-93	.5	.7
1994-05	.9	1.1
Millenia		
KJ		
right side	5.0	5.2
left side	4.1	4.3
both sides	5.5	5.7
KL		
right side	3.5	3.7
left side	.7	.9
both sides	3.8	4.0
MX-3		
4 cyl.	.5	.7
V6		
right side	2.6	2.8
left side	.7	.9
both sides	2.8	3.0
Rocker Arms or Shafts, Replace (B)		
1986-94 323, Protégé	3.2	3.4
626, MX-6		
1981-87	4.2	4.4
1988-92	2.9	3.1
929		
1988-91		
SOHC		
right side	2.6	2.8
left side	3.1	3.3
both sides	3.8	4.0
DOHC		
right side	3.9	4.1
left side	4.2	4.4
both sides	5.3	5.5
1992-95		
right side	6.4	6.6
left side	6.9	7.1
both sides	7.8	8.0
GLC	3.1	3.3
Mazda6		
right side	12.3	12.5
left side	13.5	13.7
both sides	13.6	13.8
1992-93 MX-3 4 cyl.	1.9	2.1

	LABOR TIME	SEVERE SERVICE
Valve Job (A)		
323, Protégé/5		
1986-94 8V		
SOHC		
carbureted	12.0	12.2
fuel injected	10.3	10.4
DOHC	14.9	15.1
1995-98	12.7	12.9
1999-05	12.3	12.5
626, MX-6		
1981-82	9.3	9.5
1983-92	13.7	13.9
1993-02		
4 cyl.	13.1	13.3
V6		
one side	13.6	13.8
both sides	19.8	20.0
929		
1988-91		
SOHC		
right side	12.6	12.8
left side	12.8	13.0
both sides	21.6	21.8
DOHC		
right side	13.3	13.5
left side	13.5	13.7
both sides	22.8	23.0
1992-95		
right side	14.2	14.4
left side	14.6	14.8
both sides	22.4	22.6
GLC	10.0	10.2
Mazda3	14.4	14.6
Mazda6		
4 cyl.	14.8	15.0
V6		
right side	19.0	19.2
left side	20.4	20.6
both sides	25.0	25.2
Miata		
1990-93	14.3	14.5
1994-05	12.6	12.8
Millenia		
KJ		
right side	18.3	18.5
left side	17.5	17.7
both sides	25.2	25.4
KL		
right side	14.5	14.7
left side	14.3	14.5
both sides	20.2	20.4
MX-3		
4 cyl.	13.7	13.9

MX-3 : MX-6 : PROTEGE : PROTEGE5 : RX-7 : RX-8   MAZ-15

	LABOR TIME	SEVERE SERVICE
Valve Job (A)		
V6		
one side	13.4	*13.6*
both sides	19.9	*20.1*
w/AC add		
86-94 323, Protégé	.1	*.1*
95-02 323, Protégé	.1	*.1*
81-92 626, MX-6	.5	*.5*
Miata	.1	*.1*
w/PS add		
86-94 323, Protégé	.3	*.3*
81-92 626, MX-6	.2	*.2*
Miata	.2	*.2*
w/turbo 81-92 626,		
MX-6 add	.9	*.9*
Valve Lash Adjuster, Replace (B)		
323, Protégé		
1986-94 all		
8V	3.5	*3.7*
16V	4.5	*4.7*
1995-96		
one	3.1	*3.3*
all	3.7	*3.9*
626, MX-6		
1983-92		
one	3.1	*3.3*
all	3.5	*3.7*
1993-97		
4 cyl.		
one	3.5	*3.7*
all	4.1	*4.3*
V6		
right side	6.0	*6.2*
left side	4.6	*4.8*
both sides	7.1	*7.3*
929		
1988-91		
SOHC		
right side	3.0	*3.2*
left side	3.4	*3.6*
both sides	4.4	*4.6*
DOHC		
right side	4.5	*4.7*
left side	5.4	*5.6*
both sides	9.4	*9.6*
1992-95		
right side	7.3	*7.5*
left side	7.8	*8.0*
both sides	9.4	*9.6*
Mazda6 V6		
right side	12.8	*13.0*
left side	13.6	*13.8*
both sides	13.9	*14.1*
Miata		
1990-93		
one	3.1	*3.3*
all	4.1	*4.3*
1994-96		
one	2.3	*2.5*
all	3.1	*3.3*

	LABOR TIME	SEVERE SERVICE
Millenia		
right side	6.7	*6.9*
left side	4.5	*4.7*
both sides	7.6	*7.8*
MX-3		
1992-93	1.8	*2.0*
1994-95		
4 cyl.	3.6	*3.8*
V6		
right side	6.2	*6.4*
left side	5.0	*5.2*
both sides	7.1	*7.3*
w/AC 90-02 Miata add	.1	*.1*
w/PS 90-02 Miata add	.2	*.2*
Valve Springs and/or Oil Seals, Replace (B)		
323, Protégé/5		
1986-94 8V		
SOHC		
carbureted	12.0	*12.2*
fuel injected	10.5	*10.7*
DOHC	14.2	*14.4*
1995-98	12.4	*12.6*
1999-05	10.2	*10.4*
626, MX-6		
1981-82	6.8	*7.0*
1983-92	15.3	*15.5*
1993-02		
4 cyl.	12.2	*12.4*
V6		
one side	13.2	*13.4*
both sides	17.8	*18.0*
929		
1988-91		
SOHC		
right side	7.4	*7.6*
left side	7.8	*8.0*
both sides	11.2	*11.4*
DOHC		
right side	5.9	*6.1*
left side	6.9	*7.1*
both sides	11.9	*12.1*
1992-95		
right side	10.2	*10.4*
left side	10.8	*11.0*
both sides	15.2	*15.7*
GLC	9.1	*9.3*
Mazda3	12.1	*12.3*
Mazda6		
4 cyl. all	12.7	*12.9*
V6		
right side	17.8	*18.0*
left side	18.9	*19.`*
both sides	22.2	*22.4*
Miata		
1990-93	8.9	*9.1*
1994-05	10.4	*10.6*

	LABOR TIME	SEVERE SERVICE
Millenia		
KJ		
right side	18.3	*18.5*
left side	17.9	*18.0*
both sides	23.8	*24.0*
KL		
one side	14.2	*14.4*
both sides	18.8	*19.0*
MX-3		
1992-93	9.9	*10.1*
1994-95		
4 cyl.	10.0	*10.2*
V6		
one side	10.5	*10.7*
both sides	14.6	*14.8*
w/AC add		
86-94 323, Protégé	.1	*.1*
95-02 323, Protégé	.1	*.1*
81-92 626, MX-6	.5	*.5*
Miata	.1	*.1*
w/PS add		
86-94 323, Protégé	.3	*.3*
81-92 626, MX-6	.2	*.2*
Miata	.2	*.2*
w/turbo 81-92 626,		
MX-6 add	1.2	*1.2*
CAMSHAFT		
Camshaft and/or Bearings, Replace (B)		
323, Protégé/5		
1986-94		
SOHC	3.4	*3.6*
DOHC		
one	4.1	*4.3*
both	4.2	*4.4*
1995-05		
one (2.3)	3.1	*3.3*
both (2.8)	3.8	*4.0*
626, MX-6		
1981-82	3.4	*3.6*
1983-92		
one	4.3	*4.4*
both	4.6	*4.8*
1993-02		
4 cyl.		
one	3.5	*3.7*
both	3.8	*4.0*
V6		
right side	6.2	*6.4*
left side	4.6	*4.8*
both sides	7.0	*7.2*
929		
SOHC		
right side	7.0	*7.2*
left side	7.3	*7.5*
both sides	8.8	*9.0*

MAZ-16 323 : 626 : 929 : GLC : MAZDA3 : MAZDA6 : MIATA : MILLENIA

	LABOR TIME	SEVERE SERVICE
Camshaft and/or Bearings, Replace (B)		
DOHC		
right side	6.5	6.7
left side	7.1	7.3
both sides	8.0	8.2
GLC	3.9	4.1
Mazda3		
one	2.9	3.1
both	3.1	3.3
Mazda6		
4 cyl.		
one	3.6	3.3
both	3.8	4.0
V6		
right side	12.4	12.6
left side	13.6	13.8
both sides	13.9	14.1
Miata		
1990-93		
one	4.1	4.3
both	4.2	4.4
1994-05 one or both	3.5	3.7
Millenia		
KJ		
right side	9.7	9.9
left side	9.4	9.6
both sides	10.9	11.1
KL		
right side	6.7	6.9
left side	4.5	4.7
both sides	7.4	7.6
MX-3		
4 cyl.	3.4	3.6
V6		
right side	5.9	6.1
left side	4.8	5.0
both sides	6.5	6.7
w/AC add		
86-94 323, Protégé	.1	.1
81-92 626, MX-6	.2	.2
Miata	.1	.1
w/PS add		
86-94 323, Protégé	.1	.1
81-92 626, MX-6	.1	.1
Miata	.2	.2
Camshaft Seal, Replace (B)		
323, Protégé/5		
1986-94		
SOHC	2.7	2.9
DOHC		
one side	3.5	3.7
both sides	3.7	3.9
1995-05		
one side (2.0)	2.7	2.9
both sides (2.2)	3.0	3.2
626, MX-6		
1983-87	1.5	1.7

	LABOR TIME	SEVERE SERVICE
1988-92		
one side	3.8	4.0
both sides	4.1	4.3
1993-97		
4 cyl.		
one side	3.1	3.3
both sides	3.4	3.6
V6		
right side	5.8	6.0
left side	4.2	4.4
both sides	6.2	6.4
1998-02		
4 cyl.		
one side	2.9	3.1
both sides	3.0	3.2
V6		
right side	5.8	6.0
left side	4.2	4.4
both sides	6.2	6.4
929		
1988-91		
right side	4.1	4.3
left side	4.6	4.8
both sides	5.3	5.5
1992-95		
right side	5.8	6.0
left side	6.3	6.5
both sides	6.8	7.0
Miata		
1990-93		
one side	3.7	3.9
both sides	3.9	4.1
1994-05		
one side	2.8	3.0
both sides	3.1	3.3
Millenia		
KJ		
right side	9.4	9.6
left side	9.1	9.3
both sides	9.7	9.9
KL		
right side	6.3	6.5
left side	4.1	4.3
both sides	6.7	6.9
MX-3		
1992-93	2.6	2.8
1994-95		
4 cyl.	3.1	3.3
V6		
right side	5.2	5.4
left side	4.2	4.4
both sides	5.8	6.0
w/AC add		
86-94 323, Protégé	.1	.1
83-87 626, MX-6	.2	.2
Miata	.1	.1
w/PS add		
86-94 323, Protégé	.1	.1
83-87 626, MX-6	.1	.1
Miata	.2	.2

	LABOR TIME	SEVERE SERVICE
Timing Belt, Replace (B)		
323, Protégé/5		
1986-94		
SOHC	2.4	2.6
DOHC	3.4	3.6
1995-05 (2.4)	3.2	3.3
626, MX-6		
1983-87	1.9	2.1
1988-92	3.5	3.7
1993-02		
4 cyl. (2.2)	3.0	3.2
V6 (3.1)	4.2	4.4
929		
1988-91	3.1	3.3
1992-95	3.9	4.1
Miata	3.1	3.2
Millenia		
KJ	5.8	6.0
KL	4.1	4.3
MX-3		
4 cyl.		
1992-93	2.3	2.5
1994-95	3.1	3.3
V6	3.9	4.1
w/AC add	.2	.2
w/PS add	.1	.1
Timing Belt Sprocket/Pulley, Replace (B)		
323, Protégé/5		
1986-98	2.5	2.7
1999-05		
1.6L	2.8	3.0
1.8L, 2.0L	2.4	2.6
626, MX-6		
1983-87	2.3	2.5
1988-92	3.1	3.3
1993-02		
4 cyl.	2.5	2.7
V6	3.4	3.6
929		
1988-91	3.4	3.6
1992-95	4.1	4.3
Miata		
1990-92	3.5	3.7
1993-05	2.6	2.8
Millenia		
KJ	5.3	5.5
KL	3.3	3.5
MX-3		
1992-93		
4 cyl.	2.7	2.9
V6	4.2	4.4
1994-95		
4 cyl.	3.0	3.2
V6	4.2	4.4
w/AC add	.2	.2
w/PS add	.1	.1

	LABOR TIME	SEVERE SERVICE
Timing Belt Tensioner and/or Spring, Replace (B)		
323, Protégé/5		
1986-94		
SOHC	2.6	2.8
DOHC	3.5	3.7
1995-05	2.7	2.9
626, MX-6		
1983-87	1.7	1.9
1988-92	3.5	3.7
1993-02		
4 cyl.	2.7	2.9
V6	4.1	4.3
929		
1988-91	3.0	3.2
1992-95	3.9	4.1
Miata		
1990-93	3.3	3.5
1994-05 (1.9)	2.6	2.8
Millenia		
KJ	5.9	6.1
KL	3.8	4.0
MX-3		
1992-93	2.3	2.5
1994-95		
4 cyl.	3.0	3.2
V6	3.9	4.1
w/AC add	.2	.2
w/PS add	.1	.1
Timing Chain or Sprocket, Replace (B)		
1981-82 626	8.2	8.4
Mazda3	4.5	4.7
Mazda6		
2.3L	5.0	5.2
3.0L	13.1	13.3
MX-3	9.2	9.4
w/AC 626 add	1.2	1.4
Timing Chain Housing Gasket, Replace (B)		
1981-82 626	7.1	7.3
Mazda3	4.3	4.5
Mazda6		
2.3L	4.7	4.9
3.0L	11.2	11.4
MX-3	8.2	8.4
w/AC 626 add	1.2	1.4
Replace cover add	.5	.7
Timing Chain Tensioner, Replace (B)		
1981-82 626	2.2	2.4
Mazda3	4.5	4.7
Mazda6		
2.3L	4.8	5.0
3.0L	12.4	12.6
MX-3	.9	1.1

	LABOR TIME	SEVERE SERVICE
Timing Cover and/or Gasket, Replace (B)		
323, Protégé/5		
1986-94 each	.5	.7
1995-05		
upper	.7	.9
center	1.4	1.6
lower	1.7	1.9
626, MX-6		
1983-92		
upper	.5	.7
lower	.8	1.0
1993-97		
4 cyl.	1.4	1.6
V6		
right side	3.1	3.3
left side	2.9	3.1
1998-02		
4 cyl.		
upper	1.6	1.8
lower	1.7	1.9
V6		
right side	2.8	3.0
left side	2.0	2.2
929		
1988-91		
SOHC		
right side	2.3	2.5
left side	2.8	3.0
DOHC		
right side	2.3	2.5
left side each		
upper	2.8	3.0
lower	2.9	3.1
1992-95		
right side	2.6	2.8
left side		
upper	3.3	3.5
lower	3.5	3.7
Miata		
upper	.9	1.1
center	1.7	1.9
lower	2.2	2.4
Millenia		
KJ		
upper	3.8	4.0
right side	3.1	3.3
left side	2.6	2.8
KL		
right side	2.6	2.8
left side	1.8	2.0
MX-3		
1992-93 4 cyl.		
upper	1.5	1.7
center	1.7	1.9
1994-95		
4 cyl.	2.3	2.5

	LABOR TIME	SEVERE SERVICE
V6		
right side	3.0	3.2
left side	2.7	2.9
w/AC Miata add	.1	.1
w/DOHC 83-92 626, MX-6 add	.3	.3
w/PS Miata add	.1	.1
CRANK & PISTONS		
Connecting Rod Bearings, Replace (A)		
Includes: Check bearing oil clearance.		
323, Protégé/5		
1986-94	3.8	4.0
1995-98	15.2	15.4
1999-05	14.6	14.8
626, MX-6		
1981-82	3.4	3.6
1983-92	5.2	5.4
1993-02		
4 cyl.	14.8	15.0
V6	18.9	19.1
929		
1988-91	5.5	5.7
1992-95	16.2	16.4
GLC	3.9	4.1
Mazda3	15.4	15.6
Mazda6	16.5	16.7
Miata		
1990-93	5.4	5.6
1994-05	12.9	13.41
Millenia		
KJ	21.3	21.5
KL	19.6	19.8
MX-3		
1992-93	13.0	13.2
1994-95		
4 cyl.	15.2	15.4
V6	16.2	16.4
w/AC add		
323, Protégé	.3	.3
626, MX-6	.3	.3
90-93 Miata	.1	.1
94-02 Miata	.5	.5
MX-3	.3	.3
w/AT 94-02 Miata add	1.3	1.3
w/PS 90-02 Miata add	.2	.2
Crankshaft and Main Bearings, Replace (A)		
Includes: Engine R&R.		
323, Protégé/5		
1986-94		
AT	16.4	16.6
MT	15.9	16.1
1995-98	17.5	17.7
1999-05	16.4	16.6

MAZ-18 323 : 626 : 929 : GLC : MAZDA3 : MAZDA6 : MIATA : MILLENIA

	LABOR TIME	SEVERE SERVICE
Crankshaft and Main Bearings, Replace (A)		
626, MX-6		
1981-82	18.3	18.5
1983-92		
AT	25.2	25.4
MT	20.1	20.3
1993-02		
4 cyl.	16.9	17.1
V6	19.1	19.3
929		
1988-91		
AT		
SOHC	21.2	21.4
DOHC	22.7	22.9
MT	19.7	19.9
1992-95	19.4	19.6
GLC	20.5	20.7
Mazda3	16.6	16.8
Mazda6		
2.3L	17.7	17.9
3.0L	24.2	24.7
Miata		
1990-93	18.8	19.0
1994-05	16.2	16.4
Millenia		
KJ	21.3	21.5
KL	19.6	19.8
MX-3		
1992-93	14.4	14.6
1994-95		
4 cyl.	16.3	16.5
V6	17.5	17.7
w/4WD 86-94 323, Protégé add	.5	.5
w/AC add		
86-94 323, Protégé	.5	.5
81-92 626, MX-6	.5	.5
93-97 626, MX-6	.3	.3
Miata	.2	.2
MX-3	.3	.3
95-02 Protégé	.3	.3
w/AT add		
93-97 626, MX-6	.1	.1
Miata	1.3	1.3
MX-3	.1	.1
w/DOHC add		
88-91 929	.5	.5
86-94 323, Protégé	.5	.5
w/PS add		
86-94 323, Protégé	.5	.5
81-92 626, MX-6	.2	.2
Miata	.2	.2
w/turbo 81-92 626, MX-6 add	1.2	1.2
Crankshaft Front Oil Seal, Replace (A)		
1986-94 323, Protégé	2.7	2.9
1981-82 626	1.5	1.7
Mazda3	2.3	2.5

	LABOR TIME	SEVERE SERVICE
Mazda6		
2.3L	3.0	3.2
3.0L	1.9	2.1
1990-94 Miata	3.4	3.6
MX-3	1.5	1.7
w/AC 626 add	.2	.2
Crankshaft Pulley, Replace (B)		
323, Protégé/5		
1986-98	.8	1.0
1999-05	1.1	1.3
626, MX-6		
1981-82	.9	1.1
1983-02	1.2	1.4
929		
1988-91	1.7	1.9
1992-95	2.0	2.2
GLC	.9	1.1
Mazda3	2.1	2.3
Mazda6		
2.3L	2.9	3.1
3.0L	1.7	1.9
Miata	1.1	1.3
Millenia	1.2	1.4
MX-3	1.5	1.7
w/AC add		
86-98 323, Protégé	.1	.1
81-92 626, MX-6	.2	.2
93-02 626, MX-6	.2	.2
w/PS 93-02 626, MX-6 add	.1	.1
Main Bearing Cap Side Seal, Replace (B)		
GLC	2.9	3.1
Pistons or Connecting Rods, Replace (A)		
Includes: Ridge reaming, cylinder wall deglazing, installing new rings and rod bearings, engine tune-up.		
323, Protégé/5		
1986-94		
SOHC	11.6	11.8
DOHC	12.8	13.0
1995-98	20.7	20.9
1999-05	19.2	19.4
626, MX-6		
1981-82	10.2	10.4
1983-92	16.2	16.4
1993-02		
4 cyl.	19.9	20.1
V6	26.3	26.5
929		
1988-91		
SOHC	16.2	16.4
DOHC	17.3	17.5
1992-95	26.2	26.4
GLC	11.4	11.6
Mazda3	18.8	19.0
Mazda6		
2.3L	20.1	20.3
3.0L	24.5	24.7

	LABOR TIME	SEVERE SERVICE
Miata		
1990-93	13.0	13.2
1994-05	18.2	18.4
Millenia		
KJ	31.2	31.4
KL	27.0	27.2
MX-3		
4 cyl.	19.7	19.9
V6	24.5	25.7
w/AC add		
86-94 323, Protégé	.1	.2
95-02 323, Protégé	.1	.1
81-92 626, MX-6	.5	.5
Miata	.1	.1
w/PS add		
86-94 323, Protégé	.3	.3
81-92 626, MX-6	.2	.2
Miata	.2	.2
w/turbo 81-92 626, MX-6 add	.9	.9
Rear Main Oil Seal, Replace (A)		
Includes: R&R trans.		
323, Protégé/5		
1986-94		
2WD	4.8	5.0
4WD	6.3	6.5
1995-98	6.4	6.6
1999-05		
AT	6.6	6.6
MT	5.8	6.0
626, MX-6		
1981-82		
AT	5.4	5.6
MT	4.2	4.4
1983-92		
AT	5.4	5.6
MT	5.0	5.2
1993-97		
4 cyl.		
AT	7.0	7.2
MT	6.3	6.5
V6		
AT	8.3	8.5
MT	7.6	7.8
1998-02		
AT	6.4	6.6
MT	7.1	7.3
929		
1988-91		
AT	4.8	5.0
MT	3.6	3.8
1992-95	5.8	6.0
GLC		
AT	4.8	5.0
MT	4.2	4.4
Mazda3		
AT	6.6	6.8
MT	5.4	5.6

MX-3 : MX-6 : PROTEGE : PROTEGE5 : RX-7 : RX-8 MAZ-19

	LABOR TIME	SEVERE SERVICE
Mazda6		
2.3L		
AT	8.2	8.3
MT	7.7	7.9
3.0L	9.9	10.1
Miata		
1990-93		
AT	5.9	6.1
MT	3.8	4.0
1994-02		
AT	6.3	6.5
MT	5.0	5.2
Millenia		
KJ	10.0	10.2
KL	8.3	8.5
MX-3		
4 cyl.		
AT	6.0	6.2
MT	5.2	5.4
V6		
AT	6.8	7.0
MT	6.3	6.5
w/DOHC 88-91 929 add	.5	.5

Rings, Replace (A)
Includes: Ridge reaming, cylinder wall deglazing, installing new rings, engine tune-up.

	LABOR TIME	SEVERE SERVICE
323, Protégé/5		
1986-94		
SOHC	10.9	11.1
DOHC	11.9	12.1
1995-98	20.1	20.3
1999-05	18.5	18.7
626, MX-6		
1981-82	9.3	9.5
1993-02		
4 cyl.	19.4	19.6
V6	25.3	25.5
929		
1988-91		
SOHC	14.4	14.6
DOHC	15.4	15.6
1992-95	24.2	24.4
GLC	10.0	10.2
Mazda3	18.8	19.0
Mazda6		
2.3L	20.1	20.3
3.0L	23.5	23.7
Miata		
1990-93	11.8	12.0
1994-05	17.6	17.8
Millenia		
KJ	29.9	30.1
KL	26.0	26.2

	LABOR TIME	SEVERE SERVICE
MX-3		
4 cyl.	18.6	18.8
V6	22.8	23.0
w/AC add		
86-94 323, Protégé	.1	.1
95-02 323, Protégé	.1	.1
81-92 626, MX-6	.5	.5
Miata	.1	.1
w/AT 96-05 Miata add	.9	.9
w/PS add		
86-94 323, Protégé	.3	.3
81-92 626, MX-6	.2	.2
Miata	.2	.2
w/turbo 81-92 626, MX-6 add	.9	.9

ENGINE LUBRICATION
Engine Oil Cooler, Replace (B)

	LABOR TIME	SEVERE SERVICE
323, Protégé/5		
1981-94	2.0	2.2
1995-98	2.6	2.8
1999-05 (1.0)	1.4	1.6
626, MX-6		
1981-87	.8	1.0
1988-92	1.2	1.4
1993-97	2.5	2.7
1998-02		
4 cyl.	2.0	2.2
V6	1.6	1.8
Mazda3, Mazda6 (1.0)	1.4	1.6
Miata	1.7	1.9
MX-3	1.2	1.4
RX-7		
1986-91	1.4	1.6
1993-95	1.7	1.9

Engine Oil Pressure Switch (Sending Unit), Replace (B)

	LABOR TIME	SEVERE SERVICE
Exc. below	.7	.9
Mazda3, Mazda6	.6	.8
Miata	.5	.7
MX-3	.5	.7

Oil Jets, Oil Jet Gaskets and/or Jet Valve, Replace (B)
Includes: Engine R&R if needed.

	LABOR TIME	SEVERE SERVICE
323, Protégé		
1988-94	2.0	2.2
1995-02	11.6	11.8
1993-97 626, MX-6		
4 cyl.	21.5	21.7
V6	17.8	18.0
929		
1990-91	22.0	22.2
1992-95	19.1	19.3
Miata		
1990-93	4.1	4.3
1994-02		
AT	10.7	10.9
MT	9.9	10.1

	LABOR TIME	SEVERE SERVICE
Millenia		
KJ	20.3	20.5
KL	19.0	19.2
MX-3		
1992-93	11.2	11.4
1994-95		
4 cyl.	13.3	13.5
V6	16.7	16.9
w/AC add		
323, Protégé	.3	.3
93-97 626, MX-6	.3	.3
MX-3	.3	.3
Miata	.5	.5
w/PS Miata add	.2	.2

Oil Pan and/or Gasket, Replace (B)

	LABOR TIME	SEVERE SERVICE
323, Protégé/5		
1986-94	1.7	1.9
1995-98		
1.5L	1.8	2.0
1.6L	2.2	2.4
1999-05	2.2	2.4
626, MX-6		
1981-82	2.3	2.5
1983-92	2.9	3.1
1993-02		
4 cyl.	2.0	2.2
V6	2.5	2.7
929		
1988-91	2.4	2.6
1992-95	1.9	2.1
GLC	2.3	2.5
Mazda3		
exc. 1.5L	2.1	2.3
1.5L	5.3	5.5
Mazda6		
2.3L	6.0	6.2
3.0L	2.1	2.3
Miata		
exc. below	3.8	4.0
1999-05 w/variable valve timing	4.3	4.5
Millenia	2.4	2.6
MX-3	2.5	2.7
w/AC add		
323, Protégé	.3	.3
626, MX-6	.3	.3
MX-3	.3	.3
90-93 Miata	.1	.1
94-05 Miata	.5	.5
w/AT 94-05 Miata add	1.3	1.3
w/PS 90-05 Miata add	.2	.2

Oil Pressure Gauge (Dash), Replace (B)

	LABOR TIME	SEVERE SERVICE
1993-95 RX-7	.9	.9

MAZ-20   323 : 626 : 929 : GLC : MAZDA3 : MAZDA6 : MIATA : MILLENIA

	LABOR TIME	SEVERE SERVICE
Oil Pump, R&R and Recondition (B)		
Includes: Engine R&R if needed.		
323, Protégé		
1986-94		
2WD	4.2	4.4
4WD	5.3	5.5
1995-98	15.4	15.6
1999-02	14.9	15.1
626, MX-6		
1981-82	3.1	3.3
1983-92	5.3	5.5
1993-02		
4 cyl.	13.1	13.3
V6	13.8	14.0
929		
1988-91		
positive displace	3.2	3.4
trochoid type	6.4	6.6
1992-95	19.0	19.2
GLC	3.0	3.2
Miata		
1990-93	6.0	6.2
1994-02		
AT	12.8	13.0
MT	11.5	11.7
Millenia		
KJ	17.5	17.7
KL	14.6	14.8
MX-3		
4 cyl.	13.0	13.2
V6	13.1	13.3
w/AC add		
86-94 323, Protégé	.1	.1
95-02 323, Protégé	.3	.3
626, MX-6	.3	.3
90-93 Miata	.1	.1
94-02 Miata	.5	.5
MX-3	.3	.3
w/PS add		
86-94 323, Protégé	.1	.1
90-02 Miata	.2	.2
Oil Pump, Replace (B)		
Includes: Engine R&R if needed.		
323, Protégé/5		
1986-94		
2WD	4.1	4.3
4WD	4.4	4.6
1995-98		
1.5L	6.3	6.5
1.8L	5.2	5.4
1999-05		
1.6L (4.3)	5.8	6.0
1.8L, 2.0L		
AT (7.8)	10.5	10.7
MT (7.3)	9.9	10.1
626, MX-6		
1981-82	2.7	2.9
1983-92	5.3	5.5
1993-02	13.0	13.2

	LABOR TIME	SEVERE SERVICE
929		
1988-91		
positive displace	2.7	2.9
trochoid type	5.9	6.1
1992-95	17.9	18.1
GLC	2.6	2.8
Mazda3	5.9	6.4
Mazda6		
2.3L	3.6	3.8
3.0L	14.3	14.5
Miata		
1990-93	3.8	4.0
1994-95		
AT	12.4	12.6
MT	11.2	11.4
1996-97	7.2	7.4
1999-05		
w/variable valve timing	7.6	7.8
w/o variable valve timing	7.2	7.4
Millenia		
KJ	17.2	17.4
KL	15.0	15.2
MX-3		
4 cyl.	12.7	12.9
V6	13.0	13.2
w/AC add		
86-94 323, Protégé	.1	.1
95-02 323, Protégé	.3	.3
626, MX-6	.3	.3
90-93 Miata	.1	.1
94-02 Miata	.5	.5
MX-3	.3	.3
w/PS add		
86-94 323, Protégé	.1	.1
90-02 Miata	.2	.2
Oil Pump Drive Sprocket, Replace (B)		
GLC	4.8	5.0
Oil Pump Seal, Replace (B)		
323, Protégé/5		
1986-94		
2WD	3.4	3.6
4WD	2.7	2.9
1995-05 (2.2)	3.0	3.2
626, MX-6		
1983-85	1.5	1.7
1986-87	2.5	2.7
1988-92	3.7	3.9
1993-02		
4 cyl. (2.2)	3.0	3.2
V6	4.1	4.3
929		
1988-91	3.8	4.0
1992-95	4.4	4.6
Miata		
1990-93	3.5	3.7
1994-05 (2.2)	3.0	3.2

	LABOR TIME	SEVERE SERVICE
Millenia		
KJ	6.2	6.4
KL	3.7	3.9
MX-3		
1992-93	2.4	2.6
1994-95		
4 cyl.	2.9	3.1
V6	4.1	4.3
w/AC add		
86-94 323, Protégé	.1	.1
95-02 323, Protégé	.3	.3
626, MX-6	.3	.3
90-93 Miata	.1	.1
94-02 Miata	.5	.5
MX-3	.3	.3
w/PS add		
86-94 323, Protégé	.1	.1
90-02 Miata	.2	.2

ROTARY ENGINE

Eccentric Shaft, Replace (B)

	LABOR TIME	SEVERE SERVICE
RX-7		
1981-85		
AT	21.3	21.5
MT	20.7	20.9
1986-91		
AT	21.9	22.1
MT	21.4	21.6
1993-95		
AT	28.8	29.0
MT	28.0	28.2
RX-8		
AT	23.0	23.2
MT	22.8	23.0
w/ABS 81-91 add	.1	.1
w/AC 81-95 add	.5	.5
w/PS 81-91 add	.1	.1
w/turbo 81-91 add	.5	.5

Eccentric Shaft Rear Oil Seal, Replace (B)

	LABOR TIME	SEVERE SERVICE
RX-7		
1981-85	4.2	4.4
1986-91	3.7	3.9
1993-95	4.6	4.8
RX-8	4.3	4.5

Apex Corner & Side or Springs Seals, Replace (A)

	LABOR TIME	SEVERE SERVICE
RX-7		
1981-85		
AT	21.3	21.5
MT	20.7	20.9
1986-91		
AT	21.9	22.1
MT	21.4	21.6
1993-95		
AT	28.8	29.0
MT	28.0	28.2

	LABOR TIME	SEVERE SERVICE
RX-8		
AT	23.0	*23.2*
MT	22.8	*23.0*
w/ABS 81-91 add	.1	*.1*
w/AC 81-95 add	.5	*.5*
w/PS 81-91 add	.1	*.1*
w/turbo 81-91 add	.5	*.5*
Engine Assy., R&I (B)		
RX-7		
1981-85		
AT	6.2	*6.4*
MT	5.9	*6.1*
1986-91		
AT	6.9	*7.1*
MT	6.4	*6.6*
1993-95		
AT	12.6	*12.9*
MT	11.8	*12.0*
RX-8		
AT	11.1	*11.3*
MT	10.9	*11.1*
w/ABS 81-91 add	.2	*.2*
w/AC 81-91 add	.7	*.7*
w/AC 93-95 add	.5	*.5*
w/PS 81-91 add	.2	*.2*
w/turbo 81-91 add	.2	*.2*
Engine Assy. (Complete), R&R and Recondition (A)		
RX-7		
1981-85		
AT	21.3	*21.5*
MT	20.8	*21.0*
1986-91		
AT	21.9	*22.1*
MT	21.5	*21.7*
1993-95		
AT	28.8	*29.0*
MT	28.0	*28.2*
RX-8		
AT	24.8	*25.0*
MT	24.6	*24.8*
w/ABS 81-91 add	.2	*.2*
w/AC 81-91 add	.7	*.7*
w/AC 93-95 add	.5	*.5*
w/PS 81-91 add	.2	*.2*
w/turbo 81-91 add	.2	*.2*
Engine Assy. (Short Block), Replace (B)		
RX-7		
1981-85		
AT	15.3	*15.5*
MT	14.6	*14.8*
1986-91	15.4	*15.6*
1993-95	18.9	*19.1*
RX-8	16.4	*16.6*
w/ABS 81-91 add	.2	*.2*
w/AC 81-91 add	.7	*.7*
w/AC 93-95 add	.5	*.5*
w/PS 81-91 add	.2	*.2*
w/turbo 81-91 add	.2	*.2*

	LABOR TIME	SEVERE SERVICE
Engine Mounts, Replace (B)		
RX-7		
1981-91		
front		
one side	.5	*.7*
both sides	.8	*1.0*
rear	.7	*.9*
1993-95	4.8	*5.0*
RX-8 one	1.2	*1.4*
Front Cover Oil Seal, Replace (B)		
RX-7		
1981-85	6.7	*6.9*
1986-91	7.3	*7.5*
1993-95		
AT	15.0	*15.2*
MT	14.2	*14.4*
RX-8		
AT	13.3	*13.5*
MT	13.1	*13.3*
w/ABS 81-91 add	.1	*.1*
w/AC 81-91 add	.1	*.1*
w/AC 93-95 add	.5	*.5*
w/PS 81-91 add	.1	*.1*
Jet Valves and/or Gaskets, Replace (B)		
RX-7		
1981-85		
AT	19.4	*19.6*
MT	18.8	*19.0*
1986-91	19.7	*19.9*
1993-95		
AT	28.8	*29.0*
MT	28.0	*28.2*
w/ABS 86-91 add	.1	*.1*
w/AC 81-85 add	.7	*.7*
w/AC 86-95 add	.5	*.5*
w/PS 86-91 add	.2	*.2*
w/turbo 86-91 add	.5	*.5*
Oil Metering Pump, Replace (B)		
RX-7		
1981-89	1.4	*1.6*
1990-91	3.6	*3.8*
1993-95	5.2	*5.4*
RX-8	.8	*1.0*
Oil Pan and/or Gasket, Replace (B)		
RX-7		
1981-85	1.6	*1.8*
1986-91	2.5	*2.7*
1993-95	5.9	*6.1*
RX-8	1.7	*1.9*
Oil Pump Assy., R&R and Recondition (B)		
RX-7		
1981-85	7.0	*7.2*
1986-91	7.5	*7.7*

	LABOR TIME	SEVERE SERVICE
1993-95		
AT	15.6	*15.8*
MT	15.0	*15.2*
w/ABS 81-91 add	.1	*.1*
w/AC 81-95 add	.5	*.5*
w/PS 81-91 add	.2	*.2*
Oil Pump Assy., Replace (B)		
RX-7		
1981-85	6.8	*7.0*
1986-91	7.5	*7.7*
1993-95		
AT	15.4	*15.6*
MT	14.8	*15.0*
RX-8		
AT	15.7	*15.9*
MT	15.5	*15.7*
w/ABS 81-91 add	.1	*.1*
w/AC 81-95 add	.5	*.5*
w/PS 81-91 add	.2	*.2*
Rear Stationary Gear Oil Seal, Replace (B)		
RX-7		
1981-85		
AT	5.6	*5.8*
MT	4.8	*5.0*
1986-91		
AT	5.0	*5.2*
MT	4.2	*4.4*
1993-95		
AT	5.8	*6.0*
MT	5.2	*5.4*
RX-8		
AT	6.8	*7.0*
MT	6.6	*6.8*

CLUTCH

	LABOR TIME	SEVERE SERVICE
Bleed Clutch Hydraulic System (B)		
All Models	.5	*.5*
Clutch Pedal Free Play, Adjust (B)		
All Models	.3	*.5*
Clutch Assy., Replace (B)		
323, Protégé/5		
1986-94		
2WD	3.7	*3.9*
4WD	5.9	*6.1*
1995-98	5.8	*6.0*
1999-05 (4.1)	5.5	*5.7*
626, MX-6		
1981-82	3.5	*3.7*
1983-92	4.9	*5.1*
1993-97		
4 cyl.	5.5	*5.7*
V6	7.1	*7.3*
1998-02	6.2	*6.4*
1988-89 929	3.5	*3.7*
GLC	3.7	*3.9*
Mazda3	5.0	*5.2*

MAZ-22 323 : 626 : 929 : GLC : MAZDA3 : MAZDA6 : MIATA : MILLENIA

	LABOR TIME	SEVERE SERVICE
Clutch Assy., Replace (B)		
Mazda6		
2.3L (5.6)	7.6	7.8
3.0L (7.0)	9.4	9.6
Miata		
1990-93	3.6	3.8
1994-05 (3.4)	4.6	4.8
MX-3		
4 cyl.	4.8	5.0
V6	6.0	6.2
RX-7		
1981-85	3.9	4.1
1986-91	3.5	3.7
1993-95	4.5	4.7
RX-8	4.1	4.3
Clutch Control Cable, Replace (B)		
1981-87 323, 626, GLC	.7	.9
Clutch Master Cylinder, R&R and Reconditon (B)		
Includes: System bleeding.		
323, Protégé		
1988-94	1.2	1.4
1995-02	1.6	1.8
626, MX-6		
1983-92	1.2	1.4
1993-02	1.6	1.8
1988-89 929	1.2	1.4
Miata	1.2	1.4
MX-3	1.8	2.0
RX-7	1.2	1.4
Clutch Master Cylinder, Replace (B)		
Includes: System bleeding.		
323, Protégé/5		
1988-94	.7	.9
1995-05 (.9)	1.2	1.4
626, MX-6		
1983-92	.8	1.0
1993-02	1.3	1.5
1988-89 929	.7	.9
Mazda3	1.7	1.9
Mazda6	1.0	1.2
Miata	.6	.8
MX-3	1.4	1.6
RX-7	.8	1.0
RX-8 (.4)	.6	.8
Clutch Release Bearing or Fork, Replace (B)		
323, Protégé/5		
1986-94		
2WD	3.1	3.3
4WD	5.0	5.2
1995-05 (4.1)	5.5	5.7
626, MX-6		
1981-82	3.1	3.3
1983-92	4.2	4.4
1993-97		
4 cyl.	4.9	5.1
V6	6.2	6.4
1998-02	5.2	5.4

	LABOR TIME	SEVERE SERVICE
929 1988-89	3.1	3.3
GLC	3.5	3.7
Mazda3	4.8	5.0
Mazda6		
2.3L (5.4)	7.3	7.5
3.0L (6.9)	9.3	9.5
Miata		
1990-93	3.3	3.5
1994-05 (3.2)	4.3	4.5
MX-3		
4 cyl.	4.4	4.6
V6	5.6	5.8
RX-7		
1981-85	3.5	3.7
1986-91	3.1	3.3
1993-95	4.2	4.4
RX-8 (3.0)	4.1	4.3
Clutch Slave Cylinder, R&R and Recondition (B)		
Includes: System bleeding.		
323, Protégé		
1988-94	.7	.9
1995-98	1.3	1.5
1999-02	.8	1.0
626, MX-6		
1983-92	.7	.9
1993-97	1.6	1.8
1998-02	.8	1.0
1988-89 929	.9	1.1
Miata	.9	1.1
MX-3		
4 cyl.	1.3	1.5
V6	2.2	2.4
RX-7	1.4	1.6
Clutch Slave Cylinder, Replace (B)		
Includes: System bleeding.		
323, Protégé/5		
1988-94	.5	.7
1995-05	1.2	1.4
626, MX-6		
1983-92	.5	.7
1993-97	1.4	1.6
1998-02	.7	.9
1988-89 929	.5	.7
Mazda3 (.5)	.7	.9
Mazda6 (.8)	1.1	1.3
Miata	.6	.8
MX-3		
4 cyl.	.8	1.0
V6	1.8	2.0
RX-7	1.2	1.4
RX-8 (.6)	.8	1.0
Flywheel, Replace (B)		
323, Protégé/5		
1986-94		
2WD	4.2	4.4
4WD	6.3	6.5
1995-05	6.2	6.4
626, MX-6		
1981-82	3.7	3.9

	LABOR TIME	SEVERE SERVICE
1983-92	5.0	5.2
1993-97		
4 cyl.	6.2	6.4
V6	7.4	7.6
1998-02	6.3	6.5
929	4.1	4.3
GLC	4.4	4.6
Mazda3	5.1	5.3
Mazda6		
2.3L	7.8	8.0
3.0L	9.8	10.0
Miata	5.0	5.2
Millenia		
KJ	9.8	10.0
KL	7.4	7.6
MX-3		
4 cyl.	5.2	5.4
V6	6.2	6.4
RX-7		
1981-85	4.2	4.4
1986-91	3.8	4.0
1993-95	4.9	5.1
RX-8 (3.7)	5.0	5.2

MANUAL TRANSAXLE

	LABOR TIME	SEVERE SERVICE
Extension Housing and/or Gasket, Replace (B)		
626, GLC	3.0	3.2
Miata	5.7	5.9
Extension Housing Oil Seal, Replace (B)		
626, GLC	.8	1.0
Miata	1.4	1.6
Speed Sensor, Replace (B)		
Mazda3 sensor	.6	.8
Mazda6 sensor		
2.3L	.8	.9
3.0L	11.5	11.7
Speedometer Driven Gear and/or Seal, Replace (B)		
323, Protégé/5	.6	.8
626, MX-6	.7	.9
GLC	.5	.7
Miata	.8	1.0
MX-3	.7	.9
Transaxle Assy., R&I (B)		
Includes: Reset toe.		
323, Protégé		
1986-94		
2WD	3.5	3.7
4WD	5.5	5.7
1995-02	5.4	5.6
626, MX-6		
1983-92	3.9	4.1
1993-97		
4 cyl.	5.4	5.6
V6	6.8	7.0
1998-02	5.8	6.0

MX-3 : MX-6 : PROTEGE : PROTEGE5 : RX-7 : RX-8 **MAZ-23**

	LABOR TIME	SEVERE SERVICE
GLC		
sedan	3.1	3.3
wagon	3.0	3.2
Mazda3	4.7	4.9
Mazda6		
2.3L	6.9	7.1
3.0L	8.8	9.0
MX-3		
4 cyl.	4.4	4.6
V6	5.6	5.8

Transaxle Assy. R&R and Reconditioning (A)
323, Protégé/5		
1986-89		
2WD		
4-Speed	6.8	7.0
5-Speed	7.0	7.2
4WD	10.5	10.7
1990-94		
2WD	11.9	12.1
4WD	15.2	15.4
1995-05	13.4	13.6
626, MX-6		
1983-92	8.6	8.8
1993-97		
4 cyl.	13.4	13.6
V6	15.0	15.2
1998-02	14.1	14.3
GLC		
4-Speed		
sedan	6.8	7.0
wagon	7.6	7.8
5-Speed		
sedan	7.0	7.2
wagon	8.4	8.6
Mazda3	13.2	13.4
Mazda6		
2.3L	15.6	15.8
3.0L	17.9	18.1
MX-3		
4 cyl.	11.9	12.1
V6	13.3	13.5

Transaxle Rear Cover, R&R or Replace (B)
323, Protégé/5		
1988-94	1.2	1.4
1995-05	5.7	5.9
626, MX-6		
1983-92	1.2	1.4
1993-97	1.7	1.9
1998-02	5.5	5.7
GLC	1.2	1.4
Mazda3	5.6	5.8
Mazda6	7.8	8.0
MX-3	1.7	1.9

MANUAL TRANSMISSION

Extension Housing and/or Gasket, Replace (B)
1981-82 626	2.6	2.8
1988-89 929	3.6	3.8
Miata		
1990-93	3.7	3.9
1994-97	5.6	5.8
RX-7		
1981-85	3.7	3.9
1986-91		
w/turbo	5.5	5.7
w/o turbo	4.4	4.6
1993-95	6.5	6.7
RX-8	5.1	5.3

Extension Housing Oil Seal, Replace (B)
1981-82 626	.8	1.0
1988-89 929	1.6	1.8
Miata		
1990-93	.9	1.1
1994-97	1.5	1.7
RX-7		
1981-85	.7	.9
1986-95	1.6	1.8
RX-8	1.8	2.0

Speedometer Driven Gear, Replace (B)
1981-82 626	.5	.7
1988-89 929	.5	.7
Miata	.8	1.0
RX-7		
1981-91	.5	.7
1993-95	1.2	1.4
RX-8 speed sensor	1.1	1.3

Transmission Assy., R&I (B)
1981-82 626	3.2	3.4
1988-89 929	3.1	3.3
Miata		
1990-96	3.3	3.5
1994-97	4.1	4.3
RX-7		
1981-85	3.5	3.7
1986-91	3.1	3.3
1993-95	4.2	4.4
RX-8	3.8	4.2

Transmission Assy., R&R and Recondition (A)
1981-82 626		
4-Speed	7.5	7.7
5-Speed	8.3	8.5
1988-89 929	7.0	7.2
Miata		
1990-93	8.9	9.1
1994-97	9.3	9.5
RX-7		
1981-85	8.9	9.2
1986-95		
w/turbo	9.9	10.2
w/o turbo	8.6	8.8
RX-8	10.1	10.3

Transmission Mount, Replace (B)
1988-89 929	.7	.9
1986-95 RX-7	.5	.7

AUTOMATIC TRANSAXLE

SERVICE TRANSAXLE INSTALLED

Check Unit For Oil Leaks (C)
All Models FWD	.9	1.1

Drain & Refill Unit (B)
1988-02 323, Protégé	.5	.7
626, MX-6	1.7	1.9
GLC	.8	1.0
Millenia	1.4	1.6
MX-3	.3	.5

Electronic Transmission Diagnosis (B)
1987-05 FWD	1.7	1.9

Oil Pressure Check (B)
All Models FWD	1.2	1.4

Control Valve Cover and/or Gasket, Replace (B)
1988-91 323, Protégé	1.7	1.9
626, MX-6		
1987-92	1.8	2.0
1993-97		
exc. LA4A-EL	2.7	2.9
LA4A-EL	1.4	1.6
Mazda3 (.8)	1.1	1.3
Mazda6 (4.1)	5.3	5.5
Millenia	2.6	2.8

Downshift Solenoid, Replace (B)
1981-89 FWD	.9	1.1

ECAT Switch, Replace (B)
1987-92 626, MX-6	.7	.9
Mazda6	.6	.8

Governor Assy., R&R and Recondition (B)
1986-89 323	1.3	1.5
1983-87 626	1.5	1.7
1983-85 MX-3	1.5	1.7

Governor Assy., R&R or Replace (B)
1986-89 323	.8	1.0
1983-87 626	.9	1.1
1983-85 MX-3	.9	1.1

Inhibitor Switch, Replace (B)
323, Protégé	.7	.9
626, MX-6		
1983-92	.7	.9
1993-02	1.3	1.5
Millenia	1.2	1.4
MX-3	1.2	1.4

Kickdown Relay/Switch, Replace (B)
All Models FWD	.7	.9

MAZ-24 323 : 626 : 929 : GLC : MAZDA3 : MAZDA6 : MIATA : MILLENIA

	LABOR TIME	SEVERE SERVICE
Kickdown Switch or Solenoid, Replace (B)		
All Models FWD	.5	.7
Oil Pan and/or Gasket, Replace (B)		
323, Protégé/5		
1986-94	1.4	1.6
1995-05 (1.2)	1.6	1.8
626, MX-6		
1983-92	1.4	1.6
1993-02	1.7	1.9
GLC	1.5	1.7
Mazda3 (.8)	1.1	1.3
Mazda6 (3.0)	4.1	4.3
Millenia	1.4	1.6
MX-3	1.7	1.9
Overdrive Switch, Replace (B)		
323, Protégé/5		
1988-94	.7	.9
1995-05	1.1	1.3
626, MX-6		
1988-92	.7	.9
1993-97	1.5	1.7
1998-02	1.2	1.4
Millenia	1.4	1.6
MX-3	.9	1.1
Pulse Generator, Replace (B)		
323, Protégé	.5	.7
626, MX-6	.7	.9
Millenia	.9	1.1
MX-3	.8	1.0
Speedometer Driven Gear and/or Seal, Replace (B)		
323, Protégé/5		
1986-94	7.5	7.7
1995-96	.6	.8
1997-05 sensor	.6	.8
626, MX-6		
1988-92	7.5	7.7
1993-97		
4 cyl.	10.0	10.2
V6	11.3	11.5
LA4A-EL	1.6	1.8
1998-02		
4 cyl.	1.6	1.8
V6	.6	.8
Millenia		
KJ	10.5	10.7
KL	9.7	9.9
Speed Sensor, Replace (B)		
Mazda3 sensor	.8	1.0
Mazda6 sensor		
2.3L	.8	1.0
3.0L	11.5	11.7
Throttle Control Cable, Replace (B)		
323, Protégé		
1988-94	2.0	2.2
1995-96	3.1	3.3
1987-92 626, MX-6	2.2	2.4

	LABOR TIME	SEVERE SERVICE
MX-3		
4 cyl.	2.8	3.0
V6	2.9	3.1
Transmission Control Module, Replace (B)		
1990-94 323, Protégé	.7	.9
1986-97 626, MX-6	.7	.9
Mazda6 ECAT	.6	.8
Millenia	.7	.9
MX-3	.7	.9
Vacuum Modulator, Replace (B)		
1983-87 FWD	.7	.9
Valve Body, R&R and Recondition (B)		
323, Protégé		
1986-87	2.2	2.4
1988-94		
2WD	3.2	3.4
4WD	7.9	8.1
1995-98	5.9	6.1
1999-02	4.5	4.7
626, MX-6		
1981-92		
w/ECAT	3.4	3.6
w/o ECAT	2.9	3.1
1993-97		
exc. LA4A-EL	6.5	6.7
LA4A-EL	3.7	3.9
1998-02		
4 cyl.	2.5	2.7
V6	4.2	4.4
GLC	3.1	3.3
Millenia		
KJ	5.3	5.5
KL	6.5	6.7
MX-3	5.9	6.1
Flush oil cooler and lines add	.5	.5
Valve Body, Replace (B)		
323, Protégé/5		
1986-87	1.4	1.6
1988-94		
2WD	2.5	2.7
4WD	6.3	6.5
1995-98	4.1	4.3
1999-05	3.6	3.8
626, MX-6		
1981-92		
w/ECAT	2.8	3.0
w/o ECAT	2.0	2.2
1993-97		
exc. LA4A-EL	3.8	4.0
LA4A-EL	2.5	2.7
1998-02		
4 cyl.	2.5	2.7
V6	4.2	4.4
GLC	1.7	1.9
Mazda3 (2.6)	3.5	3.7

	LABOR TIME	SEVERE SERVICE
Mazda6		
2.3L (4.7)	6.3	6.5
3.0L (4.3)	5.8	6.0
Millenia		
KJ	2.9	3.1
KL	3.8	4.0
MX-3	3.2	3.4
Flush oil cooler and lines add	.5	.7
SERVICE TRANSAXLE REMOVED		
Transaxle R&R included unless otherwise noted.		
Front Pump Oil Seal, Replace (B)		
323, Protégé		
1986-98		
2WD	3.9	4.1
4WD	6.0	6.2
1999-05	7.8	8.0
626, MX-6		
1983-92		
w/ECAT	5.2	5.4
w/o ECAT	4.3	4.5
GLC		
sedan	4.3	4.5
wagon	5.4	5.6
Mazda3	7.7	7.9
Mazda6		
2.3L (7.1)	9.6	9.8
3.0L (8.8)	11.9	12.1
Millenia (8.8)	11.9	12.1
MX-3		
4 cyl.	6.2	6.4
V6	7.2	7.4
Torque Converter, Replace (B)		
323, Protégé/5		
1988-94		
2WD	4.6	4.8
4WD	6.3	6.5
1995-96	6.3	6.5
1997-98	5.9	6.1
1999-05 (5.3)	7.2	7.4
626, MX-6		
1983-92		
w/ECAT	5.3	5.5
w/o ECAT	4.4	4.6
1993-97		
4 cyl.	6.7	6.9
V6	7.9	8.1
LA4A-EL	6.2	6.4
1998-02		
4 cyl.	6.2	6.4
V6	7.0	7.2
GLC		
sedan	4.3	4.5
wagon	4.1	4.3
Mazda3	7.2	7.7

MX-3 : MX-6 : PROTEGE : PROTEGE5 : RX-7 : RX-8 **MAZ-25**

	LABOR TIME	SEVERE SERVICE
Mazda6		
2.3L	9.1	9.3
3.0L	11.9	12.1
Millenia		
KJ	9.9	10.1
KL	7.4	7.6
MX-3		
4 cyl.	6.2	6.4
V6	7.0	7.2
Transaxle Assy., R&I (B)		
323, Protégé		
1986-94		
2WD	4.6	4.8
4WD	6.3	6.5
1995-98	6.3	6.5
1999-02	6.7	6.9
626, MX-6		
1983-92		
w/ECAT	5.2	5.4
w/o ECAT	4.2	4.4
1993-97		
4 cyl.	6.8	7.0
V6	8.0	8.2
LA4A-EL	6.2	6.4
1998-02		
4 cyl.	6.2	6.4
V6	6.8	7.0
GLC		
sedan	4.2	4.4
wagon	3.9	4.1
Mazda3	7.1	7.3
Mazda6		
2.3L (6.3)	8.5	8.7
3.0L (6.5)	8.8	9.0
Millenia		
KJ	9.7	9.9
KL	7.9	8.1
MX-3		
4 cyl.	5.9	6.1
V6	6.8	7.0
Flush oil cooler and lines add	.5	.5
Transaxle Assy., R&R and Recondition (A)		
323, Protégé/5		
1986-87	9.3	9.5
1988-94		
2WD	18.6	18.8
4WD	24.2	24.4
1995-96	20.7	20.9
1997-98	20.2	20.4
1999-05	19.8	20.0
626, MX-6		
1983-92		
w/ECAT	20.7	20.9
w/o ECAT	9.7	9.9
1993-97		
4 cyl.	22.3	22.5
V6	23.4	24.6
LA4A-EL	13.6	13.8

	LABOR TIME	SEVERE SERVICE
1998-02		
4 cyl.	13.6	13.8
V6	22.5	22.7
GLC		
sedan	9.7	9.9
wagon	9.5	9.7
Mazda3	19.3	19.5
Mazda6	21.8	22.0
Millenia		
KJ	22.2	22.4
KL	23.4	23.6
MX-3		
4 cyl.	19.3	19.5
V6	20.0	20.2
Flush oil cooler and lines add	.5	.7

AUTOMATIC TRANSMISSION

SERVICE TRANSMISSION INSTALLED

Diagnosis Electronic Transmission System Component Each (B)
1988-05 (.8)	1.1	1.1

Check Unit for Oil Leaks (C)
All Models	.9	.9

Drain and Refill Unit (B)
1981-82 626	.8	.8
929		
1988-91	.9	.9
1992-95	1.2	1.2
Miata	.9	.9
RX-7	.9	.9

Oil Pressure Check (B)
All Models	1.2	1.4

Downshift Solenoid, Replace (B)
1981-88 RX-7	.7	.9

Electronic Control Unit, Replace (B)
929	.8	.8
Miata	.5	.5
1989-95 RX-7	.5	.5

Extension Housing and/or Gasket, Replace (B)
1981-82 626	2.5	2.7
929	2.4	2.6
Miata		
1996-97	2.9	3.1
1999-05 (2.7)	3.6	3.8
RX-7		
1981-85	5.3	5.5
1986-91	2.0	2.2
1993-95	2.5	2.7
RX-8 (2.3)	3.1	3.3

Extension Housing Rear Oil Seal, Replace (B)
929		
1988-91	.8	1.0
1992-95	2.0	2.2

	LABOR TIME	SEVERE SERVICE
Miata, RX-7	1.6	1.8
1981-82 626	1.6	1.8
RX-8	1.8	2.0

Governor Assy., R&R and Recondition (B)
1981-82 626	2.8	3.0
1990-93 Miata	2.9	3.1
RX-7		
1981-85	5.5	5.7
1986-88	4.9	5.1

Governor Assy., R&R or Replace (B)
1981-82 626	2.7	2.9
1990-93 Miata	2.5	2.7
RX-7		
1981-85	5.2	5.4
1986-88	4.5	4.7

Inhibitor Switch, Replace (B)
929		
1988-91	.5	.7
1992-95	1.2	1.4

Kickdown Relay, Replace (B)
All Models	.3	.3

Kickdown Switch, Replace (B)
1981-88 RX-7	.7	.9

Neutral Safety Switch, Replace (B)
1981-82 626	.5	.7
929	.5	.7
Miata	.5	.7
RX-7		
1986-91	.5	.7
1993-95	3.4	3.6
RX-8	.5	.7

Oil Pan Gasket, Replace (B)
1981-82 626	.8	1.0
929		
1988-91	.9	1.1
1992-95	1.2	1.4
Miata	1.3	1.6
RX-7	.9	1.1
RX-8 (.8)	1.1	1.3

Overdrive Control Switch, Replace (B)
1988-05	.7	.9

Overdrive or Hold Switch, Replace (B)
1996-05 Miata	.7	.9
1986-91 RX-7	.7	.9

Overdrive Solenoid, Replace (B)
1986-88 RX-7	.7	.9

Selector Indicator, Replace (B)
1981-82 626	.7	.9
929	.5	.7
Miata	.7	.9
RX-7		
1981-91	.3	.5
1993-95	.7	.9
RX-8 (.7)	.9	1.1

MAZ-26   323 : 626 : 929 : GLC : MAZDA3 : MAZDA6 : MIATA : MILLENIA

	LABOR TIME	SEVERE SERVICE
Speed Input Sensor, Replace (B)		
1993-95 929	4.9	5.1
RX-8	.8	1.0
Speed Sensor, Replace (B)		
929		
1988-91	.5	.7
1992-95	1.5	1.7
1989-95 RX-7	.5	.7
RX-8 (.7)	.9	1.1
Speedometer Driven Gear and/or Seal, Replace (B)		
929	.5	.7
Miata	.7	.9
RX-7		
1981-91	.5	.7
1993-95	1.6	1.8
Transmission Mounts, Replace (B)		
929		
1988-91	.7	.9
1992-95	1.6	1.8
1986-91 RX-7	.5	.7
Vacuum Control Unit, Replace (B)		
1981-91 RWD	.5	.7
Valve Body Assy., R&R and Recondition (A)		
1981-82 626	3.3	3.5
929		
1988-91	5.9	6.1
1992-95	3.2	3.4
Miata		
1990-93	4.5	4.7
1994-02	3.8	4.0
RX-7		
1986-91	2.8	3.0
1993-95	5.3	5.5
Valve Body Assy., Replace (B)		
1981-82 626	2.0	2.2
929		
1988-91	5.0	5.2
1992-95	2.2	2.4
Miata		
1990-97	3.0	3.2
1999-05	3.8	4.0
RX-7		
1981-91	1.6	1.8
1993-95	2.7	2.9
RX-8 (2.7)	3.6	3.8

SERVICE TRANSMISSION REMOVED
Transmission R&R included unless otherwise noted.

	LABOR TIME	SEVERE SERVICE
Front Pump Oil Seal, Replace (B)		
1981-82 626	4.9	5.1
929		
1988-91	5.6	5.8
1992-95	6.3	6.5
1990-97 Miata	7.6	7.8

	LABOR TIME	SEVERE SERVICE
RX-7		
1981-85	5.6	5.8
1986-95	5.2	5.4
RX-8 (5.9)	8.0	8.2
Pulse Generator, Replace (B)		
1992-95 929	4.4	4.6
1993-95 RX-7	4.9	5.2
Torque Converter, Replace (B)		
1981-82 626	6.2	6.4
929		
1988-91	4.3	4.5
1992-95	5.2	5.4
Miata	6.7	6.8
RX-7		
1981-91	4.6	4.8
1993-95	5.2	5.4
RX-8 (5.0)	6.8	7.0
Transmission Assy., R&I (B)		
1981-82 626	4.3	4.5
929		
1988-91	4.2	4.4
1992-95	4.9	5.1
Miata		
1990-93	6.4	6.6
1994-02	6.0	6.2
RX-7		
1981-85	4.4	4.6
1986-91	3.7	3.9
1993-95	4.9	5.1
RX-8 (4.7)	6.3	6.9
Transmission Assy., R&R and Recondition (A)		
1981-82 626	13.6	13.8
929		
1988-91	10.8	11.0
1992-95	13.7	13.9
Miata		
1990-93	16.7	16.9
1994-02	16.2	16.4
RX-7		
1981-85	11.2	11.4
1986-91	10.3	10.5
1993-95	15.0	15.2
RX-8 (10.7)	14.4	14.5

TRANSFER CASE

	LABOR TIME	SEVERE SERVICE
Transfer Case, R&I (B)		
AT		
1990-91 323, Protégé	8.4	8.6
MT		
1988-89 323	6.3	6.5
1990-91 323, Protégé	8.2	8.4
Transfer Case, R&R and Recondition (B)		
AT		
1990-91 323, Protégé	10.5	10.7

	LABOR TIME	SEVERE SERVICE
MT		
1988-89 323	10.9	11.1
1990-91 323, Protégé	10.2	10.4

SHIFT LINKAGE
AUTOMATIC TRANS

	LABOR TIME	SEVERE SERVICE
Shift Linkage, Adjust (B)		
All Models	.5	.7
Gearshift Control Cable or Rod, Replace (B)		
323, Protégé/5		
1986-94	.7	.9
1995-05	4.8	5.0
626, MX-6		
1981-92	.8	1.0
1993-97	4.1	4.3
1998-02	4.4	4.6
929		
1988-91	1.2	1.4
1992-95	.5	.7
GLC	.7	.9
Mazda3	2.1	2.3
Mazda6	4.0	4.2
Miata	.5	.7
Millenia		
cable	6.2	6.4
rod	.5	.7
MX-3		
4 cyl.	1.4	1.6
V6	1.7	1.9
RX-7		
1981-91	1.2	1.4
1993-95	1.7	1.9
RX-8 (.6)	.8	1.0
Gear Selector Lever, Replace (B)		
323, Protégé/5		
1986-94	.7	.9
1995-02	1.2	1.4
626, MX-6		
1981-97	1.4	1.6
1998-05	1.1	1.3
929, GLC	.8	1.0
Mazda3	.7	.9
Mazda6, Miata	1.0	1.2
Millenia	1.4	1.6
MX-3	1.3	1.5
RX-7		
1981-85	1.2	1.4
1986-91	1.4	1.6
1993-95	.7	.9
RX-8 (.6)	.8	1.0

MANUAL TRANS

	LABOR TIME	SEVERE SERVICE
Gearshift Lever, Replace (B)		
323, Protégé/5		
1986-87	.5	.7
1988-94	.7	.9
1995-98	1.3	1.5
1999-05	1.6	1.8

	LABOR TIME	SEVERE SERVICE
626, MX-6		
1981-92	.7	.9
1993-97	1.3	1.5
1998-02	1.7	1.9
1988-89 929	.7	.9
GLC	.5	.7
Mazda3	.7	.9
Mazda6	.8	1.0
Miata	.6	.8
MX-3	.9	1.1
RX-8 (.3)	.5	.7

DRIVELINE

Differential, Drain and Refill (B)
All Models front or rear .. .7   .9

Axle Shaft and/or Bearing, Replace (B)
1981-82 626
  one 1.8   2.0
  both 3.1   3.3
1981-85 RX-7
  one 2.0   2.2
  both 3.7   3.9
RX-8
  one (1.6) 2.2   2.4
  both (2.6) 3.5   3.7

Axle Shaft Oil Seal, Replace (B)
Rear axle
  1988-89 323 4WD
    one side 8   1.0
    both sides 1.6   1.8
  1981-82 626
    one side 1.6   1.8
    both sides 2.8   3.0
  1988-91 929
    one side 2.7   2.9
    both sides 4.2   4.4
  Miata
    one side 1.7   1.9
    both sides 2.5   2.7
  RX-7
    1983-85
      one side 1.5   1.7
      both sides 2.6   2.8
    1986-91
      one side 1.4   1.6
      both sides 2.2   2.4
  RX-8 (1.2) 1.6   1.8

Center Support Bearing, Replace (B)
1988-91 323, Protégé
  one 1.7   1.9
  both 1.9   2.1
1981-82 626 .... 1.8   2.0
929
  1988-91 9   1.1
  1992-95 .... 2.2   2.4

Differential Assy., R&R and Recondition (B)
323, Protégé
  1986-89
    front
      AT 10.7   10.9
      MT
        2WD 7.3   7.5
        4WD 9.6   9.8
    rear 9.6   9.8
  1990-94
    front
      AT
        2WD 8.6   88
        4WD 12.5   12.7
      MT
        2WD 8.6   8.8
        4WD 10.5   10.47
    rear 6.2   6.4
626, MX-6
  1981-82 .... 6.3   6.5
  1983-92
    AT
      w/ECAT 8.6   8.8
      w/o ECAT 7.6   7.8
    MT 7.5   7.7
929
  1988-91 .... 4.5   4.7
  1992-95 .... 8.0   8.2
GLC
  AT 8.2   8.3
  MT 7.3   7.5
Miata 5.0   5.2
  1996-97 .... 5.3   5.5
  1999-05 .... 6.4   6.6
MX-3
  AT
    4 cyl. 10.8   11.0
    V6 .... 11.9   12.1
  MT
    4 cyl. 9.7   9.9
    V6 .... 11.2   11.4
RX-7
  1981-85 .... 5.5   5.7
  1986-91 .... 4.6   4.8
  1993-95 .... 8.2   8.4
RX-8 .... 8.2   8.4

Differential Assy., R&R or Replace (B)
323, Protégé
  1986-89
    front
      AT 7.4   7.6
      MT
        2WD 6.8   7.0
        4WD 9.3   9.5
    rear 9.2   9.4

1990-94
  front
    AT
      2WD 8.0   8.2
      4WD 11.7   11.9
    MT
      2WD 8.0   8.2
      4WD 10.0   10.2
  rear 8.2   8.4
626, MX-6
  1981-82 .... 3.3   3.5
  1983-92
    AT
      w/ECAT 8.0   8.2
      w/o ECAT 7.1   7.3
    MT 7.0   7.2
929
  1988-91 .... 1.8   2.0
  1992-95 .... 3.2   3.4
GLC
  AT 7.4   7.6
  MT 5.8   6.0
Miata
  1996-97 .... 2.4   2.6
  1999-05 .... 4.3   4.5
MX-3
  AT
    4 cyl. 10.2   10.4
    V6 .... 11.4   11.6
  MT
    4 cyl. 9.2   9.4
    V6 .... 10.2   10.4
RX-7
  1981-85 .... 4.2   4.4
  1986-91 .... 2.3   2.5
  1993-95 .... 3.6   3.8
RX-8 (3.3) 4.5   4.7

Drive Flange Shaft Seal, Replace (B)
Rear axle
  1988-91 323, Protégé 4WD
    one side 1.4   1.6
    both sides 1.8   2.0
  929
    1988-91
      one side 8   1.0
      both sides 1.2   1.4
    1992-95
      one side 2.3   2.5
      both sides 3.1   3.3
  Miata
    one side 1.5   1.7
    both sides 1.8   2.0
  RX-7
    1986-91
      one side 7   .9
      both sides 1.3   1.5
    1993-95
      one side 1.2   1.4
      both sides 1.6   1.8

MAZ-28 323 : 626 : 929 : GLC : MAZDA3 : MAZDA6 : MIATA : MILLENIA

	LABOR TIME	SEVERE SERVICE		LABOR TIME	SEVERE SERVICE		LABOR TIME	SEVERE SERVICE
Inboard CV Joint and/or Boot, Replace (B)			Miata 1.3		*1.5*	1993-97		
Rear axle			RX-7			AT inner and outer		
1988-91 323, Protégé 4WD			1981-85 9		*1.1*	right side 1.8		*2.0*
one side 1.6		*1.8*	1986-91 . . . . 1.7		*1.9*	left side 2.3		*2.5*
both sides 2.6		*2.8*	1993-95 . . . . 2.7		*2.9*	both sides		
929			RX-8 (1.3) 1.8		*2.0*	4 cyl. 3.1		*3.3*
1988-91			**Rear Axle Companion Flange, Replace (B)**			V6 . . . . 3.4		*3.6*
one side 1.4		*1.6*				MT inner and outer		
both sides 2.0		*2.2*	929			one side 1.8		*2.0*
1992-95			one side 1.8		*2.0*	both sides		
one side 2.3		*2.5*	both sides 3.2		*3.4*	4 cyl. 2.5		*2.7*
both sides 3.8		*4.0*	**Torque Tube Companion Flange, Replace (B)**			V6 . . . . 2.9		*3.1*
Miata						1998-02		
one side 1.6		*1.8*	1988-91 323, Protégé 4WD 1.2		*1.4*	inner		
both sides 2.5		*2.7*				right side 1.8		*2.0*
1999-05			1981-82 626 . . . . 1.2		*1.4*	left side 2.3		*2.5*
one side 2.4		*2.6*	929			both sides 3.1		*3.3*
both sides 4.0		*4.2*	1988-91 8		*1.0*	outer		
RX-7			1992-95 . . . . 1.4		*1.6*	right side 2.3		*2.5*
1986-91			Miata 1.2		*1.4*	left side 2.7		*2.9*
one side 1.4		*1.6*	RX-7			both sides 4.1		*4.3*
both sides 2.0		*2.2*	1981-85 8		*1.0*	GLC		
1993-95			1986-91 . . . . 1.5		*1.7*	both one side 1.6		*1.8*
one side 1.5		*1.7*	1993-95 . . . . 2.5		*2.7*	all both sides 2.5		*2.7*
both sides 2.3		*2.5*	**U-Joint, Replace or Recondition (B)**			Mazda6		
Outboard CV Joint and/or Boot, Replace (B)			1988-91 323, Protégé			2.3L		
			one 1.5		*1.7*	inner one side 2.4		*2.6*
Rear axle			two 1.8		*2.0*	outer one side . . . 3.0		*3.2*
1988-91 323, Protégé 4WD			all 1.9		*2.1*	3.0L		
one side 1.6		*1.8*	1981-82 626			right side inner . . . 2.9		*3.1*
both sides 2.6		*2.8*	one 1.3		*1.5*	left side inner 2.4		*2.6*
929			two 1.8		*2.0*	right side outer . . . 3.3		*3.5*
1988-91			all 2.0		*2.2*	left side outer 3.0		*3.2*
one side 1.4		*1.6*	1992-95 929 . . . . 1.8		*2.0*	Millenia		
both sides 2.0		*2.2*	1981-91 RX-7			inner one side 1.8		*2.0*
1992-95			one 1.2		*1.4*	outer one side 2.5		*2.7*
one side 2.3		*2.5*	two 1.5		*1.7*	inner both sides 3.0		*3.2*
both sides 3.8		*4.0*				outer both sides 3.7		*3.9*
Miata			**HALFSHAFTS**			MX-3		
one side 1.6		*1.8*	**CV Joint, Replace/Recondition (B)**			AT		
both sides 2.5		*2.7*	323, Protégé/5			right side inner . . . 1.6		*1.8*
1999-05			1986-94			left side inner 2.3		*2.5*
one side 2.4		*2.6*	both one side 1.4		*1.6*	right side outer . . . 1.8		*2.0*
both sides 4.0		*4.2*	all both sides 2.5		*2.7*	left side outer 2.9		*3.1*
RX-7			1995-98			MT		
1986-91			inner one side 1.7		*1.9*	inner one side 1.6		*1.8*
one side 1.4		*1.6*	outer			outer one side . . . 1.8		*2.0*
both sides 2.0		*2.2*	right side 1.8		*2.0*	inner both sides . . 2.3		*2.5*
1993-95			left side 2.2		*2.4*	outer both sides . . 2.9		*3.1*
one side 1.5		*1.7*	inner both sides . 2.5		*2.7*	**Halfshaft, R&R or Replace (B)**		
both sides 2.3		*2.5*	outer both sides . 3.1		*3.3*	1981-87		
Pinion Shaft Oil Seal, Replace (B)			1999-05			one side 1.2		*1.4*
Rear axle			inner one side 2.1		*2.3*	both sides 1.8		*2.0*
1988-91 323, Protégé 4WD 1.3		*1.5*	outer one side 2.7		*2.9*	323, Protégé/5		
			inner both sides . . 3.0		*3.2*	1988-98		
1981-82 626 . . . . 1.3		*1.5*	outer both sides . . 4.3		*4.5*	one side 1.4		*1.6*
929			626, MX-6			both sides 1.8		*2.0*
1988-91 9		*1.1*	1983-92			1999-05		
1992-95 . . . . 1.7		*1.9*	both one side 1.7		*1.9*	one side (1.2) 1.6		*1.8*
			all both sides 2.6		*2.8*	both sides (1.8) . . . 2.4		*2.6*

MX-3 : MX-6 : PROTEGE : PROTEGE5 : RX-7 : RX-8 **MAZ-29**

	LABOR TIME	SEVERE SERVICE
626, MX-6		
1983-92		
one side	1.2	*1.6*
both sides	1.6	*1.8*
1993-97		
4 cyl.		
AT		
right side	1.3	*1.5*
left side	1.8	*2.0*
both sides	2.5	*2.7*
MT		
one side	1.4	*1.6*
both sides	1.8	*2.0*
V6		
AT		
right side	1.4	*1.6*
left side	1.8	*2.0*
both sides	2.9	*3.1*
MT		
right side	1.4	*1.6*
left side	1.8	*2.0*
both sides	2.2	*2.4*
1998-02		
right side	1.6	*1.8*
left side	2.0	*2.2*
both sides	2.7	*2.9*
Mazda3		
right side	1.1	*1.3*
left side	1.2	*1.4*
both sides	2.0	*2.2*
Mazda6		
2.3L		
one side (1.3)	1.8	*2.0*
both sides (2.1)	2.8	*3.0*
3.0L		
one side (1.6)	2.2	*2.4*
both sides (2.5)	3.4	*3.6*
Millenia		
one side	1.8	*2.0*
both sides	2.5	*2.7*
MX-3		
AT		
right side	1.3	*1.5*
left side	1.8	*2.0*
both sides	2.4	*2.6*
MT		
one side	1.3	*1.5*
both sides	1.8	*2.0*
RX-8 rear (1.3)	1.8	*2.0*
Halfshaft Boot, Replace (B)		
Includes: Clean and lubricate CV-joint.		
Standard type		
323, Protégé/5		
1986-92		
both one side	1.4	*1.6*
all both sides	2.5	*2.7*

	LABOR TIME	SEVERE SERVICE
1993-98		
inner or outer		
right side	1.8	*2.0*
left side	2.2	*2.4*
both sides	3.1	*3.3*
1999-05		
inner or outer		
one side	2.5	*2.7*
both sides	4.1	*4.3*
626, MX-6		
1983-92		
both one side	1.8	*2.0*
all both sides	2.7	*2.9*
1993-97		
AT		
both one side	2.6	*2.8*
all both sides	4.1	*4.3*
MT		
both one side	2.2	*2.4*
all both sides	3.3	*3.5*
1998-02		
inner or outer		
right side	2.3	*2.5*
left side	2.6	*2.8*
both sides	3.8	*4.0*
GLC		
both one side	1.8	*2.0*
all both sides	2.6	*2.8*
Mazda3		
both one side	2.3	*2.5*
all both sides	3.5	*3.7*
Mazda6		
2.3L		
inner or outer		
one side	3.0	*3.2*
both sides	4.6	*4.8*
3.0L		
inner or outer		
one side	3.2	*3.4*
both sides	5.0	*5.2*
Millenia		
inner one side	2.0	*2.2*
outer one side	2.3	*2.5*
inner both sides	3.1	*3.3*
outer both sides	3.6	*3.8*
MX-3		
AT		
right side inner	1.8	*2.0*
left side inner	2.3	*2.5*
right side outer	1.9	*2.1*
left side outer	3.1	*3.3*
MT		
inner one side	1.8	*2.0*
outer one side	1.9	*2.1*
inner both sides	2.4	*2.6*
outer both sides	3.0	*3.2*
Split type		
one side	.8	*1.0*
both side	1.3	*1.5*

	LABOR TIME	SEVERE SERVICE
BRAKES		
ANTI-LOCK		
The following operations do not include testing. Add time as required.		
Diagnose Anti-Lock Brake System Component Each (A)		
1987-05 (.7)	1.1	*1.1*
Control Unit, Replace (B)		
323, Protégé/5		
1995-98	.5	*.7*
1999-05	2.5	*2.7*
626, MX-6		
1988-97	.5	*.7*
1998-02	2.6	*3.1*
929		
1988-91	.5	*.7*
1992-95	.8	*1.0*
Miata	.3	*.5*
Millenia	.5	*.7*
MX-3	.7	*.9*
RX-7		
1987-91	.7	*.9*
1993-95	.5	*.7*
Front Sensor Assy., Replace (B)		
323, Protégé/5		
1995-98		
one	.5	*.7*
both	.7	*.9*
1999-05		
one	.7	*.9*
both	1.2	*1.4*
626, MX-6		
1988-97		
one	.7	*.9*
both	.8	*1.0*
1998-05		
one	.8	*1.0*
both	1.1	*1.3*
929		
1988-91		
one	.5	*.7*
both	.7	*.9*
1992-95		
one	.7	*.9*
both	1.2	*1.4*
Mazda3		
one (.6)	.8	*1.0*
both (.8)	1.1	*1.3*
Mazda6		
one (.9)	1.2	*1.4*
both (1.3)	1.8	*2.0*
Miata		
one	.9	*1.1*
both	1.2	*1.4*
Millenia		
one	.5	*.7*
both	.7	*.9*

MAZ-30   323 : 626 : 929 : GLC : MAZDA3 : MAZDA6 : MIATA : MILLENIA

	LABOR TIME	SEVERE SERVICE
Front Sensor Assy., Replace (B)		
MX-3		
one	.7	.9
both	.9	1.1
RX-7		
1987-91		
one	.8	1.0
both	1.5	1.7
1993-95		
one	.5	.7
both	.8	1.0
RX-8		
one (.9)	1.2	1.4
both (1.3)	1.8	2.0
Hydraulic Unit, Replace (B)		
1995-05 323, Protégé/5	2.8	3.0
626, MX-6		
1988-92	2.0	2.2
1993-02	2.5	2.7
929	1.6	1.8
Mazda3	2.6	2.8
Mazda6	2.0	2.2
Miata		
1996-97	2.3	1.5
1999-05	1.7	1.9
Millenia	2.4	2.6
MX-3	2.8	3.0
RX-7	1.7	1.9
RX-8	1.5	1.8
Pump or Main Relay, Replace (B)		
1995-98 323, Protégé	.3	.5
1992-97 626, MX-6	.5	.7
929	.7	.9
Miata		
1991-94	.5	.7
1995-05		
exc. smaller unit	1.7	1.9
smaller ABS unit	.5	.7
Millenia	.5	.7
MX-3	.5	.7
RX-7		
1987-94	.7	.9
1995	1.7	1.9
Rear Sensor Assy., Replace (B)		
323, Protégé/5		
1995-98 one or both	.7	.9
1999-05		
one	1.3	1.5
both	1.6	1.7
626, MX-6		
1987-92		
one	.7	.9
both	.8	1.0
1993-97		
one	.8	1.0
both	1.6	1.8
1998-02		
one	1.2	1.4
both	1.4	1.6

	LABOR TIME	SEVERE SERVICE
929		
one	.5	.7
both	.8	1.0
Mazda3		
one	.6	.8
both	.8	1.0
Mazda6		
one	1.1	1.3
both	1.6	1.8
Miata		
one	1.1	1.3
both	1.4	1.6
Millenia		
one	.5	.7
both	.7	.9
MX-3		
one	.8	1.0
both	1.6	1.8
RX-7		
one	.7	.9
both	1.4	1.6
RX-8		
one	.8	1.0
both	1.0	1.2
SYSTEM		
Bleed Brakes (B)		
Includes: Add fluid.		
All Models	.8	.8
w/ABS add	.3	.3
Brakes, Adjust (B)		
Includes: System bleeding.		
All Models	.7	.7
Brake Hose (Flexible), Replace (B)		
Includes: System bleeding.		
323, Protégé/5		
1986-94	.7	.9
1995-05	1.1	1.3
626, MX6	.9	1.1
929		
1988-91	.7	.9
1992-95	1.2	1.4
GLC	.7	.9
Mazda3	.8	1.0
Mazda6	1.3	1.5
Miata	1.1	1.3
Millenia	1.2	1.4
MX-3	.7	.9
RX-7, RX-8	.8	1.0
each addl. add	.2	.4
Brake Pressure Switch, Replace (B)		
1988-91 929	.9	1.1
1986-91 RX-7	1.2	1.4
Brake Proportioning Valve, Replace (B)		
All Models single	1.4	1.6

	LABOR TIME	SEVERE SERVICE
Dual		
323, Protégé/5		
1986-94		
front	1.3	1.5
rear	.7	.9
1995-98		
front	2.2	2.4
rear	1.3	1.5
1999-05 front	1.6	1.8
626, MX-6		
1981-92		
front	1.4	1.6
rear	.7	.9
1993-02		
front		
4 cyl.	1.6	1.8
V6	2.2	2.4
Millenia front	1.7	1.9
Master Cylinder, R&R and Rebuild (B)		
Includes: System bleeding.		
323, Protégé		
1986-94	1.8	2.0
1995-98	2.2	2.4
1999-02	1.8	2.0
626, MX-6		
1981-82	2.3	2.5
1983-92	1.7	1.9
1993-97	2.2	2.4
1998-02	1.7	1.9
929		
1988-91	2.4	2.6
1992-95	2.9	3.1
GLC	2.6	2.8
Miata	2.3	2.5
Millenia	2.2	2.4
MX-3	2.5	2.7
RX-7		
1981-85	2.6	2.9
1986-95	2.3	2.5
w/ABS add	.3	.5
Master Cylinder, Replace (B)		
Includes: System bleeding.		
323, Protégé/5		
1986-94	1.3	1.5
1995-98	2.2	2.4
1999-05	1.3	1.5
626, MX-6		
1981-92	1.3	1.5
1993-97	1.6	1.8
1998-02	1.3	1.5
929		
1988-91	1.2	1.4
1992-95	1.7	1.9
GLC	2.2	2.4
Mazda3	1.1	1.3
Mazda6	1.4	1.6
Miata	1.4	1.6

MX-3 : MX-6 : PROTEGE : PROTEGE5 : RX-7 : RX-8   MAZ-31

	LABOR TIME	SEVERE SERVICE
Millenia	1.6	*1.8*
MX-3	2.0	*2.2*
RX-7		
1981-85	2.2	*2.4*
1986-91	1.8	*2.0*
1993-95	3.1	*3.3*
RX-8	.8	*1.0*
w/ABS add	1.7	*1.9*

Power Booster Unit, Replace (B)

323, Protégé/5		
1986-94	1.6	*1.8*
1995-98	2.0	*2.2*
1999-05	2.3	*2.5*
626, MX-6		
1981-92	2.2	*2.4*
1993-97	2.8	*3.0*
1998-02	2.0	*2.2*
929		
1988-91	1.7	*1.9*
1992-95	3.1	*3.3*
GLC	2.3	*2.5*
Mazda3	2.4	*2.6*
Mazda6	2.9	*3.1*
Miata	2.3	*2.5*
Millenia	2.4	*2.6*
MX-3	3.0	*3.2*
RX-7		
1981-85	2.3	*2.5*
1986-91	1.2	*1.4*
1993-95	1.7	*1.9*
RX-8 (1.1)	1.5	*1.7*

Power Booster Vacuum Check Valve, Replace (B)

| Exc. below | .3 | *.3* |
| 1999-05 Miata | 1.8 | *2.0* |

SERVICE BRAKES
Brake Drum, Replace (B)

323, Protégé/5		
1986-89		
one	1.2	*1.4*
both	2.2	*2.5*
1990-05		
one	.5	*.7*
both	.8	*1.0*
626, MX-6		
1981-87		
one	.7	*.9*
both	.9	*1.1*
1988-92		
one	.8	*1.0*
both	1.6	*1.8*
1993-02		
one	.5	*.7*
both	.7	*.9*
GLC		
one	.7	*.9*
both	1.2	*1.4*

	LABOR TIME	SEVERE SERVICE
MX-3		
one	.5	*.7*
both	.7	*.9*
1981-85 RX-7		
one	.7	*.9*
both	.9	*1.1*

Caliper Assy., R&R and Reconditon (B)
Includes: System bleeding.

One	1.6	*1.8*
Both	2.5	*2.7*
w/ABS add	.3	*.5*

Caliper Assy., Replace (B)
Includes: System bleeding.

One	1.2	*1.4*
Both	1.6	*1.8*
w/ABS add	.3	*.5*

Disc Brake Rotor, Replace (B)

Front		
323, Protégé/5		
1986-95		
one	1.4	*1.6*
both	2.5	*2.7*
1996-05		
one	.6	*.8*
both	1.1	*1.3*
626, MX-6		
1981-82		
one	1.3	*1.5*
both	2.0	*2.2*
1983-92		
one	1.5	*1.7*
both	2.5	*2.7*
1993-02		
one	.9	*1.1*
both	1.8	*2.0*
GLC		
one	1.3	*1.5*
both	2.0	*2.2*
Mazda3		
one	.6	*.8*
both	.8	*1.0*
Mazda6		
one	.8	*1.0*
both	1.1	*1.3*
Miata		
one	.8	*1.0*
both	1.0	*1.2*
Millenia		
one	.9	*1.1*
both	1.8	*2.0*
MX-3		
one	.8	*1.0*
both	1.3	*1.5*
RX-7		
1981-85		
one	1.5	*1.7*
both	2.5	*2.7*

	LABOR TIME	SEVERE SERVICE
1986-91		
one	.9	*1.1*
both	1.7	*1.9*
1993-95		
one	.7	*.9*
both	1.2	*1.4*
RX-8		
one (.4)	.6	*.8*
both (.6)	.8	*1.0*
Rear		
323, Protégé/5		
2WD		
one	.8	*1.0*
both	1.1	*1.3*
4WD		
one	1.4	*1.6*
both	2.4	*2.6*
626, MX-6		
1986-87		
one	1.2	*1.4*
both	2.0	*2.2*
1988-02		
one	.8	*1.0*
both	1.4	*1.6*
Mazda3		
one	.6	*.8*
both	.8	*1.0*
Mazda6		
one (.4)	.6	*.8*
both (.8)	1.1	*1.3*
Miata		
one	.6	*.8*
both	1.1	*1.3*
Millenia		
one	.7	*.9*
both	.8	*1.0*
MX-3		
one	.7	*.9*
both	1.2	*1.4*
RX-8		
one	.6	*.8*
both	.8	*1.0*

Pads and/or Shoes, Replace (B)

323, Protégé/5		
front or rear disc	.9	*1.1*
rear drum	1.3	*1.5*
four wheels		
disc brakes	1.0	*1.2*
drum brakes	1.6	*1.8*
626, MX-6		
front or rear disc	.8	*1.0*
rear drum	1.2	*1.4*
four wheels		
disc brakes	1.2	*1.4*
drum brakes	1.7	*1.9*

MAZ-32 323 : 626 : 929 : GLC : MAZDA3 : MAZDA6 : MIATA : MILLENIA

	LABOR TIME	SEVERE SERVICE
Pads and/or Shoes, Replace (B)		
929		
front or rear disc	.8	1.0
rear drum	1.2	1.4
four wheels		
disc brakes	1.2	1.4
drum brakes	1.7	1.9
GLC		
front	.8	1.0
rear	1.2	1.4
Mazda3, Mazda6 front or rear disc	1.0	1.2
Miata		
front or rear disc	.8	1.0
rear drum	1.2	1.4
four wheels		
disc brakes	1.2	1.4
drum brakes	1.7	1.9
Millenia		
front or rear	.8	1.0
four wheels	1.7	1.9
MX-3		
front disc	.8	1.0
rear disc or drum	1.3	1.5
four wheels	2.2	2.4
RX-7		
front disc	.8	1.0
rear disc or drum	1.2	1.4
four wheels	1.9	2.1
RX-8 front or rear disc (.7)	.9	1.1
COMBINATION ADD-ONS		
Repack wheel bearings		
two wheels add	.7	.7
Replace		
brake drum add each	.1	.1
caliper add	.3	.3
disc rotor add each	.2	.2
wheel cylinder add	.3	.3
Resurface		
brake drum add	.5	.5
brake rotor add	.5	.5
Wheel Cylinder, R&R and Rebuild (B)		
Includes: System bleeding.		
One	1.6	1.8
Both	2.2	2.4
Wheel Cylinder, Replace (B)		
Includes: System bleeding.		
One	1.3	1.5
Both	2.0	2.2
PARKING BRAKE		
Parking Brake Cable, Adjust (C)		
All Models	.3	.7
Parking Brake Apply Actuator, Replace (B)		
323, Protégé/5	.8	1.0

	LABOR TIME	SEVERE SERVICE
626, MX-6		
1981-92	.5	.7
1993-02	.9	1.1
929		
1988-91	.7	.9
1992-95	1.2	1.4
GLC	1.2	1.4
Mazda3	.8	1.0
Mazda6	.9	1.1
Miata, Millenia	.8	1.0
MX-3	.7	.9
RX-7	.8	1.0
RX-8 (.7)	.9	1.1
Parking Brake Shoes, Replace (B)		
929	1.7	1.9
Millenia	1.7	1.9
Parking Brake Cable, Replace (B)		
Front		
323, Protégé/5	1.0	1.2
626, MX-6		
1981-92	.8	1.0
1993-97	1.7	1.9
1998-02	1.2	1.4
929		
1988-91	1.2	1.4
1992-95	3.2	3.4
GLC	1.2	1.4
Mazda3	1.5	1.7
Mazda6	1.1	1.3
Miata	1.2	1.4
Millenia	.8	1.0
MX-3	1.2	1.4
RX-7		
1986-91	.9	1.1
1993-95	2.5	2.7
RX-8 (1.5)	2.0	2.2
Rear		
323, Protégé/5		
1986-94		
one	.7	.9
both	.8	1.0
1995-05		
one	.8	1.0
both	1.2	1.4
626, MX-6		
1981-92		
w/disc brakes		
one	.7	.9
both	1.2	1.4
w/drum brakes		
one	1.2	1.4
both	2.0	2.2
1993-97		
w/disc brakes		
one	1.2	1.4
both	1.4	1.6
w/drum brakes		
one	1.4	1.6
both	1.7	1.9

	LABOR TIME	SEVERE SERVICE
1998-02		
w/disc brakes		
one	.7	.9
both	.8	1.0
w/drum brakes		
one	.7	.9
both	1.2	1.4
929		
1988-91		
one	1.7	1.9
both	2.4	2.6
1992-02		
one	2.2	2.4
both	3.2	3.4
Mazda3		
one	1.5	1.6
both	1.8	2.0
Mazda6		
one	2.1	2.3
both	2.3	2.5
Miata		
1996-97		
one	.8	1.0
both	1.2	1.4
1999-05		
one	1.3	1.5
both	1.6	1.8
Millenia		
one	.7	.9
both	1.2	1.4
MX-3		
one	.9	1.1
both	1.4	1.6
RX-7		
1981-85		
w/disc brakes		
one	.7	.9
both	1.2	1.4
w/drum brakes		
one	1.3	1.5
both	1.9	2.1
1986-91		
one	.5	.7
both	.7	.9
1993-95		
one	2.3	2.5
both	2.5	2.7
RX-8		
one	1.7	1.9
both	1.8	2.0

FRONT SUSPENSION
Unless otherwise noted, time given does not include alignment.

Align Front End (B)

	LABOR TIME	SEVERE SERVICE
Exc. below	2.3	2.5
626, MX-6	1.7	1.9
GLC	1.6	1.8

MX-3 : MX-6 : PROTEGE : PROTEGE5 : RX-7 : RX-8 **MAZ-33**

	LABOR TIME	SEVERE SERVICE
Front & Rear Alignment, Check & Adjust (B)		
All Models FWD	2.5	2.7
929, RX-7	2.3	2.5
Front Toe, Adjust (B)		
All Models	.8	1.0
Rear Toe, Adjust (B)		
1988-02 323, Protégé	.7	.9
1988-02 626, MX-6	.7	.9
929		
1988-91	.7	.9
1992-95	1.2	1.4
1992-02 Miata	.7	.9
Millenia	.7	.9
MX-3	1.2	1.4
1986-95 RX-7	1.2	1.4
Ball Joint, Replace (B)		
Miata		
one	.6	.8
both	1.0	1.2
MX-3		
one	1.4	1.6
both	1.9	2.1
Front Crossmember, Replace (B)		
Includes: Alignment charges.		
323, Protégé/5		
1986-98	3.6	3.8
1999-05	6.7	6.8
626, MX-6		
1993-97	6.9	7.1
1998-02		
4 cyl.	6.4	6.6
V6	7.2	7.4
Mazda3	3.3	3.6
Mazda6	3.6	3.8
1996-05 Miata	8.2	8.4
Millenia		
KJ	5.0	5.2
KL	4.5	4.7
MX-3	4.9	5.1
RX-8 (4.9)	6.6	6.8
Coil Spring, Replace (B)		
FWD		
323, Protégé/5		
1986-94		
one side	1.3	1.5
both sides	2.0	2.2
1995-98		
one side	2.2	2.4
both sides	2.6	2.8
1999-05		
one side	1.2	1.4
both sides	1.9	2.1
626, MX-6		
1983-92		
one side	1.4	1.6
both sides	2.0	2.2
1993-97		
one side	1.4	1.6
both sides	2.6	2.8

	LABOR TIME	SEVERE SERVICE
1998-02		
one side	1.2	1.4
both sides	1.7	1.9
GLC		
one side	1.5	1.7
both sides	2.4	2.6
Mazda3		
one side (.9)	1.2	1.4
both sides (1.3)	1.8	2.0
Mazda6		
one side (.8)	1.1	1.3
both sides (1.3)	1.8	2.0
Millenia		
one side	1.8	2.0
both sides	3.2	3.4
MX-3		
one side	1.2	1.4
both sides	1.7	1.9
RWD		
1981-82 626		
one side	1.8	2.0
both sides	3.5	3.7
929		
1988-91		
one side	1.3	1.5
both sides	1.8	2.0
1992-95		
one side	3.6	3.8
both sides	5.2	5.4
Miata		
one side	2.5	2.7
both sides	3.1	3.3
RX-7		
1981-85		
one side	1.2	1.4
both sides	2.3	2.5
1986-91		
one side	.8	1.0
both sides	1.6	1.8
1993-95		
one side	1.6	1.8
both sides	2.9	3.1
RX-8		
one side (1.0)	1.4	1.6
both sides (1.5)	2.0	2.2
Front Hub Assy., Replace (B)		
323, Protégé/5		
1986-94		
one side	2.2	2.4
both sides	3.1	3.3
1995-05		
one side	1.9	2.1
both sides	3.5	3.7
626, MX-6		
1983-87		
one side	1.5	1.7
both sides	2.6	2.8
1988-92		
one side	2.2	2.4
both sides	3.1	3.3

	LABOR TIME	SEVERE SERVICE
1993-02		
one side	1.7	1.9
both sides	3.0	3.2
GLC		
one side	2.2	2.4
both sides	3.6	3.8
Mazda3		
one side	2.3	2.5
both sides	3.5	3.7
Mazda6		
one side	2.2	2.4
both sides	3.8	4.0
Miata		
one side	1.1	1.3
both sides	1.4	1.6
Millenia		
one side	2.0	2.2
both sides	3.3	3.5
MX-3		
one side	1.7	1.9
both sides	3.1	3.3
Front Hub Bearings, Replace (B)		
1981-82 626		
one side	.9	1.1
both sides	1.9	2.1
1992-95 929		
one side	.8	1.0
both sides	1.4	1.6
Front Strut Shock Absorber, Replace (B)		
1981-82 626		
one side	2.0	2.2
both sides	3.7	3.9
929		
1988-91		
one side	1.2	1.4
both sides	1.7	1.9
1992-95		
one side	3.5	3.7
both sides	5.0	5.2
Mazda3		
One (1.3)	1.8	2.0
both (1.9)	2.6	2.8
Mazda6, RX-8		
one	1.3	1.5
both	1.9	2.1
Miata		
one side	2.7	2.9
both sides	3.3	3.5
RX-7		
1981-85		
one side	1.7	1.9
both sides	2.6	2.8
1986-91		
one side	.8	1.0
both sides	1.6	1.8
1993-95		
one side	1.5	1.7
both sides	2.7	2.9

MAZ-34 323 : 626 : 929 : GLC : MAZDA3 : MAZDA6 : MIATA : MILLENIA

	LABOR TIME	SEVERE SERVICE
Hub, Bearing or Seal, Replace (B)		
1981-82 626		
one side	1.3	1.5
both sides	2.3	2.5
1992-95 929		
one side	.9	1.1
both sides	1.6	1.8
RX-7		
1983-91		
one side	1.4	1.6
both sides	2.4	2.6
1993-95		
one side	.8	1.0
both sides	1.4	1.6
RX-8		
one side (.9)	1.2	1.4
both sides (1.4)	1.9	2.1
Hub Bearings, Replace (B)		
323, Protégé/5		
1986-94		
one side	2.2	2.4
both sides	3.1	3.3
1995-98		
one side	1.6	1.8
both sides	2.9	3.1
1999-05		
one side	1.8	2.0
both sides	3.3	3.5
626, MX-6		
1983-87		
one side	1.4	1.6
both sides	2.4	2.6
1988-92		
one side	2.2	2.4
both sides	3.1	3.3
1993-02		
one side	1.7	1.9
both sides	3.1	3.3
GLC		
one side	1.4	1.6
both sides	2.6	2.8
Mazda3		
one side (1.6)	2.2	2.4
both sides (2.5)	3.4	3.6
Mazda6		
one side (1.6)	2.2	2.4
both sides (3.0)	4.1	4.3
Millenia		
one side	2.2	2.4
both sides	3.4	3.6
MX-3		
one side	1.6	1.8
both sides	2.8	3.0
RX-8		
one side (.9)	1.2	1.4
both sides (1.3)	1.8	2.0
Lateral Link Replace (B)		
Millenia		
one side	3.3	3.5
both sides	3.6	3.8

	LABOR TIME	SEVERE SERVICE
Leading Link Replace (B)		
Millenia		
one side	.5	.7
both sides	.7	.9
Lower Control Arm, Replace (B)		
323, Protégé/5		
1986-94		
one side	2.2	2.4
both sides	2.9	3.1
1995-05		
one side	.8	1.0
both sides	1.3	1.5
626, MX-6		
1981-82 626		
one side	1.8	2.0
both sides	3.1	3.3
1983-92		
one side	1.6	1.8
both sides	2.2	2.4
1993-97		
one side	1.2	1.4
both sides	1.7	1.9
1998-02		
one side	.7	.9
both sides	1.4	1.6
929		
1988-91		
one side	1.7	1.9
both sides	2.3	2.5
1992-95		
one side	1.4	1.6
both sides	2.3	2.5
GLC		
one side	2.0	2.2
both sides	3.1	3.3
Mazda3		
one side (1.1)	1.5	1.7
both sides (1.7)	2.3	2.5
Mazda6		
front side		
2.3L		
one side	1.0	1.2
both sides	1.3	1.5
3.0L		
one side	1.3	1.5
both sides	1.9	2.1
rear side		
one side	2.5	2.7
both sides	2.9	3.1
Miata		
one side	2.9	3.1
both sides	3.6	3.8
Millenia		
one side	2.6	2.8
both sides	3.0	3.2
MX-3		
one side	1.2	1.4
both sides	1.7	1.9

	LABOR TIME	SEVERE SERVICE
RX-7		
1981-85		
one side	1.6	1.8
both sides	3.0	3.2
1986-91		
one side	1.3	1.5
both sides	1.8	2.0
1993-95		
one side	2.5	2.7
both sides	3.2	3.4
RX-8		
one side (1.5)	2.0	2.2
both sides (2.4)	3.2	3.4
Lower Control Arm Bushings, Replace (B)		
323, Protégé/5		
1986-94		
one side	2.2	2.4
both sides	2.9	3.1
1995-05		
one side	1.0	1.2
both sides	1.4	1.6
626, MX-6		
1983-92		
one side	1.6	1.8
both sides	2.6	2.8
1993-97		
one side	1.2	1.4
both sides	1.8	2.0
1998-02		
one side	.9	1.1
both sides	1.6	1.8
Mazda6		
front side		
2.3L		
one side	1.0	1.2
both sides	1.9	2.1
3.0L		
one side	1.9	2.1
both sides	2.5	2.7
rear side		
one side	2.7	2.9
both sides	3.0	3.2
Millenia		
one side	2.7	2.9
both sides	3.1	3.3
MX-3		
one side	1.4	1.6
both sides	2.2	2.4
RX-8		
right side	3.1	3.3
left side	4.8	5.0
both sides	4.9	5.1
Lower Control Arm Bushings and/or Shaft, Replace (B)		
1981-82 626		
one side	2.0	2.2
both sides	3.4	3.6

MX-3 : MX-6 : PROTEGE : PROTEGE5 : RX-7 : RX-8 **MAZ-35**

	LABOR TIME	SEVERE SERVICE
1988-91 929		
one side	1.8	2.0
both sides	2.7	2.9
Miata		
one side	3.2	3.4
both sides	4.1	4.3
RX-7		
1981-85		
one side	2.0	2.2
both sides	3.6	3.8
1986-91		
one side	1.5	1.7
both sides	2.3	2.5
1993-95		
one side	3.1	3.3
both sides	4.2	4.4

Stabilizer Bar, Replace (B)

	LABOR TIME	SEVERE SERVICE
323, Protégé/5		
1990-94	2.2	2.4
1995-98	3.1	3.3
1999-05	6.3	6.5
626, MX-6		
1981-82	1.2	1.4
1983-92		
2WS	1.3	1.5
4WS	2.2	2.4
1993-97	6.2	6.4
1998-02		
4 cyl.	6.2	6.4
V6	6.8	7.0
929		
1988-91	1.2	1.4
1992-95	1.7	1.9
GLC	.8	1.0
Mazda3	2.8	3.0
Mazda6	3.3	3.5
Miata	1.1	1.3
Millenia	2.8	3.0
MX-3	3.4	3.6
RX-7		
1981-85	2.9	3.1
1986-95	.8	1.0
RX-8 (1.1)	1.5	1.7

Stabilizer Bar Bushings, Replace (B)

Body side both

	LABOR TIME	SEVERE SERVICE
323, Protégé/5		
1986-94	.7	.9
1995-98	3.1	3.3
1999-05	6.7	6.9
626, MX-6		
1981-82	.7	.9
1983-92		
2WS	1.4	1.6
4WS	2.0	2.2
1993-97	6.2	6.4
1998-02		
4 cyl.	6.2	6.4
V6	6.8	7.0

	LABOR TIME	SEVERE SERVICE
929	.7	.9
Mazda3	2.8	3.0
Mazda6	3.3	3.5
Miata	.8	1.0
Millenia	.8	1.0
MX-3	.9	1.1
RX-7		
1981-85	1.2	1.4
1986-95	.8	1.0
RX-8 (1.1)	1.5	1.7

Stabilizer Control Bolt and/or Bushings, Replace (B)

	LABOR TIME	SEVERE SERVICE
Exc. below		
one or both	.8	1.0
323, Protégé/5		
one	.5	.7
both	.8	1.0
Millenia one or both	.5	.7

Steering Knuckle, Replace (B)

	LABOR TIME	SEVERE SERVICE
323, Protégé/5		
1986-94		
one side	2.2	2.4
both sides	3.1	3.3
1995-98		
one side	1.6	1.8
both sides	2.9	3.1
1999-05		
one side	1.9	2.1
both sides	3.6	3.8
626, MX-6		
1981-82		
one side	1.5	1.6
both sides	1.9	2.1
1983-87		
one side	1.5	1.7
both sides	2.6	2.8
1988-92		
one side	2.2	2.4
both sides	3.1	3.3
1993-97		
one side	1.7	1.9
both sides	2.8	3.0
1998-02		
one side	1.8	2.0
both sides	3.1	3.3
929		
1988-91		
one side	1.5	1.7
both sides	1.9	2.1
1992-95		
one side	1.7	1.9
both sides	3.2	3.4
GLC		
one side	1.7	1.9
both sides	2.9	3.1
Mazda3		
one side	1.8	2.0
both sides	2.8	3.0

	LABOR TIME	SEVERE SERVICE
Mazda6		
one side	2.5	2.7
both sides	4.3	4.5
Miata		
1996-97		
one side	2.0	2.2
both sides	2.7	2.9
1999-05		
one side	1.3	1.5
both sides	2.2	2.4
Millenia		
one side	1.8	2.0
both sides	3.0	3.2
MX-3		
one side	1.6	1.8
both sides	2.6	2.8
1986-95 RX-7		
one side	1.4	1.6
both sides	2.4	2.6
RX-8		
one side	1.7	1.9
both sides	2.8	3.0

Steering Knuckle Arm, Replace (B)

Includes: Reset toe.

	LABOR TIME	SEVERE SERVICE
1988-91 929		
one side	.8	1.0
both sides	1.4	1.6

Steering Knuckle Oil Seal, Replace (B)

	LABOR TIME	SEVERE SERVICE
323, Protégé/5		
one side	1.3	1.5
both sides	2.2	2.4
626, MX-6		
1988-92		
one side	2.4	2.6
both sides	3.3	3.5
1993-97		
one side	1.4	1.6
both sides	2.4	2.6
1998-02		
one side	1.2	1.4
both sides	2.0	2.2
GLC		
one side	1.5	1.7
both sides	2.4	2.6
Millenia		
one side	1.6	1.8
both sides	2.6	2.8
MX-3		
one side	1.4	1.6
both sides	2.3	2.5

Strut, R&R or Replace (B)

	LABOR TIME	SEVERE SERVICE
323, Protégé/5		
1986-94		
one side	1.3	1.5
both sides	2.0	2.2
1995-96		
one side	2.0	2.2
both sides	3.3	3.5

MAZ-36   323 : 626 : 929 : GLC : MAZDA3 : MAZDA6 : MIATA : MILLENIA

	LABOR TIME	SEVERE SERVICE
Strut, R&R or Replace (B)		
1997-98		
one side	2.0	2.2
both sides	2.5	2.7
1999-05		
one side	1.0	1.2
both sides	1.6	1.8
626, MX-6		
1983-92		
one side	1.4	1.6
both sides	2.0	2.2
1993-97		
one side	1.4	1.6
both sides	2.6	2.8
1998-02		
one side	.8	1.0
both sides	1.4	1.6
GLC		
one side	1.7	1.9
both sides	2.6	2.8
Mazda3		
one side	1.8	2.0
both sides	2.6	2.8
Mazda6, RX-8		
one side (1.0)	1.4	1.8
both sides (1.5)	2.0	2.2
Millenia		
one side	2.0	2.2
both sides	3.3	3.5
MX-3		
one side	.8	1.0
both sides	1.7	1.9
Suspension Arm or Ball Joint, Replace (B)		
MX-3		
one	1.5	1.7
both	2.8	3.0
Suspension Crossmember, Replace (B)		
Includes: Alignment charges.		
929		
1988-91	5.2	5.4
1992-95	7.8	8.0
Miata	5.6	5.8
RX-7		
1981-85	4.2	4.4
1986-91	3.8	4.0
1993-95	5.2	5.4
RX-8	6.6	6.8
Torsion Bar Connecting Rods, Replace (B)		
1992-95 929		
one	1.2	1.4
both	1.9	2.1

	LABOR TIME	SEVERE SERVICE
Transverse Links, Replace (B)		
Includes: Alignment charges.		
1992-95 929		
lateral		
one side	2.8	3.0
both sides	3.2	3.4
leading		
one side	.9	1.0
both sides	1.2	1.4
Upper Control Arm Assy., Replace (B)		
Mazda6		
one side	1.4	1.6
both sides	2.1	2.3
Miata		
one side	3.2	3.4
both sides	4.3	4.5
1993-95 RX-7		
one side	1.2	1.4
both sides	1.7	1.9
RX-8		
one side (1.0)	1.4	1.6
both sides (1.6)	2.2	2.4
Replace bushings add each	.2	.3
Upper or Lower Ball Joints, Replace (B)		
1996-05 exc. Miata		
one	.6	.8
one both sides	1.0	1.2
Miata		
one	1.7	1.9
one side	2.3	2.5
Wheel Bearing Grease Seal, Replace (B)		
1988-91 929		
one side	.7	.9
both sides	1.2	1.4
1981-84 RX-7		
one side	.7	.9
both sides	1.5	1.7
Wheel Bearings, Replace (B)		
1988-91 929		
one side	1.3	1.5
both sides	1.8	2.0
1981-84 RX-7		
one side	1.3	1.5
both sides	2.3	2.5
Wheel Bearings, Clean & Pack (B)		
1986-89 323		
one side	1.5	1.7
both sides	2.8	3.0
1981-84 RX-7		
one side	.8	1.0
both sides	1.7	1.9

	LABOR TIME	SEVERE SERVICE
REAR SUSPENSION		
Rear Suspension Toe, Adjust (B)		
323, Protégé	.7	.9
1986-02 626, MX-6	.7	.9
929		
1988-91	.7	.9
1992-95	1.2	1.4
Miata	1.7	1.9
Millenia	.7	.9
MX-3	.8	1.0
RX-7	.8	1.0
RX-8	1.8	2.0
Coil Spring, R&R or Replace (B)		
323, Protégé/5		
1986-94		
one side	1.8	2.0
both sides	2.9	3.1
1995-98		
one side	1.3	1.5
both sides	1.7	1.9
1999-05		
one side	1.6	1.8
both sides	2.4	2.6
626, MX-6		
1981-82		
one side	.9	1.1
both sides	1.6	1.8
1983-97		
one side	1.8	2.0
both sides	3.1	3.3
1998-02		
one side	1.4	1.6
both sides	2.0	2.2
929		
1988-91		
one side	1.5	1.7
both sides	2.3	2.5
1992-95		
one side	2.0	2.2
both sides	3.2	3.4
GLC		
one side	1.8	2.0
both sides	2.8	3.0
Mazda3		
one side	1.5	1.7
both sides	2.0	2.2
Mazda6		
one side	1.4	1.6
both sides	2.1	2.3
Miata		
one side	1.7	1.9
both sides	3.1	3.3
Millenia		
one side	1.6	1.8
both sides	2.2	2.4
MX-3		
one side	1.4	1.6
both sides	1.8	2.0

MX-3 : MX-6 : PROTEGE : PROTEGE5 : RX-7 : RX-8 — MAZ-37

	LABOR TIME	SEVERE SERVICE
RX-7		
1981-91		
one side	.8	*1.0*
both sides	1.4	*1.6*
1993-95		
one side	1.3	*1.5*
both sides	2.4	*2.6*
RX-8		
one side	2.3	*2.5*
both sides	3.2	*3.4*
Lateral Link, Replace (B)		
Front		
323, Protégé/5		
1986-94	2.8	*3.0*
1995-98	1.2	*1.4*
1999-05		
one side	1.4	*1.6*
both sides	1.6	*1.8*
626, MX-6		
1983-92		
one side	1.4	*1.6*
both sides	1.7	*1.9*
1993-97		
one side	.8	*1.0*
both sides	1.3	*1.5*
1998-02		
one side	1.4	*1.6*
both sides	1.6	*1.8*
929		
1988-91		
lower		
one side	.8	*1.0*
both sides	1.2	*1.4*
upper		
one side	.5	*.7*
both sides	.8	*1.0*
1992-95		
lower		
one side	.8	*1.0*
both sides	1.2	*1.4*
upper		
one side	2.0	*2.2*
both sides	2.5	*2.7*
GLC	.8	*1.0*
Mazda3 lower		
one side	1.8	*2.0*
both sides	2.4	*2.6*
Mazda6 lower		
one side	1.4	*1.6*
both sides	1.9	*2.1*
Millenia one or both		
upper	1.8	*2.0*
lower	.7	*.9*
MX-3 front or rear		
one side	2.2	*2.4*
both sides	2.8	*3.0*

	LABOR TIME	SEVERE SERVICE
Rear		
323, Protégé/5		
1986-94		
one side	1.2	*1.4*
both sides	1.3	*1.5*
1995-98		
one side	1.7	*1.9*
both sides	1.8	*2.0*
1999-05		
one side	1.9	*2.1*
both sides	2.2	*2.4*
626, MX-6		
1983-92		
one side	.8	*1.0*
both sides	1.4	*1.6*
1993-98		
one side	1.6	*1.8*
both sides	1.9	*2.1*
1988-91 929		
one side	.8	*1.0*
both sides	1.2	*1.4*
GLC	.9	*1.1*
Millenia		
one side	1.8	*2.0*
both sides	2.2	*2.4*
MX-3		
one side	1.5	*1.7*
both sides	1.8	*2.0*
RX-8		
one side (1.0)	1.4	*1.7*
both sides (1.2)	1.6	*2.0*
Lower Control Arm, Replace (B)		
Includes: Alignment charges.		
Mazda3, Mazda6		
one side	1.7	*1.9*
both sides	2.2	*2.4*
Miata		
one side	2.7	*2.9*
both sides	3.5	*3.7*
MX-3		
one side	1.2	*1.4*
both sides	1.8	*2.0*
RX-7		
1986-91		
one side	1.3	*1.5*
both sides	2.3	*2.5*
1993-95		
one side	1.8	*2.0*
both sides	2.0	*2.2*
Lower Control Arms & Links, Replace (B)		
1981-85 RX-7		
one	.8	*1.0*
both	1.2	*1.4*
Lower Control Arm Bushing, Replace (B)		
Includes: Alignment charges.		
Mazda3, Mazda6		
one side	1.9	*2.1*
both sides	2.7	*2.9*

	LABOR TIME	SEVERE SERVICE
Miata		
one side	2.7	*2.9*
both sides	3.5	*3.7*
MX-3		
one side	1.4	*1.6*
both sides	1.9	*2.1*
1993-95 RX-7		
one side	2.0	*2.2*
both sides	2.5	*2.7*
Lower Link, Replace (B)		
1981-85 RX-7		
one	.5	*.7*
both	.7	*.9*
Rear Hub or Wheel Spindle, Replace (B)		
323, Protégé/5		
1986-94		
one side	1.2	*1.4*
both sides	1.6	*1.8*
1995-98		
disc brakes		
one side	.8	*1.0*
both sides	1.6	*1.8*
drum brakes		
one side	1.7	*1.9*
both sides	2.4	*2.6*
1999-05		
disc brakes		
one side	1.4	*1.6*
both sides	2.5	*2.7*
drum brakes		
one side	1.9	*2.1*
both sides	2.7	*2.9*
626, MX-6		
1983-97		
one side	1.3	*1.5*
both sides	2.0	*2.2*
1998-02		
drum brakes		
one side	1.8	*2.0*
both sides	2.6	*2.8*
disc brakes		
one side	1.4	*1.6*
both sides	2.5	*2.7*
929		
1988-91		
one side	2.8	*3.0*
both sides	4.2	*4.4*
1992-95		
one side	2.3	*2.5*
both sides	3.6	*3.8*
Mazda3		
one side	1.1	*1.3*
both sides	1.7	*1.9*
Mazda6		
one side (1.2)	1.6	*1.8*
both sides (1.8)	2.4	*2.6*
Miata		
one side	1.4	*1.6*
both sides	1.7	*1.9*

MAZ-38 323 : 626 : 929 : GLC : MAZDA3 : MAZDA6 : MIATA : MILLENIA

	LABOR TIME	SEVERE SERVICE
Rear Hub or Wheel Spindle, Replace (B)		
Millenia		
one side	1.7	1.9
both sides	2.9	3.1
MX-3		
disc brakes		
one side	1.2	1.4
both sides	1.7	1.9
drum brakes		
one side	1.9	2.1
both sides	2.6	2.8
1986-95 RX-7		
one side	1.3	1.5
both sides	1.9	2.1
RX-8		
one side	2.1	2.3
both sides	3.5	3.7
Rear Shock Absorber and/or Bushing, R&R or Replace (B)		
323, Protégé/5		
1986-94		
one side	1.7	1.9
both sides	2.9	3.1
1995-98		
one side	1.3	1.5
both sides	1.7	1.9
1999-05		
one side	1.6	1.8
both sides	2.2	2.4
626, MX-6		
1981-82		
one side	.5	.7
both sides	.8	1.0
1983-97		
one side	1.8	2.0
both sides	3.1	3.3
1998-02		
one side	1.3	1.5
both sides	1.8	2.0
929		
1988-91		
one side	1.4	1.6
both sides	2.2	2.4
1992-95		
one side	1.9	2.1
both sides	3.1	3.3
GLC		
one side	1.8	2.0
both sides	3.0	3.2
Mazda3		
one side	.6	.8
both sides	1.1	1.4
Mazda6		
one side (.6)	.8	.9
both sides (1.0)	1.4	1.6
Miata		
one side	1.7	1.9
both sides	3.1	3.3

	LABOR TIME	SEVERE SERVICE
Millenia		
one side	1.6	1.8
both sides	1.4	1.6
MX-3		
one side	1.3	1.5
both sides	1.7	1.9
RX-7		
1981-91		
one side	.8	1.0
both sides	1.4	1.6
1993-95		
one side	1.2	1.4
both sides	2.3	2.5
RX-8		
one side	2.1	2.3
both sides	2.9	3.1
Stabilizer Bar, Replace (B)		
323, Protégé/5		
1986-94	.8	1.0
1995-02	.5	.7
626, MX-6		
1983-97	.8	1.0
1998-05	.5	.7
929		
1988-91	.9	1.1
1992-95	1.4	1.6
Mazda3, Mazda6	.8	1.0
Miata	.8	1.0
Millenia	.5	.7
MX-3	.7	.9
RX-7	.8	1.0
RX-8 (1.3)	1.8	2.0
Stabilizer Bar Bushings, Replace (B)		
323, Protégé/5		
1986-89		
one side	.5	.7
both sides	.7	.9
1990-93		
one side	.7	.9
both sides	1.2	1.4
1994-05		
one side	.3	.5
both sides	.5	.7
1983-02 626, MX-6		
one side	.5	.7
both sides	.7	.9
929, GLC		
one side	.5	.7
both sides	.7	.9
Mazda3 one	.7	.9
Mazda6		
one side	.3	.5
both sides	.5	.7
Miata, Millenia		
one side	.5	.7
both sides	.7	.9

	LABOR TIME	SEVERE SERVICE
RX-7		
1983-91		
one side	.5	.7
both sides	.7	.9
1993-95	.5	.7
RX-8 one or both (1.3)	1.7	1.9
Trailing Link, Replace (B)		
323, Protégé/5		
one side	.7	.9
both sides	.8	1.0
626, MX-6		
1983-92		
one side	.7	.9
both sides	1.3	1.5
1993-02 one or both sides	.7	.9
929		
1988-91		
one side	.9	1.1
both sides	1.6	1.8
1992-95		
one side	.9	1.1
both sides	1.3	1.5
GLC		
one side	.7	.9
both sides	.9	1.1
Mazda3, Mazda6		
one side	2.8	3.0
both sides	4.3	4.5
Millenia, RX-8		
upper		
one side	1.8	2.0
both sides	2.0	2.2
lower		
one side	.7	.9
both sides	.8	1.0
Upper Control Arm, Replace (B)		
Mazda3, Mazda6		
one side	1.9	2.1
both sides	2.9	3.1
Miata		
one	2.4	2.6
both	2.7	2.9
1993-95 RX-7		
one	.7	.9
both	1.2	1.4
Wheel Bearings, Replace (B)		
1987-91 323, Protégé		
2WD		
one side	1.4	1.6
both sides	2.2	2.4
4WD		
one side	1.6	1.8
both sides	2.6	2.8
1983-92 626, MX6		
one side	.9	1.1
both sides	1.5	1.7

MX-3 : MX-6 : PROTEGE : PROTEGE5 : RX-7 : RX-8 — MAZ-39

	LABOR TIME	SEVERE SERVICE
929		
1988-91		
one side	3.1	3.3
both sides	6.0	6.2
1992-95		
one side	2.9	3.1
both sides	5.2	5.4
GLC		
one side	1.2	1.4
both sides	1.7	1.9
Miata		
1990-97		
one side	1.9	2.1
both sides	2.6	2.8
1999-05		
one side	1.5	1.7
both sides	2.3	2.5
RX-7		
1986-91		
one side	1.4	1.6
both sides	2.2	2.4
1993-95		
one side	2.2	2.4
both sides	3.1	3.3
Wheel Hub, Replace (B)		
323, Protégé/5		
1988-92		
disc brakes		
one side	.7	.9
both sides	1.3	1.5
drum brakes		
one side	1.7	1.9
both sides	2.6	2.8
1993-98		
disc brakes		
one side	.7	.9
both sides	1.3	1.5
drum brakes		
both sides	.8	1.0
1999-05		
disc brakes		
one side	.9	1.1
both sides	1.4	1.6
drum brakes		
one side	.6	.8
both sides	.9	1.1
626, MX-6		
1993-97		
one side	.9	1.1
both sides	1.4	1.6
1998-02		
w/drum brakes		
one side	.7	.9
both sides	.8	1.0
w/disc brakes		
one side	.8	1.0
both sides	1.4	1.6

	LABOR TIME	SEVERE SERVICE
929		
1988-91		
one side	3.0	3.2
both sides	5.3	5.5
1992-95		
one side	2.6	2.8
both sides	4.6	4.8
Mazda3		
one side	1.1	1.3
both sides	1.7	1.9
Mazda6		
one side	.8	1.0
both sides	1.2	1.4
Miata		
one side	1.8	2.0
both sides	2.6	2.8
Millenia		
one side	.7	.9
both sides	1.2	1.4
MX-3		
one side	.7	.9
both sides	1.2	1.4
RX-7		
1986-91		
one side	1.2	1.4
both sides	1.7	1.9
1993-95		
one side	1.4	1.6
both sides	2.5	2.7
RX-8		
one side	2.1	2.3
both sides	3.5	3.7

LEVEL CONTROL
Actuators, Replace (B)
1986-91 RX-7		
front	.7	.9
rear		
one side	.7	.9
both sides	.8	1.0
Control Box, Replace (B)		
626, MX-6		
1983-87	.8	1.0
1988-92	.5	.7
1988-91 929	.7	.9
1988-91 RX-7	.8	1.0
Push Switch, Replace (B)		
626, MX-6	.7	.9
1988-91 929	.5	.7
1986-91 RX-7	.3	.5

STEERING
AIR BAGS
Diagnose Air Bag System Component Each (A)
1990-05 (.7)	.9	1.1

	LABOR TIME	SEVERE SERVICE
Air Bag Electronic Control Unit, Replace (B)		
323, Protégé		
1995-96	3.4	3.6
1997-05	.7	.9
626, MX-6		
1993-97		
w/power seat	1.2	1.4
w/o power seat	.5	.7
1998-02	.7	.9
1992-95 929	.7	.9
Mazda3, Mazda6	.6	.8
Miata		
1990-94	.5	.7
1995-05	3.1	3.3
Millenia		
1995-96	4.1	4.3
1997-02	.7	.9
MX-3	.7	.9
RX-8	3.2	3.4
Air Bag Module Assy., Replace (B)		
Exc. below each	.7	.9
323, Protégé/5		
one side	.9	1.1
both sides	1.5	1.7
929		
driver side	.5	.7
passenger side	1.2	1.4
Mazda3, Mazda6, RX-8 curtain		
one side	2.3	2.5
both sides	2.6	2.8
Backup Battery, Replace (B)		
1990-94 Miata	.5	.7
1990-91 RX-7	.5	.7
Crash Sensor, Replace (B)		
626, MX-6		
right or left each	2.5	2.7
center	.5	.7
S sensor	3.7	3.9
929		
exc. S sensor	.8	1.0
S sensor	4.9	5.1
Mazda3 each	.8	1.0
Mazda6 side sensor	.5	.7
Miata		
right	2.3	2.5
left	2.8	3.0
center	.5	.7
S sensor	2.8	3.0
crash sensor	.8	1.0
MX-3		
right or left each	1.6	1.8
center	.5	.7
S sensor	2.9	3.1
Protégé/5 side sensor	.5	.7

MAZ-40 323 : 626 : 929 : GLC : MAZDA3 : MAZDA6 : MIATA : MILLENIA

	LABOR TIME	SEVERE SERVICE
Crash Sensor, Replace (B)		
RX-7		
1990-91		
right or left	.9	1.1
center	.5	.7
S sensor	2.8	3.0
1993-95		
right or left	.8	1.0
center	1.2	1.4
S sensor	2.8	3.0
RX-8 each	.4	.6
Diagnostic Air Bag Module, Replace (B)		
1990-95 RX-7	.5	.7
MANUAL RACK & PINION		
Horn Contact, Replace (B)		
1981-95	.3	.3
Idler Arm, Replace (B)		
1981-82 626	.8	1.0
1981-85 RX-7	.8	1.0
Replace bushings		
RX-7 add	.2	.3
Inner Tie Rod Assy., Replace (B		
Includes: Reset toe.		
1986-94 323, Protégé		
one side	2.8	3.0
both sides	3.1	3.3
1983-92 626, MX-6		
2WS		
one side	2.8	3.0
both sides	3.2	3.4
4WS		
one side	5.6	5.8
both sides	6.0	6.2
GLC		
one side	2.8	3.0
both sides	3.1	3.3
Miata		
one side	2.2	2.4
both sides	2.6	2.8
1986-88 RX-7		
one side	2.3	2.5
both sides	2.7	2.9
Outer Tie Rod End, Replace (B)		
Includes: Reset toe.		
1986-94 323, Protégé		
one side	.9	1.1
both sides	1.3	1.5
1983-92 626, MX-6		
one side	.9	1.1
both sides	1.5	1.7
GLC		
one side	.9	1.1
both sides	1.3	1.5
Miata		
one side	.9	1.1
both sides	1.5	1.7

	LABOR TIME	SEVERE SERVICE
1986-88 RX-7		
one side	.7	.9
both sides	1.4	1.6
Pitman Arm, Replace (B)		
1981-82 626	.8	1.0
1981-85 RX-7	.8	1.0
Rack & Pinion Assy., R&I (B)		
Includes: Reset toe.		
1986-94 323, Protégé	2.0	2.2
1983-92 626, MX-6		
2WS	2.2	2.4
4WS	5.2	5.4
GLC	2.3	2.5
Miata	1.7	1.9
1986-88 RX-7	3.3	3.5
Rack & Pinion, R&R and Recondition (A)		
1986-94 323, Protégé	3.4	3.6
1983-92 626, MX-6		
2WS	4.2	4.4
4WS	7.8	8.1
GLC	3.4	3.6
Miata	3.1	3.3
1986-88 RX-7	4.4	4.6
Rack & Pinion Assy. (Short), Replace (B)		
Includes: Reset toe.		
1986-94 323, Protégé	3.4	3.6
1983-92 626, MX-6		
2WS	3.2	3.4
4WS	6.2	6.4
GLC	4.2	4.4
Miata	3.4	3.6
1986-88 RX-7	4.4	4.6
Rack Boots, Replace (B)		
Includes: Reset toe.		
1986-94 323, Protégé		
one side	1.8	2.0
both sides	2.0	2.2
1983-92 626, MX-6		
one side	1.3	1.5
both sides	2.2	2.4
GLC		
one side	1.2	1.4
both sides	2.2	2.4
Miata		
one side	1.3	1.5
both sides	2.0	2.2
Steering Center Link (Relay Rod), Replace (B)		
1981-82 626	1.4	1.6
1981-85 RX-7	1.4	1.6
Steering Column Support Bushings, Replace (B)		
All Models	.9	1.1
Steering Knuckle Arm, Replace (B)		
Includes: Reset toe.		
1981-85 RX-7		
one side	.8	1.0
both sides	1.6	1.8

	LABOR TIME	SEVERE SERVICE
Steering Wheel, Replace (B)		
All Models	.5	.7
Tie Rod or Tie Rod Ends, Replace (B)		
Includes: Reset toe.		
1981-82 626		
one side	1.2	1.4
both sides	1.7	1.9
1981-85 RX-7		
one side	1.2	1.4
both sides	1.7	1.9
POWER RACK & PINION		
Control Unit, Replace (B)		
1986-91 RX-7	.5	.7
Inner Tie Rod Assy., Replace (B)		
Includes: Reset toe.		
323, Protégé/5		
1986-94		
2WD		
one side	4.1	4.3
both sides	4.5	4.7
4WD		
one side	4.5	4.7
both sides	5.0	5.2
1995-98		
one side	3.4	3.6
both sides	3.7	3.9
1999-05		
one side	4.9	5.1
both sides	5.2	5.4
626, MX-6		
1983-92		
2WS		
one side	3.8	4.0
both sides	4.2	4.4
4WS		
one side	5.3	5.5
both sides	5.5	5.7
1993-97		
4 cyl.		
one side	5.5	5.7
both sides	5.8	6.0
V6		
one side	6.2	6.4
both sides	6.3	6.5
1998-02		
4 cyl.		
one side	5.0	5.2
both sides	5.2	5.4
V6		
one side	6.3	6.5
both sides	6.5	6.7
929		
one side	3.1	3.3
both sides	3.8	4.0
GLC		
one side	3.1	3.3
both sides	3.7	3.9

MX-3 : MX-6 : PROTEGE : PROTEGE5 : RX-7 : RX-8 **MAZ-41**

	LABOR TIME	SEVERE SERVICE
Mazda3		
one side (2.7)	3.6	*3.8*
both sides (3.0)	4.1	*4.3*
Mazda6		
2.3L		
one side (2.8)	3.8	*4.0*
both sides (3.0)	4.1	*4.3*
3.0L		
one side (3.5)	4.7	*4.9*
both sides (3.7)	5.0	*5.2*
Miata		
1996-97		
one side	4.0	*4.2*
both sides	4.4	*4.6*
1999-05		
one side	4.4	*4.6*
both sides	4.7	*4.9*
Millenia		
KJ		
one side	3.6	*3.8*
both sides	4.2	*4.4*
KL		
one side	4.5	*4.7*
both sides	5.0	*5.2*
MX-3		
one side	3.6	*3.8*
both sides	3.7	*3.9*
RX-7		
1986-91		
one side	3.4	*3.6*
both sides	3.7	*3.9*
1993-95		
one side	3.7	*3.9*
both sides	4.2	*4.4*
w/ECPS, add	.5	*.7*
Outer Tie Rod Ends, Replace (B)		
Includes: Reset toe.		
323, Protégé/5		
one side	1.8	*2.0*
both sides	2.0	*2.2*
626, MX-6		
1983-92		
one side	1.7	*1.9*
both sides	2.4	*2.6*
1993-02		
one side	1.8	*2.0*
both sides	2.0	*2.2*
929		
1988-91		
one side	.8	*1.0*
both sides	1.3	*1.5*
1992-95		
one side	1.7	*1.9*
both sides	2.0	*2.2*
GLC		
one side	.8	*1.0*
both sides	1.2	*1.4*
Mazda3, Mazda6, RX-8		
one side (.8)	1.1	*1.3*
both sides (1.0)	1.4	*1.6*

	LABOR TIME	SEVERE SERVICE
Miata		
one side	1.8	*2.0*
both sides	1.9	*2.1*
Millenia		
one side	1.7	*1.9*
both sides	2.0	*2.2*
MX-3		
one side	1.8	*2.0*
both sides	2.0	*2.2*
RX-7		
1986-91		
one side	.8	*1.0*
both sides	1.4	*1.6*
1993-95		
one side	1.7	*1.9*
both sides	2.0	*2.2*
Pressure Switch, Replace (B)		
323, Protégé	.5	*.7*
626, MX-6		
1988-92	1.2	*1.4*
1993-02	.8	*1.0*
929		
1988-91	.8	*1.0*
1992-95	1.2	*1.4*
Miata	.5	*.7*
Millenia	1.7	*1.9*
MX-3		
4 cyl.	.5	*.7*
V6	2.0	*2.2*
RX-7		
1984-91	.7	*.9*
1993-95	1.5	*1.7*
Rack & Pinion Assy., R&I (B)		
Includes: Reset toe.		
323, Protégé		
1986-94		
2WD	3.4	*3.6*
4WD	3.8	*4.*
1995-98	3.1	*3.4*
1999-02	4.2	*4.4*
626, MX-6		
1983-92		
2WS	3.3	*3.5*
4WS	4.5	*4.7*
1993-97	5.8	*6.0*
1998-02		
4 cyl.	4.8	*5.0*
V6	7.8	*8.0*
929	2.3	*2.5*
GLC	2.8	*3.0*
Mazda3 (3.3)	4.5	*4.7*
Mazda6		
2.3L (3.5)	4.7	*4.9*
3.0L (4.2)	5.7	*5.9*
Miata	1.7	*1.9*
Millenia	4.1	*4.3*
RX-7		
1986-91	3.2	*3.4*
1993-95	4.2	*4.4*

	LABOR TIME	SEVERE SERVICE
RX-8	2.9	*3.1*
w/ECPS add	.5	*.7*
Rack & Pinion Assy., R&R and Recondition (A)		
Includes: Reset toe.		
323, Protégé		
1986-94		
2WD	5.6	*5.8*
4WD	5.9	*6.1*
1995-98	4.8	*5.0*
1999-02	5.6	*5.8*
626, MX-6		
1983-92		
2WS	5.8	*6.0*
4WS	7.1	*7.3*
1993-97	7.9	*8.1*
1998-02		
4 cyl.	6.3	*6.5*
V6	7.8	*8.0*
929	4.5	*4.7*
GLC	4.8	*5.0*
Miata	3.6	*3.8*
Millenia	6.4	*6.6*
MX-3	5.6	*5.8*
RX-7		
1986-91	4.5	*4.7*
1993-95	6.2	*6.4*
w/ECPS add	1.7	*1.9*
Rack Assy. (Short), Replace (A)		
Add for alignment.		
323, Protégé/5		
1986-94	5.6	*5.8*
1995-98	4.4	*4.6*
1999-05	5.7	*5.9*
626, MX-6		
1983-92		
2WS	5.8	*6.0*
4WS	7.1	*7.3*
1993-97	7.9	*8.1*
1998-02		
4 cyl.	6.0	*6.2*
V6	7.5	*7.7*
929	4.5	*4.7*
GLC	4.8	*5.0*
Mazda3	3.3	*3.4*
Mazda6		
2.3L	3.2	*3.4*
3.0L	4.3	*4.5*
Miata	4.3	*4.5*
Millenia	6.0	*6.2*
MX-3	3.7	*3.9*
RX-7		
1986-91	4.8	*5.0*
1993-95	5.8	*6.0*
w/ECPS add	1.7	*1.9*

MAZ-42 323 : 626 : 929 : GLC : MAZDA3 : MAZDA6 : MIATA : MILLENIA

	LABOR TIME	SEVERE SERVICE
Rack Boots, Replace (B)		
Includes: Reset toe.		
323, Protégé/5		
1986-94		
one side	2.2	2.4
both sides	2.5	2.7
1995-05		
one side	1.9	2.1
both sides	2.2	2.4
626, MX-6		
1983-92		
one side	1.8	2.0
both sides	2.8	3.0
1993-02		
one side	2.0	2.2
both sides	2.3	2.5
929		
1988-91		
one side	1.3	1.5
both sides	1.6	1.8
1992-95		
one side	2.0	2.2
both sides	2.4	2.6
GLC		
one side	1.2	1.4
both sides	2.2	2.4
Mazda3, Mazda6		
one side	1.4	1.6
both sides	1.7	1.9
Miata		
one side	2.1	2.3
both sides	2.4	2.6
Millenia		
one side	1.8	2.0
both sides	2.3	2.5
MX-3		
one side	2.0	2.2
both sides	2.3	2.5
RX-7		
1986-91		
one side	1.3	1.5
both sides	1.5	1.7
1993-95		
one side	2.2	2.4
both sides	2.7	2.9
RX-8		
one side	1.4	1.6
both sides	1.7	1.9
Steering Flex Coupling, Replace (B)		
1995-98 323, Protégé	1.2	1.4
1993-97 626, MX-6	2.0	2.2
1993-95 RX-7	1.2	1.4
Steering Pump, R&R and Recondition (B)		
323, Protégé/5		
1988-94	2.2	2.4
1995-98	1.8	2.0
1999-05	2.7	2.9
626, MX-6		
1983-92	2.2	2.4

	LABOR TIME	SEVERE SERVICE
1993-97		
4 cyl.	2.0	2.2
V6	2.7	2.9
1998-02		
4 cyl.	1.8	2.0
V6	2.9	3.1
929	2.3	2.5
GLC	1.7	1.9
Mazda3	2.2	2.4
Mazda6		
2.3L	2.3	2.5
3.0L	4.4	4.6
Miata	2.2	2.4
Millenia		
KJ	2.6	2.8
KL	2.3	2.5
MX-3		
4 cyl.	2.3	2.5
V6	2.7	2.9
Steering Pump, R&R or Replace (B)		
323, Protégé/5	1.6	1.8
626, MX-6		
1983-92	1.6	1.8
1993-97		
4 cyl.	1.2	1.4
V6	1.8	2.0
1998-02		
4 cyl.	1.3	1.5
V6	2.2	2.4
929	1.7	1.9
GLC	1.2	1.4
Mazda3	1.3	1.5
Mazda6		
2.3L	1.4	1.6
3.0L	3.5	3.7
Miata	1.4	1.6
Millenia	1.7	1.9
MX-3		
4 cyl.	1.6	1.8
V6	2.0	2.2
RX-7		
1986-91	.9	1.1
1993-95	1.4	1.6
Steering Pump Hoses, Replace (B)		
Does not include purging.		
Pressure		
323, Protégé/5		
1988-94	1.2	1.4
1995-98	.7	.9
1999-05	1.1	1.3
626, MX-6		
1983-92	1.2	1.4
1993-97	1.6	1.8
1998-02		
4 cyl.	1.3	1.5
V6	2.7	2.9
929		
1988-91	1.2	1.4
1992-95	1.6	1.8

	LABOR TIME	SEVERE SERVICE
GLC	1.2	1.4
Mazda3	1.1	1.3
Mazda6	1.2	1.4
Miata	1.2	1.4
Millenia	2.0	2.2
MX-3	1.5	1.7
RX-7		
1986-91	1.2	1.4
1993-95	.9	1.1
Return		
323, Protégé	.9	1.1
626, MX-6		
1983-92	.7	.9
1993-97	2.9	3.0
1998-02		
4 cyl.	1.4	1.6
V6	1.7	1.9
929		
1988-91	.8	1.0
1992-02	1.6	1.8
GLC	.7	.9
Mazda3	1.0	1.2
Mazda6	1.1	1.3
Miata		
1990-97	.7	.9
1999-05	1.2	1.4
Millenia		
KJ	.8	1.0
KL	1.7	1.9
MX-3	2.2	2.4
RX-7		
1986-91	.7	.9
1993-95	2.0	2.2
Steering Wheel, Replace (B)		
All Models	.6	.8
w/air bags add	.5	.7
MANUAL WORM & SECTOR		
Gear, Steering, Adjust (B)		
626, MX-3, RX-7	.5	.7
Gear Assy., R&R and Recondition (B)		
626, MX-3, RX-7	3.7	3.9
Gear Assy., Replace (B)		
626, MX-3, RX-7	2.8	3.0
Sector Shaft and/or Oil Seal, Replace (B)		
626, MX-3, RX-7	1.9	2.1
POWER WORM & SECTOR		
Gear Assy., R&R and Recondition (B)		
1981-85 626, RX-7	3.4	3.6
Gear Assy., Replace (B)		
1981-85 626, RX-7	3.1	3.3
Sector Shaft and/or Oil Seal, Replace (B)		
1981-85 626, RX-7	2.3	2.4

MX-3 : MX-6 : PROTEGE : PROTEGE5 : RX-7 : RX-8 **MAZ-43**

	LABOR TIME	SEVERE SERVICE
Steering Gear Side Cover and/or O-Ring, Replace (B)		
1981-85 RX-7 . . . . 1.2		1.4
Steering Pump, R&R and Recondition (B)		
1981-85 626, RX-7 . . . 2.3		2.5
Steering Pump, R&R or Replace (B)		
1981-85 626, RX-7 . . . 1.4		1.6
Steering Pump Hoses, Replace (B)		
1981-85 626, RX-7		
one 7		.9
all 9		1.1

4 WHEEL STEERING
Control Unit (4WS), Replace (B)
1988-90 626, MX-6 . . . 3.1   3.3
Shaft Boot (4WS), Replace (B)
1988-90 626, MX-6 . . . 1.3   1.5
Steering Sensor Assy. (4WS), Replace (B)
1988-90 626, MX-6 . . . 3.5   3.7
Stepper Motor (4WS), Replace (B)
1988-90 626, MX-6 . . . 3.7   3.9

HEATING & AIR CONDITIONING

When more than one component requires replacement where evacuation/recovery and recharging is already included, deduct 1.0 hour for each additional component from the time given.

Evacuate/Recover and Recharge System (B)
1981-05 . . . . 1.0   1.2
AC Hoses, Replace (B)
Add for evacuate/recover and recharge.
323, Protégé/5
  1994 each 3   .5
  1995-98 each 8   1.0
  1999-05 each 5   .7
1992-02 626, MX-6
  suction 3   .5
  discharge 8   1.0
Mazda3, Mazda6
  suction 9   1.1
  discharge 3   .5
Miata each 8   1.0
Millenia
  suction 3   .5
  discharge 8   1.0
RX-8 each 5   .7
Ambient Temperature Sensor, Replace (B)
929 5   .7
Mazda6 5   .7
Millenia 3   .5
RX-8 8   1.0

	LABOR TIME	SEVERE SERVICE
Blower Motor, Replace (B)		
323, Protégé/5		
1986-89 7		.9
1990-94 . . . . 1.6		1.8
1995-05 7		.9
626, MX-6		
1981-82 . . . . 1.5		1.7
1983-02 7		.9
929		
1988-91 5		.7
1992-95 . . . . 1.4		1.6
GLC 9		1.1
Mazda3 . . . . 1.8		2.0
Mazda6 5		.7
Miata		
1990-97 9		1.1
1999-05 5		.7
Millenia 9		1.1
MX-3 . . . . 1.2		1.4
RX-7 . . . . 1.3		1.5
RX-8 5		.7
Blower Motor Relay, Replace (B)		
1988-02 626, MX-6 7		.9
929, RX-7 5		.7
Mazda3, Mazda6, Miata, Millenia, RX-8 . . .3		.5
1995-05 Protégé/5 . . .3		.5
Blower Motor Resistor, Replace (B)		
Exc. below 5		.7
GLC 7		.9
Mazda3 . . . . 4.3		4.5
Miata		
1990-97 7		.9
1999-05 3		.5
Blower Motor Switch, Replace (B)		
323, Protégé/5		
1986-94 9		1.1
1995-96 . . . . 1.3		1.5
1997-05 6		.8
626, MX-6 8		1.0
1988-91 929 5		.7
GLC 5		.7
Mazda3 9		1.1
Miata 8		1.0
MX-3 . . . . 1.2		1.4
1981-85 RX-7 8		1.0
Blower Motor Transistor, Replace (B)		
1998-02 626 3		.5
Mazda3, Mazda6, RX-8 5		.7
Millenia 7		.9
Compressor Assy., Replace (B)		
Includes: Parts transfer, evacuate/recover and recharge.		
323, Protégé/5 . . . . . 2.5		2.7
626, MX-6		
1981-87 . . . . 1.8		2.0
1988-92 . . . . 1.9		2.1
1993-02 . . . . 2.4		2.6

	LABOR TIME	SEVERE SERVICE
929		
1988-91 . . . . 2.5		2.7
1992-95 . . . . 3.1		3.3
GLC 2.9		3.1
Mazda3, RX-8 . . . . . 1.1		1.3
Mazda6		
2.3L 2.2		2.4
3.0L 3.2		3.4
Miata 2.3		2.5
Millenia 3.4		3.6
MX-3		
4 cyl. 1.9		2.1
V6 . . . . 2.7		2.9
RX-7		
1981-85 . . . . 1.9		2.1
1986-95 . . . . 2.4		2.6
w/turbo add 5		.7
Compressor Shaft Seal, Replace (B)		
Includes: Parts transfer, evacuate/recover and recharge.		
1986-89 323 . . . . 2.6		2.8
626, MX-6		
1981-85 . . . . 2.2		2.4
1986-92 . . . . 2.0		2.2
1993-02 . . . . 2.4		2.6
929		
1988-91 . . . . 2.8		3.0
1992-95 . . . . 3.5		3.7
GLC 3.3		3.5
Miata 2.4		2.6
Millenia 2.5		2.7
MX-3		
4 cyl. 2.5		2.7
V6 . . . . 3.2		3.4
RX-7		
1981-85 . . . . 2.5		2.7
1986-95 . . . . 2.4		2.6
w/turbo add 5		.5
Condenser Assy., Replace (B)		
Includes: Evacuate/recover and recharge.		
323, Protégé/5 . . . . . 1.2		1.4
626, MX-6 . . . . 1.7		1.9
929 . . . . 1.8		2.0
GLC 2.2		2.4
Mazda3 . . . . 2.0		2.2
Mazda6 . . . . 3.1		3.3
Miata 2.3		2.5
Millenia 3.2		3.4
MX-3 . . . . 1.7		1.9
RX-7 . . . . 2.2		2.4
RX-8 8		1.0
Evaporator Coil, Replace (B)		
Includes: Evacuate/recover and recharge.		
323, Protégé/5 . . . . . 2.1		2.3
626, MX-6		
1981-92 . . . . 2.0		2.2
1993-97 . . . . 2.4		2.6
1998-02 . . . . 2.7		2.9

MAZ-44 323 : 626 : 929 : GLC : MAZDA3 : MAZDA6 : MIATA : MILLENIA

	LABOR TIME	SEVERE SERVICE
Evaporator Coil, Replace (B)		
929	2.5	2.7
GLC	2.2	2.4
Mazda3	5.5	5.7
Mazda6	5.0	5.2
Miata	1.7	1.9
Millenia	2.6	2.8
MX-3	2.3	2.5
RX-7		
1981-85	2.4	2.6
1986-91	2.0	2.2
1993-95	2.3	2.5
RX-8	3.5	3.7
Expansion Valve, Replace (B)		
Includes: Evacuate/recover and recharge.		
323, Protégé/5		
1986-94	2.2	2.4
1995-98	2.5	2.7
1999-05	2.1	2.3
626, MX-6		
1981-87	1.7	1.9
1988-92	2.0	2.2
1993-97	2.4	2.6
1998-02 orifice tube	1.7	1.9
929		
1988-91	1.7	1.9
1992-95	2.8	3.0
GLC	2.2	2.4
Mazda3	1.1	1.3
Mazda6	3.5	3.7
Miata	1.7	1.9
Millenia	2.5	2.7
MX-3	2.3	2.5
RX-7		
1981-85	2.4	2.6
1986-91	2.0	2.2
1993-95	2.3	2.5
RX-8	3.4	3.6
Heater Control Assy., Replace (B)		
1990-05 323, Protégé/5	.8	1.0
626, MX-6		
1981-82	1.6	1.8
1983-92		
w/logic control	.5	.7
w/o logic control	.8	1.0
1993-02	.7	.9
929	.9	1.1
GLC	1.3	1.5
Mazda3, Mazda6	.8	1.0
Miata	.6	.8
Millenia	.7	.9
MX-3	.9	1.1
RX-7		
1981-85	1.2	1.4
1986-95	.8	1.0
RX-8	.8	1.0

	LABOR TIME	SEVERE SERVICE
Heater Core, R&R or Replace (B)		
323, Protégé/5		
1986-94	3.1	3.3
1995-98	4.3	4.5
1999-05	3.6	3.8
626, MX-6		
1981-82	6.3	6.5
1983-85		
w/cluster switch	4.2	4.4
w/o cluster switch	3.7	3.9
1986-92	3.3	3.5
1993-02	3.9	4.1
929	6.2	6.4
GLC		
w/AC	5.8	6.0
w/o AC	2.8	3.0
Mazda3	5.1	5.3
Mazda6	3.6	3.8
Miata	3.6	3.8
Millenia	4.4	4.6
MX-3		
1992-93	6.2	6.4
1995	4.2	4.4
RX-7		
1981-91	4.2	4.4
1993-95	4.8	5.0
RX-8	3.2	3.4
w/AC exc. GLC add	.2	.2
Heater Water Shut-Off Valve, Replace (B)		
w/AC		
MX-3	3.1	3.3
1981-91 RX-7	4.1	4.3
w/o AC		
exc. below	.7	.9
1983-85 626		
w/cluster switch	4.2	4.4
w/o cluster switch	3.7	3.9
High or Low Pressure Switch, Replace (B)		
Includes evacuate/recover & recharge.		
323, Protégé/5	1.4	1.6
626, 929, MX-6	1.5	1.7
GLC	1.3	1.5
Mazda3, RX-8	.8	1.0
Mazda6	1.1	1.3
Miata	1.2	1.4
Millenia	1.5	1.7
MX-3	1.6	1.8
RX-7	1.7	1.9
In-Vehicle Sensor, Replace (B)		
All Models	.7	.9
Magnetic Clutch Assy., Replace (B)		
323, Protégé/5		
1986-89	1.2	1.4
1990-98	1.7	1.9
1999-05	1.4	1.6
626, MX-6		
1981-92	.8	1.0
1993-02	1.4	1.6

	LABOR TIME	SEVERE SERVICE
929		
1988-91	2.7	2.9
1992-95	3.4	3.6
GLC	1.7	1.9
Mazda3	1.0	1.2
Mazda6		
2.3L	1.5	1.7
3.0L	2.8	3.0
Miata	1.2	1.4
Millenia	1.6	1.8
MX-3		
4 cyl.	2.3	2.5
V6	3.0	3.2
RX-7		
1981-85	2.2	2.4
1986-91	1.4	1.6
1993-95	1.7	1.9
RX-8	1.0	1.2
Receiver/Drier Assy., Replace (B)		
Includes: Evacuate/recover and recharge.		
323, Protégé/5	1.4	1.6
626, MX-6	1.9	2.1
929	1.6	1.8
GLC	2.2	2.4
Mazda3	.5	.7
Miata		
1990-97	1.2	1.4
1999-05	2.0	2.2
Millenia	3.2	3.4
MX-3	1.7	1.9
RX-7	1.8	2.0
Sun/Photo Sensor, Replace (B)		
929, Millenia	.5	.7

WIPERS & SPEEDOMETER

	LABOR TIME	SEVERE SERVICE
Antenna, Replace (B)		
323, Protégé/5		
1986-94	2.7	2.9
1995-05		
assembly	.6	.8
mast only	.3	.5
626, MX-6		
1981-85	2.4	2.6
1986-87	.8	1.0
1988-92		
manual	2.5	2.7
power	2.8	3.0
1993-97		
assembly	.8	1.0
mast only	.3	.5
1998-02		
assembly	.5	.7
mast only	.3	.5
929	.7	.9
GLC	.7	.9
Mazda3		
assembly	1.8	2.0
mast only	.5	.7

MX-3 : MX-6 : PROTEGE : PROTEGE5 : RX-7 : RX-8 MAZ-45

	LABOR TIME	SEVERE SERVICE
Mazda6	.3	.5
Miata	.5	.7
Millenia		
assembly	.7	.9
mast only	.3	.5
MX-3	.9	1.1
RX-7		
1981-91	.7	.9
1993-95	1.7	1.9
Radio, R&R (B)		
Exc. Mazda6	.5	.5
Mazda6	.8	.8
Rear Window Washer Motor, Replace (B)		
1981-96	.5	.5
Rear Window Wiper Motor, Replace (B)		
323, Protégé/5	.7	.9
1981-91 626, MX-6	.7	.9
GLC	.7	.9
Mazda3	.6	.8
MX-3	.5	.7
RX-7	.5	.7
Rear Window Wiper & Washer Switch, Replace (B)		
1981-96	.5	.7
Speedometer Cable & Casing, Replace (B)		
1986-94 323, Protégé	.7	.9
1981-92 626, MX-6	.7	.9
1988-95 929	1.4	1.6
GLC	.8	1.0
Miata	.8	1.0
MX-3	1.6	1.8
1981-91 RX-7	.8	1.0
Speedometer Head, R&R or Replace (B)		
323, Protégé/5		
1985-96	1.3	1.3
1997-05	.8	.8
626, MX-6		
1981-82	1.2	1.2
1983-85		
analog	1.7	1.7
electronic	2.3	2.3
1986-87	1.3	1.3
1988-97	.8	.8
1998-02	1.2	1.2
GLC	1.2	1.2
Miata		
1990-97	1.2	1.2
1999-05	.8	.8
Millenia	.8	.8
MX-3	1.4	1.4
RX-7		
1981-85	1.6	1.6
1986-91	.9	.9
1993-95	1.2	1.2

	LABOR TIME	SEVERE SERVICE
Windshield Washer Motor and/or Reservoir, Replace (B)		
323, Protégé/5		
1986-94	.3	.3
1995-05	.7	.7
626, MX-6	.7	.7
929		
1988-91	.5	.5
1992-95	1.4	1.4
GLC	.3	.3
Mazda3, Mazda6	.8	.9
Miata	.3	.3
Millenia	.8	.8
MX-3	.5	.5
RX-7		
1981-85	.5	.5
1986-91	1.2	1.2
1993-95	.9	.9
RX-8	.8	1.0
Windshield Wiper & Washer Switch, Replace (B)		
323, Protégé/5		
1986-94	.7	.7
1995-98	1.2	1.2
1999-05	.5	.5
626, MX-6		
1981-97	.9	.9
1998-02	.5	.5
929	.9	.9
GLC	.7	.7
Mazda3, Mazda6	.5	.5
Miata	.8	.8
Millenia	1.3	1.3
MX-3		
1992-93	.8	.8
1994-95	1.4	1.4
RX-7, RX-8	.8	.8
Windshield Wiper Linkage, Replace (B)		
323, Protégé/5	.7	.9
626, MX-6	.8	1.0
929	.9	1.1
GLC, Mazda6	.8	1.0
Mazda3, RX-8	.5	.7
Miata, RX-7	.8	1.0
Millenia	.7	.9
MX-3	1.2	1.4
Windshield Wiper Motor, Replace (B)		
Exc. below	.8	1.0
Miata	.6	.8
626, MX-6	.6	.8

LAMPS & SWITCHES

	LABOR TIME	SEVERE SERVICE
Back-Up Lamp Assy., Replace (B)		
1993-95 RX-7	.5	.5
Back-Up Lamp Bulb, Replace (C)		
1981-98	.3	.3
1999-05	.5	.5

	LABOR TIME	SEVERE SERVICE
Back-Up Lamp Switch, Replace (B)		
Exc. below	.7	.7
929	.5	.5
GLC 5-Speed MT	2.8	2.8
Mazda6	.6	.6
Miata, MX-3	.5	.5
5-Speed requires R&R of transaxle extension housing.		
Back-Up, Park & Neutral Switch, Replace (B)		
1994-02 323, Protégé/5	.7	.7
1988-02 626, MX-6	.7	.7
1988-89 929	.7	.7
Mazda3, Mazda6	.6	.6
MX-3	.5	.5
RX-7		
1986-91	.7	.7
1993-95	2.3	2.3
RX-8	.6	.6
Clutch Start Switch, Replace (B)		
1981-92	.5	.5
1993-97	.7	.7
1998-05	.5	.5
Combination Switch Assy., Replace (B)		
All Models	.8	.9
Daylight Running Lamp Control Unit, Replace (B)		
323, Protégé/5		
1994-98	.7	.7
1999-05	.3	.3
1985-97 626	.3	.3
Mazda6	.3	.3
RX-8	.4	.4
Halogen Headlamp Bulb, Replace (B)		
1985-05 each	.3	.3
Hazard Warning Switch, Replace (B)		
Exc. below	.5	.5
1994-98 323, Protégé	.8	.8
626, MX-6	.7	.7
Headlamp Motor or Actuator, Replace (B)		
Miata one side	.5	.7
RX-7		
1981-85		
right side	.9	1.1
left side	1.2	1.4
1986-91		
one side	1.3	1.5
both sides	2.3	2.5
1993-95		
right side	1.3	1.5
left side	.9	1.1
both sides	2.2	2.4
Headlamp Switch, Replace (B)		
All Models	.8	.9
Headlamps, Aim (B)		
One side	.5	.5
Both sides	.6	.7

MAZ-46 323 : 626 : 929 : GLC : MAZDA3 : MAZDA6 : MIATA : MILLENIA

	LABOR TIME	SEVERE SERVICE
High Mount Stop Lamp Assy., Replace (B)		
Exc. below	.5	.5
626, MX-6		
1986-92		
standard	.3	.3
touring	.7	.7
1993-97	.7	.7
1998-02	.3	.3
Mazda6, Miata	.3	.3
Millenia	.3	.3
Horn, Replace (B)		
Exc. below one	.5	.5
323, Protégé/5	.9	1.0
Mazda6	.8	.9
each addl. add	.2	.2
Horn Relay, Replace (B)		
All Models	.5	.5
License Lamp Assy., Replace (B)		
Exc. below	.5	.5
1988-02 626, MX-6	.3	.3
929		
1988-91	1.2	1.2
1992-95	.8	.8
Millenia	.8	.8
1981-85 RX-7	.3	.3
Neutral Safety Switch, Replace (B)		
All Models FWD	.7	.9
Parking Brake Warning Lamp Switch, Replace (B)		
Exc. below	.5	.5
Mazda6	.8	.8
RX-7	1.0	1.0
Rear Combination Lamp Bulb, Replace (C)		
All Models each	.3	.3
Retractable Headlamp Switch, Replace (B)		
1981-85 RX-7	.7	.7
Sealed Beam Headlamp, Replace (B)		
One side	.5	.5
Side Marker Lamp Lens, Replace (B)		
All Models	.3	.3
Stop Lamp Switch, Replace (B)		
All Models	.3	.3
Turn Signal Lamp or Lens, Front, Replace (B)		
All Models	.3	.3
Turn Signal or Hazard Warning Flasher, Replace (B)		
All Models	.5	.5

BODY

	LABOR TIME	SEVERE SERVICE
Door Lock, Replace (B)		
One side	.7	.9
Both sides	1.2	1.4

	LABOR TIME	SEVERE SERVICE
Door Window Regulator, Replace (B)		
323, Protégé/5		
1986-87	1.2	1.4
1988-05	.8	1.0
626, MX-6		
1981-82	1.2	1.4
1983-02	.8	1.0
929		
1988-91		
manual	.8	1.0
power	1.2	1.4
1992-95	1.7	1.9
GLC	1.2	1.4
Mazda3, Mazda6	.7	.9
Miata		
manual	.9	1.1
power	1.2	1.4
Millenia	1.2	1.4
MX-3		
manual	.9	1.1
power	1.2	1.4
RX-7		
1981-85		
manual	.8	1.0
power	1.2	1.4
1986-91	1.4	1.6
1993-95	1.2	1.4
RX-8	.8	1.0
Replace motor add	.5	.7
Hood Hinge, Replace (B)		
323, Protégé/5		
1986-89		
one side	.7	.9
both sides	.8	1.0
1990-94		
one side	1.6	1.8
both sides	2.3	2.5
1995-98		
one side	2.6	2.8
both sides	2.9	3.1
1999-05		
one side	1.8	2.0
both sides	2.8	3.0
626, MX-6		
1981-82	.7	.9
1983-87		
one side	.8	1.0
both sides	1.2	1.4
1988-92		
one side	1.8	2.0
both sides	2.6	2.6
1993-97		
one side	3.0	3.2
both sides	3.4	3.6
1998-02		
one side	2.2	2.4
both sides	2.7	2.9

	LABOR TIME	SEVERE SERVICE
929		
1988-91		
one side	1.8	2.0
both sides	2.7	2.9
1992-95		
one side	3.1	3.3
both sides	3.7	3.9
GLC		
one side	.5	.7
both sides	.7	.9
Mazda3		
one side	1.6	1.8
both sides	2.3	2.5
Mazda6		
one side	1.7	1.9
both sides	2.6	2.8
carrier bar		
one side	2.0	2.2
both sides	3.1	3.3
Miata		
1996-97		
one side	2.8	3.0
both sides	3.2	3.5
1999-05		
one side	2.0	2.2
both sides	2.8	3.0
Millenia		
one side	3.3	3.5
both sides	4.1	4.3
MX-3		
one side	2.0	2.2
both sides	3.1	3.3
RX-7		
1981-86		
one side	.7	.9
both sides	.8	1.0
1987-95		
one side	2.5	2.7
both sides	2.8	3.0
RX-8		
one side	1.6	1.8
both sides	2.4	2.6
Hood Lock, Replace (B)		
323, Protégé/5		
1981-89	.7	.7
1990-94	.3	.3
1995-05	.6	.7
929	.5	.5
GLC, Mazda6	.5	.5
Mazda3, RX-8	.6	.6
Miata, MX-3	.5	.5
Millenia		
one	.7	.7
two	.8	.8
Hood Release Cable, Replace (B)		
323, Protégé/5		
1986-89	.5	.5
1990-05	1.2	1.4

MX-3 : MX-6 : PROTEGE : PROTEGE5 : RX-7 : RX-8 **MAZ-47**

	LABOR TIME	SEVERE SERVICE
626, MX-6		
1981-92	.7	.7
1993-02	1.2	1.2
929		
1988-91	.7	.7
1992-95	1.4	1.4
GLC	.7	.7
Mazda3, RX-8	.7	.8

	LABOR TIME	SEVERE SERVICE
Mazda6	1.1	1.2
Miata	.6	.7
Millenia	1.6	1.6
MX-3	.7	.7
RX-7		
1981-91	.5	.5
1993-95	1.2	1.2

	LABOR TIME	SEVERE SERVICE
Liftgate Lock, Replace (B)		
Exc. below	.5	.5
1988-91 626, MX-6	.7	.7
1993-95 RX-7	.7	.7
Lock Striker Plate, Replace (B)		
All Models	.3	.3

NOTES

MAZDA

B-Series : MPV : Navajo : Tribute

SYSTEM INDEX

MAINTENANCE	MAZ-50
Maintenance Schedule	MAZ-50
CHARGING	MAZ-51
STARTING	MAZ-51
CRUISE CONTROL	MAZ-51
IGNITION	MAZ-52
EMISSIONS	MAZ-52
FUEL	MAZ-53
EXHAUST	MAZ-54
ENGINE COOLING	MAZ-55
ENGINE	MAZ-56
Assembly	MAZ-56
Cylinder Head	MAZ-57
Camshaft	MAZ-59
Crank & Pistons	MAZ-61
Engine Lubrication	MAZ-63
CLUTCH	MAZ-64
MANUAL TRANSMISSION	MAZ-65
AUTO TRANSMISSION	MAZ-66
TRANSFER CASE	MAZ-68
SHIFT LINKAGE	MAZ-68
DRIVELINE	MAZ-68
BRAKES	MAZ-70
FRONT SUSPENSION	MAZ-72
REAR SUSPENSION	MAZ-74
STEERING	MAZ-75
HEATING & AC	MAZ-77
WIPERS & SPEEDOMETER	MAZ-78
LAMPS & SWITCHES	MAZ-79
BODY	MAZ-79

OPERATIONS INDEX

A
AC Hoses	MAZ-77
Air Bags	MAZ-75
Air Conditioning	MAZ-77
Alignment	MAZ-72
Alternator (Generator)	MAZ-51
Antenna	MAZ-78
Anti-Lock Brakes	MAZ-70

B
Back-Up Lamp Switch	MAZ-79
Ball Joint	MAZ-72
Battery Cables	MAZ-51
Bleed Brake System	MAZ-70
Blower Motor	MAZ-77
Brake Disc	MAZ-71
Brake Drum	MAZ-71
Brake Hose	MAZ-70
Brake Pads and/or Shoes	MAZ-71

C
Camshaft	MAZ-60
Camshaft Sensor	MAZ-52
Catalytic Converter	MAZ-54
Coolant Temperature (ECT) Sensor	MAZ-52
Crankshaft	MAZ-61
Crankshaft Sensor	MAZ-52
Cruise Control	MAZ-51
CV Joint	MAZ-69
Cylinder Head	MAZ-57

D
Differential	MAZ-70
Distributor	MAZ-52
Drive Belt	MAZ-50
Driveshaft	MAZ-68

E
EGR	MAZ-52
Electronic Control Module (ECM/PCM)	MAZ-53
Engine	MAZ-56
Engine Lubrication	MAZ-63
Engine Mounts	MAZ-57
Evaporator	MAZ-77
Exhaust	MAZ-54
Exhaust Manifold	MAZ-55
Expansion Valve	MAZ-77

F
Flexplate	MAZ-66
Flywheel	MAZ-65
Fuel Injection	MAZ-54
Fuel Pump	MAZ-53
Fuel Vapor Canister	MAZ-53

G
Gear Selector Lever	MAZ-68
Generator	MAZ-51
Glow Plug	MAZ-52

H
Halfshaft	MAZ-69
Headlamp	MAZ-79
Heater Core	MAZ-78
Horn	MAZ-79

I
Idle Air Control (IAC) Valve	MAZ-54
Ignition Coil	MAZ-52
Ignition Module	MAZ-52
Ignition Switch	MAZ-52
Injection Pump	MAZ-53
Inner Tie Rod	MAZ-75
Intake Air Temperature (IAT) Sensor	MAZ-53
Intake Manifold	MAZ-54

L
Lower Control Arm	MAZ-73

M
Maintenance Schedule	MAZ-50
Manifold Absolute Pressure (MAP) Sensor	MAZ-53
Mass Air Flow (MAF) Sensor	MAZ-53
Master Cylinder	MAZ-71
Muffler	MAZ-55
Multifunction Lever/Switch	MAZ-79

N
Neutral Safety Clutch Switch	MAZ-79
Neutral Safety Switch	MAZ-79

O
Oil Pan	MAZ-63
Oil Pump	MAZ-64
Outer Tie Rod	MAZ-75
Oxygen Sensor	MAZ-53

P
Parking Brake	MAZ-72
Pistons	MAZ-62
Positive Crankcase Ventilation (PCV) Valve	MAZ-53

R
Radiator	MAZ-55
Radiator Hoses	MAZ-55
Radio	MAZ-78
Rear Main Oil Seal	MAZ-62

S
Shock Absorber/Strut, Front	MAZ-72
Shock Absorber/Strut, Rear	MAZ-74
Spark Plug Cables	MAZ-52
Spark Plugs	MAZ-52
Spring, Front Coil	MAZ-72
Spring, Leaf	MAZ-74
Spring, Rear Coil	MAZ-74
Starter	MAZ-51
Steering Wheel	MAZ-76

T
Thermostat	MAZ-55
Throttle Body	MAZ-54
Throttle Position Sensor (TPS)	MAZ-53
Timing Belt	MAZ-60
Timing Chain	MAZ-60
Torque Converter	MAZ-67

U
U-Joint	MAZ-68
Upper Control Arm	MAZ-74

V
Valve Body	MAZ-66
Valve Cover Gasket	MAZ-58
Valve Job	MAZ-59

W
Water Pump	MAZ-55
Wheel Balance	MAZ-50
Wheel Cylinder	MAZ-72
Window Regulator	MAZ-79
Windshield Washer Pump	MAZ-78
Windshield Wiper Motor	MAZ-79

MAZ-50 B-SERIES : MPV : NAVAJO : TRIBUTE

	LABOR TIME	SEVERE SERVICE
MAINTENANCE		
Air Cleaner Filter Element, Replace (C)		
All Models	.3	.4
Chassis Lubrication, Change Oil & Filter (C)		
Includes: Correct fluid levels.		
All Models	.7	.7
Drive Belt, Adjust (B)		
All Models	.5	.6
Drive Belt Replace (B)		
Air pump		
1981-84 B2000	.3	.5
Alternator		
B-Series, Navajo	.7	.8
MPV	.8	.9
Tribute	.6	.7
Compressor		
B-Series		
1981-93	.3	.5
1994-04	.7	.9
1989-98 MPV	.5	.7
Navajo	.7	.9
Fan		
1981-93	.3	.4
Serpentine		
B2300, B2500	.7	.8
B3000, B4000	.8	.9
2000-05 MPV each	.6	.7
Navajo, Tribute	.7	.8
Steering pump		
B-Series	.7	.9
MPV		
1989-98	.5	.7
2000-05	.6	.7
Navajo	.7	.9
w/PS compressor belt only add	.1	.2
Drive Belt Tensioner, Replace (B)		
1994-04 B-Series	.5	.7
2000-05 MPV (.4)	.6	.8
Navajo	.5	.7
Tribute	.7	.8
Fuel Filter, Replace (B)		
B-Series, Navajo	.5	.7
MPV		
1989-98	.5	.6
2000-05 (1.4)	1.9	2.1
Tribute	.6	.7
w/diesel add	.2	.2
Halogen Headlamp Bulb, Replace (B)		
1986-05		
one side	.3	.3
both sides	.5	.5
Oil & Filter, Change (C)		
Includes: Correct all fluid levels.		
Exc. MPV	.5	.5
MPV	.7	.7

	LABOR TIME	SEVERE SERVICE
Sealed Beam Headlamp, Replace (B)		
1981-85		
one side	.7	.7
both sides	.8	.8
Timing Belt, Replace (B)		
B2200	3.1	3.3
B2300		
1994	2.4	2.6
1995-04	2.8	3.0
B2500	1.4	1.6
1989-98 MPV V6	3.4	3.6
Tribute 2.0L, 2.3L	4.4	4.6
w/AC add		
B2200	.2	.2
B2300, B2500		
94	.3	.3
95-04	1.2	1.2
w/PS add		
B2200	.2	.2
95-04 B2300, B2500	.5	.5
Tire, Replace (C)		
Includes: Dismount old tire and mount new tire to rim.		
One	.5	.5
Wheel, Balance (B)		
One	.5	.5
each addl. add	.3	.3
SCHEDULED MAINTENANCE INTERVALS		
If necessary, refer to appropriate Chilton maintenance service information.		
5,000 Mile Service (C)		
Exc. below	1.3	1.3
1998-04 B-Series	1.1	1.1
w/4WD short bed 96-97 add	.1	.1
w/AT 96-97 add	.1	.1
7,500 Mile Service (C)		
1996-98 MPV	.4	.4
10,000 Mile Service (C)		
Exc. below	.4	.4
1998-04 B-Series	.8	.8
15,000 Mile Service (C)		
Exc. below	1.9	1.9
1998-04 B-Series	1.4	1.4
1996-98 MPV	.6	.6
w/4WD short bed 96-97 add	.1	.1
w/AT 96-97 B-Series add	.1	.1
20,000 Mile Service (C)		
Exc. below	.4	.4
1998-04 B-Series	.8	.8
22,500 Mile Service (C)		
1996-98 MPV	.4	.4
25,000 Mile Service (C)		
Exc. below	1.3	1.3

	LABOR TIME	SEVERE SERVICE
1998-04 B-Series	1.1	1.1
w/4WD 96-97 add	.1	.1
w/AT 96-97 add	.1	.1
30,000 Mile Service (B)		
Exc. below	1.8	1.8
1998-04 B-Series		
AT		
2WD	3.0	3.0
4WD	3.9	3.9
MT		
2WD	2.4	2.4
4WD	3.2	3.2
1996-98 MPV	2.4	2.4
w/4WD 96-97 B-Series add	.6	.6
w/AT 96-97 B-Series add	.5	.5
35,000 Mile Service (C)		
Exc. below	1.3	1.3
1998-04 B-Series	1.1	1.1
w/4WD 96-97 add	.1	.1
w/AT 96-97 add	.1	.1
37,500 Mile Service (C)		
1996-98 MPV	.4	.4
40,000 Mile Service (C)		
Exc. below	.4	.4
1998-04 B-Series	.8	.8
45,000 Mile Service (C)		
Exc. below	1.9	1.9
1998-04 B-Series	1.4	1.4
1996-98 MPV	1.1	1.1
w/4WD 96-97 add	.1	.1
w/AT 96-97 add	.1	.1
50,000 Mile Service (C)		
Exc. below	.9	.9
1998-04 B-Series	1.1	1.1
52,500 Mile Service (C)		
1996-98 MPV	.4	.4
55,000 Mile Service (C)		
Exc. below	1.3	1.3
1998-04 B-Series	1.1	1.1
w/4WD 96-97 add	.1	.1
w/AT 96-97 add	.1	.1
60,000 Mile Service (B)		
Exc. below	2.3	2.3
1998-04 B-Series		
AT		
2WD	3.1	3.1
4WD	4.2	4.2
MT		
2WD	2.8	2.8
4WD	4.1	4.1
1996-98 MPV	6.4	6.4
w/4WD 96-04 B-Series add	.6	.6
w/AT 96-04 B-Series add	.6	.6
Replace spark plugs Calif. engs.		
96-97 B-Series add	.8	.8
timing belt 2.3L add	2.4	2.6

B-SERIES : MPV : NAVAJO : TRIBUTE MAZ-51

	LABOR TIME	SEVERE SERVICE
65,000 Mile Service (C)		
Exc. below	1.3	1.3
1998-04 B-Series	1.1	1.1
w/4WD 96-97 add	.1	.1
w/AT 96-97 add	.1	.1
67,500 Mile Service (C)		
1996-98 MPV	.4	.4
75,000 Mile Service (C)		
1996-98 MPV	1.1	1.1
80,000 Mile Service (C)		
1996-04 B-Series	1.1	1.1
82,500 Mile Service (C)		
1996-98 MPV	.4	.4
85,000 Mile Service (C)		
1996-04 B-Series	1.1	1.1
90,000 Mile Service (B)		
1996-98 MPV	2.4	2.4
1998-04 B-Series		
AT		
2WD	3.1	3.1
4WD	3.8	3.8
MT		
2WD	2.4	2.4
4WD	2.6	2.6
w/4WD 96-98 MPV add	.1	.1
Replace timing belt Calif. engs. 96-98 MPV add	3.1	3.3
95,000 Mile Service (C)		
1998-04 B-Series	1.1	1.1
97,500 Mile Service (C)		
1996-98 MPV	.4	.4

CHARGING

	LABOR TIME	SEVERE SERVICE
Alternator Circuits, Test (B)		
Includes: Test component output.		
All Models	.7	.7
Alternator, R&R and Recondition (B)		
1981-84 B2000	2.3	2.4
B2200, B2600		
1982-84	2.8	3.0
1987-93	2.3	2.4
MPV		
1989-98		
4 cyl.	2.3	2.5
V6	2.8	3.0
2000-05 (3.2)	4.3	4.5
Navajo	2.3	2.5
Alternator Assy., Replace (B)		
Includes: Pulley transfer.		
1982-84 B2200	1.3	1.5
1986-04 B-Series	.7	.9
MPV		
1989-98		
4 cyl.	.7	.9
V6	1.4	1.6
2000-05 (2.1)	2.8	3.0
Navajo	.7	.9

	LABOR TIME	SEVERE SERVICE
Tribute		
2.0L, 2.3L	1.1	1.1
3.0L	3.2	3.4
Alternator Voltage Regulator, Replace (B)		
1981-84	2.3	2.5
1986-88	1.8	2.0
MPV		
1989-98		
4 cyl.	1.7	1.9
V6	2.3	2.5
2000-05 (3.2)	4.3	4.5

STARTING

	LABOR TIME	SEVERE SERVICE
Starter Draw Test (On Car) (B)		
All Models	.3	.3
Battery, Replace (C)		
All Models	.4	.5
Battery Cables, Replace (C)		
B2000		
positive	.7	.8
negative	.2	.3
B2200		
positive	.5	.6
negative	.2	.3
B2300, B2500, B3000, B4000		
positive	.5	.6
negative	.7	.8
B2600		
positive	.5	.6
negative	.2	.3
MPV		
positive	.5	.6
negative	.3	.4
Navajo each	.5	.6
Tribute	.6	.7
Starter, R&R and Recondition (B)		
B2000	2.8	3.0
B2200, B2600		
1982-84	2.8	3.0
1987-93	2.5	2.7
B2300	2.2	2.4
B2500	1.7	1.9
B3000, B4000	2.2	2.4
1989-98 MPV		
2WD	2.6	2.8
4WD	5.0	5.2
Navajo	2.5	2.7
Starter Assy., Replace (B)		
B2000		
1981-84	1.5	1.7
1986-87	.7	.9
B2200, B2600	.8	1.0
B2300	.5	.7
B2500	.7	.9

	LABOR TIME	SEVERE SERVICE
MPV		
1989-98		
2WD	1.2	1.4
4WD	3.6	3.8
2000-05 (1.0)	1.4	1.6
Navajo	.8	1.0
Tribute		
2.0L, 2.3L		
2WD	.9	1.1
4WD	1.5	1.7
3.0L	1.8	2.0
Starter Drive Assy., Replace (B)		
Includes: Starter R&R.		
B-Series		
1981-84	2.2	2.4
1986-93	1.8	2.0
1994-04	.9	1.1
MPV		
1989-98		
2WD	1.4	1.6
4WD	3.8	4.0
2000-05 (2.1)	2.8	3.0
Navajo	1.2	1.4
Starter Solenoid, Replace (B)		
B-Series		
1981-84	1.4	1.6
1986-93	.9	1.1
1994-04	.8	1.0
MPV		
1989-98		
2WD	1.3	1.5
4WD	3.8	4.0
2000-05 (1.1)	1.5	1.7
Navajo	.8	1.0

CRUISE CONTROL

	LABOR TIME	SEVERE SERVICE
Diagnose Cruise Control System (A)		
1988-05	1.2	1.2
Actuator Assy., Replace (B)		
1986-05	.7	.7
Control Actuator Cable, Replace (B)		
1994-04 B-Series	.7	.9
1989-98 MPV	.5	.7
Navajo	.5	.7
Control Main Switch, Replace (B)		
1994-04 B-Series	.3	.3
1996-05 MPV	.5	.5
Tribute	.8	.9
Control Safety (Cut-Out) Switch, Replace (B)		
1986-93	.5	.5
Control Switch, Replace (B)		
1988-93 B-Series	.9	.9
2000-05 MPV	.5	.5
1991-94 Navajo	.3	.3
Tribute	.8	.9

MAZ-52 B-SERIES : MPV : NAVAJO : TRIBUTE

	LABOR TIME	SEVERE SERVICE
Control Unit, Replace (B)		
B-Series		
1988-93	.8	.8
1994	.3	.3
MPV		
1989-98	.8	.8
2000-05	.5	.5
Navajo	.5	

IGNITION

	LABOR TIME	SEVERE SERVICE
Diagnose Ignition System (A)		
1985-05		
w/distributor	.7	.7
w/o distributor	1.8	1.8
Ignition Timing, Reset (B)		
All Models	.8	1.0
Camshaft Sensor, Replace (B)		
1994-04 B-Series	.7	.9
2000-05 MPV (.6)	.8	1.0
Tribute		
2.0L, 2.3L	.5	.5
3.0L	.4	.4
Crankshaft Angle Sensor, Replace (B)		
B2300, B2500	2.7	2.9
B3000, B4000	.7	.9
MPV (.7)	.9	1.1
Navajo	.7	.9
Tribute		
2.0L, 2.3L	2.8	3.0
3.0L	.9	1.1
w/AC B2300, B2500 add	.3	.3
Distributor, R&R and Recondition (B)		
Includes: Reset base ignition timing.		
1981-93 B-Series	2.2	2.4
1989-91 MPV	2.2	2.4
1991-94 Navajo	2.5	2.7
Distributor, Replace (B)		
Includes: Reset base ignition timing.		
B-Series		
1981-93	.8	1.0
1994	1.2	1.4
MPV	.8	1.0
Navajo	1.3	1.5
Distributor Cap and/or Rotor, Replace (B)		
1981-98	.4	.4
Distributor Igniter Set, Replace (B)		
1981-84 B2000	.5	.7
1986-94	1.3	1.5
Distributor Points & Condenser, Replace (B)		
1981-82 B-Series		
one	.7	.9
both	.7	.9
Fast Glow Control Unit, Replace (B)		
1982 B2200	.5	.7

	LABOR TIME	SEVERE SERVICE
Glow Plug, Replace (B)		
1982-84 B2200		
one	.5	.7
all	.7	.9
Glow Plug Relay, Replace (B)		
1982-84 B2200	.3	.5
Ignition Coil, Replace (B)		
B-Series		
1981-93		
one	.3	.5
two	.5	.7
1994-04		
one	.5	.7
two	.7	.9
MPV		
1989-01 each	.5	.7
2002-05		
right side (1.2)	1.6	1.8
left side (.3)	.5	.7
all (1.3)	1.8	2.0
Navajo	.5	.7
Tribute		
2.0L, 2.3L	.6	.8
3.0L		
right side	1.4	1.6
left side	.5	.7
all	1.5	1.7
Ignition Module Assy., Replace (B)		
1994 B2300	.5	.7
1994 B3000, B4000	.7	.9
MPV		
1989-91		
4 cyl.	.5	.7
V6	1.4	1.6
1992-98	.5	.7
Navajo	.7	.9
Ignition Switch, Replace (B)		
B-Series		
1981-87	1.2	1.4
1988-04	.5	.7
MPV (.5)	.7	.9
Navajo	.5	.7
Tribute	.6	.8
Pickup Coil, Replace (B)		
All Models	1.3	1.5
Spark Plug (Ignition) Cables, Replace (B)		
B2000, B2200, B2600	.5	.7
B2300, B2500	.8	1.0
B3000, B4000	.7	.9
MPV		
1989-95	.5	.7
1996-05 (.5)	.7	.9
Navajo	.7	.9
Tribute	.8	1.0
Spark Plugs, Replace (B)		
B2000, B2200, B2600	.5	.7
B2300, B2500	1.2	1.4
B3000, B4000	.8	1.0

	LABOR TIME	SEVERE SERVICE
MPV		
1989-98	.7	.9
2000-01	1.1	1.3
2002-05 (1.5)	2.0	2.2
Navajo	.7	.9
Tribute		
2.0L, 2.3L	.4	.6
3.0L	1.6	1.8
Vacuum Advance Unit, Replace (B)		
Includes: Reset base ignition timing.		
B2000, B2200, B2600	.5	.7
1989-91 MPV	.7	.9

EMISSIONS

The following operations do not include testing. Add time as required.

	LABOR TIME	SEVERE SERVICE
Diagnose Emission Control System (A)		
All Models	1.0	1.0
Dynamometer Test (A)		
All Models	.5	.5
Air Flow Sensor, Replace (B)		
1989-96	.7	.9
Air Injection Nozzles, Replace (B)		
1981-84 B2000	.7	.9
Altitude Compensator, Replace (B)		
1986-94	.5	.7
Charge Temperature Sensor, Replace (B)		
Navajo	.2	.4
Closed Loop Control Module, Replace (B)		
1987-96	.5	.7
Coolant Temperature (ECT) Sensor, Replace (B)		
B2000	.5	.7
1987-93 B2200	.5	.7
B2300, B2500, B3000, B4000	.7	.9
B2600		
1987-89	.7	.9
1989-93	.9	1.1
MPV	.8	1.0
Navajo	.8	1.0
Tribute		
2.0L, 2.3L	1.5	1.7
3.0L	.9	1.1
EGR Control Valve, Replace (B)		
1983-84	1.2	1.4
1986-05	.8	1.0
EGR Modulator, Replace (B)		
1987-93 B2200, B2600	.5	.7
EGR Vacuum Control Solenoid, Replace (B)		
1994-04 B-Series	.3	.5

B-SERIES : MPV : NAVAJO : TRIBUTE **MAZ-53**

	LABOR TIME	SEVERE SERVICE
Electronic Control Module (ECM/PCM), Replace (B)		
B-Series	.5	.7
MPV		
1989-95	.7	.9
1996-05 (.4)	.6	.8
Navajo	.9	1.1
Tribute	.5	.7
Reflash MPV add	.9	.9
Electronic Control Relay, Replace (B)		
Navajo	.2	.4
Evap Check Valve, Replace (B)		
1994-04 B-Series	.3	.5
MPV	.5	.7
Fuel Vapor Canister, Replace (B)		
B-Series		
1981-90	.3	.5
1991-04	.5	.7
MPV		
1996-98	.7	.9
2000-05 (.7)	.9	1.1
Navajo	.5	.7
Tribute	.8	1.0
Heated Inlet Air Cleaner Temperature Sensor, Replace (B)		
1986-93	.5	.7
Heated Inlet Air Idle Compensator, Replace (B)		
1986-93	.5	.7
Heated Inlet Air Vacuum Motor, Replace (B)		
1986-93	.5	.7
Inertia Switch Assy., Replace (B)		
Navajo	.2	.4
Intake Air Temperature (IAT) Sensor, Replace (B)		
1994-04 B-Series	.5	.7
MPV	.5	.7
Navajo	.7	.9
Manifold Absolute Pressure (MAP) Sensor, Replace (B)		
1983-04	.5	.7
Mass Air Flow (MAF) Sensor, Replace (B)		
1994-04 B-Series	.7	.9
2000-05 MPV (.6)	.8	1.0
Navajo	.7	.9
Tribute	.7	.9
Oxygen Sensor, Replace (B)		
B2300 one or both	.7	.9
B2500		
one	.5	.7
both	.7	.9
B3000, B4000		
front		
one	.7	.9
both	.8	1.0
rear	.7	.9

	LABOR TIME	SEVERE SERVICE
MPV		
1989-95	.5	.7
1996-98	.7	.9
2000-05		
one or two (.7)	.9	1.1
three or all (1.0)	1.4	1.6
Navajo	.7	.9
Tribute		
2.0L, 2.3L		
one or two	.7	.8
three or four	1.2	1.4
3.0L		
one or two	.8	1.0
three or four	1.2	1.4
Positive Crankcase Ventilation (PCV) Valve, Replace (B)		
All Models	.3	.5
Processor Assy., Replace (B)		
Navajo	.3	.5
Throttle Air Bypass Valve, Replace (B)		
Navajo	.5	.7
Tribute		
2.0L, 2.3L	2.9	3.1
3.0L	.6	.8
Throttle Position Sensor (TPS), Replace (B)		
B2300, B3000, B4000	.7	.9
B2500	.5	.7
1989-05 MPV (.6)	.8	1.0
Navajo	.8	1.0
Tribute	.7	.9

FUEL
CARBURETOR

	LABOR TIME	SEVERE SERVICE
Carburetor, Adjust (On Car) (B)		
1981-87	.5	.7
1988-93	1.9	2.1
Float Level, Adjust (B)		
B2000, B2200	1.7	1.9
B2600	2.3	2.5
Carburetor, R&R and Clean or Recondition (B)		
Includes: Adjustments.		
1981-87	2.7	2.9
1988-93	3.1	3.3
Carburetor, Replace (B)		
Includes: Adjustments.		
1981-87	1.3	1.5
1988-93	1.9	2.1
EFE Heater Insulator, Replace (B)		
B2000	1.7	1.9
B2200	1.7	1.9
B2600	1.9	2.1

DELIVERY

	LABOR TIME	SEVERE SERVICE
Fuel Pump, Test (B)		
All Models	.5	.7

	LABOR TIME	SEVERE SERVICE
Diesel Fuel Injection Pump, Replace (B)		
1982-84 B2200	2.0	2.2
Fuel Filter, Replace (B)		
B-Series, Navajo	.5	.7
MPV		
1989-98	.5	.7
2000-05	1.9	2.1
Tribute	.6	.8
w/diesel add	.2	.3
Fuel Gauge (Dash), Replace (B)		
B-Series		
1981-93	1.6	1.8
1994-04	.8	1.0
Navajo	1.4	1.6
Fuel Gauge (Tank), Replace (B)		
B-Series		
1981-93	1.3	1.5
1994-04	1.8	2.0
MPV		
1989-98	1.3	1.5
2000-05	1.6	1.8
Navajo w/pump	1.8	2.0
Tribute	.6	.8
Fuel Pump, R&R and Recondition (B)		
1982-84 B2200 diesel	4.8	5.0
Fuel Pump, Replace (B)		
B2000	1.2	1.4
B2200		
AT	1.2	1.4
MT	.5	.7
B2300, B2500, B3000, B4000	1.8	2.0
B2600	1.2	1.4
MPV		
1989-98	1.2	1.4
2000-05 (1.4)	1.9	2.1
Navajo	1.8	2.0
Tribute	.7	.9
Fuel Pump Control Unit, Replace (B)		
1989-93 B2600	.5	.7
Fuel Pump Relay, Replace (B)		
1989-93 B2600	.8	.8
1994-04 B-Series	.5	.5
MPV		
1989-98	.5	.5
2000-05	.3	.3
Navajo	.5	.5
Tribute	.4	.4
Fuel Tank, Replace (B)		
Includes: Drain and refill.		
B-Series		
1981-93	1.6	1.8
1994-04		
reg. cab	1.6	1.8
cab plus 4WD	2.7	2.9

MAZ-54 B-SERIES : MPV : NAVAJO : TRIBUTE

	LABOR TIME	SEVERE SERVICE
Fuel Tank, Replace (B)		
MPV		
1989-98	1.7	1.9
2000-05	3.3	3.5
Navajo	1.9	2.1
Tribute		
2WD	3.1	3.3
4WD	4.9	5.1
Intake Manifold and/or Gasket, Replace (B)		
Includes: Adjustments.		
B2000, B2200		
gasoline	2.3	2.5
diesel	2.2	2.4
B2300, B2500		
upper	1.6	1.8
complete	3.3	3.5
B2600	2.6	2.8
B3000		
upper	1.6	1.8
complete	3.4	3.6
B4000		
upper	1.7	1.9
complete	3.0	3.2
MPV		
1989-98	3.1	3.3
2000-05	2.1	2.3
Navajo		
upper	1.3	1.5
complete	3.8	4.0
Tribute		
2.0L, 2.3L	2.9	3.1
3.0L	2.1	2.3
Replace manifold add	.5	.7
Sub or Main Fuel Pump, Replace (B)		
1989-93		
external	1.2	1.4
in tank	2.0	2.2

INJECTION

	LABOR TIME	SEVERE SERVICE
Air Flow Meter, Replace (B)		
Navajo	.8	1.0
Tribute	.7	.9
Bypass Air Control Valve, Replace (B)		
B2600	1.5	1.7
MPV		
1989-98		
4 cyl.	1.7	1.9
V6	1.5	1.7
2000-05 (.7)	.9	1.1
Tribute		
2.0L, 2.3L	2.9	3.1
3.0L	.6	.8
Diesel Fuel Governor Assy., Replace (B)		
1982-84 B2200	2.9	3.1
Diesel Fuel Governor Cover and/or Seal, Replace (B)		
1982-84 B2200	.7	.9

	LABOR TIME	SEVERE SERVICE
Fuel Cut-Off Switch, Replace (B)		
Navajo	.7	.9
Fuel Distributor, Replace (B)		
1982-84 B2200	3.7	3.9
Tribute		
2.0L, 2.3L	2.0	2.2
3.0L	1.2	1.4
Fuel Injector Resistor, Replace (B)		
MPV	.5	.7
Fuel Injectors, Replace (B)		
B2300		
one	1.4	1.6
all	1.6	1.8
B2500		
one	.9	1.1
all	1.3	1.5
B2600		
one side	2.0	2.2
both sides	2.3	2.5
B3000, B4000		
one	1.7	1.9
all	1.8	2.0
MPV		
1989-98		
4 cyl.	1.6	1.8
V6		
one side	2.3	2.5
both sides	2.6	2.8
2000-05 both sides	2.1	2.3
Navajo		
one	1.7	1.9
all	2.0	2.2
Tribute		
2.0L, 2.3L	1.1	1.3
3.0L	1.8	2.0
Fuel Pressure Regulator, Replace (B)		
1994-97 B2300	.7	.9
1989-93 B2600	.5	.7
1994-97 B3000, B4000	.5	.7
MPV		
1989-95		
4 cyl.	.5	.7
V6	1.2	1.4
1996-98	2.8	3.0
2000-05 (1.3)	1.8	2.0
Navajo	.7	.9
Tribute	.7	.9
Fuel Rail, Replace (B)		
B2600	1.4	1.6
MPV		
1989-95		
4 cyl.	1.4	1.6
V6	2.8	3.0
1996-98	2.7	2.9
2000-05	2.3	2.5
Tribute		
2.0L, 2.3L	2.0	2.2
3.0L	1.2	1.4

	LABOR TIME	SEVERE SERVICE
Idle Air Control (IAC) Valve, Replace (B)		
1994-04 B-Series	.5	.7
2000-04 MPV	.6	.8
Injection Nozzles (Diesel), Replace (B)		
1982-84 B2200		
one	.5	.7
all	1.7	1.9
Load Limiter (Diesel), Replace (B)		
1982-84 B2200	3.2	3.4
Pulsation Damper, Replace (B)		
1989-93 B2600	.9	1.1
MPV		
1989-95	.9	1.1
2000-05		
No.1 (.7)	.9	1.1
No.2 on fuel distributor (1.3)	1.8	2.0
Tribute		
2.0L, 2.3L	.7	.9
3.0L	1.4	1.6
Solenoid Valve, Replace (B)		
1989-05 MPV (.3)	.5	.7
System Relay, Replace (B)		
1989-98 MPV	.5	.7
Navajo	.8	.9
Throttle Body Assy., Replace (B)		
B2300, B2500	.8	1.0
B2600	1.4	1.6
B3000	1.6	1.8
B4000	.8	1.0
MPV		
1989-98	1.4	1.6
2000-05 (.8)	1.1	1.3
Navajo	.8	1.0
Tribute	.8	1.0
Throttle Body Lever Set, Replace (B)		
1989-98 MPV	.7	.9

EXHAUST

	LABOR TIME	SEVERE SERVICE
Catalytic Converter, Replace (B)		
B2000		
1981-84	1.2	1.4
1986-87		
monolith	1.6	1.8
3-way		
front	.7	.9
rear	.7	.9
B2200, B2600		
1982-84	1.2	1.4
1987-93		
monolith	1.6	1.8
3-way		
front	.7	.9
rear	.7	.9
B2300, B2500	1.2	1.4
B3000, B4000	1.7	1.9
MPV		
1989-98	.9	1.1

B-SERIES : MPV : NAVAJO : TRIBUTE MAZ-55

	LABOR TIME	SEVERE SERVICE
2000-05		
right side (1.6)	2.2	2.4
left side (1.3)	1.8	2.0
both sides (2.2)	3.0	3.3
Navajo	1.3	1.5
Tribute		
2.0L, 2.3L	1.5	1.7
3.0L		
right side	4.0	4.2
left side	1.8	2.0
both sides	4.8	5.0
warm up	.8	1.0
Center Exhaust Pipe, Replace (B)		
1981-93 B-Series	.9	1.1
1989-98 MPV	1.9	2.1
Exhaust Manifold or Gasket, Replace (B)		
B2000	1.8	2.0
B2200	1.5	1.7
B2300	1.7	1.9
B2500	2.5	2.7
B2600	.9	1.1
B3000, B4000		
right side	2.8	3.0
left side	2.3	2.5
both sides	4.2	4.4
MPV		
1989-98		
4 cyl.	.8	1.0
V6		
right side	1.5	1.7
left side	1.3	1.5
both sides	2.4	2.6
2000-05		
right side (1.8)	2.4	2.6
left side (1.3)	1.8	2.0
both sides (2.2)	3.0	3.2
Navajo		
right side	2.8	3.0
left side	2.3	2.5
both sides	4.2	4.4
Tribute		
2.0L, 2.3L	2.0	2.2
3.0L		
right side	4.0	4.2
left side	1.8	2.0
both sides	4.8	5.0
Front Exhaust Pipe, Replace (B)		
B2000	1.2	1.4
B2200, B2600		
1982-84	1.2	1.4
1987-93	.9	1.1
B2300, B2500	.8	1.0
B3000	1.6	1.8
B4000	1.7	1.9
MPV	1.0	1.2
Navajo	1.7	1.9
Tribute		
3.0L	.9	1.1

	LABOR TIME	SEVERE SERVICE
Muffler, Replace (B)		
B2000	.9	1.1
B2200, B2600		
1982-84	.9	1.1
1987-93	.8	1.0
1994-04 B-Series	.7	.9
1989-05 MPV (.5)	.7	.9
Navajo	.8	1.0
Tribute		
2.0L, 2.3L	.7	.9
3.0L	.8	1.0
Pre-Muffler, Replace (B)		
B2000	1.5	1.7
Tail Pipe, Replace (B)		
B2200, B2600	.5	.7
MPV	.6	.8
Tribute	.7	.9

ENGINE COOLING

Pressure Test Cooling System (C)
All Models	.5	.7
Bypass Hoses, Replace (B)		
1989-98 MPV	.7	.9
Tribute pipe or O-ring		
2.0L, 2.3L	2.4	2.6
3.0L	2.0	2.2
Coolant Temperature Sensor Switch, Replace (B)		
1987-93 B2200, B2600	.5	.7
Coolant Thermostat, Replace (B)		
B2000	.7	.9
B2200, B2600	.7	.9
B2300	.7	.9
B2500, B3000, B4000	1.2	1.4
MPV		
1989-98	.9	1.1
2000-05 (.9)	1.2	1.4
Navajo	1.2	1.4
Tribute		
2.0L, 2.3L	1.1	1.3
3.0L	1.5	1.7
Cooling Fan Relay, Replace (B)		
1989-05 MPV	.5	.7
Tribute	.3	.5
Electric Cooling Fan Assy., Replace (B)		
1983-93	.7	.9
MPV		
1996-98	.6	.8
2000-05	1.2	1.4
Tribute		
main	1.2	1.4
additional	1.1	1.3
Engine Coolant Temp. Sending Unit, Replace (B)		
Exc. 1989-93 B2600	.8	1.0
1989-93 B2600	.9	1.1
Fan Blade, Replace (B)		
All Models	.7	.9

	LABOR TIME	SEVERE SERVICE
Freeze Plugs (Water Jacket), Replace (B)		
Add access time as required.		
All Models each	.8	1.0
Radiator Assy., R&R or Replace (B)		
Exc. below	1.3	1.5
2000-05 MPV (1.8)	2.4	2.6
Tribute	2.6	2.8
w/AT add	.2	.4
Radiator Hoses, Replace (B)		
Includes: Refill with proper coolant mix.		
B-Series		
1981-93		
upper	.3	.5
lower	.5	.7
1994-04		
upper	.5	.7
lower	.7	.9
MPV		
upper	1.2	1.4
lower	1.4	1.6
Navajo		
upper	.5	.7
lower	.7	.9
Tribute		
upper	1.2	1.4
lower		
2.0L, 2.3L	1.8	2.0
3.0L	1.3	1.5
Temperature Gauge (Dash), Replace (B)		
1981-84 B2000	1.3	1.5
B2200, B2600		
1982-84	1.6	1.8
1987-93	1.3	1.5
B2300, B2500, B3000, B4000	.8	1.0
Navajo	1.4	1.6
Water Pump and/or Gasket, Replace (B)		
Includes: Refill with proper coolant mix.		
B2000		
1981-84	1.9	2.1
1986-87	2.8	3.0
B2200		
1982-84	1.7	1.9
1987-93	2.8	3.0
B2300	1.8	2.0
B2500	1.2	1.4
B2600	.8	1.0
B3000	2.0	2.2
B4000	2.2	2.4
MPV		
1989-98		
4 cyl.	1.7	1.9
V6	3.8	4.0
2000-05 (1.7)	2.3	2.5
Navajo	2.2	2.4

MAZ-56 B-SERIES : MPV : NAVAJO : TRIBUTE

	LABOR TIME	SEVERE SERVICE
Water Pump and/or Gasket, Replace (B)		
Tribute		
2.0L, 2.3L	1.9	2.0
3.0L	2.3	2.4

ENGINE
ASSEMBLY

Times shown are for OEM assemblies. Time to replace assemblies from aftermarket rebuilders may vary.

Engine Assy., R&I (B)
Does not include parts or component transfer.

	LABOR TIME	SEVERE SERVICE
B2000, B2200		
1981-84	5.2	5.4
1986-93		
2WD	5.9	6.1
4WD	6.3	6.5
B2300		
AT	6.3	6.5
MT		
2WD	7.4	7.6
4WD	7.6	7.8
B2500		
AT	6.3	6.5
MT	7.4	7.6
B2600		
AT		
2WD	6.8	7.0
4WD	7.7	7.9
MT	6.2	6.4
B3000		
2WD		
AT	7.6	7.8
MT	7.4	7.6
4WD	7.1	7.3
B4000		
AT	7.6	7.8
MT	7.1	7.3
MPV		
1989-92		
AT		
2WD		
4 cyl.	10.3	10.5
V6	12.0	12.2
4WD	12.6	12.8
MT		
2WD		
4 cyl.	9.6	9.8
V6	11.6	11.8
4WD	12.2	12.4
1993-95		
2WD		
4 cyl.	10.3	10.5
V6	12.0	12.2
4WD	12.6	12.8

	LABOR TIME	SEVERE SERVICE
1996-98		
2WD	10.9	11.1
4WD	11.6	11.8
2000-04	13.6	13.8
Navajo		
AT	7.6	7.8
MT	7.1	7.3
Tribute		
AT		
2WD		
2.0L, 2.3L	13.5	13.7
3.0L	14.6	14.8
4WD		
2.0L, 2.3L	14.7	14.9
3.0L	15.6	15.8
MT		
2WD		
2.0L, 2.3L	13.0	13.2
4WD		
2.0L, 2.3L	13.9	14.1
w/AC add		
81-93	.5	.7
94-05		
4 cyl.	1.2	1.4
3.0L	.5	.7
4.0L	.8	1.0
Navajo	.8	1.0
w/PS 81-93 add	.5	.7

Engine Assy., R&R and Recondition (A)
Includes: Replacing rings, rod and main bearings, cylinder head reconditioning and engine tune-up.

	LABOR TIME	SEVERE SERVICE
B2000		
1981-84	21.8	22.0
1986-87		
2WD	22.2	22.4
4WD	22.8	23.0
B2200		
1982-84	26.2	26.4
1987-93		
2WD	22.2	22.4
4WD	22.8	23.0
B2300		
AT	22.8	23.0
MT	23.3	23.5
B2500		
AT	22.8	23.0
MT	23.1	23.3
B2600		
AT		
2WD	23.8	24.0
4WD	24.8	25.0
MT	23.3	23.5
B3000		
2WD		
AT	26.7	26.9
MT	26.5	26.7
4WD	26.2	26.4

	LABOR TIME	SEVERE SERVICE
B4000		
AT	26.2	26.4
MT	25.4	25.6
MPV		
1989-92		
AT		
2WD		
4 cyl.	28.3	28.5
V6	28.9	29.0
4WD	29.3	29.5
MT		
2WD		
4 cyl.	27.4	27.6
V6	28.4	28.6
4WD	28.9	29.1
1993-95		
2WD		
4 cyl.	28.3	28.5
V6	28.9	29.1
4WD	29.3	29.5
1996-98		
2WD	28.0	28.2
4WD	28.4	28.6
2000-05 (30.8)	41.6	41.8
Navajo		
AT	26.7	26.9
MT	26.2	26.4
Tribute		
AT		
2WD		
2.0L, 2.3L	30.0	30.2
3.0L	36.2	36.4
4WD		
2.0L, 2.3L	33.1	33.3
3.0L	36.9	37.1
MT		
2WD		
2.0L, 2.3L	29.4	29.6
4WD		
2.0L, 2.3L	30.2	30.4
w/AC add		
81-93	.5	.7
94-04		
4 cyl.	1.2	1.4
3.0L	.5	.7
4.0L	.8	1.0
Navajo	.8	1.0
w/PS 81-93 add	.5	.7
Replace cylinder liners		
82-84 B2200 add	.8	1.0

B-SERIES : MPV : NAVAJO : TRIBUTE   **MAZ-57**

	LABOR TIME	SEVERE SERVICE
Engine Assy. (Short Block), Replace (B)		
Assembly consists of cylinder block, piston assemblies, crankshaft, camshaft, timing chain and gears. Does not include cylinder heads. Operation Includes: R&R engine, transfer necessary parts and all necessary adjustments.		
B2000		
1981-84	15.2	15.4
1986-87		
2WD	14.6	14.8
4WD	15.3	15.5
B2200		
1982-84	21.4	21.6
1987-93		
2WD	14.6	14.8
4WD	15.3	15.5
B2300		
AT	13.4	13.6
MT		
2WD	13.9	14.1
4WD	14.1	14.3
B2500		
AT	9.9	10.1
MT	10.8	11.0
B2600		
AT		
2WD	14.1	14.3
4WD	15.0	15.2
MT	13.5	13.7
B3000		
2WD	15.3	15.5
4WD	14.9	15.1
B4000		
AT	14.9	15.1
MT	14.3	14.5
MPV		
1989-92		
AT		
2WD		
4 cyl.	19.4	19.6
V6	20.7	20.9
4WD	21.4	21.6
MT		
2WD		
4 cyl.	18.7	18.9
V6	20.3	20.5
4WD	20.7	20.9
1993-95		
2WD		
4 cyl.	19.4	19.6
V6	20.7	20.9
4WD	21.4	21.6
1996-98		
2WD	19.8	20.0
4WD	20.3	20.5
2000-05 (17.1)	23.1	23.3

	LABOR TIME	SEVERE SERVICE
Navajo		
AT	15.2	15.4
MT	14.9	15.1
Tribute		
AT		
2WD		
2.0L, 2.3L	19.1	19.3
3.0L	20.5	20.7
4WD		
2.0L, 2.3L	20.0	20.2
3.0L	30.0	30.2
MT		
2WD		
2.0L, 2.3L	18.8	19.0
4WD		
2.0L, 2.3L	19.4	19.6
w/AC add		
81-93	.5	.7
94-04		
4 cyl.	1.2	1.4
3.0L	.5	.7
4.0L	.8	.1.0
Navajo	.8	1.0
w/PS 81-93 add	.5	.7
Replace cylinder liners		
82-84 B2200 add	.8	1.0
Engine Mounts, Replace (B)		
1981-93 B-Series		
front		
one	.7	.9
both	1.2	1.4
rear	.7	.9
B2300		
one side	2.2	2.4
both sides	2.4	2.6
B2500		
one side	1.8	2.0
both sides	2.5	2.7
B3000, B4000		
right side	.7	.9
left side	1.2	1.4
both sides	1.4	1.6
MPV		
1989-98		
front		
one	.8	1.0
both	1.2	1.4
rear	.7	.9
2000-01		
No. 1	2.1	2.3
Nos. 2, 3, 4 each	1.0	1.2
2002-05		
No. 1 (2.2)	3.0	3.2
Nos. 2 (.6)	.8	1.0
Nos. 3, 4 ea (1.3)	1.8	2.0
Navajo		
right side	.8	1.0
left side	1.2	1.4
both sides	.7	.9

	LABOR TIME	SEVERE SERVICE
Tribute		
2.0L, 2.3L		
No. 1	1.4	1.6
Nos. 2, 4	1.2	1.4
No. 3	1.1	1.3
3.0L		
Nos. 1, 2	1.2	1.4
Nos. 3, 4	1.4	1.6
CYLINDER HEAD		
Compression Test (B)		
4 cyl.	.8	1.0
V6		
1989-93	.8	1.0
1994-04		
exc. MPV	1.3	1.5
MPV		
1994-01	1.4	1.6
2002-05	2.2	2.4
Diesel	1.6	1.8
Valves, Adjust (B)		
1981-93	1.2	1.4
Cylinder Head, Replace (B)		
Includes: Transfer parts, adjustments.		
B2000		
1981-84	9.8	10.0
1986-87	11.8	12.0
B2200		
1982-84	10.1	10.3
1987-93	11.5	11.7
B2300		
1994	7.1	7.3
1995-04	7.4	7.6
B2500	6.3	7.0
B2600		
1987-88	10.5	10.7
1989-93	12.2	12.4
B3000		
right side	6.3	6.5
left side	5.9	6.1
both sides	8.4	8.6
B4000		
right side		
1994-97	6.3	6.5
1998-04	7.4	7.6
left side		
1994-97	5.9	6.1
1998-04	7.7	7.9
both sides		
1994-97	8.3	8.4
1998-04	11.3	11.5
MPV		
1989-98		
4 cyl.	12.2	12.4
V6		
right side	11.6	11.8
left side	11.9	12.1
both sides	19.8	20.0

MAZ-58 B-SERIES : MPV : NAVAJO : TRIBUTE

	LABOR TIME	SEVERE SERVICE
Cylinder Head, Replace (B)		
2000-05		
right side (12.6)	17.0	17.2
left side (13.2)	17.8	18.0
both sides (16.8)	22.7	22.9
Navajo		
right side	6.3	6.5
left side	5.9	6.1
both sides	8.4	8.6
Tribute		
2.0L, 2.3L	15.9	16.1
3.0L		
right side	18.2	18.4
left side	19.2	19.4
both sides	24.3	24.5
w/AC add		
81-93	.2	.2
94-04		
4 cyl.	.3	.3
3.0L	.5	.5
4.0L	.8	.8
Navajo	.8	.8
w/PS 81-93 add	.2	.2
Cylinder Head, Retorque (B)		
1986-04	1.5	1.7
Cylinder Head Gasket, Replace (B)		
B2000		
1981-84	5.5	5.7
1986-87	6.8	7.0
B2200		
1982-84	7.0	7.2
1987-93	6.8	7.0
B2300		
1994	6.8	7.0
1995-04	7.1	7.3
B2500	6.7	6.9
B2600		
1987-88	6.5	6.7
1989-93	5.5	5.7
B3000		
right side	6.2	6.4
left side	5.5	5.7
both sides	7.9	8.1
B4000		
right side		
1994-97	6.2	6.4
1998-04	7.1	7.3
left side		
1994-97	5.5	5.7
1998-04	7.4	7.6
both sides		
1994-97	7.9	8.1
1998-04	10.8	11.0
MPV		
1989-98		
4 cyl.	5.5	5.7
V6		
one side	7.6	7.8
both sides	8.5	8.7

	LABOR TIME	SEVERE SERVICE
2000-01		
right side	12.6	12.8
left side	14.2	14.4
both sides	15.8	16.0
2002-05		
right side (9.6)	13.0	13.2
left side (10.8)	14.6	14.8
both sides (11.8)	15.9	16.1
Navajo		
right side	6.2	6.4
left side	5.5	5.7
both sides	7.9	8.1
Tribute		
2.0L, 2.3L	11.2	11.4
3.0L		
right side	14.7	14.9
left side	15.7	15.9
both sides	17.9	18.1
w/AC add		
81-93	.2	.4
94-04		
4 cyl.	.3	.5
3.0L	.5	.7
4.0L	.8	1.0
Navajo	.8	1.0
w/PS 81-93 add	.2	.4
Jet Valves, Replace (B)		
B2600		
one	2.2	2.4
all	2.4	2.6
Rocker Arm Cover Gasket, Replace or Reseal (B)		
B2000, B2200, B2600	.5	.7
B2300, B2500		
1994	1.2	1.4
1995-04	1.8	2.0
B3000		
right side	1.7	1.9
left side	1.4	1.6
both sides	2.6	2.8
B4000		
right side	1.8	2.0
left side	2.4	2.6
both sides	3.1	3.3
MPV		
1989-95		
4 cyl.	.7	.9
V6		
right side	1.4	1.6
left side	2.3	2.5
both sides	2.8	3.0
1996-98		
one side	2.3	2.5
both sides	2.7	2.9
2000-01		
right side	2.3	2.5
left side	2.0	2.2
both sides	3.1	3.3

	LABOR TIME	SEVERE SERVICE
2002-05		
right side (1.9)	2.6	2.8
left side (1.0)	1.4	1.6
both sides (2.6)	3.5	3.7
Navajo		
one side	1.8	2.0
both sides	3.0	3.3
Tribute		
2.0L, 2.3L	.6	.8
3.0L		
right side	2.1	2.3
left side	1.5	1.7
both sides	3.2	3.4
w/AC add		
B-Series, Navajo	.3	.5
89-98 MPV	.3	.5
Rocker Arms or Shafts, Replace (B)		
B2000		
1981-84	1.9	2.1
1986-87	2.7	2.8
B2200		
1982-84	2.4	2.6
1987-93	2.7	2.9
B2600		
1987-88	2.4	2.6
1989-93	2.0	2.2
B2300, B2500		
1994	1.8	2.0
1995-04	2.5	2.7
B3000		
right side	2.2	2.4
left side	1.8	2.0
both sides	3.1	3.3
B4000		
right side	2.5	2.7
left side	2.9	3.1
both sides	4.4	4.6
MPV		
1989-95		
4 cyl.	2.3	2.5
V6		
right side	2.3	2.5
left side	2.8	3.0
both sides	3.4	3.6
1996-98		
one side	2.8	3.0
both sides	3.3	3.5
2000-01		
one side	3.8	4.0
both sides	5.4	5.6
2002-05		
right side	4.0	4.2
left side	3.8	4.0
both sides	5.7	5.9
Navajo		
one side	2.5	2.7
both sides	4.2	4.4

B-SERIES : MPV : NAVAJO : TRIBUTE MAZ-59

	LABOR TIME	SEVERE SERVICE
Tribute		
3.0L		
right side	3.2	3.4
left side	3.1	3.3
both sides	5.3	5.5
w/AC add		
B-Series, Navajo	.3	.5
89-98 MPV	.3	.5
Valve Job (A)		
B2000		
1981-84	11.5	11.7
1986-87	13.1	13.3
B2200		
1982-84	10.7	10.9
1987-93	13.1	13.3
B2300, B2500		
right side	7.4	7.6
left side	7.8	8.0
B2600		
1987-88	12.3	12.5
1989-93	13.6	13.8
B3000		
right side	6.9	7.1
left side	6.4	6.6
both sides	8.9	9.1
B4000		
one side	6.4	6.6
both sides	8.9	9.1
MPV		
1989-98		
4 cyl.	13.0	13.2
V6		
right side	12.2	12.4
left side	12.5	12.7
both sides	20.7	20.9
2000-05		
right side	18.2	18.4
left side	19.8	20.0
both sides	24.9	25.1
Navajo		
one side	6.4	6.6
both sides	8.9	9.1
Tribute		
2.0L, 2.3L	15.9	16.1
3.0L		
right side	18.2	19.4
left side	19.2	19.4
both sides	24.3	24.5
w/AC add		
81-93	.2	.4
94-04		
4 cyl.	.3	.3
3.0L	.5	.5
4.0L	.8	.8
Navajo	.8	.8
w/PS 81-93 add	.2	.2
Valve Lash Adjuster, Replace (B)		
1987-93 B2200		
one	2.9	3.1
all	3.1	3.3

	LABOR TIME	SEVERE SERVICE
B2300, B2500	2.6	2.8
B2600		
1987-88		
one	2.0	2.2
all	3.0	3.2
1989-93	2.8	3.0
B3000		
one side	6.4	6.6
both sides	8.3	8.5
B4000		
one side	7.7	7.9
both sides	8.3	8.5
MPV		
1989-98		
4 cyl.	2.9	3.1
V6		
right side	2.8	3.0
left side	3.2	3.4
both sides	4.2	4.4
2000-04		
one side	4.1	4.3
both sides	6.0	6.2
Navajo		
one side	7.8	8.0
both sides	8.3	8.5
Tribute		
3.0L		
right side	3.3	3.5
left side	3.2	3.4
both sides	5.5	5.7
Valve Lifters, Replace (B)		
1982-84 B2200	6.2	6.4
Valve Pushrods, Replace (B)		
1982-84 B2200	2.6	2.8
Valve Springs and/or Seals, Replace (B)		
Cylinder head installed		
B2300, B2500		
1994	3.1	3.3
1995-04	3.8	4.0
B3000		
one side	3.1	3.3
both sides	5.5	5.7
B4000		
right side	3.3	3.5
left side	3.8	4.0
both sides	6.3	6.5
Navajo		
one side	3.0	3.2
both sides	5.2	5.4
Cylinder head removed (includes head R&R)		
B2000		
1981-84	6.8	7.0
1986-87	8.3	8.5
B2200		
1982-84	8.6	8.8
1987-93	8.3	8.5

	LABOR TIME	SEVERE SERVICE
MPV		
1989-98		
4 cyl.	8.6	8.8
V6		
one side	9.8	10.0
both sides	12.6	12.8
2000-01		
right side	14.5	14.7
left side	16.2	16.4
both sides	20.6	20.8
2002-05		
right side (11.1)	15.0	15.2
left side (12.1)	16.3	16.5
both sides (15.3)	20.7	20.9
Tribute		
2.0L, 2.3L	13.9	14.1
3.0L		
right side	16.6	16.8
left side	17.6	17.8
both sides	21.4	21.6
w/AC head installed add		
B-Series, Navajo	.3	.3
89-98 MPV	.3	.3
w/AC 81-93 head removed add	.2	.2
w/PS 81-93 head removed add	.2	.2
CAMSHAFT		
Balance Shaft, Replace (B)		
B2600		
1987-88		
2WD		
AT		
one	9.4	9.6
two	9.9	10.1
MT		
one	9.2	9.4
two	9.3	9.5
4WD		
AT		
one	10.2	10.4
two	10.5	10.7
MT		
one	9.3	9.5
two	9.5	9.7
1989-93		
2WD		
AT		
one	9.2	9.4
two	9.4	9.6
MT		
one	8.3	8.5
two	8.6	8.8
4WD		
AT		
one	9.5	9.7
two	9.8	10.0

MAZ-60 B-SERIES : MPV : NAVAJO : TRIBUTE

	LABOR TIME	SEVERE SERVICE
Balance Shaft, Replace (B)		
MT		
one	8.6	8.8
two	8.9	9.1
1989-95 MPV 4 cyl.	13.1	13.3
w/AC B2600 add	.5	.5
w/PS B2600 add	.5	.5
Camshaft, Replace (A)		
B2000		
1981-84	3.4	3.6
1986-87	3.8	4.0
B2200		
1982-84	6.2	6.4
1987-93	3.8	4.0
B2300		
1994	6.7	6.9
1995-04	5.5	5.7
B2500	3.7	3.9
B2600		
1987-88	3.8	4.0
1989-93	2.3	2.5
B3000	13.7	13.9
B4000	12.9	13.1
MPV		
1989-98		
4 cyl.	2.3	2.5
V6		
one side	7.1	7.3
both sides	8.8	9.0
2000-01		
right side both	8.6	8.8
left side both	9.8	10.0
both sides all	9.9	10.1
2002-05		
right side		
both (6.7)	9.0	9.2
left side		
both (7.6)	10.3	10.5
both sides		
all (7.7)	10.4	10.6
Navajo	13.9	14.1
Tribute		
2.0L, 2.3L		
one	5.1	5.3
both	5.3	5.5
3.0L		
right side	10.9	11.1
left side	11.8	12.0
both sides	12.3	12.5
w/AC add		
81-93	.2	.2
94-04		
2.3L	1.2	1.2
3.0L	.5	.5
4.0L	.8	.8
B2500	.3	.3
Navajo	.8	.8
w/PS 81-93 add	.2	.2

	LABOR TIME	SEVERE SERVICE
Camshaft Seal, Replace (B)		
B2300 one	2.9	3.1
B2500 one	1.9	2.1
B3000, B4000	1.7	1.9
1989-98 MPV		
one	3.6	3.8
both	3.8	4.0
w/AC B2300, B2500 add	1.2	1.4
Camshaft Timing Gear, Replace (B)		
1982-84 B2200	4.2	4.4
Timing Belt, Replace (B)		
B2200	3.1	3.3
B2300		
1994	2.4	2.6
1995-04	2.8	3.0
B2500	1.4	1.6
1989-98 MPV V6	3.4	3.6
Tribute 2.0L, 2.3L	4.4	4.6
w/AC add		
B2200	.2	.2
B2300, B2500		
94	.3	.3
95-04	1.2	1.2
w/PS add		
B2200	.2	.2
95-04 B2300, B2500	.5	.5
Timing Belt Cover Oil Seal, Replace (B)		
1986-93	2.3	2.5
Tribute		
2.0L, 2.3L		
upper or lower	.8	1.0
w/AC add	.2	.2
w/PS add	.2	.2
Timing Belt Tensioner, Replace (B)		
B2200	1.7	1.9
B2300		
1994	2.0	2.2
1995-04	2.4	2.6
B2500	1.6	1.8
1989-98 MPV V6	3.5	3.7
Tribute		
2.0L, 2.3L	4.4	4.6
w/AC add		
B2200	.2	.2
B2300, B2500		
94	.3	.3
95-04	1.2	1.4
w/PS add		
B2200	.2	.2
95-04 B2300, B2500	.5	.5
Timing Belt Tensioner Pulley, Replace (B)		
B2200	3.4	3.6
1989-98 MPV V6	3.5	3.7
w/AC B2200 add	.2	.2
w/PS B2200 add	.2	.2

	LABOR TIME	SEVERE SERVICE
Timing Chain, Replace (B)		
B2000	8.2	8.4
B2200	8.2	8.4
B2600		
1987-88		
2WD	4.2	4.4
4WD	8.3	8.5
1989-93		
2WD	10.0	10.2
4WD	10.5	10.7
B3000	9.2	9.4
B4000		
AT	10.0	10.2
MT	9.7	9.9
MPV		
1989-95 4 cyl.	10.8	11.0
2000-05 (7.5)	10.1	10.3
Navajo		
AT	9.9	10.1
MT	9.8	10.0
Tribute		
3.0L	10.9	11.1
w/AC add		
3.0L	.5	.5
4.0L	.8	.8
89-93	.5	.5
w/PS 89-93 add	.5	.5
Timing Chain Housing Gasket, Replace (B)		
B2000	7.1	7.3
B2200	7.1	7.3
B2600		
1987-88		
2WD	3.6	3.8
4WD	7.7	7.9
1989-93		
2WD	9.4	9.6
4WD	9.8	10.0
MPV		
2000-01	7.6	7.8
2002-05 (5.9)	8.0	8.2
Navajo		
AT	9.0	9.2
MT	8.6	8.8
Tribute		
3.0L	9.6	9.8
w/AC add		
89-93	.5	.5
Navajo	.8	.8
w/PS 89-93 add	.5	.5
Replace cover add	.5	.7
Timing Chain Sprocket, Replace (B)		
1989-94 MPV 4 cyl.		
upper	2.2	2.4
lower	10.8	11.0
2000-05 (6.6)	8.9	9.1

B-SERIES : MPV : NAVAJO : TRIBUTE — MAZ-61

	LABOR TIME	SEVERE SERVICE
Timing Chain Tensioner, Replace (B)		
B2000, B2200	2.2	2.4
B2600		
1987-88		
2WD	4.2	4.4
4WD	8.3	8.5
1989-93	.9	1.1
1994-98 B3000	9.2	9.4
1994-98 B4000		
AT	10.0	10.2
MT	9.7	9.9
MPV		
1989-94 4 cyl.	.8	1.0
2000-01	8.3	8.5
2002-05 (6.5)	8.8	9.0
Navajo	9.7	9.9
Tribute		
3.0L	10.9	11.1
w/AC add		
3.0L	.5	.5
4.0L	.8	.8
89-93	.5	.5
w/PS 89-93 add	.5	.5
Timing Cover and/or Gasket, Replace (B)		
B2200		
one piece	3.1	3.3
two piece		
upper	.7	.9
lower	1.6	1.8
B2300, B2500		
1994	1.8	2.0
1995-04	2.0	2.2
B3000	8.6	8.8
B4000		
AT	9.8	10.0
MT	9.6	9.8
1989-98 MPV		
4 cyl.	9.8	10.0
V6	1.5	1.7
Tribute		
3.0L	9.6	9.8
w/AC add		
95-04 4 cyl.	1.2	1.4
94-04		
3.0L	.5	.5
4.0L	.8	.8
B2200	.2	.2
w/PS B2200 add	.2	.2
Vibration Damper and/or Chain Guide, Replace (B)		
B2000	7.6	7.8
B2200	7.6	7.8
B2600		
2WD	4.2	4.4
4WD	8.3	8.5
w/AC add	.5	.5
w/PS add	.5	.5

	LABOR TIME	SEVERE SERVICE
CRANK & PISTONS		
Connecting Rod Bearings, Replace (A)		
Includes: Check bearing oil clearance.		
B2000, B2200		
1981-84	3.4	3.6
1986-93	4.8	5.0
B2600		
1987-88		
2WD	4.9	5.1
4WD	7.8	8.0
1989-93		
2WD	4.4	4.6
4WD	8.0	8.2
B2300		
AT	9.4	9.6
MT		
2WD	9.7	9.9
4WD	9.9	10.1
B2500		
AT	9.4	9.6
MT	9.7	9.9
B3000		
AT	9.9	10.1
MT	9.8	10.0
B4000		
AT	10.3	10.5
MT	9.9	10.1
MPV		
1989-98		
4 cyl.	4.5	4.7
V6		
2WD	5.9	6.1
4WD	6.2	6.4
2000-01	17.5	17.7
2002-05	18.0	18.2
Navajo		
AT	10.9	11.1
MT	10.5	10.7
Tribute		
AT		
2WD		
2.0L, 2.3L	15.6	15.8
3.0L	17.1	17.3
4WD		
2.0L, 2.3L	16.9	17.1
3.0L	17.9	18.1
MT		
2WD		
2.0L, 2.3L	15.3	15.5
4WD		
2.0L, 2.3L	15.9	16.1
w/AC add		
2.3L, 2.5L 4 cyl.	1.2	1.4
3.0L	.5	.5
4.0L	.8	.8

	LABOR TIME	SEVERE SERVICE
Crankshaft and Main Bearings, Replace (A)		
Includes: Engine R&R, check bearing oil clearance.		
B2000		
1981-84	10.5	10.7
1986-87		
2WD	19.9	20.1
4WD	20.1	20.3
B2200		
1982-84	19.8	20.0
1987-93		
2WD	19.9	20.1
4WD	20.1	20.3
B2300		
AT	12.6	12.8
MT	13.2	13.4
B2500		
AT	9.3	9.5
MT	10.2	10.4
B2600		
1987-88		
2WD		
AT	19.8	20.0
MT	19.4	19.6
4WD		
AT	20.7	20.9
MT	19.8	20.0
1989-93		
2WD		
AT	19.8	20.0
MT	19.4	19.6
4WD		
AT	17.1	17.3
MT	16.4	16.6
B3000		
2WD	13.9	14.1
4WD	13.4	13.6
B4000		
AT	13.9	14.1
MT	13.4	13.6
MPV		
1989-95		
4 cyl.		
AT	20.8	21.0
MT	20.0	20.2
V6		
2WD		
AT	24.2	24.4
MT	23.7	23.9
4WD	24.5	24.7
1996-98		
2WD	23.1	23.3
4WD	23.7	23.9
2000-01	24.5	24.7
2002-05 (18.5)	25.0	25.2
Navajo		
AT	14.3	14.5
MT	13.9	14.1

MAZ-62 B-SERIES : MPV : NAVAJO : TRIBUTE

	LABOR TIME	SEVERE SERVICE
Crankshaft and Main Bearings, Replace (A)		
Tribute		
AT		
2WD		
2.0L, 2.3L	18.2	18.4
3.0L	24.2	24.4
4WD		
2.0L, 2.3L	19.8	20.0
3.0L	24.8	25.0
MT		
2WD		
2.0L, 2.3L	17.8	18.0
4WD		
2.0L, 2.3L	18.5	18.7
w/AC add		
81-93	.5	.5
94-04 4 cyl.	1.2	1.2
94-04		
3.0L	.5	.5
4.0L	.8	.8
Navajo	.8	.8
w/PS 81-93 add	.5	.5
Crankshaft Front Oil Seal, Replace (B)		
B2000	1.5	1.7
B2200	2.5	2.7
B2600		
1987-88	1.2	1.4
1989-93	1.6	1.8
MPV		
1989-98	1.8	2.0
2000-05 (.8)	1.1	1.3
Navajo	1.7	1.9
Tribute	1.2	1.4
Crankshaft Pulley, Replace (B)		
B2000	.8	1.0
B2200, B2600		
1982-84	2.3	2.5
1987-93	.7	.9
B2300	.7	.9
B2500	.5	.7
B3000, B4000	1.3	1.5
MPV		
1989-98	1.6	1.8
2000-05 (.7)	.9	1.1
Navajo		
AT	1.2	1.4
MT	1.8	1.9
Tribute		
2.0L, 2.3L	.8	1.0
3.0L	.9	1.1
w/AC add	.5	.5
Main Bearing Cap Side Seal, Replace (B)		
1981-1984 B2000, B2200	2.9	3.1

	LABOR TIME	SEVERE SERVICE
Main Bearings, Replace (A)		
B2000, B2200	3.7	3.9
B2300	9.9	10.1
B2500	9.7	9.9
B3000	10.0	10.2
B4000		
AT	10.8	11.0
MT	10.3	10.5
MPV		
1989-95		
4 cyl.		
AT	20.8	21.0
MT	20.0	20.2
V6		
2WD		
AT	24.2	24.4
MT	23.7	23.9
4WD	24.5	24.7
1996-98		
2WD	23.1	23.6
4WD	23.7	23.9
2000-01	24.5	24.7
2002-05 (18.5)	25.0	25.2
Navajo		
AT	11.7	11.9
MT	11.3	11.5
Tribute		
AT		
2WD		
2.0L, 2.3L	18.2	18.4
3.0L	24.2	24.4
4WD		
2.0L, 2.3L	19.8	20.0
3.0L	24.8	25.0
MT		
2WD		
2.0L, 2.3L	17.8	18.0
4WD		
2.0L, 2.3L	18.5	18.7
w/AC add		
4 cyl.	1.2	1.4
3.0L	.8	.8
4.0L	.5	.5
Pistons or Connecting Rods, Replace (A)		
Includes: Ridge reaming, cylinder wall deglazing, installing new rings and rod bearings, engine tune-up.		
B2000		
1981-84	10.2	10.4
1986-87	13.1	13.3
B2200		
1982-84	13.0	13.2
1987-93	13.1	13.3
B2600		
1987-88		
2WD	12.2	13.4
4WD	15.3	15.5

	LABOR TIME	SEVERE SERVICE
1989-93		
2WD	11.7	11.9
4WD	15.2	15.4
B2300		
AT	19.8	20.0
MT	20.2	20.4
B2500		
AT	9.9	10.1
MT	20.1	20.3
B3000		
AT	20.5	20.7
MT	20.2	20.4
B4000		
AT	24.2	24.4
MT	23.8	24.0
MPV		
1989-98		
4 cyl.	11.9	12.1
V6		
2WD	16.5	16.7
4WD	17.0	17.2
2000-01	24.8	25.0
2002-05 (18.8)	25.4	25.6
Navajo		
2WD	22.8	23.0
4WD	22.3	22.4
Tribute		
AT		
2WD		
2.0L, 2.3L	20.7	20.9
3.0L	24.4	24.6
4WD		
2.0L, 2.3L	21.8	22.0
3.0L	25.0	25.2
MT		
2WD		
2.0L, 2.3L	20.3	20.5
4WD		
2.0L, 2.3L	21.1	21.3
w/AC add		
81-93	.2	.2
94-04 4 cyl.	.3	.3
94-04		
3.0L	.5	.5
4.0L	.8	.8
Navajo	.8	.8
w/PS 81-93 add	.2	.2
Rear Main Oil Seal, Replace (B)		
Includes: R&R trans. if necessary.		
B2000		
1981-84		
AT	5.6	5.8
MT	4.2	4.4
1986-87		
2WD		
AT	4.2	4.4
MT	4.8	5.0
4WD		
AT	6.0	6.2
MT	7.0	7.2

B-SERIES : MPV : NAVAJO : TRIBUTE — MAZ-63

	LABOR TIME	SEVERE SERVICE
B2200, B2600		
1982-84		
AT	5.6	5.8
MT	3.7	3.9
1987-93		
2WD		
AT	4.2	4.4
MT	4.8	5.0
4WD		
AT	6.0	6.2
MT	5.2	5.4
B2300		
2WD		
AT	3.8	4.0
MT	4.9	5.1
4WD	5.5	5.7
B2500	4.4	4.6
B3000		
1994-97		
2WD		
AT	4.9	5.1
MT	5.2	5.4
4WD	5.5	5.7
1998-04		
2WD		
AT	5.0	5.2
MT	5.2	5.4
4WD		
AT	7.6	7.8
MT	6.7	6.9
B4000		
AT	7.7	7.9
MT	8.0	8.2
MPV		
1989-95		
4 cyl.		
AT	5.3	5.5
MT	4.2	4.4
V6		
2WD		
AT	5.2	5.4
MT	5.4	5.6
4WD		
AT	7.7	7.9
MT	6.3	6.5
1996-98		
2WD	5.4	5.6
4WD	7.8	8.0
2000-01	8.5	8.7
2002-05 (6.7)	9.0	9.2
Navajo		
2WD		
AT	5.8	6.0
MT	5.5	5.7
4WD		
AT	7.0	7.2
MT	6.8	7.0

	LABOR TIME	SEVERE SERVICE
Tribute		
AT		
2WD		
2.0L, 2.3L	9.8	10.0
3.0L	9.0	9.2
4WD		
2.0L, 2.3L	10.7	10.9
3.0L	9.9	10.1
MT		
2WD		
2.0L, 2.3L	6.9	7.1
4WD		
2.0L, 2.3L	7.6	7.8
w/electronic trans.		
86-93 add	.8	.8
w/full carpet add	1.2	1.4
Rings, Replace (A)		
Includes: Ridge reaming, cylinder wall deglazing, installing new rings, engine tune-up.		
B2000		
1981-84	9.3	9.5
1986-87	11.8	12.0
B2200		
1982-84	11.7	11.9
1987-93	11.8	12.0
B2300		
2WD		
one cyl.	13.1	13.3
all cyls.	14.2	14.4
4WD		
one cyl.	13.3	13.5
all cyls.	14.5	14.7
B2500		
one cyl.	9.3	9.5
all cyl.	10.2	10.4
B2600		
1987-88		
2WD	10.7	10.9
4WD	14.2	14.4
1989-93		
2WD	10.3	10.5
4WD	14.1	14.3
B3000		
one cyl.	10.7	10.9
all cyls.	17.0	17.2
B4000		
2WD		
one cyl.	10.7	10.9
all cyls.	17.0	17.2
4WD		
one cyl.	12.3	12.5
all cyls.	18.4	18.6
MPV		
1989-98		
4 cyl.	10.5	10.7
V6		
2WD	14.8	15.0
4WD	15.3	15.5

	LABOR TIME	SEVERE SERVICE
2000-01	24.0	24.2
2002-05 (18.2)	24.6	24.8
Navajo		
one cyl.	12.3	12.5
all cyls.	18.4	18.6
Tribute		
AT		
2WD		
2.0L, 2.3L	20.2	20.4
3.0L	23.9	24.1
4WD		
2.0L, 2.3L	21.3	21.5
3.0L	24.5	24.7
MT		
2WD		
2.0L, 2.3L	19.8	20.0
4WD		
2.0L, 2.3L	20.6	20.8
w/AC add		
81-93	.2	.2
94-04 4 cyl.	.3	.3
94-04		
3.0L	.5	.5
4.0L	.8	.8
Navajo	.8	.8
w/PS 81-93 add	.2	.2
ENGINE LUBRICATION		
Engine Oil Cooler, Replace (B)		
B2200	1.2	1.4
B2300, B2500, B3000, B4000	.7	.9
2000-05 MPV (.8)	1.1	1.3
Navajo	1.4	1.6
Oil Pan and/or Gasket, Replace (B)		
B2000		
1981-84	2.3	2.5
1986-87		
2WD	2.9	3.1
4WD	6.2	6.4
B2200, B2600		
1982-84	2.3	2.5
1987-93		
2WD	2.9	3.1
4WD	6.2	6.4
B2300		
2WD		
AT	8.3	8.5
MT	8.6	8.8
4WD	8.8	9.0
B2500		
AT	8.3	8.5
MT	8.6	8.8
B3000		
2WD		
AT		
1994-95	7.4	7.6
1996-04	9.1	9.3

MAZ-64 B-SERIES : MPV : NAVAJO : TRIBUTE

	LABOR TIME	SEVERE SERVICE
Oil Pan and/or Gasket, Replace (B)		
MT		
1994-95	7.2	7.4
1996-04	8.9	9.1
4WD	8.6	8.8
B4000		
AT	9.1	9.3
MT	8.6	8.8
MPV		
1989-98		
4 cyl.	2.7	2.9
V6		
2WD	2.4	2.6
4WD	2.8	3.0
2000-05 (1.6)	2.2	2.4
Navajo		
AT	9.0	9.2
MT	8.6	8.8
Tribute		
2.0L, 2.3L	2.0	2.4
3.0L	2.6	2.9
w/AC add		
3.0L	.5	.5
B2300, B2500	1.2	1.2
B4000, Navajo	.8	.8
Oil Pressure Gauge (Dash), Replace (B)		
B-Series	.8	.8
1989-98 Navajo	1.4	1.4
Oil Pressure Warning Switch, Replace (B)		
All Models B-Series	.5	.5
MPV	.8	.9
Navajo	.5	.5
Tribute		
2.0L, 2.3L		
2WD	1.1	1.3
4WD	1.3	1.5
3.0L	1.2	1.4
Oil Pump, R&R and Recondition (B)		
B2000, B2200		
1981-84	3.0	3.2
1986-93	4.5	4.7
1994 B2300		
2WD		
AT	8.8	9.0
MT	9.1	9.3
4WD	9.2	9.4
B2600		
1987-88		
2WD	4.8	5.0
4WD	9.1	9.3
1989-93		
2WD	9.8	10.0
4WD	10.0	10.2
1994-04 B4000		
AT	9.4	9.6
MT	9.1	9.3

	LABOR TIME	SEVERE SERVICE
1989-98 MPV		
4 cyl.	10.3	10.5
V6		
2WD	5.0	5.2
4WD	5.4	5.6
Navajo		
AT	9.1	9.3
MT	9.4	9.6
w/AC add		
B2300, B2500	1.2	1.4
B2600	.5	.5
B3000, B4000, Navajo	.8	.8
w/PS B2600 add	.5	.5
Oil Pump, Replace (B)		
B2000, B2200		
1981-84	2.6	2.8
1986-93	4.2	4.4
B2300		
1994		
2WD		
AT	8.4	8.6
MT	8.8	9.0
4WD	8.9	9.1
1995-98	3.2	3.4
B2500	2.7	2.9
B2600		
1987-88		
2WD	4.4	4.6
4WD	8.8	9.0
1989-93		
2WD	9.6	9.8
4WD	9.8	10.0
B3000		
2WD		
AT	7.6	7.8
MT	7.4	7.6
4WD	8.8	9.0
B4000		
AT	9.2	9.4
MT	8.8	9.0
MPV		
1989-98		
4 cyl.	10.0	10.2
V6		
2WD	4.6	4.8
4WD	5.3	5.5
2000-01	11.0	11.2
2002-05 (8.5)	11.5	11.7
Navajo		
AT	9.2	9.4
MT	8.8	9.0
Tribute		
2.0L, 2.3L		
exc. AT 4WD	7.4	7.6
AT 4WD	9.9	10.1

	LABOR TIME	SEVERE SERVICE
3.0L	11.6	11.8
w/AC add		
B2300, B2500	1.2	1.4
B2600	.5	.5
B3000, B4000, Navajo	.8	.8
w/PS B2600 add	.5	.5
Oil Pump Drive Sprocket, Replace (B)		
1981-84 B2000	7.7	7.9
1982-84 B2200	7.7	7.9

CLUTCH

	LABOR TIME	SEVERE SERVICE
Bleed Clutch Hydraulic System (B)		
All Models	.3	.5
Clutch Pedal Free Play, Adjust (B)		
All Models	.3	.5
Clutch Assy., Replace (B)		
B2000, B2200		
1981-84	3.5	3.7
1986-93		
2WD	4.2	4.4
4WD	4.6	4.8
B2300		
2WD	5.0	5.2
4WD	6.8	7.0
B2500	5.0	5.2
B3000		
2WD	5.2	5.4
4WD	6.8	7.0
B4000		
2WD	5.6	5.8
4WD	6.8	7.0
1989-92 MPV		
2WD	3.5	3.7
4WD	5.5	5.7
Navajo		
2WD	5.6	5.8
4WD	6.8	7.0
Tribute		
2WD	6.3	6.5
4WD	7.1	7.3
w/full carpet add	1.2	1.4
Clutch Master Cylinder, R&R and Recondition (B)		
Includes: System bleeding.		
1983-04	1.2	1.4
Clutch Master Cylinder, Replace (B)		
Includes: System bleeding.		
Exc. below	.8	1.0
Tribute	.7	.9
Clutch Release Bearing or Fork, Replace (B)		
B2000, B2200, B2600		
1981-84	3.1	3.3
1986-93		
2WD	3.8	4.0
4WD	4.3	4.5

B-SERIES : MPV : NAVAJO : TRIBUTE   **MAZ-65**

	LABOR TIME	SEVERE SERVICE
B2300		
2WD	4.9	*5.1*
4WD	6.7	*6.9*
B2500	4.9	*5.1*
B3000, B4000, Navajo		
2WD	5.5	*5.7*
4WD	6.7	*6.9*
1989-92 MPV		
2WD	3.1	*3.3*
4WD	5.2	*5.4*
Tribute		
2WD	6.2	*6.4*
4WD	6.9	*7.1*
w/full carpet add	1.2	*1.4*

Clutch Slave Cylinder, R&R and Reconditon (B)
Includes: System bleeding.
| 1983-84 | .9 | *1.1* |
| 1986-93 | .8 | *1.0* |

Clutch Slave Cylinder, Replace (B)
Includes: System bleeding.
Transmission installed
| 1983-93 | .5 | *.7* |
| Tribute | .6 | *.8* |

Transmission removed (includes trans. R&R)
B2300, B2500		
2WD	4.5	*4.7*
4WD	6.3	*6.5*
B3000		
2WD	4.9	*5.1*
4WD	6.3	*6.5*
B4000		
2WD	5.2	*5.4*
4WD	6.3	*6.5*
Navajo		
2WD	4.9	*5.1*
4WD	5.9	*6.1*
w/full carpet transmission removed add	1.2	*1.2*

Flywheel, Replace (B)
B2000		
1981-84	3.8	*4.0*
1986-87		
2WD	4.4	*4.6*
4WD	5.0	*5.2*
B2200, B2600		
1982-84	3.4	*3.6*
1987-93		
2WD	4.4	*4.6*
4WD	5.0	*5.2*
B2300 4WD	5.5	*5.7*
B2500	5.0	*5.2*
B3000, B4000		
2WD	5.2	*5.4*
4WD	6.2	*6.4*
1989-92 MPV		
4 cyl.	3.8	*4.0*
V6	5.9	*6.1*

	LABOR TIME	SEVERE SERVICE
Navajo		
2WD	5.8	*6.0*
4WD	6.5	*6.7*
Tribute		
2WD	6.8	*7.0*
4WD	7.5	*7.7*
w/full carpet add	1.2	*1.4*
Replace pilot bearing add	.2	*.3*

MANUAL TRANSMISSION

Extension Housing and/or Gasket, Replace (B)
B2000		
1981-84	3.0	*3.2*
1986-87		
2WD	3.0	*3.2*
4WD	5.2	*5.4*
B2200, B2600		
1982-84	2.7	*2.9*
1987-93		
2WD	3.0	*3.2*
4WD	5.2	*5.4*
B2300, B2500, B3000, B4000		
2WD	2.0	*2.2*
4WD	3.1	*3.3*
1989-92 MPV		
2WD	3.5	*3.7*
4WD	6.0	*6.2*
Navajo		
2WD	3.2	*3.4*
4WD	3.5	*3.7*

Extension Housing Oil Seal, Replace (B)
B-Series		
1981-93	.9	*1.1*
1994-04	1.4	*1.6*
1989-92 MPV		
2WD	.8	*1.0*
4WD	5.4	*5.6*
Navajo	1.4	*1.6*
w/4WD B2000, B2200, B2600 add	.2	*.2*

Front Transmission Mount or Bushing, Replace (B)
| 1994-04 B-Series | .7 | *.9* |
| Navajo | .7 | *.9* |

Transmission Assy., R&I (B)
B2000		
1981-84	3.2	*3.4*
1986-87		
2WD	3.5	*3.7*
4WD	4.2	*4.4*
B2200, B2600		
1982-84	2.8	*3.0*
1987-93		
2WD	3.5	*3.7*
4WD	4.2	*4.4*

	LABOR TIME	SEVERE SERVICE
B2300, B2500		
2WD	4.5	*4.7*
4WD	6.3	*6.5*
B3000		
2WD	4.9	*5.1*
4WD	6.3	*6.5*
B4000		
2WD	5.2	*5.4*
4WD	6.3	*6.5*
1989-92 MPV		
2WD	3.1	*3.3*
4WD	5.2	*5.4*
Navajo		
2WD	5.2	*5.4*
4WD	6.3	*6.5*
Tribute		
2WD	6.2	*6.4*
4WD	6.9	*7.2*
w/full carpet add	1.2	*1.2*
Replace front cover or seal add	.5	*.5*

Transmission Assy., R&R and Recondition (A)
B2000		
1981-84		
4-Speed	7.3	*7.5*
5-Speed	8.3	*8.5*
1986-87		
4-Speed	8.3	*8.5*
5-Speed	9.3	*9.5*
B2200		
4-Speed	8.3	*8.5*
5-Speed	7.8	*8.0*
B2300, B2500		
2WD	10.0	*10.2*
4WD	12.0	*12.2*
B2600		
2WD	7.4	*7.6*
4WD	9.3	*9.5*
B3000		
2WD	10.3	*10.5*
4WD	12.0	*12.2*
B4000		
2WD	10.8	*11.0*
4WD	12.0	*12.2*
1989-92 MPV		
2WD	6.8	*7.0*
4WD	9.8	*10.0*
Navajo	11.6	*11.8*
Tribute		
2WD	13.9	*14.1*
4WD	14.8	*15.0*
w/full carpet add	1.2	*1.4*

Transmission Mount, Replace (B)
| 1986-04 one | .5 | *.7* |

Transmission Rear Cover or Gasket, R&R (B)
Tribute		
2WD	6.6	*6.8*
4WD	7.4	*7.6*

AUTOMATIC TRANSMISSION

SERVICE TRANSMISSION INSTALLED

	LABOR TIME	SEVERE SERVICE
Diagnose Electronic Transmission (A)		
Exc. Tribute	1.5	1.7
Tribute	1.7	1.9
Performance Test (A)		
B-Series		
1981-93	1.4	1.6
1994-04	1.7	1.9
MPV		
1989-98	1.6	1.8
2000-05 (1.0)	1.4	1.6
Navajo	1.4	1.6
Tribute	1.6	1.8
Check Unit for Oil Leaks (C)		
All Models	.9	.9
Drain and Refill Unit (B)		
B-Series		
1981-93	.8	1.0
1994-04	1.2	1.4
MPV		
1989-98	1.0	1.2
2000-05 (1.0)	1.4	1.6
Navajo	1.4	1.6
Oil Pressure Check (B)		
All Models	1.0	1.2
Downshift Solenoid, Replace (B)		
B-Series		
1981-84	.9	1.1
1987-93	.7	.9
1994-04	1.6	1.8
1989-95 MPV	.7	.9
Electronic Control Unit, Replace (B)		
MPV	.8	.9
Extension Housing Rear Oil Seal, Replace (B)		
MPV		
1989-93		
2WD	.7	.9
4WD	6.5	6.7
1994-98		
2WD	.7	.9
4WD	4.2	4.4
Extension Housing Gasket, Replace (B)		
B-Series		
1981-84	4.8	5.0
1986-87		
2WD	3.2	3.4
4WD	7.0	7.2
1994-04		
2WD	3.2	3.4
4WD	4.1	4.3

	LABOR TIME	SEVERE SERVICE
MPV		
1989-95	2.6	2.8
1996-98		
2WD	2.6	2.8
4WD	4.6	4.8
Navajo		
2WD	3.2	3.4
4WD	4.1	4.3
Governor Assy., R&R or Replace (B)		
1994 B-Series		
2WD	2.7	2.9
4WD	3.8	4.0
Navajo		
2WD	2.0	2.2
4WD	3.5	3.7
Inhibitor Switch, Replace (B)		
B-Series		
1981-93	.7	.9
1994-04	.5	.9
Navajo	.5	.7
Kickdown Switch, Replace (B)		
1987-93 B-Series	.5	.7
1989-95 MPV	.5	.7
Oil Pan and/or Gasket, Replace (B)		
B-Series		
1981-93	.8	1.0
1994-04	1.2	1.4
MPV		
1989-98	.8	1.0
2000-05 (1.0)	1.4	1.6
Navajo	1.4	1.6
Overdrive or Hold Switch, Replace (B)		
1989-93	.5	.7
Overdrive Solenoid, Replace (B)		
1989-95 MPV	.7	.9
Speed Output Sensor, Replace (B)		
MPV		
1989-98	.8	1.0
2002-04 intermediate sensor	11.0	11.2
Tribute	.7	.9
Transmission Oil Cooler, Replace (B)		
1994-04 B-Series	.5	.7
2000-05 MPV		
air cooled (1.3)	1.8	2.0
water cooled (2.6)	3.5	3.7
Navajo	.5	.7
Transmission Speed Sensor, Replace (B)		
1990-04 B-Series	.5	.7
MPV		
1989-98	.5	.7
2000-05 (.8)	1.1	1.3
Navajo	.5	.7
Tribute	.9	1.1
Vacuum Control Unit, Replace (B)		
1981-95	.5	.7

	LABOR TIME	SEVERE SERVICE
Vacuum Diaphragm, Replace (B)		
B-Series		
1981-93	.8	1.0
1994	2.0	2.2
1989-95 MPV	.7	.9
Navajo	2.3	2.5
Valve Body Assy., R&R and Recondition (A)		
1994-04 B-Series	5.0	5.2
Navajo	4.3	4.5
Valve Body Assy., Replace (B)		
B-Series		
1981-93		
electronic	5.0	5.2
hydraulic	1.9	2.1
1994-04	2.4	2.6
MPV		
1989-98	1.6	1.8
2000-05	4.9	5.1
Navajo	2.8	3.0
Tribute	4.1	4.3

SERVICE TRANSMISSION REMOVED

Transmission R&R included unless otherwise noted.

	LABOR TIME	SEVERE SERVICE
Extension Housing Gasket, Replace (B)		
1989-93 MPV 4WD	7.6	7.8
Flywheel (Flexplate), Replace (B)		
B2000		
1981-84	5.3	5.5
1986-87		
2WD	3.8	4.0
4WD	5.6	5.8
B2200, B2600		
1982-84	5.5	5.7
1987-93		
2WD	3.8	4.0
4WD	5.6	5.8
B2300, B2500	5.0	5.2
B3000		
2WD	5.0	5.2
4WD	7.6	7.8
B4000		
2WD	6.3	6.5
4WD	8.3	8.5
MPV		
1989-99		
4 cyl.	3.8	4.0
V6	5.9	6.1
2000-05 (6.0)	8.1	8.3
Navajo		
2WD	6.8	7.0
4WD	7.1	7.3
Tribute		
2.0L, 2.3L		
2WD	9.6	9.8
4WD	10.6	10.8

B-SERIES : MPV : NAVAJO : TRIBUTE   MAZ-67

	LABOR TIME	SEVERE SERVICE
3.0L		
2WD	8.9	9.1
4WD	9.8	10.0
w/electronic trans.		
86-93 add	.8	1.0
w/full carpet add	1.2	1.4
Replace pilot bearing add	.2	.4
Front Oil Pump, Replace (B)		
B2000, B2200, B2600		
2WD	5.3	5.5
4WD	7.0	7.2
B2300, B2500	4.9	5.1
B3000		
2WD	4.9	5.1
4WD	7.4	7.6
B4000		
2WD	6.8	7.0
4WD	8.3	8.5
MPV		
1989-98		
4 cyl.	5.8	6.0
V6		
2WD	6.0	6.2
4WD	7.6	7.8
2000-01	10.2	10.4
2002-05	11.6	11.8
Navajo		
2WD	6.8	7.0
4WD	8.3	8.5
Tribute		
2.0L, 2.3L		
2WD	12.3	12.5
4WD	13.2	13.4
3.0L		
2WD	11.6	11.8
4WD	12.6	12.8
w/electronic trans.		
81-93 add	.7	.9
Flush oil cooler and lines add	.5	.7
Front Pump Oil Seal, Replace (B)		
B2000, B2200, B2600		
1981-84	4.9	5.1
1986-93		
2WD	5.3	5.5
4WD	6.4	6.6
B2300, B2500	4.9	5.1
B3000		
2WD	4.9	5.1
4WD	7.4	7.6
B4000		
2WD	6.8	7.0
4WD	8.3	8.5
MPV		
4 cyl.		
1989-98	5.0	5.2
V6		
2WD	5.2	5.4
4WD	7.1	7.3

	LABOR TIME	SEVERE SERVICE
2000-01	9.4	9.6
2002-05 (7.9)	10.7	10.9
Navajo		
2WD	6.8	7.0
4WD	8.3	8.5
w/electronic trans.		
81-93 add	1.6	1.8
Flush oil cooler and lines add	.5	.5
Governor Assy., Replace or Recondition (B)		
B2000, B2200, B2600		
1981-84	5.0	5.2
1986-93		
2WD	4.4	4.6
4WD	7.3	7.5
1989-95 MPV	5.3	5.5
Torque Converter, Replace (B)		
B2000, B2200, B2600		
1981-84	4.8	5.0
1986-93		
2WD	3.5	3.7
4WD	5.3	5.5
B2300, B2500	4.5	4.7
B3000		
2WD	4.5	4.7
4WD	7.1	7.3
B4000		
2WD	6.4	6.6
4WD	7.9	8.1
MPV		
1989-98		
4 cyl.	4.2	4.4
V6		
2WD	4.8	5.0
4WD	7.0	7.2
2000-01	9.4	9.6
2002-05 (7.6)	10.3	10.5
Navajo		
2WD	5.9	6.1
4WD	7.3	7.5
Tribute		
2.0L, 2.3L		
2WD	9.9	10.1
4WD	10.9	11.1
3.0L		
2WD	9.2	9.4
4WD	10.1	10.3
w/electronic trans.		
81-93 add	1.6	1.8
Flush oil cooler and lines add	.5	.5
Transmission Assy., R&I (B)		
B2000, B2200, B2600		
1981-84	4.3	4.5
1986-93		
2WD	3.5	3.7
4WD	5.3	5.5
B2300, B2500	4.3	4.5

	LABOR TIME	SEVERE SERVICE
B3000		
2WD	4.3	4.5
4WD	7.0	7.2
B4000		
2WD	6.3	6.5
4WD	7.7	7.9
MPV		
1989-98		
4 cyl.	4.2	4.4
V6		
2WD	4.8	5.0
4WD	6.5	6.7
2000-05 (6.8)	9.2	9.4
Navajo		
2WD	6.3	6.5
4WD	7.7	7.9
Tribute		
2.0L, 2.3L		
2WD	9.9	10.1
4WD	10.9	11.1
3.0L		
2WD	9.2	9.4
4WD	10.1	10.3
w/electronic trans.		
81-93 add	1.6	1.8
Flush oil cooler and lines add	.5	.7
Replace assembly add	1.2	1.4
Transmission Assy., R&R and Recondition (A)		
B2000, B2200, B2600		
1981-84	13.6	13.8
1986-93		
2WD	10.2	10.4
4WD	13.1	13.3
B2300, B2500		
1994	10.3	10.5
1995-04	11.4	11.6
B3000		
1994		
2WD	10.3	10.5
4WD	13.2	13.4
1995-04		
2WD	11.4	11.6
4WD	14.1	14.3
B4000		
1994		
2WD	12.6	12.8
4WD	13.9	14.1
1995-04		
2WD	13.4	13.6
4WD	14.9	15.1
MPV		
1989-98		
4 cyl.	11.4	11.6
V6		
2WD	13.3	13.5
4WD	16.9	17.1
2000-01	23.9	24.1
2002-04	21.3	21.5

MAZ-68 B-SERIES : MPV : NAVAJO : TRIBUTE

	LABOR TIME	SEVERE SERVICE
Transmission Assy., R&R and Reconditioning (A)		
Navajo		
2WD	16.9	17.1
4WD	18.3	18.5
Tribute		
2.0L, 2.3L		
2WD	17.0	17.2
4WD	18.2	18.4
3.0L		
2WD	16.4	16.6
4WD	17.3	17.5
w/electronic trans.		
81-93 add	3.8	4.0
Flush oil cooler and lines add	.5	.7

TRANSFER CASE

	LABOR TIME	SEVERE SERVICE
Control Unit, Replace (B)		
1994 B-Series	1.2	1.4
Navajo	1.9	2.1
Rear Output Shaft Seal, Replace (B)		
Navajo	1.5	1.7
Transfer Case Rear Shaft Seal, Replace (B)		
B2600	.8	1.0
Transfer Case, R&R and Reconditioning (A)		
B-Series		
1982-93		
AT	8.3	8.5
MT	7.1	7.2
1994-04	7.7	7.9
MPV		
1989-93		
AT	9.4	9.6
MT	8.0	8.2
1994-97	7.1	7.3
Navajo	8.2	8.4
Tribute		
MT	6.3	6.5
Transfer Case, R&R or Replace (B)		
B-Series		
1982-93		
AT	5.9	6.1
MT	4.8	5.0
1994-04	3.0	3.2
MPV		
1989-93		
AT	6.5	6.7
MT	5.2	5.4
1994-97	4.4	4.6
Navajo	5.9	6.1
Tribute		
MT	2.1	2.3
AT		
2.0L, 2.3L	3.6	3.8
3.0L	5.3	5.5

	LABOR TIME	SEVERE SERVICE
Transfer Case Chain Cover, R&R or Replace (B)		
B-Series		
1982-93	2.8	3.0
1994-04	2.9	3.1
1989-97 MPV	3.1	3.3
Navajo	5.9	6.1
Transfer Case Oil Seal, Replace (B)		
1994-04 B-Series	1.4	1.6
Navajo	5.9	6.1
Transfer Companion Flange Seal, R&R (B)		
Tribute	1.7	1.9

ELECTRONIC SHIFT

The following operations do not include testing. Add time as required.

	LABOR TIME	SEVERE SERVICE
Electronic Shift Control Circuit Test (B)		
Navajo	.7	.9
Electronic Shift Transfer Case Coil, Replace (B)		
Navajo	3.8	4.0
Electronic Shift Control Module, Replace (B)		
Navajo	.3	.5
Electronic Shift Control Motor, Replace (B)		
Navajo	.7	.9
Electronic Shift Control Speed Sensor, Replace (B)		
Navajo	.5	.7
Electronic Shift Control Switch, Replace (B)		
Navajo	.5	.7

SHIFT LINKAGE
AUTOMATIC TRANSMISSION

	LABOR TIME	SEVERE SERVICE
Shift Linkage, Adjust (B)		
All Models	.5	.7
Gear Selector Lever, Replace (B)		
B-Series		
1981-88	.8	1.0
1994-04	.5	.7
MPV		
1989-95	1.4	1.6
1996-05	1.2	1.4
Navajo	.5	.7
Tribute	1.4	1.6
Gearshift Control Cable or Rod Replace (B)		
B-Series	1.2	1.4
MPV		
1989-98	.8	1.0
2000-05	1.7	1.9
Navajo	1.2	1.4
Tribute	1.2	1.4
Lower Selector Rod, Replace (B)		
1981-84 B-Series	1.2	1.4

	LABOR TIME	SEVERE SERVICE
MANUAL TRANSMISSION		
Gearshift Lever, Replace (B)		
B-Series	.5	.7
1989-92 MPV	.5	.7
Navajo	.5	.7
Tribute	.7	.9
Gearshift Lever Housing or Base, Replace (B)		
1981-84	.7	.9
1986-93		
2WD	.7	.9
4WD	4.2	4.4

DRIVELINE
DRIVESHAFT

	LABOR TIME	SEVERE SERVICE
Center Bearing or Seals, Replace (B)		
1981-93 B-Series	1.8	2.0
MPV	1.2	1.4
w/4WD B-Series add	.2	.4
Driveshaft, R&R or Replace (B)		
B-Series		
1981-93		
front	.5	.7
rear	.9	1.1
1994-04 front or rear	.7	.9
1989-98 MPV		
2WD	.7	.9
4WD front or rear	.8	1.0
Navajo front or rear	.7	.9
Tribute	1.2	1.3
U-Joint, Replace (B)		
B-Series		
1981-84		
one	1.4	1.6
two	1.7	1.9
three	2.2	2.4
1986-93		
one	1.5	1.7
three	2.2	2.4
1994-04		
front		
one	1.3	1.5
both	1.6	1.8
rear		
one	1.2	1.4
both	1.4	1.6
1989-98 MPV		
one	1.2	1.4
two	1.5	1.7
Navajo		
front		
one	1.3	1.5
both	1.6	1.8
rear		
one	1.2	1.4
both	1.4	1.6

B-SERIES : MPV : NAVAJO : TRIBUTE   MAZ-69

	LABOR TIME	SEVERE SERVICE
DRIVE AXLE		
Differential, Drain & Refill (B)		
All Models rear axle	.7	*.9*
Differential Backlash, Adjust (B)		
Front axle		
Navajo	4.6	*4.8*
Actuator, Replace (B)		
B-Series	.7	*.7*
1989-98 MPV	.8	*.8*
Navajo	.7	*.7*
Axle Housing, Replace (B)		
Front axle		
1987-93 B-Series	4.5	*4.7*
Rear axle		
B-Series		
1981-84	5.8	*6.0*
1986-93	4.5	*4.7*
1994-04	6.4	*6.6*
MPV		
1989-93	4.5	*4.7*
1994-98	4.8	*5.0*
Navajo	6.4	*6.6*
Axle Shaft, Replace (B)		
Front axle		
B-Series		
1987-93		
one side	1.5	*1.7*
both sides	2.3	*2.5*
1994-04		
one side	1.7	*1.9*
both sides	2.7	*2.9*
MPV		
1989-98		
one side	1.4	*1.6*
both sides	2.5	*2.7*
2000-05		
right side	2.1	*2.3*
left side	1.8	*2.0*
both sides	3.2	*3.4*
Navajo		
one side	1.3	*1.5*
both sides	2.9	*3.1*
Tribute		
right side	1.5	*1.7*
left side	1.6	*1.8*
both sides	2.6	*2.8*
Rear axle		
B-Series		
1981-84		
one side	1.9	*2.1*
both sides	3.1	*3.3*
1986-93		
one side	1.8	*2.0*
both sides	2.9	*3.1*
1994-04		
one side	1.3	*1.5*
both sides	1.6	*1.8*

	LABOR TIME	SEVERE SERVICE
MPV		
1989-93		
one side	1.8	*2.0*
both sides	2.8	*3.0*
1994-98		
one side	2.0	*2.2*
both sides	3.3	*3.5*
Navajo		
one side	1.3	*1.5*
both sides	1.6	*1.8*
Tribute		
one side	3.1	*3.3*
both sides	4.7	*4.9*
Replace U-joints		
add each	.3	*.5*
Axle Shaft Bearings, Replace (B)		
Rear axle		
B-Series		
1981-84		
one side	2.0	*2.2*
both sides	3.4	*3.6*
1986-93		
one side	1.8	*2.0*
both sides	3.1	*3.3*
1994-04		
one side	1.3	*1.5*
both sides	1.6	*1.8*
MPV		
1989-93		
one side	1.7	*1.9*
both sides	2.8	*3.0*
1994-98		
one side	2.0	*2.2*
both sides	3.3	*3.5*
2000-05		
one side (1.6)	2.2	*2.4*
both sides (2.0)	2.7	*2.9*
Navajo		
one side	1.3	*1.5*
both sides	1.6	*1.8*
Axle Shaft Oil Seal, Replace (B)		
Front axle		
1987-93 B-Series		
each	.8	*1.0*
2000-05 MPV		
one side	1.2	*1.4*
both sides	2.3	*2.5*
Rear axle		
B-Series		
1981-84		
one side	1.6	*1.8*
both sides	2.8	*3.0*
1986-93		
one side	1.4	*1.6*
both sides	2.5	*2.7*
1994-04		
one side	1.2	*1.4*
both sides	1.4	*1.6*

	LABOR TIME	SEVERE SERVICE
MPV		
1989-93		
one side	1.6	*1.8*
both sides	2.6	*2.8*
1994-98		
one side	2.0	*2.2*
both sides	3.1	*3.3*
2000-05		
one side (1.4)	1.9	*2.1*
both sides (1.9)	2.6	*2.8*
Navajo		
one side	1.2	*1.4*
both sides	1.5	*1.7*
CV Joint, Replace (B)		
Front axle		
1998-04 B-Series		
one side	2.0	*2.2*
both sides	3.8	*4.0*
MPV		
1989-98		
one side	1.9	*2.1*
both sides	3.3	*3.5*
2000-05		
inner		
right side	2.5	*2.7*
left side	2.0	*2.2*
both sides	4.0	*4.2*
outer		
right side	3.3	*3.5*
left side	3.0	*3.2*
both sides	5.0	*5.2*
CV-Joint Boot, Replace (B)		
Front axle		
MPV		
1989-98		
one side	1.9	*2.1*
both sides	3.0	*3.2*
2000-05		
one side	3.2	*3.4*
both sides	4.9	*5.1*
Tribute		
one side		
split type	.8	*1.0*
standard type	2.8	*3.0*
both sides		
split type	1.1	*1.3*
standard type	4.4	*4.6*
Rear axle		
Tribute		
one side	4.3	*4.5*
both sides	6.5	*6.7*
Differential Carrier Assy., R&R and Recondition (A)		
Front axle		
B-Series		
1987-93	8.3	*8.5*
1994-04	8.0	*8.2*
1989-98 MPV	6.2	*6.4*
Navajo	8.0	*8.2*

MAZ-70 B-SERIES : MPV : NAVAJO : TRIBUTE

	LABOR TIME	SEVERE SERVICE
Differential Carrier Assy., R&R and Reconditioning (A)		
Rear axle		
B-Series		
1981-84	6.2	6.4
1986-93	5.3	5.5
1994-04	6.8	7.0
1989-97 MPV	5.2	5.4
Navajo	6.8	7.0
Tribute	11.3	11.5
Replace differential bearings add	.5	.8
Differential Carrier, R&R or Replace (B)		
Front axle		
B-Series		
1987-88	6.7	6.9
1989-93	7.0	7.2
1994-04	4.4	4.6
MPV		
1989-98	4.8	5.0
2000-05 (9.4)	12.7	12.9
Navajo	3.9	4.1
Rear axle		
exc. below	3.1	3.3
Tribute	7.8	8.0
Differential Side Gear Set, Replace (B)		
Front axle		
B-Series		
1987-93	6.4	6.6
1994-04	4.8	5.0
1989-98 MPV	4.8	5.0
Rear axle		
B-Series		
1987-93	4.2	4.4
1994-04	3.6	3.8
1989-98 MPV	3.8	4.0
Front Drive Shaft, Replace (B)		
Navajo	.7	.9
Recondition U-joint add		
one	.5	.7
all	.7	.9
Locking Hub, Replace (B)		
Navajo		
one side	.8	1.0
both sides	1.5	1.7
Output Shaft, Replace (B)		
1989-98 MPV	2.8	3.0
Pinion Bearings, Replace (B)		
Front axle		
B-Series		
1987-93	7.6	7.8
1994-04	5.8	6.0
1989-98 MPV	5.3	5.5
Navajo	5.8	6.0
Rear axle		
B-Series	4.4	4.6
MPV	4.4	4.6
Navajo	4.2	4.4

	LABOR TIME	SEVERE SERVICE
Pinion Shaft Oil Seal, Replace (B)		
Front axle		
1987-04 B-Series	1.2	1.4
MPV	1.2	1.4
Navajo	1.2	1.4
All Models rear axle	1.2	1.4
Remote Control Switch, Replace (B)		
Navajo	.5	.7
Ring Gear & Pinion Set, Replace (B)		
Front axle		
B-Series		
1987-88	7.4	7.6
1989-93	7.7	7.9
1994-04	6.2	6.4
1989-98 MPV	5.2	5.4
Navajo	6.2	6.4
Rear axle		
B-Series		
1981-93	8.0	8.2
1994-04	4.6	4.8
1989-98 MPV	4.8	5.0
Navajo	4.8	5.0
Tribute	11.0	11.2
U-Joint, Replace or Recondition (B)		
1994-04 front axle		
right side		
one	2.0	2.2
both	2.5	2.7
left side	2.0	2.2
all	3.8	4.0

BRAKES
ANTI-LOCK

The following operations do not include testing. Add time as required.

	LABOR TIME	SEVERE SERVICE
Diagnose Anti-Lock Brake System (A)		
1990-05	1.2	1.4
Anti-Lock Relay, Replace (B)		
1990-95 MPV	.7	.9
Control Module, Replace (B)		
1994-04 B-Series	.7	.9
Navajo	.5	.7
Tribute	1.5	1.7
Front Sensor Assy., Replace (B)		
1994-04 B-Series		
one side	.7	.9
both sides	1.2	1.4
MPV		
1996-98		
one side	.5	.7
both sides	.7	.9
2000-05		
one side (.8)	1.1	1.3
both sides (1.0)	1.4	1.6
1993-94 Navajo		
one side	.7	.9
both sides	1.2	1.4

	LABOR TIME	SEVERE SERVICE
Tribute		
one side	.9	1.1
both sides	1.1	1.3
Hydraulic Unit Assy., Replace (B)		
B2300, B2500, B3000, B4000		
2 wheel ABS	.5	.7
4 wheel ABS	.7	.9
1990-93 B2600	1.4	1.6
MPV		
1990-95	1.4	1.6
1996-98	1.8	2.0
2000-05	2.8	3.0
Navajo		
1991-92	.8	1.0
1993-94	2.0	2.2
Tribute	2.8	3.0
Rear Sensor Assy., Replace (B)		
B2300, B2500, B3000, B4000	.5	.7
1990-93 B2600 each	.5	.7
MPV		
1990-95 each	.5	.7
1996-98	.7	.9
2000-05		
one side (1.0)	1.4	1.6
both sides (1.3)	1.8	2.0
Navajo each	.5	.7
Tribute		
one side	.9	1.1
both sides	1.1	1.3

SYSTEM

	LABOR TIME	SEVERE SERVICE
Bleed Brakes (B)		
Includes: Add fluid.		
All Models	1.1	1.3
Brakes, Adjust (B)		
Includes: Refill master cylinder.		
All Models	.8	1.0
Brake Hose (Flexible), Replace (B)		
Includes: System bleeding.		
One	1.1	1.3
each addl. add	.5	.7
Brake Proportioning Valve, Replace (B)		
1981-93	1.3	1.5
2002-04	2.8	3.0
2000-05 MPV	2.0	2.2
Tribute	1.7	1.9
Master Cylinder, R&R and Recondition (B)		
Includes: System bleeding.		
B-Series		
1981-93	2.0	2.2
1994-04	1.2	1.4
1989-98 MPV	1.7	1.9
Navajo	1.2	1.4

B-SERIES : MPV : NAVAJO : TRIBUTE   **MAZ-71**

	LABOR TIME	SEVERE SERVICE
Master Cylinder, Replace (B)		
Includes: System bleeding.		
B-Series		
1981-93	1.5	1.7
1994-04	.7	.9
MPV	1.4	1.6
Navajo	.7	.9
Tribute	1.7	1.9
Power Booster Unit, Replace (B)		
B-Series		
1981-93	1.5	1.7
1994-04	.8	1.0
MPV		
1989-98		
4 cyl.	1.7	1.9
V6	1.9	2.1
2000-05 (1.9)	2.6	2.8
Navajo	.8	1.0
Tribute		
AT	1.0	1.2
MT	1.1	1.3
Power Booster Vacuum Check Valve, Replace (B)		
All Models	.3	.5
SERVICE BRAKES		
Brake Drum, Replace (B)		
B-Series		
one side	.7	.9
both sides	.9	1.1
MPV		
one side	.5	.7
both sides	.9	1.1
Navajo		
one side	.7	.9
both sides	1.2	1.4
Tribute		
one side	.5	.7
both sides	.6	.8
Caliper Assy., R&R and Recondition (B)		
Includes: System bleeding.		
B-Series		
1981-93		
one side	1.6	1.8
both sides	2.6	2.8
1994-04		
one side	1.3	1.5
both sides	1.7	2.0
MPV		
1989-98		
front		
one side	1.3	1.5
both sides	2.0	2.2
rear		
one side	1.8	2.0
both sides	2.5	2.7
2000-05		
one side	1.8	2.0
both sides	2.0	2.2

	LABOR TIME	SEVERE SERVICE
Navajo		
one side	1.7	1.9
both sides	2.2	2.4
Caliper Assy., Replace (B)		
Includes: System bleeding.		
B-Series		
1981-93		
one side	1.2	1.4
both sides	1.7	1.9
1994-04		
one side	.8	1.0
both sides	1.3	1.5
MPV		
1989-98		
front		
one side	.7	.9
both sides	1.2	1.4
rear		
one side	1.4	1.6
both sides	1.6	1.8
2000-05		
one side	1.1	1.3
both sides	1.3	1.5
Navajo		
one	1.4	1.6
both	1.7	1.9
Tribute		
one	1.2	1.4
both	1.5	1.7
Disc Brake Rotor, Replace (B)		
B-Series		
1981-84		
one side	1.6	1.8
both sides	2.6	2.8
1986-93		
2WD		
one side	.9	1.1
both sides	1.9	2.1
4WD		
one side	1.4	1.6
both sides	2.5	2.7
1994-04		
2WD		
one side	.7	.9
both sides	1.3	1.5
4WD		
one side	1.2	1.4
both sides	1.7	1.9
MPV		
1989-98		
front		
one side	.8	1.0
both sides	1.5	1.7
rear		
one side	1.2	1.4
both sides	2.0	2.2
2000-05		
one side (.4)	.6	.8
both sides (.8)	1.1	1.3

	LABOR TIME	SEVERE SERVICE
Navajo		
2WD		
one side	.8	1.0
both sides	1.4	1.6
4WD		
one side	1.3	1.5
both sides	1.8	2.0
Tribute		
one	.7	.9
both	.9	1.1
Pads and/or Shoes, Replace (B)		
Includes: Adjust service and parking brake. System bleeding.		
B-Series		
1981-93		
front disc	.7	.9
rear drum	1.3	1.5
four wheels	1.8	2.0
1994-04		
front disc	1.2	1.4
rear drum	1.6	1.8
four wheels	2.7	2.9
MPV		
1989-98		
front disc	.8	1.0
rear disc	1.0	1.2
rear drum	1.3	1.5
four wheels		
disc brakes	1.7	1.9
drum brakes	2.0	2.2
2000-05		
front disc	1.0	1.2
rear drum	1.3	1.5
four wheels	2.5	2.7
Navajo		
front disc	1.2	1.4
rear drum	1.6	1.8
four wheels	2.7	2.9
Tribute		
front disc	.9	1.1
rear drum	1.2	1.4
COMBINATION ADD-ONS		
Repack wheel bearings		
two wheels add	.7	.7
Replace		
brake drum add	.1	.1
brake hose add	.3	.3
caliper add	.3	.3
disc rotor add	.3	.3
wheel cylinder add	.2	.2
Resurface		
brake drum add	.5	.5
brake rotor add	.5	.5

MAZ-72 B-SERIES : MPV : NAVAJO : TRIBUTE

	LABOR TIME	SEVERE SERVICE
Wheel Cylinder, R&R and Rebuild (B)		
Includes: System bleeding.		
B-Series		
1981-93		
one side	1.4	1.6
both side	2.6	2.8
1994-04		
one side	1.4	1.6
both sides	2.2	2.4
1989-93 MPV		
one side	1.9	2.1
both sides	3.1	3.3
Navajo		
one side	1.7	1.9
both sides	2.2	2.4
Wheel Cylinder, Replace (B)		
Includes: System bleeding.		
B-Series		
1981-84		
one side	1.2	1.4
both sides	2.2	2.4
1986-93		
one side	1.4	1.6
both sides	2.6	2.8
1994-04		
one side	1.4	1.6
both sides	2.2	2.4
MPV		
1989-93		
one side	1.7	1.9
both sides	2.7	2.9
2000-05		
one side	1.6	1.8
both sides	1.9	2.1
Navajo		
one side	1.4	1.6
both sides	2.2	2.4
Tribute		
one side	1.5	1.6
both sides	1.7	1.8

PARKING BRAKE

	LABOR TIME	SEVERE SERVICE
Parking Brake Cable, Adjust (C)		
All Models	.5	.7
Parking Brake Apply Actuator, Replace (B)		
B-Series	.7	.9
MPV		
1989-98	.5	.7
2000-04	1.7	1.9
Navajo	.7	.9
Parking Brake Apply Warning Indicator Switch, Replace (B)		
B-Series	.5	.7
MPV		
1989-98	.3	.5
2000-05	1.6	1.8
Navajo, Tribute	.5	.7

	LABOR TIME	SEVERE SERVICE
Parking Brake Cable, Replace (B)		
B-Series		
1981-93		
front	1.2	1.4
rear		
one side	1.3	1.5
both sides	1.9	2.1
1994-04		
front	.8	1.0
rear		
one side	.7	.9
both sides	1.2	1.4
MPV		
1989-98		
front	.9	1.1
rear		
one side	1.5	1.7
both sides	2.3	2.5
2000-05		
one side (.8)	1.1	1.3
both sides (1.0)	1.4	1.6
Navajo		
front	.9	1.1
rear		
one side	.7	.9
both sides	1.2	1.4
Tribute		
front	1.2	1.4
rear		
one side	1.1	1.3
both sides	1.5	1.7

FRONT SUSPENSION

Unless otherwise noted, time given does not include alignment.

	LABOR TIME	SEVERE SERVICE
Align Front End (A)		
B-Series, MPV	1.7	1.9
Navajo	2.8	3.0
Tribute	2.1	2.32
Front Toe, Adjust (B)		
All Models	.7	.9
Ball Joint, Replace (B)		
B-Series		
1981-93		
one	.9	1.1
both one side	1.8	2.0
all both sides	3.2	3.4
1994-04		
2WD		
one side	2.6	2.8
both sides	4.9	5.1
4WD		
one side	3.7	3.9
both sides	6.3	6.5
Navajo		
2WD		
one side	3.0	3.2
both sides	5.2	5.4

	LABOR TIME	SEVERE SERVICE
4WD		
one side	4.2	4.4
both sides	6.8	7.0
Coil Spring, Replace (B)		
B-Series		
1981-93		
one side	1.6	1.8
both sides	2.9	3.1
1994-97		
one side	.8	1.0
both sides	1.3	1.5
1998-04		
2WD		
one side	1.4	1.6
both sides	2.5	2.7
4WD		
one side	2.1	2.3
both sides	2.6	2.8
MPV		
one side	1.5	1.7
both sides	2.3	2.5
Navajo, Tribute		
one side	1.3	1.5
both sides	1.7	1.9
Free Wheeling Hub, Replace (B)		
1994-97 B-Series		
one side	.8	1.0
both sides	1.5	1.7
Front Axle Arm, Replace (B)		
1994-97 B-Series		
2WD		
one side	3.7	3.9
both sides	5.9	6.1
4WD		
right side	3.9	4.1
left side	5.0	5.2
both sides	7.2	7.4
Navajo		
2WD		
one side	3.8	4.0
both sides	6.2	6.4
4WD		
right side	4.3	4.5
left side	5.3	5.5
both sides	7.6	7.8
Replace bushings add each side	.5	.7
Front Shock Absorbers, Replace (B)		
B-Series		
1981-93		
one	.7	.9
both	.9	1.1
1994-04		
one	.7	.9
both	.8	1.0
Navajo		
one	1.2	1.4
both	1.3	1.5

B-SERIES : MPV : NAVAJO : TRIBUTE — MAZ-73

	LABOR TIME	SEVERE SERVICE
Tribute		
one	1.1	1.3
both	1.7	1.9

Front Strut Assembly, R&R or Replace (B)

	LABOR TIME	SEVERE SERVICE
MPV		
one side	1.2	1.4
both sides	1.7	1.9

Hub, Bearing or Seal, Replace (B)

	LABOR TIME	SEVERE SERVICE
B-Series		
1981-84		
one	1.3	1.5
both	2.3	2.5
1986-93		
2WD		
one	1.3	1.5
both	2.3	2.5
4WD		
one	2.0	2.2
both	3.0	3.2
1994-04		
one side	1.2	1.4
both sides	1.7	1.9
MPV		
1989-98		
2WD		
one side	.8	1.0
both sides	1.4	1.6
4WD		
one side	1.5	1.7
both sides	1.9	2.1
2000-05		
one side (1.6)	2.2	2.4
both sides (2.8)	3.8	4.0
Navajo 4WD		
one side	1.2	1.4
both sides	1.7	1.9
Tribute		
one side	2.0	2.2
both sides	3.5	3.7

Lower Control Arm, Replace (B)

Includes: Alignment.

	LABOR TIME	SEVERE SERVICE
B-Series		
1981-93		
2WD		
one side	1.9	2.1
both sides	3.3	3.5
4WD		
one side	2.2	2.4
both sides	3.5	3.7
1998-04		
one side	1.8	2.0
both sides	2.9	3.1
MPV		
1989-98		
one side	1.8	2.0
both sides	2.4	2.6
2000-05		
one side (.8)	1.1	1.3
both sides (1.3)	1.8	2.0

	LABOR TIME	SEVERE SERVICE
Navajo		
2WD		
one side	3.6	3.8
both sides	5.9	6.1
4WD		
right side	4.4	4.6
left side	5.6	5.8
both sides	8.5	8.7
Tribute		
one side	.9	1.1
both sides	1.4	1.6

Lower Control Arm Bushings and/or Spindle, Replace (B)

Includes: Alignment.

	LABOR TIME	SEVERE SERVICE
B-Series		
1981-84		
one side	2.8	3.0
both sides	4.3	4.5
1986-93		
2WD		
one side	1.5	1.7
both sides	2.6	2.8
4WD		
one side	1.8	2.0
both sides	2.9	3.1
MPV		
1989-98		
one side	1.9	2.1
both sides	2.7	2.9
2000-05		
one side (1.0)	1.4	1.6
both sides (1.6)	2.2	2.5
Navajo		
2WD		
one side	3.5	3.7
both sides	5.5	5.7
4WD		
right side	3.9	4.1
left side	6.2	6.4
both sides	6.9	7.1
Tribute		
one side	1.1	1.3
both sides	1.6	1.8

Radius Arm, Replace (B)

Includes: Alignment.

	LABOR TIME	SEVERE SERVICE
1994-97 B-Series		
one side	2.6	2.8
both sides	3.8	4.0
Navajo		
one side	2.8	3.0
both sides	3.7	3.9

Stabilizer Bar, Replace (B)

	LABOR TIME	SEVERE SERVICE
B-Series	.8	1.0
MPV		
1989-98	.8	1.0
2000-01	7.2	7.3
2002-04	8.1	8.3

	LABOR TIME	SEVERE SERVICE
Navajo	.9	1.1
Tribute	1.2	1.4
w/4WD 98-04		
B-Series add	.1	.2

Stabilizer Bar Bushings, Replace (B)

	LABOR TIME	SEVERE SERVICE
1994-04 B-Series		
one side	.5	.7
both sides	.7	.9
MPV		
1989-98	.7	.9
2000-01	7.2	7.4
2002-05	8.1	8.3
Navajo, Tribute		
one	.6	.8
both	.8	1.0

Steering Knuckle, Replace (B)

	LABOR TIME	SEVERE SERVICE
B-Series		
1981-93		
2WD		
one side	2.7	2.9
both sides	4.2	4.4
4WD		
one side	3.1	3.3
both sides	4.4	4.6
1994-04		
2WD		
one side	3.6	3.8
both sides	4.1	4.3
4WD		
one side	2.3	2.5
both sides	3.1	3.3
MPV		
1989-98		
2WD		
one side	1.6	1.8
both sides	2.0	2.2
4WD		
one side	1.9	2.1
both sides	2.8	3.0
2000-05		
one side	2.2	2.4
both sides	3.8	4.0
Navajo		
2WD		
one side	3.0	3.2
both sides	4.1	4.3
4WD		
one side	2.3	2.5
both sides	3.1	3.3
Tribute		
one side	1.7	1.9
both sides	3.5	3.7

Steering Knuckle Spindle Bearings, Replace (B)

	LABOR TIME	SEVERE SERVICE
B2600 4WD		
one side	1.7	1.9
both sides	3.0	3.2

MAZ-74 B-SERIES : MPV : NAVAJO : TRIBUTE

	LABOR TIME	SEVERE SERVICE
Suspension Crossmember, Replace (B)		
Includes: Alignment.		
MPV		
1989-98	5.2	5.4
2000-01	7.5	7.7
2002-05	8.4	8.6
Tribute		
2WD	5.9	6.1
4WD		6.9
7.1		
Torsion Bar/Arms, Replace (B)		
Includes: Alignment.		
1986-93		
one side	1.2	1.4
both sides	1.8	2.1
1998-04 B-Series		
one side	1.2	1.4
both sides	1.9	2.1
Torsion Springs, Replace (B)		
Includes: Adjust camber and toe-in.		
1986-93		
one side	1.8	2.0
both sides	2.9	3.1
Upper Control Arm Assy., Replace (B)		
B-Series		
1981-84		
one side	2.9	3.1
both sides	4.3	4.5
1986-93		
one side	1.7	1.9
both sides	2.7	2.9
1998-04		
one side	.9	1.1
both sides	1.7	1.9
Wheel Bearing Grease Seal, Replace (B)		
B-Series		
1981-93		
2WD		
one side	.7	.7
both sides	1.5	1.5
4WD		
one side	1.2	1.2
both sides	2.2	2.2
1994-04		
one side	.7	.7
both sides	1.3	1.3
1989-98 MPV		
one side	1.5	1.5
both sides	1.9	1.9
Navajo		
one side	.7	.7
both sides	1.3	1.3
Wheel Bearings, Clean & Pack (B)		
B-Series		
1981-84		
one side	.8	1.0
both sides	1.6	1.8

	LABOR TIME	SEVERE SERVICE
1986-93		
2WD		
one side	.8	1.0
both sides	1.6	1.8
4WD		
one side	1.5	1.7
both sides	2.5	2.7
1994-04		
one side	1.4	1.6
both sides	1.8	2.0
1989-98 MPV 2WD		
one side	2.3	2.5
both sides	3.1	3.3
Navajo		
2WD	1.7	1.9
4WD	2.8	3.0
Wheel Bearings, Replace (B)		
B-Series		
1981-93		
2WD		
one side	.7	.9
both sides	1.9	2.1
4WD		
one side	1.7	1.9
both sides	2.8	3.0
1994-04		
one side	1.2	1.4
both sides	1.7	1.9
MPV		
1989-98		
one side	2.0	2.2
both sides	2.9	3.1
2000-05		
one side	2.2	2.4
both sides	3.8	4.0
Navajo		
one side	1.2	1.4
both sides	1.7	1.9
Tribute		
one side	2.0	2.2
both sides	3.5	3.7

REAR SUSPENSION

	LABOR TIME	SEVERE SERVICE
Align Rear End (A)		
Tribute	2.1	2.3
Air Compressor Relay, Replace (B)		
1989-98 MPV	.3	.5
Air Suspension Control Unit, Replace (B)		
1989-98 MPV	.5	.7
Compressor Assy., Replace (B)		
1989-98 MPV	.7	.9
Height Sensor, Replace (B)		
1989-98 MPV	.7	.9
Lower Arms & Links, Replace (B)		
1989-98 MPV		
one	.8	1.0
both	1.3	1.5
Tribute		
lower arm		

	LABOR TIME	SEVERE SERVICE
one side	4.8	5.0
both sides	7.2	7.4
lateral link upper or lower		
one side	.6	.8
both sides	.9	1.1
Rear Spring, Replace (B)		
B-Series		
1981-93		
one side	1.5	1.7
both sides	2.4	2.6
1994-97		
one side	1.4	1.6
both sides	2.5	2.7
1998-04		
2WD		
one side	1.3	1.5
both sides	1.8	2.0
4WD		
one side	1.2	1.4
both sides	1.6	1.8
MPV		
1989-98		
one side	.8	1.0
both sides	1.7	1.9
2000-05 one or both	2.1	2.3
Navajo		
one side	2.0	2.2
both sides	3.1	3.3
Tribute		
one side	.9	1.1
both sides	1.5	1.7
Recondition spring		
add each	.5	.7
Rear Stabilizer Control Links, Replace (B)		
1994-04 B-Series		
one side	.5	.7
both sides	.7	.9
1989-98 MPV		
one side	.5	.7
both sides	.7	.9
Navajo		
one side	.5	.7
both sides	.7	.9
Shock Absorbers or Bushings, Replace (B)		
B-Series		
one side	.7	.9
both sides	.9	1.1
MPV		
one side	.6	.8
both sides	.9	1.1
Navajo		
one	.8	1.0
both	1.2	1.4
Tribute		
one side	.9	1.1
both sides	1.3	1.5

B-SERIES : MPV : NAVAJO : TRIBUTE **MAZ-75**

	LABOR TIME	SEVERE SERVICE
Spring Shackle, Replace (B)		
B-Series, Navajo		
one side	.7	.9
both sides	.8	1.0
Spring Shackle Bushing, Replace (B)		
B-Series		
1981-93		
one side	.5	.7
both sides	.8	1.0
1994-04		
one side	.7	.9
both sides	.8	1.0
Navajo		
one side	.7	.9
both sides	1.2	1.4
Stabilizer Bar, Replace (B)		
1991-05	.8	1.0
Stabilizer Bar Bushings, Replace (B)		
1994-04 B-Series		
one side	.7	.9
both sides	.8	1.0
1989-98 MPV		
one side	.3	.5
both sides	.5	.7
Navajo		
one side	.7	.9
both sides	.8	1.0
Suspension Crossmember, Replace (B)		
Includes: Alignment.		
Tribute		
2WD	2.4	2.6
4WD	9.5	9.7

STEERING
AIR BAGS

	LABOR TIME	SEVERE SERVICE
Air Bag Assy., Replace (B)		
1994-04 B-Series		
driver side	.5	.7
passenger side	.7	.9
both sides	.7	.9
MPV		
1993-98		
driver side	.5	.7
passenger side	.8	1.0
2000-05		
front impact		
driver side	.5	.7
passenger side	.6	.8
side impact		
one side	1.2	1.4
both sides	2.2	2.4
Tribute		
front impact		
driver side	.8	1.0
passenger side	.8	.1.0
side impact		
one side	1.1	1.3
both sides	1.7	1.9

	LABOR TIME	SEVERE SERVICE
Air Bag (Crash) Sensor, Replace (B)		
1994-04 B-Series		
one	.5	.7
both	.7	.9
MPV		
1993-95		
right or left	.7	.9
center	.8	1.0
S sensor	2.8	3.0
2000-05		
side impact		
one side	.5	.7
both sides	.6	.8
Tribute		
one side	1.1	1.3
both sides	1.5	1.7
Air Bag Control Unit, Replace (B)		
1995-04 B-Series	.7	.9
MPV		
1993-95	.7	.9
1996-98	1.2	1.4
2000-05	.6	.8
Tribute	1.0	1.2
Air Bag Deactivation Switch, Replace (B)		
1995-04 B-Series	.7	.9
MPV		
1993-95	.7	.9
1996-98	1.2	1.4

LINKAGE

	LABOR TIME	SEVERE SERVICE
Inner Tie Rod or End, Replace (B)		
Includes: Reset toe.		
B-Series		
1981-93		
one side	1.0	1.2
both sides	1.4	1.6
1994-97		
one side	1.3	1.5
both sides	1.7	1.9
1998-04		
one side	1.6	1.8
both sides	1.9	2.1
MPV		
1989-98		
2WD		
one side	3.3	3.5
both sides	3.8	4.0
4WD		
one side	4.6	4.8
both sides	5.0	5.2
2000-05		
one side	6.0	6.2
both sides	6.1	6.3
Navajo		
one side	1.3	1.5
both sides	1.6	1.8

	LABOR TIME	SEVERE SERVICE
Tribute		
2.0L, 2.3L		
one side	5.3	5.5
both sides	5.6	5.8
3.0L		
one side	5.9	6.1
both sides	6.3	6.5
Outer Tie Rod End, Replace (B)		
Includes: Reset toe.		
B-Series		
1981-93		
one side	1.0	1.2
both sides	1.4	1.6
1994-04		
one side	1.3	1.5
both sides	1.7	1.9
MPV		
1989-98		
one side	1.0	1.2
both sides	1.4	1.6
2000-05		
one side (1.2)	1.6	1.8
both sides (1.6)	2.2	2.4
Navajo		
one side	1.3	1.5
both sides	1.6	1.8
Tribute		
one side	1.7	1.9
both sides	2.0	2.2
Replace seal add each	.1	.1
Pitman Arm, Replace (B)		
1981-97	.8	1.0

POWER RACK & PINION
Unless otherwise noted, time given does not include alignment.

	LABOR TIME	SEVERE SERVICE
Horn Contact or Canceling Cam, Replace (B)		
1989-04	.7	.9
Intermediate Shaft, Replace (B)		
MPV		
1989-95	1.2	1.4
1996-98	1.7	1.9
Pressure Switch, Replace (B)		
1989-98 MPV	.5	.7
Rack & Pinion Assy., R&I (B)		
Includes: Reset toe.		
B-Series		
2WD	1.9	2.1
4WD	2.7	2.9
MPV		
1989-98		
2WD	2.3	2.5
4WD	3.3	3.5
2000-05 (3.4)	4.6	4.8
Tribute		
2.0L, 2.3L	5.0	5.2
3.0L	5.6	5.8

MAZ-76 B-SERIES : MPV : NAVAJO : TRIBUTE

	LABOR TIME	SEVERE SERVICE
Rack & Pinion Assy., R&R and Reconditon (A)		
Includes: Reset toe.		
1989-98 MPV		
2WD	5.2	5.4
4WD	6.2	6.4
Rack Assy. (Short), Replace (A)		
MPV		
1989-98		
2WD	4.4	4.6
4WD	5.5	5.7
2000-05 (5.2)	7.0	7.2
Rack Boots, Replace (B)		
Includes: Reset toe.		
MPV		
1989-98		
one side	1.3	1.5
both sides	1.6	1.8
2000-05		
one side	2.1	2.3
both sides	2.5	2.7
Tribute		
one side	2.0	2.2
both sides	2.4	2.6
Steering Pump, R&R and Reconditon (B)		
B-Series	2.1	2.3
MPV		
1989-98	1.8	2.0
2000-01	3.2	3.4
2002-05	4.0	4.2
Tribute		
3.0L	3.6	3.8
Steering Pump, R&R or Replace (B)		
B-Series	1.4	1.6
MPV		
1989-98	1.3	1.5
2000-01	2.1	2.3
2002-05	2.9	3.1
Tribute		
2.0L, 2.3L	1.8	2.0
3.0L	2.6	2.8
Steering Pump Hoses, Replace (B)		
Does not include purging.		
B-Series each	.6	.8
MPV		
1989-98		
pressure hose	.7	.9
return hose	.7	.9
pressure pipe	2.0	2.2
return pipe	.9	1.1
2000-05		
pressure hose	1.4	1.6
return hose	1.0	1.2
return pipe	6.6	6.8
Tribute		
pressure hose		
2.0L, 2.3L	1.8	2.0
3.0L	1.1	1.3

	LABOR TIME	SEVERE SERVICE
return hose	1.6	1.8
pressure pipe		
3.0L	2.6	2.8
return pipe		
2.0L, 2.3L	1.5	1.7
3.0L	3.5	3.7
Steering Wheel, Replace (B)		
B-Series, Tribute	.8	1.0
MPV		
1989-95	.5	.7
1996-98	.8	1.0
2000-05	.6	.8
MANUAL WORM & SECTOR		
Steering Gear, Adjust (On Truck) (B)		
1981-93	.5	.7
Gear Assy., R&R and Reconditon (A)		
B-Series		
1981-93	2.3	2.5
1994-97	2.0	2.2
Gear Assy., Replace (B)		
B-Series		
1981-93	.9	1.1
1994-97	.8	1.0
Idler Arm, Replace (B)		
1981-93	.8	1.0
Replace bushing add	.2	.4
Steering Center Link (Relay Rod), Replace (B)		
1981-93	1.4	1.6
Steering Column Upper Bushing, Replace (B)		
1981-93	.7	.9
Steering Knuckle Arm, Replace (B)		
Includes: Reset toe-in.		
1986-93		
one side	1.3	1.5
both sides	2.0	2.2
Steering Intermediate Shaft, Replace (B)		
B2000, B2200	1.3	1.5
B2600	1.5	1.7
Steering Wheel, Replace (B)		
B-Series		
1981-93	.5	.7
1994-97	.7	.9
POWER WORM & SECTOR		
Steering Gear, Adjust (On Truck) (B)		
1986-97	.5	.7
Gear Assy., R&R and Reconditon (A)		
B-Series		
1986-93	2.7	2.9
1994-97	4.2	4.4
Navajo	4.2	4.4

	LABOR TIME	SEVERE SERVICE
Gear Assy., Replace (B)		
B-Series		
1986-93	1.3	1.5
1994-97	1.7	1.9
Navajo	1.7	1.9
Horn Contact, Replace (B)		
1994-97 B-Series	.5	.7
Navajo	.5	.7
Power Steering Hoses, Replace (B)		
Includes: System bleeding.		
B-Series		
1986-93		
pressure hose		
w/AC	1.7	1.9
w/o AC	1.2	1.4
return hose	1.2	1.4
return pipes each	.8	1.0
1994-97		
pressure or return	.7	.9
Navajo		
pressure or return	.7	.9
Power Steering Switch Assy., Replace (B)		
1994-97 B-Series	.5	.7
Steering Center Link (Relay Rod), Replace (B)		
1981-97	1.4	1.6
Steering Intermediate Shaft, Replace (B)		
1994-97 B-Series	.7	.9
Navajo	.7	.9
Steering Knuckle Arm, Replace (B)		
Includes reset toe-in.		
1986-97		
one side	1.3	1.5
both sides	2.0	2.2
Steering Pump, R&R and Reconditon (B)		
B-Series		
1986-93		
w/o AC	2.3	2.5
w/AC	3.0	3.2
1994-97	3.6	3.8
Navajo	3.6	3.8
Steering Pump, R&R or Replace (B)		
B-Series		
1986-93		
w/AC	2.2	2.4
w/o AC	1.4	1.6
1994-97		
4 cyl.	1.6	1.8
V6	1.7	1.9
Navajo	2.6	2.8
Steering Wheel, Replace (B)		
1986-97	.7	.9

B-SERIES : MPV : NAVAJO : TRIBUTE   MAZ-77

	LABOR TIME	SEVERE SERVICE
Upper Mast Jacket Bearing, Replace (B)		
Includes: Replace insulators if necessary.		
1986-97	.9	*1.1*

HEATING & AIR CONDITIONING

When more than one component requires replacement where evacuation/recovery and recharging is already included, deduct 1.0 hour for each additional component from the time given.

	LABOR TIME	SEVERE SERVICE
Diagnose Climate Control System (B)		
All Models	1.2	*1.2*
Evacuate/Recover and Recharge System (B)		
All Models	1.0	*1.0*
AC Hoses, Replace (B)		
Includes: Evacuate/recover and recharge.		
B-Series		
1981-84 each	1.7	*1.9*
1986-93		
one	1.7	*1.9*
each addl. add	.5	*.7*
1994-04 each	1.6	*1.8*
MPV		
one	1.7	*1.9*
each addl. add	.5	*.7*
Navajo		
one	1.9	*2.1*
each addl. add	.5	*.7*
Tribute each	.4	*.6*
AC On/Off Control Switch, Replace (B)		
MPV		
1989-95	.5	*.7*
1996-98	.9	*1.1*
AC Relay, Replace (B)		
1989-98 MPV		
front	.3	*.5*
rear	.5	*.7*
Tribute	.4	*.6*
Blower Motor, Replace (B)		
B-Series		
1981-84	1.9	*2.1*
1986-93	1.2	*1.4*
1994-04	.8	*1.0*
MPV		
1989-98		
front	.7	*.9*
rear	1.2	*1.4*
2000-05		
front	.5	*.7*
rear	.8	*1.0*

	LABOR TIME	SEVERE SERVICE
Navajo	.8	*1.0*
Tribute	.4	*.6*
Blower Motor Resistor, Replace (B)		
1986-04 front or rear	.6	*.8*
Tribute	.3	*.5*
Compressor Assy., Replace (B)		
Includes: Parts transfer, evacuate/recover and recharge.		
B-Series		
1981-84	1.9	*2.1*
1986-93	2.3	*2.5*
1994-04		
4 cyl.	2.9	*3.1*
V6	2.4	*2.6*
MPV		
1989-98	2.3	*2.5*
2000-05	1.9	*2.1*
Navajo	2.5	*2.7*
Tribute	1.0	*1.2*
Compressor Clutch Coil, Replace (B)		
B-Series		
1981-84	.9	*1.1*
1986-93	1.5	*1.7*
1994-04		
4 cyl.	1.9	*2.1*
V6	1.5	*1.7*
MPV		
1989-98	1.5	*1.7*
2000-05	1.2	*1.4*
Navajo	1.5	*1.7*
Tribute	1.0	*1.2*
Add time to evacuate & recharge.		
Compressor Shaft Seal, Replace (B)		
Includes: Compressor R&R, evacuate/recover and recharge.		
B-Series		
1981-84	2.3	*2.5*
1986-93	2.6	*2.8*
1994		
4 cyl.	3.4	*3.6*
V6	2.7	*2.9*
1995-04		
4 cyl.	3.2	*3.4*
V6	2.7	*2.9*
1989-98 MPV	2.6	*2.8*
Navajo	2.7	*2.9*
Condenser Assy., Replace (B)		
Includes: Evacuate/recover and recharge.		
B-Series		
1981-84	1.9	*2.1*
1986-93	1.7	*1.9*
1994-04	2.2	*2.4*
MPV		
1989-95	1.7	*1.9*
1996-98	1.9	*2.1*
2000-05	1.5	*1.7*
Navajo	1.8	*2.0*
Tribute	.9	*1.1*

	LABOR TIME	SEVERE SERVICE
Condenser Cooling Fan, Replace (B)		
1989-98 MPV		
one	.5	*.7*
all	.7	*.9*
Evaporator Coil, Replace (B)		
Includes: Evacuate/recover and recharge.		
B-Series		
1981-93	2.3	*2.5*
1994-04	3.1	*3.3*
MPV		
1989-98		
front or rear	2.3	*2.5*
2000-05		
front	1.8	*2.0*
rear	2.5	*2.7*
Navajo	3.1	*3.3*
Tribute	4.3	*4.5*
Expansion Valve, Replace (B)		
Includes: Evacuate/recover and recharge.		
B-Series		
1981-84	1.8	*2.0*
1986-93	1.6	*1.8*
MPV		
1989-98	1.9	*2.1*
2000-05		
front	1.6	*1.8*
rear	1.9	*2.1*
Heater Blower Motor, Replace (B)		
B-Series		
1981-84	1.9	*2.1*
1986-93	1.2	*1.4*
1994-04	.7	*.9*
MPV		
1989-98		
front	.7	*.9*
rear	1.2	*1.4*
2000-05		
front	.5	*.7*
rear	1.0	*1.2*
Navajo	.7	*.9*
Tribute	.4	*.6*
Heater Blower Motor Relay, Replace (B)		
1989-05 MPV		
front	.3	*.5*
rear	.6	*.8*
Tribute	.3	*.5*
Heater Blower Motor Resistor, Replace (B)		
B-Series	.5	*.7*
MPV		
1989-98		
front	.3	*.5*
rear	1.2	*1.4*
2000-05		
front	.3	*.5*
rear	.6	*.8*

MAZ-78 B-SERIES : MPV : NAVAJO : TRIBUTE

	LABOR TIME	SEVERE SERVICE
Heater Blower Motor Resistor, Replace (B)		
Navajo	.5	.7
Tribute	.4	.6
Heater Blower Motor Switch, Replace (B)		
B-Series		
1981-93	2.0	2.2
1994-04	1.2	1.4
MPV		
1989-95		
front	.9	1.1
rear	.5	.7
1996-98		
front	1.3	1.5
rear	.5	.7
instrument panel	.9	1.1
2000-05		
front	.6	.8
rear	.3	.5
Navajo	.9	1.1
Tribute	.6	.8
Heater Control Valve, Replace (B)		
1981-87	.7	.9
1988-93	.5	.7
Heater Core, R&R or Replace (B)		
B-Series		
1981-84	2.2	2.4
1986-93	4.8	5.0
1994	1.2	1.4
1995-97	4.5	4.7
1998-04	5.2	5.4
MPV		
1989-98		
front	5.2	5.4
rear	1.9	2.1
2000-05		
front	3.5	3.7
rear	1.1	1.3
Navajo	1.2	1.4
Tribute	4.4	4.6
Heater Control Assy., Replace (B)		
B-Series		
1981-84	1.9	2.1
1986-93	.7	.9
1994	1.2	1.4
1995-04	.7	.9
1989-98 MPV		
electric	.9	1.1
manual	1.2	1.4
Navajo	1.2	1.4
Tribute	.6	.8
Heater Hoses, Replace (B)		
Front		
one side	.7	.9
both sides	.7	.9
Rear	.8	1.0
1996-98	.8	1.0
2000-05	1.6	1.8
Tribute one or both	1.0	1.2

	LABOR TIME	SEVERE SERVICE
High or Low Pressure Switch, Replace (B)		
1981-93 B-Series	1.6	1.8
Navajo	1.5	1.7
Add time to evacuate & recharge.		
Receiver/Drier Assy., Replace (B)		
Includes: Evacuate/recover and recharge.		
B-Series		
1981-93	1.3	1.5
1994-04	1.6	1.8
1981-98 MPV	1.6	1.8
Navajo	1.6	1.8
Tribute	.4	.6

WIPERS & SPEEDOMETER

	LABOR TIME	SEVERE SERVICE
Antenna, Replace (B)		
B-Series		
1981-84	.5	.7
1986-93	1.8	2.0
1994-04	1.3	1.5
MPV		
1989-95	.9	1.1
1996-98	1.2	1.4
2000-05 (.4)	.6	.8
Navajo	1.3	1.5
Tribute		
manual or power	.7	.9
Radio, R&R (B)		
B-Series		
1981-84	.5	.7
1986-04	1.2	1.4
MPV		
1989-98	.5	.7
2000-05 (.5)	.7	.9
Navajo	1.4	1.6
Tribute	.4	.6
Rear Window Washer Motor, Replace (B)		
Navajo	.5	.7
Rear Window Wiper Motor, Replace (B)		
MPV, Navajo, Tribute	.7	.9
Speedometer Cable & Casing, Replace (B)		
B-Series		
1981-93	.8	1.0
1994	1.3	1.5
MPV		
1989-95	.8	1.0
1996-98	1.3	1.5
Navajo	1.6	1.8
Speedometer Drive Gear, Replace (B)		
B-Series		
AT		
1981-84	4.4	4.6
1986-93		
2WD	3.1	3.3
4WD	7.3	7.5

	LABOR TIME	SEVERE SERVICE
1994-04		
4 cyl.	.8	1.0
V6		
2WD	.8	1.0
4WD	2.3	2.5
MT		
1981-93		
2WD	2.8	3.0
4WD	3.2	3.4
1994-04		
2WD	2.8	3.0
4WD	1.5	1.7
MPV		
1989-93		
AT		
2WD	2.6	2.8
4WD	7.4	7.6
MT		
2WD	3.5	3.7
4WD	3.2	3.4
1994-98		
2WD	2.6	2.8
4WD	3.2	3.4
2000-05 (10.1)	13.6	13.8
Navajo		
2WD	3.2	3.4
4WD	.8	1.0
Speedometer Driven Gear and/or Seal, Replace (B)		
AT	.8	1.0
MT	.5	.7
Speedometer Head, R&R or Replace (B)		
B2000	1.3	1.5
B2200, B2600		
1982-84	1.6	1.8
1987-93	1.3	1.5
B2300, B2500, B3000, B4000	1.3	1.5
1989-98 MPV	1.3	1.5
Navajo	1.7	1.9
Tribute	.8	1.0
Washer Motor and/or Reservoir, Replace (B)		
1994-04 B-Series	.5	.7
MPV		
1989-05		
front	.5	.7
rear	.5	.7
Navajo	.5	.7
Tribute	.6	.8
Windshield Washer Pump, Replace (B)		
B-Series	.3	.5
Navajo	.5	.7
Windshield Wiper and Washer Switch, Replace (B)		
B-Series		
1986-93	.8	1.0
1994-04	.3	.5

B-SERIES : MPV : NAVAJO : TRIBUTE   **MAZ-79**

	LABOR TIME	SEVERE SERVICE
MPV		
1989-95		
switch	.8	1.0
lever	.9	1.1
1996-98		
front		
switch	.9	1.1
lever	1.2	1.4
rear	.9	1.1
2000-05	.5	.7
Navajo each	.5	.7
Tribute	.5	.7
Windshield Wiper Linkage, Replace (B)		
All Models	.9	1.1
Windshield Wiper Motor, Replace (B)		
B-Series, Navajo	1.0	1.2
MPV		
1989-98	.9	1.1
2000-05 (.8)	1.1	1.3
Tribute	.7	.9

LAMPS & SWITCHES

	LABOR TIME	SEVERE SERVICE
Back-Up Lamp Switch, Replace (B)		
B-Series		
1981-93	.7	.7
1994-04	.5	.5
1989-92 MPV	.7	.7
Navajo	.5	.5
Tribute	.8	.8
Cluster Switch, Replace (B)		
B-Series		
1986-93	.3	.3
1994-04	.8	.8
1989-98 MPV one or all	.8	.8
Navajo one or all	.8	.8
Combination Switch Assy., Replace (B)		
Exc. below	.9	.9
Tribute	1.1	1.1
Daylight Running Lamp Module, Replace (B)		
1994-04 B-Series	.5	.5
MPV		
1990-98	.5	.5
2000-05	.3	.3
Front Combination Lamp Bulb, Replace (C)		
B-Series		
one side	.5	.5
both sides	.7	.7
1989-98 MPV		
one side	.3	.3
both sides	.5	.5
Navajo		
one side	.5	.5
both sides	.7	.7
Tribute		
one or both	.3	.3

	LABOR TIME	SEVERE SERVICE
Halogen Headlamp Bulb, Replace (B)		
1986-05		
one side	.3	.3
both sides	.5	.5
Hazard Warning Switch, Replace (B)		
1996-98 MPV	.9	.9
2000-05	.5	.5
Headlamp Dimmer Switch, Replace (B)		
Navajo	.9	.9
Headlamp Switch, Replace (B)		
B-Series, MPV	.7	.7
Navajo	1.2	1.2
Tribute	.5	.5
Headlamps, Aim (B)		
One side	.5	.5
Both sides	.7	.7
High Mount Stop Lamp Assy., Replace (B)		
B-Series	.5	.5
MPV		
1994-98	.3	.3
2000-05	.5	.5
Navajo, Tribute	.5	.5
Horn, Replace (B)		
B-Series		
one	.5	.5
both	.7	.7
MPV		
1989-98		
one	.5	.5
both	.7	.7
2000-05 one or both	.8	.9
Navajo		
one	.3	.3
both	.5	.5
Tribute		
one or both	.7	.7
Horn Relay, Replace (B)		
All Models	.3	.3
License Lamp Assy., Replace (B)		
1981-94 each	.3	.3
Multifunction Switch, Replace (B)		
Navajo	.9	.9
Neutral or Clutch Start Switch, Replace (B)		
B-Series	.7	.7
1989-98 MPV	.7	.7
Navajo	.5	.5
Tribute	.8	.8
Parking Brake Warning Lamp Switch, Replace (B)		
B-Series		
1981-93	.3	.3
1994-04	.5	.5
MPV		
1989-98	.3	.3
2000-05 (1.0)	1.4	1.4
Navajo	.5	.5
Tribute	.6	.6

	LABOR TIME	SEVERE SERVICE
Rear Combination Lamp Bulb, Replace (C)		
B-Series		
1981-93		
one side	.2	.2
both sides	.3	.3
1994-04		
one side	.5	.5
both sides	.7	.7
MPV		
1989-98		
one side	.3	.3
both sides	.5	.5
2000-05		
one side	.5	.5
both sides	.6	.6
Navajo		
one	.5	.5
both	.7	.7
Tribute		
one	.3	.3
both	.5	.5
Rear Combination Lamp Lens, Replace (B)		
B-Series		
one side	.3	.3
both sides	.5	.5
1989-98 MPV		
one side	1.7	1.7
both sides	3.1	3.1
Navajo		
one side	.3	.3
both sides	.5	.5
Tribute		
one	.3	.3
both	.5	.5
Sealed Beam Headlamp, Replace (B)		
1981-85		
one side	.7	.7
both sides	.8	.8
Side Marker Lamp Lens, Replace (B)		
Exc. below	.3	.3
1981-93	.2	.2
Stop Lamp Switch, Replace (B)		
All Models	.5	.5
Turn Signal Lamp or Lens, Front, Replace (B)		
All Models	.3	.3
Turn Signal or Hazard Warning Flasher, Replace (B)		
All Models	.3	.3

BODY

	LABOR TIME	SEVERE SERVICE
Door Lock, Replace (B)		
All Models	.8	1.0
Door Window Regulator, Replace (B)		
1981-84	1.2	1.4
1986-05		
manual	1.2	1.4
power	1.4	1.6

MAZ-80 B-SERIES : MPV : NAVAJO : TRIBUTE

	LABOR TIME	SEVERE SERVICE
Hood Hinge, Replace (B)		
B-Series		
1981-93		
one	.5	.7
both	.7	.9
1994-04		
one	.7	.9
both	.8	1.0
MPV		
1989-95		
one	.5	.7
both	.7	.9
1996-98		
one	1.7	1.9
both	2.3	2.5
2000-05		
one (1.0)	1.4	1.6
both (1.3)	1.8	2.0

	LABOR TIME	SEVERE SERVICE
Navajo		
one	.7	.9
both	.8	1.0
Tribute		
one	.9	1.1
both	1.0	1.2
Hood Lock, Replace (B)		
1981-97	.5	.7
1998-04	.7	.9
Hood Release Cable, Replace (B)		
B-Series		
1981-97	.8	1.0
1998-04	1.2	1.4
MPV		
1989-98	.6	.8
2000-05 (.8)	1.1	1.3
Navajo, Tribute	.7	.9

	LABOR TIME	SEVERE SERVICE
Lock Striker Plate, Replace (B)		
All Models	.3	.5
Window Regulator Motor, Replace (B)		
1994-04 B-Series	.8	1.0
MPV		
front (.8)	1.1	1.3
rear (1.0)	1.4	1.6
Navajo	.8	1.0
Tribute	1.1	1.3

MITSU

3000GT : Cordia : Diamante : Eclipse : Galant : Lancer : Lancer Evolution : Mirage : Precis : Sigma : Starion : Tredia

SYSTEM INDEX

MAINTENANCE	MITSU-2
Maintenance Schedule	MITSU-2
CHARGING	MITSU-3
STARTING	MITSU-3
CRUISE CONTROL	MITSU-3
IGNITION	MITSU-4
EMISSIONS	MITSU-5
FUEL	MITSU-6
EXHAUST	MITSU-8
ENGINE COOLING	MITSU-9
ENGINE	MITSU-10
Assembly	MITSU-10
Cylinder Head	MITSU-11
Camshaft	MITSU-13
Crank & Pistons	MITSU-15
Engine Lubrication	MITSU-17
CLUTCH	MITSU-18
MANUAL TRANSAXLE	MITSU-19
MANUAL TRANSMISSION	MITSU-19
AUTO TRANSAXLE	MITSU-19
AUTO TRANSMISSION	MITSU-21
TRANSFER CASE	MITSU-21
SHIFT LINKAGE	MITSU-21
DRIVELINE	MITSU-21
BRAKES	MITSU-23
FRONT SUSPENSION	MITSU-25
REAR SUSPENSION	MITSU-26
STEERING	MITSU-27
HEATING & AC	MITSU-29
WIPERS & SPEEDOMETER	MITSU-31
LAMPS & SWITCHES	MITSU-32
BODY	MITSU-32

OPERATIONS INDEX

A
AC Hoses	MITSU-29
Air Bags	MITSU-27
Air Conditioning	MITSU-29
Alignment	MITSU-25
Alternator (Generator)	MITSU-3
Antenna	MITSU-31
Anti-Lock Brakes	MITSU-23

B
Back-Up Lamp Switch	MITSU-32
Ball Joint	MITSU-25
Battery Cables	MITSU-3
Bleed Brake System	MITSU-23
Blower Motor	MITSU-30
Brake Disc	MITSU-24
Brake Drum	MITSU-24
Brake Hose	MITSU-23
Brake Pads and/or Shoes	MITSU-24

C
Camshaft	MITSU-13
Camshaft Sensor	MITSU-4
Catalytic Converter	MITSU-8
Coolant Temperature (ECT) Sensor	MITSU-5
Crankshaft	MITSU-15
Crankshaft Sensor	MITSU-4
Cruise Control	MITSU-3
CV Joint	MITSU-22
Cylinder Head	MITSU-11

D
Differential	MITSU-22
Distributor	MITSU-4
Drive Belt	MITSU-2
Driveshaft	MITSU-21

E
EGR	MITSU-5
Electronic Control Module (ECM/PCM)	MITSU-5
Engine	MITSU-10
Engine Lubrication	MITSU-17
Engine Mounts	MITSU-10
Evaporator	MITSU-30
Exhaust	MITSU-8
Exhaust Manifold	MITSU-8
Expansion Valve	MITSU-30

F
Flexplate	MITSU-20
Flywheel	MITSU-18
Fuel Injection	MITSU-7
Fuel Pump	MITSU-6
Fuel Vapor Canister	MITSU-5

G
Generator	MITSU-3

H
Halfshaft	MITSU-22
Headlamp	MITSU-32
Heater Core	MITSU-30
Horn	MITSU-32

I
Idle Air Control (IAC) Valve	MITSU-7
Ignition Coil	MITSU-4
Ignition Switch	MITSU-4
Inner Tie Rod	MITSU-28
Intake Air Temperature (IAT) Sensor	MITSU-5
Intake Manifold	MITSU-7

K
Knock Sensor	MITSU-5

L
Lower Control Arm	MITSU-25

M
Maintenance Schedule	MITSU-2
Manifold Absolute Pressure (MAP) Sensor	MITSU-5
Master Cylinder	MITSU-24
Muffler	MITSU-9
Multifunction Lever/Switch	MITSU-32

N
Neutral Safety Switch	MITSU-32

O
Oil Pan	MITSU-17
Oil Pump	MITSU-17
Outer Tie Rod	MITSU-28
Oxygen Sensor	MITSU-5

P
Parking Brake	MITSU-25
Pistons	MITSU-16
Positive Crankcase Ventilation (PCV) Valve	MITSU-6

R
Radiator	MITSU-9
Radiator Hoses	MITSU-9
Radio	MITSU-31
Rear Main Oil Seal	MITSU-16

S
Shock Absorber/Strut, Front	MITSU-26
Shock Absorber/Strut, Rear	MITSU-27
Spark Plug Cables	MITSU-4
Spark Plugs	MITSU-5
Spring, Front Coil	MITSU-26
Spring, Rear Coil	MITSU-27
Starter	MITSU-3
Steering Wheel	MITSU-28

T
Thermostat	MITSU-9
Throttle Body	MITSU-8
Throttle Position Sensor (TPS)	MITSU-6
Timing Belt	MITSU-14
Timing Chain	MITSU-15
Torque Converter	MITSU-20

U
U-Joint	MITSU-22
Upper Control Arm	MITSU-26

V
Valve Body	MITSU-20
Valve Cover Gasket	MITSU-12
Valve Job	MITSU-12
Vehicle Speed Sensor	MITSU-6

W
Water Pump	MITSU-9
Wheel Balance	MITSU-2
Wheel Cylinder	MITSU-24
Window Regulator	MITSU-32
Windshield Washer Pump	MITSU-31
Windshield Wiper Motor	MITSU-32

MITSU-1

MITSU-2   3000GT : CORDIA : DIAMANTE : ECLIPSE : GALANT : LANCER

	LABOR TIME	SEVERE SERVICE
MAINTENANCE		
Air Cleaner Filter Element, Replace (C)		
All Models	.5	.7
Chassis Lubrication, Change Oil & Filter (C)		
Includes: Correct all fluid levels.		
All Models	.4	.4
Drive Belt, Adjust (B)		
One	.3	.5
each addl. add	.2	.3
Drive Belt, Replace (B)		
Alternator		
exc. Lancer (.5)	.7	.9
Lancer (.6)	.8	1.0
Compressor	.6	.8
Fan		
exc. below	.8	1.0
Lancer (.7)	.9	1.1
Lancer Evolution (.9)	1.2	1.4
PS		
exc. below	.8	1.0
3000GT	.9	1.1
1997-04 Diamante	1.0	1.2
Starion	.5	.7
Water pump Lancer		
Evolution (.8)	1.1	1.3
Fuel Filter, Replace (B)		
3000GT		
in line		
w/turbo	1.6	1.8
w/o turbo	1.1	1.3
in tank	1.0	1.2
Cordia, Tredia	.8	1.0
Diamante		
1992-96		
in line	.6	.8
in tank		
sedan	3.3	3.5
wagon	1.3	1.5
1997-04	1.1	1.3
Eclipse		
1990-94		
in line	.5	.7
in tank	1.0	1.2
1995-99		
in line	1.0	1.2
in tank	1.4	1.6
2000-05 (.8)	1.1	1.3
Galant		
1985-88		
in line	.5	.7
in tank	1.5	1.7
1989-93	.7	.9

	LABOR TIME	SEVERE SERVICE
1994-98		
in line	.8	1.0
in tank	1.4	1.6
1999-05 (.9)	1.2	1.4
Lancer (.8)	1.1	1.3
Lancer Evolution		
in line	1.1	1.3
in tank	.6	.8
Mirage		
1985-88	.8	1.0
1989-92		
in line	.7	.9
in tank	1.2	1.4
1993-96		
in line	1.0	1.2
in tank	2.4	2.6
1997-03		
in line	1.1	1.3
in tank	1.7	1.9
Precis, Starion	.5	.5
Sigma		
in line	.5	.5
in tank	1.7	1.9
Halogen Headlamp Bulb, Replace (B)		
Exc. below	.5	.5
3000GT	.6	.6
Lancer Evolution	1.5	1.5
Oil & Filter, Change (C)		
All Models	.4	.4
Sealed Beam Headlamp, Replace (B)		
One	.3	.3
Timing Belt, Replace (B)		
4 cyl.		
Cordia, Tredia	2.9	3.1
Eclipse		
1990-94		
SOHC	3.5	3.7
DOHC	4.5	4.8
1995-99		
SOHC	3.5	3.8
DOHC	3.9	4.1
2000-05 (2.7)	3.6	3.8
Galant		
1985-98		
SOHC	3.5	3.8
DOHC	4.7	4.9
1999-04	3.6	3.8
2005 (2.5)	3.4	3.6
Lancer (2.9)	3.9	4.0
Mirage		
1985-88	2.8	3.0
1989-92		
SOHC	2.8	3.0
DOHC	3.8	4.0
1993-96	2.9	3.1
1997-03		
1.5L	1.9	2.1
1.8L	2.6	2.8
Precis	1.9	2.1

	LABOR TIME	SEVERE SERVICE
V6		
3000GT	4.7	4.9
Diamante		
1992-96		
SOHC	4.6	4.8
DOHC	5.7	5.9
1997-04	3.9	4.1
Eclipse	3.6	3.8
Galant	3.6	3.8
Sigma	4.6	4.8
w/AC add		
3000GT	1.0	1.2
90-99 Eclipse	.5	.7
88-98 Galant	.5	.7
Mirage	.5	.7
w/PS Mirage add	.5	.7
w/turbo add	.3	.5
Replace tensioner add	.2	.4
Tire, Replace (C)		
Includes: Dismount old tire and mount new tire to rim.		
One	.5	.5
each addl. add	.3	.3
Tires, Rotate (C)		
All Models	.5	.5
Wheel, Balance (B)		
One	.4	.4
each addl. add	.2	.2
SCHEDULED MAINTENANCE SERVICES		
If necessary, refer to appropriate Chilton maintenance service information.		
7,500 Mile Service (C)		
All Models	.4	.6
15,000 Mile Service (C)		
All Models	.9	1.1
w/AT add	.2	.2
22,500 Mile Service (C)		
All Models	.4	.6
30,000 Mile Service (B)		
All Models	3.1	3.3
w/AT add	.1	.2
w/AWD add	.1	.2
Inspect rear brake lining and wheel cyls. Eclipse, Galant, Mirage add	.2	.4
37,500 Mile Service (C)		
All Models	.4	.6
45,000 Mile Service (C)		
All Models	.9	1.1
w/AT add	.1	.2
52,500 Mile Service (C)		
All Models	.4	.6

LANCER EVOLUTION : MIRAGE : PRECIS : SIGMA : STARION : TREDIA   MITSU-3

	LABOR TIME	SEVERE SERVICE
60,000 Mile Service (B)		
All Models	6.8	7.0
w/AT add	.1	.2
w/AWD add	.1	.2
Inspect		
dist. cap and rotor add	.2	.4
rear brake lining and wheel cyls. Eclipse, Galant, Mirage add	.2	.4
67,500 Mile Service (C)		
All Models	.4	.6
75,000 Mile Service (C)		
All Models	.9	1.1
w/AT add	.1	.2
82,500 Mile Service (C)		
All Models	.4	.6
90,000 Mile Service (B)		
All Models	3.1	3.3
w/AT add	.1	.2
w/AWD add	.1	.2
Inspect rear brake lining and wheel cyls. Eclipse, Galant, Mirage add	.2	.4
97,500 Mile Service (C)		
All Models	.4	.6

CHARGING

	LABOR TIME	SEVERE SERVICE
Alternator Circuits, Test (B)		
Includes: Test component output.		
All Models	.4	.4
Alternator, R&R and Recondition (B)		
3000GT	2.9	3.1
Cordia, Tredia	2.7	2.9
Diamante		
1992-96	2.8	3.0
1997-04	2.6	2.8
Eclipse (2.1)	2.8	3.0
Galant		
1985-98	2.8	3.0
1999-05 (2.1)	2.8	3.0
Lancer (1.9)	2.6	2.8
Lancer Evolution (2.8)	3.8	4.0
Mirage		
1985-96	2.8	3.0
1997-03		
1.5L	2.4	2.6
1.8L	2.8	3.0
Precis	2.9	3.1
Sigma	3.8	4.0
Starion	2.8	3.0
w/ABS Lancer add	.6	.8
w/warm-up 3-way catalytic converter 92-96 Diamante SOHC add	1.8	2.0
Alternator Assy., Replace (B)		
Includes: Pulley transfer.		
3000GT, Diamante	1.4	1.6
Eclipse (.8)	1.1	1.3

	LABOR TIME	SEVERE SERVICE
Galant		
1985-98	1.1	1.3
1999-05 (1.0)	1.4	1.6
Lancer (.7)	.9	1.1
Lancer Evolution (1.7)	2.3	2.5
Mirage		
1985-88	.8	1.0
1989-96	1.2	1.4
1997-03		
1.5L	.8	1.0
1.8L	1.2	1.4
Precis	1.2	1.4
Sigma	2.3	2.5
Starion	.8	1.0
w/ABS Lancer add	.6	.8
w/warm-up 3-way catalytic converter 92-96 Diamante SOHC add	1.8	2.0
Alternator Voltage Regulator, Replace (B)		
3000GT	2.3	2.5
Cordia, Tredia	1.9	2.1
Diamante	2.3	2.5
Eclipse (1.7)	2.3	2.5
Galant		
1985-98	1.9	2.1
1999-05 (1.6)	2.2	2.4
Lancer (1.3)	1.8	2.0
Lancer Evolution (2.3)	3.1	3.3
Mirage		
1985-88	1.9	2.1
1989-96	2.3	2.5
1997-03		
1.5L	1.7	1.9
1.8L	1.9	2.1
Precis	1.5	1.7
Sigma	3.2	3.4
Starion	1.7	1.9
w/ABS Lancer add	.6	.8
w/warm-up 3-way catalytic converter 92-96 Diamante SOHC add	1.8	2.0
Voltmeter and/or Ammeter Gauge, Replace (B)		
3000GT	3.1	3.3

STARTING

	LABOR TIME	SEVERE SERVICE
Starter Draw Test (On Car) (B)		
All Models	.3	.3
Battery, Replace (C)		
All Models	.5	.6
Battery Cables, Replace (C)		
Exc. below	.8	1.0
Diamante	1.0	1.2
2000-05 Eclipse (1.0)	1.4	1.6
Galant		
1994-98	1.1	1.3
1999-05 (1.2)	1.6	1.8

	LABOR TIME	SEVERE SERVICE
Lancer (1.0)	1.4	1.6
Lancer Evolution (1.5)	2.0	2.2
1997-03 Mirage	1.0	1.2
w/turbo add	.3	.3
Starter, R&R and Recondition (B)		
Exc. below (2.1)	2.8	3.0
Lancer (2.2)	3.0	3.2
Lancer Evolution (2.2)	3.0	3.2
Mirage 1.8L	2.9	3.1
Precis	1.3	1.5
Sigma	2.2	2.4
Starter Assy., Replace (B)		
Exc. below (.7)	.9	1.1
Lancer (1.0)	1.4	1.6
Lancer Evolution (.8)	1.1	1.3
Mirage 1.8L	1.4	1.6
Starter Drive Assy., Replace (B)		
Includes: Starter R&R.		
3000GT, Diamante	1.5	1.7
Cordia, Starion, Tredia	1.8	2.0
Eclipse, Galant (1.3)	1.8	2.0
Lancer (1.6)	2.2	2.4
Lancer Evolution (1.4)	1.9	2.1
Mirage		
exc. 1.8L	1.8	2.0
1.8L	2.2	2.4
Precis		
1987-89	1.4	1.6
1990-94	.9	1.1
Sigma	1.5	1.7
Starter Relay, Replace (B)		
Exc. below	.5	.7
1995-99 Eclipse	.7	.9
Starter Solenoid and/or Switch, Replace (B)		
Includes: Starter R&R.		
Exc. below (1.1)	1.5	1.7
Lancer (1.4)	1.9	2.1
Lancer Evolution (1.2)	1.6	1.8
Mirage 1.8L	1.9	2.1
Precis	.7	.9
Sigma	.9	1.1

CRUISE CONTROL

	LABOR TIME	SEVERE SERVICE
Diagnose Cruise Control System Component Each (A)		
All Models	.5	.5
Actuator Assy., Replace (B)		
3000GT	.7	.9
Cordia, Sigma, Starion, Tredia	.5	.7
Diamante	.6	.8
Eclipse		
1990-99	.6	.8
2000-05 (.8)	1.1	1.3
Galant		
1988-98	.6	.8
1999-05 (.8)	1.1	1.3
Lancer, Mirage	.7	.9

MITSU-4   3000GT : CORDIA : DIAMANTE : ECLIPSE : GALANT : LANCER

	LABOR TIME	SEVERE SERVICE
Brake or Clutch Release Switch, Replace (B)		
All Models	.5	.7
Control Cable & Core, Replace (B)		
Exc. Eclipse	.6	.8
Eclipse	.7	.9
Control Main Switch, Replace (B)		
All Models	.6	.8
Control Module, Replace (B)		
All Models	.8	1.0
Control Vacuum Control or Check Valve, Replace (B)		
Cordia, Starion, Tredia	.3	.5
Control Vacuum Pump, Replace (B)		
Exc. Eclipse	.5	.7
Eclipse	.6	.8
Control Vacuum Pump Relay, Replace (B)		
Starion	.5	.7
Control Vacuum Release Switch, Replace (B)		
Cordia, Starion, Tredia	.3	.5

IGNITION

	LABOR TIME	SEVERE SERVICE
Diagnose Ignition System Component Each (A)		
All Models	.8	.8
Ignition Timing, Reset (B)		
All Models	.5	.7
Camshaft Position Sensor, Replace (B)		
3000GT	1.9	2.1
Diamante		
1992-96	1.9	2.1
1997-04	1.1	1.3
Eclipse		
1995-99		
w/turbo	3.0	3.2
w/o turbo	.6	.8
2000-05 (.3)	.5	.7
Galant		
1994-98	3.6	3.8
1999-05 (.3)	.5	.7
Lancer, Lancer Evolution	.5	.7
Mirage	.6	.8
w/AC add		
95-99 Eclipse w/turbo	.5	.7
94-98 Galant	.5	.7
Crankshaft Position Sensor, Replace (B)		
3000GT	3.1	3.3
Diamante		
1992-96	3.5	3.7
1997-04	1.8	2.0

	LABOR TIME	SEVERE SERVICE
Eclipse		
1995-99		
2.0L		
w/turbo	2.1	2.3
w/o turbo	.5	.7
2.4L	2.7	2.9
2000-05		
4 cyl. (2.4)	3.2	3.4
V6 (1.1)	1.5	1.7
Galant		
1994-98		
SOHC	2.7	2.9
DOHC	1.5	1.7
1999-05		
exc. 2.4L (1.1)	1.5	1.7
2.4L (2.4)	3.2	3.4
Lancer (1.2)	1.6	1.8
Lancer Evolution (2.8)	3.8	4.0
Mirage		
1993-96	1.6	1.8
1997-03		
1.5L	1.4	1.6
1.8L	1.6	1.8
w/AC add		
3000GT	.5	.7
95-99 Eclipse exc. 2.0L w/o turbo	.5	.7
00-04 Eclipse	.5	.7
94-98 Galant SOHC	1.8	2.0
Mirage	.5	.7
Distributor, R&R and Recondition (B)		
Includes: Reset base ignition timing.		
Cordia, Tredia	1.9	2.1
1997-04 Diamante	2.4	2.6
Eclipse		
1990-94	2.4	2.6
2000-05 (2.1)	2.8	3.0
Galant		
1985-98	1.9	2.1
1999-03	2.9	3.1
Mirage	1.9	2.1
Precis		
1987-89	1.5	1.7
1990-94	1.9	2.1
Sigma	1.8	2.0
Starion	1.9	2.1
Distributor, Replace (B)		
Includes: Reset base ignition timing.		
3000GT	.6	.8
Diamante	.9	1.1
Eclipse		
1990-94	.6	.8
2000-05 (.8)	1.1	1.3
Galant		
1988-98	.6	.8
1999-03	1.1	1.3
Mirage	.6	.8
Distributor Cap and/or Rotor, Replace (B)		
Diamante	.3	.5

	LABOR TIME	SEVERE SERVICE
Eclipse		
1990-94	.3	.5
2000-05 (.7)	.9	1.1
Galant		
1988-98	.3	.5
1999-03	.9	1.1
Mirage	.5	.7
Distributor Igniter Set, Replace (B)		
Cordia, Starion, Tredia	.5	.7
Distributor Reluctor, Replace (B)		
Cordia, Starion, Tredia	1.2	1.4
Engine Spark Control Igniter, Replace (B)		
Cordia, Starion, Tredia	.5	.7
1985-88 Mirage	.5	.7
Ignition Coil, Replace (B)		
Exc below	.7	.9
Galant		
1994-98	1.0	1.2
1999-03		
2.4L	.6	.8
3.5L		
right side	2.3	2.5
left side	.5	.7
Ignition Switch, Replace (B)		
Exc. below	.7	.9
Cordia, Starion, Tredia	.8	1.0
1997-04 Diamante	1.2	1.4
1989-03 Mirage	.8	1.0
Spark Plug (Ignition) Cables, Replace (B)		
4 cyl.	.6	.8
V6		
exc. below		
SOHC	.6	.8
DOHC	1.6	1.8
3000GT		
right side	.6	.8
left side	1.5	1.7
both sides	1.7	1.9
Diamante		
1992-96		
SOHC		
one side	.5	.7
both sides	.6	.8
DOHC		
right side	.6	.8
left side	1.5	1.7
both sides	1.7	1.9
1997-04		
right side	1.2	1.4
left side	.5	.7
both sides	1.5	1.7
2000-05 Eclipse		
right side (1.8)	2.4	2.6
left side (.3)	.5	.7
both sides (1.9)	2.6	2.8

LANCER EVOLUTION : MIRAGE : PRECIS : SIGMA : STARION : TREDIA — MITSU-5

	LABOR TIME	SEVERE SERVICE
1999-03 Galant		
right side (1.7)	2.3	2.5
left side (.3)	.5	.7
both sides (1.8)	2.4	2.6
Spark Plugs, Replace (B)		
4 cyl.	.8	1.0
V6		
exc. below		
SOHC	1.0	1.2
DOHC	2.2	2.4
3000GT		
right side	.6	.8
left side	1.5	1.7
both sides	1.7	1.9
Diamante		
1992-96		
SOHC		
one side	.5	.7
both sides	.6	.8
DOHC		
right side	.6	.8
left side	1.5	1.7
both sides	1.7	1.9
1997-04		
right side	1.2	1.4
left side	.5	.7
both sides	1.5	1.7
2000-05 Eclipse		
right side (1.9)	2.6	2.8
left side (.3)	.5	.7
both sides (2.1)	2.8	3.0
1999-05 Galant		
right side (1.8)	2.4	2.6
left side (.3)	.5	.7
both sides (1.9)	2.6	2.8
Vacuum Advance Unit, Replace (B)		
Includes: Reset base timing.		
Cordia, Starion, Tredia	1.2	1.4
1985-87 Galant	1.1	1.3
Precis	.7	.9

EMISSIONS

The following operations do not include testing. Add time as required.

	LABOR TIME	SEVERE SERVICE
Dynamometer Test (A)		
All Models	.4	.4
Air Flow Sensor, Replace (B)		
Exc. below (.4)	.6	.8
1997-04 Diamante (.6)	.8	1.0
Coolant Temperature (ECT) Sensor, Replace (A)		
Exc. below	.5	.7
Diamante	.7	.9
Lancer (.4)	.6	.8
Lancer Evolution (.6)	.8	1.0
ECI Boost Sensor, Replace (B)		
All Models	.5	.7
ECI Control Relay, Replace (B)		
Exc. below	.6	.8
1995-99 Eclipse	.8	1.0
ECI Resistor, Replace (B)		
1983-99	.6	.8
EGR Control Solenoid, Replace (B)		
Exc. below	.5	.7
Lancer (.4)	.6	.8
Lancer Evolution (1.1)	1.5	1.7
EGR Thermal Vacuum Valve, Replace (B)		
1989-04	.5	.7
EGR Valve, Replace (B)		
Exc. Lancer Evolution	.6	.8
Lancer Evolution	.8	1.0
w/4G6 engine add		
00-04 Eclipse	.5	.7
99-04 Galant	.5	.7
EGR Valve and Sensor Assy., Replace (B)		
Precis	.3	.5
EGR Temperature Sensor, Replace (B)		
1989-04	.3	.5
Electronic Control Module (ECM/PCM), Replace (B)		
Exc. below	1.3	1.5
3000GT	.6	.8
1992-96 Diamante	.8	1.0
Eclipse	1.0	1.2
Galant	.9	1.1
Lancer, Lancer Evolution	.5	.7
Mirage	.6	.8
Evaporative Purge Solenoid, Replace (B)		
Exc. below	.5	.7
Lancer, Lancer Evolution	.7	.9
Fuel Vapor Canister, Replace (B)		
3000GT	.6	.8
Diamante	.5	.7
Eclipse		
1990-99	.7	.9
2000-05 (2.1)	2.8	3.0
Galant		
1988-98	.5	.7
1999-05 (1.7)	2.3	2.5
Lancer, Lancer Evolution	.8	1.0
Mirage	.8	1.0
Fuel Tank Pressure Control Valve, Replace (B)		
3000GT	.6	.8
Diamante	.5	.7
Eclipse (.5)	.7	.9
Galant		
1988-93	.5	.7
1994-98	1.7	1.9

	LABOR TIME	SEVERE SERVICE
Mirage		
1989-92	.5	.7
1993-04	1.9	2.1
Intake Air Temperature (IAT) Sensor, Replace (B)		
All Models	.3	.5
Knock Sensor, Replace (B)		
3000GT	1.9	2.1
Diamante		
1992-96		
sedan	1.9	2.1
wagon	.3	.5
1997-04	2.6	2.8
Eclipse		
1990-99	.5	.7
2000-05		
4 cyl. (.8)	1.1	1.3
V6 (2.6)	3.5	3.7
Galant		
1988-98	.5	.7
1999-05 (2.5)	3.4	3.6
Lancer (.7)	.9	1.1
Lancer Evolution (1.5)	2.0	2.2
Mirage		
1989-92	.5	.7
1997-03	1.0	1.2
Manifold Absolute Pressure (MAP) Sensor, Replace (B)		
All Models (.3)	.5	.6
Manifold Differential Pressure Sensor, Replace (B)		
Exc. Lancer (.2)	.3	.4
Lancer (.3)	.5	.6
Oxygen Sensor, Replace (B)		
Exc. below		
w/turbo one	2.1	2.3
w/o turbo one	.7	.9
3000GT		
heated		
one piece	.5	.7
right	.5	.7
left	.6	.8
before catalytic converter		
w/turbo		
right		
w/AC	3.2	3.4
w/o AC	2.9	3.1
left	.8	1.0
w/o turbo		
right	1.0	1.2
left	.5	.7
after catalytic converter		
right	.6	.8
left		
w/turbo	2.7	2.9
w/o turbo	1.0	1.2
Diamante		
heated		
1992-96	.5	.7
1997-04	1.1	1.3

MITSU-6 3000GT : CORDIA : DIAMANTE : ECLIPSE : GALANT : LANCER

	LABOR TIME	SEVERE SERVICE
Oxygen Sensor, Replace (B)		
before catalytic converter		
right	1.0	1.2
left	.5	.7
after catalytic converter		
right	.6	.8
left	2.7	2.9
Eclipse		
heated		
1990-99	.3	.5
2000-05 (.8)	1.1	1.3
before catalytic		
converter each	.6	.8
after catalytic		
converter one		
1995-99	.8	1.0
2000-05 (.8)	1.1	1.3
Galant		
heated		
1994-98	.3	.5
1999-05 (.6)	.8	1.0
before catalytic		
converter each	.6	.8
after catalytic		
converter each	.8	1.0
Lancer, Lancer		
Evolution each (.4)	.6	.8
1993-03 Mirage		
heated	.5	.7
before catalytic con.	1.0	1.2
after catalytic con.	.6	.8
Positive Crankcase Ventilation (PCV) Valve, Replace (B)		
All Models	.4	.6
Purge Control Valve, Replace (B)		
Exc. Galant (.4)	.6	.8
Galant		
1987-93	.3	.5
1994-98	.8	1.0
Throttle Position Sensor (TPS), Replace (B)		
Exc. below (.4)	.6	.8
1997-04 Diamante	1.0	1.2
Vapor Separator, Replace (B)		
Cordia, Tredia	.5	.7
1985-88 Galant,		
Mirage	.5	.7
Starion	1.3	1.5
Vehicle Speed Sensor, Replace (B)		
Exc. 3000GT	.6	.8
3000GT	1.0	1.2

FUEL
CARBURETOR

	LABOR TIME	SEVERE SERVICE
Carburetor, Adjust (On Car) (B)		
1983-94	.7	.9
Automatic Choke Coil Housing, Replace (B)		
1983-94	.8	1.0
Carburetor, R&R and Clean or Recondition (B)		
Cordia, Tredia	2.9	3.1
1985-88 Galant,		
Mirage	2.7	2.9
Precis	2.5	2.7
w/feedback		
carburetor add	.2	.4
Carburetor, Replace (B)		
Includes: Adjustments.		
Cordia, Tredia	1.5	1.7
1985-88 Galant,		
Mirage	1.7	1.9
Precis	.8	1.0
w/feedback		
carburetor add	.2	.4
Carburetor Accelerator Pump, Replace (B)		
Cordia, Tredia	2.9	3.1
1985-88 Galant,		
Mirage	2.8	3.0
Precis	.7	.9
Carburetor Needle & Seat, Replace (B)		
Cordia, Tredia	2.0	2.2
1985-88 Galant,		
Mirage	2.0	2.2
Precis	1.3	1.5
Dash Pot, Replace (B)		
1983-94	.7	.9

DELIVERY

	LABOR TIME	SEVERE SERVICE
Fuel Pump, Test (B)		
All Models	.3	.5
Combination Gauge, Replace (B)		
3000GT	3.1	3.3
Fuel Filter, Replace (B)		
3000GT		
in line		
w/turbo	1.6	1.8
w/o turbo	1.1	1.3
in tank	1.0	1.2
Cordia, Tredia	.8	1.0
Diamante		
1992-96		
in line	.6	.8
in tank		
sedan	3.3	3.5
wagon	1.3	1.5
1997-04	1.1	1.3
Eclipse		
1990-94		
in line	.5	.7
in tank	1.0	1.2
1995-99		
in line	1.0	1.2
in tank	1.4	1.6
2000-05 (.8)	1.1	1.3

	LABOR TIME	SEVERE SERVICE
Galant		
1985-88		
in line	.5	.7
in tank	1.5	1.7
1989-93	.7	.9
1994-98		
in line	.8	1.0
in tank	1.4	1.6
1999-05 (.9)	1.2	1.4
Lancer (.8)	1.1	1.3
Lancer Evolution		
in line	1.1	1.3
in tank	.6	.8
Mirage		
1985-88	.8	1.0
1989-92		
in line	.7	.9
in tank	1.2	1.4
1993-96		
in line	1.0	1.2
in tank	2.4	2.6
1997-03		
in line	1.1	1.3
in tank	1.7	1.9
Precis, Starion	.5	.5
Sigma		
in line	.5	.5
in tank	1.7	1.9
Fuel Gauge (Dash), Replace (B)		
Exc. below (.6)	.8	1.0
3000GT	1.0	1.3
1997-04 Diamante	1.1	1.3
Sigma	.9	1.1
Fuel Gauge (Tank), Replace (B)		
Exc. below (.6)	.8	1.0
3000GT	1.0	1.2
Cordia, Tredia	1.7	1.9
1990-94 Precis	1.5	1.7
w/4WS Diamante add	.3	.3
Fuel Pump, Replace (B)		
3000GT	1.0	1.2
Cordia, Tredia	.8	1.0
Diamante		
1992-96		
sedan	3.0	3.2
wagon	1.0	1.2
1997-04	1.4	1.6
Eclipse		
1990-94	1.0	1.2
1995-99	1.4	1.6
2000-05 (.8)	1.1	1.3
Galant		
1985-87	1.6	1.8
1988-93	.8	1.0
1994-98	1.4	1.6
1999-05 (.9)	1.2	1.4
Lancer (.6)	.8	1.0
Lancer Evolution (.9)	1.2	1.4

LANCER EVOLUTION : MIRAGE : PRECIS : SIGMA : STARION : TREDIA　　MITSU-7

	LABOR TIME	SEVERE SERVICE
Mirage		
1985-88	.8	1.0
1989-92	1.2	1.4
1993-96	2.4	2.6
1997-03	1.7	1.9
Precis	1.5	1.7
Sigma, Starion	1.7	1.9
Fuel Pump Relay, Replace (B)		
Exc. below	.3	.3
3000GT	.8	.8
Diamante, Mirage	.6	.6
Fuel Tank, Replace (B)		
Includes: Drain and refill.		
3000GT	1.9	2.1
Cordia, Tredia	1.8	2.0
Diamante		
1992-96		
sedan	3.2	3.4
wagon	1.7	1.9
1997-04	2.2	2.4
Eclipse		
1990-94	1.7	1.9
1995-99		
2WD	2.2	2.4
AWD	3.5	3.7
2000-05 (2.6)	3.5	3.7
Galant		
1985-87	1.9	2.1
1989-93	1.4	1.6
1994-98	2.2	2.4
1999-05 (2.2)	3.0	3.2
Lancer (1.5)	2.0	2.2
Lancer Evolution (2.0)	2.7	2.9
Mirage		
1985-88	.8	1.0
1989-92	1.4	1.6
1993-03	2.4	2.6
Precis		
1987-89	1.3	1.5
1990-94	1.7	1.9
Sigma	2.0	2.2
Starion	1.4	1.6
Intake Manifold and/or Gasket, Replace (B)		
Includes: Adjustments.		
4 cyl.		
Cordia, Tredia	2.4	2.6
Eclipse		
1990-99	2.4	2.6
2000-05 (2.6)	3.5	3.7
Galant		
1985-93	2.3	2.5
1994-98	1.3	1.5
1999-05 (2.6)	3.5	3.7
Lancer (2.3)	3.1	3.3
Lancer		
Evolution (3.2)	4.3	4.5
Mirage	2.3	2.5

	LABOR TIME	SEVERE SERVICE
Precis		
1987-89	3.3	3.5
1990-94	2.9	3.1
Starion	2.6	2.8
V6		
3000GT	2.2	2.4
Diamante		
1992-96	2.1	2.3
1997-04	2.8	3.0
Eclipse	4.0	4.2
Galant	3.8	4.0
Sigma	2.6	2.8
w/turbo 95-99 Eclipse add	.5	.7
Replace manifold add	.8	1.0
INJECTION		
Fuel Injectors, Replace (B)		
MPI		
4 cyl.		
Eclipse		
1990-99	.8	1.0
2000-05 (1.1)	1.5	1.7
Galant		
1988-98	.8	1.0
1999-05 (1.1)	1.5	1.7
Lancer (.7)	.9	1.1
Lancer		
Evolution (.9)	1.2	1.4
Mirage	.8	1.0
V6		
3000GT		
right side	.8	1.0
left or both sides	1.8	2.0
Diamante		
1992-96		
right side	.8	1.0
left or both sides	1.8	2.0
1997-04		
right side	2.0	2.2
left side	1.1	1.3
both sides	2.1	2.3
Eclipse		
one side (2.1)	2.8	3.0
both sides (2.2)	3.0	3.2
Galant		
right side (1.9)	2.6	2.8
left side (1.2)	1.6	1.8
both sides (2.1)	2.8	3.0
TBI		
one	.7	.9
each addl. add	.1	.1
Fuel Pressure Regulator, Replace (B)		
MPI		
exc. below	.6	.8
1990-94 Eclipse	.9	1.1
TBI	.5	.7

	LABOR TIME	SEVERE SERVICE
Fuel Rail, Replace (B)		
Exc. below		
4 cyl.	.8	1.0
V6	2.9	3.1
3000GT	1.2	1.4
Diamante		
1992-96	1.2	1.4
1997-04	2.0	2.2
2000-05 Eclipse		
4 cyl. (1.1)	1.5	1.7
V6 (2.1)	2.8	3.0
1999-05 Galant		
4 cyl. (1.1)	1.5	1.7
V6 (1.9)	2.6	2.8
Lancer (.7)	.9	1.1
Lancer Evolution (.9)	1.2	1.4
Idle Air Control (IAC) Valve, Replace (B)		
3000GT	1.1	1.3
Diamante		
1992-96	1.1	1.3
1997-04	1.7	1.9
Eclipse		
1990-94		
SOHC	.6	.8
DOHC	1.4	1.6
1995-05 (.8)	1.1	1.3
Galant		
1985-93	.6	.8
1994-05 (.8)	1.1	1.3
Lancer (.7)	.9	1.1
Lancer Evolution (1.1)	1.5	1.7
Mirage	.9	1.1
Sigma	1.3	1.5
Injection Mixer Assy., Replace (B)		
1983-89	1.4	1.6
Injection Mixer Boost Sensor, Replace (B)		
1983-89	.5	.7
Injection Mixer Computer, Replace (B)		
1983-89	.7	.9
Injection Mixer Control Relay, Replace (B)		
1983-89	.5	.7
Injection Mixer Resistor, Replace (B)		
1983-89	.5	.7
Injection Mixer Throttle Switch, Replace (B)		
1983-89	.5	.7
Injection Mixer Throttle Valve, Replace (B)		
1983-89	.5	.7
Injection Mixer Vacuum Switch, Replace (B)		
1983-89	.5	.7

MITSU-8   3000GT : CORDIA : DIAMANTE : ECLIPSE : GALANT : LANCER

	LABOR TIME	SEVERE SERVICE
Throttle Body Assy., Replace (B)		
MPI		
exc. Diamante	1.0	1.2
Diamante	1.1	1.3
TBI	1.5	1.7
w/DOHC 90-94 Eclipse add	.3	.3
TURBOCHARGER		
Oil Feed Pipe, Replace (B)		
Exc. Lancer Evolution	.8	1.0
Lancer Evolution	2.9	3.1
Intercooler, Replace (B)		
3000GT		
front side	2.6	2.8
rear side	3.3	3.5
Eclipse		
1990-94	1.4	1.6
1995-99	3.0	3.2
Galant	1.4	1.6
Starion	.7	.9
Turbocharger Assy., R&R or Replace (B)		
3000GT		
front	3.3	3.5
rear	4.3	4.5
Cordia, Starion, Tredia	2.7	2.9
Eclipse		
1990-94	2.9	3.1
1995-99	3.6	3.8
Galant, Mirage (1.9)	2.6	2.8
Lancer Evolution (3.5)	4.7	4.9
w/warm-up three way converter 3000GT add	2.1	2.3
Wastegate Actuator, Replace (B)		
Exc. Lancer Evolution	.7	.9
Lancer Evolution (2.0)	2.7	2.9
EXHAUST		
Catalytic Converter, Replace (B)		
3000GT		
front		
right side	1.1	1.3
left side	2.7	2.9
rear	.8	1.0
Cordia, Tredia	.8	1.0
Diamante		
1992-96		
front		
right side	1.1	1.3
left side	2.7	2.9
rear	.8	1.0
1997-04 each	1.1	1.3
Eclipse		
1990-99	1.0	1.2
2000-05		
4 cyl. (.8)	1.1	1.3
V6		
right side (1.4)	1.9	2.1
left side (1.2)	1.6	1.8

	LABOR TIME	SEVERE SERVICE
Galant		
1985-93	.7	.9
1994-98 each	1.1	1.3
1999-05		
front		
4 cyl. (.8)	1.1	1.3
V6		
right side (1.6)	2.2	2.4
left side (1.2)	1.6	1.8
rear (.8)	1.1	1.3
Lancer, Lancer Evolution	.6	.8
Mirage		
1985-92	.7	.9
1993-03		
front	1.1	1.3
rear	1.3	1.5
Precis		
1987-89		
front	.9	1.1
rear	.5	.7
1990-94	.5	.7
Sigma	.7	.9
Starion		
front	1.1	1.3
rear	.7	.9
w/turbo 3000GT add	1.0	1.2
Center Exhaust Pipe, Replace (B)		
3000GT	.8	1.0
Diamante	1.1	1.3
Eclipse		
1990-99	1.0	1.2
2000-05 (1.0)	1.4	1.6
Galant (.8)	1.1	1.3
Lancer, Lancer Evolution	.8	1.0
Mirage	1.1	1.3
Exhaust Manifold or Gasket, Replace (B)		
4 cyl.		
Cordia, Tredia	1.8	2.0
Eclipse		
1990-94		
SOHC	1.8	2.0
DOHC	2.4	2.6
1995-99		
exc. 4G6 engines	.9	1.1
4G6 engines	1.6	1.8
2000-05 (.9)	1.2	1.4
Galant		
1985-93	1.6	1.8
1994-05 (.9)	1.2	1.4
Lancer (1.8)	2.4	2.6
Lancer Evolution (3.5)	4.7	4.9
Mirage		
1985-88	1.9	2.1
1989-92	1.6	1.8
1993-96	1.3	1.5

	LABOR TIME	SEVERE SERVICE
1997-03		
1.5L	.9	1.1
1.8L	1.3	1.5
Precis		
1987-89	1.6	1.8
1990-94	1.7	1.9
Starion	2.9	3.1
V6		
3000GT		
w/turbo		
right side	4.5	4.7
left side	4.7	4.9
w/o turbo		
right side	2.7	2.9
left side	2.4	2.6
Diamante		
1992-96		
right side		
SOHC	1.6	1.8
DOHC	2.7	2.9
left side	2.4	2.6
1997-04		
right side	2.1	2.3
left side	1.6	1.8
Eclipse		
right side (2.1)	2.8	3.0
left side (1.2)	1.6	1.8
Galant		
right side (1.6)	2.2	2.4
left side (1.3)	1.8	2.0
Sigma		
right side	1.8	2.0
left side	6.3	6.4
w/3-way catalytic converter add		
3000GT left side only		
w/turbo	1.4	1.6
w/o turbo	.8	1.
92-96 Diamante sedan left side only	.8	1.0
00-04 Eclipse V6 left side only	.5	.7
Galant V6	.5	.7
Front Exhaust Pipe, Replace (B)		
3000GT	1.1	1.3
Cordia, Starion, Tredia	.7	.9
Diamante		
1992-96	.8	1.0
1997-03	1.3	1.5
Eclipse		
1990-94	.7	.9
1995-05 (1.0)	1.4	1.6
Galant		
1985-93	.7	.9
1994-05 (.9)	1.2	1.4
Lancer (.6)	.8	1.0
Lancer Evolution (1.0)	1.4	1.6
Mirage		
1985-92	.7	.9
1993-03	1.3	1.5
Sigma	.7	.9

LANCER EVOLUTION : MIRAGE : PRECIS : SIGMA : STARION : TREDIA   **MITSU-9**

	LABOR TIME	SEVERE SERVICE
Muffler, Replace (B)		
3000GT	1.0	1.2
Diamante		
1992-96	.6	.8
1997-04	1.1	1.3
Eclipse (.8)	1.1	1.3
Galant (.8)	1.1	1.3
Lancer, Lancer		
Evolution (.5)	.7	.9
Mirage	.7	.9
Precis each	.6	.8
Resonator, Replace (B)		
Exc. Mirage	.6	.8
Mirage		
1989-92	.3	.5
1993-03	1.4	1.6
ENGINE COOLING		
Cooling System Pressure Test (C)		
All Models	.3	.3
Combination Gauge Assy., Replace (B)		
3000GT	3.1	3.3
Coolant Thermostat, Replace (B)		
Exc. below	.6	.8
Diamante	.8	1.0
Eclipse	.9	1.1
Galant		
1994-03 (.7)	.9	1.1
2004-05 (.9)	1.2	1.4
Lancer Evolution	1.4	1.6
Engine Coolant Temp. Sending Unit, Replace (B)		
Exc. below	.5	.7
Lancer (.4)	.6	.8
Lancer Evolution (.8)	1.1	1.3
Fan Motor Resistor, Replace (B)		
Precis	.3	.5
Fluid Fan Drive or Fan Blade, Replace (B)		
Starion	.7	.9
Radiator Assy., R&R or Replace (B)		
3000GT	1.5	1.7
Cordia, Tredia	.9	1.1
Diamante	1.2	1.4
Eclipse (.8)	1.1	1.3
Galant, Lancer (.6)	.8	1.0
Lancer Evolution (1.3)	1.8	2.0
Mirage		
1985-92	.7	.9
1993-03	1.1	1.3
Precis		
1987-89	1.2	1.4
1990-94	.9	1.1
Sigma	1.4	1.6
Starion	1.7	1.9
w/AT add	.5	.7

	LABOR TIME	SEVERE SERVICE
Radiator Fan and/or Fan Motor, Replace (B)		
3000GT	1.1	1.3
Cordia, Tredia	.7	.9
Diamante	.9	1.1
Eclipse (.7)	.9	1.1
Galant		
1988-93	.6	.8
1994-05 (.8)	1.1	1.3
Lancer (.6)	.8	1.0
Lancer Evolution (1.3)	1.8	2.0
Mirage	.8	1.0
Precis	.7	.9
Radiator Fan Motor (Coolant Temp.) Switch, Replace (B)		
3000GT	1.0	1.2
Diamante	.6	.8
Eclipse	.7	.9
Galant	.6	.8
Mirage	.8	1.0
Radiator Fan Motor Relay, Replace (B)		
All Models	.5	.7
Radiator Hoses, Replace (B)		
Includes: Refill with proper coolant mix.		
Upper		
exc. below	.6	.8
Lancer Evolution	.8	1.0
Lower		
3000GT	1.1	1.3
Diamante		
1992-96	.6	.8
1997-04	1.1	1.3
Eclipse, Galant, Lancer (.8)	1.1	1.3
Lancer Evolution (1.5)	2.0	2.2
Mirage		
1989-92	.6	.8
1993-96	1.4	1.6
1997-03	.8	1.0
w/turbo add	.3	.5
Temperature Gauge (Dash), Replace (B)		
3000GT	3.1	3.3
Cordia, Precis, Tredia	.7	.9
Diamante		
1992-96	.7	.9
1997-04	1.1	1.3
Eclipse (.5)	.7	.9
Galant (.5)	.7	.9
Lancer, Lancer Evolution, Sigma (.7)	.9	1.1
Mirage	.8	1.0
Starion	.8	1.0

	LABOR TIME	SEVERE SERVICE
Water Control Valve, Replace (B)		
Cordia	2.9	3.1
Galant, Mirage, Sigma	.9	1.1
Precis	.7	.9
Starion	3.4	3.6
Tredia	3.6	3.8
Water Pump and/or Gasket, Replace (B)		
Includes: Refill with proper coolant mix.		
3000GT	4.8	5.0
Cordia, Tredia	3.5	3.7
Diamanate		
1992-96		
SOHC	4.7	4.9
DOHC	5.9	6.1
1997-04	3.6	3.8
Eclipse		
1990-94		
SOHC	3.6	3.8
DOHC	4.7	4.9
1995-99		
exc. below	3.5	3.7
2.0L 420A	4.3	4.5
2000-05		
4 cyl. (3.2)	4.3	4.5
V6 (2.7)	3.6	3.8
Galant		
1985-93	3.2	3.4
1994-98		
SOHC	3.5	3.7
DOHC	4.6	4.8
1999-05		
4 cyl. (3.0)	4.1	4.3
V6 (2.6)	3.5	3.7
Lancer, Lancer Evolution (3.2)	4.3	4.5
Mirage		
1985-96	3.2	3.4
1997-03		
1.5L	2.3	2.5
1.8L	2.9	3.1
Precis	2.6	2.8
Sigma	4.8	5.0
Starion	2.5	2.7
w/AC add		
3000GT	1.1	1.3
92-99 Eclipse	.5	.7
94-98 Galant	.5	.7
Mirage	.5	.7
w/PS Mirage	.5	.7
w/turbo add		
3000GT	.5	.7
95-99 Eclipse	.3	.5

MITSU-10   3000GT : CORDIA : DIAMANTE : ECLIPSE : GALANT : LANCER

ENGINE

ASSEMBLY

Times shown are for OEM assemblies. Time to replace assemblies from aftermarket rebuilders may vary.

Engine Assy., R&I (B)
Does not include parts or component transfer.

	LABOR TIME	SEVERE SERVICE
4 cyl.		
Cordia, Tredia	10.2	10.4
Eclipse		
1990-94	6.9	7.1
1995-99	7.5	7.7
2000-05 (9.2)	12.4	12.6
Galant		
1985-93	7.8	8.0
1994-03	9.7	9.9
2004-05 (10.6)	14.3	14.5
Mirage	7.1	7.3
Precis	4.8	5.0
Starion	10.9	11.1
V6		
3000GT	10.7	10.9
Diamante		
SOHC	11.2	11.4
DOHC	12.7	12.9
Galant (12.7)	17.1	17.3
Sigma	9.8	10.0

Engine Assy., R&R and Recondition (A)
Includes: Replacing rings, rod and main bearings, cylinder head reconditioning and engine tune-up.

	LABOR TIME	SEVERE SERVICE
4 cyl.		
Cordia, Tredia	24.3	24.5
Eclipse		
1990-94		
SOHC	23.7	23.9
DOHC	25.7	25.9
1995-99	30.7	30.9
2000-05 (23.3)	31.5	31.7
Galant		
1985-93		
SOHC		
2V	23.7	23.9
4V	25.8	26.0
DOHC	25.9	26.1
1994-03	27.8	28.0
2004-05 (21.9)	29.6	29.8
Mirage		
1985-92		
SOHC	23.8	24.0
DOHC	25.7	25.9
w/12V	24.5	24.7
1993-96		
1.5L	25.3	25.5
1.8L	25.8	26.0
1997-03	24.6	24.8

	LABOR TIME	SEVERE SERVICE
Precis	22.4	22.6
Starion	27.4	27.6
V6		
3000GT	38.7	38.9
Diamante		
1992-96		
SOHC	34.2	34.4
DOHC	38.6	38.8
1997-03	35.5	35.7
Galant	32.6	32.8
Sigma	30.5	30.7

Engine Assy. (Short Block), Replace (B)
Assembly consists of cylinder block, piston assemblies, crankshaft, camshaft, timing chain and gears. Does not include cylinder heads. Operation Includes: R&R engine, transfer necessary parts and all necessary adjustments.

	LABOR TIME	SEVERE SERVICE
4 cyl.		
Cordia, Tredia	17.0	17.2
Eclipse		
1990-94		
FWD	11.8	12.0
AWD	12.0	12.2
1995-99	12.8	13.0
2000-05 (10.1)	13.6	13.8
Galant		
1985-87	11.9	12.1
1989-93	11.8	12.0
1994-03	13.8	14.0
2004-05 (11.7)	15.8	16.0
Lancer (9.9)	13.4	13.6
Lancer		
Evolution (13.2)	17.8	18.0
Mirage		
1985-88		
SOHC	11.9	12.1
DOHC	12.8	13.0
1989-92	11.8	12.0
1993-03	12.3	12.5
Precis	12.9	13.1
Starion	17.4	17.6
V6		
3000GT		
FWD	16.3	16.5
AWD	16.7	16.9
Diamante		
1992-96	15.8	16.0
1997-04	18.2	18.4
Eclipse (11.4)	15.4	15.6
Galant (11.0)	14.9	15.1
Sigma	21.2	21.4
w/4WS 92-96 Diamante add	.6	.8
w/ABS 90-94 Eclipse add	.5	.7

	LABOR TIME	SEVERE SERVICE
w/AC add		
3000GT	.5	.7
Eclipse		
exc. 90-94 DOHC	.6	.8
90-94 DOHC	.8	1.0
Galant		
exc. 88-93 DOHC	.5	.7
88-93 DOHC	.8	1.0
Lancer	.3	.5
Mirage		
exc. 89-92 DOHC	.6	.8
89-92 DOHC	.8	1.0
w/AT add		
3000GT, Eclipse	.8	1.0
Galant, Mirage	.8	1.0
Lancer	1.0	1.2
w/cruise control 90-94		
Eclipse add	.3	.5
w/ECS add		
92-96 Diamante	.3	.5
88-93 Galant	.3	.5
w/PS add		
90-94 Eclipse	.3	.5
Galant, Mirage	.5	.7
w/turbo add		
3000GT	1.6	1.8
90-94 Eclipse	.8	1.0
95-99 Eclipse	2.4	2.6
88-93 Galant	.8	1.0
89-92 Mirage	.8	1.0

Engine Mounts, Replace (B)

	LABOR TIME	SEVERE SERVICE
Engine		
exc. below	.8	1.0
1993-96 Mirage	1.1	1.3
Starion		
front each	1.2	1.4
rear	1.3	1.5
Transmission		
3000GT	1.4	1.6
Cordia, Tredia	.8	1.0
Diamante		
1992-96	.8	1.0
1997-04	2.3	2.5
Eclipse		
1990-94	1.2	1.4
1995-99	2.8	3.0
2000-05 (1.6)	2.2	2.4
Galant		
1985-93	1.0	1.2
1994-98	2.6	2.8
1999-05 (1.3)	1.8	2.0
Lancer (1.3)	1.8	2.0
Lancer		
Evolution (2.3)	3.1	3.3
Mirage		
1985-96	1.1	1.3
1997-03	1.6	1.8
Sigma	.8	1.0

LANCER EVOLUTION : MIRAGE : PRECIS : SIGMA : STARION : TREDIA MITSU-11

	LABOR TIME	SEVERE SERVICE		LABOR TIME	SEVERE SERVICE		LABOR TIME	SEVERE SERVICE
CYLINDER HEAD			Diamante			1993-96		
Compression Test (B)			1992-96			1.5L	3.7	3.9
4 cyl.			SOHC			1.8L	4.8	5.0
SOHC	.8	1.0	front	8.5	8.7	1997-03		
DOHC	1.1	1.3	rear	9.7	9.9	1.5L	2.9	3.1
V6			DOHC			1.8L	5.6	5.8
SOHC	1.6	1.8	front	10.9	11.1	Precis	3.8	4.0
DOHC	2.8	3.0	rear	12.4	12.6	Starion	3.7	3.9
Cylinder Head, Replace (B)			1997-04			V6		
Includes: Parts transfer, adjustments.			front	10.9	11.1	3000GT		
4 cyl.			rear	10.8	11.0	SOHC		
Cordia, Tredia	6.6	6.8	Eclipse			front	5.9	6.1
Eclipse			right side (9.2)	12.4	12.6	rear	7.1	7.3
1990-94			left side (8.8)	11.9	12.1	DOHC		
SOHC	6.8	7.0	Galant			front	6.8	7.0
DOHC	10.9	11.1	right side (8.9)	12.0	12.2	rear	8.8	9.0
1995-99			left side (8.8)	11.9	12.1	turbo		
exc. below	8.8	9.0	Sigma			front	7.8	8.0
2.0L 420A	9.9	10.1	front	7.8	8.0	rear	9.9	10.1
turbo	11.9	12.1	rear	9.1	9.3	Diamante		
2000-05 (8.0)	10.8	12.0	w/3-way catalytic converter add			1992-96		
Galant			3000GT left head only	.8	1.0	SOHC		
1985-93			92-96 Diamante left			front	5.9	6.1
SOHC	6.8	7.0	head only	.8	1.0	rear	7.2	7.4
DOHC	10.9	11.1	w/AC add			DOHC		
1994-98			3000GT	.5	.7	front	7.4	7.6
SOHC	9.8	10.0	90-99 Eclipse DOHC	.5	.7	rear	9.2	9.4
DOHC	12.4	12.6	94-98 Galant DOHC	.5	.7	1997-04 one side	7.2	7.4
1999-03	10.9	11.1	97-02 Mirage 1.8L	.3	.5	Eclipse		
2004-05 (8.8)	11.9	12.1	**Cylinder Head Gasket, Replace (B)**			right side (5.8)	7.9	8.1
Lancer (8.3)	11.2	11.4	4 cyl.			left side (5.5)	7.4	7.6
Lancer			Cordia, Tredia	3.8	4.0	Galant		
Evolution (11.7)	15.8	16.0	Eclipse			right side (5.7)	7.7	7.9
Mirage			1990-94			left side (5.3)	7.2	7.4
1985-92			SOHC	3.7	3.9	Sigma		
1.5L	6.8	7.0	DOHC	6.9	7.1	front	5.9	6.1
1.6L	10.8	11.0	1995-99			rear	6.9	7.1
1993-96			exc. below	4.3	4.5	w/3-way catalytic converter add		
1.5L	7.7	7.9	2.0L 420A	8.4	8.6	3000GT left head only	.8	1.0
1.8L	9.6	9.8	turbo	6.8	7.0	92-96 Diamante left		
1997-03			2000-05 (3.2)	4.3	4.5	head only	.8	1.0
1.5L	6.9	7.1	Galant			w/AC add		
1.8L	11.2	11.5	1985-93			3000GT	.5	.7
Precis	5.9	6.1	SOHC	3.6	3.8	90-99 Eclipse DOHC	.5	.7
Starion	6.8	7.0	DOHC	6.8	7.0	94-98 Galant DOHC	.5	.7
V6			1994-98			97-03 Mirage 1.8L	.3	.5
3000GT			SOHC	4.8	5.0	Replace timing belt add		
SOHC			DOHC	7.9	8.1	4 cyl.	.5	.7
front	8.5	8.7	1999-03	4.4	4.6	V6		
rear	9.7	9.9	2004-05 (3.9)	5.3	5.5	SOHC		
DOHC			Lancer (3.6)	4.9	5.1	front	.5	.7
front	10.8	11.0	Lancer			rear	1.3	1.5
rear	11.8	12.0	Evolution (6.7)	9.0	9.2	DOHC		
turbo			Mirage			front	.7	.9
front	11.5	11.7	1985-92			rear	1.6	1.8
rear	13.2	13.5	1.5L	3.7	3.9	**Jet Valves, Replace (B)**		
			1.6L	6.5	6.7	Cordia, Starion, Tredia	2.3	2.5
						1985-93 Galant	2.3	2.5
						1985-92 Mirage	2.5	2.7
						Precis	.8	1.0

MITSU-12 3000GT : CORDIA : DIAMANTE : ECLIPSE : GALANT : LANCER

	LABOR TIME	SEVERE SERVICE
Rocker Arms or Shafts, Replace (B)		
4 cyl.		
Cordia, Tredia	1.7	1.9
Eclipse		
1990-94	2.1	2.3
1995-99		
exc. below	.9	1.1
2.0L 420A	4.8	5.0
turbo	2.1	2.3
2000-05 (1.0)	1.4	1.6
Galant		
1985-93	1.8	2.0
1994-98		
SOHC	1.1	1.3
DOHC	2.1	2.3
1999-05 (1.1)	1.5	1.7
Lancer, Lancer		
Evolution (.8)	1.2	1.4
Mirage		
1985-96	1.8	2.0
1997-03	1.5	1.7
Precis	1.3	1.5
Starion	1.9	2.1
V6		
3000GT		
right	1.8	2.0
left		
SOHC	1.8	2.0
DOHC	2.7	2.9
turbo	3.4	3.6
Diamante		
1992-96		
SOHC		
right side	1.5	1.7
left side	1.9	2.1
DOHC		
right side	1.8	2.0
left side	2.7	2.9
1997-04		
right side	2.7	2.9
left side	1.3	1.5
Eclipse		
right side (2.7)	3.6	3.8
left side (1.1)	1.5	1.7
Galant		
right side (2.5)	3.4	3.6
left side (1.1)	1.5	1.7
Sigma		
front	1.8	2.0
rear	2.5	2.7
w/AC 95-99 Eclipse		
2.0L 420A add	.5	.7
Valve/Cam Cover and/or Gasket, Replace (B)		
4 cyl.		
Cordia, Tredia	.5	.7
Eclipse, Mirage (.6)	.8	1.0
Galant, Lancer (.4)	.6	.8
Lancer Evolution (.7)	.9	1.1
Precis, Starion	.5	.7

	LABOR TIME	SEVERE SERVICE
V6		
3000GT		
front	.8	1.0
rear	1.4	1.6
Diamante		
1992-96		
front	.8	1.0
rear	1.4	1.6
1997-04		
front	1.7	1.9
rear	.6	.8
Eclipse		
right side (2.1)	2.8	3.0
left side (.4)	.6	.8
Galant		
right side (1.9)	2.6	2.8
left side (.4)	.6	.8
Sigma		
front	.5	.7
rear	1.1	1.3
Valve Job (A)		
4 cyl.		
Cordia, Tredia	7.8	8.0
Eclipse		
1990-94		
SOHC	6.2	6.4
DOHC	10.8	11.0
1995-99		
exc. below	9.6	9.8
2.0L 420A	10.5	10.7
turbo	11.8	12.0
2000-05 (8.0)	10.8	11.0
Galant		
1985-93		
SOHC	6.5	6.7
DOHC	10.8	11.0
1994-98		
SOHC	9.9	10.1
DOHC	12.6	12.8
1999-03	9.8	10.0
2004-05 (8.8)	11.9	12.1
Lancer (8.1)	10.9	11.1
Lancer		
Evolution (11.6)	15.7	15.9
Mirage		
1985-92		
SOHC	6.5	6.7
DOHC	10.8	11.0
1993-96		
1.5L	6.9	7.1
1.8L	9.8	10.0
1997-03		
1.5L	7.4	7.6
1.8L	12.2	12.5
Precis		
1987-89	7.8	8.0
1990-94	7.2	7.4
Starion	7.8	8.0

	LABOR TIME	SEVERE SERVICE
V6		
3000GT		
SOHC	11.3	11.5
DOHC	15.1	15.3
turbo	16.8	17.0
Diamante		
1992-96		
SOHC	11.6	11.8
DOHC	15.9	16.1
1997-04	15.6	15.8
Eclipse each		
side (11.0)	14.9	15.1
Galant		
right side (9.9)	13.4	13.6
left side (9.0)	12.2	12.4
Sigma	13.9	14.1
w/3-way cat. converter add		
3000GT		
w/turbo	1.4	1.6
w/o turbo	.8	1.0
92-96 Diamante	.8	1.0
w/AC add		
3000GT	.5	.7
90-99 Eclipse DOHC	.5	.7
94-98 Galant DOHC	.5	.7
97-03 Mirage 1.8L	.3	.5
w/PS 90-94 Eclipse		
DOHC add	.3	.5
Valve Lash Adjuster, Replace (B)		
4 cyl.		
Cordia, Tredia	1.6	1.8
Eclipse		
1990-94	2.0	2.2
1995-02		
w/turbo	1.7	1.9
w/o turbo	3.2	3.4
Galant		
1985-93	1.7	1.9
1994-02	1.7	1.9
1989-92 Mirage	2.0	2.2
Precis	.9	1.1
Starion	1.6	1.8
V6		
3000GT		
SOHC		
front side	1.4	1.6
rear side	1.7	1.9
DOHC		
front side	1.7	1.9
rear side	2.5	2.7
Diamante		
1992-96		
SOHC		
front	1.4	1.6
rear	1.7	1.9
DOHC		
front	1.7	1.9
rear	2.5	2.7

LANCER EVOLUTION : MIRAGE : PRECIS : SIGMA : STARION : TREDIA — MITSU-13

	LABOR TIME	SEVERE SERVICE
1997-03		
right side	2.8	3.0
left side	1.6	1.8
Eclipse		
right side (2.7)	3.6	3.8
left side (1.2)	1.6	1.8
Galant		
right side (2.1)	2.8	3.0
left side (1.2)	1.6	1.8
Sigma		
front	1.6	1.8
rear	2.5	2.7
Valve Springs and/or Oil Seals, Replace (B)		
4 cyl.		
Cordia, Tredia	3.8	4.0
Eclipse		
1990-94		
SOHC	3.8	4.0
DOHC	8.9	9.1
1995-99		
SOHC	5.2	5.4
DOHC	8.6	8.8
2000-05 (3.9)	5.3	5.5
Galant		
1985-93		
SOHC		
2V	3.8	4.0
4V	4.8	5.0
DOHC	8.4	8.6
1994-98		
SOHC	4.9	5.1
DOHC	9.3	9.5
1999-03	5.2	5.4
2004-05 (4.4)	5.9	6.1
Lancer (3.6)	4.9	5.1
Lancer		
Evolution (6.8)	9.2	9.4
Mirage		
1985-92		
SOHC	3.8	4.0
DOHC	7.9	8.1
1993-96		
1.5L	4.8	5.0
1.8L	5.6	5.8
1997-03		
1.5L	4.3	4.5
1.8L	5.5	5.7
Precis	6.2	6.4
Starion	3.7	3.9
V6		
3000GT		
SOHC	6.7	6.9
DOHC	12.8	13.0
turbo	13.5	13.7
Diamante		
1992-96		
SOHC	6.7	6.9
DOHC	13.7	13.9
1997-04	11.4	11.6

	LABOR TIME	SEVERE SERVICE
Eclipse		
each side (8.5)	11.5	11.7
Galant		
each side (8.1)	10.9	11.1
Sigma	6.3	6.5
w/AC add		
3000GT	1.1	1.3
90-99 Eclipse DOHC	.5	.7
CAMSHAFT		
Timing Belt, Adjust (B)		
Precis	.3	.5
Balance Shaft, Replace (B)		
Cordia, Tredia	9.6	9.8
Eclipse		
1990-94		
exc. below	9.2	9.4
AWD	10.2	10.4
turbo	9.6	9.8
1995-05 (5.1)	6.9	7.1
Galant		
1985-93	9.4	9.6
1994-98		
SOHC	6.7	6.9
DOHC	7.6	7.8
1999-03	6.2	6.4
2004-05 (4.0)	5.4	5.6
Lancer Evolution (5.1)	6.9	7.1
Starion	8.3	8.5
w/ABS 90-94 Eclipse add	.5	.7
w/AC add		
90-99 Eclipse	.5	.7
90-94 Eclipse DOHC	.9	1.1
94-98 Galant	.5	.7
w/AT 90-94 Eclipse add	.8	1.0
w/cruise control		
90-94 Eclipse add	.3	.5
Camshaft, Replace (A)		
4 cyl.		
Cordia, Tredia	1.6	1.8
Eclipse		
1990-94		
SOHC	1.8	2.0
DOHC	4.5	4.7
1995-99		
exc. below	1.8	2.0
2.0L 420A	4.8	5.0
2000-05 (1.8)	2.4	2.6
Galant		
1985-93		
SOHC	1.8	2.0
DOHC	4.5	4.7
1994-98		
SOHC	1.7	1.9
DOHC	3.9	4.1
1999-05 (2.0)	2.7	2.9
Lancer (1.4)	1.9	2.1
Lancer		
Evolution (4.0)	5.4	5.6

	LABOR TIME	SEVERE SERVICE
Mirage		
1985-92		
SOHC	1.8	2.0
DOHC		
one	4.2	4.4
both	4.5	4.7
1993-96	1.8	2.0
1997-03		
1.5L	1.8	2.0
1.8L	2.4	2.6
Precis		
1987-89	3.6	3.8
1990-94	2.9	3.1
Starion	1.7	1.9
V6		
3000GT		
SOHC		
right side	1.8	2.0
left side	4.7	4.9
DOHC		
right side	5.7	5.9
left side	6.3	6.5
Diamante		
1992-96		
SOHC		
right side	1.8	2.0
left side	4.7	4.9
DOHC		
right side	6.6	6.8
left side	7.4	7.6
1997-03		
right side	6.3	6.5
left side	6.1	6.3
2004		
right side	5.6	5.8
left side	5.2	5.4
Eclipse		
one side (4.8)	6.4	6.6
Galant		
one side (4.6)	6.2	6.4
Sigma	1.9	2.1
w/AC add		
3000GT	1.1	1.3
90-99 Eclipse DOHC	.5	.7
w/PS 90-94 Eclipse		
DOHC add	.3	.5
w/turbo add		
3000GT	.5	.7
95-99 Eclipse	2.8	3.0
Camshaft Seal, Replace (B)		
4 cyl.		
Cordia, Tredia	1.7	1.9
Eclipse		
1990-94		
SOHC	1.7	1.9
DOHC	4.4	4.6

MITSU-14 3000GT : CORDIA : DIAMANTE : ECLIPSE : GALANT : LANCER

	LABOR TIME	SEVERE SERVICE
Camshaft Seal, Replace (B)		
1995-99		
exc. below	1.1	1.3
2.0L 420A	4.1	4.3
turbo	3.8	4.0
2000-05 (.8)	1.1	1.3
Galant		
1985-93		
SOHC	1.7	1.9
DOHC	4.4	4.6
1994-05 (1.6)	2.2	2.5
Lancer (.6)	.8	1.0
Lancer		
Evolution (3.0)	4.1	4.3
Mirage		
1985-92		
SOHC	1.7	1.9
DOHC	4.4	4.6
1993-03	1.7	1.9
Precis	1.1	1.3
V6		
3000GT		
SOHC one side	4.6	4.8
DOHC		
right side	5.5	5.7
left side	6.1	6.3
Diamante		
1992-96		
SOHC one side	4.6	4.8
DOHC		
right side	6.4	6.6
left side	7.1	7.3
1997-04 one side	4.0	4.2
Eclipse		
one side (2.6)	3.5	3.7
Galant		
one side (2.9)	3.9	4.1
Sigma		
front	1.7	1.9
rear	4.5	4.7
w/AC add		
3000GT	1.1	1.1
90-99 Eclipse DOHC	.5	.7
w/PS 90-94 Eclipse		
DOHC add	.3	.5
w/turbo add		
3000GT	.5	.7
95-99 Eclipse	2.8	3.0
Front Case Assy., Replace (B)		
Includes: R&I timing belt and oil pan assembly.		
Cordia, Tredia	4.5	4.7
Eclipse		
1990-94		
SOHC	4.9	5.1
DOHC	7.2	7.4
AWD	7.7	7.9

	LABOR TIME	SEVERE SERVICE
1995-99		
2.0L 420A	5.8	6.0
4G6	6.6	6.8
AWD	6.9	7.1
2000-05 (4.9)	6.6	6.8
Galant		
1985-93	4.9	5.1
1994-98		
SOHC	5.3	5.5
DOHC	6.4	6.6
1999-05 (4.6)	6.2	6.4
Lancer (4.2)	5.7	5.9
Lancer Evolution (5.1)	6.9	7.1
Mirage		
1985-92	4.4	4.6
1993-96		
1.5L	4.4	4.6
1.8L	5.1	5.3
1997-03		
1.5L	4.0	4.2
1.8L	4.7	4.9
Precis	4.3	4.5
w/AC add		
90-99 Eclipse	.5	.7
94-98 Galant	.5	.7
Mirage	.5	.7
w/PS Mirage add	.5	.7
Silent Shaft Oil Seal, Replace (B)		
Cordia, Tredia	3.3	3.5
Eclipse, Galant	3.6	3.8
Timing Belt, Replace (B)		
4 cyl.		
Cordia, Tredia	2.9	3.1
Eclipse		
1990-94		
SOHC	3.5	3.7
DOHC	4.5	4.8
1995-99		
SOHC	3.5	3.8
DOHC	3.9	4.1
2000-05 (2.7)	3.6	3.8
Galant		
1985-98		
SOHC	3.5	3.8
DOHC	4.7	4.9
1999-04	3.6	3.8
2005 (2.5)	3.4	3.6
Lancer, Lancer		
Evolution (2.9)	3.9	4.0
Mirage		
1985-88	2.8	3.0
1989-92		
SOHC	2.8	3.0
DOHC	3.8	4.0
1993-96	2.9	3.1
1997-03		
1.5L	1.9	2.1
1.8L	2.6	2.8
Precis	1.9	2.1
V6		
3000GT	4.7	4.9

	LABOR TIME	SEVERE SERVICE
Diamante		
1992-96		
SOHC	4.6	4.8
DOHC	5.7	5.9
1997-04	3.9	4.1
Eclipse	3.6	3.8
Galant	3.6	3.8
Sigma	4.6	4.8
w/AC add		
3000GT	1.0	1.2
90-99 Eclipse	.5	.7
88-98 Galant	.5	.7
Mirage	.5	.7
w/PS Mirage add	.5	.7
w/turbo add	.3	.5
Replace tensioner add	.2	.4
Timing Belt Tensioner, Replace (B)		
4 cyl.		
Cordia, Tredia	2.9	3.1
Eclipse		
1990-94		
timing belt		
automatic	4.1	4.3
manual	3.1	3.3
balance shaft belt		
SOHC	2.1	2.3
DOHC	3.2	3.4
1995-99		
timing belt	3.3	3.5
balance		
shaft belt	2.3	2.5
2000-05		
timing belt (2.5)	3.4	3.6
balance		
shaft belt (2.1)	2.8	3.0
Galant		
1985-93	3.0	3.2
1994-98		
timing belt		
SOHC	3.1	3.3
DOHC	4.1	4.3
balance shaft belt		
SOHC	2.3	2.5
DOHC	3.3	3.5
1999-05		
timing belt each	3.3	3.5
balance		
shaft belt	2.9	3.1
Lancer (2.6)	3.5	3.7
Lancer		
Evolution (2.4)	3.2	3.4
Mirage		
1985-96	2.8	3.0
1997-03		
1.5L	1.7	1.9
1.8L	2.3	2.5
Precis	1.6	1.8
V6		
3000GT	4.4	4.6

LANCER EVOLUTION : MIRAGE : PRECIS : SIGMA : STARION : TREDIA — MITSU-15

	Labor Time	Severe Service
Diamante		
1992-96		
automatic	5.4	5.6
manual		
SOHC	4.3	4.5
DOHC	5.4	5.6
1997-04 each	3.8	4.0
Eclipse each (2.5)	3.3	3.5
Galant each (2.7)	3.6	3.8
Sigma	4.2	4.4
w/AC add		
3000GT	1.1	1.3
90-99 Eclipse	.5	.7
94-98 Galant	.5	.7
Mirage	.5	.7
w/PS Mirage add	.5	.7
w/turbo 3000GT add	.5	.7
Timing Belt Cover, Replace (B)		
4 cyl.		
front cover	2.8	3.0
upper	.5	.7
center	1.1	1.3
lower		
Cordia, Tredia	1.7	1.9
Eclipse		
1990-94		
SOHC	2.0	2.2
DOHC	3.1	3.3
1995-99	2.2	2.4
2000-05 (2.1)	2.8	3.0
Galant		
1985-93	2.0	2.2
1994-98		
SOHC	1.6	1.8
DOHC	3.1	3.3
1999-05 (2.1)	2.8	3.0
Lancer (1.3)	1.8	2.0
Lancer		
Evolution (1.9)	2.6	2.8
Mirage		
1985-96	1.7	1.9
1997-03	1.5	1.7
Precis	1.2	1.4
V6		
3000GT		
upper		
SOHC one side	.8	1.0
DOHC		
right side	.8	1.0
left side	2.0	2.2
lower	4.0	4.2
Diamante		
1992-96		
SOHC		
upper one side	.8	1.0
lower	3.3	3.5

	Labor Time	Severe Service
DOHC		
upper		
right side	.8	1.0
left side	2.0	2.2
lower	5.0	5.2
1997-04		
upper		
right side	1.3	1.5
left side	.7	.9
lower	2.0	2.2
Eclipse		
upper		
right side (1.0)	1.4	1.6
left side (.3)	.5	.7
lower (1.2)	1.6	1.8
Galant		
upper		
right side (1.0)	1.4	1.6
left side (.3)	.5	.7
lower (1.2)	1.6	1.8
Sigma		
upper		
front	.5	.7
rear	.7	.9
lower	3.3	3.5
w/AC lower only add		
3000GT	.5	.7
92-99 Eclipse	.5	.7
94-98 Galant	.5	.7
Mirage	.5	.7
w/PS Mirage lower only add	.3	.5
Timing Case Cover Seal, Replace (B)		
Starion	2.6	2.8
w/AC add	.3	.5
Timing Chain, Replace (B)		
Starion	7.9	8.1
w/AC add	1.2	1.4
Timing Chain Housing Gasket, Replace (B)		
Starion	8.2	8.4
w/AC add	.7	.9
Timing Chain Tensioner, Replace (B)		
Starion	7.5	7.7
w/AC add	.2	.2

CRANK & PISTONS

Connecting Rod Bearings, Replace (A)
Includes: Check bearing oil clearance.

	Labor Time	Severe Service
4 cyl.		
Cordia, Tredia	3.4	3.6
Eclipse		
1990-94		
SOHC	3.8	4.0
DOHC	4.9	5.1
AWD	5.3	5.5

	Labor Time	Severe Service
1995-99		
SOHC 4V	4.8	5.0
DOHC	3.9	4.1
AWD, turbo	4.8	5.0
2000-05 (3.2)	4.3	4.6
Galant		
1985-93		
SOHC	3.8	4.0
DOHC	4.9	5.1
1994-05 (3.0)	4.1	4.3
Lancer (3.0)	4.1	4.3
Lancer		
Evolution (3.6)	4.9	5.1
Mirage		
1985-92		
SOHC	3.5	3.7
DOHC	4.9	5.1
1993-03		
1.5L	3.8	4.0
1.8L	4.4	4.6
Precis	4.1	4.3
Starion	3.6	3.8
V6		
3000GT		
AWD	6.3	6.5
FWD	5.9	6.1
Diamante		
1992-96	5.9	6.1
1997-04	5.4	5.6
Eclipse (4.6)	6.2	6.4
Galant (4.2)	5.7	5.9
Sigma	4.7	4.9
w/aluminum alloy oil pan		
92-96 Diamante add		
w/4WS	.5	.7
w/o 4WS	1.1	1.3
w/turbo 3000GT w/3-way catalytic converter add	1.4	1.6

Crankshaft and Main Bearings, Replace (A)
Includes: Engine R&R.

	Labor Time	Severe Service
4 cyl.		
Cordia, Tredia	13.7	13.9
Eclipse		
1990-94	13.3	13.5
1995-99	16.4	16.6
2000-05 (11.1)	15.0	15.2
Galant		
1985-93	14.0	14.2
1994-03	15.1	15.3
2004-05 (12.5)	16.9	17.1
Lancer (10.2)	13.8	14.0
Lancer		
Evolution (14.2)	19.2	19.4
Mirage	13.7	13.9
Precis		
1987-89	14.2	14.4
1990-94	12.6	12.8
Starion	14.7	14.9

MITSU-16 3000GT : CORDIA : DIAMANTE : ECLIPSE : GALANT : LANCER

	LABOR TIME	SEVERE SERVICE
Crankshaft and Main Bearings, Replace (A)		
V6		
3000GT	17.6	17.8
Diamante		
1992-96	18.0	18.2
1997-04	19.8	20.0
Eclipse (12.8)	17.3	17.5
Galant		
1999-03	16.9	17.1
2004-05 (13.9)	18.8	19.0
Sigma	16.3	16.5
w/4WS 92-96 Diamante add	.6	.8
w/ABS 90-94 Eclipse add	.5	.7
w/AC add		
3000GT	.5	.7
Eclipse		
exc. below	.6	.8
90-94 DOHC	.9	1.1
Galant	.5	.7
Lancer	.3	.5
Mirage	.6	.8
w/AT add		
3000GT	.8	1.0
Eclipse	.8	1.0
Galant	.8	1.0
Lancer	1.1	1.3
Mirage	.8	1.0
w/AWD add		
3000GT	.5	.7
Eclipse	.6	.8
w/cruise control		
90-94 Eclipse add	.3	.5
w/ECS 92-96 Diamante add	.3	.5
w/PS Mirage add	.5	.7
w/turbo add		
3000GT	1.8	2.0
Eclipse	.5	.7
Crankshaft Front Oil Seal, Replace (B)		
4 cyl.		
Cordia, Tredia	3.2	3.4
Eclipse		
1990-94		
SOHC	3.3	3.5
DOHC	4.4	4.6
1995-05 (2.7)	3.6	3.8
Galant		
SOHC (2.5)	3.4	3.6
DOHC (3.3)	4.5	4.7
Lancer, Lancer Evolution (2.7)	3.6	3.8
Mirage		
1985-96	2.9	3.1

	LABOR TIME	SEVERE SERVICE
1997-03		
1.5L	1.9	2.1
1.8L	2.5	2.7
Starion	1.7	1.9
V6		
3000GT	4.6	4.8
Diamante		
1992-96		
sedan		
SOHC	4.4	4.6
DOHC	5.6	5.8
wagon	4.8	5.0
1997-03	4.0	4.2
Eclipse, Galant (2.5)	3.4	3.6
Sigma	4.4	4.6
w/AC add		
3000GT	1.1	1.3
90-99 Eclipse	.6	.8
94-98 Galant	.5	.7
Mirage	.5	.7
w/PS Mirage add	.5	.7
w/turbo 3000GT add	.5	.7
Crankshaft Pulley, Replace (B)		
Exc. below (.7)	.9	1.1
1995-04 Eclipse (.8)	1.1	1.3
Lancer (.9)	1.2	1.4
Lancer Evolution (1.0)	1.4	1.6
Sigma	1.3	1.5
w/AC add		
90-94 Eclipse	.2	.4
00-05 Eclipse	.2	.4
Galant	.2	.4
Mirage	.2	.4
Pistons or Connecting Rods, Replace (A)		
Includes: Ridge reaming, cylinder wall deglazing, installing new rings and rod bearings, engine tune-up.		
4 cyl.		
Cordia, Tredia	11.9	12.1
Eclipse		
1990-94	10.2	10.4
1995-99	13.1	13.3
2000-05 (8.6)	11.6	11.8
Galant		
1985-93	10.2	10.4
1994-03	11.8	12.0
2004-05 (10.0)	13.5	13.7
Lancer (7.7)	10.4	10.6
Lancer Evolution (11.8)	15.9	16.1
Mirage		
1985-92	9.9	10.1
1993-96	10.5	10.7
1997-03	9.7	9.9
Precis	8.8	9.0
Starion	12.1	13.0

	LABOR TIME	SEVERE SERVICE
V6		
3000GT	14.2	14.4
Diamante		
1992-96		
SOHC	13.8	14.0
DOHC	15.3	15.5
1997-04	16.9	17.1
Eclipse (12.1)	16.3	16.5
Galant (11.6)	15.7	15.9
Sigma	14.1	14.3
w/4WS 92-96 Diamante add	.6	.8
w/ABS 90-94 Eclipse add	.5	.7
w/AC add		
3000GT	.5	.7
Eclipse		
exc. below	.6	.8
90-94 DOHC	.9	1.1
Galant	.5	.7
Lancer	.3	.5
Mirage	.6	.8
w/AT add		
3000GT	.8	1.0
Eclipse	.8	1.0
Galant	.8	1.0
Lancer	1.1	1.3
Mirage	.8	1.0
w/AWD add		
3000GT	.5	.7
Eclipse	.6	.8
w/cruise control 90-94 Eclipse add	.3	.5
w/ECS 92-96 Diamante add	.3	.5
w/PS Mirage add	.5	.7
w/turbo add		
3000GT	1.8	2.0
90-94 Eclipse	.5	.7
Rear Main Oil Seal, Replace (B)		
4 cyl.		
Cordia, Tredia	6.1	6.3
Eclipse		
1990-94	4.6	4.8
1995-05 (6.2)	8.4	8.6
Galant		
1985-93	5.8	6.0
1994-98	9.0	9.2
1999-05 (6.0)	8.1	8.3
Lancer (4.7)	6.3	6.5
Lancer Evolution (7.9)	10.7	10.9
Mirage		
1985-92	4.6	4.8
1993-96		
1.5L	5.0	5.2
1.8L	7.4	7.6
1997-03		
1.5L	5.3	5.5
1.8L	6.8	7.0

LANCER EVOLUTION : MIRAGE : PRECIS : SIGMA : STARION : TREDIA — MITSU-17

	LABOR TIME	SEVERE SERVICE
Precis		
1987-89	3.1	3.3
1990-94	3.0	3.2
Starion	5.0	5.2
V6		
3000GT	6.8	7.0
Diamante		
1992-96	6.5	6.7
1997-04	9.1	9.3
Eclipse (5.7)	7.7	7.9
Galant (5.3)	7.2	7.4
Sigma	5.5	5.7
w/4WS 92-96 Diamante add	.6	.8
w/ABS 90-94 Eclipse add	.5	.7
w/AT add		
3000GT	.5	.7
Eclipse	.5	.7
Galant	.5	.7
Lancer	1.2	1.4
Mirage	.5	.7
w/AWD add		
3000GT	.5	.7
Eclipse	.6	.8
w/cruise control 90-94		
Eclipse add	.3	.5
w/ECS 92-96 Diamante add	.3	.5
w/turbo 3000GT add	.9	1.1

Rings, Replace (A)
Includes: Ridge reaming, cylinder wall deglazing, installing new rings, engine tune-up.

	LABOR TIME	SEVERE SERVICE
4 cyl.		
Cordia, Tredia	11.9	12.1
Eclipse		
1990-94	10.2	10.4
1995-99	13.1	13.3
2000-05 (8.6)	11.6	11.8
Galant		
1985-93	10.2	10.4
1994-03	11.8	12.0
2004-05 (10.0)	13.5	13.7
Lancer (7.7)	10.4	10.6
Lancer		
Evolution (11.8)	15.9	16.1
Mirage		
1985-92	9.9	10.1
1993-96	10.5	10.7
1997-03	9.8	10.0
Precis	8.7	8.9
Starion	12.1	12.3
V6		
3000GT	14.2	14.4
Diamante		
1992-96		
SOHC	13.8	14.0
DOHC	15.3	15.5
1997-04	16.9	17.1
Eclipse (12.0)	16.2	16.4
Galant (11.6)	15.7	15.9
Sigma	14.1	14.3
w/4WS 92-96 Diamante add	.6	.8
w/ABS 90-94 Eclipse add	.5	.7
w/AC add		
3000GT	.5	.7
Eclipse		
exc. below	.6	.8
90-94 DOHC	.9	1.1
Galant	.5	.7
Lancer	.3	.5
Mirage	.6	.8
w/AT add		
3000GT	.8	1.0
Eclipse	.8	1.0
Galant	.8	1.0
Lancer	1.1	1.3
Mirage	.8	1.0
w/AWD add		
3000GT	.5	.7
Eclipse	.6	.8
w/cruise control 90-94		
Eclipse add	.3	.5
w/ECS 92-96 Diamante add	.3	.5
w/PS Mirage add	.5	.7
w/turbo add		
3000GT	1.8	2.0
90-94 Eclipse	.5	.7

ENGINE LUBRICATION

Combination Gauge Assy., Replace (B)

	LABOR TIME	SEVERE SERVICE
3000GT	3.1	3.3

Engine Oil Cooler, Replace (B)

	LABOR TIME	SEVERE SERVICE
3000GT	1.3	1.5
Cordia, Tredia	1.2	1.4
Diamante	1.7	1.9
Eclipse		
1990-94	1.6	1.8
1995-99	.8	1.0
Lancer Evolution (1.2)	1.6	1.8
1985-92 Mirage	.7	.9
Starion	.7	.9

Engine Oil Pressure Switch (Sending Unit), Replace (B)

	LABOR TIME	SEVERE SERVICE
Exc. below	.5	.7
Lancer, Lancer Evolution	.6	.8

Oil Pan and/or Gasket, Replace (B)

	LABOR TIME	SEVERE SERVICE
4 cyl.		
Cordia, Tredia	1.5	1.7
Eclipse		
1990-94		
SOHC	1.6	1.8
DOHC	3.0	3.2
AWD	3.6	3.8
1995-99		
exc. below	1.7	1.9
SOHC 16V	3.0	3.2
turbo	3.0	3.2
2000-05 (1.8)	2.4	2.6
Galant		
1985-93	1.6	1.8
1994-05 (1.6)	2.2	2.4
Lancer		
upper (1.9)	2.6	2.8
lower (1.0)	1.4	1.6
Lancer		
Evolution (2.2)	3.0	3.2
Mirage		
1985-92	1.6	1.8
1993-03		
1.5L	1.9	2.1
1.8L	2.5	2.7
Precis	1.3	1.5
Starion	1.6	1.8
V6		
3000GT		
2WD	3.3	3.5
AWD	5.2	5.4
Diamante		
1992-96		
one-piece pan	3.3	3.5
two-piece pan		
upper	4.4	4.6
lower	1.7	1.9
1997-04		
upper	2.9	3.1
lower	1.1	1.3
Eclipse		
upper (2.6)	3.5	3.7
lower (1.0)	1.4	1.6
Galant		
upper (2.4)	3.2	3.4
lower (1.0)	1.4	1.6
Sigma	2.4	2.6

Oil Pressure Gauge (Dash), Replace (B)

	LABOR TIME	SEVERE SERVICE
Exc. below	.7	.9
3000GT	3.1	3.3
1995-99 Eclipse	1.0	1.2

Oil Pressure Gauge (Engine), Replace (B)
Includes: Pressure test.

	LABOR TIME	SEVERE SERVICE
All Models	.5	.7

Oil Pump, Replace (B)

	LABOR TIME	SEVERE SERVICE
Cordia, Tredia	4.6	4.8
Eclipse		
1990-94		
SOHC	3.8	4.0
DOHC	7.5	7.7
1995-99		
2.0L 420A	6.0	6.2
4G6	6.9	7.1
2000-05 4 cyl. (5.2)	7.0	7.2

MITSU-18   3000GT : CORDIA : DIAMANTE : ECLIPSE : GALANT : LANCER

	LABOR TIME	SEVERE SERVICE
Oil Pump, Replace (B)		
Galant		
1985-93		
SOHC	4.8	5.0
DOHC	7.2	7.4
1994-98		
SOHC	5.8	6.0
DOHC	6.9	7.1
1999-03	6.6	6.8
2004-05 (4.0)	5.4	5.6
Lancer (3.8)	5.1	5.3
Lancer Evolution (4.7)	6.3	6.5
Mirage		
1985-88	3.2	3.4
1989-92	6.9	7.1
Precis	4.1	4.3
Starion	8.5	8.7
w/AC add		
92-99 Eclipse	.5	.7
94-98 Galant	.5	.7
w/AWD Eclipse add	.6	.8
Recondition pump add	.3	.5
Oil Pump Cover and/or Gear, Replace (B)		
4 cyl.		
Mirage		
1993-96	4.6	4.8
1997-03		
1.5L	4.3	4.5
1.8L	5.0	5.2
V6		
3000GT	7.6	7.8
Diamante		
1992-96		
SOHC	7.8	8.0
DOHC	9.0	9.2
1997-04	6.7	6.9
Eclipse (5.0)	6.8	7.0
Galant, Sigma (4.7)	6.3	6.5
w/AC add		
3000GT	1.1	1.3
Mirage	.5	.7
w/aluminum alloy oil pan		
92-96 Diamante add		
w/4WS	.5	.7
w/o 4WS	1.1	1.3
w/AWD 3000GT add	.5	.7
w/PS Mirage add	.5	.7
w/turbo 3000GT add		
w/3-way cat. convert.	1.4	1.6
w/o 3-way cat. convert.	.5	.7

CLUTCH

	LABOR TIME	SEVERE SERVICE
Bleed Clutch Hydraulic System (B)		
All Models	.3	.7
Clutch Pedal Free Play, Adjust (B)		
Exc. 3000GT	.3	.5
3000GT	.8	1.0

	LABOR TIME	SEVERE SERVICE
Clutch Assy., Replace (B)		
3000GT	6.7	6.9
Cordia, Tredia	4.7	4.9
Eclipse		
1990-94	4.4	4.6
1995-99	8.1	8.3
2000-05		
4 cyl. (4.6)	6.2	6.4
V6 (5.6)	7.6	7.8
Galant		
1985-93	4.3	4.5
1994-98	7.0	7.2
1999-05 (4.6)	6.2	6.4
Lancer (4.1)	5.5	5.7
Lancer Evolution (7.9)	10.7	10.9
Mirage		
1985-92	4.4	4.6
1993-96		
1.5L	4.8	5.0
1.8L	5.2	5.4
1997-03	5.1	5.3
Precis		
1987-89	4.4	4.6
1990-94	4.6	4.8
Sigma	5.4	5.6
Starion	3.2	3.4
w/AWD add		
3000GT	.5	.7
90-99 Eclipse	.6	.8
w/ABS 90-94 Eclipse add	.5	.7
w/cruise control 90-94 Eclipse add	.3	.5
w/turbo 3000GT add	1.0	1.2
Clutch Master Cylinder, Replace (B)		
Includes: System bleeding.		
Exc. below	.8	1.0
3000GT	2.7	2.9
1995-05 Eclipse (1.0)	1.4	1.6
1994-05 Galant (.8)	1.1	1.3
Lancer (1.0)	1.4	1.6
Lancer Evolution (1.4)	1.9	2.1
1993-03 Mirage	1.0	1.2
w/turbo 3000GT add	.5	.7
Clutch Release Bearing, Replace (B)		
3000GT	6.6	6.8
Cordia, Tredia	4.4	4.6
Eclipse		
1990-94	4.3	4.5
1995-99	7.9	8.1
2000-05		
4 cyl. (4.5)	6.1	6.3
V6 (5.6)	7.6	7.8
Galant		
1985-93	4.1	4.3
1994-98	6.9	7.1
1999-05 (4.6)	6.2	6.4
Lancer (4.1)	5.5	5.7
Lancer Evolution (7.9)	10.7	10.9

	LABOR TIME	SEVERE SERVICE
Mirage		
1985-92	4.1	4.3
1993-96		
1.5L	4.6	4.8
1.8L	5.2	5.4
1997-03	5.0	5.2
Precis		
1987-89	3.8	4.0
1990-94	4.2	4.4
Sigma	5.0	5.2
Starion	3.2	3.4
w/AWD add		
3000GT	.5	.7
90-99 Eclipse	.6	.8
w/ABS 90-94 Eclipse add	.5	.7
w/cruise control 90-94 Eclipse add	.3	.5
w/ECS 90-94 Eclipse add	.3	.5
w/turbo 3000GT add	1.0	1.2
Clutch Release Cable, Replace (B)		
1985-96 Mirage	.7	.9
Clutch Slave Cylinder, Replace (B)		
Includes: System bleeding.		
Exc. 1995-99 Eclipse	1.2	1.4
1995-99 Eclipse	1.4	1.6
w/turbo 3000GT add	.5	.5
Power Clutch Booster, Replace (B)		
3000GT	2.9	3.1
Flywheel, Replace (B)		
3000GT	7.0	7.2
Cordia, Tredia	5.0	5.2
Eclipse		
1990-94	4.7	4.9
1995-99	8.4	8.6
2000-05		
4 cyl. (4.9)	6.6	6.8
V6 (5.8)	7.8	8.0
Galant		
1985-93	4.9	5.1
1994-98	7.3	7.5
1999-05 (4.9)	6.6	6.8
Lancer (4.9)	6.6	6.8
Lancer Evolution (8.2)	11.1	11.3
Mirage		
1985-92	4.7	4.9
1993-96		
1.5L	5.3	5.5
1.8L	7.6	7.8
1997-03		
1.5L	5.5	5.7
1.8L	7.0	7.2
Precis	4.9	5.1

MITSU-19

LANCER EVOLUTION : MIRAGE : PRECIS : SIGMA : STARION : TREDIA

	LABOR TIME	SEVERE SERVICE
Sigma	5.3	5.5
Starion	3.6	3.8
w/AWD add		
3000GT	.5	.7
90-99 Eclipse	.6	.8
w/ABS 90-94 Eclipse add	.5	.7
w/cruise control 90-94 Eclipse add	.3	.5
w/turbo 3000GT add	1.0	1.2

MANUAL TRANSAXLE

Differential Pinion Bearings, Replace (B)
- 4-Speed Precis
  - 1987-89 .... 4.9 / 5.1
  - 1990-94 .... 5.2 / 5.4
- 5-Speed Precis
  - 1987-89 .... 5.1 / 5.3
  - 1990-94 .... 6.0 / 6.2

Speedometer Driven Gear and/or Seal, Replace (B)
- Exc. Mirage .... .5 / .7
- Mirage .... .6 / .8

Transaxle Assy., R&I or Replace (B)
- 4-Speed
  - Cordia, Tredia ..... 4.3 / 4.5
  - Mirage ..... 4.1 / 4.3
  - Precis ..... 3.8 / 4.0
- 5-Speed
  - 3000GT ..... 6.3 / 6.5
  - Cordia, Tredia ..... 4.2 / 4.4
  - Eclipse
    - 1990-94 ..... 4.0 / 4.2
    - 1995-99 ..... 7.6 / 7.8
    - 2000-04
      - 4 cyl. ..... 5.8 / 6.0
      - V6 ..... 7.2 / 7.4
  - Galant
    - 1985-93 ..... 3.8 / 4.0
    - 1994-98 ..... 6.6 / 6.8
    - 1999-05 (4.3) ..... 5.8 / 6.0
  - Lancer (3.9) ..... 5.3 / 5.5
  - Lancer Evolution (7.6) ..... 10.3 / 10.5
  - Mirage
    - 1985-92 ..... 3.8 / 4.0
    - 1993-96
      - 1.5L ..... 4.4 / 4.6
      - 1.8L ..... 4.9 / 5.1
    - 1997-03 ..... 4.7 / 4.9

	LABOR TIME	SEVERE SERVICE
Precis	3.6	3.8
Sigma	4.9	5.1
w/AWD add		
3000GT	.5	.7
90-99 Eclipse	.6	.8
w/ABS 90-94 Eclipse add	.5	.7
w/cruise control 90-94 Eclipse add	.3	.5
w/turbo 3000GT add	1.0	1.2

Transaxle Assy., R&R and Recondition (A)
Includes: Recondition transmission only.
- 4-Speed
  - Cordia, Tredia ..... 9.0 / 9.2
  - 1985-96 Mirage ..... 9.0 / 9.2
  - Precis ..... 8.4 / 8.6
- 5-Speed
  - 3000GT ..... 10.8 / 11.0
  - Cordia, Tredia ..... 7.2 / 7.4
  - Eclipse
    - 1990-94 ..... 8.9 / 9.1
    - 1995-99
      - exc. 4G6 ..... 11.7 / 11.9
      - 4G6 ..... 12.5 / 12.7
    - 2000-05
      - 4 cyl. (7.9) ..... 10.7 / 10.9
      - V6 (8.8) ..... 11.9 / 12.1
  - Galant
    - 1985-93 ..... 8.8 / 9.0
    - 1994-98 ..... 11.4 / 11.6
    - 1999-05 (7.9) ..... 10.7 / 10.9
  - Lancer (7.3) ..... 9.9 / 10.1
  - Lancer Evolution (11.0) ..... 14.9 / 15.1
  - Mirage
    - 1985-92 ..... 8.8 / 9.0
    - 1993-96
      - 1.5L ..... 9.4 / 9.6
      - 1.8L ..... 9.8 / 10.0
    - 1997-03 ..... 9.7 / 9.9
  - Precis ..... 8.8 / 9.0
  - Sigma ..... 9.4 / 9.6
- w/AWD add
  - 90-94 Eclipse ..... 1.4 / 1.6
  - 95-99 Eclipse ..... .3 / .5
- w/ABS 90-94 Eclipse add ..... .5 / .7
- w/cruise control 90-94 Eclipse add ..... .3 / .5

Transaxle Rear Cover, R&R or Replace (B)
- Precis ..... .5 / .7

	LABOR TIME	SEVERE SERVICE

MANUAL TRANSMISSION

Extension Housing Oil Seal, Replace (B)
- Starion ..... 1.3 / 1.5

Transmission Assy., R&R and Recondition (A)
- Starion ..... 4.9 / 5.1

Transmission Assy., R&R or Replace (B)
- Starion ..... 3.2 / 3.4
- Replace assembly add ..... .5 / .7

AUTOMATIC TRANSAXLE

SERVICE TRANSAXLE INSTALLED

Check Unit for Oil Leaks (C)
- All Models ..... .9 / 1.1

Drain & Refill Unit (B)
- Exc. below ..... .8 / 1.0
- 1997-03 Diamante ..... 2.6 / 2.8
- 2000-05 Eclipse ..... 2.6 / 2.8
- 1999-05 Galant ..... 2.6 / 2.9
- 1997-03 Mirage ..... 2.9 / 3.1

Oil Pressure Check (B)
- 3-Speed
  - Cordia, Tredia ..... .9 / 1.1
  - 1985-96 Mirage ..... .9 / 1.1
  - Precis ..... .5 / .7
- 1985-02 4-Speed ..... .9 / 1.1

Accumulator Assy., R&R or Replace (B)
- Eclipse
  - 1990-94 ..... 1.4 / 1.6
  - 1995-99 ..... 1.9 / 2.1
- Galant
  - 1987-93 ..... 1.4 / 1.6
  - 1994-98 ..... 1.9 / 2.1
- 1989-96 Mirage ..... 1.4 / 1.6

Accumulator Piston, Replace or Recondition (B)
- 1985-99 4-Speed ..... 1.8 / 2.0

Converter Clutch Solenoid, Replace (B)
- Eclipse, Galant ..... 2.0 / 2.2

Electronic Control Module, Replace (B)
- 3000GT ..... 1.0 / 1.2
- Diamante ..... 1.4 / 1.6
- Eclipse (.8) ..... 1.1 / 1.3
- Galant
  - 1987 ..... 1.4 / 1.6
  - 1988-05 (.8) ..... 1.1 / 1.3
- Lancer (.3) ..... .5 / .7
- Mirage ..... .8 / 1.0
- Sigma ..... 1.5 / 1.7

Input Speed Sensor, Replace (B)
- Eclipse, Galant ..... .5 / .7

MITSU-20   3000GT : CORDIA : DIAMANTE : ECLIPSE : GALANT : LANCER

	LABOR TIME	SEVERE SERVICE
Lock-up Solenoid, Replace (B)		
Eclipse, Galant	2.0	2.2
Oil Pan and/or Gasket, Replace (B)		
Exc. below	.8	1.0
1997-04 Diamante	2.7	2.9
Eclipse		
1995-99	1.1	1.3
2000-05 (2.0)	2.7	2.9
Galant		
1994-98	1.1	1.3
1999-03	2.7	2.9
2004-05 (2.4)	3.2	3.4
Lancer (2.0)	2.7	2.9
1997-03 Mirage	3.0	3.2
Output Speed Sensor, Replace (B)		
Eclipse, Galant	.5	.7
Pulse Generator, Replace (B)		
Exc. below	.5	.7
1997-04 Diamante	1.6	1.8
2000-05 Eclipse	1.1	1.3
1999-05 Galant	.8	1.0
Lancer (.8)	1.1	1.3
1997-03 Mirage	.8	1.0
Valve Body, R&R and Recondition (A)		
3-Speed		
Cordia, Tredia	3.1	3.3
Mirage		
1985-92	3.1	3.3
1993-96	4.7	4.9
Precis	3.3	3.5
4-Speed		
exc. below	4.9	5.1
1997-04 Diamante	6.8	7.0
2000-05 Eclipse	6.8	7.0
1999-05 Galant	6.8	7.0
Lancer	7.0	7.2
1997-03 Mirage	7.1	7.3
Valve Body, Replace (B)		
3-Speed		
Cordia, Tredia	1.7	1.9
Mirage		
1985-92	1.9	2.1
1993-96	1.3	1.5
Precis		
1987-89	1.6	1.8
1990-94	1.4	1.6
4-Speed		
exc. below	1.3	1.5
1997-04 Diamante	3.3	3.5
2000-05 Eclipse	3.3	3.5
1999-05 Galant	3.3	3.5
Lancer	3.5	3.7
1997-03 Mirage	3.6	3.8

	LABOR TIME	SEVERE SERVICE
SERVICE TRANSAXLE REMOVED		
Transaxle R&R included unless otherwise noted.		
Flywheel (Flexplate), Replace (B)		
3000GT	7.1	7.3
Cordia, Tredia	5.1	5.3
Diamante		
1992-96	6.4	6.6
1997-04	8.2	8.4
Eclipse		
1990-94	4.9	5.1
1995-99	8.5	8.7
2000-05		
4 cyl. (4.8)	6.5	6.7
V6 (5.8)	7.8	8.0
Galant		
1985-93	4.9	5.1
1994-98	7.4	7.6
1999-05		
4 cyl. (4.8)	6.5	6.7
V6 (5.6)	7.6	7.8
Lancer (5.6)	7.6	7.8
Mirage		
1985-92	4.9	5.1
1993-96		
1.5L	5.3	5.5
1.8L	7.8	8.0
1997-03		
1.5L	5.3	5.5
1.8L	6.9	7.1
Precis		
1987-89	4.5	4.7
1990-94	3.4	3.6
Sigma	5.7	5.9
w/4WS 92-96 Diamante add	.6	.8
w/ABS 90-94 Eclipse add	.5	.7
w/AWD add		
3000GT	.5	.7
90-99 Eclipse	.6	.8
w/cruise control 90-94 Eclipse add	.3	.5
w/ECS 92-96 Diamante add	.3	.5
Governor Assy., Replace or Recondition (B)		
1987-89 Precis	8.3	8.5
Recondition add	.2	.4
Torque Converter, Replace (B)		
3000GT	7.2	7.4
Cordia, Tredia	5.1	5.3
Diamante		
1992-96	6.4	6.6
1997-04	8.2	8.4
Eclipse		
1990-94	4.9	5.1
1995-99	8.5	8.7

	LABOR TIME	SEVERE SERVICE
2000-05		
4 cyl. (4.9)	6.6	6.8
V6 (5.8)	7.8	8.0
Galant		
1985-93	4.8	5.0
1994-98	7.5	7.7
1999-05		
4 cyl. (4.9)	6.6	6.8
V6 (5.6)	7.6	7.8
Lancer (4.9)	6.6	6.8
Mirage		
1985-92	4.8	5.0
1993-96		
1.5L	5.3	5.5
1.8L	5.8	6.0
1997-03	5.3	5.5
Precis		
1987-89	4.3	4.5
1990-94	3.2	3.4
Sigma	5.5	5.7
w/4WS 92-96 Diamante add	.6	.8
w/ABS 90-94 Eclipse add	.5	.7
w/AWD 90-99 Eclipse add	.6	.8
w/cruise control 90-94 Eclipse add	.3	.5
w/ECS 92-96 Diamante add	.3	.5
Transaxle Assy., R&I or Replace (B)		
3000GT	7.0	7.2
Cordia, Tredia	4.8	5.0
Diamante		
1992-96	6.1	6.3
1997-04	7.9	8.1
Eclipse		
1990-94	4.6	4.8
1995-99	8.2	8.4
2000-05		
4 cyl. (4.6)	6.2	6.4
V6 (5.6)	7.6	7.8
Galant		
1985-93	4.6	4.8
1994-98	7.1	7.3
1998-05		
4 cyl. (4.6)	6.2	6.4
V6 (5.4)	7.3	7.5
Lancer (4.8)	6.5	6.7
Mirage		
1985-92	4.6	4.8
1993-96		
1.5L	5.2	5.4
1.8L	5.7	5.9
1997-03	5.2	5.4
Precis		
1987-89	4.1	4.3
1990-94	3.0	3.2

LANCER EVOLUTION : MIRAGE : PRECIS : SIGMA : STARION : TREDIA — **MITSU-21**

	LABOR TIME	SEVERE SERVICE
Sigma	5.3	5.5
w/4WS 92-96 Diamante add	.6	.8
w/ABS 90-94 Eclipse add	.5	.7
w/AWD add		
3000GT	.5	.7
90-99 Eclipse	.6	.8
w/cruise control 90-94 Eclipse add	.3	.5
w/ECS 92-96 Diamante add	.3	.5
Flush oil cooler and lines add	.6	.8

Transaxle Assy., R&R and Recondition (A)

	LABOR TIME	SEVERE SERVICE
3000GT	16.0	16.2
Cordia, Tredia	12.6	12.8
Diamante		
1992-96	15.2	15.4
1997-04	17.6	17.8
Eclipse		
1990-94		
3-Speed	13.7	13.9
4-Speed	13.1	13.3
1995-99		
exc. 2.0L 420A	17.0	17.2
2.0L 420A	17.4	17.6
2000-05		
4 cyl. (11.9)	16.1	16.3
V6 (12.8)	17.3	17.5
Galant		
1985-93		
3-Speed	16.4	16.6
4-Speed	13.1	13.3
1994-98	15.6	15.8
1999-05		
4 cyl. (11.9)	16.1	16.3
V6 (12.6)	17.0	17.2
Lancer	15.8	16.0
Mirage		
1985-92	13.1	13.3
1993-96		
1.5L	13.5	13.7
1.8L	13.9	14.1
1997-03	14.7	14.9
Precis		
3-Speed	10.5	10.7
4-Speed	12.5	12.7
Sigma	13.2	13.4
w/4WS 92-96 Diamante add	.6	.8
w/ABS 90-94 Eclipse add	.5	.7
w/AWD add		
90-94 Eclipse	2.9	3.1
95-99 Eclipse	2.4	2.6

	LABOR TIME	SEVERE SERVICE
w/cruise control 90-94 Eclipse add	.3	.5
w/ECS 92-96 Diamante add	.3	.5
Flush oil cooler and lines add	.6	.8
Recondition diff. add	2.2	2.4

AUTOMATIC TRANSMISSION
SERVICE TRANSMISSION INSTALLED

Check Unit for Oil Leaks (C)
Starion	.7	.9

Drain and Refill Unit (B)
Starion	.6	.8

Kickdown Band, Adjust (B)
Starion	.5	.7

Oil Pressure Check (B)
Starion	.9	1.1

Overdrive Band, Adjust (B)
Starion	.5	.7

Accumulator Piston, Replace or Recondition (B)
Starion	1.7	1.9

Driveshaft Seal, Replace (B)
Starion	.7	.9

Extension Housing Oil Seal, Replace (B)
Starion	.7	.9

Kickdown Servo, Replace or Recondition (B)
Includes: R&R oil pan.
Starion	1.3	1.5

Kickdown Switch, Replace (B)
Starion	.5	.7

Oil Pan Gasket, Replace (B)
Starion	.8	1.0

Overdrive Cancel Switch, Replace (B)
Starion	.5	.7

Parking Lock Sprag Control Rod, Replace (B)
Starion	1.9	2.1

Throttle & Manual Seals and/or Levers, Replace (B)
Starion	1.5	1.7

Vacuum Modulator, Replace (B)
Starion	.5	.7

Valve Body Assy., Replace (B)
Starion	1.4	1.6

Valve Body Manual Lever Shaft Seal, Replace (B)
Starion	1.5	1.7

SERVICE TRANSMISSION REMOVED
Transmission R&R included unless otherwise noted.

	LABOR TIME	SEVERE SERVICE
Front Oil Pump, Replace (B)		
Starion	4.6	4.8
Front Pump Oil Seal, Replace (B)		
Starion	3.8	4.0
Overdrive Case, Replace (B)		
Starion	4.2	4.4
Speedometer Drive Gear, Replace (B)		
Starion 4-Speed	5.5	5.7
Torque Converter, Replace (B)		
Starion	3.7	3.9
Transmission Assy., R&R (B)		
Starion	3.3	3.5
Flush converter and cooler lines add	.5	.7
Transmission Assy., R&R and Recondition (A)		
Includes: Replace torque converter.		
Starion	9.4	9.6
Transmission Assy., Reseal (B)		
Starion	6.1	6.3

TRANSFER CASE

Transfer Case, R&R and Recondition (A)
3000GT	4.3	4.5
Eclipse (3.0)	4.1	4.3
Galant (2.8)	3.8	4.0

Transfer Case, R&I or Replace (B)
3000GT	1.6	1.8
Eclipse (1.2)	1.6	1.8
Galant (1.0)	1.4	1.6
Lancer Evolution (4.9)	6.6	6.8

SHIFT LINKAGE
AUTOMATIC TRANSAXLE

Throttle Control Cable, Replace (B)
3-Speed
Cordia, Mirage, Tredia	.7	.9
Precis	1.2	1.4
4-Speed Mirage	1.0	1.2

AUTOMATIC TRANSMISSION

Throttle Linkage, Adjust (B)
Starion	.5	.7

Throttle Cable, Replace (B)
Starion	.7	.9

DRIVELINE
DRIVESHAFT

Propeller Shaft, R&R or Replace (B)
Exc. Starion	1.1	1.3
Starion	.5	.7

Propeller Shaft Center Support Bearing, Replace (B)
Exc. below	1.7	1.9
Lancer Evolution	1.1	1.3

MITSU-22   3000GT : CORDIA : DIAMANTE : ECLIPSE : GALANT : LANCER

	LABOR TIME	SEVERE SERVICE
Propeller Shaft Sliding Yoke, Replace (B)		
Eclipse (1.3)	1.8	2.0
Galant AWD (1.5)	2.0	2.2
Starion	1.3	1.5
Torque Tube Assembly, Replace (B)		
Rear axle Starion	1.1	1.3
Torque Tube Bearing, Replace (B)		
Rear axle Starion	1.3	1.5
Torque Tube Companion Flange, Replace (B)		
Rear axle Starion	.8	1.0
U-Joint, Replace or Recondition (B)		
Eclipse	1.7	1.9
Galant AWD	1.7	1.9
Starion one	1.2	1.4
DRIVE AXLE		
Axle Shaft Oil Seals, Replace (B)		
Rear axle		
3000GT	1.6	1.8
Eclipse, Galant, Starion (1.1)	1.5	1.7
Axle Shaft or Bearing, Replace (B)		
Rear axle		
3000GT	1.6	1.8
Eclipse, Galant, Starion (1.5)	2.0	2.2
Differential Carrier Assy., R&R and Recondition (A)		
Rear axle		
3000GT	9.6	9.8
Eclipse		
1990-94	9.2	9.4
1995-99	10.8	11.0
Galant, Starion (6.8)	9.2	9.4
Lancer		
Evolution (9.8)	13.2	13.4
Differential Carrier, R&R or Replace (B)		
Rear axle		
3000GT	2.6	2.8
Eclipse		
1990-94	2.3	2.5
1995-99	3.8	4.0
Galant (1.7)	2.3	2.5
Lancer		
Evolution (4.8)	6.5	6.7
Starion	2.5	2.7
Differential Side Bearings, Replace (B)		
Rear axle		
3000GT	7.1	7.3
Eclipse		
1990-94	6.8	7.0
1995-99	8.4	8.6
Galant, Starion (5.0)	6.8	7.0
Lancer		
Evolution (8.1)	10.9	11.1

	LABOR TIME	SEVERE SERVICE
Pinion Bearings, Replace (B)		
Rear axle		
3000GT	7.3	7.5
Eclipse		
1990-94	7.0	7.2
1995-99	8.5	8.7
Galant, Starion (5.2)	7.0	7.2
Lancer		
Evolution (8.2)	11.1	11.3
Pinion Shaft Oil Seal, Replace (B)		
Rear axle		
exc. Starion	1.3	1.5
Starion	2.5	2.7
Rear Axle Companion Flange, Replace (B)		
Rear axle		
exc. Starion	1.3	1.5
Starion	1.4	1.6
Rear Axle Cover Gasket, Replace (B)		
3000GT	1.9	2.1
Eclipse		
1990-94	1.4	1.6
1995-99	4.7	4.9
Galant AWD	1.3	1.5
Lancer Evolution (5.3)	7.2	7.4
Ring Gear & Pinion Set, Replace (B)		
Rear axle		
3000GT	8.1	8.3
Eclipse		
1990-94	7.7	7.9
1995-99	9.3	9.5
Galant	7.7	7.9
Lancer		
Evolution (8.8)	11.9	12.1
Starion	7.1	7.3
Viscous Coupler, Replace (B)		
Rear axle		
3000GT	4.7	4.9
Eclipse		
1990-94	4.4	4.6
1995-99	5.9	6.1
HALFSHAFTS		
Birfield Joint & Shaft Assy., Replace (B)		
Precis	1.3	1.5
CV Joint, Replace or Recondition (B)		
3000GT		
right side	2.0	2.2
left side	2.5	2.7
Cordia, Tredia	1.7	1.9
Diamante		
right side	2.5	2.7
left side	2.8	3.0
Eclipse		
1990-94	1.8	2.0
1995-99		
right side	2.4	2.6
left side	2.6	2.8

	LABOR TIME	SEVERE SERVICE
2000-05		
right side (1.7)	2.3	2.5
left side (1.5)	2.0	2.2
Galant		
1985-93	1.8	2.0
1994-98		
right side	2.4	2.6
left side	2.6	2.8
1999-05		
right side (1.7)	2.3	2.5
left side (1.5)	2.0	2.2
Mirage		
1985-92	1.8	2.0
1993-96		
right side	2.3	2.5
left side	2.4	2.6
1997-03 each side	2.1	2.3
Sigma	1.9	2.1
w/AT add	.3	.5
DO Joint Assy., Replace (B)		
Precis	1.1	1.3
DO Joint Boot (Seal), Replace (B)		
Precis	1.2	1.4
Front Halfshaft, R&R or Replace (B)		
3000GT		
right side	1.4	1.6
left side	1.9	2.1
Cordia, Tredia	1.3	1.5
Diamante one side	2.2	2.4
Eclipse		
1990-94	1.1	1.3
1995-99 one side	1.9	2.1
2000-05 one side (1.2)	1.6	1.8
Galant		
1985-93	1.1	1.3
1994-98 one side	1.9	2.1
1999-05 one side (1.2)	1.6	1.8
Lancer, Lancer Evolution one side (1.0)	1.4	1.6
Mirage		
1985-92	1.1	1.3
1993-96 one side	1.8	2.0
1997-03 one side	1.5	1.7
Precis	.9	1.1
Sigma	1.3	1.5
w/AT exc. Diamante add	.5	.7
w/AWD 95-99 Eclipse add	.5	.7
Front Halfshaft Boot, Replace (B)		
3000GT		
left side	2.3	2.5
right side	1.9	2.1
Cordia, Tredia	1.5	1.7
Diamante one side	2.6	2.8
Eclipse		
1990-94	1.6	1.8
1995-99 one side	2.3	2.5

LANCER EVOLUTION : MIRAGE : PRECIS : SIGMA : STARION : TREDIA   **MITSU-23**

	LABOR TIME	SEVERE SERVICE
2000-05 one		
side (1.5) **2.0**		*2.2*
Galant		
1985-93 .... **1.6**		*1.8*
1994-98 one side ... **2.3**		*2.5*
1999-05 one		
side (1.5) **2.0**		*2.2*
Lancer, Lancer Evolution		
one side (1.4) **1.9**		*2.1*
Mirage		
1985-92 .... **1.6**		*1.8*
1993-96 one side .. **2.2**		*2.4*
1997-03 one side ... **1.9**		*2.1*
Sigma **1.8**		*2.0*
w/AT exc. Diamante add .. **.5**		*.7*
w/AWD 95-99 Eclipse		
add .... **.5**		*.7*
Front Halfshaft Oil Seal, Replace (B)		
3000GT		
right side **1.6**		*1.8*
left side **1.9**		*2.1*
Cordia, Tredia **1.1**		*1.3*
Diamante		
1992-96		
right side **1.6**		*1.8*
left side **2.3**		*2.5*
1997-03		
right side **2.2**		*2.4*
left side **2.0**		*2.2*
Eclipse		
1990-94 .... **1.0**		*1.2*
1995-99		
AT		
right side **2.0**		*2.2*
left side **2.2**		*2.4*
MT		
right side **1.6**		*1.8*
left side **1.9**		*2.1*
2000-05		
AT		
right side (1.6) .. **2.2**		*2.4*
left side (1.5) ... **2.0**		*2.2*
MT		
right side (1.3) .. **1.8**		*2.0*
left side (1.2) ... **1.6**		*1.8*
Galant		
1985-93 .... **1.0**		*1.2*
1994-98		
AT		
right side **2.0**		*2.2*
left side **2.2**		*2.4*
MT		
right side **1.6**		*1.8*
left side **1.9**		*2.1*
1999-05		
AT		
right side (1.6) .. **2.2**		*2.4*
left side (1.5) ... **2.0**		*2.2*

	LABOR TIME	SEVERE SERVICE
MT		
right side (1.4) .. **1.9**		*2.1*
left side (1.2) ... **1.6**		*1.8*
Mirage		
1985-92 .... **1.0**		*1.2*
1993-96		
AT		
right side **1.9**		*2.1*
left side **2.2**		*2.4*
MT		
right side **1.6**		*1.8*
left side **1.9**		*2.1*
1997-03		
AT each **1.9**		*2.1*
MT each **1.4**		*1.6*
Sigma **1.2**		*1.4*
Rear Halfshaft, R&R or Replace (B)		
Rear axle		
3000GT, Galant (.8) **1.1**		*1.3*
Eclipse		
1990-94 .... **1.0**		*1.2*
1995-99 .... **1.9**		*2.1*
Lancer		
Evolution (8.7) ... **11.7**		*11.9*
Starion .... **.7**		*.9*
Replace CV joints or boots		
add each .... **.5**		*.7*
Rear Halfshaft Boot, Replace (B)		
Rear axle		
3000GT **1.6**		*1.8*
Eclipse		
1990-94 .... **1.6**		*1.8*
1995-99 .... **2.3**		*2.5*
Galant, Starion (1.2) **1.6**		*1.8*
Lancer		
Evolution (1.6) **2.2**		*2.4*

BRAKES
ANTI-LOCK

Control Unit, Replace (B)		
Exc. below (.8) **1.1**		*1.2*
Diamante wagon **1.3**		*1.5*
Starion .... **.3**		*.5*
G Sensor, Replace (B)		
3000GT **1.1**		*1.3*
Diamante wagon ...... **.6**		*.8*
Eclipse (.4) .... **.6**		*.8*
Galant, Mirage (.3) .... **.5**		*.7*
Lancer Evolution (.3) ... **.5**		*.7*
Starion .... **.3**		*.5*
Hydraulic Assy., Replace (B)		
3000GT **3.5**		*3.7*
Diamante		
1992-96 .... **2.5**		*2.7*
1997-05 (2.8) **3.8**		*4.0*
Eclipse		
1990-94 .... **2.3**		*2.5*

	LABOR TIME	SEVERE SERVICE
1995-99		
exc. 2.0L 420A ... **4.1**		*4.3*
2.0L 420A **4.7**		*4.9*
2000-05 (1.9) **2.6**		*2.8*
Galant		
exc. 2004 .... **2.5**		*2.7*
2004-05 (2.1) **2.8**		*3.0*
Lancer, Lancer		
Evolution (1.6) **2.2**		*2.4*
Mirage **2.8**		*3.0*
Sigma **2.6**		*2.8*
w/TCL 92-96 Diamante add		
w/AC **1.9**		*2.1*
w/o AC .... **.8**		*1.0*
Hydraulic Control Unit Relay, Replace (B)		
Exc. below .... **.5**		*.7*
1995-99 Eclipse **1.1**		*1.3*
Modulator, Replace (B)		
1988-89 Starion ...... **.7**		*.9*
Power Relay Valve, Replace (B)		
Exc. below .... **.3**		*.5*
1995-99 Eclipse .... **.6**		*.8*
Pulse Generator, Replace (B)		
1988-89 Starion ...... **.7**		*.9*
Wheel Speed Sensor, Replace (B)		
One piece		
Galant .... **.8**		*1.0*
Sigma **1.3**		*1.5*
Two piece		
front		
exc. below **1.3**		*1.5*
3000GT **1.5**		*1.7*
2000-04 Eclipse .. **1.5**		*1.7*
Lancer, Lancer		
Evolution .... **.9**		*1.1*
1993-96 Mirage .. **1.4**		*1.6*
Rear **1.1**		*1.3*

SYSTEM
Bleed Brakes (B)		
Includes: Add fluid.		
All Models .... **.6**		*.8*
w/ABS add .... **.6**		*.8*
Brake System, Flush and Refill (B)		
All Models **1.2**		*1.4*
Brakes, Adjust (B)		
Includes: Refill master cylinder.		
All Models rear wheels .. **.5**		*.7*
Brake Hose (Flexible), Replace (B)		
Includes: System bleeding.		
Exc. Precis one **1.1**		*1.3*
Precis one .... **.5**		*.7*
each addl. add .... **.3**		*.5*

MITSU-24 3000GT : CORDIA : DIAMANTE : ECLIPSE : GALANT : LANCER

	LABOR TIME	SEVERE SERVICE
Master Cylinder, R&R and Reconditon (B)		
Includes: System bleeding.		
3000GT 2.8		3.0
Cordia, Tredia 2.6		2.8
Diamante		
1992-96 . . . . 2.8		3.0
1997-04 . . . . 3.1		3.3
Eclipse		
1990-94 . . . . 2.3		2.5
1995-05 (2.1) 2.8		3.0
Galant		
1985-98 . . . . 2.6		2.8
1999-05 (2.1) 2.8		3.0
Lancer (1.9) 2.6		2.8
Lancer Evolution (2.2) . 3.0		3.2
Mirage		
1985-96 . . . . 2.8		3.0
1997-03 . . . . 2.9		3.1
Precis 2.1		2.3
Sigma 2.3		2.5
Starion 2.3		2.5
Master Cylinder, Replace (B)		
Includes: System bleeding.		
3000GT 1.6		1.8
Cordia, Tredia 1.2		1.4
Diamante 1.9		2.1
Eclipse		
1990-94 . . . . 1.2		1.4
1995-05 (1.4) 1.9		2.1
Galant		
1985-88 . . . . 1.2		1.4
1989-93 . . . . 1.4		1.6
1994-05 (1.2) 1.6		1.8
Lancer (1.1) 1.5		1.7
Lancer Evolution (1.3) . 1.8		2.0
Mirage		
1985-88 9		1.1
1989-92 . . . . 1.6		1.8
1993-96 . . . . 1.4		1.6
1997-03 . . . . 1.8		2.0
Precis 1.3		1.5
Sigma 1.2		1.4
Starion 1.4		1.6
Power Booster Unit, Replace (B)		
3000GT 2.3		2.5
Cordia, Tredia 1.9		2.1
Diamante		
1992-96 . . . . 2.5		2.7
1997-04 . . . . 4.1		4.3
Eclipse		
1990-94 . . . . 2.2		1.4
1995-05 (2.1) 2.8		3.0
Galant		
1985-87 . . . . 1.9		2.1
1988-05 (2.1) 2.8		3.0
Lancer (1.5) 2.0		2.2
Lancer Evolution (1.9) . 2.6		2.8

	LABOR TIME	SEVERE SERVICE
Mirage		
1985-88 . . . . 1.3		1.5
1989-04 . . . . 2.3		2.5
Precis		
1987-89 . . . . 1.6		1.8
1990-94 9		1.1
Sigma 1.9		2.1
Starion 1.9		2.1
Power Booster Vacuum Check Valve, Replace (B)		
All Models 5		.7
Proportioning Valve, Replace (B)		
3000GT 1.6		1.8
Cordia, Starion, Tredia . .7		.9
Diamante 1.6		1.8
Eclipse		
1990-94 . . . . 1.6		1.8
1995-99 . . . . 2.5		2.7
2000-05 (1.3) 1.8		2.0
Galant		
1985-88 8		1.0
1989-05 (1.2) 1.6		1.8
Lancer (1.2) 1.6		1.8
Mirage		
1985-88 7		.9
1989-92 . . . . 2.0		2.2
1993-96 . . . . 1.7		1.9
1997-03 . . . . 2.3		2.5
Precis 9		1.1
Sigma 8		1.0
SERVICE BRAKES		
Brake Drum, Replace (B)		
All Models 8		1.0
Caliper Assy., R&R and Reconditon (B)		
Includes: System bleeding.		
Front one side		
exc. below 1.6		1.8
3000GT 2.1		2.3
Lancer Evolution . . . 2.2		2.5
1993-96 Mirage 1.8		2.0
Rear one side		
exc. below 1.8		2.0
1985-87 Galant 1.4		1.6
Lancer Evolution . . . 2.2		2.4
1993-04 Mirage 2.2		2.4
Caliper Assy., Replace (B)		
Includes: System bleeding.		
Front one side		
exc. below (.8) 1.1		1.3
3000GT 1.3		1.5
1993-96 Mirage 1.3		1.5
Rear one side		
exc. below (.8) 1.1		1.3
Diamante 1.3		1.5
Lancer		
Evolution (1.0) 1.4		1.6
1993-04 Mirage 1.3		1.5

	LABOR TIME	SEVERE SERVICE
Disc Brake Rotor, Replace (B)		
Front		
exc. below 8		1.0
3000GT 1.0		1.2
Cordia, Tredia 2.0		2.2
1985-92 Mirage 2.0		2.2
Precis 1.4		1.6
Rear		
exc. Diamante 1.1		1.3
Diamante 1.3		1.5
Pads and/or Shoes, Replace (B)		
Includes: Service and parking brake adjustment, system bleeding.]		
Front or rear disc 1.3		1.5
Rear drum 1.9		2.1
COMBINATION ADD-ONS		
Repack wheel bearings		
two wheels add 7		.9
Replace		
brake drum add 1		.2
brake hose add 3		.5
caliper add 3		.5
master cylinder add 7		.9
wheel cylinder add 2		.4
Resurface		
brake drum add 5		.7
rotor add		
FWD 5		.7
RWD 9		1.1
Wheel Cylinder, R&R and Reconditon (B)		
Includes: System bleeding.		
Cordia, Tredia 9		1.1
Eclipse, Lancer 1.6		1.8
Galant		
1985-93 . . . . 1.2		1.4
1994-03 (1.2) 1.6		1.8
Mirage		
1985-92 . . . . 1.2		1.4
1993-04 (1.2) 1.6		1.8
Precis 1.4		1.6
Wheel Cylinder, Replace (B)		
Includes: System bleeding.		
Cordia, Tredia 8		1.0
Eclipse, Lancer 1.4		1.6
Galant		
1985-93 . . . . 1.0		1.2
1994-03 (1.0) 1.4		1.6
Mirage		
1985-92 . . . . 1.0		1.2
1993-04 (1.0) 1.4		1.6
Precis		
1987-89 . . . . 1.4		1.6
1990-94 9		1.1

LANCER EVOLUTION : MIRAGE : PRECIS : SIGMA : STARION : TREDIA **MITSU-25**

	LABOR TIME	SEVERE SERVICE
PARKING BRAKE		
Parking Brake Cable, Adjust (C)		
All Models	.5	.7
Parking Brake Apply Actuator, Replace (B)		
Exc. below (.6)	.8	1.0
3000GT	1.4	1.6
Diamante	1.0	1.2
1985-93 Galant	1.1	1.3
Precis	.5	.7
Parking Brake Apply Warning Indicator Switch, Replace (B)		
3000GT	1.2	1.4
Cordia, Tredia	.5	.7
Diamante		
1992-96	1.0	1.2
1997-04	.6	.8
Eclipse (.5)	.7	.9
Galant (.6)	.8	1.0
Lancer, Lancer Evolution (.4)	.6	.8
Mirage	.5	.7
Precis, Starion	.5	.7
Parking Brake Cable, Replace (B)		
1983-92 front	1.1	1.3
Rear one side		
3000GT	1.8	2.0
Cordia, Tredia	1.4	1.6
Diamante		
1992-96	1.7	1.9
1997-04	2.4	2.6
Eclipse		
1990-94	1.4	1.6
1995-05 (1.8)	2.4	2.6
Galant		
1985-87	1.7	1.9
1989-93	2.3	2.5
1994-05 (1.4)	1.9	2.1
Lancer (1.3)	1.8	2.0
Lancer Evolution (1.4)	1.9	2.1
Mirage		
1985-88	1.3	1.5
1989-04	1.7	1.9
Precis, Sigma	1.7	1.9
Starion	.9	1.1
w/AWD 95-99 Eclipse add	.3	.5
w/rear disc brakes 95-04 Eclipse add	.5	.7
Parking Brake Shoes, Replace (B)		
One side		
3000GT	1.3	1.5
Diamante		
1992-96	1.1	1.3
1997-04	1.4	1.6
Eclipse		
1995-99	1.1	1.3
2000-05 (1.2)	1.6	1.8

	LABOR TIME	SEVERE SERVICE
Galant		
1985-98	1.1	1.3
1999-05 (1.2)	1.6	1.8
Lancer		
Evolution (1.2)	1.6	1.8
Sigma	1.2	1.4
FRONT SUSPENSION		
Unless otherwise noted, time given does not include alignment.		
Align Front End (A)		
All Models (.9)	1.2	1.4
Front Toe, Adjust (B)		
All Models (.4)	.6	.8
Front Axle Hub Oil Seal, Replace (B)		
Starion	.7	.9
Front Crossmember, Replace (B)		
1987-89 Precis		
No.1	2.3	2.5
No.2	3.1	3.3
Hub Bearings, Replace (B)		
3000GT	3.3	3.5
Cordia, Tredia	2.0	2.2
Diamante	2.4	2.6
Eclipse (1.6)	2.2	2.4
Lancer (1.6)	2.2	2.4
Mirage		
1985-92	1.8	2.0
1993-03	2.7	2.9
Precis	1.9	2.1
Sigma	2.0	2.2
Starion one side	.8	1.0
Lower Ball Joint, Replace (B)		
Precis	.8	1.0
Starion	1.3	1.5
Lower Control Arm, Replace (B)		
Includes: Alignment		
3000GT		
right side	1.8	2.0
left side	2.3	2.5
Cordia, Tredia	1.3	1.5
Diamante	1.6	1.8
Eclipse, Galant (1.2)	1.6	1.8
Lancer, Lancer Evolution (1.2)	1.6	1.8
Mirage		
1985-92	1.4	1.6
1993-96	1.1	1.3
1997-03	1.6	1.8
Precis	.9	1.1
Sigma	1.6	1.8
Starion	1.2	1.4
Lower Control Arm Bushings, Replace (B)		
Includes: Alignment		
3000GT	1.7	1.9
Cordia, Tredia	1.4	1.6
Diamante	1.3	1.5

	LABOR TIME	SEVERE SERVICE
Eclipse		
1990-94	1.3	1.5
1995-05 (1.3)	1.8	2.0
Galant		
1985-93	1.3	1.5
1994-05 (1.3)	1.8	2.0
Lancer, Lancer Evolution (1.4)	1.9	2.1
Mirage		
1985-96	1.3	1.5
1997-03	1.7	1.9
Precis	1.2	1.4
Sigma	1.4	1.6
Lower Control Arm Bushings and/or Shaft, Replace (B)		
Starion	1.6	1.8
Lower Control Arm Strut, Replace (B)		
Starion	1.3	1.5
Stabilizer Bar, Replace (B)		
3000GT	2.7	2.9
Cordia, Tredia	1.4	1.6
Diamante		
1992-96	2.3	2.5
1997-04	3.7	3.9
Eclipse		
1990-94	1.6	1.8
1995-99	1.1	1.3
2000-05 (1.7)	2.3	2.5
Galant		
1985-93	1.4	1.6
1994-98	1.1	1.3
1999-03	2.3	2.5
2004-05 (3.6)	4.9	5.1
Lancer (1.9)	2.6	2.8
Lancer Evolution (2.2)	3.0	3.2
Mirage		
1985-88	1.6	1.8
1989-92	.8	1.0
1993-96	3.6	3.8
1997-03	3.1	3.3
Sigma	1.3	1.5
Starion	1.6	1.8
Stabilizer Bar Bushings, Replace (B)		
3000GT	1.4	1.6
Diamante		
1992-96	.7	.9
1997-04	1.0	1.2
Eclipse		
1990-99	.8	1.0
2000-05 (1.1)	1.5	1.7
Galant		
1985-98	.8	1.0
1999-03	1.0	1.2
2004-05 (1.1)	1.5	1.7
Lancer, Lancer Evolution (1.3)	1.8	2.0

MITSU-26 3000GT : CORDIA : DIAMANTE : ECLIPSE : GALANT : LANCER

	LABOR TIME	SEVERE SERVICE
Stabilizer Bar Bushings, Replace (B)		
Mirage		
1985-92	.7	.9
1993-96	3.6	3.8
1997-03	3.1	3.3
Precis		
1987-89	.5	.7
1990-94	1.3	1.5
Sigma	.7	.9
Starion	.7	.9
Stabilizer Link, Replace (B)		
All Models one side	.8	1.0
Steering Knuckle, Replace (B)		
3000GT	3.1	3.3
Diamante	2.4	2.6
Eclipse		
1990-99	2.1	2.3
2000-05 (1.0)	1.4	1.6
Galant		
1988-98	2.0	2.2
1999-05 (1.0)	1.4	1.6
Lancer (1.6)	2.2	2.4
Lancer Evolution (1.0)	1.4	1.6
Mirage		
1985-92	2.0	2.2
1993-04	2.7	2.9
Precis		
1987-89	1.8	2.0
1990-94	1.6	1.9
Sigma	1.9	2.1
Steering Knuckle Arm, Replace (B)		
Includes: Reset toe.		
Starion	1.4	1.6
Steering Knuckle Oil Seal, Replace (B)		
3000GT		
inner	2.3	2.5
outer	3.1	3.3
Cordia, Tredia		
inner	1.4	1.6
outer	2.1	2.3
Diamante		
inner	1.4	1.6
outer	2.3	2.5
Eclipse		
inner (1.0)	1.4	1.6
outer (1.5)	2.0	2.2
Mirage		
1985-92		
inner	1.3	1.5
outer	2.0	2.2
1993-96		
inner	1.4	1.6
outer	2.6	2.8
1997-03		
inner	1.4	1.6
outer	2.3	2.5
Precis		
inner	1.2	1.4
outer	2.0	2.2

	LABOR TIME	SEVERE SERVICE
Sigma		
inner	1.9	2.1
outer	2.0	2.2
Strut (Shock Absorber), Replace (B)		
3000GT	1.6	1.8
Cordia, Tredia	1.1	1.3
Diamante	1.3	1.5
Eclipse		
1990-94	1.1	1.3
1995-05 (1.0)	1.4	1.6
Galant		
1985-93	1.1	1.3
1994-05 (1.0)	1.4	1.6
Lancer, Lancer Evolution (1.3)	1.8	2.0
Mirage		
1985-92	1.1	1.3
1993-04	1.4	1.6
Precis	1.2	1.4
Sigma	1.1	1.3
Starion	1.4	1.6
w/ECS 92-96 Diamante add	.3	.5
Strut (Shock Absorber) Coil Spring, Replace (B)		
3000GT	1.9	2.1
Cordia, Tredia	1.5	1.7
Diamante	1.6	1.8
Eclipse		
1990-94	1.4	1.6
1995-05 (1.4)	1.9	2.1
Galant		
1985-93	1.4	1.6
1994-05 (1.4)	1.9	2.1
Lancer, Lancer Evolution (1.0)	1.4	1.6
Mirage		
1985-92	1.4	1.6
1993-04	1.7	1.9
Precis	1.4	1.6
Sigma	1.5	1.7
Starion	1.6	1.8
w/ECS 92-96 Diamante add	.3	.5
Upper Control Arm, Replace (B)		
Includes: Alignment		
1995-99 Eclipse	1.4	1.6
1994-98 Galant	1.4	1.6
Upper Control Arm Bushings, Replace (B)		
Includes: Alignment		
1995-99 Eclipse	1.6	1.8
1994-98 Galant	1.6	1.8
ELECTRONIC SUSPENSION (ECS)		
Air Compressor, Replace (B)		
All Models	.7	.9
Air Compressor Relay, Replace (B)		
All Models	.5	.7

	LABOR TIME	SEVERE SERVICE
ECS Indicator, Replace (B)		
All Models	.5	.7
Electronic Control Unit, Replace (B)		
All Models	.8	1.0
Front Actuator, Replace (B)		
Exc. Galant (.3)	.5	.7
Galant (.6)	.8	1.0
Front Height Sensor, Replace (B)		
1992-96 Diamante	1.0	1.2
1985-93 Galant	1.3	1.5
Sigma	1.2	1.4
Front Height Sensor Rod, Replace (B)		
1985-87 Galant	.7	.9
Front Solenoid Valve, Replace (B)		
1992-96 Diamante	1.1	1.3
Galant (.5)	.7	.9
Sigma	.5	.7
G Sensor, Replace (B)		
1992-96 Diamante	.5	.7
Galant (.5)	.7	.9
Rear Actuator, Replace (B)		
3000GT, Diamante	.7	.9
Galant		
1985-88	1.3	1.5
1989-93	.8	1.0
Sigma	1.2	1.4
Rear Height Sensor, Replace (B)		
1992-96 Diamante	.8	1.0
1985-93 Galant	.7	.9
Sigma	.7	.9
Rear Height Sensor Rod, Replace (B)		
1985-93 Galant	.7	.9
Sigma	.7	.9
Rear Solenoid Valve, Replace (B)		
Diamante	1.3	1.5
Galant		
1985-88	1.6	1.8
1989-93	.7	.9
Sigma	1.7	1.9
Steering Wheel Angular Velocity Sensor, Replace (B)		
3000GT, Diamante	1.4	1.6
Galant	.8	1.0
Sigma	.8	1.0
Tank Reservoir, Replace (B)		
Front		
Diamante	1.9	2.1
Galant		
1985-88	.7	.9
1989-93	1.4	1.6
Sigma	.7	.9
Rear Diamante	.7	.9

REAR SUSPENSION

Rear Suspension (Complete), Align (A)
All Models 1.6   1.8

LANCER EVOLUTION : MIRAGE : PRECIS : SIGMA : STARION : TREDIA — MITSU-27

	LABOR TIME	SEVERE SERVICE
Coil Spring, R&R or Replace (B)		
3000GT	1.3	1.5
Diamante		
1992-96		
sedan	1.6	1.8
wagon	.7	.9
1997-03	2.0	2.2
Eclipse (.9)	1.2	1.4
Galant		
1985-03	1.3	1.5
2004-05 (1.2)	1.6	1.8
Lancer, Lancer Evolution (1.0)	1.4	1.6
Mirage	1.3	1.5
Precis	.5	.7
Sigma	1.5	1.7
Starion	1.7	1.9
w/ECS 92-96 Diamante sedan add	.3	.5
w/open type coil 00-04 Eclipse add	.3	.5
Lower Arms and Links, Replace (B)		
Precis		
right side	3.6	3.8
left side	3.3	3.5
Replace bushings add	.3	.5
Lower Control Arm, Replace (B)		
Includes: Alignment.		
3000GT	1.9	2.1
Cordia, Tredia	3.3	3.5
Diamante		
1992-96		
sedan	1.9	2.1
wagon	1.0	1.2
1997-04	1.3	1.5
Eclipse (1.4)	1.9	2.1
Galant		
1985-96	1.7	1.9
1997-03	1.3	1.5
2004-05 (1.3)	1.8	2.0
Lancer, Lancer Evolution (1.4)	1.9	2.1
Mirage	2.2	2.4
Starion	1.3	1.5
w/ABS add		
97-04 Diamante	.3	.5
Diamante wagon	.3	.5
Replace bushings add	.2	.4
Rear Hub Bearings, Replace (B)		
All Models	.8	1.0
Rear Knuckle, Replace (B)		
1997-04 Diamante	2.3	2.5
1995-05 Eclipse		
FWD (1.4)	1.9	2.1
AWD (1.9)	2.6	2.8
Galant (1.4)	1.9	2.1
Lancer Evolution (2.1)	2.8	3.0
w/rear disc brakes add		
Eclipse	.5	.7
Galant	.5	.7

	LABOR TIME	SEVERE SERVICE
Rear Shock Absorber or Bushing, Replace (B)		
3000GT	1.1	1.3
Diamante		
sedan	1.9	2.1
wagon	.6	.8
Eclipse		
1990-99	1.1	1.3
2000-05 (1.0)	1.4	1.6
Galant (.9)	1.2	1.4
Lancer, Lancer Evolution (1.0)	1.4	1.6
Mirage	1.3	1.5
Precis	.5	.7
Rear Stabilizer Bar & Bushings, Replace (B)		
3000GT		
trailing arm susp.	.8	1.0
wishbone susp.	2.7	2.9
Cordia, Tredia	2.3	2.5
Diamante	.8	1.0
Eclipse		
1990-94	1.6	1.8
1995-05 (.9)	1.2	1.4
Galant		
1989-93	1.4	1.6
1994-05 (.9)	1.2	1.4
Lancer (.8)	1.1	1.3
Lancer Evolution (1.0)	1.4	1.6
1997-03 Mirage	1.0	1.2
Precis	3.2	3.4
Starion	1.5	1.7
Rear Stabilizer End Kit, Replace (B)		
All Models	.8	1.0
Torsion Axle Assy., Replace (B)		
All Models	2.6	2.8
w/ABS add	.5	.7
Trailing Arm and/or Bushings, Replace (B)		
Includes: Alignment.		
3000GT		
trailing arm susp.	2.9	3.1
wishbone susp.	3.6	3.8
Diamante		
1992-96 sedan	2.9	3.1
1997-04	1.3	1.5
Eclipse		
1990-94	2.7	2.9
1995-05 (1.0)	1.4	1.6
Galant		
1989-93	2.3	2.5
1994-05 (1.0)	1.4	1.6
Lancer (2.0)	2.7	2.9
Lancer Evolution (1.4)	1.9	2.1
Mirage	3.3	3.5
w/ABS add		
92-96 Diamante sedan	.3	.5
3000GT	.3	.5

	LABOR TIME	SEVERE SERVICE
w/drum brakes Mirage add	.5	.7
Replace bushings add	.2	.4
Upper Control Arm, Replace (B)		
Includes: Alignment.		
3000GT	1.9	2.1
Diamante		
1992-96		
sedan	2.2	2.4
wagon	.6	.8
1997-03	1.3	1.5
Eclipse (1.0)	1.4	1.6
Galant		
1985-	1.4	1.6
2004-05 (1.2)	1.6	1.8
Lancer (1.3)	1.8	2.0
Lancer Evolution (1.6)	2.2	2.4
Mirage	1.9	2.1
Replace bushings add	.2	.4
Wheel Hub, Replace (B)		
Exc. below	1.0	1.2
1997-04 Diamante	1.3	1.5
1995-99 Eclipse AWD	2.1	2.3
w/rear disc brakes add		
95-04 Eclipse	.5	.7
Galant	.5	.7
Mirage	.5	.7

STEERING

AIR BAGS

	LABOR TIME	SEVERE SERVICE
Diagnose Air Bag System Component Each (A)		
Sigma	.8	.8
1991-05	.9	.9
Air Bag Assy., Replace (B)		
Driver side	.6	.8
Passenger side		
3000GT, Eclipse	.8	1.0
Diamante	.9	1.1
Galant (.9)	1.2	1.4
Lancer, Lancer Evolution (1.3)	1.8	2.0
Mirage	.6	.8
Air Bag Module, Replace (B)		
3000GT	.7	.9
Diamante		
1992-96	.7	.9
1997-03	1.5	1.7
Eclipse (.6)	.8	1.0
Galant (.5)	.7	.9
Lancer (1.0)	1.4	1.6
Lancer Evolution (.6)	.8	1.0
Mirage	.7	.9
Sigma	1.3	1.5
Front Sensor, Replace (B)		
3000GT, Galant (.6)	.8	1.0
Diamante, Sigma (.5)	.7	.9
Eclipse (.8)	1.1	1.3
Lancer Evolution (.4)	.6	.8

MITSU-28 3000GT : CORDIA : DIAMANTE : ECLIPSE : GALANT : LANCER

	LABOR TIME	SEVERE SERVICE
MANUAL RACK & PINION		
Inner Tie Rod Assy., Replace (B)		
Includes: Reset toe.		
Cordia, Tredia	2.6	2.8
Eclipse (1.0)	1.4	1.6
Mirage		
1985-88	2.3	2.5
1989-92	1.3	1.5
1993-03	1.9	2.1
Precis	2.6	2.8
Rack & Pinion Assy., R&R or Replace (B)		
Includes: Adjust steering and toe-in.		
Cordia, Tredia	1.8	2.0
Eclipse (1.4)	1.9	2.1
Mirage		
1985-88	2.0	2.2
1989-92	1.8	2.0
1993-03	2.6	2.8
Precis	1.5	1.7
Rack Boots, Replace (B)		
Includes: Reset toe.		
Cordia, Tredia	2.4	2.6
Eclipse (1.0)	1.4	1.6
Mirage		
1985-88	.8	1.0
1989-92	1.3	1.5
1993-03	1.6	1.8
Precis		
1987-89	.8	1.0
1990-94	.5	.7
Steering Column, R&R and Recondition (B)		
1987-89 Precis	2.6	2.8
Steering Shaft U-Joint, Replace (B)		
Cordia, Precis, Tredia	1.4	1.6
Eclipse (1.2)	1.6	1.8
Steering Wheel, Replace (B)		
Cordia, Tredia	.5	.7
Eclipse (.3)	.5	.7
Mirage		
1985-96	.5	.7
1997-03	.6	.8
Precis	.3	.5
Tie Rods or Tie Rod Ends, Replace (B)		
Includes: Reset toe.		
Cordia, Tredia	1.1	1.3
Eclipse	1.1	1.3
1993-03 Mirage	1.3	1.5
Precis		
1987-89	1.4	1.6
1990-94	.7	.9
POWER RACK & PINION		
Troubleshoot Power Steering (A)		
All Models	.5	.7

	LABOR TIME	SEVERE SERVICE
Outer Tie Rod Ends, Replace (B)		
Includes: Reset toe.		
Exc. below	1.3	1.5
Cordia, Precis	.9	1.1
Sigma, Tredia	.9	1.1
Rack & Pinion Assy., R&R and Recondition (A)		
3000GT	5.3	5.5
Cordia, Tredia	3.8	4.0
Diamante		
1992-96	5.3	5.5
1997-04	6.5	6.7
Eclipse		
1990-94	4.3	4.5
1995-99		
exc. 2.0L 420A	5.1	5.3
2.0L 420A	5.6	5.8
2000-05 (4.4)	5.9	6.1
Galant		
1985-93	4.3	4.5
1994-98	4.7	4.9
1999-03	5.9	6.1
2004-05 (5.1)	6.9	7.1
Lancer (5.1)	6.9	7.1
Lancer Evolution (5.3)	7.2	7.4
Mirage		
1985-92	4.1	4.3
1993-03	5.3	5.5
Precis	4.1	4.3
Sigma	4.1	4.3
w/4WS add		
3000GT	.9	1.1
92-96 Diamante	1.2	1.4
w/AWD add		
3000GT	.3	.5
90-99 Eclipse	.5	.7
w/turbo 3000GT add	.5	.7
Rack & Pinion Assy., Replace (B)		
Includes: Reset toe.		
3000GT	3.6	3.8
Cordia, Tredia	3.0	3.2
Diamante		
1992-96	3.8	4.0
1997-04	4.9	5.1
Eclipse		
1990-94	2.8	3.0
1995-99		
exc. 2.0L 420A	3.6	3.8
2.0L 420A	3.8	4.0
2000-05 (3.0)	4.1	4.3
Galant		
1985-93	2.8	3.0
1994-98	3.5	3.8
1999-03	4.4	4.6
2004-05 (3.4)	4.6	4.8
Lancer (3.6)	4.9	5.1
Lancer Evolution (4.0)	5.4	5.6

	LABOR TIME	SEVERE SERVICE
Mirage		
1985-92	2.3	2.5
1993-03	3.8	4.0
Precis	2.9	3.1
Sigma	2.8	3.0
w/4WS add		
3000GT	.9	1.1
92-96 Diamante	1.2	1.4
w/AWD add		
3000GT	.3	.5
90-99 Eclipse	.5	.7
w/turbo 3000GT add	.5	.7
Rack Boots, Replace (B)		
Includes: Reset toe.		
3000GT	1.3	1.5
Cordia, Tredia	2.3	2.5
Diamante	1.3	1.5
Eclipse		
1990-99	1.6	1.8
2000-05 (1.0)	4.4	4.6
Galant		
1985-87	2.2	2.4
1989-98	1.6	1.8
1999-03	4.4	4.6
2004-05 (4.1)	5.5	5.7
Lancer (3.8)	5.1	5.3
Lancer Evolution (4.3)	5.8	6.0
Mirage		
1985-88	.8	1.0
1989-92	1.3	1.5
1993-96	1.6	1.8
1997-03	3.8	4.0
Precis		
1987-89	.8	1.0
1990-94	.6	.8
Sigma	2.3	2.5
Steering Pump, R&R or Replace (B)		
3000GT	1.7	1.9
Cordia, Precis, Tredia	1.2	1.4
Diamante		
1992-96	1.2	1.4
1997-04	1.9	2.1
Eclipse		
1990-94	.8	1.0
1995-99	1.3	1.5
2000-05		
4 cyl. (1.3)	1.8	2.0
V6 (1.6)	2.2	2.4
Galant		
1985-88	1.2	1.4
1989-93	.8	1.0
1994-05 (1.5)	2.0	2.2
Lancer (1.0)	1.4	1.6
Lancer Evolution (1.6)	2.2	2.4
Mirage		
1985-92	.8	1.0
1993-03	1.4	1.6
Sigma	1.3	1.5
w/AC 97-02 Mirage 1.8L add	.3	.5

LANCER EVOLUTION : MIRAGE : PRECIS : SIGMA : STARION : TREDIA MITSU-29

	LABOR TIME	SEVERE SERVICE
Steering Pump Hoses, Replace (B)		
Does not include purging.		
Pressure		
3000GT		
w/turbo	1.5	1.7
w/o turbo	1.0	1.2
Cordia, Tredia	.7	.9
Diamante		
1992-96	1.0	1.2
1997-04	1.5	1.7
Eclipse		
1990-99	1.0	1.2
2000-05 (1.0)	1.4	1.6
Galant		
1985-98	1.0	1.2
1999-05 (1.0)	1.4	1.6
Lancer, Lancer Evolution (1.9)	2.6	2.8
Mirage	.9	1.1
Precis	.7	.9
Sigma	1.1	1.3
Return		
3000GT	.6	.8
Cordia, Tredia	.7	.9
Diamante	.8	1.0
Eclipse		
1990-99	.7	.9
2000-05 (.9)	1.2	1.4
Galant		
1985-98	.7	.9
1999-05 (.8)	1.1	1.3
Lancer (1.1)	1.5	1.7
Lancer Evolution (1.6)	2.2	2.4
Mirage	.8	1.0
Precis	.5	.7
Sigma	.9	1.1
Suction		
3000GT	.5	.7
Cordia, Tredia	.5	.7
Diamante	.8	1.0
Eclipse		
1990-94	.5	.7
2000-05 (.8)	1.1	1.3
Galant		
1985-98	.6	.8
1999-05 (.9)	1.2	1.4
Lancer, Lancer Evolution (1.3)	1.8	2.0
Mirage	.7	.9
Sigma	.5	.7
w/4WS add		
3000GT	.8	1.0
92-96 Diamante	.5	.7
Steering Shaft U-Joint, Replace (B)		
3000GT	1.9	2.1
Cordia, Tredia	1.6	1.8
Diamante		
1992-96	2.2	2.4
1997-04	1.9	2.1

	LABOR TIME	SEVERE SERVICE
Eclipse		
1990-94	1.4	1.6
1995-99		
exc. 2.0L 420A	2.3	2.5
2.0L 420A	2.9	3.1
Galant		
1985-88	1.9	2.1
1989-93	1.1	1.3
1994-98	2.3	2.5
Precis	1.2	1.4
Sigma	1.8	2.0
w/turbo 3000GT add	.5	.7
Steering Wheel, Replace (B)		
Exc. Precis	.7	.9
Precis	.3	.5
Tie Rod Assy., Replace (B)		
Includes: Reset toe.		
3000GT, Diamante	1.3	1.5
Cordia, Tredia	2.3	2.5
Eclipse		
1990-99	1.7	1.9
2000-05 (3.3)	4.5	4.7
Galant		
1985-87	2.3	2.5
1989-93	1.3	1.5
1994-98	1.7	1.9
1999-05 (3.1)	4.2	4.4
Lancer (3.9)	5.3	5.5
Lancer Evolution (4.7)	6.3	6.5
Mirage		
1985-88	1.9	2.1
1989-92	1.3	1.5
1993-96	1.7	1.9
1997-03	4.1	4.3
Precis	2.6	2.8
Sigma	2.2	2.4

POWER WORM & SECTOR

	LABOR TIME	SEVERE SERVICE
Gear Assy., R&R and Recondition (A)		
Starion	3.4	3.6
Gear Assy., Replace (B)		
Starion	1.5	1.7
Hoses, Steering Pump, Replace (B)		
Does not include purging.		
Starion		
pressure or return	.7	.9
suction	.5	.7
Idler Arm, Replace (B)		
Starion	.5	.7
Pitman Arm, Replace (B)		
Starion	.5	.7
Steering Center Link (Relay Rod), Replace (B)		
Starion	.7	.9
Steering Column, R&I (B)		
Starion	1.4	1.6
Steering Pump, R&R and Recondition (B)		
Starion	1.9	2.1

	LABOR TIME	SEVERE SERVICE
Steering Pump, R&R or Replace (B)		
Starion	.8	1.0
Steering Wheel, Replace (B)		
Starion	.5	.7
Tie Rod End, Replace (B)		
Starion	1.2	1.4

4 WHEEL STEERING

	LABOR TIME	SEVERE SERVICE
Control Valve (4WS), Replace (B)		
3000GT	2.9	3.1
Diamante	2.2	2.4
Galant	1.9	2.1
Power Cylinder (4WS), Replace (B)		
3000GT	3.2	3.4
Diamante	2.2	2.4
Galant	2.2	2.4
Pressure Tubes (4WS), Replace (B)		
3000GT one side	2.0	2.2
Diamante one side	2.2	2.4
Galant		
right side	2.2	2.4
left side	1.7	1.9
Rear Oil Pump (4WS), Replace (B)		
3000GT	2.6	2.9
Diamante	2.3	2.6
Galant	2.0	2.3
Recondition pump add	.8	1.0
Tie Rod Ends (4WS), Replace (B)		
3000GT, Galant	.8	1.0
Diamante	1.3	1.5
Replace dust cover add	.1	.1

HEATING & AIR CONDTIONING

When more than one component requires replacement where evacuation, recovery and recharging is already included, deduct 1.0 hour for each additional component from the time given.

	LABOR TIME	SEVERE SERVICE
Evacuate, Recover and Recharge System (A)		
All Models	.8	.8
AC Hoses, Replace (B)		
Includes: Evacuate, recover and recharge.		
Discharge		
exc. below	1.4	1.6
1997-04 Diamante	1.3	1.5
1995-99 Eclipse	1.6	1.8
2004-05 Galant (1.3)	1.8	2.0
Sigma	1.5	1.7
Starion	1.7	1.9
Suction		
exc. below	1.4	1.6
Cordia, Tredia	1.6	1.8
1997-03 Diamante	1.7	1.9
1985-88 Galant	1.7	1.9

MITSU-30 3000GT : CORDIA : DIAMANTE : ECLIPSE : GALANT : LANCER

	LABOR TIME	SEVERE SERVICE
AC Hoses, Replace (B)		
1997-03 Mirage	1.6	1.8
Sigma	1.8	2.0
Starion	1.6	1.8
AC Low Pressure Switch, Replace (B)		
Cordia, Tredia	3.1	3.3
1985-87 Galant	1.4	1.6
1990-94 Precis	.9	1.1
Starion	2.9	3.1
Accumulator Assy., Replace (B)		
Includes: Evacuate, recover and recharge.		
1990-94 Precis	1.7	1.9
Blower Motor, Replace (B)		
Exc. below	.8	1.0
1995-99 Eclipse	.9	1.1
Galant		
1994-98	.9	1.1
2004-05 (5.5)	7.4	7.6
Blower Motor Relay, Replace (B)		
Exc. below	.6	.8
1985-87 Galant	.9	1.1
Sigma	.9	1.1
Blower Motor Resistor, Replace (B)		
All Models	.7	.9
Blower Motor Switch, Replace (B)		
3000GT	2.1	2.3
Cordia, Starion, Tredia	.5	.7
Diamante	1.0	1.2
Eclipse		
1990-94	.5	.7
1995-99	1.4	1.6
2000-05 (.6)	.8	1.0
Galant		
1989-98	1.1	1.3
1999-05 (.5)	.7	.9
Lancer, Lancer Evolution (.4)	.6	.8
Mirage		
1985-92	.7	.9
1993-03	1.1	1.3
Precis	.8	1.0
Compressor Assy., Replace (B)		
Includes: Evacuate, recover and recharge.		
3000GT	2.3	2.5
Diamante	2.4	2.6
Eclipse		
1990-94	1.7	1.9
1995-05 (1.8)	2.4	2.6
Galant (2.1)	2.8	3.0
Lancer (1.3)	1.8	2.0
Lancer Evolution (2.1)	2.8	3.0
Mirage		
1993-96	2.2	2.4
1997-03	1.7	1.9
Precis	1.3	1.5
w/turbo 3000GT add	1.5	1.5

	LABOR TIME	SEVERE SERVICE
Compressor Clutch Assy., Replace (B)		
Includes: Evacuate, recover and recharge.		
3000GT	2.6	2.8
Cordia, Tredia	2.1	2.3
Diamante	2.6	2.8
Eclipse		
1990-94	1.9	2.1
1995-05 (1.7)	2.3	2.5
Galant		
1985-88	1.8	2.0
1989-98	2.5	2.7
1999-05 (2.1)	2.8	3.0
Lancer (1.3)	1.8	2.0
Lancer Evolution (2.1)	2.8	3.0
Mirage		
1985-88	1.6	1.8
1989-96	2.5	2.7
1997-03	2.0	2.2
Precis		
1987-89	1.9	2.1
1990-94	2.5	2.7
Sigma	1.8	2.0
Starion	1.4	1.6
w/turbo 3000GT add	1.5	1.7
Condenser Assy., Replace (B)		
Includes: Evacuate, recover and recharge.		
3000GT	2.0	2.2
Cordia, Tredia	2.2	2.4
Diamante		
1992-96	1.9	2.1
1997-04	2.5	2.7
Eclipse		
1990-99	1.9	2.1
2000-05 (2.8)	3.8	4.0
Galant		
1985-88	2.3	2.5
1989-93	1.9	2.1
1994-98	2.3	2.5
1999-05 (1.8)	2.4	2.6
Lancer (1.6)	2.2	2.4
Lancer Evolution (2.6)	3.5	3.7
Mirage		
1985-88	1.4	1.6
1989-92	1.7	1.9
1993-03	1.9	2.1
Precis		
1987-89	1.2	1.4
1990-94	.8	1.0
Sigma	2.2	2.4
Starion	2.0	2.2
w/turbo 3000GT add	1.1	1.1
Condenser Fan Motor, Replace (B)		
3000GT	.8	1.0
Diamante		
1992-96	.6	.8
1997-04	1.0	1.2

	LABOR TIME	SEVERE SERVICE
Eclipse (.6)	.8	1.0
Galant (.5)	.7	.9
Lancer (.4)	.6	.8
Lancer Evolution (1.3)	1.8	2.0
Mirage	1.0	1.2
Condenser Fan Motor Relay, Replace (B)		
All Models	.5	.5
Evaporator Core, Replace (B)		
Includes: Evacuate, recover and recharge.		
3000GT	2.5	2.7
Cordia, Tredia	3.2	3.4
Diamante		
1992-96	2.3	2.5
1997-04	3.2	3.4
Eclipse		
1990-94	2.2	2.4
1995-99	4.4	4.6
2000-05 (1.9)	2.6	2.8
Galant		
1985-88	2.8	3.0
1989-03	2.5	2.7
2004-05 (5.5)	7.4	7.6
Lancer (1.7)	2.3	2.5
Lancer Evolution (1.9)	2.6	2.8
Mirage	2.5	2.7
Precis	2.8	3.0
Sigma	2.8	3.0
Starion	3.0	3.2
Expansion Valve, Replace (B)		
Includes: Evacuate, recover and recharge.		
3000GT	2.5	2.7
Cordia, Tredia	3.1	3.3
Diamante		
1992-96	2.3	2.5
1997-04	3.1	3.3
Eclipse		
1990-94	2.2	2.4
1995-99	4.4	4.6
2000-05 (1.0)	1.4	1.6
Galant		
1985-88	2.8	3.0
1989-98	2.3	2.5
1999-03	1.4	1.6
2004-05 (1.5)	2.0	2.2
Lancer (1.1)	1.5	1.7
Lancer Evolution (1.3)	1.8	2.0
Mirage	2.3	2.5
Precis	2.6	2.8
Sigma	2.8	3.0
Starion	2.8	3.0
Heater Core, R&R or Replace (B)		
Includes: Evacuate, recover and recharge when necessary.		
3000GT	5.7	5.9
Cordia	3.2	3.4
Diamante	5.8	6.0

LANCER EVOLUTION : MIRAGE : PRECIS : SIGMA : STARION : TREDIA **MITSU-31**

	LABOR TIME	SEVERE SERVICE
Eclipse		
1990-94	4.7	*4.9*
1995-99	5.7	*5.9*
2000-04		
w/AC	5.8	*6.0*
w/o AC	4.9	*5.1*
Galant		
1985-88	3.8	*4.0*
1989-93	4.7	*4.9*
1994-98	5.0	*5.2*
1999-03		
w/AC	5.8	*6.0*
w/o AC	4.9	*5.1*
2004-05 (5.2)	7.0	*7.2*
Lancer		
w/AC (3.8)	5.1	*5.3*
w/o AC (3.2)	4.3	*4.5*
Lancer Evolution (4.2)	5.7	*5.9*
Mirage		
1985-92	3.4	*3.6*
1993-96	3.8	*4.0*
1997-03		
w/AC	4.1	*4.3*
w/o AC	3.3	*3.5*
Precis		
1987-89		
w/AC	4.2	*4.4*
w/o AC	1.7	*1.9*
1990-94		
w/AC	5.2	*5.4*
w/o AC	4.2	*4.4*
Sigma	3.7	*3.9*
Starion	4.8	*5.0*
Tredia	4.1	*4.3*
w/AC w/ABS 95-99		
Eclipse add	.3	*.5*
w/AT add		
95-99 Eclipse	.3	*.5*
94-98 Galant	.3	*.5*
w/cruise control 95-99		
Eclipse 4G6 add	.6	*.8*
w/turbo 3000GT add	.5	*.7*
Heater Hoses, Replace (B)		
All Models	.5	*.7*
High Pressure Cut-Off Switch, Replace (B)		
1985-88 Galant	1.5	*1.7*
Receiver/Drier Assy., Replace (B)		
Includes: Evacuate, recover and recharge.		
Exc. below	1.6	*1.8*
Lancer, Lancer Evolution	1.8	*2.0*
Temperature Control Assy., Replace (B)		
3000GT	1.8	*2.0*
Diamante		
1992-96	.8	*1.0*
1997-03	1.5	*1.7*

	LABOR TIME	SEVERE SERVICE
Eclipse		
1990-94	.8	*1.0*
1995-99	1.2	*1.4*
2000-05 (.6)	.8	*1.0*
Galant (.6)	.8	*1.0*
Lancer, Lancer Evolution (.4)	.6	*.8*
Mirage		
1985-92	.7	*.9*
1993-96	.8	*1.0*
1997-03	1.0	*1.2*

WIPERS & SPEEDOMETER

Antenna Assy., Replace (B)
3000GT	1.0	*1.2*
Cordia, Starion, Tredia	.7	*.9*
Diamante		
1992-96		
sedan	.7	*.9*
wagon	1.2	*1.4*
1997-04	1.0	*1.2*
Eclipse		
1990-99	.8	*1.0*
2000-05 (.8)	1.1	*1.3*
Galant (.5)	.7	*.9*
Lancer, Lancer Evolution		
exc. pole (1.2)	1.8	*2.0*
pole (.3)	.5	*.7*
Mirage		
1985-88	.7	*.9*
1989-92	1.6	*1.8*
1993-03	1.1	*1.3*
Precis	.5	*.7*
Sigma	.8	*1.0*

Instrument Cluster, R&I (B)
1991-05	1.3	*1.5*

Pulse Generator, Replace (B)
3000GT	.5	*.7*
Diamante		
1992-96	.5	*.7*
1997-04	1.4	*1.6*
Eclipse		
1990-99	.5	*.7*
2000-05 (.8)	1.1	*1.3*
Galant, Mirage	.7	*.9*
Lancer	.6	*.8*

Radio, R&R (B)
All Models	.8	*1.0*

Rear Window Wiper Motor, Replace (B)
Exc. below	.6	*.8*
1994-96 Diamante wagon	1.0	*1.2*

Rear Windshield Washer Pump & Motor, Replace (B)
3000GT	1.0	*1.2*
Cordia	.9	*1.1*
Diamante wagon	.8	*1.0*

	LABOR TIME	SEVERE SERVICE
Eclipse		
1990-99	.9	*1.1*
2000-05 (1.1)	1.5	*1.7*
1994-98 Galant	.7	*.9*
Lancer (.8)	1.1	*1.3*
Lancer Evolution (.6)	.8	*1.0*
1985-92 Mirage	.5	*.7*
Precis	.3	*.5*
Starion	.8	*1.0*

Rear Windshield Wiper Interval Relay, Replace (B)
3000GT	.8	*1.0*
Eclipse	.5	*.7*
1994-98 Galant	.3	*.5*
1989-92 Mirage	.5	*.7*

Rear Windshield Wiper Switch, Replace (B)
3000GT, Eclipse	.5	*.7*
Cordia, Starion	.3	*.5*
1994-96 Diamante wagon	.5	*.7*
1994-98 Galant	.5	*.7*
1985-92 Mirage	.3	*.5*

Speedometer Cable & Casing, Replace (B)
All Models	1.1	*1.3*

Speedometer Cable (Inner), Replace or Lubricate (B)
All Models	1.1	*1.3*

Speedometer Head, R&R or Replace (B)
Does not include reset odometer.
3000GT	1.0	*1.2*
Cordia, Tredia	1.4	*1.6*
Diamante		
1992-96	.7	*.9*
1997-04	1.1	*1.3*
Eclipse, Galant (.5)	.7	*.9*
Mirage		
1985-88	1.8	*2.0*
1989-02	.8	*1.0*
Precis	.9	*1.1*
Sigma	1.2	*1.4*
Starion	1.5	*1.7*
Replace speed sensor add	.3	*.5*

Turn Signal & Washer, Wiper Lever Assy., Replace (B)
Cordia, Tredia	.7	*.9*
1985-88 Mirage	.7	*.9*
Starion	1.3	*1.5*

Windshield Washer Pump & Motor, Replace (B)
3000GT, Diamante	.7	*.9*
Cordia, Tredia	.3	*.5*
Eclipse		
1990-99	.8	*1.0*
2000-04	1.4	*1.6*

MITSU-32 3000GT : CORDIA : DIAMANTE : ECLIPSE : GALANT : LANCER

	LABOR TIME	SEVERE SERVICE
Windshield Washer Pump & Motor, Replace (B)		
Galant		
1985-98	.6	.8
1999-03	1.5	1.7
2004-05 (1.4)	1.9	2.1
Lancer (.8)	1.1	1.3
Lancer Evolution (.6)	.8	1.0
Mirage		
1985-96	.6	.8
1997-03	1.0	1.2
Precis	.5	.7
Sigma	.3	.5
Starion	.8	1.0
Windshield Wiper & Washer Switch, Replace (B)		
Exc. below	.7	.9
Precis, Sigma	.8	1.0
Windshield Wiper Interval Relay, Replace (B)		
All Models	.5	.7
Windshield Wiper Linkage, Replace (B)		
3000GT, Diamante	1.0	1.2
Cordia, Tredia	.8	1.0
Eclipse	1.1	1.3
Galant		
1985-88	1.5	1.7
1989-05 (.8)	1.1	1.3
Lancer, Lancer Evolution (.6)	.8	1.0
Mirage	.8	1.0
Precis	.8	1.0
Sigma	1.4	1.6
Starion	1.4	1.6
Windshield Wiper Motor, Replace (B)		
All Models	.9	1.1
Windshield Wiper Motor Relay, Replace (B)		
All Models	.5	.7

LAMPS & SWITCHES

	LABOR TIME	SEVERE SERVICE
Back-Up Lamp Assy., Replace (B)		
Exc. below	.3	.3
1992-96 Diamante sedan	.6	.6
Back-Up Lamp Bulb, Replace (C)		
All Models	.3	.3
Back-Up Lamp/Neutral Safety Switch, Replace (B)		
Exc. below	.8	.8
1997-04 Diamante	1.4	1.4
Combination Switch Assy., Replace (B)		
Exc. below	.9	.9
3000GT	1.1	1.1
Diamante	1.1	1.1
1989-92 Mirage	1.0	1.0

	LABOR TIME	SEVERE SERVICE
Halogen Headlamp Bulb, Replace (C)		
Exc. below	.5	.5
3000GT	.6	.6
Lancer Evolution	1.5	1.5
Hazard Warning Switch, Replace (B)		
Exc. below	.8	.8
Cordia, Tredia	.3	.3
1985-88 Galant	1.4	1.4
1985-92 Mirage	.3	.3
Sigma	1.4	1.4
Headlamp Dimmer Switch, Replace (B)		
1985-88 Galant	.9	.9
Sigma	.9	.9
Headlamp Switch, Replace (B)		
1985-88 Galant	.8	.8
1985-88 Mirage	.7	.7
Precis	.5	.5
Sigma	.8	.8
Headlamps, Aim (B)		
Two	.5	.5
Four	.6	.6
High Mount Stop Lamp Assy., Replace (B)		
Exc. below	.7	.7
1992-96 Diamante sedan	.8	.8
Horn, Replace (B)		
3000GT	.3	.3
Cordia, Tredia	.3	.3
Diamante	.6	.6
Eclipse	.7	.7
Galant		
1985-87	.3	.3
1988-05 (.6)	.8	.8
Lancer	.3	.3
Lancer Evolution (1.1)	1.5	1.5
Mirage		
1985-88	.3	.3
1989-02	.8	.8
Precis	.3	.3
Starion	.3	.3
Horn Relay, Replace (B)		
All Models	.3	.3
Horn Switch, Replace (B)		
All Models	.5	.5
License Lamp Assy., Replace (B)		
Exc. below	.5	.5
Cordia, Starion	.8	.8
1985-88 Galant	.7	.7
Lancer	1.8	1.8
Lancer Evolution	1.1	1.1
1985-88 Mirage	.7	.7
Sigma	.7	.7
Park & Turn Signal Lamp Assy., Replace (B)		
All Models	.5	.5

	LABOR TIME	SEVERE SERVICE
Parking Brake Warning Lamp Switch, Replace (B)		
Exc. below	.8	.8
3000GT	1.3	1.3
1992-96 Diamante	1.0	1.0
Sealed Beam Headlamp, Replace (B)		
One	.3	.3
Side Marker Lamp Assy., Replace (B)		
All Models	.5	.5
Stop, Tail & Turn Signal Lamp Assy., Replace (B)		
Exc. below	.7	.7
3000GT	.9	.9
1995-99 Eclipse	.8	.8
1988-93 Galant	1.0	1.0
Stop Lamp Switch, Replace (B)		
All Models	.5	.5
Turn Signal & Headlamp Dimmer Switch, Replace (B)		
All Models	.7	.7
Turn Signal Switch, Replace (B)		
3000GT, Diamante sedan	.7	.7
1985-88 Galant	1.3	1.3
Sigma	1.1	1.1
Turn Signal or Hazard Warning Flasher, Replace (B)		
Exc. below	.3	.3
1995-99 Eclipse	.9	.9

SEAT BELTS

	LABOR TIME	SEVERE SERVICE
Seat Belt Drive & Rail Assy., Replace (B)		
Eclipse	.9	.9
1989-93 Galant	1.5	1.5
1989-96 Mirage		
exc. below	1.4	1.4
2 door, hatchback	1.9	1.9
1988-89 Starion	1.6	1.6

BODY

	LABOR TIME	SEVERE SERVICE
Deck Lid Hinge, Replace (B)		
Cordia, Tredia	1.5	1.7
Eclipse	.9	1.1
Galant		
1985-87	.7	.9
1999-05 (1.0)	1.4	1.6
Lancer, Lancer Evolution (1.2)	1.8	2.0
Mirage		
1985-88	.7	.9
1989-96	1.0	1.2
1997-03	1.4	1.6
Sigma	.7	.9
Door Window Regulator, Replace (B)		
Front		
3000GT	2.8	3.0
Cordia, Tredia	.9	1.1

LANCER EVOLUTION : MIRAGE : PRECIS : SIGMA : STARION : TREDIA MITSU-33

	LABOR TIME	SEVERE SERVICE
Diamante		
1992-96		
sedan	2.3	2.5
wagon	1.4	1.6
1997-04	2.5	2.7
Eclipse		
1990-94	1.0	1.2
1995-99	2.3	2.5
2000-05 (1.6)	2.2	2.4
Galant		
1985-93	.8	1.0
1994-05 (.9)	1.2	1.4
Lancer, Lancer Evolution (.9)	1.2	1.4
Mirage		
1985-92	.7	.9
1993-03	1.1	1.3
Precis	.7	.9
Sigma	.8	1.0
Starion	.9	1.1
Rear		
Diamante		
1992-96		
sedan	2.3	2.5
wagon	1.2	1.4
1997-04	2.1	2.3
Galant		
1985-88	1.4	1.6
1989-93	.8	1.0
1994-05 (.9)	1.2	1.4
Lancer, Lancer Evolution (.9)	1.2	1.4
Mirage		
1985-92	.7	.9
1993-03	1.1	1.3
Sigma	1.2	1.4
Tredia	.8	1.0
Front Door Lock, Replace (B)		
3000GT	1.0	1.2
Cordia, Starion, Tredia	.9	1.1
Diamante	1.2	1.4

	LABOR TIME	SEVERE SERVICE
Eclipse		
1990-94	.8	1.0
1995-05 (.8)	1.1	1.3
Galant (.7)	.9	1.1
Lancer, Lancer Evolution (.7)	.9	1.1
Mirage	.7	.9
Precis	.8	1.0
Sigma	.8	1.0
Hood Hinge, Replace (B)		
3000GT	.8	1.0
Cordia, Tredia	.8	1.0
Diamante	1.1	1.3
Eclipse		
1990-94	.7	.9
1995-99	.8	1.0
2000-05 (.9)	1.2	1.4
Galant		
1988-98	.8	1.0
1999-05 (.9)	1.2	1.4
Lancer, Lancer Evolution (.8)	1.1	1.3
Mirage		
1989-92	.6	.8
1993-03	.8	1.0
Precis	.8	1.0
Sigma	.9	1.1
Starion	1.6	1.8
Hood Release Cable, Replace (B)		
3000GT	1.0	1.2
Cordia, Starion, Tredia	.7	.9
Diamante	1.9	2.1
Eclipse		
1990-94	.7	.9
1995-99	3.2	3.4
2000-05 (2.2)	3.0	3.2
Galant		
1985-88	.9	1.1
1989-93	1.2	1.4
1994-98	2.3	2.5
1999-05 (1.0)	1.4	1.6
Lancer (1.0)	1.4	1.6
Lancer Evolution (1.7)	2.3	2.5
Mirage		

	LABOR TIME	SEVERE SERVICE
1985-92	.8	1.0
1993-03	1.2	1.4
Precis	.7	.9
Sigma	.9	1.1
Lock Striker Plate, Replace (B)		
All Models	.3	.5
Window Regulator Motor, Replace (B)		
Front		
Cordia, Tredia	.9	1.1
Diamante		
1992-96		
sedan	2.3	2.5
wagon	1.4	1.6
1997-03	1.4	1.6
Eclipse (1.7)	2.3	2.5
Galant		
1985-88	.8	1.0
1994-05 (.9)	1.2	1.4
Lancer, Lancer Evolution (.9)	1.2	1.4
Mirage		
1989-92	.7	.9
1993-03	1.1	1.3
1990-94 Precis	.8	1.0
Sigma	.8	1.0
Starion	.7	.9
Rear		
Diamante		
1992-96		
sedan	2.2	2.4
wagon	1.2	1.4
1997-04	2.1	2.3
Galant (1.0)	1.4	1.6
Lancer, Lancer Evolution (.9)	1.2	1.4
Mirage		
1989-92	.7	.9
1993-03	1.1	1.3
1990-94 Precis	.7	.9
Sigma	1.3	1.5
Tredia	.9	1.1

NOTES

MITSU

Endeavor : Expo : Expo LRV : Mighty Max : Montero : Montero Sport : Outlander : Van

SYSTEM INDEX

MAINTENANCE	MITSU-36
Maintenance Schedule	MITSU-36
CHARGING	MITSU-36
STARTING	MITSU-37
CRUISE CONTROL	MITSU-37
IGNITION	MITSU-37
EMISSIONS	MITSU-38
FUEL	MITSU-39
EXHAUST	MITSU-40
ENGINE COOLING	MITSU-40
ENGINE	MITSU-41
Assembly	MITSU-41
Cylinder Head	MITSU-42
Camshaft	MITSU-43
Crank & Pistons	MITSU-45
Engine Lubrication	MITSU-46
CLUTCH	MITSU-47
MANUAL TRANSAXLE	MITSU-47
MANUAL TRANSMISSION	MITSU-47
AUTO TRANSAXLE	MITSU-48
AUTO TRANSMISSION	MITSU-48
TRANSFER CASE	MITSU-49
SHIFT LINKAGE	MITSU-49
DRIVELINE	MITSU-49
BRAKES	MITSU-51
FRONT SUSPENSION	MITSU-53
REAR SUSPENSION	MITSU-54
STEERING	MITSU-54
HEATING & AC	MITSU-56
WIPERS & SPEEDOMETER	MITSU-57
LAMPS & SWITCHES	MITSU-57
BODY	MITSU-58

OPERATIONS INDEX

A
AC Hoses	MITSU-56
Air Bags	MITSU-54
Air Conditioning	MITSU-56
Alignment	MITSU-53
Alternator (Generator)	MITSU-36
Antenna	MITSU-57
Anti-Lock Brakes	MITSU-51

B
Back-Up Lamp Switch	MITSU-57
Ball Joint	MITSU-53
Battery Cables	MITSU-37
Bleed Brake System	MITSU-52
Blower Motor	MITSU-56
Brake Disc	MITSU-52
Brake Drum	MITSU-52
Brake Hose	MITSU-52
Brake Pads and/or Shoes	MITSU-52

C
Camshaft	MITSU-43
Camshaft Sensor	MITSU-37
Catalytic Converter	MITSU-40
Coolant Temperature (ECT) Sensor	MITSU-38
Crankshaft	MITSU-45
Crankshaft Sensor	MITSU-37
Cruise Control	MITSU-37
Cylinder Head	MITSU-42

D
Differential	MITSU-50
Distributor	MITSU-37
Drive Belt	MITSU-36
Driveshaft	MITSU-49

E
EGR	MITSU-38
Engine	MITSU-41
Engine Lubrication	MITSU-46
Engine Mounts	MITSU-42
Evaporator	MITSU-56
Exhaust	MITSU-40
Exhaust Manifold	MITSU-40
Expansion Valve	MITSU-56

F
Flywheel	MITSU-47
Fuel Injection	MITSU-39
Fuel Pump	MITSU-39
Fuel Vapor Canister	MITSU-38

G
Generator	MITSU-36
Glow Plug	MITSU-38

H
Halfshaft	MITSU-51
Headlamp	MITSU-57
Heater Core	MITSU-56
Horn	MITSU-57

I
Ignition Coil	MITSU-38
Ignition Module	MITSU-38
Ignition Switch	MITSU-38
Injection Pump	MITSU-39
Inner Tie Rod	MITSU-55
Intake Manifold	MITSU-39

K
Knock Sensor	MITSU-38

L
Lower Control Arm	MITSU-53

M
Maintenance Schedule	MITSU-36
Manifold Absolute Pressure (MAP) Sensor	MITSU-38
Mass Air Flow (MAF) Sensor	MITSU-38
Master Cylinder	MITSU-52
Muffler	MITSU-40
Multifunction Lever/Switch	MITSU-57

N
Neutral Safety Switch	MITSU-57

O
Oil Pan	MITSU-46
Oil Pump	MITSU-46
Outer Tie Rod	MITSU-55
Oxygen Sensor	MITSU-38

P
Parking Brake	MITSU-53
Pistons	MITSU-46
Positive Crankcase Ventilation (PCV) Valve	MITSU-38

R
Radiator	MITSU-40
Radiator Hoses	MITSU-40
Radio	MITSU-57
Rear Main Oil Seal	MITSU-46

S
Shock Absorber/Strut, Front	MITSU-53
Shock Absorber/Strut, Rear	MITSU-54
Spark Plug Cables	MITSU-38
Spark Plugs	MITSU-38
Spring, Front Coil	MITSU-53
Spring, Leaf	MITSU-54
Spring, Rear Coil	MITSU-54
Starter	MITSU-37
Steering Wheel	MITSU-55

T
Thermostat	MITSU-40
Throttle Body	MITSU-40
Throttle Position Sensor (TPS)	MITSU-38
Timing Belt	MITSU-44
Timing Chain	MITSU-45
Torque Converter	MITSU-48

U
U-Joint	MITSU-50
Upper Control Arm	MITSU-54

V
Valve Body	MITSU-48
Valve Cover Gasket	MITSU-43
Valve Job	MITSU-43

W
Water Pump	MITSU-40
Wheel Balance	MITSU-36
Wheel Cylinder	MITSU-52
Window Regulator	MITSU-58
Windshield Washer Pump	MITSU-57
Windshield Wiper Motor	MITSU-57

MITSU-36 ENDEAVOR : EXPO : EXPO LRV : MIGHTY MAX : MONTERO

	LABOR TIME	SEVERE SERVICE
MAINTENANCE		
Air Cleaner Filter Element, Service (B)		
Exc. below	.7	.9
Endeavor (.9)	1.2	1.4
2001-04 Montero	1.2	1.4
Chassis Lubrication, Change Oil and Filter (C)		
Includes: Correct all fluid levels.		
All Models (.4)	.7	.7
Composite Headlamp Bulb, Replace (C)		
All Models	.5	.5
Drive Belt, Adjust (B)		
All Models	.3	.5
Drive Belt, Replace (B)		
Alternator		
exc. Outlander (.4)	.6	.8
Outlander (.6)	.8	1.0
Compressor (.3)	.5	.7
Fan		
serpentine belt	.7	.9
V belt	.5	.7
Power steering		
exc. below (.4)	.6	.8
Van, Wagon	.7	.9
Fuel Filter, Replace (B)		
Endeavor (.8)	1.1	1.3
Expo, LRV		
in line	.6	.8
in tank		
2WD	1.9	2.1
4WD	2.4	2.6
Montero		
1983-00	.8	1.0
2001-04	1.3	1.5
Montero Sport		
in line	.8	1.0
in tank	2.1	2.3
Outlander (.8)	1.1	1.3
Pickup		
in line	.6	.8
in tank	1.3	1.5
Van, Wagon		
in line	.5	.7
in tank	1.2	1.4
Oil and Filter, Change (C)		
Includes: Correct all fluid levels.		
All Models (.4)	.6	.7
Sealed Beam Headlamp, Replace (C)		
1983-96	.3	.5
Timing Belt, Replace (B)		
Endeavor (2.1)	2.8	3.0
Expo, LRV	2.9	3.1
Montero		
4 cyl.	3.8	4.3
V6		
1989-00		
SOHC	3.8	4.3
DOHC	4.7	5.2
2001-04	4.4	4.9
Montero Sport		
4 cyl.	2.9	3.4
V6	3.2	3.7
Outlander (2.3)	3.1	3.6
Pickup		
4 cyl.		
gasoline	2.5	3.0
diesel	1.9	2.4
V6	3.5	4.0
Van, Wagon	3.0	3.5
w/AC add		
Expo, LRV	.5	.5
83-00 Montero	.5	.5
Montero Sport V6	.3	.3
Pickup	.5	.5
Van, Wagon	.5	.5
w/PS add		
Expo, LRV	.3	.3
Pickup	.3	.3
Van, Wagon	.3	.3
Tire, Replace (C)		
Includes: Dismount old tire and mount new tire to rim.		
One	.5	.5
each addl. add	.3	.3
Tires, Rotate (C)		
All Models	.5	.5
Wheel, Balance (B)		
One	.8	1.0
each addl. add	.2	.4
SCHEDULED MAINTENANCE INTERVALS		
If necessary, refer to appropriate Chilton maintenance service information.		
7,500 Mile Service (C)		
All Models	.4	.4
15,000 Mile Service (C)		
All Models	.7	.7
w/4WD add	.1	.1
w/AT add	.1	.1
22,500 Mile Service (C)		
All Models	.4	.4
30,000 Mile Service (B)		
All Models	2.2	2.2
w/4WD add	1.0	1.0
w/AT add	.2	.2
Replace spark plugs		
Montero add	.6	.6
37,500 Mile Service (C)		
All Models	.4	.4
45,000 Mile Service (C)		
All Models	.7	.7
w/4WD add	.1	.1
w/AT add	.1	.1
52,500 Mile Service (C)		
All Models	.4	.4
60,000 Mile Service (B)		
All Models	6.2	6.2
w/4WD add	.5	.5
w/AT add	.6	.6
Replace spark plugs		
Montero add	.6	.6
67,500 Mile Service (C)		
All Models	.4	.4
75,000 Mile Service (C)		
All Models	.7	.7
w/4WD add	.1	.1
w/AT add	.1	.1
82,500 Mile Service (C)		
All Models	.4	.4
90,000 Mile Service (B)		
All Models	2.3	2.3
w/4WD add	.4	.4
w/AT add	.6	.6
Replace spark plugs		
Montero add	.6	.6
97,500 Mile Service (C)		
All Models	.4	.4
CHARGING		
Alternator Circuits, Test (B)		
Includes: Test component output.		
All Models	.7	.7
Alternator, R&R and Recondition (B)		
Exc. below	2.8	3.0
Montero		
1992-00		
SOHC	2.6	2.8
DOHC	3.1	3.3
2001-04	2.4	2.6
Montero Sport	2.4	2.6
Outlander	3.3	3.5
Pickup	2.6	2.8
Alternator Assy., Replace (B)		
Includes: Pulley transfer.		
Exc. below	1.4	1.6
Montero		
1992-00		
SOHC	.8	1.0
DOHC	1.8	2.0
2001-04	.8	1.0
Montero Sport, Pickup	.8	1.0
Outlander (1.3)	1.8	2.0
Alternator Relay, Replace (B)		
Exc. Van, Wagon	.3	.5
Van, Wagon	2.4	2.6
Alternator Voltage Regulator, Replace (B)		
Exc. below	2.3	2.5
Endeavor (1.5)	2.0	2.2
Montero		
1992-00		
SOHC	1.9	2.1
DOHC	2.6	2.8
2001-04	1.8	2.0

MONTERO SPORT : OUTLANDER : VAN — MITSU-37

	LABOR TIME	SEVERE SERVICE
Montero Sport, Pickup	1.8	2.0
Outlander	2.8	3.0
Voltmeter, Replace (B)		
Montero	.9	1.1

STARTING

Starter Draw Test (On Car) (B)
- All Models ... .3 / .3

Battery Cables, Replace (C)
- Expo, LRV ... .8 / 1.0
- Montero
  - 1983-91 ... .7 / .9
  - 1992-00 ... .9 / 1.1
  - 2001-04 ... 2.2 / 2.4
- Montero Sport
  - positive
    - 4 cyl. ... .8 / 1.0
    - V6 ... 1.9 / 2.1
  - negative ... 1.7 / 1.9
- Outlander (1.2) ... 1.6 / 1.8
- Pickup ... .6 / .8
- Van, Wagon ... .5 / .7

Starter Assy., R&R and Reconditon (B)
- Endeavor (1.8) ... 2.4 / 2.6
- Expo, LRV ... 2.6 / 2.8
- Montero
  - 1983-91
    - 4 cyl. ... 2.6 / 2.8
    - V6 ... 3.7 / 3.9
  - 1992-00
    - SOHC
      - exc. 4V ... 2.4 / 2.6
      - 4V ... 3.1 / 3.3
    - DOHC ... 3.3 / 3.5
  - 2001-04 ... 2.1 / 2.3
- Montero Sport
  - 4 cyl. ... 2.4 / 2.6
  - V6 ... 3.4 / 3.6
- Outlander (1.9) ... 2.6 / 2.8
- Pickup ... 2.4 / 2.6
- Van, Wagon ... 2.1 / 2.3

Starter Assy., Replace (B)
- Endeavor, Expo, LRV ... .8 / 1.0
- Montero
  - 1983-91
    - 4 cyl. ... .8 / 1.0
    - V6 ... 2.3 / 2.5
  - 1992-00
    - SOHC
      - exc. 4V ... .9 / 1.1
      - 4V ... 1.5 / 1.7
    - DOHC ... 1.8 / 2.0
  - 2001-04 ... .6 / .8
- Montero Sport
  - 4 cyl. ... .9 / 1.1
  - V6 ... 1.7 / 1.9
- Outlander (.8) ... 1.1 / 1.3
- Pickup ... .8 / 1.0
- Van, Wagon ... .5 / .7

Starter Drive Assy., Replace (B)
Includes: Starter R&R.
- Endeavor (1.1) ... 1.5 / 1.7
- Expo, LRV ... 1.6 / 1.8
- Montero
  - 1983-91
    - 4 cyl. ... 1.6 / 1.8
    - V6 ... 3.1 / 3.3
  - 1992-00
    - SOHC
      - exc. 4V ... 1.4 / 1.6
      - 4V ... 2.0 / 2.2
    - DOHC ... 2.3 / 2.5
  - 2001-04 ... 1.4 / 1.6
- Montero Sport
  - 4 cyl. ... 1.8 / 2.0
  - V6 ... 2.4 / 2.6
- Outlander (1.3) ... 1.8 / 2.0
- Pickup ... 1.8 / 2.0
- Van, Wagon ... 1.3 / 1.5

Starter Relay, Replace (B)
- Montero Sport ... .3 / .5
- Pickup ... .6 / .8

Starter Solenoid and/or Switch, Replace (B)
Includes: Starter R&R.
- Endeavor (.9) ... 1.2 / 1.4
- Expo, LRV ... 1.4 / 1.6
- Montero
  - 1983-91
    - 4 cyl. ... 1.4 / 1.6
    - V6 ... 2.8 / 3.0
  - 1992-00
    - SOHC
      - exc. 4V ... 1.4 / 1.6
      - 4V ... 2.0 / 2.2
    - DOHC ... 2.3 / 2.5
  - 2001-04 (.8) ... 1.1 / 1.3
- Montero Sport
  - 4 cyl. ... 1.5 / 1.7
  - V6 ... 2.1 / 2.3
- Outlander (1.1) ... 1.5 / 1.7
- Pickup ... 1.4 / 1.6
- Van, Wagon ... .9 / 1.1

CRUISE CONTROL

Diagnose Cruise Control System Component Each (A)
- All Models ... .8 / .8

Actuator Assy., Replace (B)
- All Models ... .8 / 1.0

Brake or Clutch Release Switch, Replace (B)
- All Models ... .5 / .7

Control Chain, Cable or Rod, Replace (B)
- All Models ... .8 / 1.0

Control Controller Module, Replace (B)
- Expo, LRV, Pickup ... .7 / .9

Control Main Switch, Replace (B)
- Montero Sport ... .3 / .5
- Montero ... .8 / 1.0
- Montero Sport ... .5 / .7
- Outlander ... .8 / 1.0
- Van, Wagon ... .3 / .5

Control Vacuum Control or Check Valve, Replace (B)
- Pickup, Van, Wagon ... .3 / .5

Control Vacuum Hoses, Replace (B)
- 1983-96 ... .5 / .7

Control Vacuum Pump, Replace (B)
- Expo, LRV ... .6 / .8
- Montero ... .8 / 1.0
- Montero Sport, Van, Wagon ... .5 / .7
- Pickup ... .9 / 1.1

Control Vacuum Pump Relay, Replace (B)
- Van, Wagon ... .5 / .7

Control Vacuum Release Switch, Replace (B)
- 1983-96 ... .3 / .5

IGNITION

Diagnose Ignition System Component Each (A)
- All Models ... .8 / .8

Ignition Timing, Reset (B)
- All Models ... .5 / .7

Camshaft Sensor, Replace (B)
- Endeavor (.4) ... .6 / .8
- Montero ... .7 / .9
- Montero Sport ... .9 / 1.1
- Outlander (.5) ... .7 / .9

Crankshaft Angle Sensor, Replace (B)
- Endeavor (1.3) ... 1.8 / 2.0
- Montero
  - 1992-00
    - SOHC ... 3.3 / 3.5
    - DOHC ... 4.0 / 4.2
  - 2001-04 (2.6) ... 3.5 / 3.7
- Montero Sport
  - 4 cyl. ... 2.4 / 2.6
  - V6 ... 3.2 / 3.4
- Outlander (2.4) ... 3.2 / 3.4
- w/AC add
  - 92-00 Montero ... .8 / 1.0
  - Montero Sport V6 ... .3 / .5

Distributor, R&R and Reconditon (B)
Includes: Reset base ignition timing.
- Expo, LRV, Montero Sport ... 2.3 / 2.5
- Montero ... 2.0 / 2.2
- Pickup, Van, Wagon ... 1.8 / 2.0

Distributor, Replace (B)
Includes: Reset base ignition timing.
- Exc. Montero ... .7 / .9
- Montero ... .9 / 1.1

MITSU-38 ENDEAVOR : EXPO : EXPO LRV : MIGHTY MAX : MONTERO

	LABOR TIME	SEVERE SERVICE
Distributor Cap and/or Rotor, Replace (B)		
Exc. Montero Sport	.3	.5
Montero Sport	.6	.8
Distributor Igniter Set, Replace (B)		
All Models	.5	.7
Distributor Pickup Set, Replace (B)		
Includes: Distributor R&R.		
1983-89 Montero	1.2	1.4
1983-89 Pickup	1.4	1.6
Distributor Reluctor, Replace (B)		
1983-89 Montero	1.3	1.5
1983-89 Pickup	1.4	1.6
ECI Control Relay, Replace (B)		
1987-05	.5	.7
Electronic Ignition Control Unit, Replace (B)		
All Models (.5)	.7	.9
Glow Plug, Replace (B)		
Diesel one or all	.5	.7
Glow Plug Control Unit, Replace (B)		
Diesel	.5	.9
Glow Plug Relay, Replace (B)		
Diesel	.5	.7
Ignition Coil, Replace (B)		
All Models	.7	.7
Ignition Switch, Replace (B)		
Exc. Montero Sport	.8	1.0
Montero Sport	1.1	1.3
Spark Plug (Ignition) Cables, Replace (B)		
4 cyl.	.5	.7
V6		
Montero		
1983-91		
SOHC	2.2	2.4
DOHC	3.8	4.0
1992-00		
SOHC exc. 4V		
one side	.5	.7
both sides	.6	.8
SOHC 4V		
right side	.8	1.0
left or both sides	1.8	2.0
DOHC		
one side	3.1	3.3
both sides	3.2	3.4
2001-04		
right side	1.4	1.6
left or both sides	2.1	2.3
Montero Sport		
right side	.8	1.0
left or both sides	1.7	1.9
Pickup		
one side	.5	.7
both sides	.6	.8
Spark Plugs, Replace (B)		
4 cyl. (.4)	.6	.8
V6		

	LABOR TIME	SEVERE SERVICE
Endeavor (1.7)	2.3	2.5
Montero		
1983-91		
SOHC	2.6	2.8
DOHC	3.1	3.3
1992-00		
SOHC exc. 4V		
one side	.5	.7
both sides	.6	.8
SOHC 4V		
right side	.8	1.0
left or both sides	2.0	2.2
DOHC		
right side	3.1	3.3
left side	.8	1.0
both sides	3.5	3.7
2001-04		
right side	1.4	1.6
left or both sides	2.3	2.6
Montero Sport		
right side	.8	1.0
left or both sides	1.8	2.0
Pickup		
one side	.5	.7
both sides	.6	.8
Vacuum Advance Unit, Replace (B)		
Includes: Reset timing.		
Montero, Pickup	1.4	1.4

EMISSIONS

The following operations do not include testing. Add time as required.

	LABOR TIME	SEVERE SERVICE
Dynamometer Test (A)		
All Models	.5	.5
Air Flow Sensor, Replace (B)		
Exc. Expo, LRV	.6	.8
Expo, LRV	.7	.9
Canister Purge Control Valve, Replace (B)		
All Models	.6	.8
Coolant Temperature (ECT) Sensor, Replace (B)		
Exc. Outlander (.3)	.5	.7
Outlander (.4)	.6	.8
EGR Control Solenoid, Replace (B)		
All Models	.6	.8
EGR Coolant Control Valve, Replace (B)		
1983-91 Montero	.5	.7
Pickup, Van, Wagon	.5	.7
EGR Temperature Sensor, Replace (B)		
Exc. below	.5	.7
1989-90 Van, Wagon	.7	.9
EGR Time Delay Solenoid, Replace (B)		
All Models	.5	.7

	LABOR TIME	SEVERE SERVICE
EGR Valve, Replace (B)		
All Models	.6	.8
Fuel Tank Pressure Control Valve, Replace (B)		
Expo, LRV	.8	1.0
Montero		
1983-91	.3	.5
1992-00	.8	1.0
Montero Sport, Pickup	1.0	1.2
Fuel Vapor Canister, Replace (B)		
Exc. below	.5	.7
2001-04 Montero	5.0	5.2
Endeavor (.7)	.9	1.1
Outlander (.6)	.8	1.0
Fuel Vapor Pressure Sensor, Replace (B)		
Montero	.7	.9
High Altitude Compensator, Replace (B)		
1983-91 Montero	.5	.7
Pickup	.5	.7
Knock Sensor, Replace (B)		
Endeavor (2.3)	3.1	3.3
Montero		
1992-00		
SOHC	2.1	2.3
DOHC	3.3	3.5
2001-04	2.6	2.8
Manifold Pressure Sensor, Replace (B)		
All Models	.5	.7
Mass Air Flow (MAF) Sensor, Replace (B)		
All Models	.7	.9
Oxygen Sensor, Replace (B)		
Exc. below each	.7	.9
Montero before or after catalytic converter each	.8	1.0
Montero Sport after catalytic converter each	.8	1.0
Positive Crankcase Ventilation (PCV) Valve, Replace (B)		
Exc. below	.3	.5
1992-00 Montero	2.3	2.5
Outlander	.5	.7
Purge Control Solenoid Valve, Replace (B)		
Exc. below	.5	.7
2001-04 Montero	.6	.8
Throttle Position Sensor (TPS), Replace (B)		
Exc. Van, Wagon	.7	.9
Van, Wagon	.8	1.0

MONTERO SPORT : OUTLANDER : VAN — MITSU-39

FUEL

CARBURETOR

	LABOR TIME	SEVERE SERVICE
Carburetor, Adjust (On Pickup) (B)		
Montero, Pickup	.7	.9
Carburetor, R&R and Clean or Reconditon (B)		
Includes: Adjustments.		
Montero, Pickup	3.3	3.5
w/Closed loop add	.2	.4
Carburetor, Replace (B)		
Includes: Adjustments.		
Montero, Pickup	1.6	1.8
w/Closed loop add	.2	.4
Carburetor Accelerator Pump, Replace (B)		
Montero, Pickup	2.7	2.9
Dash Pot, Replace (B)		
Montero, Pickup	.5	.7
Sub-EGR Valve, Replace (B)		
Montero, Pickup	.7	.9

DELIVERY

	LABOR TIME	SEVERE SERVICE
Fuel Pump, Test (B)		
All Models	.3	.5
Fuel Cut-off Solenoid, Replace (B)		
1983-94	.5	.7
Fuel Filter, Replace (B)		
Endeavor (.8)	1.1	1.3
Expo, LRV		
in line	.6	.8
in tank		
2WD	1.9	2.1
4WD	2.4	2.6
Montero		
1983-00	.8	1.0
2001-04	1.3	1.5
Montero Sport		
in line	.8	1.0
in tank	2.1	2.3
Outlander (.8)	1.1	1.3
Pickup		
in line	.6	.8
in tank	1.3	1.5
Van, Wagon		
in line	.5	.5
in tank	1.2	1.4
Fuel Gauge (Dash), Replace (B)		
Exc. Montero	.8	.8
Montero	.9	.9
Fuel Gauge (Tank), Replace (B)		
Exc. below	.8	1.0
Expo, LRV		
2WD	1.7	1.9
4WD	2.3	2.5
2001-04 Montero	1.4	1.6
Pickup	1.4	1.6
Fuel Pump, Replace (B)		
Electric		
Endeavor (.7)	.9	1.1
Expo, LRV		
2WD	1.7	1.9
4WD	2.2	2.4
Montero	1.3	1.5
Montero Sport	2.1	2.3
Outlander (.6)	.8	1.0
Pickup	1.4	1.6
Van, Wagon	1.6	1.8
Mechanical	.8	1.0
Fuel Tank, Replace (B)		
Includes: Drain and refill.		
Endeavor (2.5)	3.4	3.6
Expo, LRV		
2WD	2.1	2.3
4WD	2.5	2.7
Montero		
1983-91	1.2	1.4
1992-04	3.0	3.2
Montero Sport	2.2	2.4
Outlander		
2WD (2.2)	3.0	3.2
4WD (2.7)	3.6	3.8
Pickup	1.6	1.8
Van, Wagon	1.5	1.7
Intake Manifold and/or Gasket, Replace (B)		
Includes: Adjustments.		
4 cyl.		
Expo, LRV	2.4	2.6
Montero	2.4	2.6
Montero Sport	2.6	2.8
Outlander (2.4)	3.2	3.4
Pickup	2.8	3.0
Van, Wagon	3.0	3.2
V6		
Endeavor (2.1)	2.9	3.1
Montero		
1989-00		
SOHC	2.4	2.6
DOHC	3.6	3.8
2001-04	2.9	3.2
Montero Sport	2.6	2.8
Pickup	3.5	3.7
w/AC Pickup 2.6L add	1.5	1.7

INJECTION

	LABOR TIME	SEVERE SERVICE
Diesel Fuel Injection Timing, Check & Adjust (B)		
1983-85 Pickup	.7	.9
Fuel Injection Pump, Replace (B)		
1983-85 Pickup	2.2	2.4
Fuel Injectors, Replace (B)		
Endeavor (1.7)	2.3	2.5
Expo, LRV	.8	1.0
Montero		
1989-91	2.5	2.7
1992-00		
SOHC	2.0	2.2
DOHC	3.5	3.7
2001-04	2.4	2.6
Montero Sport		
4 cyl.	.9	1.1
V6	2.1	2.3
Outlander (.7)	.9	1.1
Pickup	2.5	2.7
Van, Wagon	2.4	2.6
Fuel Pressure Regulator, Replace (B)		
Endeavor (.5)	.7	.9
Expo, LRV	.6	.8
Montero		
1989-91	2.9	3.1
1992-00		
SOHC	.8	1.0
DOHC	2.9	3.1
2001-04	.6	.8
Montero Sport	.6	.8
Outlander (.3)	.5	.7
Pickup		
4 cyl.	.8	1.0
V6	2.3	2.5
Van, Wagon	.9	1.1
Fuel Rail, Replace (B)		
Endeavor (1.8)	2.4	2.6
Expo, LRV	.8	1.0
Montero		
1989-00		
SOHC	2.0	2.2
DOHC	3.5	3.7
2001-04	2.4	2.6
Montero Sport		
4 cyl.	1.1	1.3
V6	2.3	2.5
Outlander (.8)	1.1	1.3
Pickup	2.3	2.5
Van, Wagon	2.3	2.5
Fuel Shut-Off Valve, Replace (B)		
Expo, LRV		
2WD	1.7	1.9
4WD	2.2	2.4
Montero	2.2	2.4
Montero Sport	2.1	2.3
Outlander		
2WD (1.8)	2.4	2.6
4WD (2.2)	3.0	3.2
Idle Air Control Motor, Replace (B)		
Expo, LRV	.6	.8
Montero, Pickup	1.1	1.3
Montero Sport	.9	1.1
Outlander	1.2	1.4
Van, Wagon	.7	.9
Injection Nozzles Diesel, Replace (B)		
1983-85 Pickup one	.7	.9
each addl. add	.5	.6
Intake Manifold Plenum and/or Gasket, Replace (B)		
Exc. below	1.7	1.9
Expo	.8	1.0
Pickup	2.0	2.2

MITSU-40 ENDEAVOR : EXPO : EXPO LRV : MIGHTY MAX : MONTERO

	LABOR TIME	SEVERE SERVICE
Throttle Body Assy., R&R or Replace (B)		
Endeavor (.9)	1.2	1.4
Expo, LRV	.9	1.1
Montero		
SOHC	.9	1.1
DOHC		
1994-00 (1.0)	1.4	1.6
2001-04 (.6)	.8	1.0
Montero Sport	.9	1.1
Outlander (.8)	1.1	1.3
1994-96 Pickup	1.4	1.6
Turbocharger Assy., R&R or Replace (B)		
1983-86 Pickup	1.6	1.8

EXHAUST

	LABOR TIME	SEVERE SERVICE
Catalytic Converter, Replace (B)		
Expo, LRV each	1.3	1.5
Montero		
1983-91	.8	1.0
1992-00 each	1.6	1.8
2001-04 each	1.3	1.5
Montero Sport		
4 cyl.	1.4	1.6
V6		
front	2.6	2.8
rear	1.1	1.3
Outlander (.8)	1.1	1.3
Pickup	.8	1.0
Van, Wagon	.7	.9
Center Exhaust Pipe, Replace (B)		
Exc. below (.8)	1.1	1.3
Montero (1.0)	1.4	1.6
Exhaust Manifold or Gasket, Replace (B)		
4 cyl.		
exc. below	1.8	2.0
Expo, LRV 2.4L 4V	2.3	2.5
Outlander (1.0)	1.4	1.6
V6		
Endeavor		
right side (2.7)	3.6	3.8
left side (1.1)	1.5	1.7
Montero		
1992-00		
SOHC		
right side	1.8	2.0
left side		
w/4V	3.6	3.8
w/o 4V	1.5	1.7
DOHC		
right	1.8	2.0
left	3.1	3.3
2001-04		
right side	1.8	2.0
left side	2.3	2.5

	LABOR TIME	SEVERE SERVICE
Montero Sport		
right side	2.4	2.6
left side	3.1	3.3
Pickup		
right side	1.5	1.7
left side	1.9	2.1
Front Exhaust Pipe, Replace (B)		
4 cyl.		
exc. below	.7	.9
Expo/LRV	1.3	1.5
Montero Sport	1.3	1.5
Outlander (.8)	1.1	1.3
V6		
Endeavor (.5)	.7	.9
Montero	1.7	1.9
Montero Sport	2.4	2.6
Pickup	1.9	2.1
Muffler, Replace (B)		
Exc. below (.6)	.8	1.0
2001-04 Montero	1.3	1.5
Montero Sport	1.0	1.2
Tail Pipe, Replace (B)		
Montero		
1992-00	.6	.8
2001-04	1.1	1.3
Montero Sport	.8	1.0

ENGINE COOLING

	LABOR TIME	SEVERE SERVICE
Pressure Test Cooling System (B)		
All Models	.3	.3
Combination Meter Assy., Replace (B)		
Exc. below	.7	.9
Montero	.8	1.0
Montero Sport	.7	.9
Outlander, Pickup	.5	.7
Replace speed sensor add	.3	.5
Coolant Thermostat, Replace (B)		
All Models (.8)	1.1	1.3
Electric Cooling Fan Motor, Replace (B)		
Endeavor (.8)	1.1	1.3
Expo, LRV	.5	.7
Outlander (.7)	.9	1.1
Engine Coolant Temp. Sending Unit, Replace (B)		
Exc. Outlander (.4)	.6	.8
Outlander (.5)	.7	.9
Fluid Fan Drive or Fan Blade, Replace (B)		
Montero, Pickup	.9	1.1
Montero Sport	1.3	1.5
Van, Wagon	.8	1.0
w/charged air cooler add	.3	.5
Radiator Assy., R&R or Replace (B)		
Endeavor (1.5)	2.0	2.2
Expo, LRV	1.4	1.6

	LABOR TIME	SEVERE SERVICE
Montero		
1983-91	.8	1.0
1992-00		
SOHC	1.2	1.4
DOHC	1.7	1.9
2001-04	1.7	1.9
Montero Sport	1.2	1.4
Outlander (.9)	1.2	1.4
Pickup, Van, Wagon	1.4	1.6
w/AT add		
92-00 Montero	.3	.5
Montero Sport	.3	.5
Radiator Fan Motor Relay, Replace (B)		
Expo, LRV	.3	.5
Radiator Fan Motor (Coolant Temp.) Switch, Replace (B)		
Expo, LRV (.5)	.7	.9
Radiator Hoses, Replace (B)		
Includes: Refill with proper coolant mix.		
Endeavor		
upper (.4)	.6	.8
lower (.7)	.9	1.1
Expo, LRV		
upper (.5)	.7	.9
lower (1.0)	1.4	1.6
Montero		
1983-91 each	.7	.9
1992-03 each	1.1	1.3
Montero Sport each	1.0	1.2
Outlander		
upper (.4)	.6	.8
lower (.8)	1.1	1.3
Pickup each	.7	.9
Van, Wagon each	.8	.9
Temperature Gauge (Dash), Replace (B)		
Expo, LRV, Pickup	.7	.9
Montero	.9	1.1
Montero Sport, Van, Wagon	.8	1.0
Outlander	.7	.9
Water Pump and/or Gasket, Replace (B)		
Includes: Refill with proper coolant mix.		
Endeavor (2.4)	3.2	3.4
Expo, LRV	3.2	3.4
Montero		
1983-91		
4 cyl.	1.6	1.8
V6	3.7	3.9
1992-00		
SOHC	4.1	4.3
DOHC	5.3	5.5
2001-04	5.1	5.3
Montero Sport		
4 cyl.	3.2	3.4
V6	3.7	3.9
Outlander (2.8)	3.8	4.0

MONTERO SPORT : OUTLANDER : VAN — MITSU-41

	LABOR TIME	SEVERE SERVICE
Pickup		
2.0L, 2.4L	3.2	3.4
2.6L	1.6	1.8
V6	3.9	4.1
Van, Wagon	3.4	3.6
w/AC add	.3	.5
w/PS add	.2	.4

ENGINE
ASSEMBLY

Times shown are for OEM assemblies. Time to replace assemblies from aftermarket rebuilders may vary.

Engine Assy., R&I (B)
Does not include parts or component transfer.

	LABOR TIME	SEVERE SERVICE
4 cyl.		
Expo, LRV		
2WD	10.6	10.8
4WD	11.7	11.9
Montero	11.8	12.0
Montero Sport	8.3	8.5
Outlander (9.3)	12.6	12.8
Pickup		
gasoline		
2WD	9.8	10.0
4WD	11.8	12.0
diesel	5.3	5.5
Van, Wagon	10.1	10.3
V6		
Endeavor (12.9)	17.4	17.6
Montero		
1989-91	18.2	18.4
1992-02	22.2	22.4
Montero Sport		
2WD	12.3	12.5
4WD	13.5	13.7
Pickup		
2WD	14.4	14.6
4WD	16.2	16.4
w/AC add		
Expo	.7	.9
Montero	.5	.7
Montero Sport	.3	.5
Pickup		
gasoline	.5	.7
diesel	.7	.9
Van, Wagon	.5	.7
w/AT add		
Expo, LRV	.5	.7
Montero	.8	1.0
Montero Sport	.8	1.0
Pickup	.5	.7
w/PS add		
Endeavor	1.0	1.2
Expo, LRV	.2	.4
Montero	.3	.5
Pickup	.3	.5
Van, Wagon	.2	.4

Engine Assy., R&R and Recondition (A)
Includes: Replacing rings, rod and main bearings, cylinder head reconditioning and engine tune-up.

	LABOR TIME	SEVERE SERVICE
4 cyl.		
Expo, LRV		
2WD	28.8	29.0
4WD	29.7	29.9
Montero	32.9	33.1
Montero Sport	26.1	26.3
Outlander (18.9)	25.5	25.7
Pickup		
2WD	28.1	28.3
4WD	32.7	32.9
Van, Wagon	29.0	29.2
V6 SOHC		
Endeavor (25.0)	33.8	34.0
Montero		
1989-91	27.1	27.3
1992-02	32.0	32.2
Montero Sport	29.9	30.1
Pickup		
2WD	24.8	25.0
4WD	27.0	27.2
V6 DOHC Montero	33.4	33.6
w/4V DOHC add		
Expo	1.6	1.8
LRV	1.7	1.9
w/4WD Montero Sport add	1.5	1.7
w/AC add		
Expo	.7	.9
Montero	.5	.7
Montero Sport	.3	.5
Pickup		
gasoline	.5	.7
diesel	.7	.9
Van, Wagon	.5	.7
w/AT add		
Expo, LRV	.5	.7
Montero	.8	1.0
Montero Sport	.8	1.0
Pickup	.5	.7
w/PS add		
Endeavor	1.0	1.2
Expo, LRV	.2	.4
Montero	.3	.5
Pickup	.3	.5
Van, Wagon	.2	.4

Engine Assy. (Short Block), Replace (B)
Assembly consists of cylinder block, piston assemblies, crankshaft, camshaft, timing chain and gears. Does not include cylinder heads. Operation Includes: R&R engine, transfer necessary parts and all necessary adjustments.

	LABOR TIME	SEVERE SERVICE
4 cyl.		
Expo, LRV		
2WD	12.0	12.2
4WD	13.0	13.2
Montero Sport	11.8	12.0
Outlander (11.1)	15.0	15.2
Pickup		
gasoline		
2WD	11.5	11.7
4WD	13.2	13.4
diesel	14.9	15.1
Van, Wagon	11.8	12.0
V6		
Endeavor (13.6)	18.4	18.6
Montero		
1989-91	23.0	23.2
1992-00	18.9	19.1
2001-04	18.2	18.4
Montero Sport		
2WD	15.1	15.3
4WD	16.7	17.2
Pickup		
2WD	14.4	14.6
4WD	16.0	16.2
w/AC add		
Expo	.8	1.0
LRV		
1.8L	.3	.5
2.4L	.8	1.0
83-00 Montero	.6	.8
Montero Sport	.5	.7
Pickup		
gasoline	.6	.8
diesel	.7	.9
Van, Wagon	.6	.8
w/AT add		
Expo, LRV	.8	1.0
83-00 Montero	.8	1.0
Montero Sport	1.1	1.3
Pickup	.8	1.0
w/PS add		
Endeavor	1.0	1.2
Expo, LRV	.3	.5
Pickup	.5	.7
Van, Wagon	.3	.5

MITSU-42 ENDEAVOR : EXPO : EXPO LRV : MIGHTY MAX : MONTERO

	LABOR TIME	SEVERE SERVICE
Engine Mounts, Replace (B)		
Includes: Engine lifting when necessary.		
Endeavor		
mount (.4)	.6	.8
stopper each (2.3)	3.1	3.3
Expo/LRV		
mount	.6	.8
stopper each	1.2	1.4
Montero		
1983-91		
one	1.2	1.4
both	1.7	1.9
1992-00		
front	13.6	13.8
rear	1.2	1.4
2001-04		
front	7.4	7.6
rear	1.2	1.4
Montero Sport		
front		
4 cyl.	6.9	7.1
V6		
2WD	9.6	9.8
4WD	11.3	11.5
rear	1.1	1.3
Outlander		
mount (.4)	.6	.8
stopper		
front (.7)	.9	1.1
rear (1.0)	1.4	1.6
Pickup		
gasoline		
one	1.3	1.5
both	1.6	1.6
diesel		
one	.5	.7
both	.7	.9
w/AC add		
92-00 Montero	.6	.8
Montero Sport	.5	.7
w/AT add		
92-00 Montero	.8	1.0
Montero Sport	1.2	1.4
CYLINDER HEAD		
Compression Test (B)		
4 cyl.	.8	1.0
V6		
Endeavor (1.0)	1.4	1.6
Montero		
SOHC	2.9	3.1
DOHC	3.6	3.8
Montero Sport	2.3	2.5
Valve Clearance, Adjust (B)		
All Models	1.6	1.8

	LABOR TIME	SEVERE SERVICE
Cylinder Head, Replace (B)		
Includes: Parts transfer, adjustments.		
4 cyl.		
Expo		
exc. SOHC 4V	6.5	6.7
SOHC 4V	9.3	9.5
LRV	9.3	9.5
Montero	6.3	6.5
Montero Sport	9.6	9.8
Outlander (7.6)	10.3	10.5
Pickup		
gasoline		
carbureted	6.4	6.6
MPI	6.8	7.0
diesel	6.3	6.5
Van, Wagon	7.0	7.2
V6		
Endeavor		
right side (10.3)	13.9	14.1
left side (9.6)	13.0	13.2
both sides (14.7)	19.8	20.0
Montero		
1989-91		
right side	11.5	11.7
left side	11.7	11.9
both sides	12.9	13.1
1992-00		
SOHC		
right side	10.9	11.1
left side	11.3	11.5
SOHC 4V		
right side	11.9	12.1
left side	12.6	12.8
DOHC one side	15.3	15.5
2001-04 one side	13.8	14.0
Montero Sport		
right side	11.5	11.7
left side	12.3	12.5
Pickup		
right side	11.8	12.0
left side	12.1	12.3
w/AC add		
92-00 Montero	.9	1.1
Montero Sport	.3	.5
Pickup 2.6L	.3	.5
w/PS add		
Pickup 2.4L	.3	.5
Van, Wagon	.2	.4
Cylinder Head Gasket, Replace (B)		
4 cyl.		
Expo		
exc. SOHC 4V	3.6	3.8
SOHC 4V	4.7	4.9
LRV	4.7	4.9
Montero	3.5	3.7
Montero Sport	4.8	5.0
Outlander (3.3)	4.5	4.7

	LABOR TIME	SEVERE SERVICE
Pickup		
gasoline		
carbureted	3.7	3.9
MPI	4.0	4.2
diesel	3.8	4.0
Van, Wagon	4.4	4.6
V6		
Endeavor		
right side (7.2)	9.7	9.9
left side (6.4)	8.6	8.8
both sides (8.8)	11.9	12.1
Montero		
1989-91		
right side	9.8	10.0
left side	9.9	10.1
both sides	10.3	10.5
1992-00		
SOHC		
right side	7.7	7.9
left side	8.6	8.8
SOHC 4V		
right side	9.1	9.3
left side	9.6	9.8
DOHC		
right side	10.5	10.7
left sides	10.8	11.0
2001-04 one side	9.6	9.8
Montero Sport		
right side	7.9	8.1
left side	8.7	8.9
Pickup		
right side	9.9	10.1
left side	10.5	10.7
w/AC add		
92-00 Montero	.9	1.1
Montero Sport	.3	.5
Pickup 2.6L	.3	.5
w/PS add		
Pickup 2.4L	.3	.5
Van, Wagon	.2	.4
Jet Valves, Replace (B)		
Montero, Pickup	2.8	3.0
Van, Wagon	2.9	3.1
Rocker Arms or Shafts, Replace (B)		
4 cyl.		
Expo, LRV	1.9	2.1
Montero	1.8	2.0
Montero Sport	1.5	1.7
Outlander (1.1)	1.5	1.7
Pickup		
gasoline	1.7	1.9
diesel	1.2	1.4
Van, Wagon	2.4	2.6
V6		
Endeavor		
right side (2.4)	3.2	3.4
left side (1.0)	1.4	1.6

MONTERO SPORT : OUTLANDER : VAN — MITSU-43

	LABOR TIME	SEVERE SERVICE
Rocker Arms or Shafts, Replace (B)		
Montero		
1989-91		
right side	2.3	2.5
left side	3.4	3.6
1992-00		
right side		
SOHC	1.8	2.0
SOHC 4V	2.4	2.6
DOHC	4.6	4.8
left side	2.8	3.0
2001-04		
right side	2.3	2.5
left side	2.9	3.1
Montero Sport		
each side	2.6	2.8
Pickup right side	1.8	2.0
Valve/Cam Cover Gasket, Replace or Reseal (B)		
4 cyl.		
Expo, LRV, Montero	.5	.7
Montero Sport, Van, Wagon	.9	1.1
Outlander (.4)	.6	.8
Pickup		
gasoline		
carbureted	.5	.7
MPI	.9	1.1
diesel	.7	.9
V6		
Endeavor		
right side (1.5)	2.0	2.2
left side (.4)	.6	.8
Montero, Pickup		
SOHC		
right side		
2V	1.2	1.4
4V	2.1	2.3
left side	2.5	2.7
DOHC one side	3.8	4.0
Montero Sport		
one side	2.2	2.4
Valve Job (A)		
4 cyl.		
Expo		
2V	6.2	6.4
4V	9.8	10.0
LRV	9.8	10.0
Montero	7.8	8.0
Montero Sport	9.9	10.1
Outlander	9.8	10.0
Pickup		
gasoline		
carbureted	6.3	6.5
MPI	6.8	7.0
diesel	6.9	7.1
Van, Wagon	8.8	9.0
V6		
Endeavor	18.4	18.6

	LABOR TIME	SEVERE SERVICE
Montero		
1983-91		
right side	12.9	13.1
left side	12.9	13.1
both sides	14.9	15.1
1992-00		
SOHC	12.4	12.6
SOHC 4V	16.9	17.1
DOHC	19.8	20.0
2001-04	17.9	18.1
Montero Sport	15.9	16.1
Pickup	12.9	13.1
w/AC add		
92-00 Montero	.9	1.1
Montero Sport V6	.3	.5
Pickup 2.6L	.3	.5
w/PS add		
Pickup 4G6	.3	.5
Van, Wagon	.2	.4
Valve Lash Adjuster, Replace (B)		
4 cyl.		
Expo, LRV	1.9	2.1
Montero, Pickup	2.1	2.3
Van, Wagon	2.6	2.8
V6		
Endeavor		
right side	3.5	3.7
left side	1.6	1.8
Montero, Pickup		
SOHC		
right side	2.5	2.7
left side	3.7	3.9
DOHC		
right side	3.1	3.3
left side	3.5	3.7
Montero Sport		
right side	2.9	3.1
left side	3.6	3.8
Valve Springs and/or Oil Seals, Replace (B)		
4 cyl.		
Expo		
2V	3.9	4.1
4V	5.7	5.9
LRV	5.6	5.9
Montero	3.9	4.1
Montero Sport	5.3	5.5
Outlander	5.3	5.5
Pickup		
gasoline		
carbureted	3.9	4.1
MPI	4.4	4.6
diesel	4.6	4.8
Van, Wagon	4.4	4.6
V6		
Endeavor	9.8	10.0
Montero		
1983-91		
right side	3.9	4.1
left side	4.9	5.1

	LABOR TIME	SEVERE SERVICE
1992-00		
SOHC	7.7	7.9
SOHC 4V	9.3	9.5
DOHC	15.4	15.6
2001-04	9.9	10.1
Montero Sport	8.9	9.1
Pickup	7.9	8.1
w/AC 92-00 Montero		
DOHC add	.9	1.1
CAMSHAFT		
Balance Shaft, Replace (B)		
Expo, LRV		
2WD	9.5	9.7
4WD	10.5	10.7
Montero	7.3	7.5
Montero Sport	6.4	6.6
Outlander (4.6)	6.2	6.4
Pickup		
exc. 2.6L	4.8	5.0
2.6L	8.4	8.6
Van, Wagon	4.4	4.6
w/AC add		
Expo, LRV	.5	.7
Pickup		
exc. 2.6L	.5	.7
2.6L	1.5	1.7
w/aluminum alloy oil pan Outlander add	.8	1.0
w/AT Expo, LRV add	.8	1.0
w/PS Expo, LRV, Pickup add	.3	.5
Balance Shaft Chain and/or Sprockets, Replace (B)		
4 cyl.		
Montero	6.9	7.1
Pickup	7.9	8.1
w/AC add		
Montero	.3	.5
Pickup	1.5	1.7
w/PS add	.3	.5
Balance Shaft Sprocket, Replace (B)		
Expo, LRV	3.0	3.2
Montero	7.0	7.2
Montero Sport	2.9	3.1
Outlander (2.4)	3.2	3.4
Pickup	2.8	3.0
Van, Wagon	3.2	3.4
w/AC add	.5	.7
w/PS add	.3	.5
Camshaft, Replace (A)		
4 cyl.		
Expo, LRV	2.0	2.2
Montero	1.8	2.0
Montero Sport	5.1	5.3
Outlander (3.2)	4.3	4.5

MITSU-44 ENDEAVOR : EXPO : EXPO LRV : MIGHTY MAX : MONTERO

	LABOR TIME	SEVERE SERVICE
Pickup		
gasoline		
carbureted	1.9	2.1
MPI	2.1	2.3
diesel	2.2	2.4
Van, Wagon	2.5	2.7
V6		
Endeavor		
right side (5.3)	7.2	7.4
left side (4.2)	5.7	5.9
Montero		
1992-00		
SOHC		
right side	2.4	2.6
left side	3.6	3.8
SOHC 4V		
right side	8.2	8.4
left side	8.9	9.1
DOHC one side	8.3	8.5
2001-04 one side	10.5	10.7
Montero Sport		
right side	8.3	8.5
left side	9.4	9.6
Pickup		
right side	2.3	2.5
left side	3.5	3.7
w/AC add		
92-00 Montero	.9	1.1
Montero Sport	.3	.5
Camshaft Seal, Replace (B)		
4 cyl.		
Expo, LRV	1.7	1.9
Montero	1.9	2.1
Montero Sport	1.1	1.3
Outlander (2.1)	2.8	3.0
Pickup		
gasoline		
carbureted	1.7	1.9
MPI	2.1	2.3
diesel	2.2	2.4
Van, Wagon	2.4	2.6
V6		
Endeavor one side	2.8	3.1
Montero		
1992-00		
SOHC		
right side	2.3	2.5
left side	3.5	3.7
SOHC 4V		
one side	3.8	4.0
DOHC one side	5.0	5.2
2001-04 one side	4.6	4.8
Montero Sport		
one side	3.3	3.5
Pickup		
right side	2.3	2.5
left side	3.5	3.7
w/AC add		
92-00 Montero	.9	1.1
Montero Sport	.3	.5

	LABOR TIME	SEVERE SERVICE
Front Case Gasket, Replace (B)		
4 cyl.		
Expo, LRV	5.4	5.6
Montero Sport	6.2	6.4
Outlander (4.6)	6.2	6.4
Pickup	4.8	5.0
Van, Wagon	4.3	4.5
w/AC Expo, LRV,		
Pickup add	.5	.7
w/aluminum alloy oil		
pan Outlander add	.8	1.0
w/PS Expo, LRV,		
Pickup add	.3	.5
Silent Shaft, Replace (B)		
2.0L, 2.4L		
1983-88		
right	6.8	7.0
left	7.2	7.4
both	7.8	8.0
1989-96 one or both	8.8	9.0
2.3L diesel		
right side	2.9	3.1
left side	4.2	4.4
both sides	4.8	5.0
1983-90 2.6L		
right	8.1	8.3
left	7.6	7.8
both	8.3	8.5
Montero Sport	2.8	3.0
Silent Shaft Oil Seal, Replace (B)		
1983-90 4 cyl.	4.7	4.9
Diesel	2.2	2.4
Timing Belt, Replace (B)		
Endeavor (2.1)	2.8	3.0
Expo, LRV	2.9	3.1
Montero		
4 cyl.	3.8	4.0
V6		
1989-00		
SOHC	3.8	4.0
DOHC	4.9	5.1
2001-04	4.4	4.6
Montero Sport		
4 cyl.	2.9	3.1
V6	3.2	3.4
Outlander (2.3)	3.1	3.3
Pickup		
4 cyl.		
gasoline	2.8	3.0
diesel	1.9	2.1
V6	3.5	3.7
Van, Wagon	3.0	3.2
w/AC add		
Expo, LRV	.5	.7
83-00 Montero	.5	.7
Montero Sport V6	.3	.5
Pickup	.5	.7
Van, Wagon	.5	.7

	LABOR TIME	SEVERE SERVICE
w/PS add		
Expo, LRV	.3	.5
Pickup	.3	.5
Van, Wagon	.3	.5
Replace balance shaft		
belt add	.3	.5
Timing Belt Tensioner, Replace (B)		
4 cyl.		
Expo, LRV		
timing blt each	2.8	3.0
balance shaft blt	1.8	2.0
Montero Sport each	2.6	2.8
Outlander		
each (2.1)	2.8	3.0
Pickup		
gasoline		
timing belt	2.5	2.7
balance shaft blt	1.5	1.7
diesel		
cam drive	1.9	2.1
silent shaft drive	2.3	2.5
Van, Wagon	3.1	3.3
V6		
Endeavor (1.9)	2.6	2.8
Montero		
1989-00		
automatic		
SOHC	3.8	4.0
DOHC	4.7	4.9
manual	3.8	4.0
Montero Sport each	3.2	3.4
Pickup	3.5	3.7
w/AC add		
Expo, LRV	.5	.7
92-00 Montero		
AT	.9	1.1
MT	.5	.7
Montero Sport V6	.3	.5
Pickup	.5	.7
Van, Wagon	.3	.5
w/PS add		
Expo, LRV, Pickup	.3	.5
Van, Wagon	.2	.4
Timing Belt/Chain Cover, Replace (B)		
4 cyl.		
gasoline		
upper		
exc. Outlander	.5	.7
Outlander (.4)	.6	.8
lower		
Expo, LRV	1.8	2.0
Montero Sport	1.7	1.9
Outlander (1.5)	2.0	2.2
Pickup	1.5	1.7
Van, Wagon	1.9	2.1
diesel		
upper	.3	.5
lower	1.2	1.4

MONTERO SPORT : OUTLANDER : VAN — MITSU-45

	LABOR TIME	SEVERE SERVICE
V6		
Endeavor		
upper		
right side (.4)	.6	.8
left side (.3)	.5	.7
lower (1.3)	1.8	2.0
Montero		
1989-91		
upper one side	1.6	1.8
lower	3.0	3.2
1992-00		
upper		
right		
SOHC	1.6	1.8
SOHC 4V	2.3	2.5
DOHC	4.0	4.2
left		
exc. below	.5	.7
SOHC 2V	1.5	1.7
lower		
SOHC	3.0	3.2
DOHC	4.4	4.6
2001-04 each	3.8	4.0
Montero Sport		
upper		
right side	2.6	2.8
left side	.7	.9
lower	2.4	2.6
Pickup		
upper		
right	1.1	1.3
left	1.5	1.7
lower	3.1	3.3
w/AC add		
Montero		
89-91	.3	.5
92-00		
upper		
SOHC 4V	.9	1.1
DOHC	1.5	1.7
lower	.9	1.1
Montero Sport V6	.3	.5
Pickup	.5	.7
Van, Wagon	.3	.5
w/PS Expo, LRV,		
Pickup add	.3	.5
Timing Chain, Replace (B)		
Montero	7.0	7.2
Pickup	7.8	8.0
w/AC add	.3	.5
w/PS add	.2	.4
Timing Chain Housing Gasket, Replace (B)		
Montero	7.1	7.3
Pickup	7.9	8.1
w/AC add	.3	.5
w/PS add	.2	.4
Timing Chain Tensioner, Replace (B)		
Montero	7.0	7.2
Pickup	7.7	7.9
w/AC add	.3	.5
w/PS add	.2	.4
CRANK & PISTONS		
Connecting Rod Bearings, Replace (A)		
Includes: Check bearing oil clearance.		
4 cyl.		
Expo, LRV	4.4	4.6
Montero, Van, Wagon	3.1	3.3
Montero Sport	4.8	5.0
Outlander (3.3)	4.5	4.7
Pickup		
gasoline	4.0	4.2
diesel	3.4	3.6
V6		
Endeavor (3.9)	5.3	5.5
Montero		
1989-91	8.2	8.4
1992-00		
exc. SOHC 2V	9.8	10.0
SOHC 2V	8.3	8.5
2001-04	8.3	8.5
Montero Sport	9.3	9.5
Pickup	6.9	7.1
w/aluminum alloy oil pan Outlander add	.6	.8
Crankshaft and Main Bearings, Replace (A)		
Includes: Engine R&R.		
4 cyl.		
Expo, LRV	13.6	13.8
Montero	14.9	15.1
Montero Sport	13.9	14.1
Outlander (11.4)	15.4	15.6
Pickup		
2.0L, 2.4L	13.1	13.3
2.6L	13.6	13.8
diesel	7.9	8.1
Van, Wagon	13.8	14.0
V6		
Endeavor (14.7)	19.8	20.0
Montero		
1989-91	17.3	17.5
1992-00	22.4	22.6
2001-04	22.9	23.1
Montero Sport	16.7	16.9
Pickup	15.6	15.8
w/4WD add		
Expo, LRV	1.1	1.3
Montero Sport	1.8	2.0
Outlander	1.4	1.6
Pickup	2.0	2.2
w/AC add		
Expo	.9	1.1
LRV	.3	.5
83-00 Montero	.6	.8
Montero Sport	.5	.7
Pickup		
gasoline	.6	.8
diesel	.7	.9
Van, Wagon	.5	.7
w/AT add		
Expo, LRV	.8	1.0
83-00 Montero	.8	1.0
Montero Sport	1.2	1.4
Pickup	.8	1.0
w/PS add		
Expo, LRV	.3	.5
83-91 Montero	.3	.5
Pickup	.5	.7
Van, Wagon	.2	.4
Crankshaft Front Oil Seal, Replace (B)		
Endeavor (2.1)	2.8	3.0
Expo, LRV	3.1	3.3
Montero		
1983-91		
4 cyl.	1.3	1.5
V6	3.5	3.7
1992-00		
SOHC	3.9	4.1
DOHC	4.8	5.0
2001-04	4.6	4.8
Montero Sport		
4 cyl.	2.9	3.1
V6	3.4	3.6
Outlander (2.4)	3.2	3.4
Pickup		
4 cyl.		
gasoline		
exc. 4G6	1.2	1.4
4G6	2.8	3.0
diesel	1.8	2.0
V6	3.7	3.9
Van, Wagon	3.1	3.3
w/AC add		
Expo, LRV 2.4L	.5	.7
Montero		
83-91	.3	.5
92-00	.9	1.1
Montero Sport V6	.3	.5
Pickup	.5	.7
Van, Wagon	.3	.5
w/PS add		
83-91 Montero	.2	.4
Pickup	.3	.5
Van, Wagon	.2	.4

MITSU-46   ENDEAVOR : EXPO : EXPO LRV : MIGHTY MAX : MONTERO

	LABOR TIME	SEVERE SERVICE
Crankshaft Pulley, Replace (B)		
4 cyl.		
1983-96		
gasoline	.7	.9
diesel	.7	.9
Montero Sport	1.3	1.5
Outlander	1.1	1.3
V6	1.2	1.4
w/AC add	.2	.4
Pistons or Connecting Rods, Replace (A)		
Includes: Ridge reaming, cylinder wall deglazing, installing new rings and rod bearings, engine tune-up.		
4 cyl.		
Expo, LRV	10.2	10.4
Montero	11.8	12.0
Montero Sport	9.9	10.1
Outlander (8.9)	12.0	12.2
Pickup		
gasoline	9.8	10.0
diesel	8.7	8.9
Van, Wagon	10.8	11.0
V6		
Endeavor (12.5)	16.9	17.1
Montero		
1983-91	13.9	14.1
1992-00		
SOHC	16.4	16.6
DOHC	17.9	18.1
2001-04	15.8	16.0
Montero Sport	13.2	13.4
Pickup	11.2	11.3
w/4WD add		
Expo, LRV	1.1	1.3
Montero Sport	1.8	2.0
Outlander	1.4	1.6
Pickup	2.0	2.2
w/AC add		
Expo	.9	1.1
LRV	.3	.5
92-94 Montero	.6	.8
Montero Sport	.5	.7
Pickup	.6	.8
w/AT add		
Expo, LRV, Pickup	.8	1.0
92-00 Montero	.8	1.0
Montero Sport	1.2	1.4
w/PS add		
Expo, LRV	.3	.5
Pickup	.5	.7
Van, Wagon	.2	.4
Rear Main Oil Seal, Replace (B)		
Includes: Trans. R&R. when necessary.		
4 cyl.		
Expo		
exc. below	6.3	6.5
2WD SOHC 4V	6.8	7.0

	LABOR TIME	SEVERE SERVICE
LRV	5.0	5.2
Montero	6.7	6.9
Montero Sport	6.3	6.5
Outlander (4.9)	6.6	6.8
Pickup	5.1	5.3
Van, Wagon	4.8	5.0
V6		
Endeavor (5.2)	7.0	7.2
Montero		
1983-00	7.9	8.1
2001-04	6.9	7.1
Montero Sport	6.3	6.5
Pickup	5.3	5.5
w/4WD add		
Expo	1.1	1.3
LRV	.6	.8
Montero Sport	1.8	2.0
Outlander	1.5	1.7
Pickup	2.6	2.8
w/AT add		
Expo, LRV	.5	.7
83-00 Montero	.5	.7
Montero Sport	.9	1.1
Pickup, Van, Wagon	.5	.7
Rings, Replace (A)		
Includes: Ridge reaming, cylinder wall deglazing, installing new rings, engine tune-up.		
4 cyl.		
Expo, LRV	10.2	10.4
Montero	11.2	11.4
Montero Sport	9.9	10.1
Outlander (8.9)	12.0	12.2
Pickup		
gasoline	9.8	10.0
diesel	7.5	7.7
Van, Wagon	10.2	10.4
V6		
Endeavor (12.2)	16.5	16.7
Montero		
1983-91	13.2	13.4
1992-00		
SOHC	16.4	16.6
DOHC	17.9	18.1
2001-04	15.8	16.0
Montero Sport	13.2	13.4
Pickup	11.2	11.4
w/4WD add		
Expo, LRV	1.1	1.3
Montero Sport	1.8	2.0
Outlander	1.4	1.6
Pickup	2.0	2.2
w/AC add		
Expo	.9	1.1
LRV	.3	.5
92-94 Montero	.6	.8
Montero Sport	.5	.7
Pickup	.6	.8

	LABOR TIME	SEVERE SERVICE
w/AT add		
Expo, LRV, Pickup	.8	1.0
92-00 Montero	.8	1.0
Montero Sport	1.2	1.4
w/PS add		
Expo, LRV	.3	.5
Pickup	.5	.7
Van, Wagon	.2	.4
ENGINE LUBRICATION		
Engine Oil Cooler, Replace (B)		
Montero, Pickup	1.1	1.3
Engine Oil Pressure Switch (Sending Unit), Replace (B)		
Exc. below	.5	.7
Montero, Montero Sport, Outlander (.5)	.7	.9
Oil Pan and/or Gasket, Replace (B)		
4 cyl.		
Expo, LRV		
2WD	2.2	2.4
4WD	2.6	2.8
Montero	1.5	1.7
Montero Sport	3.0	3.2
Outlander (1.8)	2.4	2.6
Pickup		
gasoline	2.0	2.2
diesel	1.7	1.9
Van, Wagon	1.4	1.6
V6		
Endeavor		
upper (2.3)	3.1	3.3
lower (.8)	1.1	1.3
Montero		
one-piece pan	5.7	5.9
two-piece pan		
upper	7.2	7.4
lower	2.4	2.6
Montero Sport		
upper	6.8	7.0
lower	5.0	5.2
Pickup	4.0	4.2
Oil Pressure Gauge (Dash), Replace (B)		
All Models	.9	1.1
Oil Pump, Replace (B)		
4 cyl.		
Expo, LRV	5.3	5.5
Montero	8.0	8.2
Montero Sport	6.6	6.8
Outlander (5.1)	6.9	7.1
Pickup		
2.0L, 2.4L	4.9	5.1
2.6L	8.7	8.9
diesel	3.6	3.8
Van, Wagon	7.4	7.6

MONTERO SPORT : OUTLANDER : VAN — MITSU-47

	LABOR TIME	SEVERE SERVICE
V6		
Endeavor (4.4)	5.9	6.1
Montero		
1989-00		
SOHC	9.9	10.1
SOHC 4V	11.0	11.2
DOHC	11.8	12.0
2001-04	10.5	10.7
Montero Sport	10.0	10.2
Pickup	8.1	8.3
w/4WD add		
Expo, LRV	.3	.5
Montero Sport 4 cyl.	.6	.8
w/AC add		
Expo, LRV	.5	.7
83-00 Montero	.5	.7
Montero Sport	.8	1.0
Pickup		
exc. 2.6L	.5	.7
2.6L	1.6	1.8
Van, Wagon	.3	.5
w/aluminum alloy oil pan Outlander add	.8	1.0
w/PS add		
Expo, LRV	.3	.5
Pickup	.3	.5
Van, Wagon	.2	.4

CLUTCH

Bleed Clutch Hydraulic System (B)
All Models .... .3 / .5

Clutch Cable, Adjust (B)
Pickup 4 cyl. .... .5 / .7

Clutch Pedal Free Play, Adjust (B)
All Models .... .3 / .5

Clutch Assy., Replace (B)
	LABOR	SEVERE
4 cyl.		
Expo		
exc. below	4.3	4.5
2WD SOHC 4V	4.7	4.9
LRV	4.7	4.9
Montero	5.5	5.7
Montero Sport	4.0	4.2
Outlander (4.0)	5.4	5.6
Pickup	3.2	3.4
V6		
Montero		
1989-91	6.8	7.0
1992-00	8.8	9.0
Montero Sport	6.0	6.2
Pickup	4.7	4.9
w/4WD add		
Expo	1.1	1.3
LRV	.6	.8
Montero Sport	1.9	2.1
Pickup	2.7	2.9

Clutch Master Cylinder, Replace (B)
Includes: System bleeding.
	LABOR	SEVERE
Expo, LRV, Outlander	1.0	1.2
Montero, Pickup	.8	1.0
Montero Sport	1.3	1.5
w/ABS 92-00 Montero add	.6	.8

Clutch Release Bearing, Replace (B)
	LABOR	SEVERE
4 cyl.		
Expo		
exc. below	4.3	4.5
2WD SOHC 4V	4.7	4.9
LRV	4.7	4.9
Montero	5.2	5.4
Montero Sport	4.0	4.2
Outlander (4.2)	5.7	5.9
Pickup	3.2	3.4
V6		
Montero		
1989-91	6.6	6.8
1992-00	8.8	9.0
Montero Sport	6.0	6.2
Pickup	4.7	4.9
w/4WD add		
Expo	1.1	1.3
LRV	.6	.8
Montero Sport	1.9	2.1
Pickup	2.7	2.9
Replace fork add	.3	.5

Clutch Release Cable, Replace (B)
Pickup 4 cyl. .... .5 / .7

Clutch Slave Cylinder, Replace (B)
Includes: System bleeding.
All Models .... 1.0 / 1.2

Flywheel, Replace (B)
	LABOR	SEVERE
4 cyl.		
Expo		
exc. below	4.7	4.9
2WD SOHC 4V	5.2	5.4
LRV		
1.8L	7.0	7.2
2.4L	5.2	5.4
Montero	5.8	6.0
Montero Sport	4.4	4.6
Outlander (4.4)	5.9	6.1
Pickup	3.6	3.8
V6		
Montero		
1989-91	7.2	7.4
1992-00	9.3	9.5
Montero Sport	6.4	6.6
Pickup	5.2	5.4
w/4WD add		
Expo	1.1	1.3
LRV	.6	.8
Montero Sport	1.9	2.1
Pickup	2.7	2.9

MANUAL TRANSAXLE

Transaxle Assy. R&R and Recondition (A)
Includes: Reset toe, if required.
	LABOR	SEVERE
Expo		
exc. below	8.9	9.1
2WD SOHC 4V	9.4	9.6
LRV	9.4	9.6
w/4WD add		
Expo	1.9	2.1
LRV	1.4	1.6

Transaxle Assy., R&R or Replace (B)
	LABOR	SEVERE
Expo		
exc. below	4.0	4.2
2WD SOHC 4V	4.4	4.6
LRV	4.4	4.6
w/4WD add		
Expo	1.1	1.3
LRV	.6	.8

MANUAL TRANSMISSION

Extension Housing Oil Seal, Replace (B)
	LABOR	SEVERE
Exc. Montero Sport	.7	.9
Montero Sport	1.0	1.2

Transmission Assy., R&R and Recondition (A)
	LABOR	SEVERE
4 cyl.		
Montero	9.5	9.7
Montero Sport	8.8	9.0
Outlander (7.8)	10.5	10.7
Pickup	7.8	8.0
V6		
Montero		
1989-91	13.4	13.6
1992-00	15.6	15.8
Montero Sport	14.9	15.1
Pickup	11.2	11.4
w/4WD Pickup add	2.7	2.9

Transmission Assy., R&R or Replace (B)
	LABOR	SEVERE
4 cyl.		
Montero	5.2	5.4
Montero Sport	3.6	3.8
Outlander (3.8)	5.1	5.3
Pickup	2.9	3.1
V6		
Montero		
1989-91	6.4	6.6
1992-00	9.5	9.7
Montero Sport	5.7	5.9
Pickup	4.4	4.6
w/4WD add		
Montero Sport	2.5	2.5
Pickup	2.7	2.7

MITSU-48 — ENDEAVOR : EXPO : EXPO LRV : MIGHTY MAX : MONTERO

	LABOR TIME	SEVERE SERVICE
AUTOMATIC TRANSAXLE		
SERVICE TRANSAXLE INSTALLED		
Check Unit for Oil Leaks (C)		
All Models	.9	1.1
Drain & Refill Unit (B)		
All Models (.9)	1.2	1.4
Oil Pressure Check (B)		
All Models	.9	1.1
Accumulator Assy., R&R or Replace (B)		
Expo, LRV	1.4	1.6
Electronic Controller, Replace (B)		
All Models	.7	.9
Lock-Up Control Solenoid Valve, Replace (B)		
Exc. below	2.1	2.3
Outlander	3.0	3.2
Oil Pan Gasket, Replace (B)		
Exc. below	.8	1.0
Outlander	2.7	2.9
Pulse Generator, Replace (B)		
Endeavor	1.0	1.3
Expo, LRV	.5	.7
Outlander	.8	1.0
Solenoid Valve, Replace (B)		
Endeavor	3.1	3.3
Expo, LRV	2.1	2.3
Outlander	3.0	3.2
Speedometer Driven Gear and/or Seal, Replace (B)		
Expo, LRV	.5	.7
Valve Body, Replace (B)		
Endeavor	3.3	3.5
Expo, LRV	1.6	2.1
Outlander	3.5	3.7
SERVICE TRANSAXLE REMOVED		
Transaxle R&R included unless otherwise noted.		
Front Pump and/or Oil Seal, Replace (B)		
Endeavor (6.5)	8.8	9.0
Expo		
2WD		
exc. SOHC 4V	6.6	6.8
SOHC 4V	7.0	7.2
4WD	6.9	7.1
LRV		
2WD	7.0	7.2
4WD	6.9	7.1
Outlander		
2WD (7.3)	9.9	10.1
4WD (8.4)	11.3	11.5

	LABOR TIME	SEVERE SERVICE
Torque Converter, Replace (B)		
Endeavor (6.1)	8.2	8.4
Expo		
2WD		
exc. SOHC 4V	4.9	5.1
SOHC 4V	5.3	5.5
4WD	6.0	6.2
LRV		
2WD	5.3	5.5
4WD	6.0	6.2
Outlander		
2WD (6.3)	8.5	8.7
4WD (7.4)	10.0	10.2
Transaxle Assy., R&R and Recondition (A)		
Endeavor (12.0)	16.2	16.4
Expo		
2WD		
exc. SOHC 4V	13.1	13.3
SOHC 4V	13.5	13.7
4WD	16.6	16.8
LRV		
2WD	13.5	13.7
4WD	16.6	16.8
Outlander		
2WD (11.7)	15.8	16.0
4WD (12.7)	17.1	17.3
Flush oil cooler and lines add	.6	.8
Transaxle Assy., R&R or Replace (B)		
Endeavor	7.4	7.6
Expo		
2WD		
exc. SOHC 4V	4.7	4.9
SOHC 4V	5.2	5.4
4WD	5.8	6.0
LRV		
2WD	5.2	5.4
4WD	5.8	6.0
Outlander		
2WD	6.4	6.6
4WD	7.9	8.1
Transaxle Assy., Reseal (B)		
Expo		
2WD	5.0	5.2
4WD	5.8	6.0
LRV		
2WD	5.1	5.3
4WD	5.9	6.1
AUTOMATIC TRANSMISSION		
SERVICE TRANSMISSION INSTALLED		
Check Unit for Oil Leaks (C)		
All Models	.9	1.1
Drain and Refill Unit (B)		
Exc. Montero Sport	.8	1.0
Montero Sport	2.4	2.6

	LABOR TIME	SEVERE SERVICE
Oil Pressure Check (B)		
All Models	.9	1.9
Accumulator Assy., Replace (B)		
Exc. Montero Sport	1.7	1.9
Montero Sport	3.2	3.4
Electronic Control Unit, Replace (B)		
Montero	.8	1.0
Montero Sport, Pickup	.5	.7
Extension Housing Gasket, Replace (B)		
All Models	2.0	2.2
Extension Housing Oil Seal, Replace (B)		
Exc. Montero Sport	.7	.9
Montero Sport	.9	1.1
Governor Assy., R&R or Replace (B)		
1983-93		
2WD	2.6	2.8
4WD	3.0	3.2
Governor Support & Parking Gear, Replace (B)		
1983-93	2.6	2.8
Oil Pan Gasket, Replace (B)		
Exc. below	.8	1.0
2001-04 Montero	3.0	3.2
Montero Sport		
3-Speed	2.4	2.6
4-Speed	3.0	3.2
Parking Pawl, Shaft, Rod or Spring, Replace (B)		
Exc. Montero Sport	1.8	2.0
Montero Sport	3.2	3.4
Solenoid, Lock-Up Torque Converter, Replace (B)		
Exc. Montero Sport	2.0	2.2
Montero Sport	3.0	3.2
Speedometer Driven Gear and/or Seal, Replace (B)		
All Models	.5	.7
Valve Body Assy., Replace (B)		
Exc. below	1.8	2.0
2001-04 Montero	3.8	4.0
Montero Sport	3.6	3.8
SERVICE TRANSMISSION REMOVED		
Transmission R&R included unless otherwise noted.		
Front Oil Pump, Replace (B)		
4 cyl.		
Montero	6.3	6.5
Montero Sport	9.7	9.9
Pickup		
exc. 2.4L	5.7	5.9
2.4L	10.8	11.0
Van, Wagon	3.5	3.7

MONTERO SPORT : OUTLANDER : VAN **MITSU-49**

	LABOR TIME	SEVERE SERVICE
V6		
Montero		
1989-91	7.9	8.1
1992-00		
3.0L	11.9	12.1
3.5L	12.5	12.7
2001-04	10.6	10.8
Montero Sport	9.7	9.9
w/4WD add		
Montero Sport	2.5	2.5
Pickup	2.7	2.7
Front Pump Oil Seal, Replace (B)		
4 cyl.		
Montero	6.4	6.6
Montero Sport	9.7	9.9
Pickup		
exc. 2.4L	5.7	5.9
2.4L	10.8	11.0
Van, Wagon	3.4	3.6
V6		
Montero		
1989-91	7.5	7.7
1992-00		
3.0L	11.9	12.1
3.5L	12.5	12.7
2001-04	10.6	10.8
Montero Sport	9.7	9.9
w/4WD add		
Montero Sport	2.5	2.7
Pickup	2.7	2.9
Torque Converter, Replace (B)		
4 cyl.		
Montero	6.1	6.3
Montero Sport	7.0	7.2
Pickup		
2WD	3.8	4.0
4WD	6.4	6.6
Van, Wagon	3.2	3.4
V6		
Montero		
1989-91	7.4	7.6
1992-00	9.3	9.5
2001-04	6.6	6.8
Montero Sport		
2WD	7.0	7.2
4WD	8.9	9.1
Transmission Assy., R&R and Recondition (A)		
4 cyl.		
Montero	15.1	15.3
Montero Sport	16.4	16.6
Pickup		
2WD	13.4	13.6
4WD	15.8	16.0
Van, Wagon	12.3	12.5
V6		
Montero		
1989-91	16.2	16.4
1992-00	18.9	19.1

	LABOR TIME	SEVERE SERVICE
2001-04		
4-Speed	14.8	15.0
5-Speed	15.2	15.4
Montero Sport		
1997-98		
2WD	16.4	16.6
4WD	18.5	18.7
1999-03		
2WD	13.7	13.9
4WD	15.9	16.1
Flush oil cooler and lines add	.5	.7
Transmission Assy., Replace (B)		
4 cyl.		
Montero	6.0	6.2
Montero Sport	6.9	7.1
Pickup		
2WD	3.5	3.7
4WD	6.3	6.5
Van, Wagon	3.0	3.2
V6		
Montero		
1989-91	7.2	7.4
1992-00	9.7	9.9
2001-04	7.0	7.2
Montero Sport		
2WD	6.9	7.1
4WD	9.3	9.5
Transmission Assy., Reseal (B)		
4 cyl.		
Montero	6.3	6.5
Montero Sport	9.9	10.1
Pickup	3.8	4.0
Van, Wagon	3.5	3.7
V6		
Montero		
1989-91	7.7	7.9
1992-02	9.6	9.8
Montero Sport		
2WD	9.9	10.1
4WD	12.2	12.4

TRANSFER CASE
Rear Output Shaft Seal, Replace (B)
All Models	1.1	1.3

Speedometer Drive Pinion, Replace (B)
All Models	.5	.7

Transfer Case, R&I or Replace (B)
4 cyl.		
Expo, LRV	1.6	1.8
Montero	5.3	5.5
Outlander (3.6)	4.9	5.1
Pickup		
AT	6.3	6.5
MT	5.5	5.7
V6		
Endeavor (3.6)	4.9	5.1

	LABOR TIME	SEVERE SERVICE
Montero		
1989-91	6.4	6.6
1992-00		
AT	9.7	9.9
MT	9.5	9.7
2001-04	7.0	7.2
Montero Sport		
AT	9.4	9.6
MT	8.1	8.3
Pickup	7.0	7.2
Replace		
drive chain add	2.7	2.9
front housing add	3.0	3.2
rear housing add	2.1	2.3
Transfer Case, R&R and Recondition (A)		
4 cyl.		
Expo, LRV	4.3	4.5
Montero	8.3	8.5
Pickup		
AT	9.7	9.9
MT	9.0	9.2
V6		
Montero		
1989-91	9.8	10.0
1992-00	14.0	14.2
2001-04	11.5	11.7
Montero Sport		
AT	12.9	13.1
MT	11.8	12.0
Pickup	10.5	10.7
Transfer Case Lever and/or Boot, Replace (B)		
AT	1.1	1.3
MT	.5	.7

SHIFT LINKAGE
AUTOMATIC TRANSAXLE
Throttle Linkage, Adjust (B)
Expo, LRV	.3	.5

AUTOMATIC TRANSMISSION
Throttle Linkage, Adjust (B)
All Models	.5	.7

Throttle Cable, Replace (B)
Exc. Montero Sport	1.2	1.4
Montero Sport	2.3	2.5

DRIVELINE
DRIVESHAFT
Driveshaft, Replace (B)
Endeavor (.9)	1.2	1.4
Expo, LRV	1.0	1.2
Montero		
1983-91		
front shaft	.7	.9
rear shaft	.5	.7
1992-00 each	1.1	1.3
Montero Sport each	1.0	1.2
Outlander (.8)	1.1	1.3

MITSU-50　ENDEAVOR : EXPO : EXPO LRV : MIGHTY MAX : MONTERO

	LABOR TIME	SEVERE SERVICE
Driveshaft, Replace (B)		
Pickup		
front shaft	.8	1.0
rear shaft long bed		
trans. case		
to bearing	.7	.9
bearing to		
rear axle	.5	.7
rear shaft short bed	.7	.9
Van, Wagon	.5	.7
Driveshaft Center Bearing or Seals, Replace (B)		
All Models	1.6	1.8
U-Joint, Replace or Recondition (B)		
Expo, LRV	1.6	1.8
Montero		
1983-91		
front axle	1.4	1.6
rear axle	1.1	1.3
1992-03	1.7	1.9
Montero Sport	1.6	1.8
Outlander (1.3)	1.8	2.0
Pickup		
front axle	1.4	1.6
rear axle	1.1	1.3
Van, Wagon	.9	1.1

DRIVE AXLE

	LABOR TIME	SEVERE SERVICE
Differential, Drain & Refill (B)		
Front axle		
Montero	.7	.9
1983-94 Pickup	.7	.9
Axle Housing, Replace (B)		
Front axle		
Montero		
1983-91	7.7	7.9
1992-02	8.5	8.7
Montero Sport	9.4	9.6
Pickup	7.9	8.1
Rear axle		
Montero		
1983-91	3.6	3.8
1992-00	6.9	7.1
Montero Sport	6.6	6.8
Pickup	3.7	3.9
w/disc brakes Montero		
Sport add	.6	.8
Axle Shaft, Replace (B)		
Front axle		
Montero		
1983-91	1.6	1.7
1992-00	2.1	2.3
Montero Sport	2.3	2.5
Pickup	1.9	2.1
Replace bearing or		
seal add	.2	.4

	LABOR TIME	SEVERE SERVICE
Axle Shaft and/or Bearing, Replace (B)		
Rear axle		
Montero		
1983-91	1.6	1.8
1992-00	3.2	3.4
Montero Sport	3.8	4.0
Pickup	1.6	1.8
Van, Wagon	1.7	1.9
Axle Shaft Oil Seal, Replace (B)		
Rear axle		
Montero		
1983-91		
inner	1.2	1.4
outer	1.5	1.7
1992-00	2.0	2.2
Montero Sport	2.1	2.3
Pickup		
inner	1.1	1.3
outer	1.6	1.8
Van, Wagon		
inner	1.3	1.5
outer	1.6	1.8
Differential Carrier, R&R or Replace (B)		
Front axle		
Montero		
1983-91	4.1	4.3
1992-00	4.3	4.5
2001-04	4.7	4.9
Montero Sport	5.9	6.1
1983-94 Pickup	4.4	4.6
Rear axle		
Endeavor (2.3)	3.1	3.3
Expo, LRV	1.9	2.1
Montero		
1983-91	3.1	3.3
1992-00	4.7	4.9
2001-04	3.0	3.2
Montero Sport	4.7	4.9
Outlander (2.0)	2.7	2.9
Pickup, Van, Wagon	3.0	3.2
w/rear disc brakes Montero		
Sport add	.6	.8
Differential Carrier Assy., R&R and Recondition (A)		
Front axle		
Montero		
1983-91	10.5	10.7
1992-00	11.2	11.4
2001-04	11.6	11.8
Montero Sport	12.5	12.7
Pickup	11.3	11.5
Rear axle		
Endeavor (7.3)	9.9	10.1
Expo, LRV	9.0	9.2
Montero		
1983-91	8.7	8.9
1992-00	10.8	11.0
2001-04	10.0	10.2

	LABOR TIME	SEVERE SERVICE
Montero Sport	10.8	11.0
Outlander (7.2)	9.7	9.9
Pickup, Van, Wagon	9.1	9.3
w/rear disc brakes Montero		
Sport add	.6	.8
Differential Case, Replace (B)		
Front axle		
Montero		
1983-91	9.3	9.4
1992-00	9.9	10.1
2001-04	10.3	10.5
Montero Sport	10.7	10.9
Pickup	9.6	10.1
Rear axle		
Endeavor (6.5)	8.8	9.0
Expo, LRV	7.9	8.1
Montero		
1983-91	7.9	8.1
1992-00	9.9	10.1
2001-04		
limited slip	8.7	8.9
standard	5.7	5.9
Montero Sport	9.9	10.1
Outlander (6.2)	8.4	8.6
Pickup, Van, Wagon		
limited slip	7.6	7.8
standard	8.2	8.4
w/rear disc brakes Montero		
Sport add	.6	.8
Differential Cover Gasket, Replace (B)		
Front axle		
Montero		
1983-91	.8	1.0
1992-00	2.7	2.9
2001-04	4.4	4.6
Montero Sport	2.4	2.6
Pickup	1.0	1.2
Rear axle		
Endeavor (2.8)	3.8	4.0
Expo, LRV	1.6	1.8
Montero		
1983-91	2.8	3.0
1992-00	4.7	4.9
2001-04	3.8	4.0
Montero Sport	4.7	4.9
Outlander (2.5)	3.4	3.6
Pickup	3.0	3.2
Van, Wagon	2.8	3.0
w/rear disc brakes Montero		
Sport add	.6	.8
Differential Side Bearings, Replace (B)		
Front axle		
Montero		
1983-91	8.1	8.3
1992-00	8.5	8.7
2001-04	9.0	9.2
Montero Sport	9.9	10.1
Pickup	8.7	8.9

MONTERO SPORT : OUTLANDER : VAN — MITSU-51

	LABOR TIME	SEVERE SERVICE
Rear axle		
Endeavor (5.6)	7.6	7.8
Expo, LRV	6.6	6.8
Montero		
1983-91	5.4	5.6
1992-03	7.8	8.0
Montero Sport	7.8	8.0
Outlander (5.4)	7.3	7.5
Pickup, Van, Wagon	6.1	6.3
w/rear disc brakes Montero Sport add	.6	.8
Differential Side Gear Set, Replace (B)		
Front axle		
Montero		
1992-00	6.4	6.6
2001-04	6.9	7.1
Montero Sport	7.8	8.0
Pickup	6.6	6.8
Rear axle		
Endeavor (3.8)	5.1	5.3
Expo, LRV	4.1	4.3
Montero		
1983-91	4.2	4.4
1992-00	6.3	6.5
2001-04	5.2	5.4
Montero Sport	6.3	6.5
Outlander (3.6)	4.9	5.1
Pickup, Van, Wagon	4.6	4.8
w/rear disc brakes Montero Sport add	.6	.8
Pinion Shaft Oil Seal, Replace (B)		
Front axle		
Montero	1.4	1.6
Montero Sport	1.3	1.5
Pickup	1.3	1.5
Rear axle		
Endeavor (.9)	1.2	1.4
Expo, LRV	1.1	1.3
Montero		
1983-91	.8	1.0
1992-03	1.3	1.5
Montero Sport	1.1	1.3
Outlander (1.0)	1.4	1.6
Pickup, Van, Wagon	1.0	1.2
Rear Companion Shaft, Replace (B)		
Rear axle Expo, LRV	1.3	1.5
Rear Wheel Bearing, Replace (B)		
Rear axle		
Expo, LRV	3.6	3.8
Outlander (2.1)	2.8	3.0
w/ABS Expo, LRV add	.3	.5
Rear Wheel Hub Assy., Replace (B)		
Rear axle		
Endeavor (.9)	1.2	1.4
Expo, LRV		
FWD	1.2	1.4
AWD	3.6	3.8
2001-04 Montero	2.5	2.7

	LABOR TIME	SEVERE SERVICE
Outlander		
FWD (.6)	.8	1.0
AWD (2.1)	2.8	3.0
w/ABS Expo, LRV add	.3	.5
Ring Gear & Pinion Set, Replace (B)		
Front axle		
Montero		
1983-91	9.5	9.7
1992-00	9.8	10.0
2001-04	10.5	10.7
Montero Sport	11.3	11.5
Pickup	9.9	10.1
Rear axle		
Endeavor (3.9)	5.3	5.5
Expo, LRV	7.8	8.0
Montero		
1983-91	6.9	7.1
1992-00	9.6	9.8
2001-04	8.8	9.0
Montero Sport	9.6	9.8
Outlander (6.3)	8.5	8.7
Pickup	7.9	8.1
Van, Wagon	4.4	4.6
w/rear disc brakes Montero Sport add	.6	.8

HALFSHAFTS

	LABOR TIME	SEVERE SERVICE
Halfshaft, R&R or Replace (B)		
Front axle		
Endeavor (1.2)	1.6	1.8
Expo, LRV		
AT	1.9	2.1
MT	1.5	1.7
Montero		
1992-00		
right side	1.8	2.0
left side	1.5	1.7
2001-04 one side	1.7	1.9
Montero Sport		
right side	1.9	2.1
left side	1.5	1.7
Outlander (1.3)	1.9	2.1
Pickup	1.8	2.0
Rear axle		
Endeavor (1.0)	1.4	1.6
Expo, LRV	1.1	1.3
2001-04 Montero	1.5	1.7
Outlander (1.0)	1.4	1.6
Halfshaft Boot, Replace (B)		
Includes: Halfshaft R&R.		
Front axle		
Endeavor (1.9)	2.8	3.0
Expo, LRV		
AT	2.6	2.8
MT	2.2	2.4
Montero		
1992-00		
right side	2.5	2.7
left side	1.8	2.0
2001-04 one side	2.2	2.4

	LABOR TIME	SEVERE SERVICE
Montero Sport		
right side	2.5	2.7
left side	1.9	2.1
Outlander (1.9)	2.8	3.0
Pickup one side	2.3	2.5
Rear axle		
Endeavor (1.3)	1.8	2.0
Expo, LRV	1.5	1.7
2001-04 Montero	1.9	2.1
Outlander (1.3)	1.8	2.0
Intermediate Shaft Support Bearing, Replace (B)		
Expo, LRV		
AT	2.6	2.8
MT	2.2	2.4
Outlander (1.9)	2.6	2.8
Rear Halfshaft Oil Seal, Replace (B)		
Rear axle Expo, LRV	1.7	1.9

BRAKES
ANTI-LOCK

	LABOR TIME	SEVERE SERVICE
Diagnose Anti-Lock Brake System Component Each (A)		
1991-05	1.0	1.2
Control Module, Replace (B)		
Expo, LRV	.5	.7
Montero	1.3	1.5
Pickup	.7	.9
G Sensor, Replace (B)		
Endeavor (.7)	.9	1.1
Expo, LRV, Outlander (.4)	.6	.8
Montero	1.1	1.3
Montero Sport	.5	.7
Pickup	.6	.8
Hydraulic Assy., Replace (B)		
Endeavor (1.9)	2.6	2.8
Expo, LRV	2.3	2.5
Montero	1.7	1.9
Montero Sport	1.9	2.1
Outlander (1.6)	2.2	2.4
Hydraulic Control Unit Relay, Replace (B)		
1991-05	.5	.7
Modulator, Replace (B)		
1991-96 Pickup	1.3	1.5
Power Modulator Relay, Replace (B)		
Expo, LRV	.3	.5
Montero	.5	.7
Valve Relay, Replace (B)		
1991-04	.5	.7
Wheel Speed Sensor, Replace (B)		
Exc. below	1.4	1.6
Outlander rear (.8)	1.1	1.3
1991-96 Pickup	.8	1.0

MITSU-52 ENDEAVOR : EXPO : EXPO LRV : MIGHTY MAX : MONTERO

SYSTEM	LABOR TIME	SEVERE SERVICE
Bleed Brakes (B)		
Includes: Add fluid.		
Exc. below	.6	.8
2001-04 Montero	1.1	1.3
Brake System, Flush and Refill (B)		
All Models	1.2	1.4
Brakes, Adjust (B)		
Includes: Adjust brakes, fill master cylinder.		
Two wheels	.5	.7
Brake Hose (Flexible), Replace (B)		
Includes: System bleeding.		
Endeavor (.7)	.9	1.1
Expo, LRV,		
Outlander (1.0)	1.4	1.6
Montero		
1983-91	.7	.9
1992-03	1.3	1.5
Montero Sport	1.3	1.5
Pickup, Van, Wagon	.8	1.0
Dual Proportioning Valve, Replace (B)		
Includes: System bleeding.		
LRV, Montero,		
Montero Sport	1.9	2.1
Van, Wagon	.7	.9
Load Sensing Valve, Replace (B)		
Includes: System bleeding.		
Expo	2.8	3.0
Montero		
1983-91	.7	.9
1992-00	1.8	2.0
Pickup, Van, Wagon	1.0	1.2
Master Cylinder, R&R and Recondition (B)		
Includes: System bleeding.		
Endeavor (2.1)	2.8	3.0
Expo, LRV,		
Outlander (2.2)	3.0	3.2
Montero		
1983-91	2.3	2.5
1992-00	2.5	2.7
Montero Sport	3.2	3.4
Pickup	2.2	2.4
Van, Wagon	2.6	2.8
Master Cylinder, Replace (B)		
Includes: System bleeding.		
Endeavor, Expo, LRV,		
Outlander (1.4)	1.9	2.0
Montero		
1983-91	1.3	1.5
1992-00	1.5	1.7
Montero Sport	2.2	2.4
Pickup	1.2	1.4
Van, Wagon	1.6	1.8

	LABOR TIME	SEVERE SERVICE
Master Cylinder Reservoir, Replace (B)		
Expo, LRV	1.5	1.7
Outlander (.6)	.8	1.0
Montero		
1983-91	.5	.7
1992-00	1.4	1.6
Montero Sport	.6	.8
Pickup	.5	.7
Van, Wagon	1.4	1.6
Power Booster Unit, Replace (B)		
Endeavor (1.6)	2.2	2.4
Expo, LRV,		
Outlander (1.8)	2.4	2.6
Montero		
1983-91	1.7	1.9
1992-00	2.0	2.2
2001-04	2.9	3.1
Montero Sport		
w/ABS	2.9	3.1
w/o ABS	2.4	2.6
Pickup	1.8	2.0
Van, Wagon	2.6	2.8
Power Booster Vacuum Check Valve, Replace (B)		
Expo, LRV	.6	.8
Montero	.3	.5
Proportioning Valve, Replace (B)		
Includes: System bleeding.		
LRV	1.8	2.0
Montero		
1983-91	.7	.9
1992-00	1.7	1.9
Montero Sport	1.8	2.0
SERVICE BRAKES		
Brake Drum, Replace (B)		
Exc. below	.5	.7
Montero Sport	.8	1.0
Outlander (.6)	.8	1.0
Caliper Assy., R&R and Recondition (B)		
Includes: System bleeding.		
Front each		
exc. below	1.8	2.0
Montero		
1983-91	1.6	1.8
2001-04	2.0	2.2
Montero Sport	2.0	2.2
Pickup	1.4	1.6
Van, Wagon	1.6	1.8
Rear each	1.7	1.9
Caliper Assy., Replace (B)		
Includes: System bleeding.		
Front		
exc. below (.9)	1.2	1.4
1983-91 Montero	.8	1.0
Pickup	.9	1.0
Van, Wagon	.8	1.0
Rear	1.2	1.4

	LABOR TIME	SEVERE SERVICE
Disc Brake Rotor, Replace (B)		
Front one		
exc. below (.6)	.8	1.0
Montero		
1983-91	1.1	1.3
1992-00	1.5	1.7
2001-04	.9	1.1
Montero Sport	1.7	1.9
Pickup	1.1	1.3
Van, Wagon	.8	1.0
Rear one	1.2	1.4
Front Hub, Replace (B)		
Endeavor, Expo, LRV,		
Outlander (1.9)	2.6	2.8
Montero		
1983-91	2.0	2.2
1992-03	1.8	2.0
Montero Sport	2.2	2.4
Pickup	2.2	2.4
Pads and/or Shoes, Replace (B)		
Includes: Service and parking brake adjustment, system bleeding.		
Front or rear disc (1.0)	1.4	1.6
Rear drum		
exc. below	1.9	2.1
Pickup	2.1	2.3
Van, Wagon	2.5	2.6
COMBINATION ADD-ONS		
Repack wheel bearings		
two wheels add	.7	.9
Replace		
brake drum add	.2	.4
brake hose add	.3	.5
caliper add	.3	.5
disc rotor add	.3	.5
master cylinder add	.8	1.0
wheel cylinder add	.2	.4
Resurface		
brake drum add	.5	.7
disc rotor add	.5	.7
Wheel Cylinder, R&R and Recondition (B)		
Includes: System bleeding.		
Expo, LRV,		
Outlander (1.2)	1.6	1.8
1983-91 Montero	1.3	1.5
Montero Sport	1.8	2.0
Pickup	1.3	1.5
Van, Wagon	1.4	1.6
Wheel Cylinder, Replace (B)		
Includes: System bleeding.		
Expo, LRV,		
Outlander (.8)	1.1	1.3
1983-91 Montero	.9	1.1
Montero Sport	1.6	1.8
Pickup	1.1	1.3
Van, Wagon	.9	1.1

MONTERO SPORT : OUTLANDER : VAN — MITSU-53

	LABOR TIME	SEVERE SERVICE
PARKING BRAKE		
Parking Brake Cable, Adjust (C)		
All Models	.5	.7
Parking Brake Apply Actuator, Replace (B)		
Endeavor, Expo, LRV, Outlander, Pickup (.8)	1.1	1.3
Montero		
1988-91	.7	.9
1992-00	1.1	1.3
2001-04	1.6	1.8
Montero Sport	.8	1.0
Van, Wagon	.5	.7
Parking Brake Apply Warning Indicator Switch, Replace (B)		
Endeavor, Expo, LRV, Outlander (.5)	.7	.9
Montero		
1983-91	.3	.5
1992-00	.8	1.0
2001-04	1.4	1.6
Montero Sport, Pickup	.5	.7
Van, Wagon	.3	.5
Parking Brake Cable, Replace (B)		
Front (.6)	.8	1.0
Rear		
Endeavor, Expo, LRV (2.1)	2.8	3.0
Outlander (1.0)	1.4	1.6
Montero		
1983-91	1.5	1.7
1992-00	1.9	2.1
2001-04	2.8	3.0
Montero Sport	1.9	2.1
Pickup	1.4	1.6
Van, Wagon	2.3	2.5

FRONT SUSPENSION

Unless otherwise noted, time given does not include alignment.

	LABOR TIME	SEVERE SERVICE
Align Front End (A)		
All Models	1.6	1.8
Front Toe, Adjust (B)		
All Models	.7	.9
Free Wheeling Hub, Replace (B)		
Montero, Pickup	.5	.7
Front Actuator Assy., Replace (B)		
1992-00 Montero	.9	1.1
Front Axle Hub Oil Seal, Replace (B)		
Montero Sport 2WD	1.3	1.5
Pickup	1.0	1.2
Front Coil Spring, Replace (B)		
Endeavor (1.5)	2.0	2.2
Expo, LRV, Outlander, Pickup 2WD (1.3)	1.8	2.0
2001-04 Montero	2.7	2.9

	LABOR TIME	SEVERE SERVICE
Front Hub Assy., Replace (B)		
Endeavor, Expo, LRV, Outlander (2.0)	2.7	2.9
Montero		
1983-91	1.9	2.1
1992-03	1.8	2.0
Montero Sport	2.1	2.3
Pickup		
2WD	1.9	2.1
4WD	2.1	2.3
Hub Bearings, Replace (B)		
Endeavor, Expo, LRV, Outlander (2.0)	2.7	2.9
Montero		
1983-91	2.0	2.2
1992-00	1.6	1.8
Montero Sport	1.6	1.8
Pickup		
2WD	1.0	1.2
4WD	2.1	2.3
Lower Ball Joint, Replace (B)		
Montero		
1983-91	1.7	1.9
1992-00	3.2	3.4
2001-04	2.7	2.9
Montero Sport	3.4	3.6
Pickup		
2WD	1.3	1.5
4WD	1.8	2.0
Lower Control Arm, Replace (B)		
Exc. below (1.2)	1.6	1.8
Montero		
1983-91	2.3	2.5
1992-00	3.9	4.1
2001-04	2.9	3.1
Montero Sport	4.6	4.8
Pickup		
2WD	2.3	2.5
4WD	2.8	3.0
Lower Control Arm Bushings, Replace (B)		
Endeavor (1.1)	1.5	1.7
Expo, LRV	1.3	1.5
Outlander (1.3)	1.8	2.0
Lower Control Arm Bushings and/or Shaft, Replace (B)		
Montero		
1983-91	2.4	2.6
1992-00	3.8	4.0
Montero Sport	4.8	5.0
Pickup		
2WD	3.4	3.6
4WD	4.4	4.6
Van Wagon	2.1	2.3
Lower Control Arm Strut, Replace (B)		
1983-94 Pickup 4WD	1.2	1.4
Van, Wagon 4WD	1.3	1.5

	LABOR TIME	SEVERE SERVICE
Shock Absorber, Replace (B)		
Exc. below (.5)	.7	.9
Montero		
1992-00	1.1	1.3
2001-04	2.2	2.4
Montero Sport	1.0	1.2
Stabilizer Bar, Replace (B)		
Endeavor (3.0)	4.1	4.3
Expo, LRV		
2WD	1.6	1.8
4WD	1.9	2.1
Montero, Montero Sport		
1983-91	1.5	1.7
1992-03	1.3	1.4
Outlander (2.2)	3.0	3.2
Pickup		
2WD	2.2	2.4
4WD	2.9	3.1
Van, Wagon	1.2	1.4
Stabilizer Bar Bushings, Replace (B)		
Exc. Below (.6)	.8	1.0
Endeavor (.9)	1.2	1.4
Montero Sport	1.4	1.6
Stabilizer Bar Link, Replace (B)		
Endeavor, Expo, LRV (.4)	.6	.8
1992-03 Montero	1.0	1.2
Montero Sport	1.3	1.5
Outlander (.8)	1.1	1.3
Pickup	.5	.7
Steering Knuckle, Replace (B)		
Endeavor (1.1)	1.5	1.7
Expo, LRV, Outlander (2.0)	2.7	2.9
Montero	1.9	2.1
Montero Sport		
2WD	1.8	2.0
4WD	2.1	2.3
Pickup		
2WD	1.8	2.0
4WD	2.1	2.3
Van, Wagon	1.6	1.8
Steering Knuckle Inner Oil Seal, Replace (B)		
Montero 4WD	1.7	1.9
Montero Sport 4WD	1.7	1.9
Pickup 2WD	1.8	2.0
Steering Knuckle Oil Seal, Replace (B)		
Expo, LRV, Outlander		
inner (1.1)	1.5	1.7
outer (1.9)	2.6	2.8
Steering Knuckle Spindle Bearings, Replace (B)		
Montero, Montero Sport, Pickup	2.1	2.3
Strut Assy., Replace (B)		
Endeavor (1.0)	1.4	1.6
Expo, LRV, Outlander (.8)	1.1	1.3

MITSU-54 ENDEAVOR : EXPO : EXPO LRV : MIGHTY MAX : MONTERO

	LABOR TIME	SEVERE SERVICE
Torsion Bars, Replace (B)		
Montero		
1983-91	.8	1.0
1992-00	1.3	1.5
Montero Sport	1.6	1.8
Pickup	1.0	1.2
Van, Wagon	1.2	1.4
Torsion Bar Anchor/Adjusting Bolt, Replace (B)		
Montero		
1983-91	.8	1.0
1992-00	1.3	1.5
Montero Sport	1.6	1.8
Pickup	1.0	1.2
Van, Wagon	1.4	1.6
Upper Ball Joint, Replace (B)		
Montero		
1983-91	1.5	1.7
1992-00	2.5	2.7
2001-04	2.1	2.3
Montero Sport	2.4	2.6
Pickup	1.1	1.3
Van, Wagon	.9	1.1
Upper Control Arm Assy., Replace (B)		
Montero		
1983-91	1.4	1.6
1992-00	3.2	3.4
2001-04	2.4	2.6
Montero Sport	3.0	3.2
Pickup	1.4	1.6
Van, Wagon	1.6	1.8
Upper Control Arm Bushings and/or Shaft, Replace (B)		
Montero		
1983-91	1.7	1.9
1992-00	3.2	3.4
Pickup	1.6	1.8
Van, Wagon	2.0	2.2
Wheel Bearing & Cups, Replace (B)		
Montero 4WD		
1983-91	1.9	2.1
1992-00	1.6	2.0
Montero Sport	1.6	1.8
Pickup one side		
2WD	1.0	1.2
4WD	2.1	2.3
Van, Wagon 2WD		
one side	.8	1.0
both sides	1.8	2.0
Wheel Bearing Grease Seal, Replace (B)		
Pickup 2WD	1.3	1.5
Van, Wagon	1.3	1.5

REAR SUSPENSION

Unless otherwise noted, time given does not include alignment.

	LABOR TIME	SEVERE SERVICE
Rear Suspension, Align (Complete) (A)		
All Models (1.2)	1.6	1.8
Lower Control Arm Bushing, Replace (B)		
Montero		
1989-91	.9	1.1
1992-00	1.4	1.6
Montero Sport	1.6	1.8
Outlander (1.2)	1.6	1.8
Lower Control Arms, Replace (B)		
Endeavor (1.2)	1.6	1.8
Expo, LRV		
2WD	3.0	3.2
4WD	3.3	3.5
Montero		
1989-91	.8	1.0
1992-00	1.3	1.5
2001-04	2.2	2.4
Montero Sport, Outlander (1.0)	1.4	1.6
Rear Coil Spring, Replace (B)		
Endeavor (1.5)	2.0	2.2
Expo, LRV	1.3	1.5
Montero		
1989-91	.8	1.0
1992-00	1.6	1.8
Montero Sport	2.5	2.7
Rear Leaf Spring, Replace (B)		
Montero	1.6	1.8
Montero Sport	1.4	1.6
Pickup	1.6	1.8
Van, Wagon	.9	1.1
Rear Shock Actuator, Replace (B)		
1992-00 Montero	.7	.9
Rear Stabilizer Bar Bushings, Replace (B)		
Endeavor, Expo, LRV, Outlander (.6)	.8	1.0
1983-00 Montero	.8	1.0
Montero Sport	1.1	1.3
Rear Strut Shock Absorber, Replace (B)		
Expo, LRV	1.1	1.3
Outlander		
right side (1.2)	1.6	1.8
left side (1.6)	2.2	2.4
Rear Strut Assy., Replace (B)		
Expo, LRV (.6)	.8	1.0
Outlander		
right side (1.0)	1.4	1.6
left side (1.3)	1.8	2.0

	LABOR TIME	SEVERE SERVICE
Shackle Bushing, Replace (B)		
1983-91 Montero	.7	.9
Montero Sport	.8	1.0
Pickup	.6	.8
Van, Wagon	.7	.9
Shock Absorbers or Bushings, Replace (B)		
Endeavor (1.2)	1.6	1.8
Montero		
1983-91	.5	.7
1992-00	.8	1.0
Montero Sport	.7	.9
Pickup, Van, Wagon	.6	.8
Spring Shackle, Replace (B)		
Montero	.7	.9
Montero Sport	.8	1.0
Pickup, Van, Wagon	.7	.9
Stabilizer Bar, Replace (B)		
Endeavor (.6)	.8	1.0
Expo, LRV, Montero Sport, Outlander (1.0)	1.4	1.6
Montero		
1983-91	1.2	1.4
1992-00	1.8	2.0
2001-04	1.1	1.3
Stabilizer End Kit, Replace (B)		
Endeavor (.4)	.6	.8
Expo, LRV, Outlander (.5)	.7	.9
1992-00 Montero	.7	.9
Montero Sport	1.0	1.2

STEERING
AIR BAGS

	LABOR TIME	SEVERE SERVICE
Air Bag Assy., Replace (B)		
Endeavor, Outlander		
driver side (.4)	.6	.8
passenger side (2.8)	3.8	4.0
Montero, Montero Sport		
driver side	.6	.8
passenger side	.7	.8
Diagnostic Air Bag Module, Replace (B)		
All Models	.9	1.1
Front Sensor, Replace (B)		
Endeavor, Outlander (.4)	.6	.8
Montero		
1992-00 front	.5	.7
2001-04 front or side	.5	.7
Montero Sport		
right side	.7	.9
left side	.8	1.0

LINKAGE

	LABOR TIME	SEVERE SERVICE
Center Link (Relay Rod), Replace (B)		
1983-00 Montero	1.6	1.8
Montero Sport, Pickup	1.1	1.3

MONTERO SPORT : OUTLANDER : VAN   MITSU-55

	LABOR TIME	SEVERE SERVICE
Idler Arm, Replace (B)		
Montero		
1983-91	.8	1.0
1992-00	1.0	1.2
Montero Sport, Pickup	.8	1.0
Outer Tie Rod End, Replace (B)		
Includes: Reset toe.		
Exc. below (1.0)	1.4	1.6
Pickup	1.1	1.3
Pitman Arm, Replace (B)		
Montero		
1983-91	.8	1.0
1992-00	1.6	1.8
Montero Sport	1.0	1.2
Pickup	.8	1.0
Tie Rod, Replace (B)		
Includes reset toe-in.		
Montero, Montero Sport	1.4	1.6
Pickup	1.1	1.3

MANUAL RACK & PINION

	LABOR TIME	SEVERE SERVICE
Bevel Gear Assy., Replace (B)		
Van, Wagon	.7	.9
Inner Tie Rod Assy., Replace (B)		
Includes: Reset toe.		
Van, Wagon	1.2	1.4
Outer Tie Rod End, Replace (B)		
Includes: Reset toe.		
Van, Wagon	1.2	1.4
Rack & Pinion Assy., R&R and Recondition (A)		
Includes: Reset toe.		
Van, Wagon	3.0	3.2
Rack & Pinion Assy., Replace (B)		
Includes: Reset toe.		
Van, Wagon	1.6	1.8
Steering Column, Replace (B)		
Van, Wagon	1.2	1.4
Steering Intermediate Shaft, Replace (B)		
Van, Wagon	.8	1.0
Steering Wheel, Replace (B)		
Van, Wagon	.5	.7
Tie Rod End Boot, Replace (B)		
Van, Wagon	1.7	1.9

POWER RACK & PINION

	LABOR TIME	SEVERE SERVICE
Bevel Gear Assy., Replace (B)		
Van, Wagon	.7	.9
Inner Tie Rod Assy., Replace (B)		
Includes: Reset toe.		
Exc. below (1.0)	1.4	1.6
Endeavor (3.2)	4.3	4.5
2001-04 Montero	3.8	4.0
Outlander (2.9)	3.9	4.1
Intermediate Shaft, Replace (B)		
Van, Wagon	.8	1.0

	LABOR TIME	SEVERE SERVICE
Outer Tie Rod Ends, Replace (B)		
Includes: Reset toe.		
Endeavor (.9)	1.2	1.4
Expo, LRV, Outlander (.8)	1.1	1.3
2001-04 Montero (1.0)	1.4	1.6
Van, Wagon (.8)	1.1	1.3
Power Steering Hoses, Replace (B)		
Endeavor		
pressure or return (1.0)	1.4	1.6
suction (.9)	1.2	1.4
Expo, LRV		
pressure or return	.9	1.1
suction	.7	.9
2001-04 Montero		
pressure	2.2	2.4
return	1.8	2.0
suction	1.3	1.5
Outlander		
pressure (1.2)	1.6	1.8
return (1.6)	2.2	2.4
suction (.6)	.8	1.0
Van, Wagon		
pressure	.8	1.0
return	1.0	1.2
suction	.6	.8
Rack & Pinion Assy., R&R and Recondition (A)		
Includes: Reset toe.		
Expo, LRV		
2WD	4.8	5.0
4WD	4.5	4.7
2001-04 Montero	5.6	5.8
Outlander (4.9)	6.6	6.8
Van, Wagon	3.7	3.9
Rack & Pinion Assy., Replace (B)		
Includes: Reset toe.		
Endeavor (3.2)	4.3	4.5
Expo, LRV		
2WD	2.9	3.1
4WD	2.6	2.8
2001-04 Montero	3.3	3.5
Outlander (3.4)	4.6	4.8
Van, Wagon	1.8	2.0
Steering Column, Replace (B)		
Endeavor, Expo, LRV, Outlander (1.2)	1.6	1.8
2001-04 Montero	1.9	2.1
Van, Wagon	1.4	1.6
Steering Pump, R&R and Recondition (B)		
Exc. below (1.4)	1.9	2.1
2001-04 Montero	2.2	2.4
Outlander (1.9)	2.6	2.8
Steering Pump, R&R or Replace (B)		
Exc. below (.7)	.9	1.1
2001-04 Montero	2.4	2.6
Endeavor (1.6)	2.2	2.4
Outlander (1.3)	1.8	2.0

	LABOR TIME	SEVERE SERVICE
Steering Wheel, Replace (B)		
Endeavor, Expo, LRV, Outlander (.3)	.5	.7
2001-04 Montero	.6	.8
Van, Wagon	.5	.7
Tie Rod End Boot, Replace (B)		
Endeavor (3.5)	4.7	4.9
Expo, LRV	1.2	1.4
2001-04 Montero	1.4	1.6
Outlander (2.8)	3.8	4.0
Van, Wagon	1.9	2.1

MANUAL WORM & SECTOR

	LABOR TIME	SEVERE SERVICE
Steering Gear, Adjust (On Pickup) (B)		
1983-91 Montero	.5	.7
Montero Sport, Pickup	.5	.7
Gear Assy., R&R and Recondition (A)		
1983-91 Montero	2.3	2.5
Montero Sport	5.2	5.4
Pickup	2.1	2.3
Gear Assy., Replace (B)		
1983-91 Montero	1.2	1.4
Montero Sport	3.2	3.4
Pickup	1.2	1.4
Steering Column, R&R or Replace (B)		
1983-91 Montero	1.4	1.6
Montero Sport	1.9	2.1
Pickup	1.6	1.8
Steering Shaft Joint Assy., Replace (B)		
1983-91 Montero	.7	.9
Montero Sport	1.6	1.8
Pickup	.7	.9
Steering Wheel, Replace (B)		
All Models	.5	.7

POWER WORM & SECTOR

	LABOR TIME	SEVERE SERVICE
Steering Gear, Adjust (B)		
All Models	.5	.7
Gear Assy., R&R and Recondition (A)		
1983-91 Montero	3.1	3.3
Montero Sport	5.2	5.4
Pickup	3.5	3.7
Gear Assy., Replace (B)		
1983-91 Montero	1.3	1.5
Montero Sport	3.2	3.4
Pickup	1.4	1.6
Power Steering Hoses, Replace (B)		
1983-91 Montero	.5	.7
Montero Sport		
pressure	.9	1.1
suction	.8	1.0
return	.8	1.0
Pickup		
pressure or return	.5	.7
suction	.5	.7

MITSU-56 ENDEAVOR : EXPO : EXPO LRV : MIGHTY MAX : MONTERO

	LABOR TIME	SEVERE SERVICE
Steering Column, R&R or Replace (B)		
1983-91 Montero	1.5	1.7
Montero Sport	2.3	2.5
Pickup	1.7	1.9
Steering Pump, R&R and Recondition (B)		
1983-91 Montero	1.8	2.0
Montero Sport, Pickup	1.7	1.9
Steering Pump, R&R or Replace (B)		
1983-91 Montero	.8	1.0
Montero Sport, Pickup	1.0	1.2
Steering Pump Reservoir or Seals, Replace (B)		
All Models	.7	.9
Steering Shaft Joint Assy., Replace (B)		
1983-91 Montero	.7	.9
Montero Sport	1.6	1.8
Pickup	.7	.9
Steering Wheel, Replace (B)		
All Models	.5	.7

HEATING & AIR CONDITIONING

When more than one component requires replacement where evacuation/recovery and recharging is already included, deduct 1.0 hour for each additional component from the time given.

Evacuate/Recover and Recharge System (B)
All Models .... 1.0 / 1.0

AC Hoses, Replace (B)
Includes: Evacuate/recover and recharge.
Exc. below one (1.3) .. 1.8 / 2.0
each addl. add .... .5 / .7
Endeavor, Expo, LRV
  each (1.0) .... 1.4 / 1.6
1992-03 Montero each 1.4 / 1.6
Montero Sport
  suction .... 1.4 / 1.6
  discharge .... 2.0 / 2.2
Outlander
  suction (1.2) ..... 1.6 / 1.8
  discharge (.9) ... 1.2 / 1.9
Pickup each .... 1.5 / 1.7

AC On/Off Control Switch, Replace (B)
Expo, LRV .... 1.2 / 1.4
Montero, Pickup .... .7 / .9
Montero Sport ...... .9 / 1.1
Van
  front .... .7 / .9
  rear .... 1.6 / 1.8

Blower Motor, Replace (B)
Exc. below .... .6 / .8
Endeavor .... 8.3 / 8.5
2001-04 Montero rear .. 1.5 / 1.7
Van, Wagon rear .... 1.4 / 1.6

Blower Motor Relay, Replace (B)
All Models .... .6 / .8

Blower Motor Resistor, Replace (B)
All Models .... .7 / .9

Blower Motor Switch, Replace (B)
Exc. below .... .7 / .9
Expo, LRV .... 1.1 / 1.3
Montero Sport ...... 1.4 / 1.6

Compressor Assy., Replace (B)
Includes: Parts transfer, evacuate/recover and recharge.
Endeavor, Expo,
  Outlander (1.8) ... 2.4 / 2.6
LRV .... 1.8 / 2.0
Montero
  1983-91 .... 2.0 / 2.2
  1992-03 .... 2.5 / 2.7
Montero Sport ..... 2.1 / 2.3
Pickup .... 1.6 / 1.8
Van, Wagon .... 2.4 / 2.6

Compressor Clutch Assy., Replace (B)
Endeavor, Expo,
  Outlander (1.5) ... 2.0 / 2.2
LRV .... 1.8 / 2.0
Montero
  1983-91 .... 1.7 / 1.9
  1992-03 .... 2.7 / 2.9
Montero Sport ..... 2.2 / 2.4
Pickup .... 1.7 / 1.9
Van, Wagon .... 2.8 / 3.0

Condenser Assy., Replace (B)
Includes: Evacuate/recover and recharge.
Expo, Endeavor, LRV
  Montero Sport (1.8) .. 2.4 / 2.6
Montero
  1983-91 .... 2.3 / 2.5
  2001-04 .... 1.7 / 1.9
Outlander (1.1) .... 1.5 / 1.7
Pickup .... 1.1 / 1.3
Van, Wagon .... 1.5 / 1.7

Condenser Fan Motor Relay, Replace (B)
All Models .... .3 / .5

Evaporator Core, Replace (B)
Includes: Evacuate/recover and recharge.
Expo, LRV .... 1.6 / 1.8
Montero
  1983-00 .... 1.9 / 2.1
  2001-04 .... 7.4 / 7.6
Montero Sport ..... 2.8 / 3.0

Endeavor (5.6) ..... 7.6 / 7.8
Outlander (1.5) .... 2.0 / 2.2
Pickup .... 1.7 / 1.9
Van, Wagon each ... 2.4 / 2.6

Expansion Valve, Replace (B)
Includes: Evacuate/recover and recharge.
Exc. below (1.2) ... 1.6 / 1.8
Montero, Pickup .... 2.0 / 2.2
Montero Sport ..... 2.8 / 3.0

Front Heater Blower Motor, Replace (B)
All Models .... .7 / .9

Front Heater Blower Motor Resistor, Replace (B)
All Models .... .7 / .9

Front Heater Blower Motor Switch, Replace (B)
Exc. below .... .7 / .9
Expo, LRV .... 1.2 / 1.4
1992-00 Montero ... .9 / 1.1
Montero Sport ..... 1.6 / 1.8

Front Heater Core, R&R or Replace (B)
Endeavor (6.2) .... 8.4 / 8.6
Expo, LRV .... 3.6 / 4.0
1983-00 Montero ... 4.3 / 4.5
Montero Sport, Pickup 3.2 / 3.4
Outlander (3.6) ... 4.9 / 5.1
Van, Wagon .... 4.2 / 4.4
w/AC Montero Sport add .. .9 / 1.1

Front Heater Water Valve, Replace (B)
1983-91 Montero ... 3.2 / 3.4
Montero Sport ..... 1.2 / 1.4
Pickup .... .7 / .9

Heater Blower Motor Relay, Replace (B)
All Models .... .5 / .7

Heater Control Assy., Replace (B)
Endeavor,
  Outlander (.4) ... .6 / .8
Expo, LRV .... 1.5 / 1.7
Montero
  1983-91 .... .8 / 1.0
  1992-03 .... 1.1 / 1.3
Montero Sport ..... 1.2 / 1.4
Pickup .... 1.3 / 1.5
Van, Wagon .... .8 / 1.0

Heater Hoses, Replace (B)
All Models .... .8 / 1.0

Rear Heater Blower Motor, Replace (B)
Montero, Montero Sport ..... 1.5 / 1.7
Van, Wagon .... 1.2 / 1.4

MONTERO SPORT : OUTLANDER : VAN — MITSU-57

	LABOR TIME	SEVERE SERVICE
Rear Heater Blower Motor Resistor, Replace (B)		
Montero Sport	.9	1.1
Van, Wagon	.5	.7
Rear Heater Blower Motor Switch, Replace (B)		
Montero Sport	.5	.7
Van, Wagon	.5	.7
Rear Heater Core, R&R or Replace (B)		
Montero Sport	1.4	1.6
Van, Wagon	1.3	1.5
Rear Heater Water Valve, Replace (B)		
Montero Sport	1.3	1.5
Van, Wagon	1.7	1.9
Receiver/Drier Assy., Replace (B)		
Includes: Evacuate/recover and recharge.		
Exc. below (1.2)	1.6	1.8
Endeavor (1.6)	2.2	2.4
1983-91 Montero	1.7	1.9
Pickup	1.7	1.9

WIPERS & SPEEDOMETER

	LABOR TIME	SEVERE SERVICE
Antenna Assy., Replace (B)		
Endeavor (1.1)	1.5	1.7
Expo, LRV	1.8	2.0
Montero		
1983-91	.9	1.1
1992-03	1.5	1.7
Montero Sport	2.0	2.2
Outlander (2.7)	3.6	3.8
Pickup	.9	1.1
Van, Wagon	.7	.9
Combination Gauge Assy., Replace (B)		
Exc. below (.7)	.9	1.1
Pickup	.5	.7
Replace speed sensor add	.3	.5
Radio, R&R (B)		
Exc. below (.5)	.7	.9
Montero, Pickup	.8	1.0
Montero Sport	1.1	1.3
Rear Window Wiper Motor, Replace (B)		
Exc. below (.5)	.7	.9
1992-03 Montero	.8	1.0
Rear Windshield Wiper Switch, Replace (B)		
All Models (.6)	.8	1.0
Rear Window Washer Pump, Replace (B)		
Endeavor, Expo, LRV, Outlander (.6)	.8	1.0
Montero		
1983-91	.5	.7
1992-03	.9	1.1
Montero Sport	1.5	1.7
Van, Wagon	.5	.7

	LABOR TIME	SEVERE SERVICE
Speedometer Cable & Casing, Replace (B)		
All Models	1.2	1.4
Speedometer Head, R&R or Replace (B)		
Expo, LRV, Pickup	.7	.9
Montero, Van, Wagon	.9	1.1
Montero Sport	.7	.9
Endeavor, Outlander (.7)	.9	1.1
Replace speed sensor add	.3	.5
Windshield Washer Pump, Replace (B)		
Endeavor, Expo, LRV, Outlander (.6)	.8	1.0
Montero		
1983-91	.3	.5
1992-00	1.7	1.9
2001-04	.6	.8
Montero Sport	.5	.7
Pickup, Van, Wagon	.5	.7
Windshield Wiper Linkage, Replace (B)		
Endeavor, Expo, LRV, Outlander (1.0)	1.4	1.6
Montero		
1983-91	.7	.9
1992-03	1.2	1.4
Montero Sport	1.3	1.5
Pickup	.8	1.0
Van, Wagon	2.5	2.7
Windshield Wiper Motor, Replace (B)		
Exc. below	.8	1.0
Endeavor (.8)	1.1	1.3
Expo, LRV	1.5	1.7
1992-03 Montero	1.2	1.4
Montero Sport	.8	1.0
Windshield Wiper & Washer Switch, Replace (B)		
Endeavor (.4)	.6	.8
Expo, LRV, Outlander (.7)	.9	1.1
1992-03 Montero	.7	.9
Montero Sport	.7	.9
Pickup	.8	1.0
Van, Wagon	.5	.7

LAMPS & SWITCHES

	LABOR TIME	SEVERE SERVICE
Back-Up Lamp Assy., Replace (B)		
All Models	.3	.3
Back-Up Lamp Switch, Replace (B)		
All Models MT	.5	.5
Back-Up Lamp/Neutral Safety Switch, Replace (B)		
All Models AT	.7	.7

	LABOR TIME	SEVERE SERVICE
Combination Switch Assy., Replace (B)		
Endeavor, Expo, LRV, Outlander (1.0)	1.3	1.3
Montero, Montero Sport	.8	.8
Pickup	.9	.9
Van, Wagon	.7	.7
Composite Headlamp Bulb, Replace (C)		
All Models	.5	.5
Hazard Warning Switch, Replace (B)		
Endeavor, Expo, LRV, Montero Sport	.5	.5
Headlamps, Aim (B)		
Two	.5	.5
Four	.6	.6
High Mount Stop Lamp Assy., Replace (B)		
All Models	.5	.5
Horn, Replace (B)		
All Models	.5	.5
Horn Relay, Replace (B)		
Endeavor, Montero, Montero Sport, Outlander	.3	.3
Horn Switch, Replace (B)		
1983-96	.5	.5
License Lamp Assy., Replace (B)		
Endeavor	.4	.4
Expo, LRV, Outlander	1.1	.9
Montero	.7	1.2
Montero Sport	.8	.8
Pickup, Van, Wagon	.5	.5
Park/Neutral Position Switch, Replace (B)		
Automatic transmission		
Exc. Montero	.7	.9
Montero	.9	1.1
Park & Turn Signal Lamp Assy., Replace (B)		
All Models	.5	.5
Parking Brake Warning Lamp Switch, Replace (B)		
Endeavor, Expo, LRV	.7	.7
Montero		
1983-91	.3	.3
1992-02	.8	.8
Montero Sport, Pickup	.5	.5
Van, Wagon	.3	.3
Sealed Beam Headlamp, Replace (C)		
1983-96	.3	.5
Side Marker Lamp Assy., Replace (B)		
1983-96	.5	.5
Stop, Tail & Turn Signal Lamp Assy., Replace (B)		
All Models	.5	.5
Stop Lamp Switch, Replace (B)		
All Models	.5	.5
w/cruise control add	.1	.1

MITSU-58 ENDEAVOR : EXPO : EXPO LRV : MIGHTY MAX : MONTERO

	LABOR TIME	SEVERE SERVICE
Turn Signal & Headlamp Dimmer Switch, Replace (B)		
1992-00 Montero	.7	.7
Turn Signal & Washer Wiper Combination Lever Assy., Replace (B)		
Endeavor, Expo, LRV, Outlander (.7)	.9	.9
Montero	.7	.7
Pickup	.8	.8
Van, Wagon	.5	.5
Turn Signal or Hazard Warning Flasher, Replace (B)		
Exc. below	.3	.3
Expo, LRV, Outlander	.5	.5
1983-91 Montero	.5	.5

BODY

	LABOR TIME	SEVERE SERVICE
Door Lock, Replace (B)		
Front		
Endeavor, Expo, LRV, Outlander (.8)	1.1	1.3
Montero		
1983-00	.9	1.1
2001-04	1.1	1.3
Montero Sport	.8	1.0
Pickup	.8	1.0
Van, Wagon	1.2	1.4
Rear		
Endeavor, Expo (.8)	1.1	1.3
LRV	.6	.8
Montero		
1983-00	.9	1.1
2001-04	1.2	1.4
Montero Sport	1.0	1.2
Outlander (.9)	1.2	1.4
Door Window Regulator, Replace (B)		
Electric		
front		
Endeavor, Expo, LRV (.7)	.9	1.1
Montero		
1983-91	.8	1.0
1992-03	1.4	1.6
Montero Sport	1.5	1.7
Outlander (1.2)	1.6	1.8
Pickup	1.1	1.3
Van, Wagon	.8	1.0
rear		
Endeavor, Expo (.8)	1.1	1.3
Montero		
1983-00	.9	1.1
2001-04	1.2	1.4
Montero Sport	1.5	1.7
Outlander (1.2)	1.6	1.8

	LABOR TIME	SEVERE SERVICE
Manual front		
Endeavor, Expo, LRV	.9	1.1
Montero		
1983-91	.8	1.0
1992-03	1.3	1.5
Montero Sport	1.3	1.5
Outlander (.8)	1.1	1.3
Pickup	1.1	1.3
Van, Wagon	.8	1.0
Hood Hinge, Replace (B)		
Endeavor, Expo, LRV, Outlander (.8)	1.0	1.2
Montero		
1983-91	1.3	1.5
2001-04	1.1	1.3
Montero Sport	.9	1.1
Hood Release Cable, Replace (B)		
Endeavor, Outlander (.9)	1.2	1.4
Expo, LRV	2.3	2.5
Montero		
1983-91	.7	.9
1992-00	1.5	1.7
2001-04	.8	1.0
Montero Sport	1.9	2.1
Pickup	.6	.8
Lock Striker Plate, Replace (B)		
All Models	.3	.5

NI

200SX : 210 : 240SX : 280ZX : 300ZX : 310 : 350Z : 510 : 810 : Altima : Maxima : NX : Pulsar : Pulsar NX : Sentra : Stanza

SYSTEM INDEX

- **MAINTENANCE** NI-2
  - Maintenance Schedule NI-2
- **CHARGING** NI-2
- **STARTING** NI-3
- **CRUISE CONTROL** NI-4
- **IGNITION** NI-4
- **EMISSIONS** NI-4
- **FUEL** NI-6
- **EXHAUST** NI-8
- **ENGINE COOLING** NI-9
- **ENGINE** NI-10
  - Assembly NI-10
  - Cylinder Head NI-12
  - Camshaft NI-15
  - Crank & Pistons NI-17
  - Engine Lubrication NI-19
- **CLUTCH** NI-21
- **MANUAL TRANSAXLE** NI-22
- **MANUAL TRANSMISSION** .. NI-23
- **AUTO TRANSAXLE** NI-23
- **AUTO TRANSMISSION** NI-25
- **TRANSFER CASE** NI-26
- **SHIFT LINKAGE** NI-27
- **DRIVELINE** NI-27
- **BRAKES** NI-29
- **FRONT SUSPENSION** NI-33
- **REAR SUSPENSION** NI-35
- **STEERING** NI-37
- **HEATING & AC** NI-40
- **WIPERS & SPEEDOMETER** . NI-42
- **LAMPS & SWITCHES** NI-43
- **SEAT BELTS** NI-43
- **BODY** NI-44

OPERATIONS INDEX

A
- AC Hoses NI-40
- Air Bags NI-37
- Air Conditioning NI-40
- Alignment NI-33
- Alternator (Generator) NI-2
- Antenna NI-42
- Anti-Lock Brakes NI-29

B
- Back-Up Lamp Switch NI-43
- Ball Joint NI-33
- Battery Cables NI-3
- Bleed Brake System NI-30
- Blower Motor NI-40
- Brake Disc NI-32
- Brake Drum NI-31
- Brake Hose NI-30
- Brake Pads and/or Shoes . NI-32

C
- Camshaft NI-15
- Catalytic Converter NI-8
- Coolant Temperature (ECT) Sensor NI-5
- Crankshaft NI-17
- Crankshaft Sensor NI-4
- Cruise Control NI-4
- CV Joint NI-29
- Cylinder Head NI-12

D
- Differential NI-28
- Distributor NI-4
- Drive Belt NI-2
- Driveshaft NI-27

E
- EGR NI-5
- Electronic Control Module (ECM/PCM) NI-5
- Engine NI-11
- Engine Lubrication NI-19
- Engine Mounts NI-12
- Evaporator NI-41
- Exhaust NI-8
- Exhaust Manifold NI-9
- Expansion Valve NI-41

F
- Flexplate NI-24
- Flywheel NI-21
- Fuel Injection NI-7
- Fuel Pump NI-7
- Fuel Vapor Canister NI-5

G
- Gear Selector Lever NI-27
- Generator NI-2
- Glow Plug NI-4

H
- Halfshaft NI-29
- Headlamp NI-2
- Heater Core NI-41
- Horn NI-43

I
- Idle Air Control (IAC) Valve ... NI-8
- Ignition Coil NI-4
- Ignition Module NI-4
- Ignition Switch NI-4
- Injection Pump NI-6
- Inner Tie Rod NI-38
- Intake Manifold NI-7

K
- Knock Sensor NI-5

L
- Lower Control Arm NI-34

M
- Maintenance Schedule NI-2
- Manifold Absolute Pressure (MAP) Sensor NI-5
- Mass Air Flow (MAF) Sensor .. NI-5
- Master Cylinder NI-30
- Muffler NI-9
- Multifunction Lever/Switch NI-43

N
- Neutral Safety Switch NI-43

O
- Oil Pan NI-19
- Oil Pump NI-20
- Outer Tie Rod NI-38
- Oxygen Sensor NI-5

P
- Parking Brake NI-32
- Pistons NI-18
- Positive Crankcase Ventilation (PCV) Valve NI-6

R
- Radiator NI-10
- Radiator Hoses NI-10
- Radio NI-42
- Rear Main Oil Seal NI-18

S
- Shock Absorber/Strut, Front .. NI-33
- Shock Absorber/Strut, Rear .. NI-36
- Spark Plug Cables NI-4
- Spark Plugs NI-4
- Spring, Front Coil NI-33
- Spring, Leaf NI-36
- Spring, Rear Coil NI-36
- Starter NI-3
- Steering Wheel NI-38

T
- Thermostat NI-9
- Throttle Body NI-8
- Throttle Position Sensor (TPS) NI-6
- Timing Belt NI-16
- Timing Chain NI-16
- Torque Converter NI-25

U
- U-Joint NI-27
- Upper Control Arm NI-37

V
- Valve Body NI-24
- Valve Cover Gasket NI-14
- Valve Job NI-14
- Vehicle Speed Sensor NI-6

W
- Water Pump NI-10
- Wheel Balance NI-2
- Wheel Cylinder NI-32
- Window Regulator NI-44
- Windshield Washer Pump ... NI-43
- Windshield Wiper Motor ... NI-43

NI-1

200SX : 210 : 240SX : 280ZX : 300ZX : 310 : 350Z : 510 : 810

	LABOR TIME	SEVERE SERVICE
MAINTENANCE		
Air Cleaner Filter Element, Replace (C)		
Gasoline		
carbureted	.2	.3
fuel injected	.5	.6
Diesel	.3	.4
Drive Belt, Adjust (B)		
One	.5	.6
each addl. add	.1	.2
Drive Belt, Replace (B)		
Air pump		
1981-90	.7	.8
Fan		
exc. Maxima	.7	.8
Maxima	.8	.9
each addl. add	.3	.3
Injection pump		
810/Maxima	1.7	1.8
1982-85 Sentra	.9	1.0
Power steering		
exc. below	.7	.8
1981-83 200SX	.5	.6
300ZX	.8	.9
w/AC air pump add	.3	.3
Replace		
drive gear injection pump add	.3	.3
tensioner pulley injection pump add	.5	.5
Fuel Filter, Replace (B)		
All Models	.8	.8
Headlamp Bulb, Replace (B)		
Halogen		
exc. below each	.3	.3
350Z	1.5	1.5
Altima		
1993-01	1.1	1.1
2002-05	.7	.7
Xenon		
exc. below each	.4	.4
Altima	1.5	1.5
Oil and Filter, Change (C)		
Includes: Correct fluid levels.		
All Models	.4	.5
Sealed Beam Headlamp, Replace (C)		
Exc. 280ZX	.3	.3
280ZX	.5	.5
Timing Belt, Replace (B)		
200SX		
4 cyl.	2.6	3.1
V6	5.8	6.3
300ZX		
1984-89	7.0	7.5
1990-96	5.2	5.7
1981 310	7.0	7.5
Maxima		
1985-88	5.3	5.8
1989-94 SOHC	3.8	4.3

	LABOR TIME	SEVERE SERVICE
Pulsar		
CA	4.5	5.0
E	2.2	2.7
Sentra		
1982-86 each	4.3	4.8
1987-90	2.3	2.8
Stanza	4.9	5.4
Tire, Replace (C)		
Includes: Dismount old tire and mount new tire to rim.		
All Models	.5	.5
Tires, Rotate (C)		
All Models	.5	.5
Wheel, Balance (C)		
One	.3	.3
each addl. add	.2	.2
SCHEDULED MAINTENANCE INTERVALS		
If necessary, refer to appropriate Chilton maintenance service information.		
7,500 Mile Service (C)		
All Models	.4	.4
15,000 Mile Service (C)		
200SX, 240SX	1.0	1.0
300ZX	1.1	1.1
Altima, Maxima	.9	.9
Sentra	.9	.9
22,500 Mile Service (C)		
All Models	.4	.4
30,000 Mile Service (B)		
200SX		
w/GA16DE	2.3	2.3
w/o GA16DE	2.1	2.1
240SX	1.9	1.9
300ZX	2.4	2.4
Altima, Maxima	1.7	1.7
Sentra	2.0	2.0
37,500 Mile Service (C)		
All Models	.4	.4
45,000 Mile Service (C)		
200SX, 240SX	1.0	1.0
300ZX	1.1	1.1
Altima, Maxima	.9	.9
Sentra	.9	.9
52,500 Mile Service (C)		
All Models	.4	.4
60,000 Mile Service (B)		
200SX		
w/GA16DE	2.9	2.9
w/o GA16DE	2.7	2.7
300ZX	3.1	3.1
Altima, Maxima	2.9	2.9
Sentra	3.1	3.1
67,500 Mile Service (C)		
All Models	.4	.4
75,000 Mile Service (C)		
200SX, 240SX	1.0	1.0
300ZX	1.1	1.1

	LABOR TIME	SEVERE SERVICE
Altima, Maxima	.9	.9
Sentra	.9	.9
82,500 Mile Service (C)		
All Models	.4	.4
90,000 Mile Service (B)		
200SX		
w/GA16DE	2.3	2.3
w/o GA16DE	2.1	2.1
240SX	1.9	1.9
300ZX	2.4	2.4
Altima, Maxima	1.7	1.7
Sentra	2.0	2.0
97,500 Mile Service (C)		
All Models	.4	.4
CHARGING		
Alternator Circuits, Test (B)		
Includes: Test component output.		
All Models	.6	.6
Alternator, R&R and Recondition (B)		
200SX	3.0	3.5
210, 310	2.8	3.3
240SX		
SOHC	2.8	3.3
DOHC	3.3	3.8
280ZX	2.7	3.2
300ZX	4.0	4.5
350Z	1.1	1.6
510	2.9	3.4
810/Maxima	2.9	3.4
Altima		
1993-97	2.8	3.3
1998-01	2.0	2.5
2000-05		
2.5L QR	2.4	2.9
3.5L VQ	3.5	4.0
Maxima		
1985-94	3.1	3.6
1995-99	3.8	4.3
2000-01	3.2	3.7
2002-05	3.6	4.1
NX	2.9	3.4
Pulsar	3.2	3.7
Sentra		
1982-94	3.1	3.6
1995-05	2.7	3.2
Stanza		
1982-89	3.1	3.6
1990-92	3.3	3.8
Stanza Wagon	3.7	4.2
Alternator Assy., Replace (B)		
Includes: Pulley transfer.		
Exc. below	.9	1.1
240SX DOHC	1.2	1.4
300ZX	1.7	1.9
350Z	.6	.8
2000-05 Altima		
2.5L QR	1.1	1.3
3.5L VQ	1.8	2.0

Altima : Maxima : NX : Pulsar : Pulsar NX : Sentra : Stanza

	LABOR TIME	SEVERE SERVICE
1995-05 Maxima	2.0	2.2
Sentra		
2000-01	1.5	1.7
2002-05	1.2	1.4
1990-92 Stanza	.5	.7
Stanza Wagon	1.2	1.4
Front Alternator Bearing, Replace (B)		
200SX	1.5	1.7
210, 310	1.6	1.8
240SX		
SOHC	1.2	1.4
DOHC	1.7	1.9
280ZX	1.6	1.8
300ZX	2.2	2.4
510	1.4	1.6
810/Maxima	1.5	1.7
Altima	1.5	1.7
Maxima		
1985-94	1.7	1.9
1995-99	2.3	2.5
2000-05	3.2	3.4
NX	1.6	1.8
Pulsar	1.4	1.6
Sentra		
1982-99	1.6	1.8
2000-05	2.2	2.4
Stanza	1.4	1.6
Stanza Wagon	1.6	1.8
Voltage Regulator, Replace (B)		
200SX		
1981-88	.7	.7
1995-99	1.7	1.9
210, 310	.7	.7
240SX		
SOHC	.9	.9
DOHC	1.4	1.6
280ZX	.7	.7
300ZX	1.4	1.6
510	.7	.7
810/Maxima	.7	.7
Altima		
1993-97	.9	.9
1998-05	1.5	1.7
Maxima		
1985-94	1.4	1.6
1995-99	2.4	2.6
2000-05	2.2	2.4
NX	1.2	1.4
Sentra		
1982-87	.7	.7
1988-94	1.2	1.4
1995-05	1.9	2.1
Stanza		
1982-85	.7	.7
1986-92	1.3	1.5

	LABOR TIME	SEVERE SERVICE
Voltmeter and/or Ammeter Gauge, Replace (B)		
200SX		
1981-83	2.3	2.3
1984-88	1.6	1.6
1995-97	1.5	1.5
210	2.4	2.4
240SX		
1989-94	2.2	2.2
1995-97	1.6	1.6
280ZX	1.5	1.5
300ZX	1.9	1.9
310	.7	.7
350Z	.9	.9
510	1.6	1.6
810/Maxima	1.2	1.2
Altima	1.8	1.8
Maxima	1.9	1.9
NX, Sentra	1.6	1.6
Pulsar	1.4	1.4
Stanza		
1982-86	1.5	1.5
1987-92	.9	.9
Stanza Wagon	1.4	1.4

STARTING

	LABOR TIME	SEVERE SERVICE
Starter Draw Test (On Car) (B)		
All Models	.3	.3
Battery, Replace (C)		
All Models	.5	.7
Battery Cables, Replace (C)		
Each		
1981-88 200SX	.5	.7
210, 310	.5	.7
240SX	.5	.7
280ZX	.5	.7
300ZX	.5	.7
510	.5	.7
All		
1995-98 200SX	.7	.9
350Z	1.2	1.4
810/Maxima	.5	.7
Altima		
1993-01	2.3	2.5
2002-05	.9	1.1
Maxima		
1985-94	.5	.7
1995-04	1.5	1.7
2005	.8	.9
NX	.8	1.0
Pulsar	.5	.7
Sentra		
1982-99	.8	1.0
2000-05		
QG, SR	1.1	1.3
QR	.6	.8

	LABOR TIME	SEVERE SERVICE
Stanza		
1982-86	.5	.7
1987-89	2.2	2.4
1990-92	1.4	1.6
Stanza Wagon	.5	.7
Starter, R&R and Recondition (B)		
Exc. below	2.9	3.4
1995-98 200SX		
GA	2.3	2.8
SR	3.1	3.6
350Z	1.2	1.7
2002-05 Altima	2.1	2.6
Maxima		
1985-01	2.6	3.1
2002-05	2.1	2.6
NX	3.0	3.5
Sentra		
1987-90		
E		
2WD	2.7	3.2
4WD	7.0	7.5
GA		
2WD	2.1	2.6
4WD	2.6	3.1
1991-94	3.0	3.5
1995-99		
GA	2.3	2.8
SR	3.1	3.6
2000-05	2.3	2.8
Stanza		
1982-86	2.7	3.2
1987-89	7.1	7.6
1990-92	3.2	3.7
Stanza Wagon		
2WD	10.7	11.2
4WD	12.1	12.6
Starter Assy., Replace (B)		
Exc. below	.8	1.0
1995-98 200SX SR	2.0	2.2
350Z	1.1	1.3
Altima		
1993-97	.9	1.1
1998-05	1.4	1.6
Maxima	1.1	1.3
NX SR	1.3	1.5
Sentra		
1987-90		
E 4WD	5.2	5.4
GA 4WD	1.1	1.3
1991-93 SR	1.3	1.5
1994-99 SR	2.0	2.2
2000-05	.9	1.1
Stanza		
1987-89	5.2	5.4
1990-92	.9	1.1
Stanza Wagon		
2WD	5.2	5.4
4WD	6.2	6.4

NI-3

200SX : 210 : 240SX : 280ZX : 300ZX : 310 : 350Z : 510 : 810

	LABOR TIME	SEVERE SERVICE
Starter Drive Assy., Replace (B)		
Includes: Starter R&R.		
Exc. below	1.4	1.8
1995-98 200SX SR	2.2	2.6
280ZX	1.5	1.9
Altima	1.8	2.2
1994-99 Maxima SR	1.6	2.0
NX SR	1.6	2.0
Sentra		
1987-90		
E 4WD	5.7	6.1
GA 4WD	1.5	1.9
1991-93 SR	1.6	2.0
1994-99 SR	2.2	2.6
Stanza		
1982-86	1.5	1.9
1987-89	5.9	6.3
1990-92	1.6	2.0
Stanza Wagon		
2WD	5.9	6.3
4WD	6.6	7.0
Starter Relay, Replace (B)		
1990-96 300ZX	.3	.3
350Z	.7	.7
810/Maxima	.5	.5
Starter Solenoid and/or Switch, Replace (B)		
Includes: Starter R&R.		
Exc. below	1.4	1.8
1995-98 200SX SR	2.1	2.5
Altima		
1993-01	1.6	2.0
2002-05	.9	1.3
Pulsar, GA	.9	1.3
Sentra		
1987-90		
E 4WD	5.5	5.9
GA 2WD	.9	1.3
1994-99 SR	2.1	2.5
2000-05	1.1	1.5
Stanza		
1987-89	5.7	6.1
1990-92	1.5	1.9
Stanza Wagon		
2WD	5.5	5.9
4WD	6.2	6.6

CRUISE CONTROL

	LABOR TIME	SEVERE SERVICE
Diagnose Cruise Control System Component Each (A)		
All Models	1.1	1.1
Circuits, Test (B)		
All Models	1.0	1.0
Actuator Assy., Replace (B)		
All Models	.5	.5
Control Cable & Core, Replace (B)		
All Models	.7	.7
Control Clutch Switch, Replace (B)		
All Models	.5	.5

	LABOR TIME	SEVERE SERVICE
Control Controller Module and/or Relay, Replace (B)		
Exc. below	.7	.7
1995-98 200SX	.3	.3
1985-88 Maxima	1.2	1.2
Control On/Off Switch, Replace (B)		
All Models	.6	.6
Control Safety Switch (Cut-Out), Replace (B)		
Exc. 300ZX	.5	.5
300ZX	.6	.6
Control Sensor, Replace (B)		
200SX		
1986-88	1.2	1.2
1995-98	.5	.5
1989-94 240SX	.9	.9
1986-89 300ZX	.8	.8
Altima	.5	.5
Maxima	.7	.7
NX	.5	.5
Sentra	.5	.5
Stanza		
1986-89	1.3	1.3
1990-92	1.7	1.7
Stanza Wagon	1.4	1.4
Control Set Switch, Replace (B)		
All Models	.7	.7
Control Vacuum Pump, Replace (B)		
All Models	.5	.5
Solenoid or Servo Valve, Replace (B)		
1981-91	.5	.5

IGNITION

	LABOR TIME	SEVERE SERVICE
Diagnose Ignition System Component Each (A)		
All Models	.8	.8
Ignition Timing, Reset (B)		
All Models	.5	.5
Crank Angle Sensor, Replace (B)		
Exc. below	.5	.7
2002-05 Maxima	.8	1.0
Crankshaft Position Sensor, Replace (B)		
Exc. below	.7	.9
2002-05 Maxima		
w/OBD II	1.5	1.7
w/o OBD II	1.1	1.3
Distributor, R&R and Recondition (B)		
Includes: Reset base ignition timing.		
1981-89	2.0	2.5
Distributor, Replace (B)		
Includes: Reset base ignition timing.		
All Models	.9	1.1
Distributor Cap and/or Rotor, Replace (B)		
All Models	.5	.5
Distributor Reluctor and/or Pick-up Coil, Replace (B)		
1981-89	.8	1.0

	LABOR TIME	SEVERE SERVICE
Distributor Vacuum Control Unit, Replace (B)		
1981-90	.7	.7
Distributorless Ignition Module, Replace (B)		
1990-96 300ZX	.5	.6
1993-99 Maxima	.5	.6
1987-89 Pulsar CA	.3	.4
Glow Plug, Replace (B)		
810/Maxima	.7	.9
Glow Plug Control Unit, Replace (B)		
810/Maxima	.5	.5
Glow Plug Relay, Replace (B)		
810/Maxima	.5	.5
Ignition Coil, Replace (B)		
Exc. below	.5	.5
350Z	1.5	1.7
Maxima VQ	3.2	3.4
1995-99 200SX, Sentra GA	.7	.9
Ignition Lock Switch Assy., Replace (B)		
Exc. below	1.3	1.3
350Z	1.9	1.9
2002-05 Altima	.8	.8
2000-05 Sentra	1.4	1.4
Ignition Switch, Replace (B)		
All Models	.8	.8
Spark Plug (Ignition) Cables, Replace (B)		
4 cyl.	.5	.7
6 cyl.	.7	.9
V6	.9	1.1
w/DOHC 4 cyl. add	.3	.3
Spark Plugs, Replace (B)		
4 cyl.	.5	.7
6 cyl.	.7	.9
V6		
exc. below	1.1	1.3
350Z	2.0	2.2
Maxima VQ	3.3	3.5
w/DOHC 4 cyl. add	.3	.3
Transistorized Ignition Control Unit, Replace (B)		
All Models	.7	.7

EMISSIONS

The following operations do not include testing. Add time as required.

	LABOR TIME	SEVERE SERVICE
Diagnose Emission Control System Component Each (A)		
All Models	1.0	1.0
Dynamometer Test (B)		
All Models	.4	.4
Retrieve Fault Codes (A)		
1981-05	.3	.3
Air Cleaner Temperature Sensor, Replace (B)		
1981-88	.3	.3

Altima : Maxima : NX : Pulsar : Pulsar NX : Sentra : Stanza NI-5

	LABOR TIME	SEVERE SERVICE
Air Flow Meter, Replace (B)		
200SX, 280ZX	.8	.8
240SX, 300ZX	.7	.7
350Z	.5	.5
Altima		
1993-97	.7	.7
1998-05	.5	.5
Maxima		
1985-88	1.6	1.6
1989-94	.8	.8
1995-04	.7	.7
2005	.3	.3
NX	.5	.5
Pulsar		
CA	.5	.5
E	.7	.7
GA	.8	.8
Sentra		
1987-90	.8	.8
1991-05	.5	.5
Stanza		
1987-89	1.5	1.5
1990-92	.5	.5
Air-Fuel Control Solenoid, Replace (B)		
1981-90	.7	.7
Air Injection Check Valve, Replace (B)		
1981-90	.4	.4
Air Injection Valve or Filter, Replace (B)		
1981-90	.4	.4
Air Pump, R&R and Recondition (B)		
1981-90	2.4	2.9
w/AC add	.5	.5
Air Pump Assy., Replace (B)		
1981-90	1.8	2.0
w/AC add	.5	.5
Altitude Switch or Sensor, Replace (B)		
1984-88 200SX turbo	.7	.7
1985-88 Maxima	.4	.4
Pulsar	.7	.7
Anti-Backfire Valve, Replace (B)		
310	.7	.7
Back Pressure Variable Transducer Valve, Replace (B)		
All Models	.4	.4
Canister Purge Control Valve, Replace (B)		
All Models	.4	.4
Coolant Temperature (ECT) Sensor, Replace (B)		
All Models	.7	.7
Dash Pot, Replace (B)		
1983-89	.4	.4
ECC Relay, Replace (B)		
1981-90	.5	.5

	LABOR TIME	SEVERE SERVICE
EFE Valve, Replace (B)		
210	1.5	1.5
310	4.2	4.2
EGR Control Unit, Replace (B)		
1990-96 300ZX	.7	.7
1990-92 Stanza	.7	.7
EGR Control Valve, Replace (B)		
200SX		
1981-88	.7	.9
1998-98		
GA	.7	.9
SR	1.4	1.6
210, 310	.7	.9
240SX		
SOHC	.8	1.0
DOHC	1.6	1.8
280ZX	.7	.9
300ZX		
1984-89	.7	.9
1990-94	9.9	10.1
1995-96	3.1	3.3
510	.7	.9
810/Maxima	.7	.9
Altima		
1993-97	.4	.6
1998-05	.3	.5
Maxima		
1985-94	.7	.9
1995-01		
electric	1.5	1.7
mechanical	1.1	1.3
2002-04	1.8	2.0
2005	1.1	1.2
NX	.7	.9
Sentra		
1982-93	.7	.9
1994-99		
GA	.7	.9
SR	1.4	1.6
2000-05	.8	1.0
Stanza	.7	.9
EGR Controller, Replace (B)		
810/Maxima	.7	.7
1982-85 Sentra	.7	.7
EGR Solenoid Valve, Replace (B)		
1981-87	.3	.3
1988-02	.7	.7
EGR Temperature Sender, Replace (B)		
All Models	.7	.7
EGR Temperature Sensor, Replace (B)		
Exc. below	.7	.7
1994-96 300ZX	4.9	4.9
2005 Maxima	.5	.5
Sentra QG	.3	.3
Electronic Control Module (ECM/PCM), Replace (B)		
1981-90	.4	.4

	LABOR TIME	SEVERE SERVICE
Fuel Cut-Off Solenoid, Replace (B)		
1987-90	.5	.5
Fuel Shut-Off Clutch Switch, Replace (B)		
1987-90	.4	.4
Fuel Temperature Sensor, Replace (B)		
All Models	.9	.9
Fuel Vapor Canister, Replace (B)		
Exc. below	.5	.5
1990-96 300ZX	.9	.9
350Z	.6	.6
Altima		
1993-01	.7	.7
2002-05	1.5	1.5
Maxima	1.6	1.6
Intake Air Temperature (IAT) Sensor, Replace (B)		
All Models	.5	.5
Knock Sensor, Replace (B)		
200SX		
1984-88 turbo	.9	1.1
1995-98		
GA	1.2	1.4
SR	1.7	1.9
1995-98 240SX	.7	.9
300ZX		
1984-89 turbo	.7	.9
1990-96	10.8	11.0
350Z	2.4	2.6
Altima		
1993-97	4.1	4.3
1998-01	.5	.7
2002-05		
QR	.5	.7
VQ	2.6	2.8
Maxima		
1994		
VG	1.6	1.8
VE	6.5	6.7
1995-01	.9	1.1
2002-04	3.8	4.0
2005	2.9	3.0
Pulsar	.7	.9
Sentra		
1994	2.8	3.0
1995-99		
GA	1.2	1.4
SR	1.7	1.9
2000-05	.5	.7
Manifold Absolute Pressure (MAP) Sensor, Replace (B)		
All Models	.5	.5
Oxygen Sensor, Replace (B)		
1983-93	.7	.9
1995-98 200SX	.5	.7
240SX	.5	.7

200SX : 210 : 240SX : 280ZX : 300ZX : 310 : 350Z : 510 : 810

	LABOR TIME	SEVERE SERVICE
Oxygen Sensor, Replace (B)		
300ZX		
1992-93	1.6	*1.8*
1995-96		
front		
one	1.8	*2.0*
both	2.4	*2.6*
rear	1.9	*2.1*
350Z		
front one or both	1.6	*1.8*
rear one or both	.5	*.7*
Altima		
1993-97 one or both	.7	*.9*
1998-01 one or both	.5	*.7*
2001-05		
QR		
front one or both	.6	*.8*
rear		
one	.5	*.7*
both	1.0	*1.2*
VQ one or both	1.0	*1.2*
Maxima		
1994	.5	*.7*
1995-01		
front one or both	1.1	*1.3*
rear one or both	.9	*1.1*
2002-04		
front		
one	.8	*1.0*
both	1.9	*2.1*
rear one or both	1.1	*1.3*
2005 front or rear	.6	*.8*
Sentra		
1994-99	.5	*.7*
2000-05		
front one or both	.6	*.8*
rear		
QG one or both	1.0	*1.2*
SR	.5	*.7*
Positive Crankcase Ventilation (PCV) Valve, Replace (B)		
Exc. below	.5	*.5*
300ZX	.9	*.9*
350Z	.7	*.7*
Starting Enrichment Solenoid Valve, Replace (B)		
1981-90	.4	*.4*
Thermal Vacuum Valve, Replace (B)		
1981-90	.8	*.8*
Throttle Control Solenoid, Replace (B)		
810/Maxima	.5	*.5*
Throttle Position Sensor (TPS)/Valve Switch, Replace (B)		
200SX, 300ZX	.5	*.5*
240SX		
SOHC	.5	*.5*
DOHC	1.7	*1.7*

	LABOR TIME	SEVERE SERVICE
Altima		
1993-97	1.2	*1.2*
1998-01	.5	*.5*
2002-05	1.0	*1.0*
Maxima, Stanza	.8	*.8*
NX, Pulsar	.8	*.8*
Sentra		
1987-93	.8	*.8*
1994-05	.6	*.6*
Stanza Wagon	.5	*.5*
Throttle Positioner, Replace (B)		
1981-90	.5	*.5*
Vacuum Control Valve, Replace (B)		
Exc. below	.4	*.4*
1984-88 200SX	.5	*.5*
1984-89 300ZX	.7	*.7*
Vacuum Sensor, Replace (B)		
1981-90	.5	*.5*
Vapor Separator, Replace (B)		
1981-97	1.1	*1.1*
Vehicle Speed Sensor, Replace (B)		
1989-97 240SX	.5	*.5*
300ZX	.5	*.5*
Altima	.5	*.5*
Maxima		
1985-88	1.3	*1.3*
1989-05	.6	*.6*
NX, Pulsar	.6	*.6*
Sentra		
1987-90	1.4	*1.4*
1991-94	.6	*.6*

FUEL
CARBURETOR

	LABOR TIME	SEVERE SERVICE
Air Cleaner Vacuum Motor, Replace (B)		
1981-88	.4	*.4*
Anti-Dieseling Solenoid, Replace (B)		
1981-88	.7	*.7*
Automatic Choke, Replace (B)		
1981-88	.7	*.7*
Carburetor, R&R and Clean or Recondition (B)		
200SX	3.6	*3.6*
210, 310	3.8	*3.8*
510	3.8	*3.8*
810/Maxima	3.7	*3.7*
Maxima	3.7	*3.7*
Pulsar	3.8	*3.8*
Sentra	3.3	*3.3*
Stanza	3.6	*3.6*
Stanza Wagon	3.6	*3.6*
Carburetor, Replace (B)		
Includes: Adjustments.		
200SX	1.6	*1.6*
210, 310	1.6	*1.6*
510	1.7	*1.7*
810/Maxima	1.6	*1.6*
Maxima	1.6	*1.6*

	LABOR TIME	SEVERE SERVICE
Pulsar	1.7	*1.7*
Sentra	2.5	*2.5*
Stanza	2.2	*2.2*
Stanza Wagon	1.6	*1.6*

DELIVERY

	LABOR TIME	SEVERE SERVICE
Fuel Pump, Test (B)		
All Models	.3	*.3*
Diesel Fuel Injection Pump, Replace (B)		
810/Maxima	6.7	*7.0*
Sentra	3.6	*3.9*
Fuel Filter, Replace (B)		
All Models	.5	*.5*
Fuel Gauge (Dash), Replace (B)		
200SX		
1981-83	2.3	*2.3*
1984-88	1.4	*1.4*
1995-98	1.3	*1.3*
210	3.2	*3.2*
240SX		
1989-94	1.7	*1.7*
1995-98	1.2	*1.2*
280ZX	1.4	*1.4*
300ZX	1.6	*1.6*
310	2.2	*2.2*
350Z	1.1	*1.1*
510	1.6	*1.6*
810/Maxima	1.5	*1.5*
Altima	1.4	*1.4*
Maxima		
1985-88	1.6	*1.6*
1989-05	1.2	*1.2*
NX	1.4	*1.4*
Pulsar	1.6	*1.6*
Sentra		
1982-90	.9	*.9*
1991-99	1.4	*1.4*
2000-05	.8	*.8*
Stanza		
1982-89	.9	*.9*
1990-92	1.4	*1.4*
Stanza Wagon	1.4	*1.4*
Fuel Gauge (Tank), Replace (B)		
200SX		
1981-88	1.2	*1.4*
1995-98	.8	*1.0*
210	.9	*1.1*
240SX	.7	*.9*
280ZX	1.2	*1.4*
300ZX		
1984-89	.9	*1.1*
1990-96	.5	*.7*
310	1.5	*1.7*
350Z, Altima	1.0	*1.2*
510	1.1	*1.3*
810/Maxima	1.6	*1.8*

Altima : Maxima : NX : Pulsar : Pulsar NX : Sentra : Stanza

	LABOR TIME	SEVERE SERVICE
Maxima		
1985-91	1.6	*1.8*
1992-05	1.0	*1.2*
NX	.7	*.9*
Pulsar	1.4	*1.6*
Sentra	.9	*1.1*
Stanza		
1982-86	.7	*.9*
1987-89	1.1	*1.3*
1990-92	.5	*.7*
Stanza Wagon	1.4	*1.6*
Fuel Pump, Replace (B)		
Electrical		
200SX		
1981-88	1.3	*1.5*
1995-98	.7	*.9*
240SX	.8	*1.0*
280ZX	1.3	*1.5*
300ZX		
1984-89	1.7	*1.9*
1990-96	.8	*1.0*
350Z	1.0	*1.2*
Altima		
1993-01	1.5	*1.7*
2002-05	.8	*1.0*
Maxima		
1985-88	1.7	*1.9*
1989-05	1.0	*1.2*
NX, Sentra	.9	*1.1*
Stanza		
1987-89	1.3	*1.5*
1990-92	.7	*.9*
Stanza Wagon	1.5	*1.7*
Mechanical		
exc. below	.7	*.9*
1981 310	1.7	*1.9*
Fuel Pump Control Module, Replace (B)		
Exc. below	.7	*.7*
1990-96 300ZX	1.5	*1.7*
Fuel Pump Relay, Replace (B)		
Exc. below	.5	*.5*
350Z	.7	*.7*
Maxima	.6	*.6*
Fuel Tank, Replace (B)		
Includes: Drain and refill.		
200SX		
1981-88	1.7	*1.9*
1995-98	2.1	*2.3*
210		
exc. hatchback	1.6	*1.8*
hatchback	1.9	*2.1*
240SX		
1989-94	1.3	*1.5*
1995-98	4.1	*4.3*
280ZX	1.5	*1.7*
300ZX		
1984-89	1.3	*1.5*
1990-96	4.8	*5.0*

	LABOR TIME	SEVERE SERVICE
310	.9	*1.1*
350Z	7.3	*7.5*
510	1.2	*1.4*
810/Maxima	1.4	*1.6*
Altima		
1993-97	2.8	*3.0*
1998-05	2.1	*2.3*
Maxima		
1985-88	2.5	*2.7*
1989-94	2.1	*2.3*
1995-05	1.9	*2.1*
NX	2.4	*2.6*
Pulsar	1.8	*2.0*
Sentra		
1982-86	.9	*1.1*
1987-90	1.5	*1.7*
1991-05	2.4	*2.6*
Stanza		
1982-86	1.4	*1.6*
1987-89	2.1	*2.3*
1990-92	1.3	*1.5*
Stanza Wagon	1.4	*1.6*
w/2+2 300ZX add	1.2	*1.4*
Intake Manifold and/or Gaskets, Replace (B)		
4 cyl.		
200SX		
1981-83	3.7	*4.0*
1984-88		
w/turbo	3.6	*3.9*
w/o turbo	2.7	*3.0*
1995-98		
GA	5.7	*6.0*
SR	7.8	*8.1*
210	2.5	*2.8*
240SX	4.4	*4.7*
310		
1981	4.3	*4.6*
1982	3.0	*3.3*
510	3.0	*3.3*
Altima		
1993-01	4.3	*4.6*
2002-05	1.8	*2.1*
NX		
GA	4.2	*4.5*
SR	1.7	*2.0*
Pulsar		
CA	6.3	*6.6*
E	2.0	*2.3*
GA	3.1	*3.4*
Sentra		
1982-86	3.0	*3.3*
1987-90		
E	2.0	*2.3*
GA	3.2	*2.5*
1991-93		
GA	4.2	*4.5*
SR	1.7	*2.0*

	LABOR TIME	SEVERE SERVICE
1994		
GA	4.3	*4.6*
SR	7.9	*8.2*
1995-99		
GA	5.7	*6.0*
SR	7.8	*8.1*
2000-05		
QG, SR	3.8	*4.1*
QR	2.0	*2.3*
Stanza		
1982-86	3.1	*3.4*
1987-92	4.6	*4.9*
L6		
280ZX	5.1	*5.4*
810/Maxima	2.6	*2.9*
V6		
1987-88 200SX	2.9	*3.2*
300ZX		
1984-89	3.8	*4.1*
1990-96	10.2	*10.5*
350Z	2.1	*2.4*
Altima	2.6	*2.9*
Maxima		
1985-88	3.8	*4.1*
1989-94		
SOHC upper or lower	1.8	*2.1*
DOHC upper or lower	3.7	*4.0*
1995-01 upper or lower	3.2	*3.5*
2002-04	4.1	*4.4*
2005	2.9	*3.4*

INJECTION

Diagnose Fuel Injection System Component Each (A)

	LABOR TIME	SEVERE SERVICE
1984-05	1.1	*1.1*

Diesel Fuel Injection Timing, Check & Adjust (B)

	LABOR TIME	SEVERE SERVICE
810/Maxima	1.3	*1.5*
1982-85 Sentra	1.1	*1.3*

Air Regulator, Replace (B)

	LABOR TIME	SEVERE SERVICE
1995-98 200SX		
GA	.8	*.8*
SR	1.8	*1.8*
240SX		
SOHC	.8	*.8*
DOHC	1.9	*1.9*
300ZX	.8	*.8*
Altima		
1993-97	4.3	*4.3*
1998-05	.7	*.7*
NX	.8	*.8*
1987-90 Pulsar CA	.5	*.5*
Sentra		
1991-93	.8	*.8*
1994-05		
GA	.8	*.8*
SR	1.8	*1.8*

NI-8
200SX : 210 : 240SX : 280ZX : 300ZX : 310 : 350Z : 510 : 810

	LABOR TIME	SEVERE SERVICE
Anti-Stall Dashpot, Replace (B)		
Stanza Wagon	.7	.7
EFI Fuel Damper, Replace (B)		
1984-88 200SX	.8	.8
280ZX	.8	.8
1984-89 300ZX	1.6	1.6
Pulsar	1.1	1.1
EFI Relay, Replace (B)		
1984-05	.7	.7
Electronic Control Unit, Replace (B)		
Exc. below	.7	.7
350Z	.8	.8
2002-05 Maxima	1.6	1.6
Fuel Injection Lines, Replace (B)		
810/Maxima	1.5	1.7
1982-85 Sentra	.7	.9
Fuel Injectors, Replace (B)		
200SX		
1984-88	2.5	2.7
1995-98		
GA	1.2	1.4
SR	2.0	2.2
1994-98 240SX	1.7	1.9
280ZX		
three	1.4	1.6
all	2.4	2.6
300ZX		
1984-89	2.5	2.7
1990-96	5.3	5.5
350Z	1.7	1.9
Altima		
1993-97	1.8	2.0
1998-01	1.1	1.3
2002-05	1.8	2.0
Maxima		
1985-94		
SOHC	2.2	2.4
DOHC	5.3	5.5
1995-01	3.1	3.3
2002-04	3.5	3.7
2005	2.8	3.0
NX	1.2	1.4
1987-90 Pulsar CA	2.9	3.1
Sentra		
1991-93	1.2	1.4
1994	2.3	2.5
1995-99		
GA	1.2	1.4
SR	2.0	2.2
2000-05		
QG	1.1	1.3
QR	1.8	2.0
SR	1.4	1.6
Stanza		
1987-89	1.5	1.7
1990-92	3.2	3.4
Fuel Pressure Regulator, Replace (B)		
1995-98 200SX	.8	1.0

	LABOR TIME	SEVERE SERVICE
240SX		
SOHC	.7	.9
DOHC	1.9	2.1
300ZX	.7	.9
Altima		
1993-97	1.6	1.8
1998-05	.5	.7
Maxima		
1985-01	.7	.9
2002-04	1.2	1.4
2005	.7	.9
NX	.7	.9
Pulsar CA	2.6	2.8
Sentra		
1987-99	.8	1.0
2000-05		
QG	1.1	1.3
SR	.6	.8
Stanza		
1987-89	1.1	1.3
1990-92	.5	.7
Stanza Wagon	.9	1.1
Idle Speed Control Actuator, Replace (B)		
1981-90	.4	.4
Injection Nozzles (Diesel), Replace (B)		
810/Maxima	1.9	2.1
1982-85 Sentra	1.8	2.0
Clean and test add each	.2	.3
Mixture Heater Relay, Replace (B)		
Pulsar	.5	.5
1987-90 Sentra	.5	.5
Throttle Body Diesel Fuel Injection, Replace (B)		
1981-85	.7	.7
Throttle Body Injection Control Unit, Replace (B)		
Pulsar	.5	.5
1987-90 Sentra	.5	.5
Throttle Body Injection Unit, Replace (B)		
Pulsar	1.8	1.8
1987-90 Sentra	1.6	1.6
Replace injectors add	.3	.3
Throttle Chamber, Replace (B)		
200SX	.9	1.1
240SX		
SOHC	1.3	1.5
DOHC	1.8	2.0
300ZX	1.8	2.0
350Z	.5	.7
Altima	1.3	1.5
Maxima		
1985-01	1.7	1.9
2002-04	1.9	2.1
2005	.8	1.0
NX		
GA	.7	.9
SR	1.5	1.7

	LABOR TIME	SEVERE SERVICE
Pulsar CA	1.7	1.9
Sentra		
1991-93		
GA	.7	.9
SR	1.5	1.7
1994-05	1.1	1.3
Stanza Wagon	1.7	1.9
Throttle Control Solenoid, Replace (B)		
1981-85	.5	.5
Vacuum Modulator, Replace (B)		
1981-84	.5	.5
Vacuum Pump, Replace (B)		
Includes: Vacuum test.		
1981-85	1.1	1.3
Recondition pump add	.5	.5

TURBOCHARGER

	LABOR TIME	SEVERE SERVICE
Boost Sensor, Replace (B)		
200SX	.3	.3
300ZX	.3	.3
Pulsar	.3	.3
Emergency Relief Valve, Replace (B)		
200SX	.3	.3
Pulsar	.3	.3
Turbocharger Assy., R&R or Replace (B)		
200SX	6.7	7.0
280ZX	6.7	7.0
300ZX		
1984-89	6.2	6.5
1990-96		
right	6.8	7.1
left	8.6	8.9
both	15.5	15.8
Pulsar	3.5	3.8

EXHAUST

	LABOR TIME	SEVERE SERVICE
Catalytic Converter, Replace (B)		
200SX		
1981-88	1.2	1.3
1995-98	.8	.9
210, 310	1.2	1.3
240SX	.7	.8
280ZX	1.3	1.4
300ZX		
1984-89	.7	.8
1990-96	1.2	1.3
350Z		
one	1.1	1.2
both	1.4	1.6
510	1.1	1.2
810/Maxima	1.2	1.3
Altima		
1993-01	.8	.9
2002-05		
4 cyl.	1.7	1.8

Altima : Maxima : NX : Pulsar : Pulsar NX : Sentra : Stanza   **NI-9**

	LABOR TIME	SEVERE SERVICE
V6		
one	1.6	1.7
both	2.4	2.5
Maxima		
1985-88	.8	.9
1989-94	.7	.8
1995-05	1.2	1.3
NX	.8	.9
Pulsar	1.3	1.4
Sentra		
1982-90	1.1	1.2
1991-05		
exc. QR	.8	.9
QR	2.1	2.2
Stanza		
1982-86	.8	.9
1987-89	1.2	1.3
1990-92	.5	.6
Stanza Wagon	1.1	1.2
Exhaust Manifold, Replace (B)		
200SX		
1981-83	3.2	3.5
1984-88		
4 cyl.		
w/turbo	6.9	7.2
w/o turbo	1.7	2.0
V6		
right side	.9	1.2
left side	1.9	2.2
1995-98	1.8	2.1
210	3.0	3.3
240SX		
SOHC	1.6	1.9
DOHC	1.8	2.1
280ZX	5.5	5.8
300ZX		
1984-89		
right side	2.9	3.2
left side		
w/turbo	5.5	5.8
w/o turbo	3.7	4.0
1990-96		
w/turbo		
right side	7.1	7.4
left side	8.2	8.5
w/o turbo		
right side	2.2	2.5
left side	2.6	2.9
310		
1981	4.6	4.9
1982	3.1	3.4
350Z one side	3.6	3.9
510	3.5	3.8
810/Maxima	1.7	2.0
Altima		
4 cyl.	1.8	2.1
V6		
right side	3.2	3.5
left side	2.7	3.0

	LABOR TIME	SEVERE SERVICE
Maxima		
1985-88		
right side	3.2	3.5
left side	2.7	3.0
1989-94		
right side	1.9	2.2
left side	2.3	2.6
1995-01		
right side	6.6	6.3
left side	4.0	4.3
2002-05		
right side	6.7	7.0
left side	3.3	3.6
NX	1.6	1.9
Pulsar		
CA	1.6	1.9
E	1.4	1.7
GA	.9	1.2
Sentra		
1982-86	3.2	3.5
1987-90		
E	1.6	1.9
GA	.9	1.2
1991-94	1.6	1.9
1995-99	1.8	2.1
2000-05		
QG		
exc. CA EZEV	2.4	2.7
CA EZEV	2.9	3.2
QR	2.1	2.4
SR	1.3	1.6
Stanza		
1982-86	3.7	4.0
1987-92	1.3	1.6
Stanza Wagon	1.3	1.6
Front Exhaust Pipe, Replace (B)		
200SX		
1981-88	1.2	1.3
1995-98		
GA		
one	.8	.9
both	1.6	1.7
SR	1.2	1.3
210, 310	1.2	1.3
240SX	.8	.9
280ZX	1.3	1.4
300ZX		
1984-89	1.7	1.8
1990-96		
one side	.9	1.0
both sides	1.5	1.6
350Z	.6	.7
510	1.3	1.4
810/Maxima	1.2	1.3
Altima		
1993-97	1.3	1.4
1998-05	.8	.9

	LABOR TIME	SEVERE SERVICE
Maxima		
1985-94	1.2	1.3
1995-04	2.1	2.2
2005	1.1	1.2
NX	1.3	1.4
Pulsar	1.3	1.4
Sentra		
1982-94	1.4	1.5
1995-05		
GA		
one	.8	.9
both	1.6	1.7
QG		
exc. CA EZEV	.8	.9
CA EZEV	2.9	3.0
QR	.6	.7
SR	1.2	1.3
Stanza		
1982-89	1.2	1.3
1990-92	.8	.9
Muffler & Rear Pipe, Replace (B)		
Exc. below	.8	.9
1981-83 200SX	1.3	1.4
1990-96 300ZX	1.1	1.2
350Z	1.1	1.2
Pre-Muffler & Center Pipe, Replace (B)		
Exc. below	1.3	1.4
1995-98 240SX	.7	.8
300ZX	.8	.9
350Z	.6	.7
Maxima	.8	.9
NX	.7	.8
1991-05 Sentra	1.0	1.1
Stanza Wagon	.8	.9
Tail Pipe, Replace (B)		
280ZX	1.2	1.3

ENGINE COOLING

Coolant Thermostat, Replace (B)

	LABOR TIME	SEVERE SERVICE
200SX		
1981-83	1.1	1.3
1984-88		
4 cyl.	.5	.7
V6	1.6	1.6
1995-99	1.3	1.5
210, 310, 510, 810	.8	1.0
240SX	1.3	1.5
280ZX	.7	.9
300ZX		
1984-89	1.6	1.8
1990-96	2.5	2.7
350Z	1.5	1.7
Altima	1.1	1.3
Maxima		
1985-88	.8	1.0
1989-94		
SOHC	1.1	1.3
DOHC	2.0	2.2

NI-10 200SX : 210 : 240SX : 280ZX : 300ZX : 310 : 350Z : 510 : 810

	LABOR TIME	SEVERE SERVICE
Coolant Thermostat, Replace (B)		
1995-04	1.5	1.7
2005	.9	1.0
NX	1.1	1.3
Pulsar	.8	1.0
Sentra		
1982-86	.7	.9
1987-05	1.3	1.5
Stanza		
1982-86	.7	.9
1987-89	1.2	1.4
1990-92	.8	1.0
Stanza Wagon	1.1	1.3
Engine Coolant Temp. Sending Unit, Replace (B)		
Exc. 350Z	.5	.5
350Z	1.2	1.2
Radiator Assy., R&R or Replace (B)		
200SX		
1981-83	.8	1.0
1984-88	1.5	1.7
1995-98	1.4	1.6
210	.9	1.1
240SX		
1989-94	2.5	2.7
1995-98	1.9	2.1
280ZX	1.5	1.7
300ZX	1.9	2.1
310	1.2	1.4
350Z	2.0	2.2
510	.9	1.1
810/Maxima	1.5	1.7
Altima	1.8	2.0
Maxima	1.8	2.0
NX	1.4	1.6
Pulsar	1.3	1.5
Sentra		
1982-99	1.5	1.7
2000-05		
QG, SR	1.4	1.6
QR	1.7	1.9
Stanza	1.5	1.7
Radiator Fan and/or Fan Motor, Replace (B)		
Exc. below	.9	1.1
1995-98 200SX		
one	.7	.9
both	1.2	1.4
350Z one or both	2.0	2.2
2002-05 Altima one or both	1.7	1.9
1989-05 Maxima one or both	1.5	1.7
NX		
GA	.5	.7
SR	1.1	1.3
Sentra		
1991-94		
GA	.5	.7
SR	1.1	1.3

	LABOR TIME	SEVERE SERVICE
1995-05		
one	.8	1.0
both	1.2	1.4
1990-92 Stanza	1.4	1.6
Radiator Fan Motor Relay, Replace (B)		
Exc. 350Z	.5	.5
350Z	.7	.7
Radiator Fan Motor (Coolant Temp.) Switch, Replace (B)		
Exc. below	.5	.5
350Z	1.2	1.2
2002-05 Maxima	.8	.8
Radiator Hoses, Replace (B)		
Includes: Refill with proper coolant mix.		
Exc. below each	.7	.8
350Z		
upper	.5	.5
lower	1.4	1.5
2002-05 Altima		
upper	.5	.5
lower	.9	1.0
Maxima		
2000-04	2.3	2.5
2005	1.8	2.0
2000-05 Sentra		
upper	.5	.5
lower	.9	1.0
Temperature Gauge (Dash), Replace (B)		
1995-98 200SX	.8	.8
240SX		
1989-94	1.7	1.7
1995-98	1.6	1.6
300ZX	1.6	1.6
350Z	1.1	1.1
Altima	1.5	1.5
Maxima		
1985-88	1.6	1.6
1989-05	1.2	1.2
NX	1.6	1.6
Sentra	.8	.8
Water Pump and/or Gasket, Replace (B)		
Includes: Refill with proper coolant mix.		
200SX		
1981-83	2.1	2.6
1984-88		
4 cyl.	2.6	3.1
V6	3.5	4.0
1995-98		
GA	3.4	3.9
SR	3.0	3.5
210	1.3	1.8
240SX		
1989-94	1.8	2.3
1995-98	2.2	2.7
280ZX	2.0	2.5
300ZX	3.7	4.2

	LABOR TIME	SEVERE SERVICE
310		
1981	1.4	1.9
1982	1.9	2.4
350Z	2.1	2.6
510	1.7	2.2
810/Maxima		
gasoline	1.3	1.8
diesel	3.9	4.4
Altima		
1993-01	2.4	2.9
2002-05	1.8	2.3
Maxima		
1985-04	4.4	4.9
2005	2.8	3.1
NX		
GA		
w/ABS	3.8	4.3
w/o ABS	2.6	3.1
SR	3.1	3.6
Pulsar		
CA	2.4	2.9
E	1.8	2.3
GA	4.8	5.3
Sentra		
1982-86		
gasoline	1.5	2.0
diesel	4.0	4.5
1987-90		
E	2.0	2.5
GA	2.5	3.0
1991-94		
GA		
w/ABS	3.8	4.3
w/o ABS	2.6	3.1
SR	3.1	3.6
1995-99		
GA	3.4	3.9
SR	3.0	3.5
2000-05	1.7	2.2
Stanza		
1982-86	4.5	5.0
1987-89	3.8	4.3
1990-92	2.5	3.0
Stanza Wagon		
2WD	5.0	5.5
4WD	5.4	5.9

ENGINE
ASSEMBLY

Times shown are for OEM assemblies. Time to replace assemblies from after-market rebuilders may vary.

Engine Assy., R&R and Recondition (A)
Includes: Replacing rings, rod and main bearings, cylinder head reconditioning and engine tune-up.
200SX
1981-83 .... 21.7   22.2

Altima : Maxima : NX : Pulsar : Pulsar NX : Sentra : Stanza NI-11

	LABOR TIME	SEVERE SERVICE
1984-88		
4 cyl.		
w/turbo	19.9	20.4
w/o turbo	19.5	20.0
V6	27.2	27.7
1995-98		
GA	25.9	26.4
SR	32.5	33.0
210	22.9	23.4
240SX		
SOHC	22.5	23.0
DOHC	30.7	31.2
280ZX	29.3	29.8
300ZX		
1984-89		
w/turbo	42.4	42.9
w/o turbo	27.3	27.8
1990-96		
w/turbo	42.6	43.1
w/o turbo	39.2	39.7
310		
1981	23.8	24.3
1982	24.2	24.7
510	21.8	22.3
810/Maxima		
gasoline	29.1	29.6
diesel	27.6	28.1
Altima		
1993-97	26.1	26.6
1998-05	31.2	31.7
Maxima		
1985-88	28.7	29.2
1989-94		
SOHC	27.2	27.7
DOHC	30.5	31.0
1995-99	31.9	32.4
2000-04	32.4	32.9
2005	26.7	27.1
NX		
GA	22.0	22.5
SR	27.3	27.8
Pulsar		
CA	25.6	26.1
E	19.7	20.2
GA	21.2	21.7
Sentra		
1982-86		
gasoline	17.9	18.4
diesel	22.2	22.7
1987-90		
E		
2WD	19.4	19.9
4WD	20.8	21.3
GA		
2WD	21.1	21.6
4WD	22.9	23.4
1991-94		
GA	22.0	22.5
SR	27.3	27.8
1995-99		
GA	25.9	26.4
SR	32.5	33.0
2000-05		
QG	18.5	19.0
SR	26.1	26.6
Stanza		
1982-86	26.9	27.4
1987-89	21.9	22.4
1990-92	22.4	22.9
Stanza Wagon		
2WD	19.8	20.3
4WD	21.7	22.2

Engine (New or Rebuilt Complete) Replace (B)
Includes: Component transfer and engine tune-up.

	LABOR TIME	SEVERE SERVICE
200SX		
1981-83	9.5	10.0
1984-88		
4 cyl.		
w/turbo	9.4	9.9
w/o turbo	9.0	9.5
V6	10.6	11.1
1995-98		
GA	13.0	13.5
SR	14.5	15.0
210	8.8	9.3
240SX		
SOHC	9.8	10.3
DOHC	14.4	14.9
280ZX	13.0	13.5
300ZX		
1984-89		
w/turbo	11.9	12.4
w/o turbo	11.1	11.6
1990-96		
w/turbo	18.3	18.8
w/o turbo	15.2	15.7
310		
1981	8.5	9.0
1982	8.9	9.4
350Z	14.5	15.0
510	9.8	10.3
810/Maxima		
gasoline	12.5	13.0
diesel	10.2	10.7
Altima		
1993-01	11.4	11.9
2002-05		
4 cyl.	9.9	10.4
V6	13.5	14.0
Maxima		
1985-91	12.0	12.5
1992-01		
SOHC	12.4	12.9
DOHC	16.7	17.2
2002-04	22.2	22.7
2005	14.5	15.0

	LABOR TIME	SEVERE SERVICE
NX		
GA	10.7	11.2
SR	12.4	12.9
Pulsar		
CA	12.3	12.8
E	9.4	9.9
GA	9.8	10.3
Sentra		
1982-86		
gasoline	9.3	9.8
diesel	6.8	7.3
1987-90		
E		
2WD	8.9	9.4
4WD	10.2	10.7
GA		
2WD	9.8	10.3
4WD	11.6	12.1
1991-94		
GA	10.7	11.2
SR	12.4	12.9
1995-99		
GA	13.0	13.5
SR	14.5	15.0
2000-05		
QG, QR	11.2	11.7
SR	15.6	16.1
Stanza		
1982-86	13.6	14.1
1987-89	13.2	13.7
1990-92	10.5	11.0
Stanza Wagon		
2WD	12.4	12.9
4WD	14.5	15.0
w/AC add		
Altima 4 cyl.	.5	.5
Sentra		
GA	.6	.6
QG	.5	.5
91-94 SR	.6	.6
w/PS add		
93-01 Altima	.5	.5
Sentra		
GA	.5	.5
91-94 SR	.5	.5

Engine (Short Block), Replace (B)
Assembly may consist of engine block, piston assemblies, crankshaft, camshaft, timing chain and gears. Does not includes cylinder heads. Operation includes: R&R engine, transfer necessary parts and all necessary adjustments.

	LABOR TIME	SEVERE SERVICE
200SX		
1981-83	14.4	14.9
1984-88		
4 cyl.		
w/turbo	16.8	17.3
w/o turbo	14.2	14.7
V6	18.3	18.8

200SX : 210 : 240SX : 280ZX : 300ZX : 310 : 350Z : 510 : 810

	LABOR TIME	SEVERE SERVICE		LABOR TIME	SEVERE SERVICE		LABOR TIME	SEVERE SERVICE
Engine (Short Block), Replace (B)			2000-05			**CYLINDER HEAD**		
1995-98			QG	13.8	14.3	**Compression Test (B)**		
GA	16.3	16.8	QR	16.3	16.8	Gasoline		
SR	19.8	20.3	SR	18.8	19.3	4 cyl.	.7	1.0
210	14.5	15.0	Stanza			6 cyl.	.9	1.2
240SX			1982-86	18.4	18.9	V6	1.4	1.7
1989-94			1987-89	16.3	16.8	Diesel		
SOHC	16.9	17.4	1990-92	17.1	17.6	810/Maxima	1.6	1.9
DOHC	20.2	20.7	Stanza Wagon			Sentra	1.1	1.4
1995-98	19.0	19.5	2WD	14.5	15.0	w/DOHC add	.3	.3
280ZX	19.9	20.4	4WD	18.7	19.2	**Valve Clearance, Adjust (B)**		
300ZX			w/AC add			Exc. below	1.2	1.5
1984-89			300ZX	.7	.7	Altima V6	4.0	4.3
w/turbo	20.0	20.5	Altima 4 cyl.	.3	.3	2002-05 Maxima	4.0	4.3
w/o turbo	19.5	20.0	Sentra			**Cylinder Head, Replace (B)**		
1990-96			GA	.6	.6	Includes: Parts transfer, adjustments.		
w/turbo	33.5	34.0	QG	.5	.5	200SX		
w/o turbo	30.0	30.5	91-94 SR	.6	.6	1981-83	5.1	5.6
310	15.6	16.1	w/PS add			1984-88		
350Z	18.9	19.4	300ZX	.6	.6	4 cyl.		
510	13.4	13.9	93-01 Altima	.5	.5	w/turbo	10.9	11.4
810/Maxima			Sentra			w/o turbo	10.0	10.5
gasoline	19.9	20.4	GA	.5	.5	V6		
diesel	18.4	18.9	91-94 SR	.5	.5	one	11.8	12.3
Altima			**Engine Mounts, Replace (B)**			both	17.0	17.5
1993-01	15.1	15.6	Front			1995-98		
2002-05			exc. below	1.4	1.6	GA	16.5	17.0
4 cyl.	9.9	10.4	1984-88 200SX	1.9	2.1	SR	18.3	18.8
V6	13.5	14.0	1982 310	1.7	1.9	210	7.6	8.1
Maxima			350Z	4.1	4.3	240SX		
1985-91	21.3	21.8	Altima			SOHC	13.0	13.5
1992-04			1993-97	1.7	1.9	DOHC	16.2	16.7
VE	21.7	22.2	1998-05	2.2	2.4	280ZX	7.1	7.6
VG	16.4	16.9	Maxima			300ZX		
VQ	23.3	23.8	1985-88	1.7	1.9	1984-89		
2005	26.7	27.2	1992-94 VE	1.9	2.1	w/turbo		
NX			1995-99	1.1	1.2	right side	12.2	12.7
GA	18.9	19.4	2000-04	1.6	1.7	left side	16.6	17.1
SR	22.2	22.7	2005	.6	.7	both sides	19.7	20.2
Pulsar			NX SR	.5	.7	w/o turbo		
CA	20.0	20.5	Pulsar	1.4	1.6	right side	12.2	12.7
E	14.2	14.7	Sentra			left side	15.1	15.6
GA	15.7	16.2	1982-90	1.7	1.9	both sides	19.0	19.5
Sentra			1991-94 SR	.5	.7	1990-96		
1982-86			2002-05 QR	1.7	1.9	w/turbo		
gasoline	12.3	12.8	Stanza	1.6	1.8	one side	21.3	21.8
diesel	14.5	15.0	Rear			both sides	29.3	29.8
1987-90			exc. below	.8	1.0	w/o turbo		
E			1995-98 200SX	1.5	1.7	one side	18.7	19.2
2WD	13.9	14.4	1981 310	1.5	1.7	both sides	25.9	26.4
4WD	15.2	15.7	Altima	1.4	1.6	310		
GA			Maxima			1981	7.7	8.2
2WD	15.7	16.2	1985-94	1.6	1.8	1982	8.6	9.1
4WD	17.4	17.9	1995-04	1.9	2.1	350Z		
1991-94			2005	.8	.9	one side	18.5	19.0
GA	18.9	19.4	NX GA	1.2	1.4	both sides	24.5	25.0
SR	22.2	22.7	Sentra			510	5.1	5.6
1995-99			1991-94 GA	1.2	1.4	810/Maxima	7.0	7.5
GA	16.3	16.8	1995-05	1.5	1.7	Altima		
SR	19.8	20.3	Stanza Wagon 4WD	1.2	1.4	1993-97	15.4	15.9

Altima : Maxima : NX : Pulsar : Pulsar NX : Sentra : Stanza NI-13

	LABOR TIME	SEVERE SERVICE
1998-01	13.9	14.4
2002-05		
4 cyl.	15.9	16.4
V6		
one side	14.8	15.3
both sides	19.7	20.2
Maxima		
1985-88		
right side	14.3	14.8
left side	13.9	14.4
both sides	19.0	19.5
1989-94		
SOHC		
right side	10.7	11.5
left side	11.6	12.1
both sides	16.2	16.7
DOHC		
right side	15.0	15.5
left side	14.2	14.7
both sides	23.0	23.5
1995-04		
right side	23.8	24.3
left side	22.1	22.6
both sides	28.6	29.1
2005		
right side	15.4	15.9
left side	15.5	16.0
both sides	20.3	20.8
NX		
GA	16.5	17.0
SR	18.3	18.8
Pulsar		
CA	16.8	17.4
E	9.5	10.0
GA	10.8	11.3
Sentra		
1982-86		
gasoline	5.4	5.9
diesel	7.1	7.6
1987-90		
E	9.4	9.9
GA	10.6	11.1
1991-99		
GA	16.5	17.0
SR	18.3	18.8
2000-05		
QG	12.8	13.3
QR	18.2	18.7
SR	13.2	13.7
Stanza		
1982-86	7.3	7.8
1987-89	11.7	12.2
1990-92	13.5	14.0
Stanza Wagon	11.9	12.4
Cylinder Head Gasket, Replace (B)		
200SX		
1981-83	4.7	5.0

	LABOR TIME	SEVERE SERVICE
1984-88		
4 cyl.		
w/turbo	6.3	6.6
w/o turbo	5.1	5.4
V6		
one side	8.7	9.0
both sides	10.3	10.6
1995-98		
GA	9.2	9.5
SR	11.2	11.5
210	4.0	4.3
240SX		
SOHC	5.9	6.2
DOHC	10.9	11.2
280ZX	6.5	6.8
300ZX		
1984-89		
w/turbo		
right side	8.3	8.6
left side	11.9	12.2
both sides	13.3	13.6
w/o turbo		
right side	9.1	9.4
left side	10.0	10.3
both sides	12.7	13.0
1990-96		
w/turbo		
one side	15.8	16.1
both sides	18.4	18.7
w/o turbo		
one side	13.0	13.3
both sides	15.1	15.4
310		
1981	4.2	4.5
1982	4.9	5.2
350Z		
one side	15.6	15.9
both sides	18.3	18.6
510	4.6	4.9
810/Maxima	6.4	6.7
Altima		
1993-97	9.8	10.1
1998-01	11.7	12.0
2002-05		
4 cyl.	10.4	10.7
V6		
one side	12.1	12.4
both sides	14.5	14.8
Maxima		
1985-88		
right side	10.8	11.1
left side	10.5	10.8
both sides	12.5	12.8
1989-94		
SOHC		
right side	7.9	8.1
left side	8.6	8.9
both sides	9.7	10.0

	LABOR TIME	SEVERE SERVICE
DOHC		
right side	11.7	12.0
left side	11.5	11.8
both sides	16.0	16.3
1995-04		
right side	20.7	21.0
left side	19.3	19.6
both sides	23.0	23.3
2005		
right or left side	12.7	13.2
both sides	15.1	15.6
NX		
GA	9.2	9.5
SR	11.2	11.5
Pulsar		
CA	9.5	9.8
E	4.7	5.0
GA	4.5	4.8
Sentra		
1982-86		
gasoline	5.0	5.3
diesel	6.3	6.6
1987-90	4.6	4.9
1991-99		
GA	9.2	9.5
SR	11.2	11.5
2000-05		
QG	8.1	8.4
QR	12.6	12.9
SR	9.3	9.6
Stanza	6.8	7.1
Stanza Wagon	7.0	7.3
w/AC Altima 4 cyl. add	.2	.2
Rocker Arms or Shafts, Replace (B)		
200SX		
1981-83	1.8	2.1
1984-88		
4 cyl.	1.2	1.5
V6		
right side	1.9	2.2
left side	3.3	3.6
both sides	4.2	4.5
1995-98		
GA	5.2	5.5
SR	4.1	4.4
210, 310	2.0	2.3
240SX		
SOHC	1.1	1.4
DOHC	3.7	4.0
280ZX	3.8	4.1
300ZX		
1984-89		
right side	1.7	2.0
left side	3.4	3.7
both sides	4.4	4.7

NI-14 200SX : 210 : 240SX : 280ZX : 300ZX : 310 : 350Z : 510 : 810

	LABOR TIME	SEVERE SERVICE
Rocker Arms or Shafts, Replace (B)		
1990-96		
w/turbo		
one side	15.5	15.8
both sides	18.2	18.5
w/o turbo		
one side	14.2	14.5
both sides	16.6	16.9
350Z		
one side	14.5	14.8
both sides	16.4	16.7
510	3.2	3.5
810/Maxima		
gasoline	3.7	4.0
diesel	1.6	1.9
Altima		
1993-97	3.0	3.3
1998-01	4.1	4.4
2002-05		
4 cyl.	5.7	6.0
V6		
right side	9.8	10.1
left side	8.6	8.9
both sides	11.8	12.1
Maxima		
1985-88		
one side	1.4	1.7
both sides	2.1	2.4
1989-94		
SOHC		
right side	1.9	2.2
left side	2.9	3.2
both sides	4.4	4.7
DOHC		
right side	5.4	5.7
left side	2.8	3.1
both sides	7.4	7.7
1995-01		
right side	13.5	13.8
left side	11.3	11.6
both sides	15.0	15.3
2002-04		
right side	13.6	13.9
left side	12.0	12.3
both sides	15.1	15.4
2005		
right side	10.5	11.0
left side	8.9	9.4
both sides	12.4	12.9
NX		
GA	5.2	5.5
SR	4.1	4.4
Pulsar		
CA	7.1	7.4
E, GA	1.5	1.8
Sentra		
1982-86	3.3	3.6
1987-90	1.4	1.7

	LABOR TIME	SEVERE SERVICE
1991-99		
GA	5.2	5.5
SR	4.1	4.4
2000-05		
QG	4.1	4.4
QR	5.9	6.2
SR	3.6	3.9
Stanza		
1982-86	3.1	3.4
1987-89	1.4	1.7
1990-92	.7	1.0
Stanza Wagon	1.4	1.7
Valve Cover Gasket, Replace (B)		
Exc. below	.9	1.1
1984-88 200SX V6		
right side	1.4	1.6
left side	2.9	3.1
both sides	3.0	3.2
300ZX		
1984-89		
right side	.7	.9
left side	2.8	3.0
both sides	3.7	3.9
1990-96		
one side	5.5	5.7
both sides	6.3	6.5
Altima V6		
right side	2.1	2.3
left side	1.2	1.4
both sides	2.8	3.0
Maxima		
1989-94		
SOHC		
right side	1.4	1.6
left side	2.4	2.6
both sides	3.4	3.6
DOHC		
right side	3.8	4.0
left side	1.3	1.5
both sides	4.1	4.3
1995-01		
right side	2.8	3.0
left side	.7	.9
both sides	3.1	3.3
2002-04		
right side	4.0	4.2
left side	1.3	1.5
both sides	4.9	5.1
2005		
right side	2.6	2.9
left side	1.1	1.5
both sides	3.2	3.7
Valve Job (A)		
200SX		
1981-83	6.2	6.7
1984-88		
4 cyl.		
w/turbo	12.3	12.8
w/o turbo	11.1	11.6

	LABOR TIME	SEVERE SERVICE
V6		
one side	12.9	13.4
both sides	18.0	18.5
1995-98		
GA	17.5	18.0
SR	19.1	19.6
210	8.7	9.2
240SX		
SOHC	13.5	14.0
DOHC	17.1	17.6
280ZX	8.7	9.2
300ZX		
1984-89		
w/turbo	32.7	33.2
w/o turbo	21.5	22.0
1990-96		
w/turbo	31.4	31.9
w/o turbo	27.5	28.0
310		
1981	8.6	9.1
1982	9.3	9.8
510	6.1	6.6
810/Maxima		
gasoline	8.2	8.7
diesel	12.5	13.0
Altima		
1993-97	15.4	15.9
1998-05	15.0	15.5
Maxima		
1985-89	20.7	21.2
1990-94		
SOHC	19.3	19.8
DOHC	25.1	25.6
1995-99	26.9	27.4
2000-04	31.8	32.3
2005	20.0	21.0
NX		
GA	17.5	18.0
SR	19.1	19.6
Pulsar		
SOHC		
CA	18.6	19.1
E	11.3	11.8
GA	12.7	13.2
DOHC	12.3	12.8
Sentra		
1982-86		
gasoline	6.4	6.9
diesel	9.7	10.2
1987-90		
E	11.2	11.7
GA	12.4	12.9
1991-99		
GA	17.5	18.0
SR	19.1	19.6
2000-05	13.8	14.3

Altima : Maxima : NX : Pulsar : Pulsar NX : Sentra : Stanza   NI-15

	LABOR TIME	SEVERE SERVICE
Stanza		
1982-86	8.3	8.8
1987-89	12.7	13.2
1990-92	14.8	15.3
Stanza Wagon	12.6	13.1
Valve Lifters, Replace (B)		
210	7.8	8.3
1981 310	8.3	8.8
Valve Pushrods, Replace (B)		
210	1.7	2.0
1981 310	2.3	2.6
Valve Springs and/or Oil Seals, Replace (B)		
200SX		
1981-83	3.1	3.6
1984-88		
4 cyl.	3.2	3.7
V6		
right side	3.3	3.8
left side	4.4	4.9
both sides	7.1	7.6
1995-98		
GA	6.9	7.4
SR	7.2	7.7
210	3.3	3.8
240SX		
SOHC	3.6	4.1
DOHC	5.4	5.9
280ZX	3.7	4.2
300ZX		
1984-89		
right side	2.6	3.1
left side	4.2	4.7
both sides	6.3	6.8
1990-96		
one side	13.3	13.8
both sides	17.6	18.1
310	3.5	4.0
350Z		
right side	8.3	8.8
left side	7.6	8.1
both sides	8.8	9.3
510	3.1	3.6
810/Maxima		
gasoline	3.9	4.4
diesel		
one cyl.	1.3	1.8
each addl. cyl.	.3	.5
Altima		
1993-01	5.3	5.8
2002-05		
4 cyl.	7.2	7.7
V6		
right side	10.4	10.9
left side	9.5	10.0
both sides	13.0	13.5
Maxima		
1985-88		
one side	2.7	3.2
both sides	5.3	5.8

	LABOR TIME	SEVERE SERVICE
1989-94		
SOHC		
right side	3.5	4.0
left side	4.2	4.7
both sides	6.8	7.3
DOHC		
right side	8.6	9.1
left side	5.3	5.8
both sides	11.7	12.2
1995-01		
right side	14.7	15.2
left side	12.7	13.2
both sides	18.0	18.5
2002-04		
right side	14.8	15.3
left side	13.2	13.7
both sides	18.0	18.5
2005		
right side	11.0	11.5
left side	9.5	10.0
both sides	13.5	14.0
NX		
GA	6.9	7.4
SR	7.2	7.7
Pulsar		
CA	4.6	5.1
E	3.3	3.8
Sentra		
1982-86		
gasoline	3.1	3.6
diesel	4.8	5.3
1987-90		
E	3.2	3.7
GA	4.4	4.9
1991-99		
GA	6.9	7.4
SR	7.2	7.7
2000-05		
QG	5.6	6.1
QR	7.5	8.0
SR	5.2	5.7
Stanza		
1982-89	3.3	3.8
1990-92	4.1	4.6
CAMSHAFT		
Camshaft, Replace (A)		
4 cyl. w/lifters		
200SX		
1981-83	3.4	3.7
1984-88	5.8	6.1
1995-98		
GA		
one	4.6	4.9
both	5.2	5.5
SR		
one	3.0	3.3
both	3.8	4.1
210	8.4	8.7

	LABOR TIME	SEVERE SERVICE
240SX		
SOHC	1.8	2.1
DOHC		
one	2.5	2.8
both	3.2	3.5
310	3.0	3.3
510	3.2	3.5
Altima		
1993-97		
one	2.1	2.4
both	2.9	3.2
1998-01		
one or both	3.8	4.1
2002-05		
one	4.1	4.4
both	5.6	5.9
NX		
GA		
one	4.6	4.9
both	5.2	5.5
SR		
one	3.0	3.3
both	3.8	4.1
Pulsar		
CA		
one	5.8	6.1
both	6.3	6.6
E	5.1	5.4
GA	6.1	6.4
Sentra		
1982-86		
gasoline	5.1	5.4
diesel	4.5	4.8
1987-89		
E	5.1	5.4
GA	6.1	6.4
1991-99		
GA		
one	4.6	4.9
both	5.2	5.5
SR		
one	3.0	3.3
both	3.8	4.1
2000-05		
QG		
one	2.9	3.2
both	3.8	4.1
QR		
one	4.4	4.7
both	5.7	6.0
SR		
one	2.5	2.8
both	3.1	3.4
Stanza	7.6	7.9
Stanza Wagon	7.5	7.8
V6		
200SX		
one side	9.9	10.2
both sides	12.3	12.6

200SX : 210 : 240SX : 280ZX : 300ZX : 310 : 350Z : 510 : 810

	LABOR TIME	SEVERE SERVICE
Camshaft, Replace (A)		
300ZX		
1984-89		
w/turbo		
right side	10.8	*11.1*
left side	13.4	*13.7*
both sides	16.3	*16.6*
w/o turbo		
right side	10.7	*11.0*
left side	12.7	*13.0*
both sides	15.5	*15.8*
1990-96		
w/turbo		
one side	16.9	*17.2*
both sides	20.4	*20.7*
w/o turbo		
one side	14.3	*14.6*
both sides	17.5	*17.8*
350Z		
one side	14.5	*14.8*
both sides	16.7	*17.0*
Altima		
right side	9.8	*10.1*
left side	8.6	*8.9*
both sides	11.8	*12.1*
Maxima		
1985-88		
right side	12.3	*12.6*
left side	11.7	*12.0*
both sides	14.4	*14.7*
1989-94		
SOHC		
right side	8.7	*9.0*
left side	9.4	*9.7*
both sides	11.9	*12.2*
DOHC		
right side, both	5.6	*5.9*
left side, both	2.8	*3.1*
both sides, all	7.4	*7.7*
1995-01		
right side	13.8	*14.1*
left side	11.6	*11.9*
both sides	15.4	*15.7*
2002-04		
right side	13.8	*14.1*
left side	12.3	*12.5*
both sides	15.6	*15.9*
2005		
right side	10.4	*10.9*
left side	8.6	*9.1*
both sides	12.1	*12.6*
Camshaft Seal, Replace (B)		
300ZX		
1984-89		
one side	4.4	*4.6*
both sides	7.1	*7.3*
1990-96		
one side	6.0	*6.2*
both sides	6.3	*6.5*

	LABOR TIME	SEVERE SERVICE
Maxima		
1985-88		
one side	3.8	*4.0*
both sides	4.2	*4.4*
1989-94 SOHC		
one side	4.3	*4.5*
both sides	5.3	*5.5*
Timing Belt, Replace (B)		
200SX		
4 cyl.	2.6	*3.1*
V6	5.8	*6.3*
300ZX		
1984-89	7.0	*7.5*
1990-96	5.2	*5.7*
1981 310	7.0	*7.5*
Maxima		
1985-88	5.3	*5.8*
1989-94 SOHC	3.8	*4.3*
Pulsar		
CA	4.5	*5.0*
E	2.2	*2.7*
Sentra		
1982-86 each	4.3	*4.8*
1987-90	2.3	*2.8*
Stanza	4.9	*5.4*
Timing Chain, Replace (B)		
200SX		
1981-83	6.1	*6.6*
1995-98		
GA	9.0	*9.5*
SR	10.1	*10.6*
210	3.8	*4.3*
240SX		
SOHC	7.5	*8.0*
DOHC	15.3	*15.8*
280ZX	6.1	*6.6*
1982 310	7.4	*7.9*
350Z	9.8	*10.3*
510	6.3	*6.8*
810/Maxima		
gasoline	6.1	*6.6*
diesel	8.8	*9.3*
Altima		
1993-97	10.5	*11.0*
1998-01	7.5	*8.0*
2002-05		
4 cyl.	7.4	*7.9*
V6	9.6	*10.1*
Maxima		
1992-94 DOHC	20.0	*20.5*
1995-04	12.4	*12.9*
2005	17.9	*18.4*
NX		
GA	9.0	*9.5*
SR	10.1	*10.6*
Pulsar GA	5.4	*5.9*

	LABOR TIME	SEVERE SERVICE
Sentra		
1987-90 GA	5.7	*6.2*
1991-99		
GA	9.0	*9.5*
SR	10.1	*10.6*
2000-05		
QG	6.4	*6.9*
QR	9.8	*10.3*
SR	7.0	*7.5*
1990-92 Stanza	8.3	*8.8*
Timing Chain Tensioner, Replace (B)		
1995-98 200SX		
GA	2.8	*3.1*
SR	.7	*1.0*
240SX	10.5	*10.8*
350Z	1.4	*1.7*
Altima		
1993-97	8.9	*9.2*
1998-01	6.5	*6.8*
2002-05	2.0	*2.3*
Maxima		
Exc. 2005	1.3	*1.6*
2005	1.6	*1.8*
NX		
GA	2.8	*3.1*
SR	.7	*1.0*
Sentra		
1991-99		
GA	2.8	*3.1*
SR	.7	*1.0*
2000-05		
QG	1.4	*1.7*
QR	2.4	*2.7*
SR	.7	*1.0*
Timing Cover, Reseal (B)		
1993-97 Altima	9.0	*9.5*
Timing Cover and/or Gasket, Replace (B)		
200SX		
1981-83	5.7	*6.0*
1984-88		
4 cyl.	2.4	*2.7*
V6	5.3	*5.6*
1995-98	8.2	*8.5*
210	3.3	*3.6*
240SX		
SOHC	7.4	*7.7*
DOHC	14.8	*15.1*
280ZX	5.9	*6.2*
300ZX		
1984-89	6.1	*6.4*
1990-96	4.8	*5.1*
1981 310	6.9	*7.2*
350Z	5.4	*5.7*
510	5.6	*5.9*
810/Maxima		
gasoline	5.6	*5.9*
diesel	8.3	*8.6*

Altima : Maxima : NX : Pulsar : Pulsar NX : Sentra : Stanza NI-17

	LABOR TIME	SEVERE SERVICE
Altima		
1993-97	8.8	9.1
1998-05	6.5	6.8
Maxima		
1985-88	5.2	5.5
1989-94		
SOHC	3.4	3.7
DOHC	19.2	19.5
1995-01	9.0	9.3
2002-04	9.6	9.9
2005	6.5	7.0
NX		
GA	11.5	11.8
SR	10.9	11.2
Pulsar		
CA	4.2	4.5
E	1.6	1.9
GA	5.1	5.4
Sentra		
1982-86	4.0	4.3
1987-90		
E	1.9	2.2
GA	5.1	5.4
1991-94		
GA	11.5	11.8
SR	10.9	11.2
1995-99	8.2	8.5
2000-05		
QG, SR	5.8	6.1
QR	8.9	9.2
Stanza		
1982-89	4.3	4.6
1990-92	8.3	8.6

CRANK & PISTONS
Connecting Rod Bearings, Replace (A)
Includes: Check bearing oil clearance.

	LABOR TIME	SEVERE SERVICE
200SX		
1981-83	4.5	5.0
1984-88		
4 cyl.		
w/turbo	14.7	15.2
w/o turbo	13.8	14.3
V6	18.1	18.6
1995-98		
GA	17.0	17.5
SR	20.1	20.6
210	3.5	4.0
240SX		
SOHC	15.3	15.8
DOHC	17.8	18.3
280ZX	4.3	4.8
300ZX		
1984-89		
w/turbo	19.3	19.8
w/o turbo	18.5	19.0
1990-96		
w/turbo	28.6	29.1
w/o turbo	25.1	25.6

	LABOR TIME	SEVERE SERVICE
310		
1981	8.6	9.1
1982	5.0	5.5
350Z	19.5	20.0
510	4.2	4.7
810/Maxima	4.3	4.8
Altima		
1993-97	17.6	18.1
1998-01	19.9	20.4
2002-05		
4 cyl.	16.2	16.7
V6	22.1	22.6
Maxima		
1985-88	19.4	19.9
1989-94		
SOHC	17.7	18.2
DOHC	20.7	21.2
1995-04	24.2	24.7
2005	21.9	22.6
NX		
GA	17.0	17.5
SR	20.1	20.6
Pulsar		
CA	17.2	17.7
E	14.3	14.8
GA	14.7	15.2
Sentra		
1982-86	3.4	3.9
1987-90		
E		
2WD	13.7	14.2
4WD	15.2	15.7
GA		
2WD	14.7	15.2
4WD	16.1	16.6
1991-99		
GA	17.0	17.5
SR	20.1	20.6
2000-05		
QG	14.7	15.2
QR	16.9	17.4
SR	20.0	20.5
Stanza		
1982-86	3.5	4.0
1987-89	16.3	16.8
1990-92	15.2	15.7
Stanza Wagon		
2WD	14.3	14.8
4WD	16.3	16.8

Crankshaft and Main Bearings, Replace (A)
Includes: Engine R&R. Check bearing oil clearance.

	LABOR TIME	SEVERE SERVICE
200SX		
1981-83	12.4	13.3
1984-88		
4 cyl.		
w/turbo	14.9	15.8
w/o turbo	14.1	15.0
V6	18.5	19.4

	LABOR TIME	SEVERE SERVICE
1995-98		
GA	17.9	18.8
SR	21.1	22.0
210	9.1	10.0
240SX		
SOHC	15.5	16.4
DOHC	20.1	21.0
280ZX	15.4	16.3
300ZX		
1984-89		
w/o turbo	19.7	20.6
w/turbo	19.1	20.0
1990-96		
w/o turbo	25.5	26.4
w/turbo	29.2	30.1
310		
1981	9.3	10.2
1982	9.6	10.5
350Z	21.5	22.4
510	9.2	10.1
810/Maxima		
gasoline	11.6	12.5
diesel	11.5	12.4
Altima		
1993-97	18.7	19.6
1998-01	20.7	21.6
2002-05		
4 cyl.	18.3	19.2
V6	22.5	23.4
Maxima		
1985-88	19.8	20.7
1989-94		
SOHC	18.2	19.1
DOHC	19.7	20.6
1995-04	26.0	26.9
2005	23.6	24.2
NX		
GA	17.9	18.8
SR	21.1	22.0
Pulsar		
CA	17.6	18.5
E	14.3	15.2
GA	14.9	15.8
Sentra		
1982-86	8.2	9.1
1987-90		
E		
2WD	14.0	14.9
4WD	15.7	16.6
GA		
2WD	14.9	15.8
4WD	16.4	17.3
1991-99		
GA	17.9	18.8
SR	21.1	22.0
2000-05		
QG	15.7	16.6
QR	18.9	19.8
SR	21.8	22.7

NI-18 200SX : 210 : 240SX : 280ZX : 300ZX : 310 : 350Z : 510 : 810

	LABOR TIME	SEVERE SERVICE
Crankshaft and Main Bearings, Replace (A)		
Stanza		
1982-86	12.4	13.3
1987-89	16.4	17.3
1990-92	15.5	16.4
Stanza Wagon		
2WD	14.4	15.3
4WD	16.3	17.2
Crankshaft Front Oil Seal, Replace (B)		
200SX		
1981-83	1.7	1.9
1984-88		
4 cyl.	4.5	4.7
V6	7.9	8.1
1995-98	1.2	1.4
210	1.5	1.7
240SX	1.7	1.9
280ZX	2.4	2.6
300ZX		
1984-89	10.8	11.0
1990-96	5.4	5.6
310	1.6	1.8
350Z	1.2	1.4
510	1.9	2.1
810/Maxima		
gasoline	2.2	2.4
diesel	2.6	2.8
Altima		
1993-97	1.2	1.4
1998-01	.8	1.0
2002-05		
4 cyl.	.6	.8
V6	1.4	1.6
Maxima		
1985-88	7.5	7.7
1989-94		
SOHC	6.1	6.3
DOHC	1.3	1.5
1999-05	1.4	1.6
NX	1.2	1.4
Pulsar		
CA	4.9	5.1
E	2.2	2.4
GA	1.1	1.3
Sentra		
1982-86		
gasoline	1.3	1.5
diesel	6.3	6.5
1987-89		
E	2.2	2.4
GA	1.2	1.4
1990-99	1.2	1.4
2000-05		
QG, SR	1.6	1.8
QR	.9	1.1

	LABOR TIME	SEVERE SERVICE
Stanza		
1982-86	1.2	1.4
1987-89	5.3	5.5
1990-92	1.3	1.5
Stanza Wagon	5.3	5.5
Pistons and/or Connecting Rods, Replace (A)		
Includes: Ridge reaming, cylinder wall deglazing, installing new rings and rod bearings, engine tune-up.		
200SX		
1981-83	11.5	12.4
1984-88		
4 cyl.		
w/turbo	17.1	18.0
w/o turbo	16.3	17.2
1995-98		
GA	18.9	19.8
SR	23.3	24.2
210	9.3	10.2
240SX		
SOHC	18.4	19.3
DOHC	19.8	20.7
280ZX	13.6	14.5
300ZX		
1984-89		
w/turbo	22.6	23.5
w/o turbo	21.8	22.7
1990-96		
w/turbo	32.0	32.9
w/o turbo	28.9	29.8
310	10.1	11.0
350Z	22.0	22.9
510	9.4	10.3
810/Maxima	13.5	14.4
Altima		
1993-97	19.6	20.5
1998-01	21.3	22.2
2002-05		
4 cyl.	17.9	18.8
V6	23.0	23.9
Maxima		
1985-88	22.8	23.7
1989-94		
SOHC	21.2	22.1
DOHC	23.7	24.6
1995-04	26.5	27.4
2005	24.1	25.0
NX		
GA	18.9	19.8
SR	23.3	24.2
Pulsar		
CA	19.7	20.6
E	16.5	17.4
GA	17.0	17.9
Sentra		
1982-86		
gasoline	8.2	9.1
diesel	10.3	11.2

	LABOR TIME	SEVERE SERVICE
1987-90		
E		
2WD	16.3	17.2
4WD	17.6	18.5
GA		
2WD	16.9	17.8
4WD	18.6	19.5
1991-99		
GA	18.9	19.8
SR	23.3	24.2
2000-05		
QG	16.0	16.9
QR	18.5	19.4
SR	21.5	22.4
Stanza		
1982-86	8.6	9.5
1987-89	18.7	19.6
1990-92	17.3	18.2
Stanza Wagon		
2WD	16.5	17.4
4WD	18.5	19.4
Rear Main Oil Seal, Replace (B)		
Includes: Trans. R&R when necessary.		
200SX		
1981-83	4.3	4.9
1984-88		
4 cyl.		
w/turbo		
AT	4.7	5.3
MT	4.0	4.6
w/o turbo	4.3	4.9
V6		
AT	4.6	5.2
MT	4.1	4.7
1995-98		
GA		
AT	5.9	6.5
MT	5.5	6.1
SR	6.2	6.8
210	4.9	5.5
240SX		
AT	4.8	5.4
MT	4.5	5.1
280ZX	4.8	5.4
300ZX		
1984-89	6.1	6.7
1990-96		
AT	4.8	5.4
MT	4.0	4.6
310	9.3	9.9
350Z	5.3	5.9
510	9.9	10.5
810/Maxima	5.3	5.9
Altima		
1993-97		
AT	6.6	7.2
MT	4.7	5.3
1998-01		
AT	6.5	7.1
MT	5.2	5.8

Altima : Maxima : NX : Pulsar : Pulsar NX : Sentra : Stanza — NI-19

	LABOR TIME	SEVERE SERVICE
2002-05		
4 cyl.	4.4	5.0
V6	8.3	8.9
Maxima		
1985-94		
AT	5.4	6.0
MT		
SOHC	4.7	5.3
DOHC	5.0	5.6
1995-01		
AT	11.2	11.8
MT	10.7	11.3
2002-04	11.8	12.4
2005		
AT	8.6	9.1
MT	8.3	8.8
NX		
GA		
AT	5.9	6.5
MT	5.5	6.1
SR	6.2	6.8
Pulsar		
CA, GA	5.2	5.8
E		
AT	5.2	5.8
MT	4.5	5.1
Sentra		
1982-86		
AT	5.4	6.0
MT	4.9	5.4
1987-90		
E		
AT	5.3	5.9
MT		
2WD	4.5	5.1
4WD	7.6	8.2
GA		
AT		
2WD	5.1	5.7
4WD	8.1	8.7
MT		
2WD	5.1	5.7
4WD	8.6	9.2
1991-99		
GA		
AT	5.9	6.5
MT	5.5	6.1
SR	6.2	6.8
2000-05		
QG	5.1	5.7
QR, SR	5.0	5.6
Stanza		
1982-86	4.5	5.1
1987-89		
AT	5.4	6.0
MT	4.5	5.1
1990-92		
AT	5.7	6.3
MT	4.6	5.2

	LABOR TIME	SEVERE SERVICE
Stanza Wagon		
AT		
2WD	5.2	5.8
4WD	7.0	7.6
MT		
2WD	4.5	5.1
4WD	6.2	6.8

Rings, Replace (A)
Includes: Ridge reaming, cylinder wall deglazing, installing new rings, engine tune-up.

	LABOR TIME	SEVERE SERVICE
200SX		
1981-83	12.9	13.8
1984-88		
4 cyl.		
w/turbo	15.8	16.7
w/o turbo	15.3	16.2
V6	19.6	20.5
1991-99		
GA	17.6	18.5
SR	21.8	22.7
210	10.0	10.9
240SX		
SOHC	16.4	17.3
DOHC	18.7	19.6
280ZX	15.3	16.2
300ZX		
1984-89		
w/turbo	20.7	21.6
w/o turbo	19.9	20.8
1990-96		
w/turbo	30.0	30.9
w/o turbo	26.8	27.7
310		
1981	10.5	11.4
1982	11.4	12.3
350Z	22.0	22.9
510	10.3	11.2
810/Maxima		
gasoline	15.2	16.1
diesel	12.7	13.3
Altima		
1993-97	18.5	19.4
1998-01	19.9	20.8
2002-05		
4 cyl.	17.9	18.8
V6	23.0	23.9
Maxima		
1985-88	24.4	25.3
1989-94		
SOHC	19.4	20.3
DOHC	21.9	22.8
1995-04	26.5	27.4
2005	24.1	25.0
NX		
GA	17.6	18.5
SR	21.8	22.7

	LABOR TIME	SEVERE SERVICE
Pulsar		
CA	18.5	19.4
E	15.1	16.0
GA	15.9	16.5
Sentra		
1982-86		
gasoline	9.6	10.5
diesel	9.0	9.9
1987-90		
E		
2WD	14.9	15.8
4WD	16.3	17.2
GA		
2WD	15.7	16.6
4WD	17.4	18.3
1991-99		
GA	17.6	18.5
SR	21.8	22.7
2000-05		
QG	16.0	16.9
QR	18.5	19.4
SR	21.5	22.4
Stanza		
1982-86	9.6	10.5
1987-89	17.3	18.2
1990-92	16.2	17.1
Stanza Wagon		
2WD	15.4	16.3
4WD	17.3	18.2

ENGINE LUBRICATION

Engine Oil Cooler, Replace (B)

	LABOR TIME	SEVERE SERVICE
300ZX	.7	.9
810/Maxima	.7	.9
Altima VQ	1.4	1.6
1995-05 Maxima	1.4	1.6

Engine Oil Pressure Switch (Sending Unit), Replace (B)

	LABOR TIME	SEVERE SERVICE
Exc. below	.5	.5
350Z	.6	.6
1982-94 Sentra, NX SR	1.2	1.2
Stanza Wagon, 4WD	.7	.7

Oil Pan and/or Gasket, Replace (B)

	LABOR TIME	SEVERE SERVICE
200SX		
1981-83	3.0	3.2
1984-88		
4 cyl.	4.0	4.2
V6	3.1	3.3
1995-98		
GA	1.9	2.1
SR		
upper	4.1	4.3
lower	1.7	1.9
210	2.3	2.5
240SX		
SOHC	2.6	2.8
DOHC	3.1	3.3
280ZX	2.6	2.8

200SX : 210 : 240SX : 280ZX : 300ZX : 310 : 350Z : 510 : 810

	LABOR TIME	SEVERE SERVICE
Oil Pan and/or Gasket, Replace (B)		
300ZX		
1984-87	10.5	10.7
1988-89	4.1	4.3
1990-96		
w/turbo	3.0	3.2
w/o turbo	4.6	4.8
310		
1981	7.2	7.4
1982	3.8	4.0
350Z		
510	3.0	3.2
810/Maxima	2.8	3.0
upper	4.6	4.8
lower	1.1	1.3
Altima		
1993-01		
upper	4.2	4.4
lower	1.8	2.0
2002-05		
upper		
4 cyl.	3.2	3.4
V6	6.1	6.3
lower	1.0	1.2
Maxima		
1985-88	3.2	3.4
1989-94		
SOHC	2.2	2.4
DOHC	3.1	3.3
1995-01		
upper	5.0	5.2
lower	.8	1.0
2002-05		
upper	6.1	6.3
lower	1.3	1.5
NX		
GA	1.9	2.1
SR		
upper	4.1	4.3
lower	1.7	1.9
Pulsar		
CA, GA	1.9	2.1
E	2.4	2.6
Sentra		
1982-86	2.4	2.6
1987-90		
E	2.2	2.4
GA	2.0	2.2
1991-99		
GA	1.9	2.1
SR		
upper	4.1	4.3
lower	1.7	1.9
2000-05		
QG	1.7	1.9
QR		
upper	4.1	4.3
lower	1.0	1.2

	LABOR TIME	SEVERE SERVICE
SR		
upper	3.0	3.2
lower	1.3	1.5
Stanza		
1982-86	2.4	2.6
1987-89	2.9	3.1
1990-92	2.4	2.6
Stanza Wagon	2.8	3.0
Oil Pump, R&R and Recondition (B)		
200SX		
1981-83	1.6	2.1
1984-88		
4 cyl.	4.9	5.1
V6	8.2	8.4
1995-98	11.8	12.0
210	1.2	1.7
240SX		
SOHC	.8	1.0
DOHC	15.6	15.8
280ZX	1.5	1.7
300ZX		
1984-89	11.4	11.6
1990-96	8.4	8.6
310		
1981	1.8	2.3
1982	1.3	1.8
350Z	7.8	8.0
510	1.5	2.0
810/Maxima	1.6	2.1
Altima		
1993-97	9.8	10.0
1998-01	13.9	14.1
2002-05 V6	11.9	12.1
Maxima		
1985-88	7.7	7.9
1989-01		
SOHC	6.3	6.5
DOHC	11.9	12.1
2002-04	13.1	13.3
2005	11.9	12.4
NX	11.8	12.0
Pulsar		
CA	3.0	3.2
E	.7	.9
GA	6.2	6.4
Sentra		
1982-86		
gasoline	1.1	1.3
diesel	6.4	6.6
1987-90		
E	.7	.9
GA	3.0	3.2
SR	11.7	11.9
1991-99	11.8	12.0
2000-05		
QG	6.1	6.3
SR	4.7	4.9
Stanza	8.4	8.6

	LABOR TIME	SEVERE SERVICE
Stanza Wagon		
2WD	1.3	1.5
4WD	6.2	6.4
Oil Pump, Replace (B)		
200SX		
1981-83	1.4	1.8
1984-88		
4 cyl.	4.4	4.8
V6	7.9	8.3
1995-98	11.6	12.0
210	.7	1.1
240SX		
SOHC	.8	1.2
DOHC	15.2	15.6
280ZX	1.4	1.8
300ZX		
1984-89	11.1	11.5
1990-96	8.1	8.5
310		
1981	1.6	2.0
1982	.7	1.1
350Z	7.6	8.0
510	1.4	1.8
810/Maxima	1.3	1.7
Altima		
1993-97	9.3	9.7
1998-01	13.5	13.9
2002-05		
4 cyl.	6.9	7.3
V6	11.5	11.9
Maxima		
1985-88	7.5	7.9
1989-01		
SOHC	6.1	6.5
DOHC	11.4	11.8
2002-04	12.6	13.0
2005	11.5	12.0
NX	11.6	12.0
Pulsar		
CA	2.8	3.2
E	.5	.9
GA	6.2	6.4
Sentra		
1982-86		
gasoline	1.2	1.6
diesel	6.1	6.5
1987-90		
E	.5	.9
GA	5.3	5.7
1991-99	11.6	12.0
2000-05		
QG	5.8	6.2
QR	9.4	9.8
SR	4.4	4.8
Stanza	7.5	7.9
Stanza Wagon	6.4	6.8

Altima : Maxima : NX : Pulsar : Pulsar NX : Sentra : Stanza **NI-21**

CLUTCH

	LABOR TIME	SEVERE SERVICE
Bleed Clutch Hydraulic System (B)		
Exc. Altima	.5	.7
Altima	.8	1.0
Clutch Pedal Free Play, Adjust (B)		
All Models	.5	.6
Clutch Assy., Replace (B)		
200SX		
1981-83	4.8	5.0
1984-88		
4 cyl.	3.5	3.7
V6	3.9	4.1
1995-98	4.9	5.1
210, 510	4.5	4.7
240SX, 280ZX	4.1	4.3
300ZX		
1984-89	5.4	5.6
1990-96	4.7	4.9
310		
1981	3.3	3.5
1982	5.2	5.4
350Z	4.9	5.1
810/Maxima	4.4	4.6
Altima		
4 cyl.	4.3	4.5
V6	4.9	5.1
Maxima		
1985-94		
SOHC	4.1	4.3
DOHC	4.6	4.8
1995-04	6.6	6.8
2005	4.8	5.3
NX	5.4	5.6
Pulsar		
CA, GA	4.7	4.9
E	4.1	4.3
Sentra		
1982-86	4.3	4.5
1987-90		
E		
2WD	4.2	4.4
4WD	7.3	7.5
GA		
2WD	4.7	4.9
4WD	7.9	8.1
1991-99	5.4	5.6
2000-05	5.0	5.2
Stanza		
1982-86	4.7	4.9
1987-92	4.1	4.3
Stanza Wagon		
2WD	4.1	4.3
4WD	6.0	6.2
Clutch Control Cable, Replace (B)		
Exc. below	.8	1.0
1991-94 NX, Sentra		
GA	1.2	1.4
SR	1.7	1.9

	LABOR TIME	SEVERE SERVICE
Clutch Interlock Switch, Replace (B)		
All Models	.5	.5
Clutch Master Cylinder, Replace (B)		
Includes: System bleeding.		
Exc. below	1.3	1.5
350Z	1.6	1.8
2002-04 Altima	1.4	1.6
Clutch Release Bearing, Replace (B)		
200SX		
1981-83	4.4	4.6
1984-88	3.6	3.8
1995-98	4.5	4.7
210	4.1	4.3
240SX, 280ZX	3.9	4.1
300ZX		
1984-89	5.3	5.5
1990-96	4.2	4.4
310		
1981	2.9	3.1
1982	4.8	5.0
350Z	5.2	5.4
510	4.1	4.3
810/Maxima	4.3	4.5
Altima		
4 cyl.	4.6	4.8
V6	5.2	5.4
Maxima		
1985-94	4.1	4.3
1995-04	6.7	6.9
2005	5.0	5.5
NX		
GA	4.5	4.7
SR	5.3	5.5
Pulsar		
CA, GA	4.2	4.4
E	3.8	4.0
Sentra		
1982-86	3.8	4.0
1987-90		
E		
2WD	3.6	3.8
4WD	7.0	7.2
GA		
2WD	4.4	4.6
4WD	7.6	7.8
1991-94		
GA	4.5	4.7
SR	5.3	5.5
1995-99	5.8	6.0
2000-05	5.3	5.5
Stanza		
1982-86	4.3	4.5
1989-92	3.9	4.1
Stanza Wagon		
2WD	3.9	4.1
4WD	5.5	5.7

	LABOR TIME	SEVERE SERVICE
Clutch Slave Cylinder, Replace (B)		
Includes: System bleeding.		
Exc. below	.8	1.1
1995-04 Maxima	1.4	1.7
Flywheel, Replace (B)		
200SX		
1981-83	5.2	5.4
1984-88	4.2	4.4
1995-98	6.0	6.3
210	4.9	5.1
240SX, 280ZX	4.5	4.7
300ZX		
1984-89	5.9	6.1
1990-96	4.9	5.1
310		
1981	3.3	3.5
1982	5.2	5.4
350Z	5.2	5.4
510	4.7	4.9
810/Maxima	4.8	5.0
Altima		
1993-97	4.1	4.3
1998-01	5.2	5.4
2002-05	4.4	4.6
Maxima		
1985-94		
SOHC	4.6	4.8
DOHC	5.0	5.2
1995-01	6.6	6.8
2002-04	7.2	7.4
2005	4.4	4.9
NX		
GA	5.3	5.5
SR	5.8	6.0
Pulsar		
CA, GA	4.9	5.1
E	4.4	4.6
Sentra		
1982-86	4.4	4.6
1987-90		
E		
2WD	4.3	4.5
4WD	7.7	7.9
GA		
2WD	5.3	5.5
4WD	8.3	8.5
1991-94		
GA	5.3	5.5
SR	5.8	6.0
1995-99	6.0	6.2
2000-05	5.2	5.4
Stanza		
1982-86	5.3	5.5
1987-92	4.5	4.7
Stanza Wagon		
2WD	4.3	4.5
4WD	6.3	6.5

MANUAL TRANSAXLE

Differential Case, Recondition (B)

	LABOR TIME	SEVERE SERVICE
1995-98 200SX		
GA	6.8	7.3
SR	7.3	7.8
310	9.3	9.8
350Z	9.3	9.8
Altima		
1993-97	7.5	8.0
1998-01	7.0	7.5
2002-05		
4 cyl.	7.3	7.8
V6	8.1	8.6
Maxima		
1989-94		
SOHC	6.8	7.3
DOHC	8.8	9.3
1995-01	9.0	9.5
2002-04	9.7	10.2
2005	7.7	8.0
NX		
GA	7.8	8.3
SR	8.2	8.7
Pulsar		
CA, GA	6.9	7.4
E	6.4	6.9
Sentra		
1982-90		
E		
2WD	6.3	6.8
4WD	9.8	10.3
GA		
2WD	6.9	7.4
4WD	10.1	10.6
1991-94		
GA	7.8	8.3
SR	8.2	8.7
1995-99	8.4	8.9
2000-05	7.5	8.0
Stanza		
1982-86	6.1	6.6
1987-92	6.8	7.3
Stanza Wagon		
2WD	6.4	6.9
4WD	8.2	8.7

Differential Case, Replace (B)

	LABOR TIME	SEVERE SERVICE
1995-98 200SX	7.2	7.7
310		
1981	7.6	8.1
1982	8.9	9.4
Altima		
1993-01	7.0	7.5
2002-05		
4 cyl.	6.9	7.4
V6	7.6	8.1
Maxima		
1989-94		
SOHC	6.2	6.7
DOHC	7.9	8.4
1995-01	8.7	9.2
2002-04	9.3	9.8
2005	7.4	7.9
NX		
GA	6.3	6.8
SR	7.8	8.3
Pulsar		
CA, GA	6.7	7.2
E	6.3	6.8
Sentra		
1982-90		
2WD		
GA	6.6	7.1
E	6.0	6.5
4WD	9.7	10.2
1991-94		
GA	6.3	6.8
SR	7.8	8.3
1995-99	7.9	8.4
2000-05	7.2	7.7
Stanza		
1982-86	5.5	6.0
1987-92	6.1	6.6
Stanza Wagon		
2WD	6.2	6.7
4WD	7.8	8.3

Differential Oil Seal, Replace (B)

	LABOR TIME	SEVERE SERVICE
1995-98 200SX		
one	1.3	1.5
both	2.4	2.6
Altima		
one	1.3	1.5
both		
1993-97	2.3	2.5
1998-01	1.9	2.1
2002-05	1.4	1.6
Maxima		
1985-94		
one	1.1	1.3
both	2.0	2.2
1995-04		
one	1.6	1.8
both	2.7	2.9
2005		
one	1.0	1.2
both	1.5	1.7
NX		
one	1.3	1.5
both	2.4	2.6
Pulsar		
one	1.4	1.6
both	2.1	2.3
Sentra		
one	1.3	1.5
both		
1982-90	1.9	2.1
1991-99	2.4	2.6
2000-05	2.1	2.3

	LABOR TIME	SEVERE SERVICE
Stanza		
one	1.1	1.3
both	1.9	2.1

Speedometer Pinion, Replace (B)

	LABOR TIME	SEVERE SERVICE
Exc. below	.5	.7
1995-05 Maxima	.7	.9

Transaxle Assy. R&R and Recondition (A)

	LABOR TIME	SEVERE SERVICE
1995-98 200SX	11.6	12.2
310		
4-Speed	11.6	12.2
5-Speed	12.3	12.9
Altima		
1993-97	10.0	10.6
1998-01	11.0	11.6
2002-05		
4 cyl.	10.8	11.4
V6	11.5	12.1
Maxima		
1989-01		
SOHC	9.5	10.1
DOHC	12.3	12.9
2002-04	13.1	13.7
2005	11.0	11.6
NX		
GA	10.5	11.1
SR	10.9	11.5
Pulsar		
CA, GA	9.8	10.4
E	9.1	9.7
Sentra		
1982-90		
E		
2WD	8.6	9.2
4WD	12.3	12.9
GA		
2WD	9.3	9.9
4WD	12.9	13.5
1991-94		
GA	10.5	11.1
SR	10.9	11.5
1995-99	11.6	12.2
2000-05	12.5	13.1
Stanza	9.4	10.0
Stanza Wagon		
2WD	9.3	9.9
4WD	12.3	12.9

Transaxle Assy., R&R or Replace (B)

	LABOR TIME	SEVERE SERVICE
1995-98 200SX	5.3	5.5
310	4.8	5.0
Altima		
1993-01	4.7	4.9
2002-05		
4 cyl.	4.1	4.3
V6	4.9	5.1
Maxima		
1989-94		
SOHC	4.0	4.2
DOHC	5.6	5.8

Altima : Maxima : NX : Pulsar : Pulsar NX : Sentra : Stanza

	LABOR TIME	SEVERE SERVICE
1995-01	6.2	6.4
2002-04	6.6	6.8
2005	4.6	5.1
NX		
GA	4.6	4.8
SR	5.1	5.3
Pulsar		
CA, GA	4.3	4.5
E	3.6	3.8
Sentra		
1982-90		
E		
2WD	3.7	3.9
4WD	7.0	7.2
GA		
2WD	4.3	4.5
4WD	7.8	8.0
1991-94		
GA	4.6	4.8
SR	5.1	5.3
1995-99	5.3	5.5
2000-05	4.9	5.1
Stanza	3.8	4.0
Stanza Wagon		
2WD	3.6	3.8
4WD	5.4	5.6

MANUAL TRANSMISSION

Extension Housing and/or Gasket, Replace (B)

	LABOR TIME	SEVERE SERVICE
4-Speed		
210	4.3	4.5
510	2.6	2.8
5-Speed		
200SX		
1981-83	4.2	4.4
1984-88		
4 cyl.	3.2	3.4
V6	3.7	3.9
210	4.4	4.6
240SX	4.7	4.9
280ZX	4.3	4.5
300ZX		
1984-89	5.3	5.5
1990-96	4.3	4.5
810/Maxima	4.0	4.2

Extension Housing Oil Seal, Replace (B)

	LABOR TIME	SEVERE SERVICE
200SX	.8	1.0
210	.8	1.0
240SX	.8	1.0
280ZX	1.1	1.3
300ZX	1.9	2.1
350Z	1.3	1.5
810/Maxima	.8	1.0

Transmission Assy., R&I (B)

	LABOR TIME	SEVERE SERVICE
350Z	4.6	4.8

Transmission Assy., R&R and Recondition (A)

	LABOR TIME	SEVERE SERVICE
4-Speed		
210	7.2	7.8
510	7.8	8.4
5-Speed		
200SX		
1981-83	8.8	9.4
1984-88		
4 cyl.	6.9	7.5
V6	7.2	7.8
210	8.3	8.9
240SX	7.7	8.3
280ZX	9.4	10.0
300ZX	10.0	10.6
810/Maxima	7.7	8.3

Transmission Assy., R&R or Replace (B)

	LABOR TIME	SEVERE SERVICE
4-Speed		
210	3.6	3.8
510	4.2	4.4
5-Speed		
200SX		
1981-83	4.4	4.6
1984-88		
4 cyl.	3.4	3.6
V6	3.5	3.7
210	4.0	4.2
240SX	3.7	3.9
280ZX	3.8	4.0
300ZX		
1984-89	5.2	5.4
1990-96	5.2	5.4
810/Maxima	4.2	4.4

AUTOMATIC TRANSAXLE

SERVICE TRANSAXLE INSTALLED

Check Unit For Oil Leaks (C)

	LABOR TIME	SEVERE SERVICE
All Models	.9	.9

Drain & Refill Unit (B)

	LABOR TIME	SEVERE SERVICE
Exc. below	1.4	1.4
1982 310	1.8	1.8
Pulsar, Stanza	.8	.8

Oil Pressure Check (B)

	LABOR TIME	SEVERE SERVICE
All Models	.7	.7

Accumulator Assy., R&R or Replace (B)

	LABOR TIME	SEVERE SERVICE
Exc. below	2.5	2.7
2000-04 Sentra	1.9	2.1

Differential Oil Seal (Halfshaft), Replace (B)

	LABOR TIME	SEVERE SERVICE
1995-98 200SX		
one	1.5	1.8
both	2.8	3.1
Altima		
1993-97		
one	1.4	1.7
both	2.8	3.1
1998-05		
one	1.2	1.5
both	1.7	2.0
Maxima		
1985-88		
one	3.3	3.6
both	5.1	5.4
1989-99		
SOHC		
one	1.8	2.1
DOHC		
one	1.4	1.7
both	2.3	2.6
2000-04		
one	1.6	1.9
both	2.3	2.6
2005		
one	1.0	1.2
both	1.5	1.7
NX		
one	1.5	1.8
both	2.8	3.1
Sentra		
1982-90		
one	2.7	3.0
both	5.1	5.4
1991-99		
one	1.5	1.8
both	2.8	3.1
2000-05		
one	1.4	1.7
both	2.2	2.5
Stanza		
1982-89		
one	1.8	2.1
both	3.1	3.4
1990-92		
one	1.7	2.0
both	3.2	3.5
Stanza Wagon		
one	2.4	2.7
both	5.1	5.4

Governor Assy., R&R or Replace (B)

	LABOR TIME	SEVERE SERVICE
1995-98 200SX, GA	.8	1.3
1982 310	.7	1.2
Maxima		
1985-88	2.7	3.2
1989-94	1.3	1.8
Pulsar	.7	1.2
NX	.7	1.2
Sentra	.8	1.3
Stanza		
1982-86	.7	1.2
1987-92	1.4	1.9
Stanza Wagon	1.6	2.1

Inhibitor Switch, Replace (B)

	LABOR TIME	SEVERE SERVICE
1988-04	.8	.8

	LABOR TIME	SEVERE SERVICE
Lock-up Solenoid, Replace (B)		
1995-98 200SX	2.5	2.7
Altima		
1993-97	1.8	2.0
1998-04	2.7	2.9
1989-05 Maxima	2.2	2.4
NX	2.5	2.7
Sentra		
1991-99	2.5	2.7
2000-05		
QG, SR	2.1	2.3
QR	3.0	3.2
1990-92 Stanza	1.8	2.0
Oil Pan and/or Gasket, Replace (B)		
1995-98 200SX	1.3	1.5
1982 310	1.8	2.0
Altima	1.6	1.8
Maxima		
Exc. 2005	1.2	1.4
2005	1.5	1.8
NX, Sentra	1.3	1.5
Pulsar, Stanza	.8	1.0
Overdrive Cancel Solenoid, Replace (B)		
Exc. below	.7	.7
1995-98 200SX	2.5	2.7
1998-04 Altima	2.7	2.9
1995-05 Maxima	2.2	2.4
Sentra		
1995-99	2.5	2.7
2000-05		
QG, SR	2.1	2.3
QR	3.0	3.2
Parking Pawl, Replace (B)		
1995-98 200SX	2.1	2.4
Altima		
1993-97	7.2	7.5
1998-01	4.3	4.6
2002-05	1.6	1.9
Maxima		
Exc. 2005		
SOHC	5.4	5.7
DOHC	8.1	8.4
2005	1.5	1.8
NX	1.6	1.9
Pulsar	2.2	2.5
Sentra		
1982-90	2.1	2.4
1991-94	1.6	1.9
1995-99	2.1	2.4
2000-05	2.7	3.0
Stanza	5.5	5.8
Stanza Wagon	5.1	5.4
Speed Sensor, Replace (B)		
Exc. below	.7	.7
1998-01 Altima	.9	.9
Speedometer Drive Pinion and/or Sensor, Replace (B)		
All Models	.6	.7

	LABOR TIME	SEVERE SERVICE
Transaxle Control Unit, Replace (B)		
Exc. below	.7	.7
2000-04 Sentra	.3	.3
Valve Body, R&R and Recondition (A)		
1995-98 200SX	4.9	5.4
1982 310	4.2	4.7
Altima	4.4	4.9
Maxima		
1985-87	3.6	4.1
1988-05	4.6	5.1
NX	4.9	5.4
Pulsar	3.6	4.1
Sentra		
1982-90	3.7	4.2
1991-99	4.9	5.4
2000-05	4.3	4.8
Stanza	3.8	4.3
Stanza Wagon	3.6	4.1
Valve Body, Replace (B)		
1995-98 200SX	2.5	2.8
1982 310	2.5	2.8
Altima		
1993-97	1.9	2.2
1998-01	3.0	3.3
2002-05	2.1	2.4
Maxima, NX	2.2	2.5
Pulsar	2.0	2.3
Sentra		
1982-86	1.6	1.9
1987-90	2.0	2.3
1991-99	2.5	2.8
2000-05	1.9	2.2
Stanza	1.9	2.2
Stanza Wagon	1.9	2.2

SERVICE TRANSAXLE REMOVED
Transaxle R&R included unless otherwise noted.

Differential Case, R&R and Recondition (A)		
1995-98 200SX	8.4	8.9
Altima		
1993-01	9.3	9.8
2002-05	7.3	7.8
Maxima		
1989-94	9.2	9.7
1995-04	9.7	10.2
2005	8.2	8.7
NX		
GA	7.2	7.7
SR	7.6	8.1
Pulsar	7.3	7.8
Sentra		
1982-86	6.1	6.6
1987-90		
E	7.1	7.6
GA		
2WD	7.3	7.8
4WD	9.8	10.3

	LABOR TIME	SEVERE SERVICE
1991-94		
GA	7.2	7.7
SR	7.6	8.1
1995-99	8.4	8.9
2000-05	7.5	8.0
Stanza		
1982-86	5.9	6.4
1987-89	7.2	7.7
1990-92	7.6	8.1
Stanza Wagon		
2WD	7.2	7.7
4WD	8.9	9.4
Differential Case, Replace (B)		
1995-98 200SX		
GA	7.8	8.3
SR	8.2	8.7
Altima		
1993-01	9.0	9.5
2002-05	7.2	7.7
Maxima		
1989-94		
SOHC	6.1	6.6
DOHC	8.6	9.1
1995-04	9.5	10.0
2005	8.0	8.5
NX	7.2	7.7
Pulsar	6.9	7.4
Sentra		
1982-86	4.5	5.0
1987-90		
E	6.4	6.9
GA		
2WD	7.1	7.6
4WD	9.9	10.4
1991-94	7.2	7.7
1995-99		
GA	7.8	8.3
SR	8.2	8.7
2000-05	7.3	7.8
Stanza		
1982-86	4.6	5.1
1987-89	6.9	7.4
1990-92	7.4	7.9
Stanza Wagon		
2WD	6.9	7.4
4WD	8.5	9.0
Flywheel (Flexplate), Replace (B)		
1995-98 200SX	6.1	6.3
Altima		
1993-01	6.8	7.0
2002-05	4.7	4.9
Maxima		
1989-94		
SOHC	7.0	7.2
DOHC	7.5	7.7
1995-04	7.2	7.4
2005	5.0	5.5
NX	5.4	5.6

Altima : Maxima : NX : Pulsar : Pulsar NX : Sentra : Stanza

	LABOR TIME	SEVERE SERVICE
Pulsar		
CA, GA	5.4	5.6
E	4.9	5.1
Sentra		
1982-86	4.9	5.1
1987-90		
E	5.1	5.3
GA		
2WD	5.3	5.5
4WD	8.1	8.3
1991-94	5.4	5.6
1995-99	6.1	6.3
2000-05	5.2	5.4
Stanza		
1982-86	4.2	4.4
1987-89	5.1	5.3
1990-92	5.7	5.9
Stanza Wagon		
2WD	5.2	5.4
4WD	6.9	7.1
Front Oil Pump, Replace or Reconditon (B)		
1995-98 200SX		
GA	6.9	7.1
SR	7.3	7.5
Altima		
1993-01	8.1	8.3
2002-05	6.6	6.8
Maxima		
1989-94		
SOHC	7.8	8.0
DOHC	8.2	8.4
1995-04	9.0	9.2
2005	7.2	7.7
NX	6.4	6.6
Pulsar	7.4	7.6
Sentra		
1982-86	6.5	6.7
1987-90		
E	7.0	7.2
GA		
2WD	7.2	7.4
4WD	9.8	10.0
1991-94	6.4	6.6
1995-99		
GA	6.9	7.1
SR	7.3	7.5
2000-05	6.7	6.9
Stanza		
1982-86	5.4	5.6
1987-89	6.3	6.5
1990-92	6.9	7.1
Stanza Wagon		
2WD	6.4	6.6
4WD	8.1	8.3
Torque Converter, Replace (B)		
1995-98 200SX		
GA	5.5	5.7
SR	6.0	6.2
1982 310	5.1	5.3

	LABOR TIME	SEVERE SERVICE
Altima		
1993-01	6.5	6.7
2002-05	4.7	4.9
Maxima		
exc. 2005		
SOHC	6.7	6.9
DOHC	7.3	7.5
2005	5.9	6.2
NX	5.3	5.5
Pulsar		
CA, GA	5.3	5.5
E	4.8	5.0
Sentra		
1982-86	4.0	4.2
1987-90		
E	4.9	5.1
GA		
2WD	5.2	5.4
4WD	8.1	8.3
1991-94	5.3	5.5
1995-99		
GA	5.5	5.7
SR	6.0	6.2
2000-05	5.0	5.2
Stanza		
1982-86	4.2	4.4
1987-92	5.5	5.7
Stanza Wagon		
2WD	5.1	5.3
4WD	6.8	7.0
Transaxle, Reconditon (A)		
Includes: Replace torque converter.		
1995-98 200SX	17.9	18.8
1982 310	14.2	15.1
Altima		
1993-01	18.6	19.5
2002-05	17.1	18.0
Maxima		
1989-94		
SOHC	13.0	13.9
DOHC	18.1	19.0
1995-04	19.3	20.2
2005	17.5	18.1
NX		
GA	14.3	15.2
SR	14.8	15.7
Pulsar	14.4	15.3
Sentra		
1982-86	13.2	14.1
1987-90		
E	14.3	15.2
GA		
2WD	14.5	15.4
4WD	17.3	18.2
1991-94		
GA	14.3	15.2
SR	14.8	15.7
1995-99	17.9	18.8

	LABOR TIME	SEVERE SERVICE
2000-05		
QG	17.4	18.3
QR, SR	16.8	17.7
Stanza		
1982-86	13.0	13.9
1987-89	14.4	15.3
1990-92	15.0	15.9
Stanza Wagon		
2WD	14.5	15.4
4WD	15.9	16.8
Transaxle & Torque Converter, R&I (B)		
1995-98 200SX		
GA	5.3	5.5
SR	5.8	6.0
Altima		
1993-01	6.3	6.5
2002-05	4.6	4.8
Maxima		
1989-94		
SOHC	6.8	7.0
DOHC	7.3	7.5
1995-04	7.0	7.2
2005	5.6	5.6
NX		
GA	4.9	5.1
SR	5.3	5.5
Pulsar	4.8	5.0
Sentra		
1982-86	3.7	3.9
1987-90		
E	4.5	4.7
GA		
2WD	5.0	5.2
4WD	7.6	7.8
1991-94		
GA	4.9	5.1
SR	5.3	5.5
1995-99		
GA	5.3	5.5
SR	5.8	6.0
2000-05	4.9	5.1
Stanza		
1982-86	3.8	4.0
1987-89	4.9	5.1
1990-92	5.2	5.4
Stanza Wagon		
2WD	4.9	5.1
4WD	7.6	7.8

AUTOMATIC TRANSMISSION

SERVICE TRANSMISSION INSTALLED

Check Unit for Oil Leaks (C)
1981-05 .9 .9
Drain and Refill Unit (B)
1981-05 .8 .8

NI-25

200SX : 210 : 240SX : 280ZX : 300ZX : 310 : 350Z : 510 : 810

	LABOR TIME	SEVERE SERVICE
Oil Pressure Check (B)		
All Models	.8	.8
Downshift Solenoid, Replace (B)		
1984-88 200SX	.8	.8
1989-98 240SX	.8	.8
300ZX	.8	.8
1984 810/Maxima	.8	.8
Electronic Control Unit, Replace (B)		
1986-88 200SX		
4 cyl.	1.5	1.5
V6	.7	.7
240SX		
1989-94	.5	.5
1995-98	.7	.7
300ZX		
1986-89	2.0	2.0
1990-96	.7	.7
Extension Housing Oil Seal Replace (B)		
200SX		
1981-83	.9	1.1
1984-88	.8	1.0
210	1.3	1.5
240SX	1.2	1.4
280ZX	1.4	1.6
300ZX		
1984-89	1.8	2.0
1990-96	1.5	1.7
510	1.4	1.6
810/Maxima	1.5	1.7
Governor Assy., Replace or Recondition (B)		
1984-88 200SX	3.0	3.5
210	3.1	3.6
280ZX	3.5	4.0
1984-89 300ZX	3.7	4.2
510	3.2	3.7
810/Maxima	3.2	3.7
Kickdown Switch, Replace (B)		
1984-98	.7	.9
Lock-up Solenoid, Replace (B)		
240SX		
1989-94	1.8	2.1
1995-98	1.7	2.0
1990-96 300ZX	1.3	1.6
Oil Pan Gasket, Replace (B)		
1981-98	.8	.8
Overdrive Cancel Solenoid, Replace (B)		
1984-87	.7	.7
1988-98	.8	.8
Overdrive Cancel Switch, Replace (B)		
1984-88 200SX	.5	.5
1989-94 240SX	.5	.5
300ZX	.7	.7
1984 810 Maxima	.7	.7

	LABOR TIME	SEVERE SERVICE
Overdrive Indicator Switch, Replace (B)		
1981-98	.7	.7
Speedometer Drive Pinion, Replace (B)		
1981-98	.5	.5
Vacuum Diaphragm, Replace (B)		
1981-85	.5	.5
Valve Body Assy., R&R and Recondition (A)		
200SX	3.9	4.4
210	3.5	4.0
240SX	4.3	4.8
280ZX	2.8	3.3
300ZX	3.9	4.4
510	3.5	4.0
810, Maxima	3.1	3.6
Valve Body Assy., Replace (B)		
200SX	1.8	2.1
210	1.9	2.2
240SX	1.7	2.0
280ZX	1.5	1.8
300ZX	1.6	1.9
510	1.9	2.2
810, Maxima	1.5	1.8

SERVICE TRANSMISSION REMOVED

Transmission R&R included unless otherwise noted.

	LABOR TIME	SEVERE SERVICE
Flywheel (Flexplate), Replace (B)		
200SX		
1981-83	4.1	4.3
1984-88	4.6	4.8
210	3.9	4.1
240SX	5.4	5.6
280ZX	4.1	4.3
300ZX		
1984-89		
w/turbo	6.4	6.6
w/o turbo	5.8	6.0
1990-96	4.8	5.0
350Z	5.3	5.5
510	4.3	4.5
810/Maxima	4.1	4.3
Torque Converter, Replace (B)		
200SX		
1981-83	4.0	4.2
1984-88 4 cyl.	4.4	4.6
210	3.6	3.8
240SX	5.3	5.5
280ZX	4.0	4.2
300ZX		
1984-89		
w/turbo	6.2	6.4
w/o turbo	5.8	6.0
1990-96	4.5	4.7
510	4.2	4.4
810/Maxima	4.1	4.3

	LABOR TIME	SEVERE SERVICE
Transmission, Recondition (A)		
Includes: Replace torque converter.		
200SX		
1981-83	14.0	14.9
1984-88 4 cyl.	15.2	15.1
210	13.8	14.7
240SX		
1989-94	13.2	14.1
1995-98	13.7	14.6
280ZX	14.3	15.2
300ZX		
1984-89		
w/turbo	17.0	17.9
w/o turbo	16.6	17.5
1990-96	15.4	16.3
510	14.1	15.0
810/Maxima	14.1	15.0
Transmission & Torque Converter, R&I (B)		
200SX		
1981-83	3.7	3.9
1984-88		
4 cyl.		
w/turbo	4.2	4.4
w/o turbo	3.8	4.0
210	3.4	3.6
240SX	5.3	5.5
280ZX	3.8	4.0
300ZX		
1984-89		
w/turbo	5.9	6.1
w/o turbo	5.5	5.7
1990-96	4.3	4.5
350Z	5.2	5.4
510	3.9	4.1
810/Maxima	3.9	4.1

TRANSFER CASE

	LABOR TIME	SEVERE SERVICE
Actuator Assy., Replace (B)		
Sentra	.7	.7
Stanza Wagon	1.5	1.5
Control Unit, Replace (B)		
Sentra	.5	.5
Stanza Wagon	.5	.5
Companion Flange, Replace (B)		
Sentra	.7	.9
Stanza Wagon	.7	.9
Drive Pinion Seal, Replace (B)		
Stanza Wagon	2.3	2.5
4WD Indicator Switch, Replace (B)		
Stanza Wagon	.5	.5
Speedometer Driven Gear and/or Seal, Replace (B)		
Sentra	.5	.5
Stanza Wagon	1.7	1.7
Transfer Case, R&R and Recondition (A)		
Sentra		
E	7.6	8.2

Altima : Maxima : NX : Pulsar : Pulsar NX : Sentra : Stanza NI-27

	LABOR TIME	SEVERE SERVICE
GA		
AT	9.0	9.6
MT	8.5	9.1
Stanza Wagon		
AT	9.3	9.9
MT	8.8	9.4
Transfer Case, R&R or Replace (B)		
Sentra		
E	5.0	5.2
GA		
AT	6.2	6.4
MT	5.8	6.0
Stanza Wagon		
AT	6.5	6.7
MT	6.1	6.3
Transfer Case Relay, Replace (B)		
Sentra	.5	.5
Stanza Wagon	.5	.5

SHIFT LINKAGE
AUTOMATIC TRANSMISSION

	LABOR TIME	SEVERE SERVICE
Gear Selector, Replace (B)		
200SX		
1981-83	1.7	1.7
1984-88		
4 cyl.	1.6	1.6
V6	.7	.7
1995-98	1.2	1.2
210	1.8	1.8
240SX	.7	.7
280ZX	2.1	2.1
300ZX		
1984-89	2.0	2.0
1990-96	.7	.7
350Z	.7	.7
510	1.3	1.3
810/Maxima	1.7	1.7
Altima		
1993-01	1.5	1.5
2002-05	.5	.5
Maxima		
1985-94	.7	.7
1995-01	2.4	2.4
2002-04	1.6	1.6
2005	.4	.6
NX, Pulsar	1.2	1.2
Sentra		
1982-90	1.6	1.6
1991-05	1.2	1.2
Stanza	1.5	1.5
Gear Selector Indicator, Replace (B)		
All Models	1.2	1.2
Manual Shaft and/or Parking Rod, Replace (B)		
All Models	.7	.9

MANUAL TRANSMISSION
Gearshift Control Cable, Replace (B)

	LABOR TIME	SEVERE SERVICE
2005 Maxima	6.5	6.9

	LABOR TIME	SEVERE SERVICE
Gearshift Control Rod, Replace (B)		
Exc. below	.7	.8
1995-04 Maxima	1.3	1.4
Gearshift Lever, Replace (B)		
200SX		
1981-88	.8	.8
1995-98	1.1	1.1
210, 310, 510	.5	.5
240SX	.8	.8
280ZX	.5	.5
300ZX		
1984-89	.7	.7
1985-96	1.1	1.1
350Z	.8	.8
810/Maxima	.5	.5
Altima		
1993-97	.8	.8
1998-04	1.2	1.2
Maxima		
1985-94	.7	.7
1995-04	1.4	1.4
2005	.5	.7
NX, Pulsar	.9	.9
Sentra		
1982-90	.7	.7
1991-05	1.1	1.1
Stanza	.7	.7
Stanza Wagon	.7	.7
Gearshift Support Rod, Replace (B)		
200SX, NX, Sentra	1.1	1.1
Altima	.9	.9
Maxima		
1985-94	.8	.8
1995-05	1.5	1.5
Pulsar, Stanza	.8	.8

DRIVELINE
DRIVESHAFT
Center Support Bearing, Replace (B)

	LABOR TIME	SEVERE SERVICE
1984-88 200SX	1.3	1.6
240SX	1.1	1.4
1990-96 300ZX	2.0	2.3
810/Maxima	1.4	1.7
Sentra 4WD	1.3	1.6
Stanza Wagon 4WD	1.3	1.6

Driveshaft, Replace (B)
2-joint shaft

	LABOR TIME	SEVERE SERVICE
1981-83 200SX	.5	.7
210, 510	.5	.7
280ZX	1.3	1.5
1984-89 300ZX	1.7	1.9
350Z	1.4	1.6

3-joint shaft

	LABOR TIME	SEVERE SERVICE
1984-88 200SX	.5	.7
240SX	.7	.9
1990-96 300ZX	1.8	2.0
810/Maxima	.7	.9
Sentra 4WD	.7	.9
Stanza Wagon	.5	.7

	LABOR TIME	SEVERE SERVICE
Intermediate Shaft Support Bearing, Replace (B)		
1995-98 200SX	1.8	2.1
Altima		
1993-97	2.2	2.5
1998-01	1.7	2.0
2002-05	1.3	1.6
Maxima, NX	2.5	2.8
Sentra		
1987-91	1.7	2.0
1992-94	2.2	2.5
1995-05	1.9	2.2
1987-90 Stanza	2.3	2.6
Stanza Wagon	2.1	2.4
U-Joint, Replace or Recondition (B)		
2-joint shaft		
1981-83 200SX	.9	1.1
210	.9	1.1
280ZX	1.6	1.8
1984-89 300ZX	1.9	2.1
510	.9	1.1
3-joint shaft		
1984-88 200SX	1.3	1.5
240SX	1.2	1.4
1990-96 300ZX	1.9	2.1
810/Maxima	1.3	1.5
Sentra 4WD	1.3	1.5
Stanza Wagon	1.2	1.4

DRIVEAXLE
Axle Shaft Bearings, Replace (B)

	LABOR TIME	SEVERE SERVICE
200SX		
one side	1.8	2.1
both sides	3.2	3.5
1984-89 300ZX		
one side	1.7	2.0
both sides	3.3	3.6
Altima		
1998-01		
one side	.5	.8
both sides	.8	1.1
2002-05		
one side	.8	1.1
both sides	1.4	1.7
1982-88 Sentra		
one side	1.9	2.2
both sides	3.5	3.8
Stanza Wagon		
one side	1.9	2.2
both sides	3.6	3.9

Axle Shaft Oil Seal, Replace (B)

	LABOR TIME	SEVERE SERVICE
200SX		
one	1.7	2.0
both	3.1	3.4
300ZX		
one	1.6	1.9
both	2.9	3.2

200SX : 210 : 240SX : 280ZX : 300ZX : 310 : 350Z : 510 : 810

	LABOR TIME	SEVERE SERVICE
Axle Shaft Oil Seal, Replace (B)		
Altima		
1998-01		
one side	.5	.8
both sides	.8	1.1
2002-05		
one side	.8	1.1
both sides	1.4	1.7
1982-88 Sentra		
one	1.6	1.9
both	3.2	3.5
Stanza Wagon		
one	1.7	2.0
both	3.1	3.4
Carrier Side Oil Seals, Replace (B)		
Rear axle		
240SX		
1990-94		
one side	.7	.9
both sides	1.4	1.6
1995-98		
one side	1.2	1.4
both sides	1.7	1.9
280ZX		
one side	1.9	2.1
both sides	3.3	3.5
300ZX		
1984-89		
one side	1.4	1.6
both sides	2.9	3.1
1990-96		
one side	.7	.9
both sides	1.4	1.6
350Z		
one side	1.1	1.3
both sides	1.9	2.1
810/Maxima		
one side	.8	1.0
both sides	2.0	2.2
Stanza Wagon		
one side	1.4	1.6
both sides	2.8	3.0
Differential Carrier, R&R or Replace (B)		
Rear axle		
1984-88 200SX	2.7	3.2
240SX		
1989-94	1.8	2.3
1995-98	2.1	2.6
300ZX		
1984-89	3.1	3.6
1990-96	1.7	2.2
350Z	2.7	3.2
810/Maxima	2.7	3.2
Sentra	1.9	2.4
Stanza Wagon	3.6	4.1

	LABOR TIME	SEVERE SERVICE
Differential Carrier, R&R and Recondition (A)		
Includes: Transaxle R&R.		
Rear axle		
1984-88 200SX	6.9	7.5
240SX		
1989-94	6.2	6.8
1995-98	6.4	7.0
300ZX		
1984-89	7.1	7.7
1990-96	6.3	6.9
350Z	5.5	6.1
810/Maxima	7.0	7.6
Sentra	6.2	6.8
Stanza Wagon	7.9	8.5
Differential Case, R&R and Recondition (A)		
310		
1981		
4-Speed	9.1	9.6
5-Speed	9.2	9.7
1982	9.4	9.9
350Z	4.4	4.9
Differential Side Bearing, Replace (B)		
Rear axle		
1984-88 200SX	3.7	4.0
240SX		
1989-94	2.5	2.8
1995-98	3.1	3.4
300ZX		
1984-89	4.0	4.3
1990-96	2.7	3.0
350Z	1.6	1.9
810/Maxima	3.9	4.2
Sentra	3.2	3.5
Stanza Wagon	5.0	5.3
Final Drive Assy., R&R or Replace (B)		
310		
1981		
4-Speed	7.3	7.8
5-Speed	7.7	8.2
1982	4.1	4.6
Final Drive Flange, Replace (B)		
310		
one	1.4	1.6
both	2.2	2.4
Final Drive Side Bearing, Replace (B)		
310		
1981		
4-Speed	7.8	8.1
5-Speed	8.1	8.4
1982	9.3	9.6
Pinion Shaft Oil Seal, Replace (B)		
Rear axle		
1984-88 200SX	1.4	1.6

	LABOR TIME	SEVERE SERVICE
240SX		
1989-94	.7	.9
1995-98	.8	1.0
300ZX	1.3	1.5
350Z	1.4	1.6
810/Maxima	.8	1.0
Sentra		
coll. type	2.7	2.9
solid type	.7	.9
Stanza Wagon	.7	.9
Rear Axle Cover Gasket, Replace (B)		
1984-88 200SX	1.8	1.8
240SX		
1989-94	1.8	1.8
1995-98	2.3	2.3
280ZX	1.6	1.6
300ZX	4.1	4.1
810/Maxima	.8	.8
Rear Axle Housing, Replace (B)		
240SX		
1989-94		
one	2.3	2.8
both	4.2	4.7
1995-98		
one	2.3	2.8
both	3.5	4.0
1990-96 300ZX		
one	2.4	2.9
both	4.2	4.7
Rear Axle Shaft, Replace (B)		
200SX		
one	1.4	1.7
both	2.5	2.8
1984-89 300ZX		
one	1.5	1.8
both	2.5	2.8
Sentra		
one	1.5	1.8
both	2.9	3.2
Stanza Wagon		
one	1.5	1.8
both	2.8	3.1
Rear Wheel Bearing, Replace (B)		
240SX		
1989-94		
one side	2.9	3.1
both sides	5.0	5.2
1995-98		
one side	2.6	2.8
both sides	4.2	4.4
1990-96 300ZX		
one side	2.8	3.0
both sides	5.0	5.2
Rear Wheel Hub Assy., Replace (B)		
240SX		
1989-94		
one side	2.8	3.3
both sides	5.0	5.5

Altima : Maxima : NX : Pulsar : Pulsar NX : Sentra : Stanza — NI-29

	LABOR TIME	SEVERE SERVICE
1995-98		
one side	2.7	3.2
both sides	4.1	4.6
1990-96 300ZX		
one side	2.8	3.3
both sides	4.9	5.4
350Z		
one side	.5	1.0
both sides	1.0	1.5
Altima		
1998-01		
one side	.5	1.0
both sides	.8	1.3
2002-05		
one side	.8	1.3
both sides	1.4	1.9
Maxima		
one side	1.1	1.6
both sides	1.9	2.4
Sentra		
1995-99		
one side	1.1	1.6
both sides	1.9	2.4
2000-05		
one side	.8	1.3
both sides	1.1	1.6

HALFSHAFTS
CV Joint and Boot, Replace or Recondition (B)

	LABOR TIME	SEVERE SERVICE
Front		
1995-98 200SX		
one side	1.8	2.1
both sides	3.0	3.3
Altima		
1993-97		
one side	2.5	2.8
both sides	4.2	4.5
1998-05		
one side	1.9	2.2
both sides	3.6	3.9
Maxima		
one side	2.5	2.8
both sides	4.1	4.4
NX		
one side	2.3	2.6
both sides	4.3	4.6
Pulsar		
one side	2.2	2.5
both sides	3.9	4.2
Sentra		
1982-86		
one side	1.9	2.2
both sides	3.4	3.7
1987-94		
one side	2.3	2.6
both sides	4.3	4.6
1995-05		
one side	2.0	2.3
both sides	3.3	3.6

	LABOR TIME	SEVERE SERVICE
Stanza		
1982-86		
one side	1.8	2.1
both sides	3.6	3.9
1987-89		
one side	1.8	2.1
both sides	3.8	4.1
1990-92		
one side	2.2	2.5
both sides	4.0	4.3
Stanza Wagon		
one side	2.1	2.4
both sides	3.4	3.7
Rear		
240SX		
1989-94		
one side	1.7	2.0
both sides	3.4	3.7
1995-98		
one side	2.8	3.1
both sides	4.5	4.8
1990-96 300ZX		
one side	2.6	2.9
both sides	4.9	5.2
350Z		
one side	1.6	1.9
both sides	2.7	3.0

Halfshaft, R&R and Recondition (B)

	LABOR TIME	SEVERE SERVICE
Rear axle		
240SX		
1989-94		
one side	1.7	2.0
both sides	3.4	3.7
1995-98		
one side	2.8	3.1
both sides	4.9	5.2
1990-96 300ZX		
one side	2.5	2.8
both sides	4.9	5.2
350Z		
one side	2.2	2.5
both sides	4.1	4.4

Halfshaft, R&R or Replace (B)

	LABOR TIME	SEVERE SERVICE
Front		
1995-98 200SX		
one side	1.2	1.4
both sides	1.6	1.8
Altima		
1993-97		
one side	1.5	1.7
both sides	2.9	3.1
1998-01		
one side	1.5	1.7
both sides	2.0	2.2
2002-05		
one side	.6	.8
both sides	1.1	1.3
Maxima, NX		
one side	1.6	1.8
both sides	2.7	2.9

	LABOR TIME	SEVERE SERVICE
Pulsar		
one side	1.5	1.7
both sides	2.6	2.8
Sentra		
1982-86		
one side	1.3	1.5
both sides	2.1	2.3
1987-94		
one side	1.6	1.8
both sides	2.7	2.9
1995-05		
one side	1.2	1.4
both sides	2.1	2.3
Stanza		
1982-89		
one side	1.3	1.5
both sides	2.4	2.6
1990-92		
one side	1.5	1.7
both sides	2.7	2.9
Stanza Wagon		
one side	1.3	1.5
both sides	2.3	2.5
Rear		
240SX		
1989-94		
one side	.7	.9
both sides	1.3	1.5
1995-98		
one side	1.7	1.9
both sides	3.1	3.3
1990-96 300ZX		
one side	1.4	1.6
both sides	2.7	2.9
350Z		
one side	.8	1.0
both sides	1.4	1.6

BRAKES
ANTI-LOCK
Diagnose Anti-Lock Brake System Component Each (A)

	LABOR TIME	SEVERE SERVICE
1990-05	1.0	1.0

Retrieve Fault Codes (A)

	LABOR TIME	SEVERE SERVICE
1990-05	.3	.3

Actuator Assy., Replace (B)
Includes: System bleeding.

	LABOR TIME	SEVERE SERVICE
1995-98 200SX	3.0	3.0
240SX	1.8	1.8
1990-96 300ZX	1.5	1.5
350Z	3.5	3.5
Altima	2.5	2.5
Maxima		
1985-04	2.2	2.2
2005	2.6	2.6
NX	4.2	4.2
Sentra		
1991-94	4.2	4.2
1995-99	3.0	3.0

NI-30 200SX : 210 : 240SX : 280ZX : 300ZX : 310 : 350Z : 510 : 810

	LABOR TIME	SEVERE SERVICE
Actuator Assy., Replace (B)		
2000-05	2.2	2.2
1991-92 Stanza	1.4	1.4
Anti-Lock Relay, Replace (B)		
Exc. below	.3	.3
1994-96 300ZX	.9	.9
Maxima		
1990-04	.7	.7
2005	2.6	2.6
Control Unit, Replace (B)		
All Models	.7	.7
Front Rotor Sensor Assy., Replace (B)		
1995-98 200SX		
one	1.5	1.7
both	2.4	2.6
240SX		
1990-94		
one	.9	1.1
both	1.7	1.9
1995-98		
one	1.4	1.6
both	2.2	2.4
1990-96 300ZX		
one	.7	.9
both	1.5	1.7
350Z		
one	1.1	1.3
both	2.1	2.3
Altima		
1993-97		
one	1.6	1.8
both	2.8	3.0
1998-05		
one	.8	1.0
both	1.3	1.5
Maxima		
1990-01		
one	1.5	1.7
both	2.5	2.7
2002-04		
one	1.6	1.8
both	3.2	3.4
2005		
one	.8	1.0
both	1.2	1.4
NX		
one	1.4	1.6
both	2.7	2.9
Sentra		
1991-99		
one	1.5	1.7
both	2.7	2.9
2000-05		
one	1.1	1.3
both	1.4	1.6
1990-92 Stanza		
one	.8	1.0
both	1.7	1.9

	LABOR TIME	SEVERE SERVICE
Rear Rotor Sensor Assy., Replace (B)		
1995-98 200SX		
one	1.1	1.3
both	1.7	1.9
240SX	.8	1.0
1990-96 300ZX		
one	.7	.9
both	1.2	1.4
350Z		
one	1.1	1.3
both	1.9	2.1
Altima		
1993-97		
one	.8	1.0
both	1.5	1.7
1998-01		
one	.5	.7
both	.8	1.0
2002-05		
one	1.0	1.2
both	1.6	1.8
Maxima		
1990-94		
one	.7	.9
both	1.6	1.8
1995-04		
one	1.1	1.3
both	1.9	2.1
2005		
one	.8	1.0
both	1.2	1.4
NX		
one	1.3	1.5
both	2.0	2.2
Sentra		
1991-99		
one	1.3	1.5
both	2.0	2.2
2000-05		
one	.8	1.0
both	1.3	1.5
1990-92 Stanza		
one	.8	1.0
both	1.4	1.6
Wheel Speed Sensor, Replace (B)		
Front or rear		
one	.7	.7
both	.9	.9

SYSTEM

	LABOR TIME	SEVERE SERVICE
Bleed Brakes (B)		
Includes: Add fluid.		
All Models	.9	.9
Brake Hose (Flexible), Replace (B)		
Includes: System bleeding.		
One	.8	.9
each addl. add	.3	.4

	LABOR TIME	SEVERE SERVICE
Brake Proportioning Valve, Replace (B)		
Includes: System bleeding.		
Exc. below	.7	.9
1995-98 200SX	1.3	1.5
240SX	1.5	1.7
Altima		
1993-97	.9	1.1
1998-05	1.5	1.7
Maxima		
1995-01	4.1	4.3
2002-04	2.6	2.8
2005	1.3	1.5
NX	1.3	1.5
Pulsar	1.4	1.6
Sentra		
1982-86	1.3	1.5
1987-90	.8	1.0
1991-05	1.5	1.7
Master Cylinder, R&R and Rebuild (B)		
Includes: System bleeding.		
Exc. below	1.8	2.1
240SX	1.2	1.5
Maxima		
1985-94	1.4	1.7
2005	2.0	2.2
NX	2.1	2.4
Pulsar	1.4	1.7
Sentra		
1982-86	2.1	2.4
1987-90	1.4	1.7
1991-94	2.1	2.4
Stanza Wagon	1.5	1.8
Master Cylinder, Replace (B)		
Includes: System bleeding.		
Exc. below	.9	1.1
200SX	1.2	1.4
350Z	1.7	1.9
Altima		
1993-97	1.2	1.4
1998-05	1.9	2.1
1995-04 Maxima	1.4	1.6
2005 Maxima	1.7	1.9
NX	1.6	1.8
Pulsar	1.2	1.4
Sentra		
1982-86	1.3	1.5
1987-90	.7	.9
1991-94	1.6	1.8
1995-05	1.4	1.6
Power Booster Unit, Replace (B)		
Includes: System bleeding.		
200SX	1.7	1.9
210, 310, 510	1.3	1.5
240SX	1.6	1.8
280ZX	.8	1.0
300ZX		
1984-89	1.5	1.7
1990-96	1.1	1.3

Altima : Maxima : NX : Pulsar : Pulsar NX : Sentra : Stanza   NI-31

	LABOR TIME	SEVERE SERVICE
350Z	2.9	3.1
810/Maxima	1.4	1.6
Altima		
1993-01	4.0	4.2
2002-05	2.0	2.2
Maxima		
1985-88	1.3	1.5
1989-94	.9	1.1
1995-05	2.1	2.3
NX	2.3	2.5
Pulsar	1.7	1.9
Sentra		
1982-86	1.3	1.5
1987-90	1.7	1.9
1991-94	2.3	2.5
1995-99	1.7	1.9
2000-05	2.1	2.3
Stanza		
1982-86	1.3	1.5
1987-89	2.9	3.1
1990-92	2.0	2.2
Stanza Wagon	1.2	1.4
Power Booster Vacuum Check Valve, Replace (B)		
Exc. below	.5	.5
1998-05 Altima	.7	.7
Power Booster Vacuum Pump, Replace (B)		
1981-85	1.1	1.3
SERVICE BRAKES		
Brake Drum, Replace (B)		
All Models	.8	.9
Caliper Assy., R&R and Recondition (B)		
Includes: System bleeding.		
200SX		
1981-88		
front		
one	1.5	1.8
both	3.1	3.4
rear		
one	1.2	1.5
both	2.3	2.6
1995-98		
one	1.6	1.9
both	2.7	3.0
210, 310		
one	1.7	2.0
both	3.1	3.4
240SX		
one	1.5	1.8
both	2.6	2.9
280ZX		
one	1.7	2.0
both	3.3	3.6
300ZX		
one	1.7	2.0
both	2.7	3.0
350Z		

	LABOR TIME	SEVERE SERVICE
front		
one	2.4	2.7
both	3.3	3.6
rear		
one	1.4	1.7
both	2.5	2.8
510		
one	1.7	2.0
both	2.9	3.2
810/Maxima		
one	1.5	1.8
both	3.0	3.3
Altima		
front		
one	1.9	2.2
both	3.5	3.8
rear		
one	1.4	1.7
both	2.5	2.8
Maxima		
1985-94		
one	1.7	2.0
both	2.9	3.2
1995-05		
one	1.7	2.0
both	3.0	3.3
NX		
one	1.9	2.2
both	3.2	3.5
Pulsar		
one	1.5	1.8
both	2.9	3.2
Sentra		
1982-90		
one	1.5	1.8
both	2.9	3.2
1991-94		
one	1.9	2.2
both	3.2	3.5
1995-05		
one	1.6	1.9
both	2.4	2.7
Stanza		
one	1.7	2.0
both	3.1	3.4
Caliper Assy., Replace (B)		
Includes: System bleeding.		
200SX		
1981-88		
one	1.3	1.5
both	2.3	2.5
1995-98		
one	1.1	1.3
both	1.5	1.7
210, 310		
one	1.1	1.3
both	2.2	2.4
240SX		
one	.8	1.0
both	1.7	1.9

	LABOR TIME	SEVERE SERVICE
280ZX		
front		
one	.7	.9
both	1.6	1.8
rear		
one	1.3	1.5
both	2.1	2.3
300ZX		
1984-89		
one	1.3	1.5
both	2.2	2.4
1990-96		
one	1.1	1.3
both	1.9	2.1
350Z		
one	.8	1.0
both	1.3	1.5
510		
one	1.3	1.5
both	2.3	2.5
810/Maxima		
one	1.3	1.5
both	2.3	2.5
Altima		
1993-97		
front		
one	1.4	1.6
both	2.8	3.0
rear		
one	1.0	1.2
both	1.4	1.6
1998-05		
one	.9	1.1
both	1.4	1.6
Maxima		
1985-88		
one	1.3	1.5
both	2.2	2.4
1989-05		
one	1.1	1.3
both	2.1	2.3
NX		
one	1.3	1.5
both	1.8	2.0
Pulsar		
one	1.3	1.5
both	2.3	2.5
Sentra		
1982-90		
one	1.1	1.3
both	2.3	2.5
1991-99		
one	1.3	1.5
both	1.8	2.0
1991-05		
one	.8	1.0
both	1.3	1.5
Stanza		
one	1.2	1.4
both	2.3	2.5

200SX : 210 : 240SX : 280ZX : 300ZX : 310 : 350Z : 510 : 810

	LABOR TIME	SEVERE SERVICE
Front Disc Brake Rotor, Replace (B)		
200SX		
1981-87		
one	1.1	1.3
both	2.3	2.5
1988-98		
one	.7	.9
both	1.3	1.5
210		
one	1.2	1.4
both	2.2	2.4
240SX		
1994		
one	.7	.9
both	1.3	1.5
1995-98		
one	.5	.7
both	.9	1.1
280ZX		
one	1.3	1.5
both	2.1	2.3
300ZX		
1984-87		
one	1.1	1.3
both	2.1	2.3
1988-96		
one	.8	1.0
both	1.4	1.6
310		
one	1.5	1.7
both	2.6	2.8
350Z		
one	.8	1.0
both	1.0	1.2
510		
one	1.3	1.5
both	2.3	2.5
810/Maxima		
one	1.1	1.3
both	2.3	2.5
Altima		
one	.8	1.0
both	1.3	1.5
Maxima		
1985-94		
one	1.2	1.4
both	2.3	2.5
1995-05		
one	.8	1.0
both	1.4	1.6
NX		
one	.7	.9
both	1.2	1.4
Pulsar		
1983-87		
one	1.1	1.3
both	2.2	2.4

	LABOR TIME	SEVERE SERVICE
1988-90		
one	.7	.9
both	1.1	1.3
Sentra		
front		
1982-87		
one	1.5	1.7
both	2.7	2.9
1988-99		
one	.8	1.0
both	1.4	1.6
2000-05		
one	.6	.8
both	.8	1.0
rear		
one	.8	1.0
both	1.3	1.5
Stanza		
1982-87		
one	1.1	1.3
both	2.3	2.5
1988-92		
one	.7	.9
both	1.2	1.4
Stanza Wagon		
1986-87		
one	1.1	1.3
both	2.2	2.4
1988		
one	.7	.9
both	1.2	1.4
Pads and/or Shoes, Replace (B)		
Front disc	1.1	1.2
Rear		
disc		
exc. Maxima	1.1	1.2
Maxima	1.3	1.4
drum		
200SX		
1981-88	1.5	1.6
1995-98	2.1	2.2
210	1.4	1.5
310	1.8	1.9
510	1.3	1.4
810/Maxima	1.4	1.5
Altima		
1993-97	1.7	1.8
1998-05	1.1	1.2
Maxima	1.4	1.5
NX	1.3	1.4
Pulsar	1.8	1.9
Sentra		
1982-94	1.6	1.7
1995-99	2.1	2.2
2000-05	1.1	1.2

	LABOR TIME	SEVERE SERVICE
Stanza	1.7	1.8
Repack wheel bearings		
add both	.7	.8
Replace		
caliper add	.3	.4
rotor add	.2	.3
Resurface		
drum add	.6	.7
rotor add	.5	.6
Wheel Cylinder, R&R and Rebuild (B)		
Includes: System bleeding.		
1981-87		
one	1.2	1.5
both	2.4	2.7
1988-99		
one	1.7	2.0
both	3.2	3.5
2000-05		
one	1.7	2.0
both	2.4	2.7
Wheel Cylinder, Replace (B)		
Includes: System bleeding.		
1981-87		
one	.9	1.1
both	1.9	2.1
1988-05		
one	1.4	1.6
both	2.2	2.4
PARKING BRAKE		
Parking Brake Apply Actuator, Replace (B)		
Exc. below	.8	.8
240SX	1.2	1.2
280ZX	1.4	1.4
1984-89 300ZX	1.8	1.8
350Z	.9	.9
2002-05 Altima	1.1	1.1
Maxima		
1985-94	1.1	1.1
1995-04	1.4	1.4
2005	2.3	2.5
Sentra		
1982-86	1.4	1.4
1987-05	1.0	1.0
Parking Brake Cable, Adjust (C)		
All Models	.5	.7
Parking Brake Cable, Replace (B)		
Front		
300ZX	1.4	1.6
350Z, Sentra	1.1	1.3
Altima, Maxima	1.3	1.5
Rear		
200SX		
1981-88	.7	.9
1995-98		
disk type	.8	1.0
non-disk type	1.3	1.5

	LABOR TIME	SEVERE SERVICE
210, 310, 510	.8	1.0
240SX, 280ZX	.9	1.1
300ZX		
1984-89	.9	1.1
1990-96	1.4	1.6
350Z	2.9	3.1
810/Maxima	.9	1.1
Altima		
1993-97		
disk type	2.0	2.2
non-disk type	.8	1.0
1998-05	1.6	1.8
Maxima		
exc. 2005	1.1	1.3
2005	2.6	2.8
NX	1.4	1.6
Pulsar, Stanza	.7	.9
Sentra		
1982-86	.7	.9
1987-90	2.6	2.8
1991-94	1.4	1.6
1995-99		
disk type	.8	1.0
non-disk type	1.3	1.5
2000-05	1.6	1.8
Parking Brake Equalizer, Replace (B)		
Exc. below	.5	.7
2002-05 Altima	1.3	1.5
1987-89 Stanza	.7	.9
2005 Maxima	2.3	2.5

FRONT SUSPENSION

Unless otherwise noted, time given does not include alignment.

Align Front End (A)

	LABOR TIME	SEVERE SERVICE
All Models	1.1	1.3
Front & Rear Alignment, Check Adjust (A)		
FWD		
1995-97 200SX	1.7	1.9
1993-05 Altima	1.3	1.5
1985-05 Maxima	1.6	1.8
1988-95 NX, Sentra	1.7	1.9
1988-89 Pulsar	1.6	1.8
Stanza		
1988-89	1.6	1.8
1990-92	2.2	2.4
RWD		
1988 200SX	1.6	1.8
240SX	1.8	2.0
1988-96 300ZX	1.7	1.9
350Z	2.7	2.9
Front Toe, Adjust (B)		
All Models	.8	1.0
Front Shock Absorbers, Replace (B)		
1990-96 300ZX		
one	1.5	1.6
both	2.7	2.8

	LABOR TIME	SEVERE SERVICE
350Z		
one	1.4	1.5
both	2.1	2.2
Front Strut Assembly, R&R or Replace (B)		
FWD		
Altima		
1993-97		
one side	2.5	2.7
both sides	4.3	4.5
1998-05		
one side	2.4	2.6
both sides	3.2	3.4
Maxima		
1985-94		
one side	2.5	2.7
both sides	4.7	4.9
1995-04		
one side	1.7	1.9
both sides	2.5	2.7
2005		
one side	2.1	2.3
both sides	2.9	3.1
Pulsar		
one side	1.6	1.8
both sides	3.0	3.2
Sentra		
1982-86		
one side	1.6	1.8
both sides	3.0	3.2
1987-90		
one side	1.1	1.3
both sides	2.7	2.9
1991-05		
one side	1.9	2.1
both sides	3.0	3.2
Stanza		
1982-89		
one side	1.6	1.8
both sides	3.0	3.2
1990-92		
one side	2.5	2.7
both sides	4.7	4.9
RWD		
200SX		
one side	2.8	2.9
both sides	5.2	5.3
210, 510		
one side	2.1	2.2
both sides	3.8	3.9
240SX		
1989-94		
one side	2.5	2.6
both sides	4.2	4.3
1995-98		
one side	2.4	2.5
both sides	3.2	3.3
280ZX		
one side	2.8	2.9
both sides	3.8	3.9

	LABOR TIME	SEVERE SERVICE
1984-89 300ZX		
one side	2.9	3.0
both sides	5.2	5.3
810/Maxima		
one side	2.1	2.2
both sides	3.7	3.8
Replace coil spring add each	.2	.3
Front Strut Coil Spring, Replace (B)		
1982-87 FWD		
one side	1.5	1.7
both sides	2.9	3.1
1988-05 FWD		
one side	2.5	2.7
both sides	4.4	4.6
Front Strut Shock Absorber, Replace (B)		
1982-87 FWD		
one side	1.7	1.8
both sides	2.9	3.0
1988-05 FWD		
one side	2.4	2.5
both sides	4.2	4.3
Lower Ball Joint, Replace (B)		
1981-83 200SX		
one side	1.5	1.8
both sides	2.5	2.8
280ZX		
one side	1.4	1.7
both sides	2.7	3.0
210		
one side	1.3	1.6
both sides	2.5	2.8
1984-89 300ZX		
one side	1.4	1.7
both sides	2.2	2.5
310		
1981		
one side	1.8	2.1
both sides	3.2	3.5
1982		
one side	1.7	2.0
both sides	3.1	3.4
510		
one side	1.3	1.6
both sides	2.6	2.9
810/Maxima		
one side	1.3	1.6
both sides	2.6	2.9
Maxima		
1985-05		
one side	1.6	1.9
both sides	2.4	2.7
Pulsar		
one side	1.2	1.5
both sides	2.3	2.6
Sentra		
one side	1.5	1.8
both sides	3.0	3.3

NI-34   200SX : 210 : 240SX : 280ZX : 300ZX : 310 : 350Z : 510 : 810

	LABOR TIME	SEVERE SERVICE
Lower Ball Joint, Replace (B)		
Stanza		
1982-86		
one side	1.7	2.0
both sides	2.9	3.2
1987-92		
one side	1.3	1.6
both sides	2.1	2.4
Stanza Wagon		
one side	1.3	1.6
both sides	2.1	2.4
Stabilizer Bar Rods, Replace (B)		
One	.8	1.0
Both	1.1	1.3
Stabilizer Bar, Replace (B)		
Includes: Replace bushings.		
1995-98 200SX	1.1	1.3
240SX		
1989-94	.8	1.0
1995-98	1.2	1.4
300ZX, 350Z	1.0	1.2
Altima		
1993-97	.8	1.0
1998-05	1.5	1.7
Maxima		
1985-94	.8	1.0
1995-04	2.7	2.9
2005	1.4	1.6
NX	1.1	1.3
Pulsar	1.4	1.6
Sentra		
1982-99	1.4	1.6
2000-05	1.6	1.8
Stanza		
1982-89	1.3	1.5
1990-92	.8	1.0
Stanza Wagon	.8	1.0
Stabilizer Bar Bushings, Replace (B)		
All Models	1.3	1.5
Steering Knuckle, Replace (B)		
1982-93 FWD		
one side	2.8	3.3
both sides	4.9	5.4
1995-98 200SX		
one side	2.7	3.2
both sides	4.6	5.1
Altima		
1994-97		
one side	2.4	2.9
both sides	4.0	4.5
1998-05		
one side	2.9	3.4
both sides	4.3	4.8
Maxima		
1994-01		
one side	3.1	3.6
both sides	4.7	5.2
2002-04		
one side	3.2	3.7
both sides	5.2	5.7

	LABOR TIME	SEVERE SERVICE
2005		
one side	2.1	2.3
both sides	2.9	3.2
NX		
one side	3.2	3.7
both sides	6.3	6.8
Sentra		
1994		
one side	3.2	3.7
both sides	6.3	6.8
1995-05		
one side	2.7	3.2
both sides	4.6	5.1
Steering Knuckle Arm, Replace (B)		
Includes: Reset toe.		
200SX		
one side	.8	1.0
both sides	1.3	1.5
210		
one side	.8	1.0
both sides	1.7	1.9
280ZX		
one side	.8	1.0
both sides	1.8	2.0
300ZX		
one side	1.2	1.4
both sides	1.6	1.8
510		
one side	.8	1.0
both sides	1.6	1.8
810/Maxima		
one side	.8	1.0
both sides	1.6	1.8
Steering Knuckle Extension and/or Bearings, Replace (B)		
1990-96 300ZX		
one	2.7	3.2
both	3.2	3.7
Steering Spindle, Replace (B)		
Includes: Alignment.		
RWD		
240SX		
1984-94		
one side	1.4	1.9
both sides	2.3	2.8
1995-98		
one side	2.3	2.8
both sides	3.3	3.8
300ZX		
one side	1.3	1.8
both sides	2.3	2.8
Tension Rod and/or Bushing, Replace (B)		
Includes: Reset toe.		
200SX		
1981-83		
one side	.7	.9
both sides	1.2	1.4

	LABOR TIME	SEVERE SERVICE
1984-88		
one side	.5	.7
both sides	.9	1.1
210		
one side	.7	.9
both sides	1.2	1.4
240SX		
1989-94		
one side	.8	1.0
both sides	1.2	1.4
1995-98		
one side	.9	1.1
both sides	1.5	1.7
280ZX		
one side	.7	.9
both sides	1.2	1.4
300ZX		
one side	.7	.9
both sides	1.3	1.5
510		
one side	.7	.9
both sides	1.3	1.5
810/Maxima		
one side	.7	.9
both sides	1.4	1.6
Transverse Links, Replace (B)		
Includes: Alignment (FWD) or reset toe (RWD).		
1995-98 200SX		
one side	1.4	1.6
both sides	2.2	2.4
1994-98 240SX		
one side	1.2	1.4
both sides	1.6	1.8
1990-96 300ZX		
one side	1.1	1.3
both sides	1.9	2.1
350Z		
one side	1.4	1.6
both sides	1.7	1.9
Altima		
1993-97		
one side	2.3	2.5
both sides	3.7	3.9
1998-01		
one side	2.2	2.4
both sides	3.0	3.2
2002-05		
one side	1.9	2.1
both sides	2.2	2.4
Maxima		
1985-94		
one side	1.7	1.9
both sides	2.8	3.0
1995-04		
one side	2.9	3.1
both sides	4.4	4.6
2005		
one side	1.8	2.0
both sides	2.1	2.3

Altima : Maxima : NX : Pulsar : Pulsar NX : Sentra : Stanza   **NI-35**

	LABOR TIME	SEVERE SERVICE
NX		
one side	2.6	2.8
both sides	4.1	4.3
Pulsar		
one side	2.4	2.6
both sides	3.8	4.0
Sentra		
1982-94		
one side	2.6	2.8
both sides	4.1	4.3
1995-99		
one side	1.4	1.6
both sides	2.2	2.4
2000-05		
one side	2.1	2.3
both sides	3.5	3.7
Stanza		
1982-86		
one side	2.3	2.5
both sides	3.8	4.0
1987-89		
one side	1.7	1.9
both sides	2.8	3.0
1990-92		
one side	1.7	1.9
both sides	2.1	2.3
Stanza Wagon		
one side	1.8	2.0
both sides	2.9	3.1
Wheel Hub, Bearing or Seal, Replace (B)		
200SX		
1981-86		
one side	1.3	1.7
both sides	2.6	3.0
1995-98		
one side	2.7	3.1
both sides	4.6	5.0
210		
one side	1.7	2.1
both sides	3.2	3.6
240SX		
1989-94		
one side	3.0	3.4
both sides	5.2	5.6
1995-98		
one side	.9	1.3
both sides	1.5	1.9
280ZX		
one side	1.6	2.0
both sides	3.3	3.7
300ZX		
one side	1.4	1.8
both sides	2.7	3.1
350Z		
one side	1.1	1.5
both sides	2.1	2.5
510		
one side	1.8	2.2
both sides	3.2	3.6

	LABOR TIME	SEVERE SERVICE
810/Maxima		
one side	1.7	2.1
both sides	2.1	2.5
Altima		
1993-97		
one side	2.4	2.8
both sides	4.0	4.4
1998-01		
one side	2.9	3.3
both sides	4.3	4.7
2002-05		
one side	2.2	2.6
both sides	2.9	3.3
Maxima		
1985-01		
one side	3.1	3.5
both sides	4.8	5.2
2002-04		
one side	3.2	3.6
both sides	5.2	5.6
2005		
one side	2.1	2.3
both sides	2.9	3.2
NX		
one side	3.2	3.6
both sides	6.3	6.7
Pulsar		
one side	2.7	3.1
both sides	4.8	5.2
Sentra		
1982-86		
one side	1.8	2.2
both sides	3.2	3.6
1987-90		
one side	2.8	3.2
both sides	5.3	5.7
1991-94		
one side	3.2	3.6
both sides	6.3	6.7
1995-05		
one side	2.7	3.1
both sides	4.6	5.0
Stanza		
1982-86		
one side	1.7	2.2
both sides	3.1	3.6
1987-89		
one side	2.6	3.0
both sides	5.3	5.7
Stanza Wagon		
one side	2.6	3.0
both sides	5.0	5.4

REAR SUSPENSION
Leaf Spring Eye Mount and/or Bushings, Replace (B)

	LABOR TIME	SEVERE SERVICE
210, 510		
one side	1.1	1.3
both sides	2.3	2.5

	LABOR TIME	SEVERE SERVICE
Lower Control Arm and/or Bushings, Replace (B)		
200SX		
one side	2.2	2.5
both sides	3.7	4.0
240SX		
1989-94		
one side	1.8	2.1
both sides	3.2	3.5
1995-98		
one side	1.4	1.7
both sides	1.8	2.1
280ZX		
one side	2.5	2.8
both sides	4.1	4.4
300ZX		
1984-89		
one side	2.9	3.2
both sides	5.0	5.3
1990-96		
one side	1.0	1.3
both sides	1.7	2.0
310		
one side	2.4	2.7
both sides	4.3	4.6
810/Maxima		
one side	2.7	3.0
both sides	4.7	5.0
Altima		
one side	3.3	3.6
both sides	3.6	3.9
Pulsar		
1983-86		
one side	2.4	2.7
both sides	4.3	4.6
1987-90		
one side	2.4	2.7
both sides	3.7	4.0
Sentra		
one side	2.3	2.6
both sides	4.2	4.5
Stanza Wagon 4WD		
one side	1.7	2.0
both sides	3.1	3.4
Lower Link, Replace (B)		
1982-93		
one side	.7	.9
both sides	1.3	1.5
240SX		
1994		
one side	.7	.9
both sides	1.3	1.5
1995-98		
one side	.9	1.1
both sides	1.7	1.9
300ZX		
one side	.7	.9
both sides	1.3	1.5

200SX : 210 : 240SX : 280ZX : 300ZX : 310 : 350Z : 510 : 810

	LABOR TIME	SEVERE SERVICE
Lower Link, Replace (B)		
350Z		
one side	1.7	*1.9*
both sides	1.9	*2.1*
Altima		
1998-01		
one side	.6	*.8*
both sides	.8	*1.0*
2002-05		
one side	3.0	*3.2*
both sides	3.2	*3.4*
2005 Maxima		
one side	2.8	*3.0*
both sides	2.9	*3.1*
Parallel Rods, Replace (B)		
Altima		
1993-97		
one side	.7	*.9*
both sides	1.3	*1.5*
1998-01		
one side	2.1	*2.3*
both sides	2.7	*2.9*
2002-05 one or		
both sides	1.6	*1.8*
Maxima		
one side	1.8	*2.0*
both sides	3.4	*3.6*
1987-94 NX, Sentra 2WD		
one side	1.2	*1.4*
both sides	1.6	*1.8*
Pulsar		
one side	1.3	*1.5*
both sides	1.8	*2.0*
Stanza		
1987-89		
one side	1.8	*2.0*
both sides	3.3	*3.5*
1990-92		
one side	1.5	*1.7*
both sides	2.4	*2.6*
Rear Control Arm, Replace (B)		
Stanza Wagon 2WD	2.0	*2.3*
Rear Leaf Spring, Replace (B)		
210, 510		
one	2.0	*2.4*
both	3.8	*4.2*
Rear Shock Absorbers and/or Springs, Replace (B)		
200SX		
1984-88		
one	.7	*.8*
both	1.3	*1.4*
1995-98		
one	1.4	*1.5*
both	2.2	*2.3*
210, 510		
one	.7	*.8*
both	1.2	*1.3*

	LABOR TIME	SEVERE SERVICE
240SX		
one	1.5	*1.6*
both	2.3	*2.4*
1990-96 300ZX		
one	1.4	*1.5*
both	2.9	*3.0*
350Z		
one	1.7	*1.8*
both	2.1	*2.2*
1995-05 Maxima		
one	1.1	*1.2*
both	1.9	*2.0*
Sentra		
1987-90 4WD		
one	1.4	*1.5*
both	2.8	*2.9*
1995-99		
one	1.4	*1.5*
both	2.2	*2.3*
2000-05		
one	.8	*.9*
both	1.4	*1.5*
Stanza Wagon		
2WD		
one	.5	*.6*
both	.8	*.9*
4WD		
one	1.1	*1.2*
both	2.1	*2.3*
Rear Struts, R&R or Replace (B)		
Altima		
1993-97		
one side	1.8	*1.9*
both side	2.7	*2.8*
1998-01		
one side	2.7	*2.8*
both side	3.2	*3.3*
2002-05		
one side	.5	*.6*
both side	.8	*.9*
Maxima		
1985-88		
one side	2.1	*2.2*
both sides	3.7	*3.8*
1989-94		
one side	1.7	*1.8*
both sides	3.0	*3.1*
2000-2005		
one side	.5	*.7*
both sides	.8	*1.0*
1991-93 NX		
one side	2.7	*2.8*
both sides	4.1	*4.2*
Pulsar		
one side	2.0	*2.1*
both sides	3.3	*3.4*
Sentra		
1982-88		
one side	1.9	*2.0*
both sides	3.7	*3.8*

	LABOR TIME	SEVERE SERVICE
1987-90 2WD		
one side	1.9	*2.0*
both sides	3.3	*3.4*
1991-05		
one side	2.7	*2.8*
both sides	4.1	*4.2*
Stanza		
1987-89		
one side	2.1	*2.2*
both sides	3.8	*3.9*
1990-92		
one side	1.9	*2.0*
both sides	3.4	*3.5*
Rear Wheel Knuckle Spindle, Replace (B)		
Altima		
1993-97		
one side	1.4	*1.9*
both sides	2.3	*2.8*
1998-01		
one side	2.1	*2.6*
both sides	2.7	*3.2*
2002-05		
one side	2.5	*3.0*
both sides	3.5	*4.0*
Maxima		
2005	2.4	*2.6*
1991-93 NX		
one side	1.8	*2.3*
both sides	3.1	*3.6*
Pulsar		
one side	1.8	*2.3*
both sides	3.2	*3.7*
Sentra		
1987-90		
2WD		
one side	1.6	*2.1*
both sides	2.7	*3.2*
4WD one	3.6	*4.1*
1991-05		
one side	1.8	*2.3*
both sides	3.1	*3.6*
Stanza Wagon 4WD	3.6	*4.1*
Rear Wheel Oil Seal, Replace (B)		
240SX		
one side	1.2	*1.4*
both sides	2.7	*2.9*
1990-96 300ZX		
one side	1.3	*1.5*
both sides	2.5	*2.7*
Altima		
1993-01		
one side	.5	*.7*
both sides	.8	*1.0*
2002-05		
one side	.8	*1.0*
both sides	1.4	*1.6*

Altima : Maxima : NX : Pulsar : Pulsar NX : Sentra : Stanza NI-37

	LABOR TIME	SEVERE SERVICE
Maxima		
1985-88		
one side	1.2	*1.4*
both sides	2.1	*2.3*
1989-94		
one side	.8	*1.0*
both sides	1.6	*1.8*
1991-93 NX		
one side	1.3	*1.5*
both sides	2.2	*2.4*
Pulsar		
one side	1.2	*1.4*
both sides	2.1	*2.3*
1987-94 Sentra 2WD		
one side	1.3	*1.5*
both sides	2.2	*2.4*
Stanza		
1987-89		
one side	1.2	*1.4*
both sides	2.3	*2.5*
1990-92		
one side	.9	*1.1*
both sides	1.9	*2.1*
Stanza Wagon 4WD		
one side	1.8	*2.0*
both sides	2.5	*2.7*
Stabilizer Bar, Replace (B)		
Exc. Altima	.7	*.9*
Altima	.8	*1.0*
Stabilizer Bar & Bushings, Replace (B)		
2WD	.8	*1.0*
4WD	1.3	*1.5*
Spring Shackle, Replace (B)		
210, 510		
one side	.9	*1.1*
both side	1.9	*2.1*
Torsion Bar, Replace (B)		
Stanza Wagon 2WD		
one side	2.4	*2.9*
both sides	4.2	*4.7*
Upper Control Arm and/or Bushings, Replace (B)		
1984-88 200SX		
one	.9	*1.2*
both	1.8	*2.1*
350Z		
one	2.3	*2.6*
both	3.4	*3.8*
2005 Maxima		
one	2.6	*2.9*
both	3.0	*3.3*
Upper Link, Replace (B)		
1989-05 front or rear		
one	.6	*.9*
both	1.1	*1.4*

	LABOR TIME	SEVERE SERVICE
Wheel Hub, Replace (B)		
Includes: Rear wheel oil seal.		
1995-98 200SX		
one side	1.1	*1.6*
both sides	1.7	*2.2*
Altima		
1993-01		
one side	.5	*.7*
both sides	.8	*1.0*
2002-05		
one side	.8	*1.0*
both sides	1.4	*1.6*
1995-05 Maxima		
one side	1.1	*1.6*
both sides	1.9	*2.4*
Sentra		
1995-99		
one side	1.1	*1.6*
both sides	1.7	*2.2*
2000-05		
one	.7	*1.2*
both	1.0	*1.5*

STEERING
AIR BAGS

	LABOR TIME	SEVERE SERVICE
Diagnose Air Bag System Component Each (A)		
1990-05	1.0	*1.0*
Retrieve Fault Codes (A)		
1990-05	.3	*.3*
Air Bag Assy., Replace (B)		
Driver side	.6	*.6*
Passenger side		
200SX	.8	*.8*
240SX	2.5	*2.5*
300ZX	4.3	*4.3*
350Z	.8	*.8*
Altima		
1993-97	.9	*.9*
1998-05	1.3	*1.3*
Maxima		
exc. 2005	.8	*.8*
2005	4.4	*4.8*
Sentra	.8	*.8*
Side Impact		
350Z		
one	1.2	*1.2*
both	2.2	*2.2*
Altima		
1998-01		
one	1.1	*1.1*
both	1.6	*1.6*
2002-05		
one	.8	*.8*
both	1.2	*1.2*
Maxima		
one	.8	*.8*
both	1.0	*1.0*

	LABOR TIME	SEVERE SERVICE
Sentra		
one	1.0	*1.0*
both	1.4	*1.4*
Air Bag On/Off Switch, Replace (B)		
1990-05	.9	*.9*
Air Bag Spiral Cable, Replace (B)		
1995-98 200SX	.7	*.7*
1995-98 240SX	.7	*.7*
1990-96 300ZX	.8	*.8*
350Z	1.4	*1.4*
Altima		
1993-97	.5	*.5*
1998-05	1.1	*1.1*
Maxima		
exc. 2005	.8	*.9*
2005	1.1	*1.3*
NX	.5	*.5*
Pulsar	.9	*.9*
Sentra		
1991-94	.5	*.5*
1995-05	1.0	*1.0*
Crash Sensor, Replace (B)		
Center		
1995-98 240SX	2.7	*2.7*
1990-96 300ZX	2.0	*2.0*
350Z	.5	*.5*
Altima		
1993-97	1.7	*1.7*
2002-05	.6	*.6*
2002-05 Maxima	.5	*.5*
1991-93 NX	.5	*.5*
Pulsar	.5	*.5*
1991-94 Sentra		
coupe	.5	*.5*
sedan	1.4	*1.4*
Side		
1990-96 300ZX		
one	3.8	*3.8*
both	4.3	*4.3*
350Z, Altima		
one	.8	*.8*
both	1.1	*1.1*
Maxima		
one	.8	*.8*
both	1.4	*1.4*
NX		
one	2.5	*2.5*
both	2.7	*2.7*
Sentra		
1991-94		
one	2.5	*2.5*
both	2.7	*2.7*
2000-05		
one	.8	*.8*
both	1.0	*1.0*
Electronic Control Unit, Replace (B)		
1995-98 240SX	.7	*.7*
1990-96 300ZX	.8	*.8*
1993-97 Altima	1.2	*1.2*

200SX : 210 : 240SX : 280ZX : 300ZX : 310 : 350Z : 510 : 810

	LABOR TIME	SEVERE SERVICE
Electronic Control Unit, Replace (B)		
1991-94 Maxima	2.9	2.9
1991-93 NX	1.1	1.1
Pulsar	.9	.9
1991-94 Sentra		
coupe	1.1	1.1
sedan	1.9	1.9
Sensing and Diagnostic Module(SDM), Replace (B)		
200SX	.7	.7
1995-98 240SX	.7	.7
Altima	.8	.8
Maxima		
1995-99	.5	.5
2000-05	1.1	1.1
Sentra		
1995-99	.7	.7
2000-05	1.0	1.0
Tunnel Sensor, Replace (B)		
1995-98 240SX	2.4	2.4
1990-96 300ZX	.8	.8
1993-97 Altima	3.0	3.0
1995-01 Maxima	3.1	3.1
1991-93 NX	.5	.5
Pulsar	.5	.5
1991-94 Sentra		
coupe	.5	.5
sedan	.8	.8
MANUAL RACK & PINION		
Inner Tie Rod, Replace (B)		
Includes: Reset toe.		
200SX		
1984-88		
one side	1.1	1.3
both sides	1.6	1.8
1995-98		
one side	1.1	1.3
both sides	1.5	1.7
280ZX		
one side	2.4	2.6
both sides	2.9	3.1
310		
one side	2.2	2.4
both sides	2.9	3.1
810/Maxima		
one side	1.3	1.5
both sides	1.7	1.9
1991-93 NX		
one side	1.1	1.3
both sides	1.5	1.7
1983-86 Pulsar		
one side	1.3	1.5
both sides	1.7	1.9
Sentra		
1982-86		
one side	1.1	1.3
both sides	1.6	1.8

	LABOR TIME	SEVERE SERVICE
1987-99		
one side	1.1	1.3
both sides	1.5	1.7
1982-86 Stanza		
one side	1.2	1.4
both sides	1.5	1.7
Outer Tie Rod, Replace (B)		
Includes: Reset toe.		
One side	.8	1.0
Both sides	1.4	1.4
Pinion Oil Seal, Replace (B)		
200SX		
1984-88	1.9	2.1
1995-98	3.0	3.2
280ZX	2.9	3.1
310	1.9	2.1
810/Maxima	.9	1.1
1991-93 NX	3.9	4.1
1983-86 Pulsar	3.3	3.5
Sentra		
1982-86	3.4	3.6
1987-94	3.9	4.1
1995-99	3.0	3.2
1982-86 Stanza	3.5	3.7
Rack & Pinion Assy., R&R and Recondition (A)		
Includes: Alignment		
200SX		
1984-88	2.8	3.3
1995-98	3.4	3.9
280ZX	3.5	4.0
310	3.1	3.6
810/Maxima	3.3	3.8
1991-93 NX	3.8	4.3
Pulsar	2.8	3.3
Sentra		
1982-86	2.7	3.2
1987-88	2.9	3.4
1989-90	3.3	3.8
1991-94	3.8	4.3
1995-99	3.4	3.9
Stanza	2.8	3.3
Rack & Pinion Assy., Adjust (B)		
1981-82	.9	1.1
Rack & Pinion Assy., R&R or Replace (B)		
Includes: Alignment.		
200SX		
1984-88	1.4	1.7
1995-98	2.3	2.6
280ZX	2.5	2.8
310	1.4	1.7
810/Maxima	1.8	2.1
1991-93 NX	2.7	3.0
Pulsar	1.5	1.8
Sentra		
1982-86	1.6	1.9
1987-88	1.9	2.2
1989-90	2.5	2.8

	LABOR TIME	SEVERE SERVICE
1991-94	2.7	3.0
1995-99	2.3	2.6
2000-05	1.9	2.1
Stanza	1.7	2.0
Steering Column, R&I (B)		
200SX		
1984-88	1.8	1.8
1995-98	1.7	1.7
280ZX	1.6	1.6
310	1.6	1.6
810/Maxima	1.6	1.6
1991-93 NX	1.7	1.7
Maxima		
2004-05	2.2	2.4
Pulsar	1.8	1.8
Sentra		
1982-90	1.6	1.6
1991-94	1.7	1.7
1995-99	1.7	1.7
2000-05	1.4	1.6
Stanza	1.8	1.8
Steering Wheel, Replace (B)		
1981-99	.5	.5
2000-05	.8	.9
Upper Mast Jacket Bearing, Replace (B)		
Includes: Insulators.		
1984-88 200SX	2.2	2.4
280ZX	2.2	2.4
310	2.1	2.3
810/Maxima	2.3	2.5
1991-93 NX	2.1	2.3
Pulsar	2.1	2.3
Sentra		
1982-90	2.3	2.5
1991-97	2.1	2.3
Stanza	2.2	2.4
POWER RACK & PINION		
Troubleshoot Power Steering System Component Each (B)		
All Models	.5	.5
Bleed System (B)		
Exc. below	.5	.5
1998-05 Altima	.7	.7
Hose & Tube Assy., Replace (B)		
Includes: Bleed system.		
200SX		
1981-88	.8	1.0
1995-98	1.7	1.9
240SX		
1989-94 240SX	1.3	1.5
1995-98 240SX	.8	1.0
300ZX		
1984-89	.8	1.0
1990-96	1.5	1.7
310	.8	1.0
350Z	1.7	1.9
810/Maxima	.8	1.0

Altima : Maxima : NX : Pulsar : Pulsar NX : Sentra : Stanza

	LABOR TIME	SEVERE SERVICE
Altima		
1993-97	1.4	1.6
1998-05	2.1	2.3
Maxima		
1985-01	1.4	1.6
2002-05	2.3	2.5
NX	1.6	1.8
Pulsar	1.4	1.6
Sentra		
1982-86	.8	1.0
1987-90		
2WD	.7	.9
4WD	1.3	1.5
1991-94	1.6	1.8
1995-05		
exc. QR	1.2	1.4
QR	2.0	2.2
Stanza	1.3	1.5
Stanza Wagon	.8	1.0

Inner Tie Rod, Replace (B)
Includes: Reset toe.

	LABOR TIME	SEVERE SERVICE
200SX		
one side	1.3	1.5
both sides	1.7	1.9
1994-98 240SX		
one side	1.1	1.3
both sides	1.6	1.8
280ZX		
one side	2.4	2.6
both sides	3.0	3.2
300ZX		
one side	1.3	1.5
both sides	1.6	1.8
310		
one side	2.3	2.5
both sides	2.8	3.0
350Z		
one side	1.1	1.3
both sides	1.6	1.8
810/Maxima		
one side	1.3	1.5
both sides	1.5	1.7
Altima		
1993-97		
one side	1.1	1.3
both sides	1.5	1.7
1998-05		
one side	1.9	2.1
both sides	2.4	2.6
Maxima		
1985-01		
one side	1.2	1.4
both sides	1.6	1.8
2002-05		
one side	1.6	1.8
both sides	2.4	2.6
NX		
one side	1.1	1.3
both sides	1.7	1.9

	LABOR TIME	SEVERE SERVICE
Pulsar		
one side	1.1	1.3
both sides	1.6	1.8
Sentra		
1982-99		
one side	1.2	1.4
both sides	1.7	1.9
2000-05		
one side	1.4	1.6
both sides	2.1	2.3
Stanza		
1982-86		
one side	1.1	1.3
both sides	1.7	1.9
1987-92		
one side	1.2	1.4
both sides	1.7	1.9
Stanza Wagon		
one side	1.3	1.5
both sides	1.5	1.7

Oil Pressure Switch, Replace (B)

	LABOR TIME	SEVERE SERVICE
All Models	.7	.7

Outer Tie Rod, Replace (B)
Includes: Reset toe.

	LABOR TIME	SEVERE SERVICE
Exc. below		
one side	1.1	1.3
both sides	1.6	1.8
1998-05 Altima		
one side	1.6	1.8
both sides	1.8	2.0

Rack & Pinion Assy., R&R and Recondition (A)

	LABOR TIME	SEVERE SERVICE
200SX	4.5	5.0
240SX	3.8	4.3
310		
1981	5.8	6.3
1982	4.2	4.7
350Z	3.6	4.1
810/Maxima	3.3	3.8
Altima		
1993-01	6.3	6.8
2002-05	3.3	3.8
Maxima		
1985-94	4.5	5.0
1995-01	6.9	7.4
2002-04	8.6	9.1
NX	5.4	5.9
Pulsar	5.5	6.0
Sentra		
1982-86	5.5	6.0
1987-90		
E		
2WD	4.4	4.9
4WD	4.9	5.4
GA		
2WD	4.9	5.4
4WD	5.4	5.9

	LABOR TIME	SEVERE SERVICE
1991-94	5.4	5.9
1995-99	4.5	5.0
2000-05	3.2	3.7
Stanza	5.2	5.7

Rack & Pinion Assy., R&R or Replace (B)
Includes: Alignment.

	LABOR TIME	SEVERE SERVICE
200SX	2.9	3.2
240SX	2.4	2.7
300ZX		
1984-87	2.7	3.0
1988-89	3.1	3.4
1990-96	2.5	2.8
310		
1981	3.7	4.0
1982	2.5	2.8
350Z	2.3	2.6
810/Maxima	1.8	2.1
Altima		
1993-97	2.9	3.2
1998-01	4.4	4.7
2002-05	2.4	2.7
Maxima		
1985-94	3.2	3.5
1995-01	5.6	5.9
2002-04	7.2	7.5
2005	1.5	1.7
NX	3.8	4.3
Pulsar	4.1	4.4
Sentra		
1982-86	4.3	4.6
1987-90		
E		
2WD	3.2	3.5
4WD	3.5	3.8
GA		
2WD	3.7	4.0
4WD	4.2	4.5
1991-94	3.8	4.3
1995-99	2.9	3.2
2000-05	1.8	2.1
Stanza	3.8	4.1

Steering Column, R&I (B)

	LABOR TIME	SEVERE SERVICE
200SX	1.8	1.8
240SX	1.8	1.8
300ZX	1.6	1.6
310	1.8	1.8
350Z	2.2	2.2
810/Maxima	1.6	1.6
Altima		
1993-01	1.6	1.6
2002-05	.9	.9
Maxima		
1985-94	1.7	1.7
1995-01	1.3	1.3
2002-04	2.7	2.7
2005	2.4	2.4
NX, Pulsar	1.7	1.7

NI-40 200SX : 210 : 240SX : 280ZX : 300ZX : 310 : 350Z : 510 : 810

	LABOR TIME	SEVERE SERVICE
Steering Column, R&I (B)		
Sentra		
1982-94	1.8	*1.8*
1995-05	1.5	*1.5*
Stanza		
1982-89	1.7	*1.7*
1990-92	2.3	*2.3*
Steering Oil Cooler, Replace (B)		
1995-98 240SX	1.3	*1.5*
Altima	1.1	*1.3*
Maxima	.9	*1.1*
NX	.9	*1.1*
Sentra	1.1	*1.3*
Steering Pump, R&R or Replace (B)		
200SX		
1984-88	1.5	*1.7*
1995-98	2.3	*2.5*
240SX		
1989-94	1.3	*1.5*
1995-98	1.6	*1.8*
280ZX	1.5	*1.7*
300ZX	1.6	*1.8*
310		
1981	2.2	*2.4*
1982	1.3	*1.5*
350Z	2.0	*2.2*
810/Maxima	1.3	*1.5*
Altima	2.0	*2.2*
Maxima		
1985-94	1.3	*1.5*
1995-04	2.0	*2.2*
2005	1.1	*1.4*
Pulsar	1.1	*1.3*
Sentra		
1982-90	1.4	*1.6*
1991-05	2.3	*2.5*
Stanza	1.4	*1.6*
w/HICAS Altima add	.3	*.3*
Steering Wheel, Replace (B)		
All Models	.7	*.7*
Upper Mast Jacket Bearing, Replace (B)		
Includes: Insulators.		
1984-88 200SX	2.1	*2.3*
1989-94 240SX	2.2	*2.4*
300ZX	2.1	*2.3*
310	2.1	*2.3*
810/Maxima	2.1	*2.3*
1993-94 Altima	2.2	*2.4*
1985-94 Maxima	2.1	*2.3*
1991-93 NX	2.3	*2.5*
Pulsar	2.3	*2.5*
1982-94 Sentra	2.3	*2.5*
Stanza		
1982-89	2.2	*2.4*
1990-92	2.8	*3.0*

	LABOR TIME	SEVERE SERVICE
MANUAL WORM & SECTOR		
Gear Assy., R&R and Recondition (B)		
210, 510	3.0	*3.5*
Gear Assy., Replace (B)		
210, 510	1.4	*1.7*
Sector Shaft and/or Oil Seal, Replace (B)		
210, 510	2.0	*2.2*
Steering Column, R&R or Replace (B)		
210, 510	1.7	*1.7*
Steering Gear, Adjust (B)		
210, 510	.8	*1.0*
Steering Wheel, Replace (B)		
210, 510	.5	*.5*
Upper Mast Jacket Bearing, Replace (B)		
Includes: Insulators.		
210, 510	1.6	*1.8*
POWER WORM & SECTOR		
Gear Assy., Replace (B)		
1981-83 200SX	2.8	*3.0*
Hose & Tube Assy., Replace (B)		
1981-83 200SX	1.5	*1.7*
Sector Shaft and/or Oil Seal, Replace (B)		
1981-83 200SX	3.2	*3.4*
Steering Column, R&R or Replace (B)		
1981-83 200SX	1.5	*1.5*
Steering Pump, R&R or Replace (B)		
1981-83 200SX	1.4	*1.6*
Steering Wheel, Replace (B)		
1981-83 200SX	.5	*.5*
Upper Mast Jacket Bearing, Replace (B)		
Includes: Replace insulators if necessary.		
1981-83 200SX	1.6	*1.8*

HEATING & AIR CONDTIONING

When more than one component requires replacement where evacuation/recovery and recharging is already included, deduct 1.0 hour for each additional component from the time given.

Evacuate/Recover and Recharge System (B)

	LABOR TIME	SEVERE SERVICE
R12	1.2	*1.2*
R134A	.9	*.9*

AC Hoses, Replace (B)
Includes: Evacuate/recover and recharge.

	LABOR TIME	SEVERE SERVICE
Exc. below each	2.2	*2.4*
1995-98 200SX		
GA each	2.1	*2.3*
SR		
high pressure	3.1	*3.3*
low pressure	1.8	*2.0*
240SX each	2.1	*2.3*
1990-96 300ZX		
high pressure	4.4	*4.6*
low pressure	1.8	*2.0*
Altima		
1993-97 each	2.3	*2.5*
1998-01		
high pressure	2.4	*2.6*
low pressure	2.0	*2.2*
2002-05 each	1.7	*1.9*
Maxima		
1995-05		
high pressure	3.2	*3.4*
low pressure		
R12 system	2.0	*2.2*
R134A system	1.6	*1.8*
NX each	2.1	*2.3*
Sentra		
1991-94 each	2.1	*2.3*
1995-99		
GA each	2.1	*2.3*
SR		
high pressure	3.1	*3.3*
low pressure	1.8	*2.0*
2000-05 each	1.8	*2.0*
1990-92 Stanza		
high pressure	2.3	*2.5*
low pressure	1.8	*2.0*
Blower Motor, Replace (B)		
Exc. below	.6	*.6*
240SX	.7	*.7*
310	.9	*.9*
810/Maxima	.7	*.7*
Maxima	.8	*.8*
1982-89 Stanza	1.3	*1.3*
Blower Motor Relay, Replace (B)		
Exc. 350Z	.5	*.5*
350Z	1.6	*1.6*
Blower Motor Resistor, Replace (B)		
All Models	.7	*.7*
Compressor Assy., Replace (B)		
Includes: Parts transfer, evacuate/recover and recharge.		
200SX	3.0	*3.2*
210	3.7	*3.9*
240SX	2.3	*2.6*
280ZX	3.3	*3.5*
300ZX	2.9	*3.1*
310	3.8	*4.0*
350Z	2.0	*2.2*
510	3.5	*3.7*
810/Maxima	3.1	*3.3*
Altima	2.2	*2.4*
Maxima		
1985-88	2.4	*2.6*

Altima : Maxima : NX : Pulsar : Pulsar NX : Sentra : Stanza   NI-41

	LABOR TIME	SEVERE SERVICE
1989-94		
VE	2.5	2.7
VG	1.7	1.9
1995-05	2.6	2.8
NX	2.2	2.4
Pulsar	3.2	3.4
Sentra		
1982-90	3.2	3.4
1991-94	2.2	2.4
1995-99	2.8	3.0
2000-05	2.2	2.4
Stanza		
1982-86	3.2	3.4
1987-89		
w/turbo	2.8	3.0
w/o turbo	3.2	3.4
1990-92	2.7	2.9
Stanza Wagon	2.6	2.8
Compressor or Fan Relay, Replace (B)		
Exc. 350Z	.5	.5
350Z	.8	.8
Condenser Assy., Replace (B)		
Includes: Evacuate/recover and recharge.		
200SX		
1981-83	3.0	3.2
1984-88	3.7	3.9
1995-98	2.5	2.7
210, 310	2.8	3.0
240SX	4.1	4.3
280ZX	3.1	3.3
300ZX		
1984-89	3.1	3.3
1990-96	3.5	3.7
350Z	2.8	3.0
510	3.2	3.4
810/Maxima	3.2	3.4
Altima		
1993-01	2.1	2.3
2002-05	2.4	2.6
Maxima	3.2	3.4
NX	3.2	3.4
Pulsar	2.9	3.1
Sentra		
1982-94	3.2	3.4
1995-05	2.6	2.8
Stanza		
1982-86	2.7	2.9
1987-92	3.3	3.5
Stanza Wagon	3.2	3.4
Evaporator Core, Replace (B)		
Includes: Evacuate/recover and recharge.		
200SX		
1981-88	3.9	4.2
1995-98	1.9	2.2
210	3.4	3.7

	LABOR TIME	SEVERE SERVICE
240SX		
1989-94	3.3	3.6
1995-98	1.9	2.2
280ZX	4.7	5.0
300ZX		
1984-89	6.5	6.8
1990-96	4.3	4.6
310	3.9	4.2
350Z	3.5	4.1
510	3.4	3.7
810/Maxima	3.8	4.1
Altima		
1993-97	2.9	3.2
1998-01	2.1	2.4
2002-05	5.1	5.4
Maxima		
1985-88	4.3	4.6
1989-01	2.0	2.3
2002-04	3.2	3.5
2005	7.4	7.6
NX, Pulsar	3.7	4.0
Sentra		
1982-90	4.3	4.6
1991-94	3.7	4.0
1995-99	1.9	2.2
2000-05	2.5	2.8
Stanza		
1982-86	5.7	6.0
1987-89	4.1	4.4
1990-92	3.9	4.2
Stanza Wagon	4.9	5.2
Expansion Valve, Replace (B)		
Includes: Evacuate/recover and recharge.		
200SX		
1981-88	4.1	4.1
1995-98	2.8	2.8
210	3.8	3.8
240SX		
1989-94	3.9	3.9
1995-98	2.7	2.7
280ZX	4.9	4.9
300ZX		
1984-87	5.2	5.2
1988-89	5.9	5.9
1990-96	3.8	3.8
310	4.3	4.3
350Z	3.5	3.5
510	3.8	3.8
810/Maxima	4.1	4.1
Altima		
1993-01	2.0	2.0
2002-05	5.1	5.1
Maxima		
1985-88	4.2	4.2
1989-01	2.9	2.9
2002-04	3.3	3.3
2005	7.4	7.6
NX, Pulsar	3.9	3.9

	LABOR TIME	SEVERE SERVICE
Sentra		
1982-90	4.4	4.4
1991-94	3.9	3.9
1995-99	2.8	2.8
2000-05	2.5	2.5
Stanza		
1982-86	6.1	6.1
1987-89	3.8	3.8
1990-92	4.3	4.3
Stanza Wagon	5.4	5.4
Heater and AC Control Assy., Replace (B)		
200SX		
1981-83	1.4	1.4
1984-88	2.3	2.3
1995-98	.9	.9
210	2.0	2.0
240SX		
1989-94	2.2	2.2
1995-98	.8	.8
280ZX	1.8	1.8
300ZX		
1984-89	2.1	2.1
1990-96	.8	.8
310	1.2	1.2
350Z	.3	.3
510	1.8	1.8
810/Maxima	1.5	1.5
Altima	.7	.7
Maxima		
1985-88	2.3	2.3
1989-94	1.5	1.5
1995-04	.8	.8
2005	.4	.4
NX	1.8	1.8
Pulsar	1.4	1.4
Sentra		
1982-86	1.6	1.6
1987-90	1.1	1.1
1991-94	1.8	1.8
1995-05	.9	.9
Stanza	.8	.8
Stanza Wagon	2.3	2.3
Heater Core, R&R or Replace (B)		
200SX		
1981-83	4.7	5.0
1995-98		
GA	4.2	4.5
SR	5.1	5.4
210, 310	3.8	4.1
240SX		
1989-94	3.2	3.5
1995-97	4.2	4.5
280ZX	4.5	4.8
300ZX		
1984-89	5.7	6.0
1990-96	6.9	7.2
350Z	9.2	9.5
510	2.8	3.1
810/Maxima	4.7	5.0

NI-42   200SX : 210 : 240SX : 280ZX : 300ZX : 310 : 350Z : 510 : 810

	LABOR TIME	SEVERE SERVICE
Heater Core, R&R or Replace (B)		
Altima		
1993-97	5.3	5.6
1998-01	6.2	6.5
2002-05	4.7	5.0
Maxima		
1985-01	5.8	6.1
2002-05	7.4	7.7
NX	5.6	5.9
Pulsar	4.2	4.5
Sentra		
1982-86	3.1	3.4
1987-90	3.8	4.1
1991-94	5.6	5.9
1995-99		
GA	4.2	4.5
SR	5.1	5.4
2000-05		
QG	5.9	6.2
QR, SR	4.7	5.0
Stanza		
1982-89	3.1	3.4
1990-92	4.5	4.8
Stanza Wagon	4.1	4.4
w/AC 00-04 Sentra QG, SR add	.5	.5
Heater Hoses, Replace (B)		
200SX	1.7	1.9
210, 310, 510	.9	1.1
240SX	1.8	2.0
280ZX	1.7	1.9
300ZX		
1984-89	.9	1.1
1990-96	1.4	1.6
350Z	2.3	2.5
810/Maxima	.9	1.1
Altima		
1993-01	1.4	1.6
2002-05	.9	1.1
Maxima		
1985-88	.9	1.1
1989-94	1.5	1.7
1995-01	.6	.8
2002-04	2.4	2.6
2005	2.1	2.3
NX	1.4	1.6
Pulsar	.9	1.1
Sentra		
1982-86	.9	1.1
1987-99	1.6	1.8
2000-05	.9	1.1
Stanza	.9	1.1
Low Pressure Cut-Off Switch, Replace (B)		
Includes: Evacuate/recover and recharge.		
Exc. below		
R12	1.8	1.8
R134A	1.4	1.4

	LABOR TIME	SEVERE SERVICE
350Z	2.6	2.6
Maxima	1.8	1.8
Magnetic Clutch Assy., Replace (B)		
200SX	3.4	3.6
210, 310	3.7	3.9
240SX	2.9	3.1
280ZX	3.0	3.2
300ZX		
1984-89	3.2	3.4
1990-96		
w/turbo	3.5	3.7
w/o turbo	3.1	3.3
350Z	2.3	2.5
510	3.4	3.6
810/Maxima	3.1	3.3
Altima		
1993-97	2.5	2.7
1998-05	2.1	2.3
Maxima		
1985-88	2.8	3.0
1989-94	2.6	2.8
1995-04	2.1	2.3
2005	1.8	2.0
NX, Pulsar	3.1	3.3
Sentra		
1982-86	2.4	2.6
1987-99	3.4	3.6
2000-05	1.8	2.0
Stanza	3.1	3.3
Receiver/Drier Assy., Replace (B)		
Includes: Evacuate/recover and recharge.		
Exc. below	2.2	2.4
280ZX	1.4	1.6
350Z	2.8	3.0
Altima		
1998-01	1.4	1.6
2002-05	2.4	2.6
2000-05 Sentra	2.4	2.6

WIPERS & SPEEDOMETER

	LABOR TIME	SEVERE SERVICE
Intermittent Wiper Amplifier, Replace (B)		
1985-01 front or rear	.5	.5
Power Antenna, Replace (B)		
1984-88 200SX	1.4	1.4
240SX		
1989-94	1.3	1.3
1995-98	.5	.5
300ZX	.7	.7
Altima	.7	.7
Maxima		
1985-94	1.3	1.3
1995-05	.6	.6
Stanza		
1987-89	1.4	1.4
1990-92	.8	.8

	LABOR TIME	SEVERE SERVICE
Radio, R&R (B)		
200SX		
1981-88	1.3	1.3
1995-98	.8	.8
210, 310	1.3	1.3
240SX	.8	.8
280ZX	1.1	1.1
300ZX		
1984-89	2.1	2.1
1990-96	1.1	1.1
350Z	1.1	1.1
510	1.7	1.7
810/Maxima	1.7	1.7
Altima		
1993-97	1.2	1.2
1998-05	.7	.7
Maxima		
1985-94	.5	.5
1995-05	1.1	1.1
NX	.8	.8
Pulsar	1.3	1.3
Sentra	.8	.8
Stanza	1.3	1.3
Rear Window Washer Motor, Replace (B)		
Exc. below	.5	.5
1989-94 240SX	.8	.8
1990-96 300ZX	.9	.9
1991-93 NX	.8	.8
1987-94 Sentra	.8	.8
1982-89 Stanza	.7	.7
Rear Window Wiper & Washer Switch, Replace (B)		
Exc. below	.7	.7
1991-94 NX, Sentra	.9	.9
Rear Window Wiper Motor, Replace (B)		
Exc. 350Z	.8	1.0
350Z	.3	.5
Speedometer Cable & Casing, Replace (B)		
200SX	1.6	1.6
210	1.2	1.2
1992-94 240SX	1.8	1.8
280ZX	.9	.9
1984-89 300ZX	1.7	1.7
310		
upper	1.6	1.6
lower	.9	.9
510	.9	.9
810/Maxima	1.6	1.6
Altima	1.2	1.2
1985-94 Maxima	1.3	1.3
1991-93 NX	1.3	1.3
Pulsar	.8	.8
Sentra		
1982-90	.9	.9
1991-94	1.3	1.3

Altima : Maxima : NX : Pulsar : Pulsar NX : Sentra : Stanza NI-43

	LABOR TIME	SEVERE SERVICE
Stanza		
1982-89	1.2	1.2
1990-92	1.6	1.6
Stanza Wagon	1.4	1.4

Speedometer Head, R&R or Replace (B)

	LABOR TIME	SEVERE SERVICE
200SX		
1981-83	2.2	2.2
1984-88	1.4	1.4
1995-98	.8	.8
210	3.3	3.3
240SX		
1989-94	1.7	1.7
1995-98	1.2	1.2
280ZX	1.5	1.5
300ZX	1.7	1.7
310, 510	2.3	2.3
350Z	1.1	1.1
810/Maxima	1.4	1.4
Altima	1.4	1.4
Maxima		
exc. 2005	1.4	1.4
2005	.5	.5
NX, Pulsar	.9	.9
Sentra		
1982-94	1.5	1.5
1995-05	.8	.8
Stanza		
1982-89	.7	.7
1990-92	1.5	1.5
Stanza Wagon	1.4	1.4

Windshield Washer Pump, Replace (B)

	LABOR TIME	SEVERE SERVICE
Exc. below	.5	.7
200SX	.8	1.0
240SX	.7	.9
350Z	1.4	1.6
Altima, Maxima, NX	.8	1.0
Pulsar	.5	.7
1987-05 Sentra	.9	1.1
Stanza		
1982-89	.7	.9
1990-92 exc. fender well mount	1.3	1.5

Windshield Wiper & Washer Switch, Replace (B)

	LABOR TIME	SEVERE SERVICE
Exc. below	.9	.9
210	1.5	1.5
1995-05 Maxima	.5	.5

Windshield Wiper Motor, Replace (B)

	LABOR TIME	SEVERE SERVICE
Exc. below	.8	1.0
280ZX	1.8	2.0
1990-96 300ZX	1.3	1.5
1987-92 Stanza	1.3	1.5
2005 Maxima	1.1	1.3

LAMPS & SWITCHES

Back-Up Lamp Bulb, Replace (C)

	LABOR TIME	SEVERE SERVICE
All Models each	.3	.3

Back-Up Lamp Switch, Replace (B)

	LABOR TIME	SEVERE SERVICE
Exc. 350Z	.5	.5
350Z	4.1	4.1

Combination Switch Assy., Replace (B)

	LABOR TIME	SEVERE SERVICE
Exc. below	.9	.9
210	1.3	1.3
350Z	1.4	1.4
Altima		
1993-97	.5	.5
1998-05	1.2	1.2
2002-05 Maxima	1.1	1.1

Daylight Running Lamp Module, Replace (B)

	LABOR TIME	SEVERE SERVICE
Exc. below	.5	.5
2002-05 Altima	.7	.7

Daytime Running Lamp Relay, Replace (B)

	LABOR TIME	SEVERE SERVICE
1995-05	.3	.3

Front Combination Lamp Bulb, Replace (C)

	LABOR TIME	SEVERE SERVICE
Each	.2	.2

Hazard Warning Switch, Replace (B)

	LABOR TIME	SEVERE SERVICE
Exc. below	.7	.7
280ZX	.8	.8
1998-05 Altima	.8	.8

Headlamp & Turn Signal Switch, Replace (B)

	LABOR TIME	SEVERE SERVICE
Exc. below	.5	.5
350Z	.9	.9
1998-05 Altima	.7	.7
Sentra	.8	.8

Headlamp Bulb, Replace (B)

	LABOR TIME	SEVERE SERVICE
Halogen		
exc. below each	.3	.3
350Z	1.5	1.5
Altima		
1993-01	1.1	1.1
2002-05	.7	.7
Xenon		
exc. below each	.4	.4
Altima	1.5	1.5

Headlamp Switch, Replace (B)

	LABOR TIME	SEVERE SERVICE
200SX	.7	.7
210	1.5	1.5

Headlamps, Aim (B)

	LABOR TIME	SEVERE SERVICE
Two	.4	.4
Four	.6	.6

High Mount Stop Lamp Assy., Replace (B)

	LABOR TIME	SEVERE SERVICE
Exc. below	.5	.5
1990-96 300ZX	1.2	1.2
1993-97 Altima	.8	.8
Maxima	.6	.6

Horn, Replace (B)

	LABOR TIME	SEVERE SERVICE
Exc. below	.5	.5
2002-05 Altima	1.4	1.4
1987-89 Stanza	1.3	1.3
2005 Maxima	1.6	1.6

Horn Relay, Replace (B)

	LABOR TIME	SEVERE SERVICE
Exc. below	.5	.5
1981-83 200SX	1.3	1.3

Neutral Safety Switch, Replace (B)

	LABOR TIME	SEVERE SERVICE
FWD		
exc. below	.8	.8
1982 310	.5	.5
Stanza Wagon	2.0	2.0
1981-98 RWD	.5	.5

Park & Turn Signal Lamp Bulb or Lens, Replace (C)

	LABOR TIME	SEVERE SERVICE
Exc. below each	.3	.3
2002-05 Altima each	1.5	1.5
2000-05 Sentra each	.6	.6

Rear Combination Lamp Bulb, Replace (C)

	LABOR TIME	SEVERE SERVICE
Each	.3	.3

Parking Brake Warning Lamp Switch, Replace (B)

	LABOR TIME	SEVERE SERVICE
Exc. below	.7	.7
1984-89 300ZX	1.4	1.4
2002-05 Maxima	1.2	1.2

Sealed Beam Headlamp, Replace (C)

	LABOR TIME	SEVERE SERVICE
Exc. 280ZX	.3	.3
280ZX	.5	.5

Side Marker Lamp Lens, Replace (B)

	LABOR TIME	SEVERE SERVICE
Each	.3	.3

Stop & Tail Lamp Lens, Replace (B)

	LABOR TIME	SEVERE SERVICE
Each	.7	.7

Stop Lamp Switch, Replace (B)

	LABOR TIME	SEVERE SERVICE
Exc. below	.7	.7
1988-91 300ZX	.9	.9
1985-88 Maxima	.9	.9
1990-92 Stanza	.9	.9

Turn Signal or Hazard Warning Flasher, Replace (B)

	LABOR TIME	SEVERE SERVICE
Exc. below	.5	.5
310	1.3	1.3
350Z	1.6	1.6

SEAT BELTS

Automatic Shoulder Belt Control Unit, Replace (B)

	LABOR TIME	SEVERE SERVICE
Exc. below	.7	.7
1991-94 NX, Sentra	1.4	1.4

Front Limit Switch, Replace (B)

	LABOR TIME	SEVERE SERVICE
1989-94 240SX		
one	.7	.7
both	1.3	1.3
1993 Altima		
one	.5	.5
both	.7	.7

	LABOR TIME	SEVERE SERVICE
Front Limit Switch, Replace (B)		
1988-94 Maxima		
one	.7	.7
both	1.1	1.1
1991-93 NX		
one	1.2	1.2
both	2.1	2.1
Sentra		
1989-90		
one	.7	.7
both	1.4	1.4
1991-94		
one	1.2	1.2
both	2.1	2.1
1990-92 Stanza		
one	.5	.5
both	.7	.7
Rail Motor, Replace (B)		
1989-94 240SX		
one	.8	.8
both	1.8	1.8
1993 Altima		
one	.8	.8
both	1.6	1.6
Maxima		
1988		
one	.8	.8
both	1.7	1.7
1989-94		
one	1.3	1.3
both	1.9	1.9
1991-93 NX		
one	.8	.8
both	1.8	1.8
Sentra		
1989-90		
one	1.3	1.3
both	1.9	1.9
1991-94		
one	.8	.8
both	1.8	1.8
1990-92 Stanza		
one	2.0	2.0
both	3.3	3.3
Rear Limit Switch, Replace (B)		
1989-94 240SX		
one	.7	.7
both	1.4	1.4
1993 Altima		
one	.5	.5
both	.7	.7
1988-94 Maxima		
one	.7	.7
both	1.3	1.3
1991-93 NX		
one	1.1	1.1
both	1.9	1.9

	LABOR TIME	SEVERE SERVICE
Sentra		
1989-90		
one	.8	.8
both	1.8	1.8
1991-94		
one	1.1	1.1
both	1.9	1.9
1990-92 Stanza		
one	1.9	1.9
both	2.3	2.3
Seat Belt Guide Rails, Replace (B)		
1989-94 240SX		
one	1.5	1.5
both	2.5	2.5
1993 Altima		
one	1.4	1.4
both	2.4	2.4
Maxima		
1988		
one	1.6	1.6
both	2.9	2.9
1989-94		
one	1.5	1.5
both	2.7	2.7
1991-93 NX		
one	.8	.8
both	1.9	1.9
Sentra		
1989-90		
one	1.6	1.6
both	2.8	2.8
1991-94		
one	.8	.8
both	1.9	1.9
1990-92 Stanza		
one	1.7	1.7
both	2.8	2.8

BODY

	LABOR TIME	SEVERE SERVICE
Door Latch Rod, Replace (B)		
Front		
exc. below	1.4	1.4
210, 310	.8	.8
1995-98 240SX	.7	.7
280ZX	.8	.8
300ZX	1.1	1.1
350Z	1.6	1.6
Altima	.9	.9
Pulsar	1.1	1.1
2000-05 Sentra	.5	.5
Rear		
exc. below	1.3	1.3
210	.8	.8
Altima	.8	.8
Maxima		
1985-94	1.7	1.7
1995-99	1.1	1.1
2000-05	1.4	1.4
2000-05 Sentra	.7	.7

	LABOR TIME	SEVERE SERVICE
Door Lock, Replace (B)		
Front		
exc. below	.9	.9
1984-98 200SX	1.4	1.4
240SX		
1989-94	1.2	1.2
1995-98	1.8	1.8
280ZX	2.3	2.3
300ZX	1.5	1.5
350Z	1.8	1.8
Altima		
1993-97	1.4	1.4
1998-04	1.0	1.0
Maxima		
1989-01	1.4	1.4
2002-05	1.0	1.0
NX	1.3	1.3
1982-99 Sentra	1.3	1.3
Stanza	1.4	1.4
Rear		
exc. below	.9	.9
810/Maxima	1.5	1.5
Maxima		
1985-94	1.7	1.7
1995-05	1.4	1.4
NX	1.1	1.1
1991-94 Sentra	1.1	1.1
1982-86 Stanza	1.3	1.3
Front Door Window Regulator, Replace (B)		
200SX		
1981-83	1.8	2.0
1984-88	2.4	2.6
1995-98	1.2	1.4
210, 310, 510	1.5	1.7
240SX	1.5	1.7
280ZX	2.3	2.5
300ZX	1.8	2.0
350Z	1.6	1.8
810/Maxima	1.8	2.0
Altima	1.1	1.3
Maxima		
exc. 2005	1.3	1.5
2005	1.1	1.3
NX	1.4	1.6
Pulsar	.8	1.0
Sentra		
1982-90	1.6	1.8
1991-94		
coupe	1.4	1.6
sedan	.9	1.1
1995-99	1.2	1.4
2000-05	.7	.9
Stanza	1.8	2.0
Stanza Wagon	1.6	1.8
w/power windows add	.2	.2
Hood Hinge, Replace (B)		
Exc. below	.9	.9
1995-98 200SX	1.2	1.2
1995-98 240SX	1.6	1.6

	LABOR TIME	SEVERE SERVICE
1999-96 300ZX	2.0	2.0
350Z	1.2	1.2
2002-05 Altima	2.0	2.0
2005 Maxima	1.8	1.8
Sentra		
1995-99	1.2	1.2
2000-05	2.6	2.6
Hood Lock, Replace (B)		
Exc. below	.7	.7
2002-05 Maxima	1.1	1.1
Hood Release Cable, Replace (B)		
200SX		
1981-88	.7	.7
1995-98	1.2	1.2
210	.7	.7
240SX		
1989-94	1.2	1.2
1995-98	.5	.5
280ZX	.7	.7
300ZX	1.3	1.3
310	1.2	1.2
510	.7	.7
810/Maxima	.7	.7
Altima		
1993-97	.8	.8
1998-05	1.2	1.2
Maxima		
1985-94	.7	.7
1995-99	1.1	1.1
2000-04	1.9	1.9
2005	1.2	1.4
1991-93 NX	1.2	1.2
Pulsar	.7	.8

	LABOR TIME	SEVERE SERVICE
Sentra		
1982-90	.7	.8
1991-94		
sedan	.9	.9
coupe	1.2	1.2
1995-99	1.2	1.2
2000-05	.5	.5
Stanza		
1982-89	.7	.7
1990-92	1.5	1.5
Lock Striker Plate, Replace (B)		
All Models	.5	.5
Rear Door Window Regulator, Replace (B)		
210	1.7	1.9
510	1.7	1.9
810/Maxima	1.6	1.8
Altima	1.2	1.4
Maxima		
exc. 2005	1.4	1.6
2005	.8	1.0
1983-86 Pulsar	1.8	2.0
Sentra		
1982-94	1.6	1.8
1995-05	1.0	1.2
Stanza	1.7	1.9
w/power windows add	.1	.4
Trunk Lock, Replace (B)		
All Models	.7	.7
Window Regulator Motor, Replace (B)		
Front		
200SX		
1984-88	2.3	2.5
1995-98	.9	1.1

	LABOR TIME	SEVERE SERVICE
240SX		
1989-94	1.5	1.7
1995-98	.9	1.1
300ZX		
1984-89	1.7	1.9
1990-96	2.1	2.3
350Z	1.8	2.0
810/Maxima	2.1	2.3
Altima	1.1	1.3
Maxima		
exc. 2005	1.4	1.6
2005	.8	1.0
NX	1.6	1.8
Sentra		
1991-94		
coupe	1.6	1.8
sedan	1.3	1.5
1995-05	.9	1.1
Stanza		
1987-89	1.7	1.9
1990-92	1.4	1.6
Rear		
810/Maxima	2.0	2.2
Altima	.9	1.1
Maxima		
1985-94	1.6	1.8
1995-04	1.1	1.3
2005	.9	1.0
1991-05 Sentra	.8	1.0
Stanza		
1987-89	1.8	2.0
1990-92	2.1	2.3

NOTES

NI

720 Pickup : Amada : Axxess : D21 Pickup : Frontier : Murano : Pathfinder : Pickup : Quest : Titan : Van : Xterra

SYSTEM INDEX

MAINTENANCE	NI-48
Maintenance Schedule	NI-48
CHARGING	NI-48
STARTING	NI-49
CRUISE CONTROL	NI-49
IGNITION	NI-49
EMISSIONS	NI-50
FUEL	NI-51
EXHAUST	NI-52
ENGINE COOLING	NI-53
ENGINE	NI-54
Assembly	NI-54
Cylinder Head	NI-56
Camshaft	NI-58
Crank & Pistons	NI-59
Engine Lubrication	NI-61
CLUTCH	NI-61
MANUAL TRANSAXLE	NI-62
MANUAL TRANSMISSION	NI-62
AUTO TRANSAXLE	NI-62
AUTO TRANSMISSION	NI-63
CONTINUOUSLY VARIABLE TRANSMISSION	NI-64
TRANSFER CASE	NI-64
DRIVELINE	NI-65
BRAKES	NI-66
FRONT SUSPENSION	NI-69
REAR SUSPENSION	NI-71
STEERING	NI-72
HEATING & AC	NI-73
WIPERS & SPEEDOMETER	NI-74
LAMPS & SWITCHES	NI-75
BODY	NI-75

OPERATIONS INDEX

A
- AC Hoses NI-73
- Air Bags NI-72
- Air Conditioning NI-73
- Alignment NI-69
- Alternator (Generator) NI-48
- Antenna NI-74
- Anti-Lock Brakes NI-66

B
- Back-Up Lamp Switch NI-75
- Ball Joint NI-69
- Battery Cables NI-49
- Bleed Brake System NI-67
- Blower Motor NI-73
- Brake Disc NI-67
- Brake Drum NI-67
- Brake Hose NI-67
- Brake Pads and/or Shoes ... NI-68

C
- Camshaft NI-58
- Catalytic Converter NI-52
- Coolant Temperature (ECT) Sensor ... NI-50
- Crankshaft NI-59
- Crankshaft Sensor NI-49
- Cruise Control NI-49
- CV Joint NI-66
- Cylinder Head NI-56

D
- Differential NI-65
- Distributor NI-49
- Drive Belt NI-48
- Driveshaft NI-65

E
- EGR NI-50
- Electronic Control Module (ECM/PCM) ... NI-50
- Engine NI-54
- Engine Lubrication NI-61
- Engine Mounts NI-55
- Evaporator NI-74
- Exhaust NI-52
- Exhaust Manifold NI-53
- Expansion Valve NI-74

F
- Flexplate NI-62
- Flywheel NI-62
- Fuel Injection NI-52
- Fuel Pump NI-51
- Fuel Vapor Canister NI-50

G
- Generator NI-48
- Glow Plug NI-50

H
- Halfshaft NI-66
- Headlamp NI-75
- Heater Core NI-74
- Horn NI-75

I
- Idle Air Control (IAC) Valve ... NI-52
- Ignition Coil NI-50
- Ignition Module NI-50
- Ignition Switch NI-50
- Injection Pump NI-51
- Inner Tie Rod NI-72
- Intake Manifold NI-52

K
- Knock Sensor NI-51

L
- Lower Control Arm NI-69

M
- Maintenance Schedule NI-48
- Manifold Absolute Pressure (MAP) Sensor ... NI-50
- Mass Air Flow (MAF) Sensor ... NI-50
- Master Cylinder NI-67
- Muffler NI-53
- Multifunction Lever/Switch ... NI-75

N
- Neutral Safety Switch NI-62

O
- Oil Pan NI-61
- Oil Pump NI-61
- Outer Tie Rod NI-72
- Oxygen Sensor NI-51

P
- Parking Brake NI-68
- Pistons NI-59
- Positive Crankcase Ventilation (PCV) Valve ... NI-51

R
- Radiator NI-54
- Radiator Hoses NI-54
- Radio NI-75
- Rear Main Oil Seal NI-60

S
- Shock Absorber/Strut, Front ... NI-69
- Shock Absorber/Strut, Rear ... NI-71
- Spark Plug Cables NI-50
- Spark Plugs NI-50
- Spring, Leaf NI-71
- Spring, Rear Coil NI-71
- Starter NI-49
- Steering Wheel NI-72

T
- Thermostat NI-53
- Throttle Body NI-52
- Throttle Position Sensor (TPS) ... NI-51
- Timing Belt NI-58
- Timing Chain NI-58
- Torque Converter NI-62

U
- U-Joint NI-65
- Upper Control Arm NI-70

V
- Valve Body NI-62
- Valve Cover Gasket NI-57
- Valve Job NI-57
- Vehicle Speed Sensor NI-51

W
- Water Pump NI-54
- Wheel Balance NI-48
- Wheel Cylinder NI-68
- Window Regulator NI-76
- Windshield Washer Pump NI-75
- Windshield Wiper Motor NI-75

720 PICKUP : AMADA : AXXESS : D21 PICKUP : FRONTIER : MURANO

	LABOR TIME	SEVERE SERVICE
MAINTENANCE		
Air Cleaner Filter Element, Replace (B)		
All Models	.3	.4
Chassis Lubrication, Change Oil & Filter (C)		
Includes: Correct all fluid levels.		
Exc. Pathfinder	.6	.7
Pathfinder	.9	1.0
Drive Belt, Adjust (B)		
Exc. Pathfinder	.5	.6
Pathfinder	.6	.7
Drive Belt, Replace (B)		
AC	.7	.8
Air pump		
w/AC	1.0	1.0
w/o AC	.7	.8
Fan		
620, 720	1.3	1.4
Armada, Titan	.6	.7
Axxess	.8	.9
Frontier, Xterra		
one	.3	.4
two	.5	.6
three	.9	1.0
Murano		
one	.5	.6
two	.8	.9
Pathfinder, Pickup		
1986-95		
4 cyl.		
KA		
one	.5	.6
two	.7	.8
three	1.3	1.4
Z	.7	.8
V6		
one	.8	.9
two	.9	1.0
three	1.1	1.2
1996-04		
one	.8	.9
two	.9	1.0
three	1.1	1.2
Quest		
one	.7	.8
two	.8	.9
three	.9	1.0
Van	.7	.8
Power steering	.7	.8
Fuel Filter, Replace (B)		
Exc. below	.5	.5
Armada, Titan	2.0	2.0
Axxess	.9	1.1
Frontier, Xterra V6	.8	.9
Halogen Headlamp Bulb, Replace (C)		
Exc. Xterra	.3	.3
Xterra	.8	.8

	LABOR TIME	SEVERE SERVICE
Oil & Filter, Change (C)		
Includes: Correct all fluid levels.		
Exc. Pathfinder	.6	.7
Pathfinder	.9	1.0
Sealed Beam Headlamp, Replace (C)		
All Models	.3	.3
Timing Belt, Replace (B)		
V6		
Frontier, Xterra	3.5	4.0
Pathfinder		
1987-95	3.7	4.2
1996-00	3.2	3.7
Pickup	3.8	4.3
Quest	4.3	4.8
w/AC add	.3	.3
w/PS add	.2	.2
Replace crankshaft gear add	.5	.7
Tire, Replace (C)		
Includes: Dismount old tire and mount new tire to rim.		
All Models	.5	.5
Tires, Rotate (C)		
Exc. below	.4	.4
Armada, Murano, Pathfinder, Titan	1.0	1.2
Wheel, Balance (C)		
One	.3	.3
each addl. add	.2	.2
SCHEDULED MAINTENANCE INTERVALS		
If necessary, refer to appropriate Chilton maintenance service information.		
7,500 Mile Service (C)		
All Models	.6	.6
15,000 Mile Service (C)		
Frontier, Pathfinder, Pickup	1.0	1.0
Quest	1.0	1.0
w/4WD add	.3	.3
22,500 Mile Service (C)		
All Models	.5	.5
30,000 Mile Service (B)		
Frontier, Pathfinder, Pickup	3.0	3.0
Quest	3.2	3.2
w/4WD add	.5	.5
w/KA24E engine add	.2	.2
37,500 Mile Service (C)		
All Models	.6	.6
45,000 Mile Service (C)		
Frontier, Pathfinder, Pickup	1.0	1.0
Quest	1.0	1.0
w/4WD add	.3	.3
52,500 Mile Service (C)		
All Models	.6	.6

	LABOR TIME	SEVERE SERVICE
60,000 Mile Service (B)		
Frontier, Pathfinder, Pickup	3.8	3.8
Quest	3.3	3.3
w/4WD add	.5	.5
w/KA24E engine add	.2	.2
67,500 Mile Service (C)		
All Models	.6	.6
75,000 Mile Service (C)		
Frontier, Pathfinder, Pickup	1.0	1.0
Quest	1.1	1.1
w/4WD add	.3	.3
82,500 Mile Service (C)		
All Models	.5	.5
90,000 Mile Service (B)		
Frontier, Pathfinder, Pickup	3.8	3.8
Quest	3.3	3.3
w/4WD add	.5	.5
97,500 Mile Service (C)		
All Models	.6	.6
CHARGING		
Alternator Circuits, Test (B)		
Includes: Test component output.		
All Models	.6	.6
Alternator, R&R and Recondition (B)		
620, 720	2.2	2.7
Axxess	2.8	3.3
Frontier, Xterra	2.1	2.6
Murano	4.3	4.8
Pathfinder	3.1	3.6
Pickup		
4 cyl.		
KA	2.2	2.7
Z	2.6	3.1
V6	2.6	3.1
Quest		
1993-97	2.7	3.2
1998-02	2.0	2.5
Van	2.6	3.1
Alternator Assy., Replace (B)		
Includes: Pulley transfer.		
620, 720	.9	1.1
Armada, Titan	2.1	2.3
Axxess	.5	.7
Frontier, Xterra	1.9	2.1
Murano	2.2	2.4
Pathfinder		
1987-95	1.3	1.5
1996-05		
VG	1.9	2.1
VQ	2.1	2.3
Pickup		
4 cyl.		
KA	.8	1.0
Z	1.2	1.4

PATHFINDER : PICKUP : QUEST : TITAN : VAN : XTERRA   NI-49

	LABOR TIME	SEVERE SERVICE
V6	1.4	1.6
Quest	2.2	2.4
Van	1.3	1.5

Alternator Voltage Regulator, Replace (B)
620, 720	1.3	1.6
Axxess	1.5	1.8
Frontier, Xterra	1.9	2.1
Pathfinder		
1987-95	1.7	2.0
1996-05		
VG	1.2	1.5
VQ	2.2	2.5
Pickup	1.6	1.9
Quest		
1993-98	1.0	1.3
1999-05	2.2	2.4
Van	1.5	1.8

Voltmeter, Replace (B)
620, 720	.7	.7
Axxess	.7	.7
Pickup	.8	.8
Quest	1.0	1.0
Van	1.2	1.2

STARTING

Starter Draw Test (On Car) (B)
| All Models | .3 | .3 |

Battery, Replace (C)
| All Models | .5 | .7 |

Battery Cables, Replace (C)
620, 720	.5	.7
Armada, Titan	.6	.8
Axxess	1.3	1.5
Frontier, Xterra	1.2	1.4
Murano	1.1	1.2
Pathfinder, Pickup		
1986-2000	.3	.5
2001-04	.8	1.0
Quest	.8	1.0
Van	1.3	1.5

Starter, R&R and Recondition (B)
620, 720		
2WD	2.1	2.6
4WD	2.1	2.6
Axxess		
2WD	2.6	3.1
4WD	2.9	3.4
Frontier, Xterra		
4 cyl.	2.3	2.8
V6	1.7	2.2
Murano	2.3	2.8
Pathfinder		
1987-95	2.6	3.1
1996-01	2.0	2.5

Pickup		
4 cyl.		
2WD	2.1	2.6
4WD	2.6	3.1
V6	2.6	3.1
Quest		
1993-98	2.2	2.7
1999-02	1.6	2.1
Van	2.4	2.9

Starter Assy., Replace (B)
620, 720	.7	1.0
Armada, Titan	1.2	1.4
Axxess		
2WD	.8	1.1
4WD	1.4	1.7
Frontier, Xterra	1.2	1.5
Murano	1.1	1.4
Pathfinder		
1987-95	1.1	1.4
1996-05	.9	1.2
Pickup		
4 cyl.	.8	1.1
V6		
2WD	.7	1.0
4WD	1.2	1.5
1993-05 Quest	.7	1.0
Van	.7	1.0

Starter Drive Assy., Replace (B)
Includes: Starter R&R.
620, 720	1.1	1.4
Axxess	1.5	1.8
Frontier, Xterra	1.5	1.8
Pathfinder		
1987-95	1.4	1.7
1996-01	1.2	1.5
Pickup		
4 cyl.	1.4	1.7
V6	1.6	1.9
1993-02 Quest	1.1	1.4
Van	1.2	1.5

Starter Relay, Replace (B)
| All Models | .5 | .5 |

Starter Solenoid and/or Switch, Replace (B)
Includes: Starter R&R.
620, 720	.8	1.1
Axxess	1.6	1.9
Frontier, Xterra	1.4	1.7
Murano	1.2	1.5
Pathfinder		
1987-95	1.5	1.8
1996-01	1.1	1.4
Pickup		
4 cyl.		
2WD	.8	1.1
4WD	1.4	1.7
V6	1.5	1.8
1993-02 Quest	.8	1.1
Van	1.2	1.5

CRUISE CONTROL

Diagnose Cruise System Component Each (A)
| All Models | 1.0 | 1.0 |

Actuator Assy., Replace (B)
| All Models | .5 | .5 |

Control Cable & Core, Replace (B)
| All Models | .7 | .7 |

Control Clutch Switch, Replace (B)
| All Models | .5 | .5 |

Control Controller Module and/or Relay, Replace (B)
| All Models | .7 | .7 |

Control Main Switch, Replace (B)
| All Models | .5 | .5 |

Control Sensor, Replace (B)
620, 720	.8	.8
Frontier, Xterra	.3	.3
Pathfinder	.6	.6
1986-97 Pickup	.7	.7
Van	.7	.7

Control Stop Switch, Replace (B)
| All Models | .5 | .5 |

Control Vacuum Pump, Replace (B)
| All Models | .5 | .5 |

IGNITION

Diagnose Ignition System Component Each (A)
| All Models | .8 | .8 |

After Glow Timer, Replace (B)
| 720 | .5 | .5 |

Crank Angle Sensor, Replace (B)
| 1990-04 | .5 | .7 |

Crankshaft Position Sensor, Replace (B)
Armada, Titan	.4	.4
Frontier, Xterra		
4 cyl.	.5	.7
V6	3.0	3.2
1996-00 Pathfinder		
VG	.6	.8
VQ	.3	.5
1995-02 Quest	.3	.5

Distributor, R&R and Recondition (B)
Includes: Reset base ignition timing.
| 620, 720 | 1.8 | 2.1 |
| 1986-91 Pathfinder, Pickup V6 | 1.9 | 2.2 |

Distributor, Replace (B)
Includes: Reset base ignition timing.
| All Models | .7 | .9 |

Distributor Cap and/or Rotor, Replace (B)
| All Models | .5 | .5 |

Distributor Vacuum Control Unit, Replace (B)
| 1981-86 | .7 | .7 |

720 PICKUP : AMADA : AXXESS : D21 PICKUP : FRONTIER : MURANO

	LABOR TIME	SEVERE SERVICE
Glow Plug, Replace (B)		
720	.7	.8
Glow Plug Relay, Replace (B)		
720	.5	.5
Ignition Coil, Replace (B)		
4 cyl.	.5	.7
V6		
VG	.5	.7
VQ		
exc. Murano	1.0	1.2
Murano	1.5	1.7
V8		
Armada, Titan	.6	.8
Ignition Switch, Replace (B)		
Exc. Quest	.9	.9
Quest		
1993-98	1.0	1.0
1999-02	1.4	1.4
Ignition Timing, Reset (B)		
All Models	.5	.5
Spark Plug (Ignition Cables), Replace (B)		
Exc. below	.7	.9
Frontier, Xterra V6	1.2	1.4
Spark Plugs, Replace (B)		
620, 720	.8	1.0
Armada, Titan	.9	1.1
Axxess	.5	.7
Frontier, Xterra		
4 cyl.	.5	.7
V6	1.7	1.9
Murano	1.8	2.0
Pathfinder		
1987-95	1.3	1.5
1996-05		
VG	.5	.7
VQ	1.6	1.8
Pickup		
4 cyl.	.8	1.0
V6	1.2	1.4
Quest		
1993-98	1.1	1.3
1999-05	2.4	2.8
Van	.7	.9
Steering Lock Assy., Replace (B)		
All Models	1.3	1.3
Transistorized Ignition Control Unit, Replace (B)		
All Models	.7	.7
w/AC add	.2	.2

EMISSIONS

The following operations do not include testing. Add time as required.

	LABOR TIME	SEVERE SERVICE
Diagnose Emission Control System Component Each (A)		
1981-05	.7	.7
Retrieve Fault Codes (A)		
1981-05	.3	.3
Dynamometer Test (A)		
All Models	.4	.4
Air Control Valve, Replace (B)		
1981-97	.5	.5
Air Flow Meter, Replace (B)		
Armada, Titan	.4	.6
Axxess, Pathfinder	.7	.7
Frontier, Xterra	.5	.5
Murano	.3	.3
Quest	.5	.5
Van	1.3	1.3
Air-Fuel Control Solenoid, Replace (B)		
620, 720	1.3	1.3
Air Injection Control Unit, Replace (B)		
1981-97	.5	.7
Air Pump Assy., Replace (B)		
1987-95 Pathfinder	.7	.9
Pickup	.7	.9
Anti-Backfire Valve, Replace (B)		
1981-97	.5	.5
Barometric Manifold Absolute Pressure Sensor, Replace (B)		
1994-04	.3	.3
BCDD Assy., Replace (B)		
620, 720	.7	.7
BCDD Control Valve Assy., Replace (B)		
620, 720	.5	.5
Coolant Temperature (ECT) Sensor, Replace (B)		
4 cyl.	.5	.5
V6		
VG	.5	.5
VQ	3.1	3.1
V8		
Armada, Titan	.8	.8
Detonation Control Unit, Replace (B)		
620, 720	.5	.5
ECC Relay, Replace (B)		
620, 720	.5	.5
EFE Valve, Replace (B)		
1981-97	1.6	1.6
EGR Control Unit, Replace (B)		
Axxess	.5	.5
EGR Control Valve, Replace (B)		
720 diesel	.7	.9
Axxess	.5	.7
Frontier, Xterra	.9	1.1
Pathfinder		
1987-95	1.5	1.7
1996-00	.5	.7
Pickup		
4 cyl.		
KA	1.5	1.7
Z	1.2	1.4
V6	1.5	1.7

	LABOR TIME	SEVERE SERVICE
Quest		
1993-98	.6	.8
1999-05	1.1	1.3
Van	.8	1.0
EGR Pressure Feedback Transducer, Replace (B)		
All Models	.5	.5
EGR Temperature Sensor, Replace (B)		
Exc. below	.5	.5
1993-98 Quest	2.8	2.8
EGR Valve Solenoid, Replace (B)		
Exc. Quest	.5	.5
Quest	.6	.6
EGR Thermal Vacuum Valve, Replace (B)		
1981-89	.5	.5
Electronic Control Unit, Replace (B)		
620, 720	.5	.5
Armada, Titan	.6	.6
Frontier, Xterra	.8	.8
Murano	.8	.8
Pathfinder, Pickup	.7	.7
Quest	1.0	1.0
Van	.5	.5
Evaporative Purge Solenoid, Replace (B)		
Exc. Quest	.5	.5
Quest	2.7	2.7
Fuel Cut-Off Solenoid, Replace (B)		
620, 720	.5	.5
Fuel Shut-Off Clutch Switch, Replace (B)		
620, 720	.5	.5
Fuel Shut-Off Vacuum Switch, Replace (B)		
620, 720	.3	.3
Fuel Vapor Canister, Replace (B)		
Armada, Titan	1.0	1.1
Frontier, Xterra	.6	.6
Pathfinder	.6	.6
Pickup	.5	.5
Quest	.7	.7
Heated Inlet Air Cleaner Temperature Sensor, Replace (B)		
Exc. below	.5	.5
2001-05 Pathfinder	.6	.6
Heated Inlet Air Idle Compensator, Replace (B)		
1981-89	.5	.5
Heated Inlet Air Cleaner Vacuum Motor, Replace (B)		
1981-97	.5	.5
Knock Control Sensor, Replace (B)		
Pathfinder, Pickup		
1992-95	3.8	4.0
1996-97		
Pathfinder	2.5	2.7
Pickup	4.0	4.2
Quest	4.5	4.7

PATHFINDER : PICKUP : QUEST : TITAN : VAN : XTERRA

	LABOR TIME	SEVERE SERVICE
Knock Sensor, Replace (B)		
620, 720	.7	.7
Armada, Titan	1.8	2.0
Frontier, Xterra		
4 cyl.	.5	.5
V6	2.2	2.4
Murano	1.9	2.1
Pathfinder		
1987-95	4.4	4.6
1996-05		
VG	2.0	2.3
VQ	2.3	2.6
1986-97 Pickup	4.3	4.5
Quest	2.7	3.1
Mixture Ratio Control Solenoid, Replace (B)		
1994-95 Pathfinder	.5	.5
1994-97 Pickup	.5	.5
Oxygen Sensor, Replace (B)		
1987-93	.5	.7
Armada, Titan		
air/fuel ratio both	.8	1.0
rear		
one	.5	.6
both	.8	1.0
Frontier, Xterra	.6	.8
Murano		
front		
one	1.0	1.2
both	1.4	1.6
rear	.6	.8
Pathfinder		
1994-95	.8	1.0
1996-05		
front		
VG one or both	1.4	1.6
VQ		
one	2.2	2.4
both	2.7	2.9
rear	.8	1.0
1994-97 Pickup	.8	1.0
Quest		
1994-98	.8	.9
1999-05	.6	.6
Positive Crankcase Ventilation (PCV) Valve, Replace (B)		
Exc. below	.5	.5
Murano	.7	.7
2001-03 Pathfinder	2.7	2.7
Spark Delay Valve, Replace (B)		
Z engine	.3	.3
Thermal Vacuum Valve, Replace (B)		
Z engine	.5	.5
Throttle Position Sensor (TPS)/Switch, Replace (B)		
Exc. below	.5	.5
Axxess	.8	.8
Van	2.7	2.7

	LABOR TIME	SEVERE SERVICE
Vacuum Control Valve, Replace (B)		
1981-89 Z engine	.5	.5
Vacuum Solenoid Valve, Replace (B)		
Z engine	.5	.5
Vapor Canister Filter, Replace (B)		
All Models	.5	.5
Vapor Separator, Replace (B)		
All Models	.8	.8
Vehicle Speed Sensor, Replace (B)		
Armada, Titan	.5	.6
Frontier, Xterra	1.4	1.6
Murano	.3	.5
Pathfinder	.8	1.0
1994-97 Pickup	.8	1.0

FUEL
CARBURETOR

	LABOR TIME	SEVERE SERVICE
Air Cleaner Vacuum Motor, Replace (B)		
620, 720	.5	.5
Carburetor, R&R and Clean or Recondition (B)		
Includes: Adjustments.		
620, 720	3.0	3.0
Carburetor, Replace (B)		
Includes: Adjustments.		
620, 720	1.4	1.4
Carburetor Needle & Seat, Replace (B)		
620, 720	.9	1.2
Dash Pot, Replace (B)		
620, 720	.7	.7
Electric Choke Relay, Replace (B)		
620, 720	.7	.7
Fuel Shut-Off Vacuum Switch, Replace (B)		
620, 720	.5	.5

DELIVERY

	LABOR TIME	SEVERE SERVICE
Fuel Pump, Test (B)		
All Models	.3	.3
Diesel Fuel Injection Pump, Replace (B)		
1981-86 720	.5	.8
Fuel Filter, Replace (B)		
720	.5	.5
Axxess	.9	.9
Frontier, Xterra		
4 cyl.	.5	.5
V6	.7	.7
1990-05 Pathfinder, Pickup	.5	.5
1993-02 Quest	.5	.5
Fuel Gauge (Dash), Replace (B)		
Exc. below	1.0	1.0
620, 720	1.8	1.8
Pickup	1.3	1.3
Van	1.1	1.1

	LABOR TIME	SEVERE SERVICE
Fuel Gauge (Tank), Replace (B)		
620, 720	1.2	1.4
Armada, Titan	1.8	1.8
Axxess	.8	1.0
Murano	1.0	1.2
Frontier, Xterra		
4 cyl.	1.9	2.1
V6		
Frontier	1.6	1.8
Xterra	1.1	1.3
Pathfinder	1.9	2.1
Pickup	1.6	1.8
Quest	1.6	1.8
Van	.7	.9
Fuel Priming Pump, R&R and Recondition (B)		
720	1.6	1.9
Fuel Pump (Electric), Replace (B)		
Armada, Titan	1.8	2.0
Axxess	.8	1.0
Frontier	1.6	1.8
Murano	1.0	1.2
Pathfinder	.9	1.1
Pickup	1.7	1.9
Quest		
1993-98	1.9	2.1
1999-05	1.5	1.7
Van	.7	.9
Xterra	1.1	1.3
Fuel Pump (Mechanical), R&R or Replace (B)		
620, 720	.5	.5
Fuel Pump Control Unit, Replace (B)		
All Models	.5	.5
Fuel Pump Relay, Replace (B)		
Exc. below	.3	.3
Murano, Van	.5	.5
Fuel Supply Pump, Replace (B)		
720	.5	.7
Fuel Tank, Replace (B)		
Includes: Drain and refill.		
620, 720	.9	1.1
Armada, Titan	3.0	3.3
Axxess		
2WD	1.4	1.6
4WD	2.6	2.8
Frontier, Xterra	2.0	2.3
Murano		
2WD	3.0	3.3
4WD	4.3	4.8
Pathfinder	2.0	2.3
Pickup	1.3	1.5
Quest	2.2	2.4
Van	.7	.9
Intake & Exhaust Manifold or Gaskets, Replace (B)		
620, 720	2.7	3.0
Axxess	4.3	4.6
Van	4.1	4.4

NI-51

720 PICKUP : AMADA : AXXESS : D21 PICKUP : FRONTIER : MURANO

	LABOR TIME	SEVERE SERVICE
Intake Manifold and/or Gasket, Replace (B)		
Includes: Adjustments.		
Gasoline		
Armada, Titan	1.8	2.0
Frontier, Xterra		
4 cyl.	5.0	5.3
V6	4.1	4.4
Murano	2.0	2.3
Pathfinder, Pickup		
1986-95		
4 cyl.		
KA	4.5	5.0
Z	2.6	2.9
V6		
upper	3.2	3.5
lower	4.4	4.7
1996-05		
Pathfinder		
VG	2.8	3.1
VQ	3.6	3.9
Pickup		
4 cyl.	4.5	4.8
V6 upper or lower	2.6	3.0
Quest		
1993-98	3.9	4.2
1999-02	5.9	6.2
Diesel	1.6	1.9

INJECTION
	LABOR TIME	SEVERE SERVICE
Injection Timing, Check & Adjust (B)		
720	1.3	1.5
Air Cleaner Filter Element, Replace (B)		
All Models	.3	.4
Air Valve, Replace (B)		
720	.3	.3
Air Regulator, Replace (B)		
Axxess	.7	.7
1990-97 Pathfinder, Pickup	.5	.5
1993-98 Quest	.5	.5
Altitude Compensator, Replace (B)		
720	1.1	1.1
Automatic Timer, Replace (B)		
720	1.2	1.2
Control Relay, Replace (B)		
720	1.3	1.3
Diesel Injection Nozzles, Replace (B)		
720	1.5	1.7
Clean and test add each	.2	.4
EFI Relay, Replace (B)		
1994-04	.5	.5
Fuel Injection Pump Drive Gear, Replace (B)		
720 diesel	1.3	1.5
w/AC add	.2	.2

	LABOR TIME	SEVERE SERVICE
Fuel Injectors, Replace (B)		
Armada, Titan	1.4	1.6
Axxess	.9	1.1
Frontier, Xterra		
4 cyl.	.9	1.1
V6	2.2	2.4
Murano	2.0	2.2
Pathfinder, Pickup		
1990-97		
4 cyl.	.7	.9
V6	3.1	3.3
1998-05	2.2	2.4
Quest		
1993-98	4.4	4.6
1999-05	2.7	3.1
Fuel Pressure Regulator, Replace (B)		
Axxess	.5	.7
Frontier, Xterra		
4 cyl.	.6	.8
V6	2.6	2.8
Murano	.8	1.0
Pathfinder		
1987-95	3.1	3.3
1996-05		
VG	2.6	2.8
VQ	.6	.8
Pickup		
4 cyl.	.7	.9
V6	3.0	3.2
Quest		
1993-98	.5	.7
1999-05	1.8	2.0
Fuel Temperature Sensor, Replace (B)		
Frontier, Xterra	2.0	2.2
Murano	.8	1.0
1996-05 Pathfinder	.9	1.1
1994-97 Pickup	1.4	1.4
Intake Air Heater Controller, Replace (B)		
720	.7	.7
Mixture Heater Relay, Replace (B)		
1986-89 Pathfinder, Pickup	.5	.5
Van	.5	.5
Pressure Regulator Control Solenoid, Replace (B)		
Axxess	.7	.7
1990-97 Pathfinder, Pickup	.7	.7
Throttle Body Injection Unit, Replace (B)		
1986-95 Pathfinder, Pickup	1.2	1.4
Van	2.7	2.9
Replace injectors add	.2	.4

	LABOR TIME	SEVERE SERVICE
Throttle Body Pressure Regulator, Replace (B)		
Axxess	.5	.7
1986-95 Pathfinder, Pickup	.7	.9
Throttle Chamber, Replace (B)		
Armada, Titan	1.0	1.2
Axxess	.8	1.0
Frontier, Xterra		
4 cyl.	1.8	2.0
V6	1.4	1.6
Murano	.5	.7
Pathfinder		
1990-95	1.7	1.9
1996-05	1.4	1.6
Pickup	1.7	1.9
Quest	1.5	1.7
Van	2.7	2.9
Throttle Control Solenoid, Replace (B)		
1986-89 Pathfinder, Pickup	.7	.7
Van	.8	.8
Vacuum Pump, Replace (B)		
720	.7	.9

EXHAUST
	LABOR TIME	SEVERE SERVICE
Catalytic Converter, Replace (B)		
620, 720	1.5	1.6
Armada, Titan		
one	2.4	2.8
both	5.4	6.0
Axxess	.5	.6
Murano	2.0	2.3
Frontier, Xterra		
4 cyl.	.6	.7
V6		
2WD, one	2.5	2.6
4WD		
one	2.3	2.6
both	3.6	4.1
Pathfinder		
1987-95	1.5	1.6
1996-05		
2WD		
VG		
one	1.4	1.5
both	1.9	2.0
VQ		
one	2.3	2.6
both	3.6	4.1
4WD		
VG		
one	1.6	1.7
both	6.3	6.4
VQ		
one	2.4	2.5
both	3.6	4.1

PATHFINDER : PICKUP : QUEST : TITAN : VAN : XTERRA   NI-53

	LABOR TIME	SEVERE SERVICE
Pickup		
1986-95		
4 cyl.	.7	.8
V6	1.2	1.3
1996-97	.7	.8
1993-02 Quest	.7	.8
Van	.8	.9
Exhaust Manifold or Gasket, Replace (B)		
620, 720	1.6	2.0
Armada, Titan		
2WD right or left	3.8	4.3
4WD right or left	5.8	6.5
Axxess	1.3	1.7
Frontier, Xterra		
4 cyl.	1.9	2.3
V6		
right side	2.1	2.5
left side	3.3	3.7
Murano		
2WD		
right side	2.9	3.3
left side	2.4	2.8
4WD		
right side	3.6	4.0
left side	2.5	2.9
Pathfinder, Pickup		
1986-95		
4 cyl.	2.0	2.4
V6		
right side	2.6	3.0
left side	3.3	3.7
1996-05		
Pathfinder		
VG		
right side	2.8	3.2
left side		
2WD	3.1	3.5
4WD	4.1	4.5
VQ		
right side	12.4	13.3
left side		
2WD	3.2	3.6
4WD	3.7	4.1
Pickup	1.8	2.2
Quest		
1993-98 one side	2.3	2.7
1999-05		
right side	3.0	3.3
left side	2.3	2.6
Van	1.8	2.3
Front Exhaust Pipe, Replace (B)		
620, 720	1.4	1.5
Armada, Titan		
one	.9	1.1
both	1.2	1.4
Axxess	.8	.9

	LABOR TIME	SEVERE SERVICE
Frontier, Xterra		
4 cyl.	1.1	1.2
V6		
one side	1.3	1.4
both sides	2.2	2.3
Murano	.9	1.1
Pathfinder, Pickup		
1986-95		
4 cyl.	1.2	1.3
V6		
one	1.5	1.6
both	2.1	2.2
1996-05		
Pathfinder		
VG		
one	.8	.9
both	1.3	1.4
VQ		
one	.9	1.1
both	1.2	1.4
Pickup	1.3	1.4
Quest		
1993-98	1.5	1.6
1999-05		
one	.8	.9
both	1.1	1.2
Van	.8	.9
Intermediate Exhaust Pipe, Replace (B)		
All Models	1.3	1.4
Muffler & Rear Pipe, Replace (B)		
Exc. below	.8	.9
620, 720	.9	1.0
1999-02 Quest	.5	.6
Armada, Titan	.9	1.1
Van	.9	1.0
Pre-Muffler & Center Pipe, Replace (B)		
Armada, Titan	.9	1.1
Frontier, Xterra	1.0	1.1
Murano	.5	.6
Pathfinder		
1994-95	1.2	1.3
1996-05		
VG	.8	.9
VQ	1.3	1.4
1994-97 Pickup		
4 cyl.	.7	.8
V6	1.2	1.3
Quest		
1993-98	1.3	1.4
1999-05	.7	.8

	LABOR TIME	SEVERE SERVICE
ENGINE COOLING		
Coolant Thermostat, Replace (B)		
620, 720	.5	.7
Armada, Titan	.8	1.0
Axxess	.8	1.0
Frontier, Xterra		
4 cyl.	1.1	1.3
V6	1.5	1.7
Murano	1.5	1.7
Pathfinder		
1994-95	2.3	2.5
1996-05	1.7	1.9
Pickup		
4 cyl.		
KA	1.7	1.9
Z	.8	1.0
V6	2.0	2.2
Quest	1.8	2.0
Van	2.0	2.2
Engine Coolant Temp. Sending Unit, Replace (B)		
Exc. below	.5	.5
2001-05 Pathfinder	2.7	2.7
Engine Oil Cooler, Replace (B)		
Armada, Titan	1.0	1.3
Murano	.9	1.1
1996-05 Pathfinder	1.4	1.6
Fan Blade, Replace (B)		
620, 720	1.2	1.4
Armada, Titan	2.0	2.2
Axxess one or both	.7	.9
Frontier, Xterra	.9	1.1
Murano one or both	1.8	2.0
Pathfinder		
1987-95	.7	.9
1996-05		
VG	.6	.8
VQ	1.2	1.4
Pickup	.7	.9
Quest		
1993-98	1.1	1.3
1999-05	.7	.9
Van	2.7	2.9
Fan Coupling, Replace (B)		
620, 720		
2WD	.8	1.0
4WD	1.6	1.8
Armada, Titan	2.0	2.2
Frontier, Xterra	.9	1.1
Pathfinder		
1987-95	1.1	1.3
1996-05		
VG	.8	1.0
VQ	1.2	1.4
Pickup, Van	1.1	1.3

720 PICKUP : AMADA : AXXESS : D21 PICKUP : FRONTIER : MURANO

	LABOR TIME	SEVERE SERVICE
Radiator Assy., R&R or Replace (B)		
620, 720	1.2	1.4
Armada, Titan	2.1	2.3
Axxess	1.5	1.7
Frontier, Xterra	1.9	2.1
Murano	2.0	2.3
Pathfinder	1.7	1.9
Pickup	1.4	1.6
Quest		
1993-98	1.1	1.3
1999-05	1.5	1.7
Van	3.1	3.3
w/AT add	.2	.3
Radiator Fan and/or Fan Motor, Replace (B)		
Axxess		
one	.7	1.0
both	1.3	1.6
Frontier, Xterra	.6	.9
Murano one or both	1.8	2.1
Quest	1.1	1.4
Radiator Fan Motor Relay, Replace (B)		
All Models	.5	.7
Radiator Fan Motor (Coolant Temp.) Switch, Replace (B)		
620, 720	.7	.7
Axxess, Murano	.5	.5
Frontier, Xterra	.6	.6
Pathfinder, Pickup	.5	.5
1993-05 Quest	.3	.3
Van	1.6	1.6
Radiator Hoses, Replace (B)		
Includes: Refill with proper coolant mix.		
620, 720 each	.5	.6
Armada, Titan		
upper	.8	.9
lower	1.1	1.2
Axxess each	.8	.9
Frontier, Xterra		
upper	.6	.7
lower	1.1	1.2
Murano each	.9	1.0
Pathfinder		
1987-95		
upper	.8	.8
lower	1.3	1.4
1996-00 each	.8	.9
2001-05 each	1.4	1.5
Pickup		
upper	.7	.7
lower		
4 cyl.		
KA	1.3	1.4
Z	.8	.9
V6	.8	.9
Quest each	.7	.7
Van each	2.6	2.6

	LABOR TIME	SEVERE SERVICE
Temperature Gauge (Dash), Replace (B)		
620, 720	1.9	1.9
Axxess	.7	.7
Frontier, Xterra	1.0	1.0
Murano, Pathfinder	1.0	1.0
Pickup	.8	.8
Quest	1.0	1.0
Van	1.4	1.4
Water Pump and/or Gasket, Replace (B)		
Includes: Refill with proper coolant mix.		
620, 720		
gasoline	1.8	2.3
diesel	1.4	1.9
Armada, Titan	2.1	2.3
Axxess	2.3	2.8
Frontier, Xterra		
4 cyl.	2.1	2.6
V6	3.3	3.8
Murano	3.8	4.3
Pathfinder, Pickup		
1986-95		
4 cyl.		
KA	2.7	3.2
Z	3.3	3.8
V6	3.4	3.9
1996-05		
Pathfinder		
VG	3.5	4.0
VQ	2.4	2.9
Pickup	2.7	3.2
Quest		
1993-98	4.8	5.3
1999-05	2.4	2.8
Van	3.6	4.1

ENGINE
ASSEMBLY

Times shown are for OEM assemblies. Time to replace assemblies from aftermarket rebuilders may vary.

Engine Assy., R&I (B)
Does not include parts or component transfer.

	LABOR TIME	SEVERE SERVICE
4 cyl.		
620, 720		
gasoline		
2WD	9.5	10.0
4WD	10.7	11.2
diesel	6.3	6.8
Axxess		
2WD	10.2	10.7
4WD	10.8	11.3
Frontier, Xterra		
2WD	8.1	8.6
4WD	11.6	12.1

	LABOR TIME	SEVERE SERVICE
Pathfinder, Pickup		
2WD	7.2	7.7
4WD	12.7	13.2
Van	11.2	11.7
V6		
Frontier, Xterra		
2WD	8.4	8.9
4WD	11.8	12.3
Pathfinder		
1987-95		
2WD	7.7	8.2
4WD	10.8	11.3
1996-05		
VG		
2WD	8.2	8.7
4WD	10.3	10.8
VQ		
2WD	9.8	10.3
4WD	11.5	12.0
Pickup		
2WD	7.5	8.0
4WD	10.8	11.3
Quest		
1993-98	10.5	11.0
1999-05	8.4	8.9
V8		
Armada, Titan		
2WD	11.1	11.6
4WD	12.4	12.9
w/AC add	.5	.5
w/PS add	.5	.5

Engine Assy., R&R and Recondition (A)
Includes: Replacing rings, rod and main bearings, cylinder head reconditioning and engine tune-up.

	LABOR TIME	SEVERE SERVICE
4 cyl.		
620, 720		
gasoline		
2WD	20.1	20.6
4WD	21.7	22.2
diesel	25.2	25.7
Axxess		
2WD	23.0	23.5
4WD	23.3	23.8
Frontier, Xterra		
2WD	22.4	22.9
4WD	26.2	26.7
1990-97 Pathfinder, Pickup		
KA		
2WD	21.8	22.3
4WD	26.2	26.7
Z		
2WD	20.0	20.5
4WD	26.0	26.5
Van		
2WD	22.0	22.5
4WD	23.3	23.8

PATHFINDER : PICKUP : QUEST : TITAN : VAN : XTERRA NI-55

	LABOR TIME	SEVERE SERVICE
V6		
Frontier, Xterra		
2WD	22.7	23.2
4WD	27.4	27.9
Pathfinder		
1987-95		
2WD	27.9	28.4
4WD	32.5	33.0
1996-05		
VG		
2WD	18.3	18.8
4WD	21.2	21.7
VQ		
2WD	27.9	28.4
4WD	28.6	29.1
Pickup		
2WD	27.9	28.4
4WD	32.7	33.2
Quest		
1993-98	24.0	24.5
1999-05	19.0	19.5
V8		
Armada, Titan		
2WD	27.4	28.3
4WD	29.2	30.1
w/AC add	.5	.5
w/PS add	.5	.5

Engine (New or Rebuilt Complete) Replace (B)
Includes: Component transfer and engine tune-up.

	LABOR TIME	SEVERE SERVICE
Frontier, Xterra		
4 cyl.		
2WD	12.1	12.6
4WD	15.8	16.3
V6		
w/supercharger		
2WD	13.8	14.3
4WD	17.0	17.5
w/o supercharger		
2WD	13.0	13.5
4WD	16.3	16.8
Murano		
2WD	13.6	14.6
4WD	14.4	15.4
Pathfinder		
1994-95		
V6		
2WD	10.8	11.3
4WD	14.5	15.0
1996-05		
VG		
2WD	11.4	11.9
4WD	13.8	14.3
VQ		
2WD	17.2	17.7
4WD	17.9	18.4

	LABOR TIME	SEVERE SERVICE
1994-97 Pickup		
4 cyl.		
2WD	10.4	10.9
4WD	16.4	16.9
V6		
2WD	10.8	11.3
4WD	14.5	15.0
Quest		
1993-98	13.0	13.5
1999-05	13.4	14.4
V8		
Armada, Titan		
2WD	15.9	16.4
4WD	17.6	18.1
w/AC add		
exc. below	.5	.5
Frontier, Xterra V6	1.4	1.4
w/PS add	.5	.5

Engine (Short Block), Replace (B)
Assembly may consist of cylinder block, piston assemblies, crankshaft, camshaft, timing chain and gears. Does not include cylinder heads. Operation Includes: R&R engine, transfer necessary parts and all necessary adjustments.

	LABOR TIME	SEVERE SERVICE
4 cyl.		
620, 720		
gasoline		
2WD	10.7	11.2
4WD	12.4	12.9
diesel	13.5	14.0
Axxess		
2WD	17.5	18.0
4WD	17.9	18.4
Frontier, Xterra		
2WD	12.4	12.9
4WD	15.9	16.4
Pathfinder, Pickup		
KA		
2WD	16.4	16.9
4WD	21.4	21.9
Z		
2WD	14.0	14.5
4WD	19.4	19.9
Van	16.3	16.8
V6		
Frontier, Xterra		
2WD	14.8	15.3
4WD	18.3	18.8
Murano		
2WD	20.4	20.9
4WD	21.8	22.3
Pathfinder		
1987-95		
2WD	20.0	20.5
4WD	24.4	24.9

	LABOR TIME	SEVERE SERVICE
1996-05		
VG		
2WD	14.2	14.7
4WD	16.7	17.2
VQ		
2WD	22.8	23.3
4WD	23.3	23.8
Pickup		
2WD	20.0	20.5
4WD	24.4	24.9
Quest		
1993-98	18.0	18.5
1999-05	14.3	14.8
V8		
Armada, Titan		
2WD	21.0	21.6
4WD	22.6	23.1
w/AC add		
exc. below	.5	.5
Frontier, Xterra V6	1.4	1.4
w/PS add	.5	.5
w/supercharger Frontier, Xterra V6 add	.9	.9

Engine Mounts, Replace (B)

	LABOR TIME	SEVERE SERVICE
Front		
620, 720	1.4	1.6
Armada, Titan	1.2	1.4
Axxess	1.6	1.8
Frontier, Xterra		
4 cyl.	1.6	1.8
V6	2.1	2.3
Murano	1.9	2.1
Pathfinder		
1987-95		
4 cyl.	1.5	1.7
V6	2.6	2.8
1996-05		
VG	3.2	3.4
VQ		
2WD	2.1	2.3
4WD	4.8	5.0
Pickup		
4 cyl.	1.5	1.7
V6	2.7	2.9
Quest		
1993-98	1.0	1.2
1999-05	1.8	2.0
Van	1.6	1.8
Rear		
620, 720	.8	1.0
Armada, Titan	1.2	1.4
Axxess	1.1	1.3
Frontier, Xterra	1.0	1.2
Murano		
2WD	2.4	2.6
4WD	5.7	5.9
Pathfinder		
VG	.6	.8
VQ	1.9	2.1

	LABOR TIME	SEVERE SERVICE
Engine Mounts, Replace (B)		
Pickup	.9	1.1
Quest		
1993-98	.7	.9
1999-05	1.2	1.4
Van	.8	1.0
CYLINDER HEAD		
Compression Test (B)		
620, 720		
gasoline	1.2	1.5
diesel	1.7	2.0
Armada, Titan	1.5	1.6
Axxess	.7	1.0
Frontier, Xterra	.8	1.1
Pathfinder, Pickup		
1986-95		
4 cyl.	1.1	1.4
V6	1.5	1.8
1996-05		
Pathfinder		
VG	.7	1.0
VQ	1.3	1.6
Pickup	1.2	1.5
Quest		
1993-98	1.3	1.6
1999-05	.7	1.0
Van	.9	1.2
Cylinder Head, Replace (B)		
Includes: Parts transfer, adjustments.		
4 cyl.		
620, 720		
gasoline	10.2	10.7
diesel	5.5	6.0
Axxess	13.5	14.0
Frontier, Xterra	13.0	13.5
Pathfinder, Pickup		
KA	14.3	14.8
Z	10.2	10.7
Van	12.2	12.7
V6		
Frontier, Xterra		
w/supercharger		
right side	13.8	14.3
left side	14.4	14.9
both sides	19.3	19.8
w/o supercharger		
one side	13.6	14.1
both sides	18.9	19.4
Murano		
2WD		
one side	22.1	22.6
both sides	26.1	26.6
4WD		
one side	23.5	24.0
both sides	27.4	27.9

	LABOR TIME	SEVERE SERVICE
Pathfinder		
1987-95		
right side	12.7	13.2
left side	13.2	13.7
both sides	19.0	19.5
1996-05		
VG		
one side	10.2	10.7
both sides	13.6	14.1
VQ		
2WD		
one side	23.3	23.8
both sides	26.8	27.3
4WD		
one side	23.7	24.2
both sides	27.3	27.8
Pickup		
one side	13.1	13.6
both sides	18.8	19.3
Quest		
1993-98		
one side	14.5	15.0
both sides	19.0	19.5
1999-05		
right side	12.4	12.9
left side	13.5	14.0
both sides	18.5	19.0
V8		
Armada, Titan		
2WD		
right or left	22.2	22.8
both sides	26.9	27.8
4WD		
right or left	24.0	24.6
both sides	28.5	29.4
Cylinder Head Gasket, Replace (B)		
4 cyl.		
620, 720		
gasoline	5.1	5.4
diesel	3.7	4.0
Axxess	7.0	7.3
Frontier, Xterra	10.6	10.9
Pathfinder, Pickup		
KA	9.8	10.1
Z	5.3	5.6
Van	7.0	7.3
V6		
Frontier, Xterra		
w/supercharger		
right side	10.9	11.2
left side	11.2	11.5
both sides	13.6	13.9
w/o supercharger		
right side	11.0	11.3
left side	11.8	12.1
both sides	14.5	14.8

	LABOR TIME	SEVERE SERVICE
Murano		
2WD		
one side	19.2	19.9
both sides	20.7	21.3
4WD		
one side	20.7	21.3
both sides	22.1	22.6
Pathfinder		
1987-95		
one side	9.6	9.9
both sides	11.8	12.3
1996-05		
VG		
one side	8.3	8.6
both sides	9.9	10.2
VQ		
2WD		
one side	21.0	21.3
both sides	22.0	22.3
4WD		
one side	21.5	21.8
both side	22.5	22.8
Pickup		
right side	9.3	9.6
left side	9.8	10.1
both sides	11.6	11.9
Quest		
1993-98		
one side	12.0	12.3
both sides	14.3	14.6
1999-05		
right side	14.3	15.3
left side	13.5	14.5
both sides	16.2	17.1
V8		
Armada, Titan		
2WD		
right or left	19.6	20.2
both sides	21.3	22.0
4WD		
right or left	21.2	21.8
both sides	22.9	24.0
w/AC Frontier, Xterra		
4 cyl. add	.2	.2
Rocker Arm Pivots, Springs and Arms, Replace (B)		
4 cyl.		
620, 720	1.5	1.8
Axxess	.8	1.1
Frontier, Xterra	4.6	4.9
Pathfinder, Pickup		
KA	2.2	2.5
Z	1.2	1.5
Van	3.0	3.3

PATHFINDER : PICKUP : QUEST : TITAN : VAN : XTERRA

	LABOR TIME	SEVERE SERVICE
Rocker Arm Shafts, Lifters or Guides, Replace (B)		
V6		
Frontier, Xterra		
right side	2.4	2.7
left side	4.3	4.6
both sides	5.4	5.7
Murano		
one side	20.9	21.2
both sides	21.7	22.0
Pathfinder		
1987-95		
one side	2.1	2.4
both sides	3.5	3.8
1996-05		
VG		
right side	1.7	2.0
left side	3.6	3.9
both sides	4.7	5.0
VQ		
one side	17.0	17.3
both sides	18.2	18.5
Pickup		
one side	2.2	2.5
both sides	3.3	3.6
Quest		
1993-98		
right side	4.2	4.5
left side	1.7	2.0
both sides	5.3	5.6
1999-05		
right side	8.4	9.2
left side	8.2	9.0
both sides	9.4	10.4
Armada, Titan		
2WD		
right or left	19.3	19.9
both sides	21.0	21.8
4WD		
right or left	20.9	21.5
both sides	22.6	23.5
Valve Clearance, Adjust (B)		
4 cyl.		
620, 720	1.2	1.5
Frontier, Xterra	1.0	1.3
Pathfinder, Pickup	1.1	1.4
Van	2.7	3.0
V6		
exc. Pathfinder VQ	4.3	4.6
Pathfinder VQ	4.8	5.1
Valve Cover Gasket, Replace (B)		
4 cyl.		
exc. Van	.7	.9
Van	2.1	2.3
V6		
Frontier, Xterra		
right side	1.1	1.3
left side	3.2	3.4
both sides	3.8	4.0

	LABOR TIME	SEVERE SERVICE
Murano		
right side	2.3	2.5
left side	.9	1.1
both sides	2.9	3.1
Pathfinder		
1987-95		
one side	1.6	1.8
both sides	2.4	2.6
1996-05		
VG		
right side	.6	.8
left side	3.1	3.3
both sides	3.3	3.5
VQ		
one side	3.5	3.7
both sides	4.0	4.2
Pickup		
one side	1.6	1.8
both sides	2.3	2.5
Quest		
1993-98		
right side	3.3	3.5
left side	.9	1.1
both sides	3.9	4.1
1999-05		
right side	3.9	4.1
left side	.9	1.1
both sides	4.5	4.7
V8		
Armada, Titan		
right or left	.8	1.0
both sides	1.1	1.3
Valve Job (A)		
4 cyl.		
620, 720		
gasoline	11.5	12.0
diesel	7.6	8.1
Axxess	14.6	15.1
Frontier, Xterra	14.0	14.5
Pathfinder, Pickup		
KA	14.3	14.8
Z	11.3	11.8
Van	13.2	13.7
V6		
Frontier, Xterra	20.0	20.5
Pathfinder		
1987-95	20.0	20.5
1996-05		
VG	14.8	15.3
VQ	28.3	28.8
Pickup	20.0	20.5
Quest		
1993-98	20.2	20.7
1999-05	19.7	20.2
V8		
Armada, Titan		
2WD	27.7	28.5
4WD	29.4	30.2

	LABOR TIME	SEVERE SERVICE
Valve Pushrods, Replace (B)		
720 diesel	1.5	1.8
w/AC add	.5	.5
Valve Springs and/or Oil Seals, Replace (B)		
4 cyl.		
620, 720		
gasoline	3.2	3.6
diesel	2.3	2.7
Axxess	3.1	3.5
Frontier, Xterra	6.9	7.3
Pathfinder, Pickup	3.8	4.2
Van	3.8	4.2
V6		
Frontier, Xterra		
right side	3.2	3.6
left side	5.1	5.5
both sides	6.9	7.3
Murano		
one side	21.4	21.8
both sides	23.8	24.2
Pathfinder		
1987-95		
one side	2.9	3.3
both sides	5.0	5.4
1996-05		
VG		
right side	2.4	2.8
left side	4.6	5.0
both sides	6.2	6.6
VQ		
one side	18.2	18.6
both sides	21.0	21.4
Pickup		
one side	2.9	3.3
both sides	5.0	5.4
Quest		
1993-98		
right side	5.9	6.3
left side	3.6	4.0
both sides	8.2	8.6
1999-05		
right side	5.4	5.8
left side	2.3	2.7
both sides	7.2	7.6
V8		
Armada, Titan		
2WD		
right or left	20.0	20.5
both sides	22.2	22.7
4WD		
right or left	21.5	22.0
both sides	24.0	24.5

NI-58 720 PICKUP : AMADA : AXXESS : D21 PICKUP : FRONTIER : MURANO

	LABOR TIME	SEVERE SERVICE
CAMSHAFT		
Camshaft, Replace (B)		
4 cyl. (Includes lifters)		
620, 720		
gasoline	3.8	4.2
diesel	8.9	9.3
Axxess	2.0	2.4
Frontier, Xterra one or		
both sides	4.3	4.7
Pathfinder, Pickup		
KA	2.2	2.6
Z	4.0	4.4
Van	4.7	5.1
V6		
Frontier, Xterra		
2WD		
one side	11.3	11.7
both sides	14.1	14.5
4WD		
one side	12.0	12.4
both sides	15.0	15.4
supercharged		
right side	11.3	11.7
left side	11.8	12.2
both sides	14.7	15.1
Murano		
2WD		
one side	19.2	19.6
both sides	20.3	20.7
4WD		
one side	20.6	21.0
both sides	21.7	22.1
Pathfinder		
1987-95		
right side	10.5	10.9
left side	10.8	11.2
both sides	14.1	14.5
1996-05		
VG		
2WD		
right side	3.7	4.1
left side	7.8	8.2
both sides	9.6	10.0
4WD		
right side	4.2	4.6
left side	8.0	8.4
both sides	10.3	10.7
VQ		
2WD		
one side	15.6	16.0
both sides	16.9	17.3
4WD		
one side	17.2	17.6
both sides	18.5	18.9
Pickup		
one side	10.8	11.3
both sides	14.3	14.7

	LABOR TIME	SEVERE SERVICE
Quest		
1993-98		
one side	13.9	14.3
both sides	18.3	18.7
1999-05		
right side	10.7	11.1
left side	11.4	11.8
both sides	14.1	14.5
V8		
Armada, Titan		
2WD		
right or left	19.0	19.5
both sides	20.4	21.0
4WD		
right or left	20.6	21.2
both sides	22.0	22.6
Camshaft Seal, Replace (B)		
V6		
Frontier, Xterra		
one side	2.7	2.9
both sides	2.8	3.0
Pathfinder		
1987-95		
one side	5.3	5.5
both sides	5.5	5.7
1996-05 VG		
one side	3.2	3.4
both sides	3.4	3.6
Pickup		
one side	5.2	5.4
both sides	5.4	5.6
Quest		
1993-98		
one side	4.1	4.3
both sides	4.2	4.4
1999-05		
one side	3.1	3.3
both sides	3.9	4.1
Camshaft Sprocket or Gear, Replace (B)		
4 cyl.		
620, 720	.8	1.0
Axxess	.7	.9
Frontier, Xterra	2.6	2.8
Pathfinder, Pickup		
KA	1.3	1.5
Z	.8	1.0
Van	.8	1.0
Crankshaft Timing Gear or Sprocket, Replace (B)		
720 diesel	3.7	4.0
w/AC add	.2	.2
Timing Belt, Replace (B)		
V6		
Frontier, Xterra	3.5	4.0
Pathfinder		
1987-95	3.7	4.2
1996-00	3.2	3.7

	LABOR TIME	SEVERE SERVICE
Pickup	3.8	4.3
Quest	4.3	4.8
w/AC add	.3	.3
w/PS add	.2	.2
Replace crankshaft		
gear add	.5	.7
Timing Chain, Replace (B)		
4 cyl.		
620, 720	6.2	6.7
Axxess	7.9	8.4
Frontier, Xterra		
2WD	7.5	8.0
4WD	9.5	10.0
Pathfinder, Pickup		
KA		
2WD		
SOHC	7.2	7.7
DOHC	9.7	10.2
4WD	10.0	10.5
Z		
2WD	6.3	6.8
4WD	9.1	9.6
V6		
exc. below		
2WD	12.4	12.9
4WD	14.2	14.7
Murano	11.5	12.0
2001-05 Pathfinder		
2WD	13.3	13.8
4WD	15.1	15.6
Van	7.7	8.2
V8		
Armada, Titan		
2WD	16.0	16.6
4WD	17.8	18.4
w/AC Frontier,		
Xterra add	.5	.5
w/PS Frontier,		
Xterra add	.6	.6
Timing Cover and/or Gasket, Replace (B)		
4 cyl.		
620, 720	6.2	6.6
Axxess	6.4	6.8
Frontier, Xterra		
2WD	6.2	6.6
4WD	8.2	8.6
Pathfinder, Pickup		
KA		
2WD		
SOHC	6.8	7.2
DOHC	10.0	10.4
4WD	9.7	10.1
Z		
2WD	6.2	6.6
4WD	8.9	9.3
Van	7.2	7.6

PATHFINDER : PICKUP : QUEST : TITAN : VAN : XTERRA   NI-59

	LABOR TIME	SEVERE SERVICE
V6		
Frontier, Xterra	3.5	*3.9*
Murano	8.8	*9.2*
Pathfinder		
1987-95	3.7	*4.1*
1996-05		
VG	2.4	*2.8*
VQ		
2WD	10.9	*11.3*
4WD	12.7	*13.1*
Pickup	3.6	*4.0*
Quest		
1993-98	3.9	*4.3*
1999-05	2.5	*2.9*
V8		
Armada, Titan		
2WD	16.0	*16.6*
4WD	17.8	*18.4*
Diesel	3.2	*3.6*
w/AC add	.5	*.5*
w/PS add	.6	*.6*

CRANK & PISTONS
Connecting Rod Bearings, Replace (A)
Includes: Check bearing oil clearances.

	LABOR TIME	SEVERE SERVICE
4 cyl.		
620, 720		
2WD	3.0	*3.5*
4WD	10.2	*10.7*
Axxess		
2WD	15.7	*16.2*
4WD	16.3	*16.8*
Frontier, Xterra		
2WD	3.3	*3.8*
4WD	5.0	*5.5*
Pathfinder, Pickup		
KA		
2WD	16.6	*17.1*
4WD	18.7	*19.2*
Z	2.9	*3.4*
Van	16.5	*17.0*
V6		
Frontier, Xterra		
2WD	16.0	*16.5*
4WD	19.5	*20.0*
Murano		
2WD	21.4	*21.9*
4WD	22.8	*23.3*
Pathfinder		
1987-95		
2WD	20.8	*21.3*
4WD	22.6	*23.1*
1996-05		
VG		
2WD	16.9	*17.4*
4WD	17.5	*18.0*
VQ		
2WD	23.7	*24.2*
4WD	24.2	*24.7*

	LABOR TIME	SEVERE SERVICE
Pickup		
2WD	20.8	*21.3*
4WD	22.7	*23.2*
Quest		
1993-98	20.0	*20.5*
1999-05	14.3	*14.8*
V8		
Armada, Titan		
2WD	22.0	*22.6*
4WD	24.6	*25.2*
w/AC add	.5	*.5*
w/PS add	.5	*.5*
w/supercharger Frontier, Xterra V6 add	.9	*.9*

Crankshaft & Main Bearings, Replace (A)
Includes: Engine R&R.

	LABOR TIME	SEVERE SERVICE
4 cyl.		
620, 720		
gasoline		
2WD	15.0	*15.9*
4WD	16.4	*17.3*
diesel	10.8	*11.7*
Axxess		
2WD	16.0	*16.9*
4WD	16.4	*17.3*
Frontier, Xterra		
2WD	15.4	*16.3*
4WD	18.8	*19.7*
Pathfinder, Pickup		
2WD	16.9	*17.8*
4WD	22.0	*22.9*
Van	16.8	*17.7*
V6		
Frontier, Xterra		
2WD	21.6	*22.1*
4WD	22.4	*22.9*
Murano		
2WD	22.8	*23.3*
4WD	24.3	*25.7*
Pathfinder		
1987-95		
2WD	21.7	*22.6*
4WD	23.2	*24.1*
1996-05		
VG		
2WD	13.7	*14.6*
4WD	16.1	*17.0*
VQ		
2WD	25.6	*26.5*
4WD	26.1	*27.0*
Pickup		
2WD	21.6	*22.5*
4WD	23.2	*24.1*
Quest		
1993-98	20.1	*21.0*
1999-05	15.8	*16.7*

	LABOR TIME	SEVERE SERVICE
V8		
Armada, Titan		
2WD	26.1	*26.7*
4WD	27.7	*28.3*
w/AC add	.5	*.5*
w/PS add	.5	*.5*
w/supercharger Frontier, Xterra V6 add	.9	*.9*

Crankshaft Front Oil Seal, Replace (B)

	LABOR TIME	SEVERE SERVICE
720 diesel	1.5	*1.7*
4 cyl.	1.5	*1.7*
V6	5.2	*5.4*
Armada, Titan	2.8	*3.1*
Frontier, Xterra		
4 cyl.	1.0	*1.2*
V6	3.5	*3.7*
Murano	1.2	*1.4*
Pathfinder		
1994-95	5.3	*5.5*
1996-05		
VG		
2WD	7.2	*7.4*
4WD	12.3	*12.5*
VQ	1.7	*1.9*
Quest		
1993-98	4.0	*4.2*
1999-05	2.7	*2.9*
w/AC add	.3	*.3*
w/PS add	.2	*.2*

Crankshaft Pulley, Replace (B)

	LABOR TIME	SEVERE SERVICE
Exc. below	1.2	*1.4*
Armada, Titan	2.8	*3.1*
Frontier, Xterra 4 cyl.	.9	*1.1*
Pathfinder, Pickup	1.5	*1.7*
1993-98 Quest	1.6	*1.8*
Van	1.6	*1.8*

Pistons or Connecting Rods, Replace (A)
Includes: Ridge reaming, cylinder wall deglazing, installing new rings and rod bearings, engine tune-up.

	LABOR TIME	SEVERE SERVICE
4 cyl.		
620, 720		
gasoline		
2WD	15.9	*16.8*
4WD	17.4	*18.3*
diesel	9.9	*10.8*
Axxess		
2WD	18.4	*19.3*
4WD	18.9	*19.8*
Frontier, Xterra		
2WD	15.9	*16.8*
4WD	19.4	*20.3*
Pathfinder, Pickup		
KA		
2WD	19.2	*20.1*
4WD	21.1	*22.0*

720 PICKUP : AMADA : AXXESS : D21 PICKUP : FRONTIER : MURANO

	LABOR TIME	SEVERE SERVICE
Pistons or Connecting Rods, Replace (A)		
Z		
2WD	17.3	18.2
4WD	19.0	19.9
Van	19.2	20.1
V6		
Frontier, Xterra		
2WD	19.3	20.2
4WD	23.9	25.3
Murano		
2WD	23.6	24.5
4WD	24.9	25.8
Pathfinder		
1987-95		
2WD	24.9	25.8
4WD	26.4	27.3
1996-05		
VG		
2WD	19.1	20.0
4WD	22.6	23.5
VQ		
2WD	25.8	26.7
4WD	26.5	27.4
Pickup		
2WD	24.9	25.8
4WD	26.4	27.3
Quest		
1993-98	23.2	24.1
1999-05	23.5	24.9
V8		
Armada, Titan		
2WD	25.0	25.6
4WD	26.8	27.4
w/AC add	.5	.5
w/PS add	.5	.5
w/supercharger Frontier, Xterra V6 add	.9	.9
Rear Main Oil Seal, Replace (B)		
Includes: Trans. R&R. when necessary.		
620, 720		
gasoline		
2WD	4.3	4.9
4WD	6.0	6.6
diesel		
SD22	13.1	13.7
SD25	6.5	7.1
Armada, Titan	4.4	4.9
Axxess		
AT		
2WD	6.2	6.8
4WD	7.9	8.5
MT		
2WD	5.3	5.9
4WD	7.4	8.0

	LABOR TIME	SEVERE SERVICE
Frontier, Xterra		
AT		
2WD	5.3	5.9
4WD		
4 cyl.	5.1	5.7
V6	8.4	9.2
MT		
2WD	4.4	5.0
4WD		
4 cyl.	8.8	9.4
V6	9.2	9.8
Murano		
2WD	11.6	13.1
4WD	13.2	14.8
Pathfinder		
1987-95		
2WD		
AT	5.7	6.3
MT	4.8	5.4
4WD	10.1	10.7
1996-05		
VG		
AT		
2WD	5.8	6.4
4WD	10.5	11.1
MT		
2WD	4.6	5.2
4WD	9.6	10.2
VQ		
AT		
2WD	8.6	9.2
4WD	9.9	10.5
MT		
2WD	5.3	5.9
4WD	10.4	11.0
Pickup		
1986-95		
KA		
2WD		
AT	5.6	6.2
MT	4.9	5.5
4WD	10.1	10.7
Z		
2WD	4.2	4.8
4WD	9.0	9.6
1996-97		
AT	5.6	6.2
MT		
2WD	5.7	6.3
4WD	10.1	10.7
Quest		
1993-98	5.9	6.5
1999-05	8.4	9.5
Van		
AT	5.2	5.8
MT	3.7	4.3

	LABOR TIME	SEVERE SERVICE
Rings, Replace (A)		
Includes: Ridge reaming, cylinder wall deglazing, installing new rings, engine tune-up.		
4 cyl.		
620, 720		
gasoline		
2WD	15.5	16.4
4WD	17.0	17.9
diesel	9.2	10.1
Axxess		
2WD	16.8	17.7
4WD	17.4	18.3
Frontier, Xterra		
2WD	15.9	16.8
4WD	19.4	20.3
Pathfinder, Pickup		
KA		
2WD	17.8	18.7
4WD	20.6	21.5
Z		
2WD	15.8	16.7
4WD	17.3	18.2
Van	17.6	18.5
V6		
Frontier, Xterra		
2WD	19.3	20.2
4WD	22.4	23.3
Murano		
2WD	23.6	24.5
4WD	24.9	25.8
Pathfinder		
1987-95		
2WD	23.2	24.1
4WD	24.4	25.3
1996-05		
VG		
2WD	17.3	18.2
4WD	20.8	21.7
VQ		
2WD	25.8	26.7
4WD	26.5	27.4
Pickup		
2WD	23.0	23.9
4WD	24.6	25.5
Quest		
1993-98	21.4	22.3
1999-05	23.5	24.9
V8		
Armada, Titan		
2WD	25.0	25.6
4WD	26.8	27.4
w/AC add	.5	.5
w/PS add	.5	.5
w/supercharger Frontier, Xterra V6 add	.9	.9

PATHFINDER : PICKUP : QUEST : TITAN : VAN : XTERRA NI-61

	LABOR TIME	SEVERE SERVICE
ENGINE LUBRICATION		
Engine Oil Pressure Switch (Sending Unit), Replace (B)		
Exc. Pathfinder	.5	.5
Pathfinder	.8	.8
Oil Pan and/or Gasket, Replace (B)		
4 cyl.		
620, 720		
gasoline		
2WD	1.6	1.8
4WD	3.0	3.2
diesel	1.6	1.8
Axxess	2.5	2.7
Frontier, Xterra		
2WD	2.3	2.6
4WD	3.1	3.5
Pathfinder, Pickup		
KA		
2WD	2.2	2.4
4WD	6.1	6.3
Z		
2WD	1.7	1.9
4WD	5.2	5.4
Van	1.7	1.9
V6		
Frontier, Xterra		
2WD	3.8	4.0
4WD	4.7	5.3
Murano		
upper		
2WD	5.4	5.6
4WD	14.9	15.8
lower	1.4	1.6
Pathfinder		
1987-95		
2WD	2.1	2.3
4WD	6.2	6.4
1996-05		
VG		
2WD	3.1	3.3
4WD	6.5	6.7
VQ		
upper		
2WD	5.0	5.2
4WD	7.2	7.4
lower	.8	1.0
Pickup		
2WD	2.2	2.4
4WD	6.4	6.6
1993-05 Quest	3.5	3.7
V8		
Armada, Titan		
2WD	14.4	15.0
4WD	16.0	16.6

	LABOR TIME	SEVERE SERVICE
Oil Pressure Gauge (Dash), Replace (B)		
620, 720	1.7	1.7
Axxess	.7	.7
Frontier, Xterra	1.0	1.0
Murano, Pathfinder	1.0	1.0
Pickup	.8	.8
Quest	1.0	1.0
Van	1.4	1.4
Oil Pressure Regulator Valve, Replace (B)		
720 diesel	.7	.7
Oil Pump, Replace (B)		
4 cyl.		
620, 720		
gasoline		
2WD	1.2	1.7
4WD	2.1	2.6
diesel	2.5	3.0
Axxess		
2WD	6.8	7.3
4WD	6.5	7.0
Frontier, Xterra	6.5	7.2
Pathfinder, Pickup	.8	1.3
Van	.7	1.2
V6		
Frontier, Xterra	8.8	9.6
Murano		
2WD	15.5	16.4
4WD	17.1	18.0
Pathfinder		
1987-95		
2WD	7.9	8.4
4WD	12.3	12.8
1996-05		
VG		
2WD	7.2	7.7
4WD	8.2	8.7
VQ		
2WD	11.3	11.8
4WD	13.1	13.6
Pickup		
2WD	8.1	8.6
4WD	12.5	13.0
Quest		
1993-98	7.4	7.9
1999-05	5.5	6.0
V8		
Armada, Titan		
2WD	15.0	15.9
4WD	16.7	17.6
Engine Oil Pressure Switch (Sending Unit), Replace (B)		
4 cyl.	.5	.5
V6		
VG	.6	.6
VQ	.8	.8
V8	.5	.5

	LABOR TIME	SEVERE SERVICE
CLUTCH		
Bleed Clutch Hydraulic System (B)		
All Models	.5	.7
Clutch Pedal Free Play, Adjust (B)		
All Models	.3	.5
Clutch Assy., Replace (B)		
620, 720		
2WD	4.5	4.7
4WD	7.3	7.5
Axxess		
2WD	5.0	5.2
4WD	7.5	7.7
Frontier, Xterra		
4cyl.		
2WD	4.1	4.3
4WD	8.8	9.0
V6		
2WD	7.2	7.4
4WD	8.0	8.8
Pathfinder		
2WD	5.0	5.2
4WD	10.8	11.0
Pickup		
2WD	3.3	3.5
4WD	7.5	7.7
Van	3.6	3.8
Replace pilot bearing add	.3	.5
Clutch Damper, Replace (B)		
All Models	1.1	1.3
Clutch Master Cylinder, Replace (B)		
Includes: System bleeding.		
All Models	1.1	1.3
Clutch Release Bearing, Replace (B)		
620, 720		
2WD	4.3	4.5
4WD	7.0	7.2
Axxess		
2WD	4.5	4.7
4WD	7.1	7.3
Frontier, Xterra		
4 cyl.		
2WD	4.3	4.5
4WD	9.0	9.2
V6		
2WD	7.3	7.5
4WD	8.0	8.8
Pathfinder		
2WD	5.2	5.4
4WD	10.9	11.1
Pickup		
2WD	3.3	3.5
4WD	7.2	7.4
Van	3.2	3.4
Clutch Slave Cylinder, Replace (B)		
Includes: System bleeding.		
All Models	.8	1.0

720 PICKUP : AMADA : AXXESS : D21 PICKUP : FRONTIER : MURANO

	LABOR TIME	SEVERE SERVICE
Flywheel, Replace (B)		
620, 720		
2WD	5.0	5.2
4WD	7.6	7.8
Axxess		
2WD	5.3	5.5
4WD	7.7	7.9
Frontier, Xterra		
2WD	4.3	4.5
4WD		
4 cyl.	9.0	9.2
V6	8.4	9.2
Pathfinder		
2WD	5.2	5.4
4WD	10.9	11.1
Pickup		
2WD	3.8	4.0
4WD	7.9	8.1
Van	3.8	4.0
Replace rear main seal add	.3	.5

MANUAL TRANSAXLE

Differential Case, Recondition (A)
Axxess .... 10.2 / 10.7

Differential Oil Seal, Replace (B)
Axxess
  one side .... 1.3 / 1.5
  both sides .... 2.1 / 2.3

Speedometer Pinion, Replace (B)
Axxess .... .5 / .5

Transaxle Assy., R&R and Recondition (B)
Axxess
  2WD .... 8.1 / 8.7
  4WD .... 12.5 / 13.1

Transaxle Assy., R&I (B)
Axxess
  2WD .... 4.4 / 4.6
  4WD .... 7.3 / 7.5

MANUAL TRANSMISSION

Extension Housing and/or Gasket, Replace (B)
620, 720
  2WD .... 4.1 / 4.3
  4WD .... 7.1 / 7.3
Frontier, Xterra
  4 cyl.
    2WD .... 4.9 / 5.1
    4WD .... 9.4 / 9.6
  V6
    2WD .... 7.6 / 7.8
    4WD .... 9.7 / 9.9
Pathfinder
  2WD .... 5.5 / 5.7
  4WD .... 11.2 / 11.4

	LABOR TIME	SEVERE SERVICE
Pickup		
2WD	4.6	4.8
4WD	9.2	9.4
Van	3.7	3.9
Replace rear oil seal add	.3	.5

Speedometer Drive Pinion, Replace (B)
All Models .... .6 / .7

Transmission Assy., R&R and Recondition (A)
4 speed 620, 720 .... 6.8 / 7.4
5 speed
  620, 720
    2WD .... 7.4 / 8.0
    4WD .... 9.8 / 10.4
  Frontier, Xterra
    4 cyl.
      2WD .... 8.2 / 8.8
      4WD .... 12.6 / 13.2
    V6
      2WD .... 11.0 / 11.6
      4WD .... 13.1 / 13.7
  Pathfinder
    2WD .... 9.0 / 9.6
    4WD .... 14.5 / 15.1
  Pickup
    2WD .... 7.8 / 8.4
    4WD .... 12.9 / 13.5
  Van .... 6.3 / 6.9

Transmission Assy., R&R or Replace (B)
620, 720
  2WD .... 4.1 / 4.3
  4WD .... 7.1 / 7.3
Frontier, Xterra
  4 cyl.
    2WD .... 4.1 / 4.3
    4WD .... 8.7 / 8.9
  V6
    2WD .... 6.9 / 7.1
    4WD .... 9.0 / 9.2
Pathfinder
  2WD .... 4.7 / 4.9
  4WD .... 10.5 / 10.7
Pickup
  2WD .... 3.9 / 4.1
  4WD .... 8.7 / 8.9
Van .... 3.2 / 3.4

AUTOMATIC TRANSAXLE

SERVICE TRANSAXLE INSTALLED

Differential Oil Seal (Halfshaft), Replace (B)
Axxess
  one .... 1.4 / 1.6
  both .... 2.1 / 2.3
Quest
  1993-98
    one .... 1.4 / 1.6
    both .... 2.3 / 2.5

	LABOR TIME	SEVERE SERVICE
1999-05		
one	1.6	1.8
both	2.5	2.7

Lock-up Solenoid, Replace (B)
Axxess, Quest .... 2.2 / 2.5

Neutral Safety Switch, Replace (B)
Axxess, Quest .... .5 / .5

Overdrive Cancel Solenoid, Replace (B)
Axxess .... 1.8 / 2.0
Quest .... 2.2 / 2.4

Overdrive Switch, Replace (B)
Axxess, Quest .... .5 / .5

Speedometer Drive Pinion and/or Sensor, Replace (B)
Axxess
  2WD .... .5 / .5
  4WD .... 1.3 / 1.3
1993-05 Quest .... .5 / .5

Valve Body, R&R and Recondition (A)
Axxess .... 3.8 / 4.3
Quest
  1993-98 .... 4.6 / 5.2
  1999-05 .... 4.5 / 5.1

Valve Body, Replace (B)
Axxess .... 1.9 / 2.2
Quest
  1993-98 .... 2.5 / 2.8
  1999-05 .... 2.2 / 2.5

SERVICE TRANSAXLE REMOVED
Transaxle R&R included unless otherwise noted.

Differential Case, Recondition (A)
Axxess
  2WD .... 8.1 / 8.7
  4WD .... 9.7 / 10.3
Quest
  1993-98 .... 7.8 / 8.4
  1999-05 .... 6.5 / 7.1

Flywheel (Flexplate), Replace (B)
Axxess
  2WD .... 6.4 / 6.6
  4WD .... 8.2 / 8.4
Quest
  1993-98 .... 5.7 / 5.9
  1999-05 .... 4.6 / 4.8

Torque Converter, Replace (B)
Axxess
  2WD .... 6.3 / 6.5
  4WD .... 8.2 / 8.4
Quest
  1993-98 .... 5.6 / 5.8
  1999-05 .... 4.5 / 4.7

Transaxle Assy., R&R and Recondition (A)
Axxess
  2WD .... 15.2 / 16.1
  4WD .... 17.1 / 18.0

PATHFINDER : PICKUP : QUEST : TITAN : VAN : XTERRA NI-63

	LABOR TIME	SEVERE SERVICE
Quest		
1993-98	16.6	*17.5*
1999-05	15.5	*16.4*
Flush oil cooler and lines add	.5	*.5*
Transaxle & Torque Converter, R&I (B)		
Axxess		
2WD	6.2	*6.4*
4WD	7.8	*8.0*
Quest		
1993-98	5.4	*5.6*
1999-05	4.3	*4.5*
Replace assembly add	.5	*.5*

AUTOMATIC TRANSMISSION
SERVICE TRANSMISSION INSTALLED

	LABOR TIME	SEVERE SERVICE
Check Unit for Oil Leaks (C)		
All Models	.9	*.9*
Drain and Refill Unit (B)		
620, 720	.8	*.8*
Frontier, Xterra		
4 cyl.	1.0	*1.0*
V6	1.5	*1.5*
Pathfinder, Pickup		
1986-87	.8	*.8*
1988-05	1.2	*1.2*
Van	.8	*.8*
Oil Pressure Check (B)		
All Models	.9	*.9*
Accumulator Assy., Replace (B)		
Frontier, Xterra		
4 cyl.	1.9	*2.1*
V6	2.4	*2.6*
Pathfinder		
1994-95	2.2	*2.4*
1996-05	1.9	*2.1*
1994-97 Pickup	2.1	*2.3*
Electronic Control Unit, Replace (B)		
All Models	.6	*.6*
Extension Housing Oil Seal, Replace (B)		
620, 720	.9	*1.1*
Armada, Titan		
2WD	.8	*1.0*
4WD	4.1	*4.7*
Frontier, Xterra	1.1	*1.3*
Pathfinder		
1987-95	1.2	*1.4*
1996-05	.8	*1.0*
Pickup	1.2	*1.4*
Van	1.3	*1.5*
Inhibitor Switch (B)		
All Models	.7	*.7*
Lock-up Solenoid, Replace (B)		
1981-93	.7	*.9*
Armada, Titan	1.8	*2.1*

	LABOR TIME	SEVERE SERVICE
Frontier, Xterra		
4 cyl.	1.9	*2.1*
V6	2.5	*2.7*
Pathfinder	2.1	*2.3*
1994-97 Pickup	2.3	*2.5*
Oil Pan Gasket, Replace (B)		
620, 720	.8	*1.0*
Armada, Titan	1.4	*1.6*
Frontier, Xterra		
4 cyl.	1.1	*1.3*
V6	1.7	*1.9*
Pathfinder, Pickup		
1986-87	.8	*1.0*
1988-05	1.3	*1.5*
Van	.8	*1.0*
Speedometer Drive Pinion, Replace (B)		
All Models	.6	*.8*
Vacuum Diaphragm, Replace (B)		
1981-97	.8	*.8*
Valve Body Assy., R&R and Recondition (A)		
620, 720	4.1	*4.6*
Armada, Titan	1.8	*2.1*
Frontier, Xterra	4.1	*4.6*
Pathfinder		
1987-95	4.3	*4.8*
1996-05	2.7	*3.2*
Pickup	4.2	*4.7*
Van	4.1	*4.6*
Valve Body Assy., Replace (B)		
620, 720	1.8	*2.1*
Armada, Titan	1.8	*2.1*
Frontier, Xterra		
4 cyl.	1.9	*2.2*
V6	2.4	*2.7*
Pathfinder		
1987-95	2.0	*2.3*
1996-05	1.7	*2.0*
Pickup, Van	2.0	*2.3*

SERVICE TRANSMISSION REMOVED
Transmission R&R included unless otherwise noted.

	LABOR TIME	SEVERE SERVICE
Front Oil Pump, Replace or Recondition (B)		
620, 720		
2WD	5.1	*5.3*
4WD	9.0	*9.2*
Armada, Titan		
2WD	4.7	*5.3*
4WD	6.6	*7.0*
Frontier, Xterra		
4 cyl.		
recondition	6.6	*6.8*
replace	5.8	*6.0*
V6		
2WD	7.0	*7.2*
4WD	9.1	*9.3*

	LABOR TIME	SEVERE SERVICE
Pathfinder		
1987-87		
2WD	6.5	*6.7*
4WD	9.7	*9.9*
1988-95		
2WD	5.4	*5.6*
4WD	9.9	*10.1*
1996-05		
VG		
2WD	6.5	*6.7*
4WD	11.0	*11.2*
VQ		
2WD	6.7	*6.9*
4WD	8.6	*8.8*
Pickup		
1986-87		
2WD	6.5	*6.7*
4WD	9.7	*9.9*
1988-97		
2WD	5.6	*5.8*
4WD	9.9	*10.1*
Van	6.4	*6.6*
Governor Assy., Replace or Recondition (B)		
620, 720	5.1	*5.4*
Frontier, Xterra, 4 cyl.	1.9	*2.2*
1986-95 Pathfinder, Pickup	5.2	*5.5*
1996-97 Pickup	5.3	*5.6*
Van	4.4	*4.7*
Transmission, Recondition (A)		
Includes: Replace torque converter.		
620, 720		
2WD	14.9	*15.8*
4WD	18.2	*19.1*
Armada, Titan		
2WD	9.8	*10.4*
4WD	11.6	*12.2*
Frontier, Xterra		
4 cyl.	14.2	*15.1*
V6		
2WD	16.5	*17.4*
4WD	17.4	*18.3*
Pathfinder		
1987-87		
2WD	15.7	*16.6*
4WD	19.0	*19.9*
1988-95		
2WD	12.9	*13.8*
4WD	17.7	*18.6*
1996-05		
VG		
2WD	15.1	*16.0*
4WD	19.7	*20.6*
VQ		
2WD	14.6	*15.5*
4WD	16.5	*17.4*

720 PICKUP : AMADA : AXXESS : D21 PICKUP : FRONTIER : MURANO

	LABOR TIME	SEVERE SERVICE
Transmission, Recondition (A)		
Pickup		
1986-87		
2WD	15.7	16.6
4WD	19.0	19.9
1988-97		
2WD	13.0	13.9
4WD	17.6	18.5
Van	15.5	16.4
Flush oil cooler and lines add	.5	.5
Transmission & Torque Converter, R&I (B)		
620, 720		
2WD	4.2	4.4
4WD	8.3	8.5
Armada, Titan		
2WD	4.1	4.7
4WD	6.0	6.6
Frontier, Xterra		
4 cyl.	5.2	5.4
V6		
2WD	7.6	7.8
4WD	8.7	8.9
Pathfinder		
1988-95		
2WD	4.9	5.1
4WD	9.4	9.6
1996-05		
VG		
2WD	6.1	6.3
4WD	11.2	11.4
VQ		
2WD	5.7	5.9
4WD	7.8	8.0
Pickup		
2WD	5.1	5.3
4WD	8.2	8.4
Van	5.2	5.4
Transmission Assy., Reseal (B)		
Frontier, Xterra		
4 cyl.	10.3	10.9
V6		
2WD	12.6	13.2
4WD	13.7	14.3
Pathfinder		
1988-95		
2WD	9.5	10.1
4WD	14.2	14.8
1996-05		
VG		
2WD	11.2	11.8
4WD	16.1	16.7
VQ		
2WD	10.8	11.4
4WD	12.8	13.4
1994-97 Pickup		
2WD	9.4	10.0
4WD	14.0	14.6

CONTINUOUSLY VARIABLE TRANSMISSION
SERVICE TRANSMISSION INSTALLED

	LABOR TIME	SEVERE SERVICE
Oil Pressure Check (B)		
Murano	.8	.8
Transmission & Torque Converter, R&I (B)		
Murano		
2WD	11.2	12.1
AWD	13.6	14.1
Oil Pan Gasket, Replace (B)		
Murano	1.6	1.8

TRANSFER CASE

	LABOR TIME	SEVERE SERVICE
Companion Flange, Replace (B)		
620, 720	.7	.9
Armada, Titan	.8	1.0
Axxess	.7	.9
Frontier, Xterra	1.4	1.6
Murano	1.9	2.1
Pathfinder, Pickup	1.1	1.3
Front Driveshaft Oil Seal, Replace (B)		
Armada, Titan	.8	1.0
Axxess		
AT	2.7	2.9
MT	1.6	1.8
Front or Rear Cover Seals, Replace (B)		
620, 720 front or rear each	.9	1.1
Armada, Titan rear	1.0	1.2
Axxess front	2.1	2.3
Frontier, Xterra		
front	9.3	9.5
rear	.7	.9
Murano front	12.1	12.3
Pathfinder		
1987-95		
front		
AT	6.4	6.6
MT	9.1	9.3
rear	1.3	1.5
1996-05		
front		
AT	5.3	5.5
MT	8.7	8.9
rear	1.0	1.2
Pickup		
front		
AT	6.2	6.4
MT	9.0	9.2
rear	1.2	1.4
Mount & Insulator, Replace (B)		
620, 720	1.8	2.0

	LABOR TIME	SEVERE SERVICE
Transfer Case, R&R and Recondition (A)		
620, 720	9.4	10.0
Armada, Titan	7.2	7.7
Axxess		
AT	11.5	12.1
MT	13.0	13.6
Frontier, Xterra	13.4	14.0
Murano	15.7	16.6
Pathfinder		
1987-95	13.2	13.8
1996-05		
AT	12.1	12.7
MT	12.8	13.4
Pickup	13.2	13.8
Transfer Case, R&R or Replace (B)		
620, 720	5.1	5.3
Armada, Titan	3.2	3.6
Axxess		
AT	7.9	8.1
MT	9.3	9.5
Frontier, Xterra	5.8	6.0
Murano	11.1	12.0
Pathfinder		
1987-95	6.2	6.4
1996-05	3.4	3.8
Pickup	6.3	6.5
Transfer Case Drive Chain, Replace (B)		
Armada, Titan	4.8	5.4
Frontier, Xterra	8.4	8.9
Pathfinder		
1987-95	9.0	9.5
1996-05		
AT	6.0	6.5
MT	7.8	8.3
Pickup	9.0	9.5
Transfer Case Front Cover, R&R or Replace (B)		
620, 720	3.5	3.8
Armada, Titan	4.0	4.4
Frontier, Xterra	9.3	9.6
Murano	5.2	5.5
Pathfinder		
1987-95	9.5	9.8
1996-05		
AT	5.3	5.6
MT	8.7	9.0
Pickup	9.5	9.8
Replace front case add	.5	.7
Transfer Case Housing, Replace (B)		
620, 720	3.7	4.0
Frontier, Xterra		
front	9.9	10.2
center add	4.4	4.4
rear	4.0	4.3
Murano	15.1	15.4

PATHFINDER : PICKUP : QUEST : TITAN : VAN : XTERRA

	LABOR TIME	SEVERE SERVICE
1996-05 Pathfinder		
AT		
front	6.7	7.0
center add	2.4	2.4
rear	1.0	1.3
MT		
front	9.3	9.6
center add	4.4	4.4
rear	1.9	2.2

DRIVELINE
DRIVESHAFT

Driveshaft, Replace (B)
620, 720	.7	.9
Armada, Titan	.6	.8
Axxess	.7	.9
Frontier, Xterra	1.0	1.2
Murano	1.0	1.2
Pathfinder, Pickup	.8	1.0
Van	.5	.7

U-Joint, Replace or Recondition (B)
620, 720	1.2	1.4
Armada, Titan	1.1	1.3
Axxess	1.3	1.5
Frontier, Xterra	1.4	1.6
Pathfinder, Pickup	1.1	1.3
Van	.9	1.1

DRIVE AXLE

Differential, Drain & Refill (B)
Front axle	1.3	1.3
Rear axle removable carrier		
Pathfinder, Pickup	.7	.7

Axle Housing Cover or Gasket, Replace (C)
Front axle		
1981-93	1.4	1.6
Armada, Titan	2.3	2.5
Frontier	1.7	1.9
Pathfinder		
1994-95	1.6	1.8
1996-05	2.9	3.1
1994-97 Pickup	1.5	1.7
Rear axle		
Armada, Titan	2.2	2.4
Axxess	2.3	2.5
Frontier, Xterra	1.9	2.1
Murano	4.0	4.2
Pathfinder	.8	1.0
Pickup	.8	1.0

Axle Shaft and/or Bearing, Replace (B)
Rear axle		
integral carrier		
single rear wheels		
one side	1.8	2.1
both sides	3.1	3.4
dual rear wheels		
one side	2.2	2.5
both sides	3.5	3.8
removable carrier		
one side	2.1	2.4
both sides	3.3	3.6

Axle Shaft Oil Seals, Replace (B)
Rear axle		
integral carrier		
620, 720, Pickup		
single rear wheels		
one side	1.1	1.3
both sides	2.1	2.3
dual rear wheels		
one side	1.8	2.0
both sides	2.8	3.0
Pathfinder		
one side	1.5	1.7
both sides	2.5	2.7
Van		
one side	.8	1.0
both sides	1.9	2.1
removable carrier		
one side	1.6	1.8
both sides	2.4	2.6

Center Bearing or Seals, Replace (B)
620, 720	1.3	1.5
Armada, Titan	.8	1.0
Axxess	1.5	1.7
Frontier, Xterra	1.3	1.5
Murano		
2WD	1.1	1.3
4WD	2.5	2.7
Pathfinder, Pickup	1.1	1.3

Differential, R&R or Replace (B)
Front axle		
620, 720	5.8	6.3
Armada, Titan	4.3	4.8
Frontier, Xterra	3.2	3.7
Pathfinder		
1987-95	5.1	5.6
1996-05	4.7	5.2
Pickup	4.9	5.4
Rear axle		
620, 720	2.5	3.0
Armada, Titan	7.5	7.8
Axxess	2.2	2.7
Frontier, Xterra	4.9	5.4
Murano	3.5	4.0
Pathfinder		
1987-95		
integral	3.1	3.6
removable	4.5	5.0
1996-05	3.6	4.1
Pickup	5.2	5.7
Van	5.1	5.6

Differential Carrier Assy., R&R and Recondition (A)
Front axle		
620, 720	9.3	10.0
Armada, Titan	6.2	6.6
Frontier, Xterra	6.9	7.6
Pathfinder		
1987-95	9.6	10.3
1996-05	8.4	9.1
Pickup	9.7	10.4
Rear axle		
620, 720	5.2	5.9
Armada, Titan	8.5	9.0
Axxess	6.3	7.0
Frontier, Xterra		
limited slip	8.5	9.2
standard	9.5	10.2
Murano	9.1	9.8
Pathfinder		
1987-95		
integral	6.9	7.6
removable	9.4	10.1
1996-05		
limited slip	7.3	8.0
standard	6.3	7.0
Pickup	9.2	9.9
Van	8.0	8.7

Differential Side Bearings, Replace (B)
Front axle		
Armada, Titan	3.8	4.1
Frontier, Xterra	4.7	5.2
Pathfinder	6.3	6.8
1994-97 Pickup	6.4	6.9
Rear axle		
620, 720	3.7	4.2
Armada, Titan	4.9	5.4
Axxess	3.2	3.7
Frontier, Xterra	5.7	6.2
Murano	6.4	6.9
Pathfinder		
1994-95	5.6	6.1
1996-05	4.4	4.9
1994-97 Pickup	5.5	6.0
Van	4.9	5.4

Differential Side Gear, Replace (B)
Rear axle Pathfinder,		
Pickup	3.5	4.0

Pinion Shaft Oil Seal, Replace (B)
Front axle		
Armada, Titan	1.4	1.6
Frontier, Xterra	1.3	1.5
Pathfinder		
1994-95	1.6	1.8
1996-05	1.1	1.3
1994-97 Pickup	1.6	1.8

NI-65

NI-66 720 PICKUP : AMADA : AXXESS : D21 PICKUP : FRONTIER : MURANO

	LABOR TIME	SEVERE SERVICE
Pinion Shaft Oil Seal, Replace (B)		
Rear axle		
Armada, Titan	1.8	2.0
Frontier, Xterra		
crush spacer	4.7	4.9
solid spacer	1.3	1.5
Murano	1.4	1.6
Pathfinder		
1987-95		
crush spacer	5.7	5.9
solid spacer	1.3	1.5
1996-05	1.3	1.5
Pickup		
crush spacer	5.5	5.7
solid spacer	1.1	1.3
w/dual rear wheels add	.3	.5
Rear Wheel Hub Assy., Replace (B)		
Rear axle integral carrier		
620, 720, Pickup dual wheels		
one	1.2	1.7
both	2.4	2.9
Ring Gear & Pinion Set, Replace (B)		
Front axle		
620, 720	5.1	5.6
Armada, Titan	4.7	5.1
Frontier	5.8	6.3
Pathfinder		
1994-95	8.4	8.9
1996-05	7.3	7.8
1994-97 Pickup	8.2	8.7
Rear axle		
620, 720	4.3	4.8
Armada, Titan	8.0	8.4
Axxess	5.0	5.5
Frontier, Xterra		
integral	7.9	8.4
removable	6.7	7.2
Murano	7.5	8.0
Pathfinder		
1987-95		
integral	5.2	5.7
removable	7.8	8.3
1996-05	5.5	6.0
Pickup	7.6	8.1
Van	6.9	7.4
Side Flange Oil Seals, Replace (B)		
Front axle		
Armada, Titan		
one side	2.0	2.2
both sides	2.2	2.4
Frontier, Xterra		
one side	2.1	2.4
both sides	2.9	3.2
1996-05 Pathfinder		
one side	2.4	2.7
both sides	3.3	3.6
Rear axle integral carrier		
Armada, Titan		
one side	.6	.8
both sides	.8	1.0

	LABOR TIME	SEVERE SERVICE
Axxess		
one side	.8	1.1
both sides	1.6	1.9
Frontier, Xterra		
one side	1.8	2.1
both sides	2.5	2.8
1994-95 Pathfinder		
one side	2.3	2.6
both sides	3.7	4.0
1994-97 Pickup		
one side	1.9	2.2
both sides	3.6	3.9
Wheel Hub, Replace (B)		
Front axle		
620, 720		
one side	2.3	2.8
both sides	3.9	4.4
Armada, Titan		
one side	1.7	1.9
both sides	2.0	2.2
Murano		
one side	2.4	2.9
both sides	3.8	4.3
1996-05 Pathfinder		
one side	2.2	2.7
both sides	2.9	3.4

HALFSHAFTS
Halfshaft, R&R or Replace (B)

	LABOR TIME	SEVERE SERVICE
Front		
Armada, Titan		
one side	1.4	1.7
both sides	2.3	2.6
Axxess		
right side	1.2	1.5
left side		
MT	1.3	1.6
AT	1.2	1.5
both sides	2.1	2.4
Murano		
one side	1.9	2.2
both sides	2.4	2.7
Quest		
1993-98		
right side	1.4	1.7
left side	1.1	1.4
both sides	2.6	2.9
1999-05		
right side	1.4	1.7
left side	1.1	1.4
both sides	2.1	2.4
Rear		
Armada, Titan		
one side	.9	1.1
both sides	1.5	1.8
Axxess 4WD		
one side	.8	1.1
both sides	1.9	2.2

	LABOR TIME	SEVERE SERVICE
Frontier, Xterra 4WD		
one side	2.2	2.5
both sides	4.3	4.6
Murano AWD		
one side	1.7	2.0
both sides	2.2	2.5
Pathfinder 4WD		
one side	1.6	1.9
both sides	2.5	2.8
Recondition CV-joint		
add each	.7	1.0
Replace center		
bearing add	.3	.5
Halfshaft, R&R and Recondition (B)		
Armada, Titan		
one side	2.1	2.5
both sides	3.6	3.8
Axxess 4WD		
one side	1.6	2.1
both sides	2.9	3.4
Frontier, Xterra 4WD		
one side	3.0	3.5
both sides	5.7	6.2
Murano AWD		
front		
one side	3.5	4.0
both sides	4.7	5.2
rear		
one side	2.9	3.4
both sides	5.0	5.5
Pathfinder 4WD		
one side	2.5	3.0
both sides	4.4	4.9

BRAKES
ANTI-LOCK

	LABOR TIME	SEVERE SERVICE
Diagnose Anti-Lock Brake System Component Each (B)		
1991-05	1.1	1.1
Retrieve Fault Codes (A)		
1991-05	.3	.3
Actuator Assy., Replace (B)		
1991-97	1.1	1.1
Armada, Titan	1.4	1.6
Frontier, Xterra	2.0	2.0
Murano	4.1	4.1
1996-05 Pathfinder	1.5	1.5
Quest		
1993-98	1.1	1.1
1999-05	1.9	1.9
Anti-Lock Relay, Replace (B)		
Exc. Below	.3	.3
Armada, Titan	2.1	2.5
Murano	3.6	3.6
Control Module, Replace (B)		
Armada, Titan	2.1	2.5
Frontier, Xterra	.9	.9
Murano	3.6	3.6

PATHFINDER : PICKUP : QUEST : TITAN : VAN : XTERRA — NI-67

	LABOR TIME	SEVERE SERVICE
Pathfinder		
1991-95	.9	.9
1996-05	.5	.5
1991-97 Pickup	.9	.9
Quest		
1993-98	.6	.6
1999-05	1.9	1.9

Front Rotor Sensor Assy., Replace (B)

	LABOR TIME	SEVERE SERVICE
Exc. below		
one side	1.6	1.8
both sides	3.0	3.2
1996-05 Pathfinder		
one	1.1	1.3
both	1.9	2.1

Front Sensor Assy., Replace (B)

	LABOR TIME	SEVERE SERVICE
Exc. Frontier, Xterra		
one	.6	.6
both	.8	.8
Frontier, Xterra		
one side	.7	.7
both sides	1.1	1.1

Rear Rotor Sensor Assy., Replace (B)

	LABOR TIME	SEVERE SERVICE
Armada, Titan		
one side	2.1	2.5
both sides	2.4	2.8
Frontier, Xterra		
one side	1.1	1.3
both sides	2.1	2.3
Pathfinder		
1991-95	.8	1.0
1996-04		
one	1.3	1.5
both	2.2	2.4
1991-97 Pickup	.8	.8
1993-05 Quest		
one	.7	.7
both	1.2	1.2

Rear Sensor Assy., Replace (B)

	LABOR TIME	SEVERE SERVICE
Armada, Titan		
one side	1.5	1.7
both sides	2.4	2.6
Frontier, Xterra		
one side	.6	.6
both sides	1.0	1.0
Murano, Pathfinder		
1991-95	.8	.8
1996-05		
one	.5	.5
both	.6	.6
1991-97 Pickup	.9	.9
1993-05 Quest		
one	.5	.5
both	.7	.7

SYSTEM

Bleed Brakes (B)
Includes: Add fluid.

	LABOR TIME	SEVERE SERVICE
All Models	1.0	1.0

Brake Hose (Flexible), Replace (B)
Includes: System bleeding.

	LABOR TIME	SEVERE SERVICE
Exc. below	.8	.9
Frontier, Xterra	1.0	1.1
Quest	1.0	1.1
each addl. add	.3	.4

Brake Proportioning Valve, Replace (B)
Includes: System bleeding.

	LABOR TIME	SEVERE SERVICE
620, 720	.5	.7
Axxess	.7	.9
Pathfinder	.8	1.0
Pickup	.8	1.0
Quest	1.1	1.3

Load Sensing Valve, Replace (B)

	LABOR TIME	SEVERE SERVICE
Frontier, Xterra	1.7	1.9
Pathfinder	1.3	1.5
1994-97 Pickup	1.3	1.5
Quest	1.2	1.4

Master Cylinder, R&R and Rebuild (B)
Includes: System bleeding.

	LABOR TIME	SEVERE SERVICE
Exc. below	1.5	1.8
Armada, Titan	1.8	2.0
Murano	1.7	2.0
Quest	1.6	1.9

Master Cylinder, Replace (B)
Includes: System bleeding.

	LABOR TIME	SEVERE SERVICE
Exc. below	.9	1.1
Armada, Titan	1.5	1.6
Frontier, Xterra	1.6	1.8
Murano	1.5	1.7
Quest		
1993-98	1.0	1.2
1999-05	1.4	1.6

Power Booster Unit, Replace (B)

	LABOR TIME	SEVERE SERVICE
620, 720	1.3	1.5
Armada, Titan	3.2	3.4
Axxess	1.7	1.9
Frontier, Xterra	2.0	2.2
Murano	2.1	2.3
Pathfinder		
1987-95	1.3	1.5
1996-05	2.1	2.3
Pickup	1.2	1.4
Quest	2.3	2.5
Van	3.3	3.5

Power Booster Vacuum Check Valve, Replace (B)

	LABOR TIME	SEVERE SERVICE
Exc. Murano	.5	.5
Murano	.7	.7

SERVICE BRAKES

Brake Drum, Replace (B)

	LABOR TIME	SEVERE SERVICE
All Models	.7	.9
w/dual rear wheels add	.5	.7

Caliper Assy., Replace (B)
Includes: System bleeding.

	LABOR TIME	SEVERE SERVICE
Front		
620, 720		
2WD		
one	.7	.9
both	1.3	1.5
4WD		
one	1.6	1.8
both	2.8	3.0
Armada, Titan		
one	.9	1.1
both	1.4	1.6
Axxess		
one	.7	.9
both	1.5	1.7
Frontier, Xterra		
one	1.4	1.6
both	1.7	1.9
Murano		
one	1.0	1.2
both	1.4	1.6
Pathfinder		
one	1.3	1.5
both	1.7	1.9
Pickup		
one	1.2	1.4
both	1.6	1.8
Quest		
one	1.0	1.2
both	1.4	1.6
Van		
one	.7	.9
both	1.4	1.6
Rear		
Armada, Titan		
one	.9	1.1
both	1.4	1.6
Murano		
one	1.0	1.2
both	1.6	1.8
1988-97 Pathfinder, Pickup		
one	.9	1.1
both	1.8	2.0

Disc Brake Rotor, Replace (B)

	LABOR TIME	SEVERE SERVICE
Front		
620, 720		
2WD		
one	1.4	1.6
both	2.5	2.7
4WD		
one	2.2	2.4
both	4.0	4.2
Armada, Titan		
one	.9	1.1
both	1.4	1.6
Axxess		
one	.8	1.0
both	1.7	1.9

	LABOR TIME	SEVERE SERVICE
Disc Brake Rotor, Replace (B)		
Frontier, Xterra		
2WD		
one	.8	1.0
both	1.4	1.6
4WD		
one	1.1	1.3
both	2.1	2.2
Murano		
one	.5	.7
both	.8	1.0
Pathfinder		
1987-95		
one	.8	1.0
both	1.4	1.6
1996-05		
one	1.3	1.5
both	2.2	2.4
Pickup		
one	.8	1.0
both	1.8	2.0
Quest		
1993-98		
one	.7	.9
both	1.2	1.4
1999-05		
one	.6	.8
both	.8	1.0
Van		
one	.8	1.0
both	1.6	1.8
Rear		
Armada, Titan		
one	1.0	1.1
both	1.4	1.6
Murano		
one	.6	.8
both	.8	1.0
1988-97 Pathfinder, Pickup		
one	.8	1.0
both	1.3	1.5
Resurface rotor add each	.6	.6
Pads and/or Shoes, Replace (B)		
Includes: Service and parking brake adjustment, system bleeding.		
Front		
exc. below	.8	.9
Armada, Titan	1.1	1.2
Murano	1.1	1.2
Pathfinder, Pickup	1.0	1.1
Rear		
disc	1.1	1.2
drum		
Frontier, Xterra	1.1	1.2
Pickup		
single wheels	1.9	2.0
dual wheels	2.2	2.3
Pathfinder		
1987-95		
single wheels	1.9	2.0
dual wheels	2.2	2.3
1996-05	1.4	1.5
Quest	1.9	2.0
COMBINATION ADD-ONS		
Repack wheel bearings add	.6	.6
Replace		
brake drum add	.1	.1
brake hose add	.3	.3
brake rotor add	.2	.2
caliper add	.3	.3
wheel cylinder add	.3	.3
Wheel Cylinder, R&R and Rebuild (B)		
Includes: System bleeding.		
620, 720		
single rear wheels		
one	1.1	1.4
both	2.1	2.4
dual rear wheels		
one	1.7	2.0
both	3.2	3.5
Axxess		
one	1.6	1.9
both	2.8	3.1
Frontier, Xterra		
one	1.2	1.5
both	1.9	2.2
Pathfinder		
1987-95		
single rear wheels		
one	1.4	1.7
both	2.5	2.8
dual rear wheels		
one	1.8	2.1
both	2.9	3.2
1996-05		
one	1.4	1.7
both	2.2	2.5
Pickup		
single rear wheels		
one	1.4	1.7
both	2.5	2.8
dual rear wheels		
one	1.8	2.1
both	2.9	3.2
Quest		
1993-98		
one	1.8	2.1
both	3.0	3.3
1999-05		
one	1.4	1.7
both	2.7	3.0
Van		
one	1.5	1.8
both	2.8	3.1
Wheel Cylinder, Replace (B)		
Includes: System bleeding.		
620, 720		
single rear wheels		
one	.7	.9
both	1.6	1.8
dual rear wheels		
one	1.6	1.8
both	2.8	3.0
Axxess		
one	1.5	1.7
both	2.4	2.6
Frontier, Xterra		
one	1.1	1.3
both	1.7	1.9
Pathfinder		
1987-95		
one	1.4	1.6
both	2.5	2.7
1996-05		
one	1.3	1.5
both	1.9	2.1
Pickup		
single rear wheels		
one	1.5	1.7
both	2.5	2.7
dual rear wheels		
one	1.7	1.9
both	3.0	3.2
Quest		
1993-98		
one	1.6	1.8
both	2.6	2.8
1999-05		
one	1.2	1.4
both	2.3	2.5
Van		
one	1.4	1.6
both	2.5	2.7
PARKING BRAKE		
Parking Brake Apply Actuator, Replace (B)		
620, 720	.7	.7
Armada, Titan	.8	1.0
Axxess	.7	.7
Frontier, Xterra		
pedal type	.7	.7
stick type	1.1	1.1
Murano	.9	.9
Pathfinder, Pickup		
1987-95		
pedal type	1.1	1.1
stick type	.7	.7
1996-05	.7	.7
Quest		
1993-98	1.9	1.9
1999-05	1.2	1.2
Van	.8	.8

PATHFINDER : PICKUP : QUEST : TITAN : VAN : XTERRA   NI-69

	LABOR TIME	SEVERE SERVICE
Parking Brake Cable, Adjust (C)		
All Models	.5	.7
Parking Brake Cable, Replace (B)		
Front		
620, 720	.7	.9
Armada, Titan	2.8	3.0
Axxess	.8	1.0
Frontier, Xterra	1.1	1.3
Murano	3.2	3.4
Pathfinder	1.0	1.2
Pickup	.7	.9
Quest	2.1	2.3
Van	.8	1.0
Rear		
620, 720		
one	1.1	1.3
both	1.9	2.1
Armada, Titan		
one	2.1	2.2
both	3.5	3.7
Axxess		
one	1.1	1.3
both	1.4	1.6
Frontier, Xterra		
2WD	1.1	1.3
4WD	1.7	1.9
Murano		
one	1.3	1.5
both	1.9	2.1
Pathfinder, Pickup		
1987-95		
disc type		
2WD		
one	.7	.9
both	1.3	1.5
4WD		
one	1.2	1.4
both	1.9	2.1
drum type		
one	1.1	1.3
both	2.0	2.2
1996-05		
one	1.3	1.5
both	1.9	2.1
Quest		
1993-98		
one	1.4	1.6
both	2.5	2.7
1999-05		
one	.8	1.0
both	1.2	1.4
Van		
one	.7	.9
both	1.4	1.6
Parking Brake Equalizer, Replace (B)		
All Models	.7	.9
Parking Brake Shoes, Replace (B)		
Armada, Titan		
one side	2.0	2.2
both sides	3.2	3.4

	LABOR TIME	SEVERE SERVICE
Murano		
one side	1.0	1.2
both sides	1.4	1.6
Pathfinder, Pickup		
one side	1.5	1.7
both sides	3.0	3.2

FRONT SUSPENSION
FRONT WHEEL DRIVE
Unless otherwise noted, time given does not include alignment.

	LABOR TIME	SEVERE SERVICE
Front Toe, Adjust (B)		
All Models	.8	1.0
Front Hub Assy., Replace (B)		
Axxess		
one side	2.6	3.1
both sides	4.9	5.4
Murano		
2WD		
one side	2.4	2.9
both sides	3.8	4.3
4WD		
one side	1.7	2.2
both sides	2.7	3.2
Quest		
1993-98		
one side	1.8	2.3
both sides	3.0	3.5
1999-05		
one side	2.2	2.7
both sides	3.4	3.9
Lower Ball Joint, Replace (B)		
Includes: Alignment.		
Axxess		
one side	1.4	1.6
both sides	2.7	2.9
Quest		
1993-98		
one side	1.4	1.6
both sides	2.1	2.3
1999-05		
one side	1.4	1.6
both sides	1.6	1.8
Lower Control Arm, Replace (B)		
Includes: Alignment.		
Axxess		
one side	1.7	2.0
both sides	2.7	3.0
Murano		
one side	1.4	1.7
both sides	1.7	2.0
Quest		
one side	2.1	2.4
both sides	2.8	3.1
Replace bushings		
add each	.2	.5
Stabilizer Bar, Replace (B)		
Includes: Replace bushings..		
Exc. Murano	1.2	1.4
Murano	6.0	6.2

	LABOR TIME	SEVERE SERVICE
Strut, Replace (B)		
Includes: Alignment.		
Axxess		
one side	2.3	2.4
both sides	3.9	4.0
Murano		
one side	1.9	2.0
both sides	2.5	2.6
Quest		
1993-98		
one side	1.9	2.0
both sides	3.5	3.6
1999-05		
one side	2.3	2.6
both sides	3.4	3.8

REAR WHEEL DRIVE

	LABOR TIME	SEVERE SERVICE
Align Front End (A)		
All Models	1.7	1.9
Front Toe, Adjust (B)		
All Models	1.0	1.2
Free Wheeling Hub, Replace (B)		
One side	.7	1.0
Both sides	1.1	1.4
Front Hub Oil Seal, Replace (B)		
620, 720		
one side	2.2	2.2
both sides	3.9	3.9
Armada, Titan		
2WD		
one side	.9	1.1
both sides	1.5	1.7
4WD		
one side	1.7	1.9
both sides	3.1	3.3
Frontier, Xterra 4WD		
one side	1.6	1.6
both sides	3.1	3.1
Pathfinder		
1987-95		
one side	1.8	1.8
both sides	3.2	3.2
1996-05		
one side	2.2	2.2
both sides	4.2	4.2
Pickup		
one side	2.8	2.8
both sides	5.2	5.2
Front Shock Absorbers, Replace (B)		
One side	.6	.7
Both sides	1.0	1.1
Hub, Bearing or Seal, Replace (B)		
Includes: Alignment.		
620, 720		
one side	1.8	2.0
both sides	3.1	3.3
Armada, Titan		
2WD		
one side	.9	1.1
both sides	1.5	1.7

720 PICKUP : AMADA : AXXESS : D21 PICKUP : FRONTIER : MURANO

	LABOR TIME	SEVERE SERVICE
Hub, Bearing or Seal, Replace (B)		
4WD		
one side	1.7	1.9
both sides	3.1	3.3
Frontier, Xterra		
2WD		
one side	1.0	1.2
both sides	1.6	1.8
4WD		
one side	1.9	2.1
both sides	3.6	3.8
Pathfinder		
1987-95		
one side	2.7	2.9
both sides	5.2	5.4
1996-05		
2WD		
one side	1.7	1.9
both sides	3.2	3.4
4WD		
one side	2.5	2.7
both sides	4.9	5.1
Pickup		
one side	2.0	2.2
both sides	3.4	3.6
Van		
one side	1.5	1.7
both sides	2.4	2.6
Leaf Spring, Replace (B)		
Van	3.0	3.5
Recondition spring add	.5	.7
Lower Control Arm, Replace (B)		
Includes: Alignment.		
620, 720		
one side	1.9	2.2
both sides	3.1	3.4
Armada, Titan		
one side	3.3	3.6
both sides	4.0	4.4
Frontier, Xterra		
one side	3.2	3.5
both sides	4.6	4.9
Pathfinder		
1987-95		
one side	1.6	1.9
both sides	3.2	3.5
1996-05		
one side	1.1	1.4
both sides	1.7	2.0
Pickup		
one side	1.8	2.1
both sides	3.1	3.4
Van		
one side	2.1	2.4
both sides	3.7	4.0
Replace bushings add each	.5	.7

	LABOR TIME	SEVERE SERVICE
Stabilizer Bar and/or Bushings, Replace (B)		
Exc. below	.9	1.1
1996-05 Pathfinder	1.1	1.3
Steering Knuckle, Replace (B)		
Includes: Alignment.		
620, 720		
one side	2.7	3.0
both side	5.0	5.3
Van		
one side	2.5	2.8
both sides	4.4	4.7
Armada, Titan		
2WD		
one side	3.6	4.0
both sides	4.7	5.1
4WD		
one side	4.1	4.6
both sides	5.5	6.2
Frontier, Xterra		
2WD		
one side	4.0	4.3
both sides	4.9	5.2
4WD		
one side	4.2	4.7
both sides	5.7	6.3
Murano		
2WD		
one side	2.4	2.7
both sides	3.8	4.1
4WD		
one side	2.1	2.6
both sides	2.9	3.4
Pathfinder, Pickup		
2WD		
one side	2.1	2.4
both sides	3.8	4.1
4WD		
1994-95		
one side	2.9	3.4
both sides	5.2	5.7
1996-05		
Pathfinder		
one side	2.4	2.9
both sides	4.0	4.5
Pickup		
one side	2.8	3.3
both sides	5.1	5.6
Steering Knuckle Arm, Replace (B)		
Includes: Reset toe.		
620, 720		
one side	.7	.8
both sides	1.3	1.4
Frontier, Xterra		
one side	1.6	1.7
both sides	2.1	2.2
Pathfinder		
one side	1.1	1.2
both sides	1.5	1.6

	LABOR TIME	SEVERE SERVICE
Pickup		
one side	.8	.9
both sides	1.5	1.6
Strut Assy., Replace (B)		
1996-05 Pathfinder		
one side	1.9	2.1
both sides	3.3	3.5
Torsion Bar, Replace (B)		
Includes: Adjust vehicle height.		
620, 720		
one side	1.3	1.8
both side	2.6	3.1
Frontier, Xterra		
one side	2.4	2.9
both sides	2.9	3.4
Pathfinder, Pickup		
one side	1.4	1.9
both sides	2.4	2.9
Upper Control Arm, Replace (B)		
Includes: Alignment.		
620, 720		
one side	1.8	2.1
both sides	3.2	3.5
Armada, Titan		
one side	1.0	1.2
both sides	1.7	1.9
Frontier, Xterra		
one side	3.5	3.8
both sides	4.0	4.3
Pathfinder, Pickup		
one side	2.0	2.3
both sides	3.7	4.0
Van		
one side	1.8	2.1
both sides	3.6	3.9
Replace bushings add each	.5	.7
Upper or Lower Ball Joints, Replace (B)		
Includes: Reset toe.		
620, 720		
one side	1.4	1.7
both sides	2.2	2.5
Armada, Titan		
one side	3.3	3.6
both sides	4.0	4.4
Frontier, Xterra		
upper		
one side	3.3	3.6
both sides	3.6	3.9
lower		
one side	3.8	4.1
both sides	5.3	5.6
Pathfinder		
one side	1.5	1.8
both sides	2.5	2.8
Pickup		
one side	1.3	1.6
both sides	2.4	2.7

REAR SUSPENSION

	LABOR TIME	SEVERE SERVICE
Lower Link and/or Bushings, Replace (B)		
Axxess		
one	.5	.7
both	.8	1.0
Murano		
front		
one	2.4	2.6
both	3.2	3.4
rear		
one	2.0	2.3
both	2.7	3.1
Pathfinder, Van		
1987-95		
one	.7	.9
both	1.2	1.4
1996-05		
one	.6	.8
both	1.0	1.2
Parallel Rods, Replace (B)		
Axxess		
2WD		
one	.7	.9
both	1.1	1.3
4WD		
one	.5	.7
both	.8	1.0
Rear Axle Shaft Oil Seal, Replace (B)		
Armada, Titan		
one	2.1	2.3
both	3.6	3.8
Axxess		
2WD		
one	.8	.8
both	1.8	1.8
4WD		
one	1.8	1.8
both	3.2	3.2
Rear Coil Spring, Replace (B)		
Murano		
one	2.1	2.3
both	2.5	2.7
Pathfinder, Van		
1987-95		
one	1.5	1.7
both	2.9	3.1
1996-05		
one	3.0	3.2
both	3.2	3.4
Rear Leaf Spring, Replace (B)		
620, 720, Pickup		
one	1.8	2.3
both	3.3	3.8
Armada, Titan		
one	1.4	1.7
both	2.1	2.4

	LABOR TIME	SEVERE SERVICE
Frontier, Xterra		
one side	1.3	1.8
both sides	2.2	2.7
Quest		
1993-98		
one side	1.5	1.8
both sides	2.6	2.9
1999-05		
one side	1.1	1.4
both sides	1.8	2.1
Rear Shock Absorber, Replace (B)		
620, 720, Pickup		
one	.7	.8
both	1.3	1.4
Armada, Titan		
one	.5	.6
both	.6	.7
Frontier, Xterra		
one side	.5	.6
both sides	.6	.7
Murano		
one	.8	.9
both	1.3	1.4
Pathfinder, Van		
one	.6	.7
both	.8	.9
Quest		
1993-98		
one side	.6	.7
both sides	1.0	1.1
1999-05		
one	.7	.8
both	1.0	1.1
Rear Stabilizer Bar, Replace (B)		
Axxess		
2WD	.5	.7
4WD	.8	1.0
Frontier, Xterra	1.4	1.6
Murano, Pathfinder, Van	.8	1.0
Quest		
1993-98	.6	.8
1999-05	.5	.7
Rear Stabilizer Bar Bushings, Replace (B)		
Frontier, Xterra	.8	1.0
Murano, Pathfinder, Van	.8	1.0
Rear Strut Assy., Replace (B)		
Axxess		
2WD		
one side	1.8	1.9
both sides	3.6	3.7
4WD		
one side	1.1	1.2
both sides	2.2	2.3

	LABOR TIME	SEVERE SERVICE
Rear Wheel Knuckle Spindle, Replace (B)		
Axxess		
one	2.8	3.1
both	4.5	4.8
Quest		
one side	.8	1.1
both sides	1.4	1.7
Spring Eye Mount and/or Bushings, Replace (B)		
620, 720, Pickup		
one side	.8	1.0
both sides	1.6	1.8
Armada, Titan		
one	1.7	1.9
both	2.8	3.0
Frontier, Xterra		
one side	.5	.7
both sides	.8	1.0
Quest		
1993-98		
one side	1.2	1.4
both sides	2.2	2.4
1999-05		
one side	.6	.8
both sides	.8	1.0
Spring Shackle and/or Bushing, Replace (B)		
620, 720, Pickup		
single rear wheels		
one side	.7	.9
both sides	1.3	1.5
dual rear wheels		
one side	1.3	1.5
both sides	1.5	1.7
Armada, Titan		
one	1.5	1.7
both	2.4	2.6
Frontier, Xterra		
one side	.8	1.0
both sides	1.1	1.3
Quest		
1993-98		
one side	.8	1.0
both sides	1.5	1.7
1999-05		
one side	.7	.9
both sides	1.0	1.2
Upper Link and/or Bushings, Replace (B)		
Pathfinder, Van		
one	.7	.9
both	1.1	1.3
Wheel Bearings, Hub or Oil Seals, Replace (B)		
Axxess		
one	2.3	2.3
both	4.0	4.0

720 PICKUP : AMADA : AXXESS : D21 PICKUP : FRONTIER : MURANO

	LABOR TIME	SEVERE SERVICE
Wheel Hub & Bearing Assy., Replace (B)		
1993-01 Quest		
one side	.8	1.1
both sides	1.4	1.7

AUTO LEVEL SYSTEM
	LABOR TIME	SEVERE SERVICE
Replace Air Compressor Assy. (B)		
Armada, Titan	.6	.8
Replace Air Suspension Control Assy. (B)		
Armada, Titan	.9	1.1
Replace Sensor Rod Assy. (B)		
Armada, Titan	.5	.6
Replace Compressor-to-Shock Tube Assy. (B)		
Armada, Titan		
one	.9	1.1
both	1.2	1.4
Replace Air Adjustable Shock Assy. (B)		
Armada, Titan		
one	.7	.9
both	.9	1.1

STEERING
AIR BAGS
	LABOR TIME	SEVERE SERVICE
Diagnose Air Bag System Component Each (A)		
1993-05	.5	.5
Retrieve Fault Codes (A)		
1993-05	.3	.3
Air Bag Assy., Replace (B)		
Driver side	.7	.7
Passenger side		
Armada, Titan	4.0	4.0
Frontier, Xterra	.8	.8
Murano	1.1	1.1
Pathfinder	2.9	2.9
Quest	.8	.8
Side curtain		
Armada, Titan		
one	1.7	1.7
both	2.8	2.8
Frontier, Xterra		
one	3.2	3.2
both	3.8	3.8
Murano		
one	3.0	3.0
both	3.4	3.4
Side impact		
Murano		
one	1.6	1.6
both	2.8	2.8
Pathfinder		
one	.5	.5
both	.8	.8
Air Bag On/Off Switch, Replace (B)		
All Models	.8	.8

	LABOR TIME	SEVERE SERVICE
Air Bag Spiral Cable, Replace (B)		
Exc. below	.9	.9
Armada, Titan	1.5	1.5
Frontier, Xterra	1.1	1.1
1999-05 Quest	1.4	1.4
Crash Sensor, Replace (B)		
Center		
exc. Quest	.7	.7
Quest		
1993-98	1.7	1.7
1999-05	2.0	2.0
Side		
one	1.0	1.0
both	1.5	1.5
Electronic Control Unit, Replace (B)		
1996-97 Pickup	.5	.5
1993-98 Quest	.9	.9
Tunnel Sensor, Replace (B)		
1993-98 Quest	.8	.8
Sensing & Diagnostic Module (SDM), Replace (B)		
All Models	.7	.7

LINKAGE
	LABOR TIME	SEVERE SERVICE
Idler Arm, Replace (B)		
Exc. Frontier, Xterra	.8	1.0
Frontier, Xterra	1.1	1.3
Pitman Arm, Replace (B)		
Includes: Reset toe.		
Exc. Frontier, Xterra	.8	1.1
Frontier, Xterra	.9	1.2
Steering Center Link (Relay Rod), Replace (B)		
620, 720	1.2	1.5
Frontier, Xterra	1.1	1.4
Pathfinder, Pickup	1.2	1.5
Van	.8	1.1
Steering Damper, Replace (B)		
All Models	.7	.9
Steering Wheel, Replace (B)		
Exc. below	.7	.7
Armada, Titan, Frontier	.8	.9
Tie Rod or Tie Rod Ends, Replace (B)		
Includes: Reset toe.		
Exc. Frontier, Xterra		
one side	.9	1.1
both sides	1.8	2.0
Frontier, Xterra		
one side	1.6	1.8
both sides	1.9	2.1

POWER RACK & PINION
	LABOR TIME	SEVERE SERVICE
Inner Tie Rod, Replace (B)		
Includes: Reset toe.		
Armada, Titan		
one	1.8	2.0
both	2.8	3.0

	LABOR TIME	SEVERE SERVICE
Murano		
one side	6.6	6.8
both sides	7.0	7.2
1996-05 Pathfinder		
one side	1.1	1.3
both sides	1.9	2.1
Quest		
1993-98		
one side	1.0	1.2
both sides	1.4	1.6
1999-05		
one side	1.2	1.4
both sides	1.8	2.0
Outer Tie Rod, Replace (B)		
Includes: Reset toe.		
Armada, Titan		
one	1.2	1.4
both	1.5	1.7
Murano		
one side	1.3	1.5
both sides	1.6	1.8
1996-05 Pathfinder		
one side	.8	1.0
both sides	1.3	1.5
Quest		
one side	1.0	1.2
both sides	1.2	1.4
Pressure Switch, Replace (B)		
All Models	.7	.7
Rack & Pinion Assy., R&R and Recondition (A)		
Includes: Reset toe.		
Axxess		
2WD	4.3	4.8
4WD	4.9	5.4
Murano		
2WD	3.6	4.1
4WD	6.9	7.6
1996-05 Pathfinder	4.3	4.8
Quest	4.6	5.1
Rack & Pinion Assy., R&R or Replace (B)		
Includes: Adjustments and bleeding system.		
Armada, Titan	3.5	3.8
Axxess	3.3	3.6
Murano		
2WD	2.6	2.9
4WD	5.4	6.0
1996-05 Pathfinder	3.2	3.6
Quest		
1993-98	2.6	2.9
1999-05	2.0	2.3
Rack Boots, Replace (B)		
Armada, Titan		
one side	1.5	1.7
both sides	2.1	2.3
Axxess		
one side	.8	1.0
both sides	1.3	1.5

PATHFINDER : PICKUP : QUEST : TITAN : VAN : XTERRA NI-73

	LABOR TIME	SEVERE SERVICE
Murano		
one side	6.4	6.6
both sides	6.7	6.9
1996-05 Pathfinder		
one side	1.0	1.2
both sides	1.6	1.8
Quest		
one side	1.0	1.2
both sides	1.4	1.6
Steering Flex Coupling, Replace (B)		
Exc. Quest	.5	.7
Quest		
1993-98	1.5	1.7
1999-05	1.1	1.3
Steering Lock Assy., Replace (B)		
All Models	1.4	1.4
Steering Oil Cooler, Replace (B)		
Armada, Titan	.8	.9
Quest	1.0	1.2
Steering Pump Hoses, Replace (B)		
Includes: Bleeding system.		
Exc. below	1.8	1.9
Armada, Titan	2.5	2.7
Murano	2.1	2.2
1999-05 Quest	2.3	2.4
Steering Pump, R&R and Recondition (B)		
Armada, Titan	2.4	2.6
Axxess	2.2	2.7
Murano	3.1	2.6
1996-05 Pathfinder	2.6	3.1
Quest	2.1	2.6
Steering Pump, R&R or Replace (B)		
Armada, Titan	1.5	1.7
Axxess	1.1	1.4
Murano	2.0	2.3
1996-05 Pathfinder	1.5	1.8
Quest	1.1	1.4

MANUAL WORM & SECTOR

	LABOR TIME	SEVERE SERVICE
Gear Assy., R&R and Recondition (A)		
All Models	2.8	3.3
Gear Assy., Replace (B)		
All Models	1.5	1.8
Sector Shaft and/or Oil Seal, Replace (B)		
All Models	2.1	2.3
Steering Column, R&R or Replace (B)		
All Models	1.6	1.6
Steering Column Lock Assy., Replace (B)		
1981-97	2.2	2.2
1998-05	1.4	1.4
Steering Column Upper Bushing, Replace (B)		
All Models	1.7	1.9
Steering Gear, Adjust (B)		
All Models	.5	.7

	LABOR TIME	SEVERE SERVICE
POWER WORM & SECTOR		
Gear Assy., Replace (B)		
620, 720	1.7	2.0
Frontier, Xterra	1.8	2.1
Pathfinder, Pickup	2.0	2.3
Van	1.6	1.9
Power Steering Hoses, Replace (B)		
Includes: Bleeding system.		
620, 720	1.3	1.4
Frontier, Xterra	1.7	1.8
Pathfinder, Pickup		
4 cyl.	.7	.8
V6	1.3	1.4
Van	3.2	3.3
Power Steering Switch Assy., Replace (B)		
1981-97	.5	.5
Frontier	.7	.7
Steering Column Upper Bushing, Replace (B)		
620, 720	1.6	1.8
Van	1.7	1.9
Steering Column Lock Assy., Replace (B)		
620, 720	1.3	1.3
Frontier, Xterra	1.4	1.4
Pathfinder, Pickup	1.3	1.3
Van	1.4	1.4
Steering Pump, R&R and Recondition (B)		
620, 720	3.1	3.6
Frontier, Xterra	1.8	2.3
Pathfinder, Pickup	2.6	3.1
Van	2.6	3.1
Steering Pump, R&R or Replace (B)		
620, 720	2.0	2.3
Frontier, Xterra	1.2	1.5
1986-95 Pathfinder, Pickup		
4 cyl.		
KA	1.4	1.7
Z	.8	1.1
V6	1.4	1.7
1996-97 Pickup	1.5	1.8
Van	1.7	2.0

HEATING & AIR CONDITIONING

When more than one component requires replacement where evacuation, recovery and recharging is already included, deduct 1.0 hour for each additional component from the time given.

Evacuate/Recover and Recharge System (B)

	LABOR TIME	SEVERE SERVICE
R12	1.2	1.2
R134A	.9	.9

	LABOR TIME	SEVERE SERVICE
AC Hoses, Replace (B)		
Includes: Evacuate/recover and recharge.		
620, 720	1.9	2.0
Armada, Titan		
high pressure	3.6	3.7
low pressure	2.5	2.7
Axxess	3.0	3.1
Frontier, Xterra	1.7	1.8
Murano		
high pressure	2.9	3.0
low pressure	2.3	2.4
Pathfinder, Pickup		
4 cyl.		
KA	2.7	2.8
Z	2.0	2.1
V6	2.0	2.1
1996-05 Pathfinder		
VG	1.5	1.6
VQ		
high pressure	2.5	2.6
low pressure	1.7	1.8
Quest	1.9	2.0
Van	2.6	2.7
each addl. add	.5	.6
AC Relay, Replace (B)		
Exc. Murano, Van	.3	.3
Murano, Van	.5	.5
Auxiliary Heater Hoses, Replace (B)		
Quest	3.4	3.6
Axxess	1.2	1.4
Blower Motor, Replace (B)		
Exc. Van	.7	.7
Van	1.1	1.1
Compressor Assy., Replace (B)		
Includes: Parts transfer, evacuate/recover and recharge.		
620, 720	2.7	2.9
Armada, Titan	1.8	2.0
Axxess	2.8	3.0
Frontier, Xterra		
KA	1.8	2.0
VG	2.3	2.5
Murano	2.3	2.5
Pathfinder, Pickup		
1986-95		
4 cyl.		
KA	3.3	3.5
Z	2.6	2.8
V6	2.5	2.7
1996-05		
Pathfinder	2.3	2.5
Pickup		
4 cyl.	3.1	3.3
V6	2.3	2.5
Quest		
1993-98	3.1	3.3
1999-05	2.0	2.2
Van	2.5	2.7

720 PICKUP : AMADA : AXXESS : D21 PICKUP : FRONTIER : MURANO

	LABOR TIME	SEVERE SERVICE
Condenser Assy., Replace (B)		
Includes: Evacuate/recover and recharge.		
620, 720	3.2	3.4
Armada, Titan	2.0	2.2
Axxess	3.4	3.6
Frontier, Xterra	2.5	2.7
Murano	3.2	3.4
Pathfinder		
1987-95	3.4	3.6
1996-05	2.6	2.8
Pickup	3.3	3.5
Quest		
1993-98	3.1	3.3
1999-05	2.5	2.7
Van	2.7	2.9
Condenser Fan Motor, Replace (B)		
Armada, Titan	1.2	1.4
Axxess	1.4	1.6
Frontier, Xterra	1.5	1.7
Quest	1.1	1.3
Van		
one	1.8	2.0
both	2.7	2.9
Evaporator Core, Replace (B)		
Includes: Evacuate/recover and recharge.		
620, 720	3.0	3.2
Armada, Titan	7.2	7.4
Axxess	2.3	2.5
Frontier, Xterra	1.7	1.9
Murano	2.3	2.5
Pathfinder		
1987-95	2.8	3.0
1996-05	1.7	1.9
Pickup	2.6	2.8
Quest		
1993-98	2.6	2.8
1999-05	1.5	1.7
Van	2.6	2.8
Replace thermo switch add	.3	.3
Expansion Valve, Replace (B)		
Includes: Evacuate/recover and recharge.		
620, 720	3.1	3.2
Armada, Titan	7.3	7.5
Axxess	2.4	2.6
Frontier, Xterra	2.0	2.2
Murano	2.4	2.6
Pathfinder		
1987-95	3.4	3.6
1996-05	1.8	2.0
Pickup	3.4	3.6
Quest	3.3	3.5
Axxess	2.6	2.8
Heater/AC Relay, Replace (B)		
Exc. Van	.3	.3
Van	.5	.5

	LABOR TIME	SEVERE SERVICE
Heater Blower Motor Relay, Replace (B)		
All Models	.5	.5
Heater Blower Motor Resistor, Replace (B)		
All Models	.5	.5
Heater Core, R&R or Replace (B)		
620, 720	3.1	3.4
Armada, Titan	7.7	7.9
Axxess	4.7	5.0
Frontier, Xterra	5.0	5.3
Murano	6.8	7.1
Pathfinder	4.8	5.1
Pickup	4.5	4.8
Quest	4.7	5.0
Heater Hoses, Replace (B)		
620, 720	.8	1.0
Armada, Titan	1.4	1.6
Axxess	1.7	1.9
Frontier, Xterra	1.2	1.4
Murano	1.8	2.0
Pathfinder	.7	.9
Pickup		
4 cyl.		
KA	1.4	1.6
Z	.5	.7
V6	.5	.7
Quest		
1993-98	1.5	1.7
1999-05	1.1	1.3
Van	2.8	3.0
Low Pressure Cycling Switch, Replace (B)		
All Models	.5	.5
Magnetic Clutch Assy., Replace (B)		
Armada, Titan	2.0	2.2
Frontier, Xterra	2.4	2.6
Murano	2.0	2.2
Pathfinder		
1994-95	2.9	3.1
1996-00	4.4	4.6
2001-05	2.9	3.1
1994-97 Pickup	2.9	3.1
Quest	2.8	3.0
Rear Evaporator Assy., Replace (B)		
Includes: Evacuate/recover and recharge.		
Armada, Titan	2.8	3.0
Quest		
1993-98	3.1	3.1
1999-05	1.9	1.9
Rear Heater Blower Motor, Replace (B)		
Armada, Titan	4.6	4.6
1993-05 Quest	3.2	3.2
Van	1.8	1.8
Rear Heater Control Assy., Replace (B)		
Armada, Titan	.4	.4
Quest, Van	.5	.5

	LABOR TIME	SEVERE SERVICE
Rear Heater Core, R&R or Replace (B)		
Armada, Titan	6.0	6.2
1993-05 Quest	5.0	5.0
Van	3.2	3.2
Receiver/Drier Assy., Replace (B)		
Includes: Evacuate/recover and recharge.		
1981-93	2.1	2.1
Armada, Titan	2.1	2.3
Frontier, Xterra	1.4	1.4
Murano	1.7	1.7
Pathfinder		
1994-95	2.0	2.0
1996-05	1.7	1.7
1994-97 Pickup	2.0	2.0
Quest		
1993-98	2.2	2.2
1999-05	1.4	1.4
Temperature Control Assy., Replace (B)		
620, 720	1.3	1.3
Armada, Titan	.5	.5
Axxess	.8	.8
Frontier, Xterra	.7	.7
Murano	.7	.7
Pathfinder, Pickup		
1986-95	1.6	1.6
1996-05	.7	.7
Quest	.6	.6
Van	.8	.8

WIPERS & SPEEDOMETER

	LABOR TIME	SEVERE SERVICE
Antenna, Replace (B)		
620, 720	.5	.5
Armada, Titan		
manual	.6	.6
power	1.4	1.4
Axxess	.5	.5
Frontier	.3	.3
Murano	1.6	1.6
Pathfinder		
1986-95	.7	.7
1996-05	1.1	1.1
Pickup	.7	.7
Quest		
1993-98		
manual	.8	.8
power	1.5	1.5
1999-05	1.4	1.4
Van	.5	.5
Xterra	1.4	1.4
Intermittent Wiper Amplifier, Replace (B)		
All Models	.5	.5
Rear Window Washer Motor, Replace (B)		
Armada, Titan	.5	.5
Axxess	.7	.7
Pathfinder	.3	.3

PATHFINDER : PICKUP : QUEST : TITAN : VAN : XTERRA NI-75

	LABOR TIME	SEVERE SERVICE
1993-05 Quest	.7	.7
Van	.5	.5
Xterra	.8	.8

Rear Window Wiper Motor, Replace (B)
Armada, Titan	.8	.8
Axxess	.7	.9
Murano	.3	.5
Pathfinder		
1987-95	1.3	1.5
1996-05	.6	.8
1993-05 Quest	.5	.7
Van	.7	.8
Xterra	.5	.5

Radio, R&R (B)
Exc. below	.8	.8
620, 720	1.6	1.6
Murano	.9	.9
Van	1.5	1.5

Rear Window Wiper & Washer Switch, Replace (B)
All Models	.5	.5

Speedometer Cable & Casing, Replace (B)
620, 720	.7	.7
Pathfinder, Pickup	.9	.9
Van	1.5	1.5

Speedometer Head, R&R or Replace (B)
Exc. below	1.0	1.0
620, 720	1.8	1.8
Axxess	1.3	1.3
Van	1.2	1.2

Windshield Washer Pump, Replace (B)
Exc. below	.7	.7
Frontier, Xterra	.9	.9
Murano	1.1	1.1
Pathfinder	.8	.8
Quest	.8	.8

Windshield Wiper & Washer Switch, Replace (B)
Exc. below	.7	.7
Pathfinder, Quest	.8	.8

Windshield Wiper Interval Relay, Rear Relay, Replace (B)
All Models	.5	.5

Windshield Wiper Motor, Replace (B)
620, 720	.7	.9
Armada, Titan	.8	.9
Axxess	1.2	1.4
Frontier, Xterra	.8	1.0
Murano	.5	.7
Pathfinder		
1987-95	1.2	1.4
1996-05	.8	1.0
Pickup	1.1	1.3
1993-05 Quest	1.0	1.2
Van	.8	1.0

Windshield Wiper Pivot, Replace (B)
620, 720	.7	.9
Armada, Titan	.8	.9
Axxess	1.3	1.5
Frontier, Xterra	.7	.9
Murano	.5	.7
Pathfinder		
1987-95	1.3	1.5
1996-05	.8	1.0
Pickup	1.1	1.3
1993-05 Quest	1.0	1.2
Van	.8	1.0

LAMPS & SWITCHES

Back-Up Lamp Assy., Replace (B)
All Models	.5	.5

Back-Up Lamp Switch (w/MT), Replace (B)
All Models	.5	.5

Combination Switch Assy., Replace (B)
620, 720	.7	.7
Armada, Titan	1.4	1.4
Axxess	.5	.5
Frontier		
exc. crew	.5	.5
crew	1.2	1.2
Murano	.7	.7
Pathfinder, Pickup	.9	.9
Quest	1.1	1.1
Van	.7	.7
Xterra	1.0	1.0

Daylight Running Lamp Module, Replace (B)
Frontier, Xterra	.7	.7
Pathfinder, Pickup	.5	.5
Quest		
1995-98	.3	.3
1999-05	.7	.7

Halogen Headlamp Bulb, Replace (C)
Exc. Xterra	.3	.3
Xterra	.8	.8

Hazard Warning Switch, Replace (B)
Exc. Murano	.5	.5
Murano	.9	.9

Headlamps, Aim (B)
Two	.4	.4
Four	.6	.6

High Mount Stop Lamp Assy., Replace (B)
All Models	.5	.5

Horn, Replace (B)
Exc. Murano	.5	.5
Murano	.9	.9

Horn Relay, Replace (B)
All Models	.5	.5

License Lamp Assy., Replace (B)
All Models	.5	.5

Parking Brake Warning Lamp Switch, Replace (B)
All Models	.6	.6

Rear Combination Lamp Assy., Replace (B)
Exc. below	.5	.5
Pathfinder, Pickup	.7	.7

Sealed Beam Headlamp, Replace (C)
All Models	.3	.3

Side Marker Lamp Lens, Replace (B)
Each	.3	.3

Stop & Tail Lamp Lens, Replace (B)
Each	.3	.3

Stop Lamp Switch, Replace (B)
All Models	.5	.5

Turn Signal & Headlamp Dimmer Switch, Replace (B)
All Models	.9	.9

Turn Signal or Hazard Warning Flasher, Replace (B)
All Models	.5	.5

Turn Signal or Parking Lamp Lens or Bulb, Replace (C)
Exc. below each	.3	.3
Murano each	1.4	1.4
Pathfinder each	.6	.6

BODY

Door Lock, Replace (B)
620, 720	.9	.9
Armada, Titan	.8	.9
Axxess	.7	.7
Frontier	.8	.8
Murano		
front	1.0	1.0
rear	.7	.7
Pathfinder		
1987-95	.9	.9
1996-05		
front	1.2	1.2
rear	.6	.6
Pickup		
1986-95	.9	.9
1996-97	2.1	2.1
Quest		
1993-98	1.6	1.6
1999-05	1.2	1.2
Van	.7	.7
Xterra	.8	.8

Door Lock Cylinder, Replace (B)
Exc. Below	.7	.7
Armada, Titan	2.9	3.1
Quest	.8	.8

Door Lock Remote Control, Replace (B)
620, 720	.7	.7
Axxess	.9	.9
Frontier, Xterra	.7	.7

720 PICKUP : AMADA : AXXESS : D21 PICKUP : FRONTIER : MURANO

	LABOR TIME	SEVERE SERVICE
Door Lock Remote Control, Replace (B)		
Pathfinder, Pickup	.8	.8
Quest	1.1	1.1
Van	.7	.7
Hood Hinge, Replace (One) (B)		
Exc. below	.8	.8
Armada, Titan		
one	1.6	1.7
both	2.7	2.8
Murano	2.2	2.2
Quest	1.0	1.0
Hood Lock, Replace (B)		
All Models	.6	.6
Hood Release Cable, Replace (B)		
Exc. below	.6	.6

	LABOR TIME	SEVERE SERVICE
Armada, Titan	.8	.8
Axxess	.9	.9
1996-05 Pathfinder	1.3	1.3
Quest	1.4	1.4
Xterra	1.0	1.0
Lock Striker Plate, Replace (B)		
All Models	.5	.7
Window Regulator, Replace (B)		
620, 720	1.3	1.5
Armada, Titan	1.2	1.3
Axxess	.8	1.0
Frontier	1.0	1.2
Murano	.9	1.1
Pathfinder		
1987-95	1.7	1.9
1996-05	1.1	1.3

	LABOR TIME	SEVERE SERVICE
Pickup	1.7	1.9
1993-05 Quest	1.3	1.5
Van	.9	1.1
Xterra	.8	1.0
Window Regulator Motor, Replace (B)		
620, 720	1.4	1.6
Armada, Titan	1.2	1.3
Axxess	1.4	1.6
Frontier, Xterra	1.2	1.4
Murano	.9	1.1
Pathfinder, Pickup	1.9	2.1
Quest		
1993-98	1.3	1.5
1999-05	.8	1.0
Van	1.2	1.4

SC

TC : XA : XB

SYSTEM INDEX

- **MAINTENANCE** SC-2
  - Maintenance Schedule SC-2
- **CHARGING** SC-2
- **STARTING** SC-2
- **CRUISE CONTROL** SC-2
- **IGNITION** SC-2
- **EMISSIONS** SC-2
- **FUEL** SC-3
- **EXHAUST** SC-3
- **ENGINE COOLING** SC-3
- **ENGINE** SC-3
  - Assembly SC-3
  - Cylinder Head SC-3
  - Camshaft SC-3
  - Crank & Pistons SC-4
  - Engine Lubrication SC-4
- **CLUTCH** SC-4
- **MANUAL TRANSAXLE** SC-4
- **AUTO TRANSAXLE** SC-4
- **SHIFT LINKAGE** SC-4
- **DRIVELINE** SC-5
- **BRAKES** SC-5
- **FRONT SUSPENSION** SC-5
- **REAR SUSPENSION** SC-5
- **STEERING** SC-6
- **HEATING & AC** SC-6
- **WIPERS & SPEEDOMETER** . SC-6
- **LAMPS & SWITCHES** SC-6
- **BODY** SC-7

OPERATIONS INDEX

A
- AC Hoses SC-6
- Air Bags SC-6
- Air Conditioning SC-6
- Alignment SC-5
- Alternator (Generator) SC-2
- Antenna SC-6

B
- Back-Up Lamp Switch SC-6
- Battery Cables SC-2
- Bleed Brake System SC-5
- Blower Motor SC-6
- Brake Disc SC-5
- Brake Drum SC-5
- Brake Hose SC-5
- Brake Pads and/or Shoes .. SC-5

C
- Camshaft SC-4
- Camshaft Sensor SC-2
- Catalytic Converter SC-3
- Coolant Temperature (ECT) Sensor SC-3
- Crankshaft SC-4
- Cruise Control SC-2
- CV Joint SC-5
- Cylinder Head SC-3

D
- Distributor SC-2
- Drive Belt SC-2

E
- Electronic Control Module (ECM/PCM) SC-3
- Engine SC-3
- Engine Lubrication SC-4
- Evaporator SC-6
- Exhaust SC-3
- Exhaust Manifold SC-3
- Expansion Valve SC-6

F
- Flexplate SC-4
- Flywheel SC-4
- Fuel Injection SC-3
- Fuel Pump SC-3
- Fuel Vapor Canister SC-3

G
- Gear Selector Lever SC-4
- Generator SC-2

H
- Halfshaft SC-5
- Headlamp SC-6
- Heater Core SC-6
- Horn SC-6

I
- Idle Air Control (IAC) Valve .. SC-3
- Ignition Coil SC-2
- Ignition Module SC-2
- Ignition Switch SC-2
- Intake Air Temperature (IAT) Sensor SC-3
- Intake Manifold SC-3

K
- Knock Sensor SC-3

L
- Lower Control Arm SC-5

M
- Maintenance Schedule SC-2
- Manifold Absolute Pressure (MAP) Sensor SC-3
- Mass Air Flow (MAF) Sensor SC-3
- Master Cylinder SC-5
- Muffler SC-3
- Multifunction Lever/Switch .. SC-6

N
- Neutral Safety Switch SC-6

O
- Oil Pan SC-4
- Oil Pump SC-4
- Outer Tie Rod SC-6
- Oxygen Sensor SC-3

P
- Parking Brake SC-5
- Pistons SC-4

R
- Radiator SC-3
- Radiator Hoses SC-3
- Radio SC-6
- Rear Main Oil Seal SC-4

S
- Shock Absorber/Strut, Front .. SC-5
- Shock Absorber/Strut, Rear ... SC-5
- Spark Plug Cables SC-2
- Spark Plugs SC-2
- Spring, Front Coil SC-5
- Spring, Rear Coil SC-5
- Starter SC-2
- Steering Wheel SC-6

T
- Thermostat SC-3
- Throttle Body SC-3
- Timing Chain SC-4

V
- Valve Body SC-4
- Valve Cover Gasket SC-3
- Valve Job SC-3
- Vehicle Speed Sensor SC-3

W
- Water Pump SC-3
- Wheel Balance SC-2
- Wheel Cylinder SC-5
- Window Regulator SC-7
- Windshield Washer Pump .. SC-6
- Windshield Wiper Motor ... SC-6

SC-1

SC-2 TC : XA : XB

	LABOR TIME	SEVERE SERVICE
MAINTENANCE		
Air Cleaner Filter Element, Replace (C)		
XA, XB, TC (.3)	.5	.5
Drive Belt, Replace (B)		
XA, XB, TC (.4)	.6	.7
Fuel Filter, Replace (B)		
XA, XB, TC (.7)	.9	1.1
Halogen Headlamp Bulb, Replace (C)		
XA, XB, TC (.4)	.6	.7
Oil & Filter, Change (C)		
Includes: Correct fluid levels.		
XA, XB, TC (.4)	.6	.7
Tire, Replace (C)		
Includes: Dismount old tire and mount new tire to rim.		
XA, XB, TC (.4)	.6	.7
each addl. add	.5	.5
Tires, Rotate (C)		
XA, XB, TC (.4)	.6	.7
Wheel, Balance (B)		
Two (.5)	.7	.8
each addl. add	.2	.2

SCHEDULED MAINTENANCE INTERVALS

If necessary, refer to appropriate Chilton maintenance service information.

	LABOR TIME	SEVERE SERVICE
5,000 Mile Service (C)		
All Models (.4)	.6	.8
10,000 Mile Service (B)		
All Models (.4)	.6	.8
15,000 Mile Service (C)		
All Models (.6)	.8	.9
20,000 Mile Service (B)		
All Models (.4)	.6	.8
25,000 Mile Service (C)		
All Models (.4)	.6	.8
30,000 Mile Service (C)		
All Models (.8)	1.1	1.3
35,000 Mile Service (C)		
All Models (.4)	.6	.8
40,000 Mile Service (B)		
All Models (.4)	.6	.8
45,000 Mile Service (C)		
All Models (.6)	.8	.9
50,000 Mile Service (C)		
All Models (.4)	.6	.8
55,000 Mile Service (C)		
All Models (.4)	.6	.8
60,000 Mile Service (C)		
All Models (.9)	1.2	1.4
65,000 Mile Service (C)		
All Models (.4)	.6	.8
70,000 Mile Service (C)		
All Models (.4)	.6	.8
75,000 Mile Service (C)		
All Models (.6)	.8	.9
80,000 Mile Service (C)		
All Models (.4)	.6	.8
85,000 Mile Service (C)		
All Models (.4)	.6	.8
90,000 Mile Service (C)		
All Models (.9)	1.2	1.4
95,000 Mile Service (C)		
All Models (.4)	.6	.8
100,000 Mile Service (C)		
All Models (.6)	.8	.9
105,000 Mile Service (C)		
All Models (.4)	.6	.8
110,000 Mile Service (C)		
All Models (.4)	.6	.8
115,000 Mile Service (C)		
All Models (.4)	.6	.8
120,000 Mile Service (C)		
All Models (.9)	1.2	1.4

CHARGING

	LABOR TIME	SEVERE SERVICE
Alternator Circuits, Test (B)		
Includes: Test component output.		
XA, XB, TC (.2)	.3	.3
Alternator Assy., Replace (B)		
Includes: Pulley transfer.		
XA, XB, TC (.5)	.7	.8
Alternator Bearing, Replace (B)		
XA, XB, TC (.7)	.9	1.1
Alternator Voltage Regulator, Replace (B)		
XA, XB, TC (.7)	.9	1.1
Voltmeter and/or Ammeter Gauge, Replace (B)		
XA, XB, TC (.4)	.6	.7

STARTING

	LABOR TIME	SEVERE SERVICE
Starter Draw Test (On Car) (B)		
XA, XB, TC (.3)	.4	.4
Battery, Replace (C)		
XA, XB, TC (.4)	.6	.7
Battery Cables, Replace (C)		
Positive (.4)	.6	.7
Negative (.4)	.6	.7
Starter Assy., Replace (B)		
XA, XB, TC (.4)	.6	.7
Recondition starter add	1.1	1.3

CRUISE CONTROL

	LABOR TIME	SEVERE SERVICE
Diagnose Cruise Control System Component Each (A)		
XA, XB, TC (.8)	1.1	1.3
Actuator Assy., Replace (B)		
XA, XB, TC (.4)	.6	.7
Control Computer (ECU), Replace (B)		
XA, XB, TC (.4)	.6	.7
Control Main Switch, Replace (B)		
XA, XB, TC (.4)	.6	.7

IGNITION

	LABOR TIME	SEVERE SERVICE
Diagnose Ignition System Component Each (A)		
XA, XB, TC (.8)	1.1	1.3
Camshaft Sensor, Replace (B)		
XA, XB, TC (.3)	.5	.6
Distributor, Replace (B)		
Includes: Reset base ignition timing.		
XA, XB, TC (.3)	.5	.6
Ignition Coil, Replace (B)		
XA, XB, TC (.3)	.5	.6
Ignition Module, Replace (B)		
XA, XB, TC (.4)	.6	.7
Ignition Switch, Replace (B)		
XA, XB, TC (.5)	.7	.9
Spark Plug (ignition) Cables, Replace (B)		
XA, XB, TC (.6)	.8	.9
Spark Plugs, Replace (B)		
XA, XB, TC (.4)	.6	.7

EMISSIONS

The following operations do not include testing. Add time as required.

	LABOR TIME	SEVERE SERVICE
Diagnose Accelerator Pedal Position Sensor Circuit Malfunction (A)		
XA, XB, TC (.4)	.6	.7
Diagnose Coolant Temperature Insufficient for Closed Loop Fuel Control (A)		
XA, XB, TC (.4)	.6	.7
Diagnose Manifold Absolute Pressure Circuit Range/Performance Problem (A)		
XA, XB, TC (.4)	.6	.7
Diagnose Mass Air Flow Circuit Malfunction (A)		
XA, XB, TC (.4)	.6	.7
Diagnose Mass Air Flow Circuit Range/Performance Problem (A)		
XA, XB, TC (.4)	.6	.7
Diagnose Oxygen Sensor Circuit Malfunction (A)		
XA, XB, TC (.4)	.6	.7
Diagnose Oxygen Sensor Heater Circuit Malfunction (A)		
XA, XB, TC (.4)	.6	.7
Diagnose Oxygen Sensor Heater Circuit Slow Response (A)		
XA, XB, TC (.4)	.6	.7
Diagnose Throttle Pedal Position Sensor Switch "A" Circuit Performance Problem (A)		
XA, XB, TC (.4)	.6	.7
Diagnose Vehicle Speed Sensor Malfunction (A)		
XA, XB, TC (.4)	.6	.7
Dynamometer Test (A)		
XA, XB, TC (.4)	.6	.7

TC : XA : XB   SC-3

	LABOR TIME	SEVERE SERVICE
Coolant Temperature (ECT) Sensor, Replace (B)		
XA, XB, TC (.7)	.9	1.1
Electronic Control Module (ECM/PCM), Replace (B)		
XA, XB, TC (.5)	.7	.9
Fuel Vapor Canister, Replace (B)		
XA, XB (.3)	.5	.5
TC (.8)	1.1	1.2
Intake Air Temperature (IAT) Sensor, Replace (B)		
XA, XB, TC (.4)	.6	.7
Knock Sensor, Replace (B)		
XA, XB (1.4)	1.9	2.1
TC (.6)	.8	.9
Manifold Absolute Pressure (MAP) Sensor, Replace (B)		
XA, XB, TC (.4)	.6	.7
Mass Air Flow (MAF) Sensor, Replace (B)		
XA, XB, TC (.4)	.6	.7
Oxygen Sensor, Replace (B)		
XA, XB, TC (.6)	.8	.9
Vacuum Switching Valve, Replace (B)		
XA, XB, TC (.3)	.5	.5
Vehicle Speed Sensor, Replace (B)		
XA, XB, TC (.4)	.6	.7

FUEL

DELIVERY

	LABOR TIME	SEVERE SERVICE
Fuel Pump, Test (B)		
XA, XB, TC (.4)	.6	.7
Fuel Filter, Replace (B)		
XA, XB, TC (.7)	.9	1.1
Fuel Gauge (Dash), Replace (B)		
XA, XB, TC (.4)	.6	.7
Fuel Gauge (Tank), Replace (B)		
XA, XB, TC (1.9)	2.6	2.9
Fuel Pump, Replace (B)		
XA, XB, TC (.6)	.8	.9
Fuel Tank, Replace (B)		
Includes: Drain and refill.		
XA, XB, TC (1.9)	2.6	2.9
Intake Manifold Gasket, Replace (B)		
XA, XB (1.3)	1.8	2.0
TC (1.8)	2.4	2.8

INJECTION

	LABOR TIME	SEVERE SERVICE
Diagnose Air/Fuel Sensor Circuit Range/Performance Malfunction (A)		
XA, XB, TC (.4)	.6	.7
Diagnose Idle Control System Malfunction (A)		
XA, XB, TC (.4)	.6	.7
Air Flow Meter, Replace (B)		
XA, XB, TC (.3)	.5	.6
Fuel Injection Lines, Replace (B)		
XA, XB, TC (.5)	.7	.9

	LABOR TIME	SEVERE SERVICE
Fuel Injectors, Replace (B)		
XA, XB, TC (.6)	.8	.9
Fuel Pressure Regulator, Replace (B)		
XA, XB, TC (.6)	.8	.9
Idle Air Control (IAC) Valve, Replace (B)		
XA, XB, (1.1)	1.5	1.7
Throttle Body Assy., Replace (B)		
XA, XB, TC (1.1)	1.5	1.7

EXHAUST

	LABOR TIME	SEVERE SERVICE
Catalytic Converter, Replace (B)		
XA, XB, TC (.8)	1.1	1.2
Center Exhaust Pipe, Replace (B)		
XA, XB, TC (.5)	.7	.8
Exhaust Manifold and/or Gasket, Replace (B)		
XA, XB, TC (.9)	1.2	1.4
Front Exhaust Pipe, Replace (B)		
XA, XB, TC (.7)	.9	1.1
Muffler and Tailpipe Assy., Replace (B)		
XA, XB, TC (.5)	.7	.8

ENGINE COOLING

	LABOR TIME	SEVERE SERVICE
Pressure Test Cooling System (B)		
XA, XB, TC (.4)	.6	.7
Coolant Thermostat, Replace (B)		
XA, XB, TC (.7)	.9	1.1
Engine Coolant Temp. Sending Unit, Replace (B)		
XA, XB, TC (.4)	.6	.7
Heater Hose, Replace (B)		
XA, XB, TC (.5) one	.7	.9
Radiator Assy., R&R or Replace (B)		
XA, XB (1.6)	2.2	2.4
TC (1.0)	1.4	1.6
Radiator Fan and/or Fan Motor, Replace (B)		
XA, XB (1.8)	2.4	2.8
TC (1.2)	1.6	1.8
Radiator Fan Motor Relay, Replace (B)		
XA, XB, TC (.2)	.3	.3
Radiator Fan Motor Switch (Coolant Temp.), Replace (B)		
XA, XB, TC (.4)	.6	.7
Radiator Hoses, Replace (B)		
Includes: Refill with proper coolant mix.		
Upper (.7)	.9	1.1
Lower (.9)	1.2	1.4
Temperature Gauge (Dash), Replace (B)		
XA, XB, TC (.4)	.6	.7
Water Pump and/or Gasket, Replace (B)		
Includes: Refill with proper coolant mix.		
XA, XB, TC (1.1)	1.5	1.7
TC (1.4)	1.9	2.1

ENGINE

ASSEMBLY

Times shown are for OEM assemblies. Time to replace assemblies from aftermarket rebuilders may vary.

	LABOR TIME	SEVERE SERVICE
Engine Assy., R&R and Recondition (A)		
Includes: Replacing rings, rod and main bearings, cylinder head reconditioning and engine tune-up.		
XA, XB (27.6)	37.3	38.6
TC (28.7)	38.7	40.2
Engine Assy. (Short Block), Replace (B)		
Assembly may consist of cylinder block, piston assemblies, crankshaft, camshaft, timing chain and gears. Operation Includes: R&R engine, transfer necessary parts and all necessary adjustments.		
XA, XB (12.9)	17.4	18.3
TC (13.6)	18.4	19.2

CYLINDER HEAD

	LABOR TIME	SEVERE SERVICE
Compression Test (B)		
XA, XB, TC (.6)	.8	.9
Valve Clearance, Adjust (B)		
XA, XB (2.4)	3.2	3.6
TC (3.6)	4.9	5.4
Cylinder Head, Replace (B)		
Includes: Parts transfer, adjustments.		
XA, XB (13.6)	18.4	19.2
TC (14.4)	19.4	20.2
Cylinder Head Gasket, Replace (B)		
XA, XB (6.4)	8.6	9.4
TC (8.6)	11.6	12.6
Valve/Cam Cover Gasket, Replace or Reseal (B)		
XA, XB, TC (.6)	.8	.9
Valve Guides, Replace (B)		
XA, XB (14.8)	20.0	20.7
TC (17.3)	23.4	24.7
Valve Job (A)		
XA, XB (14.0)	18.9	19.7
TC (15.9)	21.5	22.0
Valve Lifters, Replace (B)		
XA, XB (1.9)	2.6	2.9
TC (3.8)	5.1	5.7
Valve Springs and/or Oil Seals, Replace (B)		
XA, XB, TC (.4)	.6	.7

CAMSHAFT

	LABOR TIME	SEVERE SERVICE
Balance Shaft, Replace (B)		
TC		
one (14.5)	19.6	21.2
both (15.0)	20.3	21.8

SC-4 TC : XA : XB

	LABOR TIME	SEVERE SERVICE
Camshaft, Replace (A)		
XA, XB all (2.8)	3.8	4.3
TC all (4.2)	5.7	6.3
Camshaft Timing Gear, Replace (B)		
XA, XB all (2.8)	3.8	4.3
TC all (4.2)	5.7	6.3
Crankshaft Gear or Sprocket, Replace (B)		
XA, XB (3.5)	4.7	5.3
TC (4.5)	6.1	6.8
Timing Chain, Replace (B)		
Includes: Engine R&R when required.		
XA, XB (3.2)	4.3	4.8
TC (4.7)	6.3	7.1
Timing Chain Tensioner, Replace (B)		
XA, XB (2.8)	3.8	4.3
TC (1.5)	2.0	2.3

CRANK & PISTONS

Connecting Rod Bearings, Replace (A)
Includes: Check bearing oil clearance.
XA, XB, TC (15.8)	21.3	22.9

Crankshaft and Main Bearings, Replace (A)
Includes: Engine R&R.
XA, XB, TC (15.9)	21.5	23.1

Crankshaft Front Oil Seal, Replace (B)
XA, XB, TC (1.2)	1.6	1.8

Crankshaft Pulley, Replace (B)
XA, XB, TC (.6)	.8	.9

Pistons and/or Connecting Rods, Replace (A)
Includes: Ridge reaming, cylinder wall deglazing, installing new rings and rod bearings, engine tune-up.
XA, XB (16.7)	22.5	24.0
TC (15.6)	21.1	22.6

Rear Main Oil Seal, Replace (B)
Includes: Trans. R&R. when necessary.
XA, XB, TC (5.6)	7.6	8.3

Rings, Replace (A)
Includes: Ridge reaming, cylinder wall deglazing, installing new rings, engine tune-up.
XA, XB, TC (15.4)	20.8	21.4

ENGINE LUBRICATION

Engine Oil Cooler, Replace (B)
XA, XB, TC (.4)	.6	.7

Engine Oil Pan Gasket Number 1, Replace (B)
XA, XB (11.2)	15.1	16.0
TC (2.3)	3.1	3.5

Engine Oil Pan Gasket Number 2, Replace (B)
XA, XB (1.5)	2.0	2.3

	LABOR TIME	SEVERE SERVICE
Oil Pressure Gauge (Dash), Replace (B)		
XA, XB, TC (.4)	.6	.7
Oil Pump, Replace (B)		
XA, XB (3.1)	4.2	4.7
TC (5.1)	6.9	7.6

CLUTCH

Bleed Clutch Hydraulic System (B)
XA, XB, TC (.8)	1.1	1.2

Clutch Assy., Replace (B)
XA, XB (5.3)	7.2	7.9
TC (4.7)	6.3	7.1

Clutch Master Cylinder, Replace (B)
Includes: System bleeding.
XA, XB, TC (2.3)	3.1	3.5

Clutch Pedal Subassembly, Replace (B)
XA, XB (.4)	.6	.7
TC (1.7)	2.3	2.6

Clutch Release Bearing, Replace (B)
XA, XB (5.2)	7.0	7.7
TC (4.5)	6.1	6.8

Clutch Slave Cylinder, Replace (B)
Includes: System bleeding.
XA, XB, TC (.7)	.9	1.1

Flywheel, Replace (B)
XA, XB, TC (5.6)	7.6	8.3
TC (5.0)	6.8	7.5

MANUAL TRANSAXLE

Transaxle Assy. R&R and Recondition (A)
XA, XB, TC (11.8)	15.9	16.9

Transaxle Assy., R&R or Replace (B)
XA, XB, TC (5.1)	6.9	7.6

Transaxle Assy. Differential Bearings or Gears, Replace (A)
XA, XB (7.7)	10.4	11.3
TC (6.8)	9.2	10.1

Transaxle Assy. Input Shaft, Replace (A)
XA, XB (7.4)	10.0	10.9
TC (6.4)	8.6	9.5

Transaxle Assy. Input Shaft Oil Seal, Replace (A)
XA, XB (7.0)	9.4	10.4
TC (6.4)	8.6	9.5

Transaxle Assy. Output Shaft, Replace (A)
XA, XB (8.2)	11.1	12.0
TC (7.6)	10.3	11.2

Transaxle Differential Case Subassembly, Replace (A)
XA, XB (8.1)	10.9	11.9
TC (5.5)	7.4	8.2

Type T Oil Seal for Side Gear Shaft, Replace (B)
XA, XB, TC (5.1)	6.9	7.6

	LABOR TIME	SEVERE SERVICE
AUTOMATIC TRANSAXLE		
SERVICE TRANSAXLE INSTALLED		
Check Unit for Oil Leaks (C)		
XA, XB, TC (.6)	.8	.9
Drain and Refill Unit (B)		
XA, XB, TC (1.1)	1.3	1.5
Oil Cooler Hose, Replace (B)		
XA, XB (.4)	.6	.7
TC (1.0)	1.4	1.6
Oil Pan and/or Gasket, Replace (B)		
XA, XB, TC (1.4)	1.9	2.1
Transmission Differential Case Subassembly, Replace (A)		
XA, XB (6.3)	8.5	9.3
TC (5.9)	8.0	8.8
Transmission Oil Seal for Bearing Cap Side, Replace (B)		
XA, XB, TC (2.1)	2.8	3.2
Transmission Oil Seal for Bearing Retainer, Replace (B)		
XA, XB, TC (2.1)	2.8	3.2
Transmission Solenoid Assy., Replace (B)		
XA, XB, TC (1.6)	2.2	2.4
Transmission Speed Sensor, Replace (B)		
XA, XB, TC (.4)	.6	.7
Valve Body Assy., Replace (B)		
XA, XB, TC (1.8)	2.4	2.8
Valve Body Manual Lever Shaft Seal, Replace (B)		
XA, XB, TC (.4)	.6	.7

SERVICE TRANSMISSION REMOVED

Transmission R&R included unless otherwise noted.

Flywheel (Flexplate), Replace (B)
XA, XB (5.8)	7.8	8.6
TC (5.6)	7.6	8.3

Torque Converter, Replace (B)
XA, XB (5.5)	7.4	8.2
TC (5.3)	7.2	7.9

Transmission Assy., R&I (B)
XA, XB (5.8)	7.8	8.6
TC (5.3)	7.2	7.9

Transmission Assy., R&R and Recondition (A)
XA, XB (12.1)	16.3	17.2
TC (11.6)	15.7	16.6

SHIFT LINKAGE
AUTOMATIC TRANSAXLE

Gear Selector Lever, Replace
XA, XB (.8)	1.1	1.2
TC (.4)	.6	.7

Gear Selector Lever Knob, Replace
XA, XB, TC (.2)	.3	.3

	LABOR TIME	SEVERE SERVICE

Shift Linkage Control Cable Assembly, Replace (A)
XA, XB (1.2) 1.6 *1.8*
TC (4.5) 6.1 *6.8*

MANUAL TRANSAXLE
Gear Selector Control Cable Assembly, Replace (A)
XA, XB (.8) 1.1 *1.2*
TC (4.7) 6.3 *7.1*
Shift Selector Lever, Replace
XA, XB (.5) .... .7 *.8*
TC (.3) .... .5 *.6*

DRIVELINE
HALFSHAFTS
Halfshaft, R&R or Replace (B)
XA, XB (1.4) 1.9 *2.1*
TC (1.9) 2.6 *2.9*
Halfshaft Boot, Replace (B)
XA, XB, TC (2.3) 3.1 *3.5*

BRAKES
SYSTEM
Bleed Brakes (B)
Includes: Add fluid.
XA, XB, TC (.6) .... .8 *.9*
Brake System, Flush and Refill (B)
XA, XB, TC (1.0) 1.4 *1.6*
Brakes, Adjust (B)
Includes: Refill master cylinder.
XA, XB (.4) .... .6 *.7*
Brake Hose (Flexible), Replace (B)
Includes: System bleeding.
XA, XB, TC (.8) 1.1 *1.3*
each addl. add .... .5 *.6*
Master Cylinder, Replace (B)
Includes: System bleeding.
XA, XB (1.0) 1.4 *1.6*
TC (1.5) 2.0 *2.3*
Power Booster Unit, Replace (B)
Includes: System bleeding.
XA, XB, TC (1.9) 2.6 *2.9*
Power Booster Vacuum Check Valve, Replace (B)
XA, XB, TC (.3) ...... .5 *.6*

SERVICE BRAKES
Brake Drum, Replace (B)
XA, XB (.4) .... .6 *.7*
Front Caliper Assy., Replace (B)
Includes: System bleeding.
XA, XB
 one (.8) 1.1 *1.3*
 both (1.2) 1.6 *1.8*
TC
 one (.9) 1.2 *1.4*
 both (1.4) 1.9 *2.1*

Rear Caliper Assy., Replace (B)
Includes: System bleeding.
TC
 one (.9) 1.2 *1.4*
 both (1.4) 1.9 *2.1*
Front Disc Brake Rotor, Replace (B)
XA, XB, TC
 both (1.2) 1.6 *1.8*
Pads and/or Shoes, Replace (B)
Includes: Service and parking brake adjustment, system bleeding.
Front or rear disc (1.2) 1.6 *1.8*
Rear drum (1.0) 1.4 *1.6*
Four wheels (1.9) ... 2.6 *2.7*

COMBINATION ADD-ONS
Replace
 brake drum add .... .2 *.2*
 brake hose add .... .3 *.3*
 caliper add .... .4 *.4*
 disc brake rotor add . .2 *.2*
 wheel cylinder add ... .2 *.2*
Resurface
 brake drum add .... .6 *.6*
Wheel Cylinder, Replace (B)
Includes: System bleeding.
XA, XB (.8) 1.1 *1.3*
each addl. add .... .7 *.8*

PARKING BRAKE
Parking Brake Cable, Adjust (C)
XA, XB, TC (.4) ...... .6 *.7*
Parking Brake Apply Actuator, Replace (B)
XA, XB, TC (.6) ...... .8 *.9*
Parking Brake Cable, Replace (B)
Front
 XA, XB (.5) .... .7 *.8*
 TC (1.2) 1.6 *1.8*
Rear both sides
 XA, XB (1.9) 2.6 *2.9*
 TC (1.5) 2.0 *2.2*

TRACTION CONTROL
Actuator Assy., Replace (B)
XA, XB, TC (.4) ...... .6 *.7*
Control Unit, Replace (B)
XA, XB, TC (.4) ...... .6 *.7*
Main Relay, Replace (B)
XA, XB, TC (.4) ...... .6 *.7*

FRONT SUSPENSION
Unless otherwise noted, time given does not include alignment.
Front End Alignment, Check (A)
XA, XB, TC (.8) 1.1 *1.3*
Coil Spring, Replace (B)
XA, XB, TC
 one (1.2) 1.6 *1.8*
 both (1.8) 2.4 *2.8*

Front Hub Assy., Replace (B)
XA, XB, TC
 one (2.0) 2.7 *3.1*
 both (3.1) 4.2 *4.7*
Front Hub Bearing, Replace (B)
XA, XB, TC
 one (1.8) 2.4 *2.8*
 both (2.8) 3.8 *4.3*
Front Shock Absorbers, Replace (B)
XA, XB, TC
 one (1.2) 1.6 *1.8*
 both (1.8) 2.4 *2.8*
Lower Control Arm, Replace (B)
XA, XB
 one (.6) .... .8 *.9*
 both (.8) 1.1 *1.3*
TC
 one (3.0) 4.1 *4.6*
 both (3.2) 4.3 *4.8*
Stabilizer Bar, Replace (B)
XA, XB (1.9) 2.6 *2.9*
TC (2.9) 3.9 *4.4*
Stabilizer Bar Bushing, Replace (B)
XA, XB (1.0) 1.4 *1.6*
TC (.4) .... .6 *.8*
Steering Knuckle, Replace (B)
XA, XB, TC (1.9) 2.6 *2.9*

REAR SUSPENSION
Rear Suspension Toe or Camber, Adjust (B)
XA, XB, TC (.8) 1.1 *1.3*
Coil Spring, R&R or Replace (B)
XA, XB
 one (2.2) 3.0 *3.3*
 both (2.4) 3.2 *3.6*
TC
 one (2.0) 2.7 *3.1*
 both (3.0) 2.0 *2.2*
Lower Control Arm, Replace (B)
TC (1.3) 1.8 *2.0*
Rear Hub and/or Bearing, Replace (B)
XA, XB one side (1.0) . 1.4 *1.6*
TC one side (2.0) ... 2.7 *3.1*
Shock Absorbers, Replace (B)
XA, XB
 one (1.2) 1.6 *1.8*
 both (1.4) 1.9 *2.1*
TC
 one (2.0) 2.7 *3.1*
 both (3.0) 2.0 *2.2*
Stabilizer Bar, Replace (B)
TC (.8) 1.1 *1.3*
Stabilizer Bar Bushing, Replace (B)
TC (.6) .... .8 *.9*

TC : XA : XB

	LABOR TIME	SEVERE SERVICE

STEERING

AIR BAGS

Diagnose Air Bag System Component Each (A)
XA, XB, TC (.8) ...... 1.1 — 1.3

Air Bag Assy., Replace (B)
Driver side
  XA, XB, TC (.6) ...... .8 — .9
Passenger side
  XA, XB, TC (1.1) ... 1.5 — 1.7
Side impact
  XA
    one side (1.4) .... 1.9 — 2.1
    both sides (1.8) .. 2.4 — 2.8
  TC
    one side (2.4) .... 3.2 — 3.6
    both sides (2.8) .. 3.8 — 4.3

Center (Crash) Sensor, Replace (B)
XA (1.6) .... 2.2 — 2.4
XB, TC (.6) .... .6 — .7

Front Sensor, Replace (B)
XA, XB, TC
  one (.6) .... .8 — .9
  both (1.0) .... 1.4 — 1.6

Rear Sensor, Replace (B)
XA, XB
  one (.6) .... .8 — .9
  both (1.0) .... 1.4 — 1.6

Side Sensor, Replace (B)
XA, XB, TC
  one (.6) .... .8 — .9
  both (1.0) .... 1.4 — 1.6

POWER RACK & PINION

Steering Pump Pressure Check (B)
XA, XB, TC (.4) .... .6 — .7

Steering Column Upper Bracket, Replace (B)
XA, XB, TC (1.4) ..... 1.9 — 2.1

Steering Column Assembly, Replace (B)
XA, XB, TC (1.2) ..... 1.6 — 1.8

Steering Column Main Shaft Assy., Replace (B)
Tilt column
  XA, XB, TC (1.3) ... 1.8 — 2.0

Horn Contact, Replace (B)
XA, XB, TC (.4) .... .6 — .7

Outer Tie Rod, Replace (B)
Includes: Reset toe.
XA, XB
  one (.8) .... 1.1 — 1.2
  both (1.0) .... 1.4 — 1.6
TC
  one (1.9) .... 2.6 — 2.9
  both (2.4) .... 3.2 — 3.6

Power Steering Hoses, Replace (B)
XA, XB, TC
  pressure (1.1) ...... 1.5 — 1.7
  return (1.0) .... 1.4 — 1.6

Rack & Pinion Assy., R&R or Replace (B)
Includes: Alignment.
XA, XB (3.2) .... 4.3 — 4.8
TC (4.1) .... 5.5 — 6.2

Steering Pump, R&R or Replace (B)
XA, XB, TC (1.6) ..... 2.2 — 2.4

Steering Wheel, Replace (B)
XA, XB, TC (.5) .... .7 — .9

HEATING & AIR CONDITIONING

When more than one component requires replacement where evacuation/recovery and recharging is already included, deduct 1.0 hour for each additional component from the time given.

Evacuate/Recover and Recharge System (B)
XA, XB, TC (.8) ...... 1.1 — 1.3

AC Hoses, Replace (B)
Includes: Evacuate/recover and recharge.
XA, XB, TC (2.0) ..... 2.7 — 3.1

Blower Motor, Replace (B)
XA, XB (.4) .... .6 — .8
TC (2.0) .... 2.7 — 3.1

Blower Motor Resistor, Replace (B)
XA, XB (.3) .... .5 — .6

Blower Motor Control Relay, Replace (B)
TC (.3) .... .5 — .6

Blower Motor Switch, Replace (B)
XA, XB, TC (.4) .... .6 — .7

Compressor Assy., Replace (B)
Includes: Parts transfer, evacuate/recover and recharge.
XA, XB, TC (2.6) ..... 3.5 — 4.0

Condenser Assy., Replace (B)
Includes: Evacuate/recover and recharge.
XA, XB, TC (1.8) ..... 2.4 — 2.8

Damper Servo Subassembly, Replace (B)
XA, XB (.9) .... 1.2 — 1.4
TC (2.1) .... 2.8 — 3.2

Evaporator Coil, Replace (B)
Includes: Evacuate/recover and recharge.
XA (5.1) .... 6.9 — 7.6
XB (4.0) .... 5.4 — 6.0
TC (4.7) .... 6.3 — 7.1

Expansion Valve, Replace (B)
Includes: Evacuate/recover and recharge.
XA, XB (1.4) .... 1.9 — 2.1
TC (4.2) .... 5.7 — 6.3

Heater Control Assy., Replace (B)
XA, XB, TC (.4) .... .6 — .7

Heater Core, R&R or Replace (B)
XB, TC
  w/AC (3.8) .... 5.1 — 5.7
  w/o AC (2.7) .... 3.6 — 4.1
XA
  w/AC (4.4) .... 5.9 — 6.6
  w/o AC (3.5) .... 4.7 — 5.3

Heater Hose, Replace (B)
XA, XB, TC (.5) one ... .7 — .9

Receiver/Drier Assy., Replace (B)
Includes: Evacuate/recover and recharge.
XA, XB, TC (2.9) ..... 3.9 — 4.4

WIPERS & SPEEDOMETER

Antenna, Replace (B)
XA, XB, TC (1.9) ..... 2.6 — 2.8

Radio, R&R (B)
XA, XB, TC (.7) .... .9 — 1.1

Speedometer Head, R&R or Replace (B)
XA, XB, TC (.4) .... .6 — .7

Windshield Washer Pump, Replace (B)
XA, XB, TC (.4) .... .6 — .7

Windshield Wiper Motor, Replace (B)
XA, XB, TC (.5) .... .7 — .9

Windshield Wiper Switch, Replace (B)
XA, XB, TC (.4) .... .6 — .7

LAMPS & SWITCHES

Back-Up Lamp Bulb, Replace (C)
XA, XB, TC (.2) .... .3 — .3

Back-Up Lamp Switch, Replace (B)
XA, XB, TC (.3) .... .5 — .6

Halogen Headlamp Bulb, Replace (C)
XA, XB, TC (.4) .... .6 — .7

Hazard Warning Switch, Replace (B)
XA, XB, TC (.4) .... .6 — .7

Headlamp Switch, Replace (B)
XA, XB, TC (.3) .... .4 — .4

Headlamps, Aim (B)
XA, XB, TC (.4) .... .6 — .7

High Mount Stop Lamp Assy., Replace (B)
XA, XB, TC (.3) .... .5 — .5

Horn, Replace (B)
XA, XB, TC (.4) .... .6 — .7

License Lamp Assy., Replace (B)
XA, XB, TC (.2) .... .3 — .3

Multifunction Switch, Replace (B)
XA, XB, TC (.4) .... .6 — .7

Neutral Safety Switch, Replace (B)
XA, XB, TC (.7) .... .9 — 1.1

Parking Brake Warning Lamp Switch, Replace (B)
XA, XB, TC (.6) .... .8 — .9

TC : XA : XB **SC-7**

	LABOR TIME	SEVERE SERVICE
Parking Lamp Lens or Bulb, Replace (C)		
XA, XB, TC (.3)	.5	.6
Power Window Switch, Replace (C)		
XA, XB, TC ea. (.3)	.5	.6
Power Window Master Switch, Replace (C)		
XA, XB, TC (.2)	.3	.4
Rear Combination Lamp Bulb, Replace (C)		
XA, XB, TC (.2)	.3	.3
Rear Combination Lamp Lens, Replace (B)		
XA, XB, TC (.4)	.6	.7
Rear Side Marker Lamp Lens, Replace (C)		
XA, XB, TC (.2)	.3	.3
Side Marker Lamp Lens or Bulb, Replace (C)		
XA, XB, TC front (.3)	.4	.4

	LABOR TIME	SEVERE SERVICE
Stop Lamp Switch, Replace (B)		
XA, XB, TC (.4)	.6	.7
Turn Signal Lamp Lens or Bulb, Replace (C)		
XA, XB, TC (.3)	.4	.4
Turn Signal or Hazard Warning Flasher, Replace (B)		
XA, XB, TC (.3)	.4	.4
Turn Signal Switch, Replace (B)		
XA, XB, TC (.3)	.4	.4

SEAT BELTS

	LABOR TIME	SEVERE SERVICE
Front Seat Belt, Replace (B)		
XA, XB, TC (.4)	.6	.7
Rear Seat Belt, Replace (B)		
XA, XB, TC (.6)	.8	.9

BODY

	LABOR TIME	SEVERE SERVICE
Door Check, Replace (B)		
XA, XB, TC (.4)	.6	.7
Door Lock, Replace (B)		
XA, XB, TC (.3)	.5	.6
Hood Hinges, Replace (B)		
XA, XB, TC (1.6)	2.2	2.4
Hood Lock, Replace (B)		
XA, XB, TC (.3)	.5	.6
Hood Release Cable, Replace (B)		
XA, XB, TC (.9)	1.2	1.4
Lock Striker Plate, Replace (B)		
XA, XB, TC (.4)	.6	.7
Window Regulator, Replace (B)		
XA, XB, TC front or rear (.7)	.9	1.0
Window Regulator Motor, Replace (B)		
XA, XB, TC front or rear (.9)	1.2	1.4

NOTES

Baja : Brat : DL : Forester : GL : GL-10 : GLF : Impreza : Justy :
Legacy : Loyale : Outback : RX : Standard : SVX : XT : XT6

SUB

SYSTEM INDEX

MAINTENANCE SUB-2
   Maintenance Schedule... SUB-2
CHARGING............... SUB-2
STARTING SUB-2
CRUISE CONTROL SUB-2
IGNITION................ SUB-3
EMISSIONS............... SUB-3
FUEL................... SUB-4
EXHAUST SUB-5
ENGINE COOLING........ SUB-6
ENGINE................. SUB-6
   Assembly............. SUB-6
   Cylinder Head......... SUB-7
   Camshaft............. SUB-9
   Crank & Pistons SUB-10
   Engine Lubrication...... SUB-11
CLUTCH SUB-12
MANUAL TRANSAXLE SUB-12
AUTO TRANSAXLE SUB-13
TRANSFER CASE SUB-16
SHIFT LINKAGE.......... SUB-16
DRIVELINE SUB-16
BRAKES SUB-17
FRONT SUSPENSION SUB-19
REAR SUSPENSION SUB-20
STEERING SUB-21
HEATING & AC.......... SUB-22
WIPERS &
   SPEEDOMETER SUB-23
LAMPS & SWITCHES SUB-24
SEAT BELTS............ SUB-24
BODY SUB-25

OPERATIONS INDEX

A
AC Hoses SUB-22
Air Bags................. SUB-21
Air Conditioning........... SUB-22
Alignment................ SUB-19
Alternator (Generator) SUB-2
Antenna................. SUB-23
Anti-Lock Brakes SUB-17

B
Back-Up Lamp Switch....... SUB-24
Ball Joint................. SUB-19
Battery Cables............. SUB-2
Bleed Brake System SUB-18
Blower Motor............. SUB-22
Brake Disc................ SUB-18
Brake Drum SUB-18
Brake Hose SUB-18
Brake Pads and/or Shoes ... SUB-19

C
Camshaft................. SUB-9
Camshaft Sensor........... SUB-3
Catalytic Converter SUB-5
Coolant Temperature (ECT)
   Sensor SUB-3
Crankshaft............... SUB-10
Crankshaft Sensor.......... SUB-3
Cruise Control SUB-2
CV Joint................. SUB-16
Cylinder Head SUB-7

D
Differential SUB-14
Distributor SUB-3
Drive Belt................ SUB-2
Driveshaft SUB-17

E
EGR..................... SUB-3
Electronic Control Module
   (ECM/PCM) SUB-4
Engine SUB-6
Engine Lubrication......... SUB-11
Engine Mounts SUB-7
Evaporator............... SUB-22
Exhaust SUB-5
Exhaust Manifold.......... SUB-6
Expansion Valve SUB-22

F
Flexplate SUB-15
Flywheel SUB-12
Fuel Injection............. SUB-5
Fuel Pump............... SUB-5
Fuel Vapor Canister........ SUB-4

G
Gear Selector Lever SUB-16
Generator SUB-2

H
Halfshaft SUB-17
Headlamp SUB-24
Heater Core.............. SUB-22
Horn.................... SUB-24

I
Idle Air Control (IAC) Valve .. SUB-5
Ignition Coil SUB-3
Ignition Module SUB-3
Ignition Switch SUB-3
Intake Air Temperature (IAT)
   Sensor SUB-4
Intake Manifold SUB-5

K
Knock Sensor SUB-4

L
Lower Control Arm......... SUB-19

M
Maintenance Schedule SUB-2
Manifold Absolute Pressure
   (MAP) Sensor SUB-4
Mass Air Flow (MAF)
   Sensor SUB-4
Master Cylinder........... SUB-18
Muffler SUB-6
Multifunction Lever/Switch... SUB-24

N
Neutral Safety Switch SUB-24

O
Oil Pan.................. SUB-11
Oil Pump SUB-12
Outer Tie Rod SUB-21
Oxygen Sensor SUB-4

P
Parking Brake SUB-19
Pistons.................. SUB-11
Positive Crankcase
   Ventilation (PCV) Valve SUB-4

R
Radiator................. SUB-6
Radiator Hoses SUB-6
Radio................... SUB-23
Rear Main Oil Seal SUB-11

S
Shock Absorber/Strut,
   Front................... SUB-20
Shock Absorber/Strut,
   Rear................... SUB-20
Spark Plug Cables SUB-3
Spark Plugs SUB-3
Spring, Rear Coil.......... SUB-20
Starter SUB-2
Steering Wheel SUB-21

T
Thermostat SUB-6
Throttle Body............. SUB-5
Throttle Position Sensor
   (TPS) SUB-4
Timing Belt............... SUB-10
Timing Chain SUB-10
Torque Converter.......... SUB-15

U
U-Joint.................. SUB-17

V
Valve Body SUB-14
Valve Cover Gasket........ SUB-8
Valve Job................ SUB-8
Vehicle Speed Sensor...... SUB-4

W
Water Pump.............. SUB-6
Wheel Balance SUB-2
Wheel Cylinder SUB-19
Window Regulator......... SUB-25
Windshield Washer Pump... SUB-24
Windshield Wiper Motor SUB-24

SUB-1

SUB-2 BAJA : BRAT : DL : FORESTER : GL : GL-10 : GLF : IMPREZA : JUSTY

	LABOR TIME	SEVERE SERVICE
MAINTENANCE		
Air Cleaner Filter Element, Replace (C)		
All Models	.3	.4
Chassis Lubrication, Change Oil and Filter (C)		
All Models	.5	.6
Drive Belt, Adjust (B)		
All Models	.3	.4
Drive Belt, Replace (B)		
All Models	.6	.7
w/AC 81-94 add	.2	.2
w/PS 81-94 add	.2	.2
Fuel Filter, Replace (B)		
1981-92 w/carburetor	.3	.3
1983-04 w/FI	.5	.5
Halogen Headlamp Bulb, Replace (C)		
All Models	.3	.3
Oil & Filter, Change (C)		
Includes: Correct all fluid levels.		
All Models	.5	.5
Sealed Beam Headlamp, Replace (C)		
1981-95 each	.3	.3
Timing Belt, Replace (B)		
3 cyl.	1.4	2.0
4 cyl.		
1985-94		
8V	3.1	3.7
16V	2.5	3.1
1995-98		
1.8L, 2.2L SOHC	2.4	2.7
2.5L DOHC	2.9	3.2
1999-04		
2.0L	2.9	3.2
2.2L, 2.5L SOHC	2.4	2.7
6 cyl.		
2.7L	3.3	3.9
3.3L	2.5	3.1
Tire, Replace (C)		
Includes: Dismount old tire and mount new tire to rim.		
All Models	.5	.5
Tires, Rotate (C)		
All Models	.6	.7
Wheel, Balance (B)		
One	.5	.5
each addl. add	.2	.2

SCHEDULED MAINTENANCE INTERVALS
If necessary, refer to appropriate Chilton maintenance service information.

	LABOR TIME	SEVERE SERVICE
7,500 Mile Service (C)		
All Models	.4	.4
15,000 Mile Service (C)		
All Models	1.0	1.0
22,500 Mile Service (C)		
All Models	.4	.4
30,000 Mile Service (B)		
All Models	6.1	6.1
SVX add	.2	.2
37,500 Mile Service (C)		
All Models	.4	.4
45,000 Mile Service (C)		
All Models	1.0	1.0
52,500 Mile Service (C)		
All Models	.4	.4
60,000 Mile Service (B)		
All Models	6.7	6.7
67,500 Mile Service (C)		
All Models	.4	.4
75,000 Mile Service (C)		
All Models	1.0	1.0
82,500 Mile Service (C)		
All Models	.4	.4
90,000 Mile Service (B)		
All Models	6.1	6.1
SVX add	.2	.2
97,500 Mile Service (C)		
All Models	.4	.4

CHARGING
	LABOR TIME	SEVERE SERVICE
Alternator Circuits, Test (B)		
Includes: Test component output.		
All Models	.6	.6
Alternator, R&R and Recondition (A)		
All Models	2.2	2.5
Alternator Assy., Replace (B)		
Includes: Pulley transfer.		
All Models	.6	.7

STARTING
	LABOR TIME	SEVERE SERVICE
Starter Draw Test (On Car) (B)		
All Models	.3	.3
Battery Cables, Replace (C)		
Each	.5	.7
Starter, R&R and Recondition (A)		
All Models	1.9	2.1
Starter Assy., Replace (B)		
All Models	1.1	1.3
Starter Interlock Switch, Replace (B)		
1988-04	.5	.5
Starter Interrupt Relay, Replace (B)		
Forester, Impreza	.6	.6
Justy, XT	.5	.5
Legacy, Loyale	.5	.5
Voltmeter, Replace (B)		
Baja, Legacy, Outback		
1990-94	1.8	1.8
1995-04	1.5	1.5
Brat	1.3	1.3
Forester	1.5	1.5
Hatchback		
1981-84	1.2	1.2
1985-87	1.6	1.6
Impreza	1.5	1.5
Justy	1.5	1.5
Loyale	1.6	1.6
Station Wagon		
1981-84	1.2	1.2
1985-87	1.7	1.7
XT	1.9	1.9
3 door, 4 door		
1981-84	1.2	1.2
1985-87	1.6	1.6

CRUISE CONTROL
	LABOR TIME	SEVERE SERVICE
Diagnose Cruise Control System (A)		
1985-04	.6	.6
Brake or Clutch Release Switch, Replace (B)		
All Models	.3	.3
Control Actuator Cable, Replace (B)		
1981-89	.5	.5
1990-04		
Forester	.3	.3
Loyale	.5	.5
Legacy, Impreza	.3	.3
Control Main Switch, Replace (B)		
1981-86		
exc. XT	.3	.3
XT	1.0	1.0
1987-04		
Forester	.3	.3
Legacy, Impreza	.3	.3
Loyale, XT	1.0	1.0
SVX	.5	.5
Control On/Off Switch, Replace (B)		
Exc. XT	.6	.6
XT	.8	.8
Control Regulator, Replace (B)		
1981-87	.3	.3
Control Relay, Replace (B)		
1990-94 Legacy	.5	.5
SVX	1.4	1.4
Control Sensor, Replace (B)		
Baja, Legacy, Outback		
1990-94	1.8	1.8
1995-04	1.2	1.2
Forester, Impreza	1.2	1.2
Loyale	1.4	1.4
SVX	.5	.5
XT	1.6	1.6
1981-89 3 door, 4 door	.9	.9
1981-89 Station Wagon	.9	.9
Control Servo, Replace (B)		
1981-87	.5	.5
Control Steering Wheel Switch, Replace (B)		
1995-04 Forester, Impreza, Legacy	.6	.6
Control Sub Switch, Replace (B)		
1985-94	.3	.3
w/ABS add	.2	.2
Deceleration Switch, Replace (B)		
1981-87	.5	.5

LEGACY : LOYALE : OUTBACK : RX : STANDARD : SVX : XT : XT6 — SUB-3

	LABOR TIME	SEVERE SERVICE
Disengagement Switch, Replace (B)		
1981-87	.5	.5
Electronic Speed Control Module, Replace (B)		
All Models	.7	.7
Engagement Switch, Replace (B)		
1981-87	.5	.5
Servo & Bracket Assy., Replace (B)		
1985-94	.7	.7
Valve Body Assy., Replace (B)		
1981-89	.5	.5
1990-04		
Forester, Impreza	.6	.6
Legacy, Loyale	.5	.5
SVX	.6	.6
w/turbo 81-89 add	.1	.1

IGNITION

	LABOR TIME	SEVERE SERVICE
Diagnose Ignition System Component Each (A)		
All Models	1.1	1.2
Camshaft Position Sensor, Replace (B)		
Baja, Legacy, Outback,		
w/OBD II	1.1	1.2
w/o OBD II	.6	.8
H6 3.0	1.4	1.5
Forester	1.1	1.2
Impreza		
w/OBD II	1.1	1.2
w/o OBD II	.6	.8
SVX, WRX	1.1	1.2
Crankshaft Position Sensor, Replace (B)		
Baja, Legacy, Outback		
w/OBD II	1.1	1.2
w/o OBD II	.6	.8
H6 3.0	1.2	1.4
Forester	1.1	1.2
Impreza, WRX		
w/OBD II	1.1	1.2
w/o OBD II	.6	.8
Loyale	.8	1.0
SVX		
one or both	1.1	1.2
Distributor, R&R and Recondition (B)		
Includes: Reset base ignition timing.		
1981-94	1.5	1.8
Distributor, Replace (B)		
Includes: Reset base ignition timing.		
1981-94	.5	.7
Distributor Cap and/or Rotor, Replace (B)		
3 cyl.	.5	.5
1981-94 4 cyl.	.5	.5
1988-91 6 cyl.	.6	.6
Electronic Control Module, Replace (B)		
Justy	.8	.8

	LABOR TIME	SEVERE SERVICE
Igniter Assy. (External), Replace (B)		
1987-94	.8	.8
Ignition Coil, Replace (B)		
Exc. below	.5	.5
SVX	.7	.8
1997-00	1.2	1.4
2001-04	.8	.9
Ignition Igniter, Replace (B)		
Baja, Forester, Impreza,		
Legacy, Outback	.7	.7
SVX	.8	.9
H6 3.0L, WRX	.8	.9
each addl. add	.3	.3
Ignition Lock (Column), Replace (B)		
Baja, Forester, Impreza,		
Legacy, Outback	.6	.6
Brat	1.5	1.5
Hatchback	1.7	1.7
Justy, Loyale	1.4	1.4
SVX	1.0	1.0
XT	.7	.7
3 door, 4 door	1.7	1.7
Ignition Switch, Replace (B)		
Baja, Forester, Impreza,		
Legacy, Outback	.7	.7
Brat	1.5	1.5
Loyale, XT	.7	.7
SVX	1.0	1.0
Hatchback	1.6	1.6
Station Wagon	1.7	1.7
3 door, 4 door	1.6	1.6
Ignition Timing, Reset (B)		
1981-92	.3	.3
Pickup Coil, Replace (B)		
1981-92	.7	.7
Spark Plug (Ignition) Cables, Replace (B)		
3 cyl.	.6	.8
4 cyl.	.6	.8
6 cyl.	.7	.9
Spark Plugs, Replace (B)		
3 cyl.	.6	.8
4 cyl.		
1981-95	.5	.7
1996-04		
1.8L, 2.2L	.6	.8
2.0L	2.9	3.2
2.5L	1.2	1.4
6 cyl.		
2.7L	.7	.9
3.0L	3.3	3.7
3.3L	2.8	3.0
Vacuum Advance Unit, Replace (B)		
Includes: Reset timing.		
1981-92	.9	.9

EMISSIONS

The following operations do not include testing. Add time as required.

	LABOR TIME	SEVERE SERVICE
Diagnose Emission Control System (A)		
1987-04	1.0	1.0
Dynamometer Test (A)		
All Models	.4	.4
Air Cleaner Temperature Sensor, Replace (B)		
All Models	.3	.3
Air Cleaner Vacuum Motor, Replace (B)		
All Models	.3	.3
Air Suction Manifold, Replace (B)		
1981-94	.5	.5
Air Suction Valve, Replace. (B)		
Loyale		
front		
one	.7	.7
both	1.4	1.4
rear	.7	.7
Justy	.7	.7
1993-94 Impreza	.7	.7
Anti-Stall Dashpot, Replace (B)		
1983-94	.5	.5
Coolant Temperature (ECT) Sensor, Replace (B)		
Baja, Legacy, Outback		
w/OBD II	1.2	1.4
w/o OBD II	.6	.6
Forester	1.2	1.4
Impreza		
w/OBD II	1.1	1.1
w/o OBD II	.6	.6
Justy	.7	.7
Loyale	.6	.6
SVX	.8	.8
WRX	1.2	1.4
XT		
4 cyl.	1.0	1.0
6 cyl.	1.2	1.2
Deceleration Valve, Replace (B)		
1981-94	.7	.7
EGR Control Valve, Replace (B)		
Forester	1.2	1.4
Loyale	.5	.7
SVX	1.3	1.5
XT		
4 cyl.	.5	.7
6 cyl.	.6	.8
1981-94	.5	.7
1995-04 Baja, Legacy,		
Outback	1.2	1.4
H6 3.0	2.9	3.0
1995-04 Impreza		
w/OBD II	1.2	1.4
w/o OBD I or II	.5	.7

SUB-4 BAJA : BRAT : DL : FORESTER : GL : GL-10 : GLF : IMPREZA : JUSTY

	LABOR TIME	SEVERE SERVICE
EGR Modulator Valve, Replace (B)		
Baja, Legacy, Outback	1.1	1.2
Forester, Impreza	1.1	1.2
SVX	.5	.5
EGR Solenoid Valve, Replace (B)		
Forester, Impreza	1.2	1.4
Justy	.7	.7
Loyale	.7	.7
SVX	2.6	2.6
1995-04 Baja, Legacy, Outback	1.2	1.4
EGR Temperature Sensor, Replace (B)		
Impreza	.5	.5
Justy	.9	.9
Loyale	.7	.7
SVX	.5	.5
Electronic Control Unit, Replace (B)		
w/OBD II	.9	1.0
w/o OBD II	.6	.7
Fuel Vapor Canister, Replace (B)		
1981-94	.3	.3
1995-04		
front	.3	.3
rear	.6	.6
Idle Compensator, Replace (B)		
1981-94	.3	.3
Intake Air Temperature (IAT) Sensor, Replace (B)		
Baja, Forester, Impreza, Legacy	1.1	1.2
Justy	.7	.7
Knock Sensor, Replace (B)		
Baja, Legacy, Outback		
w/OBD II	1.1	1.2
w/o OBD II	.7	.9
H6 3.0	2.8	3.1
Forester	1.1	1.2
1995-04 Impreza		
w/OBD II	1.1	1.2
Justy	1.1	1.2
Loyale	.8	1.0
SVX		
one or both	2.6	2.8
WRX, XT	1.1	1.2
Manifold Absolute Pressure (MAP) Sensor, Replace (B)		
Baja, Forester, Impreza, Legacy	1.2	1.4
Justy	.8	.9
Loyale	1.1	1.2
XT	1.3	1.5
Mass Air Flow (MAF) Sensor, Replace (B)		
Forester	1.2	1.4
Impreza		
w/OBD I or II	1.2	1.4
w/o OBD I or II	.9	1.1

	LABOR TIME	SEVERE SERVICE
Loyale, SVX, XT	.8	.8
WRX	1.2	1.4
1995-04 Baja, Legacy, Outback		
w/OBD I or II	1.2	1.4
w/o OBD II	.8	.8
Oxygen Sensor, Replace (B)		
Baja, Legacy, Outback		
front or rear	1.5	1.7
Forester		
front or rear	1.5	1.7
H6 3.0		
one side	1.5	1.7
both sides	2.1	2.4
rear	1.5	1.7
Impreza		
w/OBD II		
front or rear	1.5	1.7
w/o OBD II	.8	.9
Loyale, Justy	.8	1.0
SVX		
one	1.0	1.2
both	1.8	2.0
WRX		
front or rear	1.5	1.7
PAIR Manifold, Replace (B)		
1993-04 Impreza	.5	.5
PAIR Silencer, Replace (B)		
1993-04 Impreza	.5	.5
PAIR Valve, Replace (B)		
1993-04 Impreza	.6	.7
Loyale, Justy	.6	.7
Positive Crankcase Ventilation (PCV) Valve, Replace (B)		
Exc. 3.3L	.3	.3
3.3L	.6	.6
Throttle Position Sensor (TPS), Replace (B)		
Baja, Legacy, Outback		
w/OBD I or II	1.2	1.4
w/o OBD I or II	1.0	1.1
Forester	1.2	1.4
Impreza		
w/OBD I or II	1.2	1.4
w/o OBD I or II	1.1	1.2
Loyale, SVX, XT	1.1	1.2
WRX	1.2	1.4
Vacuum Control Valve, Replace (B)		
1981-94	.5	.5
Vapor Separator, Replace (B)		
Justy		
exc. 5 door	.9	.9
5 door	1.6	1.6
Loyale		
3 or 4 door	.7	.7
Station Wagon	1.4	1.4
XT	.7	.7
1981-89 Station Wagon	1.4	1.4
1981-89 Hatchback	.7	.7
1981-89 3 door, 4 door	.7	.7

	LABOR TIME	SEVERE SERVICE
1985-87 Brat	.7	.7
1990-94 Legacy		
4 door	.7	.7
Station Wagon	1.2	1.2
Vehicle Speed Sensor, Replace (B)		
Baja, Legacy, Outback		
1990-94	1.8	1.9
1995-04	1.1	1.2
Forester, Impreza	1.1	1.2
Loyale	1.4	1.5
Justy	1.2	1.3
SVX	.7	.8
WRX	1.1	1.2
XT	1.6	1.7

FUEL
CARBURETOR
Automatic Choke, R&R and Recondition (B)
1981-92	1.1	1.1
Carburetor, Adjust (On Car) (B)		
1981-92	.3	.3
Carburetor, R&R and Clean or Recondition (A)		
Includes: Adjustments.		
1981-92		
exc. Hitachi	2.9	2.9
Hitachi	2.2	2.2
w/AC add	.2	.2
Carburetor, Replace (B)		
Includes: Adjustments.		
1981-92	1.5	1.5

DELIVERY
Fuel Pump, Test (B)
All Models	.3	.3
Fuel Filter, Replace (B)		
1981-92 w/carburetor	.3	.3
1983-04 w/FI	.5	.5
Fuel Gauge (Dash), Replace (B)		
Baja, Legacy, Outback		
1990-94	1.8	1.8
1995-04	1.5	1.5
Brat	1.2	1.2
Forester	1.5	1.5
Loyale	1.6	1.6
Impreza	1.5	1.5
XT	1.9	1.9
Justy	1.4	1.4
SVX	1.6	1.6
1981-89 Hatchback	1.2	1.2
1981-89 Station Wagon	1.3	1.3
1981-89 3 door, 4 door	1.3	1.3
Fuel Gauge Sending Unit (Tank), Replace (B)		
Baja, Legacy, Outback		
one	1.0	1.1
both	1.6	1.8

LEGACY : LOYALE : OUTBACK : RX : STANDARD : SVX : XT : XT6

	LABOR TIME	SEVERE SERVICE
Brat	.7	1.0
Forester, Impreza		
one	1.0	1.1
both	1.6	1.8
Justy	.8	1.1
Loyale	.6	.9
SVX	1.4	1.7
WRX		
one	1.0	1.1
both	1.6	1.8
1985-94 XT	.6	.9
1981-89 Hatchback, Station Wagon, 3 door, 4 door	.7	1.0

Fuel Pump, Replace (B)

1981-82	.5	.5
1983-04	.8	.9
w/MFI add	.2	.2

Fuel Pump Relay, Replace (B)

Forester	.6	.7
SVX	.5	.5
1983-94	.5	.5
1995-04 Impreza	.5	.5
1995-04 Baja, Legacy, Outback	.6	.7

Fuel Tank, Replace (B)
Includes: Drain and refill.

Baja, Legacy, Outback		
w/AWD	3.8	4.3
w/o AWD	2.1	2.3
Brat	2.0	2.3
Forester	3.8	4.3
Impreza		
w/AWD	3.8	4.3
w/o AWD	2.1	2.3
Justy		
w/AWD	2.1	2.4
w/o AWD	1.7	2.0
Loyale		
w/AWD	1.8	2.1
w/o AWD	.8	1.1
SVX		
w/AWD	5.5	5.8
w/o AWD	3.4	3.7
XT		
w/AWD	1.8	2.1
w/o AWD	.8	1.1
1981-89 Hatchback		
w/AWD	2.1	2.4
w/o AWD	1.2	1.5
1981-89 Station Wagon		
w/AWD	2.1	2.4
w/o AWD	1.1	1.4
1981-89 3 door, 4 door		
w/AWD	2.1	2.4
w/o AWD	1.1	1.4

Intake Manifold and/or Gasket, Replace (B)

Baja, Legacy, Outback	1.8	2.1
Brat	2.3	2.8

	LABOR TIME	SEVERE SERVICE
Forester	1.8	2.1
Impreza		
1.8L	2.1	2.4
2.2L, 2.5L	1.8	2.1
Justy	1.9	2.4
H6 3.0	2.0	2.2
Loyale	2.1	2.6
SVX	2.1	2.6
WRX	3.2	3.6
XT		
1985-87	2.1	2.6
1988-91		
4 cyl.	1.3	1.8
6 cyl.	1.6	2.1
1981-89 Hatchback	2.4	2.9
1981-89 Station Wagon	2.3	2.8
1981-89 3 door, 4 door	2.4	2.9
w/MFI Brat, Forester, Impreza, Justy add	.2	.2
w/turbo Legacy, Outback add	.5	.5
Replace manifold add	.6	.7

INJECTION

Fuel Injectors, Replace (B)

3 cyl.		
one	1.1	1.3
each addl. add	.3	.3
4 cyl. exc. below		
w/OBD I or II		
one	1.2	1.4
each addl. add	.3	.3
w/o OBD I or II		
one side	.6	.7
both sides	.9	1.0
1995 Impreza 1.8L		
w/o OBD I or II		
one	.9	1.0
each addl. add	.3	.3
WRX 2.0L		
one side	1.2	1.4
both sides	1.7	1.9
6 cyl.		
one	1.2	1.4
each addl. add	.3	.3
H6 3.0		
w/OBD I or II		
one	1.2	1.4
each addl. add	.3	.3
w/o OBD I or II		
one side	1.1	1.2
both sides	1.2	1.4

Fuel Pressure Regulator, Replace (B)

Baja, Forester, Impreza, Legacy, Outback	.6	.7
H6 3.0	.9	1.0
Loyale	.6	.7
Justy	1.1	1.3
SVX	.7	.9

	LABOR TIME	SEVERE SERVICE
XT		
4 cyl.	1.0	1.2
6 cyl.	1.5	1.7
1983-89 Hatchback, Station Wagon, 3 door, 4 door	.7	.9

Fuel Rail Replace (B)

SVX		
one side	.6	.8
both sides	1.0	1.2

Idle Air Control (IAC) Valve, Replace (B)

Baja, Legacy, Outback		
1990-94		
w/turbo	1.0	1.0
w/o turbo	.8	.8
1995-04	1.2	1.4
Forester	1.2	1.4
Impreza		
1993-95	.8	.9
1996-04	1.2	1.4
Loyale	1.1	1.1
SVX		
auxiliary	2.6	2.6
main	1.4	1.4
WRX	.9	1.0

Throttle Body Assy., R&R or Replace (B)

1983-04 MPI	.6	.6
1985-94 SPI	.6	.6
Replace throttle body add	.3	.3

TURBOCHARGER

Emergency Relief Valve, Replace (B)

1983-94	.3	.3

Turbocharger Assy., R&R or Replace (B)

Forester, Legacy, WRX	2.2	2.5
1983-94	2.9	3.4

Wastegate Actuator, Replace (B)

Forester, Legacy, WRX	1.1	1.2
1983-94	2.7	3.0

EXHAUST

Center Catalytic Converter, Replace (B)

1983-94 turbo	1.0	1.1
SVX	.6	.7

Center Exhaust Pipe, Replace (B)

1981-94	.5	.6
w/turbo add		
exc. WRX	.5	.6
WRX	1.3	1.4

Crossover Pipe, Replace (B)

WRX	1.0	1.1
1983-94 turbo	1.2	1.3

SUB-5

SUB-6 BAJA : BRAT : DL : FORESTER : GL : GL-10 : GLF : IMPREZA : JUSTY

	LABOR TIME	SEVERE SERVICE
Exhaust Manifold or Gasket, Replace (B)		
Justy	1.0	1.5
Loyale turbo	1.4	1.9
SVX		
right side	1.0	1.5
left side	.7	1.2
WRX		
one side	1.1	1.3
both sides	1.3	1.4
1983-89 Station Wagon, 3 door, 4 door	1.6	2.1
1990-94 Legacy turbo	.6	1.1
1990-91 XT turbo	1.8	2.3
Exhaust Pipe or Crossover Pipe Flange Gasket or Seal, Replace (B)		
1981-94 one	.7	.9
WRX	1.0	1.1
Exhaust System (Complete), Replace (B)		
1981-94	1.3	1.8
w/turbo add	.5	.5
Exhaust Y Pipe, Replace (B)		
1981-86	.8	.9
1987-94 Legacy	1.0	1.1
1987-94 Loyale	.8	.9
1987-94 Impreza	.6	.7
XT		
4 cyl.	.8	.9
6 cyl.	1.0	1.1
1995-04	.6	.7
Front Catalytic Converter, Replace (B)		
All Models	1.1	1.3
H6 3.0		
right side	.8	.9
left side	1.0	1.1
Muffler, Replace (B)		
All Models	.3	.4
Rear Catalytic Converter, Replace (B)		
1981-94	.5	.6
1995-04	.7	.8
Rear Exhaust Pipe, Replace (B)		
All Models	.5	.6
w/turbo add		
exc. WRX	.1	.1
WRX	.8	.9
Resonator or Pre-Muffler, Replace (B)		
1981-85	.7	.8
Tail Pipe, Replace (B)		
All Models	.3	.4

ENGINE COOLING

	LABOR TIME	SEVERE SERVICE
Cooling System Pressure Test (B)		
All Models	.5	.5
By-Pass Hoses, Replace (B)		
Exc. below	.5	.7
1981-86 all models	.7	.9
Loyale	.7	.9
w/AC add	.3	.5
Coolant Fan Thermo Sensor, Replace (B)		
Justy	.5	.5
Coolant Thermostat, Replace (B)		
All Models	.6	.8
w/MFI Justy add	.5	.5
Electric Cooling Fan Assy., Replace (B)		
1985-04		
one	.5	.6
both	.6	.7
Replace motor, fan blade, or shroud add each	.1	.1
Mechanical Cooling Fan Blade, Replace (B)		
1981-94	.5	.7
Radiator Assy., R&R or Replace (B)		
All Models	.7	.9
w/AT add	.2	.2
Radiator Fan Motor Switch (Coolant Temp.), Replace (B)		
All Models	1.2	1.4
Radiator Hoses, Replace (B)		
Includes: Refill with proper coolant mix.		
Upper	.3	.5
Lower	.5	.7
Temperature Gauge (Dash), Replace (B)		
Baja, Legacy, Outback		
1990-94	1.8	1.8
1995-04	1.5	1.5
Brat	1.2	1.2
Forester, Impreza	1.5	1.5
Loyale	1.6	1.6
XT	1.9	1.9
Justy	1.4	1.4
SVX	1.6	1.6
1981-89 Hatchback	1.2	1.2
1981-89 Station Wagon, 3 door, 4 door	1.3	1.3
Temperature Gauge (Engine), Replace (B)		
1981-86	.5	.5
1987-04	1.1	1.1
w/6 cyl. add	.1	.1
Water Pump and/or Gasket, Replace (B)		
Includes: Refill with proper coolant mix.		
1981-99		
4 cyl.		
OHV		
w/o AC	2.9	3.4
w/AC	1.3	1.8
1.8L, 2.2L SOHC	2.8	3.3
2.5L DOHC	3.3	3.7

	LABOR TIME	SEVERE SERVICE
1999-04		
2.0L	3.5	3.9
2.2L, 2.5L SOHC	3.3	3.7
6 cyl.		
2.7L	2.7	3.0
3.0L	3.3	3.7
3.3L	3.6	3.9
w/PS 4 cyl. OHV add	.1	.1
w/turbo 4 cyl. OHV add	.5	.5
Water Pump Cover or Gasket, Replace (B)		
Justy	1.0	1.3
w/AC add	.1	.1
Replace		
impeller or seal add	.1	.3
water pump seal add	4.8	5.0

ENGINE

ASSEMBLY

Times shown are for OEM assemblies. Time to replace assemblies from aftermarket rebuilders may vary.

Engine Assy., R&I (B)
Does not include parts or component transfer.

	LABOR TIME	SEVERE SERVICE
3 cyl.		
w/AWD	6.5	6.5
w/o AWD	5.8	5.8
4 cyl.		
OHC		
AT	6.3	6.3
MT	5.7	5.7
OHV	3.5	3.5
6 cyl.		
2.7L	5.7	5.7
3.0L	4.7	4.7
3.3L	6.3	6.3
w/AC 4 cyl. OHV add	.5	.5
w/AT add		
2.7L	.2	.2
4 cyl. OHV	.2	.2
w/turbo add		
2.7L	.5	.5
4 cyl. OHC	.5	.5

Engine Assy. (Short Block), Replace (B)
Assembly consists of cylinder block, piston assemblies, crankshaft, camshaft, timing chain and gears. Does not include cylinder heads. Operation Includes: R&R engine, transfer necessary parts and all necessary adjustments.

	LABOR TIME	SEVERE SERVICE
3 cyl.		
w/AWD	10.1	10.1
w/o AWD	9.5	9.5

LEGACY : LOYALE : OUTBACK : RX : STANDARD : SVX : XT : XT6 **SUB-7**

	LABOR TIME	SEVERE SERVICE
4 cyl.		
1981-94		
OHC		
8V	13.6	*13.6*
16V		
1.8L, 2.2L	13.2	*13.2*
OHV	7.9	*7.9*
1995-98		
1.8L, 2.2L SOHC		
AT	15.5	*17.3*
MT	14.8	*16.5*
2.5L DOHC		
AT	16.6	*18.5*
MT	15.9	*17.7*
1999-04		
2.0L		
AT	15.1	*16.8*
MT	14.3	*16.1*
2.2L, 2.5L SOHC		
AT	16.6	*18.5*
MT	15.9	*17.7*
6 cyl.		
2.7L	14.3	*16.1*
3.0L	12.6	*14.1*
3.3L	13.8	*15.4*
w/AC 4 cyl. OHV	.5	*.5*
w/adjustable valves add		
1.8L, 2.2L	.6	*.7*
2.5L	1.9	*2.1*
w/AT add		
2.7L	.2	*.2*
4 cyl. OHV	.2	*.2*
w/turbo add		
2.7L	.5	*.5*
4 cyl. OHC	.5	*.5*

Engine Assy., R&R and Recondition (A)
Includes: Replacing rings, rod and main bearings, cylinder head reconditioning and engine tune-up.

	LABOR TIME	SEVERE SERVICE
3 cyl.		
w/AWD	14.7	*14.7*
w/o AWD	14.0	*14.0*
4 cyl.		
1981-94		
OHV	11.1	*11.1*
OHC		
8V	19.2	*19.2*
16V		
1.8L, 2.2L	19.2	*19.2*
2.5L	22.7	*22.7*
1995-98		
1.8L, 2.2L SOHC		
w/AT	20.9	*23.3*
w/MT	20.3	*22.5*
2.5L DOHC		
w/AT	21.9	*24.5*
w/MT	21.3	*23.7*

	LABOR TIME	SEVERE SERVICE
1999-04		
2.0L		
w/AT	18.4	*20.5*
w/MT	17.7	*19.8*
2.2L		
w/AT	19.3	*21.5*
w/MT	18.6	*20.8*
2.5L SOHC		
w/AT	21.9	*24.5*
w/MT	21.3	*23.7*
6 cyl.		
2.7L	21.6	*21.6*
3.0L	20.7	*23.1*
3.3L	22.3	*22.3*
w/AC 4 cyl. OHV	.5	*.5*
w/adjustable valves add		
1.8L, 2.2L	.6	*.7*
2.5L	1.9	*2.1*
w/AT add		
2.7L	.2	*.2*
4 cyl. OHV	.2	*.2*
w/turbo add		
2.7L	.5	*.5*
4 cyl. OHC	.5	*.5*

Engine Mounts, Replace (B)

	LABOR TIME	SEVERE SERVICE
3 cyl.		
right or left	.6	*.8*
center	.8	*1.0*
4 cyl.		
OHV		
front		
one side	.5	*.7*
both sides	.7	*.9*
rear		
one or both sides	.5	*.7*
OHC		
front		
w/turbo		
one side	.7	*.9*
both sides	1.2	*1.4*
w/o turbo		
one side	1.1	*1.3*
both sides	1.5	*1.7*
rear	.7	*.9*
6 cyl.		
2.7L		
front		
w/turbo		
one side	.7	*.9*
both sides	1.2	*1.4*
w/o turbo		
one side	1.0	*1.2*
both sides	1.5	*1.7*
rear		
AT	.7	*.9*
MT	.8	*1.0*

	LABOR TIME	SEVERE SERVICE
3.0L, 3.3L		
front		
one side	1.2	*1.4*
both sides	2.2	*2.4*
rear	1.0	*1.2*
w/AWD add		
3 cyl.	.2	*.2*
4 cyl.	.2	*.2*
2.7L	.1	*.1*
3.3L	.2	*.2*

CYLINDER HEAD

Compression Test (B)

	LABOR TIME	SEVERE SERVICE
3 cyl.	1.0	*1.3*
4 cyl.		
1981-94	.7	*1.0*
1995-04		
1.8L, 2.2L	1.4	*1.5*
2.0Lc, 2.5L	1.8	*2.0*
6 cyl.		
2.7L	1.2	*1.5*
3.0L	3.2	*3.6*
3.3L	2.1	*2.4*

Retorque Cylinder Heads (B)

	LABOR TIME	SEVERE SERVICE
3 cyl.	.8	*1.1*
4 cyl.	1.2	*1.5*
6 cyl.	1.7	*2.0*

Valve Clearance, Adjust (B)

	LABOR TIME	SEVERE SERVICE
3 cyl.	1.5	*1.8*
4 cyl.		
1981-94		
1.8L, 2.2L	1.5	*1.8*
1995-98		
1.8L, 2.2L SOHC		
one side	.9	*1.0*
both sides	1.7	*1.9*
2.5L DOHC		
one side	5.4	*6.0*
both sides	7.7	*8.6*
1999-04		
2.5L SOHC		
one side	.9	*1.0*
both sides	1.1	*1.2*

Cylinder Head and/or Gasket, Replace (B)

	LABOR TIME	SEVERE SERVICE
3 cyl.	3.9	*4.4*
4 cyl.		
1985-94		
OHV		
MT		
one side	5.4	*5.9*
both sides	5.8	*6.3*
AT		
one side	5.4	*5.9*
both sides	6.1	*6.6*
OHC		
8V		
right side	6.6	*7.1*
left side	7.3	*7.8*
both sides	9.4	*9.9*

SUB-8   BAJA : BRAT : DL : FORESTER : GL : GL-10 : GLF : IMPREZA : JUSTY

	LABOR TIME	SEVERE SERVICE
Cylinder Head and/or Gasket, Replace (B)		
16V		
1.8L, 2.2L		
one side	4.8	5.3
both sides	7.3	7.8
1995-98		
1.8L, 2.2L SOHC		
w/adjustable valves		
one side	5.6	6.2
both sides	8.6	9.6
w/o adjustable valves		
one side	5.0	5.5
both sides	7.6	8.5
2.5L DOHC		
w/adjustable valves		
one side	8.0	9.0
both sides	12.4	13.8
w/o adjustable valves		
one side	7.2	8.0
both sides	10.7	11.9
AWD 2.2L, 2.5L		
one side	8.0	9.0
both sides	12.4	13.8
1999-04		
2.0L		
right side	11.0	12.4
left side	10.6	11.9
both sides	14.6	16.3
2.2L		
w/adjustable valves		
one side	5.6	6.2
both sides	8.6	9.6
w/o adjustable valves		
one side	5.0	5.5
both sides	7.6	8.5
2.5L SOHC		
one side	5.6	6.2
both sides	8.6	9.6
6 cyl.		
2.7L		
right side	6.7	7.2
left side	7.6	8.1
both sides	9.8	10.3
3.0L		
one side	9.9	11.0
both sides	11.5	12.8
3.3L		
one side	8.6	9.1
both sides	13.4	13.9
w/AC 4 cyl OHV add	.5	.5
w/PS 4 cyl OHV add	.3	.3
w/turbo add		
2.2L		
right side	.5	.5
left or both sides	2.1	2.1
2.7L		
right side	.5	.5
left side	.3	.3
both sides	.5	.5

	LABOR TIME	SEVERE SERVICE
4 cyl. OHC 8V		
right side	.5	.5
left side	.3	.3
both sides	.5	.5
4 cyl OHV	.5	.5
Rocker Arm Assy. and/or Valve Lash Adjusters, Replace (B)		
1985-94 4 cyl. OHC 8V		
one side	.9	1.1
both sides	1.7	2.0
1995-98 4 cyl.		
1.8L, 2.2L SOHC		
w/adjustable valves		
one side	1.5	1.7
both sides	2.8	3.1
w/o adjustable valves		
one side	.9	1.0
both sides	1.7	1.9
2.5L DOHC		
w/adjustable valves		
one side	5.9	6.6
both sides	8.6	9.6
w/o adjustable valves		
one side	5.0	5.5
both sides	6.9	7.7
1999-04		
2.0L		
right side	5.1	5.7
left side	5.4	6.0
both sides	7.6	8.5
2.2L		
w/adjustable valves		
one side	1.5	1.7
both sides	2.8	3.1
w/o adjustable valves		
one side	.9	1.0
both sides	1.7	1.9
2.5L SOHC		
Intake or Exhaust		
one side	1.1	1.2
both sides	1.8	2.1
1988-91 6 cyl. 2.7L		
right side	4.2	4.4
left side	4.6	4.8
both sides	6.3	6.7
Recondition rocker arms add each side	.5	.7
Rocker Arm Cover Gasket, Replace or Reseal (B)		
3 cyl.	.5	.7
4 cyl.		
1981-94		
OHV		
one side	.5	.7
both sides	.7	.9
OHC		
8V		
right side	.5	.7
left side	.6	.8
both sides	.8	.9

	LABOR TIME	SEVERE SERVICE
16V		
1.8L, 2.2L		
one side	.5	.7
both sides	.8	.9
1995-98		
1.8L, 2.2L SOHC		
one side	.5	.7
both sides	.8	.9
2.5L DOHC		
one side	.8	.9
both sides	1.7	1.9
1999-04		
2.0L DOHC		
one side	1.4	1.5
both sides	2.3	2.6
2.2L, 2.5L SOHC		
one side	.5	.7
both sides	.8	.9
6 cyl.		
2.7L		
right side	.6	.8
left side	.7	.9
both sides	1.1	1.3
3.0L		
one side	2.8	3.1
both sides	4.0	4.4
3.3L		
one side	1.1	1.3
both sides	2.1	2.3
Rocker Arms or Shafts, Replace (B)		
1981-89 4 cyl. OHV		
one side	.9	1.1
both sides	2.0	2.3
w/AC add	.1	.1
w/PS add	.1	.1
w/turbo add	.2	.3
Rocker Arms, Spring and/or Shaft, Replace (B)		
Justy	1.1	1.5
Valve Lash Adjuster, Replace (B)		
1995-98		
1.8L, 2.2L SOHC		
w/adjustable valves		
one side	1.5	1.7
both sides	2.8	3.1
w/o adjustable valves		
one side	.9	1.0
both sides	1.7	1.9
1992-97 6 cyl. 3.3L		
all one side	4.9	5.1
all both sides	7.4	7.7
Valve Job (A)		
3 cyl.	8.9	9.4
4 cyl.		
1985-94		
OHV		
AT		
one side	6.4	6.9
both sides	7.1	7.6

LEGACY : LOYALE : OUTBACK : RX : STANDARD : SVX : XT : XT6 — SUB-9

	LABOR TIME	SEVERE SERVICE
MT		
one side	6.1	6.6
both sides	7.0	7.5
OHC		
8V		
right side	7.7	8.2
left side	8.4	8.9
both sides	11.6	12.1
16V		
1.8L, 2.2L		
one side	5.9	6.4
both sides	9.7	10.2
1995-98		
1.8L, 2.2L SOHC		
w/adjustable valves		
one side	7.9	9.8
both sides	13.1	14.5
w/o adjustable valves		
one side	7.4	8.1
both sides	12.1	13.4
2.5L DOHC		
one side	9.5	10.6
both sides	15.1	16.7
1999-04		
2.0L		
right side	13.2	14.7
left side	12.4	13.8
both sides	16.2	17.9
2.2L SOHC		
w/adjustable valves		
one side	7.9	9.8
both sides	13.1	14.5
w/o adjustable valves		
one side	7.4	8.1
both sides	12.1	13.4
2.5L SOHC		
one side	8.3	9.3
both sides	11.3	12.6
6 cyl.		
2.7L		
right side	8.4	8.9
left side	9.2	9.7
both sides	13.1	13.6
3.0L		
one side	8.9	9.9
both sides	10.1	11.2
3.3L		
one side	11.9	12.4
both sides	19.8	20.3
w/AC 4 cyl OHV add	.5	.5
w/adjustable valves add		
1.8L, 2.2L	.5	.5
2.5L	1.4	1.4
w/PS 4 cyl OHV add	.3	.3
w/turbo add		
2.2L		
right side	.5	.5
left or both sides	2.1	2.1

	LABOR TIME	SEVERE SERVICE
2.7L		
right side	.5	.5
left side	.3	.3
both sides	.5	.5
4 cyl. OHC 8V		
right side	.5	.5
left side	.3	.3
both sides	.5	.5
4 cyl OHV	.5	.5

CAMSHAFT

Balance Shaft Chain and/or Guides, Replace (B)

	LABOR TIME	SEVERE SERVICE
Justy		
w/AWD	8.2	8.5
w/o AWD	7.6	7.9
Replace shaft and/or bearings add	.2	.2

Camshaft, Replace (A)

	LABOR TIME	SEVERE SERVICE
3 cyl.	2.5	2.8
4 cyl.		
1985-94		
OHV	10.0	10.3
OHC		
8V		
right side	4.0	4.3
left side	4.3	4.6
both sides	5.9	6.2
16V		
right side	2.7	3.0
left side	3.2	3.5
1995-98		
1.8L, 2.2L SOHC		
w/adjustable valves		
right side	3.6	4.1
left side	4.1	4.6
w/o adjustable valves		
right side	3.1	3.4
left side	3.5	3.9
2.5L DOHC		
w/adjustable valves		
one side	5.9	6.6
both sides	4.1	4.6
w/o adjustable valves		
one side	5.0	5.5
both sides	6.9	7.7
1999-04		
2.0L		
Intake and/or Exhaust		
right side	5.1	5.7
left side	5.4	6.0
both sides	7.6	8.5
2.2L SOHC		
w/adjustable valves		
right side	3.6	4.1
left side	4.1	4.6
w/o adjustable valves		
right side	3.1	3.4
left side	3.5	3.9

	LABOR TIME	SEVERE SERVICE
2.5L SOHC		
one side	4.3	4.7
both sides	5.7	6.4
6 cyl.		
2.7L		
right side	4.2	4.5
left side	4.6	4.9
both sides	6.3	6.6
3.0L		
one side both	8.3	9.3
both sides all	8.8	9.7
3.3L		
one side both	5.0	5.3
both sides all	7.6	7.9
w/AC 4 cyl. OHV add	.5	.5
w/adjustable valves add		
1.8L, 2.2L	.5	.5
2.5L	1.4	1.4
w/cruise control 4 cyl. OHV add	.2	.2
w/PS 4 cyl. OHV add	.3	.3
w/turbo add		
4 cyl. OHC	1.9	1.9
4 cyl. OHV	1.2	1.2

Camshaft Carrier (Support), Replace (B)

	LABOR TIME	SEVERE SERVICE
1985-94 4 cyl. 8V		
right side	4.2	4.5
left side	4.6	4.9
both sides	6.2	6.5
1988-91 6 cyl. 2.7L		
right side	3.9	4.2
left side	4.3	4.6
both sides	5.8	6.1

Camshaft Seal, Replace (B)

	LABOR TIME	SEVERE SERVICE
3 cyl.	2.1	2.3
4 cyl.		
1985-94		
8V		
one side	2.5	2.7
both sides	2.7	2.9
16V		
right side	2.5	2.7
left side	2.6	2.8
both sides	2.8	3.0
1995-98		
1.8L, 2.2L SOHC		
right side	2.5	2.7
left side	2.6	2.8
both sides	2.8	3.0
2.5L DOHC		
one side	3.6	4.1
both sides	4.3	4.7
1999-04		
2.0L		
one side	3.3	3.7
both sides	3.5	3.9
2.2L, 2.5L SOHC		
one side	2.9	3.2
both sides	3.2	3.6

SUB-10 BAJA : BRAT : DL : FORESTER : GL : GL-10 : GLF : IMPREZA : JUSTY

	LABOR TIME	SEVERE SERVICE
Camshaft Seal, Replace (B)		
6 cyl.		
2.7L		
one side	2.5	2.7
both sides	2.7	2.9
3.0L		
one side	2.8	3.1
both sides	4.0	4.4
3.3L		
one side	2.8	3.0
both sides	3.2	3.4
Timing Belt, Adjust (B)		
3 cyl.	.6	.8
1985-04 4 cyl.	1.2	1.4
1988-91 6 cyl.		
right side	1.9	2.1
left side	1.2	1.4
Timing Belt, Replace (B)		
3 cyl.	1.4	2.0
4 cyl.		
1985-94		
8V	3.1	3.7
16V	2.5	3.1
1995-98		
1.8L, 2.2L SOHC	2.4	2.7
2.5L DOHC	2.9	3.2
1999-04		
2.0L	2.9	3.2
2.2L, 2.5L SOHC	2.4	2.7
6 cyl.		
2.7L	3.3	3.9
3.3L	2.5	3.1
Timing Belt Case and/or Cover, Replace (B)		
3 cyl.		
inner	1.4	1.7
outer	.8	1.0
4 cyl.		
1985-94		
8V		
outer	2.2	2.4
inner both sides	2.5	2.8
16V		
outer	1.5	1.7
inner		
right side	2.5	2.7
left side	2.6	2.8
both sides	3.0	3.3
1995-98		
1.8L, 2.2L SOHC	1.7	2.0
1996-99		
2.5L DOHC	1.7	2.0
1999-04		
2.0L	1.9	2.1
2.2L, 2.5L SOHC	1.7	2.0
6 cyl.		
2.7L		
outer	2.2	2.5
inner	2.5	2.7

	LABOR TIME	SEVERE SERVICE
3.3L		
outer	1.8	2.0
Timing Belt Tensioner, Replace (B)		
3 cyl.	1.0	1.4
4 cyl.		
1985-94		
8V, each side	2.7	3.1
16V	2.2	2.6
1995-98		
1.8L, 2.2L SOHC	2.5	2.9
2.5L DOHC	3.0	3.4
1999-04		
2.0L	3.0	3.4
2.2L, 2.5L SOHC	2.5	2.9
6 cyl.	3.0	3.4
Timing Chain, Replace (B)		
H6 3.0L	2.8	3.1
Timing Chain Tensioner, Replace (B)		
H6 3.0L	3.2	3.6
CRANK & PISTONS		
Connecting Rod Bearings, Replace (A)		
Includes: Check bearing oil clearance.		
3 cyl.	3.6	4.1
4 cyl.		
1981-94		
OHV	11.1	11.6
OHC		
8V	19.2	19.7
16V		
1.8L, 2.2L	19.2	19.7
2.5L	22.7	23.2
1995-98		
1.8L, 2.2L SOHC		
w/AT	19.8	22.1
w/MT	19.2	21.3
2.5L DOHC		
w/AT	21.9	24.5
w/MT	21.3	23.7
1999-04		
2.0L		
w/AT	16.7	18.7
w/MT	16.1	17.9
2.2L, 2.5L SOHC		
w/AT	20.4	22.7
w/MT	19.7	22.0
6 cyl.		
2.7L	21.6	22.1
3.0L	18.0	20.0
3.3L	22.3	22.8
w/AT add		
2.7L	.2	.2
4 cyl. OHC	.5	.5
w/turbo add		
2.7L	.5	.5
4 cyl. OHC	.5	.5
Crankcase Front (Timing Belt Rear) Cover and/or Gasket, Replace (B)		
Justy	4.7	5.1

	LABOR TIME	SEVERE SERVICE
Crankshaft and Main Bearings, Replace (A)		
Includes: Engine R&R.		
3 cyl.		
w/AWD	9.5	10.4
w/o AWD	8.9	9.8
4 cyl.		
1981-94		
OHV	11.1	12.0
OHC		
8V	19.2	20.1
16V		
1.8L, 2.2L	19.2	20.1
2.5L	22.7	23.6
1995-98		
1.8L, 2.2L SOHC		
w/AT	19.8	22.1
w/MT	19.2	21.3
2.5L DOHC		
w/AT	21.9	24.5
w/MT	21.3	23.7
1999-04		
2.0L		
w/AT	16.7	18.7
w/MT	16.1	17.9
2.2L, 2.5L SOHC		
w/AT	20.4	22.7
w/MT	19.7	22.0
6 cyl.		
2.7L	21.6	22.5
3.0L	18.0	20.0
3.3L	22.3	23.2
w/AC 4 cyl. OHV add	.5	.5
w/adjustable valves add		
1.8L, 2.2L	.5	.5
2.5L	1.4	1.4
w/AT add		
2.7L	.2	.2
4 cyl. OHC	.5	.5
4 cyl. OHV	.2	.2
w/turbo add		
2.7L	.5	.5
4 cyl. OHC	.5	.5
Crankshaft Front Oil Seal, Replace (B)		
3 cyl.	4.9	5.1
4 cyl.		
OHV	.9	1.1
OHC	2.7	3.0
1995-98		
1.8L, 2.2L SOHC	2.7	3.0
2.5L DOHC	3.2	3.6
1999-04		
2.0L	3.2	3.6
2.2L, 2.5L SOHC	2.7	3.0
6 cyl.		
2.7L	2.3	2.5
3.0L	1.1	1.3
3.3L	2.7	2.9

LEGACY : LOYALE : OUTBACK : RX : STANDARD : SVX : XT : XT6 — SUB-11

	LABOR TIME	SEVERE SERVICE
Crankshaft Pulley, Replace (B)		
3 cyl.	.5	.7
4 cyl.		
OHV	.7	.9
OHC		
8V	1.0	1.2
16V	.6	.7
6 cyl.		
2.7L, 3.0L	1.0	1.2
3.3L	.6	.8
w/AC 4 cyl. OHV add	.2	.2
Pistons and Connecting Rods, Replace (B)		
Includes: R&R engine assy. & cylinder heads.		
3 cyl.	10.9	11.8
4 cyl.		
1981-94		
OHV	6.1	7.0
OHC		
1.8L, 2.2L	17.5	18.4
1995-98		
1.8L, 2.2L SOHC		
w/AT	15.5	16.4
w/MT	14.8	15.7
2.5L DOHC		
w/AT	18.2	19.1
w/MT	17.5	18.4
1999-04		
2.0L		
w/AT	16.7	18.7
w/MT	16.1	17.9
2.2L, 2.5L SOHC		
w/AT	20.4	22.7
w/MT	19.7	22.0
6 cyl.		
2.7L	18.2	19.1
3.0L	15.1	16.8
3.3L	22.3	23.2
w/AC 4 cyl. OHV add	.5	.5
w/adjustable valves add		
1.8L, 2.2L	.5	.5
2.5L	1.4	1.4
w/AT add		
2.5L	.5	.5
2.7L	.2	.2
4 cyl. OHV	.2	.2
w/cruise control 4 cyl. OHV add	.2	.2
w/PS 4 cyl. OHV add	.3	.3
w/turbo add		
2.5L	.5	.5
2.7L	.5	.5
4 cyl. OHV	1.3	1.3

	LABOR TIME	SEVERE SERVICE
Rear Main Oil Seal, Replace (B)		
Includes: Trans. R&R. when necessary.		
3 cyl.		
AT		
w/AWD	6.6	7.2
w/o AWD	5.4	6.0
MT		
w/AWD	3.8	4.4
w/o AWD	3.2	3.8
4 cyl.		
OHV		
AT	3.4	4.0
MT		
w/AWD	5.0	5.6
w/o AWD	4.3	4.9
OHC		
AT		
3-Speed		
w/AWD	5.1	5.7
w/o AWD	4.8	5.4
4-Speed 8V		
w/AWD	4.4	5.0
w/o AWD	4.0	4.6
4-Speed 16V		
w/AWD	4.9	5.4
w/o AWD	4.6	5.2
1999-04		
AWD 4EAT	6.1	6.9
WRX	6.6	7.4
MT		
1981-96		
w/AWD	4.7	5.3
w/o AWD	4.0	4.6
1997-04		
w/AWD	5.8	6.5
w/o AWD	5.0	5.7
WRX	6.7	7.5
6 cyl.		
2.7L		
AT		
3-Speed		
w/AWD	5.1	5.7
w/o AWD	4.8	5.4
4-Speed		
w/AWD	4.4	5.0
w/o AWD	4.0	4.6
MT		
w/AWD	4.7	5.3
w/o AWD	4.0	4.6
3.0L		
w/AWD	6.1	6.9
3.3L		
w/AWD	4.9	5.5
w/o AWD	4.6	5.2

	LABOR TIME	SEVERE SERVICE
Rings, Replace (A)		
Includes: Ridge reaming, cylinder wall deglazing, installing new rings, engine tune-up.		
3 cyl.	8.9	9.8
4 cyl.		
1981-94		
OHV	11.1	12.0
OHC		
8V	19.2	20.1
16V		
1.8L, 2.2L	19.2	20.1
2.5L	22.7	23.6
1995-98		
1.8L, 2.2L SOHC		
w/AT	19.9	20.8
w/MT	19.2	20.1
2.5L DOHC		
w/AT	23.0	23.9
w/MT	22.2	23.1
1999-04		
2.0		
w/AT	18.4	20.5
w/MT	17.7	19.8
2.2L, 2.5L SOHC		
w/AT	21.1	22.0
w/MT	18.1	19.0
6 cyl.		
2.7L	21.6	22.5
3.0L	16.7	18.7
3.3L	22.3	23.2
w/AC 4 cyl. OHV add	.5	.5
w/adjustable valves add		
1.8L, 2.2L	.5	.5
2.5L	1.4	1.4
w/AT add		
2.5L	.5	.5
2.7L	.2	.2
4 cyl. OHV	.2	.2
w/cruise control 4 cyl. OHV add	.2	.2
w/PS 4 cyl. OHV add	.3	.3
w/turbo add		
2.5L	.5	.5
2.7L	.5	.5
4 cyl. OHV	1.3	1.3

ENGINE LUBRICATION

Engine Oil Pressure Switch (Sending Unit), Replace (B)

	LABOR TIME	SEVERE SERVICE
All Models	.9	1.0

Oil Pan and/or Gasket, Replace (B)

	LABOR TIME	SEVERE SERVICE
3 cyl.	2.5	2.7
4 cyl.		
OHV	1.6	1.8
OHC		
8V	1.6	1.8
16V	1.9	2.1

SUB-12   BAJA : BRAT : DL : FORESTER : GL : GL-10 : GLF : IMPREZA : JUSTY

	LABOR TIME	SEVERE SERVICE
Oil Pan and/or Gasket, Replace (B)		
6 cyl.		
2.7L	1.9	2.1
3.0	1.0	1.1
3.3L	1.9	2.1
w/AWD add		
2.7L	.2	.2
4 cyl. OHC 8V	.2	.2
w/turbo 4 cyl. OHC 16V add	.2	.2
Oil Pick-Up Tube and/or Seal, Replace (B)		
3 cyl.	2.6	2.8
4 cyl.		
OHV	2.1	2.3
OHC		
8V	2.2	2.4
16V	2.4	2.6
6 cyl.		
2.7L	2.5	2.7
3.0	1.5	1.8
3.3L	2.5	2.7
w/AWD add		
2.7L	.2	.2
4 cyl. OHC 8V	.2	.2
w/turbo 4 cyl. OHC 16V add	.2	.2
Oil Pressure Gauge (Dash), Replace (B)		
Brat	1.2	1.2
Forester	1.5	1.5
Legacy, Outback		
1990-94	1.8	1.8
1995-04	1.5	1.5
Loyale	1.6	1.6
Impreza	1.5	1.5
XT	1.9	1.9
Justy, SVX	1.4	1.4
1981-89 Hatchback	1.2	1.2
1981-89 Station Wagon	1.3	1.3
1981-89 3 door, 4 door	1.3	1.3
Oil Pump, R&R and Recondition (B)		
1981-94		
4 cyl.		
OHV	2.6	3.1
OHC	3.0	3.5
6 cyl.	3.2	3.7
1995-98		
1.8L, 2.2L: SOHC	2.6	3.1
2.5L DOHC	3.0	3.5
1999-04		
2.0L	4.0	4.4
2.2L, 2.5L SOHC	3.5	3.9
6 cyl.		
3.0L	4.0	4.4
3.3L	2.8	3.3
w/AC 4 cyl. OHV add	.2	.2
w/turbo 4 cyl. OHV add	.2	.2

	LABOR TIME	SEVERE SERVICE
Oil Pump, Replace (B)		
1981-94		
4 cyl.		
OHV	2.4	2.9
OHC	2.6	3.1
6 cyl.	2.8	3.3
1995-98		
1.8L, 2.2L: SOHC	2.6	3.1
2.5L DOHC	3.0	3.5
1999-04		
2.0L	3.5	3.9
2.2L, 2.5L SOHC	3.0	3.4
6 cyl.		
3.0L	3.5	3.9
3.3L	2.8	3.3
w/AC 4 cyl. OHV add	.2	.2
w/turbo 4 cyl. OHV add	.2	.2
Oil Pump Cover and/or Seal, Replace (B)		
Justy	.5	.7
Replace pump rotor or gear add	.2	.2

CLUTCH

	LABOR TIME	SEVERE SERVICE
Bleed Clutch Hydraulic System (B)		
All Models	.5	.5
Clutch Assy., Replace (B)		
1981-96		
3 cyl.		
w/AWD	3.6	2.8
w/o AWD	3.1	3.3
4 cyl.		
OHV		
w/AWD	4.3	4.5
w/o AWD	3.7	3.9
OHC		
w/AWD	4.8	5.0
w/o AWD	4.1	4.3
6 cyl.		
w/AWD	4.6	4.8
w/o AWD	4.0	4.2
1997-04		
w/AWD	5.2	5.8
w/o AWD	4.4	5.0
WRX	6.1	6.9
Clutch Cable, Adjust (B)		
1995-04	.5	.7
Clutch Cable, Replace (B)		
1995-04	1.1	1.3
Clutch Master Cylinder, R&R and Recondition (A)		
Includes: System bleeding.		
All Models	1.6	1.8
Clutch Master Cylinder, Replace (B)		
Includes: Bleed system.		
All Models	1.0	1.2

	LABOR TIME	SEVERE SERVICE
Clutch Release Bearing or Fork, Replace (B)		
1981-96		
3 cyl.		
w/AWD	3.5	3.7
w/o AWD	2.9	3.1
4 cyl.		
OHV		
w/AWD	4.1	4.3
w/o AWD	3.6	3.8
OHC		
w/AWD	4.5	4.7
w/o AWD	3.8	4.0
6 cyl.		
w/AWD	4.4	4.6
w/o AWD	3.7	3.9
1997-04		
w/AWD	5.2	5.8
w/o AWD	4.4	5.0
WRX	6.1	6.9
Clutch Slave Cylinder, Replace (B)		
Includes: System bleeding.		
All Models	.8	1.0
Flywheel, Replace (B)		
1981-96		
3 cyl.		
w/AWD	4.2	4.4
w/o AWD	3.5	3.7
4 cyl.		
OHV		
w/AWD	4.5	4.7
w/o AWD	4.3	4.5
OHC		
w/AWD	5.1	5.3
w/o AWD	4.4	4.6
6 cyl.		
w/AWD	5.0	5.2
w/o AWD	4.4	4.6
1997-04		
w/AWD	5.3	6.0
w/o AWD	4.6	5.1
WRX	6.3	7.0

MANUAL TRANSAXLE

	LABOR TIME	SEVERE SERVICE
Differential Case Bearings, Replace (B)		
Baja, Legacy, Outback		
w/AWD	7.6	8.5
w/o AWD	6.1	6.9
Brat	7.2	7.5
Forester	7.6	8.5
Impreza		
w/AWD	7.6	8.5
w/o AWD	6.1	6.9
Justy		
w/AWD	5.8	6.1
w/o AWD	4.7	5.0

LEGACY : LOYALE : OUTBACK : RX : STANDARD : SVX : XT : XT6 **SUB-13**

	LABOR TIME	SEVERE SERVICE
Loyale		
w/AWD	7.3	7.6
w/o AWD	5.9	6.2
WRX	8.5	9.6
XT		
w/AWD	7.0	7.3
w/o AWD	6.5	6.8
1981-89 Hatchback		
w/AWD	7.2	7.5
w/o AWD	6.0	6.3
1981-89 Station Wagon		
w/AWD	7.2	7.5
w/o AWD	5.9	6.2
1981-89 3 door, 4 door		
w/AWD	7.3	7.6
w/o AWD	5.9	6.2
w/turbo add	.2	.2

Differential Case, Recondition (A)
Baja, Legacy, Outback		
w/AWD	9.0	10.0
w/o AWD	7.5	8.4
Brat	8.3	8.8
Forester	9.0	10.0
Impreza		
w/AWD	9.0	10.0
w/o AWD	7.5	8.4
Justy		
w/AWD	6.5	7.0
w/o AWD	5.4	5.9
Loyale		
w/AWD	8.7	9.2
w/o AWD	7.3	7.8
WRX	10.3	11.5
XT		
w/AWD	8.7	9.2
w/o AWD	7.3	7.8
1981-89 Hatchback		
w/AWD	8.3	8.8
w/o AWD	7.0	7.5
1981-89 Station Wagon		
w/AWD	8.2	8.7
w/o AWD	7.0	7.5
1981-89 3 door, 4 door		
w/AWD	8.2	8.7
w/o AWD	7.0	7.5
w/turbo add	.2	.2

Transaxle Rear Cover, Replace (B)
1995-04		
w/o AWD	2.4	2.7

Final Drive Gear, Replace (B)
Justy		
w/AWD	6.1	6.6
w/o AWD	5.0	5.6

Ring Gear & Pinion Assy., Replace (B)
Baja, Impreza, Legacy, Outback		
w/AWD	10.0	11.2
w/o AWD	8.5	9.6

	LABOR TIME	SEVERE SERVICE
Brat	9.3	9.8
Forester	10.0	11.2
Loyale		
w/AWD	9.8	10.3
w/o AWD	8.4	8.9
WRX	10.9	12.2
XT		
w/AWD	9.7	10.2
w/o AWD	8.4	8.9
1981-89 Hatchback		
w/AWD	9.3	9.8
w/o AWD	7.9	8.4
1981-89 Station Wagon		
w/AWD	9.3	9.8
w/o AWD	7.9	8.4
1981-89 3 door, 4 door		
w/AWD	9.3	9.8
w/o AWD	8.1	8.6

Extension Housing Oil Seal, Replace (B)
Baja, Forester, Impreza, Justy, Legacy, Outback	.8	1.0
Brat	.7	.9
Loyale, XT	.7	.9
1981-89 Hatchback, Station Wagon, 3 door, 4 door	.7	.9

Transaxle Assy., R&R and Recondition (A)
Baja, Legacy, Outback		
w/AWD	12.2	13.7
w/o AWD	10.5	11.7
Brat	11.2	11.8
Forester	12.2	13.7
Impreza		
w/AWD	12.2	13.7
w/o AWD	10.5	11.7
Justy		
w/AWD	8.7	9.3
w/o AWD	8.5	9.1
Loyale		
w/AWD	12.0	12.6
w/o AWD	10.3	10.9
WRX	13.1	14.6
XT		
w/AWD	11.7	12.3
w/o AWD	10.3	10.9
1981-89 Hatchback		
w/AWD	11.2	11.8
w/o AWD	9.7	10.3
1981-89 Station Wagon		
w/AWD	11.2	11.8
w/o AWD	9.7	10.3
1981-89 3 door, 4 door		
w/AWD	11.2	11.8
w/o AWD	9.9	10.5
w/turbo add	.2	.2

Transaxle Assy., R&R or Replace (B)
Baja, Legacy, Outback		
w/AWD	7.0	7.9
w/o AWD	5.5	6.2

	LABOR TIME	SEVERE SERVICE
Brat	6.5	6.7
Forester	7.0	7.9
Impreza		
w/AWD	7.0	7.9
w/o AWD	5.5	6.2
Justy		
w/AWD	4.5	4.7
w/o AWD	3.5	3.7
Loyale		
w/AWD	6.7	6.9
w/o AWD	5.1	5.3
WRX	7.9	8.9
XT		
w/AWD	6.7	6.9
w/o AWD	5.3	5.5
1981-89 Hatchback		
w/AWD	6.3	6.5
w/o AWD	5.3	5.5
1981-89 Station Wagon		
w/AWD	6.4	6.6
w/o AWD	5.2	5.4
1981-89 3 door, 4 door		
w/AWD	6.5	6.7
w/o AWD	5.1	5.3
w/turbo add	.2	.2

AUTOMATIC TRANSAXLE

SERVICE TRANSAXLE INSTALLED

Accumulator, Seal or Spring, Replace (B)
1987-98 4-Speed		
one	2.5	2.7
each addl. add	.1	.1

AWD Engagement Solenoid, Replace (B)
1981-94 3-Speed	.7	.9

Brake Band, Adjust (B)
1981-94	.5	.7

Check Unit for Oil Leaks (C)
Includes: Clean and dry outside of unit. Operate unit to determine point of leakage.
All Models	.9	.9

Drain & Refill Unit (B)
All Models	.8	.8

Electromagnetic Clutch Brush Assy., Replace (B)
1989-94 Justy	1.3	1.5

Electromagnetic Clutch Control Unit, Replace (B)
1989-94 Justy	.3	.5

Electronic Control Unit, Replace (B)
4-Speed		
Baja, Forester, Impreza, Legacy, Outback	.6	.6
1985-91 XT	1.1	1.1
1987-94 Loyale		
3 door, 4 door	.5	.5
Station Wagon	.8	.8

SUB-14 BAJA : BRAT : DL : FORESTER : GL : GL-10 : GLF : IMPREZA : JUSTY

	LABOR TIME	SEVERE SERVICE
Electronic Control Unit, Replace (B)		
1988-89 Station Wagon	.7	.7
1988-89 3 door	.5	.5
1988-89 4 door	.5	.5
1992-97 SVX	.5	.5
Electronic Transmission Diagnosis (A)		
SVX	1.0	1.0
1989-94 Justy	1.0	1.0
Extension Housing, R&R and Recondition (B)		
1981-84 3 spd AWD	2.2	2.5
1985-94 3 spd AWD	3.1	3.4
w/turbo add	.1	.1
Extension Housing and/or Gasket, Replace (B)		
1981-84 3 spd AWD	.9	1.1
1985-94 3 spd AWD	1.9	2.1
w/turbo add	.1	.1
Extension Housing Bushing, Replace (B)		
1981-84 3 spd AWD	1.6	1.8
1985-94 3 spd AWD	2.5	2.7
w/turbo add	.1	.1
Extension Housing Oil Seal, Replace (B)		
1981-00	.8	.9
1999-04 4EAT AWD	.8	.9
4WD Cover Assy. and/or Gasket, R&R or Replace (B)		
1987-98 4-Speed	2.0	2.2
4WD Transfer Case and/or Gasket, R&R or Replace (B)		
4-Speed		
1987-00	2.1	2.3
SVX	3.1	3.3
XT	2.3	2.5
1988-89 Station Wagon	2.9	3.1
1988-89 3 door	2.7	2.9
1988-89 4 door	2.9	3.1
1999-04 4EAT AWD	2.4	2.7
w/turbo add	.5	.5
Governor Assy., R&R and Recondition (A)		
1981-94 3-Speed	1.4	1.9
w/cruise control add	.2	.2
w/turbo add	.5	.5
Governor Cover Gasket, Replace (B)		
1981-94 3-Speed	.5	.7
w/turbo add	.1	.1
Governor Seal, Replace (B)		
1981-94 3-Speed	1.6	1.8
w/turbo add	.1	.1
Kickdown Relay, Replace (B)		
1988-94 3-Speed	.7	.8

	LABOR TIME	SEVERE SERVICE
Kickdown Switch or Solenoid, Replace (B)		
1981-94 3-Speed	.6	.7
Line Pressure Solenoid or Harness, Replace (B)		
1989-94 Justy	1.3	1.4
Line Pressure Test (B)		
1989-94 Justy	.7	.7
Lock-up Solenoid, Replace (B)		
1987-98 4-Speed	.8	1.0
1999-04 4-Speed	1.3	1.4
Manual Shaft and/or Seal, Replace (B)		
1981-94 3-Speed	1.8	2.0
Manual Shaft, Lever or Parking Rod, Replace (B)		
1987-94 4-Speed	2.8	3.0
1995-98 4-Speed	2.9	3.2
1999-04 4-Speed	2.4	2.7
Neutral Safety Switch, Replace (B)		
1981-94 3-Speed	.6	.7
1987-04 4-Speed	1.4	1.6
w/turbo add	.5	.5
Oil Pan and/or Gasket, Replace (B)		
1981-98	.8	.9
1999-04	1.1	1.3
Oil Pressure Check (B)		
1981-94 3-Speed	.8	.8
Oil Pump Assy., R&R and Recondition (A)		
1989-94 Justy	2.5	3.0
Selector Cable, Replace or Adjust (B)		
1989-94 Justy		
adjust	.5	.7
replace	.8	1.0
1995-04		
adjust	.5	.5
replace		
Baja, Legacy, Outback	1.3	1.4
Forester, Impreza	1.0	1.1
Selector Shaft Seal, Replace (B)		
1987-98 4-Speed	1.1	1.3
1999-04	2.4	2.7
Servo and Accumulator, Replace (B)		
1981-94 3-Speed	2.5	2.7
Servo Cover and/or Seal, Replace (B)		
1981-94 3-Speed	1.6	1.8
Speed Sensor, Replace (B)		
4-Speed		
Baja, Legacy, Outback		
1990-92	.7	.8
1993-94	3.6	3.6
Impreza		
1993-94	4.9	5.5
Loyale		
1990-94	.6	.7

	LABOR TIME	SEVERE SERVICE
SVX		
1992	2.2	2.5
1993-94	5.0	5.7
XT 1985-91	.6	.7
1988-89 Station Wagon, 3 door, 4 door	.7	.8
1995-98	5.2	5.8
1999-04	3.6	4.1
w/AWD 88-89		
4 door add	.2	.2
w/turbo add 88-89		
4 door add	.5	.5
Speedometer Driven Gear and/or Seal, Replace (B)		
1989-94 Justy	.5	.5
Switch Accelerator, Replace (B)		
1989-94 Justy	.5	.5
Throttle Position Switch, Replace (B)		
1989-94 Justy	.5	.5
Transfer Clutch Valve Assy., Replace (B)		
1981-94 3-Speed	2.1	2.1
w/turbo add	.1	.1
Transfer Drive Gear and/or Bearings, Replace (B)		
1981-84 3 spd AWD	1.6	1.9
1985-94 3 spd AWD	2.1	2.4
w/turbo add	.1	.1
Transmission Temperature Sensor, Replace (B)		
1995-98	1.0	1.0
1999-04	1.4	1.6
Vacuum Throttle Valve, Replace (B)		
1981-85 3-Speed	.6	.6
Valve Body, R&R and Recondition (A)		
1981-94 3-Speed	2.2	2.7
1987-95 4-Speed	4.7	5.2
1989-94 Justy	2.2	2.7
1996-98	5.0	5.5
1999-04	4.3	4.8
Replace solenoids add each	.1	.1
Valve Body, Replace (B)		
SVX	3.1	3.4
1981-94 3-Speed	1.0	1.3
1987-95 4-Speed	2.1	2.4
1996-98	2.3	2.6
1999-04	1.9	2.1
1989-94 Justy	1.1	1.4

SERVICE TRANSAXLE REMOVED
Transaxle R&R included unless otherwise noted.

	LABOR TIME	SEVERE SERVICE
Differential, R&R or Replace (B)		
Front axle		
4-Speed		
1988-89 Station Wagon	4.3	4.6
1988-89 3 door	4.2	4.5

LEGACY : LOYALE : OUTBACK : RX : STANDARD : SVX : XT : XT6 — SUB-15

	LABOR TIME	SEVERE SERVICE
1988-89 4 door	4.2	4.5
1990-94 Legacy, Outback	4.2	4.5
1990-94 Loyale	3.8	4.1
1993-94 Impreza	4.2	4.5
1995-02 SVX	4.2	4.5
1995-02 XT	4.3	4.6
w/AWD add	.3	.3

Differential Assy., Recondition (A)
Front axle
  1981-94 3-Speed

w/AWD turbo	8.6	9.1
w/o AWD	7.6	8.1

Differential Carrier Assy., Recondition (A)

1989-94 Justy	6.8	7.3
w/AWD	8.0	8.5

1995-97
  Impreza, Legacy

w/AWD	8.2	9.2
w/o AWD	7.8	8.7

1998-04
  Baja, Impreza, Legacy, Outback | 8.2 | 9.2
  Forester | 8.7 | 9.7
  WRX | 9.1 | 10.2

Differential Carrier Bearings, Replace (B)
Front axle
3-Speed
  1981-94

w/AWD	6.3	6.7
w/o AWD	5.3	5.7
turbo	5.9	6.3

4-Speed
  Baja, Impreza, Legacy, Outback
  1990-97

w/AWD	7.0	7.9
w/o AWD	6.6	7.4

1998-04

w/AWD	7.5	8.4
w/o AWD	7.0	7.9
Forester	7.5	8.4

Loyale
  1990-94

w/AWD	7.0	7.9
w/o AWD	6.6	7.4

SVX

1992-97	6.8	7.2
WRX	7.5	8.4

1988-89 Station
  Wagon | 5.9 | 6.3
  1988-89 3 door | 5.7 | 6.1
  1988-89 4 door | 5.9 | 6.3
1988-89 models
  w/AWD add | .3 | .3

Flywheel (Flexplate), Replace (B)
4-Speed

1987-94 Loyale	3.5	3.7

1988-89 Station
  Wagon | 3.8 | 4.0
  1988-89 3 or 4 door | 3.9 | 4.1
1990-94 Legacy,
  Outback | 4.0 | 4.2
1993-94 Impreza | 4.0 | 4.2
SVX, XT | 3.9 | 4.1
1995-04
  Baja, Forester, Impreza,
  Legacy, Outback | 5.5 | 6.2
  WRX | 6.0 | 6.7
w/AWD 81-94 add | .3 | .3

Front Oil Pump, Replace or Recondition (B)
1981-94 3-Speed

w/AWD	5.5	5.7
w/o AWD	5.3	5.5
1999-04 4EAT	8.8	9.9

w/cruise control add | .2 | .2

Oil Pump and/or Gasket, Replace (B)
4-Speed
  Baja, Impreza, Legacy, Loyal, Outback
  1990-94 | 6.3 | 6.5
  1995-98

w/AWD	8.2	9.2
w/o AWD	7.8	8.7
1999-04	8.2	9.2
SVX	5.0	5.2
XT	6.9	7.1

1988-89 Station
  Wagon | 6.8 | 7.0
  1988-89 3 or 4 door | 7.0 | 7.2
w/AWD 81-94 add | .3 | .3
Recondition pump add | .6 | .8

Primary & Secondary Pulleys, Recondition (B)
1989-94 3-Speed

w/AWD	7.2	7.5
w/o AWD	5.5	5.8

Ring & Pinion Set, Replace (B)
Front axle
  Baja, Impreza, Legacy, Outback
  1990-94 | 8.2 | 8.8
  1995-98

w/AWD	9.9	11.1
w/o AWD	9.4	10.6

1999-04

w/AWD	10.3	11.5
w/o AWD	9.9	11.1
WRX	11.2	12.5
1985-91 XT	8.4	9.0
1987-94 Loyale	7.8	8.4

1988-89 Station
  Wagon, 3 or 4 door | 8.4 | 9.0
1992-97 SVX | 8.2 | 8.8
w/AWD add | .3 | .3
Recondition differential complete add | 2.7 | 3.0

Torque Converter or Front Oil Seal, Replace (B)
3-Speed
  1981-94

w/AWD	4.7	4.9
w/o AWD	4.5	4.7

4-Speed
  Baja, Impreza, Legacy, Outback
  1990-94 | 4.0 | 4.2
  1995-98

w/AWD	5.0	5.7
w/o AWD	4.6	5.1
1999-04	5.5	6.2

Loyale

w/AWD	4.1	4.6
w/o AWD	3.8	4.3
SVX	3.9	4.1
WRX	6.0	6.7
XT	3.8	4.0

1988-89 Station
  Wagon | 3.7 | 3.9
  1988-89 3 or 4 door | 3.9 | 4.1
w/AWD 81-94 add | .3 | .3

Transaxle Assy., R&R and Recondition (A)
3-Speed
  1981-94

w/AWD	11.2	12.1
w/o AWD	10.8	11.7

4-Speed
  Baja, Impreza, Legacy, Outback, WRX
  1990-94 | 17.5 | 18.4
  1995-98

w/AWD	20.0	22.3
w/o AWD	19.2	21.3
1999-04	18.5	20.6
Loyale	17.1	18.0
SVX	17.5	18.4
XT	14.9	15.8

1988-89 Station
  Wagon | 15.1 | 16.0
1988-89 3 or 4 door | 15.1 | 16.0
1989-94 Justy | 12.2 | 13.1
w/AWD 81-94 add | .3 | .3
w/cruise control
  3-Speed add | .2 | .2
Perform electrical test add | 1.0 | 1.0
Recondition diff. 3-Speed add | 3.0 | 3.0
transfer case 89-94 add | 3.8 | 3.8

SUB-16   BAJA : BRAT : DL : FORESTER : GL : GL-10 : GLF : IMPREZA : JUSTY

	LABOR TIME	SEVERE SERVICE
Transaxle Assy., R&I (B)		
3-Speed		
1981-94		
w/AWD	4.0	4.2
w/o AWD	4.3	4.5
4-Speed		
1987-04 Baja, Legacy, Outback	3.7	3.9
1987-04 Impreza	3.7	3.9
1987-94 Loyale	3.2	3.4
1995-00 SVX	3.7	3.9
1995-00 XT	3.6	3.8
1988-89 Station Wagon	3.4	3.6
1988-89 3 or 4 door	4.2	4.4
1989-94 Justy	4.2	4.4
w/AWD 81-04 add	.3	.3
w/cruise control		
3-Speed add	.2	.2
Replace flywheel 81-94		
3-Speed add	.3	.3

TRANSFER CASE

	LABOR TIME	SEVERE SERVICE
Access Cover and/or Gasket, Replace (B)		
1981-94 w/MT		
full time	4.7	4.9
part time	3.8	4.0
Engagement Solenoid, Replace (B)		
1981-94 w/AT	.7	.8
Extension Housing and/or Gasket, Replace (B)		
1981-94 w/MT part time	4.1	4.2
Rear Pinion Shaft Drive Gear, Replace (B)		
1981-94 w/MT part time	5.4	5.6
Shift Rail & Fork, Replace (B)		
1981-94 w/MT		
full time	5.4	5.6
part time	3.8	4.0
Tailshaft Bushing or Seal, Replace (B)		
1981-94 w/AT	1.7	1.9
w/turbo add	.1	.1
Transfer Case, R&R and Recondition (A)		
1981-94 w/MT full time	6.3	6.9
Transfer Case, R&R or Replace (B)		
1981-94		
w/AT	2.1	2.3
w/MT		
full time	5.8	6.0
part time	4.7	4.9
1995-04	2.9	3.2
w/turbo add	.5	.5
Replace assembly add	.5	.8

	LABOR TIME	SEVERE SERVICE
Transfer Case Clutch Assy., Replace (B)		
1981-91 w/AT		
3 dr, 4 dr, Justy	1.6	1.9
w/turbo add	.1	.1
Transfer Case Driven Gear, Replace (B)		
1981-91 w/AT		
3 dr, 4 dr, Justy	1.5	1.8
w/turbo add	.1	.1

SHIFT LINKAGE
AUTOMATIC TRANSAXLE

	LABOR TIME	SEVERE SERVICE
Selector Handle, Replace (B)		
Baja, Legacy, Outback		
1990-94	.7	.7
1995-04	.3	.3
Brat	.5	.5
Forester	.3	.3
Impreza		
1993-94	1.0	1.0
1995-04	.3	.3
Justy	.3	.3
Loyale	.5	.5
SVX	1.2	1.2
XT	.5	.5
1981-89 Hatchback	.5	.5
1981-89 Station Wagon	.5	.5
1981-89 3 door, 4 door	.5	.5
w/air bags 90-94 Legacy, Outback add	.2	.2
w/AWD 3-Speed add	.2	.2
Gear Selector Indicator, Replace (B)		
Baja, Legacy, Outback		
1990-94	.7	.7
1995-04	.5	.5
Brat	.7	.7
Forester	.5	.5
Impreza		
1993-94	1.0	1.0
1995-01	.7	.7
2002-04	1.1	1.1
Justy	.5	.5
Loyale	.5	.5
SVX	1.2	1.2
XT	.7	.7
1981-89 Hatchback, Station Wagon, 3 door, 4 door	.7	.7
w/4EAT Loyale add	.2	.2
w/AWD 3-Speed add	.2	.2
w/air bags add	.2	.2

MANUAL TRANSAXLE

	LABOR TIME	SEVERE SERVICE
Gear Selector Lever, Replace (B)		
All Models	.3	.3
Gearshift Mechanism, R&R and Recondition (B)		
1981-86	1.4	1.4
1987-94 Impreza	1.6	1.6

	LABOR TIME	SEVERE SERVICE
1987-94 Legacy	1.6	1.6
1987-94 Loyale	.8	.8
1987-94 Justy	.7	.7
1995-04	1.6	1.6

DRIVELINE

	LABOR TIME	SEVERE SERVICE
CV Joint & Shaft Assy., Replace (B)		
Rear axle		
Baja, Legacy, Outback		
1990-99		
one side	1.4	1.6
both sides	2.5	2.9
2000-04		
one side	1.9	2.1
both sides	2.2	2.5
Brat		
one side	1.2	1.5
both sides	2.4	2.7
Forester, Impreza		
one side	1.4	1.6
both sides	2.5	2.9
Justy		
one side	1.1	1.4
both sides	1.9	2.2
Loyale		
one side	1.1	1.4
both sides	1.9	2.2
SVX		
one side	1.1	1.4
both sides	2.3	2.6
XT		
one side	1.1	1.4
both sides	1.9	2.2
1981-89 Hatchback		
one side	1.3	1.6
both sides	2.2	2.5
1981-89 Station Wagon		
one side	1.3	1.6
both sides	2.3	2.6
1981-89 3 door, 4 door		
one side	1.3	1.6
both sides	2.4	2.7
Differential Assy., R&R or Replace (B)		
Rear axle		
Baja, Forester, Impreza Justy, Legacy, Outback		
1990-99	1.6	1.8
2000-04	1.3	1.4
Brat	2.7	2.9
Loyale	2.5	2.7
SVX	4.2	4.4
XT	2.5	2.7
1981-89 Hatchback	2.9	3.1
1981-89 Station Wagon	2.9	3.1
1981-89 3 door, 4 door	2.8	3.0

	LABOR TIME	SEVERE SERVICE

Differential Cover Gasket, Replace (B)
  Rear axle
    All Models 1.9   *2.1*
DO Joint and/or CV-Joint Boot, Replace (B)
  Rear axle
    1981-94
      one side 1.0   *1.3*
      both sides 1.6   *1.9*
    1995-99
      one side 1.0   *1.1*
      both sides 1.6   *1.8*
    2000-04
      Baja, Legacy
        one side 1.7   *2.0*
        both sides 2.1   *2.3*
      Forester, Impreza
        one side 1.0   *1.1*
        both sides 1.6   *1.8*
Driveshaft, R&R or Replace (B)
  All Models 8   *.9*
  Replace
    center bearing add 3   *.3*
    U-joint add each 5   *.5*
Halfshaft, R&R or Replace (B)
  Rear axle
    Baja, Legacy, Outback
      1990-99 7   *1.0*
      2000-04
        one side 1.6   *1.8*
        both sides 1.9   *2.1*
    Brat 5   *.8*
    Forester 7   *1.0*
      one side 8   *.9*
      both sides 1.6   *1.8*
    Impreza
      one side 8   *.9*
      both sides 1.6   *1.8*
    Justy 8   *1.1*
    Loyale 5   *.8*
    SVX
      right side 7   *1.0*
      left sides 1.0   *1.3*
    XT 5   *.8*
    1981-89 Hatchback 5   *.8*
    1981-89 Station
      Wagon 5   *.8*
    1981-89 3 door,
      4 door 5   *.8*
Carrier Side Oil Seals, Replace (B)
  Rear axle
    exc. below
      one side 8   *.9*
      both sides 1.4   *1.6*
    2002-04 Baja, Legacy
      one side 1.7   *2.0*
      both sides 2.2   *2.5*

Pinion Shaft Oil Seal, Replace (B)
  Rear axle
    Baja, Legacy,
      Outback 1.1   *1.3*
    Brat 8   *1.0*
    Forester, Impreza .. 1.1   *1.3*
    Justy 8   *1.0*
    Loyale 8   *1.0*
    SVX 1.2   *1.4*
    XT 8   *1.0*
    1981-89 Hatchback 8   *1.0*
    1981-89 Station
      Wagon 8   *1.0*
    1981-89 3 door,
      4 door 8   *1.0*
Stub Axle Shaft and/or Seal, Replace (B)
  Rear axle
    one side 8   *1.2*
    both sides 1.5   *1.9*

HALFSHAFT
Axle Boot Band, Replace (B)
  One 3   *.3*
  each addl. add 2   *.2*
CV-Joint Outer Boot (Seal), Replace (B)
  Right side 1.7   *2.0*
  Left side 1.4   *1.6*
  Both sides 2.4   *2.7*
  Replace boot band
    add each 2   *.2*
DOJ, Boot and/or Seal, Replace (B)
  Right side 1.3   *1.4*
  Left side 1.6   *1.8*
  Both sides 2.1   *2.3*
Halfshaft, R&R or Replace (B)
  Baja, Legacy, Outback
    one side 8   *.9*
    both sides 1.4   *1.6*
  Brat
    one side 1.4   *1.6*
    both sides 2.5   *2.7*
  Forester
    one side 8   *.9*
    both sides 1.4   *1.6*
  Impreza
    one side 8   *.9*
    both sides 1.4   *1.6*
  Justy
    one side 1.2   *1.4*
    both sides 2.3   *2.5*
  Loyale
    one side 1.2   *1.4*
    both sides 2.1   *2.3*
  SVX
    one side 8   *1.0*
    both sides 1.4   *1.6*

XT
  4 cyl.
    one side 1.2   *1.4*
    both sides 2.1   *2.3*
  6 cyl.
    one side 1.0   *1.2*
    both sides 1.8   *2.0*
  1981-89 Hatchback
    one side 1.5   *1.7*
    both sides 2.6   *2.8*
  1981-89 Station Wagon
    one side 1.5   *1.7*
    both sides 2.7   *2.9*
  1981-89 3 door, 4 door
    one side 1.6   *1.8*
    both sides 2.6   *2.8*
  Replace CV-joint
    add each 2   *.2*
Intermediate Shaft, R&R or Replace (B)
  Justy one 1.0   *1.3*
  Recondition shaft and support
    bearing assembly add 5   *.8*

BRAKES
ANTI-LOCK
Diagnose Anti-Lock Brake System (A)
  Baja, Legacy, Outback
    1996-97 .... 1.3   *1.3*
    1998-04 8   *.8*
  Forester 8   *.8*
  SVX 8   *.8*
  1990-95 8   *.8*
  1996-04 Impreza 8   *.8*
Control Unit, Replace (B)
  Baja, Legacy, Outback
    1996-97 8   *.8*
    1998-04 5   *.5*
  Forester 5   *.5*
  SVX 5   *.5*
  1990-95 models 5   *.5*
  1996-04 Impreza 5   *.5*
G-Sensor, Replace (B)
  Baja, Legacy,
    Outback 3   *.3*
  Forester 3   *.3*
  Impreza
    1993-95 3   *.3*
    1996-04 7   *.7*
Hydraulic Assy., Replace (B)
  1990-04 .... 1.9   *2.1*
Hydraulic Control Unit Relay, Replace (B)
  1990-04
    one or both 3   *.3*

SUB-18   BAJA : BRAT : DL : FORESTER : GL : GL-10 : GLF : IMPREZA : JUSTY

	LABOR TIME	SEVERE SERVICE
Tone Wheel, Replace (B)		
Baja, Legacy, Outback		
front		
1990-94		
one side	3.5	3.5
both sides	4.9	4.9
1995-00		
one side	4.1	4.6
both sides	5.7	6.3
2000-04		
one side	4.9	5.5
both sides	7.0	7.9
rear		
1990-94		
w/AWD		
one side	1.1	1.1
both sides	1.7	1.7
w/o AWD		
one side	2.0	2.0
both sides	3.1	3.1
1995-00		
w/AWD		
one side	2.9	3.2
both sides	3.2	3.6
w/o AWD		
one side	1.1	1.3
both sides	2.1	2.3
2000-04		
one side	1.6	1.8
both sides	1.9	2.1
Forester		
front		
one side	4.1	4.6
both sides	5.7	6.3
rear		
one side	2.9	3.2
both sides	5.2	5.8
Impreza		
front		
1993-94		
one side	3.5	3.5
both sides	4.9	4.9
1995-00		
one side	4.1	4.6
both sides	5.7	6.3
2000-04		
one side	4.9	5.5
both sides	7.0	7.9
rear		
1993-94		
w/AWD		
one side	1.1	1.1
both sides	1.7	1.7
w/o AWD		
one side	2.0	2.0
both sides	3.1	3.1
1995-04		
w/AWD		
one side	2.9	3.2
both sides	3.2	3.6
w/o AWD		
one side	1.1	1.3
both sides	2.1	2.3
SVX		
front		
one side	3.5	3.5
both sides	4.9	4.9
rear		
one side	2.5	2.5
both sides	4.5	4.5
Wheel Speed Sensor, Replace (B)		
1990-04	.7	.7

SYSTEM

	LABOR TIME	SEVERE SERVICE
Bleed Brakes (B)		
Includes: Add fluid.		
All Models	.5	.5
w/ABS/TCS	.5	.5
Brakes, Adjust (B)		
Includes: Refill master cylinder.		
All Models rear wheels	.5	.5
Brake Hose (Flexible), Replace (B)		
Includes: System bleeding.		
All Models one		
front or rear	.8	.9
each addl. add	.5	.5
Brake Proportioning Valve, Replace (B)		
All Models	1.0	1.1
w/MFI add	.3	.3
Master Cylinder, Replace (B)		
Includes: System bleeding.		
All Models	1.1	1.3
Vacuum Brake Booster Assy., Replace (B)		
Baja, Forester, Impreza, Justy, Legacy, Outback	1.4	1.6
Brat	1.4	1.6
Loyale	1.4	1.6
SVX	1.9	2.1
XT	1.4	1.6
1981-89 Hatchback	1.5	1.7
1981-89 Station Wagon	1.4	1.6
1981-89 3 door, 4 door	1.3	1.5
Vacuum Check Valve, Replace (B)		
All Models	.3	.3

SERVICE BRAKES

	LABOR TIME	SEVERE SERVICE
Brake Drum, Replace (B)		
One side	.5	.7
Both sides	.8	.9
Disc Brake Rotor, Replace (B)		
Baja, Legacy, Outback		
front		
one side	.8	.9
both sides	1.3	1.4
rear		
one side	.6	.7
both sides	1.1	1.3
Brat		
one side	1.2	1.4
both sides	2.2	2.4
Forester, Impreza		
front		
one side	.8	.9
both sides	1.3	1.4
rear		
one side	.6	.7
both sides	1.1	1.3
Justy		
one side	1.2	1.4
both sides	2.2	2.4
Loyale		
front		
one side	1.2	1.4
both side	2.2	2.4
rear		
one side	.6	.8
both sides	1.0	1.2
SVX		
front		
one side	.7	.9
both sides	1.1	1.3
rear		
one side	.6	.8
both sides	1.0	1.2
XT		
front		
one side	1.2	1.4
both sides	2.2	2.4
rear		
one side	.6	.8
both sides	1.0	1.2
1981-89 Hatchback		
one side	1.3	1.5
both sides	2.3	2.5
1981-89 Station Wagon		
one side	1.4	1.6
both sides	2.2	2.4
1981-89 3 door, 4 door		
one side	1.3	1.5
both sides	2.2	2.4
Replace rear parking brake shoes add each side	.3	.3
Front Caliper Assy., R&R and Recondition (A)		
Includes: System bleeding.		
Baja, Legacy, Outback		
one side	1.7	2.0
both sides	2.4	2.7
Brat		
one side	1.4	1.7
both sides	2.2	2.5
Forester		
one side	1.9	2.1
both sides	3.0	3.4
Impreza		
one side	1.7	2.0
both sides	2.4	2.7

LEGACY : LOYALE : OUTBACK : RX : STANDARD : SVX : XT : XT6 — SUB-19

	LABOR TIME	SEVERE SERVICE
Justy		
one side	1.4	1.7
both sides	2.1	2.4
Loyale		
one side	1.6	1.9
both sides	2.6	2.9
SVX		
one side	1.2	1.5
both sides	2.2	2.5
w/ABS add	.3	.3
XT		
one side	1.6	1.9
both sides	2.6	2.9
1981-89 Hatchback		
one side	1.4	1.7
both sides	2.3	2.6
1981-89 Station Wagon		
one side	1.6	1.9
both sides	2.3	2.6
1981-89 3 door, 4 door		
one side	1.5	1.8
both sides	2.2	2.5
w/ABS/TCS Forester, Impreza, Legacy, Outback	.3	.3
w/dual piston caliper Legacy, Outback add each	.6	.6
w/turbo Legacy, Outback add	.5	.5
Front Caliper Assy., Replace (B)		
Includes: System bleeding.		
One side	1.0	1.1
Both sides	1.3	1.4
w/ABS/TCS add	.3	.3
Pads and/or Shoes, Replace (B)		
Includes: Service and parking brake adjustment.		
Front disc	.8	.9
Rear disc	.6	.7
Rear drum		
one side	1.0	1.1
both sides	1.6	1.8
Four wheels		
Drum brakes	2.4	2.7
Disc brakes	1.4	1.6
w/AWD add	.2	.2

COMBINATION ADD-ONS

	LABOR TIME	SEVERE SERVICE
Replace		
brake drum add		
w/AWD	.1	.1
w/o AWD	.3	.3
brake hose add	.3	.3
caliper add	.3	.3
disc brake rotor add	.5	.5
wheel cylinder add	.2	.2
Resurface		
drum add	.5	.5
rotor add	.5	.5

	LABOR TIME	SEVERE SERVICE
Rear Caliper Assy., R&R and Recondition (A)		
Includes: System bleeding.		
One side	1.6	1.8
Both sides	2.4	2.7
w/ABS add	.3	.3
Rear Caliper Assy., Replace (B)		
Includes: System bleeding.		
One side	1.0	1.1
Both sides	1.3	1.4
w/ABS/TCS add	.3	.3
Wheel Cylinder, R&R and Recondition (B)		
Includes: System bleeding.		
One side	1.3	1.4
Both sides	1.9	2.1
Wheel Cylinder, Replace (B)		
Includes: System bleeding.		
One side	1.1	1.3
Both sides	1.7	2.0

HILLHOLDER

	LABOR TIME	SEVERE SERVICE
Hillholder Cable, Replace (B)		
All Models	.6	.8
Pressure Holding Valve, Replace (B)		
All Models	1.3	1.4
Three-Way Connector Valve, Replace (B)		
All Models	.8	1.0

PARKING BRAKE

	LABOR TIME	SEVERE SERVICE
Parking Brake Cable, Adjust (B)		
All Models	.5	.7
Parking Brake Apply Actuator, Replace (B)		
All Models	.6	.6
Parking Brake Apply Warning Indicator Switch, Replace (B)		
Exc. below	.7	.7
Baja, Legacy, Outback	1.0	1.0
SVX	1.4	1.4
Parking Brake Cable, Replace (B)		
Baja, Legacy, Outback		
one side	1.3	1.4
both sides	1.7	2.0
Brat		
one side	1.2	1.4
both sides	2.3	2.5
Forester		
one side	1.6	1.8
both sides	2.1	2.3
Impreza		
one side	1.6	1.8
both sides	2.1	2.3
Justy	1.1	1.3
Loyale		
one side	1.1	1.3
both sides	1.5	1.7

	LABOR TIME	SEVERE SERVICE
SVX		
one side	2.3	2.5
both sides	2.7	2.9
XT		
one side	1.1	1.3
both side	1.5	1.7
1981-89 Hatchback		
one side	1.3	1.5
both sides	2.2	2.4
1981-89 Station Wagon		
one side	1.4	1.6
both sides	2.2	2.4
1981-89 3 door, 4 door		
one side	1.3	1.5
both sides	2.2	2.4

FRONT SUSPENSION

Unless otherwise noted, time given does not include alignment.

	LABOR TIME	SEVERE SERVICE
Diagnose Air Suspension System (A)		
1988-94	.8	.8
Align Front End (A)		
All Models	1.6	1.8
Front Toe, Adjust (B)		
All Models	.8	1.0
Front Air Strut, Replace (B)		
Loyale		
one	1.0	1.2
both	1.6	1.8
1988-91 2, 4 door, Wagon, XT		
one	1.3	1.5
both	1.9	2.1
1990-94 Legacy		
one	2.8	3.0
both	3.5	3.7
Front Hub Assy., Replace (B)		
Baja, Legacy, Outback	4.3	4.8
Brat	.7	.9
Forester, Impreza	4.3	4.8
Justy	.6	.9
Loyale	.6	.9
SVX	3.6	4.1
XT		
4 cyl.	1.8	2.3
6 cyl.	2.3	2.8
1981-89 Hatchback	.7	1.0
1981-89 Station Wagon	.7	1.0
1981-89 3 door, 4 door	.7	1.0
Lower Ball Joint, Replace (B)		
One side	.6	.7
Both sides	1.1	1.3
Lower Control Arm, Replace (B)		
Baja, Legacy, Outback	2.4	2.7
Brat	.8	1.1
Forester, Impreza	2.4	2.7
Justy, Loyale	.8	1.1
SVX	3.3	3.6
XT	.8	1.1

SUB-20 BAJA : BRAT : DL : FORESTER : GL : GL-10 : GLF : IMPREZA : JUSTY

	LABOR TIME	SEVERE SERVICE
Lower Control Arm, Replace (B)		
1981-89 Hatchback	.8	1.1
1981-89 Station Wagon	.8	1.1
1981-89 3 door, 4 door	.8	1.1
Replace bushings add each side	.3	.5
Lower Control Arm Strut (Tension Bar), Replace (B)		
1981-94	.5	.8
Stabilizer Bar, Replace (B)		
All Models	.6	.8
Steering Knuckle, Replace (B)		
Includes: Alignment.		
Baja, Legacy, Outback	4.3	4.8
Brat	1.8	2.3
Forester, Impreza	4.3	4.8
Justy	2.1	2.6
Loyale	2.1	2.6
XT		
4 cyl.	2.1	2.6
6 cyl.	2.2	2.7
1981-89 Hatchback	1.9	2.1
1981-89 Station Wagon	2.0	2.2
1981-89 3 door, 4 door	2.0	2.2
Strut Assy., R&R and Recondition (B)		
Includes: Alignment.		
Baja, Legacy, Outback		
1990-99	3.8	4.3
2000-04	3.2	3.6
Brat	1.8	2.1
Forester	3.8	4.3
Impreza		
1993-94	3.2	3.5
1995-00	3.8	4.3
2001-04	3.2	3.6
Justy	1.4	1.7
Loyale	1.4	1.7
SVX	3.3	3.6
XT	1.4	1.7
1981-89 Hatchback	1.8	2.1
1981-89 Station Wagon	1.6	1.9
1981-89 3 door, 4 door	1.8	2.1
Wheel Bearings, Replace (B)		
Baja, Legacy, Outback	3.8	4.3
Brat	2.3	2.5
Forester, Impreza	3.8	4.3
Justy	2.1	2.3
Loyale	2.1	2.4
SVX	3.6	3.9
XT		
4 cyl.	2.1	2.4
6 cyl.	2.5	2.8
1981-89 Hatchback	2.1	2.4
1981-89 Station Wagon	2.2	2.5
1981-89 3 door, 4 door	2.3	2.6

REAR SUSPENSION

	LABOR TIME	SEVERE SERVICE
Rear Alignment, Adjust (A)		
1987-04	2.2	2.5
Air Compressor Relay, Replace (B)		
1988-94	.2	.2
Air Control Solenoid, Replace (B)		
1988-94	.5	.5
Air Suspension Control Unit, Replace (B)		
1988-94	.5	.5
Coil Spring, R&R or Replace (B)		
Justy		
one	.5	.7
both	.7	.9
Compressor/Drier and/or Air Tank, Replace (B)		
1988-94	.7	.7
Height Control Switch, Replace (B)		
Loyale	.8	.8
XT	1.0	1.0
1990-94 Legacy	.5	.5
Lateral Link, Replace (B)		
1990-04		
one side	3.0	3.4
both sides	3.6	4.1
Lower Control Arm, Replace (B)		
Add for alignment.		
Justy	.5	.8
Rear Air Shock, Replace (B)		
Loyale		
one	.5	.7
both	.7	.9
XT		
one	.5	.7
both	.7	.9
1988-89 2 dr, 4 dr, Wagon		
one	.5	.7
both	.7	.9
1990-94 Legacy		
one	1.7	1.9
both	2.3	2.5
Rear Control Arm, R&R or Replace (B)		
1981-95		
outer	.8	1.1
inner		
w/AWD	2.2	2.5
w/o AWD	1.6	1.9
2000-04 Baja, Legacy, Outback		
one side	3.2	3.6
both sides	5.3	6.0
Rear Crossmember, R&R or Replace (B)		
Loyale XT		
w/AWD	2.3	2.8
w/o AWD	1.9	2.4
1981-89, all		
w/AWD	2.7	3.2
w/o AWD	2.4	2.9

	LABOR TIME	SEVERE SERVICE
Rear Hub and/or Bearings, Replace (B)		
1990-99		
one side	2.1	2.6
both sides	3.5	4.0
2000-04		
Baja, Legacy, Outback	1.6	1.8
Forester, Impreza	2.4	2.7
w/ABS add each side	.6	.6
w/drum brakes add	.3	.3
Rear Hub or Wheel Spindle, Replace (B)		
Justy		
w/AWD	1.7	2.2
w/o AWD	1.1	1.6
Rear Link, Replace (B)		
2001-04 Baja, Legacy, Outback		
one side	1.1	1.3
both sides	1.4	1.6
Rear Spindle Housing and/or Bushing, Replace (B)		
Baja, Legacy, Outback	1.6	1.8
Forester	2.9	3.2
Impreza	1.6	1.8
SVX	2.2	2.5
w/ABS/TCS Forester, Impreza, Legacy, Outback, SVX add	.5	.5
w/AWD Impreza, Legacy, Outback add	1.3	1.3
w/drum brakes add	.3	.3
Rear Stabilizer Control Links, Replace (B)		
Justy	.5	.7
Rear Struts, R&R or Replace (B)		
Baja, Legacy, Outback		
1990-99		
one	2.4	2.7
both	3.8	4.3
2000-04		
one	1.9	2.1
both	3.0	3.4
Forester		
one	2.4	2.7
both	3.8	4.3
Impreza		
1993-00		
one	2.4	2.7
both	3.8	4.3
2001-04		
one	1.9	2.1
both	3.0	3.4
SVX		
one	2.1	2.3
both	2.8	3.0
1981-89		
one	.5	.7
both	.8	1.0

LEGACY : LOYALE : OUTBACK : RX : STANDARD : SVX : XT : XT6   **SUB-21**

	LABOR TIME	SEVERE SERVICE
Rear Wheel Oil Seal, Replace (B)		
All Models inner or outer		
one	.5	.6
both	.8	.9
w/ABS add each side	.1	.1
Shock Absorbers, Replace (B)		
1981-89 Hatchback		
one	.6	.8
both	.8	1.0
Justy		
one	.5	.7
both	.7	.9
Stabilizer Bar, Replace (B)		
All Models	.6	.7
Stabilizer Bar Bushings, Replace (B)		
one side	.3	.5
both sides	.5	.7
Torsion Bar, Replace (B)		
1981-89 Hatchback		
one	1.3	1.8
both	2.3	2.8
Trailing Link, Replace (B)		
Baja, Legacy, Outback		
one side	2.5	2.9
both sides	2.9	3.2
Forester, Impreza		
one side	2.5	2.9
both sides	2.9	3.2
Justy	1.1	1.4
SVX		
one side	2.3	2.6
both sides	2.7	3.0
Wheel Bearings, Replace (B)		
w/AWD		
one side	2.4	2.7
both sides	3.8	4.3
w/o AWD		
one side	.7	.9
both sides	1.2	1.4
Upper Link, Replace (B)		
2001 Legacy		
one side	1.1	1.3
both sides	1.4	1.6

STEERING
AIR BAGS

	LABOR TIME	SEVERE SERVICE
Diagnose Air Bag System (A)		
1990-04	1.0	1.0
Contact Reel, Replace (B)		
Baja, Legacy, Outback	1.4	1.4
Forester, Impreza	1.4	1.4
SVX	1.1	1.1
Electronic Control Unit, Replace (B)		
Baja, Legacy, Outback	.5	.5
Forester, Impreza	.6	.6
SVX	1.6	1.6
Front Sensor, Replace (B)		
Baja, Legacy, Outback		
right side	3.8	4.2
left side	1.9	2.1
Forester, Impreza		
one side	1.9	2.1
SVX		
right or left side	2.6	2.6
Inflatable Cushion Module, Replace (B)		
Baja, Legacy, Outback		
1990-94		
right side	1.3	1.3
left side	.8	.8
1995-04		
right side	2.4	2.4
left side	.7	.7
Forester		
right side	.9	.9
left side	.7	.7
Impreza		
1993-94		
right side	1.8	1.8
left side	.8	.8
1995-04		
right side	2.4	2.4
left side	.7	.7
SVX		
left side	.8	.8
right side	1.3	1.3

MANUAL RACK & PINION

	LABOR TIME	SEVERE SERVICE
Gear Assy., Adjust (B)		
1981-94	.5	.7
Horn Contact, Replace (B)		
1981-94	.5	.5
Outer Tie Rod End, Replace (B)		
Includes: Reset toe.		
1981-94	1.2	1.4
Replace inner tie rod assembly add	.3	.5
Rack & Pinion Assy., R&R and Recondition (A)		
Includes: Alignment.		
Brat	3.5	4.0
Justy	2.9	3.4
Loyale	4.3	4.8
XT	4.3	4.8
1981-89 Hatchback	3.6	4.1
1981-89 Station Wagon	3.7	4.2
1981-89 3 door, 4 door	3.6	4.1
Rack & Pinion Assy., R&R or Replace (B)		
Includes: Alignment.		
Brat	2.5	2.8
Justy	1.6	1.9
Loyale	2.9	3.2
XT	2.9	3.2
1981-89 Hatchback	2.7	3.0
1981-89 Station Wagon	2.6	2.9
1981-89 3 door, 4 door	2.5	2.8
Rack Boots, Replace (B)		
1981-94	1.7	1.9
Rack Mounting Bushings, Replace (B)		
1981-94	.6	.8
Steering Column, Replace (B)		
1981-94	1.0	1.0
w/air bags add	.2	.2
w/tilt wheel add	.5	.5
Replace column add	.5	.5
Steering Shaft Coupler, Replace (B)		
1981-94	.5	.7
Steering Wheel, Replace (B)		
1981-94	.5	.5
w/air bags add	.5	.5
Upper Mast Jacket Bearing, Replace (B)		
Includes: Insulators.		
1981-94	1.3	1.5

POWER RACK & PINION

	LABOR TIME	SEVERE SERVICE
Horn Contact or Canceling Cam, Replace (B)		
All Models	.5	.5
Outer Tie Rod Ends, Replace (B)		
Includes: Reset toe.		
All Models	1.4	1.6
Rack & Pinion Assy., Adjust (B)		
All Models	.5	.7
Rack & Pinion Assy., R&R and Recondition (A)		
Baja, Legacy, Outback	4.4	5.0
H6, 3.0	4.7	5.3
Forester, Impreza	4.4	5.0
Loyale	4.3	4.8
SVX	4.5	5.0
WRX	5.2	5.8
XT		
w/electronic strg	4.6	5.1
w/o electronic strg	4.3	4.8
1981-89 Hatchback	4.2	4.7
1981-89 Station Wagon	4.3	4.8
1981-89 3 door, 4 door	4.2	4.7
Rack & Pinion Assy., R&R or Replace (B)		
Includes: Adjustments.		
Baja, Legacy, Outback	3.0	3.4
H6, 3.0	3.3	3.7
Forester	3.0	3.4
Impreza	3.0	3.4
Loyale	2.9	3.2
SVX	3.2	3.5
WRX	3.8	4.3

SUB-22 BAJA : BRAT : DL : FORESTER : GL : GL-10 : GLF : IMPREZA : JUSTY

	LABOR TIME	SEVERE SERVICE
Rack & Pinion Assy., R&R or Replace (B)		
XT		
w/electronic strg	3.1	3.4
w/o electronic strg	2.9	3.2
1981-89 Hatchback	3.0	3.3
1981-89 Station Wagon	2.9	3.2
1981-89 3 door, 4 door	2.8	3.1
Rack Boots, Replace (B)		
All Models	1.9	2.1
Rack Mounting Bushings, Replace (B)		
All Models	.6	.7
Steering Column, R&I (B)		
All Models	1.0	1.0
w/air bags add	.2	.2
w/tilt wheel add	.5	.5
Replace column add	.5	.5
Steering Flex Coupling, Replace (B)		
All Models	.5	.7
Steering Pump, R&R or Replace (B)		
All Models	1.4	1.6
w/turbo add	.1	.1
Steering Pump Hoses, Replace (B)		
All Models each	.6	.8
w/turbo add	.2	.2
Steering Pump Reservoir or Seals, Replace (B)		
1995-04	.5	.7
w/air bags add	.5	.5
Steering Wheel, Replace (B)		
All Models	.7	.7
w/air bags add	.5	.5
Upper Mast Jacket Bearing, Replace (B)		
Includes: Insulators.		
Exc. Below	1.4	1.4
H6 3.0L	1.6	1.6
ELECTRONIC POWER STEERING		
Power Steering System Test (A)		
SVX, XT 6 cyl.	1.0	1.0
ECM Steering, Replace (B)		
SVX	.5	.7
Fluid Heater Relay, Replace (B)		
XT 6 cyl.	.3	.3
Fluid Heater Temperature Sensor, Replace (B)		
XT 6 cyl.	.3	.3
Photo Coupler Sensor Plate, Replace (B)		
XT 6 cyl.	1.8	2.0
Power Controller, Replace (B)		
XT 6 cyl.	.5	.5
Speed Sensor, Replace (B)		
XT 6 cyl.	.8	.8
SVX	.3	.3
Signal Control Unit, Replace (B)		
XT 6 cyl.	.3	.3

	LABOR TIME	SEVERE SERVICE
Steering Reaction Solenoid, Replace (B)		
SVX	3.6	3.8
Sub Fan Diode, Replace (B)		
XT 6 cyl.	.5	.5
Sub Fan Relay, Replace (B)		
XT 6 cyl.	.3	.3

HEATING & AIR CONDITIONING

When more than one component requires replacement where evacuation/recovery and recharging is already included, deduct 1.0 hour for each additional component from the time given.

	LABOR TIME	SEVERE SERVICE
Evacuate/Recover and Recharge System (B)		
All Models	1.0	1.0
AC Hoses, Replace (B)		
Includes: Evacuate/recover and recharge.		
One	2.0	2.2
each addl. add	.3	.2
w/PS add	.2	.2
Compressor Assy., Replace (B)		
Includes: Parts transfer, evacuate/recover and recharge.		
All Models	2.6	2.8
w/turbo add	.3	.3
Compressor Clutch Coil, Replace (B)		
All Models	3.1	3.4
Compressor Clutch Pulley (w/Hub), Replace (B)		
All Models	3.1	3.4
Condenser Assy., Replace (B)		
Includes: Evacuate/recover and recharge.		
All Models	2.1	2.4
w/turbo add	.1	.1
Evaporator Coil, Replace (B)		
Includes: Evacuate/recover and recharge.		
All Models	2.7	2.7
Expansion Valve, Replace (B)		
Includes: Evacuate/recover and recharge.		
1981-94	3.1	3.3
1995-04	2.0	2.2
Heater & AC Mode Switch, Replace (B)		
Baja, Legacy, Outback		
1990-94	2.0	2.0
1995-04	1.6	1.6
Brat	1.7	1.7
Forester, Impreza	.8	.8
Justy	1.0	1.0
Loyale	1.9	1.9
SVX	2.3	2.3

	LABOR TIME	SEVERE SERVICE
XT	1.5	1.5
1981-84		
w/AC	3.1	3.1
w/o AC	3.1	3.1
1985-89 Hatchback	1.6	1.6
1985-89 Station Wagon	1.6	1.6
1985-89 3 door, 4 door	1.7	1.7
Heater Blower Motor, Replace (B)		
Baja, Legacy, Outback		
1990-94		
w/AC	1.8	1.8
w/o AC	1.1	1.1
1995-04	1.1	1.1
Forester, Impreza	1.1	1.1
Justy	1.6	1.6
Loyale	1.1	1.1
SVX	1.0	1.0
1981-84		
w/AC	1.6	1.6
w/o AC	.9	.9
1985-88	.7	.7
1989-91 XT		
w/AC	2.7	2.7
w/o AC	1.1	1.1
1985-89 Hatchback	1.2	1.2
1985-89 Station Wagon	1.3	1.3
1985-89 3 door, 4 door	1.2	1.2
Heater Blower Motor Relay, Replace (B)		
1981-84	.5	.5
1985-04	1.0	1.0
Heater Blower Motor Resistor, Replace (B)		
All Models	1.0	1.0
Heater Blower Motor Switch, Replace (B)		
Baja, Legacy, Outback		
1990-94	2.1	2.1
1995-04	1.6	1.6
Forester, Impreza	1.1	1.1
Justy	1.2	1.2
Loyale	1.6	1.6
SVX	1.5	1.5
XT	1.7	1.7
1981-84	1.5	1.5
1985-87	.8	.8
1985-89 Hatchback	1.5	1.5
1985-89 Station Wagon	1.5	1.5
1985-89 3 door, 4 door	1.6	1.6
Heater Core, R&R or Replace (B)		
w/AC		
Baja, Legacy, Outback		
1990-94	3.5	3.5
1995-04	3.6	3.6
Forester	3.6	3.6

LEGACY : LOYALE : OUTBACK : RX : STANDARD : SVX : XT : XT6   **SUB-23**

	LABOR TIME	SEVERE SERVICE
Hatchback		
1985-87	4.2	4.2
1988-89	5.3	5.3
Impreza		
1993-97	3.5	3.5
1998-04	3.4	3.4
Justy	2.6	2.6
Loyale	4.4	4.4
Station Wagon		
1985-87	4.3	4.3
1988-89	5.2	5.2
SVX	5.5	5.5
XT	3.6	3.6
3 door, 4 door		
1985-87	4.3	4.3
1988-89	5.3	5.3
1981-84 all	5.8	5.8
1985-87 Brat	4.2	4.2
w/o AC		
Baja, Legacy, Outback		
1990-94	3.5	3.5
1995-04	3.6	3.6
Forester	3.6	3.6
Hatchback		
1981-87	2.9	2.9
1988-89	5.3	5.3
Impreza	3.4	3.4
Justy	2.6	2.6
Loyale	4.4	4.4
Station Wagon		
1985-87	2.9	2.9
1988-89	5.1	5.1
SVX	6.7	6.7
XT	3.6	3.6
3 door, 4 door		
1985-87	2.8	2.8
1988-89	5.3	5.3
1981-84	4.3	4.3
1985-87 Brat	2.7	2.7

Heater Hoses, Replace (B)
| 1981-84 | .6 | .8 |
| 1985-04 | .7 | .9 |

Receiver/Drier Assy., Replace (B)
Includes: Evacuate/recover and recharge.
| All Models | 1.8 | 2.1 |

AUTOMATIC TEMPERATURE CONTROL

Diagnose ATC System (A)
| 1990-97 | 1.0 | 1.0 |

AC Cut Relay, Replace (B)
| 1990-94 Legacy | .3 | .3 |

AC Relay, Replace (B)
| 1990-97 | .3 | .3 |

Air Mix Door Motor, Replace (B)
| 1990-97 | .5 | .5 |

Ambient Sensor, Replace (B)
| 1990-97 | .5 | .5 |

Amplifier Control Unit and/or In-Vehicle Sensor, Replace (B)
| 1990-94 Legacy | 1.1 | 1.1 |
| 1992-97 SVX | .8 | .8 |

Coolant Temperature Sensor, Replace (B)
| 1990-94 Legacy | .5 | .5 |
| 1992-97 SVX | .3 | .3 |

Fan Control Amplifier, Replace (B)
| 1990-94 Legacy | .3 | .3 |
| 1992-97 SVX | .6 | .6 |

Intake Door Motor, Replace (B)
| 1990-94 Legacy | .5 | .5 |
| 1992-97 SVX | 3.5 | 3.5 |

In-Take Sensor, Replace (B)
1990-94 Legacy	.3	.3
1992-97 SVX	2.5	2.5
w/dual air bags add	.2	.2

Mode Door Motor, Replace (B)
| 1990-97 | .6 | .6 |

Sunload Sensor, Replace (B)
| 1990-94 Legacy | .3 | .3 |
| 1992-97 SVX | 3.4 | 3.4 |

Trinary Switch, Replace (B)
| 1990-94 Legacy | .7 | .7 |
| 1992-97 SVX | 1.5 | 1.5 |

WIPERS & SPEEDOMETER

Antenna, Replace (B)
Baja, Legacy, Outback
1990-96
| manual | .7 | .7 |
power
| Station Wagon | 1.0 | 1.0 |
| 4 door | .6 | .6 |
1997-04
4 door	.6	.6
Station Wagon	1.4	1.4
Brat	.7	.7
Forester	.7	.7
Impreza
manual	.7	.7
power
2 door, 4 door	.6	.6
Station Wagon	1.1	1.1
Loyale	.7	.7
SVX
manual	.5	.5
power	.6	.6
XT
manual	.5	.5
power	.6	.6
1981-89 Hatchback	.7	.7
1981-89 Station Wagon	.7	.7
1981-89 3 door, 4 door	.7	.7

Instrument Panel, Replace (B)
| All Models | 2.1 | 2.4 |

Radio, R&R (B)
1981-99	.5	.5
2000-04 Forester	.8	.8
2000-04 Impreza	.5	.5
2000-04 Legacy	1.2	1.2

Rear Window Washer Pump, Replace (B)
| 1981-94 | .6 | .6 |
| Forester | .5 | .5 |
1995-04
Impreza	.8	.8
Legacy, Outback	.5	.5
SVX	.6	.6

Rear Window Wiper & Washer Switch, Replace (B)
| Forester | .6 | .6 |
Impreza
1993-94	1.0	1.0
1995-97	1.2	1.2
1998-00	.6	.6
2001-04	1.2	1.2
Justy	.3	.3
Legacy, Outback
1990-94	1.0	1.0
1995-04	1.2	1.2
Loyale	.5	.5
SVX	1.1	1.1
1981-89 Station Wagon	.5	.5
Hatchback	.5	.5
3 door	.5	.5
w/air bags add
90-94 Legacy, Outback	.2	.2
Forester, Impreza	.2	.2

Rear Window Wiper Intermittent Control Unit, Replace (B)
| 1990-94 Legacy | .5 | .5 |

Rear Window Wiper Motor, Replace (B)
| All Models | .8 | .9 |

Speedometer Cable & Casing, Replace (B)
| All Models | 1.1 | 1.2 |

Speedometer Head, R&R or Replace (B)
Baja, Legacy, Outback
1990-94	1.8	1.8
1995-04	1.5	1.5
Brat	1.3	1.3
Forester	1.5	1.5
Hatchback
1981-84	1.2	1.2
1985-87	1.6	1.6
Justy	1.5	1.5
Loyale	1.6	1.6
Impreza	1.5	1.5
XT	1.9	1.9
Station Wagon
1981-84	1.2	1.2
1985-87	1.7	1.7

SUB-24 BAJA : BRAT : DL : FORESTER : GL : GL-10 : GLF : IMPREZA : JUSTY

	LABOR TIME	SEVERE SERVICE
Speedometer Head, R&R or Replace (B)		
SVX	1.6	1.6
3 door, 4 door		
1981-84	1.2	1.2
1985-87	1.6	1.6
Windshield Washer Pump, Replace (B)		
1981-94	.6	.6
1995-04	.5	.5
Windshield Wiper & Washer Switch, Replace (B)		
Baja, Legacy, Outback		
1990-94	1.0	1.0
1995-04	1.2	1.2
Brat, Forester	.6	.6
Impreza		
1993-94	1.0	1.0
1995-97	1.2	1.2
1998-00	.6	.6
2001-04	1.2	1.2
Justy	.7	.7
Loyale	.7	.7
SVX	1.1	1.1
XT	.8	.8
1981-89 Hatchback	.7	.7
1981-89 Station Wagon	.7	.7
1981-89 3 door, 4 door	.7	.7
w/air bags add		
90-94 Legacy, Outback	.2	.2
Forester, Impreza	.2	.2
Windshield Wiper Intermittent Control Unit, Replace (B)		
Baja, Legacy, Outback		
1990-94	1.0	1.0
1995-04	1.2	1.2
Brat	.5	.5
Forester	.6	.6
Impreza		
1993-94	1.0	1.0
1995-97	1.2	1.2
1998-00	.6	.6
2001-04	1.2	1.2
Justy	.5	.5
Loyale	.5	.5
SVX	.6	.6
XT	.8	.8
Station Wagon	.5	.5
3 door, 4 door	.5	.5
1981-89 Hatchback	.5	.5
w/air bags add		
90-94 Legacy, Outback	.2	.2
Forester, Impreza	.2	.2
Windshield Wiper Intermittent Selector Switch, Replace (B)		
XT	.6	.6

	LABOR TIME	SEVERE SERVICE
Windshield Wiper Linkage, Replace (B)		
Baja, Legacy, Outback	.6	.8
Brat, Justy	.5	.7
Forester, Impreza	.6	.8
Loyale	.5	.7
SVX	1.0	1.2
XT	.6	.8
1981-89 Hatchback	.5	.7
1981-89 Station Wagon	.5	.7
1981-89 3 door, 4 door	.5	.7
Windshield Wiper Motor, Replace (B)		
Exc. SVX, XT	.6	.8
SVX	1.0	1.2
XT	.7	.9

LAMPS & SWITCHES

	LABOR TIME	SEVERE SERVICE
Back-Up Lamp Assy., Replace (B)		
All Models	.3	.3
Back-Up Lamp Switch, Replace (B)		
All Models	.5	.5
Combination Switch Assy., Replace (B)		
Baja, Legacy, Outback		
1990-94	1.0	1.0
1995-04	1.2	1.2
Brat	.7	.7
Forester	1.2	1.2
Impreza		
1993-94	1.0	1.0
1995-04	1.2	1.2
Justy	.7	.7
Loyale	.7	.7
SVX	1.1	1.1
XT	.8	.8
Station Wagon	.7	.7
3 door, 4 door	.7	.7
1981-89 Hatchback	.7	.7
w/air bags add		
90-94 Legacy, Outback	.2	.2
Forester, Impreza	.2	.2
Halogen Headlamp Bulb, Replace (C)		
1985-04	.3	.3
Hazard Warning Switch, Replace (B)		
1981-94		
exc. below	.6	.6
Impreza	.5	.5
Baja, Legacy	.8	.8
SVX	.5	.5
XT	.7	.7
1995-04	.5	.5
Headlamp Switch, Replace (B)		
Loyale	.7	.7
XT	1.0	1.0
Headlamps, Aim (B)		
Two	.4	.4
Four	.6	.6

	LABOR TIME	SEVERE SERVICE
High Mount Stop Lamp Assy., Replace (B)		
1985-04		
rear window	.5	.5
rear spoiler	.7	.7
Horn, Replace (B)		
Exc. below	.6	.6
1987-88 Justy	.3	.3
Baja, Legacy, Outback	.3	.3
XT 6 cyl.	.3	.3
SVX	.3	.3
Horn Switch, Replace (B)		
All Models	.5	.5
License Lamp Assy., Replace (B)		
All Models	.5	.5
Neutral Safety Switch, Replace (B)		
All Models	.5	.5
Parking Brake Warning Lamp Switch, Replace (B)		
Exc. below	.7	.7
Legacy, Outback	1.0	1.0
SVX	1.4	1.4
Parking Lamp Lens or Bulb, Replace (C)		
All Models	.3	.3
Sealed Beam Headlamp, Replace (C)		
1981-95 each	.3	.3
Side Marker Lamp Bulb, Replace (C)		
1981-99	.3	.3
2000-04	.6	.6
Stop Lamp Switch, Replace (B)		
All Models	.6	.6
Tail Lamp Lens or Bulb, Replace (B)		
All Models	.3	.3
Turn Signal Combination Switch, Replace (B)		
Baja, Impreza, Forester, Legacy, Outback	.6	.6
Turn Signal or Hazard Warning Flasher, Replace (B)		
All Models	.6	.6

SEAT BELTS

	LABOR TIME	SEVERE SERVICE
Diagnose Seat Belt System (A)		
Justy	.6	.6
Loyale	.5	.5
SVX	.7	.7
1987-89 Station Wagon	.5	.5
1987-89 4 door	.5	.5
1990-94 Legacy	.7	.7
Belt Guide Switch, Replace (B)		
1987-94		
front	.5	.5
rear	1.1	1.1

LEGACY : LOYALE : OUTBACK : RX : STANDARD : SVX : XT : XT6 — SUB-25

	LABOR TIME	SEVERE SERVICE
Control Module, Replace (B)		
Justy	.5	.5
Loyale		
4 door	.6	.6
Station Wagon	.8	.8
SVX	.6	.6
XT	.8	.8
1987-89 Station Wagon	.9	.9
1987-89 4 door	.7	.7
1990-94 Legacy		
4 door	.6	.6
Station Wagon	.8	.8
Door Lock Switch, Replace (B)		
Justy	1.2	1.2
Loyale	1.4	1.4
SVX	.7	.7
1987-89 Station Wagon	1.5	1.5
1987-89 4 door	1.5	1.5
1990-94 Legacy, XT	.8	.8
Lap Belt Switch, Replace (B)		
1987-94	.8	.8
Rail Motor, Replace (B)		
1987-94	1.2	1.2

	LABOR TIME	SEVERE SERVICE
BODY		
Door Window Regulator, Replace (B)		
Manual	1.3	1.5
Power	1.4	1.6
Hood Hinges (Both), Replace (B)		
Baja, Legacy, Outback	1.2	1.4
Brat	1.2	1.4
Forester	1.8	2.1
Impreza	1.8	2.1
Justy	1.1	1.2
Loyale	1.1	1.2
SVX, XT	1.1	1.2
1981-89 Hatchback	1.3	1.5
1981-89 Station Wagon	1.2	1.4
1981-89 3 door, 4 door	1.3	1.5
Hood Lock, Replace (B)		
All Models	.5	.5
Hood Release Cable, Replace (B)		
Baja, Legacy, Outback	1.2	1.4
Brat	.8	.8
Forester	1.1	1.2

	LABOR TIME	SEVERE SERVICE
Impreza	1.1	1.2
Justy	.7	.7
Loyale	.8	.8
SVX	1.1	1.1
XT	.8	.8
1981-89 Hatchback	.8	.8
1981-89 Station Wagon	.8	.8
1981-89 3 door, 4 door	.8	.8
Lock Striker Plate, Replace (B)		
All Models	.5	.5
Window Regulator Cable, Replace (B)		
XT	1.4	1.6
SVX	1.2	1.4
Window Regulator Motor, Replace (B)		
1988-04	1.4	1.6

NOTES

SUZ

Aerio : Esteem : Forenza : Grand Vitara : Samurai : Sidekick : Swift : Verona : Vitara : X-90 : XL-7

SYSTEM INDEX

- MAINTENANCE SUZ-2
  - Maintenance Schedule... SUZ-2
- CHARGING SUZ-2
- STARTING SUZ-3
- CRUISE CONTROL SUZ-3
- IGNITION SUZ-3
- EMISSIONS SUZ-3
- FUEL SUZ-4
- EXHAUST SUZ-5
- ENGINE COOLING SUZ-5
- ENGINE SUZ-5
  - Assembly SUZ-5
  - Cylinder Head SUZ-6
  - Camshaft SUZ-7
  - Crank & Pistons SUZ-7
  - Engine Lubrication .. SUZ-8
- CLUTCH SUZ-8
- MANUAL TRANSAXLE SUZ-8
- MANUAL TRANSMISSION ... SUZ-9
- AUTO TRANSAXLE SUZ-9
- AUTO TRANSMISSION SUZ-9
- TRANSFER CASE SUZ-10
- DRIVELINE SUZ-10
- BRAKES SUZ-11
- FRONT SUSPENSION SUZ-12
- REAR SUSPENSION SUZ-13
- STEERING SUZ-14
- HEATING & AC SUZ-15
- WIPERS & SPEEDOMETER .. SUZ-16
- LAMPS & SWITCHES SUZ-16
- SEAT BELTS SUZ-16
- BODY SUZ-17

OPERATIONS INDEX

A
- AC Hoses SUZ-15
- Air Bags SUZ-14
- Air Conditioning SUZ-15
- Alignment SUZ-12
- Alternator (Generator) . SUZ-2
- Antenna SUZ-16
- Anti-Lock Brakes SUZ-11

B
- Back-Up Lamp Switch ... SUZ-16
- Ball Joint SUZ-13
- Battery Cables SUZ-3
- Bleed Brake System SUZ-11
- Blower Motor SUZ-15
- Brake Disc SUZ-11
- Brake Drum SUZ-11
- Brake Hose SUZ-11
- Brake Pads and/or Shoes . SUZ-12

C
- Camshaft SUZ-7
- Camshaft Sensor SUZ-3
- Catalytic Converter ... SUZ-5
- Coolant Temperature (ECT) Sensor SUZ-4
- Crankshaft SUZ-7
- Crankshaft Sensor SUZ-3
- Cruise Control SUZ-3
- CV Joint SUZ-10
- Cylinder Head SUZ-6

D
- Differential SUZ-10
- Distributor SUZ-3
- Drive Belt SUZ-2
- Driveshaft SUZ-10

E
- EGR SUZ-4
- Engine SUZ-5
- Engine Lubrication SUZ-8
- Engine Mounts SUZ-6
- Evaporator SUZ-15
- Exhaust SUZ-5
- Exhaust Manifold SUZ-5
- Expansion Valve SUZ-15

F
- Flexplate SUZ-9
- Flywheel SUZ-8
- Fuel Injection SUZ-4
- Fuel Pump SUZ-4
- Fuel Vapor Canister ... SUZ-4

G
- Generator SUZ-2

H
- Halfshaft SUZ-10
- Headlamp SUZ-16
- Heater Core SUZ-15
- Horn SUZ-16

I
- Idle Air Control (IAC) Valve .. SUZ-5
- Ignition Coil SUZ-3
- Ignition Module SUZ-3
- Inner Tie Rod SUZ-14
- Intake Air Temperature (IAT) Sensor .. SUZ-4
- Intake Manifold SUZ-4

L
- Lower Control Arm SUZ-12

M
- Maintenance Schedule .. SUZ-2
- Manifold Absolute Pressure (MAP) Sensor .. SUZ-4
- Mass Air Flow (MAF) Sensor ... SUZ-4
- Master Cylinder SUZ-11
- Muffler SUZ-5
- Multifunction Lever/Switch .. SUZ-16

O
- Oil Pan SUZ-8
- Oil Pump SUZ-8
- Outer Tie Rod SUZ-14
- Oxygen Sensor SUZ-4

P
- Parking Brake SUZ-12
- Pistons SUZ-7
- Positive Crankcase Ventilation (PCV) Valve SUZ-4

R
- Radiator SUZ-5
- Radiator Hoses SUZ-5
- Radio SUZ-16
- Rear Main Oil Seal SUZ-8

S
- Shock Absorber/Strut, Front SUZ-12
- Shock Absorber/Strut, Rear SUZ-13
- Spark Plug Cables SUZ-3
- Spark Plugs SUZ-3
- Spring, Leaf SUZ-13
- Spring, Rear Coil SUZ-13
- Starter SUZ-3
- Steering Wheel SUZ-14

T
- Thermostat SUZ-5
- Throttle Body SUZ-5
- Throttle Position Sensor (TPS) SUZ-4
- Timing Belt SUZ-7
- Timing Chain SUZ-7
- Torque Converter SUZ-9

U
- U-Joint SUZ-10

V
- Valve Body SUZ-9
- Valve Cover Gasket SUZ-6
- Valve Job SUZ-7

W
- Water Pump SUZ-5
- Wheel Balance SUZ-2
- Wheel Cylinder SUZ-12
- Window Regulator SUZ-17
- Windshield Washer Pump . SUZ-16
- Windshield Wiper Motor . SUZ-16

SUZ-1

SUZ-2   AERIO : ESTEEM : FORENZA : GRAND VITARA : SAMURAI

MAINTENANCE

	LABOR TIME	SEVERE SERVICE
Air Cleaner Filter Element, Replace (B)		
All Models	.3	.4
Chassis Lubrication, Change Oil & Filter (C)		
Includes: Correct all fluid levels.		
All Models	.3	.4
Drive Belt, Adjust (B)		
All Models	.3	.4
Drive Belt, Replace (B)		
Exc. below	.6	.7
Aerio, Samurai	.8	.9
Compressor		
Aerio, Verona	.8	.9
1989-94 Swift	1.4	1.5
Steering pump		
Verona	.9	1.0
Fuel Filter, Replace (B)		
Carbureted	.5	.6
Fuel injected		
Aerio, Esteem	1.5	1.5
Forenza, Verona	.5	.5
Grand Vitara, XL-7	.8	.9
Samurai, Sidekick, X-90	.7	.8
Swift		
1989-94	.8	.9
1995-01	1.1	1.2
Vitara	1.0	1.1
Halogen Headlamp Bulb, Replace (C)		
Exc. Forenza	.3	.3
Forenza	.6	.6
Oil and Filter, Change (C)		
Includes: Correct all fluid levels.		
All Models	.2	.3
Sealed Beam Headlamp, Replace (C)		
All Models	.3	.3
Timing Belt and/or Tensioner, Replace (B)		
Esteem	1.7	1.9
Forenza	1.2	1.4
Grand Vitara, Vitara, XL-7	2.4	2.7
Samurai, Sidekick, X-90	3.2	3.6
Swift		
1989-94	3.1	3.6
1995-01	2.4	2.7
Tire, Replace (C)		
Includes: Dismount old tire and mount new tire to rim.		
One	.8	.8
each addl. add	.3	.3
Tires, Rotate (C)		
All Models	.5	.5
Wheel, Balance (B)		
One	.5	.5
each addl. add	.3	.3

SCHEDULED MAINTENANCE INTERVALS

If necessary, refer to appropriate Chilton maintenance service information.

AERIO, ESTEEM, FORENZA, SWIFT, VERONA

	LABOR TIME	SEVERE SERVICE
7,500 Mile Service (C)		
All Models	1.8	1.8
w/AT add	.1	.1
15,000 Mile Service (C)		
All Models	2.0	2.0
w/AT add	.1	.1
22,500 Mile Service (C)		
All Models	1.8	1.8
w/AT add	.1	.1
30,000 Mile Service (B)		
All Models	3.7	3.7
w/AT add	.1	.1
37,500 Mile Service (C)		
All Models	1.8	1.8
w/AT add	.1	.1
45,000 Mile Service (B)		
All Models	2.5	2.5
w/AT add	.1	.1
52,500 Mile Service (C)		
All Models	1.8	1.8
w/AT add	.1	.1
60,000 Mile Service (B)		
All Models	4.1	4.1
w/AT add	.1	.1
67,500 Mile Service (C)		
All Models	1.8	1.8
w/AT add	.1	.1
75,000 Mile Service (B)		
All Models	2.2	2.2
w/AT add	.1	.1
82,500 Mile Service (C)		
All Models	1.8	1.8
w/AT add	.1	.1
90,000 Mile Service (B)		
All Models	4.0	4.0
w/AT add	.1	.1
97,500 Mile Service (C)		
All Models	1.8	1.8
w/AT add	.1	.1

GRAND VITARA, SAMURAI, SIDEKICK, VITARA, XL-7, X-90

	LABOR TIME	SEVERE SERVICE
7,500 Mile Service (C)		
All Models	.8	.8
w/4WD add	.3	.3
w/AT add	.1	.1
15,000 Mile Service (B)		
All Models	2.1	2.1
w/4WD add	.3	.3
w/AT add	.1	.1
22,500 Mile Service (C)		
All Models	.7	.7
w/4WD add	.3	.3
w/AT add	.1	.1
30,000 Mile Service (B)		
All Models	4.5	4.5
w/4WD add	.3	.3
w/AT add	.1	.1
37,500 Mile Service (C)		
All Models	.8	.8
w/4WD add	.3	.3
w/AT add	.1	.1
45,000 Mile Service (B)		
All Models	2.1	2.1
w/4WD add	.3	.3
w/AT add	.1	.1
52,500 Mile Service (C)		
All Models	.7	.7
w/4WD add	.3	.3
w/AT add	.1	.1
60,000 Mile Service (B)		
All Models	7.7	7.7
w/4WD add	.3	.3
w/AT add	.1	.1
67,500 Mile Service (C)		
All Models	.7	.7
w/4WD add	.3	.3
w/AT add	.1	.1
75,000 Mile Service (B)		
All Models	2.0	2.0
w/4WD add	.3	.3
w/AT add	.1	.1
82,500 Mile Service (C)		
All Models	.7	.7
w/4WD add	.3	.3
w/AT add	.1	.1
90,000 Mile Service (B)		
All Models	7.4	7.4
w/4WD add	.3	.3
w/AT add	.1	.1
97,500 Mile Service (C)		
All Models	.7	.7
w/4WD add	.3	.3
w/AT add	.1	.1

CHARGING

	LABOR TIME	SEVERE SERVICE
Alternator Circuits, Test (B)		
Includes: Test component output.		
All Models	.6	.6
Alternator Assy., Replace (B)		
Includes: Pulley transfer.		
Exc. below	.8	1.0
Aerio	1.7	1.7
Samurai	1.4	1.6
Alternator, R&R and Recondition (B)		
Exc. below	1.7	1.9
Aerio	2.0	2.2
Samurai	2.3	2.8

SIDEKICK : SWIFT : VERONA : VITARA : X-90 : XL-7 **SUZ-3**

	LABOR TIME	SEVERE SERVICE
Alternator Voltage Regulator, Replace (B)		
Exc. below	1.7	1.9
Aerio	2.0	2.2
Samurai	2.3	2.8
Front Alternator Bearing, Replace (B)		
Exc. below	1.7	1.9
Aerio	2.0	2.2
Samurai	2.3	2.8
Voltmeter and/or Ammeter Gauge, Replace (B)		
Sidekick, X-90	.9	.9
Swift	1.7	1.7

STARTING

Starter Draw Test (On Car) (B)		
All Models	.3	.3
Battery Cables, Replace (C)		
One	.5	.6
Starter Assy., Replace (B)		
Aerio, Forenza, Verona	.6	.7
Esteem	.8	.9
Grand Vitara, Sidekick, Vitara, XL-7, X-90		
SOHC	.8	.9
DOHC	.6	.7
Samurai	.7	.8
Swift	.8	.9
Starter, R&R and Recondition (B)		
Aerio, Forenza, Verona	.9	1.0
Esteem	1.0	1.1
Grand Vitara, Sidekick, Vitara, XL-7, X-90		
SOHC	1.1	1.2
DOHC	.9	1.0
Samurai	1.0	1.1
Swift	1.2	1.3
Replace field coils add	.5	.5
Starter Drive Assy., Replace (B)		
Includes: Starter R&R.		
Aerio, Forenza, Verona	.9	1.0
Esteem	1.0	1.1
Grand Vitara, Sidekick, Vitara, XL-7, X-90		
SOHC	1.1	1.2
DOHC	.9	1.0
Samurai	1.0	1.1
Swift	1.2	1.3
Starter Solenoid and/or Switch, Replace (B)		
Includes: Starter R&R.		
Aerio, Forenza, Verona	.9	1.0
Esteem	1.0	1.1

	LABOR TIME	SEVERE SERVICE
Grand Vitara, Sidekick, Vitara, XL-7, X-90		
SOHC	1.1	1.2
DOHC	.9	1.0
Samurai	1.0	1.1
Swift	1.2	1.3

CRUISE CONTROL

Diagnose Cruise Control System (A)		
Sidekick, X-90	.8	.8
Actuator Assy., Replace (B)		
All Models	.7	.7
Brake or Clutch Release Switch, Replace (B)		
All Models	.5	.5
Control Cable & Core, Replace (B)		
All Models	.7	.7
Control Main Switch, Replace (B)		
All Models	.5	.5
Control Module, Replace (B)		
Forenza	.3	.3
Grand Vitara, Vitara, XL-7	.6	.6
Sidekick, X-90	.7	.7
Verona	.2	.2
Control Stepper Motor, Replace (B)		
Grand Vitara, Vitara, XL-7	.5	.5
Control Stop Switch, Replace (B)		
Aerio	.7	.7
Sidekick, X-90	1.2	1.2

IGNITION

Diagnose Ignition System Component Each (A)		
All Models	.9	.9
Ignition Timing, Reset (B)		
All Models	.5	.6
Camshaft Position Sensor, Replace (B)		
Aerio, Grand Vitara, XL-7	.6	.7
Esteem, Sidekick, Swift, Vitara, X-90	.5	.6
Forenza	1.1	1.2
Crankshaft Angle Sensor, Replace (B)		
Flywheel side		
Aerio	4.8	5.0
Esteem	3.6	3.8
Grand Vitara, XL-7	.6	.7
Sidekick, Vitara, X-90		
AT	4.9	5.1
MT		
2WD	2.9	3.1
4WD	3.7	3.9
Verona	.4	.6

	LABOR TIME	SEVERE SERVICE
Pulley side		
Esteem, Sidekick, Swift, X-90	.3	.4
Forenza	.7	.8
Grand Vitara, Vitara, XL-7	.5	.6
Distributor, R&R and Recondition (B)		
Includes: Reset base ignition timing.		
Esteem, Swift	1.1	1.2
Samurai	1.5	1.6
Sidekick, X-90	1.4	1.5
Distributor, Replace (B)		
Includes: Reset base ignition timing.		
Esteem, Swift	.6	.7
Samurai, Sidekick, X-90	.9	1.0
Distributor Cap and/or Rotor, Replace (B)		
All Models	.5	.6
Distributor Vacuum Advance, Replace (B)		
Includes: Reset base ignition timing.		
Samurai, Sidekick, X-90	1.2	1.3
1989-94 Swift	1.3	1.4
Electronic Ignition Coil, Replace (B)		
Exc. below	.5	.6
Grand Vitara, Vitara, XL-7	.8	.9
Igniter Assy. (External), Replace (B)		
All Models	.5	.6
Ignition Coil, Replace (B)		
All Models	.5	.6
Signal Generator Assy., Replace (B)		
Esteem, Samurai, Swift	1.1	1.2
Sidekick, X-90	1.2	1.3
Spark Plug (Ignition) Cables Replace (B)		
All Models	.5	.6
Spark Plugs, Replace (B)		
Exc. below	.5	.6
Grand Vitara, XL-7	.9	1.0

EMISSIONS

Dynamometer Test (A)		
All Models	.4	.4
Air Cleaner Temperature Sensor, Replace (B)		
Exc. below	.3	.3
Grand Vitara, XL-7	.5	.5
Air Cleaner Vacuum Motor, Replace (B)		
1986-89 Samurai	.5	.5
1989 Sidekick	.5	.5
Altitude Compensator, Replace (B)		
1986-89 Samurai	.5	.6
Sidekick, X-90	.3	.3

SUZ-4 AERIO : ESTEEM : FORENZA : GRAND VITARA : SAMURAI

	LABOR TIME	SEVERE SERVICE
Canister Air Control Valve, Replace (B)		
Exc. below	.6	.7
1995-01 Swift	1.4	1.5
Canister Purge Control Solenoid, Replace (B)		
All Models	.6	.7
Canister Purge Vacuum Switch, Replace (B)		
Vitara	.3	.3
Computer, Replace (B)		
Exc. Aerio	.6	.7
Aerio	.9	1.0
Coolant Temperature (ECT) Sensor, Replace (B)		
All Models	.8	.9
EGR Back Pressure Transducer, Replace (B)		
Esteem, Sidekick	.3	.3
Swift	.5	.5
EGR Control Valve, Replace (B)		
Exc. below	.9	1.0
Sidekick, Vitara, X-90	.5	.6
EGR 3-Way Solenoid Valve Assy., Replace (B)		
Esteem, Sidekick, Swift, X-90	.6	.7
Fuel Vapor Canister, Replace (B)		
Aerio, Sidekick, Verona, X-90	.8	.9
Esteem, Grand Vitara, Vitara, XL-7	.9	1.0
Forenza	1.2	1.4
Samurai	.3	.3
Swift		
1989-94	.5	.6
1995-01	1.5	1.7
Intake Air Temperature (IAT) Sensor, Replace (B)		
All Models	.3	.3
Manifold Absolute Pressure (MAP) Sensor, Replace (B)		
Exc. Forenza	.3	.4
Forenza	.5	.6
Mass Air Flow (MAF) Sensor, Replace (B)		
All Models	.7	.8
Oxygen Sensor, Replace (B)		
Exc. below	.6	.7
Esteem, Swift	.8	.9
each addl. add	.1	.2
Positive Crankcase Ventilation (PCV) Valve, Replace (B)		
All Models	.5	.6
Throttle Position Sensor (TPS), Replace (B)		
All Models	.6	.7

	LABOR TIME	SEVERE SERVICE
Vacuum Switching Valve, Replace (B)		
Exc. below	.6	.7
Samurai, Sidekick	.7	.8
Vapor Tank Pressure Sensor, Replace (B)		
Exc. Aerio	1.0	1.2
Aerio	1.2	1.4
Vapor Separator or Control Valve, Replace (B)		
Exc. Swift	1.3	1.6
Swift	1.0	1.2

FUEL
CARBURETOR

	LABOR TIME	SEVERE SERVICE
Automatic Choke, Clean and Adjust (B)		
Samurai, Sidekick	.5	.5
Carburetor, R&R and Clean or Recondition (B)		
Includes: Adjustments.		
All Models	3.0	3.0
Carburetor, Replace (B)		
Includes: Adjustments.		
All Models	2.1	2.1
Carburetor Needle & Seat, Replace (B)		
Includes: Adjustments.		
All Models	2.8	3.1

DELIVERY

	LABOR TIME	SEVERE SERVICE
Fuel Pump, Test (B)		
All Models	.3	.3
Fuel Filter, Replace (B)		
Carbureted	.5	.6
Fuel injected		
Aerio, Esteem	1.5	1.5
Forenza, Verona	.5	.5
Grand Vitara, XL-7	.8	.9
Samurai, Sidekick, X-90	.7	.8
Swift		
1989-94	.8	.9
1995-01	1.1	1.2
Vitara	1.0	1.1
Fuel Gauge (Dash), Replace (B)		
Esteem, Swift	.9	1.0
Samurai	1.3	1.4
Sidekick, X-90	1.2	1.3
Grand Vitara, Verona, Vitara, XL-7	.7	.8
Fuel Gauge (Tank), Replace (B)		
Exc. below	1.0	1.2
Aerio	1.5	1.7
Esteem		
2WD	1.3	1.5
4WD	1.8	2.0
Forenza	.4	.6

	LABOR TIME	SEVERE SERVICE
Swift		
1989-94	1.3	1.6
1995-01	1.7	2.0
Fuel Pump, Replace (B)		
Carbureted		
Samurai, Sidekick	.5	.7
Fuel injected		
exc. below	1.1	1.3
Aerio	1.5	1.7
Esteem		
2WD	1.3	1.5
4WD	1.8	2.0
Forenza	.4	.6
Swift		
1989-94	1.3	1.6
1995-01	1.7	2.0
Fuel Pump Relay, Replace (B)		
1989-05	.3	.3
Fuel Tank, Replace (B)		
Includes: Drain and refill.		
Exc. below	1.2	1.4
Aerio	1.6	1.8
Esteem		
2WD	1.5	1.7
4WD	1.8	2.0
Samurai	1.8	2.1
Verona	1.0	1.2
Intake Manifold and/or Gasket, Replace (B)		
Includes: Adjustments.		
Aerio	2.8	3.1
Esteem		
SOHC	1.7	1.9
DOHC	2.4	2.7
Forenza, Verona	1.5	1.7
Grand Vitara, XL-7	2.5	2.7
Samurai	2.9	3.4
Sidekick, X-90		
SOHC	2.6	2.8
DOHC	2.1	2.3
Swift	2.4	2.7
Vitara		
SOHC	2.1	2.4
DOHC	2.6	2.9
Replace manifold add	.5	.8

INJECTION

	LABOR TIME	SEVERE SERVICE
Fuel Injectors, Replace (B)		
Aerio	.9	1.0
Esteem		
SOHC	1.6	1.7
DOHC	.8	1.0
Forenza	.7	.9
Grand Vitara, XL-7	2.3	2.5
Sidekick, Vitara, X-90		
SOHC	1.2	1.4
DOHC	.8	1.0
Swift, Verona	1.3	1.5

SIDEKICK : SWIFT : VERONA : VITARA : X-90 : XL-7   **SUZ-5**

	LABOR TIME	SEVERE SERVICE
Fuel Pressure Regulator, Replace (B)		
Exc. Verona	.6	.7
Verona	1.3	1.4
Fuel Rail, Replace (B)		
Aerio, Esteem	.9	1.0
Forenza	.7	.9
Grand Vitara, XL-7	2.3	2.5
Sidekick, Vitara, X-90		
SOHC	1.2	1.4
DOHC	.8	1.0
Swift, Verona	1.3	1.5
Idle Air Control (IAC) Valve, Replace (B)		
Exc. below	.5	.5
Aerio, Esteem, 1995-01 Swift	.8	.9
Pulsation Damper, Replace (B)		
Sidekick	.5	.5
Throttle Body Assy., R&R or Replace (B)		
Aerio	.9	1.1
Esteem, Forenza, Swift, Vitara	.9	1.0
Grand Vitara, XL-7	1.9	2.2
Samurai, Sidekick, X-90		
SOHC	1.5	1.7
DOHC	.9	1.1
Verona	.6	.8
Throttle Body Fuel Injector and/or Gaskets, Replace (B)		
Samurai	.8	.8
Sidekick, X-90	.9	1.1
Swift	.7	.9

EXHAUST

	LABOR TIME	SEVERE SERVICE
Catalytic Converter, Replace (B)		
Exc. below	.8	.9
Sidekick, X-90	1.0	1.1
Exhaust Manifold or Gasket, Replace (B)		
Aerio	1.8	2.1
Esteem		
SOHC	.7	1.0
DOHC	1.2	1.5
Forenza	.9	1.2
Grand Vitara, XL-7	3.1	3.4
Samurai	1.4	1.9
Sidekick, Swift, Vitara, X-90	1.5	1.7
Verona	1.2	1.5
Exhaust Pipe, Replace (B)		
Exc. below one	1.3	1.4
Aerio, Esteem		
No.1	1.3	1.4
No.2	.8	.9
Samurai center	.8	.9
Swift one	.6	.7

	LABOR TIME	SEVERE SERVICE
Muffler, Replace (B)		
Exc. Samurai	.8	.9
Samurai	.9	1.0
w/skid plate add	.3	.3

ENGINE COOLING

	LABOR TIME	SEVERE SERVICE
Pressure Test Cooling System (B)		
All Models	.3	.3
Bypass Hose, Replace (B)		
All Models	.6	.7
Coolant Thermostat, Replace (B)		
All Models	.7	.9
Electric Fan Thermo Switch, Replace (B)		
1989-94 Swift	.5	.6
Engine Coolant Temp. Sending Unit, Replace (B)		
All Models	.6	.7
Fan Blades or Clutch Assy., Replace (B)		
Aerio	.7	.8
Esteem, Swift, Verona	.8	.9
Forenza	.6	.8
Grand Vitara, XL-7	1.2	1.4
Samurai	1.2	1.4
Sidekick, Vitara, X-90		
SOHC	1.3	1.5
DOHC	.8	.9
Freeze Plugs (Water Jacket), Replace (B)		
Add access time as required.		
All Models each	.5	.7
Radiator Assy., R&R or Replace (B)		
Aerio, Esteem, Swift, Verona	.9	1.1
Forenza	1.1	1.4
Grand Vitara, XL-7	1.1	1.3
Samurai	1.7	1.9
Sidekick, Vitara, X-90	1.0	1.3
w/AC add	.5	.5
Radiator Fan and/or Fan Motor, Replace (B)		
Aerio	.7	.8
Esteem, Swift, Verona	.8	.9
Forenza	.6	.8
Grand Vitara, XL-7	1.0	1.2
Samurai	1.0	1.2
Sidekick, Vitara, X-90		
SOHC	1.2	1.3
DOHC	.7	.8
Radiator Hoses, Replace (B)		
Includes: Refill with proper coolant mix.		
Upper	.5	.7
Lower	.8	1.0

	LABOR TIME	SEVERE SERVICE
Temperature Gauge (Dash), Replace (B)		
Esteem, Swift	.9	.9
Forenza	.5	.5
Grand Vitara, Vitara, XL-7	.7	.7
Samurai	1.4	1.4
Sidekick, X-90	1.2	1.2
Verona	.7	.7
Water Control Valve, Replace (B)		
Samurai	.8	1.0
Water Pump and/or Gasket, Replace (B)		
Includes: Refill with proper coolant mix.		
Aerio, Verona	1.3	1.7
Esteem		
SOHC	2.4	2.6
DOHC	1.5	1.8
Forenza	1.2	1.6
Grand Vitara, XL-7	8.0	8.2
Samurai	3.5	4.0
Sidekick, Vitara, X-90		
SOHC	2.8	3.0
DOHC	1.4	1.7
Swift	3.1	3.6
w/AC add	.5	.5

ENGINE
ASSEMBLY

Times shown are for OEM assemblies. Time to replace assemblies from aftermarket rebuilders may vary.

Engine Assy., R&I (B)
Does not include parts or component transfer.

	LABOR TIME	SEVERE SERVICE
Aerio	5.5	5.7
Esteem		
SOHC	4.0	4.3
DOHC	4.7	5.0
Forenza	4.2	4.4
Grand Vitara, XL-7	5.8	6.0
Samurai	6.2	6.2
Sidekick, X-90		
SOHC	5.1	5.3
DOHC	6.3	6.5
Swift	5.0	5.2
Verona	4.7	5.0
Vitara		
SOHC	4.5	4.8
DOHC	5.6	5.9
w/AC add		
exc. below	.4	.4
Forenza	1.6	1.6
Verona	.9	.9
w/skid plate add	.3	.3

AERIO : ESTEEM : FORENZA : GRAND VITARA : SAMURAI

	LABOR TIME	SEVERE SERVICE

Engine Assy., R&R and Recondition (A)
Includes: Replacing rings, rod and main bearings, cylinder head reconditioning and engine tune-up.
- Aerio 18.6   *19.6*
- Esteem
  - SOHC 15.2   *17.0*
  - DOHC 19.6   *21.4*
- Forenza 14.0   *15.0*
- Grand Vitara, XL-7 .. 20.8   *22.7*
- Samurai 20.5   *20.5*
- Sidekick, X-90
  - SOHC 13.2   *15.0*
  - DOHC 17.8   *19.3*
- Swift
  - 1989-94
    - SOHC 17.1   *19.1*
    - DOHC 19.2   *21.3*
  - 1995-01 .... 16.7   *18.7*
- Verona 17.6   *18.8*
- Vitara
  - SOHC 13.3   *15.1*
  - DOHC 19.9   *22.3*
- w/AC add
  - exc. below 4   *.4*
  - Forenza 1.6   *1.6*
  - Verona 9   *.9*
- w/skid plate add 3   *.3*

Engine Assy. (Short Block), Replace (B)
Assembly consists of cylinder block, piston assemblies, crankshaft, camshaft, timing chain and gears. Does not include cylinder heads. Operation Includes: R&R engine, transfer necessary parts and all necessary adjustments.
- Aerio 13.7   *15.4*
- Esteem
  - AT 12.3   *13.8*
  - MT
    - SOHC 11.0   *12.4*
    - DOHC 11.6   *13.0*
- Forenza 10.2   *12.4*
- Grand Vitara, XL-7
  - AT 15.2   *17.0*
  - MT 14.5   *16.2*
- Samurai 11.2   *12.5*
- Sidekick, X-90
  - AT 12.8   *14.3*
  - MT
    - SOHC 12.8   *14.3*
    - DOHC 13.1   *14.6*
- Swift
  - AT 13.9   *15.9*
  - MT 12.6   *14.1*

- Verona 15.7   *17.5*
- Vitara 12.1   *13.5*
- w/AC add
  - exc. below 4   *.4*
  - Forenza 1.6   *1.6*
  - Verona 9   *.9*
- w/skid plate add 3   *.3*

Engine Mounts, Replace (B)
- Front
  - Aerio, Esteem, Samurai, Verona .. 1.0   *1.1*
  - Forenza 1.1   *1.2*
  - Grand Vitara, XL-7 . 1.3   *1.4*
  - Sidekick, Vitara, X-90
    - SOHC 1.7   *2.0*
    - DOHC 1.3   *1.4*
  - Swift
    - right side 8   *.9*
    - left side 1.4   *1.6*
- Rear
  - exc. Swift 8   *.9*
  - Swift 1.0   *1.1*
- Torque bushings 8   *.9*
- w/skid plate front add 3   *.3*

CYLINDER HEAD

Compression Test (B)
- All Models 7   *1.0*

Valve Clearance, Adjust (B)
- All Models 1.2   *1.4*

Cylinder Head, Replace (B)
Includes: Parts transfer and adjustments
- Aerio 14.6   *15.8*
- Esteem
  - SOHC 8.2   *9.2*
  - DOHC 12.8   *14.3*
- Forenza 10.7   *12.6*
- Grand Vitara, XL-7
  - one side 14.8   *16.8*
  - both sides 17.6   *19.8*
- Samurai 7.8   *8.3*
- Sidekick, Vitara, X-90
  - SOHC 9.7   *10.9*
  - DOHC 13.2   *14.8*
- Swift
  - 1989-94
    - SOHC 7.8   *8.3*
    - DOHC 10.9   *11.4*
  - 1995-01 .... 8.2   *9.2*
- Verona 12.8   *14.7*
- w/AC add
  - exc. Forenza 5   *.5*
  - Forenza 1.0   *1.0*

Cylinder Head Gasket, Replace (B)
- Aerio 10.6   *12.1*
- Esteem
  - SOHC 5.5   *6.2*
  - DOHC 10.2   *11.4*
- Forenza 8.3   *10.1*

- Grand Vitara, XL-7
  - one side 12.0   *13.7*
  - both sides 14.6   *15.5*
- Samurai 4.9   *5.2*
- Sidekick, Vitara, X-90
  - SOHC 6.9   *7.7*
  - DOHC 13.2   *14.8*
- Swift
  - 1989-94
    - SOHC 4.9   *5.5*
    - DOHC 6.3   *7.0*
  - 1995-01 .... 5.5   *6.2*
- Verona 10.9   *12.5*
- w/AC add
  - exc. below 5   *.5*
  - Forenza 1.0   *1.0*
  - Verona 9   *.9*

Rocker Arm Cover Gasket, Replace or Reseal (B)
- Exc. below 8   *.9*
- Grand Vitara, XL-7 ... 3.2   *3.6*
- Samurai 5   *.7*
- Verona 2.0   *2.6*

Rocker Arms or Shafts, Replace (B)
- Esteem 1.4   *1.5*
- Grand Vitara, XL-7 ... 6.5   *7.2*
- Samurai 3.3   *3.5*
- Sidekick, Vitara, X-90 . 4.1   *4.6*
- Swift
  - 1989-94 .... 2.3   *2.5*
  - 1995-01 .... 1.4   *1.5*
- Verona 9.1   *9.8*
- w/AC add
  - *Sidekick, Vitara, X-90* 2   *.2*
  - *Verona* 9   *.9*

Springs and/or Valve Seals, Replace (B)
Includes: Cylinder head R&R.
- Aerio, Verona 14.1   *15.5*
- Esteem
  - SOHC 7.4   *8.1*
  - DOHC 12.4   *13.8*
- Forenza 9.3   *11.5*
- Grand Vitara, XL-7
  - one side 15.9   *17.6*
  - both sides 19.6   *21.5*
- Samurai 6.4   *6.9*
- Sidekick, Vitara, X-90
  - SOHC 8.8   *9.7*
  - DOHC 12.8   *14.3*
- Swift
  - 1989-94
    - SOHC 7.3   *7.8*
    - DOHC 10.0   *10.5*
  - 1995-01 .... 7.4   *8.1*
- w/AC add
  - exc. below 5   *.5*
  - Forenza 1.0   *1.0*
  - Verona 9   *.9*

SIDEKICK : SWIFT : VERONA : VITARA : X-90 : XL-7   SUZ-7

	LABOR TIME	SEVERE SERVICE
Valve Job (A)		
Aerio, Verona	16.4	18.2
Esteem		
SOHC	9.6	10.7
DOHC	16.7	18.5
Forenza	12.8	14.2
Grand Vitara, XL-7		
one side	16.7	18.5
both sides	21.0	23.3
Samurai	9.0	9.5
Sidekick, Vitara, X-90		
SOHC	8.8	9.7
DOHC	12.8	14.3
Swift		
1989-94		
SOHC	9.4	9.9
DOHC	14.2	14.7
1995-01	9.6	10.7
w/AC add		
exc. below	.5	.5
Forenza	1.0	1.0
Verona	.9	.9
Valve Lash Adjuster, Replace (B)		
Aerio	11.5	12.7
Esteem	6.9	7.7
Forenza	3.3	4.5
Grand Vitara, XL-7	13.0	14.3
Sidekick, Vitara, X-90	7.7	8.6
Swift		
1989-94	4.3	4.7
1995-01	1.1	1.2
w/AC add		
Aerio, Esteem	.5	.5
Swift	.2	.2
CAMSHAFT		
Camshaft and/or Oil Seal Replace (B)		
Aerio, Verona	11.3	12.6
Esteem		
SOHC	3.8	4.2
DOHC	6.8	7.5
Forenza	3.5	4.2
Grand Vitara, XL-7	13.0	14.3
Samurai	5.2	5.4
Sidekick, X-90	7.4	8.1
Swift		
1989-94	4.2	4.4
1995-01	4.3	4.7
Vitara	6.0	6.7
w/AC add		
exc. below	.5	.5
Forenza	1.0	1.0
Verona	.9	.9
Timing Belt and/or Tensioner, Replace (B)		
Esteem	1.7	1.9
Forenza	1.2	1.4

	LABOR TIME	SEVERE SERVICE
Grand Vitara, Vitara, XL-7	2.4	2.7
Samurai, Sidekick, X-90	3.2	3.6
Swift		
1989-94	3.1	3.6
1995-01	2.4	2.7
Timing Chain & Gears, Replace (B)		
Aerio	10.2	11.3
Esteem	6.5	7.2
Grand Vitara, XL-7	12.1	13.4
Sidekick, Vitara, X-90	6.6	7.3
Verona	11.3	12.4
CRANK & PISTONS		
Connecting Rod Bearings Replace (A)		
Includes: Check bearing oil clearance.		
Aerio	4.3	4.8
Esteem, Forenza	3.5	3.9
Grand Vitara, XL-7		
2WD	5.7	6.3
4WD	7.2	8.1
Samurai	3.0	3.5
Sidekick, X-90		
SOHC		
2WD	3.3	3.7
4WD	5.5	6.2
DOHC	12.5	13.9
Swift	3.3	3.7
Verona	14.5	15.7
Vitara		
SOHC		
2WD	4.0	4.4
4WD	5.5	6.2
DOHC	12.5	13.9
Crankshaft and Main Bearings, Replace (A)		
Includes: Engine R&R, check bearing oil clearance.		
Aerio	15.6	17.5
Esteem		
AT	12.8	14.3
MT		
SOHC	11.6	13.0
DOHC	13.5	15.0
Forenza	13.2	14.4
Grand Vitara, XL-7		
AT	18.8	21.0
MT	18.1	20.1
Samurai	11.5	12.4
Sidekick, X-90		
AT	14.8	16.5
MT		
SOHC	13.2	14.8
DOHC	14.5	16.2

	LABOR TIME	SEVERE SERVICE
Swift		
AT	14.3	16.1
MT	13.2	14.8
Verona	16.4	17.8
Vitara		
AT	13.7	15.4
MT		
1.6L	12.3	13.8
2.0L	13.5	15.0
w/AC add		
exc. below	.5	.5
Forenza	1.6	1.6
Verona	.9	.9
Crankshaft Front Oil Seal, Replace (B)		
Aerio	1.1	1.3
Esteem		
SOHC	5.2	5.8
DOHC	1.1	1.3
Forenza	4.7	5.4
Grand Vitara, XL-7	11.5	12.8
Samurai	5.6	5.8
Sidekick, Vitara, X-90		
SOHC	8.2	9.2
DOHC	1.4	1.6
Swift	5.3	6.0
Verona	9.9	10.8
w/AC add		
exc. Verona	.5	.5
Verona	.9	.9
Crankshaft Pulley, Replace (B)		
Exc. below	1.3	1.4
Grand Vitara, XL-7	1.6	1.8
Samurai	.9	1.1
Verona	6.6	6.8
Pistons or Connecting Rods, Replace (A)		
Includes: Ridge reaming, cylinder wall deglazing, installing new rings and rod bearings, engine tune-up.		
Aerio	13.5	15.0
Esteem		
SOHC	8.2	9.2
DOHC	11.9	13.3
Forenza	12.1	14.7
Grand Vitara, XL-7		
2WD	18.5	20.6
4WD	19.8	22.1
Samurai	9.4	10.3
Sidekick, X-90		
SOHC		
2WD	10.2	11.4
4WD	11.9	13.3
DOHC	13.1	14.6
Swift		
1989-94		
SOHC	8.4	9.3
DOHC	9.7	10.6
1995-01	7.6	8.5

SUZ-8 AERIO : ESTEEM : FORENZA : GRAND VITARA : SAMURAI

	LABOR TIME	SEVERE SERVICE
Pistons or Connecting Rods, Replace (A)		
Verona	15.6	*18.7*
Vitara		
SOHC		
2WD	11.6	*13.0*
4WD	11.9	*13.3*
DOHC	13.1	*14.6*
w/AC add		
exc. below	.5	*.5*
Forenza	1.6	*1.6*
Verona	.9	*.9*
Rear Main Oil Seal, Replace (B)		
Aerio		
AT	10.0	*11.2*
MT	9.4	*10.6*
Esteem		
AT	9.7	*10.3*
MT		
SOHC	8.5	*9.1*
DOHC	5.5	*6.2*
Forenza	6.6	*7.0*
Grand Vitara, XL-7	7.2	*8.1*
Samurai	5.4	*6.0*
Sidekick, X-90		
SOHC	8.9	*9.5*
DOHC	5.3	*5.9*
Swift		
AT	9.8	*10.4*
MT	8.7	*9.3*
Verona	7.2	*7.6*
Vitara		
SOHC	8.9	*9.5*
DOHC	5.3	*5.9*
w/AC add		
exc. below	.5	*.5*
Forenza	1.6	*1.6*
Verona	.9	*.9*
Rings, Replace (A)		
Includes: Ridge reaming, cylinder wall deglazing, installing new rings, engine tune-up.		
Aerio	14.2	*15.9*
Esteem		
SOHC	8.2	*9.2*
DOHC	12.3	*13.8*
Forenza	12.1	*13.5*
Grand Vitara, XL-7		
2WD	19.6	*21.8*
4WD	20.9	*23.3*
Samurai	7.9	*8.8*
Sidekick, X-90		
SOHC		
2WD	10.2	*11.4*
4WD	11.9	*13.3*
DOHC	13.5	*15.0*

	LABOR TIME	SEVERE SERVICE
Swift		
1989-94		
SOHC	8.4	*9.3*
DOHC	9.7	*10.6*
1995-01	7.6	*8.5*
Verona	15.6	*17.0*
Vitara		
SOHC		
2WD	12.1	*13.5*
4WD	11.9	*13.3*
DOHC	13.5	*15.0*
w/AC add		
exc. below	.5	*.5*
Forenza	1.6	*1.6*
Verona	.9	*.9*
w/skid plate add	.3	*.3*

ENGINE LUBRICATION

	LABOR TIME	SEVERE SERVICE
Engine Oil Pressure Switch (Sending Unit), Replace (B)		
Exc. Samurai	.7	*.7*
Samurai	.5	*.5*
Oil Pan and/or Gasket, Replace (B)		
Aerio, Forenza	2.4	*2.7*
Esteem	2.2	*2.5*
Grand Vitara, XL-7		
upper pan		
2WD	3.3	*3.7*
4WD	4.9	*5.5*
lower pan		
2WD	2.2	*2.7*
4WD	3.8	*4.3*
Samurai	1.8	*2.0*
Sidekick, X-90		
2WD	1.7	*2.0*
4WD	4.3	*4.8*
Swift		
1989-94	1.5	*1.7*
1995-01	1.9	*2.1*
Verona	1.9	*2.2*
Vitara		
2WD	2.9	*3.2*
4WD	4.6	*5.1*
Oil Pressure Gauge (Dash), Replace (B)		
Sidekick, X-90	.9	*.9*
Oil Pump (Front Cover), Replace (B)		
Aerio, Verona	3.0	*3.4*
Esteem		
SOHC	5.3	*6.0*
DOHC	2.5	*2.9*
Forenza	5.0	*5.5*
Grand Vitara, XL-7		
2WD	10.3	*11.5*
4WD	11.8	*13.2*
Samurai	5.6	*6.1*

	LABOR TIME	SEVERE SERVICE
Sidekick, X-90		
SOHC		
2WD	6.0	*6.7*
4WD	8.2	*9.2*
DOHC	4.3	*4.8*
Swift	5.5	*6.2*
Vitara		
SOHC		
2WD	6.7	*7.5*
4WD	8.2	*9.2*
DOHC	4.7	*5.3*
w/AC add		
exc. below	.2	*.2*
Sidekick, Vitara, X-90	.6	*.6*
Recondition pump add	.5	*.5*

CLUTCH

	LABOR TIME	SEVERE SERVICE
Bleed Hydraulic Clutch System (B)		
1996-01	.3	*.3*
Clutch Pedal Free Play, Adjust (B)		
All Models	.5	*.5*
Clutch Assy., Replace (B)		
Exc. below	5.3	*6.0*
Grand Vitara, XL-7	6.3	*7.0*
Samurai	3.6	*3.8*
Clutch Cable, Replace (B)		
All Models	.7	*.9*
Clutch Pedal Position Switch, Replace (B)		
All Models	.5	*.5*
Clutch Master Cylinder, Replace (B)		
Includes: System bleeding.		
All Models	1.4	*1.6*
Clutch Release Bearing, Replace (B)		
Exc. below	4.2	*4.6*
Samurai	3.4	*3.6*
Sidekick, X-90		
2WD	3.9	*4.1*
4WD	5.1	*5.3*
Vitara		
2WD	3.9	*4.1*
4WD	4.4	*4.6*
Clutch Slave Cylinder, Replace (B)		
Includes: System bleeding.		
Exc. Forenza	.8	*1.0*
Forenza	4.1	*4.5*
Flywheel, Replace (B)		
Exc. below	5.8	*6.5*
Forenza	6.4	*7.0*
Samurai	4.2	*4.4*
Verona	7.2	*7.8*
Replace pilot bearing add	.1	*.1*

MANUAL TRANSAXLE

	LABOR TIME	SEVERE SERVICE
Axle Shaft Oil Seal, Replace (B)		
Aerio, Esteem, Forenza, Swift		
one side	1.9	*2.1*
both sides	2.7	*3.0*

SIDEKICK : SWIFT : VERONA : VITARA : X-90 : XL-7 **SUZ-9**

	LABOR TIME	SEVERE SERVICE
Differential Carrier, Replace or Reconditioned (A)		
Aerio, Esteem, Swift	8.8	8.9
Forenza	4.4	4.5
Speedometer Driven Gear and/or Seal, Replace (B)		
Aerio, Esteem, Swift	.7	.7
Forenza	.5	.5
Transaxle Assy., R&R and Recondition (A)		
Aerio, Esteem, Swift	10.2	11.4
Forenza	10.5	11.7
Transaxle Assy., R&R or Replace (B)		
Aerio, Esteem, Swift	5.0	5.7
Forenza	5.3	6.0

MANUAL TRANSMISSION

Extension Housing Oil Seal, Replace (B)
Samurai	3.2	3.4
Sidekick, X-90	.8	.9

Speedometer Drive Gear, Replace (B)
Grand Vitara, Sidekick, Vitara, X-90	5.0	5.7

Speedometer Driven Gear, Replace (B)
Sidekick, X-90	.8	.9

Transmission Assy., R&R and Recondition (A)
Grand Vitara, XL-7		
2WD	8.8	9.9
4WD	11.5	12.8
Samurai	8.2	8.8
Sidekick, Vitara, X-90		
2WD	8.8	9.9
4WD	10.0	11.2

Transmission Assy., R&R or Replace (B)
Grand Vitara, XL-7		
2WD	4.1	4.6
4WD	6.6	7.4
Samurai	3.5	3.7
Sidekick, Vitara, X-90		
2WD	4.1	4.6
4WD	5.3	6.0

AUTOMATIC TRANSAXLE

SERVICE TRANSAXLE INSTALLED

Check Unit for Oil Leaks (C)
All Models	.9	.9

Drain & Refill Unit (B)
All Models	1.5	1.5

Oil Pressure Check (B)
All Models	.5	.5

	LABOR TIME	SEVERE SERVICE
Electronic Control Unit/ Module, Replace (B)		
Aerio	.8	.9
Esteem, Forenza, Verona	.5	.6
Swift		
1989-94	1.0	1.1
1995-01	.5	.6
Manual Shift Shaft Oil Seal, Replace (B)		
All Models	.8	.9
Oil Pan and/or Gasket, Replace (B)		
Forenza, Verona	1.1	1.4
Swift	1.6	1.8
Oil Pressure Control Valve Cable, Replace (B)		
Swift	2.1	2.3
Speedometer Driven Gear and/or Seal, Replace (B)		
Aerio, Esteem	.3	.4
Swift	.6	.7
Transaxle Solenoid Assy., Replace (B)		
Direct clutch or second brake		
Swift	1.7	2.0
Shift all		
Aerio, Esteem	1.8	2.1
Forenza	2.4	2.8
Verona	1.1	1.5
TCC control		
Aerio, Esteem	1.8	2.1
Valve Body, R&R and Recondition (A)		
Aerio, Esteem, Forenza	2.9	3.2
Swift	3.5	3.9
Verona	1.2	1.5

SERVICE TRANSAXLE REMOVED
Transaxle R&R included unless otherwise noted.

Flywheel (Flexplate), Replace (B)
Aerio	8.1	9.0
Esteem, Verona	5.5	6.2
Forenza	6.4	7.2
Swift	9.1	10.2

Front Oil Pump, Replace (B)
Aerio	8.7	9.7
Esteem	5.8	6.5
Forenza	7.0	7.9
Swift	9.0	10.0
Verona	6.3	7.0
Recondition pump add	.3	.4

Ring & Pinion Set, Replace (B)
Aerio	9.3	10.4
Esteem, Forenza, Verona	6.4	7.2
Swift	10.8	12.0
Replace diff. carrier add	.3	.4

	LABOR TIME	SEVERE SERVICE
Torque Converter, Replace (B)		
Aerio	7.9	8.9
Esteem, Verona	5.3	6.0
Forenza	6.3	7.0
Swift	9.0	10.0
Transaxle Assy. (Complete), R&R and Recondition (A)		
Aerio	15.8	17.5
Esteem	13.1	14.6
Forenza	11.5	13.3
Swift	16.7	18.7
Verona	10.0	11.9
Transaxle Assy., R&R (B)		
Aerio	8.2	9.2
Esteem, Verona	5.3	6.0
Forenza	6.4	7.2
Swift	9.0	10.0
Replace assembly add	1.4	1.6

AUTOMATIC TRANSMISSION

SERVICE TRANSMISSION INSTALLED

Check Unit for Oil Leaks (C)
All Models	.9	.9

Drain and Refill Unit (B)
All Models	1.5	1.5

Oil Pressure Check (B)
All Models	.5	.5

Electronic Control Unit/ Module, Replace (B)
Sidekick, X-90	.7	.7
Grand Vitara, XL-7	.5	.5

Kickdown Throttle Cable, Replace (B)
All Models	6.0	6.7

Modulator, Replace (B)
Sidekick, X-90	4.9	5.5

Oil Pan and/or Gasket, Replace (B)
All Models	1.6	1.8

Transfer Case Adapter Gasket, Replace (B)
All Models	4.0	4.4

Transmission Oil Pressure Switch, Replace (B)
Sidekick, X-90	2.2	2.5

Transmission Solenoid, Replace (B)
Grand Vitara, Vitara, XL-7		
No. 1 or No. 2	3.3	3.7
Sidekick, X-90		
valve body	2.1	2.3
transmission		
No. 1 or No. 2	3.3	3.7

Transmission Speed Sensor, Replace (B)
All Models	1.3	1.4

Valve Body Assy., Replace (B)
All Models	3.0	3.4

AERIO : ESTEEM : FORENZA : GRAND VITARA : SAMURAI

	LABOR TIME	SEVERE SERVICE
SERVICE TRANSMISSION REMOVED		
Transmission R&R included unless otherwise noted.		
Flywheel (Flexplate), Replace (B)		
All Models	12.5	13.9
Front Oil Pump, Replace (B)		
All Models	8.1	9.0
Reconditon pump add	.3	.3
Torque Converter, Replace (B)		
All Models	6.7	7.5
Torque Converter Seal, Replace (B)		
All Models	7.2	8.1
Transmission Assy., R&R and Recondition (A)		
All Models	12.5	13.9
Transmission Assy., R&I (B)		
Sidekick, X-90	6.7	6.9
Vitara		
2WD	5.4	5.6
4WD	6.9	7.1
Replace assembly add	1.0	1.5

TRANSFER CASE

	LABOR TIME	SEVERE SERVICE
Front Cover or Gasket, Replace (B)		
Grand Vitara, Vitara, XL-7		
AT	4.4	5.0
MT	6.3	7.0
Samurai		
2WD	.8	1.0
4WD	2.0	2.2
Sidekick, X-90		
2WD	6.3	7.0
4WD		
AT	4.4	5.0
MT	6.3	7.0
Output Shaft Oil Seal, Replace (B)		
Exc. Samurai one	.9	1.1
Samurai one	1.2	1.4
Speed Sensor, Replace (B)		
Vitara	.6	.7
Speedometer Driven Gear Replace (B)		
Samurai	.8	.8
Sidekick, X-90	.6	.7
Transfer Case, R&R and Recondition (A)		
Exc. below		
AT	7.6	8.9
MT	9.3	10.4
Aerio	6.6	7.4
Samurai		
2WD	3.9	4.5
4WD	5.2	5.8
Transfer Case, R&I (B)		
Aerio	3.5	3.9
Samurai		
2WD	1.4	1.9
4WD	3.4	3.9

	LABOR TIME	SEVERE SERVICE
Sidekick, X-90		
AT	3.1	3.6
MT	5.4	5.9
Vitara		
AT		
3L30	2.9	3.4
4-Speed	2.8	3.3
MT	4.3	4.8
Transfer Case Mount, Replace (B)		
Samurai one side	.7	.9
4WD Indicator Switch, Replace (B)		
Grand Vitara, XL-7	.8	.9
Samurai, Sidekick, Vitara, X-90	.6	.7

DRIVELINE

DRIVESHAFT

	LABOR TIME	SEVERE SERVICE
Driveshaft, R&R or Replace (B)		
Exc. below	.8	.9
Aerio, Esteem	1.4	1.6
Samurai	.9	1.1
U-Joint, Replace or Recondition (B)		
Each	1.3	1.4

DRIVE AXLE

	LABOR TIME	SEVERE SERVICE
Axle Housing, Replace (B)		
Front axle	4.0	4.5
Rear axle		
exc. below	5.2	5.8
Samurai	5.4	5.9
Vitara	5.0	5.7
Axle Shaft, Replace (B)		
Rear axle		
one	2.7	3.0
both	4.1	4.6
Axle Shaft Bearing and/or Oil Seal, Replace (B)		
Rear axle		
exc. Samurai		
one	2.7	3.0
both	4.1	4.6
Samurai		
one	2.6	2.9
both	3.7	4.0
Carrier Side Oil Seals, Replace (B)		
Front axle		
one	1.3	1.4
both	1.9	2.1
Differential Carrier, R&I (B)		
Front axle		
Grand Vitara, Vitara, XL-7	3.3	3.8
Samurai	4.2	4.7
Sidekick, X-90	3.7	4.2
Rear axle		
Grand Vitara, Vitara, XL-7	2.6	3.1
Samurai	3.5	4.0
Sidekick, X-90	3.6	4.1

	LABOR TIME	SEVERE SERVICE
Differential Carrier Assy., R&R and Recondition (A)		
Front axle		
exc. Samurai	7.0	7.9
Samurai	7.4	8.0
Rear axle		
exc. below	7.0	7.9
Aerio, Esteem	4.7	5.3
Samurai	6.7	7.3
Output Shaft, Replace (B)		
Front axle	1.3	1.4
Pinion Shaft Oil Seal, Replace (B)		
Front axle	1.1	1.3
Rear axle		
exc. below	1.1	1.3
Aerio, Esteem	1.3	1.4
Samurai	1.2	1.4
Ring Gear & Pinion Set, Replace (B)		
Front axle	7.0	7.9
Rear axle		
exc. below	7.0	7.9
Aerio, Esteem	4.7	5.3
Samurai	6.3	6.8

HALFSHAFTS

	LABOR TIME	SEVERE SERVICE
CV Joint, Replace and/or Recondition (B)		
Aerio, Forenza, Verona		
one side	1.5	1.7
both sides	2.2	2.4
Esteem, Swift		
one side	2.1	2.3
both sides	3.6	4.1
Grand Vitara, Vitara, XL-7		
one side	2.2	2.5
both sides	3.8	4.3
Sidekick, X-90		
one side	1.9	2.1
both sides	3.2	3.6
CV Joint Boot, Replace (B)		
Exc. below		
one side	1.9	2.1
both sides	3.2	3.6
Aerio, Forenza, Verona		
one side	1.6	1.8
both sides	2.5	2.9
Esteem, Swift		
one side	2.1	2.3
both sides	3.6	4.1
Halfshaft, R&R or Replace (B)		
Aerio, Esteem, Forenza, Verona		
one side	1.1	1.3
both sides	1.9	2.1
Grand Vitara, Vitara, XL-7		
one side	1.7	2.0
both sides	2.2	2.5

SIDEKICK : SWIFT : VERONA : VITARA : X-90 : XL-7

	LABOR TIME	SEVERE SERVICE
Sidekick, X-90		
one side	1.3	*1.4*
both sides	1.9	*2.1*
Swift		
one side	1.4	*1.6*
both sides	2.5	*2.9*
Inner Halfshaft, R&R or Replace (B)		
Swift one	1.7	*2.0*
Recondition shaft add	.5	*.8*

BRAKES
ANTI-LOCK

Diagnosis Anti-Lock Brake System (A)
All Models	1.0	*1.0*

Actuator Assy., Replace (B)
Aerio, Esteem, Grand Vitara, Vitara, XL-7	1.7	*2.0*
Forenza	1.4	*1.7*
Sidekick, Swift, X-90	2.4	*2.7*
Verona	2.1	*2.4*

Brake and Traction Control Module, Replace (B)
Exc. below	.6	*.7*
Forenza	1.2	*1.3*
Grand Vitara, Vitara, XL-7		
rear wheel	.5	*.5*
four wheel	.5	*.5*
Verona	1.8	*1.9*

Front Sensor Assy., Replace (B)
Aerio, Esteem, Forenza		
one	.6	*.7*
both	.8	*.9*
Grand Vitara, Vitara, XL-7	1.0	*1.1*
1996-98 Sidekick, X-90		
one	.8	*.8*
both	1.3	*1.3*
1995-01 Swift		
one	.8	*.9*
both	1.3	*1.4*
Verona		
one	.9	*1.0*
both	1.3	*1.4*

Pressure Relief Valve, Replace (B)
Grand Vitara, Vitara	1.5	*1.5*

Pressure Modulator Valve Replace (B)
Grand Vitara, Vitara	1.4	*1.4*

Rear Sensor Assy., Replace (B)
Aerio, Esteem, Forenza, Verona		
one	.6	*.7*
both	.8	*.9*
Grand Vitara, Vitara, XL-7	.8	*.9*
Sidekick, X-90		
one	1.0	*1.1*
both	1.1	*1.3*

	LABOR TIME	SEVERE SERVICE
Swift		
one	.8	*.9*
both	1.3	*1.4*
Relay, Replace (B)		
Esteem, Sidekick, Swift, X-90	.3	*.3*

SYSTEM

Bleed Brakes (B)
Includes: Add fluid.
All Models	1.1	*1.1*
w/ABS add	.2	*.2*

Brakes, Adjust (B)
Includes: Refill master cylinder.
Rear wheels	.5	*.5*

Brake Hose (Flexible), Replace (B)
Includes: System bleeding.
Exc. below one	1.0	*1.1*
Aerio, Esteem, Swift one	1.3	*1.4*
Forenza, Verona one	.7	*.9*
each addl. add	.6	*.6*
w/ABS add	.2	*.2*

Brake Proportioning Valve, Replace (B)
Aerio	1.4	*1.6*
Esteem, Swift	2.1	*2.3*
Grand Vitara, Vitara, XL-7	1.6	*1.8*
Samurai	1.4	*1.6*
Sidekick, X-90	1.5	*1.7*
Verona	1.0	*1.2*

Master Cylinder, R&R and Recondition (B)
Includes: System bleeding.
Aerio, Esteem, Forenza, Swift, Verona	1.6	*1.8*
Grand Vitara, Vitara, XL-7	2.2	*2.5*
Samurai	1.4	*1.7*
Sidekick, X-90	1.9	*2.1*
w/ABS add	.2	*.2*

Master Cylinder, Replace (B)
Includes: System bleeding.
Aerio, Esteem, Forenza, Swift, Verona	1.4	*1.6*
Grand Vitara, Vitara, XL-7	1.9	*2.1*
Samurai	.9	*1.1*
Sidekick, X-90	1.6	*1.8*
w/ABS add	.2	*.2*

Power Booster Unit, Replace (B)
Exc. Esteem	1.9	*2.1*
Esteem	1.6	*1.8*
w/ABS Sidekick, Swift, X-90 add	.5	*.5*

Power Booster Vacuum Check Valve, Replace (B)
All Models	.3	*.3*

	LABOR TIME	SEVERE SERVICE
SERVICE BRAKES		
Brake Differential Valve, Replace (B)		
Sidekick, X-90	1.6	*1.8*
Brake Drum, Replace (B)		
Aerio	1.0	*1.1*
Esteem, Swift	1.3	*1.4*
Forenza	1.4	*1.5*
Grand Vitara, XL-7	1.1	*1.3*
Samurai, Sidekick, Vitara, X-90	1.1	*1.3*
Brake Pressure Limiting Valve, Replace (B)		
Sidekick, X-90	1.9	*2.1*
Caliper Assy., R&R and Recondition (B)		
Includes: System bleeding.		
Exc. below		
one	2.1	*2.3*
both	2.9	*3.2*
Aerio, Verona		
one	1.6	*1.8*
both	2.1	*2.3*
Esteem		
one	2.1	*2.3*
both	3.1	*3.4*
Samurai		
one	1.9	*2.2*
both	2.8	*3.1*
Swift		
one	1.9	*2.2*
both	3.2	*3.5*
w/ABS add	.2	*.2*
Caliper Assy., Replace (B)		
Includes: System bleeding.		
One	1.7	*2.0*
Both	2.2	*2.5*
w/ABS add	.2	*.2*
Disc Brake Rotor, Replace (B)		
Aerio, Verona		
one	.8	*.9*
both	1.0	*1.1*
Esteem, Forenza, Swift		
front		
one	1.4	*1.6*
both	2.5	*2.9*
rear		
one	1.1	*1.3*
both	1.7	*2.0*
Grand Vitara, Sidekick, Vitara, X-90, XL-7		
one	1.1	*1.3*
both	1.6	*1.8*
Samurai		
one	1.4	*1.6*
both	1.9	*2.1*

SUZ-12   AERIO : ESTEEM : FORENZA : GRAND VITARA : SAMURAI

	LABOR TIME	SEVERE SERVICE
Pads and/or Brake Shoes, Replace (B)		
Includes: Adjust service and parking brake. System bleeding.		
Front disc		
Aerio, Forenza	1.0	1.1
Esteem, Swift, Verona	1.4	1.6
Grand Vitara, Vitara, XL-7	1.7	2.0
Samurai	1.6	1.7
Sidekick, X-90	1.7	1.8
Rear disc		
Esteem	1.5	1.6
Forenza, Verona	1.0	1.1
Rear drum		
Aerio	1.4	1.5
Esteem, Swift	1.7	2.0
Forenza	2.9	3.2
Grand Vitara, Vitara, XL-7	1.9	2.1
Samurai	2.1	2.2
Sidekick, X-90	2.1	2.3
Verona	1.0	1.1
w/ABS add	.2	.2
COMBINATION ADD-ONS		
Repack wheel bearings two wheels add	.7	.7
Replace		
brake hose add	.3	.3
caliper add	.3	.3
wheel cyl add each	.2	.2
Resurface		
drum add	.5	.5
rotor add	.5	.5
Wheel Cylinder, R&R and Recondition (B)		
Includes: System bleeding.		
Aerio		
one	1.9	2.1
both	3.0	3.3
Esteem, Samurai, Swift		
one	2.1	2.3
both	2.9	3.2
Grand Vitara, Vitara, XL-7		
one	2.1	2.3
both	2.7	3.0
Sidekick, X-90		
one	2.2	2.5
both	3.3	3.7
w/ABS add	.2	.2
Wheel Cylinder, Replace (B)		
Exc. below		
one	1.9	2.1
both	2.5	2.9
Sidekick, X-90		
one	2.1	2.3
both	3.0	3.4
w/ABS add	.2	.2

	LABOR TIME	SEVERE SERVICE
PARKING BRAKE		
Parking Brake Apply Actuator, Replace (B)		
All Models	.6	.6
Parking Brake Cable, Adjust (C)		
All Models	.5	.7
Parking Brake Cable, Replace (B)		
Exc. below		
one or one side	1.4	1.6
both sides	2.1	2.3
Grand Vitara, Vitara, XL-7		
one	1.7	2.0
both	2.4	2.7
Samurai one	.8	1.0

FRONT SUSPENSION

Unless otherwise noted, time given does not include alignment.

FRONT WHEEL DRIVE

	LABOR TIME	SEVERE SERVICE
Align Front End (A)		
All Models	1.6	1.8
Front End Alignment, Check (B)		
All Models	.5	.7
Front Toe, Adjust (B)		
All Models	.7	.9
Front Hub Assy., Replace (B)		
Aerio		
one	1.7	2.0
both	2.9	3.2
Esteem, Swift		
one	1.6	1.8
both	2.5	2.9
Forenza, Verona		
one	.8	1.1
both	1.4	1.7
Front Hub Seal, Replace (B)		
Esteem, Swift		
one side	2.4	2.7
both sides	4.0	4.4
Hub Bearings, Replace (B)		
Aerio		
one side	1.7	2.0
both sides	2.9	3.2
Esteem, Swift		
one side	2.4	2.7
both sides	4.0	4.4
Forenza		
one	.8	1.1
both	1.4	1.7
Lower Control Arm, Replace (B)		
Aerio, Forenza		
one	1.1	1.3
both	1.4	1.6
Esteem, Swift, Verona		
one	1.6	1.8
both	2.2	2.5
Replace bushings add each side	.2	.2

	LABOR TIME	SEVERE SERVICE
Stabilizer (Sway) Bar, Replace (B)		
Aerio, Esteem, Swift	1.0	1.1
Forenza, Verona	2.5	2.7
Stabilizer Ball Joints, Replace (B)		
Aerio, Esteem, Forenza, Swift, Verona	.8	.9
Strut (Shock Absorber), Replace (B)		
Aerio, Esteem, Forenza, Swift, Verona		
one	1.4	1.6
both	2.1	2.3
Replace springs add each	.2	.2

REAR WHEEL DRIVE

	LABOR TIME	SEVERE SERVICE
Align Front End (A)		
Samurai, Sidekick, X-90	.8	.9
Front End Alignment, Check (B)		
All Models	.5	.7
Front Toe, Adjust (B)		
All Models	.6	.8
Axle Shaft Oil Seal, Replace (B)		
Samurai	2.3	2.3
Axle Shaft Spindle, Replace (B)		
Samurai		
one side	2.0	2.5
both sides	3.0	3.5
Front Springs, Replace (B)		
Leaf		
Samurai		
one side	1.4	1.7
both sides	2.9	3.2
Coil		
Grand Vitara, Vitara, XL-7		
one side	2.2	2.5
both sides	3.3	3.7
Sidekick, X-90		
one side	1.4	1.6
both sides	2.1	2.3
Front Hub, Replace (B)		
Grand Vitara, Vitara, XL-7		
one side	1.4	1.6
both sides	2.4	2.7
Samurai		
one	1.6	2.1
both	2.5	3.0
Sidekick, X-90		
2WD		
one side	1.4	1.6
both sides	2.4	2.7
4WD		
one side	.8	.9
both sides	1.3	1.4
Front Hub Bearings or Seal, Replace (B)		
Grand Vitara, Vitara, XL-7		
one side	1.4	1.6
both sides	2.4	2.7

SIDEKICK : SWIFT : VERONA : VITARA : X-90 : XL-7 SUZ-13

	LABOR TIME	SEVERE SERVICE
Sidekick, X-90		
one side	1.4	1.6
both sides	2.3	2.6
Front Shock Absorbers, Replace (B)		
Samurai		
one	.5	.6
both	.8	.9
Front Spindle, Replace (B)		
One	1.6	1.8
Both	2.7	3.0
Front Strut (Shock Absorber), Replace (B)		
Grand Vitara, Vitara, XL-7		
one	1.3	1.4
both	1.9	2.1
Sidekick, X-90		
one	1.2	1.3
both	1.8	1.9
Locking Hub, Replace or Recondition (B)		
Grand Vitara, Vitara, XL-7		
one	1.4	1.6
both	2.4	2.7
Samurai		
one	.5	.8
both	.7	1.0
Sidekick, X-90		
one	.7	.9
both	1.3	1.5
Lower Ball Joint, Replace (B)		
Sidekick, X-90		
one	1.7	2.0
both	2.7	3.0
Lower Control/Suspension Arm Assy., Replace (B)		
Grand Vitara, Vitara, XL-7		
one	2.7	3.0
both	4.3	4.8
Sidekick, X-90		
one	1.7	2.0
both	3.0	3.4
Spring Shackle Bushings, Replace (B)		
Samurai		
front		
one side	.7	.9
both sides	.9	1.1
rear		
one side	.8	1.0
both sides	1.7	1.9
Stabilizer (Sway) Bar, Replace (B)		
Exc. Samurai	.8	.9
Samurai	1.3	1.5
Stabilizer (Sway) Bar Bushings, Replace (B)		
One or both	.5	.7
Steering Knuckle, Replace (B)		
Exc. Samurai		
one	2.4	2.7
both	3.8	4.3

	LABOR TIME	SEVERE SERVICE
Samurai		
2WD	3.0	3.2
4WD	4.1	4.6
Steering Knuckle Inner Dust Seal, Replace (B)		
Grand Vitara, Vitara, XL-7		
one	1.4	1.6
both	2.4	2.7
Steering Knuckle Pivot (King Pin), Replace (B)		
Samurai		
one	2.1	2.4
both	3.5	3.8
Steering Knuckle Pivot Bearings, Replace (B)		
Samurai		
one	3.7	4.0
both	6.2	6.5
Steering Knuckle Retainer Oil Seals, Replace (B)		
Samurai		
one	.9	.9
both	1.6	1.6

REAR SUSPENSION

Unless otherwise noted, time given does not include alignment.

	LABOR TIME	SEVERE SERVICE
Lateral Rod, Replace (B)		
Grand Vitara, Vitara, XL-7	.5	.6
Lower Control Arm, Replace (B)		
Swift		
one	1.3	1.4
both	2.1	2.3
Lower Control Arm Bushing, Replace (B)		
Swift		
one side	.8	.9
both sides	1.1	1.3
Parallel Link, Replace (B)		
Forenza	1.1	1.3
Rear Coil Spring, Replace (B)		
Aerio, Esteem, Forenza, Verona		
one	1.5	1.7
both	2.1	2.3
Grand Vitara, Vitara, XL-7		
one	1.7	2.0
both	2.5	2.9
Sidekick		
one	.6	.7
both	.8	.9
Swift		
one	1.0	1.1
both	1.4	1.6
Rear Hub and/or Bearings, Replace (B)		
Aerio		
one	2.7	3.0
both	4.0	4.4

	LABOR TIME	SEVERE SERVICE
Esteem, Swift, Verona		
one	1.3	1.4
both	2.1	2.3
Forenza		
one	1.9	2.1
both	2.5	2.9
Rear Knuckle, Replace (B)		
Aerio, Forenza		
one	2.7	3.0
both	4.0	4.4
Esteem, Swift		
one	1.9	2.1
both	3.0	3.4
Verona		
one	1.3	1.4
both	2.1	2.5
Rear Leaf Spring, Replace (B)		
Samurai		
one	1.5	1.8
both	2.8	3.1
Rear Strut Assy., Replace (B)		
Aerio, Esteem, Forenza, Swift		
one	1.5	1.6
both	2.2	2.4
Rear Wheel Spindle Control Rod, Replace (B)		
Aerio, Swift		
one	1.1	1.3
both	1.4	1.6
Esteem		
one	.8	.9
both	1.1	1.3
Shock Absorbers or Bushings, Replace (B)		
Exc. Verona		
one	.6	.7
both	1.0	1.1
Verona		
one	1.6	1.7
both	2.4	2.6
Spring Front Eye Bushing, Replace (B)		
Samurai		
one side	.8	1.0
both side	1.7	1.9
Spring Shackle, Replace (B)		
Samurai		
one	.8	1.0
both	1.3	1.5
Stabilizer Bar, Replace (B)		
Aerio, Forenza, Swift, Verona	.8	1.0
Trailing Arm and/or Bushings, Replace (B)		
Aerio, Verona		
one	.5	.6
both	.6	.7
Esteem		
one	.6	.7
both	.8	.9

SUZ-14 AERIO : ESTEEM : FORENZA : GRAND VITARA : SAMURAI

	LABOR TIME	SEVERE SERVICE
Trailing Arm and/or Bushings, Replace (B)		
Grand Vitara, Vitara, XL-7		
one	.8	.9
both	1.1	1.3
Sidekick, X-90		
one	1.3	1.4
both	1.9	2.1
Upper Control Arm, Replace (B)		
Sidekick, X-90 one	1.6	1.8
Verona	.6	.8
Upper Control Arm Bushings, Replace (B)		
Sidekick, X-90 all	1.6	1.8

STEERING

AIR BAGS

	LABOR TIME	SEVERE SERVICE
Diagnosis Air Bag System (A)		
All Models	1.0	1.0
Coil, Replace (B)		
Grand Vitara, Vitara, XL-7	.7	.7
Diagnostic Unit, Replace (B)		
Exc. below	.5	.5
1989-95 Sidekick	1.2	1.2
Forward Sensor, Replace (B)		
Forenza, Verona	.6	.7
Grand Vitara, XL-7	.8	.9
Swift	.5	.5
Module Assembly, Replace (B)		
Driver side	.5	.5
Passenger side		
Aerio, Grand Vitara, Vitara, XL-7	.8	.8
Esteem, Sidekick, Swift, X-90	.5	.5
Forenza	2.3	2.5
Verona	.8	.8

LINKAGE

	LABOR TIME	SEVERE SERVICE
Center Link (Relay Rod), Replace (B)		
Samurai, Sidekick, X-90	1.4	1.6
Idler Arm, Replace (B)		
Sidekick, X-90	1.0	1.1
Steering Damper, Replace (B)		
Samurai	.5	.7
Tie Rod or Tie Rod Ends, Replace (B)		
Includes: Reset toe.		
Samurai, Sidekick, X-90		
one	1.4	1.6
each addl. add	.3	.5

MANUAL RACK & PINION

	LABOR TIME	SEVERE SERVICE
Horn Contact, Replace (B)		
Esteem	.7	.7
Swift		
1989-94	.2	.2
1995-01	.7	.7
Inner Tie Rod Ends, Replace (B)		
Includes: Rack R&R where required.		
Esteem, Forenza, Swift		
one side	3.0	3.4
both sides	3.3	3.7
Outer Tie Rod End, Replace (B)		
Esteem, Forenza, Swift		
one side	1.3	1.4
both sides	1.9	2.1
Rack & Pinion Assy., R&R or Replace (B)		
Includes: Alignment.		
Esteem, Forenza, Swift	3.3	3.7
w/skid plate add	.3	.3
Rack Boots, Replace (B)		
Esteem, Forenza, Swift one	1.7	2.0
Steering Column, Replace (B)		
Esteem, Forenza, Swift	1.8	1.8
Steering Wheel, Replace (B)		
Esteem, Forenza, Swift	.5	.5
Tie Rod End Boot, Replace (B)		
Esteem, Forenza, Swift		
one	1.4	1.6
both	1.9	2.1

POWER RACK & PINION

	LABOR TIME	SEVERE SERVICE
Horn Contact or Canceling Cam, Replace (B)		
Esteem	.7	.7
Grand Vitara, Vitara, XL-7	.5	.5
Swift		
1989-94	.2	.2
1995-01	.7	.7
Outer Tie Rod End, Replace (B)		
Aerio, Forenza, Verona		
one side	1.0	1.1
both sides	1.1	1.3
Esteem, Swift		
one side	1.3	1.4
both sides	1.9	2.1
Grand Vitara, Vitara, XL-7		
one side	1.3	1.4
both sides	1.6	1.8
Rack & Pinion Assy., R&R or Replace (B)		
Includes: Alignment.		
Aerio		
2WD	3.0	3.4
4WD	4.3	4.8

	LABOR TIME	SEVERE SERVICE
Esteem	4.4	5.0
Forenza, Verona	2.5	2.9
Grand Vitara, XL-7	2.2	2.5
Swift	3.6	4.0
Vitara	1.4	1.6
Rack Boots, Replace (B)		
Aerio, Grand Vitara, Vitara, XL-7 one	1.3	1.4
Esteem, Swift, Verona one	1.7	2.0
Forenza one	1.2	1.3
Steering Column, R&I (B)		
Aerio, Verona	1.6	1.8
Esteem, Forenza, Swift	1.3	1.4
Grand Vitara, Vitara, XL-7	1.9	2.1
Steering Pump, R&R or Replace (B)		
Exc. below	1.2	1.4
Aerio, Verona	1.6	1.8
Esteem	1.3	1.4
Forenza	.8	1.0
Swift	1.7	2.0
Steering Pump, R&R and Recondition (B)		
Aerio	2.1	2.3
Esteem, Swift	2.2	2.5
Forenza	1.6	1.9
Grand Vitara, XL-7	1.7	2.0
Sidekick, Vitara, X-90	2.4	2.7
Steering Pump Hoses, Replace (B)		
Aerio, Esteem, Forenza, Swift, Verona		
pressure or return each	.8	.9
Grand Vitara, Vitara, XL-7		
pressure	.8	.9
return		
to gear	.7	.8
to pump	.6	.7
Steering Wheel, Replace (B)		
Exc. below	.6	.6
Aerio, Forenza, Verona	.7	.7
Tie Rod End Boot, Replace (B)		
Aerio, Forenza, Verona		
one	.6	.7
both	.8	.9
Esteem, Swift		
one	1.3	1.4
both	1.9	2.1
Grand Vitara, Vitara, XL-7		
one	1.0	1.1
both	1.3	1.4

SIDEKICK : SWIFT : VERONA : VITARA : X-90 : XL-7   SUZ-15

	LABOR TIME	SEVERE SERVICE
MANUAL WORM & SECTOR		
Gear Assy., Replace (B)		
Samurai, Sidekick, X-90	1.4	1.6
Horn Contact, Replace (B)		
Samurai, Sidekick, X-90	.3	.3
Steering Column, R&R or Replace (B)		
Samurai, Sidekick, X-90	1.9	2.1
Steering Wheel, Replace (B)		
Samurai, Sidekick, X-90	.6	.6
POWER WORM & SECTOR		
Gear Assy., Replace (B)		
Sidekick, X-90	2.2	2.5
Horn Contact, Replace (B)		
Sidekick, X-90	.3	.3
Steering Column, R&I (B)		
Sidekick, X-90	1.9	2.1
Steering Pump, R&R and Recondition (B)		
Sidekick, X-90	2.4	2.7
Steering Pump, R&R or Replace (B)		
Sidekick, X-90	1.7	2.0
Steering Pump Hoses, Replace (B)		
Sidekick, X-90		
pressure	1.0	1.1
return	.6	.7
Steering Wheel, Replace (B)		
Sidekick, X-90	.6	.6

HEATING & AIR CONDITIONING

When more than one component requires replacement where evacuation/recovery and recharging is already included, deduct 1.0 hour for each additional component from the time given.

	LABOR TIME	SEVERE SERVICE
Evacuate/Recover and Recharge System (B)		
All Models	1.0	1.0
Leak Check (B)		
Includes: Check all lines and connections.		
All Models	1.2	1.4
AC Hoses, Replace (B)		
Includes: Evacuate/recover and recharge.		
Aerio one	2.8	3.1
Esteem one	2.0	2.2
Forenza	1.6	1.8
Grand Vitara, XL-7 one	3.3	3.7
Samurai one	1.9	2.0
Sidekick, Swift, Vitara, X-90 one	1.7	1.8
Verona	2.5	2.7
Blower Motor, Replace (B)		
Aerio, Esteem, Swift	.6	.7
Grand Vitara, Vitara, XL-7		
AC	2.2	2.2
heater	1.1	1.1
Samurai	4.4	4.4
Sidekick, X-90	3.0	3.0
Verona	6.6	6.8
Blower Motor Resistor, Replace (B)		
Exc. Samurai	.6	.6
Samurai	4.5	4.5
Blower Motor Switch, Replace (B)		
Samurai	1.2	1.2
Sidekick, X-90	.7	.7
Compressor Assy., Replace (B)		
Includes: Parts transfer, evacuate/recover and recharge.		
Aerio, Forenza	2.3	2.6
Esteem, Swift	3.3	3.7
Grand Vitara, XL-7	3.1	3.4
Samurai	2.3	2.5
Sidekick, Vitara, X-90	2.3	2.6
Verona	2.1	2.4
Compressor Clutch & Pulley, Replace (B)		
Includes: Evacuate/recover and recharge.		
Aerio	2.6	2.9
Esteem	3.6	4.1
Forenza, Verona	4.3	4.8
Grand Vitara, XL-7	3.3	3.7
Samurai	2.5	2.7
Sidekick, Vitara, X-90	2.8	3.1
Swift		
1989-94	3.3	3.5
1995-01	3.6	4.1
Compressor Shaft Seal, Replace (B)		
Includes: Compressor R&R. Evacuate/recover and recharge.		
Exc. below	3.5	3.9
Aerio	3.2	3.6
Esteem, Swift	4.3	4.7
Forenza, Verona	2.3	2.5
Samurai	3.1	3.3
Condenser Assy., Replace (B)		
Includes: Evacuate/recover and recharge.		
Aerio, Forenza	2.6	2.9
Esteem, Swift	3.8	4.2
Grand Vitara, Vitara, XL-7	3.4	2.7
Samurai	2.6	2.8
Sidekick, X-90	2.6	2.9
Verona	2.1	2.4

	LABOR TIME	SEVERE SERVICE
Condenser Cooling Fan, Replace (B)		
Aerio	.5	.6
Esteem, Swift	3.3	3.7
Forenza, Verona	.8	.9
Grand Vitara, Vitara, XL-7	.9	1.0
Sidekick, X-90	1.1	1.2
Condenser Fan Motor, Replace (B)		
Samurai	.9	1.2
Evaporator Core, Replace (B)		
Includes: Evacuate/recover and recharge.		
Aerio	2.3	2.3
Esteem, Swift	3.4	3.4
Forenza	6.1	6.1
Grand Vitara, XL-7		
front unit	2.8	2.8
rear unit	2.3	2.3
Samurai	2.7	2.7
Sidekick, X-90	4.2	4.2
Verona	6.4	6.4
Vitara	2.3	2.3
Expansion Valve, Replace (B)		
Includes: Evacuate/recover and recharge.		
Aerio	2.0	2.2
Esteem, Swift	4.0	4.4
Forenza, Verona	1.6	1.8
Grand Vitara, XL-7		
front unit	3.2	3.6
rear unit	2.6	2.9
Samurai	3.0	3.0
Sidekick, X-90	4.7	5.2
Vitara	2.6	2.9
Heater Control Cable, Replace (B)		
Aerio, Esteem, Forenza	.7	.7
Grand Vitara, Vitara, XL-7	.8	.8
Samurai	2.2	2.2
Sidekick, X-90	3.0	3.0
Swift	1.9	1.9
Heater Core, R&R or Replace (B)		
Aerio		
w/AC	3.6	3.6
w/o AC	2.4	2.4
Esteem, Swift	2.9	3.2
Forenza		
w/AC	6.2	6.4
w/o AC	5.3	5.5
Grand Vitara, XL-7		
w/AC	4.9	4.9
w/o AC	3.4	3.4
Samurai	4.9	4.9
Sidekick, X-90		
w/AC	5.7	5.7
w/o AC	4.1	4.1
Verona		
w/AC	7.1	7.3
w/o AC	6.1	6.3

AERIO : ESTEEM : FORENZA : GRAND VITARA : SAMURAI

	LABOR TIME	SEVERE SERVICE
Heater Core, R&R or Replace (B)		
Vitara		
w/AC	3.9	3.9
w/o AC	2.4	2.4
Heater Hoses, Replace (B)		
One	.8	.8
Both	1.5	1.5
Receiver/Drier Assy., Replace (B)		
Includes: Evacuate/recover and recharge.		
Esteem, Swift	3.3	3.7
Forenza	2.6	2.7
Grand Vitara, Vitara,		
XL-7	2.1	2.4
Samurai	1.7	1.7
Sidekick, X-90	1.7	1.9
Verona	1.6	1.6
Temperature Control Assy. Replace (B)		
Aerio, Forenza,		
Verona	.6	.6
Esteem	.7	.7
Grand Vitara, Vitara,		
XL-7	.8	.8
Samurai	2.4	2.3
Sidekick, X-90		
exc. below	3.0	3.0
push type control	.7	.7
Swift	1.9	1.9

WIPERS & SPEEDOMETER

	LABOR TIME	SEVERE SERVICE
Antenna. Replace (B)		
Exc. below	.6	.7
Aerio	1.5	1.7
Forenza	2.0	2.1
Radio, R&R (B)		
Exc. Forenza	.6	.8
Forenza	.8	.9
Rear Window Washer Pump, Replace (B)		
Exc. below	.5	.5
Aerio	.8	.9
Esteem, Swift	1.1	1.2
Rear Window Wiper Motor, Replace (B)		
Exc. below	.6	.7
Sidekick, X-90	.9	1.0
Rear Window Wiper & Washer Switch, Replace (B)		
Exc. below	.5	.5
Sidekick, X-90	.7	.7
Speedometer Cable & Casing, Replace (B)		
Esteem, Swift	1.0	1.1
Samurai	1.2	1.2
Sidekick, X-90	1.3	1.4

	LABOR TIME	SEVERE SERVICE
Speedometer Body, R&R or Replace (B)		
Aerio, Esteem	.5	.5
Forenza, Grand Vitara,		
Vitara, XL-7	.7	.7
Samurai, Sidekick,		
X-90	1.2	1.2
Swift, Verona	.9	.9
Windshield Washer Pump, Replace (B)		
Aerio, Forenza,		
Verona	.8	.9
Esteem, Swift	1.1	1.2
Grand Vitara, Vitara,		
XL-7	.5	.5
Samurai	.3	.3
Sidekick, X-90	.5	.5
Windshield Wiper Link, Replace (B)		
Aerio, Forenza, Verona	.8	.9
Esteem, Swift		
front	.9	1.0
rear	.5	.5
Grand Vitara, Vitara,		
XL-7	.9	1.0
Samurai		
hard top	3.3	3.5
soft top	1.5	1.7
Sidekick, X-90	4.0	4.4
Windshield Wiper Motor, Replace (B)		
Exc. below	.8	.9
Grand Vitara, Vitara,		
XL-7	.5	.7
Samurai		
hard top	3.2	3.6
soft top	1.2	1.4
Sidekick, X-90	1.1	1.3

LAMPS & SWITCHES

	LABOR TIME	SEVERE SERVICE
Back-Up Lamp Bulb, Replace (C)		
Exc. Forenza	.3	.3
Forenza	.6	.6
Back-Up Lamp Switch, Replace (B)		
Exc. below	.4	.6
Grand Vitara, Vitara,		
XL-7	1.3	1.4
Samurai	.5	.5
Sidekick, X-90	1.3	1.4
Clearance Lamp Assy., Replace (B)		
Aerio	.3	.3
Esteem, Forenza	.6	.7
Swift		
1989-94	1.2	1.2
1994-01	.3	.3
Clutch Start Switch, Replace (B)		
All Models	.5	.5
Combination Switch Assy., Replace (B)		
Exc. Samurai	.8	.8
Samurai	1.1	1.1

	LABOR TIME	SEVERE SERVICE
Front Combination Lamp Bulb, Replace (C)		
All Models	.3	.3
Halogen Headlamp Bulb, Replace (C)		
Exc. Forenza	.3	.3
Forenza	.6	.6
Headlamps, Aim (B)		
Two	.5	.5
Four	.7	.7
High Mount Stop Lamp Assy., Replace (B)		
Exc. below	.5	.5
Esteem	.6	.6
Swift	.3	.3
Horn, Replace (B)		
Exc. Aerio	.5	.5
Aerio	.9	1.0
Horn Relay, Replace (B)		
All Models	.5	.5
License Lamp Assy., Replace (B)		
All Models	.5	.5
Parking Brake Warning Lamp Switch, Replace (B)		
All Models	.5	.5
Rear Combination Lamp Assy., Replace (B)		
All Models	.5	.5
Sealed Beam Headlamp, Replace (C)		
All Models	.3	.3
Side Marker & Reflector Assy., Replace (B)		
All Models	.3	.3
Side Marker Lamp Lens, Replace (B)		
All Models	.3	.3
Starter Safety Switch, Replace (B)		
All Models	.5	.5
Stop Lamp Switch, Replace (B)		
All Models	.5	.5
Tail Lamp Lens or Bulb, Replace (C)		
All Models	.3	.3
Turn Signal or Hazard Warning Flasher, Replace (B)		
All Models	.5	.5
Turn Signal Lamp Assy., Replace (B)		
All Models	.5	.5
Turn Signal Lamp Lens or Bulb, Replace (C)		
All Models	.2	.2

SEAT BELTS

	LABOR TIME	SEVERE SERVICE
Automatic Shoulder Belt Control Unit, Replace (B)		
1989-94 Swift	.8	.8
Seat Belt Assy., Replace (B)		
Aerio, Forenza, Verona		
front seat one	.6	.6
rear seat one	.8	.8
Esteem one	.5	.5
Grand Vitara, XL-7 one	.7	.7

	LABOR TIME	SEVERE SERVICE
Sidekick, Vitara,		
X-90 one	.7	.7
Swift		
1989-94 trim removal		
one side	1.2	1.2
both sides	2.2	2.2
1989-01 one	.5	.5

BODY

Door Lock Cylinder, Replace (B)
Exc. below
one side	.6	.6
both sides	.9	.9
back door	.5	.5

	LABOR TIME	SEVERE SERVICE
Samurai, Sidekick,		
Verona, X-90		
one side	.8	.8
both sides	1.4	1.4
back door	.5	.5

Door Window Regulator, Replace (B)
Exc. below
one side	.8	.8
both sides	1.5	1.5

Forenza, Sidekick,
Verona, X-90
one side	.9	.9
both sides	1.8	1.8

	LABOR TIME	SEVERE SERVICE
Hood Hinges (Both), Replace (B)		
All Models	.6	.7
Hood Lock, Replace (B)		
Exc. Aerio	.5	.5
Aerio	.9	1.0
Hood Release Cable, Replace (B)		
Exc. below	.8	.9
Aerio, Forenza	1.1	1.2
Lock Striker Plate, Replace (B)		
All Models	.3	.3

NOTES

TOY

Avalon : Camry : Celica : Corolla : Cressida : Echo : Matrix : MR2 : MR2 Spyder : Paseo : Prius : Solara : Starlet : Supra : Tercel

SYSTEM INDEX

- MAINTENANCE TOY-2
  - Maintenance Schedule... TOY-2
- CHARGING TOY-3
- STARTING TOY-3
- CRUISE CONTROL TOY-4
- IGNITION TOY-4
- EMISSIONS TOY-5
- FUEL TOY-8
- EXHAUST TOY-11
- ENGINE COOLING TOY-12
- ENGINE TOY-14
  - Assembly TOY-14
  - Cylinder Head TOY-17
  - Camshaft TOY-21
  - Crank & Pistons TOY-25
  - Engine Lubrication TOY-29
- CLUTCH TOY-31
- MANUAL TRANSAXLE TOY-32
- MANUAL TRANSMISSION . TOY-33
- AUTO TRANSAXLE TOY-33
- AUTO TRANSMISSION TOY-35
- TRANSFER CASE TOY-36
- SHIFT LINKAGE TOY-36
- DRIVELINE TOY-36
- BRAKES TOY-38
- FRONT SUSPENSION TOY-43
- REAR SUSPENSION TOY-45
- STEERING TOY-47
- HEATING & AC TOY-50
- WIPERS & SPEEDOMETER TOY-53
- LAMPS & SWITCHES TOY-53
- SEAT BELTS TOY-54
- BODY TOY-55

OPERATIONS INDEX

A
- AC Hoses TOY-50
- Air Bags TOY-47
- Air Conditioning TOY-50
- Alignment TOY-43
- Alternator (Generator) ... TOY-3
- Antenna TOY-53
- Anti-Lock Brakes TOY-38

B
- Back-Up Lamp Switch TOY-54
- Battery Cables TOY-3
- Bleed Brake System TOY-39
- Blower Motor TOY-50
- Brake Disc TOY-41
- Brake Drum TOY-40
- Brake Hose TOY-39
- Brake Pads and/or Shoes ... TOY-41

C
- Camshaft TOY-21
- Camshaft Sensor TOY-5
- Catalytic Converter TOY-11
- Coolant Temperature (ECT) Sensor TOY-7
- Crankshaft TOY-25
- Cruise Control TOY-4
- CV Joint TOY-38
- Cylinder Head TOY-17

D
- Differential TOY-37
- Distributor TOY-5
- Drive Belt TOY-2
- Driveshaft TOY-36

E
- EGR TOY-7
- Electronic Control Module (ECM/PCM) TOY-7
- Engine TOY-14
- Engine Lubrication TOY-29
- Engine Mounts TOY-16
- Evaporator TOY-51
- Exhaust TOY-11
- Exhaust Manifold TOY-11
- Expansion Valve TOY-51

F
- Flexplate TOY-35
- Flywheel TOY-32
- Fuel Injection TOY-10
- Fuel Pump TOY-9
- Fuel Vapor Canister TOY-8

G
- Generator TOY-3
- Glow Plug TOY-5

H
- Halfshaft TOY-38
- Headlamp TOY-54
- Heater Core TOY-51
- Horn TOY-54

I
- Idle Air Control (IAC) Valve .. TOY-11
- Ignition Coil TOY-5
- Ignition Switch TOY-5
- Injection Pump TOY-10
- Inner Tie Rod TOY-48
- Intake Manifold TOY-9

K
- Knock Sensor TOY-8

L
- Lower Control Arm TOY-44

M
- Maintenance Schedule TOY-2
- Master Cylinder TOY-39
- Muffler TOY-12

N
- Neutral Safety Switch TOY-54

O
- Oil Pan TOY-30
- Oil Pump TOY-30
- Outer Tie Rod TOY-48
- Oxygen Sensor TOY-8

P
- Parking Brake TOY-42
- Pistons TOY-27
- Positive Crankcase Ventilation (PCV) Valve TOY-8

R
- Radiator TOY-12
- Radiator Hoses TOY-13
- Radio TOY-53
- Rear Main Oil Seal TOY-28

S
- Shock Absorber/Strut, Front TOY-44
- Shock Absorber/Strut, Rear TOY-46
- Spark Plug Cables TOY-5
- Spark Plugs TOY-5
- Spring, Front Coil TOY-43
- Spring, Leaf TOY-46
- Spring, Rear Coil TOY-45
- Starter TOY-3
- Steering Wheel TOY-48

T
- Thermostat TOY-12
- Throttle Body TOY-11
- Throttle Position Sensor (TPS) TOY-8
- Timing Belt TOY-23
- Timing Chain TOY-24
- Torque Converter TOY-34

U
- U-Joint TOY-36
- Upper Control Arm TOY-45

V
- Valve Body TOY-34
- Valve Cover Gasket TOY-19
- Valve Job TOY-19

W
- Water Pump TOY-13
- Wheel Balance TOY-2
- Wheel Cylinder TOY-42
- Window Regulator TOY-55
- Windshield Washer Pump ... TOY-53
- Windshield Wiper Motor ... TOY-53

TOY-1

TOY-2   AVALON : CAMRY : CELICA : COROLLA : CRESSIDA : ECHO : MATRIX : MR2

	LABOR TIME	SEVERE SERVICE
MAINTENANCE		
Air Cleaner Filter Element, Replace (C)		
1981-05 (.2)	.3	.4
Drive Belt, Adjust (B)		
Exc. below	.5	.6
Corolla	.7	.8
Prius, Tercel	.7	.8
each addl. add	.1	.1
Drive Belt, Replace (B)		
Exc. below	.6	.7
2002-05 Camry, Camry Solara	.9	1.0
1997-02 Corolla compressor	.8	.9
Supra		
1993-98		
w/turbo	.7	.8
w/o turbo	.3	.4
Steering pump		
exc. below	.9	1.0
1989-91 Camry V6	1.3	1.4
1990-93 Celica	1.5	1.6
1981-84 Cressida	.7	.8
1989-92 Cressida	1.3	1.4
1987-92 Supra	1.3	1.4
Fuel Filter, Replace (B)		
Exc. below (.5)	.8	.9
2002-05 Camry, Camry Solara		
3.0L (.6)	1.0	1.1
3.3L (.3)	.4	.5
Celica		
1981-85	1.2	1.3
1986-93	.7	.8
1994-99	1.4	1.6
1995-97 Corolla FWD	1.1	1.2
1991-95 MR2	1.2	1.3
MR2 Spyder (1.3)	1.9	2.0
1996-99 Paseo	1.2	1.3
Tercel		
1981-82	.3	.4
1983-90		
3E	.3	.4
3EE	.7	.8
1991-94	.7	.8
1995-99	1.2	1.3
Halogen Headlamp Bulb, Replace (C)		
Exc. below (.3)	.5	.5
2003-05 Corolla, Matrix (.6)	.8	.8
Oil & Filter, Change (C)		
Includes: Correct fluid levels.		
All Models (.6)	.8	1.0
Sealed Beam Headlamp, Replace (B)		
1981-94	.7	.7

	LABOR TIME	SEVERE SERVICE
Timing Belt, Replace (B)		
Avalon	3.0	3.4
Camry		
1983-86		
gas	3.4	3.9
diesel	2.6	3.1
1987-91		
4 cyl.	2.9	3.4
V6	3.7	4.2
1992-96	3.2	3.7
1997-01	2.9	3.4
2002-04 V6	2.5	2.9
Camry Solara		
1999-01	3.0	3.5
2002-03 V6	2.7	3.2
Celica		
1986-89		
SOHC	3.4	3.9
DOHC	2.6	3.1
turbo	3.1	3.6
1990-93		
w/turbo	4.3	4.8
w/o turbo	3.3	3.8
1994-99		
7AFE	2.7	3.2
5SFE	2.3	2.8
Corolla FWD		
1983-87		
SOHC	2.9	3.4
DOHC, diesel	2.4	2.9
1988-92	3.1	3.6
1993-97		
2WD	2.5	3.0
4WD	3.0	3.5
Corolla FX, FX16		
SOHC	3.1	3.6
DOHC	2.3	3.8
Corolla RWD	1.8	2.3
Cressida		
1981-88	3.2	3.7
1989-92	2.3	2.8
MR2		
1985-89	2.8	3.3
1991-95		
w/turbo	3.9	4.4
w/o turbo	3.2	3.7
Paseo	2.7	3.2
Sienna	6.0	6.7
Supra		
1981-92	3.3	3.8
1993-98		
w/turbo	3.1	3.6
w/o turbo	2.5	3.0

	LABOR TIME	SEVERE SERVICE
Tercel		
1981-82	1.8	2.3
1983-99	2.4	2.9
w/AC add		
83-87 Corolla FWD	.2	.2
86-93 Celica	.1	.1
94-99 Celica	.3	.3
MR2	.5	.5
Sienna	.2	.2
Tercel	.1	.1
w/PS add Tercel	.1	.1
Tire, Replace (C)		
Includes: Dismount old tire and mount new tire to rim.		
1981-05 (.5)	.8	.9
each addl. add	.5	.5
Tires, Rotate (C)		
Exc. MR2 Spyder (.4)	.6	.6
MR2 Spyder (.5)	.8	.8
Wheels, Balance (B)		
Two (.6)	1.0	1.1
each addl. add	.3	.3
SCHEDULED MAINTENANCE INTERVALS		
If necessary, refer to appropriate Chilton maintenance service information.		
7,500 Mile Service (C)		
Exc. Paseo	.4	.4
Paseo	.7	.7
15,000 Mile Service (C)		
All Models	1.2	1.2
w/AT add	.1	.1
Inspect differential oil		
Camry, Celica, Corolla, Supra add	.1	.1
Inspect drive shaft boots		
exc. Supra add	.1	.1
22,500 Mile Service (C)		
Exc. Paseo	.4	.4
Paseo	.7	.7
30,000 Mile Service (B)		
Avalon	2.2	2.2
Camry	2.2	2.2
Celica	2.1	2.1
Corolla	2.7	2.7
Paseo	3.2	3.2
Tercel	3.2	3.2
Supra	2.1	2.1
w/AT add	.1	.1
w/LSD Supra add	.2	.2
37,500 Mile Service (C)		
Exc. Paseo	.4	.4
Paseo	.7	.7

MR2 SPYDER : PASEO : PRIUS : SOLARA : STARLET : SUPRA : TERCEL **TOY-3**

	LABOR TIME	SEVERE SERVICE
45,000 Mile Service (C)		
All Models	1.2	1.2
w/AT add	.1	.1
Inspect differential oil		
Camry, Celica, Corolla, Supra add	.1	.1
Inspect driveshaft boots		
exc. Supra add	.1	.1
52,500 Mile Service (C)		
Exc. Paseo	.4	.4
Paseo	.7	.7
60,000 Mile Service (B)		
Avalon	3.4	3.4
Camry, Celica	3.4	3.4
Corolla	3.2	3.2
Paseo	3.1	3.1
Supra	3.3	3.3
Tercel	3.1	3.1
w/AT add	.1	.1
w/LSD Supra add	.2	.2
67,500 Mile Service (C)		
Exc. Paseo	.5	.5
Paseo	.8	.8
75,000 Mile Service (C)		
All Models	1.3	1.3
w/AT add	.1	.1
Inspect differential oil		
Camry, Celica, Corolla, Supra add	.1	.1
Inspect driveshaft boots		
exc. Supra add	.1	.1
82,500 Mile Service (C)		
Exc. Paseo	.5	.5
Paseo	.8	.8
90,000 Mile Service (B)		
Avalon	2.3	2.3
Camry, Celica	2.3	2.3
Corolla	2.8	2.8
Paseo, Tercel	2.8	2.8
Supra	2.2	2.2
w/AT add	.1	.1
w/LSD Supra add	.2	.2
97,500 Mile Service (C)		
Exc. Paseo	.5	.5
Paseo	.8	.8

CHARGING

Alternator Circuits, Test (B)
Includes: Test component output.
| 1981-05 | .7 | .7 |

Alternator Assy., Replace (B)
Includes: Pulley transfer.
Avalon	.6	.8
Camry		
1983-86	1.4	1.6
1987-91		
4 cyl.	1.1	1.3
V6	.8	1.0
1992-05 (.6)	.9	1.0

	LABOR TIME	SEVERE SERVICE
Camry Solara (.5)	.8	1.0
Celica		
1981-89	1.3	1.5
1990-05 (.5)	.8	1.0
Corolla FWD		
1983-91	1.3	1.5
1992-04	.6	.8
Corolla FX, FX16	1.2	1.4
Corolla RWD	1.1	1.3
Cressida	1.2	1.4
Echo, Matrix (.4)	.6	.8
MR2, MR2 Spyder (.6)	.9	1.1
Paseo	.7	.9
Prius	16.3	18.4
Starlet	.9	1.1
Supra		
1981-87	1.2	1.4
1988-98	.8	1.0
Tercel		
1981-90	.8	1.0
1991-99	.7	.9
w/AC add Prius	.3	.3

Alternator Bearing, Replace (B)
Avalon	1.5	1.7
Camry		
1983-86	1.8	2.0
1987-01	1.6	1.8
Camry Solara		
1999-03 (1.0)	1.5	1.7
Celica		
1981-85	1.6	1.8
1986-89	2.0	2.2
1990-05 (1.1)	1.6	1.8
Corolla, Cressida (1.1)	1.6	1.8
Echo (.9)	1.3	1.5
Matrix	1.7	1.9
MR2, MR2 Spyder (1.2)	1.8	2.0
Paseo, Starlet	1.4	1.6
Supra, Tercel	1.6	1.8

Alternator Voltage Regulator, Replace (B)
Avalon	1.1	1.2
Camry		
1983-86		
SV	.7	.9
CV	1.7	2.0
1987-91	1.4	1.7
1992-96		
4 cyl.	.8	1.0
V6	1.2	1.5
1997-05 (.8)	1.2	1.4
Camry Solara		
1999-03 (.5)	.8	1.0
2004-05 (.7)	1.1	1.2
Celica		
1981-89	1.4	1.7
1990-05 (.4)	.6	.7

	LABOR TIME	SEVERE SERVICE
Corolla FWD		
1983-91	1.6	1.9
1992-05 (.7)	1.1	1.2
Corolla FX, FX16	1.5	1.8
Corolla RWD		
1981-82	.5	.7
1983		
internal	1.3	1.6
external	.5	.7
1983-87	1.5	1.8
Cressida	1.5	1.8
Echo (.4)	.6	.7
Matrix (.7)	1.1	1.2
MR2		
1985-89	.8	1.0
1991-95	1.4	1.7
MR2 Spyder (.4)	.6	.7
Paseo	.8	1.0
Starlet, Supra	1.5	1.8
Tercel		
1981-82	.5	.7
1983-90		
sedan	1.4	1.7
wagon		
internal	1.3	1.6
external	.5	.7
1991-99	.8	1.0

Voltmeter and/or Ammeter Gauge, Replace (B)
1983-89 Celica	1.7	1.7
1981-82 Corolla RWD	.8	.8
1981-85 Supra	1.7	1.7

STARTING

Starter Draw Test (On Car) (B)
| 1981-05 (.3) | .5 | .5 |

Battery, Replace (C)
| 1981-05 (.3) | .5 | .6 |

Battery Cables, Replace (C)
Positive		
exc. below (.3)	.5	.7
Camry, Celica (.5)	.7	.9
Corolla RWD		
1983-87 DOHC	.8	1.0
MR2		
1991-95	2.9	3.1
Starlet	1.5	1.7
Negative		
all models (.4)	.6	.7

Starter Assy., Replace (B)
Non-reduction type		
2002-05 Camry (.5)	.8	.9
2002-05 Camry Solara		
2.4L, 3.3L (.5)	.8	.9
2003-05 Corolla (.4)	.6	.7
Echo (.5)	.8	.9
Matrix (.4)	.6	.7
Paseo		
1992-95	.7	.9
1996-99	.8	1.0

AVALON : CAMRY : CELICA : COROLLA : CRESSIDA : ECHO : MATRIX : MR2

	LABOR TIME	SEVERE SERVICE
Starter Assy., Replace (B)		
Starlet	.9	1.1
Tercel	.8	1.0
Reduction type		
Avalon	1.0	1.1
Camry		
1983-86	.7	.9
1987-91		
4 cyl.	.5	.7
V6	1.6	1.8
1992-01	.8	1.0
Camry Solara	.9	1.0
Celica		
1981-85	.9	1.1
1986-89	.5	.7
1990-93		
w/turbo	1.2	1.4
w/o turbo	.7	.9
1994-05 (.7)	1.1	1.3
Corolla FWD		
1983-97	.7	.9
1998-02	3.0	3.2
Corolla FX, FX16	.8	1.0
Corolla RWD		
1981-82	.9	1.1
1983-87		
SOHC	.7	.9
DOHC	1.4	1.6
Cressida		
1981-82	1.6	1.8
1983-92	.8	1.0
MR2		
1985-87	.7	.9
1988-89		
4AGZE	1.1	1.3
4AGE	.7	.9
1991-95	1.1	1.3
MR2 Spyder (.5)	.8	.9
Paseo, Tercel	.7	.9
Starlet	.9	1.1
Supra		
1981-85	1.4	1.6
1986-98	.8	1.0
Recondition starter add		
exc.00-05 Echo	1.0	1.1
00-05 Echo	1.6	1.8
Starter Solenoid and/or Switch, Replace (B)		
Includes: Starter R&R.		
Non-reduction type		
Corolla, Matrix		
2002-05 (.5)	.8	.9
Echo (.9)	1.3	1.5
Paseo		
1992-95	.8	.9
1996-99	1.3	1.4
Starlet, Tercel	1.4	1.6

	LABOR TIME	SEVERE SERVICE
Reduction type		
Avalon	1.3	1.4
Camry		
1983-86		
SV	.8	.9
CV	1.1	1.3
1987-91		
4 cyl.	.8	.9
V6	1.9	2.1
1992-01		
4 cyl.	.8	.9
V6	1.3	1.4
Camry Solara		
4 cyl.	.8	.9
V6	1.2	1.3
Celica		
1981-85	1.4	1.6
1986-89		
3SFE	.8	.9
3SGE, 3SGTE	1.2	1.5
1990-93		
w/turbo	1.4	1.7
w/o turbo	.8	.9
1994-99	.8	.9
2000-05 (.9)	1.4	1.6
Corolla FWD		
1983-87	1.3	1.6
1988-97	.8	.9
1998-01	3.2	3.5
Corolla FX, FX16	1.3	1.4
Corolla RWD		
1981-82	1.3	1.4
1983-87		
SOHC	1.1	1.3
DOHC	1.6	1.9
Cressida		
1981-82	1.9	2.2
1983-92	1.1	1.3
MR2		
1985-87	.8	.9
1988-89		
4AGZE	1.5	1.8
4AGE	.8	.9
1991-95	1.3	1.6
MR2 Spyder (.7)	1.1	1.3
Paseo		
1992-95	.9	1.0
1996-99	.8	.9
Starlet	1.3	1.4
Supra		
1981-85	1.7	2.0
1986-92	1.2	1.3
1993-98		
w/turbo	1.3	1.4
w/o turbo	.8	.9
Tercel		
1981-94	1.1	1.3
1995-99	.8	.9

	LABOR TIME	SEVERE SERVICE
CRUISE CONTROL		
Diagnose Cruise Control System Component Each (A)		
1981-05	.8	.8
Actuator Assy., Replace (B)		
All Models (.4)	.6	.7
Control Computer (ECU), Replace (B)		
Exc. below (.4)	.6	.7
2000-04 Avalon	.8	.9
1987-91 Camry	1.3	1.3
Celica		
1986-89	2.5	2.5
1990-93	1.4	1.4
Control Main Switch, Replace (B)		
Avalon,		
Camry Solara (.4)	.6	.6
Camry		
1983-86	.3	.3
1987-91	.7	.7
1992-96	1.2	1.2
1997-05 (.7)	.9	.9
Celica		
1981-85	1.2	1.2
1986-89	.3	.3
1990-93	1.4	1.4
1994-05 (.4)	.6	.6
Corolla FWD		
1983-91	.5	.5
1992-02		
2WD	1.4	1.4
4WD	.5	.5
2003-05 (.6)	.8	.8
Corolla FX, FX16	.5	.5
Corolla RWD, Paseo	.5	.5
Cressida	.5	.5
Matrix, Prius (.6)	.8	.8
MR2	1.2	1.2
MR2 Spyder (.4)	.6	.6
Supra		
1981-92	1.3	1.3
1993-98	.7	.7
1983-90 Tercel	.3	.3
IGNITION		
Diagnose Ignition System Component Each (A)		
1981-05	.8	.8
Diagnose Igniter Circuit Malfunction (A)		
2002-05 Camry		
2.4L (1.0)	1.4	1.4
2000-05 Celica (.3)	.5	.5
Echo (.2)	.3	.3
2001-03 Prius	.9	.9
Solara 2.4L (1.0)	1.4	1.4

MR2 SPYDER : PASEO : PRIUS : SOLARA : STARLET : SUPRA : TERCEL — TOY-5

	LABOR TIME	SEVERE SERVICE
Ignition Timing, Reset (B)		
1995 Avalon	1.3	1.3
Camry		
1983-91	.5	.5
1992-95	1.1	1.1
Celica		
1981-89	.5	.5
1990-95	1.3	1.3
Corolla FWD		
1983-91	.5	.5
1992-95	1.3	1.3
Corolla FX, FX16	.5	.5
Corolla RWD	.5	.5
Cressida, Starlet	.5	.5
MR2		
1985-89	.5	.5
1991-95	1.1	1.1
1992-95 Paseo	1.1	1.1
Supra		
1981-92	.5	.5
1994-95	1.2	1.2
1981-94 Tercel	.5	.5
Camshaft Sensor, Replace (B)		
Exc. below (.4)	.6	.7
Avalon	.3	.4
Corolla, Matrix		
2003-05 (.5)	.8	.9
Prius		
2001-03 (2.0)	3.1	3.3
2004-05 (2.6)	3.8	4.0
Distributor, Replace (B)		
Includes: Reset base ignition timing.		
Camry		
1983-86	.8	1.0
1987-96	1.5	1.7
Celica		
1981-85	.9	1.1
1986-99	1.5	1.7
Corolla FWD		
1983-87	.8	1.0
1988-97	1.7	1.9
Corolla FX, FX16	.9	1.1
Corolla RWD	.9	1.1
Cressida		
1981-88	.9	1.1
1989-92	1.4	1.6
MR2	1.5	1.7
Paseo	1.3	1.5
Starlet	.7	.9
Supra		
1981-92		
w/turbo	1.4	1.6
w/o turbo	.8	1.0
1993-98		
w/turbo	2.4	2.6
w/o turbo	1.3	1.5

	LABOR TIME	SEVERE SERVICE
Tercel		
1981-90	.9	1.1
1991-94	1.3	1.5
Distributor Cap and/or Rotor, Replace (B)		
All Models	.5	.5
Glow Plug, Replace (B)		
1983-86 Camry, Corolla		
one	.5	.7
all	.9	1.1
Glow Plug Relay, Replace (B)		
1983-86 Camry, Corolla	.5	.5
Igniter Unit, Replace (B)		
Exc. below	.7	.7
1994-99 Celica		
7AFE	1.9	1.9
5SFE	.5	.5
Cressida	.8	.8
1991-99 Tercel	.3	.3
Ignition Coil, Replace (B)		
Exc. below (.4)	.6	.7
Avalon	.3	.3
Camry		
1983-86	.9	1.0
1987-98 4 cyl.	1.6	1.8
2002-05 V6 (.6)	.9	1.0
Celica		
1990-93 4AFE	1.7	1.9
1994-99 7AFE	1.4	1.6
Corolla FWD		
1983-87	.8	.9
1988-91 4AFE	1.6	1.8
1992-97	1.5	1.7
Echo (.2)	.3	.3
1992-95 Paseo	1.3	1.5
Supra	.8	.9
1983-94 Tercel	1.4	1.6
Ignition Switch, Replace (B)		
Exc. below (.4)	.6	.6
Avalon	.8	.8
1992-96 Camry	.9	.9
1981-99 Celica	.9	.9
Corolla, Matrix		
2003-05 (.2)	.3	.3
Corolla RWD	.9	.9
1981-88 Cressida	.9	.9
1985-89 MR2	1.4	1.4
1981-92 Supra	.8	.8
Signal Generator Assy., Replace (B)		
1983-86 Camry	1.7	1.7
1981-85 Celica	.8	.8
1983-87 Corolla FWD	1.5	1.5
Corolla FX, FX16	1.5	1.5
Corolla RWD		
1981-82	.8	.8
1983-87	1.4	1.4

	LABOR TIME	SEVERE SERVICE
1981-88 Cressida	.9	.9
Starlet	.8	.8
1981-87 Supra	.8	.8
1981-90 Tercel	1.5	1.5
Spark Plug (Ignition) Cables, Replace (B)		
Exc. below	.7	.9
1986-98 Supra	1.4	1.6
Spark Plugs, Replace (B)		
Avalon		
1995-98	.8	.9
1999-04	1.8	2.1
Camry		
1983-86	.6	.7
1987-98	.8	1.0
1999-01		
4 cyl.	1.9	2.1
V6	.5	.5
2002-05		
4 cyl. (.3)	.5	.5
V6 (.8)	1.2	1.4
Camry Solara		
4 cyl. (.3)	.5	.5
V6 (1.2)	1.8	2.1
Celica (.4)	.6	.7
Corolla, Matrix (.5)	.8	.9
Cressida	.8	.9
Echo (.3)	.5	.5
MR2	.8	.9
MR2 Spyder, Prius (.4)	.6	.7
Paseo, Starlet	.5	.5
Supra		
1981-92	1.1	1.2
1993-98		
w/turbo	.8	.9
w/o turbo	1.6	1.8
Tercel	.8	.9
Vacuum Advance Unit, Replace (B)		
Includes: Reset base ignition timing.		
1983-86 Camry	1.8	2.0
1981-85 Celica	.7	.9
1983-92 Corolla FWD	1.7	1.9
Corolla FX, FX16	1.7	1.9
Corolla RWD		
1981-82	.7	.9
1983-87	1.4	1.6
Starlet	.7	.9
Tercel		
1981-82	1.1	1.3
1983-90	1.6	1.8

EMISSIONS

The following operations do not include testing. Add time as required.

Diagnose Emission Control System Component Each (A)

1981-05	.9	.9

TOY-6   AVALON : CAMRY : CELICA : COROLLA : CRESSIDA : ECHO : MATRIX : MR2

	LABOR TIME	SEVERE SERVICE
Diagnose Accelerator Pedal Position Sensor Circuit Malfunction (A)		
2002-04 Camry		
4 cyl.	.6	.6
V6	1.2	1.2
Solara	.6	.6
Supra	1.1	1.1
Diagnose Coolant Temperature Insufficient for Closed Loop Fuel Control (A)		
1997-01 Camry	1.1	1.1
1997-99 Celica	.9	.9
Corolla		
1997	.8	.8
1998-05 (.4)	.6	.6
Paseo	.9	.9
2004-05 Prius (.1)	.2	.2
Solara (.8)	1.1	1.1
Supra, Tercel	.9	.9
Diagnose EGR Valve Position Sensor Malfunction (A)		
Avalon	.6	.6
Camry		
1997-01 3.0L	.5	.5
2002-05 3.0L (.5)	.8	.8
Solara	.5	.5
Diagnose Evaporative Emission Control System Pressure Sensor Malfunction (A)		
Avalon	.8	.8
Camry		
1997-01		
2.2L	.5	.5
3.0L	.8	.8
2002-05 (.2)	.3	.3
Celica (.4)	.6	.6
Corolla		
1998-02	1.2	1.2
2003-05 (.2)	.3	.3
Echo (.2)	.3	.3
Matrix (.2)	.3	.3
MR2 (1.0)	1.4	1.4
Paseo	.5	.5
Solara		
2.2L	.3	.3
2.4L	.8	.8
3.0L	.6	.6
Supra	.6	.6
Diagnose Evaporative Emission Control System Pressure Sensor Range/Performance Problem (A)		
1997-99 Avalon	.6	.6
Camry		
1997-01		
4 cyl.	.5	.5
6 cyl.	.8	.8
2002-05 (.2)	.3	.3
Celica (.4)	.6	.6

	LABOR TIME	SEVERE SERVICE
Corolla		
1998-02	1.2	1.2
2003-04	.3	.3
Echo (.2)	.3	.3
Matrix (.2)	.3	.3
MR2 (1.0)	1.4	1.4
Paseo	.5	.5
2004 Prius	.8	.8
Solara		
2.2L	.3	.3
3.0L	.6	.6
2004-05 (.6)	.8	.8
Supra	.6	.6
Diagnose Intake Air Temperature Circuit Malfunction (A)		
Exc. below (.4)	.6	.6
Camry (.3)	.5	.5
Corolla		
1997	.5	.5
1998-02	1.1	1.1
2003-05 (.1)	.2	.2
Echo (.2)	.3	.3
Matrix (.1)	.2	.2
MR2 (1.0)	1.4	1.4
Paseo, Prius, Tercel (.2)	.3	.3
Diagnose Manifold Absolute Pressure Circuit Malfunction (A)		
1997-01 Camry	.5	.5
1997-99 Celica	.6	.6
1998-02 Corolla	.7	.7
Paseo, Solara, Tercel	.5	.5
Diagnose Manifold Absolute Pressure Circuit Range/Performance Problem (A)		
Exc. below	.5	.5
1997-99 Celica	.6	.6
1998-02 Corolla	.8	.8
Diagnose Mass Air Flow Circuit Malfunction (A)		
Avalon	.6	.6
Camry (.5)	.7	.7
2000-05 Celica (.2)	.3	.3
Corolla, Matrix (.6)	.8	.8
Echo (.2)	.3	.3
MR2 (1.0)	1.4	1.4
Prius (.3)	.5	.5
Solara 2.4L, 3.0L, 3.3L (.5)	.7	.7
Supra	.6	.6
Diagnose Mass Air Flow Circuit Range/Performance Problem (A)		
All Models (.2)	.3	.3
Diagnose Oxygen Sensor Circuit Malfunction (A)		
Bank one sensor two exc. below (.6)	.8	.8
Camry		
1997-01	.9	.9
2002-05 (.3)	.5	.5

	LABOR TIME	SEVERE SERVICE
2000-05 Celica (.9)	1.4	1.4
1998-02 Corolla	.6	.6
Prius (.9)	1.4	1.4
Solara		
2.2L, 3.0L	.9	.9
2.4L	.5	.5
Bank two sensor one		
Avalon	.7	.7
1997-01 Camry		
3.0L	.7	.7
Solara (.5)	.7	.7
Supra	.9	.9
sensor two		
2002-05 Camry	.5	.5
Supra	.8	.8
Diagnose Oxygen Sensor Heater Circuit Malfunction (A)		
Bank one		
sensor one		
Avalon	.6	.6
1997-01 Camry	.6	.6
2000-05 Celica (.3)	.5	.5
1998-05 Corolla (.6)	.9	.9
Echo (.3)	.5	.5
Matrix (.6)	.9	.9
MR2 (1.0)	1.4	1.4
Prius, Solara, Supra (.4)	.6	.6
sensor two		
Avalon	.9	.9
Camry		
1997-01	.9	.9
2002-05 (.3)	.5	.5
Celica		
1997	.8	.8
2000-05 (1.0)	1.4	1.4
Corolla		
1997	.8	.8
1998-02	.6	.6
Echo (.6)	.8	.8
Prius (1.0)	1.4	1.4
Solara		
2.2L, 3.0L, 3.3L (.6)	.9	.9
2.4L (.3)	.5	.5
Supra, Tercel	.8	.8
Bank two		
sensor one		
Avalon	.7	.7
1997-01 Camry	.6	.6
Solara (.5)	.7	.7
Supra	.9	.9
sensor two		
2002-05 Camry (.3)	.5	.5
Supra	.8	.8

MR2 SPYDER : PASEO : PRIUS : SOLARA : STARLET : SUPRA : TERCEL TOY-7

	LABOR TIME	SEVERE SERVICE
Diagnose Oxygen Sensor Heater Circuit Slow Response (A)		
Bank one sensor one		
Exc. below (.2)	.3	.3
1997-01 Camry	.5	.5
Celica		
1997-99	.5	.5
2000-05 (.8)	1.2	1.2
Corolla		
1997	.5	.5
Echo (.8)	1.2	1.2
Bank two sensor one		
Avalon	.5	.5
1997-01 Camry	.5	.5
Supra	.3	.3
Diagnose Throttle Pedal Position Sensor Switch "A" Circuit Performance Problem (A)		
Avalon	.5	.5
Camry	.3	.3
1997-99 Celica	.3	.3
Corolla		
1997	.3	.3
1998-02	.7	.7
2003-05 (.1)	.2	.2
Echo (.8)	1.2	1.2
Matrix (.1)	.2	.2
MR2 (1.0)	1.4	1.4
Paseo	.5	.5
2001-03 Prius	.5	.5
Solara (.2)	.3	.3
Supra	1.1	1.1
Tercel	.5	.5
Diagnose Vehicle Speed Sensor Malfunction (A)		
Exc. below (.3)	.5	.5
Avalon	.8	.8
Celica (.5)	.7	.7
1998-05 Corolla (.5)	.7	.7
2000-05 MR2 (1.0)	1.4	1.4
Supra (.5)	.7	.7
Dynamometer Test (A)		
1981-05	.5	.5
Air Control Valve, Replace (B)		
1981-83 Celica	.7	.7
Air Distribution Manifold, Replace (B)		
1981-83 Celica	1.7	1.9
Air Injection Tube Check Valve Assy., Replace (B)		
1981-83 Celica	1.7	1.7
Air Pump Assy., Replace (B)		
1981-83 Celica	.8	1.0
Coolant Temperature (ECT) Sensor, Replace (B)		
Avalon	.9	1.0
Camry,		
Camry Solara (.8)	1.2	1.3

	LABOR TIME	SEVERE SERVICE
Celica		
1983-85	1.2	1.4
1986-89	.9	1.0
1990-93	1.2	1.4
1994-99	.7	.8
2000-05 (1.3)	2.0	2.2
Corolla FWD,		
Matrix (.7)	1.1	1.2
Corolla FX, FX16	.9	1.0
Corolla RWD	1.3	1.4
Cressida		
1981-88	1.3	1.4
1989-92	1.6	1.8
Echo (.7)	1.1	1.2
MR2		
1985-89	.9	1.0
1991-95	1.8	2.1
MR2 Spyder (1.2)	1.8	2.1
Paseo, Tercel	.8	.9
Prius (3.3)	4.8	5.3
Supra		
1981-92	1.4	1.5
1993-98	.8	.9
EGR Modulator, Replace (B)		
All Models	.3	.4
EGR Temperature Sensor, Replace (B)		
Avalon	1.3	1.3
Camry		
1994-96	.3	.3
1997-01		
4 cyl.	.3	.3
V6	1.3	1.3
Camry Solara		
4 cyl.	.3	.3
V6	.9	.9
Paseo		
1994-95	.7	.7
1996-99	.3	.3
1993-98 Supra	.3	.3
1994-99 Tercel	.3	.3
EGR Thermal Vacuum Valve, Replace (B)		
1981-02	1.2	1.2
EGR Vacuum Switching Valve, Replace (B)		
All Models	.5	.5
EGR Valve, Replace (B)		
Avalon	.5	.7
Camry		
1983-86	.5	.7
1987-91		
4 cyl.	.5	.7
V6	.9	1.0
1992-01	.9	1.0
2002-05		
V6 (.6)	.9	1.0

	LABOR TIME	SEVERE SERVICE
Camry Solara		
1999-01	.5	.7
2002-03		
4 cyl.	.9	1.0
V6	.5	.7
Celica		
1981-85	.8	.9
1986-89		
3SGE	2.0	2.2
3SFE	.5	.7
turbo	1.6	1.8
1990-93		
2WD	.9	1.0
4WD		
w/turbo	1.6	1.8
w/o turbo	.5	.7
1994-03	.9	1.0
Corolla FWD		
1983-87		
SOHC	.9	1.0
DOHC	1.5	1.7
1988-02	.9	1.0
Corolla FX, FX16		
SOHC	.9	1.0
DOHC	1.6	1.8
Corolla RWD		
1981-82	.5	.7
1983-87		
4AC	.9	1.0
4AGE	1.7	1.9
Cressida		
1981-88	.7	.9
1989-92	.9	1.0
Echo, Matrix	.9	1.0
MR2		
1985-89	.9	1.0
1991-95	.5	.7
MR2 Spyder, Prius	.9	1.0
Paseo, Starlet	.7	.9
Supra		
1981-92	.9	1.0
1993-98		
w/turbo	1.5	1.7
w/o turbo	.7	.9
Tercel		
1981-82	.5	.7
1983-99	.9	1.0
Electronic Control Module (ECM/PCM), Replace (B)		
Exc. below (.5)	.7	.7
1983-93 Camry	1.4	1.4
Celica		
1983-85	1.6	1.6
1986-89	1.2	1.2
1990-99	1.5	1.5
1988-97 Corolla FWD	1.2	1.2
Corolla FX, FX16	1.1	1.1
Corolla RWD	1.2	1.2

TOY-8 AVALON : CAMRY : CELICA : COROLLA : CRESSIDA : ECHO : MATRIX : MR2

	LABOR TIME	SEVERE SERVICE
Electronic Control Module (ECM/PCM), Replace (B)		
Cressida		
1981-88	1.8	1.8
1989-92	1.4	1.4
MR2	1.3	1.3
MR2 Spyder (1.2)	1.6	1.6
Prius		
2004-05 (1.0)	1.4	1.4
Supra		
1982-92	1.5	1.5
1993-98	1.2	1.2
Fuel Vapor Canister, Replace (B)		
Exc. below (.4)	.6	.7
1998-04 Avalon	.8	.9
Camry		
1998-01	.8	.9
2002-05 (1.5)	2.3	2.4
Camry Solara,		
1999-03 (.6)	.8	.9
2004-05 (1.8)	2.7	2.8
2000-05 Celica (.7)	1.1	1.2
MR2 Spyder (1.2)	1.8	1.9
Prius		
2001-03	2.1	2.2
2004-05 (1.1)	1.7	1.8
Knock Sensor, Replace (B)		
Avalon	2.8	3.1
Camry, Camry Solara		
1993-01		
4 cyl.	.7	.9
V6	2.8	3.0
2002-05		
4 cyl. (1.1)	1.7	1.9
V6 (1.8)	2.8	3.1
Celica		
1993-99	.7	.9
2000-05 (1.7)	2.6	2.9
Corolla FWD, Matrix		
1993-97	.8	1.0
1998-02	2.7	2.9
2003-05 (.3)	.5	.7
Echo (1.0)	1.5	1.7
MR2		
1993-95	.7	.9
MR2 Spyder (7.0)	9.9	10.4
1996-99 Paseo	.7	.9
Prius	2.9	3.2
1993-98 Supra		
front	.8	1.0
rear		
w/turbo	1.3	1.5
w/o turbo	.8	1.0
1995-99 Tercel	.5	.7
w/AC add MR2 Spyder	.2	.2
V6 each addl. add		
00-04 Avalon	.2	.2
02-05 Camry	.3	.3
99-05 Camry Solara	.2	.2

	LABOR TIME	SEVERE SERVICE
Oxygen Sensor, Replace (B)		
Avalon	1.0	1.1
each addl. add	.2	.3
Camry		
1983-91	1.2	1.4
1992-96	.8	1.0
1997-01	.5	.7
2002-05 (.6)	1.0	1.1
each addl. 02-05 V6 add	.2	.2
Camry Solara		
1999-03 (.3)	.5	.7
2004-05 (.7)	1.1	1.2
each addl. 04-05 V6 add	.3	.3
Celica		
1981-85	.8	1.0
1986-89	1.3	1.5
1990-05 (.7)	1.1	1.3
each addl. add	.2	.2
Corolla FWD, Matrix		
1988-92	1.2	1.4
1993-02	.8	1.0
2003-05 (.8)	1.3	1.4
each addl. add	.3	.3
Corolla FX, FX16	1.3	1.5
Corolla RWD	1.2	1.4
Cressida	1.2	1.4
Echo (.6)	1.0	1.1
each addl. add	.2	.2
MR2		
1985-89	1.4	1.6
1991-95	.7	.9
MR2 Spyder (.7)	1.1	1.3
Paseo	.7	.9
each addl. add	.3	.3
Prius		
2001-03	1.3	1.4
2004-05 (.4)	.6	.7
each addl. add	.3	.3
Supra		
1981-92	1.4	1.6
1993-98		
w/turbo	1.2	1.4
w/o turbo	.7	.9
Tercel		
1990	1.3	1.5
1991-99	.7	.9
each addl. add	.3	.3
Positive Crankcase Ventilation (PCV) Valve, Replace (B)		
Exc. below (.2)	.3	.3
Avalon	.5	.5
1998-01 Camry V6	.5	.5
2004-05 Prius (.7)	1.0	1.0
2002-03 Solara V6	.5	.5

	LABOR TIME	SEVERE SERVICE
Throttle Position Sensor (TPS), Replace (B)		
Exc. below (.5)	.8	.9
2002-05 Camry		
V6 (1.1)	1.7	1.8
Celica		
1993 turbo	.9	1.0
2000-05 (1.5)	2.2	2.3
2003-05 Corolla, Matrix (1.2)	1.8	1.9
Echo (1.0)	1.5	1.6
MR2 Spyder (1.6)	2.4	2.5
2001-03 Prius	3.2	3.3

FUEL
CARBURETOR

	LABOR TIME	SEVERE SERVICE
Idle Speed, Adjust (B)		
1981-90 Corolla	.7	.7
Tercel		
1981-82	.5	.5
1983-90		
sedan	.7	.7
wagon	.5	.5
Automatic Choke Coil Housing, Replace (B)		
1981-90 Corolla, Tercel	.8	.8
Carburetor, R&R and Clean or Recondition (B)		
Includes: Adjustments.		
1981-90 Corolla, Tercel	3.3	3.3
Carburetor, Replace (B)		
Includes: Adjustments.		
1981-90 Corolla, Tercel	1.8	1.8
Carburetor Float, Replace or Adjust (B)		
1981-90 Corolla	1.7	1.7
Tercel		
1981-82	.8	.8
1983-90		
sedan	1.5	1.5
wagon	.8	.8

DELIVERY

	LABOR TIME	SEVERE SERVICE
Fuel Pump, Test (B)		
1981-05	.3	.4
Fuel Filter, Replace (B)		
Exc. below (.5)	.8	.9
2002-05 Camry, Camry Solara		
3.0L (.6)	1.0	1.1
3.3L (.3)	.4	.5
Celica		
1981-85	1.2	1.3
1986-93	.7	.8
1994-99	1.4	1.6
1995-97 Corolla FWD	1.1	1.2

MR2 SPYDER : PASEO : PRIUS : SOLARA : STARLET : SUPRA : TERCEL TOY-9

	LABOR TIME	SEVERE SERVICE
1991-95 MR2	1.2	1.3
MR2 Spyder (1.3)	1.9	2.0
1996-99 Paseo	1.2	1.3
Tercel		
1981-82	.3	.4
1983-90		
3E	.3	.4
3EE	.7	.8
1991-94	.7	.8
1995-99	1.2	1.3
Fuel Gauge (Dash), Replace (B)		
Exc. below	.9	.9
1983-86 Camry	1.3	1.3
Celica		
1981-85	1.9	1.9
1986-93	1.5	1.5
1988-93 Corolla	1.3	1.3
1981-82 Cressida	1.4	1.4
1985-88 MR2	1.1	1.1
1981-92 Supra	1.8	1.8
1983-88 Tercel	1.2	1.2
Fuel Gauge (Tank), Replace (B)		
Exc. below (.5)	.8	1.0
MR2 Spyder (1.2)	1.9	2.2
1993-98 Supra	1.4	1.6
Fuel Pump, Replace (B)		
Exc. below (.6)	1.0	1.1
1985-91 Camry	2.3	2.5
Celica		
1981-82 electric	1.1	1.3
1983-85	1.9	2.1
1986-92	3.0	3.2
1993 4WD	2.5	2.7
1994-99	1.3	1.5
Corolla FWD		
1988-92 electric	2.5	2.7
Corolla FX, FX16		
electric	2.3	2.5
Corolla RWD		
electric	2.2	2.4
Cressida		
1981-84	1.3	1.5
1985-88	2.4	2.6
1989-92	1.7	1.9
MR2	3.4	3.6
MR2 Spyder (1.3)	2.1	2.3
Paseo	1.3	1.5
1983-84 Starlet	1.1	1.3
Supra		
1981-83	1.2	1.4
1983-92	2.3	2.5
1993-98	1.6	1.8
1995-99 Tercel	1.1	1.2
Fuel Tank, Replace (B)		
Includes: Drain and refill.		
Avalon	3.3	3.5
Camry		
1983-86	1.7	1.9
1987-91	2.6	2.8
1992-05 (2.0)	3.0	3.2

	LABOR TIME	SEVERE SERVICE
Camry Solara		
1999-03 (2.1)	3.4	3.6
2004-05 (1.5)	2.2	2.5
Celica		
1981-85	2.4	2.6
1986-92	3.0	3.2
1993		
2WD	2.8	3.1
4WD	2.5	2.7
1994-99	3.0	3.2
2000-05 (1.6)	2.4	2.6
Corolla FWD		
1983-87	2.1	2.3
1988-92	2.4	2.6
1993-02	2.9	3.1
2003-05 (1.4)	2.1	2.3
Corolla FX, FX16 (1.4)	2.1	2.3
Corolla RWD (1.4)	2.1	2.3
Cressida		
1981-84	1.9	2.1
1985-88	3.0	3.2
1989-92	1.9	2.1
Echo (1.2)	1.9	2.1
Matrix		
2WD (1.5)	2.2	2.4
4WD (2.5)	3.7	3.9
MR2		
1985-89	2.9	3.1
1991-95	3.5	3.7
MR2 Spyder (1.7)	2.7	2.9
Paseo	2.4	2.6
Prius		
2001-03	3.3	3.5
2004-05 (1.3)	2.1	2.3
Starlet	1.8	2.0
Supra		
1981-86	2.7	2.9
1987-92	2.8	3.0
1993-98	2.2	2.4
Tercel		
1981-90	1.9	2.1
1991-94	2.3	2.5
1995-99	3.0	3.2
Intake Manifold Gasket, Replace (B)		
Avalon	3.1	3.4
Camry		
1983-86	3.3	3.6
1987-91		
4 cyl.	2.5	2.8
V6	3.2	3.5
1992-01	2.5	2.8
2002-05		
4 cyl. (1.1)	1.7	2.0
V6 (1.7)	2.4	2.7
Camry Solara		
1999-03		
4 cyl.	1.7	2.0
V6	2.6	2.9

	LABOR TIME	SEVERE SERVICE
2004-05		
4 cyl. (2.1)	3.0	3.3
V6 (2.7)	3.8	4.1
Celica		
1981-85		
5MGE	1.9	2.2
22RE	3.2	3.5
1986-89		
exc. turbo	3.5	3.8
turbo	12.9	13.2
1990-93		
w/turbo	12.8	13.1
w/o turbo	2.5	2.8
1994-99	2.1	2.4
2000-05 (1.6)	2.4	2.7
Corolla FWD		
1983-87	2.9	3.2
1988-92		
SOHC	2.9	3.2
DOHC	3.3	3.6
1993-97	2.0	2.3
1998-01	2.9	3.2
2003-05 (1.0)	1.5	1.8
Corolla FX, FX16		
SOHC	2.7	3.0
DOHC	3.2	3.5
Corolla RWD		
1981-82	1.8	2.1
1983-87		
SOHC	2.7	3.0
DOHC	3.0	3.3
Cressida		
1981-88	3.3	3.6
1989-92	2.8	3.1
Echo (1.0)	1.5	1.8
Matrix (1.1)	1.8	2.1
MR2		
1985-89		
w/turbo	4.5	4.8
w/o turbo	3.3	3.6
1991-95	2.4	2.7
MR2 Spyder (7.0)	9.9	10.2
w/AC add	.2	.2
w/sequential manual		
transmission add	1.0	1.1
Paseo	2.3	2.6
Prius		
2001-03 (1.9)	2.7	3.0
2004-05 (1.6)	2.3	2.6
Starlet		
w/o FI	2.1	2.4
w/FI	3.0	3.3
Supra		
1981-86	3.2	3.5
1987-98	2.7	3.0

TOY-10 AVALON : CAMRY : CELICA : COROLLA : CRESSIDA : ECHO : MATRIX : MR2

	LABOR TIME	SEVERE SERVICE
Intake Manifold Gasket, Replace (B)		
Tercel		
1981-82	3.0	3.3
1983-90		
sedan	2.3	2.6
wagon	3.0	3.3
1991-94	2.3	2.6
1995-99	1.8	2.1

INJECTION

	LABOR TIME	SEVERE SERVICE
Diagnose Air/Fuel Sensor Circuit Range/Performance Malfunction (A)		
Bank one sensor one		
Avalon	.7	.7
Camry		
1997-01	.8	.8
2002-05 (.3)	.5	.5
Solara		
2.2L	.8	.8
2.4L (.3)	.5	.5
3.0L (.5)	.7	.7
Bank two sensor one		
Avalon	.7	.7
Solara 3.0L (.5)	.7	.7
Diagnose Electronic Control Module Battery Malfunction (A)		
Exc. below	.6	.6
2002-05 Camry		
3.0L (.1)	.2	.2
MR2 (1.0)	1.4	1.4
Diagnose Idle Control System Malfunction (A)		
1997 Corolla	1.1	1.1
Supra	.6	.6
Injection Timing, Check & Adjust (B)		
1983-86 Camry, Corolla	.8	1.1
Air Flow Meter, Replace (B)		
Exc. below (.5)	.8	.9
1995-99 Avalon	1.1	1.1
Camry, Camry Solara		
1983-96	1.3	1.3
1997-05 (.2)	.3	.3
1983-85 Celica	1.3	1.3
1988-92 Corolla FWD	1.3	1.3
Corolla FX, FX16	1.4	1.4
Corolla RWD	1.5	1.5
Cressida		
1981-88	1.4	1.4
1989-92	1.1	1.1
Echo (.3)	.5	.5
MR2		
1985-89	1.6	1.6
1991-95	1.3	1.3
Prius (.3)	.5	.5
Supra		
1981-92	1.4	1.4
1993-98	.9	.9

	LABOR TIME	SEVERE SERVICE
Cold Start Injector, Replace (B)		
Camry	.7	.7
Celica		
1983-85		
w/turbo	1.8	1.8
w/o turbo	.8	.8
1986-89		
w/turbo	1.7	1.7
w/o turbo	.8	.8
1988-92 Corolla FWD	.8	.8
Corolla FX, FX16	.7	.7
1985-87 Corolla RWD	.7	.7
Cressida		
1981-88	1.3	1.3
1989-92	.7	.7
MR2		
1985-89	.8	.8
1991-95	1.4	1.4
Supra	.8	.8
1990 Tercel	1.2	1.2
Diesel Fuel Injection Pump, Replace (B)		
1983-86 Camry, Corolla	4.2	4.5
Fuel Injection Lines, Replace (B)		
1983-86 Camry, Corolla (.5)	.8	.8
Fuel Injectors, Replace (B)		
Avalon	2.8	3.1
Camry		
1983-86	1.8	2.0
1987-91	3.0	3.2
1992-01	2.7	2.9
2002-05		
4 cyl. (.9)	1.4	1.5
V6 (1.4)	2.1	2.4
Camry Solara		
1999-01	2.6	2.8
2002-03		
4 cyl.	1.4	1.5
V6	2.7	3.1
2004-05		
4 cyl. (.9)	1.4	1.5
V6 (1.3)	2.7	3.1
Celica		
1986-89		
w/turbo	2.5	2.7
w/o turbo	1.7	1.9
1990-93		
w/turbo	2.9	3.1
w/o turbo	2.4	2.6
1994-99	2.0	2.2
2000-05 (.7)	1.1	1.2
Corolla FWD		
1988-92	1.3	1.5
1993-02	1.8	2.0
2003-05 (.9)	1.4	1.5
Corolla FX, FX16	1.6	1.8
Corolla RWD	1.7	1.9

	LABOR TIME	SEVERE SERVICE
Cressida		
1981-88	3.4	3.6
1989-92	2.6	2.8
Echo (.7)	1.1	1.2
Matrix (.9)	1.4	1.5
MR2		
1985-89	1.8	2.0
1991-95	2.4	2.6
MR2 Spyder (.6)	.9	1.0
Paseo	1.7	1.9
Prius		
2001-03 (1.8)	2.6	3.0
2004-05 (1.5)	2.2	2.5
Supra		
1981	1.9	2.1
1982-92	3.0	3.2
1993-98		
w/turbo	4.6	4.8
w/o turbo	1.8	2.0
Tercel	1.7	1.9
Fuel Pressure Regulator, Replace (B)		
Avalon	.6	.7
Camry, Camry Solara		
1983-86	1.4	1.6
1987-91	1.3	1.5
1992-05 (.5)	.8	.9
Celica		
1983-85	1.4	1.6
1986-89		
w/turbo	1.9	2.1
w/o turbo	1.2	1.4
1990-93		
w/turbo	1.9	2.1
w/o turbo	.5	.7
1994-99	1.4	1.6
2000-05 (.5)	.7	.8
Corolla FWD, Matrix		
1988-02	1.1	1.3
2003-05 (.5)	.8	.9
Corolla FX, FX16	1.3	1.5
1985-87 Corolla RWD	1.2	1.4
Cressida	3.3	3.5
Echo (.4)	.6	.7
MR2		
1985-89	1.4	1.6
1991-95		
w/turbo	1.9	2.1
w/o turbo	.5	.7
MR2 Spyder (1.2)	1.8	2.1
Paseo		
1992-95	1.3	1.5
1996-99	.7	.9
Supra		
1981	1.6	1.8
1982-92	2.9	3.1
1993-98		
w/turbo	1.5	1.7
w/o turbo	.5	.7

MR2 SPYDER : PASEO : PRIUS : SOLARA : STARLET : SUPRA : TERCEL   TOY-11

	LABOR TIME	SEVERE SERVICE
Tercel		
1990	1.2	1.4
1991-99	.7	.9
Idle Air Control (IAC) Valve, Replace (B)		
Avalon	1.1	1.2
Camry		
1987-91	1.8	1.9
1992-01	1.5	1.6
Camry Solara	1.3	1.4
Celica		
1986-99	1.8	1.9
2000-05 (1.5)	2.3	2.4
Corolla FWD, Matrix		
1988-92	.9	1.0
1993-97	1.3	1.4
1998-02	1.9	2.0
2003-05 (1.1)	1.7	1.8
Corolla FX, FX16	.9	1.0
1981-88 Cressida	1.3	1.4
Echo (.9)	1.4	1.5
1991-95 MR2	1.4	1.5
MR2 Spyder (1.6)	2.4	2.5
1998-99 Paseo	1.7	1.8
Supra		
1981	.5	.6
1982-92	1.3	1.4
1993-98		
w/turbo	1.5	1.6
w/o turbo	.9	1.0
Tercel		
1990-94	1.3	1.4
1995-99	1.9	2.0
Throttle Body Assy., Replace (B)		
Avalon	1.1	1.2
Camry		
1983-01	1.3	1.4
2002-05		
4 cyl. (.6)	.9	1.0
V6 (1.1)	1.7	1.9
Camry Solara		
1999-03	1.1	1.2
2004-05		
4 cyl (.9)	1.3	1.5
V6 (1.2)	1.8	2.0
Celica		
1983-89	1.5	1.6
1990-99	1.9	2.0
2000-05 (1.5)	2.3	2.6
Corolla FWD, Matrix		
1988-92	1.3	1.4
1993-97	1.5	1.6
1998-02	1.2	1.3
2003-05 (1.1)	1.7	1.9
Corolla FX, FX16	1.8	1.9
1985-87 Corolla RWD	1.7	1.8
Cressida	1.8	1.9
Echo (.9)	1.4	1.5

	LABOR TIME	SEVERE SERVICE
MR2	1.3	1.4
MR2 Spyder (1.6)	2.4	2.7
Paseo, Supra	1.6	1.7
Prius		
2001-03	2.8	3.1
2004-05 (1.4)	2.0	2.3
Tercel		
1990	1.2	1.3
1991-99	1.6	1.7
TURBOCHARGER		
Pressure Switch, Replace (B)		
1983-87 Camry diesel	.5	.5
Turbocharger Assy., R&R or Replace (B)		
1983-86 Camry	1.6	2.0
1983-87 Camry diesel	1.7	2.2
1989 Celica	4.1	4.5
1990-93 Celica	4.5	4.9
1991-95 MR2	5.3	5.7
1987-92 Supra	2.9	3.3
1993-98 Supra	6.2	6.6
Turbo to Exhaust Manifold Gasket, Replace (B)		
1983-87 Camry diesel	1.1	1.6
SUPERCHARGER		
Supercharger Assy. and/or Gasket, Replace (B)		
1988-89 MR2	2.8	3.3
EXHAUST		
Catalytic Converter, Replace (B)		
Avalon, Prius	.8	.9
Camry	1.3	1.4
Celica		
1981-93	1.3	1.4
1994-99	.5	.6
Corolla FWD		
1983-92	1.5	1.6
1993-97	.8	.9
Corolla FX	1.5	1.6
Corolla RWD	1.7	1.8
Cressida	1.6	1.7
MR2	1.3	1.4
Paseo		
1992-95	1.3	1.4
1996-99	.8	.9
Starlet	1.3	1.4
Supra		
1981-87	.8	.9
1988-98	1.3	1.4
Tercel		
1981-90	1.4	1.5
1991-94	.8	.9
Center Exhaust Pipe, Replace (B)		
Avalon	.7	.8
Camry		
1983-86	1.2	1.3
1987-05 (.6)	1.0	1.1

	LABOR TIME	SEVERE SERVICE
Camry Solara		
1999-03 (.5)	.8	.9
2004-05		
4 cyl. (.5)	.8	.9
V6 (.8)	.7	.8
1986-97 Celica	.9	1.0
Corolla FWD		
1988-92	1.1	1.2
1993-02	.7	.8
Cressida	1.3	1.4
Matrix 4WD	.8	.9
Paseo	.9	1.0
1993-98 Supra	.9	1.0
1991-94 Tercel	.8	.9
Exhaust Manifold and/or Gasket, Replace (B)		
Avalon		
right side	8.2	9.1
left side	2.6	2.9
Camry		
1983-86	3.5	3.9
1987-91		
4 cyl.	1.6	1.8
diesel	9.4	10.6
1992-01		
4 cyl.	1.5	1.7
V6		
right side	7.9	8.9
left side	2.5	2.9
2002-05		
4 cyl. (.9)	1.4	1.6
V6		
right side (.9)	1.4	1.6
left side (.6)	.9	1.1
Camry Solara		
1999-03		
4 cyl.	1.5	1.7
V6		
right side	8.8	9.9
left side	2.7	3.0
2004-05		
4 cyl. (.7)	1.1	1.3
V6		
right side (1.1)	1.6	1.8
left side (.9)	1.4	1.6
Celica		
1981-85	1.8	2.1
1986-89		
SOHC	3.5	3.9
DOHC	1.7	2.0
turbo	4.5	5.1
1990-93		
w/turbo	4.7	5.3
w/o turbo	1.2	1.4
1994-99	.7	.8
2000-05 (.8)	1.2	1.4

TOY-12 AVALON : CAMRY : CELICA : COROLLA : CRESSIDA : ECHO : MATRIX : MR2

	LABOR TIME	SEVERE SERVICE
Exhaust Manifold and/or Gasket, Replace (B)		
Corolla FWD		
1983-87		
SOHC	3.2	*3.6*
DOHC	2.5	*2.9*
1988-92		
SOHC	1.1	*1.3*
DOHC	2.2	*2.5*
1993-97	.8	*.9*
1998-02	1.2	*1.5*
2003-05 (.5)	.8	*.9*
Corolla FX, FX16		
SOHC	3.1	*3.5*
DOHC	2.4	*2.7*
Corolla RWD		
1981-82	1.4	*1.6*
1983	3.2	*3.6*
1984-87		
4AC	3.3	*3.7*
4AGE	1.5	*1.7*
Cressida	1.7	*2.0*
1991-95 MR2	1.5	*1.7*
Echo, Matrix (.7)	1.1	*1.3*
MR2 Spyder (.9)	1.4	*1.5*
1992-99 Paseo	1.6	*1.8*
Prius		
2001-03 (.7)	1.1	*1.3*
2004-05 (3.2)	4.5	*4.7*
Supra		
1981-86	1.9	*2.1*
1987-92		
w/turbo	2.5	*2.9*
w/o turbo	2.0	*2.3*
1993-98		
w/turbo	5.2	*5.8*
w/o turbo	1.9	*2.1*
Tercel		
1983-90		
sedan	1.5	*1.7*
wagon	3.2	*3.6*
1991-94	.8	*.9*
1995-99	1.5	*1.7*
Front Exhaust Pipe, Replace (B)		
Exc. below (.5)	.8	*.9*
Camry		
2002-05 V6 (.7)	.9	*1.0*
Celica		
1981-89	1.2	*1.3*
1990-93 w/turbo	1.4	*1.5*
1983-92 Corolla FWD	1.3	*1.4*
Corolla RWD	1.3	*1.4*
1981-88 Cressida	1.4	*1.5*
1991-95 MR2	1.3	*1.4*
Paseo, Starlet	1.3	*1.4*
1981-87 Supra	1.4	*1.5*
Tercel		
1981-82	1.3	*1.4*
1983-90 wagon	1.4	*1.5*

	LABOR TIME	SEVERE SERVICE
Muffler and Tailpipe Assy., Replace (B)		
Exc. below (.4)	.6	*.7*
2002-05 Camry		
Solara (.8)	1.2	*1.3*
1981-85 Celica	1.2	*1.3*
1983-87 Corolla FWD	.9	*1.0*
Corolla FX, FX16	.9	*1.0*
Corolla RWD	1.5	*1.6*
Cressida	.9	*1.0*
MR2 Spyder, Starlet (.7)	1.1	*1.2*
Supra		
1981-87	1.3	*1.4*
1988-92	.9	*1.0*
1981-90 Tercel	1.2	*1.3*

ENGINE COOLING

Pressure Test Cooling System (B)		
1981-05	.3	*.3*
Coolant Thermostat, Replace (B)		
Avalon	1.2	*1.4*
Camry, Camry, Solara		
1983-86		
gas	1.1	*1.2*
diesel	.7	*.8*
1987-91	1.2	*1.4*
1992-96	.9	*1.0*
1997-05		
4 cyl. (.5)	.8	*.9*
V6 (1.0)	1.4	*1.5*
Celica		
1981-85	.8	*.9*
1986-89	1.2	*1.4*
1990-93		
w/turbo	1.7	*1.9*
w/o turbo	1.2	*1.4*
1994-99	.8	*.9*
2000-04	2.0	*2.2*
Corolla FWD, Matrix		
1983-87		
gas	1.2	*1.4*
diesel	.7	*.8*
1988-92	1.5	*1.7*
1993-97	1.1	*1.2*
1998-02	1.8	*2.1*
2003-05 (.8)	1.2	*1.4*
Corolla FX, FX16, Corolla RWD (.8)	1.2	*1.4*
Cressida	1.2	*1.4*
Echo (.6)	.9	*1.0*
MR2		
1985-89	1.2	*1.4*
1991-95		
w/turbo	2.1	*2.4*
w/o turbo	1.2	*1.4*
MR2 Spyder (1.6)	2.4	*2.7*
Paseo	1.1	*1.2*

	LABOR TIME	SEVERE SERVICE
Prius		
2001-03	2.4	*2.7*
2004-05 (1.2)	1.8	*2.1*
Starlet	.7	*.9*
Supra		
1981-92	1.2	*1.4*
1993-98		
w/turbo	2.0	*2.2*
w/o turbo	.9	*1.1*
Tercel		
1981-90	1.2	*1.4*
1991-99	.8	*1.0*
Engine Coolant Temp. Sending Unit, Replace (B)		
Exc. below (.6)	.8	*.8*
1987-91 Camry	1.3	*1.3*
1991-95 MR2	.9	*.9*
Freeze Plugs (Water Jacket), Replace (B)		
Add appropriate time to access plug.		
1981-05 each (.3)	.5	*.7*
Radiator Assy., R&R or Replace (B)		
Avalon	1.7	*1.9*
Camry, Camry Solara		
1983-86	1.2	*1.4*
1987-91	1.7	*1.9*
1992-96		
4 cyl.	1.7	*1.9*
V6	2.4	*2.6*
1997-05 (1.2)	1.7	*1.9*
Celica		
1981-89	1.5	*1.7*
1990-99	1.7	*1.9*
2000-05 (2.0)	3.1	*3.4*
Corolla FWD, Matrix		
1983-97	1.6	*1.8*
1998-02	1.8	*2.0*
2003-05 (.7)	1.1	*1.2*
Corolla FX, FX16	1.5	*1.7*
Corolla RWD	1.4	*1.6*
Cressida, Echo (1.1)	1.7	*1.9*
MR2		
1985-89	1.2	*1.4*
1991-95	2.3	*2.5*
MR2 Spyder (1.4)	2.1	*2.4*
Paseo	1.6	*1.8*
Prius (2.8)	4.1	*4.4*
Starlet	1.4	*1.6*
Supra		
1981-86	1.5	*1.7*
1987-92	1.9	*2.1*
1993-98		
w/turbo	2.0	*2.2*
w/o turbo	1.4	*1.6*
Tercel	1.7	*1.9*

MR2 SPYDER : PASEO : PRIUS : SOLARA : STARLET : SUPRA : TERCEL   TOY-13

	LABOR TIME	SEVERE SERVICE
Radiator Fan and/or Fan Motor, Replace (B)		
Avalon	.8	.9
Camry		
1983-86	.5	.8
1987-91		
4 cyl.	.5	.8
V6	.9	1.0
1992-94		
4 cyl.	.8	.9
V6	2.4	2.7
1995-01	.8	.9
2002-05 (1.3)	2.0	2.2
Camry Solara		
1999-03 (.5)	.8	.9
2004-05 (1.4)	2.0	2.1
Celica		
1986-89		
w/turbo	1.4	1.6
w/o turbo	.5	.6
1990-05 (.6)	.9	1.0
Corolla FWD, Matrix (8.)	1.2	1.4
Corolla FX, FX16	.8	.9
Echo (1.0)	1.5	1.7
MR2, MR2 Spyder (.5)	.8	.9
Paseo	1.4	1.6
Prius		
2001-03 (2.0)	3.0	3.2
2004-05 (2.3)	3.3	3.5
Starlet	.7	.8
Supra	1.4	1.6
Tercel		
1981-90	.8	.9
1991-99	1.2	1.4
Radiator Fan Motor Relay, Replace (B)		
Exc. below (.1)	.2	.2
1983-97 Corolla	.3	.3
MR2	.3	.3
1993-98 Supra	.5	.5
Radiator Fan Motor Switch (Coolant Temp.), Replace (B)		
Avalon	1.1	1.1
Camry		
1983-91	.5	.5
1992-01	1.1	1.1
2002-04	.3	.3
Camry Solara	1.1	1.1
Celica		
1986-89	.5	.5
1990-93	.7	.7
1994-99	.9	.9
Corolla FWD		
1983-92	.7	.7
1993-02	.8	.8
Corolla FX	.7	.7
Echo (.8)	1.2	1.2

	LABOR TIME	SEVERE SERVICE
MR2		
1985-89	.8	.8
1991-95	1.8	1.8
Paseo	.9	.9
Starlet	.5	.5
Tercel		
1981-90	.7	.7
1991-99	.9	.9
Radiator Hoses, Replace (B)		
Includes: Refill with proper coolant mix.		
Upper		
exc. below	.9	1.0
Camry,		
Camry Solara (.7)	1.1	1.2
2000-05 Celica (1.3)	2.0	2.2
2003-05 Corolla (.8)	1.2	1.4
Echo (.7)	1.1	1.2
Matrix (.8)	1.2	1.4
MR2 Spyder	4.6	5.1
Prius		
2001-03	5.1	5.6
2004-05 (2.5)	3.6	4.1
Lower		
Avalon	1.6	1.8
Camry		
1983-86	.8	.9
1987-91		
4 cyl.	.8	.9
V6	1.4	1.6
1992-01		
4 cyl.	1.1	1.3
V6	1.6	1.8
2002-05		
4 cyl. (.7)	1.1	1.3
V6 (1.0)	1.6	1.8
Celica		
1981-85	1.0	1.1
1994-99	.8	.9
2000-05 (1.2)	1.9	2.2
Corolla FWD		
1983-87	1.0	1.1
1988-05 (.8)	1.2	1.4
Corolla FX, FX16	1.0	1.1
Corolla RWD	1.0	1.1
Cressida, Paseo	1.3	1.4
Echo, Matrix (.8)	1.2	1.4
Prius		
2001-03	2.5	2.9
2004-05 (1.2)	1.9	2.1
Starlet	1.0	1.1
Supra		
1981-92	1.4	1.6
1993-98	1.1	1.3
Tercel	1.3	1.4
Temperature Gauge (Dash), Replace (B)		
Avalon	1.3	1.3
Camry		
1983-86	.9	.9
1987-01	1.3	1.3

	LABOR TIME	SEVERE SERVICE
Camry Solara	1.1	1.1
Celica		
1981-93	2.0	2.0
1994-99	1.2	1.2
Corolla FWD		
1983-87	.9	.9
1988-91	1.4	1.4
1992-97		
2WD	.9	.9
4WD	1.3	1.3
1998-01	.9	.9
Corolla FX, FX16	.9	.9
Corolla RWD		
1981-82	1.3	1.3
1983-87	.7	.7
Cressida		
1981-82	1.4	1.4
1983-92	.9	.9
MR2	1.3	1.3
Paseo	1.2	1.2
Starlet	.9	.9
Supra		
1981-85	2.0	2.0
1986-92	1.6	1.6
1994-97	.9	.9
Tercel		
1981-82	.8	.8
1983-94	1.2	1.2
1995-99	.8	.8
Water Pump and/or Gasket, Replace (B)		
Includes: Refill with proper coolant mix.		
Avalon	3.5	3.9
Camry		
1983-86		
gas	5.2	5.7
diesel	2.9	3.4
1987-91		
4 cyl.	3.7	4.2
V6	4.6	5.1
1992-96	3.4	3.9
1997-01	3.2	3.7
2002-05		
4 cyl. (.9)	1.4	1.6
V6 (2.1)	3.0	3.3
Camry Solara		
1999-01	3.2	3.7
2002-05		
4 cyl. (.9)	1.4	1.6
V6 (2.2)	3.1	3.4
Celica		
1981-85	1.7	2.2
1986-89		
w/turbo	4.2	4.7
w/o turbo	3.6	4.1
1990-93		
w/turbo	5.4	5.9
w/o turbo	3.7	4.2
1994-99	3.2	3.7
2000-05 (1.3)	1.9	2.2

TOY-14 AVALON : CAMRY : CELICA : COROLLA : CRESSIDA : ECHO : MATRIX : MR2

	LABOR TIME	SEVERE SERVICE
Water Pump and/or Gasket, Replace (B)		
Corolla FWD		
1983-87		
gas	1.8	2.3
diesel	2.5	3.0
1988-92		
SOHC	2.0	2.5
DOHC	3.1	3.6
1993-97	2.3	2.8
1998-02	3.0	3.5
2003-05 (.9)	1.3	1.4
Corolla FX, FX16		
SOHC (.9)	1.3	1.8
DOHC	2.7	3.2
Corolla RWD	2.4	2.9
Cressida		
1981-88	2.1	2.6
1989-92	1.7	2.2
Echo (.8)	1.3	1.6
Matrix (.9)	1.4	1.6
MR2		
1985-89	2.4	2.9
1991-95		
w/turbo	5.9	6.4
w/o turbo	4.2	4.7
MR2 Spyder (1.5)	2.2	2.7
Paseo, Starlet	1.7	2.2
Prius		
2001-03	2.6	2.9
2004-05 (1.4)	2.0	2.3
Supra		
1981-86	2.3	2.8
1987-92	1.7	2.2
1993-98		
w/turbo	3.7	4.2
w/o turbo	2.4	2.9
Tercel	1.8	2.3
w/ABS add Tercel	.1	.1
w/AC add		
93-97 Corolla FWD	.2	.2
98-05 Corolla FWD	.2	.2
MR2, Prius	.5	.5
Paseo, Tercel	.1	.1
w/PS add		
88-92 Corolla FWD	.3	.3
93-05 Corolla FWD	.5	.5
Tercel	.1	.1

ENGINE
ASSEMBLY
Times shown are for OEM assemblies. Time to replace assemblies from aftermarket rebuilders may vary.

Engine Assy., R&I (B)
Does not include parts or component transfer.

	LABOR TIME	SEVERE SERVICE
Avalon	10.1	10.1

	LABOR TIME	SEVERE SERVICE
Camry		
1983-86		
gas	9.0	9.0
diesel	6.3	6.3
1987-91		
4 cyl.		
2WD	9.7	9.7
4WD	11.2	11.2
V6	14.4	14.4
1992-96		
4 cyl.	9.2	9.2
V6	9.9	9.9
1997-01		
4 cyl.	12.7	12.7
V6	17.1	17.1
Camry Solara		
4 cyl.	12.7	12.7
V6	17.1	17.1
Celica	6.8	6.8
1981-85		
1986-89		
SOHC	9.0	9.0
DOHC	9.5	9.5
turbo	13.1	13.1
1990-93	9.8	9.8
1994-99		
7AFE	9.0	9.0
5SFE	7.5	7.5
Corolla FWD		
1983-87		
SOHC	6.5	6.5
DOHC	9.8	9.8
diesel	6.2	6.2
1988-92		
SOHC	9.3	9.3
DOHC	8.8	8.8
4WD	9.2	9.2
1993-97	10.3	10.3
1998-01	9.8	9.8
Corolla FX, FX16		
SOHC	8.5	8.5
DOHC	9.7	9.7
Corolla RWD		
1981-82	7.1	7.1
1983-87		
SOHC	6.1	6.1
DOHC	8.8	8.8
Cressida		
1981-88	6.9	6.9
1989-92	7.7	7.7
MR2		
1985-89		
w/super charger	11.6	11.6
w/o super charger	10.5	10.5
1991-95		
w/turbo	10.6	10.6
w/o turbo	8.8	8.8
Paseo	10.1	10.1

	LABOR TIME	SEVERE SERVICE
Starlet		
1981-82	6.9	6.9
1983-84	8.0	8.0
Supra		
1981-86	6.7	6.7
1987-92	8.4	8.4
1993-98		
w/turbo	12.9	12.9
w/o turbo	9.4	9.4
Tercel		
1981-82		
MT	6.4	6.4
AT	7.5	7.5
1983-90		
sedan	8.8	8.8
wagon	6.2	6.2
1991-94	7.2	7.2
1995-99	8.2	8.2
w/AC add		
81-90 Tercel	1.3	1.3
81-85 Celica	1.1	1.1
81-92 Corolla RWD	.5	.5
83-86 Camry diesel	.2	.2
83-86 Camry gas	.9	.9
83-87 Corolla diesel	.2	.2
83-87 Corolla FWD	1.3	1.3
83-87 Corolla RWD	1.1	1.1
85-89 MR2	.3	.3
86-89 Celica turbo	1.6	1.6
86-89 Celica	.1	.1
87-91 Camry 4 cyl	.2	.2
87-91 Camry V6	1.6	1.6
88-92 Corolla FWD	.3	.3
90-93 Celica	.2	.2
91-94 Tercel	.3	.3
91-95 MR2	.5	.5
92-96 Camry 4 cyl.	.2	.2
92-96 Camry V6	.5	.5
94-99 Celica	.2	.2
95-99 Tercel	.2	.2
98-01 Corolla FWD	.2	.2
Corolla FX, FX16	1.1	1.1
Paseo	.2	.2
Starlet	.5	.5
w/PS add		
81-85 Celica	.3	.3
81-92 Corolla RWD	.5	.5
83-86 Camry diesel	.3	.3
83-86 Camry gas	.5	.5
83-87 Corolla diesel	.3	.3
83-87 Corolla FWD	.3	.3
83-87 Corolla RWD	.3	.3
83-90 Tercel	.3	.3
86-89 Celica	.8	.8
88-92 Corolla	.2	.2
91-94 Tercel	.3	.3
93-97 Corolla FWD	.2	.2
95-99 Tercel	.2	.2
Corolla FX, FX16	.3	.3

MR2 SPYDER : PASEO : PRIUS : SOLARA : STARLET : SUPRA : TERCEL — TOY-15

	Labor Time	Severe Service
Engine Assy., R&R and Reconditioning (A)		
Includes: Replacing rings, rod and main bearings, cylinder head reconditioning and engine tune-up.		
Avalon	47.7	52.9
Camry		
1983-86		
gas	20.3	20.3
diesel	24.1	24.1
1987-91		
4 cyl.		
2WD	29.6	29.6
4WD	30.5	30.5
V6	40.3	40.3
1992-96		
4 cyl.	31.7	31.7
V6	43.9	43.9
1997-01		
4 cyl.	33.0	33.0
V6	46.1	46.1
2002-05		
4 cyl. (30.6)	39.8	40.7
V6 (39.4)	51.2	52.1
Camry Solara		
1999-01		
4 cyl.	44.1	49.9
V6	46.4	46.4
2002-05		
4 cyl. (30.6)	39.8	40.7
V6		
2002-03 (34.2)	44.5	45.4
2004-05 (38.7)	50.3	51.2
Celica		
1981-85		
SOHC	23.2	23.2
DOHC	24.4	24.4
1986-89		
SOHC	23.6	23.6
DOHC	29.8	29.8
turbo	34.7	34.7
1990-93		
2WD	30.7	30.7
4WD	32.2	32.2
turbo	35.3	35.3
1994-99	30.7	30.7
2000-05 (26.4)	34.3	35.2
Corolla FWD		
1983-87		
SOHC	20.1	20.1
DOHC	25.6	25.6
diesel	24.1	24.1
1988-92		
SOHC	28.6	28.6
DOHC	29.0	29.0
4WD	29.3	29.3

	Labor Time	Severe Service
1993-97		
2WD		
SOHC	30.2	30.2
DOHC	31.3	31.3
4WD	29.5	29.5
1998-02	36.3	36.3
2003-05 (24.5)	31.9	32.8
Corolla FX, FX16		
SOHC	20.1	20.1
DOHC	25.5	25.5
Corolla RWD		
1981-82	20.1	20.1
1983-87		
SOHC	19.7	19.7
DOHC	30.9	30.9
Cressida		
1981-88	31.8	32.7
1989-92	38.9	39.8
Echo (23.9)	31.1	32.0
Matrix		
2WD (24.5)	31.9	32.8
4WD (26.5)	34.5	35.4
XRS (29.9)	38.9	39.8
MR2		
1985-87	26.9	26.9
1985-89		
w/super charger	31.9	31.9
w/o super charger	28.6	28.6
1991-95		
w/turbo	35.1	35.1
w/o turbo	30.9	30.9
MR2 Spyder		
5-Speed MT (25.0)	32.5	33.4
sequential MT (25.6)	33.3	34.2
Paseo	31.9	31.9
Prius		
2001-03	35.0	38.8
2004-05 (30.6)	39.8	40.7
Starlet		
1981-82	20.5	20.5
1983-84	24.1	24.1
Supra		
1981-86	31.5	31.5
1987-92		
w/turbo	37.2	37.2
w/o turbo	36.6	36.6
1993-98		
w/turbo	45.7	45.7
w/o turbo	41.3	41.3
Tercel		
1981-82		
MT	22.3	22.3
AT	23.0	23.0
1983-90		
sedan	24.3	24.3
wagon	19.4	19.4

	Labor Time	Severe Service
1991-94	25.2	25.2
1995-99	31.6	31.6
w/AC add		
81-90 Tercel	1.3	1.3
81-85 Celica	1.1	1.1
81-92 Corolla RWD	.5	.5
83-86 Camry diesel	.2	.2
83-86 Camry gas	.9	.9
83-87 Corolla diesel	.2	.2
83-87 Corolla FWD	1.3	1.3
83-87 Corolla RWD	1.1	1.1
85-89 MR2	.3	.3
86-89 Celica turbo	1.6	1.6
86-89 Celica	.1	.1
87-91 Camry 4 cyl	.2	.2
87-91 Camry V6	1.6	1.6
88-92 Corolla FWD	.3	.3
90-93 Celica	.2	.2
91-94 Tercel	.3	.3
91-95 MR2	.5	.5
92-96 Camry 4 cyl.	.2	.2
92-96 Camry V6	.5	.5
94-05 Celica	.2	.2
95-99 Tercel	.2	.2
98-05 Corolla FWD	.2	.2
Corolla FX, FX16	1.1	1.1
Corolla Matrix	.3	.3
Echo, Prius	.3	.3
MR2 Spyder, Paseo	.2	.2
Starlet	.5	.5
w/PS add		
81-85 Celica	.3	.3
81-92 Corolla RWD	.5	.5
83-86 Camry diesel	.3	.3
83-86 Camry gas	.5	.5
83-87 Corolla diesel	.3	.3
83-87 Corolla FWD	.3	.3
83-87 Corolla RWD	.3	.3
83-90 Tercel	.3	.3
86-89 Celica	.8	.8
88-92 Corolla	.2	.2
91-94 Tercel	.3	.3
93-97 Corolla FWD	.2	.2
95-99 Tercel	.2	.2
00-05 Celica	.3	.3
Corolla FX, FX16	.3	.3
Echo	.5	.5
Engine Assy. (Short Block), Replace (B)		
Assembly may consist of cylinder block, piston assemblies, crankshaft, camshaft, timing chain and gears. Operation Includes: R&R engine, transfer necessary parts and all necessary adjustments.		
Avalon	21.7	24.1

TOY-16   AVALON : CAMRY : CELICA : COROLLA : CRESSIDA : ECHO : MATRIX : MR2

	LABOR TIME	SEVERE SERVICE
Camry		
1983-86		
gasoline	13.6	15.0
diesel	18.4	20.5
1987-91		
4 cyl.		
2WD	14.9	16.6
4WD	16.3	18.0
V6	21.5	23.8
1992-96		
4 cyl.	14.1	15.7
V6	19.0	20.0
1997-01		
4 cyl.	14.6	16.3
V6	20.1	22.4
2002-05		
4 cyl. (16.5)	21.5	22.4
V6 (15.9)	20.7	21.6
Camry Solara		
1999-01		
4 cyl.	14.8	16.5
V6	20.2	20.2
2002-03		
4 cyl.	22.5	22.4
V6	21.0	23.1
2004-05		
4 cyl. (16.1)	20.9	21.8
V6 (15.8)	20.5	21.4
Celica		
1981-85		
SOHC	11.8	13.2
DOHC	12.8	14.3
1986-89		
SOHC	13.9	15.6
DOHC	15.3	17.1
turbo	19.9	22.1
1990-93		
2WD	14.4	16.1
4WD	15.4	17.1
turbo	17.2	19.1
1994-99	13.8	15.4
2000-05 (12.5)	16.8	17.7
Corolla FWD		
1983-87		
SOHC	11.4	11.4
DOHC	15.8	12.7
diesel	18.4	20.5
1988-92		
SOHC	14.1	15.7
DOHC	14.6	16.3
4WD	14.2	15.9
1993-97		
4AFE	14.8	16.5
7AFE	15.9	17.6
1998-02	15.8	17.5
2003-05 (10.8)	14.7	15.6
Corolla FX, FX16		
SOHC	11.5	12.8
DOHC	15.8	17.6

	LABOR TIME	SEVERE SERVICE
Corolla RWD		
1981-82	12.1	13.5
1983-87		
SOHC	10.8	12.0
DOHC	14.4	16.1
Cressida		
1981-88	13.0	14.5
1989-92	13.5	15.1
Echo (10.6)	14.4	14.3
Matrix		
2WD (10.4)	14.2	15.1
4WD (12.4)	16.6	17.5
XRS (11.0)	15.0	15.9
MR2		
1985-89		
w/super charger	17.6	19.6
w/o super charger	16.8	18.8
1991-95		
w/turbo	17.0	18.9
w/o turbo	14.4	16.1
MR2 Spyder		
5-Speed MT (10.9)	14.8	15.7
sequential MT (11.5)	15.5	16.4
Paseo		
1992-95	15.3	17.1
1996-99	14.4	16.1
Prius		
2001-03 (12.4)	16.6	17.5
2004-05 (14.0)	18.5	19.4
Starlet		
1981-82	11.8	13.2
1983-84	12.9	14.5
Supra		
1981-86	12.8	14.5
1987-92		
w/turbo	15.1	16.8
w/o turbo	14.2	15.9
1993-98		
w/turbo	19.7	22.0
w/o turbo	15.2	17.0
Tercel		
1981-82		
MT	11.4	12.7
AT	12.5	13.9
1983-90		
sedan	13.7	15.2
wagon	11.1	12.4
1991-94	12.2	13.7
1995-99	14.1	15.7
w/AC add		
81-90 Tercel	1.3	1.3
81-85 Celica	1.1	1.1
81-92 Corolla RWD	.5	.5
83-86 Camry diesel	.2	.2
83-86 Camry gas	.9	.9
83-87 Corolla diesel	.2	.2
83-87 Corolla FWD	1.3	1.3
83-87 Corolla RWD	1.1	1.1
85-89 MR2	.3	.3

	LABOR TIME	SEVERE SERVICE
86-89 Celica turbo	1.6	1.6
86-89 Celica	.1	.1
87-91 Camry 4 cyl	.2	.2
87-91 Camry V6	1.6	1.6
88-92 Corolla FWD	.3	.3
90-93 Celica	.2	.2
91-94 Tercel	.3	.3
91-95 MR2	.5	.5
92-96 Camry 4 cyl.	.2	.2
92-96 Camry V6	.5	.5
94-05 Celica	.2	.2
95-99 Tercel	.2	.2
98-05 Corolla FWD	.2	.2
Corolla FX, FX16	1.1	1.1
Corolla Matrix	.3	.3
Echo, Prius	.3	.3
MR2 Spyder, Paseo	.2	.2
Starlet	.5	.5
w/PS add		
81-85 Celica	.3	.3
81-92 Corolla RWD	.5	.5
83-86 Camry diesel	.3	.3
83-86 Camry gas	.5	.5
83-87 Corolla diesel	.3	.3
83-87 Corolla FWD	.3	.3
83-87 Corolla RWD	.3	.3
83-90 Tercel	.3	.3
86-89 Celica	.8	.8
88-92 Corolla	.2	.2
91-94 Tercel	.3	.3
93-97 Corolla FWD	.2	.2
95-99 Tercel	.2	.2
00-05 Celica	.3	.3
Corolla FX, FX16	.3	.3
Echo	.5	.5
Hybrid Motor, Replace (A)		
Prius		
2001-03 (12.0)	16.1	17.0
2004-05 (7.7)	10.8	11.7
w/AC add	.3	.3
Engine Mounts, Replace (B)		
Avalon, Camry Solara (.5)	.8	.9
Camry		
1983-86	.9	1.1
1987-91		
4 cyl.	.7	.9
V6	1.3	1.4
1992-01	.9	1.1
2002-05		
4 cyl. (.4)	.6	.7
V6 (.8)	1.3	1.5
Celica		
1981-99	1.1	1.3
2000-05		
GT (.6)	1.0	1.1
GTS (.2)	.3	.4
Corolla FWD, Matrix (.4)	.6	.7
Corolla FX, FX16	.8	1.0

MR2 SPYDER : PASEO : PRIUS : SOLARA : STARLET : SUPRA : TERCEL TOY-17

	LABOR TIME	SEVERE SERVICE
Corolla RWD		
1981-82	.8	1.0
1983-87		
SOHC	.8	1.0
DOHC	1.8	2.0
Cressida	1.2	1.4
Echo, Prius (.4)	.6	.7
MR2, Paseo, Starlet	.9	1.1
MR2 Spyder (.6)	1.0	1.1
Supra		
1981-86	.9	1.1
1987-98	1.7	1.9
Tercel	1.1	1.3

CYLINDER HEAD
Compression Test (B)

	LABOR TIME	SEVERE SERVICE
Avalon	1.1	1.4
Camry, Camry Solara		
1983-86	.8	1.1
1987-02		
4 cyl.	.7	1.0
V6	1.3	1.6
Celica		
1981-93	.9	1.2
1994-99	.7	1.0
Corolla	.8	1.1
Cressida	1.1	1.4
Echo	.7	1.0
MR2	.8	1.1
Paseo, Starlet	.7	1.0
Supra		
1981-92	1.5	1.8
1993-98		
w/turbo	.8	1.1
w/o turbo	1.9	2.2
Tercel	.8	1.1

Valve Clearance, Adjust (B)

	LABOR TIME	SEVERE SERVICE
Avalon	5.6	6.2
Camry, Camry Solara		
1983-86 diesel	1.6	2.1
1987-91		
4 cyl.	2.8	3.3
V6	5.3	5.8
1992-01		
4 cyl.	2.6	3.1
V6	5.3	5.8
2002-05		
4 cyl. (2.1)	3.0	3.5
V6 (3.6)	5.1	5.6
Celica		
1983-85	1.5	2.0
1986-89		
DOHC	2.9	3.4
SOHC	3.7	4.2
1990-93		
w/turbo	3.3	3.8
w/o turbo	2.9	3.4
1994-99	2.5	3.0
2000-05		
GT (3.6)	5.1	5.6
GTS (5.1)	7.1	7.6

	LABOR TIME	SEVERE SERVICE
Corolla FWD		
1983-87	1.7	2.2
1988-97	3.1	3.6
1998-02	4.9	5.4
2003-05 (2.8)	4.3	4.8
Corolla FX, FX16		
SOHC	1.6	2.1
DOHC	2.3	2.8
Corolla RWD		
SOHC (1.1)	1.7	2.2
DOHC (1.5)	2.3	2.8
1989-92 Cressida	4.0	4.5
Echo (4.6)	6.6	7.1
Matrix		
exc. XRS (2.8)	4.0	4.5
XRS (2.5)	3.6	4.1
MR2		
1985-89	2.4	2.9
1991-95		
w/turbo	3.1	3.6
w/o turbo	2.7	3.2
MR2 Spyder (3.7)	5.2	5.7
Paseo	2.8	3.3
Prius		
2001-03	7.1	7.8
2004-05 (3.5)	4.9	5.4
Supra		
1987-92	4.4	4.9
1993-98		
w/turbo	7.5	8.0
w/o turbo	4.1	4.6
Tercel		
1983-94	1.5	2.0
1995-99	2.8	3.3

Cylinder Head, Replace (B)
Includes: Parts transfer, adjustments.

	LABOR TIME	SEVERE SERVICE
Avalon		
one	14.6	15.1
both	22.0	22.5
Camry		
1983-86		
gasoline	9.3	9.8
diesel	9.8	10.3
1987-91		
4 cyl.	10.0	10.5
V6, both	21.4	22.0
1992-96		
4 cyl.	15.2	15.7
V6, both	19.7	20.5
1997-01		
4 cyl.	15.6	16.1
V6		
one	13.2	13.7
both	20.5	21.0
2002-05		
4 cyl. (12.4)	16.6	17.1
V6		
one (9.9)	13.6	14.1
both (15.9)	20.7	21.2

	LABOR TIME	SEVERE SERVICE
Camry Solara		
1999-01		
4 cyl.	15.7	16.2
V6		
one	13.5	14.0
both	20.7	21.2
2002-03		
4 cyl.	18.4	18.9
V6		
one	14.9	15.4
both	23.1	23.6
2004-05		
4 cyl. (12.3)	16.5	17.0
V6		
one (9.8)	13.4	13.9
both (15.6)	20.3	20.8
Celica		
1981-85		
SOHC	8.7	9.2
DOHC	9.8	10.3
1986-89		
SOHC	9.3	9.8
DOHC	14.1	14.6
turbo	19.2	19.7
1990-93		
2WD	14.8	15.3
4WD	15.3	15.8
turbo	19.9	20.4
1994-99	15.2	15.7
2000-05 (11.9)	16.0	16.5
Corolla FWD		
1983-87		
SOHC	9.7	10.2
DOHC	13.0	13.5
diesel	9.8	10.3
1988-92		
SOHC	13.5	14.0
DOHC	14.7	15.2
1993-97		
2WD	14.3	14.8
4WD	13.3	13.8
1998-02	15.3	15.8
2003-05 (10.5)	14.3	14.8
Corolla FX, FX16		
SOHC	9.8	10.3
DOHC	13.0	13.5
Corolla RWD		
1981-82	7.3	7.8
1983-87		
SOHC	9.6	10.1
DOHC	12.3	12.8
Cressida		
1981-88	10.1	10.6
1989-92	18.3	18.8
Echo (10.4)	14.2	14.7
Matrix		
exc. XRS (10.5)	14.3	14.8
XRS (12.8)	17.1	17.6

TOY-18 AVALON : CAMRY : CELICA : COROLLA : CRESSIDA : ECHO : MATRIX : MR2

	LABOR TIME	SEVERE SERVICE
Cylinder Head, Replace (B)		
MR2		
1985-89		
w/super charger	13.6	*14.1*
w/o super charger	13.2	*13.7*
1991-95		
w/turbo	19.9	*20.4*
w/o turbo	15.3	*15.8*
MR2 Spyder		
5-Speed MT (15.9)	20.7	*21.2*
sequential		
MT (16.5)	21.5	*22.0*
Paseo	16.4	*16.9*
Prius		
2001-03 (13.1)	17.4	*17.9*
2004-05 (11.4)	15.4	*15.9*
Starlet		
1981-82	6.9	*7.4*
1983-84	7.7	*8.2*
Supra		
1981-86	10.0	*10.5*
1987-92		
w/turbo	18.1	*18.6*
w/o turbo	17.3	*17.8*
1993-98		
w/turbo	24.3	*25.8*
w/o turbo	21.2	*21.7*
Tercel		
1981-82	7.6	*8.1*
1983-90	9.7	*10.2*
1991-94	11.2	*11.7*
1995-99	15.7	*16.2*
w/AC add		
81-82 Corolla RWD	.1	*.1*
87-96 Camry	.1	*.1*
90-99 Celica	.1	*.1*
91-94 Tercel	.1	*.1*
93-97 Corolla	.1	*.1*
95-99 Tercel	.1	*.1*
98-05 Corolla	.2	*.2*
Matrix XRS	.2	*.2*
MR2	.5	*.5*
MR2 Spyder	.2	*.2*
Paseo	.1	*.1*
w/PS add		
81-82 Corolla RWD	.5	*.5*
81-85 Celica	.5	*.5*
81-90 Tercel	.2	*.2*
83-86 Camry	.5	*.5*
83-87 Corolla diesel	.2	*.2*
83-87 Corolla FWD	.1	*.1*
83-87 Corolla RWD	.3	*.3*
86-89 Celica	.2	*.2*
88-92 Corolla FWD	.2	*.2*
91-94 Tercel	.2	*.2*
93-97 Corolla	.2	*.2*
95-99 Tercel	.3	*.3*
00-05 Celica	.3	*.3*
Corolla FX, FX16	.1	*.1*
Echo	.2	*.2*

	LABOR TIME	SEVERE SERVICE
Cylinder Head Gasket, Replace (B)		
Avalon		
one	8.5	*9.4*
both	11.0	*12.2*
Camry		
1983-86		
gas	7.3	*7.6*
diesel	5.6	*5.9*
1987-91		
4 cyl.	6.7	*7.0*
V6, one	10.9	*11.2*
1992-96		
4 cyl.	7.4	*7.7*
V6, one	9.9	*10.2*
1997-01		
4 cyl.	7.6	*7.9*
V6		
one	7.9	*8.1*
both	10.1	*10.4*
2002-05		
4 cyl. (6.7)	9.5	*10.0*
V6		
one (6.2)	8.8	*9.3*
both (8.5)	11.8	*12.3*
Camry Solara		
1999-01		
4 cyl.	7.6	*7.9*
V6		
one	8.2	*8.5*
both	10.3	*10.6*
2002-03		
4 cyl.	10.3	*10.6*
V6		
one	9.5	*9.8*
both	12.5	*12.8*
2004-05		
4 cyl. (5.8)	8.3	*8.8*
V6		
one (6.1)	8.7	*9.2*
both (8.3)	11.5	*12.0*
Celica		
1981-85		
SOHC	5.6	*5.9*
DOHC	6.5	*6.8*
1986-89		
SOHC	7.7	*8.0*
DOHC	6.3	*6.6*
turbo	12.4	*12.7*
1990-93		
2WD	7.2	*7.5*
4WD	7.8	*8.1*
turbo	11.6	*11.9*
1994-99	7.3	*7.6*
2000-04	9.6	*9.9*
Corolla FWD, Matrix		
1983-87		
SOHC	5.9	*6.2*
DOHC	7.1	*7.4*
diesel	5.7	*6.0*

	LABOR TIME	SEVERE SERVICE
1988-92		
SOHC	5.9	*6.2*
DOHC	7.7	*8.0*
4WD	6.3	*6.6*
1993-97		
2WD	6.9	*7.2*
4WD	6.1	*6.4*
1998-02	8.5	*9.6*
2003-05 (4.8)	6.9	*7.4*
Corolla FX, FX16		
SOHC	5.8	*6.1*
DOHC	7.3	*7.6*
Corolla RWD		
1981-82	5.0	*5.3*
1983-87		
SOHC	5.8	*6.1*
DOHC	6.3	*6.6*
Cressida		
1981-88	5.4	*5.7*
1989-92	6.3	*6.6*
Echo (5.0)	7.2	*7.7*
MR2		
1985-89		
w/super charger	7.5	*7.8*
w/o super charger	7.1	*7.4*
1991-95		
w/turbo	11.8	*12.1*
w/o turbo	7.9	*8.2*
MR2 Spyder		
5-Speed MT (10.2)	14.0	*14.5*
sequential		
MT (10.8)	14.7	*15.2*
Paseo	8.2	*8.5*
Prius		
2001-03	11.9	*12.2*
2004-05 (7.4)	10.4	*10.9*
Starlet		
1981-82	5.3	*5.6*
1983-84	5.9	*6.2*
Supra		
1981-86	5.1	*5.4*
1987-92		
w/turbo	9.6	*9.9*
w/o turbo	8.8	*9.1*
1993-98		
w/turbo	14.2	*14.5*
w/o turbo	10.8	*11.1*
Tercel		
1981-82	5.3	*5.6*
1983-90	5.8	*6.1*
1991-94	5.9	*6.2*
1995-99	7.5	*7.8*
w/AC add		
81-82 Corolla RWD	.1	*.1*
87-96 Camry	.1	*.1*
90-99 Celica	.1	*.1*
91-94 Tercel	.1	*.1*
93-97 Corolla	.1	*.1*
95-99 Tercel	.1	*.1*
98-05 Corolla	.2	*.2*

MR2 SPYDER : PASEO : PRIUS : SOLARA : STARLET : SUPRA : TERCEL TOY-19

	LABOR TIME	SEVERE SERVICE
Matrix XRS	.2	.2
MR2	.5	.5
MR2 Spyder	.2	.2
Paseo	.1	.1
w/PS add		
81-82 Corolla RWD	.5	.5
81-85 Celica	.5	.5
81-90 Tercel	.2	.2
83-86 Camry	.5	.5
83-87 Corolla diesel	.2	.2
83-87 Corolla FWD	.1	.1
83-87 Corolla RWD	.3	.3
86-89 Celica	.2	.2
88-92 Corolla FWD	.2	.2
91-94 Tercel	.2	.2
93-97 Corolla	.2	.2
95-99 Tercel	.3	.3
00-05 Celica	.3	.3
Corolla FX, FX16	.1	.1
Echo	.2	.2
Pushrods, Replace (B)		
1981-82 Corolla RWD	2.4	2.7
Starlet		
1981-82	1.9	2.2
1983-84	2.1	2.4
Rocker Arms or Shafts, Replace (B)		
1983-86 Camry	3.6	3.9
1981-85 Celica	3.0	3.3
Celica GTS	7.6	8.5
1983-87 Corolla FWD	2.4	2.7
Corolla RWD		
1981-82	3.0	3.3
1983-87	2.2	2.5
Corolla FX, FX16	2.4	2.7
1981-88 Cressida	4.5	4.8
Matrix XRS	5.1	5.7
Starlet		
1981-82	2.6	2.9
1983-84	3.0	3.3
1981-86 Supra	4.4	4.7
Tercel		
1981-82	3.1	3.4
1983-90		
sedan	3.2	3.5
wagon	2.4	2.7
1991-94	3.3	3.6
w/AC add	.1	.1
w/PS add	.1	.1
Valve/Cam Cover Gasket, Replace or Reseal (B)		
Avalon, both	2.9	3.2
Camry, Camry Solara		
1983-86	.8	1.0
1987-05		
4 cyl. (.5)	.9	1.1
V6, both (1.7)	2.4	2.7

	LABOR TIME	SEVERE SERVICE
Celica		
1981-85	1.3	1.5
1986-89		
SOHC	.8	1.0
DOHC	1.4	1.6
turbo	2.4	2.6
1990-93		
2WD	.7	.9
4WD	.8	1.0
turbo	1.6	1.8
1994-05 (.6)	.9	1.0
Corolla FWD, Matrix		
1983-92	1.1	1.3
1993-97		
2WD	.7	.9
4WD	1.2	1.4
1998-05 (.5)	.8	.9
Corolla FX, FX16	.9	1.1
Corolla RWD (.6)	.9	1.1
Cressida	1.4	1.6
Echo (.4)	.6	.7
MR2		
1985-89	1.2	1.4
1991-95		
w/turbo	1.4	1.6
w/o turbo	.8	1.0
MR2 Spyder (.5)	.8	.9
Paseo	.5	.7
Prius (1.1)	1.7	1.9
Starlet		
1981-82	.8	1.0
1983-84	1.4	1.6
Supra		
1981-86	1.2	1.4
1987-92	1.8	2.0
1993-98	1.9	2.1
Tercel		
1981-94	1.1	1.3
1995-99	.5	.7
Valve Adjuster, Replace (B)		
1983-86 Camry	4.0	4.3
1986-89 Celica	4.0	4.3
1981-88 Cressida	4.8	5.1
1981-86 Supra	4.7	5.0
Valve Job (A)		
Avalon	27.4	29.8
Camry		
1983-86		
gas	11.2	11.7
diesel	10.7	11.2
1987-91		
4 cyl.	15.2	15.7
V6	24.4	24.9
1992-01		
4 cyl.	20.4	22.5
V6	27.4	29.8
2002-05		
4 cyl. (14.6)	18.4	19.3
V6 (17.5)	21.9	22.8

	LABOR TIME	SEVERE SERVICE
Camry Solara		
1999-01		
4 cyl.	20.6	22.6
V6	27.5	30.0
2002-05		
4 cyl. (14.7)	18.5	19.4
V6 (20.7)	25.9	26.8
Celica		
1981-85		
SOHC	9.9	10.4
DOHC	11.1	11.6
1986-89		
SOHC	10.7	11.2
DOHC	17.2	17.7
turbo	23.2	23.7
1990-93		
2WD	18.7	19.2
4WD	19.8	20.3
turbo	23.6	24.1
1994-99		
7AFE	18.3	18.8
5SFE	19.1	20.0
2000-05 (14.9)	18.7	19.6
Corolla FWD		
1983-87		
SOHC	10.5	11.0
DOHC	16.3	16.8
diesel	10.7	11.2
1988-92		
SOHC	16.3	16.8
DOHC	18.8	19.3
1993-97		
2WD	18.7	19.2
4WD	16.3	16.8
1998-02	19.4	19.9
2003-05 (13.4)	17.1	18.0
Corolla FX, FX16		
SOHC	10.5	11.0
DOHC	15.9	16.4
Corolla RWD		
1981-82	9.2	9.7
1983-87		
SOHC	11.3	11.8
DOHC	14.7	15.2
Cressida		
1981-88	13.6	14.1
1989-92	24.9	25.4
Echo, Matrix (13.5)	17.2	18.1
MR2		
1985-89		
w/supercharger	16.3	16.8
w/o supercharger	15.8	16.3
1991-95		
w/turbo	23.6	24.1
w/o turbo	19.6	20.1
MR2 Spyder		
5-Speed MT (18.8)	23.5	24.4
sequential		
MT (19.4)	24.3	25.2

TOY-20 AVALON : CAMRY : CELICA : COROLLA : CRESSIDA : ECHO : MATRIX : MR2

	LABOR TIME	SEVERE SERVICE
Valve Job (A)		
Paseo	19.3	*19.8*
Prius (16.3)	20.4	*21.3*
Starlet		
1981-82	9.3	*9.8*
1983-84	10.1	*10.6*
Supra		
1981-86	13.5	*14.0*
1987-92		
w/turbo	17.4	*17.9*
w/o turbo	16.5	*17.0*
1993-98		
w/turbo	30.0	*30.5*
w/o turbo	27.8	*28.3*
Tercel		
1981-82	10.4	*10.9*
1983-90		
sedan	12.1	*12.6*
wagon	11.1	*11.6*
1991-94	14.1	*14.6*
1995-99	18.4	*18.9*
w/AC add		
81-82 Corolla RWD	.1	*.1*
87-96 Camry	.1	*.1*
90-99 Celica	.1	*.1*
91-94 Tercel	.1	*.1*
93-97 Corolla	.1	*.1*
95-99 Tercel	.1	*.1*
98-05 Corolla	.2	*.2*
Matrix XRS	.2	*.2*
MR2	.5	*.5*
MR2 Spyder	.2	*.2*
Paseo	.1	*.1*
w/PS add		
81-82 Corolla RWD	.5	*.5*
81-85 Celica	.5	*.5*
81-90 Tercel	.2	*.2*
83-86 Camry	.5	*.5*
83-87 Corolla diesel	.2	*.2*
83-87 Corolla FWD	.1	*.1*
83-87 Corolla RWD	.3	*.3*
86-89 Celica	.2	*.2*
88-92 Corolla FWD	.2	*.2*
91-94 Tercel	.2	*.2*
93-97 Corolla	.2	*.2*
95-99 Tercel	.3	*.3*
00-05 Celica	.3	*.3*
Corolla FX, FX16	.1	*.1*
Echo	.2	*.2*
Valve Lifters, Replace (B)		
Avalon	9.8	*10.8*
Camry		
1983-86	3.5	*3.8*
1987-91		
4 cyl.	5.4	*5.7*
V6	9.6	*9.9*

	LABOR TIME	SEVERE SERVICE
1992-01		
4 cyl.	5.7	*6.0*
V6	9.2	*9.5*
2002-05		
4 cyl. (5.7)	7.9	*8.2*
V6 (6.2)	8.5	*8.8*
Camry Solara		
1999-01		
4 cyl.	5.9	*6.2*
V6	9.4	*9.7*
2002-03		
4 cyl.	8.5	*9.4*
V6	9.2	*10.2*
2004-05		
4 cyl. (7.4)	10.1	*10.4*
V6 (8.5)	11.4	*11.7*
Celica		
1986-89		
w/turbo	6.2	*6.5*
w/o turbo	5.5	*5.8*
1990-93		
w/turbo	8.5	*8.8*
w/o turbo	5.7	*6.0*
1994-99	5.3	*5.6*
2000-05 (4.9)	6.8	*7.1*
Corolla FWD, Matrix		
1983-86 diesel	3.6	*3.9*
1988-92		
SOHC	5.3	*5.6*
DOHC	6.4	*6.7*
1993-97	5.3	*5.6*
1998-02	6.1	*6.4*
2003-05 (2.8)	3.8	*4.1*
Corolla RWD (2.8)	3.8	*4.1*
1989-92 Cressida	6.9	*7.2*
Echo (4.6)	6.4	*6.9*
MR2		
1985-89	2.8	*3.1*
1991-95		
w/turbo	8.3	*8.6*
w/o turbo	6.1	*6.4*
MR2 Spyder (3.7)	5.2	*5.7*
Paseo	5.2	*5.5*
Prius		
2001-03	5.1	*5.7*
2004-05 (2.6)	3.7	*4.0*
Starlet		
1981-82	3.4	*3.7*
1983-84	2.5	*2.8*
Supra		
1987-92		
w/turbo	7.3	*7.6*
w/o turbo	6.2	*6.5*
1993-98		
w/o turbo	7.2	*7.5*
w/turbo	7.7	*8.0*

	LABOR TIME	SEVERE SERVICE
1995-99 Tercel	4.8	*5.1*
w/AC add		
83-05 Corolla FWD	.2	*.2*
86-93 Celica	.1	*.1*
Camry	.1	*.1*
MR2	.1	*.1*
MR2 Spyder	.2	*.2*
Tercel	.1	*.1*
w/PS add		
00-05 Celica	.3	*.3*
Echo	.2	*.2*
Tercel	.1	*.1*
Valve Springs and/or Oil Seals, Replace (B)		
Avalon	14.8	*16.5*
Camry		
1983-86		
gas	8.2	*8.7*
diesel	7.0	*7.5*
1987-91		
4 cyl.	9.2	*9.7*
V6	16.3	*16.8*
1992-01		
4 cyl.	10.6	*11.1*
V6	14.1	*14.6*
2002-05		
4 cyl. (9.6)	12.8	*13.3*
V6 (10.1)	13.3	*13.8*
Camry Solara		
1999-03		
4 cyl.	11.5	*12.7*
V6	15.0	*16.6*
2004-05		
4 cyl. (8.0)	10.8	*11.3*
V6 (9.0)	12.0	*12.5*
Celica		
1981-85		
SOHC	6.3	*6.8*
DOHC	7.8	*8.3*
1986-89		
SOHC	8.3	*8.8*
DOHC	9.1	*9.6*
turbo	14.8	*15.3*
1990-93		
2WD	9.9	*10.4*
4WD	10.3	*10.8*
turbo	15.2	*15.7*
1994-99	9.9	*10.4*
2000-05 (8.3)	11.1	*11.6*
Corolla FWD		
1983-87		
SOHC	7.9	*8.4*
DOHC	8.9	*9.4*
diesel	6.9	*7.4*
1988-92		
SOHC	8.4	*8.9*
DOHC	10.1	*10.6*

MR2 SPYDER : PASEO : PRIUS : SOLARA : STARLET : SUPRA : TERCEL   TOY-21

	LABOR TIME	SEVERE SERVICE
1993-97		
2WD	9.8	10.3
4WD	8.6	9.1
1998-02	10.1	10.6
2003-05 (8.1)	10.9	11.4
Corolla FX, FX16		
SOHC	7.8	8.3
DOHC	8.9	9.4
Corolla RWD		
1981-82	5.9	6.4
1983-87		
SOHC	7.7	8.2
DOHC	8.3	8.8
Cressida		
1981-88	8.8	9.3
1989-92	7.3	7.8
Echo (6.8)	9.3	9.8
Matrix		
exc. XRS (8.1)	10.9	11.4
XRS (7.8)	10.5	11.0
MR2		
1985-89	9.8	10.3
1991-95		
w/turbo	15.1	15.6
w/o turbo	10.2	10.7
MR2 Spyder		
5-Speed MT (12.2)	15.7	16.2
sequential MT (12.8)	16.4	16.9
Paseo	10.4	10.9
Prius		
2001-03 (9.8)	13.0	13.5
2004-05 (9.4)	12.5	13.0
Starlet		
1981-82	6.0	6.5
1983-84	7.2	7.7
Supra		
1981-86	8.7	9.2
1987-93		
w/turbo	11.2	11.7
w/o turbo	10.1	10.6
1994-97		
w/turbo	18.5	19.0
w/o turbo	14.5	15.0
Tercel		
1981-82	6.7	7.2
1983-90		
sedan	7.4	7.9
wagon	8.3	8.8
1991-94	8.7	9.2
1995-99	9.9	10.4
w/AC add		
81-82 Corolla RWD	.1	.1
87-96 Camry	.1	.1
90-99 Celica	.1	.1
91-94 Tercel	.1	.1
93-97 Corolla	.1	.1
95-99 Tercel	.1	.1

	LABOR TIME	SEVERE SERVICE
98-05 Corolla	.2	.2
MR2	.5	.5
Paseo	.1	.1
w/PS add		
81-82 Corolla RWD	.5	.5
81-85 Celica	.5	.5
81-90 Tercel	.2	.2
83-86 Camry	.5	.5
83-87 Corolla diesel	.2	.2
83-87 Corolla FWD	.1	.1
83-87 Corolla RWD	.3	.3
86-89 Celica	.2	.2
88-92 Corolla FWD	.2	.2
91-94 Tercel	.2	.2
93-97 Corolla	.2	.2
95-99 Tercel	.3	.3
00-05 Celica	.3	.3
Corolla FX, FX16	.1	.1

CAMSHAFT
Balance Shaft, Replace (B)

	LABOR TIME	SEVERE SERVICE
Camry, Camry Solara 4 cyl.		
1992-01	5.5	6.2
2002-05 (11.8)	15.3	15.8

Camshaft, Replace (A)

	LABOR TIME	SEVERE SERVICE
Avalon		
1995-99	10.2	11.3
2000-04	10.6	11.8
Camry		
1983-86		
gas	4.3	4.8
diesel	3.2	3.7
1987-91		
4 cyl.	4.2	4.7
V6	9.9	10.4
1992-96		
4 cyl.	6.3	6.8
V6	9.3	9.8
1997-01		
4 cyl.	6.6	7.1
V6	9.6	10.1
2002-05		
4 cyl. (1.8)	2.6	3.1
V6 (6.1)	8.4	8.9
Camry Solara		
1999-01		
4 cyl.	6.4	6.9
V6	9.9	10.4
2002-03		
4 cyl.	2.8	3.1
V6	10.4	11.5
2004-05		
4 cyl. (3.4)	4.8	5.3
V6 (6.1)	8.4	8.9
Celica		
1981-85	3.4	3.9
1986-89		
SOHC	4.3	4.8
DOHC	5.4	5.9
turbo	6.3	6.8

	LABOR TIME	SEVERE SERVICE
1990-93		
2WD	6.1	6.6
4WD	6.5	7.0
turbo	8.5	9.0
1994-99	6.3	6.8
2000-05 (5.5)	7.6	8.1
Corolla FWD, Matrix		
1983-87		
SOHC	5.0	5.5
DOHC	4.1	4.6
diesel	3.4	3.9
1988-92		
SOHC	5.3	5.8
DOHC	5.9	6.4
1993-97		
2WD	5.9	6.4
4WD	5.3	5.8
1998-02	4.9	5.4
2003-05 (3.5)	4.9	5.4
Corolla FX, FX16		
SOHC	5.1	5.6
DOHC	4.2	4.7
Corolla RWD		
1981-82	7.8	8.3
1983-87		
SOHC	5.0	5.5
DOHC	4.3	4.8
Cressida	6.5	7.0
Echo (4.4)	6.2	6.7
MR2		
1985-89		
w/super charger	5.0	5.5
w/o super charger	4.3	4.8
1991-95		
w/turbo	8.3	8.8
w/o turbo	6.8	7.3
MR2 Spyder (5.5)	7.6	8.1
Paseo		
1992-95	5.3	5.8
1996-99	5.9	6.4
Prius		
2001-03	5.0	5.5
2004-05 (4.3)	6.0	6.5
Starlet	6.9	7.4
Supra		
1981-86	6.3	4.8
1987-92		
w/turbo	8.0	8.5
w/o turbo	7.0	7.5
1993-98		
w/turbo	8.1	8.6
w/o turbo	7.4	7.9
Tercel		
1981-82		
AT	4.2	4.7
MT	3.9	4.4
1983-90		
sedan	3.7	4.2
wagon	4.8	5.3

AVALON : CAMRY : CELICA : COROLLA : CRESSIDA : ECHO : MATRIX : MR2

	LABOR TIME	SEVERE SERVICE
Camshaft, Replace (A)		
1991-94	3.4	3.9
1995-99	5.9	6.4
w/AC add		
81-82 Corolla RWD	.2	.2
81-99 Tercel	.1	.1
83-87 Corolla RWD	.1	.1
85-89 MR2	.7	.7
88-92 Corolla FWD	.1	.1
90-93 Celica	.1	.1
91-95 MR2	.1	.1
92-96 Camry	.1	.1
93-05 Corolla FWD	.2	.2
94-99 Celica	.3	.3
Corolla FX, FX16	.2	.2
MR2 Spyder	.2	.2
Starlet	.5	.5
w/PS add		
81-82 Corolla RWD	.5	.5
81-85 Celica	.5	.5
83-86 Camry gas	.5	.5
83-87 Corolla FWD	.1	.1
83-87 Corolla RWD	.3	.3
86-89 Celica	.1	.1
88-97 Corolla FWD	.2	.2
95-99 Tercel	.1	.1
Echo	.2	.2
Camshaft Seal, Replace (B)		
Avalon	5.1	5.7
Camry		
1992-96		
4 cyl.	3.5	3.7
V6	4.9	5.1
1997-01		
4 cyl.	3.3	3.5
V6	4.8	5.0
2002-05 V6 (5.8)	8.0	8.5
Camry Solara		
4 cyl.		
1999-01	3.5	3.7
V6		
1999-01	4.8	5.0
2002-03 V6 (3.4)	4.8	5.3
2004-05 V6 (4.4)	6.2	6.7
Celica		
1990-93		
2WD	3.2	3.4
4WD	3.4	3.6
turbo	5.5	5.7
1994-99		
7AFE	3.1	3.3
5SFE	2.6	2.8
1993-97 Corolla FWD	3.3	3.5
Paseo	3.0	3.2
1993-98 Supra		
w/turbo	3.9	4.1
w/o turbo	3.2	3.4

	LABOR TIME	SEVERE SERVICE
1991-99 Tercel	2.8	3.0
w/AC add		
90-93 Celica	.1	.1
92-96 Camry	.1	.1
94-99 Celica	.3	.3
Tercel	.1	.1
w/PS add		
93-97 Corolla FWD	.2	.2
Tercel	.1	.1
Camshaft Timing Gear, Replace (B)		
Avalon	2.9	3.2
Camry, Camry Solara		
1983-86		
gas	2.6	2.9
diesel	.9	1.2
1987-91		
4 cyl.	2.8	3.1
V6	3.8	4.1
1992-01	3.3	3.6
2002-05 (1.9)	2.7	3.2
Celica		
1981-85	1.6	1.9
1986-89		
w/turbo	3.2	3.5
w/o turbo	2.6	2.9
1990-93		
w/turbo	5.6	5.9
w/o turbo	3.3	3.6
1994-99	2.7	3.0
2000-05 (3.6)	5.1	5.6
Corolla FWD, Matrix		
1983-87		
gasoline	2.1	2.4
diesel	.9	1.2
1988-92	3.0	3.3
1993-97		
2WD	2.6	2.9
4WD	3.1	3.4
1998-02	2.8	3.1
2003-05 (1.3)	2.0	2.2
Corolla FX, FX16		
SOHC (1.3)	2.0	2.3
DOHC (1.9)	2.8	3.1
Corolla RWD		
1981-82	4.2	4.5
1983-87	1.7	2.0
Cressida	2.0	2.3
Echo (2.9)	4.1	4.6
MR2		
1985-89		
w/super charger	2.4	2.7
w/o super charger	1.7	2.0
1991-95		
w/turbo	5.1	5.4
w/o turbo	3.5	3.8
MR2 Spyder (3.7)	5.2	5.7
Paseo	2.7	3.0
Prius (2.3)	3.3	3.8
Starlet	4.3	4.6

	LABOR TIME	SEVERE SERVICE
Supra		
1981-86	1.6	1.9
1987-92		
w/turbo	3.7	4.0
w/o turbo	2.9	3.2
1993-98	1.8	2.1
Tercel		
1981-82	1.7	2.0
1983-90		
sedan	2.1	2.4
wagon	1.3	1.6
1991-99	2.7	3.0
w/AC add		
81-82 Corolla RWD	.5	.5
81-82 Tercel	.2	.2
83-87 Corolla RWD	.1	.1
83-99 Tercel	.1	.1
85-89 MR2	1.1	1.1
88-92 Corolla FWD	.1	.1
91-95 MR2	.5	.6
93-97 Corolla	.2	.2
94-99 Celica	.3	.3
98-05 Corolla	.2	.2
MR2 Spyder	.2	.2
Starlet	.5	.5
w/PS add		
81-82 Corolla RWD	.2	.2
81-85 Celica	.1	.1
83-86 Camry gas	.5	.5
83-87 Corolla FWD	.1	.1
83-87 Corolla RWD	.3	.3
83-99 Tercel	.1	.1
86-89 Celica	.5	.5
88-92 Corolla FWD	.2	.2
Corolla FX, FX16	.1	.1
Echo	.2	.2
Crankshaft Gear or Sprocket, Replace (B)		
Camry		
1983-86		
gas	3.1	3.3
diesel	2.3	2.5
1987-91		
4 cyl.	3.0	3.2
V6	3.7	3.9
1992-96	3.0	3.2
2002-05		
4 cyl. (4.7)	6.8	7.2
V6 (1.6)	2.4	2.7
Camry Solara		
4 cyl.		
2002-03 (4.7)	6.8	7.3
2004-05 (3.8)	5.5	6.0
Celica		
1981-85		
SOHC	7.2	7.4
DOHC	8.2	8.4

MR2 SPYDER : PASEO : PRIUS : SOLARA : STARLET : SUPRA : TERCEL — TOY-23

	LABOR TIME	SEVERE SERVICE
1986-89		
SOHC	3.3	3.5
DOHC	2.9	3.1
turbo	3.1	3.3
1990-93		
w/turbo	4.5	4.7
w/o turbo	3.2	3.4
2000-05 (3.2)	4.7	5.2
Corolla FWD		
1983-87	2.2	2.4
1988-92	3.3	3.5
1993-97		
2WD	2.3	2.5
4WD	3.0	3.2
1998-02	2.8	3.0
2003-05 (2.5)	3.7	4.2
Corolla FX, FX16	2.0	2.2
Corolla RWD		
1981-82	4.9	5.1
1983-87	2.0	2.2
Cressida		
1981-88	3.5	3.7
1989-92	2.6	2.8
Echo	4.4	4.9
Matrix (2.5)	4.3	4.7
MR2		
1985-89	2.6	2.8
1991-94		
w/turbo	4.2	4.4
w/o turbo	3.6	3.8
MR2 Spyder (3.3)	4.7	5.2
1992-95 Paseo	2.8	3.0
Prius		
2001-03	5.3	5.9
Starlet		
1981-82	4.1	4.3
1983-84	4.9	5.1
Supra		
1981-86	3.4	3.6
1987-92		
w/turbo	3.7	3.9
w/o turbo	3.1	3.3
1993-98		
w/turbo	3.3	3.5
w/o turbo	2.9	3.1
Tercel		
1981-82	3.3	3.5
1983-90		
sedan	2.4	2.6
wagon	1.7	1.9
1991-99	2.5	2.7
w/AC or PS add		
83-87 Corolla FWD	.1	.1
Camry diesel	.1	.1
Corolla FX, FX16	.1	.1
w/AC add		
81-97 Celica	.3	.3
81-94 Tercel	.1	.1
81-92 Corolla RWD	.5	.5
83-87 Corolla RWD	.1	.1

	LABOR TIME	SEVERE SERVICE
85-89 MR2	1.1	1.1
88-92 Corolla	.1	.1
91-94 MR2	.5	.5
93-05 Corolla	.2	.2
95-99 Tercel	.1	.1
03-05 Matrix XRS	.2	.2
Corolla diesel	.1	.1
Starlet	.5	.5
w/PS add		
81-89 Celica	.5	.5
81-90 Tercel	.2	.2
81-92 Corolla RWD	.2	.2
83-86 Camry	.5	.5
83-87 Corolla RWD	.3	.3
88-92 Corolla	.2	.2
95-99 Tercel	.1	.1
Corolla diesel	.1	.1
Timing Belt, Replace (B)		
Avalon	3.0	3.4
Camry		
1983-86		
gas	3.4	3.9
diesel	2.6	3.1
1987-91		
4 cyl.	2.9	3.4
V6	3.7	4.2
1992-96	3.2	3.7
1997-01	2.9	3.4
2002-04 V6	2.5	2.9
Camry Solara		
1999-01	3.0	3.5
2002-03 V6	2.7	3.2
Celica		
1986-89		
SOHC	3.4	3.9
DOHC	2.6	3.1
turbo	3.1	3.6
1990-93		
w/turbo	4.3	4.8
w/o turbo	3.3	3.8
1994-99		
7AFE	2.7	3.2
5SFE	2.3	2.8
Corolla FWD		
1983-87		
SOHC	2.9	3.4
DOHC, diesel	2.4	2.9
1988-92	3.1	3.6
1993-97		
2WD	2.5	3.0
4WD	3.0	3.5
Corolla FX, FX16		
SOHC	3.1	3.6
DOHC	2.3	3.8
Corolla RWD	1.8	2.3
Cressida		
1981-88	3.2	3.7
1989-92	2.3	2.8

	LABOR TIME	SEVERE SERVICE
MR2		
1985-89	2.8	3.3
1991-95		
w/turbo	3.9	4.4
w/o turbo	3.2	3.7
Paseo	2.7	3.2
Sienna	6.0	6.7
Supra		
1981-92	3.3	3.8
1993-98		
w/turbo	3.1	3.6
w/o turbo	2.5	3.0
Tercel		
1981-82	1.8	2.3
1983-99	2.4	2.9
w/AC add		
83-87 Corolla FWD	.2	.2
86-93 Celica	.1	.1
94-99 Celica	.3	.3
MR2	.5	.5
Sienna	.2	.2
Tercel	.1	.1
w/PS add Tercel	.1	.1
Timing Belt Idler, Replace (B)		
Avalon	3.0	3.4
Camry		
1983-86	3.3	3.7
1987-91		
4 cyl.	2.8	3.2
V6	3.7	4.1
1992-01	3.1	3.5
2002-04 V6	2.5	2.9
Camry Solara	3.0	3.4
Celica		
1986-89		
SOHC	3.7	4.1
DOHC	2.8	3.2
w/turbo	3.1	3.5
1990-93		
w/turbo	4.3	4.7
w/o turbo	3.2	3.6
1994-99		
7AFE	2.6	3.0
5SFE	1.9	2.3
Corolla FWD		
1983-87	1.9	2.3
1988-92	3.2	3.6
1993-97		
2WD	2.6	3.0
4WD	3.1	3.5
Corolla FX, FX16		
SOHC	2.7	3.1
DOHC	2.1	2.5
Corolla RWD	1.3	1.7
Cressida		
1981-88	1.6	2.0
1989-92	2.4	2.8

TOY-24 AVALON : CAMRY : CELICA : COROLLA : CRESSIDA : ECHO : MATRIX : MR2

	LABOR TIME	SEVERE SERVICE
Timing Belt Idler, Replace (B)		
MR2		
1985-89	2.6	2.9
1991-95		
w/turbo	4.2	4.6
w/o turbo	3.5	3.9
Paseo	2.7	3.1
Sienna	6.0	6.7
Supra		
1981-86	1.4	1.8
1987-92	3.2	3.6
1993-98		
w/turbo	3.1	3.5
w/o turbo	2.6	3.0
Tercel		
1981-82	1.5	1.9
1983-90		
sedan	2.6	3.0
wagon	1.4	1.8
1991-99	2.3	2.7
w/AC add		
83-87 Corolla FWD	.2	.2
86-93 Celica	.1	.1
94-99 Celica	.3	.3
MR2	.5	.5
Tercel	.1	.1
w/PS add Tercel	.1	.1
Timing Belt/Chain Tensioner, Replace (B)		
Avalon	.5	.7
Camry		
1987-01	.5	.7
2002-05		
4 cyl. (.8)	1.2	1.4
V6 (1.4)	2.1	2.4
Camry Solara		
1999-01 V6	.5	.7
2002-03		
4 cyl.	1.2	1.4
V6	.5	.7
2004-05		
4 cyl. (3.0)	4.3	4.6
V6 (1.3)	1.9	2.2
Celica		
1981-85		
SOHC	7.1	7.6
DOHC	8.3	8.8
1990-93		
w/turbo	4.2	4.4
w/o turbo	3.0	3.2
2000-05 (.6)	.9	1.0
Corolla FWD		
1998-02	.7	1.2
2003-05 (.7)	1.1	1.2
1981-82 Corolla RWD	4.7	5.2
Echo (2.5)	3.7	4.2
Matrix, MR2 Spyder (.7)	1.1	1.2
1991-95 MR2	3.6	3.8

	LABOR TIME	SEVERE SERVICE
Prius		
2001-03	5.0	5.5
2004	6.6	7.3
Starlet		
1981-82	4.0	4.5
1983-84	4.6	5.1
1993-98 Supra	3.0	3.2
w/AC add		
81-82 Celica	.3	.3
81-92 Corolla RWD	.5	.5
98-05 Corolla	.2	.2
MR2	.5	.5
Starlet	.5	.5
w/PS add		
81-82 Celica	.5	.5
81-92 Corolla RWD	.2	.2
Echo	.2	.2
Timing Chain, Replace (B)		
Includes: Engine R&R when required.		
Camry, Camry Solara		
2002-05 2.4L (4.7)	6.8	7.3
Celica		
1981-82		
SOHC	7.1	7.7
DOHC	8.1	8.7
2000-05 (3.2)	4.7	5.2
Corolla FWD, Matrix		
1998-02	2.8	3.4
2003-05 (2.5)	3.7	4.2
Corolla FX, FX16	5.1	5.7
1981-82 Corolla RWD	5.1	5.7
1981-82 Cressida	10.5	11.1
Echo (2.8)	4.1	4.6
MR2 Spyder (3.3)	4.8	5.3
Prius		
2001-03	5.3	5.9
2004-05 (4.6)	6.6	7.1
Starlet		
1981-82	4.1	4.7
1983-84	4.7	5.3
w/AC add		
81-82 Celica	.3	.3
81-92 Corolla RWD	.5	.5
98-05 Corolla	.2	.2
00-05 MR2 Spyder	.2	.2
Starlet	.5	.5
w/PS add		
81-82 Celica	.5	.5
81-92 Corolla RWD	.2	.2
Echo	.2	.2
Timing Cover and/or Gaskets, Replace (B)		
Avalon	2.6	2.9
Camry		
1987-91	3.3	3.6
1992-96		
4 cyl.	2.2	2.5
V6	3.2	3.5

	LABOR TIME	SEVERE SERVICE
1997-01		
4 cyl.	2.2	2.5
V6	2.7	3.0
2002-05		
4 cyl. (3.9)	5.7	6.2
V6 (.7)	1.1	1.2
Camry Solara		
1999-01		
4 cyl.	2.1	2.4
V6	2.7	3.0
2002-03		
4 cyl.	4.8	5.4
V6	2.6	2.9
2004-05		
4 cyl. (3.5)	5.1	5.6
V6 (.9)	1.3	1.5
Celica		
1981-85		
SOHC	8.7	9.0
DOHC	9.7	10.0
1986-93	2.5	2.8
turbo	3.6	3.9
1994-99		
7AFE	2.5	2.8
5SFE	1.9	2.2
2000-05 (2.9)	4.3	4.8
Corolla FWD		
1993-02	1.8	2.1
2003-05 (2.2)	3.3	3.8
1981-82 Corolla RWD	4.2	4.5
Cressida		
1981-88	3.1	3.4
1989-92	2.1	2.4
Matrix (2.2)	3.3	3.8
MR2 Spyder (3.0)	4.4	4.9
Paseo		
1992-95	1.8	2.1
1996-99	2.4	2.7
Starlet	3.6	3.9
Supra		
1981-92	3.2	3.5
1993-98	2.4	2.7
Tercel		
1991-94	1.6	1.9
1995-99	2.1	2.4
w/AC add		
81-82 Corolla RWD	.5	.5
81-85 Celica	1.2	1.2
86-93 Celica	.1	.1
91-99 Tercel	.1	.1
93-97 Corolla FWD	.2	.2
94-99 Celica	.3	.3
98-05 Corolla	.2	.2
Matrix 4WD	.2	.2
MR2 Spyder	.2	.2
Starlet	.5	.5
w/PS add		
81-82 Corolla RWD	.5	.5
81-85 Celica	.5	.5
91-99 Tercel	.1	.1

MR2 SPYDER : PASEO : PRIUS : SOLARA : STARLET : SUPRA : TERCEL — TOY-25

	LABOR TIME	SEVERE SERVICE
CRANK & PISTONS		
Connecting Rod Bearings, Replace (A)		
Includes: Check bearing oil clearance.		
Avalon	7.8	8.7
Camry		
1983-86		
gas	3.8	4.3
diesel	5.4	5.9
1987-91		
4 cyl.	4.1	4.6
V6	5.2	5.7
1992-93		
4 cyl.	5.7	6.4
V6	6.5	7.2
1994-01		
4 cyl.	6.4	6.4
V6	7.5	8.3
2002-05		
4 cyl. (17.2)	22.4	22.9
V6 (12.2)	16.3	16.8
Camry Solara		
1999-01		
4 cyl.	5.7	6.4
V6	7.5	8.3
2002-03		
4 cyl.	22.6	24.9
V6	7.5	8.3
2004-05		
4 cyl. (15.9)	20.7	21.2
V6 (12.1)	16.2	16.7
Celica		
1981-85	4.3	4.8
1986-89	4.4	4.9
1990-93		
w/turbo	6.0	6.5
w/o turbo	5.4	5.9
1994-99		
7AFE	6.6	7.1
5SFE	5.3	5.8
2000-05 (2.8)	4.1	4.6
Corolla FWD		
1983-87		
gasoline	4.5	5.0
diesel	5.2	5.7
1988-92		
SOHC	5.0	5.5
DOHC	6.1	6.6
1993-97		
2WD		
SOHC	5.4	5.9
DOHC	6.8	7.3
4WD	4.5	5.0
1998-02	6.1	6.6
2003-04	17.1	19.1
Corolla FX, FX16	4.6	5.1

	LABOR TIME	SEVERE SERVICE
Corolla RWD		
1981-82	4.0	4.5
1983-87		
SOHC	4.3	4.8
DOHC	5.3	5.8
Cressida	7.7	8.2
Echo (5.7)	8.2	8.7
Matrix		
2WD (11.3)	15.3	15.8
4WD (12.6)	16.9	17.4
XRS (11.9)	16.0	16.5
MR2		
1985-89	5.2	5.7
1991-95		
w/turbo	6.0	6.5
w/o turbo	5.2	5.7
MR2 Spyder (2.7)	4.0	4.5
Paseo	5.4	5.9
Prius		
2001-03	17.6	19.6
2004-05 (15.4)	20.1	20.6
Starlet		
1981-82	3.6	4.1
1983-84	4.1	4.6
Supra		
1981-86	4.9	5.4
1987-92	7.7	8.2
1993-98		
w/turbo	20.8	21.3
w/o turbo	17.2	17.7
Tercel		
1981-82	4.3	4.8
1983-90		
sedan	4.3	4.8
wagon	3.4	3.9
1991-99	5.0	5.5
w/AC add		
04-05 Corolla	.2	.2
Matrix	.3	.3
MR2	.5	.5
Paseo	.2	.2
Prius	.3	.3
Tercel	.2	.2
w/PS add		
Echo	.2	.2
Tercel	.1	.1
Crankshaft and Main Bearings, Replace (A)		
Includes: Engine R&R.		
Avalon	23.2	25.8
Camry		
1983-86		
gas	14.2	15.1
diesel	15.3	16.2
1987-91		
4 cyl.		
2WD	16.6	17.5
4WD	18.5	19.4
V6	23.5	24.4

	LABOR TIME	SEVERE SERVICE
1992-96		
4 cyl.	16.4	17.3
V6	19.1	20.0
1997-01		
4 cyl.	17.4	18.3
V6	22.0	22.9
2002-05		
4 cyl. (17.2)	22.4	22.9
V6 (18.7)	24.3	24.8
Camry Solara		
1999-01		
4 cyl.	17.5	18.4
V6	22.2	23.1
2002-03		
4 cyl.	24.6	27.3
V6	26.6	29.4
2004-05		
4 cyl. (15.6)	20.3	20.8
V6 (15.2)	19.8	20.3
Celica		
1981-85		
SOHC	13.6	14.5
DOHC	14.3	15.2
1986-89		
SOHC	14.3	15.2
DOHC	18.1	19.0
turbo	22.0	22.9
1990-93		
2WD	16.7	17.6
4WD	17.4	18.3
turbo	19.0	19.9
1994-99		
7AFE	15.9	16.8
5SFE	15.3	16.2
2000-05 (14.8)	19.4	19.9
Corolla FWD		
1983-87		
SOHC	14.1	15.0
DOHC	16.1	17.0
diesel	15.1	16.0
1988-92		
2WD	16.8	17.7
4WD	17.3	18.2
1993-97		
2WD	15.8	16.7
4WD	18.2	19.1
1998-02	17.2	18.1
2003-05 (13.5)	17.9	18.4
Corolla FX, FX16		
SOHC	14.0	14.9
DOHC	16.1	17.0
Corolla RWD		
1981-82	13.0	13.9
1983-87		
SOHC	13.8	14.7
DOHC	14.7	15.6
1981-88	18.0	18.9
1989-92	18.7	19.6
Echo (12.5)	16.8	17.3

TOY-26   AVALON : CAMRY : CELICA : COROLLA : CRESSIDA : ECHO : MATRIX : MR2

	LABOR TIME	SEVERE SERVICE
Crankshaft and Main Bearings, Replace (A)		
Matrix		
2WD (13.5)	17.9	18.4
4WD (15.4)	20.0	20.5
XRS (12.7)	17.0	17.5
MR2		
1985-89		
w/super charger	20.3	21.2
w/o super charger	17.0	17.9
1991-95		
w/turbo	19.1	20.0
w/o turbo	16.5	17.4
MR2 Spyder		
5-Speed MT (13.5)	17.9	18.4
sequential MT (14.1)	18.6	19.1
Paseo		
1992-95	17.1	18.0
1996-99	16.3	17.2
Prius		
2001-03 (14.3)	18.8	19.3
2001-03 (18.2)	23.7	24.2
Starlet		
1981-82	13.6	14.5
1983-84	14.1	15.0
Supra		
1981-86	18.0	18.9
1987-92	18.7	19.6
1993-98		
w/turbo	21.6	22.5
w/o turbo	18.3	19.2
Tercel		
1981-82		
AT	13.8	14.7
MT	13.2	14.1
1983-90		
sedan	16.3	17.2
wagon	13.4	14.3
1991-94	13.8	14.7
1995-99	16.1	17.0
w/AC add		
81-90 Tercel	1.3	1.3
81-85 Celica	1.1	1.1
81-92 Corolla RWD	.5	.5
83-86 Camry diesel	.2	.2
83-86 Camry gas	.9	.9
83-87 Corolla diesel	.2	.2
83-87 Corolla FWD	1.3	1.3
83-87 Corolla RWD	1.1	1.1
85-89 MR2	.3	.3
86-89 Celica turbo	1.6	1.6
86-89 Celica	.1	.1
87-91 Camry 4 cyl	.2	.2
87-91 Camry V6	1.6	1.6
88-92 Corolla FWD	.3	.3
90-93 Celica	.2	.2
91-94 Tercel	.3	.3
91-95 MR2	.5	.5
92-96 Camry 4 cyl.	.2	.2

	LABOR TIME	SEVERE SERVICE
92-96 Camry V6	.5	.5
94-99 Celica	.2	.2
95-99 Tercel	.2	.2
98-05 Corolla FWD	.2	.2
00-05 Celica	.2	.2
Corolla FX, FX16	1.1	1.1
Echo	.3	.3
Matrix, Prius	.3	.3
MR2 Spyder	.2	.2
Paseo	.2	.2
Starlet	.5	.5
w/PS add		
81-85 Celica	.3	.3
81-92 Corolla RWD	.5	.5
83-86 Camry diesel	.3	.3
83-86 Camry gas	.5	.5
83-87 Corolla diesel	.3	.3
83-87 Corolla FWD	.3	.3
83-87 Corolla RWD	.3	.3
83-90 Tercel	.3	.3
86-89 Celica	.8	.8
88-92 Corolla	.2	.2
91-94 Tercel	.3	.3
93-97 Corolla FWD	.2	.2
95-99 Tercel	.2	.2
00-05 Celica	.3	.3
Corolla FX, FX16	.3	.3
Echo	.5	.5
Crankshaft Front Oil Seal, Replace (B)		
Avalon	3.8	4.3
Camry		
1983-86	4.7	4.9
1987-91		
4 cyl.		
2WD	3.3	3.5
4WD	5.7	5.9
V6	7.3	7.5
1992-96		
4 cyl.	3.1	3.3
V6	6.3	6.5
1997-01	3.2	3.4
2002-05		
4 cyl. (.7)	1.1	1.2
V6 (1.6)	2.4	2.5
Camry Solara		
1999-01		
4 cyl.	3.0	3.2
V6	3.4	3.6
2002-03		
4 cyl.	1.1	1.2
V6	3.6	4.1
2004-05		
4 cyl. (.7)	1.1	1.2
V6 (5.2)	3.6	4.1
Celica		
1981-85	2.3	2.5
1986-89		
SOHC	5.2	5.4
DOHC	5.8	6.0
turbo	6.6	6.8

	LABOR TIME	SEVERE SERVICE
1990-93		
w/turbo	4.6	4.8
w/o turbo	3.2	3.4
1994-99	2.4	2.6
2000-05 (.8)	1.3	1.4
Corolla FWD		
1988-92		
SOHC	6.9	7.1
DOHC	5.7	5.9
1993-97		
2WD	2.7	2.9
4WD	5.7	5.9
1998-02	2.5	2.7
2003-04	1.6	1.8
Corolla FX, FX16	4.2	4.4
Corolla RWD		
1981-82	2.0	2.2
1983-87	4.6	4.8
Cressida		
1981-88	3.5	3.7
1989-92	2.2	2.4
Echo (.7)	1.1	1.2
Matrix		
exc. XRS (1.0)	1.5	1.6
XRS (.7)	1.1	1.2
MR2		
1985-89		
w/super charger	6.0	6.2
w/o super charger	5.5	5.7
1991-95		
w/turbo	4.1	4.3
w/o turbo	3.4	3.6
MR2 Spyder (.9)	1.4	1.6
Paseo	2.9	3.1
Prius (.8)	1.3	1.4
Starlet	1.8	2.0
Supra	3.2	3.4
Tercel		
1981-82		
MT	5.4	5.6
AT	5.0	5.2
1983-90		
sedan	5.4	5.6
wagon	4.1	4.3
1991-94	5.1	5.3
1995-99	2.3	2.5
w/AC add		
81-82 Corolla RWD	.1	.1
81-82 Tercel	.1	.1
83-91 Camry	.1	.1
83-87 Corolla RWD	.1	.1
83-90 Tercel	.2	.2
85-89 MR2	1.3	1.3
88-92 Corolla FWD	.1	.1
90-93 Celica	.1	.1
91-94 Tercel	.1	.1
91-95 MR2	.5	.5
93-97 Corolla FWD	.2	.2
94-99 Celica	.3	.3
95-99 Tercel	.1	.1

	LABOR TIME	SEVERE SERVICE
98-05 Corolla	.2	*.2*
Corolla diesel	.1	*.1*
Corolla FWD	.1	*.1*
Corolla FX, FX16	.1	*.1*
Starlet	.1	*.1*
w/PS add		
81-87 Corolla RWD	.2	*.2*
81-89 Celica	.2	*.2*
83-86 Camry	.1	*.1*
83-90 Tercel	.2	*.2*
88-97 Corolla FWD	.2	*.2*
91-99 Tercel	.1	*.1*
98-05 Corolla FWD	.1	*.1*
Corolla diesel	.1	*.1*
Corolla FX, FX16	.1	*.1*
Echo	.2	*.2*

Crankshaft Pulley, Replace (B)

	LABOR TIME	SEVERE SERVICE
Avalon	.8	*.9*
Camry		
1983-86	1.2	*1.4*
1987-91	1.7	*1.9*
1992-01	.7	*.9*
2002-05 (.7)	1.1	*1.2*
Camry Solara		
1999-03 (.6)	.9	*1.0*
2004-05		
4 cyl. (.5)	.7	*.8*
V6 (1.0)	1.5	*1.6*
Celica		
1981-85	2.0	*2.2*
1986-93	1.6	*1.8*
1994-05 (.7)	1.1	*1.2*
Corolla FWD		
1983-97	1.5	*1.7*
1998-03	.9	*1.0*
Corolla FX, FX16	1.6	*1.8*
Corolla RWD		
1981-82	1.8	*2.0*
1983-87	1.2	*1.4*
Cressida	2.7	*2.9*
Echo, Matrix, Prius (.6)	.9	*1.0*
MR2		
1985-89	1.5	*1.7*
1991-95		
w/turbo	2.2	*2.4*
w/o turbo	1.8	*2.0*
MR2 Spyder (.7)	1.1	*1.2*
Paseo	1.4	*1.6*
Starlet	1.7	*1.9*
Supra		
1981-92	2.7	*2.9*
1993-98	2.3	*2.5*
Tercel		
1981-82	2.4	*2.6*
1983-90	1.7	*1.9*
1991-99	1.3	*1.5*
w/AC add		
81-82 Corolla RWD	.1	*.1*
81-82 Tercel	.1	*.1*
83-91 Camry	.1	*.1*
83-87 Corolla RWD	.1	*.1*
83-90 Tercel	.2	*.2*
85-89 MR2	1.3	*1.3*
88-92 Corolla FWD	.1	*.1*
90-93 Celica	.1	*.1*
91-94 Tercel	.1	*.1*
91-95 MR2	.5	*.5*
93-97 Corolla FWD	.2	*.2*
94-99 Celica	.3	*.3*
95-99 Tercel	.1	*.1*
98-05 Corolla	.2	*.2*
Corolla diesel	.1	*.1*
Corolla FWD	.1	*.1*
Corolla FX, FX16	.1	*.1*
Starlet	.1	*.1*
w/PS add		
81-87 Corolla RWD	.2	*.2*
81-89 Celica	.2	*.2*
83-86 Camry	.1	*.1*
83-90 Tercel	.2	*.2*
88-97 Corolla FWD	.2	*.2*
91-99 Tercel	.1	*.1*
98-05 Corolla FWD	.1	*.1*
Corolla diesel	.1	*.1*
Corolla FX, FX16	.1	*.1*
Echo	.2	*.2*

Pistons and/or Connecting Rods, Replace (A)

Includes: Ridge reaming, cylinder wall deglazing, installing new rings and rod bearings, engine tune-up.

	LABOR TIME	SEVERE SERVICE
Avalon	18.6	*20.8*
Camry		
1983-86		
gas	11.6	*12.5*
diesel	12.3	*13.2*
1987-91		
4 cyl.	11.9	*12.8*
V6	17.8	*18.7*
1992-96		
4 cyl.	13.7	*14.6*
V6	18.8	*19.7*
1997-01		
4 cyl.	13.7	*14.6*
V6	19.2	*20.1*
2002-05		
4 cyl. (17.0)	22.1	*22.6*
V6 (18.3)	23.8	*24.2*
Camry Solara		
1999-01		
4 cyl.	13.6	*14.5*
V6	19.2	*20.1*
2002-03		
4 cyl.	24.0	*26.4*
V6	19.7	*21.7*
2004-05		
4 cyl. (16.1)	20.9	*21.4*
V6 (18.2)	23.7	*24.2*
Celica		
1981-85		
SOHC	9.9	*10.8*
DOHC	10.5	*11.4*
1986-89		
SOHC	11.8	*12.7*
DOHC	10.5	*11.4*
turbo	15.9	*16.8*
1990-93		
2WD	11.6	*12.5*
4WD	12.2	*13.1*
turbo	16.9	*17.8*
1994-99		
7AFE	14.7	*15.6*
5SFE	12.6	*13.5*
2000-05 (9.4)	13.0	*13.5*
Corolla FWD		
1983-87		
SOHC	9.9	*10.8*
DOHC	11.7	*12.6*
diesel	12.4	*13.3*
1988-92		
SOHC	10.4	*11.3*
DOHC	13.1	*14.0*
1993-97		
2WD		
SOHC	11.6	*12.5*
DOHC	13.0	*13.9*
4WD	10.5	*11.4*
1998-05 (12.2)	16.3	*16.8*
Corolla FX, FX16		
SOHC	9.9	*10.8*
DOHC	11.9	*12.8*
Corolla RWD		
1981-82	9.3	*10.2*
1983-87		
SOHC	9.9	*10.8*
DOHC	10.3	*11.1*
Cressida		
1981-88	16.7	*17.6*
1989-92	14.6	*15.5*
Echo (8.8)	12.2	*12.7*
Matrix		
2WD (12.1)	16.2	*16.7*
4WD (14.0)	18.5	*19.0*
XRS (12.4)	16.6	*17.1*
MR2		
1985-89	12.6	*13.5*
1991-95		
w/turbo	16.3	*17.2*
w/o turbo	11.7	*11.6*
MR2 Spyder		
5-Speed MT (12.6)	16.9	*17.4*
sequential MT (13.2)	17.6	*18.1*
Paseo	13.2	*14.1*
Prius		
2001-03	20.7	*23.1*
2004-05 (19.6)	25.5	*26.0*

TOY-28 AVALON : CAMRY : CELICA : COROLLA : CRESSIDA : ECHO : MATRIX : MR2

	LABOR TIME	SEVERE SERVICE		LABOR TIME	SEVERE SERVICE		LABOR TIME	SEVERE SERVICE
Pistons and/or Connecting Rods, Replace (A)			4WD			1988-92		
Starlet			AT 10.5		11.1	2WD		
1981-82 .... 9.3		10.2	MT 9.1		9.7	AT 8.5		9.1
1983-84 .... 9.8		10.7	V6 .... 8.4		9.0	MT 7.2		7.8
Supra			1992-01			4WD		
1981-86 .... 10.1		11.0	AT 9.3		9.9	AT 9.5		10.1
1987-92 .... 16.8		17.7	MT 7.8		8.4	MT 8.2		8.8
1993-98			2002-05			1993-97		
w/turbo 19.9		20.8	4 cyl.			2WD 5.8		6.4
w/o turbo 17.7		18.6	AT (8.5) 11.8		12.3	4WD		
Tercel			MT (8.2) 11.4		11.9	AT 9.4		10.0
1981-82 .... 9.7		10.6	V6 (8.9) 12.4		12.9	MT 8.3		8.9
1983-90			Camry Solara			1998		
sedan 8.0		8.9	1999-01			AT 7.5		8.1
wagon 9.8		10.7	AT 9.1		9.7	MT 6.4		7.0
1991-94 .... 9.9		10.8	MT 7.6		8.2	1999-02		
1995-99 .... 13.0		13.9	2002-03			AT 6.7		7.3
w/AC add			4 cyl.			MT 5.5		6.1
81-82 Corolla RWD 1		.1	AT 12.4		13.8	2003-05		
87-96 Camry 1		.1	MT 12.0		13.3	AT (6.8) 9.6		10.1
90-99 Celica 1		.1	V6			MT (5.3) 7.6		8.1
91-94 Tercel 1		.1	AT 9.6		10.7	Corolla FX, FX16		
93-97 Corolla 1		.1	MT 8.2		9.1	AT 6.4		7.0
95-99 Tercel 1		.1	2004-05			MT 5.7		6.3
98-05 Corolla 2		.2	4 cyl.			Corolla RWD		
Matrix, Prius 3		.3	AT (9.3) 12.8		13.3	1981-82		
MR2 5		.5	MT (8.8) 12.2		12.7	AT 8.6		9.2
MR2 Spyder 2		.2	V6			MT 7.7		8.3
Paseo 1		.1	AT (8.5) 11.8		12.3	1983-87		
w/PS add			Celica			AT 6.2		6.8
81-82 Corolla RWD 5		.5	1981-82			MT 5.4		6.0
81-85 Celica 5		.5	AT 8.6		9.2	Cressida		
81-90 Tercel 2		.2	MT 7.9		8.5	1981-82 .... 9.3		9.9
83-86 Camry 5		.5	1983-85			1983-88		
83-87 Corolla diesel 2		.2	AT 6.1		6.7	AT 6.0		6.6
83-87 Corolla FWD 1		.1	MT 5.3		5.9	MT 5.3		5.9
83-87 Corolla RWD 3		.3	1986-89			1989-92 .... 5.9		6.5
86-89 Celica 2		.2	SOHC			Echo (4.5) 6.5		7.0
88-92 Corolla FWD 2		.2	AT 7.7		8.3	Matrix		
91-94 Tercel 2		.2	MT 7.0		7.6	2WD		
93-97 Corolla 2		.2	DOHC			AT (6.8) 9.6		10.1
95-99 Tercel 3		.3	AT 8.3		8.9	MT (5.4) 7.7		8.2
00-05 Celica 3		.3	MT 7.7		8.3	4WD (8.4) 11.7		12.2
Corolla FX, FX16 1		.1	turbo 12.3		13.1	XRS		
Echo 2		.2	1990-93			AT (4.9) 7.1		7.6
Rear Main Oil Seal, Replace (B)			AT 7.3		8.1	MT (5.7) 8.2		8.7
Includes: Trans. R&R. when necessary.			MT 6.9		7.7	MR2		
Avalon 10.0		11.2	turbo 10.1		10.7	1985-89		
Camry			1994-99			w/super charger		
1983-86			AT 7.7		8.3	AT 8.1		8.7
gasoline 6.4		7.0	MT 7.2		7.8	MT 7.1		7.7
diesel			2000-05 (5.1) 7.3		7.8	w/o super charger		
AT 6.5		7.1	Corolla FWD			AT 6.9		7.5
MT 5.9		6.5	1983-87			MT 6.3		6.9
1987-91			gasoline			1991-95		
4 cyl.			AT 6.6		7.2	AT 9.0		9.6
2WD			MT 5.6		6.2	MT 8.8		9.4
AT 7.6		8.2	diesel			turbo 10.3		10.9
MT 9.1		9.7	AT 6.4		7.0			
			MT 6.1		6.7			

MR2 SPYDER : PASEO : PRIUS : SOLARA : STARLET : SUPRA : TERCEL   TOY-29

	LABOR TIME	SEVERE SERVICE		LABOR TIME	SEVERE SERVICE		LABOR TIME	SEVERE SERVICE
MR2 Spyder			2002-05			MR2		
5-Speed MT (3.9)	5.7	6.2	4 cyl. (17.5)	22.8	23.3	1985-89	13.1	14.0
sequential MT (4.5)	6.5	7.0	V6 (18.3)	23.8	24.3	1991-95		
Paseo			Camry Solara			w/turbo	14.9	15.8
1992-95			1999-01			w/o turbo	11.6	12.5
AT	6.7	7.3	4 cyl.	12.9	13.8	MR2 Spyder		
MT	6.0	6.6	V6	17.5	18.4	5-Speed MT (12.6)	16.9	17.4
1996-99			2002-03			sequential		
AT	7.4	8.0	4 cyl.	22.9	25.2	MT (13.2)	17.6	18.1
MT	5.8	6.4	V6	18.3	20.1	Paseo	10.8	11.7
Prius			2004-05			Prius		
2001-03	12.1	13.5	4 cyl. (15.8)	20.5	21.0	2001-03	20.7	23.1
2004-05 (6.1)	8.7	9.2	V6 (16.6)	21.6	22.1	2004-05 (18.1)	23.5	24.0
Starlet	5.9	6.5	Celica			Starlet		
Supra			1981-85			1981-82	9.7	10.6
1981-92			SOHC	10.6	11.5	1983-84	10.1	11.0
AT	6.0	6.6	DOHC	11.8	12.7	Supra		
MT	5.1	5.7	1986-89			1981-86	11.6	12.5
1993-98			SOHC	12.4	13.3	1987-92	17.6	18.5
AT			DOHC	10.9	11.8	1993-98		
w/turbo	6.2	6.8	turbo	16.5	17.4	w/turbo	22.3	23.2
w/o turbo	6.8	7.4	1990-93			w/o turbo	18.1	19.0
MT			2WD	11.8	12.7	Tercel		
w/turbo	4.9	5.5	4WD	12.2	13.1	1981-82	10.5	11.4
w/o turbo	4.5	5.1	turbo	16.8	17.7	1983-99	10.1	11.0
Tercel			1994-99			w/AC add		
1981-82			7AFE	13.5	14.4	81-82 Corolla RWD	.1	.1
AT	8.3	8.9	5SFE	11.8	12.7	87-96 Camry	.1	.1
MT	7.3	7.9	2000-05 (9.4)	13.0	13.5	90-99 Celica	.1	.1
1983-90 sedan			Corolla FWD			91-94 Tercel	.1	.1
AT	7.9	8.5	1983-87			93-97 Corolla	.1	.1
MT	5.7	6.3	SOHC	10.5	11.4	95-99 Tercel	.1	.1
1985-88 wagon	6.5	7.1	DOHC	12.5	13.4	98-05 Corolla	.2	.2
1991-94			diesel	10.9	11.8	Matrix, Prius	.3	.3
AT	6.8	7.4	1988-92			MR2	.5	.5
MT	5.8	6.4	SOHC	11.1	12.0	MR2 Spyder	.2	.2
1995-99			DOHC	12.3	13.2	Paseo	.1	.1
AT	7.5	8.1	1993-97			w/PS add		
MT	5.8	6.4	2WD			81-82 Corolla RWD	.5	.5
w/AC add			SOHC	11.6	12.5	81-85 Celica	.5	.5
Matrix 4WD	.2	.2	DOHC	12.9	13.8	81-90 Tercel	.2	.2
Prius	.3	.3	4WD	11.3	12.2	83-86 Camry	.5	.5
Rings, Replace (A)			1998-05 (11.7)	15.8	16.3	83-87 Corolla diesel	.2	.2
Includes: Ridge reaming, cylinder wall deglazing, installing new rings, engine tune-up.			Corolla FX, FX16			83-87 Corolla FWD	.1	.1
			SOHC	10.4	11.3	83-87 Corolla RWD	.3	.3
			DOHC	12.6	13.5	86-89 Celica	.2	.2
Avalon	18.9	21.1	Corolla RWD			88-92 Corolla FWD	.2	.2
Camry			1981-82	9.9	10.8	91-94 Tercel	.2	.2
1983-86			1983-87			93-97 Corolla	.2	.2
gas	11.5	12.4	SOHC	10.6	11.5	95-99 Tercel	.3	.3
diesel	10.9	11.8	DOHC	12.9	13.8	00-05 Celica	.3	.3
1987-91			Cressida			Corolla FX, FX16	.1	.1
4 cyl.	10.5	11.4	1981-88	17.4	18.3	Echo	.2	.2
V6	15.9	16.8	1989-92	13.4	14.3			
1992-96			Echo (8.9)	12.4	12.9	**ENGINE LUBRICATION**		
4 cyl.	12.5	13.4	Matrix			**Diagnose Oil Control Valve Circuit Malfunction (A)**		
V6	16.8	17.7	2WD (11.7)	15.8	16.3			
1997-01			4WD (13.5)	17.9	18.4	All Models (.6)	.8	.8
4 cyl.	12.7	13.6	XRS (12.4)	16.6	17.1			
V6	17.7	18.6						

TOY-30 AVALON : CAMRY : CELICA : COROLLA : CRESSIDA : ECHO : MATRIX : MR2

	LABOR TIME	SEVERE SERVICE
Engine Oil Cooler, Replace (B)		
1992-01 Camry	1.9	2.2
Camry Solara	2.1	2.4
Celica		
1989	4.1	4.3
1990-93		
w/turbo	1.9	2.2
w/o turbo	2.5	2.8
1994-99	2.1	2.4
1987-96 Corolla FWD	1.1	1.4
1988 Corolla FX, FX16	1.2	1.5
MR2		
1985-89	1.2	1.5
1991-95	2.5	2.8
Supra		
1987-92	1.3	1.6
1993-98	4.2	4.5
Engine Oil Pressure Switch (Sending Unit), Replace (B)		
Exc. below (.4)	.6	.7
Corolla RWD		
1983-87 DOHC	.8	.9
Echo (.5)	.8	.9
Matrix (.2)	.3	.3
1988-89 MR2	.8	.9
1993-98 Supra	1.7	1.7
w/AC add		
87-93 Corolla FWD	.1	.1
Corolla RWD	.5	.5
w/PS add		
87-92 Corolla FWD		
2WD	.2	.2
Corolla RWD	.5	.5
Oil Pan and/or Gasket, Replace (B)		
Avalon		
No.1	5.3	6.0
No.2	3.3	3.7
Camry		
1983-86		
gas	2.9	3.1
diesel	3.3	3.5
1987-91	3.5	3.7
1992-01		
4 cyl.	3.6	3.8
V6		
No.1	4.8	5.0
No.2	3.0	3.2
2002-05		
4 cyl. (1.9)	2.8	3.3
V6		
No.1 (3.0)	4.4	4.9
No.2 (1.9)	2.8	3.3
Camry Solara		
1999-01		
4 cyl.	3.6	3.8
V6		
No.1	5.0	5.2
No.2	3.1	3.3
2002-03		
4 cyl.	3.0	3.4
V6		
No.1	5.3	6.0
No.2	3.3	3.7
2004-05		
4 cyl. (1.9)	2.8	3.3
V6		
No.1 (4.4)	6.4	6.9
No.2 (3.1)	4.5	5.0
Celica		
1981-85	2.7	2.9
1986-89	3.3	3.5
1990-93		
SOHC	3.7	3.9
DOHC	4.5	4.7
1994-99		
7AFE		
No.1	5.3	5.5
No.2	2.9	3.1
5SFE	4.0	4.2
2000-05 (1.9)	2.8	3.3
Corolla FWD		
1983-87	3.3	3.5
1988-92	3.5	3.7
1993-97	4.1	4.3
1998-02	4.3	4.5
2003-05 (2.0)	3.0	3.5
Corolla FX, FX16		
SOHC	2.9	3.1
DOHC	3.4	3.6
Corolla RWD		
SOHC	3.1	3.3
DOHC	4.0	4.2
Cressida		
1981-88	3.2	3.4
1989-92	5.4	5.6
Echo		
No. 1 (5.0)	7.2	7.7
No. 2 (1.4)	2.1	2.6
Matrix		
exc. XRS (2.0)	3.0	3.5
XRS (1.7)	2.5	3.0
MR2		
1985-89	3.8	4.0
1991-95		
w/turbo	4.9	5.1
w/o turbo	3.7	3.9
MR2 Spyder (1.6)	2.4	2.9
Paseo	3.8	4.0
Prius		
2001-03		
No. 1	8.8	9.9
No. 2	2.7	3.0
2004-05		
No. 1 (13.8)	18.2	18.7
No. 2 (1.4)	2.1	2.6
Starlet	2.7	2.9
Supra		
1981-86	3.0	3.2
1987-92	5.4	5.6
1993-98		
w/turbo	13.0	13.2
w/o turbo	9.9	10.1
Tercel		
1981-82	3.2	3.4
1983-90		
sedan	2.5	2.7
wagon	3.2	3.4
1991-94	3.2	3.4
1995-99	3.6	3.8
w/AC Echo, Prius add	.3	.3
w/PS Echo add	.2	.2
Oil Pressure Gauge (Dash), Replace (B)		
Celica		
1981-85	1.9	1.9
1986-89	1.3	1.3
Corolla RWD	.8	.8
1985-89 MR2	1.3	1.3
Supra		
1981-85	1.8	1.8
1986-92	1.4	1.4
Oil Pump, Replace (B)		
Avalon	8.1	9.0
Camry, Camry Solara		
1983-86		
gas	5.1	5.6
diesel	5.5	6.0
1987-91		
4 cyl.	5.8	6.3
V6	7.0	7.5
1992-01		
4 cyl.	5.3	5.8
V6	7.4	7.9
2002-05		
4 cyl. (4.4)	6.4	6.9
V6 (5.2)	7.5	8.0
Celica		
1986-89		
SOHC	5.3	5.8
DOHC	5.8	6.3
turbo	8.0	8.5
1990-93		
SOHC	5.9	6.4
DOHC	8.0	8.5
1994-99		
7AFE	7.8	8.3
5SFE	5.3	5.8
2000-05 (3.3)	4.8	5.3
Corolla FWD		
1983-87		
SOHC	4.6	5.1
DOHC	5.1	5.6
diesel	4.8	5.3
1988-92		
SOHC	5.8	6.3
DOHC	6.9	7.4

MR2 SPYDER : PASEO : PRIUS : SOLARA : STARLET : SUPRA : TERCEL TOY-31

	LABOR TIME	SEVERE SERVICE		LABOR TIME	SEVERE SERVICE		LABOR TIME	SEVERE SERVICE
1993-97			V6	7.3	7.5	**Clutch Control Cable, Replace (B)**		
SOHC	5.5	6.0	1992-01	7.2	7.4	Starlet	.7	1.0
DOHC	7.0	7.5	2002-05 (7.6)	10.6	11.1	Tercel wagon	.5	.8
1998-02	3.5	4.0	Camry Solara			**Clutch Master Cylinder, Replace (B)**		
2003-05 (2.6)	3.8	4.3	1999-01	7.1	7.3	Includes: System bleeding.		
Corolla FX, FX16			2002-03			Camry		
SOHC	4.5	5.0	4 cyl.	11.6	13.0	1983-01	1.6	1.8
DOHC	5.1	5.6	V6	7.8	8.7	2002-05 (1.8)	2.7	2.9
Corolla RWD			2004-05			Camry Solara		
SOHC	4.4	4.9	4 cyl. (8.9)	12.4	12.9	1999-01	1.5	1.7
DOHC	5.3	5.8	Celica			2002-05		
Cressida			1981-85	4.5	4.7	4 cyl. (1.8)	2.7	2.9
1981-88	3.3	3.8	1986-90			V6 (.9)	1.4	1.6
1989-92	5.8	6.3	2WD	6.8	7.0	Celica		
Echo (2.4)	3.5	4.0	4WD	11.6	11.8	1981-89	1.6	1.8
Matrix (2.6)	3.8	4.3	1991-93			1990-99	1.8	2.0
MR2			2WD	5.6	5.8	2000-05 (.8)	1.2	1.4
1985-89			4WD	9.7	9.9	Corolla FWD		
w/super charger	5.8	6.3	1994-05 (4.8)	6.9	7.4	1981-87	1.4	1.6
w/o super charger	5.2	5.7	Corolla FWD			1988-91		
1991-95			1981-82	4.6	4.8	2WD	1.3	1.5
w/turbo	7.4	7.9	1983-87			4WD	3.5	3.7
w/o turbo	6.0	6.5	2WD	5.3	5.5	DOHC	3.7	3.9
MR2 Spyder (3.4)	5.0	5.5	4WD	4.7	4.9	1992-02	1.4	1.6
Paseo			1988-92			2003-05 (1.9)	2.8	3.0
1992-95	5.1	5.6	2WD	6.3	6.5	Corolla FX, FX16	1.5	1.7
1996-99	6.3	6.8	4WD	7.9	8.1	1981-87 Cressida	1.6	1.8
Prius			1993-97			Echo (.9)	1.3	1.5
2001-03 (3.1)	4.5	5.0	2WD	5.7	5.9	Matrix (1.7)	2.4	2.6
2004-05 (4.4)	6.4	6.9	4WD	8.2	8.4	1991-95 MR2	1.7	1.9
Starlet	2.9	3.4	1998-02	6.2	6.4	MR2 Spyder (.9)	1.3	1.5
Supra			2003-05 (5.6)	8.0	8.5	Paseo, Starlet, Supra	1.7	1.9
1981-86	3.3	3.8	1981-87 Cressida	4.7	4.9	Tercel		
1987-92	6.0	6.5	Echo (4.2)	6.1	6.6	1981-87	1.5	1.7
1993-98			Matrix (5.6)	8.0	8.5	1988-91	.9	1.1
w/turbo	13.7	14.2	MR2			1992-99	1.7	1.9
w/o turbo	10.7	11.2	1985-89			*w/cruise control Matrix*		
Tercel			4AGZ	6.2	6.4	*add*	.2	.2
1981-82	5.3	5.8	4AGE	5.4	5.6	**Clutch Release Bearing, Replace (B)**		
1983-90			1991-95	5.9	6.1	Camry		
sedan	5.1	5.6	MR2 Spyder			1983-86	4.9	5.1
wagon	4.4	4.9	5-Speed MT (3.6)	5.3	5.8	1987-91		
1991-99	5.0	5.5	sequential			4 cyl.		
w/AC add			MT (4.2)	6.1	6.6	2WD	7.6	7.8
Matrix XRS	.2	.2	Paseo	4.9	5.1	4WD	9.6	9.8
MR2 Spyder	.2	.2	Starlet	3.3	3.5	V6	6.8	7.0
w/PS Echo add	.2	.2	Supra			1992-01	6.9	7.1
			1981-86	4.6	4.8	2002-05 (7.2)	10.2	10.7
CLUTCH			1987-92			Camry Solara		
Bleed Clutch Hydraulic System (B)			7MGE	3.9	4.1	1999-01	6.9	7.1
1981-05 (.4)	.6	.7	7MGTE	4.4	4.6	2002-03		
Clutch Pedal Free Play, Adjust (B)			1993-98	4.1	4.3	4 cyl.	11.3	12.7
1981-05 (.3)	.5	.6	Tercel			V6	7.5	8.4
Clutch Assy., Replace (B)			1981-94			2004-05		
Camry			2WD	4.9	5.1	4 cyl. (8.5)	11.8	12.3
1983-86	5.2	5.4	4WD	5.5	5.7	Celica		
1987-91			1995-99	4.4	4.6	1981-85	4.3	4.5
4 cyl.			*Replace*			1986-90		
2WD	7.7	7.9	*pilot bearing add*	.1	.1	2WD	6.5	6.7
4WD	9.9	10.1	*throw-out bearing add*	.2	.2	4WD	11.2	11.4

TOY-32 AVALON : CAMRY : CELICA : COROLLA : CRESSIDA : ECHO : MATRIX : MR2

	LABOR TIME	SEVERE SERVICE
Clutch Release Bearing, Replace (B)		
1991-93		
2WD	5.2	5.4
4WD	9.3	9.5
1994-05 (4.6)	6.6	7.1
Corolla FWD		
1983-87	5.0	5.2
1988-92		
2WD	6.2	6.4
4WD	7.8	8.0
1993-97		
2WD	4.8	5.0
4WD	7.7	7.9
1998-02	6.2	6.4
2003-05 (5.3)	7.6	8.1
Corolla RWD	4.2	4.4
Corolla FX, FX16	4.8	5.0
1981-87 Cressida	4.4	4.6
Echo (4.0)	5.8	6.3
Matrix (5.3)	7.6	8.1
MR2		
1985-89		
4AGZ	6.3	6.5
4AGE	4.9	5.1
1991-95	5.6	5.8
MR2 Spyder		
5-Speed MT (3.7)	5.4	5.9
sequential		
MT (4.3)	6.2	6.7
Paseo	4.5	4.7
Starlet	3.2	3.4
Supra		
1981-86	4.4	4.6
1987-92		
7MGE	3.5	3.7
7MGTE	4.0	4.2
1993-98	3.7	3.9
Tercel		
1981-94		
2WD	4.6	4.8
4WD	5.3	5.5
1995-99	3.8	4.0
Clutch Slave Cylinder, Replace (B)		
Includes: System bleeding.		
Exc. below (.7)	1.1	1.3
2003-05 Corolla (.9)	1.4	1.6
1991-95 MR2	1.7	1.9
1993-98 Supra	1.3	1.5
Matrix (.9)	1.4	1.6
Flywheel, Replace (B)		
Camry		
1983-86	5.4	5.6
1987-91		
4 cyl.		
2WD	8.3	8.5
4WD	9.8	10.0
V6	7.4	7.6
1992-98	5.5	5.7
1999-01	7.1	7.3
2002-05 (8.0)	11.2	11.7

	LABOR TIME	SEVERE SERVICE
Camry Solara		
1999-01	7.2	7.4
2002-03		
4 cyl.	12.2	13.7
V6	7.9	8.9
2004-05		
4 cyl. (9.0)	12.5	13.0
Celica		
1981-85	5.1	5.3
1986-90		
2WD	7.2	7.4
4WD	11.7	11.9
1991-93		
2WD	5.9	6.1
4WD	9.7	9.9
1994-99	6.7	6.9
2000-05 (5.1)	7.3	7.8
Corolla FWD		
1983-87	5.5	5.7
1988-92		
2WD	6.9	7.1
4WD	8.3	8.5
1993-97		
2WD	5.5	5.7
4WD	8.4	8.6
1998-02	5.7	5.9
2003-05 (5.3)	7.6	8.1
Corolla RWD	5.0	5.2
Corolla FX, FX16	5.4	5.6
1981-87 Cressida	5.3	5.5
Echo (4.3)	6.2	6.7
Matrix (5.3)	7.6	8.1
MR2		
1985-89		
4AGZ	6.6	6.8
4AGE	5.3	5.5
1991-95	6.1	6.3
MR2 Spyder		
5-Speed MT (3.9)	5.7	6.2
sequential MT (4.5)	6.5	7.0
Paseo	5.2	5.4
Prius		
2001-03 (7.8)	10.9	11.4
2004-05 (6.0)	8.6	9.1
Starlet	3.7	3.9
Supra		
1981-86	5.0	5.2
1987-92		
7MGE	4.1	4.3
7MGTE	4.6	4.8
1993-98		
5-Speed	4.0	4.2
6-Speed	4.4	4.6
Tercel		
1981-94	5.7	5.9
1995-99	5.2	5.4
w/AC add		
2000-05 Celica	.2	.2
Prius	.3	.3

	LABOR TIME	SEVERE SERVICE
MANUAL TRANSAXLE		
Transaxle Assy. R&R and Recondition (A)		
Camry		
1983-86	10.2	10.8
1987-91		
4 cyl.		
2WD	13.2	13.8
4WD	19.7	20.3
V6	16.0	16.6
1992-01		
4 cyl.	12.5	13.1
V6	14.7	15.3
2002-05 (14.4)	18.9	19.8
Camry Solara		
1999-01		
4 cyl.	12.6	13.2
V6	14.6	15.2
2002-03		
4 cyl.	21.0	23.5
V6	16.1	17.9
2004-05		
4 cyl. (15.6)	20.3	21.2
Celica		
1986-90		
2WD	12.5	13.1
4WD	22.8	23.4
1991-93		
2WD	10.5	11.1
4WD	18.2	19.8
1994-99	12.8	13.4
2000-05 (9.1)	12.6	13.5
Corolla FWD		
1983-87	12.3	12.9
1988-92		
2WD	13.4	14.0
4WD	17.9	18.5
1993-97		
2WD	10.5	11.1
4WD	18.2	18.8
1998-02	10.9	11.5
2003-05 (8.8)	12.2	13.1
Corolla FX, FX16	12.4	13.0
Echo (8.1)	11.3	12.2
Matrix		
5-Speed (8.8)	13.4	15.0
6-Speed (11.8)	15.9	16.8
MR2 Spyder		
5-Speed (7.6)	11.6	13.0
sequential		
5-Speed (8.2)	12.5	13.9
6-Speed (8.8)	12.2	13.1
Paseo	10.0	10.6
Tercel		
1981-94		
2WD		
4-Speed	9.4	10.0
5-Speed	9.9	10.5
4WD	12.3	12.9
1995-99	9.9	10.5

MR2 SPYDER : PASEO : PRIUS : SOLARA : STARLET : SUPRA : TERCEL **TOY-33**

	LABOR TIME	SEVERE SERVICE
Transaxle Assy., R&R or Replace (B)		
Camry		
1983-86	5.0	5.2
1987-91		
4 cyl.		
2WD	7.6	7.8
4WD	9.5	9.7
V6	7.0	7.2
1992-01	7.2	7.4
2002-04 (7.4)	10.4	10.9
Camry Solara		
1999-01	7.0	7.2
2002-03		
4 cyl.	11.3	12.7
V6	7.8	8.7
2004-05		
4 cyl. (8.7)	12.1	12.6
Celica		
1986-90		
2WD	6.4	6.6
4WD	11.1	11.3
1991-93		
2WD	5.3	5.5
4WD	9.4	9.6
1994-99	6.4	6.6
2000-05 (4.6)	6.6	7.1
Corolla FWD		
1983-87	4.8	5.0
1988-92		
2WD	6.3	6.5
4WD	7.8	8.0
1993-97		
2WD	4.9	5.1
4WD	7.6	7.8
1998-02	5.5	5.7
2003-05 (5.5)	7.9	8.4
Corolla FX, FX16	4.9	5.1
Echo (4.0)	5.8	6.3
Matrix		
5-Speed (5.5)	7.9	8.4
6-Speed (5.1)	7.3	7.8
MR2 Spyder		
5-Speed (3.3)	4.8	5.3
sequential (3.9)	5.7	6.2
Paseo	4.6	4.8
Tercel		
1981-94		
2WD	4.5	4.7
4WD	5.3	5.5
1995-99	4.4	4.6

MANUAL TRANSMISSION
Transmission Assy., R&R and Reconditional (A)

1981-85 Celica	9.1	9.7
Corolla RWD	8.2	8.8
1981-87 Cressida	9.1	9.7

	LABOR TIME	SEVERE SERVICE
MR2		
1985-89		
4AGZ	16.1	16.7
4AGE	12.8	13.4
1991-95		
S54	11.3	11.9
E153	13.2	13.8
Starlet	6.1	6.7
Supra		
1981-86	9.2	9.8
1987-92		
7MGE	8.4	9.0
7MGTE	12.1	12.7
1993-98		
5-Speed	10.4	11.0
6-Speed	10.2	10.8
Transmission Assy., R&R or Replace (B)		
1981-85 Celica	4.2	4.4
Corolla RWD	4.4	4.6
1981-87 Cressida	4.4	4.6
MR2		
1985-89		
4AGZ	6.1	6.3
4AGE	4.9	5.1
1991-95	5.7	5.9
Starlet	3.3	3.5
Supra		
1981-86	4.3	4.5
1987-92		
7MGE	3.5	3.7
7MGTE	4.1	4.3
1993-98		
5-Speed	3.4	3.6
6-Speed	3.8	4.0

AUTOMATIC TRANSAXLE
SERVICE TRANSAXLE INSTALLED
Check Unit for Oil Leaks (C)

1981-05	.9	.9
Drain & Refill Unit (B)		
Exc. Prius (.9)	1.4	1.6
Prius (.4)	.6	.7
Oil Pressure Check (B)		
1981-05 (1.2)	1.9	1.9
Electronic Control Module, Replace (B)		
Avalon	.8	.8
Camry		
1985-86	.5	.5
1987-91	.8	.8
1992-96	1.3	1.3
1997-98	.7	.7
Camry Solara (.5)	.7	.7
Celica		
1986-89	.8	.8
1990-99	1.6	1.6
Corolla FWD		
1987	.7	.7
1993-98	1.4	1.4

	LABOR TIME	SEVERE SERVICE
Corolla FX, FX16	.7	.7
MR2 Spyder (1.2)	1.9	1.9
1992-99 Paseo	.8	.8
Governor Assy., R&R or Replace (B)		
1983-91 Camry	1.8	2.1
Celica		
1986-89	1.9	2.2
1990-93	6.1	6.4
Corolla FWD		
1983-87	1.7	2.0
1988-92	1.9	2.2
1993-02	2.6	2.9
Corolla FX, FX16	1.7	2.0
Tercel		
1981-87	6.8	7.1
1988-94	2.0	2.3
1995-99		
3-Speed	2.0	2.3
4-Speed	.7	1.0
Manual Lever Shaft Seal, Replace (B)		
Avalon, Echo (.8)	1.3	1.4
1992-05 Camry		
exc. 3VZFE (.9)	1.4	1.6
Camry Solara		
1999-03 (.9)	1.4	1.6
2004-05 (1.9)	2.8	3.0
1994-05 Celica (.8)	1.3	1.4
1993-05 Corolla		
exc. 4WD (.9)	1.4	1.6
Matrix (.8)	1.3	1.4
1991-95 MR2	2.1	2.3
Paseo		
1992-95	2.1	2.3
1996-99	1.1	1.3
Tercel		
1991-94	2.1	2.3
1995-99	1.3	1.4
Oil Pan and/or Gasket, Replace (B)		
1981-05		
exc. below (1.4)	2.2	2.4
V6 Solara (.8)	1.2	1.4
Speed Sensor, Replace (B)		
1995-99 Avalon	1.7	1.7
Camry		
1985-01	1.7	1.7
2002-05 (.4)	.6	.6
Camry Solara		
1999-03		
4 cyl.	.6	.6
V6	1.4	1.4
2004-05 (.8)	1.2	1.3
Celica		
1986-99	1.4	1.4
2000-05 (.4)	.6	.6
Corolla FWD		
1987	1.4	1.4
1993-02	1.6	1.6
Corolla FX, FX16	1.5	1.5
Matrix (.7)	1.1	1.1
Paseo	.5	.5

TOY-34 AVALON : CAMRY : CELICA : COROLLA : CRESSIDA : ECHO : MATRIX : MR2

	LABOR TIME	SEVERE SERVICE
Throttle Cable Assy., Replace (B)		
Avalon	3.8	4.3
Camry		
1983-91	2.4	2.6
1992-01		
4 cyl.	3.3	3.5
V6	3.5	3.7
Camry Solara		
4 cyl.	3.2	3.4
V6	3.8	4.3
Celica		
1986-89	2.5	2.7
1990-93	3.0	3.2
1994-99	3.4	3.6
Corolla FWD		
1983-87	2.5	2.7
1988-92		
2WD	2.4	2.6
4WD	2.7	2.9
1993-02		
2WD	3.4	3.6
4WD	2.5	2.7
Corolla FX, FX16	2.5	2.7
Echo (2.3)	3.4	3.6
Paseo	3.1	3.3
Tercel		
1988-90	2.6	2.8
1992-99	3.2	3.4
Valve Body, Replace (B)		
Avalon	3.0	3.4
Camry		
1987-01	3.0	3.4
2002-05 (1.6)	2.4	2.6
Camry Solara		
1999-01	3.0	3.4
2002-03		
4 cyl.	2.5	2.9
V6	3.0	3.4
2004-05		
4 cyl. (1.6)	2.4	2.6
V6 (1.0)	1.5	1.7
Celica (1.8)	2.7	2.9
Corolla FWD	2.8	3.1
Echo, Paseo (1.8)	2.7	3.0
Matrix (1.6)	2.4	2.6
1988-99 Tercel	2.9	3.2

SERVICE TRANSAXLE REMOVED
Transaxle R&R included unless otherwise noted.

	LABOR TIME	SEVERE SERVICE
Manual Lever Shaft Seal, Replace (B)		
Camry		
1983-86	9.2	9.4
1987-91		
2WD	9.8	10.0
4WD	11.3	11.5
1992-93 3VZFE	10.1	10.3
Celica		
1986-89	9.6	9.8
1990-93	7.5	7.7

	LABOR TIME	SEVERE SERVICE
Corolla FWD		
1983-87	7.0	7.2
1988-92		
2WD	8.6	8.8
4WD	10.2	10.4
1993 4WD	10.3	10.5
Corolla FX, FX16	6.9	7.1
1989 MR2	6.9	7.1
Tercel		
1981-87	2.0	2.2
1988-90	8.1	8.3
Torque Converter, Replace (B)		
Avalon	8.7	9.7
Camry		
1983-86	7.8	8.7
1987-91		
4 cyl.		
2WD	7.9	8.9
4WD	9.6	10.7
V6	8.2	9.2
1992-01	8.2	9.2
2002-05		
4 cyl. (8.0)	11.2	11.7
V6 (8.5)	11.8	12.3
Camry Solara		
1999-01	7.9	8.9
2002-03		
4 cyl.	11.8	13.2
V6	8.3	9.2
2004-05		
4 cyl. (8.3)	11.5	12.0
V6 (8.6)	12.0	12.5
Celica		
1986-89	8.4	8.9
1990-93	6.3	6.8
1994-99	7.4	7.9
2000-05 (5.4)	7.7	8.2
Corolla FWD		
1983-87	6.3	7.0
1988-92		
2WD	7.5	8.4
4WD	9.6	10.7
1993-97		
2WD	6.4	7.2
4WD	9.4	10.6
1998	7.0	7.9
1999-02	6.4	7.2
2003-05 (6.5)	9.2	9.7
Corolla FX, FX16	6.3	7.0
Echo (4.7)	6.8	7.3
Matrix		
2WD (6.5)	9.2	9.7
4WD (9.0)	12.5	13.0
Paseo	5.9	6.5
Tercel		
1981-82	7.4	8.2
1983-90		
sedan	7.3	8.2

	LABOR TIME	SEVERE SERVICE
wagon		
2WD	7.4	8.2
4WD	7.8	8.7
1991-99	6.2	7.0
w/AC Matrix 4WD add	.2	.2
Replace front seal add	.1	.1
Transaxle Assy., R&R and Reconditon (A)		
Avalon	22.9	25.4
Camry		
1983-86	16.2	17.1
1987-91		
4 cyl.		
2WD	19.8	20.7
4WD	21.5	22.4
V6	20.0	20.9
1992-96	20.7	21.6
1997-01	21.4	22.3
2002-05		
4 cyl. (14.5)	19.0	19.9
V6 (15.1)	19.7	20.6
Camry Solara		
1999-01	21.6	22.5
2002-03		
4 cyl.	21.2	23.5
V6	22.9	25.4
2004-05		
4 cyl. (14.5)	19.0	19.9
V6 (15.0)	19.6	20.5
Celica		
1986-89	20.1	21.0
1990-93	18.2	19.1
1994-99		
7AFE	19.2	20.1
5SFE	20.8	21.7
2000-05 (12.4)	16.6	17.5
Corolla FWD		
1983-87		
3-Speed	16.9	17.8
4-Speed	17.6	18.5
1988-92		
3-Speed	18.3	19.2
4-Speed	19.3	20.2
4WD	22.8	23.7
1993-97		
2WD	18.4	19.3
4WD	22.6	23.5
1998-02	19.2	20.1
2003-05 (11.8)	15.9	16.8
Corolla FX, FX16		
3-Speed	16.7	17.6
4-Speed	17.6	18.5
Echo (12.4)	16.6	17.5
Matrix		
2WD (11.8)	15.9	16.8
4WD (12.8)	17.1	18.0
XRS (13.4)	17.8	18.7
Paseo	18.9	19.8

MR2 SPYDER : PASEO : PRIUS : SOLARA : STARLET : SUPRA : TERCEL   TOY-35

	LABOR TIME	SEVERE SERVICE
Tercel		
1981-82	13.5	14.4
1983-90		
sedan	18.0	18.9
wagon		
2WD	13.8	14.7
4WD	15.4	16.3
1991-94	17.8	18.7
1995-99	19.1	20.0
w/AC Matrix 4WD add	.2	.2
Transaxle Assy., R&R or Replace (B)		
Avalon	8.4	9.4
Camry		
1983-86	7.7	7.9
1987-91		
4 cyl.		
2WD	7.8	8.0
4WD	9.4	9.6
V6	7.9	8.1
1992-96	8.3	8.5
1997-01	7.7	7.9
2002-05		
4 cyl. (7.4)	10.4	10.9
V6 (8.4)	11.7	12.2
Camry Solara		
1999-01	7.6	8.1
2002-03		
4 cyl.	11.3	11.8
V6	8.4	8.9
2004-05		
4 cyl. (7.7)	10.8	11.3
V6 (7.9)	11.1	11.6
Celica		
1986-89	8.3	8.5
1990-93	6.1	6.3
1994-99	7.4	7.6
2000-05 (5.2)	7.5	8.0
Corolla FWD		
1983-87	5.9	6.1
1988-92		
2WD	7.6	7.8
4WD	9.3	9.5
1993-97		
2WD	6.2	6.4
4WD	9.2	9.4
1998-02	6.9	7.1
2003-05 (6.2)	8.8	9.3
Corolla FX, FX16	5.8	6.0
Echo (4.7)	6.8	7.3
Matrix		
2WD (6.2)	8.8	9.3
4WD (8.5)	11.8	12.3
Paseo	6.1	6.3
Tercel		
1981-82	7.2	7.4
1983-90		
sedan	7.0	7.2
wagon	7.5	7.7
1991-99	5.9	6.1

AUTOMATIC TRANSMISSION
SERVICE TRANSMISSION INSTALLED

	LABOR TIME	SEVERE SERVICE
Check Unit for Oil Leaks (C)		
1981-98	.9	.9
Drain and Refill Unit (B)		
1981-98	1.0	1.0
Oil Pressure Check (B)		
1981-98	1.8	1.8
2001-04	2.0	2.0
Electronic Control Unit, Replace (B)		
Cressida	.8	.8
MR2		
1987-89	.7	.7
1991-95	2.3	2.3
Supra		
1985-92	.9	.9
1993-98	1.2	1.2
Extension Housing and/or Gasket, Replace (B)		
1981-85 Celica	2.9	3.1
1981-87 Corolla	3.0	3.2
Cressida	3.0	3.2
Supra		
1981-92	3.0	3.2
1993-98	2.9	3.1
Extension Housing Oil Seal, Replace (B)		
1981-85 Celica	1.6	1.8
1981-87 Corolla	1.2	1.4
Cressida	1.2	1.4
Supra		
1981-92	1.5	1.7
1993-98		
w/turbo	2.9	3.1
w/o turbo	2.8	3.0
Governor Assy., R&R or Replace (B)		
1981-85 Celica	3.5	3.8
1981-85 Corolla	3.5	3.8
1981-83 Cressida	3.6	3.9
1985-89 MR2	2.5	2.8
1981-83 Supra	3.6	3.9
Oil Pan and/or Gasket, Replace (B)		
1993-98 Supra	1.0	1.0
Overdrive Solenoid, Replace (B)		
1981-85 Celica	2.0	2.2
1981-87 Corolla	1.7	1.9
1981-84 Cressida	1.6	1.8
1981-86 Supra	2.0	2.2
Throttle Valve Control Cable, Replace (B)		
1981-85 Celica	2.5	2.6
1981-87 Corolla	2.3	2.4
Cressida	2.3	2.4
Supra		
1981-92	2.3	2.4
1993-98	2.6	2.7

	LABOR TIME	SEVERE SERVICE
Transmission Speed Sensor, Replace (B)		
1985-92 Cressida	.7	.7
1987-95 MR2	1.5	1.5
1985-98 Supra	.8	.8
Valve Body Assy., Replace (B)		
1988-92 Cressida	3.0	3.3
MR2		
1985-89	2.7	3.0
1991-95	2.8	3.1
Supra		
1988-92	2.9	3.2
1993-98	2.3	2.6
Valve Body Manual Lever Shaft Seal, Replace (B)		
1981-85 Celica	4.5	4.6
1981-87 Corolla	3.7	3.8
Cressida	4.3	4.4
Supra		
1981-92	4.4	4.5
1993-98	2.7	2.8

SERVICE TRANSMISSION REMOVED
Transmission R&R included unless otherwise noted.

	LABOR TIME	SEVERE SERVICE
Flywheel (Flexplate), Replace (B)		
1981-85 Celica	5.7	5.9
1981-87 Corolla	5.8	6.0
Cressida		
1981-88	5.7	5.9
1989-92	5.6	5.8
Supra		
1981-86	5.9	6.1
1987-92	6.2	6.4
1993-98	5.1	5.3
Torque Converter, Replace (B)		
1981-85 Celica	5.7	5.9
1981-87 Corolla	5.6	5.8
Cressida		
1981-88	5.7	5.9
1989-92	5.3	5.5
MR2		
1985-89	6.5	6.7
1991-95	7.0	7.2
Supra		
1981-86	5.6	5.8
1987-92	6.2	6.4
1993-98	5.3	5.5
Replace front seal add	.1	.1
Transmission Assy., R&I (B)		
1981-85 Celica	5.5	5.7
1981-87 Corolla	5.5	5.7
Cressida		
1981-88	5.3	5.5
1989-92	5.2	5.4
MR2		
1985-89	6.1	6.3
1991-95	6.8	7.0

TOY-36 AVALON : CAMRY : CELICA : COROLLA : CRESSIDA : ECHO : MATRIX : MR2

	LABOR TIME	SEVERE SERVICE
Transmission Assy., R&I (B)		
Supra		
1981-86	5.3	5.5
1987-92	5.9	6.1
1993-98	5.0	5.2
Transmission Assy., R&R and Recondition (A)		
1981-85 Celica	13.6	14.5
1981-87 Corolla	13.6	14.5
Cressida		
1981-88	13.6	14.5
1989-92	21.9	22.8
MR2		
1985-89	14.7	15.6
1991-95	18.7	19.6
Supra		
1981-86	13.8	14.7
1987-92	20.0	20.9
1993-98	15.2	16.1

HYBRID TRANSAXLE

	LABOR TIME	SEVERE SERVICE
Transaxle Assy., R&R and Recondition (A)		
Prius		
2001-03	19.8	22.1
2004-05 (9.2)	12.7	13.6
w/AC add	.3	.3
Transaxle Assy., Replace (A)		
Prius		
2001-03	11.0	12.4
2004-05 (5.9)	8.4	8.9
w/AC add	.3	.3

TRANSFER CASE

	LABOR TIME	SEVERE SERVICE
Rear Output Shaft Seal, Replace (B)		
1988-92 Camry	1.2	1.5
1988-93 Celica	1.3	1.6
1988-93 Corolla	1.3	1.6
Matrix (1.1)	1.6	1.9
Transfer Case, R&R and Recondition (A)		
1988-92 Camry	15.5	17.3
Celica		
1988-89	17.2	19.3
1990-93	12.8	14.3
Corolla FWD		
1988-92	13.5	15.1
1993	14.3	16.1
Matrix (9.5)	13.1	14.0
Transfer Case, R&R or Replace (B)		
1988-92 Camry	9.3	9.5
Celica		
1988-89	10.8	11.0
1990-93	8.8	9.0
1988-93 Corolla	7.7	7.9
Matrix (7.7)	10.8	11.3

	LABOR TIME	SEVERE SERVICE
SHIFT LINKAGE		
AUTOMATIC TRANSAXLE		
Throttle Rod or Cable, Adjust (B)		
1981-05 (.3)	.5	.7
AUTOMATIC TRANSMISSION		
Throttle Rod or Cable, Adjust (B)		
1981-98 RWD	.5	.7

DRIVELINE
DRIVESHAFT

	LABOR TIME	SEVERE SERVICE
Center Support Bearing, Replace (B)		
1989-91 Camry 4WD	2.6	2.9
Celica		
1981-85	1.9	2.2
1988-93 4WD	2.6	2.9
Corolla RWD	2.0	2.3
Corolla wagon		
1989-93	2.5	2.8
Cressida	1.8	2.1
Matrix	2.1	2.3
Supra		
1981-92	1.9	2.2
1993-98	2.3	2.6
1985-88 Tercel wagon	1.9	2.2
Rear Driveshaft, R&R or Replace (B)		
Exc. below	1.6	1.8
Starlet	.9	1.1
1993-98 Supra	1.9	2.1
U-Joint, Replace (B)		
1989-91 Camry 4WD	2.2	2.5
Celica		
1981-85	2.0	2.3
1988-93 4WD	2.3	2.6
Corolla RWD	1.6	1.9
Corolla wagon		
1989-93	2.3	2.6
Cressida	1.9	2.2
Starlet	1.4	1.7
1981-92 Supra	1.4	1.7

DRIVE AXLE

	LABOR TIME	SEVERE SERVICE
Axle Shaft, Replace (B)		
Rear axle		
Celica		
1981-85	2.0	2.3
1988-93 4WD	1.9	2.2
Corolla	1.9	2.2
1989-93 4WD		
Corolla RWD	1.9	2.2
Starlet	2.0	2.3
Tercel wagon		
1985-88	2.0	2.3
Axle Shaft Bearings, Replace (B)		
Camry		
1983-87	1.4	1.7
1988-93	1.6	1.9

	LABOR TIME	SEVERE SERVICE
Celica		
1981-85	1.8	2.1
1986-87	1.3	1.6
1988-93	1.6	1.9
Corolla		
FWD		
1988-92	1.6	1.9
1993	.8	1.1
4WD		
1989-93	1.9	2.2
RWD	2.0	2.3
Cressida	2.8	3.1
1985-89 MR2	3.3	3.6
Paseo	1.6	1.9
Starlet	1.9	2.2
1981-92 Supra	2.9	3.2
Tercel		
1981-90		
sedan	1.7	2.0
wagon	1.5	1.8
1991-99	1.7	2.0
Axle Shaft Oil Seal, Replace (B)		
Celica		
1981-85	1.2	1.4
1988-93 4WD	1.4	1.6
Corolla RWD	1.3	1.5
1989-93 4WD	1.2	1.4
Paseo	1.1	1.3
Starlet	1.4	1.6
Tercel		
1983-90	1.4	1.6
1991-99	1.3	1.5
Differential Carrier Assy., R&R and Recondition (A)		
Rear axle		
1989-91 Camry		
4WD	7.5	8.0
Celica	7.7	8.2
Corolla wagon		
1989-93	7.7	8.2
Corolla RWD	6.7	7.2
Cressida		
1981-88		
w/IRS	6.9	7.4
w/o IRS	6.4	6.9
w/limited slip	7.8	8.3
1989-92	7.8	8.3
Matrix	11.7	12.6
Starlet	6.4	6.9
Supra		
1981-86		
w/IRS	7.9	8.4
w/o IRS	6.9	7.4
1987-92	6.9	7.4
1993-98		
w/limited slip	7.7	8.2
w/o limited slip	6.2	6.7
1985-88 Tercel		
wagon	6.6	7.1

MR2 SPYDER : PASEO : PRIUS : SOLARA : STARLET : SUPRA : TERCEL TOY-37

	LABOR TIME	SEVERE SERVICE
Differential Carrier, R&R or Replace (B)		
Rear axle		
1989-91 Camry 4WD	3.0	3.5
Celica	3.0	3.5
Corolla wagon 1989-93	1.8	2.3
Corolla RWD	2.7	3.2
Cressida		
1981-88	2.9	3.4
1989-92	2.4	2.9
Matrix	7.3	7.8
Starlet, Supra	2.3	2.8
1985-88 Tercel wagon	2.9	3.4
Differential Oil Seal, Replace (B)		
Celica		
1981-85	1.6	1.8
1988-90 4WD	3.6	3.8
1991-93 4WD	1.3	1.5
Cressida	1.7	1.9
Matrix	2.4	2.7
Supra		
1981-92	1.8	2.0
1993-98	1.4	1.6
Differential Side Bearing, Replace (B)		
Rear axle 4WD		
1988-93 Celica	4.6	5.1
1993 Corolla	2.9	3.4
Matrix	10.6	11.1
Hub and Bearing, Replace (B)		
1994-99 Celica one	1.2	1.7
Pinion Bearings, Replace (B)		
Rear axle 4WD		
Celica	6.5	7.0
1993 Corolla	5.2	5.7
Matrix	9.4	9.9
Front axle		
1983-88 Tercel		
2WD		
MT	8.6	9.1
AT	9.6	10.1
4WD	9.1	9.6
Pinion Shaft Oil Seal, Replace (B)		
Rear axle 4WD		
Celica	2.2	2.3
1993 Corolla	1.7	1.9
Front axle		
1983-88 Tercel	7.0	7.2
Rear Axle Carrier and/or Gasket, Replace (B)		
Rear axle 4WD		
Celica	3.2	3.4
1993 Corolla	3.0	3.2
Ring Gear & Pinion Set, Replace (B)		
1989-91 Camry 4WD	6.7	7.2

	LABOR TIME	SEVERE SERVICE
Celica		
1981-85	6.0	6.5
1988-93 4WD	7.0	7.5
Corolla	5.9	6.4
Cressida		
1981-88	6.1	6.6
1989-92	7.1	7.6
Matrix	10.5	11.0
Starlet	5.4	5.9
Supra		
1981-86	5.9	6.4
1987-98	7.0	7.5
1985-88 Tercel wagon	5.9	6.4
Side Gear Shaft Oil Seal, Replace (B)		
Front axle		
1983-88 Tercel	2.2	2.4
Rear axle 4WD		
Matrix	4.1	4.6
Stub Axle Assy., Replace (B)		
Camry, Camry Solara		
1983-87		
one	1.5	2.0
both	2.3	2.8
1988-91		
2WD		
one	1.6	2.1
both	2.3	2.8
4WD		
one	3.4	3.9
both	4.6	5.1
1992-96		
one	.9	1.4
both	1.7	2.2
1997-01		
one	1.4	1.9
both	2.2	2.7
Celica		
1983-85		
one	2.9	3.4
both	4.7	5.2
1986-93		
2WD		
one	1.6	2.1
both	2.4	2.9
4WD		
one	3.2	3.7
both	4.8	5.3
Corolla FWD		
1983-92		
one	1.6	2.1
both	2.3	2.8
1993-02		
one	.8	1.3
both	1.3	1.8
Corolla FX, FX16		
one	1.6	2.1
both	2.3	2.8

	LABOR TIME	SEVERE SERVICE
Cressida		
1983-88		
one	2.8	3.3
both	4.6	5.1
1989-92		
one	2.5	2.8
both	3.8	4.3
1991-95 MR2		
one	2.7	3.2
both	3.9	4.4
Supra		
1982-92		
one	3.0	3.5
both	4.7	5.2
1993-98		
one	3.7	4.2
both	5.4	5.9
Stub Axle Bearing, Replace (B)		
Camry		
1983-87		
one	1.7	2.0
both	2.3	2.6
1988-91		
2WD		
one	1.5	1.8
both	2.3	2.6
4WD		
one	3.3	3.6
both	4.5	2.8
1992-96		
one	1.5	1.8
both	2.4	2.7
Celica		
1983-85		
one	3.1	3.4
both	4.6	4.9
1986-93		
2WD		
one	1.5	1.8
both	2.5	2.8
4WD		
one	3.3	3.6
both	4.9	5.2
Corolla		
one	1.6	1.9
both	2.2	2.5
Cressida		
1983-88		
one	3.0	3.3
both	5.3	5.6
1989-92		
one	3.1	3.4
both	4.6	4.9
MR2		
1985-89		
one	3.3	3.6
both	4.6	4.9
1991-95		
one	2.6	2.9
both	3.7	4.0

TOY-38 AVALON : CAMRY : CELICA : COROLLA : CRESSIDA : ECHO : MATRIX : MR2

	LABOR TIME	SEVERE SERVICE
Stub Axle Bearing, Replace (B)		
Supra		
1982-92		
one	3.1	3.4
both	4.6	4.9
1993-98		
one	3.7	4.0
both	5.4	5.7
Stub Axle Oil Seal, Replace (B)		
Camry		
1983-87		
one	1.5	1.7
both	2.5	2.7
1988-91		
2WD		
one	1.6	1.8
both	2.4	2.6
4WD		
one	2.8	3.0
both	4.2	4.4
Celica		
1983-85		
one	3.0	3.2
both	4.6	4.8
1986-93		
2WD		
one	1.5	1.7
both	2.5	2.7
4WD		
one	3.0	3.2
both	4.0	4.2
1983-92 Corolla FWD		
one	1.4	1.6
both	2.4	2.6
Corolla FX, FX16		
one	1.5	1.7
both	1.9	2.1
Cressida		
1983-88		
one	3.0	3.2
both	4.5	4.7
1989-92		
one	2.5	2.7
both	3.6	3.8
MR2		
1985-89		
one	3.0	3.2
both	4.5	4.7
1991-95		
one	2.5	2.7
both	3.9	4.1
Supra		
1982-92		
one	3.1	3.3
both	4.6	4.8
1993-98		
one	3.6	3.8
both	5.5	5.7

	LABOR TIME	SEVERE SERVICE
HALFSHAFTS		
CV Joint and Boot, Replace or Recondition (B)		
Avalon	3.1	3.5
Camry, Camry Solara		
1983-86	1.8	2.2
1987-01	3.2	3.6
Celica		
1986-89	3.2	3.6
1990-93	2.4	2.8
1994-99	3.1	3.5
Corolla FWD		
1983-87	1.8	2.2
1988-92	2.9	3.3
1993-01		
2WD	3.5	3.9
4WD	2.9	3.3
Corolla FX, FX16	1.8	2.2
Paseo	3.0	3.4
Tercel		
1981-82	1.8	2.2
1983-90		
sedan	2.6	3.0
wagon	2.1	2.5
1991-99	3.0	3.4
Halfshaft, R&R or Replace (B)		
Avalon, Camry Solara	1.7	2.0
Camry		
1983-86	1.4	1.7
1987-91	2.7	3.0
1992-01	1.9	2.2
2002-05	3.2	3.6
Celica		
1983-85	.8	1.1
1986-89		
2WD	2.9	3.2
4WD	3.3	3.7
1990-93	1.7	2.0
1994-99	2.1	2.3
2000-05	2.5	2.9
Corolla FWD		
1983-87	1.5	1.8
1988-92	2.9	3.2
1993-02		
2WD	2.4	2.7
4WD	2.9	3.2
2003-05	3.0	3.4
Corolla FX, FX16	1.5	1.8
1983-92 Cressida	1.1	1.4
Echo	2.2	2.5
Matrix		
2WD	3.0	3.4
4WD	3.5	3.9
XRS	2.9	3.2
MR2		
1985-89	2.0	2.3
1991-95	1.5	1.8
Paseo	1.8	2.1
Prius	2.4	2.7
Supra	1.3	1.6

	LABOR TIME	SEVERE SERVICE
Tercel		
1981-82	1.5	1.8
1983-90		
sedan	2.5	2.9
wagon	1.9	2.2
1991-99	1.9	2.1
Halfshaft Boot, Replace (B)		
Avalon	3.5	3.9
Camry, Camry Solara		
1988-91	3.2	3.6
1992-96		
w/5SFE	3.2	3.6
w/o 5SFE	3.8	4.3
1997-05	3.5	3.9
Celica		
1989		
2WD	3.3	3.7
4WD	3.8	4.3
1990-93	2.5	2.9
1994-99	3.2	3.6
2000-05	3.5	3.8
Corolla		
1989-92	3.2	3.6
1993-02		
2WD	3.8	4.3
4WD	3.2	3.6
2003-05	3.3	3.7
1983-92 Cressida	1.8	2.1
Echo	3.6	4.1
Matrix		
2WD	3.3	3.7
4WD	3.6	4.1
MR2		
1985-89	2.0	2.3
1991-95	1.6	1.9
Paseo, Tercel	3.2	3.6
Prius		
2001-03	3.6	4.1
2004	2.9	3.2
1982-98 Supra	1.8	2.1

BRAKES
ANTI-LOCK

	LABOR TIME	SEVERE SERVICE
Diagnose Anti-Lock Brake System Component Each (A)		
1989-05	1.0	1.0
Actuator Assy., Replace (B)		
Avalon	1.8	2.1
Camry		
1989-91	3.8	3.8
1992-04	2.0	2.2
Camry Solara	2.0	2.2
Celica		
1988-90	5.3	5.3
1991-93	3.6	3.6
1994-99	2.3	2.3
2000-05	3.1	3.4
Corolla FWD, Matrix	2.1	2.4
1989-92 Cressida	4.3	4.3
Echo, Paseo	2.4	2.7

MR2 SPYDER : PASEO : PRIUS : SOLARA : STARLET : SUPRA : TERCEL   TOY-39

	LABOR TIME	SEVERE SERVICE
1991-95 MR2	1.9	1.9
MR2 Spyder	2.8	3.1
Prius	6.0	6.7
Supra		
1987-92	3.8	3.8
1994-96	2.6	2.6
Tercel		
1993-94	1.8	1.8
1995-99	2.3	2.3

Control Unit, Replace (B)

Exc. below (.6)	.9	.9
1995-99 Avalon	1.6	1.6
Camry		
1989-91 4 cyl.	1.3	1.3
1992-96	1.3	1.3
1988-93 Celica	1.4	1.4
Echo	2.2	2.2
Paseo	1.2	1.2
1993-98 Supra	1.4	1.4

Front Sensor Assy., Replace (B)

Avalon		
one	1.0	1.1
both	1.6	1.8
Camry, Camry Solara		
one	1.1	1.3
both	1.7	2.0
1988-05 Celica		
one	.8	.9
both	1.1	1.3
Corolla, Matrix		
1994-02		
one	.7	.8
both	1.0	1.1
2003-05		
one	1.1	1.3
both	1.7	2.0
1989-92 Cressida		
one	.7	.8
both	1.1	1.3
Echo		
one	.6	.7
both	1.0	1.1
1991-95 MR2		
one	.5	.6
both	.8	.9
MR2 Spyder		
one	2.5	2.9
both	4.0	4.4
Paseo		
1992-95		
one	.5	.6
both	.8	.9
1996-99		
one	.7	.8
both	1.4	1.6
Prius		
one	1.1	1.3
both	1.6	1.8

	LABOR TIME	SEVERE SERVICE
Supra		
1987-98		
one	.9	1.0
both	1.1	1.3
Tercel		
one	.7	.8
both	1.2	1.4

Rear Sensor Assy., Replace (B)

Avalon		
one	1.0	1.1
both	1.3	1.4
Camry		
1993-01		
one	1.1	1.1
both	1.6	1.6
2002-05		
one	2.9	3.2
both	3.6	4.1
Camry Solara		
one	1.1	1.3
both	1.4	1.6
Celica		
1993		
one	.7	.8
both	1.3	1.4
1994-99		
one	1.9	2.1
both	2.9	3.2
2000-05		
GT		
one	1.3	1.4
both	1.9	2.1
GTS		
one	1.6	1.8
both	2.5	2.9
Corolla FWD		
1994-02		
one	.7	.8
both	.8	.9
2003-05		
one	2.1	2.3
both	2.5	2.9
Echo		
2000-05		
one	1.7	2.0
both	2.4	2.7
Matrix		
2WD		
one	2.2	2.5
both	2.7	3.0
4WD		
one	1.3	1.4
both	2.2	2.5
1993-95 MR2		
one	.5	.5
both	.8	.9
MR2 Spyder		
one	1.0	1.1
both	1.4	1.6

	LABOR TIME	SEVERE SERVICE
Paseo		
one	.7	.8
both	.8	.9
Prius		
2001-03		
one	1.3	1.4
both	1.6	1.8
2004-05		
one	1.9	2.1
both	2.5	2.9
1993-98 Supra		
one	2.3	2.5
both	3.0	3.4
1993-99 Tercel		
one	.7	.8
both	.8	.9

SYSTEM

Bleed Brakes (B)
Includes: Add fluid.

Two wheels	1.0	1.1
Four wheels	1.7	2.0

Brake System, Flush and Refill (B)

1981-05	1.7	2.0

Brakes, Adjust (B)
Includes: Refill master cylinder.

1981-05 rear wheels	.5	.5

Brake Hose (Flexible), Replace (B)
Includes: System bleeding.

Exc. below, one	.8	.9
2002-05 Camry, one	1.4	1.6
Corolla FWD		
1988-91, one	1.4	1.6
2002-05, one	1.4	1.6
Corolla FX, FX16, one	1.6	1.8
Cressida		
1989-92, rear, one	1.3	1.4
Matrix, one	1.4	1.6
MR2 Spyder	1.1	1.3
Tercel wagon		
rear, one	1.4	1.6
each addl. add	.3	.4

Load Sensing Valve, Replace (B)
Includes: System bleeding.

1987-96 Camry	1.7	1.9

Master Cylinder, Replace (B)
Includes: System bleeding.

Avalon	1.4	1.6
Camry, Camry Solara		
1983-91	2.5	2.7
1992-05	1.7	1.9
Celica		
1981-85	2.0	2.2
1986-89	2.4	2.6
1990-05	1.8	2.1
Corolla FWD, Matrix		
1983-91	2.3	2.5
1992-05	1.7	1.9
Corolla FX, FX16	2.0	2.2

TOY-40 AVALON : CAMRY : CELICA : COROLLA : CRESSIDA : ECHO : MATRIX : MR2

	LABOR TIME	SEVERE SERVICE
Master Cylinder, Replace (B)		
Corolla RWD	2.1	2.3
Cressida, MR2	1.9	2.1
Echo	1.5	1.7
MR2 Spyder	1.8	2.1
Paseo		
1992-95	1.4	1.6
1996-99	1.8	2.0
Prius	4.7	5.2
Starlet	1.8	2.0
Supra		
1981-92	2.3	2.5
1993-98	1.4	1.6
Tercel		
1981-90	2.2	2.4
1991-94	2.6	2.8
1995-99	1.6	1.8
Power Booster Unit, Replace (B)		
Includes: System bleeding.		
Avalon		
1995-99	3.0	3.2
2000-04	2.6	2.9
Camry		
1983-86	2.5	2.7
1987-91	3.5	3.7
1992-01	2.9	3.2
2002-05	2.4	2.7
Camry Solara		
1999-03 (1.9)	2.9	3.2
2004-05 (1.6)	2.4	2.9
Celica		
1981-85	2.8	3.0
1986-93	3.3	3.5
1994-99	2.8	3.0
2000-05	3.6	4.1
Corolla FWD, Matrix		
1983-87	2.6	2.8
1988-91	3.5	3.7
1992-02	2.9	3.2
2003-05	2.3	2.6
Corolla FX, FX16	2.5	2.7
Corolla RWD	2.6	2.8
Cressida		
1981-82	2.1	2.3
1983-88	2.9	3.1
1989-92	3.5	3.7
Echo	3.3	3.7
MR2	2.9	3.2
MR2 Spyder	2.1	2.4
Paseo	3.0	3.2
Starlet	2.5	2.7
Supra		
1981-92	2.9	3.1
1993-98		
w/TRAC	3.1	3.3
w/o TRAC	2.5	2.7

	LABOR TIME	SEVERE SERVICE
Tercel		
1981-82	2.4	2.6
1983-94		
sedan	3.1	3.3
wagon	2.5	2.7
1995-99	2.9	3.1
Power Booster Vacuum Check Valve, Replace (B)		
Exc. below (.4)	.6	.6
2003-05 Corolla, Matrix	.2	.2
Power Booster Vacuum Pump, Replace (B)		
Camry, Corolla	.5	.8
Proportioning Valve, Replace (B)		
Includes: System bleeding.		
Avalon	1.7	1.9
Camry		
1983-91	2.1	2.3
1992-01	1.6	1.8
Camry Solara	1.5	1.7
Celica		
1981-85	1.2	1.4
1986-99	1.9	2.1
2000-05	1.2	1.4
Corolla FWD, Matrix		
1983-91	1.9	2.1
1992-02	2.6	2.8
2003-05	1.5	1.7
Corolla FX, FX16	1.9	2.1
Corolla RWD	1.2	1.4
Cressida, MR2	1.2	1.4
Echo, Prius	1.4	1.5
Paseo	1.6	1.8
Starlet	1.2	1.4
1981-92 Supra	1.1	1.3
Tercel		
1981-82	1.2	1.4
1983-99	1.9	2.1
SERVICE BRAKES		
Brake Drum, Replace (B)		
Camry		
1983-01	1.4	1.6
2002-05	.8	.9
Camry Solara	1.1	1.3
Celica		
1981-85	1.3	1.4
1986-93	1.7	1.8
1994-97	1.1	1.3
2000-05	.6	.7
Corolla FWD, Matrix		
1983-02	1.4	1.6
2003-05	1.0	1.1
Corolla FX, FX16	1.3	1.4
Corolla RWD	1.3	1.4
1981-86 Cressida	1.3	1.4
Echo	.6	.7
Paseo	1.5	1.6
Prius	1.0	1.1
Starlet	.9	1.0

	LABOR TIME	SEVERE SERVICE
Tercel		
1981-82	1.1	1.2
1983-88	1.5	1.6
1989-90	2.4	2.7
1991-99	1.4	1.6
Caliper Assy., R&R and Recondition (B)		
Includes: System bleeding.		
Avalon	1.4	1.6
Camry	1.7	2.0
Camry Solara	1.4	1.7
Celica		
1981-93	1.9	2.1
1994-05	1.4	1.7
Corolla FWD		
1983-92		
front	1.8	2.1
rear	2.4	2.7
1993-02	1.4	1.7
Corolla FX, FX16		
front	1.6	1.9
rear	2.3	2.6
Corolla RWD	1.7	2.0
Cressida	1.7	2.0
Echo	1.3	1.4
MR2		
1985-89		
front	1.7	2.0
rear	2.2	2.5
1991-95	1.6	1.9
MR2 Spyder	1.3	1.4
Paseo	1.4	1.7
Starlet	1.8	2.1
Supra		
1981-92	1.8	2.1
1993-98	1.4	1.7
Tercel	1.6	1.9
Caliper Assy., Replace (B)		
Includes: System bleeding.		
Avalon	1.2	1.4
Camry		
1983-01	1.5	1.7
2002-05	1.9	2.1
Camry Solara	1.2	1.4
Celica	1.4	1.6
Corolla FWD, Matrix		
1983-92	1.6	1.8
1993-02	1.0	1.1
2003-05	1.6	1.8
Corolla FX, FX16	1.4	1.6
Corolla RWD	1.4	1.6
Cressida	1.4	1.6
Echo, Prius	1.3	1.4
MR2	1.4	1.6
MR2 Spyder	1.3	1.4
Paseo	1.2	1.4
Starlet	1.4	1.6
Supra		
1981-92	1.5	1.7
1993-98	.8	1.0
Tercel	1.2	1.4

MR2 SPYDER : PASEO : PRIUS : SOLARA : STARLET : SUPRA : TERCEL   TOY-41

	LABOR TIME	SEVERE SERVICE
Disc Brake Rotor, Replace (B)		
Avalon		
one	.6	.7
both		
front	.8	.9
rear	1.3	1.5
Camry		
one	.8	.9
both	1.3	1.4
Camry Solara		
one	.8	.9
both		
front	.8	.9
rear	1.2	1.4
Celica		
1983-85		
front		
one	1.7	1.9
both	2.4	2.6
rear		
one	1.1	1.3
both	1.5	1.7
1986-05		
one	1.1	1.3
both	1.6	1.8
Corolla FWD, Matrix		
1983-87		
front		
one	.8	1.0
both	1.4	1.6
rear		
one	1.4	1.6
both	2.5	2.7
1988-05		
one	.8	.9
both	1.3	1.4
Corolla FX, FX16		
front		
one	.8	1.0
both	1.3	1.5
rear		
one	1.5	1.7
both	2.5	2.7
Corolla RWD		
one	1.7	1.9
both	2.6	2.8
Cressida		
1981-88		
one	1.4	1.6
both		
front	2.2	2.4
rear	1.6	1.8
1989-92		
one	.7	.9
both	1.3	1.5
Echo		
one	1.0	1.1
both	1.4	1.6

	LABOR TIME	SEVERE SERVICE
MR2		
1985-89		
front		
1985-87		
one	1.7	1.9
both	2.9	3.1
1988-89		
one	.9	1.1
both	1.4	1.6
rear		
one	1.2	1.4
both	2.0	2.2
1991-95		
one	.7	.9
both	.9	1.1
MR2 Spyder		
one	1.0	1.1
both	1.4	1.6
Paseo		
one	.7	.9
both	.8	1.0
Prius		
one	1.1	1.3
both	1.6	1.8
Starlet		
one	1.4	1.6
both	1.9	2.1
Supra		
front		
1981-85		
one	1.5	1.7
both	2.4	2.6
1986-92		
one	1.2	1.4
both	1.7	1.9
rear		
1981-85		
one	1.4	1.6
both	1.9	2.1
1993-98		
one	.7	.9
both	.9	1.1
Tercel		
1981-82		
one	2.3	2.5
both	3.0	3.2
1983-90		
one	.9	1.1
both	1.5	1.7
1991-99		
one	.7	.9
both	.8	1.0
Pads and/or Shoes, Replace (B)		
Includes: Service and parking brake adjustment, system bleeding.		
Avalon		
front or rear disc	1.0	1.1
four wheels	1.7	1.8

	LABOR TIME	SEVERE SERVICE
Camry		
1983-86		
front disc	1.5	1.6
rear drum	2.1	2.2
four wheels	3.3	3.4
1987-91		
front or rear disc	1.5	1.6
four wheels	2.3	2.4
1992-05		
front or rear disc	1.0	1.1
rear drum	2.1	2.3
four wheels	2.7	2.8
Camry Solara		
front or rear disc	1.0	1.1
rear drum	1.7	2.0
four wheels	2.6	2.7
Celica		
1981-85		
front or rear disc	1.3	1.4
rear drum	2.2	2.3
four wheels	3.2	3.3
1986-89		
front or rear disc	1.4	1.5
rear drum	1.8	1.9
four wheels	3.3	3.4
1990-93		
front or rear disc	1.1	1.3
rear drum	2.1	2.2
four wheels	2.9	3.0
1994-99		
front or rear disc	1.0	1.1
rear drum	1.6	1.7
four wheels	2.6	2.7
2000-05		
front disc	1.3	1.4
rear disc	.8	.9
rear drum	1.3	1.4
Corolla FWD		
1983-91		
front or rear disc	1.5	1.6
rear drum	2.4	2.5
four wheels	3.4	3.5
1992-05		
front disc	1.3	1.4
rear drum	1.6	1.8
four wheels	2.5	2.6
Corolla FX, FX16		
front or rear disc	1.6	1.7
rear drum	2.1	2.2
four wheels	3.2	3.3
Corolla RWD		
front or rear disc	1.5	1.6
rear drum	2.1	2.2
four wheels	3.2	3.3
Cressida		
1981-88		
front disc	1.3	1.4
rear drum	2.1	2.2
four wheels	3.2	3.3

TOY-42   AVALON : CAMRY : CELICA : COROLLA : CRESSIDA : ECHO : MATRIX : MR2

	LABOR TIME	SEVERE SERVICE
Pads and/or Shoes, Replace (B)		
1989-92		
front or rear disc	1.4	1.5
four wheels	2.5	2.6
Echo		
front disc	1.3	1.4
rear drum	1.4	1.6
Matrix, MR2 Spyder		
front disc	1.3	1.4
rear disc	.8	.9
rear drum	1.7	2.0
MR2		
1985-89		
front or rear disc	1.6	1.7
rear disc	1.7	1.8
four wheels	2.9	3.0
1991-95		
front disc	.8	.9
rear disc	1.2	1.3
four wheels	2.0	2.1
Paseo		
front disc	1.1	1.3
rear drum	1.9	2.0
four wheels	2.7	2.8
Prius		
front disc	1.1	1.3
rear drum	1.4	1.6
Starlet		
front disc	1.1	1.3
rear drum	2.2	2.3
four wheels	3.2	3.3
Supra		
1981-92		
front or rear disc	1.3	1.4
four wheels	2.7	2.8
1993-98		
front or rear disc	.8	.9
four wheels	1.7	1.8
Tercel		
1981-82		
front disc	1.2	1.3
rear drum	2.7	2.8
four wheels	3.4	3.5
1983-90		
front disc	1.5	1.6
rear drum	2.2	2.3
four wheels	3.3	3.4
1991-99		
front disc	.8	.9
rear drum	2.1	2.2
four wheels	2.7	2.8
COMBINATION ADD-ONS		
Replace		
brake drum add	.2	.2
brake hose add	.3	.3
caliper add	.2	.2
disc brake rotor add	.2	.2
wheel cylinder add	.2	.2

	LABOR TIME	SEVERE SERVICE
Resurface		
brake drum add	.5	.5
disc rotor add		
w/wheel bearings	.9	.9
w/o wheel bearings	.5	.5
Wheel Cylinder, R&R and Recondition (B)		
Includes: System bleeding.		
Camry		
1983-98	1.9	2.2
1999-01	1.3	1.6
Camry Solara	1.3	1.5
Celica	1.6	1.8
Corolla FWD		
1983-98	1.8	2.1
1999-02	1.3	1.6
Corolla FX, FX16	1.7	2.0
1981-82 Corolla RWD	1.9	2.1
1981-86 Cressida	1.9	2.1
Echo	1.4	1.6
Paseo	1.5	1.8
Starlet	2.1	2.4
Tercel	1.7	2.0
each addl. add	.8	.9
Wheel Cylinder, Replace (B)		
Includes: System bleeding.		
Camry		
1983-02	1.4	1.6
2003-05	2.4	2.7
Camry Solara	1.3	1.4
Celica	1.4	1.6
Corolla	1.7	2.0
1981-86 Cressida	1.6	1.8
Echo	1.3	1.4
Matrix	1.9	2.1
Paseo	1.5	1.7
Prius		
2001-03	1.3	1.4
2004-05	1.9	2.1
Starlet	1.7	1.9
Tercel	1.5	1.7
PARKING BRAKE		
Parking Brake Cable, Adjust (C)		
Exc. below	.8	.9
1981-85 Celica		
w/drum brakes	.3	.5
1998-02 Corolla FWD	1.3	1.5
1981-82 Cressida		
w/drum brakes	.3	.5
1991-95 MR2	.3	.5
MR2 Spyder	1.0	1.1
Starlet	.3	.5
Parking Brake Apply Actuator, Replace (B)		
Avalon	1.4	1.4
Camry	.9	.9
Pedal		
2002-05	.7	.7

	LABOR TIME	SEVERE SERVICE
Camry Solara	.8	.8
Celica		
1981-92	.7	.7
1994-99	1.3	1.3
2000-05	.6	.6
Corolla FWD	.7	.7
Corolla FX, FX16	.8	.8
Corolla RWD		
1981-82	1.3	1.3
1983-87	.7	.7
Cressida	1.4	1.4
Echo, Matrix	.6	.6
MR2		
1985-89	1.3	1.3
1991-95	1.6	1.6
MR2 Spyder	.6	.6
Paseo, Prius	.8	.8
Starlet	.5	.5
Supra		
1981-85	.7	.7
1986-92	1.3	1.3
1993-98	.5	.5
Tercel	.8	.8
Parking Brake Cable, Replace (B)		
Front		
Avalon		
1995-99	1.6	1.9
2000-04	2.1	2.3
Camry		
1983-86	1.3	1.6
1987-91	.8	1.1
1992-96		
w/pedal type	1.6	1.9
w/o pedal type	.9	1.2
1997-05	1.3	1.6
Camry Solara	1.5	1.8
Celica		
1981-82	.8	1.1
1983-05	1.6	1.9
Corolla FWD		
1983-91	1.1	1.4
1992-93	.8	1.1
1994-05	1.4	1.6
Corolla FX, FX16	1.1	1.4
Corolla RWD		
1981-83	1.5	1.8
1983-87	1.1	1.4
Cressida	1.8	2.1
Echo, MR2 Spyder	.6	.9
Matrix		
2WD	1.4	1.6
4WD	.8	.9
MR2		
1985-89	1.2	1.5
1991-95	1.6	1.9
Paseo	.8	1.1
Prius		
2001-03	1.1	1.3
2004-05	7.5	8.0

MR2 SPYDER : PASEO : PRIUS : SOLARA : STARLET : SUPRA : TERCEL — TOY-43

	LABOR TIME	SEVERE SERVICE
Supra		
1981-85	1.3	1.6
1986-92	2.0	2.3
1993-98	.7	1.0
Tercel		
1983-90		
2WD	1.4	1.7
4WD	.8	1.1
1991-99	.8	1.1
Rear		
Avalon	2.5	2.9
Camry		
1983-86	1.6	1.9
1987-91		
4 cyl.		
w/drum brakes	1.5	1.8
w/disc brakes	2.7	3.0
V6	2.5	2.9
1992-01	2.5	2.9
2002-05	1.3	1.4
Camry Solara	2.7	3.0
Celica		
1981-89	1.6	1.9
1990-99	2.0	2.3
2000-05		
GT	2.4	2.7
GTS	2.9	3.2
Corolla FWD		
1983-87	1.4	1.7
1988-04	2.4	2.7
Corolla FX, FX16	1.5	1.8
Corolla RWD	1.6	1.9
Cressida		
1981-82		
w/drum brakes	1.2	1.5
w/disc brakes	1.6	1.9
1983-92	1.9	2.1
Echo	2.2	2.5
Matrix		
2WD	2.5	2.9
4WD	1.9	2.1
XRS	3.0	3.4
MR2	1.6	1.9
MR2 Spyder	3.3	3.7
Paseo	2.6	2.9
Prius		
2001-03	2.1	2.3
2004-05	2.9	3.2
Starlet	1.3	1.6
Supra		
1981-92	1.6	1.9
1993-98	.7	1.0
Tercel		
1981-90	1.8	2.1
1991-99	2.5	2.9

TRACTION CONTROL

	LABOR TIME	SEVERE SERVICE
Actuator Assy., Replace (B)		
Avalon	1.7	1.7
1997-01 Camry	1.7	1.7
Camry Solara	1.7	1.7
1993-98 Supra	1.4	1.4
Control Unit, Replace (B)		
Avalon	1.7	1.7
1997-01 Camry	.7	.7
Camry Solara	.7	.7
1993-98 Supra	.9	.9
Main Relay, Replace (B)		
Avalon	.2	.2
1997-01 Camry	.2	.2
Camry Solara	.2	.2
1995-98 Supra	.2	.2
Pump Assy., Replace (B)		
1993-98 Supra	1.8	2.1

FRONT SUSPENSION

Unless otherwise noted, time given does not include alignment.

	LABOR TIME	SEVERE SERVICE
Front End Alignment, Check (A)		
1981-05	1.9	2.2
Adjust		
camber add	.8	.9
front toe add	.2	.2
rear camber add	.3	.3
rear toe add	.3	.3
Coil Spring, Replace (B)		
Avalon	1.9	2.1
Camry		
1983-91	2.8	3.0
1992-96	2.1	2.3
1997-01	1.7	1.9
2002-05	1.4	1.6
Camry Solara	1.9	2.1
Celica		
1981-85	2.0	2.2
1986-89	2.9	3.1
1990-99	2.1	2.3
2000-05	1.6	1.8
Corolla FWD		
1983-92	2.7	2.9
1993-05		
2WD	1.9	2.1
4WD	2.9	3.1
Corolla FX, FX16	2.9	3.1
Corolla RWD	2.0	2.2
Cressida		
1981-84	1.8	2.0
1985-92	2.3	2.5
Echo, Matrix	1.9	2.1
MR2	2.6	2.8
MR2 Spyder	1.7	2.0
Paseo	2.3	2.5
Prius		
2001-03	2.1	2.3
2004-05	2.4	2.7

	LABOR TIME	SEVERE SERVICE
Starlet	2.0	2.2
Supra		
1981-92	2.0	2.2
1993-98		
one	2.3	2.5
both	3.7	3.9
Tercel		
1981-82	1.6	1.8
1983-90		
sedan	1.9	2.1
wagon	2.4	2.6
1991-99	2.1	2.3
Front Axle Hub Oil Seal, Replace (B)		
1981-85 Celica	.9	1.0
Corolla RWD	1.4	1.5
Cressida		
1981-88	1.4	1.5
1989-92	3.2	3.3
MR2		
1985-89	3.0	3.1
1991-95	2.5	2.6
Starlet	.9	1.0
Supra		
1981-86	.9	1.0
1987-92	2.8	2.9
1993-98	2.5	2.6
Front Hub Assy., Replace (B)		
Avalon	2.9	3.2
Camry		
1983-91	3.0	3.5
1992-01	2.7	3.2
2002-05	3.0	3.5
Camry Solara		
1999-03 (1.7)	2.5	3.0
2004-05 (2.0)	3.0	3.5
Celica		
1981-85	1.8	2.3
1986-89	3.2	3.7
1990-99	2.6	3.1
2000-05	2.2	2.5
Corolla RWD	1.9	2.4
Corolla FWD, Matrix		
1983-92	3.2	3.7
1993-97		
2WD	1.7	2.2
4WD	2.9	3.4
1998-02	2.7	3.2
2003-05	3.2	3.6
Corolla FX, FX16	3.1	3.5
Cressida	1.8	2.3
Echo	3.0	3.4
MR2		
1985-89	3.1	3.6
1991-95	2.6	3.1
MR2 Spyder	2.5	3.0
Paseo	2.9	3.4
Prius		
2001-03	2.9	3.2
2004-05	4.0	4.5

TOY-44 AVALON : CAMRY : CELICA : COROLLA : CRESSIDA : ECHO : MATRIX : MR2

	LABOR TIME	SEVERE SERVICE
Front Hub Assy., Replace (B)		
Starlet	1.4	*1.9*
Supra		
1981-86	1.8	*2.3*
1987-92	3.0	*3.5*
1993-98	2.4	*2.9*
Tercel	2.9	*3.4*
Front Hub Bearing, Replace (B)		
Avalon	2.5	*2.9*
Camry		
1983-91	3.4	*3.9*
1992-01	2.6	*3.0*
Camry Solara	2.6	*3.0*
Celica		
1981-85	1.6	*1.8*
1986-89	3.2	*3.6*
1990-05	2.4	*2.7*
Corolla FWD, Matrix	2.9	*3.2*
Corolla FX, FX16	3.0	*3.4*
Corolla RWD	1.7	*2.0*
Cressida	1.6	*1.8*
Echo	2.7	*3.0*
MR2	3.0	*3.4*
Paseo, Prius	2.5	*2.9*
Starlet	1.2	*1.4*
Supra		
1981-86	1.5	*1.8*
1987-92	3.0	*3.4*
1993-98	2.5	*2.9*
Tercel		
1981-82	2.6	*3.0*
1983-90	3.2	*3.6*
1991-99	2.7	*3.0*
Front Hub Seal, Replace (B)		
Camry		
1983-86	3.1	*3.1*
1987-91	3.3	*3.3*
Celica		
1986-89	3.2	*3.2*
1990-99	2.6	*2.6*
Corolla FWD		
1983-87	2.9	*2.9*
1988-92	3.1	*3.1*
1993-02		
2WD	2.7	*2.7*
4WD	3.0	*3.0*
Corolla FX, FX16	2.2	*2.2*
Paseo	2.7	*2.7*
Tercel		
1981-82	2.6	*2.6*
1983-90	3.1	*3.1*
1991-99	2.9	*2.9*
Front Shock Absorbers, Replace (B)		
1981-85 Celica	3.0	*3.2*
1981-87 Corolla FWD	2.9	*3.1*
Cressida		
1981-84	2.9	*3.1*

	LABOR TIME	SEVERE SERVICE
1985-88		
w/TEMS	3.4	*3.6*
w/o TEMS	3.0	*3.2*
1989-92	2.1	*2.3*
MR2		
1985-89		
one	2.7	*2.9*
both	3.6	*3.8*
1991-95		
one	2.3	*2.5*
both	3.2	*3.4*
Starlet	3.0	*3.2*
Supra		
1981-86	3.0	*3.2*
1987-92		
w/TEMS		
one	2.2	*2.4*
both	3.6	*3.8*
w/o TEMS		
one	2.3	*2.5*
both	3.2	*3.4*
1993-98		
one	2.4	*2.6*
both	3.5	*3.7*
Lower Control Arm, Replace (B)		
Avalon	2.1	*2.3*
Camry		
1983-86	2.0	*2.3*
1987-91	2.8	*3.1*
1992-96	2.0	*2.3*
1997-98		
right side	1.6	*1.9*
left side		
AT	2.7	*3.0*
MT	1.5	*1.8*
1999-01	1.3	*1.6*
2002-05		
4 cyl.	8.7	*9.0*
V6	9.7	*10.0*
Camry Solara	1.4	*1.7*
Celica		
1981-99	2.7	*3.0*
2000-05	3.0	*3.3*
Corolla FWD, Matrix		
1983-92	2.4	*2.7*
1993-02		
2WD		
AT	2.1	*2.3*
MT	1.4	*1.7*
4WD	2.3	*2.6*
2003-05	1.4	*1.6*
Corolla FX, FX16	1.9	*2.1*
Corolla RWD	2.8	*3.1*
Cressida		
1981-88	2.7	*2.9*
1989-92	1.3	*1.5*
Echo	1.0	*1.1*
MR2	1.9	*2.1*
MR2 Spyder	1.1	*1.3*
Paseo	1.4	*1.6*

	LABOR TIME	SEVERE SERVICE
Prius		
2001-03	2.1	*2.3*
2004-05	3.3	*3.5*
Starlet	2.7	*2.9*
Supra		
1981-92	2.8	*3.0*
1993-98	1.7	*1.9*
Tercel		
1981-82	1.9	*2.1*
1983-90		
sedan	1.6	*1.8*
wagon	2.3	*2.5*
1991-99	1.4	*1.6*
Stabilizer Bar, Replace (B)		
Avalon	1.7	*2.0*
Camry		
1995-01	1.7	*1.9*
2002-05	2.8	*3.1*
Camry Solara	1.7	*2.0*
Celica		
1995-99	1.2	*1.4*
2000-05	3.4	*3.6*
Corolla FWD, Matrix		
1995-97	.8	*1.0*
1998-02	1.3	*1.5*
2003-05	1.9	*2.1*
Echo	2.7	*3.0*
1995 MR2	.7	*.9*
MR2 Spyder	1.3	*1.4*
1995-99 Paseo	.8	*1.0*
Prius		
2001-03	2.1	*2.3*
2004-05	3.0	*3.2*
Stabilizer Bar Bushing, Replace (B)		
Avalon	.7	*.9*
Camry		
1995-01	.7	*.9*
2002-05	2.5	*2.7*
Camry Solara	.7	*.9*
1995-05 Celica	.8	*.9*
Corolla FWD, Matrix		
1995-02	.5	*.7*
2003-05	2.1	*2.3*
Echo	.5	*.7*
1995 MR2	.5	*.7*
MR2 Spyder	1.3	*1.4*
1995-99 Paseo	.5	*.7*
Prius	.8	*.9*
Steering Knuckle, Replace (B)		
Avalon	2.9	*3.2*
Camry	3.2	*3.6*
Camry Solara	2.9	*3.2*
Celica		
1981-89	3.3	*3.7*
1990-99	2.7	*3.0*
2000-05	1.3	*1.4*
Corolla FWD, Matrix		
1983-92	3.1	*3.4*

MR2 SPYDER : PASEO : PRIUS : SOLARA : STARLET : SUPRA : TERCEL — TOY-45

	LABOR TIME	SEVERE SERVICE
1993-02		
2WD	2.5	2.6
4WD	3.0	3.4
2003-05	3.0	3.4
Corolla FX, FX16	3.0	3.4
Corolla RWD	3.2	3.6
Cressida		
1981-88	3.2	3.6
1989-92	2.6	2.8
Echo	3.0	3.4
MR2		
1985-89	3.2	3.6
1991-95	2.7	3.0
MR2 Spyder	1.6	1.8
Paseo	2.9	3.2
Prius		
2001-03	2.7	3.0
2004-05	3.8	4.1
Supra		
1981-86	3.3	3.7
1987-92	1.9	2.1
1993-98	2.5	2.9
Tercel		
1983-90		
sedan	2.5	2.9
wagon	3.2	3.6
1991-99	2.7	3.0
Strut, Replace (B)		
Avalon		
one	1.9	2.1
both	2.7	3.0
Camry		
1983-91		
one	2.8	3.0
both	3.6	3.8
1992-01		
one	1.9	2.1
both	2.5	2.7
2002-05		
one	1.4	1.6
both	2.2	2.5
Camry Solara		
one	1.8	2.0
both	2.5	2.9
Celica		
1986-89		
one	2.7	2.9
both	3.4	3.6
1990-99		
one	1.9	2.1
both	2.7	2.9
2000-05		
one	1.4	1.6
both	1.9	2.1
Corolla FWD, Matrix		
1983-92		
one	2.9	3.1
both	3.6	3.8

	LABOR TIME	SEVERE SERVICE
1993-02		
2WD		
one	1.9	2.1
both	2.5	2.7
4WD		
one	2.7	2.9
both	3.6	3.8
2003-05		
one	1.7	2.0
both	2.5	2.9
Corolla FX, FX16		
one	2.8	3.0
both	3.6	3.8
Echo, MR2 Spyder		
one	1.6	1.8
both	2.5	2.9
Paseo		
one	1.8	2.0
both	2.7	2.9
Prius		
one	1.9	2.1
both	2.4	2.7
Tercel		
1981-82		
one	2.3	2.5
both	3.1	3.3
1983-90 sedan		
one	1.9	2.1
both	2.5	2.7
1985-88 wagon		
one	2.6	2.8
both	3.1	3.3
1991-99		
one	1.8	2.0
both	2.6	2.8
Strut Cartridge, Replace (B)		
1981-85 Celica	2.9	3.2
1981-84 Corolla RWD	2.6	2.9
1981-84 Cressida	3.0	3.3
1981-86 Supra	2.9	3.2
Upper Control Arm Assy., Replace (B)		
Supra		
1987-92	2.5	2.8
1993-98	1.9	2.2
Upper Control Arm Bushings and/or Shaft, Replace (B)		
Supra		
1987-92	1.4	1.6
1993-98	2.1	2.3

REAR SUSPENSION

	LABOR TIME	SEVERE SERVICE
Rear Suspension Toe or Camber, Adjust (B)		
1986-05	2.2	2.5
Coil Spring, R&R or Replace (B)		
Corolla FWD		
1981-90		
one	.9	1.1
both	1.4	1.6

	LABOR TIME	SEVERE SERVICE
1991-93		
one	1.4	1.6
both	1.7	1.9
2003-05		
one	1.9	2.1
both	2.7	3.0
Echo		
one	.5	.7
both	.6	.8
Matrix, 2WD		
one	1.7	2.0
both	2.4	2.7
Paseo		
1992-95		
one	1.2	1.4
both	1.7	1.9
1996-99		
one	1.6	1.8
both	2.3	2.5
Prius		
one	2.7	3.0
both	3.6	4.1
Tercel		
1981-90		
one	.9	1.1
both	1.4	1.6
1991-99		
one	1.7	1.9
both	1.8	2.0
Lower Control Arm, Replace (B)		
Corolla FWD, Tercel		
1981-90	.8	1.1
1991-93	1.2	1.5
Rear Control Arm Bushing, Replace (B)		
1985-98 one side	2.6	2.8
Rear Control Arm No. 1, Replace (B)		
Avalon	2.2	2.5
Camry		
1983-91	.9	1.1
1992-01	2.1	2.3
2002-05	3.0	3.4
Camry Solara	2.2	2.5
Celica		
1981-85	4.5	4.7
1986-93	1.3	1.5
1994-99	2.1	2.3
2000-05	1.3	1.4
Corolla FWD		
1983-92	.9	1.1
1993-02	2.3	2.5
Cressida		
1981-88	4.5	4.7
1989-92	2.0	2.2
Matrix, 4WD	1.4	1.6
MR2	1.5	1.7
MR2 Spyder	.8	.9

	LABOR TIME	SEVERE SERVICE
Rear Control Arm No. 1, Replace (B)		
Supra		
1981-86	4.7	4.9
1987-92	1.6	1.8
1993-98	2.2	2.4
Tercel wagon	.9	1.1
Rear Control Arm No. 2, Replace (B)		
Exc. below (.8)	1.3	1.4
1989-92 Cressida	1.9	2.1
MR2 Spyder	.8	.9
Supra		
1987-92	1.5	1.7
1993-98	2.2	2.4
Rear Hub and/or Bearings, Replace (B)		
1997-99 Avalon		
one side	1.3	1.8
both sides	1.6	2.1
1997-01 Camry		
one side	1.1	1.6
both sides	1.7	2.2
Camry Solara		
one side	1.3	1.8
both sides	1.8	2.3
2003-05 Corolla, Matrix 2WD		
one side	1.3	1.8
both sides	1.8	2.3
Echo		
one side	1.7	2.2
both sides	2.2	2.7
Prius		
one side	1.7	2.0
both sides	2.2	2.5
Rear Leaf Spring, Replace (B)		
Corolla RWD	1.5	1.8
Shock Absorbers, Replace (B)		
1981-85 Celica		
one	1.2	1.4
both	1.4	1.6
Corolla FWD, Matrix 2WD		
1981-90		
one	.7	.9
both	.8	1.0
1991-93		
one	1.2	1.4
both	1.5	1.7
2003-05		
one	1.9	2.1
both	2.7	3.0
1981-82 Corolla RWD		
one	.5	.7
both	.8	1.0
1981-88 Cressida		
one	1.3	1.5
both	1.7	1.9
Echo		
2000-05		
one	1.9	2.1
both	2.1	2.3

	LABOR TIME	SEVERE SERVICE
Paseo		
one	1.4	1.6
both	2.0	2.2
Prius		
one	2.2	2.5
both	3.5	3.9
1981-86 Supra		
one	1.1	1.3
both	1.6	1.8
Tercel		
1981-90		
one	.7	.9
both	.8	1.0
1991-99		
one	1.7	1.9
both	1.8	2.0
Spring Shackle, Replace (B)		
Corolla RWD	.8	1.0
Stabilizer Bar, Replace (B)		
Avalon	.8	.9
Camry		
1995-01	.7	.9
2002-05	1.4	1.6
Camry Solara	.8	.9
1995-05 Celica	1.0	1.1
Corolla FWD		
1995-02	2.0	2.2
2003-05	.6	.7
Matrix		
2WD	.6	.7
4WD	6.7	7.5
1995 MR2	2.5	2.7
MR2 Spyder	.8	.9
1995-98 Supra	.7	.9
Stabilizer Bar Bushing, Replace (B)		
Avalon	.5	.7
1995-05 Camry	.5	.7
Camry Solara	.5	.7
1995-05 Celica	1.0	1.1
1995-02 Corolla FWD	.7	.9
Matrix 4WD	6.7	7.5
1995 MR2	2.5	2.7
MR2 Spyder	.6	.7
1995-98 Supra	.5	.7
Strut Coil Spring, Replace (B)		
Avalon		
one	2.4	2.7
both	3.5	3.9
Camry		
1983-96		
one	2.5	2.7
both	3.4	3.6
1997-05		
one	1.9	2.1
both	2.9	3.2
Camry Solara		
one	2.2	2.5
both	3.3	3.7

	LABOR TIME	SEVERE SERVICE
Celica		
1986-89		
2WD		
one	1.9	2.1
both	2.8	3.0
4WD		
one	2.6	2.8
both	3.0	3.2
1990-99		
one	2.4	2.6
both	3.4	3.6
2000-05		
one	1.7	2.0
both	2.7	3.0
Corolla FWD		
1983-87		
one	2.2	2.4
both	3.2	3.4
1988-92		
one	2.6	2.8
both	3.5	3.7
1993-98		
one	4.4	4.6
both	7.1	7.3
1999-02		
one	4.1	4.3
both	6.7	6.9
Corolla FX, FX16		
one	2.2	2.4
both	3.2	3.4
1989-92 Cressida		
one	1.5	1.7
both	2.3	2.5
Matrix 4WD		
one	2.1	2.3
both	2.9	3.2
MR2		
one	2.4	2.6
both		
1985-89	3.0	3.2
1991-95	3.6	3.8
MR2 Spyder		
one	1.3	1.4
both	2.2	2.5
Supra		
1987-92		
w/TEMS		
one	2.7	2.9
both	3.8	4.0
w/o TEMS		
one	2.0	2.2
both	3.1	3.3
1993-98		
w/sport roof		
one	2.8	3.0
both	3.8	4.0
w/o sport roof		
one	2.1	2.3
both	2.9	3.1

MR2 SPYDER : PASEO : PRIUS : SOLARA : STARLET : SUPRA : TERCEL **TOY-47**

	LABOR TIME	SEVERE SERVICE
1985-88 Tercel wagon		
one	2.3	2.5
both	3.1	3.3
Strut Rod, Replace (B)		
Avalon	.8	.9
Camry	1.1	1.3
Camry Solara	.8	.9
Corolla FWD		
1981-98	.9	1.1
1999-02	.7	.9
Corolla FX, FX16	.9	1.1
1988-92 Cressida	1.6	1.8
MR2	1.7	1.9
MR2 Spyder	.8	.9
1987-98 Supra	1.5	1.7
Tercel wagon	.9	1.1
Strut Shock Absorber, Replace (B)		
Avalon		
one	2.2	2.5
both	3.0	3.4
Camry		
1983-96		
one	2.3	2.5
both	3.1	3.3
1997-01		
one	1.7	1.9
both	2.7	2.9
2002-05		
one	2.1	2.3
both	3.0	3.4
Camry Solara		
one	1.9	2.1
both	2.7	3.0
Celica		
1986-89		
2WD		
one	1.6	1.8
both	2.5	2.7
4WD		
one	2.4	2.6
both	3.1	3.3
1990-99		
one	2.0	2.2
both	3.0	3.2
2000-05		
one	1.6	1.8
both	2.4	2.7
Corolla FWD		
1983-92		
one	2.6	2.8
both	3.3	3.5
1993-02		
one	3.6	3.8
both	6.2	6.4
Corolla FX, FX16		
one	2.0	2.2
both	3.1	3.3
Matrix, 4WD		
one	2.1	2.3
both	2.9	3.2

	LABOR TIME	SEVERE SERVICE
MR2		
one	2.4	2.6
both	3.2	3.4
MR2 Spyder		
one	1.1	1.3
both	1.9	2.1
Supra		
1987-92		
w/TEMS		
one	2.6	2.8
both	3.6	3.8
w/o TEMS		
one	1.8	2.0
both	3.1	3.3
1993-98		
one	2.4	2.6
both	3.2	3.4
1985-88 Tercel wagon		
one	1.9	2.1
both	2.8	3.0
ELECTRONIC SUSPENSION		
Front Shock Actuator, Replace (B)		
1985-88 Cressida	.7	.7
1987-92 Supra	.7	.7
Mode Control Switch, Replace (B)		
1985-88 Cressida	.5	.5
1987-92 Supra	.5	.5
Rear Shock Actuator, Replace (B)		
1985-88 Cressida	.7	.7
1987-92 Supra	1.1	1.1
Shock Absorber Control Computer, Replace (B)		
1985-88 Cressida	.9	.9
1987-92 Supra	.9	.9
Steering Sensor, Replace (B)		
1985-88 Cressida	1.3	1.3
1987-92 Supra	1.4	1.4

STEERING
AIR BAGS

	LABOR TIME	SEVERE SERVICE
Diagnose Air Bag System Component Each (A)		
1990-05	1.0	1.0
Air Bag Assy., Replace (B)		
Driver side		
1990-05	.8	.8
Passenger side		
Avalon		
1995-99	3.2	3.2
2000-04	2.7	2.7
Camry		
1995-96	1.6	1.6
1997-01	.7	.7
2002-05	3.0	3.0
Camry Solara		
1999-03	.7	.7
2004-05	2.1	2.1

	LABOR TIME	SEVERE SERVICE
Celica		
1995-99	3.6	3.6
2000-05	.6	.6
1995-05 Corolla	.7	.7
Echo	.8	.8
Matrix	.6	.6
1995 MR2	.8	.8
MR2 Spyder	.7	.7
1996-99 Paseo	2.9	2.9
Prius	1.1	1.1
1995-98 Supra	.8	.8
1995-99 Tercel	3.0	3.0
Side impact		
1998-99 Avalon		
one side	1.3	1.3
both sides	1.8	1.8
Camry		
1998-01		
one side	2.7	2.7
both sides	4.1	4.1
2002-05		
one side	2.4	2.4
both sides	2.8	2.8
1998-02 Corolla		
one side	1.3	1.3
both sides	1.7	1.7
Center (Crash) Sensor, Replace (B)		
Exc. below	.8	.8
Avalon	1.4	1.4
2002-05 Camry	1.1	1.1
Celica		
1991-93	1.5	1.5
2000-05	.9	.9
Echo	.5	.5
1991-95 MR2	1.4	1.4
MR2 Spyder	.5	.5
1993-95 Paseo	1.2	1.2
2001-05 Prius	.5	.5
1993-94 Tercel	1.3	1.3
Front Sensor, Replace (B)		
Avalon		
1995-99	1.4	1.4
2000-04		
one	.8	.8
both	1.3	1.3
Camry		
1992-96		
one	.8	.8
both	1.3	1.3
1997-01		
one	.7	.7
both	1.4	1.4
2002-05		
one	.6	.6
both	.8	.8
Camry Solara		
one	.7	.7
both	1.4	1.4

TOY-48 AVALON : CAMRY : CELICA : COROLLA : CRESSIDA : ECHO : MATRIX : MR2

	LABOR TIME	SEVERE SERVICE
Front Sensor, Replace (B)		
Celica		
1991-93		
one	.9	.9
both	1.7	1.7
1994-99		
one	2.1	2.1
both	2.4	2.4
2000-05		
one	.5	.5
both	.6	.6
Corolla, Matrix		
1994-02		
one	.7	.7
both	1.3	1.3
2003-05		
one	.6	.6
both	.8	.8
Echo		
one	.3	.3
both	.5	.5
1991-95 MR2		
one	.7	.7
both	.9	.9
MR2 Spyder		
one	.5	.5
both	.6	.6
1992-95 Paseo		
one	.8	.8
both	1.4	1.4
Prius		
one	.8	.8
both	.9	.9
Supra		
1990-92		
one	.8	.8
both	1.4	1.4
1993-98		
one	1.6	1.6
both	2.8	2.8
1993-94 Tercel		
one	.8	.8
both	1.3	1.3
Rear Sensor, Replace (B)		
2002-05 Camry, Camry Solara		
one side	.6	.6
both sides	.8	.8
2004-05 Prius		
one side	.7	.7
both sides	.9	.9
Side Sensor, Replace (B)		
Avalon		
1995-99		
one side	.7	.7
both sides	1.5	1.5
2000-04		
one side	.5	.5
both sides	.6	.6

	LABOR TIME	SEVERE SERVICE
1997-05 Camry		
one side	.8	.8
both sides	1.5	1.5
Camry Solara		
one side	.7	.7
both sides	1.4	1.4
2000-05 Celica		
door mounted		
one side	.6	.6
both sides	.8	.8
not door mounted		
one side	1.1	1.1
both sides	1.6	1.6
1998-05 Corolla		
one side	.8	.8
both sides	1.2	1.2
Echo		
door mounted		
one side	.5	.5
both sides	.6	.6
not door mounted		
one side	.8	.8
both sides	1.1	1.1
Matrix		
one side	.9	.9
both sides	1.5	1.5
Prius		
one side	.6	.6
both sides	.8	.8
MANUAL RACK & PINION		
Horn Contact Cable or Ring, Replace (B)		
1981-98	.5	.5
Inner Tie Rod Assy., Replace (B)		
Includes: Reset toe.		
1983-86 Camry	2.7	2.9
1981-85 Celica	2.4	2.6
Corolla FWD		
1983-87	2.4	2.6
1988-92	2.7	2.9
1993-97		
2WD	2.4	2.6
4WD	2.8	3.0
Corolla RWD	2.4	2.6
Corolla FX, FX16	2.6	2.8
MR2		
1985-89	2.6	2.8
1991-95	2.5	2.7
Starlet	2.6	2.8
Tercel		
1981-86	2.5	2.7
1987-90	2.7	2.9
1991-99	2.4	2.6
Outer Tie Rod End, Replace (B)		
Includes: Reset toe.		
1981-99	1.3	1.5

	LABOR TIME	SEVERE SERVICE
Rack & Pinion Assy., R&R or Replace (B)		
Includes: Adjustments.		
1983-86 Camry	2.8	3.3
1981-85 Celica	2.4	2.9
Corolla FWD		
1983-92	2.3	2.8
1993-97		
2WD	1.9	2.4
4WD	2.7	3.2
Corolla RWD	2.3	2.8
Corolla FX, FX16	2.2	2.7
MR2		
1985-89	2.2	2.7
1991-95	1.9	2.4
Starlet	2.4	2.9
Tercel		
1981-86	2.4	2.9
1987-90	2.4	2.9
1991-99	1.8	2.3
Rack Boots, Replace (B)		
Includes: Reset toe.		
1983-86 Camry	1.4	1.6
1981-85 Celica	2.1	2.3
Corolla FWD		
1983-87	1.5	1.7
1988-92	1.7	1.9
1993-97	1.5	1.7
Corolla RWD	1.4	1.6
Corolla FX, FX16	1.6	1.8
MR2	1.4	1.6
Starlet	3.2	3.4
Tercel		
1981-86	1.9	2.1
1987-90	2.2	2.4
1991-99	1.5	1.7
Steering Wheel, Replace (B)		
1981-99	.7	.7
POWER RACK & PINION		
Steering Pump Pressure Check (B)		
1981-05	2.2	2.3
Horn Contact or Canceling Cam, Replace (B)		
1981-05	.5	.5
Inner Tie Rod Assy., Replace (B)		
Includes: Reset toe.		
Avalon	4.3	4.8
Camry		
1983-86	3.2	3.4
1987-88	4.2	4.4
1989-91		
4 cyl.		
2WD	4.1	4.3
4WD	5.2	5.4
V6	4.5	4.7
1992-96	3.7	3.9
1997-01	3.1	3.3
2002-05	5.4	5.9

MR2 SPYDER : PASEO : PRIUS : SOLARA : STARLET : SUPRA : TERCEL — TOY-49

	LABOR TIME	SEVERE SERVICE
Camry Solara		
1999-03	3.5	3.9
2004-05	4.1	4.6
Celica		
1983-85	2.9	3.1
1986-87	4.2	4.4
1988-89		
2WD	4.0	4.2
4WD	5.2	5.4
1990-93		
2WD	3.7	3.9
4WD	5.5	5.7
1994-99	3.9	4.1
2000-05	5.3	5.8
Corolla FWD, Matrix		
1983-87	3.0	3.2
1987-92	5.6	5.8
1993		
2WD	4.2	4.4
4WD	5.7	5.9
1994-02	3.6	3.8
2003-05	6.5	7.0
Corolla FX, FX16	3.5	3.7
Corolla RWD	3.3	3.5
Cressida	3.2	3.4
Echo	4.1	4.6
1991-95 MR2	2.9	3.1
MR2 Spyder	3.0	3.4
Paseo		
1992-95	3.5	3.7
1996-99	4.2	4.4
Prius	2.4	2.7
Supra	3.1	3.3
Tercel		
1983-94	3.8	4.0
1995-99	4.3	4.5
Outer Tie Rod Ends, Replace (B)		
Includes: Reset toe.		
Exc. below	1.6	1.8
2002-05 Camry	3.0	3.4
2003-05 Corolla, Matrix	3.0	3.4
Power Steering Hoses, Replace (B)		
Avalon, each	1.4	1.6
Camry		
pressure		
1983-86	1.6	1.8
1987-88	2.5	2.7
1989-91		
4 cyl.		
2WD	2.3	2.5
4WD	3.0	3.2
V6	2.4	2.6
1992-05	1.4	1.6
return		
1983-88	.7	.9
1989-91		
2WD	.7	.9
4WD	1.2	1.4
1992-05	1.2	1.4

	LABOR TIME	SEVERE SERVICE
Camry Solara, each	1.5	1.7
Celica		
pressure		
1981-82	1.2	1.4
1983-85	1.9	2.1
1986-87	2.7	2.9
1988-89		
2WD	2.5	2.7
4WD	3.0	3.2
1990-93	2.3	2.5
1994-05	1.6	1.8
return		
1981-99	.8	1.0
2000-05	2.5	2.9
Corolla FWD, Matrix		
pressure	1.7	1.9
return		
1983-02	.7	.9
2003-05	1.8	2.1
Corolla RWD		
pressure	1.9	2.1
return	.9	1.1
1985-92 Cressida		
pressure	1.4	1.6
return	.8	1.0
Echo		
pressure	1.7	1.9
return	1.4	1.5
1991-95 MR2		
pressure	1.6	1.8
return	1.3	1.5
MR2 Spyder, pressure	1.4	1.5
Paseo		
pressure	1.6	1.8
return	.7	.9
Supra		
1981-86		
pressure	1.3	1.5
return	.7	.9
1987-92		
pressure	1.8	2.0
return	.7	.9
1993-98 each	1.5	1.7
Tercel		
pressure	1.5	1.7
return	.8	1.0
Rack & Pinion Assy., R&R or Replace (B)		
Includes: Alignment.		
Avalon	3.5	3.9
Camry		
1983-86	2.9	3.4
1987-88	3.7	4.2
1989-91		
4 cyl.		
2WD	3.7	4.2
4WD	4.5	5.0
V6	4.3	4.8
1992-96	3.0	3.5

	LABOR TIME	SEVERE SERVICE
1997-01	2.6	3.1
2002-05	5.3	5.6
Camry Solara		
1999-03	2.6	3.1
2004-05 (2.5)	3.7	4.0
Celica		
1981-85	2.5	3.0
1986-87	3.8	4.3
1988-89		
2WD	3.8	4.3
4WD	4.9	5.4
1990-93		
2WD	3.2	3.7
4WD	5.1	5.6
1994-99	3.3	3.8
2000-05	4.5	4.8
Corolla FWD, Matrix		
1983-87	2.5	3.0
1988-92	5.3	5.8
1993		
2WD	3.5	4.0
4WD	5.3	5.8
1994-02	3.3	3.8
2003-05	5.9	6.2
Corolla FX, FX16	3.2	3.7
Corolla RWD	3.2	3.7
Cressida	2.8	3.3
Echo	3.7	4.0
1991-95 MR2	2.4	2.9
MR2 Spyder, Prius	2.2	2.5
Paseo	3.5	4.0
Steering Pump, R&R or Replace (B)		
Avalon	2.0	2.2
Camry		
1983-86	1.5	1.7
1987-88	2.4	2.6
1989-91		
4 cyl.		
2WD	2.5	2.7
4WD	3.0	3.2
V6	2.7	2.9
1992-96	2.4	2.6
1997-01		
4 cyl.	2.4	2.6
V6	1.9	2.1
2002-05	2.0	2.2
Camry Solara	2.3	2.5
Celica		
1981-85	1.7	1.9
1986-87	2.6	2.8
1988-93	2.9	3.1
1994-99		
7AFE	2.4	2.6
5SFE	3.5	3.7
2000-05	1.5	1.7
Corolla FWD, Matrix	1.7	1.9
Corolla FX, FX16	1.7	1.9
Corolla RWD	1.5	1.7

TOY-50 AVALON : CAMRY : CELICA : COROLLA : CRESSIDA : ECHO : MATRIX : MR2

	LABOR TIME	SEVERE SERVICE
Steering Pump, R&R or Replace (B)		
Cressida		
1985-88	1.8	2.0
1989-92	2.8	3.0
Echo	1.5	1.7
1991-95 MR2	1.6	1.8
MR2 Spyder	1.2	1.4
Paseo	1.8	2.0
Supra		
1983-86	1.5	1.7
1987-92	2.8	3.0
1993-98	1.6	1.8
Tercel		
1983-86	1.6	1.8
1987-90	1.9	2.1
1991-99	1.6	1.8
Steering Wheel, Replace (B)		
Exc. below	.7	.7
2002-05 Camry	.8	.8

POWER WORM & SECTOR

Steering Gear, Adjust (B)		
1981-84 Cressida	.5	.7
1981-83 Supra	.5	.7
Gear Assy., Replace (B)		
1981-84 Cressida	2.5	2.8
1981-83 Supra	1.9	2.2
Steering Pump, R&R or Replace (B)		
1981-84 Cressida	1.9	2.1
1981-83 Supra	1.6	1.8
Steering Pump Hoses, Replace (B)		
1981-84 Cressida		
pressure	1.4	1.6
return	.7	.9
1981-83 Supra		
pressure	1.2	1.4
return	.7	.9

HEATING & AIR CONDTIONING

When more than one component requires replacement where evacuation/recovery and recharging is already included, deduct 1.0 hour for each additional component from the time given.

Evacuate/Recover and Recharge System (B)		
1981-05	1.5	1.5
AC Hoses, Replace (B)		
Includes: Evacuate/recover and recharge.		
Exc. MR2	3.1	3.4
MR2	3.6	4.1
Blower Motor, Replace (B)		
Exc. below	.6	.7
1983-86 Camry	1.3	1.3
1981-85 Celica	1.4	1.4
Corolla FX, FX16	.8	.8

	LABOR TIME	SEVERE SERVICE
Cressida		
1981-82	1.2	1.2
1983-88	2.3	2.3
1985-89 MR2	.9	.9
Supra		
1981-87	.8	.8
1988-92	1.4	1.4
Tercel		
1983-88	1.6	1.6
1989-90	.8	.8
Blower Motor Resistor, Replace (B)		
Exc. below	.6	.7
1983-85 Celica	1.3	1.3
1989-92 Cressida	1.4	1.4
1991-95 MR2	4.8	4.8
1988-92 Supra	1.3	1.3
Tercel		
1981-90 wagon	1.3	1.3
Blower Motor Switch, Replace (B)		
Exc. below	.7	.7
Camry		
1983-86	1.9	1.9
1987-91	1.1	1.1
Celica		
1981-85	2.1	2.1
1986-93	1.3	1.3
Corolla FWD		
1983-87	1.2	1.2
1988-02	1.6	1.6
Corolla FX, FX16	1.2	1.2
Corolla RWD		
1981-82	.9	.9
1983-87	1.5	1.5
Cressida		
1981-82	1.9	1.9
1983-88	1.3	1.3
1989-92	1.6	1.6
1991-95 MR2	.8	.8
1992-95 Paseo	1.5	1.5
1981-85 Starlet	2.1	2.1
1981-94 Tercel	1.3	1.3
Compressor Assy., Replace (B)		
Includes: Parts transfer, evacuate/recover and recharge.		
Avalon	4.8	5.4
Camry		
1983-91	4.1	4.6
1992-96		
4 cyl.	3.3	3.7
V6	5.1	5.7
1997-01	3.3	3.7
2002-05	4.6	5.1
Camry Solara		
1999-01	3.3	3.7
2002-03		
4 cyl.	4.6	5.1
V6	3.3	3.7
Celica		
1981-85	4.0	4.4

	LABOR TIME	SEVERE SERVICE
1986-89		
2WD	4.3	4.7
4WD	4.7	5.2
1990-99	3.5	3.9
2000-05	3.5	3.9
Corolla FWD, Matrix		
1983-92	4.3	4.7
1993-02	3.8	4.2
2003-05	4.3	4.7
Corolla FX, FX16	4.4	4.9
Corolla RWD		
1981-82	4.3	4.7
1983-87	3.6	4.1
Cressida	3.8	4.2
Echo	3.6	4.1
MR2	4.1	4.6
MR2 Spyder	3.8	4.2
Paseo, Starlet	3.6	4.1
Prius		
2001-03	3.8	4.2
2004	4.3	4.7
Supra		
1981-92	3.8	4.2
1993-98	4.6	5.1
Tercel		
1981-90	4.1	4.6
1991-99	3.6	4.1
Recondition receiver drier add	.3	.3
Condenser Assy., Replace (B)		
Includes: Evacuate/recover and recharge.		
Avalon	4.0	4.4
Camry		
1983-91	3.6	4.1
1992-94		
4 cyl.	4.7	5.2
V6	6.2	6.8
1995-96		
4 cyl.	4.7	5.2
V6	5.7	6.4
1997-01	3.5	3.9
2002-05	4.4	4.9
Camry Solara		
1999-03	3.5	3.9
2004-05	3.8	4.2
Celica		
1981-93	4.3	4.7
1994-99	3.3	3.7
2000-05	4.4	4.9
Corolla FWD, Matrix		
1983-91	4.1	4.6
1992-02	4.6	5.1
2003-05	4.1	4.6
Corolla FX, FX16	4.1	4.6
Corolla RWD	4.1	4.6
Cressida, MR2	4.1	4.6
Echo	3.5	3.9
MR2 Spyder	4.0	4.4

MR2 SPYDER : PASEO : PRIUS : SOLARA : STARLET : SUPRA : TERCEL — TOY-51

	LABOR TIME	SEVERE SERVICE
Paseo		
1992-95	4.1	4.6
1996-99	5.1	5.7
Prius		
2004	4.6	5.1
Starlet	3.5	3.9
Supra		
1981-92	4.3	4.7
1993-98	7.1	7.8
Tercel		
1981-94	4.1	4.6
1995-99	5.1	5.7

Evaporator Coil, Replace (B)
Includes: Evacuate/recover and recharge.

	LABOR TIME	SEVERE SERVICE
Avalon		
1995-99	4.0	4.4
2000-04	7.2	8.0
Camry		
1983-96	4.4	4.9
1997-01	8.6	9.6
2002-05	6.8	7.2
Camry Solara		
1999-03	7.7	8.1
2004-05 (4.8)	6.7	7.1
Celica		
1981-85	4.4	4.9
1986-93	4.0	4.4
1994-99	3.5	3.9
2000-04	7.7	8.6
Corolla FWD		
1983-87	4.0	4.4
1988-92	4.6	5.1
1993-02	3.6	4.1
2003-05	6.6	7.0
Corolla FX, FX16	4.1	4.6
Corolla RWD	4.4	4.9
Cressida		
1981-88	4.4	4.9
1989-92	4.0	4.4
Echo	4.8	5.2
Matrix	6.6	7.0
MR2	4.3	4.7
MR2 Spyder	3.4	4.0
Paseo	4.1	4.6
Prius		
2001-03	3.6	4.1
2004-05 (5.2)	7.2	7.6
Supra	5.1	5.7
Tercel	4.1	4.6

Expansion Valve, Replace (B)
Includes: Evacuate/recover and recharge.

	LABOR TIME	SEVERE SERVICE
Avalon		
1995-99	4.0	4.3
2000-04	7.2	7.5
Camry		
1983-91	4.0	4.3
1992-96	4.4	4.7
1997-01	8.6	8.9
2002-05 (4.8)	6.7	7.0
Camry Solara		
1999-03	8.6	8.9
2004-05 (3.9)	5.5	5.8
Celica		
1981-85	4.4	4.7
1986-93	4.0	4.3
1994-99	3.5	3.8
2000-05	7.0	7.3
Corolla FWD		
1983-87	4.1	4.6
1988-92	4.6	5.1
1993-02	3.6	4.1
2003-05	4.5	4.8
Corolla FX, FX16	4.1	4.4
Corolla RWD	4.4	4.7
Cressida		
1981-88	4.6	4.9
1989-92	4.0	4.3
Echo	2.7	3.0
Matrix	4.9	5.2
MR2	4.3	4.6
MR2 Spyder	4.0	4.3
Paseo, Tercel	4.1	4.4
Prius		
2001-03	3.5	3.8
2004-05 (4.1)	5.7	6.0
Supra	5.1	5.4

Heater Control Assy., Replace (B)

	LABOR TIME	SEVERE SERVICE
Exc. below	.6	.6
Camry		
1983-86	1.6	1.6
1987-91	1.2	1.3
2002-05	.3	.3
Celica		
1981-85	2.2	2.2
1986-89	1.6	1.6
1990-93	1.2	1.2
Corolla FWD		
1983-87	.8	.8
1988-91	2.0	2.0
1992-02		
2WD	1.8	1.8
4WD	2.3	2.3
Corolla FX, FX16	.8	.8
Corolla RWD		
1981-82	.8	.8
1984-87	1.2	1.2
1981-82 Cressida	2.1	2.1
1989-92 Cressida	1.4	1.4
Echo	.8	.8
MR2		
1985-89	1.7	1.7
1991-95	.8	.8
1992-95 Paseo	1.4	1.4
Starlet		
1981-82	1.7	1.7
1983-84	.9	.9
1981-92 Supra	2.3	2.3

	LABOR TIME	SEVERE SERVICE
Tercel		
1981-90		
sedan	.8	.8
wagon	1.2	1.2
1991-94	1.4	1.4

Heater Core, R&R or Replace (B)

	LABOR TIME	SEVERE SERVICE
Avalon		
1995-99	4.4	4.9
2000-04		
w/AC	5.9	6.6
w/o AC	4.7	5.2
Camry		
1983-86	3.1	3.4
1987-91		
w/AC	5.7	6.4
w/o AC	4.9	5.5
1992-96	2.7	3.0
1997-01	1.6	1.8
2002-05	5.7	6.0
Camry Solara		
1999-03	1.6	1.8
2004-05 (3.9)	5.5	5.8
Celica		
1981-82	4.5	5.0
1983-85	5.2	5.8
1986-89		
w/AC	3.6	4.1
w/o AC	2.7	3.0
1990-99		
w/AC	5.8	6.5
w/o AC	4.8	5.4
2000-05		
w/AC	6.2	6.5
w/o AC	5.1	5.4
Corolla FWD		
1981-83	4.3	4.7
1984-87	2.8	3.1
1988-91		
w/AC	5.2	5.8
w/o AC	3.0	3.3
1992-97		
2WD		
w/AC	7.2	8.0
w/o AC	6.2	6.8
4WD		
w/AC	4.3	4.7
w/o AC	2.7	3.0
1998-02		
w/AC	6.3	7.0
w/o AC	5.2	5.8
2003-05		
w/AC	5.6	6.2
w/o AC	4.0	4.4
4DR sedan	5.3	5.9
Corolla FX, FX16		
w/AC	3.0	3.3
w/o AC	2.3	2.6
Corolla RWD		
1981-82	4.2	4.7
1983-87	2.4	2.7

AVALON : CAMRY : CELICA : COROLLA : CRESSIDA : ECHO : MATRIX : MR2

	LABOR TIME	SEVERE SERVICE
Heater Core, R&R or Replace (B)		
Cressida		
1981-82	5.3	5.9
1983-92		
w/AC	7.2	8.0
w/o AC	6.3	7.0
Echo		
w/AC	4.4	4.7
w/o AC	3.3	3.6
Matrix, MR2 Spyder		
w/AC	5.2	5.5
w/o AC	3.7	4.0
MR2		
1985-89		
w/AC	5.3	5.9
w/o AC	3.7	4.2
1991-95		
w/AC	7.8	8.7
w/o AC	6.9	7.7
Paseo		
1992-95		
w/AC	7.4	8.1
w/o AC	5.9	6.6
1996-99		
w/AC	6.1	6.8
w/o AC	5.1	5.7
Prius	6.4	6.7
Starlet		
1981-82	3.9	4.3
1983-84	4.3	4.7
Supra		
1981-87	5.3	5.9
1988-98		
w/AC	6.1	6.8
w/o AC	5.2	5.8
Tercel		
1981-82	3.0	3.3
1983-90		
sedan	2.6	2.9
wagon	3.6	4.1
1991-94		
w/AC	7.3	8.1
w/o AC	5.7	6.4
1995-99		
w/AC	6.7	7.4
w/o AC	5.8	6.5
Heater Hose, Replace (B)		
Avalon	1.2	1.4
Camry		
1983-91	.8	1.0
1992-01	1.3	1.5
2002-05	.3	.3
Camry Solara	1.4	1.6
Celica		
1981-82	1.2	1.4
1983-89	.8	1.0
1990-99	1.4	1.6
2000-05	1.8	2.1

	LABOR TIME	SEVERE SERVICE
Corolla FWD		
1983-87	1.1	1.3
1988-91	.8	1.0
1992-02	1.3	1.5
2003-05	.9	1.0
Corolla FX, FX16	1.3	1.5
Corolla RWD	1.2	1.4
Cressida		
1981-82	1.2	1.4
1983-92	.8	1.0
Echo, Matrix	1.1	1.2
MR2		
1985-89	.8	1.0
1991-95	1.4	1.6
MR2 Spyder	1.8	2.1
Paseo	.8	1.0
Prius		
2001-03	2.8	3.1
2004-05	1.1	1.2
Starlet, Tercel	.8	1.0
Supra	1.1	1.3
Heater Water Control Valve, Replace (B)		
Exc. below	1.4	1.6
1986-89 Celica	.8	1.0
Cressida	.9	1.1
Starlet	1.1	1.3
1993-98 Supra	5.5	5.7
Low Pressure Cut-Off Switch, Replace (B)		
Includes: Evacuate, recover/recycle and charge system.		
Avalon, Camry Solara	3.5	3.9
Camry		
1983-91	4.0	4.4
1992-01	3.5	3.9
Celica		
1981-85	4.4	4.9
1986-05	3.0	3.3
Corolla FWD		
1983-92	4.0	4.4
1993-02	3.5	3.9
2003-05	3.8	4.1
Corolla FX, FX16	4.0	4.4
1983-87 Corolla RWD	4.1	4.6
Cressida		
1981-82	4.4	4.9
1983-92	3.8	4.2
Echo	2.7	3.0
Matrix	3.4	3.7
MR2	3.5	3.9
MR2 Spyder, Prius	3.1	3.4
Paseo	4.0	4.4
Supra		
1981-92	5.0	5.5
1993-98	3.3	3.7
Tercel		
1981-94	4.1	4.6
1995-99	3.6	4.1

	LABOR TIME	SEVERE SERVICE
Magnetic Clutch Assy., Replace (B)		
Avalon	3.2	3.6
Camry		
1983-86	2.2	2.4
1987-91		
4 cyl.	2.0	2.2
V6	2.4	2.6
1992-96		
4 cyl.	2.0	2.2
V6	3.5	3.7
1997-01	2.1	2.3
2002-05	3.1	3.4
Camry Solara		
1999-01	1.9	2.1
2002-03		
4 cyl.	3.1	3.4
V6	2.0	2.2
2004-05	3.1	3.4
Celica		
1981-85	1.4	1.6
1986-89		
w/turbo	2.9	3.1
w/o turbo	2.4	2.6
1990-05	1.8	2.1
Corolla FWD, Matrix	2.4	2.7
Corolla FX, FX16	2.6	2.8
Corolla RWD		
1981-82	.9	1.1
1983-87		
SOHC	1.2	1.4
DOHC	2.2	2.4
Cressida		
1981-82	1.1	1.3
1983-92	2.4	2.6
Echo	2.0	2.2
MR2	2.5	2.7
MR2 Spyder	2.0	2.3
Paseo	2.3	2.5
Prius		
2001-03	2.3	2.6
Starlet	.9	1.1
Supra		
1981-87	1.4	1.6
1988-92	2.1	2.3
1993-98	2.9	3.1
Tercel		
1981-82	.9	1.1
1983-90		
sedan	2.2	2.4
wagon	1.3	1.5
1991-99	2.1	2.3
Receiver/Drier Assy., Replace (B)		
Includes: Evacuate/recover and recharge.		
1995-99 Avalon	3.5	3.8
Camry		
1983-01	3.5	3.9
2002-05	3.8	4.1
Camry Solara	3.0	3.3

MR2 SPYDER : PASEO : PRIUS : SOLARA : STARLET : SUPRA : TERCEL **TOY-53**

	LABOR TIME	SEVERE SERVICE
Celica		
1981-99	3.5	3.8
2000-05	3.8	4.1
Corolla FWD	3.6	3.9
Cressida	3.5	3.9
Echo, MR2	3.1	3.4
MR2 Spyder	3.4	3.7
Paseo, Starlet	3.5	3.8
Prius		
2001-03	6.0	6.3
2004-05	4.0	4.3
Supra		
1981-92	3.6	3.9
1993-98	6.6	6.9
Tercel	3.6	3.9

WIPERS & SPEEDOMETER

Antenna, Replace (B)
1995-99 Avalon		
front	.5	.5
rear	.7	.7
Camry		
1983-86	1.8	1.8
1987-91	.7	.7
1992-01		
sedan	.5	.5
wagon	.8	.8
Camry Solara		
1999-03	.5	.5
2004-05	1.0	1.0
Celica		
1981-85	.8	.8
1986-89	1.5	1.5
1990-05	.6	.6
Corolla FWD		
1983-97	.9	.9
1998-02	1.1	1.1
2003-05	2.0	2.0
Corolla FX, FX16	.8	.8
Corolla RWD	1.4	1.4
Cressida		
1981-88	.7	.7
1989-92	2.7	2.7
Echo		
hatchback	1.5	1.5
sedan	.8	.8
Matrix	2.0	2.0
MR2	.7	.7
MR2 Spyder	1.4	1.4
Paseo	1.3	1.3
Prius		
2001-03	1.6	1.6
2004-05	4.0	4.0
Starlet	.3	.3
Supra		
1981-85	.7	.7
1986-92	1.3	1.3

	LABOR TIME	SEVERE SERVICE
1993-98		
w/sport roof	1.2	1.2
w/o sport roof	.8	.8
Tercel		
1981-82	.5	.5
1983-90		
sedan	1.1	1.1
wagon	1.6	1.6
1991-99	1.1	1.1

Radio, R&R (B)
Exc. below	.7	.7
1986-89 Celica	1.3	1.3
1991-95 MR2	.8	.8
1993-98 Supra	.9	.9

Rear Window Washer Motor & Pump, Replace (B)
Exc. below	.6	.6
1983-86 Camry	1.2	1.2

Rear Window Wiper Motor, Replace (B)
Exc. below	.6	.6
1992-96 Camry	1.2	1.4
1981-82 Tercel	1.3	1.5

Rear Window Wiper Switch, Replace (B)
Exc. below	.6	.6
Celica		
1981-89	.8	.8
1990-93	1.4	1.4
1994-99	.8	.8
Starlet	1.3	1.3
Supra	.8	.8

Speedometer Head, R&R or Replace (B)
Exc. below	.9	.9
1981-93 Celica	1.4	1.4
2000-03 Celica	.6	.6
1989-92 Corolla 2WD	1.2	1.2
1989-93 Corolla 4WD	1.2	1.2
1981-92 Supra	1.5	1.5

Windshield Washer Motor & Pump Assy. Replace (B)
Exc. below	.7	.7
2002-05 Camry	.9	.9
1994-99 Celica	1.4	1.4
1983-87 Corolla FWD	.8	.8
Corolla FX, FX16	.8	.8
1996-99 Paseo	1.3	1.3
1991-99 Tercel	.8	.8

Windshield Wiper Linkage, Replace (B)
Exc. below	.9	1.0
1995-99 Avalon	1.1	1.2
1992-01 Camry	1.1	1.2
Camry Solara	1.1	1.2
2000-03 Celica GTS	1.1	1.2
1981-82 Corolla RWD	1.5	1.7
1981-82 Cressida	1.4	1.6

	LABOR TIME	SEVERE SERVICE
Windshield Wiper Motor, Replace (B)		
All Models	.6	.7
Windshield Wiper Switch, Replace (B)		
Avalon		
1995-99	.7	.7
2000-04	.3	.3
Camry		
1983-96	1.5	1.5
1997-01	.7	.7
2002-05	.3	.3
Camry Solara		
1999-03	.7	.7
2004-05	.3	.3
Celica		
1981-85	1.6	1.6
1986-99	1.3	1.3
2000-05	.3	.3
Corolla FWD, Matrix		
1983-87	1.4	1.4
1988-97	.9	.9
1998-05	.3	.3
Corolla FX, FX16	1.4	1.4
Corolla RWD		
1981-82	1.4	1.4
1984-87	.5	.5
Cressida		
1981-82	1.4	1.4
1983-92	.8	.8
Echo	.3	.3
MR2, Paseo	.9	.9
MR2 Spyder	.3	.3
Prius		
2001-03	.3	.3
2004-05	.6	.6
Starlet	1.2	1.2
Supra	1.5	1.5
Tercel		
1981-94	1.3	1.3
1995-99	.7	.7

LAMPS & SWITCHES

Diagnose Park/Neutral Position Switch Malfunction (A)
Avalon, Celica, Echo	.8	.8
1997-01 Camry	.6	.6
Corolla		
1997	.6	.6
1998-02	1.4	1.4
Paseo	.6	.6
Solara 2.2L	.6	.6
Supra, Tercel	.7	.7

Diagnose Stop Lamp Signal Switch Malfunction (A)
All Models	.6	.6

Back-Up Lamp Bulb, Replace (C)
All Models	.5	.5

TOY-54 AVALON : CAMRY : CELICA : COROLLA : CRESSIDA : ECHO : MATRIX : MR2

	LABOR TIME	SEVERE SERVICE
Back-Up Lamp Switch, Replace (B)		
Exc. below	.6	.6
1981-82 Celica	.7	.7
2002-05 Corolla FWD	.7	.7
Corolla RWD		
1981-82	.7	.7
1983	1.4	1.4
Starlet	2.0	2.0
Halogen Headlamp Bulb, Replace (C)		
Exc. below (.3)	.5	.5
2003-05 Corolla, Matrix (.6)	.8	.8
Hazard Warning Switch, Replace (B)		
Exc. below	.7	.7
1983-85 Celica	1.3	1.3
1983-87 Corolla FWD	1.4	1.4
Corolla FX, FX16	1.3	1.3
Corolla RWD	1.3	1.3
MR2		
1985-89	1.3	1.3
1991-95	.8	.8
Starlet	1.2	1.2
1981-85 Supra	1.3	1.3
Tercel		
1981-82	1.2	1.2
1983-94 wagon	1.1	1.1
Headlamp Switch, Replace (B)		
Avalon	.7	.7
2000-04	.3	.3
Camry		
1983-96	1.4	1.4
1997-01	.7	.7
2002-05	.3	.3
Camry Solara	.7	.7
Celica		
1981-85	1.6	1.6
1986-93	1.2	1.2
1994-99	.5	.5
2000-05	.3	.3
Corolla FWD, Matrix		
1983-91	1.4	1.4
1992-97	.9	.9
1998-02	.5	.5
2003-05	.3	.3
Corolla FX, FX16	1.2	1.2
Corolla RWD		
1981-82	1.5	1.5
1984-87	.5	.5
Cressida	1.4	1.4
Echo	.3	.3
MR2	.8	.8
MR2 Spyder, Prius	.3	.3
Paseo	.8	.8
Prius		
2001-03	.3	.3
2004-05	.7	.7
Starlet	1.1	1.1

	LABOR TIME	SEVERE SERVICE
Supra		
1981-92	1.7	1.7
1993-98	1.1	1.1
Tercel	1.1	1.1
Headlamps, Aim (B)		
Two	.5	.5
Four	.7	.7
High Mount Stop Lamp Assy., Replace (B)		
Exc. below	.5	.5
Camry		
1995-96 w/spoiler	1.5	1.5
2000-05 w/spoiler	1.2	1.2
2005 Corolla	.8	.8
Horn, Replace (B)		
Exc. below	.5	.5
1981-88 Cressida	1.2	1.2
1991-95 MR2	2.6	2.6
License Lamp Assy., Replace (B)		
Exc. below	.6	.7
1993-98 Supra		
w/sport roof	2.5	2.5
w/o sport roof	2.1	2.1
Camry, Solara		
1999-01	.8	.8
Neutral Safety Switch, Replace (B)		
Exc. below	.9	.9
1992-02 Corolla FWD		
2WD	1.3	1.3
1991-95 MR2	1.1	1.1
1992-95 Paseo	1.2	1.2
1993-98 Supra	1.2	1.2
1991-94 Tercel	1.2	1.2
Parking Brake Warning Lamp Switch, Replace (B)		
Exc. below	.7	.7
Avalon	1.3	1.3
1992-96 Camry w/pedal type	1.5	1.5
MR2		
1985-89	1.2	1.2
1991-95	.9	.9
Parking Lamp Lens or Bulb, Replace (C)		
All Models	.5	.5
Rear Combination Lamp Bulb, Replace (C)		
All Models	.5	.5
Rear Combination Lamp Lens, Replace (B)		
Exc. below	.6	.6
1987-91 Camry	1.2	1.2
Celica		
1981-85	.9	.9
1986-89	1.3	1.3
Corolla FX, FX16	1.3	1.3
1981-82 Corolla RWD	.9	.9
1981-88 Cressida	1.3	1.3
Supra	1.1	1.1

	LABOR TIME	SEVERE SERVICE
1983-90 Tercel		
sedan	.3	.3
wagon	1.4	1.4
Rear Side Marker Lamp Lens, Replace (C)		
All Models	.5	.5
Sealed Beam Headlamp, Replace (B)		
1981-94	.7	.7
Side Marker Lamp Lens or Bulb, Replace (Front) (C)		
Exc. below	.5	.5
1993-98 Supra	1.3	1.3
Stop Lamp Switch, Replace (B)		
Exc. below	.6	.6
1983-86 Camry	.3	.3
2002-05 Camry	.3	.3
1985-89 MR2	.3	.3
Turn Signal Lamp Lens or Bulb, Replace (C)		
1981-05	.3	.3
Turn Signal or Hazard Warning Flasher, Replace (B)		
Exc. below	.7	.7
Echo	.8	.8
Turn Signal Switch, Replace (B)		
Avalon	.7	.7
Camry		
1983-96	1.3	1.3
1997-01	.7	.7
Camry Solara	.6	.6
Celica		
1981-85	1.3	1.3
1986-93	1.1	1.1
1994-99	.7	.7
Corolla	1.1	1.1
Cressida		
1981-88	1.3	1.3
1989-92	.8	.8
MR2	1.1	1.1
Paseo, Starlet	.9	.9
Prius	.6	.6
Supra		
1981-92	1.4	1.4
1993-98	.8	.8
Tercel		
1981-82	.9	.9
1983-90	1.4	1.4
1991-99	.9	.9

SEAT BELTS

	LABOR TIME	SEVERE SERVICE
Door Lock Switch, Replace (B)		
Camry		
1987-88	.8	.8
1989-91	1.2	1.2
Cressida		
1983-88	.8	.8
1989-92	1.2	1.2
Front Limit Switch, Replace (B)		
1987-91 Camry	1.4	1.4

MR2 SPYDER : PASEO : PRIUS : SOLARA : STARLET : SUPRA : TERCEL **TOY-55**

	LABOR TIME	SEVERE SERVICE
Cressida		
1983-88	1.4	1.4
1989	1.5	1.5
Rear Limit Switch, Replace (B)		
1987-89 Camry	.5	.5
Cressida		
1983-88	.5	.5
1989	1.1	1.1
Release Switch, Replace (B)		
1987-89 Camry	.8	.8
Cressida		
1983	.7	.7
1985-88	.8	.8
1989	.9	.9
Seat Belt Motor, Replace (B)		
1987-91 Camry	.9	.9
Cressida		
1983	.7	.7
1985-92	.9	.9
Seat Belt Warning Computer, Replace (B)		
1987-91 Camry	.5	.5
1983-92 Cressida	.8	.8
Shoulder Belt Track, Replace (B)		
1987-91 Camry	.9	.9
Cressida		
1983	.7	.7
1985-92	.9	.9

BODY

Door Lock, Replace (B)
	LABOR TIME	SEVERE SERVICE
Avalon		
1997-99	1.4	1.4
2000-04	.9	.9
Camry		
1983-86	.7	.7
1987-96	1.4	1.4
1997-05	.9	.9
Camry Solara	.7	.7
Celica		
1981-85	.8	.8
1986-99	1.5	1.5
2000-05	.7	.7
Corolla FWD, Matrix		
1983-91	.7	.7
1992-02	1.6	1.6
2003-05	1.1	1.1
Corolla FX, FX16	.7	.7
Corolla RWD	.8	.8
Cressida	.7	.7
Echo	.8	.8
MR2		
1985-89	.7	.7
1991-95	1.6	1.6
MR2 Spyder	.7	.7
Paseo		
1992-95	1.2	1.2
1996-99	.8	.8

	LABOR TIME	SEVERE SERVICE
Prius	.9	.9
Starlet	.8	.8
Supra		
1981-85	.7	.7
1986-92	1.1	1.1
1993-98	1.6	1.6
Tercel		
1981-90	.8	.8
1991-99	1.4	1.4
Door Window Regulator, Replace (B)		
Avalon	1.2	1.4
Camry		
1983-86	.9	1.0
1987-01	1.3	1.5
2002-05	.9	1.0
Camry Solara	1.2	1.4
Celica		
1983-85	2.0	2.2
1986-89	1.5	1.7
1990-93	2.5	2.6
1994-05	1.6	1.8
Corolla FWD, Matrix		
1983-91	1.2	1.4
1992-02	1.6	1.8
2003-05	1.2	1.4
Corolla FX, FX16	.9	1.0
Corolla RWD		
front	.9	1.0
rear	1.5	1.7
Cressida		
1981-88	1.3	1.5
1989-92	.8	.9
Echo	.9	1.0
MR2		
1985-89	2.0	2.2
1991-95	1.5	1.7
MR2 Spyder, Prius	.9	1.0
Paseo	1.3	1.5
Starlet	1.1	1.2
Supra	2.2	2.4
Tercel		
1981-82	.8	.9
1983-90		
sedan	1.2	1.4
wagon	.8	.9
1991-94	1.4	1.6
1995-99		
front	1.7	1.9
rear	1.1	1.3
Hood Hinges, Replace (B)		
Avalon, Matrix	3.3	3.7
Camry		
1983-91	.8	.9
1992-96	3.5	3.9
1997-01	2.4	2.7
2002-05	2.1	2.4
Camry Solara		
1999-01	2.4	2.7
2002-05	2.0	2.2

	LABOR TIME	SEVERE SERVICE
Celica		
1981-89	.8	.9
1990-93	1.8	2.1
1994-99	3.1	3.4
2000-05	2.8	3.1
Corolla FWD		
1983-91	.9	1.0
1992-02	1.7	1.9
2003-05	2.4	2.7
Corolla FX, FX16	.7	.8
Corolla RWD	.9	1.0
Cressida		
1981-82	1.1	1.2
1983-92	.7	.8
Echo	2.6	2.9
MR2		
1985-89	1.3	1.5
1991-95	3.3	3.7
MR2 Spyder	2.6	2.9
Paseo		
1992-95	3.1	3.4
1996-99	2.7	2.9
Prius		
2001-03	2.6	2.9
2004-05	1.8	2.1
Starlet	.9	1.0
Supra		
1981-92	.7	.8
1993-98	4.1	4.6
Tercel		
1981-82	2.9	3.2
1983-90		
sedan	.9	1.0
wagon	1.4	1.6
1991-94	1.6	1.8
1995-99	2.9	3.2
Hood Lock, Replace (B)		
Exc. below	.6	.7
1993-98 Supra	3.5	3.5
Hood Release Cable, Replace (B)		
Avalon		
1995-99	1.3	1.3
2000-04	2.0	2.0
Camry		
1983-86	.9	.9
1987-91	1.4	1.4
1992-01	1.2	1.2
2002-05	.8	.8
Camry Solara	1.2	1.2
Celica		
1983-99	1.5	1.5
2000-05	1.1	1.1
Corolla	1.1	1.1
Cressida	1.4	1.4
Echo, Matrix	.9	.9
MR2	1.4	1.4
MR2 Spyder, Paseo	.8	.8
Prius		
2001-03	.9	.9
2004-05	.5	.5

TOY-56 AVALON : CAMRY : CELICA : COROLLA : CRESSIDA : ECHO : MATRIX : MR2

	LABOR TIME	SEVERE SERVICE
Hood Release Cable, Replace (B)		
Starlet	.7	.7
Supra		
1981-92	1.5	1.5
1993-98	3.6	3.6
Tercel	.8	.8
Lock Striker Plate, Replace (B)		
1981-05	.5	.5
Window Regulator Motor, Replace (B)		
Avalon, Matrix	1.2	1.4
Camry		
1983-86	.9	1.0
1987-05	1.4	1.6
Camry Solara	1.2	1.4

	LABOR TIME	SEVERE SERVICE
Celica		
1983-85	2.1	2.3
1986-89	1.3	1.5
1990-93	2.5	2.7
1994-99	1.6	1.8
Corolla FWD		
1983-87	.9	1.0
1988-05	1.2	1.4
Corolla FX, FX16	1.7	1.9
Corolla RWD	.9	1.0
Cressida	1.1	1.3
Echo	.9	1.0

	LABOR TIME	SEVERE SERVICE
MR2		
1985-89	2.4	2.4
1991-95	1.7	1.7
MR2 Spyder, Prius	1.2	1.4
1996-99 Paseo	1.1	1.1
Supra		
1981-92	2.1	2.1
1993-98	2.5	2.5
Tercel	1.5	1.7

TOY

4Runner : Highlander : Land Cruiser : Pickup : Prerunner : Previa : RAV4 : Sequoia : Sienna : T100 : Tacoma : Tundra : Van : Van Wagon

SYSTEM INDEX

MAINTENANCE	TOY-58
Maintenance Schedule	TOY-58
CHARGING	TOY-59
STARTING	TOY-59
CRUISE CONTROL	TOY-60
IGNITION	TOY-60
EMISSIONS	TOY-60
FUEL	TOY-62
EXHAUST	TOY-64
ENGINE COOLING	TOY-65
ENGINE	TOY-66
Assembly	TOY-66
Cylinder Head	TOY-68
Camshaft	TOY-71
Crank & Pistons	TOY-74
Engine Lubrication	TOY-78
CLUTCH	TOY-79
MANUAL TRANSAXLE	TOY-80
MANUAL TRANSMISSION	TOY-80
AUTO TRANSAXLE	TOY-81
AUTO TRANSMISSION	TOY-82
TRANSFER CASE	TOY-84
SHIFT LINKAGE	TOY-84
DRIVELINE	TOY-85
BRAKES	TOY-88
FRONT SUSPENSION	TOY-90
REAR SUSPENSION	TOY-92
STEERING	TOY-92
HEATING & AC	TOY-94
WIPERS & SPEEDOMETER	TOY-96
LAMPS & SWITCHES	TOY-96
BODY	TOY-97

OPERATIONS INDEX

A
AC Hoses	TOY-94
Air Bags	TOY-92
Air Conditioning	TOY-94
Alignment	TOY-90
Alternator (Generator)	TOY-59
Antenna	TOY-96
Anti-Lock Brakes	TOY-88

B
Back-Up Lamp Switch	TOY-96
Ball Joint	TOY-91
Battery Cables	TOY-59
Bleed Brake System	TOY-88
Blower Motor	TOY-94
Brake Disc	TOY-88
Brake Drum	TOY-88
Brake Hose	TOY-88
Brake Pads and/or Shoes	TOY-89

C
Camshaft Sensor	TOY-60
Catalytic Converter	TOY-64
Coolant Temperature (ECT) Sensor	TOY-61
Crankshaft	TOY-74
Cruise Control	TOY-60
CV Joint	TOY-87
Cylinder Head	TOY-68

D
Differential	TOY-86
Distributor	TOY-60
Drive Belt	TOY-58
Driveshaft	TOY-85

E
EGR	TOY-62
Electronic Control Module (ECM/PCM)	TOY-62
Engine	TOY-66
Engine Lubrication	TOY-78
Engine Mounts	TOY-68
Evaporator	TOY-95
Exhaust	TOY-64
Exhaust Manifold	TOY-64
Expansion Valve	TOY-95

F
Flexplate	TOY-82
Flywheel	TOY-79
Fuel Injection	TOY-63
Fuel Pump	TOY-63
Fuel Vapor Canister	TOY-62

G
Gear Selector Lever	TOY-84
Generator	TOY-59
Glow Plug	TOY-60

H
Halfshaft	TOY-87
Headlamp	TOY-96
Heater Core	TOY-95
Horn	TOY-97

I
Idle Air Control (IAC) Valve	TOY-64
Ignition Coil	TOY-60
Ignition Module	TOY-60
Ignition Switch	TOY-60
Injection Pump	TOY-64
Inner Tie Rod	TOY-93
Intake Manifold	TOY-63

K
Knock Sensor	TOY-62

L
Lower Control Arm	TOY-91

M
Maintenance Schedule	TOY-58
Master Cylinder	TOY-88
Muffler	TOY-65

N
Neutral Safety Switch	TOY-96

O
Oil Pan	TOY-78
Oil Pump	TOY-79
Outer Tie Rod	TOY-93
Oxygen Sensor	TOY-62

P
Parking Brake	TOY-89
Pistons	TOY-76
Positive Crankcase Ventilation (PCV) Valve	TOY-62

R
Radiator	TOY-65
Radiator Hoses	TOY-65
Radio	TOY-96
Rear Main Oil Seal	TOY-77

S
Shock Absorber/Strut, Front	TOY-90
Shock Absorber/Strut, Rear	TOY-92
Spark Plug Cables	TOY-60
Spark Plugs	TOY-60
Spring, Front Coil	TOY-90
Spring, Leaf	TOY-92
Spring, Rear Coil	TOY-92
Starter	TOY-59
Steering Wheel	TOY-93

T
Thermostat	TOY-65
Throttle Body	TOY-64
Throttle Position Sensor (TPS)	TOY-62
Timing Belt	TOY-73
Timing Chain	TOY-73
Torque Converter	TOY-83

U
U-Joint	TOY-85
Upper Control Arm	TOY-91

V
Valve Body	TOY-82
Valve Cover Gasket	TOY-70
Valve Job	TOY-70
Vehicle Speed Sensor	TOY-62

W
Water Pump	TOY-65
Wheel Balance	TOY-58
Wheel Cylinder	TOY-89
Window Regulator	TOY-97
Windshield Washer Pump	TOY-96
Windshield Wiper Motor	TOY-96

TOY-57

4RUNNER : HIGHLANDER : LAND CRUISER : PICKUP : PRERUNNER : PREVIA

	LABOR TIME	SEVERE SERVICE
MAINTENANCE		
Air Cleaner Filter Element, Replace (C)		
Exc. below	.3	.3
RAV4	.5	.5
Drive Belt, Adjust (B)		
Exc. below	.5	.5
Previa	.6	.6
1996-00 RAV4	.6	.6
Drive Belt, Replace (B)		
AC		
exc. below	.6	.7
Land Cruiser		
1981-97	.8	.9
Previa	1.7	1.8
Van	1.2	1.3
Fan		
exc. below	.5	.6
1996-00 RAV4	.7	.8
Van, Previa	.8	.9
Power steering		
exc. below	.8	.9
1984-89 Van	1.6	1.7
Supercharger		
1995-97 Previa	.8	.9
Fuel Filter, Replace (B)		
Gasoline		
exc. below	1.3	1.4
4Runner, Pickup		
1981-88 w/o FI	.3	.4
1996-02 V6	.6	.7
Land Cruiser		
1981-88	.5	.6
1989-05	.8	.9
1998-04 PreRunner	1.1	1.3
RAV4	1.4	1.6
Sequoia, Tundra	1.1	1.3
1998-03 Sienna	.8	.9
Tacoma		
1995-04	1.1	1.3
2005 Tacoma	1.6	1.7
Van	.6	.7
Diesel	.5	.6
Halogen Headlamp Bulb, Replace (B)		
All Models	.3	.3
Oil & Filter, Change (C)		
Includes: Correct all fluid levels.		
Exc. below	.7	.7
Highlander	.5	.5
RAV4	.9	.9
Sienna	1.3	1.3
Sealed Beam Headlamp, Replace (B)		
Exc. below	.3	.3
1995-96 Tacoma	.7	.7
Timing Belt, Replace (B)		
4 cyl.		
1981-87 diesel		
Pickup	1.8	2.3
1996-00 RAV4	2.9	3.4

	LABOR TIME	SEVERE SERVICE
V6		
1989-95 4Runner, Pickup	4.5	5.0
1996-02 4Runner	2.7	3.2
Sienna	6.0	6.5
Tacoma, Tundra	2.8	3.3
T100		
1993-94	4.4	4.9
1995-98	2.6	3.1
V8	3.5	4.0
w/ABS add		
4Runner, Pickup V6	.3	.3
96-00 RAV4	.9	.9
Sienna	.1	.1
93-98 T100 V6	.5	.5
Tacoma V6, Tundra	.3	.3
diesel	.7	.7
Tire, Replace (C)		
Includes: Dismount old tire and mount new tire to rim.		
One	.7	.7
each addl. add	.5	.5
Tires, Rotate (C)		
All Models	.8	.8
Wheels, Balance (B)		
Two	1.0	1.1
each addl. add	.3	.3
SCHEDULED MAINTENANCE INTERVALS		
If necessary, refer to appropriate Chilton maintenance service information.		
5,000 Mile Service (C)		
All Models	.4	.4
10,000 Mile Service (C)		
All Models	.4	.4
15,000 Mile Service (B)		
4Runner		
2WD	1.3	1.3
4WD	1.8	1.8
Land Cruiser	2.1	2.1
Previa		
2WD	1.6	1.6
4WD	1.9	1.9
RAV4	1.9	1.9
Sienna	1.7	1.7
T100		
4 cyl.	1.8	1.8
6 cyl.		
2WD	1.3	1.3
4WD	1.8	1.8
Tacoma		
2WD	1.8	1.8
4WD	1.9	1.9
20,000 Mile Service (C)		
All Models	.4	.4
22,500 Mile Service (C)		
4Runner	1.1	1.1
Land Cruiser	.9	.9

	LABOR TIME	SEVERE SERVICE
Previa	.9	.9
RAV4	.9	.9
T100	1.1	1.1
Tacoma	.9	.9
25,000 Mile Service (C)		
All Models	.4	.4
30,000 Mile Service (B)		
4Runner	4.1	4.1
Land Cruiser	3.8	3.8
Previa	2.9	2.9
RAV4	3.7	3.7
Sienna	2.7	2.7
T100	3.7	3.7
Tacoma	3.8	3.8
35,000 Mile Service (C)		
All Models	.4	.4
37,500 Mile Service (C)		
RAV4	.9	.9
40,000 Mile Service (C)		
All Models	.4	.4
45,000 Mile Service (B)		
4Runner		
2WD	1.3	1.3
4WD	1.8	1.8
Land Cruiser	2.1	2.1
Previa		
2WD	1.6	1.6
4WD	1.9	1.9
RAV4	1.9	1.9
Sienna	1.7	1.7
T100		
4 cyl.	1.8	1.8
6 cyl.		
2WD	1.3	1.3
4WD	1.8	1.8
Tacoma		
2WD	1.8	1.8
4WD	1.9	1.9
50,000 Mile Service (C)		
All Models	.4	.4
55,000 Mile Service (C)		
All Models	.3	.3
60,000 Mile Service (B)		
4Runner	4.1	4.1
Land Cruiser	3.8	3.8
Previa	2.9	2.9
RAV4	3.7	3.7
Sienna	2.7	2.7
T100	3.7	3.7
Tacoma	3.8	3.8
65,000 Mile Service (C)		
All Models	.4	.4
70,000 Mile Service (C)		
All Models	.4	.4
75,000 Mile Service (B)		
4Runner		
2WD	1.1	1.1
4WD	1.8	1.8
Land Cruiser	2.1	2.1

RAV4 : SEQUOIA : SIENNA : T100 : TACOMA : TUNDRA : VAN : VAN WAGON   TOY-59

	LABOR TIME	SEVERE SERVICE
Previa		
2WD	1.6	1.6
4WD	1.9	1.9
RAV4	1.9	1.9
Sienna	1.7	1.7
T100		
4 cyl.	1.8	1.8
6 cyl.		
2WD	1.3	1.3
4WD	1.8	1.8
Tacoma		
2WD	1.8	1.8
4WD	1.9	1.9
80,000 Mile Service (C)		
All Models	.4	.4
85,000 Mile Service (C)		
All Models	.4	.4
90,000 Mile Service (B)		
4Runner	4.1	4.1
Land Cruiser	3.8	3.8
Previa	2.9	2.9
RAV4	3.7	3.7
Sienna	2.7	2.7
T100	3.7	3.7
Tacoma	3.8	3.8
95,000 Mile Service (C)		
All Models	.4	.4

CHARGING

Alternator Circuits, Test (B)
Includes: Test component output.
1981-05	.3	.3

Alternator, R&R and Recondition (B)
4 cyl.		
4Runner		
1983-95	2.4	2.7
1996-00	1.9	2.1
Highlander	2.1	2.4
Pickup	2.4	2.7
1998-04 PreRunner	2.0	2.2
Previa	2.5	2.9
RAV4	1.8	2.1
1995-98 T100	2.3	2.6
Tacoma		
1995-04	2.1	2.4
2005 Tacoma	1.3	1.4
Van	3.0	3.4
6 cyl.		
1981-97 Land Cruiser	2.4	2.7
V6		
4Runner		
1989-95	2.4	2.7
1996-05	1.9	2.1
Highlander	2.0	2.2
Pickup	2.4	2.7
1998-04 PreRunner	2.0	2.2

	LABOR TIME	SEVERE SERVICE
Sienna		
1998-03	2.0	2.2
2004-05	1.5	1.7
T100	2.1	2.4
Tacoma, Tundra	2.0	2.2
V8	2.0	2.2

Alternator Assy., Replace (B)
Includes: Pulley transfer.
4 cyl.		
4Runner		
1983-95	1.2	1.4
1996-00	.6	.7
Highlander	.9	1.0
Pickup	1.3	1.4
1998-04 PreRunner	.8	.9
Previa	1.1	1.2
RAV4, T100	.6	.7
Tacoma		
1995-04	.8	.9
2005	.6	.7
Van	1.7	1.9
6 cyl.	.8	.9
V6 exc. below	.8	.9
4Runner		
1996-05	1.2	1.4
Highlander	.6	.7
Sienna		
2004-05	.5	.5
V8		
exc. below	.8	.9
2003-05 4Runner	1.2	1.4

Alternator Voltage Regulator, Replace (B)
4Runner		
1983-95		
external	.5	.5
internal	1.7	2.0
2003-05	1.3	1.4
Highlander	1.0	1.1
Land Cruiser		
1981-90	.5	.5
1991-97	1.2	1.3
1998-05	2.0	2.1
Pickup		
external	.5	.5
internal	1.7	2.0
1998-04 PreRunner	1.5	1.7
Previa	1.5	1.7
RAV4	.9	1.0
Sequoia, Tundra	1.5	1.7
Sienna		
1998-03	1.4	1.5
T100	1.7	2.0
Tacoma		
1995-04	1.5	1.7
2005	1.3	1.4
Van	2.0	2.3

	LABOR TIME	SEVERE SERVICE
Front Alternator Bearing, Replace (B)		
4 cyl.		
4Runner	1.3	1.5
Pickup, Previa	1.7	1.9
RAV4	1.4	1.5
T100, Tacoma	1.5	1.7
Van	2.2	2.4
6 cyl., V6, V8		
exc. below	1.5	1.7
2003-05 V6		
4Runner	1.8	2.1
Voltmeter and/or Ammeter Gauge, Replace (B)		
Land Cruiser, Pickup	.8	.8

STARTING

Starter Draw Test (On Car) (B)
1981-05	.5	.5

Battery, Replace (C)
1981-05	.5	.7

Battery Cables, Replace (C)
Exc. below each	.6	.7
Highlander, Previa, Van positive	.9	1.0

Starter Assy., R&R and Recondition (B)
Non-reduction		
1981-88 exc. below	2.4	2.7
4Runner V6		
2003-05	2.5	2.8
2004-05 Sienna	1.3	1.4
Reduction		
4Runner		
V6	2.1	2.3
V8	4.1	4.6
Highlander	2.1	2.3
Land Cruiser		
1985-97	1.8	2.0
1998-04	4.0	4.4
Pickup	1.7	2.0
1998-04 PreRunner	1.6	1.8
Previa	2.1	2.3
RAV4	1.4	1.6
Sequoia, Tundra		
V6	1.6	1.8
V8	4.0	4.4
Sienna	1.7	2.0
T100	1.5	1.7
Tacoma	1.6	1.8
Van	1.8	2.0

Starter Assy., Replace (B)
Non-reduction		
1981-88 exc. below	.8	.9
4Runner V6		
2003-05	1.4	1.6

4RUNNER : HIGHLANDER : LAND CRUISER : PICKUP : PRERUNNER : PREVIA

	LABOR TIME	SEVERE SERVICE

Starter Assy., Replace (B)
Reduction
 Exc. below . . . . .6 .7
 4Runner
  V6
   2003-05 . . . . . . 1.4 1.6
  V8
   2003-05 . . . . . . 3.0 3.4
 Highlander, Tacoma . .8 .9
 Land Cruiser
  1985-97 . . . . .9 1.1
  1998-05 . . . . 3.3 3.7
 Previa 1.1 1.3
 Sequoia, Tundra V8 3.3 3.7
 Sienna 9 1.1
Starter Relay, Replace (B)
 Diesel . . . . .5 .5
Starter Solenoid and/or Switch, Replace (B)
Includes: Starter R&R.
 Non-reduction
  1981-88 . . . . .8 .9
 Reduction
  4Runner
   1983-95 . . . . .9 1.0
   1996-02 . . . . . . 1.3 1.5
  Highlander 1.0 1.1
  Land Cruiser
   1985-97 . . . . .9 1.0
   1998-05 . . . . 3.6 4.1
  Pickup 9 1.0
  1998-04 PreRunner . 1.0 1.1
  Previa 1.3 1.5
  RAV4 . . . . .8 .9
  Sequoia, Tundra
   V6 . . . . .9 1.0
   V8 . . . . 3.5 3.9
  Sienna 1.1 1.3
  T100 . . . . .8 .9
  Tacoma 9 1.0

CRUISE CONTROL

Diagnose Cruise System Component Each (A)
 1981-05 . . . . .5 .5
Actuator Assy., Replace (B)
 Exc. Van . . . . .7 .7
 Van . . . . .8 .8
Control Computer, Replace (B)
 Exc. below . . . . .7 .7
 1989-95 4Runner . . . 1.2 1.2
 1991-97 Land Cruiser . . .8 .8
 1989-95 Pickup 1.1 1.1
 T100 . . . . 1.3 1.3
 Van . . . . .9 .9
Speed Control (Main) Switch, Replace (B)
 All Models . . . . .8 .8

IGNITION

Diagnose Ignition System Component Each (A)
 1981-05 . . . . .5 .5
Diagnose Igniter Circuit Malfunction (A)
 Highlander 1.1 1.1
 1998-05 Land Cruiser . .6 .6
 2001-05 RAV4 . . . . .8 .8
 Sequoia . . . . .8 .8
 Tundra . . . . .6 .6
Ignition Timing, Reset (B)
 1983-95 4Runner . . . . .7 .7
 Land Cruiser
  1981-90 . . . . .7 .7
  1991-95 . . . . 1.3 1.3
 Pickup . . . . .7 .7
 1991-95 Previa 1.2 1.2
 1993-95 T100 . . . . . . .7 .7
 Van 1.1 1.1
Camshaft Sensor, Replace (B)
 Exc. below . . . . .6 .7
 1998-05 Land Cruiser . 1.1 1.2
 2003-05 4Runner 1.2 1.4
 Sequoia, Tundra 1.1 1.2
 1995-98 T100 V6 . . . . 1.2 1.4
 Tacoma 1.1 1.2
Distributor, Replace (B)
Includes: Reset base ignition timing.
 4 cyl.
  4Runner
   1983-95 . . . . .8 .9
   1996-00 . . . . 1.4 1.5
  Pickup . . . . .8 .9
  1998-04 PreRunner . 1.2 1.4
  Previa 1.5 1.7
  RAV4 . . . . 1.7 1.9
  T100, Tacoma 1.4 1.5
  Van 1.6 1.8
 6 cyl.
  Land Cruiser
   1981-90 . . . . .8 .9
   1991-97 . . . . 1.4 1.5
  V6 . . . . 1.4 1.5
Distributor Cap and/or Rotor, Replace (B)
 4 cyl.
  exc. Van . . . . .7 .7
  Van 1.2 1.2
 6 cyl. . . . . .3 .3
 V6 . . . . .7 .7
Glow Plug, Replace (B)
 Diesel
  one . . . . .5 .7
  all 8 1.0
Glow Plug Relay, Replace (B)
 Diesel . . . . .5 .5

Igniter Unit, Replace (B)
 Exc. below . . . . .5 .5
 1999-02 4Runner V6 . . .7 .7
 RAV4 . . . . .3 .3
 Sienna 1.2 1.2
 Van 1.9 1.9
Ignition Coil, Replace (B)
 One
  Exc. below . . . . .5 .5
  1996-99 4Runner
   4 cyl. 1.3 1.3
  1998-03 Land Cruiser .6 .6
  Previa 1.1 1.1
  2001-05 RAV4 . . . . . . .7 .7
  Sequoia, Tundra V8 . . .6 .6
  2004-05 Sienna 1.8 2.1
  Van 2.0 2.0
Ignition Switch, Replace (B)
 Exc. below . . . . .6 .6
 2001-05 Highlander . . . .9 .9
 1991-97 Land Cruiser . .8 .8
 2004-05 Sienna . . . . . . .8 .8
Spark Plug (Ignition) Cables, Replace (B)
 Exc. below . . . . .8 .9
 1998-04 PreRunner . . . .5 .5
 Previa 9 1.1
 Tacoma, Tundra . . . . . .5 .5
Spark Plugs, Replace (B)
 Exc. below 9 1.0
 Highlander V6 . . . . . . 1.4 1.5
 Sienna
  1998-03 . . . . 2.8 3.1
  2004-05 . . . . 2.1 2.4
Vacuum Advance Unit, Replace (B)
Includes: Reset base ignition timing.
 1985-89 4Runner . . . . .7 .7
 1981-90 Land Cruiser . .7 .7
 1981-89 Pickup . . . . . .7 .7
 Van 1.3 1.3

EMISSIONS

The following operations do not include testing. Add time as required.

Retrieve Fault Codes (A)
 1981-05 . . . . .3 .3
Diagnose Coolant Temperature Insufficient for Closed Loop Fuel Control (A)
 4Runner . . . . .8 .8
 Land Cruiser
  1997 . . . . 1.4 1.4
  1998-05 . . . . .9 .9
 1998-04 PreRunner . . . .8 .8
 1997-00 RAV4 . . . . . . .6 .6
 1998-03 Sienna . . . . . .6 .6
 T-100 . . . . .8 .8
 Tacoma 2.7L, 3.4L . . . .8 .8

RAV4 : SEQUOIA : SIENNA : T100 : TACOMA : TUNDRA : VAN : VAN WAGON — TOY-61

	LABOR TIME	SEVERE SERVICE
Diagnose Evaporative Emission Control System Pressure Sensor Malfunction (A)		
1997-02 4Runner	.6	.6
Highlander	.9	.9
1998-05 Land Cruiser	.5	.5
1998-04 PreRunner	.6	.6
RAV4		
1997-00	.5	.5
2001-05	1.1	1.1
Sequoia	.8	.8
1998-03 Sienna	.5	.5
T-100, Tacoma	.6	.6
Tundra	.3	.3
Diagnose Evaporative Emission Control System Pressure Sensor Range/Performance (A)		
4Runner		
1997-02	.6	.6
2003-05	.8	.8
Highlander	.9	.9
1998-05 Land Cruiser	.5	.5
1998-04 PreRunner	.6	.6
RAV4		
1997-00	.5	.5
2001-05	1.1	1.1
Sequoia	.8	.8
Sienna		
1998-03	.5	.5
2004-05	.8	.8
T-100, Tacoma	.6	.6
Tundra	.3	.3
Diagnose Intake Air Temperature Circuit Malfunction (A)		
Exc. below	.6	.6
4Runner	.2	.2
Highlander 2.4L	.3	.3
1998-04 PreRunner	.2	.2
2004-05 Sienna	.2	.2
Diagnose Mass Air Flow Circuit Malfunction (A)		
Exc. below	.6	.6
2001-05 RAV4	.7	.7
Diagnose Mass Air Flow Circuit Range/Performance Problem (A)		
All Models	.3	.3
Diagnose Oxygen Sensor Circuit Malfunction (A)		
Bank one sensor two		
exc. below	.8	.8
Highlander 4WD	.9	.9
1997 Land Cruiser	1.1	1.1
RAV4	.9	.9
Sequoia	.5	.5

	LABOR TIME	SEVERE SERVICE
Bank Two		
sensor one		
1998-05 Land Cruiser	.8	.8
Sienna	.7	.7
sensor two		
2003-05 4Runner	.7	.7
Highlander 2.4L	1.1	1.1
1998-05 Land Cruiser	.8	.8
Sequoia	.5	.5
Tundra	.8	.8
Diagnose Oxygen Sensor Heated Circuit Slow Response (A)		
Bank one sensor one		
exc. below	.3	.3
1997 Land Cruiser	1.1	1.1
1998-05 Sienna	.5	.5
Bank two sensor one		
1997 Land Cruiser	.3	.3
1998-05 Sienna	.5	.5
Diagnose Oxygen Sensor Heater Circuit Malfunction (A)		
Bank one sensor one or two		
exc. below	.6	.6
Highlander 3.0L sensor two	.8	.8
Land Cruiser		
1997	1.2	1.2
1998-05	.7	.7
Sequoia	.3	.3
Bank two		
sensor one		
1998-05 Land Cruiser	.7	.7
Sequoia	.3	.3
1998-05 Sienna	.6	.6
Tundra	.5	.5
sensor two		
Highlander 4WD	.6	.6
1998-05 Land Cruiser	.7	.7
RAV4	.6	.6
Sequoia	.3	.3
Tundra	.5	.5
Diagnose Throttle Pedal Position Sensor Switch "A" Circuit Performance Problem (A)		
Exc. below	.5	.5
Highlander 3.0L	1.1	1.1
Land Cruiser	.9	.9
Previa	.7	.7
2001-05 RAV4	1.2	1.2
1998-03 Sienna	.3	.3

	LABOR TIME	SEVERE SERVICE
Diagnose Vehicle Speed Sensor Malfunction (A)		
Exc. below	.6	.6
1998-05 Land Cruiser	.7	.7
1998-04 PreRunner	.9	.9
T-100, Tacoma	.9	.9
Tundra	.7	.7
Dynamometer Test (A)		
1981-05	.5	.5
Air Control Valve, Replace (B)		
1993-95 Pickup, 4Runner V6	.5	.5
Air Flow Meter, Replace (B)		
4Runner		
1983-95	1.3	1.4
1996-02		
4 cyl.	.8	.8
V6	1.1	1.2
2003-05	.5	.5
Highlander	.3	.3
Land Cruiser		
1988-90	1.5	1.5
1991-97	1.3	1.4
1998-05	.3	.3
1984-95 Pickup	1.3	1.4
1998-04 PreRunner	1.4	1.5
Previa	.9	1.0
RAV4	.9	1.0
Sequoia, Tundra	.3	.3
Sienna	.5	.5
T100, Tacoma	1.4	1.5
Van	1.3	1.4
Air Injection Check Valve, Replace (B)		
1981-92 Land Cruiser	.5	.5
1981-88 Pickup	1.7	1.7
Air Pump Assy., Replace (B)		
1981-92 Land Cruiser	.9	1.1
1981-88 Pickup	.7	.9
Coolant Temperature (ECT) Sensor, Replace (B)		
Exc. below	1.4	1.5
1996-05 4Runner	.5	.5
Highlander 4 cyl.	.5	.5
1998-05 Land Cruiser	1.8	2.1
1998-04 PreRunner	.8	.9
Sequoia, Tundra V8	1.9	2.1
Sienna		
1998-03	.8	.9
2004-05	.3	.3
T100 4 cyl.	.8	.9
Tacoma		
exc. below	.8	.9
2005 4 cyl.	1.7	1.8

4RUNNER : HIGHLANDER : LAND CRUISER : PICKUP : PRERUNNER : PREVIA

	LABOR TIME	SEVERE SERVICE
EGR Control Valve, Replace (B)		
1983-95 4Runner, Pickup		
4 cyl.	.8	1.0
V6	.7	.8
1996-02 4Runner	.7	.8
Land Cruiser		
1981-92	.8	1.0
1993-02	1.2	1.4
Previa	1.4	1.6
1998-04 PreRunner	.6	.7
RAV4	.7	.8
Sequoia, Tundra	.5	.7
T100	.7	.8
Tacoma	.6	.7
Van	1.3	1.5
Exc. below	.3	.3
1981-97 Land Cruiser	.5	.5
Previa, Van	.5	.5
EGR Temperature Sender, Replace (B)		
All Models	.3	.3
EGR Thermal Vacuum Valve, Replace (B)		
4Runner, Pickup		
4 cyl.	1.1	1.1
V6	1.9	1.9
Land Cruiser		
1981-92	1.1	1.1
1993-97	1.4	1.4
Previa	1.3	1.3
1996-98 RAV4	.8	.8
1998 Sienna	1.3	1.3
1993-94 T100	1.8	1.8
Van	.7	.7
Electronic Control Module (ECM/PCM), Replace (B)		
4Runner		
1994-02	1.1	1.1
2003-05	.5	.5
Highlander	.5	.5
Land Cruiser		
1988-90	.9	.9
1991-92	.7	.7
1993-97	1.5	1.5
1998-05	.7	.7
1984-95 Pickup	.9	.9
Previa, RAV4	.7	.7
1998-04 PreRunner	.8	.8
Sequoia, Tundra	.6	.6
Sienna	.7	.7
T100	1.2	1.2
Tacoma	.8	.8
Van	.9	.9
Fuel Vapor Canister, Replace (B)		
Exc. below	.5	.6
4Runner, Highlander	.6	.7
RAV4	.8	.9
Sequoia	.6	.7

	LABOR TIME	SEVERE SERVICE
Knock Sensor, Replace (B)		
1993-95 4Runner, Pickup	3.7	3.9
4Runner		
1996-02		
4 cyl.	1.2	1.4
V6 one	3.2	3.4
2003-05		
V6		
2WD	15.4	15.9
4WD	16.4	16.9
V8	2.4	2.7
Highlander		
4 cyl.		
2WD	1.8	2.1
4WD	2.8	3.1
V6 one	2.6	2.9
Land Cruiser		
1993-97	.7	.9
1998-05	3.1	3.4
Previa		
front	5.4	5.6
rear	5.7	5.9
1998-04 PreRunner	1.5	1.7
4 cyl.	.6	.7
V6	3.2	3.4
RAV4		
1996-00	.5	.7
2001-05	1.5	1.7
Sequoia, Tundra		
exc. below		
V6 one	3.5	3.9
V8 one	2.9	3.2
2005 Tundra		
V6 one	11.5	11.7
Sienna, one	2.6	2.9
T100		
4 cyl.	.7	.9
V6	3.2	3.4
Tacoma		
exc. below		
4 cyl.	.6	.7
V6 one	3.2	3.6
2005		
4 cyl.	1.7	1.9
V6 one	16.8	17.0
each addl. add	.2	.2
Oxygen Sensor, Replace (B)		
Exc. below	.8	.9
2003-05 4Runner		
V6	.5	.6
V8	2.3	2.4
Land Cruiser		
1988-97	1.3	1.5
1998-05	.9	1.0
Highlander	1.4	1.6
1984-95 Pickup	1.3	1.5

	LABOR TIME	SEVERE SERVICE
Sienna	.5	.5
Tacoma	.9	1.0
each addl. add		
exc. below	.2	.2
03-05 4Runner V8	.9	.9
98-05 Land Cruiser	.5	.5
T100	.5	.5
PAIR Valve, Replace (B)		
1993-97 Land Cruiser	1.4	1.4
1993-95 Pickup, 4Runner	.7	.7
Positive Crankcase Ventilation (PCV) Valve, Replace (B)		
Exc. below	.3	.4
1998-03 Sienna	.5	.6
2003-05 4Runner		
V6	1.7	1.8
Pickup	.5	.6
Previa	.8	.9
2005 Tacoma	.4	.5
2005 Tundra V6	1.4	1.5
Throttle Position Sensor (TPS), Replace (B)		
Exc. below	.7	.7
Highlander V6	1.8	2.1
2001-05 RAV4	1.8	2.1
Vehicle Speed Sensor, Replace (B)		
Exc. below	.8	.8
Land Cruiser, Previa	1.0	1.0
1996-00 RAV4	1.0	1.0
Sienna	1.0	1.0

FUEL
CARBURETOR
Idle Speed, Adjust (B)
1981-88 Pickup, Land Cruiser	.5	.5

Carburetor, R&R and Clean or Recondition (B)
Includes: Adjustments.
1981-88 Pickup, Land Cruiser	3.0	3.0

Carburetor, Replace (B)
Includes: Adjustments.
1981-88 Pickup, Land Cruiser	1.6	1.6

Carburetor Needle & Seat, Replace (B)
1981-88 Pickup, Land Cruiser	1.9	1.9

DELIVERY
Fuel Pump, Test (B)
1981-05	.3	.3

Diagnose Accelerator Pedal Position Sensor Circuit Malfunction (A)
Land Cruiser	.8	.8
Sequoia, Tundra	1.4	1.4

RAV4 : SEQUOIA : SIENNA : T100 : TACOMA : TUNDRA : VAN : VAN WAGON TOY-63

	LABOR TIME	SEVERE SERVICE
Fuel Filter, Replace (B)		
Gasoline		
exc. below	1.3	1.4
4Runner, Pickup		
1981-88 w/o FI	.3	.4
1996-02 V6	.6	.7
Land Cruiser		
1981-88	.5	.6
1989-05	.8	.9
1998-04 PreRunner	1.1	1.3
RAV4	1.4	1.6
Sequoia, Tundra	1.1	1.3
1998-03 Sienna	.8	.9
Tacoma		
1995-04	1.1	1.3
2005 Tacoma	1.6	1.7
Van	.6	.7
Diesel	.5	.6
Fuel Gauge (Dash), Replace (B)		
Exc. below	.9	.9
1989-95 4Runner	1.1	1.1
1991-97 Land Cruiser	1.1	1.1
1989-95 Pickup	1.1	1.1
Previa	1.4	1.4
Fuel Gauge (Tank), Replace (B)		
4Runner		
1983-88	2.5	2.9
1989-95	1.9	2.1
1996-02	1.2	1.4
2003-05	1.9	2.1
Land Cruiser		
1981-90		
Hardtop	1.1	1.3
Wagon	.5	.7
1991-97	2.5	2.9
1998-05	.5	.7
Highlander	1.1	1.3
Pickup		
gasoline		
1981-88	2.2	2.4
1989-95	1.8	2.0
diesel	2.2	2.4
1998-04 PreRunner	1.9	2.1
Previa	2.4	2.7
RAV4		
1996-00	.8	1.0
2001-05	1.4	1.6
Sequoia	1.0	1.1
Sienna		
1998-03	3.0	3.4
2004-05	1.1	1.3
T100	1.9	2.1
Tacoma, Tundra	1.9	2.1
Van	1.4	1.6
Fuel Pump, Replace (B)		
4Runner, Pickup		
1981-88		
electric	1.1	1.3
mechanical	.7	.9
1989-95	2.0	2.2
1996-05	2.4	2.7
Highlander	1.3	1.5
Previa	2.2	2.4
Land Cruiser		
1981-87	.5	.7
1988-90	3.0	3.2
1991-97	2.0	2.2
1998-05	.8	1.0
1998-04 PreRunner	2.1	2.3
Previa	2.2	2.4
RAV4	1.4	1.6
Sequoia	1.0	1.2
Sienna		
1998-03	3.3	3.5
2004-05	1.3	1.4
T100	1.9	2.1
Tacoma, Tundra	2.1	2.3
Van	1.9	2.1
Fuel Tank, Replace (B)		
Includes: Drain and refill.		
4Runner, Pickup	2.2	2.5
Highlander	3.4	3.7
Land Cruiser		
1981-90	3.2	3.5
1991-05	3.7	4.0
1998-04 PreRunner	1.9	2.2
Previa	2.5	2.8
RAV4		
1996-00	3.2	3.5
2001-05		
2WD	3.0	3.3
4WD	3.7	4.0
Sequoia, Tundra	1.9	2.2
Sienna	2.7	3.0
Tacoma		
1995-04	1.9	2.2
2005	1.6	1.9
T100	1.7	2.0
Van	1.4	1.7
Intake Manifold Gasket, Replace (B)		
4 cyl.		
4Runner, Pickup		
gasoline		
1981-95		
w/FI or turbo	3.2	3.6
w/o FI	2.0	2.2
1996-02	2.6	2.9
diesel	1.8	2.1
Highlander		
2WD	1.8	2.1
4WD	2.8	3.1
1998-04 PreRunner	3.6	4.1
Previa		
2WD	6.4	6.9
4WD	6.8	7.5
RAV4		
1996-00	3.0	3.3
2001-05	2.3	2.6
Tacoma		
1995-04	3.6	4.1
2005	1.6	1.9
T100	4.4	4.9
Van	3.8	4.2
6 cyl.		
Land Cruiser		
1981-90	4.9	5.4
1991-92	2.5	2.7
1993-97	8.3	8.8
V6		
4Runner		
1994-02	3.4	3.8
2003-05	2.3	2.6
Highlander	2.9	3.2
1989-01 Pickup	3.4	3.8
1998-04 PreRunner	4.0	4.4
Sienna		
1998-03	3.5	3.9
2004-05	2.7	3.0
Tacoma		
1995-00	3.5	3.9
2001-04	4.0	4.4
2005	2.2	2.5
Tundra		
2000-04	4.0	4.4
2005	2.3	2.6
T100	3.8	4.2
V8		
4 Runner	2.3	2.6
Land Cruiser	3.2	3.6
Sequoia, Tundra	3.3	3.7
Replace heat riser Land Cruiser add	2.4	2.4
INJECTION		
Diagnose Electronic Control Module Battery Malfunction (A)		
Exc. below	.6	.6
Highlander 3.0L	.3	.3
Previa, Tundra	.3	.3
2001-05 RAV4	.3	.3
Diagnose Idle Control System Malfunction (A)		
Sequoia	.3	.3
Tundra		
3.4L	1.4	1.4
4.7L	.7	.7
Diagnose Oil Control Valve Circuit Malfunction (A)		
Highlander, RAV4	.8	.8
Air Valve, Replace (B)		
1984-95 4Runner, Pickup	1.5	1.5
Van	2.8	2.8

4RUNNER : HIGHLANDER : LAND CRUISER : PICKUP : PRERUNNER : PREVIA

	LABOR TIME	SEVERE SERVICE
Cold Start Injector, Replace (B)		
1983-95 4Runner	.7	.7
1984-95 Pickup	.7	.7
Previa	.9	.9
1993-94 T100	.7	.7
Van	.7	.7
Diesel Fuel Injection Pump, Replace (B)		
1981-87 Pickup	4.2	4.5
Fuel Injection Lines, Replace (B)		
Diesel		
one	.5	.5
each addl. add	.2	.2
Fuel Injectors, Replace (B)		
4Runner, Pickup		
gasoline		
1984-95	3.0	3.2
1996-02	2.7	2.9
2003-05		
V6	2.3	2.6
V8	1.8	2.1
diesel	1.4	1.5
Highlander		
4 cyl.	1.4	1.5
V6	2.1	2.3
Land Cruiser		
1988-92	2.6	2.8
1993-97	3.6	3.8
1998-05	2.3	2.5
1998-04 PreRunner		
4 cyl.	1.5	1.7
V6	3.1	3.4
Previa	1.8	2.0
RAV4		
1996-00	2.5	2.7
2001-05	.9	1.0
Sequoia, Tundra		
V6	2.8	3.1
V8	2.3	2.6
Sienna		
1998-03	2.6	2.9
2004-05	2.3	2.6
T100		
4 cyl.	1.7	1.9
V6	2.9	3.1
Tacoma		
1995-96		
4 cyl.	1.7	1.9
V6	3.1	3.4
1997-05		
4 cyl.	2.4	2.7
V6	2.8	3.1
Van	4.1	4.3
Fuel Pressure Regulator, Replace (B)		
1984-95 4Runner, Pickup		
4 cyl.	1.4	1.6
V6	2.3	2.5
1996-05 4Runner	.7	.8
Highlander	1.2	1.4

	LABOR TIME	SEVERE SERVICE
Land Cruiser		
1988-97	1.2	1.4
1998-05	.6	.7
1998-04 PreRunner		
4 cyl.	.5	.5
V6	2.4	2.7
Previa	1.4	1.5
RAV4	1.4	1.5
Sequoia	.9	1.1
Sienna		
1998-03	.7	.9
2004-05	1.2	1.3
T100, Tacoma		
1995-04		
4 cyl.	.5	.5
V6	2.4	2.7
2005 Tacoma	.7	.8
Tundra		
V6		
3.4L	2.4	2.7
4.0L	.6	.7
V8	.9	1.0
Idle Air Control (IAC) Valve, Replace (B)		
1996-02 4Runner		
4 cyl.	2.1	2.4
V6	1.7	1.9
Highlander	1.7	1.9
Land Cruiser		
1988-92	1.4	1.5
1993-97	1.9	2.2
1998-05	.9	1.0
1998-04 PreRunner	1.7	1.9
Previa	1.6	1.8
RAV4	2.1	2.4
Sienna	1.3	1.4
T100	1.8	2.1
Tacoma, Tundra V6	1.7	1.9
Start Injector Time Switch, Replace (B)		
1984-95 4Runner, Pickup		
4 cyl.	1.3	1.3
V6	.7	.7
Previa	1.4	1.4
1993-94 T100	.7	.7
Van	1.5	1.5
Throttle Body Assy., Replace (B)		
Exc. below	1.7	1.9
RAV4	2.0	2.2
Sienna		
1998-03	2.5	2.8
2004-05	1.2	1.4
TURBOCHARGER		
Turbocharger Assy., R&R or Replace (B)		
1985-88 Pickup		
gasoline	3.3	3.8
diesel	1.7	2.2

	LABOR TIME	SEVERE SERVICE
Turbo to Exhaust Manifold Gasket, Replace (B)		
1985-88 Pickup		
gasoline	3.0	3.3
diesel	1.5	1.8
SUPERCHARGER		
Intercooler, Replace (B)		
1995-97 Previa	.8	1.0
Supercharger Air Control Valve, Replace (B)		
1995-97 Previa	.3	.5
Supercharger Assy., or Gasket, Replace (B)		
1995-97 Previa	4.8	5.3
w/AC add	.1	.1
Supercharger By-Pass Valve, Replace (B)		
1995-97 Previa	2.9	3.1
w/AC add	.1	.1
Supercharger Relay, Replace (B)		
1995-97 Previa	.3	.3
# EXHAUST		
Catalytic Converter, Replace (B)		
Exc. below	.9	1.0
Highlander	1.1	1.3
Land Cruiser		
1988-97	1.9	2.0
1998-05	.7	.8
Tacoma	.3	.3
1984-88 Van	1.2	1.3
Center Exhaust Pipe, Replace (B)		
Exc. below	.8	.9
Highlander	1.1	1.2
Land Cruiser		
1981-92	1.1	1.2
1993-97	1.5	1.6
Exhaust Manifold or Gasket, Replace (B)		
4 cyl.		
4Runner	1.7	1.9
Highlander	.9	1.0
Pickup		
gasoline	1.8	2.1
diesel	1.6	1.8
1998-04 PreRunner	1.2	1.4
Previa	1.4	1.5
RAV4, Tacoma	1.2	1.4
T100	1.3	1.6
Van	3.8	4.2
6 cyl.		
Land Cruiser		
1981-87	1.8	2.1
1988-92	2.9	3.2
1993-97	1.9	2.2

RAV4 : SEQUOIA : SIENNA : T100 : TACOMA : TUNDRA : VAN : VAN WAGON — TOY-65

	LABOR TIME	SEVERE SERVICE
V6 one side		
4Runner, Pickup		
1989-95	3.5	3.9
1996-02	1.7	1.9
2003-05	1.1	1.2
Highlander		
right side	1.4	1.5
left side	.8	.9
1998-04 PreRunner	1.7	1.9
Sienna		
1998-03		
right side	9.5	10.6
left side	2.6	2.9
2004-05		
right side	1.7	1.9
left side	1.2	1.4
T100		
1993-94	3.6	4.1
1995-98	1.6	1.8
Tacoma, Tundra		
1995-04	1.7	1.9
2005	1.3	1.4
V8 one side		
2003-05 4Runner		
right side	1.4	1.5
left side	2.3	2.6
Land Cruiser	8.5	9.0
Sequoia, Tundra	7.2	7.7
w/AC add		
89-95 4Runner,		
Pickup V6	.5	.5
Sequoia, Tundra V8	.5	.5
93-94 T100	.5	.5
Front Exhaust Pipe, Replace (B)		
Exc. below	.9	1.0
Land Cruiser		
1988-92	1.5	1.6
1993-97	1.4	.5
Van	1.1	1.2
Muffler and Tailpipe Assy., Replace (B)		
Exc. below	.6	.7
Highlander	.3	.3
Land Cruiser		
1988-92	1.4	1.5
1993-97	1.0	1.1
Van	1.1	1.2

ENGINE COOLING

	LABOR TIME	SEVERE SERVICE
Pressure Test Cooling System (C)		
1981-05	.3	.3
By-Pass Hoses, Replace (B)		
4 cyl.		
exc. below	.7	.9
1983-95 4Runner	.9	1.1
Highlander	1.2	1.4
Pickup	.9	1.1
RAV4	1.1	1.2

	LABOR TIME	SEVERE SERVICE
6 cyl.		
Land Cruiser		
1981-87	.5	.7
1988-97	.8	1.0
V6, V8		
exc. below	.7	.9
2003-05 4Runner		
V6		
2WD (11.8)	15.3	15.5
4WD (12.8)	16.4	16.6
V8	1.2	1.4
Highlander	2.3	2.4
2004-05 Sienna	3.1	3.3
Coolant Thermostat, Replace (B)		
4 cyl.		
4Runner, Pickup		
1985-95	.9	1.0
1996-00	1.2	1.4
Highlander	1.1	1.2
1998-04 PreRunner	1.1	1.2
Previa, Van	1.3	1.4
RAV4		
1996-00	.8	.9
2001-05	1.5	1.7
T100	1.2	1.4
Tacoma	1.1	1.2
6 cyl.		
Land Cruiser		
1981-87	.8	.9
1988-97	1.2	1.4
V6		
4Runner	1.2	1.4
Highlander	.9	1.0
Pickup	1.1	1.2
1998-04 PreRunner	1.1	1.2
Sienna	1.5	1.7
T100	1.2	1.4
Tacoma, Tundra	1.1	1.2
V8		
exc. 4Runner	.9	1.0
4Runner	1.4	1.5
Electric Cooling Fan Motor, Replace (B)		
Highlander	1.5	1.7
RAV4		
1996-00	2.5	2.7
2001-05	2.1	2.4
Sienna	.8	1.0
Engine Coolant Temp. Sending Unit, Replace (B)		
Exc. below	.8	.8
Previa	1.4	1.4
1995-99 T100 V6	2.2	2.2
Tacoma V6	2.2	2.2
Van	1.4	1.4
Radiator Assy., R&R or Replace (B)		
4 cyl.		
4Runner	1.7	1.9
Highlander	1.4	1.5

	LABOR TIME	SEVERE SERVICE
Pickup	1.3	1.4
1998-04 PreRunner	1.5	1.7
Previa	1.7	1.9
RAV4	2.3	2.6
T100, Van	2.1	2.3
Tacoma	1.5	1.7
6 cyl.		
Land Cruiser	1.9	2.1
V6		
4Runner, Pickup	1.6	1.8
Highlander	1.4	1.6
1998-04 PreRunner	1.4	1.5
Sienna		
1998-03	1.7	1.9
2004-05	2.6	2.9
T100	1.2	1.4
Tacoma	1.4	1.5
Tundra	1.8	12.0
V8	1.7	1.9
Radiator Fan Motor Relay, Replace (B)		
All Models	.3	.3
Radiator Hoses, Replace (B)		
Includes: Refill with proper coolant mix.		
Upper	1.1	1.2
Lower	1.3	1.4
98-05 Land Cruiser		
lower hose add	.2	.2
Temperature Gauge (Dash), Replace (B)		
4Runner, Pickup		
1981-88	.8	.8
1989-95	1.3	1.3
1996-02	.7	.7
Land Cruiser		
1981-97	1.2	1.2
1998-02	.8	.8
1998-04 PreRunner	.8	.8
Previa	1.5	1.5
1996-00 RAV4	.7	.7
Sequoia, Tundra	.7	.7
Sienna, Tacoma	.8	.8
1993-98 T100	1.2	1.2
Van	1.4	1.4
Water Pump and/or Gasket, Replace (B)		
Includes: Refill with proper coolant mix.		
4 cyl.		
4Runner	1.8	2.1
Highlander	1.4	1.5
Pickup		
gasoline	1.5	1.7
diesel	2.5	2.8
1998-04 PreRunner	2.1	2.4
Previa	1.6	1.8
RAV4		
1996-00	4.1	4.6
2001-05	1.5	1.7

4RUNNER : HIGHLANDER : LAND CRUISER : PICKUP : PRERUNNER : PREVIA

	LABOR TIME	SEVERE SERVICE
Water Pump and/or Gasket, Replace (B)		
T100, Tacoma		
1993-04	2.1	2.4
2005	1.5	1.7
Van	2.7	3.0
6 cyl.		
Land Cruiser		
1981-87	1.7	1.9
1988-92	2.8	3.1
1993-97	2.3	2.6
V6		
4Runner, Pickup		
1989-95	4.7	5.2
1996-02	3.8	4.3
2003-05	3.2	3.6
Highlander	3.2	3.6
1998-04 PreRunner	3.2	3.6
Sienna		
1998-03	4.3	4.8
2004-05	3.8	4.2
T100		
1993-94	4.8	5.4
1995-98	3.2	3.6
Tacoma, Tundra	3.2	3.6
V8		
4Runner	3.8	4.2
Land Cruiser	4.3	4.7
Sequoia, Tundra	3.6	4.1
w/AC add		
03-05 4Runner	.5	.5
98-04 PreRunner	.5	.5
Sequoia, Tundra	.5	.5
Sienna	.2	.2
Tacoma	.5	.5

ENGINE ASSEMBLY

Times shown are for OEM assemblies. Time to replace assemblies from aftermarket rebuilders may vary.

Engine Assy., R&I (B)
Does not include parts or component transfer.

	LABOR TIME	SEVERE SERVICE
4 cyl.		
4Runner, Pickup		
gasoline		
1981-95		
2WD		
w/FI	6.3	6.3
w/o FI	5.4	5.4
4WD		
w/FI	8.1	8.1
w/o FI	7.6	7.6
1996-00		
2WD	10.2	10.2
4WD	14.4	14.4
diesel		
2WD	5.8	5.8
4WD	7.2	7.2
Previa	7.5	7.5
RAV4		
2WD	14.5	14.5
4WD	15.0	15.0
T100	12.2	12.2
Tacoma		
2WD	10.0	10.0
4WD	12.6	12.6
Van		
1984-87	6.8	6.8
1988-89	9.9	9.9
6 cyl.		
Land Cruiser		
1981-87	5.8	5.8
hardtop	10.1	10.1
wagon	10.5	10.5
1988-92	12.0	12.0
1993-97	13.4	13.4
V6		
4Runner, Pickup		
1989-95		
2WD	13.2	13.2
4WD	15.5	15.5
Sienna	14.3	14.3
T100		
1993-94		
2WD	13.2	13.2
4WD	15.6	15.6
1995-98		
2WD	10.2	10.2
4WD	12.6	12.6
Tacoma		
2WD	9.8	9.8
4WD	13.9	13.9
Tundra		
2WD	10.2	10.2
4WD	12.8	12.8
V8		
Land Cruiser	18.4	18.4
Sequoia, Tundra		
2WD	19.0	19.0
4WD	21.0	21.0
w/AC add		
4Runner, Pickup		
81-95 4 cyl.	.7	.7
89-95 V6	.5	.5
96-02	.2	.2
Land Cruiser 6 cyl.		
81-87	2.2	2.2
88-92	1.1	1.1
93-97	.5	.5
RAV4	.2	.2
Sienna V6	.3	.3

	LABOR TIME	SEVERE SERVICE
T100		
4 cyl.	.3	.3
V6		
93-94	.5	.5
95-98	.3	.3
Tacoma		
4 cyl.	.3	.3
V6	.3	.3
diesel	.7	.7
w/PS add		
81-95 4Runner, Pickup	.7	.7
81-87 Land Cruiser		
6 cyl.	.5	.5
Tacoma 4 cyl.	.5	.5
diesel	.7	.7
w/turbo diesel add	.2	.2

Engine Assy., R&R and Recondition (A)
Includes: Replacing rings, rod and main bearings, cylinder head reconditioning and engine tune-up.

	LABOR TIME	SEVERE SERVICE
4 cyl.		
4Runner, Pickup		
gasoline		
1981-87		
2WD		
w/FI	21.4	21.4
w/o FI	20.5	20.5
4WD		
w/FI	23.5	23.5
w/o FI	22.6	22.6
1988-95		
2WD		
w/FI	24.4	24.4
w/o FI	22.0	22.0
4WD		
w/FI	26.6	26.6
w/o FI	22.7	22.7
1996-00		
2WD	34.0	34.0
4WD	39.7	39.7
diesel		
2WD	20.2	20.2
4WD	23.0	23.0
Highlander		
2WD (29.6)	38.6	39.5
4WD (30.3)	39.4	40.3
1998-04 PreRunner	36.5	40.5
Previa		
2WD	31.4	31.4
4WD	31.7	31.7
RAV4		
1996-00		
2WD	34.4	34.4
4WD	34.8	34.8
2001-03		
2WD	35.6	38.8
4WD	36.3	39.5

RAV4 : SEQUOIA : SIENNA : T100 : TACOMA : TUNDRA : VAN : VAN WAGON — TOY-67

	LABOR TIME	SEVERE SERVICE
2004-05		
2WD (26.4)	34.3	35.2
4WD (26.9)	35.0	35.9
Tacoma		
1995-04		
2WD	36.0	39.8
4WD	42.1	46.7
2005		
2WD (30.4)	39.5	40.4
4WD (29.9)	38.9	39.7
Van		
1984-87	21.1	21.1
1988-89	22.2	22.2
6 cyl.		
Land Cruiser		
1981-87		
hardtop	26.8	26.8
wagon	30.1	30.1
1988-92	34.4	34.4
1993-97	46.7	46.7
V6		
4Runner, Pickup		
1989-95		
2WD	36.4	36.4
4WD	38.9	38.9
1996-02		
2WD	43.4	43.4
4WD	46.1	46.1
2003-05		
2WD (42.4)	55.1	56.0
4WD (43.1)	56.0	56.9
Highlander		
2WD (40.2)	52.3	53.1
4WD (41.2)	53.6	54.5
1998-04 PreRunner	44.3	48.2
Sienna		
1998-03	47.8	51.7
2004-05		
2WD (38.7)	50.3	51.2
4WD (39.3)	51.1	52.0
T100		
1993-94		
2WD	36.3	36.3
4WD	38.7	38.7
1995-98		
2WD	44.8	44.8
4WD	47.0	47.0
Tacoma		
1995-04		
2WD	45.9	50.8
4WD	48.7	53.9
2005 (40.4)	52.5	53.4
Tundra		
2000-04		
2WD	46.6	51.6
4WD	48.4	53.6
2005 (35.3)	45.9	46.8

	LABOR TIME	SEVERE SERVICE
V8		
2003-05 4Runner		
2WD (44.7)	58.1	59.0
4WD (45.8)	59.5	60.4
Land Cruiser (41.8)	54.3	55.2
Sequoia, Tundra		
2WD (40.1)	52.1	53.0
4WD (41.4)	53.8	54.7
w/AC add		
4Runner, Pickup		
81-95 4 cyl.	.7	.7
89-95 V6	.5	.5
96-02	.2	.2
Land Cruiser 6 cyl.		
81-87	2.2	2.2
88-92	1.1	1.1
93-97	.5	.5
98-04 PreRunner	.5	.5
RAV4		
96-00	.2	.2
01-05	1.2	1.4
Sequoia, Tundra	.5	.5
Sienna	.5	.5
T100		
4 cyl.	.3	.3
V6		
93-94	.5	.5
95-98	.3	.3
Tacoma	.5	.5
diesel	.7	.7
w/PS add		
81-95 4Runner, Pickup	.7	.7
81-87 Land Cruiser		
6 cyl.	.5	.5
01-05 RAV4	.3	.3
Tacoma 2.4L	.6	.6
diesel	.7	.7
w/turbo diesel add	.2	.2

Engine Assy. (Short Block), Replace (B)

Assembly may consist of cylinder block, piston assemblies, crankshaft, camshaft, timing chain and gears. Operation Includes: R&R engine, transfer necessary parts and all necessary adjustments.

	LABOR TIME	SEVERE SERVICE
4 cyl.		
4Runner, Pickup		
gasoline		
1981-95		
2WD		
w/FI	12.9	12.9
w/o FI	10.2	10.2
4WD		
w/FI	14.6	14.6
w/o FI	12.5	12.5

	LABOR TIME	SEVERE SERVICE
1996-02		
2WD	15.3	15.3
4WD	19.9	19.9
diesel		
2WD	14.3	14.3
4WD	15.4	15.4
Highlander		
2WD (17.0)	22.1	23.0
4WD (17.7)	23.0	23.9
1998-04 PreRunner	17.5	19.3
Previa	14.1	14.1
RAV4		
1996-00		
2WD	16.8	16.8
4WD	17.4	17.4
2001-03		
2WD	21.0	23.1
4WD	21.5	23.7
2004-05		
2WD (15.0)	19.6	20.5
4WD (15.5)	20.1	22.0
T100	17.8	17.8
Tacoma		
2WD	17.7	19.8
4WD	22.0	24.6
6 cyl.		
Land Cruiser		
1981-87		
hardtop	15.9	15.9
wagon	16.3	16.3
1988-92	17.6	17.6
1993-97	19.8	19.8
V6		
4Runner, Pickup		
1989-95		
2WD	19.3	19.3
4WD	21.5	21.5
1996-02		
2WD	17.6	17.6
4WD	21.7	21.7
2003-05		
2WD (17.7)	23.0	23.9
4WD (18.4)	23.9	24.8
Highlander		
2WD (16.5)	21.5	22.4
4WD (17.5)	22.8	23.7
1998-04 PreRunner	20.1	22.4
Sienna		
1998-03	22.1	24.5
2004-05		
2WD (15.9)	20.7	21.6
4WD (16.5)	21.5	22.4

4RUNNER : HIGHLANDER : LAND CRUISER : PICKUP : PRERUNNER : PREVIA

	LABOR TIME	SEVERE SERVICE
Engine Assy. (Short Block), Replace (B)		
T100		
1993-94		
2WD	19.2	19.2
4WD	21.7	21.7
1995-98		
2WD	19.3	19.3
4WD	20.9	20.9
Tacoma		
1995-04		
2WD	20.1	22.4
4WD	24.0	26.7
Tundra		
2WD	20.7	23.1
4WD	22.5	25.0
V8		
2003-05 4Runner		
2WD (16.3)	21.2	22.1
4WD (17.0)	22.1	23.0
Land Cruiser (16.1)	20.9	21.8
Sequoia, Tundra		
2WD (14.5)	19.0	19.9
4WD (15.7)	20.4	21.3
w/AC add		
4Runner, Pickup		
81-95 4 cyl.	.7	.7
89-95 V6	.5	.5
96-02	.2	.2
Land Cruiser 6 cyl.		
81-87	2.2	2.2
88-92	1.1	1.1
93-97	.5	.5
RAV4		
96-00	.2	.2
01-05	1.2	1.4
Sequoia, Tundra	.5	.5
Sienna	.5	.5
T100		
4 cyl.	.3	.3
V6		
93-94	.5	.5
95-98	.3	.3
diesel	.7	.7
w/PS add		
81-95 4Runner, Pickup	.7	.7
81-87 Land Cruiser		
6 cyl.	.5	.5
01-05 RAV4	.3	.3
Tacoma 2.4L	.6	.6
diesel	.7	.7
w/turbo diesel add	.2	.2
Engine Mounts, Replace (B)		
Front one side		
4Runner		
1983-95	.7	.9
1996-05		
4 cyl.	.9	1.0
V6	1.3	1.5

	LABOR TIME	SEVERE SERVICE
Highlander		
2WD	.6	.7
4WD	1.2	1.4
Land Cruiser	.9	1.1
Pickup	.7	.9
1998-04 PreRunner	.5	.7
Previa	.5	.7
RAV4, Sienna	1.0	1.1
Sequoia, Tundra	.5	.7
T100, Tacoma	.5	.7
Van	.8	1.0
Rear		
4Runner		
1983-95	.8	1.0
1996-98	1.4	1.6
1999-02	.8	1.0
Pickup, Previa	.8	1.0
1993-98 T100	.8	1.0
Tacoma	.8	1.0
Tundra V6	.6	.8
Van	.9	1.1
diesel all	.7	.9
CYLINDER HEAD		
Compression Test (B)		
4 cyl.		
exc. below	.7	1.0
4Runner	.8	1.1
Previa	1.3	1.6
6 cyl.		
Land Cruiser	1.2	1.5
V6		
exc. below	1.3	1.6
Sienna	3.1	3.4
T100, Tacoma	1.5	1.8
V8	1.4	1.7
Valve Clearance, Adjust (B)		
4 cyl.		
exc. below	3.2	3.5
4Runner, Pickup		
1981-95	1.7	2.0
Highlander	2.3	2.8
2001-05 RAV4	7.5	8.0
6 cyl.		
Land Cruiser		
1981-92	1.7	2.0
1993-97	6.1	6.4
V6		
4Runner		
1990-95	3.6	3.9
1999-02	5.6	5.9
2003-05	6.8	7.3
Highlander	5.1	5.6
Pickup	3.7	4.0
1998-04 PreRunner	5.9	6.2
Sienna		
2001-03	6.5	6.8
2004-05	5.5	6.0

	LABOR TIME	SEVERE SERVICE
T100		
1993-94	3.5	3.8
1995-98	5.6	5.9
Tacoma		
1995-04	5.9	6.2
2005	6.8	7.3
Tundra		
2000-04	5.9	6.2
2005	4.8	5.3
V8		
exc. below	5.9	6.4
2003-05 4Runner	6.8	7.3
Cylinder Head, Replace (B)		
Includes: Parts transfer and adjustments.		
4 cyl.		
4Runner		
1986-95		
w/FI	9.6	10.7
w/o FI	7.6	8.5
w/turbo	9.7	10.7
1996-00		
2WD	15.3	16.9
4WD	13.7	15.2
Highlander		
2WD	13.2	13.7
4WD	13.9	14.4
Pickup		
gasoline		
w/FI	9.6	10.7
w/o FI	7.6	8.5
w/turbo	9.6	10.6
diesel	8.8	9.7
1998-04 PreRunner	16.6	18.3
Previa		
2WD	18.7	20.6
4WD	18.8	20.7
RAV4		
1996-00		
2WD	16.6	17.1
4WD	17.0	17.5
2001-03	8.8	9.3
2004-05	16.2	16.7
T100	14.0	15.4
Tacoma		
1995-04		
2.4L	16.0	17.8
2.7L	16.6	18.3
2005		
2WD	19.3	19.8
4WD	18.4	18.9
Van	8.1	9.0
6 cyl.		
Land Cruiser		
1981-87		
hardtop	8.2	9.1
wagon	9.7	10.7
1988-92	12.0	13.3
1993-97	19.9	22.0

RAV4 : SEQUOIA : SIENNA : T100 : TACOMA : TUNDRA : VAN : VAN WAGON

	LABOR TIME	SEVERE SERVICE
V6		
4Runner		
1990-95	19.1	22.1
1996-02		
one side	14.1	15.6
both sides	21.7	23.9
2003-05		
2WD		
one side	20.1	20.6
both sides	27.8	28.2
4WD		
one side	21.3	21.8
both sides	28.9	29.3
Highlander		
2WD		
one side	13.0	13.5
both sides	19.8	20.3
4WD		
one side	13.3	13.8
both sides	20.6	21.1
Pickup	19.1	19.6
1998-04 PreRunner		
one side	14.6	16.2
both sides	21.8	24.0
Sienna		
one side	14.2	14.7
both sides	20.1	20.6
T100		
1993-94		
one side	13.6	15.1
both sides	19.2	21.2
1995-98		
one side	14.2	15.7
both sides	21.4	23.6
Tacoma		
1995-04		
one side	14.6	16.2
both sides	21.8	24.0
2005		
2WD		
one side	21.4	21.9
both sides	28.8	29.3
4WD		
one side	19.3	19.8
both sides	27.0	27.5
Tundra		
2000-04		
one side	14.6	16.2
both sides	21.8	24.0
2005		
one side	16.8	17.2
both sides	23.4	23.9
V8		
2003-05 4Runner		
one side	17.8	18.3
both sides	28.0	28.5
Land Cruiser		
one side	15.7	16.2
both sides	24.8	25.3

	LABOR TIME	SEVERE SERVICE
Sequoia, Tundra		
one side	15.4	15.9
both sides	24.4	24.9
w/ABS 96-00 RAV4 add	.7	.7
w/AC add		
4Runner, Pickup		
81-95 4 cyl.	.3	.3
89-95 V6	.5	.5
96-05 V6	.3	.3
81-92 Land Cruiser		
6 cyl.	.3	.3
96-00 RAV4	.2	.2
Sequoia, Tundra	.5	.5
Sienna	.1	.1
93-94 T100 V6	.5	.5
Tacoma		
4 cyl.	.2	.2
V6	.5	.5
w/PS add		
81-95 4Runner, Pickup	.5	.5
01-05 RAV4	.3	.3
Tacoma 2.4L	.6	.6
diesel	.7	.7
w/turbo add		
96-05 4Runner 4 cyl.	.5	.5
diesel	.5	.5
Cylinder Head Gasket, Replace (B)		
4 cyl.		
4Runner		
1983-95		
w/FI	6.4	6.7
w/o FI	4.9	5.2
w/turbo	6.9	7.2
1996-00	7.8	8.1
Highlander		
2WD	9.7	10.2
4WD	10.2	10.7
Pickup		
gasoline		
w/FI	6.3	6.6
w/o FI	4.9	5.2
w/turbo	6.9	7.2
diesel	6.1	6.4
1998-04 PreRunner	8.9	9.2
Previa		
2WD	10.6	10.9
4WD	11.1	11.4
RAV4		
1996-00	9.9	10.2
2001-05	9.5	10.0
T100	8.6	8.9
Tacoma		
1995-04		
2.4L	8.3	8.6
2.7L	8.9	9.2
2005		
2WD	17.9	18.4
4WD	17.0	17.5

	LABOR TIME	SEVERE SERVICE
6 cyl.		
Land Cruiser		
1981-87		
hardtop	5.2	5.5
wagon	6.2	6.5
1988-92	7.1	7.4
1993-97	9.9	10.2
V6		
4Runner		
1990-95	11.4	11.7
1996-02		
one side	7.9	8.2
both sides	10.3	10.6
2003-05		
2WD		
one side	15.3	15.8
both sides	17.5	18.0
4WD		
one side	16.3	16.8
both sides	18.5	19.0
Highlander		
2WD		
one side	8.4	8.9
both sides	11.5	12.0
4WD		
one side	8.8	9.8
both sides	11.9	12.4
1998-04 PreRunner		
one side	9.0	9.3
both sides	11.8	12.1
Pickup	11.5	11.8
Sienna		
1998-03		
one side	9.3	9.6
both sides	11.8	12.1
2004-05		
one side	9.0	9.5
both sides	11.5	12.0
T100		
1993-94		
one side	9.0	9.3
both sides	11.6	11.9
1995-98		
one side	8.8	9.1
both sides	10.7	11.0
Tacoma, Tundra		
1995-04		
one side	9.2	9.5
both sides	11.8	12.1
2005 Tundra		
one side	12.4	12.9
both sides	14.6	15.1
V8		
4Runner		
one side	7.2	7.7
both sides	9.2	9.7
Land Cruiser		
one side	9.4	9.9
both sides	12.6	13.1

TOY-69

TOY-70 4RUNNER : HIGHLANDER : LAND CRUISER : PICKUP : PRERUNNER : PREVIA

	LABOR TIME	SEVERE SERVICE
Cylinder Head Gasket, Replace (B)		
Sequoia, Tundra		
one side	9.0	9.5
both sides	12.2	12.7
w/ABS 96-00 RAV4 add	.7	.7
w/AC add		
4Runner, Pickup		
81-95 4 cyl.	.3	.3
89-95 V6	.5	.5
96-05 V6	.3	.3
81-92 Land Cruiser		
6 cyl.	.3	.3
96-00 RAV4	.2	.2
Sequoia, Tundra	.5	.5
Sienna	.2	.2
93-94 T100 V6	.5	.5
Tacoma		
4 cyl.	.2	.2
V6	.5	.5
w/PS add		
81-95 4Runner, Pickup		
4 cyl.	.5	.5
01-05 RAV4	.3	.3
Tacoma 2.4L	.6	.6
w/turbo add		
96-05 4Runner 4 cyl.	.5	.5
diesel	.5	.5
Rocker Arms, Shafts & Supports, Replace (B)		
4 cyl.		
exc. Van	2.9	3.1
Van	1.7	1.9
6 cyl.	2.9	3.1
Valve/Cam Cover Gasket, Replace or Reseal (B)		
4 cyl.		
exc. below	.9	1.1
T100, Tacoma	1.5	1.7
6 cyl.		
Land Cruiser		
1981-92	.8	1.0
1993-97	1.2	1.4
V6		
4Runner, Pickup	2.4	2.6
Highlander	2.4	2.7
1998-04 PreRunner	3.2	3.6
Sienna		
2001-03	2.8	3.0
2004-05	3.1	3.3
T100	3.0	3.2
Tacoma, Tundra		
1995-04	3.2	3.4
2005	2.2	2.4
V8		
exc. below	1.1	1.3
2003-05 4Runner	.6	.7

	LABOR TIME	SEVERE SERVICE
Valve Job (A)		
4 cyl.		
4Runner		
1983-95		
w/FI	9.7	10.6
w/o FI	9.1	10.0
w/turbo	10.8	11.7
1996-00	19.2	20.1
Highlander		
2WD	15.2	16.1
4WD	15.9	16.8
Pickup		
gasoline		
w/FI	9.7	10.6
w/o FI	9.0	9.9
w/turbo	10.8	11.7
diesel	9.8	10.7
1998-04 PreRunner	20.8	21.7
Previa		
2WD	21.3	22.2
4WD	21.8	22.7
RAV4		
1996-00		
2WD	20.5	21.4
4WD	26.2	27.1
2001-05	19.3	20.2
Tacoma		
1995-04		
2.4L	20.3	21.2
2.7L	20.8	21.7
2005		
2WD	30.8	31.7
4WD	29.8	30.7
T100	19.2	20.1
Van		
1984-87	8.4	9.3
1988-89	9.2	10.1
6 cyl.		
Land Cruiser		
1981-87		
hardtop	9.8	10.7
wagon	12.0	12.9
1988-92	15.4	16.3
1993-97	26.8	27.7
V6		
4Runner		
1990-95	22.1	23.0
1996-02		
one side	16.5	17.4
both sides	24.4	25.3
2003-05 all		
2WD	33.1	34.0
4WD	34.3	34.2
Highlander all		
2WD	25.1	26.0
4WD	25.5	26.4
Pickup	22.1	23.0

	LABOR TIME	SEVERE SERVICE
1998-04 PreRunner		
right side	18.3	19.2
left side	16.7	17.6
both sides	27.9	28.8
Sienna		
2001-03		
right side	17.9	18.8
left side	15.3	16.2
both sides	28.2	29.1
2004-05 all	25.4	26.3
T100		
1993-94		
one side	14.9	15.8
both sides	20.8	21.7
1995-98		
one side	17.0	17.9
both sides	27.0	27.9
Tacoma		
1995-04		
one side	17.1	18.0
both sides	27.9	28.8
2005 all		
2WD	34.1	35.0
4WD	32.4	33.3
Tundra		
2000-04 all	27.8	28.7
2005 all	28.3	29.2
V8 all		
4Runner	30.5	31.4
Land Cruiser	30.6	31.5
Sequoia, Tundra	31.4	32.3
w/ABS 96-00 RAV4 add	.7	.7
w/AC add		
4Runner, Pickup		
81-95 4 cyl.	.3	.3
89-95 V6	.5	.5
96-05 V6	.3	.3
81-92 Land Cruiser		
6 cyl.	.3	.3
98-04 PreRunner		
4 cyl.	.2	.2
V6	.5	.5
RAV4	.2	.2
Sequoia, Tundra	.5	.5
Sienna	.2	.2
93-94 T100 V6	.5	.5
Tacoma		
4 cyl.	.2	.2
V6	.5	.5
w/PS add		
81-95 4Runner, Pickup		
4 cyl.	.5	.5
01-05 RAV4	.3	.3
Tacoma 2.4L	.6	.6
w/turbo add		
96-05 4Runner 4 cyl.	.5	.5
diesel	.5	.5

RAV4 : SEQUOIA : SIENNA : T100 : TACOMA : TUNDRA : VAN : VAN WAGON **TOY-71**

	LABOR TIME	SEVERE SERVICE
Valve Lifters, Replace (B)		
4 cyl.		
1996-00 4Runner	5.2	5.6
Highlander	2.0	2.4
1998-04 PreRunner	6.2	6.7
Previa	11.4	11.8
RAV4		
1996-00	6.9	7.3
2001-05	7.3	7.8
T100	5.7	6.1
Tacoma	6.2	6.6
Van	1.4	2.0
6 cyl.		
Land Cruiser		
1981-87		
hardtop	3.7	4.1
wagon	2.9	3.3
1988-92	2.5	2.9
1993-97	6.7	7.1
V6		
exc. below	9.6	10.0
4Runner		
1996-02	8.8	9.2
2003-05	7.3	7.7
Highlander	8.5	8.9
1998-04 PreRunner	6.2	6.7
Sienna		
2001-03	10.6	11.0
2004-05	8.8	9.3
2005 Tacoma	7.3	7.7
2005 Tundra	6.4	6.8
V8		
4Runner	9.8	10.2
Land Cruiser	10.2	10.6
Sequoia, Tundra	9.7	10.1
w/ABS 96-00 RAV4 add	.7	.7
w/AC add		
88-95 4Runner, Pickup V6	.6	.6
81-92 Land Cruiser 6 cyl.	.5	.5
Sequoia, Tundra	.5	.5
Sienna	.2	.2
93-94 T100 V6	.5	.5
T100		
4 cyl.	.2	.2
V6	.6	.6
Tacoma		
4 cyl.	.2	.2
V6	.5	.5
w/PS 01-05 RAV4 add	.3	.3
Valve Pushrods, Replace (B)		
Land Cruiser		
1981-87	1.8	2.1
1988-92	2.3	2.6
Van	1.1	1.4

	LABOR TIME	SEVERE SERVICE
Valve Springs and/or Oil Seals, Replace (B)		
4 cyl.		
4Runner		
1983-95		
w/FI	7.8	8.3
w/o FI	5.8	6.3
w/turbo	7.9	8.4
1996-00	11.8	12.3
Highlander		
2WD	10.9	11.4
4WD	11.7	12.2
Pickup		
gasoline		
w/FI	7.7	8.2
w/o FI	5.7	6.2
w/turbo	7.7	8.2
diesel	7.4	7.9
1998-04 PreRunner	13.2	13.7
Previa		
2WD	11.9	12.4
4WD	12.5	13.0
RAV4		
1996-00		
2WD	12.4	12.9
4WD	13.0	13.5
2001-05	11.9	12.4
T100	12.0	12.5
Tacoma		
1995-04		
2.4L	12.7	13.2
2.7L	13.2	13.7
2005		
2WD	20.9	21.4
4WD	19.9	20.4
Van		
1984-87	6.3	6.8
1988-89	6.7	7.2
6 cyl.		
Land Cruiser		
1981-87		
hardtop	6.5	7.0
wagon	7.1	7.6
1988-92	10.0	10.5
1993-97	13.7	14.2
V6		
4Runner		
1988-02	14.5	15.0
2003-05		
2WD	20.0	20.5
4WD	21.1	21.6
Highlander		
2WD	13.3	13.8
4WD	13.7	14.2
Pickup	14.6	15.1
1998-04 PreRunner	15.7	17.3

	LABOR TIME	SEVERE SERVICE
Sienna		
2001-03	15.7	16.2
2004-05	16.2	16.7
T100		
1993-94	14.4	14.9
1995-98	14.8	15.3
Tacoma	15.7	16.2
Tundra		
2000-04	15.6	16.1
2005	25.0	25.5
V8		
4Runner	14.1	14.6
Land Cruiser	15.9	16.4
Sequoia, Tundra	15.4	15.9
w/ABS 96-00 RAV4 add	.7	.7
w/AC add		
4Runner, Pickup		
81-95 4 cyl.	.3	.3
89-05 V6	.3	.3
81-02 Land Cruiser	.3	.3
RAV4	.2	.2
Sequoia, Tundra	.5	.5
Sienna	.2	.2
T100 V6		
93-94	.5	.5
95-98	.3	.3
Tacoma		
4 cyl.	.2	.2
V6	.5	.5
w/PS add		
81-95 4Runner, Pickup 4 cyl.	.5	.5
01-05 RAV4	.3	.3
Tacoma 2.4L	.6	.6
w/turbo diesel add	.5	.5
CAMSHAFT		
Balance Shaft, Replace (B)		
Includes bearings.		
1996-00 4Runner 4 cyl.		
2WD one	18.9	19.0
4WD one	19.9	20.4
Highlander 4 cyl.		
2WD one	17.0	17.5
4WD one	17.9	18.4
1998-04 PreRunner		
2WD one	19.9	20.4
RAV4		
2WD one	17.9	18.4
4WD one	17.3	17.8
1994-99 T100 4 cyl.		
one	18.5	19.0
both	19.2	19.7
Tacoma 4 cyl.		
1995-04 4WD one	20.8	21.4
2005		
2WD	17.8	18.3
4WD	17.2	17.7

TOY-72   4RUNNER : HIGHLANDER : LAND CRUISER : PICKUP : PRERUNNER : PREVIA

	LABOR TIME	SEVERE SERVICE
Camshaft, Replace (B)		
4 cyl.		
4Runner		
1985-95	3.4	3.9
1996-00	4.9	5.4
Highlander	2.6	3.1
Pickup		
gasoline		
1981-87	3.0	3.5
1988-95	3.4	3.9
diesel		
1998-04 PreRunner	5.9	6.4
Previa		
2WD	12.6	13.1
4WD	13.0	13.5
RAV4		
1996-00	7.9	8.4
2001-05	8.5	9.0
T100	5.6	6.1
Tacoma		
1995-04	5.9	6.4
2005	4.0	4.5
Van		
1984-87	4.9	5.4
1988-89	5.4	5.9
6 cyl.		
Land Cruiser		
1981-87		
hardtop	5.4	5.9
wagon	7.8	8.3
1988-92	6.9	7.4
1993-97	6.2	6.7
V6		
4Runner, Pickup		
1989-95	8.8	9.3
1996-02	9.4	9.9
2003-05	8.4	8.9
Highlander	8.5	9.0
1998-04 PreRunner	10.6	11.8
Sienna		
1998-03	11.0	11.5
2004-05	8.8	9.3
T100		
1993-94	8.6	9.1
1995-98	9.7	10.2
Tacoma		
1995-04	10.6	11.1
2005	8.4	8.9
Tundra		
2000-04	10.6	11.1
2005	6.3	6.8
V8		
exc. below	9.3	9.8
Sequoia, Tundra	8.8	9.3
w/ABS 96-00 RAV4 add	.8	.8
w/AC add		
89-95 4Runner V6	.3	.3
Land Cruiser		
81-87	1.1	1.1
88-92	1.3	1.3

	LABOR TIME	SEVERE SERVICE
Pickup		
81-95 4 cyl.	.8	.8
89-95 V6	.3	.3
Sequoia, Tundra	.3	.3
Sienna	.2	.2
T100		
4 cyl.	.2	.2
V6	.5	.5
Tacoma		
4 cyl.	.2	.2
V6	.5	.5
Van 4 cyl.	2.2	2.2
w/PS add		
01-05 RAV4	.3	.3
Van 4 cyl.	.5	.5
w/turbo diesel add	.2	.2
Camshaft Seal, Replace (B)		
4Runner		
1996-02 V6	4.3	4.5
2003-05 V8	4.5	4.7
Highlander	8.0	8.2
Land Cruiser	5.5	5.7
1996-00 RAV4	4.3	4.5
Tacoma V6	4.6	4.8
Sequoia, Tundra	4.8	5.0
2004-05 Sienna	5.7	5.9
w/ABS 96-00 RAV4 add	1.4	1.4
w/AC add		
4Runner V6	.5	.5
Sequoia, Tundra	.5	.5
Tacoma V6	.5	.5
Camshaft Timing Gear, Replace (B)		
4 cyl.		
4Runner		
1985-87	1.4	1.7
1988-00	1.8	2.1
Highlander	2.8	3.1
Pickup		
gasoline		
1981-87	1.2	1.5
1988-95	1.7	2.0
diesel	1.6	1.9
1998-04 PreRunner	1.8	2.1
Previa	1.9	2.2
RAV4		
1996-00	3.5	3.8
2001-05	5.6	5.9
T100, Tacoma		
1993-04	1.8	2.1
2005	4.0	4.3
Van	3.3	3.6
6 cyl.		
Land Cruiser		
1981-87		
hardtop	5.4	5.7
wagon	5.0	5.3
1988-92	6.9	7.2
1993-97	2.1	3.3

	LABOR TIME	SEVERE SERVICE
V6		
4Runner, Pickup		
1989-95	4.2	4.5
1996-02	1.8	2.1
2003-05	4.0	4.4
Highlander	2.6	2.9
1998-04 PreRunner	1.8	2.1
Sienna		
1998-03	3.8	4.1
2004-05	3.1	3.4
T100		
1993-94	4.1	4.4
1995-98	1.8	2.1
Tacoma		
1995-04	1.8	2.1
2005	8.4	8.7
Tundra	2.4	2.7
V8		
4Runner	4.0	4.4
Land Cruiser	4.4	4.7
Sequoia, Tundra	3.8	4.1
w/ABS 96-00 RAV4 add	.9	.9
w/AC add		
4Runner		
4 cyl.	.1	.1
89-02 V6	.3	.3
81-92 Land Cruiser	1.2	1.2
Pickup		
4 cyl.	.7	.7
89-02 V6	.3	.3
Sequoia, Tundra	.5	.5
Sienna	.2	.2
T100		
4 cyl.	.1	.1
93-98 V6	.5	.5
Tacoma		
4 cyl.	.2	.2
V6	.5	.5
w/PS 01-05 RAV4 add	.3	.3
Replace crank gear add		
Highlander		
4 cyl.	2.9	2.9
V6	.3	.3
Pickup	.3	.3
96-00 RAV4	.2	.2
V8	.5	.5
Crankshaft Gear or Sprocket, Replace (B)		
4 cyl.		
4Runner		
2WD	11.4	11.7
4WD	12.2	12.5
Highlander	6.5	6.8
Pickup		
1981-87		
w/FI	7.2	7.5
w/o FI	6.3	6.6

RAV4 : SEQUOIA : SIENNA : T100 : TACOMA : TUNDRA : VAN : VAN WAGON TOY-73

	LABOR TIME	SEVERE SERVICE
1988-95		
w/FI		
2WD	9.7	10.0
4WD	12.6	12.9
w/o FI	8.3	8.6
Previa		
2WD	12.0	12.3
4WD	12.5	12.8
RAV4		
1996-00	3.8	4.1
2001-05	5.1	5.4
T100	12.3	12.6
Tacoma		
1995-04		
2WD	11.4	11.7
4WD	13.9	14.2
2005		
2WD	13.7	14.0
4WD	12.8	13.1
Van	3.3	3.6
6 cyl.		
Land Cruiser		
1981-87	2.5	2.8
1988-92	7.0	7.3
1993-97	14.9	15.2
V6		
4Runner, Pickup		
1989-95	4.2	4.5
1999	2.9	3.2
2003-05		
2WD	12.4	12.7
4WD	13.4	13.7
Highlander	2.6	2.9
Sienna		
1998-03	3.8	4.1
2004-05	2.9	3.2
T100		
1993-94	4.1	4.4
1995-98	3.2	3.5
Tacoma		
1995-04	2.9	3.2
2005		
2WD	14.1	14.4
4WD	14.7	15.0
w/ABS 96-00 RAV4 add	1.4	1.4
w/AC add		
4Runner		
4 cyl.	.1	.1
89-02 V6	.3	.3
81-92 Land Cruiser	1.2	1.2
Pickup		
4 cyl.	.7	.7
89-02 V6	.3	.3
Sienna	.2	.2
T100		
4 cyl.	.1	.1
93-98 V6	.5	.5
Tacoma		
4 cyl.	.2	.2
V6	.5	.5
w/PS 01-05 RAV4 add	.3	.3

	LABOR TIME	SEVERE SERVICE
Timing Belt, Replace (B)		
4 cyl.		
1981-87 Pickup		
diesel	1.8	2.3
1996-00 RAV4	2.9	3.4
V6		
4Runner, Pickup		
1989-95	4.5	5.0
1996-02	2.7	3.2
Sienna	6.0	6.5
T100		
1993-94	4.4	4.9
1995-98	2.6	3.1
Tacoma, Tundra	2.8	3.3
V8	3.5	4.0
w/ABS 96-00 RAV4 add	.9	.9
w/AC add		
4Runner, Pickup V6	.3	.3
Sienna	.1	.1
93-98 T100 V6	.5	.5
Tacoma V6	.3	.3
Tundra	.3	.3
diesel	.7	.7
Timing Belt Tensioner, Replace (B)		
1993-00 4 cyl.	.5	1.0
V6		
exc. below	.7	1.2
1989-95 4Runner, Pickup	1.2	1.7
1993-94 T100	1.3	1.8
V8	.8	1.3
w/AC add		
T100 V6	.3	.3
Tacoma, Tundra V6	.5	.5
Timing Chain/Gear & Tensioner, Replace (B)		
4 cyl.		
4Runner, Pickup		
1981-87		
w/FI	9.0	9.4
w/o FI	8.3	8.7
1988-95		
w/FI		
2WD	9.7	10.1
4WD	12.3	12.7
w/o FI	9.1	9.5
1996-00		
chain only		
2WD	12.3	12.7
4WD	12.8	13.2
Highlander		
chain	6.5	7.0
tensioner	1.2	1.4
Previa		
chain		
2WD	12.0	12.4
4WD	12.2	12.6
tensioner		
2WD	12.2	12.6
4WD	12.4	12.8

	LABOR TIME	SEVERE SERVICE
2001-05 RAV4		
chain	5.4	5.8
tensioner	1.1	1.3
T100 chain only	13.1	13.5
Tacoma		
1995-04 chain only		
2WD	12.3	12.7
4WD	14.5	14.9
2005 chain and/or tensioner		
2WD	14.1	14.6
4WD	13.1	13.6
Van		
1984-87	2.9	3.3
1988-89	3.2	3.6
L6, V6		
4Runner		
chain	13.0	14.4
tensioner	.6	.7
Land Cruiser	15.2	15.6
2005 Tacoma		
2WD		
chain	14.6	15.1
tensioner	.5	.6
4WD		
chain	12.5	13.0
tensioner	.5	.6
2005 Tundra		
chain	10.3	10.8
tensioner	.5	.6
w/AC add		
4Runner, Pickup		
81-95 4 cyl.	.3	.3
96-00 4 cyl.	.2	.2
93-97 Land Cruiser	.5	.5
T100 4 cyl.	.3	.3
Tacoma 4 cyl.	.3	.3
w/PS add		
81-95 4Runner, Pickup 4 cyl.	.5	.5
Tacoma		
2.4L	.5	.5
2.7L	.2	.2
Timing Cover and/or Gasket, Replace (B)		
4 cyl.		
4Runner, Pickup		
1981-87		
gasoline		
w/FI	8.6	8.9
w/o FI	7.7	8.0
diesel	2.0	2.3
1988-95		
w/FI		
2WD	9.8	10.1
4WD	12.2	12.5
w/o FI	8.8	9.1
1996-00		
2WD	10.9	11.2
4WD	11.7	12.0

TOY-74 4RUNNER : HIGHLANDER : LAND CRUISER : PICKUP : PRERUNNER : PREVIA

	LABOR TIME	SEVERE SERVICE
Timing Cover and/or Gasket, Replace (B)		
Highlander	6.0	6.7
1998-04 PreRunner	12.4	12.7
Previa		
2WD	11.6	11.9
4WD	12.0	12.3
RAV4		
1996-00	2.7	3.0
2001-05	5.0	5.3
T100	11.6	11.9
Tacoma		
1995-04		
2WD	11.9	12.4
4WD	14.2	14.7
2005		
2WD	13.7	14.2
4WD	12.7	13.2
Van		
1984-87	1.7	2.0
1988-89	2.3	2.6
6 cyl.		
Land Cruiser		
1981-87	2.5	2.8
1988-92	3.4	3.7
1993-97	14.9	15.2
V6		
4Runner, Pickup		
1989-02	2.7	3.0
2003-05		
2WD	10.6	11.1
4WD	11.8	12.3
Highlander	1.2	1.5
1998-04 PreRunner	2.6	2.9
Sienna		
1998-03	3.5	3.8
2004-05	1.1	1.2
T100	2.8	3.1
Tacoma		
1995-04	2.6	2.9
2005		
2WD	13.0	13.5
4WD	10.8	11.3
Tundra		
2000-04	2.6	2.9
2005	8.8	9.3
V8		
4Runner	3.1	3.6
Land Cruiser	4.0	4.3
Sequoia, Tundra	3.1	3.6
w/ABS 96-00 RAV4 add	1.4	1.4
w/AC add		
4Runner, Pickup		
81-95 4 cyl.	.7	.7
96-02 4 cyl.	.2	.2
V6	.3	.3
Land Cruiser		
81-87	1.1	1.1
88-97	.5	.5

	LABOR TIME	SEVERE SERVICE
01-05 RAV4	.3	.3
Sequoia, Tundra	.5	.5
Sienna	.2	.2
T100		
4 cyl.	.3	.3
93-98 V6	.3	.3
Tacoma	.5	.5
diesel	.7	.7
w/PS add		
81-95 4Runner, Pickup 4 cyl.	.5	.5
Tacoma 2.4L	.6	.6
diesel	.1	.1
CRANK & PISTONS		
Connecting Rod Bearings, Replace (A)		
Includes: Check bearing oil clearance.		
4 cyl.		
4Runner		
1983-95		
2WD	4.1	4.6
4WD	7.6	8.1
1996-00		
2WD	5.4	5.9
4WD	6.9	7.4
Highlander		
2WD	21.5	22.0
4WD	22.5	23.0
Pickup		
gasoline		
2WD	4.2	4.7
4WD	7.6	8.1
diesel	4.0	4.5
1998-04 PreRunner	6.0	6.5
Previa	5.6	6.1
RAV4		
1996-00	5.4	5.9
2001-05	16.6	17.1
T100	5.4	5.9
Tacoma		
1995-04		
2WD	6.0	6.5
4WD	8.1	8.6
2005		
2WD	17.8	18.3
4WD	17.2	17.7
Van		
1984-87 Van	3.1	3.6
1988-89 Van	3.5	4.0
6 cyl.		
Land Cruiser		
1981-87		
hardtop	3.6	4.1
wagon	4.1	4.6
1988-92	12.2	12.7
1993-97	6.7	7.2

	LABOR TIME	SEVERE SERVICE
V6		
4Runner, Pickup		
1989-95		
2WD	5.2	5.7
4WD	8.0	8.5
1996-02	7.6	8.1
2003-05		
2WD	26.1	26.6
4WD	27.0	27.5
Highlander		
2WD	16.0	16.5
4WD	17.2	17.7
1998-04 PreRunner	6.6	7.1
Sienna		
1998-03	7.5	8.0
2004-05		
2WD	15.5	16.0
4WD	16.2	16.7
T100		
1993-94		
2WD	5.1	5.6
4WD	7.8	8.3
1995-98		
2WD	6.0	6.5
4WD	8.3	8.8
Tacoma		
1995-04		
2WD	6.3	6.8
4WD	8.5	9.0
2005		
2WD	23.0	23.5
4WD	23.5	24.0
Tundra		
2000-04		
2WD	6.6	7.1
4WD	9.1	9.6
2005	6.1	6.6
V8		
4Runner		
2WD	14.7	15.2
4WD	15.5	16.0
Land Cruiser	15.1	15.6
Sequoia, Tundra		
2WD	13.1	13.6
4WD	14.6	15.1
w/AC add		
01-05 RAV4	1.2	1.4
Sequoia, Tundra	.5	.5
w/PS 01-05 RAV4 add	.3	.3
Crankshaft and Main Bearings, Replace (A)		
Includes: Engine R&R.		
4 cyl.		
4Runner, Pickup		
gasoline		
1981-87		
2WD		
w/FI	12.5	13.9
w/o FI	11.4	12.7

RAV4 : SEQUOIA : SIENNA : T100 : TACOMA : TUNDRA : VAN : VAN WAGON — TOY-75

	LABOR TIME	SEVERE SERVICE
4WD		
w/FI	14.2	15.9
w/o FI	13.6	15.1
1988-95		
2WD		
w/FI	14.1	15.7
w/o FI	12.4	13.8
4WD		
w/FI	16.3	18.2
w/o FI	13.6	15.1
1996-02		
2WD	18.0	20.0
4WD	22.3	24.8
diesel		
2WD	12.8	14.3
4WD	14.1	15.7
Highlander		
2WD	25.5	28.1
4WD	26.4	29.3
1998-04 PreRunner	19.4	21.6
Previa		
2WD	16.1	17.9
4WD	17.1	19.1
RAV4		
1996-00		
2WD	19.3	21.5
4WD	19.8	22.1
2001-03		
2WD	21.0	23.5
4WD	21.7	24.1
2004-05		
2WD	21.9	24.5
4WD	22.6	25.1
T100	19.3	21.5
Tacoma		
2WD	19.4	21.6
4WD	23.5	26.1
Van		
1984-87	16.3	17.2
1988-89	16.9	18.6
6 cyl.		
Land Cruiser		
1981-87		
hardtop	17.1	19.1
wagon	17.7	19.8
1988-92	19.9	22.1
1993-97	22.4	24.8
V6		
4Runner, Pickup		
1989-95		
2WD	20.1	22.4
4WD	22.9	25.4
1996-02		
2WD	19.7	22.0
4WD	22.6	25.1
2003-05		
2WD	29.3	32.4
4WD	30.2	33.5

	LABOR TIME	SEVERE SERVICE
Highlander		
2WD	27.3	30.2
4WD	28.5	31.5
1998-04 PreRunner	22.2	24.7
Sienna		
1998-03	23.6	26.2
2004-05		
2WD	25.8	28.6
4WD	26.6	29.4
T100		
1993-94		
2WD	20.1	22.4
4WD	22.9	25.4
1995-98		
2WD	20.9	23.3
4WD	23.2	25.8
Tacoma		
2WD	22.2	24.7
4WD	24.7	27.5
Tundra		
2WD	22.8	25.3
4WD	24.4	27.0
V8		
4Runner		
2WD	26.6	29.4
4WD	27.1	30.0
Land Cruiser	25.2	27.9
Sequoia, Tundra		
2WD	24.2	26.8
4WD	25.8	28.6
w/AC add		
4Runner, Pickup		
81-95 4 cyl.	.7	.7
89-95 V6	.5	.5
96-02	.2	.2
Land Cruiser		
81-87 6 cyl.	2.2	2.2
88-92 6 cyl.	1.1	1.1
93-97	.5	.5
RAV4		
96-00	.2	.2
01-05	1.2	1.2
T100		
4 cyl.	.3	.3
V6		
93-94	.5	.5
95-98	.3	.3
Sequoia, Tundra	.5	.5
Sienna	.5	.5
Tacoma	.5	.5
diesel	.7	.7
w/PS add		
81-95 4Runner, Pickup	.7	.7
81-87 Land Cruiser	.5	.5
01-05 RAV4	.3	.3
Tacoma 2.4L	.5	.5
diesel	.7	.7
w/turbo diesel add	.2	.2

	LABOR TIME	SEVERE SERVICE
Crankshaft Front Oil Seal, Replace (B)		
4 cyl.		
4Runner		
1983-95	2.1	2.3
1996-00	2.3	2.5
2003-04	1.9	2.1
Highlander	1.3	1.4
Pickup		
gasoline	2.3	2.5
diesel	5.8	6.0
1998-04 PreRunner	2.2	2.5
Previa		
2WD	3.0	3.2
4WD	3.7	3.9
2001-05 RAV4	.9	1.1
T100, Tacoma	2.2	2.5
Van	1.9	2.1
6 cyl.		
Land Cruiser		
1981-87	1.5	1.7
1988-97	3.2	3.4
V6		
4Runner, Pickup		
1981-02		
2WD	6.6	6.8
4WD	9.3	9.5
2003-05	1.8	1.9
Highlander	2.5	2.9
Sienna		
1998-02	4.0	4.2
2003-05	3.2	3.6
T100		
1993-94		
2WD	6.6	6.8
4WD	9.3	9.5
1995-98	2.7	2.9
Tacoma, Tundra	2.9	3.1
V8		
4Runner	3.8	4.0
Land Cruiser	4.6	4.8
Sequoia, Tundra	3.6	3.8
w/AC add		
4Runner, Pickup	.2	.2
81-97 Land Cruiser	.1	.1
Previa	.1	.1
Sequoia, Tundra	.5	.5
Sienna	.2	.2
T100		
4 cyl.	.1	.1
93-98 V6	.3	.3
Tacoma	.5	.5
Van	.1	.1
w/PS add		
81-95 4Runner, Pickup 4 cyl.	.1	.1
Tacoma 2.4L	.2	.2

TOY-76 4RUNNER : HIGHLANDER : LAND CRUISER : PICKUP : PRERUNNER : PREVIA

	LABOR TIME	SEVERE SERVICE
Crankshaft Pulley, Replace (B)		
4 cyl.		
4Runner		
1983-95	1.7	1.9
1996-00	2.1	2.3
Highlander	1.1	1.3
Pickup		
gasoline	1.9	2.1
diesel	.8	1.0
1998-04 PreRunner	1.9	2.1
Previa		
2WD	2.3	2.5
4WD	3.0	3.2
RAV4	.8	1.0
T100, Van	1.8	2.0
Tacoma	2.1	2.3
6 cyl.		
Land Cruiser		
1981-87		
hardtop	1.4	1.6
wagon	1.9	2.1
1988-97	2.7	2.9
V6		
4Runner, Pickup		
1989-02	1.9	2.1
2003-05	1.3	1.4
Highlander	1.1	1.3
1998-04 PreRunner	1.9	2.1
Sienna		
1998-03	1.8	2.0
2004-05	1.1	1.3
T100		
1993-94	2.1	2.3
1995-98	1.7	1.9
Tacoma, Tundra	2.1	2.3
V8		
4Runner	1.1	1.3
Land Cruiser	2.1	2.3
Sequoia, Tundra	1.4	1.6
w/AC add		
4Runner, Pickup	.2	.2
81-97 Land Cruiser	.1	.1
Previa	.1	.1
Sienna	.2	.2
T100		
4 cyl.	.1	.1
93-98 V6	.3	.3
Tacoma	.2	.2
Tundra V6	.2	.2
Van	.1	.1
w/PS add		
81-95 4Runner,		
Pickup 4 cyl.	.1	.1
Piston and Cylinder Liners, Replace (A)		
Diesel	17.2	18.1
w/AC add	.7	.7
w/turbo add	.2	.2

	LABOR TIME	SEVERE SERVICE
Pistons or Connecting Rods, Replace (A)		
Includes: Ridge reaming, cylinder wall deglazing, installing new rings and rod bearings, engine tune-up.		
4 cyl.		
4Runner, Pickup		
gasoline		
1981-87		
w/FI	9.6	10.5
w/o FI	8.8	9.7
w/turbo	10.3	11.2
1988-95		
w/FI		
2WD	11.4	12.3
4WD	13.4	14.3
turbo	10.3	11.2
w/o FI	9.7	10.6
1999-00		
2WD	13.0	13.9
4WD	15.0	15.9
2003-04		
2WD	29.3	30.2
4WD	31.5	32.4
diesel	11.9	12.8
Highlander		
2WD	23.7	24.2
4WD	24.7	25.2
1998-04 PreRunner	14.3	15.2
Previa		
2WD	14.9	15.8
4WD	15.3	16.2
RAV4		
1996-00	14.5	15.4
2001-05		
2WD	18.9	19.8
4WD	20.3	23.8
T100	13.8	14.7
Tacoma		
1995-04		
2WD	14.3	15.4
4WD	16.7	17.6
2005		
2WD	19.3	19.8
4WD	18.7	19.2
Van		
1984-87	9.2	10.1
1988-89	9.8	10.7
6 cyl.		
Land Cruiser		
1981-87		
hardtop	11.8	12.7
wagon	13.9	14.8
1988-92	14.4	15.3
1993-97	18.2	19.1
V6		
4Runner, Pickup		
1989-95		
2WD	16.2	17.1
4WD	18.9	19.8

	LABOR TIME	SEVERE SERVICE
1996-02		
2WD	16.3	17.2
4WD	18.5	19.4
2003-05		
2WD	26.7	27.2
4WD	27.6	28.1
Highlander		
2WD	24.1	24.6
4WD	25.4	25.9
1998-04 PreRunner	18.1	19.0
Sienna		
1998-03	20.3	21.2
2004-05		
2WD	23.0	23.5
4WD	23.8	24.3
T100		
1993-94		
2WD	18.7	19.6
4WD	20.3	21.2
1995-98		
2WD	16.4	17.3
4WD	18.9	19.8
Tacoma		
1995-04		
2WD	18.1	19.0
4WD	20.3	21.2
2005		
2WD	24.1	24.6
4WD	24.6	25.1
Tundra		
2000-04		
2WD	17.9	18.8
4WD	20.0	20.9
2005 Tundra	17.4	17.9
V8		
4Runner		
2WD	25.2	25.7
4WD	26.1	26.6
Land Cruiser	23.0	23.5
Sequoia, Tundra		
2WD	20.9	21.4
4WD	22.5	23.0
w/ABS 96-00 RAV4 add	.7	.7
w/AC add		
4Runner, Pickup		
81-95 4 cyl.	.3	.3
90-95 V6	.5	.5
96-02 V6	.3	.3
81-92 Land Cruiser		
6 cyl.	.3	.3
RAV4		
96-00	.2	.2
01-05	1.2	1.2
Sequoia, Tundra	.5	.5
Sienna	.2	.2
93-94 T100 V6	.5	.5
Tacoma		
4 cyl.	.2	.2
V6	.5	.5

RAV4 : SEQUOIA : SIENNA : T100 : TACOMA : TUNDRA : VAN : VAN WAGON TOY-77

	LABOR TIME	SEVERE SERVICE
w/PS add		
Pickup 4 cyl.	.5	.5
86-95 4Runner 4 cyl.	.5	.5
01-05 RAV4	.3	.3
Tacoma 2.4L	.6	.6
w/turbo add		
96-02 4Runner 4 cyl.	.5	.5
diesel	.5	.5
Rear Main Oil Seal Replace (B)		
Includes: Trans. R&R when necessary.		
4 cyl.		
4Runner, Pickup		
1981-95		
2WD	5.7	6.3
4WD		
AT	7.9	8.5
MT	7.2	7.8
1996-00		
AT	6.3	6.9
MT	5.7	6.3
diesel		
2WD	4.4	5.0
4WD	5.3	5.9
Highlander		
2WD	12.2	12.7
4WD	13.2	13.7
1998-04 PreRunner	7.3	8.2
Previa		
AT	5.5	6.1
MT		
2WD	3.8	4.4
4WD	4.2	4.8
RAV4		
1996-00		
2WD	12.0	12.6
4WD	12.4	13.0
2001-03		
2WD	6.7	7.5
4WD		
AT	11.6	12.2
MT	10.7	11.3
2004-05		
2WD	6.2	6.7
4WD		
AT	11.1	12.6
MT	10.3	10.8
T100		
AT	6.7	7.3
MT	5.2	5.8
Tacoma		
1995-04		
2WD	7.6	8.9
4WD		
AT	7.8	8.7
MT	5.5	6.2
2005		
2WD		
AT	5.1	5.6
MT	4.3	4.8
4WD	5.3	5.8

	LABOR TIME	SEVERE SERVICE
Van		
1984-87		
AT	3.4	4.0
MT	3.9	4.5
1988-89		
2WD		
AT	3.4	4.0
MT	3.8	4.4
4WD		
AT	4.8	5.4
MT	4.9	5.5
6 cyl.		
Land Cruiser		
1981-87		
AT	8.3	8.9
MT	5.6	6.2
1988-97	9.4	10.0
V6		
4Runner, Pickup		
1989-95		
AT		
2WD	7.3	7.9
4WD	8.1	8.7
MT	5.9	6.5
1996-02		
AT	6.3	6.9
MT	5.7	6.3
2003-05		
2WD	6.9	7.4
4WD	7.9	8.4
Highlander		
2WD	13.0	13.5
4WD	14.3	14.8
1998-04 PreRunner	7.3	8.2
Sienna		
1998-03	11.2	11.8
2004-05		
2WD	11.5	12.0
4WD	12.4	12.9
T100		
1993-94		
AT		
2WD	7.3	7.9
4WD	8.2	8.8
MT	5.8	6.4
1995-98		
AT		
2WD	6.9	7.5
4WD	7.7	8.3
MT	5.3	5.9
Tacoma		
1995-04		
2WD	8.1	8.6
4WD		
AT	7.2	7.7
MT	5.0	5.5

	LABOR TIME	SEVERE SERVICE
2005		
2WD	4.1	4.6
4WD		
AT	5.4	5.9
MT	4.4	4.9
Tundra		
2000-04		
2WD		
AT	6.0	6.7
MT	5.2	5.8
4WD		
AT	7.6	8.5
MT	6.3	7.0
2005		
AT	6.8	7.3
MT	5.0	5.5
V8		
4Runner		
2WD	6.6	7.1
4WD	7.6	8.1
Land Cruiser	7.6	8.1
Sequoia, Tundra		
2WD	5.4	5.9
4WD	6.8	7.3
w/AC add		
01-05 RAV4 4WD	.3	.3
Sienna	.5	.5
Tacoma	.5	.5
w/PS add		
01-05 RAV4 4WD	1.2	1.2
Tacoma 2.4L	.6	.6
Rings, Replace (A)		
Includes: Ridge reaming, cylinder wall deglazing, installing new rings, engine tune-up.		
4 cyl.		
4Runner, Pickup		
gasoline		
1981-87		
w/FI	10.1	11.0
w/o FI	9.5	10.4
w/turbo	11.3	12.2
1988-95		
w/FI		
2 WD	10.7	11.6
4 WD	13.2	14.1
w/o FI	8.8	9.7
w/turbo	11.2	12.1
1996-00		
2WD	12.8	13.7
4WD	14.9	15.8
diesel	10.5	11.4
Highlander		
2WD	22.6	23.1
4WD	23.7	24.2
1998-04 PreRunner	13.9	14.7
Previa		
2WD	13.9	14.8
4WD	14.5	15.4

4RUNNER : HIGHLANDER : LAND CRUISER : PICKUP : PRERUNNER : PREVIA

	LABOR TIME	SEVERE SERVICE
Rings, Replace (A)		
RAV4		
1996-00		
2WD	13.8	14.7
4WD	14.2	15.1
2001-03		
2WD	21.0	21.9
4WD	21.7	22.6
2004-05		
2WD	19.7	20.2
4WD	20.3	20.8
T100	13.2	14.1
Tacoma		
1995-04		
2WD	13.9	14.6
4WD	16.5	17.3
2005		
2WD	18.9	19.4
4WD	18.3	18.8
Van		
1984-87	9.7	10.6
1988-89	10.3	10.9
6 cyl.		
Land Cruiser		
1981-87		
hardtop	9.8	10.7
wagon	12.1	13.0
1988-92	12.6	13.5
1993-97	16.4	17.3
V6		
4Runner, Pickup		
1989-95		
2WD	16.9	17.8
4WD	19.4	20.3
1996-02 2WD	16.2	17.1
2003-05		
2WD	26.8	27.3
4WD	27.7	28.2
Highlander		
2WD	24.2	24.7
4WD	25.5	26.0
1998-04 PreRunner	16.5	17.3
Sienna		
1998-03	18.9	19.8
2004-05		
2WD	22.2	22.7
4WD	23.0	23.5
T100		
1993-94		
2WD	16.9	17.8
4WD	19.4	20.3
1995-98		
2WD	16.9	17.8
4WD	18.9	19.8
Tacoma		
1995-04		
2WD	18.2	19.0
4WD	20.4	21.2

	LABOR TIME	SEVERE SERVICE
2005		
2WD	23.7	24.2
4WD	24.2	24.7
Tundra		
2000-04		
2WD	18.1	18.9
4WD	20.3	21.1
2005	17.2	17.7
V8		
4Runner		
2WD	24.8	25.3
4WD	25.7	26.2
Land Cruiser	23.1	23.6
Sequoia, Tundra		
2WD	21.1	21.6
4WD	22.6	23.1
w/ABS 96-00 RAV4 add	.7	.7
w/AC add		
4Runner, Pickup		
4 cyl.	.3	.3
90-95 V6	.5	.5
96-02 V6	.3	.3
81-92 Land Cruiser		
6 cyl.	.3	.3
RAV4		
96-00	.2	.2
01-05	1.2	1.2
Sequoia, Tundra	.5	.5
Sienna	.2	.2
93-94 T100 V6	.5	.5
Tacoma		
4 cyl.	.2	.2
V6	.5	.5
w/PS add		
4Runner, Pickup 4 cyl.	.5	.5
01-05 RAV4	.3	.3
Tacoma 2.4L	.6	.6
w/turbo add		
96-02 4Runner 4 cyl.	.5	.5
diesel	.5	.5

ENGINE LUBRICATION

Engine Oil Cooler, Replace (B)

	LABOR TIME	SEVERE SERVICE
Exc. below	1.4	1.6
Land Cruiser	1.7	1.9
Pickup diesel	2.2	2.4
1996-00 RAV4	2.5	2.7

Engine Oil Pressure Switch (Sending Unit), Replace (B)

	LABOR TIME	SEVERE SERVICE
All Models	.6	.6

Oil Pan and/or Gasket, Replace (B)

	LABOR TIME	SEVERE SERVICE
4 cyl.		
4Runner, Pickup		
gasoline		
1981-87	2.7	2.9
1988-95		
2WD	3.8	4.0
4WD	6.0	6.2

	LABOR TIME	SEVERE SERVICE
1996-00		
2WD	3.3	3.5
4WD	5.2	5.4
diesel	2.9	3.1
Highlander	2.2	2.4
Previa	3.4	3.6
RAV4		
1996-00	3.8	4.0
2001-05	3.0	3.2
T100	3.8	4.0
Tacoma		
2WD	4.3	4.5
4WD	6.3	6.5
Van		
1984-87	2.6	2.8
1988-89		
2WD	2.6	2.8
4WD	3.3	3.5
6 cyl.		
Land Cruiser		
1981-92	3.0	3.2
1993-97		
No.1	5.2	5.4
No.2	3.1	3.3
V6		
4Runner, Pickup		
1989-02		
2WD	3.8	4.0
4WD	5.8	6.0
2003-05		
2WD	6.4	6.9
4WD	7.6	8.1
Highlander		
No. 1	4.7	4.9
No. 2	3.0	3.2
1998-04 PreRunner	4.4	4.6
Sienna		
1998-03		
No. 1	5.3	5.5
No. 2	3.3	3.5
2004-05		
No. 1	6.0	6.2
No. 2	2.9	3.2
T100		
1993-94		
2WD	3.9	4.1
4WD	5.9	6.1
1995-98		
2WD	4.2	4.4
4WD	6.3	6.5
Tacoma		
2WD	4.4	4.6
4WD	6.7	6.9
Tundra		
2WD	4.7	4.9
4WD	7.0	7.2
V8		
4Runner		
2WD	14.3	14.5
4WD	15.3	15.5

RAV4 : SEQUOIA : SIENNA : T100 : TACOMA : TUNDRA : VAN : VAN WAGON

	LABOR TIME	SEVERE SERVICE
Land Cruiser		
No. 1	13.7	13.9
No. 2	15.2	15.4
Sequoia, Tundra		
2WD		
No. 1	12.9	13.1
No. 2	11.8	12.0
4WD		
No. 1	14.6	14.8
No. 2	13.5	13.7
w/AC add		
Sequoia, Tundra V8	.5	.5
Oil Pressure Gauge (Dash), Replace (B)		
1981-88	.8	.8
1989-99	1.1	1.1
Oil Pump, Replace (B)		
4 cyl.		
4Runner		
1983-95	2.8	3.3
1996-00		
2WD	10.9	11.4
4WD	11.8	12.3
Highlander	6.8	7.3
Pickup		
gasoline	2.9	3.4
diesel	5.5	6.0
1998-04 PreRunner	12.5	13.9
Previa		
2WD	2.4	2.9
4WD	3.1	3.6
RAV4		
1996-00	7.2	7.7
2001-05	5.5	6.0
T100	12.4	13.9
Tacoma		
2.4L	12.9	13.4
2.7L 4WD	15.5	16.0
Van		
1984-87	2.9	3.4
1988-89		
2WD	2.9	3.4
4WD	3.7	4.2
6 cyl.		
Land Cruiser		
1981-87		
hardtop	7.8	8.3
wagon	9.0	9.5
1988-92	2.9	3.4
1993-97	14.3	14.8
V6		
4Runner, Pickup		
1989-95		
2WD	7.1	7.6
4WD	9.3	10.4
1996-02		
2WD	8.0	8.9
4WD	8.4	9.4
Highlander	7.5	8.0
1998-04 PreRunner	8.5	9.6

	LABOR TIME	SEVERE SERVICE
Sienna	8.6	9.1
T100		
1993-94		
2WD	7.3	8.2
4WD	9.1	10.2
1995-98		
2WD	6.2	6.9
4WD	8.3	9.2
Tacoma		
2WD	6.7	7.5
4WD	9.1	10.2
Tundra		
2WD	6.9	7.7
4WD	9.1	10.2
V8		
4Runner		
2WD	15.4	15.9
4WD	16.0	16.5
Land Cruiser	15.2	15.7
Sequoia, Tundra		
2WD	13.2	13.7
4WD	14.7	15.2
w/ABS 96-00 RAV4 add	1.4	1.4
w/AC add		
Sequoia, Tundra	.5	.5
Sienna	.5	.5
Tacoma	.5	.5
w/PS add		
01-05 RAV4	.3	.3
Tacoma 2.4L	.6	.6

CLUTCH

Bleed Clutch Hydraulic System (B)
| 1981-05 | .7 | .8 |

Clutch Pedal Free Play, Adjust (B)
| 1981-05 | .5 | .6 |

Clutch Assy., Replace (B)
4Runner, Pickup		
1981-88		
2WD	4.3	4.5
4WD	5.8	6.0
1989-95		
4 cyl.	4.5	4.7
V6		
2WD	5.2	5.4
4WD	5.6	5.8
1996-00		
4 cyl.		
2WD	4.1	4.3
4WD	5.1	5.3
V6	5.3	5.5
1981-87 Land Cruiser		
hardtop	4.7	4.9
wagon	5.2	5.4
Previa		
2WD	3.4	3.6
4WD	4.2	4.4

	LABOR TIME	SEVERE SERVICE
RAV4		
1996-00		
2WD	11.9	12.1
4WD	12.3	12.5
2001-05		
2WD	5.7	6.3
4WD	9.9	10.4
T100	5.3	5.5
Tacoma		
1995-04		
2WD	4.6	5.1
4WD	5.2	5.8
2005		
2WD	3.8	4.3
4WD	4.5	5.0
Tundra V6		
2000-04		
2WD	4.4	4.9
4WD	5.7	6.2
2005	4.5	5.0
Van		
1984-87	3.6	3.8
1988-89		
2WD	3.5	3.7
4WD	4.1	4.3
w/AC 01-05 RAV4 add	1.3	1.3
w/PS 01-05 RAV4 add	.3	.3
Replace		
pilot bearing add	.1	.3
throw out bearing add	.1	.1

Clutch Master Cylinder, Replace (B)
Includes: System bleeding.
4Runne, Pickupr		
1983-87	1.2	1.4
1988-95	1.6	1.8
1996-00	1.7	1.9
1981-87 Land Cruiser	1.1	1.3
Previa	1.4	1.6
RAV4		
1996-00	1.3	1.5
2001-05	1.7	1.9
T100	1.5	1.7
Tacoma		
1995-04	1.4	1.6
2005	1.6	1.8
Tundra V6	1.2	1.4
Van	1.6	1.8
w/cruise control add		
96-00 RAV4	.5	.5

Clutch Slave Cylinder, Replace (B)
Includes: System bleeding.
| 1981-05 | 1.2 | 1.3 |

Flywheel, Replace (B)
4Runner, Pickup		
1981-88		
2WD	5.1	5.3
4WD	6.1	6.3

TOY-79

TOY-80 4RUNNER : HIGHLANDER : LAND CRUISER : PICKUP : PRERUNNER : PREVIA

	LABOR TIME	SEVERE SERVICE
Flywheel, Replace (B)		
1989-95		
4 cyl.		
2WD	5.2	5.4
4WD	5.6	5.8
V6		
2WD	5.2	5.4
4WD	6.0	6.2
1996-98 4 cyl.		
2WD	4.3	4.5
4WD	5.4	5.6
1999-00		
4 cyl.		
2WD	3.7	3.9
4WD	4.7	4.9
V6	4.9	5.1
1981-87 Land Cruiser		
hardtop	5.2	5.4
wagon	5.8	6.0
Previa		
2WD	3.7	3.9
4WD	4.5	4.7
RAV4		
1996-00		
2WD	12.2	12.4
4WD	12.6	12.8
2001-05		
2WD	6.1	6.6
4WD	10.3	10.8
1993-98 T100		
2WD	5.3	5.5
4WD	5.9	6.1
Tacoma		
4 cyl.		
2WD	7.9	8.1
4WD	5.8	6.0
V6		
2WD	8.1	8.3
4WD	5.2	5.4
Tundra		
2WD	4.6	4.8
4WD	6.0	6.2
Van		
1984-87	3.9	4.1
1988-89		
2WD	3.9	4.1
4WD	4.7	4.9
w/AC add		
01-05 RAV4	1.3	1.3
Tacoma	.5	.5
w/PS add		
01-05 RAV4	.3	.3
Tacoma 2.4L	.6	.6

MANUAL TRANSAXLE

	LABOR TIME	SEVERE SERVICE
Transaxle Assy. R&R and Recondition (A)		
RAV4		
1996-00		
2WD	19.2	20.1
4WD	20.0	20.9
2001-05		
2WD	12.6	13.5
4WD	17.3	18.2
w/AC add		
96-00 RAV4	.3	.3
01-05 RAV4 4WD	1.3	1.3
w/PS add		
01-05 RAV4 4WD	.3	.3
Transaxle Assy., R&R or Replace (B)		
RAV4		
1996-00		
2WD	11.6	11.8
4WD	12.1	12.8
2001-05		
2WD	5.4	5.9
4WD	10.3	10.8
w/AC add		
96-00 RAV4	.3	.3
01-05 RAV4 4WD	1.3	1.3
w/PS add		
01-05 RAV4 4WD	.3	.3

MANUAL TRANSMISSION

	LABOR TIME	SEVERE SERVICE
Extension Housing and/or Gasket, Replace (B)		
4-Speed		
4Runner, Pickup		
gasoline	3.3	3.5
diesel	2.0	2.2
5-Speed		
4Runner, Pickup		
1981-88	2.9	3.1
1989-95	6.1	6.3
1996-00	5.0	5.2
Previa	5.5	5.7
T100	6.4	6.6
Tacoma		
1995-04 2WD	6.1	6.3
2005 2WD	4.0	4.2
Tundra V6	6.9	7.1
Van	2.0	2.2
Extension Housing Oil Seal, Replace (B)		
4-Speed		
4Runner, Pickup	1.6	1.7
5-Speed		
exc. below	1.4	1.5
2005 Tacoma	1.1	1.2
Van	1.1	1.2

	LABOR TIME	SEVERE SERVICE
6-Speed		
2005 Tacoma	.6	.7
2005 Tundra	1.3	1.4
Speedometer Driven Gear, Replace (B)		
4-Speed	1.4	1.6
5-Speed		
exc. below	.8	1.0
2005 Tacoma	.5	.5
6-Speed	.5	.5
Transmission Assy., R&R and Recondition (A)		
4-Speed		
1985-88 4Runner	8.1	8.7
1981-87 Land Cruiser	10.9	11.5
1981-88 Pickup		
gasoline	7.9	8.4
diesel	9.7	10.2
5-Speed		
4Runner, Pickup		
1981-88		
2WD		
gasoline	8.3	8.5
diesel	9.9	10.4
4WD	9.9	10.4
1989-95	9.9	10.4
1996-00		
4 cyl.		
2WD	8.7	9.2
4WD	9.5	10.0
V6	9.7	10.2
Previa		
2WD	8.3	8.8
4WD	9.2	9.7
T100	9.7	10.2
Tacoma		
1995-04		
2WD	9.9	11.1
4WD	10.6	11.9
2005		
2WD	8.6	9.5
4WD	9.6	10.5
Tundra V6		
2WD	10.0	10.5
4WD	11.2	11.7
Van		
1984-87	9.1	9.6
1988-89		
2WD	8.1	8.6
4WD	8.5	9.0
6-Speed		
2005 Tacoma		
2WD	9.2	10.1
4WD	9.5	10.4
2005 Tundra	10.8	11.7

RAV4 : SEQUOIA : SIENNA : T100 : TACOMA : TUNDRA : VAN : VAN WAGON TOY-81

	LABOR TIME	SEVERE SERVICE
Transmission Assy., Replace (B)		
4-Speed	4.4	4.6
5-Speed		
4Runner, Pickup		
1981-88		
2WD	4.2	4.6
4WD	5.5	5.7
1989-95		
2WD	4.8	5.0
4WD	5.4	5.6
1996-00		
4 cyl.		
2WD	3.6	3.8
4WD	4.6	4.8
V6	5.0	5.2
1981-87 Land Cruiser		
hardtop	4.5	4.7
wagon	5.0	5.2
Previa		
2WD	3.0	3.2
4WD	4.1	4.3
T100		
4 cyl.	5.1	5.3
V6		
2WD	4.5	4.7
4WD	5.1	5.3
Tacoma		
1995-04		
2WD	4.4	4.6
4WD	5.3	5.5
2005		
2WD	3.1	3.6
4WD	4.1	4.6
Tundra		
2000-04		
2WD	4.6	4.8
4WD	6.0	6.2
2005		
2WD	3.8	4.3
4WD	4.3	4.8
Van		
1984-87	3.2	3.4
1988-89		
2WD	3.2	3.4
4WD	3.8	4.0
6-Speed		
2005 Tacoma		
2WD	3.8	4.3
4WD	4.3	4.8
2005 Tundra	4.3	4.8
Transmission Mount, Replace (B)		
Exc. below	1.0	1.1
1981-87 Land Cruiser	1.2	1.3
2005 Tacoma		
4 cyl	.7	.8
V6	1.1	1.2
2005 Tundra	.6	.7
1988-89 Van	1.3	1.4

AUTOMATIC TRANSAXLE

SERVICE TRANSAXLE INSTALLED

	LABOR TIME	SEVERE SERVICE
Check Unit for Oil Leaks (C)		
All Models	.9	.9
Drain & Refill Unit (B)		
All Models	1.4	1.6
Oil Pan and/or Gasket, Replace (B)		
All Models	2.2	2.4
Throttle Control Cable, Replace (B)		
All Models	3.8	4.0
Valve Body, Replace (B)		
Highlander	2.5	2.9
RAV4	3.0	3.4
Sienna		
1998-03	3.0	3.4
2004-05	1.7	2.0

SERVICE TRANSAXLE REMOVED
Transaxle R&R included unless otherwise noted.

	LABOR TIME	SEVERE SERVICE
Flywheel (Flexplate), Replace (B)		
Highlander		
4 cyl.		
2WD	13.2	14.8
4WD	14.3	16.1
V6		
2WD	14.0	15.7
4WD	15.6	17.5
RAV4		
1996-00		
2WD	12.6	14.1
4WD	13.1	14.6
2001-05		
2WD	6.9	7.7
4WD	12.1	13.5
Sienna		
1998-03	11.5	12.8
2004-05		
2WD	12.6	14.1
4WD	13.5	15.1
w/AC add		
RAV4		
96-00	.3	.3
01-05 4WD	1.3	1.3
Sienna	.5	.5
w/PS add		
01-05 RAV4 4WD	.3	.3
Torque Converter, Replace (B)		
Highlander		
4 cyl.		
2WD	13.2	14.8
4WD	14.8	16.5
V6		
2WD	13.7	15.2
4WD	15.2	17.0

	LABOR TIME	SEVERE SERVICE
RAV4		
1996-00		
2WD	12.1	13.5
4WD	12.6	14.1
2001-05		
2WD	7.2	8.1
4WD	11.8	13.2
Sienna		
1998-03	10.3	11.5
2004-05		
2WD	12.8	14.3
4WD	13.9	15.5
w/AC add		
RAV4		
96-00	.3	.3
01-05 4WD	1.3	1.3
Sienna	.5	.5
w/PS add		
01-05 RAV4 4WD	.3	.3
Transaxle Assy., R&R and Recondition (A)		
Includes: Replace torque converter.		
Highlander		
4 cyl.		
2WD	21.2	22.1
4WD	22.5	23.4
V6		
2WD	21.9	22.8
4WD	23.8	22.7
RAV4		
1996-00		
2WD	26.8	27.7
4WD	27.1	28.0
2001-05		
2WD	22.6	23.5
4WD	26.4	27.3
Sienna		
1998-03	24.3	25.2
2004-05		
2WD	18.2	20.3
4WD	19.3	21.5
w/AC add		
96-00 RAV4	.3	.3
Sienna	.5	.5
Transaxle Assy., R&R or Replace (B)		
Highlander		
4 cyl.		
2WD	13.5	15.1
4WD	14.6	16.3
V6		
2WD	14.3	16.1
4WD	15.9	17.7
RAV4		
1996-00		
2WD	11.9	13.3
4WD	12.3	13.8
2001-05		
2WD	7.0	7.9
4WD	11.5	12.8

TOY-82 4RUNNER : HIGHLANDER : LAND CRUISER : PICKUP : PRERUNNER : PREVIA

	LABOR TIME	SEVERE SERVICE
Transaxle Assy., R&R or Replace (B)		
Sienna		
1998-03	10.0	11.2
2005		
2WD	12.3	13.8
4WD	13.4	15.0
w/AC add		
RAV4		
96-00	.3	.3
01-05 4WD	1.3	1.3
Sienna	.5	.5
w/PS add		
01-05 RAV4 4WD	.3	.3

AUTOMATIC TRANSMISSION

SERVICE TRANSMISSION INSTALLED

	LABOR TIME	SEVERE SERVICE
Check Unit for Oil Leaks (C)		
1981-05	.9	.9
Drain and Refill Unit (B)		
Exc. Tacoma, Tundra	1.3	1.4
Tacoma, Tundra	.8	.9
Oil Pressure Test (B)		
1981-05	2.0	2.0
Electronic Control Unit, Replace (B)		
1993-97 Land Cruiser	1.5	1.5
Extension Housing Gasket, Replace (B)		
4Runner, Pickup		
1981-89	2.5	2.8
1990-95	2.8	3.1
1996-02	1.6	1.9
2003-05	4.7	5.0
1998-04 PreRunner	1.6	1.9
Previa	2.5	2.8
Sequoia, Tundra		
2WD	2.9	3.2
4WD	1.6	1.8
T100	2.9	3.2
Tacoma	2.9	3.2
Van	2.3	2.6
Extension Housing Oil Seal, Replace (B)		
Exc. below		
2WD	1.4	1.6
4WD	1.0	1.2
4Runner, Pickup		
1981-95		
2WD	1.7	1.9
4WD	1.3	1.5
1996-02 4 cyl.	1.4	1.6
2003-05		
V6	2.7	3.0
V8	6.1	6.9
Previa	1.7	1.9
Van		
2WD	1.5	1.7
4WD	1.3	1.5

	LABOR TIME	SEVERE SERVICE
Governor Assy., R&R or Replace (B)		
1983-95 4Runner	3.5	3.8
1981-92 Land Cruiser	7.7	8.0
Pickup	3.6	3.9
Tacoma 2.4L, Van	3.5	3.8
Oil Pan and/or Gasket, Replace (B)		
4Runner, Pickup		
1993-95		
4 cyl.		
2WD	2.3	2.5
4WD	2.8	3.0
V6		
2WD	2.8	3.0
4WD	3.5	3.7
1996-05	2.6	2.8
1993-05 Land Cruiser	3.0	3.2
1998-04 PreRunner	2.9	3.2
Previa	2.3	2.5
Sequoia, Tundra	2.9	3.1
T100	2.7	2.9
Tacoma		
exc. 2.4L	2.9	3.1
2.4L	2.2	2.5
Overdrive Solenoid, Replace (B)		
4Runner, Pickup		
1984 Pickup	1.6	1.8
1985-88	.9	1.1
1989-95		
4 cyl.		
2WD	.9	1.1
4WD	2.5	2.7
V6	2.4	2.6
1998-04 PreRunner	2.4	2.7
T100	2.2	2.4
Tacoma		
exc. 2.4L	2.4	2.7
2.4L	.8	1.0
Van	1.4	1.6
Throttle Cable, Replace (B)		
Exc. below	2.5	2.9
Land Cruiser		
1981-92	4.1	4.6
1993-97	3.6	4.1
Previa	3.0	3.4
Transmission Speed Sensor, Replace (B)		
Exc. Highlander	.8	.9
Highlander	1.0	1.1
Valve Body Assy., Replace (B)		
4Runner, Pickup		
1989-95		
4 cyl.	3.1	3.5
V6		
2WD	3.2	3.6
4WD	4.2	4.6
1996-02	3.2	3.6
2003-05	2.9	3.2
Land Cruiser	3.6	4.1
Previa	2.7	3.1

	LABOR TIME	SEVERE SERVICE
Sequoia, Tundra	3.5	3.9
T100	3.1	3.5
Tacoma		
exc. 2.4L	3.5	3.9
2.4L	2.9	3.3
1988-89 Van	2.8	3.2
Valve Body Manual Lever Shaft Seal, Replace (B)		
4Runner, Pickup		
1983-95		
2WD	7.6	7.8
4WD	8.5	8.7
1996-02	1.1	1.3
2003-05		
2WD	2.9	3.2
4WD	4.3	4.8
Land Cruiser		
1981-87	3.4	3.6
1988-90	4.0	4.2
1991-92	4.4	4.6
1993-05	1.3	1.5
1998-04 PreRunner	1.1	1.3
Previa	1.1	1.3
Sequoia, Tundra	1.1	1.3
T100		
4 cyl.	1.0	1.2
V6		
1993-94		
2WD	7.6	7.8
4WD	8.7	8.9
1995-99	1.0	1.2
Tacoma	1.0	1.2
Van	2.9	3.1

SERVICE TRANSMISSION REMOVED

Transmission R&R included unless otherwise noted.

	LABOR TIME	SEVERE SERVICE
Flywheel (Flexplate), Replace (B)		
4Runner, Pickup		
1981-88		
2WD	5.5	5.7
4WD	9.8	10.0
1989-95		
2WD	6.8	7.0
4WD		
4 cyl.	7.5	7.7
V6	10.7	10.9
1996-02		
2WD		
4 cyl.	5.7	5.9
V6	5.1	5.3
4WD	5.9	6.1
2003-05		
2WD	8.4	8.6
4WD	7.3	7.5
Land Cruiser		
1981-87	11.5	11.7
1988-97	7.8	8.0

RAV4 : SEQUOIA : SIENNA : T100 : TACOMA : TUNDRA : VAN : VAN WAGON — TOY-83

	LABOR TIME	SEVERE SERVICE
1998-05	8.2	8.4
1998-04 PreRunner	6.7	6.9
Previa		
2WD	5.4	5.6
4WD	6.3	6.5
Sequoia, Tundra		
2WD	5.7	5.9
4WD	7.2	7.4
1993-98 T100		
2WD	6.8	7.0
4WD	10.7	10.9
Tacoma		
4 cyl.	7.9	8.1
V6		
2WD	8.2	8.4
4WD	7.2	7.4
Van		
1984-87	4.7	4.9
1988-89		
2WD	4.8	5.0
4WD	5.9	6.1

Front Pump Oil Seal, Replace (B)
4Runner, Pickup
	LABOR TIME	SEVERE SERVICE
1981-88		
2WD	5.4	5.6
4WD	9.9	10.1
1989-95		
2WD	7.1	7.3
4WD		
4 cyl.	7.5	7.7
V6	10.8	11.0
1996-02		
2WD	5.4	5.6
4WD	6.4	6.6
Land Cruiser		
1981-87	11.6	11.8
1988-05	7.8	8.0
1998-04 PreRunner	7.0	7.2
Previa		
2WD	5.4	5.6
4WD	6.3	6.5
Sequoia, Tundra		
2WD	7.3	7.5
4WD	9.1	9.3
1993-98 T100		
2WD	6.6	6.8
4WD	10.8	11.0
Tacoma		
2WD		
4 cyl.	9.1	9.3
V6	9.5	9.7
4WD	7.6	7.8
Van		
1984-87	5.0	5.2
1988-89		
2WD	5.0	5.2
4WD	5.8	6.0

Torque Converter, Replace (B)
4Runner, Pickup
	LABOR TIME	SEVERE SERVICE
1981-88		
2WD	5.2	5.4
4WD	9.6	9.8
1989-95		
2WD	6.7	6.9
4WD		
4 cyl.	7.3	7.5
V6	10.8	11.0
1996-02		
2WD	4.9	5.1
4WD	5.6	5.8
2003-05		
V6		
2WD	7.8	8.0
4WD	8.5	8.7
V8	7.2	7.4
Land Cruiser		
1981-87	11.5	11.7
1988-05	7.6	7.8
1998-04 PreRunner	6.6	6.8
Previa		
2WD	5.3	5.5
4WD	6.1	6.3
Sequoia, Tundra		
2WD	5.3	5.5
4WD	6.9	7.1
1993-98 T100		
2WD	6.4	6.6
4WD	10.8	11.0
Tacoma		
2WD		
4 cyl.	9.0	9.2
V6	9.4	9.6
4WD	7.3	7.5
Van		
1984-87	4.7	4.9
1988-89		
2WD	4.5	4.7
4WD	5.5	5.7

Transmission Assy., R&R and Recondition (A)
Includes: Replace torque converter.
4Runner, Pickup
	LABOR TIME	SEVERE SERVICE
1981-88		
2WD	13.1	14.0
4WD	22.9	23.8
1989-95		
4 cyl.		
2WD	15.3	16.2
4WD	21.9	22.8
V6		
2WD	21.4	22.3
4WD	24.5	25.4
1996-02		
4 cyl.		
2WD	19.8	20.7
4WD	20.5	21.4
V6		
2WD	20.0	20.9
4WD	20.5	21.4
2003-05		
2WD	13.2	14.1
4WD	13.5	14.4
Land Cruiser		
1981-87	27.7	28.6
1988-04	21.8	22.7
1998-04 PreRunner	22.3	23.2
Previa		
w/supercharger		
2WD	19.2	20.1
4WD	19.8	20.7
w/o supercharger		
2WD	12.5	13.4
4WD	13.2	14.1
Sequoia, Tundra		
2WD	22.8	23.7
4WD	24.3	25.2
T100		
2WD	21.4	22.3
4WD	24.4	25.3
Tacoma		
2WD		
4 cyl.	18.4	19.3
V6	24.6	25.5
4WD	22.9	23.8
Van		
1984-87	16.6	17.5
1988-89		
2WD	16.6	17.5
4WD	17.2	18.1

Transmission Assy., Replace (B)
4Runner, Pickup
	LABOR TIME	SEVERE SERVICE
1981-88		
2WD	5.2	5.4
4WD	9.5	9.7
1989-95		
2WD	6.6	6.8
4WD		
4 cyl.	7.2	7.4
V6	10.4	10.6
1996-02		
2WD	5.3	5.5
4WD	5.8	6.0
2003-05		
V6		
2WD	7.8	8.0
4WD	9.0	9.2
V8	7.2	7.4
Land Cruiser		
1981-87	11.3	11.5
1988-05	8.4	8.6
1998-04 PreRunner	6.7	6.9
Previa		
2WD	5.2	5.4
4WD	5.7	5.9

TOY-84 4RUNNER : HIGHLANDER : LAND CRUISER : PICKUP : PRERUNNER : PREVIA

	LABOR TIME	SEVERE SERVICE
Transmission Assy., Replace (B)		
Sequoia, Tundra		
2WD	7.2	7.4
4WD	9.0	9.2
T100		
2WD	6.3	6.5
4WD	10.7	10.9
Tacoma		
2WD		
4 cyl.	9.0	9.2
V6	9.4	9.6
4WD	7.3	7.5
Van		
1984-87	4.3	4.5
1988-89		
2WD	4.3	4.5
4WD	5.5	5.7

TRANSFER CASE

Transfer Case, R&R and Reconditon (A)

	LABOR TIME	SEVERE SERVICE
4Runner, Pickup		
1981-88		
AT	14.6	15.2
MT	8.5	9.1
1989-95		
AT		
4 cyl.		
G58	11.8	12.4
A304F	10.0	10.6
V6	14.9	15.5
MT		
4 cyl.	9.3	9.9
V6	10.3	10.9
1996-02	8.5	9.1
2003-05		
V6	14.6	15.2
V8	13.8	14.4
Highlander		
4 cyl.	16.4	17.0
V6	17.1	17.7
Land Cruiser		
1981-90	9.0	9.6
1991-05	12.5	13.1
Previa	6.2	6.8
RAV4		
1996-00		
AT	19.6	20.2
MT	16.5	17.1
2001-05		
AT	18.6	19.2
MT	15.8	16.4
Sequoia, Tundra		
AT	14.6	15.2
MT	12.2	12.8
Sienna	17.3	17.9

	LABOR TIME	SEVERE SERVICE
T100		
AT	14.9	15.5
MT	10.5	11.1
Tacoma		
AT	13.1	13.7
MT		
4 cyl.	10.2	10.8
V6	11.6	12.2
1988-89 Van	6.4	7.0
w/AC add		
RAV4		
96-00	.3	.3
01-05	1.3	1.3
w/PS 01-05 RAV4 add	.3	.3

Transfer Case, Replace (B)

	LABOR TIME	SEVERE SERVICE
4Runner, Pickup		
1981-88		
AT	9.7	9.9
MT	5.7	5.9
1989-95		
AT		
4 cyl.	6.9	7.1
V6	10.6	10.8
MT		
4 cyl.	4.7	4.9
V6	5.3	5.5
1996-02	2.8	3.0
2003-05		
V6	8.7	8.9
V8	7.5	7.7
Highlander		
4 cyl.	13.2	13.4
V6	14.0	14.2
Land Cruiser		
1981-90	6.3	6.5
1991-05	3.6	3.8
Previa	2.6	2.8
RAV4		
1996-00		
AT	12.8	13.0
MT	12.1	12.3
2001-05		
AT	12.1	12.3
MT	11.5	11.7
Sequoia, Tundra		
AT	9.0	9.2
MT	6.0	6.2
Sienna	13.9	14.1
T100		
AT	10.7	10.9
MT	5.3	5.5
Tacoma		
AT	7.3	7.5
MT	4.8	5.0
1988-89 Van	2.6	2.8
w/AC add		
RAV4		
96-00	.3	.3
01-05	1.3	1.3
w/PS 01-05 RAV4 add	.3	.3

SHIFT LINKAGE
AUTOMATIC TRANSAXLE/TRANSMISSION

Throttle Rod or Cable, Adjust (B)

	LABOR TIME	SEVERE SERVICE
1981-05	.6	.7

Gear Selector Lever, Replace (B)

	LABOR TIME	SEVERE SERVICE
4Runner, Pickup		
1983-95		
2WD	1.3	1.3
4WD	1.9	1.9
1996-02	1.3	1.3
Land Cruiser	.9	.9
1998-04 PreRunner	.9	.9
RAV4	1.3	1.3
Sequoia, Tundra	1.2	1.2
Sienna	1.2	1.2
Tacoma		
2WD		
column	1.6	1.6
floor	.9	.9
4WD	1.8	1.8
Van	.7	.7

Selector Lever Boot and/or Retainer, Replace (B)

	LABOR TIME	SEVERE SERVICE
4Runner, Pickup		
1983-95		
2WD	.8	.8
4WD	2.0	2.0
1996-02	.5	.5
Land Cruiser	.9	.9
1998-04 PreRunner	.5	.5
1996-00 RAV4	1.2	1.2
Tacoma		
2WD	.8	.8
4WD	1.9	1.9

Selector Lever Indicator Bulb, Replace (B)

	LABOR TIME	SEVERE SERVICE
Exc. below	.7	.7
Highlander	1.2	1.2
Sienna 2004-05	1.6	1.6

Shift Control Cable, Replace (B)

	LABOR TIME	SEVERE SERVICE
Highlander	1.4	1.6
RAV4		
1996-00	1.7	1.9
2001-05	1.1	1.3
Sienna	.5	.6

MANUAL TRANSAXLE/TRANSMISSION

Gearshift Lever, Replace (B)

	LABOR TIME	SEVERE SERVICE
Exc. RAV4	.7	.7
RAV4	.9	.9

Gearshift Lever Retainer, Replace (B)

	LABOR TIME	SEVERE SERVICE
Exc. Previa	.8	.8
Previa	4.2	4.2

RAV4 : SEQUOIA : SIENNA : T100 : TACOMA : TUNDRA : VAN : VAN WAGON TOY-85

DRIVELINE

Note: ADD – Automatic Disconnecting Differential

	LABOR TIME	SEVERE SERVICE
DRIVESHAFT		
Center Support Bearing, Replace (B)		
4Runner, Pickup		
1981-88	2.0	2.3
1989-95	1.4	1.7
Highlander 4WD	2.9	3.2
1998-04 PreRunner	1.3	1.6
Previa	1.9	2.2
RAV4 4WD		
1996-03	1.9	2.2
2004-05	1.1	1.3
Sequoia, Tundra		
2WD (Dana)	1.7	2.0
4WD	1.4	1.7
T100	1.6	1.9
Tacoma	1.3	1.6
Driveshaft, Replace (B)		
Front		
4Runner, Pickup		
1981-02	1.8	2.0
2003-05	.6	.7
Land Cruiser		
1989-97	1.4	1.6
1998-05	1.0	1.2
Previa	1.7	1.9
Sequoia, Tundra	1.3	1.7
T100	1.6	1.8
Tacoma	1.3	1.5
1988-89 Van	1.5	1.7
Rear		
exc. below	1.1	1.3
1984-02 4Runner, Pickup	1.5	1.7
Highlander	2.9	3.1
Land Cruiser	1.3	1.5
Intermediate Driveshaft, Replace (B)		
1983-95 4Runner	1.5	1.7
Pickup	1.4	1.6
1998-04 PreRunner	1.3	1.4
Previa	1.7	1.9
RAV4	1.4	1.6
Sequoia, Tundra	1.3	1.4
1993-98 T100	1.4	1.6
Tacoma	1.3	1.4
U-Joint Flange, Replace (B)		
Front axle		
exc. below	2.2	2.4
2003-05 4Runner 4WD	3.6	4.1
1981-90 Land Cruiser	1.3	1.5
Van	2.8	3.0
Rear axle		
exc. below	1.7	1.9
2003-05 4Runner	6.6	7.4

	LABOR TIME	SEVERE SERVICE
Highlander	3.8	4.0
RAV4	1.3	1.5
2004-05 Sienna	3.5	3.9
U-Joint, Replace or Recondition (B)		
Exc. Van	2.1	2.4
Van	1.4	1.6
DRIVE AXLE		
Axle Inner Shaft, Replace (B)		
Front axle		
Land Cruiser		
1981-90	2.5	2.8
1991-97	1.7	2.0
Axle Shaft, Replace (B)		
Rear axle		
4Runner		
1984-02	2.3	2.6
2003-05		
disc brakes	3.2	3.6
Highlander 4WD	4.3	4.6
Land Cruiser		
1981-97		
semi-floating	2.2	2.5
full-floating	.8	1.1
1998-05	2.5	2.8
Pickup		
1981-88		
single rear whls	1.9	2.1
dual rear whls	.8	1.1
1989-95		
single rear whls	2.4	2.7
dual rear whls	.8	1.1
1998-04 PreRunner	1.9	2.1
Previa		
disc brakes	2.5	2.8
drum brakes	1.9	2.1
RAV4		
1996-00		
2WD	1.8	2.0
4WD	2.5	2.8
2001-05 4WD	2.5	2.8
Sequoia	2.5	2.8
T100, Van	1.9	2.1
Tacoma, Tundra	2.2	2.5
Axle Shaft Bearings, Replace (B)		
Rear axle, one		
4Runner		
1984-02	1.9	2.2
2003-05	2.9	3.2
Highlander 4WD	4.3	4.6
Land Cruiser		
1981-90	2.3	2.6
1991-97		
semi-floating	1.9	2.1
full-floating	1.2	1.5
1998-05	2.2	2.5
Pickup		
1981-95		
single rear whls	2.0	2.3
dual rear whls	2.5	2.8

	LABOR TIME	SEVERE SERVICE
1998-04 PreRunner	1.9	2.1
Previa		
disc brakes	2.3	2.6
drum brakes	1.9	2.1
RAV4		
1996-00		
2WD	1.7	2.0
4WD	2.5	2.8
2001-05 4WD	2.5	2.8
Sequoia	2.2	2.5
T100	1.9	2.1
Tacoma, Tundra	1.9	2.1
Van	1.7	2.0
Axle Shaft Oil Seal, Replace (B)		
Rear axle one		
4Runner		
1984-02	1.9	2.1
2003-05	2.4	2.7
Highlander 4WD	4.3	4.5
Land Cruiser		
1981-97		
semi-floating	1.9	2.1
full-floating	.9	1.1
1998-05	2.5	2.7
Pickup		
1981-95		
single rear whls	1.9	2.1
dual rear whls	.8	1.0
1998-04 PreRunner	1.9	2.1
Previa	2.3	2.5
Sequoia	2.5	2.7
T100	1.9	2.1
Tacoma, Tundra	1.9	2.1
Van	1.3	1.5
Differential Carrier Assy., R&R and Recondition (A)		
Front axle		
4Runner, Pickup		
1981-95		
w/ADD	10.8	11.3
w/o ADD	10.2	10.7
1996-02		
w/ADD	10.3	10.8
w/o ADD	9.5	10.0
2003-05	5.8	6.4
Land Cruiser		
1981-90	9.7	10.2
1991-97	9.1	9.6
1998-05	9.7	10.2
Previa	7.3	7.8
Sequoia, Tundra	10.3	10.8
T100		
w/ADD	10.8	11.3
w/o ADD	10.0	10.5
Tacoma		
w/ADD	10.3	10.8
w/o ADD	9.5	10.0
Van	7.8	8.3

TOY-86 4RUNNER : HIGHLANDER : LAND CRUISER : PICKUP : PRERUNNER : PREVIA

	LABOR TIME	SEVERE SERVICE
Differential Carrier Assy., R&R and Recondition (A)		
Rear axle		
4Runner		
1985-95	7.1	7.6
1996-02		
standard	7.3	7.8
locking	6.5	7.0
2003-05	8.2	8.7
Highlander 4WD	11.5	12.0
Land Cruiser		
1981-90		
standard	7.3	7.8
locking	8.1	8.6
1991-97		
standard	7.1	7.6
full-floating	6.6	7.1
locking	7.6	8.1
1998-05		
standard	7.0	7.5
limited slip, locking	7.5	8.0
Pickup		
1981-88		
single rear whls	7.0	7.5
dual rear whls	6.0	6.5
1989-95		
single rear whls	7.2	7.7
dual rear whls	5.8	6.3
1998-04 PreRunner		
standard	8.1	8.6
locking	9.0	9.5
Previa		
disc brakes	8.7	9.2
drum brakes	8.2	8.7
RAV4	9.4	9.9
Sequoia	7.0	7.5
Sienna		
1998-03	8.4	8.9
2004-05	12.2	12.7
T100	7.4	7.9
Tacoma		
standard	8.1	8.6
locking	8.8	9.3
limited-slip	8.4	8.9
Tundra		
standard	8.1	8.6
limited-slip	8.4	8.9
Van	6.2	6.7
Differential Carrier Assy., R&R or Replace (B)		
Front axle		
4Runner, Pickup		
1983-95	3.5	4.0
1996-02	3.3	3.8
2003-05	2.7	3.0
Land Cruiser		
1981-90	5.3	5.8
1991-05	3.6	4.1
Previa	2.5	3.0
Sequoia, Tundra	3.3	3.8
T100	3.6	4.1
Tacoma	3.3	3.8
1988-89 Van	2.8	3.3
Rear axle		
1996-05 4Runner	2.4	2.9
Highlander 4WD	5.0	5.5
Land Cruiser		
1981-90	3.5	4.0
1991-05	2.6	3.1
Pickup		
1981-88	2.9	3.4
1989-95	2.5	3.0
1998-04 PreRunner	2.4	2.9
Previa		
disc brakes	3.9	4.4
drum brakes	3.1	3.6
RAV4 4WD	2.1	2.6
Sequoia, Tundra	2.4	2.9
2004-05 Sienna	6.7	7.5
T100	2.3	2.8
Tacoma	2.4	2.9
Van	2.8	3.3
Differential Cover Gasket, Replace (B)		
Rear axle		
4Runner		
1996-98	2.3	2.5
2003-05	3.6	4.1
Land Cruiser		
1981-97	1.3	1.5
1998-05	2.3	2.5
1998-04 PreRunner	2.4	2.7
2001-05 RAV4	1.3	1.5
Sienna		
1998-03	3.1	3.3
2004-05	2.2	2.5
T100, Tundra	2.4	2.7
Tacoma	2.5	2.7
Differential Side Bearings, Replace (B)		
Front axle		
4Runner		
1998-02	6.3	6.5
2003-05	4.7	5.3
Land Cruiser		
1981-90	6.2	6.4
1991-97	4.3	4.5
1998-05	7.8	8.0
Pickup	6.1	6.3
Previa	2.9	3.1
Sequoia, Tundra	7.2	7.4
T100	6.2	6.4
Tacoma	7.2	7.4
1988-89 Van	5.3	5.5
Rear axle		
4Runner		
1985-95	4.1	4.3
1996-02	3.5	3.7
2003-05	8.2	8.4
Highlander	11.0	11.2
Land Cruiser		
1981-90		
standard	4.4	4.6
locking	5.4	5.6
1991-97	3.7	3.9
1998-05		
standard	3.6	3.8
limited-slip, locking	4.1	4.3
Pickup		
1981-88		
single rear whls	4.0	4.2
dual rear whls	3.2	3.4
1989-95		
single rear whls	4.3	4.5
dual rear whls	3.0	3.2
1998-04 PreRunner	4.1	4.3
Previa		
disc brakes	5.5	5.7
drum brakes	4.9	5.1
RAV4	6.3	6.5
Sequoia	3.6	3.8
2004-05 Sienna	10.9	11.1
T100	4.2	4.4
Tacoma		
standard	4.1	4.3
locking	4.9	5.1
Tundra	4.1	4.3
Van	3.7	3.9
Hub and Bearing, Replace (B)		
Rear axle		
Highlander		
2WD	1.7	2.0
4WD	2.1	2.3
2001-05 RAV4 2WD	1.0	1.3
Sienna	1.7	2.0
Pinion Bearings, Replace (B)		
Front axle		
4Runner, Pickup		
1981-02	7.4	7.7
2003-05	3.6	4.1
Land Cruiser		
1981-90	7.1	7.4
1991-97	6.7	7.0
1998-05	7.6	7.9
Previa	6.1	6.4
Sequoia, Tundra	8.2	8.5
T100	7.5	7.8
Tacoma	8.2	8.5
1988-89 Van	6.4	6.7
Rear axle		
4Runner	6.9	7.7
Highlander	11.0	11.3
Land Cruiser		
standard	6.0	6.3
limited-slip, locking	6.4	6.7
1998-04 PreRunner	7.3	7.6
RAV4	7.3	7.6

RAV4 : SEQUOIA : SIENNA : T100 : TACOMA : TUNDRA : VAN : VAN WAGON — TOY-87

	LABOR TIME	SEVERE SERVICE
Sequoia	6.0	6.3
Sienna	10.9	11.2
Tacoma		
standard	7.3	7.6
locking	8.1	8.4
Tundra	7.3	7.6

Pinion Shaft Oil Seal, Replace (B)
Front axle
4Runner, Pickup		
1981-02	2.4	2.6
2003-05	3.6	4.1
Land Cruiser		
1981-90	1.5	1.7
1991-97	2.4	2.6
1998-05	1.9	2.1
Previa	2.4	2.6
Sequoia, Tundra	2.2	2.4
T100, Tacoma	2.2	2.4
1988-89 Van	2.7	2.9

Rear axle
4Runner, Pickup		
1981-88	2.0	2.2
1989-02	1.6	1.8
2003-05	1.9	2.1
Highlander	2.7	2.9
Land Cruiser		
1981-90		
Hardtop	1.8	2.0
Wagon	1.2	1.4
1991-05	1.9	2.1
1998-04 PreRunner	1.6	1.8
Previa	2.0	2.2
RAV4	1.7	1.9
Sequoia	1.9	2.1
Sienna	3.6	3.9
T100, Tundra	1.6	1.8
Tacoma	1.7	1.9
Van	1.4	1.6

Ring Gear & Pinion Set, Replace (B)
Front axle
4Runner, Pickup		
1983-95	8.4	8.9
1996-02	7.9	8.4
2003-05	5.0	5.7
Land Cruiser		
1981-90	8.4	8.9
1991-97	7.5	8.0
1998-05	8.2	8.7
Previa	5.8	6.3
Sequoia, Tundra	8.8	9.3
T100	8.4	8.9
Tacoma	8.8	9.3
1988-89 Van	6.8	7.3

Rear axle
4Runner		
1985-88	6.3	6.8
1989-95	7.8	8.3
1996-02	7.0	7.5
2003-05	7.9	8.4
Highlander	11.8	12.3
Land Cruiser		
1981-90		
Standard	6.3	6.8
Locking	7.1	7.6
1991-97	6.1	6.6
1998-05		
standard	6.3	6.8
limited-slip, locking	6.7	7.0
Pickup		
1981-88		
single rear whls	6.3	6.8
dual rear whls	5.3	5.8
1989-95		
single rear whls	7.0	7.5
dual rear whls	6.3	6.8
1998-04 PreRunner		
2WD	7.8	8.3
4WD	8.5	9.0
Previa		
disc brakes	8.1	8.6
drum brakes	7.3	7.8
RAV4	8.1	8.6
Sequoia	6.3	6.8
Sienna	11.9	12.4
T100	7.1	7.6
Tacoma		
standard	7.8	8.3
locking	8.5	9.0
Tundra	7.8	8.3
Van	5.9	6.4

HALFSHAFTS
CV Joint Boot, Replace (B)
Front
4Runner, Pickup		
1983-88	3.0	3.3
1989-95	2.4	2.7
1996-02	3.3	3.6
2003-05 4WD	2.5	2.9
Highlander		
2WD	3.6	3.9
4WD	4.1	4.4
1998-03 Land Cruiser	2.2	2.5
Previa	2.1	2.4
RAV4		
1996-00	3.3	3.6
2001-05	3.8	4.1
Sequoia, Tundra	2.4	2.7
Sienna		
1998-03	3.5	3.8
2004-05		
2WD	2.9	3.2
4WD	3.3	3.7
1995-99 T100	2.4	2.7
Tacoma	3.3	3.6

Van		
1984-87	2.8	3.1
1988-89		
2WD	2.5	2.8
4WD	2.1	2.4
Rear		
Highlander 4WD	4.3	4.8
RAV4 4WD		
1996-03	1.3	1.6
2004-05	2.5	2.9

Halfshaft, R&R or Replace (B)
Front
1989-05 4Runner, Pickup	1.7	2.0
Highlander		
2WD		
4 cyl.	3.2	3.5
V6	2.7	3.0
4WD	3.6	3.9
1998-05 Land Cruiser	1.7	2.0
Previa	1.6	1.9
RAV4		
1996-00		
one	1.9	2.2
both	2.7	3.0
2001-05 one	2.5	2.8
Sequoia, Tacoma, Tundra	1.6	1.9
Sienna		
1998-03		
one	1.7	2.0
both	2.3	2.6
2004-05		
one	3.0	3.4
both	4.0	4.4
T100, Van	1.7	2.0
Rear		
Highlander 4WD	4.3	4.8
RAV4 4WD		
1996-03	1.3	1.6
2004-05	2.5	2.9

Halfshaft Boot, Replace (B)
Front
4Runner, Pickup		
1983-88	3.0	3.3
1989-95	2.4	2.7
1996-02	3.3	3.6
2003-05 4WD	2.5	2.9
Highlander		
2WD	3.6	3.9
4WD	4.1	4.4
1998-05 Land Cruiser	2.2	2.5
Previa	2.1	2.4
RAV4		
1996-00	3.3	3.6
2001-05	3.8	4.1
Sequoia, Tundra	2.4	2.7

TOY-88 — 4RUNNER : HIGHLANDER : LAND CRUISER : PICKUP : PRERUNNER : PREVIA

	LABOR TIME	SEVERE SERVICE
Halfshaft Boot, Replace (B)		
Sienna		
1998-03	3.5	3.8
2004-05		
2WD	2.9	3.2
4WD	3.3	3.7
1995-99 T100	2.4	2.7
Tacoma	3.3	3.6
Van		
1984-87	2.8	3.1
1988-89		
2WD	2.5	2.8
4WD	2.1	2.4
Rear		
Highlander 4WD	3.5	3.8
RAV4 4WD	2.7	3.0

BRAKES
ANTI-LOCK

	LABOR TIME	SEVERE SERVICE
Diagnose Anti-Lock Brake System Component Each (A)		
All Models	.5	.5
Retrieve Fault Codes (A)		
All Models	.3	.3
Actuator Assy., Replace (B)		
Exc. below	2.4	2.7
Previa	1.6	1.8
Control Module, Replace (B)		
Exc. below	.8	.8
Sequoia, Tundra	1.9	1.9
Sienna	1.6	1.6
Wheel Speed Sensor, Replace (B)		
Exc. below		
front one	1.0	1.1
rear one	.8	.9
Highlander		
2WD		
front one	1.3	1.4
rear one	2.7	3.0
4WD each	1.3	1.4
RAV4 each	1.3	1.4
2004-05 Sienna		
front one	1.4	1.6
rear one	2.5	2.9

SYSTEM

	LABOR TIME	SEVERE SERVICE
Bleed Brakes (B)		
Includes: Add fluid.		
Two wheels	.9	1.0
Four wheels	1.7	1.9
Brakes, Adjust (B)		
Includes: Refill master cylinder.		
1981-05	.5	.5
Brake System, Flush and Refill (B)		
1981-05	1.5	1.5
Brake Hose (Flexible), Replace (B)		
Includes: System bleeding.		
Exc. below one		
front	.8	.9
rear	1.0	1.1
each addl. add	.3	.4

	LABOR TIME	SEVERE SERVICE
Highlander one	1.4	1.6
2003-05 4Runner, Sienna one	1.3	1.4
Load Sensing Valve, Replace (B)		
Includes: System bleeding.		
Exc. below	2.3	2.5
Previa	1.8	2.0
Van	1.9	2.1
Master Cylinder, Replace (B)		
Includes: System bleeding.		
Exc. below	1.4	1.6
4Runner, Pickup		
1981-88	2.0	2.2
1989-02	1.6	1.8
2003-05	3.3	3.7
Highlander	1.7	1.9
Land Cruiser		
1981-90	2.0	2.2
1991-97	1.7	1.9
1996-05 RAV4	1.7	1.9
Sienna		
1998-03	1.8	2.0
2004-05	2.1	2.4
Van	2.7	2.9
Power Booster Unit, Replace (B)		
4Runner, Pickup		
1981-88	2.8	3.0
1989-02	2.4	2.6
2003	3.3	3.7
Highlander	2.3	2.5
Land Cruiser	2.4	2.6
1998-04 PreRunner	2.0	2.2
Previa	2.9	3.1
RAV4		
1996-00	1.8	2.0
2001-05	2.4	2.6
Sequoia, Tundra	2.0	2.2
Sienna	2.6	2.8
T100, Tacoma	2.0	2.2
Power Booster Vacuum Check Valve, Replace (B)		
All Models	.5	.5
Power Booster Vacuum Pump, Replace (B)		
Diesel Pickup	.7	1.0
Proportioning & By-Pass Valve, Replace (B)		
Includes: System bleeding.		
4Runner, Pickup	1.3	1.5
1981-90 Land Cruiser	1.4	1.6
Previa	1.7	1.9
RAV4		
1996-00	2.1	2.3
2001-05	1.5	1.7
Sequoia, Tundra	.9	1.0
Sienna	2.5	2.7
1993-98 T100, Van	1.4	1.6

	LABOR TIME	SEVERE SERVICE
SERVICE BRAKES		
Brake Drum, Replace (B)		
Exc. below	1.1	1.2
1981-92 Land Cruiser	1.6	1.8
Previa	1.3	1.5
RAV4, Sienna	.8	.9
Van	1.7	1.9
Caliper Assy., R&R and Recondition (B)		
Includes: System bleeding.		
4Runner, Pickup		
1981-95	2.0	2.2
1996-02	1.5	1.7
Land Cruiser		
1981-90	2.0	2.2
1991-05 front or rear	1.6	1.8
1998-04 PreRunner	1.4	1.6
Previa	1.8	2.0
1996-00 RAV4	1.6	1.8
1998-03 Sienna	1.4	1.6
1993-98 T100	1.7	1.9
Tacoma	1.4	1.6
Tundra	1.5	1.7
Van	1.8	2.0
Caliper Assy., Replace (B)		
Includes: System bleeding.		
4Runner, Pickup		
1981-95	1.5	1.7
1996-02	1.1	1.3
2003-05 front or rear	1.4	1.6
Highlander front or rear	1.6	1.8
Land Cruiser		
1981-97	1.5	1.7
1998-05 front or rear	1.0	1.2
1998-04 PreRunner	1.1	1.3
Previa	1.5	1.7
RAV4	1.3	1.5
Sequoia front or rear	1.1	1.3
Sienna		
1998-03	1.2	1.4
2004-05 front or rear	1.6	1.8
1993-98 T100	1.6	1.8
Tacoma, Tundra	1.1	1.3
Van	1.3	1.5
Disc Brake Rotor, Replace (B)		
4Runner, Pickup		
1981-88		
2WD	1.8	2.0
4WD	2.6	2.8
1989-02	1.8	2.0
2003-05		
front	1.6	1.8
rear	1.0	1.1

RAV4 : SEQUOIA : SIENNA : T100 : TACOMA : TUNDRA : VAN : VAN WAGON TOY-89

	LABOR TIME	SEVERE SERVICE
Highlander		
front or rear	1.3	1.4
Land Cruiser		
1981-90	2.5	2.7
1991-05		
front	1.6	1.8
rear	.8	1.0
1998-04 PreRunner	1.6	1.8
Previa	.8	1.0
RAV4	1.0	1.2
Sequoia		
front or rear	.8	1.0
Sienna		
1998-03	.7	.9
2004-05 front or rear	1.1	1.3
T100	1.6	1.8
Tacoma	1.7	1.9
Tundra	.8	1.0
Van	1.5	1.7

Pads and/or Shoes, Replace (B)
Includes: Service and parking brake adjustment, system bleeding.

	LABOR TIME	SEVERE SERVICE
4Runner, Pickup		
front disc		
1984-02	1.3	1.4
2003-05	.8	.9
rear disc	1.6	1.8
rear drum	1.9	2.0
four wheels	2.8	2.9
Highlander		
front disc	1.0	1.1
rear disc	.6	.7
Land Cruiser		
1981-97		
front disc	1.2	1.3
rear disc	1.4	1.5
rear drum	2.4	2.5
four wheels		
w/rear disc	2.4	2.5
w/rear drum	3.2	3.3
1998-05		
front or rear disc	1.1	1.3
1998-04 PreRunner		
front disc	1.1	1.3
rear drum	1.9	2.0
Previa		
front or rear disc	1.4	1.5
four wheels	2.4	2.5
RAV4		
front disc	1.3	1.4
rear drum		
1996-00	1.9	2.0
2001-05	1.4	1.5
Sequoia		
front or rear disc	1.1	1.3
Sienna		
front disc	1.1	1.3
rear disc	.8	.9
rear drum	1.7	2.0

	LABOR TIME	SEVERE SERVICE
T100		
front disc	1.1	1.2
rear drum	1.8	1.9
four wheels	2.9	3.0
Tacoma		
front disc	1.1	1.3
rear drum	1.9	2.0
Tundra		
front disc	.8	.9
rear drum	1.9	2.0
Van		
front disc	1.4	1.5
rear drum	1.8	1.9
four wheels	2.7	2.8
w/dual wheels		
Pickup add	1.1	1.1

COMBINATION ADD-ONS
Replace
brake drum add		
2WD	.2	.2
4WD	.5	.5
brake hose add	.3	.3
caliper add	.2	.2
master cylinder add	.5	.5
wheel cylinder add		
each side	.2	.2
Resurface		
brake drum add	.5	.5
disc rotor add		
2WD	.9	.9
4WD	1.0	1.0

Wheel Cylinder, R&R and Recondition (B)
Includes: System bleeding.

	LABOR TIME	SEVERE SERVICE
4Runner, Pickup		
1981-95	1.9	2.2
1996-02	1.7	2.0
Land Cruiser		
1981-90	2.8	3.1
1991-97	1.8	2.1
1998-04 PreRunner	1.6	1.8
Previa	1.6	1.8
1996-00 RAV4	1.4	1.7
Sienna		
1998-03	1.6	1.9
2004-05	2.1	2.3
T100	1.5	1.8
Tacoma, Tundra	1.6	1.9
Van	2.6	2.9

Wheel Cylinder, Replace (B)
Includes: System bleeding.

	LABOR TIME	SEVERE SERVICE
4Runner, Pickup		
1981-95	1.9	2.2
1996-02	1.7	2.0
Land Cruiser		
1981-90	2.8	3.1
1991-97	1.8	2.1
1998-04 PreRunner	1.6	1.8
Previa	1.6	1.8
1996-05 RAV4	1.4	1.7
Sienna		
1998-03	1.6	1.8
2004-05	2.1	2.3
T100	1.5	1.8
Tacoma, Tundra	1.7	1.9
Van	2.3	2.5

PARKING BRAKE
Parking Brake Cable, Adjust (C)
Exc. below	.8	.9
1996-05 RAV4	1.1	1.3

Parking Brake Apply Actuator, Replace (B)
Exc. below	.9	.9
Highlander	1.1	1.1
1996-00 RAV4	1.2	1.2
Sienna	1.4	1.4

Parking Brake Cable, Replace (B)
Front
exc. below	1.1	1.4
4Runner, Pickup		
1981-88	1.3	1.6
1989-95	.7	1.0
1996-02	1.4	1.7
Land Cruiser	1.4	1.7
Previa	1.6	1.9
2004-05 Sienna	1.3	1.4
T100	.7	1.0
Van	1.3	1.6

Rear
4Runner		
1985-88	2.0	2.3
1989-95		
2WD	2.0	2.3
4WD	1.3	1.6
2003-05	6.6	7.4
Highlander	2.7	3.0
1981-97 Land Cruiser	1.6	1.9
Pickup		
1981-88	1.4	1.7
1989-95		
2WD	2.6	2.9
4WD	1.4	1.7
1998-04 PreRunner	1.1	1.3
Previa	1.5	1.8
RAV4		
2WD	4.6	4.9
4WD	5.3	5.6
Sequoia, Tundra	1.1	1.4
Sienna		
1998-03	1.5	1.8
2004-05	2.7	3.0
T100		
2WD	2.5	2.8
4WD	1.3	1.6
Tacoma		
2WD	2.6	2.9
4WD	1.2	1.5
Van	2.3	2.6

TOY-90   4RUNNER : HIGHLANDER : LAND CRUISER : PICKUP : PRERUNNER : PREVIA

	LABOR TIME	SEVERE SERVICE
FRONT SUSPENSION		

Unless otherwise noted, time given does not include alignment.

Front End Alignment, Check (A)
- All Models 1.9   2.2

Adjust
- front camber add 8   .9
- front toe add 2   .2
- rear camber add 3   .3
- rear toe add 3   .3

Coil Spring, Replace (B)
- 4Runner
  - 1996-02
    - 2WD
      - one 2.5   2.7
      - both 3.7   3.9
    - 4WD
      - one 1.3   1.5
      - both 2.2   2.4
  - 2003-05
    - one 2.1   2.3
    - both 2.7   2.9
  - 1998-04 PreRunner
    - one 2.5   2.7
    - both 4.1   4.3
- Sequoia, Tundra
  - one 1.4   1.6
  - both 2.2   2.4
- Tacoma
  - 2WD
    - one 2.5   2.7
    - both 4.1   4.3
  - 4WD
    - one 1.4   1.6
    - both 2.2   2.4

Free Wheeling Hub, Replace (B)
- 1984-97 8   1.1

Front Axle Hub Oil Seal, Replace (B)
- 4Runner, Pickup
  - 1981-87
    - 2WD 8   1.0
    - 4WD 2.0   2.2
  - 1988-95 .... 1.8   2.0
  - 1996-02
    - 2WD 1.3   1.5
    - 4WD 2.1   2.3
  - 2003-05 .... 3.6   3.8
- Land Cruiser
  - 1981-92 .... 1.9   2.1
  - 1993-97 .... 1.3   1.5
  - 1998-05 .... 2.4   2.6
- 1998-04 PreRunner .. 1.4   1.6
- Previa 3.0   3.2
- RAV4 .... 2.9   3.1
- Sequoia, Tundra 1.6   1.8
- T100 .... 1.6   1.8
- Tacoma
  - 2WD 1.4   1.6
  - 4WD 2.1   2.3

	LABOR TIME	SEVERE SERVICE

- 1984-87 Van
  - 2WD 1.1   1.3
  - 4WD 1.9   2.1

Front Hub, Replace (B)
- 4Runner, Pickup
  - 1982-87
    - 2WD 1.6   2.1
    - 4WD 2.7   3.2
  - 1988-95
    - 2WD 1.8   2.3
    - 4WD 2.0   2.5
  - 1996-02
    - 2WD 1.9   2.4
    - 4WD 2.6   3.1
  - 2003-05 .... 3.5   3.9
- Highlander 3.2   3.7
- Land Cruiser
  - 1993-97 .... 2.5   3.0
  - 1998-05 .... 3.0   3.5
- 1998-04 PreRunner .. 2.1   2.3
- Previa
  - 2WD 3.1   3.6
  - 4WD 2.8   3.3
- RAV4 .... 3.0   3.5
- Sequoia, Tundra 2.2   2.5
- Sienna 2.9   3.2
- T100 .... 2.1   2.6
- Tacoma
  - 2WD 2.1   2.3
  - 4WD 2.5   2.9
- Van
  - 2WD 1.8   2.3
  - 4WD 2.3   2.8

Front Hub Bearings, Replace (B)
- 4Runner, Pickup
  - 2WD 1.5   1.8
  - 4WD 2.1   2.4
- Highlander 3.2   3.5
- Land Cruiser
  - 1981-97 .... 1.9   2.2
  - 1998-05 .... 3.0   3.3
- 1998-04 PreRunner .. 1.7   2.0
- Previa 3.0   3.3
- RAV4 .... 3.0   3.3
- Sequoia, Tundra 1.9   2.1
- Sienna
  - 1998-03 .... 2.5   2.8
  - 2004-05 .... 2.9   3.2
- T100 .... 1.8   2.1
- Tacoma
  - 2WD 1.8   2.1
  - 4WD 3.0   3.4
- 1984-87 Van
  - 2WD 1.4   1.6
  - 4WD 1.9   2.1

Front Shock Absorbers, Replace (B)
- 4Runner, Pickup
  - 1981-95
    - one 8   1.0
    - both 1.3   1.5

	LABOR TIME	SEVERE SERVICE

- 1996-02
  - 2WD
    - one 7   .9
    - both 1.1   1.3
  - 4WD
    - one 1.2   1.4
    - both 1.9   2.1
- 2003-05
  - one 1.9   2.1
  - both 2.4   2.7
- Land Cruiser
  - 1981-97
    - one 7   .9
    - both 1.3   1.5
  - 1998-05
    - w/TEMS
      - one 1.1   1.3
      - both 1.7   1.9
    - w/o TEMS
      - one 6   .8
      - both 8   .9
- 1998-04 PreRunner
  - one 8   .9
  - both 1.1   1.3
- Sequoia, Tundra
  - one 1.3   1.4
  - both 1.9   2.1
- T100
  - one 7   .9
  - both 1.1   1.3
- Tacoma
  - 2WD
    - one 8   1.0
    - both 1.1   1.3
  - 4WD
    - one 1.3   1.4
    - both 1.9   2.1
- Van
  - one 5   .7
  - both 7   .9

Front Strut Coil Spring, Replace (B)
- Highlander
  - one 1.7   1.9
  - both 2.7   2.9
- Previa
  - one 2.7   2.9
  - both 3.9   3.1
- RAV4
  - one 2.2   2.4
  - both 3.6   3.8
- Sienna
  - one 2.1   2.3
  - both 3.0   3.2

Front Strut Absorber, Replace (B)
- Highlander
  - one 1.7   1.9
  - both 2.7   2.9
- Previa
  - one 2.5   2.7
  - both 3.6   3.8

RAV4 : SEQUOIA : SIENNA : T100 : TACOMA : TUNDRA : VAN : VAN WAGON — TOY-91

	LABOR TIME	SEVERE SERVICE
RAV4		
one	1.7	1.9
both	2.6	2.8
Sienna		
one	1.9	2.1
both	2.7	2.9
Leading Arm, Replace (B)		
1991-97 Land Cruiser	.8	1.2
Leaf Spring, Replace (B)		
Land Cruiser		
1981-90	1.7	2.2
1991-97	1.2	1.7
1981-82 Pickup	2.0	2.5
Leaf Spring Shackle, Replace (B)		
1981-88 Land Cruiser	.8	1.0
1981-82 Pickup	.8	1.0
Lower Ball Joint Replace (B)		
4Runner		
1983-95	2.5	2.8
1996-02	1.3	1.6
1982-95 Pickup	2.5	2.8
1998-04 PreRunner	1.4	1.7
Sequoia, Tundra	1.4	1.7
T100	2.6	2.9
Tacoma	1.4	1.7
Van	2.3	2.6
Lower Control Arm Assy., Replace (B)		
4Runner, Pickup		
1981-95		
2WD	3.3	3.6
4WD	2.8	3.1
1996-02		
2WD	2.1	2.4
4WD	3.1	3.4
2003-05	1.4	1.7
Highlander		
4 cyl.		
2WD	10.5	10.8
4WD	11.6	11.9
V6		
2WD	11.5	11.8
4WD	12.6	12.9
1998-05 Land Cruiser	1.8	2.1
1998-04 PreRunner	2.2	2.5
Previa	1.4	1.7
RAV4		
1996-00	1.9	2.2
2001-04	1.4	1.7
Sequoia, Tundra	3.3	3.6
Sienna		
1998-03	2.1	2.4
2004-05	10.2	10.5
T100		
2WD	3.4	3.7
4WD	2.6	2.9

	LABOR TIME	SEVERE SERVICE
Tacoma		
2WD	2.3	2.6
4WD	3.3	3.6
Van	2.6	2.9
Stabilizer Bar, Replace (B)		
4Runner, Pickup		
1989-95		
2WD	1.7	1.9
4WD	1.1	1.3
1996-05	1.0	1.2
Highlander		
4 cyl.	2.2	2.4
V6	2.9	3.1
Land Cruiser		
1993-97	.7	.9
1998-05	1.3	1.5
1998-04 PreRunner	1.0	1.2
RAV4	1.0	1.2
Previa	.8	1.0
Sequoia, Tundra	1.0	1.2
Sienna	1.7	1.9
T100	1.3	1.5
Tacoma	1.0	1.2
Stabilizer Bar Bushings, Replace (B)		
Exc. below		
one side	.5	.6
both sides	.6	.7
4Runner		
one side	.8	.9
both sides	1.0	1.1
Highlander		
4 cyl.		
one side	2.2	2.4
both sides	2.4	2.6
V6		
one side	2.9	3.1
both sides	3.0	3.2
1998-05 Land Cruiser		
one side	1.1	1.2
both sides	1.6	1.8
Steering Knuckle, Replace (B)		
4Runner, Pickup		
1981-95		
2WD	3.3	3.7
4WD	3.6	4.1
1996-02		
2WD	2.3	2.5
4WD	3.7	4.2
2003-05	3.2	3.6
Highlander	3.2	3.6
Land Cruiser		
1981-97	3.3	3.7
1998-05	2.9	3.2
1998-04 PreRunner	2.2	2.5
Previa		
2WD	3.2	3.6
4WD	2.9	3.2

	LABOR TIME	SEVERE SERVICE
RAV4, Sienna	2.9	3.2
Sequoia, Tundra	2.4	2.7
T100		
2WD	3.5	3.9
4WD	3.2	3.6
Tacoma		
2WD	2.2	2.5
4WD	4.0	4.4
Van	2.5	2.9
Torsion Bar, Replace (B)		
Includes: Adjust height.		
1982-95 4Runner, Pickup		
2WD	1.4	1.8
4WD	1.5	1.9
T100		
2WD	1.4	1.8
4WD	1.6	2.0
Van	.8	1.2
Torsion Bar Spring, Replace (B)		
1998-05 Land Cruiser		
one	1.0	1.3
both	1.3	1.6
Upper Ball Joint, Replace (B)		
4Runner, Pickup		
1982-95		
2WD	2.6	2.9
4WD	3.0	3.3
1996-02		
2WD	1.9	2.1
4WD	2.6	2.9
1998-04 PreRunner	1.9	2.1
Sequoia, Tacoma, Tundra		
2WD	1.9	2.1
4WD	2.9	3.2
T100		
2WD	2.6	2.9
4WD	2.9	3.2
Van	2.1	2.4
Upper Control Arm, Replace (B)		
4Runner, Pickup		
1981-95		
2WD	2.5	2.8
4WD	2.9	3.2
1996-02		
2WD	2.3	2.6
4WD	1.7	2.0
2003-05	1.4	1.6
Land Cruiser	1.4	1.6
1998-04 PreRunner	2.4	2.7
Sequoia, Tundra	1.9	2.1
T100		
2WD	2.6	2.9
4WD	3.0	3.3
Tacoma		
2WD	2.4	2.7
4WD	1.9	2.1
Van	3.2	3.5

TOY-92 4RUNNER : HIGHLANDER : LAND CRUISER : PICKUP : PRERUNNER : PREVIA

	LABOR TIME	SEVERE SERVICE
Upper Control Arm Bushings and/or Shaft, Replace (B)		
1982-95 4Runner, Pickup		
2WD	2.8	3.0
4WD	3.5	3.7
1998-04 PreRunner	2.4	2.6
T100		
2WD	3.0	3.2
4WD	3.7	3.9
Tacoma	2.4	2.6
Van	3.3	3.5

REAR SUSPENSION

	LABOR TIME	SEVERE SERVICE
Rear Alignment, Adjust (A)		
Camber or Toe	2.2	2.5
Lower Control Arm, Replace (B)		
Exc. below	.7	1.0
4Runner	.8	1.1
Previa	1.8	2.1
Sequoia	.5	.5
Rear Coil Spring, Replace (B)		
4Runner, Pickup		
1990-95	1.3	1.6
1996-02	.7	1.0
2003-05	1.7	2.0
Highlander	1.7	2.0
Land Cruiser	1.5	1.8
Previa	1.5	1.8
RAV4		
1996-03		
2WD	2.9	3.2
4WD	3.6	3.9
2004-05	2.9	3.2
Sienna		
1998-03		
one	3.2	3.5
both	4.3	4.6
2004-05		
2WD		
one	3.8	4.1
both	4.1	4.4
4WD		
one	5.3	5.6
both	5.7	6.0
Van	.7	1.0
Rear Control Arm, Replace (B)		
Highlander		
2WD	1.7	2.1
4WD	4.1	4.5
RAV4		
1996-00		
2WD	3.6	4.0
4WD	4.4	4.8
2001-05	3.3	3.7
Rear Leaf Spring, R&R and Recondition (B)		
1981-95 4Runner, Pickup	2.3	2.9
1981-90 Land Cruiser	2.1	2.7

	LABOR TIME	SEVERE SERVICE
1993-98 T100	2.3	2.9
Tacoma, Tundra	1.8	2.4
1987-89 Van	2.0	2.6
Rear Leaf Spring, Replace (B)		
All Models	1.7	2.1
Shock Absorbers or Bushings, Replace (B)		
Exc. below		
one	.7	.9
both	1.0	1.2
4Runner, Pickup		
one	1.0	1.2
both	1.3	1.5
Highlander		
one	1.7	1.9
both	2.7	2.9
1996-05 RAV4		
one	.8	1.0
both	1.1	1.3
Sienna		
one	1.0	1.1
both	1.5	1.7
Spring Shackle, Replace (B)		
Exc. below	.8	1.1
2005 Tacoma	2.2	2.5
Stabilizer Bar Bushings, Replace (B)		
1996-05 4Runner		
one side	.7	.9
both sides	.9	1.1
Highlander		
one side	1.4	1.6
both side	1.6	1.8
1993-05 Land Cruiser		
one side	.5	.7
both side	.7	.9
2001-05 RAV4		
one side	.7	.9
both side	.8	1.0
Sequoia		
one side	.5	.6
both sides	.7	.9
Tacoma		
one side	1.0	1.1
both sides	1.1	1.3
Stabilizer Bar, Replace (B)		
4Runner		
1996-02	.7	.9
2003-05	1.0	1.1
Highlander	1.4	1.6
Land Cruiser	.9	1.1
2001-05 RAV4		
2WD	3.5	3.7
4WD	4.3	4.5
Sequoia	.8	.9

STEERING
AIR BAGS

	LABOR TIME	SEVERE SERVICE
Diagnose Air Bag System Component Each (A)		
1992-05	.5	.5

	LABOR TIME	SEVERE SERVICE
Retrieve Fault Codes (A)		
1992-05	.3	.3
Air Bag Assy., Replace (B)		
Driver side		
exc. Highlander	.7	.7
Highlander	.9	.9
Passenger side		
4Runner		
1996-02	2.5	2.5
2003-05	2.0	2.0
Highlander	.9	.9
Land Cruiser	2.4	2.4
1998-04 PreRunner	2.0	2.0
Previa	.8	.8
RAV4		
1996-00	4.5	4.5
2001-05	3.4	3.4
Sequoia	2.8	2.8
Sienna	2.4	2.4
Tacoma	2.0	2.0
Tundra	2.2	2.2
Side curtain one		
4Runner	3.8	3.8
Land Cruiser	1.5	1.5
RAV4	1.6	1.6
Sequoia	3.1	3.1
Sienna	4.7	4.7
Airbag (Crash) Sensor, Replace (B)		
All Models	.7	.7
Center Sensor, Replace (B)		
Exc. below	.8	.8
1993-97 Land Cruiser	.5	.5
1992-97 Previa w/cool box	1.9	1.9
Sequoia	1.1	1.1
Tundra	.5	.5
Front Sensor, Replace (B)		
Exc. below		
one	.6	.6
both	.7	.7
Land Cruiser, T100		
one	.7	.7
both	1.3	1.3
Rear Sensor, Replace (B)		
2003-05 4Runner		
one	1.1	1.1
both	1.6	1.6
Land Cruiser		
one	.5	.5
both	.7	.7
2004-05 RAV4		
one	.6	.6
both	.8	.8
Side Sensor, Replace (B)		
2003-05 4Runner	.8	.8
Highlander, Sequoia		
one	1.0	1.0
both	1.5	1.5

RAV4 : SEQUOIA : SIENNA : T100 : TACOMA : TUNDRA : VAN : VAN WAGON   TOY-93

	LABOR TIME	SEVERE SERVICE
Land Cruiser, Sienna		
one	.7	.7
both	.9	.9
1998-04 PreRunner	.5	.5
2004-05 RAV4	.5	.5

LINKAGE
Idler Arm, Replace (B)
1983-95 4Runner	1.3	1.7
Pickup	.8	1.2
1993-98 T100	.8	1.2
1984-85 Van	.9	1.3

Pitman Arm, Replace (B)
| 1981-98 | 1.3 | 1.6 |

Steering Damper, Replace (B)
| All Models | .8 | 1.0 |

Tie Rod, Replace (B)
Includes: Reset toe.
| 1981-98 | 1.5 | 1.7 |

Tie Rod End, Replace (B)
Includes: Reset toe.
Exc. below	1.4	1.6
4Runner, Pickup	1.6	1.8
1981-90 Land Cruiser	1.7	1.9

MANUAL RACK & PINION
Flexible Coupling, Replace (B)
1995-96 Tacoma	.7	.9
1986-89 Van		
w/tilt wheel	.7	.9
w/o tilt wheel	1.5	1.7

Horn Contact, Replace (B)
| 1995-96 Tacoma | .5 | .5 |
| 1986-89 Van | .5 | .5 |

Inner Tie Rod, Replace (B)
Includes: Reset toe.
| 1995-96 Tacoma | 2.5 | 2.7 |
| 1986-89 Van | 2.1 | 2.3 |

Outer Tie Rod, Replace (B)
Includes: Reset toe.
| 1995-96 Tacoma | 1.3 | 1.5 |
| 1986-89 Van | 1.4 | 1.6 |

Rack & Pinion Assy., R&R and Recondition (A)
Includes: Alignment.
| 1995-96 Tacoma | 3.2 | 3.8 |
| 1986-89 Van | 3.3 | 3.9 |

Rack & Pinion Assy., R&R or Replace (B)
Includes: Alignment.
| 1995-96 Tacoma | 1.9 | 2.2 |
| 1986-89 Van | 1.8 | 2.1 |

Rack Boots, Replace (B)
| 1995-96 Tacoma | 1.5 | 1.8 |
| 1986-89 Van | 1.4 | 1.7 |

Steering Wheel, Replace (B)
| 1995-96 Tacoma | 1.5 | 1.5 |
| 1986-89 Van | .5 | .5 |

	LABOR TIME	SEVERE SERVICE
Upper Mast Jacket Bearing, Replace (B)		
1986-89 Van	1.8	2.0

POWER RACK & PINION
Center Arm Bracket, Replace (B)
| 1984-85 Van | 1.3 | 1.6 |

Horn Contact or Canceling Cam, Replace (B)
| Exc. below | .5 | .5 |
| Highlander | .7 | .7 |

Inner Tie Rod, Replace (B)
Includes: Reset toe.
1996-02 4Runner	3.2	3.4
Highlander	5.3	5.5
1998-05 Land Cruiser	3.0	3.2
1998-04 PreRunner	3.5	3.9
Previa		
2WD	3.5	3.7
4WD	4.7	4.9
RAV4		
1996-00	3.8	4.0
2001-05	4.3	4.5
Sequoia, Tundra	3.5	3.7
Sienna	4.3	4.5
T100	3.5	3.7
Tacoma	3.6	3.8
1986-89 Van		
2WD	3.0	3.2
4WD	5.3	5.5

Rack & Pinion Assy., R&R or Replace (B)
Includes: Alignment.
4Runner		
1996-02	2.3	2.6
2003-05	3.5	3.9
Highlander	5.0	5.3
1998-05 Land Cruiser	2.2	2.5
1998-04 PreRunner	2.7	3.0
Previa		
2WD	2.6	2.9
4WD	3.7	4.0
RAV4	3.8	4.1
Sequoia, Tundra	2.7	3.0
Sienna	3.5	3.8
T100		
2WD	2.6	2.9
4WD	1.9	2.2
Tacoma	2.7	3.0
1986-89 Van	2.5	2.8

Rack & Pinion Assy., Reseal (A)
Includes: Alignment.
4Runner		
1996-02	4.0	4.5
2003-05	5.2	5.8
1998-05 Land Cruiser	3.8	4.3
1998-04 PreRunner	5.2	5.7

	LABOR TIME	SEVERE SERVICE
Previa		
2WD	4.9	5.4
4WD	6.1	6.6
RAV4		
1996-00	5.5	6.0
2001-05	6.6	7.1
Sequoia, Tundra	5.0	5.5
Sienna	6.1	6.6
T100	5.0	5.5
Tacoma	5.2	5.7
1986-89 Van	3.7	4.2

Rack Boots, Replace (B)
Exc. below	1.5	1.7
4Runner		
1984-02	1.7	1.9
2003-05	3.8	4.3
Highlander	5.2	5.4
2004-05 Sienna	3.5	4.0
1986-89 Van	1.6	1.8

Steering Pump, R&R and Recondition (B)
4Runner	2.9	3.2
Highlander, Tacoma	3.0	3.4
1998-05 Land Cruiser	5.3	6.0
1998-04 PreRunner	3.0	3.4
Previa	3.8	4.3
RAV4		
1996-00	5.0	5.5
2001-05	3.2	3.7
Sequoia, Tundra		
V6	3.3	3.8
V8	5.3	5.8
2004-05 Sienna	2.7	3.0
T100	3.3	3.8
1986-89 Van	3.6	4.1

Steering Pump, R&R or Replace (B)
4Runner	1.8	2.0
Highlander	2.0	2.2
1998-05 Land Cruiser	3.3	3.5
1998-04 PreRunner	1.7	1.9
Previa	2.1	2.3
RAV4		
1996-00	3.5	3.7
2001-05	1.7	1.9
Sequoia, Tundra		
V6	1.4	1.6
V8	3.3	3.5
Sienna		
1998-03	2.1	2.3
2004-05	1.7	1.9
T100	1.4	1.6
Tacoma	1.7	1.9
1986-89 Van	1.9	2.1

Steering Pump Hoses, Replace (B)
Pressure		
exc. below	1.5	1.7
RAV4, Van	1.9	2.1

TOY-94 4RUNNER : HIGHLANDER : LAND CRUISER : PICKUP : PRERUNNER : PREVIA

	LABOR TIME	SEVERE SERVICE
Steering Pump Hoses, Replace (B)		
Return		
exc. below	.7	.9
2003-05 4Runner	2.0	2.2
Highlander	1.1	1.3
Land Cruiser	1.5	1.7
Previa	.8	1.0
RAV4		
1996-00	.9	1.1
Sienna		
1998-03	1.4	1.6
2004-05	2.0	2.2
1986-89 Van	2.8	3.0
Steering Wheel, Replace (B)		
Exc. Highlander	.7	.7
Highlander	.9	.9

MANUAL WORM & SECTOR

	LABOR TIME	SEVERE SERVICE
Steering Gear, Adjust (B)		
1981-95	.5	.7
Center Arm Bracket, Replace (B)		
1984-85 Van	1.3	1.6
Gear Assy., R&R and Recondition (B)		
1983-95 4Runner	2.3	2.8
1981-85 Land Cruiser	2.8	3.3
Pickup	2.4	2.9
1984-85 Van	1.9	2.4
Gear Assy., Replace (B)		
1983-95 4Runner	1.3	1.5
1981-85 Land Cruiser	1.6	1.8
Pickup	1.5	1.7
1984-85 Van	1.3	1.5
Horn Contact, Replace (B)		
1981-95	.5	.5
Steering Wheel, Replace (B)		
1981-95	.5	.5

POWER WORM & SECTOR

	LABOR TIME	SEVERE SERVICE
Center Arm Bracket, Replace (B)		
1984-85 Van	1.1	1.5
Gear Assy., Replace (B)		
Exc. below	2.4	2.6
1984-85 Van	1.8	2.0
Power Steering Control Module, Replace (B)		
1993-97 Land Cruiser	.8	.8
Steering Pump, R&R and Recondition (B)		
All Models	3.1	3.6
Steering Pump, R&R or Replace (B)		
4Runner, Pickup	1.8	2.0
Land Cruiser	2.3	2.5
1984-85 Van	2.0	2.2
Steering Pump Hoses, Replace (B)		
Pressure		
exc. below	1.4	1.6
1984-85 Van	1.8	2.0

	LABOR TIME	SEVERE SERVICE
Return		
4Runner, Pickup	.5	.7
1981-97 Land Cruiser	1.3	1.5
1984-85 Van	2.7	2.9

HEATING & AIR CONDITIONING

When more than one component requires replacement where evacuation/recovery and recharging is already included, deduct 1.0 hour for each additional component from the time given.

	LABOR TIME	SEVERE SERVICE
Evacuate/Recover and Recharge System (B)		
1981-05	1.6	1.6
AC Hoses, Replace (B)		
Includes: Evacuate/recover and recharge.		
Exc. below	3.3	3.5
Van	4.3	4.5
Blower Motor, Replace (B)		
4Runner, Pickup		
1985-88	.9	.9
1989-95		
w/AC	1.7	1.7
w/o AC	.5	.5
1996-05	.5	.5
Highlander, RAV4	.3	.3
Land Cruiser		
1981-87		
w/AC	2.2	2.2
w/o AC	.9	.9
1988-05	.6	.6
1998-04 PreRunner	.5	.5
Previa	1.5	1.5
Sequoia, Tundra	.3	.3
Sienna		
1998-03	.8	.8
2004-05	2.4	2.4
T100		
w/AC	1.8	1.8
w/o AC	.5	.5
Tacoma	.5	.5
Van		
1984-87	1.7	1.7
1988-89		
w/AC	3.9	3.9
w/o AC	1.5	1.5
Blower Motor Resistor, Replace (B)		
Exc. below	.3	.3
Land Cruiser		
1981-84	1.0	1.0
1985-97	.5	.5
Pickup	.5	.5
Previa	1.6	1.6
1998-03 Sienna	.5	.5
Van	.9	.9

	LABOR TIME	SEVERE SERVICE
Blower Motor Switch, Replace (B)		
4Runner		
1985-88	.5	.5
1989-95	.8	.8
1996-02	.5	.5
Highlander	.7	.7
Land Cruiser		
1981-90	.7	.7
1991-97	1.4	1.4
1998-05	.3	.3
1998-04 PreRunner	1.8	1.8
Previa	3.0	3.0
1996-00 RAV4	1.3	1.3
Sienna		
1998-03	.5	.5
2004-05	.8	.8
T100, Tacoma	1.8	1.8
Van	.5	.5
Compressor Assy., Replace (B)		
Includes: Parts transfer, evacuate/recover and recharge.		
4Runner, Pickup		
1981-95		
4 cyl.	3.3	3.5
V6	4.6	4.8
1996-02	3.6	3.8
2003-05	4.1	4.3
Highlander	4.3	4.5
Land Cruiser	4.0	4.2
1998-04 PreRunner	3.5	3.7
Previa	4.6	4.8
RAV4	3.8	4.0
Sequoia, Tundra		
V6	3.3	3.5
V8	4.0	4.2
Sienna	4.4	4.6
T100, Tacoma	3.5	3.7
Van	4.6	4.8
Compressor Clutch Assy., Replace (B)		
4Runner, Pickup		
1981-95		
4 cyl.	2.1	2.3
V6	3.1	3.3
1996-02	1.8	2.1
2003-05	2.6	2.9
Land Cruiser		
1981-90	.9	1.1
1991-05	2.3	2.6
1998-04 PreRunner	1.8	2.1
Previa	3.3	3.5
RAV4	2.1	2.3
Sequoia, Tundra		
V6	1.8	2.1
V8	2.3	2.6
Sienna		
1998-03	2.3	2.6
2004-05	2.9	3.2

	LABOR TIME	SEVERE SERVICE
T100	2.0	2.2
Tacoma	1.8	2.1
Van	2.9	3.1

Condenser Assy., Replace (B)
Includes: Evacuate/recover and recharge.

	LABOR TIME	SEVERE SERVICE
Exc. below	3.8	4.0
1996-02 4Runner	3.3	3.5
Highlander	4.1	4.3
1998-05 Land Cruiser	3.3	3.5
1996-00 RAV4	3.2	3.4
2004-05 Sierra	4.6	4.8
Van	4.7	4.9

Evaporator Coil, Replace (B)
Includes: Evacuate/recover and recharge.

	LABOR TIME	SEVERE SERVICE
4Runner, Pickup		
1981-95	3.6	3.8
1996-02	4.8	5.0
2003-05	7.1	7.2
Highlander	7.5	7.7
Land Cruiser		
1981-90	4.6	4.8
1991-92	4.0	4.2
1993-97	3.5	3.7
1998-05	6.3	7.0
1998-04 PreRunner	3.5	3.7
Previa	4.3	4.5
RAV4		
1996-00	6.8	7.0
2001-05	7.7	7.9
Sequoia	4.7	4.9
Sienna	5.7	5.9
T100, Tacoma	3.6	3.8
Tundra	4.0	4.2
Van	5.3	5.5

Expansion Valve, Replace (B)
Includes: Evacuate/recover and recharge.

	LABOR TIME	SEVERE SERVICE
4Runner, Pickup		
1983-95	3.5	3.7
1996-02	4.8	5.0
2003-05	4.4	4.9
Highlander	7.5	7.7
Land Cruiser		
1981-92	4.3	4.5
1993-97	3.5	3.7
1998-05	5.1	5.3
1998-04 PreRunner	3.3	3.5
Previa	4.1	4.3
RAV4		
1996-00	6.8	7.0
2001-05	7.7	7.9
Sequoia	4.7	4.9
Sienna		
1998-03	5.7	5.9
2004-05	6.8	7.0
T100	3.5	3.7
Tacoma	3.6	3.8
Tundra	4.0	4.2
Van	5.3	5.5

Heater Control Assy., Replace (B)

	LABOR TIME	SEVERE SERVICE
4Runner, Pickup		
1985-95	1.3	1.3
1996-02	.7	.7
2003-05	.3	.3
Land Cruiser		
1981-90		
hardtop	1.3	1.3
wagon	.9	.9
1991-97	1.4	1.4
1998-05	.3	.3
1998-04 PreRunner	1.8	1.8
Previa	3.1	3.1
RAV4		
1996-00	.8	.8
2001-05	.5	.5
Sequoia, Tundra	.5	.5
Sienna		
1998-03	1.3	1.3
2004-05	.7	.7
T100	1.8	1.8
Tacoma	2.0	2.0
Van	.8	.8

Heater Control Valve, Replace (B)

	LABOR TIME	SEVERE SERVICE
4Runner, Pickup		
1981-95	.8	1.0
1996-02	1.3	1.5
Land Cruiser		
1981-90	.9	1.1
1991-05	1.3	1.5
1998-04 PreRunner	.9	1.1
Previa	1.4	1.6
Sequoia, Tundra	1.4	1.6
1998-05 Sienna	1.2	1.4
T100	.8	1.0
Tacoma	.9	1.1
Van	1.4	1.6

Heater Core, R&R or Replace (B)

	LABOR TIME	SEVERE SERVICE
4Runner, Pickup		
1981-88	2.4	2.9
1989-95		
w/AC	4.8	5.3
w/o AC	2.4	2.9
1996-98		
w/AC	6.3	6.8
w/o AC	5.1	5.6
1999-02		
w/AC	5.1	5.6
w/o AC	4.2	4.7
2003-05	5.6	6.1
Highlander	5.6	6.1
Land Cruiser		
1981-90		
hardtop	3.2	3.7
wagon	3.8	4.3
1991-97		
w/AC	5.1	5.6
w/o AC	4.3	4.8
1998-05	4.0	4.5
1998-04 PreRunner		
w/AC	4.1	4.6
w/o AC	2.8	3.3
Previa	6.1	6.6
RAV4		
1996-00		
w/AC	6.9	7.4
w/o AC	6.0	6.5
2001-05		
w/AC	6.6	7.1
w/o AC	5.7	6.2
Sequoia		
w/AC	6.3	6.8
w/o AC	5.0	5.5
Sienna		
1998-03		
w/AC	4.7	5.2
w/o AC	3.2	3.7
2004-05	7.5	8.0
1993-98 T100		
w/AC	4.9	5.4
w/o AC	2.4	2.9
Tacoma		
w/AC	4.1	4.6
w/o AC	2.8	3.3
Tundra		
w/AC	5.6	6.1
w/o AC	4.3	4.8
Van	2.8	3.3

Heater Hoses, Replace (B)

	LABOR TIME	SEVERE SERVICE
4Runner, Pickup		
1981-95	.7	.9
1996-05	1.3	1.5
Highlander	.3	.3
Land Cruiser	1.2	1.4
1998-04 PreRunner	.7	.9
Previa, RAV4	1.3	1.5
Sequoia, Tundra	1.2	1.4
Sienna	.9	1.1
T100, Tacoma	.7	.9
Van	1.4	1.6

High or Low Pressure Switch, Replace (B)
Includes: Evacuate/recover and recharge.

	LABOR TIME	SEVERE SERVICE
4Runner, Pickup		
1983-95	3.6	3.8
1996-02	4.8	5.0
Land Cruiser		
1981-90	4.0	4.2
1991-05	3.6	3.8
1998-04 PreRunner	3.6	3.8
Previa	4.3	4.5
RAV4	3.6	3.8

TOY-96 4RUNNER : HIGHLANDER : LAND CRUISER : PICKUP : PRERUNNER : PREVIA

	LABOR TIME	SEVERE SERVICE
High or Low Pressure Switch, Replace (B)		
Sequoia, Tundra	3.2	3.4
Sienna	3.2	3.4
T100, Tacoma	3.6	3.8
Van	5.1	5.3
Receiver/Drier Assy., Replace (B)		
Includes: Evacuate/recover and recharge.		
Exc. below	3.5	3.7
Highlander	4.3	4.5
Previa	3.6	3.8
2001-05 RAV4	3.8	4.0
2004-05 Sienna	4.6	4.8
Van	4.3	4.5

WIPERS & SPEEDOMETER

	LABOR TIME	SEVERE SERVICE
Antenna, R&R (B)		
Exc. below	.9	1.0
2001-05 RAV4	1.3	1.4
2004-05 Sienna	2.6	2.7
Radio, R&R (B)		
Exc. below	.6	.6
1988-05 4Runner, Pickup	.9	.9
1988-97 Land Cruiser	.8	.8
Previa	.8	.8
Rear Window Washer Motor & Pump, Replace (B)		
Land Cruiser, Previa	.5	.5
2001-05 RAV4	.8	.8
Sequoia	.5	.5
2004-05 Sienna	.8	.8
Van	1.3	1.3
Rear Window Wiper Motor Replace (B)		
Exc. below	.8	1.0
Highlander	.5	.5
1981-97 Land Cruiser hardtop	.9	1.1
Previa	1.3	1.5
2001-05 RAV4	.5	.5
Rear Windshield Wiper Switch, Replace (B)		
4Runner	.8	.8
Land Cruiser		
1981-90	.5	.5
1991-97	1.3	1.3
Van	.5	.5
Speedometer Cable & Casing, Replace (B)		
Exc. below	1.5	1.5
1981-92 Land Cruiser		
hardtop	.7	.7
wagon	1.8	1.8

	LABOR TIME	SEVERE SERVICE
Speedometer Head, R&R or Replace (B)		
Exc. below	.9	.9
Previa	1.2	1.2
Van	1.3	1.3
Windshield Washer Pump & Motor Assy., Replace (B)		
Exc. below	.5	.5
4Runner	.7	.8
Highlander	1.4	1.5
RAV4	.8	.9
2004-05 Sienna	.8	.9
Van	.7	.8
Windshield Washer Switch, Replace (B)		
4Runner, Pickup		
1983-95	.8	.8
1996-05	.3	.3
Land Cruiser		
1981-90	.7	.7
1991-97	1.2	1.2
1998-05	.3	.3
Highlander	.3	.3
1998-04 PreRunner	.8	.8
Previa	.9	.9
RAV4		
1996-00	1.0	1.0
2001-05	.6	.6
Sequoia, Sienna, Tundra	.5	.5
T100, Tacoma	.8	.8
Van	1.4	1.4
Windshield Wiper Linkage, Replace (B)		
Exc. below	.8	.9
1981-97 Land Cruiser	1.1	1.3
Van	3.2	3.4
Windshield Wiper Motor, Replace (B)		
Exc. below	.5	.5
4Runner	.6	.7
Highlander, Previa, RAV4	.9	1.1
Sienna	.8	1.0
Van	3.0	3.2

LAMPS & SWITCHES

	LABOR TIME	SEVERE SERVICE
Diagnose Park/Neutral Position Switch Malfunction (A)		
4Runner	.7	.7
Highlander	.8	.8
Land Cruiser	.7	.7
1998-04 PreRunner	.7	.7
Previa	.5	.5
RAV4		
1997-00	.6	.6
2001-05	.9	.9
Sequoia, T-100, Tundra	.6	.6
Tacoma	.7	.7

	LABOR TIME	SEVERE SERVICE
Diagnose Stop Light Signal Switch Malfunction (A)		
Exc. below	.6	.6
Previa	.3	.3
2001-05 RAV4	.3	.3
Back-Up Lamp Bulb, Replace (C)		
Exc. below	.3	.3
RAV4	.6	.6
Sequoia, Tundra	.5	.5
2004-05 Sienna	.6	.6
Back-Up Lamp/Neutral Safety Switch, Replace (B)		
Exc. below	.7	.7
1991-97 Land Cruiser	1.3	1.3
RAV4	1.1	1.1
2004-05 Sienna	1.1	1.1
Back-Up Lamp Switch, Replace (B)		
Exc. below	.5	.5
1981-89 Land Cruiser wagon	1.1	1.1
Previa	.7	.7
Halogen Headlamp Bulb, Replace (B)		
All Models	.3	.3
Hazard Warning Switch, Replace (B)		
Exc. below	.3	.3
1996-02 4Runner	.7	.7
1981-90 Land Cruiser	1.3	1.3
Previa	.8	.8
1996-00 RAV4	.7	.7
Sequoia, Tundra	.5	.5
Van	1.1	1.1
Headlamp Switch, Replace (B)		
4Runner, Pickup		
1985-98	.9	.9
1999-02	.3	.3
Highlander	.3	.3
Land Cruiser		
1981-97	1.4	1.4
1998-05	.3	.3
1998-04 PreRunner	.6	.6
Previa	.9	.9
RAV4		
1996-00	1.2	1.2
2001-05	.6	.6
Sequoia, Tundra	.5	.5
Sienna	.5	.5
T100	.7	.7
Tacoma	.6	.6
Van	1.2	1.2
Headlamps, Aim (B)		
Two	.5	.5
Four	.8	.8
High Mount Stop Lamp Assy., Replace (B)		
Exc. below	.3	.3
1989-95 4Runner	.9	.9
1998-05 Land Cruiser	.5	.5

RAV4 : SEQUOIA : SIENNA : T100 : TACOMA : TUNDRA : VAN : VAN WAGON **TOY-97**

	LABOR TIME	SEVERE SERVICE
Horn, Replace (B)		
Exc. below	.3	.3
Highlander	.5	.5
1998-04 PreRunner	.6	.6
2001-05 RAV4	.8	.8
Tacoma	.6	.6
License Lamp Lens or Bulb, Replace (C)		
Exc. below	.3	.3
RAV4	.6	.6
2004-05 Sienna	.6	.6
Parking Brake Warning Lamp Switch, Replace (B)		
Exc. below	.5	.5
1998-05 Land Cruiser	.8	.8
RAV4	.8	.8
Parking Lamp Lens or Bulb, Replace (C)		
Exc. below	.3	.3
1998-03 Sienna	.7	.7
Rear Combination Lamp Lens, Replace (B)		
1981-05	.5	.5
Sealed Beam Headlamp, Replace (B)		
Exc. below	.3	.3
1995-96 Tacoma	.7	.7
Side Marker Lamp Lens, Replace (B)		
1981-05	.3	.3
Stop Light Switch, Replace (B)		
Exc. below	.7	.7
2003-05 4Runner	.2	.2
Highlander	.3	.3
2004-05 Sienna	.3	.3
Turn Signal Lamp Lens or Bulb, Replace (C)		
1981-05	.3	.3

	LABOR TIME	SEVERE SERVICE
Turn Signal or Hazard Warning Flasher, Replace (B)		
Exc. below	.5	.5
Previa	.9	.9
2001-05 RAV4	4.1	4.1
Sequoia, Tundra	.6	.6
Turn Signal Switch, Replace (B)		
Exc. below	.5	.5
2003-05 4Runner	.8	.8
1988-97 Land Cruiser	1.1	1.1
Sequoia, Tundra	.8	.8

BODY

	LABOR TIME	SEVERE SERVICE
Door Window Regulator, Replace (B)		
Exc. below	1.1	1.3
1998-04 PreRunner	.7	.7
Sequoia, Tundra	.7	.7
Tacoma	.7	.7
Front Door Lock, Replace (B)		
4Runner, Van	.7	.7
Highlander	1.2	1.2
1981-97 Land Cruiser	1.1	1.1
Pickup	1.3	1.3
1998-04 PreRunner	.8	.8
Previa	1.4	1.4
RAV4	.8	.8
Sienna		
1998-03	1.1	1.1
2004-05	.7	.7
Sequoia, Tundra	1.1	1.1
T100, Tacoma	.8	.8
Hood Hinges (Both), Replace (B)		
Exc. below	.8	.8
2003-05 4Runner	.5	.5
Highlander	1.8	1.8
1998-04 PreRunner	3.1	3.1
Previa	3.1	3.1
RAV4		
1996-00	2.7	2.7
2001-05	2.0	2.0

	LABOR TIME	SEVERE SERVICE
Sienna		
1998-03	1.5	1.5
2004-05	1.1	1.1
Tacoma	3.1	3.1
Hood Lock, Replace (B)		
1981-05	.7	.7
Hood Release Cable, Replace (B)		
Exc. below	.8	.8
Highlander	1.2	1.2
1996-00 RAV4	1.2	1.2
Sequoia, Tundra	1.1	1.1
Lock Striker Plate, Replace (B)		
1981-05	.5	.5
Rear Door Lock, Replace (B)		
4Runner	.9	.9
Highlander	1.1	1.1
1981-97 Land Cruiser	1.2	1.2
1998-04 PreRunner	.9	.9
Previa	1.1	1.1
RAV4		
1996-00	1.4	1.4
2001-05	.8	.8
Sequoia	1.1	1.1
Sienna	.8	.8
Tacoma	.9	.9
Tundra	.7	.7
Van	1.2	1.2
Tail Gate Window Regulator, Replace (B)		
Includes: Transfer motor.		
4Runner		
1983-95	1.3	1.3
1996-02	.7	.7
Window Regulator Motor, Replace (B)		
Exc. below	1.1	1.2
4Runner, RAV4	1.4	1.5
2004-05 Sienna	1.4	1.5

NOTES

NOTES

TOY-100 4RUNNER : HIGHLANDER : LAND CRUISER : PICKUP : PRERUNNER : PREVIA

NOTES

NOTES

NOTES

NOTES

TOY-104 4RUNNER : HIGHLANDER : LAND CRUISER : PICKUP : PRERUNNER : PREVIA

NOTES

NOTES

TOY-106 4RUNNER : HIGHLANDER : LAND CRUISER : PICKUP : PRERUNNER : PREVIA

NOTES

RAV4 : SEQUOIA : SIENNA : T100 : TACOMA : TUNDRA : VAN : VAN WAGON TOY-107

NOTES

TOY-108 4RUNNER : HIGHLANDER : LAND CRUISER : PICKUP : PRERUNNER : PREVIA

NOTES

NOTES

TOY-110 4RUNNER : HIGHLANDER : LAND CRUISER : PICKUP : PRERUNNER : PREVIA

NOTES

RAV4 : SEQUOIA : SIENNA : T100 : TACOMA : TUNDRA : VAN : VAN WAGON   TOY-111

NOTES

TOY-112 4RUNNER : HIGHLANDER : LAND CRUISER : PICKUP : PRERUNNER : PREVIA

NOTES

NOTES

TOY-114 4RUNNER : HIGHLANDER : LAND CRUISER : PICKUP : PRERUNNER : PREVIA

NOTES

NOTES

TOY-116 4RUNNER : HIGHLANDER : LAND CRUISER : PICKUP : PRERUNNER : PREVIA

NOTES

NOTES

TOY-118 4RUNNER : HIGHLANDER : LAND CRUISER : PICKUP : PRERUNNER : PREVIA

NOTES

EUROPEAN VEHICLES

Alfa Romeo	AR-1
Audi	AUD-1
BMW	BMW-1
Jaguar	JAG-1
Land Rover	LR-1
Mercedes Benz	MB-1
Mini	MIN-1
Peugeot	PEU-1
Porsche	POR-1
Renault	REN-1
Saab	SAA-1
Sterling	STR-1
Volkswagen	VW-1
Volvo	VLV-1
Yugo	YUG-1

EUROPEAN VEHICLES

Alfa Romeo	AR-1	Porsche	POR-1
Audi	AUD-1	Renault	REN-1
BMW	BMW-1	Saab	SAA-1
Jaguar	JAG-1	Sterling	STR-1
Land Rover	LR-1	Volkswagen	VW-1
Mercedes Benz	MB-1	Volvo	VLV-1
Mini	MIN-1	Yugo	YUG-1
Peugeot	PEU-1		

AR

164 : Milano

SYSTEM INDEX

MAINTENANCE	AR-2
CHARGING	AR-2
STARTING	AR-2
CRUISE CONTROL	AR-2
IGNITION	AR-2
EMISSIONS	AR-2
FUEL	AR-2
EXHAUST	AR-3
ENGINE COOLING	AR-3
ENGINE	AR-3
Assembly	AR-3
Cylinder Head	AR-3
Camshaft	AR-3
Crank & Pistons	AR-3
Engine Lubrication	AR-3
CLUTCH	AR-4
MANUAL TRANSAXLE	AR-4
MANUAL TRANSMISSION	AR-4
AUTO TRANSAXLE	AR-4
AUTO TRANSMISSION	AR-4
DRIVELINE	AR-4
BRAKES	AR-4
FRONT SUSPENSION	AR-5
REAR SUSPENSION	AR-5
STEERING	AR-6
WIPERS & SPEEDOMETER	AR-6
LAMPS & SWITCHES	AR-6

OPERATIONS INDEX

A
- Air Bags ... AR-6
- Air Conditioning ... AR-6
- Alignment ... AR-5
- Alternator (Generator) ... AR-2
- Anti-Lock Brakes ... AR-4

B
- Back-Up Lamp Switch ... AR-6
- Ball Joint ... AR-5
- Battery Cables ... AR-2
- Bleed Brake System ... AR-5
- Blower Motor ... AR-6
- Brake Disc ... AR-5
- Brake Hose ... AR-5
- Brake Pads and/or Shoes ... AR-5

C
- Camshaft ... AR-3
- Catalytic Converter ... AR-3
- Coolant Temperature (ECT) Sensor ... AR-2
- Cruise Control ... AR-2
- CV Joint ... AR-4
- Cylinder Head ... AR-3

D
- Differential ... AR-4
- Distributor ... AR-2
- Drive Belt ... AR-2
- Driveshaft ... AR-4

E
- Engine ... AR-3
- Engine Lubrication ... AR-3
- Engine Mounts ... AR-3
- Exhaust ... AR-3
- Exhaust Manifold ... AR-3
- Expansion Valve ... AR-6

F
- Flywheel ... AR-4
- Fuel Injection ... AR-2
- Fuel Pump ... AR-2

G
- Generator ... AR-2

H
- Halfshaft ... AR-4
- Headlamp ... AR-6
- Heater Core ... AR-6
- Horn ... AR-6

I
- Ignition Coil ... AR-2
- Ignition Switch ... AR-2
- Intake Manifold ... AR-2

L
- Lower Control Arm ... AR-5

M
- Mass Air Flow (MAF) Sensor ... AR-2
- Master Cylinder ... AR-5
- Multifunction Lever/Switch ... AR-6

O
- Oil Pan ... AR-3
- Oil Pump ... AR-3
- Outer Tie Rod ... AR-6

P
- Parking Brake ... AR-5
- Pistons ... AR-3

R
- Radiator ... AR-3
- Rear Main Oil Seal ... AR-3

S
- Shock Absorber/Strut, Front ... AR-5
- Shock Absorber/Strut, Rear ... AR-5
- Spark Plug Cables ... AR-2
- Spark Plugs ... AR-2
- Spring, Front Coil ... AR-5
- Spring, Rear Coil ... AR-5
- Starter ... AR-2
- Steering Wheel ... AR-6

T
- Thermostat ... AR-3
- Throttle Body ... AR-2
- Timing Belt ... AR-3
- Torque Converter ... AR-4

U
- Upper Control Arm ... AR-5

V
- Valve Body ... AR-4
- Valve Cover Gasket ... AR-3
- Valve Job ... AR-3

W
- Water Pump ... AR-3
- Wheel Balance ... AR-2
- Windshield Washer Pump ... AR-6
- Windshield Wiper Motor ... AR-6

AR-1

AR-2 164 : MILANO

	LABOR TIME	SEVERE SERVICE
MAINTENANCE		
Air Cleaner Filter Element, Replace (C)		
All Models	.2	.3
Drive Belt, Adjust (B)		
164	.3	.4
Milano	.2	.3
Drive Belt, Replace (B)		
164	.6	.7
Milano	.5	.6
Fuel Filter, Replace (B)		
164	.2	.2
Milano	.4	.4
Halogen Headlight Bulb, Replace (B)		
All Models	.3	.3
Oil and Filter, Change (C)		
Includes: Correct all fluid levels.		
All Models	.3	.4
Timing Belt, Replace (B)		
164	4.7	5.0
164 24V		
right side		
one	5.3	5.6
both	5.9	6.2
both sides all	10.2	10.5
Milano	2.9	3.2
Tire, Replace (C)		
Includes: Dismount old tire and mount new tire to rim.		
All Models	.5	.5
Tires, Rotate (C)		
All Models	.5	.5
Wheel, Balance (B)		
All Models		
one	.3	.3
each addl. add	.2	.2
CHARGING		
Alternator Circuits, Test (B)		
Includes: Test component output.		
All Models	.6	.6
Alternator, R&R and Recondition (B)		
164	2.5	2.8
164 24V, Milano	2.9	3.2
Alternator Assy., Replace (B)		
Includes: Pulley transfer.		
164	.9	1.0
164 24V	1.6	1.7
Milano	1.6	1.7
Voltage Regulator, Replace (B)		
164		
w/AC	1.1	1.3
w/o AC	.5	.7
Milano	1.6	1.8

	LABOR TIME	SEVERE SERVICE
STARTING		
Starter Draw Test (On Car) (B)		
All Models	.3	.3
Battery Cables, Replace (C)		
Positive		
164	1.8	1.9
Milano	.6	.7
Negative	.3	.4
Starter, R&R and Recondition (B)		
Milano	3.9	4.4
Starter Assy., Replace (B)		
164	2.1	2.2
Milano	2.9	3.0
Starter Drive Assy., Replace (B)		
Includes: Starter R&R.		
164	2.8	2.9
Milano	3.5	3.7
Starter Solenoid and/or Switch, Replace (B)		
Includes: Starter R&R.		
164	2.5	2.6
Milano	3.0	3.1
CRUISE CONTROL		
Actuator Assy., Replace (B)		
164	.4	.4
Milano	.9	.9
Control Cable & Core, Replace (B)		
All Models	.4	.4
Control Controller Module, Replace (B)		
164	.5	.5
Control Switch, Replace (B)		
Milano	.3	.3
IGNITION		
Distributor, Replace (B)		
Includes: Reset base ignition timing.		
164	.5	.6
Milano	.9	1.0
Distributor Cap and/or Rotor, Replace (B)		
All Models	.4	.4
Electronic Ignition Control Unit, Replace (B)		
All Models	.5	.5
Ignition Coil, Replace (B)		
164, Milano	.3	.3
164 24V		
right side	1.8	1.8
left side	.2	.2
Ignition Switch, Replace (B)		
164	1.2	1.2
Milano	.8	.8
Ignition Timing, Reset (B)		
All Models	.3	.3

	LABOR TIME	SEVERE SERVICE
Spark Plug (Ignition) Cables, Replace (B)		
164, Milano	.7	.8
Spark Plugs, Replace (B)		
164	.9	1.0
164 24V	2.5	2.6
Milano	.8	.9
EMISSIONS		
Coolant Temperature (ECT) Sensor and/or Switch, Replace (B)		
164, Milano	.3	.3
Mass Air Flow (MAF) Sensor, Replace (B)		
All Models	.5	.5
Thermo Time Switch, Replace (B)		
Milano	.3	.3
Timing Pickup Sensor, Replace (B)		
1991-93 164	.6	.6
FUEL		
DELIVERY		
Electronic Control Unit, Replace (B)		
Milano	.6	.6
Fuel Gauge (Tank), Replace (B)		
164	.7	.9
Milano	1.2	1.4
Fuel Pump, Replace (B)		
164 in tank	.7	.9
Milano	.6	.8
Fuel Pump Relay, Replace (B)		
164	.3	.3
Fuel Tank, Replace (B)		
Includes: Drain and refill.		
164	2.2	2.4
Milano	1.8	2.0
Intake Manifold and/or Gasket, Replace (B)		
Includes: Adjustments.		
164 each	1.6	1.9
164 24V each	1.8	2.1
Milano	2.6	2.9
INJECTION		
Air Valve, Replace (B)		
All Models	.5	.5
Cold Start Injector, Replace (B)		
Milano	.4	.4
Fuel Pressure Regulator, Replace (B)		
All Models	.4	.5
Throttle Body Assy., Replace (B)		
164	1.0	1.0
Milano	1.2	1.2

164 : MILANO AR-3

	LABOR TIME	SEVERE SERVICE
EXHAUST		
Catalytic Converter, Replace (B)		
164	.7	.8
Milano	.4	.5
Center Exhaust Pipe, Replace (B)		
164	.6	.7
Exhaust Manifold or Gasket, Replace (B)		
164		
front	.8	1.1
rear	.9	1.2
Milano each	1.0	1.3
Front Exhaust Pipe, Replace (B)		
164	.8	.9
ENGINE COOLING		
Coolant Thermostat, Replace (B)		
All Models	.5	.7
Radiator Assy., R&R or Replace (B)		
All Models	1.0	1.1
Radiator Fan Motor Relay, Replace (B)		
All Models	.3	.3
Temperature Gauge (Dash), Replace (B)		
Milano	1.1	1.1
Water Pump and/or Gasket, Replace (B)		
Includes: Refill with proper coolant mix.		
164	6.2	6.5
164 24V	7.5	7.8
Milano	5.3	5.6
ENGINE		
ASSEMBLY		
Times shown are for OEM assemblies. Time to replace assemblies from aftermarket rebuilders may vary.		
Engine Assy., R&I (B)		
Does not include parts or component transfer.		
164	8.5	8.5
164 24V	11.7	11.7
Milano	9.5	9.5
Engine Assy., R&R and Recondition (A)		
Includes: Replacing rings, rod and main bearings, cylinder head reconditioning and engine tune-up.		
164	35.9	35.9
164 24V	45.2	45.2
Milano	38.8	38.8
Engine Mounts, Replace (B)		
164		
front	1.1	1.3
rear	1.3	1.5
Milano		
right side	2.1	2.3
left side	1.8	2.0

	LABOR TIME	SEVERE SERVICE
CYLINDER HEAD		
Compression Test (B)		
164	1.1	1.2
164 24V	3.3	3.4
Milano	.9	1.0
Cylinder Head, Replace (B)		
Includes: Parts transfer, adjustments.		
164		
right side	8.4	8.9
left side	7.5	8.0
both sides	9.3	9.8
164 24V		
front	10.2	10.7
rear	10.9	11.4
both	15.8	16.3
Milano		
right side	5.9	6.4
left side	8.1	8.6
both sides	10.3	10.8
Cylinder Head Gasket, Replace (B)		
164		
right side	7.4	7.7
left side	6.5	6.8
both sides	8.3	8.6
164 24V		
front	9.8	10.1
rear	10.5	10.8
both	14.8	15.1
Milano		
right side	4.9	5.2
left side	7.1	7.4
both sides	9.3	9.6
Valve Cover Gasket, Replace (B)		
164		
front	.4	.5
rear	1.6	1.7
Milano		
front	.5	.6
rear	.8	.9
Valve Job (A)		
164	15.8	16.3
164 24V	17.7	18.2
Milano	16.9	17.4
Valve Springs and/or Oil Seals, Replace (B)		
164	12.3	12.5
164 24V	15.7	15.9
Milano	13.3	13.5
CAMSHAFT		
Camshaft, Replace (B)		
164		
one side	1.6	1.8
both sides	3.2	3.4
164 24V		
left side		
one	5.5	5.7
both	6.0	6.2

	LABOR TIME	SEVERE SERVICE
Milano		
one side	2.9	3.1
both sides	3.3	3.5
Camshaft Cover Gaskets, Replace (B)		
164		
front	.3	.4
rear	1.4	1.5
Milano		
front	.4	.5
rear	.7	.8
Engine Front Cover, Replace (B)		
164	5.5	5.7
164 24V	6.5	6.7
Milano	4.9	5.1
Timing Belt, Replace (B)		
164	4.7	5.0
164 24V	5.3	5.6
right side		
one	5.3	5.6
both	5.9	6.2
both sides all	10.2	10.5
Milano	2.9	3.2
Timing Belt Tensioner, Replace (B)		
164	3.6	3.9
164 24V	5.0	5.3
Milano	2.9	3.2
Timing Belt/Chain Cover, Replace (B)		
164	1.8	2.0
164 24V	1.6	1.8
Milano	.5	.6
CRANK & PISTONS		
Crankshaft Front Oil Seal, Replace (A)		
164	5.0	5.2
164 24V	5.7	5.9
Milano	3.9	4.1
Crankshaft Pulley, Replace (B)		
1991-93 164	1.6	1.7
Milano	2.4	2.5
Pistons, Rings and Connecting Rods (A)		
Includes: Cylinder liners.		
164	20.5	21.4
164 24V, Milano	22.0	22.9
Rear Main Oil Seal, Replace (A)		
164	8.4	9.0
Milano	11.3	11.9
ENGINE LUBRICATION		
Oil Pan and/or Gasket, Replace (B)		
164	9.8	10.0
Milano	11.0	11.2
Oil Pressure Gauge (Dash), Replace (B)		
Milano	1.0	1.0
Oil Pump, Replace (B)		
164	10.1	10.6
Milano	11.5	12.0

AR-4 164 : MILANO

	LABOR TIME	SEVERE SERVICE
CLUTCH		
Clutch Hydraulic System, Bleed (B)		
All Models	.5	.5
Clutch Assy., Replace (B)		
Milano	4.3	4.5
164	7.5	7.7
Clutch Master Cylinder, Replace (B)		
Does not include system bleeding.		
All Models	1.2	1.3
Clutch Release Bearing, Replace (B)		
164	7.2	7.4
Milano	4.0	4.2
Clutch Slave Cylinder, Replace (B)		
Does not include system bleeding.		
164	.5	.6
Milano	.3	.4
Flywheel, Replace (B)		
164	8.1	8.3
Milano	10.2	10.4
MANUAL TRANSAXLE		
Transaxle Assy. R&R and Recondition (A)		
Includes: Reset toe.		
164	11.0	11.9
Transaxle Assy., R&I (B)		
Includes: Reset toe.		
164	7.0	7.2
MANUAL TRANSMISSION		
Electronic Speedometer Pulse Generator, Replace (B)		
Milano	5.4	5.4
Transmission Assy., R&R and Recondition (A)		
Milano	13.6	14.5
Transmission Assy., R&R or Replace (B)		
Milano	4.9	5.1
AUTOMATIC TRANSAXLE		
SERVICE TRANSAXLE INSTALLED		
Drain & Refill Unit (B)		
164	.8	.8
Governor Assy., R&R or Replace (B)		
164	1.9	2.0
Oil Pan and/or Gasket, Replace (B)		
164	.8	.8
Speedometer Drive Gear, Replace (B)		
164	1.0	1.0
Valve Body, Replace (B)		
164	1.2	1.5

	LABOR TIME	SEVERE SERVICE
SERVICE TRANSAXLE REMOVED		
Transaxle R&R included unless otherwise noted.		
Torque Converter Front Oil Seal, Replace (B)		
164	7.3	7.5
Torque Converter, Replace (B)		
164	7.2	7.4
Transaxle Assy., R&R or Replace (B)		
164	6.9	7.1
AUTOMATIC TRANSMISSION		
SERVICE TRANSMISSION INSTALLED		
Check Unit for Oil Leaks (C)		
All Models	.5	.5
Drain and Refill Unit (B)		
All Models	.6	.6
Oil Pan and/or Gasket, Replace (B)		
All Models	.6	.6
Oil Pressure Check (B)		
All Models	.9	.9
SERVICE TRANSMISSION REMOVED		
Transmission R&R included unless otherwise noted.		
Transmission Assy., R&R (B)		
Milano	5.1	6.0
DRIVELINE		
DRIVESHAFT		
Center Support Bearing, Replace (B)		
Milano	3.5	3.7
Driveshaft, Replace (B)		
Milano	2.4	2.6
Driveshaft Flex Couplings, Replace (B)		
Milano	3.5	3.7
DRIVE AXLE		
CV Joint, Replace (B)		
1991-93 164	1.5	1.7
Replace boot add	.2	.3
Differential Assy., R&I (B)		
Milano	3.0	3.5
Differential Assy., R&R and Recondition (A)		
Milano	5.8	6.7
Differential Output Shaft, Replace (B)		
Milano	2.9	3.1
Replace inner bearing add	.3	.4

	LABOR TIME	SEVERE SERVICE
Halfshaft Boot, Replace (B)		
Milano	1.7	1.8
Intermediate Halfshaft, Replace (B)		
Milano, 164	1.2	1.4
Replace bearings add	.3	.4
Intermediate Halfshaft Support, Replace (B)		
164	2.2	2.4
Rear Halfshaft, R&R or Replace (B)		
164	1.2	1.4
Replace inner CV-joint add	.3	.4
Rear Hub Assy., Replace (B)		
Milano	1.7	1.9
Rear Wheel Bearing, Replace (B)		
164	.7	.9
Milano	2.0	2.1
BRAKES		
ANTI-LOCK		
The following operations do not include testing. Add time as required.		
Diagnose Anti-Skid Brake System (A)		
All Models	1.0	1.0
Accumulator Assy., Replace (B)		
Milano	.4	.4
Anti-Lock Relay, Replace (B)		
164	.4	.4
Brake Master Cylinder, Replace (B)		
Milano	1.2	1.3
Front Sensor Assy., Replace (B)		
One sides	.4	.4
Both sides	.6	.6
Front Sensor Rings, Replace (B)		
Milano		
one side	.8	.8
both sides	1.3	1.3
Hydraulic Assy., Replace (B)		
164	1.2	1.2
Milano	1.7	1.7
Add for system bleeding.		
Load Sensing Proportioning Valve, Replace (B)		
Milano	.8	.9
Pump Solenoid Valve, Replace (B)		
Milano	.3	.3
Rear Sensor Assy., Replace (B)		
One side	.6	.6
Both sides	.9	.9
Rear Sensor Rings, Replace (B)		
Milano		
one side	.8	.8
both sides	1.3	1.3

164 : MILANO — AR-5

SYSTEM	LABOR TIME	SEVERE SERVICE
Bleed Brakes (B)		
Includes: Add fluid.		
w/ABS	1.0	1.0
w/o ABS	.6	.6
Brake Hose (Flexible), Replace (B)		
Includes: System bleeding.		
One	.4	.5
each addl. add	.3	.4
Brake Pressure Regulator, Replace (B)		
All Models	.4	.5
Master Cylinder, R&R and Rebuild (B)		
All Models	1.3	1.6
Master Cylinder, Replace (B)		
All Models	.8	.9
Power Booster Unit, Replace (B)		
164	1.8	1.9
Milano	2.5	2.6

SERVICE BRAKES

	LABOR TIME	SEVERE SERVICE
Caliper Assy., R&R and Recondition (B)		
Does not include system bleeding.		
Front		
164		
one side	.9	1.2
both sides	1.6	1.9
Milano		
one side	1.4	1.7
both sides	2.4	2.7
Rear		
164		
one side	1.3	1.6
both sides	2.1	2.4
Milano		
one side	4.6	4.9
both sides	7.6	7.9
Caliper Assy., Replace (B)		
Does not include system bleeding.		
Front		
164		
one side	.5	.6
both sides	.9	1.0
Milano		
one side	.7	.8
both sides	1.3	1.4
Rear		
164		
one side	.8	.9
both sides	1.1	1.2
Milano		
one side	2.3	2.4
both sides	3.9	4.0
Disc Brake Rotor, Replace (B)		
Front		
164		
one side	.5	.6
both sides	.8	.9

	LABOR TIME	SEVERE SERVICE
Milano		
one side	1.1	1.2
both sides	1.8	1.9
Rear		
164		
one side	.9	1.0
both sides	1.6	1.7
Milano		
one side	2.2	2.3
both sides	3.5	3.6
Pads, Replace (B)		
Front	.8	.9
Rear		
164	1.0	1.1
Milano	1.2	1.3

PARKING BRAKE
Parking Brake Apply Actuator, Replace (B)		
164	.8	.8
Milano	1.2	1.2
Parking Brake Cable, Adjust (C)		
164	.3	.4
Milano	.5	.6

FRONT SUSPENSION
FRONT WHEEL DRIVE
Unless otherwise noted, time given does not include alignment.

	LABOR TIME	SEVERE SERVICE
Front Toe, Adjust (B)		
164	.6	.8
Control Arm, Replace (B)		
164	.9	1.2
Replace bushings add	.3	.5
Hub Bearings, Replace (B)		
164	.9	1.1
Stabilizer Bar, Replace (B)		
164	1.6	1.8
Strut, Replace (B)		
164		
one side	1.3	1.4
both sides	2.5	2.6
Strut Coil Spring, Replace (B)		
164	1.0	1.2

REAR WHEEL DRIVE
Unless otherwise noted, time given does not include alignment.

	LABOR TIME	SEVERE SERVICE
Align Front End (A)		
Milano	2.7	2.9
Front Toe, Adjust (B)		
Milano	.6	.8
Front Shock Absorbers, Replace (B)		
Milano		
one side	.7	.8
both sides	1.3	1.4

	LABOR TIME	SEVERE SERVICE
Hub Bearings or Seal, Replace (B)		
Milano		
w/ABS	1.4	1.6
w/o ABS	1.2	1.4
Lower Ball Joint, Replace (B)		
Milano	1.2	1.4
Lower Control Arm Assy., Replace (B)		
Milano	2.0	2.3
Stabilizer Bar Bushings, Replace (B)		
Milano	.6	.7
Stabilizer Bar, Replace (B)		
All Models	.5	.6
Steering Knuckle, Replace (B)		
Milano		
w/ABS	2.9	3.1
w/o ABS	2.7	2.9
Torsion Bar, Replace (B)		
Includes: Height adjustment.		
Milano	2.0	2.3
Upper Control Arm Assy., Replace (B)		
Milano	2.4	2.6

REAR SUSPENSION

	LABOR TIME	SEVERE SERVICE
Rear Shock Absorbers or Bushings, Replace (B)		
Milano		
one	.7	.7
both	1.0	1.1
Rear Spring, Replace (B)		
164	1.4	1.6
Milano	1.9	2.1
Rear Stabilizer Bar & Bushings, Replace (B)		
164	.5	.6
Milano	1.4	1.6
Rear Struts, R&R or Replace (B)		
164		
one	1.4	1.5
both	2.4	2.5

AUTOMATIC SUSPENSION
	LABOR TIME	SEVERE SERVICE
Accumulators, Replace (B)		
164, Milano		
one	.5	.6
both	.8	.9
Level Control Valve, Replace (B)		
164	.6	.6
Milano	.8	.8
Rear Shock Absorbers, Replace (B)		
164		
one	1.8	1.9
both	2.6	2.7
Milano		
one	.9	1.0
both	1.4	1.5

AR-6 164 : MILANO

STEERING

Unless otherwise noted, time given does not include alignment.

AIR BAGS

	LABOR TIME	SEVERE SERVICE
Diagnostic Module, Replace (B)		
All Models	.3	.3
Dual Pole Arming Sensor, Replace (B)		
All Models	.2	.2
Electronic Control Unit, Replace (B)		
164	.5	.5

POWER RACK & PINION

	LABOR TIME	SEVERE SERVICE
Rack & Pinion Assy., R&R or Replace (B)		
164	4.4	4.7
Milano	2.3	2.6
Steering Column, Replace (B)		
164	1.2	1.2
Milano	1.5	1.5
Steering Pump, R&R or Replace (B)		
164, Milano	1.3	1.4
164 24V	1.5	1.6
Steering Wheel, Replace (B)		
164, Milano	.5	.5
Tie Rod Ends, Replace (B)		
164, Milano	.5	.7

HEATING & AIR CONDITIONING

When more than one component requires replacement where evacuation/recovery and recharging is already included, deduct 1.0 hour for each additional component from the time given.

	LABOR TIME	SEVERE SERVICE
Evacuate/Recover and Recharge System (B)		
All Models	1.5	1.5
Compressor Assy., Replace (B)		
164	1.6	1.8
Milano	1.9	2.1

Add time to recharge system if needed.

	LABOR TIME	SEVERE SERVICE
Condenser Assy., Replace (B)		
164	.7	.8
Milano	1.9	2.0

Add time to recharge system if needed.

	LABOR TIME	SEVERE SERVICE
Evaporator Electric Fan, Replace (B)		
Milano	4.0	4.2
Expansion Valve, Replace (B)		
164	.6	.6
Milano	1.0	1.0

Add time to recharge system if needed.

	LABOR TIME	SEVERE SERVICE
Heater Blower Motor, Replace (B)		
164	2.2	2.2
Milano	4.6	4.6
Heater Blower Motor Switch, Replace (B)		
Milano	.6	.6
Heater Control Valve, Replace (B)		
Milano	3.2	3.2
Heater Core, R&R or Replace (B)		
164	2.7	2.7
Milano	4.4	4.4
Heater Hoses, Replace (B)		
164	2.3	2.4
Milano	1.2	1.3
Receiver/Drier Assy., Replace (B)		
All Models	.6	.8

Add time to recharge system if needed.

WIPERS & SPEEDOMETER

	LABOR TIME	SEVERE SERVICE
Pulse Generator, Replace (B)		
All Models	.4	.4
Windshield Washer Pump, Replace (B)		
164	.6	.6
Milano	.3	.3
Windshield Wiper Motor, Replace (B)		
All Models	.6	.8

LAMPS & SWITCHES

	LABOR TIME	SEVERE SERVICE
Back-Up Lamp Switch, Replace (B)		
All Models	.4	.4
Front Side Marker Lamp, Replace (B)		
All Models	.2	.2
Halogen Headlight Bulb, Replace (B)		
All Models	.2	.2
Hazard Warning Switch, Replace (B)		
164	.5	.5
Milano	.2	.2
Headlight Sealed Beam, Replace (B)		
All Models	.3	.3
Headlamps, Aim (B)		
All Models	.3	.3
Horn, Replace (B)		
All Models	.4	.4
Horn Relay, Replace (B)		
All Models	.3	.3
Multifunction Switch Assy., Replace (B)		
All Models	.9	.9
Parking Brake Warning Lamp Switch, Replace (B)		
Milano	.3	.3
Rear Side Marker Lamp, Replace (B)		
All Models	.3	.3
Stop Light Switch, Replace (B)		
All Models	.3	.3
Tail Lamp Lens or Bulb, Replace (B)		
One	.2	.2
each addl. add	.1	.1
Turn Signal or Hazard Warning Flasher, Replace (B)		
All Models	.2	.2
Turn Signal Switch, Replace (B)		
Milano	.5	.5
Windshield Wiper Switch, Replace (B)		
Milano	.8	.8

AUD

80 : 90 : 100 : 200 : 4000 : 5000 : A4 : A6 : A8 : Cabriolet : Coupe : S4 : S6 : TT : V8 Quattro

SYSTEM INDEX

- MAINTENANCE AUD-2
  - Maintenance Schedule .. AUD-2
- CHARGING AUD-2
- STARTING AUD-3
- CRUISE CONTROL AUD-3
- IGNITION AUD-4
- EMISSIONS AUD-4
- FUEL AUD-5
- EXHAUST AUD-6
- ENGINE COOLING AUD-7
- ENGINE AUD-8
  - Assembly AUD-8
  - Cylinder Head AUD-9
  - Camshaft AUD-10
  - Crank & Pistons AUD-11
  - Engine Lubrication AUD-12
- CLUTCH AUD-13
- MANUAL TRANSAXLE ... AUD-13
- AUTO TRANSAXLE AUD-14
- SHIFT LINKAGE AUD-14
- DRIVELINE AUD-15
- BRAKES AUD-15
- FRONT SUSPENSION AUD-16
- REAR SUSPENSION AUD-18
- STEERING AUD-19
- HEATING & AC AUD-20
- WIPERS & SPEEDOMETER AUD-21
- LAMPS & SWITCHES AUD-22
- BODY AUD-22

OPERATIONS INDEX

A
- AC Hoses AUD-20
- Air Bags AUD-19
- Air Conditioning AUD-20
- Alignment AUD-16
- Alternator (Generator) AUD-2
- Antenna AUD-21
- Anti-Lock Brakes AUD-15

B
- Back-Up Lamp Switch AUD-22
- Ball Joint AUD-16
- Battery Cables AUD-3
- Bleed Brake System AUD-15
- Blower Motor AUD-20
- Brake Disc AUD-16
- Brake Drum AUD-16
- Brake Hose AUD-15
- Brake Pads and/or Shoes .. AUD-16

C
- Camshaft AUD-10
- Camshaft Sensor AUD-4
- Catalytic Converter AUD-6
- Coolant Temperature (ECT) Sensor AUD-4
- Crankshaft AUD-11
- Cruise Control AUD-3
- CV Joint AUD-15
- Cylinder Head AUD-9

D
- Differential AUD-15
- Distributor AUD-4
- Drive Belt AUD-2

E
- EGR AUD-4
- Electronic Control Module (ECM/PCM) AUD-4
- Engine AUD-8
- Engine Lubrication AUD-12
- Engine Mounts AUD-9
- Evaporator AUD-20
- Exhaust AUD-6
- Exhaust Manifold AUD-6
- Expansion Valve AUD-20

F
- Flywheel AUD-13
- Fuel Injection AUD-6
- Fuel Pump AUD-5
- Fuel Vapor Canister AUD-5

G
- Gear Selector Lever AUD-14
- Generator AUD-2
- Glow Plug AUD-4

H
- Halfshaft AUD-15
- Headlamp AUD-22
- Heater Core AUD-21
- Horn AUD-22

I
- Idle Air Control (IAC) Valve . AUD-6
- Ignition Coil AUD-4
- Ignition Module AUD-4
- Ignition Switch AUD-4
- Injection Pump AUD-6
- Inner Tie Rod AUD-19
- Intake Air Temperature (IAT) Sensor AUD-4
- Intake Manifold AUD-5

K
- Knock Sensor AUD-4

L
- Lower Control Arm AUD-17

M
- Maintenance Schedule ... AUD-2
- Mass Air Flow (MAF) Sensor AUD-4
- Master Cylinder AUD-16
- Muffler AUD-7
- Multifunction Lever/Switch .. AUD-22

N
- Neutral Safety Switch AUD-22

O
- Oil Pan AUD-12
- Oil Pump AUD-13
- Outer Tie Rod AUD-19
- Oxygen Sensor AUD-5

P
- Parking Brake AUD-16
- Pistons AUD-11
- Positive Crankcase Ventilation (PCV) Valve ... AUD-5

R
- Radiator AUD-7
- Radiator Hoses AUD-7
- Radio AUD-21
- Rear Main Oil Seal AUD-12

S
- Shock Absorber/Strut, Front AUD-17
- Shock Absorber/Strut, Rear AUD-18
- Spark Plug Cables AUD-4
- Spark Plugs AUD-4
- Spring, Front Coil AUD-17
- Spring, Rear Coil AUD-18
- Starter AUD-3
- Steering Wheel AUD-20

T
- Thermostat AUD-7
- Throttle Body AUD-6
- Throttle Position Sensor (TPS) AUD-5
- Timing Belt AUD-11
- Torque Converter AUD-14

V
- Valve Body AUD-14
- Valve Cover Gasket ... AUD-10
- Valve Job AUD-10
- Vehicle Speed Sensor ... AUD-5

W
- Water Pump AUD-8
- Wheel Balance AUD-2
- Wheel Cylinder AUD-16
- Window Regulator AUD-22
- Windshield Washer Pump .. AUD-21
- Windshield Wiper Motor ... AUD-21

AUD-1

AUD-2
80 : 90 : 100 : 200 : 4000 : 5000 : A4 : A6 : A8 : CABRIOLET : COUPE : S4 : S6 : TT : V8 QUATTRO

	LABOR TIME	SEVERE SERVICE
MAINTENANCE		
Air Cleaner Filter Element, Service (B)		
All Models	.3	.4
Chassis Lubrication, Change Oil & Filter (C)		
Includes: Correct fluid levels.		
All Models	.3	.4
Drive Belt, Adjust (B)		
1981-92 V-belt	1.2	1.3
1987-01 V8	.5	.6
1990-01 serpentine	.8	.8
Drive Belt, Replace (B)		
1990-01 serpentine	1.4	1.5
Injection pump		
4000, 5000, Coupe	1.6	1.7
Fuel Filter, Replace (B)		
Continuous injection	.5	.5
Electronic injection	.3	.3
Halogen Headlamp Bulb, Replace (B)		
1988-01 each	.3	.3
Oil & Filter, Change (C)		
Includes: Correct fluid levels.		
All Models	.3	.4
Sealed Beam Headlamp, Replace (B)		
1981-87		
one side	.5	.5
both sides	.7	.7
Timing Belt, Replace (B)		
80, 90, A4		
1988-92		
4 cyl.	2.3	2.6
5 cyl.	2.7	3.0
1993-01	2.0	2.3
100, 200, A6		
1989-91	3.1	3.3
1992-97	2.0	2.2
1998-01	3.1	3.3
4000, 5000, Coupe		
gasoline	2.5	2.8
diesel	2.7	3.0
A8	6.5	6.7
Coupe Quattro	2.7	2.9
S4, S6, Cabriolet		
5 cyl.	2.0	2.2
V6	2.0	2.2
twin turbo	4.1	4.3
TT	2.7	2.9
V8 Quattro	5.2	5.4
Tire, Replace (C)		
Includes: Dismount old tire and mount new tire to rim.		
All Models	.5	.5
Tires, Rotate (C)		
All Models	.5	.5
Wheel, Balance (B)		
One	.3	.3
each addl. add	.2	.2

	LABOR TIME	SEVERE SERVICE
SCHEDULED MAINTENANCE INTERVALS		
If necessary, refer to appropriate Chilton maintenance service information.		
7,500 Mile Service (C)		
A4	1.1	1.1
A6, A8, S6	1.3	1.3
Cabriolet	1.1	1.1
15,000 Mile Service (B)		
A4	1.9	1.9
A6	2.3	2.3
A8	1.8	1.8
Cabriolet, S6	2.1	2.1
22,500 Mile Service (C)		
A4, A6, A8, S6	.9	.9
Cabriolet	.6	.6
30,000 Mile Service (B)		
A4	2.4	2.4
A6, S6	2.7	2.7
A8	2.8	2.8
Cabriolet	2.5	2.5
37,500 Mile Service (C)		
A4, A6, A8, S6	.9	.9
Cabriolet	.6	.6
45,000 Mile Service (B)		
A4	1.6	1.6
A6	1.8	1.8
A8	2.1	2.1
Cabriolet, S6	1.8	1.8
52,500 Mile Service (C)		
A4	1.1	1.1
A6, A8, S6	1.3	1.3
Cabriolet	1.1	1.1
60,000 Mile Service (B)		
A4	2.4	2.4
A6, S6	2.7	2.7
A8	2.8	2.8
Cabriolet	2.5	2.5
67,500 Mile Service (C)		
A4, A6, A8, S6	.9	.9
Cabriolet	.6	.6
75,000 Mile Service (B)		
A4	1.9	1.9
A6, Cabriolet	2.1	2.1
A8	1.8	1.8
S6	2.3	2.3
82,500 Mile Service (C)		
A4, A6, A8, S6	.9	.9
Cabriolet	.6	.6
90,000 Mile Service (B)		
A4	2.9	2.9
A6, S6	3.2	3.2
A8	3.3	3.3
Cabriolet	2.8	2.8
Replace timing belt		
2.8L add	1.5	1.5

	LABOR TIME	SEVERE SERVICE
97,500 Mile Service (C)		
A4	1.1	1.1
A6, A8, S6	1.3	1.3
Cabriolet	1.1	1.1
CHARGING		
Alternator Circuits, Test (B)		
Includes: Test component output.		
All Models	.5	.5
Alternator, R&R and Recondition (B)		
80, 90	2.8	3.3
4000, 5000		
gasoline	3.7	4.2
diesel	3.3	3.8
Alternator Assy., Replace (B)		
Includes: Pulley transfer.		
80, 90, A4		
1988-91		
4 cyl.	.8	.9
5 cyl.	1.3	1.4
1992	1.3	1.4
1993-97	2.3	2.4
1998-01	3.5	3.6
100, 200, A6		
1989-91	1.3	1.4
1992-97	2.3	2.4
1998-01	1.3	1.4
4000, 5000		
gasoline	2.4	2.5
diesel	1.9	2.0
A8	1.5	1.6
Cabriolet	2.3	2.4
Coupe Quattro	1.3	1.4
S4, S6	2.3	2.4
TT	1.9	2.0
V8 Quattro	2.3	2.4
Alternator Bearings, Replace (B)		
80, 90		
4 cyl.	1.2	1.4
5 cyl.	1.5	1.7
100, 200	1.5	1.7
4000, 5000, Coupe		
gasoline	2.6	2.8
diesel	2.2	2.4
Coupe Quattro	1.5	1.7
Alternator Voltage Regulator, Replace (B)		
80, 90		
1988-91		
4 cyl.	1.2	1.4
5 cyl.	1.5	1.7
1992	1.5	1.7
100, 200	1.5	1.7
4000, 5000, Coupe		
gasoline	2.9	3.1
diesel	2.5	2.7
Coupe Quattro	1.5	1.7
4000 external	.3	.3

80 : 90 : 100 : 200 : 4000 : 5000 : A4 : A6 : A8 : CABRIOLET : COUPE : S4 : S6 : TT : V8 QUATTRO **AUD-3**

	LABOR TIME	SEVERE SERVICE
Voltmeter and/or Ammeter Gauge, Replace (B)		
1988-92 80, 90	1.3	1.3

STARTING

Starter Draw Test (On Car) (B)
All Models .... .3 .3
Battery, Replace (C)
All Models .... .6 .7
w/R&R rear seat add .... .2 .2
Battery Cables, Replace (C)
Positive
  80, 90, A4
    1988-92 .... 1.0 1.1
    1993-01 .... 4.8 5.0
  100, 200, A6
    1989-91 .... 1.0 1.1
    1992-01 .... 4.8 5.0
  4000, Coupe 1.0 1.1
  5000 .... 2.8 2.9
  A8 .... 4.8 5.0
  Cabriolet 4.8 5.0
  Coupe Quattro 8   .9
  S4, S6 .... 4.8 5.0
  TT 2.1 2.2
  V8 Quattro 4.8 5.0
Negative
  1981-96 3   .4
  1997-01
    A4 3   .4
    A6 3   .4
    A8 .... 1.3 1.5
    Cabriolet 3   .4
    TT 1.1 1.2
Starter, R&R and Recondition (A)
80, 90
  1988-91
    AT 4.9 5.4
    MT 4.1 4.6
  1992
    AT 4.9 5.4
    MT 4.1 4.6
4000, Coupe
  gasoline 3.1 3.6
  diesel 3.8 4.3
5000 .... 3.5 4.0
Coupe Quattro 4.1 4.6
Starter Assy., Replace (B)
80, 90, A4
  1988-95
    AT 3.0 3.1
    MT 2.4 2.5
  1996-01
    AT 3.3 3.4
    MT 2.4 2.5

	LABOR TIME	SEVERE SERVICE
100, 200, A6		
1989-91		
AT	3.0	3.1
MT	1.3	1.4
1992-97		
AT	3.0	3.1
MT	2.4	2.5
1998-01		
AT	3.3	3.4
MT	2.4	2.5
4000, Coupe		
gasoline	1.3	1.4
diesel	1.9	2.0
5000	1.6	1.7
A8	3.5	3.6
Cabriolet	2.4	2.5
Coupe Quattro	2.4	2.5
S4, S6		
AT	3.0	3.1
MT	2.4	2.5
TT	1.4	1.5
V8 Quattro	1.3	1.4

Starter Drive Assy., Replace (A)
Includes: Starter R&R.
80, 90, A4
  1988-92
    AT 3.1 3.4
    MT 2.7 3.0
  1993-95
    AT 3.2 3.5
    MT 2.5 2.8
  1996-01
    AT 3.6 3.9
    MT 2.7 3.0
100, 200, A6
  1989-91
    AT 3.1 3.4
    MT 1.6 1.9
  1992-97
    AT 3.2 3.5
    MT 2.5 2.8
  1998-01
    AT 3.6 3.9
    MT 2.7 3.0
4000, Coupe
  gasoline 1.8 2.1
  diesel 2.5 2.8
5000 .... 2.2 2.5
Coupe Quattro 2.7 3.0
S4, S6, Cabriolet
  AT 3.2 3.5
  MT 1.4 1.7
V8 Quattro 1.6 1.9

	LABOR TIME	SEVERE SERVICE
Starter Solenoid and/or Switch, Replace (B)		
Includes: Starter R&R.		
80, 90, A4		
1988-95		
AT	3.2	3.4
MT	2.6	2.8
1996-01		
AT	3.5	3.7
MT	2.6	2.8
100, 200, A6		
1989-91		
AT	3.2	3.4
MT	1.5	1.7
1992-97		
AT	3.2	3.4
MT	2.6	2.8
1998-01		
AT	3.5	3.7
MT	2.6	2.8
4000, Coupe		
gasoline	1.5	1.7
diesel	2.2	2.4
5000	1.8	2.0
A8	3.7	3.9
Cabriolet	2.6	2.8
Coupe Quattro	2.6	2.8
S4, S6		
AT	3.2	3.4
MT	2.6	2.8
V8 Quattro	1.5	1.7
Voltmeter, Replace (B)		
4000, Coupe	.8	.8

CRUISE CONTROL

Control Combination Switch, Replace (B)
1984-01
  w/air bags 2.4 2.4
  w/o air bags 1.3 1.3
Control Controller Module, Replace (B)
1984-87 4000, Coupe . 2.4 2.4
1988-92
  100, 200, 5000 ... 1.3 1.3
1988-01 .... 2.4 2.4
Control Transmitter, Replace (B)
Includes: R&R instrument cluster.
1984-01
  w/air bags 3.0 3.0
  w/o air bags 1.8 1.8
Control Vacuum Control or Check Valve, Replace (B)
1984-01 .... 1.3 1.3
Control Vacuum Pump, Replace (B)
1984-01 .... 1.0 1.3
Control Vacuum Servo, Replace (B)
1984-01 .... 1.0 1.1

AUD-4   80 : 90 : 100 : 200 : 4000 : 5000 : A4 : A6 : A8 : CABRIOLET : COUPE : S4 : S6 : TT : V8 QUATTRO

	LABOR TIME	SEVERE SERVICE
IGNITION		
Diagnose Ignition System Component Each (A)		
All Models	.8	.8
Ignition Timing, Reset (B)		
1981-92	.5	.5
Camshaft Top Dead Center Sensor, Replace (B)		
1988-92	1.6	1.8
DIS Control Unit, Replace (B)		
1985-01	.7	.9
Distributor, Replace (B)		
Includes: Reset base ignition timing.		
1981-95 4 cyl., 5 cyl.	.8	.9
1990-94 V8 one or both	3.0	3.2
1989-95 20V turbo Quattro	2.4	2.6
Distributor Cap and/or Rotor, Replace (B)		
1981-87 all	.5	.5
80, 90, A4		
1988-92	.5	.5
1993-01	1.2	1.2
100, 200, A6		
1989-91	.5	.5
1992-01	1.2	1.2
A8, S4, S6, Cabriolet	1.2	1.2
V8 Quattro		
one	1.2	1.2
both	1.7	1.7
Glow Plug Relay, Replace (B)		
Diesel	.5	.6
Glow Plugs, Replace (B)		
Diesel		
4000	.8	.9
5000	1.0	1.1
Hall Generator, Replace (B)		
1981-87	1.0	1.1
1988-92 80, 90	1.6	1.6
Ignition Coil, Replace (B)		
All Models	.8	.8
Ignition Coil Resistor/Ballast, Replace (B)		
1981-85	.5	.5
Ignition Lock Cylinder and/or Buzzer Switch, Replace (B)		
w/air bags		
80, 90, A4		
1988-92	2.7	2.7
1993-95	2.8	2.8
1996-01	3.0	3.0
100, 200, A6		
1989-95	6.8	6.8
1996-01	2.4	2.4

	LABOR TIME	SEVERE SERVICE
A8	1.8	1.8
Cabriolet	2.4	2.4
Coupe Quattro	2.7	2.7
S4, S6	6.8	6.8
TT	2.7	2.7
V8 Quattro	6.8	6.8
w/o air bags		
1988-92 80, 90	1.4	1.4
1989-91 100, 200	1.4	1.4
4000, Coupe	1.4	1.4
5000		
1981-83	1.4	1.4
1984-88	2.3	2.3
1984-88 Quattro	1.4	1.4
Ignition Module Assy., Replace (B)		
1988-01	.7	.7
Ignition Switch, Replace (B)		
w/air bags		
80, 90, Cabriolet	2.4	2.4
A4	1.2	1.2
A6, S4, S6		
1992-97	2.4	2.4
1998-01	1.2	1.2
A8	2.4	2.4
TT	1.5	1.5
w/o air bags		
1988-92 80, 90	1.7	1.7
1989-91 100, 200	2.4	2.4
4000, Coupe	1.0	1.0
5000		
1981-88	1.3	1.3
1983-85 Quattro Coupe	1.3	1.3
1984-88 Quattro	2.3	2.3
Coupe Quattro	1.7	1.7
Spark Plug (Ignition) Cables, Replace (B)		
1981-87	.5	.7
1988-01	1.2	1.4
Spark Plugs, Replace (B)		
4 cyl.	.5	.6
5 cyl.	.7	.8
V6	1.3	1.4
V8	1.7	1.8
Vacuum Advance Unit, Replace (B)		
Includes: Reset base ignition timing.		
1981-92	1.2	1.2

EMISSIONS

The following operations do not include testing. Add time as required.

	LABOR TIME	SEVERE SERVICE
Dynamometer Test (A)		
All Models	.5	.5
Air Flow Sensor, Replace (B)		
1981-87	1.2	1.2
1988-92	.8	.8
Air Temperature Sensor, Replace (B)		
1989-01	1.4	1.4

	LABOR TIME	SEVERE SERVICE
Coolant Temperature (ECT) Sensor, Replace (B)		
w/continuous injection		
1981-92	.5	.5
w/o continuous injection		
4 cyl.	.5	.5
5 cyl.	.5	.5
V6	.5	.5
V8	1.7	1.7
EGR Control Valve, Replace (B)		
All Models	1.8	2.0
EGR Filter, Replace (B)		
1981-85	.8	.8
EGR Frequency Valve, Replace (B)		
All Models	.3	.3
EGR Temperature Sensor, Replace (B)		
All Models	.5	.5
Electronic Control Module (ECM/PCM), Replace (B)		
w/continuous injection		
1981-92	.7	.7
w/o continuous injection		
4 cyl.	1.6	1.6
5 cyl.	1.8	1.8
V6, V8	1.6	1.6
Knock Sensor, Replace (B)		
80, 90, A4		
1988-92	1.4	1.6
1993-95	1.7	1.9
1996-01	1.8	2.0
100, 200, A6		
1989-91	.8	.9
1992-95 each	1.7	1.9
1996-01	1.8	2.0
1986-87 4000, Coupe	.5	.6
1986-88 5000	.8	.9
A8		
No. 1	4.2	4.4
No. 2	4.9	5.1
Coupe Quattro	1.4	1.6
S4, S6, Cabriolet	1.8	2.0
S4 twin turbo		
No. 1	2.8	3.0
No. 2	1.5	1.7
TT each	1.4	1.6
V8 Quattro		
No.1	2.9	3.1
No.2	1.4	1.6
Mass Air Flow (MAF) Sensor, Replace (B)		
Exc. V8	1.3	1.3
V8		
1990-96	1.6	1.6
1997-01	.7	.7

80 : 90 : 100 : 200 : 4000 : 5000 : A4 : A6 : A8 : CABRIOLET : COUPE : S4 : S6 : TT : V8 QUATTRO **AUD-5**

	LABOR TIME	SEVERE SERVICE
Oxygen Sensor, Replace (B)		
w/continuous injection		
1981-92	.5	.6
w/o continuous injection		
80, 90		
1988-92	1.0	1.1
1993-95	1.3	1.4
100, 200, A6		
1989-91	1.0	1.1
1992-95	1.6	1.7
Coupe Quattro	1.0	1.1
S4, S6	1.6	1.7
V8 Quattro	1.6	1.7
w/OBD II		
front		
A4	1.0	1.1
A6, S6		
1995-97	1.0	1.1
1998-01	1.6	1.7
Cabriolet	1.0	1.1
TT	.6	.7
w/o OBD II		
front		
90, Cabriolet	.5	.6
A6, S6	.5	.6
A8	1.2	1.3
rear		
1995-01	1.7	1.8
Positive Crankcase Ventilation (PCV) Valve, Replace (B)		
1994-01		
4 cyl.	.3	.3
5 cyl.	.3	.3
V6	.3	.3
V8	.7	.7
Pulse Generator, Replace (B)		
1988-91		
100, 200, 5000	.7	.7
Thermal Vacuum Valve, Replace (B)		
1989-01	.3	.3
Throttle Position Sensor (TPS), Replace (B)		
w/continuous injection		
1981-92	1.0	1.0
w/o continuous injection		
exc. below	.8	.8
1989-91 100, 200, A6	1.2	1.2
Vapor Canister Filter, Replace (B)		
80, 90	1.0	1.0
100, 200, A4	1.0	1.0
4000, 5000, Coupe	1.0	1.0
A6	1.8	1.8
Coupe Quattro	1.1	1.1
S4, S6	1.8	1.8
V8 Quattro	1.8	1.8
Vehicle Speed Sensor, Replace (B)		
Exc. below	.5	.5
1995-01 A8	1.4	1.4

	LABOR TIME	SEVERE SERVICE
FUEL		
DELIVERY		
Fuel Pump, Test (B)		
All Models	.3	.3
Air Cleaner Thermo Valve, Replace (B)		
1981-92	.5	.5
Air Cleaner Vacuum Diaphragm, Replace (B)		
1981-92	.3	.3
Fuel Accumulator, Replace (B)		
80, 90		
1988-91	.7	.7
1992-01	.7	.7
4000, Coupe	.8	.8
5000	.7	.7
Fuel Filter, Replace (B)		
Continuous injection	.5	.5
Electronic injection	.3	.3
Fuel Gauge (Dash), Replace (B)		
80, 90, A4		
1988-91	1.3	1.3
1992	1.6	1.6
1993-95	2.8	2.8
1996-01	2.0	2.0
100, 200, A6		
1989-91		
w/air bags	2.5	2.5
w/o air bags	1.6	1.6
1992-01	2.8	2.8
4000, Coupe	.8	.8
5000		
1981-83	1.3	1.3
1984-88	2.0	2.0
A8	2.4	2.4
Coupe Quattro	1.6	1.6
S4, S6, Cabriolet	2.8	2.8
TT	2.1	2.1
V8 Quattro	2.8	2.8
Fuel Gauge (Tank), Replace (B)		
80, 90, A4		
1988-92	1.2	1.4
1993-95	1.6	1.8
1996-01	1.8	2.0
100, 200, A6		
1989-91	.7	.8
1992-95	3.0	3.2
1996-97	2.8	3.0
1998-01	2.3	2.5
1981-88 4000, 5000, Coupe	.5	.6
A8	1.8	2.0
Cabriolet	1.6	1.8
Coupe Quattro	1.2	1.4
S4, S6	3.0	3.2
TT	1.8	2.0
V8 Quattro	3.8	4.0

	LABOR TIME	SEVERE SERVICE
Fuel Pump, Replace (B)		
80, 90, A4		
1988-92	1.2	1.4
1993-01	1.4	1.6
100, 200, A6		
1989-91	1.7	1.9
1992-95	3.1	3.3
1996-97	1.8	2.0
1998-01	2.3	2.5
4000, Coupe	1.0	1.2
5000		
1981-83	1.3	1.5
1984-88	1.4	1.6
A8	3.0	3.2
S4, S6	3.1	3.3
Cabriolet	1.8	2.0
Coupe Quattro	1.2	1.4
TT	1.8	2.0
V8 Quattro	1.7	1.9
Fuel Pump Relay, Replace (B)		
All Models	.5	.5
w/V8 97-01 add	.5	.5
Fuel Tank, Replace (B)		
Includes: Drain and refill.		
80, 90, A4		
1988-92	1.7	1.9
1988-92 Quattro	2.4	2.6
1993-01	3.9	4.1
1993-01 Quattro	6.7	6.9
100, 200, A6		
1989-91	4.4	4.6
1992-01	3.9	4.1
1992-01 Quattro	6.6	6.8
4000, Coupe	2.0	2.2
5000		
1981-87	2.3	2.5
1988	2.6	2.8
A8	4.4	4.6
Coupe Quattro	1.7	1.9
S4, S6, Cabriolet	3.9	4.1
TT	4.5	4.7
V8 Quattro	4.4	4.6
Intake & Exhaust Manifold or Gaskets, Replace (B)		
1981-88 5 cyl.		
w/turbo	5.3	5.6
w/o turbo	4.4	4.7
Intake Manifold and/or Gasket, Replace (B)		
Includes: Adjustments.		
80, 90, A4		
1988-92	2.0	2.3
1993-01	4.3	4.6
100, 200, A6		
1989-91	3.5	3.8
1992-01	4.3	4.6

AUD-6

80 : 90 : 100 : 200 : 4000 : 5000 : A4 : A6 : A8 : CABRIOLET : COUPE : S4 : S6 : TT : V8 QUATTRO

	LABOR TIME	SEVERE SERVICE
4000, Coupe	1.8	2.1
A8	3.2	3.5
Coupe Quattro	2.0	2.3
S4, S6, Cabriolet	4.3	4.6
TT	1.6	1.9
V8 Quattro	3.2	3.5

INJECTION

Injection System, Check & Adjust (B)
- 4000, 5000, Coupe … 1.2 *1.2*

Sensor Plate, Adjust (B)
- 1981-92 … .7 *.7*

Anti-Backfire Valve, Replace (B)
- 1981-92 100, 200, 5000 … 1.5 *1.5*

Auxiliary Air Regulator, Replace (B)
- 1981-92 … .5 *.5*

Cold Start Injector, Replace (B)
- 1981-92 … .3 *.3*

Differential Pressure Regulator, Replace (B)
- 1981-92 … .7 *.7*

Frequency Valve, Replace (B)
- 1981-82 … .5 *.5*

Fuel Distributor, Replace (B)
- 1981-92 … 1.4 *1.6*

Fuel Injection Lines, Replace (B)
- 4000, 5000, Coupe
  - one … .3 *.4*
  - all … .7 *.8*

Fuel Injectors, Replace (B)
- w/continuous injection
  - 4000, 5000, Coupe
    - one … .5 *.6*
    - all … 1.0 *1.2*
- w/o continuous injection
  - exc. V8 … 1.4 *1.6*
  - V8 … 2.0 *2.2*
- *Clean and test add each* … .3 *.3*

Fuel Pressure Regulator, Replace (B)
- w/continuous injection
  - 1981-92 … .7 *.8*
- w/o continuous injection
  - exc. V8 … .7 *.8*
  - V8
    - 1990-94 … 1.8 *1.9*
    - 1997-01 … 1.2 *1.3*

Fuel Pump Check Valve, Replace (B)
- 1981-92 … .7 *.7*

Fuel Return Hoses, Replace (B)
- 4000, 5000, Coupe
  - one or all … .5 *.6*

Fuel Shut-Off Valve, Replace (B)
- 4000, 5000, Coupe … .5 *.5*

Idle Air Control (IAC) Valve, Replace (B)
- w/continuous injection
  - 1981-92 … .5 *.5*

	LABOR TIME	SEVERE SERVICE

- w/o continuous injection
  - 4 cyl. … .8 *.8*
  - 5 cyl. … .8 *.8*
  - V6 … 1.3 *1.3*
  - V8 … .8 *.8*

Idle Stabilizer Valve, Replace (B)
- w/continuous injection
  - 1986-87 … .3 *.3*
  - 1988-92 … .7 *.7*
- w/o continuous injection
  - 4cyl., 5 cyl.
    - 1981-87 … .3 *.3*
    - 1988-01 … .8 *.8*
  - V6 … 1.3 *1.3*
  - V8 … .8 *.8*

Impulse Relay, Replace (B)
- 1981-92 … .7 *.7*

Injection Pump Sprocket, Replace (B)
- 4000, 5000, Coupe … 1.7 *1.7*

Injector Cooling Blower Motor, Replace (B)
- 1986-87 5000 turbo … .7 *.7*
- 1988-92 … 1.0 *1.0*

Injector Cooling Blower Switch, Replace (B)
- 1986-87 5000 turbo … .5 *.5*
- 1988-92 … .7 *.7*

Microswitch, Replace (B)
- 1981-87 … .5 *.5*
- 1988-92 … 1.0 *1.0*

Pressure Regulator Valve, Replace (B)
- 1981-92 … .3 *.3*

Series Resistor, Replace (B)
- 80, 90 20V … .5 *.5*

Thermo Time Switch, Replace (B)
- 1981-87 … .3 *.3*
- 1988-92 … .5 *.5*

Throttle Body Assembly, Replace (B)
- 1994-95 90 … 1.0 *1.0*
- 1994-95 100 … 1.0 *1.0*
- 1994 V8 Quattro … 1.6 *1.6*
- A4, A6, A8 … 1.0 *1.0*
- Coupe Quattro … 1.2 *1.2*
- S4, S6, Cabriolet … 1.0 *1.0*
- TT … 1.0 *1.0*

Throttle Valve Housing, Replace (B)
- 80, 90, A4
  - 1988-92 … 1.2 *1.2*
  - 1993-97 … 1.6 *1.6*
- 100, 200, A6
  - 1989-91 … 1.7 *1.7*
  - 1992-97 … 1.6 *1.6*
- 4000, Coupe … .7 *.7*
- 5000 … .8 *.8*
- S4, S6, Cabriolet … 1.6 *1.6*
- V8 Quattro … 1.6 *1.6*

	LABOR TIME	SEVERE SERVICE

Throttle Valve Switch, Replace (B)
- w/continuous injection
  - 1981-92 … 1.6 *1.6*
- w/o continuous injection
  - 80, 90, A4
    - 1988-92 … 2.0 *2.0*
    - 1993-01 … .5 *.5*
  - 1992-01 100, A6 … .5 *.5*
  - A8, S4, S6, Cabriolet … .5 *.5*
  - Coupe Quattro … 2.0 *2.0*
  - TT … .5 *.5*
  - V8 Quattro … .5 *.5*

Voltage Supply Relay, Replace (B)
- 1981-92 … .2 *.2*

Warm-Up Regulator, Replace (B)
- 1981-92 … .3 *.3*

TURBOCHARGER

Turbocharger Assy., R&R or Replace (B)
- 100, 200
  - air-cooled … 6.8 *7.1*
  - water-cooled … 7.3 *7.6*
- 4000, 5000, Coupe … 6.8 *7.1*
- A4, S4, S6 … 6.2 *6.5*
- TT … 6.3 *6.6*

Wastegate Actuator, Replace (B)
- All Models … 1.7 *1.9*

EXHAUST

Catalytic Converter, Replace (B)
- 4 cyl. … 1.4 *1.5*
- 5 cyl. … 1.4 *1.5*
- V6
  - right … 1.8 *1.9*
  - left … 2.5 *2.6*
- V8
  - 1990-96
    - one … 1.7 *1.8*
    - both … 2.4 *2.5*
  - 1997-01
    - right … 1.3 *1.4*
    - left … 1.6 *1.7*
    - both … 1.8 *1.9*

Exhaust Manifold Gasket, Replace (B)
- 1981-87
  - gasoline
    - 4000, Coupe … 1.9 *2.2*
    - 5000 … 3.7 *4.0*
  - diesel … 4.2 *4.5*
- 1988-01
  - 4 cyl. … 3.5 *3.8*
  - 5 cyl. 10V
    - w/turbo … 4.2 *4.5*
    - w/o turbo … 3.5 *3.8*
  - 5 cyl. 20V … 4.8 *5.1*

AUD-7

80 : 90 : 100 : 200 : 4000 : 5000 : A4 : A6 : A8 : CABRIOLET : COUPE : S4 : S6 : TT : V8 QUATTRO

	LABOR TIME	SEVERE SERVICE
V6		
right side	5.8	6.1
left side	4.2	4.5
both sides	7.9	8.2
V8		
right side	4.5	4.8
left side	3.9	4.2
both sides	7.3	7.6
Exhaust Pipe Extension, Replace (B)		
All Models	.3	.4
Exhaust System (Complete), Replace (B)		
80, 90, A4		
1988-91	1.8	1.9
1992-01	2.8	2.9
100, 200, A6		
1989-91	1.8	1.9
1992-01	2.8	2.9
Cabriolet	1.8	1.9
S4, S6	2.8	2.9
TT	2.7	2.8
w/turbo add	.3	.3
Front Exhaust Pipe, Replace (B)		
80, 90		
1988-92		
4 cyl.	1.4	1.5
5 cyl.	1.6	1.7
100, 200		
1989-91	1.6	1.7
4000, Coupe	1.2	1.3
5000		
1981-88	.8	.9
1981-88 Quattro	2.3	2.4
A4, A6	2.3	2.4
A8	2.8	2.9
Coupe Quattro	1.4	1.5
TT	1.6	1.7
V8 Quattro	2.3	2.4
w/turbo add	.5	.5
Intake & Exhaust Manifold or Gaskets, Replace (B)		
5 cyl.	4.4	4.7
Turbo	5.3	5.6
Muffler, Replace (B)		
80, 90, A4		
1988-92		
4 cyl.	.8	.9
5 cyl.	1.6	1.7
1993-01	1.6	1.7
100, 200, A6		
1989-97	1.6	1.7
1998-01	1.8	1.9
4000, Coupe		
front	1.3	1.4
rear	1.2	1.3
5000		
front	1.5	1.6
rear	1.2	1.3

	LABOR TIME	SEVERE SERVICE
Coupe Quattro	1.6	1.7
S4, S6, Cabriolet	1.6	1.7
TT		
front	1.9	2.0
rear	1.4	1.5
V8 Quattro	.8	.9
Muffler & Tailpipe Assy., Replace (B)		
4 cyl.	.5	.6
5 cyl.	.7	.8
V6	1.8	1.9
V8		
1990-96	.7	.8
1997-01	2.3	2.4
One-Piece Exhaust System, Replace (B)		
1988-91 100, 200, 5000	.8	1.0

ENGINE COOLING

	LABOR TIME	SEVERE SERVICE
Pressure Test Cooling System (C)		
All Models	.3	.3
Auxiliary Radiator Assy., Replace (B)		
1989-91 100, 200	1.7	1.8
Coolant Thermostat, Replace (B)		
Gasoline		
80, 90		
1988-92	1.2	1.4
1993-95	2.3	2.5
100, 200, A6		
1989-91	1.2	1.4
1992-97	2.0	2.2
1998-01	3.2	3.4
4000, Coupe	.7	.9
5000	.7	.9
A4	2.0	2.2
A8	3.2	3.4
Cabriolet	2.3	2.5
Coupe Quattro	.8	1.0
S4, S6		
5 cyl.	2.0	2.2
V6	2.0	2.2
twin turbo	6.8	7.0
TT	3.1	3.3
V8 Quattro	1.8	2.0
Diesel	.8	1.0
Electric Cooling Fan Motor, Replace (B)		
1988-01		
w/ABS	2.3	2.6
w/o ABS	1.6	1.9
Electric Fan Thermo Switch, Replace (B)		
All Models	.3	.3
Engine Coolant Temp. Sending Unit, Replace (B)		
All Models	1.0	1.0

	LABOR TIME	SEVERE SERVICE
Fan Blade, Replace (B)		
1988-92 80, 90	.8	1.0
1989-92 100, 200	.8	1.0
4000, Coupe	.8	1.0
5000	1.6	1.8
Freeze Plugs (Water Jacket), Replace (B)		
Add proper time for plug access.		
All Models each	.5	.7
Radiator Assy., R&R or Replace (B)		
Includes: Refill with proper coolant mix.		
Gasoline		
80, 90, A4		
1988-92		
4 cyl.	2.0	2.1
5 cyl.	3.0	3.1
1993-95	3.0	3.1
1996-01	4.2	4.3
100, 200, A6		
1989-91		
4 cyl.	3.0	3.1
5 cyl.	3.3	3.4
turbo	2.7	2.8
1992-97	3.1	3.2
1998-01	4.2	4.3
4000, Coupe		
1981-87	1.3	1.4
Quattro	2.3	2.4
5000		
1981-88	2.4	2.5
Quattro	2.6	2.7
A8	3.8	3.9
Coupe Quattro	3.0	3.1
S4, S6, Cabriolet	3.6	3.7
TT	1.8	1.9
V8 Quattro	3.0	3.1
Diesel	2.4	2.5
w/AC 5000 add	.8	.8
Radiator Fan Motor Relay, Replace (B)		
1988-01	1.0	1.0
Radiator Fan Motor Resistor, Replace (B)		
1988-01		
5 cyl., V6	.7	.7
V8	.8	.8
Radiator Hoses, Replace (B)		
Includes: Refill with proper coolant mix.		
Exc. below		
upper	.7	.8
lower	1.0	1.1
5000		
one	1.0	1.1
all	1.4	1.5
A8		
one	.5	.6
all	2.3	2.4

AUD-8
80 : 90 : 100 : 200 : 4000 : 5000 : A4 : A6 : A8 : CABRIOLET : COUPE : S4 : S6 : TT : V8 QUATTRO

	LABOR TIME	SEVERE SERVICE
Radiator Hoses, Replace (B)		
Coupe Quattro		
upper	.8	.9
lower	1.1	1.2
TT	.6	.7
V8 Quattro		
one	.5	.6
all	2.3	2.4
Secondary Coolant Pump, Replace (B)		
1988-01 turbo	.7	1.0
Temperature Gauge (Dash), Replace (B)		
80, 90, A4		
1988-92	1.4	1.5
1994-95	2.5	2.5
1996-01	1.6	1.6
100, 200, A6		
1989-91		
w/air bags	2.5	2.5
w/o air bags	1.6	1.6
1992-01	2.8	2.8
4000, Coupe	1.2	1.2
5000	1.7	1.8
A8	1.4	1.4
Cabriolet	1.6	1.6
Coupe Quattro	1.4	1.4
S4, S6	2.8	2.8
TT	1.6	1.6
V8 Quattro	2.8	2.8
Temperature Gauge (Engine), Replace (B)		
All Models	.7	.7
Viscous Fan, Replace (B)		
V8 Quattro	1.3	1.5
Water Pump and/or Gasket, Replace (B)		
Includes: Refill with proper coolant mix.		
Gasoline		
80, 90, A4		
1988-92		
4 cyl.	3.1	3.4
5 cyl. 10V	3.8	4.1
5 cyl. 20V	4.2	4.5
1993-95	3.1	3.4
1996-01	3.9	4.2
100, 200, A6		
1989-91		
5 cyl. 10V	3.0	3.3
5 cyl. 20V	4.2	4.5
1992-01	3.1	3.4
4000, Coupe	1.8	2.1
5000		
1981-83	2.4	2.7
1984-88	3.2	3.5
A8	8.6	8.9

	LABOR TIME	SEVERE SERVICE
S4, S6		
5, cyl, V6	4.4	4.7
twin turbo	5.5	5.8
Cabriolet	3.1	3.4
Coupe Quattro	3.8	4.1
TT	3.9	4.2
V8 Quattro	6.5	6.8
Diesel	2.8	3.1

ENGINE
ASSEMBLY
Times shown are for OEM assemblies. Time to replace assemblies from aftermarket rebuilders may vary.

Engine Assy., R&I (B)
Does not include parts or component transfer.

	LABOR TIME	SEVERE SERVICE
80, 90, A4		
1988-92		
4 cyl.	7.3	7.3
5 cyl.	8.4	8.4
1993-95	10.5	10.5
1996-01		
4 cyl.	12.0	12.0
V6	10.3	10.3
100, 200, A6		
1989-91		
5 cyl.	8.4	8.4
5 cyl. turbo	12.3	12.3
1992-95	10.5	10.5
1996-01	12.4	12.4
4000, Coupe		
gasoline	10.0	10.0
diesel	5.8	5.8
5000		
gasoline		
AT	10.5	10.5
MT	10.0	10.0
diesel	11.7	11.7
A8	22.6	22.6
Cabriolet	10.5	10.5
Coupe Quattro	8.3	8.3
S4, S6		
5 cyl. 20V	12.3	12.3
V6	10.5	10.5
twin turbo	18.8	18.8
V8 Quattro	15.3	15.3
TT (5.8)	8.8	8.8
w/AC add		
4000, 5000	2.3	2.3
w/AT add		
80, 90, 5000	.2	.2
w/turbo add		
4000, 5000	1.7	1.7

	LABOR TIME	SEVERE SERVICE
Engine Assy., R&R and Recondition (A)		
Includes: Replacing rings, rod and main bearings, cylinder head reconditioning and engine tune-up.		
80, 90, A4		
1988-92		
4 cyl.	27.5	27.5
5 cyl.	28.4	28.4
1993-95	30.3	30.3
1996-01		
4 cyl.	32.2	32.2
V6	30.5	30.5
100, 200, A6		
1989-91		
5 cyl.	28.4	28.4
5 cyl. turbo	32.3	32.3
1992-95	30.3	30.3
1996-01		
5 cyl. 20V	32.3	32.3
V6	30.3	30.3
4000, Coupe		
gasoline	20.0	20.0
diesel	15.8	15.8
5000		
gasoline		
AT	22.3	22.3
MT	21.8	21.8
diesel	23.5	23.5
A8	34.4	34.4
Cabriolet	30.3	30.3
Coupe Quattro	28.3	28.3
S4, S6		
5 cyl. 20V	32.3	32.3
V6	30.3	30.3
V8 Quattro	32.3	32.3
w/AC add		
4000, 5000	2.3	2.3
w/AT add		
80, 90, 5000	.2	.2
w/turbo add		
4000, 5000	1.7	1.7
Engine Assy. (New or Rebuilt Complete), Replace (B)		
Includes: Component transfer and engine tune-up.		
80		
1988-90	13.1	13.1
90		
1988-91	15.5	15.5
4000, Coupe	12.2	12.2
5 cyl. turbo	19.5	19.5
5000	15.2	15.2
Twin turbo	23.1	23.1
TT	13.1	13.1
w/AC add		
4000, 5000	2.3	2.3
w/AT add		
80, 90, 5000	.2	.2
w/turbo add		
4000, 5000	1.7	1.7

80 : 90 : 100 : 200 : 4000 : 5000 : A4 : A6 : A8 : CABRIOLET : COUPE : S4 : S6 : TT : V8 QUATTRO AUD-9

	LABOR TIME	SEVERE SERVICE
Engine Assy. (Short Block), Replace (B)		
Assembly consists of cylinder block, piston assemblies, crankshaft, camshaft, timing chain and gears. Does not include cylinder heads. Operation Includes: R&R engine, transfer necessary parts and all necessary adjustments.		
80, 90, A4		
1988-92		
4 cyl.	12.3	12.3
5 cyl.	13.8	13.8
1993-95	16.3	16.3
1996-01		
4 cyl.	19.5	19.5
V6	18.0	18.0
100, 200, A6		
1989-91		
5 cyl.	13.8	13.8
5 cyl. turbo	18.7	18.7
1992-95	16.3	16.3
1996-01	19.8	19.8
4000, Coupe		
gasoline	24.2	24.2
diesel	16.2	16.2
5000		
gasoline		
AT	26.2	26.2
MT	25.8	25.8
diesel	27.3	27.3
A8	29.7	29.7
S4, S6		
5 cyl. 20V	18.7	18.7
V6	16.3	16.3
Cabriolet	16.3	16.3
Coupe Quattro	18.3	18.3
V8 Quattro	17.1	17.1
w/AC add		
4000, 5000	2.3	2.3
w/AT add		
80, 90, 5000	.2	.2
w/turbo add		
4000, 5000	1.7	1.7
Engine Mounts, Replace (B)		
80, 90, A4		
1988-92		
4 cyl.		
right	2.3	2.5
left	1.3	1.5
center	.7	.9
5 cyl.		
right or left	.8	1.0
center	1.4	1.6
1993-01		
right	2.0	2.2
left	2.4	2.6
center	3.1	3.3

	LABOR TIME	SEVERE SERVICE
100, 200, A6		
1989-91		
right or left	.8	1.0
center	1.4	1.6
1992-01		
right	2.0	2.2
left	2.4	2.6
center	3.1	3.3
4000, Coupe each	.5	.7
5000		
gasoline each	1.4	1.6
diesel		
right or left	1.8	2.0
center	1.4	1.6
A8		
right side	3.1	3.3
left side	2.8	3.0
Coupe Quattro		
right or left	1.0	1.2
both	1.4	1.6
S4, S6, Cabriolet		
5 cyl. 20V		
right or left	.8	1.0
center	1.4	1.6
V6		
right	2.0	2.2
left	2.4	2.6
center	3.1	3.3
twin turbo		
right	1.6	1.8
left	1.9	2.1
center	2.7	2.9
TT each	.8	1.0
V8 Quattro each	1.7	1.9
CYLINDER HEAD		
Compression Test (B)		
4 cyl., 5 cyl.	1.0	1.1
V6	1.5	1.6
V8	1.6	1.7
Diesel	1.4	1.5
Valve Clearance, Adjust (B)		
1981-90 4 cyl.	1.3	1.6
1981-93 5 cyl.	1.7	2.0
1990-94 V8	2.7	3.0
Cam Followers, Replace (B)		
1981-87	3.1	3.4
Cylinder Head, Replace (B)		
Includes: Parts transfer, adjustments.		
80, 90, A4		
1988-92		
4 cyl.	6.3	6.8
5 cyl. 10V	7.8	8.3
5 cyl. 20V	8.8	9.3
1993-95 both	8.2	8.7
1996-01		
4 cyl.	8.9	9.4

	LABOR TIME	SEVERE SERVICE
V6		
right side	6.8	7.3
left side	5.6	6.1
both sides	12.0	12.5
100, 200, A6		
1989-95		
5 cyl.	9.3	9.8
V6 both	10.1	10.6
1996-97		
one side	5.8	6.3
both sides	10.8	11.3
1998-01		
one side	6.5	7.0
both sides	8.9	9.4
4000, Coupe		
gasoline	6.8	7.3
diesel	6.0	6.5
5000		
gasoline	8.8	9.3
diesel	7.5	8.0
A8		
one side	8.1	8.6
both sides	15.0	15.5
Coupe Quattro	7.7	8.3
S4, S6, Cabriolet		
5 cyl. 20V	9.3	9.8
V6 both	8.2	8.7
TT	10.6	11.1
V8 Quattro both	13.3	13.8
w/AC add		
4000, 5000	1.2	1.2
w/turbo add		
4000, 5000, A4	1.7	1.7
Cylinder Head Gasket, Replace (B)		
Includes: Adjustments.		
80, 90, A4		
1988-92		
4 cyl.	5.3	5.6
5 cyl. 10V	6.3	6.6
5 cyl. 20V	5.6	5.9
1993-95 both	7.5	7.8
1996-01		
4 cyl.	7.9	8.3
V6		
one side	5.8	6.1
both sides	10.8	11.1
100, 200, A6		
1989-95		
5 cyl.	7.5	7.8
V6 both	9.0	9.3
1996-97		
one side	5.8	6.1
both sides	10.8	11.1
1998-01		
one side	6.5	6.8
both sides	8.9	9.2

AUD-10   80 : 90 : 100 : 200 : 4000 : 5000 : A4 : A6 : A8 : CABRIOLET : COUPE : S4 : S6 : TT : V8 QUATTRO

	LABOR TIME	SEVERE SERVICE
Cylinder Head Gasket, Replace (B)		
4000, Coupe	5.0	5.3
5000		
gasoline	5.4	5.7
diesel	5.0	5.3
A8		
one side	7.3	7.6
both sides	13.5	13.8
Coupe Quattro	6.3	6.6
S4, S6, Cabriolet		
5 cyl. 20V	8.8	9.1
V6 both	7.5	7.8
twin turbo		
one side	7.1	7.4
both sides	8.7	9.0
TT (5.8)	8.8	9.1
V8 Quattro both	11.5	11.8
w/AC add		
4000, 5000	1.2	1.2
w/turbo add		
4000, 5000, A4	1.7	1.7
Cylinder Head Bolts, Retorque (B)		
4 cyl.	.8	.9
5 cyl.	1.7	1.8
V6 each side	1.7	1.8
V8 each side	1.8	1.9
Valve/Cam Cover Gasket, Replace or Reseal (B)		
80, 90, A4		
1988-92		
4 cyl.	1.0	1.1
5 cyl. 10V	1.8	1.9
5 cyl. 20V	3.1	3.2
1993-01		
one	1.4	1.5
both	2.5	2.6
100, 200, A6		
5 cyl.	3.1	3.2
V6 both	2.5	2.6
4000, Coupe	.7	.8
5000		
gasoline	1.4	1.5
diesel	.8	.9
A8	2.4	2.5
Cabriolet		
one	1.4	1.5
both	2.5	2.6
Coupe Quattro	2.0	2.1
S4, S6		
5 cyl. 20V	1.7	1.8
V6 both	1.4	1.5
twin turbo		
one side	1.1	1.2
both sides	2.1	2.2
TT	2.1	2.2
V8 Quattro both	1.9	2.0

	LABOR TIME	SEVERE SERVICE
Valve Job (A)		
80, 90, A4		
1988-92		
4 cyl.	10.2	10.7
5 cyl. 10V	12.4	12.9
5 cyl. 20V	12.9	13.4
1993-95	18.0	18.5
1996-01		
4 cyl.	11.0	11.5
V6	14.3	14.8
100, 200, A6		
1989-95		
5 cyl.	12.4	12.9
V6	13.8	14.3
1996-97	14.3	14.8
1998-01	12.0	12.5
4000, Coupe		
gasoline	10.0	10.5
diesel	8.3	8.8
5000		
gasoline	11.2	11.7
diesel	11.7	12.2
A8	19.3	19.8
Coupe Quattro	12.4	12.9
S4, S6, Cabriolet		
5 cyl. 20V	13.8	14.3
V6	12.7	13.2
TT	13.5	14.0
V8 Quattro	19.3	19.8
w/AC add		
4000, 5000	1.2	1.2
w/turbo add		
4000, 5000, A4	1.7	1.7
Valve Lifters, Replace (B)		
80, 90, A4		
1988-92		
4 cyl.	2.8	3.1
5 cyl. 10V	4.2	4.5
5 cyl. 20V	5.4	5.7
1993-01	7.5	7.8
100, 200, A6		
1989-91	3.5	3.8
1992-01	7.5	7.8
4000, Coupe	2.8	3.1
5000	3.1	3.4
A8		
one side	6.7	7.0
both sides	7.8	8.1
Coupe Quattro	4.4	4.7
S4, S6, Cabriolet		
5 cyl. 20V	5.4	5.7
V6	7.5	7.8
twin turbo		
one side	6.2	6.5
both sides	7.4	7.7
TT	6.8	7.1
V8 Quattro		
one side	6.7	7.0
both sides	7.8	8.1
w/turbo A4 add	1.7	1.7

	LABOR TIME	SEVERE SERVICE
Valve Springs and, or Oil Seals, Replace (B)		
Gasoline		
4 cyl.		
one	3.1	3.3
all	5.3	5.5
5 cyl.		
one	10.8	11.1
each addl. add	.2	.3
V8		
one side	5.6	5.8
both sides	9.3	9.5
Diesel		
4000 all	4.2	4.5
5000 all	5.2	5.7
CAMSHAFT		
Timing Belt, Adjust (B)		
All Models	1.4	1.5
Camshaft, Replace (A)		
80, 90, A4		
1988-92		
4 cyl.	2.8	3.0
5 cyl.	3.1	3.3
1993-01		
one	4.3	4.5
both	4.8	5.0
100, 200, A6		
1989-91	3.1	3.3
1992-97		
one	4.3	4.5
both	4.8	5.0
1998-01	5.3	5.5
4000, Coupe	2.8	3.0
5000		
1981-83		
gasoline	2.9	3.1
diesel	3.3	3.5
1984-88	3.3	3.5
A8		
one	2.7	2.9
both	3.6	3.8
Coupe Quattro	3.3	3.5
S4, S6, Cabriolet		
5 cyl. 20V	3.1	3.3
V6		
one	4.8	5.0
both	6.5	6.7
twin turbo		
one	9.6	9.8
both	11.4	11.6
TT		
one	6.8	7.0
both	7.3	7.5
V8 Quattro		
one	2.7	2.9
both	3.6	3.8

	LABOR TIME	SEVERE SERVICE
Camshaft Seal, Replace (B)		
Gasoline		
4 cyl.	2.0	2.1
5 cyl.	2.0	2.1
V6	1.6	1.7
V8	6.0	6.1
Diesel		
rear	2.0	2.1
Camshaft Sprocket, Replace (B)		
Gasoline		
exc. V8	1.8	1.9
V8	2.3	2.4
Diesel		
5000	2.9	3.0
Intermediate Shaft, Replace (B)		
1981-90 4 cyl.	2.4	2.6
Timing Belt, Replace (B)		
80, 90, A4		
1988-92		
4 cyl.	2.3	2.6
5 cyl.	2.7	3.0
1993-01	2.0	2.3
100, 200, A6		
1989-91	3.1	3.3
1992-97	2.0	2.2
1998-01	3.1	3.3
4000, 5000, Coupe	2.7	3.0
A8	6.5	6.7
Coupe Quattro	2.7	2.9
S4, S6, Cabriolet		
exc. twin turbo	2.0	2.2
twin turbo	4.1	4.3
TT	2.7	2.9
V8 Quattro	5.2	5.4
Timing Belt Cover and/or Gasket, Replace (B)		
1981-94		
4 cyl.	1.0	1.1
5 cyl.	1.0	1.1
V8		
right	3.1	3.3
left	.3	.4
1995-01	1.6	1.9
Timing Belt Tensioner Pulley, Replace (B)		
4 cyl.	1.4	1.7
V6	1.8	2.1
V8	4.8	5.1

CRANK & PISTONS

Connecting Rod Bearings, Replace (A)
Includes: Check bearing oil clearance.

	LABOR TIME	SEVERE SERVICE
Gasoline		
4 cyl.	3.8	4.3
5 cyl.	4.3	4.8
Diesel		
4000	3.4	3.9
5000	4.3	4.8

	LABOR TIME	SEVERE SERVICE
Crankshaft and Main Bearings, Replace (A)		
Includes: Engine R&R.		
80, 90, A4		
1988-92		
4 cyl.	11.7	12.2
5 cyl.	13.8	14.3
1993-95	15.9	16.4
1996-01		
4 cyl.	17.4	17.9
V6	15.7	16.2
100, 200, A6		
1989-91		
5 cyl.	13.8	14.3
5 cyl. turbo	17.8	18.3
1992-95	16.0	16.5
1996-01	16.9	17.4
4000, Coupe		
gasoline	9.8	10.3
diesel	9.3	9.8
5000		
gasoline		
AT	13.1	13.6
MT	12.6	13.1
diesel	9.8	10.3
A8	27.1	27.6
Coupe Quattro	8.1	8.6
S4, S6, Cabriolet		
5 cyl.	19.3	19.8
V6	17.5	18.0
twin turbo	25.8	26.3
TT	14.1	14.6
V8 Quattro	22.3	22.8
w/AC add		
4000, 5000	2.3	2.3
w/AT add		
80, 90, 5000	.2	.2
w/turbo add		
4000, 5000	1.7	1.7
Crankshaft Front Oil Seal, Replace (B)		
80, 90, A4		
1988-92		
4 cyl.	2.0	2.2
5 cyl.	3.1	3.3
1993-95	3.7	3.9
1996-01	4.3	4.5
100, 200, A6		
1989-97	2.8	3.0
1992-95	3.7	3.9
1996-97	3.5	3.7
1998-01	2.3	2.5
4000, Coupe		
gasoline	1.8	2.0
diesel	3.3	3.5

	LABOR TIME	SEVERE SERVICE
5000		
gasoline		
1981-83	1.8	2.0
1984-88	2.5	2.7
diesel	3.1	3.3
A8	7.5	7.7
Coupe Quattro	3.0	3.2
S4, S6, Cabriolet		
5 cyl. 20V	3.2	3.4
V6	3.7	3.9
twin turbo	5.1	5.3
TT	4.3	4.5
Main Bearings, Replace (A)		
Gasoline		
4 cyl.	6.4	6.9
5 cyl.	7.5	8.0
Diesel	8.3	8.8
Pistons or Connecting Rods, Replace (A)		
Includes: Ridge reaming, cylinder wall deglazing, installing new rings and rod bearings, engine tune-up.		
80, 90, A4		
1988-92		
4 cyl.		
one cyl.	6.9	7.8
all cyls.	8.9	9.8
5 cyl.		
one cyl.	8.8	9.7
all cyls.	10.2	11.1
1993-01		
one cyl.		
each side	9.6	10.5
each addl. add	1.0	1.5
100, 200, A6		
1989-95		
5 cyl.		
one cyl.	8.8	9.7
all cyls.	10.2	11.1
V6		
one cyl.	8.8	9.7
all cyls.	12.3	13.2
1996-97		
one cyl.		
each side	11.2	12.1
each addl. add	1.0	1.5
1998-01		
one cyl.		
each side	9.4	10.3
each addl. add	1.0	1.5
4000, Coupe		
gasoline	16.9	17.8
diesel	7.3	8.2
5000		
gasoline	18.3	19.2
diesel	17.3	18.2

AUD-12 80 : 90 : 100 : 200 : 4000 : 5000 : A4 : A6 : A8 : CABRIOLET : COUPE : S4 : S6 : TT : V8 QUATTRO

	LABOR TIME	SEVERE SERVICE
Pistons or Connecting Rods, Replace (A)		
Coupe Quattro		
one cyl.	6.8	7.7
all cyls.	10.8	11.7
S4, S6, Cabriolet		
5 cyl. 20V		
one cyl.	8.8	9.7
all cyls.	12.3	13.2
V6		
one cyl.		
each side	9.6	10.5
each addl. add	1.0	1.5
twin turbo		
one cyl	8.5	9.4
all cyls.	16.1	17.0
TT	7.9	8.8
w/AC add		
4000, 5000	1.2	1.2
w/turbo add		
4000, 5000, A4	1.7	1.7
Rear Main Oil Seal, Replace (B)		
Includes: Trans. R&R when necessary.		
AT		
80, 90, A4		
1988-92	6.2	6.8
1988-92 Quattro	8.3	8.9
1993-95 4-Speed	9.2	9.8
1996-01		
5-Speed	12.9	13.5
5-Speed Quattro	15.3	15.9
100, 200, A6		
1989-91	7.9	8.5
1989-91 Quattro	10.0	10.6
1992-95	8.2	8.8
1992-95 Quattro	9.6	10.2
1996-97 4-Speed	9.9	10.5
1998-01		
5-Speed	12.4	13.0
5-Speed Quattro	15.3	15.9
4000, 5000, Coupe		
1981-87	7.9	8.5
1981-87 Quattro	8.9	9.5
A8 5-Speed	19.0	19.6
A8 5-Speed Quattro	21.3	21.9
Cabriolet 4-Speed	9.2	9.8
S4	8.2	8.8
S4 twin turbo	9.4	10.0
S6 4-Speed	9.9	10.5
V8 Quattro	13.5	14.0
MT		
80, 90, A4		
1988-92	6.7	7.3
1988-92 Quattro	8.8	9.4
1993-95	10.1	10.7
1993-95 Quattro	10.3	10.9
1996-01	7.8	8.4
1996-01 Quattro	11.6	12.2

	LABOR TIME	SEVERE SERVICE
100, 200, A6		
1989-91	7.9	8.5
1989-91 Quattro	10.0	10.6
1992-95	9.7	10.3
1992-95 Quattro	11.5	12.1
1996-97	8.0	8.6
1996-97 Quattro	8.6	9.2
1996-97		
20V Quattro	7.4	8.0
1998-01	9.1	9.7
1998-01 Quattro	9.7	10.3
4000, Coupe		
1981-87	6.7	7.3
1981-87 Quattro	8.8	9.4
5000		
1981-88	7.9	8.5
1987-88 Quattro	10.0	10.6
A8	16.2	16.8
Cabriolet	8.9	9.5
Coupe Quattro	8.6	9.2
S4, S6		
1994-97	8.0	8.6
1994-97 Quattro	8.6	9.2
20V Quattro	7.4	8.0
1998-01	9.1	9.7
1998-01 Quattro	9.7	10.3
S4 twin turbo	17.0	17.6
TT	6.2	6.8
TT Quattro	12.4	13.0
V8 Quattro	9.6	10.2
Rings, Replace (A)		
Includes: Ridge reaming, cylinder wall deglazing, installing new rings, engine tune-up.		
80, 90, A4		
1988-92		
4 cyl.		
one cyl.	7.2	8.1
all cyls.	9.7	10.6
5 cyl.		
one cyl.	9.1	10.0
all cyls.	11.4	12.3
1993-01		
one cyl.		
each side	10.2	11.1
each addl. add	2.0	2.5
100, 200, A6		
1989-95		
5 cyl.		
one cyl.	9.1	10.0
all cyls.	11.4	12.3
V6		
one cyl.	9.1	10.0
all cyls.	13.1	14.0
1996-97		
one cyl.		
each side	10.2	11.1
each addl. add	2.0	2.5

	LABOR TIME	SEVERE SERVICE
1998-01		
one cyl.		
each side	8.4	9.3
each addl. add	2.0	2.5
4000, Coupe		
gasoline	15.2	16.1
diesel	9.3	10.2
5000		
gasoline	16.2	17.1
diesel	19.8	20.7
Coupe Quattro		
one cyl.	9.0	9.9
all cyls	11.8	12.7
S4, S6, Cabriolet		
5 cyl. 20V		
one cyl.	9.1	10.0
all cyls.	13.1	14.0
V6		
one cyl.		
each side	10.2	11.1
each addl. add	2.0	2.5
twin turbo		
one cyl.	8.8	9.7
each addl. add	2.0	2.5
TT	9.1	10.0
w/AC add		
4000, 5000	1.2	1.2
w/turbo add		
4000, 5000, A4	1.7	1.7
Vibration Damper, Replace (B)		
1981-97	1.4	1.4
ENGINE LUBRICATION		
Engine Oil Cooler, Replace (B)		
Exc. V8	1.6	1.7
V8	1.2	1.3
Engine Oil Pressure Switch (Sending Unit), Replace (B)		
1981-92	.7	.7
1993-01	1.0	1.0
Oil Pan and, or Gasket, Replace (B)		
Gasoline		
4 cyl.		
w/turbo	4.5	4.7
w/o turbo	4.2	4.4
5 cyl.	2.8	3.0
V6, V8		
upper	3.8	4.0
lower	2.8	3.0
twin turbo		
upper	5.8	6.0
lower	3.1	3.3
Diesel	2.7	2.9
Oil Pressure Gauge (Dash), Replace (B)		
80, 90, A4		
1988-92	1.3	1.3
1993-95	2.8	2.8
1996-01	1.6	1.6

80 : 90 : 100 : 200 : 4000 : 5000 : A4 : A6 : A8 : CABRIOLET : COUPE : S4 : S6 : TT : V8 QUATTRO — AUD-13

	LABOR TIME	SEVERE SERVICE
100, 200, A6		
1989-91		
w/air bags	2.5	2.5
w/o air bags	1.6	1.6
1992-01	2.8	2.8
A8	1.2	1.2
Coupe Quattro	1.5	1.5
S4, S6, Cabriolet	2.8	2.8
TT	2.7	2.7
V8 Quattro	2.8	2.8

Oil Pump, Replace (B)

80, 90, A4		
1988-92		
4 cyl.	3.7	4.0
5 cyl.	5.5	5.8
1993-01	6.3	6.6
100, 200, A6	6.3	6.6
4000, 5000, Coupe		
gasoline	4.8	5.1
diesel		
4000	3.1	3.4
5000	5.3	5.6
A8	2.8	3.1
Coupe Quattro	5.6	5.9
S4, S6, Cabriolet		
5 cyl. 20V	4.5	4.8
V6	6.2	6.5
twin turbo	4.0	4.3
TT	4.6	4.9
V8 Quattro	2.0	2.3

CLUTCH

Bleed Clutch Hydraulic System (B)

All Models	.5	.5

Clutch Pedal Free Play, Adjust (B)

All Models	.5	.6

Clutch Assy., Replace (B)

80, 90, A4		
1988-92	6.0	6.2
1988-92 Quattro	8.4	8.6
1993-95	9.2	9.4
1993-95 Quattro	9.4	9.6
1996-01	6.9	7.1
1996-01 Quattro	10.7	10.9
100, 200, A6		
1989-91	7.2	7.4
1989-91 Quattro	9.4	9.6
1992-95	9.2	9.4
1992-95 Quattro	9.4	9.6
1996-97	6.9	7.1
1996-97 Quattro	7.5	7.7
1998-01	8.0	8.2
1998-01 Quattro	8.8	9.0
4000, Coupe		
1981-87	5.4	5.6
1981-87 Quattro	7.5	7.7
5000		
1981-88	6.5	6.7
1981-88 Quattro	8.8	9.0
Coupe Quattro	9.3	9.5
S4, S6	9.4	9.6
S4 twin turbo	9.6	9.8
TT	5.1	5.3
TT Quattro	10.7	10.9
V8 Quattro	12.8	13.0

Clutch Control Cable, Replace (B)

4000, Coupe	.8	.9

Clutch Master Cylinder, Replace (B)
Includes: System bleeding.

All Models	2.3	2.5

Clutch Release Bearing, Replace (B)

80, 90, A4		
1988-92	5.2	5.4
1988-92 Quattro	9.1	9.3
1993-95	8.3	8.5
1993-95 Quattro	9.1	9.3
1996-01	5.9	6.1
1996-01 Quattro	9.7	9.9
100, 200, A6		
1989-91	6.2	6.4
1989-91 Quattro	8.7	8.9
1992-95	8.3	8.5
1992-95 Quattro	9.1	9.3
1996-97	5.9	6.1
1996-97 Quattro	6.4	6.6
1998-01	5.9	6.1
1998-01 Quattro	6.4	6.6
4000, Coupe	5.2	5.4
Quattro	7.0	7.2
5000	6.2	6.4
Quattro	8.3	8.5
Coupe Quattro	9.0	9.2
S4, S6	9.1	9.3
S4 twin turbo	9.4	9.6
TT	4.9	5.1
TT Quattro	10.5	10.7
V8 Quattro	12.4	12.6

Clutch Slave Cylinder, Replace (B)
Includes: System bleeding.

All Models	1.4	1.5

Flywheel, Replace (B)

80, 90, A4		
1988-92	6.3	6.5
1988-92 Quattro	9.9	10.1
1993-95	9.4	9.6
1993-95 Quattro	9.9	11.1
1996-01	7.2	7.4
1996-01 Quattro	11.2	11.4
100, 200, A6		
1989-91	7.6	7.8
1989-91 Quattro	9.9	10.1
1992-95	9.4	9.6
1992-95 Quattro	9.9	11.1
1996-97	7.2	7.4
1996-97 Quattro	7.8	8.0
1998-01	8.3	8.5
1998-01 Quattro	9.1	9.3
4000, Coupe	5.8	6.0
Quattro	7.8	8.0
5000	7.6	7.8
Quattro	9.1	9.3
Coupe Quattro	9.6	9.8
S4, S6	9.9	11.1
S4 twin turbo	10.0	10.2
TT	5.5	5.7
TT Quattro	11.1	11.3
V8 Quattro	13.1	13.3

MANUAL TRANSAXLE

Transaxle Assy. R&R and Reconditon (A)

80, 90, A4		
1988-92	13.1	14.0
1988-92 Quattro	16.8	17.7
1993-95	13.1	14.0
1993-95 Quattro	16.8	16.7
1996-01	13.0	13.9
1996-01 Quattro	19.6	20.5
100, 200, A6		
1989-91	13.1	14.0
1989-91 Quattro		
01A trans.	16.8	17.7
016 trans.	17.4	18.3
1992-95	13.1	14.0
1992-95 Quattro	16.8	17.7
1996-97	13.0	13.9
1996-97 Quattro	16.3	17.2
1998-01	14.1	15.0
1998-01 Quattro	17.4	18.3
4000, Coupe	13.4	14.3
Quattro	14.1	15.0
5000	14.8	15.7
Quattro	16.5	17.4
Coupe Quattro	16.7	17.6
S4, S6	16.8	17.7
S4 twin turbo	17.3	18.2
TT	10.3	11.2
TT Quattro	16.6	17.5
V8 Quattro	15.5	16.4

Transaxle Assy., R&R or Replace (B)

80, 90, A4		
1988-92	4.9	5.1
1988-92 Quattro	8.4	8.6
1993-95	8.4	8.6
1993-95 Quattro	9.3	9.5
1996-01	5.8	6.0
1996-01 Quattro	9.9	10.1
100, 200, A6		
1989-91	6.2	6.4
1989-91 Quattro	8.4	8.6
1992-95	8.4	8.6
1992-95 Quattro	8.9	9.1
1996-97	5.8	6.0
1996-97 Quattro	6.3	6.5
1998-01	6.9	7.1
1998-01 Quattro	7.6	7.8

AUD-14 80 : 90 : 100 : 200 : 4000 : 5000 : A4 : A6 : A8 : CABRIOLET : COUPE : S4 : S6 : TT : V8 QUATTRO

	LABOR TIME	SEVERE SERVICE
Transaxle Assy., R&R or Replace (B)		
4000, Coupe	4.9	5.1
Quattro	6.8	7.0
5000	5.8	6.0
Quattro	8.0	8.2
S4, S6	8.9	9.1
S4 twin turbo	8.6	8.8
TT	4.7	4.9
TT Quattro	10.2	10.4
V8 Quattro	12.2	12.4
Transaxle Mounts, Replace (B)		
Exc. below		
one	1.0	1.2
both	1.4	1.6
4000, Coupe	1.7	1.9

AUTOMATIC TRANSAXLE
SERVICE TRANSAXLE INSTALLED

Check Unit for Oil Leaks (C)
All Models ... 1.0 ... 1.0

Drain & Refill Unit (B)
Exc. A4 ... 1.2 ... 1.2
A4 ... .7 7

Electronic Control Unit, Replace (B)
80, 90, A4
  1988-92 ... .7 8
  1993-01 ... 1.6 ... 1.7
100, 200, A6
  1989-91 ... .7 8
  1992-95 ... 1.6 ... 1.7
  1996-97 ... 1.6 ... 1.7
  1998-01 ... .8 9
A8 ... .7 8
S4, S6, Cabriolet ... 1.6 ... 1.7
V8 Quattro ... 1.0 ... 1.1

Governor Assy., R&R or Replace (B)
80, 90
  1988-92 ... 1.7 ... 1.9
  1993-95 ... .5 7
100, 200 ... .5 7
4000, Coupe ... 1.0 ... 1.2
5000 ... .5 7
S4, Cabriolet 5 7

Oil Pan and, or Gasket, Replace (B)
Exc. A4 ... 1.2 ... 1.2
A4 ... .7 7

Speed Sensor, Replace (B)
80, 90, A4
  1988-95 ... .5 5
  1996-01 ... 1.0 ... 1.0
100, 200, A6
  1989-97 ... .5 5
  1998-01 ... 1.0 ... 1.0
S4, S6 ... 1.0 ... 1.0
V8 Quattro ... 2.7 ... 2.7

Throttle Valve Control Cable, Replace (B)
4000, Coupe ... 2.3 ... 2.4
5000 ... .7 8

	LABOR TIME	SEVERE SERVICE
Transaxle Auxiliary Oil Cooler, Replace (B)		
4-Speed	1.6	1.7
5-Speed	3.5	3.6
Transaxle Mounts, Replace (B)		
80, 90, A4		
1988-92	.5	.7
1993-01	.5	.7
100, 200, A6	.8	1.0
4000, Coupe	1.6	1.8
5000	.8	1.0
A8	.8	1.0
Cabriolet	.5	.7
Coupe Quattro	.6	.8
S4, S6	.8	1.0
V8 Quattro	.8	1.0
Valve Body, R&R and Recondition (B)		
1981-92	2.8	3.2
Valve Body, Replace (B)		
Exc. 5-Speed	1.6	1.8
5-Speed	4.5	4.7

SERVICE TRANSAXLE REMOVED
Transaxle R&R included unless otherwise noted.

Torque Converter, Replace (B)
80, 90, A4
  1988-92 ... 6.4 ... 6.6
  1993-95 ... 8.2 ... 8.4
  1996-01
    5-Speed ... 10.7 ... 10.9
    5-Speed Quattro ... 13.4 ... 13.6
100, 200, A6
  1989-95 ... 8.2 ... 8.4
  1996-97 4-Speed ... 8.0 ... 8.2
  1998-01
    5-Speed ... 10.2 ... 10.4
    5-Speed Quattro ... 13.4 ... 13.6
4000, 5000, Coupe ... 7.0 ... 7.2
A8 5-Speed ... 17.0 ... 17.2
A8 5-Speed Quattro ... 19.5 ... 19.7
Coupe Quattro ... 6.4 ... 6.6
S4, S6, Cabriolet ... 8.2 ... 8.4
V8 Quattro ... 13.7 ... 13.9
w/turbo add
  100, 200, A6 ... 1.7 ... 1.7

Transaxle Assy., R&R and Recondition (A)
80, 90
  1988-92 ... 13.8 ... 14.7
  1993-95 ... 11.0 ... 11.9
100, 200, A6
  1989-95 ... 15.4 ... 16.3
  1996-01 4-Speed ... 20.7 ... 21.6
4000, Coupe ... 14.6 ... 15.5
5000 ... 14.6 ... 15.5

	LABOR TIME	SEVERE SERVICE
Cabriolet	11.0	11.9
Coupe Quattro	14.0	14.9
S4, S6	15.4	16.3
w/turbo add		
100, 200, A6	1.7	1.7

Transaxle Assy., R&R or Replace (B)
80, 90, A4
  1988-92 ... 6.2 ... 6.4
  1993-95 ... 7.9 ... 8.1
  1996-01
    5-Speed ... 10.5 ... 10.7
    5-Speed Quattro ... 13.1 ... 13.2
100, 200, A6
  1989-95 ... 7.9 ... 8.1
  1996-97 4-Speed ... 7.8 ... 8.0
  1998-01
    5-Speed ... 10.0 ... 10.2
    5-Speed Quattro ... 13.1 ... 13.3
4000, 5000, Coupe ... 6.8 ... 7.0
A8 5-Speed ... 16.8 ... 17.0
A8 5-Speed Quattro ... 19.3 ... 19.5
Cabriolet ... 7.9 ... 8.1
Coupe Quattro ... 6.2 ... 6.4
S4, S6 ... 8.0 ... 8.2
S4 twin turbo ... 15.8 ... 16.0
V8 Quattro ... 13.5 ... 13.7
w/turbo add
  100, 200, A6 ... 1.7 ... 1.7

SHIFT LINKAGE
AUTOMATIC TRANSAXLE

Gear Selector Lever, Replace (B)
Exc. below ... 3.0 ... 3.0
1988-92 80, 90, A4 ... 1.2 ... 1.2
4000, 5000, Coupe ... 1.4 ... 1.4

Gear Shift Housing, Replace (B)
Exc. below ... 2.8 ... 2.8
1988-92 80, 90, A4 ... .8 8
4000, 5000, Coupe 5 5
Coupe Quattro ... 1.0 ... 1.0

Selector Cable, Replace (B)
Exc. below ... 2.8 ... 2.9
1988-92 80, 90, A4 ... 1.4 ... 1.5
4000, Coupe ... 2.3 ... 2.4
5000 ... 1.8 ... 1.9
Coupe Quattro ... 1.4 ... 1.5

MANUAL TRANSAXLE

Gearshift Linkage, Adjust (B)
All Models 8 9

Gearshift Lever, Replace (B)
All Models ... 1.4 ... 1.4

Gearshift Lever Boot, Replace (B)
All Models 7 7

Gearshift Lever Housing or Base, Replace (B)
4000, Coupe ... 3.2 ... 3.2

80 : 90 : 100 : 200 : 4000 : 5000 : A4 : A6 : A8 : CABRIOLET : COUPE : S4 : S6 : TT : V8 QUATTRO **AUD-15**

	LABOR TIME	SEVERE SERVICE

DRIVELINE

Differential Assy., R&R and Reseal (B)
- All Models .... 13.1 / 13.6

Differential Vacuum Lock Activators, Replace (B)
- Front .... 1.3 / 1.3
- Rear .... .5 / .5
- Both .... 1.6 / 1.6

Drive Flange Shaft Seal, Replace (B)
- One side .... 1.8 / 1.9
- Both sides .... 2.8 / 2.9

Front Differential Assy., R&R and Recondition (B)
- 1988-92 80, 90 ..... 18.0 / 18.5
- 1989-91 100, 200 ... 19.5 / 20.0
- 1987-88 5000 .... 19.5 / 20.0

Rear Differential Assy., R&R or Replace (B)
- 80, 90, A4
  - 1988-92 .... 2.3 / 2.8
  - 1993-95 .... 2.7 / 3.2
  - 1996-01 .... 2.3 / 2.8
- 100, 200, A6
  - 1989-91 .... 3.9 / 4.5
  - 1992-95 .... 2.7 / 3.2
  - 1996-97 .... 2.7 / 3.2
  - 1998-01 .... 3.5 / 4.0
- 4000, 5000, Coupe .. 2.7 / 3.2
- A8 .... 3.9 / 4.4
- Coupe Quattro ...... 2.2 / 2.7
- S4, S6, Cabriolet .. 2.7 / 3.2
- V8 Quattro .... 3.9 / 4.4

Stub Axle Assy., Replace (B)
- All Models one ..... 1.7 / 1.9

HALFSHAFTS

Front Halfshaft, R&R or Replace (B)
- 80, 90, A4
  - 1988-92
    - AT
      - one .... 2.7 / 2.9
      - both .... 3.6 / 3.8
    - MT
      - one .... 1.7 / 1.9
      - both .... 2.5 / 2.7
  - 1993-01
    - AT
      - one .... 2.7 / 2.9
      - both .... 3.6 / 3.8
    - MT
      - one .... 2.0 / 2.2
      - both .... 2.8 / 3.0
- 100, 200, A6
  - AT
    - one .... 2.7 / 2.9
    - both .... 3.6 / 3.8
  - MT
    - one .... 2.0 / 2.2
    - both .... 2.8 / 3.0
- 4000, Coupe .... 1.7 / 1.9
- 5000 .... 2.0 / 2.2
- A8
  - one .... 2.8 / 3.0
  - both .... 3.7 / 3.9
- Coupe Quattro
  - one .... 2.6 / 2.8
  - both .... 3.7 / 3.9
- S4, S6, Cabriolet
  - AT
    - one .... 2.8 / 3.0
    - both .... 3.7 / 3.9
  - MT
    - one .... 1.8 / 2.0
    - both .... 2.6 / 2.8
- V8 Quattro
  - AT
    - one .... 2.8 / 3.0
    - both .... 3.7 / 3.9
  - MT
    - one .... 1.8 / 2.0
    - both .... 2.7 / 2.9

Replace
- inner or outer CV joint and/or boot add .... .7 / .9
- inner and outer CV joints and/or boots add .... 1.0 / 1.2

Rear Halfshaft, R&R or Replace (B)
- One .... 2.7 / 2.9

Replace CV joint or boot add each .... .5 / .7

BRAKES

ANTI-LOCK

The following operations do not include testing. Add time as required.

Diagnose Anti-Lock Brake System (A)
- 1988-01 .... 1.0 / 1.0

Anti-Lock Relay, Replace (B)
- 1988-01 each .... .3 / .3

Control Unit, Replace (B)
- 1988-01 .... .5 / .5

Hydraulic Assy., Replace (B)
- 80, 90, A4
  - 1988-92 .... 1.4 / 1.4
  - 1993-95 .... 1.7 / 1.7
  - 1996-01 .... 2.0 / 2.0
- 100, 200, A6
  - 1992-97 .... .7 / .7
  - 1998-01 .... 2.0 / 2.0
- A8 .... 2.0 / 2.0
- S4, S6, Cabriolet .. .7 / .7
- V8 Quattro .... 1.4 / 1.4

Add time for system bleeding.

On/Off Switch, Replace (B)
- 1988-01 .... .3 / .3

Over Voltage Protection Switch, Replace (B)
- 1988-01 .... .3 / .3

Pump or Main Relay, Replace (B)
- 1988-01 .... .3 / .3

Solenoid Valve Relay, Replace (B)
- 1988-01 .... .3 / .3

Toothed Rotor, Replace (B)
- 1988-01 one
  - front .... 2.0 / 2.0
  - rear .... 1.7 / 1.7

Wheel Speed Sensor, Replace (B)
- Front
  - one .... .5 / .5
  - both .... .8 / .8
- Rear
  - one .... .8 / .8
  - both .... 1.4 / 1.4

SYSTEM

Bleed Brakes (B)
Includes: Add fluid.
- All Models .... .5 / .5

Brake System, Flush and Refill (B)
- All Models .... 1.2 / 1.2

Brakes, Adjust (B)
Includes: Brake adjustment, refill master cylinder.
- 1981-88 .... .7 / .7

Brake Hose (Flexible), Replace (B)
Includes: System bleeding.
- 80, 90, A4
  - one .... 1.2 / 1.3
  - each addl. add ... .3 / .4
- 100, 200, A6
  - one .... 1.2 / 1.3
  - each addl. add ... .3 / .4
- 4000, Coupe
  - front each .... .5 / .6
  - rear each .... .7 / .8
- 5000 each .... 1.0 / 1.1
- A8, S4, S6, Cabriolet
  - one .... 1.2 / 1.3
  - each addl. add ... .3 / .4
- Coupe Quattro
  - one .... 1.1 / 1.2
  - each addl. add ... .4 / .5
- TT
  - front each .... 1.2 / 1.3
  - rear each .... .6 / .7
- V8 Quattro
  - one .... 1.2 / 1.3
  - each addl. add ... .3 / .4

Brake Pressure Regulator, Replace (B)
- All Models .... 1.0 / 1.1

AUD-16 80 : 90 : 100 : 200 : 4000 : 5000 : A4 : A6 : A8 : CABRIOLET : COUPE : S4 : S6 : TT : V8 QUATTRO

	LABOR TIME	SEVERE SERVICE
Master Cylinder, R&R and Reconditioning (B)		
Includes: System bleeding.		
4000, 5000, Coupe	2.2	2.5
Master Cylinder, Replace (B)		
Includes: System bleeding.		
80, 90, A4	1.6	1.8
100, 200, A6		
1989-97	1.6	1.8
1998-01	2.3	2.5
4000, 5000, Coupe	1.5	1.7
A8	2.4	2.6
Cabriolet	1.7	1.9
Coupe Quattro	1.6	1.8
S4, S6	1.6	1.8
TT	1.2	1.4
V8 Quattro	1.6	1.8
Power Booster Low Vacuum Switch, Replace (B)		
4000, Coupe	2.6	2.7
5000	1.7	1.8
Power Booster Unit, Replace (B)		
Includes: System bleeding.		
4000, Coupe	2.3	2.5
Power Booster Vacuum Pump, Replace (B)		
All Models	.8	1.0
Pressure Accumulator, Replace (B)		
All Models	1.2	1.4
Vacuum Check Valve, Replace (B)		
All Models	.3	.3

SERVICE BRAKES

	LABOR TIME	SEVERE SERVICE
Brake Drum, Replace (B)		
1981-88		
one	1.7	1.8
both	2.0	2.1
Caliper Assy., R&R and Reconditioning (B)		
Includes: System bleeding.		
Front		
1981-87		
one	1.9	2.2
both	3.1	3.4
1988-01		
one	2.0	2.3
both	3.1	3.4
Rear		
1984-87		
one	2.6	2.9
both	3.6	3.9
1988-01		
one	2.8	3.1
both	3.7	4.0

	LABOR TIME	SEVERE SERVICE
Caliper Assy., Replace (B)		
Includes: System bleeding.		
Front		
1981-87		
one	1.4	1.5
both	2.0	2.1
1988-01		
one	1.5	1.6
both	2.3	2.4
Rear		
1984-87		
one	2.0	2.1
both	2.7	2.8
1988-01		
one	2.3	2.4
both	2.8	2.9
Disc Brake Rotor, Replace (B)		
Front		
1981-87		
one	.7	.8
both	1.2	1.3
1988-01		
w/bearings		
one	1.3	1.4
both	2.3	2.4
w/o bearings		
one	.5	.6
both	1.0	1.1
Rear		
1984-87		
one	2.0	2.1
both	2.7	2.8
1988-01		
one	2.3	2.4
both	2.8	2.9
Pads and/or Shoes, Replace (B)		
Includes: Service and parking brake adjustment, system bleeding.		
Front disc	1.2	1.3
Rear disc	1.1	1.2
Rear drum	2.3	2.4
Four wheels		
Disc brakes	2.0	2.1
Drum brakes	3.2	3.3

COMBINATION ADD-ONS

Replace		
brake hose add each	.3	.3
caliper add	.3	.3
disc rotor add each	.2	.2
master cylinder add	.5	.5
Resurface		
drum add each	.5	.5
rotor add each	.5	.5
Wheel Cylinder, R&R and Reconditioning (B)		
Includes: System bleeding.		
1981-88		
one	2.9	3.2
both	3.8	4.1

	LABOR TIME	SEVERE SERVICE
Wheel Cylinder, Replace (B)		
Includes: System bleeding.		
1981-88		
one	2.5	2.6
both	3.1	3.2

PARKING BRAKE

Parking Brake Cable, Adjust (C)		
All Models	.5	.7
Parking Brake Apply Actuator, Replace (B)		
Exc. below	1.4	1.4
1996-01 80, 90, A4	2.0	2.0
1998-01 100, 200, A6	1.7	1.7
4000, 5000, Coupe	1.6	1.6
A8	2.8	2.8
Parking Brake Apply Warning Indicator Switch, Replace (B)		
All Models	1.2	1.2
Parking Brake Cable, Replace (B)		
Drum brakes		
4000 one	1.4	1.6
5000 one	1.6	1.8
Disc brakes		
exc. below		
one	1.4	1.6
both	1.7	1.9
4000, Coupe	2.6	2.8
5000	1.9	2.1
TT		
one	1.1	1.3
both	1.4	1.6

FRONT SUSPENSION

Unless otherwise noted, time given does not include alignment.

Align Front End (A)		
All Models	1.7	1.9
Front Axle Platform, Replace (B)		
80, 90, A4		
1988-92	2.4	2.9
1993-95	3.1	3.6
1996-01	3.7	4.2
100, 200, A6		
1989-91	2.8	3.3
1992-01	3.1	3.6
4000, Coupe	2.3	2.8
5000	2.8	3.3
A8	4.3	4.8
Coupe Quattro	3.2	3.7
S4, S6, Cabriolet	3.1	3.6
TT	2.2	2.7
V8 Quattro	2.8	3.3
Lower Ball Joint, Replace (B)		
80, 90, A4		
one	.7	.9
both	1.3	1.5
100, 200, A6		
one	.7	.9
both	1.4	1.6

80 : 90 : 100 : 200 : 4000 : 5000 : A4 : A6 : A8 : CABRIOLET : COUPE : S4 : S6 : TT : V8 QUATTRO **AUD-17**

	LABOR TIME	SEVERE SERVICE
4000, Coupe		
one	1.3	*1.5*
both	1.7	*1.9*
5000		
one	.8	*1.0*
both	1.5	*1.7*
A8, S4, S6, Cabriolet		
one	.7	*.9*
both	1.4	*1.6*
Coupe Quattro		
one	.8	*1.0*
both	1.4	*1.6*
TT		
one	.6	*.8*
both	1.0	*1.2*
V8 Quattro		
one	.7	*.9*
both	1.4	*1.6*
Lower Control Arm, Replace (B)		
80, 90, A4		
1988-92		
one	1.4	*1.6*
both	1.7	*1.9*
1993-01		
one	1.7	*1.9*
both	2.0	*2.2*
100, 200, A6		
one	1.7	*1.9*
both	2.0	*2.2*
4000, Coupe		
one	1.9	*2.1*
both	2.3	*2.5*
5000		
one	1.7	*1.9*
both	2.3	*2.5*
A8		
one	1.7	*1.9*
both	2.4	*2.6*
S4, S6		
one	2.0	*2.2*
both	2.4	*2.6*
Cabriolet		
one	1.7	*1.9*
both	2.4	*2.6*
Coupe Quattro		
one	1.4	*1.6*
both	2.0	*2.2*
TT		
one	1.1	*1.3*
both	1.9	*2.1*
V8 Quattro		
one	1.4	*1.6*
both	1.8	*2.0*
Shock Absorber Strut Cartridge, Replace (B)		
80, 90, A4		
1988-95		
one	2.6	*2.9*
both	3.7	*4.0*
1996-01		
one	1.9	*2.1*
both	2.6	*2.9*
100		
one	2.3	*2.6*
both	3.0	*3.3*
200		
one	2.5	*2.8*
both	3.2	*3.5*
4000, Coupe		
one	3.2	*3.5*
both	3.7	*4.0*
5000		
one	3.4	*3.7*
both	4.6	*4.9*
A6		
one	1.4	*1.7*
both	2.0	*2.3*
A8		
one	2.5	*2.8*
both	3.2	*3.5*
Cabriolet		
one	2.2	*2.5*
both	3.0	*3.3*
Coupe Quattro		
one	2.5	*2.8*
both	3.7	*4.0*
S4, S6		
one	1.4	*1.7*
both	2.0	*2.3*
TT		
one	2.7	*3.0*
both	3.4	*3.7*
V8 Quattro		
one	2.5	*2.8*
both	3.2	*3.5*
Stabilizer Bar, Replace (B)		
80, 90, A4		
1988-92	1.3	*1.4*
1993-95	2.4	*2.5*
1996-01	1.2	*1.3*
100, 200, A6		
1989-91	1.3	*1.4*
1992-97	2.4	*2.5*
1998-01	1.6	*1.7*
4000, Coupe	1.2	*1.3*
5000	1.3	*1.4*
A8	1.6	*1.7*
Cabriolet	1.2	*1.3*
Coupe Quattro	2.3	*2.4*
S4, S6	1.8	*1.9*
TT	.7	*.8*
V8 Quattro	1.3	*1.4*
Strut Assembly, R&R or Replace (B)		
80, 90, A4		
1988-95		
one	2.4	*2.5*
both	3.5	*3.6*
1996-01		
one	1.7	*1.8*
both	2.4	*2.5*
100		
one	2.0	*2.1*
both	2.8	*2.9*
200		
one	2.3	*2.4*
both	3.0	*3.1*
4000, Coupe		
one	3.0	*3.1*
both	3.5	*3.6*
5000		
one	3.1	*3.2*
both	4.4	*4.5*
A6		
one	1.7	*1.8*
both	2.4	*2.5*
A8		
one	2.3	*2.4*
both	3.0	*3.1*
Cabriolet		
one	2.0	*2.1*
both	2.8	*2.9*
Coupe Quattro		
one	2.3	*2.4*
both	3.5	*3.6*
S4, S6		
one	2.4	*2.5*
both	3.7	*3.8*
TT		
one	1.6	*1.7*
both	3.0	*3.1*
V8 Quattro		
one	2.3	*2.4*
both	3.0	*3.1*
Replace coil spring add	.5	*.7*
Strut Coil Spring, Replace (B)		
80, 90, A4		
1988-95		
one	2.4	*2.5*
both	3.5	*3.6*
1996-01		
one	1.7	*1.8*
both	2.4	*2.5*
100		
one	2.1	*2.2*
both	2.8	*2.9*
200		
one	2.3	*2.4*
both	3.0	*3.1*
4000, Coupe		
one	3.0	*3.1*
both	3.5	*3.6*
5000		
one	3.2	*3.3*
both	4.4	*4.5*

AUD-18 80 : 90 : 100 : 200 : 4000 : 5000 : A4 : A6 : A8 : CABRIOLET : COUPE : S4 : S6 : TT : V8 QUATTRO

	LABOR TIME	SEVERE SERVICE
Strut Coil Spring, Replace (B)		
A6		
one	1.2	1.3
both	1.8	1.9
A8		
one	2.3	2.4
both	3.0	3.1
Cabriolet		
one	2.0	2.1
both	2.8	2.9
Coupe Quattro		
one	2.3	2.3
both	3.5	3.6
S4, S6		
one	1.2	1.3
both	1.8	1.9
TT		
one	2.5	2.6
both	3.2	3.3
V8 Quattro		
one	2.3	2.4
both	3.0	3.1
Wheel Bearing & Hub Assy., Replace (B)		
80, 90, A4		
1988-92	3.1	3.3
1993-01	3.3	3.5
100, 200, A6	3.3	3.5
4000, Coupe		
one	2.6	2.8
both	3.7	3.9
5000		
one	3.0	3.2
both	5.2	5.4
A8, S4, S6, Cabriolet	3.3	3.5
Coupe Quattro	3.0	3.2
TT	2.8	3.0
V8 Quattro	3.3	3.5
Wheel Bearing Housing, Replace (B)		
80, 90, A4		
1988-92	3.1	3.3
1993-95	4.3	4.5
1996-01	2.4	2.6
100, 200, A6		
1989-91	3.5	3.7
1992-97	4.3	4.5
1998-01	2.8	3.0
4000, Coupe	3.1	3.3
5000	3.5	3.7
A8	2.4	2.6
S4, S6, Cabriolet	4.3	4.5
V8 Quattro	3.5	3.7

REAR SUSPENSION

Unless otherwise noted, time given does not include alignment.

	LABOR TIME	SEVERE SERVICE
Rear Suspension (Complete), Align (A)		
All Models	1.5	1.7
Rear Toe, Check and Adjust (B)		
1983-85 Quattro Coupe	.8	1.0
Rear Wheel Bearings, Adjust (B)		
1981-97		
one side	.3	.4
both sides	.5	.6
Rear Axle Assy. Carrier, Replace (B)		
80, 90, A4		
1988-92	2.0	2.5
1993-95	3.1	3.6
1996-01	4.2	4.7
100, 200, A6		
1989-91	3.6	4.1
1992-01	3.1	3.6
4000	2.5	3.0
5000	3.6	4.1
A8	4.2	4.7
Cabriolet	3.1	3.6
Rear Control Arm, Replace (B)		
80, 90, A4		
1988-92		
one	1.4	1.6
both	2.0	2.2
1993-95		
one	1.9	2.1
both	2.8	3.0
1996-01		
one	1.7	1.9
both	2.8	3.0
100, 200, A6		
1989-95		
one	1.9	2.1
both	2.8	3.0
1996-97		
one	1.7	1.9
both	2.8	3.0
1998-01		
one	1.2	1.4
both	2.4	2.6
A8		
one	2.0	2.2
both	2.8	3.0
S4, S6		
one	1.7	1.9
both	2.8	3.0
Cabriolet		
one	1.8	2.0
both	2.8	3.0
Coupe Quattro		
one	1.4	1.6
both	2.0	2.2
V8 Quattro		
one	2.0	2.2
both	2.8	3.0

	LABOR TIME	SEVERE SERVICE
Rear Shock Absorber, Replace (B)		
2000-01 TT		
one	1.2	1.3
both	1.1	1.2
Rear Strut Assy., Replace (B)		
Includes: Alignment charges.		
80, 90, A4		
one	1.4	1.5
both	1.7	1.8
100, 200, A6		
one	1.4	1.5
both	1.7	1.8
4000, Coupe		
one	1.6	1.7
both	1.9	2.0
5000		
one	.7	.8
both	1.5	1.6
A8		
one	1.4	1.5
both	1.7	1.8
S4, S6		
one	1.5	1.6
both	1.8	1.9
Cabriolet		
one	1.4	1.5
both	1.7	1.8
Coupe Quattro		
one	1.4	1.5
both	1.7	1.8
V8 Quattro		
one	1.5	1.6
both	1.8	1.9
Replace coil spring add each	.5	.7
Stabilizer Bar, Replace (B)		
1983-85 Quattro Coupe	.8	1.0
Strut Shock Absorber, Replace (B)		
Includes: Alignment charges.		
80, 90, A4		
one	1.9	2.0
both	2.3	2.4
100, 200, A6		
one	1.9	2.0
both	2.3	2.4
4000, Coupe		
one	2.2	2.3
both	2.5	2.6
5000		
one	1.4	1.5
both	2.0	2.1
A8		
one	2.0	2.1
both	2.4	2.5
S4, S6		
one	2.0	2.1
both	2.4	2.5
Cabriolet		
one	1.9	2.0
both	2.3	2.4

AUD-19
80 : 90 : 100 : 200 : 4000 : 5000 : A4 : A6 : A8 : CABRIOLET : COUPE : S4 : S6 : TT : V8 QUATTRO

	LABOR TIME	SEVERE SERVICE
Coupe Quattro		
one	1.6	1.7
both	1.9	2.0
V8 Quattro		
one	2.0	2.1
both	2.4	2.5
Stub Axle, Replace (B)		
One	1.7	1.9
Track Arm, Replace (B)		
A4 Quattro		
one side	.7	.9
both sides	1.2	1.4
A6, S6 Quattro		
one side	1.2	1.4
both sides	1.7	1.9
A8 Quattro		
one side	1.2	1.4
both sides	1.7	1.9
Transverse Suspension Rod, Replace (B)		
1981-97 one	.7	.9
Replace bearings add	.3	.5
Wheel Bearings, Replace (B)		
One side	2.0	2.0
w/Quattro add each	.1	.1

STEERING
AIR BAGS
The following operations do not include testing. Add time as required.

	LABOR TIME	SEVERE SERVICE
Diagnose Air Bag System (A)		
1988-01	1.0	1.0
Air Bag Assy., Replace (B)		
Driver side	.5	.5
Passenger side		
1988-94	3.5	3.5
1995-01		
A4	1.7	1.7
A8	1.7	1.7
Cabriolet	1.7	1.7
S4, S6, A6	3.5	3.5
Side impact		
front	2.7	2.7
rear	1.4	1.4
Contact Coil Initiator, Replace (B)		
1988-01	1.8	1.8
Voltage Converter, Replace (B)		
1988-01	.3	.3

POWER RACK & PINION
Unless otherwise noted, time given does not include alignment.

	LABOR TIME	SEVERE SERVICE
Troubleshoot Power Steering (A)		
All Models	1.0	1.0
Inner Tie Rod Ends, Replace (B)		
80, 90, A4		
1988-92		
one	1.0	1.2
both	1.7	1.9
1993-01		
one	1.3	1.5
both	1.6	1.8
100, A6		
one	1.3	1.5
both	1.6	1.8
200		
one	1.0	1.2
both	1.7	1.9
4000, 5000, Coupe		
1981-88		
one	.7	.9
both	1.3	1.5
1984-88 Quattro		
one	1.3	1.5
both	1.6	1.8
Coupe Quattro		
one	1.0	1.2
both	1.6	1.8
S4, S6, Cabriolet		
one	1.0	1.2
both	1.7	1.9
TT		
one	1.6	1.8
both	2.1	2.3
V8 Quattro		
one	1.3	1.5
both	1.6	1.8
Lower Steering Shaft, Replace (B)		
Exc. below	1.0	1.1
4000, Coupe	1.2	1.3
5000	1.2	1.3
Outer Tie Rod Ends, Replace (B)		
1981-93 one	1.0	1.2
1994-01		
90, Cabriolet		
one	.8	1.0
both	1.6	1.8
A4, S4, S6		
one	2.0	2.2
both	2.8	3.0
A8		
one	1.3	1.5
both	1.8	2.0
Power Steering Reservoir, Replace (B)		
1988-97	1.4	1.4
Rack & Pinion Assy., Replace (B)		
2000-01 TT	3.1	3.4
Rack & Pinion Assy., Reseal (A)		
Includes: Alignment.		
1988-97	6.2	6.8
Rack Assy. (Short), Replace (A)		
80, 90, A4		
1988-92		
w/ABS	4.9	5.2
w/air bags	6.2	6.5
w/o air bags	4.3	4.6
1993-95	6.8	7.1
1996-01	5.2	5.5

	LABOR TIME	SEVERE SERVICE
100, 200, A6		
1989-91	4.3	4.6
1992-97	6.2	6.5
1998-01	4.8	5.1
4000, 5000, Coupe	5.3	5.6
A8	6.0	6.3
Coupe Quattro	6.0	6.3
S4, S6, Cabriolet	6.8	7.1
V8 Quattro	5.5	5.8
w/ABS add		
100, 200, A6	1.4	1.4
4000, 5000	1.4	1.4
Rack Boots, Replace (B)		
Each	.5	.7
Servotronic Control Unit, Replace (B)		
1990-01	1.0	1.1
Steering Damper, Replace (B)		
Exc. below	.7	.8
1988-92 80, 90	1.2	1.3
Steering Pump, R&R and Recondition (A)		
4000	3.0	3.5
Quattro	3.1	3.6
Steering Pump, R&R or Replace (B)		
80, 90, A4		
1988-92	2.5	2.6
1993-95	1.4	1.5
1996-01	2.4	2.5
100, 200, A6		
1989-97	1.7	1.8
1998-01	2.4	2.5
5000		
1981-83	2.0	2.1
1984-87	2.4	2.5
A8	2.8	2.9
Coupe Quattro	2.0	2.1
S4, S6, Cabriolet	1.2	1.3
TT	1.6	1.7
V8 Quattro	2.4	2.5
Reseal pump add	1.2	1.5
Replace front seal add	.3	.5
Steering Pump Hoses, Replace (B)		
Does not include purging.		
Pressure or Return		
5000 each	.7	.8
Pressure		
80, 90, A4		
1988-92	3.2	3.3
1993-01	3.2	3.3
100, 200, A4		
1989-91	1.7	1.8
1992-01	2.0	2.1
4000, Coupe	1.0	1.1
A8, S4, S6	2.0	2.1
Cabriolet	3.2	3.3
Coupe Quattro	3.0	3.1
TT	1.1	1.2
V8 Quattro	1.0	1.1

AUD-20 80 : 90 : 100 : 200 : 4000 : 5000 : A4 : A6 : A8 : CABRIOLET : COUPE : S4 : S6 : TT : V8 QUATTRO

	LABOR TIME	SEVERE SERVICE
Steering Pump Hoses, Replace (B)		
Return		
80, 90, A4		
1988-92	3.1	3.2
1993-01	1.0	1.1
100, 200, A6	1.0	1.1
4000, Coupe	1.0	1.1
A8, S4, S6, Cabriolet	1.0	1.1
Coupe Quattro	3.1	3.2
TT	1.1	1.2
V8 Quattro	1.0	1.1
Suction	1.0	1.1
Steering Wheel, Replace (B)		
All Models	.3	.3
w/air bag add	.5	.5
Upper Steering Shaft, Replace (B)		
80, 90, A4		
1988-92	1.6	1.7
1993-01	3.0	3.1
100, 200, A6		
1989-91	3.2	3.3
1992-01	3.0	3.1
4000, Coupe	3.0	3.1
5000		
1981-83	1.5	1.6
1984-88	3.2	3.3
1987-88 Quattro	1.6	1.7
A8, S4, S6, Cabriolet	3.0	3.1
Coupe Quattro	2.7	2.8
TT	1.9	2.0
V8 Quattro	2.8	2.9

HEATING & AIR CONDITIONING

When more than one component requires replacement where evacuation, recovery and recharging is already included, deduct 1.0 hour for each additional component from the time given.

Evacuate, Recover and Recharge System (B)
All Models .... 1.1 / *1.1*

AC Hoses, Replace (B)
Includes: Evacuate, recover and recharge.
80, 90, A4
 1988-92
  discharge ..... 3.0 / *3.1*
  suction .... 3.1 / *3.2*
 1993-01 one .... 2.7 / *2.8*
100, 200, A6 one ... 2.7 / *2.8*
4000, Coupe
 discharge ...... 1.4 / *1.5*
 suction .... 1.6 / *1.7*

	LABOR TIME	SEVERE SERVICE
5000		
1981-83 one	1.7	1.8
1984-88 one	2.7	2.8
A8	2.7	2.8
Cabriolet one	2.7	2.8
Coupe Quattro		
discharge	2.9	3.0
suction	3.2	3.3
S4, S6 one	2.7	2.8
TT one	2.2	2.3
V8 Quattro one	2.7	2.8
each addl. add	.7	.8
Accumulator Assy., Replace (B)		
Includes: Evacuate, recover and recharge.		
1988-01	2.8	2.8
Compressor Assy., Replace (B)		
Includes: Transfer parts. Evacuate, recover and recharge system.		
80, 90, A4		
1988-92	3.9	4.0
1993-95	3.2	3.3
1996-01	4.8	4.9
100, 200, A6	3.2	3.3
4000, Coupe	3.1	3.2
5000	3.3	3.4
A8	1.5	1.6
Cabriolet	3.5	3.6
Coupe Quattro	3.9	4.0
S4, S6	3.0	3.1
TT	4.5	4.6
V8 Quattro	3.9	4.0
Replace clutch assy. add	.3	.5
Compressor Clutch Relay, Replace (B)		
1988-92	1.3	1.3
Compressor Shaft Seal, Replace (B)		
Includes: Compressor R&R, evacuate, recover and recharge.		
80, 90, A4		
1988-92	4.2	4.5
1993-95	3.8	4.1
1996-01	5.2	5.5
100, 200, A6		
1989-91	3.8	4.1
1992-01	3.6	3.9
4000, Coupe	3.0	3.3
5000	2.7	3.0
Cabriolet	4.9	5.1
Coupe Quattro	4.2	4.5
S4, S6	3.1	3.3
V8 Quattro	4.2	4.5
Condenser Assy., Replace (B)		
Includes: Evacuate, recover and recharge.		
80, 90, A4		
1988-92	3.7	3.8
1993-95	3.6	3.7
1996-01	3.7	3.8

	LABOR TIME	SEVERE SERVICE
100, 200, A6		
1989-91	3.1	3.2
1992-01	3.8	3.9
4000, Coupe		
large	2.2	2.3
small	1.8	1.9
5000	3.0	3.1
A8, S4, S6	3.8	3.9
Cabriolet	4.5	4.6
Coupe Quattro	3.7	3.8
TT	3.8	3.9
V8 Quattro	3.1	3.2
Electronic Servo Motor, Replace (B)		
1992-97	2.0	2.0
Evaporator Coil, Replace (B)		
Includes: Evacuate, recover and recharge.		
80, 90, A4		
1988-92	7.2	7.2
1993	6.2	6.2
1994-95	7.8	7.8
1996-01	10.8	10.8
100, 200, A6		
1989-95	3.8	3.8
1996-97	2.5	2.5
1998-01	3.8	3.8
4000, Coupe	3.5	3.5
5000	4.1	4.1
A8	3.8	3.8
Cabriolet	1.0	1.0
Coupe Quattro	7.1	7.1
S4, S6	2.5	2.5
TT	7.4	7.4
V8 Quattro	2.0	2.0
Replace expansion valve add	.2	.2
Evaporator Fan Motor, Replace (B)		
80, 90	1.3	1.6
4000, Coupe	1.0	1.3
5000	3.2	3.5
Expansion Valve, Replace (B)		
Includes: Evacuate, recover and recharge.		
4000, Coupe	3.5	3.5
5000	4.1	4.1
Heater Blower Motor, Replace (B)		
80, 90, A4		
1988-92	1.2	1.2
1994-01	1.7	1.7
100, 200, A6		
1989-91	5.3	5.3
1992-97	2.5	2.5
1994-97	3.1	3.1
1998-01	2.3	2.3
4000, Coupe		
5000		
1981-83	1.3	1.3
1984-88	5.3	5.3
1981-87 Quattro	.8	.8

80 : 90 : 100 : 200 : 4000 : 5000 : A4 : A6 : A8 : CABRIOLET : COUPE : S4 : S6 : TT : V8 QUATTRO — AUD-21

	LABOR TIME	SEVERE SERVICE
A8	2.3	2.3
Cabriolet	1.7	1.7
Coupe Quattro	1.0	1.0
S4, S6	3.1	3.1
TT	.6	.6
V8 Quattro	4.8	4.8

Heater Blower Motor Resistor, Replace (B)

80, 90, A4		
1988-92	1.4	1.4
1993-95	1.2	1.2
1996-01	1.4	1.4
100, 200, A6	.8	.8
A8, S4, S6	.8	.8
Cabriolet	1.4	1.4
Coupe Quattro	1.0	1.0

Heater Control Assy., Replace (B)

100, 200	1.4	1.4
4000, 5000, Coupe	1.4	1.4
5000	1.7	1.7
TT	1.1	1.1

Heater Control Valve, Replace (B)

100, 200	1.2	1.2

Heater Core, R&R or Replace (B)

80, 90, A4		
1988-92		
w/air bags	6.5	6.5
w/o air bags	4.8	4.8
1993	6.2	6.2
1994-95	7.1	7.1
1996-01	13.7	13.7
100, 200, A6		
1989-91	4.2	4.2
1992-93	6.2	6.2
1994-97	7.1	7.1
1998-01	8.3	8.3
4000, Coupe	2.3	2.3
5000		
1981-83	3.1	3.1
1984-88	4.8	4.8
1984-87 Quattro	3.1	3.1
A8	8.3	8.3
Cabriolet	9.7	9.7
Coupe Quattro		
w/air bags	6.5	6.5
w/o air bags	4.7	4.7
S4, S6		
1992-93	6.2	6.2
1994-01	8.3	8.3
TT	3.1	3.1
V8 Quattro	2.0	2.0

Heater Hoses, Replace (B)

One	.5	.6

High Pressure Cut-Off Switch, Replace (B)

Exc. A4	1.8	1.8
A4	2.7	2.7

Add time to evacuate & recharge.

Low Pressure Cut-Off Switch, Replace (B)
Includes: Evacuate, recover and recharge.

1988-92 80, 90	6.0	6.0
1989-91 100, 200	1.8	1.8

Receiver/Drier Assy., Replace (B)
Includes: Evacuate, recover and recharge.

All Models	2.2	2.2

Temperature Control Assy., Replace (B)

All Models	2.3	2.3

Thermostatic Control Switch, Replace (B)

All Models	.7	.7

VIR Assy., R&R and Recondition (B)

5000	3.6	3.6

VIR Assy., Replace (B)

5000	3.0	3.0

WIPERS & SPEEDOMETER

Power Antenna, Replace (B)

1984-88 5000, Quattro	1.3	1.3

Radio, R&R (B)

1981-88	1.2	1.2
1989-01	.7	.7

Rear Window Washer Pump, Replace (B)

1981-88		
4000, Coupe	.7	.7
5000	1.0	1.0
1989-01	.7	.7

Rear Window Wiper Motor, Replace (B)

1981-88		
4000, Coupe	1.4	1.6
5000	.8	1.0
1989-94	.8	1.0
1995-01	1.4	1.6

Rear Window Wiper Relay, Replace (B)

1988-01	.5	.5

Speedometer Cable & Casing, Replace (B)

Upper
4000, Coupe	1.3	1.3
5000	1.6	1.6

Lower
4000, 5000, Coupe	.8	.8

Speedometer Head, R&R or Replace (B)

80, 90, A4		
1988-92		
w/air bags	1.7	1.7
w/o air bags	1.4	1.4
1993-01	3.1	3.1
100, 200, A6		
1989-91		
w/air bags	2.7	2.7
w/o air bags	1.9	1.9
1992-01	3.1	3.1
4000, Coupe	1.7	1.7
5000		
1981-83	1.8	1.8
1984-88	2.3	2.3
Coupe Quattro	1.4	1.4
S4, S6, Cabriolet	3.1	3.1
TT	.8	.8
V8 Quattro	3.1	3.1

Windshield Washer Pump, Replace (B)

80, 90, A4		
1988-92	.5	.5
1993-01	1.4	1.4
100, 200, A6	.8	.8
4000, 5000, Quattro	.8	.8
A8	2.3	2.3
Coupe Quattro	.6	.6
S4, S6, Cabriolet	1.4	1.4
TT	1.2	1.2
V8 Quattro	.5	.5

Windshield Wiper & Washer Switch, Replace (B)

w/air bags	2.3	2.3
w/o air bags		
exc. below	1.2	1.2
1989-91 100, 200	1.0	1.0

Windshield Wiper Interval Relay, Replace (B)

1984-01	.5	.5

Windshield Wiper Linkage, Replace (B)

All Models	1.9	2.1

Windshield Wiper Motor, Replace (B)

80, 90, A4		
1988-92	1.3	1.5
1993-01	1.6	1.8
100, 200, A6		
1989-91	1.6	1.8
1992-01	.8	1.0
4000, Coupe	1.3	1.5
5000	1.6	1.8
A8	1.6	1.8
Coupe Quattro	1.3	1.5
S4, S6, Cabriolet	.8	1.0
TT	1.2	1.4
V8 Quattro	1.6	1.8

Windshield Wiper Switch, Replace (B)

1981-84	1.5	1.5

AUD-22 80 : 90 : 100 : 200 : 4000 : 5000 : A4 : A6 : A8 : CABRIOLET : COUPE : S4 : S6 : TT : V8 QUATTRO

	LABOR TIME	SEVERE SERVICE
LAMPS & SWITCHES		
Back-Up Lamp Bulb, Replace (C)		
All Models	.2	.2
Back-Up Lamp Switch, Replace (B)		
1981-88	.7	.7
1989-01		
AT	2.5	2.5
MT	.7	.7
Brake Lamp Switch, Replace (B)		
1981-88		
4000, Coupe	.7	.7
5000	.5	.5
1989-01 all	1.0	1.0
Combination Switch Assy., Replace (B)		
1981-85	1.0	1.0
Emergency Flasher Switch Assy., Replace (B)		
1981-88		
4000, Coupe	.5	.5
5000	.7	.7
1989-01 all		
w/air bags	2.5	2.5
w/o air bags	1.3	1.3
Halogen Headlamp Bulb, Replace (B)		
1988-01 each	.3	.3
Headlamp Switch, Replace (B)		
1981-88	.7	.7
1989-01	1.6	1.6
Headlamps, Aim (B)		
Two	.5	.5
Four	.7	.7
High Mount Stop Lamp Assy., Replace (B)		
1988-01	.3	.3
Horn, Replace (B)		
All Models one or both	.7	.7
License Lamp Assy., Replace (B)		
All Models	.2	.2
Neutral Safety Switch, Replace (B)		
1981-87	.7	.7
1988-01	2.5	2.5

	LABOR TIME	SEVERE SERVICE
Parking Lamp Lens or Bulb, Replace (C)		
All Models	.2	.2
Sealed Beam Headlamp, Replace (B)		
1981-87		
one side	.5	.5
both sides	.7	.7
Tail Lamp Lens or Bulb, Replace (C)		
All Models	.2	.2
Turn Signal Lamp Lens or Bulb, Replace (C)		
All Models	.2	.2
Turn Signal or Hazard Relay, Replace (B)		
1981-87	.3	.3
1988-01	1.8	1.8
Turn Signal Switch, Replace (B)		
1981-88	1.2	1.2
1989-01		
w/air bags	2.5	2.5
w/o air bags	1.3	1.3
BODY		
Door Window Regulator, Replace (B)		
Front		
80, 90, A4	1.9	2.1
100, 200, A6		
1989-91	2.5	2.7
1992-01	1.9	2.1
4000, Coupe	1.4	1.6
5000	2.3	2.5
A8, S4, S6, Cabriolet	1.9	2.1
Coupe Quattro	3.2	3.4
TT	2.3	2.5
V8 Quattro	1.9	2.1
Rear		
80, 90, A4		
1988-92	1.9	2.1
1993-01	1.6	1.8
100, 200, A6	1.6	1.8
4000, Coupe	1.4	1.6

	LABOR TIME	SEVERE SERVICE
5000	2.3	2.5
A8	2.3	2.5
S4, S6	1.6	1.8
V8 Quattro	1.6	1.8
Hood Lock, Replace (B)		
Upper	.3	.3
Lower	.8	.8
Hood Release Cable, Replace (B)		
All Models	1.2	1.2
Lock Striker Plate, Replace (B)		
All Models	.2	.2
Rear Lid or Liftgate Lock Cylinder, Replace (B)		
All Models	.5	.5
Window Regulator Motor, Replace (B)		
Front		
80, 90, A4		
1988-92	1.7	1.9
1992-01	1.9	2.1
100, 200, A4		
1989-91	2.5	2.7
1992-01	1.9	2.1
1988 5000, Quattro	2.5	2.7
A8	2.5	2.7
Coupe Quattro	3.2	3.4
S4, S6, Cabriolet	1.9	2.1
TT	1.5	1.7
V8 Quattro	2.5	2.7
Rear		
80, 90, A4		
1988-92	1.8	2.0
1993-01	1.6	1.8
100, 200, A6	1.6	1.8
1988 5000, Quattro	1.6	1.8
A8	2.3	2.5
Cabriolet	2.0	2.2
S4, S6	1.6	1.8
V8 Quattro	1.6	1.8

BMW

3 Series : 524TD : 5 Series : 6 Series : 7 Series : 8 Series : Alpina Roadster : M3 : M5 : X3 : X5 : Z3 : Z4 : Z8

SYSTEM INDEX

MAINTENANCE	BMW-2
CHARGING	BMW-2
STARTING	BMW-3
CRUISE CONTROL	BMW-3
IGNITION	BMW-3
EMISSIONS	BMW-4
FUEL	BMW-5
EXHAUST	BMW-8
ENGINE COOLING	BMW-9
ENGINE	BMW-10
Assembly	BMW-10
Cylinder Head	BMW-12
Camshaft	BMW-15
Crank & Pistons	BMW-18
Engine Lubrication	BMW-20
CLUTCH	BMW-21
MANUAL TRANSMISSION	BMW-21
AUTO TRANSMISSION	BMW-22
TRANSFER CASE	BMW-23
SHIFT LINKAGE	BMW-23
DRIVELINE	BMW-24
BRAKES	BMW-26
FRONT SUSPENSION	BMW-27
REAR SUSPENSION	BMW-28
STEERING	BMW-29
HEATING & AC	BMW-31
WIPERS & SPEEDOMETER	BMW-33
LAMPS & SWITCHES	BMW-34
BODY	BMW-34

OPERATIONS INDEX

A
AC Hoses	BMW-31
Air Bags	BMW-29
Air Conditioning	BMW-31
Alignment	BMW-27
Alternator (Generator)	BMW-2
Antenna	BMW-33
Anti-Lock Brakes	BMW-26

B
Back-Up Lamp Switch	BMW-34
Battery Cables	BMW-3
Bleed Brake System	BMW-26
Blower Motor	BMW-31
Brake Disc	BMW-27
Brake Drum	BMW-26
Brake Hose	BMW-26
Brake Pads and/or Shoes	BMW-27

C
Camshaft	BMW-15
Catalytic Converter	BMW-8
Coolant Temperature (ECT) Sensor	BMW-4
Crankshaft	BMW-18
Cruise Control	BMW-3
CV Joint	BMW-25
Cylinder Head	BMW-13

D
Differential	BMW-24
Distributor	BMW-3
Drive Belt	BMW-2
Driveshaft	BMW-24

E
EGR	BMW-5
Electronic Control Module (ECM/PCM)	BMW-5
Engine	BMW-10
Engine Lubrication	BMW-20
Engine Mounts	BMW-11
Evaporator	BMW-32
Exhaust	BMW-8
Exhaust Manifold	BMW-8
Expansion Valve	BMW-32

F
Flexplate	BMW-23
Flywheel	BMW-21
Fuel Injection	BMW-7
Fuel Pump	BMW-6
Fuel Vapor Canister	BMW-5

G
Gear Selector Lever	BMW-23
Generator	BMW-2
Glow Plug	BMW-4

H
Halfshaft	BMW-25
Headlamp	BMW-34
Heater Core	BMW-33
Horn	BMW-34

I
Idle Air Control (IAC) Valve	BMW-7
Ignition Coil	BMW-4
Ignition Module	BMW-4
Ignition Switch	BMW-4
Injection Pump	BMW-5
Inner Tie Rod	BMW-30
Intake Manifold	BMW-6

K
Knock Sensor	BMW-5

M
Mass Air Flow (MAF) Sensor	BMW-5
Master Cylinder	BMW-26
Muffler	BMW-9

N
Neutral Safety Switch	BMW-34

O
Oil Pan	BMW-20
Oil Pump	BMW-21
Outer Tie Rod	BMW-30
Oxygen Sensor	BMW-5

P
Parking Brake	BMW-27
Pistons	BMW-19

R
Radiator	BMW-10
Radiator Hoses	BMW-10
Radio	BMW-33
Rear Main Oil Seal	BMW-19

S
Shock Absorber/Strut, Front	BMW-28
Shock Absorber/Strut, Rear	BMW-29
Spark Plug Cables	BMW-4
Spark Plugs	BMW-4
Spring, Front Coil	BMW-28
Spring, Rear Coil	BMW-28
Starter	BMW-3
Steering Wheel	BMW-30

T
Thermostat	BMW-9
Throttle Body	BMW-7
Throttle Position Sensor (TPS)	BMW-5
Timing Belt	BMW-16
Timing Chain	BMW-17

V
Valve Body	BMW-23
Valve Cover Gasket	BMW-14
Valve Job	BMW-14
Vehicle Speed Sensor	BMW-5

W
Water Pump	BMW-10
Wheel Balance	BMW-2
Wheel Cylinder	BMW-27
Window Regulator	BMW-34
Windshield Washer Pump	BMW-33
Windshield Wiper Motor	BMW-33

BMW-2   3 SERIES : 524TD : 5 SERIES : 6 SERIES : 7 SERIES : 8 SERIES

	LABOR TIME	SEVERE SERVICE
MAINTENANCE		
Air Cleaner Filter Element, Replace (B)		
1981-05 (.3) 5		.7
Drive Belt, Adjust (B)		
Exc. below 3		.5
Power steering worm & sector 7		.9
Drive Belt, Replace (B)		
Exc. below		
1994-95 540i one ..1.2		1.4
1994-95 740i, iL one 1.2		1.4
1988-94 750iL one ..1.2		1.4
840Ci one 1.2		1.4
each addl. add 5		.5
Alternator		
exc. below 8		1.0
325iX 1.2		1.4
325i, iC, is (.8) 1.1		1.3
528i (.8) 1.1		1.3
540i, iA, iT 3		.5
545i (.3) 5		.7
633CSi, 635CSi ..1.2		1.4
645Ci 5		.7
1995-01 740i, iA, iL ..3		.5
750iL 1.2		1.4
M3 ....1.6		1.8
PS		
rack and pinion 5		.7
worm and sector		
exc. below 8		1.0
1994-95 540i 1.2		1.4
633CSi, 635CSi ..1.2		1.4
1988-97 750iL ..1.2		1.4
840Ci, 850Ci, i ...1.7		1.9
Water pump		
850Ci, i 1.7		1.9
Drive Belt Tensioner, Replace (B)		
Serpentine		
exc. below (.8) 1.1		1.3
1996-98 318i, iC, is ...5		.7
325, 325i, iC, is 1.5		1.7
525i 1.5		1.7
1994 740i, iA, iL, 750iL, 760i (.9) 1.2		1.4
M3 ....2.1		2.3
M5 ....1.1		1.3
850Ci, i water pump ..1.9		2.1
Engine Oil Service (C)		
Includes: Inspect belts, hoses, suspension wear. Lube door hinges.		
4 cyl. 1.2		1.4
6 cyl. 1.5		1.7
V8 ....1.7		1.9
V12 ....2.1		2.3

	LABOR TIME	SEVERE SERVICE
Fuel Filter, Replace (B)		
Exc. below 8		1.0
325i, iC, is		
1987-95 ....1.0		1.2
2001-05 7		.9
528i 1.0		1.2
540i		
1994-95 both 1.2		1.4
1997-03 ....1.0		1.2
740i, iL, 750iL both ...1.0		1.2
840Ci, 850Ci, i both ..1.3		1.5
1995 M3 ....1.0		1.2
M5 ....1.0		1.2
Halogen Headlamp Bulb, Replace (B)		
1985-05 each 5		.5
Maintenance Inspection (B)		
4 cyl. 4.8		5.0
6 cyl. 5.2		5.4
V8 ....6.2		6.4
V12		
AT 6.8		7.0
MT 5.8		6.0
Oil and Filter, Change (C)		
1981-05 5		.7
Sealed Beam Headlamp, Replace (B)		
1981-84 4		.4
Timing Belt, Replace (B)		
325, 325e 2.9		3.1
325iX, 524td 3.1		3.3
525i 2.8		3.0
528e		
1982-87 ....2.3		2.5
1988 ....3.1		3.3
Replace tensioner add 2		.2
Tire, Replace (C)		
Includes: Dismount old tire and mount new tire to rim.		
1986-05 5		.5
Tires, Rotate (C)		
1981-05 5		.5
Wheel, Balance (B)		
One 3		.3
each addl. add 2		.2

CHARGING

	LABOR TIME	SEVERE SERVICE
Alternator Circuits, Test (B)		
Includes: Test component output.		
1981-05 5		.5
Alternator Assy., R&R and Recondition (A)		
1984-85 318i 2.8		3.0
320i 2.8		3.0
325, 325e 3.2		3.4
325iX, 528e 3.2		3.4
1981 528i 3.2		3.4
633CSi, 635CSi, 733i .2.8		3.0

	LABOR TIME	SEVERE SERVICE
Alternator Assy., Replace (B)		
Includes: Pulley transfer.		
Exc. below 1.7		1.9
318i, iC, is, Ti 1.1		1.3
323Ci, i, is, iT 1.2		1.4
325Ci, iX 1.2		1.4
328Ci, 330Ci, Xi (.9) ..1.2		1.4
524td 2.3		2.5
525i (.9) 1.2		1.4
530i		
1994-95 ....2.4		2.6
2001-05 (.9) 1.2		1.4
540i, iA, iT		
1994-95 ....2.4		2.6
1997-03 ....1.9		2.1
740i, iA, iL		
1994 ....2.3		2.5
1995-01 ....1.9		2.1
545i, 745i, Li, 760Li (1.9) 2.6		2.8
750iL		
1988-94 ....3.1		3.3
1995-01 ....3.9		4.1
760i (2.4) 3.2		3.4
840Ci 3.1		3.3
850Ci, i		
5.0L		
main 3.3		3.5
auxiliary 2.0		2.2
5.4L 2.9		3.1
M coupe/roadster, Z3 .1.4		1.6
M3 ....1.9		2.1
M5 ....2.7		2.9
X3, X5, Z4 (.8) 1.1		1.3
Alternator Voltage Regulator, Replace (A)		
Exc. below 5		.7
1991-95 318i, iC, is ..1.5		1.9
318ti 1.5		1.9
323Ci, i, is, iT 1.2		1.4
325i, is 1.8		2.0
325iX 8		1.0
325Ci, 330Ci, Xi (.8) ..1.1		1.3
1996-00 328i, iC, is ..1.6		1.8
525i 1.8		2.0
528i, iT 1.4		1.6
530i 2.2		2.4
540i, iA, iT		
1994-95 ....2.2		2.4
1997-01 ....1.4		1.6
545i (1.3) 1.8		2.0
645Ci (1.1) 1.5		1.7
735i 2.3		2.5
740i, iA, iL 1.4		1.6
750iL		
1988-94 ....3.1		3.3
1995-01 ....4.3		4.5

ALPINA ROADSTER : M3 : M5 : X3 : X5 : Z3 : Z4 : Z8 — BMW-3

	LABOR TIME	SEVERE SERVICE
760i (2.6)	3.5	3.7
840Ci	3.1	3.3
850Ci, i	3.2	3.4
M coupe/roadster, Z3		
1.9L	.7	.9
2.3L, 2.8L	1.2	1.4
M3 (1.3)	1.8	2.0
M5	2.7	2.9
X3, X5, Z4 (.8)	1.1	1.3

Front Alternator Bearing, Replace (B)
Exc. below	1.9	2.1
733i, 735i	2.4	2.6

STARTING

Starter Draw Test (On Car) (B)
1981-05	.3	.3

Battery, Replace (C)
One	.5	.7
w/power rear seat add	.2	.2
w/rear CD changer add	.2	.2

Battery Cables, Replace (C)
Positive
exc. below (.9)	1.2	1.4
1991-99 318i, iC, is, ti harness	5.0	5.2
323Ci, i, is, iT	3.6	3.8
325, 325e	2.2	2.4
325i, iC, is harness	5.4	5.6
325iX	2.2	2.4
325Ci, 328Ci	2.8	3.0
330Ci, Xi (2.1)	2.8	3.0
328i, iC, is harness	5.4	5.6
524td	2.3	2.5
525i		
1989-90	2.2	2.5
1991-96 harness	15.2	15.4
530i harness	16.2	16.4
540i, iA, iT harness	16.2	16.4
545i (1.1)	1.5	1.7
740i, iA, iL harness	17.3	17.5
750iL	1.8	2.0
M coupe/roadster, Z3 harness	5.5	5.7
M3 harness (4.1)	5.5	5.7
M5 harness	16.3	16.5
X3	6.6	6.8
X5 (2.7)	3.6	3.8

Negative
exc. below (.5)	.7	.9
1984-85 318i, iC, is	1.2	1.4
325Ci, 330Ci (.3)	.5	.7
524td, 525i, 530i	1.2	1.4
540i, iA, iT	1.2	1.4
545i	.5	.7
735i, 740i, iA, iL, 750iL	1.2	1.4

	LABOR TIME	SEVERE SERVICE
745i, Li (.3)	.5	.7
760i, Li (.3)	.5	.7
M5 (.9)	1.2	1.4
X3 (1.3)	1.8	2.0
X5	.5	.7
w/power rear seat add	.2	.2
w/rear CD changer add	.2	.2

Starter Assy., R&R and Recondition (A)
320i	4.2	4.4
325, 325e	3.9	4.1
528e	3.8	4.0
1981 528i	4.8	5.0
633CSi, 635CSi, 733i	3.7	3.9

Starter Assy., Replace (B)
318i, iC, is
1984-85	2.8	3.0
1991-93	3.2	3.4
1994-95	2.4	2.6
1996-98	1.8	2.0

318ti
1995	2.4	2.6
1996-99	1.5	1.7
320i	3.4	3.6
323Ci, i, is, iT	2.5	2.7
325, 325e	2.8	3.0

325i, iC, is
1987-94	5.0	5.2
1995		
AT	5.2	5.4
MT	1.9	2.1
1996-97		
AT	4.3	4.5
MT	1.7	1.9
325Ci, iX	2.8	3.0
328i, iC, is	2.8	3.0
330Ci, Xi (2.2)	3.0	3.2
524td, 525i	3.2	3.4
528e	2.8	3.0

528i, iT
1981	3.7	3.9
1997-00	2.3	2.5
530i	2.0	2.2
533i, 535i, is	3.1	3.3

540i, iA, iT
1994-95	2.0	2.2
1997-01	1.2	1.4
545i (4.1)	5.5	5.7

633CSi, 635CSi, 733i, 735i
733i, 735i	2.8	3.0
645Ci (4.1)	5.5	5.7
740i, iA, iL	1.4	1.6
745i, Li	6.0	6.2
750iL	2.7	2.9
760Li, 760i (5.9)	8.0	8.2
840Ci	1.8	2.0

850Ci, i
5.0L	4.3	4.5
5.4L	2.9	3.1

	LABOR TIME	SEVERE SERVICE
M coupe/roadster, Z3		
1.9L	1.0	1.2
2.3L	2.3	2.5
2.8L		
AT	3.0	3.2
MT	1.2	1.4
M3, M5 (1.1)	1.5	1.7
X3, X5	1.6	1.8
Z4 (.9)	1.2	1.4

CRUISE CONTROL

Diagnose Cruise Control System Component Each (A)
1981-05	.8	.8

Brake or Clutch Release Switch, Replace (B)
1981-05	.7	.9
w/air bags add	.5	.5

Control Chain, Cable or Rod, Replace (B)
1981-04	.7	.9

Control Controller Module, Replace (B)
1981-05	.8	1.0

Control On/Off Switch, Replace (B)
1988-05	.8	1.0
w/air bags add	.5	.5

Control Servo, Replace (B)
Exc. below (.9)	1.2	1.4
530i	1.8	2.0
1994-95 540i	1.8	2.0
740i, iA, iL		
1994	2.0	2.2
1995-01	.7	.9
M coupe/roadster, Z3	.8	1.0
M5	.7	.9

IGNITION

Diagnose Ignition System Component Each (A)
1994-01	.8	.8
2002-05	1.1	1.1

Ignition Timing, Reset (B)
1981-01	.5	.7

Distributor, Replace (B)
Includes: Reset base ignition timing.
320i	1.2	1.4
1981 528i	1.2	1.4
633CSi, 635CSi	1.3	1.5
733i	1.3	1.5

Distributor Cap and/or Rotor, Replace (B)
1984-85 318i	.5	.7
320i	.5	.7
325 series all	.7	.9
1981-90 5 series all	.7	.9
633CSi, 635CSi, 733i, 735i	.7	.9

BMW-4   3 SERIES : 524TD : 5 SERIES : 6 SERIES : 7 SERIES : 8 SERIES

	LABOR TIME	SEVERE SERVICE
Distributor Cap and/or Rotor, Replace (B)		
750iL		
one side	.8	1.0
both	1.2	1.4
850Ci, i		
5.0L		
one	1.1	1.3
both	1.8	2.0
5.4L		
right side	2.7	2.9
left side	.8	1.0
both sides	3.2	3.4
Glow Plug, Replace (B)		
524td		
one	1.2	1.4
all	2.3	2.5
Glow Plug Module, Replace (B)		
524td	.3	.5
Ignition Coil, Replace (B)		
w/distributor		
exc. below (.5)	.7	.9
318i, iC, is	.8	1.0
320i	.3	.5
525i	.8	1.0
850Ci, i		
5.0L		
one	.3	.5
both	.7	.9
5.4L		
one	.3	.5
all	.6	.8
Distributorless		
exc. below	.6	.8
325Ci, 328Ci, 330Ci, Xi		
one	.3	.5
all	.8	1.0
525i		
one (.3)	.5	.7
all (.6)	.8	1.0
528i		
one	.2	.4
all	.7	.9
540i, iA, iT		
1994-95 both	1.5	1.7
1997-01		
one	.2	.4
all	.6	.8
545i, 645Ci, 745i, Li		
one (.2)	.3	.5
all (.8)	1.1	1.3
1994-97 740i, iA, iL		
both	1.3	1.5
750iL	1.0	1.2
760Li, 760i		
one (1.8)	2.4	2.6
all (2.1)	2.8	3.0
840Ci		
one (.2)	.3	.5
all (.5)	.7	.9
X3, X5, Z4		
one (.2)	.3	.5
all (.8)	1.1	1.3
Reprogram control unit	.8	.8
Ignition Switch, Replace (B)		
Exc. below (.6)	.8	1.0
1984-85 318i	1.2	1.4
318ti	1.0	1.2
325 series all (.9)	1.2	1.4
525i, 530i (.9)	1.2	1.4
528e, 533i (1.2)	1.6	1.8
1994-95 540i	1.2	1.4
1994 740i, iA, iL	1.4	1.6
745i, Li, 760i, Li (1.5)	2.0	2.2
750iL (.8)	1.1	1.3
1994-96 840Ci	1.6	1.8
850		
1991-94 (.8)	1.1	1.2
1995-97	1.6	1.8
M coupe/roadster, Z3 (.8)	1.1	1.3
w/air bags add	.5	.5
Spark Plug (Ignition) Cables, Replace (B)		
318i, iC, is		
1984-85	.5	.7
1991-98	1.4	1.6
318ti	1.1	1.3
320i	.5	.7
323Ci, i, is, iT	1.4	1.6
325 series all	.5	.7
328i, iC, is	1.4	1.6
525i	.5	.7
528e	.7	.9
528i, iT	.7	.9
533i, 535i, is	1.2	1.4
633CSi, 635CSi, 733i, 735i	1.2	1.4
750iL	2.1	2.3
850Ci, i	2.6	2.8
M3	2.5	2.7
M5	1.2	1.4
Z3 (.8)	1.1	1.3
Spark Plugs, Replace (B)		
Exc. below (.3)	.5	.7
325Ci	1.6	1.9
1992-95 325i, iC, is	1.6	1.9
328i, iC, is (1.2)	1.6	1.9
330Ci, i, Xi (1.3)	1.8	2.0
525i (1.2)	1.6	1.8
1997-00 528i	1.5	1.7
530i (1.2)	1.6	1.8
1992-93 535i (1.2)	1.6	1.8
540i, iA, iT (1.2)	1.6	1.8
545i (1.1)	1.5	1.7
645Ci (1.1)	1.5	1.7
740i, iA, iL	1.7	1.9
745i, Li (.9)	1.2	1.4
750iL		
1988-94	1.7	1.9
1995-01	2.7	2.9
760i, 760Li (2.1)	2.8	3.0
840Ci	2.0	2.2
850Ci, i		
5.0L	2.2	2.4
5.4L	2.6	2.8
M coupe/roadster, Z3		
2.3L, 2.8L	1.0	1.2
M3	.6	.8
M5, X3, X5 (1.0)	1.4	1.6
Z4 (.9)	1.2	1.4
Transistorized Ignition Control Unit, Replace (B)		
1981-04 one	.7	.9
each addl. add	.2	.2
Vacuum Advance Unit, Replace (B)		
Includes: Reset timing.		
1981-84	.8	.8

EMISSIONS

The following operations do not include testing. Add time as required.

	LABOR TIME	SEVERE SERVICE
Dynamometer Test (A)		
1994-05	.5	.5
Air Pump Assy., Replace (B)		
318ti	1.1	1.3
325Ci, 328i, iC, is	.6	.8
330Ci, Xi (.4)	.6	.8
525i (.8)	1.1	1.3
528i, iT (1.0)	1.4	1.6
530i (.8)	1.1	1.3
540i, iA, iT (.6)	.8	1.0
545i, 645Ci	1.0	1.2
740i, iA, iT	.8	1.0
745i, Li	1.0	1.2
1995-01 750iL	1.9	2.1
760i, 760Li (.9)	1.2	1.4
1994-97 850Ci	4.3	4.5
M coupe/roadster, Z3	1.1	1.3
M3, M5 (.8)	1.1	1.3
X3, X5 (.4)	.6	.8
Z4 (.6)	.8	1.0
Air Temperature Sensor, Replace (B)		
All Models	.5	.7
Coolant Temperature (ECT) Sensor and/or Switch, Replace (B)		
Exc. below (.4)	.6	.8
325Ci, 330Ci, Xi (2.1)	2.8	3.0
1997-00 528i, iT	3.2	3.4
1995-01 750iL	1.9	2.1
850Ci 5.4L	2.0	2.2
M coupe/roadster, Z3	3.2	3.4
M3	3.6	3.8
X3 (2.4)	3.2	3.4
Z4 (1.5)	2.0	2.2

ALPINA ROADSTER : M3 : M5 : X3 : X5 : Z3 : Z4 : Z8 **BMW-5**

	LABOR TIME	SEVERE SERVICE
EGR Check Valve, Replace (B)		
1981-95	.3	.5
EGR Control Valve, Replace (B)		
1981-95	1.0	1.2
EGR Switching Valve, Replace (B)		
1981-95	.3	.5
Electronic Control Module (ECM/PCM), Replace (B)		
Exc. below (.3)	.5	.7
318ti	.8	1.0
325iX	.7	.9
1992 735i	.7	.9
850Ci, CSi	.3	.5
Electronic Control Relay, Replace (B)		
All Models	.8	1.0
Fuel Vapor Canister, Replace (B)		
Exc. below (.6)	.8	1.0
320i	.5	.7
1993-95 325i, iC (.8)	1.1	1.3
525i, 530i	1.3	1.5
535i, 535is	1.2	1.4
1994-95 540i	1.3	1.5
735i, 745i, Li, 760i, Li (1.1)	1.5	1.7
1995-99 750iL	1.6	1.8
1991-97 850Ci, i	1.7	1.9
M3, X5 (.9)	1.2	1.4
Knock Sensor, Replace (B)		
318i, iC, is		
1991-93	1.8	2.0
1994-98	3.8	4.0
318ti		
1994-95	4.1	4.3
1996-99	1.0	1.2
323Ci, i, is, iT	3.8	4.0
325, 325i, iC, is	4.3	4.5
325Ci, 330Ci, Xi (2.1)	2.8	3.0
328i, iC, is (2.8)	3.8	4.0
525i	2.4	2.6
1997-00 528i, iT	3.1	3.3
530i		
right side	3.8	4.0
left side	6.9	7.1
540i, iA, iT		
1994-95		
one	3.8	4.0
both	7.5	7.7
1997-01		
right	3.5	3.7
left	4.6	4.8
both	5.1	5.3
545i, 645Ci		
one (1.8)	2.4	2.6
both (1.9)	2.6	2.8
745i, Li		
one (1.6)	2.2	2.4
both (1.8)	2.4	2.6
760i, 760Li		
one (6.0)	8.1	8.3
both (6.2)	8.4	8.6

	LABOR TIME	SEVERE SERVICE
740i, iA, iL		
1994		
one	3.8	4.0
both	7.5	7.7
1995-01		
right	3.4	3.6
left	4.5	4.7
both	5.0	5.2
750iL one or both (3.8)	5.1	5.3
840Ci		
right side (3.2)	4.3	4.5
left side (5.5)	7.4	7.6
850Ci 5.4L one side	4.4	4.6
M coupe/roadster, Z3		
1.9L	1.0	1.2
2.3L, 2.8L	3.1	3.3
M3 (2.7)	3.6	3.8
M5 one or both (2.7)	3.6	3.8
X3 both (2.2)	3.0	3.2
X5 both (1.6)	2.2	2.4
Z4 both (1.5)	2.0	2.2
Mass Air Flow (MAF) Sensor, Replace (B)		
Exc. below (.4)	.6	.8
1990-94 750iL		
one	.5	.7
both	.8	1.0
850Ci, i 5.0L		
one	1.2	1.4
both	1.4	1.6
Oxygen Sensor, Replace (B)		
1985-05 one (.4)	.6	.8
each addl. add	.2	.2
w/AC add	.2	.2
Test sensor add	.3	.3
Pulse Sensor, Replace (B)		
Exc. below	.6	.8
1994-95 318i, iC, is	2.2	2.4
318ti		
1995	2.2	2.4
1996-99	.8	1.0
325, 325i, iC, is		
1987-91	.8	1.0
1992-95	1.5	1.7
325iX, 525i	.7	.9
530i, 545i	1.0	1.2
750iL		
1988-94		
right side	1.4	1.6
left side	1.7	1.9
1995-01 each	.2	.4
850Ci, i		
5.0L		
right side	1.0	1.2
left side	2.2	2.4
5.4L each	.2	.4
M coupe/roadster, Z3		
1.9L	.8	1.0
M5	.8	1.0

	LABOR TIME	SEVERE SERVICE
Pulse Transmitter, Replace (B)		
1981-01	1.4	1.6
Throttle Position Sensor (TPS), Replace (B)		
Exc. M5	.5	.7
M5	2.5	2.7
Water Temperature Sensor, Replace (B)		
850Ci, i	1.2	1.4
Vehicle Speed Sensor, Replace (B)		
1989-05	.5	.7
DIESEL		
The following operations do not include testing. Add time as required.		
Diesel Emission Control Check (A)		
524td	1.0	1.2
Charging Valve, Replace (B)		
524td	.3	.5
EGR Control Valve, Replace (B)		
524td	.5	.7
Electric Switching Valve, Replace (B)		
524td	.3	.5
Pressure Converter, Replace (B)		
524td	.3	.5
Temperature Switch, Replace (B)		
524td	.3	.5
FUEL		
DELIVERY		
Diesel Fuel Injection Pump, Replace (B)		
524td	5.2	5.4
Fuel Filter, Replace (B)		
Exc. below (.6)	.8	1.0
325i, iC, is		
1987-95	1.0	1.2
2001-04	.7	.9
528i	1.0	1.2
540i		
1994-95 both	1.2	1.4
1997-03	1.0	1.2
740i, iL, 750iL both	1.0	1.2
840Ci, 850Ci, i		
both (1.0)	1.4	1.6
760i, M3, M5 (.9)	1.2	1.4
Fuel Gauge (Dash), Replace (B)		
Exc. below (1.2)	1.6	1.8
1991-95 318i, iC, is	1.2	1.4
320i	1.2	1.4
1996 328i, iC, is	1.2	1.4
525i	.8	1.0
528e	1.2	1.4
733i	1.4	1.6
735i	1.2	1.4
1988-95 750iL	1.3	1.5
w/air bags add	.5	.5

BMW-6 3 SERIES : 524TD : 5 SERIES : 6 SERIES : 7 SERIES : 8 SERIES

	LABOR TIME	SEVERE SERVICE
Fuel Pump, Replace (B)		
Exc. below (.9)	1.2	1.4
318i, iC, is	2.3	2.5
318ti	2.0	2.2
323Ci, i, is, iT	2.3	2.5
325i, iC, is	2.4	2.6
325Ci	1.8	2.0
328i, iC, is (1.7)	2.3	2.5
330Ci, Xi (1.3)	1.8	2.0
525i	2.4	2.6
528i, iT	2.2	2.4
530i	2.4	2.6
540i, iA, iT (1.6)	2.2	2.4
545i, 645Ci (1.6)	2.2	2.4
740i, iA, iL	2.5	2.7
745i, Li	1.0	1.2
750iL	2.5	2.7
840Ci	2.8	3.0
850Ci, i	3.1	3.3
M coupe/roadster, Z3 (1.3)	1.8	2.0
M3 (1.7)	2.3	2.5
M5 (1.6)	2.2	2.4
X3 (1.5)	2.0	2.2
X5 (.6)	.8	1.0
Z4 (2.1)	2.8	3.0
Fuel Pump Relay, Replace (B)		
1981-05	.3	.5
Fuel Tank, Replace (B)		
Includes: Drain and refill.		
318i, iC, is		
1984-85	4.2	4.4
1991-98	5.8	6.0
318ti	6.4	6.6
320i		
right side	1.8	2.0
left side	1.2	1.4
323Ci, i, is, iT	7.1	7.3
325, 325e	4.2	4.4
325Ci, 330Ci, Xi (3.0)	4.1	4.3
325i, iC, is		
1987-91	4.8	5.0
1992-95	7.6	7.8
325iX (3.1)	4.2	4.4
328i, iC, is	6.7	6.9
525i	2.8	3.0
528e	2.3	2.5
528i, iT		
1981	2.3	2.5
1997-00	6.4	6.6
530i	2.8	3.0
533i	2.3	2.5
535i, is		
1985-91	2.8	3.0
1992-93	5.8	6.0
540i, iA, iT		
1994-95	2.8	3.0
1997-01	7.3	7.5

	LABOR TIME	SEVERE SERVICE
545i (3.9)	5.3	5.5
633CSi	2.3	2.5
635CSi	3.1	3.3
645Ci (4.1)	5.5	5.7
733i, 735i	2.3	2.5
740i, iA, iL		
1994	2.7	2.9
1995-01	5.9	6.1
745i, Li	4.8	5.0
750iL		
1988-94	2.3	2.5
1995-01	6.5	6.7
760i, 760Li (5.4)	7.3	7.5
840Ci	10.0	10.2
850Ci, 850i		
5.0L	10.5	10.7
5.4L	9.8	10.0
M coupe/roadster, Z3	8.6	8.8
M3 (4.4)	5.9	6.1
M5 (6.6)	8.9	9.1
X3 (3.2)	4.3	4.5
X5 (4.2)	5.7	5.9
Z4 (4.5)	6.1	6.3
Fuel Tank Sending Unit (w/Pump), Replace (B)		
318i, iC, is		
1984-85	.8	1.0
1991-98		
one side	2.6	2.8
both sides	3.1	3.3
318ti, 323Ci, i, is, iT		
one side	2.6	2.8
both sides	3.1	3.3
320i, 325, 325e	1.2	1.4
325Ci, 330Ci, Xi (1.3)	1.8	2.0
325i, iC, is		
1987-91	1.2	1.4
1992-95		
one side	2.6	2.8
both sides	2.9	3.1
325iX (.6)	.8	1.0
328i, iC, is		
one side (1.9)	2.6	2.8
both sides (2.5)	3.4	3.6
524td	1.3	1.5
525i	1.1	1.3
528i, iT		
1981	1.3	1.5
1997-00		
right side	2.6	2.8
left side	1.9	2.1
both sides	2.6	2.8
528e, 533i	1.2	1.4
530i	1.0	1.2
535i, is		
1985-91	1.2	1.4
1992-93	1.7	1.9

	LABOR TIME	SEVERE SERVICE
540i, iA, iT		
1994-95	1.0	1.2
1997-01		
right side	2.2	2.4
left side	1.8	2.0
both sides	2.7	2.9
545i (1.9)	2.6	2.8
645Ci (1.8)	2.4	2.6
633CSi, 635CSi, 733i, 735i	1.3	1.5
740i, iA, iL, 750iL		
1988-94	1.8	2.0
1995-01		
one side	2.6	2.8
both sides	3.3	3.4
745i, Li, 760i, Li (.8)	1.1	1.3
840Ci, 850Ci, 850i		
right side (1.7)	2.3	2.5
left side (2.5)	3.5	3.7
both sides (3.3)	4.5	4.7
M coupe/roadster, Z3 (1.6)	2.2	2.4
M3		
one side (2.0)	2.7	2.9
both sides (2.4)	3.2	3.4
M5		
right side (1.9)	2.6	2.8
left side (1.4)	1.9	2.1
both sides (1.9)	2.6	2.8
X3, X5 (.8)	1.1	1.3
Z4 (2.1)	2.8	3.0
Intake Manifold and/or Gasket, Replace (B)		
Includes: Adjustments.		
Exc. below (3.2)	4.3	4.5
318i, iC, is		
1984-85	3.1	3.3
1991-93		
upper	1.2	1.4
lower	4.3	4.5
1994-95		
upper	2.2	2.4
lower	5.3	5.5
1996-98	5.1	5.3
318ti		
1995		
upper	2.2	2.4
lower	5.3	5.5
1996-99		
upper	1.0	1.2
lower	3.6	3.8
325Ci	3.8	4.0
330Ci, Xi (2.2)	3.0	3.2
524td	3.1	3.3
525i	2.6	2.8
530i, 545i, 645Ci (2.6)	3.5	3.7
740i, iA, iL	4.0	4.2
745i, Li, 760i, Li (2.5)	3.4	3.6

ALPINA ROADSTER : M3 : M5 : X3 : X5 : Z3 : Z4 : Z8   BMW-7

	LABOR TIME	SEVERE SERVICE
750iL		
1988-94		
one	3.1	3.3
both	4.8	5.0
1995-01		
right side	5.9	6.1
left side	5.9	6.1
840Ci	4.6	4.8
850Ci, 850i		
5.0L		
one	3.8	4.0
both	5.2	5.4
5.4L		
one	4.6	4.8
both	6.3	6.5
M coupe/roadster, Z3		
1.9L		
upper	1.0	1.2
lower	3.5	3.7
2.3L, 2.8L	4.6	4.8
M3 (3.9)	5.3	5.5
M5 (2.0)	2.7	2.9
X3 (2.2)	3.0	3.2
X5 (2.6)	3.5	3.7
Z4 (2.1)	2.8	3.0

INJECTION

Diagnose Injection System Component Each (A)

	LABOR TIME	SEVERE SERVICE
1981-87	1.7	1.9
1988-05	1.1	1.3

Air Flow Meter, Replace (B)

Exc. below (.5)	.7	.9
320i	1.2	1.4
1981 528i	1.2	1.4
633CSi	1.4	1.6
733i	1.5	1.7
735i	1.7	1.9

Auxiliary Air Valve, Replace (B)

1981-95	.8	1.0

Cold Start Valve and/or Seal, Replace (B)

Exc. below	.5	.7
533i, 535i, is	1.4	1.6
633CSi, 635CSi, 733i, 735i	1.4	1.6

Fuel Injection Lines, Replace (B)

524td	2.3	2.5

Fuel Injection Relay, Replace (B)

All	.5	.7

Fuel Injectors, Replace (B)

Exc. below all (1.0)	1.4	1.6
318i, iC, is, ti		
1984-85	.5	.7
1991-93	1.5	1.7
1994-95	2.7	2.9
1996-99	1.1	1.3

	LABOR TIME	SEVERE SERVICE
323Ci, i, is, iT	1.6	1.8
325, 325e	.8	1.0
325Ci, 330Ci, Xi (1.2)	1.6	1.8
325i, iC, is		
1987-91	.8	1.0
1992-97	2.7	2.9
325iX		
1988-89	.8	1.0
1990-91	1.7	1.9
328i, iC, is (.8)	1.1	1.3
525i, 528e, 530i	1.6	1.8
1997-00 528i, iT	1.5	1.7
540i, iA, iT	1.8	2.0
545i, 645Ci (1.9)	2.5	2.7
745i, Li	2.0	2.2
750iL		
1990-94	2.0	2.2
1995-01	3.8	4.0
760Li	7.8	8.0
840Ci (1.3)	1.8	2.0
850Ci, i		
5.0L (1.5)	2.0	2.2
5.4L (4.3)	5.8	6.0
M coupe/roadster, Z3		
1.9L (1.1)	1.5	1.7
2.3L, 2.8L (.7)	.9	1.1
M3 (2.2)	3.0	3.2
M5 one (2.0)	2.7	2.9
X3 (1.3)	1.8	2.0
X5 (1.6)	2.2	2.4
Z4 (1.1)	1.5	1.7
Clean nozzles add each	.5	.5

Fuel Pressure Regulator, Replace (B)

Exc. below (.3)	.5	.7
320i	1.2	1.4
325i, iC, is (1.1)	1.5	1.7
328i, iC, is (.4)	.6	.8
1981 528i	1.8	2.0
530i, 540i, iA, iT (.5)	.7	.9
633CSi, 635CSi (1.3)	1.8	2.0
733i, 740i, iA, iL (.5)	.7	.9
745i, Li, 760Li (.5)	.7	.9
840Ci	.7	.9
850Ci, i 5.0L		
one	.5	.7
both	.7	.9
M3		
1995	1.5	1.7
1996-01	.6	.8
X5 (.6)	.8	1.0

Fuel Rail, Replace (B)

Exc. below (.9)	1.2	1.4
318i, iC, is	3.1	3.3
318ti	1.9	2.1
325i, iC, is	3.2	3.4
328i, iC, is (1.8)	2.4	2.6
525i	2.0	2.2
528i, iT	1.5	1.7
530i, 540i, iA, iT	2.2	2.4

	LABOR TIME	SEVERE SERVICE
545i, 745i, Li (1.3)	1.8	2.0
740i, iL	2.2	2.4
750iL		
one (2.8)	3.8	4.0
both (3.9)	5.3	5.5
760Li		
one (1.3)	1.8	2.0
both (1.5)	2.0	2.2
840Ci	2.4	2.6
850Ci, 850i		
5.0L		
one (1.5)	2.0	2.2
both (2.1)	2.8	3.0
5.4L one or both (4.5)	6.1	6.3
M coupe/roadster, Z3 (1.4)	1.9	2.1
M3 (2.5)	3.4	3.6
M5 (1.6)	2.2	2.4

Idle Air Control (IAC) Valve, Replace (B)

Exc. below (.5)	.7	.9
1994-95 318i, iC, is	1.6	1.8
318ti	.3	.5
325, 325i, iC, is	1.3	1.5
328i, iC, is	1.3	1.5
M coupe/roadster, Z3		
1.9L	.2	.4
1995 M3	1.3	1.5
M5 (1.7)	2.3	2.5
X3 (1.1)	1.5	1.7

Idle Switch, Replace (B)

524td	.3	.5

Mixture Regulator, Replace (B)

320i	1.6	1.8

Thermo Time Switch, Replace (B)

1981-86	.5	.7

Throttle Body Assy., Replace (B)

Exc. below (.8)	1.1	1.3
1991-95 318i, iC, is	1.6	1.8
323Ci, i, is, iT	1.5	1.7
328i, iC, is	1.6	1.8
1997-00 528i, iT	1.6	1.8
530i	1.7	1.9
1994-95 540i, iA, iT	1.7	1.9
645Ci	.8	1.0
740iL	2.6	2.8
750iL both (1.0)	1.4	1.6
760Li one (.3)	.5	.7
840Ci	2.6	2.8
850Ci, 850i both (1.3)	1.8	2.0
M coupe/roadster, Z3		
1.9L (.8)	1.1	1.3
2.3L, 2.8L (1.2)	1.6	1.8
M3 (1.1)	1.5	1.7
X3 (1.3)	1.8	2.0

Throttle Body Injection Control Unit, Replace (B)

1990-94 750iL	.7	.9
850Ci, 850i (.6)	.8	1.0

BMW-8   3 SERIES : 524TD : 5 SERIES : 6 SERIES : 7 SERIES : 8 SERIES

	LABOR TIME	SEVERE SERVICE
Throttle Valve Housing, Replace (B)		
320i	3.1	3.3
Throttle Valve Switch, Replace (B)		
318 series all	.5	.7
320i	.5	.7
325 series all	.7	.9
328i, iC, is	.5	.7
525i	.7	.9
528e	.5	.7
1981 528i	.7	.9
533i, 535i, is	.5	.7
633CSi, 635CSi, 733i, 735i	.5	.7
Transmission Program Control Switch, Replace (B)		
850Ci, 850i	.7	.9
Transmission Switch, Replace (B)		
1990-94 750iL	.5	.7
850Ci, i	1.1	1.3
Warm-Up Regulator, Replace (B)		
320i	1.5	1.7

EXHAUST

	LABOR TIME	SEVERE SERVICE
Catalytic Converter, Replace (B)		
318i, iC, is		
1984-85	1.2	1.4
1991-98	1.6	1.8
318ti, 320i, 325, 325e	1.2	1.4
325i, iC, is (1.2)	1.6	1.8
325iX (.9)	1.2	1.4
328i, iC, is (1.0)	1.4	1.6
330Ci, i, Xi (.8)	1.1	1.3
525i	2.2	2.4
528e	1.2	1.4
528i, iT	1.7	1.9
530i one	2.0	2.2
533i, 535i, is	1.2	1.4
540i, iA, iT		
1994-95 one	2.0	2.2
1997-01 one	2.6	2.8
545i, 645Ci (.8)	1.1	13
633CSi, 635CSi, 733i, 735i	1.7	1.9
1994-01 740i, iA, iL one	2.0	2.2
745i, Li, 760Li (.8)	1.1	1.3
750iL		
1988-94 one	2.3	2.5
1995-01 one	2.6	2.8
840Ci one (1.6)	2.2	2.4
850Ci, 850i		
5.0L one (1.8)	2.4	2.6
5.4L one (2.5)	3.4	3.6
M coupe/roadster, Z3 (.9)	1.2	1.4
M3 (1.2)	1.6	1.8
M5 one (3.1)	4.2	4.4
X3 (.8)	1.1	1.3
X5 (1.0)	1.4	1.6

	LABOR TIME	SEVERE SERVICE
Exhaust Manifold and/or Gasket, Replace (B)		
318i, iC, is		
1984-85	1.5	1.7
1991-98		
front	4.4	4.6
rear	5.8	6.0
318ti	2.5	2.7
320i	1.2	1.4
325, 325e one	1.3	1.5
325i, iC, is		
1987-91		
front	1.2	1.4
rear	1.6	2.0
1992-95	1.6	2.0
325iX one	1.4	1.6
328i, iC, is one		
front (3.3)	4.5	4.7
rear or both (4.3)	5.8	6.0
330Ci, i		
one (2.2)	3.0	3.2
both (3.0)	4.1	4.3
330Xi		
one (2.5)	3.4	3.6
both (3.1)	4.2	4.4
524td	3.7	3.9
525i both	3.9	4.1
528e	1.7	1.9
528i, iT		
1981		
front	3.3	3.5
rear	2.9	3.1
1997-00		
front	4.4	4.6
rear or both	5.8	6.0
530i		
right	3.5	3.7
left	8.6	8.8
both	9.9	10.0
533i, 535i, is		
front	1.8	2.0
rear	1.9	2.1
540i, iA, iT		
1994-95		
right	3.5	3.7
left	8.6	8.8
both	9.8	10.0
1997-01		
right	2.7	2.9
left	3.7	3.9
both	4.8	5.0
545i		
right (3.8)	5.1	5.3
left (3.5)	4.7	4.9
both (4.2)	5.7	5.9
633CSi, 635CSi		
front	1.8	2.0
rear	1.9	2.1

	LABOR TIME	SEVERE SERVICE
645Ci		
right	5.3	5.5
left	4.9	5.1
both	5.9	6.1
733i, 735i		
front	1.8	2.0
rear	1.7	1.9
740i, iA, iL		
1994		
right side	3.6	3.8
left side	8.7	8.9
both sides	9.8	10.0
1995-01		
right side	3.2	3.4
left side	5.0	5.2
both sides	5.9	6.1
745i, Li		
right	5.6	5.8
left	4.2	4.4
both	6.5	6.7
750iL		
1988-94		
both right side	5.2	5.4
both left side	4.8	5.0
all both sides	9.3	9.5
1995-01		
front	5.9	6.1
rear		
right side	7.0	7.2
left side	7.4	7.6
all	12.9	13.1
760i, 760Li		
one (5.2)	7.0	7.2
both (6.9)	9.3	9.5
840Ci		
right	5.2	5.4
left	6.4	6.6
both sides	8.9	9.1
850Ci, 850i		
5.0L		
both right side	6.7	6.9
both left side	8.9	9.1
all both sides	14.8	15.0
5.4L		
right side		
front	6.4	6.6
rear	7.8	8.0
left side		
front	8.6	8.8
rear	9.8	10.0
M coupe/roadster, Z3		
1.9L (2.0)	2.7	2.9
2.3L, 2.8L		
front (3.1)	4.2	4.4
rear or both (4.0)	5.4	5.6
M3		
front (3.3)	4.5	4.7
rear or both (4.4)	5.9	6.1

ALPINA ROADSTER : M3 : M5 : X3 : X5 : Z3 : Z4 : Z8 **BMW-9**

	LABOR TIME	SEVERE SERVICE
M5		
right	5.4	5.6
left	6.3	6.5
all	7.1	7.3
X3	4.6	4.8
X5		
right (2.7)	3.6	3.8
left (2.9)	3.9	4.1
both (4.0)	5.4	5.6
Z4		
one (2.9)	3.9	4.1
both (3.6)	4.9	5.1
Final Muffler, Replace (B)		
1988-04 2 pc system	1.7	1.9
Front Exhaust Pipe, Replace (B)		
750iL		
1988-94		
one side	2.7	2.9
both sides	4.2	4.4
1995-01		
right	1.2	1.4
left	1.6	1.8
both	2.5	2.7
850Ci, i		
5.0L		
right	3.3	3.5
left	4.6	4.8
5.4L		
right	1.6	1.8
left	2.4	2.6
Intermediate & Final Muffler, Replace (B)		
525i	2.9	3.1
1997-00 528i	2.2	2.4
530i	1.7	1.9
540i, iA, iT		
1994-95	1.7	1.9
1997-01	2.9	3.1
633CSi, 635CSi, 733i	2.3	2.5
735i	1.2	1.3
740i, iA, iL	4.3	4.5
750iL	4.7	4.9
840Ci		
right	2.6	2.8
left	.8	1.0
850Ci	5.3	5.5
M5	6.5	6.7
Primary Muffler, Replace (B)		
318i, iC, is		
1984-85	2.3	2.5
1991-98	.8	1.0
318ti	.7	.9
320i	1.8	2.0
323Ci, i, is, iT	.8	1.0
325, 325e	2.3	2.5
325i, iC, is (.6)	.8	1.0

	LABOR TIME	SEVERE SERVICE
325iX (.9)	1.2	1.4
328i, iC, is (.6)	.8	1.0
330Ci, i, Xi (.9)	1.2	1.4
525i	3.1	3.3
528e	.8	1.0
528i, iT		
1981	2.3	2.5
1997-00	.8	1.0
530i	1.7	1.9
540i, iA, iT		
1994-95	1.7	1.9
1997-01	.8	1.0
545i (1.0)	1.4	1.6
633CSi, 635CSi	2.3	2.5
733i, 735i	2.4	2.6
740i, iA, iL		
1994	1.7	1.9
1995-01		
one	.7	.9
both	1.1	1.3
745i, Li, 760Li (1.1)	1.5	1.7
1995-01 750iL		
one	.7	.9
both	1.1	1.3
840Ci	2.9	3.1
850Ci, 850i		
5.0L	2.9	3.1
5.4L	4.9	5.1
M coupe/roadster, Z3		
1.9L (.5)	.7	.9
2.3L, 2.8L (.6)	.8	1.0
M3 (.6)	.8	1.0
M5 (1.4)	1.9	2.1
X3 (1.2)	1.6	1.8
X5, Z4 (.8)	1.1	1.3
Secondary Muffler, Replace (B)		
320i	1.4	1.6
330Ci, i, Xi, 545i (.6)	.8	1.0
633CSi, 635CSi	1.8	2.0
733i	2.3	2.5
745i, Li, 760Li (.6)	.8	1.0
840Ci		
right	2.6	2.8
left	.8	1.0
850Ci, 850i		
5.0L		
right	2.6	2.8
left	.8	1.0
5.4L		
one	.8	1.0
both	1.5	1.7
X3, X5 (.6)	.8	1.0
Z4 (1.2)	1.6	1.8

	LABOR TIME	SEVERE SERVICE
ENGINE COOLING		
Pressure Test Cooling System (C)		
1981-05	.4	.4
Auxiliary Cooling Fan Assy., Replace (B)		
Exc. below (1.0)	1.4	1.6
1984-85 318i, iC, is	2.3	2.5
1995 318ti	.5	.7
320i	3.1	3.3
325, e, iX (1.7)	2.3	2.5
524td, 528e	2.3	2.5
528i, iT		
1981	2.3	2.5
1997-00	2.5	2.7
533i, 535i, is	2.3	2.5
1997-01 540i, iA, iT	2.6	2.8
633CSi, 635CSi, 733i, 735i	2.3	2.5
1995-01 740i, iA, iL	1.9	2.1
750iL		
1988-94	2.3	2.5
1995-01	2.1	2.3
840Ci	1.3	1.5
M coupe/roadster, Z3	.8	1.0
M5 (2.2)	3.0	3.2
Z4 (.6)	.8	1.0
X5 (1.9)	2.6	2.8
Auxiliary Cooling Fan Relay, Replace (B)		
1981-05	.3	.5
Auxiliary Fan Temperature Switch, Replace (B)		
All Models	.8	1.0
Auxiliary Water Pump, Replace (B)		
525i	1.5	1.7
1996-01 528i, iT (.8)	1.1	1.3
530i, 540i, iA, iT (.8)	1.1	1.3
735i	1.2	1.4
740i, iA, iL (1.1)	1.5	1.7
750iL		
1988-94	1.8	2.0
1995-99	1.1	1.3
840Ci (1.2)	1.6	1.8
850Ci, 850i (1.5)	2.0	2.2
M5 (.8)	1.1	1.3
Coolant Thermostat, Replace (B)		
Exc. below (1.1)	1.5	1.7
1996-98 318i, iC, is	1.1	1.3
323Ci, i, is, iT	1.9	2.1
325Ci (1.2)	1.8	2.0
325i, is	2.4	2.6
325iC (1.6)	2.2	2.4
328i, iC, is, Xi (1.4)	1.9	2.1
330Ci, i (1.3)	1.8	2.0
524td	1.7	1.9
525i	1.8	2.0

BMW-10 3 SERIES : 524TD : 5 SERIES : 6 SERIES : 7 SERIES : 8 SERIES

	LABOR TIME	SEVERE SERVICE
Coolant Thermostat, Replace (B)		
535i, is	1.7	1.9
540i	1.9	2.1
545i, 645Ci (.9)	1.2	1.4
740i, iA, iL	1.9	2.1
745i, iL (1.2)	1.8	2.0
750iL		
1988-94	1.7	1.9
1995-01	3.0	3.2
760i, Li	1.8	2.0
840Ci	1.1	1.3
850Ci, 850i		
5.0L (.9)	1.2	1.4
5.4L (2.0)	2.7	2.9
M3		
1995	2.6	2.8
1996-01	3.4	3.6
2001-05 (1.9)	2.6	2.8
M5 (2.7)	3.6	3.8
M coupe/roadster, Z3		
1.9L (.8)	1.1	1.3
2.3L, 2.8L (1.2)	1.6	1.8
X3 (1.3)	1.8	2.0
X5		
3.0 (1.3)	1.8	2.0
4.4 (1.0)	1.4	1.6
Z4 (1.2)	1.6	1.8
Fan Coupling, Replace (B)		
Exc. below (.6)	.8	1.0
320i	1.3	1.5
1981 528i	2.2	2.4
633CSi, 635CSi	2.2	2.4
w/AT add	.3	.3
Fan or Clutch Assy. Blades, Replace (B)		
1981-05 (.7)	.9	1.1
Radiator Assy., R&R or Replace (B)		
Includes: Refill with proper coolant mix.		
Exc. below (1.2)	1.6	1.8
325iX	1.2	1.4
328i, Ci (.8)	1.1	1.3
525i AT, 530i AT	2.2	2.4
533i, 535i, is	1.2	1.4
1997-01 540i, iA, iT	1.9	2.1
1988-94 750iL	2.3	2.5
840Ci	2.7	2.9
850Ci, 850i		
AT	3.5	3.7
MT	3.2	3.4
M coupe/roadster, Z3		
1.9L	1.8	2.0
M3 (2.1)	2.8	3.0
M5 (1.9)	2.6	2.8
X5, Z4 (.8)	1.1	1.3

	LABOR TIME	SEVERE SERVICE
Radiator Hoses, Replace (B)		
Includes: Refill with proper coolant mix.		
Upper (.5)	.7	.9
Lower		
exc. below (.6)	.8	1.0
325Ci, 328Ci (.8)	1.1	1.3
330Ci, Xi (.8)	1.1	1.3
524td	1.2	1.4
525i	1.3	1.5
528i, iT	1.1	1.3
540i	1.3	1.5
545i (.8)	1.1	1.3
645Ci	1.2	1.4
745i, Li (1.2)	1.6	1.8
750iL	1.2	1.4
760i, Li (1.2)	1.6	1.8
840Ci (1.0)	1.4	1.6
850Ci, 850i	1.2	1.4
M3 (.8)	1.1	1.3
M5 (.6)	.8	1.0
X3, X5, Z4 (1.0)	1.4	1.5
Temperature Gauge (Dash), Replace (B)		
318i, iC, is		
1984-85	1.7	1.9
1991-95	1.5	1.7
318ti	.7	.9
325, 325e	1.7	1.9
325i, iC, is	1.5	1.7
325iX (1.2)	1.6	1.8
328i, iC, is	1.5	1.7
330Ci, i, Xi, 525i (.9)	1.2	1.4
540i, 545i, 645Ci (1.2)	1.6	1.8
735i	1.2	1.4
740i, iL, 745i, Li (1.2)	1.6	1.8
750iL	1.2	1.4
760Li (1.2)	1.6	1.8
840Ci, 850Ci, 850i		
1991-93	1.3	1.5
1994-97	1.6	1.8
M3 (1.1)	1.5	1.7
X3, X5, Z4 (.9)	1.2	1.4
w/air bags add	.5	.5
Temperature Gauge (Engine), Replace (B)		
Exc. below (.5)	.7	.9
1994-98 318i, iC, is	1.5	1.7
318ti	1.5	1.7
325 series all	1.1	1.3
1997-01 540i	1.1	1.3
1995-01 740i, iA, iL	.8	1.0
840Ci, 850Ci, 850i	.8	1.0
X3, X5, Z4 (.9)	1.2	1.4
Water Pump and/or Gasket, Replace (B)		
Includes: Refill with proper coolant mix.		
318i, iC, is		
1984-85	2.8	3.0
1991-98	1.8	2.0

	LABOR TIME	SEVERE SERVICE
318ti	1.8	2.0
320i		
w/AC	3.4	3.6
w/o AC	1.9	2.1
323Ci, i, is, iT	2.6	2.8
325, 325e	2.8	3.0
325i, iC, is	2.2	2.4
325i, Ci, 330Ci, i (1.5)	2.0	2.2
325iX (2.3)	3.1	3.3
328i, 328iC, is (1.7)	2.3	2.5
330Xi (1.4)	1.9	2.1
524td	4.2	4.4
525i	2.2	2.4
528e	2.8	3.0
528i		
1981	2.8	3.0
1997-00	1.8	2.0
530i	4.9	5.1
533i, 535i, is	3.1	3.3
540i		
1994-95	4.9	5.1
1997-01	3.6	3.8
545i (1.6)	2.2	2.4
633CSi, 635CSi, 733i, 735i	3.1	3.3
645Ci (1.6)	2.2	2.4
740i, iA, iL	3.6	3.8
745i, Li	2.3	2.5
750iL		
1988-94	3.1	3.3
1995-01	4.5	4.7
760i, 760Li (2.3)	3.1	3.3
840Ci	3.8	4.0
850Ci, 850i		
5.0L	3.5	3.7
5.4L	6.0	6.2
M coupe/roadster, Z3		
1.9L	1.8	2.0
2.3L, 2.8L	2.5	2.7
M3 (2.5)	3.4	3.6
M5 (4.4)	5.9	6.1
X3 (1.3)	1.8	2.0
X5 (1.5)	2.0	2.2
Z4 (2.1)	2.8	3.0

ENGINE
ASSEMBLY
Times shown are for OEM assemblies. Time to replace assemblies from aftermarket rebuilders may vary.

Engine Assy., R&I (B)
Does not include parts or component transfer.

	LABOR TIME	SEVERE SERVICE
318i, iC, is		
1984-85	9.9	10.1
1991-95	14.2	14.4
1996-98	13.8	14.0

ALPINA ROADSTER : M3 : M5 : X3 : X5 : Z3 : Z4 : Z8 BMW-11

	LABOR TIME	SEVERE SERVICE
1995-99 318ti		
AT	13.8	14.0
MT	11.9	12.1
320i	10.0	10.2
323Ci, i, is, iT	16.4	16.6
325, e, i, iC, is		
1987-91	18.9	19.1
1992-95	20.0	20.2
325iX	14.2	14.4
328i, iC, is		
AT	16.1	16.3
MT	14.6	14.8
330Ci, 330i (8.9)	12.0	12.2
330Xi (11.5)	15.5	15.7
524td	10.0	10.2
525i	12.5	12.7
528e	9.3	9.5
528i		
1981	13.4	13.6
1997-99	15.2	15.4
530i	14.5	14.7
533i, 535i, is	9.3	9.5
540i		
1994-95	16.8	17.0
1997-01	18.8	19.0
545i, 645Ci (9.6)	13.0	13.2
633CSi, 635CSi	14.1	14.3
733i, 735i	12.4	12.6
740i, iA, iL		
1994	17.2	17.4
1995-01	19.0	19.2
745i, Li	13.8	14.0
750iL		
1988-94	15.2	15.4
1995-01	22.5	22.7
760i, 760Li (15.5)	20.9	21.1
840Ci (14.7)	19.8	20.0
850Ci, i		
5.0L	21.8	22.0
5.4L	22.4	22.6
M coupe/roadster, Z3		
1.9L	11.9	12.1
2.3L, 2.8L	15.3	15.5
M3		
AT (12.2)	16.5	16.7
MT (11.1)	15.0	15.2
M5 (15.2)	20.4	20.6
X3 (12.5)	16.9	17.1
X5 (10.4)	14.0	14.2
Z4	9.8	10.0

Engine Assy., R&R and Recondition (A)
Includes: Replacing rings, rod and main bearings, cylinder head reconditioning and engine tune-up.

	LABOR TIME	SEVERE SERVICE
318i, iC, is		
1984-85	24.6	24.8
1991-95	31.5	31.7
1996-98	30.2	30.4
318ti		
AT	28.3	28.5
MT	26.0	26.2
320i	24.6	24.8
325, e, i, iC, is		
1987-91	28.9	29.1
1992-95	30.0	30.2
325iX	33.1	33.3
328i, iC, is		
AT	35.9	36.1
MT	34.7	34.9
330Ci, i	24.0	24.2
330Xi	28.1	28.3
524td	32.3	32.5
525i	32.8	33.0
528e	26.7	26.9
528i, iT		
1981	30.3	30.5
1997-00	34.0	34.2
530i	37.4	37.6
533i, 535i, is	26.7	26.9
540i, iA, iT		
1994-95	38.8	39.0
1997-01	44.1	44.3
545i	29.8	30.0
633CSi, 635CSi	31.0	31.2
645Ci	29.9	30.1
733i, 735i	28.9	29.1
740i, iA, iL		
1994	39.3	39.5
1995-01	45.1	45.3
745i, Li	30.4	30.6
750iL		
1988-94	40.0	40.2
1995-01	49.9	50.1
760Li	34.5	34.7
840Ci	41.4	41.6
850Ci, i		
5.0L	52.6	52.8
5.4L	51.6	51.8
M coupe/roadster, Z3		
1.9L	29.9	30.1
2.3L, 2.8L	34.6	34.8
M3		
AT	35.9	36.1
MT	34.4	34.6
M5	54.2	54.4
X3	24.2	24.4
X5	30.8	31.0
Z4	22.8	23.0

Engine Assy., Replace (B)

	LABOR TIME	SEVERE SERVICE
318i, iC, is		
1984-85	16.2	16.4
1991-98	19.2	19.4
318ti		
AT	17.0	17.2
MT	14.9	15.1
320i	15.2	15.4
323Ci, i, is, iT	22.3	22.5
325, e, i, iC, is	17.3	17.5
325iX	23.1	23.3
328i, iC, is		
AT	22.3	22.5
MT	21.1	21.3
330Ci, 330i (11.1)	15.0	15.2
330Xi (14.2)	19.2	19.4
524td	25.2	25.4
525i	19.3	19.5
528e	17.3	17.5
528i		
1981	17.3	17.5
1997-00	20.9	21.1
530i	21.8	22.0
533i, 535i, is	20.0	20.2
540i, iA, iT		
1994-95	23.4	23.6
1997-01	24.8	25.0
545i, 645Ci (12.5)	16.9	17.1
633CSi, 635CSi, 733i, 735i	20.0	20.2
740i, iA, iL		
1994	23.9	24.1
1995-01	25.8	26.0
745i, Li	17.8	18.0
750iL		
1988-94	30.0	30.2
1995-01	31.1	31.3
760Li (16.6)	22.4	22.6
840Ci	26.7	26.9
850Ci, 850i	30.7	30.9
M coupe/roadster, Z3		
1.9L	17.6	17.8
2.3L, 2.8L	21.0	21.2
M3		
AT	22.2	22.4
MT	20.9	21.1
X3 (14.5)	19.6	19.8
X5	16.4	16.6
Z4	13.4	13.6

Engine Mounts, Replace (B)

	LABOR TIME	SEVERE SERVICE
318i, iC, is		
1984-85		
one	.5	.7
both	1.2	1.4
1991-98		
one	1.0	1.2
both	1.1	1.3
318ti		
one	1.1	1.3
both	1.2	1.4
320i, 325, e, i, iC, is		
one	.5	.7
both	1.2	1.4
325iX		
one	1.9	2.1
both	2.3	2.5
328i, iC, is		
one	1.4	1.6
both	1.6	1.8

BMW-12 3 SERIES : 524TD : 5 SERIES : 6 SERIES : 7 SERIES : 8 SERIES

	LABOR TIME	SEVERE SERVICE
Engine Mounts, Replace (B)		
330Ci, i, Xi		
one (.9)	1.2	1.4
both (1.4)	1.9	2.1
524td		
one	.7	.9
both	1.0	1.2
525i		
one	1.0	1.2
both	1.3	1.5
528e, 533i, 535i, is		
one	.5	.7
both	1.2	1.4
528i, iT		
1981		
one	.5	.7
both	1.2	1.4
1997-00		
one	1.4	1.6
both	1.6	1.8
530i		
right	2.8	3.0
left	3.2	3.4
both	3.3	3.5
540i, iA, iT		
1994-95		
right side	2.8	3.0
left side	3.2	3.4
both sides	3.3	3.5
1997-01		
one side	1.4	1.6
both sides	1.6	1.8
545i		
one (2.4)	3.2	3.4
both (2.7)	3.6	3.8
733i, 735i		
one	.5	.7
both	1.2	1.4
740i, iA, iL		
1994		
right side	1.3	1.5
left side	3.2	3.4
both sides	3.3	3.5
1995-01		
right side	1.8	2.0
left side	1.5	1.7
both sides	2.2	2.4
745i, Li		
one (1.8)	2.4	2.6
both (2.1)	2.8	3.0
750iL		
1988-94		
right side	1.2	1.4
left side	1.7	1.9
both sides	1.9	2.1
1995-01		
right side	3.1	3.3
left side	1.6	1.8
both sides	2.3	2.5

	LABOR TIME	SEVERE SERVICE
760i, 760Li		
one (2.0)	2.7	2.9
both (2.2)	3.0	3.2
840Ci		
right side	2.0	2.2
left or both sides	4.3	4.5
850Ci, 850i		
right side (1.1)	1.5	1.7
left side (1.4)	1.9	2.1
M coupe/roadster, Z3		
one side (.8)	1.1	1.3
both sides (1.0)	1.4	1.6
M3		
one side (1.0)	1.4	1.6
both sides (1.2)	1.6	1.8
M5		
right side (1.0)	1.4	1.6
left or both sides (1.1)	1.5	1.7
X3		
one (1.7)	2.3	2.5
both (1.9)	2.6	2.8
X5		
one (1.8)	2.4	2.6
both (2.1)	2.8	3.0
Z4		
one (1.6)	2.2	2.4
both (1.9)	2.6	2.8
CYLINDER HEAD		
Compression Test (B)		
318i, iC, is	1.2	1.4
318ti	.8	1.0
320i	1.2	1.4
325, e, i, iC, is, iX	2.0	2.2
328i, iC, is	1.9	2.1
330Ci, i, Xi (1.2)	1.8	2.0
525i, 530i	2.2	2.4
528e, 528i, 533i	1.7	1.9
535i, is	2.3	2.5
540i, 545i, 645Ci (1.6)	2.2	2.4
633CSi, 635CSi, 733i, 735i	1.7	1.9
740i, iA, iL	2.2	2.4
745i, Li, 760Li (1.6)	2.2	2.4
750iL		
1988-94	2.8	3.0
1995-01	3.6	3.8
840Ci	2.6	2.8
850Ci, 850i		
5.0L	3.6	3.8
5.4L	3.9	4.1
M coupe/roadster, Z3		
1.9L	.7	.9
2.3L, 2.8L	1.2	1.4
M3, X3 (1.4)	1.9	2.1
M5 (1.2)	1.8	2.0
X5 (1.6)	2.2	2.4
Z4 (1.1)	1.5	1.7

	LABOR TIME	SEVERE SERVICE
Cylinder Head, Retorque (B)		
3 series all	.9	1.1
5 series 6 cyl.	.9	1.1
633CSi, 635CSi, 733i, 735i	1.2	1.4
Valve Clearance, Adjust (A)		
3 series	1.8	2.0
5, 6, 7 series		
M20	1.4	1.6
M30	2.4	2.6
M50		
1987-88	3.1	3.3
1991-99	3.8	4.0
M60	3.2	3.4
Coolant End Cover, Replace or Reseal (B)		
530i		
AT	11.4	11.6
MT	9.7	9.9
540i, iA, iT		
1994-95	11.5	11.7
1997		
AT	9.7	9.9
MT	8.8	9.0
1998-01		
rear	8.2	8.4
upper	5.3	5.5
545i		
rear (4.9)	6.6	6.8
upper (3.0)	4.1	4.3
740i, iA, iL		
rear	9.4	9.6
upper	5.9	6.1
745i, LI		
rear	8.2	8.4
upper	3.8	4.0
750iL		
1988-94		
rear	8.1	8.3
upper	6.2	6.4
1995-01		
rear	8.1	8.3
upper	11.2	11.4
760i, 760Li		
rear (6.5)	8.8	9.0
upper (7.1)	9.6	9.8
840Ci	11.9	12.1
850Ci, 850i		
5.0L		
AT	12.8	13.0
MT	10.7	10.9
5.4L	10.5	10.7
X5		
rear	7.8	8.0
upper	3.8	4.0

ALPINA ROADSTER : M3 : M5 : X3 : X5 : Z3 : Z4 : Z8   BMW-13

	LABOR TIME	SEVERE SERVICE
Cylinder Head, Replace (B)		
Includes: Adjustments.		
318i, iC, is		
1984-85	12.8	13.0
1991-98	15.7	15.9
318ti	20.7	20.9
320i	12.8	13.0
325, 325e	13.1	13.3
325i, iC, is		
1987-91	13.8	14.0
1992-95	28.6	28.8
325iX	15.8	16.0
328i, iC, is	28.4	28.6
330Ci, 330i (13.9)	18.8	19.0
330Xi (14.7)	19.8	20.0
524td, 528e	13.4	13.6
525i	27.8	28.0
528i, iT		
1981	17.8	18.0
1997-00	26.8	27.0
530i		
right side	29.2	29.4
left side	37.5	37.7
both sides	38.4	38.6
533i, 535i, is	13.8	14.0
540i, iA, iT		
1994-95		
right side	26.0	26.2
left side	29.8	30.0
both sides	39.7	39.9
1997-01		
right side	23.3	23.5
left side	24.5	24.7
both sides	35.6	35.8
545i		
right side (10.3)	13.9	14.1
left side (10.0)	13.5	13.7
both sides (15.2)	20.5	20.7
633CSi, 635CSi	17.8	18.0
645Ci		
right side	13.8	14.0
left side	13.8	14.0
both sides	20.5	20.7
733i, 735i	18.4	18.6
740i, iA, iL		
1994		
right side	26.3	26.5
left side	29.8	30.0
both sides	39.7	39.9
1995-01		
right side	22.8	23.0
left side	23.9	23.1
both sides	35.1	35.3
745i, Li		
right side	13.9	14.1
left side	12.5	12.7
both sides	19.8	20.0
750iL		
1988-94		
one	20.0	20.2
both	29.3	29.5
1995-01		
right side	28.1	28.3
left side	28.4	28.6
760i, 760Li		
right side (13.2)	17.8	18.0
left side (13.0)	17.6	17.8
both sides (19.9)	26.9	27.1
840Ci		
right side	28.5	28.7
left side	29.5	29.7
both sides	41.7	41.9
850Ci, 850i		
5.0L		
AT		
one side	40.3	40.5
both sides	48.4	48.6
MT		
one side	40.5	40.7
both sides	46.3	46.5
5.4L		
one side	37.8	38.0
both sides	43.7	43.9
M coupe/roadster, Z3		
1.9L	20.9	21.1
2.3L, 2.8L	27.0	27.2
M3 (21.1)	28.5	28.7
M5		
right side (23.6)	31.9	32.1
left side (23.9)	32.3	32.5
both sides (43.6)	58.9	59.1
X3 (14.1)	19.0	19.2
X5		
right side (8.3)	11.2	11.4
left side (8.5)	11.5	11.7
both sides (13.9)	18.8	19.0
Z4 (14.1)	19.0	19.2
Cylinder Head Gasket, Replace (B)		
318i, iC, is		
1984-85	7.3	7.5
1991-95	12.8	13.0
1996-98	10.0	10.2
318ti		
1995	12.8	13.0
1996-99	10.8	11.0
320i	6.8	7.0
325, 325e	7.4	7.6
325i, iC, is		
1987-91	7.4	7.6
1992-97	12.8	13.0
325iX (6.0)	8.1	8.3
328i, iC, is	20.5	20.7
330Ci, 330i (10.2)	13.8	14.0
330Xi (10.8)	14.6	14.8
524td	8.1	8.3
525i	13.0	13.2
528e	7.3	7.5
528i, iT		
1981	9.6	9.9
1997-00	19.0	19.2
530i		
right side	22.8	23.0
left side	19.0	19.2
both sides	25.2	25.5
533i, 535i, is	9.0	9.2
540i, iA, iT		
1994-95		
right side	19.4	19.6
left side	22.9	23.1
both sides	25.2	25.4
1997-01		
right side	16.8	17.0
left side	17.9	18.1
both sides	23.0	23.2
545i, 745i, Li		
right side (10.2)	13.8	14.0
left side (9.5)	12.8	13.0
both sides (14.6)	19.7	19.9
740i, iA, iL		
1994		
right side	19.5	19.7
left side	22.9	23.1
both sides	25.2	25.4
1995-01		
right side	16.6	16.8
left side	17.8	18.0
both sides	22.8	23.0
750iL		
1988-94		
one	15.2	15.4
both	19.3	19.5
1995-01		
right side	22.5	22.7
left side	23.0	23.2
both sides	26.0	26.2
760Li		
right side (13.0)	17.6	17.8
left side (12.5)	16.9	17.1
both sides (19.0)	25.7	25.9
840Ci		
right side	21.4	21.6
left side	22.5	22.7
both sides	28.6	28.8
850Ci, i		
5.0L		
AT		
one side	34.1	34.3
both sides	35.5	35.7
MT		
one side	31.8	32.0
both sides	33.2	33.4
5.4L		
one side	31.4	31.6
both sides	32.5	32.7

BMW-14 3 SERIES : 524TD : 5 SERIES : 6 SERIES : 7 SERIES : 8 SERIES

	LABOR TIME	SEVERE SERVICE
M coupe/roadster, Z3		
1.9L	10.9	11.1
2.3L, 2.8L	19.4	19.6
M3 (15.3)	20.7	20.9
M5		
right side (16.8)	22.7	22.9
left side (17.1)	23.1	23.3
both sides (30.3)	40.9	41.1
X3 (10.5)	14.2	14.4
X5		
one side (8.3)	11.2	11.4
both sides (13.3)	18.0	18.2
Cylinder Head Rear End Cover, Replace or Reseal (B)		
750iL		
right side	4.2	4.4
left side	8.1	8.3
850Ci, i		
5.0L		
right side	5.0	5.2
left side		
AT	27.0	27.2
MT	24.6	24.8
5.4L		
right side	6.0	6.2
left side	26.4	26.6
Rocker Arm Cover Gasket, Replace or Reseal (B)		
318i, iC, is		
1984-85	.5	.7
1991-98	1.1	1.3
318ti		
1995	2.6	2.8
1996-99	2.2	2.4
320i	.7	.9
323Ci, i, is, iT	1.9	2.1
325, e, iX	.8	1.0
325i, iC, is		
1987-91	.5	.7
1992-95	2.8	3.0
328i, iC, is	1.9	2.1
330Ci, i, Xi (1.2)	1.6	1.8
524td	.7	.9
525i	2.6	2.8
528i, iT		
1981	.8	1.0
1997-00	1.6	1.8
530i		
right side	2.7	2.9
left side	2.4	2.6
both sides	4.3	4.5
533i, 535i, is	.5	.7
540i, iA, iT		
1994-95		
right side	2.4	2.6
left side	2.7	2.9
both sides	4.3	4.5

	LABOR TIME	SEVERE SERVICE
1997-01		
right side	1.9	2.1
left side	2.2	2.4
both sides	3.5	3.7
545i		
right side (2.8)	3.8	4.0
left side (2.5)	3.4	3.6
both sides (4.2)	5.7	5.9
633CSi, 635CSi, 733i, 735i	.8	1.0
740i, iA, iL		
1994		
right side	2.4	2.6
left side	2.8	3.0
both sides	4.3	4.5
1995-01		
right side	1.8	2.0
left side	2.1	2.3
both sides	3.4	3.6
745i, Li		
right side	3.4	3.6
left side	2.5	2.7
both sides	4.9	5.1
750iL		
1988-94		
one	3.1	3.3
both	4.2	4.4
1995-01		
right side	5.9	6.1
left side	5.7	5.9
both sides	7.8	8.0
760Li		
right side (5.3)	7.2	7.4
left side (5.1)	6.9	7.1
both sides (7.0)	9.4	9.6
840Ci		
right side (2.6)	3.5	3.7
left side (1.8)	2.4	2.6
both sides (3.9)	5.3	5.5
850Ci, 850i		
5.0L		
one side	4.4	4.6
both sides	6.9	7.1
5.4L		
one side	5.4	5.6
both sides	7.8	8.0
M coupe/roadster, Z3		
1.9L	2.1	2.3
2.3L, 2.8L	1.8	2.0
M3 (1.4)	1.9	2.1
M5		
right side (2.7)	3.6	3.8
left side (2.9)	3.9	4.1
both sides (3.7)	5.0	5.2
X3 (1.1)	1.5	1.7
X5		
right side (2.4)	3.2	3.4
left side (1.8)	2.4	2.6
both sides (3.7)	5.0	5.2
Z4 (1.1)	1.5	1.7

	LABOR TIME	SEVERE SERVICE
Rocker Arms or Shafts, Replace (B)		
1984-85 318i, iC, is	8.0	8.2
318t	4.2	4.4
320i	8.0	8.2
524td	3.7	3.9
325, e, iX, 528e	9.6	9.8
325i, iC, is	8.1	8.3
528i	10.3	10.5
533i, 535i, is	13.1	13.3
545i, 645Ci	11.0	11.2
633CSi, 635CSi	12.3	12.5
733i, 735i	13.1	13.3
745i, Li	9.4	9.6
750iL	10.5	10.7
760Li	15.6	15.8
850Ci, i		
5.0L	8.6	8.8
5.4L	9.8	10.0
X5	9.4	9.6
Z3 1.9L	3.5	3.7
Valve Job (A)		
318i, iC, is, 320i	15.6	15.8
318ti	18.5	18.7
323Ci, i, is, iT	26.6	26.8
325, 325e	14.8	15.0
325i, iC, is		
1987-91	15.8	16.0
1992-95	29.8	30.0
325iX	18.4	18.6
328i, iC, is	26.6	26.8
330Ci, i, Xi (16.2)	21.9	22.1
525i	28.9	29.1
528e	15.8	16.0
528i, iT		
1981	20.2	20.4
1997-00	24.8	25.0
530i		
right side	29.8	30.0
left side	38.6	38.8
both sides	39.4	39.6
533i, 535i, is	20.0	20.2
540i, iA, iT		
1994-95		
right side	26.8	27.0
left sides	30.0	30.2
both sides	40.2	40.4
1997-01		
right side	22.8	23.0
left sides	24.0	24.2
both sides	35.1	35.3
545i, 645Ci		
right side (14.0)	18.9	19.1
left sides (13.0)	17.6	17.8
both sides (23.3)	31.5	31.7
633CSi, 635CSi	20.2	20.4
733i, 735i	21.2	21.4

ALPINA ROADSTER : M3 : M5 : X3 : X5 : Z3 : Z4 : Z8 — BMW-15

	LABOR TIME	SEVERE SERVICE
740i, iA, iL		
1994		
right side	26.9	27.1
left side	30.0	30.2
both sides	40.2	40.4
1995-01		
right side	22.6	22.8
left side	23.9	24.1
both sides	35.0	35.2
745i, Li		
right side	17.6	17.8
left side	16.8	17.0
both sides	38.6	38.8
750iL		
1988-94	29.8	30.0
1995-99		
right side	27.5	27.7
left side	28.0	28.2
both sides	36.0	36.2
760Li		
one side (16.8)	22.7	22.9
both sides (23.3)	31.5	31.7
840Ci		
right side (21.6)	29.2	29.4
left side (22.1)	29.8	30.0
both sides (32.2)	43.5	43.7
850Ci, 850i		
AT	49.3	49.5
MT	46.2	46.4
M coupe/roadster, Z3		
1.9L	18.2	18.4
2.3L, 2.8L	25.1	25.3
M3	26.3	26.5
M5		
right side (21.7)	29.3	29.5
left side (22.1)	29.8	30.0
both sides (40.4)	54.5	54.8
X3 (16.7)	22.5	22.9
X5		
right side (12.5)	16.9	17.1
left side (13.2)	17.8	18.0
both sides (23.5)	31.7	31.9
Z4 (16.8)	22.7	22.9
Valve Lifters, Replace (B)		
318i, iC, is		
1991-95	7.3	7.5
1996-98	7.1	7.3
318ti		
1995	7.3	7.5
1996-99	7.6	7.8
325i, iC, is	9.3	9.5
328i, iC, is	7.2	7.4
330Ci, i, Xi (5.3)	7.2	7.4
525i	8.1	8.3
1997-00 528i, iT	8.2	8.4
530i	13.7	13.9
540i, iA, iT		
1994-95	13.7	13.9
1997-01	12.8	13.0

	LABOR TIME	SEVERE SERVICE
545i, 645Ci (8.3)	11.2	11.4
740i, iA, iL		
1994	13.1	13.3
1995-01	12.8	13.0
745i, 745Li (9.5)	12.8	13.0
750iL		
1988-94	7.3	7.5
1995-01	10.6	10.8
760i, 760Li (12.5)	16.9	17.1
840Ci (10.8)	14.6	14.8
850Ci, 850i		
5.0L	8.9	9.1
5.4L	9.7	9.9
M coupe/roadster, Z3		
1.9L	7.3	7.5
2.3L, 2.8L	8.3	8.5
M3 (6.9)	9.3	9.5
X3 (5.3)	7.2	7.4
X5 (7.9)	10.7	10.9
Valve Springs and/or Seals, Replace (B)		
Includes: Cylinder head R&R when necessary.		
318i, iC, is		
1984-85	7.4	7.6
1991-95	19.4	19.6
1996-98	14.6	14.8
318ti		
1995	19.4	19.6
1996-99	16.9	17.1
320i	7.4	7.6
325, 325e	9.6	9.8
325i, iC, is		
1987-91	12.3	12.5
1992-95	18.1	18.3
325iX	9.8	10.0
328i, iC, is	24.2	24.4
330Ci, i, Xi (13.0)	17.6	17.8
524td	14.6	14.8
525i	21.8	22.0
528e	15.2	15.4
528i, iT		
1981	10.3	10.5
1997-00	22.6	22.8
530i		
AT	34.9	35.1
MT	32.6	32.8
533i, 535i, is	11.2	11.4
540i, iA, iT		
1994-95	32.8	33.0
1997-01	31.1	31.3
545i (24.3)	32.8	33.0
633CSi, 635CSi	10.0	10.2
645Ci	34.9	35.1
733i, 735i	11.2	11.4
740i, iA, iL		
1994	35.2	35.4
1995-01	31.9	32.1
745i, Li	22.3	22.5

	LABOR TIME	SEVERE SERVICE
750iL		
1988-94	22.3	22.5
1995-01	34.2	34.4
760Li	32.6	32.8
840Ci (28.2)	38.1	38.3
850Ci, 850i		
5.0L	44.4	44.6
5.4L	41.2	41.4
M coupe/roadster, Z3		
1.9L	16.9	17.1
2.3L, 2.8L	23.0	23.2
M3 (17.9)	24.2	24.4
X3 (13.0)	17.6	17.8
X5 (16.4)	22.1	22.3
Z4 (11.6)	15.7	15.9
CAMSHAFT		
Camshaft Timing, Check (B)		
1991-98 318i, iC, is	1.6	1.8
318ti	2.6	2.8
325i, iC, is	3.3	3.5
328i, iC, is	3.8	4.0
330Ci, i, Xi	3.8	4.0
525i	3.3	3.5
528i	2.7	2.9
530i, 540i	6.2	6.4
545i, 645Ci, 745i, Li	3.8	4.0
1994 740i	6.2	6.4
750iL		
1988-94	7.2	7.4
1995-01	10.6	10.8
840Ci	8.8	9.0
850Ci, i		
5.0L	8.8	9.0
5.4L	8.3	8.5
M coupe/roadster, Z3		
1.9L	3.3	3.5
2.3L, 2.8L	2.7	2.9
M3		
1995	3.3	3.5
1996-01	3.8	4.0
M5	7.2	7.4
X3	2.6	2.8
X5	3.6	3.8
Z4	2.7	2.9
Timing Belt, Adjust (B)		
325, 325e	.7	.9
524td	2.8	3.0
528e	.5	.7
Camshaft, Replace (A)		
318i, iC, is		
1984-85	7.6	7.8
1991-95		
one	6.3	6.5
both	7.2	7.4
1996-98		
one	7.0	7.2
both	8.1	8.3

BMW-16 3 SERIES : 524TD : 5 SERIES : 6 SERIES : 7 SERIES : 8 SERIES

	LABOR TIME	SEVERE SERVICE
318ti		
1995		
one	6.3	6.5
both	7.2	7.4
1996-99		
one	6.5	6.7
both	7.4	7.6
320i	7.6	7.8
325, 325e	9.3	9.5
325i, iC, is		
1987-91	10.0	10.2
1992-95		
one	7.3	7.5
both	8.9	9.1
325iX	9.3	9.5
328i, iC, is		
one	7.7	7.9
both	8.4	8.6
330Ci, i, Xi		
one (5.7)	7.7	7.9
both (6.2)	8.4	8.6
524td	3.7	3.9
525i		
one	7.3	7.5
both	8.1	8.3
528e	9.3	9.5
528i, iT		
1981	10.2	10.4
1997-00		
one	7.3	7.5
both	8.2	8.4
530i		
right side		
one	9.7	9.9
both	10.0	10.2
left side		
one	11.4	11.6
both	11.5	11.7
all	13.4	13.6
533i, 535i, is	10.0	10.2
540i, iA, iT		
1994-95		
right side	9.9	10.1
left side	11.4	11.6
all	13.4	13.6
1997-01		
one side		
one	11.2	11.4
both	11.5	11.7
all	12.3	12.5
545i, 645Ci, 745i, Li, X5		
one side		
one (5.6)	7.6	7.8
both (5.8)	7.8	8.0
all (9.1)	12.3	12.5
633CSi, 635CSi	11.0	11.2
733i, 735i	10.0	10.2

	LABOR TIME	SEVERE SERVICE
740i, iA, iL		
1994		
right side		
one	9.7	9.9
both	10.0	10.2
left side		
one	11.4	11.6
both	11.6	11.8
all	13.6	13.8
1995-01		
right side		
one	10.7	10.9
both	11.2	11.4
left side		
one	11.1	11.3
both	11.4	11.6
all	12.2	12.4
750iL		
1988-94		
one	9.3	9.5
both	10.5	10.7
1995-01		
one	16.4	16.6
both	17.3	17.5
760Li		
right side		
one (7.9)	10.7	10.9
both (8.3)	11.2	11.4
left side		
one (8.2)	11.1	11.3
both (8.5)	11.5	11.7
all (9.0)	12.2	12.4
840Ci		
right side		
one (8.9)	12.0	12.2
both (9.1)	12.3	12.5
left side		
one (8.5)	11.5	11.7
both (8.8)	11.9	12.1
all (10.5)	14.2	14.4
850Ci, 850i		
5.0L		
one	13.5	13.7
both	15.2	15.4
5.4L		
one side	16.2	16.4
both sides	17.2	17.4
M coupe/roadster, Z3		
1.9L		
one	6.4	6.6
all	7.1	7.3
2.3L, 2.8L		
one	7.6	7.8
all	8.2	8.4
M3		
one (6.2)	8.3	8.5
all (6.9)	9.3	9.5

	LABOR TIME	SEVERE SERVICE
M5		
right side		
one (8.2)	11.1	11.3
both (8.5)	11.5	11.7
left side		
one (10.0)	13.5	13.7
both (10.5)	14.5	14.7
X3	6.4	6.6
Camshaft Sprocket, Replace (B)		
325, e, iX	3.1	3.3
525i	19.9	20.1
530i		
AT	29.7	29.9
MT	28.0	28.2
1994-95 540i	29.7	29.9
840Ci	13.1	13.3
850Ci, i		
5.0L	12.3	12.5
5.4L	12.8	13.0
Intermediate Shaft, Replace (B)		
524td	6.2	6.4
Intermediate Shaft Sprocket, Replace (B)		
524td	3.6	3.8
Lower Timing Case Cover Seal, Replace (B)		
1991-94 318i, iC, is	1.7	1.9
525i	2.5	2.7
1988-94 750iL	2.3	2.5
w/AC 318 add	.5	.5
Timing Belt, Replace (B)		
325, 325e	2.9	3.1
325iX, 524td	3.1	3.3
525i	2.8	3.0
528e		
1982-87	2.3	2.5
1988	3.1	3.3
Replace tensioner add	.2	.2
Timing Belt Tensioner, Replace (B)		
524td	2.9	3.1
Timing Case Cover Seal, Replace (B)		
Exc. below	3.2	3.4
318i, iC, is		
1984-85	2.3	2.5
1991-98	1.9	2.1
318ti	2.2	2.4
320i, 325, 325e	2.3	2.5
325i, iC, is	5.2	5.4
328i, iC, is	1.8	2.0
330Ci, 330i (7.6)	10.3	10.5
330Xi (9.3)	12.6	12.8
524td	6.3	6.5
528i, iT		
1981	8.6	8.8
1997-00	1.8	2.0
545i, 645Ci	3.7	3.9
633CSi, 635CSi	9.4	9.6
733i	6.8	7.0

ALPINA ROADSTER : M3 : M5 : X3 : X5 : Z3 : Z4 : Z8 — BMW-17

	LABOR TIME	SEVERE SERVICE
740i, iA, iL		
1994	2.7	2.9
1995-01	2.5	2.7
745i, 745Li	3.8	4.0
750iL		
1988-94		
upper	9.3	9.5
lower	1.5	1.7
1995-01	3.4	3.6
760Li	7.8	8.0
850Ci 5.4L	3.6	3.8
M coupe/roadster, Z3		
1.9L	2.1	2.3
2.3L, 2.8L	1.5	1.7
1996-01 M3	1.8	2.0
M5 (2.0)	2.7	2.9
X3	13.2	13.4
X5	23.8	24.0
Z4	17.3	17.5
Timing Chain, Replace (B)		
318i, iC, is		
1984-85	8.9	9.1
1991-95	8.8	9.0
1996-98	6.8	7.0
318ti		
1995	8.8	9.0
1996-99	8.5	8.7
320i	7.2	7.4
325i, iC, is	22.2	22.4
328i, iC, is	14.7	14.9
330Ci, 330i (6.9)	9.3	9.5
330Xi (9.6)	13.0	13.2
528i		
1981	9.3	9.5
1997-99	14.2	14.4
530i		
AT	29.8	30.0
MT	27.3	27.5
533i, 535i, is	8.1	8.3
545i (17.3)	23.4	23.6
540i, iA, iT		
1994-95	29.8	30.0
1997-01	23.6	23.8
633CSi, 635CSi	9.8	10.0
733i, 735i	7.9	8.1
740i, iA, iL		
1994-95	31.8	32.0
1996-01	23.1	23.3
745i, Li	24.8	25.0
750iL		
1988-94	12.8	13.0
1995-01	26.9	27.1
760Li (20.9)	28.2	28.4
840Ci	32.4	32.6
850Ci, i	15.2	15.4
5.0L	19.0	19.2
5.4L	27.5	27.7

	LABOR TIME	SEVERE SERVICE
M coupe/roadster, Z3		
1.9L	7.9	8.1
2.3L, 2.8L	14.2	14.4
M3		
1995	22.5	22.7
1996-01	14.7	14.9
M5	45.6	45.8
X3 (9.9)	13.4	13.6
X5 (17.6)	23.8	24.0
Z4 (13.0)	17.6	17.8
Timing Chain Cover Oil Seal, Replace (B)		
325, e, iX	4.4	4.6
528e	4.4	4.6
Timing Chain Sprocket, Replace (B)		
318i, iC, is	6.8	7.0
320i	8.3	8.5
525i	17.6	17.8
528i, iT		
1981	10.0	10.2
1997-00	14.0	14.2
540i, iA, iT		
1994-95	29.8	30.0
1997-01	24.1	24.3
633CSi, 635CSi	10.3	10.5
733i, 735i	8.7	8.9
740i, iA, iL		
1994	28.9	29.1
1995-01	24.1	24.3
750iL		
1988-94	32.4	32.6
1995-01	27.3	27.5
840Ci	13.1	13.3
850Ci, i		
5.0L	12.3	12.5
5.4L	34.9	35.1
M coupe/roadster, Z3		
1.9L	7.9	8.1
2.3L, 2.8L	14.3	14.5
M3	14.9	15.1
M5	45.8	46.0
Timing Chain Tensioner, Replace (B)		
1992-98 318i, iC, is	6.1	6.3
318ti, 330Ci, i, Xi	2.6	2.8
525i	.8	1.0
1997-00 528i, iT	.6	.8
540i, iA, iT		
1994-95	1.5	1.7
1997-01	.7	.9
740i, iA, iL		
1994	1.5	1.7
1995-01	.7	.9
1995-01 750iL	1.1	1.3
840Ci	.6	.8
850Ci, i		
5.0L	1.1	1.3
5.4L	1.6	1.8

	LABOR TIME	SEVERE SERVICE
M coupe/roadster, Z3		
1.9L	3.6	3.8
2.3L, 2.8L	.6	.8
M3	.6	.8
M5	.8	1.0
X3, Z4	2.7	2.9
Timing Cover and/or Gasket, Replace (B)		
318i, iC, is		
1991-93	3.2	3.4
1994-95		
upper	4.5	4.7
lower	3.1	3.3
1996-98		
upper	3.5	3.7
lower	5.9	6.1
318ti		
upper	4.8	5.0
lower	3.4	3.6
320i		
upper	2.4	2.6
lower	6.9	7.1
325, 325e	8.2	8.4
325i, iC, is		
1987-91	14.3	14.5
1992-95	12.4	12.6
325iX	5.9	6.1
328i, iC, is	14.6	14.8
330Ci, 330i (7.6)	10.3	10.5
330Xi (9.5)	12.8	13.0
524td	.6	.8
525i	11.8	12.0
528i, iT		
1981		
upper	2.4	2.6
lower	8.9	9.1
1997-00	13.8	14.0
530i		
upper		
left side	5.9	6.1
right side	4.9	5.1
both sides	9.8	10.0
lower	26.6	26.8
533i, 535i, is		
upper	1.8	2.0
lower	7.9	8.1
540i, iA, iT		
1994-95		
upper		
right side	5.9	6.1
left side	4.9	5.1
both sides	9.8	10.0
lower	28.8	29.0
1997-01		
upper		
right side	4.5	4.7
left side	5.7	5.9
both sides	8.7	8.9
lower	21.1	21.3

BMW-18 3 SERIES : 524TD : 5 SERIES : 6 SERIES : 7 SERIES : 8 SERIES

	LABOR TIME	SEVERE SERVICE
545i		
upper		
right side (4.4)	5.9	6.1
left side (3.6)	4.9	5.1
both sides (6.5)	8.8	9.0
lower (16.9)	22.8	23.0
633CSi, 635CSi		
upper	2.8	3.0
lower	9.4	9.6
733i, 735i		
upper	2.1	2.3
lower	6.9	7.1
740i, iA, iL		
1994		
right side	4.9	5.1
left side	5.9	6.1
both sides	9.8	10.0
1995-01		
upper		
right side	4.4	4.6
left side	5.4	5.6
both sides	7.3	7.5
lower	20.8	21.0
745i, Li		
upper		
right side	4.8	5.0
left side	5.4	5.6
both sides	7.3	7.5
lower	23.8	24.0
1995-01 750iL		
upper	15.2	15.4
lower	26.6	26.8
760Li		
upper		
one side (6.1)	8.2	8.4
both sides (8.6)	11.6	11.8
lower (20.6)	27.8	28.0
840Ci		
upper		
right side	4.9	5.1
left side	3.3	3.5
lower	31.5	31.7
850Ci, 850i		
5.0L		
upper	12.3	12.5
lower	18.8	19.0
5.4L		
upper	14.5	14.7
lower	27.3	27.5
M coupe/roadster, Z3		
1.9L		
upper	4.6	4.8
lower	2.9	3.1
2.3L, 2.8L	14.1	14.3
M3 (10.8)	14.6	14.8
M5		
upper		
one side (7.0)	9.4	9.6
both sides (9.5)	12.8	13.0
lower (33.6)	45.4	45.6

	LABOR TIME	SEVERE SERVICE
X3 (10.8)	14.6	14.8
X5		
upper		
one side (3.4)	4.6	4.8
both sides (6.1)	8.2	8.4
lower (17.6)	23.8	24.0
Z4	18.8	19.0
Upper Timing Case Cover Seal, Replace (B)		
1991-97 318i, iC, is	4.4	4.6
1995-97 318ti	4.4	4.6
320i	1.6	1.8
1981 528i	2.9	3.1
533i, 535i, is	2.9	3.1
633CSi, 635CSi	3.0	3.2
733i, 735i	2.8	3.0
Variable Cam Control Unit, Replace (B)		
1994-95 325i, iC, is	4.4	4.6
328i, iC, is	5.3	5.5
330Ci, i, Xi	3.8	4.0
1994-96 525i	4.4	4.6
1997-00 528i	5.1	5.3
530i	4.4	4.6
1994-95 540i	4.4	4.6
M coupe/roadster, Z3		
2.3L, 2.8L	5.1	5.3
M3		
1995	4.4	4.6
1996-01	5.3	5.5
X3, Z4	3.8	4.0
CRANK & PISTONS		
Connecting Rod Bearings, Replace (A)		
Includes: Check bearing oil clearance.		
320i	4.4	4.6
325e	4.9	5.1
325i, iS	5.8	6.0
328i, iC	20.4	20.6
330Ci, i	22.4	22.6
330Xi	24.6	24.8
524td	4.1	4.3
528e	4.2	4.4
1981 528i	5.4	5.6
1994-95 540i	12.8	13.0
545i	26.2	26.4
633CSi, 635CSi, 733i, 735i	5.4	5.6
645Ci	26.4	26.6
1994-95 740i, iL	13.1	13.3
745i, Li	28.4	28.6
760Li	33.6	33.8
1994-95 840Ci	13.0	13.2
1991-95 850Ci, i		
AT	34.9	35.1
MT	32.9	33.1
X3	21.5	21.7
X5	28.0	28.2
Z4	21.3	21.5

	LABOR TIME	SEVERE SERVICE
Crankshaft and Main Bearings, Replace (A)		
Includes: Engine R&R.		
318i, iC, is		
1984-85	22.0	22.2
1991-93	16.8	17.0
1994-95		
AT	33.6	33.8
MT	31.2	31.3
1996-98		
AT	30.8	31.0
MT	29.2	29.4
318ti		
1995		
AT	33.1	33.3
MT	30.5	30.7
1996-99		
AT	28.7	28.9
MT	26.6	26.8
320i	17.9	18.1
325, 325e	22.0	22.2
325i, iC, is		
1987-91	28.1	28.3
1992-95	30.0	30.2
328i, iC, is		
AT (26.9)	36.3	36.5
MT (26.0)	35.1	35.3
330Ci, i (19.3)	26.2	26.4
330Xi (21.2)	28.6	28.8
524td	25.2	25.4
525i		
AT	35.2	35.4
MT	33.3	33.4
528e	28.1	28.3
528i, iT		
1981	21.5	21.7
1997-00		
AT	34.5	34.7
MT	33.3	33.5
530i		
AT	37.5	37.7
MT	36.2	36.4
533i, 535i, is	21.7	21.9
540i, iA, iT		
1994-95	37.4	37.6
1997-01		
AT	43.0	43.2
MT	41.3	41.5
545i (22.4)	30.2	30.4
633CSi, 635CSi	22.2	22.4
645Ci	31.2	31.4
733i, 735i	20.2	20.4
740i, iA, iL		
1994	37.9	38.1
1995-01	43.6	43.8
745i, Li	32.4	32.6
750iL		
1988-94	40.0	40.2
1995-01	47.1	47.3
760i, 760Li (35.8)	48.3	48.5

ALPINA ROADSTER : M3 : M5 : X3 : X5 : Z3 : Z4 : Z8 — BMW-19

	LABOR TIME	SEVERE SERVICE
840Ci	40.0	40.2
850Ci, i		
5.0L		
AT	52.4	52.4
MT	50.1	50.3
5.4L	49.9	50.1
M coupe/roadster, Z3		
1.9L		
AT	30.8	31.0
MT	28.8	29.0
2.3L, 2.8L		
AT	35.1	35.3
MT	33.1	33.3
M3		
1995		
AT	38.6	38.8
MT	35.6	35.8
1996-01		
AT	36.3	36.5
MT	35.1	35.3
M5	53.3	53.5
X3	26.4	26.6
X5 (23.5)	31.7	31.9
Z4 (18.1)	24.4	24.6

Crankshaft Front Oil Seal, Replace (B)

320i	1.6	1.8
528e	4.4	4.6
1981 528i	2.6	2.8
633CSi, 635CSi	3.1	3.3
733i	2.5	2.7

Crankshaft Pulley, Replace (B)

320i	.8	1.0
633CSi, 635CSi	2.5	2.7

Pistons or Connecting Rods, Replace (A)
Includes: Ridge reaming, cylinder wall deglazing, installing new rings and rod bearings, engine tune-up.

318i, iC, is		
1984-85	15.5	15.7
1991-93	27.8	28.0
1994-95		
AT	26.0	26.2
MT	23.2	23.4
1996-98		
AT	23.6	23.8
MT	21.7	21.9
318ti		
1995		
AT	27.3	27.5
MT	24.2	24.4
1996-99		
AT	21.7	21.9
MT	19.8	20.0
320i	15.5	15.7
325, 325e	17.7	17.9
325i, iC, is		
1987-91	20.0	20.2
1992-95	25.2	25.4

	LABOR TIME	SEVERE SERVICE
325iX	20.0	20.2
328i, iC, is		
AT	30.4	30.6
MT	28.8	29.0
330Ci, 330i (18.7)	25.2	25.4
524td	22.3	22.5
525i		
AT	27.8	28.0
MT	25.6	25.8
528e	17.7	17.9
528i, iT		
1981	19.9	20.1
1997-00		
AT	28.4	29.3
MT	27.2	27.4
530i		
AT	31.3	31.5
MT	29.8	30.0
533i, 535i, is	18.2	18.4
540i, iA, iT		
1994-95	31.3	31.5
1997-01		
AT	38.3	38.5
MT	31.1	31.3
545i (22.4)	30.2	30.4
633CSi, 635CSi	19.8	20.0
645Ci	31.2	31.4
733i, 735i	18.3	18.5
740i, iA, iL		
1994	36.2	36.4
1995-01	38.4	38.6
745i, Li	32.4	32.6
750iL		
1988-94	35.2	35.4
1995-01	41.9	42.1
760i, 760Li (31.9)	43.1	43.3
840Ci	34.3	34.5
850Ci, i		
5.0L		
AT	46.2	46.4
MT	43.9	44.1
5.4L	43.5	43.7
M coupe/roadster, Z3		
1.9L		
AT	23.6	23.8
MT	21.6	22.5
2.3L, 2.8L		
AT	29.9	30.1
MT	27.1	27.3
M3		
AT	30.4	30.6
MT	28.8	29.0
M5	49.8	50.0
X3 (18.8)	25.4	25.6
X5 (23.5)	31.7	31.9
Z4 (17.4)	23.5	23.7

Rear Main Oil Seal, Replace (B)

	LABOR TIME	SEVERE SERVICE
318i, iC, is		
1984-85	7.3	7.5
1991-95		
AT	9.8	10.0
MT	7.3	7.5
1996-98		
AT	7.6	7.8
MT	5.9	6.1
318ti		
AT	7.6	7.8
MT	5.6	5.8
320i, 325, 325e	7.3	7.5
1987-95 325i, iC, is	7.3	7.5
325iX	8.9	9.1
328i, 328iC, 328is		
AT	8.6	8.8
MT	7.8	8.0
330Ci, 330i (4.9)	6.6	6.8
330Xi (6.5)	8.8	9.0
524td	6.2	6.4
525i		
AT	9.9	10.1
MT	8.6	8.8
528e	7.3	7.5
528i, iT		
1981	7.3	7.5
1997-00		
AT	7.4	7.6
MT	6.8	7.0
530i		
AT	10.5	10.7
MT	8.5	8.7
533i, 535i, is	7.7	7.9
540i, iA, iT		
1994-95	10.6	10.8
1997-01		
AT	8.9	9.1
MT	7.5	7.7
545i (5.6)	7.6	7.8
633CSi, 635CSi, 733i, 735i	7.7	7.9
645Ci	7.9	8.1
740i, iA, iL		
1994	10.2	10.4
1995-01	8.3	8.5
745i, Li	8.8	9.0
750iL		
1988-94	10.5	10.7
1995-01	9.1	9.3
760i, 760Li (8.5)	11.5	11.7
840Ci	10.5	10.7
850Ci, 850i		
5.0L		
AT	12.5	12.7
MT	10.0	10.2
5.4L	9.8	10.0

BMW-20 3 SERIES : 524TD : 5 SERIES : 6 SERIES : 7 SERIES : 8 SERIES

	LABOR TIME	SEVERE SERVICE
Rear Main Oil Seal, Replace (B)		
M coupe/roadster, Z3		
1.9L		
AT	6.8	7.0
MT	5.6	5.8
2.3L, 2.8L		
AT	8.1	8.3
MT	6.4	6.6
M3		
AT	8.8	9.0
MT	7.4	7.6
X3	5.6	5.8
X5 (6.4)	8.6	8.8
Z4 (5.2)	7.0	7.3
Rings, Replace (A)		
Includes: Ridge reaming, cylinder wall deglazing, installing new rings, engine tune-up.		
318i, iC, is		
1984-85	17.3	17.5
1991-93	26.0	26.2
1994-95		
AT	37.0	37.2
MT	24.3	24.5
1996-98		
AT	24.5	24.7
MT	22.6	22.8
318ti		
1995		
AT	25.4	25.6
MT	22.5	22.7
1996-99		
AT	22.6	22.8
MT	20.1	20.3
320i	16.8	17.0
325, 325e	17.3	17.5
325i, iC, is		
1987-91	19.8	20.0
1992-95	24.1	24.3
325iX	22.0	22.2
328i, iC, is		
AT	31.7	31.9
MT	30.2	30.4
330Ci, 330i (17.3)	23.4	23.6
330Xi (18.9)	25.5	25.7
524td	24.1	24.3
525i		
AT	26.0	26.2
MT	23.8	24.0
528e	19.9	20.1
528i, iT		
1981	20.8	21.0
1997-00		
AT	29.9	30.1
MT	28.6	28.8
530i		
AT	29.8	30.0
MT	26.7	26.9

	LABOR TIME	SEVERE SERVICE
533i, 535i, is	18.7	18.9
540i, iA, iT		
1994-95	29.8	30.0
1997-01		
AT	40.1	40.3
MT	32.3	32.5
545i (23.1)	31.2	31.4
633CSi, 635CSi	20.0	20.2
645Ci	32.2	32.4
733i, 735i	18.8	19.0
740i, iA, iL		
1994	33.8	34.0
1995-01	40.9	41.1
750iL		
1988-94	37.8	38.0
1995-01	44.5	44.7
760Li (27.1)	36.6	36.8
840Ci	36.3	36.5
850Ci, 850i		
5.0L		
AT	49.9	50.1
MT	46.8	47.0
5.4L	46.4	46.6
M coupe/roadster, Z3		
1.9L		
AT	24.4	24.6
MT	22.5	22.7
2.3L, 2.8L		
AT	30.8	31.0
MT	28.4	28.6
M3		
AT	31.8	32.0
MT	30.2	30.4
X3 (19.6)	26.5	26.7
X5 (24.3)	32.8	33.0
Z4 (18.2)	24.6	24.8
Vibration Damper, Replace (B)		
Exc. below	1.9	2.1
328i, iC, is	1.4	1.6
524td	1.3	1.5
750iL		
1988-94	3.1	3.3
1995-01	3.0	3.2
850Ci 5.4L	3.2	3.4
M coupe/roadster, Z3		
1.9L	1.4	1.6
2.3L, 2.8L	1.1	1.3
1995 M3	2.2	2.4
M5	2.5	2.7

ENGINE LUBRICATION
Engine Oil Cooler, Replace (B)
325iX, 760i, 760iL (.8) .. 1.1 1.3
Engine Oil Pressure Switch (Sending Unit), Replace (B)
Exc. below 8 1.0
318i, iC, is 1.6 1.8
318ti 1.5 1.7
524td .... .5 .7

	LABOR TIME	SEVERE SERVICE
Oil Pan and/or Gasket, Replace (B)		
318i, iC, is		
1984-85		
top	2.3	2.5
center	1.7	1.9
lower	.7	.9
1991-95		
upper	3.1	3.3
lower	1.2	1.4
1996-98		
AT	6.4	6.6
MT	5.6	5.8
318ti	3.5	3.7
320i	3.1	3.3
325, 325e	2.8	3.0
325i, iC, is		
1987-92	7.0	7.2
1993-95	4.2	4.4
325iX	5.8	6.0
328i, 328iC, 328is	5.5	5.7
330Ci, 330i (2.8)	3.8	4.0
330Xi (5.0)	6.8	7.0
524td	2.3	2.5
525i		
1989-91	3.7	3.9
1992-96	4.6	4.8
528e	2.0	2.2
528i, iT		
1981	3.1	3.3
1997-00	5.8	6.0
530i		
upper	7.3	7.5
lower	2.0	2.2
533i, 535i, is	3.6	3.8
540i, iA, iT		
1994-95		
upper	7.3	7.5
lower	2.0	2.2
1997-01		
upper	8.2	8.4
lower	2.0	2.2
545i, 645Ci, 745i, Li		
upper (5.3)	7.2	7.4
lower (1.3)	1.8	2.0
633CSi, 635CSi	3.6	3.8
733i, 735i	3.1	3.3
740i, iA, iL		
upper	8.2	8.4
lower	2.0	2.2
750iL		
1988-94		
upper	13.1	13.3
lower	2.3	2.5
1995-01		
upper	9.8	10.0
lower	1.9	2.1
760i, 760Li		
upper (6.5)	8.8	9.0
lower (1.4)	1.9	2.1

ALPINA ROADSTER : M3 : M5 : X3 : X5 : Z3 : Z4 : Z8 BMW-21

	LABOR TIME	SEVERE SERVICE
840Ci		
upper	7.7	7.9
lower	2.6	2.8
850Ci, i		
5.0L		
upper		
AT	19.0	19.2
MT	16.5	16.7
lower	2.6	2.8
5.4L		
upper	10.3	10.5
lower	2.2	2.4
M coupe/roadster, Z3		
1.9L	4.1	4.3
2.3L, 2.8L	5.3	5.5
M3 (4.1)	5.5	5.7
X3 (6.1)	8.2	8.4
X5		
upper (7.3)	9.9	10.1
lower (1.4)	1.9	2.1
Z4 (2.8)	3.8	4.0
Oil Pump, R&R and Recondition (B)		
1984-85 318i	2.3	2.5
320i	3.4	3.6
525i	5.2	5.4
528e	3.0	3.2
1981 528i	4.4	4.6
533i, 535i, is	4.5	4.7
633CSi, 635CSi	4.5	4.7
733i, 735i	4.3	4.5
Oil Pump, Replace (B)		
318i, iC, is		
1984-85	1.8	2.0
1991-98	15.3	15.5
318ti		
1995	13.8	14.0
1996-99	13.2	13.4
320i	2.8	3.0
325, 325e	3.2	3.4
325i, iC, is		
1987-92	7.3	7.5
1993-95	5.4	5.6
325iX	6.3	6.5
328i, 328iC, 328is	6.4	6.6
330Ci, 330i (4.0)	5.4	5.6
330Xi (5.5)	7.4	7.6
524td	2.3	2.5
525i	5.4	5.6
528e	2.6	2.8
528i, iT		
1981	3.5	3.7
1997-00	6.8	7.0
530i	2.3	2.5
533i, 535i, is	4.2	4.4
540i, iA, iT	2.3	2.5
545i, 645Ci,		
745i, Li (5.5)	7.4	7.6
633CSi, 635CSi	4.2	4.4

	LABOR TIME	SEVERE SERVICE
733i, 735i	3.7	3.9
740i, iA, iL	2.4	2.6
750iL		
1988-94	3.2	3.4
1995-01	2.3	2.5
760i, 760Li (7.0)	9.4	9.6
840Ci	2.4	2.6
850Ci, 850i		
5.0L	3.2	3.4
5.4L	2.7	2.9
M coupe/roadster, Z3		
1.9L	12.7	12.9
2.3L, 2.8L	6.3	6.5
M3 (4.8)	6.5	6.7
M5 (3.0)	4.1	4.3
X3 (6.2)	8.4	8.6
X5 (5.5)	7.4	7.6
Z4 (3.1)	4.2	4.4
Oil Pump Chain, Replace (B)		
320i	9.9	10.1
1981 528i	12.8	13.0
633CSi, 635CSi	12.8	13.0

CLUTCH

Bleed Clutch Hydraulic System (B)
1981-05	.7	.9

Clutch Assy., Replace (B)
318i, iC, is		
1984-85	6.2	6.4
1991-98	5.3	5.5
318ti	5.0	5.2
320i, 325, 325e	6.2	6.4
325i, 325iC, 325is	8.0	8.2
325iX	8.1	8.3
328i, 328iC, 328is	6.2	6.4
330Ci, 330i, 330Xi	6.2	6.4
525i	6.7	6.9
528e, 533i, 535i, is	6.2	6.4
528i, iT		
1981	6.2	6.4
1997-00	5.3	5.5
530i	7.3	7.5
1997-01 540i, iA	6.1	6.3
545i	5.3	5.5
633CSi, 635CSi,		
733i, 735i	6.2	6.4
645Ci	6.2	6.4
1991-97 850Ci, i	9.3	9.5
M coupe/roadster, Z3		
1.9L, 2.3L	5.0	5.2
2.8L	5.2	5.4
M3, X5 (4.6)	6.2	6.4
M5 (6.5)	8.8	9.0
X3 (3.9)	5.3	5.5
Z4 (3.7)	5.0	5.2

	LABOR TIME	SEVERE SERVICE
Clutch Master Cylinder, Replace (B)		
Includes: System bleeding.		
1981-05	2.8	3.0
w/air bags add	.5	.5
Clutch Slave Cylinder, Replace (B)		
Includes: System bleeding.		
1981-05	1.2	1.4
Flywheel, Replace (B)		
318i, iC, is		
1984-85	6.8	7.0
1991-98	5.6	5.8
318ti	5.3	5.5
320i, 325, 325e	6.8	7.0
325i, iC, is	8.6	8.8
325iX	8.1	8.3
328i, iC, is	6.6	6.8
330Ci, 330i (4.6)	6.2	6.4
330Xi (5.6)	7.6	7.8
525i	7.2	7.4
528e, 533i, 535i, is	6.8	7.0
528i, iT		
1981	6.8	7.0
1997-00	5.3	5.5
530i	7.8	8.0
540i	6.7	6.9
545i (4.6)	6.2	6.4
633CSi, 635CSi,		
733i, 735i	6.8	7.0
645Ci	7.2	7.4
745i, Li	7.6	7.8
760Li	8.6	8.8
850Ci, 850i	9.8	10.0
M3 (4.8)	6.5	6.7
M5 (6.9)	9.3	9.5
X3 (3.9)	5.3	5.5
X5 (5.6)	7.6	7.8
Z3		
1.9L, 2.3L (3.9)	5.3	5.5
2.8L (4.4)	5.9	6.1
Z4 (4.2)	5.7	5.9

MANUAL TRANSMISSION

Rear Transmission Mount, Replace (B)
Exc. below	.8	1.0
325iX	2.3	2.5
330Xi	1.5	1.7
1997-01 540i, iA, iT	1.0	1.2
545i	1.2	1.4
745i, Li, 760Li	1.8	2.0
1991-97 850Ci, i	1.3	1.5
M3	.5	.7
M5	1.1	1.3
X5	2.2	2.4

Transmission Assy., R&R and Recondition (A)
5-Speed	13.8	14.0

BMW-22 3 SERIES : 524TD : 5 SERIES : 6 SERIES : 7 SERIES : 8 SERIES

	LABOR TIME	SEVERE SERVICE
Transmission Assy., R&R or Replace (B)		
318i, iC, is		
1984-85	6.2	6.4
1991-98	4.9	5.1
318ti	4.6	4.8
320i, 325, 325e	6.2	6.4
325i, iC, is	7.0	7.2
325iX	7.8	8.0
328i, iC, is	5.8	6.0
330Ci, 330i (3.4)	4.6	4.8
330Xi, 528e (4.6)	6.2	6.4
525i	5.6	5.8
528i, iT		
1981	6.2	6.4
1997-00	4.7	4.9
530i	6.4	6.6
533i, 535i, is	6.2	6.4
540i, iT, 545i (4.3)	5.8	6.0
633CSi, 635CSi, 733i, 735i	6.2	6.4
645Ci	6.2	6.4
850Ci, 850i	8.6	8.8
M coupe/roadster, Z3		
1.9L, 2.3L	4.3	4.5
2.8L	5.2	5.4
M3, X5	5.8	6.0
M5	8.4	8.6
X3, Z4	4.7	4.9
Replace assembly add	.5	.5
Transmission Rear Seal, Replace (B)		
1981-05	2.8	3.0

AUTOMATIC TRANSMISSION
SERVICE TRANSMISSION INSTALLED

	LABOR TIME	SEVERE SERVICE
Diagnose Electronic Transmission Component Each (A)		
1989-05	1.1	1.3
Bands, Adjust (B)		
1981-01	.7	.9
Check Unit for Oil Leaks (C)		
1981-05	.8	1.0
Oil Pressure Check (B)		
1981-05	1.2	1.4
Electronic Control Unit, Replace (B)		
Exc. below (.6)	.8	1.0
318i, iC, is	.3	.5
325i, iC, is	.3	.5
328i, iC, is	.3	.5
545i	3.0	3.2
745i, Li, 760i, Li (1.6)	2.2	2.4
Z3	.3	.5
Z4 91.6)	2.2	2.4

	LABOR TIME	SEVERE SERVICE
Extension Housing Oil Seal, Replace (B)		
Exc. below (2.1)	2.8	3.0
1991-98 318i, iC, is	3.1	3.3
325i, iC, is	3.6	3.8
330Xi (1.8)	2.4	2.6
525i	5.3	5.5
530i	4.2	4.4
540i, iA, iT		
1994-95	4.2	4.4
1997-01	3.3	3.5
545i, 645Ci	3.1	3.3
740i, iA, 750iL	4.4	4.6
840Ci	4.6	4.8
X3, Z4 (1.6)	2.2	2.4
X5 (1.8)	2.4	2.6
Z3 1.9L, 2.3L (1.8)	2.4	2.6
Governor Assy., R&R or Replace (B)		
1984-85 318i	2.8	3.0
320i	3.1	3.3
325, 325e	2.8	3.0
325iX	4.2	4.4
528e, 533i	2.8	3.0
1981 528i	2.8	3.0
633CSi, 635CSi, 733i	2.8	3.0
Main Case Solenoid Valve, Replace (B)		
All Models	2.2	2.4
Manual Shift Valve Seal, Replace (B)		
318i, iC, is		
1991-97	6.4	6.6
1998	1.9	2.1
318ti	1.4	1.6
325i, 325iC, 325is	6.4	6.6
328i, iC, is	1.1	1.3
330Ci, 330i, 330Xi (.8)	1.1	1.3
525i	6.2	6.4
530i	1.8	2.0
540i, iA, iT		
1994-95	1.7	1.9
1997-01	2.5	2.7
740i, iA, iL		
1994	2.9	3.1
1995-01	2.3	2.5
750iL	1.4	1.6
840i		
one	1.6	1.8
both	2.4	2.6
850Ci		
5.0L	.8	1.0
5.4L		
one	1.6	1.8
both	2.4	2.6
M3 (1.0)	1.4	1.6
X3, Z4 (.8)	1.1	1.3
Z3 (1.2)	1.6	1.8

	LABOR TIME	SEVERE SERVICE
Oil Pan and/or Gasket, Replace (B)		
Includes: Fluid and filter change.		
318i, iC, is		
1984-85	1.7	1.9
1991-98	2.2	2.4
318ti	2.0	2.2
320i, 325, e, iX	1.7	1.9
325i, iC, is		
1987-91	1.7	1.9
1992-95	2.4	2.6
328i, iC, is	2.0	2.2
330Ci, i, Xi (1.5)	2.0	2.2
524td	1.7	1.9
525i both	2.9	3.1
528i, iT		
1981	1.7	1.9
1997-00	2.0	2.2
530i	2.0	2.2
533i, 535i, is	1.7	1.9
540i, iA, iT, 545i (1.6)	2.2	2.4
633CSi, 635CSi, 733i, 735i	1.7	1.9
740i, iA, iL	2.3	2.5
745i, Li, 750iL, 760Li	2.2	2.4
840Ci	2.3	2.5
850Ci	2.6	2.8
M3, Z4 (1.5)	2.0	2.2
X3, X5 (1.2)	1.6	1.8
Z3 both (1.5)	2.0	2.2
Z4 (1.5)	2.0	2.2
Output Speed Pulse Sender, Replace (B)		
3 series all	.3	.5
525i	.3	.5
1997-00 528i, iT	2.2	2.4
540i, iA, iT	2.3	2.5
740i, iA, iL	2.9	3.1
850Ci		
5.0L	1.5	1.7
5.4L	2.7	2.9
M3		
1995	4.2	4.4
1996-01	2.3	2.5
X3 (1.1)	1.5	1.7
Z3 both (1.7)	2.3	2.5
Z4 (1.7)	2.3	2.5
Solenoid Valves, Replace (B)		
1991-98 318i, iC, is	2.9	3.1
318ti	2.9	3.1
1991-95 325i, iC, is	3.3	3.5
328i, iC, is (2.1)	2.9	3.1
330Ci, i, Xi (2.1)	2.8	3.0
525i	3.1	3.3
1997-00 528i, iT	2.9	3.1
530i	4.6	4.8
540i	3.2	3.4

ALPINA ROADSTER : M3 : M5 : X3 : X5 : Z3 : Z4 : Z8 — BMW-23

	LABOR TIME	SEVERE SERVICE
740i, iA, iL, 750iL	2.3	2.5
840i	3.2	3.4
850Ci		
5.0L	2.9	3.1
5.4L	3.2	3.4
M3 (1.6)	2.2	2.4
X3, Z4 (1.3)	1.8	2.0
Z3 (2.1)	2.8	3.0

Speedometer Drive Pinion, Replace (B)
1981-88	.7	.9

Torque Converter Lockup Solenoid Valve, Replace (B)
318i, iC, is, ti	1.2	1.4
1991-95 325i, iC, is	1.6	1.8
328i, iC, is	1.6	1.8
525i	1.7	1.9
Z3	1.6	1.8

Valve Body Pressure Regulator, Replace (B)
1991-98 318i, iC, is	1.6	1.8
318ti	1.6	1.8
1991-95 325i, iC, is	1.8	2.0
328i, 328iC, 328is	4.9	5.1
330Ci, i, Xi (2.1)	2.8	3.0
525i	1.8	2.0
1997-00 528i, iT	1.6	1.8
540i, iA, iT	3.8	4.0
740i, iA, iL, 750iL	3.1	3.3
850Ci	3.1	3.3
M3 (3.6)	4.9	5.1
X3 (1.7)	2.3	2.5
Z3 (1.2)	1.6	1.8
Z4 (1.4)	1.9	2.1

SERVICE TRANSMISSION REMOVED
Transmission R&R included unless otherwise noted.

Torque Converter Drive Plate (Flexplate), Replace (B)
318i, iC, is		
1984-91	6.8	7.0
1992-98	7.1	7.3
318ti	7.1	7.3
320i, 325, 325e	6.8	7.0
325i, iC, is		
1987-91	6.8	7.0
1992-95	10.0	10.2
325iX	7.8	8.0
328i, iC, is, 330Xi (5.6)	7.6	7.8
330Ci, 330i (5.0)	6.8	7.0
524td, 528e (5.0)	6.8	7.0
525i		
1989-91	6.8	7.0
1992-96	7.8	8.0
528i, iT		
1981	6.8	7.0
1997-00	6.4	6.6

	LABOR TIME	SEVERE SERVICE
530i	9.1	9.3
533i, 535i, is	6.8	7.0
540i, iA, iT		
1994-95	9.8	10.0
1997-01	8.1	8.3
545i (4.9)	6.6	6.8
633CSi, 635CSi, 733i, 735i	6.8	7.0
740i, iA, iL		
1994	9.8	10.0
1995-01	7.8	8.0
745i, Li, 760i, Li (6.0)	8.1	8.3
750iL		
1988-94	7.8	8.0
1995-01	8.7	8.9
840Ci	9.7	9.9
850Ci, i		
5.0L	11.4	11.6
5.4L	7.6	7.8
M3 (7.4)	10.0	10.2
X3 (4.2)	5.7	5.9
X5 (5.9)	8.0	8.2
Z3		
1.9L (4.8)	6.5	6.7
2.3L, 2.8L (5.2)	7.0	7.2
Z4 (4.2)	5.7	5.9

Transmission Assy., R&R and Recondition (A)
Exc. below	15.5	15.7
Z3		
1.9L	6.5	6.7
2.3L, 2.8L	7.1	7.3
Flush cooler lines add	.5	.5

Transmission Assy., R&R or Replace (B)
318i, iC, is		
1984-91	6.2	6.4
1992-98	7.3	7.5
318ti	7.4	7.6
320i	5.4	5.6
325, e, 330Ci, i (4.6)	6.2	6.4
325i, 325iC, 325is	9.7	9.9
325iX	7.3	7.5
328i, 328iC, 328is	7.1	7.3
330Xi (5.5)	7.4	7.6
524td, 528e	6.2	6.4
525i		
1989-91	6.2	6.4
1992-96	7.5	7.7
528i, iT		
1981	6.2	6.4
1997-00	5.9	6.1
530i	8.9	9.1
533i, 535i, is	6.2	6.4
540i, iA, iT		
1994-95	9.3	9.5
1997-01	8.1	8.3
545i (4.6)	6.2	6.4

	LABOR TIME	SEVERE SERVICE
633CSi, 635CSi, 733i, 735i	6.2	6.4
740i, iA, iL		
1994	9.3	9.5
1995-01	8.1	8.3
745i, Li, 760i, Li (5.6)	7.6	7.8
750iL		
1988-94	7.3	7.5
1995-99	8.6	8.8
840Ci	9.5	9.7
850Ci, 850i		
5.0L	10.8	11.0
5.4L	8.8	9.0
M3, M5 (6.0)	8.1	8.3
X3, Z4 (4.1)	5.5	5.7
X5 (5.8)	7.8	8.0
Z3 (5.4)	7.3	7.5

TRANSFER CASE

Output Flange Oil Seal, Replace (B)
325iX	2.3	2.5
330Xi (1.4)	1.9	2.1
X3 (1.9)	2.6	2.8
X5 (2.5)	3.4	3.6

Transfer Case, R&R or Replace (B)
325iX	3.1	3.3
330Xi (1.9)	2.6	2.8
X3 (1.7)	2.3	2.5
X5 (2.6)	3.5	3.7

Transfer Case Drive Chain, Replace (B)
325iX	4.4	4.6

SHIFT LINKAGE

AUTOMATIC TRANSMISSION Linkage, Adjust (B)
1981-04	.7	.9

Gear Selector Lever, Replace (B)
Exc. below (1.9)	2.6	2.8
325i, is	1.6	1.8
330i, Li, Xi (1.0)	1.4	1.6
525i, 530i (1.0)	1.4	1.6
540i, iA, iT		
1994-95	1.3	1.5
1997-01	1.5	1.7
545i (1.3)	1.8	2.0
733i, 735i	2.8	3.0
740i, iA, iL	1.6	1.8
1995-01 750iL	1.6	1.8
840Ci, 850Ci, i (1.3)	1.8	2.0
Z4 (1.3)	1.8	2.0
Z3 (1.4)	1.9	2.1

Kickdown Switch, Replace (B)
All Models	.5	.7

BMW-24 3 SERIES : 524TD : 5 SERIES : 6 SERIES : 7 SERIES : 8 SERIES

	LABOR TIME	SEVERE SERVICE
Selector Cable, Replace (B)		
1996-98 318i, iC, is	1.2	1.4
318ti, 330Ci, i, Xi (1.4)	1.9	2.1
328i, 328iC, 328is	1.2	1.4
1997-00 528i	2.2	2.4
535i, is	1.7	1.9
1997-01 540i, iA, iT	2.2	2.4
545i (1.8)	2.4	2.6
635CSi, 735i	1.7	1.9
740i, iA, iL	1.2	1.4
750iL		
1988-98	1.2	1.4
1999-01	3.5	3.7
840i	1.6	1.8
850Ci		
5.0L	1.6	1.8
5.4L	3.2	3.4
M3 (.9)	1.2	1.4
X3 (1.8)	2.4	2.6
X5 (1.9)	2.6	2.8
Z3 (1.3)	1.8	2.0
Z4 (1.5)	2.0	2.2
Selector Lever Bushings, Replace (B)		
1981-05	2.0	2.2

MANUAL TRANSMISSION

	LABOR TIME	SEVERE SERVICE
Gearshift Lever, Replace (B)		
3 series		
exc. 325iX (.9)	1.2	1.4
325iX (1.2)	1.8	2.0
525i	.8	1.0
528e	1.7	1.9
528i, iT		
1981	1.7	1.9
1997-00	.8	1.0
530i	.8	1.0
533i, 535i, is	1.7	1.9
540i, iA, iT	.7	.9
545i (1.0)	1.4	1.6
633CSi, 635CSi, 733i, 735i	1.7	1.9
840i, 850Ci, i	1.6	1.8
M coupe/roadster, Z3		
1.9L	1.5	1.7
2.3L, 2.8L	2.2	2.4
M3 (.8)	1.1	1.3
X3 (1.5)	2.0	2.2
Z4 (1.0)	1.4	1.6
Selector Rod, Replace (B)		
320i	.7	.9
528e	1.8	2.0
1981 528i	1.9	2.1
633CSi, 635CSi	1.7	1.9
733i	.7	.9
Shift Lever Bushing, Replace (B)		
1981-05	.7	.9

DRIVELINE
DRIVESHAFT

	LABOR TIME	SEVERE SERVICE
Center Support Bearing, Replace (B)		
3 series		
exc. below (2.1)	2.8	3.0
325i, 325iC, 325is	3.5	3.7
328i, i328C, 328is	3.5	3.7
330Ci, i, Xi (1.7)	2.3	2.5
524td	2.8	3.0
525i		
1989-91	2.8	3.0
1992-96	3.3	3.5
528e, i, iT	2.8	3.0
530i, 533i, 535i, is	2.8	3.0
540i, iA, iT		
1994-95	2.9	3.1
1997-01	4.0	4.2
545i (2.1)	2.8	3.0
633CSi, 635CSi, 733i, 735i	2.8	3.0
645Ci	3.1	3.3
740i, iA, iL		
1994	2.8	3.0
1995-01	3.3	3.5
745i, Li, 760i, Li (2.3)	3.1	3.3
750iL		
1988-94	3.1	3.3
1995-01	4.0	4.2
840Ci	4.4	4.6
850Ci, 850i		
5.0L	5.0	5.2
5.4L	4.3	4.5
M coupe/roadster, Z3		
1.9L (1.6)	2.2	2.4
2.3L, 2.8L (1.4)	1.9	2.1
M3, X3, X5 (1.7)	2.3	2.5
M5 (3.5)	4.7	4.9
Z4 (2.1)	2.8	3.0
Coupling Shaft, Replace (B)		
1981-05	2.8	3.0
Driveshaft, R&R or Replace (B)		
Exc. below (1.7)	2.3	2.5
318ti, 328i, iC, is	1.9	2.1
325, i, iC, is	2.8	3.0
1992-96 525i	3.3	3.5
530i	2.8	3.0
1992-93 535i, is	2.8	3.0
540i, iA, iT		
1994-95	2.8	3.0
1997-01	3.1	3.3
740i, iA, iL		
1994	2.8	3.0
1995-01	2.6	2.8
745i, Li, 760i, Li (2.1)	2.8	3.0
750iL	3.1	3.3
840Ci	3.9	4.1
850Ci, 850i	4.2	4.4

	LABOR TIME	SEVERE SERVICE
M3, X3, X5 (1.4)	1.9	2.1
M5 (3.4)	4.6	4.8
Z3 (1.3)	1.8	2.0
Z4 (1.3)	1.8	2.0
Driveshaft CV-Joint, Replace (B)		
318i, iC, is		
1984-85	1.8	2.0
1991-98	2.2	2.4
318ti, 328i, iC, is (1.4)	1.9	2.1
320i, 325, e, iX	1.8	2.0
325i, 325iC, 325is		
1987-91	1.8	2.0
1992-95	2.9	3.1
330Ci, 330i (1.9)	2.6	2.8
524td, 528e, 533i	1.8	2.0
525i	3.9	4.1
528i, iT		
1981	1.8	2.0
1997-00	2.6	2.8
530i, 535i, is	3.8	4.0
540i		
1994-95	3.8	4.0
1997-01	3.3	3.5
545i (1.9)	2.6	2.8
633CSi, 635CSi, 733i, 735i	1.8	2.0
645Ci	3.3	3.5
740i, iA, iL		
1994	3.3	3.5
1995-01	2.4	2.6
745i, Li, 760i, Li (2.8)	3.8	4.0
750iL		
1988-94	3.8	4.0
1995-01	2.8	3.0
840Ci	4.4	4.6
850Ci, 850i		
5.0L	4.9	5.1
5.4L	3.9	4.1
M coupe/roadster, Z3	2.3	2.5
M3 (1.3)	1.8	2.0
M5 (3.1)	4.2	4.4

DRIVE AXLE

	LABOR TIME	SEVERE SERVICE
Differential, Drain & Refill (B)		
1981-05 rear axle	.6	.8
Differential Assy., R&R or Replace (B)		
Rear		
318i, iC, is		
1984-85	3.4	3.6
1991-98	3.8	4.0
318ti, 320i	3.4	3.6
325, e, iX	3.4	3.6
325i, 325iC, 325is	3.9	4.1
328i, 328iC, 328is	3.6	3.8
330Ci, i, Xi (2.7)	3.6	3.8
525i	3.4	3.6

ALPINA ROADSTER : M3 : M5 : X3 : X5 : Z3 : Z4 : Z8   BMW-25

	LABOR TIME	SEVERE SERVICE
528i, iT		
1981	3.3	3.6
1997-00	3.8	4.0
530i, 533i, 535i, is	3.4	3.6
540i, iA, iT		
1994-95	3.4	3.6
1997-01	3.8	4.0
545i, 645Ci,		
745i, Li (2.8)	3.8	4.0
633CSi, 635CSi,		
733i, 735i	3.4	3.6
740i, iA, iL, 750iL	4.3	4.5
760i, 760Li (3.2)	4.3	4.5
840Ci	4.5	4.7
850Ci, 850i		
5.0L	4.5	4.7
5.4L	5.8	6.0
M coupe/roadster,		
Z3 (2.5)	3.4	3.6
M3, X5 (2.7)	3.6	3.8
M5 (1.2)	1.8	2.0
X3 (2.9)	3.9	4.1
Z4 (3.5)	4.7	4.9

Drive Flange Shaft Seal, Replace (B)
Front axle
325iX (1.7)	2.3	2.5
330Xi (2.1)	2.8	3.0
X3, X5 (2.5)	3.4	3.6

Final Drive Cover, Replace or Reseal (B)
Front axle 325iX	.7	.9

Final Drive Flange, Replace (B)
Front axle
325iX (1.7)	2.3	2.5
330Xi (1.3)	1.8	2.0
X3, X5 (1.9)	2.6	2.8

Rear Axle Crossmember, Replace (B)
325iX, 330Xi	4.8	5.0
X3, X5	6.8	7.0

Rear Wheel Bearing, Replace (B)
Rear axle
1981-92
one side	2.8	3.0
both sides	4.2	4.4
1993-05		
one side	3.9	4.1
both sides	6.7	6.9

Ring Gear & Pinion Set, Replace (B)
Front axle 325iX	6.9	7.1

Stub Axle Drive Flange, Replace (B)
Rear axle 325iX one	1.8	2.0

HALFSHAFTS

CV Joint and/or Boot, Replace (B)
Front one side
325iX		
one	1.8	2.0
both	3.1	3.3
330Xi		
one (1.4)	1.9	2.1
both (2.5)	3.4	3.6
X3		
one (1.5)	2.0	2.2
both (2.4)	3.2	3.4
X5		
one (1.8)	2.4	2.6
both (2.7)	3.6	3.8

Rear
318i, iC, is		
1984-85 one	1.8	2.0
1991-98 one	3.2	3.4
318ti one	2.8	3.0
320i, 325, e, iX one	1.8	2.0
325i, iC, is one	3.3	3.4
328i, 328iC		
328is one (2.3)	3.1	3.3
330Ci, 330i		
each (1.6)	2.2	2.4
330Xi each (2.1)	2.8	3.0
525i, 528e		
right side	2.4	2.6
left side	2.9	3.1
528i, iT		
1981		
right side	2.4	2.6
left side	2.9	3.1
1997-00		
one side	1.9	2.1
both sides	3.2	3.4
530i, 533i, 535i, is		
right side	2.4	2.6
left side	2.9	3.1
540i, iA, iT		
1994-95		
right side	2.4	2.6
left side	2.9	3.1
1997-01		
one side	1.9	2.1
both sides	3.2	3.4
545i		
one side (1.7)	2.3	2.5
both sides (2.5)	3.4	3.6
633CSi, 635CSi,		
733i, 735i		
right side	2.4	2.6
left side	2.9	3.1
645Ci		
one side	2.7	2.9
both sides	3.4	3.6

740i, iA, iL		
1994		
right side	2.4	2.6
left side	2.9	3.1
1995-01		
one side	3.6	3.8
both sides	5.7	5.9
745i, Li, 760Li		
one side (2.0)	2.7	2.9
both sides (2.5)	3.4	3.6
750iL		
1988-94		
right side	2.4	2.6
left side	2.9	3.1
1995-04		
one side	4.2	4.4
both sides	5.9	6.1
840Ci		
one side	3.7	3.9
both sides	5.9	6.1
850Ci, 850i		
5.0L		
one side	3.7	3.9
both sides	5.9	6.1
5.4L		
one side	4.4	4.6
both sides	6.8	7.0
M coupe/roadster		
one	2.8	3.0
M3 one (2.3)	3.1	3.3
M5 one (1.4)	1.9	2.1
X3 one side		
one (1.4)	1.9	2.1
both (2.5)	3.4	3.6
X5 one side		
one (1.7)	2.3	2.5
both (2.7)	3.6	3.8
Z3 one (2.1)	2.8	3.0
Z4 one side		
one (1.9)	2.6	2.8
both (2.5)	3.4	3.6
each addl. joint with shaft removed add	.6	.6

Halfshaft, R&R or Replace (B)
Front axle
325iX, 330Xi, X3, X5		
one (.9)	1.2	1.4
both (1.5)	2.0	2.2

Rear axle
318i, iC, is		
1984-85 one	1.2	1.4
1991-98 one	1.8	2.0
318ti one	1.4	1.6
320i, 325, 325e one	1.2	1.4
325i, iC, is one	2.4	2.6
325iX, 330Xi		
one (.9)	1.2	1.4

BMW-26 3 SERIES : 524TD : 5 SERIES : 6 SERIES : 7 SERIES : 8 SERIES

	LABOR TIME	SEVERE SERVICE
Halfshaft, R&R or Replace (B)		
328i, iC, is one	1.6	1.9
330Ci, i one (1.3)	1.8	2.0
525i, 528e one	1.8	2.0
528i, iT		
1981 one	1.8	2.0
1997-00 one side	3.6	3.8
530i, 533i, 535i, is		
one	1.8	2.0
540i, iA, iT		
1994-95 one	1.8	2.0
1997-01 one	3.6	3.8
545i, 645Ci		
one (1.9)	2.5	2.7
633CSi, 635CSi,		
733i, 735i one	1.8	2.0
740i, iA, iL		
1994 one	1.8	2.0
1995-01 one	2.4	2.6
745i, Li, 760i		
760Li (1.6)	2.2	2.4
750iL		
1988-94 one	1.8	2.0
1995-01 one	2.8	3.0
840Ci one	2.8	3.0
850Ci, 850i		
5.0L one side	2.8	3.0
5.4L one	3.6	3.8
M coupe/roadster, Z3		
one	1.5	1.7
M3 one (1.2)	1.6	1.8
X3, X5 (1.3)	1.8	2.0
Z4 (1.8)	2.4	2.6
Replace seal all front		
axle add each	.2	.2

BRAKES
ANTI-LOCK
The following operations do not include testing. Add time as required.

Diagnose Anti-Lock Brake System Component Each (A)

1986-05	1.1	1.3
w/air bags add	.5	.5
Control Unit, Replace (B)		
1986-05	.7	.9
Front Sensor Assy., Replace (B)		
Each	.7	.9
Hydraulic Assy., Replace (B)		
Exc. below (2.3)	3.1	3.3
318i, iC, is, ti	2.6	2.8
1997-00 528i, iT	2.1	2.3
540i, iA, iT		
1994-95	2.3	2.5
1997-01	2.1	2.3
740i, iA, iL	1.9	2.1
754i, Li, 760i, Li (1.7)	2.3	2.5

	LABOR TIME	SEVERE SERVICE
1995-01 750iL	1.9	2.1
840Ci, 850	3.8	4.0
1991-92		
w/traction control	8.8	9.0
w/o traction control	6.0	6.2
1993-97		
w/traction control	3.8	4.0
w/o traction control	5.9	6.1
1994-96	8.9	9.1
M coupe/roadster, Z3	2.3	2.5
M5, Z4 (1.6)	2.2	2.4
Pump or Main Relay, Replace (B)		
1986-05	.3	.5
Rear Sensor Assy., Replace (B)		
One	.7	.9
Both	1.2	1.4

SYSTEM
Bleed Brakes (B)
Includes: Add fluid.

w/ABS	1.7	1.9
w/o ABS	.7	.9
Brakes, Adjust (B)		
Includes: Refill master cylinder.		
1981-85 rear wheels	.5	.7
Brake Pressure Regulator, Replace (B)		
Exc. 633CSi, 635CSi	1.4	1.6
633CSi, 635CSi	1.7	1.9
Brake Hose (Flexible), Replace (B)		
Includes: System bleeding.		
One	.7	.9
each addl. add	.3	.3
Master Cylinder, Replace (B)		
Includes: System bleeding.		
Exc. below (1.7)	2.3	2.5
1991-98 318i, iC, is	1.9	2.1
318ti	1.6	1.8
524td, 528e	2.8	3.0
528i, iT		
1981	2.8	3.0
1997-00	5.3	5.5
530i	3.8	4.0
533i, 535i, is	2.8	3.0
540i		
1994-95	3.8	4.0
1997-01	5.3	5.5
545i (2.4)	3.2	3.3
633CSi, 635CSi,		
733i, 735i	2.8	3.0
740i, iA, iL		
1994	3.5	3.7
1995-01	1.6	1.8
745i, Li, 760i, Li (2.6)	3.5	3.7
750iL	1.6	1.8
840Ci, 850Ci, 850i	3.3	3.5

	LABOR TIME	SEVERE SERVICE
M coupe/roadster,		
Z3 (1.2)	1.6	1.8
M5 (3.2)	4.3	4.5
Z4 (2.6)	3.5	3.7
Power Booster Unit, Replace (B)		
Includes: System bleeding.		
318i, iC, is		
1984-85	2.3	2.5
1991-98	3.1	3.3
318ti	3.8	4.0
320i	3.1	3.3
325, e, i, iC, is	2.8	3.0
325iX	3.1	3.3
328i, 328iC, 328is	4.6	4.8
330Ci, i, Xi (2.3)	3.1	3.3
524td, 525i	3.1	3.3
528e, 533i	2.8	3.0
528i, iT		
1981	2.8	3.0
1997-00	6.0	6.2
530i	5.4	5.6
535i, is	3.1	3.3
540i	5.4	5.6
545i, 645Ci (2.1)	2.8	3.0
633CSi, 635CSi,		
733i, 735i	3.1	3.3
740i, iA, iL, 750iL	3.9	4.1
745i, Li, 760i, Li (2.7)	3.6	3.8
840Ci, 850Ci, 850i		
1991-97	6.3	6.5
1994-96	7.0	7.2
M coupe/roadster, Z3	4.6	4.8
M3 (3.4)	4.6	4.8
M5 (3.9)	5.3	5.5
X3 (2.8)	3.8	4.0
X5 (2.4)	3.2	3.4
Z4 (1.7)	2.3	2.5
w/air bags add	.5	.5
Vacuum Check Valve, Replace (B)		
1981-05	.3	.5

SERVICE BRAKES
Brake Drum, Replace (B)

1981-85		
one	.7	.9
both	1.3	1.5
Caliper Assy., R&R and Recondition (A)		
Includes: System bleeding.		
One	2.3	2.5
Both	3.8	4.0
Caliper Assy., Replace (B)		
Includes: System bleeding.		
Front		
one	1.7	1.9
both	2.8	3.0
Rear		
one	1.7	1.9
both	2.5	2.7

ALPINA ROADSTER : M3 : M5 : X3 : X5 : Z3 : Z4 : Z8   BMW-27

	LABOR TIME	SEVERE SERVICE
Disc Brake Rotor, Replace (B)		
One	.5	.7
Both	1.2	1.4
Four wheels	2.3	2.5
Pads and/or Shoes, Replace (B)		
Includes: Service and parking brake adjustment, system bleeding.		
Front disc	1.2	1.4
Rear drum	1.7	1.9
Four wheels	2.8	3.0
COMBINATION ADD-ONS		
Repack wheel bearings		
both add	.7	.7
Replace		
brake drum add each	.2	.2
brake hose add each	.3	.3
caliper add	.3	.3
disc rotor add each	.3	.3
wheel cylinder add	.3	.3
Resurface		
drum add each	.5	.5
rotor add each	.5	.5
Wheel Cylinder, R&R and Recondition (B)		
Includes: System bleeding.		
One	1.8	2.0
Both	3.1	3.3
Wheel Cylinder, Replace (B)		
Includes: System bleeding.		
One	1.4	1.6
Both	2.6	2.8
PARKING BRAKE		
Parking Brake Cables, Adjust (C)		
All Models	.9	1.1
Parking Brake Apply Actuator, Replace (B)		
318i, iC, is		
1984-85	.7	.9
1991-98	1.1	1.3
318ti	1.1	1.3
320i, 325, 325e	.7	.9
325i, 325iC, 325is	1.5	1.7
325iX	1.2	1.4
328i, 328iC, 328is	1.1	1.3
330Ci, i, Xi (1.2)	1.6	1.8
525i (1.2)	1.6	1.8
528e	.7	.9
528i, iT		
1981	.8	1.0
1997-00	1.5	1.7
530i, 540i	1.7	1.9
533i, 535i, is	.7	.9
545i, 645Ci (2.0)	2.7	2.9
633CSi, 635CSi, 733i	.7	.9
735i	1.4	1.6

	LABOR TIME	SEVERE SERVICE
740i, iA, iL		
1994	2.3	2.5
1995-97	1.2	1.4
745i, Li, 760i, Li (2.4)	3.2	3.4
750iL		
1988-94	1.7	1.9
1995-97	1.3	1.5
840Ci, 850Ci, i	3.6	3.8
850CSi	3.9	4.1
M coupe/roadster, Z3	1.1	1.3
M3 (.8)	1.1	1.3
M5, X3, X5 (1.1)	1.5	1.7
Z4 (.9)	1.2	1.4
Parking Brake Cable, Replace (B)		
318i, iC, is		
one	2.1	2.3
both	2.6	2.8
318ti		
one	2.2	2.4
both	3.0	3.2
320i		
one	1.4	1.6
both	1.9	2.1
325, 325e		
one	1.7	1.9
both	2.6	2.8
325i, 325iC, 325is, 325iX		
one	2.9	3.1
both	3.8	4.0
328i, 328iC, 328is		
one	2.2	2.4
both	3.0	3.2
330Ci, 330i, 330Xi		
one (1.5)	2.0	2.2
both (2.1)	2.8	3.0
524td, 528e		
one	1.6	1.8
both	2.3	2.5
525i, 530i		
one	2.3	2.5
both	3.0	3.2
528i, iT		
1981		
one	1.6	1.8
both	2.3	2.5
1997-00		
one	3.3	3.5
both	4.2	4.4
533i, 535i, is		
one	1.6	1.8
both	2.3	2.5
540i, iA, iT		
one	4.3	4.5
both	5.1	5.3
545i		
one (1.8)	2.4	2.6
both (2.4)	3.2	3.4

	LABOR TIME	SEVERE SERVICE
633CSi, 635CSi, 645i, 733i		
one	1.6	1.8
both	2.3	2.5
735i		
one	2.3	2.5
both	3.1	3.3
740i, 740iA, 740iL		
1994		
one	2.7	2.9
both	3.5	3.7
1995-01		
front	4.3	4.5
rear	3.9	4.1
745i, Li, 760i, Li		
one (2.5)	3.4	3.6
both (3.1)	4.2	4.4
750iL		
1988-94		
one	2.3	2.5
both	3.1	3.3
1995-99		
front	4.4	4.6
rear	4.3	4.5
840Ci		
one	3.5	3.7
both	4.2	4.4
850Ci, 850i		
5.0L		
one	3.5	3.7
both	4.2	4.4
5.4L		
one	4.3	4.5
both	5.0	5.2
M coupe/roadster, Z3		
one	3.1	3.3
both	3.6	3.8
M3		
one (1.2)	1.6	1.8
both (1.3)	1.8	2.0
M5		
one (4.3)	5.8	6.0
both (4.8)	6.5	6.7
X3, X5		
one (1.6)	2.2	2.4
both (2.2)	3.0	3.2
Z4		
one (1.9)	2.6	2.8
both (2.4)	3.2	3.4
Parking Brake Shoes, Replace (B)		
One side (1.0)	1.4	1.6
Both sides (1.7)	2.3	2.5

FRONT SUSPENSION

Unless otherwise noted, time given does not include alignment.

Front & Rear Alignment, Check & Adjust (A)
1981-05 .... 2.0 2.2
Front Toe, Adjust (B)
1981-05 .... 1.1 1.3

BMW-28 3 SERIES : 524TD : 5 SERIES : 6 SERIES : 7 SERIES : 8 SERIES

	LABOR TIME	SEVERE SERVICE
Control Arm Assy., Replace (B)		
3 series		
exc. below		
one	2.2	2.4
both	3.6	3.8
325, 325e, 330Ci, 330i		
one (1.3)	1.8	2.0
both (2.5)	3.4	3.6
524td		
one	2.3	2.5
both	3.7	3.9
525i		
one (1.2)	1.6	1.8
both (2.0)	2.7	2.9
528e, 533i, 535i, 535is		
one	1.8	2.0
both	3.2	3.4
528i, iT		
1981		
one	1.8	2.0
both	3.2	3.4
1997-00		
one	.8	1.0
both	1.3	1.5
530i, 545i		
one (1.2)	1.6	1.8
both (2.0)	2.7	2.9
540i, 540iA, 540iT		
1994-95		
one	1.6	1.8
both	2.7	2.9
1997-01		
one	1.1	1.3
both	1.5	1.7
633CSi, 635CSi, 733i, 735i		
one	1.8	2.0
both	3.3	3.5
740i, iA, iL		
1994		
one	1.6	1.8
both	2.7	2.9
1995-01		
one	1.1	1.3
both	1.5	1.7
745i, Li		
one (1.2)	1.6	1.8
both (2.0)	2.7	2.9
750iL		
1988-94		
one	1.8	2.0
both	3.3	3.5
1995-01		
one	1.1	1.3
both	1.5	1.7
760i, iL		
one (1.3)	1.8	2.0
both (2.1)	2.8	3.0

	LABOR TIME	SEVERE SERVICE
840Ci, 850Ci, 850i		
one (1.3)	1.8	2.0
both (2.1)	2.8	3.0
M coupe/roadster, Z3		
one (1.9)	2.6	2.8
both (2.5)	3.4	3.6
M3		
one (1.6)	2.2	2.4
both (1.9)	2.6	2.8
M5		
one (.8)	1.1	1.3
both (1.3)	1.8	2.0
X3, X5		
one (.8)	1.1	1.3
both (1.1)	1.5	1.7
Z4		
one (1.2)	1.6	1.8
both (2.9)	2.8	3.0
Front Hub Bearings or Seal, Replace (B)		
One side	1.7	1.9
Both sides	3.1	3.3
Front Strut, Replace (B)		
3 series both	3.2	3.4
5, 6 series both	3.5	3.7
7 series both (2.8)	3.8	4.0
840Ci, 850Ci, i		
both (2.1)	2.8	3.0
M coupe/roadster,		
Z3 both (2.1)	2.8	3.0
M3 both (2.4)	3.2	3.4
M5 both (2.2)	3.0	3.2
X3, X5 both (2.1)	2.8	3.0
Z4 both (1.9)	2.6	2.8
Front Strut Coil Spring, Replace (B)		
3 series both (2.5)	3.4	3.6
5, 6 series both (2.7)	3.6	3.8
7 series both (2.8)	3.8	4.0
840Ci, 850Ci, i		
both (2.1)	2.8	3.0
M coupe/roadster,		
Z3 both (2.1)	2.8	3.0
M3 both (2.5)	3.4	3.6
M5 both (2.3)	3.1	3.3
X3, X5 both (2.1)	2.8	3.0
Z4 both (1.9)	2.6	2.8
Stabilizer Bar, Replace (B)		
Exc. below (.6)	.8	1.0
330Xi (1.1)	1.5	1.7
528i, iT	2.7	2.9
540i, iA, iT	5.1	5.3
740i, iA, iL	5.5	5.7
750iL, 760i, iL (4.2)	5.7	5.9
M5 (3.9)	5.3	5.5
X3 (2.8)	3.8	4.0
X5 (2.2)	3.0	3.2
Stabilizer Bar Bushings, Replace (B)		
1981-05	1.2	1.4

	LABOR TIME	SEVERE SERVICE
Stabilizer Bar Push Rods, Replace (B)		
One side (.3)	.5	.7
Both sides (.6)	.8	1.0
Steering Knuckle, Replace (B)		
3 series, M3		
one	1.9	2.1
both	3.4	3.6
740i, iA, iL, 750iL		
one	2.3	2.5
both	3.9	4.1
M coupe/roadster, Z3		
one	2.2	2.4
both	3.9	4.1
Z4		
one	1.4	1.6
both	2.6	2.8
Wheel Bearings, Clean & Pack (B)		
Both wheels	2.1	2.3

REAR SUSPENSION

The following operations do not include testing. Add time as required. Unless otherwise noted, time given does not include alignment.

	LABOR TIME	SEVERE SERVICE
Rear Alignment, Adjust (A)		
1981-05	1.2	1.4
Level Control System, Check (B)		
1986-05	1.5	1.7
Height Lever, Adjust (B)		
1986-05	.5	.7
Air Suspension Control Unit, Replace (B)		
528iT, 540iT, 545i (.4)	.6	.8
740i, iA, iL, 750iL (.6)	.8	1.0
745i, 745Li, 760Li (.6)	.8	1.0
840Ci	.7	.9
X5 (.6)	.8	1.0
w/power rear seat add	.7	.7
Level Control Valve, Replace (B)		
1986-05	1.2	1.4
Rear Strut Coil Spring, Replace (B)		
Each	1.4	1.6
w/level control add	.5	.5
Rear Strut Shock Absorber, Replace (B)		
Each	1.2	1.4
Rear Suspension Carrier, Replace (B)		
318i, iC, is		
1984-85	6.2	6.4
1991-98	7.2	7.4
318ti	5.1	5.3
320i, 325, e, iX	6.2	6.4
325i, 325iC, 325is	7.0	7.2
328i, 328iC, 328is	7.0	7.2
330Ci, i, Xi (4.0)	5.4	5.6

ALPINA ROADSTER : M3 : M5 : X3 : X5 : Z3 : Z4 : Z8 BMW-29

	LABOR TIME	SEVERE SERVICE
525i, 528e	6.2	6.4
528i, iT		
1981	6.2	6.4
1997-00	5.4	5.6
530i, 533i, 535i, is	6.2	6.4
540i, 540iA, 540iT	6.2	6.4
545i (4.0)	5.4	5.6
633CSi, 635CSi, 733i, 735i	6.2	6.4
645Ci	6.9	7.1
740i, iA, iL		
1994	6.2	6.4
1995-01	6.9	7.1
745i, Li, 760i, Li (4.0)	5.4	5.6
750iL		
1988-94	6.2	6.4
1995-01	7.4	7.6
840Ci	7.0	7.2
850Ci, 850i		
5.0L	7.0	7.2
5.4L	8.0	8.2
M coupe/roadster, Z3	5.9	6.1
M3 (5.2)	7.0	7.2
M5 (5.8)	7.8	8.0
X3 (5.1)	6.9	7.1
X5 (4.0)	5.4	5.6
Z4 (4.4)	5.9	6.1

Rear Torque Rod, Replace (B)
5 series		
one (.8)	1.1	1.3
both (1.1)	1.5	1.7
6 series		
one (.6)	.8	1.0
both (1.0)	1.4	1.6
7 series		
one (.5)	.7	.9
both (.8)	1.1	1.3
840i, 850Ci, 850i		
one (1.1)	1.5	1.7
both (1.8)	2.4	2.6
Replace bushings add each	.5	.5

Rear Wheel Knuckle Assy., Replace (B)
528i, iT		
one	3.1	3.3
both	5.7	5.9
545i		
one (1.9)	2.6	2.8
bot (3.5)	4.7	4.9
740i, 740iA, 740iL, 750iL		
one	2.6	2.8
both	4.7	4.9
745i, 745Li, 760i, 760iL		
one (2.0)	2.7	2.9
both (3.0)	4.1	4.3
645i		
one	2.1	2.3
both	4.7	4.9

	LABOR TIME	SEVERE SERVICE
840i		
one	4.4	4.6
both	7.2	7.4
850Ci, 850i		
5.0L		
one	4.4	4.6
both	7.2	7.4
5.4L		
one	5.3	5.5
both	8.0	8.2
M5		
one (2.3)	3.1	3.3
both (4.2)	5.7	5.9
Replace ball joint add each	.3	.3

Shock Absorbers or Bushings, Replace (B)
318i, iC, is		
1884-85 both	2.3	2.6
1991-93 both	2.6	2.8
1994-98 both	1.4	1.6
318ti both	1.1	1.3
320i, 325, e, iX both	2.3	2.5
325i, is both	2.6	2.8
325iC both	1.3	1.5
328i, iC, is, 330Ci, 330i 330Xi both (1.0)	1.4	1.6
545i, 645Ci both	2.3	2.5
M coupe, Z3 2.8L coupe both	4.6	4.8
M roadster, Z3 roadster both	1.2	1.4
M3 both	1.4	1.6
X3, X5 both (2.7)	3.6	3.8
Z4 both	1.6	1.8

Stabilizer Bar, Replace (B)
1981-05	1.1	1.3

Stabilizer Bar Bushings, Replace (B)
1981-05	.5	.7

Stub Axle, Replace (B)
1981-05 one	2.8	3.0

Trailing Arm or Link, Replace (B)
Exc. below	3.2	3.4
318i, iC, is	3.6	3.8
325i, iC, is	4.4	4.6
328i, iC, is	3.5	3.7
330Ci, 330i		
330Xi (2.1)	2.9	3.1
525i, 530i	4.4	4.6
1994-95 540i	4.4	4.6
840Ci, 850Ci, i	1.4	1.6
M coupe/roadster, Z3	3.5	3.7
M3 (2.6)	3.5	3.7
X3, Z4 (2.0)	2.7	2.9
Replace bushings add each	.5	.5

	LABOR TIME	SEVERE SERVICE
Upper Control Arm, Replace (B)		
1991-98 318i, iC, is	1.9	2.1
325i, iC, is	2.8	3.0
330Ci, i, Xi (2.1)	2.8	3.0
1997-00 528i, iT	1.1	1.3
540i, iA, iT, 740i, iA, iL	1.5	1.7
545i, 645Ci, 745i, Li	1.3	1.5
750iL	1.1	1.3
760i, 760Li (1.0)	1.4	1.6
840I, 850Ci, 850i	1.6	1.8
M3 (1.8)	2.4	2.6
M5 (.8)	1.1	1.3
X3 (2.1)	2.8	3.0
X5 (1.3)	1.8	2.0
Z4 (3.6)	4.9	5.1
Replace bushings add each	.5	.5

STEERING
AIR BAGS
The following operations do not include testing. Add time as required.

Diagnose Air Bag System System Component Each (A)
1990-05	1.1	1.1

Air Bag Assy., Replace (B)
1990-93		
driver side	1.2	1.4
driver & pass side	3.8	4.0
1994-05		
driver side	1.2	1.4
driver & pass side	1.8	2.0

Contact Ring, Replace (B)
Exc. below	1.7	1.9
540i, iA, iT		
1994-95	1.4	1.6
1997-01	.6	.8
740i, iA, iL	.6	.8
750iL	.7	.9
840Ci, 850Ci, i	.8	1.0

Diagnostic Unit, Replace (B)
1991-98 318i, iC, is	1.5	1.7
325i, iC, is	2.0	2.2
328i, iC, is	1.6	1.8
1991-96 525i	1.8	2.0
530i	1.8	2.0
535i, 535is	1.2	1.4
540i, iA, iT	1.6	1.8
735I, 740i, iA, iL	1.1	1.3
750iL	1.1	1.3
840Ci	3.1	3.3
850Ci, i	1.5	1.7

Electronic Control Unit, Replace (B)
Exc. below	.4	.6
1994-98 318i, iC, is	.6	.8
318ti, 328i, iC, is	.6	.8
325i, 325i, iC, is	.7	.9
M coupe/roadster, Z3	1.8	2.0
M3	.6	.8
M5	.7	.9

BMW-30 3 SERIES : 524TD : 5 SERIES : 6 SERIES : 7 SERIES : 8 SERIES

	LABOR TIME	SEVERE SERVICE
Front Sensor, Replace (B)		
Each	.7	.9
Module Assembly, Replace (B)		
Exc. below		
front door each	.8	1.0
passenger side	.8	1.0
740i, iA, iL, 750iL		
front door each	1.3	1.5
passenger side	1.1	1.3
850Ci, i passenger side	.5	.7
Side Sensor, Replace (B)		
Exc. below		
one side	1.7	1.9
both sides	3.1	3.3
330Ci, 330i, 330Xi		
one side (.8)	1.1	1.3
both sides (1.6)	2.2	2.4
X3, X5		
one side (.8)	1.1	1.3
both sides (1.5)	2.0	2.2
Z4		
one side	.7	.9
both sides	1.4	1.6

MANUAL RACK & PINION
Unless otherwise noted, time given does not include alignment.

Gear Assy., Adjust (B)		
320i	1.4	1.6
Gear Assy., R&R (B)		
320i	3.1	3.63
Horn Contact, Replace (B)		
320i	.5	.7
Steering Wheel, Replace (B)		
320i	.3	.5

POWER RACK & PINION
Unless otherwise noted, time given does not include alignment.

Troubleshoot Power Steering (A)		
All Models	1.2	1.4
Gear Assy., Adjust (On Vehicle) (B)		
All Models	.7	.9
Gear Assy., R&I (B)		
318i, iC, is		
1984-85	2.3	2.5
1991-98	1.9	2.1
318ti	1.6	1.8
320i, 325, 325e	3.7	3.9
325i, iC, is, 325iX	2.3	2.5
328i, iC, is	1.9	2.1
330Ci, 330i (1.4)	1.9	2.1
330Xi (1.9)	2.6	2.8
545i, 645Ci	4.8	5.0
745i, Li, 760i, Li (5.8)	7.8	8.0
M coupe/roadster, Z3	1.9	2.1

	LABOR TIME	SEVERE SERVICE
M3	1.9	2.1
M5	3.2	3.4
X3	2.3	2.5
X5 (3.2)	4.3	4.5
w/air bags add	.5	.5
Horn Contact or Canceling Cam, Replace (B)		
All Models	.7	.9
w/air bags add	.5	.5
Inner Tie Rod End, Replace (B)		
All Models	1.0	1.2
Outer Tie Rod End, Replace (B)		
All Models	.8	1.0
Steering Column, Replace (B)		
318i, iC, is	2.4	2.6
318ti	2.1	2.3
320i, 325, e, i, is	2.3	2.5
325iC	2.6	2.8
325iX	2.3	2.5
328i, iC, is	2.5	2.7
330Ci, i, Xi (1.8)	2.4	2.6
545i (1.7)	2.3	2.6
645Ci	2.6	2.8
745i, Li, 760i, Li (1.7)	2.3	2.6
M coupe/roadster, Z3	2.3	2.5
M3 (1.9)	2.6	2.8
X3 (1.3)	1.8	2.0
X5 (1.8)	2.4	2.6
w/air bags add	.5	.5
Steering Pump, R&R or Replace (B)		
318i, iC, is		
1984-85	2.3	2.5
1991-93	1.8	2.0
1994-98	1.2	1.4
318ti	1.1	1.3
325, e, iX	2.3	2.5
325i, iC, is	1.5	1.7
328i, iC, is	1.2	1.4
330Ci, 330i (.8)	1.1	1.3
330Xi (1.7)	2.3	2.5
545i (1.1)	1.5	1.7
645Ci	1.1	1.3
745i, Li, 760i, Li (1.4)	1.9	2.1
M coupe/roadster, Z3		
1.9L	1.6	1.8
2.3L, 2.8L	1.2	1.4
M3	1.2	1.4
X3, X5	1.5	1.7
Reseal pump add	1.2	1.2
Steering Pump Hose, Replace (B)		
Does not include purging.		
All Models	1.2	1.4
Steering Wheel, Replace (B)		
All Models	.7	.9
w/air bags add	.5	.5
Tie Rod End Boot, Replace (B)		
3 series		
one side	.7	.9
both sides	1.1	1.3

	LABOR TIME	SEVERE SERVICE
545i, 745i, Li		
one side	1.6	1.8
both sides	2.9	3.1
M coupe/roadster, Z3		
one side	.7	.9
both sides	1.0	1.2
M3, X3, X5		
one side	.8	1.0
both sides	1.2	1.4
Upper Mast Jacket Bearing, Replace (B)		
Includes: Insulators.		
3 series		
exc. below	1.2	1.4
318i, iC, is	.5	.7
328i, iC, is	.8	1.0
M coupe/roadster, Z3	.8	1.0
M3	.5	.7
w/air bags add	.5	.5

POWER WORM & SECTOR
Unless otherwise noted, time given does not include alignment.

Troubleshoot Power Steering (A)		
1981-05	1.1	1.1
Center Tie Rod, Replace (B)		
Exc. 850Ci, i	1.8	2.0
850Ci, i	3.2	3.4
Gear Assy., Replace (B)		
524td, 528e	4.2	4.4
525i	2.4	2.6
528i, iT		
1981	4.2	4.4
1997-00	3.2	3.4
530i	6.3	6.5
533i, 535i, is	4.2	4.4
540i, iA, iT		
1994-95	6.3	6.5
1997-01	3.2	3.4
633CSi, 635CSi, 733i, 735i	4.2	4.4
740i, iA, iL	3.2	3.4
750iL	3.5	3.7
840Ci	7.7	7.9
850i, 850CSi		
1991-93		
AT	5.3	5.5
MT	5.9	6.1
1994-95	6.4	6.6
w/air bags add	.7	.7
Horn Contact Cable or Ring, Replace (B)		
1981-05	1.2	1.4
w/air bags add	.5	.5
Idler Arm, Replace (B)		
Includes: Reset toe.		
525i	1.4	1.6
530i, 540i, iA, iT	1.6	1.8
535i, 735i, 750iL	1.3	1.5
850Ci, i	1.8	2.0

ALPINA ROADSTER : M3 : M5 : X3 : X5 : Z3 : Z4 : Z8 BMW-31

	LABOR TIME	SEVERE SERVICE
Inner & Outer Tie Rod Assy., Replace (B)		
524td, 528e		
one side	1.3	1.5
both sides	2.4	2.6
525i, 530i		
one side	1.7	1.9
both sides	2.9	3.1
533i, 535i, is		
one side	1.3	1.5
both sides	2.4	2.6
540i		
1994-95		
one side	1.7	1.9
both sides	2.9	3.1
1997-01 one side	1.3	1.5
633CSi, 635CSi, 733i, 735i		
one side	1.3	1.5
both sides	2.4	2.6
740i, iA, iL, 750iL		
one side	1.3	1.5
both sides	1.6	1.8
840Ci, 850Ci, i		
one side	1.6	1.8
both sides	2.7	2.9
Pitman Arm, Replace (B)		
1981-05	1.1	1.3
Pitman Arm Bushing, Replace (B)		
1981-97	1.4	1.6
Steering Column, R&I (B)		
525i, 528e	2.3	2.5
528i, iT		
1981	2.3	2.5
1997-00		
AT	2.5	2.7
MT	2.2	2.4
530i, 533i, 535i, is	2.3	2.5
540i		
1994-95	2.3	2.5
1997-01		
AT	2.5	2.7
MT	2.2	2.4
633CS, 635CSi, 733i, 735i	2.3	2.5
740i, iA, iL		
1994	2.3	2.5
1995-01	2.6	2.8
750iL		
1988-94	2.3	2.5
1995-01	2.7	2.9
840Ci, 850CSi	4.1	4.3
850Ci	4.3	4.5
w/air bags add	.5	.5
Steering Control Unit, Replace (B)		
Exc. 850Ci, i	.7	.9
850Ci, i	.3	.5

	LABOR TIME	SEVERE SERVICE
Steering Pump, R&R or Replace (B)		
524td	2.8	3.0
525i	1.7	1.9
528e	2.3	2.5
528i, iT		
1981	2.3	2.5
1997-00	1.4	1.6
530i	1.5	1.7
533i, 535i, is	2.3	2.5
1994-01 540i	1.5	1.7
633CSi, 635CSi, 733i, 735i	2.3	2.5
740i, iA, iL		
1994	2.3	2.5
1995-01	1.6	1.8
750iL		
1988-94	3.1	3.3
1995-01	1.6	1.8
840Ci	2.2	2.4
850Ci, 850i	3.7	3.9
5.0L		
AT	3.1	3.3
MT	3.6	3.8
5.4L	2.6	2.8
850CSi	3.5	3.7
Reseal pump add	1.2	1.2
Steering Pump Hoses, Replace (B)		
Does not include purging.		
Each	1.1	1.3
Steering Wheel, Replace (B)		
1981-05	.7	.9
w/air bags add	.5	.5
Upper Mast Jacket Bearing, Replace (B)		
Includes: Insulators.		
1981-01	1.2	1.4
w/air bags add	.5	.5

HEATING & AIR CONDITIONING

When more than one component requires replacement where evacuation/recovery and recharging is already included, deduct 1.0 hour for each additional component from the time given.

Evacuate/Recover and Recharge System (B)
  1981-05 (1.0) 1.4   1.4

AC Hoses, Replace (B)
Includes: Evacuate/recover and recharge.
  Exc. below (2.1) 2.8   3.0
  735i 3.1   3.3
  740i, iA, iL
    1994-95 .... 3.7   3.9
    1996-01 .... 4.4   4.6

	LABOR TIME	SEVERE SERVICE
750iL	4.6	4.8
760i, 760Li (3.1)	4.2	4.4
840Ci, 850Ci, 850i	4.3	4.5
X3, X5 (1.6)	2.2	2.4
each addl. add	.5	.5
Blower Motor, Replace (B)		
318i, iC, is		
1984-85	1.2	1.4
1991-98		
w/AC	2.4	2.6
w/o AC	1.6	1.8
318ti		
w/AC	2.0	2.2
w/o AC	1.4	1.6
320i	1.1	1.3
325, e, iX	1.2	1.4
325i, iC, is		
1987-91	2.3	2.5
1992-95		
w/AC	3.8	4.0
w/o AC	2.3	2.5
328i, iC, is, 330Ci, i, Xi (1.5)	2.0	2.2
524td	2.3	2.5
525i	2.6	2.8
528e	.8	1.0
528i, iT		
1981	1.4	1.6
1997-00	4.1	4.3
530i	2.4	2.6
533i	.8	1.0
535i, is		
w/AC	4.8	5.0
w/o AC	.8	1.0
540i, iA, iT		
1994-95	2.4	2.6
1997-01	4.1	4.3
545i (2.7)	3.6	3.8
633CSi, 635CSi	1.1	1.3
645Ci	3.8	4.0
733i, 735i	1.7	1.9
740i, iA, iL		
1994	2.6	2.8
1995-01	4.7	4.9
745i, Li, 760i, Li (1.2)	1.6	1.8
750iL		
1988-94	2.8	3.0
1995-01	4.7	4.9
840Ci	1.5	1.7
850Ci, i	3.3	3.5
M coupe/roadster, Z3	1.6	1.8
M3 (1.5)	2.0	2.2
M5 (3.0)	4.1	4.3
X3 (1.2)	1.6	1.8
X5 (3.5)	4.7	4.9
Z4 (1.6)	2.2	2.4

BMW-32 3 SERIES : 524TD : 5 SERIES : 6 SERIES : 7 SERIES : 8 SERIES

	LABOR TIME	SEVERE SERVICE
Blower Motor Resistor, Replace (B)		
1991-98 318i, iC, is	1.1	1.3
318ti	1.2	1.4
325i, iC, is, 525i	.8	1.0
528i	.5	.7
530i	1.5	1.7
540I		
1994-95	1.5	1.7
1997-01	.5	.7
M coupe/roadster, Z3	1.1	1.3
M3 (.5)	.7	.9
M5 (.5)	.7	.9
w/air bags add	.5	.5
Blower Motor Switch, Replace (B)		
318i, iC, is	1.4	1.6
318ti	.8	1.0
325i, iC, is, 330Ci, i, Xi (.8)	1.1	1.3
M coupe/roadster, Z3	.6	.8
M3, X3, X5 (.8)	1.1	1.3
Compressor Assy., Replace (B)		
Includes: Parts transfer, evacuate/recover and recharge.		
318i, iC, is	2.8	3.0
318ti	3.1	3.3
320i	3.2	3.4
325, e, iX	2.8	3.0
325i, is	3.5	3.7
328i, iC, is, 330Ci, i, Xi (2.4)	3.2	3.4
524td	3.1	3.3
525i	3.5	3.7
528e	2.8	3.0
528i, 530i	3.7	3.9
533i, 535i, is	3.1	3.
540i, 540iA, 540iT	3.2	3.4
545, 645Ci (2.8)	3.8	4.0
633CSi, 635CSi, 733i, 735i	3.7	3.9
740i, iA, iL, 750iL	3.2	3.4
745i, Li (2.8)	3.8	4.0
760i, 760Li (3.3)	4.5	4.7
840Ci	3.8	4.0
850Ci, 850i	3.9	4.1
M coupe/roadster, Z3	3.2	3.4
M3, Z4 (2.4)	3.2	3.4
M5 (2.8)	3.8	4.0
X3, X5 (2.7)	3.6	3.8
Condenser Assy., Replace (B)		
Includes: Evacuate/recover and recharge.		
318i, iC, is		
1984-92	3.1	3.3
1993-94	4.5	4.7
1995-98	3.3	3.5
320i	4.4	4.6
325, 325e, iX	3.1	3.3
1992-97 325i, iC, is	4.5	4.7

	LABOR TIME	SEVERE SERVICE
328i, 328iC, 328is	3.8	4.0
330Ci, i, Xi (2.5)	3.4	3.6
524td, 528e	3.1	3.3
525i, 530i	4.5	4.7
528i, iT		
1981	5.3	5.5
1997-00		
AT	5.0	5.2
MT	4.6	4.8
533i, 535i, is	3.1	3.3
540i, 540iA, 540iT	4.6	4.8
545i, 645Ci (2.0)	2.7	2.9
633CSi, 635CSi, 733i	4.3	4.5
735i	3.7	3.9
740i, iA, iL	5.0	5.2
745i, Li, 750i, 760Li (2.0)	2.7	2.9
750iL		
1988-94	3.7	3.9
1995-01	5.0	5.2
840Ci	5.0	5.2
850Ci, 850i	5.3	5.5
M coupe/roadster, Z3	3.2	3.4
M3 (2.8)	3.8	4.0
M5 (3.7)	5.0	5.2
X3, X5 (2.5)	3.4	3.6
Z4 (2.2)	3.0	3.2
Evaporator Core, Replace (B)		
Includes: Evacuate/recover and recharge.		
318i, iC, is		
1984-85	4.2	4.4
1991-93	12.3	12.5
1994-98	13.6	13.8
318ti	12.5	12.7
320i	5.5	5.7
325, 325e, iX	4.2	4.4
325i, 325iC, 325is	2.3	2.5
328i, 328iC, 328is	4.2	4.4
330Ci, i, Xi (2.3)	3.1	3.3
524td	4.2	4.4
525i	5.6	5.8
528e	4.8	5.0
528i, iT		
1981	6.4	6.6
1997-00	14.6	14.8
530i	6.3	6.5
533i, 535i, is	4.8	5.0
540i, iA, iT		
1994-95	6.3	6.5
1997-01	15.3	15.5
545i (10.1)	13.6	13.8
633CSi, 635CSi	4.8	5.0
645Ci	11.8	12.0
733i	16.2	16.4
735i	3.8	4.0
740i, iA, iL	5.2	5.4
745i, Li, 760i, Li (10.1)	13.6	13.8

	LABOR TIME	SEVERE SERVICE
750iL		
1988-94	3.8	4.0
1995-01	5.2	5.4
840Ci, 850Ci, i	7.0	7.2
M3 (10.4)	14.0	14.2
M5 (11.0)	14.9	15.1
X3, X5 (7.3)	9.9	10.1
Z3 (10.5)	14.2	14.4
Z4 (6.8)	9.2	9.4
w/air bags add	.5	.5
Expansion Valve, Replace (B)		
Includes: Evacuate/recover and recharge.		
318i, iC, is		
1984-85	4.2	4.4
1991-98	3.5	3.7
318ti	3.5	3.7
320i	5.2	5.4
325, 325e, iX	4.2	4.4
325i, 325iC, 325is	5.3	5.5
328i, 328iC, 328is	3.5	3.7
524td	5.2	5.4
525i	5.3	5.5
528e	4.8	5.0
528i	5.5	5.7
530i	5.9	6.1
533i, 535i, is	4.8	5.0
540i, iA, iT	4.8	5.0
545i (6.4)	8.6	8.8
633CSi, 635CSi	4.8	5.0
645Ci	11.8	12.0
733i, 735i	3.7	3.9
740i, iA, iL	4.7	4.9
745i, Li, 760i, Li (10.1)	13.6	13.8
750iL		
1988-94	3.7	3.9
1995-01	4.7	4.9
840Ci, 850Ci, i	6.2	6.4
M coupe/roadster, Z3		
1.9L	4.1	4.3
2.3L, 2.8L	4.3	4.5
M3	3.5	3.7
M5 (3.6)	4.9	5.1
X3, X5 (7.2)	9.7	9.9
Z4 (6.8)	9.2	9.4
Heater Hoses, Replace (B)		
1981-01	2.3	2.5
Heater Control Assy., Replace (B)		
318i, iC, is		
1984-85	1.8	2.0
1991-98	1.1	1.3
318ti	.7	.9
320i	1.7	1.9
325, 325e	1.8	2.0
325i, 325iC, 325is	.7	.9
325iX	1.7	1.9
328i, iC, is, 330Ci, i, Xi (.6)	.8	1.0

	LABOR TIME	SEVERE SERVICE
525i	.7	.9
528e	1.7	1.9
528i, iT		
1981	2.5	2.7
1997-00	.6	.8
530i, 540i	.7	.9
533i, 535i, is	1.7	1.9
545i (2.4)	3.2	3.4
633CSi, 635CSi	1.8	2.0
645Ci	1.8	2.0
733i, 735i	2.5	2.7
740i, iA, iL		
1994	1.6	1.8
1995-01	.8	1.0
745i, Li, 760i, Li (1.2)	1.6	1.8
750iL		
1988-94	2.5	2.7
1995-01	.8	1.0
840Ci, 850Ci, i	1.3	1.5
M coupe/roadster, Z3	1.4	1.6
M3, M5	.6	.8
X3, X5 (.5)	.7	.9
Z4 (.9)	1.2	1.4
Recode unit add	.2	.2

Heater Core, R&R or Replace (B)

	LABOR TIME	SEVERE SERVICE
318i, iC, is		
1984-85	1.7	1.9
1991-98		
w/AC	3.6	3.8
w/o AC	6.4	6.6
318ti		
w/AC	3.5	3.7
w/o AC	2.3	2.5
320i	4.2	4.4
325, 325e	1.7	1.9
325i, 325iC, 325is		
1987-91	7.8	8.0
1992-95	5.2	5.4
325iX	2.3	2.5
328i, 328iC, 328is	3.8	4.0
330Ci, i, Xi (2.5)	3.4	3.6
525i	4.9	5.1
528e	6.2	6.4
528i	6.5	6.7
530i	5.4	5.6
533i, 535i, is	6.8	7.0
540i, iA, iT		
1994-95	5.4	5.6
1997-01	5.7	5.9
545i (6.1)	8.2	8.4
633CSi, 635CSi	8.9	9.1
645Ci	10.8	11.0
733i	16.2	16.4
735i	4.8	5.0
740i, iA, iL		
1994	6.2	6.4
1995-01	10.0	10.2
745i, Li, 760i, Li (9.3)	12.6	12.8

	LABOR TIME	SEVERE SERVICE
750iL		
1988-94	5.2	5.4
1995-01	10.0	10.2
840Ci	8.1	8.3
850Ci, i		
5.0L	8.1	8.3
5.4L	9.1	9.3
850CSi	9.5	9.7
M coupe/roadster, Z3	4.1	4.3
M3 (2.8)	3.8	4.0
M5 (4.4)	5.9	6.1
X3 (4.2)	5.7	5.9
X5 (4.9)	6.6	6.8
Z4 (6.4)	8.6	8.8
w/air bags add	.5	.5

Heater Water Shut-Off Valve, Replace (B)

	LABOR TIME	SEVERE SERVICE
318i, iC, is		
1984-85	1.7	1.9
1991-98	1.1	1.3
318ti	.8	1.0
320i	1.9	2.1
325 series all	1.7	1.9
328i, iC, is, 330Ci, i, Xi (.6)	.8	1.0
524td	1.7	1.9
525i	1.3	1.5
528e	.5	.7
528i, iT		
1981	2.6	2.8
1997-00	1.1	1.3
530i, 540i	1.3	1.5
533i, 535i, is	.5	.7
545i (.9)	1.2	1.4
633CSi, 635CSi	5.3	5.5
645Ci	1.8	2.0
733i, 735i, 740i, iA, iL	1.5	1.7
745i, Li, 760i, Li (1.0)	1.4	1.6
750iL		
1988-94	1.7	1.9
1995-01	.7	.9
840Ci, 850Ci, i	1.8	2.0
M coupe/roadster, Z3	1.1	1.3
M3 (.6)	.8	1.0
M5 (.8)	1.1	1.3
X3, X5 (.7)	.9	1.1

Receiver/Drier Assy., Replace (B)
Includes: Evacuate/recover and recharge.

	LABOR TIME	SEVERE SERVICE
1981-05	2.8	3.0

WIPERS & SPEEDOMETER

Electronic Wiper/Washer Control Unit, Replace (B)

	LABOR TIME	SEVERE SERVICE
1982-05	.8	.8

Power Antenna, Replace (B)

	LABOR TIME	SEVERE SERVICE
All Models	.8	.8

Radio, R&R (B)

	LABOR TIME	SEVERE SERVICE
Exc. 320i	.6	.6
320i	1.2	1.2

Rear Window Wiper Motor, Replace (B)

	LABOR TIME	SEVERE SERVICE
318ti	1.0	1.2
528iT	1.1	1.3

Speedometer Cable & Casing, Replace (B)

	LABOR TIME	SEVERE SERVICE
1981-88		
upper	1.2	1.4
center	.8	1.0
lower	.5	.7

Speedometer Head, R&R or Replace (B)

	LABOR TIME	SEVERE SERVICE
318i, iC, is		
1984-85	1.7	1.9
1991-97	1.5	1.7
318ti	.7	.9
320i	1.2	1.4
325, e, iX	1.7	1.9
1987-97 325i, iC, is	1.5	1.7
328i, iC, is	1.6	1.8
330Ci, i, Xi (.9)	1.2	1.4
524td	1.7	1.9
525i	.8	1.0
528e	1.1	13
528i	1.4	1.6
530i, 540i	1.6	1.8
533i, 535i, is	1.1	1.3
545i, 645Ci (1.2)	1.6	1.8
633CSi, 635CSi	1.6	1.8
733i, 735i	1.2	1.4
740i, 740iL		
1994	1.6	1.8
1995-97	1.3	1.5
745i, 745Li	1.6	1.8
750iL	1.3	1.5
760i, 760Li, 840i (1.2)	1.6	1.8
850Ci, 850i		
1991-93	1.3	1.5
1994-96	1.6	1.8
1995-97 M3	1.5	1.7
X3, X5, Z4 (.9)	1.2	1.4
w/air bags add	.5	.5

Windshield Washer Pump, Replace (B)

	LABOR TIME	SEVERE SERVICE
1981-05	.5	.7

Windshield Wiper & Washer Switch, Replace (B)

	LABOR TIME	SEVERE SERVICE
1981-05	1.2	1.4
w/air bags add	.5	.5

Windshield Wiper Motor, Replace (B)

	LABOR TIME	SEVERE SERVICE
318i, iC, is		
1984-85	1.8	2.0
1991-98	1.5	1.7
318ti	1.9	2.1
320i	.8	1.0
325, 325e	1.8	2.0

BMW-34 3 SERIES : 524TD : 5 SERIES : 6 SERIES : 7 SERIES : 8 SERIES

	LABOR TIME	SEVERE SERVICE
Windshield Wiper Motor, Replace (B)		
325i, iC, is		
1989-91	2.6	2.8
1992-95	2.8	3.0
325iX	1.4	1.6
328i, iC, is	1.5	1.7
330Ci, i, Xi (1.3)	1.8	2.0
524td	1.7	1.9
525i	3.6	3.8
528e	.8	1.0
528i, iT		
1981	1.6	1.8
1997-00	2.1	2.3
530i	4.3	4.5
533i	1.7	1.9
535i, is	2.8	3.0
540i, iA, iT		
1994-95	4.3	4.5
1997-01	2.1	2.3
545i, 645Ci (1.6)	2.2	2.4
633CSi, 635CSi	.7	.9
733i	1.8	2.0
735i	2.8	3.0
740i, iA, iL		
1994	4.2	4.4
1995-01	2.1	2.3
745i, Li, 760i, Li (2.7)	3.6	3.8
750iL		
1988-94	3.1	3.3
1995-01	2.1	2.3
840Ci	3.6	3.8
850Ci, 850i	5.3	5.5
M coupe/roadster, Z3	1.2	1.4
M3 (1.1)	1.5	1.7
M5 (1.6)	2.2	2.4
X3, X5, Z4 (1.2)	1.6	1.8

LAMPS & SWITCHES

	LABOR TIME	SEVERE SERVICE
Back-Up Lamp Switch, Replace (B)		
1981-91	.7	.7
1992-05	1.2	1.2
w/air bags add	.5	.5
Digital Electronics Master Relay, Replace (B)		
All Models	.6	.6
Electronic Body Module, Replace (B)		
Exc. 1991-93	1.8	1.8
1991-93	1.3	1.3
w/air bags add	.5	.5
General Module, Replace (B)		
Exc. below	.8	.8
1997-00 528i	.3	.3
540i, iA, iT	.3	.3
840i, 850Ci	1.5	1.5
M coupe/roadster, Z3		
1.9L	.8	.8
2.3L, 2.8L	1.5	1.5

	LABOR TIME	SEVERE SERVICE
M3	1.0	1.0
M5	.3	.3
w/air bags add	.5	.5
Halogen Headlamp Bulb, Replace (B)		
1985-05 each	.5	.5
Hazard Warning Switch, Replace (B)		
1981-05	.7	.7
Headlamp Switch, Replace (B)		
1981-05	.8	.8
w/air bags add	.5	.5
Headlamps, Aim (B)		
Two	.5	.5
Four	.7	.7
High Mount Stop Lamp Assy., Replace (B)		
318i, iC, is		
1991-97	1.5	1.5
1993-98	.7	.7
318ti	.7	.7
325, 325e	.5	.5
325i, iC, is	1.5	1.5
325iX	1.2	1.2
328i, iC, is		
525i, 530i	2.0	2.0
528i, iT	2.3	2.3
535i, is	1.7	1.7
540i, iA, iT		
1994-95	2.0	2.0
1997-01	2.3	2.3
735i	1.7	1.7
740i, iA, iL		
1994-97	1.4	1.4
1998-01	2.3	2.3
750iL		
1988-94	1.7	1.7
1995-01	2.3	2.3
840Ci	1.6	1.6
850Ci, i		
1991-93	2.2	2.2
1994-97	1.6	1.6
M coupe/roadster, Z3	.8	.8
M3	.5	.5
M5	2.3	2.3
High Mount Stop Lamp Bulb, Replace (C)		
1985-05	.2	.2
Horn, Replace (B)		
Exc. below	.5	.5
1997-00 528i, iT	.7	.7
1997-01 540i, iA, iT	.7	.7
740i, iA, iL, 750iL	.6	.6
each addl. add	.6	.6
Horn Relay, Replace (B)		
1981-05	.2	.2
License Lamp Assy., Replace (B)		
Each	.5	.5
Neutral Safety Switch, Replace (B)		
1981-05 (2.1)	2.8	3.0

	LABOR TIME	SEVERE SERVICE
Parking Brake Warning Lamp Switch, Replace (B)		
Exc. below	.8	.8
318ti	.7	.7
325i, iC, is	1.0	1.0
325iX	.5	.5
740i, iA, iL, 750iL	1.0	1.0
Z3, Z4	1.1	1.1
Parking Lamp Lens or Bulb, Replace (C)		
1981-05	.3	.3
Sealed Beam Headlamp, Replace (B)		
1981-84	.4	.4
Stop & Tail Lamp Lens, Replace (B)		
1981-05	.3	.3
Stop Lamp Switch, Replace (B)		
1981-91	.5	.5
1992-05	1.2	1.2
w/air bags add	.5	.5
Stop Lamp Test Switch, Replace (B)		
850Ci, i		
5.0L	.7	.9
5.4L	1.5	1.7
w/air bags add	.5	.5
Turn Signal & Headlamp Dimmer Switch, Replace (B)		
1981-05	1.2	1.2
w/air bags add	.5	.5
Turn Signal Lamp Lens or Bulb, Replace (C)		
1981-05	.2	.2
Turn Signal or Hazard Warning Flasher, Replace (B)		
1981-05	.3	.3

BODY

	LABOR TIME	SEVERE SERVICE
Front Door Lock, Replace (B)		
Includes: R&R trim panel.		
1981-91	1.7	1.9
1992-04	2.3	2.5
Front Door Window Regulator and/or Motor, Replace (B)		
3 series (1.4)	1.9	2.1
524td, 533i, 535i, is	1.7	1.9
525i, 530i, 545i (1.6)	2.2	2.4
528e	1.2	1.4
528i, iT		
1981	1.1	1.3
1997-00	1.5	1.7
540i, iA, iT		
1994-95	2.2	2.4
1997-01	1.5	1.7
633CSi, 635CSi	1.2	1.4
733i, 735i	1.4	1.6
740i, iA, iL		
1993-94	1.4	1.6
1995-01	1.9	2.1

ALPINA ROADSTER : M3 : M5 : X3 : X5 : Z3 : Z4 : Z8   BMW-35

	LABOR TIME	SEVERE SERVICE
745i, Li, 750iL,		
760i, Li (1.6)	2.2	2.4
840Ci	2.0	2.2
850Ci, 850i	2.2	2.4
M coupe/roadster, Z3	1.5	1.7
M3, X3 (1.3)	1.8	2.0
M5 (1.2)	1.6	1.8
X5 (1.5)	2.0	2.2
Z4 (1.0)	1.4	1.6
Hood Lock, Replace (B)		
1981-05	1.2	1.4
Hood Release Cable, Replace (B)		
318i, iC, is		
1984-85	1.2	1.4
1991-93	1.9	2.1
1994-98	4.9	5.1
318ti	5.2	5.4
320i	1.1	1.3
325, e, iX	1.2	1.4
325i, iC, is		
1987-93	1.9	2.1
1994-95	6.3	6.5
328i, iC, is,		
330Ci, i, Xi (3.6)	4.9	5.1
524td	1.2	1.4
525i	1.6	1.8
528e	1.1	1.3
528i, iT	1.4	1.6
530i, 545i	1.6	1.8
533i, 535i, is	1.2	1.4

	LABOR TIME	SEVERE SERVICE
540i, iA, iT	1.4	1.6
633CSi, 635CSi	.8	1.0
733i, 735i	1.2	1.4
740i, iA, iL, 750iL	1.4	1.6
745i, Li, 760i, Li (1.0)	1.4	1.6
840Ci	2.8	3.0
850Ci, 850i	1.2	1.4
M coupe/roadster, Z3	1.6	1.8
M3	4.9	5.1
M5, Z4 (1.0)	1.4	1.6
X3, X5 (1.3)	1.8	2.0
Lock Striker Plate, Replace (B)		
1981-05	.3	.5
Rear Door Lock, Replace (B)		
3 series		
exc. below (1.8)	2.4	2.6
325i	2.6	2.8
328i, 328is	1.2	1.4
525i	1.5	1.7
528e	1.1	1.3
528i, iT	1.2	1.4
530i, 545i (1.1)	1.5	1.7
540i, iA, iT		
1994-95	1.5	1.7
1997-01	1.2	1.4
733i	1.4	1.6
1995-01 740i, iA, iL	1.6	1.8
1995-01 750iL	1.6	1.8
745i, Li, 760i, Li (1.2)	1.6	1.8

	LABOR TIME	SEVERE SERVICE
M5 (1.0)	1.4	1.6
X3 (1.2)	1.6	1.8
X5 (1.5)	2.0	2.2
Rear Door Window Regulator and/or Motor, Replace (B)		
3 series (1.2)	1.6	1.8
525i	2.0	2.2
528e	1.1	1.3
528i, iT		
1981	1.1	1.3
1997-00	1.6	1.8
530i, 545i (1.5)	2.0	2.2
540i, iA, iT		
1994-95	2.0	2.2
1997-01	1.6	1.8
733i	1.3	1.5
740i, iA, iL		
1994		
motor	1.5	1.7
regulator	1.4	1.6
1995-01	1.9	2.1
745i, Li, 760i, Li (1.3)	1.8	2.0
1995-01 750iL	2.1	2.3
850Ci, i		
motor	3.6	3.8
regulator	3.1	3.3
M5 (1.2)	1.6	1.8
X3 (1.3)	1.8	2.0
X5 (1.6)	2.2	2.4

BMW-36 3 SERIES : 524TD : 5 SERIES : 6 SERIES : 7 SERIES : 8 SERIES

NOTES

JAG

**S-Type : Super V8 : Vanden Plas :
XJ12 : XJ6 : XJ8 : XJR : XJS : XK8 : XKR : X-Type**

SYSTEM INDEX

- MAINTENANCE JAG-2
  - Maintenance Schedule . . JAG-2
- CHARGING JAG-2
- STARTING JAG-2
- CRUISE CONTROL JAG-3
- IGNITION JAG-3
- EMISSIONS JAG-5
- FUEL JAG-6
- EXHAUST JAG-8
- ENGINE COOLING JAG-10
- ENGINE JAG-11
  - Assembly JAG-11
  - Cylinder Head JAG-12
  - Camshaft JAG-13
  - Crank & Pistons JAG-15
  - Engine Lubrication JAG-16
- CLUTCH JAG-17
- MANUAL TRANSMISSION . JAG-17
- AUTO TRANSMISSION . . . JAG-17
- SHIFT LINKAGE JAG-19
- DRIVELINE JAG-19
- BRAKES JAG-20
- FRONT SUSPENSION JAG-23
- REAR SUSPENSION JAG-25
- STEERING JAG-27
- HEATING & AC JAG-29
- WIPERS &
  SPEEDOMETER JAG-31
- LAMPS & SWITCHES JAG-31
- BODY JAG-32

OPERATIONS INDEX

A
- AC Hoses JAG-29
- Air Bags JAG-27
- Air Conditioning JAG-29
- Alignment JAG-23
- Alternator (Generator) JAG-2
- Antenna JAG-31
- Anti-Lock Brakes JAG-20

B
- Back-Up Lamp Switch JAG-31
- Ball Joint JAG-24
- Battery Cables JAG-2
- Bleed Brake System JAG-20
- Blower Motor JAG-29
- Brake Disc JAG-22
- Brake Hose JAG-21
- Brake Pads and/or Shoes . . JAG-22

C
- Camshaft JAG-14
- Camshaft Sensor JAG-3
- Catalytic Converter JAG-8
- Coolant Temperature (ECT)
  Sensor JAG-5
- Crankshaft JAG-15
- Crankshaft Sensor JAG-3
- Cruise Control JAG-3
- Cylinder Head JAG-12

D
- Differential JAG-19
- Distributor JAG-4
- Drive Belt JAG-2
- Driveshaft JAG-19

E
- EGR JAG-5
- Electronic Control Module
  (ECM/PCM) JAG-5
- Engine JAG-11
- Engine Lubrication JAG-16
- Engine Mounts JAG-12
- Evaporator JAG-30
- Exhaust JAG-8
- Exhaust Manifold JAG-9
- Expansion Valve JAG-30

F
- Flexplate JAG-18
- Flywheel JAG-17
- Fuel Injection JAG-7
- Fuel Pump JAG-6
- Fuel Vapor Canister JAG-5

G
- Generator JAG-2

H
- Halfshaft JAG-20
- Headlamp JAG-32
- Heater Core JAG-30
- Horn JAG-32

I
- Idle Air Control (IAC) Valve . JAG-8
- Ignition Coil JAG-4
- Ignition Module JAG-4
- Ignition Switch JAG-4
- Inner Tie Rod JAG-28
- Intake Air Temperature (IAT)
  Sensor JAG-5
- Intake Manifold JAG-7

K
- Knock Sensor JAG-5

L
- Lower Control Arm JAG-24

M
- Maintenance Schedule JAG-2
- Manifold Absolute Pressure
  (MAP) Sensor JAG-6
- Mass Air Flow (MAF)
  Sensor JAG-6
- Master Cylinder JAG-21
- Muffler JAG-9
- Multifunction Lever/Switch . . JAG-31

O
- Oil Pan JAG-16
- Oil Pump JAG-16
- Outer Tie Rod JAG-28
- Oxygen Sensor JAG-6

P
- Parking Brake JAG-22
- Pistons JAG-16

R
- Radiator JAG-10
- Radiator Hoses JAG-10
- Radio JAG-31
- Rear Main Oil Seal JAG-16

S
- Shock Absorber/Strut,
  Front JAG-24
- Shock Absorber/Strut,
  Rear JAG-27
- Spark Plug Cables JAG-4
- Spark Plugs JAG-4
- Spring, Front Coil JAG-24
- Spring, Rear Coil JAG-26
- Starter JAG-3
- Steering Wheel JAG-29

T
- Thermostat JAG-10
- Throttle Body JAG-8
- Throttle Position Sensor
  (TPS) JAG-6
- Timing Chain JAG-14
- Torque Converter JAG-18

U
- U-Joint JAG-20
- Upper Control Arm JAG-25

V
- Valve Body JAG-18
- Valve Cover Gasket JAG-13
- Valve Job JAG-13
- Vehicle Speed Sensor JAG-6

W
- Water Pump JAG-11
- Wheel Balance JAG-2
- Window Regulator JAG-33
- Windshield Washer Pump . . JAG-31
- Windshield Wiper Motor . . . JAG-31

JAG-1

JAG-2   S-TYPE : SUPER V8 : VANDEN PLAS : XJ12 : XJ6 : XJ8 : XJR : XJS : XK8 : XKR : X-TYPE

	LABOR TIME	SEVERE SERVICE
MAINTENANCE		
Air Cleaner Filter Element, Service (B)		
1981-87 each (.5)	.7	.9
1988-05 (.2)	.3	.4
Chassis Lubrication, Change Oil & Filter (C)		
Includes: Correct fluid levels.		
1981-05	.7	.7
Drive Belt, Adjust (B)		
Exc. below (.2)	.3	.4
1995-97 XJR (.3)	.5	.6
XJS (.3)	.5	.6
Drive Belt, Replace (B)		
Air pump		
1981-97 XJ6 (.5)	.8	.9
XJ12 (.3)	.5	.6
1981-94 XJS (.4)	.8	.9
Serpentine		
S-Type		
V6 (.6)	.8	.9
V8 (.5)	.6	.7
XJ8,		
Vanden Plas (.3)	.5	.6
XJR		
1998-03 (.4)	.6	.7
2004-05 (.8)	1.1	1.2
XK8		
4.0L (.3)	.4	.5
4.2L (.4)	.6	.7
XKR (.4)	.6	.7
X-Type (.7)	.9	1.0
Steering pump		
XJ6 (.3)	.5	.7
XJ12 (.5)	.7	.9
XJS (1.0)	1.4	1.6
Supercharger		
S-Type (.7)	.9	1.0
XJR		
1995-03 (.2)	.3	.5
2004-05 (.7)	.9	1.0
XKR (.4)	.6	.7
V-belt		
S-Type (.4)	.6	.7
1981-94 XJ6 (.3)	.5	.6
XJ8, Vanden Plas,		
XJ12 (.4)	.6	.7
XJR		
1995-97 (.6)	.8	.9
1996-2003 (.4)	.6	.7
2004-05 (.8)	1.1	1.2
XJS one (.5)	.7	.8
XK8, XKR (.4)	.6	.7
X-Type (.7)	.9	1.0

	LABOR TIME	SEVERE SERVICE
Drive Belt Tensioner, Replace (B)		
S-Type (.4)	.6	.8
XJ8, Vanden Plas (.3)	.5	.7
XJR		
1995-03 (.4)	.6	.9
2004-05 (.8)	1.1	1.2
XK8, XKR (.3)	.5	.8
X-Type (.6)	.8	1.0
Halogen Headlamp Bulb, Replace (B)		
1981-05 (.2)	.3	.4
Oil & Filter, Change (C)		
Includes: Correct fluid levels.		
1981-05 (.4)	.6	.7
Sealed Beam Headlamp, Replace (B)		
XJ6		
1981-87 each (.3)	.5	.5
1988-94 each (.5)	.7	.7
1995-97 each (.2)	.3	.3
XJ8, XJR, Vanden Plas		
2004-05 (.7)	.9	.9
XJ12, XJR each (.2)	.3	.3
XJS each	.5	.5
Tire, Replace (C)		
Includes: Dismount old tire and mount new tire to rim.		
1981-05 (.3)	.5	.5
Wheels, Balance (B)		
1981-05		
one (.3)	.5	.5
each addl. add	.3	.3

SCHEDULED MAINTENANCE INTERVALS

If necessary, refer to appropriate Chilton maintenance service information.

	LABOR TIME	SEVERE SERVICE
10,000 Mile Service (C)		
All Models (1.5)	2.0	2.2
20,000 Mile Service (B)		
All Models (1.7)	2.3	2.5
30,000 Mile Service (C)		
All Models (1.6)	2.2	2.4
40,000 Mile Service (B)		
All Models (1.7)	2.3	2.5
50,000 Mile Service (C)		
All Models (1.5)	2.0	2.2
60,000 Mile Service (C)		
All Models (1.7)	2.3	2.5
70,000 Mile Service (C)		
All Models (2.6)	3.5	3.7
80,000 Mile Service (B)		
All Models (1.7)	2.3	2.5
90,000 Mile Service (C)		
All Models (1.6)	2.2	2.4
100,000 Mile Service (B)		
All Models (1.5)	2.0	2.2

	LABOR TIME	SEVERE SERVICE
CHARGING		
Alternator Circuits, Test (B)		
Includes: Test component output.		
1981-05 (.5)	.7	.7
Alternator, R&R and Recondition (A)		
1981-94 XJ6 (2.1)	2.8	3.3
1981-94 XJS (3.2)	4.3	4.8
w/air pump XJ6 add	.9	.9
Alternator Assy., Replace (B)		
Includes: Pulley transfer.		
S-Type		
3.0L, 4.0L (.7)	.9	1.0
4.2L (1.7)	2.3	2.5
XJ6		
1981-87 (1.2)	1.7	1.9
1988-94 (1.5)	2.1	2.3
1995-97 (1.1)	1.5	1.7
XJ8, Vanden Plas		
1981-2003 (.9)	1.2	1.4
2004-05 (1.3)	1.8	2.0
XJ12 (1.3)	1.8	2.0
XJR		
1995-97 (1.3)	1.8	2.0
1998-05 (1.2)	1.7	1.9
XJS		
6 cyl. (.3)	.5	.7
V12 (2.0)	2.7	2.9
XK8, XKR (.9)	1.2	1.4
X-Type	1.4	1.5
w/air pump		
81-87 XJ6 add	.8	.8
w/supercharger		
Vanden Plas add	.5	.5
Alternator Voltage Regulator, Replace (B)		
1981-87 XJ6 (1.4)	1.9	1.9
XJS (2.3)	3.1	3.1
XK8 (1.3)	1.7	1.9
Voltmeter and/or Ammeter Gauge, Replace (B)		
1981-87 XJ6 (.5)	.8	.8
XJS (1.8)	2.5	2.5
STARTING		
Starter Draw Test (On Car) (B)		
Exc. below (.2)	.3	.3
2002-05 XJ Series (.4)	.6	.7
Battery, Replace (C)		
1981-05 (.2)	.3	.5
Battery Cables, Replace (C)		
Positive		
S-Type		
2000-01 (.8)	1.1	1.3
2002-05 (2.1)	2.8	3.0
XJ6 (.9)	1.2	1.4

S-TYPE : SUPER V8 : VANDEN PLAS : XJ12 : XJ6 : XJ8 : XJR : XJS : XK8 : XKR : X-TYPE **JAG-3**

	LABOR TIME	SEVERE SERVICE
XJ8, Vanden Plas		
1981-01 (.6)	.8	1.0
2002-03 (.2)	.3	.4
2004-05 (4.3)	5.9	6.1
XJ12 (.9)	1.2	1.4
XK8, XKR		
1997-01 (.6)	.8	1.0
2002-05 (.2)	.3	.4
X-Type (.2)	.3	.4
Negative		
exc. X-Type (.2)	.3	.4
X-Type (.3)	.5	.6
Starter to generator		
S-Type (1.0)	1.4	1.6

Starter, R&R and Recondition (A)
Includes: Turn down armature.
- 1981-87 XJ6 (2.7) 3.6   4.1
- 1981-91 XJS (4.2) 5.6   6.1

Starter Assy., Replace (B)
Add draw test if performed.
- S-Type (.8) 1.0   1.1
- XJ6
  - 1981-87 (1.7) 2.3   2.5
  - 1988-94 (.9) 1.2   1.4
  - 1995-97 (1.1) 1.5   1.7
- XJ8, Vanden Plas 1.1   1.3
- XJ12
  - 1994 (1.5) 2.1   2.3
  - 1995-96 (2.0) 2.7   2.9
- XJR (.8) 1.1   1.3
- XJS
  - 1981-91 (3.2) 4.4   4.6
  - 1992-96
    - 6 cyl. (1.2) 1.6   1.8
    - V12 (2.0) 2.7   2.9
- XK8, XKR
  - exc. below (1.9) 2.6   2.8
  - convertible (2.2) ... 3.0   3.2
- X-Type
  - 2002-03 (.3) 5   .7
  - 2004-05 (.9) 1.2   1.4

Starter Drive Assy., Replace (B)
Includes: R&R starter.
- XJ6
  - 1981-87 (2.2) 3.0   3.3
  - 1988-94 (1.2) 1.6   1.9
- 1981-94 XJS (3.6) 4.9   5.2

Starter Relay, Replace (B)
- S-Type (.2) 3   .4
- XJ6
  - 1981-87 (.3) 5   .6
  - 1988-97 (.6) 8   .9
- XJ8, Vanden Plas (.4) . .6   .7
- XJ12 (.5) 7   .8
- XJS (.3) 5   .6
- XK8, XKR, X-Type (.2) . .3   .4

	LABOR TIME	SEVERE SERVICE
Starter Solenoid and/or Switch, Replace (B)		
Includes: R&R starter.		
XJ6		
1981-87 (1.9)	2.6	2.8
1988-94 (1.4)	1.9	2.1
1995-97 (1.2)	1.6	1.8
XJR (1.2)	1.6	1.8
XJS		
1981-91 (3.6)	4.9	5.1
1992-96		
6 cyl. (1.3)	1.8	2.0
V12 (2.1)	2.8	3.0

CRUISE CONTROL

Diagnose Cruise Control System Component Each (A)
- 1988-05 (.6) 8   .8

Control Combination Switch, Replace (B)
- 1988-94 XJ6 (1.0) ... 1.4   1.4
- 1988-94 XJS (.3) 5   .5
- 1995-97 (.6) 8   .9

Control Controller Module, Replace (B)
- 1988-94 XJ6 (.6) 8   .8
- 1988-94 XJS (.3) 5   .5

Control Metering Valve (Dump), Replace (B)
- 1988-97 (.9) 1.2   1.2

Control Pedal Cancel Switch, Replace (B)
- 2000-05 S-Type (.2) . .3   .3

Control Relay, Replace (B)
- 1988-94 XJ6 (.5) 7   .7
- 1994 XJ12 (.5) 7   .7

Control Set Switch, Replace (B)
- 1988-94 XJS (.9) 1.2   1.2
- 1988-05 XJ8, XJR, Vanden Plas (.2) 3   .3
- 2000-05 S-Type (.5) . .7   .7

Control Throttle or Servo Cable, Replace (B)
- 1988-96 XJS (1.3) ... 1.8   1.9

Control Transducer, Replace (B)
- 1988-96 XJS (.6) 8   .8

Control Vacuum Pump, Replace (B)
- 1988-97 (.6) 8   .9

Control Vacuum Tank, Replace (B)
- 1988-03 XJ8, XJR (.3) . .5   .5
- 1998-05 XK8, XKR (.5) . .7   .7

Electronic Speed Control Module, Replace (B)
- 1988-97 (.9) 1.2   1.2

Throttle Actuator, Replace (B)
- 1988-97 (.1) 2   .2

	LABOR TIME	SEVERE SERVICE
IGNITION		
Diagnose Ignition System Component Each (A)		
All Models (.9)	1.2	1.2
Ignition Timing, Reset (B)		
1981-94 XJS, XJ6 (.5)	.7	.7
Amplifier Unit, Replace (B)		
1988-94 XJ6 (.3)	.5	.6
XJ8, Vanden Plas, XJR (.2)	.3	.4
1994 XJ12 (1.0)	1.4	1.6
XJS		
1981-87 (.5)	.7	.8
1988-94		
6 cyl. (.5)	.7	.8
V12 (.1)	.2	.3
XK8, XKR (.2)	.3	.4

Camshaft Top Dead Center Sensor, Replace (B)
- XJ12 (.7) 9   1.1

Camshaft Position Sensor, Replace (B)
- S-Type
  - 3.0L, 4.0L (.4) 6   .8
  - 4.2L
    - right side (.7) 9   1.1
    - left side (1.0) ... 1.4   1.6
- 1995-97 XJ6 (.5) 7   .9
- XJ8, Vanden Plas
  - 1982-03 (.5) 7   .9
  - 2004-05 (1.3) 1.8   2.0
- XJ12 (.6) 8   1.0
- XJR
  - 1995-97 (.7) 9   1.1
  - 1998-03 (.5) 7   .9
  - 2004-05 (1.2) 1.8   2.0
- XK8, XKR
  - 1997-03 (.5) 7   .9
  - 2004-05 (.2) 3   .4
- X-Type (.2) 3   .4

Crankshaft Angle Sensor, Replace (B)
- S-Type
  - 3.0L, 4.0L, 4.2L (.3) . .5   .7
- 1995-97 XJ6 (.6) 8   1.0
- XJ8, Vanden Plas (.4) . .6   .8
- XJ12 (.9) 1.2   1.4
- 1995-05 XJR 9 (.6) .. .8   1.0
- 1988-96 XJS (1.0) ... 1.4   1.6
- XK8, XKR, X-Type (.4) . .6   .7

Distributor, R&R and Recondition (A)
Includes: Reset base ignition timing.
- 1981-87 XJ6 (1.4) ... 1.9   2.2
- 1981-94 XJS (2.4) ... 3.2   3.5

JAG-4 S-TYPE : SUPER V8 : VANDEN PLAS : XJ12 : XJ6 : XJ8 : XJR : XJS : XK8 : XKR : X-TYPE

	LABOR TIME	SEVERE SERVICE
Distributor, Replace (B)		
Includes: Reset base ignition timing.		
XJ6		
1981-87 (.9)	1.2	1.3
1988-94 (.3)	.5	.6
Distributor Cap and/or Rotor, Replace (B)		
1981-92 XJ6 (.3)	.5	.5
1994 XJ12 (.6)	.8	.8
XJS		
1981-87 (.9)	1.2	1.2
1988-94		
6 cyl. (.3)	.5	.5
V12 (1.0)	1.4	1.4
Electronic Ignition Control Unit, Replace (B)		
S-Type		
2000-02 (.6)	.8	1.0
2003-05 (1.6)	2.2	2.4
XJ6		
1988-94 (.5)	.7	.9
1995-97 (1.0)	1.4	1.6
XJ8, Vanden Plas (1.6)	2.2	2.4
XJ12 (1.1)	1.5	1.7
XJR		
1995-97 (1.0)	1.4	1.6
1998-01 (1.6)	2.2	2.4
XJS (.9)	1.2	1.4
XK8, XKR (1.3)	1.8	2.0
Ignition Coil, Replace (B)		
Distributor		
XJ6		
1981-87 (.7)	.9	1.1
1988-94 (.5)	.7	.9
XJ12 each		
side (1.0)	1.4	1.6
XJS		
1981-87 (.5)	.7	.9
1988-94		
6 cyl. (.9)	1.2	1.2
V12 (.3)	.5	.5
Distributorless		
S-Type		
V6		
one		
right side (1.0)	1.4	1.6
left side (.2)	.3	.4
all both		
sides (1.3)	1.8	2.0
V8		
right side one (.4)	.6	.7
left side one		
4.0L (.5)	.8	.9
4.2L (.2)	.3	.4
all both sides		
4.0L (1.1)	1.5	1.7
4.2L (.5)	.8	.9

	LABOR TIME	SEVERE SERVICE
1995-97 XJ6		
one (.5)	.8	.9
all (1.3)	1.8	2.0
XJ8, Vanden Plas		
1982-03		
right or left		
side one (.4)	.6	.7
all both sides (1.1)	1.5	1.7
2004-05		
right or left one (.2)	.3	.4
all both sides (.9)	1.2	1.4
XJ12 one side (1.0)	1.4	1.6
XJR		
1995-97		
one (.4)	.6	.7
all (1.0)	1.4	1.6
1998-03		
right or left one (.4)	.6	.7
all both sides (1.1)	1.5	1.7
2004-05		
right or left one (.2)	.3	.4
all both sides (.9)	1.2	1.4
XJS (.6)	.8	.9
XK8, XKR		
1997-03		
right side one (.4)	.6	.7
left side one (.3)	.5	.6
all both sides (1.1)	1.5	1.7
2004-05		
right side one (.3)	.5	.6
left side one (.2)	.3	.4
all both sides (.9)	1.2	1.4
X-Type		
right side one (1.0)	1.4	1.6
left side one (.7)	.9	1.0
all both sides (1.8)	2.4	2.6
Ignition/Injection Combined-Control Module, Replace (B)		
XJ8, XJR		
1998-01		
one (.1)	.2	.2
all (.2)	.3	.3
2002-03 (.9)	1.2	1.4
2004-05 (1.3)	1.8	2.0
XJ12		
one (.6)	.8	.8
all (.8)	1.2	1.4
XJS (1.2)	1.6	1.8
XK8, XKR		
1997-03 one or		
both (.2)	.3	.3
2004-05 (.9)	1.2	1.4

	LABOR TIME	SEVERE SERVICE
Ignition Switch, Replace (B)		
1981-92 (.6)	.8	1.0
1993-97 (1.0)	1.4	1.6
1998-05		
S-Type (.3)	.5	.7
XJ8, XJR,		
Vanden Plas (.9)	1.2	1.4
XK8, XKR		
1997-01 (.9)	1.2	1.2
2002-05 (.3)	.5	.5
X-Type (.6)	.8	.8
Pickup Module & Base Plate, Replace (B)		
1981-87 XJ6 (1.0)	1.4	1.6
1981-94 XJS (1.0)	1.4	1.6
Spark Plug (Ignition) Cables, Replace (B)		
XJ6		
1981-87 (1.0)	1.4	1.5
1988-94 (.3)	.5	.6
XJS (1.0)	1.4	1.6
Spark Plugs, Replace (B)		
S-Type		
V6		
2000-01 (1.4)	1.9	2.1
2002-03 (1.9)	2.6	2.9
2004-05 (1.5)	2.0	2.3
V8		
2000-01 (1.6)	2.2	2.5
2002-05 (1.1)	1.5	1.8
XJ6		
1981-87 (.5)	.7	1.0
1988-94 (.7)	.9	1.2
1995-97 (1.1)	1.5	1.8
XJ8,		
Vanden Plas (1.3)	1.8	2.1
XJ12 (1.4)	1.9	2.2
XJR		
1995-97 (1.1)	1.5	1.8
1998-03 (1.3)	1.8	2.1
2004-05 (1.0)	1.4	1.7
XJS		
6 cyl. (.5)	.7	.9
V12 (1.4)	1.9	2.1
XK8, XKR		
1997-03 (1.3)	1.8	2.0
2004-05 (.9)	1.2	1.5
X-Type (2.1)	2.8	3.1
Speed Sensor, Replace (B)		
1988-97 XJ6 (.5)	.7	.8
XJ12 (.6)	.8	.9
XJS (.9)	1.2	1.3
Vacuum Advance Unit, Replace (B)		
Includes: Reset timing.		
1981-94 XJS (2.2)	3.0	3.3

S-TYPE : SUPER V8 : VANDEN PLAS : XJ12 : XJ6 : XJ8 : XJR : XJS : XK8 : XKR : X-TYPE **JAG-5**

	LABOR TIME	SEVERE SERVICE
EMISSIONS		

The following operations do not include testing. Add time as required.

Dynamometer Test (A)
  1981-05 (.3) .... .5 .5
Air Distribution Manifold, Replace (B)
Includes: R&R exhaust manifold when required.
  XJ6
    1981-87 (.9) 1.2 *1.4*
    1988-97
      front (1.2) 1.6 *1.8*
      rear (1.4) 1.9 *2.1*
  XJ12 (.9) 1.2 *1.4*
  XJS each (2.0) 2.7 *2.9*
Air Flow Meter, Replace (B)
  S-Type .... .3 *.3*
  XJ6
    1981-87 (.5) .... .7 *.7*
    1988-94 (1.2) 1.6 *1.6*
    1995-97 (.9) 1.2 *1.2*
  XJ8, Vanden Plas (.2) ... .3 *.3*
  XJR
    1995-97 (.7) .... .9 *.9*
    1998-05 (.2) .... .3 *.3*
  XK8, XKR, X-Type (.2) ... .3 *.3*
Air Injection Pump Relay, Replace (B)
  1995-97 (.5) .... .8 *.8*
Air Pump Assy., Replace (B)
  XJ6
    1981-87 (.6) 9 *1.1*
    1988-94 (1.2) 1.6 *1.8*
    1995-97 (.5) 8 *1.0*
  XJ12 (.9) 1.2 *1.4*
  XJR (.6) 8 *1.0*
  XJS
    1981-87 (1.3) 1.8 *2.0*
    1988-96 (.5) 8 *1.0*
Air Switching Valve, Replace (B)
  XJ12 (.3) .... .5 *.5*
Air Temperature Sensor, Replace (B)
  S-Type (.2) .... .3 *.3*
  XJ6
    1988-94 (.1) .... .2 *.2*
    1995-97 (.5) .... .7 *.7*
  XJ12 (.5) .... .7 *.7*
  XJ8, Vanden Plas (.3) ... .5 *.5*
  XJR
    1995-97 (1.0) 1.4 *1.4*
    1998-05 .... .3 *.3*
  1988-96 XJS .... .2 *.2*
  XKR
    2002-05 .... .3 *.3*
  X-Type .... .3 *.3*
Test sensor add .... 3 *.3*

Canister Purge Control Valve, Replace (B)
  S-Type, XJ8, XJR,
  XK8, X-Type .... 3 *.3*
Coolant Temperature (ECT) Sensor, Replace (B)
  Exc. below .... .3 *.4*
  XJ12, XJR, XJS .... .2 *.3*
  Test sensor add .... .3 *.4*
EGR Temperature Sender, Replace (B)
  1995-97 XJ6, XJR (.5) .. .8 *.8*
EGR Valve, Replace (B)
  S-Type
    V6 .... .3 *.5*
    V8
      4.2L (.4) .... .6 *.8*
      4.2L SC (2.4) 3.2 *3.4*
  XJ6
    1990-94 (.3) .... .5 *.7*
    1995-97 (.6) 8 *1.0*
  XJ8, Vanden Plas (.6) ... .8 *.9*
  XJR
    1996-03 (.7) 9 *1.0*
    2004-05 (1.3) 1.8 *2.0*
  XK8 (.4) .... .6 *.8*
  XKR
    4.0L (1.2.) 1.7 *1.9*
    4.2L (.7) 9 *1.0*
Electronic Control Module (ECM/PCM), Replace (B)
  S-Type
    2000-02 (.8) 1.1 *1.3*
    2003-05 (1.6) 2.2 *2.4*
  1996-05 XJ8, XJR,
  Vanden Plas (.9) 1.2 *1.3*
  XK8, XKR (1.2) 1.6 *1.8*
  X-Type (.6) .... .8 *.9*
Fuel Vapor Canister, Replace (B)
  S-Type
    2000-01 (.2) .... .3 *.3*
    2002-03 (4.9) 6.6 *6.8*
    2004-05 (3.1) 4.2 *4.4*
  XJ6
    1981-87 (.8) 1.2 *1.3*
    1988-94 (.5) .... .7 *.8*
    1995-97 (.3) .... .5 *.6*
  XJ8, Vanden Plas, XJR
    1996-03 (.2) .... .3 *.4*
    2004-05 (2.9) 3.9 *4.0*
  1995-97 XJ12 (.3) ...... .5 *.6*
  XJS
    1981-87 (.6) .... .8 *.9*
    1988-96 (.5) .... .7 *.8*
  XK8, XKR (.4) .... .6 *.7*
  X-Type
    2002-03 (4.5) 6.1 *6.3*
    2004-05 (4.9) 6.6 *6.8*

Knock Sensor, Replace (B)
  S-Type
    V6
      front
        2000-01 (1.2) ... 1.6 *1.8*
        2002-03 (2.6) ... 3.5 *3.9*
      2004-05
        right side (.4) .... .6 *.8*
        left side (1.8) .. 2.4 *2.6*
      rear
        2000-03 (.7) 9 *1.0*
      both
        2000-01 (1.5) ... 2.0 *2.2*
        2002-03 (3.2) ... 4.3 *4.7*
        2004-05 (2.1) ... 2.8 *3.0*
    V8
      4.0L (.3) .... .5 *.6*
      4.2L one
        side (2.1) 2.8 *3.1*
      4.2L SC (5.8) 7.8 *8.6*
  1995-97 XJ6
    front (1.0) 1.4 *1.6*
    rear (.9) 1.2 *1.4*
    both (1.2) 1.6 *1.8*
  XJ8, Vanden Plas
    1982-95
      one (.3) .... .5 *.7*
      all (.5) .... .7 *.9*
    1996-03
      one (.3) .... .5 *.7*
      all (.6) 8 *1.0*
    2004-05 one
      side (1.5) 2.0 *2.2*
  XJR
    1995-97
      front (1.9) 2.6 *2.9*
      rear (.9) 1.2 *1.4*
      both (2.1) 2.8 *3.1*
    1998-01
      one (2.3) 3.1 *3.4*
      all (2.4) 3.2 *3.5*
    2002-03
      one (1.1) 1.5 *1.7*
      all (2.4) 3.2 *3.5*
    2004-05 one
      side (5.3) 7.2 *7.4*
  XK8
    4.0L
      one (.3) .... .5 *.7*
      all (.5) .... .7 *.9*
    4.2L
      all (2.1) 2.8 *3.1*
  XKR (3.8) 5.1 *5.4*
  X-Type (3.0) 4.1 *4.4*

JAG-6 S-TYPE : SUPER V8 : VANDEN PLAS : XJ12 : XJ6 : XJ8 : XJR : XJS : XK8 : XKR : X-TYPE

	LABOR TIME	SEVERE SERVICE
Manifold Absolute Pressure (MAP) Sensor, Replace (B)		
2004 S-Type		
V6 (.9)	1.1	1.3
V8 4.2L (2.5)	3.4	3.6
1988-97 XJ6 each	.3	.4
2004 XJ8,		
Vanden Plas (1.3)	1.8	2.0
XJR		
1995-97 each (.2)	.3	.4
2004 (1.6)	2.2	2.4
XJ12 each (.2)	.3	.4
1988-96 XJS (.5)	.7	.8
2004 XK8, XKR (.2)	.3	.4
Test sensor add	.3	.4
Mass Air Flow (MAF) Sensor, Replace (B)		
S-Type (.3)	.5	.5
XJ8, XJR,		
XK8, XKR (.2)	.3	.3
X-Type (.2)	.3	.3
Oxygen Sensor, Replace (B)		
S-Type		
upstream (.4)	.6	.7
downstream (.2)	.3	.4
all (.6)	.8	.9
XJ6		
1981-87 (.5)	.7	.9
1988-94 (.4)	.6	.8
1995-97		
upstream (.5)	.7	.9
downstream (1.0)	1.4	1.6
XJ8, Vanden Plas		
1998-03		
upstream		
one side (.4)	.6	.8
both sides (.5)	.7	.9
downstream		
one side (.2)	.3	.5
all (.8)	1.1	1.3
2004-05		
one side (.2)	.3	.5
all both sides (.8)	1.1	1.3
XJ12		
right side (1.0)	1.4	1.6
left side (.9)	1.2	1.4
both sides (1.2)	1.6	1.8
XJR		
1995-97		
upstream (.5)	.7	.9
downstream (1.0)	1.4	1.6
1998-03		
upstream		
one side (.4)	.6	.8
both (.5)	.7	.9
downstream		
one side (.2)	.3	.5
all (.8)	1.1	1.3

	LABOR TIME	SEVERE SERVICE
2004-05		
one side (.2)	.3	.5
all both sides (.8)	1.1	1.3
XJS		
1988-91		
one (.2)	.3	.5
both (.5)	.7	.9
1992-96		
6 cyl. (.6)	.8	1.0
V12 (.8)	1.1	1.3
XK8		
1997-03		
upstream		
one side (.4)	.6	.8
both sides (.6)	.8	1.0
downstream		
one side (.2)	.3	.5
all both sides (.8)	1.1	1.3
2004-05		
one side (.2)	.3	.5
all both sides (.8)	1.1	1.3
XKR		
2002-03		
sensor one		
right side (1.1)	1.5	1.7
left side (1.3)	1.8	2.0
both sides (1.8)	2.4	2.7
sensor two		
one side (.2)	.3	.5
both sides (.3)	.5	.6
2004-05		
one side (.2)	.3	.5
all both sides (.8)	1.1	1.3
X-Type		
sensor one		
right side (2.1)	2.8	3.0
left side (.3)	.5	.6
both sides (2.4)	3.2	3.5
sensor two		
one side (.3)	.5	.6
both sides (.7)	.9	1.0
Throttle Position Sensor (TPS), Replace (B)		
S-Type V6 (.3)	.5	.7
XJ6		
1988-94 (1.3)	1.8	2.0
1995-97 (.9)	1.2	1.4
XJ12 (1.1)	1.5	1.7
1995-97 XJR (2.7)	3.6	3.9
XJS		
1988-91 (1.8)	2.4	2.7
1992-96		
6 cyl. (2.4)	3.2	3.3
V12 (2.4)	3.2	3.3
Throttle Switch, Replace or Adjust (B)		
1988-94 XJS		
adjust (.1)	.2	.2
replace (.2)	.3	.3

	LABOR TIME	SEVERE SERVICE
Vehicle Speed Sensor, Replace (B)		
XJ12 (1.0)	1.4	1.6
1984-96 XJS (.7)	.9	1.1

FUEL
DELIVERY

	LABOR TIME	SEVERE SERVICE
Fuel Gauge (Dash), Replace (B)		
S-Type (1.4)	1.9	1.9
XJ6		
1981-87 (.6)	.8	.8
1988-97 (.9)	1.2	1.2
XJ8		
Vanden Plas (1.0)	1.4	1.4
1995-97 XJR (1.7)	2.3	2.3
XJ12 (1.7)	2.3	2.3
XJS		
1981-87 (1.9)	2.6	2.8
1988-96 (1.5)	2.0	2.0
XK8, XKR		
X-Type (1.0)	1.4	1.4
Fuel Gauge (Tank), Replace (B)		
S-Type (1.0)	1.4	1.5
XJ6		
1981-87 (1.1)	1.5	1.7
1988-94 (.6)	.8	1.0
XJ8, Vanden Plas		
1982-03 (.3)	.5	.6
2004-05 (1.6)	2.2	2.4
XJR		
1995-97 (1.0)	1.4	1.6
2002-03 (.3)	.5	.6
2004-05 (1.4)	1.9	2.1
XJ12 (1.0)	1.4	1.6
XJS		
1981-89 (1.7)	2.3	2.5
1990-96 (.8)	1.2	1.4
XK8, XKR		
1997-03 (1.1)	1.5	1.7
2004-05 (2.5)	3.4	3.7
Add time to drain and refill tank.		
Fuel Pump, R&R and Recondition (A)		
1981-87 XJ6 (1.3)	1.8	2.1
Fuel Pump, Replace (B)		
S-Type (.8)	1.2	1.5
XJ6		
1981-87 (1.0)	1.4	1.7
1988-90 (1.7)	2.3	2.6
1991-94 (3.6)	4.9	5.2
1995-97 (2.8)	3.8	4.1
XJ8 Vanden Plas		
1982-97 (1.8)	2.4	2.7
1998-03 (2.1)	2.8	3.1
2004-05 (1.8)	2.4	2.7
XJ12		
one (3.3)	4.5	4.8
both (3.6)	4.9	5.2

S-TYPE : SUPER V8 : VANDEN PLAS : XJ12 : XJ6 : XJ8 : XJR : XJS : XK8 : XKR : X-TYPE — JAG-7

	LABOR TIME	SEVERE SERVICE
XJR		
1995-97		
one (3.3)	4.5	4.8
both (3.6)	4.9	5.2
1998-03		
one (2.0)	2.7	3.0
both (2.1)	2.9	3.2
2004-05 (1.5)	2.0	2.3
XJS		
1981-91 (.7)	.9	1.2
1992-96 (3.9)	5.3	5.6
XK8, XKR		
convertible		
exc. 4.2L (1.6)	2.2	2.5
4.2L (2.8)	3.8	4.1
coupe (1.8)	2.4	2.7
X-Type		
2002-03 (5.4)	7.3	7.6
2004-05 (3.2)	4.3	4.6
Fuel Tank, Replace (B)		
Includes: Drain and refill.		
S-Type		
2000-01 (2.2)	3.0	3.4
2002-05 (3.3)	4.5	4.9
XJ6		
1981-87 (2.0)	2.7	3.1
1988-94 (2.3)	3.1	3.4
1995-97 (3.0)	4.1	4.5
XJ8, Vanden Plas		
1982-97 (2.1)	2.8	3.2
1998-03 (2.5)	3.4	3.8
2004-05 (4.5)	6.1	6.5
XJ12 (3.3)	4.5	5.0
XJR		
1995-03 (2.1)	2.8	3.2
2004-05 (2.3)	3.3	3.7
XJS		
main (4.0)	5.4	5.9
auxiliary (1.3)	1.8	2.1
XK8, XKR		
convertible		
1997-01 (1.9)	2.6	3.0
2002-05 (3.2)	4.2	4.5
coupe		
1997-01 (1.5)	2.2	2.5
2002-05 (1.9)	2.6	3.0
X-Type		
2002-03 (5.4)	7.3	7.6
2004-05 (3.3)	4.5	4.9
Intake Manifold and/or Gasket, Replace (B)		
Includes: Adjustments.		
S-Type		
V6		
2000-01 (2.6)	3.5	3.8
2002-05 (2.3)	3.1	3.4
V8 (2.1)	2.8	3.2

	LABOR TIME	SEVERE SERVICE
XJ6		
1981-87 (3.5)	4.7	5.1
1988-94 (1.9)	2.6	2.9
1995-97 (1.7)	2.3	2.6
XJ8, Vanden Plas		
1998-05 (1.5)	2.0	2.3
XJ12		
left side (3.6)	4.9	5.2
right side (4.6)	4.6	4.9
XJR		
1995-97 (2.6)	3.5	3.8
1998-01 (1.2)	1.7	2.0
2002-05 (1.5)	2.0	2.3
1981-96 XJS		
left side (2.6)	3.5	3.8
right side (2.4)	3.2	3.5
XK8, XKR		
1997-01 (1.6)	2.2	2.5
2002-05 (1.9)	2.6	2.9
X-Type		
2002-05 (1.1)	1.5	1.7
Replace manifold XJ8, XJ12, XJR XK8, XKR, add	2.7	3.0
Secondary Fuel Pump Relay, Replace (B)		
1995-96 XJ12 (.7)	.9	.9
INJECTION		
Fuel Pressure Regulator, Adjust (B)		
XJS each side (.9)	1.2	1.4
Auxiliary Air Valve, Replace (B)		
1981-87 XJ6 (.5)	.7	.7
Cold Start Injector, Replace (B)		
1981-87 XJ6 (.9)	1.2	1.2
Electronic Control Unit, Replace (B)		
1981-87 XJ6 (.3)	.5	.5
XJ12 (.7)	.9	.9
XJS		
1981-91 (.3)	.5	.5
1992-96 (.7)	.9	.9
Fuel Cut-Off Switch, Replace (B)		
S-Type	.3	.4
XJ6		
1981-87 (.3)	.5	.5
1988-94 (.6)	.8	.9
1995-97 (1.0)	1.4	1.5
XJ8, Vanden Plas		
1982-03 (.6)	.8	.9
2004-05 (.2)	.3	.3
XJ12 (1.2)	1.6	1.8
XJR		
1995-01 (.7)	.9	1.0
2002-05 (.3)	.5	.6
XJS		
1988-91 (.3)	.5	.6
1992-96 (.7)	.9	1.0
XK8, XKR, X-Type (.3)	.5	.6

	LABOR TIME	SEVERE SERVICE
Fuel Filter, Replace (B)		
S-Type (.6)	.8	.9
XJ6		
1981-87 (.6)	.8	.9
1988-97 (.5)	.7	.8
XJ8, XJR, Vanden Plas		
1982-03 (.4)	.6	.7
2004-05 (.9)	1.1	1.2
XJ12 (.4)	.6	.7
XJS		
1981-91 (.5)	.7	.8
1992-96 (.9)	1.2	1.3
XK8, XKR (.4)	.6	.7
X-Type (.6)	.8	.9
Fuel Injection Relay, Replace (B)		
XJ6		
1981-94 (.3)	.5	.5
1995-97 (.5)	.7	.7
XJ8, Vanden Plas (.4)	.6	.6
1995-01 XJR (.5)	.7	.7
XJS (.3)	.5	.5
XK8, XKR (.5)	.7	.7
Fuel Injectors, Replace (B)		
S-Type		
V6		
2000-01		
one or all (1.6)	2.2	2.5
2002-05		
one or all (2.1)	2.8	3.1
V8		
4.0L, 4.2L		
one (1.0)	1.4	1.6
all (1.6)	2.2	2.4
4.2L SC one or		
all (7.0)	9.4	10.2
XJ6		
1981-87		
one (.9)	1.2	1.4
all (1.6)	2.2	2.4
1988-94		
one (1.2)	1.6	1.8
each addl. add	.2	.3
all (1.5)	2.0	2.2
1995-97		
one (1.5)	2.0	2.2
all (1.8)	2.4	2.6
XJ8, Vanden Plas		
1982-95		
one (.6)	.8	.9
all (1.7)	2.3	2.6
1996-05		
one (1.0)	1.4	1.6
all (1.2)	1.6	1.9
XJ12		
one side (1.7)	2.3	2.5
both sides (2.1)	2.8	3.1

JAG-8 S-TYPE : SUPER V8 : VANDEN PLAS : XJ12 : XJ6 : XJ8 : XJR : XJS : XK8 : XKR : X-TYPE

	LABOR TIME	SEVERE SERVICE
Fuel Injectors, Replace (B)		
XJR		
1995-97		
one (1.5)	2.0	2.2
all (1.8)	2.4	2.6
1998-03		
No. 1 injector (.5)	.7	.9
all others ea. (3.4)	4.6	5.1
all (4.3)	5.8	6.2
2004-05		
one (6.2)	8.4	8.7
all (6.5)	8.8	9.3
XJS		
6 cyl.		
one (1.2)	1.6	1.8
all (1.5)	2.0	2.2
V12		
one (2.1)	2.8	3.0
each addl. add	.3	.3
all (3.2)	4.3	4.6
XK8		
1997-03		
one (.6)	.8	.9
each addl. add	.3	.3
all (1.6)	2.1	2.3
2004-05		
one (.8)	1.2	1.4
each addl. add	.3	.3
all (1.3)	1.8	2.0
XKR		
2000-03		
No. 1 injector (.5)	.7	.9
all others ea. (3.4)	4.6	4.8
all (4.1)	5.5	5.7
2004-05		
one (4.2)	5.6	5.8
each addl. add	.3	.3
all (4.8)	6.5	6.8
X-Type		
one (1.3)	1.8	2.0
each addl. add	.3	.3
all (1.5)	2.0	2.2
Fuel Pressure Regulator, Replace (B)		
XJ12 (.5)	.7	.7
1988-97 XJ6 (.3)	.5	.5
1995-97 XJR (.3)	.5	.5
Fuel Pump Relay, Replace (B)		
1988-97 XJ6 (.6)	.8	.8
XJ8, XJR, Vanden Plas (.2)	.3	.3
1988-96 XJS (.7)	.9	.9
XK8, XKR (.2)	.3	.3
Fuel Rail, Replace (B)		
S-Type		
6 cyl. (2.5)	3.4	3.7
V8		
4.0L, 4.2L one side (1.6)	2.1	2.4
4.2L SC (7.0)	9.4	10.1

	LABOR TIME	SEVERE SERVICE
XJ6		
1988-94 (1.2)	1.6	1.8
1995-97 (1.6)	2.1	2.3
XJ8		
Vanden Plas (1.2)	1.8	2.0
XJ12		
1981-94 (2.6)	3.5	3.8
1995-96		
right side (6.3)	8.5	8.8
left side (6.2)	8.3	8.5
XJR		
1995-03		
one side (4.3)	5.8	6.1
2004-05 (6.5)	8.8	9.1
XJS		
6 cyl. (1.0)	1.4	1.6
V12 each bank (2.4)	3.2	3.4
XK8 (1.7)	2.3	2.5
XKR		
2000-02		
right side (4.6)	6.2	6.4
left side (4.7)	6.4	6.6
2003-05		
one side (5.0)	6.8	7.5
X-Type (1.8)	2.4	2.6
Idle Air Control (IAC) Valve, Replace (B)		
1995-97 XJ6 (1.5)	2.0	2.0
XJ12		
one (2.0)	2.7	2.7
both (2.4)	3.2	3.2
XJR (2.8)	3.8	3.8
Idle Speed Control Actuator, Replace (B)		
1988-94 XJ6 (1.2)	1.6	1.6
Overrun Valve, Replace (B)		
1981-87 XJ6 (.5)	.7	.7
Power Amplifier, Replace (B)		
1981-96 XJS (.5)	.7	.7
Pressure Regulator Valve, Replace (B)		
S-Type (.2)	.3	.3
XJ6		
1981-87 (.7)	.9	.9
1988-97 (.3)	.5	.5
XJS		
6 cyl. (.2)	.3	.3
V12 each side (.5)	.7	.7
Thermo-Time Switch, Replace (B)		
1981-87 XJ6 (.5)	.7	.7
Test switch add	.3	.3
Throttle Body Assy., R&R and Clean or Recondition (A)		
XJ6 (.9)	1.2	1.4
XJS each (.6)	.8	1.0

	LABOR TIME	SEVERE SERVICE
Throttle Body Assy., R&R or Replace (B)		
S-Type		
exc. 4.2L SC (.7)	.9	1.0
4.2L SC (1.1)	1.5	1.7
XJ8, Vanden Plas (.6)	.8	.9
XJR (1.0)	1.4	1.6
XK8, XKR (.5)	.7	.8
2004-05 (.8)	1.2	1.4
X-Type (.3)	.5	.6
Throttle Valve Housing, Replace (B)		
1995-97 XJ6 (3.1)	4.2	4.4
1995-97 XJR (2.3)	3.1	3.3
XJS each (1.2)	1.6	1.8
Throttle Valve Switch, Replace (B)		
1981-87 XJ6 (.5)	.7	.7
1994 XJ12 (.2)	.3	.3
1981-96 XJS (.5)	.7	.7

SUPERCHARGER

	LABOR TIME	SEVERE SERVICE
Supercharger Assy. or Gasket, Replace (B)		
S-Type (5.6)	7.6	8.1
XJR, XK8, Vanden Plas		
1995-97 (5.2)	7.0	7.5
1998-03 (2.8)	3.8	4.2
2004 (5.2)	7.0	7.5
XKR		
2000-03 (3.8)	5.1	5.6
2004-05 (3.3)	4.5	5.0

EXHAUST

	LABOR TIME	SEVERE SERVICE
Catalytic Converter, Replace (B)		
S-Type		
one side (.4)	.6	.7
both sides (.7)	.9	1.0
XJ6		
1981-87 (.9)	1.2	1.4
1988-94		
one (1.0)	1.4	1.5
both (1.5)	2.1	2.2
1995-97		
intermediate (.7)	.9	1.0
down pipe (2.1)	2.8	2.9
XJ8, Vanden Plas		
right or left side (.8)	1.1	1.3
both sides (1.0)	1.4	1.6
XJ12		
one side (1.3)	1.8	2.0
both sides (1.8)	2.4	2.5
XJR		
1995-01		
right or left side (.6)	.8	.9
both sides (1.0)	1.4	1.6
2002-05		
right or left side (1.0)	1.4	1.6
both sides (1.3)	1.8	2.0

S-TYPE : SUPER V8 : VANDEN PLAS : XJ12 : XJ6 : XJ8 : XJR : XJS : XK8 : XKR : X-TYPE   **JAG-9**

	LABOR TIME	SEVERE SERVICE
XJS		
1981-87		
left side (3.6)	4.9	5.1
right side (2.3)	3.1	3.3
1988-91		
exhaust manifold mount		
left side (1.7)	2.3	2.4
right side (1.2)	1.6	1.7
under floor		
each (.5)	.7	.8
1992-96		
6 cyl.(.9)	1.2	1.3
V12		
left (1.5)	2.0	2.1
under floor		
each (.5)	.7	.8
XK8		
1997-01		
right or left		
side (.8)	1.1	1.3
both sides (1.0)	1.4	1.6
2002-05		
right or left		
side (1.0)	1.4	1.6
both sides (1.7)	2.3	2.5
XKR		
2000-01		
right or left		
side (1.0)	1.4	1.6
both sides (1.6)	2.1	2.3
2002-05		
right or left		
side (1.3)	1.7	1.9
both sides (1.9)	2.6	2.8
Exhaust Manifold or Gasket, Replace (B)		
S-Type		
2000-03		
V6		
one side (1.1)	1.5	1.7
both sides (1.6)	2.2	2.4
V8		
right side (1.1)	1.5	1.7
left side (1.3)	1.8	2.1
both sides (1.9)	2.6	2.8
2004-05		
one side (1.3)	1.8	2.0
both sides (2.4)	3.2	3.4
XJ6		
1981-87 (1.0)	1.4	1.6
1988-94		
front (1.5)	2.0	2.2
rear (1.7)	2.3	2.5
both (1.9)	2.6	2.8
1995-97		
front (1.3)	1.8	2.0
rear (2.0)	2.7	2.9

	LABOR TIME	SEVERE SERVICE
XJ8, XJR, Vanden Plas		
1995-03		
right or		
left side (2.4)	3.2	3.4
both sides (3.7)	5.0	5.2
2004-05		
right or		
left side (1.3)	1.8	2.0
both sides (2.3)	3.1	3.3
XJ12		
right side (2.7)	3.6	3.8
left side (2.5)	3.4	3.6
both sides (3.9)	5.3	5.5
XJS		
left side (6.8)	9.2	9.4
right side (6.2)	8.4	8.6
XK8		
1997-03		
one side (2.6)	3.5	3.7
both sides (4.7)	6.3	6.5
2004-05		
right side (1.2)	1.7	1.9
left side (3.0)	4.1	4.3
both sides (3.9)	5.3	5.5
XKR		
2000-05		
right side (1.3)	1.8	2.0
left side (3.4)	4.6	5.1
both sides (4.3)	5.8	6.0
X-Type		
2002-03		
right side (4.7)	6.3	6.5
left side (2.6)	3.5	3.7
both sides (6.3)	8.5	8.7
2004-05		
right side (5.1)	6.9	7.1
left side (2.8)	3.8	4.0
both sides (6.7)	9.0	9.2
Exhaust System (Complete), Replace (B)		
S-Type (.8)	1.0	1.2
XJ6		
1988-97 (3.1)	4.2	4.4
XJ8, Vanden Plas		
1982-97 (1.9)	2.6	2.8
1998-01 (1.7)	2.3	2.4
2002-03 (2.0)	2.7	2.9
2004-05 (1.2)	1.6	1.8
XJR		
1995-97 (3.1)	4.2	4.4
1998-03 (2.4)	3.2	3.4
2004-05 (1.1)	1.6	1.8
XK8		
1997-01 (2.8)	3.8	4.0
2002-05 (2.3)	3.1	3.3
XKR		
2000-03 (3.6)	4.9	5.1
2004-05 (2.3)	3.1	3.3
X-Type (4.5)	6.1	6.3

	LABOR TIME	SEVERE SERVICE
Front Intermediate Pipe, Replace (B)		
1981-87 XJ6 (.6)	.8	1.0
XJS		
1981-87 each		
side (1.8)	2.4	2.6
1988-96 each		
side (.7)	.9	1.1
Front Muffler, Replace (B)		
S-Type (.6)	.8	1.0
XJ8, XJR		
2002-05 (.4)	.6	.8
XK8 each (.7)	.9	1.1
XKR (.4)	.6	.8
Intermediate Pipe and Muffler, Replace (B)		
XJ6		
1981-87 (1.0)	1.4	1.6
1988-97 (.7)	.9	1.1
XJ8, XJR,		
Vanden Plas (.8)	1.1	1.3
XJS		
1981-87 each		
side (1.0)	1.4	1.6
X-Type (.4)	.6	.8
Muffler, Replace (B)		
S-Type (.5)	.7	.9
XJ6		
1981-87		
one (.5)	.7	.8
both (1.2)	1.6	1.8
1988-94		
right or		
left side (1.2)	1.6	1.8
both (1.5)	2.0	2.2
1995-97		
one (.5)	.7	.9
both (.7)	.9	1.1
XJ8, Vanden Plas		
one side (.4)	.6	.8
both sides (.6)	.8	1.0
XJR, XJS		
one side (.3)	.5	.7
both sides (.5)	.7	.9
XK8, XKR		
one side (.5)	.7	.9
both sides (.8)	1.1	1.3
X-Type (.6)	.8	1.0
Muffler and Tailpipe Assy., Replace (B)		
S-Type, Vanden Plas		
one side (.2)	.3	.5
both sides (.3)	.5	.7
XJ6		
1981-87 (1.0)	1.4	1.6
1988-94		
right or left (.6)	.8	1.0
both (1.2)	1.6	1.8
1995-97		
one side (.5)	.7	.9
both sides (1.0)	1.4	1.6

JAG-10 S-TYPE : SUPER V8 : VANDEN PLAS : XJ12 : XJ6 : XJ8 : XJR : XJS : XK8 : XKR : X-TYPE

	LABOR TIME	SEVERE SERVICE
Muffler and Tailpipe Assy., Replace (B)		
XJ8, XJR		
one side (.2)	.3	.5
both sides (.3)	.5	.7
XJS		
6 cyl., V12 each (.7)	.9	1.1
XK8, XKR		
one or both sides (.2)	.3	.5
X-Type		
one side (.4)	.6	.8
both sides (.8)	1.0	1.2
Resonator or Pre-Muffler, Replace (B)		
1988-96 XJS each (.5)	.7	.8
Rear Intermediate Pipe, Replace (B)		
Each		
1981-87 XJ6 (.9)	1.3	1.5
XJS		
6 cyl. (1.2)	1.6	1.7
V12 (.9)	1.3	1.5

ENGINE COOLING

	LABOR TIME	SEVERE SERVICE
Pressure Test Cooling System (B)		
1981-05 (.3)	.4	.4
Auxiliary Cooling Assy. Fan, Replace (B)		
1995-97 XJ6 (.9)	1.3	1.5
XJ8 (.9)	1.3	1.5
XJ12 (1.8)	2.4	2.7
Auxiliary Fan Temperature Switch, Replace (B)		
1994-97 (1.8)	2.4	2.6
Coolant Thermostat, Replace (B)		
S-Type		
V6 (1.6)	2.2	2.4
V8 (1.9)	2.6	2.8
XJ6		
1981-97 (.6)	.8	1.0
XJ8, Vanden Plas		
1998-03 (.9)	1.3	1.5
2004-05 (1.5)	2.0	2.2
XJ12		
one (.4)	.6	.8
both (.9)	1.1	1.3
XJR		
1995-97 (.7)	.9	1.1
1998-03 (.9)	1.1	1.3
2004-05 (1.8)	2.4	2.6
XJS		
one (.6)	.8	1.0
both (1.0)	1.4	1.6
XK8 (1.0)	1.4	1.6
XKR (1.0)	1.7	1.9
X-Type (1.3)	1.8	2.1

	LABOR TIME	SEVERE SERVICE
Electric Cooling Fan Motor, Replace (B)		
S-Type		
2000-01		
V6 one or both (1.6)	2.1	2.3
V8 on or both (2.0)	2.7	2.9
2002-05		
V6 (2.7)	3.6	3.8
V8 (3.1)	4.2	4.4
XJ6		
1981-94 (1.2)	1.7	1.9
1995-97 (1.0)	1.4	1.6
XJ8, Vanden Plas		
1982-03		
one side (.3)	.5	.7
both sides (.6)	.8	1.0
2004-05 (2.1)	2.8	3.0
XJ12 (1.4)	1.9	2.1
XJR		
1995-97 (1.0)	1.4	1.6
1998-03		
one side (.4)	.6	.8
both sides (.6)	.8	1.0
2004-05 (2.7)	3.6	3.8
XJS (1.2)	1.7	1.9
XK8		
1997-03		
one side (.5)	.7	.9
both sides (.8)	1.1	1.3
2004-05 (1.7)	2.3	2.5
XKR		
2000-03 (1.6)	2.1	2.3
2004-05 (2.1)	2.8	3.0
X-Type (2.4)	3.2	3.4
Engine Coolant Temp. Sending Unit, Replace (B)		
S-Type (.2)	.3	.5
XJ6		
1981-87 (.9)	1.2	1.4
1988-94 (.3)	.5	.7
1995-97 (.5)	.7	.9
XJ8, Vanden Plas (.3)	.5	.7
XJR		
1995-97 (.5)	.7	.9
1998-05 (.3)	.5	.7
XJ12, XJS (.5)	.7	.9
XK8, XKR, X-Type (.2)	.3	.5
Fan Blade, Replace (B)		
XJ6 (.7)	.9	1.2
XJS (1.8)	2.4	2.7
Radiator Assy., R&R or Replace (B)		
Includes: Refill with proper coolant mix.		
S-Type		
2000-01		
V6 (2.4)	3.2	3.4
V8 (2.8)	3.8	4.0

	LABOR TIME	SEVERE SERVICE
2002-05		
V6		
AT (4.9)	6.6	6.8
MT (4.1)	5.5	5.7
V8		
4.0L (3.1)	4.2	4.4
4.2L (5.2)	7.0	7.2
4.2L SC (5.5)	7.4	7.6
XJ6		
1981-87 (3.0)	4.1	4.3
1988-94 (1.9)	2.6	2.8
1995-97 (2.3)	3.1	3.3
XJ8, Vanden Plas		
1982-03 (1.9)	2.6	2.8
2004-05 (3.6)	4.9	5.1
XJ12 (2.5)	3.4	3.6
XJR		
1995-01 (1.4)	1.9	2.1
2002-03 (1.7)	2.3	2.5
2004-05 (4.5)	6.1	6.4
XJS		
6 cyl. (2.6)	3.5	3.7
V12 (5.0)	6.8	7.0
XK8		
1997-03 (2.4)	3.2	3.4
2004-05 (3.2)	4.3	4.5
XKR (3.7)	5.0	5.2
X-Type (2.6)	3.5	3.7
Radiator Fan Module, Replace (B)		
2000 S-Type		
V6 (.2)	.3	.5
V8 (.3)	.5	.7
Radiator Fan Motor Relay, Replace (B)		
1994 XJ12 (.9)	1.3	1.5
XJ6, XJS (.6)	.8	1.0
Radiator Hoses, Replace (B)		
Includes: Refill with proper coolant mix.		
S-Type		
2000-01		
upper (.5)	.7	.9
lower (.8)	1.1	1.3
2002-05		
upper (1.5)	2.0	2.2
lower		
4.0L (1.7)	2.3	2.5
4.2L (2.3)	3.1	3.3
XJ6		
1981-87		
upper (.3)	.5	.7
lower (.5)	.7	.9
1988-97		
upper (.6)	.8	1.0
lower (1.0)	1.4	1.6
XJ8, Vanden Plas		
1998-99 one (.7)	.9	1.1
2000-03 one (.8)	1.1	1.3
2004-05		
upper (1.2)	1.8	2.0
lower (1.8)	2.4	2.6

S-TYPE : SUPER V8 : VANDEN PLAS : XJ12 : XJ6 : XJ8 : XJR : XJS : XK8 : XKR : X-TYPE   **JAG-11**

	LABOR TIME	SEVERE SERVICE
XJ12		
upper (1.2)	1.6	1.8
lower (1.5)	2.0	2.2
XJR		
1995-97		
upper (1.2)	1.6	1.8
lower (1.5)	2.0	2.2
1998-03 one (.9) . . .	1.2	1.4
2004-05		
upper (1.3)	1.7	1.9
lower (2.1)	2.8	3.0
XJS		
6 cyl.		
upper (.6)	.8	1.0
lower (1.0)	1.4	1.6
V12		
upper (.3)	.5	.7
lower (1.0)	1.4	1.6
XK8		
1998-99 one (.7)	.9	1.1
2000-04 one (.9) . . .	1.2	1.4
XKR		
2002-03		
upper (1.0)	1.4	1.6
lower (2.3)	3.1	3.33
2004-05		
upper (1.1)	1.5	1.7
lower (.9)	1.2	1.4
X-Type		
upper (2.1)	2.8	3.0
lower (1.3)	1.8	2.0
Temperature Control Assy., Replace (B)		
1981-87 (2.3)	3.1	3.3
Temperature Gauge (Dash), Replace (B)		
XJ6		
1981-94 (.6)	.8	.8
1995-97 (1.7)	2.3	2.3
XK8 (1.0)	1.4	1.4
XJ12 (1.7)	2.3	2.3
XJR		
1995-03 (1.7)	2.3	2.3
2004-05 (.5)	.5	.7
XJS (1.5)	2.0	2.0
Thermostatic Switch, Replace (B)		
XJ6, XJS (1.2)	1.6	1.6
Water Control Valve, Replace (B)		
XJ6		
1981-87 (.9)	1.2	1.4
1988-94 (.9)	1.2	1.4
1995-97 (2.0)	2.7	2.9
XJ8, XK8 (2.8)	3.8	4.0
XJ12		
1994 (2.5)	3.4	3.6
1995-96 (2.0)	2.7	2.9
XJR (2.0)	2.7	2.9

	LABOR TIME	SEVERE SERVICE
XJS		
1981-87 (1.0)	1.4	1.6
1988-96		
6 cyl. (.9)	1.2	1.4
V12 (1.2)	1.6	1.8
Water Pump and/or Gasket, Replace (B)		
Includes: Refill with proper coolant mix.		
S-Type		
2000-01 (1.6)	2.2	2.4
2002-05		
V6 (2.5)	3.4	3.6
V8		
4.0L (1.8)	2.4	2.6
4.2L (2.5)	3.4	3.6
4.2L SC (2.8) . . .	3.8	4.0
XJ6		
1981-87 (1.9)	2.6	2.8
1988-94 (2.3)	3.1	3.3
1995-97 (2.0)	2.7	2.9
XJ8, Vanden Plas		
1982-03 (1.3)	1.8	2.0
2004-05 (1.6)	2.2	2.4
XJ12 (3.2)	4.3	4.5
XJR		
1995-97 (3.3)	4.5	4.7
1998-01 (2.0)	2.7	2.9
2002-05 (1.2)	1.6	1.8
XJS		
6 cyl. (2.1)	2.8	3.0
V12 (4.7)	6.3	6.5
XK8 (1.3)	1.8	2.0
XKR (1.6)	2.2	2.4
X-Type (1.9)	2.6	2.8

ENGINE
ASSEMBLY

Times shown are for OEM assemblies. Time to replace assemblies from aftermarket rebuild may vary.

Engine Assy., R&I (B)
Does not include parts or component transfer.

	LABOR TIME	SEVERE SERVICE
S-Type		
2000-03		
V6 (9.0)	12.2	13.1
V8 (8.5)	11.5	12.4
2004-05		
V6		
AT (13.4)	18.1	19.0
MT (11.5)	15.5	16.5
4.2L (14.5)	19.6	20.3
4.2L SC (15.3) . . .	20.7	21.7
Vanden Plas		
1982-95 (7.6)	10.3	11.2
1996-03 (8.6)	11.6	12.6
2004-05 (8.1)	10.9	11.9

	LABOR TIME	SEVERE SERVICE
XJ6		
1981-87 (8.9)	12.0	13.0
1988-94 (11.4)	15.4	16.4
1995-97 (9.3)	12.6	13.5
XJ8 (7.5)	10.1	11.0
XJ12 (13.8)	18.6	19.6
XJR		
1995-97 (10.7)	14.4	15.3
1998-03 (7.7)	10.4	11.3
2004-05 (8.3)	11.2	12.2
XJS V12 (17.0)	23.0	23.4
XK8 (8.1)	10.9	11.9
X-Type		
AT (14.3)	19.3	20.3
MT (13.8)	18.6	19.6
w/AC V12 add	1.4	1.4
Engine Assy., R&R and Recondition (A)		
Includes: Replacing rings, rod and main bearings, cylinder head reconditioning and engine tune-up.		
XJ6		
1981-87 (25.7)	34.6	34.9
1988-94 (44.0)	59.3	59.6
1995-97 (46.5)	62.6	62.9
XJR (47.8)	64.4	64.7
XJS V12 (73.6)	91.9	92.2
w/AC V12 add	1.4	1.4
Engine Assy. (Complete), Replace (B)		
Includes: Component transfers and engine tune-up.		
S-Type		
2000-01		
V6 (9.1)	12.3	13.3
V8 (8.5)	11.5	12.4
2002-05		
V6		
AT (17.7)	23.9	24.9
MT (15.9)	21.5	22.5
V8		
4.0L (15.7)	21.2	22.2
4.2L (16.9)	22.8	23.9
4.2L SC (18.2) .	24.6	25.5
XJ8, Vanden Plas		
1982-95 (8.5)	11.5	12.5
1996-03 (12.4)	16.7	17.4
2004-05 (16.3)	22.0	22.9
XJ6		
1981-87 (8.9)	12.0	13.0
1988-94 (11.5)	15.5	16.3
1995-97 (9.4)	12.7	13.7
XJ12 (13.8)	18.6	19.6
XJR		
1995-97 (10.7)	14.4	15.3
1998-03 (13.8)	18.6	19.6
2004-05 (16.5)	22.3	23.3
XJS V12 (17.0)	23.0	23.4

S-TYPE : SUPER V8 : VANDEN PLAS : XJ12 : XJ6 : XJ8 : XJR : XJS : XK8 : XKR : X-TYPE

	LABOR TIME	SEVERE SERVICE
Engine Assy. (Complete), Replace (B)		
XK8		
1997-01 (8.8)	12.3	11.9
2002-03		
4.0L (13.9)	18.8	19.8
4.2L (11.5)	15.5	16.5
2004-05 (13.2)	17.8	18.6
XKR		
4.0L (13.0)	17.6	18.6
4.2L (13.2)	17.8	18.6
X-Type		
AT (14.6)	19.7	20.7
MT (13.9)	18.8	19.8
w/AC V12 add	1.4	1.4
Engine Assy. (Short Block), Replace (B)		

Assembly consists of cylinder block, piston assemblies, crankshaft, camshaft, timing chain and gears. Does not include cylinder heads. Operation Includes: R&R engine, transfer necessary parts and all necessary adjustments.

1988-97 XJ6 (16.0)	21.6	22.1
XJ12 (24.9)	33.5	34.1
XJR (17.4)	23.5	23.5
Engine Mounts, Replace (B)		
S-Type		
2000-01		
V6		
one side (.5)	.7	.9
both sides (.6)	.8	1.0
V8		
one side (.4)	.6	.8
both sides (.5)	.7	.9
2002-05		
V6		
one side (2.7)	3.6	4.1
both sides (3.0)	4.1	4.6
V8		
one side (1.0)	1.4	1.6
both sides (1.5)	2.0	2.2
Vanden Plas, XJ8, XJR		
right (1.0)	1.4	1.6
left (2.1)	2.8	3.1
both (2.7)	3.6	4.1
XJ6		
1981-87		
front, both (2.6)	3.5	3.7
rear (1.3)	1.8	2.0
1988-94		
front		
right (1.2)	1.6	1.8
left (1.3)	1.7	1.9
rear (1.8)	2.4	2.6
1995-97		
front (1.7)	2.3	2.5
all (3.9)	5.3	5.5

	LABOR TIME	SEVERE SERVICE
XJ12		
front (1.5)	2.0	2.2
all (3.4)	4.6	4.8
XJS		
6 cyl.		
front (1.2)	1.6	1.8
rear (2.1)	2.8	3.0
V12		
front both (2.1)	2.8	3.0
rear (2.4)	3.2	3.4
XK8		
front all (1.3)	1.7	2.0
rear (3)	.5	.7
X-Type		
right (2.2)	3.0	3.4
left (1.0)	1.4	1.6
both (2.8)	3.8	4.3

CYLINDER HEAD
Compression Test (B)

S-Type		
2000-01		
V6 (1.6)	2.2	2.5
V8 (1.7)	2.3	2.6
2002-05		
V6 (1.1)	1.5	1.7
V8 (1.3)	1.8	2.0
XJ6		
1981-87 (.7)	.9	1.2
1988-94 (.8)	1.1	1.4
1995-97 (1.2)	1.7	2.0
XJ8, Vanden Plas		
1982-01 (1.6)	2.2	2.4
2002-05 (2.0)	2.7	3.0
XJ12 (1.5)	2.0	2.3
XJR		
1995-97 (1.2)	1.7	2.0
1998-01 (1.6)	2.2	2.5
2002-05 (2.0)	2.7	3.0
XJS		
6 cyl. (.7)	.9	1.2
V12 (1.6)	2.2	2.4
XK8 (1.6)	2.2	2.4
X-Type (2.1)	2.8	3.1
Valve Clearance, Adjust (B)		
S-Type		
2002-05 (5.6)	7.6	8.0
XJ6		
1981-87 (3.6)	4.9	5.2
1988-94 (7.1)	9.6	10.1
1995-97 (8.0)	10.8	11.3
XJ8, Vanden Plas		
1996-03 (7.1)	9.6	10.3
2004-05 (6.6)	8.9	9.4
XJR		
1995-97 (8.0)	10.8	11.3
1998-03 (6.3)	8.5	9.0
2004-05 (3.1)	4.2	4.5

	LABOR TIME	SEVERE SERVICE
XJS		
6 cyl. (7.4)	10.0	11.5
V12 (11.3)	15.3	15.8
XK8, XKR		
1997-00 (6.3)	8.5	9.0
2002-03 (7.3)	9.9	10.4
2004-05 (6.8)	9.2	9.7
X-Type (6.3)	8.5	9.0
Cylinder Head, Replace (B)		

Includes: Parts transfer, adjustments.

XJ6		
1981-87 (7.0)	9.4	9.8
1988-94 (6.1)	8.2	8.7
1995-97 (7.2)	9.7	10.2
XJR (8.5)	11.5	11.9
XJS		
6 cyl. (8.5)	11.5	11.9
V12		
left side (14.1)	19.0	19.5
right side (12.1)	16.3	16.6
both sides (21.8)	29.4	29.7
Cylinder Head Gasket, Replace (B)		
S-Type		
V6		
2000-03		
right side (9.0)	12.2	12.5
left side (10.2)	13.8	14.1
both sides (12.5)	16.9	17.2
2004-05		
right side (10.5)	14.2	14.5
left side (10.7)	14.4	14.7
both sides (11.7)	15.8	16.1
V8		
4.0L		
right side (11.9)	16.1	16.4
left side (12.6)	17.0	17.3
both sides (14.1)	19.0	19.3
4.2L		
one side (18.8)	17.0	17.3
both sides (25.7)	34.6	34.9
4.2L SC		
right side (21.2)	28.6	28.9
left side (21.3)	28.7	29.0
both sides (25.9)	34.8	35.1
XJ6		
1981-87 (5.4)	7.3	7.6
1988-94 (6.6)	8.9	9.2
1995-97 (6.1)	8.2	8.5
XJ8		
1998-03		
one side (10.0)	13.5	13.8
both sides (12.6)	17.0	17.3
2004-05		
one side (15.4)	20.8	21.1
both sides (17.2)	23.2	23.5
XJ12		
right side (8.9)	12.0	12.3
left side (11.4)	15.4	15.7

S-TYPE : SUPER V8 : VANDEN PLAS : XJ12 : XJ6 : XJ8 : XJR : XJS : XK8 : XKR : X-TYPE   JAG-13

	LABOR TIME	SEVERE SERVICE
XJR		
1995-97 (7.9)	10.7	11.0
1998-03		
one side (10.5)	14.2	14.5
both sides (13.0)	17.6	17.9
2004-05		
one side (15.4)	20.8	21.1
both sides (16.9)	22.8	23.1
XJS		
6 cyl. (7.2)	9.7	10.0
V12		
right side (11.2)	15.2	15.5
left side (12.6)	17.0	17.3
both sides (18.9)	25.5	25.8
XK8		
1997-03		
one side (10.2)	13.8	14.1
both sides (12.6)	17.0	17.3
2004-05		
right side (12.0)	16.2	16.5
left side (12.3)	16.6	16.9
both sides (16.2)	21.9	22.2
XKR		
2000-03		
one side (11.5)	15.5	15.8
both sides (13.9)	18.8	19.0
2004-05		
one side (13.0)	17.6	17.9
both sides (17.3)	23.4	23.7
X-Type		
left or right side		
AT (20.4)	27.5	27.8
MT (19.1)	25.8	26.1
both		
AT (21.1)	28.5	28.9
MT (20.7)	27.9	28.2
Valve Cover Gasket, Replace (B)		
S-Type		
V6		
2000-01		
right side (1.6)	2.2	2.4
left side (1.0)	1.4	1.6
both sides (2.3)	3.1	3.3
2002-03		
right side (2.5)	3.4	3.6
left side (1.0)	1.4	1.6
both sides (3.3)	4.5	4.7
2004-05		
right side (1.8)	2.4	2.6
left side (.9)	1.2	1.4
both sides (2.6)	3.5	3.7
V8		
4.0L		
right side (1.1)	1.5	1.7
left side (1.4)	1.9	2.1
both sides (2.4)	3.2	3.4

	LABOR TIME	SEVERE SERVICE
4.2L		
right side (1.3)	1.8	2.0
left side (1.8)	2.4	2.6
both sides (3.0)	4.1	4.3
XJ6		
1981-94 (.5)	.7	.9
1995-97 (1.5)	2.0	2.2
XJ8, Vanden Plas		
3.6L (1.5)	2.0	2.2
4.0L		
one side (1.3)	1.8	2.0
both sides (2.3)	3.1	3.4
4.2L		
right side (1.3)	1.8	2.0
left side (1.9)	2.6	2.8
both sides (3.0)	4.1	4.3
XJ12		
right side (3.3)	4.5	4.7
left side (2.8)	3.8	4.0
both sides (4.7)	6.3	6.5
XJR		
1995-97 (1.5)	2.0	2.2
1998-05		
4.0L		
one side (1.3)	1.8	2.0
both sides (2.4)	3.2	3.6
4.2L		
right side (1.1)	1.5	1.7
left side (1.9)	2.6	2.8
both sides (3.2)	4.3	4.5
XJS		
6 cyl. (.5)	.5	.7
V12		
right side (3.2)	4.3	4.5
left side (3.2)	4.3	4.5
both sides (6.2)	8.4	8.6
XK8, XKR		
4.0L		
right side (1.1)	1.5	1.7
left side (1.2)	1.6	1.8
both sides (2.4)	3.2	3.4
4.2L		
right side (1.6)	2.2	2.4
left side (1.8)	2.4	2.6
both sides (3.2)	4.3	4.5
X-Type		
one side (1.9)	2.6	2.8
both sides (3.4)	4.6	4.8
Valve Job (A)		
XJ6		
1981-87 (22.9)	30.9	31.4
1988-94 (29.1)	39.3	39.8
1995-97 (22.9)	30.7	31.2
XJR (24.2)	32.7	33.2
XJS		
6 cyl. (34.1)	46.0	46.5
V12 (45.5)	61.4	61.9

	LABOR TIME	SEVERE SERVICE
Valve Springs and/or Oil Seals, Replace (B)		
S-Type		
V6		
2000-03		
one side (11.8)	15.9	16.5
2004-05		
one side (11.1)	15.0	15.9
V8		
4.0L		
right side (13.1)	17.7	18.6
left side (13.8)	18.6	19.5
4.2L one		
side (17.5)	23.6	24.1
4.2L SC one		
side (20.0)	27.0	27.9
XJ6		
1988-94 (21.7)	29.3	30.2
1995-97 (22.8)	30.8	31.1
XJ8, Vanden Plas		
1998-03		
one side (11.4)	15.4	16.3
both sides (16.5)	22.3	22.6
XJR		
1995-03		
right side (12.2)	16.6	16.9
left side (13.0)	17.6	18.5
XJS		
6 cyl. (14.7)	19.8	20.2
V12 (25.3)	34.2	35.1
XK8		
4.0L		
right side (13.9)	18.8	19.7
left side (12.4)	16.7	17.6
4.2L		
right side (13.5)	18.2	19.1
left side (13.2)	17.8	18.1
XKR		
4.0L		
right side (14.7)	19.8	20.7
left side (12.5)	16.9	17.5
4.2L		
right side (14.5)	19.6	20.0
left side (13.2)	17.8	18.7

CAMSHAFT
Auxiliary Shaft, Replace (B)

	LABOR TIME	SEVERE SERVICE
1988-97 XJ6 (10.5)	14.2	14.5

Cam Followers, Replace (B)

	LABOR TIME	SEVERE SERVICE
XJ6		
1988-94 (5.8)	7.8	8.1
1995-97 (6.4)	8.6	8.9
XJ8, XK8		
one side (8.8)	11.9	12.2
XJ12		
left side (8.3)	11.2	11.5
right side (6.9)	9.3	9.6
XJR (6.4)	8.6	8.9

S-TYPE : SUPER V8 : VANDEN PLAS : XJ12 : XJ6 : XJ8 : XJR : XJS : XK8 : XKR : X-TYPE

	LABOR TIME	SEVERE SERVICE
Cam Followers, Replace (B)		
XJS		
6 cyl. (6.2) 8.4		8.7
V12		
right side (5.4) 7.3		7.6
left side (8.3) 11.2		11.5
both sides (9.7) . . 13.1		13.4
Camshaft, Replace (A)		
S-Type		
2000-01		
V6 one side (8.3) . 11.2		12.5
V8 one side (5.2) . . 7.0		7.9
2002-05		
V6 one side (9.3) . 12.6		13.5
V8 one		
side (10.0) 13.5		14.4
XJ6		
1981-87 (.5) 7		.9
1988-97 (5.4) 7.3		8.2
XJ8		
1998-03		
one side (8.3) 11.2		12.5
2004-05		
one side (12.6) . . . 17.0		17.9
XJ12		
right side (6.9) 9.3		10.4
left side (8.0) 10.8		12.0
XJR		
1995-03 (5.4) 7.3		8.2
2004-05		
one side (12.6) . . . 17.0		17.9
XJS		
6 cyl.		
one (4.0) 5.4		5.9
both (5.6) 7.6		7.9
V12		
right side (5.0) 6.8		7.5
left side (7.9) 10.7		11.8
both sides (9.1) . . 12.3		13.8
XK8, XKR		
1997-01		
one side (8.3) 11.2		12.5
2002-05		
one side (9.8) 13.2		14.8
X-Type		
2002-03		
one side (19.0) . . 25.7		28.6
both sides (20.0) . 27.0		29.8
2004-05		
one side (18.2) . . . 24.6		25.5
Camshaft Cover Gaskets, Replace (B)		
S-Type		
V6		
2000-01		
right side (1.6) . . 2.2		2.4
left side (1.0) . . . 1.4		1.6
both sides (2.3) . 3.1		3.3

	LABOR TIME	SEVERE SERVICE
2002-03		
right side (2.4) . . 3.2		3.4
left side (.9) 1.2		1.4
both sides (3.2) . 4.3		4.5
2004-05		
right side (1.8) . . . 2.4		2.6
left side (.9) 1.2		1.4
both sides (2.6) . 3.5		3.7
V8		
4.0L		
right side (1.1) . . . 1.5		1.7
left side (1.4) . . . 1.9		2.1
both sides (2.4) . 3.2		3.4
4.2L, 4.2SC		
right side (1.3) . . . 1.8		2.0
left side (1.8) . . . 2.4		2.6
both sides (3.0) . 4.1		4.3
XJ6		
1981-94 (.5) 7		.9
1995-97 (1.5) 2.0		2.2
XJ8, Vanden Plas		
4.0L		
one side (1.3) 1.8		2.0
both sides (2.4) . 3.2		3.4
4.2L		
right side (1.1) . . . 1.5		1.7
left side (1.9) 2.6		2.8
both sides (3.2) . . 4.3		4.5
XJ12		
right side (3.3) 4.5		4.7
left side (2.8) 3.8		4.0
both sides (4.7) 6.3		6.5
XJR		
1995-97 (1.5) 2.0		2.2
4.0L		
one side (1.3) 1.8		2.0
both sides (2.4) . 3.2		3.4
4.2L		
right side (1.3) . . . 1.8		2.0
left side (1.9) 2.6		2.8
both sides (3.0) . . 4.1		4.3
XJS		
6 cyl. (.3) 5		.7
V12		
one side (3.2) 4.3		4.6
both sides (6.2) . . 8.4		8.7
XK8, XKR		
right side (1.6) 2.2		2.5
left side (1.8) 2.4		2.6
both sides (3.2) 4.3		4.5
X-Type		
one side (1.7) 2.3		2.5
both sides (3.0) 4.1		4.3
Camshaft Timing Adjustment Solenoid, Replace (B)		
S-Type (1.0) 1.4		1.5
XJ8, XJR, XK8, XKR, Vanden Plas		
one side (.1) 2		.2
both sides (.2) 3		.3

	LABOR TIME	SEVERE SERVICE
Intermediate Shaft Seal, Replace (B)		
1988-97 XJ6 (2.5) 3.4		3.7
Timing Chain or Gear, Replace (B)		
Includes: Oil seal.		
S-Type		
V6		
2000-01		
right side (6.2) . . 8.4		8.9
left side (6.5) . . . 8.8		9.3
both sides (8.2) . 11.1		11.6
2002-03		
right side (9.1) . 12.3		12.8
left side (9.9) . . 13.4		13.9
2004-05		
right side (8.4) . 11.3		11.8
left side (9.0) . . 12.2		12.5
V8 4.0L		
primary		
right side (4.6) . . 6.2		6.7
left side (5.0) . . . 6.8		7.3
secondary		
right side (4.8) . . 6.5		6.8
left side (5.4) . . . 7.3		7.8
all (6.2) 8.4		8.9
4.2L, 4.2SC		
primary (8.9) 12.0		12.5
secondary (9.2) . . 12.4		12.9
Vanden Plas		
primary		
right side (6.3) 8.5		9.0
left side (6.9) 9.3		9.9
secondary		
one side (6.5) 8.8		9.3
all (7.2) 9.7		10.2
XJ6		
1981-87 (14.6) 19.7		20.2
1988-97 (12.8) 17.3		17.9
XJ8		
4.0L		
primary		
right side (5.9) . . 8.0		8.5
left side (6.5) . . . 8.8		9.1
secondary		
one side (6.1) . . . 8.2		8.7
all (6.6) 8.9		9.4
4.2L		
primary		
one side (10.1) . 13.6		14.1
secondary		
one side (10.2) . 13.8		14.3
all (10.5) 14.2		14.7
XJR		
1995-97 (13.5) 18.2		19.3
4.0L		
primary		
right side (6.1) . . 8.2		8.7
left side (6.6) . . . 8.9		9.2
secondary		
one side (6.4) . . . 8.6		9.1
all (6.7) 9.0		9.5

S-TYPE : SUPER V8 : VANDEN PLAS : XJ12 : XJ6 : XJ8 : XJR : XJS : XK8 : XKR : X-TYPE JAG-15

	LABOR TIME	SEVERE SERVICE
4.2L		
primary		
one side (10.2)	13.8	14.1
secondary		
one side (10.3)	13.9	14.2
all (10.5)	14.2	14.7
XJS		
6 cyl.		
upper (9.6)	13.0	13.5
lower (10.7)	14.4	14.9
V12 (31.3)	42.3	44.8
XK8		
4.0L		
primary or secondary		
right side (5.8)	7.8	8.3
left side (6.2)	8.4	8.9
all (6.5)	8.8	9.1
4.2L		
primary		
right side (6.6)	8.9	9.4
left side (6.9)	9.3	9.8
secondary		
one side (7.1)	9.6	9.9
all (7.3)	9.9	10.2
XKR		
primary		
right side (6.6)	8.9	9.4
left side (6.8)	9.2	9.7
secondary		
one side (7.1)	9.6	9.9
all (7.4)	10.0	10.5
X-Type		
AT (19.0)	25.7	26.4
MT (18.4)	24.8	25.1
w/AC 1981-01 add	1.2	1.2
Timing Chain Cover and/or Gasket, Replace (B)		
S-Type		
2000-01 (7.6)	10.3	10.6
2002-03 (8.8)	11.9	12.2
XJ8, Vanden Plas		
1982-03 (5.5)	7.4	8.4
2004-05 (9.3)	12.6	13.6
XJ6		
1981-87 (14.4)	19.4	20.4
1988-97 (9.9)	13.7	14.2
XJR		
1995-97 (9.9)	13.7	14.7
1998-03 (5.8)	7.8	8.1
2004-05 (9.5)	12.8	13.8
XJS		
6 cyl. (9.7)	13.1	14.1
V12 (30.8)	41.6	41.9
XK8		
1997-03 (5.0)	6.8	7.8
2004-05 (5.9)	8.0	9.0
XKR (6.3)	8.5	9.5

	LABOR TIME	SEVERE SERVICE
X-Type		
AT (17.3)	23.4	23.5
MT (16.9)	22.8	23.8
w/AC 1981-01 add	1.2	1.2
Timing Chain Guide, Replace (B)		
2004 S-Type		
one side (9.0)	12.2	12.5
both sides (9.3)	12.6	12.9
1981-87 XJ6 (14.7)	19.8	20.1
XJ8, Vanden Plas		
4.0L (6.1)	8.2	8.5
4.2L (10.2)	13.8	14.1
XK8, XKR (6.9)	9.3	9.6
XJS V12 (25.8)	34.8	35.3
w/AC add	1.2	1.2
Timing Chain Tensioner, Replace (B)		
S-Type		
V6 (9.1)	12.3	12.5
V8		
2000-01		
one side (4.0)	5.4	5.7
both sides (5.4)	7.3	7.6
2002-05		
4.0L		
one side (4.8)	6.5	6.8
both sides (5.4)	7.3	7.6
4.2L, 4.2SC		
one side (8.9)	12.0	12.3
both sides (9.1)	12.3	12.6
XJ6		
1981-87 (14.5)	19.6	19.9
1988-97 (9.7)	13.1	13.4
XJ8, Vanden Plas		
4.0L (5.8)	7.8	8.1
4.2L (9.9)	13.4	13.7
XJR		
4.0L (6.1)	8.2	8.5
4.2L (10.0)	13.5	14.0
XJS		
V12 (25.4)	34.3	34.8
XK8		
4.0L		
one side (5.3)	7.2	7.5
both sides (5.6)	7.6	7.9
4.2L		
one side (6.6)	8.9	9.2
both sides (6.8)	9.2	9.5
XKR		
one side (6.9)	9.3	9.6
both sides (7.1)	9.6	9.9
X-Type		
MT (17.8)	24.0	24.3
AT (18.3)	24.7	25.0
w/AC 1981-01 add	1.2	1.2
Timing Cover, Reseal (B)		
1988-97 XJ6 (9.6)	13.0	13.5
Upper Timing Chain Tensioner, Replace (B)		
1988-97 XJ6 (.5)	.7	1.0

	LABOR TIME	SEVERE SERVICE
CRANK & PISTONS		
Connecting Rod Bearings, Replace (A)		
Includes: Check bearing oil clearance.		
XJ6		
1981-87 (4.8)	6.5	6.8
1988-94 (7.9)	10.7	11.9
1995-97 (8.5)	11.5	11.8
XJ8		
Vanden Plas (12.5)	16.9	17.2
XJR		
1995-97 (8.5)	11.5	11.8
1998-03 (13.2)	17.8	18.1
XJS		
6 cyl (7.6)	10.3	10.6
V12 (11.3)	15.3	15.6
XK8, XKR (10.8)	14.6	14.9
Crankshaft and Main Bearings, Replace (A)		
Includes: Engine R&R.		
XJ6		
1981-87 (24.3)	32.8	33.3
1988-94 (21.1)	28.5	28.8
1995-97 (22.0)	29.7	30.2
XJR (23.5)	31.7	32.2
XJS		
6 cyl. (22.6)	30.5	31.0
V12 (40.9)	55.2	55.7
w/AC add	1.2	1.2
Crankshaft Timing Gear, Replace (B)		
1988-97 XJ6 (11.6)	15.7	16.0
Crankshaft Front Oil Seal, Replace (B)		
S-Type		
V6 (2.8)	3.8	4.1
V8		
4.0L (1.2)	1.6	1.9
4.2L, 4.2SC (4.0)	5.4	5.7
XJ6		
1981-87 (4.3)	5.8	6.1
1988-94 (1.5)	2.0	2.2
1995-97 (1.9)	2.6	2.8
XJ8, Vanden Plas		
1982-03 (1.6)	2.2	2.4
2004-05 (3.4)	4.6	4.9
1994 XJ12 (1.6)	2.2	2.4
XJR		
1995-97 (1.8)	2.4	2.7
2002-03 (1.6)	2.2	2.5
2004-05 (2.6)	3.5	3.8
XJS		
6 cyl. (2.5)	3.4	3.7
V12 (3.2)	4.3	4.8
XK8		
1997-03 (.9)	1.2	1.5
2004-05 (1.9)	2.6	2.9
XKR (2.3)	3.1	3.4
X-Type (1.2)	1.6	1.9

JAG-16 S-TYPE : SUPER V8 : VANDEN PLAS : XJ12 : XJ6 : XJ8 : XJR : XJS : XK8 : XKR : X-TYPE

	LABOR TIME	SEVERE SERVICE
Crankshaft Pulley or Damper, Replace (B)		
S-Type		
2000-01		
V6 (1.1)	1.5	1.7
V8 (2.0)	2.7	2.9
2002-05		
V6 (2.8)	3.8	4.3
V8		
4.0L (1.2)	1.6	1.8
4.2L, 4.2SC (4.2)	5.7	6.3
XJ6		
1981-87 (1.0)	1.4	1.6
1988-94		
pulley (.5)	.7	.9
damper (1.3)	1.8	2.0
1995-97 (1.4)	1.9	2.1
XJ8, Vanden Plas		
1982-97 (1.4)	1.9	2.1
1998-03 (1.6)	2.2	2.4
2004-05 (3.5)	4.7	5.0
XJR		
1995-97 (1.4)	1.9	2.1
1998-05		
4.0L (1.6)	2.2	2.5
4.2L (3.5)	4.7	5.0
XJS		
V12 (2.9)	3.9	4.1
XK8		
1997-01 (.5)	.7	.9
2002-03 (1.0)	1.4	1.6
2004-05 (2.0)	2.7	3.0
XKR		
2000-01 (2.2)	3.0	3.2
2002-05 (2.6)	3.5	3.9
X-Type (1.2)	1.6	1.8
Main Bearings, Replace (A)		
Includes: Check bearing oil clearance.		
XJ6		
1981-87 (5.0)	6.8	7.1
1988-94 (8.8)	11.9	12.4
1995-97 (9.4)	12.7	13.2
XJR (10.7)	14.4	14.9
XJS		
6 cyl. (8.5)	11.5	12.0
V12 (9.8)	13.2	13.7
Piston and Cylinder Liners, Replace (A)		
XJS V12 (35.6)	48.1	48.6
Pistons or Connecting Rods, Replace (A)		
Includes: Ridge reaming, cylinder wall deglazing, installing new rings and rod bearings, engine tune-up.		
S-Type		
2002-03		
V6		
AT (17.9)	24.2	25.2
MT (17.4)	23.5	24.5
V8 4.0L (20.8)	28.1	29.1

	LABOR TIME	SEVERE SERVICE
XJ6		
1981-87 (15.2)	20.5	21.5
1988-94 (18.9)	25.5	26.5
1995-97 (20.1)	27.1	28.1
XJR (21.5)	29.0	30.0
XJS		
V12 (34.1)	46.0	47.0
w/AC add	1.2	1.2
Rear Main Oil Seal, Replace (B)		
S-Type		
V6		
AT (6.5)	8.8	9.8
MT (4.3)	5.8	6.8
V8		
4.0L (4.7)	6.3	7.0
4.2L (6.3)	8.5	9.5
XJ6		
1981-87 (22.2)	30.0	30.6
1988-94 (5.8)	7.8	8.4
1995-97 (6.9)	9.3	9.9
XJ8, XK8		
1997-01 (7.3)	9.9	10.5
2002-03 (5.6)	7.6	7.9
2004-05 (6.0)	8.1	9.1
XJ12 (7.5)	10.1	10.6
XJR		
1995-01 (6.2)	8.4	9.0
2002-03 (5.2)	7.0	7.9
2004-05 (5.8)	7.8	8.1
XJS		
6 cyl.		
AT (6.1)	8.2	8.8
MT (5.8)	7.8	8.4
XKR		
2002-05 (6.3)	8.5	9.6
X-Type		
2002-03		
AT (13.1)	17.7	19.8
MT (12.2)	16.5	18.5
2004-05		
AT (10.3)	13.9	14.9
MT (10.0)	13.5	14.5
ENGINE LUBRICATION		
Engine Oil Cooler, Replace (B)		
S-Type		
V6 (4.3)	5.8	6.1
V8		
4.0L (1.8)	2.4	2.7
4.2L (2.0)	2.7	3.0
4.2L SC (.7)	1.0	1.3
XJ12 (1.3)	1.8	2.1
XJ8		
4.0L (.6)	.8	1.1
4.2L (1.6)	2.2	2.5
XJR		
1995-97 (.3)	.5	.7
1998-03 (.5)	.7	.9
2004-05 (.8)	1.1	1.3

	LABOR TIME	SEVERE SERVICE
XK8 (.4)	.6	.7
X-Type (1.6)	2.2	2.5
Engine Oil Pressure Switch (Sending Unit), Replace (B)		
S-Type		
V6		
2000-03 (.3)	.5	.5
2004-05 (.7)	.9	.9
V8 (.4)	.6	.6
XJ6		
1981-87 (.5)	.7	.7
1988-97 (1.2)	1.6	1.6
XJ8, Vanden Plas		
1982-03 (.1)	.2	.2
2004-05 (.4)	.6	.6
XJR		
1997 (1.2)	1.6	1.8
1998-01 (.2)	.3	.3
2004-05 (.4)	.6	.6
XK8, XKR (.2)	.3	.3
Oil Pan and/or Gasket, Replace (B)		
S-Type		
V6		
2000-01 (2.5)	3.4	3.6
2002-05 (3.4)	4.6	4.8
V8		
2000-01 (1.0)	1.4	1.6
2002-05 (1.2)	1.6	1.8
XJ6		
1981-87 (3.4)	4.6	4.8
1988-97 (6.9)	9.3	9.5
XJ8.		
Vanden Plas (1.0)	1.4	1.6
XJR		
1995-97 (6.9)	9.3	9.5
1998-05 (1.1)	1.5	1.7
XJS		
6 cyl. (6.2)	8.4	8.6
V12 (1.4)	1.9	2.1
XK8, XKR (.8)	1.1	1.3
X-Type (7.3)	9.9	10.1
Oil Pressure Gauge (Dash), Replace (B)		
XJ6		
1981-94 (.6)	.8	.8
1995-97 (1.7)	2.3	2.3
XK8 (1.0)	1.4	1.4
XJ12 (1.7)	2.3	2.3
1995-97 XJR (1.7)	2.3	2.3
XJS (1.5)	2.0	2.0
Oil Pump, Replace (B)		
S-Type		
V6		
2000-01 (8.2)	11.1	11.4
2002-05 (9.9)	13.4	13.7
V8		
4.0L (5.6)	7.6	7.9
4.2L (10.2)	13.8	14.1

S-TYPE : SUPER V8 : VANDEN PLAS : XJ12 : XJ6 : XJ8 : XJR : XJS : XK8 : XKR : X-TYPE — JAG-17

	LABOR TIME	SEVERE SERVICE
XJ6		
1981-87 (4.0)	5.4	5.9
1988-97 (7.3)	9.9	10.2
XJ8,		
Vanden Plas (6.9)	9.3	9.8
XJR		
1995-97 (7.1)	9.6	10.1
1998-03 (7.0)	9.4	9.9
2004-05 (8.1)	10.9	11.2
XJS		
6 cyl.(7.2)	9.7	10.2
V12 (32.0)	43.2	44.0
XK8		
1997-03 (6.5)	8.8	9.1
2004-05		
exc. below (8.2)	11.1	11.4
convertible (8.5)	11.5	11.8
XKR		
4.0L (7.3)	9.9	10.4
4.2L		
exc. below (8.1)	10.9	11.4
convertible (8.5)	11.5	12.0
X-Type		
AT (19.3)	26.1	26.4
MT (18.8)	25.4	25.7

Oil Pump Drive Chain and Damper, Replace (B)
XJ6
  1988-94 (10.0) 13.5 14.1
  1995-97 (10.6) 14.3 14.9
XJR (11.7) 15.8 16.4

Oil Pump Drive Sprocket, Replace (B)
1988-94 XJ6 (7.0) 9.4 9.9

CLUTCH

Bleed Clutch Hydraulic System (B)
All Models (.5) .... .7 .7

Clutch Assy., Replace (B)
1993-96 XJS (4.7) 6.3 6.5
S-Type (3.9) 5.3 5.5
X-Type
  2002-03 (11.9) 16.1 17.0
  2004-05 (8.5) 11.5 11.8

Clutch Master Cylinder, Replace (B)
Add for system bleeding.
1993-96 XJS (.7) 9 1.2
S-Type (.8) 1.1 1.3
X-Type (.6) .... .8 .9

Clutch Release Bearing, Replace (B)
1993-96 XJS (4.5) 6.1 6.4
S-Type (3.9) 5.3 5.6
X-Type
  2002-03 (11.9) 16.1 16.4
  2004-05 (8.5) 11.5 11.8

Clutch Slave Cylinder, Replace (B)
Add for system bleeding.
1993-96 XJS (.5) .... .5 .7
S-Type (3.9) 5.3 5.6
X-Type
  2002-03 (11.9) 16.1 16.4
  2004-05 (8.5) 11.5 11.8

Flywheel, Replace (B)
S-Type (3.7) 5.0 5.7
1993-96 XJS (5.0) 6.8 7.0
X-Type
  2002-03 (12.1) 16.3 16.6
  2004-05 (8.7) 11.7 12.0

MANUAL TRANSMISSION

Rear Oil Seal and/or Bushing, Replace (B)
1993-96 XJS (1.0) 1.4 1.7
S-Type (1.3) 1.8 2.1
X-Type (1.8) 2.4 2.7

Transmission Assy., R&R or Replace (B)
1993-96 XJS (4.5) 6.1 6.4
S-Type (3.9) 5.3 5.6
X-Type
  2002-03 (11.8) 15.9 16.2
  2004-05 (8.4) 11.3 11.6

AUTOMATIC TRANSMISSION

SERVICE TRANSMISSION INSTALLED

Drain and Refill Unit (B)
S-Type
  2000-01 (.8) 1.1 1.2
  2002-05 (1.4) 1.9 2.1
XJ6
  1981-87 (.6) .... .8 .9
  1988-97 (.9) 1.2 1.4
XJ8,
  Vanden Plas (1.0) ... 1.4 1.6
XJ12 (.9) 1.2 1.4
XJR
  1995-2003 (.8) 1.1 1.3
  2004-05 (1.2) 1.6 1.8
XJS (.6) .... .8 .9
XK8, XKR (1.2) 1.6 1.8
X-Type (1.2) 1.6 1.8

Transmission, Service (B)
S-Type
  2000-01 (1.2) 1.6 1.8
  2002-05 (2.2) 3.0 3.2
XJ6
  1988-97 (1.2) 1.6 1.8
XJ8,
  Vanden Plas (1.6) ... 2.2 2.4

	LABOR TIME	SEVERE SERVICE
XJ12 (1.2)	1.6	1.8
XJR (1.0)	1.4	1.6
XJS (2.8)	3.8	4.0
XK8		
1997-03 (1.6)	2.2	2.4
2004-05 (2.1)	2.8	3.0
XKR (1.0)	1.4	1.6

Vacuum Pressure Control, Check & Adjust (B)
1981-87 XJ12,
  XJ6, XJS (.9) 1.2 1.4

Decoder, Replace (B)
XJ6, XJ12
  1990-94 (1.1) 1.5 1.5
  1995-97 (.9) 1.2 1.2

Electronic Control Unit Replace (B)
S-Type (3.0) 4.1 4.1
XJ8, XJR (1.3) 1.8 1.8
XJ12 (1.1) 1.5 1.5
XK8
  1997-03 (.7) .... .9 .9
  2004-05 (2.6) 3.5 3.5
X-Type
  2002-03 (.8) 1.1 1.1
  2004-05 (1.2) 1.6 1.6

Extension Housing Gasket, Replace (B)
Vanden Plas (1.9) 2.6 2.8
XJ6
  1988-89 (2.5) 3.4 3.6
  1990-97 (2.6) 3.5 3.9
XJ8 (1.7) 2.3 2.5
XJ12 (2.8) 3.8 4.3
XJR
  1995-97 (2.8) 3.8 4.3
  1998-01 (1.7) 2.3 2.5
XJS (3.1) 4.2 4.6
XK8, XKR (1.8) 2.4 2.7

Extension Housing Oil Seal, Replace (B)
S-Type (2.7) 3.6 3.9
XJ8, Vanden Plas
  1982-03 (2.1) 2.8 3.1
  2004-05 (3.2) 4.3 4.5
XJ6
  1988-97 (1.7) 2.3 2.5
XJ12 (1.9) 2.6 2.9
XJR
  1995-03 (1.8) 2.4 2.7
XJS (2.8) 3.8 4.0
XK8, XKR
  1997-03 (2.1) 2.8 3.1
  2004-05 (2.7) 3.6 3.9

S-TYPE : SUPER V8 : VANDEN PLAS : XJ12 : XJ6 : XJ8 : XJR : XJS : XK8 : XKR : X-TYPE

	LABOR TIME	SEVERE SERVICE
Oil Pan and/or Gasket, Replace (B)		
S-Type		
2000-01 (1.1)	1.5	1.7
2002-05 (2.2)	3.0	3.2
XJ6		
1981-87 (.6)	.8	.9
1988-97 (.9)	1.2	1.4
XJ8,		
Vanden Plas (1.9)	2.6	2.8
XJ12 (.9)	1.2	1.4
XJR (1.0)	1.4	1.6
XJS (.6)	.8	.9
XK8		
1997-03 (1.6)	2.2	2.4
2004-05 (2.1)	2.8	3.1
XKR (1.0)	1.4	1.6
X-Type (1.0)	1.4	1.6
Parking Pawl, Replace (B)		
1988-89 XJ6 (26)	3.5	3.7
1990-97 XJ6,		
XJ8, XJS (2.8)	3.8	4.0
XJ12, XJR (2.2)	3.0	3.2
Speed Input Sensor, Replace (B)		
S-Type (.2)	.3	.3
XJ8,		
Vanden Plas (1.6)	2.2	2.5
XJ12 (.6)	.8	.9
1995-97 XJR (.2)	.3	.3
XK8, XKR (1.6)	2.2	2.5
Speed Output Sensor, Replace (B)		
S-Type (.2)	.3	.3
XJ8,		
Vanden Plas (1.6)	2.2	2.5
XJ12 (.6)	.8	.9
1995-97 XJR (.2)	.3	.3
XK8, XKR (1.6)	2.2	2.5
Transmission Power Relay, Replace (B)		
1990-94 XJ6,		
XJS (1.0)	1.4	1.6
Transmission Speed Sensor, Replace (B)		
S-Type (.2)	.3	.5
XJ8,		
Vanden Plas (1.9)	2.6	2.8
Vacuum Modulator, Replace (B)		
1981-93 XJS (.5)	.7	.9
Valve Body Assy., R&R and Recondition (A)		
XJR (7.3)	9.9	10.2
Valve Body Assy., Replace (B)		
S-Type (3.0)	4.1	4.4
XJ6		
1981-87 (1.9)	2.6	2.9
1988-89 (1.9)	2.6	2.9
1990-97 (2.5)	3.4	3.7
XJ8,		
Vanden Plas (2.1)	2.8	3.1
XJ12 (3.6)	4.9	5.2

	LABOR TIME	SEVERE SERVICE
XJR		
1995-97 (3.3)	4.5	4.8
1998-03 (2.0)	2.7	3.0
XJS (3.6)	4.9	5.2
XK8 (2.6)	3.5	3.9
XKR (2.3)	3.1	3.4
X-Type		
2002-03 (1.4)	1.9	2.1
2004-05 (1.8)	2.4	2.7

SERVICE TRANSMISSION REMOVED
Transmission R&R included unless otherwise noted.

	LABOR TIME	SEVERE SERVICE
Front Oil Pump, Replace (B)		
XJ6		
1988-89 (6.2)	8.4	8.7
1990-97 (6.6)	8.9	9.2
XJ12 (7.4)	10.0	10.3
XJR		
1995-01 (6.6)	10.0	10.3
2002-05 (4.8)	6.5	6.8
XJS (7.9)	10.7	11.0
Front Pump Oil Seal, Replace (B)		
S-Type		
2000-01 (3.0)	4.1	4.6
2002-03 (6.5)	8.8	9.1
2004-05 (5.8)	7.8	8.1
Vanden Plas (4.5)	6.5	6.8
XJ6		
1988-89 (5.7)	7.7	8.0
1990-97 (6.4)	8.6	8.9
XJ8		
1998-03 (4.7)	6.3	6.6
2004-05 (6.1)	8.2	8.5
XJ12 (7.2)	9.8	10.1
XJR		
1995-97 (7.2)	9.7	10.0
1998-03 (4.8)	6.5	6.8
2004-05 (6.0)	8.1	8.4
XJS		
1981-93 (7.8)	10.5	10.8
1994-96 (6.4)	8.6	9.0
XK8 (5.0)	6.8	7.1
XKR		
2000-01 (5.4)	7.3	7.8
2002-05 (6.1)	8.2	8.7
Torque Converter, Replace (B)		
S-Type		
2000-03 (3.0)	4.1	4.5
2004-05		
exc. 4.2L SC (6.1)	8.2	8.6
4.2L SC (5.8)	7.8	8.2
Vanden Plas (4.9)	6.6	7.0
XJ6		
1988-89 (5.6)	7.6	8.0
1990-97 (6.2)	8.4	8.8
XJ8		
1998-03 (4.9)	6.6	7.0
2004-05 (6.2)	8.4	8.8

	LABOR TIME	SEVERE SERVICE
XJ12 (7.3)	9.9	10.3
XJR		
1995-97 (7.3)	9.9	10.3
1998-03 (4.9)	6.6	7.0
2004-05 (6.0)	8.1	8.5
1994-96 XJS (6.2)	8.4	8.8
1997-05 XK8 (5.1)	6.9	7.3
XKR		
2000-01 (5.4)	7.3	7.7
2002-05 (6.1)	8.2	8.6
X-Type		
2002-03 (12.2)	16.5	16.9
2004-05 (10.5)	14.2	14.7
Torque Converter Drive Plate (Flexplate), Replace (B)		
S-Type		
2000-01 (4.2)	5.7	6.0
2002-05 (6.3)	8.5	8.8
XJ6		
1988-89 (5.6)	7.6	7.9
1990-97 (6.5)	8.8	9.1
XJ8, Vanden Plas		
1982-03 (5.0)	6.8	7.1
2004-05 (6.0)	8.1	8.4
XJ12 (7.2)	9.7	10.0
XJR		
1995-97 (7.2)	9.7	10.0
1998-05 (5.0)	6.8	7.1
XK8		
1997-03 (5.2)	7.0	7.3
2004-05 (6.0)	8.1	8.4
XKR		
2000-01 (5.2)	7.2	7.5
2002-05 (5.8)	7.8	8.1
X-Type		
2002-03 (12.6)	17.0	18.0
2004-05 (10.0)	13.5	13.8
Transmission Assy., R&R and Recondition (A)		
XJ6		
1981-87 (15.2)	20.5	20.8
1988-89 (17.9)	24.2	24.5
1990-97 (18.8)	25.4	25.7
XJS		
1981-87 Borg		
Warner (20.2)	27.3	27.6
1981-93		
GM400 (16.9)	22.8	23.1
Transmission Assy., R&I (B)		
S-Type		
2000-01 (4.2)	5.7	6.0
2002-03 (6.5)	8.8	9.1
2004-05 (5.6)	7.6	7.9
XJ6		
1981-87 (9.7)	13.1	13.4
1988-89 (5.4)	7.3	7.6
1990-97 (6.2)	8.4	8.7
XJ8, Vanden Plas		
1998-03 (5.1)	6.9	7.2
2004-05 (5.6)	7.6	7.9

S-TYPE : SUPER V8 : VANDEN PLAS : XJ12 : XJ6 : XJ8 : XJR : XJS : XK8 : XKR : X-TYPE — JAG-19

	LABOR TIME	SEVERE SERVICE
XJ12 (7.3)	9.9	10.2
XJR		
1995-97 (7.3)	9.9	10.2
1998-05 (5.6)	7.6	7.9
XJS		
1981-87 Borg Warner (14.3)	19.3	19.6
1981-93 GM400 (7.6)	10.2	10.5
XK8 (5.2)	7.0	7.3
XKR		
2000-01 (5.5)	7.4	7.7
2002-03 (8.5)	11.5	11.8
2004-05 (5.8)	7.8	8.1
X-Type		
2002-03 (12.7)	17.1	17.4
2004-05 (10.5)	14.2	14.5

SHIFT LINKAGE

Selector Cable, Adjust (B)

	LABOR TIME	SEVERE SERVICE
S-Type (.3)	.5	.5
XJ6		
1981-87 (.5)	.7	.8
1990-96 (.5)	.7	.8
XJ8		
1998-03 (1.0)	1.4	1.5
2004-05 (.3)	.5	.5
XK8 (1.1)	1.5	1.6
XJS (.5)	.7	.8

Downshift Cable, Replace (B)

1988-89 XJ6 (1.7)	2.3	2.5

Selector Cable, Replace (B)

S-Type (.6)	.8	.9
XJ6		
1981-87 (2.6)	3.5	3.7
1988-89 (1.7)	2.3	2.5
1990-97 (1.0)	1.4	1.6
XJ8, Vanden Plas		
2000-01 (.8)	1.1	1.3
2002-03 (1.2)	1.6	1.8
2004-05 (.6)	.8	1.0
XJ12 (1.5)	2.0	2.2
XK8 (1.2)	1.6	1.8
X-Type (.8)	1.1	1.3

Selector Rod or Cable, Replace (B)

XJS		
6 cyl. (1.9)	2.6	2.8
V12 (2.5)	3.4	3.6

Selector Switch, Replace (B)

1990-97 XJ6, XJS (1.8)	2.4	2.4

Sport Mode Switch, Replace (B)

1990-97 XJ6, XJS (.5)	.7	.7
XK8 (.5)	.7	.7

DRIVELINE

DRIVESHAFT

Differential, Drain and Refill (B)

	LABOR TIME	SEVERE SERVICE
Exc. below (.3)	.5	.7
1981-87 XJ6, XJS (1.3)	1.8	2.0

Driveshaft, R&R or Replace (B)

S-Type (1.4)	1.9	2.1
XJ6		
1981-87 (1.8)	2.4	2.6
1988-94		
3.6L (1.3)	1.8	2.0
4.0L (1.5)	2.0	2.2
1995-97 (1.3)	1.8	2.0
XJ8, Vanden Plas		
1996-03 (1.2)	1.6	1.8
2004-05 (2.0)	2.7	3.0
XJ12 (1.8)	2.4	2.6
XJR (1.0)	1.4	1.6
XJS		
1981-91 (2.9)	3.9	4.1
1992-96		
6 cyl.		
MT (.9)	1.2	1.4
AT (1.3)	1.8	2.0
V12 (2.9)	3.9	4.1
XK8, XKR (1.4)	1.9	2.1
X-Type (1.2)	1.6	1.8

End Cover Gasket, Replace (B)

1981-87 XJ6 (1.1)	1.5	1.7
XJS		
1981-91 (1.2)	1.6	1.8
1992-96 (3.0)	4.1	4.3

Final Drive Assy., R&R and Recondition (A)

XJ6		
1981-87 (13.0)	17.6	17.9
1988-94		
w/Power-lok (17.3)	23.4	23.7
w/o Power-lok (15.7)	21.2	21.5
XJS (16.8)	22.7	23.0

Final Drive Assy., R&R or Replace (B)

S-Type		
2000-01 (2.7)	3.6	3.9
2002-03 (3.3)	4.5	4.8
2004-05 (2.7)	3.6	3.9
XJ6		
1981-87 (8.9)	12.0	12.3
1988-94 (7.1)	9.6	9.9
1995-97 (8.0)	10.8	11.1
XJ8, Vanden Plas		
1996-03 (5.6)	7.6	7.9
2004-05 (2.6)	3.5	3.8
XJ12 (8.0)	10.8	11.1

	LABOR TIME	SEVERE SERVICE
XJR		
1995-97 (8.0)	10.8	11.1
1998-03 (5.6)	7.6	7.9
2004-05 (2.2)	3.0	3.3
XJS (8.8)	11.9	12.2
XK8, XKR (6.5)	8.8	9.1
X-Type		
2002-03 (2.8)	3.8	4.1
2004-05 (2.5)	3.4	3.7

Output Shaft Bearings and/or Oil Seals, Replace (B)

S-Type		
2000-01		
one side (1.0)	1.4	1.7
both sides (1.7)	2.3	2.6
2002-03		
one side (1.9)	2.6	2.9
both sides (2.8)	3.8	4.3
2004-05		
one side (1.3)	1.8	2.1
both sides (2.2)	3.0	3.3
XJ6		
1981-87		
one side (6.2)	8.4	8.7
both sides (11.3)	15.3	15.6
1988-97		
one side (2.2)	3.0	3.3
both sides (3.9)	5.3	5.6
XJ12		
one side (2.2)	3.0	3.3
both sides (3.9)	5.3	5.6
XJR		
1995-03		
one side (2.2)	3.0	3.3
both sides (3.9)	5.3	5.6
2004-05		
one side (1.6)	2.2	2.5
both sides (3.0)	4.1	4.4
XJS		
1981-93		
one side (6.2)	8.4	8.7
both sides (11.3)	15.3	15.6
1994-96		
one side (2.2)	3.0	3.3
both sides (3.9)	5.3	5.6
XK8, XKR		
2002-05		
one side (2.8)	3.8	4.1
both sides (3.9)	5.3	5.6
X-Type		
2002-03		
one side (1.6)	2.2	2.5
both sides (3.0)	4.1	4.4
2004-05		
one side (1.2)	1.6	1.9
both sides (2.2)	3.0	3.3

JAG-20 S-TYPE : SUPER V8 : VANDEN PLAS : XJ12 : XJ6 : XJ8 : XJR : XJS : XK8 : XKR : X-TYPE

	LABOR TIME	SEVERE SERVICE

Pinion Shaft Oil Seal, Replace (B)
S-Type (1.6) 2.2   2.5
Vanden Plas, XJ8, XJR
  1996-03 (1.6) 2.2   2.5
  2004-05 (2.2) 3.0   3.3
XJ6
  1981-87 (1.0) 1.4   1.6
  1988-94
    w/Jurid shaft (1.5) . 2.0   2.2
    w/o Jurid shaft (.9) 1.2   1.4
  1995-97 (1.5) 2.0   2.2
XJ12 (1.5) 2.0   2.2
XJS (1.3) 1.8   2.0
XK8, XKR (1.8) 2.4   2.7
X-Type (2.2) 3.0   3.4

U-Joints, Recondition (B)
XJ6 (1.9) 2.6   2.9
XJS (4.4) 5.8   6.1

HALFSHAFTS

Halfshaft, Replace (B)
S-Type
  one side (1.2) 1.6   1.9
  both sides (1.9) 2.6   2.9
XJ6
  1981-87 each (2.3) .. 3.1   3.4
  1988-94
    one (2.3) 3.1   3.4
    both (3.4) 4.6   4.9
  1995-97
    one (2.4) 3.2   3.5
    both (3.6) 4.9   5.2
XJ8, XJR, Vanden Plas
  1995-97
    one (2.4) 3.2   3.5
    both (3.6) 4.9   5.2
  1998-03
    one (2.5) 3.5   3.8
    both (3.5) 4.7   5.3
  2004-05
    one (1.3) 1.8   2.1
    both (2.1) 2.8   3.1
XJ12
  one (2.4) 3.2   3.5
  both (3.6) 4.9   5.2
XJS
  one (2.0) 2.7   3.0
  both (3.2) 4.3   4.6
XK8, XKR
  one (1.6) 2.2   2.5
  both (2.7) 3.6   3.9
X-Type 2.4   2.7
  front
    left side
      AT (2.4) 3.2   3.6
      MT (1.8) 2.4   2.7
    right side (1.3) 1.8   2.1
  both sides
    AT (3.9) 5.3   5.6
    MT (2.8) 3.8   4.1
  rear
    one (1.6) 2.2   2.5
    both (3.0) 4.1   4.4

BRAKES
ANTI-LOCK

The following operations do not include testing. Add time as required.

Accumulator Assy., Replace (B)
XJ6
  1988-89 (1.2) 1.6   1.8
  1990-94 (.2) 3   .5
1994 XJ12 (.2) 3   .5

Accumulator Pressure Vessel, Replace (B)
1988-89 XJ6 (.7) 9   .9

Accumulator Switch, Replace (B)
1988-89 XJ6
  low charge (1.2) 1.6   1.8
  low pressure (1.2) .. 1.6   1.8

Brake Computer Module, Replace (B)
S-Type
  2000-01 (.3) 5   .7
  2002-05 (1.4) 1.9   2.1
XJ6
  1990-94 (.7) 9   1.1
  1995-97 (2.2) 3.0   3.2
XJ8, XJR,
  Vanden Plas (1.4) .. 1.9   2.1
1989-96 XJS (.6) 8   1.0
XJ12 (2.2) 3.0   3.2
XK8, XKR
  1997-03 (1.0) 1.4   1.6
  2004-05 (1.6) 2.2   2.4

Front Rotor Assy., Replace (B)
XJ8, XJR,
  Vanden Plas (1.2) .. 1.6   1.8
XK8, XKR (1.0) 1.4   1.6

Front Sensor Assy., Replace (B)
S-Type (.3) 5   .7
XJ6 (1.0) 1.4   1.6
XJ8, XJR,
  Vanden Plas (.3) 5   .5
XJ12 (1.0) 1.4   1.6
1989-96 XJS (1.0) 1.4   1.6
XJR
  1995-97 (1.0) 1.4   1.6
  1998-03 (.6) 8   1.0
  2004-05 (.3) 5   .7
XK8, XKR (.2) 3   .5
X-Type (.2) 3   .5

Modulator, Replace (B)
XJ6
  1988-89 (1.5) 2.0   2.2
  1994-97 (1.6) 2.2   2.4
XJ12, XJR (1.6) 2.2   2.4
XK8, XKR (1.0) 1.4   1.6

Over Voltage Protection Relay, Replace (B)
1989-94 XJ6, XJS (.5) .. .7   .9
1994 XJ12 (.5) 7   .9

Power Servo, Replace (B)
1988-89 XJ6 (.7) 9   1.0

Pressure Reducing Valve, Replace (B)
1989-94 XJ6, XJS (.7) .. .9   1.0
1994 XJ12 (.7) 9   1.0

Pressure Switch, Replace (B)
1989-94 XJ6, XJS (.7) .. .9   1.0
1994 XJ12 (.7) 9   1.0

Pump & Motor Assy., Replace (B)
XJ6
  1988-89 (.6) 8   .9
  1990-97 (1.7) 2.3   2.5
1989-96 XJS (1.5) 2.0   2.2
XK8 (1.2) 1.6   1.8

Pump or Main Relay, Replace (B)
1994 XJ12 (.7) 9   .9

Rear Rotor Assy., Replace (B)
XJ8, XJR,
  Vanden Plas (1.1) .. 1.5   1.7
XK8, XKR (.9) 1.2   1.4

Rear Sensor Assy., Replace (B)
S-Type (.3) 5   .7
1990-97 XJ6 (1.1) 1.5   1.7
XJ8, Vanden Plas (.2) .. .3   .5
XJ12 (1.1) 1.5   1.7
XJR
  1995-97 (1.1) 1.5   1.7
  1998-01 (.6) 8   1.0
  2002-05 (.2) 3   .5
1989-96 XJS (1.2) 1.6   1.8
XK8, XKR (.2) 3   .5
X-Type (.6) 8   1.0

Valve Block, Replace (B)
XJ6
  1988-89 (1.5) 2.0   2.2
  1990-97 (2.3) 3.1   3.3
1994 XJ12 (2.5) 3.4   3.6
1989-94 XJS (2.1) 2.8   3.0

SYSTEM

Bleed Brakes (B)
Includes: Add fluid.
w/ABS (.6) 8   1.0
w/o ABS (.4) 6   .8

Brake System, Flush and Refill (B)
1981-05 (1.0) 1.4   1.6

S-TYPE : SUPER V8 : VANDEN PLAS : XJ12 : XJ6 : XJ8 : XJR : XJS : XK8 : XKR : X-TYPE — JAG-21

	LABOR TIME	SEVERE SERVICE
Brake Hose (Flexible), Replace (B)		
Includes: System bleeding.		
One (1.0)	1.4	1.6
each addl. add	.3	.4
Master Cylinder, R&R and Rebuild (A)		
Includes: System bleeding.		
1981-89 (1.4)	1.9	2.1
Master Cylinder, Replace (B)		
Includes: System bleeding.		
S-Type		
2000-01 (.9)	1.2	1.4
2002-05 (1.3)	1.8	2.0
XJ6 (1.3)	1.8	2.0
XJ8		
1998-03 (1.5)	2.0	2.2
2004-05 (1.1)	1.5	1.7
XJR		
1995-01 (1.1)	1.5	1.7
2002-03 (1.5)	2.0	2.2
2004-05 (2.1)	2.8	3.0
XJS, Vanden Plas (1.1)	1.5	1.7
XK8, XKR (1.9)	2.6	2.8
X-Type (1.0)	1.4	1.6
Master Cylinder Reservoir, Replace (B)		
1981-89 XJS (.7)	.9	1.1
S-Type (.9)	1.2	1.4
XJ8		
1998-03 (1.5)	2.0	2.2
2004-05 (.7)	.9	1.1
XJR (1.5)	2.0	2.2
XK8, XKR (1.8)	2.4	2.6
X-Type (.7)	.9	1.0
Power Booster Unit, Replace (B)		
S-Type		
2000-01 (1.3)	1.8	2.0
2002-05 (1.7)	2.3	2.6
XJ6		
1981-87 (2.7)	3.6	3.8
1988-89 (.7)	.9	1.1
XJ8, Vanden Plas		
1982-01 (2.5)	3.4	3.6
2002-03 (1.8)	2.4	2.6
2004-05 (1.5)	2.0	2.2
XJR		
1995-03 (1.8)	2.4	2.6
2004-05 (2.4)	3.2	3.5
XJS		
1981-93 (1.0)	1.4	1.6
1994-96 (3.7)	5.0	5.2
XK8, XKR		
2002-03 (1.6)	2.2	2.4
2004-05 (2.1)	2.8	3.0
X-Type		
AT (.9)	1.2	1.4
MT (1.3)	1.8	2.0

	LABOR TIME	SEVERE SERVICE
Pressure Differential Valve, Replace (B)		
1981-05 (1.0)	1.4	1.6
Vacuum Check Valve, Replace (B)		
1981-05 (.3)	.5	.7
SERVICE BRAKES		
Caliper Assy., R&R and Recondition (A)		
XJ6, XJ12		
1981-87		
front		
one (1.6)	2.2	2.4
both (2.5)	3.4	3.6
rear		
one (2.3)	3.1	3.3
both (4.0)	5.4	5.6
four wheels (6.2)	8.4	8.6
1988-97		
front or rear		
one (1.5)	2.0	2.2
two (2.3)	3.1	3.3
four wheels (4.6)	6.2	6.4
XJS		
1981-93		
front		
one (1.5)	2.0	2.2
both (2.7)	3.6	3.8
rear		
one (2.9)	3.9	4.1
both (5.0)	6.8	7.0
four wheels (7.3)	9.9	10.1
1994-96		
front or rear		
one (1.5)	2.0	2.2
two (2.3)	3.1	3.3
four wheels (4.6)	6.2	6.4
Caliper Assy., Replace (B)		
Includes: System bleeding.		
S-Type		
front		
one (1.0)	1.4	1.6
both (1.4)	1.9	2.1
rear		
2000-01		
one (.9)	1.2	1.4
both (1.2)	1.6	1.8
2002-03		
one (1.6)	2.2	2.4
both (2.1)	2.8	3.0
2004-05		
one (1.0)	1.4	1.6
both (1.5)	2.0	2.2
XJ6		
1981-87		
front		
one (1.2)	1.6	1.8
both (1.9)	2.6	2.8

	LABOR TIME	SEVERE SERVICE
rear		
one (2.1)	2.8	3.0
both (3.3)	4.5	4.7
four wheels (4.7)	6.3	6.5
1988-94		
front or rear		
one (1.1)	1.5	1.7
two (1.7)	2.3	2.5
four wheels (2.5)	3.7	3.9
1995-97		
front or rear		
one (1.3)	1.8	2.0
two (1.9)	2.6	2.8
four wheels (2.8)	4.2	4.4
XJ8, XJR, Vanden Plas		
front		
one (.8)	1.1	1.3
both (1.2)	1.6	1.8
rear		
one (1.0)	1.4	1.6
both (1.4)	1.9	2.1
four wheels (1.9)	2.6	2.8
1995-97		
front		
one (1.5)	2.0	2.2
both (2.1)	2.8	3.0
rear		
one (1.3)	1.8	2.0
both (1.9)	2.6	2.8
four wheels (3.3)	4.5	4.7
1998-05		
front		
one (1.0)	1.4	1.6
both (1.3)	1.8	2.0
rear		
one (1.0)	1.4	1.6
both (1.4)	1.9	2.1
XJ12		
front		
one (1.5)	2.0	2.2
both (2.1)	2.8	3.0
rear		
one (1.3)	1.8	2.0
both (1.9)	2.6	2.8
four wheels (3.3)	4.5	4.7
XJS		
1981-93		
front		
one (1.1)	1.5	1.7
both (2.0)	2.7	2.9
rear		
one (2.5)	3.4	3.6
both (4.3)	5.8	6.0
four wheels (5.8)	7.8	8.0
1994-96		
front or rear		
one (1.1)	1.5	1.7
two (1.7)	2.3	2.5
four wheels (2.7)	3.6	3.8

JAG-22   S-TYPE : SUPER V8 : VANDEN PLAS : XJ12 : XJ6 : XJ8 : XJR : XJS : XK8 : XKR : X-TYPE

	LABOR TIME	SEVERE SERVICE
XK8, XKR		
front		
one (1.0) 1.4		1.6
both (1.3) 1.8		2.0
rear		
one (1.0) 1.4		1.6
both (1.6) 2.2		2.4
four wheels (1.9) . . 2.6		2.8
X-Type (1.8) 2.4		2.6
front		
one (1.0) 1.4		1.6
both (1.4) 1.9		2.1
rear		
one (1.2) 1.6		1.8
both (1.6) 2.2		2.4
four wheels (2.2) . . 3.0		3.2
Disc Brake Pads, Replace (B)		
Front (1.0) 1.4		1.6
Rear		
exc. below (.7) 9		1.1
S-Type (1.3) 1.8		2.0
X-Type (.9) 1.2		1.4
Disc Brake Rotor, Replace (B)		
S-Type		
front		
one (.8) 1.1		1.3
both (1.0) 1.4		1.6
rear		
2000-01		
one (.5) 7		.9
both (.8) 1.1		1.3
2002-05		
one (1.0) 1.4		1.6
both (1.6) 2.2		2.4
2004-05		
one (.6) 8		1.0
both (1.0) 1.4		1.6
four wheels		
2000-01 (1.3) 1.8		2.0
2002-03 (2.4) 3.2		3.4
2004-05 (1.8) 2.4		2.6
XJ6		
1981-87		
front		
one (1.3) 1.8		2.0
both (2.4) 3.2		3.4
rear		
one (3.9) 5.3		5.5
both (7.1) 9.6		9.8
four wheels (9.1) . 12.3		12.5
1988-97		
front		
one (.6) 8		1.0
both (1.2) 1.6		1.8
rear		
one (.9) 1.2		1.4
both (1.5) 2.0		2.2
four wheels (2.8) . . 3.8		4.0

	LABOR TIME	SEVERE SERVICE
XJ8, XJR, Vanden Plas		
1982-97		
front		
one (.6) 8		1.0
both (1.2) 1.6		1.8
rear		
one (.9) 1.2		1.4
both (1.5) 2.0		2.2
four wheels (2.8) . . 3.8		4.0
1998-05		
front		
one (.4) 6		.8
both (.9) 1.2		1.4
rear		
one (.5) 7		.9
both (.9) 1.2		1.4
four wheels (1.6) . . 2.2		2.4
XJ12		
front		
one (.6) 8		1.0
both (1.2) 1.6		1.8
rear		
one (.9) 1.2		1.4
both (1.5) 2.0		2.2
four wheels (2.8) . . 3.8		4.0
XJS		
1981-93		
front		
one (1.3) 1.8		2.0
both (2.4) 3.2		3.4
rear		
one (3.9) 5.3		5.5
both (7.1) 9.6		9.8
four wheels (9.1) . 12.3		12.5
1994-96		
front		
one (.6) 8		1.0
both (1.2) 1.6		1.8
rear		
one (.9) 1.2		1.4
both (1.5) 2.0		2.2
four wheels (2.8) . . 3.8		4.0
XK8, XKR		
front		
one (.6) 8		.9
both (1.0) 1.4		1.6
rear		
one (.4) 6		.8
both (.8) 1.1		1.3
four wheels (1.3) . . 1.8		2.0
X-Type		
front		
one (.6) 8		.9
both (1.2) 1.6		1.8
rear		
one (.8) 1.1		1.3
both (1.8) 2.4		2.6
four wheels (2.2) . . 3.0		3.2

	LABOR TIME	SEVERE SERVICE
PARKING BRAKE		
Parking Brake Cable, Adjust (C)		
1981-05 (.2) 3		.5
Parking Brake Apply Actuator, Replace (B)		
S-Type		
2000-01 (1.4) 1.9		2.1
2002-05 (2.0) 2.7		2.9
XJ6		
1981-87 (1.6) 2.2		2.4
1988-97 (1.2) 1.6		1.8
XJ8, XJR,		
Vanden Plas (.8) 1.1		1.3
XJ12 (1.2) 1.6		1.8
XJS (1.2) 1.6		1.8
XK8, XKR (.4) 6		.8
X-Type (1.9) 2.6		2.9
Parking Brake Cable, Replace (B)		
S-Type		
front lever and		
cable (2.4) 3.2		3.4
rear each (1.0) 1.4		1.6
XJ6		
1981-87 (.9) 1.2		1.4
1988-97		
front (.5) 7		.9
rear each (1.6) . . 2.2		2.4
intermediate (.6) . . .8		1.0
XJ8		
1998-01		
front (.8) 1.1		1.3
rear (1.4) 1.9		2.1
2002-03		
front (.3) 5		.6
rear (1.0) 1.4		1.7
2004-05		
front (.3) 5		.6
vehicle set (.8) . . . 1.1		1.3
XJ12		
front (.6) 8		1.0
rear each (1.6) 2.2		2.4
intermediate (.6) 8		1.1
XJR		
1995-03		
front (.6) 8		1.0
rear each (1.6) . . 2.2		2.4
intermediate (.6) . . .8		1.1
2004-05		
vehicle set (.8) . . . 1.1		1.3
XJS (1.5) 2.0		2.3
XK8, XKR		
front (.8) 1.1		1.3
rear each (1.2) 1.6		1.8
X-Type		
intermediate (1.8) . . 2.4		2.7
rear each (.6) 8		.9
vehicle set (2.6) . . . 3.5		3.7

S-TYPE : SUPER V8 : VANDEN PLAS : XJ12 : XJ6 : XJ8 : XJR : XJS : XK8 : XKR : X-TYPE **JAG-23**

	LABOR TIME	SEVERE SERVICE
Parking Brake Caliper, Replace (B)		
S-Type		
2000-03		
one (.8)	1.1	1.3
both (1.2)	1.6	1.8
2004-05		
one (.3)	.5	.7
both (.8)	1.1	1.3
XJ6		
one (1.2)	1.6	1.8
both (1.5)	2.0	2.2
XJ8, XJR, Vanden Plas		
one (.3)	.5	.7
both (.8)	1.1	1.3
XJS		
one (1.7)	2.3	2.5
both (2.5)	3.4	3.6
Parking Brake Caliper Pads, Replace (B)		
S-Type		
2000-03 (1.4)	1.9	2.1
2004-05 (.8)	1.1	1.3
XJ6, XJS (2.9)	3.9	4.1
XJ8, XJR		
Vanden Plas (.6)	.8	1.0
Parking Brake Shoes, Replace (B)		
XJ6		
1981-87 (1.7)	2.3	2.5
1988-94 (2.5)	3.4	3.6
1995-97 (2.4)	3.2	3.4
XJ8, XJR,		
Vanden Plas (1.6)	2.2	2.4
XJ12 (2.4)	3.2	3.4
XK8, XKR (1.6)	2.2	2.5

FRONT SUSPENSION

Unless otherwise noted, time given does not include alignment.

Align Front End (A)
S-Type		
2002-03 (1.3)	1.8	2.0
2004-05 (.8)	1.1	1.4
XJ6		
1981-93 (1.9)	2.6	2.8
1994-97 (1.5)	2.0	2.2
XJ8, XJR,		
Vanden Plas (1.1)	1.5	1.7
XJS (.9)	1.2	1.4
XK8, XKR		
1998-01 (1.1)	1.5	1.7
2002-05 (.8)	1.1	1.3
X-Type (.8)	1.1	1.3
Front Toe, Adjust (B)		
1981-05 (.8)	1.1	1.3
Lower Ball Joint, Adjust (B)		
1981-87 (1.0)	1.4	1.6

	LABOR TIME	SEVERE SERVICE
Front Axle Hub Oil Seal, Replace (B)		
XJ6		
1981-87		
one side (1.0)	1.4	1.6
both sides (1.7)	2.3	2.5
1988-97		
one side (1.1)	1.5	1.7
both sides (2.1)	2.8	3.0
XJ12, XJR, Vanden Plas		
one side (1.1)	1.5	1.7
both sides (2.1)	2.8	3.0
XJS		
1981-89		
one side (1.0)	1.4	1.6
both sides (1.7)	2.3	2.5
1990-96		
one side (1.3)	1.8	2.0
both sides (2.5)	3.4	3.6
Front Hub, Replace (B)		
S-Type		
one side (.8)	1.1	1.3
both sides (1.2)	1.6	1.8
1988-97 XJ6		
one (1.1)	1.5	1.7
both (2.1)	2.8	3.0
XJ8, Vanden Plas		
1982-01		
one (2.2)	3.0	3.2
both (3.7)	5.0	5.2
2002-03		
one side (2.6)	3.5	3.7
both sides (4.2)	5.7	6.0
2004-05		
one side (.8)	1.1	1.3
both sides (1.2)	1.6	1.8
XJ12		
one (1.1)	1.5	1.7
both (2.1)	2.8	3.0
XJR		
1995-97		
one (1.1)	1.5	1.7
both (2.1)	2.8	3.0
1998-01		
one (2.2)	3.0	3.2
both (3.7)	5.0	5.2
2002-03		
one side (2.6)	3.5	3.7
both sides (4.3)	5.8	6.0
2004-05		
one side (.8)	1.1	1.3
both sides (1.2)	1.6	1.8
XJS		
1981-89		
one (.7)	.9	1.1
both (1.5)	2.0	2.2
1990-96		
one (1.1)	1.5	1.8
both (2.1)	2.8	3.0

	LABOR TIME	SEVERE SERVICE
XK8, XKR		
1997-01		
one side (2.0)	2.7	2.9
both sides (3.3)	4.5	4.6
2000-05		
one side (2.4)	3.2	3.4
both sides (3.9)	5.3	5.5
X-Type		
one side (2.0)	2.7	2.9
both sides (3.0)	4.1	4.3
Front Hub Bearings, Replace (B)		
1988-97 XJ6		
one side (1.2)	1.8	2.0
both sides (2.5)	3.4	3.6
XJ8, Vanden Plas		
1982-95		
one (2.2)	3.0	3.2
both (3.7)	5.0	5.2
1996-03		
one side (3.0)	4.1	4.3
both sides (4.7)	6.3	6.5
XJ12		
one side (1.3)	1.8	2.0
both sides (2.5)	3.4	3.6
XJR		
1995-97		
one side (1.3)	1.8	2.0
both sides (2.5)	3.4	3.6
1998-03		
one side (3.0)	4.1	4.3
both sides (4.9)	6.6	6.8
2004-05		
one side (1.1)	1.5	1.7
both sides (1.8)	2.4	1.6
XJS		
1981-89		
one side (1.0)	1.4	1.6
both sides (1.7)	2.3	2.5
1990-96		
one side (1.3)	1.8	2.0
both sides (2.5)	3.4	3.6
XK8, XKR		
1997-01		
one side (2.0)	2.7	2.9
both sides (3.3)	4.5	4.7
2000-05		
one side (2.5)	3.4	3.6
both sides (3.9)	5.3	5.5
X-Type		
one side (2.1)	2.8	3.0
both sides (3.3)	4.5	4.7
Front Hub Stub Axle, Replace (B)		
XJ6		
1981-87		
one side (2.1)	2.8	3.0
both sides (3.7)	5.0	5.4
1995-97		
one side (2.7)	3.6	4.0
both sides (4.3)	5.8	6.2

JAG-24 S-TYPE : SUPER V8 : VANDEN PLAS : XJ12 : XJ6 : XJ8 : XJR : XJS : XK8 : XKR : X-TYPE

	LABOR TIME	SEVERE SERVICE
Front Hub Stub Axle, Replace (B)		
XJ12		
one side (2.7)	3.6	4.0
both sides (4.3)	5.8	6.2
XJS		
1981-89		
one side (2.1)	2.8	3.0
both sides (3.7)	5.0	5.2
1990-96		
one side (3.2)	4.3	4.6
both sides (5.3)	7.2	7.4
Front Shock Absorbers, Replace (B)		
S-Type		
2000-01		
one (.6)	.8	.9
both (1.1)	1.5	1.7
2002-05		
one side (1.0)	1.4	1.6
both sides (1.6)	2.2	2.4
XJ6		
1981-87		
one (1.0)	1.4	1.6
both (1.4)	1.9	2.1
1988-94		
one (1.1)	1.5	1.7
both (1.9)	2.6	2.8
1995-97		
right side (1.1)	1.5	1.7
left side (1.2)	1.6	1.8
both sides (2.1)	2.8	3.0
XJ8, Vanden Plas		
one side (.8)	1.1	1.3
both sides (1.4)	1.9	2.1
XJ12		
right side (1.1)	1.5	1.6
left side (1.2)	1.6	1.8
both sides (2.1)	2.8	3.0
XJR		
1995-97		
right side (1.1)	1.5	1.6
left side (1.2)	1.6	1.8
both sides (2.1)	2.8	3.0
1998-05		
one (.8)	1.1	1.3
both (1.4)	1.9	2.1
XJS		
1981-89		
one (.9)	1.2	1.4
both (1.4)	1.9	2.1
1990-96		
one (.5)	.7	.9
both (1.0)	1.4	1.6
XK8, XKR		
one side (1.0)	1.4	1.6
both sides (1.8)	2.4	2.6
X-Type		
one side (2.1)	2.8	3.0
both sides (3.2)	4.3	4.6

	LABOR TIME	SEVERE SERVICE
Front Spring, Replace (B)		
S-Type		
2000-01		
one (.6)	.8	1.0
both (.8)	1.1	1.3
2002-05		
one (1.2)	1.6	1.8
both (1.9)	2.6	2.8
XJ6		
1981-87		
one (1.6)	2.2	2.4
both (2.8)	3.8	4.0
1988-94		
one (1.2)	1.6	1.8
both (2.3)	3.1	3.3
1995-97		
one (1.2)	1.6	1.8
both (2.4)	3.2	3.4
XJ8, Vanden Plas		
one (1.0)	1.4	1.6
both (1.6)	2.2	2.5
XJ12		
one (1.2)	1.6	1.8
both (2.4)	3.2	3.4
XJR		
1995-97		
one (1.2)	1.6	1.8
both (2.4)	3.2	3.4
1998-05		
one (1.0)	1.4	1.6
both (1.6)	2.2	2.5
XJS		
1988-89		
one (1.6)	2.2	2.4
both (2.9)	3.9	4.1
1990-96		
one (1.2)	1.6	1.8
both (2.1)	2.8	3.0
XK8, XKR		
one (1.0)	1.4	1.6
both (2.0)	2.7	3.0
X-Type		
one side (2.5)	3.4	3.6
both sides (3.9)	5.3	6.0
Front Sub Frame, Replace (B)		
XJ6, XJ12, XJR (15.2)	20.5	20.8
XJS (19.0)	25.7	26.0
Lower Ball Joint, Replace (B)		
1981-96		
one (1.9)	2.6	2.8
both (2.8)	3.7	3.9
1997-05		
one (1.6)	2.2	2.4
both (2.3)	3.1	3.3

	LABOR TIME	SEVERE SERVICE
Lower Control Arm Assy., Recondition (B)		
XJ6		
1981-87 (5.0)	6.8	7.0
1988-94		
one (3.0)	4.1	4.3
both (4.8)	6.5	6.7
1995-97		
one (3.3)	4.5	4.7
both (5.6)	7.6	7.8
XJ12, both (7.8)	10.5	10.7
XJR		
one (3.3)	4.5	4.7
both (5.6)	7.6	7.8
XJS (9.1)	12.3	12.5
Lower Control Arm Assy., Replace (B)		
S-Type		
2000-01		
one side (1.1)	1.5	1.8
both sides (1.4)	1.9	2.2
2002-05		
forward arm (1.9)	2.6	2.8
lateral control arm (2.1)	2.8	3.0
both lateral arms (2.7)	3.6	3.8
1981-87 XJ6 (4.6)	6.2	6.4
XJ8, XJR, Vanden Plas		
1982-01		
one side (2.2)	3.0	3.2
both sides (3.1)	4.2	4.4
2002-03		
one side (1.8)	2.4	2.6
both sides (2.6)	3.5	3.7
2004-05		
forward arm (1.6)	2.2	2.4
lateral control arm (1.6)	2.2	2.4
both lateral arms (2.1)	2.8	3.0
XJS (8.7)	11.7	12.0
XK8, XKR		
1997-01		
one side (.5)	.7	1.0
both sides (.8)	1.1	1.4
2002-05		
one side (2.1)	2.8	3.0
both sides (2.7)	3.6	3.8
X-Type		
one side (3.9)	5.3	5.5
both sides (4.1)	5.5	5.7
Lower Control Arm Bushings and/or Shaft, Replace (B)		
S-Type		
2000-01		
one side (1.3)	1.8	2.0
both sides (1.9)	2.6	2.8

S-TYPE : SUPER V8 : VANDEN PLAS : XJ12 : XJ6 : XJ8 : XJR : XJS : XK8 : XKR : X-TYPE   **JAG-25**

	LABOR TIME	SEVERE SERVICE
2002-05		
lateral control		
arm (2.5)	3.5	3.7
both arms (3.4)	4.6	5.1
XJ8, XJR, Vanden Plas		
1982-01		
one side (2.4)	3.2	3.4
both sides (3.5)	4.7	4.9
2002-05		
one side (3.9)	5.3	5.5
both sides (6.1)	8.2	8.4
XK8, XKR		
1998-01		
one side (2.0)	2.7	2.9
both sides (2.8)	3.8	4.0
2002-05		
one side (3.3)	4.5	4.7
both sides (4.4)	6.1	6.3
Stabilizer Bar, Replace (B)		
S-Type		
2000-01 (3.3)	4.5	4.7
2002-05		
V6 (4.1)	5.5	5.7
V8 (3.3)	4.5	4.7
XJ6		
1981-87 (1.3)	1.8	2.0
1988-97 (.6)	.8	1.0
XJ8, XJR, Vanden Plas		
1982-03 (.6)	.8	1.0
2004-05 (3.3)	4.5	4.7
XJ12 (.6)	.8	1.0
XJS		
1981-89 (2.1)	2.8	3.0
1990-96 (2.3)	3.1	3.3
XK8, XKR (2.1)	2.8	3.0
X-Type (3.9)	5.3	5.5
Stabilizer Bar Bushings, Replace (B)		
S-Type		
2000-01 (3.0)	4.1	4.3
2002-05		
V6 (4.1)	5.5	5.7
V8		
4.2L (3.5)	4.7	4.9
4.2L SC (4.7)	6.3	7.0
1988-97 XJ6 (.5)	.7	1.1
XJ8, XJR, Vanden Plas		
1982-03 (.3)	.5	.9
2004-05 (3.3)	4.5	4.7
XJ12 (.5)	.7	.9
XJS		
1981-89 (1.0)	1.4	1.6
1990-96 (.5)	.7	.9
XK8, XKR (.8)	1.1	1.3
X-Type (3.9)	5.3	5.5
Stabilizer Bar Links and/or Grommets, Replace (B)		
S-Type		
one side (.3)	.5	.7
both sides (.6)	.8	1.0

	LABOR TIME	SEVERE SERVICE
XJ6		
1981-87 (.6)	.8	1.0
1988-94 (.2)	.3	.5
1995-97 (.3)	.5	.7
XJ8, XJR, Vanden Plas		
1982-03		
one side (.1)	.2	.4
both sides (.2)	.3	.5
2004-05		
one side (.3)	.5	.7
both sides (.6)	.8	1.0
XJ12 (.3)	.5	.7
XJS (.6)	.8	1.0
XK8, XKR		
one side (.2)	.3	.5
both sides (.3)	.5	.7
X-Type		
one side (.3)	.5	.7
both sides (.4)	.6	.8
Steering Knuckle, Replace (B)		
S-Type		
one (1.0)	1.4	1.6
both (2.0)	2.7	2.9
XJ6		
one (2.9)	3.9	4.1
both (4.7)	6.3	6.5
XJ8, XJR, Vanden Plas		
1982-01		
one (2.6)	3.5	3.7
both (4.1)	5.5	5.7
2002-03		
one (3.0)	4.1	4.3
both (4.7)	6.3	6.5
2004-05		
one (1.0)	1.4	1.6
both (2.0)	2.7	2.9
XJ12		
one (2.9)	3.9	4.1
both (4.7)	6.3	6.5
XK8, XKR		
1997-01		
one (2.4)	3.2	3.4
both (3.7)	5.0	5.2
2002-05		
one (3.0)	4.1	4.3
both (4.5)	6.1	6.4
Upper Ball Joint, Replace (B)		
XJ6		
one (1.5)	2.0	2.2
both (2.1)	2.8	3.0
XJ8, Vanden Plas		
one (1.0)	1.4	1.6
both (2.0)	2.7	3.0
XJ12, XJS		
one (1.5)	2.0	2.2
both (2.1)	2.8	3.0
XJR		
1995-97		
one (1.5)	2.0	2.2
both (2.1)	2.8	3.0

	LABOR TIME	SEVERE SERVICE
1996-03		
one (1.0)	1.4	1.6
both (2.0)	2.7	2.9
XK8, XKR		
1997-01		
one (2.0)	2.7	2.9
both (2.8)	3.8	4.0
2002-05		
one (1.6)	2.2	2.5
both (2.4)	3.2	3.6
Upper Control Arm Assy., Replace (B)		
S-Type		
one (1.3)	1.8	2.0
both (1.9)	2.6	2.8
Upper Control Arm (Wishbone), Replace (B)		
S-Type		
right side (2.4)	3.2	3.4
left side (2.7)	3.6	3.8
XJ6		
1981-87 (1.1)	1.5	2.0
1988-94		
right side (2.3)	3.1	3.6
left side (2.0)	2.7	3.2
both sides (3.2)	4.3	4.8
1995-97		
one (2.1)	2.8	3.0
both (3.0)	4.1	4.3
XJ8, Vanden Plas		
one (1.3)	1.8	2.0
both (2.4)	3.2	3.4
XJ12		
one (2.1)	2.8	3.0
both (3.0)	4.1	4.3
XJR		
1995-97		
one (2.1)	2.8	3.0
both (3.0)	4.1	4.3
1998-05		
one (1.3)	1.8	2.0
both (2.4)	3.2	3.4
XJS (2.1)	2.8	3.0
XK8, XKR		
one (1.1)	1.5	1.7
both (1.9)	2.6	2.8

REAR SUSPENSION

Rear Alignment, Adjust (A)
1981-97 (1.6)	2.2	2.4
1998-00 (.8)	1.1	1.3

Air Suspension Solenoid, Replace (B)
1988-93 XJ6		
up (2.1)	2.8	3.0
down (1.5)	2.0	2.2

Height Sensor, Replace (B)
1988-93 XJ6 (1.2)	1.6	1.6

JAG-26 S-TYPE : SUPER V8 : VANDEN PLAS : XJ12 : XJ6 : XJ8 : XJR : XJS : XK8 : XKR : X-TYPE

	LABOR TIME	SEVERE SERVICE
Hydraulic System Hoses, Replace (B)		
1988-93 XJ6 each (.7)	.9	.9
Hydraulic System Reservoir, Replace (B)		
1988-93 XJ6 (.3)	.5	.5
Hub Grease Seal, Replace (B)		
1981-87 XJ6		
one side (1.8)	2.4	2.6
both sides (2.9)	3.9	4.1
XJ8, XJR, Vanden Plas		
one side (1.8)	2.4	2.6
both sides (3.1)	4.2	4.4
XJS		
one side (1.8)	2.4	2.6
both sides (2.9)	3.9	4.1
XK8, XKR		
one side (1.6)	2.2	2.4
both sides (2.9)	3.9	4.1
Pump Assembly, Replace (B)		
1988-93 XJ6 (.5)	.8	1.0
Radius Arm, Replace (B)		
1981-87 XJ6 (1.1)	1.5	1.7
XJS (1.1)	1.5	1.7
Rear Control Arm (Wishbone), Replace (B)		
S-Type		
upper		
one (1.5)	2.0	2.2
both (2.2)	3.0	3.2
lower		
one (.4)	.6	.7
both (1.0)	1.4	1.6
XJ6		
1981-87 (2.1)	2.8	3.0
1988-97		
one (2.5)	3.4	3.6
both (3.1)	4.2	4.4
XJ8, Vanden Plas		
one (2.1)	2.8	3.0
XJ12, XJR		
one (2.5)	3.4	3.6
both (3.1)	4.2	4.4
XJS (2.1)	2.8	3.0
X-Type		
upper		
one (.4)	.6	.7
both (.8)	1.1	1.3
lower		
front (.4)	.6	.8
rear (1.3)	1.8	2.0
Rear Hub Bearing and/or Seal, Replace (B)		
1988-94 XJ6		
one side (2.8)	3.8	4.0
Rear Hub Bearings, Replace (B)		
S-Type		
2000-01		
one side (1.2)	1.6	1.8
both sides (2.4)	3.2	3.4
2002-03		
one side (2.1)	2.8	3.0
both sides (3.1)	4.2	4.4
2004-05		
one side (1.6)	2.2	2.4
both sides (2.5)	3.4	3.6
XJ6		
1981-87		
one side (2.8)	3.8	3.0
both sides (5.3)	7.2	7.4
1995-97		
one side (2.8)	3.8	4.0
both sides (5.5)	7.4	7.6
XJ8, XJR, Vanden Plas		
1982-01		
one side (2.1)	2.8	3.0
both sides (3.7)	5.0	5.7
2002-03		
one side (2.4)	3.2	3.6
both sides (4.2)	5.7	6.3
2004-05		
one side (1.5)	2.0	2.2
both sides (2.5)	3.4	3.6
XJ12		
one side (2.8)	3.8	4.0
both sides (5.5)	7.4	7.6
XJS		
one side (2.8)	3.8	4.0
both sides (5.3)	7.2	8.1
XK8, XKR		
1997-01		
one side (2.1)	2.7	3.0
both sides (3.6)	4.9	5.1
2002-03		
one side (2.4)	3.2	3.4
both sides (4.3)	5.8	6.0
2004-05		
one side (2.4)	3.2	3.4
both sides (4.3)	5.8	6.0
Rear Hub Carrier Bearing, Replace (B)		
S-Type		
one side (1.6)	2.2	2.4
both sides (3.0)	4.1	4.3
XJ6		
1981-87		
one side (1.9)	2.6	2.8
both sides (3.6)	4.9	5.1
1988-97		
one side (2.2)	3.0	3.2
both sides (4.2)	5.7	5.9
XJ8, XJR, Vanden Plas		
one side (1.8)	2.4	2.6
both sides (3.3)	4.5	4.7
XJ12		
one side (2.2)	3.0	3.2
both sides (4.2)	5.7	5.9
XJS		
one side (1.9)	2.6	2.8
both sides (3.6)	4.9	5.1
XK8, XKR		
1997-01		
one side (1.6)	2.2	2.4
both sides (2.8)	3.8	4.0
2002-05		
one side (1.8)	2.4	2.6
both sides (3.4)	4.6	4.8
X-Type		
one side (2.1)	2.8	3.0
both sides (3.2)	4.3	4.5
Rear Strut Shock Absorber, Replace (B)		
1988-93 XJ6		
one (2.1)	2.8	3.0
both (3.1)	4.2	4.4
Rear Spring, Replace (B)		
S-Type		
2000-01		
one (.8)	1.1	1.3
both (1.2)	1.6	1.8
2002-05		
one (1.2)	1.6	1.8
both (2.1)	2.8	3.0
XJ6		
1981-87		
one (1.0)	1.4	1.6
two (1.6)	2.2	2.4
all (2.8)	3.8	4.3
1988-95		
one (1.2)	1.6	1.8
two (1.9)	2.6	2.8
all (3.0)	4.1	4.3
1995-97		
one (1.8)	2.4	2.6
both (3.2)	4.3	4.5
XJ8, Vanden Plas		
1982-01		
one (2.1)	2.8	3.0
both (2.8)	3.8	4.0
2002-05		
one (1.7)	2.3	2.5
both (2.4)	3.2	3.4
XJ12		
one (1.8)	2.4	2.6
both (3.2)	4.3	4.5
XJR		
1995-97		
one (1.8)	2.4	2.6
both (3.2)	4.3	4.5
1998-01		
one (2.0)	2.7	2.9
both (3.1)	4.2	4.3
2002-03		
one (1.7)	2.3	2.5
both (2.4)	3.2	3.4
2004		
air springs		
one (1.2)	1.6	1.8
both (2.1)	2.8	3.0

S-TYPE : SUPER V8 : VANDEN PLAS : XJ12 : XJ6 : XJ8 : XJR : XJS : XK8 : XKR : X-TYPE   **JAG-27**

	LABOR TIME	SEVERE SERVICE		LABOR TIME	SEVERE SERVICE		LABOR TIME	SEVERE SERVICE
XJS			XJ12			XJ12		
one (1.0)	1.4	1.6	one (1.8)	2.4	2.6	1994		
two (1.6)	2.2	2.4	both (3.2)	4.3	4.5	driver side (.3)	.5	.5
all (2.8)	3.8	4.0	XJR			pass. side (2.1)	2.8	2.8
XK8, XKR			1995-97			1995-96		
1997-01			one (1.8)	2.4	2.6	driver side (.7)	.9	.9
one (1.1)	1.5	1.7	both (3.2)	4.3	4.5	pass. side (1.0)	1.4	1.4
both (2.1)	2.8	3.0	1998-01			XJR		
2002-05			one (2.2)	3.0	3.4	1995-97		
one (1.3)	1.8	2.0	both (3.1)	4.2	4.4	driver side (.7)	.9	.9
both (2.4)	3.2	3.4	2002-05			pass. side (1.0)	1.4	1.4
X-Type			one (2.6)	3.5	3.7	1998-05		
one (1.2)	1.6	1.8	both (4.5)	6.1	6.3	driver side (.3)	.5	.5
both (1.6)	2.2	2.4	XJS			pass. side (.5)	.7	.7
Rear Suspension Carrier, Replace (B)			one (1.0)	1.4	1.6	side impact		
			two (1.6)	2.2	2.4	1998-03		
1987-87 XJ6 (8.0)	10.8	11.3	all (2.8)	3.8	4.0	each (.8)	1.1	1.1
XK8 (2.7)	3.6	4.1	XK8, XKR			2004-05 (1.2)	1.6	1.6
XJS (8.0)	10.8	11.3	one (1.3)	1.8	2.0	1990-96 XJS (.2)	.3	.3
Rear Suspension Carrier Mounts, Replace (B)			both (2.5)	3.4	3.6	XK8, XKR		
			X-Type			driver side (.4)	.6	.6
1981-87 XJ6, XJS			one or both (.2)	.3	.5	passenger side (.5)	.7	.7
one side (1.2)	1.6	1.8	**Wheel Hub, Replace (B)**			side impact		
both sides (2.2)	3.0	3.2	1981-87 XJ6			each (.9)	1.2	1.2
Rear Wheel Hub Assy., R&R and Reconditition (B)			one (2.1)	2.8	3.0	X-Type		
			both (3.9)	5.3	5.5	driver side (.3)	.5	.5
1981-87 XJ6, XJS			XJS, XKR, XK8			passenger side (.7)	.9	.9
one side (3.2)	4.3	4.5	one (2.0)	2.7	2.9	side impact		
both sides (6.4)	8.6	8.8	both (3.9)	5.3	5.5	each (.9)	1.2	1.2
Roll Bar, Replace (B)						**Cassette & Cancel Module, Replace (B)**		
XJS (1.2)	1.6	1.8	**STEERING**					
XK8, XKR (.3)	.5	.7	**AIR BAGS**			1995-97 XJ6 (1.5)	2.0	2.0
Roll Bar Bushing, Replace (B)			**Diagnose Air Bag System Component Each (A)**			XJ8, Vanden Plas (.8)	1.1	1.1
XJS (.2)	.3	.5				1995-96 XJ12 (1.9)	2.6	2.6
XK8, XKR (.2)	.4	.6	1995-05 (.8)	1.0	1.0	XJR		
Shock Absorbers or Bushings, Replace (B)			**Air Bag Assy., Replace (B)**			1995-97 (1.5)	2.0	2.0
			S-Type			1998-02 (.8)	1.1	1.1
S-Type			driver side (.4)	.6	.6	XK8, XKR (.8)	1.1	1.1
one (1.0)	1.4	1.6	passenger side			**Front Sensor, Replace (B)**		
both (1.8)	2.4	2.6	2000-01 (3.3)	4.5	4.5	S-Type (.2)	.3	.3
XJ6			2002-05 (.4)	.6	.6	1995-97 XJ6 (1.1)	1.5	1.5
1981-87			side impact			XJ8, Vanden Plas (.3)	.5	.5
one (1.0)	1.4	1.6	each (.8)	1.1	1.3	1995-96 XJ12 (1.1)	1.5	1.5
two (1.6)	2.2	2.4	XJ6			XJR (1.1)	1.5	1.5
all (2.8)	3.8	4.0	1990-94			XK8, XKR		
1988-94			driver side (.3)	.5	.5	1997-03 (.4)	.6	.6
one (2.1)	2.8	3.0	pass. side (2.1)	2.8	2.8	2004-05 primary		
both (3.1)	4.2	4.4	1995-97			sensor (1.2)	1.6	1.6
1995-97			driver side (.7)	.9	.9	X-Type (.2)	.3	.3
one (1.8)	2.4	2.8	pass. side (1.0)	1.4	1.4	**Side Sensor, Replace (B)**		
both (3.2)	4.3	4.7	XJ8, Vanden Plas			S-Type		
XJ8, Vanden Plas			driver side (.3)	.5	.5	2000-03		
1982-01			passenger side (.6)	.8	.8	one side (.8)	1.1	1.1
one (2.2)	3.0	3.4	side impact			both sides (1.3)	1.8	1.8
both (3.1)	4.2	4.4	1990-03 each (.8)	1.1	1.1	2004-05		
2002-05			2004-05 (1.2)	1.6	1.6	one side (.8)	1.1	1.1
one (2.6)	3.5	3.7				both sides (1.0)	1.4	1.4
both (4.5)	6.1	6.3						

S-TYPE : SUPER V8 : VANDEN PLAS : XJ12 : XJ6 : XJ8 : XJR : XJS : XK8 : XKR : X-TYPE

	LABOR TIME	SEVERE SERVICE
Side Sensor, Replace (B)		
XJ8, XJR,		
Vanden Plas (.8)	1.0	1.0
XK8, XKR		
1997-03		
driver side (.3)	.5	.5
passenger side (.2)	.3	.3
2004-05 one (1.2)	1.6	1.6
X-Type (.6)	.8	.8
POWER RACK & PINION		
Troubleshoot Power Steering System (A)		
1981-05 (.8)	1.1	1.1
Control Valve Pinion Seal, Replace (B)		
XJ6 (2.8)	3.8	4.0
XJS (6.2)	8.4	8.6
Inner Tie Rod Assy., Replace (B)		
Includes: Reset toe.		
XJS (1.8)	2.4	2.6
Outer Tie Rod Ends, Replace (B)		
1981-05		
one (1.1)	1.5	1.7
both (1.5)	2.0	2.2
Pressure Switch, Replace (B)		
XJ8, XK8 (.5)	.7	.7
Pump Flow Control Valve, Replace (B)		
1988-94 XJ6 (.2)	.3	.3
Rack & Pinion Assy., R&R and Recondition (A)		
XJ6 (4.7)	6.3	6.9
XJS (8.9)	12.0	12.6
Rack & Pinion Assy., R&R or Replace (B)		
Includes: Adjustments.		
S-Type		
2000-03 (2.1)	2.8	3.0
2004-05 (2.8)	3.8	4.1
XJ6		
1981-87 (2.5)	3.4	3.7
1988-94 (2.2)	3.0	3.3
1995-97 (2.6)	3.5	3.8
XJ8, XJR, Vanden Plas		
1982-03 (1.9)	2.6	2.9
2002-05 (2.6)	3.5	3.7
XJ12		
1994 (2.5)	3.4	3.6
1995-96 (2.8)	3.8	4.0
XJS (6.7)	9.0	9.2
XK8, XKR (2.2)	3.0	3.2
X-Type (4.5)	6.1	6.3
Rack Boots, Replace (B)		
S-Type		
one side (1.0)	1.4	1.6
both sides (1.4)	1.9	2.1
1988-97 XJ6		
one (1.3)	1.8	2.0
both (1.9)	2.6	2.8

	LABOR TIME	SEVERE SERVICE
2002-05 XJ8, Vanden Plas		
one side (.8)	1.1	1.3
both sides (1.0)	1.4	1.6
XJ12		
one side (1.5)	2.0	2.2
both sides (2.1)	2.8	3.0
XJR		
1995-03		
one (1.3)	1.8	2.0
both (1.9)	2.6	2.8
2004-05		
one side (.8)	1.1	1.3
both sides (1.1)	1.5	1.7
XJS		
6 cyl. both (2.1)	2.8	3.0
V12 both (2.4)	3.2	3.4
X-Type		
one (.4)	.6	.7
both (.7)	1.1	1.3
Rack Mounting Bushings, Replace (B)		
S-Type (1.0)	1.3	1.6
XJ8, XJR		
1995-03 (.5)	.7	.9
2004-05 (2.5)	3.4	3.6
XJS (2.5)	3.4	3.6
XK8, XKR (.6)	.8	1.0
Steering Column, Replace (B)		
S-Type		
2002-05		
upper (2.1)	2.8	3.0
lower (1.2)	1.6	1.8
XJ6		
1981-87		
upper (1.9)	2.6	2.8
lower (1.0)	1.4	1.6
1988-94		
upper (1.1)	1.5	1.7
lower (.5)	.7	.9
1995-97		
upper (1.5)	2.0	2.2
lower (.5)	.7	.9
XJ8, XJR, Vanden Plas		
1996-03		
upper (1.2)	1.6	1.8
lower (.4)	.6	.6
2004-05		
upper (1.9)	2.6	2.8
lower (.8)	1.1	1.1
XJ12		
upper (1.5)	2.0	2.2
lower (.3)	.5	.7
XJS		
1981-89		
upper (2.3)	3.1	3.3
lower (4.7)	6.3	6.5
1990-96		
upper (1.7)	2.3	2.5
lower (.3)	.5	.7

	LABOR TIME	SEVERE SERVICE
XK8, XKR		
upper (1.2)	1.6	1.8
lower (.3)	.5	.7
X-Type		
upper (1.0)	1.4	1.6
lower (.2)	.3	.5
Steering Flex Coupling, Replace (B)		
XJ6		
1981-87 (.9)	1.2	1.4
1988-97 (.3)	.5	.7
XJS		
6 cyl. (.3)	.5	.7
V12 (1.0)	1.4	1.6
XJ12, XJR (.3)	.5	.7
Steering Lock Assy., Replace (B)		
S-Type (.8)	1.1	1.1
XJ6 (1.0)	1.3	1.3
XJ8, XJR (.4)	.6	.6
XJS (2.4)	3.2	3.2
XK8, XKR (.5)	.7	.7
X-Type (.6)	.8	.8
Steering Pump, R&R and Recondition (B)		
XJ6		
1981-87 (1.6)	2.2	2.4
1988-94 (2.5)	3.4	3.6
XJS		
6 cyl. (1.8)	2.4	2.8
V12 (2.1)	2.8	3.0
Steering Pump, R&R or Replace (B)		
S-Type		
2000-01 (1.6)	2.2	2.4
2002-05		
V6 (1.8)	2.4	2.6
V8 (2.4)	3.2	3.4
XJ6		
1981-87 (.9)	1.2	1.4
1988-94 (1.6)	2.2	2.4
1995-97 (.5)	.7	1.0
XJ8, XJR, Vanden Plas		
1995-03 (1.5)	2.0	2.2
2004-05 (3.0)	4.1	4.3
XJ12 (1.8)	2.4	2.6
XJS		
6 cyl. (.7)	.9	1.2
V12 (1.5)	2.0	2.3
XK8, XKR		
1997-03 (1.3)	1.8	2.1
2004-05 (2.6)	3.5	3.7
X-Type (1.5)	2.0	2.2
Steering Pump Hoses, Replace (B)		
Does not include purging.		
S-Type pressure		
and return (1.1)	1.5	1.6
XJ6 each (.6)	.8	.9
XJ8, XJR, Vanden Plas		
pressure (.4)	.6	.7
return (.5)	.7	.8
XJ12 each (.6)	.8	.9

S-TYPE : SUPER V8 : VANDEN PLAS : XJ12 : XJ6 : XJ8 : XJR : XJS : XK8 : XKR : X-TYPE   JAG-29

	LABOR TIME	SEVERE SERVICE
XJS		
6 cyl. each (.3)	.5	.6
V12 each (3.7)	5.0	5.2
XK8, XKR each (.4)	.6	.7
X-Type (.8)	1.1	1.3
pressure (3.6)	4.9	5.5
return (4.1)	5.5	6.2
Steering Wheel, Replace (B)		
Exc. below (.5)	.7	.7
2002-05 S-Type (.7)	.9	.9
1988-94 XJ6 (.1)	.2	.2
1988-96 XJS (.3)	.5	.5

HEATING & AIR CONDITIONING

When more than one component requires replacement where evacuation/recovery and recharging is already included, deduct 1.0 hour for each additional component from the time given.

Evacuate/Recover and Recharge System (B)
  1981-05 (1.1) 1.5   *1.5*

AC Hoses, Replace (B)
Includes: Evacuate/recover and recharge.
  S-Type
    compressor suction or
      discharge (1.8) 2.4   *2.6*
    condenser to expansion
      valve (1.6) 2.2   *2.4*
  XJ6
    1988-94
      compressor to
        condenser (1.9) .. 2.6   *2.8*
        evaporator (2.2) .. 3.0   *3.2*
      muffler to
        condenser (1.9) .. 2.6   *2.8*
      receiver drier to
        evaporator (1.9) .. 2.6   *2.8*
    1995-97
      compressor to
        condenser (.9) .. 1.2   *1.4*
        evaporator (.9) .. 1.2   *1.4*
      receiver drier to
        evaporator (1.9) .. 2.6   *2.8*
  XJ8, XJR, Vanden Plas
    condenser to
      receiver (2.1) 2.8   *3.0*
    evap. to union (.8) .. 1.1   *1.3*
    union to
      compressor (1.6) .. 2.2   *2.4*
    discharge (1.5) 2.0   *2.2*
    liquid line
      1982-01 (1.0) 1.4   *1.6*
      2002-05 (1.3) 1.8   *2.0*
    suction line
      2002-05 (1.5) 2.0   *2.2*

	LABOR TIME	SEVERE SERVICE
XJ12		
1994		
one (2.3)	3.1	3.3
each addl. add	.5	.7
1995-96		
compressor to		
condenser (.9)	1.2	1.4
evaporator (.9)	1.2	1.4
receiver drier to		
evaporator (1.8)	2.4	2.6
1988-96 XJS		
compressor to		
condenser (2.1)	2.8	3.0
evaporator (2.0)	2.7	2.9
XK8, XKR		
condenser to		
receiver (.6)	.8	1.0
evap. to union (.6)	.8	1.0
union to		
compressor (1.6)	2.2	2.4
discharge/suction		
1997-01 (.8)	1.1	1.3
2002-05 (.7)	.9	1.1
liquid line		
1997-05 (1.0)	1.4	1.6
jump hose (.7)	.9	1.1
X-Type		
evap. inlet (1.2)	1.6	1.8
evap. outlet (1.2)	1.6	1.8
liquid line (2.4)	3.4	3.6
discharge (1.2)	1.6	1.8
AC Control Module, Replace (B)		
S-Type (.6)	.8	1.0
1988-97 XJ6 (1.0)	1.4	1.6
XJ8, Vanden Plas (.8)	1.1	1.3
XJ12 (1.0)	1.4	1.6
XJR		
1995-97 (1.0)	1.4	1.6
1998-01 (.8)	1.1	1.3
1988-96 XJS (1.2)	1.6	1.8
XK8, XKR (.8)	1.1	1.3
X-Type	.3	.3
AC Electrical Control Module, Replace (B)		
1988-94 XJ6 (1.5)	2.0	2.2
1988-96 XJS w/Mark III		
AC (1.2)	1.6	1.8
AC Switch Control Module, Replace (B)		
1988-94 XJ6 (1.1)	1.5	1.7
Ambient Temperature Sensor, Replace (B)		
S-Type (.5)	.7	.9
XJ6		
1988-94 (2.1)	2.8	3.0
1995-97 (1.3)	1.8	2.0
XJ8, XJR,		
Vanden Plas (.2)	.3	.5
XK8, XKR, X-Type (.2)	.3	.5

	LABOR TIME	SEVERE SERVICE
Auxiliary Cooling Fan Motor, Replace (B)		
1988-96 XJS (1.2)	1.6	1.8
Blower Motor, Replace (B)		
S-Type	.5	.7
XJ6		
1981-87		
right side (1.5)	2.0	2.2
left side (2.3)	3.1	3.3
1988-94		
right side (2.0)	2.7	2.9
left side (1.1)	1.5	1.7
1995-97		
right side (2.2)	3.0	3.2
left side (2.1)	2.8	3.0
XJ8, Vanden Plas		
right side		
1982-01 (3.8)	5.1	5.8
2002-05 (3.5)	4.7	4.9
left side		
1982-01 (3.4)	4.6	4.8
2002-05 (1.5)	2.0	2.2
XJ12		
right side (2.3)	3.1	3.3
left side (2.1)	2.8	3.0
XJR		
1995-97		
right side (2.3)	3.1	3.3
left side (2.1)	2.8	3.0
1998-01		
right side (3.5)	4.7	4.9
left side (1.5)	2.0	2.0
2002-05		
right side (3.8)	5.1	5.3
left side (3.4)	4.6	4.8
XJS each (2.2)	3.0	3.2
XK8, XKR, Vanden Plas		
1997-03		
right side (4.7)	6.3	6.5
left side (1.7)	2.3	2.5
2004-05		
right side (4.3)	5.8	6.0
left side (1.0)	1.4	1.6
X-Type	.3	.3
Blower Motor Relay, Replace (B)		
XJ6		
1981-87 (1.3)	1.8	2.0
1988-94 (.6)	.8	1.0
1995-97 (1.0)	1.4	1.6
XJ12, XJR (1.0)	1.4	1.6
XJS (1.4)	1.9	2.1
XK8, XKR, Vanden Plas each		
1997-01 (.6)	.8	.8
2002-05 (.2)	.3	.3
Blower Motor Resistor, Replace (B)		
1981-87 XJ6 (1.3)	1.8	2.0
XJS		
1981-87 (1.5)	2.0	2.2
1988-96 (.7)	.9	1.1

S-TYPE : SUPER V8 : VANDEN PLAS : XJ12 : XJ6 : XJ8 : XJR : XJS : XK8 : XKR : X-TYPE

	LABOR TIME	SEVERE SERVICE
Center Ventilator Vacuum Actuator, Replace (B)		
XJ6		
1988-94 (2.7)	3.6	3.8
1995-97 (1.7)	2.3	2.5
Combination Temperature & Air Flow Switch, Replace (B)		
1988-96 XJS	.5	.5
Compressor Assy., Replace (B)		
Includes: Parts transfer, evacuate/recover and recharge.		
S-Type		
2000-01 (1.2)	1.6	1.8
2002-05		
V6 (1.6)	2.2	2.4
V8 (2.6)	3.5	3.7
XJ6		
1981-87 (2.5)	3.4	3.6
1988-94 (2.8)	3.8	4.0
1995-97 (1.9)	2.6	2.8
XJ8, XJR, Vanden Plas		
1982-03 (1.9)	2.6	2.8
2004-05 (2.8)	3.8	4.0
XJ12		
1994 (2.8)	3.8	4.0
1995-96 (1.9)	2.6	2.8
XJS		
1981-87 (2.5)	3.4	3.6
1988-96		
6 cyl. (3.0)	4.1	4.3
V12 (2.1)	2.8	3.0
XK8, XKR (1.9)	2.6	2.8
X-Type (1.1)	1.5	1.7
Compressor or Fan Relay, Replace (B)		
Exc. below	.7	.9
XJS		
1988-91	.3	.5
1992-96	.8	1.0
Compressor Shaft Seal, Replace (B)		
Includes: Parts transfer, evacuate/recover and recharge.		
XJ6		
1981-87 (3.2)	4.3	4.5
1988-97 (2.5)	3.4	3.6
XJS		
6 cyl. (2.8)	3.8	4.0
V12 (3.6)	4.9	5.1
Condenser Assy., Replace (B)		
Includes: Evacuate/recover and recharge.		
S-Type		
exc. 4.2L SC (1.2)	1.6	1.8
4.2L SC (3.4)	4.6	4.8
XJ6		
1981-87 (2.7)	3.6	3.8
1988-94 (2.1)	2.8	3.0
1995-97 (1.3)	1.8	2.0

	LABOR TIME	SEVERE SERVICE
XJ8,		
Vanden Plas (2.3)	3.1	3.3
XJ12		
1994 (2.2)	3.0	3.2
1995-96 (1.3)	1.8	2.0
XJR		
1995-97 (1.3)	1.8	2.0
1998-01 (2.3)	3.1	3.3
2002-05 (2.3)	3.1	3.3
XJS		
6 cyl. (2.3)	3.1	3.3
V12 (2.5)	3.4	3.6
XK8, XKR X-Type (2.1)	2.8	3.1
Condenser Cooling Fan Switch, Replace (B)		
1988-94 XJ6 (1.2)	1.6	1.6
1989-96 XJS		
6 cyl. (.7)	.9	.9
V12 (1.2)	1.6	1.8
Defroster Duct Servo, Replace (B)		
1988-94 XJ6 (3.1)	4.2	4.2
1995-97 (1.2)	1.6	1.6
Evaporator Core, Replace (B)		
Includes: Evacuate/recover and recharge.		
S-Type		
2000-01 (3.7)	5.0	5.2
2002-05		
exc. 4.2L SC (5.7)	7.7	7.9
4.2L SC (6.5)	8.8	9.1
XJ6		
1981-87 (9.1)	12.3	12.5
1988-97 (6.9)	9.3	9.5
XJ8,		
Vanden Plas (5.7)	7.7	7.9
XJ12 (6.9)	9.3	9.5
XJR		
1995-97 (6.9)	9.3	9.5
1998-03 (5.7)	7.7	7.9
XJS		
1981-87 (11.9)	16.0	16.2
1988-96		
6 cyl. (10.5)	14.2	14.4
V12 (11.9)	16.0	16.2
XK8, XKR (6.9)	9.3	10.0
X-Type (7.5)	10.1	10.3
Evaporator Temperature Sensor, Replace (B)		
1988-94 XJ6 (.7)	.9	.9
Expansion Valve, Replace (B)		
Includes: Evacuate/recover and recharge.		
2002-05 S-Type		
exc. 4.2L SC (5.9)	8.0	8.2
4.2L SC (6.7)	9.0	9.2
XJ6		
1981-87 (2.3)	3.1	3.3
1988-97 (2.4)	3.2	3.4
XJS (3.2)	4.3	4.5

	LABOR TIME	SEVERE SERVICE
Flap Motor & Gearbox Assy., Replace (B)		
1988-94 XJ6 (1.5)	2.0	2.0
1988-96 XJS		
upper (9.8)	13.2	13.4
lower (1.2)	1.6	1.8
Heater Blower Motor, Replace (B)		
1981-87 (2.2)	3.0	3.2
Heater Blower Motor Relay, Replace (B)		
1981-87 (1.0)	1.4	1.6
Heater Blower Motor Resistor, Replace (B)		
1981-87 (.9)	1.2	1.4
Heater Blower Motor Switch, Replace (B)		
1981-87 (1.3)	1.8	2.0
Heater Core, R&R or Replace (B)		
S-Type		
2000-01 (3.7)	5.0	5.2
2002-05		
exc. 4.2L SC (4.8)	6.5	6.7
4.2L SC (5.5)	7.4	7.6
XJ6		
1981-87 (9.7)	13.1	13.3
1988-94 (2.1)	2.8	3.0
1995-97 (2.7)	3.6	3.8
XJ8, Vanden Plas		
1982-03 (4.1)	5.5	5.7
2004-05 (2.8)	3.8	4.0
XJ12 (2.7)	3.6	3.8
XJR		
1995-97 (2.7)	3.6	3.8
1998-03 (4.1)	5.5	5.7
2004-05 (2.7)	3.6	3.8
XJS		
w/Mark III AC (2.7)	3.6	3.8
w/o Mark III AC (9.7)	13.1	13.3
XK8, XKR (2.3)	3.1	3.3
X-Type (2.2)	3.0	3.2
In-Vehicle Sensor, Replace (B)		
1988-94 XJ6 (1.4)	1.9	2.1
1995-97 (.7)	.9	1.1
1998-04	.3	.3
Lower Servo Feedback Potentiometer, Replace (B)		
1988-94 XJ6 (.7)	.9	1.1
1988-96 XJS (1.2)	1.6	1.8
Power Transistor Module, Replace (B)		
XK8		
1997-03		
right side (5.8)	7.8	7.8
left side (2.1)	2.8	3.0
2004-05		
right side (4.4)	5.9	5.9
left side (1.4)	1.9	1.9

S-TYPE : SUPER V8 : VANDEN PLAS : XJ12 : XJ6 : XJ8 : XJR : XJS : XK8 : XKR : X-TYPE

	LABOR TIME	SEVERE SERVICE
Receiver/Drier Assy., Replace (B)		
Includes: Evacuate/recover and recharge.		
S-Type (.7)	.9	1.1
XJ6		
1981-87 (2.2)	3.0	3.2
1988-94 (2.1)	2.8	3.0
1995-97 (1.1)	1.5	1.7
XJ8, Vanden Plas (.8)	1.1	1.3
XJ12		
1994 (2.1)	2.8	3.0
1995-96 (1.1)	1.5	1.7
XJR (.8)	1.1	1.3
XJS		
1981-87 (1.7)	2.3	2.5
1988-96		
6 cyl. (2.1)	2.8	3.0
V12 (1.7)	2.3	2.5
XK8, XKR (.8)	1.1	1.3
2002-05 X-Type (2.1)	2.8	3.0
Thermostatic Control Switch, Replace (B)		
1988-96 XJS (1.9)	2.6	2.8
Upper Servo Feedback Potentiometer, Replace (B)		
1988-94 XJ6 (.7)	.9	1.1
1988-96 XJS (9.8)	13.2	13.4

WIPERS & SPEEDOMETER

	LABOR TIME	SEVERE SERVICE
Antenna, Replace (B)		
1988-97 XJ6		
mast	.3	.5
motor (1.0)	1.4	1.6
XJ8, Vanden Plas		
2002-05 (.3)	.5	.7
XJ12 (1.0)	1.4	1.6
XJR (1.0)	1.4	1.6
1988-96 XJS		
mast	.3	.5
motor (.7)	.9	1.1
XK8		
1997-01 (.9)	1.2	1.4
2002-05 (.3)	.5	.5
Central Processing, Replace (B)		
1988-97 XJ6 (.7)	.9	1.1
XJ12, XJR (.7)	.9	1.1
XK8 (.8)	1.2	1.4
Instrument Cluster, R&I (B)		
S-Type (.5)	.7	.9
XJ8, XJR,		
Vanden Plas (.4)	.6	.8
XK8, XKR (.8)	1.1	1.3
Radio, R&R (B)		
S-Type	.5	.7
1988-94 XJ6 (1.2)	1.6	1.8
XJ8, XJR		
1997-01 (1.2)	1.6	1.8
2002-05 (.7)	.9	1.1

	LABOR TIME	SEVERE SERVICE
XJ12		
1994 (1.3)	1.8	2.0
1995-97 (1.5)	2.0	2.2
1988-96 XJS (.5)	.7	.7
XK8		
1997-03 (1.5)	2.0	2.0
2004-05 (.7)	.9	.9
X-Type (.8)	1.1	1.1
Speedometer Cable & Casing, Replace (B)		
1981-83 XJS (1.2)	1.6	1.6
Speedometer Cable (Inner), Replace or Lubricate (B)		
1981-83 XJ6 (.5)	.7	.7
1981-83 XJS (.8)	1.2	1.2
Speedometer Head, R&R or Replace (B)		
XJ6		
1981-94 (.5)	.7	.9
1995-97 (1.7)	2.3	2.5
XJ12, XJR (1.7)	2.3	2.5
XJS (1.5)	2.0	2.2
XK8 (1.0)	1.4	1.6
Windshield Washer Pump, Replace (B)		
S-Type (.4)	.6	.6
XJ6		
1981-87 (.3)	.5	.5
1988-97 (1.0)	1.4	1.4
XJ12, XJR (1.2)	1.6	1.6
XJS		
1981-87 (.3)	.5	.5
1988-91	.2	.2
1992-96 (1.2)	1.6	1.8
XK8, XKR, X-Type (.5)	.7	.7
Windshield Washer Reservoir or Sensor, Replace (B)		
S-Type (1.1)	1.5	1.7
1988-97 XJ6 each (1.0)	1.4	1.6
1988-96 XJS		
under hood (.1)	.2	.2
in fender (1.2)	1.6	1.8
XK8, XKR (.8)	1.1	1.2
X-Type (1.5)	2.0	2.2
Windshield Wiper Logic Module, Replace (B)		
1988-97 XJ6 (.6)	.8	.8
XJ12 (.6)	.8	.8
Windshield Wiper & Washer Switch, Replace (B)		
S-Type		
2000-01 (.8)	1.1	1.3
2002-05 (.3)	.5	.5
XJ6		
1981-87 (1.2)	1.6	1.8
1988-94 (.8)	1.2	1.4
1995-97 (1.5)	2.0	2.2
XJ8, Vanden Plas		
1996-03 (1.1)	1.5	1.5
2004-05 (.3)	.5	.7

	LABOR TIME	SEVERE SERVICE
XJ12		
1994 (.3)	.5	.7
1995-96 (1.5)	2.0	2.2
XJR		
1995-97 (1.5)	2.0	2.2
1998-01 (1.1)	1.5	1.7
XJS		
1988-91 (1.2)	1.6	1.8
1992-96 (.8)	1.2	1.4
XK8, XKR (.8)	1.1	1.3
X-Type (.5)	.7	.7
Windshield Wiper Motor, Replace (B)		
S-Type		
2000-01 (1.5)	2.0	2.2
2002-05 (.7)	.9	1.1
XJ6		
1981-87 (1.2)	1.6	1.8
1988-97 (1.3)	1.8	2.0
XJ8, Vanden Plas		
1996-03 (.8)	1.1	1.3
2004-05 (2.4)	3.2	3.4
XJ12 (1.7)	2.3	2.5
XJR		
1995-97 (1.3)	1.8	2.0
1998-03 (.8)	1.1	1.3
2004-05 (2.4)	3.2	3.4
XJS		
1981-91 (.9)	1.2	1.4
1992-96 (1.2)	1.6	1.8
XK8, XKR (.8)	1.2	1.4
X-Type (.5)	.8	.9
Windshield Wiper Motor Relay, Replace (B)		
1981-97 XJ6 (.3)	.5	.5
XK8 (.5)	.7	.7

LAMPS & SWITCHES

	LABOR TIME	SEVERE SERVICE
Back-Up Lamp Bulb, Replace (C)		
1981-05 each (.2)	.3	.3
Back-Up Lamp Switch, Replace (B)		
S-Type (.3)	.5	.5
XJ6, XJR,		
Vanden Plas (.8)	1.1	1.3
XJ12 (.7)	.9	.9
XJS (.6)	.8	.8
XK8, XKR (.3)	.5	.5
Combination Switch Assy., Replace (B)		
S-Type (.5)	.7	.7
XJ6		
1981-87 (1.3)	1.8	2.0
1988-94 (1.0)	1.4	1.6
1995-97 (1.5)	2.0	2.2
XJ8, XJR,		
Vanden Plas (1.1)	1.5	1.7
XJ12 (1.5)	2.0	2.2
XJS (12)	1.6	1.8
XK8, XKR (1.1)	1.5	1.7

JAG-31

JAG-32 S-TYPE : SUPER V8 : VANDEN PLAS : XJ12 : XJ6 : XJ8 : XJR : XJS : XK8 : XKR : X-TYPE

	LABOR TIME	SEVERE SERVICE
Front Bulb Failure Module, Replace (B)		
1988-97 XJ6 (1.0)	1.4	1.6
1995-02 (.3)	.5	.5
Halogen Headlamp Bulb, Replace (B)		
1981-05 (.2)	.3	.4
Hazard Warning Switch, Replace (B)		
XJ6 (.8)	1.2	1.2
XJS (.1)	.2	.3
Headlamp Assy., Replace (B)		
S-Type		
2000-01 (1.2)	1.6	1.8
2002-05 (1.5)	2.0	2.2
XJ6 (1.0)	1.4	1.6
XJ8, XJR, Vanden Plas		
1982-03 (.4)	.6	.8
2004-05 (.8)	1.1	1.3
XJS (1.5)	2.0	2.2
XK8, XKR (.6)	.8	1.0
X-Type (1.5)	2.0	2.2
Headlamp Switch, Replace (B)		
Exc. below (.8)	1.1	1.1
2002-05 S-Type (.3)	.5	.5
2004 XJ8, XJR, Vanden Plas (.3)	.5	.5
XJS (.6)	.8	.8
X-Type (.2)	.3	.3
Headlamps, Aim (B)		
1981-05		
two	.4	.4
four	.6	.6
High Mount Stop Lamp Assy., Replace (B)		
S-Type	.3	.3
1985-97 XJ6	.3	.3
XJ8, XJR, Vanden Plas	.2	.2
XJS, XJ12	.3	.3
XK8, XKR	.3	.3
Horn, Replace (B)		
S-Type	.3	.3
XJ6		
1981-87		
upper (.8)	1.2	1.4
lower (1.0)	1.4	1.6
1988-94 each (.3)	.5	.5
1995-97 (.7)	.9	.9
XJ8, Vanden Plas		
1982-03 (.2)	.3	.3
2004-05 (.6)	.8	.8
XJ12, XJR (.7)	.9	.9
XJS (.3)	.5	.5
XK8, XKR		
1997-03 (.2)	.3	.3
2004-05 (1.1)	1.5	1.5
X-Type	.3	.3
Horn Relay, Replace (B)		
Exc. XJS (.3)	.5	.5
XJS (.2)	.3	.3

	LABOR TIME	SEVERE SERVICE
License Lamp Assy., Replace (B)		
S-Type (.5)	.5	.5
XJ6		
1988-94		
one (.5)	.7	.7
both (.7)	.9	.9
1995-97 (1.2)	1.6	1.8
XJ8, Vanden Plas		
one side (.4)	.6	.6
both sides (.6)	.8	.8
XJR		
1997 (1.2)	1.6	1.8
1998-05		
one side (.6)	.8	.8
both sides (1.2)	1.6	1.6
XK8, XKR, X-Type	.5	.5
License & Back-up Light Assy., Replace (B)		
1988-96 XJS (1.2)	1.6	1.6
License Lamp Lens or Bulb, Replace (C)		
1981-04 each	.3	.3
Parking Lamp Lens or Bulb, Replace (C)		
1981-04 each	.3	.3
Parking Brake Warning Lamp Switch, Replace (B)		
S-Type (.5)	.7	.7
XJ6		
1981-87 (1.0)	1.4	1.4
1988-97 (.7)	.9	.9
XJ8, Vanden Plas (.5)	.7	.7
XJ12, XJR (.6)	.8	.8
XJS (.8)	1.2	1.2
XK8 (.6)	.8	.8
X-Type (.5)	.7	.7
Rear Bulb Failure Module, Replace (B)		
1988-97 XJ6 (1.0)	1.3	1.3
Rear Combination Lamp Assy., Replace (B)		
1988-97 XJ6 each (.7)	.9	.9
1995-96 XJ12 each (.7)	.9	.9
XJR each (.7)	.9	.9
1988-96 XJS each	.2	.2
2004-05 XK8, XKR	.3	.3
Sealed Beam Headlamp, Replace (B)		
XJ6		
1981-87 each (.3)	.5	.5
1988-94 each (.5)	.7	.7
1995-97 each (.2)	.3	.3
XJ8, XJR, Vanden Plas 2004-05 (.7)	.9	.9
XJ12, XJR each (.2)	.3	.3
XJS each	.5	.5
Side Marker Lamp Assy., Replace (B)		
1981-05 (.2)	.3	.3
Side Marker Lamp Bulb, Replace (C)		
1981-05 (.1)	.2	.2

	LABOR TIME	SEVERE SERVICE
Side Marker Lamp Lens, Replace (B)		
1981-05 each (.2)	.3	.3
Stop & Tail Lamp Lens, Replace (B)		
1981-05 each (.2)	.3	.3
Stop Light Switch, Replace (B)		
S-Type (.2)	.3	.3
XJ6		
1981-87 (.6)	.8	.8
1988-94 (.8)	1.1	1.1
1995-97 (.6)	.8	.8
XJ8, Vanden Plas (.2)	.3	.3
XJ12		
1994 (.8)	1.1	1.1
1995-96 (.6)	.8	.8
XJR (.2)	.3	.3
XJS		
1981-88 (.5)	.7	.7
1989-96 (.3)	.5	.5
XK8, XKR		
1997-03 (.1)	.2	.2
2004-05 (.4)	.6	.6
X-Type (.2)	.3	.3
Turn Signal or Hazard Warning Flasher, Replace (B)		
1981-05 (.2)	.3	.3

BODY

	LABOR TIME	SEVERE SERVICE
Door Hinges, Replace (B)		
S-Type (1.2)	1.6	1.8
XJS		
upper or lower (3.6)	4.9	4.9
X-Type (.6)	.8	.8
Door Lock, Replace (B)		
S-Type each set (.8)	1.2	1.2
XJ6		
1981-87		
2 door each (1.0)	1.4	1.4
4 door each (1.3)	1.8	1.8
1988-94 front (2.7)	3.6	3.6
1995-97 each (1.5)	2.0	2.0
XJ8, Vanden Plas each (.8)	1.2	1.2
XJR		
1997 (1.5)	2.0	2.0
1998-03 (.8)	1.1	1.1
2004-05 (.9)	1.2	1.2
XJS each (1.1)	1.5	1.5
XK8, XKR each (.6)	.8	.8
X-Type (.4)	.6	.6
Door Lock Control Module, Replace (B)		
1988-97 XJ6 (.7)	.9	.9
1988-96 XJS (.3)	.5	.5
Door Lock Motor, Replace (B)		
1988-94 XJ6 ea. (1.6)	2.2	2.4
1995-02 each (1.5)	2.0	2.0

S-TYPE : SUPER V8 : VANDEN PLAS : XJ12 : XJ6 : XJ8 : XJR : XJS : XK8 : XKR : X-TYPE — JAG-33

	LABOR TIME	SEVERE SERVICE
Door Lock Remote Control, Replace (B)		
XJ6		
1981-87		
2 door each (1.0)	1.4	1.6
4 door		
front (1.1)	1.4	1.6
rear (1.2)	1.6	1.8
1988-94 (.5)	.7	.7
1995-97 (.6)	.8	.8
Door Window Regulator and/or Motor (Electric), Replace (B)		
S-Type		
2002-03		
front (1.4)	1.9	2.1
rear (2.0)	2.7	2.9
2004-05		
front (.7)	.9	.9
rear (.8)	1.1	1.1
XJ6		
1988-97		
front or rear (1.3)	1.8	2.0
XJ8, Vanden Plas		
1982-03 front or rear (.8)	1.1	1.3
2004-05 (1.2)	1.6	1.6
XJ12		
front or rear (1.3)	1.8	2.0
XJR		
1995-97		
front or rear (1.3)	1.8	2.0
1998-03		
front or rear (.8)	1.1	1.3
XJS		
1981-89 front (1.2)	1.6	1.8
1990-91		
front (.8)	1.2	1.4
rear (2.2)	3.0	3.2
1992-96		
front (2.2)	3.0	3.2
rear (1.8)	2.4	2.6
XK8, XKR (1.2)	1.6	1.8
X-Type		
front (.7)	.9	.9
rear (.8)	1.1	1.1
Gas Tank Filler Door Solenoid, Replace (B)		
XJ6		
1988-94 (1.7)	2.3	2.5
1995-97 (.6)	.8	.8
Hood Hinge, Replace (B)		
S-Type (1.5)	2.0	2.2
XJ6		
1981-87 (3.7)	5.0	5.2
1988-94		
right (.3)	.5	.5
left (1.0)	1.4	1.6
1995-97 each (.6)	.8	.8
XJ8, XJR Vanden Plas		
2002-03 (1.0)	1.4	1.5
2004-05 (.6)	.8	.8
XJS (1.4)	1.9	1.9
2002-05 XK8, XKR (.8)	1.1	1.2
Hood Lock, Replace (B)		
S-Type	.5	.5
XJ6		
1981-87 (.5)	.7	.7
1988-97 each (.3)	.5	.5
XJ8, XK8 each (.3)	.5	.5
XJS each (.7)	.9	.9
X-Type hood set (.3)	.5	.5
Hood Release Cable, Replace (B)		
S-Type		
2000-01 (.9)	1.2	1.2
2002-05 (1.8)	2.4	2.7
XJ6		
1981-87 (.7)	.9	.9
1988-97		
one (.5)	.7	.7
both (.6)	.8	.8
XJ12, XJR		
one (.5)	.7	.7
both (.6)	.8	.8
XJS (.5)	.7	.7
Hood Strut, Replace (B)		
S-Type each	.3	.3
XJ8, XJR, Vanden Plas	.3	.3
XK8, XKR, X-Type	.3	.3
Lock Striker Plate, Replace (B)		
S-Type each	.3	.3
XJ6		
1981-87		
2 door each (1.7)	2.3	2.5
4 door		
front (.5)	.7	.7
rear (1.0)	1.4	1.6
1988-97	.2	.2
XJS each (1.0)	1.4	1.6
XK8, XKR each	.2	.2
X-Type front or rear	.3	.3
Rear Compartment Lid Hinge, Replace (B)		
XJ6		
1981-87 (1.5)	2.0	2.2
1988-97		
one (.5)	.7	.7
both (.6)	.8	.8
XJ12, XJR, XK8		
one (.5)	.7	.7
both (.6)	.8	.8
XJS (1.2)	1.6	1.8
Rear Compartment Strut, Replace (B)		
S-Type each	.3	.3
XK8, XKR each	.3	.3

NOTES

Defender 90 : Discovery : Freelander : LR3 : Range Rover

SYSTEM INDEX

MAINTENANCE LR-2
   Maintenance Schedule LR-2
CHARGING LR-2
STARTING LR-2
CRUISE CONTROL LR-2
IGNITION LR-2
EMISSIONS LR-3
FUEL LR-3
EXHAUST LR-4
ENGINE COOLING LR-4
ENGINE LR-4
   Assembly LR-4
   Cylinder Head LR-4
   Camshaft LR-5
   Crank & Pistons LR-5
   Engine Lubrication LR-5
CLUTCH LR-6
MANUAL TRANSMISSION ... LR-6
AUTO TRANSMISSION LR-6
TRANSFER CASE LR-6
SHIFT LINKAGE LR-6
DRIVELINE LR-7
BRAKES LR-7
FRONT SUSPENSION LR-8
REAR SUSPENSION LR-9
STEERING LR-10
HEATING & AC LR-11
WIPERS & SPEEDOMETER . LR-12
LAMPS & SWITCHES LR-12
BODY LR-13

OPERATIONS INDEX

A
AC Hoses LR-11
Air Bags LR-10
Air Conditioning LR-11
Alignment LR-8
Alternator (Generator) LR-2
Anti-Lock Brakes LR-7

B
Back-Up Lamp Switch LR-12
Ball Joint LR-9
Battery Cables LR-2
Bleed Brake System LR-8
Blower Motor LR-11
Brake Disc LR-8
Brake Drum LR-8
Brake Hose LR-8
Brake Pads and/or Shoes ... LR-8

C
Camshaft LR-5
Camshaft Sensor LR-2
Catalytic Converter LR-4
Coolant Temperature (ECT)
  Sensor LR-3
Crankshaft LR-5
Crankshaft Sensor LR-2
Cruise Control LR-2
Cylinder Head LR-4

D
Differential LR-7
Distributor LR-2
Drive Belt LR-2
Driveshaft LR-7

E
Electronic Control Module
  (ECM/PCM) LR-3
Engine LR-4
Engine Lubrication LR-5
Engine Mounts LR-4
Evaporator LR-11
Exhaust LR-4
Exhaust Manifold LR-4
Expansion Valve LR-11

F
Flexplate LR-6
Flywheel LR-6
Fuel Injection LR-3
Fuel Pump LR-3
Fuel Vapor Canister LR-3

G
Gear Selector Lever LR-7
Generator LR-2

H
Headlamp LR-12
Heater Core LR-12
Horn LR-12

I
Idle Air Control (IAC) Valve ... LR-4
Ignition Coil LR-3
Ignition Module LR-3
Ignition Switch LR-3
Inner Tie Rod LR-10
Intake Manifold LR-3

K
Knock Sensor LR-3

L
Lower Control Arm LR-9

M
Maintenance Schedule LR-2
Mass Air Flow (MAF) Sensor .. LR-3
Master Cylinder LR-8
Muffler LR-4
Multifunction Lever/Switch ... LR-12

N
Neutral Safety Switch LR-6

O
Oil Pan LR-5
Oil Pump LR-5
Oxygen Sensor LR-3

P
Parking Brake LR-8
Pistons LR-5
Positive Crankcase Ventilation
  (PCV) Valve LR-3

R
Radiator LR-4
Radiator Hoses LR-4
Radio LR-12
Rear Main Oil Seal LR-5

S
Shock Absorber/Strut, Front .. LR-9
Shock Absorber/Strut, Rear ... LR-10
Spark Plug Cables LR-3
Spark Plugs LR-3
Spring, Front Coil LR-9
Spring, Rear Coil LR-10
Starter LR-2
Steering Wheel LR-10

T
Thermostat LR-4
Throttle Position Sensor
  (TPS) LR-3
Timing Belt LR-5
Timing Chain LR-5
Torque Converter LR-6

U
U-Joint LR-7

V
Valve Body LR-6
Valve Cover Gasket LR-5
Valve Job LR-5

W
Water Pump LR-4
Wheel Balance LR-2
Wheel Cylinder LR-8
Window Regulator LR-13
Windshield Washer Pump LR-12
Windshield Wiper Motor LR-12

LR-1

LR-2 DEFENDER 90 : DISCOVERY : FREELANDER : LR3 : RANGE ROVER

	LABOR TIME	SEVERE SERVICE
MAINTENANCE		
Air Cleaner Filter Element, Replace (B)		
1987-05	.5	.6
Drive Belt, Adjust (B)		
1987-99		
exc. below	.3	.4
1987-95 PS pump	.7	.8
2003-05 Range Rover	.9	1.0
Drive Belt, Replace (B)		
Exc. below	.8	.9
Freelander	1.5	1.6
2003-05 Range Rover		
alternator	.4	.5
compressor	1.1	1.2
Drive Belt Tensioner, Replace (B)		
Serpentine		
exc. below	.8	1.0
Freelander	1.7	1.9
2003-05 Range Rover	1.4	1.6
Fuel Filter, Replace (B)		
Exc. below	1.2	1.2
Discovery w/evap. emissions	.8	.8
Freelander	.8	.8
1996-05 Range Rover	.6	.6
Halogen Headlight Bulb, Replace (B)		
1994-05	.3	.3
Oil & Filter, Change (C)		
Includes: Correct all fluid levels.		
1987-05	.8	.8
Tire, Replace (C)		
Includes: Dismount old tire and mount new tire to rim.		
1987-05	.7	.7
Wheels, Balance (B)		
One	.5	.5
each addl. add	.3	.3

SCHEDULED MAINTENANCE INTERVALS

If necessary, refer to appropriate Chilton maintenance service information.

7,500 Mile Service (C)		
All Models	2.4	2.4
15,000 Mile Service (B)		
All Models	5.0	5.0
22,500 Mile Service (C)		
All Models	3.0	3.0
30,000 Mile Service (B)		
All Models	5.9	5.9
37,500 Mile Service (C)		
All Models	2.4	2.4
45,000 Mile Service (C)		
All Models	1.1	1.1
52,500 Mile Service (C)		
All Models	2.7	2.7
60,000 Mile Service (B)		
All Models	5.9	5.9
67,500 Mile Service (C)		
All Models	2.6	2.6
75,000 Mile Service (B)		
All Models	5.2	5.2
82,500 Mile Service (C)		
All Models	2.4	2.4
90,000 Mile Service (B)		
All Models	6.0	6.0
97,500 Mile Service (C)		
All Models	2.4	2.4
105,000 Mile Service (B)		
All Models	5.2	5.2
112,500 Mile Service (C)		
All Models	2.6	2.6
120,000 Mile Service (B)		
All Models	5.9	5.9

CHARGING

Alternator Circuits, Test (B)		
Includes: Test component output.		
1987-05	.7	.7
Alternator, R&R and Recondition (B)		
1987-02	2.2	2.7
Alternator Assy., Replace (B)		
Includes: Pulley transfer.		
Exc. below	.9	1.2
Freelander	1.4	1.7
LR3	1.9	2.2
Range Rover LM	2.2	2.5
Voltmeter, Replace (B)		
Defender 90	1.8	1.9

STARTING

Starter Draw Test (On Car) (B)		
1987-05	.4	.4
Battery Cables, Replace (B)		
Defender 90		
positive	2.0	2.2
negative	.7	.9
Discovery negative	.3	.5
Freelander each	.2	.4
LR3		
positive	1.9	2.1
negative	.4	.6
Range Rover		
positive	2.3	2.5
negative	.7	.9
Starter Assy., Replace (B)		
Exc. below	1.8	2.0
Defender, Discovery	.7	.9
Freelander	.9	1.1
LR3	2.0	2.2
Starter Relay, Replace (B)		
Defender 90	.5	.5
Discovery	.5	.5
Freelander, LR3	.2	.2

Range Rover		
1987-95	.3	.3
1996-05	.7	.7
Starter Motor Solenoid, Replace (B)		
Exc. below	1.9	2.2
Defender, Discovery	.9	1.1

CRUISE CONTROL

Diagnose Cruise Control System Component Each (B)		
Exc. Freelander	1.3	1.3
Freelander	.5	.5
Actuator Assy., Replace (B)		
All Models	.5	.5
Control Brake or Clutch Switch, Replace (B)		
Exc. Discovery	.2	.2
Discovery	.8	.8
Control Engagement Switch, Replace (B)		
Exc. below	.5	.5
Range Rover		
1996-05 SE	1.2	1.2
Control Module, Replace (B)		
All Models	.8	.8
Control Relay, Replace (B)		
All Models	.5	.5
Control Set Switch, Replace (B)		
Exc. below	.9	.9
Freelander	.3	.3
1996-05 Range Rover	1.8	1.8
Control Transducer, Replace (B)		
All Models	.7	.7
Control Vacuum Pump, Replace (B)		
All Models	.5	.7

IGNITION

Diagnose Ignition System Component Each (A)		
1987-05	1.1	1.1
Camshaft Position Sensor, Replace (B)		
1995-05	.5	.7
Crankshaft Angle Sensor, Replace (B)		
Exc. below	.7	.9
Freelander	.9	1.1
2003-05 Range Rover	.8	1.0
Distributor, Replace (B)		
Includes: Reset base ignition timing.		
1987-02	1.3	1.5
Distributor Cap and/or Rotor, Replace (B)		
1987-02	.3	.3
Distributor Vacuum Advance Unit, Replace (B)		
Includes: Reset base timing.		
1987-95	1.4	1.5

DEFENDER 90 : DISCOVERY : FREELANDER : LR3 : RANGE ROVER

	LABOR TIME	SEVERE SERVICE
Ignition Coil, Replace (B)		
Distributor ignition	.3	.3
Distributorless ignition		
Defender 90 one or all	1.3	1.4
Discovery one	.5	.6
each addl. add	.1	.2
Freelander all	1.1	1.2
LR3 one	.6	.7
each addl. add	.3	.4
Range Rover		
1987-05 one	1.5	1.6
each addl. add	.2	.3
Ignition Module, Replace (B)		
On distributor	1.8	1.8
Remote	.7	.7
Ignition Switch, Replace (B)		
Defender 90, LR3	1.6	1.6
Discovery	.9	.9
Freelander	.3	.3
Range Rover		
1987-94	.9	.9
1995-05	1.7	1.7
Ignition Timing, Reset (B)		
1987-02	.7	.7
Spark Plug (Ignition) Cables Replace (B)		
Exc. below	.7	.9
Range Rover		
w/air injection	3.1	3.2
w/o air injection	1.9	2.1
Spark Plugs, Replace (B)		
Exc. below	1.7	1.9
Defender, Discovery	.8	1.0
Freelander	1.5	1.7
LR3	1.2	1.4

EMISSIONS

The following operations do not include testing. Add time as required.

	LABOR TIME	SEVERE SERVICE
Diagnose Electronic Emission System Component Each (A)		
Exc. Freelander	1.2	1.2
Freelander	.6	.6
Dynamometer Test (B)		
1987-05	.5	.5
Coolant Temperature (ECT) Sensor, Replace (B)		
Exc. Freelander	1.2	1.2
Freelander	3.4	3.4
Electronic Control Module (ECM/PCM), Replace (B)		
All Models	.3	.5
Fuel Vapor Canister, Replace (B)		
Exc. below	.5	.5
1997 Defender 90	.8	.8
Freelander	1.2	1.2
2003-05 Range Rover	3.3	3.3

	LABOR TIME	SEVERE SERVICE
Fuel Vapor Canister Purge Control Valve, Replace (B)		
1987-05	.3	.3
Knock Sensor, Replace (B)		
Exc. below each	.8	1.0
Freelander each	5.0	5.2
LR3 each	1.6	1.9
Range Rover		
1987-02		
left side	.3	.5
right side	1.6	1.8
2003-05	4.7	4.9
Maintenance Reminder, Replace (B)		
Range Rover	.3	.3
Mass Air Flow (MAF) Sensor, Replace (B)		
1987-05	.5	.5
Oxygen Sensor, Replace (B)		
Exc. below each	.8	.8
Discovery one	.9	.9
each addl. add	.3	.3
Freelander		
front each	1.0	1.2
rear each	.3	.5
Positive Crankcase Ventilation (PCV) Valve, Replace (B)		
Exc. Freelander	.3	.5
Freelander	1.2	1.4
PROM, Replace (A)		
Defender 90		
1994	1.4	1.4
1995-97	.7	.7
Discovery	1.4	1.4
Range Rover		
1987-94	.7	.7
1995 Classic	1.4	1.4
1996-05 SE	.8	.8
w/AC 94 Defender add	1.2	1.2
Throttle Position Sensor (TPS), Replace (B)		
1987-05	.6	.6

FUEL
DELIVERY

	LABOR TIME	SEVERE SERVICE
Fuel Filter, Replace (B)		
Exc. below	1.2	1.2
Discovery w/evap. emissions	.8	.8
Freelander	.8	.8
1996-05 Range Rover	.6	.6
Fuel Gauge (Dash), Replace (B)		
Exc. below	1.6	1.6
Defender 90	.7	.7
Freelander	.8	.8
Fuel Gauge (Tank), Replace (B)		
Defender 90	2.4	2.7
Discovery	1.9	2.2
Freelander	.5	.8
LR3	2.3	2.6

	LABOR TIME	SEVERE SERVICE
Range Rover		
1987-90	2.8	3.1
1991-93	1.6	1.9
1994-02	2.8	3.1
2003-05	.8	1.1
Fuel Pump, Replace (B)		
Defender 90	3.0	3.3
Discovery		
w/evap. emissions	1.4	1.7
w/o evap. emissions	1.8	2.1
Freelander	.6	.9
LR3	1.8	2.1
Range Rover		
1987-2002	2.8	3.1
2003-05	2.2	2.5
Fuel Pump Relay, Replace (B)		
All Models	.5	.5
Fuel Tank, Replace (B)		
Includes: Drain and refill.		
Defender 90	2.5	2.8
Discovery		
w/evap. emissions	2.7	3.0
w/o evap. emissions	3.6	3.9
Freelander	4.1	4.4
LR3	2.5	2.8
Range Rover		
1987-2002	2.8	3.1
2003-05	7.3	7.6
Intake Manifold and/or Gasket, Replace (B)		
Includes: Adjustments.		
Exc. below	4.5	5.0
Freelander	3.6	4.1
LR3	2.6	3.1
2003-05 Range Rover	4.7	5.2

INJECTION

	LABOR TIME	SEVERE SERVICE
Diagnose Fuel Injection System Component Each (B)		
1987-05	1.2	1.2
EFI Relay, Replace (B)		
1987-05	.5	.5
Electronic Control Unit, Replace (B)		
Exc. Freelander	.8	.8
Freelander	.3	.3
Fuel Injectors, Replace (B)		
Exc. below		
one	3.2	3.4
all	3.6	3.8
LR3		
one	1.3	2.5
all	2.0	2.3
2003-05 Range Rover		
one	2.3	2.5
all	3.0	3.3
Fuel Pressure Regulator, Replace (B)		
1987-05	1.2	1.4

LR-3

DEFENDER 90 : DISCOVERY : FREELANDER : LR3 : RANGE ROVER

	LABOR TIME	SEVERE SERVICE
Fuel Rail, Replace (B)		
Exc. below	3.7	4.0
Freelander	3.3	3.6
LR3	1.8	2.0
2003-05 Range Rover	2.3	2.6
Fuel Temperature Sensor, Replace (B)		
1987-05	.5	.5
Idle Air Control (IAC) Valve, Replace (B)		
1987-05	.5	.5
Idle Speed, Adjust (B)		
1987-05	.7	.7

EXHAUST

	LABOR TIME	SEVERE SERVICE
Catalytic Converter, Replace (B)		
Exc. below	2.2	2.3
Defender	1.5	1.6
LR3 each	1.6	1.7
2003-05 Range Rover	1.6	1.7
Exhaust Manifold or Gasket, Replace (B)		
Exc. below		
right side	2.0	2.5
left side	1.7	2.2
Defender each	1.5	2.0
Freelander each	2.5	3.0
LR3		
right side	3.8	4.2
left side	4.2	4.7
Range Rover LM		
right side	1.1	1.3
left side	3.2	3.5
Exhaust Pipe Flange Gasket, Replace (B)		
1987-05 both	1.7	1.9
Exhaust System (Complete), Replace (B)		
Exc. below	3.8	4.3
Defender, Freelander, LR3	1.4	1.9
Discovery	3.0	3.5
Range Rover		
1996-02 SE	2.4	2.9
2003-05	1.9	2.4
Intermediate Exhaust Pipe, Replace (B)		
Freelander	1.0	1.1
LR3	.8	.9
Range Rover		
1996-02 SE	.8	.9
2003-05	1.3	1.4
Muffler and Tailpipe Assy., Replace (B)		
Exc. below	2.5	2.6
Defender, LR3	.7	.8
Freelander	.5	.6
1996-05 Range Rover	.8	.9

ENGINE COOLING

	LABOR TIME	SEVERE SERVICE
Pressure Test Cooling System (C)		
1987-05	.3	.3
Coolant Thermostat, Replace (B)		
Includes: Refill with proper coolant mix.		
Defender 90	1.7	1.9
Discovery	1.2	1.4
Discovery II	1.8	2.0
Freelander	3.8	4.0
LR3	1.0	1.2
Range Rover		
1987-95	.9	1.1
1996-05	1.8	2.0
Engine Coolant Temp. Sending Unit, Replace (B)		
Exc. below	.9	.9
Defender 90, LR3	.5	.5
Freelander	.2	.2
Range Rover		
1996-02 SE	1.2	1.2
2003-05	.6	.6
Fan Blade, Replace (B)		
Exc. below	.7	.9
2003-05 Range Rover	1.4	1.6
Fan Clutch Assy., Replace (B)		
Exc. below	.7	.9
2003-05 Range Rover	1.2	1.4
Radiator Assy., R&R or Replace (B)		
Exc. below	2.8	3.0
Defender	2.3	2.5
Freelander	3.9	4.1
LR3	2.0	2.2
2003-05 Range Rover	1.9	2.1
Radiator Hoses, Replace (B)		
Includes: Refill with proper coolant mix.		
Exc. below		
upper	.9	1.0
lower	1.6	1.7
2003-05 Range Rover		
upper	1.5	1.6
lower	1.6	1.7
Temperature Gauge (Dash), Replace (B)		
Exc. below	1.6	1.6
Defender 90	.7	.7
Freelander	.8	.8
Water Pump and/or Gasket, Replace (B)		
Includes: Refill with proper coolant mix.		
Exc. below	3.1	3.6
Freelander	7.2	7.7
LR3	1.8	2.3
Range Rover		
1996-02 SE	2.5	3.0
2003-05	2.9	3.4

ENGINE

ASSEMBLY

Times shown are for OEM assemblies. Time to replace assemblies from aftermarket rebuilders may vary.

	LABOR TIME	SEVERE SERVICE
Engine Assy., R&I (B)		
Does not include parts or component transfer.		
Exc. below	12.3	12.6
Defender	9.6	9.9
Discovery	9.0	9.3
Freelander	11.2	11.5
LR3	15.0	15.3
2003-05 Range Rover	10.4	10.7
Engine Assy., Replace (B)		
Includes: Component transfer.		
Exc. below	21.2	21.7
Defender	14.2	14.7
Discovery	13.0	13.5
Freelander	13.1	13.6
LR3	23.8	24.3
Cylinder Block, Replace (Short Block) (B)		

Assembly consists of cylinder block, piston assemblies, crankshaft, camshaft, timing chain and gears. Does not include cylinder heads. Operation Includes: R&R engine, transfer necessary parts and all necessary adjustments.

	LABOR TIME	SEVERE SERVICE
Exc. below	24.2	24.7
Defender, Discovery	17.7	18.3
Range Rover LM	25.0	25.5
Engine Mounts, Replace (B)		
Exc. below each	.8	1.0
Freelander each	1.3	1.5
LR3 each	2.3	2.5
Range Rover		
1996-02 SE rear or left side	1.5	1.7
2003-05 each	2.9	3.1

CYLINDER HEAD

	LABOR TIME	SEVERE SERVICE
Compression Test (B)		
1987-05	1.7	1.9
Cylinder Head, Replace (B)		
Includes: Parts transfer, adjustments.		
Exc. below one	12.8	13.3
Freelander each	10.7	11.2
LR3 each	15.7	16.2
Range Rover		
1996-02 SE one	13.1	13.6
2003-05 each	21.8	22.3
Cylinder Head Gasket, Replace (B)		
Defender 90, Discovery		
right side	7.3	7.6
left side	6.5	6.8
both sides	10.0	10.3

DEFENDER 90 : DISCOVERY : FREELANDER : LR3 : RANGE ROVER — LR-5

	LABOR TIME	SEVERE SERVICE
Freelander w/engine removed both sides	7.4	7.7
LR3 each	14.7	15.2
Range Rover		
1987-1995		
right side	6.5	6.8
left side	7.3	7.6
both sides	10.0	10.3
1996-05 SE		
right side	7.0	7.3
left side	7.7	8.0
both sides	10.0	10.3
2003-05 w/water pump gasket both sides	23.0	23.3
Rocker Arm Cover Gasket, Replace or Reseal (B)		
Exc. below		
right side	1.4	1.6
left side	2.0	2.2
both sides	3.2	3.4
Freelander		
front	1.2	1.4
rear	2.6	2.8
both	3.2	3.4
LR3	1.2	1.4
2003-05 Range Rover		
one	1.7	1.9
both	2.8	3.0
Rocker Arm Shaft, Replace (B)		
Right side	1.6	1.8
Left side	2.3	2.5
Recondition shaft add each	.5	.7
Pushrods, Replace (B)		
1987-05	7.8	8.0
Valve Job (A)		
Exc. below		
right side	10.8	11.3
left side	10.3	10.8
both sides	17.5	18.0
LR3		
right side	20.3	20.8
left side	19.3	19.8
both sides	30.5	31.0
1996-05 Range Rover SE		
right side	12.9	13.4
left side	13.1	13.6
both sides	21.7	22.2
Valve Lifters, Replace (B)		
Exc. below	6.5	7.0
Freelander per head	9.2	9.7
2003-05 Range Rover per head	9.5	10.0
Valve Springs and/or Oil Seals, Replace (B)		
Exc. below	5.8	6.3
Freelander	17.3	17.8
2003-05 Range Rover	28.9	29.4

CAMSHAFT

	LABOR TIME	SEVERE SERVICE
Camshaft, Replace (B)		
Exc. below	14.6	15.1
Freelander all	12.4	12.9
LR3		
right bank	12.2	12.7
left bank	12.4	12.9
both banks	15.8	16.3
Range Rover		
1995-02	12.7	13.2
2003-05 all	10.6	111.1
Front Cover Gasket, Replace (B)		
Exc. below	4.8	5.0
Freelander all	1.2	1.4
LR3	4.3	4.6
Range Rover		
1995-02	4.6	4.8
2003-05	1.7	1.9
Timing Belt, Replace (B)		
Freelander all	8.0	8.5
Timing Belt Tensioner, Replace (B)		
Freelander	6.8	7.1
Timing Chain & Gears, Replace (B)		
Exc. below	5.3	5.8
LR3	10.0	10.2
Range Rover		
1995-02		
Classic	4.5	5.0
SE	4.9	5.4
2003-05 includes tensioner	20.7	21.2
Range Rover LM	18.5	19.0
Timing Chain Cover Oil Seal, Replace (B)		
Exc. below	2.8	3.0
Range Rover 1996-05 SE	1.5	1.8

CRANK & PISTONS

	LABOR TIME	SEVERE SERVICE
Connecting Rod Bearings, Replace (A)		
Includes: Check bearing oil clearance.		
Exc. below	5.2	5.8
Range Rover		
1996-02 SE	3.7	4.3
2003-05	15.0	15.6
Crankshaft and Main Bearings, Replace (A)		
Includes: Engine R&R, check bearing oil clearance.		
Exc. below	15.2	16.0
Freelander	24.0	24.8
Range Rover		
1996-02 SE	19.8	20.6
2003-05	31.7	32.5
Crankshaft Pulley, Replace (B)		
Defender 90	2.0	2.2
Discovery	2.0	2.2
Freelander	.9	1.1
LR3	1.5	1.7

	LABOR TIME	SEVERE SERVICE
Range Rover		
1987-95	2.4	2.6
1996-05	1.8	2.0
Main Bearings, Replace (A)		
Includes: Check bearing oil clearance.		
Exc. below	9.8	10.3
Freelander	23.4	23.9
Range Rover 1996-05 SE	13.3	13.8
Pistons or Connecting Rods, Replace (A)		
Includes: Ridge reaming, cylinder wall deglazing, installing new rings and rod bearings, engine tune-up.		
Exc. below	20.1	21.0
Freelander	27.5	28.4
Range Rover		
1996-02 SE	22.5	23.4
2003-05	32.0	32.9
Rear Main Bearing Oil Seal (Full Circle), Replace (B)		
Exc. below	10.5	11.1
Freelander	14.4	15.0
LR3	10.3	10.9
1996-05 Range Rover	9.3	9.9
Rings, Replace (A)		
Includes: Ridge reaming, cylinder wall deglazing, installing new rings, engine tune-up.		
Exc. below	17.4	18.3
Freelander	27.0	27.9
Range Rover		
1996-02 SE	19.0	19.9
2003-05	31.3	32.2

ENGINE LUBRICATION

	LABOR TIME	SEVERE SERVICE
Engine Oil Cooler, Replace (B)		
Exc. below	1.5	1.8
Freelander	2.8	3.1
LR3	.9	1.2
Engine Oil Pressure Switch (Sending Unit), Replace (B)		
Exc. below	.9	.9
2003-05 Range Rover	.4	.4
Oil Pan and/or Gasket, Replace (B)		
Exc. below	1.4	1.6
Freelander	2.0	2.2
LR3	16.7	16.9
Oil Pressure Gauge (Dash), Replace (B)		
Defender 90	1.8	1.8
Oil Pump, Replace (B)		
Defender 90		
1994-95	1.6	2.1
1997	5.3	5.8
Discovery	4.9	5.4
Freelander	9.0	9.5
LR3	18.0	18.5

DEFENDER 90 : DISCOVERY : FREELANDER : LR3 : RANGE ROVER

	LABOR TIME	SEVERE SERVICE
Oil Pump, Replace (B)		
Range Rover		
1987-94	1.6	2.1
1995-02	5.2	5.7
2003-05	11.6	12.1

CLUTCH

	LABOR TIME	SEVERE SERVICE
Clutch Hydraulic System, Bleed (B)		
All Models	.7	.7
Clutch Pedal Free Play, Adjust (B)		
1994-05	.3	.5
Clutch Assy., Replace (B)		
All Models	8.6	8.8
Clutch Master Cylinder, Replace (B)		
Includes: System bleeding.		
All Models	1.4	1.6
Clutch Release Bearing or Fork, Replace (B)		
Defender 90	8.4	8.6
Discovery	7.8	8.0
Clutch Slave Cylinder, Replace (B)		
Includes: System bleeding.		
All Models	.8	1.0
Flywheel, Replace (B)		
All Models	8.9	9.1

MANUAL TRANSMISSION

	LABOR TIME	SEVERE SERVICE
Extension Housing and/or Gasket, Replace (B)		
Defender 90	12.4	12.7
Output Shaft Seal, Replace (B)		
Defender 90	6.3	6.6
Shift Lever, Replace (B)		
Defender 90	1.7	1.7
Speedometer Driven Gear, Replace (B)		
Defender 90	.7	.7
Transmission Assy., R&R and Recondition (A)		
Defender 90	26.9	27.8
Recondition transfer case add	11.5	12.1
Transmission and Transfer Case Assy., R&I (B)		
Defender 90	12.2	12.4
Discovery	9.8	10.0
Replace assembly add	1.7	1.9

AUTOMATIC TRANSMISSION
SERVICE TRANSMISSION INSTALLED

	LABOR TIME	SEVERE SERVICE
Check Unit for Oil Leaks (C)		
1987-05	.9	.9
Oil Pressure Test (B)		
All Models	1.5	1.5
Drain and Refill Unit (B)		
All Models	1.2	1.2

	LABOR TIME	SEVERE SERVICE
Back-up Lamp and Park/Neutral Switch, Replace (B)		
All Models	.8	.8
Downshift Cable, Replace (B)		
Exc. Defender 90	3.1	3.3
Defender 90	4.2	4.4
Electronic Control Unit, Replace (B)		
Freelander	.2	.2
LR3	4.6	4.6
Range Rover	.7	.7
Extension Housing Rear Oil Seal, Replace (B)		
Exc. below	6.3	6.5
Defender 90	8.3	8.5
LR3	5.9	6.1
Range Rover		
1996-05 SE	4.8	5.0
Kickdown Cable, Adjust (B)		
All Models	.5	.7
Oil Pan and/or Gasket, Replace (B)		
Exc. below	2.8	2.8
LR3	4.6	4.6
1996-05 Range Rover	1.7	1.9
Sport Mode Switch, Replace (B)		
Range Rover SE	.7	.7
Transmission Oil Cooler, Replace (B)		
Discovery	1.2	1.4
LR3	2.8	3.0
1995-05 Range Rover		
Classic	1.2	1.4
SE	1.9	2.1
Valve Body Assy., Replace (B)		
Defender	3.9	4.2
Discovery	4.3	4.6
LR3	5.0	5.3
Range Rover		
1987-2002	3.1	3.4
2003-05	2.7	3.0

SERVICE TRANSMISSION REMOVED
Transmission R&R included unless otherwise noted.

	LABOR TIME	SEVERE SERVICE
Flexplate, Replace (B)		
Exc. below	9.8	10.0
Defender	12.0	12.2
Freelander	14.8	15.0
LR3	10.3	10.5
Range Rover		
1996-02 SE	7.4	7.6
2003-05	8.8	9.0
Front Pump Oil Seal, Replace (B)		
Exc. below	10.5	10.7
Defender	11.9	12.1
LR3	9.6	9.8
Range Rover		
1996-05 SE	7.3	7.5
Torque Converter, Replace (B)		
Exc. below	9.9	10.1
Defender	9.8	10.0

	LABOR TIME	SEVERE SERVICE
Freelander	14.3	14.5
LR3	9.6	9.8
Range Rover		
1996-02 SE	7.2	7.4
2003-05	8.7	8.9
Transmission and Torque Converter, Replace (B)		
Exc. below	11.0	11.2
Defender	12.9	12.1
Freelander	13.5	13.7
LR3	9.7	9.9
Range Rover		
1996-02 SE	8.3	8.5
2003-05	9.3	9.5

TRANSFER CASE

	LABOR TIME	SEVERE SERVICE
Input Shaft Cover Seal, Replace (B)		
Defender 90	8.3	8.5
Discovery	8.3	8.5
Freelander, LR3	5.0	5.2
Range Rover		
1987-95	8.9	9.1
1996-05	6.0	6.2
Output Shaft Oil Seal, Replace (B)		
Exc. below	2.3	2.5
Freelander		
one	2.4	2.6
both	3.6	3.8
LR3	5.0	5.2
Range Rover		
1987-02	1.7	1.9
2003-05		
front	1.7	1.9
rear	4.0	4.2
Transfer Case, R&R and Recondition (B)		
Exc. Range Rover	18.9	19.8
Range Rover	19.8	20.7
Transfer Case, Replace (B)		
Exc. below	6.8	7.3
Freelander	9.7	10.2
LR3	5.0	5.2
Range Rover		
1987-02	8.4	8.9
2003-05	6.6	7.1
Transfer Case Shift Module, Replace (B)		
1996-05 Range Rover	1.1	1.1
Viscous Coupling, Replace (B)		
1987-95 Range Rover	4.5	4.8

SHIFT LINKAGE
AUTOMATIC TRANSMISSION

	LABOR TIME	SEVERE SERVICE
Gear Selector Lever, Replace (B)		
Defender	1.6	1.7
Discovery	1.3	1.4
Freelander	1.4	1.5
LR3	1.1	1.2

DEFENDER 90 : DISCOVERY : FREELANDER : LR3 : RANGE ROVER — LR-7

	LABOR TIME	SEVERE SERVICE
Range Rover		
exc. Classic	1.3	1.4
Classic	.9	1.0
Selector Cable, Replace (B)		
Discovery	2.5	2.7
Freelander	1.8	2.0
LR3	1.2	1.4
Range Rover		
1987-94	1.8	2.0
1995-02		
Classic	2.3	2.5
SE	1.6	1.8
2003-05	1.7	1.9

MANUAL TRANSMISSION

Gear Selector Lever, Replace (B)

	LABOR TIME	SEVERE SERVICE
Discovery	2.0	2.1

DRIVELINE
DRIVESHAFT

Propeller Shaft, R&R or Replace (B)

	LABOR TIME	SEVERE SERVICE
Exc. below	1.2	1.5
Freelander		
front	1.6	1.9
rear	1.3	1.6
assembly	1.1	1.4
LR3	2.8	3.1
2003-05 Range Rover		
front	1.0	1.3
rear	3.3	3.6

U-Joints, Recondition (B)

	LABOR TIME	SEVERE SERVICE
1987-05	2.6	2.8

DRIVE AXLE

Axle Assy., R&R or Replace (B)

	LABOR TIME	SEVERE SERVICE
Front axle		
Defender 90	6.0	6.5
Discovery	7.7	8.2
LR3	4.7	5.2
Range Rover		
1987-94		
w/ABS	7.7	8.2
w/air susp	8.3	8.8
1995-05		
Classic	8.4	8.9
SE	6.0	6.5
Rear axle		
exc. below	7.8	8.3
Defender 90	4.5	5.0
LR3	4.1	4.6
Range Rover		
1996-05 SE	6.0	6.5

Axle Shaft, Replace (B)

	LABOR TIME	SEVERE SERVICE
Front axle		
exc. below		
one side	2.0	2.3
both sides	3.7	4.0
Defender 90	2.2	2.5
Freelander		
one side	2.4	2.7
both sides	3.6	3.9

	LABOR TIME	SEVERE SERVICE
Range Rover		
1995-02		
Classic	3.7	4.0
SE	2.4	2.7
2003-05		
one side	2.1	2.4
both sides	4.0	4.3
Rear axle		
exc. below	.7	1.0
Freelander		
one side	.8	1.1
both sides	1.4	1.7
LR3	1.8	2.1
Range Rover		
1995-02		
Classic	.8	1.1
SE	1.7	2.0
2003-05		
one side	2.1	2.4
both sides	4.7	5.0

Axle Shaft Oil Seal, Replace (B)

	LABOR TIME	SEVERE SERVICE
Front axle		
exc. below each	3.2	3.4
Defender 90		
1994-97 each	1.9	2.1
Range Rover		
1996-05 each	2.8	3.0

Differential Assy., R&R and Recondition (B)

	LABOR TIME	SEVERE SERVICE
Front axle		
exc. below	11.2	11.8
Defender 90	9.9	10.5
Range Rover		
1996-05 SE	8.3	8.9
Rear axle		
exc. Defender 90	8.2	8.8
Defender 90	7.5	8.1

Assy., R&R or Replace (B)

	LABOR TIME	SEVERE SERVICE
Front axle		
exc. below	6.7	7.2
Defender 90	4.8	5.3
LR3	4.7	5.2
Range Rover		
1996-02 SE	5.3	5.8
2003-05	8.7	9.2
Rear axle		
exc. below	2.7	3.2
Freelander	2.1	2.6
LR3	5.7	6.2
Range Rover		
2003-05	5.0	5.3

Pinion Shaft Oil Seal, Replace (B)

	LABOR TIME	SEVERE SERVICE
Front axle		
exc. below	.9	1.1
LR3	1.1	1.3
Range Rover		
1996-05	1.3	1.5

	LABOR TIME	SEVERE SERVICE
Rear axle		
exc. below	.8	1.0
Freelander	2.1	2.3
LR3	1.6	1.8
Range Rover		
2003-05	2.7	2.9

BRAKES
ANTI-LOCK

The following operations do not include testing. Add time as required.

Diagnose Anti-Lock Brake System (B)

	LABOR TIME	SEVERE SERVICE
Exc. below	1.2	1.2
Freelander	.6	.6
2003-05 Range Rover	.8	.8

Accumulator Assy., Replace (B)

	LABOR TIME	SEVERE SERVICE
Range Rover		
1987-95	3.0	3.0
1996-05 SE	1.6	1.6

Anti-Lock Relay, Replace (B)

	LABOR TIME	SEVERE SERVICE
Each	.5	.5

Control Module, Replace (B)

	LABOR TIME	SEVERE SERVICE
Exc. below	.9	.9
Freelander	1.4	1.4
Range Rover		
1996-05 SE	.5	.5

Power Booster, Replace (B)

	LABOR TIME	SEVERE SERVICE
Discovery	2.2	2.2
Range Rover		
1987-95	4.4	4.4
1996-05 SE	3.7	3.7

Pump & Motor Assy., Replace (B)

	LABOR TIME	SEVERE SERVICE
Discovery	2.2	2.4
Freelander	1.5	1.7
Range Rover		
1987-02	2.6	2.8
2003-05	1.7	1.9

Valve Block, Replace (B)

	LABOR TIME	SEVERE SERVICE
Range Rover	2.8	2.8

Wheel Speed Sensor and Harness, Replace (B)

	LABOR TIME	SEVERE SERVICE
Front		
Discovery		
one	1.8	1.8
both	2.9	2.9
Freelander		
one	1.4	1.4
both	2.3	2.3
LR3 each	1.2	1.2
Range Rover		
1987-94 each	2.7	2.7
1995-02		
Classic	2.7	2.7
SE each	.7	.7
2003-05 each	.5	.5

LR-8 DEFENDER 90 : DISCOVERY : FREELANDER : LR3 : RANGE ROVER

	LABOR TIME	SEVERE SERVICE
Wheel Speed Sensor and Harness, Replace (B)		
Rear		
Discovery		
1994-99 I		
one	1.2	1.2
both	2.2	2.2
1999-02 II		
one	.9	.9
both	1.2	1.2
Freelander		
one	.5	.5
both	.9	.9
LR3 each	.6	.6
Range Rover		
1987-94 each	.9	.9
1995-02		
Classic	1.2	1.2
SE each	.8	.8
2003-05 each	.8	.8
SYSTEM		
Bleed Brakes (B)		
Includes: Add fluid.		
Exc. below	1.7	1.7
Freelander	.8	.8
2003-05 Range Rover	1.0	1.0
Brake System, Flush and Refill (B)		
1987-05	1.5	1.5
Brake Hose (Flexible), Replace (B)		
Includes: System bleeding.		
Exc. below	2.4	2.5
Defender 90	2.0	2.1
Freelander, LR3		
front or rear		
one	.8	.9
both	1.4	1.5
all	1.9	2.0
Range Rover		
1996-02 SE	1.6	1.7
2003-05		
front or rear		
one	1.0	1.1
both	1.6	1.7
all	2.2	2.3
Master Cylinder, Replace (B)		
Includes: System bleeding.		
Exc. below	2.3	2.5
Freelander	.9	1.1
2003-05 Range Rover	1.3	1.5
Power Booster Unit, Replace (B)		
Exc. below	2.7	2.9
Defender 90, LR3	2.3	2.5
Freelander	1.5	1.7
2003-05 Range Rover	5.6	5.8
Power Booster Vacuum Check Valve, Replace (B)		
1987-05	.5	.5

	LABOR TIME	SEVERE SERVICE
Pressure Differential Valve, Replace (B)		
Defender 90, LR3	1.6	1.8
Discovery	2.0	2.2
Range Rover	1.8	2.0
SERVICE BRAKES		
Caliper Assy., R&R and Recondition (B)		
Includes: System bleeding.		
Front each	3.0	3.3
Rear each	2.5	2.8
Caliper Assy., Replace (B)		
Includes: System bleeding.		
Exc. below each	2.5	2.6
Freelander		
one	1.1	1.2
both	1.7	1.8
Range Rover		
2003-05 each	1.3	1.4
Disc Brake Rotor, Replace (B)		
Defender 90		
one	2.3	2.5
both	3.9	4.1
Discovery		
front		
one	2.8	3.0
both	3.8	4.0
rear		
one	1.5	1.7
both	2.7	2.9
Freelander		
one	.5	.7
both	1.3	1.5
Range Rover		
1987-95		
front		
one	3.1	3.3
both	3.9	4.1
rear		
one	1.6	1.8
both	2.8	3.0
1996-02 SE		
front		
one	.7	.9
both	1.5	1.7
rear		
one	1.2	1.4
both	1.7	1.9
2003-05		
one	.6	.8
both	1.0	1.2
Drum, Replace (B)		
One	.5	.7
Both	1.0	1.2
Pads/Shoes, Replace (B)		
Includes: System bleeding.		
Front	1.3	1.4
Rear disc	1.3	1.4
Rear drum	2.1	2.2

	LABOR TIME	SEVERE SERVICE
Four wheels		
rear disc	2.0	2.1
rear drum	2.9	3.0
COMBINATION ADD-ONS		
Repack wheel bearings		
add each	.5	.5
Resurface rotor		
add each	.5	.5
Wheel Cylinder, Replace (B)		
One	2.5	2.6
Both	2.9	3.0
PARKING BRAKE		
Parking Brake Apply Actuator, Replace (B)		
Defender 90	.7	.7
Discovery	1.7	1.7
Freelander	.6	.6
LR3	3.1	3.1
Range Rover		
1987-02	1.6	1.6
2003-05	1.9	1.9
Parking Brake Apply Warning Indicator Switch, Replace (B)		
Exc. below	1.5	1.5
Discovery	.9	.9
Freelander	.5	.5
2003-05 Range Rover	.8	.8
Parking Brake Cable, Adjust (C)		
1987-05	.6	.8
Parking Brake Cable, Replace (B)		
Exc. below each	1.8	2.1
LR3 cable set	2.6	2.9
Range Rover		
2003-05 each	2.5	2.8
Parking Brake Drum, Replace (B)		
1987-02	1.2	1.5
Parking Brake Shoes, Replace (B)		
1987-05	1.5	1.7
FRONT SUSPENSION		
Unless otherwise noted, time given does not include alignment.		
Front End (A)		
Exc. below	2.0	2.2
Freelander	1.3	1.5
2003-05 Range Rover	1.3	1.5
Front & Rear Alignment (A)		
2003-05 Range Rover	1.7	1.9
Front Toe, Adjust (B)		
1987-05	.9	.9
Front Height Sensor, Replace (B)		
Range Rover		
1987-95 each	2.0	2.0
1996-02 SE each	.7	.7
2003-05	1.1	1.1
Front Hub, Replace (B)		
Defender 90	2.4	2.6
Discovery	2.2	2.4

DEFENDER 90 : DISCOVERY : FREELANDER : LR3 : RANGE ROVER   **LR-9**

	LABOR TIME	SEVERE SERVICE
Freelander		
one side	2.1	2.3
both side	4.3	4.5
LR3 each	1.8	2.0
Range Rover		
1987-95	3.4	3.6
1996-02 SE	2.2	2.4
2003-05 each	1.9	2.1
Front Hub Bearings, Replace (B)		
Exc. below	3.2	3.4
Defender 90	2.6	2.8
Freelander		
one side	2.1	2.3
both side	4.3	4.5
LR3 each	1.4	1.6
2003-05 Range Rover		
one side	3.2	3.4
both side	6.3	6.5
Front Hub Oil Seal, Replace (B)		
Exc. Range Rover	1.6	1.6
Range Rover	2.8	2.8
Front Hub Stub Axle, Replace (B)		
Exc. Range Rover	2.4	2.7
Range Rover	4.3	4.6
Front Shock Absorbers, Replace (B)		
w/air suspension		
one side	.8	1.0
both sides	1.4	1.6
w/o air suspension		
exc. Freelander		
one side	1.8	2.0
both sides	2.4	2.6
Freelander		
one side	1.4	1.6
both sides	2.7	2.9
Front Spring, Replace (B)		
w/air suspension		
each side	1.7	1.9
w/o air suspension		
exc. Freelander		
right side	1.2	1.4
left side	1.4	1.6
Freelander		
one side	1.4	1.6
both sides	2.7	2.9
Front Strut, Replace (B)		
2003-05 Range Rover	1.9	2.1
Lower Ball Joint, Replace (B)		
Freelander, LR3	2.1	2.4
2003-05 Range Rover		
one side	1.0	1.3
both sides	1.6	1.9
Lower Control Arm, Replace (B)		
Freelander, LR3		
one side	1.6	1.9
both sides	2.7	3.0
2003-05 Range Rover	.5	.8
Radius Rod, Replace (B)		
One side	.8	1.0
Both sides	1.4	1.6

	LABOR TIME	SEVERE SERVICE
Radius Rod Bushings, Replace (B)		
One side	1.2	1.4
Both sides	2.3	2.5
w/stabilizer bar add	.5	.7
Stabilizer Bar, Replace (B)		
Exc. Freelander	.9	1.1
Freelander	1.4	1.6
Stabilizer Bar Bushings, Replace (B)		
Exc. Freelander	.8	1.0
Freelander	.3	.5
Stabilizer Bar Links, Replace (B)		
One side	.7	.9
Both sides	.8	1.0
Strut Rod Bushings, Replace (B)		
Exc. below	.8	1.0
Range Rover		
1996-05 SE	1.2	1.4
Steering Knuckle, Replace (B)		
Range Rover		
1996-05 SE	3.2	3.7
Swivel Pin Assembly, Recondition (B)		
Exc. Defender 90		
one side	4.5	5.0
both sides	8.3	8.8
Defender 90		
one side	4.2	4.7
both sides	7.8	8.3
Swivel Pin Bearings, Replace (B)		
Right side	3.6	4.1
Left side	6.9	7.4
Swivel Pin Bearing Housing, Replace (B)		
Each side	5.6	6.1
Swivel Pin Housing Oil Seal, Replace (B)		
Exc. Range Rover		
each side	3.7	4.0
Range Rover		
one side	3.7	4.0
both sides	6.2	6.5
Upper Ball Joint, Replace (B)		
1996-05 Range Rover SE		
one upper or lower	4.2	4.5
both one side	5.5	5.8
Upper Suspension Link Bushing, Replace (B)		
Defender 90		
one side	.5	.7
both sides	.9	1.1

REAR SUSPENSION

	LABOR TIME	SEVERE SERVICE
Rear End Alignment (A)		
Freelander	2.5	2.7
2003-05 Range Rover	1.4	1.6
Diagnose Air Suspension System (A)		
LR3	1.8	1.8
Range Rover		
1993-02	1.0	1.0
2003-05	.7	.7

	LABOR TIME	SEVERE SERVICE
Air Compressor, Replace (B)		
Range Rover		
1993-95	3.3	3.3
1996-05	1.5	1.5
Air Pressure Pipe Harness, Replace (B)		
1993-95 Range Rover	5.8	5.8
Compressor Dryer, Replace (B)		
Range Rover		
1993-95	3.1	3.1
1996-05 SE	.5	.5
Compressor Relay, Replace (B)		
Range Rover	.5	.5
Electronic Control Module, Replace (B)		
Range Rover		
1993-95	1.6	1.6
1996-02 SE	.7	.7
2003-05	1.4	1.4
Height Control Switch, Replace (B)		
Range Rover		
1993-94 each	.9	.9
1995-05		
Classic each	.5	.5
SE	1.2	1.2
Level Control Valve, Replace (B)		
Range Rover		
1993-95	3.2	3.2
1996-05 SE	1.6	1.6
Rear Height Sensor, Replace (B)		
LR3	.5	.5
Range Rover		
1993-95		
one	2.0	2.0
both	2.5	2.5
1996-05 each	.9	.9
Rear Hub, Replace (B)		
Exc. below	1.9	2.4
Freelander		
one side	2.5	3.0
both sides	6.0	6.5
2003-05 Range Rover	2.5	3.0
Rear Hub Bearing, Replace (B)		
Exc. below	1.4	1.6
Freelander		
one	2.5	2.7
both	6.0	6.2
LR3	3.1	3.3
Range Rover		
1996-02 SE	1.9	2.1
2003-05 one side	3.6	3.8
Rear Hub Grease Seal, Replace (B)		
All Models	1.5	1.5
Rear Lower Control Arm Bushings, Replace (B)		
2003-05 Range Rover		
one	2.7	2.9
pair	3.2	3.4

LR-10 DEFENDER 90 : DISCOVERY : FREELANDER : LR3 : RANGE ROVER

	LABOR TIME	SEVERE SERVICE
Rear Lower Link, Replace (B)		
1987-05	.5	.7
Replace bushing add	.2	.4
Rear Shock Absorbers, Replace (B)		
Exc. below		
one side	.7	.9
both sides	1.2	1.4
Freelander		
one side	1.9	2.1
both sides		
3 door	3.0	3.2
5 door	3.5	3.7
LR3 each	1.9	2.1
2003-05 Range Rover		
one side	1.4	1.6
both sides	1.9	2.1
Rear Spring, Replace (B)		
w/air suspension	1.7	2.2
w/o air suspension		
exc. Freelander	.9	1.4
Freelander		
one side	1.9	2.4
both sides		
3 door	3.0	3.5
5 door	3.6	4.1
Rear Stabilizer Bar, Replace (B)		
1987-02	.9	1.1
LR3	2.7	2.9
2003-05 Range Rover	2.9	3.1
Rear Stabilizer Bar Link, Replace (B)		
1987-05	.5	.7
Rear Stabilizer Bar Bushings, Replace (B)		
1987-05	.7	.9
Rear Trailing Arm, Replace (B)		
Freelander	.5	.7
Rear Trailing Arm Bushing, Replace (B)		
Freelander	.6	.8
LR3	1.5	1.7
Rear Transverse Link, Replace (B)		
Freelander		
one side	.5	.7
both sides	1.1	1.3
2003-05 Range Rover	.5	.7
Rear Transverse Link Bushing, Replace (B)		
Freelander each	.6	.8
Rear Upper Control Arm Bushing, Replace (B)		
Exc. below		
one side	.7	.9
both sides	.9	1.1
LR3 each side	1.1	1.3
2003-05 Range Rover	3.6	3.8
Stub Axle, Replace (B)		
1987-05	1.5	2.0
Stub Axle Seal, Replace (B)		
1987-05	1.5	1.5

STEERING

Unless otherwise noted, time given does not include alignment.

AIR BAGS

	LABOR TIME	SEVERE SERVICE
Diagnose Air Bag System Component Each (A)		
Exc. below	1.0	1.0
Freelander	.3	.3
2003-05 Range Rover	.6	.6
ABS Wiring Harness, Replace (B)		
Discovery	10.5	10.5
Range Rover		
1996-05 SE	7.3	7.3
Air Bag Assy., Replace (B)		
Driver side		
exc. below	.5	.5
Range Rover		
1996-05 SE	1.2	1.2
Passenger side		
exc. below	.9	.9
Freelander	.2	.2
Range Rover		
2003-05	4.2	4.2
Side curtain		
2003-05 Range Rover		
one side	8.4	8.4
both sides	8.8	8.8
Crash Sensor, Replace (B)		
Exc. below each side	.7	.7
2003-05 Range Rover		
each side	.9	.9
Electronic Control Unit, Replace (B)		
Freelander	4.6	4.3
1995-05 Range Rover	1.3	1.3
Sensor Assy., Replace (B)		
Freelander w/AC	5.8	5.8
2003-05 Range Rover	2.2	2.2

LINKAGE

	LABOR TIME	SEVERE SERVICE
Center Link (Relay Rod), Steering, Replace (B)		
1987-05	1.8	2.3
Drag Link, Replace (B)		
Each	1.2	1.7
Idler Arm, Replace (B)		
1987-05	.9	1.4
Steering Damper, Replace (B)		
1987-05	.5	.7

POWER RACK & PINION

	LABOR TIME	SEVERE SERVICE
Power Steering Hoses, Replace (B)		
Includes: System bleeding.		
Freelander each	.9	1.1
2003-05 Range Rover		
pressure	1.5	1.7
return	.5	.5
Rack & Pinion Assy., Replace (B)		
Includes: Reset toe. System bleeding.		
Freelander	3.5	3.8
2003-05 Range Rover	5.9	6.2
Rack Boots, Replace (B)		
Includes: Reset toe.		
Freelander each	3.5	3.6
2003-05 Range Rover		
one	.6	.6
set	1.1	1.1
Steering Column, Replace (B)		
Exc. below	1.7	1.8
2003-05 Range Rover	2.1	2.2
Steering Pump, Replace (B)		
Includes: System bleeding.		
All Models	1.7	1.9
Steering Wheel, Replace (B)		
All Models	.6	.6
Tie Rod Assy., Replace (B)		
Includes: Reset toe.		
Exc. below	1.1	1.7
Freelander	4.0	4.3
2003-05 Range Rover		
inner	1.4	1.7
outer	.5	.8

POWER WORM & SECTOR

	LABOR TIME	SEVERE SERVICE
Troubleshoot Power Steering (A)		
1987-05	1.2	1.2
Gear Assy., Adjust (On Vehicle) (B)		
1987-05	.5	.7
Gear Assy., R&R and Recondition (B)		
1987-05	6.9	7.5
Gear Assy., Replace (B)		
1987-05	2.7	3.0
Steering Column, R&R or Replace (B)		
Defender 90	3.9	3.9
Discovery	2.2	2.2
LR3	2.0	2.2
Range Rover	2.5	2.5
Steering Column Lock Assy., Replace (B)		
Defender 90, LR3	1.4	1.4
Discovery	1.7	1.7
Range Rover		
1987-94	1.4	1.4
1995-05	2.0	2.0
Steering Pump, R&R or Replace (B)		
1987-05	2.2	2.4
Steering Pump Hoses, Replace (B)		
Does not include purging.		
Each	1.2	1.4
Steering Wheel, Replace (B)		
Exc. below	.8	.8
Discovery, LR3	.9	.9
Range Rover		
1996-05 SE	1.9	1.9

DEFENDER 90 : DISCOVERY : FREELANDER : LR3 : RANGE ROVER LR-11

	LABOR TIME	SEVERE SERVICE
Tie Rod Ends, Replace (B)		
One	1.4	1.6
each addl. add	.3	.5

HEATING & AIR CONDITIONING

When more than one component requires replacement where evacuation/recovery and recharging is already included, deduct 1.0 hour for each additional component from the time given.

Evacuate/Recover and Recharge System (B)
- 1987-05 .... 1.5   1.5

AC Hoses, Replace (B)
Includes: Evacuate/recover and recharge.
- Defender 90 one 2.0   2.2
- each addl. add 5   .6
- Discovery
  - compressor to condenser 3.1   3.3
  - liquid line 3.1   3.3
  - compressor to evap. 5.4   5.6
- Freelander
  - compressor to condenser 4.1   4.3
  - exc. compressor to condenser 5.0   5.2
- LR3
  - compressor to expansion valve ... 9.3   9.5
  - pressure hose assy. 4.0   4.2
- Range Rover
  - 1987-94
    - compressor to evap. 4.4   4.6
    - receiver/drier to evap. 4.9   5.1
  - 1995-05
    - Classic
      - compressor to evap. 1.8   2.0
      - receiver/drier to evap. 2.4   2.6
    - SE one 3.1   3.3
      - each addl. add 5   .7
    - 2003-05 each 2.8   3.0

AC On/Off Control Switch, Replace (B)
- Defender 90 .... 1.7   1.7
- Freelander 3   .3

Blower Motor, Replace (B)
- Defender 90
  - 1994-95 .... 1.7   1.7
  - 1997 .... 2.5   2.5
- Discovery
  - front 7.0   7.0
  - rear 3.8   3.8
- Freelander 3   .3
- LR3 .... 1.2   1.2
- Range Rover
  - 1987-94 .... 4.9   4.9
  - 1995 .... 7.0   7.0
  - 1996-02 8   .8
  - 2003-05
    - main 15.1 15.1
    - auxiliary 1.4   1.4

Blower Motor Relay, Replace (B)
- Defender 90, LR3 3   .3
- Discovery
  - front 3   .3
  - rear 1.2   1.2
- Range Rover 7   .7

Blower Motor Resistor, Replace (B)
- Defender 90, LR3 5   .5
- Discovery
  - front 5.3   5.3
  - rear 3.5   3.5
- Freelander 2   .2
- Range Rover
  - 1987-94 .... 8.9   8.9
  - 2003-05
    - main 2.2   2.2
    - auxiliary 1.2   1.2

Blower Motor Switch, Replace (B)
- Defender 90, LR3 .... 1.2   1.2
- Discovery
  - front 2.2   2.2
  - rear 5   .5
- Freelander 1.2   1.2
- Range Rover 3.1   3.1

Compressor Assy., Replace (B)
Includes: Transfer parts as required. Evacuate/recover and recharge.
- Defender 90 .... 3.1   3.3
- Discovery 3.5   3.7
- Freelander 4.8   5.0
- Range Rover
  - 1987-94 .... 2.3   2.5
  - 1995-02
    - Classic 2.7   2.9
    - SE 3.1   3.3
  - 2003-05 .... 3.3   3.5

Compressor or Fan Relay, Replace (B)
- 1987-95 Range Rover .. .3   .3

Condenser Assy., Replace (B)
Includes: Evacuate/recover and recharge.
- Exc. below 4.3   4.4
- Defender 90 .... 2.8   2.9
- LR3 .... 2.4   2.5
- 2003-05 Range Rover 4.8   4.9

Condenser Fan Motor, Replace (B)
- Defender 90, LR3 .... 1.2   1.5
- Discovery 4.3   4.6
- Range Rover
  - 1987-94 .... 1.4   1.7
  - 1995-02
    - Classic 3.9   4.2
    - SE 1.4   1.7
  - 2003-05 .... 2.0   2.3

Evaporator Core, Replace (B)
Includes: Evacuate/recover and recharge.
- Defender 90 .... 5.4   5.6
- Discovery
  - front 9.7   9.9
  - rear 7.1   7.3
- Freelander 6.0   6.2
- LR3 .... 11.6 11.8
- Range Rover
  - 1987-94 .... 5.4   5.6
  - 1995-02
    - Classic 9.7   9.9
    - SE 12.8 13.0
  - 2003-05 .... 18.0 18.2

Expansion Valve, Replace (B)
Includes: Evacuate/recover and recharge.
- Defender 90
  - 1994-95 .... 2.5   2.5
  - 1997 .... 4.4   4.4
- Discovery
  - front 10.0 10.0
  - rear 6.2   6.2
- LR3 .... 6.2   6.4
- Range Rover
  - 1987-94 .... 5.4   5.4
  - 1995-02
    - Classic 9.7   9.7
    - SE 3.1   3.1
  - 2003-05 .... 5.7   5.7

Heater Blower Motor, Replace (B)
- Defender 90, LR3 .... 2.7   2.7
- Discovery 7.5   7.5
- 1987-95 Range Rover 8.9   8.9

Heater Blower Motor Relay, Replace (B)
- All Models 7   .7

Heater Blower Motor Resistor, Replace (B)
- Defender 90 .... 2.3   2.3
- Discovery, LR3 8   .8
- Range Rover
  - 1987-94 .... 8.9   8.9
  - 1995 8   .8

Heater Blower Motor Switch, Replace (B)
- Defender 90, LR3 8   .8
- Discovery 2.7   2.7
- Range Rover
  - 1987-95 front 3.1   3.1

DEFENDER 90 : DISCOVERY : FREELANDER : LR3 : RANGE ROVER

	LABOR TIME	SEVERE SERVICE
Heater Control Assy., Replace (B)		
Defender 90, LR3	.9	.9
Discovery	2.8	2.8
Freelander	1.5	1.5
Range Rover		
1987-95	3.1	3.1
1996-05	.7	.7
Heater Core, R&R or Replace (B)		
Defender 90	3.1	3.3
Discovery	8.4	8.6
Freelander, LR3	4.5	4.7
Range Rover		
1987-95	8.9	9.1
1996-02 SE	12.2	12.4
2003-05	15.7	15.9
Heater Hoses, Replace (B)		
Exc. below each	.9	1.0
Freelander, LR3 each	1.5	1.6
2003-05 Range Rover		
exc. engine to heater pump each	1.6	1.7
engine to heater pump	1.9	2.0
High Pressure Cut-Off Switch, Replace (B)		
Discovery	.5	.5
Receiver/Drier Assy., Replace (B)		
Includes: Evacuate/recover and recharge.		
Defender 90		
1994-95	1.7	1.7
1997	3.3	3.3
Discovery	4.6	4.6
Freelander	3.6	3.6
LR3	2.6	2.6
Range Rover		
1987-95	2.5	2.5
1996-02 SE	2.9	2.9
2003-05	4.7	4.7
Water Control Valve, Replace (B)		
Exc. below	.9	1.1
Defender 90	.5	.7
2003-05 Range Rover	1.5	1.7

WIPERS & SPEEDOMETER

	LABOR TIME	SEVERE SERVICE
Radio, R&R (B)		
1987-05	.8	.8
Rear Window Washer Pump, Replace (B)		
Exc. below	.7	.7
Freelander	.8	.8
2003-05 Range Rover	.9	.9
Rear Window Wiper Motor, Replace (B)		
Defender 90	.7	1.0
Discovery	1.7	2.0
Freelander	1.2	1.5
Range Rover		
1987-95	2.7	3.0
1996-05	.8	1.1
Rear Window Wiper Switch, Replace (B)		
1987-05	.5	.5
Speedometer Cable & Casing, Replace (B)		
Defender 90	1.4	1.4
Speedometer Head, R&R or Replace (B)		
Exc. below	1.4	1.4
Defender 90	.8	.8
Freelander	.9	.9
Windshield Washer Pump, Replace (B)		
Exc. below	.8	.8
1996-05 Range Rover	1.2	1.2
Windshield Wiper and Washer Switch Assy., Replace (B)		
Defender 90		
1994-95	1.8	1.8
1997	1.2	1.2
Discovery, LR3	.9	.9
Freelander	.6	.6
Range Rover		
1987-95	.7	.7
1996-05 SE	1.2	1.2
Windshield Wiper Linkage, Replace (B)		
Exc. below	3.1	3.4
Discovery	1.6	1.9
2003-05 Range Rover	1.0	1.3
Windshield Wiper Motor, Replace (B)		
Defender 90	1.8	2.1
Discovery	1.5	1.8
Freelander	.8	.8
LR3	1.2	1.5
Range Rover		
1987-95	3.0	3.3
1996-02 SE	2.3	2.6
2003-05	.8	1.2

LAMPS & SWITCHES

	LABOR TIME	SEVERE SERVICE
Back-Up Lamp Bulb, Replace (C)		
1987-05	.3	.3
Back-Up Lamp Switch, Replace (B)		
Defender 90	.5	.5
Discovery	1.7	1.7
Front Parking Lamp Assy., Replace (B)		
Defender 90	.8	.8
Range Rover	.3	.3
Halogen Headlamp Bulb, Replace (B)		
1994-05	.3	.3
Hazard Warning Switch, Replace (B)		
Defender 90		
1994-95	1.8	1.8
1997	.3	.3
Discovery	.5	.5
Range Rover		
1987-94	.9	.9
1995-02		
Classic	.5	.5
SE	1.4	1.4
2003-05	.4	.4
Headlight Switch, Replace (B)		
Exc. below	.8	.8
Range Rover		
1996-97 SE	1.4	1.4
Headlights, Aim (B)		
Two	.5	.5
Horn, Replace (B)		
Exc. Freelander	.7	.7
Freelander	.2	.2
Horn Relay, Replace (B)		
Exc. below	.7	.7
2003-05 Range Rover	1.8	1.8
High Mount Stop Lamp Assy., Replace (B)		
Exc. below	.5	.5
1996-05 Range Rover	.9	.9
High Mount Stop Lamp Bulb, Replace (C)		
All Models	.2	.2
License Lamp Lens, Replace (B)		
1987-05	.5	.5
Multifunction Switch, Replace (B)		
Exc. below	.9	.9
1994-95 Defender 90	1.6	1.6
Range Rover		
1996-05 SE	1.4	1.4
Reflectors, Replace (B)		
One	.3	.3
each addl. add	.2	.2
Rear Combination Lamp Assy., Replace (B)		
All Models	.8	.8
Sealed Beam Headlight, Replace (B)		
All Models each	.7	.7
Side Marker Lamp Assy., Replace (B)		
All Models	.3	.3
Side Marker Lamp Bulb, Replace (C)		
Each	.2	.2
Stop Light Switch, Replace (B)		
Exc. Freelander	.8	.8
Freelander	.2	.2
Tail Lamp Lens or Bulb, Replace (C)		
All Models	.2	.2
each addl. add	.2	.2
Turn Signal or Hazard Warning Flasher, Replace (B)		
Exc. Freelander	.8	.8
Freelander	.2	.2

DEFENDER 90 : DISCOVERY : FREELANDER : LR3 : RANGE ROVER — LR-13

BODY

	LABOR TIME	SEVERE SERVICE
Front Door Lock, Replace (B)		
Exc. below	2.4	2.4
Discovery	2.9	2.9
Freelander		
3 door	1.0	1.0
5 door	1.2	1.2
Hood Hinge, Replace (B)		
Discovery	1.7	1.7
Freelander	.8	.8
Range Rover	2.5	2.5
Hood Lock, Replace (B)		
Exc. below each	.7	.7
Freelander		
lock w/AC	4.1	4.1
safety catch	.3	.3

	LABOR TIME	SEVERE SERVICE
Hood Release Cable, Replace (B)		
Defender 90	.8	.8
Discovery, LR3	.9	.9
Freelander w/AC	4.3	4.3
Range Rover		
1987-95	.5	.5
1996-02 SE	1.7	1.7
2003-05 each	.6	.6
Lock Striker Plate, Replace (B)		
1987-05	.3	.3
Rear Door Lock, Replace (B)		
Exc. below	.9	.9
1987-94 Range Rover	2.2	2.2
Window Regulator, Replace (B)		
Front	1.5	1.7
Rear	1.6	1.8

	LABOR TIME	SEVERE SERVICE
Window Regulator Motor, Replace (B)		
Exc. below		
front	.9	1.1
rear	1.6	1.8
Defender 90, LR3	1.2	1.4
Freelander		
front		
3 door	1.4	1.6
5 door	1.0	1.2
1996-05 Range Rover SE		
front	1.8	2.0
rear	1.5	1.7

NOTES

190 Series : 240D : 260E : 280 Series : 300 Series : 380 Series : 400 Series : 420SEL : 500 Series : 560 Series : 600 Series : C Series : CL Series : CLK Series : E Series : S Series : SL Series : SLK Series

SYSTEM INDEX

- MAINTENANCE MB-2
  - Maintenance Schedule ... MB-2
- CHARGING MB-3
- STARTING MB-3
- CRUISE CONTROL MB-3
- IGNITION MB-4
- EMISSIONS MB-5
- FUEL MB-5
- EXHAUST MB-8
- ENGINE COOLING MB-9
- ENGINE MB-10
  - Assembly MB-10
  - Cylinder Head MB-11
  - Camshaft MB-14
  - Crank & Pistons MB-16
  - Engine Lubrication ... MB-18
- CLUTCH MB-19
- MANUAL TRANSMISSION MB-19
- AUTO TRANSMISSION MB-19
- TRANSFER CASE MB-21
- SHIFT LINKAGE MB-21
- DRIVELINE MB-21
- BRAKES MB-22
- FRONT SUSPENSION MB-25
- REAR SUSPENSION MB-27
- STEERING MB-28
- HEATING & AC MB-30
- WIPERS & SPEEDOMETER ... MB-32
- LAMPS & SWITCHES MB-33
- BODY MB-33

OPERATIONS INDEX

A
- Air Bags MB-28
- Air Conditioning MB-30
- Alignment MB-25
- Alternator (Generator) .. MB-3
- Anti-Lock Brakes MB-22

B
- Back-Up Lamp Switch MB-33
- Ball Joint MB-26
- Battery Cables MB-3
- Bleed Brake System MB-23
- Blower Motor MB-30
- Brake Disc MB-24
- Brake Hose MB-23
- Brake Pads and/or Shoes .. MB-24

C
- Camshaft MB-14
- Camshaft Sensor MB-4
- Catalytic Converter MB-8
- Coolant Temperature (ECT) Sensor ... MB-5
- Crankshaft MB-17
- Crankshaft Sensor MB-4
- Cruise Control MB-3
- Cylinder Head MB-12

D
- Distributor MB-4
- Drive Belt MB-2
- Driveshaft MB-21

E
- EGR MB-5
- Electronic Control Module (ECM/PCM) ... MB-5
- Engine MB-10
- Engine Lubrication MB-18
- Engine Mounts MB-11
- Evaporator MB-31
- Exhaust MB-8
- Exhaust Manifold MB-8

F
- Fuel Injection MB-7
- Fuel Pump MB-6
- Fuel Vapor Canister MB-5

G
- Generator MB-3
- Glow Plug MB-4

H
- Halfshaft MB-21
- Headlamp MB-33
- Heater Core MB-32
- Horn MB-33

I
- Ignition Coil MB-4
- Ignition Switch MB-4
- Injection Pump MB-6
- Inner Tie Rod MB-29
- Intake Air Temperature (IAT) Sensor ... MB-5
- Intake Manifold MB-6

L
- Lower Control Arm MB-26

M
- Maintenance Schedule ... MB-2
- Mass Air Flow (MAF) Sensor . MB-5
- Master Cylinder MB-23
- Muffler MB-9
- Multifunction Lever/Switch ... MB-33

O
- Oil Pan MB-18
- Oil Pump MB-19
- Outer Tie Rod MB-29
- Oxygen Sensor MB-5

P
- Parking Brake MB-24
- Pistons MB-17

R
- Radiator MB-9
- Radiator Hoses MB-10

S
- Shock Absorber/Strut, Front .. MB-25
- Shock Absorber/Strut, Rear .. MB-27
- Spark Plug Cables MB-5
- Spark Plugs MB-5
- Spring, Front Coil MB-25
- Spring, Rear Coil MB-27
- Starter MB-3
- Steering Wheel MB-29

T
- Thermostat MB-9
- Throttle Body MB-7
- Timing Chain MB-15

U
- Upper Control Arm MB-27

V
- Valve Body MB-20
- Valve Cover Gasket MB-13
- Vehicle Speed Sensor ... MB-32

W
- Water Pump MB-10
- Wheel Balance MB-2
- Window Regulator MB-33
- Windshield Washer Pump .. MB-33
- Windshield Wiper Motor .. MB-33

MB-1

MB-2
190 SERIES : 240D : 260E : 280 SERIES : 300 SERIES : 380 SERIES : 400 SERIES : 420SEL : 500 SERIES : 560 SERIES

	LABOR TIME	SEVERE SERVICE
MAINTENANCE		
Air Cleaner Filter Element, Replace (C)		
Each	.3	.4
Drive Belt, Adjust (B)		
Exc. below		
one	.6	.7
each addl. add	.2	.3
240D	.8	.9
300CD	.8	.9
300TD	.8	.9
380SE, 380SEC, 380SEL, 380SL	.8	.9
Drive Belt, Replace (B)		
Air pump	.6	.6
Alternator		
exc. below	1.2	1.4
190D	.5	.6
190E	.3	.4
300CD	1.7	1.8
300E	.8	.9
300SD		
1981-93	1.7	1.8
1981-85	.8	.9
1992-93	.8	.9
300TD	1.7	1.8
Compressor		
190E 2.3L	.8	.9
240D	.8	.9
280CE, 280E	.3	.4
300CD, 300D, 300SD, 300TD	.7	.8
380SEC, 380SEL	.5	.6
420SEL	.5	.6
500SEC, 500SEL	.5	.6
560SEC, 560SEL	.5	.6
Power steering		
exc. below	1.3	1.4
190E 2.3L	.7	.8
Serpentine		
exc. below	.9	1.1
600SEC, 600SEL	1.7	1.8
600SL, S600	1.8	1.9
C220	.5	.6
C280		
1994-97	1.4	1.5
1998-00	.5	.6
CL600	1.5	1.6
CLK320, CLK430	.5	.6
E300D	1.4	1.5
S320	1.4	1.5
S350TD	.7	.8
S420	.5	.6
S500, S600	.5	.6
SL600	1.1	1.2
SLK230, SLK32	.6	.7
SLK320	.5	.5

	LABOR TIME	SEVERE SERVICE
Drive Belt Tensioner, Replace (B)		
190D 2.2L	1.3	1.6
190E 2.6L	2.8	3.1
240D	1.7	2.0
260E	2.8	3.1
300CD, 300D, E300D	1.7	2.0
300CE, 300TE	1.7	2.0
300E	2.8	3.1
300SD		
1981-85	1.7	2.0
1992-93	1.3	1.6
300SE, 300SEL	3.1	3.3
300SL	1.7	2.0
300TD	3.1	3.3
380SE, 380SEC, 380SEL, 380SL	1.7	2.0
400E, 400SEL	1.6	1.9
420SEL	1.7	2.0
500E, E500	1.6	1.9
500SL	1.7	2.0
500SEC, 500SEL		
1984-85	1.7	2.0
1993	1.3	1.6
560SEC, 560SEL, 560SL	1.7	2.0
600SEC, 600SEL, 600SL	2.8	3.1
C220, C230, C240	1.1	1.4
C280		
1994-97	2.8	3.1
1998-00	1.4	1.7
C32, C320	.9	1.2
C36, C55	2.8	3.1
CL500	2.2	2.5
CL600	1.6	1.9
CLK320, CLK55	1.3	1.6
CLK430, CLK500	1.0	1.3
E320	3.0	3.3
E420, S420	1.6	1.9
E430, E500	2.0	2.3
S320, S430	2.8	3.1
S350TD, E55	1.3	1.6
S500, S55	2.2	2.5
S600	3.6	3.9
SL500, SL55	1.4	1.7
SL600	3.7	4.0
SLK230, SLK32	1.1	1.4
SLK320, SLK55	1.5	1.8
Fuel Filter, Replace (B)		
Exc. below	.9	1.1
280CE, 280E	.3	.3
300CE, 300TE	.3	.3
380SEC, 380SEL	.3	.3
420SEL	1.4	1.4
560SEC	1.4	1.4
560SL	.3	.3
E300D	.3	.3
Halogen Headlight Bulb, Replace (B)		
1985-05	.5	.5

	LABOR TIME	SEVERE SERVICE
Oil and Filter, Change (C)		
Includes: Correct all fluid levels.		
1981-05	.6	.8
Tire, Replace (C)		
Includes: Dismount old tire and mount new tire to rim.		
1981-05	.5	.5
Tires, Rotate (C)		
1981-05	.5	.5
Wheel, Balance (B)		
One	.3	.3
each addl. add	.2	.2

SCHEDULED MAINTENANCE INTERVALS

If necessary, refer to appropriate Chilton maintenance service information.

7,500 Mile Service (B)
1996-05	.9	.9
w/AT add	.1	.1

15,000 Mile Service (B)
1996-05	2.5	2.5
w/AT add	.1	.1

22,500 Mile Service (B)
1996-05	.9	.9
w/AT add	.1	.1

30,000 Mile Service (B)
1996-05	3.7	3.7
w/AT add	.1	.1
w/convertible add	.1	.1

37,500 Mile Service (B)
1996-05	.9	.9
w/AT add	.1	.1

45,000 Mile Service (B)
1996-05	2.9	2.9
w/AT add	.1	.1

52,500 Mile Service (B)
1996-05	.9	.9
w/AT add	.1	.1

60,000 Mile Service (B)
1996-05	4.9	4.9
w/AT add	.1	.1
w/convertible	.1	.1

67,500 Mile Service (B)
1996-05	.9	.9
w/AT add	.1	.1

75,000 Mile Service (B)
1996-05	2.4	2.4
w/AT add	.1	.1

82,500 Mile Service (B)
1996-05	.9	.9
w/AT add	.1	.1

90,000 Mile Service (B)
1996-05	4.2	4.2
w/AT add	.1	.1
w/convertible add	.1	.1

97,500 Mile Service (B)
1996-05	.9	.9
w/AT add	.1	.1

600 SERIES : C SERIES : CL SERIES : CLK SERIES : E SERIES : S SERIES : SL SERIES : SLK SERIES — MB-3

	LABOR TIME	SEVERE SERVICE
CHARGING		
Alternator Circuits, Test (B)		
Includes: Test component output.		
1981-05	.7	.7
Alternator Assy., Replace (B)		
Includes: Pulley transfer.		
190D, 190E		
1984-85	1.2	1.4
1986-87 16V	1.6	1.8
1986-93	.9	1.1
240D	2.8	3.0
260E	.9	1.1
280CE, 280E	2.8	3.0
300CD	2.8	3.0
300D		
1981-85	2.8	3.0
1986-93	.9	1.1
300E	.9	1.1
300SD	1.2	1.4
300SE, 300SEL	.8	1.0
300SL	1.3	1.5
380SE, 380SEL	2.8	3.0
380SEC	2.0	2.2
400E, E420	1.8	2.0
420SEL	.9	1.1
500E, E500	1.7	1.9
500SEC, 500SEL	2.0	2.2
500SL	1.6	1.8
560SEC, 560SEL, 560SL	.9	1.1
600SEC, 600SEL, 600SL	2.5	2.7
C220, C230	1.2	1.4
C240, C280	1.4	1.6
C32, C36 AMG	1.4	1.6
C320, CLK55	1.5	1.7
CLK320, CLK430	1.1	1.3
CL500, CL55	2.4	2.6
CLK500	2.7	2.9
CL600	6.2	6.4
E300D	.8	1.0
E320	.9	1.1
E430	1.1	1.3
E55, C55	2.8	3.1
S320	1.4	1.6
S350TD	1.2	1.4
S420, S430, S500	2.7	2.9
S55	2.4	2.6
S600, SL55	2.7	2.9
SL500	1.2	1.4
SL600	2.2	2.4
SLK230	.7	.9
SLK320	1.1	1.3
Alternator Voltage Regulator, Replace (B)		
Exc. below	1.0	1.2
190D, 190E		
1986-87 16V	1.3	1.5

	LABOR TIME	SEVERE SERVICE
300SL	1.6	1.8
500SL	1.8	2.0
600SL	2.8	3.0
C220, C230	1.3	1.5
C280	1.7	1.9
C36 AMG	1.7	1.9
CLK320, CLK430	1.7	1.9
E300D	1.4	1.6
E430	1.5	1.7
S420	3.0	3.2
S600	3.0	3.2
SL500, SL600	2.5	2.7
STARTING		
Starter Draw Test (On Car) (B)		
1981-05	.3	.3
Battery Cables, Replace (C)		
Positive		
exc. below	2.0	2.2
190D, 190E	1.4	1.6
240D	.9	1.1
300CD, 300D	.9	1.1
300SL	3.5	3.7
500SL, SL500	3.5	3.7
560SL	3.2	3.4
600SL, SL600	3.5	3.7
C230, C240, C32, C55	1.5	1.7
E430	.8	1.0
S320	1.8	2.0
Negative		
exc. below	.5	.7
380SE, 380SL	.8	1.0
560SL	.8	1.0
CL230	.3	.5
Starter Assy., Replace (B)		
190D	3.1	3.3
190E	4.2	4.4
190E 16V	5.2	5.4
240D	1.8	2.0
260E	1.7	1.9
280CE, 280E	2.2	2.3
300CD	2.2	2.4
300CE, 300TE	1.7	1.9
300D		
1981-85	3.0	3.2
1992-93	1.8	2.0
300E	1.7	1.9
300SE, 300SEL	2.2	2.4
300SL	1.3	1.5
300TD	2.8	3.0
380SE	1.8	2.0
380SEC, 380SEL	2.3	2.5
400E, E420	1.8	2.0
420SEL	1.9	2.1
500E, E500	1.8	2.0
500SEC, 500SEL	2.3	2.5
500SL	1.8	2.0
560SEC, 560SEL	1.9	2.1

	LABOR TIME	SEVERE SERVICE
560SL	4.2	4.4
600SEC, 600SEL, 600SL	1.8	2.0
C220, C230	1.5	1.7
C240	3.0	3.2
C280		
1994-97	3.0	3.2
1998-00	1.2	1.4
C32, C320	3.0	3.2
C36, C55	3.0	3.2
CL500	1.1	1.3
CL600	1.2	1.4
CLK320	1.2	1.4
CLK430	1.0	1.2
CLK500	3.0	3.2
CLK55	1.0	1.2
E320	1.7	1.9
E430	7.0	7.2
E500	2.6	2.8
E55	2.3	2.5
S350TD	1.8	2.0
S420	1.5	1.7
S430, S55	1.2	1.4
S500	2.8	3.0
S600	1.5	1.7
SL500, SL600	1.5	1.7
SLK230	7.0	7.2
SLK320	1.2	1.4
SLK55	2.0	2.2
CRUISE CONTROL		
Circuits, Test (B)		
Exc. below	1.2	1.2
260E	1.6	1.6
300D, 300E, 300TD	1.6	1.6
E300D	1.6	1.6
E320	1.6	1.6
Actuator Assy., Replace (B)		
Exc. below	1.6	1.8
190D, 190E		
1984-85	1.2	1.2
1986-93 2.3L	2.3	2.3
1986-87 16V	3.9	3.9
240D	1.2	1.2
280CE, 280E	1.2	1.2
300CD	1.2	1.2
300D		
1986-93	2.3	2.3
300SD		
1981-85	.8	.8
1992-93	2.3	2.3
300TD	1.2	1.2
380SEC, 380SEL	1.2	1.2
500SEC, 500SEL	.8	.8
560SEC, 560SEL	1.7	1.7
560SL	.9	.9
600SEC, 600SEL, 600SL		
right	.9	.9
left	2.5	2.5

MB-4 — 190 SERIES : 240D : 260E : 280 SERIES : 300 SERIES : 380 SERIES : 400 SERIES : 420SEL : 500 SERIES : 560 SERIES

	LABOR TIME	SEVERE SERVICE
Actuator Assy., Replace (B)		
C220	2.0	2.0
C230, C55	4.4	4.4
E300D	1.5	1.5
S350TD	2.3	2.3
Control Amplifier Assy., Replace (B)		
Exc. below	1.2	1.2
190D, 190E		
1984-85	.7	.7
1986-93	.3	.3
240D	.7	.7
260E	.3	.3
280CE, 280E	.7	.7
300CD	.7	.7
300CE, 300SE, 300SEL, 300TE	.8	.8
300D		
1981-85	.7	.7
1986-93	.3	.3
300E	.3	.3
300SD	.7	.7
300SL	.8	.8
300TD	.7	.7
380SEC, 380SEL	.7	.7
420SEL	.8	.8
500SEC, 500SEL	.7	.7
500SL	.8	.8
560SEC, 560SEL, 560SL	.8	.8
Control Switch, Replace (B)		
Exc. below	1.9	1.9
240D	.9	.9
280CE, 280E	1.2	1.2
300CD	1.2	1.2
300CE, 300SE, 300SEL, 300TE	1.4	1.4
300TD	1.2	1.2
380SEC, 380SEL	1.8	1.8
420SEL	1.4	1.4
500SL	1.4	1.4
560SEC, 560SEL, 560SL	1.4	1.4
C230, C55	2.5	2.5
CLK320	2.5	2.5
CLK430	2.5	2.5
E300D	2.5	2.5
E430	2.5	2.5
S420	2.5	2.5
S500, SL500	2.5	2.5
S600, SL600	2.5	2.5
Electronic Control Module, Replace (B)		
Exc. below	.8	.8
190D, 190E	.3	.3
260E	.3	.3
300CE, E300D, 300D turbo	.3	.3
300E, 300SL, 300TE	.5	.5
500SL	.5	.5
C220	.3	.3

	LABOR TIME	SEVERE SERVICE
Linkage, Adjust (B)		
1981-05	.5	.7
Speed Sensor, Replace (B)		
Exc. below	1.0	1.0
190D, 190E		
1984-85	.5	.5
1986-93	1.4	1.4
240D	.5	.5
280CE, 280E	.5	.5
300CD	.5	.5
300D		
1981-85	.5	.5
300SD	.5	.5
300TD	.7	.7
380SEC, 380SEL	.5	.5
500SEC, 500SEL	.5	.5

IGNITION

	LABOR TIME	SEVERE SERVICE
Diagnose Ignition System Component (B)		
1981-05	1.2	1.2
Diagnose Glow Plug System (A)		
1981-05	.8	.8
Camshaft Position Sensor, Replace (B)		
Exc. below	.7	.9
300SE, 300SL	1.4	1.6
C220	1.2	1.4
C230, C55	2.2	2.4
C280	1.4	1.6
E320	.8	1.0
Crankshaft Sensor, Replace (B)		
Exc. below	1.3	1.5
300SE	2.3	2.5
560SL	1.9	2.1
C220	2.4	2.6
C230, C55	.5	.7
C280	3.1	3.3
C36 AMG	3.1	3.3
CL600	1.8	2.0
CLK430	.6	.6
E320, E430	.6	.8
S600	1.2	1.4
Distributor, Replace (B)		
Includes: Reset base ignition timing.		
Exc. below	1.5	1.8
260E	.9	1.2
380SEC	.9	1.2
500SEC		
6 cyl.	1.3	1.6
V8	1.6	1.9
500SEL	.9	1.2
Distributor Cap and/or Rotor, Replace (B)		
Single 300SE	.5	.5
Dual each exc. below	.7	.7
600SEC, 600SEL, S600	.9	.9

	LABOR TIME	SEVERE SERVICE
Electronic Control Module, Replace (B)		
1981-04	.5	.5
Glow Plug, Replace (B)		
1984-89 190D	1.4	1.6
240D		
one	.3	.5
all	.8	1.0
300CD		
one	.3	.5
all	.9	1.1
300D		
1981-87		
one	.3	.5
all	.9	1.1
1988-93	3.0	3.2
300SD		
1981-85		
one	.3	.5
all	.9	1.1
1992-93	3.0	3.2
300TD	.9	1.1
E300D		
1995-97	3.3	3.5
1998-99	3.7	3.9
2005 E320	1.4	1.6
S350TD	3.0	3.2
Ignition Coil, Replace (B)		
1981-01	.7	.7
Ignition Switch, Replace (B)		
190E	2.5	2.5
260E	2.3	2.3
280CE, 280E	2.3	2.3
300CE, 300TE	2.3	2.3
300E	2.3	2.3
300SD	3.0	3.0
300SE	3.0	3.0
300SEL	2.5	2.5
380SE, 380SEC, 380SEL	2.5	2.5
380SL	1.4	1.4
400E, 400SEL, E420	2.3	2.3
400SE	3.0	3.0
420SEL	2.5	2.5
500E, E500	2.3	2.3
500SEC, 500SEL	2.5	2.5
500SL	3.1	3.1
560SEC, 560SEL	2.5	2.5
560SL	1.4	1.4
600SEC, 600SEL, S600	3.0	3.0
600SL	3.1	3.1
C220	2.3	2.3
C230, C240	2.7	2.7
C280, C32, C320	2.7	2.7
C36, C55	2.9	2.9
CL500, CL600	3.8	3.8
CLK320, CLK55	2.7	2.7
CLK430, CLK500	2.2	2.2

600 SERIES : C SERIES : CL SERIES : CLK SERIES : E SERIES : S SERIES : SL SERIES : SLK SERIES — MB-5

	LABOR TIME	SEVERE SERVICE
E300D	2.7	2.7
E320, E500, E55	2.3	2.2
E430	2.2	2.2
G500, G55	2.3	2.2
S320, S350TD	3.0	3.0
S420, S430, S55	4.4	4.4
S500, SL55	3.8	3.8
SL500, SL600	3.4	3.4
SLK32	2.2	2.2
SLK320, SLK55	2.7	2.7

Spark Plug (Ignition) Cables, Replace (B)

	LABOR TIME	SEVERE SERVICE
Exc. below	.8	1.1
280CE, 280E	.7	1.0
380SE, 380SEC, 380SEL, 380SL	2.0	2.3
420SEL	2.0	2.3
500SEC, 500SEL	2.0	2.3
560SEC, 560SEL, 560SL	2.0	2.3
560SEL, 600SEC w/R&R intake manifold	10.7	11.2
600SL, SL600 w/R&R intake manifold	8.8	9.3
C230, C240, C32	.3	.6
CLK320, CLK430	.3	.6
E320, E430	.3	.6
S600 w/R&R intake manifold	10.7	11.2
SLK230	.3	.6
SLK320, CLK320	.3	.6
CL600	2.0	2.3

Spark Plugs, Replace (B)

	LABOR TIME	SEVERE SERVICE
Exc. below	.8	1.1
190E 4 cyl.	.5	.8
500SEC V8	1.5	1.8
600SEC, 600SEL, 600SL	1.4	1.7
C240, C280, C55	3.0	3.3
CL600, S600	4.5	4.8
CLK320, C320, C32, C55	3.0	3.3
CLK430, CLK500	3.0	3.3
E320, CLK55, SLK55	2.5	2.8
E430, E500, G500	2.6	2.9
S420, S430	1.8	2.1
S500, SL500	1.4	1.7
S600, SL600	3.0	3.3
SLK320	3.0	3.3
SLK430, SLK55	3.3	3.6

Vacuum Advance Unit, Replace (B)
Includes: Reset timing.

	LABOR TIME	SEVERE SERVICE
Exc. 380SL	.5	.7
380SL	.7	.9

EMISSIONS

The following operations do not include testing. Add time as required.

Dynamometer Test (A)

	LABOR TIME	SEVERE SERVICE
1981-05	.8	.8

EGR System, Test (A)

1981-05	.5	.5

Air Flow Sensor, Replace (B)

1981-89	.3	.3
1990-05	.7	.7

Air Injection Check Valve, Replace (B)

1981-05	.8	.8

Air Intake Sensor, Replace (B)

Gasoline	.3	.3
Diesel	.7	.7

Air Pump Assy., Replace (B)

	LABOR TIME	SEVERE SERVICE
190E	.9	1.1
260E	.9	1.1
280CE, 280E	2.8	3.0
300CE, 300E	.9	1.1
300SE 1992-93	1.4	1.6
1984-85	3.1	3.3
300SEL, 300SL, 300TE	.9	1.1
380SE, 380SEL	3.1	3.3
380SL	4.2	4.4
400SE	1.2	1.4
400SEL	1.4	1.6
420SEL	3.1	3.3
500E, E500	1.4	1.6
500SEC 1984-85	3.1	3.3
1993	1.4	1.6
500SEL	1.4	1.6
500SL	2.4	2.6
560SEC, 560SEL	3.1	3.3
560SL	4.2	4.2
600SEC, 600SEL	3.6	3.8
600SL	2.4	2.6
C220, C55	1.2	1.4
C230	1.3	1.5
C240, C280, C320	.8	1.0
CL500	1.2	1.4
CL600	3.4	3.6
CLK320, CLK430	.8	1.0
E320, CLK500, CLK55	.9	1.1
E420	1.2	1.4
E430, E500, E55	.8	1.0
S320, SLK55	1.3	1.5
S420	1.5	1.7
S430, S500, S55	.6	.8
S500, SLK320	1.2	1.4
S600	3.9	4.1
SL500, SL600	1.9	2.1
SLK230, SLK32	.8	1.0

Air Temperature Sensor, Replace (B)

	LABOR TIME	SEVERE SERVICE
1990-05	.3	.3

Coolant Temperature (ECT) Sensor, Replace (B)

Gasoline	.5	.5
Diesel	.3	.3

EGR Back Pressure Transducer, Replace (B)

1981-99 diesel	.7	.7

EGR Control Unit, Replace (B)

1981-05	.5	.5

EGR Control Valve, Replace (B)

Gasoline	.7	1.0
Diesel	.5	.8

EGR Frequency Valve, Replace (B)

1981-85	.5	.5

EGR Temperature Sensor, Replace (B)

1984-05	.6	.6

EGR Vacuum Amplifier, Replace (B)

1981-05	.4	.4

Electronic Control Module (ECM/PCM), Replace (B)

1984-05	.5	.5

Fuel Vapor Canister, Replace (B)

1981-87	.2	.2
1988-05	.9	.9

Mass Air Flow (MAF) Sensor, Replace (B)

1990-05	.8	.8

Oxygen Sensor, Replace (B)

Exc. below	.6	.9
300SL	.9	1.2
500SL, CL500, SL500	.9	1.2
C280, C320, CLK320	.8	1.1
S420, CLK430, C55, CLK500, SLK55	.8	1.1
CLK55, S600,	.8	1.1
CL600, SL600	1.5	1.7

Pressure Relief Valve, Replace (B)

All Models	.5	.5

Purge Control Valve, Replace (B)

1981-05	.5	.5

FUEL

DELIVERY

Auxiliary Fuel Pump, Replace (B)

	LABOR TIME	SEVERE SERVICE
Exc. below	.9	1.1
300SE, S320	1.3	1.5
400SE, 400SEL	1.3	1.5
500SEC, 500SEL	1.3	1.5
600SEC, 600SEL, S600, CL600, SL600	1.3	1.5
S420, C55	1.2	1.4

Fuel Accumulator, Replace (B)

1981-05	.9	.9

Fuel Damper, Replace (B)

1981-85	.7	.7
1986-05	.5	.5

MB-6 — 190 SERIES : 240D : 260E : 280 SERIES : 300 SERIES : 380 SERIES : 400 SERIES : 420SEL : 500 SERIES : 560 SERIES

	LABOR TIME	SEVERE SERVICE
Fuel Filter, Replace (B)		
Exc. below	.9	*1.1*
280CE, 280E	.3	*.3*
300CE, 300TE	.3	*.3*
380SEC, 380SEL	.3	*.3*
420SEL	1.4	*1.4*
560SEC	1.4	*1.4*
560SL	.3	*.3*
E300D	.3	*.3*
E320, E430	.6	*.6*
Fuel Gauge (Tank), Replace (B)		
190D, 190E	.7	*.9*
260E	.7	*.9*
280CE, 280E	1.3	*1.5*
300CD, 300D	1.3	*1.5*
300E	.7	*.9*
300SD		
1981-85	1.6	*1.8*
1992-93	.8	*1.0*
300SE, 300SL	.8	*1.0*
300SEL	1.6	*1.8*
300TD	1.3	*1.5*
300TE	1.6	*1.8*
380SE, 380SEC, 380SEL	1.6	*1.8*
380SL	1.3	*1.5*
400E	.7	*.9*
400SE	1.6	*1.8*
400SEL	.8	*1.0*
420SEL	1.6	*1.8*
500E, E500	.7	*.9*
500SL	.8	*1.0*
560SEC, 560SEL	1.6	*1.8*
560SL	1.3	*1.5*
600SEC, 600SEL, 600SL	.8	*1.0*
C220	1.4	*1.6*
C230, C240	1.3	*10.5*
C280, C32, C320	1.3	*1.5*
C36, C55	1.5	*1.7*
CL500, CL600	.8	*1.0*
CLK320, CLK430	1.4	*1.6*
CLK500, CLK55	1.3	*1.5*
E300D	3.8	*4.0*
E320	1.6	*1.8*
E420, S420	.7	*.9*
E430, E55, E500	1.0	*1.0*
G500, G55	1.3	*1.5*
S320, S350TD, S55	.8	*1.0*
S430, S500, SL500	.8	*1.0*
S600, SL600	.8	*1.0*
SLK320, SLK32, SLK55	1.0	*1.0*
Fuel Pump, Replace (B)		
Exc. below	1.4	*1.6*
190D	.7	*.9*
190E	1.4	*1.6*
300D, S350TD	.5	*.7*
300SE, S320	.9	*1.1*

	LABOR TIME	SEVERE SERVICE
300TD	.7	*.9*
400SE, S420	.9	*1.1*
500SEC, 500SEL	.9	*1.1*
600SEC, 600SEL	.9	*1.1*
600SL, S600, SL600	.9	*1.1*
Fuel Pump Control Module, Replace (B)		
1981-05	.5	*.8*
Fuel Tank, Replace (B)		
Includes: Drain and refill.		
190D, 190E	1.8	*2.0*
240D	1.6	*1.8*
260E	3.0	*3.2*
280CE, 280E	1.6	*1.8*
300D	1.6	*1.8*
300E	3.0	*3.2*
300SD		
1981-85	2.0	*2.2*
1992-93	3.0	*3.2*
300SE	2.5	*2.7*
300SEL	2.8	*3.0*
300SL	5.3	*5.5*
300TD	3.9	*4.1*
380SEC, 380SEL	2.0	*2.2*
400E, E420	4.5	*4.7*
420SEL	2.8	*3.0*
500E, E500	4.4	*4.6*
500SEC, 500SEL	2.5	*2.7*
500SL	5.8	*6.0*
560SEC, 560SEL	2.8	*3.0*
560SL	3.6	*3.8*
600SEC, 600SEL	2.5	*2.7*
600SL	5.8	*6.0*
C220	8.3	*8.5*
C230	9.0	*9.2*
C240	3.0	*3.2*
C280	9.0	*9.2*
C32, C320, C55	9.4	*9.6*
C36	8.3	*8.5*
CL500, CL600	2.9	*3.1*
CLK320, CLK430	10.3	*10.5*
CLK500	7.7	*7.9*
CLK55	9.2	*9.4*
E300D	3.6	*3.8*
E320		
convertible	4.4	*4.6*
coupe, sedan	3.1	*3.3*
wagon	5.3	*5.5*
E430, E500, E55	3.5	*3.7*
S320, S350TD	3.0	*3.2*
S420, S55	3.1	*3.3*
S430, S500	2.9	*3.1*
S600	3.1	*3.3*
SL500, SL600	6.2	*6.4*
SLK230, SLK32	4.7	*4.9*
SLK320, SLK55	5.0	*5.2*
Fuel Injection Pump, Replace (B)		
190D	8.2	*8.7*
240D	5.4	*5.9*
300CD, 300D	5.6	*6.1*

	LABOR TIME	SEVERE SERVICE
300SD		
1981-85	6.5	*7.0*
1992-93	6.7	*7.2*
300TD, S350TD	6.7	*7.2*
E300D	6.7	*7.2*
Intake Manifold and/or Gasket, Replace (B)		
Includes: Adjustments.		
1984-89 190D		
w/turbo	3.1	*3.6*
w/o turbo	2.5	*3.0*
190E		
1985-93 2.3L	4.2	*4.7*
1986-87 16V	5.8	*6.3*
1987-93 2.6L	4.8	*5.3*
260E	4.8	*5.3*
280CE, 280E	7.3	*7.8*
300CE, 300TE	4.8	*5.3*
300D	4.4	*4.9*
300E	4.8	*5.3*
300SD	3.8	*4.3*
300SE	5.6	*6.1*
300SEL	3.5	*4.0*
300SL	5.2	*5.7*
300TD	4.4	*4.9*
380SE, 380SEC, 380SEL, 380SL	7.5	*8.0*
400E	6.0	*6.5*
400SE, 400SEL	5.6	*6.1*
420SEL	7.5	*8.0*
500E, E500	5.6	*6.1*
500SEC, 500SEL	7.5	*8.0*
500SL	9.3	*9.8*
560SEC, 560SEL, 560SL	7.5	*8.0*
600SEC, 600SEL, S600	9.9	*10.4*
600SL	7.6	*8.1*
C220, C230	3.9	*4.4*
C240, C32, CLK500	2.8	*3.0*
C280		
1994-97	4.4	*4.9*
1998-00	3.6	*4.1*
C55	5.1	*5.6*
CL500	5.8	*6.3*
CL600, SLK55	7.0	*7.5*
CLK320	3.6	*4.1*
CLK430	3.5	*4.0*
E300D	3.1	*3.6*
E320	3.8	*4.3*
E420, S420	6.0	*6.5*
E430, E500	3.8	*4.3*
S320, E55	3.0	*3.5*
S430, S500, S55	3.2	*3.7*
S350TD, SLK320	3.8	*4.3*
SL500, S600	6.4	*6.9*
SL600	7.6	*8.1*
SLK230, SLK32	3.1	*3.6*
Replace manifold add	1.2	*1.7*

600 SERIES : C SERIES : CL SERIES : CLK SERIES : E SERIES : S SERIES : SL SERIES : SLK SERIES — MB-7

	LABOR TIME	SEVERE SERVICE
Lower Intake Manifold and/or Gasket, Replace (B)		
1985-93 190E 2.3L	2.0	2.5
INJECTION		
EFI System, Test (B)		
1981-85 CIS	1.4	1.4
Fuel Injector Hydraulic Test (B)		
1986-2005	1.3	1.3
Fuel Injector Electronic Test (B)		
1986-2005	1.6	1.6
Injection Timing, Check & Adjust (B)		
Diesel	.9	1.2
Coolant Temperature Switch, Replace (B)		
Diesel	.5	.5
EDS Control Unit, Replace (B)		
Diesel	.5	.5
EGR Reference Resistor, Replace (B)		
Diesel	.2	.3
Electromagnetic Actuator, Replace (B)		
Diesel	.9	.9
Electronic Control Unit, Replace (B) Does not include diagnosis.		
1984-05	.5	.5
Engine Speed Sensor, Replace (B)		
1981-95 diesel	.9	.9
Fuel Accumulator, Replace (B)		
1981-85 CIS	.8	.8
1984-97 CISE	.7	.7
Fuel Distributor, Replace (B)		
1981-01	2.0	2.0
Fuel Pressure Damper, Replace (B)		
1981-85	.7	.7
Fuel Injectors, Replace (B)		
190E		
2.3L	.8	1.2
2.6L	1.3	1.7
260E	1.3	1.7
280	2.0	2.4
280CE, 280E	2.4	2.8
300CE, 300E, 300SEL	1.3	1.7
300SE	1.4	1.8
300SL	2.3	2.7
300TE	1.3	1.7
380SE, 380SEL, 380SL	2.4	2.8
400E, 400SE, E420, S420	2.0	2.4
400SEL	2.0	2.4
420SEL	2.4	2.8
500E, 500SEC, 500SEL, E500	2.0	2.4
500SL	2.4	2.8
560SEC, 560SEL, 560SL	2.4	2.8
600SEC, 600SEL, 600SL	2.8	3.2
C220, C230	1.4	1.8
C240, C280, C32	2.5	2.9
C320	2.3	2.7
CL500	1.6	2.0
CL600	3.4	3.8
CLK320, CLK500	1.4	1.8
CLK430, CLK55	2.4	2.8
E300D	6.7	7.1
E320, S320	2.0	2.4
E430, E500, E55	2.5	2.9
S430, S55, C55	2.8	3.2
S500	1.6	2.0
S600	3.9	4.3
SL500	2.0	2.4
SL600, SLK32	2.4	2.8
SLK230	1.1	1.5
SLK320, SLK55	1.4	1.8
Fuel Rail, Replace (B)		
300E, 300SE, 300TE	1.6	1.9
400E, E420, S420	3.1	3.4
400SEL	3.2	3.5
500E, E500	3.0	3.3
500SEC, 500SEL	3.2	3.5
500SL	3.0	3.3
600SEC, 600SEL	3.0	3.3
600SL	3.1	3.4
C220, C230, C240	1.6	1.9
C280, C32, C320	2.4	2.7
CL500, C55	2.7	3.0
CL600	3.4	3.7
CLK320, E320	2.2	2.5
CLK430, CLK55	1.8	2.1
E430, E500, E55	1.8	2.1
S320, S430, S55	2.4	2.7
S500, SL500, SL55	2.7	3.0
S600	3.9	4.2
SL600	3.4	3.7
SLK230, SLK32	1.6	1.9
SLK320, SLK55	2.0	2.3
Fuel Pressure Regulator, Replace (B)		
1984-04	2.4	2.7
Injector Nozzle and/or Seal, Replace (B)		
190D		
one	.7	1.0
all	1.4	1.7
240D		
one	.9	1.2
all	1.4	1.7
300CD, 300D		
one	.9	1.2
all	1.6	1.9
300SD		
1981-85		
one	.9	1.2
all	1.6	1.9
1992-93		
one	.7	1.0
all	2.4	2.7
300TD		
one	.9	1.2
all	1.6	1.9
E300D		
1995-97		
one	.8	1.1
all	2.4	2.7
1998-99		
one	2.2	2.5
all	3.6	3.9
S350TD		
one	.7	1.0
all	2.4	2.7
Clean and test injectors add each	.2	.5
Overvoltage Protection Relay Replace (B)		
Diesel	.2	.2
Thermal Valve, Replace (B)		
1981-85	.3	.3
Throttle Valve Switch, Replace (B)		
1981-85	3.0	3.0
Throttle Valve Housing, Replace (B)		
190E		
1984-85	1.4	1.6
1986-87 16V	3.6	3.8
1986-93	2.3	2.5
260E	1.6	1.8
280CE, 280E	2.8	3.0
300CE, 300E, 300SE, 300TE	1.6	1.8
300SEL	1.6	1.8
300SL	2.2	2.4
380SE, 380SEC, 380SEL, 380SL	2.8	3.0
420SEL	2.8	3.0
500SEC, 500SEL	2.8	3.0
500SL, SL500	2.7	2.9
560SEC, 560SEL, 560SL	2.8	3.0
Vacuum Control Flap Actuator, Replace (B)		
1981-94 diesel	.5	.5
Warm-Up Regulator, Replace (B)		
1981-85	.5	.5
TURBOCHARGER		
Turbocharger Assy., R&R or Replace (B)		
1987 190D 2.5L	3.8	4.3
1982-85 300CD	3.2	3.7
1982-87 300D	4.2	4.7
300SD		
1981-85	3.2	3.7
1992-93	3.9	4.4

MB-8 — 190 SERIES : 240D : 260E : 280 SERIES : 300 SERIES : 380 SERIES : 400 SERIES : 420SEL : 500 SERIES : 560 SERIES

	Labor Time	Severe Service
Turbocharger Assy., R&R or Replace (B)		
1981-87 300TD	4.2	4.7
S350TD	3.9	4.3
Replace assembly w/exchange unit add	.3	.5
wastegate actuator add	.3	.5

EXHAUST

	Labor Time	Severe Service
Catalytic Converter, Replace (B)		
1984-93 190E	2.0	2.1
260E	2.5	2.6
280CE, 280E	2.7	2.8
300CE	2.5	2.6
300E	2.8	2.9
300SE	2.4	2.5
300SEL	2.7	2.8
300SL	2.8	2.9
300TE	2.5	2.6
380SE, 380SEL, 380SL	2.8	2.9
400E	3.1	3.2
400SE, 400SEL	3.3	3.4
420SEL	3.1	3.2
500E, E500	3.1	3.2
500SEC, 500SEL	2.7	2.8
500SL	3.3	3.4
560SEC, 560SEL, 560SL	3.0	3.1
600SEC, 600SEL	4.2	4.3
600SL	3.5	3.6
C220, C230	2.5	2.6
C240, C320, C32, C55 left or right	2.5	2.6
C280	2.5	2.6
CL500, CL600 left or right	2.3	2.4
CLK320, CLK430	3.8	3.9
CLK500, S430, SLK55 left or right	2.3	2.4
CLK55	3.2	3.3
E300D	1.7	1.8
E320	2.8	2.9
E420	3.1	3.2
E430	2.5	2.6
E500 left or right	2.0	2.1
S320	3.0	3.1
S420	3.6	3.7
S500	3.1	3.2
S600, SL600	4.4	4.5
SL500	3.6	3.7
SLK230	1.1	1.2
SLK320, SLK32	2.6	2.7

	Labor Time	Severe Service
Exhaust Manifold or Gasket, Replace (B)		
190D, 190E	2.3	2.7
260E, 300E		
right front or rear	1.4	1.8
right both	2.3	2.7
240D	2.8	3.2
280CE, 280E	2.8	3.2
300CD	5.0	5.4
300CE, 300TE	2.8	3.2
300D		
1981-85	5.0	5.4
1986-93	2.8	3.2
300E	2.8	3.2
300SD		
1981-85	5.2	5.6
1992-93	4.4	4.8
300SE	4.4	4.8
300SEL	2.8	3.2
300TD	5.0	5.4
380SE, 380SL		
right side	4.2	4.6
left side	4.8	5.2
380SEC, 380SEL		
right side	2.8	3.2
left side	3.7	4.1
400E, 400SEL, E420		
right side	6.7	7.1
left side	7.0	7.4
420SEL		
right side	4.2	4.6
left side	4.8	5.2
500E, E500		
right side	6.7	7.1
left side	7.0	7.4
500SEC, 500SEL		
right side	3.5	3.9
left side	4.8	5.2
500SL		
right side	4.4	4.8
left side	7.5	7.9
560SEC, 560SEL		
right side	3.5	3.9
left side	4.8	5.2
560SL		
right side	4.2	4.6
left side	4.8	5.2
600SEC, 600SEL		
right side	13.2	13.6
left side	10.8	11.2
600SL		
right side	11.9	12.3
left side	9.7	10.1
C220, C230	2.7	3.1
C240, C320, C55		
right side	4.5	4.9
left side	4.5	4.9

	Labor Time	Severe Service
C280		
1994-97	5.2	5.6
1998-00		
right side	3.6	4.0
left side	4.5	4.9
CL500		
right side	5.1	5.5
left side	4.6	5.0
CL600		
right side	5.8	6.2
left side	8.6	9.0
CLK320		
right side	3.6	4.0
left side	4.5	4.9
CLK430		
right side	3.5	3.9
left side	4.4	4.8
CLK55, E320, S320		
right side	3.1	3.5
left side	4.0	4.4
E430		
right side	3.5	3.9
left side	4.1	4.5
E500		
right side	5.1	5.5
left side	4.0	4.4
E55		
right side	3.4	3.8
left side	3.5	3.9
S350TD	4.4	4.8
S420		
right side	3.6	4.0
left side	4.4	4.8
S430, S55		
each side	3.1	3.5
S500		
right side	3.5	3.9
left side	4.3	4.7
SL500		
right side	4.3	4.7
left side	7.3	7.7
SL55		
right side	3.8	4.2
left side	3.4	3.8
S600		
right side	5.8	6.2
left side	9.3	9.7
SL600		
right side	13.2	13.6
left side	10.8	11.2
SLK230	4.6	5.0
SLK32, SLK55		
right side	5.9	6.3
left side	5.9	6.3
SLK320		
right side	3.4	3.8
left side	4.3	4.7

600 SERIES : C SERIES : CL SERIES : CLK SERIES : E SERIES : S SERIES : SL SERIES : SLK SERIES — MB-9

	LABOR TIME	SEVERE SERVICE
Exhaust System, Replace (Complete) (B)		
190D	1.8	2.3
190E	2.4	2.9
240D	2.0	2.5
260E	3.1	3.6
280CE, 280E	2.8	3.3
300CD	2.0	2.5
300CE, 300SEL, 300TE	3.1	3.6
300D	3.0	3.5
300E	3.1	3.6
300SD		
1981-85	3.0	3.5
1992-93	2.0	2.5
300SE	3.0	3.5
1981-87 300TD	2.5	3.0
400E, 400SEL	3.6	4.1
420SEL, E420, S420	3.6	4.1
500E, E500	3.6	4.1
500SEC, 500SEL	3.3	3.8
500SL	3.6	4.1
560SEC, 560SEL	3.6	4.1
560SL	3.8	4.3
600SEC, 600SEL	4.5	5.0
600SL	3.9	4.4
C220, C55	2.7	3.2
C230, C240, CL500	2.5	3.0
C280		
1994-97	2.0	2.5
1998-00	2.5	3.0
CLK320	3.9	4.4
CLK430, CLK55	3.0	3.5
E300D		
1995-97	1.8	2.3
1998-99	1.7	2.2
E320, C32	3.1	3.6
E430, CLK500	1.9	2.4
S320, S350TD, E500	2.0	2.5
S500, E55, S430, S500	2.5	3.0
S55, SL55, SLK32	2.0	2.5
S600, CL600	4.5	5.0
SL500	3.6	4.1
SL600, S600	4.4	4.9
SLK230	1.6	2.1
SLK320, SLK55	3.0	3.5
Front Exhaust Pipe, Replace (B)		
Diesel	1.7	1.9
Front Exhaust Pipe to Converter, Replace (B)		
400SE	2.7	2.9
500SEL, 500SL	2.7	2.9
S430, S500	1.5	1.7
SL500, SL600	.8	1.0
Rear Muffler, Replace (B)		
190D, 190E	.9	1.0
1986-87 190E 16V	1.4	1.5
240D	.9	1.0

	LABOR TIME	SEVERE SERVICE
260E	2.0	2.1
280CE, 280E	1.7	1.8
300CD, 300D, E300D	.9	1.0
300CE, 300TE	2.0	2.1
300E	1.4	1.5
300SD		
1981-85	2.0	2.1
1992-93	.9	1.0
300SE, 300SEL	1.4	1.5
300SL	2.0	2.1
300TD	.9	1.0
380SE, 380SEC, 380SEL, 380SL	2.0	2.1
400E, 400SEL, E420	1.4	1.5
420SEL	2.0	2.1
500E, E500	1.4	1.5
500SEC, 500SEL	2.0	2.1
500SL, SL500	.9	1.0
560SEC, 560SEL	2.0	2.1
560SL	1.7	1.8
600SEC, 600SEL, 600SL	.9	1.0
C220, C230	.8	.9
C240, C32, C320, C55	1.6	1.8
C280, SLK320, SL55	.9	1.0
C36, SLK230, SLK32	.9	1.0
CLK430, CLK500	1.6	1.8
E320, E430	1.4	1.5
S320, S350TD	.9	1.0
S600, S430	1.4	1.5
SL600, CL600	.9	1.0
Resonator or Pre-Muffler, Replace (B)		
1981-85	2.7	2.8
Tailpipe Extension, Replace (B)		
1981-85	.5	.6

ENGINE COOLING

	LABOR TIME	SEVERE SERVICE
Pressure Test Cooling System (C)		
1981-05	.3	.3
Coolant Thermostat, Replace (B)		
Exc. below	1.2	1.4
240D	2.0	2.2
300CE, 300TE	2.0	2.2
1981-85 300SD	2.0	2.2
380SE, 380SEC, 380SEL, 380SL	2.0	2.2
400E, 400SEL	2.3	2.5
420SEL	1.7	1.9
500E, E500	2.3	2.5
500SL	1.5	1.7
500SEC, 500SEL		
1984-85	2.0	2.2
1993	1.6	1.8
560SEC, 560SEL, 560SL	1.7	1.9
600SEC, 600SEL, 600SL	2.0	2.2
CLK430, C55	1.6	1.8

	LABOR TIME	SEVERE SERVICE
E300D	1.8	2.0
E320	1.5	1.7
E420	2.3	2.5
E430	1.5	1.7
S55, SL55, SLK55	1.5	1.7
S420, S430	1.4	1.6
S600, SL600	2.0	2.2
Electric Cooling Fan Assy., Replace (B)		
Exc. below	2.0	2.2
190D	2.8	3.0
190E		
2.3L	.7	.9
2.6L	2.8	3.0
240D	.8	1.0
260E	1.3	1.5
300CD, 300D	1.4	1.6
300CE, 300E	1.3	1.5
300SD	1.4	1.6
300SE	1.4	1.6
300SEL	1.3	1.5
300SL	.9	1.1
380SE, 380SEC	1.4	1.6
400E	2.3	2.5
400SE	1.4	1.6
420SEL	1.4	1.6
500SEC, 500SEL	1.4	1.6
500SL, SL500	1.2	1.4
560SEC, 560SEL	1.4	1.6
560SL	2.3	2.5
600SEC, 600SEL, 600SL	1.4	1.6
E300D	1.2	1.4
E320, S320	1.3	1.5
S350TD	1.4	1.6
S600, SL600	1.4	1.6
Engine Coolant Temp. Sending Unit, Replace (B)		
1981-05	.4	.4
Radiator Assy., R&R or Replace (B)		
190D, 190E		
1984-85	2.3	2.5
1986-89 190D		
w/turbo	3.1	3.3
w/o turbo	2.3	2.5
1986-93 2.3L	1.9	2.1
1987-93 2.6L	3.5	3.7
240D	2.3	2.5
260E	1.7	1.9
280CE, 280E	2.3	2.5
300CD	2.3	2.5
300CE, 300SE, 300SEL, 300TE	1.7	1.9
300D		
1981-85	2.3	2.5
1986-93	1.7	1.9
300E	1.7	1.9
300SD		
1981-85	2.3	2.5
1992-93	2.0	2.2

MB-10 — 190 SERIES : 240D : 260E : 280 SERIES : 300 SERIES : 380 SERIES : 400 SERIES : 420SEL : 500 SERIES : 560 SERIES

	LABOR TIME	SEVERE SERVICE
Radiator Assy., R&R or Replace (B)		
300SL	2.5	2.7
300TD	2.3	2.5
380SE, 380SEC, 380SEL, 380SL	2.3	2.5
400E, 400SEL, E420	2.8	3.0
420SEL	1.7	1.9
500E, E500	2.8	3.0
1984-85 500SEC, 500SEL	2.3	2.5
1993 500SEC	2.5	2.7
500SEL	1.7	1.9
500SL	3.0	3.2
560SEC, 560SEL	1.7	1.9
560SL	2.0	2.2
600SEC, 600SEL, S600	2.4	2.6
600SL	2.0	2.2
C220, C230	2.2	2.4
C240, C320, C32	2.6	2.8
C280	3.0	3.2
C36	2.5	2.7
CL500	2.1	2.3
CL600	1.9	2.1
CLK320, C55	3.5	3.7
CLK430, CLK55	3.1	3.3
CLK500	2.6	2.8
E300D	2.7	2.9
E320	2.2	2.4
E430, E500	3.1	3.3
S320, S350TD	2.0	2.2
S420, S430	2.5	2.7
S500, S55	2.5	2.7
SL500	2.2	2.4
SL600, S600	1.6	1.8
SLK230, SLK32, SLK55	1.8	2.0
SLK320	2.6	2.8
Radiator Fan Motor Switch (Coolant Temp.), Replace (B)		
1984-93 190E	.3	.3
Radiator Hoses, Replace (B)		
Includes: Refill with proper coolant mix.		
Upper		
1981-85	.7	.9
1986-05	1.2	1.4
Lower		
190D, 190E		
1984-85	.7	.8
1986-93	1.3	1.4
240D	.8	.9
260E	1.3	1.4
280CE, 280E	1.2	1.3
300CD	.8	.9
300CE, 300E, 300SE, 300TE	1.3	1.4
300D		
1981-85	.8	.9
1986-93	1.3	1.4

	LABOR TIME	SEVERE SERVICE
300SD		
1981-85	.8	.9
1992-93	.9	1.0
300SEL, 300SL	1.3	1.4
300TD	.8	.9
380SE, 380SEC, 380SEL, 380SL	.8	.9
400E, 400SEL	.9	1.0
420SEL	1.3	1.4
500E	.9	1.0
500SEC, 500SEL	.9	1.0
560SEC, 560SEL, 560SL	1.3	1.4
600SEC, 600SEL, 600SL, E500	.9	1.0
E320, E420, E430	.9	.9
C220, C240, C280	.7	.7
C32, C36, C55	1.3	1.3
CL500, S600	2.2	2.4
CLK320, CLK430	.9	.9
S430, S500, S55	2.2	2.4
SLK230, SLK32, SLK55	1.8	2.0
SLK320	1.6	1.8
Water Pump and/or Gasket, Replace (B)		
Includes: Refill with proper coolant mix.		
190D		
2.2L	2.4	2.6
2.5L	3.9	4.1
190E		
1984-93	3.9	4.1
1986-87 16V	4.4	4.6
240D	2.7	3.0
260E	6.0	6.2
280CE, 280E	4.8	5.0
300CD, 300D	4.8	5.0
300E	6.0	6.2
300SD	2.4	2.6
300SE, 300SEL	6.2	6.4
300SL	6.7	6.9
300TD	4.8	5.0
380SE, 380SL	6.7	6.9
380SEC, 380SEL	5.2	5.4
400E, 400SEL, E420	6.7	6.9
420SEL	6.7	6.9
500E, E500	6.7	6.9
500SEC, 500SEL	5.3	5.5
500SL	5.8	6.0
560SEC, 560SEL	6.7	7.0
560SL	7.5	7.7
600SEC, 600SEL, 600SL	5.8	6.0
C220, C240, C32	3.5	3.7
C230	2.2	2.4
C280		
1994-97	5.3	5.5
1998-00	3.3	3.5
C36	5.3	5.5
C320, C55	3.5	3.8

	LABOR TIME	SEVERE SERVICE
CL500	5.8	6.0
CL600	5.5	5.7
CLK320, CLK430	3.3	3.4
CLK500, CLK55	2.8	3.0
E300D	2.8	3.0
E320	2.7	2.9
E430, E500, E55	3.2	3.4
S320	5.3	5.5
S350TD	2.4	2.6
S420	5.3	5.5
S430, S500, S55	3.8	4.0
SL500	5.3	5.5
SL600	4.3	4.5
SLK230	2.5	2.7
SLK320, SLK32	3.2	3.4
SLK55	3.6	3.8

ENGINE

ASSEMBLY

Times shown are for OEM assemblies. Time to replace assemblies from aftermarket rebuilders may vary.

Engine Assy., R&I (B)
Does not include parts or component transfer.

	LABOR TIME	SEVERE SERVICE
190D		
2.2L	9.9	10.1
2.5L	14.8	15.0
turbo	11.6	11.8
190E		
1986-93 2.3L	10.8	11.0
1986-87 16V	14.8	15.0
1987-93 2.6L	11.4	11.6
240D	10.8	11.0
260E	11.4	11.6
280CE, 280E	10.2	10.4
300CD, 300D, E300D	10.8	11.0
300E	11.4	11.6
300SD, 300TD	12.7	12.9
300SL	13.2	13.4
380SEC, 380SEL	15.8	16.0
400E	21.9	22.1
400SEL	16.3	16.5
420SEL	12.7	12.9
500E, E500	21.9	22.1
500SEC	15.6	15.8
500SEL	16.3	16.5
500SL	16.0	16.2
560SEC, 560SEL	12.7	12.9
560SL	19.3	19.5
600SEC, 600SEL	23.2	23.4
600SL	18.7	18.9
C220, C230	9.4	9.6
C240, C32	13.2	13.4
C280	9.9	10.1
C36	9.9	10.1
C55	10.4	10.6
CL500	15.2	15.4

600 SERIES : C SERIES : CL SERIES : CLK SERIES : E SERIES : S SERIES : SL SERIES : SLK SERIES

MB-11

	LABOR TIME	SEVERE SERVICE
CL600	23.3	23.5
CLK320, CLK55	10.7	10.9
CLK430	12.1	12.3
CLK500	14.0	14.2
E320	13.2	13.4
E420	21.9	22.1
E430, E500, E55	13.1	13.3
S320	15.2	15.4
S350TD	12.7	12.9
S420, S430	14.8	15.0
S500, S55	15.2	15.4
S600	23.2	23.4
SL500	15.8	16.0
SL600	19.6	19.8
SLK230, SLK32	11.4	11.6
SLK320	12.0	12.2
SLK55	13.0	13.2

Engine Assy. (New or Rebuilt), Replace (Complete) (B)
Includes: Component transfer and engine tune-up.

	LABOR TIME	SEVERE SERVICE
C230, C240	12.9	13.1
C280, C32, C320	15.2	15.4
C36	15.5	15.7
C55	13.5	13.7
CLK320	14.2	14.4
CLK430, CLK55	15.8	16.0
CL500	14.6	14.8
E300D	14.3	14.5
E320	13.9	14.1
S320	20.5	20.7
S350TD, S430	17.8	18.0
S420	20.7	20.9
S500, S55	15.7	15.9
S600	25.9	26.1
SL500	21.2	21.3
SL600	23.6	23.8
SLK230, SLK320	15.3	15.5
SLK55	15.6	15.8

Engine Assy. (Short Block), Replace (B)
Assembly consists of engine block, piston assemblies, crankshaft, camshaft, timing chain and gears. Does not includes cylinder heads. Operation includes: R&R engine, transfer necessary parts and all necessary adjustments.

	LABOR TIME	SEVERE SERVICE
190D		
2.2L		
AT	27.4	27.6
MT	25.3	25.5
2.5L		
AT	36.0	36.2
MT	33.9	34.1
turbo	30.7	30.9
190E		
2.3L	28.3	28.5
16V	33.9	34.1
2.6L	23.3	23.5

	LABOR TIME	SEVERE SERVICE
240D		
AT	32.5	32.7
MT	30.7	30.9
260E	23.3	23.5
280CE, 280E	31.7	31.9
300CD, 300D	35.6	35.8
300CE, 300TE	26.2	26.4
300E	23.3	23.5
300SD	33.9	34.1
300SE, 300SEL	37.0	37.2
300TD	35.6	35.8
380SE, 380SL	42.2	42.4
380SEC, 380SEL	37.3	37.5
420SEL	46.0	46.2
500SEC, 500SEL	37.3	37.5
500SL	20.5	20.7
560SEC, 560SEL	46.0	46.2
560SL	48.2	48.4
C240, C32, C320	31.1	31.3
C55	32.0	32.2
CL500, S500, S55	32.9	33.1
CLK320	27.5	27.7
CLK430, CLK55	33.5	33.7
CLK500	36.9	37.1
E320	27.3	27.5
E430	36.7	36.9
E500	31.1	31.3
E55	33.5	33.7
S320, S350TD	31.7	31.9
S420	46.4	46.6
S430	35.9	36.1
SLK320	27.5	27.7
SLK55	29.0	29.2

Engine Mounts, Replace (B)
Front

	LABOR TIME	SEVERE SERVICE
190D		
2.2L	3.1	3.3
2.5L	1.6	1.8
190E	.8	1.0
240D	1.6	1.8
260E	.8	1.0
300CD, 300D	1.6	1.8
300CE, 300TE	1.4	1.6
300E	1.4	1.6
300SD		
1981-85	2.4	2.6
1992-93	1.7	1.9
300SE, 300SEL	2.2	2.4
300SL	2.0	2.2
300TD	2.4	2.6
380SE, 380SL	2.4	2.6
380SEC, 380SEL	2.3	2.5
400E, 400SEL, E420	2.7	2.9
420SEL	1.7	1.9
500E, E500	2.7	2.9
500SEC, 500SEL	2.7	2.9
500SL	3.5	3.7
560SEC, 560SEL	1.7	1.9
560SL	2.5	2.7

	LABOR TIME	SEVERE SERVICE
600SEC, 600SEL	3.1	3.3
C220	1.4	1.6
C230	2.0	2.2
C240, C32, C320	3.6	3.8
C280, 280CE, C280E, C36, C55	2.0	2.2
CL500, CL600	3.5	3.7
CLK320, CLK430	3.2	3.4
CLK500, CLK55	3.5	3.7
E300D	1.7	1.9
E320	1.4	1.6
E430	2.4	2.6
E500	3.2	3.4
E55, S55	6.5	6.7
S420	3.0	3.2
S430	5.1	5.3
S500	2.8	3.0
S600	2.5	2.7
SL500	5.3	5.5
SL600	5.8	6.0
SLK230	2.5	2.7
SLK320	1.8	2.0
SLK55	3.0	3.2
Rear	.9	1.1

CYLINDER HEAD
Compression Test (B)

	LABOR TIME	SEVERE SERVICE
190D	2.0	2.3
190E	.7	1.0
240D	1.4	1.7
260E	.7	1.0
280CE, 280E	1.2	1.5
300CD, 300D	1.8	2.1
300CE, 300TE	.7	1.0
300E	.7	1.0
300SD		
1981-85	1.8	2.1
1992-93	2.4	2.7
300SE, 300SEL, 300SL	.7	1.0
300TD	1.8	2.1
380SE, 380SL	1.7	2.0
380SEC, 380SEL	1.4	1.7
400E, 400SEL, E420	1.4	1.7
420SEL	1.4	1.7
500E, 500SL, E500	1.4	1.7
500SEC, 500SEL	1.4	1.7
560SEC, 560SEL	1.4	1.7
560SL	1.7	2.0
600SEC, 600SEL, 600SL	2.0	2.3
C220	1.2	1.5
C230	.9	1.2
C240	3.1	3.3
C280		
1994-97	1.4	1.7
1998-00	3.1	3.3
C36, C55	1.4	1.7
CL500	1.5	1.8
CL600	1.6	1.9

MB-12 — 190 SERIES : 240D : 260E : 280 SERIES : 300 SERIES : 380 SERIES : 400 SERIES : 420SEL : 500 SERIES : 560 SERIES

	Labor Time	Severe Service
Compression Test (B)		
CLK320, C32	3.1	3.3
CLK430, CLK500	3.6	3.8
E300D		
1995-97	2.4	2.7
1998-99	4.4	4.7
E320	1.3	1.6
E430, E500	3.7	4.0
S320	1.4	1.7
S350TD	2.4	2.7
S420, S430	2.2	2.5
S500, SL500	2.3	2.6
S600, S55	3.5	3.8
SL600	3.1	3.3
SLK230	1.2	1.5
SLK320, SLK32, SLK55	3.0	3.3
Cylinder Head, Replace (B)		
Includes: Parts transfer and adjustments.		
1984-89 190D		
2.2L	14.6	15.1
2.5L	18.0	18.5
240D	13.5	14.0
300CD, 300D	15.0	15.5
300SD		
1981-85	17.7	18.2
1992-93	19.1	19.6
300TD		
1981-86	15.2	15.7
1987	20.1	20.6
C230	19.5	20.0
C240		
right side	10.3	10.8
left side	10.3	10.8
both sides	14.6	15.1
C280		
right side	14.3	14.8
left side	16.3	16.8
both sides	24.6	25.1
C32, C320		
right side	11.3	11.8
left side	11.3	11.8
both sides	15.6	16.1
CL500, CL600		
one side	28.0	28.5
both sides	40.8	41.3
CLK320		
right side	18.5	19.0
left side	20.4	20.9
both sides	30.5	31.0
CLK430		
right side	12.3	12.8
left side	11.9	12.4
both sides	17.6	18.1
E300D		
1995-97	17.1	17.6
1998-99	21.2	21.7
E320		
right side	18.5	19.0
left side	20.5	21.0
both sides	30.2	30.7
E430		
right side	19.9	20.4
left side	20.5	21.0
both sides	32.5	33.0
S350TD	19.1	19.6
SL500		
right side	23.7	24.2
left side	24.3	24.8
both sides	33.4	33.9
E500		
right side	9.9	10.4
left side	10.5	11.0
both sides	14.5	15.0
E55		
right side	11.9	12.4
left side	12.5	13.0
both sides	17.5	18.0
S430, S500, S55		
right side	11.9	12.4
left side	12.5	13.0
both sides	14.5	15.0
SL600		
right side	9.2	9.7
left side	9.2	9.7
both sides	36.0	36.5
SLK230	19.5	20.0
SLK320		
right side	18.4	18.9
left side	20.3	20.8
both sides	30.2	30.7
Cylinder Head Front Cover, Replace (B)		
190E	2.5	2.8
260E	2.5	2.8
280CE, 280E	2.5	2.8
300CE, 300E, 300SE, 300TE	2.5	2.8
300SEL	2.3	2.6
300SL	3.8	4.1
380SE, 380SEC, 380SEL, 380SL	2.5	2.8
400E		
right side	3.0	3.3
left side	3.6	3.9
400SEL	3.3	3.6
420SEL	2.5	2.8
500E, E500		
right side	3.0	3.3
left side	3.6	3.9
500SEC, 500SEL		
right side	2.4	2.7
left side	3.3	3.6
500SL		
right side	2.7	3.0
left side	3.6	3.9
560SEC, 560SEL	2.5	2.8
600SEC, 600SEL	18.4	18.7
600SL	9.6	9.9
C220	3.3	3.6
C230	3.1	3.4
C280	5.3	5.6
CL500		
right side	2.8	3.1
left side	3.1	3.4
CL600		
right side	4.8	5.1
left side	5.0	5.3
E320	4.4	4.7
E420		
right side	3.0	3.3
left side	3.6	3.9
S420		
right side	3.0	3.3
left side	3.1	3.4
S500		
right side	2.8	3.1
left side	3.5	3.8
S600	15.6	15.9
SL500		
right side	3.0	3.3
left side	3.1	3.4
SL600	13.5	13.8
SLK230	3.2	3.5
Cylinder Head Gasket, Replace (B)		
190D		
2.2L	10.9	11.3
2.5L	14.1	14.4
190E		
1984-85	8.2	8.5
1986-93		
2.3L	9.2	9.5
16V	11.7	12.0
2.6L	11.7	12.0
240D	9.1	9.4
260E	15.0	15.3
280CE, 280E	15.6	15.9
300CD, 300D	9.7	10.0
300CE	11.7	12.0
300E	15.0	15.3
300SE	10.8	11.1
300SEL	10.8	11.3
300SL	11.4	11.7
300SD		
1981-85	12.4	12.7
1992-93	13.4	13.7
300TD	15.0	15.3
300TE	10.5	10.8
380SE, 380SL		
right side	18.7	19.0
left side	19.0	19.3
both sides	27.7	28.0
380SEC, 380SEL		
one side	13.4	13.7
both sides	29.3	29.6

600 SERIES : C SERIES : CL SERIES : CLK SERIES : E SERIES : S SERIES : SL SERIES : SLK SERIES — MB-13

	LABOR TIME	SEVERE SERVICE
400E, 400SEL		
right side	18.2	18.5
left side	19.3	19.6
both sides	23.4	23.7
420SEL		
right side	18.7	19.0
left side	19.3	19.6
both sides	27.7	28.0
500E, E500		
right side	18.2	18.5
left side	19.1	19.4
both sides	23.4	23.7
500SEC		
right side	16.0	16.3
left side	17.7	18.0
both sides	23.4	23.7
500SEL		
right side	18.7	19.0
left side	19.3	19.6
both sides	27.7	28.0
500SL		
right side	19.5	19.8
left side	20.1	20.4
both sides	25.3	25.6
560SEC, 560SEL		
right side	18.7	19.0
left side	19.3	19.6
both sides	27.7	28.0
560SL		
right side	16.3	16.6
left side	15.0	15.3
both sides	24.2	24.5
600SEC, 600SEL		
right side	34.2	34.5
left side	34.4	34.7
both sides	36.2	36.5
600SL		
right side	20.5	20.8
left side	20.7	21.0
both sides	25.3	25.6
C220	13.7	14.0
C230	13.4	13.7
C240		
right side	11.3	11.8
left side	11.3	11.8
both sides	15.6	16.1
C280		
1994-97	13.7	14.0
1998-00		
right side	12.0	12.3
left side	14.1	14.4
both sides	17.5	17.8
C32, C320		
right side	12.3	12.8
left side	12.3	12.8
both sides	16.6	17.1
C36	13.7	14.0

	LABOR TIME	SEVERE SERVICE
C55 (engine removed)		
right side	6.5	6.8
left side	6.1	6.4
both sides	10.6	11.0
CL500	6.8	7.1
CL600		
one side	20.8	21.1
both sides	27.5	27.8
CLK320		
right side	12.0	12.3
left side	14.1	14.4
both sides	17.5	17.8
CLK430		
right side	13.4	13.7
left side	14.0	14.3
both sides	19.2	19.5
E300D		
1995-97	9.7	10.0
1998-99	17.1	17.4
E320		
right side	12.0	12.3
left side	14.0	14.3
both sides	17.2	17.5
E420		
right side	18.2	18.5
left side	19.1	19.4
both sides	23.4	23.7
E430		
right side	13.4	13.7
left side	14.0	14.3
both sides	19.1	19.4
E500		
right side	10.9	11.4
left side	11.5	12.0
both sides	15.5	16.0
E55		
right side	12.9	13.4
left side	13.5	14.0
both sides	18.5	19.0
S320	14.2	14.5
S350TD	13.4	13.7
S430, S500, S55		
right side	12.9	13.4
left side	13.5	14.0
both sides	15.5	16.0
S600		
right side	34.2	34.5
left side	34.4	34.7
both sides	36.2	36.5
SL500		
right side	18.2	18.5
left side	19.1	19.4
both sides	25.9	26.2
SL600		
right side	21.9	22.2
left side	22.2	22.5
both sides	27.8	28.1

	LABOR TIME	SEVERE SERVICE
SLK320		
right side	12.0	12.3
left side	14.1	14.4
both sides	17.5	17.8
Replace w/exchange head add		
190E	5.4	5.7
240D, 260E	6.0	6.3
280CE, 280E	5.4	5.7
300CD, CE, D, E, SE, SL	7.3	7.6
300SEL	5.8	6.1
350TD, 380SEC, SEL, SL	3.0	3.3
400E, SEL	4.5	4.8
420SEL	3.0	3.3
500E, E500	4.5	4.8
500SEC	4.5	4.8
500SEL	4.5	4.8
500SL	4.5	4.8
560SEC, SEL	3.0	3.3
600SEC, SEL	7.3	7.6
600SL, SL600	7.3	7.6
C220	5.4	5.7
C230	5.3	5.6
C280	6.2	6.5
C36, C55	6.2	6.5
CL500	3.0	3.3
CL600	4.7	5.0
CLK320	6.7	7.0
E320	6.0	6.3
E420	4.5	4.8
E430	4.3	4.6
S320	6.2	6.5
S500	2.9	3.2
S600	7.3	7.6
SL500	4.5	4.8
SL600	4.7	5.0
Valve Clearance, Adjust (B)		
Exc. below	1.7	2.2
190E 16V		
one side	2.4	2.9
both sides	3.1	3.6
280CE, 280E	2.0	2.5
300SE, 300SEL	4.4	4.9
560SEC, 560SEL, 560SL	2.4	2.9
Valve Cover, Replace or Reseal (B)		
1981-85		
gasoline		
4 cyl, 5 cyl	.7	.9
6 cyl	1.6	1.8
V8		
one	.8	1.0
both	1.6	1.8
diesel		
2.2L	.5	.7
2.5L	.8	1.0

MB-14

190 SERIES : 240D : 260E : 280 SERIES : 300 SERIES : 380 SERIES : 400 SERIES : 420SEL : 500 SERIES : 560 SERIES

	Labor Time	Severe Service
Valve Cover, Replace or Reseal (B)		
190E	.7	.9
240D	.5	.7
260E	.9	1.1
300CD, 300D	.5	.7
300CE, 300E, 300SE, 300TE	.9	1.1
300SD, 300TD	.9	1.1
300SEL, 300SL	.9	1.1
400E, 400SEL, E420		
one side	2.2	2.4
both sides	3.6	3.8
420SEL	.7	.9
500E, E500		
one side	2.2	2.4
both sides	3.6	3.8
500SEC, 500SEL	.9	1.1
500SL		
right side	.9	1.1
left side	1.4	1.6
both sides	2.3	2.5
560SEC, 560SEL, 560SL	.9	1.1
600SEC, 600SEL		
both	12.4	13.0
600SL, S600 both	9.5	10.0
C220, C230	1.5	1.7
C240, C32, C320		
right side	1.8	2.0
left side	1.8	2.0
both sides	2.6	2.8
C280		
1994-97	1.8	2.0
1998-00	2.8	3.0
C36	1.8	2.0
C55		
right side	1.5	1.7
left side	1.9	2.1
both sides	3.0	3.2
CL500		
right side	2.2	2.4
left side	2.3	2.5
both sides	3.4	3.6
CL600		
right side	7.2	7.4
left side	7.3	7.5
both sides	8.2	8.4
CLK320, CLK55		
one side	2.5	2.7
both sides	3.9	4.1
CLK430		
one side	2.5	2.7
both sides	4.2	4.4
E300D		
1995-97	4.4	4.6
1998-99	3.9	4.1
E320	2.0	2.2
E430		
one side	2.5	2.7
both sides	4.2	4.4

	Labor Time	Severe Service
E500		
right side	1.5	1.7
left side	2.0	2.2
both sides	3.2	3.4
E55		
right side	2.5	2.7
left side	2.5	2.7
both sides	4.2	4.4
S320	1.8	2.0
S350TD	.9	1.1
S420, S430		
right side	2.7	2.9
left side	2.8	3.0
both sides	3.9	4.1
S500, S55		
right side	2.2	2.4
left side	2.3	2.5
both sides	3.8	4.0
SL500		
right side	2.7	2.9
left side	2.8	3.0
both sides	3.9	4.1
SL600 both	9.3	9.5
SLK230	1.1	1.3
SLK320, SLK32		
one side	2.4	2.6
both sides	3.8	4.0
Replace all rocker arms add	2.3	2.5
Valve Lifters, Replace (B)		
Valve cover(s) removed		
190D		
2.2L	2.3	2.6
2.5L	2.5	2.8
300SD, 300TD	3.0	3.3
C230	6.2	6.5
C240, C32, C320	3.5	3.8
C280, C36	7.3	7.6
CL500	2.8	3.1
CL600, S600	9.0	9.3
CLK320	3.4	3.7
E300D	8.9	9.2
S320	4.8	5.1
S350TD	3.0	3.3
S430, S500, S55	2.8	3.1
Valve Springs and/or Oil Seals, Replace (B)		
190D		
2.2L	3.5	4.0
2.5L	4.2	4.7
190E		
1984-85	9.2	9.7
1986-93		
2.3L	4.2	4.7
16V	5.8	6.3
2.6L	5.3	5.8
240D	5.8	6.3
260E	5.3	5.8
280CE, 280E	9.8	10.3
300CD	6.3	6.8

	Labor Time	Severe Service
300CE, 300TE	9.2	9.7
300D	6.3	6.8
300E, 300SE, 300SEL	5.3	5.8
300SD		
1981-85	6.3	6.8
1992-93	4.8	5.3
300SL	9.9	10.4
300TD	6.3	6.8
380SE, 380SL	9.2	9.7
380SEC, 380SEL	8.6	9.1
400E, 400SEL, E420	18.4	18.9
420SEL	9.2	9.7
500E, E500	18.4	18.9
500SEL, 500SEC	8.6	9.1
500SL	19.5	20.0
560SEC, 560SEL, 560SL	9.2	9.7
600SEC, 600SEL	34.8	35.3
600SL	29.7	30.2
C220	6.2	6.7
C230, C240, C320	8.8	9.3
C280		
1994-97	10.7	11.2
1998-00	10.0	10.5
C36	10.7	11.2
CLK320	9.7	10.2
CLK430	8.8	9.3
CLK500	10.4	10.9
CL500, CLK55	17.0	17.5
CL600	25.8	26.3
E300D		
1995-97	14.1	14.6
1998-99	9.9	10.4
E320, S320	16.8	17.3
E430	12.1	12.6
E500	10.5	11.0
E55	13.2	13.7
S350TD	4.8	5.3
S420	21.6	22.1
S600	31.7	32.2
SL500	21.6	22.1
SL600	28.6	29.1
SLK230	7.1	7.6
CLK320	11.7	12.2
CAMSHAFT		
Camshaft, Replace (B)		
190D	2.5	3.0
190E		
2.3L	3.8	4.3
16V	2.7	3.2
2.6L	5.3	5.8
240D	3.9	4.4
260E	5.3	5.8
280CE, 280E		
one	10.0	10.5
both	10.8	11.3
300CD, 300D, 300TD	4.4	4.9
300E	5.3	5.8

600 SERIES : C SERIES : CL SERIES : CLK SERIES : E SERIES : S SERIES : SL SERIES : SLK SERIES — MB-15

	LABOR TIME	SEVERE SERVICE
300SD		
1981-85	4.9	5.4
1992-93	3.1	3.6
300SE, 300SEL	5.8	6.3
300SL	5.6	6.1
380SEC, 380SEL		
one	3.7	4.2
both	5.8	6.3
400E, 400SEL,		
E420 all	11.4	11.9
420SEL	3.6	4.1
500E, E500 all	11.4	11.9
500SEC, 500SEL		
1984-85		
one	3.7	4.2
both	5.8	6.3
1993 all	11.4	11.9
500SL		
right side		
one	6.5	7.0
both	8.8	9.3
left side		
one	6.2	6.7
both	8.2	8.7
560SEC, 560SEL	3.6	4.1
560SL	4.2	4.7
600SEC, 600SEL all	25.5	30.0
600SL all	20.2	20.7
C220, C230	5.4	5.9
C240, C32, C320	8.8	9.3
C280		
1994-97	8.6	9.1
1998-00		
one	6.5	7.0
all	8.2	8.7
C36	8.6	9.1
CL500		
right side		
one	7.0	7.5
both	7.9	8.4
left side		
one	8.1	8.6
both	9.0	9.5
CL600		
right side		
one	14.9	15.4
both	16.4	16.9
left side		
one	14.7	15.2
both	16.3	16.8
CLK320 all	8.2	8.7
CLK430, CLK55		
right side	5.5	6.0
left side	5.0	5.6
both sides	8.8	9.3
CLK500		
right side	5.9	6.4
left side	5.4	5.9
both sides	10.8	11.3

	LABOR TIME	SEVERE SERVICE
E300D all	7.0	7.5
E320	6.2	6.7
E430		
one side	5.5	6.0
both sides	7.3	7.8
E500, E55		
one side	6.5	7.0
both sides	8.3	8.8
S320	4.2	4.7
S350TD	3.1	4.6
S420 all	14.8	15.3
S430	12.8	13.3
S500		
right side		
one	7.3	7.8
both	8.5	9.0
left side		
one	8.1	8.6
both	9.3	9.8
S600 all	22.6	23.1
SL500		
right side		
one	8.7	9.2
both	9.6	10.1
left side		
one	9.7	10.0
both	10.6	11.1
SL600		
right side		
one	11.4	11.9
both	12.6	13.1
left side		
one	12.2	12.7
both	13.1	13.6
SLK230, SLK32		
one	4.6	5.1
both	4.2	4.7
SLK320 all	8.2	8.7
Camshaft and/or Rocker Arms, Replace (B)		
Right side		
420SEL	2.9	3.4
500SEC, 500SEL	3.3	3.8
560SEC, 560SEL	3.3	3.8
560SL	6.9	7.4
Left side		
420SEL	3.1	3.6
500SEC, 500SEL	3.3	3.8
560SEC, 560SEL	3.3	3.8
560SL	4.5	5.0
w/elec. injectors add	.7	1.0
Camshaft Timing Adjusters, Replace (B)		
300CE, 300E, 300SE, 300SL	4.4	4.7
300TE	4.4	4.7
400E, 400SE, 400SEL	10.7	11.0
500E, E500	10.7	11.0

	LABOR TIME	SEVERE SERVICE
500SEC, 500SEL,		
500SL	7.5	7.8
C220, C230, C240	4.9	5.2
C280, C36	6.2	6.5
C320, C32	5.7	6.0
CL500	6.3	6.6
CL600	8.6	8.9
CLK320	6.0	6.3
CLK430, CLK55	6.6	6.9
E320	7.2	7.5
E420, S420	10.5	10.8
E430	7.2	7.5
E500, E55	6.2	6.5
S320, S430	5.3	5.6
S500	10.6	10.9
SL500	10.5	10.8
SL600	19.1	19.4
SLK230	4.6	4.9
Camshaft Timing Adjustment Solenoid, Replace (B)		
400SE, 400SEL, E420	1.4	1.9
500E, 500SEC, 500SEL, 500SL, E500	1.4	1.9
C220, C280, C230, C36	.5	1.0
CL500	1.3	1.8
S320	.5	1.0
S420	.8	1.3
S500	1.3	1.8
S600	.8	1.3
SL500	.5	1.0
SL600	1.4	1.9
SLK230	.3	.8
Timing Chain, Replace (B)		
190D	10.5	11.0
190E		
2.3L	16.4	16.9
16V	21.2	21.7
2.6L	22.8	23.3
240D	12.7	13.2
260E	11.6	12.1
300CD, 300D	10.0	10.5
300CE, 300E, 300TE	12.0	12.5
300SD	10.0	10.5
300SEL	11.6	12.1
300SL	14.4	14.9
300TD		
1981-86	10.0	10.5
1987	12.6	13.1
380SE, 380SEC, 380SEL	24.3	24.8
380SL	29.0	29.5
400E	37.9	38.4
420SEL	24.3	24.8
500E, E500	37.9	38.4
500SEC, 500SEL	24.3	24.8
500SL	25.9	26.4
560SEC, 560SEL	24.3	24.8
560SL	29.0	29.5

MB-16 — 190 SERIES : 240D : 260E : 280 SERIES : 300 SERIES : 380 SERIES : 400 SERIES : 420SEL : 500 SERIES : 560 SERIES

	LABOR TIME	SEVERE SERVICE
Timing Chain, Replace (B)		
600SEC, 600SEL	37.5	38.0
600SL	26.3	26.8
C220, C230	3.9	4.4
C240, C32, C320	16.4	16.9
C280		
1994-97	4.5	5.0
1998-00	6.2	6.7
C36	4.5	5.0
CL500	16.1	16.6
CLK320	7.3	7.8
CLK430	19.0	19.5
CLK500	20.2	20.7
E300D	5.3	5.8
E320	4.5	5.0
E420	8.6	9.1
E430	8.1	8.6
E500, E55	16.4	16.9
S320	6.3	6.8
S350TD	10.0	10.5
S420	7.3	7.8
S430, S500, S55	16.1	16.6
S600	19.1	19.6
SL500	9.4	9.9
SL600	20.2	20.7
SLK230	4.1	4.6
SLK32	16.3	16.8
SLK320	7.3	7.8
Timing Chain or Belt Cover, Replace (B)		
Includes: Replace gasket.		
190D		
2.2L	8.6	9.0
2.5L	10.2	10.6
190E		
2.3L	14.2	14.6
16V	15.6	16.0
2.6L	9.3	9.7
240D, 260E	8.6	9.0
300CD, 300D	8.6	9.0
300CE, 300TE	9.1	9.5
300E	9.1	9.5
300SD		
1981-85	8.6	9.0
1992-93	8.8	9.2
300SEL	8.6	9.0
300SL	11.3	11.7
300TD		
1981-86	8.6	9.0
1987	9.7	10.1
380SE, 380SEC, 380SEL	20.5	20.9
380SL	25.2	25.6
400E, E420	36.7	37.1
420SEL	20.5	20.9
500E, E500	36.7	37.1
500SEC, 500SEL	20.5	20.9
500SL	24.6	25.0
560SEC, 560SEL	20.5	20.9
560SL	25.2	25.6

	LABOR TIME	SEVERE SERVICE
600SL	21.2	21.6
C220	14.3	14.7
C230	15.2	15.6
C240, C32, C320	15.4	15.9
C280		
1994-97	10.3	10.7
1998-00	17.2	17.6
C36	10.3	10.7
C55	19.3	19.7
CL500	15.1	15.6
CLK320	17.2	17.6
CLK430	17.1	17.5
CLK500	19.2	19.7
E300D	11.6	12.0
E320	9.7	10.1
E430	17.2	17.6
E500, E55	15.4	15.9
S320	10.2	10.6
S350TD	8.8	9.2
S420	28.3	28.7
S430, S500, S55	15.1	15.6
S500	26.2	16.6
S600	19.9	20.3
SL500	26.3	26.7
SL600	19.9	20.3
SLK230	15.2	15.6
SLK32	15.3	15.8
SLK320	17.2	17.6
Replace tensioner and rails add	.8	1.2
Timing Chain Tensioner, Replace (B)		
190D		
2.2L	.7	1.0
2.5L	.7	1.0
190E		
2.3L	1.4	1.7
16V	2.2	2.5
2.6L	.8	1.1
240D	2.2	2.5
260E	.8	1.1
280CE, 280E	1.3	1.6
300CD, 300D	2.2	2.5
300CE, 300TE	1.7	2.0
300E	.8	1.1
300SD		
1981-85	2.2	2.5
1992-93	.7	1.0
300SE	1.7	2.0
300SEL	.8	1.1
300SL	1.7	2.0
300TD	2.2	2.5
380SE, 380SEC, 380SEL	.9	1.2
380SL	1.6	1.9
400E, 400SEL	3.3	3.6
420SEL	.9	1.2
500E, E500	3.3	3.6
500SEC	3.3	3.6
500SL, 500SEL	3.3	3.6
560SEC, 560SEL	.9	1.2

	LABOR TIME	SEVERE SERVICE
560SL	1.7	2.0
600SEC, S600	3.5	3.8
600SEL	3.3	3.6
600SL	3.0	3.3
C220, C230	.9	1.2
C280		
1994-97	2.0	2.3
1998-00	1.2	1.5
C240, C32, C320, C36	2.4	2.7
CL500	3.2	3.5
CL600	3.3	3.6
CLK320, CLK430	2.2	2.5
CLK55	2.5	2.8
E300D	.5	.8
E320, S320	1.7	2.0
E420, S420	3.3	3.6
E430, E500, E55	1.1	1.4
S430, S500, S55	2.7	3.0
S600	1.8	2.1
SL500	3.1	3.3
SL600	2.8	3.1
SLK230, SLK32	2.5	2.8
SLK320	1.2	1.5
CRANK & PISTONS		
Connecting Rod Bearings, Replace (A)		
Includes: Check bearing oil clearance.		
190D		
2.2L	8.3	8.8
2.5L	9.1	9.6
190E		
1984-89		
AT	13.0	13.5
MT	9.7	10.2
1986-87 16V	8.6	9.1
1986-93 2.6L	6.0	6.5
240D	10.0	10.5
260E	6.9	7.4
280CE, 280E	7.5	8.0
300CD, 300D	14.4	14.9
300CE, 300E, 300TE	6.9	7.4
300SD		
1981-85	12.6	13.1
1992-93	19.9	20.4
300SE, 300SEL	18.2	18.7
300SL	8.3	8.8
300TD		
1981-86	14.4	14.9
1987	12.4	12.9
380SE, 380SL	14.8	15.3
380SEC, 380SEL	15.2	15.7
400E, 400SEL, E420	15.2	15.7
420SEL	22.9	23.4
500E, 500SEC, 500SEL, E500	15.2	15.7
500SL, SL500	13.5	14.0
560SEC, 560SEL	13.5	14.0
560SL	19.5	20.0
600SEC, 600SEL	17.4	17.9

600 SERIES : C SERIES : CL SERIES : CLK SERIES : E SERIES : S SERIES : SL SERIES : SLK SERIES — MB-17

	LABOR TIME	SEVERE SERVICE
600SL	17.5	18.0
E300D	14.1	14.6
E320	11.3	11.8
S350TD	19.9	20.4

Crankshaft and Main Bearings, Replace (A)
Includes: Engine R&R.

	LABOR TIME	SEVERE SERVICE
190D		
2.2L	33.6	34.1
2.5L	35.8	36.4
190E		
1984-85		
MT	17.5	18.0
AT	20.5	21.0
1986-93	29.8	30.3
240D	35.6	36.1
280CE, 280E	29.3	29.8
300CD, 300D	38.3	38.8
300SD		
1981-93	32.8	33.3
1981-87	38.3	38.8
380SE, 380SL	48.0	48.5
380SEC, 380SEL	41.4	41.9
400E, 400SEL, E420	41.4	41.9
420SEL	41.4	41.9
500E, 500SEC, 500SEL, E500	41.4	41.9
560SEC, 560SEL	44.4	44.9
C230	23.1	23.6
C240, C280, C32	26.3	26.8
CL500	40.2	40.7
CL600	45.1	45.6
S420, S430	25.2	25.7
S500	40.6	41.1
S600	21.8	22.3
SL500	40.7	41.2
SL600	40.9	41.4

Crankshaft Front Oil Seal, Replace (A)

	LABOR TIME	SEVERE SERVICE
190D		
2.2L	3.0	3.3
2.5L	5.3	5.6
190E		
1984-85		
MT	5.6	5.9
AT	6.2	6.5
1986-93		
2.3L	3.6	3.9
16V	4.4	4.7
2.6L	5.8	6.1
240D	3.9	4.2
260E	3.3	3.6
300CD	4.8	5.1
300CE, 300TE	3.8	4.1
300D	4.8	5.1
300E	3.3	3.6
300SD		
1981-85	6.0	6.3
1992-93	4.2	4.5
300SE, 300SEL	3.8	4.1

	LABOR TIME	SEVERE SERVICE
300SL	2.2	2.5
300TD		
1981-86	4.8	5.1
1987	3.8	4.1
380SE, 380SL	6.2	6.5
380SEC, 380SEL	4.9	5.2
400E, 400SEL, E420	6.0	6.3
420SEL	6.2	6.5
500E, E500	6.0	6.3
500SEC, 500SEL		
1984-85	4.9	5.2
1993	6.2	6.5
500SL	3.9	4.2
560SEC, 560SEL	6.2	6.5
560SL	5.6	5.9
600SL	2.8	3.1
C220, C240, C32	1.7	2.0
C230, C320	2.4	2.7
C280	4.2	4.5
C36, C55	4.2	4.5
CL500	3.8	4.1
CL600	2.6	2.9
CLK320, CLK430	1.8	2.1
CLK55, CLK500	2.1	2.4
E300D	3.9	4.2
E320, S320	3.1	3.4
E430, E500, E55	1.6	1.9
S350TD	4.2	4.5
S420	3.9	4.2
S430, S55	1.8	2.1
S500, SL500	3.8	4.1
S600	3.8	4.1
SL600	2.6	2.9
SLK230, SLK32	1.8	2.1
SLK320	2.2	2.5
w/AT add	.5	.7

Crankshaft Pulley or Damper, Replace (B)

	LABOR TIME	SEVERE SERVICE
190D		
2.2L	1.4	1.5
2.5L	1.9	2.0
190E		
1984-85	1.3	1.4
1986-93 2.3L	1.4	1.5
1987-93 2.6L	1.9	2.0
240D	1.7	1.8
260E	1.9	2.0
280CE, 280E	2.6	2.7
300CD	1.7	1.8
300CE, 300TE	1.9	2.0
300D, E300D	1.8	1.9
300E	1.9	2.0
300SD		
1981-85	1.7	1.8
1992-93	2.5	2.6
300SE, 300SEL	2.0	2.1
300SL	1.8	1.9
300TD	1.7	1.8
380SE, 380SL	2.7	2.8
380SEC, 380SEL	3.1	3.2

	LABOR TIME	SEVERE SERVICE
400E	3.6	3.7
420SEL	2.7	2.8
500E, E500	3.6	3.7
500SEC, 500SEL	2.7	2.8
500SL	1.8	1.9
560SEC, 560SEL	2.8	2.9
560SL	3.4	3.5
600SEC, 600SEL	2.8	2.9
600SL	1.5	1.6
C220, C230, C320	2.0	2.1
C240, C280, C32	1.6	1.7
C36, C55	1.7	1.8
CL500	1.6	1.7
CL600	1.3	1.4
CLK430, CLK500	2.0	2.1
CLK55	1.7	1.8
E320, E430, E500	1.8	1.9
E420	3.6	3.7
E55, S430, S500, S55	1.3	1.4
S320	2.0	2.1
S350TD	2.5	2.6
S500	1.6	1.7
S600	2.8	2.9
SL500	1.7	1.8
SL55	2.4	2.5
SL600	1.4	1.5
SLK230, SLK32	1.8	1.9

Pistons or Connecting Rods, Replace (A)
Includes: Ridge reaming, cylinder wall deglazing, installing new rings and rod bearings, engine tune-up.

	LABOR TIME	SEVERE SERVICE
190D		
2.2L	17.9	18.4
2.5L	21.4	21.9
190E	28.3	28.8
260E	32.5	33.0
280CE, 280E	38.9	39.4
300CE, 300TE	33.9	34.4
300E	34.8	35.3
1992-93 300SD	33.6	34.1
300SE	33.6	34.1
300SEL	32.8	33.3
300SL	34.2	34.7
380SE, 380SEC, 380SEL, 380SL	52.6	53.1
400E	58.3	58.8
400SEL	52.2	52.7
420SEL	52.6	53.1
500E, E500	58.3	58.8
500SEC, 500SEL	52.6	53.1
500SL	52.2	52.7
560SEC, 560SEL, 560SL	55.5	56.0
600SEC, 600SEL, S600	47.5	48.0
600SL	54.6	55.1
C220	27.7	28.2
C230	28.3	28.8
C240	32.3	32.8

MB-18 — 190 SERIES : 240D : 260E : 280 SERIES : 300 SERIES : 380 SERIES : 400 SERIES : 420SEL : 500 SERIES : 560 SERIES

	LABOR TIME	SEVERE SERVICE
Pistons or Connecting Rods, Replace (A)		
C280		
1994-97	31.4	31.9
1998-00	36.0	36.5
C32, C320	32.5	33.0
C36	31.4	31.9
CL500	23.7	24.2
CL600	34.3	34.8
CLK320	34.9	35.4
CLK430	20.5	21.0
CLK500	27.3	27.8
CLK55	28.3	28.8
E320	33.9	34.4
E420	58.3	58.8
E430	21.6	22.1
E500, E55	23.4	23.9
S320	36.2	36.7
S350TD	33.6	34.1
S420	49.3	50.8
S430	32.4	32.9
S500, S55	23.7	24.2
SL500	24.3	24.8
SL600	30.7	31.2
SLK230	17.1	17.6
SLK32	27.4	27.9
SLK320	34.9	35.4
Rings, Replace (A)		
Includes: Ridge reaming, cylinder wall deglazing, installing new rings, engine tune-up.		
190D		
2.2L	15.9	15.4
2.5L	19.5	20.0
190E	27.1	27.6
260E	31.2	31.7
280CE, 280E	37.6	38.1
300CE	32.8	33.3
300E	33.5	34.0
1992-93 300SD	35.9	36.4
300SE	32.5	33.0
300SEL	31.5	32.0
300SL	33.1	33.6
300TE	29.4	29.9
380SE, 380SEC, 380SEL	49.9	50.4
380SL	53.2	53.7
400E	55.9	56.4
400SE	49.8	50.3
400SEL	49.8	50.3
420SEL	49.9	50.4
500E, E500	55.9	56.4
500SEC, 500SEL	49.9	50.4
500SL	51.9	52.4
560SEC, 560SEL, 560SL	53.2	53.7
600SEC, 600SEL, S600	51.9	52.4
600SL	59.3	59.8
C220	29.3	29.8
C230	29.8	30.3
C240	31.3	31.8
C280		
1994-97	33.6	34.1
1998-00	38.3	38.8
C32, C320	31.5	32.0
C36	33.6	34.1
CL500	29.5	30.0
CL600	41.9	42.4
CLK320	37.2	37.7
CLK430	26.3	26.8
CLK500	25.3	25.8
CLK55	26.3	26.8
E300D	37.8	38.3
E320	35.8	36.3
E420	55.9	56.4
E430	27.4	27.9
E500, E55	21.4	21.9
S320	34.2	34.7
S350TD	35.9	36.4
S420	52.3	52.8
S430	31.4	31.9
S500	29.5	30.0
SL500	30.1	30.6
SL600	38.3	38.8
SLK230	18.7	19.2
SLK32	25.4	25.9
SLK320	37.2	37.7
ENGINE LUBRICATION		
Engine Oil Pressure Switch (Sending Unit), Replace (B)		
16V	.3	.3
Diesel	.5	.5
Lower Oil Pan, Replace or Reseal (B)		
190E	2.2	2.4
260E	2.3	2.5
280CE, 280E	2.2	2.4
300CE, 300TE	2.3	2.5
300E, 300SE, 300SEL	2.3	2.5
300SL	2.0	2.2
380SEC	2.3	2.5
400E, 400SEL	2.5	2.7
420SEL	2.3	2.5
500E, E500	2.5	2.7
1984-85 500SEC, 500SEL	2.2	2.4
1993 500SEC	3.1	3.3
500SL	1.8	2.0
560SEC, 560SEL	2.3	2.5
560SL	2.8	3.0
600SEC, 600SEL	3.0	3.2
600SL	1.8	2.0
C240, C32, C55	4.7	4.9
C280	3.1	3.3
S420	3.6	3.8
CL500	3.5	3.7
CL600	2.8	3.0
CLK320	3.0	3.2
CLK430	2.8	3.0
CLK500	4.7	4.9
CLK55	2.5	2.7
E320	2.9	3.1
E420	2.5	2.7
E430	2.8	3.0
E500, E55	5.1	5.3
S430, S500, S55	3.5	3.7
S600	2.3	2.5
SL500, SL600	1.7	1.9
SL55	8.1	8.3
SLK320	2.8	3.0
Oil Pan and/or Gasket, Replace (B)		
190D		
2.2L	5.3	5.6
2.5L	6.0	6.3
190E		
1984-85		
AT	11.7	12.0
MT	8.6	8.9
1986-87 16V	7.3	7.6
1986-93 2.6L	4.8	5.1
240D	7.3	7.6
260E	5.6	5.9
280CE, 280E	6.2	6.5
300CD	9.7	10.0
300CE, 300TE	5.6	5.9
300D	9.7	10.0
300E	5.6	5.9
300SD		
1981-85	8.0	8.3
1992-93	9.3	9.6
300SE	15.9	16.2
300SEL	15.6	15.9
300SL	5.8	6.1
300TD		
1981-86	9.7	10.0
1987	7.9	8.2
380SE, 380SL	12.4	12.7
380SEC, 380SEL	12.9	13.2
400E, 400SEL	26.3	26.6
420SEL	10.9	11.2
500E, E500	26.3	26.6
1984-85 500SEC, 500SEL	12.9	13.2
1993 500SEC	10.9	11.2
500SL	13.1	13.3
560SEC, 560SEL	10.9	12.2
560SL	17.1	17.4
600SEC, 600SEL	16.5	16.8
600SL	22.6	22.9
C220, C230	7.0	7.3
C280		
1994-97	5.4	5.7
1998-00	8.3	8.6
C36, C55	4.9	5.2
CL500, CL600	16.6	16.9
CLK320	8.3	8.6
E300D	9.3	9.6
E320	5.6	5.9
E420	26.3	26.6

600 SERIES : C SERIES : CL SERIES : CLK SERIES : E SERIES : S SERIES : SL SERIES : SLK SERIES — MB-19

	LABOR TIME	SEVERE SERVICE
E430	9.1	9.4
S320		
std. wheelbase	12.7	13.0
long wheelbase	9.7	10.0
S350TD	9.3	9.6
S420	18.3	18.6
S500	16.6	16.9
S600	16.5	16.8
SL500	13.2	13.5
SL600	23.6	23.9
SLK230	6.8	7.1
SLK320	8.3	8.6

Oil Pressure Warning Switch, Replace (B)

1981-05	.6	.6

Oil Pump, Replace (B)

190D		
2.2L	5.3	5.8
2.5L turbo	6.0	6.5
190E		
1984-85		
AT	17.6	18.1
MT	14.2	14.7
1986-93 2.3L	12.7	13.2
1986-87 16V	16.8	17.3
240D	6.0	6.5
260E	5.2	5.7
280CE, 280E	2.5	3.0
300CD	6.0	6.5
300CE, 300TE	2.8	3.3
300D	5.8	6.3
300E	5.3	5.8
300SD		
1981-85	8.2	8.7
1992-93	9.4	9.9
300SE, 300SEL	6.2	6.7
300SL	2.5	3.0
300TD	5.8	6.3
380SE, 380SL	18.7	19.2
380SEC, 380SEL	2.7	3.2
400E, 400SEL, E420	2.8	3.3
420SEL	2.7	3.2
500E, E500	2.8	3.3
500SEC, 500SEL	2.7	3.2
500SL	12.3	12.6
560SEC, 560SEL	2.7	3.2
560SL	17.2	17.7
C220	5.4	5.9
C230	7.8	8.3
C240, C32, C320	10.0	10.5
C280	5.2	5.7
C36	5.2	5.7
C55	11.2	11.7
CL500	4.3	4.8
CL600	2.4	2.9
E300D	8.3	8.8
E320	8.9	9.4
E430	2.6	3.1

	LABOR TIME	SEVERE SERVICE
S320	10.2	10.7
S350TD	9.4	9.9
S420	4.4	4.9
S430, S500, S55	2.0	2.5
S500	4.3	4.8
S600	2.4	2.9
SL500	2.9	3.4
SL55	10.7	11.2
SL600	3.3	3.8
SLK32, SLK320	2.6	3.1

Side Pan, Replace or Reseal (B)

1981-05	1.4	1.9

CLUTCH

Bleed Clutch Hydraulic System (B)

1984-05	.6	.6

Clutch Pedal Free Play, Adjust (B)

1984-05	.3	.3

Clutch Assy., Replace (B)

190D	6.2	6.4
190E		
2.3L	6.2	6.4
16V	7.9	8.1
2.6L	7.5	7.7
240D	4.8	5.0
260E	6.7	6.9
300E	6.7	6.9
300SL	8.4	8.6
C230, C240	5.8	6.0
SLK230, SLK320	12.8	13.0

Clutch Master Cylinder, Replace (B)
Does not include system bleeding.

190D, 190E	2.7	2.9
240D	2.4	2.6
1988 260E	3.5	3.7
1988 300E	3.5	3.7
C230, C240	2.8	3.0

Clutch Slave Cylinder, Replace (B)
Does not include system bleeding.

1984-85 190D, 190E	2.3	2.4
1986-87 190E 16V	2.7	2.8
1988 260E	1.6	1.7
1988 300E	1.6	1.7
300SL	1.6	1.7

MANUAL TRANSMISSION

Transmission Assy., R&R and Recondition (A)

1984-89 190D	16.0	16.9
1984-93 190E		
2.3L	16.0	16.9
16V	17.8	18.7
2.6L	17.1	18.0
240D	12.7	13.6
260E, 300E	16.5	17.4
300SL	18.8	19.7

	LABOR TIME	SEVERE SERVICE
Transmission Assy., R&R or Replace (B)		
1984-89 190D	5.4	5.6
1984-93 190E		
2.3L	5.4	5.6
16V	7.3	7.5
2.6L	6.5	6.7
240D	4.2	4.4
260E, 300E	6.2	6.4
300SL	7.8	8.0
C230, C240	5.5	5.7
SLK230, SLK320	12.0	12.2
Replace rear main seal add	.8	1.0

AUTOMATIC TRANSMISSION

SERVICE TRANSMISSION INSTALLED

Drain and Refill Unit (B)

Exc. below	1.4	1.4
190D, 190E, 260E	1.7	1.7
300CE, 300TE	1.7	1.7
300D, 300TD	1.7	1.7
300SE, 300SEL, 300SL	1.7	1.7
420SEL	1.7	1.7
500SEC, 500SEL, 500SL	1.7	1.7
560SEC, 560SEL	1.7	1.7
560SL	2.5	2.5
600SL	1.5	1.5
C230, C320, C55	2.0	2.0
CLK320	2.0	2.0
S420	2.0	2.0
S600	2.0	2.0
SL500	1.5	1.5

Control Pressure Cable, Replace (B)

190D, 190E, 260E	2.3	2.4
280CE, 280E	2.3	2.4
300CD, 300D	2.3	2.4
300CE, 300TE	3.0	3.1
300SD		
1981-85	2.5	2.6
1992-93	3.3	3.4
300SE, 300SEL	2.5	2.6
300SL	3.3	3.4
300TD	2.3	2.4
380SE, 380SL	2.8	2.9
380SEC, 380SEL	2.5	2.6
400E, 400SEL, E420	3.0	3.1
420SEL	1.4	1.5
500E, E500	3.0	3.1
500SL, SL500	3.3	3.4
1984-85 500SEC, 500SEL	3.1	3.2
1993 500SEC	2.5	2.6
560SEC, 560SEL	1.4	1.5

MB-20 — 190 SERIES : 240D : 260E : 280 SERIES : 300 SERIES : 380 SERIES : 400 SERIES : 420SEL : 500 SERIES : 560 SERIES

	LABOR TIME	SEVERE SERVICE
Control Pressure Cable, Replace (B)		
600SEC, 600SEL, 600SL	3.3	3.4
C280	3.0	3.1
E300D	2.2	2.3
E320	3.0	3.1
S320	3.3	3.4
S600, SL600	3.3	3.4
Governor Assy., R&R or Replace (B)		
1984-85 190D, 190E	5.0	5.3
1986-89 190D	3.5	3.8
240D, 260E	3.2	3.5
280CE, 280E	5.0	5.3
300D	3.2	3.5
300SD		
1981-85	3.2	3.5
1992-93	2.7	3.0
300SE, 300SEL	3.1	3.4
300SL	2.4	2.7
300TD	3.2	3.5
380SE, 380SEC, 380SEL, 380SL	3.1	3.4
400E, 400SEL	3.1	3.4
420SEL, E420	3.1	3.4
500E, 500SEC, 500SEL, E500	3.1	3.4
500SL, SL500	2.4	2.7
560SEC, 560SEL, 560SL	3.1	3.4
600SEC, 600SEL	2.7	3.0
600SL, SL600	2.4	2.7
C220	2.4	2.7
C280	2.7	3.0
E300D	3.2	3.5
E320	3.2	3.5
S320, S350TD	2.7	3.0
S600	2.7	3.0
Recondition governor add	.5	.8
Kickdown Solenoid or Valve, Replace (B)		
1981-05	.7	.7
Oil Pan Gasket, Replace (B)		
Exc. below	1.7	1.7
560SL	2.5	2.5
C230, C240, C280	1.1	1.1
CL500, CL600	1.6	1.6
CLK320, CLK430	2.0	2.0
C230, C320, C55	2.5	2.5
E320, E430	1.6	1.6
S420	2.0	2.0
S500	1.6	1.6
S600	2.0	2.0
SLK230	1.6	1.6
SLK320, SLK32	2.0	2.0
Parking Pawl, Replace (B)		
Exc. below	5.8	6.1
380SEL	6.0	6.3
380SL	7.2	7.5
1993 500SEC	6.0	6.3

	LABOR TIME	SEVERE SERVICE
560SL	7.0	7.3
600SEC, 600SEL, S600	13.7	14.0
C220	5.2	5.5
C280	5.2	5.5
C36	5.2	5.5
SL500	11.4	11.7
SL600	13.4	13.7
Valve Body Assy., Replace (B)		
Exc. below	2.2	2.5
560SL	3.1	3.4
C230	2.8	3.1
C280	2.8	3.1
CL500, CL600	2.3	2.6
CLK320, CLK430	2.5	2.8
E300D	2.8	3.1
E320, E430	2.1	2.4
S420	2.8	3.2
S500	2.3	2.7
S600	2.8	3.2
SL500	2.8	3.2
SL600	2.8	3.2
SLK230	2.1	2.4
SLK320	2.5	2.8
Recondition valve body add	1.7	2.0
SERVICE TRANSMISSION REMOVED		
Transmission Assy., R&R and Recondition (B)		
190D		
2.2L	23.9	24.8
2.5L	24.5	25.4
turbo	25.2	26.1
190E		
1984-93		
2.3L	23.9	24.8
16V	25.2	26.1
2.6L	25.2	26.1
240D	24.3	25.2
260E	22.9	23.8
280CE, 280E	20.0	20.9
300CD	19.3	20.2
300CE, 300E, 300TE	24.1	25.0
300D	22.9	23.8
300SD	24.1	25.0
300SE, 300SEL	25.3	26.2
300SL	27.3	28.2
300TD	23.3	24.2
380SE	26.2	27.1
380SEC, 380SEL	20.9	21.8
380SL	26.7	27.6
400E	30.5	31.4
400SEL	26.7	27.6
420SEL	27.7	28.6
500E, E500	30.5	31.4
500SEC		
1984-85	20.9	21.8
1993	26.7	27.6

	LABOR TIME	SEVERE SERVICE
500SEL	27.7	28.6
500SL	27.6	28.5
560SEC, 560SEL	27.7	28.6
560SL	28.6	29.5
600SEC, 600SEL	27.7	28.6
600SL	27.6	28.5
C220	23.7	24.6
C230, C240	25.5	26.4
C280		
1994-97	25.0	25.9
1998-00	16.2	17.1
C320	21.9	22.8
C36	25.0	25.9
CLK320	23.9	24.8
CLK430	31.6	32.5
CLK500, CLK55	33.1	34.0
E320, S320, S350TD	24.1	25.0
E300D		
1995-97	23.3	24.2
1998-99	23.9	24.8
E320	25.1	26.0
E420, E500	30.5	31.4
E430	27.1	28.0
S420	25.3	26.2
S500	25.2	26.1
S600	26.4	27.3
SL500	24.8	25.7
SL600	27.3	28.2
SLK230	17.8	18.7
SLK320	23.9	24.5
w/torque converter add	.7	.7
Flush oil cooler and lines add	.5	.5
Transmission Assy., R&R (B)		
190D		
2.2L	8.4	8.6
2.5L	9.1	9.3
turbo	9.7	9.9
190E		
1984-93		
2.3L	8.4	8.6
16V	9.7	9.9
2.6L	9.7	9.9
240D	6.9	7.1
260E	7.3	7.5
280CE, 280E	6.4	6.6
300CD, 300D, E300D	7.8	8.0
300CE, 300E, 300SD, 300TE	8.6	8.8
300SE, 300SEL	9.6	9.8
300SL	9.1	9.3
300TD	7.8	8.0
380SE	8.6	8.8
380SEC, 380SEL	9.3	9.5
380SL	10.8	11.0
400E	15.2	15.4
400SEL	10.8	11.0
420SEL	10.2	10.4
500E, E500	15.2	15.4

600 SERIES : C SERIES : CL SERIES : CLK SERIES : E SERIES : S SERIES : SL SERIES : SLK SERIES — MB-21

	LABOR TIME	SEVERE SERVICE
500SEC. 500SEL		
1984-85	9.3	9.5
1993	10.8	11.0
500SL	11.9	12.1
560SEC, 560SEL	10.2	10.4
560SL	10.9	11.1
600SEC, 600SEL	12.0	12.2
600SL	11.9	12.1
C220	8.3	8.5
C230	9.4	9.6
C240	8.9	9.1
C280		
1994-97	9.4	9.6
1998-00	7.8	8.0
C320	6.9	7.1
C36	9.4	9.6
CL500, CL600	7.3	7.5
CLK320	7.8	8.0
CLK430	8.7	8.9
CLK500, CLK55	7.2	7.4
E320, S320, S350TD	8.6	8.8
E420	15.2	15.4
E430	7.6	7.8
E500, E55	6.3	6.5
S420	9.1	9.3
S430, S55	7.3	7.5
S500	8.8	9.0
S600	10.0	10.2
SL500	8.4	8.6
SL600	10.9	11.1
SLK230, SLK32	7.6	7.8
SLK320	7.8	8.0
Replace		
flexplate add	.9	1.3
rear main seal add	.8	1.1

TRANSFER CASE

Bleed Hydraulic System (B)
300E, 300TE	.7	.7

Transfer Case, R&I (B)
300E, 300TE	5.8	6.3
E320, E430	4.1	4.6
Replace assembly add	.7	.9

SHIFT LINKAGE
MANUAL TRANSMISSION

Shift Linkage, Adjust (B)
1981-05	.5	.5

DRIVELINE

Axle Flange Oil Seal, Replace (B)
One side
240D, 280CE, 280E	3.9	4.1
300CD, 300D, 300SD	3.9	4.1
300SEL	4.5	4.7
380SE, 380SEC, 380SEL	5.2	5.4
380SL	3.9	4.1

	LABOR TIME	SEVERE SERVICE
420SEL	4.5	4.7
500SEC, 500SEL	5.2	5.4
560SEC, 560SEL	4.5	4.7
560SL	3.9	4.1
E320	5.3	5.5
S320	6.9	7.1

Axle Shaft Bearings, Replace (B)
One side	5.4	5.7

Axle Shaft Flange & Bearing, Replace (B)
One side
1992-93 300SD	6.2	6.5
300SE, 300SEL	4.5	4.8
420SEL	4.9	5.2
500SEC, 500SEL	6.2	6.5
560SEC, 560SEL, 560SL	4.9	5.2
600SEC, 600SEL, S600	6.7	7.0
600SL	5.3	5.6
C220, C240	5.4	5.7
C230	5.3	5.6
C280, C32	5.4	5.7
C36, C55	5.4	5.7
CL500, CL600	5.4	5.7
E300D	5.3	5.6
E320, E430, E500	4.8	5.1
S320, S350TD	6.2	6.5
S500	5.4	5.7
SL500, SL600	6.0	6.3
SLK230, SLK32	4.5	4.8

Axle Shaft Oil Seal, Replace (B)
280CE, 280E
one	3.1	3.4
both	4.4	4.7
300CD, 300D, 300SD, 300TD		
one	3.1	3.4
both	4.4	4.7
380SEC		
one	3.1	3.4
both	4.4	4.7
380SEL		
one	3.9	4.2
both	6.5	6.8
420SEL		
one	3.9	4.2
both	6.5	6.8
500SEL, 500SL, SL500		
one	3.9	4.2
both	6.5	6.8
560SEC, 560SEL		
one	3.9	4.2
both	6.5	6.8
560SL		
one	4.2	4.5
both	5.6	5.9
E300D		
one	5.3	5.6
both	5.6	5.9

	LABOR TIME	SEVERE SERVICE
Driveshaft, Replace (B)		
190D, 190E	2.4	2.6
240D, 260E	2.8	3.0
280CE, 280E	2.8	3.0
300CD, 300D	2.8	3.0
300SD	3.8	4.0
300SE, 300TD	3.8	4.0
300SEL	3.2	3.4
300SL	3.5	3.7
380SE, 380SEC, 380SEL	3.2	3.4
380SL	4.4	4.6
400E	5.8	6.0
400SEL	3.8	4.0
420SEL	3.2	3.4
500E, E500	5.8	6.0
500SEC		
1984-85	3.2	3.4
1993	3.8	4.0
500SEL	3.8	4.0
500SL	3.5	3.7
560SEC, 560SEL	3.2	3.4
560SL	6.5	6.7
600SEC, 600SEL	3.8	4.0
600SL	3.5	3.7
C220, C230, C240	2.4	2.6
C280, C32, C36	2.4	2.6
C320, C55	2.8	3.0
CL500, CL600	3.1	3.3
CLK320, CLK430	2.5	2.7
CLK500, CLK55	3.1	3.3
E300D	2.5	2.7
E320 E430	2.1	2.3
E420, E500, E55	5.8	6.0
S320, S350TD	3.8	4.0
S420, S430	3.6	3.8
S500, S55	3.1	3.3
S600	3.6	3.8
SL500, SL600	2.8	3.0
SLK230	2.3	2.5
SLK320, SLK32	2.5	2.7
Replace		
center bearing add	.7	.9
flex disc add	.7	.9
inter bearing add	.8	1.0
shaft add each	.3	.5

Front Halfshaft, R&R or Replace (B)
300E, 300TE
one	2.3	2.8
both	3.5	4.0
E320, E430		
one side	2.2	2.7
both sides	3.0	3.5

Pinion Shaft Oil Seal, Replace (B)
Rear axle
1984-85 190D, 190E	2.7	2.9
240D	3.9	4.1
260E	3.0	3.2
280CE, 280E	3.3	3.5

MB-22 190 SERIES : 240D : 260E : 280 SERIES : 300 SERIES : 380 SERIES : 400 SERIES : 420SEL : 500 SERIES : 560 SERIES

	LABOR TIME	SEVERE SERVICE
Pinion Shaft Oil Seal, Replace (B)		
300CD	3.3	3.5
300CE	3.0	3.2
300D		
1981-85	3.3	3.5
1986-93	3.9	4.1
300E	3.0	3.2
300SD		
1981-85	4.2	4.4
1992-93	6.2	6.4
300SE, 300SEL	4.2	4.4
300SL	5.3	5.5
300TD	3.3	3.5
380SEC, 380SEL	3.6	3.8
400E	5.3	5.5
400SEL	7.0	7.2
420SEL	4.2	4.4
500E, E500	5.3	5.5
500SEC		
1984-85	3.6	3.8
1993	5.2	5.4
500SEL	5.2	5.4
500SL	5.3	5.5
560SEC, 560SEL	4.2	4.4
560SL	4.9	5.1
600SEC, 600SEL	6.5	6.7
600SL	5.3	5.5
C220, C230, C240	5.4	5.6
C280, C36, C32	5.4	5.6
C320, C55	5.3	5.5
CL500, CL600	4.7	4.9
CLK320, CLK430	4.9	5.1
CLK500, CLK55	5.4	5.6
E300D	3.3	3.5
E320	3.3	3.5
E420, E500, E55	5.3	5.5
E430	3.4	3.6
S320, S350TD	6.2	6.4
S420, S430	6.9	5.1
S500	4.2	4.4
S600	6.5	6.7
SL500, SL600	4.5	4.7
SLK230	2.8	3.0
SLK320, SLK32	4.9	5.1
Rear Axle Assy., R&I (B)		
190D, 190E		
exc. 16V	6.5	7.0
16V	7.5	8.0
240D	8.4	8.9
260E	6.8	7.3
280CE, 280E	8.9	9.4
300CD	8.4	8.9
300CE	6.7	7.2
300D	6.8	7.3
300E	6.8	7.3
300SD		
1981-85	9.2	9.7
1992-93	11.7	12.2
300SE, 300SEL	8.6	9.1
300SL	9.7	10.2

	LABOR TIME	SEVERE SERVICE
300TD	7.5	8.0
300TE	7.5	8.0
380SE, 380SEC, 380SEL, 380SL	9.2	9.7
400E	7.6	8.1
400SEL	11.4	11.9
420SEL	8.6	9.1
500E, E500	7.6	8.1
500SEC, 500SEL		
1984-85	9.2	9.7
1993	13.7	14.2
500SL	10.2	10.7
560SEC, 560SEL	9.7	10.2
560SL	9.4	9.9
600SEC, 600SEL, S600	13.7	14.2
600SL	12.0	12.5
C220, C320	9.1	9.6
C230, C240, C55	8.8	9.3
C280, C36, C32	9.1	9.6
CL500, CL600	6.7	7.2
CLK320, CLK430	6.5	7.0
CLK500, CLK55	7.8	8.3
E300D	8.8	9.3
E320		
convertible	8.3	8.8
coupe, sedan	6.8	7.3
wagon	7.5	8.0
E420, E430	7.6	8.1
E500, E55	10.2	10.7
S320, S350TD	11.7	12.2
S420, S430	12.0	12.5
S500, S55	6.7	7.2
SL500	6.4	6.9
SL600, SL55	7.3	7.8
SLK230	4.9	5.4
SLK320, SLK32	6.5	7.0
Rear Axle Housing Cover, Replace or Reseal (B)		
1981-05	1.8	1.8
Rear Axle Shaft, Replace (B)		
190D, 190E one	2.4	2.7
240D one	3.1	3.4
260E one	2.4	2.7
280CE, 280E one	3.1	3.4
300CD, 300D, 300TD one	3.1	3.4
300E one	2.4	2.7
300SD		
1981-85 one	3.1	3.4
1992-93 one	3.6	3.9
300SE one	3.6	3.9
300SEL one	2.5	2.8
300SL one	2.4	2.7
380SEC one	2.5	2.8
380SEL one	3.1	3.4
400E one	2.4	2.7
400SEL one	3.6	3.9
420SEL one	2.5	2.8
500E, E500 one	2.4	2.7

	LABOR TIME	SEVERE SERVICE
500SEC one	2.5	2.8
500SEL one	4.2	4.5
500SL one	2.4	2.7
560SEC, 560SEL, 560SL one	2.5	2.8
600SEC, 600SEL, S600 one	3.7	4.0
600SL one	2.4	2.7
C220 one	2.7	3.0
C230, C240, C55 one	2.5	2.8
C280, C36, C32, C320 one	2.7	3.0
CL500, CL600 one	2.1	2.4
CLK320 one	2.5	2.8
CLK430, CLK500, CLK55 one	2.1	2.4
E300D one	2.7	3.0
E320, S320, S350TD one	2.4	2.7
E420 one	2.4	2.7
E430, E500, E55 one	2.1	2.4
S420, S430, S600 one	2.2	2.5
S500, S55 one	1.8	2.1
SL500 one	2.2	2.5
SL600 one	2.2	2.5
SLK230 one	2.1	2.4
SLK320, SLK32 one	2.5	2.8
Replace boots add each	1.7	2.0
FRONT AXLE 4WD		
Axle Shaft Bearings, Replace (B)		
300E, 300TE		
one side	3.6	3.9
both sides	6.5	6.8
E320, E430		
one side	2.6	2.9
both sides	4.8	5.1
Differential Housing, R&R or Reseal (B)		
300E, 300TE	11.9	12.4
E320, E430	6.5	7.0
Differential Oil Pan, Reseal (B)		
300E, 300TE	.9	1.2
Halfshaft Assy., R&R or Replace (B)		
300E, 300TE		
one	2.3	2.7
both	3.5	3.9
E320, E430		
one side	2.2	2.6
both sides	3.0	3.4

BRAKES
ANTI-LOCK

Diagnose Anti-Lock Brake System (A)
1987-05 .... 1.6 *1.6*

Control Unit, Replace (B)
1987-05 .... .5 *.5*

600 SERIES : C SERIES : CL SERIES : CLK SERIES : E SERIES : S SERIES : SL SERIES : SLK SERIES — MB-23

	LABOR TIME	SEVERE SERVICE
Hydraulic Assy., Replace (B)		
1987-05		
w/ASR	2.4	2.7
w/o ASR	1.6	1.8
Pump or Main Relay, Replace (B)		
1987-05	.3	.3
Wheel Speed Sensor, Replace (B)		
Front		
190D, 190E one	1.3	1.3
260E one	.9	.9
300CE, 300TE one	.9	.9
300D, 300E one	.9	.9
300SD, 300SE, 300SEL one	1.3	1.3
300SL one	.7	.7
400E, 400SEL one	.9	.9
420SEL one	1.3	1.3
500E, 500SEL one	1.3	1.3
500SL one	.7	.7
560SEC, 560SEL one	1.7	1.7
560SL one	1.3	1.3
600SEC, 600SEL one	1.3	1.3
600SL one	.7	.7
C220, C230, C240, C55 one	1.3	1.3
C280, C36, C32, C320 one	1.2	1.2
CL500, CL600 one	1.1	1.1
CLK320 one	.8	.8
CLK430, CLK500, CLK55 one	1.0	1.0
E300D one	.8	.8
E320 one	.9	.9
E420, S420 one	.9	.9
E430, E500, E55 one	.7	.7
S320, S350TD one	1.3	1.3
S430, S500, S55 one	1.1	1.1
S600 one	1.4	1.4
SL500, SL600, SL55 one	.7	.7
SLK230 one	.7	.7
SLK320, SLK32 one	.8	.8
Rear		
190D, 190E	1.6	1.6
260E	1.6	1.6
300CE, 300TE	1.6	1.6
300D	1.6	1.6
300E, 300TD	1.6	1.6
300SD	1.3	1.3
300SE, 300SEL		
center	1.3	1.3
side one	1.7	1.7
300SL one	6.2	6.2
400E, 400SEL	1.3	1.3
420SEL	1.6	1.6

	LABOR TIME	SEVERE SERVICE
500E, 500SEL	1.3	1.3
500SL one	6.2	6.2
560SEC, 560SEL, 560SL	1.6	1.6
600SEC, 600SEL one	2.0	2.0
600SL one	6.2	6.2
C220	1.4	1.4
C230, C240, C32, C320, C55 one	1.4	1.4
C280, C36	1.4	1.4
CL500, CL600 one	.7	.7
CLK320 one	.5	.5
CLK430 CLK500, CLK55 one	1.1	1.1
E300D	1.2	1.2
E320	1.6	1.6
E430, E500, E55 one	.8	.8
S420	1.2	1.2
S430, S500, S55 one	1.0	1.0
S600 one	1.3	1.3
SL500, SL600 one	5.2	5.2
SLK230, SLK32 one	.6	.6
SLK320 one	.5	.5
SYSTEM		
Bleed Brakes (B)		
Includes: Add fluid.		
1981-05	.7	.7
w/ABS add	.5	.5
Brake Hose (Flexible), Replace (B)		
Includes: System bleeding.		
1981-05 front or rear one	1.4	1.8
each addl. add	.6	.6
Master Cylinder, Replace (B)		
190D, 190E	1.4	1.6
240D	.9	1.1
260E	1.4	1.6
280CE, 280E	.9	1.1
300CD, 300D, E300D	.9	1.1
300CE, 300E, 300SD	1.4	1.6
300SE	1.8	2.0
300SEL, 300SL, 300TD, 300TE	1.4	1.6
380SC	2.0	2.2
380SE, 380SEL	1.4	1.6
400E	.9	1.1
400SE, 400SEL	2.0	2.2
420SEL	1.4	1.6
500E, 500SEC, 500SEL, E500	2.0	2.2
500SL, 560SEC, 560SEL	1.4	1.6
560SL	.9	1.1
600SEC, 600SEL, S600	3.1	3.3

	LABOR TIME	SEVERE SERVICE
00SL	.9	1.1
C220, C230, C55		
C240, C280, C36	1.4	1.6
C320, C32, CL500	1.5	1.7
CL600	2.7	2.9
CLK320, CLK500	1.5	1.7
CLK430, CLK55	1.1	1.3
E320, E430	1.4	1.6
E420	.9	1.1
E430, E500, E55	1.0	1.2
S320, S350TD	1.8	2.0
S420, S430, S600	2.2	2.4
S500, S55	1.5	1.7
SL500	1.8	2.0
SL55, SL600	2.1	2.3
SLK230	1.0	1.2
SLK320, SLK32	1.5	1.7
w/ABS add	1.7	1.9
Power Booster Unit, Replace (B)		
190D, 190E	2.0	2.2
240D, 260E	2.3	2.5
280CE, 280E	2.3	2.5
300CD, 300D, 300E	2.3	2.5
300CE, 300TE	2.3	2.5
300SD		
1981-85	2.8	3.0
1992-93	3.0	3.2
300SE, 300SL	3.0	3.2
300SEL	2.8	3.0
300TD	2.3	2.5
380SE, 380SEC, 380SEL, 380SL	2.8	3.0
400E, 400SEL, E420	4.5	4.7
420SEL	3.0	3.2
300E, E500	5.2	5.4
500SEC, 500SEL		
1984-85	2.8	3.0
1993	3.3	3.5
500SL	3.3	3.5
560SEC, 560SEL	2.8	3.0
560SL	2.3	2.5
600SEC, 600SEL	7.9	8.1
600SL, S600	3.3	3.5
C220, C240	3.6	3.8
C230, C32, C55	3.8	4.0
C280, C320, C36	3.6	3.8
CL500, CL600	2.3	2.5
CLK320, CLK55	3.6	3.8
CLK430, CLK500	1.1	1.3
E300D, E55	2.5	2.7
E320, E430, E500	2.3	2.5
S320, S350TD	3.0	3.2
S420, S430	3.1	3.3
S500, S55	2.3	2.5
S600	8.4	8.6
SL500, SL600	1.2	1.4
SLK230	1.1	1.3
SLK320, SLK32	3.6	3.8
w/ASR add	.5	.7

190 SERIES : 240D : 260E : 280 SERIES : 300 SERIES : 380 SERIES : 400 SERIES : 420SEL : 500 SERIES : 560 SERIES

	LABOR TIME	SEVERE SERVICE
Power Booster Vacuum Pump, Replace (B)		
190D	2.3	2.6
240D	1.8	2.1
300CD	2.5	2.8
300D	2.3	2.6
300SD		
1981-85	2.8	3.1
1992-93	2.3	2.6
300TD	2.8	3.1
E300D		
1995-97	1.4	1.7
1998-99	1.7	2.0
S350TD	2.3	2.6
Power Booster Vacuum Check Valve, Replace (B)		
1981-05	.8	.8
SERVICE BRAKES		
Brake Pads, Replace (B)		
Front or rear	.9	1.0
Four wheels	1.9	2.0
Resurface disc rotor add each		
w/wheel bearings	.9	.9
w/o wheel bearings	.5	.5
Disc Brake Rotor, Replace (B)		
Front both		
exc. below	2.4	2.6
240D	4.2	4.4
280CE, 280E	4.2	4.4
300CD, 300D, 300SD, 300TD	4.2	4.4
300SEL	4.2	4.4
380SE, 380SEC, 380SEL, 380SL	4.2	4.4
420SEL	4.2	4.4
500SEC, 500SEL	4.2	4.4
560SEC, 560SEL	4.2	4.4
Rear both		
exc. below	3.0	3.2
240D, 280CE, 280E	2.8	3.0
300CD, 300D, 300SEL, 300TD	2.8	3.0
380SE, 380SEC, 380SEL, 380SL	2.8	3.0
400SEL	2.8	3.0
500SEC, 500SEL	2.8	3.0
560SL	2.8	3.0
Front Caliper Assy., Replace (B)		
Does not include system bleeding.		
One	1.3	1.6
Both	2.3	2.6
Rear Caliper Assy., Replace (B)		
One	1.4	1.7
Both	2.5	2.8

	LABOR TIME	SEVERE SERVICE
PARKING BRAKE		
Parking Brake Apply Actuator, Replace (B)		
190D, 190E	2.5	2.5
190E 16V	3.3	3.3
240D, 260E	3.1	3.1
280CE, 280E	2.5	2.5
300CD, 300D	2.5	2.5
300CE, 300TE	3.1	3.1
300E	1.8	1.8
300SD		
1981-85	3.1	3.1
1992-93	3.6	3.6
300SE, 300SEL	1.3	1.3
300SL	3.0	3.0
300TD	2.5	2.5
380SE, 380SEC, 380SEL, 380SL	3.1	3.1
400E, 400SEL	1.8	1.8
420SEL	3.1	3.1
500E, E500	1.8	1.8
500SEC, 500SEL		
1984-85	3.1	3.1
1993	1.3	1.3
500SL	1.3	1.3
560SEC, 560SEL, 560SL	3.1	3.1
600SEC, 600SEL	3.6	3.6
600SL	1.3	1.3
C220, C230, C240	1.8	1.8
C280, C320, C36	1.8	1.8
CL500, CL600, C32, C55	3.0	3.0
CLK320, CLK430	1.9	1.9
CLK500, CLK55	2.3	2.3
E300D	2.0	2.0
E320, E430, E500	1.8	1.8
E420, S420, E55	1.8	1.8
E430	1.9	1.9
S320, S350TD	3.6	3.6
S430, S500	3.0	3.0
S600, S55, SL55	1.8	1.8
SL500, SL600, SLK32	2.3	2.3
SLK230, SLK320	1.4	1.4
Parking Brake Apply Warning Indicator Switch, Replace (B)		
Exc. below	1.4	1.4
190D, 190E	.5	.5
300SE, 300SEL, 300SL	.7	.7
500SL	.7	.7
560SL	1.6	1.6
600SL	.7	.7
C220, C230, C280, C32, C36, C55	.8	.8
CLK320, CLK430	.9	.9
S600	.9	.9
SLK320, SLK32	.9	.9

	LABOR TIME	SEVERE SERVICE
Parking Brake Cable, Adjust (C)		
1981-05	.6	.8
Parking Brake Cable, Replace (B)		
Front		
190D, 190E	2.3	2.6
1986-87 190E 16V	3.3	3.6
240D	4.2	4.5
260E	4.4	4.7
280CE, 280E	2.0	2.3
300D, 300E, E300D	4.4	4.7
300SD		
1981-85	7.8	8.1
1992-93	4.9	5.2
300SE, 300SEL	8.9	9.2
300SL	6.7	7.0
300TD	4.4	4.7
400E, E420	4.4	4.7
420SEL	8.6	8.9
500E, E500	4.4	4.7
500SEC, 500SEL	7.8	8.1
500SL	6.7	7.0
560SEC, 560SEL	8.6	8.9
560SL	3.8	4.1
600SEC, 600SEL	4.9	5.2
600SL	6.7	7.0
C220, C230, C280, C36, C55	5.3	5.6
C240, C32	4.9	5.2
CL500, CL600	4.3	4.6
CLK320, CLK430	4.6	4.9
CLK500, CLK55	5.0	5.3
E320	4.4	4.7
E430, E500, E55	2.7	3.0
S320, S350TD	4.9	5.2
S420, S430	5.2	5.5
S500, S55	4.3	4.6
S600	5.2	5.5
SL500, SL600	5.9	6.2
SLK230, SLK32	4.6	4.9
SLK320	4.6	4.9
Rear		
190D, 190E	2.3	2.6
240D	2.4	2.7
260E	2.3	2.6
280CE, 280E	2.4	2.7
300CD, 300D	2.3	2.6
300CE, 300E, 300TE, E300D	2.4	2.7
300SD	5.2	5.5
300SE	2.5	2.8
300SL	3.6	3.9
380SE, 380SEC, 380SEL, 380SL	2.4	2.7
400E, 400SEL, E420	3.8	4.1
420SEL	2.4	2.7
500E	3.8	4.1

600 SERIES : C SERIES : CL SERIES : CLK SERIES : E SERIES : S SERIES : SL SERIES : SLK SERIES — MB-25

	LABOR TIME	SEVERE SERVICE
500SEC, 500SEL		
1984-85	2.4	2.7
1993	4.2	4.5
500SL	3.6	3.9
560SEC, 560SEL	2.4	2.7
560SL	3.6	3.9
600SEC, 600SEL, S600	4.4	4.7
600SL	4.8	5.1
C220, C240	4.2	4.5
C230, C32, C36	3.8	4.1
C280, C320, C55	4.2	4.5
CL500, CL600	3.4	3.7
CLK320	5.2	5.5
CLK430	3.4	3.7
CLK500, CLK55	4.3	4.7
E320	2.4	2.7
E430, E500, E55	3.4	3.7
S320, S430	4.9	5.2
S350TD, S500	5.2	5.5
S420, S55, S600	5.3	5.6
SL500, SL600	4.3	4.6
SLK230, SLK320	3.4	3.7
SLK32	3.7	4.0

Parking Brake Shoes, Replace (B)

	LABOR TIME	SEVERE SERVICE
190D, 190E	3.6	3.8
240D	3.5	3.7
260E	3.6	3.8
280CE, 280E	3.5	3.7
300CD, 300CE, 300D	3.6	3.8
300E	3.8	4.0
300SD, 300SE	4.4	4.6
300SEL, 300SL, 300TE, 380SE, 380SEC, 380SEL	3.6	3.8
380SEL	3.5	3.7
400E	3.1	3.3
400SE, 400SEL	4.4	4.6
420SEL	3.5	3.7
500E, E500	3.6	3.8
500SEC		
1984-85	3.5	3.7
1993	4.4	4.6
500SEL		
1984-85	4.2	4.4
1993	4.4	4.6
500SL	3.6	3.8
560SEC, 560SEL	4.2	4.4
600SEC, 600SEL, S600	4.8	5.0
600SL	3.6	3.8
C220, C240, C32	4.2	4.4
C230	3.1	3.3
C280, C320, C36, C55	4.2	4.4
CL500, CL600	2.7	2.9
CLK320	4.4	4.6
CLK430	2.7	2.9
CLK500	3.6	3.8

	LABOR TIME	SEVERE SERVICE
E300D	4.2	4.4
E320, E500	3.8	4.0
E420, E55	3.1	3.3
E430	2.7	2.9
S320	4.2	4.4
S350TD	4.4	4.6
S420, S430, S55	3.9	4.1
S500, S600, SLK32	3.6	3.9
SL500, SL600	2.3	2.5
SLK230, SLK320	2.7	2.9

Release Button, Replace (B)

	LABOR TIME	SEVERE SERVICE
190D, 190E	.3	.3

FRONT SUSPENSION

Unless otherwise noted, time given does not include alignment.

Align Front End (A)

	LABOR TIME	SEVERE SERVICE
1981-05	1.7	2.0

Front Toe, Adjust (B)

	LABOR TIME	SEVERE SERVICE
1981-05	.8	1.0

Front & Rear Alignment, Check & Adjust (B)

	LABOR TIME	SEVERE SERVICE
Exc. below	2.4	2.7
C220	1.7	2.0
300TD	3.0	3.3
560SEC, 560SEL	3.0	3.3

Rear Toe-In, Adjust (B)

	LABOR TIME	SEVERE SERVICE
1981-05	.7	1.0

Front Axle Hub Oil Seal, Replace (B)

	LABOR TIME	SEVERE SERVICE
190D, 190E one	2.3	2.6
240D, 260E one	2.3	2.6
280CE, 280E one	2.3	2.6
300CD, 300D one	2.3	2.6
300CE, 300E, 300TE, E300D one	2.3	2.6
300SD, 300TD one	2.3	2.6
300SE, 300SL one	2.3	2.6
300SEL one	2.3	2.6
380SEC, 380SEL one	2.3	2.6
400E, 400SEL, E420 one	2.3	2.6
420SEL one	2.3	2.6
500E, 500SEC, 500SEL one	2.3	2.6
500SL one	2.3	2.6
560SEC, 560SEL one	2.3	2.6
560SL one	2.7	3.0
600SEC, 600SEL, 600SL one	2.3	2.6
C220, C280, C32, C36, C55 one	2.2	2.5
C230, C240, C320, CLK55 one	1.7	2.0
CL500, CL600 one	1.4	1.7
CLK320, CLK430, CLK500 one	1.7	2.0
E320, S320, S350TD one	2.3	2.6

Front Axle Hub Oil Seal, Replace (B)

	LABOR TIME	SEVERE SERVICE
E430, E500, E55 one	1.4	1.7
S420, S430 one	2.2	2.5
S500, SL500, S55 one	1.0	1.3
S600, SL600 one	2.2	2.5

Replace
	LABOR TIME	SEVERE SERVICE
wheel bearings add each side	.5	.7
wheel hub add	.7	.9

Front Shock Absorbers, Replace (B)

	LABOR TIME	SEVERE SERVICE
240D one	.9	1.1
280CE, 280E one	1.2	1.4
300D, 300CD, 300TD one	1.2	1.4
300SD		
1981-85 one	1.2	1.4
1992-93 one	.9	1.1
300SE, 300SEL one	1.2	1.4
380SE one	1.2	1.4
380SEC, 380SEL one	.9	1.1
400SE, 400SEL one	1.2	1.4
420SEL one	1.2	1.4
500SEC, 500SEL one	1.2	1.4
560SEC, 560SEL one	1.2	1.4
560SL one	1.2	1.4
600SEC, 600SEL, S600 one	1.4	1.6
C220, C280, C32, C36, C55 one	1.2	1.4
C230, C240, C320 one	1.3	1.5
CL500 one	1.0	1.2
CL600 one	.9	1.1
CLK320, CLK55 one	1.3	1.5
CLK430, CLK500 one	1.1	1.3
E300D one	.8	1.0
E320, E430 one	.8	1.0
E500, E55 one	1.2	1.4
S320 one	.7	.9
S350TD one	.9	1.1
S420, S430, S55 one	1.2	1.4
S500, S55, S600 one	1.0	1.2
SLK230, SLK320, SLK32, SLK55 one	.8	1.0
w/ADS add each	.3	.5

Front Spring, Replace (B)

	LABOR TIME	SEVERE SERVICE
190D, 190E one	1.2	1.4
240D one	1.7	1.9
280CE, 280E one	1.6	1.8
300CD, 300TD one	1.6	1.8
300D one	1.6	1.8
300E one	1.7	1.9
300SD		
1981-85 one	1.8	2.0
1992-93 one	1.9	2.1
300SE, 300SEL one	1.7	1.9
300SL one	1.4	1.6
380SE, 380SL one	1.7	1.9

MB-26 190 SERIES : 240D : 260E : 280 SERIES : 300 SERIES : 380 SERIES : 400 SERIES : 420SEL : 500 SERIES : 560 SERIES

	LABOR TIME	SEVERE SERVICE
Front Spring, Replace (B)		
380SEC, 380SEL one	1.9	2.1
400E, E420 one	1.2	1.4
400SEL one	1.4	1.6
420SEL one	1.7	1.9
500E, E500 one	1.2	1.4
500SEC, 500SEL		
1984-85 one	1.9	2.1
1993 one	1.7	1.9
500SL one	1.4	1.6
560SEC, 560SEL,		
560SL one	1.7	1.9
600SEC, 600SEL one	1.9	2.1
600SL one	1.3	1.5
C220, C280, C36,		
C55 one	1.3	1.5
C230, C240, C32,		
C320 one	1.4	1.6
CL500, CL600 one	1.0	1.2
CLK320, CLK500,		
CLK55 one	1.5	1.7
CLK430 one	1.0	1.2
E300D one	1.3	1.5
E320, E500, E55 one	1.7	1.9
E430 one	1.0	1.2
S320, S350TD one	1.9	2.1
S420, S430 one	2.0	2.2
S500, S55 one	1.6	1.8
S600 one	2.0	2.2
SL500, SL55,		
SL600 one	1.0	1.2
SLK230, SLK32,		
SLK320, SLK55 one	1.4	1.6
Front Torsion Bar Bushings, Replace (B)		
Exc. below	1.5	1.7
300SD	2.3	2.5
300SE, 300SEL	3.9	4.1
420SEL	3.9	4.1
500SEC, 500SEL	3.9	4.1
560SEC, 560SEL	3.9	4.1
560SL	.8	1.0
600SEC, 600SEL,		
S500, S55, S600	2.3	2.5
C220, C230, C240 C280,		
C36, C32, C55	1.8	2.0
CLK320, SLK320	1.8	2.0
S320, S350TD	2.3	2.5
Lower Ball Joint, Replace (B)		
Exc. below	1.8	2.1
300E, 300TE, E300D	1.4	1.7
C220, C230, C240 C280,		
C36, C32, C55	1.4	1.7
CL500, CL55, CL600	1.5	1.8
CLK320, CLK55	1.5	1.8
CLK430	1.2	1.5
E430, E500	1.2	1.5
SLK230, SLK320,		
SLK55	1.2	1.5

	LABOR TIME	SEVERE SERVICE
Lower Control Arm Assy., Replace (B)		
190D, 190E		
1984-85	1.4	1.6
1986-93	1.7	1.9
240D	2.7	2.9
260E	1.7	1.9
300D, 300E	1.7	1.9
300SD	2.8	3.0
300SE, 300SEL	2.7	2.9
300SL, 300TD	1.7	1.9
380SE, 380SEC,		
380SEL, 380SL	2.7	2.9
400E	1.7	1.9
400SEL	2.7	2.9
420SEL	2.7	2.9
500E, E500	1.7	1.9
500SEC, 500SEL	2.8	3.0
500SL	1.8	2.0
560SEC, 560SEL,		
560SL	2.7	2.9
600SEC, 600SEL	2.7	2.9
600SL	1.7	1.9
C220, C230, C240, C280		
C320 C32, C36, C55	1.8	2.0
CL500, CL600	1.6	1.8
CLK320, CLK500	2.0	2.2
CLK430, CLK55	1.6	1.8
E300D	2.0	2.2
E320	1.7	1.9
E420, E500, E55	1.7	1.9
E430	1.6	1.8
S320, S350TD	2.7	2.9
S420, S430	2.8	3.0
S500, S55	2.3	2.5
S600	2.8	3.0
SL500, SL55	1.5	1.7
SL600, SLK32	1.8	2.0
SLK230, SLK320,		
SLK55	1.6	1.8
Replace		
ball joint add	.5	.7
bushing add each side	.8	1.0
Steering Knuckle, Replace (B)		
190D, 190E	2.5	3.0
240D	2.3	2.8
260E	2.8	3.3
280CE, 280E	2.3	2.8
300CD	2.3	2.8
300CE, 300TE	2.3	2.8
300D, 300E, E300D	2.8	3.3
300SD		
1981-85	2.3	2.8
1992-93	3.1	3.6
300SE, 300SEL	2.7	3.2
300SL	2.5	3.0
300TD	2.8	3.3
380SE, 380SEC,		
380SEL, 380SL	2.3	2.8

	LABOR TIME	SEVERE SERVICE
400E, 400SEL, E420	2.8	3.3
420SEL	2.3	2.8
500E, E500	2.8	3.3
500SEC, 500SEL		
1984-85	2.3	2.8
1993	2.7	3.2
500SL	2.5	3.0
560SEC, 560SEL,		
560SL	2.3	2.8
600SEC, 600SEL,		
S600	3.1	3.6
600SL	3.0	3.5
C220, C280, C32, C36,		
C320, C55	3.0	3.5
C230, C240	2.8	3.3
CL500, CL600	2.6	3.1
CLK320, E320, E55	2.8	3.3
CLK430, CLK500	2.3	2.8
E430, CLK55, SLK32	2.3	2.8
S320, S350TD	3.1	3.6
S420, E500, S500	3.1	3.6
SL500, SL600, S55	2.6	3.1
SLK230, SLK320	2.3	2.8
Steering Knuckle Arm, Replace (B)		
Includes: Reset toe.		
One	.7	1.1
Both	1.3	1.7
Strut Assy., Replace (B)		
190D, 190E, 260E one	2.0	2.3
300CE, 300D, 300E,		
300TE, E300D one	2.0	2.3
300E 4-Matic one	2.2	2.5
300SL one	2.2	2.5
300TE 4-Matic one	2.7	3.0
400E, E320, E420 one	2.0	2.3
500E, 500SL one	2.0	2.3
600SL one	2.0	2.3
SL500, SL600 one	1.6	1.9
w/level control add each	.8	1.0
Torsion Bar, Replace (B)		
Includes: Adjust vehicle height.		
190D, 190E	1.4	1.9
240D	6.2	6.7
260E	1.4	1.9
280CE, 280E	6.2	6.7
300CD, 300D	6.2	6.7
300CE, 300E,		
300TE, E300D	1.4	1.9
300SD		
1981-85	18.7	19.2
1992-93	2.3	2.8
300SE, 300SEL	18.7	19.2
300SL	1.4	1.9
300TD	6.2	6.7
400E	1.4	1.9
400SEL	2.3	2.8
420SEL	18.7	19.2
500E, E500	1.4	1.9
500SEC, 500SEL	18.7	19.2
500SL, SL500	1.4	1.9

600 SERIES : C SERIES : CL SERIES : CLK SERIES : E SERIES : S SERIES : SL SERIES : SLK SERIES

	LABOR TIME	SEVERE SERVICE
560SEC, 560SEL ...	18.7	19.2
560SL	2.3	2.8
600SEC, 600SEL, S600	2.3	2.8
600SL, SL600	1.4	1.9
C220, C230, C240 C280, C32, C36, C55	1.7	2.2
E320, E500, E55	1.4	1.9
E420	1.4	1.9
S320, S350TD	2.3	2.8
S420, S430	2.2	2.7

Tie Rod/Drag Link Ball Joint Seal, Replace (B)
One	.3	.3

Torsion Bar Connecting Rods, Replace (B)
One side	1.2	1.5
Both sides	2.0	2.3

Upper Control Arm Assy., Replace (B)
190D, 190E, 240D, 260E	1.9	2.2
280CE, 280E	1.9	2.2
300CD, 300CE, 300D, 300E, 300TE	1.9	2.2
300SE, 300SEL	1.4	1.7
380SEC, 380SEL	1.9	2.2
400SEL	.8	1.1
420SEL	1.9	2.2
500SEC, 500SEL	1.9	2.2
560SEC, 560SEL	1.9	2.2
560SL	2.3	2.6
600SEC, 600SEL	.8	1.1
C220	1.2	1.5
C230, C240, C32	2.2	2.5
C280, C320, C55	2.3	2.6
C36	1.9	2.2
CL500, CL600	1.4	1.7
CLK320	1.4	1.7
CLK430, CLK500	1.8	2.1
E300D, CLK55	2.3	2.6
E320, E500, E55	1.6	1.9
E430	1.8	2.1
S320, S430, S55	1.8	2.1
S500, SL500, SL600 ..	1.5	1.8
S600, SLK32, SLK55 ..	1.8	2.1
SLK230, SLK320	1.6	1.9
Replace bushings add each side	.3	.5

REAR SUSPENSION

Rear Toe, Check and Adjust (B)
1981-05	.7	1.0

Pulling Link, Replace (B)
One	.9	1.1
Both	2.0	2.2

Pushing Link, Replace (B)
One	.9	1.1
Both	1.9	2.1

	LABOR TIME	SEVERE SERVICE
Rear Axle Tie Rod Assy., Replace (B)		
One	.9	1.1
Both	1.9	2.1
Rear Knuckle, Replace (B)		
Exc. below	4.2	4.7
300SD	4.8	5.3
300SL	3.1	3.6
S320, S350TD, S420, S430	4.8	5.3
500SEC, 500SEL	7.5	8.0
500SL	3.1	3.6
560SEC, 560SEL	7.5	8.0
600SL	3.1	3.6
C220, C280, C320 ...	4.5	5.0
S600, SL500, SL600 ..	4.8	5.3
Rear Spring, Replace (B)		
190D, 190E one	1.2	1.5
1986-87 190E 16V one	2.3	2.6
240D one	2.3	2.6
260E one	1.2	1.5
280CE, 280E one	2.3	2.6
300CD, 300D one ...	2.3	2.6
300E one	1.3	1.6
300SD one	1.9	2.2
300SE one	1.5	1.8
300SEL one	2.3	2.6
300SL one	1.7	2.0
1981-87 300TD one ..	3.9	4.2
380SEC, 380SEL one .	2.3	2.6
400E, 400SEL one ...	1.6	1.9
420SEL one	2.3	2.6
500E one	1.6	1.9
500SEC, 500SEL 1984-85 one	2.3	2.6
1993 one	1.6	1.9
500SL one	1.7	2.0
560SEC, 560SEL, 560SL one	3.0	3.3
600SEC, 600SEL one .	1.9	2.2
600SL one	1.7	2.0
C220, C230, C240, C280, C32, C320, C36 one ...	1.4	1.7
CL500, CL600 one ...	1.0	1.3
CLK320, CLK500 one ..	1.5	1.8
CLK430, CLK55 one ..	1.0	1.3
E300D one	1.5	1.8
E320 convertible one	1.9	2.2
coupe, sedan one ..	1.3	1.6
wagon one	1.8	2.1
E420, E500, E55 one .	1.6	1.9
E430 one	1.4	1.7
S320, S350TD one ...	1.9	2.2
S420, S430, S55 one .	2.0	2.3
S500 one	1.5	1.8
S600, SL55 one	1.9	2.2
SL500, SL600 one ...	1.6	1.9
SLK230, SLK320, SLK32, SLK55 one ..	1.5	1.8

	LABOR TIME	SEVERE SERVICE
Rear Strut Assy., Replace (B)		
1986-87 190E 16V one	2.3	2.5
300CE one	2.3	2.5
300SL one	2.5	2.7
1981-87 300TD one ..	2.5	2.7
300TE one	2.5	2.7
400E one	2.3	2.5
400SE, 400SEL one ..	2.5	2.7
500E one	2.3	2.5
500SEC, 500SEL one .	2.5	2.7
560SEC, 560SEL one .	2.3	2.5
560SL one	2.5	2.7
600SEC, 600SEL, S600 one	2.2	2.4
E320 wagon one	2.5	2.7
E420 one	2.3	2.5
Semi-Trailing Arm, Replace (B)		
280CE, 280E one	1.8	2.0
300CD, 300D, 300TD one	1.8	2.0
300SD, 300SEL one ..	1.3	1.5
380SEC, 380SEL one .	1.3	1.5
420SEL one	1.3	1.5
500SEC, 500SEL, 560SEC, 560SEL one	1.3	1.5
560SL one	1.8	2.0
Shock Absorbers or Bushings, Replace (B)		
190D, 190E, 260E one	1.4	1.6
240D one	1.7	1.9
280CE, 280E one	1.7	1.9
300CD, 300D one	1.7	1.9
300CE, 300SE one ...	1.4	1.6
300SD one	1.2	1.4
300SEL, 300TD one ..	1.7	1.9
300SL one	2.2	2.4
380SEC, 380SEL one .	1.7	1.9
400E, 400SEL, E420 one	1.2	1.4
420SEL one	1.7	1.9
500E, E500 one	1.2	1.4
500SEC, 500SEL one .	1.7	1.9
500SL one	2.2	2.4
560SEC, 560SEL, C55 one	1.7	1.9
C220, C230, C240, C280, C320, C32, C36 one ...	1.4	1.6
CL500, CL600 one ...	.9	1.1
CLK320 one	1.4	1.6
CLK430, CLK500, CLK55 one	1.3	1.5
E300D one	1.2	1.4
E320 convertible one	2.2	2.4
coupe, sedan one ..	1.2	1.4
E430, E500, E55 one .	1.4	1.6
S320 one	1.2	1.4

MB-27

MB-28 — 190 SERIES : 240D : 260E : 280 SERIES : 300 SERIES : 380 SERIES : 400 SERIES : 420SEL : 500 SERIES : 560 SERIES

	LABOR TIME	SEVERE SERVICE
Shock Absorbers or Bushings, Replace (B)		
S350TD, S55 one	1.2	1.4
S420, S430 one	.8	1.0
S500 one	.8	1.0
SL500, SL600 one	1.6	1.8
Spring Link, Replace (B)		
Exc. below	1.5	1.7
300SL	2.4	2.6
500SL, SL500	2.8	3.0
560SEC, 560SEL	2.8	3.0
600SL, SL600	2.8	3.0
E320 convertible	1.7	1.9
Torsion Bar, Replace (B)		
Exc. below	1.4	1.9
190D, 190E	2.2	2.7
380SEC, 380SEL	2.3	2.8
560SL	2.3	2.8

LEVEL CONTROL

	LABOR TIME	SEVERE SERVICE
Hydraulic System Reservoir, Replace (B)		
1981-05	1.3	1.3
Pump Assy., Replace (B)		
Exc. below	6.5	6.8
190E 16V	.9	1.2
300CE, 300SL	2.7	3.0
300SD	3.0	3.3
300TD	.9	1.2
300TE	2.7	3.0
560SEC, 560SEL	2.4	2.7
E320, S320, S350TD	3.0	3.3
Ride Level Control Valve, Replace (B)		
Exc. below	1.7	1.9
300CE, 300TE, E320	2.0	2.2
400E, E420, E430	2.0	2.2
1990 500E	2.0	2.2
500SL, SL500 front	2.2	2.4

STEERING

AIR BAGS

	LABOR TIME	SEVERE SERVICE
Diagnose Air Bag System (A)		
1988-05	.9	.9
Air Bag Unit, Replace (B)		
1988-05		
driver side	.5	.5
passenger side		
C220, C320, C55	1.3	1.3
C230, C240, C280	.7	.7
E300D, E320, S320, S420, S430	.8	.8
E430	1.6	1.6
S500, S600	.8	.8
SL500, SL600	.7	.7
SLK230, SLK320	1.1	1.1
side impact		
C230, C280, C320	1.7	1.7
CL500, CL600	2.1	2.1

	LABOR TIME	SEVERE SERVICE
CLK320, CLK430	1.0	1.0
E300D, E320, S320, S420	2.0	2.0
E430	1.6	1.6
S420, S500, S600	2.0	2.0
SL320, SL500, SL600	2.4	2.4
SLK230, SLK320, SLK55	1.1	1.1
Contact Ring, Replace (B)		
1988-05	1.4	1.4
Electronic Control Unit, Replace (B)		
190D, 190E	1.4	1.4
260E	2.3	2.3
300CE, 300D, 300E, 300TE	2.3	2.3
300SE	3.2	3.2
300SEL	1.4	1.4
300SL	2.5	2.5
400E	2.3	2.3
400SEL	3.2	3.2
420SEL	1.4	1.4
500E	2.3	2.3
500SEC, 500SEL	3.2	3.2
500SL	2.7	2.7
560SEC, 560SEL	1.4	1.4
560SL	2.3	2.3
600SEC, 600SEL	3.2	3.2
600SL	2.7	2.7
C220, C240, C280, C32, C36, CLK500	1.3	1.3
C230, C320, CLK55	1.7	1.7
CL500, CL600, C55	2.7	2.7
CLK320, CLK430	1.0	1.0
E300D	.7	.7
E320	3.6	3.6
E430, E500, E55	1.6	1.6
S320	3.1	3.1
S420	3.1	3.1
S500	2.7	2.7
S600	3.1	3.1
SL500, SL600	2.2	2.2
SLK230, SLK320		
SLK32, SLK55	1.0	1.0

POWER WORM & SECTOR

	LABOR TIME	SEVERE SERVICE
Pump Pressure Check (B)		
1981-05	1.6	1.6
Drag Link, Replace (B)		
Exc. below	.8	1.0
300SD, 300SL	1.2	1.4
500SEC	1.3	1.5
500SL	1.3	1.5
600SEC, 600SEL, 600SL, S600	1.4	1.6
CL500, CL600	1.1	1.3
S320, S350TD	1.2	1.4
S420, S430, S500	1.7	1.9
SL500, SL600	1.3	1.5

	LABOR TIME	SEVERE SERVICE
Flexible Coupling, Replace (B)		
Exc. below	1.9	2.1
190D, 190E	2.4	2.6
300CE, 300D, 300E, 300TE	2.4	2.6
300SD, 300SE, 300SEL	2.4	2.6
420SEL	2.4	2.6
500SEL, 500SL	2.4	2.6
560SEC, 560SEL	2.4	2.6
560SL	2.7	2.9
Gear Assy., Replace (B)		
190D, 190E	3.1	3.6
240D, 260E	2.8	3.3
280CE, 280E	3.2	3.7
300CD, 300D	3.1	3.6
300CE, 300TE	2.8	3.3
300E, E300D	2.8	3.3
300SD	3.5	4.0
300SE, 300SEL	2.9	3.4
300SL	2.7	3.2
1981-87 300TD	3.1	3.6
380SE, 380SL	2.8	3.3
380SEC, 380SEL	2.9	3.4
400E, 400SEL, E420	4.2	4.7
420SEL	2.8	3.3
500E, E500	4.2	4.7
500SEC, 500SEL		
1984-85	2.9	3.4
1993	5.3	5.8
500SL	3.1	3.6
560SEC, 560SEL	2.8	3.3
560SL	4.2	4.7
600SEC, 600SEL	8.4	8.9
C220, C230, C240, C32		
C280, C36, C55	3.1	3.6
C280, C320	3.5	4.0
CL500	4.5	5.0
CL600	7.2	7.5
CLK320, CLK430	5.3	5.8
CLK500, CLK55	5.7	6.2
E320, E430	2.8	3.3
E500, E55	4.1	4.6
S320, S350TD	3.5	4.0
S420, S430	5.3	5.8
S500, S55	4.5	5.0
S600	8.3	8.8
SL500	2.5	3.0
SL600	3.2	3.7
Recondition gear add	6.2	6.7
Horn Contact, Replace (B)		
Exc. below	1.4	1.4
190D, 190E	.7	.7
240D	.9	.9
300E	1.3	1.3
300SD		
1981-85	.9	.9
1992-93	1.7	1.7

600 SERIES : C SERIES : CL SERIES : CLK SERIES : E SERIES : S SERIES : SL SERIES : SLK SERIES — MB-29

	LABOR TIME	SEVERE SERVICE
380SE, 380SEC, 380SEL, 380SL	.9	.9
500SEC, 500SEL	.9	.9
560SL	1.7	1.7
600SEC, 600SEL, S600	1.7	1.7
E300D	1.7	1.7
E320	1.4	1.4
S320, S350TD	1.7	1.7
S420	.9	.9
S500	.7	.7
w/air bags add	.5	.5

Intermediate Rod & Bushings, Replace (B)
1981-05	1.8	2.1

Steering Column, R&I (B)
Electronic adjust column
exc. below	4.2	4.2
190D, 190E, 260E	3.0	3.0
300CE, 300D, 300TD, 300TE	3.0	3.0
300SE, 300SEL	2.8	2.8
CL500, CL600	6.0	6.0
S420	7.0	7.0
S500	6.1	6.1
S600	7.0	7.0
SL500, SL600	3.4	3.4

Non-electronic adjust column
exc. below	2.8	2.8
C220, C230, C240, C32 C280, C320, C36, C55	4.4	4.4
300SE, 300SEL	3.0	3.0
500SEL, 560SL	4.2	4.2

Steering Column Lock Assy., Replace (B)
190D, 190E	2.5	2.5
240D	2.5	2.5
260E	2.7	2.7
280CE, 280E	2.5	2.5
300CD, 300D	2.5	2.5
300CE, 300E, 300TE	2.7	2.7
300SD 1981-85	2.8	2.8
1992-93	3.9	3.9
300SE	3.0	3.0
300SEL	2.8	2.8
300SL	3.2	3.2
300TD	3.0	3.0
400E, 400SEL, E420	2.7	2.7
420SEL	3.0	3.0
500E, E500	2.7	2.7
500SEC, 500SEL	2.8	2.8
500SL	3.2	3.2
560SEC, 560SEL	2.8	2.8
560SL	3.2	3.2
600SEC, 600SEL	3.9	3.9
600SL	3.2	3.2

	LABOR TIME	SEVERE SERVICE
C220, C230, C240, C280, C320, C36	2.7	2.7
CL500, CL600, C55	3.7	3.7
CLK320, CLK430	1.8	1.8
E300D, CLK500	2.4	2.4
E320	2.7	2.7
E430, E500, E55	3.2	3.2
S320, S350TD	3.9	3.9
S420, S430, S500	4.4	4.4
S600	4.4	4.4
SL500	3.9	3.9
SL600	3.4	3.4

Steering Gear, Adjust (B)
1981-05	1.7	1.9

Steering Pump, R&R or Replace (B)
Exc. below	3.1	3.3
190D, 190E	2.8	3.0
240D, 280CE, 280E	2.0	2.2
300CD, 300D, 300TD, E300D	2.0	2.2
380SEC, 380SEL	2.0	2.2
400E, 400SEL, 420SEL, E420	6.5	6.7
500E, E500	6.5	6.7
500SEC, 500SEL 1984-85	2.0	2.2
1993	6.5	6.7
500SL	6.5	6.7
560SEC, 560SEL	2.0	2.2
600SEC, 600SEL, 600SL	6.0	6.2
C220, C230, C320	2.7	2.9
C55, CL500	2.7	2.9
CL600	5.0	5.2
S500	2.7	2.9
S600	5.8	6.0
SL500	3.6	3.8
SL600	7.0	7.2
SLK230	1.5	1.7
SLK320, SLK55	2.4	2.6
Replace front seal add	.5	.8

Steering Pump Hoses, Replace (B)
Does not include purging.
190D, 190E, 280CE, 280E, 300CD, 300D, 300TD
pressure	.8	.9
return or supply	.7	.8

240D pressure or return | .8 | .9

260E, 300SEL
pressure	.9	1.0
return	.8	.9

300CE, 300E, 300TE, E320, E500
pressure	.9	1.0
return	1.4	1.5

300SD pressure or return | .7 | .8

380SE, 380SL, 560SL, E300D
pressure	1.4	1.5
return	.9	1.0

	LABOR TIME	SEVERE SERVICE
380SEC, 380SEL, 500SEL pressure or return	1.2	1.3
supply	.8	.9
420SEL, 560SEC, 560SEL pressure	1.4	1.5
return	.8	.9
500SEC pressure	1.4	1.5
500SL pressure	1.7	1.8
return	1.8	1.9
C220, C230, C280 pressure or return	1.4	1.5
C240, C32, C320, C55 pressure	2.4	2.5
return	1.8	1.9
CL500, S500, SL500, SL600 pressure	1.2	1.3
return	1.4	1.5
CLK320, CLK430, CLK500 pressure	1.4	1.5
return	1.3	1.4
S420 pressure	1.5	1.6
return	1.7	1.8
SLK230 pressure	1.6	1.7
return	1.5	1.6
SLK320, SLK55 pressure	1.8	1.9
return	1.6	1.7

Steering Wheel, Replace (B)
1981-05	.5	.5
w/air bags add	.5	.5

Tie Rod Ends, Replace (B)
Add for alignment.
One	.7	.9
Both	1.5	1.7

Tie Rods, Replace (B)
Add for alignment.
One	.7	.9
Both	1.2	1.4

POWER RACK AND PINION
Troubleshoot Power Steering (B)
All Models	1.2	1.2

Gear Assy., Adjust (On Vehicle) (B)
All Models	.7	.9

Gear Assy., Replace (B)
CLK430, CLK500	5.0	5.5
E430, E55	4.3	4.8
SLK230, SLK320, SLK32	3.4	3.9

Horn Contact, Replace (B)
All Models	.7	.7

Inner Tie Rod End, Replace (B)
CLK430, SLK230, SLK320	2.5	2.7

MB-30   190 SERIES : 240D : 260E : 280 SERIES : 300 SERIES : 380 SERIES : 400 SERIES : 420SEL : 500 SERIES : 560 SERIES

	LABOR TIME	SEVERE SERVICE
Steering Column Lock Assy., Replace (B)		
CLK430	1.8	1.8
E430	.9	.9
SLK230, SLK320	2.6	2.6
Steering Pump, R&R or Replace (B)		
CLK430	2.5	2.7
E430, SLK230	1.4	1.6
SLK320, SLK32	1.9	2.1
Steering Pump Hoses, Replace (B)		
Does not include purging.		
CLK430	1.9	2.0
E430	1.6	1.7
Steering Wheel, Replace (B)		
All Models	.5	.5

HEATING & AIR CONDTIONING

When more than one component requires replacement where evacuation/recovery and recharging is already included, deduct 1.0 hour for each additional component from the time given.

	LABOR TIME	SEVERE SERVICE
Evacuate/Recover and Recharge System (A)		
1981-05	1.5	1.5
AC Control Module, Replace (B)		
Exc. below	.7	.9
240D, 280CE, 280E	.9	1.1
300CD, 300D	.9	1.1
300SD		
1981-85	1.3	1.5
1992-93	.9	1.1
300SE, 300TD	.9	1.1
300SEL	1.3	1.5
S350TD	.9	1.1
380SE, 380SEC, 380SEL	1.3	1.5
400SE, 400SEL	.9	1.1
420SEL	1.3	1.5
500SEC, 500SEL		
1984-85	1.3	1.5
1993	.9	1.1
560SEL	1.3	1.5
600SEC, 600SEL, S600	.9	1.1
SL600	.5	.7
Auxiliary Fan, Replace (B)		
190D, 190E	.8	1.0
240D	.8	1.0
260E	1.4	1.6
280CE, 280E	1.3	1.5
300CD, 300D, 300SD, 300TD	1.3	1.5
300CE, 300E, 300TE	3.0	3.2
300SE	1.4	1.6
300SEL	.8	1.0
300SL	3.9	4.1

	LABOR TIME	SEVERE SERVICE
400E, 400SEL, E420	3.0	3.2
420SEL	3.0	3.2
500E, E500	3.0	3.2
500SEC, 500SEL		
1984-85	.8	1.0
1993	1.4	1.6
500SL, SL500	3.9	4.1
560SEC, 560SEL		
single	.8	1.0
dual	3.0	3.2
560SL	2.4	2.6
600SEC, 600SEL, S600	1.4	1.6
600SL, SL600	3.9	4.1
E300D	1.5	1.7
E320, E500	3.0	3.2
E430	1.2	1.4
S320, S350TD	1.4	1.6
S420	1.4	1.4
Add time to evacuate and recharge system.		
Auxiliary Fan Relay, Replace (B)		
1981-05	.4	.4
Aspirator Blower Motor, Replace (B)		
Exc. below	.8	.8
C230, C280	.3	.3
S420	.3	.3
S600	.3	.3
Blower Motor, Replace (B)		
190D, 190E	3.0	3.0
240D	.9	.9
260E	3.3	3.3
280CE, 280E	.9	.9
300CD, 300D, 300TD	.9	.9
300CE, 300E	3.3	3.3
300SD		
1981-85	1.7	1.7
1992-93	1.3	1.3
300SE	1.3	1.3
300SEL	1.7	1.7
300SL	3.3	3.3
380SEC, 380SEL	1.7	1.7
400E	3.3	3.3
400SEL	1.3	1.3
420SEL	1.7	1.7
500E, E500	3.3	3.3
500SEC, 500SEL		
1984-85	1.7	1.7
1993	1.3	1.3
500SL	3.3	3.3
560SEC, 560SEL	1.7	1.7
560SL	2.3	2.3
600SEC, 600SEL	1.3	1.3
600SL	3.3	3.3
C220, C230, C240, C32 C280, C320, C36	.8	.8
CL500, CL600, C55	1.1	1.1
CLK320, CLK430	.8	.8
CLK500,	3.4	3.4

	LABOR TIME	SEVERE SERVICE
E300D	3.3	3.3
E320	3.3	3.3
E430	.8	.8
E500, E55	8.5	8.5
S320	1.4	1.4
S350TD	1.3	1.3
E420, S430	3.3	3.3
S420	1.4	1.4
S500	1.1	1.1
S600, SL55	1.4	1.4
SL500	3.6	3.6
SL600	3.6	3.6
SLK230, SLK320	.8	.8
Blower Motor Switch, Replace (B)		
Exc. below	1.4	1.4
190D, 190E	.7	.7
280CE, 280E	1.3	1.3
300CD	1.3	1.3
300D	1.3	1.3
1981-85 300SD	1.4	1.4
300TD	1.3	1.3
1984-85 380SE	1.4	1.4
1981-83 380SEL	1.4	1.4
380SL	.7	.7
1984-85 500SEC, 500SEL	1.4	1.4
560SEC, 560SEL	1.4	1.4
560SL	.7	.7
Compressor Cut-Off Switch, Replace (B)		
1981-05	.4	.4
Compressor Assy., Replace (B)		
Includes: Parts transfer, evacuate/recover and recharge.		
190D, 190E	3.0	3.2
240D	3.0	3.2
260E	2.4	2.6
280CE, 280E	2.0	2.2
300CD, 300D, 300TD	3.0	3.2
300CE, 300E, 300TE	2.5	2.7
300SD	2.0	2.2
300SE	1.7	1.9
300SEL, 300SL	2.5	2.7
380SEC, 380SEL	3.3	3.5
400E	2.5	2.7
400SEL	1.7	1.9
420SEL, S420	2.0	2.2
500E, E500	2.5	2.7
500SEC, 500SEL		
1984-85	3.3	3.5
1993	1.7	1.9
500SL	2.4	2.6
560SEC, 560SEL	2.0	2.2
560SL	2.5	2.7
600SEC, 600SEL	8.2	8.4
600SL	2.4	2.6
C220, C230, C280, C36, C55	2.7	2.9

600 SERIES : C SERIES : CL SERIES : CLK SERIES : E SERIES : S SERIES : SL SERIES : SLK SERIES

MB-31

	LABOR TIME	SEVERE SERVICE
C240, C32, C320	2.0	2.2
CL500	1.8	2.0
CL600	8.1	8.3
CLK320, CLK430		
CLK500, CLK55	3.2	3.4
E300D	1.8	2.0
E320, S320	2.5	2.7
E420	2.5	2.7
E430	2.3	2.5
E500, E55	4.8	5.0
S350TD	2.0	2.2
S500	2.5	2.7
S600	7.9	8.1
SL500	3.0	3.2
SL600, SLK55	3.8	4.0
SLK230	3.4	3.6
SLK320, SLK32	3.9	4.1

Condenser Assy., Replace (B)

	LABOR TIME	SEVERE SERVICE
190D 2.2L	1.4	1.6
190E		
2.3L	2.0	2.2
16V	2.7	2.9
2.6L	1.3	1.5
240D	2.8	3.0
260E	2.5	2.7
280CE, 280E	2.8	3.0
300CD, 300D	2.8	3.0
300CE, 300E, 300TE	2.5	2.7
300SD, E300D	2.5	2.7
300SE	1.7	1.9
300SEL	2.5	2.7
300SL	3.9	4.1
300TD	2.8	3.0
380SEC, 380SEL	2.5	2.7
400E	3.5	3.7
400SEL	1.7	1.9
420SEL	2.5	2.7
500E, E500	3.5	3.7
500SEC, 500SEL		
1984-85	2.5	2.7
1993	1.7	1.9
500SL	3.9	4.1
560SEC, 560SEL	2.5	2.7
560SL	2.8	3.0
600SEC, 600SEL, S600	1.7	1.9
600SL	3.9	4.1
C230, C240, C32, C320	2.0	2.2
CL500, CL600, C55	2.6	2.8
CLK320, CLK430, CLK500	2.4	2.6
E320	2.5	2.7
E420	3.5	3.7
E430	2.8	3.0
E500, E55	2.2	2.4
S320, S350TD	1.7	1.9
S420	1.7	1.9
S500	2.6	2.8

	LABOR TIME	SEVERE SERVICE
SL500, SL600	4.6	4.8
SLK230	2.8	3.0
SLK320, SLK32, SLK55	3.3	3.5

Add time to evacuate and charge.

Evaporator Coil, Replace (B)
Includes: Evacuate/recover and recharge system.

	LABOR TIME	SEVERE SERVICE
190D, 190E	5.3	5.3
240D	16.2	16.2
260E	16.4	16.4
280CE, 280E	18.3	18.3
300CD, 300D, 300TD	17.7	17.7
300CE, 300E, 300TE	16.8	16.8
300SD	26.3	26.3
300SE	27.7	27.7
300SEL	22.2	22.2
300SL	17.2	17.2
380SEC, 380SEL	22.2	22.2
400E	16.8	16.8
400SEL	24.5	24.5
420SEL	22.2	22.2
500E, E500	24.5	24.5
500SEC, 500SEL		
1984-85	22.2	22.2
1993	19.2	19.2
500SL	17.2	17.2
560SEC, 560SEL	22.2	22.2
560SL	21.5	21.5
600SEC, 600SEL	27.6	27.6
600SL	18.7	18.7
C220, C280, C32, C36, C55	18.7	18.7
C230, C240, C320	18.4	18.4
CL500, CL600	17.0	17.0
CLK320, CLK430	16.2	16.2
CLK500, CLK55	12.2	12.2
E300D	18.7	18.7
E320	16.8	16.8
E420	24.5	24.5
E430	17.2	17.2
E500, E55	15.2	15.2
S320, S350TD	26.3	26.3
S420	27.0	27.0
S430	23.5	23.5
S500	16.1	16.1
S55	17.2	17.2
S600	25.5	25.5
SL500, SL600, SL55	17.2	17.2
SLK230, SLK320, SLK32, SLK55	11.9	11.9

Heater Blower Motor, Replace (B)

	LABOR TIME	SEVERE SERVICE
190D, 190E	3.0	3.0
240D	.9	.9
260E	3.3	3.3
280CE, 280E	.9	.9
300CD, 300D, 300TD, E300D	.9	.9
300CE, 300E	3.3	3.3

	LABOR TIME	SEVERE SERVICE
Heater Blower Motor, Replace (B)		
300SD		
1981-85	1.7	1.7
1992-93	1.3	1.3
300SE	1.3	1.3
300SEL, 300SL	3.3	3.3
380SEC, 380SEL	1.7	1.7
400E	3.3	3.3
400SEL	1.3	1.3
420SEL	1.7	1.7
500E, E500	3.3	3.3
500SEC, 500SEL		
1984-85	1.7	1.7
1993	1.3	1.3
500SL, SL500	3.3	3.3
560SEC, 560SEL	1.7	1.7
560SL	2.3	2.3
600SEC, 600SEL	1.3	1.3
600SL	3.3	3.3
C220, C230, C240, C280, C32, C320, C36	.7	.7
CLK320, CLK430, C55	.8	.8
E320	3.3	3.3
E420, E430, E500	3.3	3.3
S320	1.4	1.4
S350TD	1.3	1.3
S420	1.4	1.4
S600	1.4	1.4
SL600	3.6	3.6

Heater Blower Motor Switch, Replace (B)

	LABOR TIME	SEVERE SERVICE
240D	1.3	1.3

Heater Control Valve, Replace (B)

	LABOR TIME	SEVERE SERVICE
190D, 190E	1.3	1.5
240D	1.4	1.6
260E	.8	1.0
280CE, 280E	.9	1.1
300CD, 300D	.9	1.1
300CE, 300E, E300D	.8	1.0
300SD		
1981-85	2.3	2.5
1992-93	1.3	1.5
300SE	1.3	1.5
300SEL	2.3	2.5
300SL	.9	1.1
300TD, 300TE	.8	1.0
380SE, 380SEC, 380SEL	2.3	2.5
380SL	.9	1.1
400E	.8	1.0
400SE	1.3	1.5
420SEL	2.3	2.5
500E, E500	.8	1.0
500SEC		
1984-85	2.3	2.5
1993	1.3	1.5
500SEL	1.3	1.5
500SL	.9	1.1
560SEC, 560SEL	2.3	2.5
560SL	.9	1.1

190 SERIES : 240D : 260E : 280 SERIES : 300 SERIES : 380 SERIES : 400 SERIES : 420SEL : 500 SERIES : 560 SERIES

	LABOR TIME	SEVERE SERVICE
Heater Control Valve, Replace (B)		
600SEC, 600SEL, S600	1.3	1.5
600SL, SL600	.9	1.1
C220, C230, C240, C280, C320, C36, C55	1.7	1.9
CLK320, CLK430, CLK500, CLK55	1.4	1.6
E320	.8	1.0
E420	.8	1.0
S320, S350TD	1.3	1.5
S420	1.3	1.5
Heater Core, R&R or Replace (B)		
190D, 190E	9.3	9.3
240D	13.7	13.7
260E	8.3	8.3
280CE, 280E	15.6	15.6
300CD, 300D	15.6	15.6
300CE, 300TE	8.3	8.3
300E, E300D	8.3	8.3
300SD		
1981-85	18.3	18.3
1992-93	22.8	22.8
300SE	22.8	22.8
300SEL	18.3	18.3
300SL	15.2	15.2
300TD	15.6	15.6
380SE, 380SL	19.5	19.5
380SEC, 380SEL	18.3	18.3
400E	8.3	8.3
400SE, 400SEL	22.8	22.8
420SEL	18.3	18.3
500E, E500	8.3	8.3
500SEC, 500SEL		
1984-85	18.3	18.3
1993	22.8	22.8
500SL	16.8	16.8
560SEC, 560SEL	18.3	18.3
560SL	14.6	14.6
600SEC, 600SEL, S600	22.8	22.8
600SL	16.8	16.8
C220, C240, C280, C32, C36, C55	16.2	16.2
C230, C320	15.9	15.9
CL500, CL600	15.8	15.8
CLK320, CLK55	16.3	16.3
CLK430, CLK500	16.1	16.1
E320	8.3	8.3
E420	8.3	8.3
E430, E500, E55	16.3	16.3
S320, S350TD	22.8	22.8
S420, S430	22.6	22.6
S500	12.6	12.6
SL500, SL600	16.7	16.7
SLK230, SLK320, SLK32, SLK55	11.2	11.2
Heater Hoses, Replace (B)		
190D, 190E	2.3	2.5
240D	1.8	2.0
260E	3.1	3.3

	LABOR TIME	SEVERE SERVICE
280CE, 280E	1.8	2.0
300CD	1.8	2.0
300CE, 300TE	3.1	3.3
300D, 300E, 300TD	3.1	3.3
1981-85 300SD	2.8	3.0
300SE, 300SEL	2.8	3.0
300SL	9.1	9.3
380SEC, 380SEL	2.8	3.0
400E, E420	2.5	2.7
420SEL	2.8	3.0
500E, E500	2.5	2.7
500SEC, 500SEL	2.8	3.0
500SL, SL500	9.1	9.3
560SEC, 560SEL	2.8	3.0
E320	3.1	3.3
Receiver/Drier Assy., Replace (B)		
Includes: Evacuate/recover and recharge system.		
Exc. below	2.3	2.3
190E 16V	2.7	2.7
280CE, 280E	2.6	2.6
300SD, 300SEL	2.6	2.6
300SL	3.0	3.0
380SEC, 380SEL	2.6	2.6
420SEL	2.6	2.6
500E, E500	2.8	2.8
500SEC, 500SEL	2.6	2.6
500SL, SL500	2.8	2.8
560SEC, 560SEL, 560SL	2.6	2.6
600SL, SL600	2.8	2.8
E300D	2.8	2.8

WIPERS & SPEEDOMETER

Instrument Cluster, R&I (B)

	LABOR TIME	SEVERE SERVICE
190D, 190E	1.7	1.7
240D	.9	.9
260E	1.4	1.4
300CD, 300D	.9	.9
300CE, 300TE	1.4	1.4
300SD, 300TD	.9	.9
300SE, 300SEL	.9	.9
300SL	1.3	1.3
400E	1.4	1.4
400SEL	.8	.8
420SEL	.9	.9
500E, E500	1.4	1.4
500SEC, 500SEL	.9	.9
500SL	1.3	1.3
560SEC, 560SEL	1.4	1.4
560SL	1.7	1.7
600SEC, 600SEL, S600	.8	.8
600SL	1.3	1.3
C220	.9	.9
C230, C240	1.3	1.3
C280, C320, C55	1.2	1.2
C32, C36	.9	.9
CL500, CL600	.6	.6
CLK320, CLK55	1.3	1.3

	LABOR TIME	SEVERE SERVICE
CLK430, CLK500	1.0	1.0
E300D	1.4	1.4
E320	1.4	1.4
E420, E500, E55	1.4	1.4
E430	1.0	1.0
S320, S350TD	.8	.8
S420	.8	.8
S500	.6	.6
SL500, SL600	1.0	1.0
SLK230, SLK320, SLK32, SLK55	1.6	1.6
Rear Window Washer Pump, Replace (B)		
1981-05	.5	.5
Rear Window Wiper and Washer Switch, Replace (B)		
1981-87 300TD	.5	.5
1988-93 300TE	.5	.5
Rear Window Wiper Motor, Replace (B)		
1981-05	2.0	2.3
Rear Windshield Wiper Interval Relay, Replace (B)		
1981-05	.7	.7
Speedometer Cable and Casing, Replace (B)		
Exc. below	1.4	1.4
190D, 190E		
1984-85	1.9	1.9
1986-93	2.5	2.5
260E	2.4	2.4
300D, E300D	2.4	2.4
300CE, 300E, 300TE	2.4	2.4
380SEC, 380SEL	1.9	1.9
500SEC, 500SEL	1.9	1.9
E320	2.4	2.4
Speedometer Head, R&R or Replace (B)		
Exc. below	1.4	1.4
190D, 190E		
1984-85	1.5	1.5
1986-93	2.0	2.0
260E	1.8	1.8
300CE, 300E, 300TE	1.8	1.8
300SD, 300TD, E300D	1.6	1.6
300SE	1.6	1.6
400E, 400SEL, E420	1.8	1.9
500E, E500	1.8	1.8
500SEC	1.6	1.6
560SL	2.0	2.0
600SEC, 600SEL, 600SL	1.6	1.6
CL500, CL600	1.1	1.1
E320	1.8	1.8
SL500, SL600	1.5	1.5
Vehicle Speed Sensor, Replace (B)		
1988-05		
w/B2 shift	4.5	4.5
w/o B2 shift	2.8	2.8

600 SERIES : C SERIES : CL SERIES : CLK SERIES : E SERIES : S SERIES : SL SERIES : SLK SERIES MB-33

	LABOR TIME	SEVERE SERVICE
Windshield Washer Check Valve, Replace (B)		
1981-05	.3	.3
Windshield Washer Pump, Replace (B)		
1981-85	.5	.5
1986-05	.7	.7
Windshield Wiper Interval Relay, Replace (B)		
1981-05	.7	.7
Windshield Wiper Linkage, Replace (B)		
1981-85	2.8	3.0
1986-05	2.4	2.6
Windshield Wiper Motor, Replace (B)		
190D, 190E		
1984-85	2.3	2.5
1986-93	3.0	3.2
240D	2.4	2.6
260E	3.0	3.2
280CE, 280E	2.4	2.6
300CD, 300D	2.4	2.6
300CE, 300E, 300TE	3.0	3.2
300SD		
1981-85	2.4	2.6
1992-93	3.1	3.3
300SE, 300SEL	2.4	2.6
300SL	2.8	3.0
300TD	2.4	2.6
380SEC, 380SEL	2.4	2.6
400E, 400SEL, E420	3.0	3.2
420SEL	2.4	2.6
500E, E500	3.0	3.2
500SEC, 500SEL	2.4	2.6
500SL	2.8	3.0
560SEC, 560SEL	2.4	2.6
560SL	1.7	1.9
600SEC, 600SEL	2.8	3.0
600SL	3.1	3.3
C220, C230, C240, C280, C320, C32, C36	2.2	2.4
CL500, CL600, C55	2.5	2.7
CLK320	2.2	2.4
CLK430, CLK500	1.4	1.6
E300D	3.0	3.2
E320, S320, S350TD	3.1	3.3
E430, E500, E55	2.1	2.3
S420, S430	3.3	3.5
S500, S55	2.5	2.7
S600	3.3	3.5
SL500, SL55, SL600	2.2	2.4
SLK230, SLK320, SLK32, SLK55	1.1	1.3

LAMPS & SWITCHES
Back-Up Lamp Switch, Replace (B)
AT
exc. below	.9	.9
260E, C220, C230, C280	1.3	1.3

	LABOR TIME	SEVERE SERVICE
300CE, 300E, 300TE	1.3	1.3
300D, 300TD	1.3	1.3
400E, 400SEL, E420	1.3	1.3
500E, E500	1.3	1.3
560SL	1.8	1.8
E320	1.3	1.3
MT		
190D, 190E	.5	.5
240D	.5	.5
260E	.5	.5
300E	.5	.5
Halogen Headlamp Bulb, Replace (B)		
1985-05	.5	.5
Hazard Warning Switch, Replace (B)		
1981-05	.5	.5
Headlight Switch, Replace (B)		
Exc. below	1.7	1.7
1986-93 190D, 190E	1.3	1.3
300SEL	2.3	2.3
380SEC, 380SEL	2.3	2.3
420SEL	2.3	2.3
500SEC, 500SEL	2.3	2.3
560SEC, 560SEL	2.3	2.3
CLK320	1.5	1.5
Headlights, Aim (B)		
1981-05	.5	.5
High Mount Stop Lamp Assy., Replace (B)		
1985-05	.3	.3
Horn, Replace (B)		
1981-05	.7	.7
License Lamp Lens or Bulb, Replace (C)		
1981-05	.3	.3
Multifunction Switch Assy., Replace (B)		
Exc. below	2.3	2.3
240D	1.6	1.6
280CE, 280E	1.6	1.6
300CD, 300D, 300TD	1.6	1.6
500SEC, 500SEL	1.8	1.8
560SL	1.6	1.6
Park & Turn Signal Lamp Assy., Replace (B)		
1981-05	.3	.3
Parking Lamp Lens or Bulb, Replace (C)		
1981-05	.5	.5
Stop Light Switch, Replace (B)		
Exc. below	.9	.9
190D, 190E, 260E	.5	.5
300CE, 300TD, 300TE	.5	.5
300D, 300E, E300D	.5	.5
300SE, 300SEL	1.4	1.4
380SEC, 380SEL	1.4	1.4
500E, E500	.5	.6
E320	.5	.5
SL500, SL600	1.4	1.4

	LABOR TIME	SEVERE SERVICE
Turn Signal Lamp Lens or Bulb, Replace (C)		
1981-05	.5	.5
Turn Signal or Hazard Warning Flasher, Replace (B)		
Exc. below	.5	.5
1984-85 190D, 190E	.7	.7
240D	.7	.7
280CE, 280E	.7	.7
300CD, 300D, 300SD, 300TD	.7	.7
380SEC, 380SEL	.7	.7
400E, 400SEL	.3	.4
420SEL, S420	.5	.5
500E, E500	.3	.3
500SEC, 500SEL	.7	.7
C220, C230, C280	.3	.3
E300D	.3	.3
E420	.3	.3

BODY
Front Door Window Regulator and/or Motor, Replace (B)
Exc. below	3.1	3.3
1984-93 190	2.4	2.6
260E	2.2	2.4
300D	2.2	2.4
300E, 300TE	2.0	2.2
400E	2.5	2.7
500E, E500	2.5	2.7
600SEC, 600SEL, 600SL	2.5	2.7
C220, C280, C320, C36, C55	2.0	2.2
C230	2.7	2.9
CL500, CL600	2.8	3.0
CLK320, CLK430	2.5	2.7
E320		
exc. wagon	3.9	4.1
wagon	2.2	2.4
E430, E500	2.3	2.5
S500	2.5	2.7
S600, SL600	3.8	4.0
SL500, SLK32	3.8	4.0
SLK230, SLK320, SLK55	2.1	2.3

Hood Release Cable, Replace (B)
190	2.2	2.2
260E	2.0	2.0
300D, 300E, E300D	2.0	2.0
300SD, 300SE, 300SEL	2.8	2.8
300SL	3.0	3.0
300TE	2.0	2.0
400E	2.0	2.0
400SE, 400SEL	3.0	3.0
420SEL	2.8	2.8
500E, E500	2.0	2.0
500SEL, 500SL	3.0	3.0
560SEC	2.0	2.0
560SEL, 560SL	2.8	2.8

MB-34 190 SERIES : 240D : 260E : 280 SERIES : 300 SERIES : 380 SERIES : 400 SERIES : 420SEL : 500 SERIES : 560 SERIES

	LABOR TIME	SEVERE SERVICE		LABOR TIME	SEVERE SERVICE		LABOR TIME	SEVERE SERVICE
Hood Release Cable, Replace (B)			SL500, SL600	2.5	2.5	500E, E500	1.6	1.8
600SEC, 600SEL,			SLK230, SLK320,			500SEL	2.4	2.6
600SL	3.0	3.0	SLK32, SLK55	1.2	1.2	560SEL	2.5	2.7
C220, C230, C240,			**Rear Door Window Regulator and/or**			600SEL, SL600	2.4	2.6
C280, C320, C36	2.5	2.5	**Motor, Replace (B)**			C220, C280, C240,		
CL500, CL600, C55	2.5	2.5	Exc. below	2.7	2.9	C36	1.5	1.7
CLK320, CLK430,			190	1.5	1.7	C230	2.0	2.2
CLK500, CLK55	2.5	2.5	260E	1.5	1.7	CL500, CL600	3.0	3.2
E320	2.0	2.0	300D, E300D	2.4	2.6	E320		
E420	2.0	2.0	300SD, 300SE,			coupe, sedan	1.5	1.7
E430, E500, E55	2.1	2.1	300SEL	2.5	2.7	wagon	1.7	1.9
S320, S350TD	3.0	3.0	300TE	1.5	1.7	E420	1.6	1.8
S420	3.0	3.0	400E	1.6	1.8	S320, S350TD	2.4	2.6
S500	2.5	2.5	400SE, 400SEL	2.4	2.6			
S600, S55	3.0	3.0	420SEL	2.5	2.7			

G Series : ML Series

SYSTEM INDEX

MAINTENANCE	MB-36
Maintenance Schedule	MB-36
CHARGING	MB-36
STARTING	MB-36
CRUISE CONTROL	MB-36
IGNITION	MB-36
EMISSIONS	MB-36
FUEL	MB-36
EXHAUST	MB-36
ENGINE COOLING	MB-36
ENGINE	MB-37
Assembly	MB-37
Cylinder Head	MB-37
Camshaft	MB-37
Crank & Pistons	MB-37
Engine Lubrication	MB-37
AUTO TRANSMISSION	MB-37
TRANSFER CASE	MB-37
DRIVELINE	MB-37
BRAKES	MB-37
FRONT SUSPENSION	MB-38
REAR SUSPENSION	MB-38
STEERING	MB-38
HEATING & AC	MB-38
WIPERS & SPEEDOMETER	MB-38
LAMPS & SWITCHES	MB-38
BODY	MB-38

OPERATIONS INDEX

A
AC Hoses	MB-38
Air Bags	MB-38
Air Conditioning	MB-38
Alignment	MB-38
Alternator (Generator)	MB-36
Anti-Lock Brakes	MB-37

B
Ball Joint	MB-38
Battery Cables	MB-36
Bleed Brake System	MB-38
Brake Hose	MB-38
Brake Pads and/or Shoes	MB-38

C
Camshaft	MB-37
Camshaft Sensor	MB-36
Coolant Temperature (ECT) Sensor	MB-36
Crankshaft Sensor	MB-36
Cruise Control	MB-36
Cylinder Head	MB-37

D
Differential	MB-37
Drive Belt	MB-36

E
EGR	MB-36
Engine	MB-37
Engine Lubrication	MB-37
Evaporator	MB-38
Exhaust	MB-36
Exhaust Manifold	MB-36
Expansion Valve	MB-38

F
Fuel Injection	MB-36
Fuel Pump	MB-36

G
Generator	MB-36

H
Halfshaft	MB-37
Headlamp	MB-38
Horn	MB-38

I
Ignition Coil	MB-36
Intake Air Temperature Sensor	MB-36
Intake Manifold	MB-36

K
Knock Sensor	MB-36

M
Maintenance Schedule	MB-36
Manifold Absolute Pressure (MAP) Sensor	MB-36
Mass Air Flow (MAF) Sensor	MB-36
Master Cylinder	MB-38
Multifunction Lever/Switch	MB-38

O
Oil Pan	MB-37
Outer Tie Rod	MB-38
Oxygen Sensor	MB-36

P
Pistons	MB-37

R
Radiator	MB-36
Radiator Hoses	MB-36
Radio	MB-38

S
Shock Absorber/Strut, Front	MB-38
Shock Absorber/Strut, Rear	MB-38
Spark Plug Cables	MB-36
Spark Plugs	MB-36
Spring, Rear Coil	MB-38
Starter	MB-36
Steering Wheel	MB-38

T
Thermostat	MB-36

U
Upper Control Arm	MB-38

V
Valve Body	MB-37
Valve Cover Gasket	MB-37

W
Water Pump	MB-37
Wheel Balance	MB-36
Window Regulator	MB-38
Windshield Washer Pump	MB-38
Windshield Wiper Motor	MB-38

G SERIES : ML SERIES

	LABOR TIME	SEVERE SERVICE
MAINTENANCE		
Air Cleaner Filter Element, Replace (C)		
1998-05	.4	.6
Drive Belt, Replace (B)		
One	.8	1.0
each addl. add	.1	.2
Drive Belt Tensioner, Replace (B)		
1998-05	1.4	1.8
Fuel Filter, Replace (B)		
1998-05	.9	1.1
Halogen Headlamp Bulb, Replace (B)		
Each	.3	.3
Tire, Replace (C)		
1998-05	.5	.5
Tires, Rotate (C)		
1998-05	.5	.5
Wheels, Balance (B)		
Includes: Dismount old tire and mount new tire to rim.		
One	.3	.3
each addl. add	.2	.2

SCHEDULED MAINTENANCE INTERVALS

If necessary, refer to appropriate Chilton maintenance service information.

15,000 Mile Service (B)		
All Models	1.3	1.3
30,000 Mile Service (B)		
All Models	2.6	2.6
45,000 Mile Service (B)		
All Models	2.8	2.8
60,000 Mile Service (B)		
All Models	5.9	5.9
75,000 Mile Service (B)		
All Models	1.3	1.3
90,000 Mile Service (B)		
All Models	4.2	4.2
105,000 Mile Service (B)		
All Models	1.3	1.3
120,000 Mile Service (B)		
All Models	5.9	5.9

CHARGING

Alternator Circuits, Test (B)		
Includes: Test component output.		
1998-05	.7	.7
Alternator Assy., Replace (B)		
Includes: Pulley transfer.		
1998-05	1.4	1.6
Alternator Voltage Regulator, Replace (B)		
1998-05	1.7	1.9

	LABOR TIME	SEVERE SERVICE
STARTING		
Starter Draw Test (On Car) (B)		
1998-05	.3	.3
Battery Cables, Replace (C)		
Positive	.6	.8
Negative	.5	.7
Starter Assy., Replace (B)		
Add draw test if performed.		
1998-05	1.5	1.8

CRUISE CONTROL

Speed Control Switch, Replace (B)		
1998-05	1.8	1.8

IGNITION

Diagnose Ignition System (B)		
1998-05	1.0	1.2
Camshaft Position Sensor, Replace (B)		
1998-05	.7	.9
Crankshaft Position Sensor, Replace (B)		
1998-05	.7	.9
Ignition Coil, Replace (B)		
One	.7	.8
All	1.7	1.9
Spark Plug (Ignition) Cables, Replace (B)		
1998-05	.3	.5
Spark Plugs, Replace (B)		
1998-05	3.0	3.3

EMISSIONS

Air Pump Assy., Replace (B)		
1998-05	1.1	1.3
Air Switching Valve, Replace (B)		
1998-05	.7	.7
Air Temperature Sensor, Replace (B)		
1998-05	.5	.5
Coolant Temperature (ECT) Sensor, Replace (B)		
1998-05	.5	.5
EGR Back Pressure Transducer, Replace (B)		
1998-05	.6	.6
EGR Valve, Replace (B)		
1998-05	.7	.9
Knock Sensor, Replace (B)		
One	3.6	3.8
Both	3.8	4.0
Manifold Absolute Pressure (MAP) Sensor, Replace (B)		
1998-05	.5	.5
Mass Air Flow (MAF) Sensor, Replace (B)		
1998-05	.5	.5
Oxygen Sensor, Replace (B)		
1998-05	.6	.8

	LABOR TIME	SEVERE SERVICE
FUEL		
DELIVERY		
Fuel Filter, Replace (B)		
1998-05	.9	1.1
Fuel Pressure Check (B)		
1998-05	.7	.7
Fuel Pump, Replace (B)		
G55, G500	1.1	1.3
ML55, 320, 350, 430, 500 exc. below	2.5	2.7
2006 ML350, 500	1.8	2.0
Fuel Pump Relay Module, Replace (B)		
1998-05	.5	.8
Intake Manifold Gasket, Replace (B)		
1998-05	3.8	4.3
Replace manifold add	.5	.8

INJECTION

Electronic Control Unit, Replace (B)		
1998-05	.5	.5
Fuel Injectors, Replace (B)		
One	1.5	1.7
each addl. add	.2	.3
All	3.2	3.7
Fuel Rail, Replace (B)		
1998-05	2.5	2.8

EXHAUST

Exhaust Manifold or Gasket, Replace (B)		
3.2L, 3.7L		
right side	3.0	3.5
left side	3.5	4.0
both sides	4.8	5.3
4.3L, 5.5L		
right side	3.8	4.3
left side	4.6	5.1
both sides	6.4	6.9

ENGINE COOLING

Pressure Test Cooling System (C)		
1998-05	.3	.3
Coolant Thermostat, Replace (B)		
1998-05	1.6	1.8
Cooling Fan, Clutch and/or Pulley, Replace (B)		
1998-05	.5	.8
Engine Coolant Temp. Sending Unit, Replace (B)		
1998-05	.5	.5
Radiator Assy., R&R or Replace (B)		
1998-05	4.5	5.0
Radiator Hoses, Replace (B)		
Includes: Refill with proper coolant mix.		
One	.9	1.0

G SERIES : ML SERIES — MB-37

	LABOR TIME	SEVERE SERVICE
Water Pump and/or Gasket, Replace (B)		
Includes: Refill with proper coolant mix.		
3.2L, 3.7L	3.6	4.1
4.3L, 5.5L	3.0	3.5

ENGINE
ASSEMBLY
Times shown are for OEM assemblies. Time to replace assemblies from aftermarket rebuilders may vary.

Engine Assy., R&I (A)
Does not include parts or component transfer.
- 3.2L, 3.7L 13.4 *13.9*
- 4.3L, 5.0L, 5.5L 14.2 *14.7*

Engine Assy., Replace (A)
Includes: Engine R&R engine, component transfer, minor tune-up.
- 3.2L, 3.7L 21.6 *22.1*
- 4.3L, 5.0L, 5.5L 22.5 *23.0*

CYLINDER HEAD
Compression Test (B)
- All Models 3.1 *3.5*

Cylinder Head, Replace (B)
Includes: Parts transfer, adjustments.
- 3.2L, 3.7L
  - right side 12.1 *12.6*
  - left side 12.9 *13.4*
  - both sides 17.1 *17.6*
- 4.3L, 5.0L, 5.5L
  - right side 14.8 *15.3*
  - left side 13.8 *14.3*
  - both sides 18.6 *19.1*

Cylinder Head Gasket, Replace (B)
- 3.2L, 3.7L
  - right side 10.6 *11.1*
  - left side 12.4 *12.9*
  - both sides 15.4 *15.9*
- 4.3L, 5.0L, 5.5L
  - right side 12.3 *12.8*
  - left side 13.0 *13.5*
  - both sides 18.3 *18.8*

Valve Springs and/or Oil Seals, Replace (B)
- 3.2L, 3.7L 11.6 *12.1*
- 4.3L, 5.0L, 5.5L 14.2 *14.7*

Valve/Cam Cover and/or Gasket, Replace (B)
- 3.2L, 3.7L
  - one side 2.4 *2.6*
  - both sides 3.6 *3.8*
- 4.3L, 5.0L, 5.5L
  - one side 2.5 *2.7*
  - both sides 4.1 *4.3*

CAMSHAFT
Balance Shaft, Replace (B)
Includes: R&R engine.
- 3.2L, 3.7L 24.4 *24.9*

Camshaft, Replace (B)
- 3.2L, 3.7L all 7.9 *8.4*
- 4.3L, 5.0L, 5.5L all .. 11.5 *12.0*

Timing Chain Tensioner, Replace (B)
- 3.2L, 3.7L, 4.3L, 5.0L, 5.5L 3.0 *3.5*

Timing Cover and/or Gasket, Replace (B)
- 3.2L, 3.7L, 4.3L, 5.0L, 5.5L 16.2 *16.7*
- *Replace tensioner and guide rails add* 8 *1.2*

CRANK & PISTONS
Crankshaft Pulley, Replace (B)
- 1998-05 .... 1.8 *2.0*

Pistons or Connecting Rods, Replace (A)
Includes: Ridge reaming, cylinder wall deglazing, installing new rings and rod bearings, engine tune-up.
- 3.2L, 3.7L 37.3 *37.8*
- 4.3L, 5.0L, 5.5L 38.5 *39.0*

Rings, Replace (A)
Includes: Ridge reaming, cylinder wall deglazing, installing new rings and rod bearings, engine tune-up.
- 3.2L, 3.7L 39.6 *40.1*
- 4.3L, 5.0L, 5.5L 40.5 *41.0*

ENGINE LUBRICATION
Engine Oil Pressure Switch (Sending Unit), Replace (B)
- 1998-05 .... .5 *.5*

Oil Pan and/or Gasket, Replace (B)
- Upper 7.9 *8.1*
- Lower 4.5 *4.7*

AUTOMATIC TRANSMISSION
SERVICE TRANSMISSION INSTALLED
Check Unit for Oil Leaks (C)
- 1998-05 .... .9 *.9*

Drain and Refill Unit (B)
- 1998-05 .... .9 *.9*

Electronic Transmission Diagnosis (A)
- 1998-05 .... 1.8 *1.8*

Oil Pan and/or Gasket, Replace (B)
- 1998-05 .... 1.6 *1.8*

Transmission Control Module, Replace (B)
- 1998-05 .... .7 *.7*

Valve Body Assy., Replace (B)
Includes: R&R oil pan.
- 1998-05 .... 2.5 *3.0*

SERVICE TRANSMISSION REMOVED
Transmission Assy., R&R and Recondition (B)
- 1998-05 .... 27.3 *27.8*
- *Flush oil cooler and lines add* .... .5 *.7*

Transmission Assy., Remove & Install (B)
- 1998-05 .... 10.4 *10.9*
- *Replace assembly add* 7 *1.0*

TRANSFER CASE
Transfer Case, R&R or Replace (B)
- 1998-05 .... 5.4 *5.9*

DRIVELINE
Rear Axle Shaft, Replace (B)
- One side 3.1 *3.6*
- Both sides 4.4 *4.9*
- *Replace boots add each* 1.5 *1.8*

FRONT AXLE 4WD
Axle Assy., R&R or Replace (B)
- Both 2.7 *3.1*
- *Replace diff. oil seal add* .. .5 *.7*

Differential Assy, R&R or Replace (B)
- 1998-05 .... 3.8 *4.3*

Halfshaft Assy., R&R or Replace (B)
- Right side 2.8 *3.1*
- Left side 3.0 *3.3*
- Both sides 4.4 *4.7*

BRAKES
ANTI-LOCK
Diagnose Anti-Lock Brake System (A)
- 1998-05 .... 1.7 *1.7*

Control Module, Replace (B)
- 1998-05 .... .3 *.3*

Hydraulic Unit Assy., Replace (B)
- 1998-05 .... 1.5 *1.8*

Wheel Speed Sensor, Replace (B)
- Front or rear
  - one .... .8 *.8*
  - both 1.4 *1.4*

G SERIES : ML SERIES

	LABOR TIME	SEVERE SERVICE

SYSTEM
Bleed Brakes (Four Wheels) (B)
Includes: Add fluid.
- 1998-05 .... .8 .8

Brake Hose (Flexible), Replace (B)
Includes: System bleeding.
- One 1.2 1.3
- each addl. add .... .3 .4

Master Cylinder, Replace (B)
- 1998-05 .... 1.9 2.1

SERVICE BRAKES
Brake Pads, Replace (B)
- Front 1.4 1.5
- Rear 1.2 1.3
- Four wheels 2.2 2.3

FRONT SUSPENSION
Front End Alignment, Check (A)
- 1998-05 .... 1.2 1.2

Front Hub Bearings, Replace (B)
- Both 2.8 3.3

Front Shock Absorbers, Replace (B)
- Both 1.5 1.7

Lower Ball Joint, Replace (B)
Add alignment charges.
- 1998-05 .... 2.3 2.8

Stabilizer Bar Bushings, Replace (B)
- 1998-05 .... 1.8 2.0

Torsion Bar Anchor/Adjusting Bolt, Replace (B)
- Each 1.2 1.7

Torsion Bars, Replace (B)
- Both 1.5 2.0

Upper Control Arm Assy., Replace (B)
- Each 1.5 2.0
- Replace bushings add .... .3 .6

REAR SUSPENSION
Rear Coil Spring, Replace (B)
- Both 3.5 3.9

Rear Shock Absorbers, Replace (B)
- 1998-05 .... 3.5 3.7

Rear Stabilizer Bar Bushings, Replace (B)
- 1998-05 .... 1.7 1.9

Stabilizer Bar, Replace (B)
- 1998-05 .... 1.7 1.9

STEERING
AIR BAGS
Air Bag Assy., Replace (B)
- Driver side .... .5 .5
- Passenger side 3.1 3.1
- Side impact .... .8 .8

Air Bag Control Unit, Replace (B)
- 1998-05 .... 1.3 1.3

POWER RACK & PINION
Pump Pressure Check (B)
- 1998-05 .... 1.4 1.4

Horn Contact or Canceling Cam, Replace (B)
- 1998-05 .... 1.4 1.4

Outer Tie Rod Ends, Replace (B)
- One .... .5 .8
- Both 9 1.2

Rack & Pinion Assy., R&I (B)
- 1998-05 .... 6.8 7.3

Steering Column, R&I (B)
- 1998-05 .... 2.0 2.0

Steering Pump, R&R or Replace (B)
- 1998-05 .... 2.4 2.6

Steering Wheel, Replace (B)
- 1998-05 .... .7 .7

HEATING & AIR CONDITIONING

When more than one component requires replacement where evacuation/recovery and recharging is already included, deduct 1.0 hour for each additional component from the time given.

Evacuate/Recover and Recharge System (B)
- 1998-05 .... 1.2 1.2

AC Hoses, Replace (B)
- 1998-05
  - condenser to drier .. 2.5 2.7
  - pressure 1.5 1.7
  - suction 1.5 1.7

Compressor Assy., Replace (B)
- 1998-05 .... 2.6 2.8

Condenser Assy., Replace (B)
- 1998-05 .... 2.5 2.7

Evaporator Coil, Replace (B)
- 1998-05 .... 1.9 1.9

Expansion Valve, Replace (B)
- 1998-05 .... 1.5 1.5

Receiver/Drier Assy., Replace (B)
- 1998-05 .... 2.1 2.1

WIPERS & SPEEDOMETER
Instrument Cluster, Replace (B)
- 1998-05 .... 1.8 1.8

Radio, R&R (B)
- 1998-05 .... .5 .5

Rear Window Washer Pump, Replace (B)
- 1998-05 .... .3 .3

Rear Windshield Wiper Motor, Replace (B)
- 1998-05 .... .8 1.0

Windshield Washer Pump, Replace (B)
- 1998-05 .... .5 .5

Windshield Wiper Motor, Replace (B)
- 1998-05 .... 1.5 1.8

LAMPS & SWITCHES
Halogen Headlamp Bulb, Replace (B)
- Each .... .3 .3

Headlights, Aim (B)
- 1998-05 .... .7 .7

High Mount Stop Lamp Bulb, Replace (C)
- 1998-05 .... .3 .3

Horn, Replace (B)
- 1998-05 .... .4 .4

License Lamp Assy., Replace (B)
- Each .... .3 .3

Multifunction Switch, Replace (B)
- 1998-05 .... 1.5 1.5

Stop Light Switch, Replace (B)
- 1998-05 .... .7 .7

Turn Signal or Hazard Warning Flasher, Replace (B)
- 1998-05 .... .4 .4

BODY
Door Window Regulator, Replace (B)
- Front one 2.3 2.6
- Rear one 1.9 2.2

Hood Lock, Replace (B)
- 1998-05 .... .8 .8

Hood Release Cable, Replace (B)
- 1998-05 .... 1.0 1.0

Hood Hinge, Replace (B)
- One 1.0 1.0

Lock Striker Plate, Replace (B)
- 1998-05 .... .5 .5

Cooper

SYSTEM INDEX

MAINTENANCE MIN-2
   Maintenance Schedule MIN-2
CHARGING MIN-2
STARTING MIN-2
CRUISE CONTROL MIN-2
IGNITION MIN-2
EMISSIONS MIN-2
FUEL MIN-2
EXHAUST MIN-3
ENGINE COOLING MIN-3
ENGINE MIN-3
   Assembly MIN-3
   Cylinder Head MIN-3
   Camshaft MIN-3
   Crank & Pistons MIN-3
   Engine Lubrication MIN-3
CLUTCH MIN-4
MANUAL TRANSAXLE MIN-4
SHIFT LINKAGE MIN-4
DRIVELINE MIN-4
BRAKES MIN-4
FRONT SUSPENSION MIN-5
REAR SUSPENSION MIN-5
STEERING MIN-5
HEATING & AC MIN-5
WIPERS & SPEEDOMETER . MIN-5
LAMPS & SWITCHES MIN-6
BODY MIN-6

OPERATIONS INDEX

A
AC Hoses MIN-5
Air Bags MIN-5
Air Conditioning MIN-5
Alignment MIN-5
Alternator (Generator) MIN-2
Antenna MIN-5
Anti-Lock Brakes MIN-4

B
Back-Up Lamp Switch MIN-6
Battery Cables MIN-2
Bleed Brake System MIN-4
Blower Motor MIN-5
Brake Disc MIN-4
Brake Hose MIN-4
Brake Pads and/or Shoes ... MIN-4

C
Camshaft MIN-3
Camshaft Sensor MIN-2
Catalytic Converter MIN-3
Crankshaft MIN-3
Crankshaft Sensor MIN-2
Cruise Control MIN-2
Cylinder Head MIN-3

D
Drive Belt MIN-2

E
Engine MIN-3
Engine Lubrication MIN-3
Engine Mounts MIN-3
Evaporator MIN-5
Exhaust MIN-3
Exhaust Manifold MIN-3
Expansion Valve MIN-5

F
Flexplate MIN-4
Flywheel MIN-4
Fuel Injection MIN-2
Fuel Pump MIN-2

G
Gear Selector Lever MIN-4
Generator MIN-2

H
Halfshaft MIN-4
Headlamp MIN-6
Heater Core MIN-5
Horn MIN-6

I
Ignition Coil MIN-2
Ignition Module MIN-2
Ignition Switch MIN-2
Inner Tie Rod MIN-5
Intake Manifold MIN-2

K
Knock Sensor MIN-2

L
Lower Control Arm MIN-5

M
Maintenance Schedule MIN-2
Master Cylinder MIN-4
Muffler MIN-3
Multifunction Lever/Switch .. MIN-6

N
Neutral Safety Switch MIN-6

O
Oil Pan MIN-3
Oil Pump MIN-3
Outer Tie Rod MIN-5
Oxygen Sensor MIN-2

P
Parking Brake MIN-4
Pistons MIN-3
Positive Crankcase Ventilation
  (PCV) Valve MIN-2

R
Radiator MIN-3
Radiator Hoses MIN-3
Radio MIN-5
Rear Main Oil Seal MIN-3

S
Shock Absorber/Strut, Front .. MIN-5
Shock Absorber/Strut, Rear ... MIN-5
Spark Plug Cables MIN-2
Spark Plugs MIN-2
Spring, Front Coil MIN-5
Spring, Rear Coil MIN-5
Starter MIN-2
Steering Wheel MIN-5

T
Thermostat MIN-3
Throttle Body MIN-2
Timing Chain MIN-3

V
Valve Cover Gasket MIN-3
Valve Job MIN-3
Vehicle Speed Sensor MIN-2

W
Water Pump MIN-3
Wheel Balance MIN-2
Window Regulator MIN-6
Windshield Washer Pump .. MIN-6
Windshield Wiper Motor ... MIN-6

MIN-1

MIN-2 COOPER

	LABOR TIME	SEVERE SERVICE
MAINTENANCE		
Air Cleaner Filter Element, Service (C)		
2002-05	.6	.8
Drive Belt, Replace (B)		
2002-05 (.5)	.7	.9
Drive Belt Tensioner, Replace (B)		
Cooper	1.4	1.6
Cooper S	1.8	2.0
Halogen Headlamp Bulb, Replace (B)		
2002-05	.4	.4
Oil & Filter, Change (B)		
Includes: Correct fluid levels.		
2002-05 (.7)	.9	1.1
Tire, Replace (C)		
Includes: Dismount old tire and mount new tire to rim.		
2002-05	.5	.5
Tires, Rotate (C)		
2002-05	.4	.4
Wheel, Balance (B)		
One	.3	.3
each addl. add	.2	.2

SCHEDULED MAINTENANCE INTERVALS

If necessary, refer to appropriate Chilton maintenance service information.

	LABOR TIME	SEVERE SERVICE
Maintenance Inspection 1 (B)		
2002-05 (2.1)	2.8	3.0
Maintenance Inspection 2 (B)		
2002-05 (2.5)	3.4	3.6

CHARGING

	LABOR TIME	SEVERE SERVICE
Alternator Circuits, Test (B)		
Includes: Test component output.		
2002-05	.4	.4
Alternator Assy., Replace (B)		
Includes: Pulley transfer.		
Cooper (1.3)	1.8	2.0
Cooper S (1.9)	2.6	2.8
Replace voltage regulator add	.2	.4
Alternator Voltage Regulator, Replace (B)		
Cooper (1.5)	2.0	2.2
Cooper S (1.9)	2.6	2.8

STARTING

	LABOR TIME	SEVERE SERVICE
Starter Draw Test (On Car) (B)		
2002-05	.3	.3
Battery, Replace (C)		
2002-05	.4	.6
Battery Cables, Replace (C)		
Cooper		
Positive (1.3)	1.8	2.0
negative (.7)	.9	1.1
Cooper S		
Positive (2.6)	3.5	3.7
negative (.5)	.7	.9
Starter Assy., Replace (B)		
2002-05	2.4	2.6

CRUISE CONTROL

	LABOR TIME	SEVERE SERVICE
Diagnose Cruise Control System Component Each (A)		
2002-05 (.8)	1.1	1.1
Control Module, Replace (B)		
2002-05 (.6)	.8	1.0
Control Switch, Replace (B)		
2002-05 (.6)	.8	1.0

IGNITION

	LABOR TIME	SEVERE SERVICE
Diagnose Ignition System Component Each (A)		
2002-05 (.8)	1.1	1.1
Camshaft Position Sensor, Replace (B)		
2002-05 (.6)	.8	1.0
Crankshaft Sensor, Replace (B)		
Cooper (.8)	1.1	1.3
Cooper S (1.4)	1.9	2.1
Ignition Coil, Replace (B)		
2002-05 (.3)	.5	.7
Ignition Module, Replace (B)		
2002-05 (.3)	.5	.7
Ignition Switch, Replace (B)		
2002-05	.8	.8
Spark Plug (Ignition) Cables, Replace (B)		
2002-05 (.6)	.8	1.0
Spark Plug Tube Replace (B)		
Cooper (.6)	.8	1.0
Cooper S (1.0)	1.4	1.6
Spark Plugs, Replace (B)		
Cooper (.4)	.6	.9
Cooper S (.6)	.8	1.0

EMISSIONS

The following operations do not include testing. Add time as required.

	LABOR TIME	SEVERE SERVICE
Dynamometer Test (A)		
2002-05	.8	.8
Intake Pressure Sensor, Replace (B)		
2002-05 (.3)	.5	.7
Knock Sensor, Replace (B)		
Cooper (.8)	1.1	1.3
Cooper S (4.1)	5.5	5.7
Oxygen Sensor, Replace (B)		
2002-05 (.4)	.6	.8
Positive Crankcase Ventilation (PCV) Valve, Replace (B)		
2002-05 (.2)	.4	.6
Throttle Valve Assembly, Replace (B)		
2002-05 (.6)	.8	.8
Vapor Canister Filter, Replace (B)		
Cooper (.4)	.6	.8
Cooper S (.6)	.8	1.0
Vehicle Speed Sensor, Replace (B)		
2002-05 (.6)	.8	.8

FUEL

DELIVERY

	LABOR TIME	SEVERE SERVICE
Fuel Pump, Test (B)		
2002-05 (.8)	1.1	1.1
Fuel Gauge (Dash) Replace (B)		
2002-05 (.6)	.8	.8
Fuel Gauge (Tank), Replace (B)		
2002-05 (1.3)	1.8	2.0
Fuel Pump, Replace (B)		
Cooper (1.2)	1.6	1.8
Cooper S (1.4)	1.9	2.1
Fuel Pump Relay, Replace (B)		
2002-05	.8	.8
Fuel Tank, Replace (B)		
Includes: Drain and refill.		
Cooper (3.1)	4.2	4.4
Cooper S (3.6)	4.9	5.1
Intake Manifold and/or Gasket, Replace (B)		
Includes: Adjustments.		
Cooper (.9)	1.2	1.4
Cooper S (1.3)	1.8	2.0

INJECTION

	LABOR TIME	SEVERE SERVICE
Electronic Control Unit, Replace (B)		
2002-05	.3	.3
Fuel Filter with Level Sensor, Replace (B)		
2002-05 (1.7)	2.3	2.5
Fuel Injectors, Replace (B)		
Cooper (.6)	.8	1.0
Cooper S (.9)	1.2	1.4
Fuel Pressure Regulator, Replace (B)		
Cooper (.3)	.5	.7
Cooper S (.6)	.8	.8
Fuel Rail, Replace (B)		
Cooper (.8)	1.1	1.3
Cooper S (1.0)	1.4	1.6
Tank Venting Valve, Replace (B)		
2002-05	.3	.3
Temperature Sensor, Replace (B)		
2002-05 (.6)	.8	.8
Throttle Body Assy., Replace (B)		
Cooper (.5)	.7	.9
Cooper S (.8)	1.1	1.3

SUPERCHARGER

	LABOR TIME	SEVERE SERVICE
Intercooler, Replace (B)		
Cooper S (1.3)	1.8	2.0
Supercharger Assy. or Gasket, Replace (B)		
Cooper S (3.6)	4.9	5.1

COOPER MIN-3

	LABOR TIME	SEVERE SERVICE
EXHAUST		
Catalytic Converter, Replace (B)		
Cooper (1.2)	1.6	1.8
Cooper S (1.4)	1.9	2.1
Complete Exhaust System, Replace (B)		
Cooper	.7	.9
Cooper S	1.2	1.4
Exhaust Manifold or Gasket, Replace (B)		
Cooper	1.6	1.8
Cooper S	1.9	2.1
Muffler & Resonator, Replace (B)		
2002-05 (.5)	.7	.9
Muffler & Tailpipe Assy., Replace (B)		
2002-05 (.5)	.7	.9
Rear Exhaust Pipe, Replace (B)		
2002-05 (.5)	.7	.9
ENGINE COOLING		
Pressure Test Cooling System (B)		
2002-05	.5	.5
Coolant Thermostat, Replace (B)		
2002-05 (1.3)	1.8	2.0
Electric Fan Thermo Switch, Replace (B)		
2002-05	.6	.8
Engine Coolant Temp. Sending Unit, Replace (B)		
2002-05	.8	.8
Radiator Assy., R&R or Replace (B)		
Cooper (1.7)	2.3	2.5
Cooper S (1.5)	2.0	2.2
Radiator Fan and/or Fan Motor, Replace (B)		
Cooper	2.3	2.5
Cooper S	1.8	2.0
Radiator Hoses, Replace (B)		
Includes: Refill system with proper coolant mix.		
Cooper		
upper	.6	.8
lower	.8	1.0
both	1.6	1.8
Cooper S		
upper	1.0	1.2
lower	1.8	2.0
both	2.1	2.3
Temperature Gauge (Dash), Replace (B)		
2002-05 (.6)	.8	1.0
Temperature Gauge (Engine), Replace (B)		
2002-05	.6	.8
Water Pump and/or Gasket, Replace (B)		
Includes: Refill with proper coolant mix.		
Cooper (2.5)	3.4	3.4
Cooper S (3.4)	4.6	4.8

	LABOR TIME	SEVERE SERVICE
ENGINE		
ASSEMBLY		
Times shown are for OEM assemblies. Time to replace assemblies from aftermarket rebuilders may vary.		
Engine Assy., R&I (B)		
Does not include parts or component transfer.		
Cooper (8.9)	12.0	12.2
Cooper S (9.5)	12.8	13.0
Engine Assy. (New or Rebuilt Complete), Replace (B)		
Includes: Component transfer and engine tune-up.		
Cooper	16.1	16.3
Cooper S	17.3	17.5
Engine Mounts, Replace (B)		
Right or left	.8	1.0
Top	.5	.7
Transaxle	1.2	1.4
CYLINDER HEAD		
Compression Test (B)		
2002-05	.8	1.0
Hydraulic Chain Tensioner, Replace (B)		
Cooper (1.0)	1.4	1.6
Cooper S (1.3)	1.8	2.0
Cylinder Head, Replace (B)		
Includes: Parts transfer, adjustments.		
Cooper	13.5	13.7
Cooper S	16.7	16.9
Cylinder Head Gasket, Replace (B)		
Cooper	8.0	8.2
Cooper S	11.2	11.4
Rocker Arm Shafts, Replace (B)		
Cooper	1.4	1.6
Cooper S	2.5	2.7
Rocker Arms, Replace (B)		
Cooper	1.4	1.6
Cooper S	2.8	3.0
Valve Cover Gasket, Replace (B)		
Cooper (.6)	.8	1.0
Cooper S (.9)	1.2	1.4
Valve Job (A)		
Cooper (11.2)	15.1	15.3
Cooper S (15.1)	20.4	20.6
CAMSHAFT		
Camshaft, Replace (A)		
Cooper	3.0	3.2
Cooper S	3.5	3.7
Timing Chain, Replace (B)		
Cooper	5.9	6.1
Cooper S	6.1	6.3
Timing Chain Sprockets, Replace (B)		
Cooper	5.9	6.1
Cooper S	6.1	6.3

	LABOR TIME	SEVERE SERVICE
Timing Cover and/or Gasket, Replace (B)		
Cooper	4.1	4.3
Cooper S	2.8	3.0
Timing Cover Seal, Replace (B)		
Cooper	1.3	1.5
Cooper S	1.8	2.0
CRANK & PISTONS		
Connecting Rod Bearings, Replace (A)		
Includes: Check bearing oil clearance.		
Cooper (15.0)	20.3	20.5
Cooper S (16.2)	21.9	22.1
Crankshaft and Main Bearings, Replace (A)		
Includes: Engine R&R.		
Cooper	24.8	25.0
Cooper S	26.6	26.8
Crankshaft Front Oil Seal, Replace (B)		
Cooper	1.5	1.7
Cooper S	1.8	2.0
Crankshaft Pulley, Replace (B)		
2002-04	1.2	1.4
Pistons or Connecting Rods, Replace (A)		
Includes: Ridge reaming, cylinder wall deglazing, installing new rings and rod bearings, engine tune-up.		
Cooper	20.3	20.5
Cooper S	21.8	22.0
Rear Main Oil Seal, Replace (B)		
Includes: Trans. R&R when necessary.		
Cooper (10.8)	14.6	14.8
Cooper S (8.2)	11.1	11.3
Rings, Replace (A)		
Includes: Ridge reaming, cylinder wall deglazing, installing new rings, engine tune-up.		
Cooper	22.3	22.5
Cooper S	23.8	24.0
Vibration Damper, Replace (B)		
2002-05 (.9)	1.2	1.4
ENGINE LUBRICATION		
Engine Oil Cooler, Replace (B)		
Cooper S (1.3)	1.8	2.0
Engine Oil Pressure Switch (Sending Unit), Replace (B)		
Cooper (.3)	.5	.7
Cooper S (.6)	.8	1.0
Oil Pan and/or Gasket, Replace (B)		
Cooper (3.0)	4.1	4.3
Cooper S (2.1)	2.8	3.0
Oil Pump, Replace (B)		
Cooper (3.3)	4.5	4.7
Cooper S (2.4)	3.2	3.4

MIN-4 COOPER

	LABOR TIME	SEVERE SERVICE
CLUTCH		
Bleed Clutch Hydraulic System (B)		
2002-05	.8	1.0
Clutch Pedal Free Play, Adjust (B)		
2002-05	.4	.6
Clutch Assy., Replace (B)		
Cooper (7.2)	9.7	9.9
Cooper S (8.0)	10.8	11.0
Clutch Master Cylinder, Replace (B)		
Includes: System bleeding.		
Cooper (1.1)	1.5	1.7
Cooper S (1.5)	2.0	2.2
Clutch Release Bearing Lever, Replace (B)		
Cooper (6.9)	9.3	9.5
Cooper S (7.7)	10.4	10.6
Clutch Slave Cylinder, Replace (B)		
2002-05	1.6	1.8
Flywheel, Replace (B)		
Cooper (7.4)	10.0	10.2
Cooper S (8.2)	11.1	11.3

MANUAL TRANSAXLE

Differential Seal Replace, R&I (B)		
2002-05 (12)	1.6	1.8
Input Shaft Seal Replace, R&I (B)		
Includes: R&I Transmission		
Cooper	9.0	9.2
Cooper S	10.5	10.7
Transaxle Assy., R&I (B)		
Cooper	9.2	9.4
Cooper S	10.3	10.5
Transaxle Mount, Replace (B)		
Cooper S (.9)	1.2	1.4

CONTINUOUSLY VARIABLE TRANSAXLE

SERVICE TRANSAXLE INSTALLED

Check Unit for Oil Leaks (C)		
Cooper	.8	1.0
Drain & Refill Unit (B)		
Cooper	1.2	1.4
Shift Lock Relay, Replace (B)		
Cooper	.6	.8
Shift Lock Solenoid Switch, Replace (B)		
Cooper	.8	1.0
Speed Sensor, Replace (B)		
Cooper	.9	1.1
Transaxle Mounts, Replace (B)		
Cooper		
left	1.4	1.6
rear	1.2	1.4
Transmission Oil Strainer, Replace (B)		
Cooper	3.6	3.8

SERVICE TRANSAXLE REMOVED

Flywheel (Flexplate), Replace (B)		
Cooper	14.3	14.5
Transaxle Assy., R&I (B)		
Cooper (10.3)	13.9	14.1

SHIFT LINKAGE

CONTINUOUSLY VARIABLE TRANSAXLE

Selector Cable, Adjust (B)		
Cooper	1.8	2.0
Selector Cable, Replace (B)		
Cooper	.8	1.0
Gear Selector Lever, Replace (B)		
Cooper	2.2	2.4

MANUAL TRANSAXLE

Gearshift Cable, Adjust (B)		
2002-05	1.9	2.1
Gearshift Console, Replace (B)		
2002-05	2.2	2.4
Gearshift Knob, Replace (B)		
2002-05	.3	.3

DRIVELINE

Halfshaft, R&R or Replace (B)		
2002-05 (.9)	1.2	1.4
Replace boot add each	.5	.8

BRAKES

ANTI-LOCK

Diagnose Anti-Lock Brake System Component Each (A)		
2002-05	1.2	1.2
Anti-Lock Relay, Replace (B)		
2002-05	.3	.5
Control Unit, Replace (B)		
2002-05	.4	.6
Front Rotor, Replace (B)		
2002-05	1.2	1.4
Front Sensor Assy., Replace (B)		
2002-05	.8	1.0
Hydraulic Assy., Replace (B)		
2002-05	1.8	2.0
Rear Rotor, Replace (B)		
2002-05	1.2	1.4
Rear Sensor Assy., Replace (B)		
2002-05	.8	1.0

SYSTEM

Bleed Brakes (B)		
Includes: Add fluid.		
2002-05	.8	1.0
Brake System, Flush and Refill (B)		
2002-05	1.2	1.4
Brake Hose (Flexible), Replace (B)		
Includes: System bleeding.		
One	1.1	1.2
each addl. add	.3	.4
Master Cylinder, Replace (B)		
Includes: System bleeding.		
2002-05 (1.3)	1.8	2.0
Power Booster Unit, Replace (B)		
Includes: System bleeding.		
Cooper (2.6)	3.5	3.7
Cooper S (2.3)	3.1	3.3
Power Booster Vacuum Check Valve, Replace (B)		
2002-05 (.4)	.6	.8

SERVICE BRAKES

Brake Pads, Replace (B)		
Includes: Service and parking brake adjustment, system bleeding.		
Front disc	1.2	1.4
Rear disc	1.3	1.5
Four wheels	2.2	2.4

COMBINATION ADD-ONS

Replace		
brake hose add	.3	.5
brake rotor add	.2	.4
caliper add	.3	.5
Resurface		
brake rotor add	.5	.7
Brake Pressure Regulator, Replace (B)		
2002-05	2.8	3.0
Caliper Assy., R&R and Recondition (B)		
Includes: System bleeding.		
Front		
one	1.9	2.1
both	2.8	3.0
Rear		
one	1.8	2.0
both	3.0	3.2
Caliper Assy., Replace (B)		
Includes: System bleeding.		
Front		
one	1.4	1.6
both	2.2	2.4
Rear		
one	1.8	2.0
both	3.2	3.5
Disc Brake Rotor, Replace (B)		
Front		
one	1.3	1.5
both	1.6	1.8
Rear		
one	.7	.9
both	1.6	1.8

PARKING BRAKE

Parking Brake Apply Warning Indicator Switch, Replace (B)		
2002-05	.6	.6
Parking Brake Cable, Replace (B)		
2002-05	2.4	2.6

COOPER MIN-5

	LABOR TIME	SEVERE SERVICE
Parking Brake Lever, Replace (B)		
2002-05	1.4	1.6

FRONT SUSPENSION

Unless otherwise noted, time given does not include alignment.

Align Front End (A)
- 2002-05 . . . . 1.2   *1.4*

Front Toe, Adjust (B)
- 2002-05 8   *1.0*

Rear Alignment, Adjust (B)
- 2002-05 . . . . 1.2   *1.2*

Axle Carrier, Replace (B)
- 2002-05 (3.9) 5.3   *5.5*

Lower Control Arm, Replace (B)
- Cooper
  - one (.8) 1.1   *1.3*
  - both (1.4) 1.9   *2.1*
- Cooper S
  - one (1.0) 1.4   *1.6*
  - both (1.9) 2.6   *2.8*

Stabilizer Bar, Replace (B)
- 2002-05 . . . . 3.5   *3.7*

Steering Knuckle, Replace (B)
- 2002-05 . . . . 1.6   *1.8*

Strut, R&R or Replace (B)
- 2002-05 (1.0) 1.4   *1.6*

Strut Coil Spring, Replace (B)
- One (1.2) 1.6   *1.8*
- Both (2.2) 3.0   *3.2*

Wheel Hub, Bearing or Seal, Replace (B)
- One side (1.0) 1.4   *1.6*
- Both sides (1.9) 2.6   *2.8*

REAR SUSPENSION

Check Front & Rear Axle Alignment (A)
- 2002-05 . . . . 2.0   *2.2*

Axle Carrier, Replace (B)
- Cooper (3.7) 5.0   *5.2*
- Cooper S (4.4) 5.9   *6.1*

Coil Spring, R&R or Replace (B)
- One 1.2   *1.6*
- Both 2.1   *2.3*

Lower Control Arm, Replace (B)
- 2002-05 (.9) 1.2   *1.4*

Rear Strut Assy., R&R or Replace (B)
- One 1.2   *1.4*
- Both 2.2   *2.4*

Rear Sway Bar, Replace (B)
- Cooper S (2.2) 3.0   *3.2*
- Cooper (1.9) 2.6   *2.8*

Shock Absorbers or Bushings, Replace (B)
- 2002-05 (1.2) 1.6   *1.8*

Upper Control Arm, Replace (B)
- 2002-05 (.9) 1.2   *1.4*

Wheel Bearing, Replace (B)
Includes: Replace bearing cups and grease seal.
- 2002-05 (1.0) 1.4   *1.6*

STEERING
AIR BAGS

Diagnose Air Bag System Component Each (A)
- 2002-05 . . . . 1.1   *1.1*

Air Bag Assy., Replace (B)
- Driver side 5   *.7*
- Passenger side 9   *1.1*

Air Bag Sensor, Replace (B)
- One 6   *.8*
- Both 1.2   *1.4*

Electronic Control Unit, Replace (B)
- 2002-05 5   *.7*

POWER RACK & PINION

Inner Tie Rod Assy., Replace (B)
Includes: Reset toe.
- 2002-05 . . . . 1.2   *1.4*

Outer Tie Rod End, Replace (B)
- One 6   *.8*
- Both 1.6   *1.8*

Rack & Pinion Assy., R&I (B)
Includes: Reset toe.
- Cooper (2.1) 2.8   *3.0*
- Cooper S (2.5) 3.4   *3.6*

Steering Pump, R&R or Replace (B)
- 2002-05 (.8) 1.1   *1.3*

Steering Pump Hoses, Replace (B)
- Pressure (1.2) 1.6   *1.8*
- Return (.8) 1.1   *1.3*

Steering Wheel, Replace (B)
- 2002-05 6   *.8*

Tie Rod End Boot, Replace (B)
- 2002-05 6   *.8*

HEATING & AIR CONDITIONING

When more than one component requires replacement where evacuation/recovery and recharging is already included, deduct 1.0 hour for each additional component from the time given.

Evacuate/Recover and Recharge System (B)
- 2002-05 (1.0) 1.4   *1.4*

AC Hoses, Replace (B)
Includes: Evacuate/recover and recharge.
- One 4.1   *4.3*
- each addl. add 8   *1.0*

AC Low Pressure Switch, Replace (B)
- 2002-05 . . . . 2.5   *2.7*

AC Relay, Replace (B)
- 2002-05 4   *.6*

Compressor Assy., Replace (B)
Includes: Parts transfer, evacuate/recover and recharge.
- 2002-05 (3.0) 4.1   *4.3*

Condenser Assy., Replace (B)
Includes: Evacuate/recover and recharge.
- 2002-05 (2.4) 3.2   *3.4*
- *Replace receiver/drier add* 5   *.5*

Condenser Cooling Fan Relay, Replace (B)
- 2002-05 4   *.4*

Evaporator Core, Replace (B)
Includes: Evacuate/recover and recharge.
- 2002-05 (1.9) 2.6   *2.8*

Expansion Valve, Replace (B)
Includes: Evacuate/recover and recharge.
- 2002-05 (3.0) 4.1   *4.3*

Heater Blower Motor, Replace (B)
- 2002-05 (5.1) 6.9   *7.1*

Heater Blower Motor Resistor, Replace (B)
- 2002-05 (1.3) 1.8   *2.0*

Heater Blower Motor Switch, Replace (B)
- 2002-05 (.6) 8   *1.0*

Heater Core, R&R or Replace (B)
- 2002-05 (5.1) 6.9   *7.1*

Heater Hoses, Replace (B)
- One (.5) 7   *.9*
- Both (.7) 9   *1.1*

High Pressure Cut-Off Switch, Replace (B)
- 2002-05 (1.9) 2.6   *2.8*

Receiver/Drier Assy., Replace (B)
Includes: Evacuate/recover and recharge.
- 2002-05 (1.3) 1.8   *2.0*

Temperature Control Assy., Replace (B)
- 2002-05 8   *1.0*

WIPERS & SPEEDOMETER

Power Antenna, Replace (B)
- 2002-05 7   *.9*

Radio, Replace (B)
- 2002-05 8   *.8*

Rear Window Washer Pump, Replace (B)
- 2002-05 6   *.8*

MIN-6 COOPER

	LABOR TIME	SEVERE SERVICE
Rear Window Wiper Motor, Replace (B)		
2002-05	1.0	*1.2*
Rear Window Wiper Relay, Replace (B)		
2002-05	.3	*.3*
Rear Window Wiper Switch, Replace (B)		
2002-05	.6	*.8*
Speedometer Head, R&R or Replace (B)		
2002-05	.8	*.8*
Windshield Washer Pump, Replace (B)		
2002-05 (.4)	.6	*.8*
Windshield Wiper Motor, Replace (B)		
2002-05 (1.3)	1.8	*2.0*
Windshield Wiper Motor Relay, Replace (B)		
2002-05	.3	*.3*
Windshield Wiper & Washer Switch, Replace (B)		
2002-05 (.6)	.8	*1.0*

LAMPS & SWITCHES

	LABOR TIME	SEVERE SERVICE
Back-Up Lamp Switch, Replace (B)		
2002-05	.8	*.8*
Brake Lamp Switch, Replace (B)		
2002-05	.4	*.4*
Emergency Flasher Switch Assy., Replace (B)		
2002-05	.8	*.8*
Emergency Fuel Pump Switch, Replace (B)		
2002-05	.8	*.8*
Halogen Headlamp Bulb, Replace (B)		
2002-05	.4	*.4*
Headlamp Switch, Replace (B)		
2002-05	.8	*.8*
Headlamps, Aim (B)		
2002-05	.8	*.8*
High Mount Stop Lamp Bulb, Replace (B)		
2002-05	.3	*.3*
Horn, Replace (B)		
2002-05	.8	*.8*
License Lamp Lens, Replace (B)		
2002-05	.3	*.3*
Multifunction Switch, Replace (B)		
Cooper	1.0	*1.2*
Neutral Safety Switch, Replace (B)		
Cooper	1.4	*1.6*
Parking Lamp Lens or Bulb, Replace (C)		
2002-05	.3	*.3*
Side Marker Lamp Lens, Replace (B)		
2002-05	.3	*.3*
Stop & Tail Lamp Lens, Replace (B)		
2002-05	.6	*.6*
Sunroof Switch, Replace (B)		
2002-05	.6	*.6*
Tire Pressure Warning Switch, Replace (B)		
2002-05	.6	*.6*
Turn Signal or Hazard Relay, Replace (B)		
2002-05	.8	*.8*
Turn Signal Switch, Replace (B)		
2002-05	.8	*.8*

BODY

	LABOR TIME	SEVERE SERVICE
Convertible Top, Adjust (A)		
2005 (2.9)	3.9	*4.1*
Convertible Top (Complete), Replace (A)		
2005 (3.1)	4.2	*4.4*
Convertible Top Fabric Cover, Replace (A)		
2005 (6.2)	8.4	*8.6*
Convertible Top Hydraulic Lines (Complete), Replace (A)		
2005 (4.6)	6.2	*6.4*
Convertible Top Hydraulic System (Complete), Replace (A)		
2005 (4.0)	5.4	*5.6*
Door Check, Replace (B)		
2002-05 (.8)	1.1	*1.3*
Door Lock, Replace (B)		
2002-05	1.4	*1.6*
Door Handle, Replace (B)		
2002-05 outside (.8)	1.1	*1.3*
Door Window Regulator, Replace (B)		
2002-05 electric	1.8	*2.0*
Hood Hinge, Replace (B)		
2002-05	.8	*1.0*
Hood Lock, Replace (B)		
2002-05	.8	*1.0*
Hood Release Cable, Replace (B)		
2002-05	3.5	*3.7*
Instrument Trim Panel (Complete), Replace (B)		
2002-05 (5.4)	7.3	*7.5*
Lock Striker Plate, Replace (B)		
2002-05	.6	*.8*
Rear Compartment Lid Lock and/or Cylinder, Replace (B)		
2002-05		

PEU

405 : 504 : 505 : 604

SYSTEM INDEX

MAINTENANCE	PEU-2
CHARGING	PEU-2
STARTING	PEU-2
CRUISE CONTROL	PEU-2
IGNITION	PEU-2
EMISSIONS	PEU-3
FUEL	PEU-3
EXHAUST	PEU-4
ENGINE COOLING	PEU-4
ENGINE	PEU-4
Assembly	PEU-4
Cylinder Head	PEU-4
Camshaft	PEU-5
Crank & Pistons	PEU-5
Engine Lubrication	PEU-5
CLUTCH	PEU-6
MANUAL TRANSAXLE	PEU-6
MANUAL TRANSMISSION	PEU-6
AUTO TRANSAXLE	PEU-6
AUTO TRANSMISSION	PEU-6
SHIFT LINKAGE	PEU-6
DRIVELINE	PEU-7
BRAKES	PEU-7
FRONT SUSPENSION	PEU-8
REAR SUSPENSION	PEU-8
STEERING	PEU-8
HEATING & AC	PEU-8
WIPERS & SPEEDOMETER	PEU-9
LAMPS & SWITCHES	PEU-9
BODY	PEU-9

OPERATIONS INDEX

A
AC Hoses	PEU-8
Air Conditioning	PEU-8
Alignment	PEU-8
Alternator (Generator)	PEU-2
Antenna	PEU-9

B
Back-Up Lamp Switch	PEU-9
Ball Joint	PEU-8
Battery Cables	PEU-2
Bleed Brake System	PEU-7
Blower Motor	PEU-9
Brake Disc	PEU-7
Brake Drum	PEU-7
Brake Hose	PEU-7
Brake Pads and/or Shoes	PEU-7

C
Camshaft	PEU-5
Catalytic Converter	PEU-4
Crankshaft	PEU-5
Crankshaft Sensor	PEU-3
Cruise Control	PEU-2
Cylinder Head	PEU-4

D
Differential	PEU-7
Distributor	PEU-2
Drive Belt	PEU-2
Driveshaft	PEU-7

E
Engine	PEU-4
Engine Lubrication	PEU-5
Engine Mounts	PEU-4
Evaporator	PEU-9
Exhaust	PEU-4
Exhaust Manifold	PEU-4
Expansion Valve	PEU-9

F
Flywheel	PEU-6
Fuel Injection	PEU-3
Fuel Pump	PEU-3
Fuel Vapor Canister	PEU-3

G
Gear Selector Lever	PEU-6
Generator	PEU-2
Glow Plug	PEU-3

H
Halfshaft	PEU-7
Headlamp	PEU-9
Heater Core	PEU-9
Horn	PEU-9

I
Ignition Coil	PEU-3
Ignition Switch	PEU-3
Injection Pump	PEU-3
Intake Manifold	PEU-3

K
Knock Sensor	PEU-3

L
Lower Control Arm	PEU-8

M
Master Cylinder	PEU-7
Muffler	PEU-4
Multifunction Lever/Switch	PEU-9

N
Neutral Safety Switch	PEU-6

O
Oil Pan	PEU-6
Oil Pump	PEU-6

P
Parking Brake	PEU-7
Pistons	PEU-5

R
Radiator	PEU-4
Radiator Hoses	PEU-4
Rear Main Oil Seal	PEU-5

S
Shock Absorber/Strut, Front	PEU-8
Shock Absorber/Strut, Rear	PEU-8
Spark Plug Cables	PEU-3
Spark Plugs	PEU-3
Spring, Front Coil	PEU-8
Spring, Rear Coil	PEU-8
Starter	PEU-2
Steering Wheel	PEU-8

T
Thermostat	PEU-4
Throttle Body	PEU-3
Timing Belt	PEU-5
Torque Converter	PEU-6

U
U-Joint	PEU-7
Upper Control Arm	PEU-8

V
Valve Body	PEU-6
Valve Cover Gasket	PEU-5
Valve Job	PEU-5

W
Water Pump	PEU-4
Wheel Balance	PEU-2
Wheel Cylinder	PEU-7
Window Regulator	PEU-9
Windshield Washer Pump	PEU-9
Windshield Wiper Motor	PEU-9

PEU-2 405 : 504 : 505 : 604

	LABOR TIME	SEVERE SERVICE

MAINTENANCE
Drive Belt, Replace (B)
Air pump
  505 .... 1.4   1.5
  604 .... 1.0   1.1
Alternator
  405 4   .5
  504
    gasoline 5   .6
    diesel 1.2   1.3
  505
    gasoline 1.2   1.3
    diesel or turbo ... 1.5   1.6
  604 .... 2.1   2.2
Brake vacuum pump
  Diesel 504, 505 5   .6
Compressor
  405 7   .8
  504, 505, 604
    exc. 505 gasoline 9   .9
    505 gasoline 1.6   1.7
Power steering
  405 5   .6
  504 8   .9
  505
    gasoline 2.9   3.0
    diesel 9   1.0
    turbo 9   1.0
  604 .... 1.4   1.5
Water pump
  504
    gasoline 6   .7
    diesel 2.6   2.7
  505
    gasoline 8   .9
    diesel 2.1   2.2
  604 .... 2.6   2.7
w/AC alternator add 5   .5
w/PS alternator add 2   .2
Replace tensioner
  compressor add 1.0   1.3
Halogen Headlamp Bulb, Replace (B)
1988-91 3   .3
Oil and Filter, Change (C)
Includes: Correct all fluid levels.
  Gasoline 3   .4
  Diesel 5   .6
Tire, Replace (C)
Includes: Dismount old tire and mount new tire to rim.
  All Models 5   .5
Tires, Rotate (C)
All Models 5   .5
Wheel, Balance (B)
All Models 3   .3
  each addl. add 2   .2

CHARGING
Alternator Circuits, Test (B)
Includes: Test component output.
  All Models 5   .5
Alternator, R&R and Recondition (A)
504, 604
  gasoline 4.2   4.7
  diesel 5.0   5.5
505
  gasoline 4.7   5.2
  diesel or turbo 5.0   5.5
w/AC add 5   .5
w/PS add 2   .2
Alternator Assy., Replace (B)
Includes: Pulley transfer.
  405 .... 1.2   1.4
  504, 604
    gasoline 1.7   1.9
    diesel 2.5   2.7
  505
    gasoline 2.2   2.4
    diesel or turbo 2.5   2.7
w/AC add 5   .5
w/PS add 2   .2
Alternator Voltage Regulator, Replace (B)
All Models 1.0   1.3

STARTING
Starter Draw Test (On Car) (B)
1981-91 3   .3
Battery Cables, Replace (C)
405, 504, 505 6   .8
604 positive 1.2   1.4
Starter, R&R and Recondition (A)
504
  gasoline 4.2   4.7
  diesel 4.4   4.9
505
  gasoline 5.2   5.7
  diesel or turbo 4.4   4.9
604 .... 5.7   6.2
Starter Assy., Replace (B)
405 .... 1.8   2.0
504
  gasoline 1.8   2.0
  diesel 2.1   2.3
505
  gasoline 2.8   3.0
  diesel or turbo 2.1   2.3
604 .... 3.3   3.5
Starter Drive Assy., Replace (B)
Includes: Starter R&R.
405 .... 2.6   2.9
504
  gasoline 2.5   2.8
  diesel 2.8   3.1

505
  gasoline 3.5   3.8
  diesel or turbo 2.9   3.2
604 .... 4.1   4.4
Starter Relay, Replace (B)
1981-91 6   .6
Starter Solenoid and/or Switch, Replace (B)
Includes: Starter R&R.
405 .... 2.1   2.4
504
  gasoline 2.3   2.6
  diesel 2.6   2.9
505
  gasoline 3.3   3.6
  diesel or turbo 2.6   2.9
604 .... 3.8   4.1

CRUISE CONTROL
Control Box, Replace (B)
1981-91 5   .5
Control Actuator Switch, Replace (B)
1981-91 .... 1.6   1.6
Control Brake or Clutch Switch, Replace (B)
405 5   .5
Control Electro Valve, Replace (B)
405 5   .5
Control Mode Switch, Replace (B)
405 6   .6
Control Module, Replace (B)
405 5   .5
Control Relay, Replace (B)
405 4   .4
Control Sensor, Replace (B)
405 6   .6
Control Servo, Replace (B)
1981-91 8   .6
Control Vacuum Pump, Replace (B)
405 5   .5
Deceleration Switch, Replace (B)
1981-91 5   .5
Control Transducer, Replace (B)
1981-91 5   .5

IGNITION
Distributor, Replace (B)
Includes: Reset base ignition timing.
405, 505, 604 .... 1.2   1.4
504 .... 1.0   1.2
Distributor Cap and/or Rotor, Replace (B)
405
  8V 5   .5
  16V 4   .4
504, 505 4   .4
604 5   .5

405 : 504 : 505 : 604   PEU-3

	LABOR TIME	SEVERE SERVICE
Distributor Points & Condenser, Replace (B)		
Includes: Distributor R&R, set dwell and base ignition timing.		
504	1.3	1.5
Electronic Control Unit, Replace (B)		
405	.4	.4
504, 604	1.6	1.6
505 diesel, turbo	.5	.5
Glow Plug Relay, Replace (B)		
504, 505	.6	.6
Glow Plug, Replace (B)		
504, 505	1.1	1.4
Ignition Coil, Replace (B)		
1981-91	.5	.5
Ignition Lock Switch Assy., Replace (B)		
1981-91	1.1	1.1
w/AC add	.6	.6
Ignition Timing, Reset (B)		
1981-91	.5	.5
Impulse Generator, Replace (B)		
405, 504, 505	.8	.8
604	1.9	1.9
Spark Plug, (Ignition) Cables, Replace (B)		
405	.5	.7
504, 505	.6	.8
604	.8	1.0
Spark Plugs, Replace (B)		
405	.6	.8
504	.7	.9
505, 604	1.0	1.2
Top Dead Center (TDC) Sensor, Adjust (B)		
405 16V	.5	.7
Vacuum Advance Unit, Replace (B)		
Includes: Reset base timing.		
405, 505, 604	1.6	1.6
504	1.4	1.4

EMISSIONS

	LABOR TIME	SEVERE SERVICE
Air Injection Tube, Replace (B)		
505	1.8	2.0
604		
one	1.8	2.0
both	3.2	3.4
Air Intake Sensor, Replace (B)		
405	1.1	1.1
505	.7	.7
Air Pump Assy., Replace (B)		
505	1.8	2.0
604	1.9	2.1
Air Pump Centrifugal Filter, Replace (B)		
604	1.0	1.0
Anti-Backfire Valve, Replace (B)		
505	.8	.8
604		
one	1.6	1.6
both	2.9	2.9

	LABOR TIME	SEVERE SERVICE
Diverter and/or Switching Valve, Replace (B)		
505	.7	.7
Fuel Vapor Canister, Replace (B)		
405	.4	.4
505	.5	.5
604	.8	.8
Knock Sensor, Replace (B)		
405 16V	.5	.7
Mixture Regulator, Replace (B)		
505	1.8	1.8
R&R air flow sensor add	.4	.4

FUEL
CARBURETOR

	LABOR TIME	SEVERE SERVICE
Carburetor, R&R and Clean or Recondition (A)		
Includes: Adjustments.		
1981-84 both	4.6	4.6
w/AT add	.2	.2
Carburetor, Replace (B)		
Includes: Adjustments.		
1981-84		
one	1.8	1.8
both	3.2	3.2
w/AT add	.2	.2

DELIVERY

	LABOR TIME	SEVERE SERVICE
Fuel Pump, Test (B)		
1981-84	.3	.3
Fuel Accumulator, Replace (B)		
504, 505	1.5	1.5
Fuel Filter, Replace (B)		
405	.5	.5
Fuel Gauge, (Tank), Replace (B)		
Gasoline		
504, 505	1.3	1.5
604	2.4	2.6
Diesel		
504 wagon	1.3	1.5
505		
sedan	2.4	2.6
wagon	1.3	1.5
Fuel Injection Pump, Replace (B)		
Diesel 504, 505	5.9	6.2
w/turbo add	.9	.9
Fuel Pump, Replace (B)		
405		
8V	.8	1.0
16V	1.1	1.3
504	.9	1.1
505	2.3	2.5
604	1.2	1.4
Fuel Return Hoses, Replace (B)		
505		
one	.6	.8
all	.8	1.0

	LABOR TIME	SEVERE SERVICE
Fuel Tank, Replace (B)		
Includes: Drain and refill.		
Gasoline		
405	2.2	2.4
504, 604	2.1	2.3
504 wagon	2.9	3.1
505	2.2	2.4
Diesel		
1981-88 504, 505	2.6	2.8
1982-83 504 wagon	3.1	3.3
Intake Manifold and/or Gasket, Replace (B)		
Includes: Adjustments.		
Gasoline		
405	2.2	2.6
504	2.4	2.8
505	3.1	3.5
604	3.1	3.5
Diesel 504, 505	2.9	3.3
w/PS add	.5	.5
Replace manifold add	.5	.7
Priming Pump and/or Gasket, Replace (B)		
505	2.7	2.9

INJECTION

	LABOR TIME	SEVERE SERVICE
Fuel Injection Timing, Check & Adjust (B)		
Diesel 504, 505	2.9	3.2
Fuel Injectors, R&R and Test (A)		
505	2.0	2.4
Auxiliary Air Regulator, Replace (B)		
505	.7	.7
Cold Start Injector, Replace (B)		
505	.7	.7
Fuel Damper, Replace (B)		
405	.5	.5
Fuel Distributor, Replace (B)		
505	1.5	1.7
Fuel Injection Lines, Replace (B)		
Diesel 504, 505	3.3	3.5
Fuel Injectors, Replace (B)		
405		
one	1.2	1.4
all	1.5	1.7
505		
one	.8	1.0
all	1.5	1.7
Fuel Pressure Regulator, Replace (B)		
405	1.5	1.7
Fuel Rail, Replace (B)		
405	1.0	1.2
Injection Nozzles, Replace (B)		
Diesel 504, 505	3.4	3.6
Throttle Body Pressure Regulator, Replace (B)		
505	1.0	1.2
Throttle Body Assy., Replace (B)		
405	2.8	2.8

PEU-4 405 : 504 : 505 : 604

	LABOR TIME	SEVERE SERVICE
TURBOCHARGER		
Turbocharger Assy., R&R or Replace (B)		
Diesel 504, 505	4.6	5.0

EXHAUST

	LABOR TIME	SEVERE SERVICE
Catalytic Converter, Replace (B)		
1981-91 one	1.5	1.5
Exhaust Manifold, Replace (B)		
405	2.0	2.4
504		
gasoline	2.7	3.1
diesel	3.8	3.2
505		
gasoline	2.5	2.9
diesel	3.8	4.2
604		
one	2.7	3.1
both	4.2	4.6
Front Exhaust Pipe, Replace (B)		
504		
gasoline	1.7	1.8
diesel	1.2	1.3
505	1.2	1.3
604		
one	2.0	2.1
both	2.8	2.9
Front Muffler, Replace (B)		
505, 604 diesel	1.4	1.5
604 gasoline	3.2	3.3
Intermediate Exhaust Pipe, Replace (B)		
1981-91	.7	.8
Rear Exhaust Pipe, Replace (B)		
604		
one	2.4	2.5
both	2.8	2.9
Rear Muffler, Replace (B)		
504	1.2	1.3
505	1.4	1.5
604	2.0	2.1
Thermal Reactor, Replace (B)		
504	1.4	1.5

ENGINE COOLING

	LABOR TIME	SEVERE SERVICE
Coolant Thermostat, Replace (B)		
405, 504, 505	.9	1.1
604	1.0	1.2
Fan Assy., Replace (B)		
405		
one	.7	.9
both	1.0	1.2
604		
AT	2.9	3.1
MT	2.4	2.6
Fan and/or Fan Motor, Replace (B)		
505	2.5	2.8

	LABOR TIME	SEVERE SERVICE
Fan Coupling, Replace (B)		
504, 604		
AT	3.3	3.5
MT	2.7	2.9
505		
gasoline		
AT	2.9	3.1
MT	2.2	2.4
diesel		
AT	3.3	3.5
MT	2.8	3.0
w/AC add	.5	.5
w/PS add	.3	.3
Radiator Assy., R&R or Replace (B)		
Gasoline		
405	1.5	1.7
505, 604		
AT	2.8	3.0
MT	2.3	2.5
Diesel 504, 505		
AT	2.6	2.8
MT	2.1	2.3
Radiator Hoses, Replace (B)		
Includes: Refill with proper coolant mix.		
405, 504, 505		
upper	.5	.5
lower	.7	.8
604	.8	.9
Temperature Gauge (Engine), Replace (B)		
405	.9	.9
504, 505, 604	.5	.5
Water Pump and/or Gasket, Replace (B)		
Includes: Refill with proper coolant mix.		
Gasoline		
405	4.8	5.3
504		
AT	4.6	5.1
MT	4.1	4.6
505		
AT	4.1	4.6
MT	3.4	3.9
604	5.0	5.5
Diesel 504, 505		
AT	5.0	5.5
MT	4.6	5.1
w/AC add	1.0	1.0
w/PS add	.3	.3

ENGINE

ASSEMBLY
Times shown are for OEM assemblies. Time to replace assemblies from aftermarket rebuilders may vary.

Engine Assy., R&I (B)
Does not include parts or component transfer.

	LABOR TIME	SEVERE SERVICE
Gasoline		
405	11.7	11.7
504	10.5	10.5

	LABOR TIME	SEVERE SERVICE
505	10.0	10.0
604	14.5	14.5
Diesel 504, 505	11.0	11.0
w/AC add	1.0	1.0
w/AT add	.5	.5
w/PS add	.5	.5
w/turbo add	1.0	1.0
Engine Mounts, Replace (B)		
Gasoline		
front		
405		
right	1.1	1.3
left	1.0	1.2
504	1.1	1.3
505, 604		
one	1.5	1.7
both	2.2	2.4
rear 405	2.5	2.7
Diesel 504, 505		
front		
one	1.5	1.7
both	2.5	2.7
rear	6.1	6.3

CYLINDER HEAD

	LABOR TIME	SEVERE SERVICE
Compression Test (B)		
1981-87		
4 cyl.	.8	1.1
V6	1.2	1.5
diesel	2.0	2.3
1988-91		
405	1.2	1.5
505	.8	1.1
diesel	2.0	2.3
Cylinder Head Gasket, Replace (B)		
Gasoline		
405		
8V	8.5	8.8
16V	9.5	9.8
504	6.0	6.3
505	7.9	8.2
604		
one side	10.0	10.3
both sides	18.0	18.3
Diesel 504, 505	8.0	8.3
w/AC add	.5	.5
w/AT add	.5	.5
w/PS add	.5	.5
w/turbo add	.5	.5
Pushrods, Replace (B)		
Gasoline		
504	2.1	2.3
505	2.4	2.7
Diesel 504, 505	1.9	2.2
Valve Clearance, Adjust (B)		
Gasoline		
405 8V	6.1	6.5
504	1.3	1.7
505	1.6	2.0
604	4.6	5.0
Diesel 504, 505	1.4	1.8

405 : 504 : 505 : 604 **PEU-5**

	LABOR TIME	SEVERE SERVICE

Valve Cover Gasket, Replace (B)
Gasoline
- 405 8   1.0
- 504 8   1.0
- 505 .... 1.1 1.3
- 604
  - one 2.4 2.6
  - both 4.0 4.2
- Diesel 504, 505 ...... .9 1.1

Valve Job (A)
Gasoline
- 405
  - 8V 14.0 14.5
  - 16V 16.0 16.5
- 504 .... 8.0   8.5
- 505 .... 11.0 11.5
- 604
  - one side 10.4 10.9
  - both sides ... 20.0 20.5
- Diesel 504, 505 ..... 17.0 17.5
- w/AC add 5   .5
- w/AT add 5   .5
- w/PS add 5   .5
- w/turbo add 5   .5

Valve Lifters, Replace (B)
Gasoline
- 504 .... 8.5   8.9
- 505 .... 8.5   8.9
- Diesel 504, 505 ..... 21.0 22.0

Valve Springs and/or Oil Seals, Replace (B)
Gasoline
- 504 one 1.9   2.4
- 505 one 2.2   2.7
- 604 one 3.2   3.7
  - each addl. add 5   .6
- Diesel 504, 505 one .. 2.0 2.4
  - each addl. add 8   .9

CAMSHAFT

Camshaft, Replace (A)
Gasoline
- 405
  - 8V 7.3   7.5
  - 16V 9.0   9.2
- 504 .... 16.0 16.5
- 505 .... 16.5 17.0
- 604
  - one 13.3 13.5
  - both 22.0 23.0
- Diesel 504, 505 ..... 23.0 24.0
- w/AC add 5   .5
- w/AT add 5   .5
- w/PS add 5   .5
- w/turbo add 5   .5

Timing Belt, Replace (B)
- 405 .... 3.9   4.4

Timing Chain Tensioner, Replace (B)
Gasoline
- 504 .... 4.0   4.5
- 505
  - 1981 .... 4.5   5.0
  - 1982-91 .... 2.3   2.8
- 604 incl. R&R assy. 15.1 16.1
Diesel
- 504 .... 4.9   5.4
- 505 .... 5.4   5.9
- 604 .... 5.4   5.9
- w/AC add 5   .5
- w/AT add 5   .5
- w/PS add 5   .5
- w/turbo add 5   .5

Timing Cover and/or Gasket, Replace (B)
Gasoline
- 405 .... 1.5   1.8
- 504 .... 3.5   3.8
- 505 .... 1.9   2.2
- 604 incl. R&R assy. 15.5 15.8
Diesel
- 504 .... 3.9   4.2
- 505 .... 4.4   4.7
- w/AC add 5   .5
- w/AT add 5   .5
- w/PS add 5   .5
- w/turbo add 5   .5

CRANK & PISTONS

Connecting Rod Bearings, Replace (A)
Includes: Check bearing oil clearance.
Gasoline
- 405 .... 22.0 22.5
- 504 .... 21.0 21.5
- 505 .... 20.5 21.0
- Diesel 504, 505 ..... 23.5 24.0
- w/AC add 1.0   1.0
- w/AT add 5   .5
- w/PS add 5   .5
- w/turbo add 1.0   1.0

Crankshaft & Main Bearings, Replace (A)
Includes: Engine R&R.
Gasoline
- 504 .... 25.0 25.9
- 505 .... 25.5 26.4
- 604 .... 28.0 28.9
- Diesel 504, 505 ..... 31.0 31.9
- w/AC add 1.0   1.0
- w/AT add 5   .5
- w/PS add 5   .5
- w/turbo add 1.0   1.0

Crankshaft Pulley, Replace (B)
Gasoline
- 405 9   1.1
- 504 .... 2.8   3.0

505
- 1981 .... 3.5   3.7
- 1982-91 .... 1.5   1.7
- 604 .... 4.1   4.3

Crankshaft Rear Main Oil Seal, Replace (B)
Gasoline
- 504 .... 13.0 13.6
- 505 .... 12.5 13.1
- 604 .... 17.0 17.6
- Diesel 504, 505 ..... 15.0 15.6
- w/AC add 5   .5
- w/AT add 5   .5
- w/PS add 5   .5
- w/turbo add 5   .5

Pistons and Cylinder Liners, Replace (A)
Includes: Cylinder head and oil pan R&R, install new pistons, rings, pins and cylinder liners. and make all adjustments.
Gasoline
- 405 .... 24.0 24.9
- 504 .... 27.5 28.4
- 405 .... 24.0 24.9
- 604
  - one side 29.5 30.4
  - both sides 36.0 36.9
- Diesel 504, 505 ..... 29.0 29.9
- w/AC add 1.0   1.0
- w/AT add 5   .5
- w/PS add 5   .5
- w/turbo add 1.0   1.0

Rings, Replace (A)
Includes: Ridge reaming, cylinder wall deglazing, installing new rings, engine tune-up.
Gasoline
- 405 .... 23.2 24.1
- 504 .... 25.5 26.4
- 505
  - 1981 .... 24.0 24.9
  - 1982-91 .... 23.5 24.4
- 604 .... 29.5 30.4
- Diesel 504, 505 ..... 26.0 26.9
- w/AC add 1.0   1.0
- w/AT add 5   .5
- w/PS add 5   .5
- w/turbo add 1.0   1.0

ENGINE LUBRICATION

Crankcase and Oil Pan, R&R or Replace (B)
Gasoline
- 504 .... 3.0   3.2
- 505 .... 4.0   4.2
- 604 incl. R&R assy. 19.5 20.0
- Diesel 504, 505 ..... 3.0   3.2
- w/turbo add 5   .5

PEU-6 405 : 504 : 505 : 604

	LABOR TIME	SEVERE SERVICE
Oil Pan and/or Gasket, Replace (B)		
Gasoline		
405	2.7	2.9
504	1.0	1.2
505	1.0	1.2
604	1.3	1.5
Diesel 504, 505	1.0	1.2
w/crankcase add	.5	.7
w/turbo add	.5	.5
Oil Pump, Replace (B)		
Gasoline		
405	3.4	3.9
504	3.9	4.4
505	4.9	5.4
604	16.5	17.0
Diesel 504, 505	3.5	3.9
w/crankcase add	.5	.5
w/turbo add	.5	.5
Engine Oil Pressure Switch (Sending Unit), Replace (B)		
All Models	.5	.5

CLUTCH

	LABOR TIME	SEVERE SERVICE
Clutch Hydraulic System, Bleed (B)		
1981-91	.5	.5
Clutch Pedal Free Play, Adjust (B)		
1981-91	.3	.5
Clutch Assy., Replace (B)		
405		
8V	7.0	7.2
16V	8.2	8.4
504, 505	8.0	8.2
604	13.5	13.7
w/turbo add	.3	.3
w/wagon add	.5	.5
Clutch Cable, Replace (B)		
405	1.1	1.3
Clutch Master Cylinder, Replace (B)		
Includes: System bleeding.		
504	2.0	2.2
505		
FI, turbo	2.2	2.4
diesel	2.0	2.2
604		
FI	3.2	3.4
turbo	2.2	2.4
Clutch Slave Cylinder, Replace (B)		
Includes: System bleeding.		
1981-91	2.0	2.2
Flywheel, Replace (B)		
405		
8V	7.3	7.5
16V	8.5	8.7
504	9.0	9.2
505	9.0	9.2
604	14.5	14.7
w/turbo add	.5	.5
w/wagon add	.5	.5
Replace pilot bushing add	.8	.8

MANUAL TRANSAXLE

	LABOR TIME	SEVERE SERVICE
Transaxle Assy., R&R and Recondition (A)		
405	10.9	11.8
Recondition differential assembly add	1.9	2.4
Transaxle Assy., R&R or Replace (B)		
405	6.9	7.1

MANUAL TRANSMISSION

	LABOR TIME	SEVERE SERVICE
Speedometer Driven Gear, Replace (B)		
All Models	.8	.8
Transmission Assy., R&R and Recondition (A)		
504, 505		
exc. below	14.5	15.1
1985-88 505 turbo	15.0	15.6
604	20.5	21.1
Transmission Assy., R&I (B)		
504, 505		
exc. below	7.5	7.7
1985-88 505 turbo	8.0	8.2
604	15.0	15.2

AUTOMATIC TRANSAXLE

SERVICE TRANSAXLE INSTALLED

	LABOR TIME	SEVERE SERVICE
Check Unit For Oil Leaks (B)		
1988-91	.5	.5
Drain & Refill Unit (B)		
1988-91	.9	.9
Oil Pan and/or Gasket, Replace (B)		
1988-91	.9	.9
Oil Pressure Check (B)		
1988-91	1.0	1.0
Valve Body, Replace (B)		
1988-91	1.7	2.0

SERVICE TRANSAXLE REMOVED

	LABOR TIME	SEVERE SERVICE
Torque Converter, Replace (B)		
405	7.0	7.2
Transaxle Assy., R&R and Recondition (A)		
405	13.0	13.9
Recondition diff. add	1.5	1.7
Transaxle Assy., R&R or Replace (B)		
405	6.8	7.0
Replace flexplate add	.3	.3

AUTOMATIC TRANSMISSION

SERVICE TRANSMISSION INSTALLED

	LABOR TIME	SEVERE SERVICE
Bands, Adjust (B)		
604	2.5	2.7
Check Unit for Oil Leaks (B)		
1981-91	.5	.5
Drain and Refill Unit (B)		
1981-91	1.0	1.0
Kickdown Cable, Adjust (B)		
504, 505	.7	.9
604	1.3	1.5
Kickdown Cable, Replace (B)		
504, 505, 604	2.1	2.3
Neutral Safety Switch, Replace (B)		
504, 505, 604	.6	.6
Oil Pan Gasket, Replace (B)		
504, 505, 604	1.0	1.0
Oil Pressure Check (B)		
1981-91	1.0	1.0
Speedometer Driven Gear and/or Seal, Replace (B)		
504, 505, 604	.8	.8
Vacuum Modulator, Replace (B)		
604	.7	.7
Valve Body Assy., R&R and Recondition (A)		
504, 505, 604	3.4	3.9
Valve Body Assy., Replace (B)		
504, 505, 604	1.7	2.0

SERVICE TRANSMISSION REMOVED

	LABOR TIME	SEVERE SERVICE
Torque Converter, Replace (B)		
504, 505		
exc. 505 turbo	8.2	8.4
505 turbo	9.2	9.4
604	15.4	15.6
Transmission Assy., R&R and Recondition (A)		
504, 505		
exc. 505 turbo	14.8	15.7
505 turbo	15.8	16.7
604	22.1	23.0
Transmission Assy., R&I (B)		
504, 505		
exc. 505 turbo	7.9	8.1
505 turbo	8.9	9.1
604	15.1	15.3
Replace		
flexplate add	.3	.3
transmission add	1.0	1.5

SHIFT LINKAGE

AUTOMATIC TRANSMISSION

	LABOR TIME	SEVERE SERVICE
Shift Linkage, Adjust (B)		
1981-91	.9	1.1
Gear Selector Lever, Replace (B)		
1981-91 504, 505	1.4	1.4
604	3.1	3.1

MANUAL TRANSMISSION

	LABOR TIME	SEVERE SERVICE
Shift Linkage, Adjust (B)		
1981-91	.8	1.0
Gearshift Lever, Replace (B)		
504	2.1	2.1
505	2.6	2.6
604	3.6	3.6

405 : 504 : 505 : 604 **PEU-7**

	LABOR TIME	SEVERE SERVICE
DRIVELINE		
DRIVESHAFT		
Torque Tube, Replace (B)		
504		
sedan	5.4	5.8
wagon	7.1	7.5
505	5.4	5.8
604	6.9	7.3
Replace center bearing add	.5	.7
U-Joint, Replace (B)		
504 wagon one	6.6	6.8
Replace oil seal add	.5	.7
DRIVE AXLE		
CV-Joint Outer Boot, Replace (B)		
504 wagon		
one	2.5	2.8
both	4.6	4.9
Differential Assy., R&I (B)		
504		
sedan	4.9	5.4
wagon	8.1	8.6
505	4.9	5.4
604	5.6	6.1
Differential Assy., R&R and Recondition (A)		
504		
sedan	9.6	10.1
wagon	14.1	14.6
505	9.6	10.1
604	11.6	12.1
Rear Axle Shaft, Replace (B)		
504 wagon		
one	1.8	2.1
both	3.6	3.9
Replace oil seal add each	.3	.3
Rear Hub Carrier, Replace (B)		
1981-91		
one	3.1	3.5
both	5.1	5.5
Rear Wheel Bearings, Clean and Repack or Replace (B)		
1981-91		
one side	3.4	3.4
both sides	5.6	5.6
HALFSHAFTS		
CV-Joint Boot, Replace (B)		
405		
one	2.4	2.7
both	3.3	3.6
Replace differential oil seal add each	.2	.3
Halfshaft, R&R or Replace (B)		
405		
one	1.4	1.7
both	2.2	2.5

	LABOR TIME	SEVERE SERVICE
BRAKES		
SYSTEM		
Bleed Brakes (B)		
Includes: Add fluid.		
1981-91	.5	.5
Brakes, Adjust (B)		
Includes: Refill master cylinder.		
Rear wheels	.3	.3
Brake Hose (Flexible), Replace (B)		
Includes: System bleeding.		
One	1.0	1.1
Master Cylinder, R&R and Rebuild (A)		
1981-91	2.5	2.8
Master Cylinder, Replace (B)		
405	1.5	1.7
504, 505, 604	2.0	2.2
Power Brake Booster, Replace (B)		
405	1.6	1.8
504, 505, 604	2.5	2.7
Power Booster Vacuum Check Valve, Replace (B)		
1981-91	.3	.3
Vacuum Pump, Replace (B)		
Diesel 504, 505	1.1	1.3
Vacuum Pump Diaphragm, Replace (B)		
Diesel 504, 505	1.5	1.7
Vacuum Pump Valve, Replace (B)		
Diesel 504, 505	.5	.5
SERVICE BRAKES		
Brake Drum, Replace (B)		
Each	.5	.7
Caliper Assy., R&R and Recondition (A)		
Includes: System bleeding.		
405		
one	2.6	2.9
both	4.2	4.5
504, 505, 604		
one	2.1	2.4
both	3.9	4.2
Caliper Assy., Replace (B)		
Includes: System bleeding.		
405		
one	1.8	2.0
both	2.8	3.0
504, 505, 604		
one	1.3	1.5
both	2.4	2.6
Disc Brake Rotor, Replace (B)		
Front		
405		
one	.9	1.1
both	1.6	1.8
504, 505, 604		
one	1.6	1.8
both	2.7	2.9

	LABOR TIME	SEVERE SERVICE
Rear		
405		
one	1.7	1.9
both	3.0	3.2
504, 505, 604		
one	2.6	2.8
both	4.7	4.9
Pads and/or Shoes, Replace (B)		
Includes: Adjust service and parking brake. System bleeding.		
405		
front or rear	1.2	1.3
four wheels	2.1	2.2
504, 505, 604		
front disc	1.1	1.2
rear disc	1.6	1.7
rear drum	2.5	2.6
four wheels		
disc	2.6	2.7
drum	3.2	3.3
COMBINATION ADD-ONS		
Replace caliper add	.4	.4
Resurface		
brake drum add	.5	.5
brake rotor add	.5	.5
Wheel Cylinder, R&R and Rebuild (B)		
Includes: System bleeding.		
One	2.4	2.7
Both	4.3	4.6
Wheel Cylinder, Replace (B)		
Includes: System bleeding.		
One	2.1	2.3
Both	3.7	3.9
PARKING BRAKE		
Parking Brake Apply Actuator, Replace (B)		
405	1.2	1.2
504, 505	1.3	1.3
604	2.5	2.5
Parking Brake Cable, Adjust (C)		
1981-91	.3	.5
Parking Brake Cable, Replace (B)		
405		
primary	1.8	2.0
secondary	1.9	2.1
504		
sedan each	1.4	1.6
wagon		
one	2.0	2.2
both	3.4	3.6
505		
one	2.0	2.2
both	2.5	2.7
604		
one	2.8	3.0
both	3.6	3.8

PEU-8   405 : 504 : 505 : 604

	LABOR TIME	SEVERE SERVICE

FRONT SUSPENSION
Unless otherwise noted, time given does not include alignment.

FRONT WHEEL DRIVE
Front Toe, Adjust (B)
  405 .... 1.0   *1.2*
Ball Joints, Replace (B)
  405 one side 3.5   *3.8*
Control Arm, Replace (B)
  405 one 2.3   *2.6*
  Replace bushings add .. 1.2   *1.4*
Front Strut Coil Spring, Replace (B)
  405
    one 3.1   *3.4*
    both 5.4   *5.7*
Hub Bearings, Replace (B)
  405
    one side 3.0   *3.0*
    both sides 5.2   *5.2*
Shock Absorber Strut Cartridge, Replace (B)
  405 one 2.9   *3.3*
Stabilizer Bar Link Bushings, Replace (B)
  405 9   *1.1*
Stabilizer Bar, Replace (B)
  405 .... 2.0   *2.2*
Steering Knuckle, Replace (B)
  405 one 2.9   *3.4*
Strut, R&R or Replace (B)
  405
    one 2.8   *3.0*
    both 4.9   *5.1*

REAR WHEEL DRIVE
Front Toe, Adjust (B)
  504, 505, 604 .... 1.0   *1.2*
Ball Joints, Replace (B)
  504
    one 2.8   *3.1*
    both 5.0   *5.3*
  505, 604
    one 3.1   *3.4*
    both 5.8   *6.1*
Coil Spring, Replace (B)
  504
    one 3.3   *3.6*
    both 5.4   *5.7*
  505
    one 4.4   *4.7*
    both 6.5   *6.8*
  604
    one 4.1   *4.4*
    both 6.2   *6.5*
Control Arm Assy., Replace (B)
  504, 505, 604 .... 3.1   *3.4*

Front Hub Bearings or Seal, Replace (B)
  504, 505, 604
    one side 2.0   *2.0*
    both sides 3.5   *3.5*
Stabilizer Bar, Replace (B)
  504, 505 .... 1.0   *1.2*
  604 .... 1.5   *1.7*
Strut Assy., Replace (B)
  504
    one 2.6   *2.8*
    both 5.0   *5.2*
  505, 604
    one 3.0   *3.2*
    both 5.7   *5.9*

REAR SUSPENSION
FRONT WHEEL DRIVE
Rear Alignment, Adjust (A)
  405 .... 1.0   *1.2*
Rear Spring and/or Shock Absorbers, Replace (B)
  405
    one 7   *.9*
    both 1.1   *1.2*
Rear Stabilizer Bar, Replace (B)
  405 .... 1.0   *1.2*
Rear Suspension Arm, Replace (B)
  405 one 5.5   *5.8*
Rear Torsion Bars, Replace (B)
  405
    one 3.4   *3.8*
    both 6.0   *6.4*
Rear Wheel Hub, Replace (B)
  405
    one side 1.4   *1.9*
    both sides 2.5   *3.0*

REAR WHEEL DRIVE
Coil Spring or Shock Absorbers, Replace (B)
  504
    one 8   *1.0*
    both 1.4   *1.6*
  505, 604
    one 1.3   *1.5*
    both 1.7   *1.9*
Rear Control Arm, Replace (B)
  504, 604 .... 6.2   *6.5*
  505 .... 5.7   *6.0*
  Replace bushings add each 5   *.7*
Rear Springs, Replace (B)
  504
    sedan
      one 2.8   *3.1*
      both 5.0   *5.3*
    wagon 2.0   *2.3*

  505
    one 3.0   *3.3*
    both 4.5   *4.8*
Stabilizer Bar and/or Bushings, Replace (B)
  504 .... 1.0   *1.2*
  505 .... 1.2   *1.4*

STEERING
MANUAL RACK & PINION
Rack & Pinion, R&R and Recondition (A)
  504 .... 5.7   *6.2*
Rack & Pinion Assy., R&I (B)
Includes: Reset toe.
  504 .... 2.7   *3.2*
Steering Wheel, Replace (B)
  1981-83 5   *.5*

POWER RACK & PINION
Rack & Pinion Assy., R&R or Replace (B)
Includes: Alignment.
  405 .... 3.3   *3.8*
  504 .... 4.7   *5.2*
  505 .... 8.0   *8.5*
  604 .... 4.8   *5.3*
Steering Pump, R&R or Replace (B)
  405 .... 2.4   *2.6*
  504 .... 3.0   *3.2*
  505
    gasoline 3.6   *3.8*
    diesel 3.0   *3.2*
    turbo 3.1   *3.3*
  604 .... 3.0   *3.2*
  w/AC add 5   *.5*
  w/AT add 5   *.5*
Steering Pump Hoses, Replace (B)
Does not include purging.
  1981-91 one 1.0   *1.3*
Steering Wheel, Replace (B)
  1981-91 5   *.5*

HEATING & AIR CONDITIONING
When more than one component requires replacement where evacuation/recovery and recharging is already included, deduct 1.0 hour for each additional component from the time given.

Evacuate/Recover and Recharge System (B)
  1981-91 .... 1.0   *1.0*
AC Hoses, Replace (B)
Includes: Evacuate/recover and recharge.
  One 3.0   *3.1*
  each addl. add 4   *.5*

405 : 504 : 505 : 604 **PEU-9**

	LABOR TIME	SEVERE SERVICE
Compressor Assy., Replace (B)		
Includes: Parts transfer. Evacuate/recover and recharge.		
405	4.0	4.2
504		
gasoline	3.5	3.7
diesel	4.0	4.2
505		
gasoline	4.0	4.2
diesel or turbo	3.5	3.7
w/AT add	.5	.5
Replace clutch add	.6	.8
Condenser Assy., Replace (B)		
Includes: Evacuate/recover and recharge.		
405	5.0	5.2
504		
sedan	3.2	3.4
wagon	4.0	4.2
505	3.0	3.2
604	4.0	4.2
w/AT add	.5	.5
Evaporator Core, Replace (B)		
Includes: Evacuate/recover and recharge.		
405	10.0	10.0
504	3.5	3.5
505	7.0	7.0
604	6.0	6.0
Expansion Valve, Replace (B)		
Includes: Evacuate/recover and recharge.		
405	4.4	4.4
504	4.2	4.2
505	3.2	3.2
604	5.8	5.8
Heater Blower Motor, Replace (B)		
405	.8	.8
504	5.6	5.6
505	3.7	3.7
604	8.5	8.5
w/AC add	.5	.5
Heater Core, R&R or Replace (B)		
405	12.1	12.1
505	2.1	2.1
604	3.8	3.8
w/AC add	1.0	1.0
Receiver/Drier Assy., Replace (B)		
Includes: Evacuate/recover and recharge.		
405	3.9	3.9
504, 505, 604	3.0	3.0

WIPERS & SPEEDOMETER

	LABOR TIME	SEVERE SERVICE
Antenna, Replace (B)		
405	.4	.4
Rear Window Wiper Motor, Replace (B)		
405 wagon	.5	.7
Speedometer Cable & Casing, Replace (B)		
405		
upper	.5	.5
lower	.4	.4
both	1.0	1.0
505		
upper or main	1.8	1.8
lower	.8	.8
Speedometer Head, R&R or Replace (B)		
1981-91	1.6	1.6
Windshield Washer Pump, Replace (B)		
1981-91	.5	.5
Windshield Wiper & Washer Switch, Replace (B)		
405	.8	.8
504	1.4	1.4
505, 604	.9	.9
w/cruise control add	.2	.2
Windshield Wiper Linkage, Replace (B)		
405	1.5	1.7
504	.8	1.0
505	1.8	2.0
604	3.9	4.1
Windshield Wiper Motor, Replace (B)		
405	.9	1.1
504	1.5	1.7
505	1.8	2.0
604	2.5	2.7
Windshield Wiper Motor Relay, Replace (B)		
504	.8	.8
604	1.2	1.2

LAMPS & SWITCHES

	LABOR TIME	SEVERE SERVICE
Back-Up Lamp/Neutral Safety Switch, Replace (B)		
1981-91	.5	.5
Halogen Headlamp Bulb, Replace (C)		
1988-91	.3	.3
Headlamp and Wiper Control Switch, Replace (B)		
504	1.5	1.5
604	.8	.8
w/cruise control add	.3	.3
Headlamp, Horn and Directional Switch, Replace (B)		
1981-91	.8	.8
w/cruise control add	.3	.3
Headlamp Sealed Beam, Replace (C)		
One	.3	.3
Headlamps, Aim (B)		
1981-91	.3	.3
High Mount Stop Lamp Assy., Replace (B)		
405	.5	.5
Horn, Replace (B)		
Each	.5	.5
License Lamp Bulb, Replace (C)		
405	.3	.3
504, 505, 604	.5	.5
Parking Brake Warning Lamp Switch, Replace (B)		
504	.6	.6
604	2.5	2.5
Side Marker Lamp Bulb, Replace (C)		
Each	.3	.3
Stop Light Switch, Replace (B)		
1981-91	.5	.5
Tail Lamp Assy., Replace (B)		
405, 504, 604	.6	.6
505	.5	.5
Turn Signal or Hazard Warning Flasher, Replace (B)		
405	.4	.4
504	.6	.6
505, 604	.5	.5
Turn Signal or Parking Lamp Lens or Bulb, Replace (C)		
1981-91	.3	.3

SEAT BELTS

	LABOR TIME	SEVERE SERVICE
Electronic Control Unit, Replace (B)		
405		
one	.5	.5
both	.8	.8
Lap Belt Assy., Replace (B)		
405	.6	.6
Seat Belt Motor, Replace (B)		
405 one	.5	.5
Seat Belt Assy., Replace (B)		
405 one	.9	.9
Seat Belt Switch, Replace (B)		
405	1.3	1.3

BODY

	LABOR TIME	SEVERE SERVICE
Door Handle (Outside), Replace (B)		
One	.8	.8
Door Latch, Replace (B)		
1981-91	2.0	2.0
Door Lock, Replace (B)		
1981-91	1.0	1.0
Door Window Regulator, Replace (B)		
1981-87		
front one	1.8	2.0
rear one		
exc. 1981-83 505	1.9	2.1
1981-83 505	2.5	2.7
1988-91	1.6	1.8
Hood Release Cable, Replace (B)		
1981-91	.5	.5
Trunk Lock, Replace (B)		
504, 604	.5	.5
Window Regulator Motor, Replace (B)		
1988-91	1.5	1.7

NOTES

POR

911 : Boxster

SYSTEM INDEX

MAINTENANCE POR-2
   Maintenance Schedule .. POR-2
CHARGING POR-2
STARTING POR-2
CRUISE CONTROL POR-2
IGNITION POR-2
EMISSIONS POR-3
FUEL POR-4
EXHAUST POR-5
ENGINE COOLING POR-5
ENGINE POR-5
   Assembly POR-5
   Cylinder Head POR-6
   Camshaft POR-7
   Crank & Pistons POR-7
   Engine Lubrication . POR-8
CLUTCH POR-8
MANUAL TRANSMISSION . POR-8
AUTO TRANSMISSION POR-9
TRANSFER CASE POR-9
SHIFT LINKAGE POR-9
DRIVELINE POR-9
BRAKES POR-10
FRONT SUSPENSION POR-11
REAR SUSPENSION POR-12
STEERING POR-12
HEATING & AC POR-13
WIPERS & SPEEDOMETER POR-14
LAMPS & SWITCHES POR-14
BODY POR-15

OPERATIONS INDEX

A
AC Hoses POR-13
Air Bags POR-12
Air Conditioning POR-13
Alignment POR-11
Alternator (Generator) . POR-2
Antenna POR-14
Anti-Lock Brakes POR-10

B
Back-Up Lamp Switch ... POR-14
Ball Joint POR-11
Battery Cables POR-2
Bleed Brake System POR-10
Blower Motor POR-13
Brake Disc POR-10
Brake Hose POR-10
Brake Pads and/or Shoes .. POR-10

C
Camshaft POR-7
Catalytic Converter ... POR-5
Coolant Temperature (ECT)
   Sensor POR-3

Crankshaft POR-7
Cruise Control POR-2
CV Joint POR-9
Cylinder Head POR-6

D
Differential POR-9
Distributor POR-2
Drive Belt POR-2

E
EGR POR-3
Electronic Control Module
   (ECM/PCM) POR-3
Engine POR-5
Engine Lubrication POR-8
Engine Mounts POR-6
Evaporator POR-13
Exhaust POR-5
Exhaust Manifold POR-5
Expansion Valve POR-13

F
Fuel Injection POR-4
Fuel Pump POR-4
Fuel Vapor Canister ... POR-3

G
Gear Selector Lever ... POR-9
Generator POR-2

H
Halfshaft POR-9
Headlamp POR-14
Heater Core POR-13
Horn POR-14

I
Ignition Coil POR-3
Ignition Module POR-3
Ignition Switch POR-3
Inner Tie Rod POR-12
Intake Air Temperature (IAT)
   Sensor POR-3

K
Knock Sensor POR-3

L
Lower Control Arm POR-11

M
Maintenance Schedule .. POR-2
Mass Air Flow (MAF)
   Sensor POR-3
Master Cylinder POR-10
Muffler POR-5
Multifunction Lever/Switch .. POR-14

O
Oil Pan POR-8
Oil Pump POR-8
Outer Tie Rod POR-12
Oxygen Sensor POR-3

P
Parking Brake POR-11
Pistons POR-8

R
Radiator POR-5
Radio POR-14
Rear Main Oil Seal POR-8

S
Shock Absorber/Strut, Front . POR-11
Shock Absorber/Strut, Rear . POR-12
Spark Plug Cables POR-3
Spark Plugs POR-3
Starter POR-2
Steering Wheel POR-13

T
Throttle Body POR-4
Throttle Position Sensor
   (TPS) POR-4
Timing Belt POR-7
Timing Chain POR-7
Torque Converter POR-9

V
Valve Body POR-9
Valve Cover Gasket POR-6
Valve Job POR-6
Vehicle Speed Sensor .. POR-4

W
Water Pump POR-5
Wheel Balance POR-2
Window Regulator POR-15
Windshield Washer Pump .. POR-14
Windshield Wiper Motor . POR-14

POR-1

POR-2 911 : BOXSTER

	LABOR TIME	SEVERE SERVICE
MAINTENANCE		
Air Cleaner Filter Element, Replace (B)		
Exc. turbo	.3	.5
Turbo	.5	.7
Chassis Lubrication, Change Oil & Filter (C)		
Includes: Correct all fluid levels.		
911	.8	.9
Boxster	.5	.5
Drive Belt, Replace (B)		
Alternator		
911, 911 Turbo		
1981-89 exc. C4	.3	.5
1989-94 C2/4	.7	.9
1995-98 911	.7	.9
1999-01 911	.5	.7
Boxster	1.0	1.2
Compressor		
911, 911 Turbo		
1981-89	.5	.6
1991-92	1.2	1.3
911 Carrera	.8	.9
Fan		
1989-98 911	1.5	1.8
Steering pump		
1991-92 911 Turbo	4.1	4.2
1989-94 911 Carrera 2/4	3.0	3.1
Fuel Filter, Replace (B)		
911		
1981-83	.8	.8
1984-89	.5	.5
911 Carrera 2/4		
1989-94	.9	.9
1995-01	1.3	1.3
911 Turbo		
1981-89	1.1	1.1
1991-92	1.7	1.7
Boxster	1.1	1.1
Halogen Headlamp Bulb, Replace (B)		
911		
1981-94	.7	.7
1995-01	.3	.3
Boxster	.3	.3
Oil & Filter, Change (C)		
Includes: Correct all fluid levels.		
911	.8	.9
Boxster	.5	.5
Tire, Replace (C)		
Includes: Dismount old tire and mount new tire to rim.		
1981-01, one	.5	.5
Tires, Rotate (C)		
1981-01	.5	.5
Wheel, Balance (B)		
1981-01		
one	.3	.3
each addl. add	.2	.2

	LABOR TIME	SEVERE SERVICE
SCHEDULED MAINTENANCE INTERVALS		
If necessary, refer to appropriate Chilton maintenance service information.		
7,500 Mile Service (C)		
911, Boxster	.4	.4
15,000 Mile Service (B)		
911, Boxster	2.5	2.5
22,500 Mile Service (C)		
911, Boxster	.4	.4
30,000 Mile Service (B)		
911, Boxster	6.2	6.2
w/AT add	.5	.5
37,500 Mile Service (C)		
911, Boxster	.4	.4
45,000 Mile Service (B)		
911, Boxster	2.5	2.5
52,500 Mile Service (C)		
911, Boxster	.3	.3
60,000 Mile Service (B)		
911, Boxster	12.9	12.9
w/AT add	.8	.8
67,500 Mile Service (C)		
911, Boxster	.4	.4
75,000 Mile Service (B)		
911, Boxster	2.5	2.5
82,500 Mile Service (C)		
911, Boxster	.3	.3
90,000 Mile Service (B)		
911, Boxster	6.2	6.2
w/AT add	.5	.5
97,500 Mile Service (C)		
911, Boxster	.4	.4
CHARGING		
Alternator Circuits, Test (B)		
Includes: Test component output.		
1981-01	1.0	1.0
Alternator Assy., Replace (B)		
Includes: Pulley transfer.		
1981-89 911	1.9	2.1
1981-89 911 Turbo	4.2	4.2
911, Carrera, C2/4, 993, 996		
1989-98	3.1	3.3
1999-01	2.2	2.4
Boxster	2.2	2.4
R&R fan housing		
81-89 911 add	.5	.5
STARTING		
Starter Draw Test (On Car) (B)		
1981-01	.5	.5
Battery Cables, Replace (C)		
1981-89 911, 911 Turbo		
positive	3.8	4.0
negative	.5	.7

	LABOR TIME	SEVERE SERVICE
Starter Assy., Replace (B)		
911		
1981-89 exc. C4	2.6	2.8
1990-01	1.1	1.3
AT	3.7	3.9
MT	3.5	3.7
Boxster	1.9	2.1
Starter Relay, Replace (B)		
Does not include starter R&R.		
All Models	.3	.3
CRUISE CONTROL		
Diagnose Cruise Control System (A)		
1981-01	.5	.5
Actuator Assy., Replace (B)		
911		
1981-83	.9	.9
1984-89	.8	.8
911 Carrera		
1989-94	3.2	3.2
1995-01		
w/turbo	5.4	5.4
w/o turbo	3.2	3.2
Boxster	2.4	2.4
Control Brake or Clutch Switch, Replace (B)		
1981-89 911	.8	.8
Control Switch, Replace (B)		
1981-89 911	.7	.7
911 Carrera		
1989-94		
w/air bags	1.6	1.6
w/o air bags	1.1	1.1
1995-01		
w/air bags	1.6	1.6
w/o air bags	1.3	1.3
Control Unit, Replace (B)		
1981-89 911	.3	.3
911 Carrera		
1989-94	1.4	1.4
1995-01	1.5	1.5
IGNITION		
Diagnose Ignition System Component Each (A)		
1981-01	.7	.7
Distributor, Replace (B)		
Includes: Reset base ignition timing.		
1981-89 911	.5	.8
911 Turbo		
1981-89	1.9	2.2
1991-92	1.5	1.8
911 Carrera		
1989-94	1.7	2.0
1995-01	1.5	1.8
Distributor Cap and/or Rotor, Replace (B)		
911, 911 Turbo		
1981-89	.5	.5
1991-92	1.8	1.8

911 : BOXSTER — POR-3

	LABOR TIME	SEVERE SERVICE
1993-94		
one	.7	.7
both	.9	.9
911 Carrera		
1989-94		
one	.7	.7
both	.9	.9
1995-01		
one	.7	.7
both	.9	.9

Distributor Points & Condenser, Replace (B)
- 1981-83 911 8   *1.0*

Ignition Coil, Replace (B)
- 911, 911 Turbo
  - 1981-89 5   *.5*
- 1991-92
  - one 1.6   *1.6*
  - both 1.9   *1.9*
- 911 RS
  - one 1.7   *1.7*
  - both 1.9   *1.9*
- 911 Carrera
  - 1989-94
    - one 1.6   *1.6*
    - both 1.9   *1.9*
  - 1995-01
    - one 7   *.7*
    - both 1.8   *1.8*
- Boxster
  - one 7   *.7*
  - all 1.2   *1.2*

Ignition Module, Reset (B)
- 1995-01 911 Carrera (993) 1.7   *1.7*

Ignition Switch, Replace (B)
- 911, 911 Turbo
  - 1981-89 .... 1.5   *1.5*
  - 1991-92 .... 3.0   *3.0*
- 911 Carrera
  - 1989-94
    - w/air bags 3.1   *3.1*
    - w/o air bags 1.6   *1.6*
  - 1995-01
    - w/air bags 3.2   *3.2*
    - w/o air bags 1.6   *1.6*

Ignition Timing, Reset (B)
- 1981-89 911, 911 Turbo 6   *.6*

Spark Plug (Ignition) Cables, Replace (B)
- 1981-89 911 exc. C4 .... 1.6   *1.9*
- 911 Turbo
  - 1981-89 .... 4.1   *4.4*
  - 1991-92 .... 5.3   *5.6*
- 911 Carrera
  - 1990-98
    - w/turbo 6.4   *6.7*
    - w/o turbo 3.8   *4.1*

Spark Plugs, Replace (B)
- 1981-89 911 exc. C4 .... 1.2   *1.5*
- 911 Turbo
  - 1981-89 .... 3.8   *4.1*
  - 1991-92 .... 3.2   *3.5*
- 1990-01 911 Carrera
  - w/turbo 5.2   *5.5*
  - w/o turbo 4.2   *4.5*
- Boxster 1.7   *2.0*

Vacuum Advance Unit, Replace (B)
Includes: Reset timing.
- 1981-83 911 .... 1.1   *1.1*

EMISSIONS

The following operations do not include testing. Add time as required.

Dynamometer Test (A)
- 1981-01 5   *.5*

Air Flow Sensor, Replace (B)
- 1991-92 911 Turbo 4.2   *4.2*

Air Injection Check Valve, Replace (B)
- 1986-89 911 Turbo
  - right 3.1   *3.1*
  - left 8   *.8*
- 1995-01 911 Carrera 1.4   *1.4*

Air Injection Pump Relay, Replace (B)
- Boxster 3   *.3*

Air Pump Assy., Replace (B)
- 1981-89 911 5   *.7*
- 1981-89 911 Turbo 2.9   *3.1*
- 1995-01 911 Carrera w/turbo 5.8   *6.0*

Coolant Temperature (ECT) Sensor, Replace (B)
- 1981-89 911
  - exc. C4 .... 1.2   *1.2*
  - w/turbo 2.5   *2.5*
- 1990-98
  - w/turbo 2.2   *2.2*
  - w/o turbo 2.0   *2.0*
- 1999-01
  - 911 .... 1.2   *1.2*
  - Boxster 1.2   *1.2*

Diverter and/or Switching Valve, Replace (B)
- 1986-89 911 Turbo
  - right 2.6   *2.8*
  - left 1.7   *1.9*

EGR Control Valve, Replace (B)
- 1981-94 911 .... 1.3   *1.5*

EGR Filter, Replace (B)
- 1981-94 911 8   *.8*

Electronic Control Unit, Replace (B)
- 911, 911 Turbo 5   *.5*
- 911 RS 7   *.7*

- 911 Carrera
  - 1989-94 7   *.7*
  - 1995-01 .... 1.7   *1.7*
- Boxster 7   *.7*

Electronic Ignition Control Unit, Replace (B)
- 1991-92 911 Turbo 1.1

Electronic Spark Control Unit, Replace (B)
- 1984-93 911 7   *.7*

Fuel Vapor Canister, Replace (B)
- 1989-01
  - 911 9   *.9*
  - Boxster 1.2   *1.2*

Intake Air Temperature (IAT) Sensor, Replace (B)
- 1991-92 911 Turbo 1.6   *1.6*

Knock Sensor, Replace (B)
- 911 Carrera
  - 1989-94
    - one 4.4   *4.7*
    - both 4.7   *5.0*
  - 1995-01
    - one 4.5   *4.8*
    - both 4.9   *5.2*
- Boxster
  - one 1.8   *2.1*
  - both 2.1   *2.4*

Mass Air Flow (MAF) Sensor, Replace (B)
- 1995-01 911 Carrera 1.4   *1.4*
- Boxster 3   *.3*

Oxygen Sensor, Replace (B)
- 1984-89 911 7   *.9*
- 911 Turbo
  - 1981-89 5   *.7*
  - 1991-92 8   *1.0*
- 911 Carrera
  - 1989-94 .... 1.8   *2.0*
  - 1995-01
    - w/turbo
      - one 9   *1.1*
      - all 6.3   *6.5*
    - w/o turbo
      - one 9   *1.1*
      - all 2.4   *2.6*
- Boxster
  - one 3   *.5*
  - all 8   *1.0*

Pulse Generator, Replace (B)
- 1981-89 911
  - one 1.5   *1.5*
  - both 1.8   *1.8*
- 1991-92 911 Turbo 2.1   *2.1*
- 911 Carrera
  - 1989-94 .... 2.1   *2.1*
  - 1995-01
    - w/air bags 4.1   *4.1*
    - w/o air bags 2.0   *2.0*

POR-4　911 : BOXSTER

	Labor Time	Severe Service
Throttle Position Sensor (TPS), Replace (B)		
1995 911 Carrera		
w/turbo	1.7	1.7
w/o turbo	.8	.8
Boxster	1.6	1.6
Vapor Canister Filter, Replace (B)		
911	.5	.5
Vehicle Speed Sensor, Replace (B)		
1981-01	1.5	1.5

FUEL
DELIVERY

	Labor Time	Severe Service
Fuel Pressure, Check (B)		
Exc. below	.9	.9
1991-92 911 Turbo	3.7	3.7
Boxster	.7	.7
Fuel Pump, Test (B)		
911	1.8	1.8
Boxster	.9	.9
Fuel Accumulator, Replace (B)		
1981-89 911	.7	.7
Fuel Filter, Replace (B)		
911		
1981-83	.8	.8
1984-89	.5	.5
911 Turbo		
1981-89	1.1	1.1
1991-92	1.7	1.7
911 Carrera		
1989-94	1.1	1.1
1995-01	1.3	1.3
Boxster	1.1	1.1
Fuel Gauge (Dash), Replace (B)		
911	.3	.3
Fuel Gauge (Tank), Replace (B)		
Exc. below	.3	.5
1981-89 911, 911 Turbo	.5	.7
Boxster	1.8	2.0
Fuel Pump, Replace (B)		
911	1.6	1.8
Boxster	2.3	2.5
Fuel Pump Relay, Replace (B)		
1981-89 911, 911 Turbo	.3	.3
Boxster	.5	.5
Fuel Tank, Replace (B)		
Includes: Drain and refill.		
911, 911 Turbo		
1981-89	2.6	2.8
1991-92	3.2	3.4
911 Carrera		
1989-94	3.2	3.4
1995-01	3.1	3.3

INJECTION

	Labor Time	Severe Service
Air Distributor Housing, R&I (B)		
911		
1981-83	5.2	5.2
1984-89	5.9	5.9
911 Turbo	6.8	6.8
911 Carrera		
1989-94		
one side	3.5	3.5
both sides	5.4	5.4
1995-01		
one side	3.5	3.5
both sides	5.2	5.2
Replace housing		
911 Turbo add	2.2	2.2
Air Flow Meter, Replace (B)		
1984-89 911	1.4	1.4
Auxiliary Air Regulator, Replace (B)		
1981-83 911	.7	.7
1981-89 911 Turbo	.7	.7
Auxiliary Air Valve, Replace (B)		
1981-83 911	.5	.5
1981-89 911 Turbo	1.6	1.6
Cold Start Valve, Replace (B)		
1981-83 911	1.3	1.3
1981-89 911 Turbo	1.3	1.3
Enrichment Switch, Replace (B)		
1991-92 911 Turbo	.5	.5
Fuel Distributor, Replace (B)		
1981-83 911	1.1	1.3
911 Turbo		
1981-89	1.6	1.8
1991-92	3.4	3.6
Fuel Injection Lines, Replace (B)		
1981-89 911 Turbo		
one	.7	.7
all	1.3	1.3
Fuel Injectors, Replace (B)		
911		
1981-83		
one	.3	.5
all	1.3	1.5
1984-89		
one side	1.6	1.8
both sides	2.3	2.5
911 Turbo		
1981-89		
one	.5	.7
all	1.2	1.4
1991-92 all	5.0	5.2
911 Carrera		
1989-94		
one side	1.9	2.1
both sides	3.0	3.2
1995-01		
right side	1.5	1.7
left side	1.9	2.1
both sides	2.6	2.8

	Labor Time	Severe Service
Fuel Pressure Regulator, Replace (B)		
1984-89 911	6.1	6.3
911 Turbo		
1981-89	.5	.7
1991-92	1.7	1.9
911 Carrera		
1989-94	3.2	3.4
1995-01		
w/turbo	2.5	2.7
w/o turbo	1.5	1.7
Boxster	.3	.5
Fuel Pressure Damper, Replace (B)		
1984-89 911	1.2	1.4
Fuel Rail, Replace (B)		
911 RS		
one side	1.9	2.1
both sides	2.6	2.8
1989-94 911 Carrera 2/4		
one side	1.9	2.1
both sides	2.8	3.0
Intake Pipe, Replace (B)		
1981-83 911		
one	.7	.7
all	4.3	4.3
Microswitch, Replace (B)		
911		
1981-83	.5	.5
1984-89	.3	.3
1981-89 911 Turbo	.5	.5
Mixture Regulator, Replace (B)		
1981-83 911	1.8	1.8
911 Turbo		
1981-89	1.7	1.7
1991-92	3.8	3.8
Pneumatic Valve, Replace (B)		
1981-83 911	.5	.5
1981-89 911 Turbo	.5	.5
Thermal Valve, Replace (B)		
1981-89 911 Turbo	.8	.8
Thermo Time Switch, Replace (B)		
1981-83 911	.7	.7
1981-89 911 Turbo	.5	.5
Throttle Body Assy., Replace (B)		
1984-89 911	1.8	1.8
911 Turbo		
1981-89	1.7	1.7
1991-92	2.4	2.4
911 Carrera		
1989-94	3.1	3.1
1995-01		
w/turbo	5.6	5.6
w/o turbo	2.7	2.7
Boxster	1.5	1.5
Throttle Regulating Valve, Replace (B)		
1981-83 911	.7	.7
1981-89 911 Turbo	.7	.7
Throttle Valve Switch, Replace (B)		
1984-89 911	1.7	1.7

911 : BOXSTER POR-5

	LABOR TIME	SEVERE SERVICE
Timing Valve, Replace (B)		
1991-92 911 Turbo	1.9	1.9
Vacuum Switch, Replace (B)		
1989-94 911		
Carrera 2/4	.5	.5
Boxster	.5	.5
Voltage Supply Relay, Replace (B)		
1984-97 911	.3	.3
Warm-Up Control Valve, Replace (B)		
1981-89 911 Turbo	.8	.8
Warm-Up Regulator, Replace (B)		
1981-89 911 Turbo	.7	.7

TURBOCHARGER

	LABOR TIME	SEVERE SERVICE
Intercooler Radiator, Replace (B)		
911		
1981-89	.7	1.0
1991-97	1.4	1.7
Turbocharger Assy., R&R or Replace (B)		
1981-89	6.8	7.3
1991-94	6.1	6.6
1996-97		
right side	9.2	9.7
left side	10.7	11.2
both sides	13.4	13.9
Wastegate Actuator, Replace (B)		
1981-89	2.4	2.7
1991-94	1.7	2.0

EXHAUST

	LABOR TIME	SEVERE SERVICE
Catalytic Converter, Replace (B)		
1981-89 911	2.2	2.3
911 Turbo		
1981-89	1.8	1.9
1991-92	2.6	2.7
911 Carrera		
1989-94	2.5	2.6
1995-01		
w/turbo		
one side	6.9	7.0
both sides	8.6	8.7
w/o turbo	4.1	4.2
Boxster		
one	1.7	1.8
both	2.5	2.6
Center Muffler, Replace (B)		
1989-94 911		
Carrera 2/4	.9	1.0
Exhaust Manifold or Gasket, Replace (B)		
Boxster		
one	1.3	1.8
both	1.8	2.3
Exhaust System (Complete), Replace (B)		
1981-89 911	2.8	3.3
1989-94 911		
Carrera 2/4	3.4	3.9

	LABOR TIME	SEVERE SERVICE
911 Carrera		
1989-94	3.4	3.9
1995-01		
w/turbo	9.2	9.7
w/o turbo	4.3	4.8
Boxster	3.1	3.6
Front Muffler, Replace (B)		
1981-89 911	1.6	1.7
1989-94 911		
Carrera 2/4	1.2	1.3
Heat Exchanger and/or Gaskets, Replace (B)		
1981-89 exc. C4		
one	2.2	2.7
both	4.0	4.5
1990-01		
exc. below		
one side	2.5	3.0
both	4.5	5.0
1995-01 turbo		
right side	9.9	10.4
left side	10.8	11.3
both sides	14.5	15.0
Rear Muffler, Replace (B)		
911		
1981-83	.9	1.0
1984-89	1.6	1.7
911 Turbo		
1986-89	2.4	2.5
1991-92	1.3	1.4
911 Carrera		
1989-94	1.2	1.3
1995-01		
one	1.5	1.6
both	2.4	2.5
Boxster	2.2	2.3
Tail Pipe, Replace (B)		
1991-92 911 Turbo	.3	.4
911 Carrera		
1989-94	.3	.4
1995-01		
one	.7	.8
both	.9	1.0

ENGINE COOLING

	LABOR TIME	SEVERE SERVICE
Auxiliary Fan Temperature Switch, Replace (B)		
1981-89 911, 911 Turbo	.3	.3
Auxiliary Heater Blower Motor, Replace (B)		
1981-89 911, 911 Turbo	.9	.9
Auxiliary Heater Relay, Replace (B)		
1981-89 911, 911 Turbo	.3	.3

	LABOR TIME	SEVERE SERVICE
Cooling Fan Assy., Replace (B)		
1991-92 911 Turbo	2.8	3.3
911 Carrera		
1989-94	1.1	1.5
1995-01	1.5	1.9
Electric Cooling Fan Motor, Replace (B)		
Boxster		
one side	1.6	2.1
both sides	2.8	3.3
Electric Fan Thermo Switch, Replace (B)		
Boxster	.8	.8
Fan Motor Resistor, Replace (B)		
Boxster		
one	2.9	3.1
both	3.5	3.7
Radiator Assy., Replace (B)		
Boxster		
one	4.2	4.4
both	5.7	5.9
Radiator Fan Motor Relay, Replace (B)		
Boxster	.5	.5
Silencers, Replace (B)		
1991-92 911 Turbo		
one	4.4	4.4
both	8.1	8.1
1989-94 911 Carrera 2/4		
one	4.5	4.5
both	8.3	8.3
Temperature Gauge (Dash), Replace (B)		
1981-01 911	.3	.3
Water Pump and/or Gasket, Replace (B)		
Includes: Refill with proper coolant mix.		
Boxster	3.6	4.1

ENGINE
ASSEMBLY

Times shown are for OEM assemblies. Time to replace assemblies from aftermarket rebuilders may vary.

Engine Assy., R&I (B)
Does not include parts or component transfer.

	LABOR TIME	SEVERE SERVICE
1981-89 911		
exc. C4	7.0	7.0
911 Turbo		
1986-89	8.7	8.7
1991-92	10.7	10.7
911 Carrera		
1989-94 Carrera 4	11.5	11.5
1990-94 Carrera 2	10.0	10.0
1995-01		
2WD	13.0	13.0
4WD	13.1	13.1
turbo	15.6	15.6
Boxster	14.1	14.1

POR-6 911 : BOXSTER

	LABOR TIME	SEVERE SERVICE

Engine Assy., R&R and Recondition (A)
Includes: Replacing rings, rod and main bearings, cylinder head reconditioning and engine tune-up.
  911
    1981-83 911 ...... 60.0   *60.0*
    1984-89 911 ...... 70.0   *70.0*
    1991-92 911 Turbo 80.0   *80.0*
  911 Carrera
    1989-94 Carrera 4 . 71.5   *71.5*
    1990-94 Carrera 2 . 68.3   *68.3*
    1995-01
      2WD 41.9   *41.9*
      4WD 42.5   *42.5*

Engine Assy., Replace (B)
Includes: Engine R&R, parts transfer, tune-up.
  1981-89 exc. C4 ..... 16.7   *16.7*
  1990-94 .... 18.5   *18.5*
    4WD 24.4   *24.4*
    turbo 31.3   *31.3*
  1995-98 911 Carrera
    2WD 23.8   *23.8*
    4WD 24.4   *24.4*
    turbo 31.3   *31.3*

Engine Mounts, Replace (B)
  911
    one 9   *1.1*
    both 1.3   *1.5*

CYLINDER HEAD

Compression Test (B)
  1981-89 exc. C4 ..... 1.0   *1.3*
  1990-98 911 .... 1.8   *2.1*
  1999-01 911 .... 1.9   *2.2*
  Boxster 1.9   *2.2*
  w/turbo add 8   *1.0*

Valves, Adjust (B)
  1981-89 911 exc. C4 .. 6.0   *6.5*
  911 Turbo
    1981-92 .... 10.0   *10.5*
  1989-94 911
    Carrera 2/4 .... 7.8   *8.3*

Camshaft Housing Cover Gasket, Replace (B)
  1981-89 911
    one 7   *1.0*
    two 1.3   *1.6*
    four 1.8   *2.1*
  911 Turbo
    1981-89
      upper
        one 2.0   *2.3*
        two 4.1   *4.4*
      lower
        one 1.5   *1.8*
        two 1.9   *2.2*
        all 5.9   *6.2*

  1991-92
    top
      right 2.1   *2.4*
      left 2.4   *2.7*
    bottom
      right 1.9   *2.2*
      left 2.4   *2.7*
  911 Carrera
    1989-94
      top
        right 1.3   *1.6*
        left 1.6   *1.9*
      bottom
        right 2.4   *2.7*
        left 3.1   *3.4*
    1995-98
      w/turbo
        top
          right 9.1   *9.4*
          left 4.5   *4.8*
        bottom
          right 9.9   *10.2*
          left 11.7   *12.0*
      w/o turbo
        top
          right 2.2   *2.5*
          left 2.8   *3.1*
        bottom each 2.3   *2.6*

Cylinder Head, Replace (B)
Includes: Parts transfer, adjustments.
  911
    1981-83
      one side 21.6   *22.1*
      both sides 29.2   *29.7*
    1984-89
      one side 20.8   *21.3*
      both sides 29.7   *30.2*
  911 Turbo
    1986-89
      right side 27.7   *28.2*
      left side 29.8   *30.3*
      both sides 39.2   *39.7*
    1991-92
      one side 34.0   *34.5*
      both sides 44.3   *44.8*
  911 Carrera
    1989-94 Carrera 4
      one side 30.0   *30.5*
      both sides 38.3   *38.5*
    1990-94 Carrera 2
      one side 27.8   *28.3*
      both sides 35.6   *36.1*
    1995-01
      w/turbo
        one sides 31.3   *31.8*
        both sides ... 39.7   *40.2*
      w/o turbo
        one sides 26.2   *26.7*
        both sides ... 33.7   *34.2*

Cylinder Head Gasket, Replace (B)
  911
    1981-83
      one side 18.3   *18.6*
      both sides 22.4   *22.7*
    1984-89
      one side 19.9   *20.2*
      both sides 24.1   *24.4*
  911 Turbo
    1986-89
      right side 26.9   *27.2*
      left side 29.0   *29.3*
      both sides 35.8   *36.1*
    1991-92
      one side 30.7   *31.0*
      both sides 38.6   *38.9*
  911 Carrera
    1989-94 Carrera 4
      one side 26.5   *26.8*
      both sides 31.4   *31.7*
    1990-94 Carrera 2
      one side 23.8   *24.1*
      both sides 28.9   *29.2*
    1995-01
      w/turbo
        one side 30.0   *30.3*
        both sides ... 37.7   *38.0*
      w/o turbo
        one sides 25.4   *25.7*
        both sides ... 31.9   *32.2*

Lower Rocker Arm Shaft (One), Replace (B)
  1981-89 911 exc. C4 .. 1.6   *1.9*
  each addl. add 6   *.8*
  1990-98 911
    intake 1.6   *1.9*
    exhaust 2.2   *2.5*
    each addl. add 7   *.9*

Valve Job (A)
  911
    1981-83
      one side 22.4   *22.9*
      both sides 29.4   *29.9*
    1984-89
      one side 22.3   *22.8*
      both sides 31.4   *31.9*
  911 Turbo
    1986-89
      right side 29.7   *30.2*
      left side 31.6   *32.1*
      both sides 41.8   *42.3*
    1991-92
      one side 35.2   *35.7*
      both sides 45.1   *45.6*
  911 Carrera
    1989-94 Carrera 4
      one side 31.5   *32.0*
      both sides 39.3   *39.8*

911 : BOXSTER — POR-7

	LABOR TIME	SEVERE SERVICE
1990-94 Carrera 2		
one side	28.7	29.2
both sides	37.0	37.5
1995-01		
w/turbo		
one side	32.7	33.2
both sides	43.0	43.5
w/o turbo		
one side	27.2	27.7
both sides	37.0	37.5

Valve Lifters, Replace (B)

	LABOR TIME	SEVERE SERVICE
1995-98 911 Carrera		
w/turbo		
right side top	9.4	9.9
left side top	4.3	4.8
right side bottom	10.3	10.8
left side bottom	12.3	12.8
w/o turbo		
right side top	3.1	3.6
left side top	3.1	3.6
right side bottom	2.4	2.9
left side bottom	2.6	3.1
each addl. add	.3	.4

CAMSHAFT

Camshaft, Replace (A)

	LABOR TIME	SEVERE SERVICE
911		
1981-83		
one	12.2	12.6
both	14.7	15.1
1984-89		
one side	13.8	14.2
both sides	16.4	16.8
911 Turbo		
1986-89		
right side	16.3	16.7
left side	19.2	19.6
both sides	22.8	23.2
1991-92		
one side	21.3	21.7
both sides	25.5	25.4
911 Carrera		
1989-94 Carrera 4		
one side	19.8	20.2
both sides	23.0	23.4
1990-94 Carrera 2		
one side	17.6	18.0
both sides	20.2	20.6
1995-01		
w/turbo		
one side	26.5	26.9
both sides	33.4	33.8
w/o turbo		
one side	21.2	21.6
both sides	24.4	24.8
Boxster		
one side	19.1	19.5
both sides	21.5	21.9

Camshaft Housing, Replace (B)

	LABOR TIME	SEVERE SERVICE
911		
1981-83		
one side	13.8	14.3
both sides	17.1	17.6
1984-89		
one side	15.4	15.9
both sides	18.8	19.3
911 Turbo		
1986-89		
right side	17.6	18.1
left side	20.2	20.7
both sides	25.4	25.9
1991-92		
one side	22.7	23.2
both sides	28.4	28.9
911 Carrera		
1989-94 Carrera 4		
one side	21.9	22.4
both sides	25.7	26.2
1990-94 Carrera 2		
one side	19.4	19.9
both sides	23.1	23.6
1995-01		
w/turbo		
one side	27.9	28.4
both sides	33.2	33.7
w/o turbo		
one side	22.8	23.3
both sides	27.6	28.1

Intermediate Shaft, Replace (B)

	LABOR TIME	SEVERE SERVICE
1981-89 911	45.1	45.6

Timing Belt Tensioner, Replace (B)

	LABOR TIME	SEVERE SERVICE
Boxster	5.1	5.6

Timing Chain Cover and/or Gasket, Replace (B)

	LABOR TIME	SEVERE SERVICE
1981-89 911		
one	2.4	2.7
both	2.8	3.1
911 Turbo		
1986-89		
right side	12.2	12.5
left side	13.1	13.4
both sides	13.6	13.9
1991-92		
one	15.1	15.4
both	15.9	16.2
911 Carrera		
1989-94 Carrera 4		
one	4.5	4.8
both	6.0	6.3
1990-94 Carrera 2		
one	4.5	4.8
both	7.0	7.3
1995-01		
w/turbo		
right side	18.8	19.1
left side	17.0	17.3
w/o turbo		
right side	6.9	7.2
left side	5.8	6.1

Timing Chain Tensioner, Replace (B)

	LABOR TIME	SEVERE SERVICE
1981-89 911		
one	4.1	4.6
both	5.1	5.6
911 Turbo		
1986-89		
right side	12.9	13.4
left side	13.4	13.9
both sides	15.0	15.5
1991-92		
one	15.9	16.4
both	17.1	17.6
911 Carrera		
1989-94		
one	1.4	1.9
both	1.7	2.2
1995-01		
w/turbo		
right side	2.1	2.6
left side	1.3	1.8
w/o turbo		
each side	1.3	1.8

CRANK & PISTONS

Connecting Rod Bearings, Replace (A)

Includes: Check bearing oil clearance.

	LABOR TIME	SEVERE SERVICE
911		
1981-83	55.1	56.0
1984-89	65.1	66.0

Crankshaft and Main Bearings, Replace (A)

Includes: Engine R&R.

	LABOR TIME	SEVERE SERVICE
911		
1981-83	55.1	56.0
1984-89	65.1	66.0
1991-92 911 Turbo	59.8	60.7
911 Carrera		
1989-94 Carrera 4	51.6	52.5
1990-94 Carrera 2	49.3	50.2

Crankshaft Front Oil Seal, Replace (A)

	LABOR TIME	SEVERE SERVICE
1981-89 911	4.3	4.8
911 Carrera		
1989-94	5.0	5.5
1995-01		
w/turbo	18.2	18.7
w/o turbo	6.9	7.4

Crankshaft Pulley, Replace (B)

	LABOR TIME	SEVERE SERVICE
1981-89 911	4.3	4.5
1991-92 911 Turbo	8.8	9.0
911 Carrera		
1989-94		
w/AC	4.4	4.6
w/o AC	4.1	4.3
1995-01		
w/turbo	18.0	18.2
w/o turbo		
w/AC	6.4	6.6
w/o AC	6.2	6.4

POR-8 911 : BOXSTER

	LABOR TIME	SEVERE SERVICE
Pistons & Cylinder Assy., Replace (B)		
Includes: Engine and Cylinder head R&R.		
1981-89 911		
one side	25.3	*26.2*
both sides	30.0	*30.9*
1991-92 911 Turbo		
one side	31.8	*32.7*
both sides	39.7	*40.6*
911 Carrera		
1989-94 Carrera 4		
one side	27.3	*28.2*
both sides	33.0	*33.9*
1990-94 Carrera 2		
one side	24.6	*25.5*
both sides	30.1	*31.0*
1995-01		
w/turbo		
one side	33.6	*34.5*
both sides	40.8	*41.7*
w/o turbo		
one side	28.1	*29.0*
both sides	34.8	*35.7*
Rear Main Oil Seal, Replace (A)		
Includes: Engine & Trans. R&R.		
911		
1981-89 exc. C4	8.4	*9.0*
1990-94		
C2	11.2	*11.8*
C4	12.7	*13.3*
1995-01		
2WD	14.2	*14.8*
4WD	14.3	*14.9*
turbo	16.8	*17.3*
Boxster		
AT	10.5	*11.0*
MT	7.8	*8.3*
w/turbo 81-89 911 add	1.7	*2.0*
Rings, Replace (A)		
911		
1981-83		
one side	30.0	*30.9*
both sides	35.3	*36.2*
1984-89		
one side	35.2	*36.1*
both sides	45.1	*46.0*
911 Turbo		
1986-89		
right side	41.5	*42.4*
left side	44.5	*45.4*
both sides	56.6	*57.5*
1991-92		
one side	48.0	*48.9*
both sides	60.3	*61.2*
911 Carrera		
1989-94 Carrera 4		
one side	40.9	*41.8*
both sides	49.8	*50.7*

	LABOR TIME	SEVERE SERVICE
1990-94 Carrera 2		
one side	37.6	*38.5*
both sides	46.4	*47.3*
1995-01		
w/turbo		
one side	32.4	*33.3*
both sides	39.8	*40.7*
w/o turbo		
one side	26.8	*27.7*
both sides	33.7	*34.6*
Vibration Damper, Replace (B)		
1981-89 911	.3	*.6*

ENGINE LUBRICATION
	LABOR TIME	SEVERE SERVICE
Auxiliary Oil Cooler, Replace (B)		
1984-89 911	3.0	*3.3*
911 Turbo		
1981-89	2.9	*3.2*
1991-92	3.0	*3.3*
Combination Gauge Assy., Replace (B)		
1981-01	.3	*.3*
Engine Oil Cooler, Replace (B)		
1981-89 911	1.4	*1.6*
1990-97 911	3.2	*3.4*
1995-98 911 Carrera (993)		
w/turbo	4.5	*4.7*
w/o turbo	3.3	*3.5*
Boxster	1.9	*2.1*
Engine Oil Pressure Switch (Sending Unit), Replace (B)		
1981-01	.8	*.8*
Oil Pan and/or Gasket, Replace (B)		
Boxster	1.9	*2.2*
Oil Pump & Gaskets, Replace (B)		
1981-89 911	54.2	*54.7*
1991-92 911 Turbo	45.1	*45.6*
911 Carrera		
1989-94 Carrera 4	38.3	*38.8*
1990-94 Carrera 2	29.8	*30.3*
1995-01		
w/turbo	46.3	*46.8*
w/o turbo	39.8	*40.3*

CLUTCH
	LABOR TIME	SEVERE SERVICE
Bleed Clutch Hydraulic System (B)		
1987-01 911	.3	*.3*
Boxster	.7	*.7*
Clutch Pedal Free Play, Adjust (B)		
1981-86 911	.5	*.7*
1981-88 911 Turbo	.5	*.7*
Clutch Assy., Replace (B)		
911		
1981-89 exc. C4	8.0	*8.2*
1990-94		
C2	11.0	*11.2*
C4	12.5	*12.7*

	LABOR TIME	SEVERE SERVICE
1995-01		
2WD	14.0	*14.2*
4WD	14.5	*14.7*
turbo		
2WD	16.6	*16.8*
4WD	17.6	*17.8*
Boxster	7.0	*7.2*
w/turbo 81-89 911 add	1.7	*1.7*
Replace flywheel add	.5	*.7*
Clutch Cable, Replace (B)		
1981-86 911	1.1	*1.3*
1981-88 911 Turbo	1.2	*1.4*
Clutch Master Cylinder, Replace (B)		
Includes: System bleeding.		
1987-89 911	3.0	*3.2*
Clutch Release Bearing, Replace (B)		
911		
1981-83	6.3	*6.5*
1984-86	7.6	*7.8*
1987-89	8.2	*8.4*
911 Turbo		
1981-89	9.7	*9.9*
1991-92	12.1	*12.3*
911 Carrera		
1989-94 Carrera 4		
w/AC	13.5	*13.7*
w/o AC	12.6	*12.8*
1990-94 Carrera 2		
w/AC	10.8	*11.0*
w/o AC	9.9	*10.1*
1995-01		
w/AC		
2WD	15.2	*15.4*
4WD	17.1	*17.3*
w/o AC		
2WD	14.5	*14.7*
4WD	15.2	*15.4*
Boxster	6.6	*6.8*
Clutch Slave Cylinder, Replace (B)		
Includes: System bleeding.		
1987-89 911	1.8	*2.0*
1989 911 Turbo	1.8	*2.0*
1990-01 911 Carrera		
w/turbo	3.0	*3.2*
w/o turbo	2.5	*2.7*
Boxster	1.4	*1.6*

MANUAL TRANSMISSION
	LABOR TIME	SEVERE SERVICE
Transmission Assy., R&I (B)		
1981-89 911 exc. C4	7.0	*7.2*
911 Turbo		
1986-89	8.7	*8.9*
1991-92	10.7	*10.9*
911 Carrera		
1989-94 Carrera 4	11.5	*11.7*
1990-94 Carrera 2	10.0	*10.2*

	LABOR TIME	SEVERE SERVICE
1995-01		
2WD	13.0	*13.2*
4WD	13.1	*13.3*
turbo	15.6	*15.8*
Boxster	14.1	*14.3*
Transmission Assy., R&R and Recondition (A)		
1981-86 911	14.0	*14.6*
911 Turbo		
1981-89	17.6	*18.2*
1991-92	19.9	*20.5*
911 Carrera		
1989-94 Carrera 4		
w/AC	23.8	*24.4*
w/o AC	23.0	*23.6*
1990-94 Carrera 2		
w/AC	19.2	*19.8*
w/o AC	18.2	*18.8*
1995-01		
w/AC		
2WD	26.5	*27.1*
4WD	29.8	*30.4*
w/o AC		
2WD	26.0	*26.6*
4WD	27.8	*28.4*
Boxster	22.7	*23.3*
Transmission Mount, Replace (B)		
911		
1981-89	1.1	*1.3*
1990-94	.5	*.7*
1995-97	.8	*1.0*
1997 Boxster		
one	.8	*1.0*
both	.9	*1.1*

AUTOMATIC TRANSMISSION

SERVICE TRANSMISSION INSTALLED

	LABOR TIME	SEVERE SERVICE
Check Unit for Oil Leaks (C)		
1990-01	.9	*.9*
Drain and Refill Unit (B)		
1990-01 911	2.2	*2.2*
Boxster	1.5	*1.5*
Electronic Control Unit, Replace (B)		
1990-94	.7	*.7*
1995-01	1.7	*1.7*
Pulse Generator, Replace (B)		
1990-94 911	2.8	*3.0*
Transmission Auxiliary Oil Cooler, Replace (B)		
911		
1990-94	3.1	*3.3*
1995-01	2.8	*3.0*
Boxster	2.6	*2.8*
Transmission Mount, Replace (B)		
Boxster		
one	1.6	*1.8*
both	1.7	*1.9*

	LABOR TIME	SEVERE SERVICE
Valve Body Assy., Replace (B)		
1990-01 911	2.8	*3.3*
Boxster	3.7	*4.2*

SERVICE TRANSMISSION REMOVED

Transmission R&R included unless otherwise noted.

	LABOR TIME	SEVERE SERVICE
Front Oil Pump, Replace (B)		
1990-94 911 Carrera 2		
w/AC	18.5	*18.7*
w/o AC	17.8	*18.0*
1995-01 911 Carrera		
w/AC	18.8	*19.0*
w/o AC	18.2	*18.4*
Torque Converter, Replace (B)		
1990-94 911 Carrera 2		
w/AC	15.4	*15.6*
w/o AC	14.4	*14.6*
1995-01 911 Carrera		
w/AC	15.9	*16.1*
w/o AC	15.2	*15.4*
Boxster	12.4	*12.6*
Replace front seal add	.2	*.2*
Transmission Assy., R&I (B)		
1981-89 911 exc. C4	7.0	*7.2*
911 Turbo		
1986-89	8.7	*8.9*
1991-92	10.7	*10.9*
911 Carrera		
1989-94 Carrera 4	11.5	*11.7*
1990-94 Carrera 2	10.0	*10.2*
1995-01		
2WD	13.0	*13.2*
4WD	13.1	*13.3*
turbo	15.6	*15.8*
Boxster	14.1	*14.3*
Transmission Assy., R&R and Recondition (A)		
1995-01 911 Carrera		
w/AC	23.1	*24.0*
w/o AC	22.9	*23.8*

TRANSFER CASE

	LABOR TIME	SEVERE SERVICE
Transfer Case, R&I (B)		
1989-94 911 Carrera 4		
w/AC	16.5	*17.0*
w/o AC	15.8	*16.3*

SHIFT LINKAGE

AUTOMATIC TRANSMISSION

	LABOR TIME	SEVERE SERVICE
Gear Selector Lever, Replace (B)		
1990-01 911	3.2	*3.2*
Gear Shift Housing, Replace (B)		
1990-01 911	2.8	*2.8*
Selector Handle, Replace (B)		
1990-01 911	.2	*.2*
Shift Control Cable, Replace (B)		
1990-01 911	1.7	*1.9*

	LABOR TIME	SEVERE SERVICE
Shifter Lock Cable, Replace (B)		
1990-01 911	3.4	*3.6*

MANUAL TRANSMISSION

	LABOR TIME	SEVERE SERVICE
Gearshift Linkage, Adjust (B)		
1981-01	.7	*.9*
Gearshift Rod, Replace (B)		
1981-88 911	1.5	*1.7*
Shift Lever, Replace (B)		
911		
1981-88	1.6	*1.6*
1989-01	2.7	*2.7*
Transmission Control Linkage Assy., Replace (B)		
1989-01 911	2.3	*2.3*

DRIVELINE

	LABOR TIME	SEVERE SERVICE
Axle Flange Oil Seal, Replace (B)		
1981-97 911, one	1.8	*2.0*
Boxster	1.3	*1.5*
Differential Side Cover, Reseal (B)		
1981-94 911	1.8	*2.1*
Final Drive Assy., R&R or Replace (B)		
1989-94 911 Carrera 4	4.4	*4.9*
1995-97 911 Carrera	4.8	*5.3*
Final Drive Cover, Replace or Reseal (B)		
911		
1981-86	8.3	*8.6*
1987-89	9.0	*9.3*
1990-94	13.6	*13.9*
1981-89 911 Turbo	10.2	*10.5*
1995-01 911 Carrera		
2WD	18.7	*19.0*
4WD	8.3	*8.6*
Boxster	17.9	*18.2*
Front Halfshaft, R&R or Replace (B)		
1989-94 Carrera 4		
one side	2.0	*2.4*
both sides	3.2	*3.6*
1995-01 911 Carrera		
one side	1.9	*2.3*
both sides	3.1	*3.5*
Replace CV joint add each	.8	*.8*
Replace CV joint boot add each	.8	*.8*
Halfshaft, R&R or Replace (B)		
911, 911 Turbo		
1981-89		
one side	.9	*1.2*
both sides	1.8	*2.1*
1991-92		
one side	1.8	*2.1*
both sides	2.1	*2.4*

POR-10 911 : BOXSTER

	LABOR TIME	SEVERE SERVICE
Halfshaft, R&R or Replace (B)		
911 Carrera		
1989-94		
one side	1.8	2.1
both sides	2.3	2.6
1995-01		
AT		
one side	3.9	4.2
both sides	6.3	6.6
MT		
one side	2.5	2.8
both sides	5.2	5.5
1997 Boxster		
one side	1.2	1.5
both sides	2.2	2.5
Torque Tube, Replace (B)		
1989-94 911 Carrera 4	7.6	8.1
1995-97 911 Carrera (993)	7.2	7.7

BRAKES
ANTI-LOCK

Diagnose Anti-Lock Brake System (A)
1989-01 . . . . 1.0   1.0

Anti-Lock Relay, Replace (B)
1989-01 911 one 5   .5

Control Unit, Replace (B)
1989-01 911 7   .7

Hydraulic Assy., Replace (B)
1989-01 911 . . . . 1.3   1.3
Boxster 1.3   1.3

Wheel Speed Sensor, Replace (B)
1989-01 911
 front or rear
  one 7   .7
  both 9   .9
Boxster
 front or rear
  one 5   .5
  both 7   .7

SYSTEM

Bleed Brakes (B)
Includes: Add fluid.
1981-89 911 . . . . 1.0   1.0
911 Turbo
 1981-89 . . . . 1.0   1.0
 1991-92 . . . . 2.3   2.3
911 RS 2.2   2.2
911 Carrera 2.3   2.3
Boxster 2.2   2.2
w/dual circuit system add .5   .5

Brake Hose (Flexible), Replace (B)
Includes: System bleeding.
911, 911 Turbo
 1981-89
  one 1.1   1.2
  each addl. add 3   .4

	LABOR TIME	SEVERE SERVICE
1991-92		
front one	3.0	3.1
rear one	3.0	3.1
each addl. add	.3	.4
911 Carrera		
1989-94		
front one	3.1	3.2
rear one	3.0	3.1
each addl. add	.3	.4
1995-01		
front or rear one	2.3	2.4
each addl. add	.3	.4
Boxster		
front or rear one	2.0	2.1
each addl. add	.3	.4

Master Cylinder, Replace (B)
911, 911 Turbo
 1981-89 . . . . 3.4   3.6
 1991-92 . . . . 2.8   3.0
911 Carrera
 1989-94 Carrera 4 . . 3.0   3.2
 1990-94 Carrera 2 . . 3.4   3.6
 1995-01
  2WD 2.8   3.0
  4WD 2.4   2.6
Boxster 2.5   2.7

Power Booster Unit, Replace (B)
911, 911 Turbo
 1981-89 . . . . 3.7   3.9
 1991-92 . . . . 2.3   2.5
911 Carrera
 1989-94 Carrera 4 . . 2.3   2.5
 1990-94 Carrera 2 . . 4.1   4.3
 1995-01 . . . . 3.3   3.5
Boxster 3.5   3.7

Power Booster Vacuum Check Valve, Replace (B)
1981-89 911, 911 Turbo 3   .3
Boxster 7   .7

Proportioning Valve, Replace (B)
Includes: System bleeding.
1995-01 911 Carrera (993)
 2WD 2.5   1.7
 4WD 2.6   2.8
Boxster 2.3   2.5

SERVICE BRAKES

Brake Pads, Replace (B)
Front 1.3   1.4
Rear 1.4   1.5
Four wheels 2.7   2.8
Resurface brake rotor add
 w/wheel bearings 9   .9
 w/o wheel bearings 5   .5

	LABOR TIME	SEVERE SERVICE
Front Caliper Assy., Replace (B)		
Includes: System bleeding.		
911, 911 Turbo		
1981-89		
one	1.7	1.9
both	2.3	2.5
1991-92		
one	3.2	3.4
both	3.7	3.9
911 Carrera		
1989-94		
one	3.0	3.2
both	3.8	4.0
1995-01		
one	2.8	3.0
both	3.1	3.3
Boxster		
one	2.7	2.9
both	3.1	3.3

Front Disc Brake Rotor, Replace (B)
911, 911 Turbo
 1981-89
  one 1.9   2.1
  both 2.7   2.9
 1991-92
  one 9   1.1
  both 1.8   2.0
911 Carrera
 1989-94
  one 9   1.1
  both 1.8   2.0
 1995-01
  one 1.3   1.5
  both 1.7   1.9
Boxster
 one 1.2   1.4
 both 2.1   2.3

Rear Caliper Assy., Replace (B)
Includes: System bleeding.
911, 911 Turbo
 1981-89
  one 1.7   1.9
  both 2.4   2.6
 1991-92
  one 3.3   3.5
  both 4.3   4.5
911 Carrera
 1989-94
  one 3.3   3.5
  both 4.2   4.4
 1995-01
  one 2.6   2.8
  both 3.1   3.3
Boxster
 one 2.6   2.8
 both 3.0   3.2

911 : BOXSTER POR-11

	LABOR TIME	SEVERE SERVICE
Rear Disc Brake Rotor, Replace (B)		
911, 911 Turbo		
1981-89		
one	1.9	2.1
both	2.4	2.6
1991-92		
one	.9	1.1
both	1.9	2.1
911 Carrera		
1989-94		
one	.9	1.1
both	1.7	1.9
1995-01		
one	1.3	1.5
both	2.8	3.0
Boxster		
one	1.5	1.7
both	2.5	2.7

PARKING BRAKE

	LABOR TIME	SEVERE SERVICE
Parking Brake Cable, Adjust (C)		
1981-89 911, 911 Turbo	.8	1.0
Parking Brake Apply Actuator, Replace (B)		
1981-89 911	1.7	1.7
911 Turbo		
1981-89	1.9	1.9
1991-92	1.3	1.3
911 Carrera	1.3	1.3
Boxster	2.2	2.2
Parking Brake Apply Warning Indicator Switch, Replace (B)		
Exc. below	.8	.8
1981-89 911, 911 Turbo	1.5	1.5
Boxster	1.4	1.4
Parking Brake Cable, Replace (B)		
1981-89 911		
w/automatic heat		
one	1.6	1.9
both	3.1	3.4
w/o automatic heat		
one	1.3	1.6
both	2.6	2.9
1991-92 911 Turbo		
one	2.3	2.6
both	3.4	3.7
911 Carrera		
1989-94		
one	2.4	2.7
both	3.4	3.7
1995-01		
one	2.3	2.6
both	3.3	3.6
Boxster		
one side	2.3	2.6
both sides	3.8	4.1

	LABOR TIME	SEVERE SERVICE
Parking Brake Cable and/or Shoes, Adjust (B)		
1991-92 911 Turbo	1.3	1.5
911 Carrera	1.2	1.4
Boxster	1.3	1.5
Parking Brake Shoes, Replace (B)		
911, 911 Turbo		
1981-89		
one side	1.8	2.0
both sides	3.2	3.4
1991-92		
one side	2.3	2.5
both sides	3.6	3.8
911 Carrera		
1989-94		
one side	2.5	2.7
both sides	3.6	3.8
1995-01		
one side	2.4	2.6
both sides	3.4	3.6
Boxster		
one side	2.4	2.6
both sides	3.7	3.9

FRONT SUSPENSION

Unless otherwise noted, time given does not include alignment.

	LABOR TIME	SEVERE SERVICE
Align Front End (A)		
1981-01	2.7	2.9
Front & Rear Alignment, Check Adjust (B)		
911, 911 Turbo		
1981-89	2.9	3.2
1991-92	3.8	4.1
911 Carrera		
1989-94	3.9	4.2
1995-01	4.1	4.4
Boxster	2.9	3.2
Front Toe, Adjust (B)		
1981-01	1.6	1.9
Vehicle Height, Adjust (B)		
1981-89 911	2.4	2.7
1981-89 911 Turbo	3.0	3.3
1995-01 911 Carrera		
front	2.8	3.1
rear	1.6	1.9
both	4.0	4.3
Front Axle Hub Oil Seal, Replace (B)		
1981-89 911, 911 Turbo		
one side	2.0	2.0
both sides	3.6	3.6
Front Strut Assembly, R&R or Replace (B)		
911, 911 Turbo		
1981-89		
one side	2.1	2.3
both sides	3.2	3.4
1991-92		
one side	1.6	1.8
both sides	2.6	2.8

	LABOR TIME	SEVERE SERVICE
911 Carrera		
1989-94		
one side	1.8	2.0
both sides	2.6	2.8
1995-01		
one side	1.8	2.0
both sides	2.6	2.8
Boxster		
one side	1.9	2.1
both sides	3.2	3.4
Replace		
coil spring add	.5	.7
strut cartridge add	.8	1.0
Hub, Bearing or Seal, Replace (B)		
911, 911 Turbo		
1981-89		
one side	1.9	2.1
both sides	3.1	3.3
1991-92		
one side	3.9	4.1
both sides	6.7	6.9
911 Carrera		
1989-94		
one side	3.9	4.1
both sides	6.8	7.0
1995-01		
one side	3.8	4.0
both sides	6.8	7.0
Boxster		
one side	3.0	3.2
both sides	5.3	5.5
Lower Ball Joint, Replace (B)		
911		
one	.8	1.1
both	1.3	1.6
Lower Control Arm Assy., Replace (B)		
1981-97 911		
one	1.6	1.9
both	2.5	2.8
Boxster		
one	.8	1.1
both	.9	1.2
Lower Trailing Arm, Replace (B)		
1995-01 911 Carrera		
one side	1.7	2.0
both sides	2.5	2.8
Boxster		
one	1.1	1.4
both	1.6	1.9
Stabilizer Bar, Replace (B)		
911, 911 Turbo		
1981-89	.9	1.0
1991-92	1.6	1.8
911 Carrera		
1989-94	1.5	1.7
1995-01	1.5	1.7
Boxster	1.7	1.9

POR-12 911 : BOXSTER

	LABOR TIME	SEVERE SERVICE
Stabilizer Bar Bushings, Replace (B)		
1981-89 911,		
911 Turbo	.9	1.1
1989-94 911 Carrera 2/4		
one	.7	.9
both	.9	1.1
Boxster	1.1	1.3
Steering Knuckle, Replace (B)		
1991-92 911 Turbo		
one side	3.8	4.3
both sides	6.7	7.2
911 Carrera		
1989-94		
one side	3.9	4.4
both sides	6.8	7.3
1995-01		
one side	3.9	4.4
both sides	6.8	7.3
Boxster		
one side	3.1	3.6
both sides	5.3	5.8
Torsion Bar, Replace (B)		
1981-89 911, 911 Turbo		
one	.8	1.2
both	1.8	2.2

REAR SUSPENSION
Unless otherwise noted, time given does not include alignment.

	LABOR TIME	SEVERE SERVICE
Rear Control Arm, R&R or Replace (B)		
1981-89 911		
one side	2.4	2.7
both sides	3.9	4.2
911 Turbo		
1981-89		
one side	4.5	4.8
both sides	9.0	9.4
1991-92		
one	3.8	4.2
both	6.2	6.6
911 Carrera		
1989-94		
one	3.7	4.1
both	6.4	6.8
1995-01		
one side	.9	1.3
both sides	1.8	2.2
Boxster		
one side	.7	1.1
both sides	.8	1.2
Rear Hub and/or Bearings, Replace (B)		
1981-89 911		
one side	3.0	3.5
both sides	4.2	4.7

	LABOR TIME	SEVERE SERVICE
911 Turbo		
1981-89		
one side	3.1	3.6
both sides	5.6	6.1
1991-92		
one side	3.9	4.4
both sides	6.2	6.7
911 Carrera		
1989-94		
one side	3.9	4.4
both sides	6.2	6.7
1995-01		
one side	3.9	4.4
both sides	6.4	6.9
Rear Stabilizer Bar & Bushings, Replace (B)		
1981-89 911	.7	.9
911 Turbo		
1981-89	.7	.9
1991-92	1.6	1.8
911 Carrera		
1989-94	1.5	1.7
1995-01	1.6	1.8
Rear Stabilizer Bar Bushings, Replace (B)		
1991-92 911 Turbo		
one	.7	.9
both	.9	1.1
1989-01 911 Carrera		
one	.7	.9
both	.9	1.1
Rear Wheel Knuckle Assy., Replace (B)		
1995-01 911 Carrera		
one side	4.0	4.5
both sides	6.4	6.9
Boxster		
one side	3.5	4.0
both sides	5.8	6.3
Shock Absorbers or Bushings, Replace (B)		
1981-89 911		
one	.5	.7
both	.8	1.0
1981-89 911 Turbo		
one	.9	1.1
both	1.3	1.5
Strut Assy., R&R or Replace (B)		
1991-92 911 Turbo		
right	2.0	2.2
left	2.0	2.2
both	3.2	3.4
911 Carrera		
1989-94		
right	1.8	2.0
left	1.8	2.0
both	3.0	3.2
1995-01		
one side	1.7	1.9
both sides	2.7	2.9

	LABOR TIME	SEVERE SERVICE
Boxster		
one	3.7	3.9
both	5.5	5.7
Replace		
coil spring add each	.8	.8
cartridge add each	.8	.8
Tie Rod End and/or Sleeve, Replace (B)		
1995-01 911 Carrera		
each side	1.5	1.7
Boxster		
one side	.8	1.0
both sides	1.5	1.7
Torsion Bar, Replace (B)		
1981-89 911, 911 Turbo		
one	3.1	3.5
both	5.2	5.6
Trailing Arm or Link, Replace (B)		
1995-01 911 Carrera		
one side	1.5	1.8
both sides	1.9	2.2
Boxster		
one side	1.5	1.8
both sides	2.0	2.3

STEERING
Unless otherwise noted, time given does not include alignment.

AIR BAGS

	LABOR TIME	SEVERE SERVICE
Diagnose Air Bag System (A)		
1995-01	.9	.9
Air Bag Assy., Replace (B)		
1995-01 911 Carrera		
driver side	.5	.5
passenger side	3.8	3.8
Boxster		
driver side	.5	.5
passenger side	.9	.9
Dual Pole Arming Sensor, Replace (B)		
1995-01 911	1.6	1.6

MANUAL RACK & PINION

	LABOR TIME	SEVERE SERVICE
Inner Tie Rod Assy., Replace (B)		
1981-89 911		
one side	1.1	1.2
both sides	2.0	2.2
1981-89 911 Turbo		
one side	1.8	2.0
both sides	2.3	2.5
Outer Tie Rod End, Replace (B)		
1981-89 911, 911 Turbo		
one	.5	.7
Rack & Pinion Assy., R&I (B)		
1981-89 911,		
911 Turbo	2.4	2.9
w/AC add	.3	.3

911 : BOXSTER POR-13

	LABOR TIME	SEVERE SERVICE
Rack Boots, Replace (B)		
1981-89 911, 911 Turbo		
one side	1.3	1.5
both sides	2.0	2.2
Steering Wheel, Replace (B)		
1981-89 911, 911 Turbo	.3	.3
POWER RACK & PINION		
Inner Tie Rod Ends, Replace (B)		
1991-92 911 Turbo		
one side	1.7	1.9
both sides	2.4	2.6
1989-94 911 Carrera 2/4		
one side	1.7	1.9
both sides	2.3	2.5
Boxster		
one side	3.2	3.4
both sides	3.6	3.8
Outer Tie Rod Ends, Replace (B)		
1991-92 911 Turbo		
one side	.7	.9
both sides	1.2	1.4
911 Carrera		
1989-94		
one side	.7	.9
both sides	1.2	1.4
1995-01		
one side	1.5	1.7
both sides	2.5	2.7
Boxster		
one side	.9	1.1
both sides	1.5	1.7
Rack & Pinion Assy., R&R or Replace (B)		
1991-92 911 Turbo	4.7	5.2
1993-94 911 RS	4.9	5.4
911 Carrera		
1989-94	4.9	5.4
1995-01	5.4	5.9
Boxster	2.8	3.3
Replace assembly add	.5	.5
Rack Boots, Replace (B)		
1991-92 911 Turbo		
one	.9	1.1
both	1.5	1.7
911 Carrera		
1989-94		
one	.9	1.1
both	1.3	1.5
1995-01		
one	.8	1.0
both	1.2	1.4
Boxster		
one side	3.1	3.3
both sides	3.5	3.7

	LABOR TIME	SEVERE SERVICE
Steering Pump, R&R or Replace (B)		
1991-92 911 Turbo	4.8	5.1
911 Carrera		
1989-94	3.6	3.9
1995-01		
w/turbo	7.9	8.2
w/o turbo	4.3	4.5
Boxster	4.7	5.0
w/AC add	.5	.5
Steering Wheel, Replace (B)		
1991-92 911 Turbo		
w/air bags	.5	.5
1989-01 911 Carrera		
w/air bags	.5	.5
w/o air bags	.3	.3
Boxster	.5	.5

HEATING & AIR CONDITIONING

When more than one component requires replacement where evacuation/recovery and recharging is already included, deduct 1.0 hour for each additional component from the time given.

Evacuate/Recover and Recharge System (B)
1981-01 .... 2.2 2.2
Flush Refrigerant System, Complete (B)
1981-01 .... 1.6 1.6
AC Hoses, Replace (B)
Includes: Evacuate/recover and recharge.
1981-89 911
  compressor to condenser 2.4 2.5
  condenser to condenser 4.1 4.2
  compressor to evaporator 3.7 3.8
  liquid line 3.1 3.2
1991-92 911 Turbo
  condenser line 4.3 4.4
  liquid line 3.5 3.6
1989-01 911 Carrera
  condenser line 4.4 4.5
  liquid line 3.6 3.7
Clutch Holding Coil, Replace (B)
1981-89 911, 911 Turbo 9 1.1
Compressor Assy., Replace (B)
Includes: Parts transfer, evacuate/recover and recharge.
911 .... 2.9 3.1
Boxster 4.8 5.0

	LABOR TIME	SEVERE SERVICE
Condenser Assy., Replace (B)		
Includes: Evacuate/recover and recharge.		
911, 911 Turbo		
1981-89 front or rear	2.7	2.9
1991-92	4.9	5.1
911 Carrera	4.8	5.0
Condenser Cooling Fan, Replace (B)		
1981-89 911	.7	1.0
Evaporator Coil, Replace (B)		
Includes: Evacuate/recover and recharge.		
911, 911 Turbo		
1981-89	3.1	3.1
1991-92	8.1	8.1
911 Carrera		
1989-94	8.1	8.1
1995-01	8.3	8.3
Expansion Valve, Replace (B)		
Includes: Evacuate/recover and recharge.		
911, 911 Turbo		
1981-89	3.0	3.0
1991-92	8.3	8.3
911 Carrera		
1989-94	8.2	8.2
1995-01	3.5	3.5
Boxster	2.7	2.7
Heater Blower Motor, Replace (B)		
1981-89 911	.5	.5
1991-92 911 Turbo	.9	.9
911 Carrera		
1989-94	.9	.9
1995-01 right or left	1.7	1.7
Boxster	1.3	1.3
Heater Blower Motor Relay, Replace (B)		
1981-89 911, 911 Turbo	.3	.3
1989-94 911 Carrera 2/4	.3	.3
1995-01 911 Carrera (993)	.3	.3
Boxster	.5	.5
Heater Blower Motor Resistor, Replace (B)		
Exc. Boxster	.3	.3
Boxster	1.4	1.4
Heater Blower Motor Switch, Replace (B)		
1995-01 911 Carrera	.3	.3
Heater Control Assy., Replace (B)		
1981-89 911, 911 Turbo	3.2	3.2
Heater Core, Replace (B)		
1997-01 Boxster	1.8	1.8
Heater Hoses, Replace (B)		
1981-89 911, 911 Turbo		
one	.7	.9
both	.9	1.1

POR-14 911 : BOXSTER

	LABOR TIME	SEVERE SERVICE
Heater Mode Control Cable, Replace (B)		
1981-89 911, 911 Turbo	.8	.8
Low Pressure and/or Cycling Switch, Replace (B)		
1991-92 911 Turbo	3.1	3.1
911 Carrera		
1989-94	3.2	3.2
1995-01	.3	.3
Boxster	.5	.5
Receiver/Drier Assy., Replace (B)		
Includes: Evacuate/recover and recharge.		
1981-01 911	3.0	3.0
Temperature Control Unit, Replace (B)		
Exc. Boxster	.3	.3
Boxster	.5	.5

WIPERS & SPEEDOMETER

Intermittent Relay, Replace (B)		
1995-01 911 Carrera	.3	.3
Interval Switch, Replace (B)		
1981-01	.3	.3
Power Antenna, Replace (B)		
1981-84 911, 911 Turbo	.5	.5
Radio, R&R (B)		
Exc. below	.3	.3
1981-82 911	.9	.9
Rear Window Wiper Motor, Replace (B)		
1981-94 911	.9	1.1
1995-01 911 Carrera		
coupe	.9	1.1
Targa	1.6	1.8
Rear Window Wiper Switch, Replace (B)		
911, 911 Turbo		
1981-89	.7	.7
1991-92	.3	.3
1989-01 911 Carrera	.3	.3
Speedometer Head, R&R or Replace (B)		
911, 911 Turbo		
1981-89	.5	.5
1991-92	.3	.3
911 Carrera		
1989-94	.3	.3
1995-01	.8	.8
Boxster	.8	.8
Tachometer, R&R or Replace (B)		
1981-89 911, 911 Turbo	.3	.3
1995-97 911 Carrera	.8	.8
Boxster	.8	.8

	LABOR TIME	SEVERE SERVICE
Windshield Washer Pump, Replace (B)		
911, 911 Turbo		
1981-89	.5	.5
1991-92	.8	.8
1989-94 911 Carrera 2/4	.8	.8
Windshield Wiper Linkage, Replace (B)		
1981-89 911, 911 Turbo	2.0	2.3
Windshield Wiper & Washer Switch, Replace (B)		
911, 911 Turbo		
1981-89	1.2	1.2
1991-92	1.5	1.5
911 Carrera		
1989-94		
w/air bags	1.5	1.5
w/o air bags	1.4	1.4
1995-01		
w/air bags	1.5	1.5
w/o air bags	1.3	1.3
Windshield Wiper Interval Relay, Replace (B)		
Exc. below	.3	.3
1981-89 911, 911 Turbo	1.6	1.6
Boxster	.5	.5
Windshield Wiper Linkage, Replace (B)		
1995-01 911 Carrera	7.1	7.3
Boxster	1.3	1.5
Windshield Wiper Motor, Replace (B)		
911, 911 Turbo		
1981-89	2.2	2.5
1991-92	3.1	3.4
911 Carrera		
1989-94	3.1	3.4
1995-01		
w/AC	4.8	5.1
w/o AC	3.0	3.3
Boxster	.7	1.0

LAMPS & SWITCHES

Back-Up Lamp Bulb, Replace (B)		
1995-01 911 Carrera	.5	.5
Back-Up Lamp Switch, Replace (B)		
911, 911 Turbo		
1981-89	.5	.5
1991-92	.8	.8
1989-94 911 Carrera 2/4	.8	.8
Brake Lamp Switch, Replace (B)		
911, 911 Turbo		
1981-89		
hydraulic	1.7	1.7
mechanical	.5	.5

	LABOR TIME	SEVERE SERVICE
1991-92	.9	.9
1989-01 911 Carrera	.9	.9
Emergency Flasher Switch Assy., Replace (B)		
1981-89 911, 911 Turbo	.5	.5
Fog Lamp Switch, Replace (B)		
1985-89 911, 911 Turbo	.7	.7
Hazard Warning Switch, Replace (B)		
Exc. below	.3	.3
1981-89 911, 911 Turbo	.5	.5
Halogen Headlamp Bulb, Replace (B)		
1981-94 911	.7	.7
1995-01 911 Carrera	.3	.3
Boxster	.3	.3
Headlamp Switch, Replace (B)		
911	.5	.5
Boxster	.8	.8
Headlamps, Aim (B)		
Two	.5	.5
Four	.7	.7
High Mount Stop Lamp and/or Lens, Replace (B)		
1985-89 911, 911 Turbo	.3	.3
1990-01 911	.9	.9
Boxster	.3	.3
Horn, Replace (B)		
911, 911 Turbo		
1981-89		
one	.7	.7
both	1.1	1.1
1991-92		
one	.8	.8
both	.9	.9
911 Carrera		
1989-01		
one	.8	.8
both	.9	.9
Boxster	.7	.7
Horn Relay, Replace (B)		
1981-01	.3	.3
License Lamp Assy., Replace (B)		
1995-01 911 Carrera	.3	.3
Boxster	.3	.3
License Lamp Lens, Replace (C)		
1981-01	.2	.2
Multifunction Switch, Replace (B)		
Cruise control		
1991-92 911 Turbo	.6	.6
AT		
1995-01 911 Carrera	1.5	1.5
Boxster	2.5	2.5
Parking Lamp Lens or Bulb, Replace (B)		
1981-94	.7	.7
1995-01 911 Carrera	.3	.3
Boxster	.3	.3

911 : BOXSTER POR-15

	LABOR TIME	SEVERE SERVICE
Side Marker Lamp Lens, Replace (B)		
1981-01	.3	.3
Stop Lamp Switch, Replace (B)		
1981-01	.7	.7
Tail Lamp Lens or Bulb, Replace (C)		
1981-01	.2	.2
Turn Signal Combination Switch, Replace (B)		
911, 911 Turbo		
1981-89	1.2	1.2
1991-92	1.6	1.6
911 Carrera		
1989-94		
w/air bags	1.7	1.7
w/o air bags	1.5	1.5
1995-01		
w/air bags	1.7	1.7
w/o air bags	1.3	1.3
Boxster	1.3	1.3
Turn Signal Lamp Lens or Bulb, Replace (C)		
1981-01	.2	.2
Turn Signal or Hazard Warning Flasher, Replace (B)		
1981-89 911, 911 Turbo	.5	.5
1990-01 911	.3	.3

BODY

	LABOR TIME	SEVERE SERVICE
Door Lock, Replace (B)		
911, 911 Turbo		
1981-89		
coupe	3.9	3.9
Targa	2.5	2.5
1991-92	2.6	2.6
911 Carrera		
1989-94	2.6	2.6
1995-01	2.5	2.5
Boxster	2.5	2.5
Door Window Regulator (Electric), Replace (B)		
911, 911 Turbo		
1981-89		
coupe	2.6	2.8
Targa	2.9	3.1
1991-92	2.1	2.3
911 Carrera		
1989-94	2.1	2.3
1995-01		
coupe	2.0	2.2
Targa	2.3	2.5
Boxster	1.9	2.1

	LABOR TIME	SEVERE SERVICE
Door Window Regulator (Manual), Replace (B)		
1981-89 911, 911 Turbo	2.3	2.5
Hood Lock, Replace (B)		
911 upper & lower	1.3	1.3
Boxster upper & lower	1.4	1.4
Hood Release Cable, Replace (B)		
Front		
911, 911 Turbo		
1981-89	1.5	1.5
1991-92	1.9	1.9
911 Carrera		
1989-94	1.8	1.8
1995-01	1.9	1.9
Boxster	2.4	2.4
Rear		
911		
1981-88	1.9	1.9
1989-01	.5	.5
Lock Striker Plate, Replace (B)		
1981-01	.5	.5

NOTES

POR

924 : 928 : 944 : 968

SYSTEM INDEX

MAINTENANCE POR-18
CHARGING POR-18
STARTING POR-18
CRUISE CONTROL POR-18
IGNITION POR-18
EMISSIONS POR-19
FUEL POR-19
EXHAUST POR-20
ENGINE COOLING POR-20
ENGINE POR-21
   Assembly POR-21
   Cylinder Head POR-21
   Camshaft POR-21
   Crank & Pistons POR-22
   Engine Lubrication .. POR-23
CLUTCH POR-23
MANUAL TRANSMISSION .. POR-23
AUTO TRANSMISSION POR-23
SHIFT LINKAGE POR-24
DRIVELINE POR-24
BRAKES POR-24
FRONT SUSPENSION POR-25
REAR SUSPENSION POR-26
STEERING POR-26
HEATING & AC POR-26
WIPERS &
  SPEEDOMETER POR-27
LAMPS & SWITCHES POR-27
BODY POR-28

OPERATIONS INDEX

A
AC Hoses POR-27
Air Bags POR-26
Air Conditioning POR-26
Alignment POR-25
Alternator (Generator) POR-18
Antenna POR-27
Anti-Lock Brakes POR-24

B
Back-Up Lamp Switch .. POR-27
Ball Joint POR-25
Battery Cables POR-18
Bleed Brake System ... POR-24
Blower Motor POR-27
Brake Disc POR-25
Brake Drum POR-25
Brake Hose POR-24
Brake Pads and/or Shoes .. POR-25

C
Camshaft POR-22
Catalytic Converter .. POR-20
Crankshaft POR-22
Crankshaft Sensor POR-18
Cruise Control POR-18

CV Joint POR-24
Cylinder Head POR-21

D
Differential POR-24
Distributor POR-18
Drive Belt POR-18
Driveshaft POR-24

E
EGR POR-19
Engine POR-21
Engine Lubrication ... POR-23
Engine Mounts POR-21
Evaporator POR-27
Exhaust POR-20
Exhaust Manifold POR-20
Expansion Valve POR-27

F
Flywheel POR-23
Fuel Injection POR-19
Fuel Pump POR-19
Fuel Vapor Canister .. POR-19

G
Gear Selector Lever .. POR-24
Generator POR-18

H
Halfshaft POR-24
Headlamp POR-27
Heater Core POR-27
Horn POR-27

I
Ignition Coil POR-18
Ignition Switch POR-19
Inner Tie Rod POR-26
Intake Manifold POR-19

K
Knock Sensor POR-19

L
Lower Control Arm POR-25

M
Master Cylinder POR-25
Muffler POR-20

O
Oil Pan POR-23
Oil Pump POR-23
Outer Tie Rod POR-26
Oxygen Sensor POR-19

P
Parking Brake POR-25
Pistons POR-22

R
Radiator POR-20
Radiator Hoses POR-20
Radio POR-27
Rear Main Oil Seal ... POR-22

S
Shock Absorber/Strut,
  Front POR-25
Shock Absorber/Strut,
  Rear POR-26
Spark Plug Cables POR-19
Spark Plugs POR-19
Spring, Rear Coil POR-26
Starter POR-18
Steering Wheel POR-26

T
Thermostat POR-20
Throttle Body POR-20
Timing Belt POR-22
Timing Chain POR-22
Torque Converter POR-24

U
Upper Control Arm POR-26

V
Valve Body POR-24
Valve Cover Gasket ... POR-21
Valve Job POR-21
Vehicle Speed Sensor . POR-19

W
Water Pump POR-21
Wheel Balance POR-18
Wheel Cylinder POR-25
Window Regulator POR-28
Windshield Washer Pump .. POR-27
Windshield Wiper Motor .. POR-27

POR-17

POR-18 924 : 928 : 944 : 968

	LABOR TIME	SEVERE SERVICE

MAINTENANCE

Air Cleaner Filter Element, Replace (B)
- 1981-89 .... .3 / .5
- 1989-95 944S2, 968 .... .8 / 1.0

Chassis Lubrication, Change Oil & Filter (C)
Includes: Correct all fluid levels.
- 1981-95 .... .6 / .7

Composite Headlamp Bulb, Replace (B)
- 928, 944, 968 one ..... .3 / .5

Drive Belt, Adjust (B)
- All Models .... .5 / .7

Drive Belt, Replace (B)
Exc. below
- 924 .... .5 / .7
- 924S, 944, 968 .... .8 / 1.0
- 928 one .... .8 / 1.0
- each addl. add .... .3 / .5

Air pump
- 924 .... .9 / 1.1
- 928 .... .9 / 1.1

Compressor
- 924 .... .8 / 1.0
- 928 .... 1.8 / 2.0
- 924S, 944, 968 .... 1.0 / 1.2

Serpentine .... .9 / 1.1

Fuel Filter, Replace (B)
- 1981-95 .... .7 / .9

Oil & Filter, Change (C)
Includes: Correct all fluid levels.
- 1981-95 .... .6 / .7

Sealed Beam Headlamp, Replace (B)
- 1981-85 924, 928 ..... .3 / .5
- 1983-89 944 .... .5 / .7

Timing Belt, Replace (B)
- 924 .... 3.3 / 3.5
- 924S, 944, 968 .... 4.0 / 4.2
- 1981-91 928
  - 2V .... 5.6 / 5.8
  - 4V .... 7.2 / 7.4
- w/turbo 924, 924S, 944, 968 add .... .5 / .7

Tire, Replace (C)
Includes: Dismount old tire and mount new tire to rim.
- 1981-95 one .... .5 / .7

Wheel, Balance (B)
- One .... .3 / .3
- each addl. add .... .2 / .2

CHARGING

Alternator Circuits, Test (B)
Includes: Test component output.
- 1981-95 .... .5 / .5

Alternator Assy., R&R and Recondition (A)
- 924 .... 4.3 / 4.5
- 1981-91 928 .... 5.0 / 5.2
- 924S, 944, 968 .... 5.1 / 5.3

Alternator Assy., Replace (B)
Includes: Pulley transfer.
- 924S, 944, 968 .... 2.5 / 2.7
- 928, 928S .... 1.8 / 2.0
- 928S4/GT/GTS .... 2.0 / 2.2
- w/turbo 924S, 944, 968 add .... .8 / 1.0

Alternator Voltage Regulator, Replace (B)
- All Models .... .8 / 1.0
Does not include alt. removal

Front Alternator Bearing, Replace (B)
- All Models .... 1.2 / 1.4
Does not include alt. removal

Voltmeter, Replace (B)
- 924S .... .9 / 1.1
- 944
  - 1981-85 .... .9 / 1.1
  - 1986-89 .... 1.4 / 1.6
- 968 .... 1.4 / 1.6
- 928
  - 1981-87 .... .8 / 1.0
  - 1988-91 .... 1.6 / 1.8

STARTING

Starter Draw Test (On Car) (B)
- 1981-95 .... .5 / .5

Battery Cables, Replace (C)
Positive
- 924S, 944, 968 ...... 2.5 / 2.7
- 928 .... 4.3 / 4.5
Negative
- 924, 944, 928 .... .5 / .7
- 968 .... 1.3 / 1.5
- w/turbo 924S, 944, 968 add .... .8 / 1.0

Starter Assy., R&R and Recondition (A)
- 924 .... 3.6 / 3.8
- 928, 944, 968 .... 3.0 / 3.2

Starter Assy., Replace (B)
- 924 .... 1.5 / 1.7
- 924S, 944, 968 ...... 1.3 / 1.5
- 928 .... 1.3 / 1.5

Starter Drive Assy., Replace (B)
Includes: Starter R&R.
- 924 .... 2.4 / 2.6
- 924S, 944, 968 ...... 2.1 / 2.3
- 928 .... 2.1 / 2.3

Starter Solenoid and/or Switch, Replace (B)
Includes: Starter R&R.
- 924 .... 2.2 / 2.4
- 924S, 944, 968 ...... 1.6 / 1.8
- 928 .... 1.6 / 1.8

CRUISE CONTROL

Actuator Assy., Replace (B)
- 924S, 944 .... .5 / .7
- 928 .... 1.3 / 1.5
- 968 .... .5 / .7

Clutch Switch, Replace (B)
- 928, 944, 968 .... .5 / .7

Control Chain, Cable or Rod, Replace (B)
- 924S, 944, 968 ...... .5 / .7
- 928 .... 2.0 / 2.2

Control Controller Module, Replace (B)
- 924S, 944 .... .5 / .7
- 1984-94 928 .... 1.2 / 1.4
- 968 .... .5 / .7

Control Switch, Replace (B)
- 924S, 944 .... 1.2 / 1.4
- 1984-94 928 .... 1.3 / 1.5
- 968 .... 1.2 / 1.4

IGNITION

Ignition Timing, Reset (B)
- 924, 928 .... .5 / .7

Crankshaft Sensor, Replace (B)
- 1987-94 928GT/GTS ... 3.2 / 3.4
- 1987-89 944S .... 1.8 / 2.0

Digital Ignition Timing Control Unit, Replace (B)
- 944 .... .5 / .7

Distributor, Replace (B)
Includes: Reset base ignition timing.
- 1981-82 .... .7 / .9
- 928 .... .9 / 1.1

Distributor Cap and/or Rotor, Replace (B)
- 1981-82 924 .... .5 / .7
- 924S, 944, 968 ...... .5 / .7
- 928 .... .7 / .9

Electronic Ignition Control Unit, Replace (B)
- 944 turbo .... .8 / 1.0

Hall Timing Sensor, Replace (B)
- 968 .... 2.0 / 2.2

Ignition Coil, Replace (B)
- 924 .... .5 / .7
- 1987-88 924S .... .7 / .9
- 928
  - one .... .7 / .9
  - both .... 1.3 / 1.5
- 944, 968 .... .7 / .9

924 : 928 : 944 : 968 POR-19

	LABOR TIME	SEVERE SERVICE
Ignition Switch, Replace (B)		
924	1.1	1.3
1987-88 924S	1.2	1.4
928	.8	1.0
944	1.4	1.6
968	1.5	1.7
Pulse Generator, Replace (B)		
1988-94 928	1.3	1.5
1990-91 944S2	.7	.9
944 turbo		
one	.7	.9
both	1.3	1.5
968	.7	.9
Spark Plug (Ignition) Cables, Replace (B)		
924S, 944, 968	.8	1.0
Spark Plugs, Replace (B)		
924	.8	1.0
924S, 944, 968	.7	.9
928		
2V	1.5	1.7
4V	1.6	1.8
Vacuum Advance Unit, Replace (B)		
Includes: Reset timing.		
924	.7	.9
1981-86 928	.9	1.1

EMISSIONS

	LABOR TIME	SEVERE SERVICE
Air Flow Meter, Replace (B)		
924S, 928, 944	.9	1.1
944S, 944 turbo, 968	.8	1.0
Air Injection Check Valve, Replace (B)		
1981-94 924, 928	.5	.7
Air Intake Sensor, Replace (B)		
928	.8	1.0
944	.7	.9
Air Pump Assy., Replace (B)		
1981-94 924, 928	2.2	2.4
Anti-Backfire Valve, Replace (B)		
924	.5	.7
Digital Timing Control Sensor, Replace (B)		
944	.7	.9
Digital Timing Thermo Valve, Replace (B)		
1987-88 924S	.7	.9
944	.7	.9
Diverter and/or Switching Valve, Replace (B)		
928	1.8	2.0
EGR Control Valve, Replace (B)		
924	.7	.9
EGR Filter, Replace (B)		
924	.8	1.0
Fuel Vapor Canister, Replace (B)		
1987-88 924S	.7	.9
944	1.6	1.8
968	1.5	1.7

	LABOR TIME	SEVERE SERVICE
Knock Sensor, Replace (B)		
924S, 944, 968		
one	3.2	3.4
two	3.7	3.9
1990-91 928S4/GT	6.8	7.0
w/turbo add 924S, 944, 968 add	.8	1.0
Oxygen Sensor, Replace (B)		
924S, 944, 968	.7	.9
928		
1985-93	1.1	1.3
1990-91	2.2	2.4
TCI Switching Unit, Replace (B)		
1981-82 924	.7	.9
Water Temperature Sensor, Replace (B)		
1988-94 928S4/GT	.8	1.0
924S, 944, 968	3.0	3.2
w/turbo add	.8	1.0
Vehicle Speed Sensor, Replace (B)		
All Models	1.2	1.4

FUEL

DELIVERY

	LABOR TIME	SEVERE SERVICE
Fuel Pump, Test (B)		
1981-95	.7	.9
Fuel Filter, Replace (B)		
1981-95	.7	.9
Fuel Gauge (Dash), Replace (B)		
924	.7	.9
928		
fuel gauge	1.5	1.7
combination gauge	1.5	1.7
944		
1981-88	.7	.9
1989-91	1.5	1.7
968	1.5	1.7
Fuel Gauge (Tank), Replace (B)		
1981-95	.8	1.0
Fuel Pump, Replace (B)		
924	.8	1.0
928	1.6	1.8
944		
1983-85	.8	1.0
1986-91	1.5	1.7
968	1.5	1.7
Fuel Tank, Replace (B)		
Includes: Drain and refill.		
924	7.2	7.4
928	3.9	4.1
944		
1983-85	7.1	7.3
1986-91	9.3	9.5
968		
MT		
cabriolet	7.2	7.4
coupe	7.9	8.1
w/Tiptronic trans. add	.9	1.1

	LABOR TIME	SEVERE SERVICE
Intake Manifold and/or Gasket, Replace (B)		
Includes: Adjustments.		
924	3.0	3.2
924S, 944, 968	4.3	4.5
928		
1981-88		
LH Jetronic	2.8	3.0
K Jetronic	7.6	7.8
1989-94	7.6	7.8
w/turbo add		
924	1.8	2.0
924S, 944, 968	.8	1.0

INJECTION

	LABOR TIME	SEVERE SERVICE
Air Distributor Housing, R&I (B)		
924	2.4	2.6
1981-86 928	3.1	3.3
Auxiliary Air Regulator, Replace (B)		
924	.5	.7
1981-86 928	.8	1.0
Auxiliary Air Valve, Replace (B)		
1981-86 928	.7	.9
Cold Start Valve, Replace (B)		
1981-86 928	.8	1.0
Electric Air Valve, Replace (B)		
928	.7	.9
Electronic Control Unit, Replace (B)		
1981-95	.7	.9
Fuel Distributor, Replace (B)		
924	.8	1.0
1981-86 928	1.7	1.9
Fuel Injection Lines, Replace (B)		
924		
one	.5	.7
all	1.3	1.5
1981-86 928		
one side	.8	1.0
both sides	1.6	1.8
Fuel Injectors, Replace (B)		
1981-82 924 continuous injection		
one	.5	.7
all	1.0	1.2
924S, 944, 968		
one	.7	.9
each addl. add	.3	.5
928		
w/continuous injection		
one	.7	.9
all	2.1	2.3
w/o continuous injection		
one	.7	.9
each addl. add	.5	.7
w/turbo 924, 944, 968 add	.4	.6
Fuel Pressure Regulator, Replace (B)		
1987-88 924S	.7	.9
1987-94 928	.8	1.0
944	.7	.9
944 turbo, 968	1.3	1.5

POR-20 924 : 928 : 944 : 968

	LABOR TIME	SEVERE SERVICE
Mixture Regulator, Replace (B)		
924	1.4	1.6
1981-86 928	3.2	3.4
Pneumatic Valve, Replace (B)		
1981-86 928	.5	.7
Pressure Limiter, Replace (B)		
924S, 928, 944	.7	.9
Pressure Regulator Valve, Replace (B)		
968	.9	1.1
Temperature Sensor, Replace (B)		
1987-88 924S	.7	.9
928	.9	1.1
944	.7	.9
944S	.9	1.1
944 turbo	1.3	1.5
968	.7	.9
Thermo Time Switch, Replace (B)		
924	.7	.9
1981-86 928	.8	1.0
Throttle Body Assy., Replace (B)		
968	1.4	1.6
Throttle Valve Housing, Replace (B)		
924	.7	.9
924S	1.5	1.7
928		
1981-93 928		
w/continuous injection	3.9	4.1
w/o continuous injection		
2V	3.4	3.6
4V	2.6	2.8
1990-91 928S4/GT/GTS	7.1	7.3
944	1.6	1.8
944S	3.3	3.5
Throttle Valve Switch, Replace (B)		
924S	1.5	1.7
928		
2V	.9	1.1
4V	2.7	2.9
944	1.6	1.8
944S	3.6	3.8
944S2	.7	.9
944 turbo	1.1	1.3
Warmup Control Valve, Replace (B)		
924	.5	.7
1981-86 928	.5	.7

TURBOCHARGER

	LABOR TIME	SEVERE SERVICE
Turbocharger Assy., R&R or Replace (B)		
924	8.0	8.2
944	12.1	12.3
Wastegate Actuator, Replace (B)		
924	2.2	2.4

EXHAUST

	LABOR TIME	SEVERE SERVICE
Catalytic Converter, Replace (B)		
924	1.5	1.7
924S	2.8	3.0
928	1.4	1.6
944	2.9	3.1
944 turbo	2.1	2.3
968	1.7	1.9
Exhaust Manifold or Gasket, Replace (B)		
924	2.2	2.4
924S	3.2	3.4
928		
2V		
one	18.1	18.3
both	18.2	18.4
4V		
one	20.5	20.7
both	20.7	20.9
Includes engine R&R.		
944		
MT	3.3	3.5
AT	4.2	4.4
944 turbo	5.9	6.1
968	3.2	3.4
Front Exhaust Pipe, Replace (B)		
924	.9	1.1
924S	2.4	2.6
928		
one	1.3	1.5
both	1.6	1.8
944, 968	3.0	3.2
w/turbo add		
1 piece pipe	1.5	1.7
2 piece pipe	.5	.7
Front Muffler, Replace (B)		
924	1.4	1.6
924S	2.7	2.9
928	2.4	2.6
944	2.8	3.0
944 turbo	1.9	2.1
968	1.9	2.1
Rear Muffler, Replace (B)		
928		
1981-85	3.1	3.3
1986-94	3.0	3.2
968	1.6	1.8
Tail Pipe, Replace (B)		
968	.7	.9

ENGINE COOLING

	LABOR TIME	SEVERE SERVICE
Coolant Thermostat, Replace (B)		
924, 928	.7	.9
924S, 944	2.1	2.3
944 turbo	3.7	3.9
944S, 944S2	2.2	2.4
968	1.8	2.0

	LABOR TIME	SEVERE SERVICE
Electric Cooling Fan Assy., Replace (B)		
924	.8	1.0
924S, 944, 968		
one	1.7	1.9
both	2.0	2.2
928		
1981-84	.7	.9
1985-94		
2V	2.0	2.2
4V	2.8	3.0
w/turbo 924, 924S, 944, 968 add	1.2	1.2
Electric Fan Thermo Switch, Replace (B)		
924	.7	.9
924S, 944	1.8	2.0
968	1.5	1.7
w/turbo 924, 924S, 944 add	1.2	1.4
Engine Coolant Temp. Sending Unit, Replace (B)		
924, 944, 968	.7	.9
928	.8	1.0
Fan Motor Resistor, Replace (B)		
1988-94 928	.9	1.1
968		
one	.7	.9
both	.7	.9
Magnetic Fan Clutch, Replace (B)		
928	.9	1.1
Radiator Assy., R&R or Replace (B)		
924	2.3	2.5
924S, 944, 968		
AT	3.3	3.5
MT	2.8	3.0
1981-87 928	3.0	3.2
1988-94 928, 928S	2.2	2.4
1988-94 928S4/GT	3.4	3.6
w/AT add	.5	.7
w/turbo 924, 924S, 944 add	1.5	1.7
Radiator Hoses, Replace (B)		
Includes: Refill with proper coolant mix.		
924		
one	.5	.7
all	1.4	1.6
928, 944, 968		
1981-87 928		
one	.7	.9
all	1.9	2.1
1983-87 944		
one	.5	.7
all	.7	.9
1988-95 one	1.9	2.1
Temperature Gauge (Dash), Replace (B)		
1988-91 944	1.5	1.7
1988-95 928, 968	1.7	1.9

924 : 928 : 944 : 968 **POR-21**

	LABOR TIME	SEVERE SERVICE
Temperature Gauge (Engine), Replace (B)		
924	.9	1.1
924S, 944, 968	.7	.9
928	.8	1.0
Water Pump and/or Gasket, Replace (B)		
Includes: Refill with proper coolant mix.		
924	2.4	2.6
924S, 944, 968	9.0	9.2
928		
1981-84	8.6	8.8
1985-94		
2V	9.5	9.7
4V	11.0	11.2
w/turbo 924, 924S, 944, 968 add	1.5	1.7
update belt housing 924, 924S, 944, 968 add	.5	.7

ENGINE
ASSEMBLY
Times shown are for OEM assemblies. Time to replace assemblies from after-market rebuilders may vary.

Engine Assy., R&I (B)
Does not include parts or component transfer.

924	9.2	9.4
924S, 944, 968 MT	14.0	14.2
928		
2V	18.4	18.6
4V	19.9	20.1
w/AC add	.5	.7
w/AT add		
924, 924S, 944, 968	1.5	1.7
928	1.3	1.5
w/Tiptronic trans. add	.9	1.1

Engine Assy., R&R and Recondition (A)
Includes: Replacing rings, rod and main bearings, cylinder head reconditioning and engine tune-up.

924	29.4	29.6
924S, 944 MT	39.6	39.8
928		
2V	49.2	49.4
4V	59.6	59.8
944S, 968	44.4	44.6
w/AC add	.5	.7
w/AT add		
924, 924S, 944, 968	1.5	1.7
928	1.3	1.5
w/Tiptronic trans. add	.9	1.1

Engine Assy. (Short Block), Replace (B)
Assembly consists of cylinder block, piston assemblies, crankshaft, camshaft, timing chain and gears. Does not include cylinder heads. Operation Includes: R&R engine, transfer necessary parts and all necessary adjustments.

924	25.7	25.9
924S, 944 MT	39.6	39.8
928		
2V	34.7	34.9
4V	40.0	40.2
944S, 968	44.4	44.6
w/AC add	.5	.7
w/AT add		
924, 924S, 944, 968	1.5	1.7
928	1.3	1.5
w/Tiptronic trans. add	.9	1.1

Engine Mounts, Replace (B)

924		
one	.8	1.0
both	1.6	1.8
924S, 944, 968		
right	2.7	2.9
left	1.7	1.9
both	3.8	4.0
928		
one	6.3	6.5
both	6.9	7.1
w/turbo add	.5	.7

CYLINDER HEAD
Compression Test (B)

924	1.4	1.6
924S, 944, 968	.9	1.1
928	1.9	2.1

Cylinder Head, Replace (B)
Includes: Parts transfer, adjustments.

924	7.9	8.1
924S	10.3	10.5
928		
2V		
one side	15.1	15.3
both sides	21.7	21.9
4V one side	26.2	26.4
944	12.9	13.1
944S	12.8	13.0
944S2	15.0	15.2
944 turbo	14.9	15.1
968	14.4	14.6

Cylinder Head Gasket, Replace (B)

924	5.8	6.0
924S	9.8	10.0
928		
2V		
one side	14.3	14.5
both sides	19.8	20.0
4V one side	25.4	25.6
944	9.9	10.1
944S	11.8	12.0
944S2	14.0	14.2
944 turbo	13.7	13.9
968	13.8	14.0

Valve/Cam Cover and/or Gasket, Replace (B)

924	.7	.9
924S, 944	5.0	5.2
928		
2V one	6.5	6.7
4V		
right	4.2	4.4
left	3.1	3.3
944S, 944S2	1.6	1.8
968	1.9	2.1

Valve Job (A)

924	12.6	12.8
924S	14.8	15.0
928		
2V		
one side	17.9	18.1
both sides	26.3	26.5
4V one side	31.4	31.6
944, 944S	18.2	18.4
944S2	20.9	21.1
944 turbo	18.4	18.6
968	23.3	23.5

Valve Lifters, Replace (B)

924	3.0	3.2
924S, 944	5.3	5.5
1981-93 928, each side	6.0	6.2

Valve Springs and/or Oil Seals, Replace (B)

924	3.4	3.6
924S	7.0	7.2
928 2V	16.3	16.5
944	7.0	7.2
944S2	9.9	10.1
944 turbo	9.2	9.4
968	11.1	11.3

CAMSHAFT
Balance Shaft Drive Belt, Adjust (B)

924S, 944, 968	2.0	2.2
w/turbo add	.5	.7

Timing Belt, Adjust (B)

924	1.2	1.4
924S, 944	2.3	2.5
928		
2V	3.0	3.2
4V	3.5	3.7
968	1.8	2.0
w/turbo 924, 924S, 944 add	.5	.7

POR-22 924 : 928 : 944 : 968

	LABOR TIME	SEVERE SERVICE
Balance Shaft, Replace (A)		
944S2		
right side	9.9	10.1
left side	10.5	10.7
944 turbo		
one side	21.7	21.9
both sides	27.3	27.5
968		
right side	9.1	9.3
left side	9.8	10.0
Balance Shaft Drive Belt, Replace (B)		
944	2.5	2.7
944S2	2.8	3.0
944 turbo	3.8	4.0
968	2.6	2.8
Replace tensioner 944, 944S2, 944 turbo add	.5	.7
Balance Shaft Seals, Replace (B)		
924S, 944, 968		
right side	3.3	3.5
left side	3.8	4.0
w/turbo add	.5	.7
Camshaft, Replace (A)		
924	2.9	3.1
924S, 944	5.0	5.2
928		
2V		
right	7.6	7.8
left	6.3	6.5
4V		
one	12.9	13.1
both	19.8	20.0
944S	6.7	6.9
968	8.1	8.3
Camshaft Housing, Replace (B)		
928		
K-Jetronic		
right	5.9	6.1
left	4.9	5.1
LH-Jetronic		
right	8.2	8.4
left	8.2	8.4
944 turbo	7.2	7.4
Timing Belt, Replace (B)		
924	3.3	3.5
924S, 944, 968	4.0	4.2
1981-91 928		
2V	5.6	5.8
4V	7.2	7.4
w/turbo 924, 924S, 944, 968 add	.5	.7
Timing Belt Tensioner, Replace (B)		
924	1.9	2.1
924S, 944	3.0	3.2
944S	4.0	4.2

	LABOR TIME	SEVERE SERVICE
928		
1981-89		
right	4.2	4.4
left	3.0	3.2
both	6.0	6.2
1990-94		
right	4.0	4.2
left	3.6	3.8
both	6.6	6.8
1990-91 944S2	4.5	4.7
944 turbo	4.0	4.2
968	3.3	3.5
Timing Chain, Replace (B)		
1985-91 928		
one side	12.9	13.1
both sides	13.1	13.3
944S, 944S2		
1987-88	7.0	7.2
1989-91	6.9	7.1
Timing Chain Tensioner, Replace (B)		
944S	2.0	2.2
968	2.0	2.2
Timing Cover and/or Gasket, Replace (B)		
924	.9	1.1
924S, 944	5.3	5.5
928	4.2	4.5
944S, 944S2	5.3	5.5
944 turbo	6.5	6.7
968	5.1	5.3
CRANK & PISTONS		
Connecting Rod Bearings Replace (A)		
924	5.5	5.7
924S, 944, 968	6.8	7.0
928	8.8	9.0
924S, 944, 968	6.8	7.0
w/turbo add	1.5	1.7
w/Tiptronic trans. add	.9	1.1
Crankshaft and Main Bearings, Replace (A)		
Includes: Engine R&R.		
924	9.9	10.1
924S, 944, 968 MT	21.5	21.7
928		
2V	39.8	40.0
4V	45.3	45.5
944S	23.5	23.7
968	24.9	25.1
w/AC add	.5	.7
w/AT add		
924, 924S, 944, 968	1.5	1.7
928	1.3	1.5
w/Tiptronic trans. add	.9	1.1
w/turbo 924, 924S, 944, 968 add	15.5	15.7

	LABOR TIME	SEVERE SERVICE
Crankshaft Front Oil Seal Replace (A)		
924	2.8	3.0
924S, 944	5.1	5.3
928		
1981-89		
2V	6.6	6.8
4V	7.1	7.3
1990-94	9.8	10.0
944S2, 944 turbo	6.3	6.5
968	5.3	5.4
Crankshaft Pulley, Replace (B)		
924	.8	1.0
924S	1.2	1.4
928	2.1	2.3
944	1.2	1.4
944S2, 944 turbo	1.6	1.8
968	2.2	2.3
Pistons and Connecting Rods, Replace (A)		
Includes: Ridge reaming, cylinder wall deglazing, installing new rings and rod bearings, engine tune-up. Time is for engine in vehicle; includes cylinder head and oil pan R & R.		
924	11.9	12.1
924S, 944, 968	16.9	17.1
928		
2V	35.0	35.2
4V	39.8	40.0
w/Tiptronic trans. 924, 924S, 944, 968 add	2.0	2.2
w/turbo 924, 924S, 944, 968 add	2.0	2.2
Rear Main Oil Seal, Replace (B)		
924	7.0	7.2
924S, 944, 968		
w/turbo	20.5	20.7
w/o turbo	12.7	12.9
924S, 944, 968	16.9	17.1
928		
1981-82		
AT		
A22	10.9	11.1
A28	13.5	13.7
MT	5.2	5.4
1983-86		
AT	13.3	13.5
MT	7.0	7.2
1987-94		
AT	13.3	13.5
MT	6.3	6.5
w/Tiptronic trans. 924, 924S, 944, 968 add	2.0	2.2

924 : 928 : 944 : 968 **POR-23**

	LABOR TIME	SEVERE SERVICE

Rings, Replace (A)
Includes: Ridge reaming, cylinder wall deglazing, installing new rings, engine tune-up. Time is for engine in vehicle; includes cylinder head and oil pan R & R.
- 924 .... 10.7 *10.9*
- 1983-89 924S, 944 .. 18.0 *18.2*
- 928
  - 2V 36.3 *36.5*
  - 4V 40.8 *41.0*
- 944S 20.7 *20.9*
- 944S2 .... 23.0 *23.2*
- 968
  - w/AC 21.9 *22.1*
  - w/o AC 21.8 *22.0*
- w/Tiptronic trans. 924, 924S, 944, 968 add 2.0 *2.2*
- w/turbo 924, 924S, 944, 968 add 2.0 *2.2*

ENGINE LUBRICATION
Engine Oil Cooler, Replace (B)
- 928 .... 2.3 *2.5*
- 924S, 944 .... 2.6 *2.8*
- 944 turbo 5.0 *5.2*
- 968 .... 1.9 *2.1*
- w/cruise control 928 add .. .5 *.7*

Engine Oil Pressure Switch (Sending Unit), Replace (B)
- 1981-82 .... .5 *.7*
- 1983-91 .... 1.1 *1.3*

Oil Pan and/or Gasket, Replace (B)
- 924 .... 3.7 *3.9*
- 1981-91 924S, 944 .. 8.4 *8.6*
- 928 .... 9.4 *9.6*
- 944 turbo 15.4 *15.6*
- 968 .... 6.9 *7.1*

Oil Pressure Gauge (Dash), Replace (B)
- 924 .... .7 *.9*
- 924S 9 *1.1*
- 928 .... 1.7 *1.9*
- 944 .... 1.6 *1.8*
- 968 .... 1.8 *2.0*

Oil Pump, Replace (B)
- 924 .... 3.8 *4.0*
- 924S, 944 .... 13.9 *14.1*
- 928
  - 2V 5.5 *5.7*
  - 4V 6.4 *6.6*
  - 928S4/GT 8.3 *8.5*
- 944 turbo 20.2 *20.4*
- 968 .... 11.8 *12.0*

Oil Temperature Sensor, Replace (B)
- 928, 944 .... .5 *.7*

CLUTCH
Bleed Clutch Hydraulic System (B)
- 1981-91 .... .5 *.7*

Clutch Pedal Free Play, Adjust (B)
- 1981-91 .... .5 *.9*

Clutch Assy., Replace (B)
- 924 .... 6.3 *6.5*
- 924S 11.9 *12.1*
- 928
  - 1981-86 .... 6.8 *7.0*
  - 1987-94 .... 7.4 *7.6*
- 944
  - w/turbo 20.3 *20.5*
  - w/o turbo 12.5 *12.7*
- 968 .... 6.8 *7.0*

Clutch Cable, Replace (B)
- 924 .... 1.1 *1.3*

Clutch Master Cylinder, Replace (B)
Includes: System bleeding.
- 924S 2.4 *2.6*
- 928 .... 5.3 *5.5*
- 944 .... 2.2 *2.4*
- 968 .... 2.5 *2.7*

Clutch Release Bearing, Replace (B)
- 924 .... 5.9 *6.1*
- 924S 11.8 *12.0*
- 928
  - 1981-86 .... 6.5 *6.7*
  - 1987-94 .... 7.3 *7.5*
- 944
  - w/turbo 20.2 *20.4*
  - w/o turbo 12.5 *12.7*
- 968 .... 6.3 *6.5*

Clutch Slave Cylinder, Replace (B)
Includes: System bleeding.
- 924S 1.8 *2.0*
- 928
  - 1981-86 .... 1.3 *1.5*
  - 1987-94 .... 1.6 *1.8*
- 944, 968 .... 1.7 *1.9*

Flywheel, Replace (B)
- 924
  - 4-Speed 5.2 *5.4*
  - 5-Speed 6.1 *6.3*
- 924S 12.4 *12.6*
- 928
  - 1981-86 .... 7.2 *7.4*
  - 1987-94 .... 7.6 *7.8*
- 944
  - w/turbo 20.7 *20.9*
  - w/o turbo 12.8 *13.0*
- 968 .... 7.1 *7.3*
- *Replace rear main seal add* .... .7 *.9*

MANUAL TRANSMISSION
Input Shaft Seal, Replace (B)
- 924 .... 5.0 *5.2*
- 924S, 944, 968 .... 5.8 *6.0*
- 928 .... 9.1 *9.3*
- w/turbo 924, 924S, 944, 968 add .... .7 *.9*

Transmission Assy., R&R and Recondition (A)
- 924 .... 11.7 *11.9*
- 924S, 944, 968 ... 18.1 *18.3*
- 928 .... 15.4 *15.6*
- w/turbo 924, 924S, 944, 968 add 1.0 *1.2*

Transmission Assy., R&R or Replace (B)
- 924 .... 6.2 *6.4*
- 924S, 944, 968 .... 5.8 *6.0*
- 928 .... 9.3 *9.5*
- w/turbo 924, 924S, 944, 968 add 1.0 *1.2*

Transmission Mount, Replace (B)
- 1981-91 924, 944 each .. .7 *.9*
- 968 each 1.7 *1.9*

AUTOMATIC TRANSMISSION
SERVICE TRANSMISSION INSTALLED

Transmission Performance Test (A)
- 1981-95 .... 1.0 *1.0*

Check Unit for Oil Leaks (C)
- 1981-95 .... .5 *.7*

Drain and Refill Unit (B)
- 924, 924S, 944, 968 .. 2.0 *2.2*
- 928 .... 2.0 *2.2*

Electronic Control Unit, Replace (B)
- 968 .... .7 *.9*

Governor Assy., R&R or Replace (B)
- 924 .... 1.1 *1.3*
- 924S, 944 .... 3.0 *3.2*
- *Recondition governor add* .... .3 *.5*

Oil Cooler, Replace (B)
- 924S, 944 .... 2.7 *2.9*

Oil Pan and/or Gasket, Replace (B)
- 924, 924S, 944 .... 1.1 *1.3*
- 928, 968 .... 1.4 *1.6*

Selector Cable, Replace (B)
- 928 .... 6.3 *6.5*
- 968 .... 2.3 *2.5*

Transmission Speed Sensor, Replace (B)
- All Models 1.2 *1.4*

Valve Body Assy., R&R and Recondition (A)
- 924, 944 .... 3.1 *3.3*

POR-24 924 : 928 : 944 : 968

	LABOR TIME	SEVERE SERVICE
Valve Body Assy., Replace (B)		
924, 944	1.8	2.0
928	2.6	2.8
968	2.5	2.7

SERVICE TRANSMISSION REMOVED
Transmission R&R included unless otherwise noted.

Solenoid Valve, Replace (B)
968	8.8	9.0

Torque Converter, Replace (B)
924	4.2	4.4
924S, 944	5.4	5.6
928		
A22	10.3	10.5
A28	13.4	13.6
968	6.9	7.1
Replace seal add	.2	.4

Transmission Assy., R&I (B)
924	4.2	4.4
924S, 944	5.3	5.5
928		
A22	10.2	10.4
A28	13.5	13.7
968	6.6	6.8
Replace assembly add	.5	.7

Transmission Assy., R&R and Recondition (A)
924	16.3	16.5
924S, 944	17.1	17.3
928		
A22	23.6	23.9
A28	26.4	26.6

SHIFT LINKAGE
AUTOMATIC TRANSMISSION
Gear Selector Lever, Replace (B)
924, 924S, 944	1.8	2.0
928	6.3	6.5

Selector Cable, Replace (B)
924, 924S, 944	2.8	3.0

MANUAL TRANSMISSION
Gearshift Linkage, Adjust (B)
1981-87	.8	1.0
1988-95	.5	.7

Gearshift Lever, Replace (B)
1981-95	.7	.9

DRIVELINE
DRIVESHAFT
Driveshaft, R&R or Replace (B)
924	5.9	6.1
928	12.8	13.0
944S2	7.7	7.9
944 turbo	8.6	8.8
968	10.5	10.7
w/turbo 924 add	1.2	1.4

DRIVE AXLE
Carrier Side Bearings, Replace (B)
924		
AT	7.1	7.3
MT	7.4	7.6
924S, 944	9.8	10.0
928		
A22	14.8	15.0
A28	16.8	17.0

Differential Assy., R&R or Replace (B)
924		
AT	15.7	15.9
MT	18.1	18.3
924S, 944	18.1	18.3
928		
A22	15.0	15.2
A28	17.4	17.6
968		
AT	19.2	19.4
MT	19.2	19.4

Drive Flange Shaft Seal, Replace (B)
928		
right	1.3	1.5
left	1.6	1.8
944, 968		
one	1.2	1.4
both	2.0	2.2

Final Drive Cover, Replace or Reseal (B)
968	4.9	5.1

Rear Axle Shaft, Replace (B)
924	1.0	1.2
924S, 944, 968	1.0	1.2
928	1.5	1.7

Rear Wheel Bearings and/or Hub, Replace (B)
924		
one side	2.1	2.3
both sides	3.6	3.8
928		
one side	3.1	3.3
both sides	5.9	6.1
924S, 944, 968		
one side	4.3	4.5
both sides	8.0	8.2

Ring Gear & Pinion Set, Replace (B)
924		
AT	18.2	18.4
MT	18.4	18.6
924S, 944		
AT	12.9	13.1
MT	18.5	18.7
928		
A22	15.0	15.2
A28	17.4	17.6
968	19.1	19.3

Stub Axle Assy., Replace (B)
924, 924S, 944, 968		
one	1.5	1.7
both	2.8	3.0

HALFSHAFTS
Halfshaft, R&R or Replace (B)
928		
one side	1.4	1.6
both sides	2.4	2.6
944, 968		
one side	1.4	1.6
both sides	2.4	2.6
Replace		
CV joint add each	.5	.7
CV joint boot add each	.5	.7

BRAKES
ANTI-LOCK
Diagnose Anti-Lock Brake System Component Each (A)
1987-95	1.0	1.0

Anti-Lock Relay, Replace (B)
1987-95	.5	.7

Control Unit, Replace (B)
1987-95	.5	.7

Front Sensor Assy., Replace (B)
One	.9	1.1
Both	1.6	1.8

Hydraulic Assy., Replace (B)
1988-94 928	2.9	3.1
1987-91 944	2.5	2.7
968	1.8	2.0

Pump & Motor Relay, Replace (B)
1987-95 928, 968	.5	.7

Rear Sensor Assy., Replace (B)
One	.7	.9
Both	1.3	1.5

Valve Relay, Replace (B)
1987-94 928	.5	.7

SYSTEM
Bleed Brakes (B)
Includes: Add fluid.
1981-95	.5	.7

Brake System, Flush and Refill (B)
1981-95	1.3	1.5

Brakes, Adjust (B)
Includes: Refill master cylinder.
1981-82 rear wheels	.5	.7

Brake Hose (Flexible), Replace (B)
Includes: System bleeding.
Front
1981-87		
one	1.9	2.1
both	2.2	2.4
1988-96		
one	1.4	1.6
both	1.9	2.1
Rear one	1.4	1.6

924 : 928 : 944 : 968 **POR-25**

	LABOR TIME	SEVERE SERVICE
Master Cylinder, R&R and Reconditioning (B)		
924, 924S	1.9	2.1
928	4.2	4.4
944, 944S	1.8	2.0
944 turbo	6.2	6.4
968	1.9	2.1
Master Cylinder, Replace (B)		
924, 924S	1.5	1.7
928	3.6	3.8
944, 944S	1.4	1.6
944 turbo	6.0	6.2
968	1.3	1.5
Power Booster Assy., Replace (B)		
924, 924S	2.3	2.5
928	5.2	5.4
944, 944S	2.4	2.6
944 turbo	7.4	7.6
968	3.4	3.6
Power Booster Vacuum Check Valve, Replace (B)		
1981-95	.3	.5

SERVICE BRAKES

	LABOR TIME	SEVERE SERVICE
Brake Drum, Replace (B)		
924		
one	.5	.7
both	.8	1.0
Caliper Assy., Replace (B)		
Includes: System bleeding.		
1981-95 one	1.7	1.9
Disc Brake Rotor, Replace (B)		
Front		
924		
one	1.9	2.1
both	3.3	3.5
924S		
one	1.4	1.6
both	3.0	3.2
928		
one	1.4	1.6
both	2.5	2.7
944		
one	1.5	1.7
both	2.6	2.8
968		
one	.9	1.1
both	1.9	2.1
Rear		
one	1.0	1.2
both	1.8	2.0
Pads and/or Shoes, Replace (B)		
Includes: Adjust service and parking brake. System bleeding.		
Front disc	1.4	1.6
Rear disc	1.4	1.6
Rear drum	2.4	2.6
Four wheels		
Rear drum	3.2	3.4
Rear disc	2.6	2.8

	LABOR TIME	SEVERE SERVICE
COMBINATION ADD-ONS		
Resurface		
brake rotor add	.5	.7
brake drum add	.5	.7
Wheel Cylinder, R&R and Reconditioning (B)		
Includes: System bleeding.		
924		
one	2.0	2.2
both	3.6	3.8
Wheel Cylinder, Replace (B)		
Includes: System bleeding.		
924		
one	1.8	2.0
both	3.0	3.2

PARKING BRAKE

	LABOR TIME	SEVERE SERVICE
Parking Brake Cable, Adjust (C)		
1981-95	.5	.7
Parking Brake Apply Actuator, Replace (B)		
924, 944, 968	.7	.9
928	1.8	2.0
Parking Brake Apply Warning Indicator Switch, Replace (B)		
928	.9	1.1
944, 968	.5	.7
Parking Brake Cable, Replace (B)		
924	2.1	2.3
924S		
one	1.5	1.7
both	2.2	2.4
928	2.5	2.7
944	2.7	2.9
968		
right	1.4	1.6
left	2.1	2.3
Parking Brake Shoes, Replace (B)		
928		
1981-87	2.5	2.7
1988-94	3.1	3.3
968		
one side	2.0	2.2
both sides	3.7	3.9

FRONT SUSPENSION

Unless otherwise noted, time given does not include alignment.

	LABOR TIME	SEVERE SERVICE
Align Front End (A)		
1981-95	2.3	2.5
Front & Rear Alignment, Check & Adjust (A)		
928, 944	4.2	4.4
968	3.8	4.0
Front Toe, Adjust (B)		
1981-95	1.2	1.4
Ball Joint, Replace (B)		
924, 924S, 944, 968 one	1.8	2.0
928 one	1.3	1.5

	LABOR TIME	SEVERE SERVICE
Front Axle Hub Oil Seal, Replace (B)		
968		
one side	1.6	1.8
both sides	3.1	3.3
Front Hub, Replace (B)		
924, 944 one	1.8	2.0
968		
one side	1.8	2.0
both sides	3.5	3.7
Front Hub Bearings, Replace (B)		
924, 944		
1981-87 one side	1.4	1.6
1988-91 one side	1.4	1.6
928		
1981-91 one side	1.4	1.6
968		
one side	1.6	1.8
both sides	3.0	3.2
Front Strut Assy., R&R or Replace (B)		
924, 944		
one	1.2	1.4
both	2.0	2.2
928		
one	3.6	3.8
both	5.4	5.6
968		
one	1.2	1.4
both	2.0	2.2
Front Strut Cartridge, Replace (B)		
924, 944		
one	1.5	1.7
both	2.9	3.1
928		
one	4.5	4.7
both	7.5	7.7
968		
one	1.7	1.9
both	3.2	3.4
Lower Control Arm Assy., Replace (B)		
924, 924S, 944, 968	1.6	1.8
928		
1981-89	1.2	1.4
1990-94	1.5	1.7
Replace bushings add each side	.5	.7
Stabilizer Bar, Replace (B)		
1981-91 924, 944	.9	1.1
928		
1981-87	1.3	1.5
1988-94 928, 928S	.9	1.1
1988-94 928S4/GT	1.5	1.7
968	.9	1.1

924 : 928 : 944 : 968

	LABOR TIME	SEVERE SERVICE
Steering Knuckle, Replace (B)		
924, 944	1.4	1.6
928		
1981-87	3.0	3.2
1988-94	2.4	2.6
968		
one side	1.5	1.7
both sides	2.7	2.9
Upper Control Arm Assy., Replace (B)		
928, 928S	3.0	3.3
1990-94 928S4/GT	3.8	4.0
Wheel Bearings, Clean & Pack (B)		
1981-95 both wheels	1.8	2.0

REAR SUSPENSION

Unless otherwise noted, time given does not include alignment.

	LABOR TIME	SEVERE SERVICE
Rear Toe, Check and Adjust (B)		
928, 944, 968	2.4	2.6
Lower Control Arm, Replace (B)		
928		
one side	2.3	2.5
both sides	4.3	4.5
944, 968		
one	2.8	3.0
both	4.9	5.1
Rear Coil Spring, Replace (B)		
928		
one	2.8	3.0
both	5.2	5.4
Rear Strut Assy., R&R or Replace (B)		
928		
one	1.8	2.0
both	3.0	3.2
Rear Strut Cartridge, Replace (B)		
928		
one	2.7	2.9
both	5.0	5.2
Shock Absorber or Bushings, Replace (B)		
924, 944, 968		
one	.5	.7
both	.9	1.1
Spring Supports and/or Torsion Bars, Replace (B)		
924		
one	5.1	5.3
both	7.5	7.7
928, 944, 968		
one	7.5	7.7
both	9.0	9.2
Stabilizer Bar & Bushings, Rear, Replace (B)		
924	2.3	2.5
928	5.3	5.5
944, 968	1.5	1.7

STEERING

AIR BAGS

	LABOR TIME	SEVERE SERVICE
Diagnose Air Bag System Component Each (A)		
1990-95	1.0	1.0
Air Bag Unit, Replace (B)		
928	.5	.7
944S2, 944 turbo, 968		
driver side	.5	.7
passenger side	2.4	2.6
Contact Coil Initiator, Replace (B)		
928	1.1	1.3
944S2, 944 turbo	1.4	1.6
Crash Sensor, Replace (B)		
944S2, 944 turbo, 968		
driver side	1.4	1.6
passenger side	1.0	1.2
Electronic Control Unit, Replace (B)		
944S2, 944 turbo	5.4	5.6
968	1.9	2.1

MANUAL RACK & PINION

Unless otherwise noted, time given does not include alignment.

	LABOR TIME	SEVERE SERVICE
Outer Tie Rod End, Replace (B)		
1981-86		
one side	1.2	1.4
both sides	2.1	2.3
Rack & Pinion Assy., Replace (B)		
924, 944	2.1	2.3
Rack Boots, Replace (B)		
924, 944	1.7	1.9
Steering Shaft U-Joint, Replace (B)		
924, 944		
1981-86	1.2	1.4
1987-91	1.6	1.8
Steering Wheel, Replace (B)		
924, 944	.5	.7
w/air bags add	.2	.2
Upper Mast Jacket Bearing, Replace (B)		
Includes: Insulators.		
924	.9	1.1
944		
w/air bags	1.9	2.1
w/o air bags	1.7	1.9

POWER RACK & PINION

Unless otherwise noted, time given does not include alignment.

	LABOR TIME	SEVERE SERVICE
Inner Tie Rod Ends, Replace (B)		
1988-94 928		
one side	1.8	2.0
both sides	2.6	2.8
1988-91 944		
one side	1.8	2.0
both sides	2.6	2.8
968		
one side	1.6	1.8
both sides	2.5	2.7
Outer Tie Rod Ends, Replace (B)		
1988-94 928		
one side	.8	1.0
both sides	1.5	1.7
1988-91 944		
one side	.8	1.0
both sides	1.5	1.7
968		
one side	.8	1.0
both sides	1.5	1.7
Rack & Pinion Assy., R&R or Replace (B)		
924, 944	4.1	4.3
928	3.3	3.5
968	3.6	3.8
Rack Assy. (Short), Replace (A)		
968	4.2	4.4
Rack Boots, Replace (B)		
One	1.0	1.2
Both	1.8	2.0
Steering Pump, R&R or Replace (B)		
928		
1981-87	1.6	1.8
1988-94	2.2	2.4
1990-94 928S4/GT	2.3	2.5
944	2.2	2.4
968	1.8	2.0
Steering Shaft U-Joint, Replace (B)		
924, 944, 968		
1981-86	1.5	1.7
1987-95	1.8	2.0
1988-94 928	4.3	4.5
Steering Wheel, Replace (B)		
1981-95	.5	.7
w/air bags add	.2	.4
Upper Mast Jacket Bearing, Replace (B)		
Includes: Insulators.		
924, 928		
1981-87	.7	.9
1988-94	1.5	1.7
944		
w/air bags	1.8	2.0
w/o air bags	1.4	1.6
968	1.8	2.0

HEATING & AIR CONDITIONING

When more than one component requires replacement where evacuation/recovery and recharging is already included, deduct 1.0 hour for each additional component from the time given.

	LABOR TIME	SEVERE SERVICE
Evacuate/Recover and Recharge System (B)		
1981-95	1.3	1.5

924 : 928 : 944 : 968   POR-27

	LABOR TIME	SEVERE SERVICE
AC Hoses, Replace (B)		
Includes: Evacuate/recover and recharge.		
1981-87		
discharge	2.0	2.2
suction	2.7	2.9
1988-95		
discharge	3.3	3.5
suction	3.4	3.6
Blower Motor, Replace (B)		
1987-88 924S	4.2	4.4
968	1.7	1.9
Compressor Assy., Replace (B)		
Includes: Parts transfer. Evacuate/recover and recharge.		
924	2.8	3.0
924S	3.7	3.9
928	3.5	3.7
944, 968	3.6	3.8
944 turbo	3.8	4.0
Compressor Pressure Switch, Replace (B)		
968	2.9	3.1
Add time to evacuate/recover & recharge.		
Compressor Shaft Seal, Replace (B)		
Includes: Compressor R&R. Evacuate/recover and recharge.		
924	3.5	3.7
924S	4.5	4.7
928, 944	4.0	4.2
968	3.6	3.8
Condenser Assy., Replace (B)		
Includes: Evacuate/recover and recharge.		
924	2.4	2.6
924S	3.2	3.4
928	4.2	4.4
944, 944S	3.5	3.7
944 turbo	3.8	4.0
968	3.4	3.6
w/fog lamps 944 add	.5	.7
Condenser Cooling Fan, Replace (B)		
944, 944S	1.3	1.5
944 turbo	3.3	3.5
Evaporator Coil, Replace (B)		
Includes: Evacuate/recover and recharge.		
924	3.1	3.3
924S	4.4	4.6
928	12.4	12.6
944	4.5	4.7
968	12.8	13.0
Expansion Valve, Replace (B)		
Includes: Evacuate/recover and recharge.		
924	2.4	2.6
924S	5.1	5.3

	LABOR TIME	SEVERE SERVICE
928	4.0	4.2
944	4.8	5.0
944S2, 944 turbo	3.0	3.2
968	4.1	4.3
Heater Blower Motor, Replace (B)		
924, 924S, 944	1.4	1.6
928	2.5	2.7
944 turbo	1.8	2.0
968	1.6	1.8
Heater Blower Motor Relay, Replace (B)		
928, 944, 968	.5	.7
Heater Blower Motor Resistor, Replace (B)		
928GTS	1.7	1.9
944, 968	.5	.7
Heater Blower Motor Switch, Replace (B)		
1988-91 944	.5	.7
Heater Control Assy., Replace (B)		
924	2.4	2.6
928	3.3	3.5
944		
1983-87	7.2	7.4
1988-91	.7	.9
Heater Control Valve, Replace (B)		
924S, 944	2.4	2.6
928, 944S, 968	1.7	1.9
944 turbo	2.3	2.5
Heater Core, R&R or Replace (B)		
924	2.0	2.2
924S, 944	9.5	9.7
928		
1981-87	8.9	9.1
1988-94	10.0	10.2
968	7.9	8.1
Heater Hoses, Replace (B)		
928 one	2.5	2.7
944 one	2.1	2.3
968 one	1.7	1.9
Receiver/Drier Assy., Replace (B)		
Includes: Evacuate/recover and recharge.		
1981-95	3.3	3.5

WIPERS & SPEEDOMETER

	LABOR TIME	SEVERE SERVICE
Power Antenna, Replace (B)		
1987-88 924S	1.8	2.0
1988-94 928	1.4	1.6
Radio, R&R (B)		
924	1.3	1.5
1981-87 928	1.7	1.9
1983-87 944	1.4	1.6
1988-95 928, 944, 968	.5	.7
Rear Window Wiper Motor, Replace (B)		
1981-95	.5	.7

	LABOR TIME	SEVERE SERVICE
Rear Window Wiper Switch, Replace (B)		
All Models	.7	.7
Speedometer Cable & Casing, Replace (B)		
924	1.5	1.7
924S	1.4	1.6
Speedometer Head, R&R or Replace (B)		
924	.8	1.0
924S	.9	1.1
928	1.5	1.7
944	1.4	1.6
968	1.6	1.8
Windshield Washer Pump, Replace (B)		
924, 944, 968	.5	.7
928	2.3	2.5
Windshield Wiper & Washer Switch, Replace (B)		
924, 924S, 944, 928	1.4	1.6
968	1.5	1.7
Windshield Wiper Motor, Replace (B)		
924	.8	1.0
924S, 928	1.5	1.7
944	2.2	2.4
968	2.3	2.5

LAMPS & SWITCHES

	LABOR TIME	SEVERE SERVICE
Back-Up Lamp Assy., Replace (B)		
1981-95	.5	.5
Back-Up Lamp Switch, Replace (B)		
1981-95	.5	.5
Brake Lamp Switch, Replace (B)		
1981-87	1.3	1.3
1988-95	.9	.9
Composite Headlamp Bulb, Replace (C)		
928, 944, 968 one	.3	.3
Hazard Warning Switch, Replace (B)		
944, 968	.9	.9
Headlamp Switch, Replace (B)		
All Models	.5	.5
Headlamps, Aim (B)		
Two	.5	.5
Four	.7	.7
Horn, Replace (B)		
924, 944	.5	.5
928	.5	.5
968		
one	1.3	1.3
both	1.5	1.5
Horn Relay, Replace (B)		
1981-95	.5	.5
License Lamp Assy., Replace (B)		
968 one or all	.2	.2
Parking Lamp Lens or Bulb, Replace (C)		
1981-95	.2	.2

POR-28 924 : 928 : 944 : 968

	LABOR TIME	SEVERE SERVICE
Sealed Beam Headlamp, Replace (B)		
1981-85 924, 928	.3	.5
1983-89 944	.5	.7
Stop Lamp Switch, Replace (B)		
968	.5	.5
Tail Lamp Lens or Bulb, Replace (C)		
1981-95	.2	.2
Turn Signal Lamp Lens or Bulb, Replace (B)		
1981-95	.2	.2
Turn Signal or Hazard Warning Flasher, Replace (B)		
1981-95	.5	.5
Turn Signal Switch, Replace (B)		
All Models	1.4	1.4

BODY

	LABOR TIME	SEVERE SERVICE
Door Lock, Replace (B)		
968	3.3	3.5
Front Door Window Regulator and/or Motor, Replace (B)		
924	1.8	2.0
924S, 944		
electric	2.8	3.0
manual	2.8	3.0
928	3.1	3.3
968		
electric	2.7	2.9
manual	2.5	2.7

	LABOR TIME	SEVERE SERVICE
Hood Lock, Replace (B)		
1981-95	2.0	2.2
Hood Release Cable, Replace (B)		
924	.7	.9
928	3.8	4.0
944 turbo	2.5	2.7
944S, 968	1.7	1.9
Lock Striker Plate, Replace (B)		
1981-95	.5	.7

REN

Alliance : Encore : Fuego : Lecar : R181

SYSTEM INDEX

MAINTENANCE	REN-2
CHARGING	REN-2
STARTING	REN-2
CRUISE CONTROL	REN-2
IGNITION	REN-2
EMISSIONS	REN-2
FUEL	REN-3
EXHAUST	REN-3
ENGINE COOLING	REN-3
ENGINE	REN-4
Assembly	REN-4
Cylinder Head	REN-4
Camshaft	REN-4
Crank & Pistons	REN-5
Engine Lubrication	REN-5
CLUTCH	REN-5
MANUAL TRANSAXLE	REN-5
AUTO TRANSAXLE	REN-6
SHIFT LINKAGE	REN-6
DRIVELINE	REN-6
BRAKES	REN-6
FRONT SUSPENSION	REN-7
REAR SUSPENSION	REN-7
STEERING	REN-7
HEATING & AC	REN-8
WIPERS & SPEEDOMETER	REN-8
LAMPS & SWITCHES	REN-8
BODY	REN-8

OPERATIONS INDEX

A
AC Hoses	REN-8
Air Conditioning	REN-8
Alignment	REN-7
Alternator (Generator)	REN-2

B
Back-Up Lamp Switch	REN-8
Ball Joint	REN-7
Battery Cables	REN-2
Bleed Brake System	REN-6
Blower Motor	REN-8
Brake Disc	REN-6
Brake Drum	REN-6
Brake Hose	REN-6
Brake Pads and/or Shoes	REN-6

C
Camshaft	REN-4
Catalytic Converter	REN-3
Coolant Temperature (ECT) Sensor	REN-3
Crankshaft	REN-5
Cruise Control	REN-2
Cylinder Head	REN-4

D
Distributor	REN-2
Drive Belt	REN-2

E
EGR	REN-3
Electronic Control Module (ECM/PCM)	REN-3
Engine	REN-4
Engine Lubrication	REN-5
Engine Mounts	REN-4
Evaporator	REN-8
Exhaust	REN-3
Exhaust Manifold	REN-3

F
Flexplate	REN-6
Flywheel	REN-5
Fuel Injection	REN-3
Fuel Pump	REN-3
Fuel Vapor Canister	REN-3

G
Gear Selector Lever	REN-6
Generator	REN-2

H
Halfshaft	REN-6
Headlamp	REN-8
Heater Core	REN-8
Horn	REN-8

I
Ignition Coil	REN-2
Ignition Module	REN-2
Ignition Switch	REN-2
Inner Tie Rod	REN-7
Intake Air Temperature (IAT) Sensor	REN-3
Intake Manifold	REN-3

L
Lower Control Arm	REN-7

M
Manifold Absolute Pressure (MAP) Sensor	REN-3
Master Cylinder	REN-6
Muffler	REN-3

O
Oil Pan	REN-5
Oil Pump	REN-5
Outer Tie Rod	REN-7
Oxygen Sensor	REN-3

P
Parking Brake	REN-7
Pistons	REN-5
Positive Crankcase Ventilation (PCV) Valve	REN-3

R
Radiator	REN-3
Radiator Hoses	REN-3
Rear Main Oil Seal	REN-5

S
Shock Absorber/Strut, Front	REN-7
Shock Absorber/Strut, Rear	REN-7
Spark Plug Cables	REN-2
Spark Plugs	REN-2
Spring, Front Coil	REN-7
Spring, Rear Coil	REN-7
Starter	REN-2
Steering Wheel	REN-7

T
Thermostat	REN-3
Throttle Body	REN-3
Throttle Position Sensor (TPS)	REN-3
Timing Belt	REN-5
Timing Chain	REN-5
Torque Converter	REN-6

U
Upper Control Arm	REN-7

V
Valve Body	REN-6
Valve Cover Gasket	REN-4
Valve Job	REN-4

W
Water Pump	REN-4
Wheel Balance	REN-2
Wheel Cylinder	REN-7
Window Regulator	REN-8
Windshield Washer Pump	REN-8
Windshield Wiper Motor	REN-8

REN-2   ALLIANCE : ENCORE : FUEGO : LECAR : R181

	LABOR TIME	SEVERE SERVICE
MAINTENANCE		
Air Cleaner Filter Element, Replace (B)		
1981-87	.2	.3
Chassis Lubrication, Change Oil & Filter (C)		
1981-87	.4	.4
Drive Belt, Adjust (B)		
Serpentine	.3	.4
V belt		
one	.2	.3
each addl. add	.2	.3
Drive Belt, Replace (B)		
Serpentine	.6	.7
V belt		
one	.5	.6
each addl. add	.2	.3
Oil & Filter, Change (C)		
1981-87	.4	.4
Sealed Beam Headlamp, Replace (C)		
1981-87	.3	.3
Tire, Replace (C)		
Includes: Dismount old tire and mount new tire to rim.		
All Models	.5	.5
Wheel, Balance (B)		
One	.3	.3
each addl. add	.2	.2
CHARGING		
Alternator Circuits, Test (B)		
Includes: Test component output.		
1981-87	.5	.5
Alternator Assy., R&R and Recondition (A)		
Alliance, Encore, LeCar	2.8	3.0
Fuego, R18i	3.0	3.2
2.2L or turbo add	.2	.2
Alternator Assy., Replace (B)		
Includes: Pulley transfer.		
Alliance, Encore, LeCar	.8	1.0
Fuego, R18i	1.3	1.5
2.2L or turbo add	.2	.2
Alternator Bearings, Replace (B)		
Includes: Alternator R&R.		
Alliance, Encore, LeCar	2.1	2.3
Fuego, R18i	2.5	2.7
2.2L or turbo add	.2	.2
Alternator Voltage Regulator, Replace (B)		
Alliance, Encore, LeCar	.7	.9
Fuego, R18i	.9	1.1

	LABOR TIME	SEVERE SERVICE
STARTING		
Starter Draw Test (On Car) (B)		
1981-87	.5	.5
Battery Cables, Replace (C)		
Each	.3	.5
Starter Assy., R&R and Recondition (A)		
Alliance, Encore		
1.4L	2.8	3.0
1.7L	4.1	4.3
Fuego, R18i		
1.6L		
carburetor	3.8	4.0
FI or turbo	5.4	5.6
2.2L	3.8	4.0
LeCar	4.8	5.0
Starter Assy., Replace (B)		
Alliance, Encore		
1.4L	.8	1.0
1.7L	1.9	2.1
Fuego, R18i		
1.6L		
carburetor	1.3	1.5
FI or turbo	2.9	3.1
2.2L	1.6	1.8
LeCar	3.3	3.5
Starter Drive Assy., Replace (B)		
Includes: Starter R&R.		
Alliance, Encore		
1.4L	1.3	1.5
1.7L	2.4	2.6
Fuego, R18i		
1.6L		
carburetor	1.8	2.0
FI or turbo	3.4	3.6
2.2L	2.1	2.3
LeCar	3.9	4.1
Starter Relay, Replace (B)		
Fuego, R18i	.8	.8
Starter Solenoid and/or Switch, Replace (B)		
Includes: Starter R&R.		
Alliance, Encore		
1.4L	1.3	1.5
1.7L	2.4	2.
Fuego, R18i		
1.6L		
carburetor	1.8	2.0
FI or turbo	3.4	3.6
2.2L	2.1	2.3
LeCar	3.8	4.0
CRUISE CONTROL		
Control Regulator, Replace (B)		
Alliance, Encore	.6	.6
Fuego, R18i	.5	.5
Control Sensor, Replace (B)		
Alliance, Encore	.5	.7
Fuego, R18i	1.0	1.2

	LABOR TIME	SEVERE SERVICE
Control Servo, Replace (B)		
Alliance, Encore	.5	.7
Fuego, R18i	.8	1.0
Control Switch, Replace (B)		
Alliance, Encore	.5	.7
Fuego, R18i	.8	1.0
Control Throttle or Servo Cable, Replace (B)		
1981-87	.5	.7
Control Vacuum Release Switch, Replace (B)		
Alliance, Encore	.5	.7
IGNITION		
Ignition Timing, Reset (B)		
1981-87	.3	.3
Distributor, Replace (B)		
Includes: Reset base ignition timing.		
1981-87	.6	.8
Distributor Cap and/or Rotor, Replace (B)		
1981-87	.3	.3
Distributor Vacuum Control Unit, Replace (B)		
1981-87	.8	1.0
Electronic Ignition Control Unit, Replace (B)		
1981-87	.4	.6
Ignition Coil, Replace (B)		
Alliance, Encore, LeCar	.4	.6
Fuego, R18i	.6	.8
Ignition Module Assy., Replace (B)		
Exc. Fuego	.5	.7
Fuego	.7	.9
Ignition Switch, Replace (B)		
Exc. LeCar	1.0	1.2
LeCar	1.5	1.7
Speed Sensor, Replace (B)		
1983-87	.5	.7
Spark Plug (Ignition) Cables, Replace (B)		
1981-87	.4	.6
Spark Plugs, Replace (B)		
1981-87	.6	.8
TDC Pickup, Replace (B)		
1981-87	.4	.6
EMISSIONS		
Air Control Valve, Replace (B)		
1981-87	.5	.5
Air Manifold Pipe, Replace (B)		
1981-87	2.5	2.7
Air Pump Assy., Replace (B)		
1981-87	1.3	1.5
Air Pump Centrifugal Filter, Replace (B)		
1981-87	.4	.6
Anti-Backfire Valve, Replace (B)		
1981-87	.3	.5

ALLIANCE : ENCORE : FUEGO : LECAR : R181   REN-3

	LABOR TIME	SEVERE SERVICE
Coolant Temperature (ECT) Sensor and/or Switch, Replace (B)		
1981-87	.5	.7
EGR Control Valve, Replace (B)		
1981-87	.7	.9
EGR Vacuum Amplifier, Replace (B)		
1981-87	.3	.3
EGR Valve Solenoid, Replace (B)		
1981-87	.5	.7
Electronic Control Module (ECM/PCM), Replace (B)		
1981-87	.5	.7
Fuel Vapor Canister, Replace (B)		
1981-87	.6	.8
Injection Tube Check Valve Assy., Replace (B)		
1981-87	.3	.5
Manifold Air Temperature Sensor, Replace (B)		
1981-87	.5	.7
Manifold Pressure Sensor, Replace (B)		
1981-87	.5	.7
Oxygen Sensor, Replace (B)		
1981-87	.5	.7
Positive Crankcase Ventilation (PCV) Valve, Replace (B)		
1981-87	.3	.3
Reed Valve, Replace (B)		
1981-87	.5	.7
Thermo Valve, Replace (B)		
1981-87	.7	.9
Throttle Position Sensor (TPS), Replace (B)		
1981-87	.8	1.0
Transmission Switch, Replace (B)		
1981-87	.5	.7
Vacuum Delay Valve, Replace (B)		
1981-87	.4	.6
Vapor Separator, Replace (B)		
1981-87	.4	.6

FUEL

CARBURETOR

	LABOR TIME	SEVERE SERVICE
Carburetor, R&R and Clean or Recondition (A)		
Includes: Adjustments.		
1981-87	2.0	2.2
Carburetor, Replace (B)		
Includes: Adjustments.		
1981-87	1.2	1.4

DELIVERY

	LABOR TIME	SEVERE SERVICE
Fuel Pump, Test (B)		
1981-87	.3	.3
Fuel Gauge (Dash), Replace (B)		
Alliance	1.6	1.8
Fuego, R18i	1.8	2.0
LeCar	1.2	1.4
Fuel Gauge (Tank), Replace (B)		
Alliance, Encore	1.2	1.4
Fuego, R18i	1.1	1.3
LeCar	2.2	2.4
Fuel Pump, Replace (B)		
In gas tank	1.3	1.5
On engine	.7	.9
Fuel Tank, Replace (B)		
Includes: Drain and refill.		
Alliance, Encore	1.1	1.3
Fuego	2.0	2.2
LeCar	1.5	1.7
R18i		
sedan	1.5	1.7
station wagon	1.7	1.9
Intake and Exhaust Manifold or Gaskets, Replace (B)		
Alliance, Encore	2.3	2.5
LeCar	3.5	3.7
R18i 1.6L	2.6	2.8
Replace manifold add each	.5	.8
Intake Manifold and/or Gasket, Replace (B)		
Includes: Adjustments.		
Fuego, R18i	2.8	3.0
Replace manifold add	.5	.7

INJECTION

	LABOR TIME	SEVERE SERVICE
Diagnose Injection System Component Each (A)		
1981-87	1.0	1.0
Injection Timing, Check & Adjust (B)		
Alliance, Encore	.8	1.0
Fuego, R18i	1.7	1.9
Air Flow Meter, Replace (B)		
Does not include testing.		
1981-87	.6	.8
Auxiliary Air Regulator, Replace (B)		
1981-87	.8	1.0
Cold Start Injector, Replace (B)		
1.6L	.5	.7
2.2L	2.0	2.2
Fuel Injectors, Replace (B)		
One	.5	.7
All	1.0	1.2
Fuel Pressure Regulator, Replace (B)		
1981-87	.7	.9
Idle Speed Motor and/or Actuator, Replace (B)		
Alliance, Encore	.9	1.1
Thermo Time Switch, Replace (B)		
1981-87	.5	.7
Throttle Body Assy., Replace (B)		
Alliance, Encore	.8	1.0
Wide Open Throttle Switch, Replace (B)		
Alliance, Encore	.6	.8

	LABOR TIME	SEVERE SERVICE
TURBOCHARGER		
Intercooler, Replace (B)		
Fuego	1.0	1.2
Turbocharger Assy., R&R or Replace (B)		
Fuego	2.8	3.0

EXHAUST

	LABOR TIME	SEVERE SERVICE
Catalytic Converter, Replace (B)		
Alliance, Encore	.9	1.1
Fuego, LeCar, R18i	1.5	1.7
Exhaust Manifold, Replace (B)		
Fuego, R18i		
w/turbo	2.9	3.1
w/o turbo	1.3	1.5
Front Exhaust Pipe, Replace (B)		
Alliance, Encore	.5	.7
Intake and Exhaust Manifold or Gaskets, Replace (B)		
Alliance, Encore	2.0	2.2
LeCar	3.2	3.4
R18i 1.6L	2.3	2.5
Replace manifold add each	.5	.7
Intermediate Exhaust Pipe, Replace (B)		
Alliance, Encore	.7	.9
Fuego, LeCar, R18i	1.2	1.4
Muffler, Replace (B)		
1981-87	1.2	1.4
Rear Exhaust Pipe, Replace (B)		
1981-87	1.1	1.3
Resonator, Replace (B)		
1981-87	1.2	1.4
Tail Pipe, Replace (B)		
Alliance, Encore	.6	.8

ENGINE COOLING

	LABOR TIME	SEVERE SERVICE
Coolant Temperature Sensor Switch, Replace (B)		
LeCar	.6	.8
Coolant Thermostat, Replace (B)		
1981-87	.5	.7
Radiator Assy., R&R or Replace (B)		
Alliance, Encore	1.0	1.2
Fuego, R18i	1.6	1.8
LeCar	2.6	2.8
w/AC add	.5	.5
Radiator Fan and/or Fan Motor, Replace (B)		
One	1.1	1.3
Both	1.4	1.6
Radiator Fan Motor Relay, Replace (B)		
1981-87	.5	.7
Radiator Hoses, Replace (B)		
Includes: Refill with proper coolant mix.		
Upper	.4	.6
Lower	.6	.8
Both	.9	1.1

REN-4 ALLIANCE : ENCORE : FUEGO : LECAR : R181

	LABOR TIME	SEVERE SERVICE
Temperature Sending Unit (Engine), Replace (B)		
1981-87	.5	.7
Water Pump and/or Gasket, Replace (B)		
Includes: Refill with proper coolant mix.		
Alliance, Encore		
1.4L	1.7	1.9
1.7L	3.1	3.3
Fuego, R18i	2.6	2.8
LeCar	3.9	4.1

ENGINE
ASSEMBLY

Times shown are for OEM assemblies. Time to replace assemblies from aftermarket rebuilders may vary.

Engine Assy., R&I (B)
Does not include parts or component transfer.

	LABOR TIME	SEVERE SERVICE
Alliance, Encore		
1.4L, 1.7L	6.4	6.6
Fuego		
1.6L		
w/o turbo	8.8	9.0
w/turbo	10.3	10.5
2.2L	10.3	10.5
LeCar	8.1	8.3
R18i		
1.6L		
carburetor	6.4	6.6
FI	9.5	9.7
2.2L	7.8	8.0

Engine Assy., R&R and Recondition (A)
Includes: Replacing rings, rod and main bearings, cylinder head reconditioning and engine tune-up.

	LABOR TIME	SEVERE SERVICE
Alliance, Encore		
1.4L	22.4	22.6
1.7L	20.9	21.1
Fuego		
1.6L		
w/turbo	30.4	30.6
w/o turbo	29.0	29.2
2.2L	27.0	27.2
LeCar	22.9	23.1
R18i		
1.6L		
carburetor	25.8	26.0
FI	30.1	30.3
2.2L	27.0	27.2

Engine Assy. (Complete), Replace (B)
Includes: Component transfer and engine tune-up.

	LABOR TIME	SEVERE SERVICE
Alliance, Encore		
1.4L, 1.7L	10.8	11.0
Fuego		
1.6L		
w/turbo	14.6	14.8
w/o turbo	13.0	13.2
2.2L	14.6	14.8
LeCar	14.3	14.5
R18i		
1.6L		
carburetor	11.8	12.0
FI	15.0	15.2
2.2L	13.2	13.4

Engine Assy. (Short) Replace (B)
Assembly consists of cylinder block, piston assemblies, crankshaft, camshaft, timing chain and gears. Does not include cylinder heads. Operation Includes: R&R engine, transfer necessary parts and all necessary adjustments.

	LABOR TIME	SEVERE SERVICE
Alliance, Encore		
1.4L	18.1	18.3
1.7L	20.6	20.8
Fuego		
1.6L		
w/turbo	21.2	21.4
w/o turbo	19.7	19.9
2.2L	21.2	21.4
LeCar	19.7	19.9
R18i		
1.6L		
carburetor	16.9	17.1
FI	19.7	19.9

Engine Mounts, Replace (B)

	LABOR TIME	SEVERE SERVICE
Alliance, Encore	1.4	1.6
LeCar		
right side	.9	1.1
left side	4.0	4.2
both sides	4.8	5.0

CYLINDER HEAD

Compression Test (B)
1981-87 8   1.0

Valve Clearance, Adjust (B)
	LABOR TIME	SEVERE SERVICE
1.4L, 1.6L, 2.2L	1.3	1.5
1.7L	1.9	2.1

Cylinder Head, Replace (B)
Includes: Parts transfer and adjustments.

	LABOR TIME	SEVERE SERVICE
Alliance, Encore, LeCar		
1.4L	8.9	9.1
1.7L	7.9	8.1
Fuego		
1.6L		
w/turbo	10.0	10.2
w/o turbo	9.8	10.0
2.2L	9.3	9.5
R18i		
1.6L		
carburetor	8.7	8.9
FI	9.8	10.0
2.2L	9.3	9.5

Cylinder Head Gasket, Replace (B)

	LABOR TIME	SEVERE SERVICE
Alliance, Encore, LeCar		
1.4L	7.2	7.4
1.7L	4.7	4.9
Fuego		
1.6L		
w/turbo	8.3	8.5
w/o turbo	7.8	8.0
2.2L	7.1	7.3
R18i		
1.6L		
carburetor	6.7	6.9
FI	7.8	8.0
2.2L	7.1	7.3

Rocker Arm Cover Gasket, Replace or Reseal (B)
1981-87 9   1.1

Valve Job (A)

	LABOR TIME	SEVERE SERVICE
Alliance, Encore, LeCar		
1.4L	9.9	10.1
1.7L	8.9	9.1
Fuego		
1.6L		
w/turbo	11.4	11.6
w/o turbo	10.8	11.0
2.2L	9.7	9.9
R18i		
1.6L		
carburetor	9.7	9.9
FI	10.8	11.0
2.2L	9.9	10.1

Valve Springs and/or Oil Seals, Replace (B)
	LABOR TIME	SEVERE SERVICE
One	1.8	2.0
each addl. add	.4	.6

CAMSHAFT

Camshaft, Replace (A)

	LABOR TIME	SEVERE SERVICE
Alliance, Encore		
1.4L	11.3	11.6
1.7L	3.1	3.3
Fuego		
1.6L		
w/turbo	11.9	12.1
w/o turbo	11.4	11.6
2.2L	5.9	6.1
LeCar	15.8	16.0
R18i		
1.6L		
carburetor	9.9	10.1
FI	11.1	11.3
2.2L	5.8	6.0

ALLIANCE : ENCORE : FUEGO : LECAR : R181 **REN-5**

	LABOR TIME	SEVERE SERVICE
Camshaft Seal, Replace (B)		
Alliance, Encore 1.7L	2.5	2.7
Fuego, R18i 2.2L	2.5	2.7
Camshaft Timing Gear, Replace (B)		
LeCar	1.3	1.5
Timing Belt, Replace (B)		
1.7L, 2.2L	2.2	2.7
Replace tensioner add	.3	.5
Timing Belt/Chain Cover, Replace (B)		
Alliance, Encore		
1.4L	1.4	1.6
1.7L	.7	.9
Fuego, R18i		
1.6L	1.4	1.6
2.2L	1.7	1.9
w/AC add	.2	.2
Timing Chain, Replace (B)		
Includes: Engine R&R when necessary, adjustments.		
Alliance, Encore 1.4L	3.9	4.1
Fuego, R18i 1.6L	13.3	13.5
LeCar	12.1	12.3
Timing Chain Tensioner, Replace (B)		
Alliance, Encore 1.4L	3.1	3.3
Fuego, R18i 1.6L	4.9	5.1
LeCar	10.9	11.1
CRANK & PISTONS		
Connecting Rod Bearings, Replace (A)		
Includes: Check bearing oil clearance.		
Alliance, Encore		
1.4L	3.8	4.0
1.7L	3.9	4.1
Fuego, R18i		
1.6L	3.8	4.0
2.2L	4.7	4.9
LeCar	4.8	5.0
Crankshaft and Main Bearings, Replace (A)		
Includes: Engine R&R, check bearing oil clearance.		
Alliance, Encore		
1.4L	12.9	13.1
1.7L	11.1	11.3
Fuego		
1.6L		
w/turbo	20.8	21.0
w/o turbo	19.8	20.0
2.2L	20.8	21.0
LeCar	13.8	14.0
R18i		
1.6L		
carburetor	16.9	17.1
FI	19.8	20.0
2.2L	20.9	21.1

	LABOR TIME	SEVERE SERVICE
Crankshaft Front Oil Seal, Replace (A)		
Alliance, Encore		
1.7L	2.5	2.7
Fuego, R18i		
2.2L	4.5	4.7
Cylinder Liner Bottom Seals, Replace (A)		
Alliance, Encore	11.1	11.3
Fuego		
1.6L		
w/turbo	14.1	14.3
w/o turbo	13.2	13.4
2.2L	14.1	14.3
LeCar	11.7	11.9
R18i		
1.6L		
carburetor	11.9	12.1
FI	13.2	13.4
2.2L	13.2	13.4
Pistons and Cylinder Liners, Replace (A)		
Alliance, Encore	14.0	14.2
Fuego		
1.6L		
w/turbo	17.1	17.3
w/o turbo	16.0	16.2
2.2L	17.1	17.3
LeCar	15.2	15.4
R18i		
1.6L		
carburetor	14.8	15.0
FI	16.0	16.2
2.2L	16.0	16.2
Rear Main Oil Seal, Replace (B)		
Alliance, Encore		
1.4L	5.9	6.1
1.7L	8.7	8.9
Fuego, R18i		
1.6L	7.4	7.6
2.2L	8.7	8.9
ENGINE LUBRICATION		
Engine Oil Pressure Switch (Sending Unit), Replace (B)		
1981-87	.5	.7
Oil Pan and/or Gasket, Replace (B)		
Alliance, Encore		
1.4L	1.8	2.0
1.7L	2.2	2.4
Fuego, R18i		
1.6L	1.8	2.0
2.2L	2.7	2.9
Oil Pump, Replace (B)		
Alliance, Encore		
1.4L	2.0	2.2
1.7L	2.4	2.6
Fuego, R18i		
1.6L	2.0	2.2
2.2L	3.1	3.3

	LABOR TIME	SEVERE SERVICE
CLUTCH		
Clutch Pedal Free Play, Adjust (B)		
1981-87	.5	.7
Clutch Assy., Replace (B)		
Alliance, Encore	6.6	6.8
Fuego, R18i		
1.6L		
w/turbo	7.1	7.3
w/o turbo	6.6	6.8
2.2L	7.6	7.8
LeCar		
4-Speed	9.1	9.3
5-Speed	9.6	9.8
Replace		
flywheel add	.3	.3
release bearing add	.2	.2
Clutch Control Cable, Replace (B)		
Exc. LeCar	.9	1.1
LeCar	1.4	1.6
MANUAL TRANSAXLE		
Transaxle Assy. R&R and Recondition (A)		
Alliance, Encore		
4-Speed	15.2	15.4
5-Speed	16.2	16.4
Fuego, R18i		
4-Speed	15.2	15.4
5-Speed		
1341-51, 1348-58	16.2	16.4
135B, 1368-136A	16.7	16.9
136B	17.0	17.2
LeCar	17.7	17.9
Transaxle Assy., R&R or Replace (B)		
Alliance, Encore	6.2	6.4
Fuego, R18i		
4-Speed	6.2	6.4
5-Speed		
1341-51, 1348-58	8.6	8.8
135B	9.1	9.3
1368-136A	8.6	8.8
136B	8.9	9.1
Transaxle Shift Cover and/or Gasket, Replace (B)		
Alliance, Encore	2.1	2.3
Fuego, R18i		
4-Speed	2.2	2.4
5-Speed	2.4	2.6
LeCar	2.6	2.8
Transaxle Mounts, Replace (B)		
Alliance, Encore		
front	1.3	1.5
rear	2.9	3.1
Fuego, LeCar, R18i	1.1	1.3

REN-6 ALLIANCE : ENCORE : FUEGO : LECAR : R181

	LABOR TIME	SEVERE SERVICE
AUTOMATIC TRANSAXLE		
SERVICE TRANSAXLE INSTALLED		
Check Unit for Oil Leaks (C)		
1981-87	.5	.5
Drain & Refill Unit (B)		
1981-87	1.0	1.2
Oil Pressure Check (B)		
1981-87	.5	.7
Oil Pan and/or Gasket, Replace (B)		
Alliance, Encore	1.0	1.2
Fuego, R18i	1.0	12
Solenoid Valve, Replace (B)		
Alliance, Encore	2.3	2.5
Fuego, R18i	2.1	2.3
Speed Sensor, Replace (B)		
Alliance, Encore	2.1	2.3
Vacuum Modulator, Replace (B)		
1981-87	1.3	1.5
Valve Body, Replace (B)		
Alliance, Encore	2.8	3.0
Fuego, R18i	3.2	3.4
SERVICE TRANSAXLE REMOVED		
Transaxle R&R included unless otherwise noted.		
Torque Converter Flexplate, Replace (B)		
Alliance, Encore	7.5	7.7
Fuego, R18i		
type 4139	8.0	8.2
type MJ3	7.5	7.7
Torque Converter or Seal, Replace (B)		
Alliance, Encore	7.4	7.6
Fuego, R18i		
type 4139	7.9	8.1
type MJ3	7.4	7.6
Transaxle Assy., R&R and Recondition (A)		
Alliance, Encore	12.1	12.3
Fuego, R18i		
type 4139	14.9	15.1
type MJ3	12.1	12.3
Recondition differential add		
Alliance, Encore	4.1	4.3
type 4139	8.3	8.5
type MJ3	6.6	6.8
Transaxle Assy., Replace (B)		
Alliance, Encore	7.6	7.8
Fuego, R18i		
type 4139	8.1	8.3
type MJ3	7.6	7.8

	LABOR TIME	SEVERE SERVICE
SHIFT LINKAGE		
AUTOMATIC TRANSAXLE		
Shift Linkage, Adjust (B)		
1981-87	.5	.7
Shift Control Cable, Replace (B)		
1981-87	.7	.9
Gear Selector Lever, Replace (B)		
1981-87	1.1	1.1
MANUAL TRANSAXLE		
Shift Lever, Replace (B)		
1981-87	1.5	1.5
DRIVELINE		
HALFSHAFTS		
Halfshaft, R&R or Replace (B)		
Alliance, Encore		
right side	1.0	1.2
left side	1.3	1.5
both sides	2.1	2.3
Fuego, R18i		
one side	1.3	1.5
both sides	2.3	2.5
LeCar		
one side	1.6	1.8
both sides	2.5	2.7
Replace boots add	.5	.7
BRAKES		
SYSTEM		
Bleed Brakes (B)		
Includes: Add fluid.		
1981-87	.5	.5
Brake System, Flush and Refill (B)		
1981-87	1.2	1.2
Brakes, Adjust (B)		
Includes: Refill master cylinder.		
1981-87	.3	.3
Brake Hose (Flexible), Replace (B)		
Includes: System bleeding.		
One	.7	.9
each addl. add	.3	.4
Master Cylinder, R&R and Recondition (A)		
Alliance, Encore	2.0	2.2
Fuego, R18i	2.0	2.2
LeCar	2.5	2.7
Master Cylinder, Replace (B)		
Alliance, Encore	1.4	1.6
Fuego, R18i	1.4	1.6
LeCar	1.9	2.1
Power Booster Unit, Replace (B)		
Alliance, Encore	1.2	1.4
Fuego, R18i	2.1	2.3
LeCar	2.4	2.6
Power Booster Vacuum Check Valve, Replace (B)		
1981-87	.3	.3

	LABOR TIME	SEVERE SERVICE
Pressure Limiting Valve, Replace (B)		
Includes: System bleeding.		
1981-87	1.0	1.2
SERVICE BRAKES		
Brake Drum, Replace (B)		
One	.5	.7
Both	.9	1.1
Caliper Assy., R&R and Recondition (A)		
Includes: System bleeding.		
Alliance, Encore, LeCar		
one	1.6	1.8
both	2.5	2.7
Fuego, R18i		
one	2.2	2.4
both	4.0	4.2
Caliper Assy., Replace (B)		
Includes: System bleeding.		
Alliance, Encore, LeCar		
one	1.1	1.3
both	1.6	1.8
Fuego, R18i		
one	1.7	1.9
both	3.0	3.2
Disc Brake Rotor, Replace (B)		
Alliance, Encore		
one	.9	1.1
both	1.5	1.7
Fuego, R18i		
one	1.3	1.5
both	2.1	2.3
LeCar		
one	1.5	1.7
both	2.8	3.0
Pads and/or Brake Shoes, Replace (B)		
Includes: Adjust service and parking brake. System bleeding.		
Front disc	.9	1.1
Rear drum	1.5	1.7
Four wheels	2.2	2.4
COMBINATION ADD-ONS		
Repack wheel bearings add	.6	.8
Replace		
caliper add	.4	.6
wheel cylinder add	.4	.6
Resurface		
brake drum add	.5	.7
brake rotor add	.5	.7
Wheel Cylinder, R&R and Recondition (B)		
Includes: System bleeding.		
One	1.8	2.0
Both	3.0	3.2

ALLIANCE : ENCORE : FUEGO : LECAR : R181 **REN-7**

	LABOR TIME	SEVERE SERVICE
Wheel Cylinder, Replace (B)		
Includes: System bleeding.		
One	1.5	1.7
Both	2.8	3.0

PARKING BRAKE
Parking Brake Cable, Adjust (C)
1981-87 3   .5
Parking Brake Apply Actuator, Replace (B)
Alliance, Encore 1.0   1.1
Fuego, R18i 9   1.1
LeCar 1.5   1.7
Parking Brake Cable, Replace (B)
1981-87
 front 1.5   1.7
 rear 1.7   1.9

FRONT SUSPENSION
Unless otherwise noted time given does not include alignment.
Align Front End (A)
1981-87 .... 1.0   1.2
Front Toe, Adjust (B)
1981-87 6   .8
Engine Cradle Frame, Replace (B)
Includes: Reset toe.
Alliance, Encore 14.9   15.1
Front Strut Coil Spring, Replace (B)
Alliance, Encore 1.8   2.0
Fuego, R18i 2.6   2.8
Lower Ball Joint, Replace (B)
Alliance, Encore 9   1.1
Fuego, LeCar, R18i ... 2.1   2.3
Lower Control Arm, Replace (B)
Includes: Reset toe.
Alliance, Encore 2.3   2.5
Fuego, R18i 1.6   1.8
LeCar 3.5   3.7
Lower Control Arm Bushings, Replace (B)
Includes: Reset toe.
Alliance, Encore 2.7   2.9
Fuego, R18i one side . 2.0   2.2
LeCar 3.9   4.1
Stabilizer Bar Rods, Replace (B)
LeCar 1.8   2.0
Stabilizer Bar, Replace (B)
Alliance, Encore,
 LeCar 7   .9
Fuego, R18i 1.0   1.2
Stabilizer Bar Bushing and/or Bracket, Replace (B)
Fuego, LeCar, R18i 5   .7
Stabilizer Bar Link Bushings, Replace (B)
Fuego, R18i 8   1.0

	LABOR TIME	SEVERE SERVICE
Strut, Replace (B)		
Alliance, Encore		
one	1.1	1.3
both	2.0	2.2
Fuego, R18i		
one	1.0	1.2
both	1.7	1.9
LeCar		
one	1.0	1.2
both	1.7	1.9

Strut Rod Bushings, Replace (B)
Includes: Reset toe.
Fuego, R18i 1.1   1.3
Stub Axle Carrier, Replace (B)
Includes: Reset toe.
Alliance, Encore 2.4   2.6
LeCar 3.3   3.5
Upper Ball Joints, Replace (B)
Includes: Reset toe.
Fuego, LeCar, R18i ... 2.0   2.2
Upper Control Arm, Replace (B)
Includes: Reset toe.
Fuego, R18i 2.1   2.3
LeCar 2.6   2.8
Wheel Hub Bearing or Seal, Replace (B)
Alliance, Encore
 one side 1.8   2.0
 both sides 3.5   3.7
Fuego, R18i
 one side 1.7   1.9
 both sides 2.7   2.7
LeCar
 one side 2.0   2.2
 both sides 3.8   4.0

REAR SUSPENSION
Rear Control Arm, Replace (B)
Includes: Alignment.
Alliance, Encore 3.9   4.1
LeCar 4.9   5.1
Replace bushing add 8   1.0
Rear Shock Absorbers or Bushings, Replace (B)
Alliance, Encore
 one 7   .9
 both 1.1   1.3
Fuego, R18i
 one 1.0   1.2
 both 1.8   2.0
LeCar
 one 1.0   1.2
 both 1.8   2.0
Rear Spring, Replace (B)
Fuego, R18i
 one 1.0   1.2
 both 1.8   2.0

	LABOR TIME	SEVERE SERVICE
Rear Stabilizer Bar & Bushings, Replace (B)		
1981-87	.8	1.0

Rear Torsion Bar, Replace (B)
Alliance, Encore, LeCar
 one 1.8   2.0
 both 3.5   3.7
Rear Wheel Bearings & Seals, Replace (B)
Alliance, Encore
 one side 9   1.1
 both sides 1.7   1.9
Fuego, R18i
 one side 1.0   1.2
 both sides 1.9   2.1

STEERING
Unless otherwise noted, time given does not include alignment.
MANUAL RACK & PINION
Flexible Coupling, Replace (B)
1981-87 .... 2.2   2.4
Gear Assy., R&R and Recondition (A)
Alliance, Encore 3.8   4.0
Fuego, R18i 5.4   5.6
LeCar 6.9   7.1
Gear Assy., R&R or Replace (B)
Alliance, Encore 2.4   2.6
Fuego, R18i 3.0   3.2
LeCar 4.5   4.7
Rack Boots, Replace (B)
1981-87 .... 1.6   1.8
Steering Shaft U-Joint, Replace (B)
1981-87 .... 1.6   1.8
Steering Wheel, Replace (B)
1981-87 5   .5
Tie Rod Ball Joints, Replace (B)
Includes: Alignment.
1981-87 .... 1.9   2.1
Tie Rods or Tie Rod Ends, Replace (B)
Includes: Reset toe.
Alliance, Encore 1.9   2.1
Fuego, R18i 1.9   2.1
LeCar 2.3   2.5

POWER RACK & PINION
Pump Pressure Check (B)
1981-87 5   .5
Rack Boots, Replace (B)
1981-87 .... 1.6   1.8
Rack & Pinion Assy., R&R and Recondition (A)
Alliance, Encore 4.6   4.8
Fuego, R18i 7.1   7.3

REN-8 ALLIANCE : ENCORE : FUEGO : LECAR : R181

	LABOR TIME	SEVERE SERVICE
Rack & Pinion Assy., R&R or Replace (B)		
Alliance, Encore	2.4	2.6
Fuego, R18i	4.8	5.0
Steering Pump, R&R or Replace (B)		
Alliance, Encore, LeCar	.8	1.0
Fuego, R18i	1.3	1.5

HEATING & AIR CONDITIONING

When more than one component requires replacement where evacuation/recovery and recharging is already included, deduct 1.0 hour for each additional component from the time given.

	LABOR TIME	SEVERE SERVICE
Evacuate/Recover and Recharge System (B)		
1981-87	1.0	1.0
AC Hoses, Replace (B)		
Includes: Evacuate/recover and recharge.		
Discharge	2.0	2.2
Suction	1.1	1.3
Blower Motor, Replace (B)		
LeCar	.7	.9
Compressor Assy., Replace (B)		
Includes: Parts transfer, evacuate/recover and recharge.		
Alliance, Encore	3.6	3.8
Fuego, R18i	4.7	4.9
LeCar	2.6	2.8
Compressor Clutch Field Coil, Replace (B)		
LeCar	1.3	1.5
Compressor Shaft Seal, Replace (B)		
Includes: Compressor R&R. Evacuate/recover and recharge.		
Alliance, Encore	4.1	4.3
Fuego, R18i	5.2	5.4
LeCar	3.1	3.3
Condenser Assy., Replace (B)		
Includes: Evacuate/recover and recharge.		
Alliance, Encore	2.3	2.5
Fuego, R18i	3.5	3.7
LeCar	2.5	2.7
Evaporator Coil, Replace (B)		
Includes: Evacuate/recover and recharge.		
Alliance, Encore	3.6	3.8
Fuego, R18i	5.5	5.7
LeCar	3.1	3.3

	LABOR TIME	SEVERE SERVICE
Heater Blower Motor, Replace (B)		
Alliance, Encore	.6	.8
Fuego, R18i		
w/AC	5.4	5.6
w/o AC	3.1	3.3
LeCar	1.4	1.6
Heater Control Valve, Replace (B)		
Alliance	.6	.8
Fuego, R18i	1.1	1.3
LeCar	2.3	2.5
Heater Core, R&R or Replace (B)		
Alliance, Encore	3.9	4.1
Fuego, R18i		
w/AC	5.4	5.6
w/o AC	3.1	3.3
LeCar	2.3	2.5
Heater Hoses, Replace (B)		
1981-87	1.5	1.7
Pressure Relief Valve, Replace (B)		
Includes: Evacuate/recover and recharge.		
1981-87	1.8	2.0
Receiver/Drier Assy., Replace (B)		
Includes: Evacuate/recover and recharge.		
1981-87	1.7	1.9
Thermostatic Control Switch, Replace (B)		
LeCar	.8	1.0

WIPERS & SPEEDOMETER

	LABOR TIME	SEVERE SERVICE
Front Windshield Wiper Switch, Replace (B)		
Alliance, Encore, LeCar	.7	.9
Fuego, R18i	1.1	1.3
Rear Window Wiper Motor, Replace (B)		
Exc. R18i	.6	.8
R18i	1.0	1.2
Speedometer Cable & Casing, Replace (B)		
Governor to trans.	.5	.7
Speedo to governor	1.0	1.2
Speedometer Head, R&R or Replace (B)		
Alliance, Encore	1.2	1.4
Fuego, R18i	1.5	1.7
Windshield Washer Motor and/or Reservoir, Replace (B)		
LeCar front or rear	.5	.7
Windshield Washer Pump, Replace (B)		
Alliance, Encore, LeCar	.5	.7
Fuego, R18i	.7	.9

	LABOR TIME	SEVERE SERVICE
Windshield Wiper Linkage, Replace (B)		
Alliance, Encore	.9	1.1
Fuego, R18i	1.2	1.4
LeCar	1.7	1.9
Windshield Wiper Motor, Replace (B)		
Alliance, Encore	1.5	1.7
Fuego, R18i	1.0	1.2

LAMPS & SWITCHES

	LABOR TIME	SEVERE SERVICE
Back-Up Lamp Assy., Replace (B)		
1981-87	.5	.5
Back-Up Lamp Switch, Replace (B)		
1981-87	.5	.5
Headlamps, Aim (B)		
1981-87	.4	.4
Headlamp Switch, Replace (B)		
Alliance, Encore	1.1	1.1
Fuego, R18i	1.1	1.1
LeCar	.8	.8
Horn, Replace (B)		
1981-87 one	.4	.4
License Lamp Lens, Replace (B)		
1981-87	.2	.2
Sealed Beam Headlamp, Replace (C)		
1981-87	.3	.3
Stop Lamp Switch, Replace (B)		
1981-87	.5	.5
Tail Lamp Lens or Bulb, Replace (B)		
1981-87	.2	.2
Turn Signal and Parking Lamp Lens, Replace (B)		
1981-87	.3	.3
Turn Signal or Hazard Warning Flasher, Replace (B)		
1981-87	.3	.3
Turn Signal Switch, Replace (B)		
Alliance, Encore	.9	.9
Fuego, R18i	.9	.9
LeCar	.8	.8

BODY

	LABOR TIME	SEVERE SERVICE
Door Lock, Replace (B)		
1981-87	1.1	1.3
w/power locks add	.2	.4
Hood Lock, Replace (B)		
1981-87	.7	.9
Hood Release Cable, Replace (B)		
Alliance, Encore	1.1	1.3
Fuego, R18i	1.1	1.3
LeCar	.5	.7
Window Regulator Motor, Replace (B)		
Alliance, Encore	1.1	1.3
Fuego, R18i	1.6	1.8
LeCar	1.3	1.5
w/power window add	.5	.7

SAA

900 : 9000 : 9-2X : 9-3 : 9-5

SYSTEM INDEX

MAINTENANCE	SAA-2
CHARGING	SAA-2
STARTING	SAA-2
CRUISE CONTROL	SAA-2
IGNITION	SAA-3
EMISSIONS	SAA-3
FUEL	SAA-4
EXHAUST	SAA-5
ENGINE COOLING	SAA-5
ENGINE	SAA-6
Assembly	SAA-6
Cylinder Head	SAA-6
Camshaft	SAA-7
Crank & Pistons	SAA-8
Engine Lubrication	SAA-9
CLUTCH	SAA-9
MANUAL TRANSAXLE	SAA-10
AUTO TRANSAXLE	SAA-10
SHIFT LINKAGE	SAA-11
DRIVELINE	SAA-11
BRAKES	SAA-12
FRONT SUSPENSION	SAA-13
REAR SUSPENSION	SAA-14
STEERING	SAA-15
HEATING & AC	SAA-16
WIPERS & SPEEDOMETER	SAA-17
LAMPS & SWITCHES	SAA-17
BODY	SAA-17

OPERATIONS INDEX

A
AC Hoses	SAA-16
Air Bags	SAA-15
Air Conditioning	SAA-16
Alignment	SAA-13
Alternator (Generator)	SAA-2
Antenna	SAA-17
Anti-Lock Brakes	SAA-12

B
Back-Up Lamp Switch	SAA-17
Ball Joint	SAA-14
Battery Cables	SAA-2
Bleed Brake System	SAA-12
Blower Motor	SAA-16
Brake Disc	SAA-13
Brake Hose	SAA-12
Brake Pads and/or Shoes	SAA-12

C
Camshaft	SAA-8
Catalytic Converter	SAA-5
Coolant Temperature (ECT) Sensor	SAA-3
Crankshaft	SAA-8
Crankshaft Sensor	SAA-3
Cruise Control	SAA-2
CV Joint	SAA-12
Cylinder Head	SAA-7

D
Differential	SAA-11
Distributor	SAA-3
Drive Belt	SAA-2

E
EGR	SAA-3
Electronic Control Module (ECM/PCM)	SAA-3
Engine	SAA-6
Engine Lubrication	SAA-9
Engine Mounts	SAA-6
Evaporator	SAA-16
Exhaust	SAA-5
Exhaust Manifold	SAA-5
Expansion Valve	SAA-16

F
Flexplate	SAA-11
Flywheel	SAA-10
Fuel Injection	SAA-4
Fuel Pump	SAA-4
Fuel Vapor Canister	SAA-3

G
Gear Selector Lever	SAA-11
Generator	SAA-2

H
Halfshaft	SAA-11
Headlamp	SAA-17
Heater Core	SAA-16
Horn	SAA-17

I
Idle Air Control (IAC) Valve	SAA-4
Ignition Coil	SAA-3
Ignition Module	SAA-3
Ignition Switch	SAA-3
Inner Tie Rod	SAA-15
Intake Manifold	SAA-4

K
Knock Sensor	SAA-3

L
Lower Control Arm	SAA-14

M
Mass Air Flow (MAF) Sensor	SAA-4
Master Cylinder	SAA-12
Muffler	SAA-5

N
Neutral Safety Switch	SAA-17

O
Oil Pan	SAA-9
Oil Pump	SAA-9
Outer Tie Rod	SAA-15
Oxygen Sensor	SAA-4

P
Parking Brake	SAA-13
Pistons	SAA-9
Positive Crankcase Ventilation (PCV) Valve	SAA-4

R
Radiator	SAA-5
Radiator Hoses	SAA-6
Radio	SAA-17
Rear Main Oil Seal	SAA-9

S
Shock Absorber/Strut, Front	SAA-14
Shock Absorber/Strut, Rear	SAA-14
Spark Plug Cables	SAA-3
Spark Plugs	SAA-3
Spring, Front Coil	SAA-13
Spring, Rear Coil	SAA-14
Starter	SAA-2
Steering Wheel	SAA-16

T
Thermostat	SAA-5
Throttle Body	SAA-4
Throttle Position Sensor (TPS)	SAA-4
Timing Belt	SAA-8
Timing Chain	SAA-8
Torque Converter	SAA-10

U
Upper Control Arm	SAA-14

V
Valve Body	SAA-10
Valve Cover Gasket	SAA-7
Valve Job	SAA-7
Vehicle Speed Sensor	SAA-4

W
Water Pump	SAA-6
Wheel Balance	SAA-2
Window Regulator	SAA-17
Windshield Washer Pump	SAA-17
Windshield Wiper Motor	SAA-17

SAA-2 900 : 9000 : 9-2x : 9-3 : 9-5

	LABOR TIME	SEVERE SERVICE
MAINTENANCE		
Air Cleaner Filter Element, Replace (B)		
1981-05	.4	.6
Composite Headlamp Bulb, Replace (B)		
1981-05	.3	.5
Chassis Lubrication, Change Oil & Filter (C)		
Includes: Correct all fluid levels.		
1981-05	.4	.6
Drive Belt, Replace (B)		
Exc. below		
900		
4 cyl.	.5	.7
V6	.7	.9
9000		
4 cyl.	.7	.9
V6	1.2	1.4
9-2X, 9-5	1.4	1.6
9-3	.8	1.0
Serpentine		
9-2X, 9-3	.6	.8
9-5	1.2	1.4
Serpentine Drive Belt Tensioner, Replace (B)		
900, 9000	.5	.7
9-2X	1.2	1.4
9-3	.6	.8
9-5		
4 cyl.	.8	1.0
V6	1.4	1.6
Fuel Filter, Replace (B)		
900		
1981-93	.3	.5
1994-98	.7	.9
9000, 9-5	.5	.7
9-3	.6	.8
Oil & Filter, Change (C)		
Includes: Correct all fluid levels.		
1981-05	.4	.6
Sealed Beam Headlamp, Replace (B)		
900 each	.2	.4
Timing Belt, Replace (B)		
900 V6	2.8	3.0
9000 V6	2.9	3.1
9-5 V6	3.1	3.3
Tire, Replace (C)		
Includes: Dismount old tire and mount new tire to rim.		
1981-05	.5	.5
Tires, Rotate (C)		
1981-05	.5	.5
Wheel, Balance (B)		
One	.3	.3
each addl. add	.2	.2

	LABOR TIME	SEVERE SERVICE
CHARGING		
Alternator Circuits, Test (B)		
Includes: Test component output.		
1981-98	.7	.7
1999-05	.3	.3
Alternator, R&R and Recondition (A)		
900	3.6	3.8
Alternator Assy., Replace (B)		
Includes: Pulley transfer.		
900		
1981-93	.9	1.1
1994-98	1.6	1.8
9000		
4 cyl.	1.8	2.0
V6	2.3	2.5
9-2X	.9	1.1
9-3	1.5	1.7
9-5	2.4	2.6
Alternator Voltage Regulator, Replace (B)		
900		
1981-93	.5	.7
1994-98	1.5	1.7
9000		
4 cyl.	2.0	2.2
V6	2.6	2.8
9-2X, 9-3	1.5	1.7
9-5	2.6	2.8
Front Alternator Bearing, Replace (B)		
900	2.3	2.5
9000	3.4	3.6
STARTING		
Starter Draw Test (On Car) (B)		
1981-05	.3	.3
Battery Cables, Replace (C)		
Positive	.5	.7
Negative	.3	.5
Starter Assy., R&R and Recondition (A)		
900	3.4	3.6
w/turbo add	.7	.9
Starter Assy., Replace (B)		
900		
1981-93	1.7	1.9
1994-98	.8	1.0
9000		
2.0L	.8	1.0
2.3L	.7	.9
V6	2.6	2.8
9-2X	.9	1.1
9-3	.6	.8
9-5		
4 cyl.	.5	.7
V6	2.1	2.3
w/turbo add		
81-93 900	.7	.9
9000 2.3L	.7	.9

	LABOR TIME	SEVERE SERVICE
Starter Drive Assy., Replace (B)		
Includes: Starter R&R.		
900		
1981-93	3.1	3.3
1994-98		
4 cyl.	2.8	3.0
V6	3.1	3.3
9000 4 cyl.	2.9	3.1
w/turbo add		
81-93 900	.7	.9
9000 2.3L	.4	.6
Starter Solenoid and/or Switch, Replace (B)		
Includes: Starter R&R.		
900		
1981-93	2.0	2.2
1994-98		
4 cyl.	.8	1.0
V6	1.2	1.4
9000		
2.0L	1.1	1.3
2.3L	.8	1.0
w/turbo add		
81-93 900	.7	.9
9000 2.3L	.6	.8
Starter Relay, Replace (B)		
1994-98 900	.5	.7
9-3, 9-5	.3	.5
CRUISE CONTROL		
Control Chain, Cable or Rod, Replace (B)		
1994-98 900	.3	.5
Control Controller (Module), Replace (B)		
900, 9000	.5	.7
9-3	.6	.8
Control Relay, Replace (B)		
900	.5	.7
Control Sensor, Replace (B)		
900	1.8	2.0
9000	.5	.7
Control Transmitter, Replace (B)		
9000	1.6	1.8
Control Vacuum Boost Switch, Replace (B)		
1985-93 900	.3	.5
9000	.9	1.1
Control Vacuum Regulator, Replace (B)		
900, 9000	.5	.7
Control Vacuum Reservoir, Replace (B)		
1981-01	.3	.5
Disengagement Switch, Replace (B)		
All Models	.5	.7
Engagement Switch, Replace (B)		
900	.7	.9
9000	.5	.7
9-2X, 9-3, 9-5	.4	.6

900 : 9000 : 9-2x : 9-3 : 9-5 — SAA-3

	LABOR TIME	SEVERE SERVICE
IGNITION		
Diagnose Cruise Control System Component Each (A)		
All Models	1.1	1.1
Ignition Timing, Reset (B)		
1981-98	.4	.6
Crankshaft Angle Sensor, Replace (B)		
900		
1989-93	2.6	2.8
1994-98	.8	1.0
9000		
1990-91	2.2	2.4
1992-98	1.4	1.6
9-3	1.2	1.4
9-5	.8	1.0
Distributor, R&R and Recondition (A)		
Includes: Reset base ignition timing.		
1981-85 900	2.3	2.5
Distributor, Replace (B)		
Includes: Reset base ignition timing.		
900		
1981-93	.7	.9
1994-98	.5	.7
1986-90 9000	.9	1.1
Distributor Cap and/or Rotor, Replace (B)		
900	.3	.5
1986-90 9000	.3	.5
Electronic Ignition Control Unit, Replace (B)		
1986-93 900	.5	.7
9000		
1986-87	.5	.7
1988-91	1.2	1.4
1992-98	.5	.7
9-2X, 9-3	.6	.8
9-5	1.1	1.3
Ignition Coil, Replace (B)		
1981-98 900	.4	.6
Ignition Coil and Module (Distributorless), Replace (B)		
1994-98 900	.5	.7
9000		
1990-92	.7	.9
1993-98	.2	.4
9-2X	1.2	1.4
9-3	.8	1.0
9-5		
4 cyl.	.3	.5
V6		
one	.8	1.0
both	.9	1.1
Ignition Lock Cylinder and/or Buzzer Switch, Replace (B)		
900		
1981-93	2.4	2.6
1994-98	.5	.7
1986-98 9000	.5	.7
9-2X, 9-3, 9-5	.8	1.0

	LABOR TIME	SEVERE SERVICE
Ignition Switch, Replace (B)		
900		
1981-82	1.6	1.8
1983-93	2.0	2.2
1994-98	3.0	3.2
9000		
1986-89	.3	.5
1989-98	.5	.7
9-2X	1.4	1.6
9-3	1.8	2.0
9-5	2.0	2.2
Pulse Transmitter, Replace (B)		
900	2.0	2.2
1986-90 9000	1.8	2.0
Spark Plug (Ignition) Cables, Replace (B)		
900		
exc. V6 rear	.5	.7
V6 rear	1.4	1.6
9000	.8	1.0
Spark Plugs, Replace (B)		
900		
1981-93	.7	.9
1994-98		
4 cyl.	.5	.7
V6	1.2	1.4
9000		
1986-90	.7	.9
1991-98		
4 cyl.	1.2	1.4
V6	1.4	1.6
9-2X, 9-3	.8	1.0
9-5		
4 cyl.	.5	.7
V6	.8	1.0
Transistorized Ignition Control Unit, Replace (B)		
1986-93 900	.5	.7
1987-90 9000	.5	.7
Vacuum Advance Unit, Replace (B)		
Includes: Reset base timing.		
900	.7	.9
1986-90 9000	.7	.9

EMISSIONS

	LABOR TIME	SEVERE SERVICE
Air Control Valve, Replace (B)		
1994-98 900	.3	.5
1994-98 9000	.5	.7
9-3	.6	.8
Air Flow Sensor, Replace (B)		
1981-88 900	2.3	2.5
Air Injection Tube Check Valve Assy., Replace (B)		
1994-04	.3	.5
Air Pump Assy., Replace (B)		
1994-98 900	.5	.7
1994-98 9000	.9	1.1
9-3	1.1	1.3
9-5 V6	.6	.8

	LABOR TIME	SEVERE SERVICE
Air Switching Valve, Replace (B)		
1994-98 900	.5	.7
1994-98 9000	.3	.5
1999-01 9-3	.5	.7
Air Temperature Sensor, Replace (B)		
All Models	.5	.7
Canister Purge Control Valve, Replace (B)		
1994-04	.3	.5
9-3	.6	.8
9-5	.8	1.0
Coolant Temperature (ECT) Sensor, Replace (B)		
1985-98 900	.5	.7
9000		
2.0L, 3.0L	.5	.7
2.3L	.9	1.1
9-3	.8	1.0
9-2X, 9-5	1.1	1.3
EGR Check Valve, Replace (B)		
1990-98 9000	.5	.7
EGR Control Valve, Replace (B)		
1990-98 9000	.5	.7
EGR Modulator Valve, Replace (B)		
1990-98 9000	.5	.7
EGR Temperature Sensor, Replace (B)		
1990-92 9000	.5	.7
Electronic Control Module (ECM/PCM), Replace (B)		
900		
1985-89	.5	.7
1990-93	.9	1.1
1994-98	.5	.7
9000	.5	.7
9-3	.9	1.1
9-2X, 9-5	1.1	1.3
Electronic Spark Control Module, Replace (B)		
1990-98 9000	.5	.7
Fuel Vapor Canister, Replace (B)		
900	.5	.7
9000	.7	.9
9-3	.6	.8
9-5		
exc. CA	.5	.7
CA	1.8	2.0
Knock Sensor, Replace (B)		
900		
1982-93	.5	.7
1994-98		
4 cyl.	.5	.7
V6		
front	.5	.7
rear	1.1	1.3
9000	.5	.7
9-2X, 9-3	1.1	1.3

SAA-4   900 : 9000 : 9-2x : 9-3 : 9-5

	LABOR TIME	SEVERE SERVICE
Mass Air Flow (MAF) Sensor, Replace (B)		
1994-05	1.1	1.3
Oxygen Sensor, Replace (B)		
900, 9000	.7	.9
9-2x front or rear	1.4	1.6
9-3	.8	1.0
9-5		
front	1.1	1.3
rear	.8	1.0
Positive Crankcase Ventilation (PCV) Valve, Replace (B)		
1981-98	.5	.7
Throttle Position Sensor (TPS), Replace (B)		
1994-98	.5	.7
1999-05	1.1	1.3
Vehicle Speed Sensor, Replace (B)		
1988-93 900	2.0	2.2
9000	.5	.7

FUEL

DELIVERY

	LABOR TIME	SEVERE SERVICE
Fuel Gauge (Dash), Replace (B)		
900		
1981-93	2.0	2.2
1994-98	.7	.9
9000	2.0	2.2
9-3, 9-5	1.1	1.3
Fuel Gauge (Tank), Replace (B)		
900		
1981-93	.8	1.0
1994-98	1.9	2.1
9000		
1986-89	.5	.7
1990-98	1.3	1.5
9-2X, 9-3	1.8	2.0
9-5	.5	.7
Fuel Pump, Replace (B)		
900		
1981-89	1.3	1.5
1990-93	.9	1.1
1994-98	2.1	2.3
9000		
1986-89		
main or feed		
Bosch	.8	1.0
main Walbro	1.2	1.4
feed Walbro	1.6	1.8
1990-98 main or feed	1.1	1.3
9-3	2.6	2.8
9-2X, 9-5	1.8	2.0
Fuel Pump Relay, Replace (B)		
900		
1985-93	1.2	1.4
1994-98	.5	.7
9000	.8	1.0
9-2X, 9-3, 9-5	.5	.7

	LABOR TIME	SEVERE SERVICE
Fuel Tank, Replace (B)		
Includes: Drain and refill.		
1981-98	2.1	2.3
9-3	2.2	2.4
9-5	2.0	2.2
Intake Manifold and/or Gasket, Replace (B)		
Includes: Adjustments.		
900		
1981-93	2.3	2.5
1994-98		
4 cyl.	2.7	2.9
V6		
upper	1.1	1.3
center	1.6	1.8
lower	2.4	2.6
9000		
single piece	2.6	2.8
two piece		
2.0L	3.3	3.5
2.3L	3.0	3.2
V6		
upper	1.1	1.3
lower	1.5	1.7
9-2X	2.8	3.0
9-3	4.3	4.5
9-5		
4 cyl.	4.4	4.6
V6		
upper	.9	1.1
center	1.1	1.3
lower	1.5	1.7

INJECTION

	LABOR TIME	SEVERE SERVICE
Diagnose Fuel Injection System Component Each (A)		
1981-88 900 CIS	1.3	1.5
1985-89 900 LH	1.1	1.3
1994-98 900	.8	1.0
9000	1.1	1.3
9-2X, 9-3, 9-5	1.1	1.3
Auxiliary Air Valve, Replace (B)		
1981-88 900	.7	.9
Cold Start Injector, Replace (B)		
1981-88 900	.5	.7
Fuel Accumulator, Replace (B)		
900	.8	1.0
Fuel Filter, Replace (B)		
900		
1981-93	.3	.5
1994-98	.7	.9
9000, 9-5	.5	.7
9-3	.6	.8
Fuel Injectors, Replace (B)		
900		
1981-93	.5	.7
1994-98		
4 cyl.	.8	1.0
V6	1.3	1.5

	LABOR TIME	SEVERE SERVICE
9000		
4 cyl.	.7	.9
V6		
one	1.3	1.5
all	2.0	2.2
9-2X, 9-3	1.8	2.0
9-5		
4 cyl.	2.1	2.3
V6	2.3	2.5
Fuel Pressure Regulator, Replace (B)		
900		
1985-93	.5	.7
1995-98	3.1	3.3
9000		
4 cyl.	.5	.7
V6	1.9	2.1
9-2X 9-3	1.1	1.3
9-5	.9	1.1
Fuel Rail, Replace (B)		
900		
1985-93	.5	.7
1994-98		
4 cyl.	1.2	1.4
V6	1.5	1.7
9000		
4 cyl.	.7	.9
V6	1.8	2.0
9-2X, 9-3	1.2	1.4
9-5	1.8	2.0
High Idle Solenoid, Replace (B)		
1981-88 900	.5	.7
Idle Air Control (IAC) Valve, Replace (B)		
1994-98 900	.7	.9
9000	.5	.7
9-3	.6	.8
Temperature Sensor, Replace (B)		
1985-98 900	.5	.7
9000		
2.0L	.5	.7
2.3L	.9	1.1
9-2X, 9-3, 9-5	.8	1.0
Thermo Time Switch, Replace (B)		
1981-88 900	.3	.5
Throttle Body Assy., R&R or Replace (B)		
900		
1981-93	1.1	1.3
1994-98	.9	1.1
9000	.9	1.1
9-3, 9-5	1.2	1.4
Throttle Valve Switch, Replace (B)		
1981-05	.5	.7
Warm-Up Regulator, Replace (B)		
1981-88 900	.5	.7

SAA-5
900 : 9000 : 9-2x : 9-3 : 9-5

	LABOR TIME	SEVERE SERVICE
TURBOCHARGER		
Charging Pressure, Adjust (B)		
1981-05	.7	.9
Intercooler, Replace (B)		
1985-93 900	.5	.7
9000	2.2	2.4
9-3	1.7	1.9
9-5		
4 cyl.		
AT	2.6	2.8
MT	2.1	2.3
V6	2.8	3.0
Pressure Sensor, Replace (B)		
1983-93 900	.7	.9
9000		
1986-89	.8	1.0
1990-98	.5	.7
9-3, 9-5	.5	.7
Pressure Switch, Replace (B)		
900	.5	.7
9000	.7	.9
Regulator Diaphragm, Replace (B)		
900	.9	1.1
9000	3.3	3.5
Turbocharger Assy., R&R or Replace (B)		
900		
1981-87	3.2	3.4
1988-93	4.4	4.6
9000		
1986	4.4	4.6
1987-98		
2.0L	4.8	5.0
2.3L	2.8	3.0
9-2x	1.9	2.1
9-3	2.8	3.0
9-5		
4 cyl.	2.8	3.0
V6	3.2	3.4
EXHAUST		
Catalytic Converter, Replace (B)		
900		
each	.7	.9
dual catalyst	1.3	1.5
9000		
1986-92	.5	.7
1993-98		
4 cyl.		
w/turbo	1.6	1.8
w/o turbo	.9	1.1
V6	1.1	1.3
9-2X, 9-3, 9-5	1.1	1.3
Center Exhaust Pipe/Muffler, Replace (B)		
900	.8	1.0
9000, 9-3	.5	.7

	LABOR TIME	SEVERE SERVICE
Exhaust Manifold and/or Gasket, Replace (B)		
900		
1981-93		
w/turbo	3.1	3.3
w/o turbo		
8V	1.6	1.8
16V	.8	1.0
1994-98		
4 cyl.		
upper	.8	1.0
lower	1.3	1.5
V6		
front	1.9	2.1
rear	2.8	3.0
9000		
2.0L	1.8	2.0
2.3L	.9	1.1
V6		
front	1.9	2.1
rear	2.4	2.6
9-2X	1.2	1.4
9-3	3.8	4.0
9-5		
4 cyl.	4.0	4.2
V6		
front	3.2	3.4
rear	1.3	1.5
w/turbo add		
94-98 900	1.1	1.2
9000 2.3L	3.0	3.2
Exhaust System (Complete), Replace (B)		
900		
1981-93		
w/turbo	1.8	2.0
w/o turbo	1.4	1.6
1994-98	.8	1.0
9000	1.4	1.6
9-2X, 9-3, 9-5	1.6	1.8
Front Exhaust Pipe, Replace (B)		
900		
1981-93		
w/turbo	1.3	1.5
w/o turbo	.7	.9
1994-98	.7	.9
9000		
4 cyl.		
w/turbo	1.8	2.0
w/o turbo	.8	1.0
V6	.9	1.1
9-2X, 9-3, 9-5	1.1	1.3
Front Muffler, Replace (B)		
900		
1985-93	.8	1.0
1994-98	.5	.7
9000, 9-5	.5	.7

	LABOR TIME	SEVERE SERVICE
Muffler and Rear Pipe, Replace (B)		
900		
1981-93	1.3	1.5
1994-98	.5	.7
9000, 9-2X	.7	.9
9-3	.3	.5
9-5		
sport system	1.1	1.3
standard system	.3	.5
Rear Exhaust Pipe, Replace (B)		
900	.7	.9
9-3	.2	.4
ENGINE COOLING		
Pressure Test Cooling System (C)		
1981-05	.4	.4
Coolant Thermostat, Replace (B)		
900		
1981-93	.7	.9
1994-98		
4 cyl.	1.3	1.5
V6	2.0	2.2
9000		
4 cyl.	.9	1.1
V6	2.6	2.8
9-2X, 9-3	1.2	1.4
9-5		
4 cyl.	1.2	1.4
V6	2.4	2.6
Fan Blade, Replace (B)		
900	.5	.7
Radiator Assy., R&R or Replace (B)		
900		
1981-93	1.6	1.8
1994-98		
4 cyl.	1.4	1.6
V6	2.0	2.2
9000		
4 cyl.	2.0	2.2
V6	1.8	2.0
9-2X	.8	1.0
9-3	2.8	3.0
9-5		
4 cyl.	1.5	1.7
V6	2.1	2.3
w/AT add		
900	.2	.4
9000 2.0L	.3	.5
9-5 4 cyl.	.3	.5
w/turbo add		
94-98 900	.6	.8
9000	1.1	1.3
Radiator Fan and/or Fan Motor, Replace (B)		
900	.8	1.0
9000	.9	1.1
9-2X, 9-3	1.2	1.4
9-5		
4 cyl.	1.5	1.7
V6	2.1	2.3

	LABOR TIME	SEVERE SERVICE

Radiator Fan Motor Switch (Coolant Temp.), Replace (B)
900 .... .5 .7
9000 .... .7 .9
9-2X, 9-3, 9-5 .... .5 .7

Radiator Hoses, Replace (B)
Includes: Refill with proper coolant mix.
900, 9-3 each .... .5 .7
9000
  upper .... .5 .7
  lower .... .6 .8
9-2X, 9-5
  upper 8 1.0
  lower 1.2 1.4

Temperature Gauge (Dash), Replace (B)
900
  1981-93 .... 2.2 2.4
  1994-98 .... .7 .9
9000 .... 2.2 2.4
9-2X, 9-3, 9-5 .... 1.1 1.3

Temperature Gauge (Engine), Replace (B)
900 .... .5 .7
9000
  2.0L .... .7 .9
  2.3L, 3.0L .... .5 .7
9-2X, 9-3, 9-5 .... .5 .7

Water Control Valve, Replace (B)
900 .... 2.1 2.3

Water Pump and/or Gasket, Replace (B)
Includes: Refill with proper coolant mix.
900
  1981-93 .... 1.4 1.6
  1994-98
    4 cyl. 1.8 2.0
    V6 .... 1.9 2.1
9000
  2.0L 2.9 3.1
  2.3L 1.9 2.1
  V6 .... 2.5 2.7
9-3 .... 4.8 5.0
9-5 .... 2.7 2.9

ENGINE
ASSEMBLY
Times shown are for OEM assemblies. Time to replace assemblies from aftermarket rebuilders may vary.

Engine Assy., R&I (B)
Does not include parts or component transfer.
900
  1981-93
    w/turbo
      8V 6.3 6.5
      16V 7.9 8.1
    w/o turbo 4.9 5.1

1994-98
  4 cyl. 6.9 7.1
  V6 .... 7.8 8.0
9000
  1986-91 .... 8.8 9.0
  1992-98
    4 cyl. 8.5 8.7
    V6 .... 7.7 7.9
9-3
  1999 .... 7.9 8.1
  2000-05 .... 7.8 8.0
9-5
  4 cyl.
    AT 8.9 9.1
    MT 8.8 9.0
  V6 .... 9.4 9.6
w/turbo add
  94-98 900 .... .9 1.1
  9000 .... 1.1 1.3

Engine Assy., R&R and Recondition (A)
Includes: Replacing rings, rod and main bearings, cylinder head reconditioning and engine tune-up.
900
  1981-93
    8V 25.8 26.0
    16V 27.9 28.1
  1994-98
    4 cyl. 24.8 25.0
    V6 .... 29.9 30.1
9000
  2.0L 27.7 27.9
  2.3L 26.8 27.0
w/turbo add
  900 .... 1.2 1.4
  9000 .... .7 .9

Engine Assy. (Short Block), Replace (B)
Assembly consists of cylinder block, piston assemblies, crankshaft, camshaft, timing chain and gears. Does not include cylinder heads. Operation Includes: R&R engine, transfer necessary parts and all necessary adjustments.
900
  1981-93
    8V 13.9 14.1
    16V 15.7 15.9
  1994-98
    4 cyl. 15.9 16.1
    V6 .... 16.8 17.0
9000 .... 13.8 14.0
  2.0L 15.0 15.2
  2.3L 13.9 14.1
  V6 .... 13.8 14.0
9-3
  1999 .... 14.8 15.0
  2000-04 .... 14.9 15.1

9-5
  4 cyl.
    AT 15.8 16.0
    MT 15.9 16.1
  V6 .... 16.8 17.0
w/turbo add
  900 .... 1.2 1.4
  9000 .... .7 .9

Engine Mounts, Replace (B)
900
  1981-93
    front 8 1.0
    right side .... .7 .9
    left side 16V 2.2 2.4
  1994-98
    front one 8 1.0
    rear 1.6 1.8
9000
  2.0L
    front 1.4 1.6
    left 1.8 2.0
    rear 9 1.1
  2.3L
    front 1.2 1.4
    left 2.0 2.2
    rear 1.3 1.5
9-3
  left 1.0 1.2
  rear .... .6 .8
9-5
  4 cyl.
    right .... .6 .8
    left 1.1 1.3
    rear
      1999
        AT 1.9 2.1
        MT 8 1.0
      2000-04 .... 2.2 2.4
  V6
    right 8 1.0
    left 1.1 1.3
    rear 1.7 1.9
w/turbo 81-93 900 add ... .7 .9

CYLINDER HEAD
Compression Test (B)
900
  1981-93 .... .8 1.0
  1994-98
    4 cyl. 8 1.0
    V6 .... 1.8 2.0
9000
  1986-90 .... .9 1.1
  1991-98
    4 cyl. 1.4 1.6
    V6 .... 1.9 2.1
9-3 .... .6 .8
9-5
  4 cyl. 9 1.1
  V6 .... 1.1 1.3

900 : 9000 : 9-2x : 9-3 : 9-5   SAA-7

	LABOR TIME	SEVERE SERVICE
Valve Clearance, Adjust (B)		
1981-88 900 8V	1.8	*2.0*
Cam Followers, Replace (B)		
900		
1981-93		
8V	2.4	*2.6*
16V		
intake side	3.5	*3.7*
exhaust side	4.3	*4.5*
1994-98		
4 cyl. one side	3.3	*3.5*
V6 one side	4.4	*4.6*
9000		
intake side	3.6	*3.8*
exhaust side	4.4	*4.6*
Cylinder Head, Replace (B)		
Includes: Parts transfer and adjustments.		
900		
1981-93		
8V	9.9	*10.1*
16V	14.3	*14.5*
1994-98		
4 cyl.	6.4	*6.6*
V6 one	7.9	*8.1*
9000		
2.0L	13.8	*14.0*
2.3L	12.7	*12.9*
V6		
one	7.7	*7.9*
both	9.9	*10.1*
9-3	7.8	*8.0*
9-5		
4 cyl.	8.7	*8.9*
V6		
front	9.8	*10.0*
rear	8.5	*8.7*
both	11.5	*11.7*
w/turbo add		
900		
81-93	.7	*.9*
94-98	1.1	*1.3*
9000		
2.0L	.7	*.5*
2.3L	2.8	*3.0*
1999 9-3 2.0L	.2	*.4*
Cylinder Head Gasket, Replace (B)		
900		
1981-93		
8V	5.2	*5.4*
16V	9.4	*9.6*
1994-98		
4 cyl.	5.3	*5.5*
V6 one	6.9	*7.1*
9000		
2.0L	8.4	*8.6*
2.3L	6.5	*6.7*
V6 one	6.9	*7.1*
9-2X, 9-3	7.6	*7.8*

	LABOR TIME	SEVERE SERVICE
9-5		
4 cyl.	8.5	*8.7*
V6		
front	9.9	*10.1*
rear	8.3	*8.5*
both	11.3	*11.5*
w/turbo add		
900		
81-93	.7	*.9*
94-98	1.1	*1.3*
9000		
2.0L	.7	*.9*
2.3L	2.8	*3.0*
Valve Cover Gasket, Replace (B)		
900		
1981-93	.7	*.9*
1994-98		
4 cyl.	.5	*.7*
V6		
right side	.5	*.7*
left side	1.3	*1.5*
9000	.7	*.9*
9-2X, 9-3, 9-5 each	.8	*1.0*
Valve Job (A)		
900		
1981-93		
8V	12.0	*12.2*
16V	15.8	*16.0*
1994-98 4 cyl.	11.8	*12.0*
9000 intake or exhaust		
2.0L	12.9	*13.1*
2.3L	11.8	*12.0*
9-3 intake or		
exhaust all	12.9	*13.1*
9-5 intake or exhaust		
4 cyl.	10.4	*10.6*
V6		
front	11.3	*11.5*
rear	9.8	*10.0*
w/turbo add		
900		
81-93	.7	*.9*
94-98	1.2	*1.4*
9000		
2.0L	1.1	*1.3*
2.3L	.9	*1.1*
Valve Springs, Replace (B)		
900 8V	3.3	*3.5*
Valve Springs and/or Oil Seals, Replace (B)		
900 16V	6.8	*7.0*
9000	6.3	*6.5*
9-3	6.8	*7.0*
9-5	4.9	*5.1*

	LABOR TIME	SEVERE SERVICE
CAMSHAFT		
Balance Shaft, Replace (B)		
Includes: R&I engine and timing cover on 900, 9-3, and 9-5. R&I balance shaft chain on 1990-93 9000. R&I cylinder heads and timing cover on 1994-98 9000.		
1994-98 900 4 cyl.		
w/turbo	14.2	*14.4*
w/o turbo	13.1	*13.3*
9000 2.3L		
1990-93	7.7	*7.9*
1994-98		
w/turbo	12.9	*13.1*
w/o turbo	11.9	*12.1*
9-3		
1999	14.2	*14.4*
2000-04	13.7	*13.9*
9-5 4 cyl.		
one	11.8	*12.0*
both	12.1	*12.3*
Balance Shaft Chain and/or Sprockets, Replace (B)		
Includes: R&I engine and timing cover on 900 and 9-3. R&I cylinder heads and timing cover on 1994-98 9000. R&I engine, timing cover, gears, idler pulley and chain controls on 9-5.		
1994-98 900 4 cyl.		
w/turbo	13.9	*14.1*
w/o turbo	12.8	*13.0*
9000 2.3L		
1990-93	7.7	*7.9*
1994-98		
w/turbo	12.8	*13.0*
w/o turbo	11.9	*12.1*
9-3		
1999	13.9	*14.1*
2000-04	13.4	*13.6*
9-5 4 cyl.	11.9	*12.1*
Balance Shaft Chain Tensioner and/or Guide, Replace (B)		
Includes: R&I engine and timing cover on 900 AND 9-3. R&I cylinder heads and timing cover on 1994-98 9000. R&I engine, timing cover, gears, idler pulley and chain controls on 9-5.		
1994-98 900 4 cyl.		
w/turbo	13.9	*14.1*
w/o turbo	12.8	*13.0*
9000 2.3L		
1990-93	7.7	*7.9*
1994-98		
w/turbo	12.8	*13.0*
w/o turbo	11.9	*12.1*
9-3		
1999	13.9	*14.1*
2000-04	13.4	*13.6*
9-5 4 cyl.	11.7	*11.9*

SAA-8 900 : 9000 : 9-2x : 9-3 : 9-5

	LABOR TIME	SEVERE SERVICE
Camshaft, Replace (A)		
900		
1981-93		
8V	2.5	2.7
16V		
intake	3.3	3.5
exhaust	3.8	4.0
1994-98		
4 cyl. each	2.2	2.4
V6 each	4.7	4.9
9000		
4 cyl.		
intake	3.1	3.3
exhaust	3.7	3.9
V6 each	4.4	4.6
9-3 each	2.2	2.4
9-5		
4 cyl. each	1.8	2.0
V6 front each	6.9	7.1
Camshaft Seal, Replace (B)		
900 V6 one or both	3.7	3.9
9000 V6 each	4.2	4.4
Timing Belt, Replace (B)		
900 V6	2.8	3.0
9000 V6	2.9	3.1
9-5 V6	3.1	3.3
Timing Belt Tensioner, Replace (B)		
900 V6	2.4	2.6
9000 V6	2.8	3.0
9-5 V6	2.9	3.1
Timing Chain, Replace (B)		
Includes: R&I engine, timing cover, and replace balance shaft chain, all chain controls, timing gear, balance shaft gear with idler pulley on 1994-98 900. R&I engine on 1986-93 9000. R&I cylinder heads and timing cover on 1994-98 9000. R&I cylinder head also included on 9-3 and 9-5.		
900		
1981-93		
8V	9.8	10.0
16V	12.9	13.1
1994-98 4 cyl.	14.7	14.9
9000		
1986-93 w/engine removal	10.9	11.1
1994-98 4 cyl.	13.3	13.5
9-2X	6.3	6.5
9-3		
1999	14.7	14.9
2000-04	13.9	14.1
9-5 4 cyl.	15.7	15.9
w/R12 86-93 9000		
lower add	.5	.7
w/turbo add		
81-93 900	1.5	1.7
86-93 9000 lower	1.1	1.3

	LABOR TIME	SEVERE SERVICE
Timing Chain Guide, Replace (B)		
Includes: R&I engine and timing cover on lower 900, 9-3, and 9-5. R&I engine on 1986-93 9000. R&I cylinder heads and timing cover on 1994-98 9000.		
900		
1981-93		
8V all	9.8	10.0
16V		
upper	.5	.7
lower	12.9	13.1
1994-98 4 cyl.		
upper	.6	.8
lower	13.9	14.1
9000		
1986-93		
upper	.5	.7
lower	10.1	10.3
1994-98 4 cyl.		
upper	.5	.7
lower	13.0	13.2
9-2X	6.6	6.8
9-3		
upper	.6	.8
lower		
1999	13.9	14.1
2000-04	13.8	14.0
9-5 4 cyl.		
upper	.6	.8
lower	11.8	12.0
w/R12 86-93 9000		
lower add	.5	.5
w/turbo add		
81-93 900 8V	1.7	1.9
86-93 9000 lower	1.0	1.2
Timing Chain Housing Gasket, Replace (B)		
Includes: R&I engine on 1994-98 900, 9000 2.0L, 9-3, and 9-5. R&I cylinder head on 1994-98 9000.		
900		
1981	6.3	6.5
1982-93		
8V	8.8	9.0
16V	11.8	12.0
1994-98 4 cyl.	12.3	12.5
9000		
1986-93		
2.0L	10.5	10.7
2.3L	7.8	8.0
1994-98 4 cyl.	12.4	12.6
9-2X	2.4	2.6

	LABOR TIME	SEVERE SERVICE
9-3		
1999	13.9	14.1
2000-04	13.2	13.4
9-5 4 cyl.	11.8	12.0
w/turbo add		
900		
81-93 8V	1.6	1.8
94-98	1.0	1.2
86-93 9000		
2.0L	1.1	1.3
2.3L	.7	.9
Timing Chain Tensioner, Replace (B)		
900		
1981-93		
8V	9.5	9.7
16V	1.4	1.6
1994-98	.8	1.0
9000	.8	1.0
9-2X, 9-3	1.1	1.3
9-5	1.5	1.7
w/turbo 81-93 900		
8V add	1.7	1.9
CRANK & PISTONS		
Connecting Rod Bearings, Replace (A)		
Includes: Check bearing oil clearance.		
900		
1981-93		
8V	10.8	11.0
16V	12.6	12.8
1994-96	11.8	12.0
9000	11.9	12.1
w/turbo add	.7	.9
Crankshaft and Main Bearings, Replace (A)		
Includes: R&I engine. Check bearing oil clearance.		
900		
1981-93		
8V	14.3	14.5
16V	16.4	16.6
1994-98		
4 cyl.	9.8	10.0
V6	9.9	10.1
9000	16.7	16.9
w/turbo add		
900	1.2	1.4
9000	.7	.9
Crankshaft Front Oil Seal, Replace (B)		
900		
1981-83	1.5	1.7
1984-93	3.0	3.2
1994-98		
4 cyl.	.7	.9
V6	2.7	2.9

	LABOR TIME	SEVERE SERVICE
9000		
2.0L	1.1	1.3
2.3L	.7	.9
V6	3.2	3.4
9-3	.8	1.0
9-5		
4 cyl.	1.6	1.8
V6	3.5	3.7

Main & Rod Bearings, Replace (A)
Includes: Check bearing oil clearance.
900		
8V	11.3	11.5
16V	13.0	13.2
9000	13.3	13.5
w/turbo add	.7	.9

Main Bearings, Replace (A)
Includes: Check bearing oil clearance.
900		
8V	10.7	10.9
16V	12.4	12.6
9000	12.4	12.6
w/turbo add	.7	.9

Pistons or Connecting Rods, Replace (A)
Includes: Remove cylinder top ridge, deglaze cylinder walls, replace connecting rod bearings, install new rings. Minor tune-up.
900		
1981-93		
8V	15.4	15.6
16V	17.7	17.9
1994-98	14.3	14.5
9000	17.1	17.3
9-3	13.9	14.1
9-5		
4 cyl.	15.5	15.7
V6	12.5	12.7
w/turbo 900, 9000 add	.7	.9

Rear Main Oil Seal, Replace (B)
900		
1981-93		
AT		
8V	9.9	10.1
16V	10.9	11.1
MT	2.9	3.1
1994-98		
AT	5.8	6.0
MT	4.8	5.0
9000		
4 cyl.		
AT	8.9	9.1
MT	6.9	7.1
V6		
AT	8.4	8.6
MT	6.7	6.9
9-3		
AT	7.5	7.7
MT	7.2	7.4

	LABOR TIME	SEVERE SERVICE
9-5		
4 cyl.		
AT	9.9	10.1
MT	9.8	10.0
V6		
AT	10.8	11.0
MT	5.6	5.8
w/turbo 900, 9000 add	.7	.9

Rings, Replace (A)
Includes: Remove cylinder top ridge, deglaze cylinder walls, clean carbon. Minor tune-up.
900		
1981-93		
8V	13.8	14.0
16V	16.2	16.4
1994-98	13.1	13.3
9000	15.9	16.1
9-3	14.4	14.6
9-5		
4 cyl.	14.1	14.3
V6	15.5	15.7
w/turbo 900, 9000 add	.7	.9

ENGINE LUBRICATION

Engine Oil Cooler, Replace (B)
900		
1981-93		
8V	.9	1.1
16V	1.9	2.1
1994-98		
4 cyl.	.7	.9
V6	3.7	3.9
9000		
2.0L	.5	.7
2.3L		
w/turbo	.8	1.0
w/o turbo	1.2	1.4
V6	3.8	4.0
9-2X, 9-3	.8	1.0
9-5		
4 cyl.	.8	1.0
V6	4.1	4.3

Engine Oil Pressure Switch (Sending Unit), Replace (B)
900	.5	.7
9000 4 cyl.	.5	.7
9-2X, 9-3	.8	1.0
9-5		
4 cyl.	.8	1.0
V6	.3	.5

Oil Pan Gasket, Replace (B)
900		
1981-93		
8V	7.6	7.8
16V	10.8	11.0
1994-98 4 cyl.	4.3	4.5

	LABOR TIME	SEVERE SERVICE
9000		
2.0L	5.8	6.0
2.3L	3.8	4.0
V6	3.6	3.8
9-2X, 9-3	4.3	4.5
9-5		
4 cyl.	2.1	2.3
V6	2.4	2.6
w/turbo add		
81-93 900 8V	1.8	2.0
9000 2.3L	.8	1.0

Oil Pump, R&R and Recondition (B)
900	3.9	4.1

Oil Pump, Replace (B)
900		
1981-93	4.1	4.3
1994-98		
4 cyl.	1.0	1.2
V6	9.4	9.6
9000		
1986-93	1.9	2.1
1994-98		
4 cyl.	.8	1.0
V6	8.8	9.0
9-3	1.0	1.2
9-5		
4 cyl.	1.6	1.8
V6	7.1	7.3

CLUTCH

Bleed Clutch Hydraulic System (B)
1981-05	.5	.7

Clutch Assy., Replace (B)
900		
1981-93	2.6	2.8
1994-98		
4 cyl.	6.0	6.2
V6	5.7	5.9
9000	6.4	6.6
9-2X, 9-3	6.7	6.9
9-5	8.8	9.0
w/turbo 81-93 900		
16V add	.7	.9
Replace pilot		
bearing add	.3	.5

Clutch Cable, Replace (B)
1994-98 900	1.2	1.4

Clutch Master Cylinder, R&R and Recondition (B)
Includes: System bleeding.
900	1.4	1.6
9000	1.5	1.7

Clutch Master Cylinder, Replace (B)
Includes: System bleeding.
900	.8	1.0
9000		
1986-89	.8	1.0
1990-98	1.4	1.6
9-2X, 9-3	1.4	1.6

SAA-10 900 : 9000 : 9-2x : 9-3 : 9-5

	LABOR TIME	SEVERE SERVICE
Clutch Master Cylinder, Replace (B)		
9-5		
1999-01	1.1	1.3
2002-05	2.4	2.6
Clutch Release Bearing, Replace (B)		
900		
1981-93	2.4	2.6
1994-98		
4 cyl.	5.5	5.7
V6	5.3	5.5
9000	6.2	6.4
9-2X, 9-3	6.7	6.9
9-5	8.8	9.0
w/turbo 81-93 900 16V add	.7	.9
Clutch Slave Cylinder, R&R and Reconditon (B)		
Includes: System bleeding.		
900	2.8	3.0
w/turbo 81-93 900 16V add	.7	.9
Clutch Slave Cylinder, Replace (B)		
Includes: Bleed system.		
900	2.2	2.4
9000	6.2	6.4
9-2X, 9-3	6.4	6.6
9-5	8.0	8.2
w/turbo 81-93 900 16V add	.7	.9
Flywheel, Replace (B)		
900		
1981-93	3.0	3.2
1994-98	5.3	5.5
9000	6.8	7.0
9-2X, 9-3	6.9	7.1
9-5	8.8	9.0

MANUAL TRANSAXLE

	LABOR TIME	SEVERE SERVICE
Chain Drive Gear Cover (In Chassis), R&R or Replace (B)		
900	1.8	2.0
w/turbo add	.7	.9
Drive Sprockets, Chain or Upper Bearing (In Chassis), Replace (B)		
900	3.2	3.4
w/turbo add	.7	.9
Replace tensioner add	.8	1.0
Ring Gear & Pinion Assy., Replace (B)		
900		
8V	17.7	17.9
16V	19.9	20.1
9-2X, 9-3	12.5	12.7
9-5	14.8	15.0
w/turbo 900 add	.7	.9

	LABOR TIME	SEVERE SERVICE
Transaxle Assy. R&R and Recondition (A)		
900		
1981-93		
8V	17.1	17.3
16V	19.4	19.6
1994-98	12.5	12.7
9000	15.1	15.3
9-2X, 9-3	10.2	10.4
9-5	17.9	18.1
w/turbo 81-93 900 add	.7	.9
Recondition differential 81-93 900 add	1.2	1.4
Transaxle Assy., R&I (B)		
900		
1981-93		
8V	9.7	9.9
16V	11.4	11.6
1994-98	5.3	5.5
9000	6.8	7.0
9-2X, 9-3	7.5	7.7
9-5	8.5	8.7
w/turbo 81-93 900 add	.7	.9
Replace assembly add	.5	.7

AUTOMATIC TRANSAXLE

SERVICE TRANSAXLE INSTALLED

	LABOR TIME	SEVERE SERVICE
Check Unit for Oil Leaks (C)		
1981-05	.7	.9
Drain & Refill Unit (B)		
1981-05	.9	1.1
Pressure Test (B)		
1981-05	.9	1.1
Kickdown Band, Adjust (B)		
900		
front	1.9	2.1
rear	.7	.9
9000	.5	.7
Throttle Control Cable, Adjust (B)		
900, 9000	.5	.7
Differential Cover Gasket, Replace (B)		
1981-05	1.3	1.5
Electronic Control Unit, Replace (B)		
1994-98	.5	.7
9-2X, 9-3	.9	1.1
9-5		
1999-01	.7	.9
2002-04	1.5	1.7
Governor Assy., R&R and Recondition (B)		
900	2.1	2.3
9000	2.6	2.8
Input Speed Sensor, Replace (B)		
1994-98 900	.7	.9
Lock-up Solenoid, Replace (B)		
1994-98 900	2.8	3.0
9-3	3.5	3.7
1999-05 9-5	2.4	2.6

	LABOR TIME	SEVERE SERVICE
Oil Pan Gasket, Replace (B)		
900	1.4	1.6
9000	1.3	1.5
9-2X, 9-3, 9-5	1.1	1.3
Output Speed Sensor, Replace (B)		
1994-98 900	.5	.7
9-2X, 9-3	.4	.6
9-5	.3	.5
Shift Solenoid, Replace (B)		
1994-98 900 one or all	3.6	3.8
9-3 one or all	3.6	3.8
9-5 one or all	2.5	2.7
Throttle Control Cable, Replace (B)		
900	2.6	2.8
9000	2.2	2.4
Transmission Drive Chain and/or Sprockets, Replace (B)		
900	3.3	3.5
w/turbo add	.2	.4
Transmission Front Cover, Replace (B)		
900	1.6	1.8
w/turbo 900 add	.3	.5
Valve Body, R&R and Recondition (B)		
900	4.8	5.0
Valve Body, Replace (B)		
900		
1981-93	3.3	3.5
1994-98	3.0	3.2
9000	1.5	1.7
9-2X, 9-3	5.0	5.2
9-5		
1999-01	3.8	4.0
2002-05		
4 cyl.	4.6	4.8
V6	5.8	6.0

SERVICE TRANSAXLE REMOVED

	LABOR TIME	SEVERE SERVICE
Ring & Pinion Set, Replace (B)		
900		
8V	24.3	24.5
16V	26.7	26.9
Torque Converter, Replace (B)		
900		
1981-93		
8V	14.0	14.2
16V	17.9	18.1
1994-98	7.3	7.5
9000	8.8	9.0
9-2X, 9-3	7.3	7.5
9-5		
1999-01		
4 cyl.	9.5	9.7
V6	10.2	10.4
2002-05		
4 cyl.	10.0	10.2
V6	10.6	10.8
w/turbo 81-93 900 add	2.4	2.6

900 : 9000 : 9-2x : 9-3 : 9-5 SAA-11

	LABOR TIME	SEVERE SERVICE
Transaxle Assy., R&R and Reconditon (A)		
900		
8V	21.9	22.1
16V	24.1	24.3
w/turbo 81-93 900 add	.7	.9
Transaxle Assy., Replace (B)		
900		
1981-93		
8V	10.5	10.7
16V	14.3	14.5
1994-98	8.2	8.4
9000	12.2	12.4
9-2X, 9-3	8.2	8.4
9-5		
4 cyl.	10.6	10.8
V6	11.0	11.2
w/turbo 81-93 900 add	2.4	2.6
Replace flexplate add	.3	.5

SHIFT LINKAGE
AUTOMATIC TRANSMISSION

	LABOR TIME	SEVERE SERVICE
Selector Cable, Adjust (B)		
1983-93 900	.5	.7
1986-96 9000	.3	.5
9-2X, 9-3, 9-5	.3	.5
Gear Shift Housing, Replace (B)		
900		
1983-93	2.8	3.0
1994-96	2.3	2.5
1986-96 9000	.9	1.1
9-2X, 9-3	1.6	1.8
9-5	1.8	2.0
Selector Cable, Replace (B)		
1983-93 900	2.0	2.2
1986-96 9000	1.7	1.9
9-2X, 9-3	1.9	2.1
9-5	2.5	2.7

MANUAL TRANSMISSION

	LABOR TIME	SEVERE SERVICE
Gearshift Linkage, Adjust (B)		
1981-05	.8	1.0
Gear Selector Lever, Replace (B)		
900		
1983-93	.5	.7
1994-98	1.8	2.0
9000	.5	.7
9-2X, 9-3, 9-5	2.0	2.2
Gearshift Lever Housing or Base, Replace (B)		
900		
1983-93	2.5	2.7
1994-98	2.2	2.4
9000	1.2	1.4
9-2X, 9-3, 9-5	2.0	2.2

DRIVELINE

	LABOR TIME	SEVERE SERVICE
Differential Cover Gasket, Replace (B)		
900		
AT	1.4	1.6
MT	.8	1.0
9000	1.4	1.6
Differential Housing and/or Bearings, R&R or Replace (B)		
900		
1981-93		
w/turbo		
8V	12.6	12.8
16V	14.2	14.4
w/o turbo	11.6	11.8
1994-98		
AT	11.5	11.7
MT	13.1	13.3
9000		
AT	13.8	14.0
MT	14.6	14.8
9-2X, 9-3		
AT	11.5	11.7
MT	13.2	13.4
9-5		
AT		
4 cyl.	14.0	14.2
V6		
1999-01	14.2	14.4
2002-05	14.6	14.8
MT	15.6	15.8
Differential Side & Pinion Gear, Replace (B)		
1981-93 900		
w/turbo		
8V	9.4	9.6
16V	11.0	11.2
w/o turbo	8.4	8.6
9000	11.8	12.0
Differential Side Bearings, Replace (B)		
900	12.3	12.5
9000		
AT	13.1	13.3
MT	12.7	12.9
Inner Driveshaft Bearing or Seal, Replace (B)		
900		
1981-93		
w/turbo		
8V	10.2	10.4
16V	11.8	12.0
w/o turbo	9.1	9.3
1994-98	13.5	13.7
9000		
right side		
2.0L	2.2	2.4
2.3L	2.9	3.1
left side	2.1	2.3

	LABOR TIME	SEVERE SERVICE
9-2X, 9-3	1.7	1.9
9-5 each	1.8	2.0
Inner Shaft, R&R or Replace (B)		
9000	2.2	2.4
Intermediate Shaft, R&R or Replace (B)		
1994-98 900		
4 cyl.	1.3	1.5
V6	.9	1.1
9000	1.8	2.0
Intermediate Shaft Support Bearing, Replace (B)		
1994-98 900		
4 cyl.	1.4	1.6
V6	.9	1.1
9000	2.1	2.3
Ring Gear & Pinion Set, Replace (B)		
900		
1981-93		
AT		
w/turbo		
8V	25.9	26.1
16V	26.9	27.1
w/o turbo	24.5	24.7
4-Speed MT		
w/turbo	18.1	18.3
w/o turbo	17.1	17.3
5-Speed MT		
w/turbo		
8V	19.8	20.0
16V	20.9	21.1
w/o turbo	18.5	18.7
1994-98		
AT	11.8	12.0
MT	13.4	13.6
9000		
AT	13.8	14.0
MT	14.6	14.8
9-2X, 9-3		
AT	11.5	11.8
MT	13.2	13.4
9-5		
AT		
4 cyl.	14.0	14.2
V6		
1999-01	14.2	14.4
2002-04	14.6	14.8
MT	15.6	15.8

HALFSHAFTS

	LABOR TIME	SEVERE SERVICE
Halfshaft, R&R or Replace (B)		
900		
1981-93		
one side	1.4	1.6
both sides	2.9	3.1
1994-98		
right side	.6	.8
left side	1.3	1.5
9000 one side	1.3	1.5

SAA-12 900 : 9000 : 9-2x : 9-3 : 9-5

	LABOR TIME	SEVERE SERVICE
Halfshaft, R&R or Replace (B)		
9-2X, 9-3		
right side	.8	1.0
left side	1.3	1.5
9-5		
right side	1.0	1.2
left side	1.4	1.6
Inner CV Joint, Replace (B)		
900		
1981-93	1.8	2.0
1994-98		
right side	1.8	2.0
left side	2.1	2.3
9000		
right side	2.1	2.3
left side	1.4	1.6
9-2X, 9-3		
right side	1.4	1.6
left side	1.9	2.1
9-5		
right side	1.4	1.6
left side	2.1	2.3
Inner CV Joint Boot, Replace (B)		
900	1.6	1.8
9000	.8	1.0
9-2X, 9-3		
right side	1.4	1.6
left side	1.9	2.1
9-5		
right side	1.4	1.6
left side	2.1	2.3
Outer CV Joint, Replace (B)		
900	1.9	2.1
9000	.9	1.1
9-2X, 9-3		
right side	1.4	1.6
left side	1.9	2.1
9-5		
right side	1.4	1.6
left side	2.1	2.3
Outer CV Joint Boot, Replace (B)		
900		
1981-93	1.4	1.6
1994-98		
right side	1.8	2.0
left side	1.9	2.1
9000	1.2	1.4
9-2X, 9-3		
right side	1.4	1.6
left side	1.9	2.1
9-5		
right side	1.4	1.6
left side	2.1	2.3

BRAKES
ANTI-LOCK

	LABOR TIME	SEVERE SERVICE
Diagnose Anti-Lock Brake System Component Each (A)		
900		
1990-93	2.2	2.4
1994-98	.5	.7
1988-98 9000	2.1	2.3
9-2X, 9-3, 9-5	.9	1.1
Accumulator Assy., Replace (B)		
1990-93 900	.5	.7
1988-98 9000	.7	.9
Control Unit, Replace (B)		
900		
1990-93	.5	.7
1994-95	1.6	1.8
1996-98	1.3	1.5
1988-98 9000	.5	.7
9-2X, 9-3, 9-5	1.7	1.9
Front Sensor Assy., Replace (B)		
900	.5	.7
9000		
1988-89	1.2	1.4
1990-98	.9	1.1
9-2X, 9-3	1.1	1.3
9-5	1.3	1.4
High Pressure Hoses, Replace (B)		
1990-93 900	1.5	1.7
1988-98 9000	4.6	4.8
Pump & Motor Assy., Replace (B)		
1990-93 900	.5	.7
1988-98 9000	4.6	4.8
Rear Sensor Assy., Replace (B)		
900		
1990-93	.5	.7
1994-98	.8	1.0
9000	.9	1.1
9-2X, 9-3, 9-5	1.1	1.3
Reservoir, Replace (B)		
900		
1990-93	1.3	1.5
1994-98	.5	.7
1988-98 9000		
w/MKIV ABS	.5	.7
w/o MKIV ABS	4.4	4.6
9-2X, 9-3	.6	.8
9-5	.8	1.0
Valve Block, Replace (B)		
1990-93 900	1.4	1.6
9-3, 9-5	1.7	1.9
Warning Switch, Replace (B)		
1990-93 900	.5	.7
1988-98 9000	1.2	1.4
9-2X, 9-3, 9-5	.6	.8

SYSTEM

	LABOR TIME	SEVERE SERVICE
Bleed Brakes (B)		
Includes: Add fluid.		
1981-05	.6	.8
w/ABS 900, 9000 add	.3	.5
Brake Hose (Flexible), Replace (B)		
Includes: System bleeding.		
One	.8	1.0
each addl. add	.5	.7
w/ABS 900, 9000 add	.3	.5
Master Cylinder, R&R and Recondition (B)		
Includes: System bleeding.		
900	2.4	2.6
w/ABS 900, 9000 add	.3	.5
Master Cylinder, Replace (B)		
Includes: System bleeding.		
900	1.3	1.5
9000	1.4	1.6
9-2X, 9-3	.8	1.0
9-5	1.2	1.4
w/ABS 900, 9000 add	.3	.5
Power Booster Unit, Replace (B)		
Includes: System bleeding.		
900		
1981-89	1.4	1.6
1994-98	1.8	2.0
9000	2.7	2.9
9-2X, 9-3	1.8	2.0
9-5	4.5	4.7
Power Booster Vacuum Check Valve, Replace (B)		
1981-05	.3	.5
Traction Control Unit, Replace (B)		
1994-96 900	.7	.9

SERVICE BRAKES

	LABOR TIME	SEVERE SERVICE
Brake Pads, Replace (B)		
Front		
900		
1981-87	1.2	1.4
1988-98	.9	1.1
9000	.9	1.1
9-3, 9-5	1.0	1.2
Rear		
900		
1981-87	.8	1.0
1988-93	1.3	1.5
1994-98	.7	.9
9000	1.3	1.5
9-2X, 9-3, 9-5	1.1	1.3
Caliper Assy. R&R and Recondition (A)		
Includes: System bleeding.		
900		
1981-87		
front		
one side	2.3	2.5
both sides	4.2	4.4

	LABOR TIME	SEVERE SERVICE
rear		
one side	1.9	2.1
both sides	3.0	3.2
1988-93		
one side	2.6	2.8
both sides	3.4	3.6
1994-98		
one side	1.2	1.4
both sides	1.7	1.9
9000		
front		
one side	2.3	2.5
both sides	2.9	3.1
rear		
one side	2.8	3.0
both sides	3.2	3.4
9-2X, 9-3		
one side	1.4	1.6
both sides	2.9	3.1
9-5		
front		
one side	.8	1.0
both sides	1.4	1.6
rear		
one side	1.1	1.3
both sides	2.2	2.4
w/ABS 900, 9000 add	.3	.5
w/turbo 900, 9000 add	.2	.2
Caliper Assy., Replace (B)		
Includes: System bleeding.		
900		
1981-87		
front		
one side	1.4	1.6
both sides	2.6	2.8
rear		
one side	1.4	1.6
both sides	1.9	2.1
1988-93		
front		
one side	1.6	1.8
both sides	2.0	2.2
rear		
one side	2.0	2.2
both sides	2.6	2.8
1994-98		
one side	.7	.9
both sides	.9	1.1
9000		
front		
one side	1.7	1.9
both sides	2.1	2.3
rear		
one side	2.0	2.2
both sides	2.6	2.8
9-2X, 9-3		
one side	1.3	1.5
both sides	1.7	1.9

	LABOR TIME	SEVERE SERVICE
9-5		
front		
one side	.8	.9
both sides	1.3	1.4
rear		
one side	1.0	1.2
both sides	1.8	2.0
w/ABS 900, 9000 add	.3	.5
w/turbo 900, 9000 add	.2	.4
Disc Brake Rotor, Replace (B)		
900		
1981-87		
front		
one side	.5	.7
both sides	.8	1.0
rear		
one side	.5	.7
both sides	1.4	1.6
1988-93		
one side	1.3	1.5
both side	1.7	1.9
1994-98		
one side	.5	.7
both sides	.8	1.0
9000		
front		
w/turbo		
one side	1.8	2.0
both sides	1.9	2.1
w/o turbo		
one side	1.4	1.6
both sides	1.7	1.9
rear		
one side	1.7	1.9
both sides	2.3	2.5
9-2X, 9-3		
one side	1.3	1.5
both sides	1.6	1.8
9-5		
one side	1.4	1.6
both sides	1.8	2.0

PARKING BRAKE

Parking Brake Cable, Adjust (C)		
900		
1988-90	.5	.7
1991-98	.3	.5
9000	.3	.5
9-2X, 9-3, 9-5	.5	.7
Parking Brake Apply Actuator, Replace (B)		
900		
1981-93	1.2	1.4
1994-98	1.6	1.8
9000	1.1	1.3
9-2X, 9-3	1.6	1.8
9-5	1.2	1.4

	LABOR TIME	SEVERE SERVICE
Parking Brake Apply Warning Indicator Switch, Replace (B)		
900		
1981-93	.5	.7
1994-98	.9	1.1
9000	.3	.5
9-2X, 9-3, 9-5	.7	.9
Parking Brake Cable, Replace (B)		
900		
1981-87		
one side	2.8	3.0
both sides	3.4	3.6
1988-93		
one side	1.8	2.0
both sides	2.4	2.6
1994-98		
right side	.9	1.1
left side	.6	.8
9000		
one side	1.9	2.1
both sides	2.5	2.7
9-2X, 9-3		
right side	1.0	1.2
left side	.6	.8
both sides	1.4	1.6
9-5		
one side	.8	1.0
rear	1.4	1.6
Parking Brake Shoes, Replace (B)		
900	1.3	1.5
9-5	1.4	1.6

FRONT SUSPENSION

Unless otherwise noted, time given does not include alignment.

Align Front End (A)		
1981-05	1.1	1.3
Front & Rear Toe, Adjust (B)		
9000	.9	1.1
Front Toe, Adjust (B)		
1981-05	.7	.9
Rear Toe, Adjust (B)		
900	3.0	3.2
9000	.5	.7
9-2X, 9-3	.8	.9
9-5	1.0	1.1
Front Coil Spring, Replace (B)		
900		
one side	.9	1.1
both sides	1.8	2.0
Front Stabilizer Link, Replace (B)		
9000	.7	.9
Front Strut Coil Spring, Replace (B)		
1994-98 900	1.6	1.8
9000	1.5	1.7
9-2X, 9-3		
one side	1.6	1.8
both sides	2.9	3.1

SAA-14   900 : 9000 : 9-2x : 9-3 : 9-5

	LABOR TIME	SEVERE SERVICE
Front Strut Coil Spring, Replace (B)		
9-5 both		
one side	1.3	*1.5*
both sides	2.4	*2.6*
Front Strut Shock Absorber, Replace (B)		
1994-98 900	1.7	*1.9*
9000	1.1	*1.3*
9-2X, 9-3		
one side	1.6	*1.8*
both sides	2.9	*3.1*
9-5 both		
one side	1.1	*1.3*
both sides	1.9	*2.1*
Hub Bearings, Replace (B)		
1994-98 900	1.6	*1.8*
9-2X, 9-3	1.6	*1.8*
9-5	1.1	*1.3*
Lower Ball Joint, Replace (B)		
900	.8	*1.0*
9000	.6	*.8*
9-2X, 9-3, 9-5	.8	*1.0*
Lower Control Arm, Replace (B)		
900		
1981-93	2.8	*3.0*
1994-98	.7	*.9*
9000	.8	*1.0*
9-2X, 9-3	.6	*.8*
9-5		
one side	2.9	*3.1*
both sides	4.3	*4.5*
Lower Control Arm Bushings, Replace (B)		
900	1.9	*2.1*
9000		
front bushing	2.5	*2.7*
rear bearing	.7	*.9*
9-2X, 9-3		
one side	1.6	*1.8*
all	2.5	*2.7*
9-5		
one side	2.9	*3.1*
all	3.8	*4.0*
Shock Absorber, Replace (B)		
900		
one	.5	*.7*
both	.7	*.9*
Stabilizer Bar, Replace (B)		
900	1.2	*1.4*
9000	2.0	*2.2*
9-2X, 9-3	1.6	*1.8*
9-5	2.5	*2.7*
Stabilizer Bar Bushings, Replace (B)		
900	.3	*.5*
9000	.7	*.9*
9-2X, 9-3	1.4	*1.6*
9-5	.3	*.5*
Steering Arm, Replace (B)		
900	2.0	*2.2*

	LABOR TIME	SEVERE SERVICE
Steering Knuckle, Replace (B)		
900		
1981-87	4.8	*5.0*
1988-98	1.9	*2.1*
9000	1.9	*2.1*
9-2X, 9-3, 9-5	2.4	*2.6*
Upper Ball Joints, Replace (B)		
900	.9	*1.1*
9-5	.7	*.9*
Upper Control Arm, Replace (B)		
900 one side	3.0	*3.2*
Upper Control Arm Bushings, Replace (B)		
900		
right side	2.1	*2.3*
left side	2.7	*2.9*
Wheel Hub, Bearing or Seal, Replace (B)		
1994-98 900	1.6	*1.8*
9-2X, 9-3	1.6	*1.8*
9-5	1.1	*1.3*

REAR SUSPENSION

	LABOR TIME	SEVERE SERVICE
Rear Suspension Toe, Adjust (B)		
900	3.2	*3.4*
Rear Wheel Camber & Toe-In, Check (B)		
900		
1981-93	1.2	*1.4*
1994-98	.5	*.7*
Rear Axle Assy., Replace (B)		
900		
1981-93	4.0	*4.2*
1994-98	2.9	*3.1*
9000	3.0	*3.2*
9-3	3.0	*3.2*
9-5	3.6	*3.8*
Rear Coil Spring, Replace (B)		
900, 9000		
one	.5	*.7*
both	.9	*1.1*
9-2X, 9-3		
one	.5	*.7*
both	.6	*.8*
9-5		
one	.8	*1.0*
both	1.6	*1.8*
Rear Control Arm, Replace (B)		
900		
one	.8	*1.0*
both	1.4	*1.6*
9000		
one	.8	*1.0*
both	1.2	*1.4*
9-5	1.4	*1.6*
Rear Cross Bar, Replace (B)		
900	.7	*.9*
9000	.5	*.7*

	LABOR TIME	SEVERE SERVICE
Rear Hub and/or Bearings, Replace (B)		
1994-98 900		
one side	.7	*.9*
both sides	1.3	*1.5*
9000		
one side	.7	*.9*
both sides	1.3	*1.5*
9-2X, 9-3		
one side	.8	*1.0*
both sides	1.6	*1.8*
9-5		
one side	1.1	*1.3*
both sides	2.2	*2.4*
Rear Side Link, Replace (B)		
900, 9000		
one side	.5	*.7*
both sides	.8	*1.0*
Rear Stabilizer Bar & Bushings, Replace (B)		
900		
1981-93	.7	*.9*
1994-98	.5	*.7*
9000	.8	*1.0*
9-2X, 9-5	1.1	*1.3*
9-3		
front bar	.5	*.7*
rear bar	.2	*.4*
Rear Stabilizer Bar Bushings, Replace (B)		
Exc. 9-5	.5	*.7*
9-5	1.1	*1.3*
Rear Struts, Replace (B)		
9-5		
one side	.6	*.8*
both sides	1.1	*1.3*
Rear Wheel Bearing & Hub Assy., Clean, Repack or Replace (B)		
1982-93 900		
one side	.7	*.9*
both sides	1.2	*1.4*
9-2X, 9-3		
one side	.8	*1.0*
both sides	1.6	*1.8*
9-5		
one side	1.1	*1.3*
both sides	2.2	*2.4*
Shock Absorbers or Bushings, Replace (B)		
900		
1981-93		
one side	.7	*.9*
both sides	1.2	*1.4*
1994-98		
one side	.5	*.7*
both sides	.7	*.9*

	LABOR TIME	SEVERE SERVICE
9000		
4 door		
one side	1.9	2.1
both sides	2.6	2.8
5 door		
one side	.9	1.1
both sides	2.1	2.3
CS		
one side	.7	.9
both sides	1.1	1.3
9-2X, 9-3		
one side	.5	.7
both sides	1.0	1.2
Stub Axle, Replace (B)		
1988-93 900	3.8	4.0

STEERING
AIR BAGS

	LABOR TIME	SEVERE SERVICE
Diagnose Air Bag System Component Each (A)		
1990-05	1.1	1.1
Air Bag Assy., Replace (B)		
900		
driver side	.5	.7
passenger side	1.3	1.5
9000 each side	.5	.7
9-2X, 9-3		
driver side	.7	.9
passenger side	1.5	1.7
9-5		
driver side	.7	.9
passenger side	4.1	4.3
Air Bag Clock Spring, Replace (B)		
1990-05	.7	.9
Air Bag Sliding Contact, Replace (B)		
1990-98 900	.7	.9
1989-98 9000	1.8	2.0
Capacitor Pack, Replace (B)		
1989-98 9000	1.3	1.5
Electronic Control Unit, Replace (B)		
900		
1990-93	2.2	2.4
1994-98	.5	.7
9000		
1989-91	1.9	2.1
1992-98	.7	.9
9-2X, 9-3	.8	1.0
9-5	1.2	1.4
Front Sensor, Replace (B)		
1990-93 900	.3	.5
9000		
1989-91	.5	.7
1992-98	.7	.9

	LABOR TIME	SEVERE SERVICE
POWER RACK & PINION		
Horn Contact or Canceling Cam, Replace (B)		
900	.5	.7
9000		
1986-89	.3	.5
1990-98	2.0	2.2
9-2X, 9-3	.5	.7
9-5	.7	.9
Inner Tie Rod Assy., Replace (B)		
Includes: Reset toe.		
900		
1981-93		
one side	3.4	3.6
both sides	4.2	4.4
1994-98 one	1.8	2.0
9000		
one side	4.2	4.4
both sides	4.9	5.1
9-2X, 9-3		
one side	1.1	1.3
both sides	1.6	1.8
9-5		
one side	5.7	5.9
both sides	6.0	6.2
Intermediate Shaft, Replace (B)		
900		
1981-93	2.1	2.3
1994-98	1.3	1.5
9000		
1986-89	.8	1.0
1990-98	1.5	1.7
Outer Tie Rod Ends, Replace (B)		
Includes: Reset toe.		
900		
1981-93		
one side	.7	.9
both sides	.9	1.1
1994-98		
one side	1.3	1.5
both sides	1.6	1.8
9000		
one side	.8	1.0
both sides	1.2	1.4
9-2X, 9-3, 9-5		
one side	1.1	1.3
both sides	1.6	1.8
Rack & Pinion Assy., R&R and Recondition (A)		
Includes: Reset toe.		
900	7.2	7.4
9000		
1986-89	6.9	7.1
1990-98		
AT	6.9	7.1
MT	7.6	7.8

	LABOR TIME	SEVERE SERVICE
Rack & Pinion Assy., R&R or Replace (B)		
Includes: Alignment check.		
900		
1981-93	3.7	3.9
1994-98	3.2	3.4
9000		
1986-89	3.4	3.6
1990-98		
AT	3.3	3.5
MT	4.3	4.5
9-2X, 9-3	3.0	3.2
9-5		
4 cyl.	5.2	5.4
V6	5.7	5.9
Rack Boots, Replace (B)		
900		
1981-93		
one side	1.3	1.5
both sides	1.8	2.0
1994-98 one	3.1	3.3
9000		
one side	.8	1.0
both sides	1.5	1.7
9-2X, 9-3	2.9	3.1
9-5 one side	1.0	1.2
Steering Column, Replace (B)		
900		
1981-89	1.9	2.1
1990-93	2.4	2.6
1994-98	1.4	1.6
9000		
1986-89	4.1	4.3
1989-98	10.9	11.1
9-2X, 9-3, 9-5	1.2	1.4
Steering Pump, R&R or Replace (B)		
900		
1981-93		
8V	1.6	1.8
16V	2.0	2.2
1994-98	.7	.9
9000	1.7	1.9
9-2X, 9-3	1.1	1.3
9-5		
4 cyl.	1.0	1.2
V6	2.2	2.4
Steering Pump Hoses, Replace (B)		
900		
1981-93		
return	2.3	2.5
suction	.8	1.0
1994-98		
pressure		
front	1.5	1.7
rear	.7	.9
suction	1.2	1.4

SAA-16 900 : 9000 : 9-2x : 9-3 : 9-5

	LABOR TIME	SEVERE SERVICE
Steering Pump Hoses, Replace (B)		
9000		
pump to reservoir	.5	.7
suction or return	1.3	1.5
upper	.5	.7
9-2X, 9-3		
pressure		
front	1.7	1.9
center	.8	1.0
rear	1.1	1.3
suction	.6	.8
9-5		
pressure		
front, center	1.4	1.6
rear		
4 cyl.	2.1	2.3
V6	2.6	2.8
suction	.8	1.0
Steering Wheel, Replace (B)		
900, 9000	.5	.7
9-2X, 9-3, 9-5	.6	.8
Upper Mast Jacket Bearing, Replace (B)		
900		
1981-89	1.9	2.1
1990-93	2.8	3.0
1994-98	.7	.9
9000	.9	1.1

HEATING & AIR CONDTIONING

Note: If more than one item requires replacement where evacuation and discharging the system is already included in the operation, deduct 1.0 hour for each additional item from the time listed.

	LABOR TIME	SEVERE SERVICE
Evacuate/Recover and Recharge System (B)		
1981-05	1.2	1.2
AC Hoses, Replace (B)		
Includes: Evacuate/recover and recharge.		
900		
1981-93		
compressor to condenser	1.6	1.8
each addl. add	1.3	1.5
1994-98		
compressor to condenser	1.8	2.0
9000		
compressor to condenser	.9	1.1
each addl. add	.7	.9
9-2X, 9-3		
compressor to condenser	2.8	3.0
each addl. add	1.0	1.2

	LABOR TIME	SEVERE SERVICE
9-5		
compressor to condenser	3.1	3.4
each addl. add	1.0	1.2
Compressor Assy., Replace (B)		
Includes: Parts transfer. Evacuate/recover and recharge.		
900		
1981-93	1.6	1.8
1994-98	2.3	2.5
9000		
2.0L	2.8	3.0
2.3L	2.0	2.2
9-2X, 9-3	3.5	3.7
9-5	4.1	4.3
Compressor Clutch & Pulley, Replace (B)		
Includes: Evacuate/recover and recharge if needed.		
900		
1981-93	2.7	2.9
1994-98	.9	1.1
1986-90 9000	3.5	3.7
9-2X, 9-3	1.1	1.2
9-5		
4 cyl.	1.1	1.3
V6	2.1	2.3
Compressor Shaft Seal, Replace (B)		
Includes: Compressor R&R. Evacuate/recover and recharge.		
900	2.8	3.0
1986 9000	3.7	3.9
Condenser Assy., Replace (B)		
Includes: Evacuate/recover and recharge.		
900		
1981-93	1.6	1.8
1994-98		
w/turbo	2.9	3.1
w/o turbo	2.1	2.3
9000		
w/turbo	2.2	2.4
w/o turbo	1.8	2.0
9-2X, 9-3	2.8	3.0
9-5	3.3	3.5
Evaporator Core, Replace (B)		
Includes: Evacuate/recover and recharge.		
900		
1981-93		
8V	2.3	2.5
16V	1.3	1.5
1994-98		
4 cyl.	2.9	3.1
V6	2.5	2.7

	LABOR TIME	SEVERE SERVICE
9000		
1986-90	2.3	2.5
1991	2.6	2.8
1992-98		
front	2.4	2.6
rear	.9	1.1
9-2X, 9-3	3.6	3.8
9-5	4.8	5.0
Expansion Valve, Replace (B)		
Includes: Evacuate/recover and recharge.		
900		
1981-93	1.3	1.5
1994-98		
4 cyl.	1.8	2.0
V6	2.1	2.3
9000		
1986-91	1.3	1.5
1992-98		
front	1.8	2.0
rear	.7	.9
9-2X, 9-3	2.6	2.8
9-5	3.1	3.3
Heater Blower Motor, Replace (B)		
900		
1981-93	2.5	2.7
1994-98	.8	1.0
9000		
1986-91	4.3	4.5
1992-98	2.6	2.8
9-2X, 9-5	1.4	1.6
9-3	.8	1.0
Heater Blower Motor Resistor, Replace (B)		
1981-05	.5	.7
Heater Blower Motor Switch, Replace (B)		
900		
1981-93	1.4	1.6
1994-98	.5	.7
9000	.5	.7
9-2X, 9-3, 9-5	.5	.7
Heater Core, R&R or Replace (B)		
900	2.8	3.0
9000		
1986-89	4.3	4.5
1990-98	2.7	2.9
9-3	1.7	1.9
9-5	1.1	1.3
Heater Hoses, Replace (B)		
1981-05	.6	.8
Low Pressure and/or Cycling Switch, Replace (B)		
1994-98 900	.5	.7
9000	.5	.7
Pressure Relief Valve, Replace (B)		
Includes: Evacuate/recover and recharge.		
1990-93 900	.5	.7
1990-98 9000	.5	.7

	LABOR TIME	SEVERE SERVICE
Receiver/Drier Assy., Replace (B)		
Includes: Evacuate/recover and recharge.		
900		
1981-93	1.4	1.6
1994-98	1.7	1.9
9000		
1986-91	1.4	1.6
1992-98	1.7	1.9
9-2X, 9-3	2.1	2.3
9-5	3.1	3.3

WIPERS & SPEEDOMETER

	LABOR TIME	SEVERE SERVICE
Power Antenna, Replace (B)		
900, 9000	.5	.7
Radio, R&R (B)		
1981-05	.5	.7
Rear Window Wiper Switch, Replace (B)		
1994-05	.4	.6
Rear Window Wiper Motor, Replace (B)		
1994-05	.8	1.0
Speedometer Cable & Casing, Replace (B)		
900	.7	.9
9000	1.7	1.9
Speedometer Driven Gear, Replace (B)		
900	.5	.7
9000	.7	.9
Speedometer Head, R&R or Replace (B)		
900		
1981-93	1.9	2.1
1994-98	.7	.9
9000	2.1	2.3
9-2X, 9-3, 9-5	1.6	1.8
Windshield Washer Motor, Replace (B)		
1981-05	.8	1.0
Windshield Wiper & Washer Switch, Replace (B)		
1981-05	.9	1.1
w/ABS add	.2	.4
Windshield Wiper Motor, Replace (B)		
900	.8	1.0
9000	.9	1.1
9-2X, 9-3	.6	.8
9-5	.8	1.0
Windshield Wiper Pivot, Replace (B)		
900		
1981-93 each	.7	.9
1994-98	.5	.7
9000		
1986-89		
right side	1.5	1.7
left side	.5	.7
1990-98	1.8	2.0

	LABOR TIME	SEVERE SERVICE
9-2X, 9-3	.5	.7
9-5	.8	1.0

LAMPS & SWITCHES

	LABOR TIME	SEVERE SERVICE
Back-Up Lamp Switch, Replace (B)		
900	.5	.7
9000	.7	.9
9-2X, 9-3, 9-5	.2	.4
Composite Headlamp Bulb, Replace (B)		
1981-05	.3	.5
Hazard Warning Switch, Replace (B)		
900	.3	.5
9000	.5	.7
9-2X, 9-3, 9-5	.3	.5
Headlamp Switch, Replace (B)		
1981-05	.5	.7
Headlamps, Aim (B)		
Two	.4	.6
Four	.6	.8
High Mount Stop Lamp Assy., Replace (B)		
1985-05	.4	.6
Horn, Replace (B)		
900, 9000 each	.5	.7
9-2X, 9-3, 9-5	.2	.4
Horn Relay, Replace (B)		
1994-01 900	.3	.3
License Lamp Bulb, Replace (C)		
Each	.2	.2
Neutral Safety Switch, Replace (B)		
900	.7	.9
9000	1.1	1.3
9-2X, 9-3	.8	1.0
9-5	1.0	1.2
Parking Brake Warning Lamp Switch, Replace (B)		
900	.5	.5
9000	.3	.3
9-2X, 9-3	1.1	1.1
9-5	.6	.6
Parking Lamp Lens or Bulb, Replace (C)		
1981-05	.2	.2
Sealed Beam Headlamp, Replace (B)		
900 each	.2	.4
Side Marker Lamp Assy., Replace (B)		
Each	.2	.2
Stop Lamp Switch, Replace (B)		
900	.7	.7
9000	.5	.5
9-2X, 9-3, 9-5	.3	.3
Tail Lamp Lens or Bulb, Replace (C)		
Exc. below	.2	.2
1981-93 900 2, 4 door	.5	.5

	LABOR TIME	SEVERE SERVICE
Turn Signal & Headlamp Dimmer Switch, Replace (B)		
900	.7	.7
9000	.5	.5
9-2X, 9-3, 9-5	.2	.2
Turn Signal or Hazard Warning Flasher, Replace (B)		
900	.5	.5
9000	.3	.3
9-2X, 9-3, 9-5	.2	.2

BODY

	LABOR TIME	SEVERE SERVICE
Door Window Regulator, Replace (B)		
Front		
900		
1981-93	1.4	1.6
1994-98	.7	.9
9000	.9	1.1
9-2X, 9-3, 9-5	.9	1.1
Rear		
900		
1981-93	.9	1.1
1994-98	.5	.7
9000	.9	1.1
9-2X, 9-3, 9-5	.6	.8
Front Door Lock, Replace (B)		
900		
1981-93		
exc. convertible	.8	1.0
convertible	1.4	1.6
1994-98	.8	1.0
9000	.9	1.1
9-2X, 9-5	1.4	1.6
9-3	.6	.8
Hood Lock, Replace (B)		
1981-98	.5	.7
9-2X, 9-3, 9-5	.2	.4
Hood Release Cable, Replace (B)		
900	.9	1.1
9000	1.3	1.5
9-2X, 9-3, 9-5	.6	.8
Lock Striker Plate, Replace (B)		
1981-05	.4	.6
Rear Door Lock, Replace (B)		
1981-05	.9	1.1
Trunk Lock, Replace (B)		
1981-05	.6	.8
Window Regulator Motor, Replace (B)		
900		
1981-93		
front	1.5	1.7
rear	1.2	1.4
1994-98	.7	.9
9000	1.0	1.2
9-2X, 9-3, 9-5	.7	.9

NOTES

STR

825 : 827

SYSTEM INDEX

MAINTENANCE	STR-2
CHARGING	STR-2
STARTING	STR-2
CRUISE CONTROL	STR-2
IGNITION	STR-2
EMISSIONS	STR-2
FUEL	STR-2
EXHAUST	STR-2
ENGINE COOLING	STR-3
ENGINE	STR-3
Assembly	STR-3
Cylinder Head	STR-3
Camshaft	STR-3
Crank & Pistons	STR-3
Engine Lubrication	STR-3
CLUTCH	STR-3
MANUAL TRANSAXLE	STR-3
AUTO TRANSAXLE	STR-4
SHIFT LINKAGE	STR-4
DRIVELINE	STR-4
BRAKES	STR-4
FRONT SUSPENSION	STR-5
REAR SUSPENSION	STR-5
STEERING	STR-5
HEATING & AC	STR-5
WIPERS & SPEEDOMETER	STR-6
LAMPS & SWITCHES	STR-6
BODY	STR-6

OPERATIONS INDEX

A
AC Hoses	STR-5
Air Bags	STR-5
Air Conditioning	STR-5
Alignment	STR-5
Alternator (Generator)	STR-2
Antenna	STR-6
Anti-Lock Brakes	STR-4

B
Back-Up Lamp Switch	STR-6
Ball Joint	STR-5
Bleed Brake System	STR-4
Blower Motor	STR-6
Brake Disc	STR-4
Brake Pads and/or Shoes	STR-4

C
Camshaft	STR-3
Catalytic Converter	STR-2
Coolant Temperature (ECT) Sensor	STR-2
Crankshaft	STR-3
Crankshaft Sensor	STR-2
Cruise Control	STR-2
CV Joint	STR-4
Cylinder Head	STR-3

D
Differential	STR-4
Distributor	STR-2
Drive Belt	STR-2

E
EGR	STR-2
Electronic Control Module (ECM/PCM)	STR-2
Engine	STR-3
Engine Lubrication	STR-3
Engine Mounts	STR-3
Evaporator	STR-6
Exhaust	STR-2
Exhaust Manifold	STR-2
Expansion Valve	STR-6

F
Flywheel	STR-3
Fuel Injection	STR-2
Fuel Pump	STR-2
Fuel Vapor Canister	STR-2

G
Gear Selector Lever	STR-4
Generator	STR-2

H
Halfshaft	STR-4
Headlamp	STR-6
Heater Core	STR-6

I
Ignition Coil	STR-2
Ignition Switch	STR-2
Inner Tie Rod	STR-5
Intake Air Temperature (IAT) Sensor	STR-2
Intake Manifold	STR-2

L
Lower Control Arm	STR-5

M
Master Cylinder	STR-4
Muffler	STR-3

O
Oil Pan	STR-3
Oil Pump	STR-3
Outer Tie Rod	STR-5
Oxygen Sensor	STR-2

P
Parking Brake	STR-5
Pistons	STR-3

R
Radiator	STR-3
Radiator Hoses	STR-3
Radio	STR-6
Rear Main Oil Seal	STR-3

S
Shock Absorber/Strut, Front	STR-5
Shock Absorber/Strut, Rear	STR-5
Spark Plug Cables	STR-2
Spark Plugs	STR-2
Spring, Front Coil	STR-5
Spring, Rear Coil	STR-5
Starter	STR-2
Steering Wheel	STR-5

T
Thermostat	STR-3
Throttle Body	STR-2
Throttle Position Sensor (TPS)	STR-2
Timing Belt	STR-3
Torque Converter	STR-4

U
Upper Control Arm	STR-5

V
Valve Cover Gasket	STR-3
Valve Job	STR-3
Vehicle Speed Sensor	STR-2

W
Water Pump	STR-3
Wheel Balance	STR-2
Window Regulator	STR-6
Windshield Washer Pump	STR-6
Windshield Wiper Motor	STR-6

STR-2   825 : 827

	LABOR TIME	SEVERE SERVICE

MAINTENANCE

Air Cleaner Filter Element, Replace (C)
1987-91 . . . . .2 .3
Drive Belt, Adjust (B)
1987-91 . . . . .2 .3
Drive Belt, Replace (B)
1987-91 . . . . .5 .6
Fuel Filter, Replace (B)
1987-91 . . . . .6 .6
Sealed Beam Headlamp, Replace (C)
One . . . . .3 .3
Wheel, Balance (B)
One . . . . .3 .3
each addl. add . . . . .2 .2

CHARGING

Alternator Circuits, Test (B)
Includes: Test component output.
1987-91 . . . . .5 .5
Alternator, R&R and Recondition (A)
1987-91 . . . . 2.0 2.5
Alternator Assy., Replace (B)
Includes: Pulley transfer.
1987-91 . . . . 1.4 1.6
Alternator Voltage Regulator, Replace (B)
1987-91 . . . . 1.7 2.0

STARTING

Starter Draw Test (On Car) (B)
1987-91 . . . . .5 .5
Starter, R&R and Recondition (A)
1987-91 . . . . 2.4 2.9
Starter Assy., Replace (B)
1987-91 . . . . 1.2 1.5

CRUISE CONTROL

Diagnose Cruise Control System (A)
1987-91 . . . . 1.0 1.0
Actuator Assy., Replace (B)
1987-91 . . . . .6 .6
Control Actuator Cable, Replace (B)
1987-91 . . . . .5 .5
Control Actuator Valve, Replace (B)
1987-91 . . . . .4 .4
Control Controller Module, Replace (B)
1987-91 . . . . .4 .4
Control Sensor, Replace (B)
1987-91 . . . . 1.9 1.9
Control Switch, Replace (B)
Brake, clutch or main . . .5 .5
Steering wheel . . . . .4 .4

IGNITION

Ignition Timing, Reset (B)
1987-91 . . . . .5 .5
Crank Angle Sensor, Replace (B)
1987-89 . . . . 2.0 2.2
1990-91 . . . . 3.9 4.1
Distributor, R&R or Replace (B)
Includes: Reset base ignition timing.
1987-91 . . . . 1.0 1.2
Distributor Cap and/or Rotor, Replace (B)
1987-91 . . . . .3 .3
Distributor Coil, Replace (B)
1987-91 . . . . 1.0 1.0
Distributor Vacuum Advance, Replace (B)
Includes: Reset base ignition timing.
1987-91 . . . . 1.2 1.3
EFI Control Unit, Replace (B)
1987-91 . . . . .5 .5
Igniter Unit, Replace (B)
1987-91 . . . . 1.0 1.0
Ignition Coil, Replace (B)
1987-91 . . . . .6 .6
Ignition Switch, Replace (B)
1987-91 . . . . 1.0 1.0
Spark Plug (Ignition) Cables, Replace (B)
1987-91 . . . . .5 .7
Spark Plugs, Replace (B)
1987-91 . . . . .6 .8
TDC Sensor, Replace (B)
1987-91 . . . . 1.0 1.2

EMISSIONS

Diagnose Engine Electronic Controls (A)
1987-91 . . . . 1.0 1.0
Atmospheric Pressure Valve, Replace (B)
1987-91 . . . . .5 .5
Coolant Temperature (ECT) Sensor, Replace (B)
1987-91 . . . . .5 .5
EGR Control Valve, Replace (B)
1987-91 . . . . .9 1.1
EGR Frequency Solenoid Control Valve, Replace (B)
1987-91 . . . . .4 .4
Electronic Control Module (ECM/PCM), Replace (B)
1987-91 . . . . .5 .5
Electronic Idle Control Valve, Replace (B)
1987-91 . . . . .9 .9
Fuel Vapor Canister, Replace (B)
1987-91 . . . . .5 .5
Igniter Unit, Replace (B)
1987-91 . . . . .5 .5

Manifold Air Temperature Sensor, Replace (B)
1987-91 . . . . .5 .5
Oxygen Sensor, Replace (B)
Front . . . . .7 .9
Rear . . . . .4 .6
Throttle Angle Sensor, Replace (B)
1987-91 . . . . 1.4 1.4
Timing Adjuster, Replace (B)
1987-91 . . . . .5 .7
Two-Way Valve, Replace (B)
1987-91 . . . . 1.5 1.5
Vehicle Speed Sensor, Replace (B)
1987-89 . . . . 1.5 1.5
1990-91 . . . . .9 .9

FUEL

DELIVERY

Fuel Filter, Replace (B)
1987-91 . . . . .6 .6
Fuel Gauge (Dash), Replace (B)
1987-91 . . . . 1.4 1.4
Fuel Pump, Replace (B)
In gas tank 1.5 1.8
Fuel Pump Relay, Replace (B)
1987-91 . . . . .5 .5
Fuel Tank, Replace (B)
1987-91 . . . . 1.4 1.7
Fuel Tank Sending Unit, Replace (B)
1987-91 . . . . .5 .8
Intake Manifold and/or Gasket, Replace (B)
1987-91 . . . . 2.9 3.4
Replace manifold add . . . .5 .7

INJECTION

Fast Idle Valve, Replace (B)
1987-91 . . . . .6 .6
Fuel Injector Resistor, Replace (B)
1987-91 . . . . .5 .5
Fuel Injectors, Replace (B)
One bank 2.0 2.4
Both banks 3.5 3.9
Fuel Pressure Regulator, Replace (B)
1987-91 . . . . .7 .9
Throttle Body Assy., Replace (B)
1987-91 . . . . 1.7 1.7

EXHAUST

Catalytic Converter, Replace (B)
1987-91 . . . . 1.1 1.2
Exhaust Manifold or Gasket, Replace (B)
One 1.2 1.5
Both 2.0 2.3
Exhaust Pipe, Replace (B)
One 1.0 1.1
each addl. add . . . . .4 .5

825 : 827 **STR-3**

	LABOR TIME	SEVERE SERVICE
Exhaust Pipe Flange Gasket, Replace (B)		
One	.8	1.0
Muffler, Replace (B)		
1987-91	1.0	1.1
Tailpipe Extension, Replace (B)		
1987-91	.5	.6

ENGINE COOLING

Coolant Thermostat, Replace (B)
1987-91 .... 1.0   1.2
Radiator Assy., R&R or Replace (B)
1987-91 .... 2.1   2.3
w/AT add 2   .2
Radiator Fan and/or Fan Motor, Replace (B)
1987-91 .... 1.3   1.6
Radiator Fan Motor Switch (Coolant Temp.), Replace (B)
1987-91 6   .6
Radiator Hoses, Replace (B)
Includes: Refill with proper coolant mix.
Upper or lower 5   .6
Both 9   1.0
Temperature Gauge (Dash), Replace (B)
1987-91 .... 1.6   1.6
Temperature Gauge (Engine), Replace (B)
1987-91 7   .7
Water Pump and/or Gasket, Replace (B)
Includes: Refill with proper coolant mix.
1987-91 .... 4.1   4.6

ENGINE

ASSEMBLY
Times shown are for OEM assemblies. Time to replace assemblies from aftermarket rebuilders may vary.
Engine Assy., R&I (B)
Does not include parts or component transfer.
1987-91 .... 9.5   9.5
Engine Assy., R&R and Recondition (A)
Includes: Replacing rings, rod and main bearings, cylinder head reconditioning and engine tune-up.
1987-91 .... 34.5   34.5
Engine Assy. (Complete), Replace (B)
Includes: Component transfer and engine tune-up.
1987-91 .... 13.0   13.0

	LABOR TIME	SEVERE SERVICE

Engine Assy. (Short), Replace (B)
Assembly consists of cylinder block, piston assemblies, crankshaft, camshaft, timing chain and gears. Does not include cylinder heads. Operation Includes: R&R engine, transfer necessary parts and all necessary adjustments.
1987-91 .... 21.5   21.5
Engine Mounts, Replace (B)
One 1.1   1.3
Both 2.2   2.4

CYLINDER HEAD
Compression Test (B)
1987-91 8   1.1
Valves, Adjust (B)
1987-91 .... 1.7   2.1
Cylinder Head, Replace (B)
Includes: Adjustments.
One 11.3   11.8
Both 14.2   14.7
Cylinder Head Gasket, Replace (B)
One 8.5   8.8
Both 11.1   11.4
Rocker Arms, Replace (B)
One head 2.5   2.8
Both heads 3.8   4.1
Valve Cover Gasket, Replace (B)
Each 5   .7
Valve Job (A)
1987-91 .... 19.2   19.7

CAMSHAFT
Timing Belt, Adjust (B)
1987-91 .... 1.1   1.3
Camshaft Replace (B)
One 2.5   2.7
Both 3.7   3.9
Camshaft Seal, Replace (B)
One 1.5   1.7
Both 1.8   2.0
Timing Belt and/or Drive Pulley, Replace (B)
1987-91 .... 4.5   5.0

CRANK & PISTONS
Connecting Rod Bearings, Replace (A)
Includes: Check bearing oil clearance.
1987-91 .... 3.9   4.4
Crankshaft and Main Bearings, Replace (A)
Includes: Engine R&R, check bearing oil clearance.
1987-91 .... 16.9   17.8
Crankshaft Front Oil Seal, Replace (A)
1987-91 .... 4.1   4.3

	LABOR TIME	SEVERE SERVICE

Main Bearings, Replace (A)
1987-91 .... 5.9   6.4
Pistons or Connecting Rods, Replace (A)
Includes: Ridge reaming, cylinder wall deglazing, installing new rings and rod bearings, engine tune-up.
One cyl. 10.8   11.7
All cyls. 16.8   17.7
Rear Main Oil Seal, Replace (A)
AT 5.8   6.4
MT 5.1   5.7
Rings, Replace (A)
Includes: Ridge reaming, cylinder wall deglazing, installing new rings, engine tune-up.
1987-91 .... 15.0   15.9

ENGINE LUBRICATION
Engine Oil Cooler, Replace (B)
1987-91 .... 1.5   1.7
Engine Oil Pressure Switch (Sending Unit), Replace (B)
1987-91 5   .5
Oil Pan and/or Gasket, Replace (B)
1987-91 .... 1.6   1.8
Oil Pump, Replace (B)
1987-91 .... 3.6   4.1

CLUTCHCLUTCH
Clutch Hydraulic System, Bleed (B)
1987-91 5   .5
Clutch Pedal Free Play, Adjust (B)
1987-91 3   .5
Clutch Assy., Replace (B)
1987-91 .... 6.9   7.1
Replace rear main seal add 1.6   1.6
Clutch Master Cylinder, Replace (B)
Includes: System bleeding.
1987-91 .... 1.4   1.6
Clutch Release Bearing or Fork, Replace (B)
1987-91 .... 6.6   6.2
Clutch Slave Cylinder, Replace (B)
Includes: System bleeding.
1987-91 .... 1.4   1.6
Flywheel, Replace (B)
1987-91 .... 7.2   7.4
Replace pilot bearing add 3   .3

MANUAL TRANSAXLE
Transaxle Assy., R&R and Recondition (A)
1987-91 .... 10.3   10.9
Recondition differential assembly add 1.6   1.8

STR-4 825 : 827

	LABOR TIME	SEVERE SERVICE

Transaxle Assy., R&R or Replace (B)
1987-91 .... 6.6 *6.8*
Transaxle Assy., R&R and Reseal (B)
1987-91 .... 8.8 *9.4*

AUTOMATIC TRANSAXLE
Diagnose Electronic Transmission (A)
1987-91 .... 1.0 *1.0*
3-4 Shift Switch, Replace (B)
1987-91 .... .5 *.5*
EAT Computer, Replace (B)
1987-91 .... .5 *.5*
Governor Assy., R&R or Replace (B)
1987-91 .... 7.6 *8.0*
Lock-Up Control Solenoid Valve, Replace (B)
1987-91 .... .6 *.9*
Shift Control Solenoid Valve, Replace (B)
1987-91 .... .6 *.9*
Speed Pulser, Replace (B)
1987-91 .... .5 *.5*
Transaxle and Torque Converter, R&R and Recondition (A)
1987-91 .... 14.1 *15.0*
Recondition differential assembly add 1.7 *1.9*
Transaxle and Torque Converter, R&R or Replace (B)
1987-91 .... 6.6 *6.8*
Replace assembly add ... .5 *.5*

SHIFT LINKAGE
AUTOMATIC TRANSAXLE
Shift Control Cable, Adjust (B)
1987-91 .... .4 *.6*
Throttle Lever Control Cable, Adjust (B)
1987-91 .... .5 *.7*
Shift Control Cable, Replace (B)
1987-91 .... 1.6 *1.8*
Throttle Lever Control Cable, Replace (B)
1987-91 .... 1.3 *1.5*

MANUAL TRANSAXLE
Gearshift Lever, Replace (B)
1987-91 .... 1.6 *1.6*
Gearshift Lever Boot, Replace (B)
1987-91 .... .5 *.5*
Shift Lever Bushing, Replace (B)
1987-91 .... 1.1 *1.3*
Shift Rod, Replace (B)
1987-91 .... .6 *.8*
Torque Rod Bushing, Replace (B)
1987-91 .... .5 *.7*

DRIVELINE
DRIVE AXLE
Differential Assy., R&R and Recondition (A)
1987-91 .... 11.5 *12.0*
Differential Oil Seal, Replace (B)
Right 1.6 *1.8*
Left 1.4 *1.6*
Both 2.6 *2.8*

HALFSHAFTS
Halfshaft, R&R or Replace (B)
Right 1.4 *1.7*
Left 1.2 *1.5*
Both 2.5 *2.8*
Intermediate 1.6 *1.9*
Replace support bearing add .... .2 *.4*
Halfshaft Oil Seal, Replace (B)
Right 1.6 *1.8*
Left 1.4 *1.6*
Both 2.6 *2.8*
Inboard CV Joint and/or Boot, Replace (B)
Right 1.5 *1.8*
Left 1.7 *2.0*
Both 2.8 *3.1*
Intermediate Shaft Support Bearing, Replace (B)
1987-91 .... 1.9 *2.1*
Outboard CV Joint and/or Boot, Replace (B)
Right 1.8 *2.1*
Left 2.0 *2.3*
Both 3.1 *3.4*

BRAKES
ANTI-LOCK
Diagnose Anti-Lock Brake System (A)
1987-91 .... 1.0 *1.0*
ABS System, Bleed (B)
1987-91 .... .8 *.8*
Accumulator Assy., Replace (B)
1987-91 .... 1.2 *1.2*
Control Module, Replace (B)
Does not include testing.
1987-91 .... 1.0 *1.0*
High Pressure Hoses, Replace (B)
Upper or lower 1.0 *1.0*
Low Pressure Hoses, Replace (B)
1987-91 .... 1.3 *1.3*
Modulator, Replace (B)
1987-91 .... 1.5 *1.5*
Power Unit, Replace (B)
1987-91 .... 1.5 *1.5*
Wheel Speed Sensor, Replace (B)
Front one .... .7 *.7*
Rear one 1.3 *1.3*

SYSTEM
Bleed Brakes (B)
Includes: Add fluid.
1987-91 .... .5 *.5*
Brake Failure Relay, Replace (B)
1987-91 .... .2 *.2*
Brake Warning Switch, Replace (B)
1987-91 .... .2 *.2*
Master Cylinder, R&R and Rebuild (A)
1987-91 .... 2.6 *2.9*
Master Cylinder, Replace (B)
1987-91 .... 1.6 *1.8*
Power Booster Unit, Replace (B)
1987-91 .... 1.8 *2.0*
Power Booster Vacuum Check Valve, Replace (B)
1987-91 .... .3 *.3*

SERVICE BRAKES
Caliper Assy., R&R and Recondition (A)
Includes: System bleeding.
Front
  one 2.1 *2.4*
  both 3.5 *3.8*
Rear
  one 1.6 *1.9*
  both 2.8 *3.1*
Four wheels 6.0 *6.3*
Replace brake hose add .. .3 *.3*
Caliper Assy., Replace (B)
Includes: System bleeding.
Front
  one 1.7 *1.9*
  both 2.2 *2.4*
Rear
  one 1.2 *1.4*
  both 1.7 *1.9*
Four wheels 3.6 *3.8*
Replace brake hose add .. .2 *.2*
Disc Brake Rotor, Replace (B)
Front
  one 1.4 *1.6*
  both 2.6 *2.8*
Rear
  one 1.2 *1.4*
  both 1.9 *2.1*
Four wheels 4.2 *4.4*
Pads, Replace (B)
Front or rear 1.0 *1.1*
Four wheels 1.8 *1.9*

COMBINATION ADD-ONS
Replace disc brake rotor add .... .4 *.4*
Resurface disc brake rotor add .... .5 *.5*

825 : 827 **STR-5**

	LABOR TIME	SEVERE SERVICE
PARKING BRAKE		

Parking Brake Cable, Adjust (B)
  1987-91 3   .5
Parking Brake Cable, Replace (B)
  One 1.7   2.0
  each addl. add 5   .6

FRONT SUSPENSION

Unless otherwise noted, time given does not include alignment.

Front Toe, Adjust (B)
  1987-91 9   1.1
Front Crossmember, Replace (B)
  1987-91 .... 3.5   4.0
Front Radius Arm Bushings, Replace (B)
  One side 1.3   1.5
  Both sides 1.7   2.0
Front Strut Coil Spring, Replace (B)
  One side 1.8   2.0
  Both sides 3.2   3.4
Lower Control Arm Bushings, Replace (B)
  Includes: Reset toe.
  One side 2.5   2.7
  Both sides 3.7   3.9
Rear Crossmember, Replace (B)
  1987-91 .... 4.9   5.4
Stabilizer Bar, Replace (B)
  1987-91 9   1.1
Steering Knuckle, Replace (B)
  One side 1.9   2.4
  Both sides 3.2   3.7
  w/ABS add each side 3   .3
Strut, R&R or Replace (B)
  One side 1.6   1.8
  Both sides 3.0   3.2
Tie Rod Ball Joint Boot, Replace (B)
  Each 7   .9
Upper Control Arm, Replace (B)
  One 1.6   1.8
  Both 2.2   2.4
Wheel Bearings and Cups, Replace (B)
  Includes: Oil seal.
  One side 1.6   1.6
  Both sides 3.0   3.0
  w/ABS add each side 3   .3

REAR SUSPENSION

Unless otherwise noted, time given does not include alignment.

Rear Alignment, Adjust (A)
  1987-91 .... 1.0   1.2

Lower Control Arm Bushing, Replace (B)
  Includes: Alignment.
  One side 1.5   1.7
  Both sides 2.8   3.0
  w/ABS add each side 3   .3
Radius Rod or Bushings, Replace (B)
  One side 1.4   1.6
  Both sides 2.0   2.2
Rear Spring, Replace (B)
  One 8   1.0
  Both 1.5   1.7
Rear Stabilizer Bar & Bushings, Replace (B)
  1987-91 9   1.1
Rear Wheel Knuckle Assy., Replace (B)
  One side 1.8   2.3
  Both sides 3.4   3.9
Shock Absorbers or Bushings, Replace (B)
  One 9   1.1
  Both 1.6   1.8
Stabilizer Bar Bushings, Replace (B)
  1987-91 8   1.0
Trailing Arm or Link, Replace (B)
  One side 1.9   2.3
  Both sides 3.0   3.4

STEERING

AIR BAGS

Diagnose Air Bag System (A)
  1987-91 .... 1.0   1.0
Air Bag Assy., Replace (B)
  1987-91 6   .6
Cable Reel, Replace (B)
  1987-91 9   .9
Electronic Control Unit, Replace (B)
  1987-91 5   .5
Sensor or Harness, Replace (B)
  Right 3.8   3.8
  Left 4.8   4.8
  Main 1.5   1.5

POWER RACK & PINION

Pump Pressure Check (B)
  1987-91 5   .5
Rack & Pinion Assy., Adjust (B)
  1987-91 8   1.0
Control Unit, Replace (B)
  1987-91 .... 1.4   1.4
Pump Shaft Seal, Replace (B)
  1987-91 .... 1.4   1.6
Rack & Pinion Assy., R&R or Replace (B)
  1987-91 .... 3.3   3.8

Rack & Pinion Assy., Reseal (A)
  Includes: Alignment.
  1987-91 .... 4.6   5.2
Steering Column, Replace (B)
  1987-91 .... 1.4   1.4
  w/air bags add 3   .3
Steering Pump, R&R or Replace (B)
  1987-91 .... 1.6   1.8
Steering Pump, R&R and Reseal (B)
  1987-91 .... 2.1   2.6
Steering Pump Hoses, Replace (B)
  Pressure 2.5   2.6
  Return 7   .8
Steering Pump Reservoir, Replace (B)
  1987-91 4   .6
Steering Shaft Coupler, Replace (B)
  1987-91 8   1.0
Steering Wheel, Replace (B)
  1987-91 5   .5
  w/air bags add 3   .3
Tie Rods or Tie Rod Ends, Replace (B)
  Includes: Reset toe.
  One side 1.1   1.3
  Both sides 1.4   1.6

HEATING & AIR CONDITIONING

When more than one component requires replacement where evacuation/recovery and recharging is already included, deduct 1.0 hour for each additional component from the time given.

Evacuate/Recover and Recharge System (B)
  1987-91 .... 1.0   1.0
AC Hoses, Replace (B)
  Includes: Evacuate/recover and recharge.
  Suction or discharge
    one 1.5   1.7
    each addl. add 3   .5
AC On/Off Control Switch, Replace (B)
  1987-91 .... 1.2   1.2
Compressor Assy., Replace (B)
  Includes: Parts transfer, evacuate/recover and recharge.
  1987-91 .... 3.5   3.7
Compressor or Fan Relay, Replace (B)
  1987-91 3   .3
Compressor Clutch Assy., Replace (B)
  1987-91 .... 1.5   1.7

	LABOR TIME	SEVERE SERVICE

Compressor Shaft Seal, Replace (B)
Includes: Compressor R&R, evacuate/recover and recharge.
1987-91 .... 4.4 4.6

Condenser Assy., Replace (B)
Includes: Evacuate/recover and recharge.
1987-91 .... 2.4 2.6

Evaporator Coil, Replace (B)
Includes: Evacuate/recover and recharge.
1987-91 .... 3.4 3.4

Expansion Valve, Replace (B)
Includes: Evacuate/recover and recharge.
1987-91 .... 2.8 2.8

Fan Motor, Replace (B)
1987-91 .... .9 1.2

Heater Blower Motor, Replace (B)
1987-91 .... 1.9 1.9

Heater Blower Motor Resistor, Replace (B)
1987-91 .... .5 .5

Heater Control Assy., Replace (B)
1987-91 .... 1.2 1.2

Heater Core, R&R or Replace (B)
1987-91 .... 5.3 5.3

Heater Hoses, Replace (B)
1987-91 .... .7 .9

Receiver/Drier Assy., Replace (B)
Includes: Evacuate/recover and recharge.
1987-91 .... 1.5 1.5

Temperature Control Assy., Replace (B)
1987-91 .... 1.2 1.2

Thermostat, Replace (B)
1987-91 .... 2.5 2.5

WIPERS & SPEEDOMETER

Antenna, Replace (B)
1987-91 .... .7 .7

Electronic Wiper/Washer Control Unit, Replace (B)
1987-91 .... .9 .9

Radio, R&R (B)
1987-89 .... 1.0 1.0
1990-91 .... 1.5 1.5

Rear Window Washer Motor, Replace (B)
1987-91 .... .6 .8

Speedometer Cable & Casing, Replace (B)
1987-91 .... 1.5 1.5

Speedometer Head, R&R or Replace (B)
1987-91 .... 1.9 1.9

Windshield Washer Pump, Replace (B)
1987-91 .... .8 .8

Windshield Wiper & Washer Switch, Replace (B)
1987-91 .... .9 .9

Windshield Wiper Motor, Replace (B)
1987-91 .... 1.0 1.3

LAMPS & SWITCHES

Back-Up Lamp Bulb, Replace (C)
One .... .2 .2
All .... .3 .3

Back-Up Lamp Switch, Replace (B)
1987-91 w/MT .... .4 .4

Headlamps, Aim (B)
1987-91 .... .4 .4

High Mount Stop Lamp Bulb, Replace (C)
1987-91 .... .3 .3

Horn, Replace (B)
1987-91 .... .3 .3

License Lamp Bulb, Replace (C)
One or all .... .2 .2

Parking Brake Warning Lamp Switch, Replace (B)
1987-91 .... .4 .4

Sealed Beam Headlamp, Replace (C)
One .... .3 .3

Stop Light Switch, Replace (B)
1987-91 .... .5 .5

Tail Lamp Lens or Bulb, Replace (C)
One .... .2 .2
each addl. add .... .1 .1

Turn Signal and Headlight Dimmer Switch, Replace (B)
1987-91 .... .9 .9

Turn Signal and Parking Lamp Lens, Replace (B)
One .... .3 .3
each addl. add .... .2 .2

Turn Signal or Hazard Warning Flasher, Replace (B)
1987-91 .... .2 .2

BODY

Front Door Latch, Replace (B)
1987-91 .... .9 .9

Front Door Window Regulator and/or Motor, Replace (B)
1987-91 .... 1.6 1.9

Hood Release Cable, Replace (B)
1987-91 .... 1.0 1.0

Rear Door Latch, Replace (B)
1987-91 .... .8 .8

Rear Door Window Regulator and/or Motor, Replace (B)
1987-91 .... 1.0 1.3

Trunk Lock, Replace (B)
1987-91 .... .5 .5

VW

Beetle : Cabrio : Cabriolet : Corrado : Dasher : Fox : Golf : Jetta : Passat : Quantum : Rabbit : Rabbit Convertible : Rabbit Pickup : Scirocco

SYSTEM INDEX

- MAINTENANCE VW-2
  - Maintenance Schedule... VW-2
- CHARGING VW-3
- STARTING VW-3
- CRUISE CONTROL VW-4
- IGNITION VW-4
- EMISSIONS VW-5
- FUEL VW-5
- EXHAUST VW-7
- ENGINE COOLING VW-8
- ENGINE VW-9
  - Assembly VW-9
  - Cylinder Head VW-11
  - Camshaft VW-12
  - Crank & Pistons VW-13
  - Engine Lubrication ... VW-15
- CLUTCH VW-15
- MANUAL TRANSAXLE VW-16
- AUTO TRANSAXLE VW-17
- SHIFT LINKAGE VW-18
- DRIVELINE VW-18
- BRAKES VW-19
- FRONT SUSPENSION VW-21
- REAR SUSPENSION VW-22
- STEERING VW-23
- HEATING & AC VW-25
- WIPERS & SPEEDOMETER VW-27
- LAMPS & SWITCHES VW-27
- BODY VW-28

OPERATIONS INDEX

A
- AC Hoses VW-25
- Air Bags VW-23
- Air Conditioning VW-25
- Alignment VW-21
- Alternator (Generator) .. VW-3
- Antenna VW-27
- Anti-Lock Brakes VW-19

B
- Back-Up Lamp Switch VW-27
- Ball Joint VW-21
- Battery Cables VW-3
- Bleed Brake System VW-20
- Blower Motor VW-25
- Brake Disc VW-21
- Brake Drum VW-20
- Brake Hose VW-20
- Brake Pads and/or Shoes . VW-21

C
- Camshaft VW-12
- Camshaft Sensor VW-4
- Catalytic Converter VW-7
- Coolant Temperature (ECT) Sensor VW-8
- Crankshaft VW-13
- Crankshaft Sensor VW-4
- Cruise Control VW-4
- CV Joint VW-19
- Cylinder Head VW-11

D
- Differential VW-18
- Distributor VW-4
- Drive Belt VW-2

E
- EGR VW-5
- Electronic Control Module (ECM/PCM) VW-5
- Engine VW-9
- Engine Lubrication VW-15
- Engine Mounts VW-10
- Evaporator VW-26
- Exhaust VW-7
- Exhaust Manifold VW-7
- Expansion Valve VW-26

F
- Flexplate VW-18
- Flywheel VW-16
- Fuel Injection VW-6
- Fuel Pump VW-6
- Fuel Vapor Canister ... VW-5

G
- Gear Selector Lever ... VW-18
- Generator VW-3
- Glow Plug VW-4

H
- Halfshaft VW-19
- Headlamp VW-28
- Heater Core VW-26
- Horn VW-28

I
- Idle Air Control (IAC) Valve .. VW-7
- Ignition Coil VW-4
- Ignition Module VW-4
- Ignition Switch VW-5
- Injection Pump VW-6
- Inner Tie Rod VW-24
- Intake Manifold VW-6

K
- Knock Sensor VW-5

L
- Lower Control Arm VW-21

M
- Maintenance Schedule .. VW-2
- Mass Air Flow (MAF) Sensor VW-5
- Master Cylinder VW-20
- Muffler VW-8
- Multifunction Lever/Switch ... VW-28

N
- Neutral Safety Switch .. VW-28

O
- Oil Pan VW-15
- Oil Pump VW-15
- Outer Tie Rod VW-24
- Oxygen Sensor VW-5

P
- Parking Brake VW-21
- Pistons VW-14
- Positive Crankcase Ventilation (PCV) Valve ... VW-5

R
- Radiator VW-8
- Radiator Hoses VW-9
- Radio VW-27
- Rear Main Oil Seal VW-14

S
- Shock Absorber/Strut, Front VW-22
- Shock Absorber/Strut, Rear VW-23
- Spark Plug Cables VW-5
- Spark Plugs VW-5
- Spring, Front Coil VW-22
- Spring, Leaf VW-22
- Spring, Rear Coil VW-22
- Starter VW-3
- Steering Wheel VW-24

T
- Thermostat VW-8
- Throttle Body VW-7
- Throttle Position Sensor (TPS) VW-5
- Timing Belt VW-13
- Timing Chain VW-13
- Torque Converter VW-17

U
- Upper Control Arm VW-22

V
- Valve Body VW-17
- Valve Cover Gasket VW-11
- Valve Job VW-12
- Vehicle Speed Sensor .. VW-5

W
- Water Pump VW-9
- Wheel Balance VW-2
- Wheel Cylinder VW-21
- Window Regulator VW-28
- Windshield Washer Pump ... VW-27
- Windshield Wiper Motor ... VW-27

VW-1

VW-2 — BEETLE : CABRIO : CABRIOLET : CORRADO : DASHER : FOX : GOLF : JETTA : PASSAT

	LABOR TIME	SEVERE SERVICE
MAINTENANCE		
Air Cleaner Filter Element, Service (C)		
1981-04	.6	.7
Chassis Lubrication, Change Oil & Filter (C)		
Includes: Correct fluid levels.		
Gasoline		
exc. W8	.7	.8
W8	.9	1.0
Diesel	.8	.8
Drive Belt, Replace or Adjust (B)		
Exc. below		
1981-87	.3	.4
1988-97	.7	.8
1998-04	.3	.4
Air pump		
1981-93	.3	.4
Compressor		
1981-84		
gasoline	.9	1.0
diesel	.8	.9
1985-02	1.1	1.2
Serpentine		
Beetle		
gasoline	.8	.9
diesel	1.4	1.5
Cabrio	.9	1.0
Corrado		
4 cyl.	1.2	1.3
6 cyl.	1.4	1.5
Golf, Jetta		
1981-97		
4 cyl.	.9	1.0
6 cyl.	1.3	1.4
1998-04		
gasoline		
w/AC	1.1	1.2
w/o AC	.6	.7
diesel	1.1	1.2
Passat		
1990-96	1.4	1.5
1997-04		
4 cyl.	1.5	1.6
6 cyl.	.8	.9
8 cyl.	1.8	1.9
Steering pump		
1981-97	1.1	1.2
1998-04		
w/AC	1.1	1.2
w/o AC	.6	.7
Drive Belt Tensioner, Replace (B)		
Corrado	.5	.8
1994-97		
4 cyl.	1.7	2.0
6 cyl.	.5	.8

	LABOR TIME	SEVERE SERVICE
1998-04		
Beetle	.8	1.1
Golf		
gasoline	.5	.8
diesel	1.1	1.4
Passat		
4 cyl., 6 cyl.	2.0	2.3
8 cyl. each	2.9	3.2
Fuel Filter, Replace (B)		
1981-97	.5	.5
1998-04		
gasoline	.6	.6
diesel	.7	.7
Halogen Headlamp Bulb, Replace (B)		
1985-04	.6	.6
Oil & Filter, Change (C)		
Includes: Correct fluid levels.		
Gasoline		
exc. W8	.7	.8
W8	.9	1.0
Diesel	.8	.8
Sealed Beam Headlamp, Replace (B)		
1981-93 each	.3	.3
Timing Belt, Replace (B)		
Beetle		
1.8L	3.3	3.8
2.0L	2.9	3.4
diesel	5.0	5.5
Cabrio	2.6	3.1
Cabriolet	2.4	2.9
Dasher	1.6	2.1
Fox	1.5	2.0
Golf, Jetta		
1981-84		
gasoline	2.0	2.5
diesel	2.6	3.1
1985-92		
8V	2.8	3.3
16V	3.0	3.5
diesel	2.8	3.3
1993-97		
gasoline	2.5	3.0
diesel	2.8	3.3
1998-04		
1.8L	3.3	3.8
diesel	5.0	5.5
Passat		
1990-97		
gasoline	2.5	3.0
diesel	2.8	3.3
1998-04		
4 cyl.	3.8	4.3
6 cyl.	5.0	5.5
diesel	4.3	4.8
Quantum		
4 cyl.	1.7	2.2
5 cyl.	1.9	2.4
diesel	2.7	3.2

	LABOR TIME	SEVERE SERVICE
Rabbit, Scirocco		
1981-84		
gasoline	2.1	2.6
diesel	2.7	3.2
1985-89		
8V	2.4	2.9
16V	2.8	3.3
w/turbo diesel add	.9	.9
Replace water pump add	.3	.5
Tire, Replace (C)		
Includes: Dismount old tire and mount new tire to rim.		
One	.6	.6
each addl. add	.3	.3
Tires, Rotate (B)		
1981-02	.5	.5
Wheel, Balance (B)		
One	.3	.3
each addl. add	.2	.2
SCHEDULED MAINTENANCE INTERVALS		
If necessary, refer to appropriate Chilton maintenance service information.		
7,500 Mile Service (C)		
All Models	.4	.4
w/diesel add	.2	.2
15,000 Mile Service (B)		
All Models	2.1	2.1
w/16V engine add	.2	.2
w/AT add	.1	.1
w/diesel add	.2	.2
22,500 Mile Service (C)		
All Models	.4	.4
w/diesel add	.2	.2
30,000 Mile Service (B)		
All Models	3.2	3.2
w/16V engine add	.2	.2
w/AT add	.1	.1
w/diesel add	.7	.7
37,500 Mile Service (C)		
All Models	.4	.4
w/diesel add	.2	.2
45,000 Mile Service (B)		
All Models	2.1	2.1
w/16V engine add	.2	.2
w/AT add	.1	.1
w/diesel add	.2	.2
52,500 Mile Service (C)		
All Models	.4	.4
w/diesel add	.2	.2
60,000 Mile Service (B)		
All Models	3.2	3.2
w/AT add	.1	.1
w/Cabrio add	.1	.1
w/diesel add	.7	.7
w/supercharger add	.3	.3

QUANTUM : RABBIT : RABBIT CONVERTIBLE : RABBIT PICKUP : SCIROCCO — VW-3

	LABOR TIME	SEVERE SERVICE
67,500 Mile Service (C)		
All Models	.4	.4
w/diesel add	.2	.2
75,000 Mile Service (B)		
All Models	2.1	2.1
w/16V engine add	.2	.2
w/AT add	.1	.1
w/diesel add	.2	.2
82,500 Mile Service (C)		
All Models	.4	.4
w/diesel add	.2	.2
90,000 Mile Service (B)		
All Models	3.2	3.2
w/16V engine add	.2	.2
w/AT add	.1	.1
w/diesel add	.7	.7
97,500 Mile Service (C)		
All Models	.4	.4
w/diesel add	.2	.2

CHARGING

	LABOR TIME	SEVERE SERVICE
Alternator Circuits, Test (B)		
Includes: Test component output.		
1981-04	.7	.7
Alternator, R&R and Recondition (B)		
Beetle	2.4	2.9
Dasher	2.5	3.0
Fox	2.4	2.9
Golf, Jetta		
gasoline	2.3	2.8
diesel	2.7	3.2
1990-94 Passat		
4 cyl.	2.4	2.9
6 cyl.	4.3	4.8
Quantum	3.1	3.6
Rabbit		
gasoline	2.2	2.7
diesel	2.7	3.2
Scirocco	2.2	2.7
w/AC add	.8	.8
w/turbo add	.7	.7
Alternator Assy., Replace (B)		
Includes: Pulley transfer.		
Beetle		
gasoline	1.1	1.3
diesel	4.3	4.5
Cabrio, Cabriolet	.8	1.0
Corrado		
4 cyl.	.8	1.0
6 cyl.	3.2	3.4
Dasher	.9	1.1
Fox	.8	1.0
Golf, Jetta		
1981-84		
gasoline	.8	1.0
diesel	1.2	1.4

	LABOR TIME	SEVERE SERVICE
1985-97		
4 cyl.	.8	1.0
6 cyl.	3.3	3.5
diesel	1.2	1.4
1998-04		
1.8L	1.1	1.3
2.0L	1.4	1.6
diesel	1.7	1.9
Passat		
1990-97		
4 cyl.	.8	1.0
6 cyl.	3.0	3.2
1998-04		
4 cyl.	1.4	1.6
6 cyl.	2.9	3.1
8 cyl.	3.3	3.5
diesel	1.6	1.8
Quantum	1.8	2.0
Rabbit, Scirocco		
gasoline	.8	1.0
diesel	1.4	1.6
w/AC add		
81-96	.8	.8
Beetle 1.8L	.3	.3
98-04 Golf		
1.8L	.3	.3
diesel	.3	.3
98-04 Passat 6 cyl.	.4	.4
w/turbo 81-96 add	.7	.7
Replace voltage regulator add	.1	.1
Alternator Voltage Regulator, Replace (B)		
4 cyl.		
exc. below	1.5	1.8
Beetle 1.8L	1.1	1.4
1998-04 Golf		
1.8L	1.1	1.4
2.0L	.8	1.1
5 cyl.	1.6	1.9
6 cyl.		
VR6		
exc. 1998-04 Golf	3.4	3.7
1998-04 Golf	.8	1.1
V6		
w/AC	3.5	3.8
w/o AC	3.1	3.4
8 cyl.	3.5	3.8
Diesel		
exc. below	1.5	1.8
Beetle	4.4	4.7
1998-04 Golf	1.1	1.4

STARTING

	LABOR TIME	SEVERE SERVICE
Starter Draw Test (On Car) (B)		
1981-04	.3	.3
Battery, Replace (C)		
1981-04	.4	.6

	LABOR TIME	SEVERE SERVICE
Battery Cables, Replace (C)		
1981-02 each	.3	.5
Starter, R&R and Recondition (B)		
1981-87		
gasoline		
AT	3.9	4.4
MT	3.1	3.6
diesel	3.2	3.7
turbo diesel	3.4	3.9
1988-94		
4WD	5.0	5.5
gasoline		
4 cyl.		
AT	4.2	4.7
MT	3.3	3.8
6 cyl.	3.2	3.7
diesel	3.5	4.0
turbo diesel	3.7	4.2
Starter Assy., Replace (B)		
1981-84		
AT	1.3	1.5
MT	.9	1.1
diesel	1.3	1.5
turbo diesel	1.6	1.8
1985-93		
4 cyl.		
AT		
Cabriolet	1.5	1.7
Corrado	.9	1.1
Scirocco	1.6	1.8
MT	.8	1.0
6 cyl.	.9	1.1
Quantum 4WD	3.0	3.2
diesel	1.5	1.7
1994-04		
Beetle	1.7	1.9
Cabrio	.9	1.1
Corrado	.9	1.1
Golf, Jetta		
1994-97		
4 cyl.	1.1	1.3
6 cyl.	.9	1.1
diesel	1.5	1.7
1998-04		
1.8L	1.7	1.9
2.0L	1.1	1.3
diesel	1.4	1.6
Passat		
1994-97		
4 cyl.	.9	1.1
6 cyl.	1.1	1.3
diesel	1.5	1.7
1998-04		
4 cyl.	1.7	1.9
6 cyl.	3.3	3.5
diesel	1.6	1.8

VW-4 — BEETLE : CABRIO : CABRIOLET : CORRADO : DASHER : FOX : GOLF : JETTA : PASSAT

	LABOR TIME	SEVERE SERVICE
Starter Drive Assy., Replace (B)		
Includes: Starter R&R.		
1981-84		
AT	2.0	2.3
MT	1.5	1.8
diesel	1.6	1.9
turbo diesel	2.3	2.6
1985-95		
4 cyl.		
AT		
Cabrio, Cabriolet	1.8	2.1
Corrado	1.4	1.7
Golf, Jetta	2.5	2.8
Passat	.9	1.2
Scirocco	1.8	2.1
MT	1.3	1.6
6 cyl.	1.4	1.7
Quantum 4WD	3.0	3.3
diesel	1.8	2.1
1996-02		
Beetle		
gasoline	2.4	2.7
diesel	2.6	2.9
Cabrio, Golf, Jetta		
gasoline	1.7	2.0
diesel	2.1	2.4
Passat		
gasoline		
4 cyl.		
AT	2.8	3.1
MT	1.8	2.2
6 cyl.	3.8	4.1
diesel	2.1	2.4
Starter Solenoid and/or Switch, Replace (B)		
Includes: Starter R&R.		
1981-84		
AT	1.6	1.9
MT	1.2	1.5
diesel	1.8	2.1
turbo diesel	2.3	2.6
1985-97		
4 cyl.		
AT		
Cabrio	1.7	2.0
Cabriolet	1.6	1.9
Corrado	1.2	1.5
Golf, Jetta	2.4	2.7
Passat	.8	1.1
Scirocco	1.6	1.9
MT	1.4	1.7
6 cyl.	1.4	1.7
Quantum 4WD	3.3	3.6
diesel	1.8	2.1
1998-04		
4 cyl.		
Beetle	2.0	2.3
Cabrio	1.7	2.0
Golf		
1.8L	2.1	2.4
2.0L	2.4	1.7
Jetta	2.4	2.7
Passat	1.7	2.0
6 cyl.		
Jetta	1.4	1.7
Passat		
w/AC	3.6	3.9
w/o AC	3.2	3.5
diesel	2.0	2.3

CRUISE CONTROL

	LABOR TIME	SEVERE SERVICE
Diagnose Cruise Control System (A)		
1985-02	.8	.8
Control Module, Replace (B)		
1985-97	1.3	1.3
1998-04		
exc. Passat	1.3	1.3
Passat	.8	.8
Control Switch, Replace (B)		
1985-97	1.3	1.3
1998-04	1.1	1.1
Control Transmitter, Replace (B)		
Cabrio, Cabriolet	1.7	1.7
Corrado	1.8	1.8
Golf, Jetta		
1985-92	2.3	2.3
1993-02	1.6	1.6
1990-94 Passat	1.1	1.1
1985-89 Scirocco	1.6	1.6
Control Vacuum Pump, Replace (B)		
1985-97	.9	1.1
1998-04		
exc. Passat	.9	1.1
Passat	1.4	1.6
Control Vacuum Servo, Replace (B)		
Cabrio, Cabriolet	.9	.9
Corrado	1.4	1.4
Golf, Jetta	.9	.9
Passat	.9	.9
1985-89 Scirocco	.9	.9

IGNITION

	LABOR TIME	SEVERE SERVICE
Diagnose Ignition System Component Each (A)		
1981-02	.8	.8
Ignition Timing, Reset (B)		
1981-04	.6	.6
Camshaft Position Sensor, Replace (B)		
1988-92		
8V	1.6	1.8
16V	1.2	1.4
1993-02	.5	.7
Crankshaft Sensor, Replace (B)		
1990-97	.5	.7
1998-04		
exc. below	.5	.7
Golf 1.8L	1.1	1.3
Passat	1.0	1.2
Distributor, Replace (B)		
Includes: Reset base ignition timing.		
8V	.9	1.1
16V	.5	.7
Distributor Cap and/or Rotor, Replace (B)		
All Models	.3	.3
Distributor Points & Condenser, Replace (B)		
1981	1.1	1.3
Glow Plug, Replace (B)		
1981-97		
one	1.2	1.3
all	1.4	1.5
1998-04		
Beetle		
one	.8	.9
all	1.2	1.5
Golf, Jetta, Passat		
one	.8	1.3
all	1.4	1.6
Glow Plug Relay, Replace (B)		
1981-97	.5	.5
1998-04		
Beetle	.7	.7
Golf, Jetta, Passat	.5	.5
Hall Generator, Replace (B)		
1981-87	.7	.8
Ignition Coil, Replace (B)		
1981-97	.7	.8
1998-04		
Beetle		
1.8L one	.6	.7
each addl. add	.2	.2
2.0L	.7	.8
Golf		
1.8L, 6 cyl.		
one	.6	.7
all	.8	.9
2.0L	.6	.7
Jetta	.7	.8
Passat		
1.8L, 6 cyl. one	.6	.7
each addl. add	.2	.2
8 cyl. one	1.1	1.2
each addl. add	.3	.3
Ignition Module, Replace (B)		
1985-02	.7	.8

QUANTUM : RABBIT : RABBIT CONVERTIBLE : RABBIT PICKUP : SCIROCCO VW-5

	LABOR TIME	SEVERE SERVICE
Ignition Switch, Replace (B)		
1981-97		
w/air bags	3.1	3.1
w/o air bags	2.3	2.3
w/tilt wheel	2.5	2.5
w/o tilt wheel	1.1	1.1
1998-04	1.4	1.4
Spark Plug (Ignition) Cables, Replace (B)		
1981-97	.5	.7
1998-04		
Beetle		
1.8L	.7	.9
2.0L	1.7	1.9
Cabrio, Jetta		
4 cyl.	.7	.9
6 cyl.	.8	1.0
Golf	.8	1.0
Passat		
4 cyl., 8 cyl.	.6	.8
6 cyl.	.8	1.0
Spark Plugs, Replace (B)		
1981-97		
4 cyl.	.5	.7
5 cyl.	.5	.7
6 cyl.	.8	1.0
1998-04		
Beetle		
1.8L	1.1	1.3
2.0L	1.7	1.9
Cabrio, Jetta		
4 cyl.	.5	.7
6 cyl.	.8	1.0
Golf	.8	1.0
Passat		
4 cyl.	.9	1.1
6 cyl.	1.1	1.3
8 cyl.	1.7	1.9
Vacuum Advance Unit, Replace (B)		
Includes: Reset base ignition timing.		
1981-87	.9	1.0
1988-02	1.3	1.4

EMISSIONS

The following operations do not include testing. Add time as required.

	LABOR TIME	SEVERE SERVICE
Dynamometer Test (A)		
1981-97	.5	.5
1998-04	1.0	1.0
Air Flow Sensor, Replace (B)		
w/continuous inj	1.5	1.5
w/o continuous inj	.7	.7
Adjust sensor plate add	.3	.3
Air Injection Check Valve, Replace (B)		
1981-93	.3	.3

	LABOR TIME	SEVERE SERVICE
Air Pump Assy., Replace (B)		
1981-93	.7	.9
1998-04 Beetle, Golf	.8	1.0
Passat		
1994-97	2.7	2.9
1998-04 8 cyl.	1.8	2.0
Air Pump Centrifugal Filter, Replace (B)		
1981-93	.3	.3
Anti-Backfire Valve, Replace (B)		
1981-93	.3	.3
EGR Control Valve, Replace (B)		
1981-97	.7	.9
1998-04		
Beetle diesel	1.2	1.4
Cabrio, Jetta	.5	.7
Golf 6 cyl.	.8	1.0
Passat	1.9	2.1
EGR Delay Valve, Replace (B)		
1981-02	.3	.3
EGR Filter, Replace (B)		
1981-02	.7	.7
Electronic Control Unit, Replace (B)		
1981-97		
w/continuous inj	.3	.3
w/o continuous inj	.8	.8
1998-04		
exc. below		
w/continuous inj	.3	.3
w/o continuous inj	.8	.8
gasoline		
Beetle	1.0	1.0
Golf	.8	.8
Passat	.6	.6
diesel	1.1	1.1
Knock Sensor, Replace (B)		
1988-04		
single	.9	1.1
dual		
one	1.1	1.3
two	1.7	1.9
Maintenance Reminder, Replace (B)		
1988-93	.5	.5
Mass Air Flow (MAF) Sensor, Replace (B)		
1983-97	.7	.7
1998-04		
exc. 1.8L	.7	.7
1.8L	1.1	1.1
Oxygen Sensor, Replace (B)		
4 cyl.		
exc. 1.8L	.8	1.0
1.8L	1.4	1.6
6 cyl.		
one	.8	1.0
both	1.1	1.3
After catalyst	.8	1.0
Positive Crankcase Ventilation (PCV) Valve, Replace (B)		
1981-02	.3	.3

	LABOR TIME	SEVERE SERVICE
Throttle Position Sensor (TPS), Replace (B)		
1988-04	.8	.8
Vapor Canister Filter, Replace (B)		
Exc. Passat	.8	.8
Passat	1.3	1.3
Vehicle Speed Sensor, Replace (B)		
Beetle	.8	.8
Cabrio	1.3	1.3
Golf, Jetta		
1984-97	1.5	1.5
1998-04	.8	.8
Passat		
1994-97	.5	.5
1998-04	.8	.8

FUEL
CARBURETOR

	LABOR TIME	SEVERE SERVICE
Carburetor, R&R and Clean or Recondition (B)		
Jetta	3.3	3.3
Rabbit, Scirocco	3.3	3.3
Carburetor, Replace (B)		
Includes: Adjustments.		
Jetta	1.3	1.3
Rabbit, Scirocco	1.4	1.4

DELIVERY

	LABOR TIME	SEVERE SERVICE
Fuel Pump, Test (B)		
1981-02	.3	.3
Fuel Gauge (Dash) Replace (B)		
Beetle	1.8	1.8
Cabrio	1.7	1.7
Cabriolet	1.9	1.9
Corrado		
1990	.9	.9
1991-94	1.4	1.4
Dasher	.9	.9
Fox	2.4	2.4
Golf, Jetta		
1981-84	.7	.7
1985-92	2.4	2.4
1993-02	1.6	1.6
Passat		
1990-97	.9	.9
1998-02	2.3	2.3
Quantum	2.0	2.0
Rabbit, Scirocco		
1981-84	.7	.7
1985-89	1.9	1.9
Fuel Gauge (Tank), Replace (B)		
Beetle	1.1	1.4
Cabrio, Cabriolet	.9	1.2
Corrado	2.1	2.4
Dasher	.5	.8
Fox	.7	1.0
Golf, Jetta		
1981-92	.8	1.1
1993-04	1.1	1.4

VW-6 BEETLE : CABRIO : CABRIOLET : CORRADO : DASHER : FOX : GOLF : JETTA : PASSAT

	LABOR TIME	SEVERE SERVICE
Fuel Gauge (Tank), Replace (B)		
Passat		
1990-97	1.8	2.1
1998-04	1.4	1.7
Quantum	.8	1.1
Rabbit, Scirocco		
1981-84	.8	1.1
1985-89	.9	1.2
Fuel Pump, Replace (B)		
Beetle	1.3	1.6
Cabrio	1.4	1.7
Cabriolet	.8	1.1
Corrado		
early	.7	1.0
late	2.0	2.3
Fox	.7	1.0
Golf, Jetta		
1981-84	.7	1.0
1985-04	1.4	1.7
Passat		
1990-97	1.3	1.6
1998-04	1.7	2.0
Quantum	1.4	1.7
Rabbit, Scirocco		
1981-84	.7	1.0
1985-89	.8	1.1
Fuel Pump Relay, Replace (B)		
1988-04	.6	.6
Fuel Tank, Replace (B)		
Includes: Drain and refill.		
Beetle	3.5	3.8
Cabrio	1.7	2.0
Cabriolet	2.5	2.8
Corrado	2.0	2.3
Dasher	1.2	1.5
Fox	1.5	1.8
Golf, Jetta		
1981-84	2.7	3.0
1985-04	3.5	3.8
Passat		
2WD	2.1	2.4
4Motion	5.5	5.8
Quantum	1.9	2.2
Rabbit, Scirocco	2.6	2.9
Intake Manifold and/or Gasket, Replace (B)		
Includes: Adjustments.		
Gasoline		
1981-87	1.6	2.1
1988-97		
one piece	2.8	3.3
two piece		
upper	1.7	2.2
lower	2.4	2.9
1998-04		
4 cyl.		
Beetle		
1.8L	1.7	2.2
2.0L upper	1.4	1.9
Cabrio, Jetta	3.8	4.3

	LABOR TIME	SEVERE SERVICE
Golf		
1.8L	1.7	2.2
2.0L	2.4	2.9
Passat	1.7	2.2
6 cyl.		
exc. Passat		
upper	3.2	3.7
lower	2.2	2.7
Passat	2.4	2.9
Diesel		
1981-92	1.7	2.2
1993-97	2.4	2.9
1998-04		
exc. below	2.4	2.9
Beetle	3.3	3.8
Golf	1.7	2.2
Mechanical Fuel Pump, R&R or Replace (B)		
1981-84 Jetta, Rabbit, Scirocco	.7	.8
INJECTION		
Fuel Injection System, Bleed (B)		
1981-02	.3	.3
Idle Speed, Adjust (B)		
Diesel	.5	.7
Injection Timing, Check & Adjust (B)		
Diesel	1.1	1.3
Throttle Body Sensor, Adjust (B)		
1988-02	.5	.7
Cold Start Cable, Replace (B)		
1981-84 Jetta	.7	.9
Quantum	1.3	1.5
Rabbit	.7	.9
Differential Pressure Regulator, Replace (B)		
1981-02	1.2	1.4
Fuel Accumulator, Replace (B)		
1988-04	.7	.9
Fuel Distributor, Replace (B)		
1981-97	1.5	1.8
1998-04		
exc. below	1.5	1.8
Beetle 2.0L	.8	1.1
Passat		
4 cyl.	.8	1.1
6 cyl.	1.4	1.7
Fuel Filter, Replace (B)		
1981-97	.5	.5
1998-04		
gasoline	.6	.6
diesel	.7	.7
Fuel Injection Lines, Replace (B)		
Diesel		
1981-89		
one	.5	.7
all	.8	1.0
1990-04		
one	.3	.5
all	.7	.9

	LABOR TIME	SEVERE SERVICE
Fuel Injection Pump, Replace (B)		
Gasoline	.9	1.1
Diesel		
1981-89	3.1	3.3
Beetle	3.8	4.0
1998-04 Golf	3.8	4.0
Passat	2.9	3.1
w/turbo diesel 81-89 add	.7	.7
w/AC 81-89 add	.5	.5
Fuel Injectors, Replace (B)		
Gasoline		
1981-97 one or all		
w/continuous inj	1.4	1.7
w/o continuous inj	1.8	2.1
1998-04		
exc. below		
w/continuous inj	1.4	1.7
w/o continuous injection	1.8	2.1
Beetle		
one	1.7	2.0
all	2.0	2.3
Golf 4 cyl.		
one	1.4	1.7
all	1.7	2.0
Passat		
4 cyl.		
one	.8	1.1
all	1.4	1.7
6 cyl.		
one	1.4	1.7
all	1.7	2.0
8 cyl.		
one	1.4	1.7
all	2.0	2.3
Diesel		
1981-89		
one	.5	.8
all	1.2	1.5
1990-97		
one	.3	.6
all	.8	1.1
1998-04		
exc. below		
one	.3	.6
all	.8	1.1
Beetle all	1.1	1.4
Golf		
one	.8	1.1
all	1.1	1.4
w/turbo add	.3	.3
Clean and test add each	.3	.3
Fuel Pressure Regulator, Replace (B)		
1988-97	1.3	1.5
1998-04	.8	1.0
Fuel Rail, Replace (B)		
Beetle	1.8	2.0
Cabrio	1.3	1.5
Cabriolet	.9	1.1
Corrado	.9	1.1

QUANTUM : RABBIT : RABBIT CONVERTIBLE : RABBIT PICKUP : SCIROCCO   VW-7

	LABOR TIME	SEVERE SERVICE
Fox	.9	1.1
Golf, Jetta		
1985-92	.9	1.1
1993-02		
4 cyl.	1.3	1.5
6 cyl.	1.8	2.0
Passat	.9	1.1
Scirocco	.9	1.1
Fuel Shut-Off Valve, Replace (B)		
1981-89	.5	.5
1990-92	.7	.7
1993-94	.8	.8
Idle Air Control (IAC) Valve, Replace (B)		
1988-04	.8	.9
Impulse Relay, Replace (B)		
1981-02	.5	.5
Injection Pump Sprocket, Replace (B)		
Diesel		
1981-89	1.9	2.1
1990-02	1.8	2.0
Microswitch, Replace (B)		
1981-97	.9	.9
Pressure Regulator Valve, Replace (B)		
1981-04	.6	.7
Temperature Sensor, Replace (B)		
1981-97		
w/continuous inj	.9	.9
w/o continuous inj	.3	.3
1998-04		
exc. below		
w/continuous inj	.9	.9
w/o continuous inj	.3	.3
Beetle, Golf	.5	.5
Passat		
4 cyl.	.6	.6
6 cyl.	1.0	1.0
Throttle Body Assy., Replace (B)		
1988-97	1.5	1.7
1998-04		
Beetle		
1.8L	.6	.8
2.0L	1.4	1.6
Cabrio, Jetta	.8	1.0
Golf		
4 cyl.	.8	1.0
6 cyl.	1.1	1.3
Passat		
4 cyl.	.6	.8
6 cyl.	1.5	1.7
Throttle Body Switches, Replace (B)		
1988-04		
exc. below	1.9	2.1
1998-04 Passat		
4 cyl., 8 cyl.	.5	.7
6 cyl.	1.1	1.3

	LABOR TIME	SEVERE SERVICE
Voltage Supply Relay, Replace (B)		
1981-97	.5	.5
1998-04	.3	.3
SUPERCHARGER		
Intercooler, Replace (B)		
Corrado	1.7	2.0
Supercharger Assy. or Gasket, Replace (B)		
Corrado	2.0	2.5
TURBOCHARGER		
Turbocharger Assy., R&R or Replace (B)		
Gasoline		
Beetle, Golf	5.6	6.1
1998-04 Passat	3.4	3.9
Diesel		
Beetle	3.8	4.3
1996-04 Golf, Jetta	3.8	4.3
1996-98 Passat	2.6	3.1
Quantum	2.3	2.8
Wastegate Actuator, Replace (B)		
1983-88	.9	1.1
EXHAUST		
Catalytic Converter, Replace (B)		
Beetle		
gasoline		
1.8L	1.1	1.2
2.0L w/front pipe	1.4	1.5
diesel w/front pipe	1.4	1.5
Cabrio	1.1	1.2
Cabriolet	1.3	1.4
Corrado	.9	1.0
Fox	1.3	1.4
Golf, Jetta		
1981-84	1.1	1.2
1985-92	1.7	1.8
1993-97	1.3	1.4
1998-04		
gasoline	1.4	1.5
diesel	2.1	2.2
Passat		
1990-97	1.3	1.4
1998-04		
single	1.7	1.8
dual		
one side	1.7	1.8
both sides	2.5	2.6
Rabbit, Scirocco		
1981-84	1.3	1.4
1985-89		
8V	1.4	1.5
16V	2.4	2.5

	LABOR TIME	SEVERE SERVICE
Exhaust Manifold or Gasket, Replace (B)		
Beetle		
gasoline		
1.8L	3.5	4.0
2.0L	3.0	3.5
diesel	4.0	4.5
Cabrio	2.8	3.3
Cabriolet		
AT	2.7	3.2
MT	2.2	2.7
Corrado		
4 cyl.	2.1	2.6
6 cyl. one or both	3.6	4.1
Dasher	1.9	2.4
Fox	2.7	3.2
Golf, Jetta		
1981-84		
gasoline	2.0	2.5
diesel	1.6	2.1
turbo diesel	4.0	4.5
1985-97		
8V		
AT	5.3	5.8
MT	4.5	5.0
16V	2.4	2.9
6 cyl. one or both	4.5	5.0
diesel	2.9	3.4
turbo diesel	4.0	4.5
1998-04		
4 cyl.	3.3	3.8
6 cyl.	2.9	3.4
diesel	4.4	4.9
Passat		
1990-97	4.5	5.0
1998-04		
4 cyl.	3.0	3.5
6 cyl.		
one side	4.6	5.1
both sides	5.8	6.3
diesel	4.2	4.7
Quantum		
4 cyl.	2.4	2.9
5 cyl.	3.2	3.7
turbo diesel	3.3	3.8
Rabbit, Scirocco		
1981-84		
gasoline	1.9	2.4
diesel	1.5	2.0
turbo diesel	4.0	4.5
1985-89		
8V		
AT	2.8	3.3
MT	2.3	2.8
16V	2.6	3.1

VW-8 — BEETLE : CABRIO : CABRIOLET : CORRADO : DASHER : FOX : GOLF : JETTA : PASSAT

Operation	Labor Time	Severe Service
Front Exhaust Pipe, Replace (B)		
1981-87	1.6	1.7
1988-97		
4 cyl.		
8V	1.3	1.4
16V	1.7	1.8
5 cyl.	1.5	1.6
6 cyl.	2.4	2.5
1998-04		
exc. below		
4 cyl. 8V	1.3	1.4
6 cyl.	2.4	2.5
Beetle w/catalytic converter	1.4	1.5
Golf		
1.8L	1.1	1.2
2.0L	1.4	1.5
diesel	2.0	2.1
Passat		
4 cyl.	1.1	1.2
V6 w/catalytic converter		
one	1.4	1.5
both	2.1	2.2
Muffler, Replace (B)		
1981-97		
center	1.1	1.2
muffler and resonator		
4 cyl.		
1981-87	1.4	1.5
1988-97	1.2	1.3
5 cyl.	1.7	1.8
muffler and tailpipe		
4 cyl.		
1981-87	.9	1.0
1988-97	.5	.6
5 cyl.	.7	.8
1998-04		
Beetle		
gasoline		
center		
convertible	1.4	1.5
hard top	1.7	1.8
rear	1.1	1.2
diesel		
w/intermediate pipe	1.7	1.8
w/o intermediate pipe	.8	.9
Golf		
center	1.1	1.2
rear	.8	.9
Passat		
center	1.4	1.5
rear		
one	.8	.9
both	1.4	1.4
Rear Exhaust Pipe, Replace (B)		
Dasher, Quantum	.7	.8

ENGINE COOLING

Operation	Labor Time	Severe Service
Pressure Test Cooling System (B)		
1981-04	.5	.5
Coolant Thermostat, Replace (B)		
Beetle		
gasoline	1.9	2.1
diesel	1.4	1.6
Cabrio	1.5	1.7
Cabriolet		
w/AC	1.4	1.6
w/o AC	.8	1.0
Corrado, Fox		
w/AC	1.8	2.0
w/o AC	1.4	1.6
Golf, Jetta		
1981-92		
w/AC	1.9	2.1
w/o AC	1.5	1.7
1993-97		
4 cyl.	1.5	1.7
6 cyl.	1.9	2.1
diesel	1.5	1.7
1998-04		
1.8L	3.3	3.5
2.0L	1.4	1.6
6 cyl.	2.4	2.6
diesel	1.7	1.9
Passat		
1990-97		
w/AC	1.7	1.9
w/o AC	1.4	1.6
1998-04		
4 cyl.	1.4	1.6
6 cyl.	3.3	3.5
Quantum		
w/AC	1.9	2.1
w/o AC	1.3	1.5
Rabbit, Scirocco		
w/AC	1.7	1.9
w/o AC	1.3	1.5
Electric Fan Thermo Switch, Replace (B)		
1981-87	.5	.5
1988-97	1.3	1.3
1998-04		
Beetle, Passat		
gasoline	1.5	1.5
diesel	.8	.8
Golf	1.1	1.1
Engine Coolant Temp. Sending Unit, Replace (B)		
1981-97	.7	.7
1998-04		
Beetle		
gasoline	.9	.9
diesel	1.2	1.2

Operation	Labor Time	Severe Service
Golf		
gasoline	.5	.5
diesel	1.4	1.4
Passat	.3	.3
Freeze Plugs (Water Jacket), Replace (B)		
Add access time as required.		
1981-02 each	.5	.7
Radiator Assy., R&R or Replace (B)		
Beetle	3.3	3.5
Cabrio	2.9	3.1
Cabriolet	1.2	1.4
Corrado		
4 cyl.	2.7	2.9
6 cyl.	3.2	3.4
Dasher	1.5	1.7
Fox	1.5	1.7
Golf, Jetta		
1981-84	1.4	1.6
1985-92	1.9	2.1
1993-97	3.1	3.3
1998-04 4 cyl.		
w/AC	2.9	3.1
w/o AC	2.4	2.6
Passat		
1990-96		
4 cyl.	2.0	2.2
6 cyl.	3.0	3.2
1997-04		
4 cyl.		
w/AC	2.4	2.6
w/o AC	2.0	2.2
6 cyl.	4.0	4.2
Quantum		
4 cyl.	1.5	1.7
5 cyl.	2.3	2.5
Rabbit, Scirocco		
1981-84	1.3	1.5
1985-89	1.4	1.6
Radiator Fan and/or Fan Motor, Replace (B)		
Beetle		
1.8L	1.1	1.3
2.0L	.6	.8
Cabrio, Cabriolet	1.7	1.9
Corrado	1.9	2.1
Fox	.9	1.1
Golf, Jetta		
1981-84	.9	1.1
1985-92	1.5	1.7
1993-97	3.2	3.4
1998-04 4 cyl.	1.1	1.3
Passat		
1990-96		
4 cyl.	2.4	2.6
6 cyl.	2.8	3.0
1997-04		
4 cyl., 8 cyl.	2.5	2.7
6 cyl.	2.1	2.3

QUANTUM : RABBIT : RABBIT CONVERTIBLE : RABBIT PICKUP : SCIROCCO

	LABOR TIME	SEVERE SERVICE
Quantum	1.4	1.6
Rabbit, Scirocco		
1981-84	.9	1.1
1985-89	1.5	1.7

Radiator Hoses, Replace (B)
Includes: Refill system with proper coolant mix.

1981-87		
one	.5	.7
both	.7	.9
1988-04		
exc. below		
one	.7	.9
both	.8	1.0
1997-04 Passat 4 cyl.		
upper	1.7	1.9
lower or both	1.8	2.0

Temperature Gauge (Dash), Replace (B)

Beetle	2.0	2.0
Cabrio	1.3	1.3
Cabriolet	1.9	1.9
Corrado	1.2	1.2
Fox	2.1	2.1
Golf, Jetta		
1981-84	1.5	1.5
1985-92	2.7	2.7
1993-02	1.3	1.3
Passat		
1990-97	1.8	1.8
1998-02	2.6	2.6
Quantum	1.4	1.4
Rabbit, Scirocco		
1981-84	1.3	1.3
1985-89	1.9	1.9

Temperature Gauge (Engine), Replace (B)

1981-84	.3	.3
1985-97	.7	.7
1998-02 Beetle	1.9	1.9
1998-02 Cabrio, Golf, Jetta	1.8	1.8
1998-02 Passat	2.5	2.5

Water Pump and/or Gasket, Replace (B)
Includes: Refill with proper coolant mix.

Beetle		
gasoline		
1.8L	3.5	4.0
2.0L	3.1	3.6
diesel	4.6	5.1
Cabrio	2.2	2.7
Cabriolet	2.7	3.2
Corrado		
4 cyl.	3.7	4.2
6 cyl.	2.8	3.3

	LABOR TIME	SEVERE SERVICE
Fox		
w/AC	2.8	3.3
w/o AC	1.5	2.0
Golf, Jetta		
1981-84		
gasoline		
w/AC	4.1	4.6
w/o AC	2.5	3.0
diesel	2.3	2.8
1985-92		
8V		
w/AC	3.9	4.4
w/AC & PS	4.9	5.4
w/o AC	3.1	3.6
16V	3.0	3.5
diesel		
w/AC	3.7	4.2
w/AC & PS	5.0	5.5
w/o AC	3.1	3.6
1993-97		
gasoline		
4 cyl.		
w/AC	2.8	3.3
w/o AC	2.4	2.9
6 cyl.	2.7	3.2
diesel		
w/AC	5.0	5.5
w/o AC	3.2	3.7
1998-04		
gasoline		
1.8L	3.3	3.8
2.0L, 6 cyl.	2.9	3.4
diesel	4.4	4.9
Passat		
1990-96		
4 cyl.	2.7	3.2
6 cyl.	2.2	2.7
1997-04		
4 cyl.	4.0	4.5
6 cyl.	5.8	6.3
auxiliary	1.4	1.9
Quantum		
4 cyl.		
w/AC	2.7	3.2
w/o AC	1.7	2.2
5 cyl.	2.2	2.7
Rabbit, Scirocco		
gasoline		
w/AC	4.2	4.7
w/o AC	2.5	3.0
diesel	2.5	3.0

ENGINE
ASSEMBLY

Times shown are for OEM assemblies. Time to replace assemblies from aftermarket rebuilders may vary.

Engine Assy., R&I (B)
Does not include parts or component transfer.

	LABOR TIME	SEVERE SERVICE
Beetle		
1.8L	9.2	9.7
2.0L	8.5	9.0
diesel	9.2	9.7
Cabrio	5.7	6.2
Cabriolet	6.4	6.9
Corrado		
4 cyl.	9.2	9.7
6 cyl.	9.9	10.4
Dasher	6.3	6.8
Fox	7.4	7.9
Golf, Jetta		
1981-84		
gasoline	6.5	7.0
diesel	4.9	5.4
1985-97		
4 cyl.	5.5	6.0
6 cyl.	8.3	8.8
diesel	6.8	7.3
1998-04		
4 cyl.	8.5	9.0
6 cyl.	10.9	11.4
diesel	8.5	9.0
Passat		
1990-96		
4 cyl.	7.2	7.7
6 cyl.	9.0	9.5
diesel	7.6	8.1
1997-04		
4 cyl., 6 cyl.	8.8	9.3
8 cyl.	14.2	14.7
Quantum		
4 cyl.	6.9	7.4
5 cyl.	8.0	8.5
diesel	8.3	8.8
Rabbit, Scirocco		
1981-84		
gasoline	6.5	7.0
diesel	6.9	7.4
1985-89	6.4	6.9
w/AC add	1.4	1.4
w/AT add	.3	.3
w/turbo diesel add	1.0	1.0

Engine Assy. (New or Rebuilt Complete), Replace (B)
Includes: Component transfer and engine tune-up.

Beetle		
1.8L	9.9	10.4
2.0L	9.2	9.7
diesel	10.2	10.7

VW-10 BEETLE : CABRIO : CABRIOLET : CORRADO : DASHER : FOX : GOLF : JETTA : PASSAT

	LABOR TIME	SEVERE SERVICE
Engine Assy. (New or Rebuilt Complete), Replace (B)		
Cabrio	13.0	13.5
Cabriolet	8.6	9.1
Corrado		
4 cyl.	13.2	13.7
6 cyl.	15.1	15.6
Dasher	8.2	8.7
Fox	9.0	9.5
Golf, Jetta		
1981-84		
gasoline	8.6	9.1
diesel	9.1	9.6
1985-97		
4 cyl.	12.8	12.8
6 cyl.	14.3	14.3
diesel	15.1	15.1
1998-04		
4 cyl.	9.2	9.7
6 cyl.	11.6	12.1
diesel	9.5	10.0
Passat		
1990-97		
4 cyl.	14.2	14.2
6 cyl.	15.3	15.3
1998-04		
4 cyl., 6 cyl.	9.2	9.7
8 cyl.	14.4	14.9
Rabbit, Scirocco		
1981-84		
gasoline	8.6	9.1
diesel	9.4	9.5
1985-89	8.5	9.0
w/AC add	1.4	1.4
w/AT add	.3	.3
w/turbo diesel add	1.0	1.0

Engine Assy. (Short Block), Replace (B)
Assembly consists of engine block, piston assemblies, crankshaft, camshaft, timing chain and gears. Does not include cylinder heads. Operation includes: R&R engine, transfer necessary parts and all necessary adjustments.

	LABOR TIME	SEVERE SERVICE
Beetle		
1.8L	15.7	16.6
2.0L	17.8	18.7
diesel	15.7	16.6
Cabrio	10.2	11.1
Cabriolet	10.9	11.8
Corrado		
4 cyl.	13.7	14.6
6 cyl.	17.7	18.6
Dasher	10.0	10.9
Fox	11.6	12.5

	LABOR TIME	SEVERE SERVICE
Golf, Jetta		
1981-84		
gasoline	10.5	11.4
diesel	12.9	13.8
1985-97		
4 cyl.	10.3	11.2
6 cyl.	12.4	13.3
diesel	13.1	14.0
1998-04		
1.8L	15.0	15.9
2.0L	13.6	14.5
6 cyl.	12.4	13.3
diesel	15.0	15.9
Passat		
1990-97		
4 cyl.	11.9	12.8
6 cyl.	13.1	14.0
diesel	13.7	14.6
1998-04		
4 cyl.	15.0	15.9
6 cyl.	17.0	17.0
8 cyl.	18.0	18.9
Quantum		
4 cyl.	11.6	12.5
5 cyl.	13.1	14.0
diesel	14.0	14.9
Rabbit, Scirocco		
1981-84		
gasoline	10.9	11.8
diesel	12.8	13.7
1985-89	14.1	15.0
w/AC add	1.4	1.4
w/AT add	.3	.3
w/turbo diesel add	1.0	1.0

Engine Mounts, Replace (B)
Includes: R&I noise dampening pan.
Includes: Loosen and tighten anti-roll bar for Passat 6 cyl. only.

	LABOR TIME	SEVERE SERVICE
Beetle		
2.0L		
carrier	1.1	1.3
rubber mount	1.7	1.9
diesel rubber mount		
right	1.4	1.6
left	.8	1.0
Cabrio		
right	.9	1.1
center	.5	.7
Cabriolet		
right	2.8	3.0
center	.5	.7
Corrado		
right or left	1.7	1.9
center	1.4	1.6
Dasher		
right or left	.5	.7
center	.5	.7
rear	1.2	1.4

	LABOR TIME	SEVERE SERVICE
Fox		
right	.5	.7
left	.8	1.0
Golf, Jetta		
1981-84		
gasoline		
right side	2.3	2.5
front center	.5	.7
rear each	.5	.7
diesel		
right	3.8	4.0
left	.5	.7
center	.5	.7
1985-97		
gasoline		
right or left	1.5	1.7
center	.5	.7
diesel		
right	.7	.9
left	1.4	1.6
center	.7	.9
1998-04		
1.8L	1.7	1.9
2.0L		
right	1.7	1.9
left	.8	1.0
6 cyl.		
right	1.1	1.3
left	1.4	1.6
diesel		
right	1.1	1.3
left	.8	1.0
center	.7	.9
Passat		
1990-97		
right or left	1.4	1.6
center	.5	.7
1998-04		
4 cyl.		
one	1.4	1.6
both	1.7	1.9
6 cyl.		
one	1.7	1.9
both	2.1	2.3
Quantum		
gasoline		
right or left	.7	.9
center	1.2	1.4
diesel		
right	4.2	4.4
left	.5	.7
rear each	.5	.7
Rabbit, Scirocco		
1981-84		
gasoline		
right side	2.2	2.4
front center	.5	.7
rear each	.5	.7

QUANTUM : RABBIT : RABBIT CONVERTIBLE : RABBIT PICKUP : SCIROCCO — VW-11

	LABOR TIME	SEVERE SERVICE
diesel		
right side	3.8	4.0
front center	.5	.7
rear each	.5	.7
1985-89		
right	2.6	2.8
center	.5	.7
Right Engine/Injection Pump Mount, Replace (B)		
1981-84 diesel	3.8	4.0

CYLINDER HEAD

Compression Test (B)
	LABOR TIME	SEVERE SERVICE
1981-87		
4 cyl.	.7	1.0
5 cyl.	.7	1.0
6 cyl.	.8	1.1
diesel	1.3	1.6
1998-04		
Beetle		
gasoline	1.9	2.2
diesel	.9	1.2
Cabrio, Golf, Jetta		
4 cyl.	1.1	1.4
6 cyl.	.8	1.1
diesel	1.1	1.4
Passat	1.5	1.8

Valve Clearance, Adjust (B)
4 cyl.	1.2	1.5
5 cyl.	1.6	1.9
Diesel	1.2	1.5

Cam Followers, Replace (B)
1981-88 Golf, Jetta	2.8	3.2
Quantum		
gasoline	3.0	3.4
diesel	2.8	3.2
1981-84 Rabbit, Scirocco	2.8	3.2
w/turbo diesel add	.5	.5

Cylinder Head, R&I (B)
Includes: Parts transfer, adjustments.
Beetle		
gasoline	6.3	6.8
diesel	7.7	8.2
Cabrio	9.8	10.3
Cabriolet	8.4	8.9
Corrado		
4 cyl.	8.3	8.8
6 cyl.	9.9	10.4
Dasher	5.9	6.4
Fox	8.4	8.9
Golf, Jetta		
1981-84		
gasoline	5.3	5.8
diesel	5.3	5.8
1985-92		
gasoline	8.3	8.8
diesel	5.6	6.1

	LABOR TIME	SEVERE SERVICE
1993-97		
4 cyl.	9.8	10.3
6 cyl. both	26.3	26.8
diesel	11.3	11.8
1998-04		
4 cyl.	6.3	6.8
6 cyl.	9.2	9.7
diesel	7.7	8.2
Passat		
1990-97		
4 cyl.	14.2	14.7
6 cyl. both	26.3	26.8
1998-04		
4 cyl.	7.7	8.2
6 cyl.		
right side	7.1	7.6
left side	7.7	8.2
both sides	10.1	10.6
Quantum		
4 cyl.	5.4	5.9
5 cyl.	6.5	7.0
diesel	5.7	6.2
Rabbit, Scirocco		
1981-84		
gasoline	5.2	5.7
diesel	5.5	6.0
1985-89		
8V	8.3	8.8
16V	9.8	10.3
w/turbo diesel add	1.8	1.8

Cylinder Head Bolts, Retorque (B)
1981-02	.8	1.1

Cylinder Head Gasket, Replace (B)
Beetle		
gasoline	6.5	6.8
diesel	7.9	8.2
Cabrio	4.9	5.2
Cabriolet	4.8	5.1
Corrado		
4 cyl.	4.7	5.0
6 cyl.	7.2	7.5
Dasher	4.5	4.8
Fox	4.8	5.1
Golf, Jetta		
1981-84		
gasoline	5.1	5.4
diesel	5.1	5.4
1985-97		
gasoline		
4 cyl.	4.9	5.2
6 cyl.	10.3	10.6
diesel	6.1	6.4
1998-04		
4 cyl.	6.5	6.8
6 cyl.	9.3	9.6
diesel	7.9	8.2
Passat		
1990-97		
4 cyl.	9.1	9.4
6 cyl.	10.3	10.6

	LABOR TIME	SEVERE SERVICE
1998-04		
4 cyl.	7.9	8.2
6 cyl.		
right side	7.2	7.5
left side	8.2	8.5
both sides	10.3	10.6
Quantum		
4 cyl.	3.8	4.1
5 cyl.	5.7	6.0
diesel	5.3	5.6
Rabbit, Scirocco		
1981-84	5.1	5.4
1985-89		
8V	4.9	5.2
16V	5.9	6.2
w/turbo diesel add	1.8	1.8

Valve Cover Gasket, Replace (B)
1981-87	.5	.7
1988-92		
4 cyl.		
8V	.7	.9
16V	1.5	1.7
5 cyl.	1.1	1.3
diesel	.7	.9
Beetle		
1.8L	2.4	2.6
2.0L	1.7	1.9
diesel	.8	1.0
Cabrio	1.5	1.7
1993 Cabriolet		
8V	.7	.9
16V	1.4	1.6
1993-94 Corrado	1.4	1.6
Fox	.7	.9
Golf, Jetta		
1993-97		
4 cyl.	1.6	1.8
6 cyl. both	1.9	2.1
diesel	.5	.7
1998-04		
1.8L	2.6	2.8
2.0L	1.8	2.0
diesel	.9	1.1
Passat		
1993-97		
4 cyl.	1.7	1.9
6 cyl. both	1.8	2.0
diesel	.5	.7
1998-04		
4 cyl.	1.2	1.4
6 cyl.		
one side	1.5	1.7
both sides	2.5	2.7
8 cyl.		
right side	2.6	2.8
left side	2.1	2.3
both sides	3.1	3.3

VW-12 BEETLE : CABRIO : CABRIOLET : CORRADO : DASHER : FOX : GOLF : JETTA : PASSAT

	LABOR TIME	SEVERE SERVICE
Valve Job (A)		
Beetle		
gasoline	7.9	8.4
diesel	11.3	11.8
Cabrio	10.8	11.3
Cabriolet	9.4	9.9
Corrado		
4 cyl.	9.7	10.2
6 cyl.	11.3	11.8
Dasher	7.7	8.2
Fox	9.7	10.2
Golf, Jetta		
1981-84		
gasoline	9.7	10.2
diesel	10.0	10.5
1985-92		
4 cyl.	11.6	12.1
diesel	10.0	10.5
1993-97		
4 cyl.	11.5	12.0
6 cyl.		
right side	16.4	16.9
left side	17.6	18.1
both sides	29.7	30.2
diesel	13.0	13.5
1998-04		
1.8L	10.4	10.9
2.0L	9.9	10.4
6 cyl.	14.7	15.2
diesel	11.3	11.8
Passat		
1990-97		
8V	9.7	10.2
16V	14.0	14.5
1998-04		
4 cyl.	10.6	11.1
6 cyl.		
right side	9.5	10.0
left side	10.2	10.7
both sides	14.7	15.2
Quantum		
4 cyl.	9.4	9.9
5 cyl.	11.8	12.3
diesel	10.0	10.5
Rabbit, Scirocco		
1981-84		
gasoline	9.7	10.2
diesel	10.0	10.5
1985-89		
8V	9.7	10.2
16V	13.1	13.6
Valve Lifters, Replace (B)		
1988-95		
4 cyl.		
8V	2.5	2.8
16V	2.8	3.1
6 cyl.	3.1	3.4

	LABOR TIME	SEVERE SERVICE
1996-02 Beetle	4.3	4.6
1996-02 Cabrio	3.1	3.4
1996-02 Golf, Jetta		
gasoline		
4 cyl.	3.0	3.3
6 cyl.	3.6	3.9
diesel	3.2	3.5
Passat		
1996-97		
gasoline		
4 cyl.	3.2	3.5
6 cyl.	3.4	3.7
diesel	3.3	3.6
1998-02		
gasoline		
4 cyl.	3.4	3.7
6 cyl.	4.7	5.0
diesel	3.3	3.6
Valve Stem Oil Seal, Replace (B)		
Beetle		
1.8L	7.7	8.1
2.0L	4.4	4.8
diesel	5.0	5.4
Cabrio	3.2	3.6
Cabriolet	4.9	5.3
Fox	4.1	4.5
Golf, Jetta		
1981-84		
gasoline	3.2	3.6
diesel	3.7	4.1
1985-92		
8V	3.1	3.5
16V	4.8	5.2
diesel	3.6	4.0
1993-97		
4 cyl.	3.3	3.7
diesel	5.2	5.6
1998-04		
1.8L	8.5	8.9
6 cyl.	9.2	9.6
diesel	5.0	5.4
Passat		
1990-97	4.8	5.2
1998-04		
4 cyl.	8.5	8.9
diesel	5.2	5.6
Quantum		
4 cyl.	5.2	5.6
5 cyl.	6.4	6.8
diesel	3.7	4.1
Rabbit, Scirocco		
1981-84		
gasoline	3.1	3.5
diesel	3.5	3.9
1985-89		
8V	3.0	3.4
16V	4.8	5.2

	LABOR TIME	SEVERE SERVICE
CAMSHAFT		
Camshaft, Replace (A)		
Gasoline		
1981-87	2.7	3.0
1988-97		
4 cyl.	2.5	2.8
6 cyl.		
one	4.2	4.5
both	4.8	5.1
1998-04		
Beetle		
1.8L		
one	3.8	4.1
both	5.0	5.3
2.0L	3.3	3.6
Cabrio, Jetta		
4 cyl.	2.5	2.8
6 cyl.		
one	4.2	4.5
both	4.8	5.1
Golf		
1.8L		
one	3.8	4.1
both	4.4	4.7
2.0L	3.3	3.6
6 cyl.	5.6	5.9
Passat	3.1	3.4
4 cyl.		
one	3.8	4.1
both	5.0	5.3
6 cyl.		
one	3.3	3.6
both	4.7	5.0
8 cyl. one	2.9	3.2
Diesel	3.3	3.6
Camshaft Seal, Replace (B)		
Gasoline		
1994-97	1.8	2.0
1998-04		
Beetle		
1.8L		
one	2.0	2.2
both	2.4	2.6
2.0L		
one	1.7	1.9
both	3.8	4.0
Cabrio, Jetta	1.8	2.0
Golf		
1.8L	2.0	2.2
2.0L	1.7	1.9
6 cyl.	3.3	3.5
Passat	3.1	3.4
4 cyl.	2.4	2.6
6 cyl. each	4.4	4.6
Diesel		
1981-89	2.1	2.3
1990-04	3.0	3.2

QUANTUM : RABBIT : RABBIT CONVERTIBLE : RABBIT PICKUP : SCIROCCO VW-13

	LABOR TIME	SEVERE SERVICE
Camshaft Sprocket, Replace (B)		
8V	1.8	2.0
16V	1.8	2.0
Intermediate Shaft Sprocket, Replace (B)		
Gasoline		
1981-87	2.8	3.0
1988-02	2.7	2.9
Diesel	3.0	3.2
w/turbo diesel add	.3	.3
Timing Belt, Adjust (B)		
Gasoline		
8V	.7	.8
16V	.8	.9
Diesel		
1981-02	1.1	1.2
Timing Belt, Replace (B)		
Beetle		
1.8L	3.3	3.8
2.0L	2.9	3.4
diesel	5.0	5.5
Cabrio	2.6	3.1
Cabriolet	2.4	2.9
Dasher	1.6	2.1
Fox	1.5	2.0
Golf, Jetta		
1981-84		
gasoline	2.0	2.5
diesel	2.6	3.1
1985-92		
8V	2.8	3.3
16V	3.0	3.5
diesel	2.8	3.3
1993-97		
gasoline	2.5	3.0
diesel	2.8	3.3
1998-04		
1.8L	3.3	3.8
diesel	5.0	5.5
Passat		
1990-97		
gasoline	2.5	3.0
diesel	2.8	3.3
1998-04		
4 cyl.	3.8	4.3
6 cyl.	5.0	5.5
diesel	4.3	4.8
Quantum		
4 cyl.	1.7	2.2
5 cyl.	1.9	2.4
diesel	2.7	3.2
Rabbit, Scirocco		
1981-84		
gasoline	2.1	2.6
diesel	2.7	3.2
1985-89		
8V	2.4	2.9
16V	2.8	3.3
w/turbo diesel add	.9	.9
Replace water pump add	.3	.5

	LABOR TIME	SEVERE SERVICE
Timing Belt Tensioner, Replace (B)		
4 cyl.		
exc. below	.8	1.2
1998-04 Golf 1.8L	3.5	3.9
6 cyl.	1.7	2.1
Diesel	5.1	5.5
Timing Belt Tensioner Pulley, Replace (B)		
Gasoline		
1981-87	1.7	1.9
1988-97		
4 cyl.	1.4	1.6
5 cyl.	1.8	2.0
6 cyl.	1.7	1.9
1998-04		
Beetle		
1.8L	3.5	3.7
2.0L	1.7	1.9
Cabrio, Jetta		
4 cyl.	1.4	1.6
6 cyl.	1.7	1.9
Golf	3.5	3.7
Passat		
4 cyl.	4.0	4.2
6 cyl.	5.1	5.3
Diesel		
1981-97	1.8	2.0
1998-04	5.1	5.3
w/turbo diesel 81-97 add	.3	.3
Timing Chain, Replace (B)		
1988-02 16V	2.4	2.9
1990-97 Corrado, Passat 6 cyl.		
front		
AT	7.5	8.0
MT	8.9	9.4
rear		
AT	7.7	8.2
MT	9.3	9.8
1992-02 Golf, Jetta 6 cyl.		
front	7.5	8.0
rear	7.7	8.2
Timing Chain Tensioner, Replace (B)		
1998-04 Golf 6 cyl.	1.4	1.9
Timing Cover and/or Gasket, Replace (B)		
Gasoline		
1981-87	.7	1.0
1988-97		
4 cyl.		
upper	.3	.5
lower	1.7	2.0
6 cyl.	2.1	2.4
1998-04 4 cyl.		
upper	.3	.5
all	1.1	1.3
Diesel	.7	1.0

	LABOR TIME	SEVERE SERVICE
CRANK & PISTONS		
Connecting Rod Bearings, Replace (A)		
Includes: Check bearing oil clearance.		
Gasoline		
1981-84	3.3	3.8
1985-02	3.0	3.5
Diesel	3.1	3.6
Crankshaft and Main Bearings, Replace (A)		
Includes: Engine R&R.		
Beetle		
1.8L	15.7	16.6
2.0L	14.4	15.3
diesel	16.3	17.2
Cabrio	9.8	10.7
Cabriolet	9.3	10.2
Corrado		
4 cyl.	7.3	8.2
6 cyl.	14.9	15.8
Dasher	7.3	8.2
Fox	8.7	9.6
Golf, Jetta		
1981-84		
gasoline	10.0	10.9
diesel	11.2	12.1
1985-97		
4 cyl.		
AT	10.0	10.9
MT	9.8	10.7
6 cyl.	14.3	15.2
diesel	10.8	11.7
1998-04		
1.8L	15.0	15.9
2.0L	14.4	15.3
diesel	15.7	16.6
Passat		
1990-97		
4 cyl.	10.2	11.1
6 cyl.	15.2	16.1
1998-04		
4 cyl.	15.0	15.9
6 cyl.	15.7	16.6
8 cyl.	19.2	20.1
diesel	11.7	12.6
Quantum		
4 cyl.	9.7	10.6
5 cyl.	14.8	15.7
diesel	11.3	12.2
Rabbit, Scirocco		
1981-84		
gasoline	10.0	10.9
diesel	11.2	12.1
1985-89	9.2	10.1
w/AC add	1.4	1.4
w/AT add	.3	.3
w/turbo diesel add	1.0	1.0

VW-14 — BEETLE : CABRIO : CABRIOLET : CORRADO : DASHER : FOX : GOLF : JETTA : PASSAT

	LABOR TIME	SEVERE SERVICE
Crankshaft Front Oil Seal, Replace (B)		
Beetle		
1.8L	2.4	2.6
2.0L	3.3	3.5
diesel	3.8	4.0
Cabrio	2.7	2.9
Cabriolet	2.9	3.1
Corrado	2.9	3.1
Dasher	2.0	2.2
Fox	2.5	2.7
Golf, Jetta		
1981-84	2.2	2.4
1985-92	2.7	2.9
1993-97		
gasoline	3.0	3.2
diesel	3.2	3.4
1998-04		
1.8L	2.5	2.7
2.0L	3.5	3.7
6 cyl.	2.5	2.7
diesel	4.0	4.2
Passat		
1990-97	2.9	3.1
1998-97		
4 cyl.	2.7	2.9
6 cyl.	4.2	4.4
1998-04	3.8	4.0
Quantum		
4 cyl.	1.9	2.1
5 cyl	3.6	3.8
diesel	2.2	2.4
Rabbit, Scirocco		
1981-84		
gasoline	2.5	2.7
diesel	2.2	2.4
1985-89	2.7	2.9
w/turbo diesel add	1.7	1.7
Crankshaft Pulley, Replace (B)		
Gasoline		
1981-87	.5	.7
1988-02		
exc. below	1.4	1.6
1998-04 Passat		
8 cyl. w/AC	2.9	3.1
Diesel	1.3	1.5
Pistons or Connecting Rods, Replace (A)		
Includes: Ridge reaming, cylinder wall deglazing, installing new rings and rod bearings, engine tune-up.		
Beetle		
1.8L	7.7	8.6
2.0L	8.5	9.4
diesel	7.1	8.0
Cabrio	7.8	8.7
Cabriolet	9.1	10.0

	LABOR TIME	SEVERE SERVICE
Corrado		
4 cyl.	9.0	9.9
6 cyl.	13.6	14.5
Dasher	7.8	8.7
Fox	9.0	9.9
Golf, Jetta		
1981-84		
gasoline	10.1	11.0
diesel	6.4	7.3
1985-97		
4 cyl.	10.0	10.9
6 cyl.	14.4	15.3
diesel	7.4	8.3
1998-04		
1.8L	7.7	8.6
2.0L	8.5	9.4
diesel		
connecting rod	8.5	9.4
piston	7.1	8.0
Passat		
1990-97		
4 cyl.	7.9	8.8
6 cyl.	14.2	15.1
1998-04		
4 cyl. w/engine removed	16.3	17.2
6 cyl.	19.6	20.5
Quantum		
4 cyl.	8.6	9.5
5 cyl.	11.7	12.2
diesel	9.7	10.6
Rabbit, Scirocco		
8V	9.9	10.8
16V	10.0	10.9
diesel	9.8	10.7
Rear Main Oil Seal, Replace (B)		
Includes: Trans. R&R when necessary.		
Beetle		
1.8L	12.3	12.9
2.0L		
AT	7.2	7.8
MT	6.6	7.2
diesel	12.3	12.9
Cabrio		
AT	6.0	6.6
MT	6.8	7.4
Cabriolet		
AT	6.0	6.6
MT	6.2	6.8
Corrado		
4 cyl.	9.1	9.7
6 cyl.	9.7	10.3
Dasher	6.4	7.0
Fox	6.2	6.8
Golf, Jetta		
1981-84		
AT	6.4	7.0
MT	5.3	5.9

	LABOR TIME	SEVERE SERVICE
1985-92		
AT	7.5	8.1
MT	6.9	7.5
1993-97		
4 cyl.		
AT	6.5	7.1
MT	6.2	6.8
6 cyl.	8.4	9.0
diesel	7.5	8.1
1998-04		
AT	8.2	8.8
MT	7.6	8.2
Passat		
1990-97		
4 cyl.		
2WD	6.9	7.5
4WD	9.1	9.7
6 cyl.		
2WD	8.3	8.9
4WD	9.3	9.9
diesel	7.6	8.2
1998-04		
2WD		
AT		
4 cyl.	15.2	15.8
6 cyl.	12.1	12.7
8 cyl.	11.5	12.1
MT		
4 cyl.	10.8	11.4
6 cyl.	8.2	8.8
8 cyl.	6.9	7.5
4WD		
AT		
4 cyl.	16.1	16.7
6 cyl.	12.9	13.5
8 cyl.	12.3	12.9
MT		
4 cyl.	13.7	14.3
6 cyl.	11.2	11.8
8 cyl.	9.9	10.5
diesel	7.2	7.8
Quantum		
gasoline		
AT.	6.3	6.9
MT		
2WD	5.5	6.1
4WD	8.6	9.2
diesel		
AT	6.2	6.8
MT	5.5	6.1
Rabbit, Scirocco		
1981-84		
AT	6.3	6.9
MT	5.3	5.9
1985-89		
AT	6.4	7.0
MT	6.2	6.8

QUANTUM : RABBIT : RABBIT CONVERTIBLE : RABBIT PICKUP : SCIROCCO

VW-15

	LABOR TIME	SEVERE SERVICE
Rings, Replace (A)		
Includes: Ridge reaming, cylinder wall deglazing, installing new rings, engine tune-up.		
Beetle		
1.8L	8.0	8.9
2.0L	9.2	10.1
diesel	7.4	8.3
Cabrio	8.8	9.7
Cabriolet	8.2	9.1
Corrado		
4 cyl.	8.1	9.0
6 cyl.	10.8	11.7
Dasher	6.9	7.8
Fox	8.1	9.0
Golf, Jetta		
1981-84		
gasoline	10.4	11.3
diesel	6.8	7.7
1985-97		
4 cyl.	10.4	11.3
6 cyl.	14.8	15.7
diesel	7.8	8.7
1998-04		
4 cyl.	9.2	10.1
diesel	7.8	8.7
Passat		
1990-97		
4 cyl.	9.1	10.0
6 cyl.	15.5	16.4
diesel	16.3	17.2
1998-04		
4 cyl. w/engine removed	17.0	17.9
6 cyl.	14.1	15.0
diesel	7.9	8.8
Quantum		
4 cyl.	7.3	8.2
5 cyl.	10.2	11.1
diesel	8.7	9.6
Rabbit, Scirocco		
gasoline	9.3	10.2
diesel	8.5	9.4
Recondition cylinder head add		
4 cyl.		
8V	3.6	3.9
16V	5.3	5.6
6 cyl.	7.7	8.0
Vibration Damper, Replace (B)		
Beetle		
gasoline	1.1	1.3
diesel	1.7	1.9
Golf, Jetta		
1985-92	.3	.5
1998-04		
gasoline	1.1	1.3
diesel	1.4	1.6

	LABOR TIME	SEVERE SERVICE
Passat		
1990-97	.3	.5
1998-04		
4 cyl.	2.4	2.6
diesel	3.3	3.5
ENGINE LUBRICATION		
Engine Oil Cooler, Replace (B)		
Beetle	1.1	1.3
Cabrio	.8	1.0
Golf, Jetta		
1994-97		
4 cyl.	.8	1.0
6 cyl.	2.4	2.6
1998-04		
4 cyl.	1.1	1.3
6 cyl.	1.7	1.9
diesel	1.1	1.3
Passat		
1994-97		
4 cyl.	.8	1.0
6 cyl.	2.4	2.6
1998-04		
4 cyl., 6 cyl.	1.1	1.3
8 cyl.	1.7	1.9
Engine Oil Pressure Switch (Sending Unit), Replace (B)		
Gasoline		
1981-87	.5	.5
1988-04	.8	.8
Diesel	.7	.7
Oil Pan and/or Gasket, Replace (B)		
Beetle	2.1	2.3
Cabrio	1.9	2.1
Cabriolet	1.6	1.8
Corrado		
4 cyl.	1.7	1.9
6 cyl.	1.8	2.0
Dasher	2.3	2.5
Fox	1.8	2.0
Golf, Jetta	2.1	2.3
Passat		
1990-97		
4 cyl.	1.6	1.8
6 cyl.	2.3	2.5
1998-04		
4 cyl.		
w/AC	4.6	4.8
w/o AC	4.0	4.2
6 cyl.		
upper	4.4	4.6
lower	2.4	2.6
Quantum		
4 cyl	2.4	2.6
5 cyl.	2.7	2.9
diesel	2.4	2.6
Rabbit, Scirocco	1.6	1.8
w/turbo diesel add	.7	.7

	LABOR TIME	SEVERE SERVICE
Oil Pump, Replace (B)		
Beetle		
2.0L	4.6	5.1
diesel	2.5	3.0
Cabrio	2.1	2.6
Cabriolet	1.9	2.4
Corrado	1.8	2.3
Dasher	2.4	2.9
Fox	2.0	2.5
Golf, Jetta		
1981-84	1.9	2.4
1985-92	2.1	2.6
1993-97		
gasoline		
4 cyl.	2.1	2.6
6 cyl.	2.7	3.2
diesel	2.0	2.5
1998-04		
1.8L	2.5	3.0
2.0L	4.6	5.1
diesel	2.5	3.0
Passat		
1990-97		
4 cyl.	1.7	2.2
6 cyl.	2.7	3.2
1998-04		
4 cyl.		
w/AC	5.2	5.7
w/o AC	4.0	4.5
6 cyl.	8.8	9.3
Quantum		
4 cyl.	3.2	3.7
5 cyl.	4.6	5.1
diesel	2.6	3.1
Rabbit, Scirocco	2.1	2.6
w/turbo diesel add	.7	.7

CLUTCH

	LABOR TIME	SEVERE SERVICE
Bleed Clutch Hydraulic System (B)		
1988-04	.5	.5
Clutch Pedal Free Play, Adjust (B)		
1981-02	.3	.5
Clutch Control Cable, Replace (B)		
1981-02	.8	1.0
Clutch Assy., Replace (B)		
Beetle	5.2	5.4
Cabrio	6.6	6.8
Cabriolet	4.8	5.0
Corrado	7.8	8.0
Dasher	5.0	5.2
Fox	4.5	4.7
Golf, Jetta		
1981-84	5.2	5.4
1985-92	4.1	4.3
1993-97		
gasoline		
4 cyl.	6.6	6.8
6 cyl.	7.7	7.9
diesel	7.1	7.3

VW-16 BEETLE : CABRIO : CABRIOLET : CORRADO : DASHER : FOX : GOLF : JETTA : PASSAT

	LABOR TIME	SEVERE SERVICE
Clutch Assy., Replace (B)		
1998-04		
gasoline		
4 cyl.	5.2	5.4
6 cyl.	7.2	7.4
diesel	5.2	5.4
Passat		
1990-97		
4 cyl.		
2WD	6.4	6.6
4WD	8.7	8.9
6 cyl.		
2WD	7.8	8.0
4WD	8.9	9.1
diesel	7.3	7.5
1998-04		
4 cyl.		
2WD	6.4	6.6
4WD	9.4	9.6
6 cyl.		
2WD	7.2	7.4
4WD	10.2	10.4
diesel	7.3	7.5
Quantum		
4-Speed		
2WD	5.7	5.9
4WD	8.0	8.2
5-Speed	5.5	5.7
Rabbit, Scirocco	4.7	4.9
w/turbo diesel add	.7	.7
Clutch Master Cylinder, Replace (B)		
Includes: System bleeding.		
1990-97	2.3	2.5
1998-04		
exc. Passat	1.8	2.0
Passat	3.3	3.5
Clutch Release Bearing, Replace (B)		
Beetle	4.7	4.9
Cabrio	6.5	6.7
Cabriolet	4.7	4.9
Corrado	7.6	7.8
Dasher	4.9	5.1
Fox	4.4	4.6
Golf, Jetta		
1981-84	5.1	5.3
1985-92	4.0	4.2
1993-97		
gasoline		
4 cyl.	6.5	6.7
6 cyl.	7.6	7.8
diesel	7.0	7.2
1998-04		
gasoline		
4 cyl.	4.7	4.9
6 cyl.	6.7	6.9
diesel	4.7	4.9

	LABOR TIME	SEVERE SERVICE
Passat		
1990-97		
4 cyl.		
2WD	6.3	6.5
4WD	8.6	8.8
6 cyl.		
2WD	7.6	7.8
4WD	8.8	9.0
diesel	7.2	7.4
1998-04		
4 cyl.		
2WD	6.0	6.2
4WD	9.0	9.2
6 cyl.		
2WD	6.7	6.9
4WD	9.7	9.9
diesel	7.2	7.4
Quantum		
4-Speed		
2WD	5.6	5.8
4WD	7.9	8.1
5-Speed	5.4	5.6
Rabbit, Scirocco	4.5	2.7
w/turbo diesel add	.7	.7
Clutch Self-Adjusting Cable, Replace (B)		
1988-93	.8	1.0
Clutch Slave Cylinder, Replace (B)		
1981-97		
exc. Corrado	.8	1.0
Corrado	1.3	1.5
1998-04	1.5	1.7
Flywheel, Replace (B)		
Beetle		
gasoline	5.4	5.6
diesel	5.0	5.2
Cabrio	7.6	7.8
Cabriolet	5.3	5.5
Corrado	8.5	8.7
Dasher	5.4	5.6
Fox	4.8	5.0
Golf, Jetta		
1981-84	5.6	5.8
1985-92	4.4	4.6
1993-97		
gasoline		
4 cyl.	7.7	7.9
6 cyl.	8.0	8.2
diesel	6.0	6.2
1998-04		
gasoline		
4 cyl.	5.7	5.9
6 cyl.	8.2	8.4
diesel	5.7	5.9
Passat		
1990-97		
4 cyl.		
2WD	7.5	7.7
4WD	9.6	9.8

	LABOR TIME	SEVERE SERVICE
6 cyl.		
2WD	8.9	9.1
4WD	9.8	10.0
diesel	8.2	8.4
1998-04		
4 cyl.		
2WD	6.9	7.1
4WD	9.9	10.1
6 cyl.		
2WD	7.3	7.5
4WD	10.3	10.5
diesel	8.2	8.4
Quantum		
4-Speed		
2WD	6.0	6.2
4WD	8.3	8.5
5-Speed	5.9	6.1
Rabbit, Scirocco	5.0	5.2
w/turbo diesel add	.7	.7

MANUAL TRANSAXLE

	LABOR TIME	SEVERE SERVICE
Transaxle Assy., R&I (B)		
Beetle	4.6	4.8
Cabrio	6.3	6.5
Cabriolet	4.5	4.7
Corrado	7.5	7.7
Dasher	4.6	4.8
Fox	4.2	4.4
Golf, Jetta		
1981-84	4.8	5.0
1985-92	3.7	3.9
1993-97		
gasoline		
4 cyl.	6.3	6.5
6 cyl.	7.4	7.6
diesel	6.7	6.9
1998-04		
gasoline		
4 cyl.	6.4	6.6
6 cyl.	8.4	8.6
diesel	6.4	6.6
Passat		
1990-97		
4 cyl.		
2WD	6.0	6.2
4WD	8.4	8.6
6 cyl.		
2WD	7.5	7.7
4WD	8.5	8.7
diesel	7.0	7.2
1998-04		
4 cyl.		
2WD	6.0	6.2
4WD	8.8	9.0
6 cyl.		
2WD	6.6	6.8
4WD	9.5	9.7
diesel	7.0	7.2

VW-17
QUANTUM : RABBIT : RABBIT CONVERTIBLE : RABBIT PICKUP : SCIROCCO

	LABOR TIME	SEVERE SERVICE
Quantum		
4-Speed		
2WD	5.4	5.6
4WD	7.6	7.8
5-Speed	5.1	5.3
Rabbit, Scirocco	4.4	4.6
w/turbo diesel add	.7	.7
Transaxle Assy. R&R and Recondition (A)		
Beetle	9.5	10.1
Cabrio, Cabriolet	10.1	10.7
Corrado	12.9	13.5
Dasher, Fox	9.6	10.1
Golf, Jetta		
1981-84	8.5	9.1
1985-92	9.5	10.1
1993-97		
gasoline		
4 cyl.	12.7	13.3
6 cyl.	14.2	14.8
diesel	13.0	13.6
1998-04		
gasoline		
4 cyl.	9.5	10.1
6 cyl.	9.9	10.5
diesel	9.5	10.1
Passat		
1990-97		
4 cyl.		
2WD	12.7	13.3
4WD	14.9	15.5
6 cyl.		
2WD	14.0	14.6
4WD	15.0	15.6
diesel	13.1	13.7
1998-04		
4 cyl.		
2WD	10.8	11.4
4WD	14.9	15.5
6 cyl.		
2WD	11.5	12.1
4WD	15.6	16.2
diesel	13.1	13.7
Quantum		
4-Speed		
2WD	9.8	10.4
4WD	14.4	15.0
5-Speed	11.1	11.7
Rabbit, Scirocco		
1981-84	8.4	9.0
1985-89	10.1	10.7
w/turbo diesel add	.7	.7
Transaxle Mounts, Replace (B)		
Exc. below	.8	1.0
Golf, Jetta		
1981-84		
left	.3	.5
center	.7	.9
1993-02	1.9	2.1

	LABOR TIME	SEVERE SERVICE
Quantum, rear	.5	.7
Rabbit, Scirocco		
left	.3	.5
center	.7	.9

AUTOMATIC TRANSAXLE
SERVICE TRANSAXLE INSTALLED

	LABOR TIME	SEVERE SERVICE
Check Unit for Oil Leaks (C)		
1981-02	.9	.9
Drain & Refill Unit (B)		
1981-97	1.1	1.1
1998-04		
exc. Cabrio, Jetta	1.1	1.1
Cabrio, Jetta	1.3	1.3
Electro Magnet, Replace (B)		
1981-87	2.3	2.5
Electronic Control Unit, Replace (B)		
1990-04	.9	.9
Governor Assy., R&R or Replace (B)		
1981-93	.5	.8
Kickdown Switch or Solenoid, Replace (B)		
1981-89	.9	1.1
Oil Pan and/or Gasket, Replace (B)		
Exc. below	1.1	1.1
1998-04 Passat		
gasoline	3.5	3.5
Shift Lock Relay, Replace (B)		
1985-02	.3	.3
Shift Lock Solenoid Switch, Replace (B)		
1990-02	1.5	1.7
Speed Sensor, Replace (B)		
1990-04	.8	.8
Transaxle Mounts, Replace (B)		
Beetle	.8	1.0
Cabrio		
left	1.8	2.0
rear	1.9	2.1
Cabriolet		
left	.9	1.1
rear	.8	1.0
Dasher	.8	1.0
Golf, Jetta		
1981-84		
left	.3	.5
center	.7	.7
1985-02	1.9	2.1
Passat		
1990-97	1.8	2.0
1998-02	.8	1.0
Quantum	1.7	1.9
Rabbit, Scirocco		
1981-84		
left	.3	.5
center	.7	.9
1985-89	.9	1.1

	LABOR TIME	SEVERE SERVICE
Vacuum Modulator, Replace (B)		
1981-89	.7	.7
Valve Body, R&R and Recondition (B)		
3-Speed	2.8	3.3
4-Speed	3.0	3.5
Valve Body, Replace (B)		
3-Speed	1.5	1.8
4-Speed	2.1	2.4
5-Speed	4.6	4.9

SERVICE TRANSAXLE REMOVED
Transaxle R&R included unless otherwise noted.

	LABOR TIME	SEVERE SERVICE
Torque Converter, Replace (B)		
Beetle	5.3	5.5
Cabrio	5.0	5.2
Cabriolet	5.6	5.8
Corrado		
4 cyl.	5.6	5.8
6 cyl.	6.5	6.7
Dasher	5.5	5.7
Golf, Jetta		
1981-84	5.6	5.8
1985-92	6.2	6.4
1993-97		
4 cyl.	5.8	6.0
6 cyl.	7.2	7.4
1998-04 4-Speed	5.3	5.5
Passat		
2WD		
1990-97		
4 cyl.	5.4	5.6
6 cyl.	6.0	6.2
1998-04 5-Speed	10.6	10.8
4WD		
1990-97	13.4	13.6
1998-04 5-Speed	11.5	11.7
Quantum		
1982-84	5.5	5.7
1985-88	6.9	7.1
Rabbit, Scirocco	5.7	5.9
Transaxle Assy., R&R and Recondition (A)		
Beetle	10.7	11.6
Cabrio	17.0	17.9
Cabriolet	9.8	10.7
Corrado		
4 cyl.	13.9	14.8
6 cyl.	15.0	15.9
Dasher	9.8	10.7
Golf, Jetta		
1985-92	14.0	14.9
1993-02		
4 cyl.	17.0	17.9
6 cyl.	18.3	19.2

VW-18   BEETLE : CABRIO : CABRIOLET : CORRADO : DASHER : FOX : GOLF : JETTA : PASSAT

	LABOR TIME	SEVERE SERVICE
Transaxle Assy., R&R and Reconditioning (A)		
Passat		
1990-97		
4 cyl.	16.2	*17.1*
6 cyl.	18.2	*19.1*
1998-02	14.9	*15.8*
Quantum	15.3	*16.2*
Rabbit, Scirocco	9.7	*10.6*
Transaxle Assy., R&I (B)		
Beetle	5.2	*5.4*
Cabrio, Cabriolet	5.3	*5.5*
Corrado		
4 cyl.	5.3	*5.5*
6 cyl.	6.2	*6.4*
Dasher	5.2	*5.4*
Golf, Jetta		
1981-84	5.3	*5.5*
1985-92	5.9	*6.1*
1993-97		
4 cyl.	5.4	*5.6*
6 cyl.	6.9	*7.1*
1998-04 4-Speed	5.2	*5.4*
Passat		
2WD		
1990-97		
4 cyl.	4.8	*5.0*
6 cyl.	6.8	*7.0*
1998-04	10.6	*10.8*
4Motion		
1990-97	13.0	*13.2*
1998-04	11.3	*11.5*
Quantum		
1982-84	5.2	*5.4*
1985-88	6.5	*6.7*
Rabbit, Scirocco	5.4	*5.6*
Replace flexplate add	.4	*.4*

SHIFT LINKAGE
AUTOMATIC TRANSAXLE
Selector Cable, Adjust (B)
1981-04	.5	*.7*

Gear Selector Lever, Replace (B)
Beetle	1.2	*1.2*
Cabrio	1.2	*1.2*
Cabriolet	1.1	*1.1*
Dasher	1.3	*1.3*
Golf, Jetta	1.2	*1.2*
Passat		
1990-97	2.1	*2.1*
1998-04	1.7	*1.7*
Quantum	1.5	*1.5*
Rabbit, Scirocco	1.1	*1.1*

Selector Cable, Replace (B)
Beetle		
convertible	3.2	*3.4*
hard top	2.5	*2.7*
Cabrio, Cabriolet	1.8	*2.0*
Corrado	2.1	*2.3*
Dasher	2.3	*2.5*
Golf, Jetta	2.2	*2.4*
Passat		
1990-97	2.1	*2.3*
1998-02	1.8	*2.0*
Quantum	1.6	*1.8*
Rabbit, Scirocco	2.1	*2.3*

MANUAL TRANSAXLE
Gearshift Cable/Linkage, Adjust (B)
4-Speed	.8	*1.0*
5-Speed	1.0	*1.2*
4WD	.9	*1.1*

Gearshift Lever, Replace (B)
Beetle	2.3	*2.3*
Cabriolet	1.2	*1.2*
Corrado	1.7	*1.7*
Dasher, Fox	1.3	*1.3*
Golf, Jetta		
1981-84	.7	*.7*
1985-02	2.5	*2.5*
Passat		
1990-97	1.8	*1.8*
1998-02	1.3	*1.3*
Quantum	1.4	*1.4*
Rabbit, Scirocco		
1981-84	.7	*.7*
1985-89	1.2	*1.2*

Gearshift Cable/Rod, Replace (B)
Beetle	2.1	*2.1*
Cabrio	1.6	*1.6*
Cabriolet	1.7	*1.7*
Corrado	1.6	*1.6*
Dasher	1.3	*1.3*
Fox	1.4	*1.4*
Golf, Jetta		
1981-84	1.4	*1.4*
1985-92	2.4	*2.4*
1993-97	1.7	*1.7*
1998-04 4 cyl.	3.0	*3.0*
Passat		
1990-97	1.6	*1.6*
1998-04 5-Speed	1.7	*1.7*
Quantum	1.4	*1.4*
Rabbit, Scirocco		
1981-84	1.5	*1.5*
1985-89	1.6	*1.6*

DRIVELINE
Differential Assy., R&R or Replace (B)
Rear axle		
Passat 4WD	2.8	*3.3*
Passat 4Motion	2.1	*2.6*
Quantum 4WD	2.6	*3.1*

Differential Bearings, Replace (B)
Includes: R&R differential. Adjust ring gear and pinion and bearing preload.
Front axle		
Cabrio	9.2	*9.5*
Cabriolet	7.8	*8.1*
Corrado		
4 cyl.	7.6	*7.9*
6 cyl.	8.8	*9.1*
Dasher		
AT	7.7	*8.0*
MT	6.4	*6.7*
Fox	5.3	*5.6*
Golf, Jetta		
1981-84	6.2	*6.5*
1985-92		
AT	8.3	*8.6*
MT	5.2	*5.5*
1993-02		
4 cyl.	9.3	*9.6*
6 cyl.	9.1	*9.4*
Passat		
1990-97		
2WD	9.1	*9.4*
4WD	10.9	*11.2*
1998-02		
4 cyl.	9.2	*9.5*
6 cyl.		
2WD	10.2	*10.5*
4WD	13.5	*13.8*
diesel	9.8	*10.1*
Quantum		
AT	7.9	*8.2*
MT		
2WD	6.1	*6.4*
4WD	8.9	*9.2*
Rabbit, Scirocco		
1981-84	6.3	*6.6*
1985-89		
AT	7.8	*8.1*
MT	6.3	*6.6*

Differential Case, Replace (B)
1981-84	11.6	*12.1*

Differential Lock Valve/Switch, Replace (B)
Passat	1.3	*1.5*
Quantum	1.4	*1.6*

Differential Oil Seal, Replace (B)
Rear axle		
Passat 4WD	.7	*.9*
Quantum 4WD	.7	*.9*

Differential Side Gear Set, Replace (B)
Front axle		
Cabrio	8.3	*8.6*
Cabriolet	6.2	*6.5*
Corrado	7.9	*8.2*

QUANTUM : RABBIT : RABBIT CONVERTIBLE : RABBIT PICKUP : SCIROCCO — VW-19

	LABOR TIME	SEVERE SERVICE
Golf, Jetta		
1985-92	5.3	5.6
1993-02		
gasoline		
4 cyl.	8.3	8.6
6 cyl.	9.5	9.8
diesel	8.9	9.2
Passat		
1990-97		
2WD	7.8	8.1
4WD	9.2	9.5
1998-02		
4 cyl.	8.1	8.4
6 cyl.		
2WD	9.5	9.8
4WD	12.7	13.0
diesel	8.7	9.0
1985-89 Scirocco	6.1	6.4

Differential Vacuum Lock Activators, Replace (B)

Passat	2.8	3.2
Quantum		
front	1.5	1.9
rear	.5	.9

Pinion Bearings, Replace (B)

Front axle		
Dasher	10.0	10.5
1981-84 Jetta	10.0	10.5
Quantum		
AT	11.4	11.9
MT	10.0	10.5
Rabbit, Scirocco	10.0	10.5

Pinion Shaft Oil Seal, Replace (B)

Front axle		
Dasher	6.7	6.9
1981-84 Jetta, Rabbit, Scirocco	5.3	5.5

Rear Axle Cover Gasket, Replace (B)

1985-02	3.9	3.9

Ring Gear & Pinion Set, Replace (B)
Includes: R&R axle shafts, drain and refill unit. Road test.

Front axle		
Cabrio		
AT	12.3	12.8
MT	12.0	12.5
Cabriolet		
AT	11.9	12.4
MT	12.4	12.9
Corrado		
AT	11.7	12.2
MT	14.3	14.8
Dasher	12.4	12.9
Fox	12.0	12.5

	LABOR TIME	SEVERE SERVICE
Golf, Jetta		
1981-84		
AT	9.3	9.8
MT	10.5	11.0
1985-92		
AT	10.8	11.3
MT	11.3	11.8
1993-02		
AT	12.3	12.8
MT		
gasoline		
4 cyl.	12.4	12.9
6 cyl.	13.4	13.9
diesel	12.9	13.4
Passat		
1990-97		
AT		
4 cyl.	11.7	12.2
6 cyl.	13.6	14.1
MT		
2WD	12.4	12.9
4WD	14.1	14.6
1998-02		
AT		
4 cyl.	14.2	14.7
6 cyl.	17.4	17.9
MT		
4 cyl.	11.9	12.4
6 cyl.		
2WD	13.6	14.1
4WD	16.6	17.1
diesel	12.8	13.3
Quantum		
AT	11.9	12.4
MT		
2WD	11.4	11.9
4WD	14.0	14.5
Rabbit, Scirocco		
1981-84		
AT	9.2	9.7
MT	10.5	11.0
1985-89		
AT	11.9	12.4
MT	12.4	12.9

Stub Axle, Replace (B)

1981-97		
one	1.6	2.1
both	2.7	3.2
1998-04		
Beetle, Golf		
disc brakes		
one side	1.4	1.9
both sides	2.4	2.9
drum brakes		
one side	1.1	1.6
both sides	1.7	2.2
Cabrio, Jetta		
one side	1.6	2.1
both sides	2.5	3.0
Passat one side	1.1	1.6

	LABOR TIME	SEVERE SERVICE
HALFSHAFTS		
Front Halfshaft, R&R or Replace (B)		
Beetle		
one		
AT	1.7	2.0
MT	1.4	1.7
both	2.1	2.4
Cabrio	1.8	2.1
Cabriolet	1.1	1.4
Corrado	1.2	1.5
Fox	1.2	1.5
Golf, Jetta		
1981-84	.7	1.0
1985-97	1.7	2.0
1998-04		
one		
AT	1.7	2.0
MT	1.4	1.7
both	2.1	2.4
Passat		
1990-97	1.3	1.6
1998-04		
one	1.4	1.7
both	2.1	2.4
Quantum	1.4	1.7
Rabbit, Scirocco		
1981-84	.7	1.0
1981-89	1.1	1.4
Replace boot add each	.5	.8
Replace or recondition CV joint add		
one	.7	.9
both one side	.9	1.1

Rear Halfshaft, R&R or Replace (B)

Passat 4WD	1.4	1.9
Passat 4Motion		
one	1.1	1.6
both	1.7	2.2
Quantum 4WD	1.3	1.8
Replace CV-joint or boot add each	.5	.8

Rear Halfshaft Oil Seal, Replace (B)

Passat 4WD	1.7	1.9
Passat 4Motion	2.8	3.0
Quantum 4WD	1.6	1.8

BRAKES
ANTI-LOCK

The following operations do not include testing. Add time as required.

Diagnose Anti-Lock Brake System (A)

Beetle	1.5	1.5
Cabrio	1.3	1.3
Corrado	1.9	1.9
Golf, Jetta	1.3	1.3
Passat	1.8	1.8

Anti-Lock Relay, Replace (B)

1988-04	.3	.3

VW-20 BEETLE : CABRIO : CABRIOLET : CORRADO : DASHER : FOX : GOLF : JETTA : PASSAT

	LABOR TIME	SEVERE SERVICE
Control Unit, Replace (B)		
1988-97	.5	.5
Beetle	1.1	1.1
1998-04 Golf	1.1	1.1
1998-04 Jetta, Passat	.5	.5
Front Sensor Assy., Replace (B)		
1988-04 each	.5	.5
Front Toothed Rotor, Replace (B)		
Exc. below one	2.5	2.5
Beetle		
right	2.1	2.1
left	1.4	1.4
both	2.5	2.5
1998-04 Golf		
right	2.1	2.1
left	1.4	1.4
both	2.5	2.5
Hydraulic Assy., Replace (B)		
1990-97	2.7	2.9
Beetle	1.1	1.3
1998-04 Golf, Jetta	1.1	1.3
1998-04 Passat	1.4	1.6
Over Voltage Protection Relay, Replace (B)		
1988-02	.3	.3
Rear Sensor Assy., Replace (B)		
1988-04 each	.8	.8
Rear Toothed Rotor, Replace (B)		
One		
Cabrio	.8	.8
Corrado	1.7	1.7
Golf, Jetta	.8	.8
Passat	1.7	1.7
Return Flow Pump, Replace (B)		
1988-02	1.6	1.6
Solenoid Valve Relay, Replace (B)		
1988-04	.3	.3

SYSTEM

	LABOR TIME	SEVERE SERVICE
Bleed Brakes (B)		
Includes: Add fluid.		
1981-04	.8	.8
Brake System, Flush and Refill (B)		
1981-02	1.2	1.2
Brakes, Adjust (B)		
Includes: Brake adjustment, filling master cylinder.		
1981-02	.5	.5
Brake Hose (Flexible), Replace (B)		
Includes: System bleeding.		
One	1.4	1.5
each addl. add	.3	.4
Master Cylinder, Replace (B)		
Includes: System bleeding.		
Beetle	1.8	2.0
Cabrio	1.7	1.9
Cabriolet, Dasher, Fox	1.5	1.7
Corrado		
w/ABS	2.5	2.7
w/o ABS	1.4	1.6

	LABOR TIME	SEVERE SERVICE
Golf, Jetta		
1981-94	1.7	1.9
1995-97		
w/ABS	2.5	2.7
w/o ABS	1.6	1.8
1998-04	1.8	2.0
Passat		
1995-97	2.7	2.9
1998-04	1.9	2.1
Quantum, Rabbit, Scirocco	1.6	1.8
Power Booster Unit, Replace (B)		
Includes: System bleeding.		
Beetle	2.9	3.1
Cabrio	2.1	2.3
Cabriolet, Dasher, Fox	1.9	2.1
Corrado		
w/ABS	3.0	3.2
w/o ABS	1.8	2.0
Golf, Jetta		
1981-94		
w/ABS	2.2	2.4
w/o ABS	1.9	2.1
1995-97		
w/ABS	3.0	3.2
w/o ABS	2.0	2.2
1998-04	3.0	3.2
Passat		
1995-97	3.1	3.3
1998-04	2.4	2.6
Quantum, Rabbit, Scirocco	2.0	2.2
Power Booster Vacuum Check Valve, Replace (B)		
1981-02	.3	.3
Power Booster Vacuum Pump, Replace (B)		
1981-02	.9	1.1

SERVICE BRAKES

	LABOR TIME	SEVERE SERVICE
Brake Drum, Replace (B)		
Exc. Beetle		
one	.9	1.1
both	1.6	1.8
Beetle		
one	.5	.7
both	.6	.8
Brake Pressure Regulator, Replace (B)		
Cabrio	1.6	1.8
Cabriolet	1.3	1.5
Corrado	1.3	1.5
Fox		
sedan	.8	1.0
wagon	1.3	1.5
Golf, Jetta	1.5	1.7
Passat	1.6	1.8
Quantum	1.3	1.5
Rabbit, Scirocco	1.5	1.7

	LABOR TIME	SEVERE SERVICE
Caliper Assy., R&R and Recondition (B)		
Includes: System bleeding.		
1981-84		
one	1.7	2.0
two	3.2	3.5
1985-97		
front		
one	1.9	2.2
both	3.2	3.5
rear		
one	1.7	2.0
both	3.2	3.5
1998-04 Beetle, Golf		
front		
one	2.7	3.0
both	3.6	3.6
rear one	2.5	2.8
1998 Cabrio, Jetta		
front		
one	2.0	2.3
both	3.2	3.5
rear		
one	1.7	2.0
both	3.2	3.5
1998-04 Passat		
front		
one	2.7	3.0
both	4.0	4.3
rear		
one	2.6	2.9
both	2.8	3.1
w/ABS 88-97 add	.1	.1
Caliper Assy., Replace (B)		
Includes: System bleeding.		
1981-84		
one	1.2	1.4
two	2.0	2.2
1985-97		
front		
one	1.4	1.6
both	2.1	2.3
rear		
one	1.6	1.8
both	2.5	2.7
1998-04 Beetle, Golf		
front		
one	1.9	2.1
both	2.2	2.4
rear		
one	1.9	2.1
both	2.5	2.7
1998 Cabrio, Jetta		
front		
one	1.5	1.7
both	3.2	3.4
rear		
one	1.1	1.3
both	1.6	1.8

QUANTUM : RABBIT : RABBIT CONVERTIBLE : RABBIT PICKUP : SCIROCCO VW-21

	LABOR TIME	SEVERE SERVICE
1998-04 Passat		
one	2.1	2.3
both	2.6	2.8
w/ABS 88-97 add	.1	.1
Brake Disc Rotor, Replace (B)		
Front		
1981-97		
one	1.3	1.5
both	1.6	1.8
1998-04		
one	.8	1.0
both	1.1	1.3
Rear		
1981-97		
one	1.3	1.5
both	1.7	1.9
1998-04		
exc. Cabrio, Jetta		
one	.8	1.0
both	1.1	1.3
Cabrio, Jetta		
one	1.2	1.4
both	1.9	2.1
Pads and/or Shoes, Replace (B)		
Includes: Service and parking brake adjustment, system bleeding.		
Front disc	1.2	1.3
Rear disc	1.3	1.4
Rear drum	1.9	2.0
Four wheels		
disc brakes	2.2	2.3
drum brakes	2.9	3.0
COMBINATION ADD-ONS		
Replace		
brake drum add	.2	.2
brake hose add	.3	.3
brake rotor add	.2	.2
caliper add	.3	.3
wheel cylinder add	.2	.2
Resurface		
brake drum add	.5	.5
brake rotor add	.5	.5
Wheel Cylinder, R&R and Recondition (B)		
Includes: System bleeding.		
One	2.2	2.5
Both	2.8	3.1
Wheel Cylinder, Replace (B)		
Includes: System bleeding.		
1981-97		
one	1.7	1.9
both	2.3	2.5
1998-04		
one	2.1	2.3
both	2.7	2.9

	LABOR TIME	SEVERE SERVICE
PARKING BRAKE		
Parking Brake Apply Actuator, Replace (B)		
Beetle	1.1	1.1
Cabrio, Cabriolet, Corrado	.8	.8
Dasher	1.2	1.2
Fox	1.8	1.8
Golf, Jetta	1.1	1.1
Passat		
1990-97	.8	.8
1998-04	1.9	1.9
Quantum	1.1	1.1
Rabbit, Scirocco	.7	.7
Parking Brake Apply Warning Indicator Switch, Replace (B)		
Exc. below	.5	.5
1998-04 Passat	1.1	1.1
Parking Brake Cable, Replace (B)		
1981-84		
one side	1.6	1.9
both sides	2.5	2.8
1985-97		
one side	1.3	1.6
both sides	2.4	2.7
1998-04		
Beetle one or both sides	1.1	1.4
Cabrio, Jetta		
one side	1.3	1.6
both sides	2.4	2.7
Golf		
one side	1.1	1.4
both sides	1.7	2.0
Passat		
one side	1.7	2.0
both sides	2.1	2.4

FRONT SUSPENSION

Unless otherwise noted, time given does not include alignment.

	LABOR TIME	SEVERE SERVICE
Align Front End (A)		
1981-97	1.6	1.8
1998-04		
exc. Passat	1.9	2.1
Passat	2.2	2.4
Front Toe, Adjust (B)		
1981-97	1.3	1.5
1998-04		
exc. Cabrio, Jetta	1.9	2.1
Cabrio, Jetta	1.2	1.4
Rear Alignment, Adjust (B)		
1981-04	1.7	1.9
Lower Ball Joint, Replace (B)		
One	1.1	1.4
Both	2.1	2.4

	LABOR TIME	SEVERE SERVICE
Lower Control Arm, Replace (B)		
Beetle		
AT		
one	1.4	1.7
both	2.6	2.9
MT		
one	1.2	1.5
both	1.9	2.2
Cabrio		
one	1.8	2.1
both	2.5	2.8
Cabriolet		
AT	1.5	1.8
MT	1.2	1.5
Corrado		
AT		
one	1.8	2.1
both	2.3	2.6
MT		
one	1.3	1.6
both	2.0	2.3
Dasher		
AT		
one	1.5	1.8
both	1.9	2.2
MT		
one	.8	1.1
both	1.7	2.0
Fox		
one	1.5	1.8
both	1.9	2.2
Golf, Jetta		
1981-84		
AT		
one	1.4	1.7
both	1.9	2.2
MT		
one	.8	1.1
both	1.6	1.9
1985-97		
one	1.7	2.0
both	2.4	2.7
1998-04		
AT		
one	1.4	1.7
both	2.5	2.8
MT		
one	1.1	1.4
both	1.7	2.0
Passat		
1990-97		
one	1.8	2.1
both	2.4	2.7
1998-04		
one	1.4	1.7
both	2.1	2.4
Quantum one	1.3	1.6

VW-22   BEETLE : CABRIO : CABRIOLET : CORRADO : DASHER : FOX : GOLF : JETTA : PASSAT

	LABOR TIME	SEVERE SERVICE
Lower Control Arm, Replace (B)		
Rabbit, Scirocco		
1981-84		
AT		
one	1.5	1.8
both	2.0	2.3
MT		
one	.8	1.1
both	1.7	2.0
1985-89		
AT		
one	1.8	2.1
both	2.2	2.5
MT		
one	1.5	1.8
both	2.1	2.4
w/turbo diesel add	.2	.2
Stabilizer Bar, Replace (B)		
Beetle	2.1	2.3
Cabrio, Cabriolet	.8	1.0
Dasher	1.2	1.4
Fox	1.3	1.5
Golf, Jetta		
1981-84	.5	.7
1985-97	.8	1.0
1998-04	2.1	2.3
Passat		
1990-97	2.4	2.6
1998-04	1.7	1.9
Quantum	1.4	1.6
Rabbit, Scirocco		
1981-84	.5	.7
1985-89	.8	1.0
Steering Knuckle, Replace (B)		
Beetle		
one side	2.5	3.0
both sides	4.1	4.6
Cabrio	2.4	2.9
Cabriolet	2.3	2.8
Corrado	2.5	3.0
Golf, Jetta		
one side	2.5	3.0
both sides	4.1	4.6
Passat		
1990-97	2.3	2.8
1998-04		
one side	2.7	3.2
both sides	4.6	5.1
Quantum	3.3	3.8
1981-84 Rabbit, Scirocco		
one side	2.3	2.8
both sides	3.7	4.2
Strut, R&R or Replace (B)		
Beetle		
right side	1.7	1.9
left side	1.1	1.3
both sides	2.1	2.3
Cabrio, Cabriolet	.9	1.1

	LABOR TIME	SEVERE SERVICE
Dasher		
one	1.9	2.1
both	3.4	3.6
Fox	2.1	2.3
Golf, Jetta		
1981-97		
one	.9	1.1
both	2.0	2.2
1998-04		
right side	1.7	1.9
left side	1.1	1.3
both sides	2.1	2.3
Passat		
1990-97	.9	1.1
1998-04		
one	1.5	1.7
both	2.1	2.3
Quantum		
one	2.0	2.2
both	3.5	3.7
Rabbit, Scirocco		
one	.9	1.1
both	1.9	2.1
Replace cartridge add each	.8	.8
Strut Coil Spring, Replace (B)		
Beetle		
right side	2.2	2.4
left side	1.6	1.8
both sides	2.5	2.7
Cabrio, Cabriolet	1.5	1.7
Corrado	1.5	1.7
Dasher		
one	2.3	2.5
both	4.3	4.5
Fox	2.5	2.7
Golf, Jetta		
1981-97		
one	1.6	1.8
both	2.8	3.0
1998-04		
right side	2.2	2.4
left side	1.6	1.8
both sides	2.5	2.7
Passat		
1990-97	1.5	1.7
1998-04	2.3	2.5
Quantum	2.5	2.7
one	1.8	2.0
both	3.3	3.5
Rabbit, Scirocco		
one	1.4	1.6
both	2.7	2.9
Upper Control Arm, Replace (B)		
Passat		
one	1.4	1.9
both	2.4	2.9

	LABOR TIME	SEVERE SERVICE
Wheel Hub, Bearing or Seal, Replace (B)		
Beetle		
bearing		
one	2.2	2.7
both	3.8	4.3
hub		
right side	2.2	2.7
left side	1.6	2.1
both sides	2.9	3.4
Cabrio, Cabriolet, Corrado	2.5	3.0
Dasher		
one	2.9	3.4
both	4.7	5.2
Fox	2.3	2.8
Golf, Jetta		
1981-97		
one side	2.5	3.0
both sides	4.1	4.6
1998-04		
w/VAG 1459 B		
one side	1.7	2.2
both sides	3.3	3.8
w/o VAG 1459 B		
one side	2.2	2.7
both sides	3.8	4.3
Passat		
1990-97		
one	1.2	1.7
both	2.2	2.7
1998-04		
w/VAG 1459 B		
one side	1.7	2.2
both sides	3.3	3.8
w/o VAG 1459 B		
one side	2.7	3.2
both sides	4.6	5.1
Quantum	2.7	3.2
Rabbit, Scirocco		
one side	2.4	2.9
both sides	4.3	4.8

REAR SUSPENSION

	LABOR TIME	SEVERE SERVICE
Front & Rear Axle Alignment, Check (A)		
1981-87	1.1	1.3
1988-04	1.7	1.9
Coil Spring, R&R or Replace (B)		
Dasher		
one	.8	1.0
both	.9	1.1
Rear Leaf Spring, Replace (B)		
1981-84 Rabbit Pickup		
each	1.4	1.9
Rear Strut Coil Spring, Replace (B)		
Beetle		
one	.8	1.0
both	1.4	1.6

QUANTUM : RABBIT : RABBIT CONVERTIBLE : RABBIT PICKUP : SCIROCCO — VW-23

	LABOR TIME	SEVERE SERVICE
Cabrio, Cabriolet, Corrado		
one	1.3	1.5
both	2.3	2.5
Fox		
one	1.0	1.2
both	2.0	2.2
Golf, Jetta		
1981-84		
one	.8	1.0
both	1.4	1.6
1985-97		
one	1.4	1.6
both	2.4	2.6
1998-04		
one	.6	.8
both	.8	1.0
Passat		
1990-97		
one	1.2	1.4
both	2.1	2.3
1998-04		
2WD one or both	.8	1.0
4WD	1.4	1.6
Quantum one	1.2	1.4
Rabbit, Scirocco		
1981-84		
one	.8	1.0
both	1.5	1.7
1985-89		
one	1.0	1.2
both	1.9	2.1
Rear Strut Shock Absorber, Replace (B)		
Cabrio, Cabriolet, Corrado		
one	1.0	1.2
both	1.8	2.0
Fox		
one	1.3	1.5
both	2.0	2.2
Golf, Jetta		
1981-84		
one	.8	1.0
both	1.4	1.6
1985-97		
one	1.1	1.3
both	1.8	2.0
Passat		
one	1.6	1.8
both	2.4	2.6
Quantum one	1.2	1.4
Rabbit, Scirocco		
1981-84		
one	.8	1.0
both	1.4	1.6
1985-89		
one	1.0	1.2
both	1.8	2.0

	LABOR TIME	SEVERE SERVICE
Rear Strut Assy., R&R or Replace (B)		
Cabrio, Cabriolet, Corrado		
one	.7	.9
both	1.4	1.6
Fox		
one	.7	.9
both	1.6	1.8
Golf, Jetta		
1981-84		
one	.5	.7
both	.9	1.1
1985-97		
one	.7	.9
both	1.5	1.7
Passat		
one	1.1	1.3
both	1.8	2.0
Quantum	.8	1.0
Rabbit, Scirocco		
1981-84		
one	.5	.7
both	.9	1.1
1985-89		
one	.7	.9
both	1.3	1.5
Rear Wheel Oil Seal, Replace (B)		
1981-87 one side	.7	.9
1988-04 one side	1.2	1.4
Shock Absorbers or Bushings, Replace (B)		
Beetle		
one	.6	.8
both	.8	1.0
Dasher		
one	.5	.7
both	.8	1.0
1998-04 Golf, Jetta		
2WD		
one	.6	.8
both	.8	1.0
4WD		
one	1.1	1.3
both	1.7	1.9
Passat		
1990-97		
one	2.4	2.6
both	3.2	3.4
1998-04		
one	.6	.8
both	.8	1.0
Quantum		
one	1.7	1.9
both	3.2	3.4
Rabbit Pickup		
each	.3	.5
Sway Bar, Replace (B)		
1981-04	.9	1.1

	LABOR TIME	SEVERE SERVICE
Wheel Bearing, Replace (B)		
Includes: Replace bearing cups and grease seal.		
Exc. below		
2WD		
disc brakes		
one side	1.4	1.6
both sides	2.4	2.6
drum brakes		
one side	1.6	1.8
both sides	2.9	3.1
4WD		
one side	2.4	2.6
both sides	3.2	3.4
1998-04 Beetle, Passat		
one side	1.1	1.3
both sides	1.4	1.6
1998-04 Golf w/knuckle off vehicle		
one side	1.4	1.6
both sides	2.4	2.6

STEERING
AIR BAGS

	LABOR TIME	SEVERE SERVICE
Diagnose Air Bag System (A)		
1988-02	1.0	1.0
Air Bag Assy., Replace (B)		
Beetle each	.5	.5
Cabrio		
driver side	.5	.5
passenger side	.9	.9
Cabriolet	.5	.5
Golf, Jetta		
driver side	.5	.5
passenger side	1.1	1.1
Passat		
driver side	.5	.5
passenger side	.9	.9
Scirocco	.5	.5
Air Bag Clock Spring, Replace (B)		
Beetle	1.0	1.0
Cabrio	.5	.5
Cabriolet	1.9	1.9
Golf, Jetta, Passat		
1988-97	.5	.5
1998-04	1.0	1.0
Scirocco	1.8	1.8
Dual Pole Arming Sensor, Replace (B)		
Cabriolet	.9	.9
Scirocco	.9	.9
Electronic Control Unit, Replace (B)		
Beetle		
convertible	1.5	1.5
hard top	1.0	1.0
Cabrio	1.7	1.7
Cabriolet	4.4	4.4
Golf, Jetta		
1988-97	1.5	1.5
1998-04	.8	.8

VW-24 BEETLE : CABRIO : CABRIOLET : CORRADO : DASHER : FOX : GOLF : JETTA : PASSAT

	LABOR TIME	SEVERE SERVICE
Electronic Control Unit, Replace (B)		
Passat		
1990-97	1.7	1.7
1998-04	1.1	1.1
Scirocco	4.3	4.3

MANUAL RACK & PINION

Tie Rod End Boot, Replace (B)		
1981-94	.5	.7
Upper Mast Jacket Bearing, Replace (B)		
1981-94	.9	1.1
Rack & Pinion Assy., R&R or Replace (B)		
Includes: Adjustments.		
1981-92	3.0	3.3
1993-94	2.8	3.1
Steering Damper, Replace (B)		
Dasher	.5	.7
Steering Wheel, Replace (B)		
1981-94	.3	.3
w/air bag add	.5	.5

POWER RACK & PINION

Inner Tie Rod Assy., Replace (B)		
Includes: Reset toe.		
Exc. below	2.1	2.3
1998-04 Beetle, Golf	6.0	6.2
1998-04 Passat	3.0	3.2
Outer Tie Rod End, Replace (B)		
Beetle	.5	.7
Cabrio		
one	.7	.9
both	1.2	1.4
Cabriolet		
one	.7	.9
both	.8	1.0
Corrado		
one	.7	.9
both	.9	1.1
Dasher		
one	1.3	1.5
both	1.7	1.9
Fox		
one	.9	1.1
both	1.6	1.8
Golf, Jetta		
1981-84		
one	.5	.7
both	.8	1.0
1985-04		
one	.7	.9
both	.9	1.1
Passat		
1990-97		
one	.5	.7
both	.9	1.1

	LABOR TIME	SEVERE SERVICE
1998-04		
one	1.3	1.5
both	1.7	1.9
Quantum		
one	1.3	1.5
both	1.7	1.9
Rabbit, Scirocco		
1981-84		
one	.5	.7
both	.8	1.0
1985-89		
one	.7	.9
both	.9	1.1
Rack & Pinion Assy., R&I (B)		
Includes: Reset toe.		
Beetle	5.8	6.3
Cabrio	3.8	4.3
Cabriolet		
AT	5.3	5.8
MT	2.9	3.4
Golf, Jetta		
1981-84	2.9	3.4
1985-97	3.9	4.4
1998-04	5.8	6.3
Passat		
1990-97		
4 cyl.	3.6	4.1
6 cyl.	5.2	5.7
1998-04		
4 cyl.	5.8	6.3
6 cyl.	7.2	7.7
Quantum		
1982-85	3.1	3.6
1986-88	4.3	4.8
Rabbit, Scirocco		
1981-84	2.9	3.4
1985-89		
AT	5.2	5.7
MT	3.0	3.5
Steering Pump, R&R or Replace (B)		
Beetle	1.9	2.1
Cabrio	2.1	2.3
Cabriolet	2.0	2.2
Golf, Jetta		
1981-84	1.5	1.7
1985-04	2.0	2.2
Passat		
1990-97	2.8	3.0
1998-04		
4 cyl.	2.7	2.9
6 cyl.	2.2	2.4
Quantum		
1982-85	2.3	2.5
1986-88	2.7	2.9
Rabbit, Scirocco		
1981-84	1.5	1.7
1985-89	2.1	2.3

	LABOR TIME	SEVERE SERVICE
Steering Pump Hoses, Replace (B)		
Beetle		
pressure	1.1	1.2
return	1.2	1.3
Cabriolet		
pressure	1.2	1.3
return	.9	1.0
Corrado		
pressure	1.4	1.5
return	.9	1.0
Golf, Jetta		
1981-84		
pressure	1.2	1.3
return	.8	.9
1985-97		
pressure	1.5	1.6
return	.9	1.0
1998-04 each	1.1	1.2
Passat		
1990-97		
pressure	1.3	1.4
return	.9	1.0
1998-04		
pressure	2.0	2.1
return		
exc. 1.8L	.6	.7
1.8L	1.4	1.5
Quantum		
pressure	.9	1.0
suction	.7	.8
Rabbit, Scirocco		
1981-84		
pressure	1.4	1.5
return	.8	.9
1985-89		
pressure	1.2	1.3
return	.9	1.0
Steering Wheel, Replace (B)		
1981-04	.5	.5
w/air bag 88-97 add	.5	.5
Tie Rod End Boot, Replace (B)		
Beetle	4.3	4.5
Cabrio, Cabriolet, Corrado	.5	.7
Golf, Jetta		
1981-84	1.5	1.7
1985-97	.5	.7
1998-04	4.3	4.5
Passat		
1990-97	.5	.7
1998-04		
exc. 6 cyl.	4.1	4.3
6 cyl.	5.9	6.1
Quantum		
1982-84	3.4	3.6
1985-88	2.9	3.1
Rabbit	1.5	1.7
Scirocco	1.3	1.5

QUANTUM : RABBIT : RABBIT CONVERTIBLE : RABBIT PICKUP : SCIROCCO

VW-25

	LABOR TIME	SEVERE SERVICE
HEATING & AIR CONDITIONING		

When more than one component requires replacement where evacuation/recovery and recharging is already included, deduct 1.0 hour for each additional component from the time given.

Evacuate/Recover and Recharge System (B)
- 1981-04 .... 1.1 — 1.1

AC Condenser Cooling Fan Relay, Replace (B)
- 1988-02 .... .9 — .9

AC Hoses, Replace (B)
Includes: Evacuate/recover and recharge.
- Beetle
  - compressor-condenser ...... 1.8 — 2.0
  - compressor-evaporator ...... 2.4 — 2.6
  - evaporator-condenser
    - gasoline .... 4.1 — 4.3
    - diesel .... 4.6 — 4.8
  - evaporator-receiver/drier ..... 5.0 — 5.2
- Cabrio, Cabriolet
  - suction .... 2.0 — 2.2
  - discharge .... 2.1 — 2.3
- Corrado
  - suction .... 1.8 — 2.0
  - discharge .... 1.8 — 2.0
- Dasher
  - suction
    - integrated or factory .... 2.5 — 2.7
    - recirculating ..... 1.9 — 2.1
  - discharge .... 2.4 — 2.6
- Fox
  - suction .... 1.8 — 2.0
  - discharge .... 2.1 — 2.3
- Golf, Jetta
  - 1981-97
    - suction .... 1.9 — 2.1
    - discharge ...... 2.0 — 2.2
  - 1998-04
    - suction .... 2.4 — 2.6
    - discharge ...... 1.7 — 1.9
- Passat
  - 1990-97
    - suction .... 1.8 — 2.0
    - discharge ...... 1.9 — 2.1

- 1998-04
  - exc. 6 cyl.
    - exc. compressor-condenser ..... 2.1 — 2.3
    - compressor-condenser ..... 2.6 — 2.8
  - 6 cyl.
    - compressor-condenser, compressor-evaporator ..... 3.3 — 3.5
    - evaporator-condenser, evaporator-receiver/drier ... 2.1 — 2.3
- Quantum
  - 1982-87
    - suction
      - integrated or factory .... 2.5 — 2.7
      - recirculating .... 2.7 — 2.9
    - discharge ...... 2.3 — 2.5
  - 1988
    - suction or discharge ...... 2.5 — 2.7
    - evaporator ...... 2.8 — 3.0
- Rabbit, Scirocco
  - suction .... 1.9 — 2.2
  - discharge .... 1.8 — 2.0

AC Low Pressure Switch, Replace (B)
- 1985-97 .... 1.8 — 1.8
- 1998-04 .... .5 — .5

AC Relay, Replace (B)
- 1985-97 .... 1.2 — 1.2
- 1998-04
  - exc. Passat .... .8 — .8
  - Passat .... .3 — .3

Blower Motor, Replace (B)
- Dasher
  - integrated .... .9 — .9
  - recirculating .... 3.9 — 3.9
- 1981-87 Jetta .... 1.7 — 1.7
- Quantum
  - integrated .... .9 — .9
  - recirculating .... 3.9 — 3.9

Compressor Assy., Replace (B)
Includes: Parts transfer, evacuate/recover and recharge.
- Beetle .... 1.8 — 2.0
- Cabrio .... 3.2 — 3.4
- Cabriolet .... 4.0 — 4.2
- Corrado .... 5.1 — 5.3
- Dasher .... 3.4 — 3.6
- Fox .... 4.1 — 4.3
- Golf, Jetta
  - 1981-87
    - gasoline .... 3.1 — 3.3
    - diesel .... 2.6 — 2.8

	LABOR TIME	SEVERE SERVICE

- 1988-97
  - gasoline .... 3.6 — 3.8
  - diesel .... 2.9 — 3.1
- 1998-04 .... 1.5 — 1.7
- Passat
  - 1990-97 .... 3.5 — 3.7
  - 1998-04 .... 1.4 — 1.6
- Quantum
  - 1982-87 .... 4.5 — 4.7
  - 1988 .... 3.6 — 3.8
- Rabbit, Scirocco
  - 1981-87
    - gasoline .... 3.2 — 3.4
    - diesel .... 2.8 — 3.0
  - 1988-89 .... 4.0 — 4.2

Compressor Clutch & Pulley, Replace (B)
Includes: Evacuate, recover/recycle and recharge system.
- Beetle .... 1.8 — 2.0
- Cabrio .... 3.9 — 4.1
- Cabriolet .... 4.6 — 4.8
- Corrado .... 5.6 — 5.8
- Dasher .... 2.9 — 3.1
- Fox .... 4.7 — 4.9
- Golf, Jetta
  - 1981-84
    - gasoline .... 2.5 — 2.7
    - diesel .... 2.3 — 2.5
  - 1985-97
    - gasoline .... 4.1 — 4.3
    - diesel .... 3.3 — 3.5
  - 1998-04 .... 2.1 — 2.3
- Passat
  - 1990-97 .... 4.3 — 4.5
  - 1998-04
    - exc. 6 cyl. ..... 1.4 — 1.6
    - 6 cyl. .... 2.6 — 2.8
- Quantum .... 4.2 — 4.4
- Rabbit, Scirocco
  - 1981-84
    - gasoline .... 2.6 — 2.8
    - diesel .... 2.3 — 2.5
  - 1985-89 .... 4.4 — 4.6

Compressor Shaft Seal, Replace (B)
Includes: Parts transfer, evacuate/recover and recharge.
- Beetle .... 3.3 — 3.5
- Cabriolet .... 4.6 — 4.8
- Corrado .... 5.9 — 6.1
- Fox .... 4.9 — 5.1
- Golf, Jetta
  - 1981-87
    - gasoline .... 3.3 — 3.5
    - diesel .... 2.9 — 3.1
  - 1988-92 .... 4.2 — 4.4

VW-26 BEETLE : CABRIO : CABRIOLET : CORRADO : DASHER : FOX : GOLF : JETTA : PASSAT

	LABOR TIME	SEVERE SERVICE
Compressor Shaft Seal, Replace (B)		
Passat		
1990-94	5.0	5.2
1995-97	3.5	3.7
1998-02		
4 cyl.	3.6	3.8
6 cyl.	3.2	3.4
Quantum		
1982-87	2.1	2.3
1988	4.2	4.4
Rabbit, Scirocco		
1981-87		
gasoline	3.3	3.5
diesel	2.7	2.9
1988-89	4.6	4.8
Condenser Assy., Replace (B)		
Includes: Evacuate/recover and recharge.		
Beetle	5.3	5.5
Cabrio	3.1	3.3
Cabriolet	2.7	2.9
Corrado	4.0	4.2
Dasher		
large	2.3	2.5
small	2.3	2.5
Fox	3.1	3.3
Golf, Jetta		
1981-84 large or small	2.4	2.6
1985-97	3.0	3.2
1998-04	3.8	4.0
Passat		
1990-97	3.2	3.4
1998-04	2.6	2.8
Quantum	4.1	4.3
Rabbit, Scirocco		
1981-84		
large	2.4	2.6
small	2.2	2.4
1985-89	2.7	2.9
Replace receiver/drier add	.2	.2
Evaporator Core, Replace (B)		
Includes: Evacuate/recover and recharge.		
Beetle	7.6	7.8
Cabrio	7.9	8.1
Cabriolet	4.2	4.4
Corrado	8.4	8.6
Dasher	3.0	3.2
Fox	6.1	6.3
Golf, Jetta		
1981-84		
factory	3.8	4.0
recirculating	3.1	3.3
1985-92	7.3	7.5
1993-97	9.3	9.5
1998-04	7.5	7.7

	LABOR TIME	SEVERE SERVICE
Passat		
1990-97	9.1	9.3
1998-04	3.3	3.5
Quantum		
factory or Integrated	3.5	3.7
recirculating	3.2	3.4
Rabbit, Scirocco		
1981-84		
factory	3.8	4.0
recirculating	3.2	3.4
1985-89	4.2	4.4
Evaporator Fan Motor, Replace (B)		
Fox, Corrado	1.7	2.0
Golf, Jetta		
1985-92	1.4	1.7
1993-95	.8	1.1
1990-95 Passat	1.8	2.1
Expansion Valve, Replace (B)		
Includes: Evacuate/recover and recharge.		
Beetle	2.3	2.5
Cabrio	2.2	2.4
1988-93 Cabriolet	2.5	2.7
Corrado	2.1	2.3
Dasher	3.8	4.0
Fox	6.8	7.0
Golf, Jetta		
1981-87		
factory	3.1	3.3
recirculating	3.9	4.1
1988-92	2.5	2.7
1993-97	2.2	2.4
1998-04	1.2	1.4
Passat	2.3	2.5
Quantum		
1982-87	3.8	4.0
1988	2.8	3.0
Rabbit, Scirocco		
1981-87		
factory	3.0	3.2
recirculating	3.9	4.1
1988-89	2.4	2.6
Heater Blower Motor, Replace (B)		
Beetle	.7	.7
Cabrio, Corrado	.5	.5
Cabriolet	1.8	1.8
Dasher	2.6	2.6
Fox	1.5	1.5
Golf, Jetta		
1981-84	2.3	2.3
1985-92	1.2	1.2
1993-04	.8	.8
Passat		
1990-97	1.5	1.5
1998-04	.6	.6
Quantum	1.2	1.2
Rabbit, Scirocco		
1981-84	2.3	2.3
1985-89	1.7	1.7

	LABOR TIME	SEVERE SERVICE
Heater Blower Motor Resistor, Replace (B)		
Beetle	.7	.7
Cabrio, Cabriolet	1.9	1.9
Corrado	.5	.5
Fox	1.5	1.5
Golf, Jetta		
1985-97	.5	.5
1998-04	.9	.9
Passat	.7	.7
Quantum	.3	.3
1985-89 Scirocco	1.8	1.8
Heater Blower Motor Switch, Replace (B)		
1981-97	.7	.7
1998-04	.3	.3
Heater Control Valve, Replace (B)		
1981-84	.5	.7
1985-97	.7	.9
1998-04	1.4	1.4
Heater Core, R&R or Replace (B)		
Beetle	3.4	3.6
Cabrio	7.1	7.3
Cabriolet		
w/AC	6.3	6.5
w/o AC	1.8	2.0
Corrado	6.2	6.4
Dasher		
w/AC	3.1	3.3
w/o AC	2.4	2.6
Fox		
w/AC	6.3	6.5
w/o AC	4.8	5.0
Golf, Jetta		
1981-84	2.3	2.5
1985-92		
w/AC	6.2	6.4
w/o AC	2.9	3.1
1993-97	7.1	7.3
1998-04		
w/AC	6.3	6.5
w/o AC	5.7	5.9
Passat		
1990-97	7.7	7.9
1998-04	4.4	4.6
Quantum		
w/AC	3.2	3.2
w/o AC	2.4	2.4
Rabbit, Scirocco		
1981-84		
w/AC	6.3	6.5
w/o AC	2.2	2.4
1985-89		
w/AC	6.4	6.6
w/o AC	1.8	2.0
Heater Hoses, Replace (B)		
One	.7	.9
Both	.9	1.1

QUANTUM : RABBIT : RABBIT CONVERTIBLE : RABBIT PICKUP : SCIROCCO — VW-27

	LABOR TIME	SEVERE SERVICE
High Pressure Cut-Off Switch, Replace (B)		
1985-97	2.0	2.0
1998-04	1.5	1.5
Receiver/Drier Assy., Replace (B)		
Includes: Evacuate/recover and recharge.		
Beetle	1.5	1.5
Cabrio	1.8	1.8
Cabriolet	2.3	2.3
Corrado		
4 cyl.	2.3	2.3
6 cyl.	1.7	1.7
Dasher	2.6	2.6
Fox	2.9	2.9
Golf, Jetta		
1981-87	2.0	2.0
1988-89	1.5	1.5
1990-92	2.4	2.4
1993-04	1.8	1.8
Passat		
1990-97	1.6	1.6
1998-04	2.0	2.0
Quantum		
1982-87	2.6	2.6
1988	2.7	2.7
Rabbit, Scirocco		
1981-87	1.8	1.8
1988-89	2.5	2.5
Temperature Control Assy, Replace (B)		
Beetle	.9	.9
Cabrio	.8	.8
Golf, Jetta		
1994-97	.8	.8
1998-04	1.6	1.6
Passat		
1994-97	1.7	1.7
1998-02	.8	.8

WIPERS & SPEEDOMETER

	LABOR TIME	SEVERE SERVICE
Power Antenna, Replace (B)		
1985-02	3.0	3.0
Radio, R&R (B)		
Beetle	.7	.7
Cabrio	.5	.5
Cabriolet	.9	.9
Dasher	.9	.9
Golf, Jetta		
1981-84	.7	.7
1985-92	.9	.9
1993-04	.5	.5
Passat		
1990-94	.9	.9
1995-04	.5	.5
Quantum	.9	.9
Rabbit, Scirocco		
1981-84	.7	.7
1985-89	.9	.9

	LABOR TIME	SEVERE SERVICE
Rear Window Washer Pump, Replace (B)		
1990-02	.7	.7
Rear Window Wiper Motor, Replace (B)		
Cabriolet	1.3	1.5
Corrado	2.3	2.5
Fox	1.3	1.5
Golf, Jetta		
1981-92	1.4	1.6
1993-97	1.2	1.4
1998-04	.8	1.0
Passat		
1990-97	1.9	2.1
1998-04	.8	1.0
Quantum	.8	1.0
Rabbit, Scirocco	1.3	1.5
Rear Window Wiper Relay, Replace (B)		
1985-04	.5	.5
Rear Windshield Wiper Switch, Replace (B)		
1981-84 Jetta	.5	.5
1981-84 Rabbit, Scirocco	.5	.5
Speedometer Cable & Casing, Replace (B)		
1981-87		
upper	1.4	1.4
lower	.5	.5
1988-02		
upper	.8	.8
lower	.7	.7
one piece	.9	.9
Speedometer Driven Gear and/or Seal, Replace (B)		
1981-02 MT	.5	.5
Speedometer Driven Gear, Replace (B)		
1981-84	.5	.5
1985-02	.7	.7
Speedometer Head, R&R or Replace (B)		
Beetle	1.9	1.9
Cabrio	1.6	1.6
Cabriolet	1.9	1.9
Corrado	1.3	1.3
Dasher		
Fox	2.5	2.5
Golf, Jetta		
1981-84	.7	.7
1985-92	2.6	2.6
1993-02		
electronic	1.8	1.8
mechanical	1.9	1.9
Passat	1.4	1.4
Quantum	2.1	2.1
Rabbit, Scirocco		
1981-84	.7	.7
1985-89	2.1	2.1

	LABOR TIME	SEVERE SERVICE
Vehicle Speed Sensor, Replace (B)		
1990-02	.5	.5
Windshield Washer Pump, Replace (B)		
1981-84	.5	.5
1985-97	.7	.7
1998-04		
exc. Beetle	.3	.3
Beetle	1.1	1.1
Windshield Wiper & Washer Switch, Replace (B)		
Beetle, Cabrio	.9	.9
Cabriolet		
1985-89	.9	.9
1990-93	2.8	2.8
Corrado	1.8	1.8
Dasher	.9	.9
Fox	.9	.9
Golf, Jetta		
1981-84	.5	.5
1985-92	1.4	1.4
1993-04	1.0	1.0
Passat	1.1	1.1
Quantum	.7	.7
Rabbit, Scirocco		
1981-84	.5	.5
1985-89	.9	.9
Windshield Wiper Linkage, Replace (B)		
1981-84	1.1	1.3
1985-93	.8	1.0
1998-04	1.0	1.2
Windshield Wiper Motor, Replace (B)		
1981-84	.8	1.0
1985-04		
exc. Passat	1.3	1.5
Passat	1.5	1.7
Windshield Wiper Motor Relay, Replace (B)		
Exc. Beetle	.5	.5
Beetle	.8	.8

LAMPS & SWITCHES

	LABOR TIME	SEVERE SERVICE
Back-Up Lamp Switch, Replace (B)		
Beetle	.8	.8
Cabrio	.7	.7
Cabriolet	.5	.5
Corrado		
AT	.7	.7
MT	.5	.5
Dasher		
AT	.7	.7
MT	1.8	1.8
Fox	.5	.5
Golf, Jetta, Passat		
1981-97	.7	.7
1998-04	.6	.6
Quantum each	.5	.5

VW-28 — BEETLE : CABRIO : CABRIOLET : CORRADO : DASHER : FOX : GOLF : JETTA : PASSAT

Labor Time	Severe Service

Back-Up Lamp Switch, Replace (B)
Rabbit, Scirocco
- AT .... .7 / .7
- MT .... .5 / .5

Brake Lamp Switch, Replace (B)
- 1981-84 .... 1.2 / 1.2
- 1985-02 one or both .... .6 / .6

Emergency Flasher Switch Assy., Replace (B)
- Exc. below .... .7 / .7
- 1989-92 Golf, Jetta ... 1.2 / 1.2
- Passat
  - 1994-97 .... .9 / .9
  - 1998-04 .... .3 / .3

Halogen Headlamp Bulb, Replace (B)
- 1985-04 .... .6 / .6

Headlamp Switch, Replace (B)
- 1981-84 .... .5 / .5
- 1985-04 .... .7 / .7

Headlamps, Aim (B)
- Two .... .4 / .4
- Four .... .6 / .6

High Mount Stop Lamp Bulb, Replace (B)
- 1985-04 .... .5 / .5

Horn, Replace (B)
- Exc. Passat each .... .5 / .5
- Passat
  - 1990-00 .... .6 / .6
  - 2001-04 .... 1.2 / 1.2

License Lamp Lens, Replace (B)
- 1981-04 .... .3 / .3

Multifunction Switch, Replace (B)
- 1990-97 .... 1.8 / 1.8
- 1998-04 Beetle, Golf .... .5 / .5
- 1998-02 Cabrio, Jetta . 1.8 / 1.8
- 1998-04 Passat .... .9 / .9

Neutral Safety Switch, Replace (B)
- 1981-84 .... .5 / .5
- 1985-02 .... .7 / .7

Parking Lamp Lens or Bulb, Replace (C)
- 1981-02 .... .2 / .2

Sealed Beam Headlamp, Replace (B)
- 1981-93 each .... .3 / .3

Side Marker Lamp Lens, Replace (B)
- 1981-04 .... .3 / .3

Stop & Tail Lamp Lens, Replace (B)
- Exc. 1998-04 Passat .... .5 / .5
- 1998-04 Passat .... .6 / .6

Turn Signal or Hazard Relay, Replace (B)
- 1981-84 .... .3 / .3
- 1985-04 .... .8 / .8

Turn Signal Switch, Replace (B)
- Beetle .... .9 / .9
- Cabrio .... 1.4 / 1.4

- Cabriolet .... .8 / .8
- Dasher .... .9 / .9
- Fox .... .8 / .8
- Golf, Jetta
  - 1981-84 .... .8 / .8
  - 1985-88 .... 1.1 / 1.1
  - 1989-97 .... 1.3 / 1.3
  - 1998-04 .... 1.0 / 1.0
- Passat
  - 1990-97 .... 1.5 / 1.5
  - 1998-04 .... 1.1 / 1.1
- Quantum .... .9 / .9
- Rabbit, Scirocco .... .8 / .8

BODY

Door Lock Remote Control, Replace (B)
- 1981-87 front or rear .... .7 / .7
- 1988-04 front or rear .. 1.4 / 1.4

Door Lock, Replace (B)
Front
- Beetle .... 1.4 / 1.4
- Cabrio .... 1.1 / 1.1
- Cabriolet .... .5 / .5
- Corrado .... .3 / .3
- Dasher .... .5 / .5
- Fox .... .5 / .5
- Golf, Jetta
  - 1981-92 .... .7 / .7
  - 1993-97 .... 1.3 / 1.3
  - 1998-04 .... .8 / .8
- Passat .... 1.5 / 1.5
- Quantum .... 1.3 / 1.3
- Rabbit, Scirocco
  - 1981-84 .... .7 / .7
  - 1985-89 .... .5 / .5

Rear
- exc. below .... .3 / .3
- 1993-04 Golf, Jetta . 1.6 / 1.6
- 1998-04 Passat .... 1.4 / 1.4
- 1981-84 Rabbit ...... .5 / .5

Door Window Regulator (Electric), Replace (B)
Front
- Beetle .... 1.7 / 1.9
- Cabrio .... 2.6 / 2.8
- Cabriolet .... 1.5 / 1.7
- Corrado .... 2.5 / 2.7
- Golf, Jetta
  - 1985-97 .... 1.3 / 1.5
  - 1998-04 .... .9 / 1.1
- Passat
  - 1990-97 .... 1.9 / 2.1
  - 1998-04 .... .8 / 1.0
- 1985-89 Scirocco .. 1.7 / 1.9
- *Replace motor add* .... .3 / .5

Door Window Regulator (Manual), Replace (B)
- Exc. below .... 1.3 / 1.5
- Beetle .... 1.5 / 1.7
- Golf, Jetta
  - 1993-97 .... 1.6 / 1.8
  - 1998-04 .... .3 / .5

Hood Hinge, Replace (B)
- Cabrio .... .9 / .9
- Cabriolet .... 1.3 / 1.3
- Corrado .... 1.5 / 1.5
- Fox .... 1.4 / 1.4
- Golf, Jetta .... 1.2 / 1.2
- Passat
  - 1990-97 .... 1.4 / 1.4
  - 1998-04 .... 1.1 / 1.1
  - each addl. add .... .2 / .2
- Quantum .... .5 / .5
- Rabbit, Scirocco .... 1.4 / 1.4

Hood Lock, Replace (B)
- Exc. below .... .9 / .9
- Beetle .... .3 / .3
- 1998-04 Golf, Passat ... .5 / .5

Hood Release Cable, Replace (B)
- Beetle .... .6 / .6
- Cabrio .... .7 / .7
- Cabriolet .... .5 / .5
- Dasher .... .8 / .8
- Fox .... .5 / .5
- Golf, Jetta
  - 1981-84 .... .9 / .9
  - 1985-92 .... .5 / .5
  - 1993-04 .... .7 / .7
- Passat .... .7 / .7
- Quantum .... .8 / .8
- Rabbit, Scirocco
  - 1981-84 .... .9 / .9
  - 1985-89 .... .5 / .5

Lock Striker Plate, Replace (B)
- 1981-87 .... .2 / .2
- 1988-02 .... .3 / .3

Rear Compartment Lid Lock and/or Cylinder, Replace (B)
- 1981-84 .... .5 / .5
- 1985-97 .... 1.3 / 1.3
- 1998-04 .... .8 / .8

Rear Door Window Regulator and/or Motor, Replace (B)
- Cabrio .... 2.1 / 2.3
- Golf, Jetta
  - 1985-92 .... 1.4 / 1.6
  - 1993-97 .... 1.6 / 1.8
  - 1998-04 .... .8 / 1.0
- Passat
  - 1990-97 .... 1.9 / 2.1
  - 1998-04 .... 1.5 / 1.7

Eurovan : Vanagon

SYSTEM INDEX

MAINTENANCE	**VW-30**
CHARGING	**VW-30**
STARTING	**VW-30**
CRUISE CONTROL	**VW-30**
IGNITION	**VW-30**
EMISSIONS	**VW-31**
FUEL	**VW-31**
EXHAUST	**VW-32**
ENGINE COOLING	**VW-32**
ENGINE	**VW-32**
Assembly	VW-32
Cylinder Head	VW-32
Camshaft	VW-33
Crank & Pistons	VW-33
Engine Lubrication	VW-34
CLUTCH	**VW-34**
MANUAL TRANSAXLE	**VW-34**
MANUAL TRANSMISSION	**VW-34**
AUTO TRANSAXLE	**VW-34**
AUTO TRANSMISSION	**VW-35**
SHIFT LINKAGE	**VW-35**
DRIVELINE	**VW-35**
BRAKES	**VW-36**
FRONT SUSPENSION	**VW-37**
REAR SUSPENSION	**VW-37**
STEERING	**VW-38**
HEATING & AC	**VW-38**
WIPERS & SPEEDOMETER	**VW-39**
LAMPS & SWITCHES	**VW-39**
BODY	**VW-39**

OPERATIONS INDEX

A
AC Hoses	VW-38
Air Conditioning	VW-38
Alignment	VW-37
Alternator (Generator)	VW-30
Antenna	VW-39
Anti-Lock Brakes	VW-36

B
Back-Up Lamp Switch	VW-39
Ball Joint	VW-37
Battery Cables	VW-30
Bleed Brake System	VW-36
Blower Motor	VW-38
Brake Disc	VW-36
Brake Drum	VW-36
Brake Hose	VW-36
Brake Pads and/or Shoes	VW-37

C
Camshaft	VW-33
Camshaft Sensor	VW-30
Catalytic Converter	VW-32
Crankshaft	VW-33
Cruise Control	VW-30
CV Joint	VW-36
Cylinder Head	VW-33

D
Differential	VW-35
Distributor	VW-30
Drive Belt	VW-30
Driveshaft	VW-35

E
EGR	VW-31
Engine	VW-32
Engine Lubrication	VW-34
Engine Mounts	VW-32
Evaporator	VW-38
Exhaust	VW-32
Exhaust Manifold	VW-32
Expansion Valve	VW-38

F
Flexplate	VW-35
Flywheel	VW-34
Fuel Injection	VW-31
Fuel Pump	VW-31
Fuel Vapor Canister	VW-31

G
Gear Selector Lever	VW-35
Generator	VW-30
Glow Plug	VW-30

H
Halfshaft	VW-36
Headlamp	VW-39
Heater Core	VW-38
Horn	VW-39

I
Idle Air Control (IAC) Valve	VW-31
Ignition Coil	VW-30
Ignition Switch	VW-30
Injection Pump	VW-31
Inner Tie Rod	VW-38
Intake Manifold	VW-31

L
Lower Control Arm	VW-37

M
Mass Air Flow (MAF) Sensor	VW-31
Master Cylinder	VW-36
Muffler	VW-32

N
Neutral Safety Switch	VW-39

O
Oil Pan	VW-34
Oil Pump	VW-34
Outer Tie Rod	VW-38
Oxygen Sensor	VW-31

P
Parking Brake	VW-37
Pistons	VW-34
Positive Crankcase Ventilation (PCV) Valve	VW-31

R
Radiator	VW-32
Radio	VW-39
Rear Main Oil Seal	VW-34

S
Shock Absorber/Strut, Front	VW-37
Shock Absorber/Strut, Rear	VW-37
Spark Plug Cables	VW-30
Spark Plugs	VW-31
Spring, Front Coil	VW-37
Spring, Rear Coil	VW-37
Starter	VW-30
Steering Wheel	VW-38

T
Thermostat	VW-32
Throttle Body	VW-31
Throttle Position Sensor (TPS)	VW-31
Timing Belt	VW-33
Torque Converter	VW-35

U
Upper Control Arm	VW-37

V
Valve Body	VW-35
Valve Cover Gasket	VW-33
Valve Job	VW-33
Vehicle Speed Sensor	VW-31

W
Water Pump	VW-32
Wheel Balance	VW-30
Wheel Cylinder	VW-37
Window Regulator	VW-39
Windshield Washer Pump	VW-39
Windshield Wiper Motor	VW-39

VW-30 EUROVAN : VANAGON

	LABOR TIME	SEVERE SERVICE

MAINTENANCE

Air Cleaner Filter Element, Replace (C)
- Gasoline ....3 .5
- Diesel ....5 .7

Chassis Lubrication, Change Oil & Filter (C)
Includes: Correct all fluid levels.
- 1981-94 ....4 .6

Composite Headlamp Bulb, Replace (C)
- 1993-94 ....3 .3

Drive Belt, Replace (B)
- Eurovan 9 1.1
- 1988-91 Vanagon 9 1.1

Fuel Filter, Replace (B)
- Gasoline ....5 .7
- Diesel ....7 .9

Oil & Filter, Change (C)
Includes: Correct all fluid levels.
- 1981-94 ....4 .6

Sealed Beam Headlamp, Replace (C)
- 1981-91 ....3 .3

Timing Belt, Replace (B)
- Gasoline
  - Eurovan 3.8 4.0
- Diesel
  - 1982-84 Vanagon ...3.2 3.4

Tire, Replace (C)
Includes: Dismount old tire and mount new tire to rim.
- One ....5 .5
- each addl. add ....3 .3

Wheel, Balance (B)
- One ....3 .5
- each addl. add ....1 .3

CHARGING

Alternator Circuits, Test (B)
Includes: Test component output.
- 1981-94 ....5 .5

Alternator Assy., R&R and Recondition (A)
- Eurovan 2.8 3.0
- Vanagon
  - 1981-87
    - air-cooled 3.0 3.2
    - water-cooled ...2.6 2.8
    - diesel 3.0 3.2
  - 1988-91 ....2.4 2.6

Alternator Assy., Replace (B)
Includes: Pulley transfer.
- Eurovan 1.6 1.8
- Vanagon
  - 1981-87
    - air-cooled 1.6 1.8
    - water-cooled 8 1.0
    - diesel 1.5 1.7
  - 1988-91 ....1.5 1.7

Alternator Voltage Regulator, Replace (B)
- Eurovan 1.8 2.0
- Vanagon
  - 1981-87 ....9 1.1
  - 1988-91 ....1.8 2.

Front Alternator Bearing, Replace (B)
- Eurovan 1.5 1.7
- Vanagon
  - 1981-87
    - air-cooled 1.8 2.0
    - water-cooled ...1.1 1.3
    - diesel 1.6 1.8
  - 1988-91 ....1.6 1.8

STARTING

Starter Draw Test (On Car) (B)
- 1981-94 ....3 .3

Battery Cables, Replace (C)
- Exc. below
  - positive ....7 .9
  - negative ....5 .7
- 1988-91 Vanagon
  - positive 1.1 1.3
  - negative ....5 .7

Starter Assy., R&R and Recondition (A)
- Eurovan 2.6 2.8
- Vanagon
  - 1981-87
    - gasoline 3.3 3.5
    - diesel 3.9 4.1
  - 1988-91 ....3.3 3.5

Starter Assy., Replace (B)
- Eurovan 1.5 1.7
- Vanagon
  - 1981-87
    - gasoline 1.3 1.5
    - diesel 1.8 2.0
  - 1988-91 ....1.3 1.5

Starter Drive Assy., Replace (B)
Includes: Starter R&R.
- Eurovan
  - AT 2.2 2.4
  - MT 1.9 2.1
- Vanagon
  - 1981-87
    - gasoline 1.9 2.1
    - diesel 2.5 2.7
  - 1988-91 ....1.8 2.0

Starter Solenoid and/or Switch, Replace (B)
Includes: Starter R&R.
- Eurovan
  - AT 2.3 2.5
  - MT 1.8 2.0

Vanagon
- 1981-87
  - gasoline 1.7 1.9
  - diesel 2.2 2.4
- 1988-91 ....1.4 1.6

CRUISE CONTROL

Diagnose Cruise Control System Component Each (A)
- Eurovan ....8 .8

Control Controller Module, Replace (B)
- 1981-94 ....1.1 1.3

Control Switch, Replace (B)
- 1981-94 ....1.3 1.5

Control Transmitter, Replace (B)
- 1981-94 ....1.8 2.0

Control Vacuum Clutch Valve, Replace (B)
- 1981-94 ....1.2 1.4

Control Vacuum Pump, Replace (B)
- 1981-94 ....1.3 1.5

Control Vacuum Servo, Replace (B)
- 1981-94 ....1.1 1.3

IGNITION

Diagnose Ignition System Component Each (A)
- 1981-94 ....8 .8

Ignition Timing, Reset (B)
- 1981-94 ....3 .3

Camshaft Position Sensor, Replace (B)
- Eurovan 1.5 1.7

DIS Control Unit, Replace (B)
- Vanagon ....7 .9

Distributor, Replace (B)
Includes: Reset base ignition timing.
- 1981-94 ....8 1.0

Distributor Cap and/or Rotor, Replace (B)
- All ....3 .3

Glow Plug, Replace (B)
- 1982-84 Vanagon
  - one ....7 .9
  - all 9 1.1

Glow Plug Relay, Replace (B)
- 1982-84 Vanagon ......5 .7

Hall Generator, Replace (B)
- 1988-91 Vanagon 1.6 1.8

Ignition Coil, Replace (B)
- 1981-94 ....5 .7

Ignition Switch, Replace (B)
- All Models 1.3 1.5

Spark Plug (Ignition) Cables, Replace (B)
- Eurovan 8 1.0
- Vanagon
  - 1981-87 ....7 .9
  - 1988-91 ....8 1.0

EUROVAN : VANAGON VW-31

	LABOR TIME	SEVERE SERVICE
Spark Plugs, Replace (B)		
Eurovan	.7	.9
Vanagon	.5	.7
TCI Control Unit, Replace (B)		
Vanagon	.7	.9
Vacuum Advance Unit, Replace (B)		
Includes: Reset base ignition timing.		
Eurovan	1.3	1.5
Vanagon		
1981-87	.9	1.1
1988-91	1.4	1.6

EMISSIONS

Air Intake Sensor, Replace (B)		
Vanagon	.5	.7
Anti-Backfire Valve, Replace (B)		
Vanagon	.5	.7
Cold Start Valve, Replace (B)		
Vanagon	1.3	1.5
Deceleration Valve, Replace (B)		
Vanagon	.5	.7
EGR Check Valve, Replace (B)		
Vanagon	.5	.7
EGR Control Valve, Replace (B)		
Vanagon	.7	.9
EGR Filter, Replace (B)		
Vanagon	.7	.9
Fuel Vapor Canister, Replace (B)		
Eurovan	.7	.9
Vanagon	.5	.7
Fuel Vapor Canister Filter, Replace (B)		
Vanagon	.2	.4
Mass Air Flow (MAF) Sensor, Replace (B)		
1981-94	.8	1.0
Oxygen Sensor, Replace (B)		
1981-94	1.3	1.5
Positive Crankcase Ventilation (PCV) Valve, Replace (B)		
Eurovan	.3	.5
Pressure Sensor, Replace (B)		
Vanagon	.5	.7
Temperature Sensor, Replace (B)		
Eurovan	.7	.9
Vanagon	.5	.7
Throttle Position Sensor (TPS), Replace (B)		
Eurovan	.9	1.1
Thermo Time Switch, Replace (B)		
Vanagon	.5	.7
Vehicle Speed Sensor, Replace (B)		
Eurovan	.7	.9

FUEL
DELIVERY

	LABOR TIME	SEVERE SERVICE
Fuel Filter, Replace (B)		
Gasoline	.5	.7
Diesel	.7	.9
Fuel Gauge (Dash), Replace (B)		
Eurovan	1.9	2.1
Vanagon		
1981-87	.9	1.1
1988-91	1.5	1.7
Fuel Gauge (Tank), Replace (B)		
Eurovan	1.6	1.8
Vanagon		
1981-87	2.8	3.0
1988-91		
2WD	2.9	3.1
4WD	2.1	2.3
Fuel Pump, Replace (B)		
Eurovan	2.1	2.3
Vanagon	1.3	1.5
Fuel Tank, Replace (B)		
Includes: Drain and refill. R&R transmission.		
Eurovan	2.2	2.4
Vanagon		
1981-87	2.6	2.8
1988-91		
2WD	2.3	2.5
4WD	7.1	7.3
Intake Manifold and/or Gasket, Replace (B)		
Includes: Adjustments.		
Gasoline		
Eurovan	3.9	4.1
Diesel		
Vanagon	1.9	2.1

INJECTION

Diagnose Fuel Injection System Component Each (A)		
1981-94	1.0	1.0
Diesel Fuel Injection Timing, Check & Adjust (B)		
1982-84 Vanagon	.9	1.1
Idle Speed, Adjust (B)		
Vanagon diesel	.7	.9
Auxiliary Air Regulator, Replace (B)		
Vanagon		
1981-87	.9	1.1
1988-91	.7	.9
Electronic Control Unit, Replace (B)		
Eurovan	.7	.9
Vanagon		
1981-87	.5	.7
1988-91	.7	.9
Fuel Accumulator, Replace (B)		
Eurovan	.7	.9
Fuel Distributor, Replace (B)		
Eurovan	1.7	1.9

	LABOR TIME	SEVERE SERVICE
Fuel Injection Lines, Replace (B)		
Vanagon diesel	.7	.9
Fuel Injection Pump, Replace (B)		
Gasoline		
Eurovan	1.7	1.9
Diesel		
Vanagon	4.0	4.2
Fuel Injectors, Replace (B)		
Gasoline		
Eurovan		
one	1.6	1.8
each addl.	.5	.7
Vanagon		
one side	.7	.9
both sides	1.1	1.3
Diesel		
Vanagon		
one	.5	.7
all	1.3	1.5
Fuel Pressure Regulator, Replace (B)		
Eurovan	1.4	1.6
Vanagon	1.5	1.7
Fuel Pump Relay, Replace (B)		
Eurovan	.7	.9
Fuel Shut-Off Valve, Replace (B)		
Vanagon diesel	.5	.7
Idle Air Control (IAC) Valve, Replace (B)		
Eurovan	.9	1.1
Idle Speed Motor and/or Actuator, Replace (B)		
Eurovan	1.6	1.8
Intake Pipe, Replace (B)		
Vanagon		
1981-87		
one side	1.6	1.8
both sides	2.5	2.7
1988-91		
one side	1.7	1.9
both sides	2.7	2.9
Throttle Body Air Flow Meter, Replace (B)		
Vanagon		
1981-87	1.7	1.9
1988-91	2.3	2.5
Throttle Body Assy., Replace (B)		
Eurovan	1.6	1.8
Throttle Body Switches, Replace (B)		
Eurovan	1.8	2.0
Throttle Valve Housing, Replace (B)		
1981-91	1.1	1.3
Throttle Valve Switch, Replace or Adjust (B)		
Eurovan		
adjust	.7	.9
replace	2.2	2.4
Vanagon	.7	.9
Voltage Supply Relay, Replace (B)		
Vanagon	.5	.7

EUROVAN : VANAGON

	LABOR TIME	SEVERE SERVICE

EXHAUST

Catalytic Converter, Replace (B)
- Eurovan 1.5   *1.7*
- Vanagon
  - 1981-87 .... 1.3   *1.5*
  - 1988-91 .... 1.5   *1.7*

Exhaust Manifold, Replace (B)
- Eurovan 5.3   *5.5*
- Vanagon
  - 1981-87
    - air-cooled
      - one 1.8   *2.0*
      - both 2.5   *2.7*
    - water-cooled
      - one 1.5   *1.7*
      - both 1.9   *2.1*
    - diesel 1.8   *2.0*
  - 1988-91
    - one 1.6   *1.8*
    - both 2.9   *3.1*

Front Exhaust Pipe, Replace (B)
- Eurovan 1.7   *1.9*
- Vanagon
  - 1981-87 9   *1.1*
  - 1988-91 .... 1.6   *1.8*

Muffler, Replace (B)
- Eurovan
  - front or intermediate 8   *1.0*
  - rear 5   *.7*
- Vanagon 1.4   *1.6*
- w/diesel add 3   *.5*

Tail Pipe, Replace (B)
- Eurovan 7   *.9*
- Vanagon 3   *.5*
- w/diesel add 2   *.4*

ENGINE COOLING

Pressure Test Cooling System (C)
- 1982-93 3   *.3*

Coolant Thermostat, Replace (B)
- Eurovan 9   *1.1*
- Vanagon
  - 1982-85
    - gas 2.2   *2.4*
    - diesel 1.7   *1.9*
  - 1986-91 9   *1.1*

Electric Cooling Fan Assy., Replace (B)
- Eurovan 1.5   *1.7*
- Vanagon
  - 1981-87 .... 1.6   *1.8*
  - 1988-91 .... 2.3   *2.5*

Electric Fan Relay, Replace (B)
- Eurovan 7   *.9*
- Vanagon
  - 1984-87 5   *.7*
  - 1988-91 8   *1.0*

Fan Housing, R&I (B)
- 1981-87 Vanagon 4.9   *5.1*
- w/AC add 1.3   *1.5*

Heat Exchanger, Replace (B)
- 1981-84 Vanagon
  - gasoline 1.9   *2.1*
  - diesel 3.6   *3.8*

Radiator Assy., R&R or Replace (B)
- Eurovan
  - AT 2.7   *2.9*
  - MT 2.5   *2.7*
- 1982-91 Vanagon 2.9   *3.1*

Temperature Gauge (Dash), Replace (B)
- Eurovan 1.9   *2.1*
- Vanagon 1.7   *1.9*

Temperature Gauge (Engine), Replace (B)
- Eurovan 7   *.9*
- Vanagon
  - 1981-87 7   *.9*
  - 1988-91 .... 1.3   *1.5*

Water Pump and/or Gasket, Replace (B)
Includes: Refill with proper coolant mix.
- Eurovan 4.0   *4.2*
- Vanagon
  - 1982-87
    - gas 2.9   *3.1*
    - diesel 2.4   *2.6*
  - 1988-91 .... 3.1   *3.3*
- w/AC add 5   *.7*
- w/PS add 2   *.4*

ENGINE

ASSEMBLY

Times shown are for OEM assemblies. Time to replace assemblies from aftermarket rebuilders may vary.

Engine Assy., R&I (B)
Does not include parts or component transfer.
- Gasoline
  - Eurovan 10.5   *10.7*
  - Vanagon
    - 1981-83
      - AT 4.5   *4.7*
      - MT 3.1   *3.3*
    - 1984-87 .... 6.8   *7.0*
    - 1988-91 .... 10.8   *11.0*
- Diesel
  - Vanagon 5.3   *5.5*
- w/AC add
  - Eurovan 1.2   *1.4*
  - 81-83 Vanagon ... 1.2   *1.4*
- w/AT Eurovan add 1.1   *1.3*

Engine Assy., R&R and Recondition (A)
Includes: Replacing rings, rod and main bearings, cylinder head reconditioning and engine tune-up.
- 1981-83 Vanagon
  - AT 22.2   *22.4*
  - MT 20.5   *20.7*
- w/AC add 1.2   *1.4*

Engine Assy., Replace (B)
Includes: Component transfer and engine tune-up.
- Eurovan 13.8   *14.0*
- Vanagon
  - 1984-87 .... 6.9   *7.1*
  - 1988-91 .... 10.8   *11.0*
- w/AC add
  - Eurovan 1.2   *1.4*
  - 81-83 Vanagon ... 1.2   *1.4*
- w/AT Eurovan add 1.1   *1.3*

Engine Assy. (Short), Replace (B)
Assembly consists of cylinder block, piston assemblies, crankshaft, camshaft, timing chain and gears. Does not include cylinder heads. Operation Includes: R&R engine, transfer necessary parts and all necessary adjustments.
- Gasoline
  - Eurovan
    - AT 20.9   *21.1*
    - MT 19.7   *19.9*
  - Vanagon
    - AT 10.2   *10.4*
    - MT 9.2   *9.4*
- Diesel
  - Vanagon 19.1   *19.3*
- w/AC add
  - Eurovan 1.2   *1.4*
  - 81-83 Vanagon ... 1.2   *1.4*
- w/AT Eurovan add 1.1   *1.3*

Engine Carrier, Replace (B)
- 1984-91 Vanagon 2.9   *3.1*

Engine Mounts, Replace (B)
- Gasoline
  - Eurovan one side ... 1.2   *1.4*
  - Vanagon 3.0   *3.2*
- Diesel
  - Vanagon 1.4   *1.6*

CYLINDER HEAD

Compression Test (B)
- Eurovan 8   *1.0*
- Vanagon
  - gas 7   *.9*
  - diesel 1.1   *1.3*

Cylinder Head, Retorque (B)
- 1984-94
  - water-cooled 9   *1.1*
  - diesel 1.8   *2.0*

EUROVAN : VANAGON

	LABOR TIME	SEVERE SERVICE
Valve Clearance, Adjust (B)		
Gasoline		
Eurovan	3.4	3.6
1988-91 Vanagon	1.3	1.5
Diesel		
Vanagon	1.4	1.6
Cylinder Head, Replace (B)		
Includes: Adjustments.		
Gasoline		
Eurovan	8.2	8.4
Vanagon		
1981-83		
one side	9.4	9.6
both sides	11.6	11.8
1984-87 one	9.1	9.3
1988-91		
one side	7.1	7.3
both sides	9.8	10.0
Diesel		
Vanagon	5.4	5.6
w/AC Vanagon add	1.3	1.5
Cylinder Head Gasket, Replace (B)		
Gasoline		
Eurovan	6.4	6.6
Vanagon		
1981-83		
one side	9.0	9.2
both sides	10.2	10.4
1984-87 one	7.4	7.6
1988-91		
one side	6.3	6.5
both sides	8.8	9.0
Diesel		
Vanagon	5.4	5.6
w/AC Vanagon add	1.3	1.5
Pushrod Tubes, Reseal (B)		
Vanagon		
1981-83		
one side	1.3	1.5
both sides	2.9	3.1
1984-87 one side	2.6	2.8
1988-91 one side	1.8	2.0
Pushrods, Replace (B)		
1981-83 Vanagon		
one side	1.9	2.1
both sides	2.4	2.6
Rocker Arms, Replace (B)		
1981-83 Vanagon		
one side	1.3	1.5
both sides	1.9	2.1
Valve Cover Gasket, Replace (B)		
Gasoline		
Eurovan	2.2	2.4
Vanagon		
one side	.5	.7
both sides	.7	.9
Diesel		
Vanagon	.7	.9

	LABOR TIME	SEVERE SERVICE
Valve Job (A)		
Gasoline		
Eurovan	9.9	10.1
Vanagon		
1981-83		
one side	9.8	10.0
both sides	13.0	13.2
1984-87 one side	9.8	10.0
1988-91		
one side	7.8	8.0
both sides	10.7	10.9
Diesel		
Vanagon all	10.6	10.8
w/AC Vanagon add	1.3	1.5
Valve Lifters, Replace (B)		
Eurovan all	3.7	3.9
1988-91 Vanagon all	2.8	3.0
Valve Springs and/or Oil Seals, Replace (B)		
Gasoline		
Eurovan	6.8	7.0
1981-83 Vanagon		
one	1.6	1.8
all	3.0	3.2
Diesel		
Vanagon	3.8	4.0

CAMSHAFT

	LABOR TIME	SEVERE SERVICE
Timing Belt, Adjust (B)		
Gasoline		
Eurovan	1.5	1.7
Diesel		
Vanagon	1.2	1.4
Cam Followers, Replace (B)		
Vanagon		
1981-83		
one side	1.9	2.1
both sides	2.9	3.1
1984-91	2.1	2.3
Vanagon diesel	2.8	3.0
Camshaft, Replace (A)		
Gasoline		
Vanagon		
1981-83		
AT	16.9	17.1
MT	15.3	15.5
1984-91	15.7	15.9
w/AC 81-83 Vanagon add	1.3	1.5
Camshaft Seal, Replace (B)		
Gasoline		
Eurovan front	2.7	2.9
Diesel		
Vanagon rear	2.3	2.5
Camshaft Timing Gear or Sprocket, Replace (B)		
Eurovan	2.5	2.7
1984-91 Vanagon	1.9	2.1

	LABOR TIME	SEVERE SERVICE
Timing Belt, Replace (B)		
Gasoline		
Eurovan	3.8	4.0
Diesel		
1982-84 Vanagon	3.2	3.4
Timing Belt Tensioner, Replace (B)		
Eurovan	1.4	1.6
Timing Belt/Chain Cover, Replace (B)		
Gasoline		
Eurovan	.8	1.0
Diesel		
Vanagon	.7	.9

CRANK & PISTONS

	LABOR TIME	SEVERE SERVICE
Connecting Rod Bearings, Replace (A)		
Includes: Check bearing oil clearance.		
1981-94	3.3	3.5
Crankshaft and Main Bearings, Replace (A)		
Includes: Engine R&R, check bearing oil clearance.		
Gasoline		
Eurovan	18.3	18.5
Vanagon		
1981-83		
AT	19.8	20.0
MT	18.7	18.9
1984-91	18.0	18.2
Diesel		
Vanagon	9.9	10.1
w/AC add		
Eurovan	1.2	1.4
81-83 Vanagon	1.2	1.4
w/AT Eurovan add	1.1	1.3
Crankshaft Front Oil Seal, Replace (B)		
Gasoline		
Eurovan	2.9	3.1
Vanagon		
1981-83	2.1	2.3
1984-87	2.1	2.3
1988-91	2.8	3.0
Diesel		
Vanagon	3.4	3.6
w/AC 81-87 Vanagon add	.9	1.1
Crankshaft Pulley, Replace (B)		
Eurovan	1.6	1.8
Vanagon		
1981-83	.9	1.1
1984-91	2.4	2.6
w/AC 81-83 Vanagon add	.9	1.1

VW-33

VW-34 EUROVAN : VANAGON

	LABOR TIME	SEVERE SERVICE
Piston and Cylinder Assy., Replace (A)		
Gasoline		
Eurovan		
one cyl.	10.2	*10.4*
all cyls.	12.3	*12.5*
Vanagon		
1981-83		
AT		
one side	9.0	*9.2*
both sides	10.9	*11.1*
MT		
one side	8.4	*8.6*
both sides	10.2	*10.4*
1984-91		
one side	8.9	*9.1*
both sides	12.0	*12.2*
w/AC 81-91 Vanagon add	1.3	*1.5*
Pistons or Connecting Rods, Replace (A)		
Includes: Ridge reaming, cylinder wall deglazing, installing new rings and rod bearings, engine tune-up.		
Diesel		
Vanagon	9.9	*10.1*
Rear Main Oil Seal, Replace (B)		
Gasoline		
Eurovan		
AT	9.8	*10.0*
MT	9.1	*9.3*
Vanagon		
1981-83		
AT	5.0	*5.2*
MT	3.6	*3.8*
1984-91		
AT	6.2	*6.4*
MT	5.7	*5.9*
Diesel		
Vanagon	5.1	*5.3*
w/AC 81-91 Vanagon add	1.3	*1.5*
Rings, Replace (A)		
Includes: Ridge reaming, cylinder wall deglazing, installing new rings, engine tune-up.		
Gasoline		
Eurovan		
one cyl.	9.3	*9.5*
all cyls.	11.8	*12.0*
Vanagon		
1981-83		
AT		
one side	9.8	*10.0*
both sides	11.8	*12.0*
MT		
one side	8.3	*8.5*
both sides	10.2	*10.4*

	LABOR TIME	SEVERE SERVICE
1984-91		
one side	9.8	*10.0*
both sides	13.1	*13.3*
Diesel		
Vanagon	9.5	*9.7*
w/AC 81-91 Vanagon add	1.3	*1.5*
ENGINE LUBRICATION		
Engine Oil Pressure Switch (Sending Unit), Replace (B)		
Gasoline		
Eurovan	.7	*.9*
Vanagon		
1981-83		
AT	6.7	*6.9*
MT	5.5	*5.7*
1984-87	.7	*.9*
1988-91	1.3	*1.5*
Diesel		
Vanagon	.7	*.9*
w/AC 81-83 Vanagon add	1.3	*1.5*
Engine Oil Cooler, Replace (B)		
Eurovan	1.5	*1.7*
1981-83 Vanagon		
AT	6.1	*6.3*
MT	4.8	*5.0*
w/AC add	1.3	*1.5*
Oil Pan and/or Gasket, Replace (B)		
Gasoline		
Eurovan	3.0	*3.2*
Vanagon	1.3	*1.5*
Diesel		
Vanagon	3.2	*3.4*
Oil Pump, Replace (B)		
Gasoline		
Eurovan	4.9	*5.1*
Vanagon	4.1	*4.3*
Diesel		
Vanagon	3.6	*3.8*
CLUTCH		
Clutch Hydraulic System, Bleed (B)		
1981-94	.5	*.7*
Clutch Pedal Free Play, Adjust (B)		
1981-94	.3	*.5*
Clutch Assy., Replace (B)		
Eurovan	7.8	*8.0*
Vanagon		
1981-87		
gasoline	3.4	*3.6*
diesel	4.6	*4.8*
1988-91		
2WD	3.6	*3.8*
4WD	4.9	*5.1*
Clutch Master Cylinder, Replace (B)		
Includes: System bleeding.		
Eurovan	2.3	*2.5*
Vanagon	1.5	*1.7*

	LABOR TIME	SEVERE SERVICE
Clutch Release Bearing, Replace (B)		
Eurovan	7.7	*7.9*
Vanagon		
1981-87		
gasoline	3.2	*3.4*
diesel	4.1	*4.3*
1988-91		
2WD	3.3	*3.5*
4WD	4.3	*4.5*
Clutch Slave Cylinder, Replace (B)		
Includes: System bleeding.		
Eurovan	1.5	*1.7*
Vanagon	1.6	*1.8*
Flywheel, Replace (B)		
Eurovan	8.4	*8.6*
Vanagon		
1981-87		
gasoline	3.8	*4.0*
diesel	5.1	*5.3*
1988-91		
2WD	4.1	*4.3*
4WD	5.1	*5.3*
MANUAL TRANSAXLE		
Differential Assembly, R&R and Recondition (A)		
Eurovan	16.7	*16.9*
Transaxle Assy., R&R and Recondition (A)		
Eurovan	14.9	*15.1*
Transaxle Assy., R&R or Replace (B)		
Eurovan	7.6	*7.8*
MANUAL TRANSMISSION		
Transmission Assy., R&I (B)		
Vanagon		
1981-87		
gasoline	3.1	*3.3*
diesel	4.1	*4.3*
1988-91		
2WD	3.3	*3.5*
4WD	4.5	*4.7*
Transmission Assy., R&R and Recondition (A)		
Vanagon		
1981-87		
gasoline	8.9	*9.1*
diesel	9.5	*9.7*
1988-91		
2WD	12.5	*12.7*
4WD	13.5	*13.7*
AUTOMATIC TRANSAXLE		
SERVICE TRANSAXLE INSTALLED		
Performance Test (B)		
Eurovan	1.3	*1.5*
Check Unit for Oil Leaks (C)		
Eurovan	.5	*.7*
Drain & Refill Unit (B)		
Eurovan	.9	*1.1*

EUROVAN : VANAGON — VW-35

	LABOR TIME	SEVERE SERVICE
Governor Assy., R&R or Replace (B)		
Eurovan	.5	.7
Oil Pan and/or Gasket, Replace (B)		
Eurovan	.9	1.1
Shift Lock Relay, Replace (B)		
Eurovan	.5	.7
Shift Lock Solenoid & Switch, Replace (B)		
Eurovan	1.6	1.8
Speed Sensor, Replace (B)		
Eurovan	.7	.9
Transaxle Control Unit, Replace (B)		
Eurovan	.7	.9
Valve Body, Replace (B)		
Eurovan	1.2	1.4

SERVICE TRANSAXLE REMOVED
Transaxle R&R included unless otherwise noted.

	LABOR TIME	SEVERE SERVICE
Differential Assy., Recondition (A)		
Eurovan	15.5	15.7
Flywheel & Ring Gear Assy., Replace (B)		
Eurovan	9.2	9.4
Oil Pump and/or Gasket, Replace (B)		
Eurovan	9.5	9.7
Torque Converter Assy., Replace (B)		
Eurovan	9.1	9.3
Torque Converter Front Oil Seal, Replace (B)		
Eurovan	9.1	9.3
Transaxle Assy. (Complete), R&R and Recondition (A)		
Eurovan	17.6	17.8
Transaxle Assy., R&R or Replace (B)		
Eurovan	8.9	9.1

AUTOMATIC TRANSMISSION

SERVICE TRANSMISSION INSTALLED

	LABOR TIME	SEVERE SERVICE
Performance Test (B)		
Vanagon	1.1	1.3
Check Unit for Oil Leaks (C)		
Vanagon	.5	.7
Drain & Refill Unit (B)		
Vanagon	.9	1.1
Governor Assy., R&R or Replace (B)		
Vanagon	.7	.9
Kickdown Switch, Replace (B)		
Vanagon	.7	.9
Oil Pan and/or Gasket, Replace (B)		
Vanagon	.9	1.1
Vacuum Modulator, Replace (B)		
Vanagon	.7	.9
Valve Body Assy., R&R and Recondition (A)		
Vanagon		
1981-87	1.8	2.0
1988-91	2.9	3.1
Valve Body Assy., Replace (B)		
Vanagon	1.4	1.6

SERVICE TRANSMISSION REMOVED
Transmission R&R included unless otherwise noted.

	LABOR TIME	SEVERE SERVICE
Torque Converter, Replace (B)		
Vanagon		
1981-87		
air-cooled	2.9	3.1
water-cooled	5.0	5.2
1988-91	4.9	5.1
Replace seal add	.3	.5
Transmission Assy., R&I (B)		
Vanagon		
1981-87		
air-cooled	2.7	2.9
water-cooled	4.7	4.9
1988-91	4.6	4.8
Transmission Assy., R&R and Recondition (A)		
Vanagon		
1981-87		
air-cooled	10.7	10.9
water-cooled	13.1	13.3
1988-91	13.0	13.2

SHIFT LINKAGE

AUTOMATIC TRANSMISSION

	LABOR TIME	SEVERE SERVICE
Selector Cable, Adjust (B)		
1981-94	.7	.9
Gear Selector Lever, Replace (B)		
Eurovan	1.4	1.6
Vanagon	1.5	1.7
Selector Cable, Replace (B)		
Eurovan	2.2	2.4
Vanagon	1.3	1.5

MANUAL TRANSMISSION

	LABOR TIME	SEVERE SERVICE
Gearshift Linkage, Adjust (B)		
Eurovan	.7	.9
Vanagon		
1981-87	.7	.9
1988-91	.9	1.1
Gearshift Control Rod, Replace (B)		
Eurovan	1.7	1.9
Vanagon		
front	1.7	1.9
rear	.7	.9
Gearshift Lever Boot, Replace (B)		
1981-93	.5	.7
Gearshift Lever, Replace (B)		
Vanagon		
1981-87	1.1	1.3
1988-91	1.5	1.7

DRIVELINE

DRIVESHAFT

	LABOR TIME	SEVERE SERVICE
Driveshaft, R&R or Replace (B)		
Vanagon Syncro	1.2	1.4
Driveshaft Shaft Seal, Replace (B)		
Vanagon Syncro	1.5	1.7

DRIVE AXLE

	LABOR TIME	SEVERE SERVICE
Axle Housing & Differential Assy., Replace (B)		
AT		
Eurovan	15.0	15.2
Vanagon		
1981-87		
air-cooled	6.9	7.1
water-cooled	9.0	9.2
1988-91	9.1	9.3
MT		
Eurovan	13.3	13.5
Vanagon		
2WD	8.0	8.2
4WD	8.3	8.5
Differential Carrier Bearings, Replace (B)		
Eurovan	10.7	10.9
Vanagon		
2WD	7.4	7.6
4WD	8.0	8.2
Differential Pinion Seal, Replace (B)		
Vanagon		
1981-87		
air-cooled	4.3	4.5
water-cooled	6.4	6.6
1988-91	4.9	5.1
Differential Side Gear Set, Replace (B)		
MT		
Eurovan	11.4	11.6
Vanagon		
2WD	6.8	7.0
4WD	7.5	7.7
Drive Flange Shaft Seal, Replace (B)		
Eurovan		
one	3.3	3.5
both	4.0	4.2
Vanagon	2.1	2.3
Final Drive Assy., R&R or Replace (B)		
Vanagon Syncro	2.9	3.1
Final Drive Flange, Replace (B)		
Vanagon	1.9	2.1

EUROVAN : VANAGON

	LABOR TIME	SEVERE SERVICE
Rear Wheel Bearings and/or Seals, Replace (B)		
One side	2.0	2.2
Both sides	3.2	3.4
Ring Gear & Pinion Set, Replace (A)		
AT		
Eurovan	15.4	15.6
Vanagon		
1981-87		
air-cooled	7.9	8.1
water-cooled	9.7	9.9
1988-91	9.8	10.0
MT		
Eurovan	14.1	14.3
Vanagon		
2WD	10.1	10.3
4WD	10.8	11.0
Stub Axle Shaft, Replace (B)		
Vanagon		
one	2.5	2.7
both	4.5	4.7

HALFSHAFTS

	LABOR TIME	SEVERE SERVICE
CV Joint, Replace or Recondition (B)		
Eurovan		
one side	2.8	3.0
Front Halfshaft, R&R or Replace (B)		
Front		
Vanagon		
1981-87		
one	.7	.9
both	1.4	1.6
1988-91		
4WD one	2.5	2.7
Replace or recondition CV joint/boot add		
inner or outer one	.5	.7
inner and outer one side	.7	.9
Halfshaft, R&R or Replace (B)		
Eurovan	1.4	1.6
Replace CV joint boot add	.7	.9
Rear CV Joint, Replace or Recondition (B)		
Vanagon Syncro		
one side	3.1	3.3
both sides	4.0	4.2
Rear Halfshaft, R&R or Replace (B)		
Vanagon		
1981-87		
one	1.4	1.6
both	2.5	2.7
1988-91		
one	1.2	1.4
both	2.0	2.2
Recondition shafts add each	.7	.9

BRAKES
ANTI-LOCK

	LABOR TIME	SEVERE SERVICE
Diagnose Anti-Lock Brake System (A)		
Eurovan	1.0	1.0
Anti-Lock Relay, Replace (B)		
Eurovan each	.5	.7
Control Unit, Replace (B)		
Eurovan	.7	.9
Hydraulic Assy., Replace (B)		
Eurovan	2.4	2.6
Toothed Rotor, Replace (B)		
Eurovan		
right front	2.1	2.3
left front	2.6	2.8
rear each	1.7	1.9
Wheel Speed Sensor, Replace (B)		
Eurovan		
front		
one	.7	.9
both	.9	1.1
rear each	.9	1.1

SYSTEM

	LABOR TIME	SEVERE SERVICE
Bleed Brakes (B)		
Includes: Add fluid.		
1981-94	.5	.5
Brake System, Flush and Refill (B)		
1981-94	1.2	1.4
Brakes, Adjust (B)		
Includes: Refill master cylinder.		
1981-94 rear wheels	.3	.5
Brake Hose (Flexible), Replace (B)		
Includes: System bleeding.		
Eurovan		
one	.7	.9
each addl. add	.3	.5
Vanagon		
1981-87	.9	1.1
1988-91		
one	.7	.9
each addl. add	.3	.5
Brake Pressure Regulator, Replace (B)		
Eurovan		
w/ABS	1.2	1.4
w/o ABS	1.9	2.1
Vanagon	1.8	2.0
Master Cylinder, Replace (B)		
Eurovan		
w/ABS	2.2	2.4
w/o ABS	.9	1.1
Vanagon		
1981-87	2.8	3.0
1988-91	2.8	3.0

	LABOR TIME	SEVERE SERVICE
Power Booster Unit, Replace (B)		
Eurovan		
w/ABS	3.1	3.3
w/o ABS	1.9	2.1
Vanagon		
1981-87	3.3	3.5
1988-91	3.0	3.2
Power Booster Vacuum Check Valve, Replace (B)		
1981-94	.3	.3
Pressure Limiting Valve, Replace (B)		
Includes: System bleeding.		
Vanagon		
1981-87	1.6	1.8
1988-91	1.4	1.6
Vacuum Pump, Replace (B)		
Eurovan	.8	1.0
Vanagon	.8	1.0

SERVICE BRAKES

	LABOR TIME	SEVERE SERVICE
Brake Drum, Replace (B)		
Eurovan		
one	.5	.7
both	.7	.9
Vanagon		
1981-87	.7	.9
1988-91	1.2	1.4
Caliper Assy., R&R and Recondition (A)		
Includes: System bleeding.		
Eurovan		
one	2.2	2.4
both	3.8	4.0
Vanagon		
1981-87		
one	2.2	2.4
both	3.7	3.9
1988-91		
one	1.7	1.9
both	3.3	3.5
Caliper Assy., Replace (B)		
Includes: System bleeding.		
Exc. below		
one	1.3	1.5
both	2.3	2.5
1988-91 Vanagon		
one	1.4	1.6
both	2.4	2.6
Disc Brake Rotor, Replace (B)		
Eurovan		
one	.9	1.1
both	1.8	2.0
Vanagon		
1981-87		
one	1.6	1.8
both	3.0	3.2
1988-91		
one	1.1	1.3
both	2.1	2.3

EUROVAN : VANAGON **VW-37**

	LABOR TIME	SEVERE SERVICE
Pads and/or Shoes, Replace (B)		
Includes: Adjust service and parking brake. System bleeding.		
Eurovan		
front disc	1.1	*1.3*
rear drum	1.9	*2.1*
four wheels	2.9	*3.1*
Vanagon		
1981-87		
front disc	1.3	*1.5*
rear drum	1.4	*1.6*
four wheels	2.4	*2.6*
1988-91		
front disc	1.3	*1.5*
rear drum	1.9	*2.1*
four wheels	2.8	*3.0*
COMBINATION ADD-ONS		
Repack wheel bearings		
two wheels add	.5	*.7*
Replace		
brake drum add	.3	*.5*
brake hose add	.5	*.7*
caliper add	.5	*.7*
disc rotor add each	.3	*.5*
wheel cylinder add	.5	*.7*
Resurface		
brake rotor add	.5	*.7*
brake drum add	.5	*.7*
Wheel Cylinder, R&R and Recondition (B)		
Includes: System bleeding.		
Eurovan		
one	2.0	*2.2*
both	3.6	*3.8*
Vanagon		
1981-87		
one	2.0	*2.2*
both	3.4	*3.6*
1988-91		
one	2.0	*2.2*
both	3.5	*3.7*
Wheel Cylinder, Replace (B)		
Includes: System bleeding.		
Eurovan		
one	1.7	*1.9*
both	3.2	*3.4*
Vanagon		
one	1.7	*1.9*
both	3.0	*3.2*

PARKING BRAKE

	LABOR TIME	SEVERE SERVICE
Parking Brake Cable, Adjust (C)		
1981-94	.7	*.9*
Parking Brake Apply Actuator, Replace (B)		
1981-94	1.3	*1.5*
Parking Brake Apply Warning Indicator Switch, Replace (B)		
Eurovan	.7	*.9*
Parking Brake Cable, Replace (B)		
Eurovan		
one	1.4	*1.6*
both	2.2	*2.4*
Vanagon		
1981-87		
one	1.2	*1.4*
both	2.0	*2.2*
1988-91		
one	1.4	*1.6*
both	1.8	*2.0*

FRONT SUSPENSION

Unless otherwise noted, time given does not include alignment.

	LABOR TIME	SEVERE SERVICE
Align Front End (A)		
Eurovan	1.5	*1.7*
Vanagon	1.4	*1.6*
Front Toe, Adjust (B)		
All Models	.7	*.9*
Coil Spring, Replace (B)		
Includes: Alignment.		
Vanagon	2.3	*2.5*
Front Hub & Bearing Assy., Replace (B)		
Eurovan		
one side	1.4	*1.6*
both sides	2.5	*2.7*
Front Hub Bearing or Seal, Replace (B)		
Vanagon		
one	1.4	*1.6*
both	2.4	*2.6*
Front Shock Absorbers, Replace (B)		
Eurovan		
one	.8	*1.0*
both	1.3	*1.5*
Vanagon		
2WD		
one	1.1	*1.3*
both	2.1	*2.3*
4WD		
one	1.7	*1.9*
both	3.1	*3.3*
Lower Ball Joint, Replace (B)		
Eurovan		
one side	.8	*1.0*
both sides	1.1	*1.3*
Vanagon		
one side	2.0	*2.2*
both sides	3.5	*3.7*
Lower Control Arm, Replace (B)		
Eurovan		
AT		
one side	2.0	*2.2*
both sides	2.8	*3.0*
MT		
one side	1.8	*2.0*
both sides	1.9	*2.1*

	LABOR TIME	SEVERE SERVICE
Vanagon		
one side	2.8	*3.0*
both sides	4.7	*4.9*
w/AT add	.5	*.7*
Radius Rod, Replace (B)		
Vanagon	1.2	*1.4*
Stabilizer Bar, Replace (B)		
Eurovan	.9	*1.1*
Vanagon	1.2	*1.4*
Steering Knuckle, Replace (B)		
Eurovan	2.4	*2.6*
Vanagon		
one side	2.5	*2.7*
both sides	4.3	*4.5*
Upper Ball Joint, Replace (B)		
Eurovan		
one side	.8	*1.0*
both sides	1.7	*1.9*
Vanagon		
one side	1.1	*1.3*
both sides	1.7	*1.9*
Upper Control Arm Assy., Replace (B)		
Eurovan		
one side	3.1	*3.3*
both sides	4.0	*4.2*
Vanagon		
one side	1.2	*1.4*
both sides	2.2	*2.4*
Wheel Bearings, Clean & Pack (B)		
Vanagon both wheels		
1981-87	1.5	*1.7*
1988-91		
2WD	1.6	*1.8*
4WD	2.3	*2.5*
Wheel Bearings, Replace (B)		
Vanagon		
1981-87	1.4	*1.6*
1988-91		
2WD	1.5	*1.7*
4WD	2.4	*2.6*
Wheel Hub, Replace (C)		
1992-94	1.9	*2.1*

REAR SUSPENSION

	LABOR TIME	SEVERE SERVICE
Rear Alignment, Adjust (A)		
1981-93	1.3	*1.5*
Rear Control Arm, Replace (B)		
Vanagon		
one	2.5	*2.7*
both	4.4	*4.6*
Rear Spring, Replace (B)		
Eurovan		
one side	.8	*1.0*
both sides	1.7	*1.9*
Vanagon	1.4	*1.6*
Shock Absorbers or Bushings, Replace (B)		
One	.7	*.9*
Both	1.5	*1.7*

EUROVAN : VANAGON

	LABOR TIME	SEVERE SERVICE
Trailing Arm, Replace (B)		
Eurovan		
one side	1.9	2.1
both sides	3.7	3.9
Wheel Hub, Replace (B)		
Vanagon	1.2	1.4

STEERING

MANUAL RACK & PINION

Unless otherwise noted, time given does not include alignment.

Coupling Disc, Replace (B)
- 1988-91 Vanagon 7   .9

Inner Tie Rod Assy., Replace (B)
- Vanagon
  - 1981-87
    - one side 1.3   1.5
    - both sides 1.7   1.9
  - 1988-91
    - one side 8   1.0
    - both sides 1.5   1.7

Inner Tie Rod End Boot, Replace (B)
- 1981-91 each 7   .9

Outer Tie Rod End, Replace (B)
- Vanagon
  - one side 7   .9
  - both sides 1.1   1.3

Rack & Pinion Assy., Replace (B)
- Vanagon
  - 1981-87 .... 2.6   2.8
  - 1988-91 .... 2.3   2.5

Steering Wheel, Replace (B)
- 1981-91 5   .7

Upper Mast Jacket Bearing, Replace (B)
- 1988-91 Vanagon 1.2   1.4

POWER RACK & PINION

Unless otherwise noted, time given does not include alignment.

Outer Tie Rod Ends, Replace (B)
- Eurovan
  - one 9   1.1
  - both 1.3   1.5

Rack & Pinion Assy., R&R or Replace (B)
- Eurovan 4.9   5.1
- Vanagon 2.9   3.1

Steering Pump, R&R or Replace (B)
- Eurovan 1.6   1.8
- Vanagon 1.2   1.4

Steering Pump Hoses, Replace (B)
- Eurovan
  - pressure 1.4   1.6
  - return 8   1.0
- Vanagon
  - pressure 8   1.0
  - return 8   1.0

Steering Wheel, Replace (B)
- Eurovan
  - w/air bag 8   1.0
  - w/o air bag 3   .5
- Vanagon 5   .7

Upper Mast Jacket Bearing Replace (B)
- Eurovan 1.7   1.9
- Vanagon 1.4   1.6

HEATING & AIR CONDITIONING

When more than one component requires replacement where evacuation/recovery and recharging is already included, deduct 1.0 hour for each additional component from the time given.

Evacuate/Recover and Recharge System (B)
- 1981-94 .... 1.1   1.1

AC Hoses, Replace (B)
Includes: Evacuate/recover and recharge.
- Eurovan
  - suction 1.9   2.1
  - discharge 2.0   2.2
- Vanagon
  - suction 4.5   4.7
  - discharge 3.2   3.4

AC Low Pressure Switch, Replace (B)
Includes: Evacuate/recover and recharge.
- Eurovan 1.4   1.6

AC Relay, Replace (B)
- Eurovan 8   1.0

Compressor Assy., Replace (B)
Includes: Parts transfer. Evacuate/recover and recharge.
- Eurovan 3.8   4.0
- Vanagon
  - 1981-87 .... 3.7   3.9
  - 1988-91 .... 3.8   4.0

Compressor Clutch Assy., Replace (B)
Includes: Evacuate/recover and recharge.
- Eurovan 3.8   4.0
- Vanagon 4.4   4.6

Compressor Shaft Seal, Replace (B)
Includes: Compressor R&R. Evacuate/recover and recharge.
- Eurovan 3.9   4.1
- Vanagon
  - 1981-87 .... 2.5   2.7
  - 1988-91 .... 4.2   4.4

Condenser Assy., Replace (B)
Includes: Evacuate/recover and recharge.
- Eurovan 3.3   3.5
- Vanagon
  - 1981-87 .... 2.8   3.0
  - 1988-91 .... 2.8   3.0

Condenser Fan Motor, Replace (B)
- Vanagon
  - 1981-87 one 1.9   2.1
  - 1988-91 one 8   1.0

Evaporator Coil, Replace (B)
Includes: Evacuate/recover and recharge.
- Eurovan
  - front 5.0   5.2
  - rear 5.4   5.6
- Vanagon
  - 1981-87 .... 3.9   4.1
  - 1988-91 .... 5.4   5.6

Evaporator Fan Motor, Replace (B)
- Eurovan 7   .9
- Vanagon 3.4   3.6

Expansion Valve, Replace (B)
Includes: Evacuate/recover and recharge.
- Vanagon
  - 1981-87 .... 2.9   3.1
  - 1988-91 .... 3.9   4.1

Front Heater Blower Motor, Replace (B)
- Eurovan 7   .9
- Vanagon 4.4   4.6

Heater Control Valve, Replace (B)
- Vanagon diesel 7   .9

Heater Core, R&R or Replace (B)
- Eurovan 5.4   5.6
- 1984-91 Vanagon ... 4.8   5.0

High Pressure Cut-Off Switch, Replace (B)
Includes: Evacuate/recover and recharge.
- Eurovan 1.8   2.0

Mode Door Actuator or Damper, Replace (B)
- Vanagon 2.6   2.8

Rear Heater Blower Motor, Replace (B)
- Eurovan 7   .9

Rear Heater Blower Motor Switch, Replace (B)
- Eurovan 5   .7

Rear Heater Core, R&R or Replace (B)
- Eurovan 1.9   2.1
- 1984-91 Vanagon ... 1.7   1.9

Temperature Control Assy., Replace (B)
- Eurovan 1.4   1.6

EUROVAN : VANAGON VW-39

	LABOR TIME	SEVERE SERVICE
Temperature Control Cable, Replace (B)		
1981-87 Vanagon	1.8	2.0
Thermostatic Switch, Replace (B)		
Eurovan	1.4	1.6
1982-91 Vanagon	1.8	2.0

WIPERS & SPEEDOMETER

Antenna Assy., Replace (B)		
Eurovan	1.3	1.5
Radio, R&R (B)		
1981-94	.8	1.0
Rear Window Washer Pump, Replace (B)		
Eurovan	.9	1.1
Vanagon	.7	.9
Rear Window Wiper Interval Relay, Replace (B)		
Eurovan	.7	.9
Rear Window Wiper Motor, Replace (B)		
Eurovan	1.9	2.1
1988-91 Vanagon	1.9	2.1
Speedometer Cable & Casing, Replace (B)		
Vanagon		
1981-87	1.5	1.7
1988-91 each	.9	1.1
Speedometer Head, R&R or Replace (B)		
Eurovan	1.7	1.9
Vanagon		
1981-87	.9	1.1
1988-91	1.6	1.8
Windshield Washer Pump, Replace (B)		
Eurovan	.9	1.1
Vanagon		
1981-87	.7	.9
1988-91	1.2	1.4
Windshield Wiper & Washer Switch, Replace (B)		
Eurovan	1.2	1.4
Vanagon	1.4	1.6
Windshield Wiper Interval Relay, Replace (B)		
Eurovan	.7	.9
Vanagon		
1981-87	.5	.7
1988-91	.7	.9

	LABOR TIME	SEVERE SERVICE
Windshield Wiper Linkage, Replace (B)		
Eurovan	.9	1.1
Vanagon		
1981-87	2.7	2.9
1988-91	2.8	3.0
Windshield Wiper Motor, Replace (B)		
Eurovan	1.4	1.6
Vanagon		
1981-87		
w/AC	2.5	2.7
w/o AC	1.4	1.6
1988-91		
w/AC	2.8	3.0
w/o AC	1.6	1.8

LAMPS & SWITCHES

Back-Up Lamp Bulb, Replace (B)		
1981-94	.2	.2
Back-Up Lamp Switch, Replace (B)		
All Models	.5	.5
Brake Lamp Switch, Replace (B)		
1981-94	.7	.7
Composite Headlamp Bulb, Replace (C)		
1993-94	.3	.3
Emergency Flasher Switch Assy., Replace (B)		
Eurovan	.5	.5
Vanagon		
1981-87	.3	.3
1988-91	.5	.5
Headlamp Dimmer Relay, Replace (B)		
Vanagon	.5	.5
Headlamp Switch, Replace (B)		
All	.7	.7
Headlamps, Aim (B)		
1981-94	.5	.5
High Mount Stop Lamp Bulb, Replace (C)		
Eurovan	.2	.2
Horn, Replace (B)		
All Models	.5	.5
Horn Relay, Replace (B)		
Vanagon	.3	.3
License Lamp Bulb, Replace (C)		
Eurovan one or all	.2	.2

	LABOR TIME	SEVERE SERVICE
Neutral Safety Switch, Replace (B)		
Vanagon	1.6	1.8
Parking Brake Warning Lamp Switch, Replace (B)		
Eurovan	.5	.5
Vanagon	.2	.2
Parking Lamp Lens or Bulb, Replace (C)		
1981-94 each	.2	.2
Sealed Beam Headlamp, Replace (C)		
1981-91	.3	.3
Stop & Tail Lamp Bulb, Replace (C)		
1981-94	.2	.2
Turn Signal & Windshield Wiper Switch, Replace (B)		
All Models	1.1	1.1
Turn Signal or Hazard Relay, Replace (B)		
Eurovan	.7	.7
Vanagon		
1981-87	.5	.5
1988-91	.7	.7
Turn Signal Switch, Replace (B)		
All Models	.9	.9

BODY

Front Door Lock, Replace (B)		
All Models	.5	.7
Hood Hinge, Replace (B)		
Eurovan		
one	1.5	1.7
both	1.6	1.8
Hood Release Cable, Replace (B)		
Eurovan	.5	.7
Lock Striker Plate, Replace (B)		
1981-94	.5	.7
Window Regulator and/or Motor, Replace (B)		
Eurovan		
electric	1.9	2.1
manual	1.5	1.7
Vanagon		
1981-87	.8	1.0
1988-91		
electric	1.9	2.1
manual	1.6	1.8

NOTES

VLV

240 : 242 : 244 : 245 : 262 : 264 : 265 : 740 : 745 : 760 : 780 : 850 : 940 : 960 : C70 : S40 : S60 : S70 : S80 : S90 : V40 : V70 : V90 : XC70 : XC90

SYSTEM INDEX

MAINTENANCE	VLV-2
Maintenance Schedule	VLV-2
CHARGING	VLV-3
STARTING	VLV-3
CRUISE CONTROL	VLV-4
IGNITION	VLV-4
EMISSIONS	VLV-5
FUEL	VLV-6
EXHAUST	VLV-7
ENGINE COOLING	VLV-8
ENGINE	VLV-9
Assembly	VLV-9
Cylinder Head	VLV-10
Camshaft	VLV-11
Crank & Pistons	VLV-12
Engine Lubrication	VLV-13
CLUTCH	VLV-13
MANUAL TRANSAXLE	VLV-14
MANUAL TRANSMISSION	VLV-14
AUTO TRANSAXLE	VLV-14
AUTO TRANSMISSION	VLV-15
SHIFT LINKAGE	VLV-15
DRIVELINE	VLV-15
BRAKES	VLV-16
FRONT SUSPENSION	VLV-18
REAR SUSPENSION	VLV-19
STEERING	VLV-21
HEATING & AC	VLV-22
WIPERS & SPEEDOMETER	VLV-23
LAMPS & SWITCHES	VLV-24
BODY	VLV-25

OPERATIONS INDEX

A
AC Hoses	VLV-22
Air Bags	VLV-21
Air Conditioning	VLV-22
Alignment	VLV-18
Alternator (Generator)	VLV-3
Antenna	VLV-23
Anti-Lock Brakes	VLV-16

B
Back-Up Lamp Switch	VLV-24
Ball Joint	VLV-18
Battery Cables	VLV-3
Bleed Brake System	VLV-16
Blower Motor	VLV-23
Brake Disc	VLV-17
Brake Hose	VLV-16
Brake Pads and/or Shoes	VLV-17

C
Camshaft	VLV-11
Camshaft Sensor	VLV-4
Catalytic Converter	VLV-7
Coolant Temperature (ECT) Sensor	VLV-5
Crankshaft	VLV-12
Crankshaft Sensor	VLV-4
Cruise Control	VLV-4
CV Joint	VLV-16
Cylinder Head	VLV-10

D
Differential	VLV-15
Distributor	VLV-4
Drive Belt	VLV-2
Driveshaft	VLV-15

E
EGR	VLV-5
Electronic Control Module (ECM/PCM)	VLV-5
Engine	VLV-9
Engine Lubrication	VLV-13
Engine Mounts	VLV-9
Evaporator	VLV-22
Exhaust	VLV-7
Exhaust Manifold	VLV-8
Expansion Valve	VLV-23

F
Flexplate	VLV-14
Flywheel	VLV-14
Fuel Injection	VLV-6
Fuel Pump	VLV-6
Fuel Vapor Canister	VLV-5

G
Generator	VLV-3
Glow Plug	VLV-4

H
Halfshaft	VLV-16
Headlamp	VLV-24
Heater Core	VLV-23
Horn	VLV-24

I
Idle Air Control (IAC) Valve	VLV-7
Ignition Coil	VLV-4
Ignition Switch	VLV-4
Injection Pump	VLV-7
Inner Tie Rod	VLV-21
Intake Air Temperature (IAT) Sensor	VLV-5
Intake Manifold	VLV-6

K
Knock Sensor	VLV-5

L
Lower Control Arm	VLV-18

M
Maintenance Schedule	VLV-2
Manifold Absolute Pressure (MAP) Sensor	VLV-5
Mass Air Flow (MAF) Sensor	VLV-5
Master Cylinder	VLV-17
Muffler	VLV-8

N
Neutral Safety Switch	VLV-24

O
Oil Pan	VLV-13
Oil Pump	VLV-13
Outer Tie Rod	VLV-21
Oxygen Sensor	VLV-5

P
Parking Brake	VLV-17
Pistons	VLV-12

R
Radiator	VLV-8
Radiator Hoses	VLV-8
Radio	VLV-23
Rear Main Oil Seal	VLV-12

S
Shock Absorber/Strut, Front	VLV-19
Shock Absorber/Strut, Rear	VLV-20
Spark Plug Cables	VLV-4
Spark Plugs	VLV-5
Spring, Front Coil	VLV-18
Spring, Rear Coil	VLV-20
Starter	VLV-3
Steering Wheel	VLV-22

T
Thermostat	VLV-8
Throttle Body	VLV-7
Throttle Position Sensor (TPS)	VLV-5
Timing Belt	VLV-11
Timing Chain	VLV-11
Torque Converter	VLV-15

U
U-Joint	VLV-15

V
Valve Body	VLV-15
Valve Cover Gasket	VLV-10
Valve Job	VLV-11
Vehicle Speed Sensor	VLV-6

W
Water Pump	VLV-9
Wheel Balance	VLV-2
Window Regulator	VLV-25
Windshield Washer Pump	VLV-24
Windshield Wiper Motor	VLV-24

VLV-2 240 : 242 : 244 : 245 : 262 : 264 : 265 : 740 : 745 : 760 : 780 : 850 : 940

MAINTENANCE

Air Cleaner Filter Element, Replace (C)
	LABOR TIME	SEVERE SERVICE
Exc. XC90	.3	.5
XC90	.6	.8

Chassis Lubrication, Change Oil & Filter (C)
Includes: Correct fluid levels.
Exc. XC90	.3	.5
XC90	.8	1.0

Drive Belt, Adjust (B)
Exc. below one	.3	.5
each addl. add	.1	.2
Injection pump	.8	1.0

Drive Belt, Replace (B)
Exc. below one	.7	.9
each addl. add	.1	.2

Compressor
740, 760, 780 V6	.3	.5
960, C70, S70, V70	.3	.5
S90, V90	.3	.5

Fan
S80	.9	1.1
XC90 B6294T	.8	1.0
Injection pump	2.2	2.4

Serpentine
1992-97 850, 960	.4	.6

1998-05
S80 (.7)	.9	1.1
XC90 B6294T (.7)	.9	1.1

Steering pump
1986-92 740	.3	.5
1986-90 760	.3	.5
780	.3	.5
940	.3	.5

Drive Belt Tensioner, Replace (B)
All Models	.5	.7

Fuel Filter, Replace (B)
1981-97 (.5)	.7	.9

1998-05
5 cyl.
 exc. XC90
w/AWD (.8)	1.1	1.3
w/o AWD (.4)	.6	.8
XC90 (.3)	.5	.7
6 cyl.	.6	.8

Halogen Headlamp Bulb, Replace (C)
Exc. below	.3	.3
1988-93 240	.5	.5
850	.5	.5

Oil & Filter, Change (C)
Includes: Correct fluid levels.
Exc. XC90	.3	.5
XC90	.8	1.0

Sealed Beam Headlamp, Replace (B)
	LABOR TIME	SEVERE SERVICE
240 each	.5	.5
260 each	.3	.3

740
 1985-87
one side	.7	.7
both sides	1.3	1.3
1988-89 each	.5	.5

760
 1983-87
one side	.7	.7
both sides	1.3	1.3
1988 each	.5	.5
1987-89 780 each	.5	.6

Timing Belt, Replace (B)
Gasoline
B21F, FT	2.6	2.8
B23F	2.8	3.0
B230F, FTL	2.9	3.1
B4914T, B4204T	3.8	4.0

B5234FT
1994-98	2.7	2.9
1999-01	2.0	2.2
B5234T3	2.8	3.0
B5244T	2.0	2.2
B5244S, B5244TC	3.5	3.7
B5254, S, T	2.8	3.0
B6244, B6254	1.9	2.1
B6284	3.6	3.8

B6284T
1999	5.3	5.5
2000-02	4.2	4.4
B6294	3.6	3.8
B6294F	2.8	3.0
B6294S	3.2	3.4
B6294T (3.0)	4.1	4.3
B6304	2.7	2.9
B6304F	3.9	4.1
B6304S	4.4	4.6

Diesel
240, 360, 740	3.4	3.6
760	3.7	3.9

Tire, Replace (C)
Includes: Dismount old tire and mount new tire to rim.
1981-05	.4	.4

Tires, Rotate (C)
All Models	.5	.5

Wheel, Balance (B)
One	.5	.5
each addl. add	.2	.2

SCHEDULED MAINTENANCE INTERVALS

If necessary, refer to appropriate Chilton maintenance service information.

5,000 Mile Service (C)
	LABOR TIME	SEVERE SERVICE
All Models	.5	.7

10,000 Mile Service (C)
All Models	1.7	1.9

20,000 Mile Service (C)
All Models	1.8	2.0

30,000 Mile Service (B)
850	3.5	3.7
940	2.9	3.1
960	2.8	3.0
S40, V40, V50	2.5	2.7
S90, V90	2.8	3.0
C70, S70, V70, XC90	3.5	3.7

40,000 Mile Service (B)
All Models	2.3	2.5

50,000 Mile Service (B)
850	1.8	2.0
940	1.8	2.0
960	3.8	4.0
C70, S70, V70, XC90	1.8	2.0
S90, V90	3.8	4.0

Replace timing belt add ... 2.0 / 2.2

60,000 Mile Service (B)
850	4.8	5.0
940	3.8	4.0
960	4.2	4.4
S40, V40, V50	3.5	3.7
S90, V90	4.2	4.2
C70, S70, V70, XC90	4.8	5.0

Inspect timing belt add ... .2 / .4

70,000 Mile Service (B)
850	4.1	4.3
940	1.8	2.0
960	3.8	4.0
S90, V90	3.8	4.0
C70, S70, V70, XC90	4.1	4.3

80,000 Mile Service (B)
850	2.4	2.6
940	2.2	2.4
960	2.1	2.3
S90, V90	2.1	2.3
C70, S70, V70, XC90	2.4	2.6

90,000 Mile Service (B)
850	3.9	4.1
940	3.1	3.3
960	3.2	3.4
S40, V40, V50	3.0	3.2
S90, V90	3.2	3.4
C70, S70, V70, XC90	3.9	4.1

960 : C70 : S40 : S60 : S70 : S80 : S90 : V40 : V70 : V90 : XC70 : XC90 VLV-3

	LABOR TIME	SEVERE SERVICE
100,000 Mile Service (B)		
850	2.6	2.8
940	2.4	2.6
960	2.3	2.5
S90, V90	2.5	2.7
C70, S70, V70, XC90	2.6	2.8
Replace timing belt add	2.0	2.2
110,000 Mile Service (C)		
850	2.1	2.3
940	1.8	2.0
960	1.8	2.0
S90, V90	1.8	2.0
C70, S70, V70, XC90	2.1	2.3
Inspect timing belt add	.2	.4
120,000 Mile Service (B)		
850	4.4	4.6
940	3.7	3.9
960	3.8	4.0
S40, V40, V50	3.5	3.7
S90, V90	3.8	4.0
C70, S70, V70, XC90	4.4	4.6

CHARGING

Alternator Circuits, Test (B)
Includes: Test component output.

	LABOR TIME	SEVERE SERVICE
1981-04	.7	.7
Alternator, R&R and Recondition (B)		
B21F, FT	3.7	3.9
B23F, FT, FTL	3.3	3.5
B230F, FT, B234F	3.8	4.0
B27, B28	3.8	4.0
B28F, B280F	4.1	4.3
B5234FT, B5244T, B5234T3		
w/AC	4.7	4.9
w/o AC	4.0	4.2
B5254	2.3	2.5
B5254S, T (2.7)	3.6	3.8
B6304	3.6	3.8
D24	3.4	3.6
D24T	3.6	3.8

Alternator Assy., Replace (B)
Includes: Pulley transfer.

	LABOR TIME	SEVERE SERVICE
B21F, FT	1.5	1.7
B23F, FT, FTL	1.3	1.5
B230F, FT	1.1	1.3
B234F	1.3	1.5
B27, B28, B28F, B280F	1.2	1.4
B4192T3, T4	2.0	2.2
B4194T, B4204T		
w/AC	2.0	2.2
w/o AC	1.4	1.6
B5234T3, B5244T3	2.6	2.8
B5244S, B5254	2.0	2.2
B5254FS, T (1.0)	1.4	1.6
B6244, B6254	1.0	1.2

	LABOR TIME	SEVERE SERVICE
B6284T, B6294S, T, B6304F	1.5	1.7
B6304	1.2	1.4
D24, D24T	1.7	1.9
XC90		
B5254T	2.3	2.5
B6294T	3.8	4.0

Alternator Voltage Regulator, Replace (B)

	LABOR TIME	SEVERE SERVICE
Exc. below	.8	1.0
850	1.5	1.7
C70, S70		
w/AC	2.3	2.5
w/o AC	1.4	1.6
S40, V40	1.5	1.7
S60, V70		
w/turbo (2.1)	2.8	3.0
w/o turbo (1.1)	1.5	1.7
S80 turbo (1.3)	1.8	2.0
XC90		
5 cyl. B5254T (2.1)	2.8	3.0
6 cyl. B6294T (3.0)	4.1	4.3

Voltmeter and/or Ammeter Gauge, Replace (B)

	LABOR TIME	SEVERE SERVICE
240	.5	.7
260	.5	.7
740		
1985-87	.8	1.0
1988-92	1.5	1.7
760		
1983-87	.8	1.0
1988-90	1.3	1.5
780		
1987	.8	1.0
1988-91	1.3	1.5
850	2.0	2.2
940	1.5	1.7
960	1.4	1.6
S40, V40, V50	1.4	1.6
S80	1.0	1.2
S90, V90	.8	1.0

STARTING

Starter Draw Test (On Car) (B)

	LABOR TIME	SEVERE SERVICE
1981-05	.3	.3

Battery Cables, Replace (C)

	LABOR TIME	SEVERE SERVICE
Positive		
exc. below	.5	.7
260	.7	.9
C70, S70, V70 (1.0)	1.4	1.6
S40, V40, V50 (1.6)	2.2	2.4
Negative		
exc. below	.5	.7
850	.7	.9
C70, S70, V70 (.5)	.7	.9

Starter Assy., R&R and Recondition (B)

	LABOR TIME	SEVERE SERVICE
B21F, FT		
AT	3.0	3.2
MT	2.8	3.0

	LABOR TIME	SEVERE SERVICE
B23F		
1983-84		
AT	3.1	3.3
MT	2.8	3.0
1984		
AT	2.9	3.1
MT	2.6	2.8
B230		
B230F		
AT	3.1	3.3
MT	2.9	3.1
B230FTL	2.6	2.8
B27, B28	2.7	2.9
B28F, B280F	2.5	2.7
B4194T, B4204T	3.5	3.7
B5234T3, B5244T3	3.5	3.7
B5244S, T	3.6	3.8
B5254	2.9	3.1
B6284T, B6294F, S, B6304F	3.5	3.7
B6304	3.6	3.8
D24, D24T	2.6	2.8
XC90		
B5254T	3.2	3.4
B6294T	4.0	4.2

Starter Assy., Replace (B)

	LABOR TIME	SEVERE SERVICE
B21F, FT, B23F, FT, B230F, FTL		
AT	.8	1.0
MT	.5	.7
B27, B28	.8	1.0
B28F, B280F	.7	.9
B4194T, B4204T	1.6	1.8
B5234FT, B5234T3	1.6	1.8
B5244S, T, T3	1.6	1.8
B5254, B5254S, T	1.5	1.7
B6244, B6254	1.6	1.8
B6284T, B6294F, B6304F	1.2	1.4
B6304	1.7	1.9
B6294T	1.9	2.1
D24, D24T	.5	.7

Starter Drive Assy., Replace (B)
Includes: R&R starter.

	LABOR TIME	SEVERE SERVICE
B21F, FT		
AT	1.1	1.3
MT	.8	1.0
B23F, FT		
AT	1.3	1.5
MT	.8	1.0
B230F, FTL		
AT	1.2	1.4
MT	.8	1.0
B28	1.1	1.3
B28F, B280F	.8	1.0
B4194T, B4204T	1.8	2.0
B5234FT, B5234T3	1.8	2.0
B5244S, T, B5244T3	1.8	2.0
B5254FS, FT, T	1.6	1.8
B5254S, T (1.2)	1.6	1.8

VLV-4   240 : 242 : 244 : 245 : 262 : 264 : 265 : 740 : 745 : 760 : 780 : 850 : 940

	LABOR TIME	SEVERE SERVICE
Starter Drive Assy., Replace (B)		
B6284T, B6294S, B6304	1.9	2.1
B6294T (1.6)	2.2	2.4
D24, D24T	.8	1.0
Starter Solenoid and/or Switch, Replace (B)		
Includes: R&R starter.		
B21FT		
AT	1.4	1.6
MT	.8	1.0
B23F		
AT	1.6	1.8
MT	1.4	1.6
B23FT		
AT	1.4	1.6
MT	.8	1.0
B230F		
AT	1.8	2.0
MT	1.5	1.7
B230FTL		
AT	1.4	1.6
MT	.8	1.0
B27, B28	1.6	1.8
B28F, B280F	1.2	1.4
B4194T, B4204T	1.9	2.1
B5234FT, B5234T3	2.1	2.3
B5244S, B5244T	2.1	2.3
B5244T3	1.9	2.1
B5254FS, FT, T	1.8	2.0
B6284T, B6294S, B6304F	1.9	2.1
B6294T	2.5	2.7
D24	.9	1.1

CRUISE CONTROL

	LABOR TIME	SEVERE SERVICE
Diagnose Cruise Control System (A)		
1988-05	1.1	1.1
Control Controller (Module), Replace (B)		
1988-05	.5	.7
Control Main Switch, Replace (B)		
1981-05		
keypad	.9	1.1
switch	.5	.7
Control Regulator, Replace (B)		
1981-85 240	.5	.7
260	.5	.7
Control Servo, Replace (B)		
1981-04	.6	.8
Control Throttle or Servo Cable, Replace (B)		
Exc. below	.5	.7
240, 260	.8	1.0
Control Vacuum Pump, Replace (B)		
1981-04	.6	.8
Control Vacuum Release Switch, Replace (B)		
1981-04	.3	.5

	LABOR TIME	SEVERE SERVICE
Control Vacuum Release Valve, Replace (B)		
1981-85 240	.5	.7
260	.5	.7
Vacuum Motor, Replace (B)		
Exc. below	.7	.9
240, 260	.5	.7
C70, S70, V70	.5	.7
S40, V40	.5	.7
S90, V90	.3	.5

IGNITION

	LABOR TIME	SEVERE SERVICE
Diagnose Ignition System Component Each (A)		
1981-05	1.0	1.0
Ignition Timing, Reset (B)		
1981-95	.3	.5
Camshaft Position Sensor, Replace (B)		
850, 960	.5	.7
C70, S70, V70		
w/turbo	.9	1.1
w/o turbo	.5	.7
S40, V40, V50	1.2	1.4
S60, S80		
w/turbo	.9	1.1
w/o turbo	.5	.7
XC90	.9	1.1
Crankshaft Angle Sensor, Replace (B)		
1990-98	.5	.7
1999-02		
exc. below	.8	1.0
S40, V40, V50	1.1	1.3
S60, V70 B5244S		
2001	1.3	1.5
2002	.3	.5
S80		
1999-01 B6284T, B6294S, B6294T, B6304S	1.0	1.2
2002 B6294S	.3	.5
XC90		
B5254T	.3	.5
B6294T	.6	.8
Distributor, R&R and Recondition (B)		
Includes: Reset base ignition timing.		
1981-87 240	1.9	2.1
260	2.1	2.3
1985 740		
4 cyl.	1.8	2.0
V6	2.3	2.5
1983-85 760		
4 cyl.	1.8	2.0
V6	2.3	2.5
1986-95		
4 cyl.	1.7	1.9
V6	2.4	2.6

	LABOR TIME	SEVERE SERVICE
Distributor, Replace (B)		
Includes: Reset base ignition timing.		
240	.7	.9
260	.9	1.1
1985 740		
4 cyl.	1.1	1.3
V6	1.4	1.6
1983-85 760		
4 cyl.	1.3	1.5
V6	1.2	1.4
1986-95		
4 cyl., 5 cyl.	.7	.9
6 cyl.	1.3	1.5
V6	1.4	1.6
Replace impulse sender add	.3	.5
Distributor Cap and/or Rotor, Replace (B)		
1981-85	.3	.5
1986-95 4 cyl.	.3	.5
1993-02 5 cyl.	.7	.9
1992-02 6 cyl.	.7	.9
1986-90 V6	.7	.9
Glow Plug, Replace (B)		
One	.5	.7
each addl. add	.1	.2
Ignition Coil, Replace (B)		
Exc. below one	.6	.8
B5234FT, B5234T3	.9	1.1
B6284T, B6294T	1.4	1.6
XC90		
B5254T (.7)	.9	1.1
B6294T (1.0)	1.4	1.6
each addl. add	.2	.4
Ignition Switch, Replace (B)		
1981-85	.8	1.0
1986-05	.7	.9
Ignition Timing Magnetic Pickup, Replace (B)		
1984-87 240	.5	.7
260	.7	.9
1988-95	.7	.9
Spark Plug (Ignition) Cables, Replace (B)		
240	.5	.7
260	.7	.9
740		
1985-87		
4 cyl.	.3	.5
V6	.7	.9
1988-92	.5	.7
760		
4 cyl.	.3	.5
V6	.7	.9
780	.7	.9
850, 940	.5	.7
960	.7	.9
C70, S70, V70	.5	.7
S40, V40	.3	.5
XC90 (1.5)	2.0	2.2

960 : C70 : S40 : S60 : S70 : S80 : S90 : V40 : V70 : V90 : XC70 : XC90 VLV-5

	LABOR TIME	SEVERE SERVICE
Spark Plugs, Replace (B)		
1981-98		
4 cyl.	.5	.7
5 cyl.	.7	.9
6 cyl.	2.1	2.3
V6	.8	1.0
1999-04		
C70, S70	1.1	1.3
S40, V40	.5	.7
S60, V70	1.4	1.6
S80		
w/turbo	1.8	2.0
w/o turbo	1.4	1.6
XC90		
5 cyl. (1.0)	1.4	1.6
6 cyl. (1.3)	1.8	2.0

EMISSIONS

The following operations do not include testing. Add time as required.

	LABOR TIME	SEVERE SERVICE
Dynamometer Test (A)		
1981-05	.5	.5
EGR System, Test (A)		
1981-04	.8	.8
Air Control Valve, Replace (B)		
B230FD, B5254	.3	.5
C70, S70, V70	.3	.5
Air Distribution Manifold, Replace (B)		
240		
1981-85	1.5	1.7
1986-87	1.3	1.5
260		
right side	1.2	1.4
left side	1.3	1.5
Air Diverter and/or Switching Valve, Replace (B)		
1981-87	.3	.5
Air Pump Assy., Replace (B)		
1981-87	.5	.7
850, C70, S70, V70	.8	1.0
Anti-Backfire Valve, Replace (B)		
1981-85	.3	.5
1986-87	.5	.7
Charge Pressure Sensor Replace (B)		
940	.5	.7
Coolant Temperature (ECT) Sensor, Replace (A)		
B230F	.5	.7
B6304, B5254	.7	.9
1998-99	.7	.9
2000-03		
C70, S70	1.3	1.5
S60, V70		
exc. below	1.3	1.5
2002 B5244S	.5	.7
S40, V40, V50	.8	1.0

	LABOR TIME	SEVERE SERVICE
S80		
w/turbo	3.1	3.3
w/o turbo	2.1	2.3
XC90		
5 cyl. (1.3)	1.8	2.0
6 cyl. (2.1)	2.8	3.0
EGR Control Valve, Replace (B)		
Gasoline		
1981-87 240	1.1	1.3
260	.5	.7
1988-97		
4 cyl.	1.2	1.4
5 cyl.		
w/turbo	2.2	2.4
w/o turbo	1.8	2.0
1994 6 cyl.	1.9	2.1
1995-02		
5 cyl. B5254FT	2.3	2.5
6 cyl.	2.5	2.7
Diesel		
740, 760	.5	.7
EGR Micro Switch, Replace (B)		
740, 760 diesel	.5	.7
EGR Solenoid Valve, Replace (B)		
C70, S70, V70	.2	.4
D24T	.2	.4
EGR Temperature Sender, Replace (B)		
B230F	.3	.5
B5254, 5234T3	.5	.7
B6304	1.1	1.3
C/S/V70 EGR	.3	.5
EGR Thermostat, Replace (B)		
1981-87 240	1.2	1.4
260	2.9	3.1
EGR Time Relay, Replace (B)		
740, 760 diesel	.2	.4
EGR Vacuum Booster, Replace (B)		
1985-95	.3	.5
Electronic Control Module (ECM/PCM), Replace (A)		
1981-99		
exc. below	.7	.9
B21F, FT	.5	.7
B28F 260, 760	.5	.7
2000-04	.4	.6
Frequency Valve, Replace (B)		
B21F, FT	.8	1.0
B27F, B28F 260, 760	.7	.9
Fuel Vapor Canister, Replace (B)		
1981-98	.5	.7
1999-04		
S40, V40		
w/fog lamp	1.3	1.5
w/o fog lamp	.6	.8
C70, S70		
RULOEVAP	.6	.8
RULOORVR	.3	.5
S60, S80, V70 (1.2)	1.6	1.8
XC90 (.6)	.8	1.0

	LABOR TIME	SEVERE SERVICE
Intake Air Temperature (IAT) Sensor, Replace (B)		
940	.3	.5
1998-05	.2	.4
Knock Sensor, Replace (B)		
1984 B23FT	.5	.7
B230F, FTL	.5	.7
1987-91 B280	2.7	2.9
B4194T, B4204T	2.0	2.2
B5234FT, B5234T3, B5244T3	1.3	1.5
B5244S	3.3	3.5
B5244T	2.1	2.3
B5254S, T	2.8	3.0
B6294S, B6304S	.8	1.0
B6284T, B6294T	1.3	1.5
B6304		
front	.5	.7
rear	1.8	2.0
XC90 (1.0)	1.4	1.6
Manifold Absolute Pressure (MAP) Sensor, Replace (B)		
1994-97	.2	.4
Manifold Pressure Sensor, Replace (B)		
940	.3	.5
Mass Air Flow (MAF) Sensor, Replace (B)		
1991-97 240, 850, 940	.5	.7
1998-04	.5	.7
Oxygen Sensor, Replace (B)		
850, 940, 960	.5	.7
1998-05		
C70, S70	.5	.7
S40, V40, V50 each	.5	.7
S60 each	.8	1.0
S80 each	1.4	1.6
V70		
1998-00 each	.5	.7
2001-02 each	.8	1.0
XC90 each (1.0)	1.4	1.6
Oxygen Sensor Thermostat, Replace (B)		
1981-86 240	1.2	1.4
260	2.6	2.8
1983-86 760	.9	1.1
Pressure Differential Switch, Replace (B)		
760	.3	.5
Thermo Valve, Replace (B)		
740, 760 diesel	.2	.4
Throttle Position Sensor (TPS), Replace (B)		
850	.5	.7
940, 960	.3	.5
C70, S70, V70	.5	.7
S40, V40, S80	.3	.5
S90, V90	.3	.5

VLV-6   240 : 242 : 244 : 245 : 262 : 264 : 265 : 740 : 745 : 760 : 780 : 850 : 940

	LABOR TIME	SEVERE SERVICE
Throttle Position Switch, Replace (B)		
1991-95 240, 850, 940	.5	.7
Vehicle Speed Sensor, Replace (B)		
850	.5	.7
1994-98 960, S90, V90	.5	.7
1998-02 C70, S70	.8	1.0
S40, V40, V50 (.8)	1.1	1.3
S60		
exc. below	.3	.5
2001 B5244S	1.2	1.4
S80		
exc. below	.9	1.1
2002 B6294S	.3	.5
V70		
exc. below	.5	.7
2001 B5244S	1.2	1.4
XC90 (.4)	.6	.8

FUEL

DELIVERY

	LABOR TIME	SEVERE SERVICE
Fuel Filter, Replace (B)		
1981-97 (.5)	.7	.9
1998-05		
5 cyl.		
exc. XC90		
w/AWD (.8)	1.1	1.3
w/o AWD (.4)	.6	.8
XC90 (.3)	.5	.7
6 cyl.	.6	.8
Fuel Gauge (Dash), Replace (B)		
240, 260	.9	1.1
740, 760	1.4	1.6
850	1.6	1.8
940	1.4	1.6
960	1.6	1.8
C70, S70	1.4	1.6
S40, V40, V50	1.2	1.4
S60, S80	.6	.8
S90, V90	1.5	1.7
V70		
1998-00	1.4	1.6
2001-05	.6	.8
XC90 (.4)	.6	.8
Fuel Gauge (Tank), Replace (B)		
1981-85	1.3	1.5
1986-87		
sedan	1.3	1.5
wagon	1.5	1.7
1988-93 240		
sedan	1.4	1.6
wagon	1.7	1.9
1988-91 740		
sedan	1.7	1.9
wagon	1.8	2.0

	LABOR TIME	SEVERE SERVICE
760		
sedan	1.9	2.1
wagon	1.6	1.8
780		
sedan	1.7	1.9
wagon	1.8	2.0
850	.8	1.0
940		
sedan	1.9	2.1
wagon	1.8	2.0
960		
sedan	1.9	2.1
wagon	1.7	1.9
C70, S70	.8	1.0
S40, V40, V50	1.6	1.8
S60, S80	2.4	2.7
S90, V90	1.8	2.0
V70		
1998-00	1.8	2.0
2001-05	2.5	2.9
XC90 (2.2)	3.0	3.4
Fuel Pump, Replace (B)		
1981-95 external	.9	1.1
In tank		
240		
sedan	1.4	1.6
wagon	1.5	1.7
740	1.9	2.1
760		
sedan	1.9	2.1
wagon	1.8	2.0
780	1.7	1.9
850	1.3	1.5
940		
sedan	1.9	2.1
wagon	2.0	2.2
960		
sedan	1.7	1.9
wagon	1.8	2.0
C70, S70		
w/AWD	8.8	9.0
w/o AWD	1.3	1.5
S40, V40	1.4	1.6
S60, S80	2.5	2.7
S90, V90, V50	1.8	2.0
V70		
1998-00		
w/AWD	8.8	9.0
w/o AWD	1.3	1.5
2001-05 (2.1)	2.8	3.0
XC90 (2.4)	3.2	3.4
Fuel Pump Relay, Replace (B)		
1981-05	.3	.5
Fuel Tank, Replace (B)		
Includes: Drain and refill.		
240	1.7	1.9
260	1.6	1.8
740	2.7	2.9

	LABOR TIME	SEVERE SERVICE
760		
1983-84	1.5	1.7
1985-90	2.5	2.7
780	2.9	3.1
850	2.8	3.0
940		
coupe, GLE, turbo	2.7	2.9
SE	2.9	3.1
960	2.8	3.0
C70, S70		
w/AWD	9.8	10.0
w/o AWD	3.5	3.7
S40, V40, V50	3.2	3.4
S60		
w/AWD	8.1	8.3
w/o AWD	5.2	5.4
S80	6.3	6.5
S90, V90	3.0	3.2
V70		
1998-00		
w/AWD	9.9	10.1
w/o AWD	3.5	3.7
2001-05		
w/AWD (6.1)	8.2	8.4
w/o AWD (3.9)	5.3	5.5
XC90 (5.8)	7.8	8.0
Intake Manifold and/or Gasket, Replace (B)		
B21F, FT	1.1	1.3
B23F	1.2	1.4
B23FT, B230F, FTL	1.7	1.9
B27, B28, B28F	2.1	2.3
B280F	1.8	2.0
F254F	2.1	2.3
B4194T, B4204T	2.6	2.8
B5234FT, B5234T3, B5244T	2.9	3.1
B5244S	2.9	3.1
B5254	2.2	2.4
B5254T	3.3	3.5
B6284T, B6294T	3.8	4.0
B6294F	2.6	2.8
B6294S, B6304S	3.5	3.7
D24	.8	1.0
D24T	1.4	1.6
Variable Intake Manifold Control Actuator, Replace (B)		
850	2.9	3.1
C70, S70, V70	2.8	3.0

INJECTION

	LABOR TIME	SEVERE SERVICE
Fuel Injection System, Test (B)		
1981-05	1.1	1.1
Air Flow Sensor Plate, Adjust (B)		
1981-86 B28F	.3	.5
Diesel Fuel Injection Timing, Check & Adjust (B)		
1981-85	1.1	1.3

960 : C70 : S40 : S60 : S70 : S80 : S90 : V40 : V70 : V90 : XC70 : XC90   VLV-7

	LABOR TIME	SEVERE SERVICE
Idle Speed, Adjust (B)		
1981-01	.3	.5
Throttle Valve Switch, Adjust (B)		
1981-87 240	.2	.4
260	.2	.4
Air/Fuel Control Unit, Replace (B)		
B21F, FT	1.9	2.1
B27F, B28F	.8	1.0
Air Flow Meter, Replace (B)		
Exc. B23F	.5	.7
B23F	.7	.9
Auxiliary Air Regulator, Replace (B)		
1981-87 240	.7	.9
260	.7	.9
Ballast Resistor, Replace (B)		
940	.3	.5
Cold Start Injector, Replace (B)		
B21F, T B27F, B28F	.5	.7
B230F	.7	.9
Diesel Fuel Injection Pump, Replace (B)		
1981-85	3.6	3.8
Diesel Injectors, R&R and Recondition (B)		
1981-85	4.9	5.1
Electronic Control Unit, Replace (B)		
B23F, FT	.5	.7
B5254	.3	.5
Fuel Distributor, Replace (B)		
B27F, B28F	.7	.9
Fuel Injection Lines, Replace (B)		
Diesel	1.6	1.8
Fuel Injectors, Replace (B)		
Gasoline one		
B28F	1.8	2.0
B280	.5	.7
B230F, FT	1.3	1.5
B21F, FT, B23F, FT	1.2	1.4
B234F	1.2	1.4
B4194T, B4204T	1.1	1.3
B5234FT, B5234T3, B5244T, B5254S	.8	1.0
B5244S	.9	1.1
B5254T	1.1	1.3
B6284T, B6294T	1.7	1.9
B6294F, B6294S, B6304S	1.4	1.6
each addl. add	.2	.4
Diesel one	1.1	1.3
each addl. add	.2	.4
Fuel Pressure Regulator, Replace (B)		
B21F	.5	.7
B230F	.3	.5
B4194T, B4204T		
2000-02 engine compartment	.3	.5
fuel tank	.6	.8

	LABOR TIME	SEVERE SERVICE
B5202, B5252 B5234F, B5254F		
w/MO4.4	1.9	2.1
B5254	.5	.7
B5254T	.2	.4
B6304	.3	.5
1999-00 S70, V70		
w/AWD	6.4	6.6
w/o AWD	1.9	2.1
w/fuel system draining add	.3	.5
Fuel Rail, Replace (B)		
B23F, B230F	.5	.7
B27F, B28F, B28, B280	.7	.9
B4194T, B4204T	1.2	1.4
B5234T3	1.5	1.7
B5244S, B5244T	1.5	1.7
B5254, 5254S, T	1.5	1.7
B6284T, B6294T	2.1	2.3
B6294S, B6304S	1.8	2.0
Idle Air Control (IAC) Valve, Replace (B)		
1994-97 850	.5	.7
1994-97 940, 960	.3	.5
1998-04 C/S/V70, S40, V40, V50	.5	.7
S80, S90, V90	.3	.5
Idle Air Control Valve Module, Replace (B)		
B21F, FT	.7	.9
B23F, FT, B28F	.8	1.0
Impulse Relay, Replace (B)		
1981-86 B28F	.3	.5
Pressure Accumulator, Replace (B)		
B21F, FT	.8	1.0
B27F, B28F		
260	.7	.9
760	.8	1.0
Thermo Time Switch, Replace (B)		
1981-87 240	.3	.5
260	.3	.5
Throttle Body Pressure Regulator, Replace (B)		
B21F, FT, B27F, B28F	.5	.7
Vacuum Switch, Replace (B)		
1986-87 240	.5	.7
TURBOCHARGER		
Intercooler, Replace (B)		
1988-01	.5	.7
Turbocharger Assy., R&R or Replace (B)		
240		
1981-85	4.3	4.5
1986-87	4.3	4.5

	LABOR TIME	SEVERE SERVICE
1988-93		
air cooled	6.0	6.2
water cooled	6.1	6.3
740		
air cooled	6.3	6.5
water cooled	5.6	5.8
760		
1983-85		
gasoline	3.6	3.8
diesel	1.9	2.1
1986-87	4.3	4.5
1988-90		
air cooled	6.3	6.5
water cooled	5.6	5.8
850	4.9	5.1
940		
air cooled	6.2	6.4
water cooled	5.5	5.7
C70, S70	3.8	4.0
S40, V40	4.7	4.9
S60	3.3	3.5
S80 B6284T, B6294T		
one	7.4	7.6
both	8.8	9.0
V70	3.3	3.5
XC90		
B5254T (3.0)	4.1	4.3
B6294T		
one (7.1)	9.6	9.8
both (8.4)	11.3	11.5
Replace		
comp. housing add	.5	.7
turbine housing add	.9	1.1
Wastegate Actuator, Replace (B)		
1981-85	1.1	1.3
1986-04		
exc. below	.9	1.1
S40, V40	2.3	2.5
S60, V70	1.1	1.3
S80 B6284T, B6294T	2.8	3.0
XC90 B6294T	2.6	2.8
EXHAUST		
Catalytic Converter, Replace (B)		
B21FT	1.2	1.4
B230F	.8	1.0
B5254	.8	1.0
C70, S70		
w/AWD	2.2	2.4
w/o AWD	.8	1.0
S40, V40, V50	1.4	1.6
S60, V70		
1998-00		
w/AWD	2.2	2.4
w/o AWD	.8	1.0
2001-02		
w/AWD or turbo	2.4	2.6
w/o turbo	1.7	1.9

240 : 242 : 244 : 245 : 262 : 264 : 265 : 740 : 745 : 760 : 780 : 850 : 940

	LABOR TIME	SEVERE SERVICE
Catalytic Converter, Replace (B)		
S80	3.0	3.2
XC90		
B5254T (1.8)	2.4	2.6
B6294T (5.2)	7.0	7.2
Exhaust Manifold and/or Gasket, Replace (B)		
B21F, FT, B23F	.7	.9
1984 B23FT	3.2	3.4
B230F	.8	1.0
B230FTL	3.1	3.3
B27, B28		
1981-82 B27, B28		
one side	.9	1.1
both sides	1.8	2.0
1981-86 B28F		
one side	1.5	1.7
both sides	2.6	2.8
B280F		
one side	1.4	1.6
both sides	2.0	2.2
B4194T, B4204T	5.1	5.3
B5234FT, B5234T3	3.3	3.5
B5244S	2.6	2.8
B5244T, B5244T3	3.5	3.7
5254	2.0	2.2
B5254FS	1.8	2.0
B5254FT, B5254T	3.3	3.5
B6284T	8.6	8.8
B6294S B6304S	3.5	3.7
B6304	2.0	2.2
D24		
one side	1.2	1.4
both sides	1.7	1.9
D24T	2.3	2.5
XC90		
B5254T (2.1)	2.8	3.0
B6294T (5.0)	6.8	7.0
Front Exhaust Pipe, Replace (B)		
240		
1981-85	1.6	1.8
1986-93	.8	1.0
260	1.8	2.0
740, 760		
gasoline	.8	1.1
diesel	.7	.9
780	.8	1.0
940, 960	.8	1.0
1998-02	.8	1.0
Front Muffler/Pipe, Replace (B)		
1983-85	.8	1.0
1986-90	.7	.9
1986-93 240	.9	1.1
740	.7	.9
760		
780	.7	.9
S40, V40	.8	1.0
S60, S80, V70, XC90	1.0	1.2

	LABOR TIME	SEVERE SERVICE
Intermediate Exhaust Pipe, Replace (B)		
1985-87 740	.7	.9
1983-87 760	.7	.9
Muffler and Tailpipe Assy., Replace (B)		
C70, S70, V70, S80 AWD	.7	.9
Rear Exhaust Pipe, Replace (B)		
1981-85	.9	1.1
1986-01	1.3	1.5
Rear Muffler, Replace (B)		
1981-02		
exc. below	1.1	1.3
S40, S60, S80, V40, V70	.5	.7
Tail Pipe, Replace (B)		
1981-02		
exc. below	.7	.9
C70, S70, V70		
w/AWD	.5	.7
w/o AWD	.2	.4

ENGINE COOLING

	LABOR TIME	SEVERE SERVICE
Pressure Test Cooling System (B)		
1981-05	.4	.4
Coolant Fan Thermo Sensor, Replace (B)		
1988-05	.5	.7
Coolant Thermostat, Replace (B)		
Exc. below	.9	1.1
S60 gas turbo	1.4	1.6
S80	2.0	2.2
XC90		
5 cyl. (1.0)	1.4	1.6
6 cyl. (1.5)	2.0	2.2
Electric Cooling Fan Assy., Replace (B)		
Exc. below	.6	.8
1986-93 240	.8	1.0
1983-87 740, 760	.8	1.0
S40, V40		
w/turbo	1.8	2.0
w/o turbo	.8	1.0
2001-05 S60, V70		
w/turbo (1.0)	1.4	1.6
w/o turbo (.8)	1.1	1.3
S80		
w/turbo	1.2	1.4
w/o turbo	.9	1.1
XC90 (2.1)	2.8	3.0
Engine Coolant Temp. Sending Unit, Replace (B)		
1981-98	.7	.9
1999-02		
S40, V40, V50	.8	1.0
C70, S70	1.2	1.4

	LABOR TIME	SEVERE SERVICE
S60, V70		
exc. below	1.2	1.4
2002 B5244S	.5	.7
S80		
w/turbo	3.1	3.3
w/o turbo	2.1	2.3
XC90		
B5254T	1.7	1.9
B6294T	2.9	3.1
Fan or Clutch Assy. Blades, Replace (B)		
Exc. below	.7	.9
260	1.4	1.6
1985-87 740	.5	.7
1983-87 760	.5	.7
Radiator Assy., R&R or Replace (B)		
1981-97		
AT	1.5	1.7
MT	1.4	1.6
1998-03		
C70, S70		
AT	2.3	2.5
MT	2.0	2.2
S40, V40, V50	2.9	3.1
S60	3.7	3.9
S80		
w/turbo	3.9	4.1
w/o turbo	2.9	3.1
V70		
1998-00		
AT	2.3	2.5
MT	2.0	2.2
2001-02	3.5	3.7
XC90		
5 cyl.	5.7	5.9
6 cyl.	6.2	6.4
w/AC add	.3	.5
w/diesel add	.1	.2
Radiator Fan Motor Relay, Replace (B)		
1981-05		
exc. below	.5	.7
S40, V40, C70, S70, V70	.2	.4
Radiator Fan Motor Switch (Coolant Temp.), Replace (B)		
1981-05	.5	.7
Radiator Hoses, Replace (B)		
Includes: Refill with proper coolant mix.		
Upper	.6	.8
Lower		
exc. below	.5	.7
S60, S80	1.1	1.3
2001-05 V70	1.1	1.3

960 : C70 : S40 : S60 : S70 : S80 : S90 : V40 : V70 : V90 : XC70 : XC90

	LABOR TIME	SEVERE SERVICE
Temperature Control Assy., Replace (B)		
240		
1981-85	1.7	1.9
1986-93	1.2	1.4
260	1.7	1.9
1994 700, 900	.5	.7
850		
ECC	.3	.5
MCC	1.3	1.5
1995-97 960 ECC	.8	1.0
Temperature Gauge (Dash), Replace (B)		
240, 360	.9	1.1
740, 760	1.3	1.5
780	1.2	1.4
850	1.6	1.8
940, 960	1.4	1.6
C70, S70	1.4	1.6
S40, V40, V50	1.2	1.4
S60, S80	.6	.8
S90, V90	1.4	1.6
V70		
1998-00	1.4	1.6
2001-02	.6	.8
XC90	.6	.8
Water Pump and/or Gasket, Replace (B)		
Includes: Refill with proper coolant mix.		
B21F, FT	2.3	2.5
B23F, FT	2.1	2.3
B230F, FT, B234F	2.4	2.6
1981-82 B27F, B28F	4.5	4.7
B280F	3.0	3.2
B4194T, B4204T	4.8	5.0
B5234FT, B5234T3, B5244T	2.9	3.1
B5244S, B5254, T	3.2	3.4
B6284T		
1999	5.9	6.1
2000-02	5.0	5.2
B6294S	4.3	4.5
B6294T	5.0	5.2
B6304F, B6304S	5.0	5.2
D24, D24T	3.1	3.3
Replace thermostat add	.3	.7

ENGINE
ASSEMBLY

Times shown are for OEM assemblies. Time to replace assemblies from aftermarket rebuilders may vary.

Engine Assy., R&I (B)
Does not include parts or component transfer.

	LABOR TIME	SEVERE SERVICE
Gasoline		
B21F	5.8	6.0
B21FT	6.2	6.4
B23F	5.7	5.9

	LABOR TIME	SEVERE SERVICE
B23FT	6.2	6.4
B230F	5.7	5.9
B230FTL	5.9	6.1
B234F	5.3	5.5
B27, B28	6.4	6.6
B28F, B280F	6.0	6.2
B4194T, B4204T	10.2	10.4
B5234FT, B5234T3, B5244T, B5244T3	9.3	9.5
B5244S	9.3	9.5
B5254, 5254S, T	13.3	13.5
B6284T, B6294T	12.6	12.8
B6294S, B6304S	11.5	11.7
B6304	8.8	9.0
Diesel		
1981-84	6.2	6.4
1985 240	6.3	6.5
1985 740	4.3	4.5
1985 760	4.1	4.3

Engine Assy., R&R and Recondition (A)
Includes: Replacing rings, rod and main bearings, cylinder head reconditioning and engine tune-up.

	LABOR TIME	SEVERE SERVICE
B21F	19.1	19.3
B21FT	20.0	20.2
B23F	19.2	19.4
B23FT	23.7	23.9
B230F	19.7	19.9
B230FTL	22.1	22.3
B234F	19.7	19.9
B27, B28	36.9	37.1
B28F, B280F	37.3	37.5
B4194T, B4204T	27.1	27.3
B5202, B5252	25.7	25.9
B5234FT, 5234T3	28.3	28.5
B5244T, B5254, 5254S, T	28.3	28.5
B6304	28.9	29.1

Engine Assy. (Complete), Replace (B)
Includes: Engine R&R and parts transfer.

	LABOR TIME	SEVERE SERVICE
Gasoline		
B21F, FT	9.2	9.4
B23F	7.8	8.0
B23FT	7.9	8.1
B230F	8.7	8.9
B230FT, FTL	8.0	8.2
B234F	8.1	8.3
B27, B28, B28F, B280F	10.2	10.4
B4194T, B4204T	15.5	15.7
B5234FT, B5234T3, B5244T, B5244T3	15.3	15.5
B5244S	14.5	14.7
B5254, B5254FS, T	14.5	14.7
B6284T, B6294T	12.2	12.4

	LABOR TIME	SEVERE SERVICE
B6294S, B6304S	20.9	21.1
B6304F	12.2	12.4
XC90		
B5254T	15.3	15.5
B6294T	24.0	24.2
Diesel		
1981-84	10.0	10.2
1985 240	9.7	9.9
1985 740	7.2	7.4
1985 760	7.1	7.3

Engine Assy. (Short Block), Replace (B)
Assembly consists of engine block, piston assemblies, crankshaft, camshaft, timing chain and gears. Does not includes cylinder heads. Operation includes: R&R engine, transfer necessary parts and all necessary adjustments.

	LABOR TIME	SEVERE SERVICE
Gasoline		
B21F	10.2	10.4
B21FT	12.3	12.5
B23F, FT	11.3	11.5
B230F	10.2	10.4
B230FT, FTL	12.4	12.6
B234F	9.9	10.1
B27, B28	10.5	10.7
B28F	12.3	12.5
B280F	11.9	12.1
B4194T, B4204T	15.3	15.5
B5234FT, B5234T3, B5244T	14.5	14.7
B5254, 5254S, T	13.7	13.9
B6294S, B6304S	20.9	21.1
B6284T, B6294T	23.8	24.0
B6304	15.4	15.6
XC90		
B5254T	15.3	15.5
B6294T	24.2	24.4
Diesel		
1981-84	19.4	19.6
1985 240	19.8	20.0
1985 740	16.6	16.8
1985 760	16.4	16.6

Engine Mounts, Replace (B)

	LABOR TIME	SEVERE SERVICE
Gasoline		
1981-84 240		
right side	.9	1.1
left side	.8	1.0
both sides	1.6	1.8
260		
right side	.9	1.1
left side	.8	1.0
both sides	1.6	1.8
740		
right side	.9	1.1
left side	1.3	1.5

VLV-9

VLV-10   240 : 242 : 244 : 245 : 262 : 264 : 265 : 740 : 745 : 760 : 780 : 850 : 940

	LABOR TIME	SEVERE SERVICE
Engine Mounts, Replace (B)		
760		
right side	.9	1.1
left side	.8	1.0
both sides	1.6	1.8
780		
one side	.9	1.1
both sides	1.8	2.0
850		
front	1.1	1.3
rear	1.4	1.6
right	.7	.9
upper	1.3	1.5
940, 960		
right side	.9	1.1
left side	1.1	1.3
both sides	1.9	2.1
C70, S70		
front		
AT	1.3	1.5
MT	1.0	1.2
right front	.6	.8
rear	1.1	1.3
S40, V40, V50		
front	.6	.8
right front	1.0	1.2
rear	.8	1.0
S60		
front	1.8	2.0
right front	1.1	1.3
rear	2.1	2.3
S80		
front/right front	1.6	1.8
rear		
exc. B5244S	2.9	3.1
B5244S	2.1	2.3
S90, V90		
right side	.9	1.1
left side	1.1	1.3
both sides	1.9	2.1
V70		
1998-00		
front		
AT	1.5	1.7
MT	.9	1.1
right front	.6	.8
rear	1.3	1.5
2001-05		
front (1.3)	1.8	2.0
right front (.8)	1.1	1.3
rear (1.5)	2.0	2.2
XC90		
front (1.2)	1.6	1.8
right front		
exc. B6294T (.8)	1.1	1.3
B6294T (1.2)	1.6	1.8
rear (2.1)	2.8	3.0

	LABOR TIME	SEVERE SERVICE
Diesel		
right	.7	.9
left	.9	1.1
rear	.7	.9
CYLINDER HEAD		
Compression Test (B)		
4 cyl.	.8	1.0
5 cyl.	.9	1.1
6 cyl.	2.5	2.7
V6	1.1	1.3
Cylinder Head, Retorque (B)		
Diesel	1.3	1.5
Valves, Adjust (B)		
1981-95 gasoline		
4 cyl.	1.2	1.4
V6	3.5	3.7
Diesel		
1981-84	2.1	2.3
1985 240	1.9	2.1
1985 740	1.6	1.8
1985 760	1.7	1.9
w/AC add	.5	.7
Cylinder Head, Replace (B)		
Includes: Parts transfer, adjustments.		
Gasoline		
B21F	6.8	7.0
B21FT	8.2	8.4
B23F	6.7	6.9
B23FT	8.7	8.9
B230F	6.7	6.9
B230FTL	8.9	9.1
B234F	10.5	10.7
B27, B28, B28F		
right side	10.8	11.0
left side	10.0	10.2
both sides	17.2	17.4
B234F	10.5	10.7
B280F		
right side	10.7	10.9
left side	9.9	10.1
both sides	15.4	15.6
B4194T, B4204T	15.0	16.2
B5234FT, B5234T3,		
B5244T	21.7	21.9
B5244F, B5244S	20.9	21.1
B5254, B5254S	13.2	13.4
B6284T		
1999	22.6	22.8
2000-02	22.1	22.3
B6294S	17.9	18.1
B6294T	22.1	22.3
B6304	12.6	12.8
B6304S	18.5	18.7
XC90		
B5254T	21.7	21.9
B6294T	28.6	28.8

	LABOR TIME	SEVERE SERVICE
Diesel		
1981-84	12.2	12.4
1985 240	14.4	14.6
1985 740, 760	14.6	14.8
Cylinder Head Gasket, Replace (B)		
Gasoline		
B21F	5.4	5.6
B21FT	5.2	5.4
B23F	5.3	5.5
B23FT	7.4	7.6
B230F	5.4	5.6
B230FTL	7.6	7.8
B234F	10.0	10.2
B27, B28		
right side	9.9	10.1
left side	8.3	8.5
both sides	14.9	15.1
B28F		
right side	7.9	8.1
left side	7.2	7.4
B280F		
right side	9.0	9.2
left side	7.4	7.6
both sides	13.8	14.0
B4194T, B4204T	12.1	12.3
B5234FT, B5234T3,		
B5244T	15.6	15.8
B5244F, B5244S	14.7	14.9
B5254	10.0	10.2
5254S	12.0	12.2
B5254T	15.7	15.9
B6294S	17.6	17.8
B6284T		
1999	22.4	22.6
2000-02	21.8	22.0
B6294F	19.9	20.1
B6304	11.6	11.8
B6304F, B6304S	18.2	18.4
B6294T	21.8	22.0
Diesel		
1981-84	7.3	7.5
1985 240	7.5	7.7
1985 740	7.3	7.5
1985 760	7.4	7.6
Valve Cover Gasket, Replace (B)		
Gasoline		
B21F, FT, B23F, FT,		
B230F, FTL	.7	.9
B234F	1.7	1.9
B27, B28, B28F, B280F		
right side	1.8	2.0
left side	1.2	1.4
both sides	2.9	3.1
B4194T, B4204T	7.4	7.6
B5234FT, B5234T3,		
B5244T	7.4	7.6
B5254, 5254S, T	7.4	7.6
B6304	7.9	8.1

	LABOR TIME	SEVERE SERVICE
Diesel		
1981-84	1.1	1.3
1985 240	.8	1.0
1985 740, 760	.7	.9
Valve Job (A)		
Gasoline		
B21F	10.9	11.1
B21FT	12.8	13.0
B23F	10.9	11.1
B23FT	12.8	13.0
B230F	10.9	11.1
B230FTL	13.1	13.3
B234F	15.8	16.0
B27, B28		
one side	13.3	13.5
both sides	21.8	22.0
B28F		
one side	12.4	12.6
both sides	22.0	22.2
B280F		
one side	12.3	12.5
both sides	13.2	13.4
B4194T, B4204T	17.9	18.1
B5234T3, B5234FT, B5244T	20.0	20.2
B5254, B5254S, T	19.5	19.7
B6304	20.0	20.2
Diesel		
1981-84	11.1	11.3
1985 240	13.7	13.9
1985 740	14.3	14.5
1985 760	14.2	14.4
Valve Lifters, Replace (B)		
Includes: Camshaft R&R.		
B234F	7.2	7.4
B4194T, B4204T	8.8	9.0
B5234FT, B5234T3	8.6	8.8
B5254, 5254S, T	7.9	8.1
B6304	8.7	8.9
CAMSHAFT		
Timing Belt, Adjust (B)		
Gasoline		
1981-95 SOHC	.3	.5
1990-95 DOHC	1.3	1.5
1981-85 diesel	.8	1.0
Balance Shaft Housing, Replace (A)		
B234F	4.3	4.5
Camshaft Seal, Replace (B)		
Gasoline		
B21F, FT	2.7	2.9
B23F, FT	2.5	2.7
B230F, FTL	2.7	2.9
B234F	3.7	3.9
B4194T, B4204T	4.6	4.8

	LABOR TIME	SEVERE SERVICE
B5234T3, B5234FT, B5244T, B5244T3	4.3	4.5
B5244S	3.6	3.8
B5254	3.3	3.5
B5254S, T	3.7	3.9
B6284T		
1999	6.0	6.2
2000-02	5.3	5.5
B6294S	4.1	4.3
B6294F	3.4	3.6
B6294T	5.3	5.5
B6304	3.5	3.7
B6304F	4.1	4.3
B6304S	4.8	5.0
XC90		
B5254T	4.4	4.6
B6294T	5.7	5.9
Diesel		
front	1.4	1.6
rear	2.4	2.6
both	3.2	3.4
Camshaft, Replace (A)		
Gasoline		
B21F, FT, B23F, FT	4.3	4.5
B230F	4.2	4.4
B230FTL	4.4	4.6
B234F	7.1	7.3
B27, B28		
right	9.7	9.9
left	9.4	9.6
B28F		
right	9.4	9.6
left	8.5	8.7
B280F		
right	9.2	9.4
left	8.7	8.9
B4194T, B4204T	8.5	8.7
B5234T3, B5244T		
one or both	7.4	7.6
B5254		
one	5.4	5.6
both	6.2	6.4
B5254S, T one or both	7.2	7.4
B6304 one or both	8.2	8.4
Diesel	4.3	4.5
Timing and Balance Shaft Belt, Replace (B)		
B234F	3.8	4.0
Replace sealing ring for balance shafts add	.5	.7
Timing Belt, Replace (B)		
Gasoline		
B21F, FT	2.6	2.8
B23F	2.8	3.0
B230F, FTL	2.9	3.1
B4914T, B4204T	3.8	4.0

	LABOR TIME	SEVERE SERVICE
B5234FT		
1994-98	2.7	2.9
1999-01	2.0	2.2
B5234T3	2.8	3.0
B5244T	2.0	2.2
B5244S, B5244TC	3.5	3.7
B5254, S, T	2.8	3.0
B6244, B6254	1.9	2.1
B6284	3.6	3.8
B6284T		
1999	5.3	5.5
2000-02	4.2	4.4
B6294	3.6	3.8
B6294F	2.8	3.0
B6294S	3.2	3.4
B6294T (3.0)	4.1	4.3
B6304	2.7	2.9
B6304F	3.9	4.1
B6304S	4.4	4.6
Diesel		
240, 360, 740	3.4	3.6
760	3.7	3.9
Timing Belt Tensioner, Replace (B)		
C70, S70		
1998	2.9	3.1
1999-02	2.2	2.4
S40, V40, V50	3.9	4.1
S60	2.7	2.9
S80		
B6284T		
1999	5.5	5.7
2000-02	4.4	4.6
B6294S	3.4	3.6
B6294T, B6304S	4.5	4.7
V70		
1998	2.9	3.1
1999-00	2.2	2.4
2001-02	2.7	2.9
XC90		
B6254T (2.1)	2.8	3.0
B6294T (3.3)	4.5	4.7
Timing Chain or Sprocket, Replace (B)		
B28F	7.5	7.7
B27, B28, B280F	7.6	7.8
Replace sprocket add	.2	.4
Timing Cover and/or Gasket, Replace (B)		
Diesel		
240		
1981-84	2.7	2.9
1985	.5	.7
260	2.8	3.0
1985 740	.5	.7
760		
1983-84	.9	1.1
1985	.5	.7

VLV-12 240 : 242 : 244 : 245 : 262 : 264 : 265 : 740 : 745 : 760 : 780 : 850 : 940

	LABOR TIME	SEVERE SERVICE
Timing Cover Front Seal, Replace (B)		
B28F	2.2	2.4
B27, B28, B280F	2.4	2.6
Timing Gear Cover and/or Gasket, Replace (B)		
Gasoline		
B21F, FT, B23F, FT, B230F, FTL	.5	.7
B27, B28, B28F	5.9	6.1
B280F	6.0	6.2
B4194T, B4204T	.5	.7
B5244S	.6	.8
B5254, 5254S, T	.6	.8
B6284T, B6294F, B6294S, B6294T	.8	1.0
B6304	.5	.7
B6304F, B6304S	.8	1.0

CRANK & PISTONS

Connecting Rod Bearings, Replace (A)
Includes: Check bearing oil clearance.

	LABOR TIME	SEVERE SERVICE
Diesel	9.8	10.0
Crankshaft and Main Bearings, Replace (A)		
Includes: Engine R&R.		
Gasoline		
B21F	11.3	11.5
B21FT	12.9	13.1
B23F, FT, B230F	11.1	11.3
B230FTL	11.6	11.8
B27, B28	12.7	12.9
B28F	12.3	12.5
B280F	12.4	12.6
B4194T, B4204T	19.7	19.9
B5234T3, B5244T	19.3	19.5
B5254, B5254S, T	19.8	20.0
Diesel		
240		
1981-84	12.7	12.9
1985	15.4	15.6
260	12.9	13.1
1985 740	14.0	14.2
760		
1983-84	15.3	15.5
1985	13.9	14.1
Crankshaft Front Oil Seal, Replace (B)		
Gasoline		
B21F	2.6	2.8
B21FT	3.0	3.2
B23F	2.8	3.0
B230F, FTL	3.4	3.6
B4194T, B4204T	4.5	4.7
B5234FT, B5234T3, B5244T	2.9	3.1
B5244S, B5244T3	3.4	3.6
B5254	1.9	2.1
5254S, T	3.5	3.7

	LABOR TIME	SEVERE SERVICE
B6284T		
1999	5.9	6.1
2000-02	4.7	4.9
B6294F	3.4	3.6
B6294S	3.8	4.0
B6294T, B6304S	4.8	5.0
B6304	4.2	4.4
B6304F	4.5	4.7
Diesel		
1981-84	3.1	3.3
1985 240	3.5	3.7
1985 740	3.8	4.0
1985 760	3.6	3.8
Crankshaft Pulley, Replace (B)		
Gasoline		
B21F, FT, B23F, FT	.7	.9
B230F, FTL	1.6	1.8
B27, B28	2.2	2.4
B280F	2.4	2.6
B234F	1.6	1.8
B4194T, B4204T	1.1	1.3
B5254, 5254S, T	.9	1.1
B5234FT, B5244T	.9	1.1
B6304	.9	1.1
Diesel		
240		
1981-84	2.4	2.6
1985	1.7	1.9
260	2.2	2.4
1985 740	1.7	1.9
760		
1983-84	2.0	2.2
1985	1.9	2.1
Liners, Piston and Cylinder, Replace (A)		
B27, B28	13.8	14.0
B28F	13.6	13.8
B280F	13.7	13.9
Pistons or Connecting Rods, Replace (A)		
Includes: Ridge reaming, cylinder wall deglazing, installing new rings and rod bearings, engine tune-up.		
Gasoline		
B21F	12.6	12.8
B21FT	13.0	13.2
B23F, FT	13.6	13.8
B230F	12.7	12.9
B230FT, FTL	13.8	14.0
B234F	16.8	17.0
B27, B28	19.8	20.0
B28F, B280F	20.2	20.4
B4194T, B4204T	18.4	18.6
B5254	24.9	25.1
B5234T3, B5244T, B5254S, T	20.0	20.2
B6304	27.7	27.9
1981-85 diesel	18.4	18.6

	LABOR TIME	SEVERE SERVICE
Rear Main Oil Seal, Replace (B)		
Includes: Trans R&R when necessary.		
Gasoline		
240		
AT	5.9	6.1
MT	4.2	4.4
260		
AT	5.7	5.9
MT	4.2	4.4
740		
AT	5.1	5.3
MT	3.5	3.7
760		
AT	5.3	5.5
MT	3.3	3.5
780	5.2	5.4
850		
AT	9.5	9.7
MT	8.8	9.0
940	6.7	6.9
960	4.5	4.7
C70, S70, V70		
AT		
w/AWD	11.9	12.1
w/o AWD	9.9	10.1
MT		
w/AWD	10.9	11.1
w/o AWD	8.2	8.4
S40, V40, V50	10.6	11.8
S60		
w/AWD	11.9	12.1
w/o AWD	10.9	11.1
S80		
exc. below	11.0	11.2
4T65EV trans.	12.3	12.5
turbo	12.8	13.0
S90, V90	4.3	4.5
XC90		
B5254T (8.8)	11.9	12.1
B6294T (9.3)	12.6	12.8
w/OD 240, 260 add	.5	.7
w/turbo 740, 760, 780, 850 add	.5	.7
Rings, Replace (A)		
Includes: Ridge reaming, cylinder wall deglazing, installing new rings, engine tune-up.		
Gasoline		
B21F	11.1	11.3
B21FT	11.9	12.1
B23F	11.3	11.5
B23FT	12.6	12.8
B230F	11.3	11.5
B230FT, FTL	12.7	12.9
B234F	15.3	15.5
B27, B28	17.7	17.9
B28F	18.2	18.4
B280F	18.3	18.5
B4194T, B4204T	18.4	18.6
B5254	24.9	25.1

	LABOR TIME	SEVERE SERVICE
B5234T3, B5244T,		
B5254S, T	20.0	20.2
B6304	27.7	27.9
Diesel		
1981-84	16.4	16.6
1985 240	16.7	16.9
1985 740, 760	14.6	14.8

ENGINE LUBRICATION

Engine Oil Cooler, Replace (B)
Exc. D24T	.8	1.0
D24T	1.3	1.5

Engine Oil Pressure Switch (Sending Unit), Replace (B)
1981-04	.5	.7

Oil Pan and/or Gasket, Replace (B)
Gasoline
B21F, FT	2.4	2.6
B23F, FT, B230F, FT, B234F	3.1	3.3
B27, B28	2.4	2.6
B28F	2.2	2.4
B280F	4.4	4.6
B5254	2.3	2.5
5254S, T	2.9	3.1
B6304	2.1	2.3
B6304F	3.1	3.3
C70, S70, V70 B5234FT, B5234T3, B5244T	3.5	3.7
S40, V40, V50	3.3	3.5
S60, V70	2.9	3.1
S80		
B6284T, B6294T	7.3	7.5
B6294S, B6304S	6.9	7.1
V70		
1999-00	3.5	3.7
2001-02	2.9	3.1
XC90		
B5254T (2.2)	3.0	3.2
B6294T (5.4)		7.3
	7.5	
Diesel		
240	8.9	9.1
260	8.7	8.9
740	8.9	9.1
760	3.4	3.6

Oil Pressure Gauge (Dash), Replace (B)
Exc. below	.7	.9
1988-93 240	.5	.7

Oil Pump, R&R and Recondition (B)
Diesel
240, 260	9.2	9.4
740	6.5	6.7
760		
1983-84	9.2	9.4
1985	6.5	6.7

Oil Pump, Replace (B)
Gasoline
B21F, FT	2.9	3.1
B23F	2.8	3.0
B23FT	2.5	2.7
B230F	2.8	3.0
B230FTL	2.4	2.6
B234F	3.4	3.6
B27, B28	5.9	6.1
B28F, B280F	6.3	6.5
1998-04		
C70, S70	4.0	4.2
S40, V40, V50, B4194T, B4204T	6.1	6.3
S60, V70	3.0	3.2
S80		
B6284T		
1999	5.8	6.0
2000-02	4.7	4.9
B6294S	3.8	4.0
B6294T, B6304S	4.9	5.1
S90, V90	4.9	5.1
V70		
1998-00	4.0	4.2
2001-05 (2.2)	3.0	3.2
XC90		
B5254T (2.4)	3.2	3.4
B6294T (3.4)	4.6	4.8
Diesel		
240, 260	8.7	8.9
740	5.9	6.1

Oil Pump Seal, Replace (B)
B234F	3.2	3.4
1998-03		
C70, S70	4.0	4.2
S40, V40 B4194T, B4204T	6.1	6.3
S60, V70	3.0	3.2
S80		
B6284T		
1999	5.7	5.9
2000-02	4.7	4.9
B6294S	3.8	4.0
B6294T, B6304S	4.9	5.1
S90, V90	4.7	4.9
V70		
1998-00	4.0	4.2
2001-05 (2.2)	3.0	3.2
XC90		
B5254T (2.4)	3.2	3.4
B6294T (3.4)	4.6	4.8

CLUTCH

Bleed Clutch Hydraulic System (B)
1981-05	.3	.3

Free Play, Clutch Pedal, Adjust (B)
1981-05	.3	.5

Clutch Assy., Replace (B)
240	3.8	4.0
260	3.6	3.8
740, 760	3.1	3.3
850	6.8	7.0
C70, S70	8.5	8.7
w/AWD	10.6	10.8
w/o AWD	8.5	8.7
S40, V40, V50, B4194T, B4204T	8.4	8.6
S60		
w/AWD	11.2	11.4
w/o AWD	9.8	10.0
V70		
1998-00		
w/AWD	10.6	10.8
w/o AWD	8.5	8.7
2001-02		
w/AWD	11.2	11.4
w/o AWD	9.4	9.6
w/OD 240, 260 add	.5	.7
w/turbo add		
740, 760, 850	.5	.7
C70, S70	.5	.7
98-00 V70	.5	.7
Replace pilot bearing add	.3	.5

Clutch Cable, Replace (B)
1981-97	.7	.9

Clutch Master Cylinder, Replace (B)
240, 260, 740, 760	.8	1.0
850	2.2	2.4
C70, S70	2.2	2.4
S40, V40, V50	1.2	1.4
S60, S80	1.6	1.8
V70		
1998-00	2.2	2.4
2001-02	1.6	1.8

Clutch Slave Cylinder, Replace (B)
Includes: System bleeding.
240, 260, 740, 760	.7	.9
850	1.3	1.5
C70, S70		
1994-97	.6	.8
1998-02	7.9	8.1
S40, V40, V50	8.1	8.3
S60		
w/AWD	10.9	11.1
w/o AWD	9.4	9.6
V70		
1998-00		
w/AWD	10.2	10.4
w/o AWD	7.9	8.1
2001-02		
w/AWD	10.9	11.1
w/o AWD	9.3	9.5
w/turbo add		
C70, S70	.6	.8
98-00 V70	.6	.8

VLV-14   240 : 242 : 244 : 245 : 262 : 264 : 265 : 740 : 745 : 760 : 780 : 850 : 940

	LABOR TIME	SEVERE SERVICE
Flywheel, Replace (B)		
240	4.1	4.3
260	4.2	4.4
740	3.0	3.2
760	3.2	3.4
850	8.0	8.2
C70, S70		
w/AWD	10.5	10.7
w/o AWD		
1998	9.1	9.3
1999-02	8.4	8.6
S40, V40, V50	9.1	9.3
S60		
w/AWD	11.5	11.7
w/o AWD	9.9	10.1
V70		
1998-00		
w/AWD	10.5	10.7
w/o AWD	7.8	8.0
2001-05		
w/AWD	11.5	11.7
w/o AWD	9.9	10.1
w/OD 240, 260 add	.5	.7
w/turbo add		
740, 760, 850	.5	.7
C70, S70	.2	.4
98-00 V70	.2	.4

MANUAL TRANSAXLE

	LABOR TIME	SEVERE SERVICE
Axle/Output Shaft Oil Seal, Replace (B)		
850, C70, S70		
right side	1.4	1.6
left side	.8	1.0
S40, V40, V50 each	2.2	2.4
S60, S80 one side	1.6	1.8
V70		
1998-00		
right side	1.4	1.6
left side	.8	1.0
2001-05 one side	1.6	1.8
Transaxle Assy., R&I (B)		
850	7.2	7.4
C70, S70		
w/AWD	9.9	10.1
w/o AWD	7.6	7.8
S60		
w/AWD	10.8	11.0
w/o AWD	9.1	9.3
S40, V40, V50, B4194T, B4204T	7.9	8.1
V70		
1998-99		
w/AWD	9.9	10.1
w/o AWD	7.6	7.8
2000-05		
w/AWD	10.8	11.0
w/o AWD	9.1	9.3
Replace assembly add	.5	.7

MANUAL TRANSMISSION

	LABOR TIME	SEVERE SERVICE
Extension Housing Oil Seal, Replace (B)		
1981-05	.9	1.1
Transmission Assy., R&I (B)		
240	3.3	3.5
260	3.5	3.7
740	2.8	3.0
760	2.7	2.9
w/OD 240, 260 add	.5	.7
Transmission Assy., R&R and Recondition (A)		
240		
4-Speed	9.3	9.5
5-Speed	10.0	10.2
260		
4-Speed	9.2	9.4
5-Speed	10.0	10.2
740, 760		
4-Speed	7.7	7.9
5-Speed	8.9	9.1
w/OD 240, 260 add	.5	.7

OVERDRIVE

	LABOR TIME	SEVERE SERVICE
Oil Pressure Check (B)		
1981-93	.7	.9
Engage Switch Assy., Replace (B)		
1981-93	.5	.7
Oil Filter, Replace (B)		
1981-93	.8	1.0
One Way Clutch, Replace (B)		
1981-93	1.4	1.6
Overdrive Assy., R&I (B)		
1981-93	1.8	2.0
Overdrive Assy., R&R and Recondition (A)		
1981-93	4.8	5.0
Relay, Replace (B)		
1981-93	.5	.7
Solenoid, Replace (B)		
1981-93	.5	.7

AUTOMATIC TRANSAXLE

SERVICE TRANSAXLE INSTALLED

	LABOR TIME	SEVERE SERVICE
Check Unit for Oil Leaks (C)		
All Models	.9	1.1
Drain & Refill Unit (B)		
850	3.8	4.0
C70, S70, V70	1.6	1.8
S40, V40, V50, S60	1.4	1.6
S80	2.9	3.1
XC90		
w/AWD	2.5	2.7
w/o AWD	1.1	1.3
Oil Pressure Check (B)		
All Models	.8	1.0
Electronic Control Unit, Replace (B)		
All Models	.5	.7

	LABOR TIME	SEVERE SERVICE
Oil Pan and/or Gasket, Replace (B)		
850	3.8	4.0
C70, S70	4.7	4.9
S40, V40, V50	3.3	3.5
S60	6.1	6.3
S80		
exc. 4T65EV	3.6	3.8
4T65EV	2.9	3.1
V70		
1998-00	4.7	4.9
2000-02	6.1	6.3
XC90		
4T65AWD (2.1)	2.8	3.0
AW55-20 (4.7)	6.3	6.5
Output Shaft Seal, Replace (B)		
850		
right side	1.4	1.6
left side	.8	1.0
C70, S70		
right side	1.2	1.4
left side	.8	1.0
S40, V40, V50 each	2.2	2.4
S60, S80 one side	1.6	1.8
V70		
1998-00		
right side	1.4	1.6
left side	.8	1.0
2001-02 one side	1.6	1.8
XC90 one side	1.8	2.0
Shift Solenoid Switch, Replace (B)		
850	.7	.9
C70, S70, V70	.6	.8
S40, V40, V50	.7	.9
S60, S80	1.1	1.3
XC90	.9	1.1
Speed Sensor, Replace (B)		
850	1.7	1.9
S40, V40, V50, S80	1.0	1.2
C70, S70, S60, V70, XC90 (1.0)	1.4	1.6

SERVICE TRANSAXLE REMOVED

Transaxle R&R included unless otherwise noted.

	LABOR TIME	SEVERE SERVICE
Flywheel (Flexplate), Replace (B)		
850	9.2	9.4
C70, S70		
w/AWD	10.9	11.1
w/o AWD	9.9	10.1
S40, V40, V50	9.9	10.1
S60		
w/AWD	11.9	12.1
w/o AWD	10.9	11.1
S80		
4T65EV	12.2	12.4
B6284T, B6294T	12.8	13.0
V70		
1998-00		
w/AWD	10.9	11.1
w/o AWD	9.7	9.9

960 : C70 : S40 : S60 : S70 : S80 : S90 : V40 : V70 : V90 : XC70 : XC90   **VLV-15**

	LABOR TIME	SEVERE SERVICE
2001-05		
w/AWD	11.9	12.1
w/o AWD	10.9	11.1
XC90		
w/AWD	12.6	12.8
w/o AWD	11.8	12.0
w/turbo 850 add	.5	.7
Torque Converter Oil Seal, Replace (B)		
850	7.7	7.9
Transaxle Assy., R&I (B)		
850	8.7	8.9
C70, S70, V70		
w/AWD	11.2	11.4
w/o AWD	9.0	9.2
S40, V40, V50	9.3	9.5
S60		
w/AWD	11.2	11.4
w/o AWD	10.2	10.4
S80		
4T65EV	11.5	11.7
B6284T, B6294T	12.1	12.3
XC90		
w/AWD	11.9	12.1
w/o AWD	11.0	11.2
w/turbo 850 add	.5	.7
Replace		
converter add	.2	.4
front seal add	.3	.5

AUTOMATIC TRANSMISSION
SERVICE TRANSMISSION INSTALLED

	LABOR TIME	SEVERE SERVICE
Check Unit for Oil Leaks (C)		
1981-05	.9	1.1
Drain and Refill Unit (B)		
1981-05	1.7	1.9
Oil Pressure Check (B)		
1981-05	.8	1.0
Electronic Control Unit, Replace (B)		
1992-05	.5	.7
Extension Housing and/or Gasket, Replace (B)		
1981-05	2.9	3.1
Replace extension housing bushing add	.2	.4
Extension Housing Oil Seal, Replace (B)		
1981-05	1.1	1.3
Governor Assy., R&R and Recondition (B)		
1981-94	2.2	2.4
Kickdown Throttle Cable, Replace (B)		
1981-01	3.2	3.4
Oil Pan and/or Gasket, Replace (B)		
1981-05	1.8	2.0

	LABOR TIME	SEVERE SERVICE
Transmission Mounts, Replace (B)		
1981-05	.8	1.0
Valve Body Assy., R&R and Recondition (A)		
1981-05	7.0	7.2
Valve Body Assy., Replace (B)		
1981-05	3.2	3.4

SERVICE TRANSMISSION REMOVED
Transmission R&R included unless otherwise noted.

	LABOR TIME	SEVERE SERVICE
Flywheel (Flexplate), Replace (B)		
240, 260	5.3	5.5
740	5.0	5.2
760, 780	4.9	5.1
940, 960	3.7	3.9
S90, V90	3.9	4.1
Front Pump Oil Seal, Replace (B)		
240, 260	4.4	4.6
740, 760, 780	3.5	3.7
940	3.8	4.0
960	5.1	5.3
S90, V90	5.1	5.3
Replace front pump add	.5	.7
Torque Converter, Replace (B)		
240	4.4	4.6
260, 740	4.3	4.5
760	4.1	4.3
780, 940	4.2	4.4
960	4.0	4.2
S90, V90	4.2	4.4
Transmission Assy., R&I (B)		
240, 260	5.1	5.3
740	4.9	5.1
760	4.7	4.9
780	4.9	5.1
940	3.5	3.7
960	3.4	3.6
S90, V90	3.3	3.5
Transmission Assy., R&R and Recondition (A)		
Includes: Replace torque converter.		
240	14.8	15.0
260	14.9	15.1
740	14.5	14.7
760	14.6	15.8
780	14.5	14.7
940	14.6	14.8
960	16.8	17.0
S90, V90	16.7	16.9

SHIFT LINKAGE
AUTOMATIC TRANSAXLE

	LABOR TIME	SEVERE SERVICE
Shift Control Cable, Replace (B)		
Exc. below	1.5	1.7
C70, S70, V70	1.9	2.1
S40, V40, V50	1.9	2.1
S60, S80	2.5	2.7
XC90	2.4	2.6

	LABOR TIME	SEVERE SERVICE
AUTOMATIC TRANSMISSION		
Linkage, Adjust (B)		
1981-04	.5	.7

DRIVELINE
DRIVESHAFT

	LABOR TIME	SEVERE SERVICE
Center Bearing, Replace (Driveshaft Removed) (B)		
1981-01	.7	.9
Driveshaft, Replace (B)		
1981-87 front or rear		
one	.5	.7
both	.8	1.0
1988-04		
exc. below		
front	.5	.7
rear	.7	.9
both	.9	1.1
1998-00 S70, V70 AWD	1.4	1.6
2001-02 S60, V70 AWD	2.5	2.7
XC90	2.5	2.7
U-Joint, Recondition (B)		
1981-04 each w/driveshaft removed	.5	.7

DRIVE AXLE

	LABOR TIME	SEVERE SERVICE
Differential, Drain & Refill (B)		
1981-05	.6	.8
Axle Shaft, Seal and/or Bearing, Replace (B)		
1981-04 w/o multi-link		
one side	1.6	1.8
both sides	2.8	3.0
Axle Shaft Oil Seal, Replace (B)		
1981-94	1.5	1.7
1995-97 960	2.3	2.5
1998-01 AWD	1.9	2.1
2002-05		
S60, V70 AWD one	5.8	6.0
XC90 one	2.5	2.7
Differential Carrier, Replace (B)		
240	3.8	4.0
260	3.6	3.8
740, 760	3.3	3.5
780	3.4	3.6
940	3.5	3.7
960	3.3	3.5
S80	4.3	4.5
S90, V90	3.8	4.0
Inspection Cover Gasket, Replace (B)		
1981-05	.8	1.0

VLV-16   240 : 242 : 244 : 245 : 262 : 264 : 265 : 740 : 745 : 760 : 780 : 850 : 940

	LABOR TIME	SEVERE SERVICE
Pinion Shaft Oil Seal, Replace (B)		
w/multi-link		
exc. below	1.6	1.8
1999-01 S70, V70 AWD	2.2	2.4
2002 S60, V70 AWD	2.8	3.0
2003 S60, V70 AWD	2.3	2.5
w/o multi-link	.9	1.1
Rear Axle Assy., R&I (B)		
240	3.7	3.9
260	3.8	4.0
740	3.0	3.2
760, 780	2.8	3.0
940, 960	2.8	3.0
1998-04		
w/AWD	6.7	6.9
w/o AWD	3.0	3.2
Rear Axle Assy., R&R and Recondition (B)		
240, 260	9.7	9.9
740, 760	7.9	8.1
780	7.7	7.9
940, 960	7.9	8.1
S90, V90	7.9	8.1
Rear Axle Assy., Replace (B)		
240, 260	5.6	5.8
740	4.1	4.3
760	4.2	4.4
780	4.3	4.5
940	4.1	4.3
960	4.2	4.4
S90, V90	4.1	4.3
Rear CV Joint, Replace or Recondition (B)		
One	2.3	2.5
each addl. add	.5	.7

HALFSHAFTS

	LABOR TIME	SEVERE SERVICE
CV Joint, Replace or Recondition (B)		
1993-05		
one	1.6	1.8
one each side	2.8	3.0
both one side	1.9	2.1
all both sides	3.0	3.2
Halfshaft, Replace (B)		
1988-91		
760, 780, 940	1.4	1.6
960	1.9	2.1
1998-05		
C70, S70		
front	1.0	1.2
rear AWD	1.6	1.8
S60		
front	1.4	1.6
rear AWD	5.5	5.7
S80	1.6	1.8

	LABOR TIME	SEVERE SERVICE
V70		
1998-00		
front	1.0	1.2
rear AWD	1.6	1.8
2000-02		
front	1.4	1.6
rear AWD	5.5	5.7
XC90		
front (1.0)	1.4	1.6
rear (1.8)	2.4	2.6
both (3.2)	4.3	4.5
Halfshaft Boot, Replace (B)		
1988-90 760	2.1	2.3
1988-91 780	1.9	2.1
940	2.1	2.3
960	1.9	2.1
1998-05		
exc. below	1.8	2.0
C70, S70		
front	1.4	1.6
rear AWD	2.4	2.6
S60		
front	1.9	2.1
rear AWD	6.3	6.5
S80	1.9	2.1
S90, V90	1.9	2.1
V70		
1998-00		
front	1.4	1.6
rear AWD	2.4	2.6
2000-02		
front	1.9	2.1
rear AWD	6.3	6.5
XC90		
front (1.6)	2.2	2.4
rear (2.1)	2.8	3.0

BRAKES
ANTI-LOCK

	LABOR TIME	SEVERE SERVICE
Diagnose Anti-Lock Brake System Component Each (A)		
1988-05	1.0	1.0
Anti-Lock Relay, Replace (B)		
1988-05	.3	.5
Control Unit, Replace (B)		
1988-95	.3	.5
1996-97 850	.7	.9
1996-97 960	.3	.5
1998-00		
C70, S70, V70	.7	.9
S80	1.2	1.4
2001-04		
C70	.7	.9
S40, V40, V50	1.4	1.6
S60, S80, V70	1.7	1.9
XC90 (1.2)	1.6	1.8
Converter, Replace (B)		
1988-04	.3	.5

	LABOR TIME	SEVERE SERVICE
Front Pulse Wheel, Replace (B)		
1988-05 one (1.0)	1.4	1.6
Front Sensor Assy., Replace (B)		
Exc. below one	.5	.7
S40, V40	1.0	1.2
Hydraulic Modulator Assy., R&R or Replace (B)		
Exc. below	2.4	2.6
C70, S70	1.4	1.6
S40, V40, V50	1.8	2.0
S60, S80, V70, XC90	2.5	2.7
Pressure Reducing Valve, Replace (B)		
1991-05	.9	1.1
Rear Sensor Assy., Replace (B)		
1988-04 one		
740, 760, 780, 940, 960	.5	.7
850	.9	1.1
C70, S70	1.4	1.6
S40, V40, V50	1.1	1.3
S60		
left side	1.1	1.3
right side	1.8	2.0
S80	1.3	1.5
V70		
1998-00	1.4	1.6
2001-02		
left side	1.1	1.3
right side	1.8	2.0
XC90 (1.3)	1.8	2.0
Surge Protector, Replace (B)		
1988-05	.3	.5

SYSTEM

	LABOR TIME	SEVERE SERVICE
Bleed Brakes (B)		
Includes: Add fluid.		
240, 260	.8	1.0
740, 760, 780	.7	.9
850	.7	.9
940, 960	.7	.9
1998-05	1.1	1.3
w/ABS add	.2	.4
Brake Hose (Flexible), Replace (B)		
Includes: System bleeding.		
Exc. below one (1.0)	1.4	1.6
S60, S80, XC90 (1.2)	1.8	2.0
2001-05 V70 (1.2)	1.8	2.0
each addl. add	.3	.5
w/ABS add	.2	.4
Brake Proportioning Valve, Replace (B)		
1988-93 240	1.7	1.9
Brake Warning Valve Assy., Replace (B)		
Includes: System bleeding.		
1981-04	1.7	1.9

VLV-17

960 : C70 : S40 : S60 : S70 : S80 : S90 : V40 : V70 : V90 : XC70 : XC90

	LABOR TIME	SEVERE SERVICE
Master Cylinder, R&R and Recondition (B)		
Includes: System bleeding.		
Exc. below	3.6	3.8
S40	3.2	3.4
Master Cylinder, Replace (B)		
Includes: System bleeding.		
Exc. below	1.1	1.3
S60, S80	2.9	3.1
S40, V50	2.6	2.8
V70 01-02	2.9	3.1
XC90 (2.1)	2.8	3.0
Power Booster Unit, Replace (B)		
240	2.1	2.3
260	2.2	2.4
740	1.2	1.4
760	1.3	1.5
780	1.1	1.3
850	2.5	2.7
940	1.3	1.5
960	1.2	1.4
C70, S70	2.8	3.0
S40	1.4	1.6
V40, V50	3.4	3.6
S60	3.8	4.0
S80	4.1	4.3
S90, V90	1.1	1.3
V70		
1998-00	2.8	3.0
2001-05 (3.0)	4.1	4.3
XC90 (3.2)	4.3	4.5
Power Booster Vacuum Check Valve, Replace (B)		
1981-04	.3	.5
Power Booster Vacuum Pump, Replace (B)		
1981-85 240	.5	.7
260	.5	.7
S40, V40	.5	.7
Rear Wheel Limiter Valve, Replace (B)		
240, 260		
one	.8	1.0
both	1.3	1.5
SERVICE BRAKES		
Brake Pads, Replace (B)		
Exc. XC90		
front or rear	1.2	1.4
four wheels	2.2	2.4
XC90		
front or rear	1.6	1.8
four wheels	3.0	3.2

	LABOR TIME	SEVERE SERVICE
COMBINATION ADD-ONS		
Repack wheel bearings		
add each wheel	.3	.5
Replace		
brake hose add	.3	.5
caliper add	.2	.4
disc rotor add	.3	.5
Resurface disc rotor add	.9	1.1
Caliper Assy., R&R and Recondition (B)		
Includes: System bleeding.		
Exc. below		
front		
one	1.8	2.0
both	2.9	3.1
rear		
one	1.6	1.8
both	2.5	2.7
S60, S80, V70 rear		
one	2.1	2.3
both	3.0	3.2
XC90		
front		
one	2.2	2.4
both	3.3	3.5
rear		
one	2.1	2.3
both	3.0	3.2
w/ABS thru 1998		
add each	.2	.4
Caliper Assy., Replace (B)		
Includes: System bleeding.		
Exc. below		
front		
one	1.2	1.4
both	1.5	1.7
rear		
one	.9	1.1
both	1.5	1.7
four wheels	2.9	3.1
C70, S70, V70		
front		
one	1.8	2.0
both	2.2	2.4
rear		
one	1.4	1.6
both	1.8	2.0
S40, V40, V50		
front		
one	1.8	2.0
both	2.9	3.1
rear		
one	2.1	2.3
both	2.4	2.6

	LABOR TIME	SEVERE SERVICE
XC90		
front		
one	2.2	2.4
both	2.8	3.0
rear		
one	1.8	2.0
both	2.8	3.0
w/ABS thru 1999		
add each	.2	.4
Disc Brake Rotor, Replace (B)		
Exc. below		
front		
one	.7	.9
both	1.0	1.2
rear		
one	1.0	1.2
both	1.3	1.5
S40, V40, V50 rear		
one	1.3	1.5
both	1.6	1.8
XC90 rear		
one	1.9	2.1
both	2.5	2.7
Front Hub, Replace (B)		
1985-87 740		
one	.8	1.0
both	1.9	2.1
1983-87 760		
one	.8	1.0
both	1.6	1.8
1987 780		
one	.8	1.0
both	1.7	1.9
Front Pads & Rotors, Replace (B)		
1981-05	1.1	1.3
PARKING BRAKE		
Parking Brake Cable, Adjust (C)		
1981-05	.5	.7
Parking Brake Apply Actuator, Replace (B)		
Exc. below	.7	.9
740, 780	.5	.7
Parking Brake Cable, Replace (B)		
240, 260		
one	2.1	2.3
two	3.2	3.4
1985-87 740		
front or rear	1.3	1.5
one piece	1.3	1.5
1988-92 740		
front	1.8	2.0
rear		
right	.8	1.0
left	1.5	1.7

VLV-18 240 : 242 : 244 : 245 : 262 : 264 : 265 : 740 : 745 : 760 : 780 : 850 : 940

	LABOR TIME	SEVERE SERVICE
Parking Brake Cable, Replace (B)		
760		
1983-87		
front	1.4	1.6
rear	1.3	1.5
one piece	1.3	1.5
1988-90		
front	1.9	2.1
rear		
right	.8	1.0
left	1.4	1.6
780		
front	1.8	2.0
rear		
right	.8	1.0
left	1.3	1.5
850		
right	2.0	2.2
left	1.9	2.1
940, 960		
front	1.9	2.1
rear		
right	.8	1.0
left	1.5	1.7
C70, S70		
one	2.4	2.6
both	3.3	3.5
S40, V40, V50		
front	1.3	1.5
rear	2.4	2.6
S60, S80		
front	1.3	1.5
rear		
one	2.4	2.6
both	3.3	3.5
S90, V90		
front	1.2	1.4
rear		
right	.8	1.0
left	1.7	1.9
V70		
1998-00		
one	2.4	2.6
both	3.3	3.5
2001-05		
front	1.3	1.5
rear		
one	2.4	2.6
both	3.3	3.5
XC90		
front (1.3)	1.2	1.4
rear		
right (1.7)	2.3	2.5
left (1.7)	2.3	2.5
Parking Brake Shoes, Replace (B)		
One side	.8	1.0
Both sides	1.6	1.8

FRONT SUSPENSION

Unless otherwise noted, time given does not include alignment.

	LABOR TIME	SEVERE SERVICE
Align Front End (A)		
1981-05	1.1	1.3
w/air bags add	.2	.4
Front & Rear Alignment, Check Adjust (A)		
850	1.8	2.0
1988-92 740	1.9	2.1
1988-90 760	1.8	2.0
780	2.0	2.2
940, 960	1.9	2.1
C70, S70, V70	2.0	2.2
S40, V40, V50	1.5	1.7
S60, S80, XC90	1.7	2.0
w/air bags add	.2	.4
Front Toe, Adjust (B)		
1981-05	.7	.9
Front Coil Spring, Replace (B)		
850		
one	1.5	1.7
both	2.6	2.8
1995-97 960		
one	1.5	1.7
both	3.0	3.2
C70, S70, V70		
one	1.4	1.6
both	2.5	2.7
S40, V40, V50		
one	1.4	1.6
both	2.8	3.0
S60, S80		
one	1.4	1.6
both	2.7	2.9
S90, V90		
one	1.8	2.0
both	2.9	3.1
XC90		
one (1.4)	1.9	2.1
both (2.6)	3.5	37
Front Crossmember, Replace (B)		
850	4.9	5.1
C70, S70, V70 (3.9)	5.3	5.5
S40, V40, V50	3.6	3.8
S60, S80	5.8	6.0
XC90		
B5254T (4.9)	6.6	6.8
B6294T (5.3)	7.2	7.4
Front Hub Assy., Replace (B)		
1993-98		
one	1.8	2.0
both	3.5	3.7
1999-03		
C70, S70		
one	1.1	1.3
both	2.2	2.4

	LABOR TIME	SEVERE SERVICE
S60, S80		
one	2.2	2.4
both	4.3	4.5
V70		
1999-00		
one	1.1	1.3
both	2.2	2.4
2001-05		
one (1.6)	2.2	2.4
both (3.2)	4.3	4.5
XC90		
One (1.3)	1.8	2.0
both (2.4)	3.2	3.4
Lower Ball Joint, Replace (B)		
Exc. below one	.7	.9
850, C70, S70		
one side	.5	.7
both sides	.8	1.0
S60, S80		
one side	1.8	2.0
both sides	3.0	3.2
2000-02 V70		
one side	1.8	2.0
both sides	3.0	3.2
XC90		
one side	1.6	1.8
both sides	3.0	3.2
Lower Control Arm, Replace (B)		
240, 260		
one side	1.4	1.6
both sides	2.5	2.7
740, 760, 780		
one side	.8	1.0
both sides	1.2	1.4
850		
one side	1.2	1.4
both sides	1.8	2.0
940, 960		
one side	.8	1.0
both sides	1.2	1.4
C70, S70, V70		
w/AWD both sides	2.8	3.0
w/o AWD		
one side	1.6	1.8
both sides	2.5	2.7
S40, V40, V50		
one side	1.3	1.5
both sides	1.9	2.1
S60		
one side	1.6	1.8
both sides	2.8	3.0
S80		
exc. below		
one side	1.6	1.8
both sides	3.4	3.6
B6284T, B6294S, B6304S		
one side	2.4	2.6
both sides	4.3	4.5

	LABOR TIME	SEVERE SERVICE
S90, V90		
one side	.8	1.0
both sides	1.6	1.8
XC90		
one side (1.0)	1.4	1.6
both sides (1.8)	2.4	2.6
Lower Control Arm Bushings, Replace (B)		
850, C70, S70, V70		
one side	1.3	1.5
both sides	2.2	2.4
S40, V40, V50		
one side	1.6	1.8
both sides	2.9	3.1
Lower Control Arm Bushings and/or Shaft, Replace (B)		
240, 260		
one side	1.7	1.9
both sides	2.9	3.1
740, 760, 780		
one side	.8	1.0
both sides	1.7	1.9
940, 960		
one side	.8	1.0
both sides	1.6	1.8
Lower Control Arm Strut Bushings, Replace (B)		
1981-92 one side	.5	.7
Stabilizer Bar, Replace (B)		
850	1.9	2.1
940, 960		
1994	.5	.7
1995-97	.9	1.1
C70, S70	1.8	2.0
S40, V40, V50	2.8	3.0
S60, S80,	2.9	3.1
S90, V90	.9	1.1
V70		
1998-00	1.8	2.0
2001-05 (2.1)	2.8	3.0
XC90 (3.2)	4.3	4.5
Stabilizer Bar Bushing and/or Bracket, Replace (B)		
850, C70, S70, V70	.9	1.1
S40, V40, V50	2.9	3.1
Stabilizer Bar Links and/or Grommets, Replace (B)		
One	.5	.7
Both	.7	.9
Strut Assembly, R&R or Replace (B)		
240, 260		
one side	1.3	1.5
both sides	2.3	2.5
740, 760, 780		
one side	1.5	1.7
both sides	2.8	3.0
850		
one side	1.2	1.4
both sides	2.4	2.6

	LABOR TIME	SEVERE SERVICE
940, 960		
one side	1.4	1.6
both sides	2.8	3.0
C70, S70, V70		
one side (1.0)	1.4	1.6
both sides (1.8)	2.4	2.6
S40, V40, V50		
one side	1.4	1.6
both sides	2.7	2.9
S60, S80		
one side	1.4	1.6
both sides	2.7	2.9
S90, V90		
one side	1.4	1.6
both sides	2.8	3.0
XC90		
one side	1.7	2.0
both sides	3.3	3.5
Replace		
cartridge add each	.2	.4
coil spring add each	.2	.4
Wheel Bearing and Cups, Replace (B)		
One side	.8	1.0
Both sides	1.7	1.9
Wheel Bearing and/or Grease Seal, Replace (B)		
Exc. below		
one side	.7	.9
both sides	1.1	1.3
S40, V40, V50		
one side	2.5	2.7
both sides	4.9	5.5
Wheel Bearings, Clean & Pack (B)		
1981-05 both wheels	1.6	1.8

REAR SUSPENSION

	LABOR TIME	SEVERE SERVICE
Rear Alignment, Adjust (A)		
Multi-Link Suspension		
1988-92 740	1.7	1.9
1988-90 760	1.5	1.7
850	1.8	2.0
940	1.5	1.7
960	1.6	1.8
C70, S70, V70	2.1	2.3
S40, V40, V50	2.1	2.3
S60, S80	2.1	2.3
S90, V90	1.6	1.8
XC90 (1.6)	2.2	2.4
Bearing Housing Support Arm Bushing, Replace (B)		
1988-90 760	2.4	2.6
1988-91 780	2.4	2.6
940	2.6	2.8
960	2.4	2.6
S90, V90	2.6	2.8

	LABOR TIME	SEVERE SERVICE
Lower Control/Link Arm, Replace (B)		
S40, V40, V50		
one side	.6	.8
both sides	1.3	1.5
S60, S70, V70		
1998-00 AWD		
one side	2.5	2.7
both sides	4.1	4.3
2001-02		
w/AWD		
one side	1.0	1.2
both sides	1.9	2.1
w/o AWD		
one side	1.9	2.1
both sides	2.7	2.9
S80		
one side	1.7	1.9
both sides	2.7	2.9
XC90		
one side (1.6)	2.2	2.4
both sides (2.6)	3.5	3.7
Rear Control/Link Arm Bushing, Replace (B)		
1988-90 760		
inner	.7	.9
upper	2.3	2.5
lower	2.7	2.9
1988-91 780		
inner	.7	.9
upper	2.2	2.4
lower	2.8	3.0
940		
inner	.7	.9
upper	2.2	2.4
lower	2.7	2.9
960		
inner	.7	.9
upper	2.4	2.6
lower	2.9	3.1
S90, V90		
inner	.7	.9
upper	2.4	2.6
lower	2.9	3.1
1999-05		
S60, S80, V70, V40, V50		
inner upper		
w/AWD (5.6)	7.6	7.8
w/o AWD (5.2)	7.0	7.2
inner lower (1.6)	2.2	2.4
upper (2.4)	3.2	3.4
lower (1.5)	2.0	2.2
XC90		
inner upper	7.6	7.8
inner lower	2.1	2.3
upper	3.2	3.4
lower	2.1	2.3

VLV-20   240 : 242 : 244 : 245 : 262 : 264 : 265 : 740 : 745 : 760 : 780 : 850 : 940

	LABOR TIME	SEVERE SERVICE
Rear Hub & Bearings, Replace (B)		
1981-94	2.2	2.4
1995-03		
exc. below		
one side	2.9	3.1
both sides	4.7	4.9
S40, V40, V50		
one side	1.0	1.2
both sides	1.7	1.9
S60, S80, V70	1.6	1.8
XC90	1.8	2.0
Rear Spring, Replace (B)		
240		
1981-85		
one	.7	.9
both	1.4	1.6
1986-93		
one	.5	.7
both	1.3	1.5
260		
one	.7	.9
both	1.3	1.5
740		
1985-87		
one	.5	.7
both	.7	.9
1988-92		
multi-link		
one	.7	.9
both	1.5	1.7
solid axle		
one	.8	1.0
both	1.8	2.0
760		
1983-87		
one	.5	.7
both	.7	.9
1988-90		
solid axle		
one	.8	1.0
both	1.8	2.0
multi-link		
one	.7	.9
both	1.5	1.7
780		
solid axle		
one	.8	1.0
both	1.7	1.9
multi-link		
one	.7	.9
both	1.5	1.7
850		
one	.5	.7
both	.8	1.0
940, 960		
multi-link		
one	.7	.9
both	1.6	1.8

	LABOR TIME	SEVERE SERVICE
solid axle		
one	.8	1.0
both	1.8	2.0
C70, S70, V70 99-00		
w/AWD		
one	2.1	2.3
both	3.5	3.7
w/o AWD		
one	.5	.7
both	1.0	1.2
S40, V40, V50		
one	1.6	1.8
both	2.8	3.0
2001-05 S60, V70		
w/AWD		
one (1.2)	1.6	1.8
both (2.1)	2.8	3.0
w/o AWD		
one	1.8	2.0
both	2.5	2.7
S80		
one	1.6	1.8
both	2.2	2.4
S90, V90	4.9	5.1
XC90		
one (1.4)	1.9	2.1
both (1.8)	2.4	2.6
Rear Torque Rod, Replace (B)		
1981-85 240	.5	.7
260	.5	.7
Rear Torque Rod Bushing, Replace (B)		
240, 260		
one side	.7	.9
both sides	1.6	1.8
740, 760, 780		
one side	.7	.9
both sides	1.4	1.6
940, 960		
one side	.7	.9
both sides	1.6	1.8
S90, V90		
one side	.7	.9
both sides	1.6	1.8
Rear Track Bar and/or Bushings, Replace (B)		
All Models	.8	1.0
Shock Absorbers or Bushings, Replace (B)		
240, 260		
one	.5	.7
both	.7	.9
740, 760, 780		
one	.3	.5
both	.5	.7
850		
one	.7	.9
both	.5	.7

	LABOR TIME	SEVERE SERVICE
940, 960		
one	.3	.5
both	.5	.7
C70		
cabriolet		
one	1.4	1.6
both	2.2	2.4
coupe		
one	.7	.9
both	1.2	1.4
S60		
one	1.3	1.5
both	2.2	2.4
S70, V70		
w/AWD		
one (1.0)	1.4	1.6
both (1.3)	1.8	2.0
w/o AWD		
one (.5)	.7	.9
both (1.0)	1.4	1.6
S40, V40, V50		
one	1.4	1.6
both	2.5	2.7
S80		
one	1.4	1.6
both	1.6	1.8
S90, V90		
one	.3	.5
both	.5	.7
XC90		
one (1.2)	1.6	1.8
both (1.6)	2.2	2.4
Stabilizer Bar, Replace (B)		
1981-05		
exc. below	.7	.9
S60, V70, S80 (1.0)	1.4	1.6
XC90 (2.1)	2.8	3.0
Stabilizer Bar Bushings, Replace (B)		
1981-04	.5	.7
Stub Axle, Replace (B)		
850, C70, S70, V70	2.2	2.4
Trailing Arm and/or Bushings, Replace (B)		
240	1.6	1.8
260	1.8	2.0
740	1.5	1.7
760	1.8	2.0
780	1.6	1.8
850	3.8	4.0
940	1.6	1.8
960	1.8	2.0
C70, S70	3.8	4.0
S40, V40, V50	1.6	1.8
S60, S80	2.1	2.3
S90, V90	1.5	1.7
V70		
1998-00	3.8	4.0
2001-05 (1.6)	2.2	2.4
XC90 (1.2)	1.6	1.8

960 : C70 : S40 : S60 : S70 : S80 : S90 : V40 : V70 : V90 : XC70 : XC90 **VLV-21**

	LABOR TIME	SEVERE SERVICE
STEERING		
AIR BAGS		
Diagnose Air Bag System Component Each (A)		
1987-05	.9	.9
Air Bag Assy., Replace (B)		
Driver side		
1988-02		
exc. below	.5	.7
2001-05 S80, V70	.8	1.0
Passenger side		
850	1.7	1.9
940	.9	1.1
960	.8	1.0
C70, S70	1.8	2.0
S40, V50	2.5	2.7
V40	3.9	4.1
S80	4.7	4.9
S90, V90	.8	1.0
V70		
1999-00	1.8	2.0
2001-05 (3.4)	4.6	4.8
XC90 (.9)	1.2	1.4
Side Impact each	1.6	1.8
Inflatable curtain		
S60, V70, XC90	3.6	3.8
S40, V40, V50	2.2	2.4
S80	4.1	4.3
Cable Reel, Replace (B)		
All	.9	1.1
Contact Reel, Replace (B)		
1988-05	1.3	1.5
Crash Sensor, Replace (B)		
1990-93 240	.8	1.0
1991-92 740	4.6	4.8
780	1.1	1.3
850	.7	.9
940		
1992	4.5	4.7
1993-95	1.2	1.4
960		
1992	3.9	4.1
1993-97	1.2	1.4
C70, S70, V70 (.5)	.7	.9
S40, V40, V50, S80 (.5)	.7	.9
S90, V90 (.8)	1.1	1.3
XC90 (.5)	.7	.9
Energy Reserve Module, Replace (B)		
1990-93 240	.5	.7
1990-92 740	1.4	1.6
1990 760	1.2	1.4
850	.5	.7
940	1.3	1.5
960	1.4	1.6
C70, S70	.6	.8
S90, V90	1.2	1.4

	LABOR TIME	SEVERE SERVICE
V70		
1999-00	.6	.8
2001-04	1.1	1.3
XC90	.9	1.1
Standby Power Unit, Replace (B)		
1988-05	.8	1.0
POWER RACK & PINION		
Pump Pressure Check & Flow Test (B)		
1981-05		
exc. below (.5)	.7	.9
S60, S80, V70 (1.1)	1.5	1.7
XC90 (1.1)	1.5	1.7
Gear Assy., Adjust (On Vehicle) (B)		
1981-05	.8	1.0
Inner Tie Rod Ends, Replace (B)		
Includes: Reset toe.		
Exc. below		
one	1.7	1.9
both	2.9	3.1
C70, S70		
one	1.3	1.5
both	1.8	2.0
S40, V50		
one	1.1	1.3
both	1.5	1.7
V40		
left side	.5	.7
right side	1.1	1.5
S60, S80, XC90		
one	1.3	1.5
both	2.4	2.6
V70		
1998-00		
one	1.3	1.5
both	1.8	2.0
2001-05		
one (1.0)	1.4	1.6
both (1.8)	2.4	2.6
Outer Tie Rod Ends, Replace (B)		
Does not include reset toe-in.		
1981-97		
one	.8	1.0
both	1.1	1.3
1998-05		
one	.5	.7
both	.6	.8
Rack & Pinion Assy., R&R and Recondition (A)		
1981-93 240	5.1	5.3
260	5.1	5.3
740, 760, 780, 940	4.9	5.1
940, 960	4.8	5.0
C70, S70, V70 (4.9)	6.6	6.8
V40,	8.1	8.3
S80	7.0	7.2
S90, V90 (3.5)	4.7	4.9

	LABOR TIME	SEVERE SERVICE
Rack & Pinion Assy., R&R or Replace (B)		
240, 260	2.4	2.6
740	2.3	2.5
1985-90 760	2.3	2.5
780	1.9	2.1
850	2.9	3.1
940, 960	2.1	2.3
C70, S70	3.2	3.4
S40, V50	3.6	3.8
V40	5.2	5.4
S60	4.4	4.6
S80	4.9	5.1
S90, V90	1.9	2.1
V70		
1998-00	3.2	3.4
2001-05 (3.4)	4.6	4.8
XC90		
5 cyl (4.8)	6.5	6.7
6 cyl (5.2)	7.0	7.2
Steering Column, Replace (B)		
w/air bags		
1990-93 240	3.2	3.4
1987-92 740	3.0	3.2
1987-90 760	3.2	3.4
780	3.2	3.4
850	2.2	2.4
940, 960	3.2	3.4
C70, S70, V70	2.4	2.6
S40, V40, V50	1.9	2.1
S60, S80	2.5	2.7
S90, V90	2.2	2.4
XC90	2.8	3.0
w/o air bags		
1981-89 240	2.4	2.6
260	2.4	2.6
1985-86 740	1.9	2.1
1983-86 760	2.0	2.2
Steering Flex Coupling, Replace (B)		
240, 260		
upper	.9	1.1
lower	.8	1.0
740, 760, 780	.5	.7
850	.8	1.0
940, 960	.5	.7
C70, S70, V70	.9	1.1
S60, S80	1.1	1.3
S90, V90	.5	.7
XC90	.8	.9
Steering Pump, R&R and Recondition (B)		
240	3.0	3.2
260	2.4	2.6
1986-92 740		
4 cyl.	1.9	2.1
V6	2.5	2.7

VLV-22 240 : 242 : 244 : 245 : 262 : 264 : 265 : 740 : 745 : 760 : 780 : 850 : 940

	LABOR TIME	SEVERE SERVICE
Steering Pump, R&R and Reconditioning (B)		
760		
1983-85	2.4	2.6
1986-90		
4 cyl.	2.1	2.3
V6	2.5	2.7
780		
4 cyl.	2.0	2.2
V6	2.5	2.7
940	1.9	2.1
960	2.3	2.5
S40, V40		
exc. below	1.6	1.8
B4194T, B4204T	2.9	3.1
S90, V90	2.3	2.5
Steering Pump, R&R or Replace (B)		
240	1.6	1.8
260	1.1	1.3
1986-92 740		
4 cyl.	.8	1.0
V6	1.2	1.4
760		
1983-85		
4 cyl.	1.5	1.7
V6	1.2	1.4
diesel	.9	1.1
1986-90		
4 cyl.	.8	1.0
V6	1.2	1.4
780		
4 cyl	.8	1.0
V6	1.1	1.3
850	1.8	2.0
940	.8	1.0
960	1.2	1.4
C70, S70, V70 (1.3)	1.8	2.0
S40, V40, V50	1.6	1.8
S60, S80	1.8	2.0
S90, V90	1.1	1.3
XC90		
5 cyl (1.1)	1.5	1.7
6 cyl (1.6)	2.2	2.4
Steering Pump Hoses, Replace (B)		
Exc. below	.8	1.0
850	.9	1.1
C70, S70, S40, V50	1.5	1.7
S60, V70		
1998-00	1.5	1.7
2001-05		
pressure hose		
w/AWD (2.5)	3.4	3.6
w/o AWD (1.9)	2.6	2.8
return hose (1.3)	1.8	2.0
S80		
pressure hose	2.6	2.7
return hose	1.8	2.0
XC90		
pressure hose (4.7)	6.3	6.5
return hose (4.5)	6.1	6.3

	LABOR TIME	SEVERE SERVICE
Steering Wheel, Replace (B)		
Exc. below	.5	.7
S60, S80, V70, XC90	.7	.9
Tie Rod End Boot, Replace (B)		
One	1.1	1.3
Both	1.6	1.8

HEATING & AIR CONDTIONING

When more than one component requires replacement where evacuation/recovery and recharging is already included, deduct 1.0 hour for each additional component from the time given.

	LABOR TIME	SEVERE SERVICE
Evacuate/Recover and Recharge System (B)		
1981-05	1.1	1.3
AC Hoses, Replace (B)		
Includes: Evacuate/recover and recharge.		
Exc. below one (1.4)	1.9	2.1
each addl. add	.5	.7
S60, S80, V70		
pressure side (2.1)	2.8	3.0
Compressor Assy., Replace (B)		
Includes: Parts transfer, evacuate/recover and recharge.		
240		
1981-85	3.2	3.4
1986-93	3.5	3.7
260	2.9	3.1
740		
4 cyl.	3.2	3.4
V6	2.7	2.9
760		
1983-85		
gasoline	2.2	2.4
diesel	3.6	3.8
1986-90		
4 cyl.	3.2	3.4
V6	2.7	2.9
780		
4 cyl.	3.2	3.4
V6	2.7	2.9
850	3.7	3.9
940	3.1	3.3
960	2.5	2.7
C70, S70	3.3	3.5
S40, V40, V50	2.7	2.9
S60		
w/turbo	5.1	5.3
w/o turbo	4.6	4.8
S80	3.5	3.7
S90, V90	2.5	2.7

	LABOR TIME	SEVERE SERVICE
V70		
1998-00	3.3	3.5
2001-05		
w/turbo (3.4)	4.6	4.8
w/o turbo (3.0)	4.1	4.3
XC90		
5 cyl (3.2)	4.3	4.5
6 cyl (4.2)	5.7	5.9
Compressor Clutch Assy., Replace (B)		
240		
1981-85	.8	1.0
1986-93	1.1	1.3
260	.9	1.1
740, 760		
4 cyl.	1.2	1.4
V6	.9	1.1
diesel	3.8	4.0
780		
4 cyl.	1.3	1.5
V6	.9	1.1
940	1.3	1.5
960	1.9	2.1
S40, V40, V50	1.7	1.9
S60, V70		
w/turbo (3.0)	4.1	4.3
w/o turbo (2.6)	3.5	3.7
S80	2.4	2.6
S90, V90	2.1	2.3
XC90		
5 cyl (2.7)	3.6	3.8
6 cyl (3.8)	5.1	5.3
Compressor or Fan Relay, Replace (B)		
740, 760, 780	.5	.7
850	.5	.7
940, 960	.5	.7
1998-02	.5	.7
Condenser Assy., Replace (B)		
Includes: Evacuate/recover and recharge.		
240	2.4	2.6
260, 740	2.6	2.8
760, 780	2.5	2.7
850	3.2	3.4
940, 960	2.6	2.8
C70, S70, V70 (2.1)	2.8	3.0
S40, V50	3.8	4.0
V40	2.0	2.2
S60, S80	3.3	3.7
S90, V90	2.6	2.8
XC90 (2.6)	3.5	3.7
Evaporator Core, Replace (B)		
Includes: Evacuate/recover and recharge.		
240		
1981-90	4.1	4.3
1991-93	4.9	5.1
260	4.3	4.5

960 : C70 : S40 : S60 : S70 : S80 : S90 : V40 : V70 : V90 : XC70 : XC90   **VLV-23**

	LABOR TIME	SEVERE SERVICE
740		
ACC or MCC	5.3	5.5
ECC	4.3	4.5
760, 780		
ACC or MCC	5.3	5.5
ECC	4.1	4.3
850	7.7	7.9
940, 960		
ACC or MCC	5.1	5.3
ECC	4.2	4.4
C70, S70	7.7	7.9
S40, V50	4.9	5.1
V40	5.0	5.2
S60, S80	9.8	10.0
S90, V90		
ACC or MCC	5.2	5.4
ECC	4.2	4.4
V70		
1998-00	7.7	7.9
2001-05 (7.5)	10.1	10.3
XC90 (7.3)	9.9	10.1
Expansion Valve/Tube, Replace (B)		
Includes: Evacuate/recover and recharge.		
240		
1981-90	2.4	2.6
1991-93	1.7	1.9
260	2.2	2.4
740, 760, 780	1.9	2.1
850, 940, 960	1.8	2.0
C70, S70, V70	1.8	2.0
S40, V40, V50	2.2	2.4
S60, S80	2.3	2.5
V70		
1998-00	1.8	2.0
2001-05 (1.7)	2.3	2.5
XC90 (1.7)	2.3	2.5
Heater Blower Motor, Replace (B)		
Exc. below	.9	1.1
240	6.3	6.5
260	4.3	4.5
S40, V50	5.9	6.1
XC90	7.2	7.4
Heater Blower Motor Switch, Replace (B)		
1981-05		
exc. below	.5	.7
S60, S80, V70, XC90	.9	1.1
Heater Control Valve, Replace (B)		
Exc. below	.8	1.0
240, 260	1.7	1.9
Heater Core, R&R or Replace (B)		
240, 260	7.9	8.1
740, 760, 780		
w/EEC	12.3	12.5
w/o EEC	10.0	10.2
850	2.8	3.0

	LABOR TIME	SEVERE SERVICE
940, 960		
w/EEC	12.3	12.5
w/o EEC	10.0	10.2
C70, S70, V70	1.8	2.0
S40, V40		
2000		
w/EEC	6.8	7.0
w/o EEC	5.0	5.2
2001-05		
w/EEC (4.8)	6.5	6.7
w/o EEC (3.5)	4.7	4.9
S60, S80	1.2	1.4
S90, V90		
w/EEC	12.3	12.5
w/o EEC	10.0	10.2
XC90 (2.3)	3.1	3.3
Heater Hoses, Replace (B)		
1981-04 each	.8	1.0
High Pressure Cut-Off Switch, Replace (B)		
850	.5	.7
940, 960	.5	.7
1998-04		
exc. below	.5	.7
S60, S80, V70	1.4	1.6
Receiver/Drier Assy., Replace (B)		
Includes: Evacuate/recover and recharge.		
240, 260	2.3	2.5
740, 760	2.4	2.6
780	2.3	2.5
850	3.2	3.4
940, 960	2.4	2.6
C70, S70	2.4	2.6
S40, V40, V50	2.1	2.3
S60, S80	3.8	4.0
S90, V90	2.2	2.4
V70		
1998-00	2.4	2.6
2000-05 (2.8)	3.8	4.0
XC90 (1.7)	2.3	2.5

AUTOMATIC TEMPERATURE CONTROL (ATC)

Air Mix Servo Motor, Replace (B)

	LABOR TIME	SEVERE SERVICE
1988-90 760	.8	1.0
1988-91 780	.8	1.0
850	.8	1.0
940, 960	.8	1.0
S90, V90	.8	1.0

Ambient Sensor, Replace (B)

	LABOR TIME	SEVERE SERVICE
1988-90 760	.5	.7
1988-91 780	.5	.7
850	.7	.9
940, 960	.5	.7
C70, S70, S40, V40	.3	.5
S60, S80, V70, V50	.7	.9
S90, V90	.5	.7
XC90	.7	.9

	LABOR TIME	SEVERE SERVICE
Climate Control Solenoid Valve, Replace (B)		
1988-90 760	.7	.9
1988-91 780	.7	.9
940, 960, S90, V90	.7	.9
Control Assy., Replace (B)		
1988-90 760	.7	.9
1988-91 780	.7	.9
850	.5	.7
940, 960	.7	.9
S40, V40, V50	.5	.7
S80, S90, V90	.7	.9
In-Vehicle Sensor, Replace (B)		
1988-97	.7	.9
1998-04	.3	.5
Power Servo Assy., R&R or Replace (B)		
1988-90 760	.8	1.0
1988-91 780	.8	1.0
940, 960	.8	1.0
S90, V90	.8	1.0
Programmer, Replace (B)		
1988-90 760	1.3	1.5
1988-91 780	1.3	1.5
940, 960	1.3	1.5
S90, V90, V50	1.1	1.3
Thermostat, Replace (B)		
1988-93 240	1.9	2.1
Vacuum Motor, Replace (B)		
1988-90 760	3.2	3.4
1988-91 780	3.1	3.3
940, 960	3.3	3.5
S90, V90	3.1	3.3
Water Temperature Sensor Replace (B)		
1988-90 760	.5	.7
1988-91 780	.5	.7
940, 960	.5	.7
S90, V90	.5	.7

WIPERS & SPEEDOMETER

Power Antenna, Replace (B)

	LABOR TIME	SEVERE SERVICE
Exc. below	.8	1.0
240, 260, S40	.5	.7
C70, S70, V70	.5	.7

Radio, R&R (B)

	LABOR TIME	SEVERE SERVICE
Exc. below	.5	.7
240		
1981-87	.8	1.0
1988-93	1.1	1.3
260, 740, 760	.8	1.0
S60, S80, V70	.9	1.1

Rear Window Washer Motor, Replace (B)

	LABOR TIME	SEVERE SERVICE
1981-85	.7	.9
1986-97 200, 700, 900 Series	.5	.7
850	.7	.9
S40, V40, V50	.6	.8
V70	.7	.9

	LABOR TIME	SEVERE SERVICE
Rear Window Washer Pump, Replace (B)		
1981-05	.5	.7
Rear Window Wiper Motor, Replace (B)		
1981-05	.7	.9
Rear Window Wiper Switch, Replace (B)		
1981-05	.6	.8
Speedometer Cable & Casing, Replace (B)		
1981-87	.7	.9
S40, V40	.8	1.0
Speedometer Head, R&R or Replace (B)		
Includes: Cluster R&R.		
240, 260	1.3	1.5
740	1.3	1.5
760		
1983-84	.8	1.0
1985-90	1.4	1.6
780	1.5	1.7
850	2.3	2.5
940, 960	1.4	1.6
C70, S70, V70	1.4	1.6
S40, V40, V50	1.5	1.7
S60, S80	1.0	1.2
S90, V90	1.4	1.6
XC90	.6	.8
Speedometer/ABS Sender, Replace (B)		
1988-05	.5	.7
Windshield Washer Motor, Replace (B)		
1981-05		
exc. below	.6	.8
S60, S80, V70 (.7)	.9	1.1
XC90 (.6)	.8	1.0
Windshield Washer Pump, Replace (B)		
1981-05	.5	.7
Windshield Wiper Interval Relay, Replace (B)		
1981-85	.5	.7
1986-05	.3	.5
Windshield Wiper Motor, Replace (B)		
240		
1981-87	.9	1.1
1988-93	.5	.7
260	.9	1.1
740	.9	1.1
760		
1983-87	.9	1.1
1988-90	1.2	1.4
780	1.2	1.4
850	.9	1.1
940, 960	1.3	1.5

	LABOR TIME	SEVERE SERVICE
C70, S70, V70 (.7)	.9	1.1
S40, V40, V50	.8	1.0
S60, S80	1.4	1.6
S90, V90, XC90 (1.0)	1.4	1.6
Windshield Wiper Switch (Combination), Replace (B)		
1981-85	.7	.9
1986-04	.8	1.0

LAMPS & SWITCHES

	LABOR TIME	SEVERE SERVICE
Back-Up Lamp Switch, Replace (B)		
240, 260	.9	.9
740		
1985	.3	.3
1986-92	.5	.5
760		
1983-85	.3	.3
1986-90	.5	.5
780	.5	.5
850	.5	.5
940, 960	.5	.5
1998-03		
exc. below	.5	.5
S40, V40, V50	1.1	1.3
XC90	.6	.7
Brake Lamp Switch, Replace (B)		
1981-04	.5	.5
Halogen Headlamp Bulb, Replace (C)		
Exc. below	.3	.3
1988-93 240	.5	.5
850	.5	.5
Hazard Warning Switch, Replace (B)		
1985-04	.5	.5
Headlamp Switch, Replace (B)		
1981-04	.8	.8
Headlamps, Aim (B)		
Two	.4	.4
Four	.6	.6
High Mount Stop Lamp Assy., Replace (B)		
1985-97	.5	.5
1998-04		
C70	.8	.8
S40, V40, S60, S80	.5	.5
S70, V50	.7	.7
S90, V90	.5	.5
V70	.5	.5
XC90	.3	.3
Horn, Replace (B)		
1981-04 each	.5	.5
Horn Relay, Replace (B)		
1981-04 each	.3	.3
Horn Switch, Replace (B)		
1981-04	.3	.3
License Lamp Assy., Replace (B)		
1998-04		
C70	.5	.5
S40, V40, V50	.5	.5
S60, S80	.3	.3

	LABOR TIME	SEVERE SERVICE
S70	.7	.7
V70	.8	.8
XC90	.3	.3
License Lamp Lens, Replace (B)		
1981-04	.2	.2
Neutral Safety Switch, Replace (B)		
1981-05	.8	1.0
Sealed Beam Headlamp, Replace (B)		
240 each	.5	.5
260 each	.3	.3
740		
1985-87		
one side	.7	.7
both sides	1.3	1.3
1988-89 each	.5	.5
760		
1983-87		
one side	.7	.7
both sides	1.3	1.3
1988 each	.5	.5
1987-89 780 each	.5	.6
Side Marker Lamp Lens, Replace (B)		
1981-04		
exc. below	.2	.2
S40, V40, V50		
front	.6	.6
rear	.5	.5
Stop Lamp Switch, Replace (B)		
1981-04	.5	.5
Tail Lamp Lens or Bulb, Replace (C)		
1981-04		
exc. below each	.4	.4
XC90	.6	.6
Turn Signal and Headlamp Dimmer Switch, Replace (B)		
1981-04	.8	.8
Turn Signal and Parking Lamp Lens, Replace (B)		
1981-04 each	.3	.3
Turn Signal or Hazard Warning Flasher, Replace (B)		
1981-04	.3	.3

SEAT BELTS

	LABOR TIME	SEVERE SERVICE
Lap Belt Assy., Replace (B)		
1990-93 240	.8	1.0
1990-91 780	1.7	1.9
850	.9	1.1
940, 960	1.8	2.0
C70, S70, V70 (1.0)	1.4	1.6
S40, V40, V50 (.7)	.9	1.1
S60, S80	.8	1.0
S90, V90 (1.3)	1.8	2.0
XC90 (.6)	.8	1.0

960 : C70 : S40 : S60 : S70 : S80 : S90 : V40 : V70 : V90 : XC70 : XC90   **VLV-25**

BODY

	LABOR TIME	SEVERE SERVICE
Door Window Regulator, Replace (B)		
Electric		
front		
1981-87	1.6	1.8
1988-97	1.1	1.3
C70	2.0	2.2
S40, V40, V50	1.1	1.3
S80	1.2	1.4
S60, S70, V70	.9	1.1
S80	1.4	1.6
XC90	1.2	1.4
rear		
exc. below	1.1	1.3
S80, XC90	1.4	1.6
Manual		
exc. below	.9	1.1
S40, V40	1.1	1.3
Front Door Latch, Replace (B)		
1981-05	.8	1.0
Hood Hinge, Replace (B)		
Exc. below	.8	1.0
C70, S70, V70	2.8	3.0
Hood Lock, Replace (B)		
Exc. below	1.5	1.7
S60, V70		

	LABOR TIME	SEVERE SERVICE
one	2.8	3.0
both	2.9	3.1
S80		
one	3.3	3.5
both	3.5	3.7
Hood Release Cable, Replace (B)		
240, 260	.7	.9
740, 760, 780	1.2	1.4
850	1.3	1.5
940, 760	1.3	1.5
C70, S70	1.1	1.3
S40, V40, V50	1.0	1.2
S60, V70	3.3	3.5
S80	4.0	4.2
XC90	1.2	1.4
Lock Striker Plate, Replace (B)		
1981-05	.3	.5
Rear Door Latch, Replace (B)		
Exc. below	.8	1.0
S40, V40, V50	1.5	1.7
Trunk or Liftgate Hinges, Replace (B)		
240, 260		
sedan	.9	1.1
wagon	1.7	1.9
740, 760		
sedan	2.1	2.3
wagon	1.2	1.4

	LABOR TIME	SEVERE SERVICE
780	2.3	2.5
850	2.0	2.2
940, 960	1.3	1.5
C70, S70	1.8	2.0
S40, V40, V50	.9	1.1
S60, S80	1.1	1.3
V70	1.2	1.3
XC90	.8	1.0
Window Regulator Motor, Replace (B)		
Front		
240	1.2	1.4
260	1.4	1.6
740, 760, 780	1.3	1.5
850	1.3	1.5
940, 960	1.3	1.5
C70	2.3	2.5
S40, V40, V50	1.2	1.4
S60, S80	1.1	1.3
S70, V70	.9	1.1
XC90	1.2	1.4
Rear		
exc. below	.9	1.1
XC90	1.4	1.6

NOTES

YUG

Cabrio : GV : GVL : GVS : GVX

SYSTEM INDEX

MAINTENANCE	YUG-2
CHARGING	YUG-2
STARTING	YUG-2
IGNITION	YUG-2
EMISSIONS	YUG-2
FUEL	YUG-2
EXHAUST	YUG-2
ENGINE COOLING	YUG-2
ENGINE	YUG-3
Assembly	YUG-3
Cylinder Head	YUG-3
Camshaft	YUG-3
Crank & Pistons	YUG-3
Engine Lubrication	YUG-3
CLUTCH	YUG-3
MANUAL TRANSAXLE	YUG-3
DRIVELINE	YUG-3
BRAKES	YUG-3
FRONT SUSPENSION	YUG-4
REAR SUSPENSION	YUG-4
STEERING	YUG-4
HEATING & AC	YUG-4
WIPERS & SPEEDOMETER	YUG-5
LAMPS & SWITCHES	YUG-5
BODY	YUG-5

OPERATIONS INDEX

A
- AC Hoses YUG-4
- Air Conditioning YUG-4
- Alternator (Generator) YUG-2

B
- Back-Up Lamp Switch YUG-5
- Battery Cables YUG-2
- Bleed Brake System YUG-3
- Blower Motor YUG-4
- Brake Disc YUG-4
- Brake Drum YUG-4
- Brake Hose YUG-3
- Brake Pads and/or Shoes YUG-4

C
- Camshaft YUG-3
- Catalytic Converter YUG-2
- Crankshaft YUG-3
- CV Joint YUG-3
- Cylinder Head YUG-3

D
- Differential YUG-3
- Distributor YUG-2
- Drive Belt YUG-2

E
- EGR YUG-2
- Electronic Control Module (ECM/PCM) YUG-2
- Engine YUG-3
- Engine Lubrication YUG-3
- Engine Mounts YUG-3
- Evaporator YUG-4
- Exhaust YUG-2
- Exhaust Manifold YUG-2
- Expansion Valve YUG-4

F
- Flywheel YUG-3
- Fuel Pump YUG-2
- Fuel Vapor Canister YUG-2

G
- Generator YUG-2

H
- Halfshaft YUG-3
- Headlamp YUG-5
- Heater Core YUG-4
- Horn YUG-5

I
- Ignition Coil YUG-2
- Ignition Switch YUG-2
- Inner Tie Rod YUG-4
- Intake Air Temperature (IAT) Sensor YUG-2
- Intake Manifold YUG-2

L
- Lower Control Arm YUG-4

M
- Master Cylinder YUG-4
- Muffler YUG-2

O
- Oil Pan YUG-3
- Oil Pump YUG-3
- Outer Tie Rod YUG-4

P
- Parking Brake YUG-4
- Pistons YUG-3
- Positive Crankcase Ventilation (PCV) Valve YUG-2

R
- Radiator YUG-2
- Radiator Hoses YUG-3
- Radio YUG-5
- Rear Main Oil Seal YUG-3

S
- Shock Absorber/Strut, Front YUG-4
- Shock Absorber/Strut, Rear YUG-4
- Spark Plug Cables YUG-2
- Spark Plugs YUG-2
- Spring, Front Coil YUG-4
- Spring, Leaf YUG-4
- Starter YUG-2
- Steering Wheel YUG-4

T
- Thermostat YUG-2
- Timing Belt YUG-3

V
- Valve Job YUG-3

W
- Water Pump YUG-3
- Wheel Balance YUG-2
- Wheel Cylinder YUG-4
- Window Regulator YUG-5
- Windshield Washer Pump YUG-5
- Windshield Wiper Motor YUG-5

YUG-1

YUG-2   CABRIO : GV : GVL : GVS : GVX

	LABOR TIME	SEVERE SERVICE
MAINTENANCE		
Air Cleaner Filter Element, Replace (C)		
1987-91	.2	.3
Drive Belt, Adjust (B)		
1987-91	.2	.3
Drive Belt, Replace (B)		
1987-91		
exc. compressor	.4	.5
compressor	.8	.9
Fuel Filter, Replace (B)		
1987-91	.4	.4
Sealed Beam Headlamp, Replace (C)		
Each	.3	.3
Oil & Filter, Change (C)		
Includes: Correct all fluid levels.		
1987-91	.3	.4
Timing Belt, Replace (B)		
1987-91	1.8	2.3
Replace tensioner add	.3	.5
Tire, Replace (C)		
Includes: Dismount old tire and mount new tire to rim.		
One	.5	.5
Wheel, Balance (B)		
One	.5	.5
CHARGING		
Alternator Circuits, Test (B)		
Includes: Test battery, regulator and alternator output.		
1987-91	.5	.5
Alternator Assy., R&R and Recondition (A)		
1987-91	2.1	2.6
Alternator Assy., Replace (B)		
Includes: Transfer pulley, if required.		
1987-91	1.6	1.8
Alternator Voltage Regulator, Replace (B)		
1987-91	.4	.4
Front Alternator Bearing, Replace (B)		
1987-91	1.6	1.9
Replace rear bearing add	.2	.4
STARTING		
Starter Draw Test (On Car) (B)		
1987-91	.3	.3
Battery Cables, Replace (C)		
Positive	.5	.7
Negative	.4	.6
Starter, R&R and Recondition (A)		
1987-91	2.3	2.9
Starter Assy., Replace (B)		
1987-91	1.1	1.3

	LABOR TIME	SEVERE SERVICE
Starter Drive Assy., Replace (B)		
Includes: Starter R&R.		
1987-91	1.3	1.6
Starter Solenoid and/or Switch, Replace (B)		
Includes: Starter R&R.		
1987-91	1.3	1.6
IGNITION		
Ignition Timing, Reset (B)		
1987-91	.3	.3
Distributor, Replace (B)		
Includes: Reset base ignition timing.		
1987-91	.5	.7
Distributor Cap and/or Rotor, Replace (B)		
1987-91	.4	.4
Ignition Coil, Replace (B)		
1987-91	.5	.5
Ignition Switch, Replace (B)		
1987-91	1.0	1.0
Pickup Coil, Replace (B)		
1987-91	.9	1.1
Spark Plug (Ignition) Cables, Replace (B)		
1987-91	.4	.5
Spark Plugs, Replace (B)		
1987-91	.4	.5
EMISSIONS		
Air Cleaner Temperature Sensor, Replace (B)		
1987-91	.3	.3
Air Pump Assy., Replace (B)		
1987-91	.9	1.1
Altitude Compensator, Replace (B)		
1987-91	.5	.5
Crankcase Vent Cleaner Filter Element, Replace (B)		
1987-91	.2	.2
Diverter and/or Switching Valve, Replace (B)		
1987-91	.4	.4
EGR Tube, Replace (B)		
1987-91	.5	.5
EGR Valve, Replace (B)		
1987-91	.6	.8
Electronic Control Module (ECM/PCM), Replace (B)		
1987-91	.4	.4
Fuel Vapor Canister, Replace (B)		
1987-91	.5	.5
Injection Tube Check Valve Assy., Replace (B)		
1987-91	.5	.5
Positive Crankcase Ventilation (PCV) Valve, Replace (B)		
1987-91	.3	.3
Vacuum Motor, Replace (B)		
1987-91	.4	.4

	LABOR TIME	SEVERE SERVICE
FUEL		
CARBURETOR		
Carburetor, Adjust (B)		
1987-91	.4	.4
Carburetor, R&R and Clean or Recondition (A)		
Includes: Adjustments.		
1987-91	2.0	2.0
Carburetor, Replace (B)		
Includes: Adjustments.		
1987-91	1.1	1.1
Electric Choke Relay, Replace (B)		
1987-91	.4	.4
Vacuum Kicker Assy., Replace (B)		
1987-91	.4	.4
DELIVERY		
Fuel Pump, Test (B)		
1987-91	.3	.3
Air Cleaner Filter Element, Replace (C)		
1987-91	.2	.3
Fuel Filter, Replace (B)		
1987-91	.4	.4
Fuel Gauge (Dash), Replace (B)		
1987-91	.8	.8
Fuel Gauge (Tank), Replace (B)		
1987-91	1.3	1.5
Fuel Pump, Replace (B)		
1987-91	.6	.8
Fuel Tank, Replace (B)		
Includes: Drain and refill tank.		
1987-91	1.4	1.6
Intake & Exhaust Manifold or Gaskets, Replace (B)		
1987-91	2.1	2.6
EXHAUST		
Catalytic Converter, Replace (B)		
1987-91	.9	1.0
Exhaust Pipe, Replace (B)		
1987-91	1.0	1.1
Intake & Exhaust Manifold or Gaskets, Replace (B)		
1987-91	2.1	2.6
Muffler, Replace (B)		
1987-91	1.0	1.1
Resonator, Replace (B)		
1987-91	.6	.7
ENGINE COOLING		
Coolant Thermostat, Replace (B)		
1987-91	.9	1.1
Fan Blade, Replace (B)		
1987-91	.5	.7
Radiator Assy., R&R or Replace (B)		
1987-91	1.3	1.5

CABRIO : GV : GVL : GVS : GVX YUG-3

	LABOR TIME	SEVERE SERVICE
Radiator Fan Motor Switch (Coolant Temp.), Replace (B)		
1987-91	.6	.6
Radiator Hoses, Replace (B)		
Includes: Refill with proper coolant mix.		
Upper	.5	.6
Lower	.7	.9
Temperature Gauge (Engine Unit), Replace (B)		
1987-91	.4	.4
Water Pump and/or Gasket, Replace (B)		
Includes: Refill with proper coolant mix.		
1987-91	3.7	4.2
w/AC add	1.0	1.0

ENGINE
ASSEMBLY

Times shown are for OEM assemblies. Time to replace assemblies from aftermarket rebuilders may vary.

Engine Assy., R&I (B)
Does not include parts or component transfer.
1987-91 .... 7.8 7.8

Engine Assy., R&R and Recondition (A)
Includes: Replacing rings, rod and main bearings, cylinder head reconditioning and engine tune-up.
1987-91 .... 19.4 19.4

Engine Assy. (Short) Replace (B)
Assembly consists of cylinder block, piston assemblies, crankshaft, camshaft, timing chain and gears. Does not include cylinder heads. Operation Includes: R&R engine, transfer necessary parts and all necessary adjustments.
1987-91 .... 11.3 11.3

Engine Mounts, Replace (B)
Front 1.7 1.9
Rear 1.0 1.2

CYLINDER HEAD
Compression Test (B)
1987-91 .... .6 .8
Valve Clearance, Adjust (B)
1987-91 .... 1.4 1.6
Camshaft Housing Cover and/or Gasket, Replace (B)
1987-91 .... .6 .8
Cylinder Head, Replace (B)
Includes: Adjustments.
1987-91 .... 7.7 8.2
Cylinder Head Gasket, Replace (B)
1987-91 .... 5.6 5.9

	LABOR TIME	SEVERE SERVICE
Rocker Arms, Pushrods and/or Pivots, Replace (B)		
1987-91	.6	.8
Valve Job (A)		
1987-91	8.7	9.2
Valve Lifters, Replace (B)		
1987-91	2.6	2.9

CAMSHAFT
Auxiliary Shaft, Replace (B)
1987-91 .... 1.5 1.7
Auxiliary Shaft Cover Gasket, Replace (B)
1987-91 .... 2.3 2.5
Camshaft, Replace (A)
1987-91 .... 2.7 3.2
Replace oil seal add .....2 .2
Camshaft Housing, Replace (B)
1987-91 .... 2.4 2.9
Timing Belt, Replace (B)
1987-91 .... 1.8 2.3
Replace tensioner add ....3 .5

CRANK & PISTONS
Connecting Rod Bearings, Replace (A)
Includes: Check bearing oil clearance.
1987-91 .... 3.2 3.7
Crankshaft and Main Bearings, Replace (A)
Includes: Engine R&R.
1987-91 .... 12.9 13.8
Crankcase Front Cover and/or Gasket, Replace (B)
1987-91 .... 2.9 3.1
Crankshaft Pulley, Replace (B)
1987-91 .... 1.3 1.5
Pistons or Connecting Rods, Replace (A)
Includes: Ridge reaming, cylinder wall deglazing, installing new rings and rod bearings, engine tune-up.
1987-91 .... 12.6 13.5
Rear Main Oil Seal, Replace (B)
1987-91 .... 6.5 7.1
Rings, Replace (A)
Includes: Ridge reaming, cylinder wall deglazing, installing new rings, engine tune-up.
1987-91 .... 11.4 12.3

ENGINE LUBRICATION
Engine Oil Pressure Switch (Sending Unit), Replace (B)
1987-91 .... .5 .5
Oil Pan and/or Gasket, Replace (B)
1987-91 .... 2.0 2.2
Oil Pump, Replace (B)
1987-91 .... 2.7 3.2

	LABOR TIME	SEVERE SERVICE
## CLUTCH		
Clutch Pedal Free Play, Adjust (B)		
1987-91	.3	.5
Clutch Assy., Replace (B)		
1987-91	5.7	5.9
Clutch Cable, Replace (B)		
1987-91	.9	1.1
Clutch Release Bearing, Replace (B)		
1987-91	5.8	6.0
Flywheel, Replace (B)		
1987-91	6.0	6.2

MANUAL TRANSAXLE
Transaxle Assy. R&R and Recondition (A)
1987-91 .... 10.5 11.1
Transaxle Assy., R&R or Replace (B)
1987-91 .... 5.5 5.7
Transaxle Rear Cover, R&R or Replace (B)
1987-91 .... 1.0 1.2

DRIVELINE
DRIVE AXLE
Differential Assy., R&R and Recondition (A)
1987-91 .... 7.8 8.4
Differential Assy., R&R or Replace (B)
1987-91 .... 5.1 5.6

HALFSHAFTS
CV Joint, Replace or Recondition (B)
One 1.5 1.8
CV-Joint Boot, Replace (B)
One 1.2 1.5
Halfshaft, R&R or Replace (B)
One 1.0 1.2
Halfshaft Boot, Replace (B)
Includes: Clean and lubricate CV-joint.
One 1.2 1.5

BRAKES
SYSTEM
Bleed Brakes (B)
Includes: Add fluid.
1987-91 .... .5 .5
Brakes, Adjust (B)
Includes: Refill master cylinder.
1987-91 .... .3 .3
Brake Compensator Valve, Replace (B)
1987-91 .... 1.2 1.4
Brake Hose (Flexible), Replace (B)
Includes: System bleeding.
One .... .7 .9
each addl. add ....2 .3

YUG-4 CABRIO : GV : GVL : GVS : GVX

	LABOR TIME	SEVERE SERVICE
Master Cylinder, R&R and Rebuild (A)		
1987-91	1.8	2.1
Master Cylinder, Replace (B)		
1987-91	1.0	1.2
Power Booster Unit, Replace (B)		
1987-91	1.5	1.5
Power Booster Vacuum Check Valve, Replace (B)		
1987-91	.3	.3

SERVICE BRAKES

	LABOR TIME	SEVERE SERVICE
Brake Drum, Replace (B)		
One	.5	.6
Both	.7	.8
Disc Brake Rotor, Replace (B)		
One	.8	1.0
Both	1.5	1.7
Caliper Assy., R&R and Recondition (A)		
Includes: System bleeding.		
One	1.7	2.0
Both	3.3	3.6
Caliper Assy., Replace (B)		
Includes: System bleeding.		
One	1.5	1.7
Both	2.8	3.0
Pads and/or Shoes, Replace (B)		
Includes: Adjust service and parking brake. System bleeding.		
Front disc	.9	1.0
Rear drum	1.0	1.1
Four wheels	1.7	1.8

COMBINATION ADD-ONS

	LABOR TIME	SEVERE SERVICE
Replace		
brake drum add	.2	.2
caliper add	.3	.3
wheel cylinder add	.3	.3
Resurface		
brake drum add	.5	.5
disc rotor add	.5	.5
Wheel Cylinder, R&R and Rebuild (B)		
Includes: System bleeding.		
One	1.0	1.3
Both	1.8	2.1
Wheel Cylinder, Replace (B)		
Includes: System bleeding.		
One	.8	1.0
Both	1.4	1.6

PARKING BRAKE

	LABOR TIME	SEVERE SERVICE
Parking Brake Apply Actuator, Replace (B)		
1987-91	1.0	1.0
Parking Brake Cable, Adjust (C)		
1987-91	.3	.5
Parking Brake Cable, Replace (B)		
1987-91	1.0	1.3

FRONT SUSPENSION

Unless otherwise noted, time given does not include alignment.

	LABOR TIME	SEVERE SERVICE
Align Front End (A)		
1987-91	1.1	1.3
Front Toe, Adjust (B)		
1987-91	.7	.9
Control Arm, Replace (B)		
One side	2.0	2.3
Both sides	3.9	4.2
Replace bushings add each	.3	.5
Front Strut Coil Spring, Replace (B)		
One side	1.1	1.4
Both sides	2.0	2.3
Stabilizer Bar, Replace (B)		
1987-91	1.1	1.3
Strut, Replace (B)		
One side	1.0	1.2
Both sides	1.8	2.0
Wheel Hub Bearing or Seal, Replace (B)		
One	1.8	1.8
Both	3.4	3.4

REAR SUSPENSION

	LABOR TIME	SEVERE SERVICE
Rear Alignment, Adjust (A)		
1987-91	1.0	1.2
Rear Control Arm, Replace (B)		
1987-91	1.5	1.8
Replace bushings add	.3	.5
Rear Leaf Spring, Replace (B)		
1987-91	1.5	1.9
Shock Absorbers or Bushings, Replace (B)		
One	.8	1.0
Both	1.4	1.6
Stub Axle, Replace (B)		
One	1.1	1.6
Both	1.9	2.4
Wheel Hub and Bearing Assy., Replace (B)		
One	.8	1.1
Both	1.4	1.7

STEERING

Unless otherwise noted, time given does not include alignment.

MANUAL RACK & PINION

	LABOR TIME	SEVERE SERVICE
Horn Contact, Replace (B)		
1987-91	.5	.5
Rack & Pinion Assy., R&R and Recondition (A)		
1987-91	3.8	4.3
Rack & Pinion Assy., Replace (B)		
1987-91	2.2	2.5
Steering Column, Replace (B)		
1987-91	1.0	1.0
Steering Wheel, Replace (B)		
1987-91	.5	.5
Tie Rod End Boot, Replace (B)		
1987-91	1.0	1.2
Tie Rods or Tie Rod Ends, Replace (B)		
Includes: Reset toe.		
One side	1.0	1.2

HEATING & AIR CONDITIONING

When more than one component requires replacement where evacuation/recovery and recharging is already included, deduct 1.0 hour for each additional component from the time given.

	LABOR TIME	SEVERE SERVICE
Evacuate/Recover and Recharge System (B)		
1987-91	1.0	1.0
AC Hoses, Replace (B)		
Includes: Evacuate/recover and recharge.		
High pressure	1.5	1.7
Pressure line condenser to drier	1.6	1.8
drier to expansion valve	2.1	2.3
Suction line	1.9	2.1
Compressor Assy., Replace (B)		
Includes: Parts transfer. Evacuate/recover and recharge.		
1987-91	2.5	2.7
Condenser Assy., Replace (B)		
Includes: Evacuate/recover and recharge.		
1987-91	1.9	2.1
Condenser Cooling Fan, Replace (B)		
1987-91	.9	2.1
Condenser Cooling Fan Switch, Replace (B)		
1987-91	.5	.5
Evaporator Coil, Replace (B)		
Includes: Evacuate/recover and recharge.		
1987-91	2.3	2.3
Expansion Valve, Replace (B)		
Includes: Evacuate/recover and recharge.		
1987-91	1.3	1.3
Heater Blower Motor, Replace (B)		
1987-91	1.1	1.1
Heater Blower Motor Switch, Replace (B)		
1987-91	.4	.4
Heater Core, R&R or Replace (B)		
1987-91	1.6	1.6

CABRIO : GV : GVL : GVS : GVX **YUG-5**

	LABOR TIME	SEVERE SERVICE
High or Low Pressure Switch, Replace (B)		
1987-91	1.3	1.3
Add time to recharge system if needed.		
Mode Door Actuator or Damper, Replace (B)		
1987-91	2.5	2.5
Receiver/Drier Assy., Replace (B)		
Includes: Evacuate/recover and recharge.		
1987-91	1.5	1.5
Thermostat, Replace (B)		
1987-91	.5	.7
Thermostatic Control Switch, Replace (B)		
1987-91	.6	.6
Water Control Valve, Replace (B)		
1987-91	.7	.9

WIPERS & SPEEDOMETER

	LABOR TIME	SEVERE SERVICE
Radio, R&R (B)		
1987-91	.8	.8
Rear Window Washer Pump, Replace (B)		
1987-91	.3	.3
Rear Window Wiper Motor, Replace (B)		
1987-91	.5	.7
Rear Window Wiper & Washer Switch, Replace (B)		
1987-91	.4	.4
Speedometer Cable & Casing, Replace (B)		
1987-91	.5	.5

	LABOR TIME	SEVERE SERVICE
Speedometer Head, R&R or Replace (B)		
Includes: Cluster R&R.		
1987-91	.8	.8
Windshield Washer Pump, Replace (B)		
1987-91	.4	.4
Windshield Wiper & Washer Switch, Replace (B)		
1987-91	.8	.8
Windshield Wiper Intermittent Selector Switch, Replace (B)		
1987-91	.8	.8
Windshield Wiper Linkage, Replace (B)		
1987-91	1.1	1.3
Windshield Wiper Motor, Replace (B)		
1987-91	.8	1.0

LAMPS & SWITCHES

	LABOR TIME	SEVERE SERVICE
Back-Up Lamp Assy., Replace (B)		
1987-91	.3	.3
Back-Up Lamp Switch, Replace (B)		
1987-91	.4	.4
Combination Switch Assy., Replace (B)		
1987-91	.8	.8
Headlamp Switch, Replace (B)		
1987-91	.4	.4
Headlamps, Aim (B)		
Two	.4	.4
Four	.7	.7
Horn, Replace (B)		
1987-91	.5	.5
Horn Relay, Replace (B)		
1987-91	.3	.3

	LABOR TIME	SEVERE SERVICE
License Lamp Assy., Replace (B)		
1987-91	.2	.2
Parking Brake Warning Lamp Switch, Replace (B)		
1987-91	.2	.2
Parking Lamp Assy., Front, Replace (B)		
1987-91	.2	.2
Sealed Beam Headlamp, Replace (C)		
Each	.3	.3
Side Marker Lamp Assy., Replace (B)		
1987-91	.2	.2
Stop Light Switch, Replace (B)		
1987-91	.4	.4
Tail Lamp Assy., Replace (B)		
1987-91	.3	.3
Turn Signal or Hazard Warning Flasher, Replace (B)		
1987-91	.3	.3

BODY

	LABOR TIME	SEVERE SERVICE
Door Lock, Replace (B)		
1987-91	.9	.9
Door Lock Remote Control, Replace (B)		
1987-91	.8	.8
Door Window Regulator, Replace (B)		
1987-91	1.1	1.3
Hood Lock, Replace (B)		
1987-91	.4	.4
Hood Release Cable, Replace (B)		
1987-91	.7	.7
Lock Striker Plate, Replace (B)		
1987-91	.4	.4

NOTES

NOTES

NOTES

NOTES

YUG-10 CABRIO : GV : GVL : GVS : GVX

NOTES

NOTES

YUG-12 CABRIO : GV : GVL : GVS : GVX

NOTES

NOTES

NOTES

NOTES

YUG-16 CABRIO : GV : GVL : GVS : GVX

NOTES

NOTES

YUG-18 CABRIO : GV : GVL : GVS : GVX

NOTES

NOTES

YUG-20 CABRIO : GV : GVL : GVS : GVX

NOTES

CHILTON LABOR GUIDES

Domestic Manual ISBN 1-4180-0606-8/Part No. 130606
Import Manual ISBN 1-4180-1537-7/Part No. 131537

Chilton has added so much to its labor guide manual that we've had to put it in two volumes! We've added hundreds of new labor operations—including maintenance services and electronic system diagnosis—to the *Chilton® 2006 Import and Domestic Labor Guide* manuals. All labor times for 1981 through 2006 vehicles consider the real world environment in which technicians work: worn, rusted or dirty components, being serviced with tools commonly used in the aftermarket. Chilton labor times are accepted by most insurance and extended warranty companies. Vehicle makes and models conform to current Automotive Aftermarket Industry Association standards.

Labor Guide Manual Benefits:

- parts terminology is more standardized across different OEMs to simplify reference.
- a total of more than 2,500 pages of updated Chilton labor times appear in these two volumes
- make sure your students have this latest edition because our experts have updated hundreds of labor times for earlier models

Hardcover Manuals are 8 7/8" x 11", ©2006

Labor Guide CD-ROM Benefits:

- easy-to-use software to create and print professional-quality estimates and invoices
- three user-defined levels of labor rates correspond to different types of job scenarios, for "real-world" application
- functions as a database of aftermarket labor times for monitoring warranty and insurance claims
- software keeps track of customers and prior estimates for time-saving recall
- customizable application allows service writers to add labor operations and times, and parts companies to add labor times to existing parts ordering systems

CD ISBN 1-4180-0605-X/Part No. 130605
©2006

Previous Year Editions:
Chilton 2005 Labor Guide Manual, ISBN 1-4018-7412-6/Part No. 27412
Chilton 2005 Labor Guide CD-ROM, ISBN 1-4018-7818-0/Part No. 27818

FOR CUSTOMER SUPPORT CALL **1-800-477-3692**

CHILTON 2006 MECHANICAL SERVICE MANUALS

The *Chilton® 2006 Mechanical Service Manuals* provides updated coverage through 2005 models and even many 2006 models, as made available from original equipment manufacturers (OEMs). Chilton is still your reliable source for fast, accurate repairs and reassembly and it still provides the lowest-priced professional repair manuals on the market! These manuals are organized by make, model and system so information gathering is easier. Now with even more illustrations and a streamlined index, it's no wonder more automotive professionals turn to Chilton Professional Manuals for their mechanical service and repair information.

Mechanical Service Manual Benefits:

- access up-to-date service and repair information covering model years 2002-2006, all logically arranged by manufacturer
- follow clear, step-by-step procedures—from drive train to chassis —to yield fast, accurate results
- service more mechanical systems, including brakes, engines, suspensions, steering and related components
- know what special tools are required for specific jobs, as Chilton editors describe and illustrate them to make repair work go more smoothly

2006 Editions

Chilton 2006 DaimlerChrysler Mechanical Service Manual—ISBN 1-4180-0600-9/Part No. 130600
Chilton 2006 Ford Mechanical Service Manual—ISBN 1-4180-0601-7/Part No. 130601
Chilton 2006 General Motors Mechanical Service Manual—ISBN 1-4180-0602-5/Part No. 130602
Chilton 2006 Asian Mechanical Service Manual—Volume I—ISBN 1-4180-0947-4/Part No. 130947
Chilton 2006 Asian Mechanical Service Manual—Volume II—ISBN 1-4180-0948-2/Part No. 130948
Chilton 2006 Asian Mechanical Service Manual—Volume III—ISBN 1-4180-0949-0/Part No. 130949
Chilton 2006 Asian Mechanical Service Manual—3 Volume Set—ISBN 1-4180-0603-3/Part No. 130603
Chilton 2006 European Mechanical Service Manual—ISBN 1-4180-0604-1/Part No. 130604

Manuals are 8 1/2" x 11", ©2006

2005 Editions

Chilton 2005 General Motors Mechanical Service Manual—ISBN 1-4018-7146-1/Part No. 27146
Chilton 2005 Chrysler Mechanical Service Manual—ISBN 1-4018-6718-9/Part No. 26718
Chilton 2005 Ford Mechanical Service Manual—ISBN 1-4018-6719-7/Part No. 26719
Chilton 2005 European Mechanical Service Manual—ISBN 1-4018-6720-0/Part No. 126720
Chilton 2005 Asian Mechanical Service Manual – Volume I—(Acura-Mazda) ISBN 1-4018-6716-2/Part No. 26716
Chilton 2005 Asian Mechanical Service Manual – Volume II—(Mitsubishi-Toyota)
      ISBN 1-4018-6717-0/Part No. 26717
Chilton 2005 Asian Mechanical Service Manual – Set of Volumes I & II—ISBN 1-4018-7180-1/Part No. 27180

Manuals are 8 1/2" x 11", ©2005

ONLINE www.trainingbay.com   TO PLACE AN ORDER CALL **1-800-347-7707**

CHILTON 2006 DIAGNOSTIC SERVICE MANUALS

Chilton Timing Belts, 1985-2005

Chilton
ISBN 1-4018-9880-7/Part No. 129880

Timing belt procedures can represent increased profits for automotive repair shops and service stations, and this manual contains all the information automotive technicians need to properly service timing belts on domestic and imported cars, vans, and light trucks through 2005 models. Clear, straightforward procedures, illustrations, and specifications help to communicate 20 years of vehicle applications for fast, accurate inspection, replacement, and tensioning of timing belts. Readers will learn step-by-step how to perform key procedures both quickly and safely, while learning the correct labor time to charge for the service. OEM-recommended replacement intervals for proper maintenance of customer's vehicles are also featured.

BENEFITS
- detailed illustrations clearly demonstrate important concepts, such as how to correctly align camshaft and crankshaft timing marks, and how to simplify serpentine belt installation
- readers are made aware of potential hazards and time-wasting practices that can impede safe and profitable service procedures
- special tools are identified so that completing the service is as easy and quick as possible

544 pp, 8 1/2" x 11", softcover, ©2006

The *Chilton® 2006 Diagnostic Service Manuals* provide technicians with the critical diagnostic information they need to accurately identify and solve engine performance problems. Clear explanations, specifications, and illustrations help technicians diagnose second generation on-board diagnostic (OBD-II) systems. Chilton Diagnostic Service Manuals, when used with an engine analyzer, scan tool, or lab scope, allow diagnosticians to understand functions of engine performance components and systems, simplify testing procedures, and diagnose trouble codes.

Diagnostic Service Manual Benefits:
- provide training information in addition to reference material
- explain engine performance components and system operation
- function as exceptional diagnostic companions when analyzing automotive drive-train performance problems
- provide a comprehensive list of trouble code titles, conditions, and possible causes
- reduce diagnostic and repair time using expert testing procedures and troubleshooting hints

2006 Editions
Chilton 2006 DaimlerChrysler Diagnostic Service Manual
   ISBN 1-4180-2118-0/Part No. 132118
Chilton 2006 Ford Diagnostic Service Manual
   ISBN 1-4180-2119-9/Part No. 132119
Chilton 2006 General Motors Diagnostic Service Manual
   ISBN 1-4180-2120-2/Part No. 132120
Chilton 2006 Asian Diagnostic Service Manual, Volume 1
   ISBN 1-4180-2913-0/Part No. 132913
Chilton 2006 Asian Diagnostic Service Manual, Volume 2
   ISBN 1-4180-2914-9/Part No. 132914
Chilton 2006 Asian Diagnostic Service Manual, Volume 3
   ISBN 1-4180-2915-7/Part No. 132915
Chilton 2006 European Diagnostic Service Manual
   ISBN 1-4180-2924-6/Part No. 132924
Manuals are 8 1/2" x 11", ©2006

2005 Editions
Chilton 2005 General Motors Diagnostic Service Manual
   ISBN 1-4180-0552-5/Part No. 130552
Chilton 2005 Chrysler Diagnostic Service Manual
   ISBN 1-4180-0550-9/Part No. 130550
Chilton 2005 Ford Diagnostic Service Manual
   ISBN 1-4180-0551-7/Part No. 130551
Chilton 2005 Asian Diagnostic Service Manual
   ISBN 1-4180-0553-3/Part No. 130553
Manuals are 8 1/2" x 11", ©2005

FOR CUSTOMER SUPPORT CALL **1-800-477-3692**

CHILTON SERVICE MANUALS—PERENNIAL EDITIONS

The *Chilton® Perennial Editions* contain repair and maintenance information for popular mechanical systems that may not be available elsewhere. They offer a wide range of repair information on cars, trucks, vans, and SUVs dating back to the early 1960s, and as current as 2002. Information for 1993 and later model years includes scheduled maintenance interval charts.

Benefits:
- covers the most common vehicle models found in the repair aftermarket today
- gain quick understanding of systems using exploded-view illustrations, diagrams, and charts
- simplify tough jobs with easy-to-follow removal and installation instructions for heater core and other components
- obtain complete coverage of repair procedures from drive train to chassis and associated components

Auto Repair Manual, 1998-2002, 1,426 pages
ISBN 0-8019-9362-8/Part No. 9362
Auto Repair Manual, 1993-1997, 2,064 pages
ISBN 0-8019-7919-6/Part No. 7919
Auto Repair Manual, 1988-1992, 1,284 pages
ISBN 0-8019-7906-4/Part No. 7906
Auto Repair Manual, 1980-1987, 1,344 pages
ISBN 0-8019-7670-7/Part No. 7670

Import Car Repair Manual, 1998-2002, 1,792 pps
ISBN 0-8019-9363-6/Part No. 9363
Import Car Repair Manual, 1993-1997, 2,080 pps
ISBN 0-8019-7920-X/Part No. 7920
Import Car Repair Manual, 1988-1992, 1,632 pages
ISBN 0-8019-7907-2/Part No. 7907
Import Car Repair Manual, 1980-1987, 1,488 pages
ISBN 0-8019-7672-3/Part No. 7672

Truck & Van Repair Manual, 1998-2002, 1,408 pages
ISBN 0-8019-9364-4/Part No. 9364
Truck & Van Repair Manual, 1993-1997, 2,096 pages
ISBN 0-8019-7921-8/Part No. 7921
Truck & Van Repair Manual, 1991-1995, 1,664 pages
ISBN 0-8019-7911-0/Part No. 7911
Truck & Van Repair Manual, 1986-1990, 1,536 pages
ISBN 0-8019-7902-1/Part No. 7902
Truck & Van Repair Manual, 1979-1986, 1,440 pages
ISBN 0-8019-7655-3/Part No. 7655

SUV Repair Manual, 1998-2002, 1,292 pages
ISBN 0-8019-9365-2/Part No. 9365

Hardcover manuals are 8 1/2" x 11".

Chilton Collector's Editions—*Reference Manuals for Vintage Vehicles*
Auto Repair Manual, 1964-1971, ISBN 0-8019-5974-8/Part No. 5974,
Truck & Van Repair Manual, 1961-1971, ISBN 0-8019-6198-X/Part No. 6198
Truck & Van Repair Manual, 1971-1978, ISBN 0-8019-7012-1/Part No. 7012

ONLINE www.trainingbay.com   TO PLACE AN ORDER CALL **1-800-347-7707**